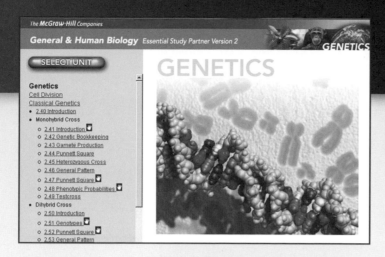

Essential Study Partner

McGraw-Hill's Essential Study
Partner offers art activities, animations, and quizzes that are embedded
within tutorials designed to walk you
through the key concepts in *Biology*.

Test Yourself

Take a chapter quiz on the
Biology website. Each quiz is
specially constructed to test
your comprehension of key
concepts. Feedback on your
responses help you gauge your
mastery of the material.

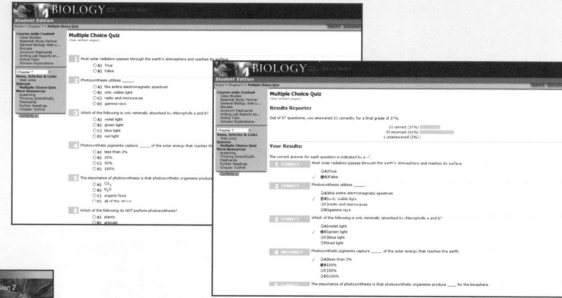

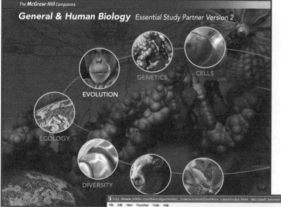

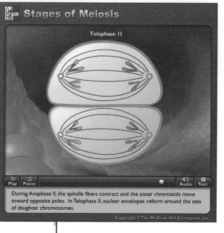

Access to Premium Learning Materials

The *Biology* website is your
portal to exclusive study tools
such as McGraw-Hill's
Essential Study Partner,
animations, and case studies.

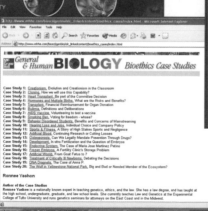

The Next Generation of Digital Assets

This collection of visual resources allows instructors to utilize artwork from the text, full-color animations, digitized video clips, and other resources in multiple formats to create customized course tools.

NEW!

The **Introductory General Biology Animations** bring the next generation of animations to the classroom. These full-color animations of biological concepts and processes offer total flexibility in classroom presentations or online course materials. Designed to be used in lectures, these full-screen animations are available on both the Digital Content Manager and the *Biology* ARIS website and feature pause, rewind, fast-forward, and audio on/off options. A Spanish version is also provided for many of the animations.

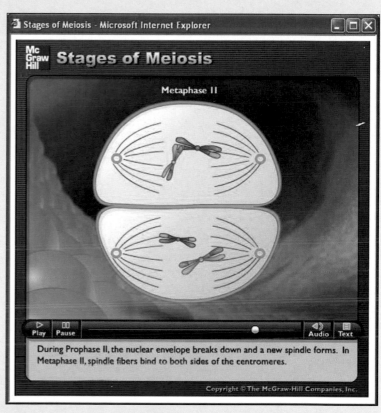

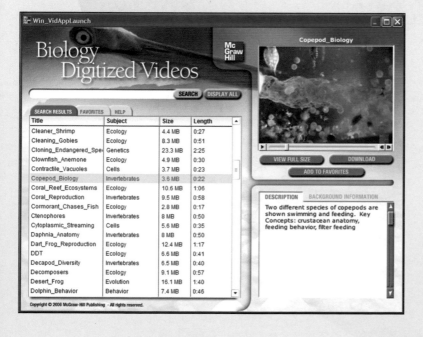

McGraw-Hill's Biology Digitized Video Clips

These video clips are an exciting new presentation tool for adopting instructors to use in lecture or lab! Licensed from some of the highest-quality life science video producers in the world, these brief segments range in length from about five seconds to just under three minutes and cover all areas of general biology, from cells to ecosystems. Engaging and informative, McGraw-Hill's digitized biology video clips will help capture students' interest while illustrating key biological concepts and processes through clips on mitosis, cytoplasmic streaming, stem cell research, Darwin's finches, poison dart frogs, maggot therapy, brain surgery, parasites, deforestation, and much, much more.

A Powerful Lecture Resource

Active Art

Key art pieces can be customized in almost any way imaginable inside of PowerPoint®. Each piece can be broken down to its core elements, grouped or ungrouped, and edited to create customized illustrations.

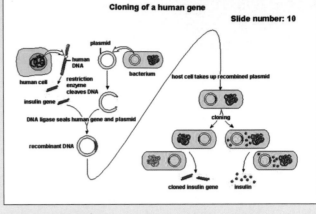

Change Colors – Colors can be removed and/or changed from any Active Art slide or object. Black and white images can be created for use in lecture supplements or exams.

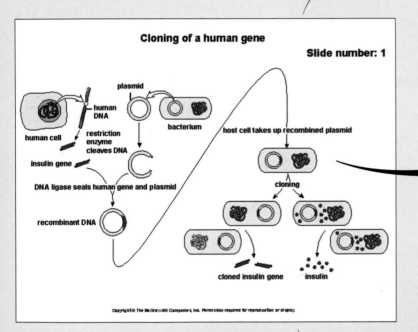

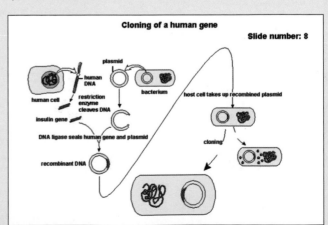

Remove Labels – Labels and leader lines can be easily repositioned, edited, or removed. Your customized images can then be used for course assignments or additional quizzing for your students.

Resize Objects – The entire image or parts of the image can be made larger or smaller depending on what you choose to emphasize.

Digital Content Manager

The Digital Content Manager provides all illustrations, photos, and tables from the textbook, in addition to supplemental media materials. With the Digital Content Manager, instructors will have access to:

Art Libraries
Color-enhanced, digital files of all the illustrations in the book, plus the same art saved in unlabeled and gray scale versions, are included in the Art Libraries.

Photo Library
Digital files of all photographs from the book are available.

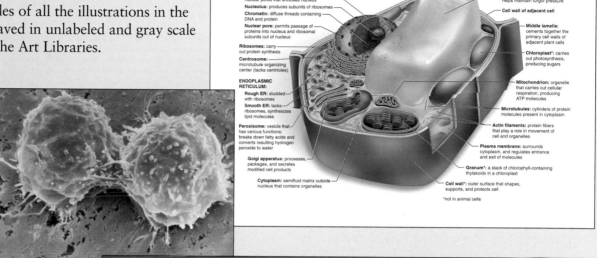

Table Library
Every table that appears in the book is provided in electronic form.

Additional Photo Library
Over 700 photos not found in the text are available for use in creating lecture presentations.

TABLE 20.3

Major Distinctions Among the Three Domains of Life

	Bacteria	Archaea	Eukarya
Unicellularity	Yes	Yes	Some, many multicellular
Membrane lipids	Phospholipids, unbranched	Varied branched lipids	Phospholipids, unbranched
Cell wall	Yes (contains peptidoglycan)	Yes (no peptidoglycan)	Some yes, some no
Nuclear envelope	No	No	Yes
Membrane-bounded organelles	No	No	Yes
Ribosomes	Yes	Yes	Yes
Introns	No	Some	Yes

PowerPoint® Lecture Outlines
These ready-made presentations combine art and lecture notes for each chapter in the book. The presentations can be used as they are, or can be customized to reflect your preferred lecture topics and organization.

All line art, photos, and tables are also pre-inserted into blank PowerPoint® slides for ease of lecture preparation.

TextEdit Art Library
Every line art piece is placed into a PowerPoint® presentation that allows the user to revise, move, or delete labels as desired for creation of customized presentations and/or for testing purposes.

Animations Library
The next generation of general biology animations bring key processes to life! They are designed to be used in lectures to create dynamic presentations.

Making Our Instructors' Lives Easier!

ARIS

McGraw-Hill's ARIS–Assessment, Review, and Instruction System–for *Biology* ninth edition is a complete online tutorial, electronic homework, and course management system designed for greater ease of use than any other system available.   Free with adoption of McGraw-Hill's *Biology* ninth edition text, instructors can create and share course materials and assignments with colleagues with a few clicks of the mouse. ARIS has easy-to-assign home-work, quizzing, and testing. All student activity within ARIS is automatically recorded and available to the instructor through a fully integrated grade book that can be downloaded to Excel. ARIS also provides access to a wealth of opportunities for the instructor, such as the Instructor's Manual, Case Studies, Classroom Performance System questions, PageOut, and much more.

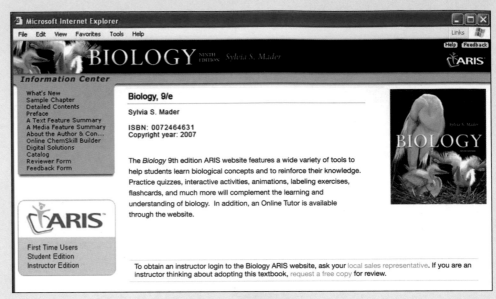

The Instructor's Testing and Resource CD

This cross-platform CD-ROM contains the Instructor's Manual and Test Item File, both available in Word and PDF formats. The manual contains chapter outlines, lecture enrichment ideas, critical thinking questions and answers, and details of changes from the eighth edition of *Biology*. The Test Item File offers questions that can be used for homework assignments or the preparation of exams. The computerized test bank utilizes flexible testing software, which allows the user to quickly create customized exams. This user-friendly program allows instructors to search questions by format or difficulty level; edit existing questions or add new ones; and scramble questions and answer keys for multiple versions of the same test.

eInstruction Classroom Performance System (CPS)

Wireless technology brings interactivity into the classroom or lecture hall. Instructors and students receive immediate feedback through wireless response pads that engage students and are easy to use. eInstruction can assist instructors by:

- taking attendance
- administering quizzes and tests
- creating a lecture with intermittent questions
- using the CPS grade book to manage lectures and student comprehension
- integrating interactivity into their PowerPoint® presentations

Sylvia S. Mader

BIOLOGY

NINTH EDITION

with significant contributions by

Murray P. Pendarvis
Southeastern Louisiana University

Boston   Burr Ridge, IL   Dubuque, IA   Madison, WI   New York   San Francisco   St. Louis
Bangkok   Bogotá   Caracas   Kuala Lumpur   Lisbon   London   Madrid   Mexico City
Milan   Montreal   New Delhi   Santiago   Seoul   Singapore   Sydney   Taipei   Toronto

The McGraw-Hill Companies

Higher Education

BIOLOGY, NINTH EDITION

Published by McGraw-Hill, a business unit of The McGraw-Hill Companies, Inc., 1221 Avenue of the Americas, New York, NY 10020. Copyright © 2007 by The McGraw-Hill Companies, Inc. All rights reserved. No part of this publication may be reproduced or distributed in any form or by any means, or stored in a database or retrieval system, without the prior written consent of The McGraw-Hill Companies, Inc., including, but not limited to, in any network or other electronic storage or transmission, or broadcast for distance learning.

Some ancillaries, including electronic and print components, may not be available to customers outside the United States.

 This book is printed on recycled, acid-free paper containing 10% postconsumer waste.

1 2 3 4 5 6 7 8 9 0 QPD/QPD 0 9 8 7 6 5

ISBN-13 978–0–07–246463–4
ISBN-10 0–07–246463–1

Publisher: *Janice Roerig-Blong*
Sponsoring Editor: *Thomas C. Lyon*
Director of Development: *Kristine Tibbetts*
Senior Developmental Editor: *Margaret B. Horn*
Marketing Manager: *Tamara Maury*
Senior Project Manager: *Jayne Klein*
Lead Production Supervisor: *Sandy Ludovissy*
Senior Media Project Manager: *Tammy Juran*
Media Producer: *Eric A. Weber*
Design Manager: *K. Wayne Harms*
Cover/Interior Designer: *Kaye Farmer*
(USE) Cover Image: *B. Moose Peterson/WRP*
Senior Photo Research Coordinator: *Lori Hancock*
Photo Research: *Connie Mueller*
Supplement Producer: *Melissa M. Leick*
Compositor: *Electronic Publishing Services Inc., NYC*
Typeface: *10/12 Palatino*
Printer: *Quebecor World Dubuque, IA*

The credits section for this book begins on page C-1 and is considered an extension of the copyright page.

Library of Congress Cataloging-in-Publication Data

Mader, Sylvia S.
    Biology / Sylvia S. Mader. —— 9th ed.
        p.   cm.
    Includes bibliographical references and index.
    ISBN 978–0–07–246463–4 — ISBN 0–07–246463–1 (hard copy : alk. paper)
    1. Biology. I. Title.

QH308.2.M23   2007
570—dc22   2005027781
                                                 CIP

www.mhhe.com

BRIEF CONTENTS

CONTENTS

1

A VIEW OF LIFE   1

PART I: THE CELL   18

2

BASIC CHEMISTRY   19

3

THE CHEMISTRY OF ORGANIC MOLECULES   35

4

CELL STRUCTURE AND FUNCTION   57

5

MEMBRANE STRUCTURE AND FUNCTION   83

6

METABOLISM: ENERGY AND ENZYMES   101

7

PHOTOSYNTHESIS   115

8

CELLULAR RESPIRATION   131

READINGS

ECOLOGY FOCUS

HEALTH FOCUS

SCIENCE FOCUS

PREFACE

Biology was born out of my desire for students to develop a particular view of the world—a biological view. It seemed to me that a thorough grounding in biological principles would bring about an appreciation of the structure and function of individual organisms, how they have evolved, and how they interact in the biosphere. This led me to use the levels of biological organization as my guide; thus, the book begins with chemistry and ends with the biosphere.

Students need to be aware that our knowledge of biology is built on scientific discovery. The first chapter explains the process of science and thoroughly reviews examples of how this process works. Throughout the text, biologists are introduced and their experiments are explained. An appreciation of the scientific process should include the perception that without it, the study of biology would not exist.

Evolution of *Biology*

The ninth edition of *Biology* is the result of a dramatic evolutionary change. Previously, the text gradually improved with each edition; in comparison, this edition represents a giant leap forward. As soon as the eighth edition left the presses, we started working on the next edition. Murray P. "Pat" Pendarvis, a talented biology professor much beloved by his students, assisted in updating the text and improving the illustrations. Pat's many additions to the text and choice of photographs increased its beauty and, in particular, its relevancy. I, too, worked diligently from cover to cover refining all that went before and making additions to improve content and pedagogy. My work was greatly assisted by the talented staff of EPS (Electronic Publishing Services Inc.) who laid out the pages and reworked every illustration to produce the most detailed, refined, and pedagogically sound figures ever developed for an introductory biology book.

Pedagogy

Pages xxi-xxiii of this preface review "The Learning System" of *Biology*. As explained, each chapter opening page provides an outline and lists the concepts that are discussed and reinforced within the chapter. Opening vignettes capture the interest of students, and at the close of each chapter, "Connecting the Concepts" discusses the relationships between various biological principles. The end matter of the chapter gives students an opportunity to test themselves on their progress.

It has been my privilege to develop a style and methodology that appeals to students because it meets them where they are and brings them along to a thorough understanding of the concept being presented. Concepts are only grasped if a student comes away with "take-home messages." The interweaving of concepts allows the student to develop a biological view of the world that is essential in the twenty-first century.

OVERVIEW OF CHANGES TO *BIOLOGY*, NINTH EDITON

VISUALS
A brilliant new visuals program combined with innovative page layouts enhances the pedagogical value and visual appeal.

GENETICS
Reproductive and therapeutic cloning are illustrated. There is an improved emphasis on regulation of gene activity, expansion of genomics to include proteomics and bioinformatics, and much more.

EVOLUTION AND CLASSIFICATION
Micro- and macroevolution are better explained and illustrated. Fungal and animal classifications are reorganized based on molecular data.

BOTANY
Discussion now centers on a generalized flowering plant, and plant anatomy is more expansive.

ANIMAL PHYSIOLOGY
New homeostasis art, contrast of nervous system with hormone system, reorganization of the development chapter, and the importance of chronic inflammatory response to general health are now included.

ECOLOGY
Population ecology is more comprehensive and understandable.

Changes in *Biology*, Ninth Edition

Perhaps the first significant enhancement that readers will notice in the ninth edition of *Biology* is the brilliant new visuals program. Virtually every illustration is either completely new or significantly revised to convey basic concepts and processes as effectively as possible. In addition to new artwork, hundreds of new photos grace the pages of *Biology*. Finally, we employed an innovative page layout process that combines text, art, and photos in a seamless manner to enhance pedagogical value and visual appeal.

Other significant content updates of special interest include:

- *Chapter 9, The Cell Cycle and Cellular Reproduction*, was reorganized and updated. The descriptions of stages now applies to both plant and animal cells, with differences still clearly designated. A new section, "The Functions of Mitosis," includes a Science Focus reading on Reproductive and Therapeutic Cloning. The cancer section has been completely rewritten to include the origin of cancer and how it relates to the regulation of the cell cycle. The action of oncogenes and tumor repressor genes is stressed.

- *Chapter 15, Regulation of Gene Activity and Gene Mutations*, has taken on new significance because we now know that humans have far fewer genes than was estimated before the sequencing of the human genome. This chapter was revised to reflect the importance of chromatin organization, transcription factors, and activators to the control of gene activity within the nucleus. Translational control within the cytoplasm, including the possible role of RNA to expand each gene's functions, is discussed, as is the importance of gene mutations to the development of cancer.

- *Chapter 16, Biotechnology and Genomics*. This chapter was updated and the topic of genomics was expanded to include a discussion of a genomic profile, proteomics, and bioinformatics. The importance of all these advances for improved health care is explored.

- *Chapter 18, Process of Evolution*, was reorganized to include a section on microevolution and macroevolution. Under macroevolution a more thorough discussion of speciation due to reproductive isolating mechanisms precedes real-life examples of allopatric speciation.

- *Chapter 20, Classification of Living Things*, was rewritten and now includes a better explanation of the phylogenetic tree and its connection to the classification and evolutionary relationships between organisms. The utilization of molecular data to guide classification from domain to species is stressed. A new Science Focus reading describes the proposal to use DNA differences as a basis to develop bar codes for all living species.

- *Chapter 23, The Fungi*, was reorganized to reflect the classification of fungi based on DNA sequencing. Fungi previously classified as imperfect fungi have been incorporated into the ascomycetes, and this chapter now has an expanded discussion of the sac fungi and their relationship to human beings.

- *Chapter 25, Structure and Organization of Plants*. A generalized flowering plant has been developed to present the basics of plant anatomy. New additional structural information permeates this chapter, which seeks to have students understand the overall functioning of a flowering plant. The discussion of primary versus secondary growth has been expanded to provide a better explanation for plant growth.

- *Part VI, Animal Evolution*. This part has been reorganized to be consistent with molecular data regarding the relationship of groups of animals. Traditional classification is the backbone of this part, but new hypotheses regarding the classification of animals are introduced. To accommodate the new hypotheses, Chapter 30 now includes molluscs, annelids, arthropods and echinoderms. Chapter 31 is devoted exclusively to the vertebrates.

- *Chapter 33, Animal Organization and Homeostasis*, has been reorganized to lead students to a better understanding of tissues, organs, and organ systems. Professors will particularly appreciate the improved homeostasis diagrams that explain negative feedback mechanisms. A new Health Focus regarding nerve regeneration stresses advances in this field and touches on the possible use of stem cells to cure paralysis.

- *Chapter 35, Lymph Transport and Immunity*, has been revised to include updated explanations of nonspecific and specific defenses. New data regarding the role of chronic inflammatory response to human illnesses is included. This chapter also has a new Health Focus reading regarding Opportunistic Infections and HIV.

- *Chapter 42, Hormones and Endocrine Systems*, now begins with an overview of the endocrine system, which includes a contrast between hormone and nervous signaling. An in-depth look at hormone signaling follows. The review of endocrine glands and their hormones includes an updated discussion of diabetes mellitus.

- *Chapter 44, Animal Development*, has been reorganized to present a more logical progression of animal developmental stages before developmental processes are explained. The discussion of developmental processes places an emphasis on experimental data to explain the orderliness of development. As before, the chapter ends with a look at the stages of human development.

- *Chapter 46, Ecology of Populations*, was reorganized and rewritten to better present the modern principles of population ecology. The sections now include demographics of populations, population growth models, and regulation of population size before life history patterns and human population growth are considered.

ACKNOWLEDGMENTS

The hard work of many dedicated and talented individuals helped to vastly improve this edition of *Biology*. Let me begin by thanking the people who guided this revision at McGraw-Hill. I am very grateful for the help of so many professionals who were involved in bringing this book to fruition. In particular, let me thank Margaret Horn, the developmental editor who lent her talents and advice to all those who worked on this edition of *Biology*. The biology editor was Thomas Lyon, who was also intimately involved in putting *Biology* through its paces. The project manager, Jayne Klein, faithfully and carefully steered the book through the publication process. Tamara Maury, the marketing manager, tirelessly promoted the text and educated the sales reps on its message.

The design of the book is the result of the creative talents of Wayne Harms and many others who assisted in deciding the appearance of each element in the text. EPS followed their guidelines as they created and reworked each illustration, emphasizing pedagogy and beauty to arrive at the best presentation on the page. Lori Hancock and Connie Mueller did a superb job of finding just the right photographs and micrographs.

My staff, consisting of Evelyn Jo Hebert and Beth Butler, worked faithfully as they helped proof the chapters and made sure all was well before the book went to press. As always, my family was extremely patient with me as I remained determined to make every deadline on the road to publication. My husband, Arthur Cohen, is also a teacher of biology. The many discussions we have about the minutest detail to the gravest concept are invaluable to me.

As stated previously, the content of the ninth edition of *Biology* is not due to my efforts alone. I want to thank the many specialists who were willing to share their knowledge to improve *Biology*. Also, this edition was enriched by Pat Pendarvis, who went through every chapter improving the presentation, making relevant additions, and helping to seek and/or select photographs to enhance the text. I am extremely grateful to Pat for his dedicated efforts. The ninth edition of *Biology* would not have the same excellent quality without his suggested changes and those of the many reviewers who are listed here.

David Adegboye
  Southern University at New Orleans
Patricia Adumanu Ahanotu
  Georgia Perimeter College
Nancy C. Aiello
  Northern Virginia Community College
Felix Akojie
  West Kentucky Community and Technical College
Nurul Alam
  Jarvis Christian College
Mark Albrecht
  University of Nebraska at Kearney
Jorge E. Arriagada
  St. Cloud State University
Judy Awong-Taylor
  Armstrong Atlantic State University
Jessica Baack
  Montgomery College
Dave S. Bachoon
  Georgia College & State University
Ellen Baker
  Santa Monica College
LaQuetta Ballard-Anderson
  Grambling State

Richard R. Barkosky
  Minot State University
Sarah F. Barlow
  Middle Tennessee State University
Michael C. Bell
  Richland College
Karen S. Benoit
  Kankakee Community College
Bradley J. Bergstrom
  Valdosta State University
Kimberly A. Bjorgo
  West Virginia University
Lois Brewer Borek
  Georgia State University
Robert Boyd
  Auburn University
Ruby L. Broadway
  Dillard University
Bob Broyles
  Butler County Community College
Evelyn K. Bruce
  University of North Alabama
Arthur L. Buikema
  Virginia Tech
Sharon K. Bullock
  Virginia Commonwealth University
Matthew Rex Burnham
  Jones County Junior College

Carol T. Burton
  Bellevue Community College
Stefan Cairns
  Central Missouri State University
Jane Caldwell
  University of West Virginia
Pam Cole
  Shelton State Community College
Alexander Collier
  Armstrong Atlantic State University
Donald L. Collins
  Orange Coast College
Ruth Conley
  Shepherd College
Jerry L. Cook
  Sam Houston State University
William Wade Cooper
  Shelton State Community College
David T. Corey
  Midlands Technical College
David Cox
  Lincoln Land Community College
Kathryn Stephenson Craven
  Armstrong Atlantic State University
Jean DeSaix
  University of North Carolina–Chapel Hill

Heather Dickinson-Anson
  University of California, Irvine
Kevin E. Dixon
  East Central College
Sondra Dubowsky
  Allen County Community College
Terese Dudek
  Kishwaukee College
P. K. Duggal
  Maplewoods Community College
James F. Duke
  Calhoun Community College
Howard B. Duncan
  Norfolk State University
Carmen Eilertson
  Georgia State University
Peter Ekechukwu
  Horry-Georgetown Technical College
Clarence A. Elkins
  Baton Rouge Community College
Lee F. Famino
  Cuyahoga Community College
Gerald Farr
  Texas State University–San Marcos
Jill E. Feinstein
  Kishwaukee College

Lynn Firestone
Brigham Young University
Teresa G. Fischer
Indian River Community College
Jeff D. Foster
Southern State Community College
Christine A. Fredrich
Milwaukee Area Technical College
Mitchell A. Freymiller
University of Wisconsin–Eau Claire
Anne Galbraith
University of Wisconsin–La Crosse
Raul Galvan
South Texas College
Ric A. Garcia
Clemson University
John R. Geiser
Western Michigan University
N. Ghosh
West Texas A&M University
Shashuna J. Gray
Alabama State University
Melvin H. Green
University of California, San Diego
Paige Guilliams
Piedmont Virginia Community College
Bonnie S. Gunn
South Texas College
Paul H. Gurn
Bristol Community College
Karen Guzman
Campbell University
Fred E. Halstead
Wallace State Community College
Mijitaba Hamissou
Jacksonville State University
Laszlo Hanzely
Northern Illinois University
Bob Harms
Saint Louis Community College at Meramec
Joseph W. Hayes
Southern Union State Community College
Christopher Haynes
Shelton State Community College—Tuscaloosa
Kate He
Murray State University
Tom Heebner
Milwaukee Area Technical College
Wiley J. Henderson
Alabama A&M University

Sally G. Hornor
Anne Arundel Community College
Adam W. Hrincevich
Louisiana State University
Robert D. Hunter, Jr.
Trident Technical College
Allison B. Jablonski
Lynchburg College
Dawn Janich
Community College of Philadelphia
Scott T. Johnson
Brookdale Community College
Kenneth H. Jones
Dyersburg State Community College
Beverly Joseph
Bishop State Community College
Walter S. Judd
University of Florida
Ragupathy Kannan
University of Arkansas–Fort Smith
Joanne M. Kilkpatrick
Auburn University at Montgomery
Michael Koban
Morgan State University
Pramod Kumar
The University of Texas at San Antonio
Joseph L. Kyle
Kennedy-King College
David E. Lemke
Texas State University–San Marcos
Gina Crowder Levesque
Northeastern State University
Roger Lightner
University of Arkansas–Fort Smith
Tammy J. Liles
Lexington Community College
Michael D. Marlen
Southwestern Illinois College
Craig E. Martin
University of Kansas
Mark E. McCallum
Pfeiffer University
Joseph E. McCauley
Shorter College
Gabrielle Lynn McLemore
Morgan State University
Alice Mills
University of Tennessee at Martin
Thomas H. Milton
Richard Bland College

Lance G. Morris
Arkansas Northeastern College
David H. Niebuhr
College of William and Mary
Juliet K. F. Noor
Louisiana State University
John C. Osterman
University of Nebraska–Lincoln
Mary O'Sullivan
Elgin Community College
Vanessa Passler
Wallace Community College
David L. Pindel
Corning Community College
Fiona J. Qualls
Jones County Junior College
Shannon F. Quick
Wallace State Community College
Talitha T. Rajah
Indiana University Southeast
Darrell L. Ray
The University of Tennessee at Martin
Jill D. Reid
Virginia Commonwealth University
Anthony A. Rothering
Lincoln Land Community College
Clay Runck
Benedictine University
Lisa L. Rutledge
Columbia State Community College
Linda A. Salicce
Community College of Allegheny
Judith A. Schneidewent
Milwaukee Area Technical College
Betty B. Schroeder
Southeastern Louisiana University
Brian W. Schwartz
Columbus State University
Roger Seeber, Jr.
West Liberty State College
Beatrice Sirakaya
Pennsylvania State University
A. Denny Smith
Clemson University
Marc A. Smith
Sinclair Community College
Linda Smith-Staton
Pellissippi State Technical Community College
Dianne C. Snyder
Augusta State University

Robert R. Speed
Wallace Community College
Judith Stewart
Community College of Southern Nevada
Walter J. Strength
Mountain Empire Community College
Gregory W. Stunz
Texas A&M University
Eileen M. Synnott
Bristol Community College
Sharon Thoma
University of Wisconsin–Madison
Kip R. Thompson
Ozarks Technical Community College
Ronald Toth
Northern Illinois University
Dorothy Kraeuter Trevelyan
Anne Arundel Community College
Jagan Valluri
Marshall University
Staria Vanderpool
Arkansas State University
Neal J. Voelz
St. Cloud State University
Emily R. Watkinson
Virginia Commonwealth University
Catherine Weinstein
Nassau Community College
Annette N. Wells
Durham Technical Community College
James Wetzel
Presbyterian College
Allison Widmaier
University of Missouri–Columbia
Robert R. Wise
University of Wisconsin, Oshkosh
Cosima B. Wiese
College Misericordia
J.D. Wilhide
Arkansas State University
Michael Windelspecht
Appalachian State University
David B. Wing
Shepherd College
Michelle D. Withers
Louisiana State University
Tony Yates
Seminole State College
Brenda Zink
Northeastern Junior College

I am also grateful to the following for their contributions to this edition of *Biology*:

Tammy Atchison
Pitt Community College
David Huffman
Texas State University–San Marcos
Kimberly Lyle-Ippolito
Anderson University

David Pindel
Corning Community College
William Rogers
Ball State University
Betty Schroeder
Southeastern Louisiana University

Stephanie Songer
North Georgia College & State University
Kent Thomas
Wichita State University

Wendy Vermillion
Columbus State Community College
Michelle Zurawski
Moraine Valley Community College

Reviewers of the previous edition:

John Alcock
Arizona State University
Lawrence A. Alice
Western Kentucky University
Saad Al-Jassabi
Yarmouk University
Irbid, Jordan
Karl Aufderheide
Texas A&M University
Robert Beason
University of Louisiana
at Monroe
Gerald Bergtrom
University of Wisconsin–
Milwaukee
James Enderby Bidlack
University of Central Oklahoma
George B. Biggs
New Mexico Junior College
Benjie Blair
Jacksonville State University
David Boehmer
Volunteer State Community
College
Robert S. Boyd
Auburn University
Randall M. Brand
Southern Union State
Community College
Chantae Calhoun
Lawson State Community
College
Richard W. Cheney, Jr.
Christopher Newport University
Andrew N. Clancy
Georgia State University
George R. Cline
Jacksonville State University
Donald Collins
Orange Coast College
Jerry L. Cook
Sam Houston University
David T. Corey
Midlands Technical College
Don C. Dailey
Austin Peay State University
W. Marshall Darley
University of Georgia
Kristie Deramus
Odessa College
Jean DeSaix
University of North Carolina
at Chapel Hill
Jean Dickey
Clemson University
Jessica Boyce Doiron
Coastal Carolina Community
College
David Eldridge
Baylor University
Harold W. Elmore
Marshall University
Thomas C. Emmel
University of Florida

Laurie A. Folgate
Kishwaukee College
Katherine A. Foreman
Moraine Valley Community
College
Lawton Fox
Spokane Falls Community
College
Stephen Gallik
Mary Washington University
Doug Gayou
University of Missouri–
Columbia
Nandini Ghosh-Choudhury
University of Texas
at San Antonio
Marcia Gillette
Indiana University Kokomo
Andrew Goliszek
North Carolina A&T State
University
David J. Grisé
Southwest Texas State University
Peggy Guthrie
University of Central Oklahoma
Fred E. Halstead
Wallace State Community
College
Blanche C. Haning
North Carolina State University
Rosemary K. Harkins
Langston University
John P. Harley
Eastern Kentucky University
Victoria S. Hennessy
Sinclair Community College
Eva Ann Horne
Kansas State University
John C. Inman
Presbyterian College
Beverly Joseph
Bishop State Community College
Geeta S. Joshi
Southeastern Community College
Mark E. Knauss
Shorter College
William Kroll
Loyola University of Chicago
Harry D. Kurtz, Jr.
Clemson University
Mary V. Lipscomb
Virginia Polytechnic Institute
and State University
Douglas Lyng
Indiana University–
Purdue University
Bill Mathena
Kaskaskia College
John Mathwig
College of Lake County
Jerry W. McClure
Miami University
Greg McCormac
American River College

Bonnie McCormick
University of the Incarnate Word
Scott Murdoch
Moraine Valley Community
College
Joseph Murray
Blue Ridge Community College
Hao Nguyen
University of Tennessee–Martin
William M. Olivero
Cumberland County College
Vanessa Passler
Wallace Community College–
Dothan
Robert P. Patterson
North Carolina State University
William Joseph Pegg
Frostburg State University
Matthew K. Pelletier
Houghton College
Scott Porteous
Fresno City College
Charles Pumpuni
Northern Virginia Community
College
John Raasch
University of Wisconsin–
Madison
Katherine Rasmussen
University of South Dakota
Darrell Ray
University of Tennessee
at Martin
James Rayburn
Jacksonville State University
Mary Ann Reihman
California State University,
Sacramento
Bill Rogers
Ball State University
Donald J. Roufa
University of Arkansas
Kathleen W. Roush
University of North Alabama
Lisa Rutledge
Columbia State Community
College
Connie E. Rye
Bevill State Community
College
Judith A. Schneidewent
Milwaukee Area Technical
College
Pat Selelyo
College of Southern Idaho
Doris Shoemaker
Dalton State College
Thomas E. Snowden
Florida Memorial College
Eric P. Spaziani
Pasco-Hernando Community
College
Robert R. Speed
Wallace Community College

Amy C. Sprinkle
Jefferson Community College
Southwest
Bruce Stallsmith
University of Alabama
in Huntsville
Frederick E. Stemple, Jr.
Tidewater Community College
John Sternfeld
SUNY Cortland
John D. Story
NorthWest Arkansas
Community College
David L. Swanson
University of South Dakota
Beth Thornton
Abraham Baldwin Agricultural
College
Gene R. Trapp
California State University,
Sacramento
Renn Tumlison
Henderson State University
David Turnbull
Lake Land College
Paul Twigg
University of Nebraska
Jagan V. Valluri
Marshall University
Thomas Vance
Navarro College
Brenda Boyd Vaughn
Calhoun Community
College
Otelia S. Vines
Virginia State University
Tracy L. Wacker
University of Michigan–
Flint
O. Eugene Walton
Tallahassee Community
College
Susan Weinstein
Marshall University
Jennifer L. Wells
SUNY Cortland
George Williams, Jr.
Southern University
Mike Woller
University of Wisconsin–
Whitewater
Tony Yates
Seminole State College
Robert W. Yost
Indiana University-Purdue
University Indianapolis
Henry H. Ziller
Southeastern Louisiana
University
Michael Zimmerman
University of Wisconsin–
Oshkosh

TEACHING SUPPLEMENTS FOR THE INSTRUCTOR

McGraw-Hill offers a variety of tools and technology products to support the ninth edition of *Biology*. Instructors can obtain teaching aids by calling the Customer Service Department at (800) 338-3987 or by contacting their local McGraw-Hill sales representative.

Biology Laboratory Manual

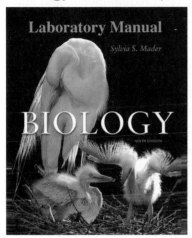

The *Biology Laboratory Manual*, Ninth Edition, is written by Dr. Sylvia Mader. With few exceptions, each chapter in the text has an accompanying laboratory exercise in the manual. Every laboratory has been written to help students learn the fundamental concepts of biology and the specific content of the chapter to which the lab relates, as well as gain a better understanding of the scientific method.

ISBN-13: 978-0-07-298955-7 (ISBN-10: 0-07-298955-6)

Digital Content Manager

This collection of multimedia resources provides tools for rich visual support of your lectures. You can utilize artwork from the text in multiple formats to create customized classroom presentations, visually based tests and quizzes, dynamic course website content, or attractive printed support materials. The following digital assets are available either on a cross-platform CD-ROM or on a DVD and are grouped by chapters:

Art Libraries. Full-color digital files of all illustrations in the book, plus the same art saved in unlabeled and grayscale version, can be readily incorporated into lecture presentations, exams, or custom-made classroom materials.

TextEdit Art Library. Every illustration is available in a PowerPoint®-compatible art file that allows the user to revise, move, or delete labels and leader lines as desired for creation of customized presentations and/or for testing purposes.

Active Art Library. Illustrations depicting key processes have been converted to a format that allows the artwork to be edited inside of PowerPoint.® Each piece can be broken down to its core elements, grouped or ungrouped, and edited to create customized illustrations.

Animations Library. The next generation of biology animations is now available! These new animations bring key processes to life and offer total flexibility. Designed to be used in lectures, you can pause, rewind, fast-forward, and turn the audio on or off to create dynamic lecture presentations. Many of the animations are also available with Spanish narration and audio.

Tables Library. Every table that appears in the text is provided in electronic format.

Photos Library. All photos from the text are available in digital format.

Additional Photos Library. Over 700 photos not found in *Biology* are available for use in creating lecture presentations.

PowerPoint® Lecture Outlines. A ready-made presentation that combines lecture notes and illustrations is written for each chapter. They can be used as they are, or the instructor can customize them to preferred lecture topics and organization.

PowerPoint® Art Slides. Art, photos, and tables from each chapter have been pre-inserted into blank PowerPoint® slides to which you can add your own notes.

CD-ROM ISBN-13: 978-0-07-295427-2
(ISBN-10: 0-07-295427-2)

DVD ISBN-13: 978-0-07-326197-3
(ISBN-10: 0-07-326197-1)

Instructor's Testing and Resource CD-ROM

This cross-platform CD-ROM provides these resources for the instructor:

Instructor's Manual contains learning objectives, extended lecture outlines, lecture enrichment and student activities suggestions, and critical thinking questions. In addition, there is an explanation of text changes and reorganization as well as information on new and revised illustrations and tables.

Test Bank offers questions that can be used for homework assignments or the preparation of exams.

Computerized Test Bank utilizes testing software to quickly create customized exams. This user-friendly program allows instructors to sort questions by format or level of difficulty; edit existing questions or add new ones; and scramble questions and answer keys for multiple versions of the same test.
ISBN-13: 978-0-07-296753-1 (ISBN-10: 0-07-296753-6)

eInstruction Classroom Performance System (CPS)

Wireless technology brings interactivity into the classroom or lecture hall. Instructors and students receive immediate feedback through wireless response pads that are easy to use and engage students. eInstruction can be used by instructors to:

- Take attendance
- Administer quizzes and tests
- Create a lecture with intermittent questions
- Manage lectures and student comprehension through use of the CPS grade book
- Integrate interactivity into their PowerPoint® presentations

Transparencies

This set of overhead transparencies includes every piece of line art in the textbook plus every table. The images are printed with better visibility and contrast than ever before, and labels are large and bold for clear projection.
ISBN-13: 978-0-07-296752-4 (ISBN-10: 0-07-296752-8)

ARIS

McGraw-Hill's ARIS—Assessment, Review, and Instruction System—for *Biology* Ninth Edition is a complete online tutorial, electronic homework, and course management system designed for greater ease of use than any other system available. Free with adoption of McGraw-Hill's *Biology* Ninth Edition text, instructors can create and share course materials and assignments with colleagues with a few clicks of the mouse. All PowerPoint® lectures, assignments, quizzes, tutorials, and interactives are directly tied to text-specific materials in *Biology*, but instructors can also edit questions, import their own content, and create announcements and due dates for assignments. ARIS has automatic grading and reporting of easy-to-assign homework, quizzing, and testing. All student activity within McGraw-Hill's ARIS is automatically recorded and available to the instructor through a fully integrated grade book that can be downloaded to Excel.

The *Biology* Ninth Edition ARIS site at www.mhhe.com/maderbiology9 offers access to a vast array of premium online content to fortify the learning and teaching experience for students and instructors.

Instructor Edition. In addition to all of the resources for students, the Instructor Edition of ARIS has these assets:

- *eInstruction Classroom Performance System (CPS) Question Bank* A set of questions for use with the CPS is provided for every textbook chapter to assist instructors in quickly assessing student comprehension of the concepts.

- *Animations* The next generation of biology animations is available with the ninth edition of *Biology*. Key biological processes in full color have been brought to life via animation. These animations offer flexibility for instructors. Designed to be used in lectures, you can pause, rewind, fast-forward, and turn the audio on or off. Many of the animations are also available with Spanish narration and audio.

- *Laboratory Resource Guide* A preparation guide that provides set-up instructions, sources for materials and supplies, time estimates, special requirements, and suggested answers to all questions in the *Biology Laboratory Manual*, Ninth Edition.

- *PageOut* McGraw-Hill's exclusive tool for creating your own website for your general biology course. It requires no knowledge of coding and is hosted by McGraw-Hill.

- *Active Art Demo* Teaches you how to use the Active Art that is on the Digital Content Manager CD-ROM.

- *Case Studies* Offers suggestions on how to use Case Studies in your classroom.

McGraw-Hill: Biology Digitized Video Clips

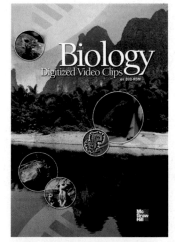

McGraw-Hill is pleased to offer adopting instructors a new presentation tool—digitized biology video clips on DVD! Licensed from some of the highest-quality science video producers in the world, these brief segments range from about five seconds to just under three minutes in length and cover all areas of general biology from cells to ecosystems. Engaging and informative, McGraw-Hill's Biology Digitized Video Clips will help capture students' interest while illustrating key biological concepts and processes such as mitosis, how cilia and flagella work, and how some plants have evolved into carnivores.
ISBN-13: 978-0-07-312155-0 (ISBN-10: 0-07-312155-X)

Mader Micrograph Slides

This set contains one hundred 35mm slides of many of the photomicrographs and electron micrographs in the text.
ISBN-13: 978-0-07-239977-6 (ISBN-10: 0-07-239977-5)

LEARNING SUPPLEMENTS FOR THE STUDENT

Student Study Guide

Dr. Sylvia Mader has written the *Student Study Guide* that accompanies *Biology*, thereby ensuring close coordination with the text. Each text chapter has a corresponding study guide chapter that includes a chapter review, learning objectives and study questions for each section of the chapter, and a chapter test. Answers to all questions are provided to give students immediate feedback. Students who make use of the *Student Study Guide* should find that performance increases dramatically.
ISBN-13: 978-0-07-297671-7 (ISBN-10: 0-07-297671-3)

ARIS

McGraw-Hill's ARIS—Assessment, Review, and Instruction System—for *Biology*, Ninth Edition at www.mhhe.com/maderbiology9 offers access to a vast array of premium online content to fortify the learning experience.

Student Edition. The Student Edition of ARIS features a wide variety of tools to help students learn biological concepts and to reinforce their knowledge:

- *Interactive Activities* These online study aids, organized by chapter, include **practice quizzes, animations, labeling exercises, flashcards,** and much more.
- *Online Tutoring* The tutorial service is moderated by qualified instructors. Help with difficult concepts is only an email away!

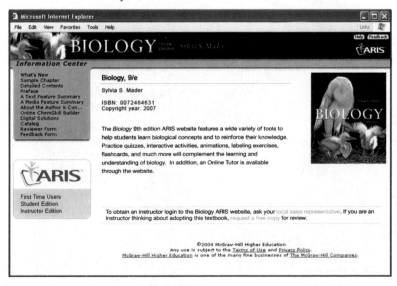

- *Essential Study Partner* This collection of interactive study modules contains hundreds of animations, learning activities, and quizzes designed to help students grasp complex concepts.

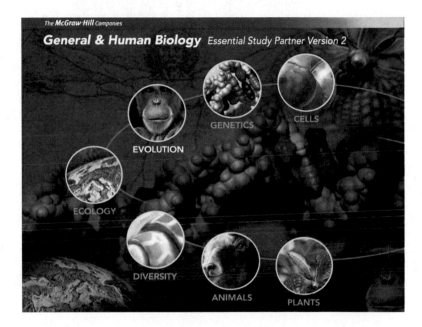

- *Animations* Full-color presentations of key biological processes have been brought to life via animation. You can pause, rewind, fast-forward, and turn the audio on or off. Many of the animations are also available with Spanish narration and audio.
- *Animation Quizzes* Quizzes based on the new animations will help you assess your understanding of the concepts.

Student Interactive CD-ROM

This interactive CD-ROM is an indispensable resource for studying topics covered in the text. It includes chapter outlines, chapter-based quizzes, animations of complex processes, flashcards, PowerPoint® lecture outlines, and PowerPoint® slides of all art and photos found in the textbook. All of the material is organized chapter-by-chapter. Direct links to the text's ARIS website and to the Essential Study Partner are also provided.
ISBN-13: 978-0-07-326525-4 (ISBN-10: 0-07-326525-X)

BIOLOGY, Ninth Edition
Guided Tour

A brilliant new visuals program brings Biology *to life!*

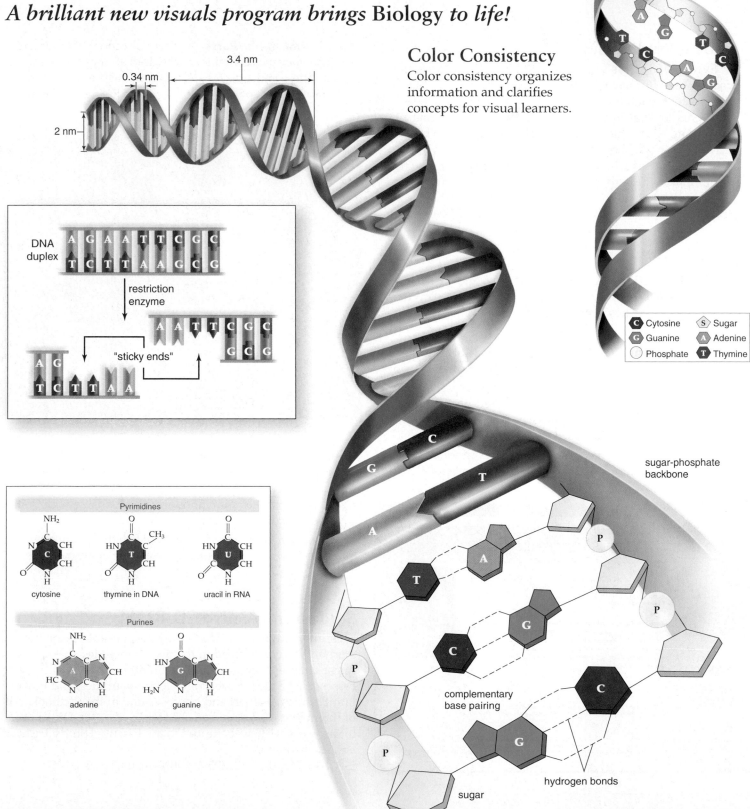

Color Consistency

Color consistency organizes information and clarifies concepts for visual learners.

3.4 nm

0.34 nm

2 nm

DNA duplex

A G A A T T C G C
T C T T A A G C G

restriction enzyme

A A T T C G C C
G C G

A G
T C T T A A

"sticky ends"

C Cytosine	**S** Sugar
G Guanine	**A** Adenine
Phosphate	**T** Thymine

Pyrimidines

cytosine

thymine in DNA

uracil in RNA

Purines

adenine

guanine

sugar-phosphate backbone

complementary base pairing

hydrogen bonds

sugar

nuclear envelope

nucleolus

nuclear pore

chromatin

nucleoplasm

Nuclear envelope:
inner membrane
outer membrane
nuclear pore

phospholipid

Multi-Level Perspective

Illustrations depicting complex structures connect macroscopic and microscopic views to help students connect the two levels.

Small intestine

Section of intestinal wall

lumen

villus

lacteal

blood capillaries

goblet cell

lymph nodule

venule

lymphatic vessel

arteriole

villus

microvilli

100 μm

Villi

Combination Art

Drawings of structures are often paired with micrographs to enhance visualization.

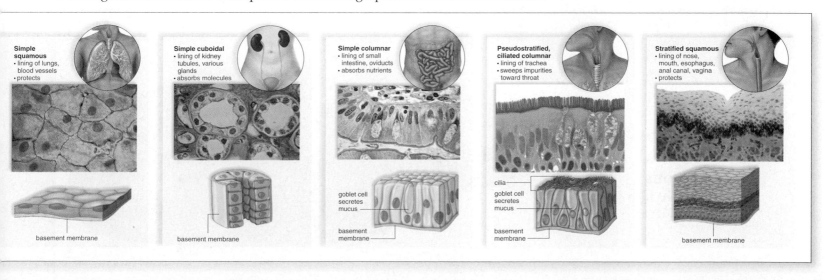

Simple squamous
- lining of lungs, blood vessels
- protects

basement membrane

Simple cuboidal
- lining of kidney tubules, various glands
- absorbs molecules

basement membrane

Simple columnar
- lining of small intestine, oviducts
- absorbs nutrients

goblet cell secretes mucus

basement membrane

Pseudostratified, ciliated columnar
- lining of trachea
- sweeps impurities toward throat

cilia

goblet cell secretes mucus

basement membrane

Stratified squamous
- lining of nose, mouth, esophagus, anal canal, vagina
- protects

basement membrane

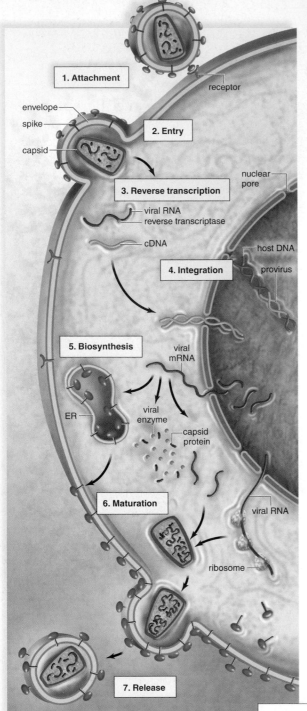

1. Attachment

receptor

envelope
spike

2. Entry

capsid

nuclear pore

3. Reverse transcription

viral RNA
reverse transcriptase

cDNA

host DNA

4. Integration

provirus

5. Biosynthesis

viral mRNA

ER

viral enzyme

capsid protein

6. Maturation

viral RNA

ribosome

7. Release

Process Figures

These figures break down processes into a series of smaller steps and organize them in an easy-to-follow format.

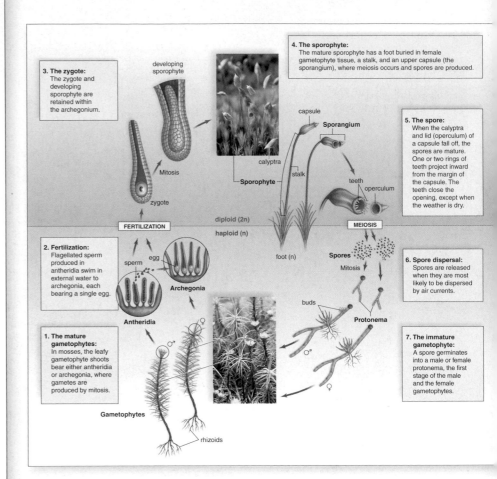

3. The zygote: The zygote and developing sporophyte are retained within the archegonium.

developing sporophyte

4. The sporophyte: The mature sporophyte has a foot buried in female gametophyte tissue, a stalk, and an upper capsule (the sporangium), where meiosis occurs and spores are produced.

Mitosis

capsule

Sporangium

calyptra

Sporophyte

stalk

5. The spore: When the calyptra and lid (operculum) of a capsule fall off, the spores are mature. One or two rings of teeth project inward from the margin of the capsule. The teeth close the opening, except when the weather is dry.

teeth

operculum

zygote

diploid (2n)

FERTILIZATION

haploid (n)

MEIOSIS

foot (n)

2. Fertilization: Flagellated sperm produced in antheridia swim in external water to archegonia, each bearing a single egg.

sperm

egg

Archegonia

Spores

Mitosis

6. Spore dispersal: Spores are released when they are most likely to be dispersed by air currents.

buds

Protonema

Antheridia

1. The mature gametophytes: In mosses, the leafy gametophyte shoots bear either antheridia or archegonia, where gametes are produced by mitosis.

7. The immature gametophyte: A spore germinates into a male or female protonema, the first stage of the male and the female gametophytes.

Gametophytes

rhizoids

Icons

Icons help orient the student.

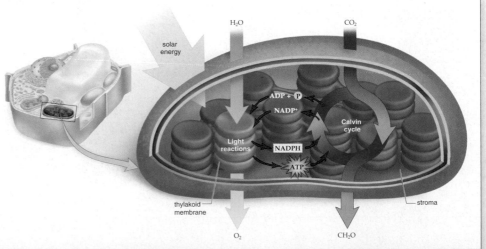

H₂O

CO₂

solar energy

ADP + P

NADP⁺

Calvin cycle

Light reactions

NADPH

ATP

thylakoid membrane

stroma

O₂

CH₂O

The Learning System

Proven Pedagogical Features That Will Facilitate Your Understanding of Biology

Chapter Concepts

The chapter begins with an integrated outline that numbers the major topics of the chapter and lists the concepts for each topic.

PHOTOSYNTHESIS

The fate of life on Earth literally hinges on a star 93 million miles away because this star provides photosynthesizers with solar energy. Only 42% of the solar energy directed towards Earth reaches the planet; the remainder is absorbed by or reflected into the atmosphere and becomes heat. Of this, only 1–2% is captured by photosynthesizers and, only a portion of this is incorporated into plant materials.

Yet all living things are dependent on the amount of solar energy that photosynthesizers transform into chemical energy. Exceptions do exist. In rare hydrothermal environments, some prokaryotes acquire energy by oxidizing inorganic molecules and are the producers of food for others. In the majority of ecosystems, photosynthesizers are the producers that take in inorganic molecules and produce food by using solar energy.

When photosynthesis occurs, carbon dioxide is absorbed and oxygen is released. Oxygen is required by organisms when they carry on cellular respiration. The collective action of algae and plants is responsible for placing copious amounts of oxygen into the atmosphere. It rises high into the atmosphere and forms an ozone layer that makes terrestrial life possible. Accordingly, this layer that protects us against damaging ultraviolet rays of the sun is called the ozone shield.

The products of photosynthesis are critical to humankind in a number of ways. They provide our food, to be sure, but they also are a source of building materials, fabrics, paper, fuel, and pharmaceuticals. Even plants that existed hundreds of millions of years ago are important as a source of fossil fuels. And while we are thanking green plants for their many services, let's not forget the simple beauty of a magnolia blossom or the majesty of the Earth's forests.

Photosynthesizers use solar energy to produce organic nutrients for themselves and all other organisms.

CONCEPTS

7.1 PHOTOSYNTHETIC ORGANISMS
- Plants, algae, and cyanobacteria are photosynthetic organisms that produce most of the carbohydrate used as an energy source by the living world. 116
- In flowering plants, photosynthesis takes place within membrane-bounded chloroplasts, organelles that contain membranous thylakoids surrounded by a fluid called stroma. 116–117

7.2 PLANTS AS SOLAR ENERGY CONVERTERS
- Plants use solar energy in the visible light range when they carry on photosynthesis. 118
- Photosynthesis has two sets of reactions: Solar energy is captured by the pigments in thylakoids, and carbon dioxide is reduced by enzymes in the stroma. 119

7.3 LIGHT REACTIONS
- Solar energy energizes electrons and permits a buildup of ATP and NADPH molecules. 120–21

7.4 CALVIN CYCLE REACTIONS
- Carbon dioxide reduction requires ATP and NADPH from the light reactions. 124–25

7.5 OTHER TYPES OF PHOTOSYNTHESIS
- Plants use C_3 or C_4 or CAM photosynthesis, which are distinguishable by the manner in which CO_2 is fixed. 126–27

Phases of Cellular Respiration

The oxidation of glucose by removal of hydrogen atoms involves four phases (Fig. 8.2). Glycolysis takes place outside the mitochondria and does not require the presence of oxygen. Therefore, glycolysis is **anaerobic.** The other phases of cellular respiration take place inside the mitochondria, where oxygen is the final acceptor of electrons.

- During **glycolysis** [Gk *glycos*, sugar, and *lysis*, splitting], glucose is broken down in the cytoplasm to two molecules of pyruvate. Oxidation by removal of hydrogen atoms results in NADH and provides enough energy for the net yield of two molecules of ATP.
- During the **preparatory (prep) reaction,** pyruvate enters a mitochondrion and is oxidized to a 2-carbon acetyl group carried by CoA; NADH is formed; and the waste product CO_2 is removed. Since glycolysis ends with two molecules of pyruvate, the prep reaction occurs twice per glucose molecule.
- The **citric acid cycle** is a cyclical series of oxidation reactions in the matrix of a mitochondrion that result in NADH and $FADH_2$. CO_2 is given off and one ATP is produced. The citric acid cycle turns twice because two acetyl CoA molecules enter the cycle per glucose molecule. Altogether, the citric acid cycle accounts for two immediate ATP molecules per glucose molecule.

- The **electron transport chain** is a series of carriers in the inner mitochondrial membrane that accept the electrons removed from glucose and pass them along from one carrier to the next until they are finally received by O_2, which then combines with hydrogen ions and becomes water. As the electrons pass from a higher-energy to a lower-energy state, energy is released and later used for ATP synthesis by chemiosmosis. The electrons from one glucose result in 32 or 34 ATP, depending on certain conditions.

  Pyruvate is a pivotal metabolite in cellular respiration. If oxygen is not available to the cell, fermentation occurs in the cytoplasm (see Fig. 8.10). During **fermentation,** glucose is incompletely metabolized to lactate or to carbon dioxide and alcohol, depending on the organism. As we shall see on page 142, fermentation results in a net gain of only two ATP per glucose molecule.

  Cellular respiration involves the oxidation of glucose to carbon dioxide and water. As glucose breaks down, energy is made available for ATP synthesis. A total of 36 or 38 ATP molecules are produced per glucose molecule in cellular respiration (2 from glycolysis, 2 from the citric acid cycle, and 32 or 34 from the electron transport chain).

Internal Summary Statements

A summary statement appears at the end of each major section of the chapter to help students focus on the key concepts.

FIGURE 8.2 The four phases of complete glucose breakdown.
The complete breakdown of glucose consists of four phases. Glycolysis in the cytoplasm produces pyruvate, which enters mitochondria if oxygen is available. The preparatory reaction and the citric acid cycle that follow occur inside the mitochondria. Also, inside mitochondria, the electron transport chain receives the electrons that were removed from glucose breakdown products. The result of glucose breakdown is 36 or 38 ATP, depending on the particular cell.

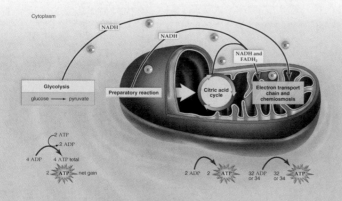

Readings

Biology offers three types of boxed readings:

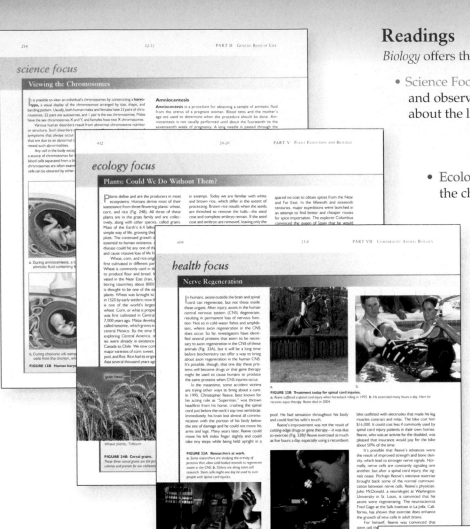

- Science Focus readings describe how experimentation and observations have contributed to our knowledge about the living world.

- Ecology Focus readings show how the concepts of the chapter can be applied to ecological concerns.

- Health Focus readings review procedures and technology that can contribute to our well-being.

Connecting the Concepts

These appear at the close of the text portion of the chapter, and they stimulate critical thinking by showing how the concepts of the chapter are related to other concepts in the text.

Chapter Summary

The summary is organized according to the major sections in the chapter and helps students review the important topics and concepts.

End-of-Chapter Study Tools

Reviewing the Chapter
These page-referenced study questions follow the sequence of the chapter.

Testing Yourself
These objective questions allow you to test your ability to answer recall-based questions. Answers to *Testing Yourself* questions are given in Appendix A.

Thinking Scientifically
Critical thinking questions give you an opportunity to reason as a scientist. Detailed answers to these questions are found on ARIS, the *Biology*, Ninth Edition website.

Bioethical Issue
A *Bioethical Issue* is found at the end of most chapters. These short readings discuss a variety of controversial topics that confront our society. Each reading ends with appropriate questions to help you fully consider the issue and arrive at an opinion.

Understanding the Terms
The boldface terms in the chapter are page referenced, and a matching exercise allows you to test your knowledge of the terms.

Website Reminder
Located at the end of the chapter is this reminder that additional study questions and other learning activities are on ARIS, the *Biology*, Ninth Edition website.

Reviewing the Chapter

1. What is the overall chemical equation for the complete breakdown of glucose to CO_2 and H_2O? Explain how this is an oxidation-reduction reaction. Why is the reaction able to drive ATP synthesis? 132
2. What are NAD^+ and FAD? What are their functions? 132
3. What are the three pathways involved in the complete breakdown of glucose to carbon dioxide (CO_2) and water (H_2O)? What reaction is needed to join two of these pathways? 132–33
4. What are the main events of glycolysis? How is ATP formed? 134–35
5. Give the substrates and products of the prep reaction. Where does it take place? 136–37
6. What are the main events of the citric acid cycle? 137
7. What is the electron transport chain, and what are its functions? 138
8. Describe the organization of protein complexes within the cristae. Explain how the complexes are involved in ATP production. 138–39
9. Calculate the energy yield of glycolysis and complete glucose breakdown. Distinguish yields between substrate-level phosphorylation and oxidative phosphorylation. 140
10. What is fermentation, and how does it differ from glycolysis? Mention the benefit of pyruvate reduction during fermentation. What types of organisms carry out lactic acid fermentation, and what types carry out alcoholic fermentation? 141–43
11. Give examples to support the concept of the metabolic pool. 144

Testing Yourself

Choose the best answer for each question. For questions 1–8, identify the pathway involved by matching each description to the terms in the key.

KEY:
 a. glycolysis
 b. citric acid cycle
 c. electron transport chain
1. carbon dioxide (CO_2) given off
2. water (H_2O) formed
3. G3P
4. NADH becomes NAD^+
5. oxidative phosphorylation
6. cytochrome carriers
7. pyruvate
8. FAD becomes $FADH_2$
9. The prep reaction
 a. connects glycolysis to the citric acid cycle.
 b. gives off CO_2.
 c. uses NAD^+.
 d. results in an acetyl group.
 e. All of these are correct.
10. The greatest contributor of electrons to the electron transport chain is
 a. oxygen.
 b. glycolysis.
 c. the citric acid cycle.
 d. the prep reaction.
 e. fermentation.
11. Substrate-level phosphorylation takes place in
 a. glycolysis and the citric acid cycle.
 b. the electron transport chain and the prep reaction.
 c. glycolysis and the electron transport chain.
 d. the citric acid cycle and the prep reaction.
 e. Both b and d are correct.
12. Which of these is not true of fermentation?
 a. net gain of only two ATP
 b. occurs in cytoplasm
 c. NADH donates electrons to electron transport chain
 d. begins with glucose
 e. carried on by yeast
13. Fatty acids are broken down to
 a. pyruvate molecules, which take electrons to the electron transport chain.
 b. acetyl groups, which enter the citric acid cycle.
 c. amino acids, which excrete ammonia.
 d. glycerol, which is found in fats.
 e. All of these are correct.
14. How many ATP molecules are usually produced per NADH?
 a. 1   c. 36
 b. 3   d. 10
15. How many NADH molecules are produced during the complete breakdown of one molecule of glucose?
 a. 5   c. 10
 b. 30   d. 6
16. What is the name of the process that adds the third phosphate to an ADP molecule using the flow of hydrogen ions?
 a. substrate-level phosphorylation
 b. fermentation
 c. reduction
 d. chemiosmosis
17. Which are possible products of fermentation?
 a. lactic acid
 b. alcohol
 c. CO_2
 d. All of these are possible.
18. The metabolic process that produces the most ATP molecules is
 a. glycolysis.   c. electron transport chain.
 b. citric acid cycle.   d. fermentation.
19. Which of these is not true of citric acid cycle? The citric acid cycle
 a. includes the prep reaction.
 b. produces ATP by substrate-level phosphorylation.
 c. occurs in the mitochondria.
 d. is a metabolic pathway, as is glycolysis.
20. Which of these is not true of the electron transport chain? The electron transport chain
 a. is located on the cristae.
 b. produces more NADH than any metabolic pathway.
 c. contains cytochrome molecules.
 d. ends when oxygen accepts electrons.
21. Which of these is not true of the prep reaction? The prep reaction
 a. begins with pyruvate and ends with acetyl CoA.
 b. produces more NADH than does glycolysis.
 c. occurs in the mitochondria.
 d. occurs after glycolysis and before the citric acid cycle.

Thinking Scientifically

1. A certain flower generates heat. This heat attracts pollinating insects to the flower. While the evolutionary benefit of attracting insects is obvious, the metabolic cost of this particular adaptation is high. What metabolic mechanism(s) might a plant use to generate heat, and under what circumstances would the metabolic cost be high?
2. The free energy of carbon dioxide and water is considerably less than the free energy of sucrose (table sugar). However, the conversion of sucrose to carbon dioxide and water is never spontaneous under normal conditions. How would you explain this observation?

Bioethical Issue: Greenhouse Effect and Emerging Diseases

Today, we are very much concerned about emerging diseases caused by pathogens. Examples of emerging diseases are AIDS and Ebola, which emerge from their natural hosts to cause illness in humans. In 1993, the hantavirus strain emerged from the common deer mouse and killed about 60 young people in the Southwest. In the case of hantavirus, we know that climate was involved. An unusually mild winter and wet spring caused piñon trees to bloom well and provide pine nuts to the mice. The increasing deer mouse population came into contact with humans, and the hantavirus leaped easily from mice to humans.

The prediction is that global warming, caused in large part by the burning of fossil fuels, will upset normal weather cycles and result in outbreaks of hantavirus as well as malaria, dengue and yellow fevers, filariasis, encephalitis, schistosomiasis, and cholera. Clearly, any connection between global warming and emerging diseases offers another reason that greenhouse gases should be curtailed when fossil fuels such as gasoline are consumed. Examples of greenhouse gases are carbon dioxide and methane, which allow the sun's rays to pass through but then trap the heat from escaping.

In December 1997, 159 countries met in Kyoto, Japan, to work out a protocol that would reduce greenhouse gases worldwide. This protocol, called the Kyoto Protocol, entered into force February 16, 2005. It is believed that the emission of greenhouse gases, especially from power plants, will cause Earth's temperature to rise 1.5°–4.5° by 2060. The U.S. Senate still does not want to ratify the agreement because it does not include a binding emissions commitment from the developing countries, which are only now becoming industrialized. While the United States presently emits a large proportion of the greenhouse gases, China is expected to surpass that amount in about 2020 to become the biggest source of greenhouse emissions.

Negotiations with the developing countries are still going on, and some creative ideas have been put forward. Why not have a trading program that allows companies to buy and sell emission credits across international boundaries? Accompanying that would be a market in greenhouse reduction techniques. If it became monetarily worth their while, companies in developed countries would have an incentive to reduce greenhouse emissions. If you were a CEO, would you be willing to reduce greenhouse emissions simply because they cause a deterioration of the environment and probably cause human illness? Why or why not? Instead, would you approve of giving companies monetary incentives to reduce greenhouse emissions? Why or why not?

Understanding the Terms

active site 106
ADP (adenosine diphosphate) 104
ATP (adenosine triphosphate) 104
ATP synthase complex 111
chemical energy 102
chemiosmosis 111
coenzyme 108
cofactor 108
competitive inhibition 109
coupled reactions 105
denatured 108
electron transport chain 104
endergonic reaction 104
energy 102
energy of activation 106
entropy 103
enzyme 106
enzyme inhibition 109
exergonic reaction 104
feedback inhibition 109

free energy 104
heat 102
induced fit model 106
kinetic energy 102
laws of thermodynamics 102
mechanical energy 102
metabolic pathway 106
metabolism 104
NAD^+ (nicotinamide adenine dinucleotide) 110
$NADP^+$ (nicotinamide adenine dinucleotide phosphate) 110
noncompetitive inhibition 109
oxidation 110
phosphorylation 109
potential energy 102
product 104
reactant 104
reduction 110
substrate 106
vitamin 109

Match the terms to these definitions:

a. _____ All of the chemical reactions that occur in a cell during growth and repair.
b. _____ Stored energy as a result of location or spatial arrangement.
c. _____ Essential requirement in the diet, needed in small amounts. They are often part of coenzymes.
d. _____ Measure of disorder or randomness.
e. _____ Nonprotein organic molecule that aids the action of the enzyme to which it is loosely bound.
f. _____ Loss of one or more electrons from an atom or molecule; in biological systems, generally the loss of hydrogen atoms.

ARIS, the *Biology* Website

ARIS, the website for *Biology*, provides a wealth of information organized and integrated by chapter. You will find practice quizzes, interactive activities, labeling exercises, flashcards, and much more that will complement your learning and understanding of general biology.

www.mhhe.com/maderbiology9

1

A VIEW OF LIFE

From bacteria to bats, toadstools to trees, whip-poor-wills to whales—the diversity of the living world boggles the mind. Yet all organisms are united by a common heritage that began during the early years of our planet. Just as you are descended from your parents, grandparents, and so forth from generation to generation back in time, all forms of life can trace their history to the original living thing.

At first glance, humans and irises seem very different in their appearance and way of life. However, closer examination reveals that they share a common chemistry, genetic code, and basic life processes as would be expected if their ancestry is joined. But why are they different? Because animals, such as humans, and plants, such as irises, are the products of evolutionary forces which shaped them differently over eons of time.

Living things are linked in another way. Their existence depends on one another, and collectively they form a giant web of life that stretches about the globe. Members of this web, from orchids in tropical rain forests to the big cats in Africa, are threatened with extinction due to human activities. The diversity accomplished by evolution over billions of years can be destroyed in a much shorter length of time. With every extinction, the web of life is threatened with collapse. What do we want to do—preserve diversity and the web of life, or destroy it? We are in the driver's seat, but which road will we choose? The diversity of living things and the future of the planet are in our hands.

Humans and irises, like all living things, have many characteristics in common.

1.1 HOW TO DEFINE LIFE

Life on Earth takes on a staggering variety of forms, often functioning and behaving in ways strange to humans. For example, gastric-brooding frogs swallow their embryos and give birth to them later by throwing them up! Some species of puffballs, a type of fungus, are capable of producing trillions of spores when they reproduce. Fetal sand sharks kill and eat their siblings while still inside their mother. Some *Ophrys* orchids look so much like female bees that male bees try to mate with them. Octopi and squid have remarkable problem-solving abilities despite a small brain. Some bacteria can live out their entire life in 15 minutes, while bristlecone pine trees outlive ten generations of humans. Simply put, from the deepest oceanic trenches to the reaches of the atmosphere, life abounds.

Figure 1.1 illustrates the major groups of living things, also called **organisms.** From left to right, bacteria are widely distributed, tiny, microscopic organisms with a very simple structure. A *Paramecium* is an example of a microscopic protist. Protists are larger in size and more complex than bacteria. The other organisms in Figure 1.1 are quite complex and easily seen with the naked eye. They can be distinguished by how they get their food. A morel is a fungus that digests its food externally. A sunflower is a photosynthetic plant that makes its own food and a snow goose is an animal that ingests its food.

Because life is so diverse, it seems reasonable that it cannot be defined in a straight-forward manner. Instead, life is best defined by several basic characteristics shared by all organisms. Like nonliving things, organisms are composed of chemical elements. Also, organisms obey the same laws of chemistry and physics that govern everything within the universe. The characteristics of life, however, will provide great insight into the unique nature of life and will help us distinguish living things from nonliving things.

Living Things Are Organized

The complex organization of living things begins with atoms, which make up basic building blocks known as elements (Fig. 1.2). Elements combine with themselves or other elements to form molecules. The **cell,** which is composed of a variety of molecules working together, is the basic unit of structure and function of all living things. Some cells, notably **unicellular** paramecia, live independently. Other cells, for example, the colonial alga *Volvox,* cluster together in microscopic colonies. An elephant is a **multicellular** organism in which similar cells combine to form a tissue; nerve tissue is a common tissue in animals. Tissues make up organs, as when various tissues combine to form the brain. Organs work together in systems; for example, the brain works with the spinal cord and a network of nerves to form the nervous system. Organ systems are joined together to form a complete living thing, or organism.

The levels of biological organization extend beyond the individual organism. All the members of one species in a particular area belong to a population. A nearby forest may have a population of gray squirrels and a population of white oaks, for example. The populations of various animals and plants in the forest make up a community. The community of populations interacts with the physical environment and forms an ecosystem. Finally, all the Earth's ecosystems make up the biosphere.

Emergent Properties

Biological organization includes the following levels: cell, tissues, organs, organ systems, organisms, populations, communities, ecosystems, and the biosphere. Each level of organization is more complex than the level preceding it and has properties beyond those of the former level. For example, when cells are broken down into bits of membrane and liquids, these parts themselves cannot carry out the business of living. Slice a nonliving lump of coal and fit the pieces together, and you still have a lump of coal. Slice a living plant and arrange the slices, and you have a collection of plant parts that can no longer function as a complete plant.

In the living world, the whole is indeed more than the sum of its parts. Each new level of biological organization has **emergent properties** that are due to interactions between the parts making up the whole. All properties, even emergent properties, are governed by the laws of physics and chemistry.

Bacteria   *Paramecium*   Morel   Sunflower   Snow goose

FIGURE 1.1 Diversity of life.
Biology is the scientific study of life. Many diverse forms of life are found on planet Earth.

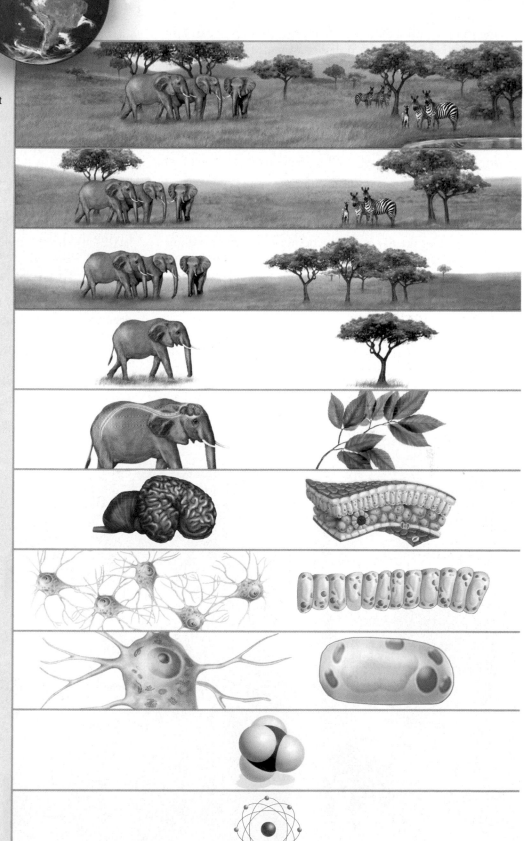

Biosphere
Regions of the Earth's crust, waters, and atmosphere inhabited by living things

Ecosystem
A community plus the physical environment

Community
Interacting populations in a particular area

Population
Organisms of the same species in a particular area

Organism
An individual; complex individuals contain organ systems

Organ System
Composed of several organs working together

Organ
Composed of tissues functioning together for a specific task

Tissue
A group of cells with a common structure and function

Cell
The structural and functional unit of all living things

Molecule
Union of two or more atoms of the same or different elements.

Atom
Smallest unit of an element composed of electrons, protons, and neutrons

FIGURE 1.2 Levels of biological organization.

Living Things Acquire Materials and Energy

Living things cannot maintain their organization or carry on life's activities without an outside source of nutrients and energy (Fig. 1.3). Food provides nutrients, which are used as building blocks or for energy. **Energy** is the capacity to do work, and it takes work to maintain the organization of the cell and the organism. When cells use nutrient molecules to make their parts and products, they carry out a sequence of chemical reactions. The term **metabolism** [Gk. *meta*, implying change] encompasses all the chemical reactions that occur in a cell.

The ultimate source of energy for nearly all life on Earth is the sun. Plants and certain other organisms are able to capture solar energy and carry on **photosynthesis,** a process that transforms solar energy into the chemical energy of organic nutrient molecules. All life on Earth acquires energy by metabolizing nutrient molecules made by photosynthesizers. This applies even to plants.

Remaining Homeostatic

To survive, it is imperative that an organism maintain a state of biological balance or **homeostasis** [Gk. *homoios*, like, resembling; and *stasis*, standing]. For life to continue, temperature, moisture level, acidity, and other physiological factors must remain within the tolerance range of the organism.

Homeostasis is maintained by systems that monitor internal conditions and make routine and necessary adjustments.

Organisms have intricate feedback and control mechanisms that do not require any conscious activity. When a student is so engrossed in her textbook that she forgets to eat lunch, her liver releases stored sugar to keep blood sugar levels within normal limits. In this case, hormones regulate sugar storage and release, but in other instances the nervous system is involved in maintaining homeostasis. Many organisms depend on behavior to regulate their internal environment. The same student may realize that she is hungry and decide to visit the local diner. A lizard may raise its internal temperature by basking in the sun or cool down by moving into the shade.

Living Things Respond

Living things interact with the environment as well as with other living things. Even unicellular organisms can respond to their environment. In some, the beating of microscopic hairs and, in others, the snapping of whiplike tails move them toward or away from light or chemicals. Multicellular organisms can manage more complex responses. A vulture can detect a carcass a mile away and soar toward dinner. A monarch butterfly can sense the approach of fall and begin its flight south where resources are still abundant.

The ability to respond often results in movement: the leaves of a plant turn toward the sun, and animals dart to-

a.

b.

c.

d.

e.

f.

FIGURE 1.3 Acquiring nutrient materials and energy.
a. An eagle ingesting fish. **b.** A human eating an apple. **c.** A cypress tree capturing sunlight. **d.** An amoeba engulfing debris. **e.** A fungus feeding on a tree. **f.** A bison eating grass.

ward safety. Appropriate responses help ensure survival of the organism and allow it to carry on its daily activities. All together, these activities are termed the behavior of the organism. Organisms display a variety of behaviors as they search and compete for energy, nutrients, shelter, and mates. Many organisms display complex communication, hunting, and defense behaviors.

Living Things Reproduce and Develop

Life comes only from life. Every type of living thing can **reproduce,** or make another organism like itself (Fig. 1.4). Bacteria, protists, and other unicellular organisms simply split in two. In most multicellular organisms, the reproductive process begins with the pairing of a sperm from one partner and an egg from the other partner. The union of sperm and egg, followed by many cell divisions, results in an immature stage, which grows and develops through various stages to become the adult.

An embryo develops into a humpback whale or a purple iris because of a blueprint inherited from its parents. The instructions, or blueprint, for an organism's metabolism and organization are encoded in genes. The **genes,** which contain specific information for how the organism is to be ordered, are made of long molecules of DNA (deoxyribonucleic acid). DNA has a shape resembling a spiral staircase with millions of steps. Housed in this spiral staircase is the genetic code that is shared by all living things.

Living Things Have Adaptations

Adaptations [L. *ad*, toward, and *aptus*, fit, suitable] are modifications that make organisms suited to their way of life. For example, penguins are adapted to an aquatic existence in the Antarctic. An extra layer of downy feathers is covered by short, thick feathers that form a waterproof coat. Layers of blubber also keep the birds warm in cold water. Most birds have forelimbs proportioned for flying, but penguins have stubby, flattened wings suitable for swimming. Their feet and tails serve as rudders in the water, but the flat feet also allow them to walk on land. Rockhopper penguins have a bill adapted to eating small shellfish.

Charles Darwin illustrated that organisms become modified over time by a process called **natural selection.** Certain members of a **species** [L. *species*, model, kind], defined as a group of interbreeding individuals, may inherit a genetic change that causes them to be better suited to a particular environment. These members can be expected to produce more surviving offspring who also have the favorable characteristic. In this way, the typical attributes of a species change over time.

Descent with Modification

All living things share the same basic characteristics. They are all composed of cells organized in a similar manner. Their genes are composed of DNA, and they carry out the same

FIGURE 1.4 Rockhopper penguins with their offspring.
Rockhopper penguins, which are named for their skill in leaping from rock to rock, produce one or two offspring at a time. Both male and female have a brood patch, a feather-free area of vascularized skin where the egg(s) are kept warm when either parent sits on the nest.

metabolic reactions to acquire energy and maintain their organization. This unity suggests that all living things are descended from a common ancestor—the first cell or cells. **Evolution** [L. *evolutio*, an unrolling] is descent with modification over time. One species can serve as a common ancestor to several species, each adapted to a particular set of environmental conditions. Specific adaptations increase the chances for survival and allow an organism to play a particular role in an ecosystem. Through the eons of time, evolution is responsible for the great diversity of life on Earth.

Descent from a common ancestor explains the unity of life. Adaptations to different ways of life account for the great diversity of life-forms.

1.2 HOW THE BIOSPHERE IS ORGANIZED

The organization of life extends beyond the individual to the **biosphere,** the zone of air, land, and water at the surface of the Earth where organisms exist. Individual organisms belong to a **population,** which is all the members of a species within a particular area. The populations of a **community** interact among themselves and with the physical environment (e.g., soil, atmosphere, and chemicals), thereby forming an **ecosystem.**

Figure 1.5 depicts a grassland inhabited by populations of rabbits, mice, snakes, hawks, and various types of plants. These populations exchange gases with and give off heat to the atmosphere. They also take in water from and give off water to the physical environment. In addition, populations interact with each other by forming food chains in which one population feeds on another. Mice feed on plants and seeds, snakes feed on mice, and hawks feed on rabbits and snakes, for example. The interactions between the various food chains make up a food web.

Ecosystems are characterized by chemical cycling and energy flow, both of which begin when photosynthetic plants, algae, and some bacteria take in solar energy and inorganic nutrients to produce food in the form of organic nutrients. The gray arrows in Figure 1.5 represent chemical cycling—chemicals move from one population to another in a food chain, until with death and decomposition, inorganic nutrients are returned to living plants once again. The yellow to red arrows represent energy flow. Energy flows from the sun through plants and other members of the food chain as one population feeds on another. With each transfer some energy is lost as heat. Eventually, all the energy taken in by photosynthesizers has dissipated into the atmosphere. Because energy flows and does not cycle, ecosystems could not stay in existence without a constant input of solar energy and the ability of photosynthesizers to absorb it.

The Human Population

Humans tend to modify existing ecosystems for their own purposes. Humans clear forests or grasslands to grow crops; later, they build houses on what was once farmland; and finally, they convert small towns into cities. As coasts are developed, humans send sediments, sewage, and other pollutants into the sea. Human activities destroy valuable coastal wetlands, which serve as protection against storms and as nurseries for a myriad of invertebrates and vertebrates.

The two most biologically diverse ecosystems—tropical rain forests and coral reefs—are home to many organisms. The canopy of the tropical rain forest alone supports a variety of organisms including orchids, insects, and monkeys. Coral reefs, which are found just offshore of the continents and islands of the Southern Hemisphere, are built up from calcium carbonate skeletons of sea animals called corals. Reefs provide a habitat for many animals, including jellyfish, sponges, snails, crabs, lobsters, sea turtles, moray eels, and some of the world's most colorful fishes (Fig. 1.6). Like tropical rain forests, coral reefs are severely threatened as the human population increases in size. Some reefs are 50 million years old, and yet in just a

FIGURE 1.5 Grassland, a terrestrial ecosystem.
In an ecosystem, chemical cycling (gray arrows) and energy flow (yellow to red arrows) begin when plants use solar energy and inorganic nutrients to produce food for themselves and directly or indirectly for all other populations in the ecosystem. As one population feeds on another, chemicals and energy are passed along a food chain. With each transfer, some energy is lost as heat. Eventually, all the energy dissipates. With the death and decomposition of organisms, inorganic nutrients are returned to the environment and eventually may be used by plants.

heat

solar energy

heat

heat

heat

heat

heat

WASTE MATERIAL, DEATH, AND DECOMPOSITION

Chemical cycling
Energy flow

few decades, human activities have destroyed 10% of all coral reefs and seriously degraded another 30%. At this rate, nearly three-quarters could be destroyed within 50 years. Similar statistics are available for tropical rain forests.

It has long been clear that human beings depend on healthy ecosystems for food, medicines, and various raw materials. We are only now beginning to realize that we depend on them even more for the services they provide. Just as chemical cycling occurs within an ecosystem, so ecosystems keep chemicals cycling throughout the entire biosphere. The workings of ecosystems ensure that the environmental conditions of the biosphere are suitable for the continued existence of humans. And many ecologists (scientists who study ecosystems) believe that ecosystems cannot function properly unless they remain biologically diverse.

Biodiversity

Biodiversity is the total number and relative abundance of species, the variability of their genes, and the different ecosystems in which they live. The present biodiversity of our planet has been estimated to be as high as 15 million species, and so far, less than 2 million have been identified and named. **Extinction** is the death of a species or larger classification category. It is estimated that presently we are losing as many as 400 species per day due to human activities. For example, several species of fishes have all but disappeared

from the coral reefs of Indonesia and along the African coast because of overfishing. Many biologists are alarmed about the present rate of extinction and believe it may eventually rival the rates of the five mass extinctions that have occurred during our planet's history. The last mass extinction, about 65 million years ago, caused many plant and animal species, including the dinosaurs, to become extinct.

FIGURE 1.6 Coral reef, a marine ecosystem.
Coral reefs, a type of ecosystem found in tropical seas, contain many diverse forms of life, a few of which are shown here. Coral reefs are now threatened because of many adverse human activities. Saving biodiversity is a modern-day challenge of great proportions.

It has been suggested that the primary bioethical issue of our time is preservation of ecosystems. Just as a native fisherman who assists in overfishing a reef is doing away with his own food source, so are we as a society contributing to the destruction of our home, the biosphere. If instead we adopt a conservation ethic that preserves the biosphere, we would help ensure the continued existence of our own species.

Living things belong to ecosystems, where populations interact among themselves in communities and with the physical environment. Preservation of ecosystems is of primary importance because they perform services that ensure our continued existence.

1.3 HOW LIVING THINGS ARE CLASSIFIED

Because life is so diverse, it is helpful to have a classification system to group organisms into categories (see Appendix B). **Taxonomy** [Gk. *tasso,* arrange, classify, and *nomos,* usage, law] is the discipline of identifying and classifying organisms according to certain rules. Taxonomy makes sense out of the bewildering variety of life on Earth and provides valuable insight into evolution. As more is learned about living things, including the evolutionary relationships between species, taxonomy changes. Taxonomists are constantly making observations and performing experiments that will one day bring about changes in the classification system adopted by this text.

Categories of Classification

Several of the basic classification categories, or taxa, going from least inclusive to most inclusive, are **species, genus, family, order, class, phylum, kingdom,** and **domain** (Table 1.1). Each successive classification category above species contains more types of organisms than the preceding one. Species placed within one genus share many specific characteristics and are the most closely related, while species placed in the same king-

dom share only general characteristics with one another. For example, all species in the genus *Pisum* look pretty much the same—that is, like pea plants—but species in the plant kingdom can be quite varied, as is evident when we compare grasses to trees. By the same token, only modern humans are in the genus *Homo,* but many types of species, from tiny hydras to huge whales, are members of the animal kingdom. Species placed in different domains are the most distantly related.

Domains

Biochemical evidence suggests that there are only three domains: **domain Bacteria, domain Archaea,** and **domain Eukarya.** Both domain Bacteria and domain Archaea contain unicellular prokaryotes, which lack the membrane-bounded nucleus found in the eukaryotes of domain Eukarya.

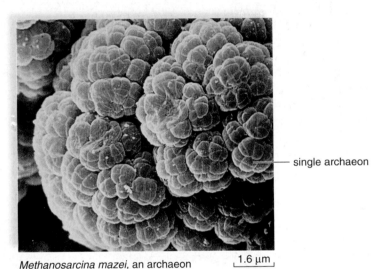

Methanosarcina mazei, an archaeon   1.6 µm

FIGURE 1.7 Domain Archaea.
Archaea are prokaryotes capable of surviving in extreme environments, such as those with high salinity and temperature and low pH.

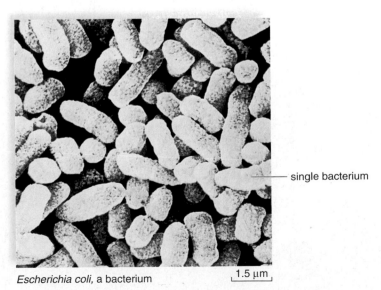

Escherichia coli, a bacterium   1.5 µm

FIGURE 1.8 Domain Bacteria.
Bacteria are metabolically diverse prokaryotes widely distributed in various environments.

TABLE 1.1

Levels of Classification

Category	Human	Corn
Domain	Eukarya	Eukarya
Kingdom	Animalia	Plantae
Phylum	Chordata	Anthophyta
Class	Mammalia	Monocotyledones
Order	Primates	Commelinales
Family	Hominidae	Poaceae
Genus	*Homo*	*Zea*
Species*	*H. sapiens*	*Z. mays*

* To specify an organism, you must use the full binomial name, such as *Homo sapiens.*

Prokaryotes are structurally simple but metabolically complex (Figs. 1.7 and 1.8). Archaea can live in aquatic environments that lack oxygen or are too salty, too hot, or too acidic for most other organisms. Perhaps these environments are similar to those of the primitive Earth, and archaea are representative of the first cells that evolved. Bacteria are variously adapted to living almost anywhere—in the water, soil, and atmosphere, as well as on our skin and in our mouths and large intestines. Although some bacteria cause dreaded diseases, others perform many valuable services, both environmentally and commercially. They are used to conduct genetic research in our laboratories, to produce innumerable products in our factories, and to purify water in our sewage treatment plants, for example.

Kingdoms

Taxonomists are in the process of deciding how to categorize archaea and bacteria into kingdoms. Domain Eukarya, on the other hand, contains four kingdoms (Fig. 1.9). **Protists** (kingdom Protista) range from unicellular forms to a few multicellular ones. Some are photosynthesizers, and some must acquire their food. Common protists include algae, the protozoans, and the water molds. Among the **fungi** (kingdom Fungi) are the familiar molds and mushrooms that,

DOMAIN EUKARYA

KINGDOM PROTISTA (protists)

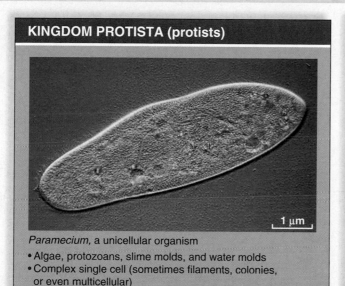

Paramecium, a unicellular organism

- Algae, protozoans, slime molds, and water molds
- Complex single cell (sometimes filaments, colonies, or even multicellular)
- Absorb, photosynthesize, or ingest food

KINGDOM PLANTAE (plants)

Passiflora, passion flower, a flowering plant

- Mosses, ferns, conifers, and flowering plants (both woody and nonwoody)
- Multicellular with specialized tissues containing complex cells
- Photosynthesize food

KINGDOM FUNGI

Coprinus, a shaggy mane mushroom

- Molds, mushrooms, yeasts, and ringworms
- Mostly multicellular fillaments with specialized, complex cells
- Absorb food

KINGDOM ANIMALIA (animals)

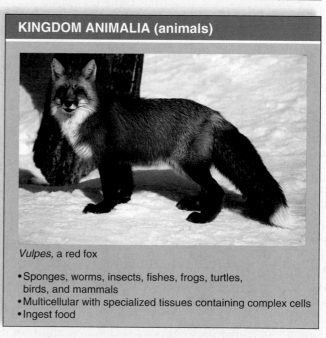

Vulpes, a red fox

- Sponges, worms, insects, fishes, frogs, turtles, birds, and mammals
- Multicellular with specialized tissues containing complex cells
- Ingest food

FIGURE 1.9 The four kingdoms in domain Eukarya.

along with bacteria, help decompose dead organisms. **Plants** (kingdom Plantae) are multicellular photosynthetic organisms. Example plants include azaleas, zinnias, and pines. **Animals** (kingdom Animalia) are multicellular organisms that must ingest and process their food. Aardvarks, jellyfish, and zebras are representative animals.

Scientific Name

Biologists use **binomial nomenclature** [L. *bi*, two, and *nomen*, name] to assign each living thing a two-part name called a scientific name. For example, the scientific name for mistletoe is *Phoradendron tomentosum*. The first word is the genus, and the second word is the specific epithet of a species within a genus. The genus may be abbreviated (e.g., *P. tomentosum*) and the species may simply be indicated if it is unknown (e.g., *Phoradendron* sp.). Scientific names are universally used by biologists to avoid confusion. Common names tend to overlap and often are in the language of a particular country. But scientific names are based on Latin, a universal language that not too long ago was well known by most scholars.

Taxonomy places species into classification categories. The categories are species, genus, family, order, class, phylum, kingdom, and domain (the most inclusive).

1.4 THE PROCESS OF SCIENCE

The process of science pertains to **biology,** the scientific study of life. Biology consists of many disciplines and areas of specialty because life has numerous aspects. Some biological disciplines are cytology, the study of cells; anatomy, the study of structure; physiology, the study of function; botany, the study of plants; zoology, the study of animals; genetics, the study of heredity; and ecology, the study of the interrelationships between organisms and their environment.

Religion, aesthetics, ethics, and science are all ways in which human beings seek order in the natural world. Science differs from other ways of knowing and learning by its process, which can be quite varied because it can be adjusted to where and how a study is being conducted. Still, the **scientific process** often involves the use of the scientific method, which begins with observation (Fig. 1.10).

Observation

Scientists believe that nature is orderly and measurable—that natural laws, such as the law of gravity, do not change with time, and that a natural event, or **phenomenon,** can be understood more fully through observation. Scientists use all of their senses in making **observations.** The behavior of chimps can be observed through visual means, the disposition of a skunk can be observed through olfactory means, and the warning rattles of a rattlesnake provide auditory information of imminent danger. Scientists also extend the ability of their senses by using instruments; for example, the microscope enables us to see objects that could never be seen by the

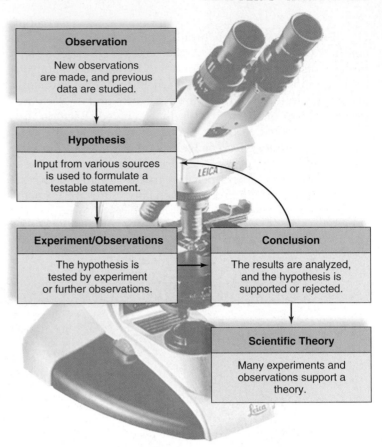

FIGURE 1.10 Flow diagram for the scientific method.
On the basis of new and/or previous observations, a scientist formulates a hypothesis. The hypothesis is tested by further observations and/or experiments, and new data either support or do not support the hypothesis. The return arrow indicates that a scientist often chooses to retest the same hypothesis or to test a related hypothesis. Conclusions from many different but related experiments may lead to the development of a scientific theory. For example, studies pertaining to development, anatomy, and fossil remains all support the theory of evolution.

naked eye. Finally, scientists may expand their understanding even further by taking advantage of the knowledge and experiences of other scientists. For instance, they may look up past studies at the library or on the Internet, or they may write or speak to others who are researching similar topics.

Nevertheless, chance alone can help a scientist get an idea. The most famous case pertains to penicillin. When examining a petri dish, Alexander Fleming observed an area around a mold that was free of bacteria. Upon investigating, Fleming found that the mold, a *Penicillium* species, produced an antibacterial substance he called penicillin, and he thought that perhaps penicillin would be useful in humans. This discovery changed medicine and has saved countless lives.

Hypothesis

After making observations and gathering knowledge about a phenomenon, a scientist uses inductive reasoning. **Inductive reasoning** occurs whenever a person uses creative thinking to combine isolated facts into a cohesive whole. In this way, a scientist comes up with a **hypothesis,** a possible explanation for a natural event. The scientist presents the hypothesis as an actual statement.

All of a scientist's past experiences, no matter what they might be, will most likely influence the formation of a hypothesis. But a scientist only considers hypotheses that can be tested. Moral and religious beliefs, while very important to the lives of many people, differ between cultures and through time and may not be testable.

Experiments/Further Observations

Testing a hypothesis involves either conducting an **experiment** or making further observations. To determine how to test a hypothesis, a scientist uses deductive reasoning. **Deductive reasoning** involves "if, then" logic. For example, a scientist might reason, if organisms are composed of cells, then microscopic examination of any part of an organism should reveal cells. We can also say that the scientist has made a **prediction** that the hypothesis can be supported by doing microscopic studies. Making a prediction helps a scientist know what to do next.

The manner in which a scientist intends to conduct an experiment is called the **experimental design.** A good experimental design ensures that scientists are testing what they want to test and that their results will be meaningful. It is always best for an experiment to include a control group. Often, a control group, or simply the **control,** goes through all the steps of an experiment but lacks the factor (is not exposed to the factor) being tested.

Scientists often use a **model,** a representation of an actual object when doing an experiment. Later in this section, a scientist uses bluebird models because it would have been impossible to get live birds to cooperate. Another type of modeling occurs when scientists use software to decide how human activities will affect climate, or when they use mice instead of humans for medical research. Nevertheless, a medicine that is effective in mice should still be tested in humans. And whenever it is impossible to study the actual phenomenon, a model remains a hypothesis in need of testing. Someday, a scientist might devise a way to test it.

Data

The results of an experiment are referred to as the **data.** Data should be observable and objective, rather than subjective. Mathematical data are often displayed in the form of a graph or table. Many studies rely on statistical data. Let's say an investigator wants to know if eating onions can prevent women from getting osteoporosis (weak bones). The scientist conducts a survey asking women about their onion-eating habits and then correlates this data with the condition of their bones. Other scientists critiquing this study would want to know: How many women were surveyed? How old were the women? What were their exercise habits? What proportion of the diet consisted of onions? And what criteria were used to determine the condition of their bones? Should the investigators conclude that eating onions does protect a woman from osteoporosis, other scientists would want to know the statistical probability of error. If the results are significant at a 0.30 level, then the probability that the correlation is incorrect is 30% or less. (This would be considered a high probability of error.) The greater the variance in the data, the greater the probability of error. And in the end, statistical data of this sort would only be suggestive until we learn of some ingredient in onions that has a direct biochemical or physiological effect on bones. Therefore, scientists must be skeptics who always pressure one another to keep on investigating a particular topic.

Conclusion

Scientists must analyze the data in order to reach a **conclusion** as to whether the hypothesis is supported or not. Because science progresses, the conclusion of one experiment can lead to the hypothesis for another experiment, as represented by the return arrow in Figure 1.10. Results that do not support one hypothesis can often help a scientist formulate another hypothesis to be tested. Scientists report their findings in scientific journals so that their methodology and data are available to other scientists for critique. Experiments and observations must be repeatable—that is, the reporting scientist and any scientist who repeats the experiment must get the same results, or else the data are suspect.

Scientific Theory

The ultimate goal of science is to understand the natural world in terms of **scientific theories,** which are concepts that join together well-supported and related hypotheses. In ordinary speech, the word *theory* refers to a speculative idea. In contrast, a scientific theory is supported by a broad range of observations, experiments, and data often from a variety of disciplines. Some of the basic theories of biology are:

Theory	Concept
Cell	All organisms are composed of cells, and new cells only come from preexisting cells.
Homeostasis	The internal environment of an organism stays relatively constant—within a range that is protective of life.
Gene	Organisms contain coded information that dictates their form, function, and behavior.
Ecosystem	Organisms are members of populations, which interact with each other and the physical environment within a particular locale
Evolution	All living things have a common ancestor, but each is adapted to a particular way of life.

The theory of evolution is the unifying concept of biology because it pertains to many different aspects of living things. For example, the theory of evolution enables scientists to understand the history of life, the diversity of living things, and the anatomy, physiology, and embryological development of organisms. Even behavior can be described through evolution, as we shall see in a study discussed later in this chapter.

The theory of evolution has been a fruitful scientific theory, meaning that it has helped scientists generate new hypotheses. Because this theory has been supported by so many observations and experiments for over 100 years, some biologists refer to the **principle** of evolution, a term sometimes used for theories that are generally accepted by an overwhelming number of scientists. The term **law** instead of principle is preferred by some. For instance, in a subsequent chapter concerning energy relationships, we will examine the laws of thermodynamics.

Scientists carry out studies in which they test hypotheses. The conclusions of many different types of related experiments eventually enable scientists to arrive at a scientific theory that is generally accepted by all.

A Controlled Study

Most investigators do controlled studies in which all groups get the same treatment. Controlled studies ensure that the outcome is due to the **experimental variable** or independent variable, the component or factor being tested. The result is called the **responding variable** or dependent variable because it is due to the experimental variable:

Experimental Variable (Independent Variable)	Responding Variable (Dependent Variable)
Factor of the experiment being tested	Result or change that occurs due to the experimental variable

In the study we discuss here, researchers perform an experiment in which nitrogen fertilizer is the experimental variable and enhanced yield is the responding variable. Nitrogen fertilizer in the short run has long been known to enhance yield and increase food supplies. However, excessive nitrogen fertilizer application can cause pollution by adding toxic levels of nitrates to water supplies. Also, applying nitrogen fertilizer year after year may alter soil properties to the point that crop yields may decrease instead of increase. Then the only solution is to let the land remain unplanted for several years until the soil recovers naturally.

An alternative to the use of nitrogen fertilizers is the use of legumes, plants such as peas and beans, that increase soil nitrogen. Legumes provide a home for bacteria that convert atmospheric nitrogen to a form usable by plants. The bacteria live in nodules on the roots (Fig. 1.11). The products of photosynthesis move from the leaves to the root nodules; in turn, the nodules supply the plant with nitrogen compounds the plant can use to make proteins.

There are numerous legume crops that can be rotated (planted every other season) with any number of cereal crops. The nitrogen added to the soil by the legume crop is a natural fertilizer that increases the yield of cereal crops. The particular rotation used by farmers tends to depend on the location, climate, and market demand.

The Experiment

The investigators doing this study knew that the pigeon pea plant is a legume with a high rate of atmospheric nitrogen conversion. This plant is widely grown as a food crop in India, Kenya, Uganda, Pakistan, and other subtropical countries. Researchers formulated the hypothesis that a pigeon pea/winter wheat rotation would be a reasonable alternative to the use of nitrogen fertilizer to increase the yield of winter wheat.

> HYPOTHESIS: A pigeon pea/winter wheat rotation will cause winter wheat production to increase as well as or better than the use of nitrogen fertilizer.

> PREDICTION: Wheat biomass following the growth of pigeon peas will surpass wheat biomass following nitrogen fertilizer treatment.

In this study, the investigators decided on the following experimental design (Fig. 1.12a):

CONTROL POTS
■ Winter wheat was planted in pots of soil that received no fertilization treatment, i.e., no nitrogen fertilizer and no preplanting of pigeon peas.

TEST POTS
■ Winter wheat was grown in clay pots in soil treated with nitrogen fertilizer equivalent to 45 kilograms (kg)/hectare (ha).
■ Winter wheat was grown in clay pots in soil treated with nitrogen fertilizer equivalent to 90 kg/ha.
■ Pigeon pea plants were grown in clay pots in the summer. The pigeon pea plants were then tilled into the soil and winter wheat was planted in the same pots.

To ensure a controlled experiment, the conditions for the control pots and the test pots were identical; the plants were exposed to the same environmental conditions and watered equally. During the following spring, the wheat plants were

FIGURE 1.11 Root nodules.
Bacteria that live in nodules on the roots of legumes, such as pea plants, convert nitrogen in the air to a form that plants can use to make proteins and other nitrogen-containing molecules.

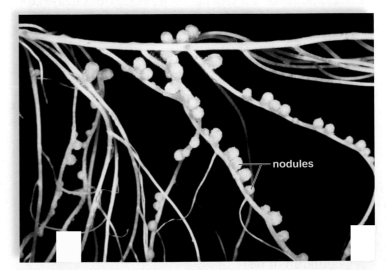

nodules

FIGURE 1.12 Pigeon pea/winter wheat rotation study.
a. Experiment involves control pots and test pots of three types: test pots that
received 45 kg/ha of nitrogen; test pots that received 90 kg/ha of nitrogen; and test
pots in which pigeon peas rotated with winter wheat. **b.** The graph compares wheat
biomass for each of three years. Wheat biomass in test pots that received the most
nitrogen fertilizer (green) declined while wheat biomass in test pots with pigeon
pea/winter wheat rotation (brown) increased dramatically.

b. Results

dried and weighed to determine wheat biomass production
in each of the pots.

The Results.   After the first year, wheat biomass was higher
in certain test pots than in the control pots (Fig. 1.12*b*). Specifically, test pots with 45 kg/ha of nitrogen fertilizer (orange) had only slightly more wheat biomass production
than the control pots, but test pots which received 90 kg/ha
treatment (green), demonstrated nearly twice the biomass
production of the control pots. This suggested that nitrogen fertilizer application has the potential to increase wheat
biomass. To the surprise of investigators, wheat production
following summer planting of pigeon peas did not demonstrate as high a biomass production as the control pots.

> CONCLUSION: The hypothesis is not supported. Wheat
> biomass following the growth of pigeon peas is not as
> great as that obtained with nitrogen fertilizer treatments.

Continuing the Experiment

The researchers decided to continue the experiment using the same design and the same pots as before, to see
if the buildup of residual soil nitrogen from pigeon peas

would eventually increase wheat biomass. This was their
new hypothesis.

> HYPOTHESIS: A sustained pigeon pea/winter wheat
> rotation will eventually cause an increase in winter
> wheat production.

> PREDICTION: Wheat biomass following two years of pigeon
> pea/winter wheat rotation will surpass wheat biomass
> following nitrogen fertilizer treatment.

After two years, the yield following 90 kg/ha nitrogen treatment (green) was not as much as it was the first year (Fig.
1.12*b*). Indeed, wheat biomass following summer planting
of pigeon peas (brown) was the highest of all treatments,
suggesting that buildup of residual nitrogen from pigeon
peas had the potential to provide fertilization for winter
wheat growth.

> CONCLUSION: The hypothesis is supported. At the end of two
> years, the yield of winter wheat following a pigeon pea/
> winterwheat rotation was better than for the other type pots.

The researchers continued their experiment for still another
year. After three years, winter wheat biomass production had

decreased in the control pots and in the pots treated with nitrogen fertilizer. Pots treated with nitrogen fertilizer still had increased wheat biomass production compared with the control pots but not nearly as much as pots following summer planting of pigeon peas. Compared to the first year, wheat biomass increased almost fourfold in pots having a pigeon pea/winter wheat rotation (brown, Fig. 1.12b). The researchers suggested that the soil was improved by the organic matter as well as the addition of nitrogen from the pigeon peas. The researchers published their results in a scientific journal.[1]

A Field Study

A scientist, David Barash, while observing the mating behavior of mountain bluebirds (Fig. 1.13a,b), formulated the hypothesis that aggression of the male varies during the reproductive cycle. To test this hypothesis, he reasoned that he should eval-

uate the intensity of male aggression at three stages: after the nest is built, after the first egg is laid, and after the eggs hatch.

HYPOTHESIS: Male bluebird aggression varies during the reproductive cycle.

PREDICTION: Aggression intensity will change after the nest is built, after the first egg is laid, and after hatching.

Testing the Hypothesis

For his experiment, Barash decided to measure aggression intensity by recording the "number of approaches per minute" a male made toward a rival male and his own female mate. To provide a rival, Barash posted a male bluebird model near the nests while resident males were out foraging. The aggressive behavior (approaches) of the resident male was noted during the first 10 minutes of the male's return (Fig. 1.13c). To give his results validity, Barash included a control group. For his control, Barash posted a male robin model instead of a male bluebird near certain nests.

FIGURE 1.13 A field study.
Observation of normal male bluebird behavior (**a** and **b**) allowed David Barash to formulate a testable hypothesis. He (**c**) collected data, which was (**d**) displayed in a graph. Then, he came to a conclusion.

a. Scientist making observations

b. Normal mountain bluebird nesting behavior

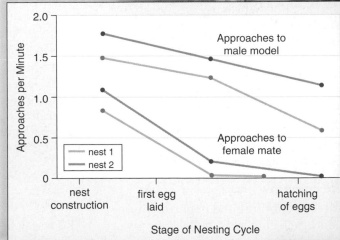
c. Resident male attacking a male model near nest

d. Observation of two experimental nests provided data for graph.

CONNECTING THE CONCEPTS

What we know about biology and what we'll learn in the future result from objective observation and testing of the natural world. The ultimate goal of science is to understand the natural world in terms of theories—conceptual schemes supported by abundant research. Evolution is a theory that accounts for the differences that divide and the unity that joins all living things. All living things have the same levels of organization and function similarly because they share a common evolution extending back through time to the first cells on Earth.

Scientific creationism, which states that God created all species, and, of late, the intelligent design theory, which states that an "intelligent agent" is responsible for life on Earth, are not considered science by the majority of biologists because they hypothesize a super-natural cause rather than a natural cause for events. When faith is involved, a hypothesis cannot be tested in a purely objective way.

Just as science does not test religious beliefs, it does not make ethical or moral decisions. The general public may want scientists to label certain research as "good" or "bad" and to predict whether any resulting technology will primarily benefit or harm our society. Yet science, by its very nature, is impartial and simply attempts to study natural phenomena.

Scientists should provide the public with as much information as possible when such issues as recombinant DNA technology or environmental preservation are being debated. Then they, along with other citizens, can help make intelligent decisions about what is most likely best for society. All men and women have a responsibility to decide how to use scientific knowledge so that it benefits all living things, including the human species.

This textbook was written to help you understand the scientific process and learn the basic concepts of general biology so that you will be better informed. This chapter has introduced you to the levels of biological organization, from the cell to the biosphere. The cell, the simplest of living things, is composed of nonliving molecules. Therefore, we must begin our study of biology with a brief look at cellular chemistry. In the next two chapters, you will study some important inorganic and organic molecules as they relate to cells. Then, you will learn how the cell makes use of energy and materials to maintain itself and to reproduce.

Resident males of the control group did not exhibit any aggressive behavior, but resident males of the experimental groups did exhibit aggressive behavior. Barash graphed his mathematical data (Fig. 1.13d). By examining the graph, you can see that the resident male was more aggressive toward the rival male model than toward his female mate, and that he was most aggressive while the nest was under construction, less aggressive after the first egg was laid, and least aggressive after the eggs hatched.

The Conclusion

The results allowed Barash to conclude that aggression in male bluebirds is related to their reproductive cycle. Therefore, his hypothesis was supported. If male bluebirds were always aggressive, even toward male robin models, his hypothesis would not have been supported.

> CONCLUSION: The hypothesis is supported. Male bluebird aggression does vary during the reproductive cycle.

Barash reported his experiment in the *American Naturalist*.[2] In this article, Barash gave an evolutionary interpretation to his results. It was adaptive, he said, for male bluebirds to be less aggressive after the first egg is laid because by then the male bird is "sure the offspring is his own." It was maladaptive for the male bird to waste energy being aggressive after hatching because his offspring are already present.

[2] Barash, D. P. 1976. The male responds to apparent female adultery in the mountain bluebird, *Sialia currucoides:* An evolutionary interpretation. *American Naturalist* 110:1097–101.

Summary

1.1 HOW TO DEFINE LIFE

Although living things are diverse, they have certain characteristics in common. Living things (a) are organized, and their levels of organization extend from the cell to ecosystems, (b) need an outside source of materials and energy, (c) respond to external stimuli, (d) reproduce and develop, passing on genes to their offspring, and (e) have adaptations suitable to their way of life in a particular environment.

The process of evolution explains both the unity and the diversity of life. Descent from a common ancestor explains why all organisms share the same characteristics, and adaptation to various ways of life explains the diversity of life-forms.

1.2 HOW THE BIOSPHERE IS ORGANIZED

Within an ecosystem, populations interact with one another and with the physical environment. Nutrients cycle within and between ecosystems, but energy flows unidirectionally and eventually becomes heat. Adaptations of organisms allow them to play particular roles within an ecosystem.

1.3 HOW LIVING THINGS ARE CLASSIFIED

Each living thing is given an italicized binomial name that consists of the genus and the specific epithet. For example, *Pisum sativum* is the name of the garden pea. From the least inclusive to the most inclusive category, each species belongs to genus, family, order, class, phylum, kingdom, and finally domain.

The three domains of life are Archaea, Bacteria, and Eukarya. The first two domains contain prokaryotic organisms that are structurally simple but metabolically complex. Domain Eukarya contains the kingdoms Protista, Fungi, Plantae, and Animalia. Protists range from unicellular to multicellular organisms and include the protozoans and algae. Among the fungi are the familiar molds and mushrooms. Plants are well known as the multicellular photosynthesizers of the world, while animals are multicellular and ingest their food.

1.4 THE PROCESS OF SCIENCE

When studying the natural world, scientists use the scientific process. Observations, along with previous data, are used to formulate a hypothesis. New observations and/or experiments are carried out in order to test the hypothesis. A good experimental design includes an experimental variable and a control group. The experimental and observational results are analyzed, and the scientist comes to a conclusion as to whether the results support the hypothesis or do not support the hypothesis.

Several conclusions in a particular area may allow scientists to arrive at a theory, such as the cell theory, the gene theory, or the theory of evolution. The theory of evolution is a unifying concept of biology.

Reviewing the Chapter

1. What are the common characteristics of life listed in the chapter?   2–5
2. What evidence can you cite to show that living things are organized?   2
3. Why do living things require an outside source of materials and energy? Describe these sources.   4
4. What is passed from generation to generation when organisms reproduce? What has to happen to the hereditary material DNA for evolution to occur?   5
5. How does evolution explain both the unity and the diversity of life?   5
6. What is an ecosystem, and why should human beings preserve ecosystems?   5–8
7. What are the categories of classification? What four kingdoms are in the domain Eukarya? Explain the scientific name of an organism.   8–10
8. Describe the series of steps involved in the scientific method.   10–11
9. What is the ultimate goal of science? Give an example that supports your answer.   11
10. Give an example of a scientific experiment. Name the experimental variable and the responding variable.   12–14

Testing Yourself

Choose the best answer for each question. For questions 1–3, match the statements with the characteristics of life in the key.

KEY:

   a. Living things are organized.
   b. Living things are homeostatic.
   c. Living things respond to stimuli.
   d. Living things reproduce.
   e. Living things have adaptations.

1. Genes made up of DNA are passed from parent to child.
2. A herd of zebra will scatter when a lion approaches.
3. The long, sharp talons of a hawk can hold onto a mouse.
4. Which of these is mismatched?
   a. domain Bacteria—mosses, ferns, pine trees
   b. kingdom Protista—protozoans, algae, water molds
   c. kingdom Fungi—molds, mushrooms, and ringworms
   d. kingdom Plantae—woody and nonwoody flowering plants
   e. kingdom Animalia—fish, frogs, birds, humans
5. The level of organization that includes cells of similar structure and function would be
   a. an organ.
   b. a tissue.
   c. an organ system.
   d. an organism.
6. In which kingdom are you most likely to find unicellular organisms?
   a. kingdom Protista
   b. kingdom Fungi
   c. kingdom Plantae
   d. kingdom Animalia
7. After performing an experiment and collecting data, the next step in the scientific method would be to
   a. propose a theory.
   b. design a model.
   c. form a hypothesis.
   d. come to a conclusion.

8. An example of chemical cycling occurs when
   a. plants absorb solar energy and make their own food.
   b. energy flows through an ecosystem and becomes heat.
   c. hawks soar and nest in trees.
   d. death and decay make inorganic nutrients available to plants.
   e. we eat food and use the nutrients to grow or repair tissues.
9. Science always studies a phenomenon that
   a. has previously been published.
   b. lends itself to experimentation.
   c. is observable with the eye or with instruments.
   d. fits in with an already existing theory.
   e. Both b and c are correct.
10. After formulating a hypothesis, a scientist
   a. proves the hypothesis true or false.
   b. tests the hypothesis.
   c. decides how to best avoid having a control.
   d. makes sure environmental conditions are just right.
   e. formulates a scientific theory.
11. The experimental variable in the bluebird experiment was the
   a. use of a model male bluebird.
   b. observations of the experimenter.
   c. various behavior of the males.
   d. identification of what bluebirds to study.
   e. All of these are correct.
12. The control group in the pigeon pea/winter wheat experiment was the pots that were
   a. planted with pigeon peas.
   b. treated with nitrogen fertilizer.
   c. not treated.
   d. not watered.
   e. Both c and d are correct.
13. Evolution from the first cell(s) best explains why
   a. ecosystems have populations of organisms.
   b. photosynthesizers produce food.
   c. human activities are threatening the biosphere.
   d. diverse organisms share common characteristics.
   e. All of these are correct.
14. Adaptation to a way of life best explains why living things
   a. display homeostasis.
   b. are diverse.
   c. began as single cells.
   d. are classified into three domains.
   e. mate with their own kind.
15. The second word of a scientific name, such as *Homo sapiens*, is the
   a. genus.
   b. phylum.
   c. specific epithet.
   d. species.
   e. family.
16. Energy is brought into ecosystems by which of the following?
   a. fungi and other decomposers
   b. cows and other organisms that graze on grass
   c. meat-eating animals
   d. organisms that photosynthesize, such as plants
   e. All of these are correct.
17. A scientist cannot
   a. make value judgments like everyone else.
   b. prove a hypothesis true.
   c. contribute to a long-standing scientific theory.
   d. make use of preexisting mathematical data.
   e. be as objective as possible.

For questions 18–21, fill in the blanks.

18. To reproduce is to make a _____ of one's self.

19. _____ has a responsibility to decide how scientific knowledge should be used.

20. The _____ population is now causing a mass extinction of living species.

21. The _____ and the bacteria are both prokaryotes.

22. An investigator spills dye on a culture plate and then notices that the bacteria live despite exposure to sunlight. He hypothesizes that the dye protects bacteria against death by ultraviolet (UV) light. To test this hypothesis, he decides to expose 200 culture plates to UV light. One hundred plates contain bacteria and dye; the other 100 plates contain only bacteria. Result: After exposure to UV light, the bacteria on both plates die. Fill in the right-hand portion of this diagram.

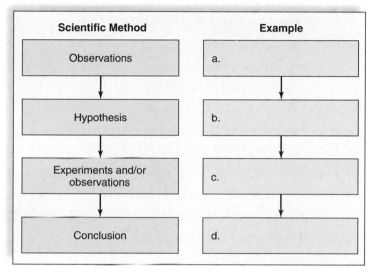

Thinking Scientifically

1. In a study testing a new drug to treat migraine headaches, 25% of the control group, which had been given a placebo (looks like but does not contain medication), and 46% of the experimental group reported an improvement in their symptoms. Would you conclude that this is an effective drug? What variables in the study might have influenced whether participants benefited from the new drug?

2. Viruses are small, infectious particles. Outside a living cell they can be stored just like chemical substances can be stored. Viruses can reproduce, but only inside a living cell. They do contain genes and can evolve, thereby avoiding destruction by the host's immune system. Should a virus be considered alive? What are the criteria on which you base your answer?

Bioethical Issue: Oil Drilling in the Arctic

The Arctic National Wildlife Refuge is home to a diverse array of wildlife, including migratory birds, caribou, grizzly bears, Dall sheep, polar bears, and musk oxen. The nearby continental shelf provides the coastal waters with a rich nutrient base that supports a wide variety of marine mammals during the summer months.

Those who favor oil drilling in the Arctic National Wildlife Refuge suggest that it would affect only an area the size of an airport within a state. They contend that the effect would mainly be underground because new techniques allow us to go lower and spread out beneath the surface to get the oil. Acquiring the oil, advocates say, would also protect jobs and security

in the United States by lessening dependence on foreign countries for oil. If these countries were to stop supplying the United States with oil, it would cause economic hardships, including a high price of gasoline.

Those who do not favor oil drilling in the Arctic National Wildlife Refuge believe that U.S. citizens can reduce their need for energy by adopting simple efficiency measures. They suggest that this would save many times the oil that could come from drilling in the Arctic refuge and that, by using a renewable energy resource, the environment in the lower 48 United States would be protected, in addition to protecting the wildlife in the Arctic National Wildlife Refuge.

Should citizens decide this matter on the basis of party politics? If not, how should they go about making a decision?

Understanding the Terms

adaptation 5	genus 8
animal 10	homeostasis 4
binomial nomenclature 10	hypothesis 10
biodiversity 7	inductive reasoning 10
biology 10	kingdom 8
biosphere 5	law 12
cell 2	metabolism 4
class 8	model 11
community 5	multicellular 2
conclusion 11	natural selection 5
control 11	observation 10
data 11	order 8
deductive reasoning 11	organism 2
domain 8	phenomenon 10
domain Archaea 8	photosynthesis 4
domain Bacteria 8	phylum 8
domain Eukarya 8	plant 10
ecosystem 5	population 5
emergent property 2	prediction 11
energy 4	principle 12
evolution 5	protist 9
experiment 11	reproduce 5
experimental design 11	responding variable 12
experimental variable 12	scientific process 10
extinction 7	scientific theory 11
family 8	species 5, 8
fungus 9	taxonomy 8
gene 5	unicellular 2

Match the terms to these definitions:

a. _____ All of the chemical reactions that occur in a cell during growth and repair.

b. _____ Changes that occur among members of a species with the passage of time, often resulting in increased adaptation to the prevailing environment.

c. _____ Component in an experiment that is manipulated as a means of testing it.

d. _____ Process by which plants use solar energy to make their own organic food.

e. _____ Sample that goes through all the steps of an experiment but lacks the factor being tested.

ARIS, the *Biology* Website

ARIS, the website for *Biology*, provides a wealth of information organized and integrated by chapter. You will find practice quizzes, interactive activities, labeling exercises, flashcards, and much more that will complement your learning and understanding of general biology.

www.mhhe.com/maderbiology9

PART I

THE CELL

Our study of the cell begins with the atoms and molecules necessary to its structure and function. Cellular organization requires an ongoing input of matter and energy. Plant cells and other types of photosynthetic cells capture solar energy and store it in the form of nutrient molecules that can be used later as a source of matter and energy for all living things. Metabolic pathways carry out cellular respiration and other energy conversions needed to keep the cell operational.

This part provides a foundation for all the other parts of the text because a cell is the basic unit of life, and all living things are composed of cells. Our knowledge of the structure and function of the cell can be applied directly to other disciplines of biology. Genes control cell structure and function and determine the characteristics of the organism. The study of evolution will include the origin of the cell and thereafter the history of life. Adaptation to the environment is an essential part of the evolutionary history of the various species that make up the living world. Knowledge of chemistry, energy transformations, and metabolic pathways increases our understanding of plants and animals, and it increases our capacity to keep ourselves healthy and the world capable of sustaining living things.

2

BASIC CHEMISTRY

T he difference between the nonliving and the living is beautifully illustrated in the photograph of a bottle-nosed dolphin jumping up out of the water. Most anyone would be able to say that the dolphin was alive while the water was not alive. Throughout history, scientists have attempted to determine the difference between the composition of nonliving and living things. At one time, it was believed that organisms contained a vital force and this force accounted for their "vitality." Such a hypothesis has never been supported, and, instead, today we know that living things are composed of the same elements as inanimate objects. It's true, though, as we shall see, that some types of molecules are unique to living things.

A knowledge of chemistry is also necessary to our understanding of the particular organism. Chemistry plays a role in the life of the bottle-nosed dolphin playing in the Gulf of Mexico or performing at SeaWorld. After all, a dolphin has a certain salinity tolerance, can only stay underwater for so long, and must have a particular diet to keep its complex organ systems functioning. A dolphin can't jump unless its nervous system is able to direct its muscles to contract. These and all aspects of a dolphin's biology involve molecular chemistry.

Charles Darwin told us many years ago that only an appreciation of every note leads to an understanding of a musical composition. In the same way we can only understand an organism when we have a fundamental knowledge and respect for its chemistry, which is necessary to its existence.

Bottle-nosed dolphin, *Tursiops truncatus*.

2.1 CHEMICAL ELEMENTS

Turn the page, throw a ball, pat your dog, rake leaves; everything that we touch—from the water we drink to the air we breathe—is composed of matter. **Matter** refers to anything that takes up space and has mass. Although matter has many diverse forms—anything from molten lava to kidney stones—it only exists in three distinct states: solid, liquid, and gas.

All matter, both nonliving and living, is composed of certain basic substances called **elements.** An element is a substance that cannot be broken down to simpler substances with different properties (a property is a physical or chemical characteristic, such as density, solubility, melting point, and reactivity) by ordinary chemical means. It is quite remarkable that there are only 92 naturally occurring elements that serve as the building blocks of matter. Other elements have been "human-made" and are not biologically important.

Both the Earth's crust and all organisms are composed of elements, but they differ as to which ones are predominant (Fig. 2.1). Only six elements—carbon, hydrogen, nitrogen, oxygen, phosphorus, and sulfur—are basic to life and make up about 95% of the body weight of organisms. The acronym CHNOPS helps us remember these six elements. The properties of these elements are essential to the uniqueness of cells and organisms. Other elements are also important to living things, including sodium, potassium, calcium, iron, and magnesium.

All living and nonliving things are matter composed of elements. Six elements (CHNOPS) in particular are basic to life.

Atomic Structure

In the early 1800s, the English scientist John Dalton championed the atomic theory, which says that elements consist of tiny particles called **atoms** [Gk. *atomos,* uncut, indivisible]. An atom is the smallest part of an element that displays the properties of the element. An element and its atoms share the same name. One or two letters create the **atomic symbol,** which stands for this name. For example, the symbol H means a hydrogen atom, the symbol Rn stands for radon, and the symbol Na (for *natrium* in Latin) is used for a sodium atom.

Each atom has its own specific mass. The **atomic mass,** or mass number, of an atom depends on the presence of certain subatomic particles. Physicists have identified a number of subatomic particles that make up atoms. The three best known subatomic particles include positively charged **protons,** uncharged **neutrons,** and negatively charged **electrons** [Gk. *elektron,* electricity]. Protons and neutrons are located within the nucleus of an atom, and electrons move about the nucleus. Figure 2.2 shows the arrangement of the subatomic particles in a helium atom, which has only two electrons. In Figure 2.2*a,* the stippling shows the probable location of electrons, and in Figure 2.2*b,* the circle represents an **electron shell,** the average location of electrons.

The concept of an atom has changed greatly since Dalton's day. If an atom could be drawn the size of a football field, the nucleus would be like a gumball in the center of the field, and the electrons would be tiny specks whirling about in the upper stands. Most of an atom is empty space. We should also realize that we can only indicate

FIGURE 2.1 Elements that make up the Earth's crust and its organisms.
Scarlet and red-blue-green macaws gather on a salt lick in South America. The graph inset shows the Earth's crust primarily contains the elements silicon (Si), aluminum (Al), and oxygen (O). Organisms primarily contain the elements oxygen, nitrogen (N), carbon (C), and hydrogen (H). Along with sulfur (S) and phosphorus (P), these elements make up biological molecules.

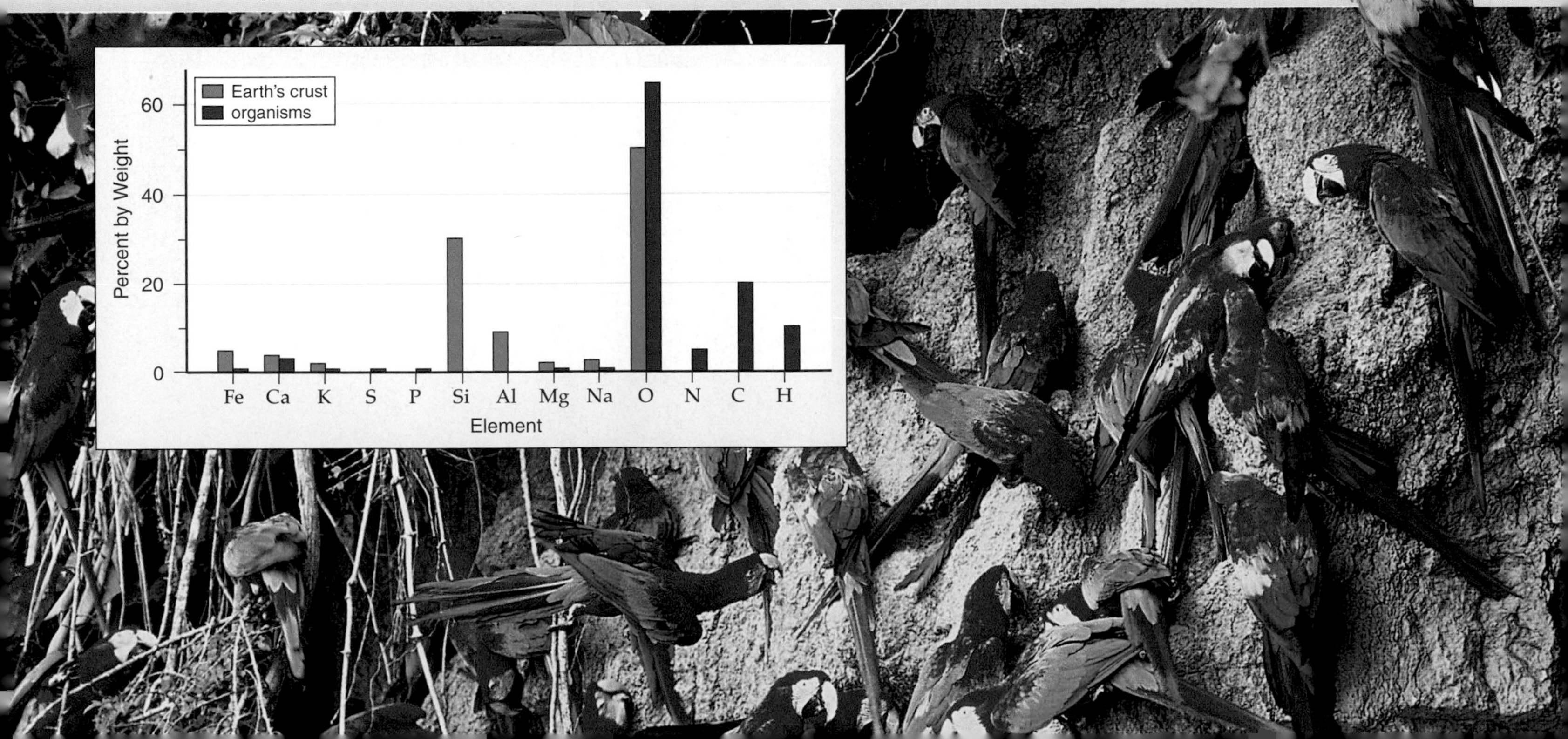

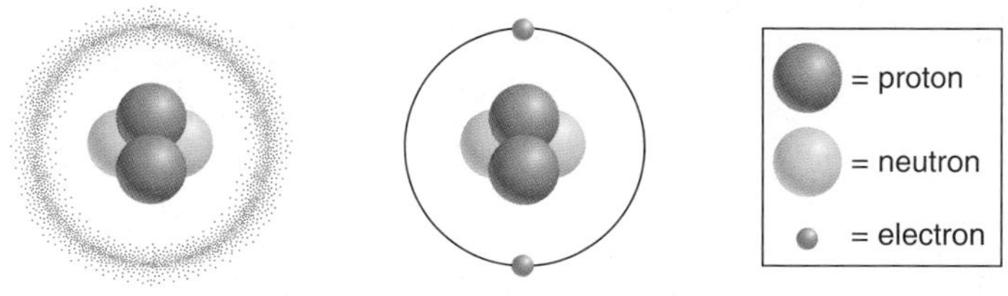

a. b.

Subatomic Particles			
Particle	**Electric Charge**	**Atomic Mass**	**Location**
Proton	+1	1	Nucleus
Neutron	0	1	Nucleus
Electron	−1	0	Electron shell

c.

FIGURE 2.2 Model of helium (He).
Atoms contain subatomic particles, which are located as shown. Protons and neutrons are found within the nucleus, and electrons are outside the nucleus. **a.** The stippling shows the probable location of the electrons in the helium atom. **b.** The average location of an electron is sometimes represented by a circle termed an electron shell. **c.** The electric charge and the atomic mass units of the subatomic particles vary as shown.

Groups / Periods

I	II	III	IV	V	VI	VII	VIII
1 **H** 1.008							2 **He** 4.003
3 **Li** 6.941	4 **Be** 9.012	5 **B** 10.81	6 **C** 12.01	7 **N** 14.01	8 **O** 16.00	9 **F** 19.00	10 **Ne** 20.18
11 **Na** 22.99	12 **Mg** 24.31	13 **Al** 26.98	14 **Si** 28.09	15 **P** 30.97	16 **S** 32.07	17 **Cl** 35.45	18 **Ar** 39.95
19 **K** 39.10	20 **Ca** 40.08	31 **Ga** 69.72	32 **Ge** 72.59	33 **As** 74.92	34 **Se** 78.96	35 **Br** 79.90	36 **Kr** 83.60

FIGURE 2.3 A portion of the periodic table.
In the periodic table, the elements, and therefore atoms, are in the order of their atomic numbers but arranged so that they are placed in groups (vertical columns) and periods (horizontal rows). All the atoms in a particular group have certain chemical characteristics in common. These four periods contain the elements that are most important in biology; the complete periodic table is in Appendix D.

where the electrons are expected to be most of the time. In our analogy, the electrons might very well stray outside the stadium at times.

All atoms of an element have the same number of protons. This is called the **atomic number.** The number of protons housed in the nucleus makes each atom unique. The atomic number is often written as a subscript to the lower left of the atomic symbol.

The atomic mass of an atom is essentially the sum of its protons and neutrons. Protons and neutrons are assigned one atomic mass unit each. Electrons are so small that their mass is considered zero in most calculations (Fig. 2.2*c*). The term *atomic mass* is used, and not *atomic weight,* because mass is constant while weight changes according to the gravitational force of a body. The gravitational force of the Earth is greater than that of the moon; therefore, substances weigh less on the moon even though their mass has not changed. The atomic mass is often written as a superscript to the upper left of the atomic symbol. For example, the carbon atom can be noted in this way:

atomic mass —— $^{12}_{6}\mathrm{C}$ —— atomic symbol
atomic number ——

The Periodic Table

Once chemists discovered a number of the elements, they began to realize that even though each element consists of a different atom, certain chemical and physical characteristics recur. The periodic table, developed by the Russian chemist Dmitri Mendeleev (1834–1907), was constructed as a way to group the elements, and therefore atoms, according to these characteristics. Notice that the periodic table is arranged according to increasing atomic number. The vertical columns in the table are groups; the horizontal rows are periods, which cause each atom to be in a particular group. For example, all the atoms in group 7 react with one atom at a time, for reasons we will soon explore. The atoms in group 8 are called the noble gases because they are inert and rarely react with another atom. Notice that helium and krypton are noble gases.

In Figure 2.3, the atomic number is above the atomic symbol and the atomic mass is below the atomic symbol. The atomic number tells you the number of positively charged protons, and also the number of negatively charged electrons if the atom is electrically neutral. To determine the number of neutrons, subtract the number of protons from the atomic mass, and take the closest whole number.

Atoms have an atomic symbol, an atomic number, and an atomic mass. The subatomic particles (protons, neutrons, and electrons) determine the characteristics of atoms.

Isotopes

Isotopes [Gk. *isos*, equal, and *topos*, place] are atoms of the same element that differ in the number of neutrons. Isotopes have the same number of protons, but they have different atomic masses. For example, the element carbon has three common isotopes:

$$^{12}_{6}C \qquad\qquad ^{13}_{6}C \qquad\qquad ^{14}_{6}C*$$

*radioactive

Carbon 12 has six neutrons, carbon 13 has seven neutrons, and carbon 14 has eight neutrons. Unlike the other two isotopes of carbon, carbon 14 is unstable; it changes over time into nitrogen 14, which is a stable isotope of the element nitrogen. As carbon 14 decays, it releases various types of energy in the form of rays and subatomic particles, and therefore it is a radioactive isotope. The radiation given off by radioactive isotopes can be detected in various ways. The Geiger counter is an instrument that is commonly used to detect radiation. In 1860, the French physicist Antoine-Henri Becquerel discovered that a sample of uranium would produce a bright image on a photographic plate because it was radioactive. A similar method of detecting radiation is still in use today. Marie Curie, who worked with Becquerel, contributed much to the study of radioactivity, as she named it. Today, radiation is used by biologists to date objects, create images, and trace the movement of substances.

Low Levels of Radiation

The chemical behavior of a radioactive isotope is essentially the same as that of the stable isotopes of an element. This means that you can put a small amount of radioactive isotope in a sample and it becomes a **tracer** by which to detect molecular changes. Melvin Calvin and his co-workers used carbon 14 to detect all the various reactions that occur during the process of photosynthesis.

The importance of chemistry to medicine is nowhere more evident than in the many medical uses of radioactive isotopes. Specific tracers are used in imaging the body's organs and tissues. For example, after a patient drinks a solution containing a minute amount of ^{131}I, it becomes concentrated in the thyroid—the only organ to take it up. A subsequent image of the thyroid indicates whether it is healthy in structure and function (Fig. 2.4a). Positron-emission tomography (PET) is a way to determine the comparative activity of tissues. Radioactively labeled glucose, which emits a subatomic particle known as a positron, is injected into the body. The radiation given off is detected by sensors and analyzed by a computer. The result is a color image that shows which tissues took up glucose and are metabolically active (Fig. 2.4b). A PET scan of the brain can help diagnose a brain tumor, Alzheimer disease, epilepsy, or whether a stroke has occurred.

High Levels of Radiation

Radioactive substances in the environment can harm cells, damage DNA, and cause cancer. When Marie Curie was studying radiation, its harmful effects were not known,

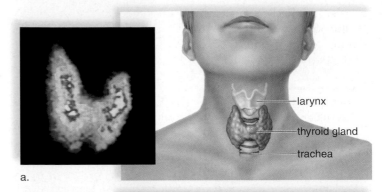
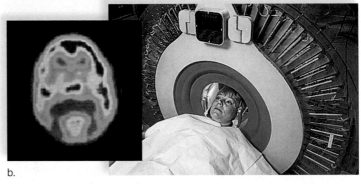

—larynx

—thyroid gland

—trachea

a.

b.

FIGURE 2.4 Low levels of radiation.
a. Incomplete scan of the thyroid gland on the left indicates the presence of cancer. **b.** A PET (positron-emission tomography) scan reveals which portions of the brain are most active.

and she and many of her co-workers developed cancer. The release of radioactive particles following a nuclear power plant accident can have far-reaching and long-lasting effects on human health. The harmful effects of radiation can be put to good use, however (Fig. 2.5). Radiation from radioactive isotopes has been used for many years to sterilize medical and dental products. Now it can be used to sterilize the U.S. mail and other packages to free them of possible pathogens, such as anthrax spores. The ability of radiation to kill cells

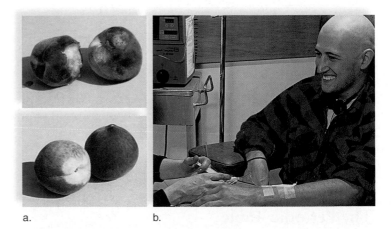

a.   b.

FIGURE 2.5 High levels of radiation.
a. Radiation kills bacteria and fungi. After irradiation, peaches spoil less quickly and can be kept for a longer length of time. **b.** Physicians use radiation therapy to kill cancer cells.

is often applied to cancer cells. Targeted radioisotopes can be introduced into the body so that the subatomic particles emitted destroy only cancer cells, with little risk to the rest of the body.

Isotopes of an element have the same atomic number but differ in mass due to a different number of neutrons. Radioactive isotopes, which emit radiation, have the potential to do harm but also have many beneficial uses.

Electrons and Energy

In an electrically neutral atom, the positive charges of the protons in the nucleus are balanced by the negative charges of electrons moving about the nucleus. Various models in years past have attempted to illustrate the precise location of electrons. Figure 2.6 uses the Bohr model, which is named after the physicist Niels Bohr. The Bohr model is useful, but we need to realize that today's physicists tell us it is not possible to determine the precise location of any individual electron at any given moment. An **orbital** is defined as a particular volume of space where an electron is most apt to be found most of the time. Orbitals near the nucleus are circular or dumbbell-shaped. Other shapes occur in more distant orbitals. All orbitals, regardless of their shape, contain no more than two electrons.

It seems reasonable to suggest that negatively charged electrons are attracted to the positively charged nucleus, and that it takes energy to push them away and keep them in their orbitals. Further, the more distant the orbital, the more energy it takes. Therefore, it is proper to speak of electrons as being at particular energy levels in relation to the nucleus. When you study photosynthesis, you will learn that when atoms absorb the energy of the sun, electrons are boosted to a higher energy level. Later, as the electrons return to their original energy level, the energy is transformed into chemical energy, and this chemical energy supports all life on Earth. Our very existence is dependent on the energy of electrons.

Just now, we want to consider that each energy level contains a certain number of orbitals, and therefore a certain number of electrons. In the diagrams shown in Figure 2.6, the energy levels (also termed electron shells) are drawn as concentric rings about the nucleus. For atoms up through number 20 (i.e., calcium), the first shell (closest to the nucleus) can contain two electrons; thereafter, each additional shell can contain eight electrons. For these atoms, each lower level is filled with electrons before the next higher level contains any electrons.

The sulfur atom, with an atomic number of 16, has two electrons in the first shell, eight electrons in the second shell, and six electrons in the third, or outer, shell. Revisit the periodic table (see Fig. 2.3), and note that sulfur is in the third period. In other words, the horizontal row tells you how many shells an atom has. Also note that sulfur is in group 6. The group tells you how many electrons an atom has in its outer shell.

If an atom has only one shell, the outer shell is complete when it has two electrons. Otherwise, the **octet rule,** which states that the outer shell is most stable when it has eight electrons, holds. As mentioned previously, atoms in group 8 of the periodic table are called the noble gases because they do not ordinarily react. Atoms with fewer than eight electrons in the outer shell react with other atoms in such a way that after the reaction, each has a stable outer shell. Atoms can give up, accept, or share electrons in order to have eight electrons in the outer shell.

The number of electrons in the outer shell determines whether an atom reacts with other atoms.

FIGURE 2.6 Bohr models of atoms. Electrons orbit the nucleus at particular energy levels (electron shells): the first shell contains up to two electrons, and each shell thereafter can contain up to eight electrons as long as we consider only atoms with an atomic number of 20 or below. Each shell is filled before electrons are placed in the next shell. Why does carbon have only two shells while phosphorus and sulfur have three shells?

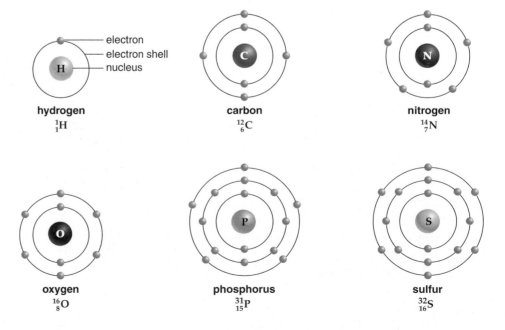

hydrogen
$_1^1H$

carbon
$_6^{12}C$

nitrogen
$_7^{14}N$

oxygen
$_8^{16}O$

phosphorus
$_{15}^{31}P$

sulfur
$_{16}^{32}S$

2.2 ELEMENTS AND COMPOUNDS

Atoms, except for noble gases, routinely bond with one another. For example, oxygen does not exist in nature as a single atom, O; instead, two oxygen atoms are joined to form a diatomic molecule (O_2). Other naturally occurring diatomic molecules include hydrogen (H_2) and nitrogen (N_2). When atoms of two or more different elements bond together, the product is called a **compound**. Water (H_2O) is a compound that contains atoms of hydrogen and oxygen. We can also speak of molecules of water because a **molecule** [L. *moles*, mass] is the smallest part of a compound that still has the properties of that compound. Some molecules can be very large, such as one of the types of chlorophyll ($C_{55}H_{72}O_5N_4Mg$).

Electrons possess energy, and the bonds that exist between atoms also contain energy. Organisms are directly dependent on chemical-bond energy to maintain their organization. When a chemical reaction occurs, electrons shift in their relationship to one another, and energy may be given off or absorbed. This same energy is used to carry on our daily lives.

Ionic Bonding

Ions form when electrons are transferred from one atom to another. For example, sodium (Na), with only one electron in its third shell, tends to be an electron donor (Fig. 2.7a). Once it gives up this electron, the second shell, with eight electrons, becomes its outer shell. Chlorine (Cl), on the other hand, tends to be an electron acceptor. Its outer shell has seven electrons, so if it acquires only one more electron it has a completed outer shell. When a sodium atom and a chlorine atom come together, an electron is transferred from the sodium atom to the chlorine atom. Now both atoms have eight electrons in their outer shells.

This electron transfer, however, causes a charge imbalance in each atom. The sodium atom has one more proton than it has electrons; therefore, it has a net charge of +1 (symbolized by Na^+). The chlorine atom has one more electron than it has protons; therefore, it has a net charge of −1 (symbolized by Cl^-). Such charged particles are called **ions**. Sodium (Na^+) and chloride (Cl^-) are not the only biologically important ions. Some, such as potassium (K^+), are formed by the transfer of a single electron to another atom; others, such as calcium (Ca^{2+}) and magnesium (Mg^{2+}), are formed by the transfer of two electrons.

Ionic compounds are held together by an attraction between negatively and positively charged ions called an **ionic bond**. When sodium reacts with chlorine, an ionic compound called sodium chloride (NaCl) results. Sodium chloride is a salt, commonly known as table salt, because it is used to season our food (Fig. 2.7b). **Salts** can exist as a dry solid, but when salts are placed in water, they release ions as they dissolve. NaCl separates into Na^+ and Cl^-. Ionic compounds are most commonly found in this dissociated (ionized) form in biological systems because these systems are 70–90% water.

The transfer of electron(s) between atoms results in ions that are held together by an ionic bond, the attraction of negative and positive charges.

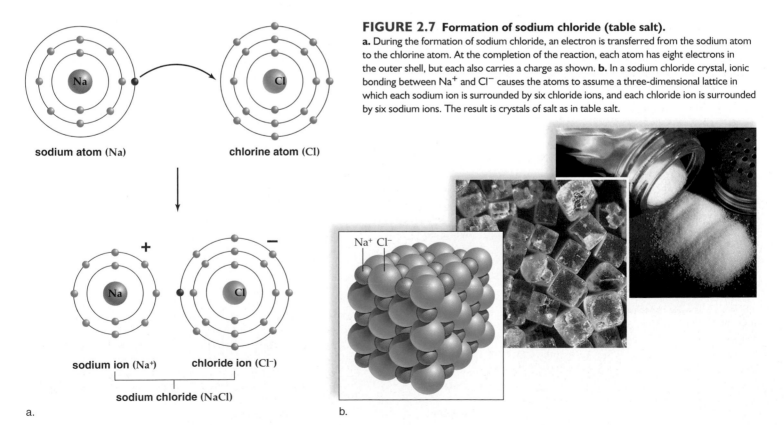

FIGURE 2.7 Formation of sodium chloride (table salt).
a. During the formation of sodium chloride, an electron is transferred from the sodium atom to the chlorine atom. At the completion of the reaction, each atom has eight electrons in the outer shell, but each also carries a charge as shown. **b.** In a sodium chloride crystal, ionic bonding between Na^+ and Cl^- causes the atoms to assume a three-dimensional lattice in which each sodium ion is surrounded by six chloride ions, and each chloride ion is surrounded by six sodium ions. The result is crystals of salt as in table salt.

sodium atom (Na)   chlorine atom (Cl)

+   −

sodium ion (Na^+)   chloride ion (Cl^-)

sodium chloride (NaCl)

Na^+ Cl^-

a.   b.

Covalent Bonding

A **covalent bond** [L. *co*, together, with, and *valens*, strength] results when two atoms share electrons in such a way that each atom has an octet of electrons in the outer shell (or two electrons, in the case of hydrogen). In a hydrogen atom, the outer shell is complete when it contains two electrons. If hydrogen is in the presence of a strong electron acceptor, it gives up its electron to become a hydrogen ion (H^+). But if this is not possible, hydrogen can share with another atom and thereby have a completed outer shell. For example, one hydrogen atom will share with another hydrogen atom. Their two orbitals overlap and the electrons are shared between them (Fig. 2.8a). Because they share the electron pair, each atom has a completed outer shell.

A more common way to symbolize that atoms are sharing electrons is to draw a line between the two atoms, as in the structural formula H—H. In a molecular formula, the line is omitted and the molecule is simply written as H_2.

Sometimes, atoms share more than one pair of electrons to complete their octets. A double covalent bond occurs when two atoms share two pairs of electrons (Fig. 2.8b). To show that oxygen gas (O_2) contains a double bond, the molecule can be written as O=O.

It is also possible for atoms to form triple covalent bonds, as in nitrogen gas (N_2), which can be written as N≡N. Single covalent bonds between atoms are quite strong, but double and triple bonds are even stronger.

Shape of Molecules

Structural formulas make it seem as if molecules are one-dimensional, but actually molecules have a three-dimensional shape that often determines their biological function. Molecules consisting of only two atoms are always linear, but a molecule such as methane with five atoms (Fig. 2.8c) has a tetrahedral shape. Why? Because, as shown in the ball-and-stick model, each bond is pointing to the corners of a tetrahedron (Fig. 2.8d, *left*). The space-filling model comes closest to the actual shape of the molecule. In space-filling models, each type of atom is given a particular color—carbon is always black and hydrogen is always off-white (Fig. 2.8d, *right*).

The shapes of molecules are necessary to the structural and functional roles they play in living things. For example, hormones have specific shapes that allow them to be recognized by the cells in the body. We can stay well only when antibodies combine with disease-causing agents, like a key fits a lock. Similarly, homeostasis is only maintained when enzymes have the proper shape to carry out their particular reactions in cells.

In a covalent molecule, atoms share electrons; the final shape of the molecule often determines the role it plays in cells and organisms.

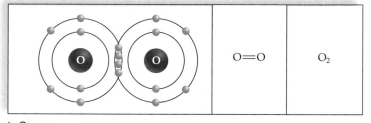

a. Hydrogen gas

b. Oxygen gas

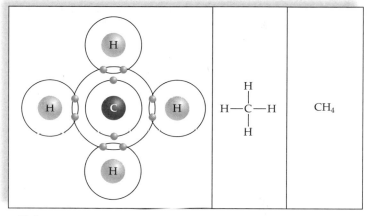

c. Methane

d. Methane—continued

FIGURE 2.8 Covalently bonded molecules.
In a covalent bond, atoms share electrons, allowing each atom to have a completed outer shell. **a.** A molecule of hydrogen (H_2) contains two hydrogen atoms sharing a pair of electrons. This single covalent bond can be represented in any of the three ways shown. **b.** A molecule of oxygen (O_2) contains two oxygen atoms sharing two pairs of electrons. This results in a double covalent bond. **c.** A molecule of methane (CH_4) contains one carbon atom bonded to four hydrogen atoms. **d.** When carbon binds to four other atoms, as in methane, each bond actually points to one corner of a tetrahedron. Ball-and-stick models and space-filling models are three-dimensional representations of the molecule.

Nonpolar and Polar Covalent Bonds

When the sharing of electrons between two atoms is fairly equal, the covalent bond is said to be a **nonpolar covalent bond.** All the molecules in Figure 2.8, including methane (CH_4), are nonpolar. In the case of water (H_2O), however, the sharing of electrons between oxygen and each hydrogen is not completely equal. The attraction of an atom for the electrons in a covalent bond is called its **electronegativity.** The larger oxygen atom, with the greater number of protons, is more electronegative than the hydrogen atom. The oxygen atom can attract the electron pair to a greater extent than each hydrogen atom can. In a water molecule, this causes the oxygen atom to assume a slightly negative charge (δ^-), and it causes the hydrogen atoms to assume slightly positive charges (δ^+). The unequal sharing of electrons in a covalent bond creates a **polar covalent bond,** and in the case of water, the molecule itself is a polar molecule (Fig. 2.9a).

The water molecule is a polar molecule and has an asymmetrical distribution of charge: one end of the molecule (the oxygen atom) carries a slightly negative charge, and the other ends of the molecule (the hydrogen atoms) carry slightly positive charges.

Hydrogen Bonding

Polarity within a water molecule causes the hydrogen atoms in one molecule to be attracted to the oxygen atoms in other water molecules (Fig. 2.9b). This attraction, although weaker than an ionic or covalent bond, is called a **hydrogen bond.** Because a hydrogen bond is easily broken, it is often represented by a dotted line. Hydrogen bonding is not unique to water. Many biological molecules have polar covalent bonds involving an electropositive hydrogen and usually an electronegative oxygen or nitrogen. In these instances, a hydrogen bond can occur within the same molecule or between different molecules.

Although a hydrogen bond is more easily broken than a covalent bond, many hydrogen bonds taken together are quite strong. Hydrogen bonds between cellular molecules help maintain their proper structure and function. For example, hydrogen bonds hold the two strands of DNA together. When DNA makes a copy of itself, each hydrogen bond easily breaks, allowing the DNA to unzip. On the other hand, the hydrogen bonds acting together add stability to the DNA molecule. As we shall see, many of the important properties of water are the result of hydrogen bonding.

A hydrogen bond occurs between a slightly positive hydrogen atom of one molecule and a slightly negative atom of another molecule, or between atoms of the same molecule.

Electron Model	Ball-and-stick Model	Space-filling Model
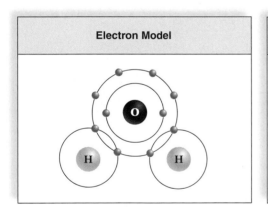	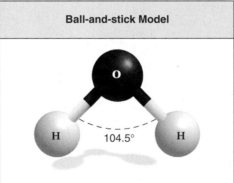	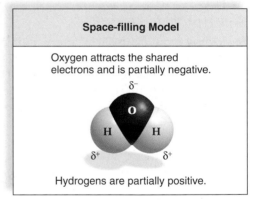

a. Water (H_2O)

FIGURE 2.9 Water molecule.
a. Three models for the structure of water. The electron model does not indicate the shape of the molecule. The ball-and-stick model shows that the two bonds in a water molecule are angled at 104.5°. The space-filling model also shows the V shape of a water molecule. **b.** Hydrogen bonding between water molecules. A hydrogen bond is the attraction of a slightly positive hydrogen to a slightly negative atom in the vicinity. Each water molecule can hydrogen-bond to four other molecules in this manner. When water is in its liquid state, some hydrogen bonds are forming and others are breaking at all times.

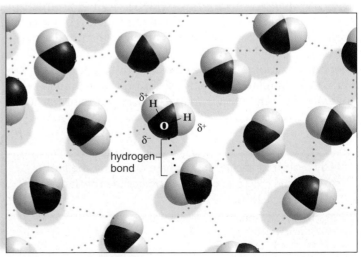

b. Hydrogen bonding between water molecules

2.3 CHEMISTRY OF WATER

The first cell(s) evolved in water, and all living things are 70–90% water. Water is a polar molecule, and water molecules are hydrogen-bonded to one another (Fig. 2.9b). Due to hydrogen bonding water molecules cling together. Without hydrogen bonding between molecules, water would melt at −100°C and boil at −91°C, making most of the water on Earth steam, and life unlikely. But because of hydrogen bonding, water is a liquid at temperatures typically found on the Earth's surface. It melts at 0°C and boils at 100°C. These and other unique properties of water make it essential to the existence of life.

Properties of Water

Water has a high heat capacity. A **calorie** is the amount of heat energy needed to raise the temperature of 1 g of water 1°C. In comparison, other covalently bonded liquids require input of only about half this amount of energy to rise in temperature 1°C. The many hydrogen bonds that link water molecules help water absorb heat without a great change in temperature.

Converting 1 g of the coldest liquid water to ice requires the loss of 80 calories of heat energy (Fig. 2.10a). Water holds onto its heat, and its temperature falls more slowly than that of other liquids. This property of water is important not only for aquatic organisms but also for all living things. Because the temperature of water rises and falls slowly, organisms are better able to maintain their normal internal temperatures and are protected from rapid temperature changes.

Water has a high heat of vaporization. Converting 1 g of the hottest water to a gas requires an input of 540 calories of heat energy. Water has a high heat of vaporization because hydrogen bonds must be broken before water boils and water molecules vaporize—that is, evaporate into the environment. Water's high heat of vaporization gives animals in a hot environment an efficient way to release excess body heat. When an animal sweats, or gets splashed, body heat is used to vaporize water, thus cooling the animal (Fig. 2.10b).

Because of water's high heat capacity and high heat of vaporization, temperatures along coasts are moderate. During the summer, the ocean absorbs and stores solar heat, and during the winter, the ocean releases it slowly. In contrast, the interior regions of continents experience abrupt changes in temperatures.

Water is a solvent. Due to its polarity, water facilitates chemical reactions, both outside and within living systems. It dissolves a great number of substances. A **solution** contains dissolved substances, which are then called **solutes.** When ionic salts—for

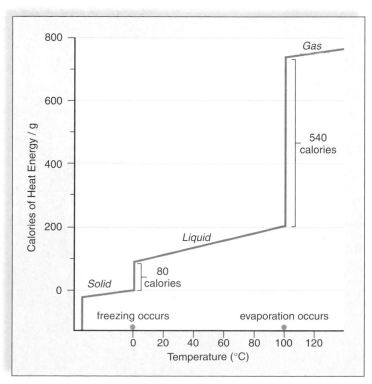

a. Calories lost when 1 g of liquid water freezes and calories required when 1 g of liquid water evaporates.

b. Bodies of organisms cool when their heat is used to evaporate water.

FIGURE 2.10 Temperature and water.
a. Water can be a solid, a liquid, or a gas at naturally occurring environmental temperatures. At room temperature and pressure, water is a liquid. When water freezes and becomes a solid (ice), it gives off heat, and this heat can help keep the environmental temperature higher than expected. On the other hand, when water vaporizes, it takes up a large amount of heat as it changes from a liquid to a gas. **b.** This means that splashing water on the body will help keep body temperature within a normal range. Can you also see why water's properties help keep the coasts moderate in both winter and summer?

example, sodium chloride (NaCl)—are put into water, the negative ends of the water molecules are attracted to the sodium ions, and the positive ends of the water molecules are attracted to the chloride ions. This causes the sodium ions and the chloride ions to separate, or dissociate, in water.

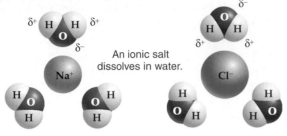

An ionic salt dissolves in water.

Water is also a solvent for larger molecules that contain ionized atoms or are polar molecules.

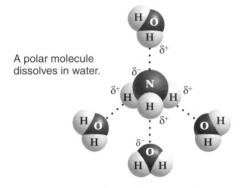

A polar molecule dissolves in water.

Those molecules that can attract water are said to be **hydrophilic** [Gk. *hydrias,* of water, and *phileo,* love]. When ions and molecules disperse in water, they move about and collide, allowing reactions to occur. Nonionized and nonpolar molecules that cannot attract water are said to be **hydrophobic** [Gk. *hydrias,* of water, and *phobos,* fear]. Gasoline contains nonpolar molecules, and therefore it does not mix with water and is hydrophobic.

Water molecules are cohesive and adhesive.   Cohesion is apparent because water flows freely, yet water molecules do not separate from each other. They cling together because of hydrogen bonding. Water's positive and negative poles allow it to adhere to polar surfaces; therefore, water exhibits adhesion. Cohesion and adhesion allow water to fill a tubular vessel. Therefore, water is an excellent transport system, both outside of and within living organisms. Unicellular organisms rely on external water to transport nutrient and waste molecules, but multicellular organisms often contain internal vessels in which water serves to transport nutrients and wastes. For example, the liquid portion of our blood, which transports dissolved and suspended substances throughout the body, is 90% water.

Cohesion and adhesion also contribute to the transport of water in plants. Plants have their roots anchored in the soil, where they absorb water, but the leaves are uplifted and exposed to solar energy. How is it possible for water to rise to the top of even very tall trees (Fig. 2.11)? A plant contains a system of vessels that reaches from the roots to the leaves. Water

evaporating from the leaves is immediately replaced with water molecules from the vessels. Because water molecules are cohesive, a tension is created that pulls a water column up from the roots. Adhesion of water to the walls of the vessels also helps prevent the water column from breaking apart.

Water has a high surface tension.   The stronger the force between molecules in a liquid, the greater the surface tension. As with cohesion, hydrogen bonding causes water to have a high surface tension. This property makes it possible for humans to skip rocks on water. Water striders, a common insect, can even walk on the surface of a pond without breaking the surface.

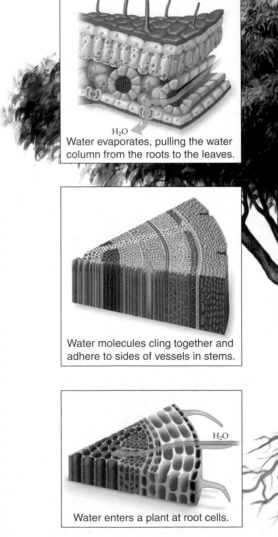

H₂O
Water evaporates, pulling the water column from the roots to the leaves.

Water molecules cling together and adhere to sides of vessels in stems.

H₂O
Water enters a plant at root cells.

FIGURE 2.11 Water as a transport medium.
How does water rise to the top of tall trees? Vessels are water-filled pipelines from the roots to the leaves. When water evaporates from the leaves, this water column is pulled upward due to the cohesion of water molecules with one another and the adhesion of water molecules to the sides of the vessels.

Frozen water (ice) is less dense than liquid water. As liquid water cools, the molecules come closer together. They are densest at 4°C, but they are still moving about (Fig. 2.12). At temperatures below 4°C, there is only vibrational movement, and hydrogen bonding becomes more rigid but also more open. This means that water expands as it freezes, which is why cans of soda burst when placed in a freezer or why frost heaves make northern roads bumpy in the winter. It also means that ice is less dense than liquid water, and therefore ice floats on liquid water.

If ice did not float on water, it would sink, and ponds, lakes, and perhaps even the ocean would freeze solid, making life impossible in the water and also on land. Instead, bodies of water always freeze from the top down. When a body of water freezes on the surface, the ice acts as an insulator to prevent the water below it from freezing. This protects aquatic organisms so that they can survive the winter. As ice melts in the spring, it draws heat from the environment, helping to prevent a sudden change in temperature that might be harmful to life.

Water has unique properties that allow cellular activities to occur and that make life on Earth possible.

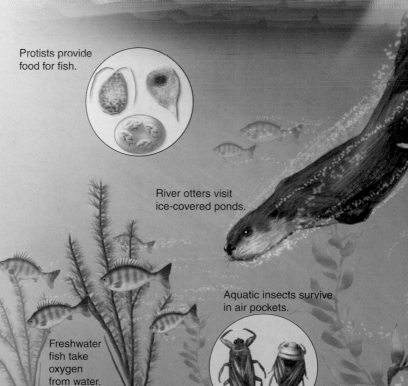

FIGURE 2.12 A pond in winter.
Above: Remarkably, water is more dense at 4°C than at 0°C. Most substances contract when they solidify, but water expands when it freezes because in ice, water molecules form a lattice in which the hydrogen bonds are farther apart than in liquid water. *Below:* The layer of ice that forms at the top of a pond shields the water and protects the protists, plants, and animals so that they can survive the winter. These animals, except for the otter, are ectothermic, which means that they take on the temperature of the outside environment. This might seem disadvantageous until you realize that water remains relatively warm because of its high heat capacity. During the winter, frogs and turtles hibernate and in this way, lower their oxygen needs. Insects survive in air pockets. Fish, as you will learn later in this text, have an efficient means of extracting oxygen from the water and they need less oxygen than the endothermic otter, which depends on muscle activity to warm its body.

Protists provide food for fish.

River otters visit ice-covered ponds.

Aquatic insects survive in air pockets.

Freshwater fish take oxygen from water.

ice lattice

liquid water

Density (g/cm³)

Temperature (C)

ice layer

Common frogs and pond turtles hibernate.

Acids and Bases

When water ionizes, it releases an equal number of **hydrogen ions (H^+)** and **hydroxide ions (OH^-)**:

$$H—O—H \rightleftharpoons H^+ + OH^-$$
water   hydrogen   hydroxide
                 ion   ion

Only a few water molecules at a time dissociate, and the actual number of H^+ and OH^- is very small (1×10^{-7} moles/liter).[1]

Acidic Solutions (High H^+ Concentrations)

Lemon juice, vinegar, tomatoes, and coffee are all acidic solutions. What do they have in common? **Acids** are substances that dissociate in water, releasing hydrogen ions (H^+).[2] For example, hydrochloric acid (HCl) is an important inorganic acid that dissociates in this manner:

$$HCl \longrightarrow H^+ + Cl^-$$

Dissociation is almost complete; therefore, HCl is called a strong acid. If hydrochloric acid is added to a beaker of water, the number of hydrogen ions (H^+) increases greatly.

Basic Solutions (Low H^+ Concentration)

Milk of magnesia and ammonia are common basic solutions familiar to most people. **Bases** are substances that either take up hydrogen ions (H^+) or release hydroxide ions (OH^-). For example, sodium hydroxide (NaOH) is an important inorganic base that dissociates in this manner:

$$NaOH \longrightarrow Na^+ + OH^-$$

Dissociation is almost complete; therefore, sodium hydroxide is called a strong base. If sodium hydroxide is added to a beaker of water, the number of hydroxide ions increases.

pH Scale

The **pH scale** is used to indicate the acidity or basicity (alkalinity) of a solution.[3] The pH scale (Fig. 2.13) ranges from 0 to 14. A pH of 7 represents a neutral state in which the hydrogen ion and hydroxide ion concentrations are equal. A pH below 7 is an acidic solution because the hydrogen ion concentration [H^+] is greater than the hydroxide concentration [OH^-]. A pH above 7 is basic because [OH^-] is greater than [H^+]. Further, as we move down the pH scale from pH 14 to pH 0, each unit has ten times the [H^+] of the previous unit. As we move up the scale from 0 to 14, each unit has ten times the [OH^-] of the previous unit.

The pH scale was devised to eliminate the use of cumbersome numbers. For example, the possible hydrogen ion

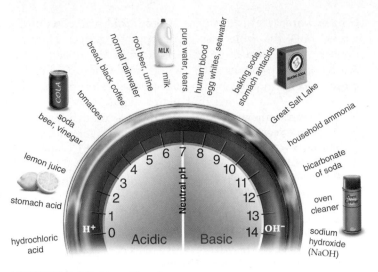

FIGURE 2.13   The pH scale.
The dial of this pH meter indicates that pH ranges from 0 to 14 with 0 being the most acidic and 14 being the most basic. pH 7 (neutral pH) has equal amounts of hydrogen ions (H^+) and hydroxide ions (OH^-). An acidic pH has more H^+ than OH^- and a basic pH has more OH^- than H^+.

concentrations of a solution are on the left of this listing and the pH is on the right:

	[H^+] (moles per liter)	pH
0.000001	$= 1 \times 10^{-6}$	6
0.0000001	$= 1 \times 10^{-7}$	7
0.00000001	$= 1 \times 10^{-8}$	8

To further illustrate the relationship between hydrogen ion concentration and pH, consider the following question. Which of the pH values listed indicates a higher hydrogen ion concentration [H^+] than pH 7, and therefore would be an acidic solution? A number with a smaller negative exponent indicates a greater quantity of hydrogen ions than one with a larger negative exponent. Therefore, pH 6 is an acidic solution.

The Ecology Focus on page 31 describes detrimental environmental consequences to nonliving and living things as rain and snow have become more acidic. In humans, pH needs to be maintained with a narrow range or there are health consequences. The pH of blood is around 7.4, and blood is buffered in the manner described next to keep the pH within a normal range.

Buffers and pH

A **buffer** is a chemical or a combination of chemicals that keeps pH within normal limits. Many commercial products such as Bufferin, shampoos, or deodorants, are buffered as an added incentive for us to buy them. Buffers resist pH changes because they can take up excess hydrogen ions (H^+) or hydroxide ions (OH^-).

In living things, the pH of body fluids is maintained within a narrow range, or else health suffers. The pH of

---

[1] In chemistry, a mole is defined as the amount of matter that contains as many objects (atoms, molecules, ions) as the number of atoms in exactly 12 g of ^{12}C.
[2] A hydrogen atom contains one electron and one proton. A hydrogen ion has only one proton, so it is often simply called a proton.
[3] pH is defined as the negative log of the hydrogen ion concentration [H^+]. A log is the power to which ten must be raised to produce a given number.

ecology focus

The Harm Done by Acid Deposition

Normally, rainwater has a pH of about 5.6 because the carbon dioxide in the air combines with water to give a weak solution of carbonic acid. Rain falling in the northeastern United States and southeastern Canada now has a pH between 5 and 4. We have to remember that a pH of 4 is ten times more acidic than rain with a pH of 5 to comprehend the increase in acidity this represents.

Strong evidence indicates that this observed increase in rainwater acidity is a result of the burning of fossil fuels such as coal, oil, and gasoline derived from oil. When fossil fuels are burned, sulfur dioxide and nitrogen oxides are produced, and they combine with water vapor in the atmosphere to form sulfuric and nitric acids. These acids return to Earth dissolved in rain or snow, a process properly called wet deposition but more often called acid rain. Dry particles of sulfate and nitrate salts descend from the atmosphere during dry deposition.

Unfortunately, regulations that require the use of tall smokestacks to reduce local air pollution only cause pollutants to be carried farther from their place of origin. For example, acid deposition in southeastern Canada results from the burning of fossil fuels in factories and power plants in the midwestern United States. Acid deposition adversely affects lakes, particularly in areas where the soil is thin and lacks limestone (calcium carbonate, $CaCO_3$), a buffer to acid deposition. Acid deposition leaches aluminum from the soil, carries aluminum into the lakes, and converts mercury deposits in lake bottom sediments to soluble and toxic methyl mercury. Lakes not only become more acidic, but they also accumulate toxic substances. The increasing deterioration of thousands of lakes and rivers in southern Norway and Sweden during the past two decades has been attributed to acid deposition. Some lakes contain no fish, and others have decreasing numbers of fish. The same phenomenon has been observed in Canada and the United States (mostly in the Northeast and upper Midwest).

In forests, acid deposition weakens trees because it leaches away nutrients and releases aluminum. By 1988, most spruce, fir, and other conifers atop North Carolina's Mt. Mitchell were dead from being bathed in ozone and acid fog for years. The soil was so acidic that new seedlings could not survive. Many countries in northern Europe have also reported woodland and forest damage most likely due to acid deposition (Fig. 2A).

Lake and forest deterioration aren't the only effects of acid deposition. Reduction of agricultural yields, damage to marble and limestone monuments and buildings, and even illnesses in humans have been reported. Acid deposition has been implicated in the increased incidence of lung cancer and possibly colon cancer in residents of the East Coast. Tom McMillan, former Canadian Minister of the Environment, says that acid rain is "destroying our lakes, killing our fish, undermining our tourism, retarding our forests, harming our agriculture, devastating our heritage, and threatening our health."

a.

b.

FIGURE 2A Effects of acid deposition.
The burning of gasoline derived from oil, a fossil fuel, leads to acid deposition, which causes (a) statues to deteriorate and (b) trees to die.

our blood when we are healthy is always about 7.4—that is, just slightly basic (alkaline). If the blood pH drops to about 7, acidosis results. If the blood pH rises to about 7.8, alkalosis results. Both conditions can be life threatening; the blood pH must be kept around 7.4. Normally, pH stability is possible because the body has built-in mechanisms to prevent pH changes. Buffers are the most important of these mechanisms. Buffers help keep the pH within normal limits because they are chemicals or combinations of chemicals that take up excess hydrogen ions (H^+) or hydroxide ions (OH^-). For example, carbonic acid (H_2CO_3) is a weak acid that minimally dissociates and then re-forms in the following manner:

$$\underset{\text{carbonic acid}}{H_2CO_3} \underset{\text{re-forms}}{\overset{\text{dissociates}}{\rightleftharpoons}} H^+ + \underset{\text{bicarbonate ion}}{HCO_3^-}$$

Blood always contains a combination of some carbonic acid and some bicarbonate ions. When hydrogen ions (H^+) are added to blood, the following reaction occurs:

$$H^+ + HCO_3^- \longrightarrow H_2CO_3$$

When hydroxide ions (OH^-) are added to blood, this reaction occurs:

$$OH^- + H_2CO_3 \longrightarrow HCO_3^- + H_2O$$

These reactions prevent any significant change in blood pH.

A pH value is the hydrogen ion concentration [H^+] of a solution. Buffers act to keep the pH within normal limits.

CONNECTING THE CONCEPTS

The methods of science applied to the structure of matter have revealed that all substances consist of various combinations of the same 92 elements. Living things consist primarily of just six of these elements—carbon, hydrogen, nitrogen, oxygen, phosphorus, and sulfur (CHNOPS for short). These elements combine to form the unique types of molecules found in living cells. In organisms, many other elements exist in smaller amounts as ions, and their functions are dependent on their charged nature. Cells consist largely of water, a molecule that contains only hydrogen and oxygen. Polar covalent bonding between the atoms and hydrogen bonding between the molecules give water the properties that make life possible. Presently, no other planet we are aware of has liquid water.

In the next chapter, we will learn that a carbon atom combines covalently with CHNOPS to form the organic molecules of cells. It is these unique molecules that set living forms apart from nonliving objects. Carbon-containing molecules can be modified in numerous ways, and this accounts for life's diversity, such as differences between a bottle-nosed dolphin and a black shoulder peacock. Varying molecular compositions in plants can also tell us, for example, why some trees have leaves that change color in the fall.

It is difficult for us to visualize that a bottle-nosed dolphin, a kangaroo, or a pine tree is a combination of molecules and ions, but later in this text we will learn that even our thoughts about these organisms are simply the result of molecules flowing from one brain cell to another. An atomic, ionic, and molecular understanding of the variety of processes unique to life provides a deeper understanding of the definition of life and offers tools for the improvement of its quality, preservation of its diversity, and appreciation of its beauty.

Summary

2.1 CHEMICAL ELEMENTS

Both living and nonliving things are composed of matter consisting of elements. The acronym CHNOPS stands for the most significant elements (atoms) found in living things: carbon, hydrogen, nitrogen, oxygen, phosphorus, and sulfur. Elements contain atoms, and atoms contain subatomic particles. Protons and neutrons in the nucleus determine the atomic mass of an atom. The atomic number indicates the number of protons and the number of electrons in electrically neutral atoms. Protons have positive charges, neutrons are uncharged, and electrons have negative charges. Isotopes are atoms of a single element that differ in their numbers of neutrons. Radioactive isotopes have many uses, including serving as tracers in biological experiments and medical procedures.

Electrons occupy energy levels (electron shells) at discrete distances from the nucleus. The number of electrons in the outer shell determines the reactivity of an atom. The first shell is complete when it is occupied by two electrons. In atoms up through calcium, number 20, every shell beyond the first shell is complete with eight electrons. The octet rule states that atoms react with one another in order to have a completed outer shell. Most atoms, including those common to living things, do not have filled outer shells and this causes them to react with one another to form compounds and/or molecules. Following the reaction, the atoms have completed outer shells.

2.2 ELEMENTS AND COMPOUNDS

Ions form when atoms lose or gain one or more electrons to achieve a completed outer shell. An ionic bond is an attraction between oppositely charged ions. When covalent compounds form, atoms share electrons. A covalent bond is one or more shared pairs of electrons. There are single, double, and triple covalent bonds.

When carbon combines with four other atoms, the resulting molecule has a tetrahedral shape. The shape of a molecule is important to its biological role. Hormones and other molecules, for example, are recognized by a cell's receptors because of their specific shapes.

In polar covalent bonds, the sharing of electrons is not equal; one of the atoms exerts greater attraction for the shared electrons than the other, and a slight charge results on each atom. A hydrogen bond is a weak attraction between a slightly positive hydrogen atom and a slightly negative oxygen or nitrogen atom within the same or a different molecule. Hydrogen bonds help maintain the structure and function of cellular molecules.

2.3 CHEMISTRY OF WATER

Water is a polar molecule. The polarity of water molecules allows hydrogen bonding to occur between water molecules. Water's polarity and hydrogen bonding account for its unique properties. These features allow living things to exist and carry on cellular activities.

A small fraction of water molecules dissociate to produce an equal number of hydrogen ions and hydroxide ions. Solutions with equal numbers of H^+ and OH^- are termed neutral. In acidic solutions, there are more hydrogen ions than hydroxide ions; these solutions have a pH less than 7. In basic solutions, there are more hydroxide ions than hydrogen ions; these solutions have a pH greater than 7. Cells are sensitive to pH changes. Biological systems often contain buffers that help keep the pH within a normal range.

Reviewing the Chapter

1. Name the kinds of subatomic particles studied. What is their atomic mass unit, charge, and location in an atom?   20–21
2. What is an isotope? A radioactive isotope? What are some uses of radioactive isotopes?   22–23
3. Using the Bohr model, draw an atomic structure for a carbon that has six protons and six neutrons.   23
4. Draw an atomic representation for $MgCl_2$. Using the octet rule, explain the structure of the compound.   24
5. Explain whether CO_2 (O=C=O) is an ionic or a covalent compound. Why does this arrangement satisfy all atoms involved?   25
6. Of what significance is the shape of molecules in organisms?   25
7. Explain why water is a polar molecule. What is the relationship between the polarity of the molecule and the hydrogen bonding between water molecules?   26
8. Name six properties of water, and relate them to the structure of water, including its polarity and hydrogen bonding between molecules.   27–29
9. Define an acid and a base. On the pH scale, which numbers indicate a solution is acidic? Basic? Neutral?   30
10. What are buffers, and why are they important to life?   30–31

Testing Yourself

Choose the best answer for each question.

1. Which of the subatomic particles contributes almost no weight to an atom?
   a. protons in the electron shells
   b. electrons in the nucleus
   c. neutrons in the nucleus
   d. electrons at various energy levels

2. The atomic number tells you the
   a. number of neutrons in the nucleus.
   b. number of protons in the atom.
   c. atomic mass of the atom.
   d. number of its electrons if the atom is neutral.
   e. Both b and d are correct.

3. An atom that has two electrons in the outer shell, such as calcium, would most likely
   a. share to acquire a completed outer shell.
   b. lose these two electrons and become a negatively charged ion.
   c. lose these two electrons and become a positively charged ion.
   d. bind with carbon by way of hydrogen bonds.
   e. bind with another calcium atom to satisfy its energy needs.

4. Radioactive elements differ in their
   a. number of protons.
   b. atomic number.
   c. number of neutrons.
   d. number of electrons.

5. When an atom gains electrons, it
   a. forms a negatively charged ion.
   b. forms a positively charged ion.
   c. forms covalent bonds.
   d. forms ionic bonds.
   e. gains atomic mass.

6. A covalent bond is indicated by
   a. plus and minus charges attached to atoms.
   b. dotted lines between hydrogen atoms.
   c. concentric circles about a nucleus.
   d. overlapping electron shells or a straight line between atomic symbols.
   e. the touching of atomic nuclei.

7. The shape of a molecule
   a. is dependent in part on the angle of bonds between its atoms.
   b. influences its biological function.
   c. is dependent on its electronegativity.
   d. is dependent on its place in the periodic table.
   e. Both a and b are correct.

8. In which of these are the electrons always shared unequally?
   a. double covalent bond
   b. triple covalent bond
   c. hydrogen bond
   d. polar covalent bond
   e. ionic and covalent bonds

9. In the molecule

$$H-\overset{\displaystyle H}{\underset{\displaystyle H}{C}}-H$$

   a. all atoms have eight electrons in the outer shell.
   b. all atoms are sharing electrons.

   c. carbon could accept more hydrogen atoms.
   d. the bonds point to the corners of a square.
   e. All of these are correct.

10. Which of these properties of water cannot be attributed to hydrogen bonding between water molecules?
    a. Water stabilizes temperature inside and outside the cell.
    b. Water molecules are cohesive.
    c. Water is a solvent for many molecules.
    d. Ice floats on liquid water.
    e. Both b and c are correct.

11. H_2CO_3/$NaHCO_3$ is a buffer system in the body. What effect will the addition of an acid have on the pH of this buffer?
    a. The pH will rise.
    b. The pH will lower.
    c. The pH will not change.
    d. All of these are correct.

12. Rainwater has a pH of about 5.6; therefore, rainwater is
    a. a neutral solution.
    b. an acidic solution.
    c. a basic solution.
    d. It depends if it is buffered.

13. Acids
    a. release hydrogen ions in solution.
    b. cause the pH of a solution to rise above 7.
    c. take up hydroxide ions and become neutral.
    d. increase the number of water molecules.
    e. Both a and b are correct.

14. Which type of bond results from the sharing of electrons between atoms?
    a. covalent   c. hydrogen
    b. ionic   d. neutral

15. Complete this diagram of a nitrogen atom by placing the correct number of protons and neutrons in the nucleus and electrons in the shells. Explain why the correct formula for ammonia is NH_3, not NH_4.

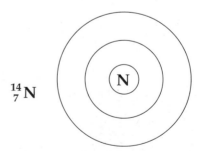

$$^{14}_{7}N$$

16. In the molecule CH_4,
    a. all atoms have eight electrons in the outer shell.
    b. all atoms are sharing electrons.
    c. carbon could accept more hydrogen atoms.
    d. All of these are correct.

17. If a chemical accepted H^+ from the surrounding solution, the chemical could be
    a. a base.
    b. an acid.
    c. a buffer.
    d. None of the above are correct.
    e. Both a and c are correct.

18. Which of these best describes the changes that occur when a solution goes from pH 5 to pH 8?
    a. The hydrogen ion concentration decreases as the solution goes from acidic to basic.
    b. The hydrogen ion concentration increases as the solution goes from basic to acidic.
    c. The hydrogen ion concentration decreases as the solution goes from basic to acidic.

For questions 19–21, indicate whether the statement is true (T) or false (F).

19. The higher the pH, the higher the H^+ concentration. _____

20. Ionic bonds share electrons, and covalent bonds are an attraction between charges. _____

21. Protons are located in the nucleus, while electrons are located in shells about the nucleus. _____

22. Complete this diagram by placing an O for oxygen or an H for hydrogen on the appropriate atoms. Place partial charges where they belong.

Thinking Scientifically

1. Consider the following bar graph showing the proportions of oxygen and carbon in humans versus pumpkins. Knowing that pumpkins contain a higher percentage of water than humans, suggest reasons for the similarities and differences between the two organisms.

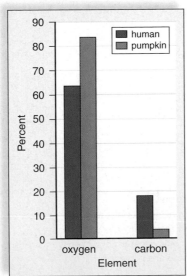

2. Natural antifreeze molecules allow many animals to exist in conditions cold enough to freeze the blood (or equivalent fluid) of animals without these additives. Knowing the role of hydrogen bonding in the transition from liquid water to ice, how might these "natural antifreeze" molecules interact with ice/water to prevent crystal growth?

Bioethical Issue: The Right to Refuse an IV

When a person gets sick or endures physical stress—as, for example, during childbirth—pH levels may dip or rise too far, endangering that person's life. In most U.S. hospitals, doctors routinely administer IVs, or intravenous infusions, of certain fluids to maintain a patient's pH level. Some people who oppose IVs for philosophical reasons may refuse an IV. That's relatively safe, as long as the person is healthy.

Problems arise when hospital policy dictates an IV, even though a patient does not want one. Should a patient be allowed to refuse an IV? Or does a hospital have the right to insist, for health reasons, that patients accept IV fluids? And what role should doctors play—patient advocates or hospital representatives?

Understanding the Terms

acid 30	hydroxide ion (OH⁻) 30
atom 20	ion 24
atomic mass 20	ionic bond 24
atomic number 21	isotope 22
atomic symbol 20	matter 20
base 30	molecule 24
buffer 30	neutron 20
calorie 27	nonpolar covalent bond 26
compound 24	octet rule 23
covalent bond 25	orbital 23
electron 20	pH scale 30
electronegativity 26	polar covalent bond 26
electron shell 20	proton 20
element 20	salt 24
hydrogen bond 26	solute 27
hydrogen ion (H^+) 30	solution 27
hydrophilic 28	tracer 22
hydrophobic 28	

Match the terms to these definitions:
a. _____ Bond in which the sharing of electrons between atoms is unequal.
b. _____ Charged particle that carries a negative or positive charge(s).
c. _____ Molecules tending to raise the hydrogen ion concentration in a solution and to lower its pH numerically.
d. _____ The smallest part of a compound that still has the properties of that compound.
e. _____ A chemical or a combination of chemicals that maintains a constant pH upon the addition of small amounts of acid or base.

ARIS, the *Biology* Website

ARIS, the website for *Biology*, provides a wealth of information organized and integrated by chapter. You will find practice quizzes, interactive activities, labeling exercises, flashcards, and much more that will complement your learning and understanding of general biology.

www.mhhe.com/maderbiology9

3

THE CHEMISTRY OF ORGANIC MOLECULES

ormal red blood cells are biconcave disks that easily change shape as they squeeze along tiny blood vessels. A person with sickle cell disease has sickle-shaped red blood cells that aren't as flexible. Their inflexibility causes them to either break down or clog small blood vessels. Typically, the individual suffers from anemia, poor circulation, and lack of resistance to infection. Internal hemorrhaging leads to jaundice, episodic pain in the abdomen and joints, and damage to internal organs.

     As with many humans ills, it is necessary to look to chemistry to tell us the cause of this particular condition. Red blood cells contain the respiratory pigment hemoglobin. The chemistry of sickle cell hemoglobin is different from that of normal hemoglobin. Sickle cell hemoglobin isn't as soluble as normal hemoglobin. It stacks up into long, semirigid rods that distort the red blood cells into a sickle shape. The molecule acts this way because a different chemical (amino acid) is found at two tiny spots, compared to normal hemoglobin. This difference accounts for abnormally shaped red blood cells and the symptoms of sickle cell disease. Sickle cell disease reminds us that the structure and shape of the organic compounds making up our bodies can influence our health, and determine if we are well or suffering from an illness. Only a knowledge of chemistry offers the hope that one day sickle cell disease and other diseases will be curable.

A sickled red blood cell compared to a normal red blood cell.

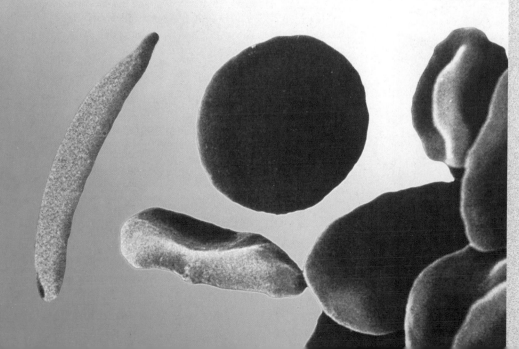

3.1 ORGANIC MOLECULES

Because chemists of the nineteenth century thought that the molecules of cells must contain a vital force, they divided chemistry into **organic chemistry,** the chemistry of organisms, and **inorganic chemistry,** the chemistry of the nonliving world. This terminology is still with us even though many types of organic molecules can now be synthesized in the laboratory. Today, we simply define **organic molecules** as those that contain both carbon and hydrogen atoms (Table 3.1).

There are only four classes of organic compounds in any living thing: carbohydrates, lipids, proteins, and nucleic acids. Despite the limited number of classes, the so-called biomolecules in cells are quite diverse. A bacterial cell contains some 5,000 different organic molecules, and a plant or animal cell has twice that number. This diversity of organic molecules makes the diversity of life possible (Fig. 3.1). It is quite remarkable that the variety of organic molecules can be traced to the unique chemical properties of the carbon atom.

The Carbon Atom

What is there about carbon that makes organic molecules so dissimilar and complex? Carbon is quite small, with only a

FIGURE 3.1 Carbohydrates as structural materials.
a. Plants are held erect partly by the incorporation of the polysaccharide cellulose into the cell wall, which surrounds each cell. **b.** The shell of a blue crab contains chitin, a different type of polysaccharide. **c.** Like plant cells, most bacterial cells are enclosed by a cell wall, but in this case the wall is strengthened by another type of polysaccharide known as peptidoglycan.

a.   c.

TABLE 3.1	
Inorganic Versus Organic Molecules	
Inorganic Molecules	*Organic Molecules*
Usually contain positive and negative ions	Always contain carbon and hydrogen
Usually ionic bonding	Always covalent bonding
Always contain a small number of atoms	Often quite large, with many atoms
Often associated with nonliving matter	Usually associated with living organisms

total of six electrons: two electrons in the first shell and four electrons in the outer shell. In order to acquire four electrons to complete its outer shell, a carbon atom almost always shares electrons with—you guessed it—CHNOPS, the elements that make up most of the weight of living things (see Fig. 2.1).

Because carbon needs four electrons to complete its outer shell, it can share with as many as four other elements, and this spells diversity. But even more significant to the shape, and therefore the function, of biomolecules, carbon often shares electrons with another carbon atom. The C—C bond is quite stable, and the result is carbon chains that can be quite long. Hydrocarbons are chains of carbon atoms bonded exclusively to hydrogen atoms.

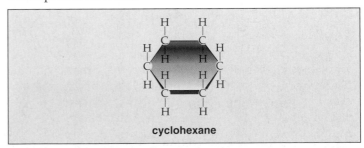

octane

Branching at any carbon atom is possible, and also a hydrocarbon can turn back on itself to form a ring compound when placed in water:

cyclohexane

Carbon can form double bonds with itself and other atoms. Double bonds restrict the movement of attached atoms, and in that way contribute to the shape of the molecule. As in acetylene, H—C≡C—H, carbon is also capable of forming a triple bond with itself.

The Carbon Skeleton and Functional Groups

The carbon chain of an organic molecule is called its skeleton or backbone. The terminology is appropriate because just as a skeleton accounts for your shape, so does the carbon skeleton

Functional Groups			
Group	**Structure**	**Compound**	**Significance**
Hydroxyl	R—OH	Alcohol as in ethanol	Polar, forms hydrogen bond Present in sugars, some amino acids
Carbonyl	R—C⟨O, H	Aldehyde as in formaldehyde	Polar Present in sugars
	R—C(=O)—R	Ketone as in acetone	Polar Present in sugars
Carboxyl (acidic)	R—C⟨O, OH	Carboxylic acid as in acetic acid	Polar, acidic; Present in fatty acids, amino acids
Amino	R—N⟨H, H	Amine as in tryptophan	Polar, basic, forms hydrogen bonds Present in amino acids
Sulfhydryl	R—SH	Thiol as in ethanethiol	Forms disulfide bonds Present in some amino acids
Phosphate	R—O—P(=O)(OH)—OH	Organic phosphate as in phosphorylated molecules	Polar, acidic; Present in nucleotides, phospholipids

R = remainder of molecule

FIGURE 3.2 Functional groups.
Molecules with the same carbon skeleton can still differ according to the type of functional group attached to the carbon skeleton. Many of these functional groups are polar, helping to make the molecule soluble in water. In this illustration, the remainder of the molecule (does not include the functional groups) is represented by an *R*.

of an organic molecule account for its shape. The reactivity of an organic molecule is largely dependent on the attached functional groups. A **functional group** is a specific combination of bonded atoms that always reacts in the same way, regardless of the particular carbon skeleton. As in Figure 3.2, it is even acceptable to use an *R* to stand for the remainder of the molecule because only the functional group is involved in the reaction.

Notice that functional groups with a particular name and structure are in certain types of compounds. For example, the addition of an —OH (hydroxyl group) to a carbon skeleton turns that molecule into an alcohol. When an —OH replaces one of the hydrogens in ethane, a 2-carbon hydrocarbon, it becomes ethanol, a type of alcohol that is familiar because it is consumable by humans. Whereas ethane, like other hydrocarbons, is **hydrophobic** (not soluble in water), ethanol is **hydrophilic** (soluble in water) because the —OH functional group is polar. Since cells are 70–90% water, the ability to interact with and be soluble in water profoundly affects the function of organic molecules in cells.

Organic molecules containing carboxyl (acidic) groups (—COOH) are polar. They tend to ionize and release hydrogen ions in solution:

$$—COOH \longrightarrow —COO^- + H^+$$

Functional groups determine the polarity of an organic molecule and also the types of reactions it will undergo. We will see that alcohols react with carboxyl groups when a fat forms, and that carboxyl groups react with amino groups during protein formation.

Isomers

Isomers [Gk. *isos*, equal, and *meros*, part, portion] are organic molecules that have identical molecular formulas but a different arrangement of atoms. In essence, isomers are variations in the molecular architecture of a molecule. Isomers are another example of how the chemistry of carbon leads to variations in organic molecules.

The two molecules in Figure 3.3 are isomers of one another; they have the same molecular formula but different functional groups. Therefore, we would expect them to react differently in chemical reactions.

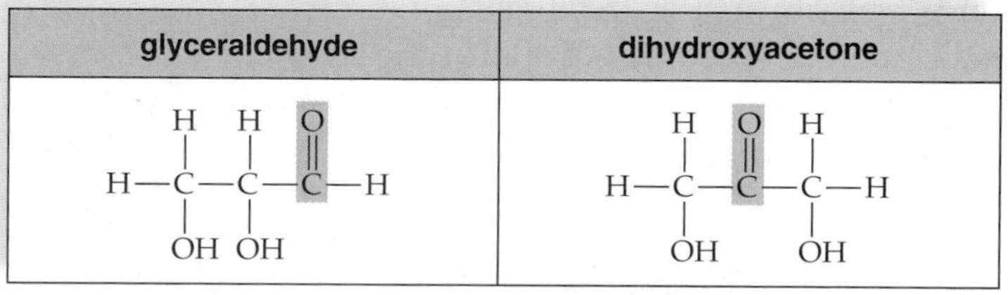

glyceraldehyde	dihydroxyacetone
H—C(H)—C(H)—C(=O)—H with OH OH below	H—C(H)—C(=O)—C(H)—H with OH OH below

FIGURE 3.3 Isomers.
Isomers have the same molecular formula but different atomic configurations. Both of these compounds have the formula $C_3H_6O_3$. In glyceraldehyde, oxygen is double-bonded to an end carbon. In dihydroxyacetone, oxygen is double-bonded to the middle carbon.

Organic molecules are diverse and complex. Carbon skeletons vary in size and shape; functional groups have various chemical characteristics, and isomers occur.

FIGURE 3.4 Common foods.
Lipids such as oils are digested to glycerol and fatty acids; proteins such as those in meat are digested to amino acids; and carbohydrates such as those in bread and pasta are digested to sugars. Cells use these subunit molecules to build their own macromolecules and as a source of energy.

The Macromolecules of Cells

Carbohydrates, lipids, proteins, and nucleic acids are called **macromolecules** because of their large size. You are very familiar with these molecules because certain foods are known to be rich in them, as illustrated in Figure 3.4. Even a midnight pizza, a quick hamburger, or an afternoon snack contains many of these essential molecules. When you digest these foods, they get broken down into the subunit molecules listed in the last column of Table 3.2. Your body then takes these subunits and builds from them the macromolecules that make up your cells. Many different foods also contain nucleic acids, the type of molecule that forms your genes.

The largest of the macromolecules are called **polymers** because they are constructed by linking together a large number of the same type of subunits, called **monomers.** A protein can contain hundreds of amino acids, and a nucleic acid can contain hundreds of nucleotides. How can polymers get so large? Cells use the modular approach when constructing polymers. Just as a train increases in length when boxcars are hitched together one by one, so a polymer gets longer as monomers bond to one another.

A cell uses the same type of condensation reaction to synthesize any type of macromolecule. It's called a **dehydra-**

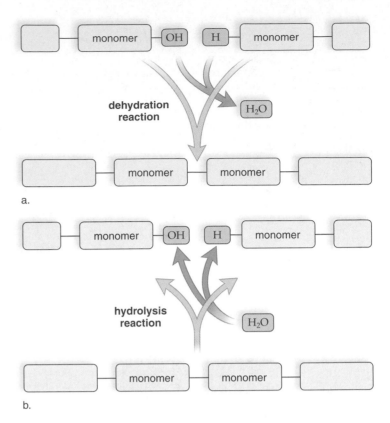

FIGURE 3.5 Synthesis and degradation of polymers.
a. In cells, synthesis often occurs when monomers bond during a dehydration reaction (removal of H_2O). **b.** Degradation occurs when the monomers in a polymer separate during a hydrolysis reaction (the addition of H_2O).

tion reaction because the equivalent of a water molecule, that is, an —OH (hydroxyl group) and an —H (hydrogen atom), is removed as the reaction occurs (Fig. 3.5a). To degrade a macromolecule, our digestive tract, or any cell, uses an opposite type of reaction. During a **hydrolysis** [Gk. *hydro*, water, and *lyse*, break] **reaction,** an —OH group from water attaches to one subunit, and an —H from water attaches to the other subunit (Fig. 3.5b). In other words, water is used to break the bond holding subunits together.

For these reactions or, for that matter, almost any other type of reaction to occur in a cell, an enzyme must be present. An **enzyme** is a molecule that speeds a reaction by bringing reactants together, and the enzyme may even participate in the reaction but it is unchanged by it. Frequently, monomers must be activated before they will react.

> Organic molecules are routinely built up in cells by the removal of water (H_2O) during a dehydration reaction. They are degraded in cells by the addition of water during a hydrolysis reaction.

3.2 CARBOHYDRATES

Carbohydrates are almost universally used as an immediate energy source in living things, but they also play structural

TABLE 3.2

Macromolecules

Category	Example	Subunit(s)
Lipids	Fat	Glycerol and fatty acids
Carbohydrates*	Polysaccharide	Monosaccharide
Proteins*	Polypeptide	Amino acid
Nucleic acids*	DNA, RNA	Nucleotide

*Polymers

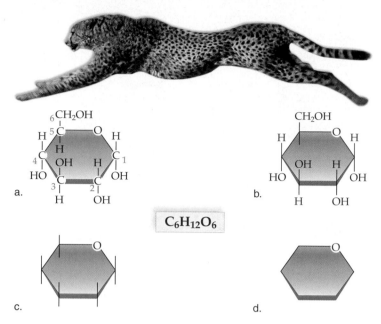

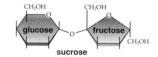

FIGURE 3.6 Glucose.

Glucose provides energy for organisms, such as this cheetah. Each of these structural formulas is glucose. **a.** The carbon skeleton and all attached groups are shown. **b.** The carbon skeleton is omitted. **c.** The carbon skeleton and attached groups are omitted. **d.** Only the ring shape, which includes one oxygen atom, remains.

Glucose, with six carbon atoms, is a **hexose** [Gk. *hex*, six] and has a molecular formula of $C_6H_{12}O_6$. Despite the fact that glucose has several isomers, such as fructose and galactose, we usually think of $C_6H_{12}O_6$ as glucose. This signifies that glucose has a special place in the chemistry of organisms. This simple sugar is the major source of cellular fuel for all living things. Glucose is transported in the blood of animals, and it is the molecule that is broken down in nearly all types of organisms during cellular respiration, with the resulting buildup of ATP molecules.

Ribose and **deoxyribose,** with five carbon atoms, are **pentoses** [Gk. *pent*, five] of significance because they are found respectively in the nucleic acids RNA and DNA. RNA and DNA are discussed later in the chapter.

Disaccharides: Varied Uses

A **disaccharide** contains two monosaccharides that have joined during a dehydration reaction. Figure 3.7 shows how the disaccharide maltose (an ingredient used in brewing) arises when two glucose molecules bond together. Note the position of the bond that results when the —OH groups participating in the reaction project below the ring. When our hydrolytic digestive juices break this bond, the result is two glucose molecules.

Sucrose or table sugar (structure shown at right) is another disaccharide of special interest because it is the form in which sugar is transported in plants. Sucrose is also the sugar we use to sweeten our food. We acquire the sugar from plants such as sugarcane and sugar beets. You may also have heard of lactose, a disaccharide found in milk. Lactose is glucose combined with galactose. Individuals that are lactose intolerant cannot break this disaccharide down and have subsequent medical problems.

roles in a variety of organisms (see Fig. 3.1). The majority of carbohydrates have a carbon to hydrogen to oxygen ratio of 1:2:1. The term *carbohydrate* includes single sugar molecules and also chains of sugars. Chain length varies from a few sugars to hundreds of sugars. The long chains are thus called polymers.

Monosaccharides: Ready Energy

Monosaccharides [Gk. *monos*, single, and *sacchar*, sugar], consisting of only a single sugar molecule, are called simple sugars. A simple sugar can have a carbon backbone of three to seven carbons. The molecular formula for a simple sugar is some multiple of CH_2O, suggesting that every carbon atom is bonded to an H and an —OH. This is not strictly correct, as you can see by examining the structural formula for glucose (Fig. 3.6). Still, sugars do have many hydroxyl groups, and this polar functional group makes them soluble in water.

Polysaccharides: Energy Storage Molecules

Polysaccharides are polymers of monosaccharides. Some types of polysaccharides function as short-term energy storage molecules. They serve as storage molecules because they are not as soluble in water and are much larger than a sugar. Therefore, polysaccharides cannot easily pass through the plasma membrane, a sheetlike structure that encloses the cell.

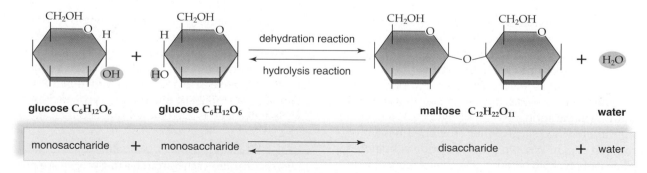

FIGURE 3.7 Synthesis and degradation of maltose, a disaccharide.

Synthesis of maltose occurs following a dehydration reaction when a bond forms between two glucose molecules, and water is removed. Degradation of maltose occurs following a hydrolysis reaction when this bond is broken by the addition of water.

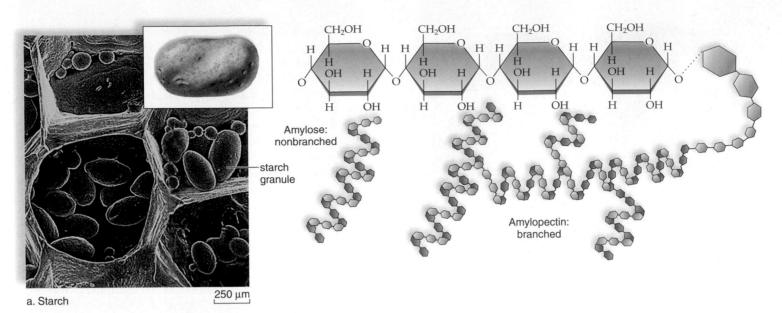

a. Starch   250 μm

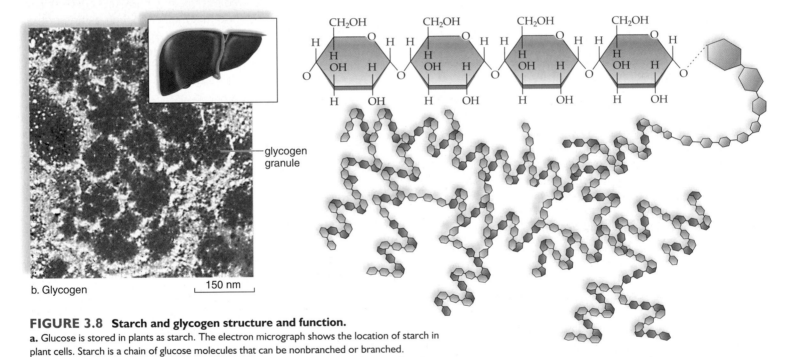

b. Glycogen   150 nm

FIGURE 3.8 Starch and glycogen structure and function.
a. Glucose is stored in plants as starch. The electron micrograph shows the location of starch in plant cells. Starch is a chain of glucose molecules that can be nonbranched or branched.
b. Glucose is stored in animals as glycogen. The electron micrograph shows glycogen deposits in a portion of a liver cell. Glycogen is a highly branched polymer of glucose molecules.

When an organism requires energy, the polysaccharide is broken down to release sugar molecules. The helical shape of the polysaccharides in Figure 3.8 exposes the sugar linkages to the hydrolytic enzymes that can break them down.

Plants store glucose as **starch.** The cells of a potato contain granules where starch resides during winter until energy is needed for growth in the spring. Notice in Figure 3.8*a* that amylose is a simple unbranched starch, while amylopectin is a more complex branched starch. When a polysaccharide is branched, there is no main carbon chain because new chains occur at regular intervals, always at the sixth carbon of the monomer. Both starches serve as glucose reservoirs in plants. The plant as well as animals can tap into these reservoirs for energy.

Animals store glucose as **glycogen.** In our bodies and those of other vertebrates, liver cells contain granules where glycogen is being stored until needed. The storage and release of glucose from liver cells is under the control of hormones. After we eat, the release of the hormone insulin from the pancreas promotes the storage of glucose as glycogen. Notice in Figure 3.8*b* that glycogen is even more branched than starch.

Polysaccharides such as starch and glycogen are polymers of glucose. Plant cells use starch, and animal cells use glycogen for short-term energy storage.

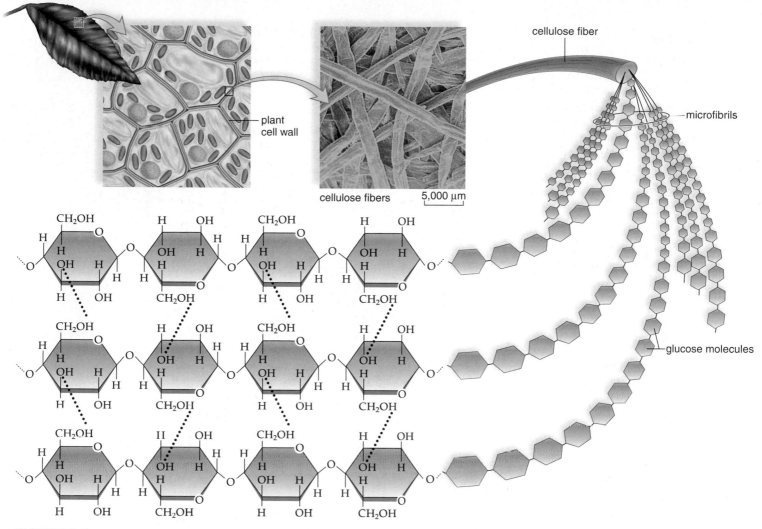

FIGURE 3.9 Cellulose fibrils.
Cellulose fibers criss-cross in plant cell walls for added strength. A cellulose fiber contains several microfibrils, each a polymer of glucose molecules—notice that the linkage bonds differ from those of starch. Every other glucose is flipped, permitting hydrogen bonding between the microfibrils.

Polysaccharides: Structural Molecules

Structural polysaccharides include **cellulose** in plants, **chitin** in animals and fungi, and **peptidoglycan** in bacteria (see Fig. 3.1). In all three, monomers are joined by the type of bond shown for cellulose in Figure 3.9. The cellulose monomer is simply glucose, but in chitin, the monomer has an attached amino group. The structure of peptidoglycan is more complex because each monomer also has an amino acid chain.

Cellulose is the most abundant carbohydrate and, indeed, the most abundant organic molecule on Earth—over 100 billion tons of cellulose is produced by plants each year. Wood, a cellulose plant product, is used for construction, and cotton is used for cloth. Microorganisms, but not animals, are able to digest the bond between glucose monomers in cellulose. The protozoans in the gut of termites allows termites to digest wood. In cows and other ruminants, microorganisms break down cellulose in a special pouch before the "cud" is returned to the mouth for more chewing and reswallowing. In rabbits, microorganisms digest cellulose in

a pouch where it is packaged into pellets. In order to make use of these nutrient pellets, rabbits have to reswallow them as soon as they pass out at the anus. For animals that have no means of digesting cellulose, cellulose is dietary fiber, which maintains regularity of elimination.

Chitin [Gk. *chiton*, tunic], is found in fungal cell walls and in the exoskeletons of crabs and related animals, such as lobsters, scorpions, and insects. Chitin, like cellulose, cannot be digested by animals; however, humans have found many other good uses for chitin. Seeds are coated with chitin, and this protects them from attack by soil fungi. Because chitin also has antibacterial and antiviral properties, it is processed and used in medicine as a wound dressing and suture material. Chitin is even useful during the production of cosmetics and various foods.

Plant cell walls contain cellulose. The shells of crabs and related animals, and cell walls of fungi, contain chitin.

3.3 LIPIDS

A variety of organic compounds are classified as **lipids** [Gk. *lipos*, fat] (Table 3.3). These compounds are insoluble in water due to their hydrocarbon chains. Hydrogens bonded only to carbon have no tendency to form hydrogen bonds with water molecules. Fat, a well-known lipid, is used for both insulation and long-term energy storage by animals. Fat below the skin of marine mammals is called blubber (Fig. 3.10); in humans, it is given slang expressions such as "spare tire" and "love handles." Plants use oil instead of fat for long-term energy storage. We are familiar with fats and oils because we use them as foods and for cooking.

Phospholipids and steroids are also important lipids found in living things. They serve as major components of the plasma membrane in cells. Waxes, which are sticky, not greasy like fats and oils, tend to have a protective function in living things.

Triglycerides: Long-term Energy Storage

Fats and **oils** contain two types of subunit molecules: glycerol and fatty acids. **Glycerol** is a compound with three —OH groups. The —OH groups are polar; therefore, glycerol is soluble in water. When a fat or oil forms, the acid portions of three fatty acids react with the —OH groups of glycerol during a dehydration reaction. In addition to a fat molecule, three molecules of water result. Fats and oils are degraded following a hydrolysis reaction (Fig. 3.11a). Because there are three fatty acids attached to each glycerol molecule, fats and oils are sometimes called **triglycerides.** Notice that triglycerides have many C—H bonds; therefore, they do not mix with water. Despite the liquid nature of both cooking oils and water, cooking oils separate out of water even after shaking.

Each **fatty acid** consists of a long hydrocarbon chain with a —COOH (carboxyl) group at one end. Most of the fatty acids in cells contain 16 or 18 carbon atoms per molecule, although

FIGURE 3.10 Blubber.
The fat (blubber) beneath the skin of marine mammals protects them well from the cold. Blubber accounts for about 25% of their body weight.

smaller ones are also found. Fatty acids are either saturated or unsaturated. **Saturated fatty acids** have no double bonds between the carbon atoms. The carbon chain is saturated, so to speak, with all the hydrogens that can be held. **Unsaturated fatty acids** have double bonds in the carbon chain wherever the number of hydrogens is less than two per carbon atom.

Triglycerides containing fatty acids with unsaturated bonds melt at a lower temperature than those containing only saturated fatty acids. This is because a double bond creates a kink in the fatty acid chain that prevents close packing between the hydrocarbon chains (Fig. 3.11b, c). We can reason that butter, a fat that is solid at room temperature, must contain primarily saturated fatty acids, while corn oil, which is a liquid even when placed in the refrigerator, must contain primarily unsaturated fatty acids. This difference is useful to living things. For example, the feet of reindeer and penguins contain unsaturated triglycerides, and this helps protect those exposed parts from freezing.

In general, however, fats, which are most often of animal origin, are solid at room temperature, and oils, which are liquid at room temperature, are of plant origin. Diets high in animal fat have been associated with circulatory disorders. Replacement of fat whenever possible with oils such as peanut oil and sunflower oil has been suggested.

Nearly all animals use fat in preference to glycogen for long-term energy storage. Gram per gram, fat stores more energy than glycogen. The C—H bonds of fatty acids make them a richer source of chemical energy than glycogen, because glycogen has many C—OH bonds. Also, fat droplets, being nonpolar, do not contain water. Small birds, like the broad-tailed hummingbird, store a great deal of fat before they start their long spring and fall migratory flights. About 0.15 g of fat per gram of body weight is accumulated each day. If the same amount of energy were stored as glycogen, a bird would be so heavy it would not be able to fly.

Fats and oils are triglycerides (one glycerol plus three fatty acids). They are used as long-term energy storage compounds in plants and animals.

TABLE 3.3

Lipids

Type	Functions	Human Uses
Fats	Long-term energy storage and insulation in animals	Butter, lard
Oils	Long-term energy storage in plants and their seeds	Cooking oils
Phospholipids	Component of plasma membrane	—
Steroids	Component of plasma membrane (cholesterol), sex hormones	Medicines
Waxes	Protection, prevent water loss (cuticle of plant surfaces), beeswax, earwax	Candles, polishes

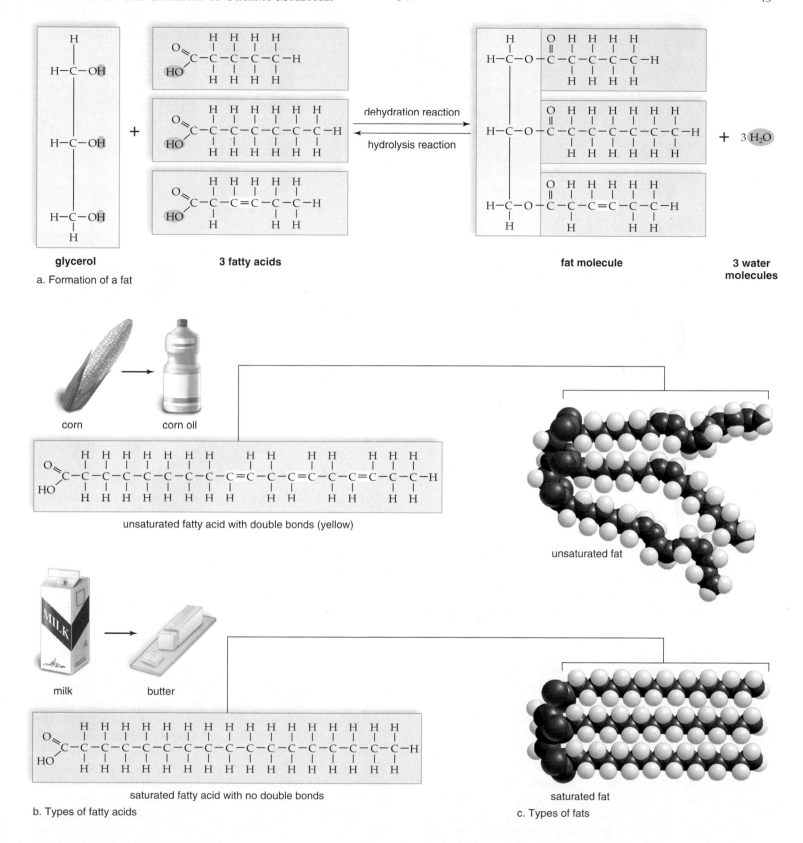

a. Formation of a fat

corn → corn oil

unsaturated fatty acid with double bonds (yellow)

unsaturated fat

milk → butter

saturated fatty acid with no double bonds

saturated fat

b. Types of fatty acids

c. Types of fats

FIGURE 3.11 Fat and fatty acids.

a. Following a dehydration reaction, glycerol is bonded to three fatty acid molecules as fat forms and water is given off. Following a hydrolysis reaction, the bonds are broken due to the addition of water. **b.** A fatty acid has a carboxyl group attached to a long hydrocarbon chain. If there are double bonds between some of the carbons in the chain, the fatty acid is unsaturated. If there are no double bonds, the fatty acid is saturated. **c.** Space-filling models of an unsaturated fat and a saturated fat.

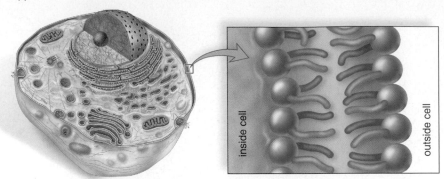

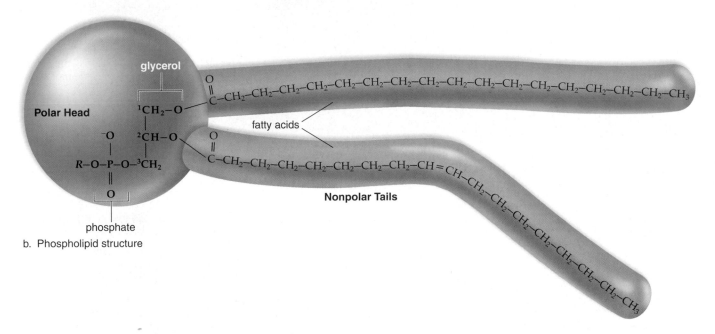

FIGURE 3.12 Phospholipids form membranes.
a. Phospholipids arrange themselves as a bilayer in the plasma membrane that surrounds cells because (**b**) phospholipids are constructed like fats, except that in place of the third fatty acid, they have a polar phosphate group. The hydrophilic (polar) head is soluble in water, whereas the two hydrophobic (nonpolar) tails are not.

a. Plasma membrane of a cell

b. Phospholipid structure

Phospholipids: Membrane Components

The bulk of the plasma membrane that surrounds cells consists of a phospholipid bilayer (Fig. 3.12*a*). **Phospholipids** [Gk. *phos*, light, and *lipos*, fat], have hydrophilic heads and hydrophobic tails; they tend to arrange themselves so that only the polar heads are adjacent to a watery medium. Between two compartments of water, such as the outside and inside of a cell, phospholipids become a bilayer in which the hydrophilic heads project outward and the hydrophobic tails project inward. A plasma membrane is absolutely essential to the structure and function of a cell, and this signifies the importance of phospholipids to living things.

Phospholipids, as implied by their name, contain a phosphate group. Essentially, a phospholipid is constructed like a fat, except that in place of the third fatty acid attached to glycerol, there is a polar phosphate group. The phosphate group is usually bonded to another organic group, indicated by *R* in Figure 3.12*b*. This portion of the molecule becomes the polar head, while the hydrocarbon chains of the fatty acids become the nonpolar tails.

Phospholipids have polar heads and nonpolar tails; therefore, they are suited to forming a membrane that surrounds the cell.

Steroids: Four Fused Rings

Steroids are lipids that have entirely different structures from those of fats. Steroid molecules have skeletons of four fused carbon rings (Fig. 3.13). Each type of steroid differs primarily by the types of functional groups attached to the carbon skeleton.

Cholesterol is an essential component of an animal cell's plasma membrane, where it provides physical stability. Cholesterol is the precursor of several other steroids, such as the sex hormones estrogen and testosterone (Fig. 3.13). The male sex hormone, testosterone, is formed primarily in the testes, and the female sex hormone, estrogen, is formed primarily in the ovaries. Testosterone and estrogen differ only by the functional groups attached to the same carbon skeleton, and yet they each have their own profound effect on the body and the sexuality of an animal.

We know also that a diet high in saturated fats and cholesterol can contribute to circulatory disorders. When fatty material accumulates inside the lining of blood vessels, blood flow is reduced and high blood pressure results.

Steroids are a type of lipid because of their insolubility in water. Steroids are derived from cholesterol, a component of the plasma membrane. The sex hormones are steroids.

OH
CH₃
CH₃

O

b. Testosterone

CH₃
HC — CH₃
(CH₂)₃
HC — CH₃
CH₃

a. Cholesterol

CH₃

HO

OH
CH₃

HO

c. Estrogen

FIGURE 3.13 Steroid diversity.
a. Like cholesterol (**b**) testosterone and (**c**) estrogen have different effects on the body due to different functional groups attached to the same carbon skeleton. An important component of the plasma membrane, all steroids have four adjacent rings, but vary by their attached groups.

Waxes

In **waxes,** long-chain fatty acids bond with long-chain alcohols:

fatty acid

$$O=C-CH_2-CH_2-CH_2-CH_2-CH_2-CH_2-CH_2-CH_2-CH_2-CH_3$$
O H
$$C-CH_2-CH_2-CH_2-CH_2-CH_2-CH_2-CH_2-CH_2-CH_3$$
H   **long-chain alcohol**

Waxes are solid at normal temperatures because they have a high melting point. Being hydrophobic, they are also water-proof and resistant to degradation. In many plants, waxes, along with other molecules, form a protective cuticle (covering) that re-tards the loss of water for all exposed parts (Fig. 3.14*a*). In many animals, waxes are involved in skin and fur maintenance. In hu-

mans, wax is produced by glands in the outer ear canal. Earwax contains cerumin, an organic compound that at the very least repels insects, and in some cases even kills them. It also traps dust and dirt, preventing them from reaching the eardrum.

A honeybee produces beeswax in glands on the under-side of its abdomen. Beeswax is used to make the six-sided cells of the comb where honey is stored (Fig. 3.14*b*). Honey contains the sugars fructose and glucose, breakdown products of the sugar sucrose.

Humans have found a myriad of uses for waxes, from making candles to polishing cars, furniture, and floors (see Table 3. 3).

Waxes contain fatty acids attached to long-chain alcohols. Waxes have many different types of functions, primarily in plants but also in animals.

FIGURE 3.14 Waxes.
Waxes are a type of lipid. **a.** Fruits are protected by a waxy coating that is visible on these plums. **b.** Bees secrete the wax that allows them to build a comb where they store honey. This bee has collected pollen (yellow) to feed growing larvae.

a.

b.

3.4 PROTEINS

Proteins [Gk. *proteios,* first place], as their Greek derivation implies, are of primary importance to the structure and function of cells. As much as 50% of the dry weight of most cells consists of proteins. Presently, over 100,000 proteins have been identified. Here are some of their many functions:

Support   Some proteins—such as keratin, which makes up hair and nails, and collagen, which lends support to ligaments, tendons, and skin—are structural proteins.

Enzymes   Enzymes bring reactants together and thereby speed chemical reactions in cells. They are specific for one particular type of reaction and can function at body temperature.

Transport   Channel and carrier proteins in the plasma membrane allow substances to enter and exit cells. Some other proteins transport molecules in the blood of animals; **hemoglobin** is a complex protein that transports oxygen.

Defense   Antibodies are proteins. They combine with foreign substances, called antigens. In this way, they prevent antigens from destroying cells and upsetting homeostasis.

Hormones   Hormones are regulatory proteins. They serve as intercellular messengers that influence the metabolism of cells. The hormone insulin regulates the content of glucose in the blood and in cells; the presence of growth hormone determines the height of an individual.

Motion   The contractile proteins actin and myosin allow parts of cells to move and cause muscles to contract. Muscle contraction accounts for the movement of animals from place to place.

The structures and functions of vertebrate cells and tissues differ according to the type of protein they contain. For example, muscle cells contain actin and myosin; red blood cells contain hemoglobin; and support tissues contain collagen.

Amino Acids: Building Blocks of Proteins

The so-called α carbon atom of an **amino acid** bonds to a hydrogen atom, and also to three other groups of atoms (Fig. 3.15). The name of the amino acid molecule is appropriate because one of these groups is an —NH_2 (amino group) and another is a —COOH (acidic or car-

boxyl group). The third group is called an *R* group because it is the *R*emainder of the molecule.

Amino acids differ according to their particular *R* group. The *R* groups range in complexity from a single hydrogen atom to a complicated ring compound. The unique chemical properties of an amino acid depend on those of the *R* group. For example, some *R* groups are polar and some are not. Also, the amino acid cysteine has an *R* group that ends with a sulfhydryl (—SH) group that often serves to connect one chain of amino acids to another by a disulfide bond, —S—S—. The 20 different amino acids commonly found in cells are shown in Figure 3.16.

Peptides

Figure 3.15 shows how two amino acids join by a dehydration reaction between the carboxyl group of one and the amino group of another. The resulting covalent bond between two amino acids is called a **peptide bond.** The atoms associated with the peptide bond share the electrons unevenly because oxygen is more electronegative than nitrogen. Therefore, the hydrogen attached to the nitrogen has a slightly positive charge, while the oxygen has a slightly negative charge.

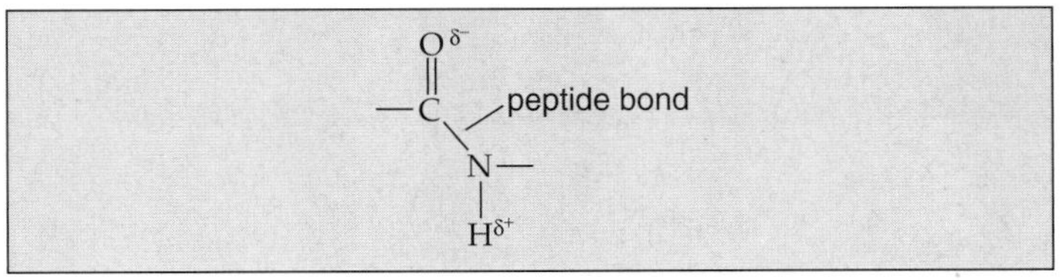

The polarity of the peptide bond means that hydrogen bonding is possible between the —CO of one amino acid and the —NH of another amino acid in a polypeptide.

A **peptide** is two or more amino acids bonded together, and a **polypeptide** is a chain of many amino acids joined by peptide bonds. A protein may contain more than one polypeptide chain; therefore, you can see why a protein could have a very large number of amino acids. In 1953, Frederick Sanger developed a method to determine the sequence of amino acids in a polypeptide. Now that we know the sequences of thousands of polypeptides, it is clear that each polypeptide has its own normal sequence. This sequence influences the final three-dimensional shape of the protein. Proteins that have an abnormal sequence have the wrong shape and cannot function properly.

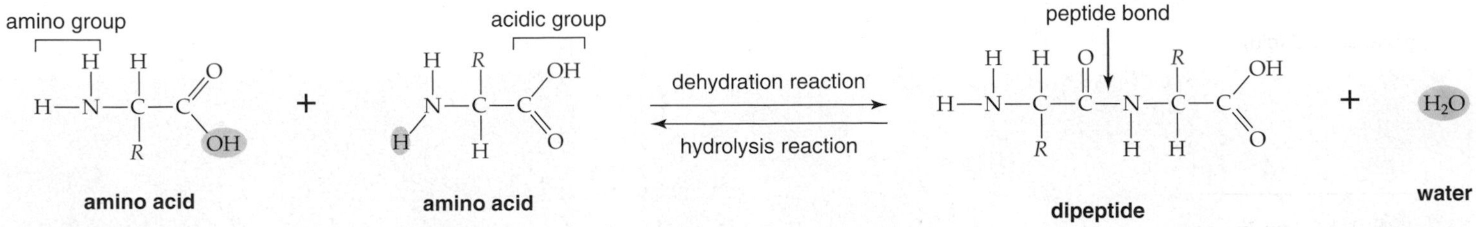

FIGURE 3.15 Synthesis and degradation of a peptide.
Following a dehydration reaction, a peptide bond joins two amino acids and a water molecule is released. Following a hydrolysis reaction, the bond is broken due to the addition of water.

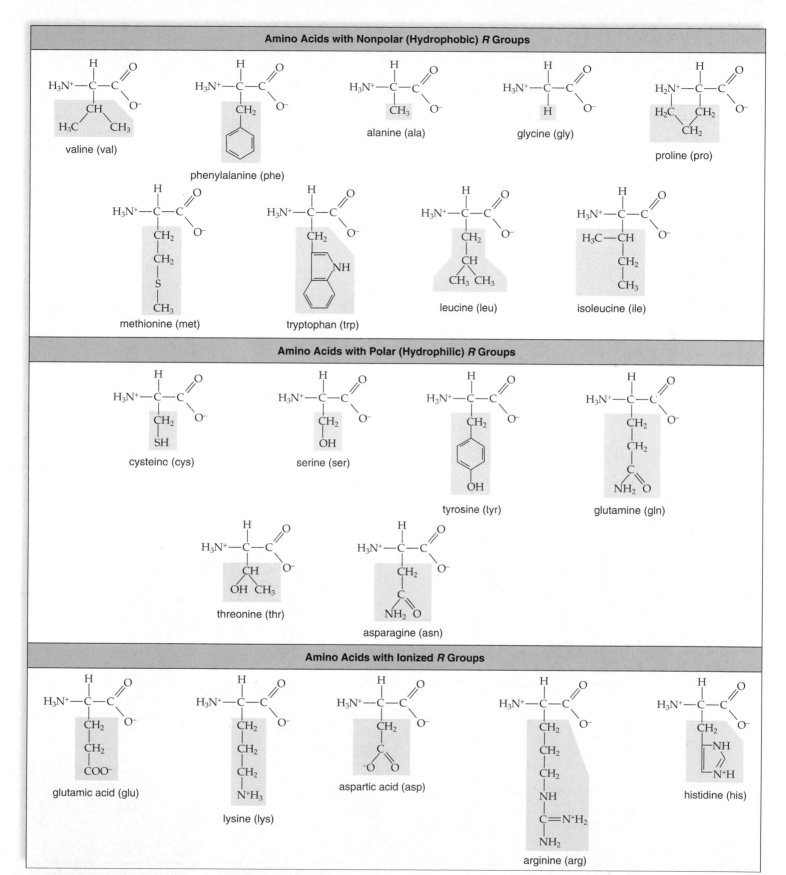

FIGURE 3.16   Amino acids.
Polypeptides contain 20 different kinds of amino acids, which are shown here. Amino acids differ by the particular R group (blue) attached to the central carbon. Some R groups are nonpolar and hydrophobic, some are polar and hydrophilic, and some are ionized and hydrophilic. The amino acids are shown in ionized form.

Shape of Proteins

The final shape of a protein determines its function in the cells and body of an organism. A protein can have up to four levels of structure, but not all proteins have all four levels.

Primary Structure

The primary structure of one protein is its own particular sequence of amino acids. The following analogy can help you see that hundreds of thousands of different polypeptides can be built from just 20 amino acids: the English alphabet contains only 26 letters, but an almost infinite number of words can be constructed by varying the number and sequence of these few letters. In the same way, many different proteins can result by varying the number and sequence of just 20 amino acids.

Secondary Structure

The secondary structure of a protein occurs when the polypeptide coils or folds in a particular way (Fig. 3.17).

Linus Pauling and Robert Corey, who began studying the structure of amino acids in the late 1930s, concluded that a coiling they called an α (alpha) helix and a pleated sheet they called the β (beta) sheet were two basic patterns of amino acids within a polypeptide. The names came from the fact that the α helix was the first pattern they discovered, and the β sheet was the second pattern they discovered.

Hydrogen bonding often holds the secondary structure of a polypeptide in place. Hydrogen bonding between every fourth amino acid accounts for the spiral shape of the helix. In a β sheet, the polypeptide turns back upon itself, and hydrogen

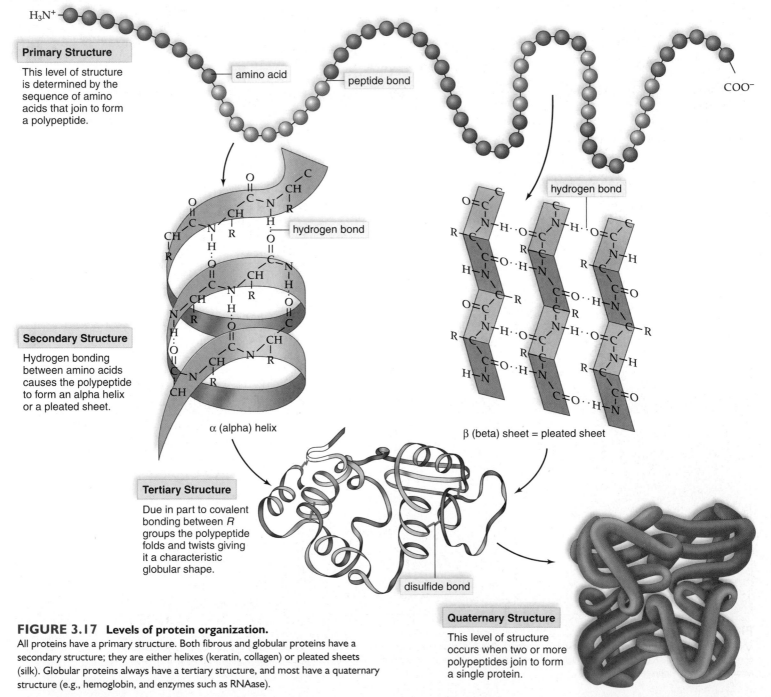

Primary Structure

This level of structure is determined by the sequence of amino acids that join to form a polypeptide.

amino acid

peptide bond

hydrogen bond

Secondary Structure

Hydrogen bonding between amino acids causes the polypeptide to form an alpha helix or a pleated sheet.

α (alpha) helix

β (beta) sheet = pleated sheet

Tertiary Structure

Due in part to covalent bonding between *R* groups the polypeptide folds and twists giving it a characteristic globular shape.

disulfide bond

Quaternary Structure

This level of structure occurs when two or more polypeptides join to form a single protein.

FIGURE 3.17 Levels of protein organization.
All proteins have a primary structure. Both fibrous and globular proteins have a secondary structure; they are either helixes (keratin, collagen) or pleated sheets (silk). Globular proteins always have a tertiary structure, and most have a quaternary structure (e.g., hemoglobin, and enzymes such as RNAase).

bonding occurs between extended lengths of the polypeptide. **Fibrous proteins,** which are structural proteins, exist as helices or pleated sheets that hydrogen-bond to each other. Keratin, a protein found in hair and silk, the protein that forms spider webs, are examples of secondary structure proteins (Fig. 3.18).

Tertiary Structure

The tertiary structure is a folding that results in the final three-dimensional shape of a polypeptide. So-called **globular proteins,** which tend to ball up into rounded shapes, have a tertiary structure. Various types of bonding between the R groups of the amino acids bring about the tertiary structure. Hydrogen bonds, ionic bonds, and covalent bonds all contribute to the tertiary structure of a polypeptide. Strong disulfide linkages in particular help maintain the tertiary shape. Hydrophobic R groups do not bond with other R groups, and they tend to collect in a common region where they are not exposed to water. These are called hydrophobic interactions. Although hydrophobic interactions are not as strong as hydrogen bonds, they are important in creating and stabilizing the tertiary structure.

Enzymes are globular proteins. Enzymes work best at body temperature, and each one also has an optimal pH at which the rate of the reaction is highest. At this temperature and pH, the enzyme has its normal shape. A high temperature and change in pH can disrupt the interactions that maintain the shape of the enzyme. When a protein loses its natural shape, it is said to be **denatured.**

Quaternary Structure

Some proteins have a quaternary structure because they consist of more than one polypeptide. Hemoglobin is a much-studied globular protein that consists of four polypeptides, and there-fore it has a quaternary structure. Each polypeptide in hemoglobin has a primary, secondary, and tertiary structure.

Protein-Folding Diseases

Proteins cannot function properly unless they fold into their correct shape. In recent years it has been shown that the cell contains **chaperone proteins,** which help new proteins fold into their normal shape. At first it seemed as if chaperone proteins ensured that proteins folded properly, but now it seems that they might correct any misfolding of a new protein. In any case, without fully functioning chaperone proteins, a cell's proteins may not be functional because they have misfolded. Several diseases in humans such as cystic fibrosis and Alzheimer disease are associated with misshapen proteins. The possibility exists that the diseases are due to missing or malfunctioning chaperone proteins.

Other diseases in humans are due to misfolded proteins, but the cause may be different. For years, investigators have been studying fatal brain diseases, known as TSEs,[1] that have no cure because no infective agent can be found. Mad cow disease is a well-known example of a TSE disease. Now it appears that TSE diseases could be due to misfolded proteins, called **prions,** that cause other proteins of the same type to fold the wrong way too. A possible relationship between prions and the functioning of chaperone proteins is now under investigation.

The sequence of amino acids in a polypeptide determines its final shape because it determines which R groups interact. The function of a protein is dependent on its shape.

---

[1] TSE (transmissible spongiform encephalopathies)

a.   b.   c.

FIGURE 3.18 Fibrous proteins.
Fibrous proteins are structural proteins. **a.** Keratin—found, for example, in hair, horns, and hoofs—exemplifies fibrous proteins that are helical for most of their length. Keratin is a hydrogen-bonded triple helix. In this photo, Drew Barrymore has straight hair. **b.** In order to give her curly hair, water was used to disrupt the hydrogen bonds, and when the hair dried, new hydrogen bonding allowed it to take on the shape of a curler. A permanent-wave lotion induces new covalent bonds within the helix. **c.** Silk made by spiders and silkworms exemplifies fibrous proteins that are pleated sheets for most of their length. Hydrogen bonding between parts of the molecule occurs as the pleated sheet doubles back on itself.

3.5 NUCLEIC ACIDS

Nucleic acids are polymers of nucleotides with very specific functions in cells. **DNA (deoxyribonucleic acid)** is the genetic material that stores information regarding its own replication and the order in which amino acids are to be joined to make a protein. **RNA (ribonucleic acid)** is another type of nucleic acid. One type of RNA molecule called messenger RNA (mRNA) is an intermediary in the process of protein synthesis, conveying information from DNA regarding the amino acid sequence in a protein.

Some nucleotides have independent metabolic functions in cells. For example, some are components of **coenzymes,** nonprotein organic molecules that facilitate enzymatic reactions. **ATP (adenosine triphosphate)** is a nucleotide that supplies energy for synthetic reactions and for various other energy-requiring processes in cells.

Structure of DNA and RNA

Every **nucleotide** is a molecular complex of three types of molecules: phosphate (phosphoric acid), a pentose sugar, and a nitrogen-containing base (Fig. 3.19a). In DNA, the pentose sugar is deoxyribose, and in RNA the pentose sugar is ribose. A difference in the structure of these 5-carbon sugars accounts for their respective names because deoxyribose lacks an oxygen atom found in ribose (Fig. 3.19b).

There are four types of nucleotides in DNA and four types of nucleotides in RNA (Fig. 3.19c). The base of a nucleotide can be a pyrimidine with a single ring or a purine with a double ring. In DNA, the pyrimidine bases are cytosine and thymine; in RNA, the pyrimidine bases are cytosine and uracil. In both DNA and RNA, the purine bases are adenine or guanine. These molecules are called bases because their presence raises the pH of a solution.

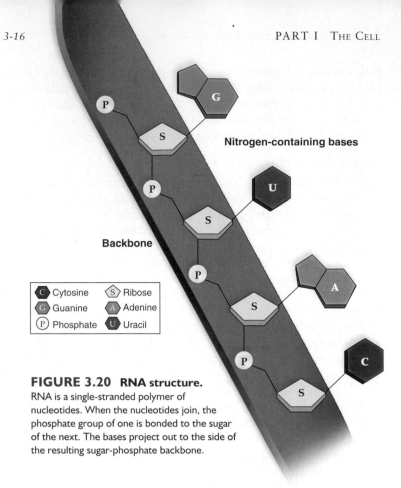

FIGURE 3.20 RNA structure.
RNA is a single-stranded polymer of nucleotides. When the nucleotides join, the phosphate group of one is bonded to the sugar of the next. The bases project out to the side of the resulting sugar-phosphate backbone.

Nucleotides join in a definite sequence by a series of dehydration reactions when DNA and RNA form. The polynucleotide is a linear molecule called a strand in which the backbone is made up of a series of sugar-phosphate-sugar-phosphate molecules. The bases project to one side of the backbone. Since the nucleotides occur in a definite order, so do the bases. RNA is single stranded (Fig. 3.20).

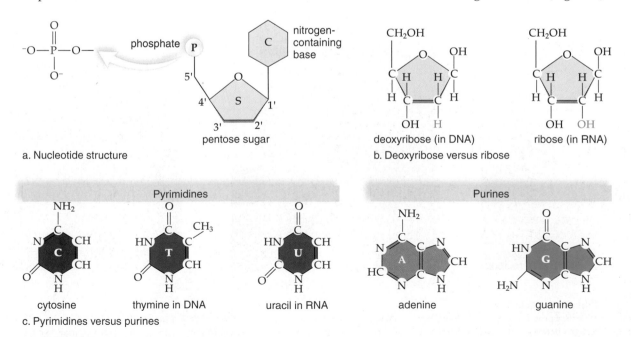

a. Nucleotide structure

b. Deoxyribose versus ribose

deoxyribose (in DNA)   ribose (in RNA)

c. Pyrimidines versus purines

Pyrimidines — cytosine, thymine in DNA, uracil in RNA

Purines — adenine, guanine

FIGURE 3.19 Nucleotides.
a. A nucleotide consists of a phosphate molecule, a pentose sugar, and a nitrogen-containing base. **b.** DNA contains the sugar deoxyribose, and RNA contains the sugar ribose. **c.** RNA contains the pyrimidines C and U and the purines A and G. DNA contains the pyrimidines C and T and the purines A and G.

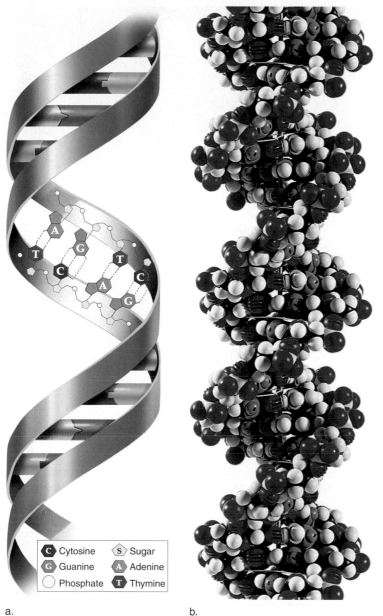

C Cytosine   S Sugar
G Guanine   A Adenine
○ Phosphate T Thymine

a.   b.

FIGURE 3.21   DNA structure.
DNA is a double helix in which the two polynucleotide strands twist about each other. **a.** Hydrogen bonds (dotted lines) occur between the complementarily paired bases: C is always paired with G, and A is always paired with T. **b.** Space-filling model of DNA.

TABLE 3.4

DNA Structure Compared to RNA Structure

	DNA	RNA
Sugar	Deoxyribose	Ribose
Bases	Adenine, guanine, thymine, cytosine	Adenine, guanine, uracil, cytosine
Strands	Double stranded with base pairing	Single stranded
Helix	Yes	No

The nucleic acids DNA and RNA are polymers of nucleotides. DNA is the genetic material, and RNA is an intermediary during the process of protein synthesis.

DNA is double stranded, with the two strands usually twisted about each other in the form of a double helix. The two strands are held together by hydrogen bonds between pyrimidine and purine bases. The bases can be in any order within a strand, but between strands, thymine (T) is always paired with adenine (A), and guanine (G) is always paired with cytosine (C) (Fig. 3.21). This is called **complementary base pairing.** Therefore, regardless of the order or the quantity of any particular base pair, the number of purine bases (A + G) always equals the number of pyrimidine bases (T + C).

Table 3.4 summarizes the differences between DNA and RNA.

ATP (Adenosine Triphosphate)

ATP is a nucleotide in which **adenosine** is composed of adenine and ribose. Triphosphate stands for the three phosphate groups that are attached together and to ribose, the pentose sugar (Fig. 3.22). ATP is a high-energy molecule because the last two phosphate bonds are unstable and are easily broken. In cells the terminal phosphate bond is usually hydrolyzed to give the molecule **ADP (adenosine diphosphate)** and a phosphate molecule Ⓟ.

The energy that is released by ATP breakdown is coupled to energy-requiring processes in cells, such as the synthesis of macromolecules such as carbohydrates and proteins.

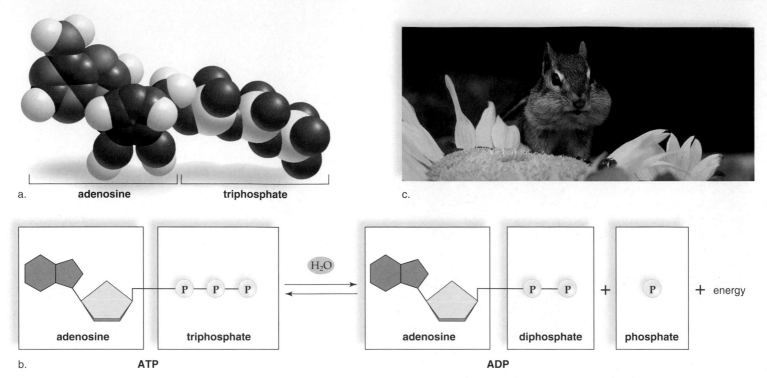

a.   **adenosine**   **triphosphate**

c.

b.   **ATP**   **ADP**

FIGURE 3.22 ATP.

ATP, the universal energy currency of cells, is composed of adenosine and three phosphate groups. **a.** Space-filling model of ATP. **b.** When cells require energy, ATP becomes ADP + Ⓟ, and energy is released. **c.** The breakdown of ATP provides the energy that an animal, such as a chipmunk, needs to acquire food and make more ATP.

In muscle cells, the energy is used for muscle contraction, and in nerve cells, it is used for the conduction of nerve impulses. Just as you spend money when you pay for a product or a service, cells "spend" ATP when they need something. Therefore, ATP is called the energy currency of cells.

Because energy is released when the last phosphate bond of ATP is hydrolyzed, it is sometimes called a high-energy bond, symbolized by a wavy line. But this terminology is misleading—the breakdown of ATP releases energy because the products of hydrolysis (ADP and Ⓟ) are more stable than the original reactant ATP. It is the entire molecule that releases energy, not a particular bond.

> ATP is a common high-energy molecule in cells. ATP breaks down to ADP + Ⓟ, releasing energy, which is used for all the metabolic work done in a cell or organism.

CONNECTING THE CONCEPTS

What does the term *organic* mean? For some, organic means that food products have been grown without the use of chemicals or have been minimally processed. Biochemically speaking, organic refers to molecules containing carbon and hydrogen. In biology, organic also refers to living things or anything that has been alive in the past. Therefore, the food we eat and the wood we burn are organic substances. Fossil fuels (coal and oil) formed over 300 million years ago from plant and animal life that, by chance, did not fully decompose are also organic. When burned, they release carbon dioxide into the atmosphere just as we do when we breathe!

Although living things are very complex, their macromolecules are simply polymers of small organic molecules. Simple sugars are the monomers of complex carbohydrates; amino acids are the monomers of proteins; nucleotides are the monomers of nucleic acids. Fats are composed of fatty acids and glycerol.

This system of forming macromolecules still allows for diversity. Monomers exist in modified forms and can combine in slightly different ways; therefore, a variety of macromolecules can come about. In cellulose, a plant product, glucose monomers are linked in a slightly different way than glucose monomers in glycogen, an animal product. One protein differs from another by the number and/or sequence of the same 20 amino acids.

There is no doubt that the chemistry of carbon is the chemistry of life. The groups of molecules discussed in this chapter, as well as other small molecules and ions, are assembled into structures that make up cells. As discussed in Chapter 4, each structure has a specific function necessary to the life of a cell.

Summary

3.1 ORGANIC MOLECULES

The chemistry of carbon accounts for the diversity of organic molecules found in living things. Carbon can bond with as many as four other atoms. It can also bond with itself to form both chains and rings. Differences in the carbon skeleton and attached functional groups cause organic molecules to have different chemical properties. The chemical properties of a molecule determine how it interacts with other molecules and the role the molecule plays in the cell. Some functional groups are hydrophobic and others are hydrophilic.

There are four classes of organic compounds in cells: carbohydrates, lipids, proteins, and nucleic acids (Table 3.5). Polysaccharides, the largest of the carbohydrates, are polymers of simple sugars called monosaccharides. The polypeptides of proteins are polymers of amino acids, and nucleic acids are polymers of nucleotides. Polymers are formed by the joining together of monomers. For each bond formed during a dehydration reaction, a molecule of water is removed, and for each bond broken during a hydrolysis reaction, a molecule of water is added.

3.2 CARBOHYDRATES

Monosaccharides, disaccharides, and polysaccharides are all carbohydrates. Therefore, the term *carbohydrate* includes both the monomers (e.g., glucose) and the polymers (e.g., starch, glycogen, and cellulose). Glucose is the immediate energy source of cells. Polysaccharides such as starch, glycogen, and cellulose are polymers of glucose. Starch in plants and glycogen in animals are energy storage compounds, but cellulose in plants and chitin in crabs and related animals, as well as fungi, have structural roles. Chitin's monomer is glucose with an attached amino group.

3.3 LIPIDS

Lipids include a wide variety of compounds that are insoluble in water. Fats and oils, which allow long-term energy storage, contain one glycerol and three fatty acids. Both glycerol and fatty acids have polar groups, but fats and oils are nonpolar, and this accounts for their insolubility in water. Fats tend to contain saturated fatty acids, and oils tend to contain unsaturated fatty acids. Saturated fatty acids do not have carbon–carbon double bonds, but unsaturated fatty acids do have double bonds in their hydrocarbon chain.

In a phospholipid, one of the fatty acids is replaced by a phosphate group. In the presence of water, phospholipids form a double layer because the head of each molecule is hydrophilic and the tails are hydrophobic.

Waxes are composed of a long fatty acid bonded to an alcohol with a long hydrocarbon chain. Steroids have the same four-ring structure as cholesterol, but each differs by the groups attached to these rings.

TABLE 3.5

Organic Compounds in Cells

	Categories	Elements	Examples	Functions
Carbohydrates	Monosaccharides 6-carbon sugar 5-carbon sugar	C, H, O	Glucose Deoxyribose, ribose	Immediate energy source Structure of DNA, RNA
	Disaccharides 12-carbon sugar	C, H, O	Sucrose	Transport sugar in plants
	Polysaccharides Polymer of glucose	C, H, O	Starch, glycogen Cellulose	Energy storage in plants, animals Plant cell wall structure
Lipids	Triglycerides 1 glycerol + 3 fatty acids	C, H, O	Fats, oils	Long-term energy storage
	Phospholipids Like triglyceride except the head group contains phosphate	C, H, O, P	Lecithin	Plasma membrane component
	Steroids Backbone of 4 fused rings	C, H, O	Cholesterol Testosterone, estrogen	Plasma membrane component Sex hormones
	Waxes Fatty acid + alcohol	C, H, O	Cuticle Earwax	Protective covering in plants Protective wax in ears
Proteins	Polypeptides Polymer of amino acids	C, H, O, N, S	Enzymes Myosin and actin Insulin Hemoglobin Collagen	Speed cellular reactions Muscle cell components Regulates sugar content of blood Oxygen carrier in blood Fibrous support of body parts
Nucleic Acids	Nucleic acids Polymer of nucleotides	C, H, O, N, P	DNA RNA	Genetic material Protein synthesis
	Nucleotides		ATP Coenzymes	Energy carrier Assist enzymes

3.4 PROTEINS

Proteins are polymers of amino acids. Proteins carry out many diverse functions in cells and organisms, including support, enzymes, transport, defense, hormones, and motion.

A polypeptide is a long chain of amino acids joined by peptide bonds. There are 20 different amino acids in cells, and they differ only by their R groups. Polarity and nonpolarity are important aspects of the R groups. A polypeptide has up to four levels of structure: the primary level is the sequence of the amino acids; the secondary level contains α helices and β (pleated) sheets held in place by hydrogen bonding between amino acids along the polypeptide chain; and the tertiary level is the final folding of the polypeptide, which is held in place by bonding and hydrophobic interactions between R groups. Proteins that contain more than one polypeptide have a quaternary level of structure as well.

The shape of an enzyme is important to its function. Both high temperatures and a change in pH can cause proteins to denature and lose their shape.

3.5 NUCLEIC ACIDS

The nucleic acids DNA and RNA are polymers of nucleotides. Each nucleotide has three components: a phosphate (phosphoric acid), a 5-carbon sugar, and a nitrogen-containing base.

DNA, which contains the sugar deoxyribose, is the genetic material that stores information for its own replication and for the order in which amino acids are to be sequenced in proteins. DNA, with the help of mRNA, specifies protein synthesis. DNA, which contains phosphate, the sugar deoxyribose, and the bases A, T, C, and G, is a double-stranded helix. RNA, containing phosphate, the sugar ribose, and the bases A, U, C, and G, is single stranded.

ATP, with its unstable phosphate bonds, is the energy currency of cells. Hydrolysis of ATP to ADP + ℗ releases energy, which is used by the cell to make a product or do any other type of metabolic work.

Reviewing the Chapter

1. How do the chemical characteristics of carbon affect the characteristics of organic molecules? 36–37
2. Give examples of functional groups, and discuss the importance of their being hydrophobic or hydrophilic. 37
3. What molecules are monomers of the polymers studied in this chapter? How do monomers join to produce polymers, and how are polymers broken down to monomers? 38
4. Name several monosaccharides, disaccharides, and polysaccharides, and give a function of each. How are these molecules structurally distinguishable? 39–40
5. What is the difference between a saturated and an unsaturated fatty acid? Explain the structure of a fat molecule by stating its components and how they join together. 42–43
6. How does the structure of a phospholipid differ from that of a fat? How do phospholipids form a bilayer in the presence of water? 44
7. Describe the structure of a generalized steroid. How does one steroid differ from another? 44–45
8. Draw the structure of an amino acid and a peptide, pointing out the peptide bond. 46
9. Discuss the four possible levels of protein structure, and relate each level to particular bonding patterns. 48–49
10. How do nucleotides bond to form nucleic acids? State and explain several differences between the structure of DNA and that of RNA. 50–51
11. Discuss the structure and function of ATP. 52

Testing Yourself

Choose the best answer for each question.

1. Which of these is not a characteristic of carbon?
   a. forms four covalent bonds
   b. bonds with other carbon atoms
   c. is sometimes ionic
   d. can form long chains
   e. sometimes shares two pairs of electrons with another atom

2. The functional group —COOH is
   a. acidic.
   b. basic.
   c. never ionized.
   d. found only in nucleotides.
   e. All of these are correct.

3. A hydrophilic group is
   a. attracted to water.
   b. a polar and/or ionized group.
   c. found at the end of fatty acids.
   d. the opposite of a hydrophobic group.
   e. All of these are correct.

4. Which of these is an example of a hydrolysis reaction?
   a. amino acid + amino acid $\longrightarrow$ dipeptide + H_2O
   b. dipeptide + H_2O $\longrightarrow$ amino acid + amino acid
   c. denaturation of a polypeptide
   d. Both a and b are correct.
   e. Both b and c are correct.

5. Which of these makes cellulose nondigestible in humans?
   a. a polymer of glucose subunits
   b. a fibrous protein
   c. the linkage between the glucose molecules
   d. the peptide linkage between the amino acid molecules
   e. The carboxyl groups ionize.

6. A fatty acid is unsaturated if it
   a. contains hydrogen.
   b. contains carbon–carbon double bonds.
   c. contains a carboxyl (acidic) group.
   d. bonds to glycogen.
   e. bonds to a nucleotide.

7. Which of these is not a lipid?
   a. steroid
   b. fat
   c. polysaccharide
   d. wax
   e. phospholipid

8. The difference between one amino acid and another is found in the
   a. amino group.
   b. carboxyl group.
   c. R group.
   d. peptide bond.
   e. carbon atoms.

9. The shape of a polypeptide is
   a. maintained by bonding between parts of the polypeptide.
   b. important to its function.
   c. ultimately dependent on the primary structure.
   d. necessary to its function.
   e. All of these are correct.

10. Which of these illustrates a peptide bond?

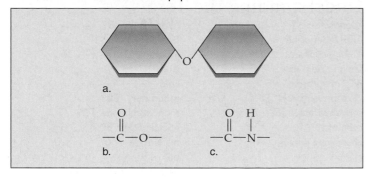

a.

b.   c.

11. Nucleotides
    a. contain a sugar, a nitrogen-containing base, and a phosphate group.
    b. are the monomers of fats and polysaccharides.
    c. join together by covalent bonding between the bases.
    d. are present in both DNA and RNA.
    e. Both a and d are correct.

12. ATP
    a. is an amino acid.
    b. has a helical structure.
    c. is a high-energy molecule that can break down to ADP and phosphate.
    d. provides enzymes for metabolism.
    e. is most energetic when in the ADP state.

13. Label the following diagram using the terms H_2O and *monomer* (as often as necessary), *hydrolysis reaction, dehydration reaction,* and *polymer.*

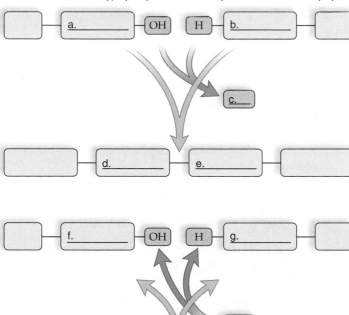

14. The monomer of a carbohydrate is
    a. an amino acid.
    b. a nucleic acid.
    c. a monosaccharide.
    d. a fatty acid.

15. The joining of two adjacent amino acids is called
    a. a peptide bond.
    b. a dehydration reaction.
    c. a covalent bond.
    d. All of these are correct.

16. The characteristic globular shape of a protein is the
    a. primary structure.
    b. secondary structure.
    c. tertiary structure.
    d. quaternary structure.

17. The shape of a polypeptide
    a. is maintained by bonding between parts of the polypeptide.
    b. is ultimately dependent on the primary structure.
    c. involves hydrogen bonding.
    d. All of these are correct.

18. Which of the following pertains to an RNA nucleotide and not to a DNA nucleotide?
    a. contains the sugar ribose
    b. contains a nitrogen-containing base
    c. contains a phosphate molecule
    d. becomes bonded to other nucleotides following a dehydration reaction

19. Nucleotides
    a. contain a sugar, a nitrogen-containing base, and a phosphate molecule.
    b. are the monomers for fats and polysaccharides.
    c. join together by covalent bonding between the bases.
    d. are found in DNA, RNA, and proteins.

For questions 20–27, match the items to those in the key. Some answers are used more than once.

KEY:
    a. carbohydrate
    b. fats and oils
    c. protein
    d. nucleic acid

20. contains the bases adenine, guanine, cytosine, and thymine

21. the 6-carbon sugar, glucose

22. polymer of amino acids

23. glycerol and fatty acids

24. enzymes

25. long-term energy storage

26. genes

27. plant cell walls

For questions 28–34, indicate whether the statement is true (T) or false (F).

28. Cholesterol is a type of phospholipid. _____

29. Fibrous proteins have a globular shape. _____

30. Sucrose is a type of monosaccharide. _____

31. Both proteins and nucleic acids contain nitrogen. _____

32. Both ribose and deoxyribose are 5-carbon sugars. _____

33. Polypeptides that contain fewer amino acids are said to be unsaturated. _____

34. ATP contains a sugar, adenine, and three phosphates. _____

Thinking Scientifically

1. You are studying the fat content of different types of seeds. You have discovered that some types of seeds have a much higher percentage of saturated fatty acids than others. You know the property difference (solid versus liquid) and the structural difference (more hydrogen versus less) between saturated and unsaturated fatty acids. How might these fatty acid differences correlate with climate (tropical compared to temperate), the size of seeds (small compared to large), and environmental conditions for germination (favorable compared to unfavorable)?
2. You are investigating molecules that inhibit a bacterial enzyme. You discover that the addition of several phosphate groups to an inhibitor improves its effectiveness. Why would knowledge of the three-dimensional structure of the bacterial enzyme help you understand why the phosphate groups improve the inhibitor's effectiveness?

Bioethical Issue: Organic Pollutants

Organic compounds include the carbohydrates, proteins, lipids, and nucleic acids that make up our bodies. Modern industry also uses all sorts of organic compounds that are synthetically produced. Indeed, our modern way of life wouldn't be possible without synthetic organic compounds.

Pesticides, herbicides, disinfectants, plastics, and textiles contain organic substances that are termed pollutants when they enter the natural environment and cause harm to living things. Global use of pesticides has increased dramatically since the 1950s, and modern pesticides are ten times more toxic than those of the 1950s. The Centers for Disease Control and Prevention in Atlanta reports that 40% of children working in agricultural fields now show signs of pesticide poisoning. The U.S. Geological Survey estimates that 32 million people in urban areas and 10 million people in rural areas are using groundwater that contains organic pollutants. J. Charles Fox, an official of the Environmental Protection Agency, says that "over the life of a person, ingestion of these chemicals has been shown to have adverse health effects such as cancer, reproductive problems, and developmental effects."

At one time, people failed to realize that everything in the environment is connected to everything else. In other words, they didn't know that an organic chemical can wander far from the site of its entry into the environment and that eventually these chemicals can enter our own bodies and cause harm. Now that we are aware of this outcome, we have to decide as a society how to proceed. We might decide to do nothing if the percentage of people dying from exposure to organic pollutants is small. Or we might decide to regulate the use of industrial compounds more strictly than has been done in the past. We could also decide that we need better ways of purifying public and private water supplies so that they do not contain organic pollutants.

Understanding the Terms

adenosine   52
ADP (adenosine diphosphate)   52
amino acid   46
ATP (adenosine triphosphate)   50
carbohydrate   38
cellulose   41
chaperone protein   49
chitin   41
coenzyme   50
complementary base pairing   51
dehydration reaction   38
denatured   49
deoxyribose   39
disaccharide   39
DNA (deoxyribonucleic acid)   50
enzyme   38
fat   42
fatty acid   42
fibrous protein   49
functional group   37
globular protein   49
glucose   39
glycerol   42
glycogen   40
hemoglobin   46
hexose   39
hydrolysis reaction   38
hydrophilic   37
hydrophobic   37
inorganic chemistry   36
isomer   37
lipid   42
monomer   38
monosaccharide   39
nucleic acid   50
nucleotide   50
oil   42
organic chemistry   36
organic molecule   36
pentose   39
peptide   46
peptide bond   46
peptidoglycan   41
phospholipid   44
polymer   38
polypeptide   46
polysaccharide   39
prion   49
protein   46
ribose   39
RNA (ribonucleic acid)   50
saturated fatty acid   42
starch   40
steroid   44
triglyceride   42
unsaturated fatty acid   42
wax   45

Match the terms to these definitions:
a. _____ Class of organic compounds that includes monosaccharides, disaccharides, and polysaccharides.
b. _____ Class of organic compounds that tend to be soluble in nonpolar solvents such as alcohol.
c. _____ Macromolecule consisting of covalently bonded monomers.
d. _____ Molecules that have the same molecular formula but a different structure and, therefore, shape.
e. _____ Two or more amino acids joined together by covalent bonding.

ARIS, the *Biology* Website

ARIS, the website for *Biology*, provides a wealth of information organized and integrated by chapter. You will find practice quizzes, interactive activities, labeling exercises, flashcards, and much more that will complement your learning and understanding of general biology.

www.mhhe.com/maderbiology9

CELL STRUCTURE AND FUNCTION

4

C ells, the fundamental building blocks of organisms, are so tiny they weren't discovered until the development of the microscope. The Dutch shopkeeper Antonie van Leeuwenhoek (1632–1723) was probably the first person to see cells. After using a magnifying glass to evaluate cloth, he taught himself to grind more powerful lenses and built a microscope. He looked at everything possible including the teeming life in a drop of pond water. Perhaps he even saw the unicellular protist, Stentor, featured in the accompanying micrograph. Leeuwenhoek's reports of innumerable microscopic "wee beasties and animalcules" were enthusiastically received by the Royal Society of London.

The English scientist Robert Hooke (1635–1703) was a microscopist who confirmed Leeuwenhoek's observations and was the first to use the word cell. The tiny chambers he observed in the honeycomb structure of cork reminded him of the rooms, or cells, in a monastery. Naturally, he referred to the boundaries of these chambers as walls. Today, we know that all tissues from the cork of a tree to the nervous tissue of humans are composed of cells.

A light micrograph does not reveal much cellular detail; a cell's often centrally placed nucleus is the only highly visible structure. The other parts of a cell are largely indistinguishable. Electron microscopy and biochemical analysis of the last century led to the discovery that plant and animal cells contain a variety of organelles, each specialized to perform a particular function. Even the content of organelles is now known. The nucleus may contain numerous chromosomes and thousands of genes!

Micrograph of a freshwater protozoan, Stentor.

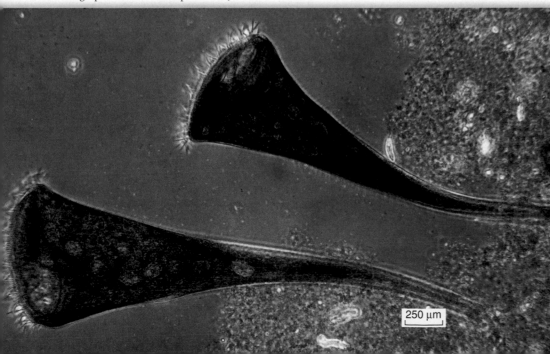

250 μm

4.1 CELLULAR LEVEL OF ORGANIZATION

The 1830s were exciting times in the history of our knowledge of the cell. In 1831, the English botanist Robert Brown described the nucleus of cells. In 1838, the German botanist Matthais Schleiden stated that all plants are composed of cells. A year later, the German zoologist Theodor Schwann declared that all animals are composed of cells (Fig. 4.1). As a result of their work, the field of cytology (study of cells) began, and we can conclude that a **cell** is the smallest unit of living matter.

In the 1850s, the German physician Rudolph Virchow viewed the human body as a state in which each cell was a citizen. Today, we know that various illnesses of the body, such as diabetes and prostate cancer, are due to a malfunctioning of cells rather than the organ itself. It also means that a cell is the basic unit of function as well as structure in organisms.

Virchow was the first to tell us that cells reproduce and "every cell comes from a preexisting cell." When unicellular organisms reproduce, a single cell divides, and when multicellular organisms grow, many cells divide. Cells are also involved in the sexual reproduction of multicellular organisms. In reality, there is a continuity of cells from generation to generation even back to the very first cell (or cells) in the history of life. Due to countless investigations that began with the work of Virchow, it is evident that cells are capable of self-reproduction.

The **cell theory** is based upon the work of Schleiden, Schwann, and Virchow. It states that (1) all organisms are composed of cells, (2) cells are the basic units of structure and function in organisms, and (3) cells come only from preexisting cells because cells are self-reproducing.

All organisms are made up of cells, and a cell is the structural and functional unit of organs and, ultimately, of organisms. Cells are capable of self-reproduction, and cells come only from preexisting cells.

FIGURE 4.1 Organisms and cells.
All organisms, including plants and animals, are composed of cells. This is not readily apparent because a microscope is usually needed to see the cells. **a.** Lilac plant. **b.** Light micrograph of a cross section of a lilac leaf showing many individual cells. **c.** Rabbit. **d.** Light micrograph of a rabbit's intestinal lining showing that it, too, is composed of cells. The dark-staining bodies are nuclei.

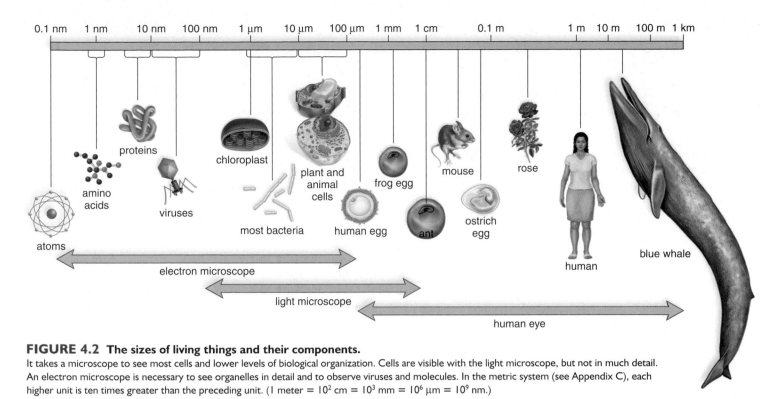

FIGURE 4.2 The sizes of living things and their components.
It takes a microscope to see most cells and lower levels of biological organization. Cells are visible with the light microscope, but not in much detail. An electron microscope is necessary to see organelles in detail and to observe viruses and molecules. In the metric system (see Appendix C), each higher unit is ten times greater than the preceding unit. (1 meter $= 10^2$ cm $= 10^3$ mm $= 10^6$ μm $= 10^9$ nm.)

Cell Size

Cells are quite small. A frog's egg, at about 1 millimeter (mm) in diameter, is large enough to be seen by the human eye. But most cells are far smaller than 1 mm; some are even as small as 1 micrometer (μm)—one thousandth of a millimeter. Cell inclusions and macromolecules are smaller than a micrometer and are measured in terms of nanometers (nm). Figure 4.2 outlines the visual range of the eye, light microscope, and electron microscope, and the discussion of microscopy in the Science Focus on pages 60 and 61 explains why the electron microscope allows us to see so much more detail than the light microscope does.

Why are cells so small? To answer this question, consider that a cell needs a surface area large enough to allow adequate nutrients to enter and to rid itself of wastes. Small cells, not large cells, are likely to have an adequate surface area for exchanging wastes for nutrients. For example, Figure 4.3 visually demonstrates that cutting a large cube into smaller cubes provides a lot more surface area per volume. The calculations show that a 4-cm cube has a **surface-area-to-volume ratio** of only 1.5:1, whereas a 1-cm cube has a surface-area-to-volume ratio of 6:1.

We would expect, then, that actively metabolizing cells would be small. A chicken's egg is several centimeters in diameter, but the egg is not actively metabolizing. Once the egg is incubated and metabolic activity begins, the egg divides repeatedly without growth. Cell division restores the amount of surface area needed for adequate exchange of materials. Further, cells that specialize in absorption have modifications that greatly increase the surface area per volume of the cell. The columnar epithelial cells along the surface of the intestinal wall have surface foldings called microvilli (sing., microvillus) that increase their surface area. Nerve cells and some large plant cells are long and thin in order to keep the cytoplasm near the plasma membrane.

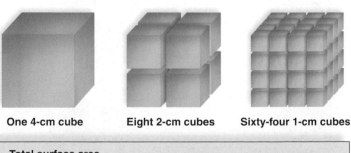

One 4-cm cube	Eight 2-cm cubes	Sixty-four 1-cm cubes
Total surface area (height × width × number of sides × number of cubes)		
96 cm^2	192 cm^2	384 cm^2
Total volume (height × width × length × number of cubes)		
64 cm^3	64 cm^3	64 cm^3
Surface area: Volume per cube (surface area ÷ volume)		
1.5:1	3:1	6:1

FIGURE 4.3 Surface-area-to-volume relationships.
As cell size decreases from 4 cm^3 to 1 cm^3, the surface-area-to-volume ratio increases.

A cell needs a surface area that can adequately exchange materials with the environment. Surface-area-to-volume considerations require that cells stay small.

---

science focus

Microscopy Today

Cells were not discovered until the seventeenth century (when the microscope was invented). Since that time, various types of microscopes have been developed for the study of cells and their components.

In the *compound light microscope,* light rays passing through a specimen are brought into focus by a set of glass lenses, and the resulting image is then viewed by the human eye. In the *transmission electron microscope (TEM),* electrons passing through a specimen are brought into fo-

cus by a set of magnetic lenses, and the resulting image is projected onto a fluorescent screen or photographic film. In the *scanning electron microscope (SEM),* a narrow beam of electrons is scanned over the surface of the specimen, which is coated with a thin metal layer. The metal gives off secondary electrons that are collected by a detector to produce an image on a television screen. The SEM permits the development of three-dimensional images. Figure 4A shows these three types of microscopic images.

Magnification, Resolution, and Contrast

The magnifying capability of a transmission electron microscope is greater than that of a compound light microscope. A light microscope can magnify objects about a thousand times, but an electron microscope can magnify them hundreds of thousands of times. The difference lies in the means of illumination. The path of light rays and electrons moving through space is wavelike, but the wavelength of elec-

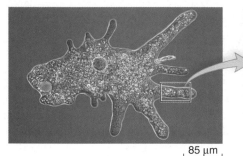

85 μm

amoeba, light micrograph

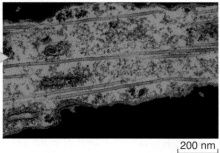

200 nm

pseudopod segment, transmission electron micrograph

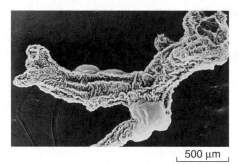

500 μm

amoeba, scanning electron micrograph

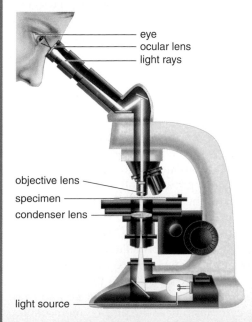

a. Compound light microscope

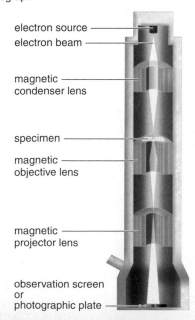

b. Transmission electron microscope

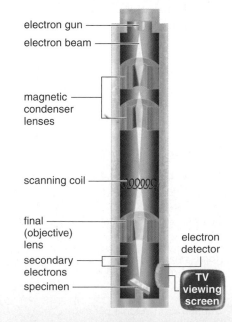

c. Scanning electron microscope

FIGURE 4A Diagram of microscopes with accompanying micrographs of *Amoeba proteus.*
The compound light microscope and the transmission electron microscope provide an internal view of an organism. The scanning electron microscope provides an external view of an organism.

trons is much shorter than the wavelength of light. This difference in wavelength accounts for the electron microscope's greater magnifying capability and its greater resolving power. The greater the resolving power, the greater the detail eventually seen. *Resolution* is the minimum distance between two objects at which they can still be seen, or resolved, as two separate objects. If oil is placed between the sample and the objective lens of the compound light microscope, the resolving power is increased, and if ultraviolet light is used instead of visible light, it is also increased. But typically, a light microscope can resolve down to 0.2 μm, while the transmission electron microscope can resolve down to 0.0002 μm. If the resolving power of the average human eye is set at 1.0, then that of the typical compound light microscope is about 500, and that of the transmission electron microscope is 100,000. This means that this type of electron microscope distinguishes much greater detail (Fig. 4A*b*).

Some microscopes view living specimens, but often specimens are treated prior to observation. Cells are killed, fixed so that they do not decompose, and embedded into a matrix. The matrix strengthens the specimen so that it can be thinly sliced. These sections are often stained with colored dyes (light microscopy) or with electron-dense metals (electron microscopy) to provide contrast. Another way to increase contrast is to use optical methods such as phase contrast and differential interference contrast (Fig. 4B). In addition to optical and electronic methods for contrasting transparent cells, a third very prominent research tool is called *immunofluorescence microscopy*, because it uses fluorescent antibodies to reveal the location of a protein in the cell (see Fig. 4.18). The importance of this method is that the cellular distribution of a single type of protein can be examined.

Illumination, Viewing, and Recording

Light rays can be bent (refracted) and brought to focus as they pass through glass lenses, but electrons do not pass through glass. Electrons have a charge that allows them to be brought into focus by magnetic lenses. The human eye uses light to see an object but cannot use electrons for the same purpose. Therefore, electrons leaving the specimen in the electron microscope are directed toward a screen or a photographic plate that is sensitive to their presence. Humans can view the image on the screen or photograph.

A major advancement in illumination has been the introduction of *confocal microscopy,* which uses a laser beam scanned across the specimen to focus on a single shallow plane within the cell. The microscopist can "optically section" the specimen by focusing up and down, and a series of optical sections can be combined in a computer to create a three-dimensional image, which can be displayed and rotated on the computer screen.

An image from a microscope may be recorded by replacing the human eye with a television camera. The television camera converts the light image into an electronic image, which can be entered into a computer. In *video-enhanced contrast microscopy,* the computer makes the darkest areas of the original image much darker and the lightest areas of the original much lighter. The result is a high-contrast image with deep blacks and bright whites. Even more contrast can be introduced by the computer if shades of gray are replaced by colors.

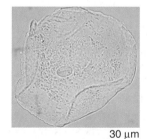 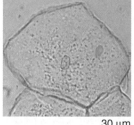 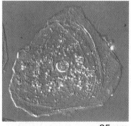 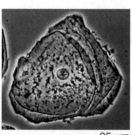 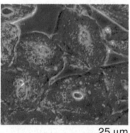

30 μm   30 μm   25 μm   25 μm   25 μm

Bright-field. Light passing through the specimen is brought directly into focus. Usually, the low level of contrast within the specimen interferes with viewing all but its largest components.

Bright-field (stained). Dyes are used to stain the specimen. Certain components take up the dye more than other components, and therefore contrast is enhanced.

Differential interference contrast. Optical methods are used to enhance density differences within the specimen so that certain regions appear brighter than others. This technique is used to view living cells, chromosomes, and organelle masses.

Phase contrast. Density differences in the specimen cause light rays to come out of "phase." The microscope enhances these phase differences so that some regions of the specimen appear brighter or darker than others. The technique is widely used to observe living cells and organelles.

Dark-field. Light is passed through the specimen at an oblique angle so that the objective lens receives only light diffracted and scattered by the object. This technique is used to view organelles, which appear quite bright against a dark field.

FIGURE 4B Photomicrographs of cheek cells.
Bright-field microscopy is the most common form used with a compound light microscope. Other types of microscopy include differential interference contrast, phase contrast, and dark-field.

4.2 PROKARYOTIC CELLS

Fundamentally, two different types of cells exist. **Prokaryotic cells** [Gk. *pro,* before, and *karyon,* kernel, nucleus] are so named because they lack a membrane-bounded nucleus. The other type of cell, called a **eukaryotic cell,** has a nucleus (see Figs. 4.6 and 4.7). Prokaryotic cells are simpler and much smaller than eukaryotic cells, and they are present in great numbers in the air, in bodies of water, in the soil, and they also live in and on other organisms. Prokaryotes are an extremely successful group of organisms whose evolutionary history dates back to the first cells on Earth.

All prokaryotic cells are structurally simple, and they can be divided into two groups, largely based on nucleic and base sequence differences. These two groups are so biochemically different that they have been placed in separate domains, called domain Bacteria and domain Archaea. Bacteria are well known because they cause some serious diseases, such as tuberculosis, anthrax, tetanus, throat infections, and gonorrhea. But many species of bacteria are important to the environment because they decompose the remains of dead organisms and contribute to ecological cycles. Bacteria also assist humans in another way—we use them to manufacture all sorts of products, from industrial chemicals to foodstuffs and drugs.

The Structure of Bacteria

Bacteria are quite small; an average size is 1.1–1.5 μm wide and 2.0–6.0 μm long. These different shapes are common.

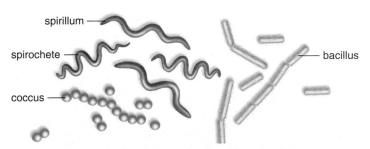

A rod-shaped bacterium is called a **bacillus,** while a spherical-shaped bacterium is a **coccus.** Both of these can occur as pairs or chains, and in addition, cocci can occur as clusters. Some long rods are twisted into spirals, in which case they are **spirilla** if they are rigid or **spirochetes** if they are flexible.

Figure 4.4 shows the generalized structure of a bacterium. This example is a bacillus because of its rod shape. For the sake of discussion, we will divide the organization of bacteria into the cell envelope, the cytoplasm, and the appendages.

Cell Envelope

The **cell envelope** includes the plasma membrane, the cell wall, and the glycocalyx. The **plasma membrane** of a bacterial cell has the same composition as that of a eukaryotic cell—it is a phospholipid bilayer with both embedded and peripheral proteins. The plasma membrane has the important function of regulating the entrance and exit of substances into and out of the cytoplasm. After all, the cytoplasm has a normal composition that needs to be maintained.

The plasma membrane can form internal pouches called mesosomes. **Mesosomes** most likely increase the internal surface area for the attachment of enzymes that are carrying on metabolic activities.

The **cell wall** maintains the shape of the cell even if the cytoplasm should happen to take up an abundance of water. You may recall that the cell wall of a plant cell is strengthened by the presence of cellulose, while the cell wall of a bacterium contains peptidoglycan, a complex molecule containing a unique amino disaccharide and peptide fragments.

The **glycocalyx** is a layer of polysaccharides lying outside the cell wall. When the layer is well organized and not easily washed off, it is called a **capsule.** A slime layer, on the other hand, is not well organized and is easily removed. The glycocalyx aids against drying out and helps bacteria resist a host's immune system. It also helps bacteria attach to almost any surface.

Cytoplasm

The **cytoplasm** is a semifluid solution composed of water and inorganic and organic molecules encased by a plasma membrane. Among the organic molecules are a variety of enzymes, which speed the many types of chemical reactions involved in metabolism.

The DNA of a bacterium is in a single chromosome that coils up and is located in a region called the **nucleoid.** Many bacteria also have an extrachromosomal piece of circular DNA called a **plasmid.** Plasmids are routinely used in biotechnology laboratories as vectors to transport DNA into a bacterium—even human DNA can be put into a bacterium by using a plasmid as a vector. This technology is important in the production of new medicines.

The many proteins specified for by bacterial DNA are synthesized on tiny particles called **ribosomes.** A bacterial cell contains thousands of ribosomes that are smaller than eukaryotic ribosomes. However, bacterial ribosomes still contain RNA and protein in two subunits, as do eukaryotic ribosomes. The **inclusion bodies** found in the cytoplasm are stored granules of various substances. Some are nutrients that can be broken down when needed.

The **cyanobacteria** are bacteria that photosynthesize in the same manner as plants. These organisms live in water, in ditches, on buildings, and on the bark of trees. Their cytoplasm contains extensive internal membranes called **thylakoids** [Gk. *thylakon,* and *eidos,* form] where chlorophyll and other pigments absorb solar energy for the production of carbohydrates. Cyanobacteria are called the blue-green bacteria because some have a pigment that adds a shade of blue to the cell, in addition to the green color of chlorophyll. The cyanobacteria release oxygen as a side product of photosynthesis, and perhaps ancestral cyanobacteria were the first types of organisms on Earth to do so. The addition of oxygen changed the composition of the Earth's atmosphere.

Appendages

The appendages of a bacterium, namely the flagella, fimbriae, and sex pili, are made of protein. Motile bacteria can propel themselves in water by the means of appendages called

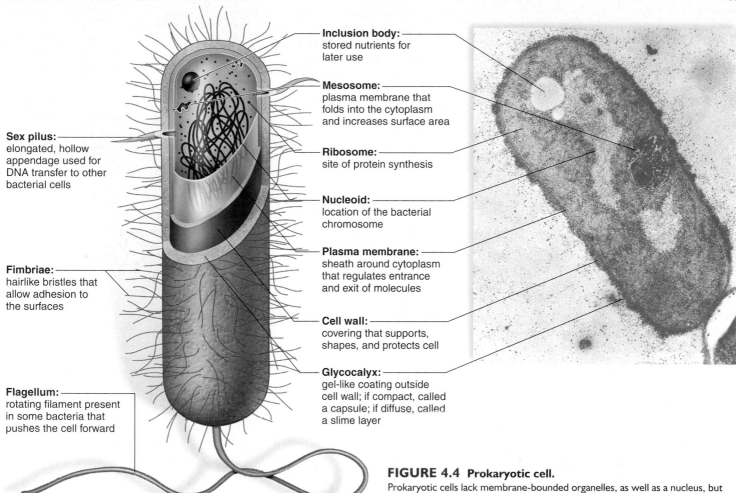

Inclusion body: stored nutrients for later use

Mesosome: plasma membrane that folds into the cytoplasm and increases surface area

Ribosome: site of protein synthesis

Nucleoid: location of the bacterial chromosome

Plasma membrane: sheath around cytoplasm that regulates entrance and exit of molecules

Cell wall: covering that supports, shapes, and protects cell

Glycocalyx: gel-like coating outside cell wall; if compact, called a capsule; if diffuse, called a slime layer

Sex pilus: elongated, hollow appendage used for DNA transfer to other bacterial cells

Fimbriae: hairlike bristles that allow adhesion to the surfaces

Flagellum: rotating filament present in some bacteria that pushes the cell forward

FIGURE 4.4 Prokaryotic cell.
Prokaryotic cells lack membrane-bounded organelles, as well as a nucleus, but they possess a nucleoid region that houses DNA.

flagella (usually 20 nm in diameter and 1–70 nm long). The bacterial flagellum has a filament, a hook, and a basal body. The basal body is a series of rings anchored in the cell wall and membrane. The hook rotates 360° within the basal body, and this motion propels bacteria—the bacterial flagellum does not move back and forth like a whip. Sometimes flagella occur only at the two ends of a cell, and sometimes they are dispersed randomly over the surface. The number and location of flagella are important in distinguishing different types of bacteria.

Fimbriae are small, bristlelike fibers that sprout from the cell surface. They are not involved in locomotion; instead, fimbriae attach bacteria to a surface. **Sex pili** are rigid tubular structures used by bacteria to pass DNA from cell to cell. Bacteria reproduce asexually by binary fission, but they can exchange DNA by way of the sex pili. They can also take up DNA from the external medium or by way of viruses.

The Structure of Archaea

Like bacteria, archaea are prokaryotes. Archaea are more diverse in shape than bacteria because, in addition to the shapes illustrated on page 62, they can be lobed, platelike, or irregular in shape.

The cell wall of archaea does not contain peptidoglycan; it contains polysaccharides and proteins arranged in different ways in various archaea. The membrane lipids are composed of glycerol bonded to hydrocarbons, not fatty acids.

The base sequences of DNA and RNA in archaea match that of eukaryotes better than that of bacteria! Therefore, it is thought that archaea may be more closely related to eukaryotes than to bacteria.

The archaea are well known for living in extreme habitats. They abound in extremely salty and/or hot, aqueous environments. These conditions are thought to resemble the earliest environments on Earth, and archaea may have been the first type of cell to evolve. They are able to adapt to various environments, however and they are even prevalent in the waters off the coast of Antarctica.

Bacteria and archaea are prokaryotic cells. Bacterial cells have these features:

Cell envelope	Glycocalyx
	Cell wall
	Plasma membrane
Cytoplasm	Nucleoid
	Ribosomes
	Thylakoids (cyanobacteria)
Appendages	Flagella
	Sex pili
	Fimbriae

4.3 EUKARYOTIC CELLS

Organisms with eukaryotic cells, namely protists, fungi, plants, and animals, are members of domain Eukarya, the third domain of living things. Unlike prokaryotic cells, eukaryotic cells do have a membrane-bounded **nucleus** [L. *nucleus*, kernel], which houses their DNA. It has been suggested by some scientists that the nucleus evolved as the result of the invagination of the plasma membrane (Fig. 4.5).

Eukaryotic cells are much larger than prokaryotic cells, and therefore they have less surface area per volume than prokaryotic cells (see Fig. 4.2). This difference in surface-area-to-volume ratio is not detrimental to the cells' existence because, unlike prokaryotic cells, eukaryotic cells are compartmentalized. They have small structures called **organelles** that are specialized to perform specific functions.

Eukaryotic cells, like prokaryotic cells, have a plasma membrane that separates the contents of the cell from the environment and regulates the passage of molecules into and out of the cytoplasm. The plasma membrane is a phospholipid bilayer with embedded proteins:

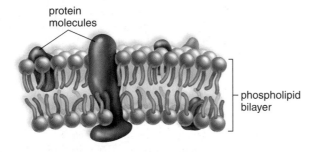

Some eukaryotic cells, notably plant cells, also have an outer boundary called a cell wall. A plant cell wall contains cellulose fibrils and therefore has a different composition than the cell wall of bacteria. A cell wall supports and protects the cell but does not interfere with the movement of molecules across the plasma membrane. The plasmodesmata are channels in a cell wall that allow cytoplasmic strands to extend between adjacent cells.

The Structure of Eukaryotic Cells

Figures 4.6 and 4.7 illustrate cellular anatomy and types of structures and organelles found in animal and plant cells. In this chapter, our discussion of the structures found in eukaryotic cells will be divided into these categories: the nucleus and ribosomes; the organelles of the endomembrane system; the peroxisomes and vacuoles; the energy-related organelles; and the cytoskeleton.

The nucleus communicates with ribosomes in the cytoplasm, and the organelles of the endomembrane system communicate with one another. Each organelle has its own particular set of enzymes and produces its own products, and the products move from one organelle to the other. The products are carried between organelles by little transport vesicles, membranous sacs that enclose the molecules

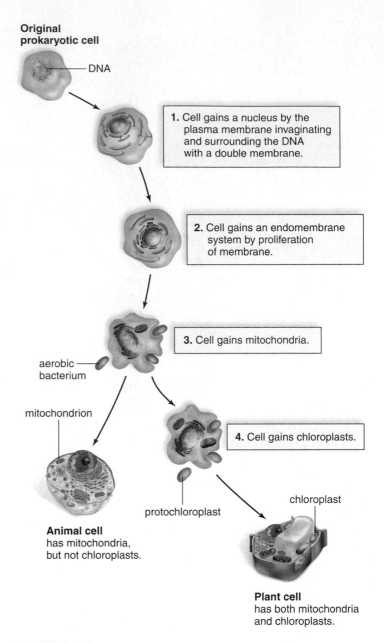

FIGURE 4.5 Origin of organelles.
Invagination of the plasma membrane could have created the nuclear envelope and an endomembrane system. The endosymbiotic hypothesis suggests that mitochondria and chloroplasts could have been independent prokaryotes that took up residence in a eukaryotic cell.

and keep them separate from the cytoplasm. In contrast, the energy-related organelles—the mitochondria in plant and animal cells, and the chloroplasts in plant cells—do not communicate with the other organelles of the cell. Except for importing certain proteins, these organelles are self-sufficient. They even have their own genetic material, and their ribosomes resemble those of prokaryotic cells. This and other evidence suggest that the mitochondria and chloroplasts are derived from prokaryotes that took up residence in an early eukaryotic cell (see Fig. 4.5). Notice that an animal cell has only mitochondria, while a plant cell has both mitochondria and chloroplasts.

science focus

Cell Fractionation and Differential Centrifugation

Modern microscopic techniques can be counted on to reveal the structure and distribution of organelles in a cell. But how do researchers isolate the different types of organelles from a cell so that they can determine their function? Suppose, for example, you wanted to study the function of ribosomes. How would you acquire some ribosomes? First, researchers remove cells from an organism or cell culture and place them in a sugar or salt solution. Then they fractionate (break open) the cells in a homogenizer.

A process called *differential centrifugation* allows researchers to separate the parts of a cell by size and density. A centrifuge works like the spin cycle of a washing machine. Only when the centrifuge spins do cell components come out of suspension and form a sediment. The faster the centrifuge spins, the smaller the components that settle out.

Figure 4C shows that the slowest spin cycle separates out the nuclei, and then progressively faster cycles separate out ever smaller components. In between spins, the fluid portion of the previous cycle must be poured into a clean tube. Why? If you didn't start with a fresh tube, all the different cell parts would pile up in one tube.

By using different salt solutions and different centrifuge speeds, researchers can obtain essentially pure preparations of almost any cell component. Biochemical analysis and manipulation then allow them to determine the function of that component.

FIGURE 4C Cell fractionation and differential centrifugation.
Cells are broken open mechanically by the action of the pestle against the side of a homogenizer. Then a centrifuge spins the tubes, and this action separates out the contents of the cell. The notations above the arrows indicate the force of gravity (g) and length of time necessary to separate out the structures listed. With ever-increasing speed, first the larger and then the smaller components of the cell are in the sediment. Under the proper conditions, an organelle will continue to work in isolation so that its functions can be determined.

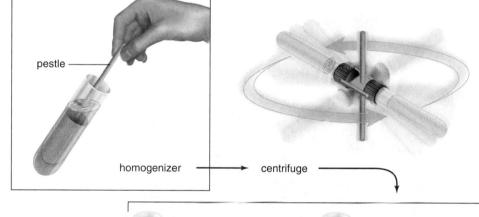

pestle

homogenizer ⟶ centrifuge

600 g × 10 min   15,000 g × 5 min   100,000 g × 60 min

nuclei in sediment   mitochondria and lysosomes in sediment   ribosomes and endoplasmic reticulum in sediment   soluble portion of cytoplasm

The cytoskeleton is a lattice of protein fibers that maintains the shape of the cell and assists in the movement of organelles. The protein fibers serve as tracks for the transport vesicles that are taking molecules from one organelle to another. In other words, the tracks direct and speed them on their way. The manner in which vesicles and other types of organelles move along these tracks will be discussed in more detail later in the chapter. Without a cytoskeleton, a eukaryotic cell would not have an efficient means of moving organelles and their products within the cell and possibly could not exist.

Each structure in an animal or plant cell (Figs. 4.6 and 4.7) has been given a particular color that will be used for this structure throughout the text.

Compartmentalization is seen in eukaryotic cells, and they are larger than prokaryotic cells. We will discuss the nucleus and ribosomes; the organelles of the endomembrane system; the energy-related organelles; and the cytoskeleton. Each of these has a specific structure and function.

FIGURE 4.6
Animal cell anatomy.
Micrograph of liver cell and drawing of a generalized animal cell.

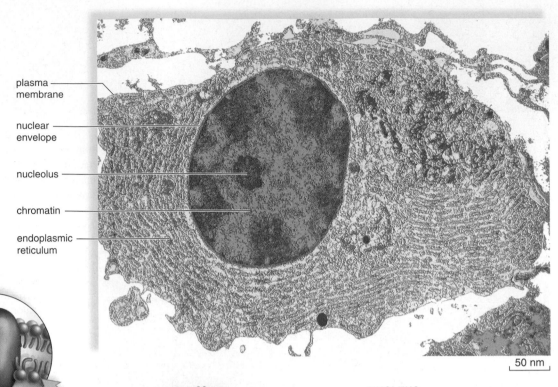

plasma membrane

nuclear envelope

nucleolus

chromatin

endoplasmic reticulum

50 nm

Plasma membrane: outer surface that regulates entrance and exit of molecules

protein

phospholipid

CYTOSKELETON: maintains cell shape and assists movement of cell parts:

Microtubules: cylinders of protein molecules present in cytoplasm, centrioles, cilia, and flagella

Intermediate filaments: protein fibers that provide support and strength

Actin filaments: protein fibers that play a role in movement of cell and organelles

Centrioles*: short cylinders of microtubules of unknown function

Centrosome: microtubule organizing center that contains a pair of centrioles

Lysosome*: vesicle that digests macromolecules and even cell parts

Vesicle: membrane-bounded sac that stores and transports substances

Cytoplasm: semifluid matrix outside nucleus that contains organelles

NUCLEUS:

Nuclear envelope: double membrane with nuclear pores that encloses nucleus

Chromatin: diffuse threads containing DNA and protein

Nucleolus: region that produces subunits of ribosomes

ENDOPLASMIC RETICULUM:

Rough ER: studded with ribosomes

Smooth ER: lacks ribosomes, synthesizes lipid molecules

Ribosomes: particles that carry out protein synthesis

Peroxisome: vesicle that has various functions; breaks down fatty acids and converts resulting hydrogen peroxide to water

Polyribosome: string of ribosomes simultaneously synthesizing same protein

Mitochondrion: organelle that carries out cellular respiration, producing ATP molecules

Golgi apparatus: processes, packages, and secretes modified cell products

*not in plant cells

peroxisome

mitochondrion

nucleus

ribosomes

central vacuole

plasma membrane

cell wall

chloroplast

1 μm

FIGURE 4.7
Plant cell anatomy.
False-colored micrograph
of a young plant cell and
drawing of a generalized
plant cell.

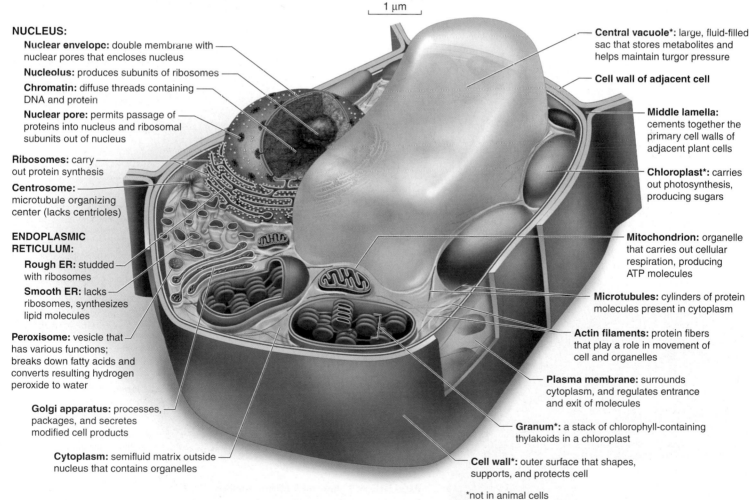

NUCLEUS:

Nuclear envelope: double membrane with
nuclear pores that encloses nucleus

Nucleolus: produces subunits of ribosomes

Chromatin: diffuse threads containing
DNA and protein

Nuclear pore: permits passage of
proteins into nucleus and ribosomal
subunits out of nucleus

Ribosomes: carry
out protein synthesis

Centrosome:
microtubule organizing
center (lacks centrioles)

**ENDOPLASMIC
RETICULUM:**

Rough ER: studded
with ribosomes

Smooth ER: lacks
ribosomes, synthesizes
lipid molecules

Peroxisome: vesicle that
has various functions;
breaks down fatty acids and
converts resulting hydrogen
peroxide to water

Golgi apparatus: processes,
packages, and secretes
modified cell products

Cytoplasm: semifluid matrix outside
nucleus that contains organelles

Central vacuole*: large, fluid-filled
sac that stores metabolites and
helps maintain turgor pressure

Cell wall of adjacent cell

Middle lamella:
cements together the
primary cell walls of
adjacent plant cells

Chloroplast*: carries
out photosynthesis,
producing sugars

Mitochondrion: organelle
that carries out cellular
respiration, producing
ATP molecules

Microtubules: cylinders of protein
molecules present in cytoplasm

Actin filaments: protein fibers
that play a role in movement of
cell and organelles

Plasma membrane: surrounds
cytoplasm, and regulates entrance
and exit of molecules

Granum*: a stack of chlorophyll-containing
thylakoids in a chloroplast

Cell wall*: outer surface that shapes,
supports, and protects cell

*not in animal cells

The Nucleus and Ribosomes

The nucleus is essential to the life of a cell. It contains the genetic information that is passed on from cell to cell and from generation to generation. The ribosomes use this information to carry out protein synthesis.

The Nucleus

The nucleus, which has a diameter of about 5 μm, is a prominent structure in the eukaryotic cell (Fig. 4.8). It generally appears as an oval structure located near the center of most cells. A cell can have more than one nucleus. The nucleus contains **chromatin** [Gk. *chroma,* color, and *teino,* stretch] in a semifluid matrix called the **nucleoplasm.** Chromatin looks grainy, but actually it is a network of strands that condenses and undergoes coiling into rodlike structures called **chromosomes** [Gk. *chroma,* color, and *soma,* body], just before the cell divides. All of the cells of an individual contain the same number of chromosomes, and the mechanics of nuclear division ensure that each daughter cell receives the normal number of chromosomes, except for the egg and sperm, which have half this

number. This alone suggested to early investigators that the chromosomes are the carriers of genetic information.

Chromatin, and therefore chromosomes, contains DNA, protein, and some RNA (ribonucleic acid). Genes, composed of DNA, are units of heredity located on the chromosomes.

RNA, of which there are several forms, is produced in the nucleus. A **nucleolus** is a dark region of chromatin where a type of RNA, called ribosomal RNA (rRNA), is produced and where rRNA joins with proteins to form the subunits of ribosomes. Ribosomes are small bodies in the cytoplasm where protein synthesis occurs. Another type of RNA, called messenger RNA (mRNA), acts as an intermediary for DNA and specifies the sequence of amino acids during protein synthesis. Transfer RNA (tRNA) is used in the assembly of amino acids during protein synthesis. The proteins of a cell determine its structure and functions; therefore, the nucleus is the command center for a cell.

The nucleus is separated from the cytoplasm by a double membrane known as the **nuclear envelope.** Even so, the nucleus communicates with the cytoplasm. The nuclear envelope has **nuclear pores** of sufficient size (100 nm) to per-

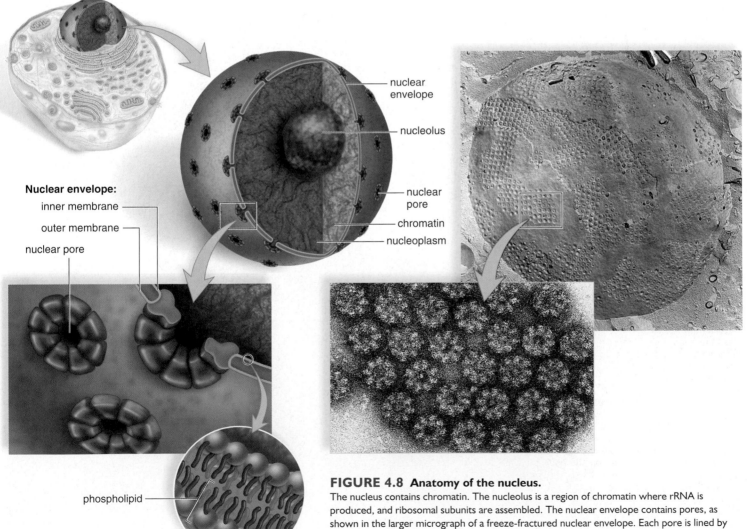

Nuclear envelope:
inner membrane
outer membrane
nuclear pore

phospholipid

nuclear envelope
nucleolus
nuclear pore
chromatin
nucleoplasm

FIGURE 4.8 Anatomy of the nucleus.
The nucleus contains chromatin. The nucleolus is a region of chromatin where rRNA is produced, and ribosomal subunits are assembled. The nuclear envelope contains pores, as shown in the larger micrograph of a freeze-fractured nuclear envelope. Each pore is lined by a complex of eight proteins, as shown in the smaller micrograph and drawing. Nuclear pores serve as passageways for substances to pass into and out of the nucleus.

mit the passage of ribosomal subunits and mRNA out of the nucleus into the cytoplasm and the passage of proteins from the cytoplasm into the nucleus. High-power electron micrographs show that nonmembranous components associated with the pores form a nuclear pore complex.

Ribosomes

Ribosomes are non-membrane-bounded particles where protein synthesis occurs. In eukaryotes, ribosomes are 20 nm by 30 nm, and in prokaryotes they are slightly smaller. In both types of cells, ribosomes are composed of two subunits, one large and one small. Each subunit has its own mix of proteins and rRNA. The number of ribosomes in a cell varies depending on its functions. For example, pancreatic cells and those of other glands have many ribosomes because they produce secretions that contain proteins.

In eukaryotic cells, some ribosomes occur freely within the cytoplasm, either singly or in groups called **polyribosomes,** and others are attached to the endoplasmic reticulum (ER), a membranous system of flattened saccules (small sacs) and tubules, which is discussed more fully on the next page. Ribosomes receive mRNA from the nucleus, and this nucleic acid carries a coded message from DNA indicating the correct sequence of amino acids in a protein. Proteins synthesized by cytoplasmic ribosomes are used in the cytoplasm, and those synthesized by attached ribosomes end up in the ER.

What causes a ribosome to bind to the endoplasmic reticulum? Binding occurs only if the protein being synthesized by a ribosome begins with a signal peptide. The signal peptide binds to a signal recognition particle (SRP), which then binds to an SRP receptor on the endoplasmic reticulum. Once the protein enters the ER, a peptidase cleaves off the signal peptide, and the protein ends up within the lumen (interior) of the ER (Fig. 4.9).

The nucleus is in constant communication with the cytoplasm. The nucleus is the command center of the cell because it contains DNA, the genetic material. DNA, which is located in the chromosomes, specifies the sequence of amino acids in a protein through an intermediary called mRNA. Protein synthesis occurs in the cytoplasm, at the ribosomes, whose subunits are made in the nucleolus. Ribosomes occur singly and in groups (i.e., polyribosomes). Numerous ribosomes become attached to the endoplasmic reticulum.

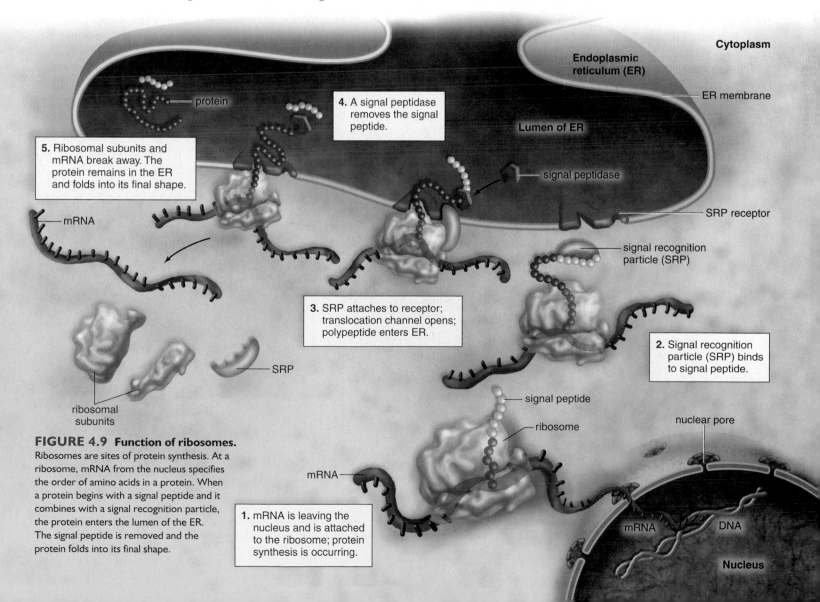

FIGURE 4.9 Function of ribosomes.
Ribosomes are sites of protein synthesis. At a ribosome, mRNA from the nucleus specifies the order of amino acids in a protein. When a protein begins with a signal peptide and it combines with a signal recognition particle, the protein enters the lumen of the ER. The signal peptide is removed and the protein folds into its final shape.

Cytoplasm

Endoplasmic reticulum (ER)

ER membrane

Lumen of ER

4. A signal peptidase removes the signal peptide.

protein

signal peptidase

5. Ribosomal subunits and mRNA break away. The protein remains in the ER and folds into its final shape.

SRP receptor

mRNA

signal recognition particle (SRP)

3. SRP attaches to receptor; translocation channel opens; polypeptide enters ER.

2. Signal recognition particle (SRP) binds to signal peptide.

SRP

signal peptide

ribosomal subunits

ribosome

nuclear pore

mRNA

1. mRNA is leaving the nucleus and is attached to the ribosome; protein synthesis is occurring.

mRNA   DNA

Nucleus

The Endomembrane System

The **endomembrane system** consists of the nuclear envelope, the membranes of the endoplasmic reticulum, the Golgi apparatus, and several types of vesicles. This system compartmentalizes the cell so that particular enzymatic reactions are restricted to specific regions. The vesicles transport molecules from one part of the system to another.

Endoplasmic Reticulum

The **endoplasmic reticulum (ER)** [Gk. *endon,* within; *plasma,* something molded; L. *reticulum,* net], consisting of a complicated system of membranous channels and saccules (flattened vesicles), is physically continuous with the outer membrane of the nuclear envelope (Fig. 4.10). **Rough ER** is studded with ribosomes on the side of the membrane that faces the cytoplasm; therefore, it is correct to say that rough ER synthesizes proteins. It also modifies proteins after they have entered the ER lumen (see Fig. 4.9). Certain ER enzymes add carbohydrate (sugar) chains to proteins, and then these proteins are called glycoproteins. Other proteins assist the folding process that results in the final shape of the protein. The rough ER forms **vesicles** in which large molecules are transported to other parts of the cell. Often these vesicles are on their way to the plasma membrane or the Golgi apparatus.

Smooth ER, which is continuous with rough ER, does not have attached ribosomes. It is more abundant in gland cells, which synthesize lipids, such as phospholipids and steroids.

The specific function of smooth ER is dependent on the particular cell. In the testes, it produces testosterone, and in the liver, it helps detoxify drugs. Smooth ER increases in quantity when a person consumes alcohol or takes barbiturates on a regular basis. Regardless of any specialized function, rough and smooth ER also form vesicles that transport molecules to other parts of the cell, notably the Golgi apparatus.

The Golgi Apparatus

The **Golgi apparatus** is named for Camillo Golgi, who discovered its presence in cells in 1898. The Golgi apparatus typically consists of a stack of three to twenty slightly curved, flattened saccules whose appearance can be compared to a stack of pancakes (Fig. 4.11). In animal cells, one side of the stack (the cis or inner face) is directed toward the ER, and the other side of the stack (the trans or outer face) is directed toward the plasma membrane. Vesicles can frequently be seen at the edges of the saccules.

Protein-filled vesicles that bud from the rough ER and lipid-filled vesicles that bud from the smooth ER are received by the Golgi apparatus at its inner face. Thereafter, the apparatus alters these substances as they move through its saccules. For example, the Golgi apparatus contains enzymes that modify the carbohydrate chains first attached to proteins in the rough ER. It can change one sugar for another sugar. In some cases, the modified carbohydrate chain serves as a signal molecule that determines the protein's final destination in the cell.

The Golgi apparatus sorts and packages proteins and lipids in vesicles that depart from the outer face. In animal cells, some of these vesicles are lysosomes, which are discussed next. Other vesicles proceed to the plasma membrane, where they become part of the membrane as they discharge their contents during **secretion.** Secretion is termed exocytosis because the substance exits the cytoplasm.

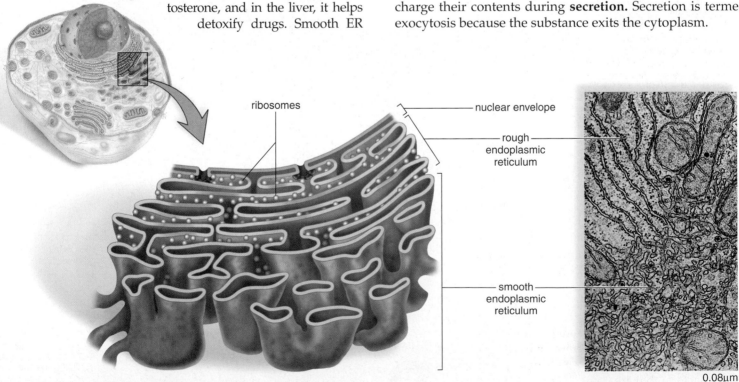

ribosomes   nuclear envelope

rough endoplasmic reticulum

smooth endoplasmic reticulum

0.08µm

FIGURE 4.10 Endoplasmic reticulum (ER).
Ribosomes are present on rough ER, which consists of flattened saccules, but not on smooth ER, which is more tubular. Proteins are synthesized and modified by rough ER, whereas smooth ER is involved in lipid synthesis, detoxification reactions, and several other possible functions.

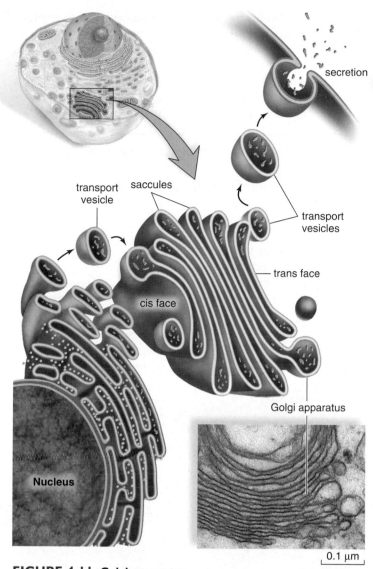

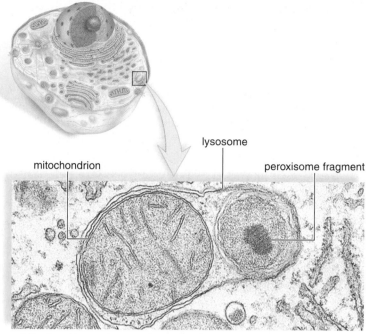

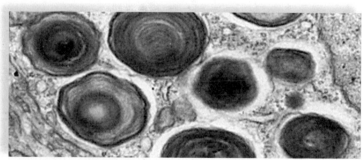

a. Mitochondrion and a peroxisome in a lysosome

b. Storage bodies in a cell with defective lysosomes

0.1 µm

FIGURE 4.11 Golgi apparatus.
The Golgi apparatus is a stack of flattened, curved saccules. It modifies proteins and lipids and packages them in vesicles that distribute these molecules to various locations.

FIGURE 4.12 Lysosomes.
a. Lysosomes, which bud off the Golgi apparatus in cells, are filled with hydrolytic enzymes that digest molecules and parts of the cell. Here a lysosome digests a worn mitochondrion and a peroxisome. **b.** The nerve cells of a person with Tay-Sachs disease are filled with membranous cytoplasmic bodies storing a fat that lysosomes are unable to digest.

Lysosomes

Lysosomes [Gk. *lyo,* loose, and *soma,* body] are membrane-bounded vesicles produced by the Golgi apparatus. They have a very low pH and contain powerful hydrolytic digestive enzymes. Lysosomes are important in recycling cellular material, and destroying nonfunctional organelles and portions of cytoplasm (Fig. 4.12).

Sometimes macromolecules are brought into a cell by vesicle formation at the plasma membrane. When a lysosome fuses with such a vesicle, its contents are digested by lysosomal enzymes into simpler subunits that then enter the cytoplasm. Some white blood cells defend the body by engulfing bacteria that are then enclosed within vesicles. When lysosomes fuse with these vesicles, the bacteria are digested.

A number of human lysosomal storage diseases are due to a missing lysosomal enzyme. In Tay-Sachs disease, the missing enzyme digests a fatty substance that helps insulate nerve cells and increases their efficiency. Because the enzyme is missing, the fatty substance accumulates, literally smothering nerve cells. Affected individuals appear normal at birth but begin to develop neurological problems at four to six months of age. Eventually, the child suffers cerebral degeneration, slow paralysis, blindness, and loss of motor function. Children with Tay-Sachs disease live only about three to four years. In the future, it may be possible to provide the missing enzyme and in that way prevent lysosomal storage diseases.

Lysosomes also participate in **apoptosis,** or programmed cell death, which is a normal part of development. When a tadpole becomes a frog, lysosomes digest away the cells of the tail. The fingers of a human embryo are at first webbed, but they are freed from one another as a result of lysosomal action.

Labels (Figure 4.11): secretion, transport vesicle, saccules, transport vesicles, trans face, cis face, Golgi apparatus, Nucleus

Labels (Figure 4.12): mitochondrion, lysosome, peroxisome fragment

Endomembrane System Summary

We have seen that the endomembrane system is a series of membranous organelles that work together and communicate by means of transport vesicles. The endoplasmic reticulum (ER) and the Golgi apparatus are essentially flattened saccules, and lysosomes are specialized vesicles.

Figure 4.13 shows how the components of the endomembrane system work together. Proteins produced in rough ER and lipids produced in smooth ER are carried in transport vesicles to the Golgi apparatus, where they are further modified before being packaged in vesicles that leave the Golgi. Using signaling sequences, the Golgi apparatus sorts proteins and packages them into vesicles that transport them to various cellular destinations. Secretory vesicles take the proteins to the plasma membrane, where they exit the cell when the vesicles fuse with the membrane. This is called secretion by exocytosis. For example, secretion into ducts occurs when the mammary glands produce milk or the pancreas produces digestive enzymes.

In animal cells, the Golgi apparatus also produces specialized vesicles called lysosomes that contain hydrolytic enzymes. Lysosomes fuse with incoming vesicles from the plasma membrane and digest macromolecules and/or even debris brought into a certain cell. White blood cells are well known for engulfing pathogens (e.g., disease-causing viruses and bacteria) that are then broken down in lysosomes.

The organelles of the endomembrane system are as follows:

Endoplasmic reticulum (ER): series of tubules and saccules
Rough ER: ribosomes are present
Smooth ER: ribosomes are not present
Golgi apparatus: stack of curved saccules
Lysosomes: specialized vesicles
Vesicles: membranous sacs

FIGURE 4.13 Endomembrane system.
The organelles in the endomembrane system work together to carry out the functions noted. Plant cells do not have lysosomes.

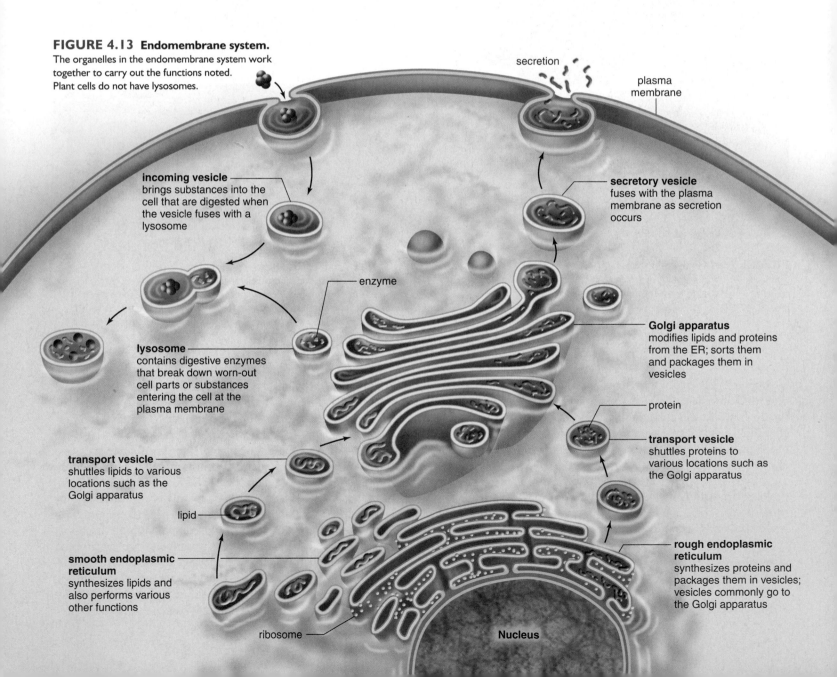

secretion

plasma membrane

incoming vesicle
brings substances into the cell that are digested when the vesicle fuses with a lysosome

secretory vesicle
fuses with the plasma membrane as secretion occurs

enzyme

lysosome
contains digestive enzymes that break down worn-out cell parts or substances entering the cell at the plasma membrane

Golgi apparatus
modifies lipids and proteins from the ER; sorts them and packages them in vesicles

protein

transport vesicle
shuttles proteins to various locations such as the Golgi apparatus

transport vesicle
shuttles lipids to various locations such as the Golgi apparatus

lipid

smooth endoplasmic reticulum
synthesizes lipids and also performs various other functions

rough endoplasmic reticulum
synthesizes proteins and packages them in vesicles; vesicles commonly go to the Golgi apparatus

ribosome

Nucleus

Peroxisomes and Vacuoles

Peroxisomes and the vacuoles of cells do not communicate with the organelles of the endomembrane system and therefore are not part of it.

Peroxisomes

Peroxisomes, similar to lysosomes, are membrane-bounded vesicles that enclose enzymes. However, the enzymes in peroxisomes are synthesized by free ribosomes and transported into a peroxisome from the cytoplasm. All peroxisomes contain enzymes whose action results in hydrogen peroxide (H_2O_2).

$$RH_2 + O_2 \longrightarrow R + H_2O_2$$

Hydrogen peroxide, a toxic molecule, is immediately broken down to water and oxygen by another peroxisomal enzyme called catalase. When hydrogen peroxide is applied to a wound, bubbling occurs as peroxisomal enzymes break it down.

The enzymes in a peroxisome depend on the function of a particular cell. However, peroxisomes are especially prevalent in cells that are synthesizing and breaking down lipids. In the liver, some peroxisomes produce bile salts from cholesterol, and others break down fats. In a 1992 movie *Lorenzo's Oil,* the peroxisomes in a boy's cells lack a membrane protein needed to import a specific enzyme from the cytoplasm. As a result, long chain fatty acids accumulate in his brain, and he suffers neurological damage. This disorder is known as adrenoleukodystrophy.

Plant cells also have peroxisomes (Fig. 4.14). In germinating seeds, they oxidize fatty acids into molecules that can be converted to sugars needed by the growing plant. In leaves, peroxisomes can carry out a reaction that is opposite to photosynthesis—the reaction uses up oxygen and releases carbon dioxide.

Vacuoles

Like vesicles, **vacuoles** are membranous sacs, but vacuoles are larger than vesicles. The vacuoles of some protists are quite specialized; they include contractile vacuoles for ridding the cell of excess water and digestive vacuoles for breaking down nutrients. Vacuoles usually store substances. Plant vacuoles contain not only water, sugars, and salts but also water-soluble pigments and toxic molecules. The pigments are responsible for many of the red, blue, or purple colors of flowers and some leaves. The toxic substances help protect a plant from herbivorous animals.

Plant Cell Central Vacuole.   Typically, plant cells have a large **central vacuole** that may take up to 90% of the volume of the cell. The vacuole is filled with a watery fluid called cell sap that gives added support to the cell (Fig. 4.15). Animals must produce more cytoplasm, including organelles, in order to grow, but a plant cell can rapidly increase in size by enlarging its vacuole. Eventually, a plant cell also produces more cytoplasm. The central vacuole maintains hydrostatic pressure or turgor pressure in plant cells.

The central vacuole functions in storage of both nutrients and waste products. A system to excrete wastes never evolved in plants; instead, metabolic waste products are pumped across the vacuole membrane and stored permanently in the central vacuole. As organelles age and become nonfunctional, they fuse with the vacuole, where digestive enzymes break them down. This is a function carried out by lysosomes in animal cells.

FIGURE 4.14 Peroxisomes.
Peroxisomes contain one or more enzymes that can oxidize various organic substances. Peroxisomes also contain the enzyme catalase, which breaks down hydrogen peroxide (H_2O_2), which builds up after organic substances are oxidized.

100 nm

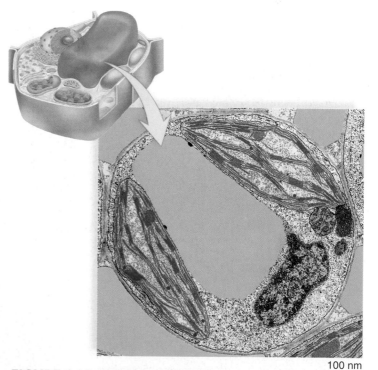

FIGURE 4.15 Plant cell central vacuole.
The large central vacuole of plant cells has numerous functions, from storing molecules to helping the cell increase in size.

100 nm

Energy-Related Organelles

Life is possible only because a constant input of energy maintains the structure of cells. Chloroplasts and mitochondria are the two eukaryotic membranous organelles that specialize in converting energy to a form that can be used by the cell. **Chloroplasts** use solar energy to synthesize carbohydrates, and carbohydrate-derived products are broken down in **mitochondria** (sing., mitochondrion) to produce ATP molecules. *Photosynthesis*, which usually occurs in chloroplasts [Gk. *chloros*, green, and *plastos*, formed, molded], is the process by which solar energy is converted to chemical energy within carbohydrates. Photosynthesis can be represented by this equation:

solar energy + carbon dioxide + water → carbohydrate + oxygen

Plants, algae, and cyanobacteria are capable of carrying on photosynthesis in this manner, but only plants and algae have chloroplasts.

Cellular respiration is the process by which the chemical energy of carbohydrates is converted to that of ATP (adenosine triphosphate). Cellular respiration can be represented by this equation:

carbohydrate + oxygen → carbon dioxide + water + energy

Here the word *energy* stands for ATP molecules. When a cell needs energy, ATP supplies it. The energy of ATP is used for synthetic reactions, active transport, and all energy-requiring processes in cells. In eukaryotes, mitochondria are necessary to the process of cellular respiration, which produces ATP.

Chloroplasts

Chloroplasts are a type of **plastid**, plant and algal organelles that are bounded by a double membrane and contain a series of internal membranes and/or vesicles. Plastids have DNA and are produced by division of existing plastids. Chloroplasts contain chlorophyll and carry on photosynthesis, while the other types of plastids have a storage function.

Some algal cells have only one chloroplast, while some plant cells have as many as a hundred. Chloroplasts can be quite large, being twice as wide and as much as five times the length of a mitochondrion. Their structure is shown in Figure 4.16. The double membrane encloses a large space called the **stroma**, which contains **thylakoids**, disklike sacs formed from a third chloroplast membrane. A stack of thylakoids is a **granum**. The lumens of the thylakoids are believed to form a large internal compartment called the thylakoid space. Chlorophyll and the other pigments that capture solar energy are located in the thylakoid membrane, and the enzymes that synthesize carbohydrates are located outside the thylakoid in the fluid of the stroma.

The structure of chloroplasts and the discovery that chloroplasts also have their own DNA and ribosomes support the endosymbiotic hypothesis that chloroplasts are derived from cyanobacteria that entered a eukaryotic cell. It cannot be said too often that, as shown in Figure 4.5, plant and algal cells contain both mitochondria and chloroplasts.

Other Types of Plastids. Other plastids are different from chloroplasts in color, form, and function. **Chromoplasts** contain pigments that result in a yellow, orange, or red color. Chromoplasts are responsible for the color of autumn leaves, fruits, carrots, and some flowers. **Leucoplasts** are generally colorless plastids that synthesize and store starches and oils. A microscopic examination of potato tissue yields a number of leucoplasts.

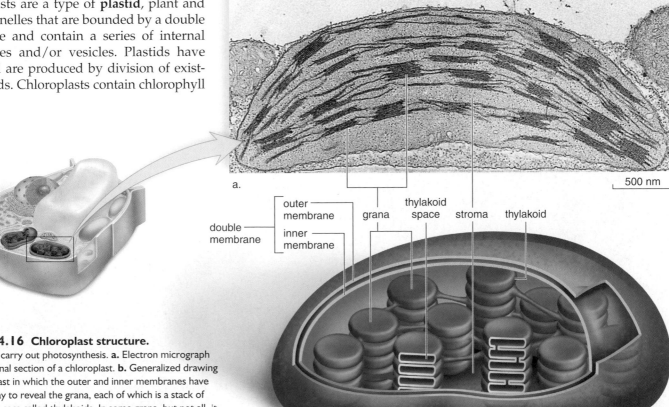

a.

500 nm

double membrane
outer membrane
inner membrane
grana
thylakoid space
stroma
thylakoid

b.

FIGURE 4.16 Chloroplast structure.
Chloroplasts carry out photosynthesis. **a.** Electron micrograph of a longitudinal section of a chloroplast. **b.** Generalized drawing of a chloroplast in which the outer and inner membranes have been cut away to reveal the grana, each of which is a stack of membranous sacs called thylakoids. In some grana, but not all, it is obvious that thylakoid spaces are interconnected.

Mitochondria

Even though mitochondria are smaller than chloroplasts, they can be seen in some cells by using a light microscope. The number of mitochondria can vary in cells depending on their activity. Some cells such as liver cells may have as many as 1,000 mitochondria. We think of mitochondria as having a shape like that shown in Figure 4.17, but actually they often change shape to be longer and thinner or shorter and broader. Mitochondria can form long, moving chains, or they can remain fixed in one location—often where energy is most needed. For example, they are packed between the contractile elements of cardiac cells and wrapped around the interior of a sperm's flagellum.

Mitochondria have two membranes, the outer membrane and the inner membrane. The inner membrane is highly convoluted into **cristae** that project into the matrix. These cristae increase the surface area of the inner membrane so much that in a liver cell they account for about one-third the total membrane in the cell. The inner membrane encloses a **matrix,** which contains mitochondrial DNA and ribosomes. The presence of a double membrane and mitochondrial genes is consistent with the endosymbiotic hypothesis regarding the origin of mitochondria, which was illustrated in Fig. 4.5. This figure suggests that mitochondria are derived from bacteria that took up residence in an early eukaryotic cell.

Mitochondria are often called the power-houses of the cell because they produce most of the ATP used by the cell through cellular respiration. Cell fractionation and centrifugation, which is described in the Science Focus on page 65, allowed investigators to separate the inner membrane, the outer membrane, and the matrix from each other. Then they discovered that the matrix is a highly concentrated mixture of enzymes that break down carbohydrates and other nutrient molecules. These reactions supply the chemical

energy that permits a chain of proteins on the inner membrane to create the conditions that allow ATP synthesis to take place. The entire process, which also involves the cytoplasm, is called cellular respiration because oxygen is used and carbon dioxide is given off, as shown on the previous page.

Mitochondrial Diseases.   So far, more than 40 different mitochondrial diseases that affect the brain, muscles, kidneys, heart, liver, eyes, ears, or pancreas have been identified. The common factor among these genetic diseases is that the patient's mitochondria are unable to completely metabolize organic molecules to produce ATP. As a result, toxins accumulate inside the mitochondria and the body. The toxins can be free radicals (sustances that readily form harmful compounds when they react with other molecules), and they damage mitochondria over time. In the United States, between 1,000 and 4,000 children per year are born with a mitochondrial disease. In addition, it is possible that many diseases of aging are due to malfunctioning mitochondria.

Chloroplasts and mitochondria are organelles that transform energy. Chloroplasts capture solar energy and produce carbohydrates. Mitochondria convert the energy within carbohydrates to that of ATP molecules.

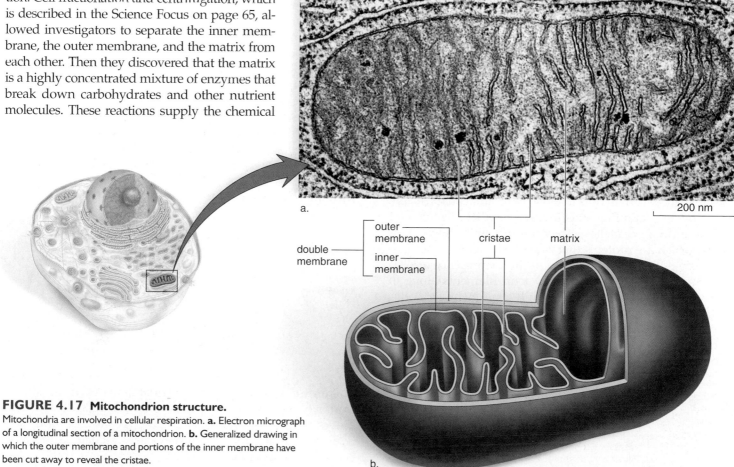

a.

200 nm

double membrane
outer membrane
inner membrane
cristae
matrix

FIGURE 4.17 Mitochondrion structure.
Mitochondria are involved in cellular respiration. **a.** Electron micrograph of a longitudinal section of a mitochondrion. **b.** Generalized drawing in which the outer membrane and portions of the inner membrane have been cut away to reveal the cristae.

b.

The Cytoskeleton

The protein components of the **cytoskeleton** [Gk. *kytos,* cell, and *skeleton,* dried body] interconnect and extend from the nucleus to the plasma membrane in eukaryotic cells. Prior to the 1970s, it was believed that the cytoplasm was an unorganized mixture of organic molecules. Then, high-voltage electron microscopes, which can penetrate thicker specimens, showed instead that the cytoplasm was instead highly organized. The technique of immunofluorescence microscopy identified the makeup of the protein components within the cytoskeletal network (Fig. 4.18).

The cytoskeleton contains actin filaments, intermediate filaments, and microtubules, which maintain cell shape and allow the cell and its organelles to move. Therefore, the cytoskeleton is often compared to the bones and muscles of an animal. However, the cytoskeleton is dynamic, especially because its protein components can assemble and disassemble as appropriate. Apparently a number of different mechanisms regulate this process, including protein kinases that phosphorylate proteins. Phosphorylation leads to disassembly, and dephosphorylation causes assembly.

Actin Filaments

Actin filaments (formerly called microfilaments) are long, extremely thin, flexible fibers (about 7 nm in diameter) that occur in bundles or meshlike networks. Each actin filament contains two chains of globular actin monomers twisted about one another in a helical manner.

Actin filaments play a structural role when they form a dense, complex web just under the plasma membrane, to which they are anchored by special proteins. They are also seen in the microvilli that project from intestinal cells, and their presence most likely accounts for the ability of microvilli to alternately shorten and extend into the intestine. In plant cells, actin filaments apparently form the tracks along which chloroplasts circulate in a particular direction; doing so is called cytoplasmic streaming. Also, the presence of a network of actin filaments lying beneath the plasma membrane accounts for the formation of **pseudopods** (false feet), extensions that allow certain cells to move in an amoeboid fashion.

How are actin filaments involved in the movement of the cell and its organelles? They interact with **motor molecules,** which are proteins that can attach, detach, and reattach farther along an actin filament. In the presence of ATP, the motor molecule myosin pulls actin filaments along in this way. Myosin has both a head and a tail. In muscle cells, the tails of several muscle myosin molecules are joined to form a thick filament. In nonmuscle cells, cytoplasmic myosin tails are bound to membranes, but the heads still interact with actin:

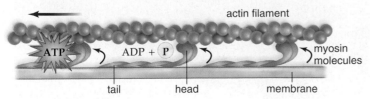

During animal cell division, the two new cells form when actin, in conjunction with myosin, pinches off the cells from one another.

Intermediate Filaments

Intermediate filaments (8–11 nm in diameter) are intermediate in size between actin filaments and microtubules. They are a ropelike assembly of fibrous polypeptides, but the specific type varies according to the tissue. Some intermediate filaments support the nuclear envelope, whereas others support the plasma membrane and take part in the formation of cell-to-cell junctions. In the skin, intermediate filaments, made of the protein keratin, give great mechanical strength to skin cells. We now know that intermediate filaments are also highly dynamic and will disassemble when phosphate is added by a kinase.

Microtubules

Microtubules [Gk. *mikros,* small, little; L. *tubus,* pipe] are small, hollow cylinders about 25 nm in diameter and from 0.2–25 µm in length.

Microtubules are made of a globular protein called tubulin, which is of two types called α and β. There is a slightly different amino acid sequence in α tubulin compared to β tubulin. When assembly occurs, α and β tubulin molecules come together as dimers, and the dimers arrange themselves in rows. Microtubules have 13 rows of tubulin dimers, surrounding what appears in electron micrographs to be an empty central core.

The regulation of microtubule assembly is under the control of a microtubule organizing center (MTOC). In most eukaryotic cells, the main MTOC is in the **centrosome** [Gk. *centrum,* center, and *soma,* body], which lies near the nucleus. Microtubules radiate from the centrosome, helping to maintain the shape of the cell and acting as tracks along which organelles can move. Whereas the motor molecule myosin is associated with actin filaments, the motor molecules kinesin and dynein are associated with microtubules:

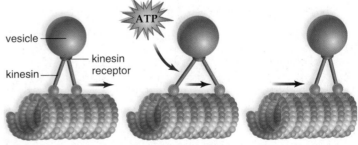

vesicle moves, not microtubule

There are different types of kinesin proteins, each specialized to move one kind of vesicle or cellular organelle. Kinesin moves vesicles or organelles in an opposite direction from dynein. Cytoplasmic dynein is closely related to the molecule dynein found in flagella.

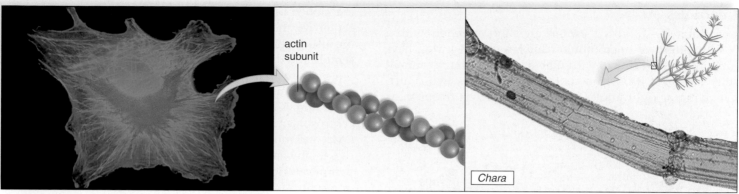

a. Actin filaments

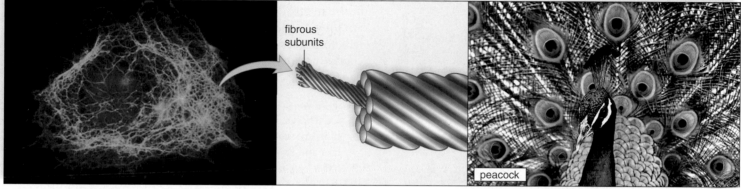

b. Intermediate filaments

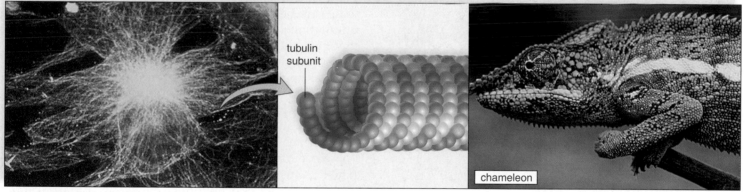

c. Microtubules

FIGURE 4.18 The cytoskeleton.
The cytoskeleton maintains the shape of the cell and allows its parts to move. Three types of protein components make up the cytoskeleton. They can be detected in cells by using a special fluorescent technique that detects only one type of component at a time. **a.** *Left to right:* Fibroblasts in animal tissue have been treated so that actin filaments can be microscopically detected; the drawing shows that actin filaments are composed of a twisted double chain of actin subunits. The giant cells of the green alga *Chara* rely on actin filaments to move organelles from one end of the cell to another. **b.** *Left to right:* Fibroblasts in an animal tissue have been treated so that intermediate filaments can be microscopically detected; the drawing shows that fibrous proteins account for the ropelike structure of intermediate filaments. A peacock's colorful feathers are strengthened by the presence of intermediate filaments. **c.** *Left to right:* Fibroblasts in an animal tissue have been treated so that microtubules can be microscopically detected; the drawing shows that microtubules are hollow tubes composed of tubulin subunits. The skin cells of a chameleon rely on microtubules to move pigment granules around so that they can take on the color of their environment.

Before a cell divides, microtubules disassemble and then reassemble into a structure called a spindle that distributes chromosomes in an orderly manner. At the end of cell division, the spindle disassembles, and microtubules reassemble once again into their former array. In the arms race between plants and herbivores, plants have evolved various types of poisons that prevent them from being eaten. Colchicine is a plant poison that binds tubulin and blocks the assembly of microtubules.

The cytoskeleton is an internal skeleton composed of actin filaments, intermediate filaments, and microtubules that maintain the shape of the cell and assist movement of its parts.

Centrioles

Centrioles [Gk. *centrum,* center] are short cylinders with a 9 + 0 pattern of microtubule triplets—that is, a ring having nine sets of triplets with none in the middle. In animal cells and most protists, a centrosome contains two centrioles lying at right angles to each other. A centrosome, as

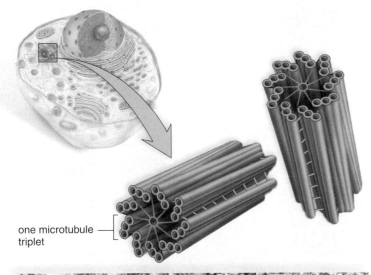

one microtubule triplet

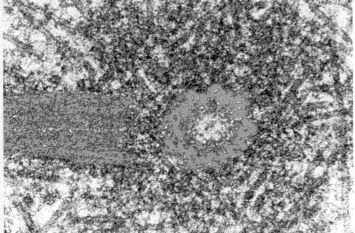

one pair of centrioles

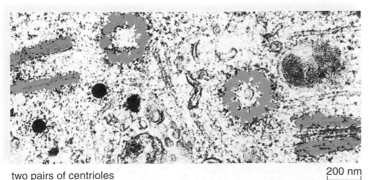

two pairs of centrioles

200 nm

FIGURE 4.19 Centrioles.
In a nondividing animal cell, there is a single pair of centrioles in the centrosome located just outside the nucleus. Just before a cell divides, the centrioles replicate, producing two pairs of centrioles. During cell division, centrioles in their respective centrosomes separate so that each new cell has one centrosome containing one pair of centrioles.

mentioned previously, is the major microtubule-organizing center for the cell. Therefore, it is possible that centrioles are also involved in the process by which microtubules assemble and disassemble.

Before an animal cell divides, the centrioles replicate, and the members of each pair are at right angles to one another (Fig. 4.19). Then each pair becomes part of a separate centrosome. During cell division, the centrosomes move apart and most likely function to organize the mitotic spindle. In any case, each new cell has its own centrosome and pair of centrioles. Plant and fungal cells have the equivalent of a centrosome, but this structure does not contain centrioles, suggesting that centrioles are not necessary to the assembly of cytoplasmic microtubules.

In cells with cilia and flagella, centrioles are believed to give rise to **basal bodies** that direct the organization of microtubules within these structures. In other words, a basal body may do for a cilium or flagellum what the centrosome does for the cell.

Centrioles, which are short cylinders with a 9 + 0 pattern of microtubule triplets, may give rise to the basal bodies of cilia and flagella.

Cilia and Flagella

Cilia [L. *cilium,* eyelash, hair] and **flagella** [L. *flagello,* whip] are hairlike projections that can move either in an undulating fashion, like a whip, or stiffly, like an oar. Cells that have these organelles are capable of movement. For example, unicellular paramecia move by means of cilia, whereas sperm cells move by means of flagella. The cells that line our upper respiratory tract have cilia that sweep debris trapped within mucus back up into the throat, where it can be swallowed. This action helps keep the lungs clean.

In eukaryotic cells, cilia are much shorter than flagella, but they have a similar construction. Both are membrane-bounded cylinders enclosing a matrix area. In the matrix are nine microtubule doublets arranged in a circle around two central microtubules; this is called the 9 + 2 pattern of microtubules (Fig. 4.20). Cilia and flagella move when the microtubule doublets slide past one another.

As mentioned, each cilium and flagellum has a basal body lying in the cytoplasm at its base. Basal bodies have the same circular arrangement of microtubule triplets as centrioles and are believed to be derived from them. It is possible that basal bodies organize the microtubules within cilia and flagella, but this is not supported by the observation that cilia and flagella grow by the addition of tubulin dimers to their tips.

Cilia and flagella, which have a 9 + 2 pattern of microtubules, are involved in the movement of cells.

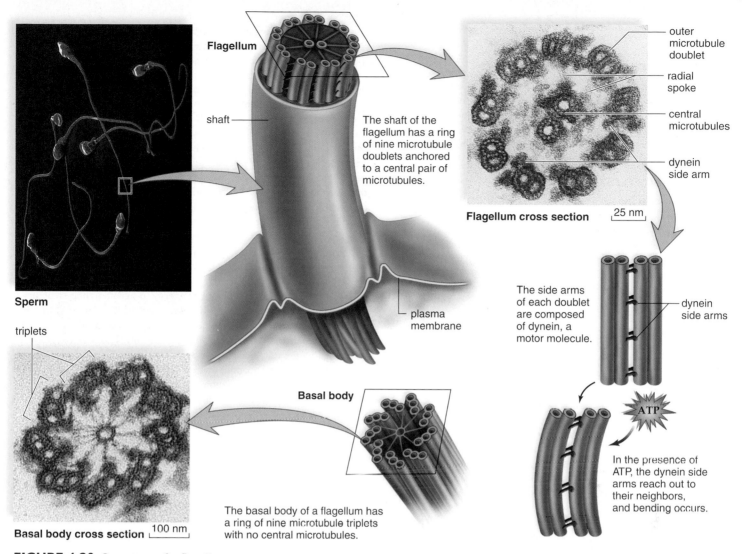

FIGURE 4.20 Structure of a flagellum.

A flagellum has a basal body with a 9 + 0 pattern of microtubule triplets. (Notice the ring of nine triplets, with no central microtubules.) The shaft of the flagellum has a 9 + 2 pattern (a ring of nine microtubule doublets surrounds a central pair of microtubules). Compare the cross section of the basal body to the cross section of the flagellum shaft, and note that in place of the third microtubule, a flagellum's outer doublets have side arms of dynein, a motor molecule. In the presence of ATP, the dynein side arms reach out and attempt to move along their neighboring doublet. Because of the radial spokes connecting the doublets to the central microtubules, bending occurs.

CONNECTING THE CONCEPTS

Cell biology is fundamental to the other fields of biology studied in this text. Many of the organelles of a cell are involved in producing, modifying, and transporting proteins. Each type of cell is characterized by the presence of particular proteins, which are specified by active genes in that cell. The process of protein synthesis makes cell biology directly applicable to the field of genetics, which is the topic of the next part in this text.

Next, we will study the classification of living things, which is dependent on cell type. The organisms in two domains—domain Archaea and domain Bacteria—have prokaryotic cells. The organisms in domain Eukarya—the protists, fungi, plants, and animals—have eukaryotic cells. We will learn that differences in

DNA (and RNA) base sequences resulted in recognition of these three domains and that classification based on DNA base sequences are increasingly influencing classification, even down to the species level.

The evolution of life began with the first cell or cells, and we will study how the evolution of the eukaryotic cell led to multicellularity and the forms of life with which we are the most familiar. Without genetic and, consequently, cellular protein changes, adaptations to the environment that resulted in the diversity of life on our planet would never have occurred. In addition to structural differences, DNA sequence data are now helping us trace the relatedness of organisms through time.

Knowledge of the anatomy and physiology of organisms is based on understanding the structure and function of the cells making up different organs. This is never more apparent than in the field of medicine. Almost any human illness can be traced back to malfunctioning cells and metabolic processes that involve cellular enzymes.

This text concludes with ecology, and you might think that cell biology is not pertinent to ecology. However, preservation of organisms within an ecosystem is dependent in part on our knowledge of how the cells of organisms are affected by environmental pollutants. This leads us back to the necessity of understanding the structure and function of cells.

Summary

4.1 CELLULAR LEVEL OF ORGANIZATION

All organisms are composed of cells, the smallest units of living matter. Cells are capable of self-reproduction, and existing cells come only from preexisting cells. Cells are very small and are measured in micrometers. The plasma membrane regulates exchange of materials between the cell and the external environment. Cells must remain small in order to have an adequate amount of surface area to volume.

4.2 PROKARYOTIC CELLS

There are two major groups of prokaryotic cells: the bacteria and the archaea. Prokaryotic cells lack the nucleus of eukaryotic cells. The cell envelope of bacteria includes a plasma membrane, a cell wall, and an outer glycocalyx. The cytoplasm contains ribosomes, inclusion bodies, and a nucleoid that is not bounded by a nuclear envelope. The cytoplasm of cyanobacteria also includes thylakoids. The appendages of a bacterium are the flagella, the fimbriae, and the sex pili.

4.3 EUKARYOTIC CELLS

Eukaryotic cells are much larger than prokaryotic cells, but they are compartmentalized by the presence of organelles, each with a specific structure and function (Table 4.1). The nuclear envelope most likely evolved through invagination of the plasma membrane, but mitochondria and chloroplasts may have arisen when a eukaryotic cell took up bacteria and algae in separate events. Perhaps this accounts for why the mitochondria and chloroplasts function independently. Other membranous organelles are in constant communication by way of transport vesicles.

The nucleus of eukaryotic cells is bounded by a nuclear envelope containing pores. These pores serve as passageways between the cytoplasm and the nucleoplasm. Within the nucleus, chromatin, which contains DNA, undergoes coiling into chromosomes at the time of cell division. The nucleolus is a special region of the chromatin where rRNA is produced and ribosomal subunits are formed.

Ribosomes are organelles that function in protein synthesis. When protein synthesis occurs, mRNA leaves the nucleus with a coded message from DNA that specifies the sequence of amino acids in that protein. After mRNA attaches to a ribosome, it binds to the ER if it has a signal peptide. The signal peptide attaches to a signal recognition particle that, in turn, binds to an SRP receptor on the ER. When completed, the protein enters the lumen of the ER.

The endomembrane system includes the ER (both rough and smooth), the Golgi apparatus, the lysosomes (in animal cells), and transport vesicles. Newly produced proteins are modified in the ER before they are packaged in transport vesicles, many of which go to the Golgi apparatus. The smooth ER has various metabolic functions, depending on the cell type, but it also forms vesicles that carry lipids to different locations, particularly to the Golgi apparatus. The Golgi apparatus modifies, sorts, and repackages proteins. Some proteins are packaged into lysosomes, which carry out intracellular digestion, or into vesicles that fuse with the plasma membrane. Following fusion, secretion occurs.

Cells require a constant input of energy to maintain their structure. Chloroplasts capture the energy of the sun and carry on photosynthesis, which produces carbohydrates. Carbohydrate-derived products are broken down in mitochondria as ATP is produced. This is an oxygen-requiring process called cellular respiration.

The cytoskeleton contains actin filaments, intermediate filaments, and microtubules. These maintain cell shape and allow it and the organelles to move. Actin filaments, the thinnest filaments, interact with the motor molecule myosin in muscle cells to bring about contraction; in other cells, they pinch off daughter cells and have other dynamic functions. Intermediate filaments support the nuclear envelope and the plasma membrane and probably participate in cell-to-cell junctions. Microtubules radiate out from the centrosome and are present in centrioles, cilia, and flagella. They serve as tracks along which vesicles and other organelles move, due to the action of specific motor molecules.

TABLE 4.1

Comparison of Prokaryotic Cells and Eukaryotic Cells

	Prokaryotic Cells	Eukaryotic Cells	
		Animal	*Plant*
Size	*Smaller* *(1–20 µm in diameter)*	*Larger* *(10–100 µm in diameter)*	
Cell wall	Usually (peptidoglycan)	No	Yes (cellulose)
Plasma membrane	Yes	Yes	Yes
Nucleus	No	Yes	Yes
Nucleolus	No	Yes	Yes
Ribosomes	Yes (smaller)	Yes	Yes
Endoplasmic reticulum	No	Yes	Yes
Golgi apparatus	No	Yes	Yes
Lysosomes	No	Yes	No
Mitochondria	No	Yes	Yes
Chloroplasts	No	No	Yes
Peroxisomes	No	Usually	Usually
Cytoskeleton	No	Yes	Yes
Centrioles	No	Yes	No
9 + 2 cilia or flagella	No	Often	No (in flowering plants) Yes (sperm of bryophytes, ferns, and cycads)

Reviewing the Chapter

1. What are the three basic principles of the cell theory? 58
2. Why is it advantageous for cells to be small? 59
3. Roughly sketch a bacterial (prokaryotic) cell, label its parts, and state a function for each of these. 63
4. How do eukaryotic and prokaryotic cells differ? 64
5. Describe how the nucleus, the chloroplast, and the mitochondrion may have become a part of the eukaryotic cell. 64
6. What does it mean to say that the eukaryotic cell is compartmentalized? 64–65
7. Describe the structure and the function of the nuclear envelope and the nuclear pores. 68
8. Distinguish between the nucleolus, rRNA, and ribosomes. 68–69
9. Name organelles that are a part of the endomembrane system and explain the term. 70
10. Trace the path of a protein from rough ER to the plasma membrane. 72
11. Give the overall equations for photosynthesis and cellular respiration, contrast the two, and tell how they are related. 74
12. Describe the structure and function of chloroplasts and mitochondria. How are these two organelles related to one another? 74–75
13. What are the three components of the cytoskeleton? What are their structures and functions? 76–77
14. Relate the structure of flagella (and cilia) to centrioles, and discuss the function of both. 78–79

Testing Yourself

Choose the best answer for each question.

1. The small size of cells best correlates with
   a. the fact that they are self-reproducing.
   b. their prokaryotic versus eukaryotic nature.
   c. an adequate surface area for exchange of materials.
   d. the fact that they come in multiple sizes.
   e. All of these are correct.

2. Which of these is not a true comparison of the compound light microscope and the transmission electron microscope?

LIGHT / **ELECTRON**
a. Uses light to "view" object / Uses electrons to "view" object
b. Uses glass lenses for focusing / Uses magnetic lenses for focusing
c. Specimen must be killed and stained / Specimen may be alive and nonstained
d. Magnification is not as great / Magnification is greater
e. Resolution is not as great / Resolution is greater

3. Which of these best distinguishes a prokaryotic cell from a eukaryotic cell?
   a. Prokaryotic cells have a cell wall, but eukaryotic cells never do.
   b. Prokaryotic cells are much larger than eukaryotic cells.
   c. Prokaryotic cells have flagella, but eukaryotic cells do not.
   d. Prokaryotic cells do not have a membrane-bounded nucleus, but eukaryotic cells do have such a nucleus.
   e. Prokaryotic cells have ribosomes, but eukaryotic cells do not have ribosomes.

4. Which of these is not found in the nucleus?
   a. functioning ribosomes
   b. chromatin that condenses to chromosomes
   c. nucleolus that produces rRNA
   d. nucleoplasm instead of cytoplasm
   e. all forms of RNA

5. Vesicles from the ER most likely are on their way to
   a. the rough ER.
   b. the lysosomes.
   c. the Golgi apparatus.
   d. the plant cell vacuole only.
   e. the location suitable to their size.

6. Lysosomes function in
   a. protein synthesis.
   b. processing and packaging.
   c. intracellular digestion.
   d. lipid synthesis.
   e. production of hydrogen peroxide.

7. Mitochondria
   a. are involved in cellular respiration.
   b. break down ATP to release energy for cells.
   c. contain grana and cristae.
   d. are present in animal cells but not plant cells.
   e. All of these are correct.

8. Which organelle releases oxygen?
   a. ribosome
   b. Golgi apparatus
   c. mitochondrion
   d. chloroplast
   e. smooth ER

9. Label these parts of the cell that are involved in protein synthesis and modification. Give a function for each structure.

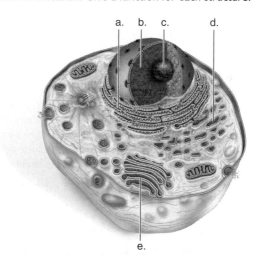

10. Which of these is not true?
    a. Actin filaments are found in muscle cells.
    b. Microtubules radiate out from the ER.
    c. Intermediate filaments sometimes contain keratin.
    d. Motor molecules use microtubules as tracks.

11. Cilia and flagella
    a. have a 9 + 0 pattern of microtubules.
    b. contain myosin that pulls on actin filaments.
    c. are organized by basal bodies derived from centrioles.
    d. are constructed similarly in prokaryotes and eukaryotes.
    e. Both a and c are correct.

12. Which of the following organelles contains its (their) own DNA, which suggests they were once independent prokaryotes?
    a. Golgi apparatus
    b. mitochondria
    c. chloroplasts
    d. ribosomes
    e. Both b and c are correct.

13. Which organelle most likely originated by invagination of the plasma membrane?
    a. mitochondria
    b. flagella
    c. nucleus
    d. chloroplasts
    e. All of these are correct.

14. Which structures are found in a prokaryotic cell?
    a. cell wall, ribosomes, thylakoids, chromosome
    b. cell wall, plasma membrane, nucleus, flagellum
    c. nucleoid, ribosomes, chloroplasts, capsule
    d. plasmid, ribosomes, enzymes, DNA, mitochondria
    e. chlorophyll, enzymes, Golgi apparatus, plasmids

15. Study the example given in (a) below. Then for each other organelle listed, state another that is structurally and functionally related. Tell why you paired these two organelles.
    a. The nucleus can be paired with nucleoli because nucleoli are found in the nucleus. Nucleoli occur where chromatin is producing rRNA.
    b. mitochondria
    c. centrioles
    d. ER

Thinking Scientifically

1. Protists of the phylum Apicomplexa cause malaria and contribute to infections associated with AIDS. These parasites are unusual because they contain plastids. (Chloroplasts are a type of plastid.) Scientists have discovered that an antibiotic that inhibits prokaryotic enzymes will kill the parasite because it is effective against the plastids contained in the cell. What can be concluded about the plastids?

2. In a cell biology laboratory, students are examining sections of plant cells. Some students report seeing the nucleus; others do not. Some students see a large vacuole; others see a small vacuole. Some see evidence of an extensive endoplasmic reticulum; others see almost none. How can these observations be explained if the students are looking at the same cells?

Bioethical Issue: Stem Cells

Embryonic stem cells are cells from young embryos that divide indefinitely. They are of interest to researchers because they are undifferentiated and have the ability to develop into a variety of cell types, including brain, heart, bone, muscle, and skin cells. They can help researchers to understand how cells change as they mature. In addition, they offer the hope of curing cell-based diseases such as diabetes, Parkinson disease, and heart disease. Embryonic stem cell research involves the destruction of human embryos. Opponents of this research argue that embryos should not be destroyed because they have the same rights as all human beings. Proponents counter that the embryos used for research are excess embryos from in vitro fertilization procedures, and they would be destroyed anyway. Therefore, they should be used to alleviate human suffering. Do you think that human embryos should be used for stem cell research?

Understanding the Terms

actin filament 76	leucoplast 74
apoptosis 71	lysosome 71
bacillus 62	matrix 75
basal body 78	mesosome 62
capsule 62	microtubule 76
cell 58	mitochondrion 74
cell envelope	motor molecule 76
(of prokaryotes) 62	nuclear envelope 68
cell theory 58	nuclear pore 68
cell wall 62	nucleoid 62
central vacuole	nucleolus 68
(of plant cell) 73	nucleoplasm 68
centriole 78	nucleus 64
centrosome 76	organelle 64
chloroplast 74	peroxisome 73
chromatin 68	plasma membrane 62
chromoplast 74	plasmid 62
chromosome 68	plastid 74
cilium 78	polyribosome 69
coccus 62	prokaryotic cell 62
cristae 75	pseudopod 76
cyanobacteria 62	ribosome 62, 69
cytoplasm 62	rough ER 70
cytoskeleton 76	secretion 70
endomembrane system 70	sex pili 63
endoplasmic reticulum	smooth ER 70
(ER) 70	spirillum 62
eukaryotic cell 62	spirochete 62
fimbriae 63	stroma 74
flagellum (pl., flagella) 62, 78	surface-area-to-volume
glycocalyx 62	ratio 59
Golgi apparatus 70	thylakoid 62, 74
granum 74	vacuole 73
inclusion body 62	vesicle 70
intermediate filament 76	

Match the terms to these definitions:
a. _____ Organelle, consisting of saccules and vesicles, that processes, packages, and distributes molecules about or from the cell.
b. _____ Especially active in lipid metabolism; always produces H_2O_2.
c. _____ Dark-staining, spherical body in the cell nucleus that produces ribosomal subunits.
d. _____ Internal framework of the cell, consisting of microtubules, actin filaments, and intermediate filaments.
e. _____ Allows prokaryotic cells to attach to other cells.

ARIS, the *Biology* Website

ARIS, the website for *Biology*, provides a wealth of information organized and integrated by chapter. You will find practice quizzes, interactive activities, labeling exercises, flashcards, and much more that will complement your learning and understanding of general biology.

www.mhhe.com/maderbiology9

5

MEMBRANE STRUCTURE AND FUNCTION

A t first glance, an African pygmy, an overweight diabetic, and a young child with cystic fibrosis seem to have little in common. In reality, however, each suffers from a defect in the cells' plasma membrane. Growth hormone does not bind to the pygmy's plasma membrane, the diabetic's does not respond properly to insulin, and the membrane does not transport chloride from the cells of a child who has cystic fibrosis. A plasma membrane was essential to the origin of the first cell(s), and its proper functioning is essential to our good health today.

A plasma membrane encloses every cell, whether the cell is a unicellular amoeba or one of many from the body of a squid, carnation, mushroom, or human. Universally, a plasma membrane protects a cell by acting as a barrier between its living contents and the surrounding environment. It regulates what goes into and out of the cell and marks the cell as being unique to the organism. In multicellular organisms, cell junctions requiring specialized features of the plasma membrane connect cells together in specific ways and pass on information to neighboring cells so that the activities of tissues and organs are coordinated. Inside these cells, membrane compartmentalizes the cell so that specific enzymes for particular functions are isolated from one another. Communication is achieved by membrane-bounded vesicles that carry molecules from one part of the cell to another.

A eukaryotic cell is surrounded by a plasma membrane, and membrane compartmentalizes the cell into regions specific in structure and function.

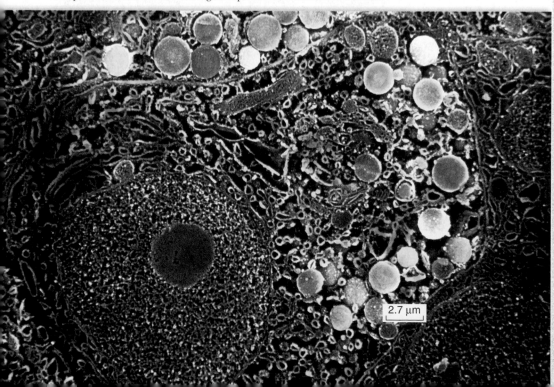

2.7 μm

5.1 MEMBRANE MODELS

At the turn of the 20th century, investigators noted that lipid-soluble molecules entered cells more rapidly than water-soluble molecules. This prompted them to suggest that lipids are a component of the plasma membrane. Later, chemical analysis disclosed that the plasma membrane contains phospholipids. In 1925, E. Gorter and G. Grendel measured the amount of phospholipids extracted from red blood cells and determined that there where just enough to form a bilayer around the cells. They further suggested that the nonpolar (hydrophobic) tails are directed inward and the polar (hydrophilic) heads are directed outward, forming a phospholipid bilayer:

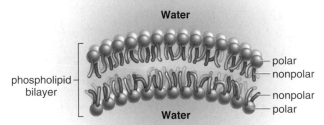

In the 1940s, J. Danielli and H. Davson suggested that proteins are also a part of the membrane. They proposed a sandwich model in which the phospholipid bilayer is a filling between two continuous layers of proteins. By the late 1950s, electron microscopy had advanced to allow viewing of the plasma membrane, and indeed a sandwich appearance was observed (Fig. 5.1a). At that time, J. D. Robertson modified the sandwich model and proposed that the outer dark layer (stained with heavy metals) contained protein *plus* the hydrophilic heads of the phospholipids. The interior was assumed to be the hydrophobic tails of these molecules. Robertson went on to suggest that all membranes in various cells have basically the same composition. Therefore, his proposal was called the unit membrane model (Fig. 5.1b *left*).

After some years, investigators began to doubt the accuracy of the unit membrane model. Not all membranes have the same appearance in electron micrographs, and they do not have the same function. For example, the inner membrane of a mitochondrion, which is coated with rows of particles, functions in cellular respiration, and it has a far different appearance than the plasma membrane.

Fluid-Mosaic Model

Finally, in 1972, S. Singer and G. Nicolson introduced the **fluid-mosaic model** of membrane structure, which proposes that the membrane is a fluid phospholipid bilayer in which protein molecules are either partially or wholly embedded (Fig. 5.1b,c right). The proteins are scattered throughout the membrane in an irregular pattern that can vary from membrane to membrane. The mosaic distribution of proteins is supported especially by electron micrographs of freeze-fractured membranes (Fig. 5.1d).

The fluid-mosaic model of membrane structure consists of a fluid phospholipid bilayer in which embedded proteins form a mosaic pattern.

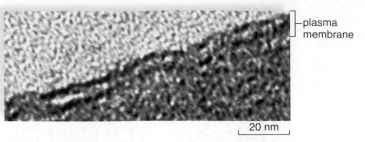

a. Electron micrograph of red blood cell plasma membrane

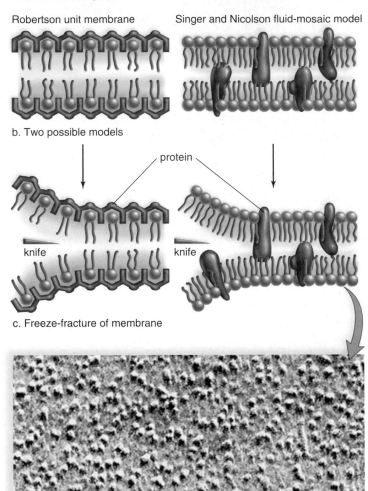

b. Two possible models

c. Freeze-fracture of membrane

d. Electron micrograph of freeze-fractured membrane shows presence of particles

FIGURE 5.1 Membrane structure.
a. The red blood cell plasma membrane, like all cells, typically has a three-layered appearance in electron micrographs. **b.** The Robertson unit membrane model proposed that the outer, dark layers in electron micrographs were made up of protein and polar heads of phospholipid molecules, while the inner, light layer was composed of the nonpolar tails of the phospholipid molecules. The Singer and Nicolson fluid-mosaic model put protein molecules within the lipid bilayer. **c.** A technique called freeze-fracture allows an investigator to view the interior of the membrane. Cells are rapidly frozen in liquid nitrogen and then fractured with a special knife. **d.** Platinum and carbon are applied to the fractured surface to produce a faithful replica that is observed by electron microscopy. A fracture in the middle of the bilayer shows the presence of particles, consistent with the fluid-mosaic model.

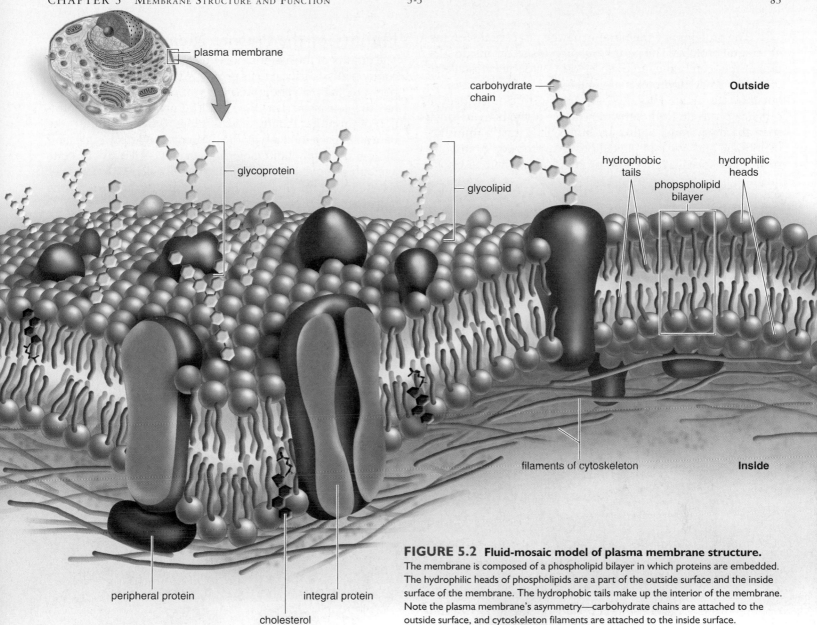

plasma membrane

carbohydrate chain

Outside

glycoprotein

glycolipid

hydrophobic tails

hydrophilic heads

phopspholipid bilayer

filaments of cytoskeleton

Inside

peripheral protein

integral protein

cholesterol

FIGURE 5.2 Fluid-mosaic model of plasma membrane structure.
The membrane is composed of a phospholipid bilayer in which proteins are embedded. The hydrophilic heads of phospholipids are a part of the outside surface and the inside surface of the membrane. The hydrophobic tails make up the interior of the membrane. Note the plasma membrane's asymmetry—carbohydrate chains are attached to the outside surface, and cytoskeleton filaments are attached to the inside surface.

5.2 PLASMA MEMBRANE STRUCTURE AND FUNCTION

The plasma membrane separates the internal environment of the cell from the external environment. It regulates the entrance and exit of molecules into the cell. In this way, it helps the cell and the organism maintain a steady internal environment. Figure 5.2 illustrates the membrane as a **phospholipid bilayer** with embedded proteins. Some of the proteins span the membrane and others do not. Some are on the inside surface of the membrane. All together, the proteins form a *mosaic* pattern.

A phospholipid is a molecule that has both a hydrophilic (water-loving) region and a hydrophobic (water-fearing) region. The hydrophilic polar heads of the phospholipid molecules face the outside and inside of the cell, where water is found. The hydrophobic nonpolar tails face

each other. **Cholesterol** is another lipid found in animal plasma membrane; related steroids are found in the plasma membrane of plants. Cholesterol stiffens and strengthens the membrane, thereby helping to regulate its fluidity:

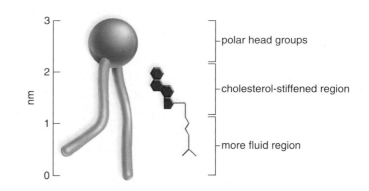

3

nm

2

1

0

polar head groups

cholesterol-stiffened region

more fluid region

The proteins in a membrane may be peripheral proteins or integral proteins. The peripheral proteins on the inside surface of the membrane are often held in place by cytoskeletal filaments. Integral proteins are embedded in the membrane, but they can move laterally back and forth. Some integral proteins protrude from only one surface of the bilayer. Most span the membrane, with a hydrophobic region within the membrane and hydrophilic regions that protrude from both surfaces of the bilayer. These proteins are known as transmembrane proteins:

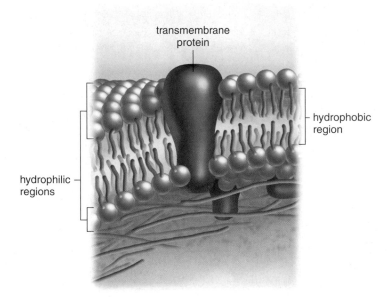

Both phospholipids and proteins can have attached carbohydrate (sugar) chains. If so, these molecules are called **glycolipids** and **glycoproteins,** respectively. Since the carbohydrate chains occur only on the outside surface and peripheral proteins occur asymmetrically on one surface or the other, the two sides of the membrane are not identical.

Carbohydrate Chains

In animal cells, the carbohydrate chains of proteins give the cell a "sugar coat," more properly called the glycocalyx. The glycocalyx protects the cell and has various other functions. For example, it facilitates adhesion between cells, reception of signal molecules, and cell-to-cell recognition.

The possible diversity of the carbohydrate (sugar) chains is enormous. The chains can vary by the number (15 is usual, but there can be several hundred) and sequence of sugars and by whether the chain is branched. Each cell within the individual has its own particular "fingerprint" because of these chains. As you probably know, transplanted tissues are often rejected by the recipient. This is because the immune system is able to recognize that the foreign tissue's cells do not have the appropriate carbohydrate chains. In humans, carbohydrate chains are also the basis for the A, B, and O blood groups.

Fluidity of the Plasma Membrane

The fluidity of the membrane, which is dependent on its lipid components, is critical to its proper functioning. At body temperature, the phospholipid bilayer of the plasma membrane has the consistency of olive oil. The greater the concentration of unsaturated fatty acid residues, the more fluid is the bilayer. In each monolayer, the hydrocarbon tails wiggle, and the entire phospholipid molecule can move sideways at a rate averaging about 2 μm—the length of a prokaryotic cell—per second. (Phospholipid molecules rarely flip-flop from one layer to the other, because this would require the hydrophilic head to move through the hydrophobic center of the membrane.) The fluidity of a phospholipid bilayer means that cells are pliable. Imagine if they were not—the long nerve fibers in your neck would crack whenever you nodded your head!

Although some proteins are often held in place by cytoskeletal filaments, in general, proteins are free to drift laterally in the fluid lipid bilayer. This has been demonstrated by fusing mouse and human cells, and watching the movement of tagged proteins (Fig. 5.3). Forty minutes after fusion, the proteins are completely mixed. The fluidity of the membrane is needed for the functioning of some proteins, such as enzymes, which become inactive when the membrane solidifies.

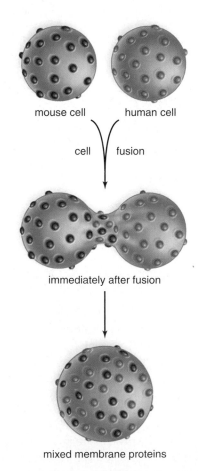

FIGURE 5.3 Experiment to demonstrate lateral drifting of plasma membrane proteins.
After human and mouse cells fuse, the plasma membrane proteins of the mouse (purple circles) and of the human cell (orange circles) mix within a short time.

The Functions of the Proteins

The plasma membranes of various cells and the membranes of various organelles each have their own unique collections of proteins. The proteins form different patterns according to the particular membrane and also within the same membrane at different times. When you consider that the plasma membrane of a red blood cell contains over 50 different types of proteins, it is apparent why the membrane is said to be a mosaic.

The integral proteins largely determine a membrane's specific functions. The integral proteins can be:

Channel proteins   Channel proteins are involved in the passage of molecules through the membrane. They have a channel that allows a substance to simply move across the membrane (Fig. 5.4*a*). For example, a channel protein allows hydrogen ions to flow across the inner mitochondrial membrane. Without this movement of hydrogen ions, ATP would never be produced.

Carrier proteins   Carrier proteins are also involved in the passage of molecules through the membrane. They combine with a substance and help it move across the membrane (Fig. 5.4*b*). A carrier protein transports sodium and potassium ions across a nerve cell membrane. Without this carrier protein, nerve conduction would be impossible.

Cell recognition proteins   Cell recognition proteins are glycoproteins (Fig. 5.4*c*). Among other functions, these proteins help the body recognize when it is being invaded by pathogens so that an immune reaction can occur. Without this recognition, pathogens would be able to freely invade the body.

Receptor proteins   Receptor proteins have a shape that allows a specific molecule to bind to it (Fig. 5.4*d*). The binding of this molecule causes the protein to change its shape and thereby bring about a cellular response. The coordination of the body's organs is totally dependent on such signal molecules. For example, the liver stores glucose after it is signaled to do so by insulin.

Enzymatic proteins   Some plasma membrane proteins are enzymatic proteins that carry out metabolic reactions directly (Fig. 5.4*e*). Without the presence of enzymes, some of which are attached to the various membranes of the cell, a cell would never be able to perform the metabolic reactions necessary to its proper function.

The peripheral proteins often have a structural role in that they help stabilize and shape the plasma membrane.

The plasma membrane consists of a phospholipid bilayer that has the consistency of olive oil and accounts for the fluidity of the membrane. The integral proteins, which are either partially or wholly embedded in the membrane, have specific functions. Some are involved in the passage of molecules through the membrane, others are receptors for signal molecules, and still others play an enzymatic role.

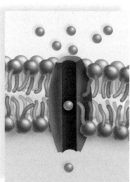

Channel Protein
Allows a particular molecule or ion to cross the plasma membrane freely. Cystic fibrosis, an inherited disorder, is caused by a faulty chloride (Cl^-) channel; a thick mucus collects in airways and in pancreatic and liver ducts.

a.

Carrier Protein
Selectively interacts with a specific molecule or ion so that it can cross the plasma membrane. The inability of some persons to use energy for sodium-potassium (Na^+–K^+) transport has been suggested as the cause of their obesity.

b.

Cell Recognition Protein
The MHC (major histocompatibility complex) glycoproteins are different for each person, so organ transplants are difficult to achieve. Cells with foreign MHC glycoproteins are attacked by white blood cells responsible for immunity.

c.

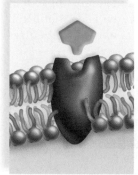

Receptor Protein
Is shaped in such a way that a specific molecule can bind to it. Pygmies are short, not because they do not produce enough growth hormone, but because their plasma membrane growth hormone receptors are faulty and cannot interact with growth hormone.

d.

Enzymatic Protein
Catalyzes a specific reaction. The membrane protein, adenylate cyclase, is involved in ATP metabolism. Cholera bacteria release a toxin that interferes with the proper functioning of adenylate cyclase; sodium ions and water leave intestinal cells, and the individual may die from severe diarrhea.

e.

FIGURE 5.4 Membrane protein diversity.
These are some of the functions performed by proteins found in the plasma membrane.

TABLE 5.1

Passage of Molecules into and out of the Cell

	Name	Direction	Requirement	Examples
Energy Not Required	Diffusion	Toward lower concentration	Concentration gradient	Lipid-soluble molecules, water, and gases
	Facilitated transport	Toward lower concentration	Channels or carrier and concentration gradient	Some sugars, and amino acids
Energy Required	Active transport	Toward higher concentration	Carrier plus energy	Sugars, amino acids, and ions
	Exocytosis	Toward outside	Vesicle fuses with plasma membrane	Macromolecules
	Endocytosis	Toward inside	Vesicle formation	Macromolecules

5.3 PERMEABILITY OF THE PLASMA MEMBRANE

The plasma membrane regulates the passage of molecules into and out of the cell. This function is critical because the life of the cell depends on maintenance of its normal composition. The plasma membrane can carry out this function because it is **differentially** (selectively) **permeable,** meaning that certain substances can move across the membrane while others cannot.

Table 5.1 lists, and Figure 5.5 illustrates, which types of molecules can freely (i.e., passively) cross a membrane and which may require transport by a carrier protein and/or an expenditure of energy. In general, water and small, noncharged molecules, such as carbon dioxide, oxygen, glycerol, water, and alcohol, can freely cross the membrane. They are able to slip between the hydrophilic heads of the phospholipids and pass through the hydrophobic tails of the membrane. These molecules are said to follow their **concentration gradient** as they move from an area where their concentration is high to an area where their concentration is low. Consider that a cell is always using oxygen when it carries on cellular respiration. Therefore, the concentration of oxygen is always lower inside a cell than outside a cell, and so oxygen has a tendency to enter a cell. Carbon dioxide, on the other hand, is produced when a cell carries on cellular respiration. Therefore, carbon dioxide is also following a concentration gradient when it moves from inside the cell to outside the cell.

Ions and polar molecules, such as glucose and amino acids, can cross a membrane but slowly. Therefore, they are often assisted across the plasma membrane by carrier proteins. The carrier protein must combine with an ion, such as sodium (Na^+) or a molecule, such as glucose, before transporting it across the membrane. Therefore, carrier proteins are specific for the substances they transport across the plasma membrane.

Vesicle formation is another way a molecule can exit a cell (called exocytosis) or enter a cell (called endocytosis). This method of crossing a plasma membrane is reserved for macromolecules or even for something larger than a macromolecule, such as a virus. You might think that endocytosis is not specific, but we will see that a cell does have a means to be selective about what enters by endocytosis.

FIGURE 5.5 How molecules cross the plasma membrane.
The curved arrows indicate that these substances cannot diffuse across the plasma membrane, and the long back-and-forth arrows indicate that these substances can diffuse across the plasma membrane.

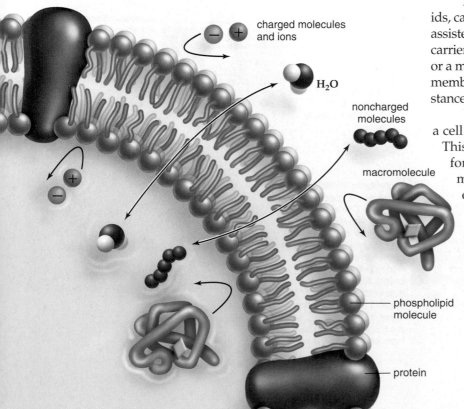

charged molecules and ions

H_2O

noncharged molecules

macromolecule

phospholipid molecule

protein

The plasma membrane is differentially permeable. Certain substances can freely pass through the membrane, and others cannot. Those that cannot freely cross the membrane may be transported across either by carrier proteins or by vesicle formation.

FIGURE 5.6 Process of diffusion.
Diffusion is spontaneous, and no chemical energy is required to bring it about. **a.** When a dye crystal is placed in water, it is concentrated in one area. **b.** The dye dissolves in the water, and there is a net movement of dye molecules from a higher to a lower concentration. There is also a net movement of water molecules from a higher to a lower concentration. **c.** Eventually, the water and the dye molecules are equally distributed throughout the container.

water molecules
(solvent)

dye molecules
(solute)

a. Crystal of dye is placed
   in water

b. Diffusion of water and
   dye molecules

c. Equal distribution of
   molecules results

Diffusion and Osmosis

Diffusion is the movement of molecules from a higher to a lower concentration—that is, down their concentration gradient—until equilibrium is achieved and they are distributed equally. Diffusion is a physical process that can be observed with any type of molecule. For example, when a crystal of dye is placed in water (Fig. 5.6), the dye and water molecules move in various directions, but their net movement, which is the sum of their motion, is toward the region of lower concentration. Eventually, the dye is dissolved in the water, resulting in equilibrium and a colored solution.

A **solution** contains both a solute, usually a solid, and a solvent, usually a liquid. In this case, the **solute** is the dye and the **solvent** is the water molecules. Once the solute and solvent are evenly distributed, they continue to move about, but there is no net movement of either one in any direction.

The chemical and physical properties of the plasma membrane allow only a few types of molecules to enter and exit a cell simply by diffusion. Gases can diffuse through the lipid bilayer; this is the mechanism by which oxygen enters cells and carbon dioxide exits cells. Also, consider the movement of oxygen from the alveoli (air sacs) of the lungs to the blood in the lung capillaries (Fig. 5.7). After inhalation (breathing in), the concentration of oxygen in the alveoli is higher than that in the blood; therefore, oxygen diffuses into the blood.

Several factors influence the rate of diffusion. Among these factors are temperature, pressure, electrical currents, and molecular size. For example, as temperature increases, the rate of diffusion increases.

Molecules diffuse down their concentration gradients. A few types of small molecules can simply diffuse through the plasma membrane.

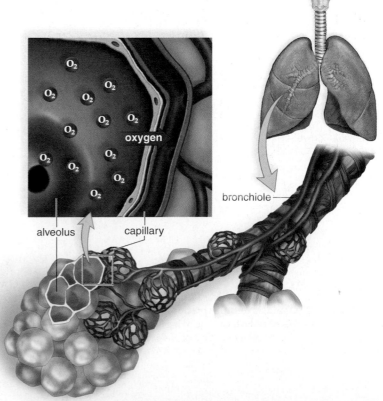

O_2 O_2 O_2 O_2 O_2 O_2 O_2 O_2 O_2 O_2 O_2 O_2

oxygen

alveolus   capillary

bronchiole

FIGURE 5.7 Gas exchange in lungs.
Oxygen (O_2) diffuses into the capillaries of the lungs because there is a higher concentration of oxygen in the alveoli (air sacs) than in the capillaries.

Osmosis

The diffusion of water across a differentially (selectively) permeable membrane due to concentration differences is called **osmosis**. To illustrate osmosis, a thistle tube containing a 10% solute solution[1] is covered at one end by a differentially permeable membrane and then placed in a beaker containing a 5% solute solution (Fig. 5.8). The beaker has a higher concentration of water molecules (lower percentage of solute), and the thistle tube has a lower concentration of water molecules (higher percentage of solute). Diffusion always occurs from higher to lower concentration. Therefore, a net movement of water takes place across the membrane from the beaker to the inside of the thistle tube.

The solute does not diffuse out of the thistle tube. Why not? Because the membrane is not permeable to the solute. As water enters and the solute does not exit, the level of the solution within the thistle tube rises (Fig. 5.8c). In the end, the concentration of solute in the thistle tube is less than 10%. Why? Because there is now less solute per unit volume. And the concentration of solute in the beaker is greater than 5%. Why? Because there is now more solute per unit volume.

Water enters the thistle tube due to the osmotic pressure of the solution within the thistle tube. **Osmotic pressure** is the pressure that develops in a system due to osmosis.[2] In other words, the greater the possible osmotic pressure, the more likely it is that water will diffuse in that direction. Due to osmotic pressure, water is absorbed by the kidneys and taken up by capillaries in the tissues. Osmosis also occurs across the plasma membrane, as we shall now see (Fig. 5.9).

Isotonic Solution. In the laboratory, cells are normally placed in **isotonic solutions**—that is, the solute concentration and the water concentration both inside and outside the cell are equal, and therefore there is no net gain or loss of water. The prefix *iso* means "the same as," and the term **tonicity** refers to the strength of the solution. A 0.9% solution of the salt sodium chloride (NaCl) is known to be isotonic to red blood cells. Therefore, intravenous solutions medically administered usually have this tonicity. Terrestrial animals can usually take in either water or salt as needed to maintain the tonicity of their internal environment. Many animals living in an estuary, such as oysters, blue crabs, and some fishes, are able to cope with changes in the salinity (salt concentrations) of their environment. Their kidneys, gills, and other structures help them do this.

Hypotonic Solution. Solutions that cause cells to swell, or even to burst, due to an intake of water are said to be **hypotonic solutions.** The prefix *hypo* means "less than" and refers to a solution with a lower concentration of solute (higher concentration of water) than inside the cell. If a cell is placed in a hypotonic solution, water enters the cell; the net movement of water is from the outside to the inside of the cell.

---

[1] Percent solutions are grams of solute per 100 ml of solvent. Therefore, a 10% solution is 10 g of sugar with water added to make 100 ml of solution.

[2] Osmotic pressure is measured by placing a solution in an osmometer and then immersing the osmometer in pure water. The pressure that develops is the osmotic pressure of a solution.

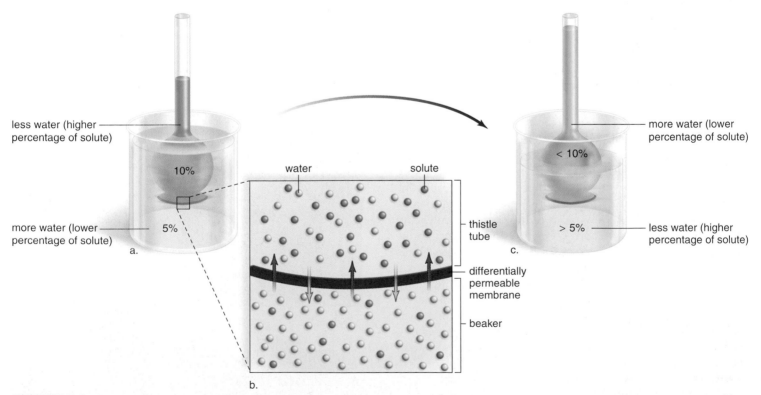

less water (higher percentage of solute)

more water (lower percentage of solute)

10%

5%

a.

water   solute

thistle tube

differentially permeable membrane

beaker

b.

more water (lower percentage of solute)

< 10%

> 5%

less water (higher percentage of solute)

c.

FIGURE 5.8 Osmosis demonstration.
a. A thistle tube, covered at the broad end by a differentially permeable membrane, contains a 10% solute solution. The beaker contains a 5% solute solution. **b.** The solute (green circles) is unable to pass through the membrane, but the water (blue circles) passes through in both directions. There is a net movement of water toward the inside of the thistle tube, where there is a lower percentage of water molecules. **c.** Due to the incoming water molecules, the level of the solution rises in the thistle tube.

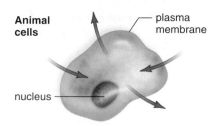
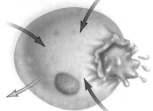

Animal cells — plasma membrane, nucleus

In an isotonic solution, there is no net movement of water.

In a hypotonic solution, water enters the cell, which may burst (lysis).

In a hypertonic solution, water leaves the cell, which shrivels (crenation).

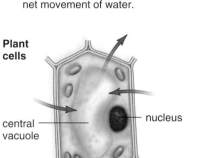
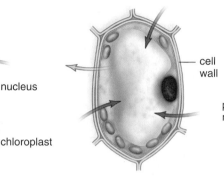
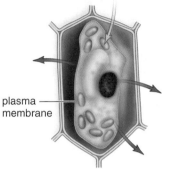

Plant cells — central vacuole, nucleus, chloroplast, cell wall, plasma membrane

In an isotonic solution, there is no net movement of water.

In a hypotonic solution, vacuoles fill with water, turgor pressure develops, and chloroplasts are seen next to the cell wall.

In a hypertonic solution, vacuoles lose water, the cytoplasm shrinks (plasmolysis), and chloroplasts are seen in the center of the cell.

FIGURE 5.9 Osmosis in animal and plant cells. The arrows indicate the movement of water molecules. To determine the net movement of water, compare the number of arrows that are taking water molecules into the cell versus the number that are taking water out of the cell. In an isotonic solution, a cell neither gains nor loses water; in a hypotonic solution, a cell gains water; and in a hypertonic solution, a cell loses water.

Any concentration of a salt solution lower than 0.9% is hypotonic to red blood cells. Animal cells placed in such a solution expand and sometimes burst due to the buildup of pressure. The term *cytolysis* is used to refer to disrupted cells; hemolysis, then, is disrupted red blood cells.

The swelling of a plant cell in a hypotonic solution creates **turgor pressure.** When a plant cell is placed in a hypotonic solution, we observe expansion of the cytoplasm because the large central vacuole gains water and the plasma membrane pushes against the rigid cell wall. The plant cell does not burst because the cell wall does not give way. Turgor pressure in plant cells is extremely important to the maintenance of the plant's erect position. If you forget to water your plants, they wilt due to decreased turgor pressure.

Organisms that live in fresh water have to prevent their internal environment from becoming hypotonic. Many protozoans, such as paramecia, have contractile vacuoles that rid the body of excess water. Freshwater fishes have well-developed kidneys that excrete a large volume of dilute urine. Even so, they have to take in salts at their gills. Even though fresh-water fishes are good osmoregulators, they would not be able to survive in either distilled water or a marine environment.

Hypertonic Solution. Solutions that cause cells to shrink or shrivel due to loss of water are said to be **hypertonic solutions.** The prefix *hyper* means "more than" and refers to a solution with a higher percentage of solute (lower concentration of water) than the cell. If a cell is placed in a hypertonic solution, water leaves the cell; the net movement of water is from the inside to the outside of the cell.

Any concentration of a salt solution higher than 0.9% is hypertonic to red blood cells. If animal cells are placed in this solution, they shrink. The term **crenation** refers to red blood cells in this condition. Meats are sometimes preserved by salting them. The bacteria are not killed by the salt but by the lack of water in the meat.

When a plant cell is placed in a hypertonic solution, the plasma membrane pulls away from the cell wall as the large central vacuole loses water. This is an example of **plasmolysis,** a shrinking of the cytoplasm due to osmosis. The dead plants you may see along a salted roadside died because they were exposed to a hypertonic solution during the winter. Also, when salt water invades coastal marshes due to storms and human activities, coastal plants die. Without roots to hold the soil, it washes into the sea, doing away with many acres of valuable wetlands.

Marine animals cope with their hypertonic environment in various ways that prevent them from losing water to the environment. Sharks increase or decrease urea in their blood until their blood is isotonic with the environment and in this way do not lose excessive water. Marine fishes and other types of animals excrete salts across their gills. Have you ever seen a marine turtle cry? It is ridding its body of salt by means of glands near the eye.

In an isotonic solution, a cell neither gains nor loses water. In a hypotonic solution, a cell gains water. In a hypertonic solution, a cell loses water and the cytoplasm shrinks.

Transport by Carrier Proteins

The plasma membrane impedes the passage of all but a few substances. Yet, biologically useful molecules are able to enter and exit the cell at a rapid rate because of carrier proteins in the membrane. Carrier proteins are specific; each can combine with only a certain type of molecule or ion, which is then transported through the membrane. It is not completely understood how carrier proteins function, but after a carrier combines with a molecule, the carrier is believed to undergo a change in shape that moves the molecule across the membrane. Carrier proteins are required for both facilitated transport and active transport (see Table 5.1).

Some of the proteins in the plasma membrane are carriers. They transport biologically useful molecules into and out of the cell.

Facilitated Transport

Facilitated transport explains the passage of such molecules as glucose and amino acids across the plasma membrane even though they are not lipid-soluble. The passage of glucose and amino acids is facilitated by their reversible combination with carrier proteins, which in some manner transport them through the plasma membrane. These carrier proteins are specific. For example, various sugar molecules of identical size might be present inside or outside the cell, but glucose can cross the membrane hundreds of times faster than the other sugars. As stated earlier, this is the reason the membrane can be called differentially permeable.

A model for facilitated transport (Fig. 5.10) shows that after a carrier has assisted the movement of a molecule to the other side of the membrane, it is free to assist the passage of other similar molecules. Neither diffusion nor facili-

tated transport requires an expenditure of energy because the molecules are moving down their concentration gradient in the same direction they tend to move anyway.

Active Transport

During **active transport,** molecules or ions move through the plasma membrane, accumulating either inside or outside the cell. For example, iodine collects in the cells of the thyroid gland; glucose is completely absorbed from the gut by the cells lining the digestive tract; and sodium can be almost completely withdrawn from urine by cells lining the kidney tubules. In these instances, molecules have moved to the region of higher concentration, exactly opposite to the process of diffusion.

Both carrier proteins and an expenditure of energy are needed to transport molecules against their concentration gradient. In this case, chemical energy (ATP molecules usually) is required for the carrier to combine with the substance to be transported. Therefore, it is not surprising that cells involved primarily in active transport, such as kidney cells, have a large number of mitochondria near membranes where active transport is occurring.

Proteins involved in active transport often are called pumps because, just as a water pump uses energy to move water against the force of gravity, proteins use energy to move a substance against its concentration gradient. One type of pump that is active in all animal cells, but is especially associated with nerve and muscle cells, moves sodium ions (Na^+) to the outside of the cell and potassium ions (K^+) to the inside of the cell. These two events are linked, and the carrier protein is called a **sodium-potassium pump.** A change in carrier shape after the attachment and again after the detachment

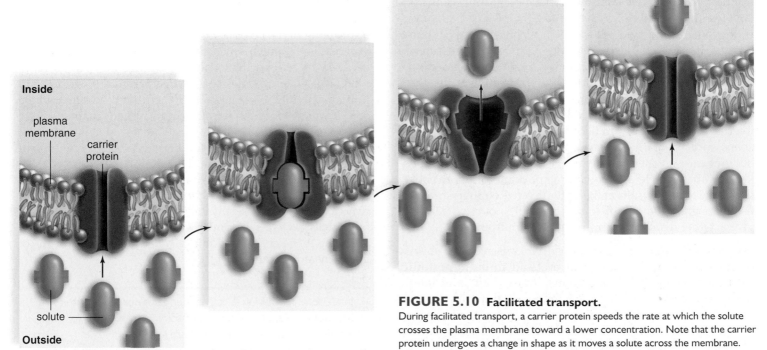

FIGURE 5.10 Facilitated transport.
During facilitated transport, a carrier protein speeds the rate at which the solute crosses the plasma membrane toward a lower concentration. Note that the carrier protein undergoes a change in shape as it moves a solute across the membrane.

of a phosphate group allows it to combine alternately with sodium ions and potassium ions (Fig. 5.11). The phosphate group is donated by ATP when it is broken down enzymatically by the carrier. The sodium-potassium pump results in both a solute concentration gradient and an electrical gradient for these ions across the plasma membrane.

The passage of salt (NaCl) across a plasma membrane is of primary importance to most cells. The chloride ion (Cl⁻) usually crosses the plasma membrane because it is attracted by positively charged sodium ions (Na⁺). First sodium ions are pumped across a membrane, and then chloride ions simply diffuse through channels that allow their passage.

As noted in Figure 5.4a, the genetic disorder cystic fibrosis results from a faulty chloride channel. Ordinarily, after chloride ions have passed through the membrane, sodium ions (Na⁺) and water follow. It is believed that the lack of water causes abnormally thick mucus in the bronchial tubes and pancreatic ducts, thus interfering with the function of the lungs and pancreas.

During facilitated transport, small molecules follow their concentration gradient. During active transport, small molecules and ions move against their concentration gradient.

FIGURE 5.11 The sodium-potassium pump.
The same carrier protein transports sodium ions (Na⁺) to the outside of the cell and potassium ions (K⁺) to the inside of the cell because it undergoes an ATP-dependent change in shape. Three sodium ions are carried outward for every two potassium ions carried inward; therefore, the inside of the cell is negatively charged compared to the outside.

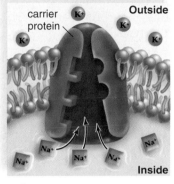

1. Carrier has a shape that allows it to take up 3 Na⁺.

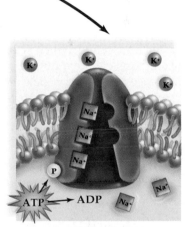

2. ATP is split, and phosphate group attaches to carrier.

6. Change in shape results and causes carrier to release 2 K⁺ inside the cell.

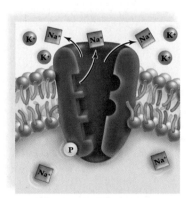

3. Change in shape results and causes carrier to release 3 Na⁺ outside the cell.

5. Phosphate group is released from carrier.

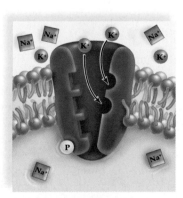

4. Carrier has a shape that allows it to take up 2 K⁺.

Vesicle Formation

How do macromolecules such as polypeptides, polysaccharides, or polynucleotides enter and exit a cell? Because they are too large to be transported by carrier proteins, macromolecules are transported into and out of the cell by vesicle formation. Vesicle formation is membrane-assisted transport because membrane is needed to form the vesicle. Vesicle formation requires an expenditure of cellular energy, but vesicle formation has the added benefit that the vesicle membrane keeps the contained macromolecules from mixing with molecules within cytoplasm. Exocytosis is a way substances can enter a cell, and exocytosis is a way substances can exit a cell.

Exocytosis

During **exocytosis,** a vesicle fuses with the plasma membrane as secretion occurs (Fig. 5.12). Hormones, neurotransmitters, and digestive enzymes are secreted from cells in this manner. The Golgi body often produces the vesicles that carry these cell products to the membrane. Notice that during exocytosis, the membrane of the vesicle becomes a part of the plasma membrane, which is thereby enlarged. For this reason, exocytosis can be a normal part of cell growth. The proteins released from the vesicle adhere to the cell surface or become incorporated in an extracellular matrix.

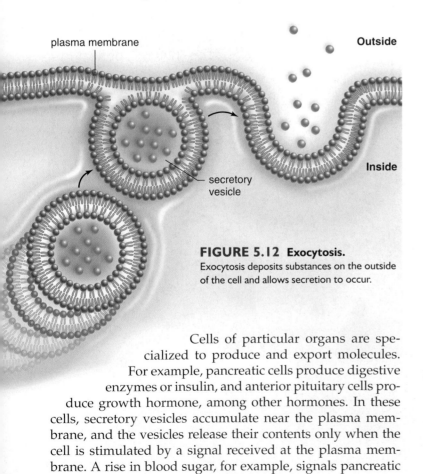

plasma membrane

Outside

secretory vesicle

Inside

FIGURE 5.12 Exocytosis.
Exocytosis deposits substances on the outside of the cell and allows secretion to occur.

Cells of particular organs are specialized to produce and export molecules. For example, pancreatic cells produce digestive enzymes or insulin, and anterior pituitary cells produce growth hormone, among other hormones. In these cells, secretory vesicles accumulate near the plasma membrane, and the vesicles release their contents only when the cell is stimulated by a signal received at the plasma membrane. A rise in blood sugar, for example, signals pancreatic cells to release the hormone insulin. This is called regulated secretion, because vesicles fuse with the plasma membrane only when it is appropriate to the needs of the body.

Endocytosis

During **endocytosis,** cells take in substances by vesicle formation. A portion of the plasma membrane invaginates to envelop the substance, and then the membrane pinches off to form an intracellular vesicle. Endocytosis occurs in one of three ways, as illustrated in Figure 5.13. Phagocytosis transports large substances, such as a virus, and pinocytosis transports small substances, such as a macromolecule, into a cell. Receptor-mediated endocytosis is a special form of pinocytosis.

Phagocytosis.   When the material taken in by endocytosis is large, such as a food particle or another cell, the process is called **phagocytosis** [Gk. *phagein,* to eat]. Phagocytosis is common in unicellular organisms such as amoebas (Fig. 5.13*a*). It also occurs in humans. Certain types of human white blood cells are amoeboid—that is, they are mobile like an amoeba, and they are able to engulf debris such as worn-out red blood cells or viruses. When an endocytic vesicle fuses with a lysosome, digestion occurs. We will see that this process is a necessary and preliminary step toward the development of immunity to bacterial diseases.

Pinocytosis.   Pinocytosis [Gk. *pinein,* to drink] occurs when vesicles form around a liquid or around very small particles (Fig. 5.13*b*). Blood cells, cells that line the kidney tubules or the intestinal wall, and plant root cells all use pinocytosis to ingest substances.

Whereas phagocytosis can be seen with the light microscope, the electron microscope must be used to observe pinocytic vesicles, which are no larger than 0.1–0.2 μm. Still, pinocytosis involves a significant amount of the plasma membrane because it occurs continuously. The loss of plasma membrane due to pinocytosis is balanced by the occurrence of exocytosis, however.

Receptor-Mediated Endocytosis.   Receptor-mediated endocytosis is a form of pinocytosis that is quite specific because it uses a receptor protein shaped in such a way that a specific molecule such as a vitamin, peptide hormone, or lipoprotein can bind to it (Fig. 5.13*c*). The receptors for these substances are found at one location in the plasma membrane. This location is called a coated pit because there is a layer of protein on the cytoplasmic side of the pit. Once formed, the vesicle is uncoated and may fuse with a lysosome. When an empty, used vesicle fuses with the plasma membrane, the receptors return to their former location.

Receptor-mediated endocytosis is selective and much more efficient than ordinary pinocytosis. It is involved in uptake and also in the transfer and exchange of substances between cells. Such exchanges take place when substances move from maternal blood into fetal blood at the placenta, for example.

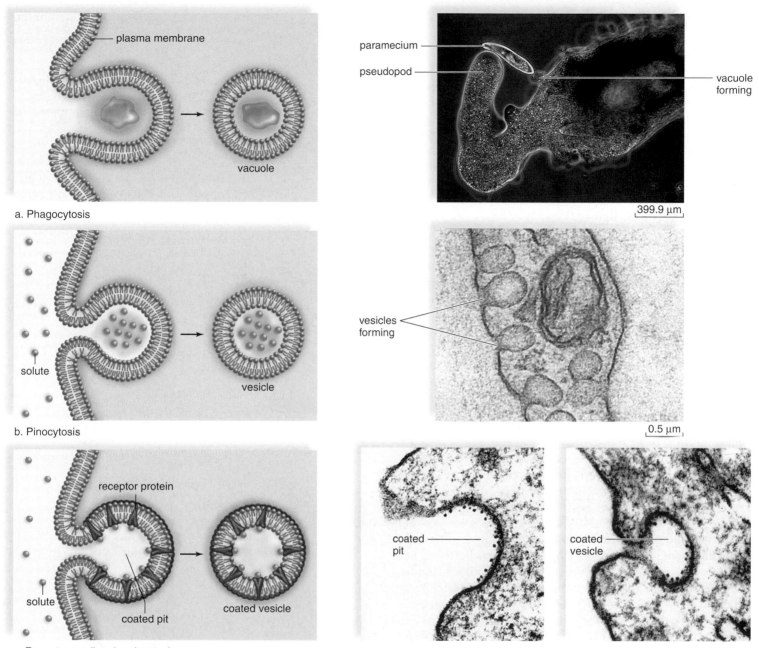

a. Phagocytosis

b. Pinocytosis

c. Receptor-mediated endocytosis

FIGURE 5.13 Three methods of endocytosis.

a. Phagocytosis occurs when the substance to be transported into the cell is large; amoebas ingest by phagocytosis. Digestion occurs when the resulting vacuole fuses with a lysosome. **b.** Pinocytosis occurs when a macromolecule such as a polypeptide is transported into the cell. The result is a vesicle (small vacuole). **c.** Receptor-mediated endocytosis is a form of pinocytosis. Molecules first bind to specific receptor proteins, which migrate to or are already in a coated pit. The vesicle that forms contains the molecules and their receptors.

The importance of receptor-mediated endocytosis is demonstrated by a genetic disorder called familial hypercholesterolemia. Cholesterol is transported in blood by a complex of lipids and proteins called low-density lipoprotein (LDL). Ordinarily, body cells take up LDL when LDL receptors gather in a coated pit. But in some individuals, the LDL receptor is unable to properly bind to the coated pit, and the cells are unable to take up cholesterol. Instead, cholesterol accumulates in the walls of arterial blood vessels, leading to high blood pressure, occluded (blocked) arteries, and heart attacks.

Substances are secreted from a cell by exocytosis. Substances enter a cell by endocytosis. Receptor-mediated endocytosis allows cells to take up specific kinds of molecules and then they are released within the cell.

5.4 MODIFICATION OF CELL SURFACES

Now that we have completed our discussion of the plasma membrane, you might think we have finished our tour of the cell. However, most cells have extracellular structures that take shape from materials the cell produces and transports across its plasma membrane. In plants, prokaryotes, fungi, and most algae, the extracellular component of the cell is a fairly rigid cell wall. A cell wall occurs in organisms that have a rather inactive lifestyle. Animals that have an active way of life have a more varied extracellular anatomy appropriate to the particular tissue type.

Cell Surfaces in Animals

We will consider two different types of animal cell surface features: (1) junctions that occur between some types of cells and (2) the extracellular matrix that is observed outside cells.

Junctions Between Cells

Certain organs of vertebrate animals are well known to have junctions between their cells that allow them to behave in a coordinated manner. These junctions are of the three types shown in Figure 5.14.

Anchoring junctions serve to mechanically attach adjacent cells. Two types of anchoring junctions are described here. In **adhesion junctions,** internal cytoplasmic plaques, firmly attached to the cytoskeleton within each cell, are joined by intercellular filaments. The result is a sturdy but flexible sheet of cells. In some organs—such as the heart, stomach, and bladder, where tissues get stretched—adhesion junctions hold the cells together. At a **desmosome,** a single point of attachment between adjacent cells connects the cytoskeletons of adjacent cells. Desmosomes are the most common type of intercellular junction between skin cells. Neither adhesion junctions nor desmosomes affect the movement of substances between adjacent cells.

Adjacent cells are even more closely joined by **tight junctions,** in which plasma membrane proteins actually attach to each other, producing a zipperlike fastening. The cells of tissues that serve as barriers are held together by tight junctions; in the intestine, the digestive juices stay out of the body, and in the kidneys the urine stays within kidney tubules, because the cells are joined by tight junctions. Tight junctions are important in the blood-brain barrier.

A **gap junction** allows cells to communicate. A gap junction is formed when two identical plasma membrane channels join. The channel of each cell is lined by six plasma membrane proteins. A gap junction lends strength to the cells, but it also allows small molecules and ions to pass between them. Gap junctions are important in heart muscle and smooth muscle because they permit a flow of ions that is required for the cells to contract as a unit.

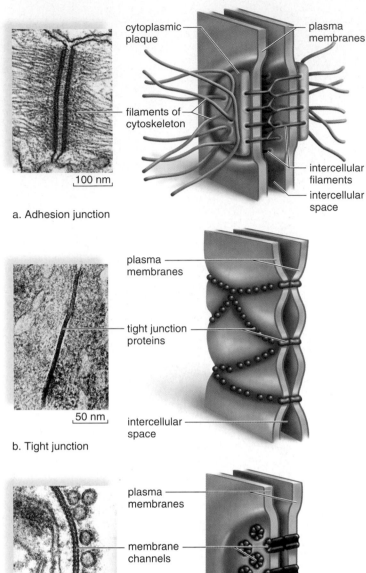

a. Adhesion junction

b. Tight junction

c. Gap junction

FIGURE 5.14 Junctions between cells of the intestinal wall.
a. In adhesion junctions, intercellular filaments run between two cells. **b.** Tight junctions between cells form an impermeable barrier because their adjacent plasma membranes are joined. **c.** Gap junctions allow communication between two cells because adjacent plasma membrane channels are joined.

Extracellular Matrix

An extracellular matrix is a nonliving meshwork of polysaccharides and proteins in close association with the cell that produced them (Fig. 5.15). Collagen and elastin fibers are two well-known structural proteins in the extracellular matrix. Collagen gives the matrix strength, and elastin gives it resilience. Fibronectins and laminins are two adhesive proteins that seem

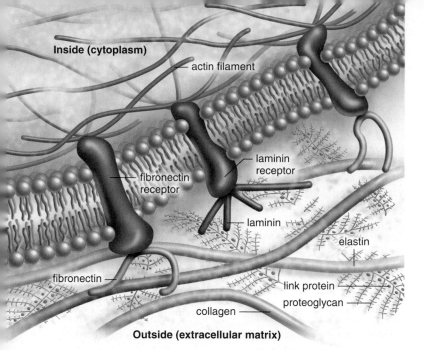

Inside (cytoplasm)

— actin filament

laminin receptor

fibronectin receptor

laminin

elastin

fibronectin —

link protein

proteoglycan

collagen —

Outside (extracellular matrix)

FIGURE 5.15 Animal cell extracellular matrix.
The extracellular matrix supports an animal cell and also affects its behavior. Collagen and elastin have a support function, while fibronectins and laminins bind to receptors in the plasma membrane and most likely assist cell communication processes.

to play a dynamic role in influencing the behavior of cells. For example, fibronectin and laminin form "highways" that direct the migration of cells during development. Recently, laminins were found to be necessary for the production of milk by the mammary gland cells of mice. Fibronectins and laminins also bind to receptors in the plasma membrane and permit communication between the extracellular matrix and the cytoplasm of the cell, perhaps via cytoskeletal connections.

The polysaccharides in the extracellular matrix contain amino sugars, and when they join to proteins, they are called proteoglycans. The polysaccharides and proteoglycans provide a rigid packing gel for the various matrix proteins. More work will be needed to determine the functions of proteoglycans in the extracellular matrix, but for now we know that the gel they help create permits rapid diffusion of nutrients, metabolites, and hormones between blood and tissue cells. Most likely, they regulate the activity of signal molecules that bind to receptor proteins in the plasma membrane.

The extracellular matrix of various tissues varies between being quite flexible, as in cartilage, and being rock solid, as in bone. The extracellular matrix of bone is hard because in addition to the components mentioned, mineral salts, notably calcium salts, are deposited outside the cell.

Plant Cell Walls

In addition to a plasma membrane, plant cells are surrounded by a porous **cell wall** that varies in thickness, depending on the function of the cell. All plant cells have a primary cell wall. The primary cell wall contains cellulose fibrils in which microfibrils are held together by noncellulose substances. Pectins allow the wall to stretch when the cell is growing, and noncellulose polysaccharides harden the wall when the cell is mature. Pectins are especially abundant in the middle lamella,

which is a layer of adhesive substances that holds the cells together. Some cells in woody plants have a secondary wall that forms inside the primary cell wall. The secondary wall has a greater quantity of cellulose fibrils than the primary wall, and layers of cellulose fibrils are laid down at right angles to one another. Lignin, a substance that adds strength, is a common ingredient of secondary cell walls in woody plants.

In a plant, the cytoplasm of living cells is connected by **plasmodesmata** (sing., plasmodesma), numerous narrow, membrane-lined channels that pass through the cell wall (Fig. 5.16). Cytoplasmic strands within these channels allow direct exchange of some materials between adjacent plant cells and eventually all the cells of a plant. The plasmodesmata are large enough to allow only water and small solutes to pass freely from cell to cell. This limitation means that plant cells can maintain their own concentrations of larger substances and differentiate into particular cell types.

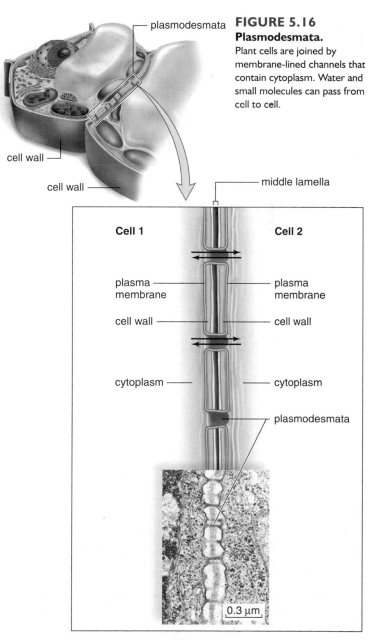

plasmodesmata

FIGURE 5.16 Plasmodesmata.
Plant cells are joined by membrane-lined channels that contain cytoplasm. Water and small molecules can pass from cell to cell.

cell wall —

cell wall —

middle lamella

Cell 1

Cell 2

plasma membrane

plasma membrane

cell wall —

cell wall

cytoplasm —

cytoplasm

plasmodesmata

0.3 μm

CONNECTING THE CONCEPTS

The plasma membrane is quite appropriately called the gatekeeper of the cell because it maintains the integrity of the cell and stands guard over what enters and leaves. But we have seen that the plasma membrane also does much more than this. Its glycoproteins and glycolipids mark the cell as belonging to the organism. Its numerous proteins allow communication between cells and enable tissues to function as a whole. Now it appears that the extracellular material secreted by

cells assists the plasma membrane in its numerous functions.

The progression in our knowledge about the plasma membrane illustrates how science works. The concepts and techniques of science evolve and change, and the knowledge we have today will be amended and expanded by new investigative work. Basic science has applications that promote the health of human beings. To know that the plasma membrane is malfunctioning in a person who has diabetes

or cystic fibrosis or in someone who has a high cholesterol count is a first step toward curing these conditions. Even cancer is sometimes due to receptor proteins that signal the cell to divide even when no growth factor is present.

Our ability to understand the functioning of the plasma membrane is dependent on a thorough understanding of the molecules and ions that make up the cell. Today, it is impossible to deny the premise that biology and medicine have a biochemical basis.

Summary

5.1 MEMBRANE MODELS

The fluid-mosaic model of membrane structure developed by Singer and Nicolson was preceded by several other models. Electron micrographs of freeze-fractured membranes support the fluid-mosaic model, rather than Robertson's unit membrane model based on the Danielli and Davson sandwich model.

5.2 PLASMA MEMBRANE STRUCTURE AND FUNCTION

Two components of the plasma membrane are lipids and proteins. In the lipid bilayer, phospholipids are arranged with their hydrophilic (polar) heads at the surfaces and their hydrophobic (nonpolar) tails in the interior. The lipid bilayer has the consistency of oil but acts as a barrier to the entrance and exit of most biological molecules. Membrane glycolipids and glycoproteins are involved in marking the cell as belonging to a particular individual and tissue.

The hydrophobic portion of an integral protein lies in the lipid bilayer of the plasma membrane, and the hydrophilic portion lies at the surfaces. Proteins act as receptors, carry on enzymatic reactions, join cells together, form channels, or act as carriers to move substances across the membrane.

5.3 PERMEABILITY OF THE PLASMA MEMBRANE

Some molecules (lipid-soluble compounds, water, and gases) simply diffuse across the membrane from the area of higher concentration to the area of lower concentration. No metabolic energy is required for diffusion to occur.

The diffusion of water across a differentially permeable membrane is called osmosis. Water moves across the membrane into the area of higher solute (less water) content per volume. When cells are in an isotonic solution, they neither gain nor lose water. When cells are in a hypotonic solution, they gain water, and when they are in a hypertonic solution, they lose water (Table 5.2).

Other molecules are transported across the membrane by carrier proteins that span the membrane. During facilitated transport, a carrier

protein assists the movement of a molecule down its concentration gradient. No energy is required.

During active transport, a carrier protein acts as a pump that causes a substance to move against its concentration gradient. The sodium-potassium pump carries Na^+ to the outside of the cell and K^+ to the inside of the cell. Energy in the form of ATP molecules is required for active transport to occur.

Larger substances can enter and exit a membrane by exocytosis and endocytosis. Exocytosis involves secretion. Endocytosis includes phagocytosis, pinocytosis, and receptor-mediated endocytosis. Receptor-mediated endocytosis makes use of receptor proteins in the plasma membrane. Once a specific solute binds to receptors, a coated pit becomes a coated vesicle. After losing the coat, the vesicle can join with the lysosome, or after discharging the substance, the receptor-containing vesicle can fuse with the plasma membrane.

5.4 MODIFICATION OF CELL SURFACES

Some animal cells have anchoring junctions. Adhesion junctions and tight junctions help hold cells together; gap junctions allow passage of small molecules between cells. Other animal cells have an extracellular matrix that holds their shape and influences their behavior.

Plant cells have a freely permeable cell wall, with cellulose as its main component. Plant cells are joined by narrow, membrane-lined channels called plasmodesmata that span the cell wall and contain strands of cytoplasm that allow materials to pass from one cell to another.

TABLE 5.2

Effect of Osmosis on a Cell

Tonicity of Solution	Concentrations		Net Movement of Water	Effect on Cell
	Solute	Water		
Isotonic	Same as cell	Same as cell	None	None
Hypotonic	Less than cell	More than cell	Cell gains water	Swells, turgor pressure
Hypertonic	More than cell	Less than cell	Cell loses water	Shrinks, plasmolysis

Reviewing the Chapter

1. Describe the fluid-mosaic model of membrane structure as well as the models that preceded it. Cite the evidence that either disproves or supports these models.   84
2. Tell how the phospholipids are arranged in the plasma membrane. What other lipids are present in the membrane, and what functions do they serve?   85
3. Describe how proteins are arranged in the plasma membrane. What are their various functions? Describe an experiment indicating that proteins can laterally drift in the membrane.   86–87
4. Define diffusion. What factors can influence the rate of diffusion? What substances can diffuse through a differentially permeable membrane?   89
5. Define osmosis. Describe verbally and with drawings what happens to an animal cell when placed in isotonic, hypotonic, and hypertonic solutions.   90–91
6. Describe verbally and with drawings what happens to a plant cell when placed in isotonic, hypotonic, and hypertonic solutions.   90–91
7. Why do most substances have to be assisted through the plasma membrane? Contrast movement by facilitated transport with movement by active transport.   92–93
8. Draw and explain a diagram that shows how the sodium-potassium pump works.   92–93
9. Describe and contrast three methods of endocytosis.   94–95
10. Give examples to show that cell surface modifications help plant and animal cells communicate.   96–97

Testing Yourself

Choose the best answer for each question.

1. Write hypotonic solution or hypertonic solution beneath each cell. Justify your conclusions.

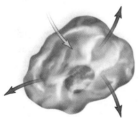

a. _____

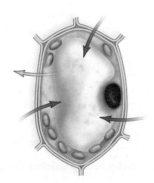

b. _____

2. Electron micrographs following freeze-fracture of the plasma membrane indicate that
    a. the membrane is a phospholipid bilayer.
    b. some proteins span the membrane.
    c. protein is found only on the surfaces of the membrane.
    d. glycolipids and glycoproteins are antigenic.
    e. there are receptors in the membrane.

3. A phospholipid molecule has a head and two tails. The tails are found
    a. at the surfaces of the membrane.
    b. in the interior of the membrane.
    c. spanning the membrane.
    d. where the environment is hydrophilic.
    e. Both a and b are correct.

4. During diffusion,
    a. solvents move from the area of higher to lower concentration, but solutes do not.
    b. there is a net movement of molecules from the area of higher to lower concentration.
    c. a cell must be present for any movement of molecules to occur.
    d. molecules move against their concentration gradient if they are small and charged.
    e. All of these are correct.

5. When a cell is placed in a hypotonic solution,
    a. solute exits the cell to equalize the concentration on both sides of the membrane.
    b. water exits the cell toward the area of lower solute concentration.
    c. water enters the cell toward the area of higher solute concentration.
    d. solute exits and water enters the cell.
    e. Both c and d are correct.

6. When a cell is placed in a hypertonic solution,
    a. solute exits the cell to equalize the concentration on both sides of the membrane.
    b. water exits the cell toward the area of lower solute concentration.
    c. water exits the cell toward the area of higher solute concentration.
    d. solute exits and water enters the cell.
    e. Both a and c are correct.

7. Active transport
    a. requires a carrier protein.
    b. moves a molecule against its concentration gradient.
    c. requires a supply of chemical energy.
    d. does not occur during facilitated transport.
    e. All of these are correct.

8. The sodium-potassium pump
    a. helps establish an electrochemical gradient across the membrane.
    b. concentrates sodium on the outside of the membrane.
    c. uses a carrier protein and chemical energy.
    d. is present in the plasma membrane.
    e. All of these are correct.

9. Receptor-mediated endocytosis
    a. is no different from phagocytosis.
    b. brings specific solutes into the cell.
    c. helps concentrate proteins in vesicles.
    d. results in high osmotic pressure.
    e. All of these are correct.

10. Plant cells
   a. always have a secondary cell wall, even though the primary one may disappear.
   b. have channels between cells that allow strands of cytoplasm to pass from cell to cell.
   c. develop turgor pressure when water enters the nucleus.
   d. do not have cell-to-cell junctions like animal cells.
   e. All of these are correct.

11. Label this diagram of the plasma membrane.

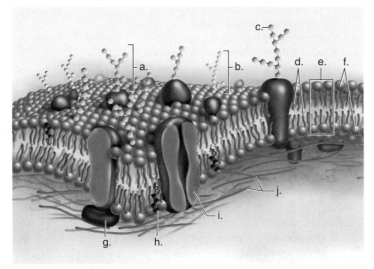

12. The fluid-mosaic model of membrane structure refers to
   a. the fluidity of proteins and the pattern of phospholipids in the membrane.
   b. the fluidity of phospholipids and the pattern of proteins in the membrane.
   c. the fluidity of cholesterol and the pattern of carbohydrate chains outside the membrane.
   d. the lack of fluidity of internal membranes compared to the plasma membrane, and the ability of the proteins to move laterally in the membrane.
   e. the fluidity of hydrophobic regions, proteins, and the mosaic pattern of hydrophilic regions.

13. Which of the following is not a function of proteins present in the plasma membrane? Proteins
   a. assist the passage of materials into the cell.
   b. interact and recognize other cells.
   c. bind with specific hormones.
   d. carry out specific metabolic reactions.
   e. produce lipid molecules.

14. The carbohydrate chains projecting from the plasma membrane are involved in
   a. adhesion between cells.
   b. reception of molecules.
   c. cell-to-cell recognition.
   d. All of these are correct.

15. Plants wilt on a hot summer day because of a decrease in
   a. turgor pressure.
   b. evaporation.
   c. condensation.
   d. diffusion.

Thinking Scientifically

1. The mucus in bronchial tubes must be thin enough for cilia to move bacteria and viruses up into the throat away from the lungs. Which way would Cl^- normally cross the plasma membrane of bronchial tube cells in order for mucus to be thin (see Fig. 5.4a)? Use the concept of osmosis to explain your answer.

2. Winter wheat is planted in the early fall, grows over the winter when the weather is colder, and is harvested in the spring. As the temperature drops, the makeup of the plasma membrane of winter wheat changes. Unsaturated fatty acids replace saturated fatty acids in the phospholipids of the membrane. Why is this a suitable adaptation?

Understanding the Terms

active transport 92	hypertonic solution 91
adhesion junction 96	hypotonic solution 90
anchoring junction 96	isotonic solution 90
carrier protein 87	osmosis 90
cell recognition protein 87	osmotic pressure 90
cell wall 97	phagocytosis 94
channel protein 87	phospholipid bilayer 85
cholesterol 85	pinocytosis 94
concentration gradient 88	plasmodesmata 97
crenation 91	plasmolysis 91
desmosome 96	receptor-mediated
differentially permeable 88	endocytosis 94
diffusion 89	receptor protein 87
endocytosis 94	sodium-potassium pump 92
enzymatic protein 87	solute 89
exocytosis 94	solution 89
facilitated transport 92	solvent 89
fluid-mosaic model 84	tight junction 96
gap junction 96	tonicity 90
glycolipid 86	turgor pressure 91
glycoprotein 86	

Match the terms to these definitions:

a._____ Characteristic of the plasma membrane due to its ability to allow certain molecules but not others to pass through.

b._____ Diffusion of water through the plasma membrane of cells.

c._____ Higher solute concentration (less water) than the cytoplasm of a cell; causes cell to lose water by osmosis.

d._____ Protein in plasma membrane that bears a carbohydrate chain.

e._____ Process by which a cell engulfs a substance, forming an intracellular vacuole.

ARIS, the *Biology* Website

ARIS, the website for *Biology*, provides a wealth of information organized and integrated by chapter. You will find practice quizzes, interactive activities, labeling exercises, flashcards, and much more that will complement your learning and understanding of general biology.

www.mhhe.com/maderbiology9

6

METABOLISM: ENERGY AND ENZYMES

*L*iving things cannot maintain their organization or carry on life's other activities without a source of organic nutrients. Green plants use solar energy, carbon dioxide, and water to make organic nutrients for themselves and other living things. Animals, including caterpillars and human beings, feed on plants or other animals that have eaten plants.

Nutrient molecules are used by organisms as a source of energy and cellular building blocks. Energy is the capacity to do work, and it takes work to maintain the organization of a cell and the organism, including a growing caterpillar. When nutrients are broken down, they provide the necessary energy to make ATP (adenosine triphosphate). ATP fuels chemical reactions in cells, such as the synthetic reactions that produce cell parts and products in a caterpillar. The universal use of ATP in cells is substantial evidence for the relatedness of all life-forms, be they bacteria, protists, fungi, plants, or animals.

The term metabolism encompasses all the chemical reactions that occur in a cell. Enzymes are protein molecules that speed metabolic reactions in cells. This chapter deals with energy and enzymes, two essential requirements for cellular metabolism and life.

A caterpillar feeding on a leaf.

6.1 CELLS AND THE FLOW OF ENERGY

Energy is the ability to do work or bring about a change. In order to maintain their organization and carry out metabolic activities, cells as well as organisms need a constant supply of energy. This energy allows living things to carry on the processes of life, including growth, development, locomotion, metabolism, and reproduction.

All organisms depend on organic nutrients as a source of energy after they are produced by photosynthesizers (algae, plants, and some bacteria). Photosynthesizers use solar energy to produce organic nutrients, and therefore life on Earth is ultimately dependent on solar energy.

Forms of Energy

Energy occurs in two forms: kinetic and potential energy. **Kinetic energy** is the energy of motion, as when a ball rolls down a hill or a moose walks through grass. **Potential energy** is stored energy—its capacity to accomplish work is not being used at the moment. The food we eat has potential energy because it can be converted into various types of kinetic energy. Food is specifically called **chemical energy** because it is composed of organic molecules such as carbohydrates, proteins, and fat. When a moose walks, it has converted chemical energy into a type of kinetic energy called **mechanical energy** (Fig. 6.1).

Two Laws of Thermodynamics

Figure 6.1 illustrates the flow of energy in a terrestrial ecosystem. Plants capture only a small portion of solar energy, and much of it dissipates as **heat.** When plants photosynthesize and then make use of the food they produce, more heat results. Still, there is enough remaining to sustain a moose and the other organisms in an ecosystem. As they metabolize nutrient molecules, all the captured solar energy eventually dissipates as heat. Therefore, energy flows and does not cycle. Two **laws of thermodynamics** explain why energy flows in ecosystems and in cells. These laws were formulated by early researchers who studied energy relationships and exchanges.

The first law of thermodynamics—the law of conservation of energy—states energy cannot be created or destroyed, but it can be changed from one form to another.

When leaf cells photosynthesize, they use solar energy to form carbohydrate molecules from carbon dioxide and water. (Carbohydrates are energy-rich molecules, while carbon dioxide and water are energy-poor molecules.) Not all of the captured solar energy becomes carbohydrates; some becomes heat:

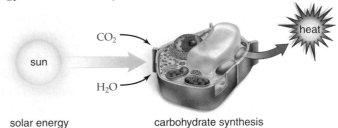

solar energy   carbohydrate synthesis

Obviously, plant cells do not create the energy they use to produce carbohydrate molecules; that energy comes from the sun. Is any energy destroyed? No, because heat is also a

FIGURE 6.1

Flow of energy.
The plant converts solar energy to the chemical energy of nutrient molecules. The moose converts a portion of this chemical energy to the mechanical energy of motion. Eventually, all solar energy absorbed by the plant dissipates as heat.

form of energy. Similarly, a moose uses the energy derived from carbohydrates to power its muscles. And as its cells use this energy, none is destroyed, but some becomes heat, which dissipates into the environment:

carbohydrate   muscle contraction

The second law of thermodynamics therefore applies to living systems.

The second law of thermodynamics states energy cannot be changed from one form to another without a loss of usable energy.

In our example, this law is upheld because some of the solar energy taken in by the plant and some of the chemical energy within the nutrient molecules taken in by the moose become heat. When heat dissipates into the environment, it is no longer usable—that is, it is not available to do work. With transformation upon transformation, eventually all usable forms of energy become heat that is lost to the environment. Heat that dissipates into the environment cannot be captured and converted to one of the other forms of energy.

As a result of the second law of thermodynamics, no process requiring a conversion of energy is ever 100% efficient. Much of the energy is lost in the form of heat. In automobiles, the gasoline engine is between 20% and 30% efficient in converting chemical energy into mechanical energy. The majority of energy is obviously lost as heat. Cells are capable of about 40% efficiency, with the remaining energy being given off to the surroundings as heat.

Cells and Entropy

The second law of thermodynamics can be stated another way: Every energy transformation makes the universe less organized and more disordered. The term **entropy** [Gk. *entrope,* a turning inward] is used to indicate the relative amount of disorganization. Since the processes that occur in cells are energy transformations, the second law means that every process that occurs in cells always does so in a way that increases the total entropy of the universe. Then, too, any one of these processes makes less energy available to do useful work in the future.

Figure 6.2 shows two processes that occur in cells. The second law of thermodynamics tells us that glucose tends to break apart into carbon dioxide and water. Why? Because glucose is more organized, and therefore less stable, than its breakdown products. Also, hydrogen ions on one side of a membrane tend to move to the other side unless they are prevented from doing so. Why? Because when they are distributed randomly, entropy has increased. As an analogy, you know from experience that a neat room is more organized

but less stable than a messy room, which is disorganized but more stable. How do you know a neat room is less stable than a messy room? Consider that a neat room always tends to become more messy.

On the other hand, you know that some cells can make glucose out of carbon dioxide and water, and all cells can actively move ions to one side of the membrane. How do they do it? These cellular processes obviously require an input of energy from an outside source. This energy ultimately comes from the sun. Living things depend on a constant supply of energy from the sun because the ultimate fate of all solar energy in the biosphere is to become randomized in the universe as heat. A living cell is a temporary repository of order purchased at the cost of a constant flow of energy.

Energy exists in several different forms. When energy transformations occur, energy is neither created nor destroyed. However, there is always a loss of usable energy. For this reason, living things are dependent on an outside source of energy that ultimately comes from the sun.

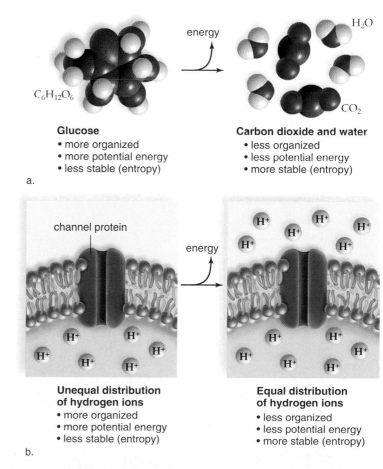

Glucose
• more organized
• more potential energy
• less stable (entropy)

Carbon dioxide and water
• less organized
• less potential energy
• more stable (entropy)

a.

Unequal distribution of hydrogen ions
• more organized
• more potential energy
• less stable (entropy)

Equal distribution of hydrogen ions
• less organized
• less potential energy
• more stable (entropy)

b.

FIGURE 6.2 Cells and entropy.
The second law of thermodynamics tells us that (**a**) glucose, which is more organized, tends to break down to carbon dioxide and water, which are less organized. **b.** Similarly, hydrogen ions (H^+) on one side of a membrane tend to move to the other side so that the ions are randomly distributed. Both processes result in a loss of potential energy and an increase in entropy.

6.2 METABOLIC REACTIONS AND ENERGY TRANSFORMATIONS

Metabolism is the sum of all the chemical reactions that occur in a cell. **Reactants** are substances that participate in a reaction, while **products** are substances that form as a result of a reaction. In the reaction A + B → C + D, A and B are the reactants while C and D are the products. How would you know that this reaction will occur spontaneously—that is, without an input of energy? Using the concept of entropy, it is possible to state that a reaction will occur spontaneously if it increases the entropy of the universe. But in cell biology, we do not usually wish to consider the entire universe. We simply want to consider this reaction. In such instances, cell biologists use the concept of free energy instead of entropy. **Free energy** is the amount of energy available—that is, energy that is still "free" to do work—after a chemical reaction has occurred. Free energy is denoted by the symbol G after Josiah Gibbs, who first developed the concept. The change (Δ) in free energy (G) after a reaction occurs (ΔG) is calculated by subtracting the free energy content of the reactants from that of the products. A negative ΔG means that the products have less free energy than the reactants, and the reaction will occur spontaneously. In our reaction, if C and D have less free energy than A and B, then the reaction will "go."

Exergonic reactions are ones in which ΔG is negative and energy is released, while **endergonic reactions** are ones in which ΔG is positive and the products have more free energy than the reactants. Endergonic reactions can only occur if there is an input of energy. In the body, many reactions, such as protein synthesis, nerve conduction, or muscle contraction, are endergonic, and they occur because the energy released by exergonic reactions is used to drive endergonic reactions. ATP is a carrier of energy between exergonic and endergonic reactions.

ATP: Energy for Cells

ATP (adenosine triphosphate) is the common energy currency of cells; when cells require energy, they "spend" ATP. A sedentary oak tree as well as a flying bat requires vast amounts of ATP. The more active the organism, the greater the demand for ATP. However, the amount on hand at any one moment is minimal because ATP is constantly being generated from **ADP (adenosine diphosphate)** and a molecule of inorganic phosphate $\textcircled{P}$ (Fig. 6.3). A cell is assured of a supply of ATP, because glucose breakdown during cellular respiration provides the energy for the buildup of ATP in mitochondria. Only 39% of the free energy of glucose is transformed to ATP; the rest is lost as heat.

There are many biological advantages to the use of ATP as an energy carrier in living systems. ATP provides a common and universal energy currency because it can be used in many different types of reactions. Also, when ATP is converted to energy, ADP, and $\textcircled{P}$, the amount of energy released is sufficient for a particular biological function, and there is little waste of energy. In addition, ATP breakdown can be coupled to endergonic reactions in such a way that it minimizes energy loss.

Structure of ATP

ATP is a nucleotide composed of the nitrogen-containing base adenine and the 5-carbon sugar ribose (together called adeno-

adenosine triphosphate

Energy from exergonic reactions (e.g., cellular respiration)

ATP

Energy for endergonic reactions (e.g., protein synthesis, nerve conduction, muscle contraction)

ADP + P

P — P + P

adenosine diphosphate   +   phosphate

a.

FIGURE 6.3 The ATP cycle.
a. In cells, ATP carries energy between exergonic reactions and endergonic reactions. When a phosphate group is removed by hydrolysis, ATP releases the appropriate amount of energy for most metabolic reactions. **b.** In order to produce light, a firefly breaks down ATP.

b.   2.25×

sine) and three phosphate groups. ATP is called a "high-energy" compound because a phosphate group can be easily removed. Under cellular conditions, the amount of energy released when ATP is hydrolyzed to ADP + $\textcircled{P}$ is about 7.3 kcal per mole.[1]

Coupled Reactions

In **coupled reactions,** the energy released by an exergonic reaction is used to drive an endergonic reaction. ATP breakdown is often coupled to cellular reactions that require an input of energy. Coupling, which requires that the exergonic reaction and the endergonic reaction be closely tied, can be symbolized like this:

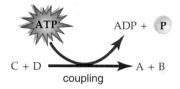

Notice that the word *energy* does not appear following ATP breakdown. Why not? Because this energy was used to drive forward the coupled reaction. Figure 6.4 tells us that ATP breakdown provides the energy necessary for muscular contraction to occur. During muscle contraction, myosin filaments pull actin filaments to the center of the muscle cells, and a muscle shortens. The coupling of energy from ATP breakdown

to muscle contraction occurs like this. First, myosin combines with both ATP and an actin filament. Then ATP breaks down to ADP + $\textcircled{P}$. The release of ADP + $\textcircled{P}$ from myosin causes it to change shape and pull on the actin filament. The transfer of energy is not complete, and some energy is lost as heat.

Function of ATP

In living systems, three obvious uses of ATP are:

Chemical work ATP supplies the energy needed to synthesize macromolecules that make up the cell, and therefore the organism.

Transport work ATP supplies the energy needed to pump substances across the plasma membrane.

Mechanical work ATP supplies the energy needed to permit muscles to contract, cilia and flagella to beat, chromosomes to move, and so forth.

In most cases, ATP is the immediate source of energy for these processes.

ATP is a carrier of energy in cells. It is the common energy currency because it supplies energy for many different types of reactions.

---

[1] A mole is the number of molecules present in the molecular weight of a substance (in grams).

FIGURE 6.4 Coupled reactions.
Muscle contraction occurs only when it is coupled to ATP breakdown.

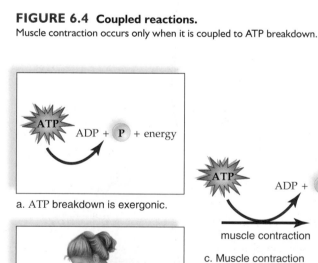

a. ATP breakdown is exergonic.

b. Muscle contraction is endergonic and cannot occur without an input of energy.

c. Muscle contraction becomes exergonic and can occur when it is coupled to ATP breakdown.

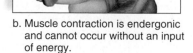

6.3 METABOLIC PATHWAYS AND ENZYMES

Reactions do not occur haphazardly in cells; they are usually part of a **metabolic pathway,** a series of linked reactions. Metabolic pathways begin with a particular reactant and terminate with an end product. While it is possible to write an overall equation for a pathway as if the beginning reactant went to the end product in one step, actually many specific steps occur in between. In the pathway, one reaction leads to the next reaction, which leads to the next reaction, and so forth in an organized, highly structured manner. This arrangement makes it possible for one pathway to lead to several others, because various pathways have several molecules in common. Also, metabolic energy is captured and used more easily if it is released in small increments rather than all at once.

A metabolic pathway can be represented by the following diagram:

$$A \xrightarrow{E_1} B \xrightarrow{E_2} C \xrightarrow{E_3} D \xrightarrow{E_4} E \xrightarrow{E_5} F \xrightarrow{E_6} G$$

In this diagram, the letters A–F are reactants and the letters B–G are products in the various reactions. In other words, the products from the previous reaction become the reactants of the next reaction. The letters E_1–E_6 are enzymes.

An **enzyme** is a protein molecule that functions as an organic catalyst to speed a chemical reaction without itself being affected by the reaction. Enzymes can only speed reactions that are possible. In the cell, an enzyme brings together particular molecules and causes them to react with one another. Ribozymes, which are made of RNA instead of proteins, can also serve as biological catalysts. Ribozymes are involved in the synthesis of RNA and the synthesis of proteins at the ribosomes.

The reactants in an enzymatic reaction are called the **substrates** for that enzyme. In the first reaction, A is the substrate for E_1, and B is the product. Now B becomes the substrate for E_2, and C is the product. This process continues until the final product G forms.

Any one of the molecules (A–G) in this linear pathway could also be a substrate for an enzyme in another pathway. A diagram showing all the possibilities would be highly branched.

Energy of Activation

Molecules frequently do not react with one another unless they are activated in some way. In the lab, for example, in the absence of an enzyme, activation is very often achieved by heating a reaction flask to increase the number of effective collisions between molecules. The energy that must be added to cause molecules to react with one another is called the **energy of activation** (E_a). Figure 6.5 compares E_a when an enzyme is not present to when an enzyme is present, illustrating that enzymes lower the amount of energy required for activation to occur. Nevertheless, the addition of the enzyme does not change ΔG of the reaction.

Enzymes lower the energy of activation by bringing the substrates into contact with one another and even by participating in the reaction at times.

Enzyme-Substrate Complex

The following equation, which is pictorially shown in Figure 6.6, is often used to indicate that an enzyme forms a complex with its substrate:

$$\underset{\text{enzyme}}{E} + \underset{\text{substrate}}{S} \longrightarrow \underset{\substack{\text{enzyme-substrate} \\ \text{complex}}}{ES} \longrightarrow \underset{\text{enzyme}}{E} + \underset{\text{product}}{P}$$

In most instances, only one small part of the enzyme, called the **active site,** binds with the substrate(s). It is here that the enzyme and substrate fit together, seemingly like a key fits a lock; however, it is now known that the active site undergoes a slight change in shape to accommodate the substrate(s). This is called the **induced fit model** because the enzyme is induced to undergo a slight alteration to achieve optimum fit (Fig. 6.7).

The change in shape of the active site facilitates the reaction that now occurs. After the reaction has been completed, the product(s) is released, and the active site returns to its original state, ready to bind to another substrate molecule. Only a small amount of enzyme is actually needed in a cell because enzymes are not used up by the reaction.

Some enzymes do more than simply complex with their substrate(s); they participate in the reaction. Trypsin digests protein by breaking peptide bonds. The active site of trypsin contains three amino acids with R groups that

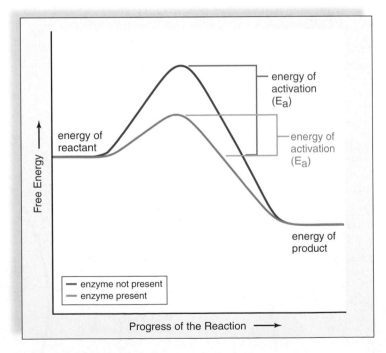

FIGURE 6.5 Energy of activation (E_a).
Enzymes speed the rate of reactions because they lower the amount of energy required for the reactants to react. Even reactions like this one, in which the energy of the product is less than the energy of the reactant (ΔG is negative), speed up when an enzyme is present.

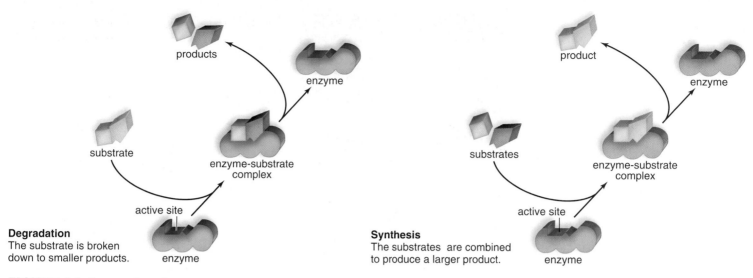

FIGURE 6.6 Enzymatic action.
An enzyme has an active site where the substrates and enzyme fit together in such a way that the substrates react. Following the reaction, the products are released, and the enzyme is free to act again.

actually interact with members of the peptide bond—first to break the bond and then to introduce the components of water. This illustrates that the formation of the enzyme-substrate complex is very important in speeding the reaction.

Sometimes it is possible for a particular reactant(s) to produce more than one type of product(s). The presence or absence of an enzyme determines which reaction takes place. If a substance can react to form more than one product, then the enzyme that is present and active determines which product is produced.

Every reaction in a metabolic pathway requires its specific enzyme. Because enzymes only complex with their substrates, they are sometimes named for their substrates, and usually end in *ase* as shown in Table 6.1. However, some enzymes, such as lysozyme, trypsin, and pepsin, have retained their traditional names.

Enzymes are protein molecules that speed chemical reactions by lowering the energy of activation. They do this by forming an enzyme-substrate complex.

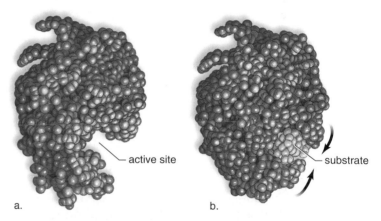

FIGURE 6.7 Induced fit model.
These computer-generated images show an enzyme called lysozyme that hydrolyzes its substrate, a polysaccharide that makes up bacterial cell walls. **a.** Configuration of enzyme when no substrate is bound to it. **b.** After the substrate binds, the configuration of the enzyme changes so that hydrolysis can better proceed.

Factors Affecting Enzymatic Speed

Generally, enzymes work quickly, and in some instances they can increase the reaction rate more than 10 million times. The rate of a reaction is the amount of product produced per unit time. To achieve maximum product per unit time, there should be enough substrate to fill active sites most of the time. Increasing the amount of substrate and providing an adequate temperature and optimal pH also increase the rate of an enzymatic reaction.

Substrate Concentration

Molecules must collide to react. Generally, enzyme activity increases as substrate concentration increases because there are more collisions between substrate molecules and the enzyme. As more substrate molecules fill active sites, more product

TABLE 6.1

Enzymes Named for Their Substrate

Substrate	Enzyme
Lipid	Lipase
Urea	Urease
Maltose	Maltase
Ribonucleic acid	Ribonuclease
Lactose	Lactase
Sucrose	Sucrase

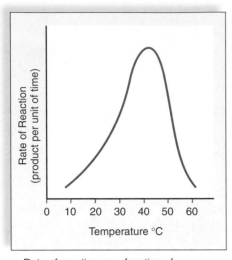

a. Rate of reaction as a function of temperature.

b. Body temperature of ectothermic animals often limits rates of reactions.

c. Body temperature of endothermic animals promotes rates of reactions.

FIGURE 6.8 The effect of temperature on rate of reaction.
a. Usually, the rate of an enzymatic reaction doubles with every 10°C rise in temperature. This enzymatic reaction is maximum at about 40°C; then it decreases until the reaction stops altogether, because the enzyme has become denatured. **b.** The body temperature of ectothermic animals, which take on the temperature of their environment, often limits rates of reactions. **c.** The body temperature of endothermic animals promotes rates of reaction.

results per unit time. But when the enzyme's active sites are filled almost continuously with substrate, the enzyme's rate of activity cannot increase any more. Maximum rate has been reached.

Temperature and pH

Typically, as temperature rises, enzyme activity increases (Fig. 6.8 a). This occurs because warmer temperatures cause more effective collisions between enzyme and substrate. The body temperature of an animal seems to affect whether it is normally active or inactive (Fig. 6.8 b,c). It has been suggested that mammals are more prevalent today than reptiles because they maintain a warm internal temperature that allows their enzymes to work at a rapid rate. In the laboratory, if the temperature rises beyond a certain point, enzyme activity eventually levels out and then declines rapidly because the enzyme is **denatured**. An enzyme's shape changes during denaturation, and then it can no longer bind its substrate(s) efficiently. Many organisms cannot survive in extremely hot temperatures. Some prokaryotes that live in hot springs are an exception to the temperature barrier. These organisms thrive in the hot water and are responsible for the brilliant colors of the hot springs.

Another exception involves the coat color of animals. Siamese cats have inherited a mutation that causes an enzyme to be active only at cooler body temperatures! Therefore, only cooler regions of the body—the face, ears, legs, and tail—are dark in color (Fig. 6.9). The coat color pattern in several other animals can be explained similarly.

Each enzyme also has an optimal pH at which the rate of the reaction is highest. Figure 6.10 shows the optimal pH for the enzymes pepsin and trypsin. At this pH value, these enzymes have their normal configurations. The globular structure of an enzyme is dependent on interactions, such as hydrogen bonding, between R groups. A change in pH can alter the ionization of these side chains and disrupt normal interactions, and under extreme conditions of pH, denaturation eventually occurs. Again, the enzyme has an altered shape and is then unable to combine efficiently with its substrate.

Enzyme Concentration

Just as the amount of substrate can limit the rate of an enzymatic reaction, so the amount of active enzyme can also limit the rate of an enzymatic reaction. Cells regulate which enzymes are present and/or active at any one time. Gene expression is the first way to regulate the amount of enzyme present, and cells have various ways to control gene activity (see Chapter 15). A few metabolic considerations are discussed here.

Enzyme Cofactors. Many enzymes require the presence of an inorganic ion or nonprotein organic molecule in order to be active; these necessary ions or molecules are called **cofactors.** The inorganic ions are metals such as copper, zinc, or iron. The nonprotein organic molecules are called **coenzymes.** These cofactors assist the enzyme and may

FIGURE 6.9

The effect of temperature on enzymes.

Siamese cats have inherited a mutation that causes an enzyme to be active only at cooler body temperatures. Therefore, only certain regions of the body are dark in color.

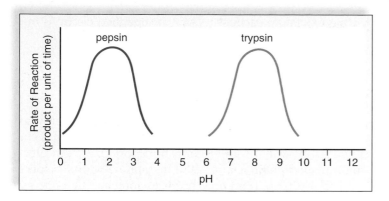

FIGURE 6.10 The effect of pH on rate of reaction.
The preferred pH for pepsin, an enzyme that acts in the stomach, is about 2, while the preferred pH for trypsin, an enzyme that acts in the small intestine, is about 8. The preferred pH of an enzyme maintains its shape so that it can bind with its substrates.

even accept or contribute atoms to the reactions. In the next section, we will discuss two coenzymes that play significant roles in photosynthesis and cellular respiration, respectively.

Vitamins are relatively small organic molecules that are required in trace amounts in our diet and in the diets of other animals for synthesis of coenzymes. The vitamin becomes part of a coenzyme's molecular structure. If a vitamin is not available, enzymatic activity will decrease, and the result will be a vitamin-deficiency disorder: Niacin deficiency results in a skin disease called pellagra, and riboflavin deficiency results in cracks at the corners of the mouth.

Phosphorylation is one way to activate an enzyme. Signal molecules received by membrane receptors often turn on enzymes known as kinases, which then activate enzymes by phosphorylating them. Some hormones act in this manner:

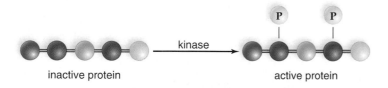

Enzyme Inhibition. It is imperative that the production of substances in biochemical pathways be coordinated and regulated. Once sufficient product is present, it may be necessary to inhibit further production to conserve raw materials and energy. **Enzyme inhibition** occurs when a substance known as an inhibitor binds to an enzyme and decreases its activity. Normally, enzyme inhibition is *reversible,* and the enzyme is not damaged by being inhibited, as in competitive and noncompetitive inhibition. In **competitive inhibition,** the substrate and the inhibitor are both able to bind to the enzyme's active site, and each spends time complexing with the enzyme. Only when the substrate, not the inhibitor, is at the active site will product form. In this way, the amount of product is regulated.

In **noncompetitive inhibition,** the inhibitor binds to an enzyme at a location other than the active site. The site is called an *allosteric site* [Gk. *allo,* other; *steric,* space, structure]. When the inhibitor is at the allosteric site, the shape of an enzyme changes, it is unable to bind its substrate, and no product forms. Both competitive inhibition and noncompetitive inhibition are also examples of **feedback inhibition** in which the final product of a pathway can inhibit an earlier reaction in a sequence of reactions (Fig. 6.11).

When enzyme inhibition is *irreversible,* the inhibitor permanently inactivates or destroys an enzyme. Many poisons are irreversible enzyme inhibitors. Cyanide is an inhibitor for an essential enzyme (cytochrome *c* oxidase) in all cells, which accounts for its lethal effect on humans. Penicillin blocks the active site of an enzyme unique to bacteria. Therefore, penicillin is a poison for bacteria. When penicillin is administered, bacteria die, but humans are unaffected. Mercury and lead poisoning result from these heavy metals, irreversibly destroying enzymes. Certain nerve gases that could possibly be used by terrorists interfere with the enzyme acetylcholinesterase, which is essential to proper nerve and muscle function.

Enzymes speed reactions by forming a complex with their substrates. Various factors affect enzyme speed, including substrate and enzyme concentrations, the temperature and pH of the medium, and the presence or absence of cofactors and inhibitors.

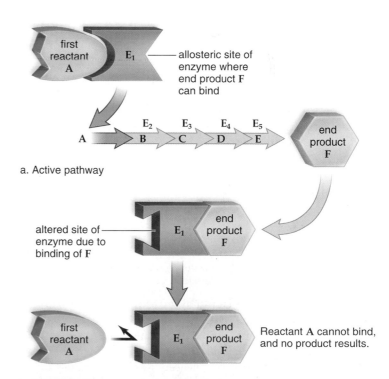

FIGURE 6.11 Feedback inhibition.
In feedback inhibition, an end product of a pathway inhibits an earlier reaction.
a. First reactant (substrate) can bind to its enzyme and the pathway is active.
b. Noncompetitive inhibition occurs when the end product of a metabolic pathway binds to the allosteric site of an enzyme. The enzyme changes shape and is unable to bind its substrate.

6.4 OXIDATION-REDUCTION AND THE FLOW OF ENERGY

In oxidation-reduction (redox) reactions, electrons pass from one molecule to another. **Oxidation** is the loss of electrons and **reduction** is the gain of electrons. Oxidation and reduction always take place at the same time because one molecule accepts the electrons given up by another molecule. Oxidation-reduction reactions occur during photosynthesis and cellular respiration.

Photosynthesis

In living things, hydrogen ions often accompany electrons, and oxidation is a loss of hydrogen atoms, while reduction is a gain of hydrogen atoms. For example, the overall reaction for photosynthesis can be written like this:

$$\underset{\substack{\text{carbon}\\\text{dioxide}}}{6\,CO_2} + \underset{\text{water}}{6\,H_2O} + \text{energy} \longrightarrow \underset{\text{glucose}}{C_6H_{12}O_6} + \underset{\text{oxygen}}{6\,O_2}$$

This equation shows that when hydrogen atoms are transferred from water to carbon dioxide, glucose is formed. Water has been oxidized and carbon dioxide has been reduced. Since glucose is a high-energy molecule, an input of energy is needed to make the reaction go. Chloroplasts are able to capture solar energy and convert it by way of an electron transport chain (discussed next) to the chemical energy of ATP molecules. ATP is then used along with hydrogen atoms to reduce glucose.

A coenzyme of oxidation-reduction called **NADP$^+$ (nicotinamide adenine dinucleotide phosphate)** is active during photosynthesis. This molecule carries a positive charge, and therefore is written as NADP$^+$. During photosynthesis, NADP$^+$ accepts electrons and a hydrogen ion derived from water and later passes them by way of a metabolic pathway to carbon dioxide, forming glucose. The reaction that reduces NADP$^+$ is:

$$NADP^+ + 2\,e^- + H^+ \longrightarrow NADPH$$

Cellular Respiration

The overall equation for cellular respiration is opposite that of photosynthesis:

$$\underset{\text{glucose}}{C_6H_{12}O_6} + \underset{\text{oxygen}}{6\,O_2} \longrightarrow \underset{\substack{\text{carbon}\\\text{dioxide}}}{6\,CO_2} + \underset{\text{water}}{6\,H_2O} + \text{energy}$$

In this reaction, glucose has lost hydrogen atoms (been oxidized), and oxygen has gained hydrogen atoms (been reduced). When oxygen gains hydrogen atoms, it becomes water. Glucose is a high-energy molecule, while carbon dioxide and water are low-energy molecules; energy is released. Mitochondria use the energy released from glucose breakdown to build ATP molecules by way of an electron transport chain that passes electrons to oxygen. Oxygen then becomes water.

In metabolic pathways, most oxidations such as those that occur during cellular respiration involve a coenzyme called **NAD$^+$ (nicotinamide adenine dinucleotide).** This molecule carries a positive charge, and therefore it is represented as NAD$^+$.

During oxidation reactions, NAD$^+$ accepts two electrons but only one hydrogen ion. The reaction that reduces NAD$^+$ is:

$$NAD^+ + 2\,e^- + H^+ \longrightarrow NADH$$

Electron Transport Chain

As previously mentioned, chloroplasts use solar energy to generate ATP, and mitochondria use glucose energy to generate ATP by way of an electron transport chain. An **electron transport chain** is a series of membrane-bound carriers that pass electrons from one carrier to another. High-energy electrons are delivered to the chain, and low-energy electrons leave it. Every time electrons are transferred to a new carrier, energy is released; this energy is ultimately used to produce ATP molecules (Fig. 6.12).

In certain redox reactions, the result is release of energy, and in others, energy is required. In an electron transport chain, each carrier is reduced and then oxidized in turn. The overall effect of oxidation-reduction as electrons are passed from carrier to carrier of the electron transport chain is the release of energy for ATP production.

ATP Production

For many years, it was known that ATP synthesis was somehow coupled to the electron transport chain, but the exact mechanism could not be determined. Peter Mitchell, a British

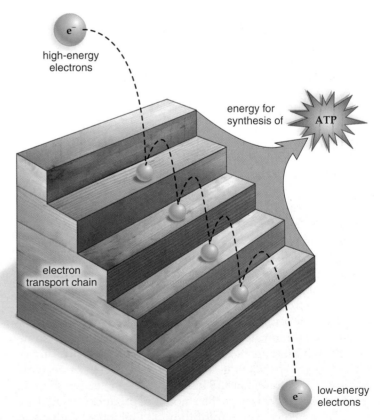

FIGURE 6.12 Electron transport chain.
High-energy electrons are delivered to the chain and, with each step as they pass from carrier to carrier, energy is released and used for ATP production.

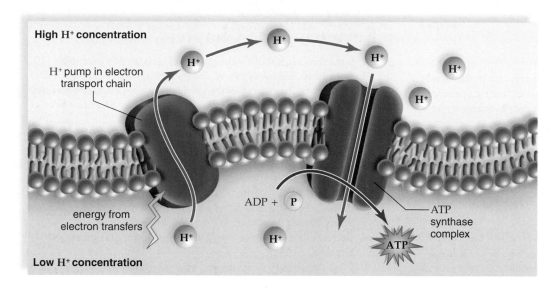

High H+ concentration

H+ pump in electron transport chain

energy from electron transfers

ADP + P

ATP synthase complex

ATP

Low H+ concentration

FIGURE 6.13 Chemiosmosis.
Carriers in the electron transport chain pump hydrogen ions (H+) across a membrane. When the hydrogen ions flow back across the membrane through an ATP synthase complex, ATP is synthesized by an enzyme called ATP synthase. Chemiosmosis occurs in mitochondria and chloroplasts.

biochemist, received a Nobel Prize in 1978 for his theory of ATP production in both mitochondria and chloroplasts.

In mitochondria and chloroplasts, the carriers of the electron transport chain are located within a membrane: thylakoid membranes in chloroplasts and cristae in mitochondria. Hydrogen ions (H+), which are often referred to as protons in this context, tend to collect on one side of the membrane because they are pumped there by certain carriers of the electron transport chain. This establishes an electrochemical gradient across the membrane that can be used to provide energy for ATP production. Enzymes and their carrier proteins, called **ATP synthase complexes,** span the membrane. Each complex contains a channel that allows hydrogen ions to flow down their electrochemical gradient. The flow of hydrogen ions through the channel provides the energy for the ATP synthase enzyme to produce ATP from ADP + P (Fig. 6.13). The production of ATP due to a hydrogen ion gradient across a membrane is called **chemiosmosis** [Gk. *osmos*, push].

Consider this analogy to understand chemiosmosis. The sun's rays evaporate water from the seas and help create the winds that blow clouds to the mountains, where water falls in the form of rain and snow. The water in a mountain reservoir has a higher potential energy than water in the ocean. The po-

tential energy is converted to electrical energy when water is released and used to turn turbines in an electrochemical dam before it makes its way to the ocean. The continual release of water results in a continual production of electricity.

Similarly, during photosynthesis, solar energy collected by chloroplasts continually leads to ATP production. Energized electrons lead to the pumping of hydrogen ions across a thylakoid membrane, which acts like a dam to retain them. The hydrogen ions flow through the channel of an ATP synthase complex. This complex couples the flow of hydrogen ions to the formation of ATP, as the turbines in a hydroelectric dam system couple the flow of water to the formation of electricity.

Similarly, during cellular respiration, glucose breakdown provides the energy to establish a hydrogen ion gradient across the cristae of mitochondria. And again, hydrogen ions flow through the channel within an ATP synthase complex that couples the flow of hydrogen ions to the formation of ATP.

The oxidation-reduction pathways of photosynthesis in chloroplasts and cellular respiration in mitochondria permit a flow of energy from the sun through all living things.

CONNECTING THE CONCEPTS

All cells use energy. Energy is the ability to do work, to bring about change, and to make things happen, whether it's a leaf growing or a human running. The metabolic pathways inside cells use the chemical energy of ATP to synthesize molecules, cause muscle contraction, and even allow you to read these words.

A metabolic pathway consists of a series of individual chemical reactions, each with its own enzyme. The cell can regulate the activity of the many hundreds of different enzymes taking part in cellular metabolism. Enzymes are proteins, and as such they are sensitive to

environmental conditions, including pH, temperature, and even certain pollutants, as will be discussed in later chapters.

ATP is called the universal energy "currency" of life. This is an apt analogy—before we can spend currency (e.g., money), we must first make some money. Similarly, before the cell can spend ATP molecules, it must make them. Cellular respiration in mitochondria transforms the chemical energy of carbohydrates into that of ATP molecules. ATP is spent when it is hydrolyzed, and the resulting energy is coupled to an endergonic reaction. All cells are continually making and

breaking down ATP. If ATP is lacking, the organism dies.

What is the ultimate source of energy for ATP production? In Chapter 7, we will see that, except for a few deep ocean vents and certain cave communities, the answer is the sun. Photosynthesis inside chloroplasts transforms solar energy into the chemical energy of carbohydrates. And then in Chapter 8 we will discuss how carbohydrate products are broken down in mitochondria as ATP is built up. Chloroplasts and mitochondria are the cellular organelles that permit a flow of energy from the sun through all living things.

Summary

6.1 CELLS AND THE FLOW OF ENERGY

Two energy laws are basic to understanding energy-use patterns at all levels of biological organization. The first law of thermodynamics states that energy cannot be created or destroyed, but can only be transferred or transformed. The second law of thermodynamics states that one usable form of energy cannot be completely converted into another usable form. As a result of these laws, we know that the entropy of the universe is increasing and that only a constant input of energy maintains the organization of living things.

6.2 METABOLIC REACTIONS AND ENERGY TRANSFORMATIONS

The term *metabolism* encompasses all the chemical reactions occurring in a cell. Considering individual reactions, only those that result in a negative free-energy difference—that is, the products have less usable energy than the reactants—occur spontaneously. Such reactions, called exergonic reactions, release energy. Endergonic reactions, which require an input of energy, occur only in cells because it is possible to couple an exergonic process with an endergonic process. For example, glucose breakdown is an exergonic metabolic pathway that drives the buildup of many ATP molecules. These ATP molecules then supply energy for cellular work. Thus, ATP goes through a cycle in which it is constantly being built up from, and then broken down to, ADP + $\textcircled{P}$.

6.3 METABOLIC PATHWAYS AND ENZYMES

A metabolic pathway is a series of reactions that proceed in an orderly, step-by-step manner. Each reaction requires a specific enzyme. Enzymes speed reactions by lowering the energy of activation when they form a complex with their substrates. Generally, enzyme activity increases as substrate concentration increases; once all active sites are filled, maximum rate has been achieved. Any environmental factor, such as temperature or pH, affects the shape of a protein and therefore also affects the ability of an enzyme to do its job.

Gene expression regulates how much of an enzyme is present. The activity of enzymes is controlled in various ways. Many enzymes need cofactors or coenzymes to carry out their reactions. The activity of most metabolic pathways is regulated by feedback inhibition.

6.4 OXIDATION-REDUCTION AND THE FLOW OF ENERGY

Photosynthesis is a metabolic pathway in chloroplasts that transforms solar energy to the chemical energy within carbohydrates (e.g., glucose). Cellular respiration is a metabolic pathway completed in mitochondria that transforms this energy into that of ATP molecules.

The overall equation for photosynthesis is the opposite of that for cellular respiration. The coenzyme $NADP^+$ is active during photosynthesis, while NAD^+ is active during cellular respiration. During photosynthesis, NADPH reduces substrates, while during cellular respiration, NAD^+ oxidizes substrates. Redox reactions are a major way in which energy transformation occurs in cells.

Both processes make use of an electron transport chain in which electrons are transferred from one carrier to the next with the release of energy that is ultimately used to produce ATP molecules. Chemiosmosis explains how the electron transport chain produces ATP. The carriers of this system deposit hydrogen ions (H^+) on one side of a membrane. When the ions flow down an electrochemical gradient through an ATP synthase complex, an enzyme uses the release of energy to make ATP from ADP and $\textcircled{P}$.

Reviewing the Chapter

1. State the first law of thermodynamics, and give an example.   102
2. State the second law of thermodynamics, and give an example.   102–3
3. Explain why the entropy of the universe is always increasing and why an organized system such as an organism requires a constant input of useful energy.   103
4. What is the difference between exergonic reactions and endergonic reactions? Why can exergonic but not endergonic reactions occur spontaneously?   104
5. Why is ATP called the energy currency of cells? What is the ATP cycle?   104
6. Define coupling, and write an equation that shows an endergonic reaction being coupled to ATP breakdown.   105
7. Diagram a metabolic pathway. Label the reactants, products, and enzymes.   106
8. Why is less energy needed for a reaction to occur when an enzyme is present?   106
9. Why are enzymes specific, and why can't each one speed many different reactions?   106–7
10. Name and explain the manner in which at least three factors can influence the speed of an enzymatic reaction. How do cells regulate the activity of enzymes?   107–9
11. What are cofactors and coenzymes?   108
12. Compare and contrast competitive and noncompetitive inhibition.   109
13. Compare and contrast reversible and irreversible inhibition.   109
14. Describe how oxidation-reduction occurs in cells, and discuss the overall equations for photosynthesis and cellular respiration in terms of oxidation-reduction.   110
15. Describe an electron transport chain.   110
16. Tell how cells form ATP during chemiosmosis.   110–11

Testing Yourself

Choose the best answer for each question.

1. A form of potential energy is
   a. a boulder at the top of a hill.
   b. the bonds of a glucose molecule.
   c. a starch molecule.
   d. stored fat tissue.
   e. All of these are correct.

2. A lit lightbulb can be used as an example of
   a. the creation of heat energy.
   b. the second law of thermodynamics.
   c. the conversion of electrical energy into heat energy.
   d. the first law of thermodynamics.
   e. All of the above except (a) are correct.

3. Consider this reaction: A + B $\longrightarrow$ C + D + energy.
   a. This reaction is exergonic.
   b. An enzyme could still speed the reaction.
   c. ATP is not needed to make the reaction go.
   d. A and B are reactants; C and D are products.
   e. All of these are correct.

4. The active site of an enzyme
   a. is similar to that of any other enzyme.
   b. is the part of the enzyme where its substrate can fit.
   c. can be used over and over again.
   d. is not affected by environmental factors, such as pH and temperature.
   e. Both b and c are correct.

5. If you want to increase the amount of product per unit time of an enzymatic reaction, do not increase
   a. the amount of substrate.
   b. the amount of enzyme.
   c. the temperature somewhat.
   d. the pH.
   e. All of these are correct.

6. An allosteric site on an enzyme is
   a. the same as the active site.
   b. nonprotein in nature.
   c. where ATP attaches and gives up its energy.
   d. often involved in feedback inhibition.
   e. All of these are correct.

7. During photosynthesis, carbon dioxide
   a. is oxidized to oxygen.
   b. is reduced to glucose.
   c. gives up water to the environment.
   d. is a coenzyme of oxidation-reduction.
   e. All of these are correct.

8. The difference between NAD^+ and $NADP^+$ is that
   a. only NAD^+ production requires niacin in the diet.
   b. one is an organic molecule, and the other is inorganic because it contains phosphate.
   c. one carries electrons to the electron transport chain, and the other carries them to synthetic reactions.
   d. one is involved in cellular respiration, and the other is involved in photosynthesis.
   e. Both c and d are correct.

9. Chemiosmosis is dependent on
   a. the diffusion of water across a differentially permeable membrane.
   b. an outside supply of phosphate and other chemicals.
   c. the establishment of an electrochemical hydrogen ion (H^+) gradient.
   d. the ability of ADP to join with Ⓟ even in the absence of a supply of energy.
   e. All of these are correct.

10. Use these terms to label the following diagram: substrates, enzyme (used twice), active site, product, and enzyme-substrate complex. Explain the importance of an enzyme's shape to its activity.

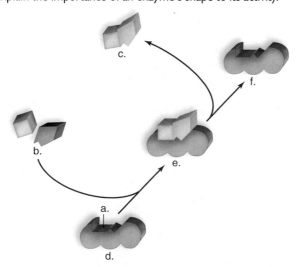

11. Electron transport chains
    a. are found in both mitochondria and chloroplasts.
    b. release energy as electrons are transferred.
    c. are involved in the production of ATP.
    d. are located in a membrane.
    e. All of these are correct.

12. Label this diagram describing chemiosmosis.

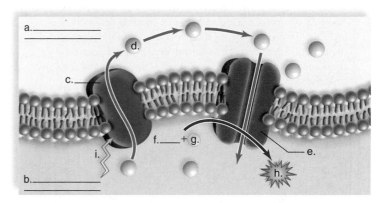

13. NAD^+ is the _____ form, and when it later becomes NADH, it is said to be _____.
    a. reduced, oxidized
    b. neutral, a coenzyme
    c. oxidized, reduced
    d. active, denatured

14. Coenzymes
    a. have specific functions in reactions.
    b. have an active site just as enzymes do.
    c. can be carriers for proteins.
    d. always have a phosphate group.
    e. are used in photosynthesis, but not in cellular respiration.

For questions 15–20, match each pair to a description in the key. Choose more than one answer if correct.

KEY:

    a. first includes the other
    b. first breaks down to the other
    c. have nothing to do with each other
    d. work together

15. metabolic pathway, enzyme

16. allosteric site, reduction

17. kinetic energy, mechanical energy

18. ATP, ADP + Ⓟ

19. enzyme, coenzyme

20. chemiosmosis, electron transport chain

21. Oxidation
    a. is the opposite of reduction.
    b. sometimes uses NAD^+.
    c. is involved in cellular respiration.
    d. occurs when ATP goes to ADP + Ⓟ.
    e. All of these but d are correct.

Thinking Scientifically

1. A certain flower generates heat. This heat attracts pollinating insects to the flower. While the evolutionary benefit of attracting insects is obvious, the metabolic cost of this particular adaptation is high. What metabolic mechanism(s) might a plant use to generate heat, and under what circumstances would the metabolic cost be high?

2. The free energy of carbon dioxide and water is considerably less than the free energy of sucrose (table sugar). However, the conversion of sucrose to carbon dioxide and water is never spontaneous under normal conditions. How would you explain this observation?

Bioethical Issue: Greenhouse Effect and Emerging Diseases

Today, we are very much concerned about emerging diseases caused by pathogens. Examples of emerging diseases are AIDS and Ebola, which emerge from their natural hosts to cause illness in humans. In 1993, the hantavirus strain emerged from the common deer mouse and killed about 60 young people in the Southwest. In the case of hantavirus, we know that climate was involved. An unusually mild winter and wet spring caused piñon trees to bloom well and provide pine nuts to the mice. The increasing deer mouse population came into contact with humans, and the hantavirus leaped easily from mice to humans.

The prediction is that global warming, caused in large part by the burning of fossil fuels, will upset normal weather cycles and result in outbreaks of hantavirus as well as malaria, dengue and yellow fevers, filariasis, encephalitis, schistosomiasis, and cholera. Clearly, any connection between global warming and emerging diseases offers another reason that greenhouse gases should be curtailed when fossil fuels such as gasoline are consumed. Examples of greenhouse gases are carbon dioxide and methane, which allow the sun's rays to pass through but then trap the heat from escaping.

In December 1997, 159 countries met in Kyoto, Japan, to work out a protocol that would reduce greenhouse gases worldwide. This protocol, called the Kyoto Protocol, entered into force February 16, 2005. It is believed that the emission of greenhouse gases, especially from power plants, will cause Earth's temperature to rise 1.5°–4.5° by 2060. The U.S. Senate still does not want to ratify the agreement because it does not include a binding emissions commitment from the developing countries, which are only now becoming industrialized. While the United States presently emits a large proportion of the greenhouse gases, China is expected to surpass that amount in about 2020 to become the biggest source of greenhouse emissions.

Negotiations with the developing countries are still going on, and some creative ideas have been put forward. Why not have a trading program that allows companies to buy and sell emission credits across international boundaries? Accompanying that would be a market in greenhouse reduction techniques. If it became monetarily worth their while, companies in developed countries would have an incentive to reduce greenhouse emissions. If you were a CEO, would you be willing to reduce greenhouse emissions simply because they cause a deterioration of the environment and probably cause human illness? Why or why not? Instead, would you approve of giving companies monetary incentives to reduce greenhouse emissions? Why or why not?

Understanding the Terms

active site 106	free energy 104
ADP (adenosine diphosphate) 104	heat 102
	induced fit model 106
ATP (adenosine triphosphate) 104	kinetic energy 102
	laws of thermodynamics 102
ATP synthase complex 111	mechanical energy 102
chemical energy 102	metabolic pathway 106
chemiosmosis 111	metabolism 104
coenzyme 108	NAD^+ (nicotinamide adenine dinucleotide) 110
cofactor 108	
competitive inhibition 109	$NADP^+$ (nicotinamide adenine dinucleotide phosphate) 110
coupled reactions 105	
denatured 108	noncompetitive inhibition 109
electron transport chain 110	oxidation 110
endergonic reaction 104	phosphorylation 109
energy 102	potential energy 102
energy of activation 106	product 104
entropy 103	reactant 104
enzyme 106	reduction 110
enzyme inhibition 109	substrate 106
exergonic reaction 104	vitamin 109
feedback inhibition 109	

Match the terms to these definitions:

a. _____ All of the chemical reactions that occur in a cell during growth and repair.

b. _____ Stored energy as a result of location or spatial arrangement.

c. _____ Essential requirement in the diet, needed in small amounts. They are often part of coenzymes.

d. _____ Measure of disorder or randomness.

e. _____ Nonprotein organic molecule that aids the action of the enzyme to which it is loosely bound.

f. _____ Loss of one or more electrons from an atom or molecule; in biological systems, generally the loss of hydrogen atoms.

ARIS, the *Biology* Website

ARIS, the website for *Biology*, provides a wealth of information organized and integrated by chapter. You will find practice quizzes, interactive activities, labeling exercises, flashcards, and much more that will complement your learning and understanding of general biology.

www.mhhe.com/maderbiology9

7

PHOTOSYNTHESIS

T he fate of life on Earth literally hinges on a star 93 million miles away because this star provides photosynthesizers with solar energy. Only 42% of the solar energy directed towards Earth reaches the planet; the remainder is absorbed by or reflected into the atmosphere and becomes heat. Of this, only 1–2% is captured by photosynthesizers and, only a portion of this is incorporated into plant materials.

Yet all living things are dependent on the amount of solar energy that photosynthesizers transform into chemical energy. Exceptions do exist. In rare hydrothermal environments, some prokaryotes acquire energy by oxidizing inorganic molecules and are the producers of food for others. In the majority of ecosystems, photosynthesizers are the producers that take in inorganic molecules and produce food by using solar energy.

When photosynthesis occurs, carbon dioxide is absorbed and oxygen is released. Oxygen is required by organisms when they carry on cellular respiration. The collective action of algae and plants is responsible for placing copious amounts of oxygen into the atmosphere. It rises high into the atmosphere and forms an ozone layer that makes terrestrial life possible. Accordingly, this layer that protects us against damaging ultraviolet rays of the sun is called the ozone shield.

The products of photosynthesis are critical to humankind in a number of ways. They provide our food, to be sure, but they also are a source of building materials, fabrics, paper, fuel, and pharmaceuticals. Even plants that existed hundreds of millions of years ago are important as a source of fossil fuels. And while we are thanking green plants for their many services, let's not forget the simple beauty of a magnolia blossom or the majesty of the Earth's forests.

Photosynthesizers use solar energy to produce organic nutrients for themselves and all other organisms.

7.1 PHOTOSYNTHETIC ORGANISMS

Photosynthesis converts solar energy into the chemical energy of a carbohydrate. Photosynthetic organisms, including plants, algae, and cyanobacteria, are called **autotrophs** because they produce their own food (Fig. 7.1). Photosynthesis produces an enormous amount of carbohydrate. So much that, if it were instantly converted to coal and the coal were loaded into standard railroad cars (each car holding about 50 tons), the photosynthesizers of the biosphere would fill more than 100 cars per second with coal.

No wonder photosynthetic organisms are able to sustain themselves and all other living things on Earth. With few exceptions it is possible to trace any food chain back to plants and algae. In other words, producers, which have the ability to synthesize carbohydrates, feed not only themselves but also consumers, which must take in preformed organic molecules. Collectively, consumers are called **heterotrophs.** Both autotrophs and heterotrophs use organic molecules produced by photosynthesis as a source of building blocks for growth and repair and as a source of chemical energy for cellular work.

Our analogy about photosynthetic products becoming coal is apt because the bodies of many ancient plants did become the coal we burn today in large part to produce electricity. This happened several hundred million years ago, and that is why coal is called a fossil fuel. The wood of trees is also commonly used as fuel. Then, too, the fermentation of plant materials produces alcohol, which can be used directly to fuel automobiles or as a gasoline additive.

Flowering Plants as Photosynthesizers

The green portions of plants, particularly the leaves, carry on photosynthesis. The leaf of a flowering plant contains mesophyll tissue in which cells are specialized for photosynthesis (Fig. 7.2). The raw materials for photosynthesis are water and carbon dioxide. The roots of a plant absorb water, which then moves in vascular tissue up the stem to a leaf by way of the leaf veins. Carbon dioxide in the air enters a leaf through small openings called **stomata** (sing., stoma). After entering a leaf, carbon dioxide and water diffuse into **chloroplasts** [Gk. *chloros,* green, and *plastos,* formed, molded], the organelles that carry on photosynthesis.

A double membrane surrounds a chloroplast and its fluid-filled interior called the **stroma** [Gk. *stroma,* bed, mattress]. A different membrane system within the stroma

FIGURE 7.1
Photosynthetic organisms.

Photosynthetic organisms include (**a**) cyanobacteria such as *Oscillatoria,* which are a type of bacterium; (**b**) algae such as kelp, which typically live in water and can range in size from microscopic to macroscopic; and (**c**) plants such as the sequoia, which typically live on land.

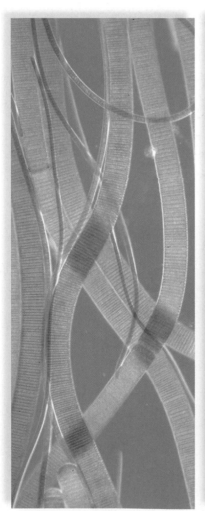

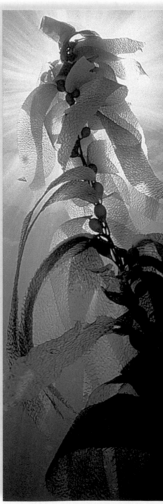

a. *Oscillatoria*   100×   b. Kelp   c. Sequoia

forms flattened sacs called **thylakoids** [Gk. *thylakos*, sack, and *eides*, like, resembling], which in some places are stacked to form **grana** (sing., granum), so called because they looked like piles of seeds to early microscopists. The space of each thylakoid is thought to be connected to the space of every other thylakoid within a chloroplast, thereby forming an inner compartment within chloroplasts called the thylakoid space.

Chlorophyll and other pigments that are part of a thylakoid membrane are capable of absorbing solar energy. This is the energy that drives photosynthesis. The stroma is an enzyme-rich solution where carbon dioxide is first attached to an organic compound and is then reduced to a carbohydrate.

Therefore, it is proper to associate the absorption of solar energy with the thylakoid membranes making up the grana and to associate the reduction of carbon dioxide to a carbohydrate with the stroma of a chloroplast.

Human beings, and indeed nearly all organisms, release carbon dioxide into the air. This is some of the same carbon dioxide that enters a leaf through the stoma and is converted to carbohydrate. Carbohydrate, in the form of glucose, is the chief energy source for most organisms.

Photosynthesis, which occurs in chloroplasts, is critically important because photosynthetic organisms are able to use solar energy to produce carbohydrate, an organic nutrient. Almost all organisms depend either directly or indirectly on these organic nutrients to sustain themselves.

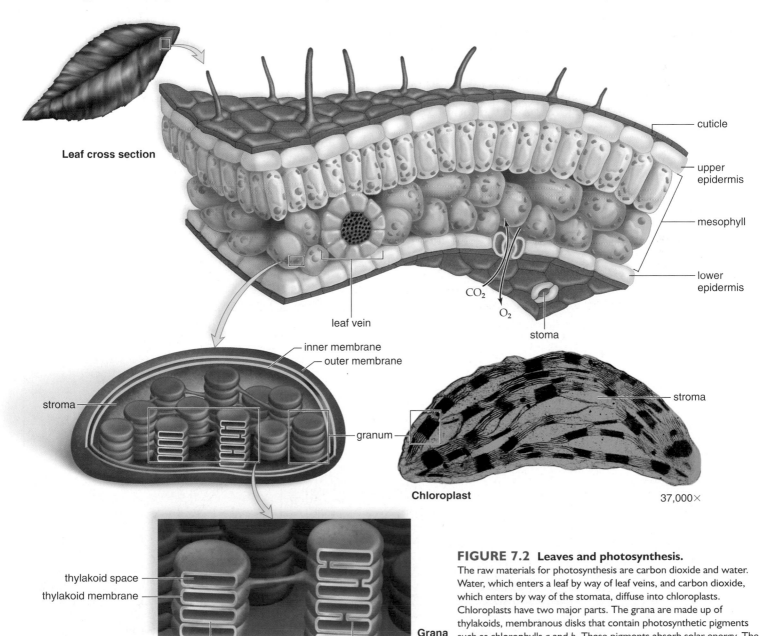

Leaf cross section

cuticle

upper epidermis

mesophyll

lower epidermis

CO_2

O_2

stoma

leaf vein

inner membrane

outer membrane

stroma

granum

stroma

Chloroplast   37,000×

thylakoid space

thylakoid membrane

independent thylakoid in a granum

overlapping thylakoid in a granum

Grana

FIGURE 7.2 Leaves and photosynthesis.
The raw materials for photosynthesis are carbon dioxide and water. Water, which enters a leaf by way of leaf veins, and carbon dioxide, which enters by way of the stomata, diffuse into chloroplasts. Chloroplasts have two major parts. The grana are made up of thylakoids, membranous disks that contain photosynthetic pigments such as chlorophylls *a* and *b*. These pigments absorb solar energy. The stroma is a fluid-filled space where carbon dioxide is enzymatically reduced to a carbohydrate such as glucose.

7.2 PLANTS AS SOLAR ENERGY CONVERTERS

Only about 42% of solar radiation passes through the Earth's atmosphere and reaches its surface. Most of this radiation is within the visible-light range. Higher-energy wavelengths are screened out by the ozone layer in the atmosphere, and lower-energy wavelengths are screened out by water vapor and carbon dioxide before they reach the Earth's surface. The conclusion is, then, that organic molecules and processes within organisms, such as vision and photosynthesis, are chemically adapted to the radiation that is most prevalent in the environment—**visible light.**

Photosynthetic Pigments

Pigments are molecules that absorb wavelengths of light. Most pigments absorb only some wavelengths; they reflect or transmit the other wavelengths. The pigments found in chloroplasts are capable of absorbing various portions of visible light. This is called their **absorption spectrum.** Photosynthetic organisms differ by the type of chlorophyll they contain. In plants, chlorophyll *a* and chlorophyll *b* play prominent roles in photosynthesis. **Carotenoids** play an accessory role. Both chlorophylls *a* and *b* absorb violet, blue, and red light better than the light of other colors. Because green light is transmitted and reflected by chlorophyll, plant leaves appear green to us. The carotenoids, which are shades of yellow and orange, are able to absorb light in the violet-blue-green range. These pigments become noticeable in the fall when chlorophyll breaks down.

How do you determine the absorption spectrum of pigments? To identify the absorption spectrum of a particular pigment, a purified sample is exposed to different wavelengths of light inside an instrument called a spec-

trophotometer. A spectrophotometer measures the amount of light that passes through the sample, and from this it is possible to calculate how much was absorbed. The amount of light absorbed at each wavelength is plotted on a graph, and the result is a record of the pigment's absorption spectrum (Fig. 7.3*a*). How do we know that the peaks shown in the absorption spectrum for chlorophylls *a* and *b* and also the carotenoids indicate which wavelengths plants use for photosynthesis? Because photosynthesis gives off oxygen, we can use the production rate of oxygen as a means to measure the rate of photosynthesis at each wavelength of light. When such data are plotted, the resulting graph is a record of the **action spectrum** for photosynthesis in plants (Fig. 7.3*b*). An action spectrum shows which wavelengths are used to perform a function, in this case, photosynthesis. Because the sum of the absorption spectrums for chlorophylls *a* and *b* plus the carotenoids matches the action spectrum for plant photosynthesis, we are confident that the light absorbed by these pigments does contribute extensively to photosynthesis in plants.

Photosynthetic Reaction

In 1930, C. B. van Niel of Stanford University found that oxygen given off by photosynthesis comes from water, not from carbon dioxide as had been originally thought. This was proven by two separate experiments. When an isotope of oxygen, namely ^{18}O, was a part of carbon dioxide, the O_2 given off by a plant did not contain the isotope. On the other hand, when the isotope (red color) was a part of water, the isotope did appear in the O_2 given off by the plant.

$$CO_2 + 2\,H_2O \xrightarrow{\text{solar energy}} (CH_2O) + H_2O + O_2$$

FIGURE 7.3

Photosynthetic pigments and photosynthesis.
Visible light contains forms of energy that differ according to wavelength and color. **a.** The photosynthetic pigments in chlorophylls *a* and *b* and the carotenoids absorb certain wavelengths within visible light. This is their absorption spectrum. **b.** The action spectrum for photosynthesis in plants—the wavelengths that are used when photosynthesis is taking place—matches well the sum of the absorption spectrums for chlorophylls *a* and *b* and the carotenoids.

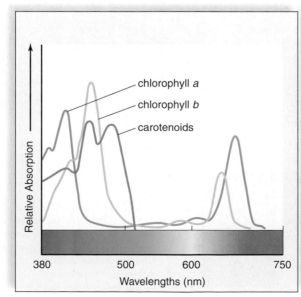

a. The absorption spectrums for chlorophylls *a* and *b* and the carotenoids.

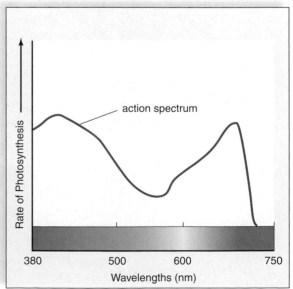

b. The action spectrum for photosynthesis.

The following equation for photosynthesis is sometimes preferred because it shows only the net consumption of water and gives glucose as the end product. Glucose is the molecule most often broken down during cellular respiration:

$$6\,CO_2 + 6\,H_2O \xrightarrow{\text{solar energy}} C_6H_{12}O_6 + 6\,O_2$$

You can arrive at still a third equation for photosynthesis by dividing all reactants and products by 6:

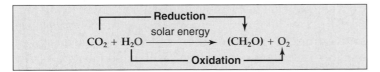

$$CO_2 + H_2O \xrightarrow{\text{solar energy}} (CH_2O) + O_2$$

Some prefer this equation because it shows a generalized carbohydrate (CH_2O) as the end product of photosynthesis. Regardless of the particular equation used, we need to realize that photosynthesis is an oxidation-reduction (redox) reaction. During photosynthesis, carbon dioxide is reduced and water is oxidized.

Two Sets of Reactions

In 1905, F. F. Blackman suggested that two sets of reactions are involved in photosynthesis because enzymes are needed to produce carbohydrate. The two sets of reactions are called the **light reactions** and the **Calvin cycle reactions** (Fig 7.4).

Light Reactions. During the light reactions, chlorophyll located within the thylakoid membranes absorbs solar energy and energizes electrons. When these energized electrons move down an **electron transport chain,** energy is captured and later used for ATP production.

Energized electrons are also taken up by $NADP^+$ (nicotinamide adenine dinucleotide phosphate), an electron carrier. After $NADP^+$ accepts electrons, it becomes NADPH.

During the light reactions:
solar energy $\longrightarrow$ chemical energy
(ATP, NADPH)

Calvin Cycle Reactions. During the Calvin cycle reactions in the stroma, CO_2 is taken up and reduced to a carbohydrate. The ATP and NADPH formed during the light reactions carries out this reduction.

After the production of a carbohydrate, ADP + ⓅP and $NADP^+$ return to the light reactions, where they become ATP and NADPH once more.

During the Calvin cycle reactions:
chemical energy $\longrightarrow$ chemical energy
(ATP, NADPH)   (carbohydrate)

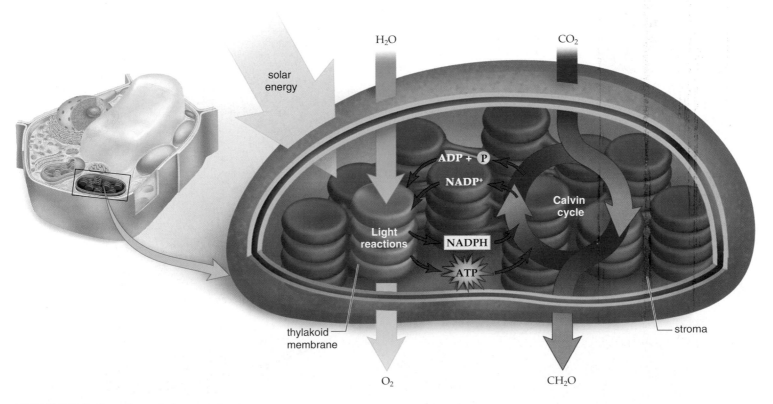

FIGURE 7.4 Overview of photosynthesis.
The process of photosynthesis consists of the light reactions and the Calvin cycle reactions. The light reactions, which produce ATP and NADPH, occur in the thylakoid membrane. These molecules are used in the Calvin cycle reactions to reduce carbon dioxide to a carbohydrate.

7.3 LIGHT REACTIONS

Photosynthesis takes place in chloroplasts. The light reactions that occur in the thylakoid membrane consist of two electron pathways called the noncyclic electron (e^-) pathway and the cyclic electron pathway. In both pathways, solar energy is transformed to the chemical energy that will be used to reduce carbon dioxide in the stroma of chloroplasts.

Both electron pathways produce ATP, but only the noncyclic pathway also produces NADPH. ATP production during photosynthesis is sometimes called photophosphorylation because light powers the process. The production of ATP during the cyclic electron pathway is called cyclic photophosphorylation, whereas ATP production during the noncyclic pathway is called noncyclic photophosphorylation.

Noncyclic Electron Pathway

The **noncyclic electron pathway** is so named because the electron flow can be traced from water to a molecule of $NADP^+$ (Fig. 7.5). This pathway uses two photosystems, called photosystem I (PS I) and photosystem II (PS II). The photosystems are named for the order in which they were discovered, not for the order in which they occur in the thylakoid membrane or participate in the photosynthetic process. A **photosystem** consists of a pigment complex (molecules of chlorophyll a, chlorophyll b, and the carotenoids) and electron acceptor molecules within the thylakoid membrane. The pigment complex serves as an "antenna" for gathering solar energy.

The noncyclic pathway begins with photosystem II. The pigment complex absorbs solar energy, which is then passed from one pigment to the other until it is concentrated in a particular pair of chlorophyll a molecules, called the *reaction center*. Electrons (e^-) in the reaction center become so energized that they escape from the reaction center and move to nearby electron acceptor molecules.

PS II would disintegrate without replacement electrons, and these are removed from water, which splits, re-

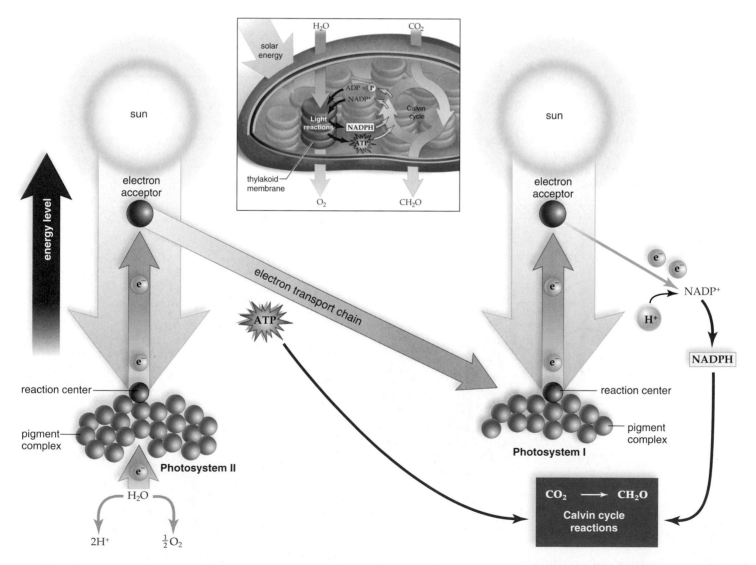

FIGURE 7.5 Noncyclic electron pathway: Electrons move from water to $NADP^+$.
Energized electrons (replaced from water, which splits, releasing oxygen) leave photosystem II and pass down an electron transport chain, leading to the formation of ATP. Energized electrons (replaced by photosystem II) leave photosystem I and pass to $NADP^+$, which then combines with H^+, becoming NADPH.

leasing oxygen to the atmosphere. Notice that with the loss of electrons, water has been oxidized, and that indeed, the oxygen released during photosynthesis does come from water. Many organisms, including plants and even ourselves, use this oxygen within their mitochondria. The hydrogen ions (H^+) stay in the thylakoid space and contribute to the formation of a hydrogen ion gradient.

An electron acceptor sends energized electrons, received from the reaction center, down an electron transport chain, a series of carriers that pass electrons from one to the other. As the electrons pass from one carrier to the next, energy is captured and stored in the form of a hydrogen ion (H^+) gradient. When these hydrogen ions flow down their electrochemical gradient through ATP synthase complexes, ATP production occurs (see page 122). Notice that this ATP will be used in the Calvin cycle reactions in the stroma to reduce carbon dioxide to a carbohydrate.

When the PS I pigment complex absorbs solar energy, energized electrons leave its reaction center and are captured by different electron acceptors. (Low-energy electrons from the electron transport chain adjacent to PS II replace those lost by PS I.) The electron acceptors in PS I pass their electrons to $NADP^+$ molecules. Each one accepts two electrons and an H^+ to become a reduced form of the molecule, that is, NADPH. This NADPH will also be used by the Calvin cycle reactions in the stroma to reduce carbon dioxide to a carbohydrate.

Results of noncyclic electron flow in the thylakoid membrane: water is oxidized (split), yielding H^+, e^-, and O_2, which is released. Also, ATP is produced; and $NADP^+$ becomes NADPH. In the stroma, ATP and NADPH reduce CO_2 to CH_2O, a carbohydrate.

Cyclic Electron Pathway

The **cyclic electron pathway** (Fig. 7.6) begins when the PS I pigment complex absorbs solar energy and it is passed from one pigment to the other until it is concentrated in a reaction center. Electrons (e^-) become so energized that they escape from the reaction center and move to nearby electron-acceptor molecules.

Energized electrons (e^-) taken up by electron acceptors are sent down an electron transport chain. As the electrons pass from one carrier to the next, energy is captured and stored in the form of a hydrogen (H^+) gradient. When these hydrogen ions flow down their electrochemical gradient through ATP synthase complexes, ATP production occurs (see page 122).

This time, instead of the electrons moving on to $NADP^+$, they return to PS I; this is how PS I receives replacement electrons and why this electron pathway is called cyclic. It is also why the cyclic pathway produces ATP but does not produce NADPH.

In plants, the reactions of the Calvin cycle can use any extra ATP produced by the cyclic pathway because

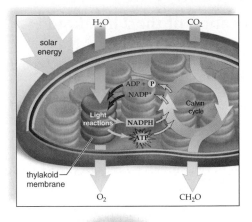

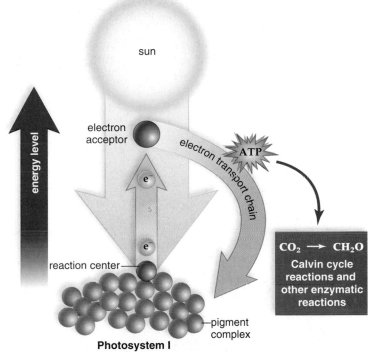

FIGURE 7.6 Cyclic electron pathway: Electrons leave and return to photosystem I.
Energized electrons leave photosystem I reaction-center chlorophyll *a* and are taken up by an electron acceptor, which passes them down an electron transport chain before they return to photosystem I. Only ATP production occurs as a result of this pathway.

the Calvin cycle reaction requires more ATP molecules than NADPH molecules. Also, other enzymatic reactions, aside from those involving photosynthesis, are occurring in the stroma and can use this ATP. It's possible that the cyclic flow of electrons is used alone when carbohydrate is not being produced. At this time, there would be no need for NADPH, which is produced only by the noncyclic electron pathway.

The cyclic electron pathway, from PS I back to PS I, has only one effect: production of ATP.

The Organization of the Thylakoid Membrane

As we have discussed, the following molecular complexes are in the thylakoid membrane (Fig. 7.7):

PS II, which consists of a pigment complex and an electron-acceptor molecule, receives electrons from water and it splits, releasing oxygen.

The electron transport chain, consisting of Pq (plastoquinone) and cytochrome complexes, carries electrons from PS II to PS I. Pq also pumps H⁺ from the stroma into the thylakoid space.

PS I, which also consists of a pigment complex and an electron-acceptor molecule, is adjacent to NADP reductase, which reduces $NADP^+$ to NADPH.

The ATP synthase complex has a channel and a protruding ATP synthase, an enzyme that joins ADP + $\circled{P}$.

ATP Production

The thylakoid space acts as a reservoir for hydrogen ions (H^+). First, each time water is oxidized, two H^+ remain in the thylakoid space. Second, as the electrons move from carrier to carrier along the electron transport chain, the electrons give up energy, which is used to pump H^+ from the stroma into the thylakoid space. Therefore, there are more H^+ in the thylakoid space than in the stroma. The flow of H^+ (often referred to as protons in this context) from high to low concentration across the thylakoid membrane provides the energy that allows an **ATP synthase** enzyme to enzymatically produce ATP from ADP + $\circled{P}$. This method of producing ATP is called **chemiosmosis** because ATP production is tied to the establishment of an H^+ gradient.

Chemiosmosis is the production of ATP due to an H^+ gradient.

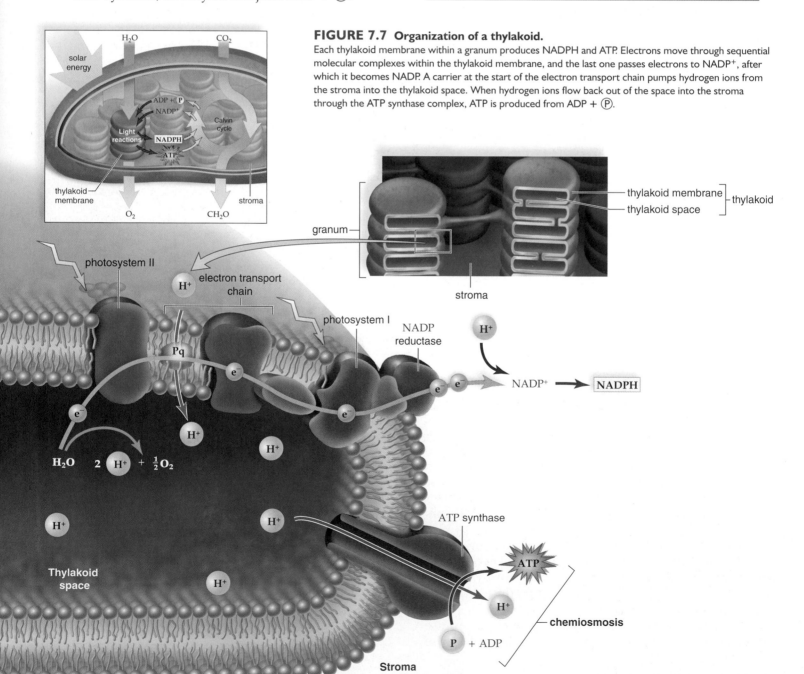

FIGURE 7.7 Organization of a thylakoid.
Each thylakoid membrane within a granum produces NADPH and ATP. Electrons move through sequential molecular complexes within the thylakoid membrane, and the last one passes electrons to $NADP^+$, after which it becomes NADP. A carrier at the start of the electron transport chain pumps hydrogen ions from the stroma into the thylakoid space. When hydrogen ions flow back out of the space into the stroma through the ATP synthase complex, ATP is produced from ADP + $\circled{P}$.

ecology focus

Tropical Rain Forests

Tropical rain forests occur near the equator. They can exist wherever temperatures are above 26°C and rainfall is heavy (from 100–200 cm) and regular. Huge trees with buttressed trunks and broad, undivided, dark green leaves predominate. Nearly all plants in a tropical rain forest are woody, and woody vines are also abundant. There is no undergrowth except at clearings. Instead, orchids, ferns, and bromeliads live in the branches of the trees.

Despite their reduction in size from an original 14% to 6% of land surface today, tropical rain forests make a substantial contribution to global carbon dioxide (CO_2) fixation. Taking into account all ecosystems, marine and terrestrial, photosynthesis produces organic matter that is 300 to 600 times the mass of people currently on Earth this year. Tropical rain forests contribute greatly to the uptake of CO_2 and the productivity of photosynthesis because they are the most efficient of all terrestrial ecosystems.

We have learned that organic matter produced by photosynthesizers feeds all living things and that photosynthesis releases oxygen (O_2), a gas that is needed to complete the process of cellular respiration. Does photosynthesis by tropical rain forests provide any other service that has significant worldwide importance? Figure 7A projects an expected rise in the average global temperature during the twenty-first century due to introduction of certain gases, chiefly carbon dioxide, into the atmosphere. The process of photosynthesis and also the oceans act as a sink for carbon dioxide. For at least a thousand years prior to 1850, atmospheric CO_2 levels remained fairly constant at 0.028%. Since the 1850s, when industrialization began, the amount of CO_2 in the atmosphere increased to 0.036%.

In much the same way as the panes of a greenhouse, carbon dioxide in our atmosphere traps radiant heat from the sun and warms the world. Therefore, CO_2 and other gases that act similarly are called greenhouse gases. Without the greenhouse gases, the Earth's temperature would be about 33°C cooler than it is now. Likewise, increasing the concentration of these gases causes an increase in global tempatures.

Burning fossil fuels adds CO_2 to the atmosphere. Are there any other factors that have contributed to the increases in CO_2 in the atmosphere? Between 10 and 30 million hectares of rain forests are lost every year to ranching, logging, mining, and otherwise developing areas of the forest for human needs. The clearing of forests often involves burning them. Each year, deforestation in tropical rain forests accounts for 20–30% of all carbon dioxide in the atmosphere. The consequence of burning forests is double trouble for global warming because burning a forest adds CO_2 to the atmosphere and at the same time removes trees that would ordinarily absorb CO_2.

It might be hypothesized that an increased amount of CO_2 in the atmosphere will cause photosynthesis to increase in the remaining portion of the forest. To study this possibility, investigators measured atmospheric CO_2 levels, daily temperature levels, and tree girth in La Selva, Costa Rica, for 16 years. The data collected demonstrated relatively lower forest productivity at higher temperatures. These findings suggest that, as temperatures rise, tropical rain forests may add to ongoing atmospheric CO_2 accumulation and accelerated global warming rather than the reverse. All the more reason to do what we can now to slow the process of global warming before it gets out of hand.

Some countries have programs to combat the problem of deforestation. In the mid-1970s, Costa Rica established a system of national parks and reserves to protect 12% of the country's land area from degradation. The current Costa Rican government wants to expand the goal by increasing protected areas to 25% in the near future. Similar efforts in other countries may help slow the ever-increasing threat of global warming.

a.

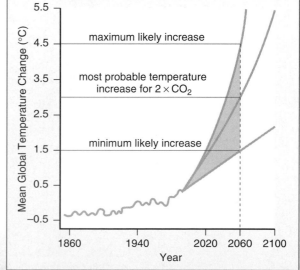

b.

FIGURE 7A Global warming.
a. The burning of tropical rain forests adds CO_2 to the atmosphere and at the same time removes a sink for CO_2. b. Mean global temperature change is expected to rise due to the introduction of greenhouse gases into the atmosphere.

7.4 CALVIN CYCLE REACTIONS

The Calvin cycle reactions follow the light reactions. The Calvin cycle is a series of reactions that produce carbohydrate before returning to the starting point once more (Fig. 7.8). Therefore, this set of reactions is called a cycle. The cycle is named for Melvin Calvin, who, with colleagues, used the radioactive isotope ^{14}C as a tracer to discover the reactions making up the cycle.

This series of reactions uses carbon dioxide from the atmosphere to produce carbohydrate. How does carbon dioxide get into the atmosphere? We and most other organ-isms take in oxygen from the atmosphere and release carbon dioxide to the atmosphere. The Calvin cycle includes (1) carbon dioxide fixation, (2) carbon dioxide reduction, and (3) regeneration of RuBP (ribulose-1, 5-bisphosphate).

Fixation of Carbon Dioxide

Carbon dioxide (CO_2) fixation is the first step of the Calvin cycle. During this reaction, carbon dioxide from the atmosphere is attached to RuBP, a 5-carbon molecule. The result is one 6-carbon molecule, which splits into two 3-carbon molecules.

The enzyme that speeds this reaction, called **RuBP carboxylase,** is a protein that makes up about 20–50% of the protein content in chloroplasts. The reason for its abundance may be that it is unusually slow (it processes only a few molecules of substrate per second compared to thousands per second for a typical enzyme), and so there has to be a lot of it to keep the Calvin cycle going.

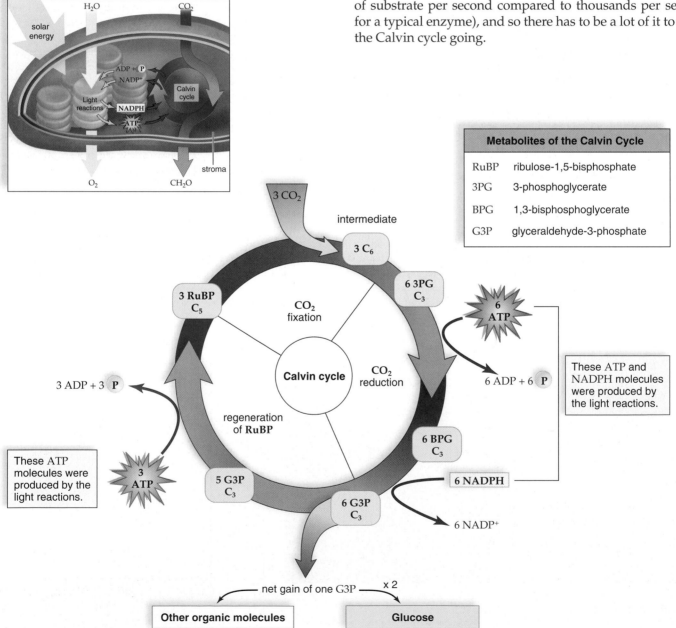

FIGURE 7.8 The Calvin cycle reactions.
The Calvin cycle is divided into three portions: CO_2 fixation, CO_2 reduction, and regeneration of RuBP. Because five G3P are needed to re-form three RuBP, it takes three turns of the cycle to have a net gain of one G3P. Two G3P molecules are needed to form glucose.

Reduction of Carbon Dioxide

The first 3-carbon molecule in the Calvin cycle is called 3PG (3-phosphoglycerate). Each of two 3PG molecules undergoes reduction to G3P in two steps:

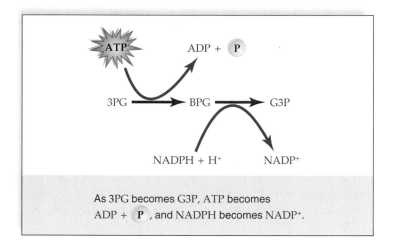

As 3PG becomes G3P, ATP becomes ADP + P, and NADPH becomes NADP$^+$.

This is the sequence of reactions that uses some ATP and NADPH from the light reactions. This sequence signifies the reduction of carbon dioxide to a carbohydrate because R—CO_2 has become R—CH_2O. Energy and electrons are needed for this reduction reaction, and these are supplied by ATP and NADPH.

Regeneration of RuBP

Notice that the Calvin cycle reactions in Figure 7.8 are multiplied by three because it takes three turns of the Calvin cycle to allow one G3P to exit. Why? Because, for every three turns of the Calvin cycle, five molecules of G3P are used to re-form three molecules of RuBP and the cycle continues. Notice that 5 × 3 (carbons in G3P) = 3 × 5 (carbons in RuBP):

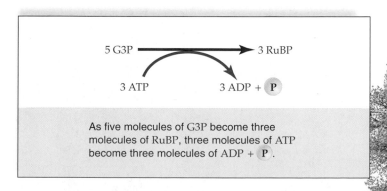

As five molecules of G3P become three molecules of RuBP, three molecules of ATP become three molecules of ADP + P.

This reaction also uses some of the ATP produced by the light reactions.

The carbohydrate produced by the Calvin cycle is the ultimate nutrient source for most living things on Earth.

The Importance of the Calvin Cycle

G3P (glyceraldehyde-3-phosphate) is the product of the Calvin cycle that can be converted to all sorts of organic molecules. Compared to animal cells, algae and plants have enormous biochemical capabilities. They use G3P for the purposes described in Figure 7.9.

Notice that glucose phosphate is among the organic molecules that result from G3P metabolism. This is of interest to us because glucose is the molecule that plants and animals most often metabolize to produce the ATP molecules they require for their energy needs. Glucose is blood sugar in human beings.

Glucose phosphate can be combined with fructose (and the phosphate removed) to form sucrose, the molecule that plants use to transport carbohydrates from one part of the plant to the other.

Glucose phosphate is also the starting point for the synthesis of starch and cellulose. Starch is the storage form of glucose. Some starch is stored in chloroplasts, but most starch is stored in amyloplasts in roots. Cellulose is a structural component of plant cell walls and becomes fiber in our diet because we are unable to digest it.

A plant can use the hydrocarbon skeleton of G3P to form fatty acids and glycerol, which are combined in plant oils. We are all familiar with corn oil, sunflower oil, or olive oil used in cooking. Also, when nitrogen is added to the hydrocarbon skeleton derived from G3P, amino acids are formed.

FIGURE 7.9 Fate of G3P.
G3P is the first reactant in a number of plant cell metabolic pathways. Two G3Ps are needed to form glucose phosphate; glucose is often considered the end product of photosynthesis. Sucrose is the transport sugar in plants; starch is the storage form of glucose; and cellulose is a major constituent of plant cell walls.

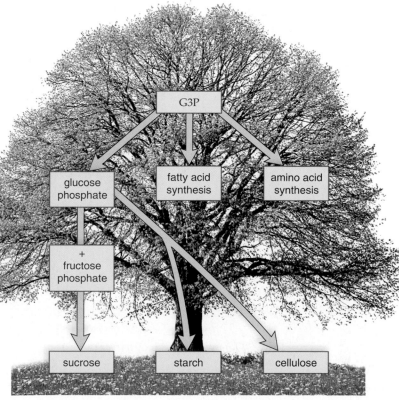

7.5 OTHER TYPES
OF PHOTOSYNTHESIS

The majority of green plants such as azaleas, maples, and tulips carry on photosynthesis as described and are called **C$_3$ plants.** C$_3$ plants use the enzyme RuBP carboxylase to fix CO$_2$ to RuBP in mesophyll cells. The first detected molecule following fixation is the 3-carbon molecule 3PG:

$$\text{RuBP carboxylase}$$
$$\text{RuBP} + \text{CO}_2 \longrightarrow 2 \text{ 3PG}$$

As shown in Figure 7.2, leaves have small openings called stomata through which water can leave and carbon dioxide (CO$_2$) can enter. If the weather is hot and dry, the stomata close, conserving water. (Water loss might cause the plant to wilt and die.) Now the concentration of CO$_2$ decreases in leaves, while oxygen, a by-product of photosynthesis, increases. When oxygen rises in C$_3$ plants, it combines with RuBP instead of CO$_2$. The result is one molecule of 3PG and the eventual release of CO$_2$. This is called **photorespiration** because in the presence of light *(photo)*, oxygen is taken up and CO$_2$ is released *(respiration)*.

An adaptation called C$_4$ photosynthesis enables some plants to avoid photorespiration.

C$_4$ Photosynthesis

In a C$_3$ plant, the mesophyll cells contain well-formed chloroplasts and are arranged in parallel layers. In a C$_4$ leaf, the bundle sheath cells, as well as the mesophyll cells, contain chloroplasts. Further, the mesophyll cells are arranged concentrically around the bundle sheath cells:

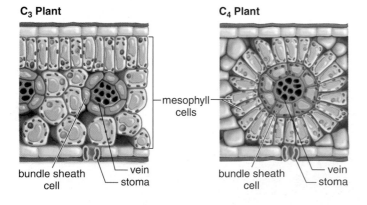

C$_4$ plants use the enzyme PEP carboxylase (PEPCase) to fix CO$_2$ to PEP (phosphoenolpyruvate, a C$_3$ molecule). The result is oxaloacetate, a C$_4$ molecule:

$$\text{PEPCase}$$
$$\text{PEP} + \text{CO}_2 \longrightarrow \text{oxaloacetate}$$

In a C$_4$ plant, CO$_2$ is taken up in mesophyll cells, and then malate, a reduced form of oxaloacetate, is pumped

into the bundle sheath cells (Fig. 7.10). Here, and only here, does CO$_2$ enter the Calvin cycle. It takes energy to pump molecules, and you would think that the C$_4$ pathway would be disadvantageous. Yet in hot, dry climates, the net photosynthetic rate of C$_4$ plants such as sugarcane, corn, and Bermuda grass is about two to three times that of C$_3$ plants such as wheat, rice, and oats. Why do C$_4$ plants enjoy such an advantage? The answer is that they can avoid photorespiration, discussed previously. Photorespiration is wasteful because it is not part of the Calvin cycle. Photorespiration does not occur in C$_4$ leaves because PEP, unlike RuBP, does not combine with O$_2$. Even when stomata are closed, CO$_2$ is delivered to the Calvin cycle in the bundle sheath cells.

When the weather is moderate, C$_3$ plants ordinarily have the advantage, but when the weather becomes hot and dry, **C$_4$ plants** have the advantage, and we can expect them to predominate. In the early summer, C$_3$ plants such as Kentucky bluegrass and creeping bent grass predominate

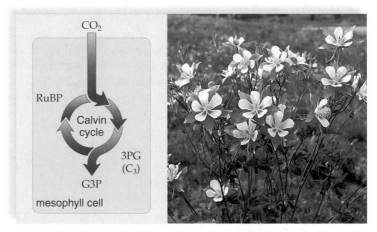

a. CO$_2$ fixation in a C$_3$ plant, blue columbine, *Aquilegia caerulea*

b. CO$_2$ fixation in a C$_4$ plant, corn, *Zea mays*

FIGURE 7.10 Carbon dioxide fixation in C$_3$ and C$_4$ plants.
a. In C$_3$ plants, CO$_2$ is taken up by the Calvin cycle directly in mesophyll cells.
b. C$_4$ plants form a C$_4$ molecule in mesophyll cells prior to releasing CO$_2$ to the Calvin cycle in bundle sheath cells.

in lawns in the cooler parts of the United States, but by mid-summer, crabgrass, a C_4 plant, begins to take over.

CAM Photosynthesis

CAM stands for crassulacean-acid metabolism; the Crassulaceae is a family of flowering succulent (water-containing) plants that live in warm, dry regions of the world. CAM was first discovered in these plants, but now it is known to be prevalent among most succulent plants that grow in desert environments, including cacti.

Whereas a C_4 plant represents partitioning in space—carbon dioxide fixation occurs in mesophyll cells and the Calvin cycle occurs in bundle sheath cells—CAM is partitioning by the use of time. During the night, CAM plants use PEPCase to fix some CO_2, forming C_4 molecules, which are stored in large vacuoles in mesophyll cells. During the day, C_4 molecules (malate) release CO_2 to the Calvin cycle when NADPH and ATP are available from the light reactions (Fig. 7.11). The primary advantage for this partitioning again has to do with the conservation of water. CAM plants open their stomata only at night, and therefore only at that time is atmospheric CO_2 available. During the day, the stomata close. This conserves water, but CO_2 cannot enter the plant.

Photosynthesis in a CAM plant is minimal because a limited amount of CO_2 is fixed at night, but it does allow CAM plants to live under stressful conditions.

Photosynthesis and Adaptation to the Environment

The different types of photosynthesis give us an opportunity to consider that organisms are metabolically adapted to their environment. Each method of photosynthesis has its advantages and disadvantages, depending on the climate.

C_4 plants most likely evolved in, and are adapted to, areas of high light intensities, high temperatures, and

CO_2 fixation in a CAM plant, pineapple, *Ananas comosus*

FIGURE 7.11 Carbon dioxide fixation in a CAM plant.
CAM plants, such as pineapple, fix CO_2 at night, forming a C_4 molecule.

limited rainfall. C_4 plants, however, are more sensitive to cold, and C_3 plants do better than C_4 plants below 25°C. CAM plants, on the other hand, compete well with either type of plant when the environment is extremely arid. Surprisingly, CAM is quite widespread and has evolved in 23 families of flowering plants, including some lillies and orchids! And it is found among nonflowering plants, including some ferns and cone-bearing trees.

In C_3 plants, the Calvin cycle fixes carbon dioxide (CO_2) directly, and the first detectable molecule following fixation is 3PG, a C_3 molecule. C_4 plants fix CO_2 by forming a C_4 molecule prior to the involvement of the Calvin cycle. CAM plants open their stomata at night and also fix CO_2 by forming a C_4 molecule.

CONNECTING THE CONCEPTS

"Have You Thanked a Green Plant Today?" is a bumper sticker that you may have puzzled over until now. Plants, you now know, capture solar energy and store it in carbon-based organic nutrients that are passed to other organisms when they feed on plants and/or on other organisms. In this context, plants are called autotrophs because they make their own organic food. Heterotrophs are organisms that take in preformed organic food.

Both autotrophs and heterotrophs contribute to the global carbon cycle. In the carbon cycle, organisms in both terrestrial and aquatic ecosystems exchange carbon dioxide with the atmosphere. Autotrophs such as plants take in carbon dioxide when

they photosynthesize. Carbon dioxide is returned to the atmosphere when autotrophs and heterotrophs carry on cellular respiration. In this way, the very same carbon atoms cycle from the atmosphere to autotrophs, then to heterotrophs, and then back to autotrophs again.

Living and dead organisms contain organic carbon and serve as a reservoir of carbon in the carbon cycle. Some 300 million years ago, a host of plants died and did not decompose. These plants were compressed to form the coal that we mine and burn today. (Oil has a similar origin, but it most likely formed in marine sedimentary rocks that included animal remains.)

The amount of carbon dioxide in the atmosphere is increasing steadily because we humans burn fossil fuels to run our modern industrial society. This buildup of carbon dioxide will contribute to global warming, because carbon dioxide is a greenhouse gas. Like the glass of a greenhouse, carbon dioxide allows sunlight to pass through but then traps the resulting heat. Yet while we are pumping carbon dioxide into the atmosphere due to fossil fuel burning, we are destroying vast tracts of tropical rain forests, which soak up carbon dioxide like a sponge. Clearly, these actions are not in our self-interest, and it would behoove us to reduce the burning of fossil fuels and preserve tropical rain forests.

Summary

7.1 PHOTOSYNTHETIC ORGANISMS

Photosynthesis produces carbohydrate and releases oxygen, both of which are used by the majority of living things. Cyanobacteria, algae, and plants carry on photosynthesis. In plants, photosynthesis takes place in chloroplasts. A chloroplast is bounded by a double membrane and contains two main components: the liquid stroma and the membranous grana made up of thylakoids.

7.2 PLANTS AS SOLAR ENERGY CONVERTERS

Photosynthesis uses solar energy in the visible-light range. Specifically, chlorophylls a and b absorb violet, blue, and red wavelengths best. This causes chlorophyll to appear green to us. The carotenoids absorb light in the violet-blue-green range and are shades of yellow to orange. During photosynthesis, the light reactions take place in the thylakoids, and the Calvin cycle reactions take place in the stroma.

7.3 LIGHT REACTIONS

The noncyclic electron pathway of the light reactions begins when solar energy enters PS II. In PS II, energized electrons are picked up by electron acceptors. The oxidation (splitting) of water replaces these electrons in the reaction-center chlorophyll a molecules. Oxygen is released to the atmosphere, and hydrogen ions (H^+) remain in the thylakoid space. An electron acceptor molecule passes electrons to PS I by way of an electron transport chain. When solar energy is absorbed by PS I, energized electrons leave and are ultimately received by $NADP^+$, which also combines with H^+ from the stroma to become NADPH.

In the cyclic electron pathway, electrons energized by the sun leave PS I. They pass down an electron transport chain and back to PS I again. In keeping with its name, PS I probably evolved first and is the only photosynthetic pathway present in certain bacteria today. Most photosynthetic cells regulate the activity of the cyclic and noncyclic pathways to suit the needs of the cell.

Chemiosmosis requires an organized membrane. The thylakoid membrane is highly organized: PS II is associated with an enzyme that oxidizes (splits) water, the cytochrome complexes transport electrons and pump H^+, PS I is associated with an enzyme that reduces $NADP^+$, and ATP synthase produces ATP.

The energy made available by the passage of electrons down the electron transport chain allows carriers to pump H^+ into the thylakoid space. The buildup of H^+ establishes an electrochemical gradient. When H^+ flows down this gradient through the channel present in ATP synthase complexes, ATP is synthesized from ADP and Ⓟ by ATP synthase. This method of producing ATP is called chemiosmosis.

7.4 CALVIN CYCLE REACTIONS

The energy yield of the light reactions is stored in ATP and NADPH. These molecules are used by the Calvin cycle reactions to reduce CO_2 to carbohydrate, namely G3P, which is then converted to all the organic molecules a plant needs.

During the first stage of the Calvin cycle, the enzyme RuBP carboxylase fixes CO_2 to RuBP, producing a 6-carbon molecule that immediately breaks down to two C_3 molecules. During the second stage, CO_2 (incorporated into an organic molecule) is reduced to carbohydrate (CH_2O). This step requires the NADPH and some of the ATP from the light reactions. For every three turns of the Calvin cycle, the net gain is one G3P molecule; the other five G3P molecules are used to re-form three molecules of RuBP. This step also requires ATP for energy. It takes two G3P molecules to make one glucose molecule.

7.5 OTHER TYPES OF PHOTOSYNTHESIS

In C_4 plants, as opposed to the C_3 plants just described, the enzyme PEPCase fixes carbon dioxide to PEP to form a 4-carbon molecule, oxaloacetate, within mesophyll cells. A reduced form of this molecule is pumped into bundle sheath cells where CO_2 is released to the Calvin cycle. C_4 plants avoid photorespiration by a partitioning of pathways in space: carbon dioxide fixation occurs in mesophyll cells, and the Calvin cycle occurs in bundle sheath cells.

During CAM photosynthesis, PEPCase fixes CO_2 to PEP at night. The next day, CO_2 is released and enters the Calvin cycle within the same cells. This represents a partitioning of pathways in time: carbon dioxide fixation occurs at night, and the Calvin cycle occurs during the day. The plants that carry on CAM are desert plants, in which the stomata only open at night, thereby conserving water.

Reviewing the Chapter

1. Why is it proper to say that almost all living things are dependent on solar energy?   116
2. Name the two major components of chloroplasts, and associate each portion with the two sets of reactions that occur during photosynthesis. How are the two sets of reactions related?   116, 119
3. Discuss the absorption spectrum of chlorophylls a and b and the carotenoids. Why is chlorophyll a green pigment, and the carotenoids a yellow-orange pigment?   118
4. Trace the noncyclic electron pathway, naming and explaining all the events that occur as the electrons move from water to $NADP^+$.   120
5. Trace the cyclic electron pathway, naming and explaining the main events that occur as electrons cycle.   121
6. How is the thylakoid membrane organized? Name the main complexes in the membrane. Give a function for each.   122
7. Explain what is meant by chemiosmosis, and relate this process to the electron transport chain present in the thylakoid membrane.   122
8. Describe the three stages of the Calvin cycle. Which stage uses the ATP and NADPH from the light reactions?   124
9. Explain C_4 photosynthesis, contrasting the actions of RuBP carboxylase and PEPCase.   126
10. Explain CAM photosynthesis, contrasting it to C_4 photosynthesis in terms of partitioning a pathway.   127

Testing Yourself

Choose the best answer for each question.

1. The absorption spectrum of chlorophyll
   a. is not the same as that of carotenoids.
   b. approximates the action spectrum of photosynthesis.
   c. explains why chlorophyll is a green pigment.
   d. shows that some colors of light are absorbed more than others.
   e. All of these are correct.
2. The final acceptor of electrons during the noncyclic electron pathway is
   a. PS I.   d. $NADP^+$.
   b. PS II.   e. water.
   c. ATP.

3. A photosystem contains
   a. pigments, a reaction center, and electron acceptors.
   b. ADP, (P), and hydrogen ions (H$^+$).
   c. protons, photons, and pigments.
   d. cytochromes only.
   e. Both b and c are correct.

For questions 4–8, match each item to those in the key. Use an answer more than once, if possible.

KEY:
   a. solar energy
   b. chlorophyll
   c. chemiosmosis
   d. Calvin cycle

4. light energy

5. ATP synthase

6. thylakoid membrane

7. green pigment

8. RuBP

For questions 9–13, indicate whether the statement is true (T) or false (F).

9. RuBP carboxylase is the enzyme that fixes carbon dioxide to RuBP in the Calvin cycle. _____

10. Oxygen is given off during the cyclic pathway of the light reactions. _____

11. When 3PG becomes G3P during the light reactions, carbon dioxide is reduced to carbohydrate. _____

12. NADPH and ATP cycle between the Calvin cycle and the light reactions constantly. _____

13. The action spectrum for photosynthesis matches the combined absorption spectrums of chlorophylls *a* and *b* and the carotenoids. _____

14. The NADPH and ATP from the light reactions are used to
   a. split water.
   b. cause RuBP carboxylase to fix CO_2.
   c. re-form the photosystems.
   d. cause electrons to move along their pathways.
   e. convert 3PG to G3P.

15. Chemiosmosis
   a. depends on complexes in the thylakoid membrane.
   b. depends on an electrochemical gradient.
   c. depends on a difference in H$^+$ concentration between the thylakoid space and the stroma.
   d. results in ATP formation.
   e. All of these are correct.

16. Under what conditions might plants use the cyclic electron pathway instead of the noncyclic pathway?
   a. hot, dry conditions   c. periods of darkness
   b. low CO_2 levels   d. low oxygen levels

17. The function of the light reactions is to
   a. obtain CO_2.
   b. make carbohydrate.
   c. convert light energy into a usable form of chemical energy.
   d. regenerate RuBP.

18. Label the following diagram of a chloroplast:

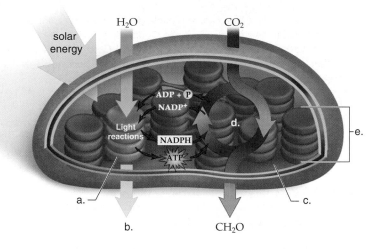

f. The light reactions occur in which part of a chloroplast?
g. The Calvin cycle reactions occur in which part of a chloroplast?

19. Label the following diagram using these labels: water, carbohydrate, carbon dioxide, oxygen, ATP, ADP + (P), NADPH, and NADP$^+$.

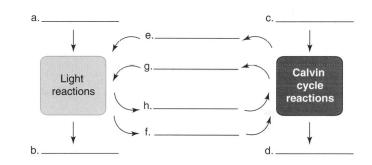

20. The oxygen given off by photosynthesis comes from
   a. H_2O.   c. glucose.
   b. CO_2.   d. RuBP.

21. The glucose formed by photosynthesis can be used by plants to make
   a. starch.   d. proteins.
   b. cellulose.   e. All of these are correct.
   c. lipids and oils.

22. The Calvin cycle reactions
   a. produce carbohydrate.
   b. convert one form of chemical energy into a different form of chemical energy.
   c. regenerate more RuBP.
   d. use the products of the light reactions.
   e. All of these are correct.

23. CAM photosynthesis
   a. is the same as C_4 photosynthesis.
   b. is an adaptation to cold environments in the Southern Hemisphere.
   c. is prevalent in desert plants that close their stomata during the day.
   d. occurs in plants that live in marshy areas.
   e. stands for chloroplasts and mitochondria.

24. Compared to RuBP carboxylase, PEPCase has the advantage that
    a. PEPCase is present in both mesophyll and bundle sheath cells, but RuBP carboxylase is not.
    b. RuBP carboxylase fixes carbon dioxide (CO_2) only in C_4 plants, but PEPCase does it in both C_3 and C_4 plants.
    c. RuBP carboxylase combines with O_2, but PEPCase does not.
    d. PEPCase conserves energy, but RuBP carboxylase does not.
    e. Both b and c are correct.
25. C_4 photosynthesis
    a. is the same as C_3 photosynthesis because it takes place in chloroplasts.
    b. occurs in plants whose bundle sheath cells contain chloroplasts.
    c. takes place in plants such as wheat, rice, and oats.
    d. is an advantage when the weather is hot and dry.
    e. Both b and d are correct.

Thinking Scientifically

1. In the fall of the year, the leaves of many trees change from green to red or yellow. Two hypotheses can explain this color change: (1) In the fall, chlorophyll degenerates, and red or yellow pigments that were earlier masked by chlorophyll become apparent. (2) In the fall, red or yellow pigments are synthesized, and they mask the color of chlorophyll. How could you test these two hypotheses?
2. You have discovered a type of bacterium that seems to be capable of photosynthesis but has no chloroplasts. However, there are invaginations of the plasma membrane inside the bacterium. Would photosynthesis be possible with such an arrangement? What would you expect to find in the invaginated membrane if these bacteria were photosynthetic?

Bioethical Issue: Feeding the Expanding Population

Whether there will be enough food to feed the increase in population expected by the middle of the twenty-first century is unknown. Over the past 40 years, the world's food supply has expanded faster than the population, due to the development of high-yielding plants and the increased use of irrigation, pesticides, and fertilizers. Unfortunately, modern farming techniques result in pollution of the air, water, and land. One of the most worrisome threats to food production is an increasing degradation of agricultural land. Soil erosion in particular is robbing the land of its topsoil and reducing its productivity.

In 1986, it was estimated that humans already use nearly 40% of the Earth's terrestrial photosynthetic production, and therefore we should reach maximum capacity in the middle of the twenty-first century, when the population is projected to double its 1986 size. Most population growth will occur in the developing countries, which are countries in Africa, Asia, and Latin America that are only now becoming industrialized. Even the United States will face an increased drain on its economic resources and increased pollution problems due to population growth. In the developed countries, the technical gains needed to prevent a disaster will be enormous.

Some people feel that technology will continue to make great strides for many years to come. They maintain that technology hasn't begun to reach the limits of performance and therefore will be able to solve the problems of increased population growth. Others feel that technology's successes are self-defeating. The newly-developed hybrid crops that led to enormous increases in yield per acre can also cause pollution problems that degrade the environment. Some are in favor of limiting human population growth as quickly as possible. What do you feel is the best approach to meeting the world population's growing food production needs?

Understanding the Terms

absorption spectrum 118	electron transport
action spectrum 118	chain 119
ATP synthase 122	grana (sing., granum) 117
autotroph 116	heterotroph 116
C_3 plant 126	light reactions 119
C_4 plant 126	noncyclic electron
Calvin cycle reactions 119	pathway 120
CAM 127	photorespiration 126
carbon dioxide (CO_2)	photosynthesis 116
fixation 124	photosystem 120
carotenoid 118	RuBP carboxylase 124
chemiosmosis 122	stomata 116
chlorophyll 117	stroma 116
chloroplast 116	thylakoid 117
cyclic electron pathway 121	visible light 118

Match the terms to these definitions:
a. _____ Energy-capturing portion of photosynthesis that takes place in thylakoid membranes of chloroplasts and cannot proceed without solar energy; it produces ATP and NADPH.
b. _____ Photosynthetic unit where solar energy is absorbed and high-energy electrons are generated; contains an antenna complex and an electron acceptor.
c. _____ Passage of electrons along a series of carrier molecules from a higher to a lower energy level; the energy released is used for the synthesis of ATP.
d. _____ Process usually occurring within chloroplasts whereby chlorophyll traps solar energy and carbon dioxide is reduced to a carbohydrate.
e. _____ Series of photosynthetic reactions in which carbon dioxide is fixed and reduced to G3P.

ARIS, the *Biology* Website

ARIS, the website for *Biology*, provides a wealth of information organized and integrated by chapter. You will find practice quizzes, interactive activities, labeling exercises, flashcards, and much more that will complement your learning and understanding of general biology.

www.mhhe.com/maderbiology9

8

CELLULAR RESPIRATION

Whether you run with the bulls in Spain, take an aerobics class, or just sit under a shady tree, ATP molecules provide the energy necessary for your muscles to contract. ATP molecules are produced during cellular respiration, a process that includes reactions in both the cytoplasm and the mitochondria. Muscle cells, in particular, contain numerous mitochondria.

The glucose and oxygen for cellular respiration are delivered to cells by the cardiovascular system, and the end products—water and carbon dioxide—are removed by the cardiovascular system. Cellular respiration consists of many small steps mediated by specific enzymes. High-energy electrons are removed from glucose breakdown products and passed down an electron transport chain located on the cristae of mitochondria. As the electrons move from one carrier to the next, energy is released and captured for the production of ATP molecules.

One form of chemical energy (glucose) cannot be transformed completely into another (ATP molecules) without the loss of usable energy in the form of heat. When ATP is produced, heat is given off. Since exercise requires plentiful ATP, your body's internal temperature rises, and you begin sweating when you exercise.

Effort requires energy.

8.1 CELLULAR RESPIRATION

In order to function, living things must acquire energy by breaking down nutrient molecules produced by photosynthesizers. Cellular respiration includes various metabolic pathways that break down carbohydrates and other metabolites, with the concomitant buildup of ATP. **Cellular respiration,** as implied by its name, is a cellular process that requires oxygen and gives off carbon dioxide (CO_2). Most often it involves the complete breakdown of glucose to carbon dioxide and water (H_2O).

Glucose is a high-energy molecule, and its breakdown products, CO_2 and H_2O, are low-energy molecules. Therefore, we expect the process to be exergonic and release energy. As breakdown occurs, electrons are removed from substrates and are eventually received by oxygen atoms, which then combine with H^+ to become H_2O.

The following equation illustrating the overall reaction for cellular respiration shows changes in regard to hydrogen atom (H) distribution. But remember that a hydrogen atom consists of a hydrogen ion plus an electron ($H^+ + e^-$). Therefore, when hydrogen atoms are removed from glucose, so are electrons. Since oxidation is the loss of electrons, and reduction is the gain of electrons, glucose breakdown is an oxidation-reduction reaction. Glucose is oxidized and O_2 is reduced:

$$\text{Oxidation}$$

$$C_6H_{12}O_6 \ + \ 6\,O_2 \longrightarrow 6\,CO_2 \ + \ 6\,H_2O \ + \ \text{energy}$$

glucose

$$\text{Reduction}$$

The buildup of ATP is an endergonic reaction, which requires energy. The pathways of cellular respiration allow the energy within a glucose molecule to be released slowly so that ATP can be produced gradually. Cells would lose a tremendous amount of energy if glucose breakdown occurred all at once—much energy would become nonusable heat. The step-by-step breakdown of glucose to carbon dioxide and water usually realizes a maximum yield of 36 or 38 ATP molecules, dependent on conditions to be discussed later. The energy in these ATP molecules is equivalent to about 39% of the energy that was available in glucose. This conversion is more efficient than many others; for example, only between 20% and 30% of the energy within gasoline is converted to the motion of a car.

NAD$^+$ and FAD

Cellular respiration involves many individual metabolic reactions, each one catalyzed by its own enzyme. Enzymes of particular significance are those that use **NAD$^+$** (nicotinamide adenine dinucleotide), a coenzyme of oxidation-reduction sometimes called a redox coenzyme. When a metabolite is oxidized, NAD$^+$ accepts two electrons plus a hydrogen ion (H^+), and NADH results. The electrons re-

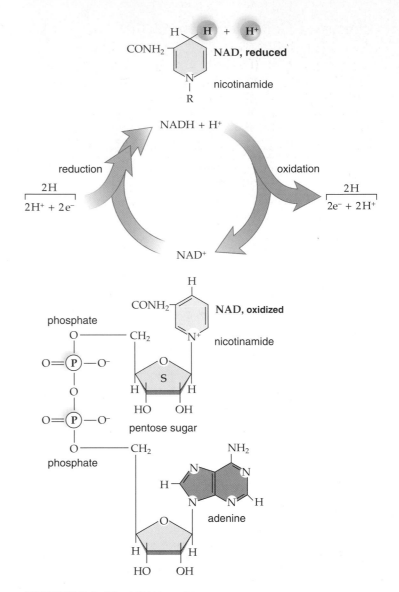

FIGURE 8.1 The NAD$^+$ cycle.
The coenzyme NAD$^+$ is a dinucleotide (two nucleotides joined by bonding between their phosphates). NAD$^+$ accepts two electrons (e^-) plus a hydrogen ion (H^+), and NADH + H^+ results. When NADH passes the electrons to another substrate or carrier, NAD$^+$ is reformed.

ceived by NAD$^+$ are high-energy electrons that are usually carried to the electron transport chain. Figure 8.1 illustrates how NAD$^+$ carries electrons.

NAD$^+$ is called a coenzyme of oxidation-reduction because it can oxidize a metabolite by accepting electrons and can reduce a metabolite by giving up electrons. Only a small amount of NAD$^+$ need be present in a cell, because each NAD$^+$ molecule is used over and over again. **FAD** (flavin adenine dinucleotide), another coenzyme of oxidation-reduction, is sometimes used instead of NAD$^+$. FAD accepts two electrons and two hydrogen ions (H^+) to become FADH$_2$.

NAD$^+$ and FAD are two coenzymes of oxidation-reduction that are active during cellular respiration.

Phases of Cellular Respiration

The oxidation of glucose by removal of hydrogen atoms involves four phases (Fig. 8.2). Glycolysis takes place outside the mitochondria and does not require the presence of oxygen. Therefore, glycolysis is **anaerobic.** The other phases of cellular respiration take place inside the mitochondria, where oxygen is the final acceptor of electrons.

- During **glycolysis** [Gk. *glycos,* sugar, and *lysis,* splitting], glucose is broken down in the cytoplasm to two molecules of pyruvate. Oxidation by removal of hydrogen atoms results in NADH and provides enough energy for the net yield of two molecules of ATP.
- During the **preparatory (prep) reaction,** pyruvate enters a mitochondrion and is oxidized to a 2-carbon acetyl group carried by CoA; NADH is formed; and the waste product CO_2 is removed. Since glycolysis ends with two molecules of pyruvate, the prep reaction occurs twice per glucose molecule.
- The **citric acid cycle** is a cyclical series of oxidation reactions in the matrix of a mitochondrion that result in NADH and $FADH_2$. CO_2 is given off and one ATP is produced. The citric acid cycle turns twice because two acetyl CoA molecules enter the cycle per glucose molecule. Altogether, the citric acid cycle accounts for two immediate ATP molecules per glucose molecule.

- The **electron transport chain** is a series of carriers in the inner mitochondrial membrane that accept the electrons removed from glucose and pass them along from one carrier to the next until they are finally received by O_2, which then combines with hydrogen ions and becomes water. As the electrons pass from a higher-energy to a lower-energy state, energy is released and later used for ATP synthesis by chemiosmosis. The electrons from one glucose result in 32 or 34 ATP, depending on certain conditions.

Pyruvate is a pivotal metabolite in cellular respiration. If oxygen is not available to the cell, fermentation occurs in the cytoplasm (see Fig. 8.10). During **fermentation,** glucose is incompletely metabolized to lactate or to carbon dioxide and alcohol, depending on the organism. As we shall see on page 142, fermentation results in a net gain of only two ATP per glucose molecule.

Cellular respiration involves the oxidation of glucose to carbon dioxide and water. As glucose breaks down, energy is made available for ATP synthesis. A total of 36 or 38 ATP molecules are produced per glucose molecule in cellular respiration (2 from glycolysis, 2 from the citric acid cycle, and 32 or 34 from the electron transport chain).

FIGURE 8.2 The four phases of complete glucose breakdown.
The complete breakdown of glucose consists of four phases. Glycolysis in the cytoplasm produces pyruvate, which enters mitochondria if oxygen is available. The preparatory reaction and the citric acid cycle that follow occur inside the mitochondria. Also, inside mitochondria, the electron transport chain receives the electrons that were removed from glucose breakdown products. The result of glucose breakdown is 36 or 38 ATP, depending on the particular cell.

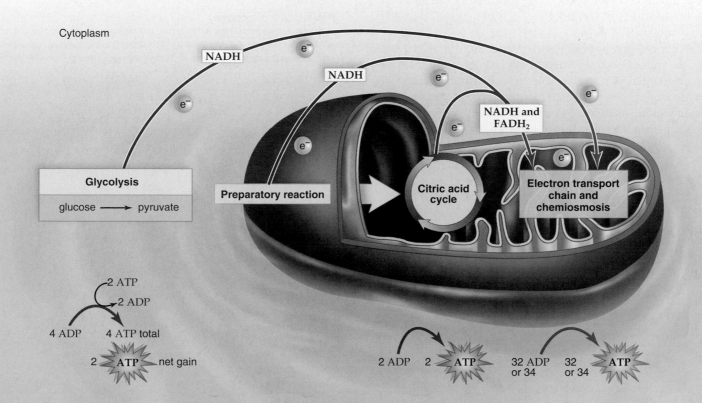

8.2 OUTSIDE THE MITOCHONDRIA: GLYCOLYSIS

Glycolysis, which takes place within the cytoplasm outside the mitochondria, is the breakdown of glucose to two pyruvate molecules. Since glycolysis is universally found in organisms, it most likely evolved before the citric acid cycle and the electron transport chain. This may be why glycolysis occurs in the cytoplasm and does not require the presence of oxygen.

Energy-Investment Steps

As glycolysis begins, two ATP are used to activate glucose, a C_6 (6-carbon) molecule that splits into two C_3 molecules known as G3P. Each G3P has a phosphate group. From this point on, each C_3 molecule undergoes the same series of reactions.

Energy-Harvesting Steps

Oxidation of G3P now occurs by the removal of electrons, which are accompanied by hydrogen ions. In two duplicate reactions, hydrogen atoms ($e^- + H^+$) are picked up by coenzyme NAD^+.

$$2\,NAD^+ + 4\,H \longrightarrow 2\,NADH + 2\,H^+$$

Later, when the NADH molecules pass two electrons on to another **electron carrier,** they become NAD^+ again. Only a small amount of NAD^+ need be present in a cell, because like other coenzymes, it is used over and over again.

The oxidation of G3P and subsequent substrates results in four high-energy phosphate groups, which are used to synthesize four ATP. This is **substrate-level phosphorylation** in which an enzyme passes a high-energy phosphate to ADP, and ATP results (Fig. 8.3). Subtracting the two ATP that were used to get started, there is a net gain of two ATP from glycolysis (Fig. 8.4).

When oxygen is available, the end product, pyruvate, enters the mitochondria, where it undergoes further breakdown. If oxygen is not available, fermentation occurs and pyruvate undergoes reduction. In humans, pyruvate is reduced to lactate, as discussed on page 142.

Altogether, the inputs and outputs of glycolysis are as follows:

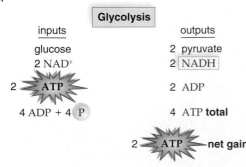

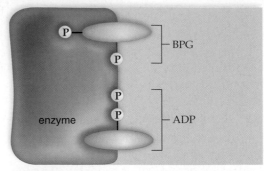

a. The enzyme has a shape that accommodates both BPG and ADP.

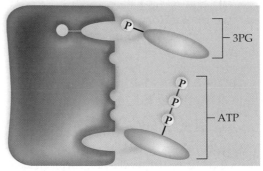

b. A phosphate has been transferred from BPG to ADP, forming 3PG and ATP.

FIGURE 8.3 Substrate-level phosphorylation.
a. Substrates participating in the reaction are oriented on the enzyme. **b.** Then a phosphate group is transferred to ADP, producing one ATP molecule. During glycolysis (see Fig. 8.4), BPG is a substrate that gives up a phosphate group to ADP, forming 3PG and ATP.

FIGURE 8.4 Glycolysis.
This metabolic pathway begins with glucose and ends with pyruvate. Net gain of two ATP molecules can be calculated by subtracting those expended during the energy-investment steps from those produced during the energy-harvesting steps.

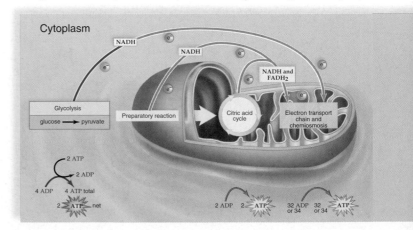

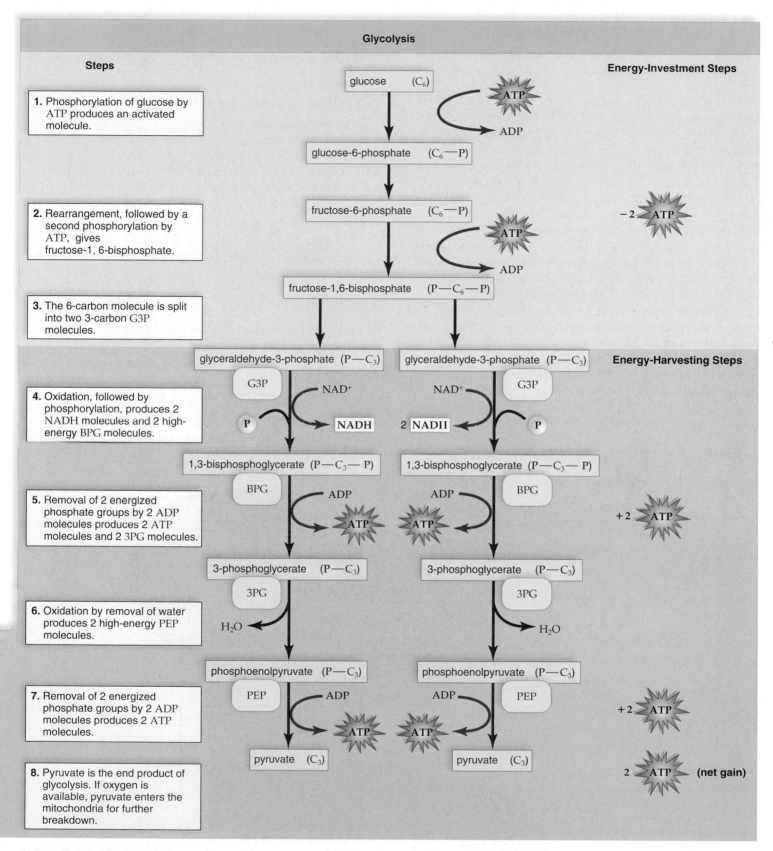

Glycolysis

Steps

Energy-Investment Steps

1. Phosphorylation of glucose by ATP produces an activated molecule.

2. Rearrangement, followed by a second phosphorylation by ATP, gives fructose-1, 6-bisphosphate.

3. The 6-carbon molecule is split into two 3-carbon G3P molecules.

Energy-Harvesting Steps

4. Oxidation, followed by phosphorylation, produces 2 NADH molecules and 2 high-energy BPG molecules.

5. Removal of 2 energized phosphate groups by 2 ADP molecules produces 2 ATP molecules and 2 3PG molecules.

6. Oxidation by removal of water produces 2 high-energy PEP molecules.

7. Removal of 2 energized phosphate groups by 2 ADP molecules produces 2 ATP molecules.

8. Pyruvate is the end product of glycolysis. If oxygen is available, pyruvate enters the mitochondria for further breakdown.

FIGURE 8.4 continued

8.3 INSIDE THE MITOCHONDRIA

It is interesting to think about how our bodies provide the reactants for cellular respiration. The air we breathe contains oxygen, and the food we eat contains glucose. These enter the bloodstream, which carries them about the body, and they move into each and every cell. The end product of glycolysis, pyruvate, enters the mitochondria, where it is oxidized to CO_2 during the prep reaction and the citric acid cycle. Its hydrogen atoms, which are carried to the electron transport chain, result in the reduction of oxygen to H_2O as ATP is produced. The CO_2 and ATP are transported out of mitochondria into the cytoplasm. The ATP is used in the cell for energy-requiring processes. Carbon dioxide diffuses out of the cell and enters the bloodstream. The bloodstream takes the CO_2 to the lungs, where it is exchanged. The H_2O can remain in the mitochondria or in the cell, or it can enter the blood and be excreted by the kidneys as need be.

Just exactly where are the prep reaction, the citric acid cycle, and the electron transport chain located in a mitochondrion? A **mitochondrion** has a double membrane with an intermembrane space (between the outer and inner membrane). Cristae are folds of inner membrane that jut out into the matrix, the innermost compartment, which is filled with a gel-like fluid (Fig. 8.5). The prep reaction and the citric acid cycle enzymes are located in the matrix, and the electron transport chain is located in the cristae. Most of the ATP produced during cellular respiration is produced in mitochondria; therefore, mitochondria are often called the powerhouses of the cell.

The Preparatory Reaction

The **preparatory (prep) reaction** is so called because it occurs before the citric acid cycle. In this reaction, pyruvate is converted to a 2-carbon (C_2) *acetyl group* attached to *coenzyme A*, or CoA, and CO_2 is given off. This is an oxidation reaction in which electrons are removed from pyruvate by NAD^+, and NAD^+ goes to $NADH + H^+$ as **acetyl CoA** forms. This reaction occurs twice per glucose molecule:

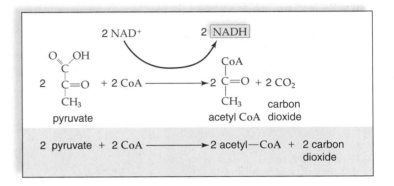

$$2\ \text{pyruvate} + 2\ \text{CoA} \longrightarrow 2\ \text{acetyl—CoA} + 2\ \text{carbon dioxide}$$

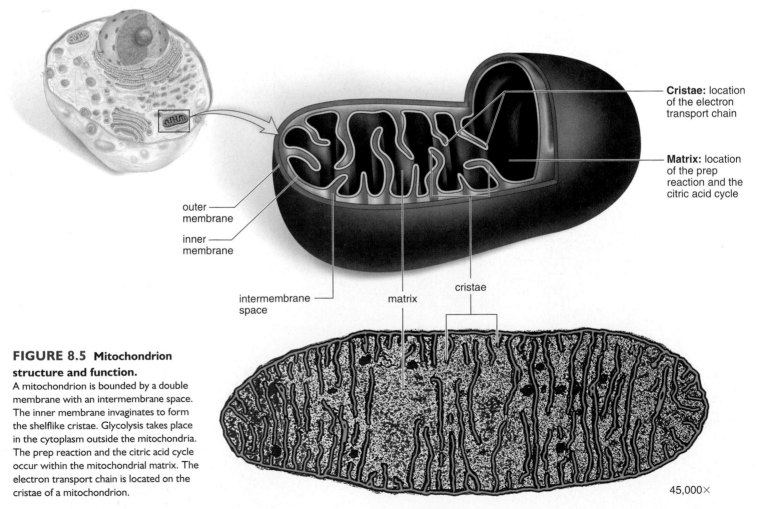

FIGURE 8.5 Mitochondrion structure and function.
A mitochondrion is bounded by a double membrane with an intermembrane space. The inner membrane invaginates to form the shelflike cristae. Glycolysis takes place in the cytoplasm outside the mitochondria. The prep reaction and the citric acid cycle occur within the mitochondrial matrix. The electron transport chain is located on the cristae of a mitochondrion.

outer membrane
inner membrane
intermembrane space
matrix
cristae

Cristae: location of the electron transport chain

Matrix: location of the prep reaction and the citric acid cycle

45,000×

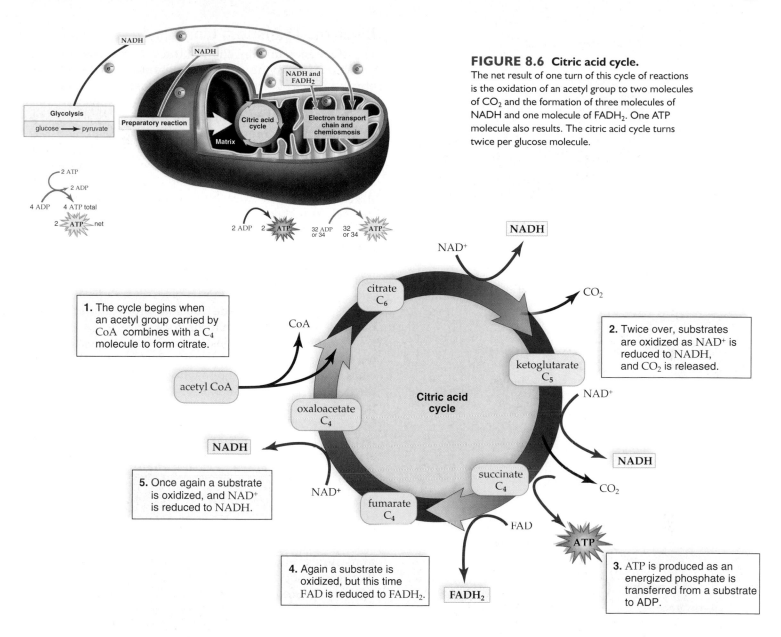

FIGURE 8.6 Citric acid cycle.
The net result of one turn of this cycle of reactions is the oxidation of an acetyl group to two molecules of CO_2 and the formation of three molecules of NADH and one molecule of $FADH_2$. One ATP molecule also results. The citric acid cycle turns twice per glucose molecule.

1. The cycle begins when an acetyl group carried by CoA combines with a C_4 molecule to form citrate.

2. Twice over, substrates are oxidized as NAD^+ is reduced to NADH, and CO_2 is released.

3. ATP is produced as an energized phosphate is transferred from a substrate to ADP.

4. Again a substrate is oxidized, but this time FAD is reduced to $FADH_2$.

5. Once again a substrate is oxidized, and NAD^+ is reduced to NADH.

Citric Acid Cycle

The **citric acid cycle** is a cyclical metabolic pathway located in the matrix of mitochondria (Fig. 8.6). The citric acid cycle is also known as the Krebs cycle, after Hans Krebs, the chemist who worked out the fundamentals of the process in the 1930s. At the start of the citric acid cycle, the (C_2) acetyl group carried by CoA joins with a 4-carbon (C_4) molecule, and a 6-carbon (C_6) citrate molecule results. During the citric acid cycle, each acetyl group received from the prep reaction is oxidized to two CO_2 molecules.

During the cycle, oxidation occurs when electrons are accepted by NAD^+ in three instances and by FAD in one instance. On each occasion, NAD^+ accepts two electrons and one hydrogen ion to become NADH. FAD accepts two electrons and two hydrogen ions to become $FADH_2$.

Substrate-level phosphorylation is also an important event of the citric acid cycle. In substrate-level phosphoryla-

tion, you will recall, an enzyme passes a high-energy phosphate to ADP, and ATP results.

Because the citric acid cycle turns twice for each original glucose molecule, the inputs and outputs of the citric acid cycle per glucose molecule are as follows:

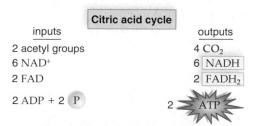

Citric acid cycle	
inputs	outputs
2 acetyl groups	4 CO_2
6 NAD^+	6 NADH
2 FAD	2 $FADH_2$
2 ADP + 2 P	2 ATP

The six carbon atoms originally located in a glucose molecule have now become CO_2. The prep reaction produces two CO_2, and the citric acid cycle produces four CO_2 per glucose molecule.

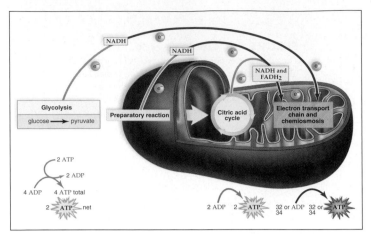

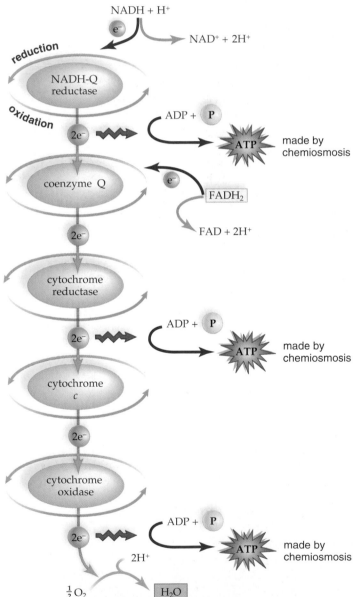

FIGURE 8.7 The electron transport chain.
NADH and FADH$_2$ bring electrons to the electron transport chain. As the electrons move down the chain, energy is captured and used to form ATP. For every pair of electrons that enters by way of NADH, three ATP result. For every pair of electrons that enters by way of FADH$_2$, two ATP result. Oxygen, the final acceptor of the electrons, becomes a part of water.

Electron Transport Chain

The **electron transport chain** located in the cristae of the mitochondria and the plasma membrane of aerobic prokaryotes is a series of carriers that pass electrons from one to the other. The electrons that enter the electron transport chain are carried by NADH and FADH$_2$. Figure 8.7 is arranged to show that high-energy electrons enter the chain, and low-energy electrons leave the chain. When NADH gives up its electrons, it becomes NAD$^+$; the next carrier gains the electrons and is reduced. This oxidation-reduction reaction starts the process, and each of the carriers in turn becomes reduced and then oxidized as the electrons move down the chain. Many of the carriers are cytochrome molecules. A **cytochrome** is a protein that has a tightly bound heme group with a central atom of iron, the same as hemoglobin does. The iron becomes reduced when it accepts electrons and becomes oxidized when it gives them up. A number of poisons, such as cyanide, cause death by binding to and blocking the function of cytochromes.

As the pair of electrons is passed from carrier to carrier, energy is captured and eventually used to form ATP molecules. The term **oxidative phosphorylation** refers to the production of ATP as a result of energy released by the electron transport chain. Oxygen receives the energy-spent electrons from the last of the carriers (i.e., cytochrome oxidase). After receiving electrons, oxygen combines with hydrogen ions, and water forms:

$$\tfrac{1}{2} O_2 + 2\,e^- + 2\,H^+ \longrightarrow H_2O$$

The critical role of oxygen as the final acceptor of electrons during cellular respiration is exemplified by noting that if oxygen is not present, the chain does not function, and no ATP is produced by mitochondria. The limited capacity of the body to form ATP in a way that does not involve the electron transport chain means that death will eventually result.

When NADH delivers electrons to the first carrier of the electron transport chain, enough energy is captured by the time the electrons are received by O$_2$ to permit the production of three ATP molecules. When FADH$_2$ delivers electrons to the electron transport chain, only two ATP are produced.

Once NADH has delivered electrons to the electron transport chain, it is "free" to return and pick up more hydrogen atoms. In the same manner, the components of ATP are recycled in cells. Energy is required to join ADP + Ⓟ and then when ATP is used to do cellular work, ADP and Ⓟ result once more. The recycling of coenzymes and ADP increases cellular efficiency since it does away with the necessity to synthesize NAD$^+$, FAD, and ADP anew.

The Cristae of a Mitochondrion

The carriers of the electron transport chain and the proteins concerned with ATP synthesis are spatially arranged in a particular manner in the cristae of mitochondria. Essentially, the electron transport chain consists of three protein complexes and two carriers. The three protein complexes include NADH-Q reductase complex, the cytochrome reductase complex, and cytochrome oxidase complex. The two other carriers that transport electrons between the complexes are coenzyme Q and cytochrome *c* (Fig. 8.8).

The members of the electron transport chain accept electrons, which they pass from one to the other. What happens to the hydrogen ions (H^+) carried by NADH and $FADH_2$? The complexes of the electron transport chain use the released energy to pump these hydrogen ions from the matrix into the intermembrane space of a mitochondrion. The vertical arrows in Figure 8.8 show that all the protein complexes of the electron transport chain all pump H^+ into the intermembrane space. This establishes a strong electrochemical gradient; there are about ten times as many hydrogen ions in the intermembrane space as there are in the matrix.

ATP Production. The cristae contain an ATP synthase complex through which hydrogen ions flow down a gradient from the intermembrane space into the matrix. As hydrogen ions flow from high to low concentration, the enzyme ATP synthase synthesizes ATP from ADP + Ⓟ. This process is called **chemiosmosis** because ATP production is tied to the establishment of an H^+ gradient.

The finding of respiratory poisons has lent support to the chemiosmotic model of ATP synthesis. When one poison inhibits ATP synthesis, the H^+ gradient subsequently becomes larger than usual. When another poison makes the membrane leaky so that an H^+ gradient does not form, no ATP is made.

Once formed, ATP moves out of mitochondria and is used to perform cellular work, during which it breaks down to ADP and Ⓟ. Then these molecules are returned to mitochondria for recycling. At any given time, the amount of ATP in a human would sustain life for only about a minute; therefore, ATP synthase must constantly produce ATP. It is estimated that mitochondria produce our body weight in ATP every day.

Active tissues require greater amounts of ATP and have more mitochondria than less active cells. The dark meat of chickens, the legs, contains more mitochondria than the white meat of the breast. This suggests that chickens mainly walk or run, rather than fly about the barnyard. The color of dark meat is also due to the presence of blood vessels and a respiratory pigment called myoglobin that is found in muscles.

Rigor mortis is the hardening and stiffening of the body three to four hours after death. This phenomenon is also related to ATP production. To relax, muscle fibers require ATP. After death, no more ATP is produced and the muscle fibers remain contracted and stiff. Usually rigor mortis peaks 12 hours after death and begins to diminish over the next 48 to 60 hours.

> Mitochondria synthesize ATP by chemiosmosis. ATP production is dependent on an H^+ gradient established by the pumping of H^+ into the intermembrane space.

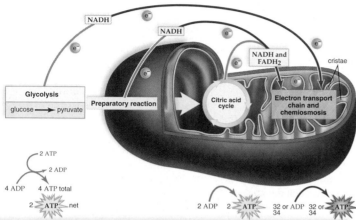

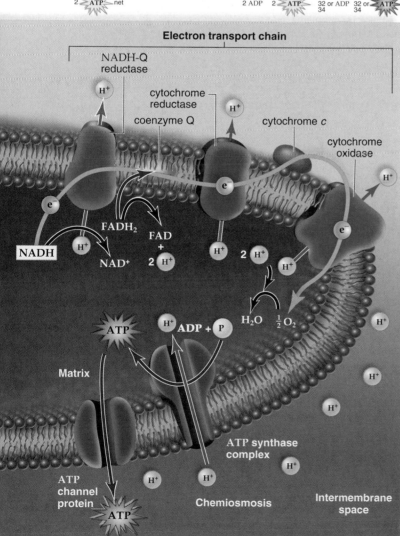

FIGURE 8.8 Organization and function of cristae.
The electron transport chain is located in the cristae. As electrons move from one protein complex to the other, hydrogen ions (H^+) are pumped from the matrix into the intermembrane space. As hydrogen ions flow down a concentration gradient from the intermembrane space into the mitochondrial matrix, ATP is synthesized by the enzyme ATP synthase. ATP leaves the matrix by way of a channel protein.

ecology focus

Carbon Monoxide: A Deadly Poison

Carbon monoxide (CO) is an air pollutant that comes primarily from the incomplete combustion of natural gas and gasoline. However, it can be generated by other sources, including the gaseous emissions of grazing animals. Figure 8A shows that transportation contributes most of the carbon monoxide to our cities' air. But power plants, factories, waste incineration, and home heating also contribute to the carbon monoxide level. Cigarette smoke contains carbon monoxide and is delivered directly to the smoker's blood and also to nonsmokers nearby.

Because carbon monoxide is a colorless, odorless gas, people can be unaware that it is affecting their systems. But it binds to iron 200 times more tightly than oxygen. Hemoglobin contains iron, and so does cytochrome oxidase, the carrier in the electron transport chain that passes electrons on to oxygen. When these molecules bind to carbon monoxide preferentially, they cannot perform their

usual functions. The result is that delivery of oxygen to mitochondria is impaired, and so is the functioning of the mitochondria.

Flushed red skin, especially on facial cheeks, is a first sign of carbon monoxide poisoning, because hemoglobin bound to carbon monoxide is brighter red than oxygenated hemoglobin. Marked euphoria, then sleepiness, coma, and death follow. Removing a person from the carbon monoxide source is not sufficient treatment because carbon monoxide, unlike oxygen, remains tightly bound to iron for many hours. A transfusion of red blood cells will help increase the carrying capacity of the blood, and pure oxygen given under pressure will displace some carbon monoxide. Despite good medical care, some people still die each year from CO poisoning.

The level of carbon monoxide in polluted air may not be sufficient to kill people, but it does interfere with the body's ability to function properly. Older people and those with

cardiovascular disease are especially at risk. One study found that even levels once thought to be safe can cause angina patients to experience chest pains, because oxygen delivery to the heart is reduced.

We can all lessen air pollution and reduce the amount of carbon monoxide in the air by doing the following:

- Don't smoke (especially indoors).
- Walk or bicycle instead of driving.
- Use public transportation instead of driving.
- Heat your home with solar energy—not a furnace.
- Support the development of more efficient automobiles, factories, power plants, and home furnaces.
- Support the development of alternative fuels. When hydrogen gas is burned, for example, the result is water, not carbon monoxide and carbon dioxide.
- Support better science and environmental education.

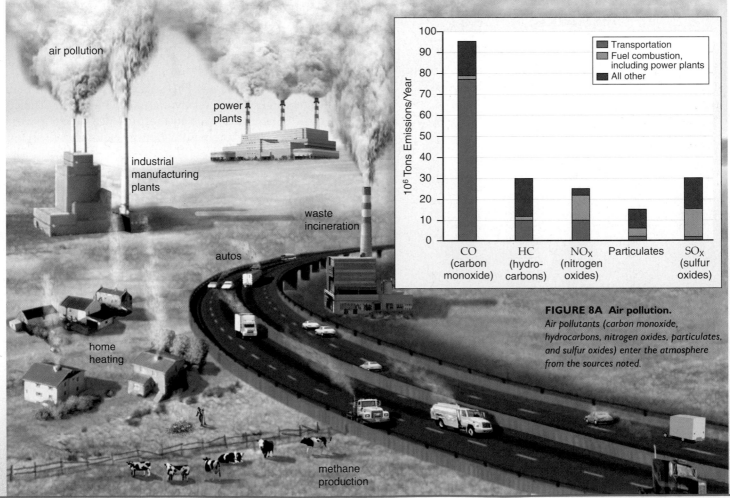

FIGURE 8A   Air pollution.
Air pollutants (carbon monoxide, hydrocarbons, nitrogen oxides, particulates, and sulfur oxides) enter the atmosphere from the sources noted.

Energy Yield from Glucose Metabolism

Figure 8.9 calculates the ATP yield for the complete breakdown of glucose to CO_2 and H_2O during cellular respiration. Notice that the diagram includes the number of ATP produced directly by glycolysis and the citric acid cycle, as well as the number produced as a result of electrons passing down the electron transport chain. Thirty-two or 34 molecules of ATP are produced by the electron transport chain.

Per glucose molecule, there is a net gain of two ATP from glycolysis, which takes place in the cytoplasm. The citric acid cycle, which occurs in the matrix of mitochondria, accounts for two ATP per glucose molecule. This means that a total of four ATP are formed by substrate-level phosphorylation outside the electron transport chain.

Most ATP is produced by the electron transport chain and chemiosmosis. Per glucose molecule, ten NADH and two $FADH_2$ take electrons to the electron transport chain. For each NADH formed *inside* the mitochondria by the citric acid cycle, three ATP result, but for each $FADH_2$, only two ATP are produced. Figure 8.7 explains the reason for this difference: $FADH_2$ delivers its electrons to the transport chain after NADH, and therefore these electrons cannot account for as much ATP production.

What about the ATP yield of NADH generated *outside* the mitochondria by the glycolytic pathway? NADH cannot cross mitochondrial membranes, but a "shuttle" mechanism allows its electrons to be delivered to the electron transport chain inside the mitochondria. The shuttle consists of an organic molecule, which can cross the outer membrane, accept the electrons, and in most but not all cells, deliver them to a FAD molecule in the inner membrane. If FAD is used, only two ATP result because the electrons have not entered the start of the electron transport chain.

Efficiency of Cellular Respiration

It is interesting to calculate how much of the energy in a glucose molecule eventually becomes available to the cell. The difference in energy content between the reactants (glucose and O_2) and the products (CO_2 and H_2O) is 686 kcal. An ATP phosphate bond has an energy content of 7.3 kcal, and 36 of these are usually produced during glucose breakdown; 36 phosphates are equivalent to a total of 263 kcal. Therefore, 263/686, or 39%, of the available energy is usually transferred from glucose to ATP. The rest of the energy is lost in the form of heat.

8.4 FERMENTATION

Some forms of respiration are **anaerobic.** They occur in the absence of oxygen and do not use oxygen as a final electron receptor. Anaerobic bacteria that live in the intestines of certain herbivorous animals and in other anaerobic environments can use inorganic molecules such as nitrate and sulfate as electron acceptors. Other organisms carry on **fermentation,** which also occurs anaerobically.

In fermentation, pyruvate—the end product of glycolysis—is reduced by NADH to either alcohol and carbon dioxide or to lactate (Fig. 8.10), depending on an organism's particular enzymes. During both types of fermentation, NADH

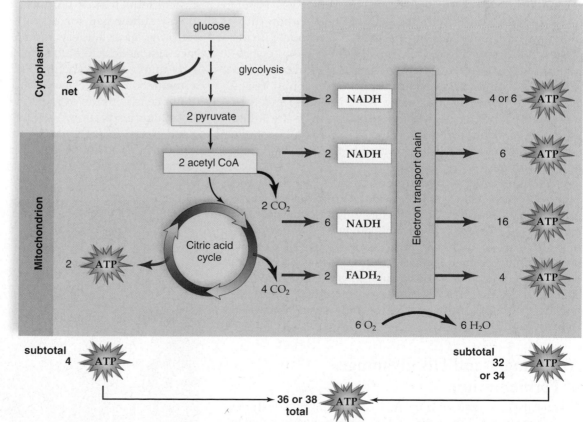

FIGURE 8.9 Accounting of energy yield per glucose molecule breakdown. Substrate-level phosphorylation during glycolysis and the citric acid cycle accounts for 4 ATP. Oxidative phosphorylation accounts for 32 or 34 ATP, and the grand total of ATP is therefore 36 or 38 ATP. Cells differ as to the delivery of the electrons from NADH generated outside the mitochondria. If they are delivered by a shuttle mechanism to the start of the electron transport chain, 6 ATP result; otherwise, 4 ATP result.

FIGURE 8.10 Fermentation.
Fermentation consists of glycolysis followed by a reduction of pyruvate. This "frees" NAD$^+$ and it returns to the glycolytic pathway to pick up more electrons.

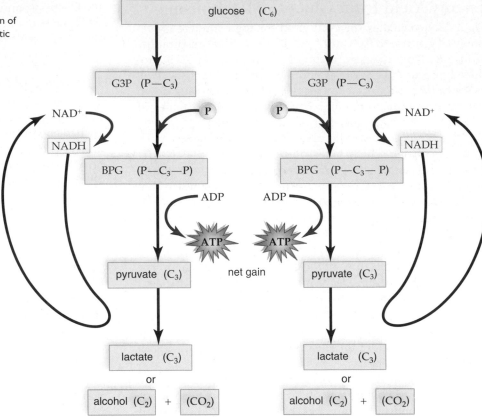

reduces pyruvate, and then it is "free" to return and pick up more electrons during an earlier reaction of glycolysis.

Yeasts are unicellular fungi that carry on **alcoholic fermentation** when oxygen is absent. When yeasts ferment, they produce both ethyl alcohol and carbon dioxide. Yeast is used in baking because the carbon dioxide makes bread rise (Fig. 8.11). Usually, the alcohol evaporates; however, once in a while fresh bread or pizza dough may have a slight taste of unevaporated alcohol. Yeasts are used to ferment sugar when wine is made, and then the ethyl alcohol is desired. Eventually yeasts are killed by the very alcohol they produce. In addition, alcoholic fermentation is utilized to produce beer and other alcoholic beverages.

Animals, including humans, certain bacteria, and fungi carry on **lactic acid fermentation,** during which NADH reduces pyruvate to lactate (Fig. 8.10). Typically, muscle tissue ferments in animals when oxygen is not being delivered to cells at a rapid enough rate. Anaerobic bacteria that produce lactate are used in the production of cheese, yogurt, and sauerkraut. Other bacteria produce chemicals of industrial importance, including isopropanol, butyric acid, proprionic acid, and acetic acid when they ferment.

Advantages and Disadvantages of Fermentation

Despite its low yield of only two ATP made by substrate-level phosphorylation, lactic acid fermentation is essential to

certain animals and/or tissues. Typically, animals use lactic acid fermentation for only a short period of time. However, in a common pond turtle known as a red-eared slider, lactic acid fermentation produces enough ATP for the animal to stay submerged for up to two weeks. Physicians, who perform cancer surgery, report that they observe the presence of a lactic acid halo surrounding new tumors. During early tumor development, ATP is formed only by lactic acid fermentation. Later, new blood vessel formation provides tumor cells with the oxygen and glucose they need to carry on cellular respiration. New medicines designed to prevent blood vessel formation near tumors are being developed. Without a blood supply, the tumor shrinks and disappears. Normally, lactic acid fermentation provides a rapid burst of

FIGURE 8.11
Products of fermentation.
Humans use a number of the by-products of fermentation. Fermentation is important in the bakery, cheese, and wine industries.

health focus

Exercise: A Test of Homeostatic Control

Exercise is a dramatic test of the body's homeostatic control systems—there is a large increase in muscle oxygen (O_2) requirement, and a large amount of carbon dioxide (CO_2) is produced. These changes must be countered by increases in breathing and blood flow to increase oxygen delivery and removal of the metabolically produced carbon dioxide. Also, heavy exercise can produce a large amount of lactate due to the use of fermentation, an anaerobic process. The accumulation of both carbon dioxide and lactate can lead to an increase in intracellular and extracellular acidity. Further, during heavy exercise, working muscles produce large amounts of heat that must be removed to prevent overheating. In a strict sense, the body rarely maintains true homeostasis while performing intense exercise or during prolonged exercise in a hot or humid environment. However, better maintenance of homeostasis is observed in those who have had endurance training.

The number of mitochondria increases in the muscles of persons who train; therefore, their bodies rely more on the citric acid cycle and the electron transport chain to generate energy. Muscle cells with few mitochondria must have a high ADP concentration to stimulate the limited number of mitochondria to start consuming oxygen. After an endurance training program, the large number of mitochondria start consuming oxygen as soon as the ADP concentration starts ris-

In athletes, there is:

- a smaller oxygen debt due to a more rapid increase in oxygen uptake at the onset of work

- an increase in fat metabolism, which spares blood glucose

- a reduction in lactate and hydrogen ion (H^+) formation

- an increase in ability to remove and process lactate

ing due to muscle contraction and subsequent breakdown of ATP. Therefore, a steady state of oxygen intake by mitochondria is achieved earlier in the athlete. This faster rise in oxygen uptake at the onset of work means that oxygen debt is less, and the formation of lactate due to fermentation is less. Further, any lactate that is produced is removed and processed more quickly.

Training also results in greater reliance on the citric acid cycle and increased fatty acid metabolism, because fatty acids are broken down to acetyl CoA, which enters the citric acid cycle. This preserves plasma glucose concentration and also helps the body maintain homeostasis.

ATP when immediate muscle contraction is needed. Also, when muscles are working vigorously over a short period of time, lactic acid fermentation provides them with ATP, even though oxygen is temporarily in limited supply.

Lactate, however, is toxic to cells. At first, blood carries away all the lactate formed in muscles. But eventually, lactate begins to build up, changing the pH and causing the muscles to fatigue and perhaps cramp. After running for awhile, our bodies are in **oxygen debt,** a term that refers to the amount of oxygen needed to rid the body of lactate. Oxygen debt is evidenced when we continue breathing heavily for a time after exercise. Recovery involves transporting most of the lactate to the liver, where it is converted back to pyruvate. Some of the pyruvate is respired completely, and the rest is converted back to glucose.

Efficiency of Fermentation

The two ATP produced per glucose during alcoholic fermentation and lactic acid fermentation are equivalent to 14.6 kcal. Complete glucose breakdown to CO_2 and H_2O represents a possible energy yield of 686 kcal per molecule. Therefore, the efficiency of fermentation is only 14.6 kcal/686 kcal $\times$ 100, or 2.1%. This is much less efficient than the complete breakdown of glucose. The inputs and outputs of fermentation are shown here:

Fermentation

inputs	outputs
glucose	2 lactate or 2 alcohol and 2 CO_2

2 ADP + 2 P   2 ATP net gain

8.5 Metabolic Pool

Degradative reactions, which participate in **catabolism,** break down molecules and tend to be exergonic. Synthetic reactions, which participate in **anabolism,** tend to be endergonic. It is correct to say that catabolism drives anabolism because catabolism results in an ATP synthesis that is used by anabolism.

Catabolism

We already know that glucose is broken down during cellular respiration. However, other molecules can also undergo catabolism. When a fat is used as an energy source, it breaks down to glycerol and three fatty acids. As Figure 8.12 indicates, glycerol can enter glycolysis. The fatty acids are converted to acetyl CoA, which enters the citric acid cycle. An 18-carbon fatty acid results in nine acetyl CoA molecules. Calculation shows that respiration of these can produce a total of 108 ATP molecules. For this reason, fats are an efficient form of stored energy—there are three long fatty acid chains per fat molecule.

The carbon skeleton of amino acids can enter glycolysis, be converted to acetyl CoA, or enter the citric acid cycle. The carbon skeleton is produced in the liver when an amino acid undergoes **deamination,** or the removal of the amino group. The amino group becomes ammonia (NH_3), which enters the urea cycle and becomes part of urea, the primary excretory product of humans. Just where the carbon skeleton begins degradation depends on the length of the R group, since this determines the number of carbons left after deamination.

Anabolism

We have already mentioned that the ATP produced during catabolism drives anabolism. But catabolism is also related to anabolism in another way. The substrates making up the pathways in Figure 8.12 can be used as starting materials for synthetic reactions. In other words, compounds that enter the pathways are oxidized to substrates that can be used for biosynthesis. This is the cell's **metabolic pool,** in which one type of molecule can be converted to another. In this way, carbohydrate intake can result in the formation of fat. G3P of glycolysis can be converted to glycerol, and acetyl groups can be joined to form fatty acids. Fat synthesis follows. This explains why you gain weight from eating too much candy, ice cream, or cake.

Some substrates of the citric acid cycle can be converted to amino acids through transamination, the transfer of an amino group to an organic acid, forming a different amino acid. Plants are able to synthesize all of the amino acids they need. Animals, however, lack some of the enzymes necessary for synthesis of all amino acids. Adult humans, for example, can synthesize 11 of the common amino acids, but they cannot synthesize the other 9. The amino

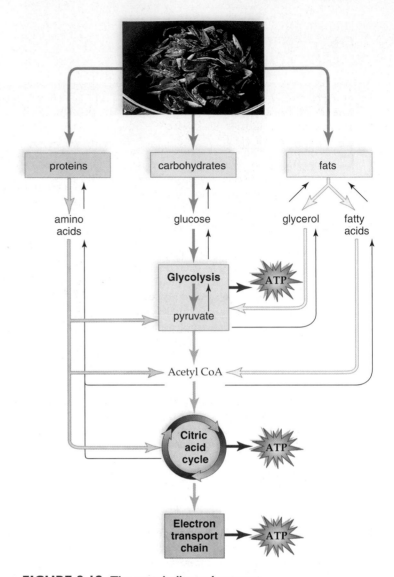

FIGURE 8.12 The metabolic pool concept.
Carbohydrates, fats, and proteins can be used as energy sources, and they enter degradative pathways at specific points. Catabolism produces molecules that can also be used for anabolism of other compounds.

acids that cannot be synthesized must be supplied by the diet; they are called the essential amino acids. (The amino acids that can be synthesized are called nonessential.) It is quite possible for animals to suffer from protein deficiency if their diets do not contain adequate quantities of all the essential amino acids.

> All the reactions involved in cellular respiration are part of a metabolic pool, and their substrates can be used for catabolism or for anabolism.

CONNECTING THE CONCEPTS

The organelles of eukaryotic cells have a structure that suits their function. Cells didn't arise until they had a membranous covering, and membrane is also absolutely essential to the organization of chloroplasts and mitochondria. In a chloroplast, membrane forms the grana, which are stacks of interconnected, flattened membranous sacs called thylakoids. The inner membrane of a mitochondrion invaginates to form the convoluted cristae.

The detailed structure of chloroplasts and mitochondria is different, but essentially they operate similarly: An assembly line of particles in the thylakoid membrane and cristae carry out functions necessary to photosynthesis and cellular respiration, respectively.

In both organelles, electron transport chain carriers pump hydrogen ions into an enclosed space, establishing an electrochemical gradient. When hydrogen ions flow down this gradient through an ATP synthase complex, the energy released is used to produce ATP. In chloroplasts, the electrons sent to the electron transport chain are energized by the sun; in mitochondria, energized electrons are removed from the substrates of glycolysis and the citric acid cycle.

In chloroplasts, the gel-like fluid of the stroma contains the enzymes of the Calvin cycle, which reduce carbon dioxide to a carbohydrate, and in mitochondria, the gel-like fluid of the matrix contains the enzymes of the citric acid cycle, which oxidize carbohydrate products received from the cytoplasm. In chloroplasts, reduction of carbon dioxide requires ATP produced by chemiosmosis, while in mitochondria, oxidation of substrates releases carbon dioxide; the ATP produced by chemiosmosis is made available to the cell.

According to the endosymbiotic hypothesis, chloroplasts and mitochondria were independent prokaryotic organisms at one time. Indeed, each contains genes (DNA) not found in the eukaryotic nucleus. Through evolution, all organisms are related, and the similar organization of these organelles suggests that they may be related also.

Summary

8.1 CELLULAR RESPIRATION

The oxidation of glucose to CO_2 and H_2O is an exergonic reaction that drives ATP synthesis, an endergonic reaction. Four phases are required: glycolysis, the prep reaction, the citric acid cycle, and passage of electrons along the electron transport chain. Oxidation of substrates involves the removal of hydrogen atoms ($H^+ + e^-$) from the substrate molecules, usually by redox coenzymes. NAD becomes NADH, and FAD becomes $FADH_2$.

8.2 OUTSIDE THE MITOCHONDRIA: GLYCOLYSIS

Glycolysis, the breakdown of glucose to two molecules of pyruvate, is a series of enzymatic reactions that occur in the cytoplasm. Breakdown releases enough energy to immediately give a net gain of two ATP by substrate-level phosphorylation. Two NADH are formed. Glycolysis is anaerobic.

8.3 INSIDE THE MITOCHONDRIA

When oxygen is available, pyruvate from glycolysis enters the mitochondrion, where the prep reaction takes place. During this reaction, oxidation occurs as CO_2 is removed from pyruvate. NAD^+ is reduced, and CoA receives the C_2 acetyl group that remains. Since the reaction must take place twice per glucose molecule, two NADH result.

The acetyl group enters the citric acid cycle, a cyclical series of reactions located in the mitochondrial matrix. Complete oxidation follows, as two CO_2 molecules, three NADH molecules, and one $FADH_2$ molecule are formed. The cycle also produces one ATP molecule. The entire cycle must turn twice per glucose molecule.

The final stage of glucose breakdown involves the electron transport chain located in the cristae of the mitochondria. The electrons received from NADH and $FADH_2$ are passed down a chain of carriers until they are finally received by oxygen, which combines with H^+ to produce water. As the electrons pass down the chain, energy is captured and stored for ATP production. The term *oxidative phosphorylation* is sometimes used for ATP production associated with the electron transport chain.

The cristae of mitochondria contain complexes of the electron transport chain that not only pass electrons from one to the other but also pump H^+ into the intermembrane space, setting up an electrochemical gradient. When H^+ flows down this gradient through an ATP synthase complex, energy is captured and used to form ATP molecules from ADP and $\textcircled{P}$. This is ATP synthesis by chemiosmosis.

Of the 36 or 38 ATP formed by complete glucose breakdown, four are the result of substrate-level phosphorylation and the rest are produced by oxidative phosphorylation. The energy for the latter comes from the electron transport chain. For most NADH molecules that donate electrons to the electron transport chain, three ATP molecules are produced. However, in some cells each NADH formed in the cytoplasm results in only two ATP molecules. This occurs when the hydrogen atoms are shuttled across the mitochondrial membrane by a carrier that passes them to FAD. $FADH_2$ results in the formation of only two ATP because its electrons enter the electron transport chain at a lower energy level.

8.4 FERMENTATION

Fermentation involves glycolysis followed by the reduction of pyruvate by NADH either to lactate or to alcohol and carbon dioxide (CO_2). The reduction process "frees" NAD^+ so that it can accept more hydrogen atoms from glycolysis.

Although fermentation results in only two ATP molecules, it still serves a purpose. Many of the products of fermentation are used in the baking and brewing industries. In vertebrates, it provides a quick burst of ATP energy for short-term, strenuous muscular activity. The accumulation of lactate puts the individual in oxygen debt because oxygen is needed when lactate is completely metabolized to CO_2 and H_2O.

8.5 METABOLIC POOL

Carbohydrate, protein, and fat can be metabolized by entering the degradative pathways at different locations. These pathways also provide metabolites needed for the anabolism of various important substances. Therefore, catabolism and anabolism both use the same pools of metabolites.

Reviewing the Chapter

1. What is the overall chemical equation for the complete breakdown of glucose to CO_2 and H_2O? Explain how this is an oxidation-reduction reaction. Why is the reaction able to drive ATP synthesis?   132
2. What are NAD^+ and FAD? What are their functions?   132
3. What are the three pathways involved in the complete breakdown of glucose to carbon dioxide (CO_2) and water (H_2O)? What reaction is needed to join two of these pathways?   132–33
4. What are the main events of glycolysis? How is ATP formed?   134–35
5. Give the substrates and products of the prep reaction. Where does it take place?   136–37
6. What are the main events of the citric acid cycle?   137
7. What is the electron transport chain, and what are its functions?   138
8. Describe the organization of protein complexes within the cristae. Explain how the complexes are involved in ATP production.   138–39
9. Calculate the energy yield of glycolysis and complete glucose breakdown. Distinguish yields between substrate-level phosphorylation and oxidative phosphorylation.   140
10. What is fermentation, and how does it differ from glycolysis? Mention the benefit of pyruvate reduction during fermentation. What types of organisms carry out lactic acid fermentation, and what types carry out alcoholic fermentation?   141–43
11. Give examples to support the concept of the metabolic pool.   144

Testing Yourself

Choose the best answer for each question. For questions 1–8, identify the pathway involved by matching each description to the terms in the key.

KEY:
   a. glycolysis
   b. citric acid cycle
   c. electron transport chain

1. carbon dioxide (CO_2) given off
2. water (H_2O) formed
3. G3P
4. NADH becomes NAD^+
5. oxidative phosphorylation
6. cytochrome carriers
7. pyruvate
8. FAD becomes $FADH_2$
9. The prep reaction
   a. connects glycolysis to the citric acid cycle.
   b. gives off CO_2.
   c. uses NAD^+.
   d. results in an acetyl group.
   e. All of these are correct.
10. The greatest contributor of electrons to the electron transport chain is
   a. oxygen.
   b. glycolysis.
   c. the citric acid cycle.
   d. the prep reaction.
   e. fermentation.

11. Substrate-level phosphorylation takes place in
   a. glycolysis and the citric acid cycle.
   b. the electron transport chain and the prep reaction.
   c. glycolysis and the electron transport chain.
   d. the citric acid cycle and the prep reaction.
   e. Both b and d are correct.
12. Which of these is not true of fermentation?
   a. net gain of only two ATP
   b. occurs in cytoplasm
   c. NADH donates electrons to electron transport chain
   d. begins with glucose
   e. carried on by yeast
13. Fatty acids are broken down to
   a. pyruvate molecules, which take electrons to the electron transport chain.
   b. acetyl groups, which enter the citric acid cycle.
   c. amino acids, which excrete ammonia.
   d. glycerol, which is found in fats.
   e. All of these are correct.
14. How many ATP molecules are usually produced per NADH?
   a. 1   c. 36
   b. 3   d. 10
15. How many NADH molecules are produced during the complete breakdown of one molecule of glucose?
   a. 5   c. 10
   b. 30   d. 6
16. What is the name of the process that adds the third phosphate to an ADP molecule using the flow of hydrogen ions?
   a. substrate-level phosphorylation
   b. fermentation
   c. reduction
   d. chemiosmosis
17. Which are possible products of fermentation?
   a. lactic acid
   b. alcohol
   c. CO_2
   d. All of these are possible.
18. The metabolic process that produces the most ATP molecules is
   a. glycolysis.   c. electron transport chain.
   b. citric acid cycle.   d. fermentation.
19. Which of these is not true of citric acid cycle? The citric acid cycle
   a. includes the prep reaction.
   b. produces ATP by substrate-level phosphorylation.
   c. occurs in the mitochondria.
   d. is a metabolic pathway, as is glycolysis.
20. Which of these is not true of the electron transport chain? The electron transport chain
   a. is located on the cristae.
   b. produces more NADH than any metabolic pathway.
   c. contains cytochrome molecules.
   d. ends when oxygen accepts electrons.
21. Which of these is not true of the prep reaction? The prep reaction
   a. begins with pyruvate and ends with acetyl CoA.
   b. produces more NADH than does glycolysis.
   c. occurs in the mitochondria.
   d. occurs after glycolysis and before the citric acid cycle.

22. The oxygen required by cellular respiration is reduced and becomes part of which molecule?
    a. ATP
    b. H_2O
    c. pyruvate
    d. CO_2

For questions 23–26, match each pathway to metabolite in the key. Choose more than one if correct.

KEY:
   a. pyruvate
   b. acetyl CoA
   c. G3P
   d. NADH
   e. None of these are correct.

23. electron transport chain

24. glycolysis

25. citric acid cycle

26. prep reaction

27. Which of these is not true of glycolysis? Glycolysis
    a. is anaerobic.
    b. occurs in the cytoplasm.
    c. is a part of fermentation.
    d. evolved after the citric acid cycle.

28. Label this diagram of a mitochondrion, and state a function for each portion indicated.

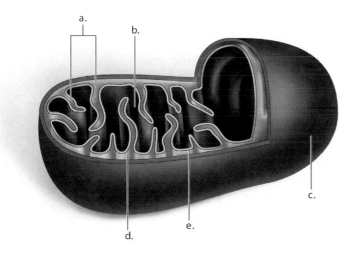

Thinking Scientifically

1. In mitochondria, NADH supplies electrons to the electron transport chain. Molecular oxygen (with the highest affinity for electrons) is the final receiver. In photosynthesis, however, the oxygen in water (H_2O) is the source of electrons and the final receiver is $NADP^+$. How can oxygen act as both a donor and receiver of electrons?

2. You are working with pyruvate molecules that contain radioactive carbon. They are incubated with all the components of the citric acid cycle long enough for one turn of the cycle. Carbon dioxide is produced, but only one-third of it is radioactive. How can this observation be explained?

Bioethical Issue: Alternative Medicine

Feeling tired and run-down? Want to jump-start your mitochondria? If you seem to have no specific ailment, you might be tempted to turn to what is now called alternative medicine. Alternative medicine includes such nonconventional therapies as herbal supplements, acupuncture, chiropractic therapy, homeopathy, osteopathy, and therapeutic touch (e.g., laying on of hands).

Advocates of alternative medicine have made some headway in having alternative medicine practices accepted by almost anyone. For example, Congress has established the National Center for Complementary and Alternative Medicine. It has also passed the Dietary Supplement Health and Education Act, which allows vitamins, minerals, and herbs to be marketed without first being approved by the Food and Drug Administration (FDA).

But is this a mistake? Many physicians believe control studies are needed to test the efficacy of alternative medications and practices. Do you agree, or is word of mouth good enough?

Understanding the Terms

acetyl CoA 136	fermentation 133, 141
alcoholic fermentation 142	glycolysis 133, 134
anabolism 144	lactic acid fermentation 142
anaerobic 133, 141	metabolic pool 144
catabolism 144	mitochondrion 136
cellular respiration 132	NAD^+ 132
chemiosmosis 139	oxidative
citric acid cycle 133, 137	phosphorylation 138
cytochrome 138	oxygen debt 142
deamination 144	preparatory (prep)
electron carrier 134	reaction 133, 136
electron transport	pyruvate 133
chain 133, 138	substrate-level
FAD 132	phosphorylation 134

Match the terms to these definitions:
a. _____ A metabolic pathway that begins with glucose and ends with two molecules of pyruvate.
b. _____ Occurs due to the accumulation of lactate following vigorous exercise.
c. _____ Metabolic process that degrades molecules and tends to be exergonic.
d. _____ Type of ATP production that uses oxygen as the final acceptor for electrons.
e. _____ Flow of hydrogen ions down their concentration gradient through an ATP synthase and resulting in ATP production.
f. _____ Metabolic pathway described by the equation:
$$C_6H_{12}O_6 + 6\,O_2 \longrightarrow 6\,CO_2 + 6\,H_2O$$

ARIS, the *Biology* Website

ARIS, the website for *Biology*, provides a wealth of information organized and integrated by chapter. You will find practice quizzes, interactive activities, labeling exercises, flashcards, and much more that will complement your learning and understanding of general biology.

www.mhhe.com/maderbiology9

PART II

GENETIC BASIS OF LIFE

Hereditary information is stored in DNA, molecules that compose the genes located within chromosomes. When cells reproduce, chromosomes are distributed to daughter cells. In this way, DNA is passed on to all body cells or to the next generation of organisms. Knowing how cellular reproduction is regulated has contributed greatly to our knowledge of cancer and inherited disorders.

Principles of inheritance include those that allow us to predict the chances that an offspring will inherit a particular characteristic from one of the parents. These principles have been applied to the breeding of plants and animals and to the study of human genetic disorders. But to go further and control an organism's characteristics, it is necessary to understand how DNA and RNA function in protein synthesis. The human endeavor known as biotechnology is based on our newfound knowledge of nucleic acid structure and function.

The principles of inheritance are central to understanding many other topics in biology—from the evolution and diversity of life to the reproduction and development of organisms.

9

THE CELL CYCLE AND CELLULAR REPRODUCTION

Consider the development of a human being. A new life begins as one cell—an egg fertilized by a sperm. Yet in nine short months, a human becomes a complex organism consisting of trillions of cells. How is such a feat possible? Cell division enables a single cell to produce many cells, allowing an organism to grow and develop. Cell division also occurs when repair is needed and worn-out tissues need to be replaced. Adult humans have over 100 trillion cells working together in harmony.

The instructions for cell division lie in the genes. During the first part of an organism's life, the genes instruct all cells to divide. When adulthood is reached, however, only specific cells—human blood and skin cells, for example—continue to divide daily. Other cells, such as nerve cells, no longer routinely divide and produce new cells.

Since all types of adult cells contain the full complement of genetic material in their nuclei, why don't they all reproduce routinely? Such questions are being intensely studied. Cell biologists have recently discovered that specific signaling proteins regulate the cell cycle, the period extending from the time a new cell is produced until it, too, completes division. The presence or absence of signaling proteins ensures the regulation of cell division. Cancer results when the genes that code for these signaling proteins mutate and cell division occurs nonstop.

Cells come from preexisting cells.

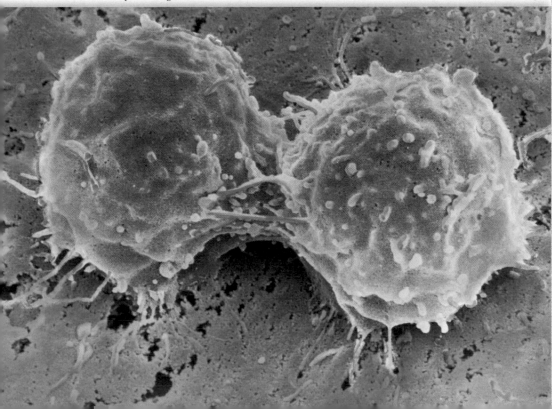

9.1 THE CELL CYCLE

The **cell cycle** is an orderly set of stages that take place between the time a eukaryotic cell divides and the time the resulting daughter cells also divide. When a cell is going to divide, it grows larger, the number of organelles doubles, and the amount of DNA doubles as DNA replication occurs. The two portions of the cell cycle are interphase, which includes a number of stages, and the mitotic stage, when mitosis and cytokinesis occur.

Interphase

As Figure 9.1a shows, most of the cell cycle is spent in **interphase.** This is the time when a cell performs its usual functions, depending on its location in the body. The amount of time the cell takes for interphase varies widely. Some cells, such as nerve and muscle cells, typically do not complete the cell cycle and are permanently arrested. These cells are said to have entered a G_0 stage. Embryonic cells complete the entire cell cycle in just a few hours. For adult mammalian cells, interphase lasts for about 20 hours, which is 90% of the cell cycle. In the past, interphase was known as the resting stage. However, today it is known that interphase is very busy and consists of three stages referred to as G_1, S, and G_2.

G_1 Stage

Cell biologists named the stage before DNA replication G_1, and they named the stage after DNA replication G_2. G stood for "gap," but now that we know how metabolically active the cell is, it is better to think of G as standing for "growth." Protein synthesis is very much a part of these growth stages.

During G_1, the cell recovers from the previous division. Then, the cell increases in size, doubles its organelles (such as mitochondria and ribosomes), and accumulates materials that will be used for DNA synthesis.

S Stage

Following G_1, the cell enters the S stage, when DNA synthesis or replication occurs. At the beginning of the S stage, each chromosome is composed of one DNA double helix. Following DNA replication, each chromosome is composed of two identical DNA double helix molecules. Each double helix is called a **chromatid.** Another way of expressing these events is to say that DNA replication has resulted in duplicated chromosomes.

G_2 Stage

Following the S stage, G_2 is the stage from the completion of DNA replication to the onset of mitosis. During this stage, the cell synthesizes proteins that will assist cell division. For example, it makes the proteins that form microtubules. The role of microtubules in cell division is described in a later section.

M (Mitotic) Stage

Following interphase, the cell enters the M (for mitotic) stage. This cell division stage includes **mitosis** (nuclear division) and **cytokinesis** (division of the cytoplasm). During mitosis, daughter chromosomes are distributed to two daughter nuclei. When division of the cytoplasm is complete, two daughter cells are present.

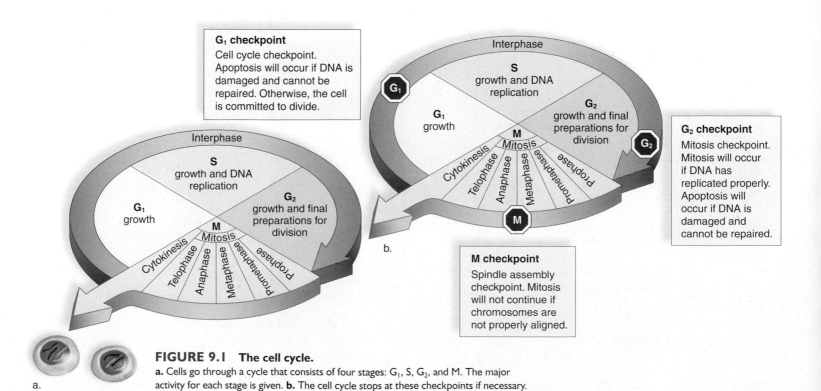

FIGURE 9.1   The cell cycle.
a. Cells go through a cycle that consists of four stages: G_1, S, G_2, and M. The major activity for each stage is given. **b.** The cell cycle stops at these checkpoints if necessary.

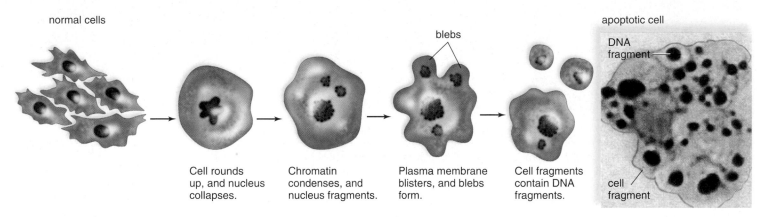

normal cells

blebs

apoptotic cell

DNA fragment

cell fragment

Cell rounds up, and nucleus collapses.

Chromatin condenses, and nucleus fragments.

Plasma membrane blisters, and blebs form.

Cell fragments contain DNA fragments.

FIGURE 9.2 Apoptosis.
Apoptosis is a sequence of events that results in a fragmented cell. The fragments are phagocytized by white blood cells and neighboring tissue cells.

Control of the Cell Cycle

The cell cycle is controlled by internal and external signals. A **signal** is a molecule that stimulates or inhibits a metabolic event. These signals ensure that the stages follow one another in the normal sequence and that each stage is properly completed before the next stage begins. **Growth factors** are external signals received at the plasma membrane. Even cells that are arrested in G_0 will finish the cell cycle if stimulated to do so by growth factors. For example, when blood platelets release a growth factor, skin fibroblasts in the vicinity are stimulated to finish the cell cycle, thereby repairing an injury.

Cell Cycle Checkpoints

The red barriers in Figure 9.1b represent three checkpoints when the cell cycle either stops or continues on, depending on the internal signal it receives. Researchers have identified a family of internal signaling proteins called **cyclins** that increase and decrease as the cell cycle continues. Cyclin has to be present for the cell to proceed from the G_2 stage to the M stage and for the cell to proceed from the G_1 stage to the S stage.

The cell cycle stops at the G_2 checkpoint if DNA has not finished replicating. This prevents the initiation of the M stage before completion of the S stage. Also, if DNA is damaged, such as from exposure to solar radiation or X-rays, stopping the cell cycle at this checkpoint allows time for the damage to be repaired so that it is not passed on to daughter cells.

Another cell cycle checkpoint occurs during the mitotic stage. The cycle stops if the chromosomes are not going to be distributed accurately to the daughter cells.

DNA damage can also stop the cell cycle at the G_1 checkpoint. In mammalian cells, the signaling protein p53 stops the cycle at the G_1 checkpoint when DNA is damaged. First, p53 attempts to initiate DNA repair, but if that is not possible, it brings about apoptosis.

Apoptosis

Apoptosis is often defined as programmed cell death because the cell progresses through a usual series of events that bring about its destruction (Fig. 9.2). The cell rounds up and loses contact with its neighbors. The nucleus fragments, and the plasma membrane develops blisters. Finally, the cell fragments, and its bits and pieces are engulfed by white blood cells and/or neighboring cells. A remarkable finding of the past few years is that cells routinely harbor the enzymes, now called caspases, that bring about apoptosis. The enzymes are ordinarily held in check by inhibitors but can be unleashed by either internal or external signals.

Apoptosis and Cell Division. In living systems, opposing events keep the body in balance and maintain homeostasis. There are many examples to support this statement. For now, consider that some carrier proteins transport molecules into the cell, and others transport molecules out of the cell. Some hormones increase the level of blood glucose, and others decrease the level. Similarly, two opposing processes keep the number of cells in the body at an appropriate level. In other words, cell division increases and apoptosis decreases the number of **somatic** (body) **cells.** Both mitosis and apoptosis are normal parts of growth and development. An organism begins as a single cell that repeatedly divides to produce many cells, but eventually some cells must die for the organism to take shape. For example, when a tadpole becomes a frog, the tail disappears as apoptosis occurs. The fingers and toes of a human embryo are at first webbed, but then they are usually freed from one another as a result of apoptosis.

Cell division occurs during your entire life. Even now, your body is producing thousands of new red blood cells, skin cells, and cells that line your respiratory and digestive tracts. Also, if you suffer a cut, cell division repairs the injury. Apoptosis occurs all the time too, particularly if an abnormal cell that could become cancerous appears. Death through apoptosis prevents a tumor from developing.

The cell cycle consists of interphase (G_1, S, and G_2) plus cell division (mitosis and cytokinesis). During the life of an organism, both cell division and apoptosis occur.

---

science focus

What's in a Chromosome?

When early investigators decided that the genes are contained in the chromosomes, they had no idea of chromosome composition. By the mid-1900s, it was known that chromosomes are made up of both DNA and protein. Only in recent years, however, have investigators been able to produce models suggesting how chromosomes are organized.

A eukaryotic chromosome is more than 50% protein. Some of these proteins are concerned with DNA and RNA synthesis, but a large proportion, termed histones, seem to play primarily a structural role. The five primary types of histone molecules are designated H1, H2A, H2B, H3, and H4. Remarkably, the amino acid sequences of H3 and H4 vary little between organisms. For example, the H4 of peas is only two amino acids different from the H4 of cattle. This similarity suggests that few mutations in the histone proteins have occurred during the course of evolution and that the histones therefore have important functions.

A human cell contains at least 2 m of DNA. Yet all of this DNA is packed into a nucleus that is about 5 μm in diameter. The histones are responsible for packaging the DNA so that it can fit into such a small space. First the DNA double helix is wound at intervals around a core of eight histone molecules (two copies each of H2A, H2B, H3, and H4), giving the appearance of a string of beads (Fig. 9A*a, b*). Each bead is called a nucleosome, and the nucleosomes are said to be joined by "linker" DNA. This string is coiled tightly into a fiber that has six nucleosomes per turn (Fig. 9A*c*). The H1 histone appears to mediate this coiling process. The fiber loops back and forth (Fig. 9A*d, e*) and can condense to produce a highly compacted form (Fig. 9A*f*) characteristic of metaphase chromosomes. No doubt, compact chromosomes are easier to move about than extended chromatin. During interphase, extended chromatin makes DNA available for RNA synthesis and subsequent protein synthesis. Many of these proteins are enzymes, and this accounts for ongoing metabolic activity during interphase.

Some regions of chromatin remain tightly coiled and compacted throughout the cell cycle and appear as dark stained fibers known as heterochromatin (Fig. 9A*e*). Heterochromatin is considered inactive chromatin. Euchromatin (Fig. 9A*d*), the active chromatin, is condensed only during cell division. The condensation allows for the movement of chromosomes. Most of the time, euchromatin is loosely coiled, allowing for gene activity.

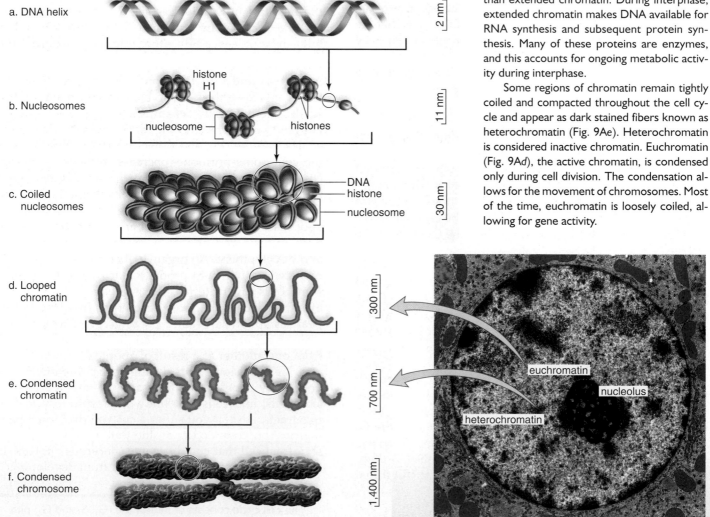

a. DNA helix   2 nm

histone H1

b. Nucleosomes   11 nm

nucleosome   histones

c. Coiled nucleosomes   30 nm

DNA
histone
nucleosome

d. Looped chromatin   300 nm

e. Condensed chromatin   700 nm

f. Condensed chromosome   1,400 nm

euchromatin

nucleolus

heterochromatin

2,730×

FIGURE 9A Levels of chromosome structure. *The level of euchromatin and heterochromatin condensation in the nucleus is noted.*

TABLE 9.1

Diploid Chromosome Numbers of Some Eukaryotes

Type of Organism	Name of Organism	Chromosome Number
Fungi	*Aspergillus nidulans* (mold)	8
	Neurospora crassa (mold)	14
	Saccharomyces cerevisiae (yeast)	32
Plants	*Vicia faba* (broad bean)	12
	Pisum sativum (garden pea)	14
	Zea mays (corn)	20
	Solanum tuberosum (potato)	48
	Nicotiana tabacum (tobacco)	48
	Ophioglossum vulgatum (Southern adder's tongue fern)	1,320
Animals	*Drosophila melanogaster* (fruit fly)	8
	Rana pipiens (frog)	26
	Felis domesticus (cat)	38
	Homo sapiens (human)	46
	Pan troglodytes (chimp)	48
	Carassius auratus (goldfish)	94

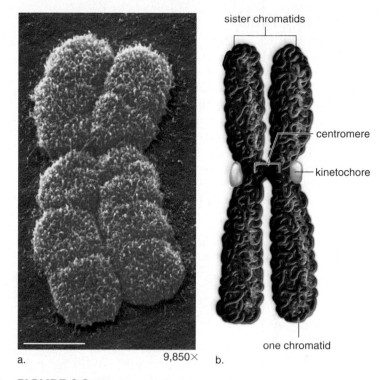

a.   9,850×   b.

FIGURE 9.3 Duplicated chromosomes.
A duplicated chromosome contains two sister chromatids, each with copies of the same genes. **a.** Electron micrograph of a highly coiled and condensed chromosome, typical of a nucleus about to divide. **b.** Diagrammatic drawing of a condensed chromosome. The chromatids are held together at a region called the centromere.

9.2 MITOSIS AND CYTOKINESIS

As mentioned, cell division in eukaryotes involves mitosis, which is nuclear division (also called karyokinesis), and cytokinesis, which is division of the cytoplasm. During mitosis, chromosomes are distributed to two daughter cells.

Eukaryotic Chromosomes

The DNA in the chromosomes of eukaryotes is associated with various proteins, including *histone proteins* that are especially involved in organizing chromosomes. When a eukaryotic cell is not undergoing division, the DNA (and associated proteins) within a nucleus is a tangled mass of thin threads called **chromatin** [Gk. *chroma*, color, and *teino*, stretch]. Before mitosis begins, as discussed in the Science Focus on the previous page, chromatin becomes highly coiled and condensed, and it is easy to see the individual chromosomes.

When the chromosomes are visible, it is possible to photograph and count them. Each species has a characteristic chromosome number (Table 9.1); for instance, human cells contain 46 chromosomes, corn has 20 chromosomes, and a goldfish has 94. This is the full or **diploid (2n) number** [Gk. *diplos*, twofold, and *-eides*, like] of chromosomes that is found in all cells of the body. The diploid number includes two chromosomes of each kind. Half the diploid number, called the **haploid (n) number** [Gk. *haplos*, single, and *-eides*, like] of chromosomes, contains only one chromosome of

each kind. Typically, only sperm and eggs have the haploid number of chromosomes in the life cycle of animals.

During mitosis, a 2n nucleus divides to produce daughter nuclei that are also 2n. The dividing cell is called the parent cell, and the resulting cells are called the daughter cells. Before nuclear division takes place, DNA replicates, duplicating the chromosomes in the parent cell. Each chromosome now has two identical double helix molecules; each double helix is a chromatid, and the two identical chromatids are called **sister chromatids** (Fig. 9.3). Sister chromatids are constricted and attached to each other at a region called the **centromere.** Protein complexes called *kinetochores* develop on either side of the centromere during cell division.

During nuclear division, the two sister chromatids separate at the centromere, and in this way each duplicated chromosome gives rise to two daughter chromosomes. Each daughter chromosome has only one double helix molecule. The daughter chromosomes are distributed equally to the daughter cells. In this way, each daughter nucleus gets a copy of each chromosome that was in the parent cell.

Each type of eukaryote has a characteristic number of chromosomes in the nucleus of each cell. The chromosomes duplicate prior to mitosis, and this allows the chromosome number to stay constant despite nuclear division.

Phases of Mitosis

The **centrosome** [Gk. *centrum*, center, and *soma*, body], the main microtubule-organizing center of the cell, divides before mitosis begins. Each centrosome in an animal cell—but not a plant cell—contains a pair of barrel-shaped organelles called **centrioles.**

The centrosomes organize the mitotic **spindle**, which contains many fibers, each composed of a bundle of microtubules. Microtubules are hollow cylinders made up of the protein tubulin. They assemble when tubulin subunits join, and when they disassemble, tubulin subunits become free once more. The microtubules of the cytoskeleton disassemble when spindle fibers begin forming. Most likely, this provides tubulin for the formation of the spindle fibers.

Mitosis is a continuous process that is arbitrarily divided into five phases for convenience of description: prophase, prometaphase, metaphase, anaphase, and telophase (Fig. 9.4).

Prophase

It is apparent during **prophase** that nuclear division is about to occur because chromatin has condensed and the chromosomes are visible. Recall that DNA replication occurred during interphase, and therefore the *parental chromosomes are already duplicated and composed of two sister chromatids held together at a centromere.* Counting the number of centromeres in diagrammatic drawings gives the number of chromosomes for the cell depicted.

FIGURE 9.4 Phases of mitosis in animal and plant cells.

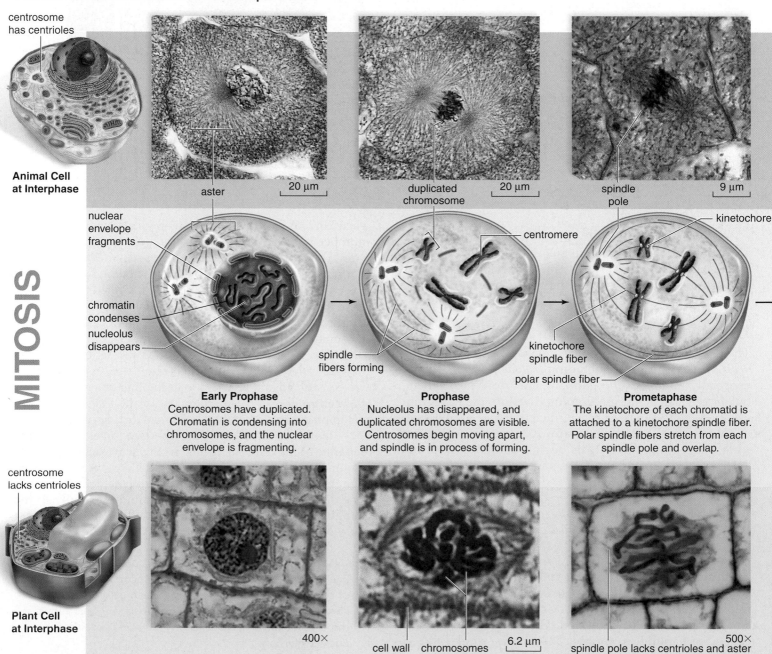

MITOSIS

Animal Cell at Interphase — centrosome has centrioles

aster   20 μm

duplicated chromosome   20 μm

spindle pole   9 μm

nuclear envelope fragments — chromatin condenses — nucleolus disappears — spindle fibers forming

centromere — kinetochore — kinetochore spindle fiber — polar spindle fiber

Early Prophase
Centrosomes have duplicated. Chromatin is condensing into chromosomes, and the nuclear envelope is fragmenting.

Prophase
Nucleolus has disappeared, and duplicated chromosomes are visible. Centrosomes begin moving apart, and spindle is in process of forming.

Prometaphase
The kinetochore of each chromatid is attached to a kinetochore spindle fiber. Polar spindle fibers stretch from each spindle pole and overlap.

Plant Cell at Interphase — centrosome lacks centrioles

400×

cell wall   chromosomes   6.2 μm

500×
spindle pole lacks centrioles and aster

During prophase, the nucleolus disappears and the nuclear envelope fragments. The spindle begins to assemble as the two centrosomes migrate away from one another. In animal cells, an array of microtubules radiates toward the plasma membrane from the centrosomes. These structures are called **asters.** It is thought that asters serve to brace the centrioles during later stages of cell division. Notice that the chromosomes have no particular orientation because the spindle has not formed as yet.

Prometaphase (Late Prophase)

During **prometaphase,** preparations for sister chromatid separation are evident. Kinetochores appear on each side of the centromere, and these attach sister chromatids to the so-called kinetochore spindle fibers. These fibers extend from the poles to the chromosomes, which will soon be located at the center of the spindle.

The kinetochore fibers attach the sister chromatids to opposite poles of the spindle, and the chromosomes are pulled first toward one pole and then toward the other before the chromosomes come into alignment. Notice that even though the chromosomes are attached to the spindle fibers in prometaphase, they are still not in alignment.

Metaphase

During **metaphase,** the centromeres of chromosomes are now in alignment at the center of the cell. When viewed with a light microscope, the chromosomes appear to be in

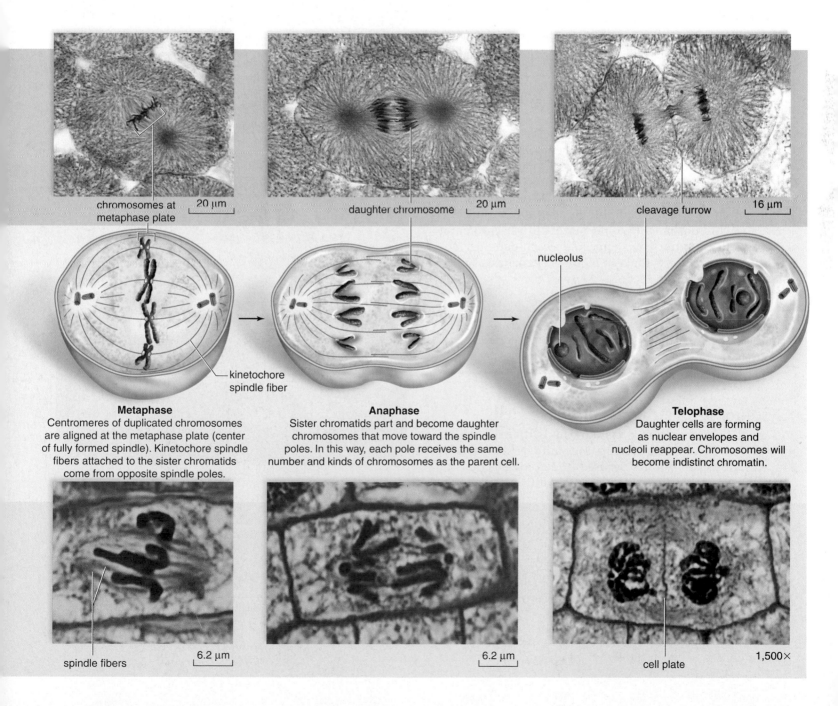

chromosomes at metaphase plate   20 μm

daughter chromosome   20 μm

cleavage furrow   16 μm

nucleolus

kinetochore spindle fiber

Metaphase
Centromeres of duplicated chromosomes are aligned at the metaphase plate (center of fully formed spindle). Kinetochore spindle fibers attached to the sister chromatids come from opposite spindle poles.

Anaphase
Sister chromatids part and become daughter chromosomes that move toward the spindle poles. In this way, each pole receives the same number and kinds of chromosomes as the parent cell.

Telophase
Daughter cells are forming as nuclear envelopes and nucleoli reappear. Chromosomes will become indistinct chromatin.

spindle fibers   6.2 μm

6.2 μm

cell plate   1,500×

a circle that encompasses the circumference of the cell. An imaginary plane that is perpendicular and passes through this circle is called the **metaphase plate.** It indicates the future axis of cell division. Many nonattached spindle fibers called *polar spindle fibers* are apparent in metaphase. Some of these fibers reach beyond the metaphase plate and overlap.

Anaphase

At the start of **anaphase,** the two sister chromatids of each duplicated chromosome separate at the centromere, giving rise to two daughter chromosomes. Daughter chromosomes, each with a centromere and single chromatid composed of a single double helix, appear to move toward opposite poles. Actually, the daughter chromosomes are being pulled to the opposite poles as the kinetochore spindle fibers disassemble at the region of the kinetochores. Even as the daughter chromosomes move toward the spindle poles, the poles themselves are moving farther apart because the polar spindle fibers are sliding past one another. Microtubule-associated proteins such as the motor molecules kinesin and dynein are involved in the sliding process. Anaphase is the shortest phase of mitosis.

Telophase

During **telophase,** the spindle disappears as new nuclear envelopes form around the daughter chromosomes. Each daughter nucleus contains the same number and kinds of chromosomes as the original parent cell. Remnants of the polar spindle fibers are still visible between the two nuclei.

The chromosomes become more diffuse chromatin once again, and a nucleolus appears in each daughter nucleus. Division of the cytoplasm requires cytokinesis, which is discussed in the next section.

During mitosis, daughter chromosomes go into daughter nuclei by a mechanism that ensures each daughter nucleus has a full set of chromosomes.

Cytokinesis in Animal and Plant Cells

As mentioned previously, cytokinesis is division of the cytoplasm. Cytokinesis accompanies mitosis in most cells but not all. When mitosis occurs but cytokinesis doesn't occur, the result is a multinucleated cell. For example, we will see that the embryo sac in flowering plants are multinucleated.

Division of the cytoplasm begins in anaphase, continues in telophase, but does not reach completion until the following interphase begins. By the end of mitosis each newly forming cell has received a share of the cytoplasmic organelles that duplicated during interphase.

Cytokinesis in Animal Cells

In animal cells, a cleavage furrow, which is an indentation of the membrane between the two daughter nuclei, begins as anaphase draws to a close. By that time, the newly forming cells have received a share of the cytoplasmic organelles that duplicated during the previous interphase.

The cleavage furrow deepens when a band of actin filaments, called the contractile ring, slowly forms a circular constriction between the two daughter cells. The action of the contractile ring can be likened to pulling a drawstring ever tighter about the middle of a balloon. As the drawstring is pulled tight, the balloon constricts in the middle as the material on either side of the constriction gathers in folds. These folds are represented by the longitudinal lines in Figure 9.5.

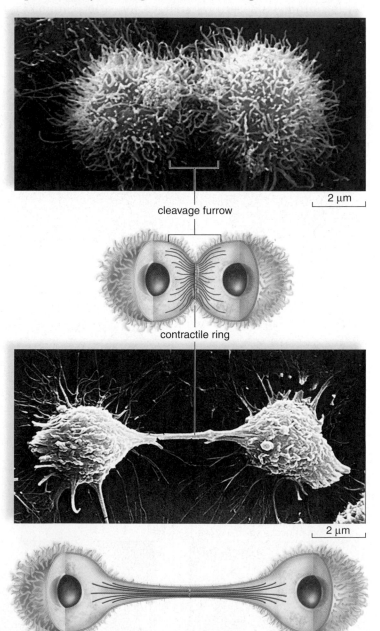

2 μm

cleavage furrow

contractile ring

2 μm

FIGURE 9.5 Cytokinesis in animal cells.
A single cell becomes two cells by a furrowing process. A contractile ring composed of actin filaments gradually gets smaller, and the cleavage furrow pinches the cell into two cells.
Copyright by R. G. Kessel and C. Y. Shih, *Scanning Electron Microscopy in Biology: A Students' Atlas on Biological Organization*, Springer-Verlag, 1974.

A narrow bridge between the two cells can be seen during telophase, and then the contractile ring continues to separate the cytoplasm until there are two independent daughter cells (Fig. 9.5).

Cytokinesis in Plant Cells

Cytokinesis in plant cells occurs by a process different from that seen in animal cells (Fig. 9.6). The rigid cell wall that surrounds plant cells does not permit cytokinesis by furrowing. Instead, cytokinesis in plant cells involves the building of new cell walls between the daughter cells.

Cytokinesis is apparent when a small, flattened disk appears between the two daughter plant cells. In electron micrographs, it is possible to see that the disk is at right angles to a set of microtubules. The Golgi apparatus produces vesicles, which move along the microtubules to the region of the disk. As more vesicles arrive and fuse, a cell plate can be seen. The **cell plate** is simply newly formed plasma membrane that expands outward until it reaches the old plasma membrane and fuses with this membrane. The new membrane releases molecules that form the new plant cell walls. These cell walls, known as primary cell walls, are later strengthened by the addition of cellulose fibrils. The space between the daughter cells becomes filled with middle lamella, which cements the primary cell walls together.

> Cytokinesis in animal cells is accomplished by a furrowing process. Cytokinesis in plant cells begins with the formation of a cell plate, which eventually becomes new plasma membrane between the daughter cells.

The Functions of Mitosis

Mitosis permits growth and repair. In both plants and animals, mitosis is required during development as a single cell develops into an individual. In plants, the individual could be a fern or daisy, while in animals, the individual could be a grasshopper or a human being.

In flowering plants, meristematic tissue retains the ability to divide throughout the life of a plant. Meristematic tissue at the shoot tip accounts for an increase in the height of a plant for as long as it lives. Then, too, lateral meristem accounts for the ability of trees to increase their girth each growing season.

In human beings and other mammals, mitosis is necessary as a fertilized egg becomes an embryo and as the embryo becomes a fetus. Mitosis also occurs after birth as a child becomes an adult. Throughout life, mitosis allows a cut to heal or a broken bone to mend.

Stem Cells

Many mammalian organs contain stem cells (often called adult stem cells), which retain the ability to divide. In

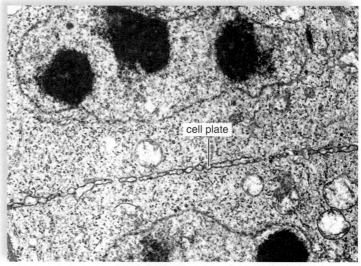
Vesicles containing cell wall components fusing to form cell plate

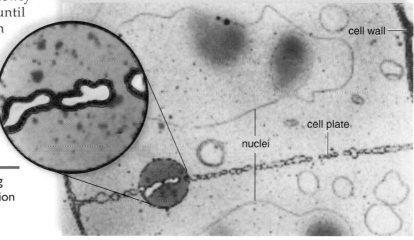

FIGURE 9.6 Cytokinesis in plant cells.
During cytokinesis in a plant cell, a cell plate forms midway between two daughter nuclei and extends to the plasma membrane.

the body, red bone marrow stem cells repeatedly divide to produce millions of cells that go on to become various types of blood cells. The possibility exists that researchers can learn to manipulate the production of various types of tissues from red bone marrow stem cells in the laboratory. If so, these tissues could be used to cure illnesses. As discussed in the Science Focus on page 158, **therapeutic cloning** to produce human tissues can begin with either adult stem cells or embryonic stem cells. Embryonic stem cells can also be used for **reproductive cloning,** the production of a new individual.

> Mitosis is critical to the development and repair of organisms. The meristem cells of flowering plants and the stem cells of human tissues are capable of dividing throughout the lifetime of the individual.

science focus

Reproductive and Therapeutic Cloning

Reproductive cloning and therapeutic cloning are done for different purposes. In reproductive cloning, the desired end is an individual that is genetically identical to the original individual. At one time, it was thought that the cloning of adult animals would be impossible because investigators found it difficult to have the nucleus of an adult cell "start over," even when it was placed in an enucleated egg cell. (An enucleated egg cell has had its own nucleus removed.)

In March 1997, Scottish investigators announced they had cloned a sheep called Dolly. How was their procedure different from all the others that had been attempted? Unlike other attempts, the donor cells were starved before the cell's nucleus was placed in an enucleated egg. Starving the donor cells caused them to stop dividing and go into a resting stage, and this made the nuclei amenable to cytoplasmic signals for initiation of development (Fig. 9Ba). Now, it is common practice to clone all sorts of farm animals that have desirable traits and

even to clone rare animals that might otherwise become extinct.

In the United States, no federal funds can be used in experiments to clone human beings. Cloning is wasteful—even in the case of Dolly, out of 29 clones, only one was successful. Also, there is concern that cloned animals may not be healthy. Dolly was euthanized in 2003 because she was suffering from lung cancer and crippling arthritis. She had lived only half the normal life span for a Dorset sheep.

In therapeutic cloning, the desired end is not an individual, rather, it is mature cells of various cell types. The purpose of therapeutic cloning is (1) to learn more about how specialization of cells occurs and (2) to provide cells and tissues that could be used to treat human illnesses such as diabetes, spinal cord injuries, Parkinson disease, and so forth.

There are two possible ways to carry out therapeutic cloning. The first way is to use the exact same procedure as reproductive cloning, except embryonic cells, called *embryonic*

stem cells, are separated and each is subjected to a treatment that causes it to develop into a particular type of cell, such as red blood cells, muscle cells, or nerve cells (Fig. 9Bb). Some have ethical concerns about this type of therapeutic cloning, which is still experimental, because if the embryo were allowed to continue development, it would become an individual.

The second way to carry out therapeutic cloning is to use *adult stem cells.* Stem cells are found in many organs of the adult's body; for example, the skin has stem cells that constantly divide and produce new skin cells. The bone marrow has stem cells that produce new blood cells as does the umbilical cord of newborns. It has already been possible to use stem cells from the brain to regenerate nerve tissue for the treatment of Parkinson disease. However, the goal is to develop techniques that would allow scientists to turn any adult stem cell into any type of specialized cell. Many investigators are engaged in this endeavor. To do this, scientists need to know how to control gene expression.

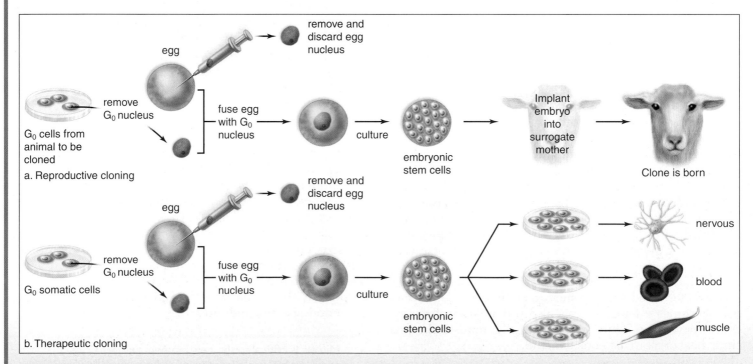

FIGURE 9B Two types of cloning. *a.* *The purpose of reproductive cloning is to produce an individual that is genetically identical to the one that donated a nucleus. The nucleus is placed in an enucleated egg, and, after several mitotic divisions, the embryo comes to term in a surrogate mother.* ***b.*** *The purpose of therapeutic cloning is to produce specialized tissue cells. A nucleus is placed in an enucleated egg, and, after several mitotic divisions, the embryonic cells (called embryonic stem cells) are separated and treated to become specialized cells.*

9.3 THE CELL CYCLE AND CANCER

A **neoplasm** is an abnormal growth of cells. A **benign** neoplasm is not cancerous. A **malignant** neoplasm is cancerous. **Cancer** is a cellular growth disorder that results from the mutation of the genes that regulate the cell cycle. Essentially, cancer results from a loss of control and a disruption of the cell cycle. As medical science advances, many questions will be answered regarding the characteristics of cancer cells, causes of cancer, and potential methods of prevention and cures for cancer.

Characteristics of Cancer Cells

Carcinogenesis, the development of cancer, is gradual; it may be decades before a cell has the characteristics of a cancer cell (Table 9.2 and Fig. 9.7).

Cancer cells lack differentiation. Cancer cells are nonspecialized and do not contribute to the functioning of a body part. A cancer cell does not look like a differentiated epithelial, muscle, nervous, or connective tissue cell; instead, it looks distinctly abnormal. Normal cells can enter the cell cycle about 50 times, and then they die. Cancer cells can enter the cell cycle repeatedly, and in this way seem immortal.

Cancer cells have abnormal nuclei. The nuclei of cancer cells are enlarged and may contain an abnormal number of chromosomes. The chromosomes have mutated; some parts may be duplicated and some may be deleted. In addition, gene amplification (extra copies of specific genes) is seen much more frequently than in normal cells. Ordinarily, cells with damaged DNA undergo apoptosis, or programmed cell death. Cancer cells fail to undergo apoptosis even though they are abnormal cells.

Cancer cells form tumors. Normal cells anchor themselves to a substratum and/or adhere to their neighbors. They exhibit contact inhibition—in other words, when they come in contact with a neighbor, they stop dividing. Cancer cells have lost all restraint and do not exhibit contact inhibition. The abnormal cancer cells pile on top of one another and grow in multiple layers, forming a **tumor.** During carcinogenesis, the most aggressive cell becomes the dominant cell of the tumor.

Cancer cells undergo metastasis and angiogenesis. A *benign tumor* is usually encapsulated and, therefore, will

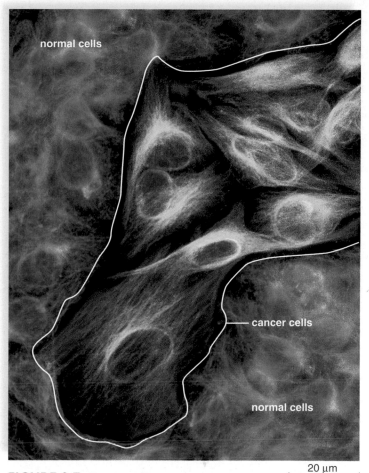

FIGURE 9.7 Cancer cells.
In this micrograph, the orange cells are cancer cells, and the other cells are normal cells. These cancer cells were detected by using a special fluorescent technique that detects abnormal proteins.

never invade adjacent tissue. *Cancer in situ* is a tumor in its place of origin, but it is not encapsulated and will eventually invade surrounding tissues. Many types of cancer can undergo **metastasis** and spread, forming new tumors distant from the primary tumor. Cancer cells produce enzymes that allow tumor cells to invade underlying tissues. Then, they travel through the blood and lymph, to start tumors elsewhere in the body.

Angiogenesis, the formation of new blood vessels, is required to bring nutrients and oxygen to a cancerous tumor whose growth is not contained within a capsule. Some modes of cancer treatment are aimed at preventing angiogenesis from occurring.

The patient's prognosis (probable outcome) is dependent on (1) whether the tumor has invaded surrounding tissues, (2) if so, whether there is any lymph node involvement, and (3) whether there are metastatic tumors in distant parts of the body. Therefore, early detection and treatment of cancer are of critical importance.

TABLE 9.2	
Cancer Cells Versus Normal Cells	
Cancer Cells	*Normal Cells*
Nondifferentiated cells	Differentiated cells
Abnormal nuclei	Normal nuclei
Do not undergo apoptosis	Undergo apoptosis
No contact inhibition	Contact inhibition
Disorganized, multilayered	One organized layer
Undergo metastasis and angiogenesis	

Cancer cells grow and divide uncontrollably, and then they metastasize, forming new tumors wherever they relocate.

Origin of Cancer

Mutations due to influences such as those illustrated in Figure 9.8 are associated with the development of cancer. The following mutations are of particular interest:

1. During the cell cycle, a DNA repair system ordinarily corrects mutations occurring when DNA replicates. There are enzymes that read a newly forming DNA strand, making sure it has the same sequence of bases as the template strand. Any mistakes are then corrected by DNA repair enzymes. Mutations that arise in the genes encoding these various enzymes can contribute to carcinogenesis.
2. **Proto-oncogenes** specify proteins that directly and indirectly promote the cell cycle, and **tumor suppressor genes** specify proteins that directly and indirectly inhibit the cell cycle. When cancer develops, normal regulation of the cell cycle no longer occurs due to mutations in these two types of genes.
3. DNA segments called **telomeres** occur at the ends of chromosomes and protect them from damage. Telomeres usually get shorter every time chromosomes replicate prior to cell division. Shortened telomeres eventually signal the cell to enter senescence and stop dividing. In cancer cells, an enzyme called telomerase keeps telomeres at a constant length so that cancer cells keep on dividing over and over again.

Regulation of the Cell Cycle

Proto-oncogenes are at the end of a *stimulatory pathway* that extends from the plasma membrane to the nucleus. In the stimulatory pathway, a growth factor is received by a receptor protein in the plasma membrane. This sets in motion a whole series of enzymatic reactions that lead to the turning on of proto-oncogenes (Fig. 9.8). Proto-oncogene products promote the cell cycle both directly and indirectly.

Tumor suppressor genes are at the start of an *inhibitory pathway.* Tumor suppressor gene products directly or indirectly inhibit the cell cycle. They are often likened to the brakes of a car because they inhibit acceleration.

For our purposes, we can consider that the balance between stimulatory signals associated with proto-oncogenes and inhibitory signals associated with tumor suppressor genes determines whether cell division occurs or does not occur.

Oncogenes

Proto-oncogenes are so called because a mutation can cause them to become **oncogenes** (cancer-causing genes). They are often likened to the gas pedal of a car because they cause acceleration of the cell cycle. An oncogene may code for a faulty receptor in the stimulatory pathway. A faulty receptor may be able to start the process that stimulates the cell cycle even when no growth factor is present! Or an on-cogene may specify either an abnormal protein product or abnormally high levels of a normal product that stimulates the cell cycle to begin or to go to completion. For example, oncogenes bring about the presence of excess cyclin, the protein that promotes the cell cycle, and also excess inhibitors of p53 so that apoptosis does not occur. In either case, uncontrolled growth threatens.

Researchers have identified perhaps 100 oncogenes that can cause increased growth and lead to tumors. The oncogenes most frequently involved in human cancers belong to the *ras* gene family. An alteration of only a single nucleotide pair is sufficient to convert a normally functioning *ras* proto-oncogene to an oncogene. The *ras*K oncogene is associated with lung cancer, colon cancer, and pancreatic cancer. The *ras*N oncogene is associated with **leukemias** (cancers of blood-forming cells) and lymphomas (cancers of lymphoid tissue), and both *ras* oncogenes are frequently found in thyroid cancers. Another gene, *BRCA1* (breast cancer predisposition gene 1), is associated with inheriting certain forms of breast and ovarian cancer.

Tumor Suppressor Genes

When a tumor suppressor gene undergoes a mutation, inhibitory proteins fail to be active, and the cell cycle is unchecked. For example, mutated tumor suppressor genes no longer code for effective inhibitors of cyclin nor for effective promoters of p53 and apoptosis. Researchers have identified about a half-dozen tumor suppressor genes. The *RB* tumor suppressor gene was discovered when the inherited condition retinoblastoma was being studied. If a child receives only one normal *RB* gene and that gene mutates, eye tumors develop in the retina by the age of three. The *RB* gene has now been found to malfunction in cancers of the breast, prostate, and bladder, among others. Loss of the *RB* gene through chromosome deletion is particularly frequent in a type of lung cancer called small-cell lung carcinoma. How the RB protein fits into the inhibitory pathway is known; an active RB protein turns off the expression of a proto-oncogene.

Another major tumor suppressor gene is *p53,* a gene that is more frequently mutated in human cancers than any other known gene. It has been found that the p53 protein is involved in turning on the expression of other genes whose products are cell cycle inhibitors. *p53* can also stimulate apoptosis, programmed cell death. It is estimated that over half of human cancers involve an abnormal or deleted *p53* gene. In recent years, cancers of the colon, bladder, lung, liver, esophagus, and skin have shown distinct types of *p53* mutations.

When a proto-oncogene mutates and becomes an oncogene, or if a tumor suppressor gene mutates, the cell cycle is not properly regulated, and tumor formation is more likely to occur.

FIGURE 9.8 Causes of cancer.
a. Mutated genes that cause cancer can be due to the influences noted. **b.** A growth factor that binds to a receptor protein initiates a reaction that triggers a stimulatory pathway. **c.** A stimulatory pathway that begins at the plasma membrane turns on proto-oncogenes. The products of these genes promote the cell cycle and double back to become part of the stimulatory pathway. When proto-oncogenes become oncogenes, they are turned on all the time. An inhibitory pathway begins with tumor suppressor genes whose products inhibit the cell cycle. When tumor suppressor genes mutate, the cell cycle is no longer inhibited. **d.** Cancerous skin cell.

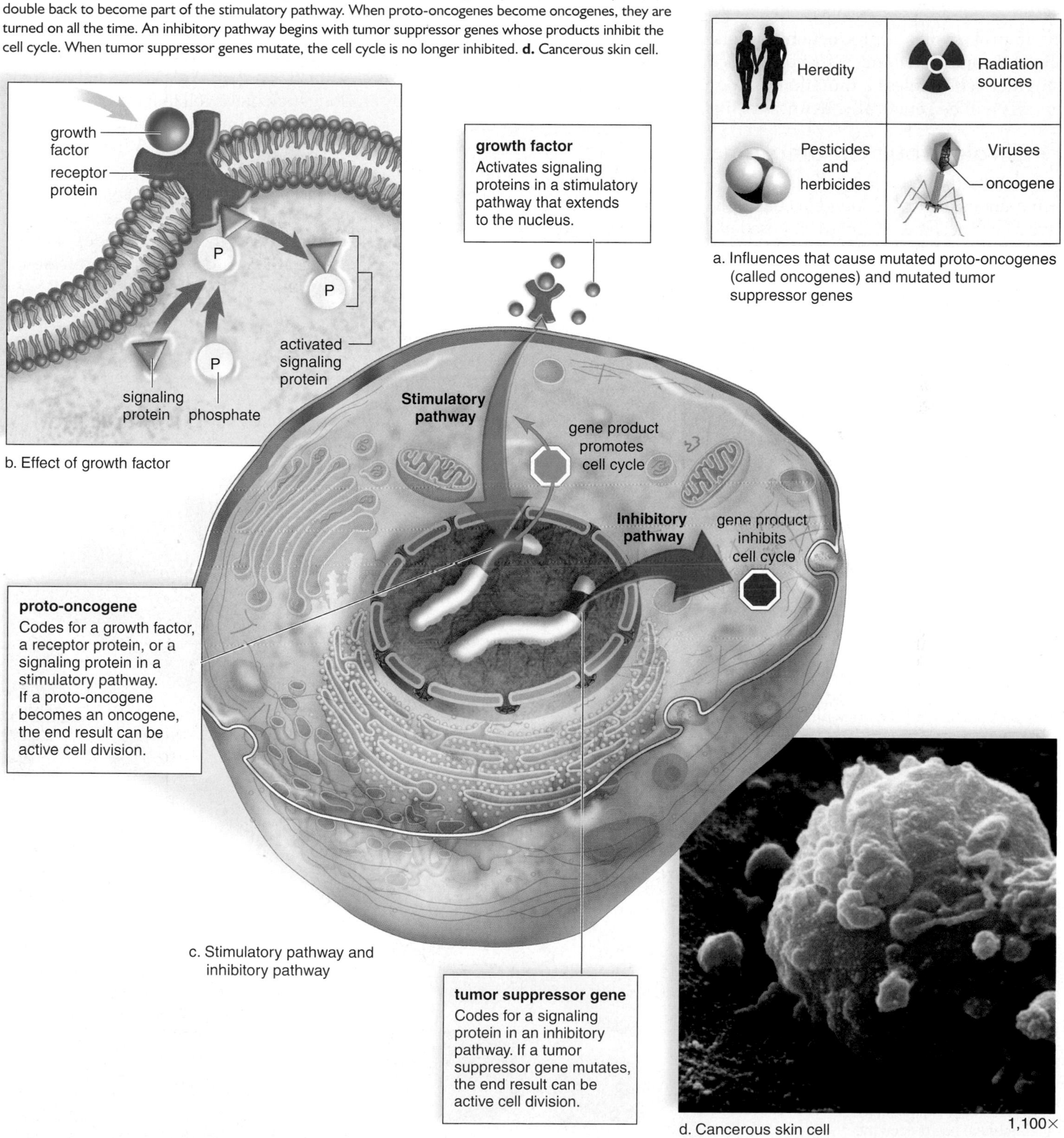

growth factor

receptor protein

P

P

P

signaling protein   phosphate

activated signaling protein

b. Effect of growth factor

growth factor
Activates signaling proteins in a stimulatory pathway that extends to the nucleus.

Heredity

Radiation sources

Pesticides and herbicides

Viruses

oncogene

a. Influences that cause mutated proto-oncogenes (called oncogenes) and mutated tumor suppressor genes

Stimulatory pathway

gene product promotes cell cycle

Inhibitory pathway

gene product inhibits cell cycle

proto-oncogene
Codes for a growth factor, a receptor protein, or a signaling protein in a stimulatory pathway. If a proto-oncogene becomes an oncogene, the end result can be active cell division.

c. Stimulatory pathway and inhibitory pathway

tumor suppressor gene
Codes for a signaling protein in an inhibitory pathway. If a tumor suppressor gene mutates, the end result can be active cell division.

d. Cancerous skin cell

1,100×

9.4 PROKARYOTIC CELL DIVISION

Cell division in unicellular organisms, such as prokaryotes, produces two new individuals. This is **asexual reproduction** in which the offspring are genetically identical to the parent. In prokaryotes, reproduction consists of duplicating the single chromosome and distributing a copy to each of the daughter cells. Unless a mutation has occurred, the daughter cells will be genetically identical to the parent cell.

The Prokaryotic Chromosome

Prokaryotes (bacteria and archaea) lack a nucleus and other membranous organelles found in eukaryotic cells. Still, they do have a chromosome, which is composed of DNA and a limited number of associated proteins. The single chromosome of prokaryotes contains just a few proteins and is organized differently from eukaryotic chromosomes. A eukaryotic chromosome has many more proteins than a prokaryotic chromosome.

In electron micrographs, the bacterial chromosome appears as an electron-dense, irregularly shaped region called the **nucleoid** [L. *nucleus,* nucleus, kernel; Gk. *-eides,* like], which is not enclosed by membrane. When stretched out, the chromosome is seen to be a circular loop with a length that is up to about a thousand times the length of the cell. No wonder it is folded when inside the cell.

Binary Fission

Prokaryotes reproduce asexually by binary fission. The process is termed **binary fission** because division (fission) produces two (binary) daughter cells that are identical to the original parent cell. Before division takes place, the cell enlarges, and after DNA replication occurs, there are two chromosomes. These chromosomes attach to a special plasma membrane site and separate by an elongation of the cell that pulls them apart. During this period, new plasma membrane and cell wall develop and grow inward to divide the cell. When the cell is approximately twice its original length, the new cell wall and plasma membrane for each cell are complete (Fig. 9.9).

Escherichia coli, which lives in our intestines, has a generation time (the time it takes the cell to divide) of about 20 minutes under favorable conditions. In about seven hours, a single cell can increase to over 1 million cells! The dividing rate of other bacteria varies depending on the species and conditions.

Prokaryotes reproduce asexually by binary fission. DNA replicates, and the two resulting chromosomes separate as the cell elongates.

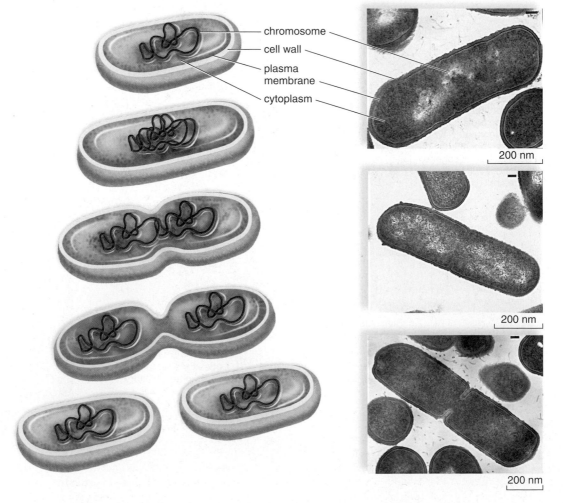

1. Attachment of chromosome to a special plasma membrane site indicates that this bacterium is about to divide.

2. The cell is preparing for binary fission by enlarging its cell wall, plasma membrane, and overall volume.

3. DNA replication has produced two identical chromosomes. Cell wall and plasma membrane begin to grow inward.

4. As the cell elongates, the chromosomes are pulled apart. Cytoplasm is being distributed evenly.

5. New cell wall and plasma membrane has divided the daughter cells.

chromosome
cell wall
plasma membrane
cytoplasm

200 nm

200 nm

200 nm

FIGURE 9.9 Binary fission.
First, DNA replicates, and as the cell lengthens, the two chromosomes separate, and the cells become divided. The two resulting bacteria are identical.

Comparing Prokaryotes and Eukaryotes

Both binary fission and mitosis ensure that each daughter cell is genetically identical to the parent cell. The genes are portions of DNA found in the chromosomes.

Prokaryotes (bacteria and archaea), protists (many algae and protozoans), and some fungi (yeasts) are unicellular. Cell division in unicellular organisms produces two new individuals:

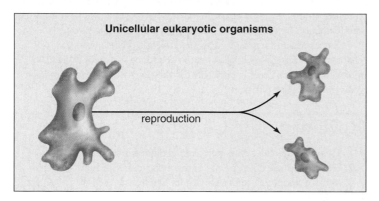

Unicellular eukaryotic organisms

reproduction

This is a form of asexual reproduction because one parent has produced identical offspring (Table 9.3).

In multicellular fungi (molds and mushrooms), plants, and animals, cell division is part of the growth process. It produces the multicellular form we recognize as the mature organism. Cell division is also important in multicellular forms for renewal and repair:

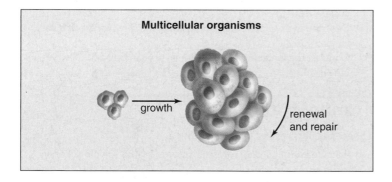

Multicellular organisms

growth

renewal and repair

TABLE 9.3

Functions of Cell Division

Type of Organism	Cell Division	Function
Prokaryotes Bacteria and archaea	Binary fission	Asexual reproduction
Eukaryotes Protists, and some fungi (yeast)	Mitosis and cytokinesis	Asexual reproduction
Other fungi, plants, and animals	Mitosis and cytokinesis	Development, growth, and repair

In prokaryotes, the single chromosome consists largely of DNA with some associated proteins. During binary fission, this chromosome duplicates, and each daughter cell receives one copy as the parent cell elongates, and a new cell wall and plasma membrane form between the daughter cells. No spindle apparatus is involved in binary fission.

The chromosomes of eukaryotic cells are composed of DNA and associated proteins. The protein histone organizes a chromosome, allowing it to extend as chromatin during interphase and to coil and condense just prior to mitosis. Each species of multicellular eukaryotes has a characteristic number of chromosomes in the nuclei. As a result of mitosis, each daughter cell receives the same number and kinds of chromosomes as the parent cell. The spindle, which appears during mitosis, is involved in distributing the daughter chromosomes to the daughter nuclei. Cytokinesis either by the formation of a cell plate (plant cells) or by furrowing (animal cells) is division of the cytoplasm.

Due to binary fission and mitosis, daughter cells are genetically identical to the parent cell. Cell division allows unicellular organisms to reproduce and is necessary to growth and repair in multicellular organisms

---

CONNECTING THE CONCEPTS

Cells have an elaborate internal structure that is not revealed by the light microscope, as well as complex enzymatic pathways that can only be discovered by biochemical analyses. Similarly, the process of cell division is more complicated than it first appears. We now know that cell division is part of a tightly regulated cell cycle and there are negative consequences if the cell cycle should come out of synchronization. For example, in humans, overproduction of skin cells due to an overstimulated cell cycle produces a chronic inflammatory condition known as psoriasis. In a

rare condition called progeria, the reproductive capacity of all the body's cells is severely diminished, and young people grow old and die at an early age. Many different types of cancer can result when the signals that keep the cell cycle in check are not transmitted or received properly. The hope is that if we learn to control the cell cycle, these conditions will be curable one day.

The newfound ability to reproductively clone farm animals shows that we are now able to initiate and control the process of cell division, even in mammals. The possibilities

of therapeutic cloning and also tissue engineering, during which organs are formed in the laboratory (see Chapter 16), are even better examples that we can now manipulate the cell cycle.

The end product of ordinary cell division (i.e., mitosis) is two new cells, each with the same number and kinds of chromosomes as the parent cell. But what about the special type of cell division (i.e., meiosis) that produces the sex cells—sperm and egg? As we'll see in the next chapter, following meiosis, the sex cells have only half the chromosome number.

Summary

9.1 THE CELL CYCLE

Eukaryotic cells go through a cell cycle that includes (1) interphase and (2) a mitotic stage that consists of mitosis and cytokinesis. Interphase, in turn, includes G_1 (growth as certain organelles double), S (DNA synthesis, which is DNA replication), and G_2 (growth as the cell prepares to divide). During the mitotic stage (M), daughter cells receive a full complement of chromosomes.

The cell cycle is controlled, and there are three well-known checkpoints—one in G_1 prior to the S stage, one in G_2 prior to the M stage, and another near the end of mitosis. DNA damage is a reason the cell cycle stops; the *p53* gene is active at the G_1 checkpoint, and if DNA is damaged and can't be repaired, this gene initiates apoptosis. During apoptosis, enzymes called caspases bring about destruction of the nucleus and the rest of the cell. Cell division and apoptosis are two opposing processes that keep the number of healthy cells in balance.

9.2 MITOSIS AND CYTOKINESIS

Between nuclear divisions, the chromosomes are not distinct and are collectively called chromatin. Each eukaryotic species has a characteristic number of chromosomes. The total number is called the diploid number, and half this number is the haploid number.

Among eukaryotes, cell division involves nuclear division (karyokinesis) and division of the cytoplasm (cytokinesis). Mitosis is nuclear division in which the chromosome number stays constant because each chromosome is duplicated and gives rise to two daughter chromosomes.

Mitosis has five phases. Recall that the chromosomes were duplicated during the S stage of the cell cycle.

Prophase—The nucleolus disappears, the nuclear envelope fragments, and the spindle forms between centrosomes. In animal cells, asters radiate from the centrioles within the centrosomes. Plant cells lack centrioles and, therefore, asters. Even so, the mitotic spindle forms, and the same five mitotic phases are observed.

Prometaphase (late prophase)—The kinetochores of sister chromatids attach to kinetochore spindle fibers extending from opposite poles. The chromosomes move back and forth until they are aligned at the metaphase plate.

Metaphase—The spindle is fully formed, and the duplicated chromosomes are aligned at the metaphase plate. The spindle consists of polar spindle fibers that overlap at the metaphase plate and kinetochore spindle fibers that are attached to chromosomes.

Anaphase—Sister chromatids separate, becoming daughter chromosomes that move toward the poles. The polar spindle fibers slide past one another, and the kinetochore spindle fibers disassemble. Cytokinesis by furrowing begins.

Telophase—Nuclear envelopes re-form, chromosomes begin changing back to chromatin, the nucleoli reappear, and the spindle disappears.

Cytokinesis in animal cells is a furrowing process that divides the cytoplasm. Cytokinesis in plant cells involves the formation of a cell plate from which the plasma membrane and cell wall are completed.

9.3 THE CELL CYCLE AND CANCER

The development of cancer is primarily due to the mutation of genes involved in control of the cell cycle. The failure of the DNA repair system and the presence of telomerase are also involved. Cancer cells lack differentiation, have abnormal nuclei, form tumors, and undergo metastasis and angiogenesis.

Proto-oncogenes stimulate the cell cycle after they are turned on by environmental signals such as growth factors. Oncogenes are mutated proto-oncogenes that stimulate the cell cycle without need of environmental signals. Tumor suppressor genes inhibit the cell cycle. Mutated tumor suppressor genes no longer inhibit the cell cycle, and cell division continues to occur.

9.4 PROKARYOTIC CELL DIVISION

The prokaryotic chromosome has a few proteins and a single, long loop of DNA. When binary fission occurs, the chromosome attaches to the inside of the plasma membrane and replicates. As the cell elongates, the chromosomes are pulled apart. Inward growth of the plasma membrane and formation of new cell wall material divide the cell in two.

Binary fission (in prokaryotes) and mitosis (in cellular eukaryotic protists and fungi) allow organisms to reproduce asexually. Mitosis in multicellular eukaryotes is primarily for the purpose of development, growth, and repair of tissues.

Reviewing the Chapter

1. Describe the cell cycle, including its different stages.   150
2. Describe three checkpoints of the cell cycle.   151
3. What is apoptosis, and what are its functions?   151
4. Define the following words: chromosome, chromatin, chromatid, centriole, cytokinesis, centromere, and kinetochore.   150, 152–55
5. Describe the events that occur during the phases of mitosis.   154–56
6. How does plant cell mitosis differ from animal cell mitosis?   154–57
7. Contrast cytokinesis in animal cells and plant cells.   156
8. List and discuss four characteristics of cancer cells that distinguish them from normal cells.   159
9. Mutations of what types of genes decrease regulation of the cell cycle?   160–61
10. Describe the prokaryotic chromosome and the process of binary fission.   162
11. Contrast the function of cell division in prokaryotic and eukaryotic cells.   163

Testing Yourself

Choose the best answer for each question.

1. What feature in prokaryotes substitutes for the spindle action in eukaryotes?
   a. centrioles with asters
   b. fission instead of cytokinesis
   c. elongation of plasma membrane
   d. looped DNA
   e. presence of one chromosome

2. How does a prokaryotic chromosome differ from a eukaryotic chromosome? A prokaryotic chromosome
   a. is shorter and fatter.
   b. has a single loop of DNA.
   c. never replicates.
   d. contains many histones.
   e. All of these are correct.

3. The diploid number of chromosomes
   a. is the 2n number.
   b. is in a parent cell and therefore in the two daughter cells following mitosis.
   c. varies according to the particular organism.
   d. is in every somatic cell.
   e. All of these are correct.

For questions 4–6, match the descriptions that follow to the terms in the key.

KEY:

    a. centriole   d. centromere
    b. chromatid   e. cyclin
    c. chromosome

4. Point of attachment for sister chromatids

5. Found at a spindle pole in the center of an aster

6. Coiled and condensed chromatin

7. If a parent cell has 14 chromosomes prior to mitosis, how many chromosomes will each daughter cell have?
    a. 28 because each chromatid is a chromosome
    b. 14 because the chromatids separate
    c. only 7 after mitosis is finished
    d. any number between 7 and 28
    e. 7 in the nucleus and 7 in the cytoplasm, for a total of 14

8. In which phase of mitosis are the chromosomes moving toward the poles?
    a. prophase   d. anaphase
    b   prometaphase   e. telophase
    c. metaphase

9. Interphase
    a. is the same as prophase, metaphase, anaphase, and telophase.
    b. includes stages G_1, S, and G_2.
    c. requires the use of polar spindle fibers and kinetochore spindle fibers.
    d. is a part of the cell cycle.
    e. Both b and d are correct.

10. Cytokinesis
    a. is mitosis in plants.
    b. requires the formation of a cell plate in plant cells.
    c. is the longest part of the cell cycle.
    d. is half a chromosome.
    e. is a form of apoptosis.

11. At the metaphase plate during metaphase of mitosis, there are
    a. single chromosomes.
    b. unpaired duplicated chromosomes.
    c. G_1 stage chromosomes.
    d. always 23 chromosomes.

12. A cell plate
    a. is characteristic of plant cells.
    b. will become the cell wall.
    c. is necessary to the movement of chromosomes.
    d. helps make the spindle fibers.
    e. Both a and b are correct.

13. During which mitotic phases are duplicated chromosomes present?
    a. all but telophase
    b. prophase and anaphase
    c. all but anaphase and telophase
    d. only during metaphase at the metaphase plate
    e. Both a and b are correct.

14. Which of these is paired incorrectly?
    a. prometaphase—the kinetochore become attached to spindle fibers
    b. anaphase—daughter chromosomes are located at the spindle poles
    c. prophase—the nucleolus disappears and the nuclear envelope disintegrates
    d. metaphase—the chromosomes are aligned in the metaphase plate
    e. telophase—a resting phase between cell division cycles

15. When cancer occurs,
    a. mutations have occurred.
    b. the *p53* gene is operational.
    c. apoptosis has occurred.
    d. the cells can no longer enter the cell cycle.
    e. All of these are correct.

16. Which of the following is not characteristic of cancer cells?
    a. Cancer cells often undergo angiogenesis.
    b. Cancer cells tend to be nonspecialized.
    c. Cancer cells undergo apoptosis.
    d. Cancer cells often have abnormal nuclei.
    e. Cancer cells can metastasize.

17. If a cell is cancerous, you might find an abnormality in
    a. a receptor protein.   c. genes.
    b. a signaling protein.   d. All of these are correct.

For questions 18–21, match the descriptions to a stage in the key.

KEY:

    a. G_1 stage
    b. S stage
    c. G_2 stage
    d. M (mitotic) stage

18. At the end of this stage, each chromosome consists of two identical DNA molecules.

19. During this stage, daughter chromosomes are distributed to two daughter nuclei.

20. The cell doubles its organelles and accumulates the materials needed for DNA synthesis.

21. The cell synthesizes the proteins needed for cell division.

22. Which is not true of the cell cycle?
    a. The cell cycle is controlled by internal/external signals.
    b. During the cell cycle, cyclin increases and decreases as the cycle continues.
    c. DNA damage can stop the cell cycle at the G_1 checkpoint.
    d. Apoptosis occurs frequently during the cell cycle.

23. In human beings, mitosis is necessary to
    a. growth and repair of tissues.
    b. formation of the gametes.
    c. maintaining the chromosome number in all body cells.
    d. the death of unnecessary cells
    e. Both a and c are correct.

24. Label this diagram of a cell in early prophase of mitosis.

Thinking Scientifically

1. After DNA is duplicated in eukaryotes, it must be bound to histone proteins. This requires the synthesis of hundreds of millions of new protein molecules, a process that the cell most likely regulates. With reference to Figure 9.1, when in the cell cycle would histones be made? At what point might histone synthesis be switched on and off?

2. When animal cells are grown in dishes in the laboratory, most divide a few dozen times and then die. It would seem likely that the cell cycle is suddenly not functioning properly. With the help of Figure 9.1*b*, hypothesize what could have happened to turn the cell cycle off.

Bioethical Issue: Paying for Cancer Treatment

Cancer is more likely to develop in tissues whose cells frequently divide, such as the blood-forming cells in the bone marrow. A cancer drug called Gleevec is particularly helpful in treating a deadly form of blood cancer called myeloid leukemia. Gleevec is a pill taken by mouth, and it doesn't cause hair loss! For some patients, the results have been dramatic. Within weeks, some have gone from being bedridden and about to die to returning to work, doing volunteer work, and enjoying life as before.

Gleevec is expensive. It costs something like $2,400 a month, or nearly $30,000 a year, and treatment may have to continue for life. How should the cost of treatment be met? Drug companies claim that it costs them between $500 million and $1 billion to bring a single new medicine to market. This cost may seem overblown, especially when you consider that the National Cancer Institute funds basic research into cancer biology. But the drug companies tell us that they need one successful drug to pay for the many drugs they try to develop that do not pay off. Still, it does seem as if successful drug companies try to keep lower-cost competitors out of the market.

Now that Gleevec has been taken off the experimental list, insurance companies will probably pick up the cost, but of course, this may increase the cost of insurance for everyone. Cancer most often strikes older people. In the future, Medicare may pay for cancer drugs as well as all drugs needed by older people. In that case, the cost of cancer treatment will be borne by everyone who pays taxes.

The question of how much drug companies can charge for drugs and who should pay for them is a thorny one. If drug companies don't show a profit, they may go out of business and there will be no new drugs. The same is true for insurance companies if they can't raise the cost of insurance to pay for expensive drugs. If the government buys drugs for Medicare patients, taxes may go up dramatically.

Understanding the Terms

anaphase   156
angiogenesis   159
apoptosis   151
asexual
   reproduction   162
aster   155
benign   159
binary fission   162
cancer   159
carcinogenesis   159
cell cycle   150
cell plate   157
centriole   154
centromere   153
centrosome   154
chromatid   150
chromatin   153
cyclin   151
cytokinesis   150
diploid (2n) number   153
growth factor   151
haploid (n) number   153
interphase   150
kinetochore   153

leukemia   160
malignant   159
metaphase   155
metaphase plate   156
metastasis   159
mitosis   150
neoplasm   159
nucleoid   162
oncogene   160
p53 (gene)   160
prometaphase   155
prophase   154
proto-oncogene   160
reproductive cloning   157
signal   151
sister chromatid   153
somatic cell   151
spindle   154
telomere   160
telophase   156
therapeutic cloning   157
tumor   159
tumor suppressor gene   160

Match the terms to these definitions:

a. _____ Central microtubule organizing center of cells, consisting of granular material. In animal cells, it contains two centrioles.

b. _____ Constriction where sister chromatids of a chromosome are held together.

c. _____ Microtubule structure that brings about chromosome movement during nuclear division.

d. _____ One of two genetically identical chromosome units that are the result of DNA replication.

e. _____ Programmed cell death that is carried out by enzymes routinely present in the cell.

ARIS, the *Biology* Website

ARIS, the website for *Biology*, provides a wealth of information organized and integrated by chapter. You will find practice quizzes, interactive activities, labeling exercises, flashcards, and much more that will complement your learning and understanding of general biology.

www.mhhe.com/maderbiology9

10

MEIOSIS AND SEXUAL REPRODUCTION

Take a moment to think about what sexual reproduction accomplishes—a sperm fertilizes an egg, and a new life begins. During the production of egg and sperm, the chromosome number is reduced, and the gametes have one-half the number of chromosomes. Otherwise, the chromosome number would double with each new generation. Also, a gamete is genetically unique, and this along with fertilization ensures that every new individual has a combination of genes that is different from the parents and from any other of its own kind.

One of the most significant events in the evolutionary history of life was the development of sexual reproduction, the mode of reproduction that is prevalent across the world of multicellular eukaryotes. Because sexual reproduction brings together new combinations of genes, it offers an opportunity to fine-tune adaptations to an environment that is always changing. Then, too, sexually reproducing individuals have two genes instead of one for every possible characteristic. If one gene has mutated and doesn't work, the other one may very well work. In this way, sexual reproduction ensures survival until just the right combination of genes for the environment can occur.

One sperm out of many fertilizes an egg.

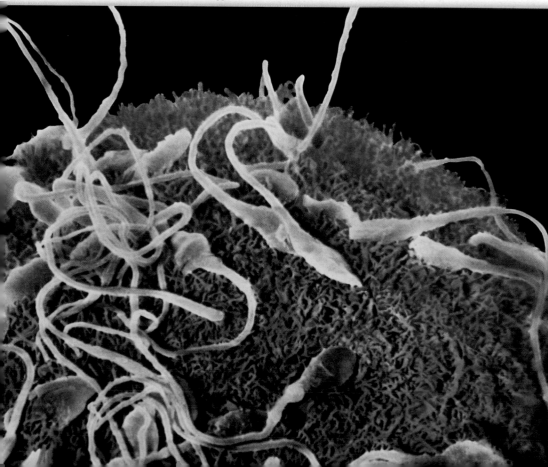

10.1 HALVING THE CHROMOSOME NUMBER

In sexually reproducing organisms, **meiosis** [Gk. *mio,* less, and *-sis,* act or process of] is the type of nuclear division that reduces the chromosome number from the diploid (2n) number [Gk. *diplos,* twofold, and *-eides,* like] to the haploid (n) number [Gk. *haplos,* single, and *-eides,* like]. The **diploid (2n) number** refers to the total number of chromosomes. The **haploid (n) number** of chromosomes is half the diploid number. In humans, the diploid number of 46 is reduced to the haploid number of 23. **Gametes** (reproductive cells, often the sperm and egg) usually have the haploid number of chromosomes. Gamete formation and then fusion of gametes to form a cell called a zygote are integral parts of **sexual reproduction.** A **zygote** always has the full or diploid (2n) number of chromosomes. In plants and animals, the zygote undergoes development to become an adult organism.

Obviously, if the gametes contained the same number of chromosomes as the body cells, the number of chromosomes would double with each new generation. Within a few generations, the cells of an animal would be nothing but chromosomes! For example, in humans with a diploid number of 46 chromosomes, in five generations the chromosome number would increase to 1,472 chromosomes (46×2^5). In 10 generations this number would increase to a staggering 47,104 chromosomes (46×2^{10}). The early cytologists (biologists who study cells) realized this, and Pierre-Joseph van Beneden, a Belgian, was gratified to find in 1883 that the sperm and the egg of the roundworm *Ascaris* each contain only two chromosomes, while the zygote and subsequent embryonic cells always have four chromosomes.

Homologous Pairs of Chromosomes

In diploid body cells, the chromosomes occur in pairs. Figure 10.1*a,* a pictorial display of human chromosomes, shows the chromosomes arranged according to pairs. The members of each pair are called homologous chromosomes. **Homologous chromosomes** or **homologues** [Gk. *homologos,* agreeing, corresponding] look alike; they have the same length and centromere position. When stained, homologues have a similar banding pattern because they contain genes for the same traits in the same order. But while homologous chromosomes have genes for the same traits, such as finger length, the gene on one homologue may code for short fingers and the gene at the same location on the other homologue may code for long fingers. Alternate forms of a gene (as for long fingers and short fingers) are called **alleles.**

The chromosomes in Figure 10.1*a* are duplicated as they would be just before nuclear division. Recall that during the S stage of the cell cycle, DNA replicates and the chromosomes become duplicated. The results of the duplication process are depicted in Figure 10.1*b.* When duplicated, a chromosome is composed of two identical parts called sister chromatids, each containing one DNA double helix molecule. The sister chromatids are held together at a region called the centromere.

Why does the zygote have two chromosomes of each kind? One member of a homologous pair was inherited from the male parent, and the other was inherited from the female parent by way of the gametes. In Figure 10.1*b* and throughout the chapter, the paternal chromosome is colored blue, and the maternal chromosome is colored red. *Therefore, you should use size and centromere location, not color, to recognize homologues.* We will see how meiosis reduces the chromosome number. Whereas the zygote and body cells have homologous pairs of chromosomes, the gametes have only one chromosome of each kind—derived from either the paternal or maternal homologue.

Overview of Meiosis

Meiosis requires two nuclear divisions and produces four haploid daughter cells, each having one of each kind of chromosome. Therefore, the daughter cells have half the total number of chromosomes as were in the diploid parent nucleus. The daughter cells receive one of each kind of parental chromosome, but in different combinations. Therefore, the daughter cells are not genetically identical to the parent cell.

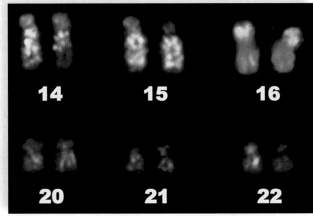

a.

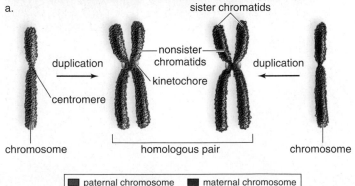

b.

FIGURE 10.1 Homologous chromosomes.
In diploid body cells, the chromosomes occur in pairs called homologous chromosomes. **a.** In this micrograph of stained chromosomes from a human cell, the pairs have been numbered. **b.** These chromosomes are duplicated, and each one is composed of two chromatids. The sister chromatids contain the exact same genes; the nonsister chromatids contain genes for the same traits (e.g., type of hair, color of eyes), but they may differ in that one could "call for" dark hair and eyes and the other for light hair and eyes.

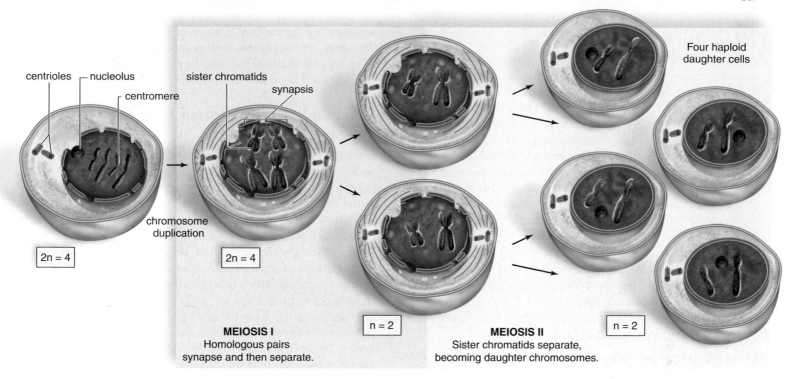

FIGURE 10.2 Overview of meiosis.
Following DNA replication, each chromosome is duplicated and consists of two chromatids. During meiosis I, homologous chromosomes
pair and separate. During meiosis II, the sister chromatids of each duplicated chromosome separate. At the completion of meiosis, there
are four haploid daughter cells. Each daughter cell has one of each kind of chromosome.

Figure 10.2 presents an overview of meiosis, indicating the two cell divisions, meiosis I and meiosis II. Prior to meiosis I, DNA (deoxyribonucleic acid) replication has occurred; therefore, each chromosome has two sister chromatids. During meiosis I, something new happens that does not occur in mitosis. The homologous chromosomes come together and line up side by side due to a means of attraction still unknown. This so-called **synapsis** [Gk. *synaptos*, united, joined together] results in a **bivalent** [L. *bis*, two, and *valens,* strength]—that is, two homologous chromosomes that stay in close association during the first two phases of meiosis I. Sometimes the term tetrad [Gk. *tetra,* four] is used instead of bivalent because, as you can see, a bivalent contains four chromatids.

Following synapsis, homologous pairs align at the metaphase plate, and then the members of each pair separate. This separation means that only one duplicated chromosome from each homologous pair reaches a daughter nucleus. It is important for daughter nuclei to have a member from each pair of homologous chromosomes because only in that way can there be a copy of each kind of chromosome in the daughter nuclei. Notice in Figure 10.2 that two possible combinations of chromosomes in the daughter cells are shown: short red with long blue and short blue with long red. Knowing that all daughter cells have to have one short chromosome and one long chromosome, what are the other two possible combinations of chromosomes for these particular cells?

During meiosis I, homologous chromosomes pair and separate. The daughter cells, therefore, have one copy of each kind of chromosome in various combinations.

No replication of DNA is needed between meiosis I and meiosis II because the chromosomes are still duplicated; they already have two sister chromatids. During meiosis II, the sister chromatids separate, becoming daughter chromosomes that move to opposite poles. The chromosomes in the four daughter cells contain only one DNA double helix molecule because they are not duplicated.

The number of centromeres can be counted to verify that the parent cell has the diploid number of chromosomes. Each daughter cell that forms has the haploid number of chromosomes.

Following meiosis II, there are four haploid daughter cells, and each chromosome consists of one chromatid.

In the plant life cycle, the daughter cells become haploid spores that germinate to become a haploid generation. This generation produces the gametes by mitosis. The plant life cycle is studied in Chapter 24. In the animal life cycle, the daughter cells become the gametes, either sperm or eggs. The body cells of an animal normally contain the diploid number of chromosomes due to the fusion of sperm and egg during fertilization.

Occasionally, meiotic events go wrong. Several chromosomal abnormalities can occur in humans that are due to meiotic errors. For example, the birth of an individual with trisomy 21 (often called Down syndrome) is due to a meiotic error. Most often the egg had two chromosomes 21 instead of one. Sometimes the sperm has the extra chromosome 21. People with trisomy 21 have abnormalities but can often lead fairly normal lives.

10.2 GENETIC VARIATION

We have seen that meiosis provides a way to keep the chromosome number constant generation after generation. Without meiosis, the chromosome number of the next generation would continually increase. The events of meiosis also help ensure that genetic variation occurs with each generation. Asexually reproducing organisms, such as the prokaryotes, depend primarily on mutations to generate variation among offspring. This is sufficient because they produce great numbers of offspring within a limited amount of time. Mutation also occurs among sexually reproducing organisms, but the reshuffling of genetic material during sexual reproduction contributes greatly to the possibility that offspring will have a different combination of genes from their parents. Meiosis brings about genetic variation in two key ways: crossing-over and independent assortment of homologous chromosomes.

Genetic Recombination

Crossing-over is an exchange of genetic material between nonsister chromatids of a bivalent during meiosis I. It is estimated that an average of two or three crossovers occur per human chromosome. At synapsis, homologues line up side by side, and a nucleoprotein lattice appears between them (Fig. 10.3). This lattice holds the bivalent together in such a way that the DNA of the nonsister chromatids is aligned. Now crossing-over may occur. As the lattice breaks down, homologues are temporarily held together by *chiasmata* (sing., chiasma), regions where the nonsister chromatids are attached due to crossing-over. Then homologues separate and are distributed to different daughter cells.

To appreciate the significance of crossing-over, it is necessary to remember that the members of a homologous pair can carry slightly different instructions for the same genetic traits. In the end, due to a swapping of genetic material during crossing-over, the chromatids held together by a centromere are no longer identical. Therefore, when the chromatids

separate during meiosis II, some of the daughter cells receive daughter chromosomes with recombined genes. Due to **genetic recombination,** the offspring have a different sequence of alleles and therefore genes than their parents.

Independent Assortment of Homologous Chromosomes

During **independent assortment,** the homologous chromosomes separate independently or in a random manner. When homologues align at the metaphase plate, the maternal or paternal homologue may be oriented toward either pole. Figure 10.4 shows the possible orientations for a cell that contains only three pairs of homologous chromosomes. Once all possible alignments are considered, the result will be 2^3, or eight, combinations of maternal and paternal chromosomes in the resulting gametes from this cell, simply due to independent assortment of homologues.

In humans, who have 23 pairs of chromosomes, the possible chromosomal combinations in the gametes is a staggering 2^{23}, or 8,388,608. And this does not even consider the genetic variations that are introduced due to crossing-over.

Fertilization

The variation that results from meiosis is enhanced by **fertilization,** the union of the male and female gametes. Each person has genes for the same traits, but again, each gene's specific instructions can vary. Therefore, the gametes produced by one person are expected to be genetically different from the gametes produced by another person. *When the gametes fuse at fertilization,* the chromosomes donated by the parents are combined, and in humans, this means that $(2^{23})^2$, or 70,368,744,000,000, chromosomally different zygotes are possible, even assuming no crossing-over. If crossing-over occurs once, then $(4^{23})^2$, or 4,951,760,200,000,000,000,000,000,000, genetically different zygotes are possible for every couple. Keep in mind that crossing-over can occur several times in each chromosome!

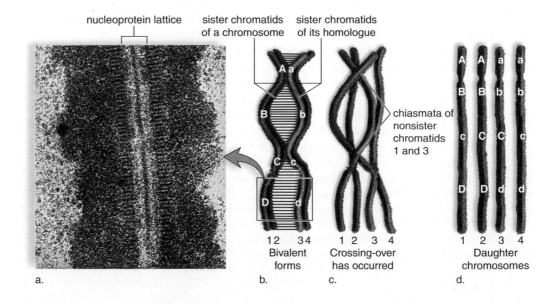

FIGURE 10.3 Crossing-over occurs during meiosis I.
a. The homologous chromosomes pair up, and a nucleoprotein lattice develops between them. This is an electron micrograph of the lattice. It zippers the members of the bivalent together so that corresponding genes are in alignment. **b.** This diagrammatic representation shows only two places where nonsister chromatids 1 and 3 have come into contact. Actually, the other two nonsister chromatids most likely are also crossing-over. **c.** Chiasmata indicate where crossing-over has occurred. The exchange of color represents the exchange of genetic material. **d.** Following meiosis II, daughter chromosomes have a new combination of genetic material due to crossing-over, which occurred between nonsister chromatids during meiosis I.

nucleoprotein lattice

sister chromatids of a chromosome

sister chromatids of its homologue

chiasmata of nonsister chromatids 1 and 3

12   34
Bivalent forms

1 2 3 4
Crossing-over has occurred

1 2 3 4
Daughter chromosomes

a.   b.   c.   d.

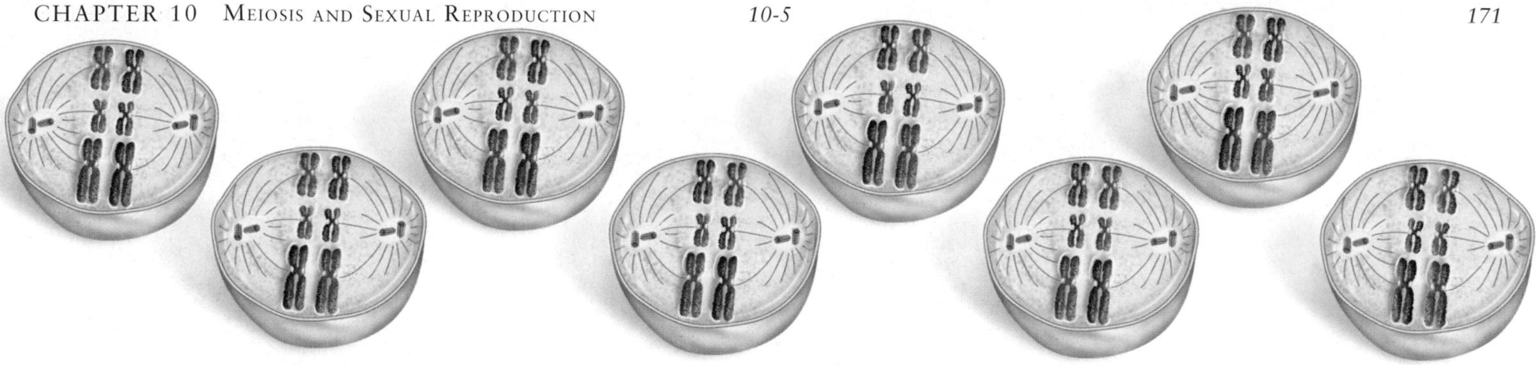

FIGURE 10.4 Independent assortment.
When a parent cell has three pairs of homologous chromosomes, there are 2³, or 8, possible chromosome alignments at the metaphase plate due to independent assortment. Among the 16 daughter nuclei resulting from these alignments, there are 8 different combinations of chromosomes.

Significance of Genetic Variation

Asexual reproduction passes on exactly the same combination of chromosomes and genes that the parent possesses. The process of sexual reproduction brings about genetic recombinations among members of a population (Fig. 10.5). If a parent is already successful in a particular environment, is asexual reproduction advantageous? It would seem so as long as the environment remains unchanged. However, if the environment changes, genetic variability among offspring, introduced by sexual reproduction, may be advantageous. Under these conditons, some offspring may have a better chance of survival and reproductive success than others in a population. For example, suppose the ambient temperature were to rise due to global warming. Perhaps a dog with genes for the least amount of fur may have an advantage over other dogs of its generation.

In a changing environment, asexual reproduction might saddle an offspring with a parent's disadvantageous gene combination. In contrast, sexual reproduction, with its reshuffling of genes due to meiosis and fertilization, might give a few offspring a better chance of survival when environmental conditions change.

> Genetic variation is achieved during sexual reproduction for three reasons: (1) Due to crossing-over, the nonsister chromatids carry recombined genes; (2) due to independent assortment, the gametes have various recombined chromosomes; and (3) due to fertilization, the zygote receives recombined chromosomes from both parents.

FIGURE 10.5 Genetic variation.
Why do the puppies in this litter have a different appearance even though they have the same two parents? Because crossing-over and independent assortment occurred during meiosis, and fertilization brought different gametes together.

10.3 THE PHASES OF MEIOSIS

Meiosis consists of two unique cell divisions, meiosis I and meiosis II. The phases of both meiosis I and meiosis II—prophase, metaphase, anaphase, and telophase—are described.

Prophase I

It is apparent during prophase I that nuclear division is about to occur because a spindle forms as the centrosomes migrate away from one another. The nuclear envelope fragments, and the nucleolus disappears.

The homologous chromosomes, each having two sister chromatids, undergo synapsis to form bivalents. As depicted in Figure 10.3 by the exchange of color, crossing-over between the nonsister chromatids may occur at this time. After crossing-over, the sister chromatids of a duplicated chromosome are no longer identical.

Throughout prophase I, the chromosomes have been condensing so that by now they have the appearance of metaphase chromosomes.

Metaphase I

During metaphase I, the bivalents held together by chiasmata (see Fig. 10.3) have moved toward the metaphase plate (equator of the spindle). Metaphase I is characterized by a fully formed spindle and alignment of the bivalents at the metaphase plate. **Kinetochores,** protein complexes just outside the centromeres, are seen, and these are attached to spindle fibers called kinetochore spindle fibers.

Bivalents independently align themselves at the metaphase plate of the spindle. The maternal homologue of each bivalent may be oriented toward either pole, and the paternal homologue of each bivalent may be oriented toward either pole. This means that all possible combinations of chromosomes can occur in the daughter cells.

Anaphase I

During anaphase I, the homologues of each bivalent separate and move to opposite poles. Notice that each chromosome still has two chromatids (see Fig. 10.6 p. 172).

Telophase I

Telophase I need not go to completion during meiosis. That is, the spindle disappears, but new nuclear envelopes need

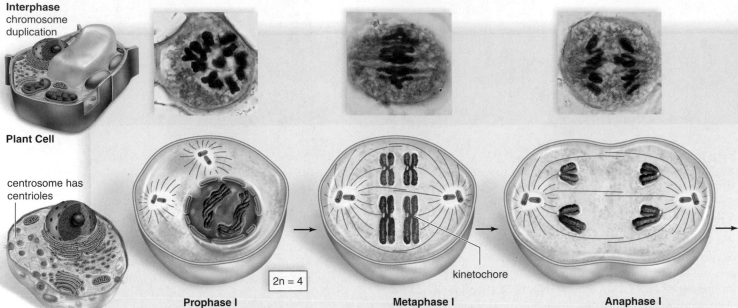

Interphase
chromosome
duplication

Plant Cell

centrosome has
centrioles

**Animal Cell
at Interphase**

2n = 4

Prophase I
Chromosomes have duplicated. Homologous
chromosomes pair during synapsis and
crossing-over occurs.

kinetochore

Metaphase I
Homologous pairs align independently
at the metaphase plate.

Anaphase I
Homologous chromosomes separate
and move toward the poles.

MEIOSIS I

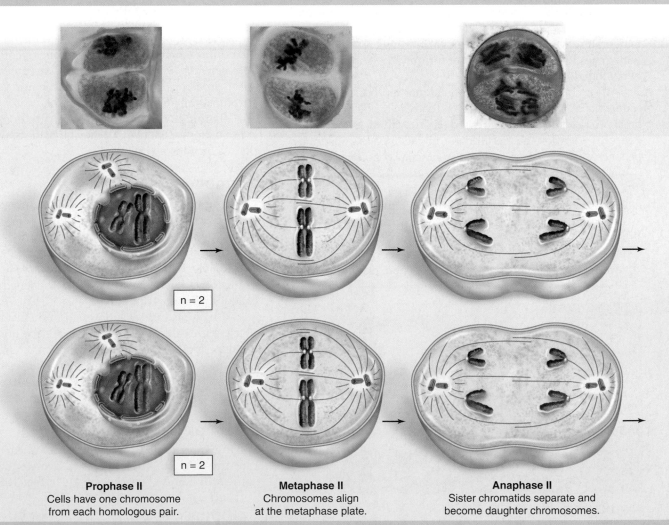

**FIGURE
10.6**

**Meiosis I and
II in plant cell
micrographs
and animal
cell drawings.**
When
homologous
chromosomes
pair during
meiosis I,
crossing-over
occurs as
represented by
the exchange
of color. Pairs
of homologous
chromosomes
separate during
meiosis I, and
chromatids
separate,
becoming
daughter
chromosomes
during meiosis II.
Following
meiosis II, there
are four haploid
daughter cells.

n = 2

n = 2

Prophase II
Cells have one chromosome
from each homologous pair.

Metaphase II
Chromosomes align
at the metaphase plate.

Anaphase II
Sister chromatids separate and
become daughter chromosomes.

MEIOSIS II

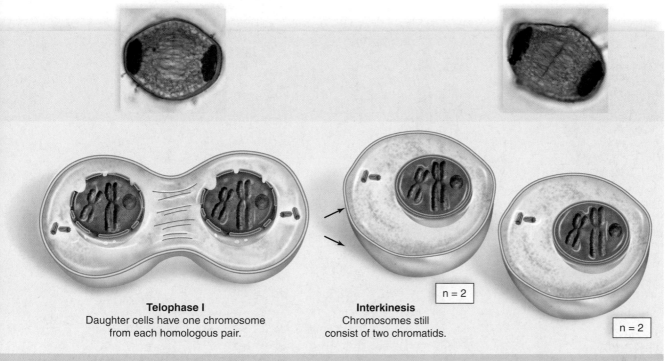

Telophase I
Daughter cells have one chromosome
from each homologous pair.

Interkinesis
Chromosomes still
consist of two chromatids.

n = 2

n = 2

MEIOSIS I cont'd

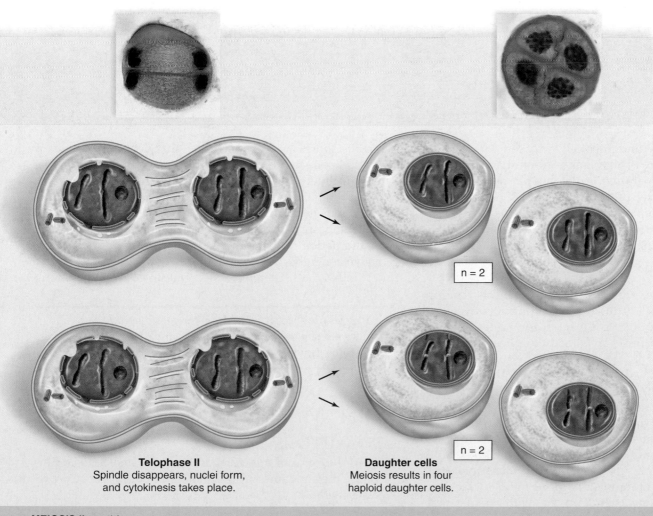

Telophase II
Spindle disappears, nuclei form,
and cytokinesis takes place.

Daughter cells
Meiosis results in four
haploid daughter cells.

n = 2

n = 2

MEIOSIS II cont'd

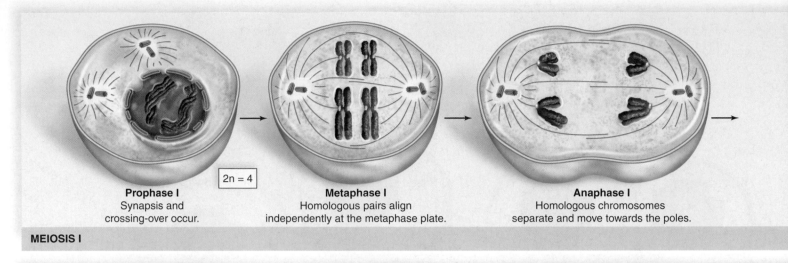

Prophase I
Synapsis and
crossing-over occur.

Metaphase I
Homologous pairs align
independently at the metaphase plate.

Anaphase I
Homologous chromosomes
separate and move towards the poles.

2n = 4

MEIOSIS I

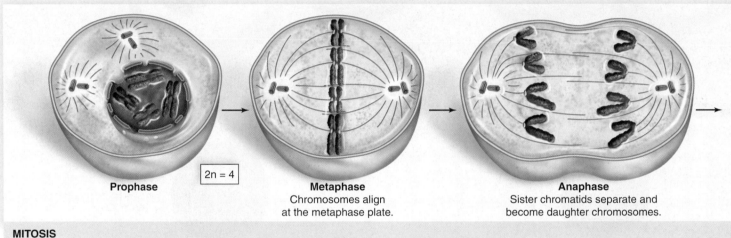

Prophase

Metaphase
Chromosomes align
at the metaphase plate.

Anaphase
Sister chromatids separate and
become daughter chromosomes.

2n = 4

MITOSIS

not form before the daughter cells proceed to meiosis II. Also, this phase may or may not be accompanied by cytokinesis, which is separation of the cytoplasm. Figure 10.6, p. 173, shows only two of the four possible combinations of haploid chromosomes when the parent cell has two homologous pairs of chromosomes. Can you determine what the other two possible combinations of chromosomes are?

Interkinesis

Following telophase, the cells enter interkinesis. The process of **interkinesis** is similar to interphase between mitotic divisions except that DNA replication does not occur because the chromosomes are already duplicated.

Meiosis II and Gamete Formation

During metaphase II, the haploid number of chromosomes, which are still duplicated, align at the metaphase plate (see Fig. 10.6, p. 172). During anaphase II, the sister chromatids separate, becoming daughter chromosomes that are not duplicated. These daughter chromosomes move toward the poles. At the end of telophase II and cytokinesis, there are four haploid cells. Due to crossing-over of chromatids during meiosis I, each gamete will most likely contain chromosomes with varied genes.

As mentioned, following meiosis II, the haploid cells become gametes in animals (see Section 10.5). In plants, they

become **spores,** a reproductive cell that develops into a new multicellular structure without the need to fuse with another reproductive cell. This structure is the haploid generation, which produces gametes. The resulting zygote develops into a diploid generation. Therefore, plants have both haploid and diploid phases in their life cycle, and plants are said to exhibit an **alternation of generations.** In most fungi and algae, the zygote undergoes meiosis, and the daughter cells develop into new individuals. Therefore, the organism is always haploid.

Meiosis produces haploid daughter cells that have a different fate depending on the organism. In animals, they become gametes.

10.4 MEIOSIS COMPARED TO MITOSIS

Figure 10.7 graphically compares meiosis and mitosis. Several of the fundamental differences between the two processes include:

• Meiosis requires two nuclear divisions, but mitosis requires only one nuclear division.

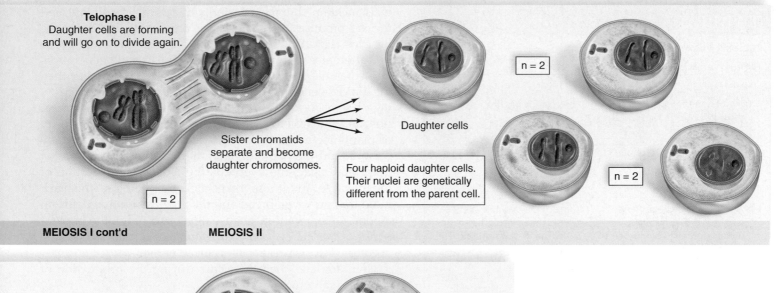

Telophase I
Daughter cells are forming
and will go on to divide again.

Sister chromatids
separate and become
daughter chromosomes.

n = 2

n = 2

Daughter cells

Four haploid daughter cells.
Their nuclei are genetically
different from the parent cell.

n = 2

n = 2

MEIOSIS I cont'd   MEIOSIS II

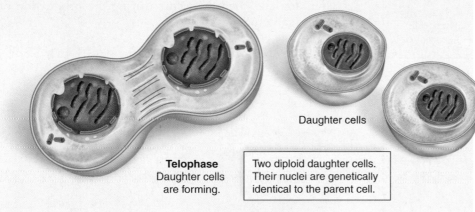

Daughter cells

Telophase
Daughter cells
are forming.

Two diploid daughter cells.
Their nuclei are genetically
identical to the parent cell.

MITOSIS cont'd

FIGURE 10.7 Meiosis compared to mitosis.
Why does meiosis produce daughter cells with half
the number while mitosis produces daughter cells
with the same number of chromosomes as the parent
cell? Compare metaphase I of meiosis to metaphase
of mitosis. Only in metaphase I are the homologous
chromosomes paired at the metaphase plate. Members
of homologous chromosome pairs separate during
anaphase I, and therefore the daughter cells are
haploid. The blue chromosomes were inherited from
the paternal parent, and the red chromosomes were
inherited from the maternal parent. The exchange of
color between nonsister chromatids represents the
crossing-over that occurred during meiosis I.

- Meiosis produces four daughter nuclei. Following cytokinesis there are four daughter cells. Mitosis followed by cytokinesis results in two daughter cells.
- Following meiosis, the four daughter cells are haploid and have half the chromosome number as the diploid parent cell. Following mitosis, the daughter cells have the same chromosome number as the parent cell.
- Following meiosis, the daughter cells are not genetically identical to each other or to the parent cell. Following mitosis, the daughter cells are genetically identical to each other and to the parent cell.

In addition to the fundamental differences between meiosis and mitosis, two specific differences between nuclear divisions can be categorized. These differences involve occurrence and process.

Occurrence

Meiosis occurs only at certain times in the life cycle of sexually reproducing organisms. In humans, meiosis occurs only in the reproductive organs and produces the gametes. Mitosis is more common because it occurs in all tissues during growth and repair.

Process

To summarize the process differences between meiosis and mitosis, see Tables 10.1 and 10.2, which separately compare meiosis I and meiosis II to mitosis.

Meiosis I Compared to Mitosis

Notice that these events distinguish meiosis I from mitosis:

- During meiosis I, bivalents form and crossing-over occurs during prophase I. These events do not occur during mitosis.
- During metaphase I of meiosis, bivalents independently align at the metaphase plate. The paired chromosomes have a total of four chromatids each. During metaphase in mitosis, individual chromosomes align at the metaphase plate. They each have two chromatids.
- During anaphase I of meiosis, homologues of each bivalent separate and duplicated chromosomes (with centromeres intact) move to opposite poles. During anaphase of mitosis, sister chromatids separate, becoming daughter chromosomes that move to opposite poles.

Meiosis II Compared to Mitosis

The events of meiosis II are similar to those of mitosis except in meiosis II, the nuclei contain the haploid number of

TABLE 10.1

Meiosis I Compared to Mitosis

Meiosis I	Mitosis
Prophase I	Prophase
Pairing of homologous chromosomes	No pairing of chromosomes
Metaphase I	Metaphase
Bivalents at metaphase plate	Duplicated chromosomes at metaphase plate
Anaphase I	Anaphase
Homologues of each bivalent separate and duplicated chromosomes move to poles	Sister chromatids separate, becoming daughter chromosomes that move to the poles
Telophase I	Telophase
Two haploid daughter cells not identical to the parent cell	Two diploid daughter cells, identical to the parent cell

TABLE 10.2

Meiosis II Compared to Mitosis

Meiosis II	Mitosis
Prophase II	Prophase
No pairing of chromosomes	No pairing of chromosomes
Metaphase II	Metaphase
Haploid number of duplicated chromosomes at metaphase plate	Diploid number of duplicated chromosomes at metaphase plate
Anaphase II	Anaphase
Sister chromatids separate, becoming daughter chromosomes that move to the poles	Sister chromatids separate, becoming daughter chromosomes that move to the poles
Telophase II	Telophase
Four haploid daughter cells, not genetically identical	Two diploid daughter cells, identical to the parent cell

chromosomes. In mitosis, the original number of chromosomes is maintained.

Meiosis reduces the chromosome number during the production of gametes. Mitosis keeps the chromosome number constant during growth and repair of tissues.

10.5 THE HUMAN LIFE CYCLE

The term **life cycle** refers to all the reproductive events that occur from one generation to the next similar generation. In ani-

mals, including humans, the individual is always diploid, and meiosis produces the gametes, the only haploid phase of the life cycle (Fig. 10.8). In contrast, plants have a haploid phase that alternates with a diploid phase. The haploid generation, known as the **gametophyte,** may be larger or smaller than the diploid generation, called the **sporophyte.** Mosses growing on bare rocks and forest floors are the haploid generation, and the diploid generation is short-lived. The majority of plants including pine, corn, and sycamores are usually diploid, and the haploid generation is short-lived. In most fungi and algae, the zygote is the only diploid portion of the life cycle, and it undergoes meiosis. Therefore, the black mold that grows on bread and the green scum that floats on a pond are haploid. In plants, algae, and fungi, the haploid phase of the life cycle produces gamete nuclei without the need for meiosis because it occurred earlier.

In animals, meiosis occurs during the production of gametes (**gametogenesis).**

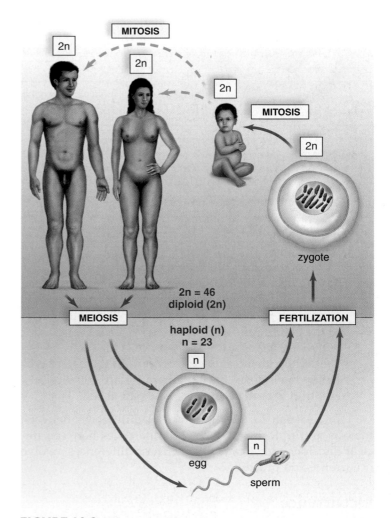

FIGURE 10.8 Life cycle of humans.
Meiosis in males is a part of sperm production, and meiosis in females is a part of egg production. When a haploid sperm fertilizes a haploid egg, the zygote is diploid. The zygote undergoes mitosis as it develops into a newborn child. Mitosis continues throughout life during growth and repair.

Animals are diploid and meiosis occurs during the production of gametes (**gametogenesis**). In males, meiosis is a part of **spermatogenesis** [Gk. *sperma*, seed; L. *genitus*, producing], which occurs in the testes and produces sperm. In females, meiosis is a part of **oogenesis** [Gk. *oon*, egg; L. *genitus*, producing], which occurs in the ovaries and produces eggs. A sperm and egg join at fertilization, and the resulting zygote undergoes mitosis during development of the fetus, which is the stage before birth. After birth, mitosis is involved in the continued growth of the child and repair of tissues at any time. As a result of mitosis, which occurs during most of the life cycle, each somatic cell in the body has the same number of chromosomes.

Spermatogenesis and Oogenesis in Humans

In humans, as in most animals, the sexes are separate. In males, spermatogenesis occurs within the testes, and in females, oogenesis occurs within the ovaries. The testes contain stem cells called spermatogonia, and these cells keep the testes supplied with primary spermatocytes that will undergo spermatogenesis as described in Figure 10.9, *top*. Primary spermatocytes with 46 chromosomes undergo meiosis I to form two secondary spermatocytes, each with 23 duplicated chromosomes. Secondary spermatocytes undergo meiosis II to produce four spermatids with 23 daughter chromosomes. Spermatids then differentiate into four viable sperm (spermatozoa). Upon sexual arousal, the sperm enter ducts and exit the penis upon ejaculation.

The ovaries contain stem cells called oogonia that actively produce primary oocytes during fetal development. Primary oocytes with 46 chromosomes undergo oogenesis once the female is sexually mature. Primary oocytes divide through meiosis I into two cells, each having 23 chromosomes (Fig. 10.9, *bottom*). One of these cells, termed the **secondary oocyte** [Gk, *oon*, egg, and *kytos*, cell], receives almost all the cytoplasm. The other is a **polar body** that may either disintegrate or divide again. The secondary oocyte begins meiosis II but stops at metaphase II. Then the secondary oocyte leaves the ovary and enters an oviduct, where sperm may be present. If no sperm are in the oviduct or one does not enter the secondary oocyte, it eventually disintegrates. If a sperm does enter the oocyte, it is activated to continue meiosis II, and another polar body forms. At the completion of oogenesis, following entrance of a sperm, there is one egg and two to three polar bodies. The polar bodies are a way to dispose of chromosomes while retaining much of the cytoplasm in the egg. Cytoplasmic molecules are needed by a developing embryo following fertilization.

The mature egg has 23 chromosomes but the zygote that results when the sperm and egg nuclei fuse has 46 chromosomes. In other words, fertilization restores the diploid (full number) of chromosomes.

In animals, including humans, meiosis is a part of gametogenesis, that is, oogenesis and spermatogenesis. Spermatogenesis produces four sperm, but oogenesis (completed only if fertilization occurs) produces one egg and at least two polar bodies.

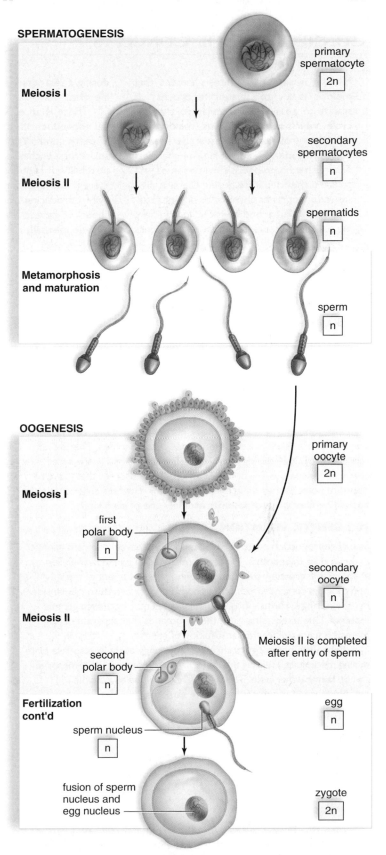

FIGURE 10.9 Spermatogenesis and oogenesis in mammals.
Spermatogenesis produces four viable sperm, whereas oogenesis produces one egg and at least two polar bodies. In humans, both sperm and egg have 23 chromosomes each; therefore, following fertilization, the zygote has 46 chromosomes.

CONNECTING THE CONCEPTS

Meiosis is similar to mitosis except that meiosis is a more elaborate process. It stands to reason that similar to the cell cycle, which involves mitosis, meiosis is a tightly controlled event. Researchers tell us that during meiosis, controlling mechanisms ensure that homologous chromosomes first pair and then separate during the first division and that sister chromatids do not separate until the second division. In addition, meiosis only occurs in certain types of cells during a restricted period of an organism's life span.

There is an evolutionary cost to sexual reproduction: The increased number of genes controlling the process can lead to an increased chance of mutations and the possibility of faulty or inviable gametes. But there is also an evolutionary advantage: Sexually reproducing species have a greater likelihood of genetic diversity among offspring than do asexually reproducing species.

Understanding the behavior of chromosomes during meiosis is critical to understanding the manner in which genes segregate during gamete formation. Chapter 11 reviews the fundamental laws of genetics established by Gregor Mendel. Although Mendel had no knowledge of chromosome behavior, modern students have the advantage of applying their knowledge of meiosis to their understanding of Mendel's laws.

Summary

10.1 HALVING THE CHROMOSOME NUMBER

Meiosis ensures that the chromosome number in offspring stays constant generation after generation. The nucleus contains pairs of chromosomes, called homologous chromosomes (homologues).

Meiosis requires two cell divisions and results in four daughter cells. Replication of DNA takes place before meiosis begins. During meiosis I, the homologues undergo synapsis (resulting in a bivalent) and align independently at the metaphase plate. The daughter cells receive one member of each pair of homologous chromosomes. There is no replication of DNA during interkinesis. During meiosis II, the sister chromatids separate, becoming daughter chromosomes that move to opposite poles as they do in mitosis. The four daughter cells contain the haploid number of chromosomes and only one of each kind.

10.2 GENETIC VARIATION

Sexual reproduction ensures that the offspring have a different genetic makeup than the parents. Meiosis contributes to genetic variability in two ways: crossing-over and independent assortment of the homologous chromosomes. When homologous chromosomes lie side by side during synapsis, nonsister chromatids may exchange genetic material. Due to crossing-over, the chromatids that separate during meiosis II have a different combination of genes.

When the homologous chromosomes align at the metaphase plate during metaphase I, either the maternal or the paternal chromosome can be facing either pole. Therefore, there will be all possible combinations of chromosomes in the gametes.

10.3 THE PHASES OF MEIOSIS

Meiosis I is divided into four phases:

Prophase I—Bivalents form, and crossing-over occurs as chromosomes condense; the nuclear envelope fragments.

Metaphase I—Bivalents independently align at the metaphase plate.

Anaphase I—Homologous chromosomes separate, and duplicated chromosomes move to poles.

Telophase I—Nuclei become haploid, having received one duplicated chromosome from each homologous pair.

Meiosis II is divided into four phases:

Prophase II—Chromosomes condense, and the nuclear envelope fragments.

Metaphase II—The haploid number of still duplicated chromosomes align at the metaphase plate.

Anaphase II—Sister chromatids separate, becoming daughter chromosomes that move to the poles.

Telophase II—Four haploid daughter cells are genetically different from the parent cell.

10.4 MEIOSIS COMPARED TO MITOSIS

Mitosis and meiosis can be compared in this manner:

Meiosis I	Mitosis
Prophase	
Pairing of homologous chromosomes	No pairing of chromosomes
Metaphase	
Bivalents at metaphase plate	Duplicated chromosomes at metaphase plate
Anaphase	
Homologous chromosomes separate and move to poles	Sister chromatids separate, becoming daughter chromosomes that move to the poles
Telophase	
Daughter nuclei have the haploid number of chromosomes	Daughter nuclei have the parent cell chromosome number

Meiosis II is like mitosis except the nuclei are haploid.

10.5 THE HUMAN LIFE CYCLE

Meiosis occurs in any life cycle that involves sexual reproduction. In the animal life cycle, only the gametes are haploid; in plants, meiosis produces spores that develop into a multicellular haploid adult that produces the gametes. In unicellular protists and fungi, the zygote undergoes meiosis, and spores become a haploid adult that gives rise to gametes.

During the life cycle of humans and other animals, meiosis is involved in spermatogenesis and oogenesis. Whereas spermatogenesis produces four sperm per meiosis, oogenesis produces one egg and two to three nonfunctional polar bodies. Spermatogenesis occurs in males, and oogenesis occurs in females. When a sperm fertilizes an egg, the zygote has the diploid number of chromosomes. Mitosis, which is involved in growth and repair, also occurs during the life cycle of all animals.

Reviewing the Chapter

1. Why did early investigators predict that there must be a reduction division in the sexual reproduction process?   168
2. What are homologous chromosomes? Contrast the genetic makeup of sister chromatids with that of nonsister chromatids.   168
3. What is synapsis? Contrast in general meiosis I with meiosis II.   169
4. Draw and explain a diagram that illustrates synapsis and another that shows all possible results from independent assortment of homologous pairs. How do these events ensure genetic variation among the gametes?   170–71
5. Draw and explain a series of diagrams that illustrate the stages of meiosis I.   171–74
6. Draw and explain a series of diagrams that illustrate the stages of meiosis II.   172–73
7. Construct a chart to describe the many differences between meiosis and mitosis.   171–75
8. What accounts for (1) the genetic similarity between daughter cells and the parent cell following mitosis, and (2) the genetic dissimilarity between daughter cells and the parent cell following meiosis?   174–76
9. Explain the human (animal) life cycle and the roles of meiosis and mitosis.   176–77
10. Compare spermatogenesis in males to oogenesis in females.   177

Testing Yourself

Choose the best answer for each question.

1. A bivalent is
   a. a homologous chromosome.
   b. the paired homologous chromosomes.
   c. a duplicated chromosome composed of sister chromatids.
   d. the two daughter cells after meiosis I.
   e. the two centrioles in a centrosome.

2. If a parent cell has 12 chromosomes, then each of the daughter cells following meiosis will have
   a. 48 chromosomes.
   b. 24 chromosomes.
   c. 12 chromosomes.
   d. 6 chromosomes.
   e. Any one of these could be correct.

3. At the metaphase plate during metaphase I of meiosis, there are
   a. chromosomes consisting of one chromatid.
   b. unpaired duplicated chromosomes.
   c. bivalents.
   d. homologous pairs of chromosomes.
   e. Both c and d are correct.

4. At the metaphase plate during metaphase II of meiosis, there are
   a. chromosomes consisting of one chromatid.
   b. unpaired duplicated chromosomes.
   c. bivalents.
   d. homologous pairs of chromosomes.
   e. Both c and d are correct.

5. Gametes contain one of each kind of chromosome because
   a. the homologous chromosomes separate during meiosis.
   b. the chromatids separate during meiosis.
   c. only one replication of DNA occurs during meiosis.
   d. crossing-over occurs during prophase I.
   e. the parental cell contains only one of each kind of chromosome.

6. Crossing-over occurs between
   a. sister chromatids of the same chromosome.
   b. two different kinds of bivalents.
   c. two different kinds of chromosomes.
   d. nonsister chromatids of a bivalent.
   e. two daughter nuclei.

7. During which phase of meiosis do homologous chromosomes separate?
   a. prophase II
   b. telophase I
   c. metaphase I
   d. anaphase I
   e. anaphase II

8. Fertilization
   a. is a source of variation during sexual reproduction.
   b. is fusion of the gametes.
   c. occurs in both animal and plant life cycles.
   d. restores the diploid number of chromosomes.
   e. All of these are correct.

9. Which of these is not a difference between spermatogenesis and oogenesis in humans?

Spermatogenesis	**Oogenesis**
a. Occurs in males	Occurs in females
b. Produces four sperm per meiosis	Produces one egg per meiosis
c. Produces haploid cells	Produces diploid cells
d. Always goes to completion	Does not always go to completion

For questions 10–15, indicate whether the statement is true (T) or false (F).

10. Normally, a person has 46 chromosomes in his or her karyotype. _____

11. Nondisjunction can occur during meiosis I or meiosis II. _____

12. During meiosis I, chromatids separate, and during meiosis II the members of homologous pairs separate. _____

13. The chromosomes are aligned at the metaphase plate of the spindle during anaphase. _____

14. Half of your chromosomes were inherited from your father, and half were inherited from your mother. _____

15. Homologous chromosomes pair during prophase of mitosis. _____

For questions 16–20, fill in the blanks.

16. If the parent cell has 24 chromosomes, the daughter cells following mitosis will have _____ chromosomes and following meiosis will have _____ chromosomes.

17. Meiosis in males is a part of _____, and meiosis in females is a part of _____.

18. Oogenesis will not go to completion unless _____ occurs.

19. In humans, meiosis produces _____, and in plants, meiosis produces _____.

20. During oogenesis, the primary oocyte is _____ and the secondary oocyte is _____.

For questions 21–27, match the statements that follow to the items in the key. Answers may be used more than once, and more than one answer may be used.

KEY:
    a. mitosis
    b. meiosis I
    c. meiosis II
    d. All of these are correct.
    e. None of these are correct.

21. Spindle fibers are attached to kinetochores.

22. The parent cell has ten duplicated chromosomes, and the daughter cells have five duplicated chromosomes.

23. Consists of a number of phases.

24. The parent cell has five duplicated chromosomes, and the daughter cells have five chromosomes, consisting of one chromatid each.

25. In humans, occurs only in the sexual organs.

26. The parent cell has ten duplicated chromosomes, and the daughter cells have ten duplicated chromosomes.

27. Involved in growth and repair of tissues.

28. Which of these drawings represents metaphase I? How do you know?

Thinking Scientifically

1. A population of lizards does not appear to contain any males, yet young are being born. If analysis shows that the offspring are diploid and have the same genetic traits as their mothers, what is the most likely source of maternal DNA to fertilize the eggs? How would you test this hypothesis?

2. In the nineteenth century, physicians noticed that people with Down syndrome were often the youngest children in large families. Some physicians suggested that the disorder is due to "maternal reproductive exhaustion." How would you disprove this hypothesis? What might be a reasonable explanation for the relationship between maternal age and incidence of Down syndrome?

Understanding the Terms

allele 168	independent assortment 170
alternation of generations 174	interkinesis 174
bivalent 169	kinetochore 171
crossing-over 170	life cycle 176
diploid (2n) number 168	meiosis 168
fertilization 170	oogenesis 177
gamete 168	polar body 177
gametogenesis 176	secondary oocyte 177
gametophyte 176	sexual reproduction 168
genetic recombination 170	spermatogenesis 177
haploid (n) number 168	spore 174
homologous	sporophyte 176
chromosome 168	synapsis 169
homologue 168	zygote 168

Match the terms to these definitions:
a. _____ Production of sperm in males by the process of meiosis and maturation.
b. _____ Pair of homologous chromosomes at the metaphase plate during meiosis I.
c. _____ A nonfunctional product of oogenesis.
d. _____ The functional product of meiosis I in oogenesis becomes the egg.
e. _____ Member of a pair of chromosomes in which both members carry genes for the same traits.

ARIS, the *Biology* Website

ARIS, the website for *Biology*, provides a wealth of information organized and integrated by chapter. You will find practice quizzes, interactive activities, labeling exercises, flashcards, and much more that will complement your learning and understanding of general biology.

www.mhhe.com/maderbiology9

11

MENDELIAN PATTERNS OF INHERITANCE

"*S he has her mother's eyes.*" "*Do you think that I will be bald like Dad?*" "*How can two apparently normal parents have a child with cystic fibrosis?*" "*Is it possible to increase the chances that my next litter of puppies will be champions?*" "*Why is there so much variety in parakeet coloration?*" *Humans have always been curious about how traits are passed from one generation to the next generation. This accounts for why comments and questions are endless when people begin considering inheritance. Genetics, the study of heredity, provides answers to their many questions.*

Today, we live in the age of genetics with its vast potential to map the human genome and engineer the genes themselves. Therefore, it is hard for us to believe that the science of genetics began humbly in an Austrian monastery during the 1800s. Our fundamental knowledge of genetics can be attributed to the experiments conducted by a monk, Gregor Mendel (1822–84), who studied the inheritance patterns of the garden pea plant. Mendel's work was ignored during his lifetime, and not published until early in the twentieth century when scientists revisited his research. Today, Gregor Mendel is known as the father of genetics.

Our knowledge of genetics has grown tremendously since those early days of pea cross-breeding at a monastery. Our modern knowledge of genetics has been acquired from studying a variety of other model organisms such as bacteria, bread mold, nematodes, and fruit flies. Extensive studies have also broadened our knowledge of human genetics. In addition, molecular genetics has blossomed over the past century. In 1953, Francis Crick and James Watson ushered in the modern era of genetics when they determined the molecular configuration of DNA. Because of their findings, biology has changed forever, and we are better able to answer questions pertaining to inheritance, health, and evolution.

Genetic inheritance accounts for the diversity of parakeet colorations.

11.1 GREGOR MENDEL

Like begets like—zebras always produce zebras, never camels; pumpkins always produce seeds for pumpkins, never watermelons. It is apparent to anyone who observes such phenomenon that parents pass hereditary information to their offspring. However, an offspring can be markedly different from either parent. For example, black-coated mice occasionally produce white-coated mice. The science of genetics provides explanations about not only the stability of inheritance but also the variations that are observed between generations and organisms.

Virtually every culture in history has attempted to explain inheritance patterns. An understanding of these patterns has always been important to agriculture and animal husbandry, the science of breeding animals. However, it was not until the 1860s that the Austrian monk Gregor Mendel developed the fundamental laws of heredity after performing a series of ingenious experiments (Fig. 11.1). Previously, he had studied science and mathematics at the University of Vienna, and at the time of his genetic research he was a substitute natural science teacher at a local high school. Various hypotheses about heredity had been proposed before Mendel began his experiments. In particular, investigators had been trying to support a blending concept of inheritance.

FIGURE 11.1 Gregor Mendel, 1822–84.
Mendel grew and tended the pea plants he used for his experiments. For each experiment, he observed as many offspring as possible. For a cross that required him to count the number of round seeds to wrinkled seeds, he observed and counted a total of 7,324 peas!

Blending Concept of Inheritance

When Mendel began his work, most plant and animal breeders acknowledged that both sexes contribute equally to a new individual. They felt that parents of contrasting appearance always produce offspring of intermediate appearance. Therefore, according to this concept, a cross between plants with red flowers and plants with white flowers would yield only plants with pink flowers. When red and white flowers reappeared in future generations, the breeders mistakenly attributed this to an instability in the genetic material.

The blending concept of inheritance offered little help to Charles Darwin, the father of evolution, whose treatise on natural selection lacked a strong genetic basis. If populations contained only intermediate individuals and normally lacked variations, how could diverse forms evolve? However, the theory of inheritance eventually proposed by Mendel did account for the presence of discrete variations (differences) among the members of a population, generation after generation. Although Darwin was a contemporary of Mendel, Darwin never learned of Mendel's work because it went unrecognized until 1900. Therefore, Darwin was never able to make use of Mendel's research to support his theory of evolution.

Mendel's Experimental Procedure

Most likely his background in mathematics prompted Mendel to use a statistical basis for his breeding experiments. He prepared for his experiments carefully and conducted preliminary studies with various animals and plants. He then chose to work with the garden pea, *Pisum sativum* (Fig. 11.2*a*).

The garden pea was a good choice. The plants were easy to cultivate and had a short generation time. Although peas normally self-pollinate (pollen only goes to the same flower), they could be cross-pollinated by hand by transferring pollen from the anther to the stigma. Many varieties of peas were available, and Mendel chose 22 for his experiments. When these varieties self-pollinated, they were *true-breeding*—meaning that the offspring were like the parent plants and like each other. In contrast to his predecessors, Mendel studied the inheritance of relatively simple and discrete traits, such as seed shape, seed color, and flower color, and he observed no intermediate characteristics among the offspring (Fig. 11.2*b*).

As Mendel followed the inheritance of individual traits, he kept careful records, and he used his understanding of the mathematical laws of probability to interpret his results and to arrive at a theory that has been supported by innumerable experiments. It's called a *particulate theory of inheritance* because it is based on the existence of minute particles or hereditary units we now call genes. Inheritance involves the reshuffling of the same genes from generation to generation.

At the time Mendel began his study of heredity, the blending concept of inheritance was popular. Mendel carefully designed his experiments and gathered mathematical data to arrive at a particulate theory of inheritance rather than a blending theory of inheritance.

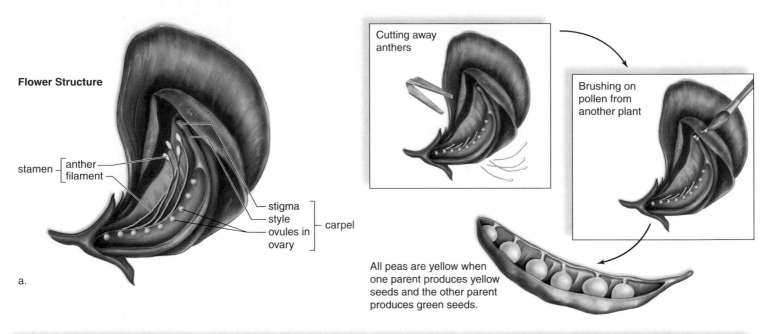

Flower Structure

a.

Cutting away anthers

Brushing on pollen from another plant

All peas are yellow when one parent produces yellow seeds and the other parent produces green seeds.

Trait	Characteristics			F₂ Results*	
	*Dominant		*Recessive	Dominant	Recessive
Stem length	Tall		Short	787	277
Pod shape	Inflated		Constricted	882	299
Seed shape	Round		Wrinkled	5,474	1,850
Seed color	Yellow		Green	6,022	2,001
Flower position	Axial		Terminal	651	207
Flower color	Purple		White	705	224
Pod color	Green		Yellow	428	152

*All of these produce approximately a 3:1 ratio. For example, $\frac{787}{277} = \frac{3}{1}$.

b.

FIGURE 11.2 Garden pea anatomy and a few traits.
a. In the garden pea, *Pisum sativum,* pollen grains produced in the anther contain sperm, and ovules in the ovary contain eggs. When Mendel performed crosses, he brushed pollen from one plant onto the stigma of another plant. After sperm fertilized eggs, the ovules developed into seeds (peas). The open pod shows the results of a cross between plants with yellow seeds and plants with green seeds. **b.** Mendel selected traits like these for study. He made sure his parent (P generation) plants bred true, and then he cross-pollinated the plants. The offspring called F₁ (first filial) generation always resembled the parent with the dominant characteristic (*left*). Mendel then allowed the F₁ plants to self-pollinate. In the F₂ (second filial) generation, he always achieved a 3:1 (dominant to recessive) ratio. The text explains how Mendel went on to interpret these results.

11.2 MENDEL'S LAW OF SEGREGATION

After ensuring that his pea plants were true-breeding—for example, that his tall plants always had tall offspring and his short plants always had short offspring—Mendel was ready to perform a cross-pollination experiment between two strains. For these initial experiments, Mendel chose varieties that differed in only one trait. If the blending theory of inheritance were correct, the cross should yield offspring with an intermediate appearance compared to the parents. For example, the offspring of a cross between a tall plant and a short plant should be intermediate in height.

Mendel called the original parents the *P generation* and the first generation the *F₁*, or filial [L. *filius*, sons and daughters], *generation* (Fig. 11.3). He performed *reciprocal crosses:* First he dusted the pollen of tall plants onto the stigmas of short plants, and then he dusted the pollen of short plants onto the stigmas of tall plants. In both cases, all F₁ offspring resembled the tall parent.

Certainly, these results were contrary to those predicted by the blending theory of inheritance. Rather than being intermediate, the F₁ plants were tall and resembled only one parent. Did these results mean that the other characteristic (i.e., shortness) had disappeared permanently? Apparently not, because when Mendel allowed the F₁ plants to self-pollinate, ¾ of the *F₂ generation* were tall and ¼ were short, a 3:1 ratio (Fig. 11.3). Therefore, the F₁ plants were able to pass on a factor for shortness—it didn't just disappear. Perhaps the F₁ plants were tall because tallness was dominant to shortness?

Mendel counted many plants. For this particular cross, called a **monohybrid cross,** he counted a total of 1,064 plants, of which 787 were tall and 277 were short. In all crosses that he performed, he found a 3:1 ratio in the F₂ generation. The characteristic that had disappeared in the F₁ generation reappeared in ¼ of the F₂ offspring.

His mathematical approach led Mendel to interpret his results differently from previous breeders. He knew that the same ratio was obtained among the F₂ generation time and time again for the same type of cross for the seven traits that he studied. Eventually Mendel arrived at this explanation: A 3:1 ratio among the F₂ offspring was possible if the F₁ parents contained two separate copies of each hereditary factor, one of these being dominant and the other recessive; the factors separated when the gametes were formed, and each gamete carried only one copy of each factor; and random fusion of all possible gametes occurred upon fertilization. Only in this way would shortness reoccur in the F₂ generation.

After doing monohybrid crosses, Mendel arrived at the first of his laws of inheritance—the law of segregation, which is a cornerstone of his particulate theory of inheritance.

The law of segregation states the following:

- Each individual has two factors for each trait.
- The factors segregate (separate) during the formation of the gametes.
- Each gamete contains only one factor from each pair of factors.
- Fertilization gives each new individual two factors for each trait.

As Viewed by Modern Genetics

Figure 11.3 also shows how we now interpret the results of Mendel's experiments on inheritance of stem length in peas. Each trait in a pea plant is controlled by two **alleles** [Gk. *allelon*, reciprocal, parallel], alternate forms of a gene that in this case control the length of the stem. The **dominant allele** is so named because of its ability to mask the expression of the other allele, called the **recessive allele.** The dominant allele is identified by an uppercase (capital) letter and the recessive allele by the same but lowercase (small) letter. Usually, the first letter designating a trait is chosen to identify the allele. With reference to the cross being

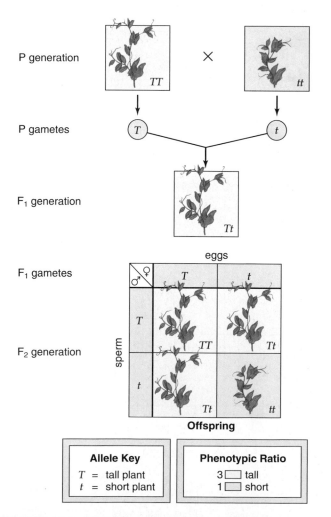

P generation

P gametes

F₁ generation

F₁ gametes

F₂ generation

Allele Key	Phenotypic Ratio
T = tall plant t = short plant	3 ☐ tall 1 ☐ short

FIGURE 11.3 Monohybrid cross done by Mendel.
The P generation plants differ in one regard—length of the stem. The F₁ generation plants are all tall, but the factor for short has not disappeared because ¼ of the F₂ generation plants are short. The 3:1 ratio allowed Mendel to deduce that individuals have two discrete and separate genetic factors for each trait.

discussed, there is an allele for tallness (*T*) and an allele for short-ness (*t*). Alleles occur on a homologous pair of chromosomes at a particular location that is called the **gene locus** (Fig. 11.4).

Meiosis is the type of cell division that reduces the chro-mosome number. During meiosis I, bivalents (homologous chromosomes each having sister chromatids) separate. There-fore, as discussed in the Science Focus on page 190, the process of meiosis gives an explanation for Mendel's law of segrega-tion, and why only one allele for each trait is in a gamete.

In Mendel's cross, the original parents (P generation) were true-breeding; therefore, the tall plants had two alleles for tall-ness (*TT*), and the short plants had two alleles for shortness (*tt*). When an organism has two identical alleles, as these had, we say it is **homozygous** [Gk. *homo*, same, and *zygos*, balance, yoke]. Because the parents were homozygous, all gametes produced by the tall plant contained the allele for tallness (*T*), and all gametes produced by the short plant contained an allele for shortness (*t*).

After cross-pollination, all the individuals of the resulting F$_1$ generation had one allele for tallness and one for shortness (*Tt*). When an organism has two different alleles at a gene locus, we say that it is **heterozygous** [Gk. *hetero*, different, and *zygos*, bal-ance, yoke]. Although the plants of the F$_1$ generation had one of each type of allele, they were all tall. The allele that is expressed in a heterozygous individual is the dominant allele. The allele that is not expressed in a heterozygote is the recessive allele.

Genotype Versus Phenotype

It is obvious from our discussion that two organisms with dif-ferent allelic combinations for a trait can have the same out-ward appearance. (*TT* and *Tt* pea plants are both tall.) For this reason, it is necessary to distinguish between the alleles present in an organism and the appearance of that organism.

The word **genotype** [Gk. *genos*, birth, origin, race, and *typos*, image, shape] refers to the alleles an individual receives at fertilization. Genotype may be indicated by letters or by short, descriptive phrases. Genotype *TT* is called homozygous

dominant, and genotype *tt* is called homozygous recessive. Genotype *Tt* is called heterozygous.

The word **phenotype** [Gk. *phaino*, appear, and *typos*, image, shape] refers to the physical appearance of the indi-vidual. The homozygous dominant (*TT*) individual and the heterozygous (*Tt*) individual both show the dominant pheno-type and are tall, while the homozygous recessive individual shows the recessive phenotype and is short (Table 11.1).

One-Trait Genetics Problems

When solving genetics problems, it is first necessary to know which allele is dominant. For example, the following allele key indicates that unattached earlobes are dominant over attached earlobes:

Allele key: *E* = unattached earlobes
            e = attached earlobes

If a man, homozygous for unattached earlobes, reproduces with a woman who has attached earlobes, what type of ear-lobe will their child have? In row 1 of the following diagram, P represents the parent generation, and the letters in this row are the genotypes of the parents. Row 2 shows that the gametes of each parent have only one type of allele for ear-lobes; therefore, the child (F$_1$ generation, row 3) will have a heterozygous genotype and unattached earlobes.

P: *EE*   ×   *ee*
Gametes: *E*   *e*
F$_1$:   *Ee*

If two heterozygotes reproduce with one another, will the child have unattached or attached earlobes? In the P gen-eration, there was only one possible type of gamete for each parent because they were homozygous. Heterozygotes, however, can produce two types of gametes; ½ of the gam-etes will contain an *E*, and ½ will contain an *e*.

P: *Ee*   ×   *Ee*
Gametes: *E, e*   *E, e*

When determining the gametes, it is necessary to keep in mind that although an individual is diploid (has two alleles for each trait), *each gamete is haploid—that is, has only one allele for each trait.* This is true of single-trait crosses as well as multiple-trait crosses. When solving genetics problems, first decide on the appropriate allele key, and then determine the genotype of both parents and the various types of gametes for both parents. Practice Problems 11.1 (p. 186) will help you learn to designate the gametes.

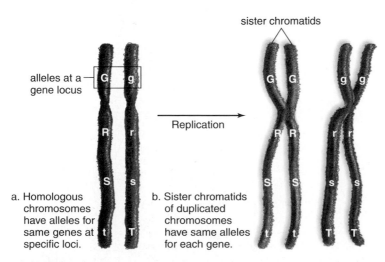

FIGURE 11.4 Homologous chromosomes.
a. The letters represent alleles; that is, alternate forms of a gene. Each allelic pair, such as *Gg* or *Tt*, is located on homologous chromosomes at a particular gene locus. **b.** Sister chromatids carry the same alleles in the same order.

TABLE 11.1		
Genotype Versus Phenotype		
Genotype	Genotype	Phenotype
TT	Homozygous dominant	Tall plant
Tt	Heterozygous	Tall plant
tt	Homozygous recessive	Short plant

Practice Problems 11.1*

1. For each of the following genotypes, give all genetically different gametes, noting the proportion of each for the individual.
   a. WW
   b. Ww
   c. Tt
   d. TT

2. For each of the following, state whether a genotype (genetic makeup of an organism) or a type of gamete is represented.
   a. D
   b. GG
   c. P

*Answers to Practice Problems appear in Appendix A.

Laws of Probability

The next step in doing a genetics problem is to decide the expected results of a cross. When we calculate the expected results of the cross under consideration, or any genetic cross, we use the laws of probability. Imagine flipping a coin; each time you flip the coin, there is a 50% chance of heads and a 50% chance of tails. In like manner, if the parent has the genotype *Ee*, what is the chance of any child inheriting either an *E* or *e* from that parent?

$$\text{The chance of } E = \tfrac{1}{2}$$

$$\text{The chance of } e = \tfrac{1}{2}$$

But in the cross *Ee* × *Ee*, a child will inherit an allele from each parent. Therefore, we are observing two separate, independent events. How likely is it that an offspring will inherit a specific set of two genes, one from each parent? The multiplicative law of probability answers this question by stating that *the chance, or probability, of two or more independent events occurring together is the product (multiplication) of their chance of occurring separately.* Therefore, the probability of receiving these genotypes is as follows:

1. The chance of $EE = \tfrac{1}{2} \times \tfrac{1}{2} = \tfrac{1}{4}$
2. The chance of $Ee = \tfrac{1}{2} \times \tfrac{1}{2} = \tfrac{1}{4}$
3. The chance of $eE = \tfrac{1}{2} \times \tfrac{1}{2} = \tfrac{1}{4}$
4. The chance of $ee = \tfrac{1}{2} \times \tfrac{1}{2} = \tfrac{1}{4}$

Now we have to consider the additive law of probability: *The chance of an event that can occur in two or more independent ways is the sum (addition) of the individual chances.* Therefore:

The chance of a child with unattached earlobes (*EE*, *Ee*, or *eE*) is ¾ (add **1, 2,** and **3**), or 75%.
The chance of a child with attached earlobes (*ee*) is ¼ (only **4**), or 25%.

Many students use the Punnett square discussed next to help use the laws of probability and calculate the expected results of a cross.

The Punnett Square

The **Punnett square** was introduced by a prominent poultry geneticist, R. C. Punnett, in the early 1900s as a simple way to figure the probable results of a genetic cross. Figure 11.5 shows how the results of the cross under consideration (*Ee* × *Ee*) can be determined using a Punnett square. In a Punnett square, all possible kinds of sperm are lined up vertically and all possible kinds of eggs are lined up horizontally (or vice versa), and every possible combination of alleles is placed within the squares. In our cross, each parent has two possible types of gametes for a trait (*E* or *e*), so these two types of gametes are lined up vertically and horizontally. To determine the number of boxes to place in a Punnett square, use the formula 4^N (where N represents the number of traits) if both parents are heterozygous for all traits involved. Therefore, a Punnett square for a monohybrid cross will have 4 boxes, a Punnett square for a dihybrid cross will have 16 boxes, and one for a trihybrid cross will have 64 boxes.

The results of the Punnett square calculations show that the expected genotypes of offspring are ¼ *EE*, ½ *Ee*, and ¼ *ee*, giving a 1:2:1 genotypic ratio. Since ¾ have unattached earlobes and ¼ have attached earlobes, this is a 3:1 phenotypic ratio.

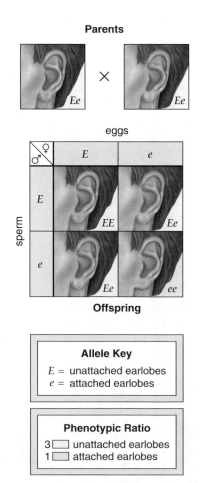

FIGURE 11.5 Genetic inheritance in humans.
When the parents are heterozygous, each child has a 75% chance of having the dominant phenotype and a 25% chance of having the recessive phenotype.

The gametes combine at random, and in practice it is usually necessary to observe a very large number of offspring before a 3:1 ratio can be verified. Only if many offspring are counted can it be ensured that all genetically different sperm have had a chance to fertilize all genetically different eggs. If a number of heterozygotes produced 200 offspring, approximately 150 of these would have unattached earlobes and approximately 50 would have attached earlobes (a 3:1 ratio). In terms of genotypes, approximately 50 would be *EE,* about 100 would be *Ee,* and the remaining 50 would be *ee.*

We cannot arrange crosses between humans in order to count a large number of offspring. Therefore, in humans the phenotypic ratio is used to estimate the chances any child has for a particular characteristic. In the cross under consideration, each child has a ¾ (or 75%) chance of having unattached earlobes and a ¼ (or 25%) chance of having attached earlobes. And we must remember that *chance has no memory:* if two heterozygous parents already have a child with attached earlobes, the next child still has a 25% chance of having attached earlobes. Each conception is an independent event.

The Punnett square makes use of the laws of probability mentioned in the previous section. First, there is a ½ chance of a child getting the *E* allele or the *e* allele from a parent. In this way it is possible to designate the possible gametes. What operation is in keeping with the multiplicative law of probability? The operation of combining the alleles and placing them in the squares. What operation is in keeping with the additive law of probability? The operation of adding the results and arriving at the phenotypic ratio.

Mendel was quite familiar with the laws of probability and was able to use them to see that inheritance of traits depended on the passage of individual factors from generation to generation.

The laws of probability allow us to calculate the probable results of one-trait genetic crosses.

Practice Problems 11.2*

1. In rabbits, if *B* = dominant black allele and *b* = recessive white allele, which of these genotypes (*Bb, BB, bb*) could a white rabbit have?
2. In pea plants, yellow seed color is dominant over green seed color. When two heterozygous plants are crossed, what percentage of plants would have yellow seeds? Green seeds?
3. In humans, freckles is dominant over no freckles. A man with freckles reproduces with a woman with freckles, but the children have no freckles. What chance did each child have for freckles?
4. In horses, trotter (*T*) is dominant over pacer (*t*). A trotter is mated to a pacer, and the offspring is a pacer. Give the genotype of all the horses.

Answers to Practice Problems appear in Appendix A.

One-Trait Testcross

Mendel's experimental use of simple dominant and recessive traits allowed him to test the law of segregation. To confirm that the F_1 was heterozygous, Mendel crossed his F_1 generation tall plants with true-breeding, short (homozygous recessive) plants. He reasoned that half the offspring should be tall and half should be short, producing a 1:1 phenotypic ratio (Fig. 11.6*a*). Indeed, those were the results he obtained; therefore, his hypothesis that alleles segregate when gametes are formed was supported.

In Figure 11.6*a*, the homozygous recessive parent can produce only one type of gamete—*t*—and so the Punnett square has only one column. The use of one column signifies that all the gametes carry a *t*.

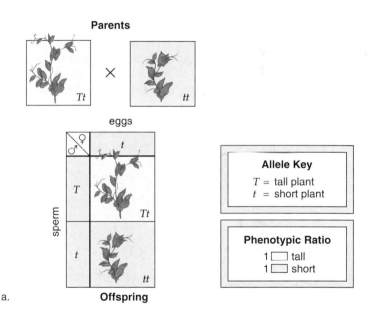

a.

Allele Key
T = tall plant
t = short plant

Phenotypic Ratio
1 ☐ tall
1 ☐ short

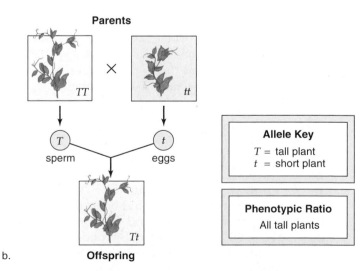

b.

Allele Key
T = tall plant
t = short plant

Phenotypic Ratio
All tall plants

FIGURE 11.6 One-trait testcross.
Crossing an individual with the dominant phenotype with a recessive individual indicates the genotype. **a.** If a parent with the dominant phenotype is heterozygous, the phenotypic ratio among the offspring is 1:1. **b.** If a parent with the dominant phenotype is homozygous, all offspring have the dominant phenotype.

Today, a one-trait **testcross** is used to determine if an individual with the dominant phenotype is homozygous dominant or heterozygous for a particular trait. Since both of these genotypes produce the dominant phenotype, it is not possible to determine the genotype by observation. Figure 11.6*b* shows that if the individual is homozygous dominant, all the offspring will be tall. Each parent has only one type of gamete and therefore a Punnett square is not required to determine the results.

The results of a testcross indicate whether an individual with the dominant phenotype is heterozygous or homozygous dominant.

Practice Problems 11.3*

1. In horses, *B* = black coat and *b* = brown coat. What type of cross should be done to best determine whether a black-coated horse is homozygous dominant or heterozygous?
2. In fruit flies, *L* = long wings and *l* = short wings. When a long-winged fly is crossed with a short-winged fly, the offspring exhibit a 1:1 ratio. What is the genotype of the parent flies?
3. In the garden pea, round seeds are dominant over wrinkled seeds. An investigator crosses a plant having round seeds with a plant having wrinkled seeds. He counts 400 offspring. How many of the offspring have wrinkled seeds if the plant having round seeds is a heterozygote?

Answers to Practice Problems appear in Appendix A.

11.3 Mendel's Law of Independent Assortment

Mendel performed a second series of crosses in which true-breeding plants differed in two traits. This F₁ cross is known as a **dihybrid cross.** For example, he crossed tall plants having green pods with short plants having yellow pods (Fig. 11.7). The F₁ plants showed both dominant characteristics. As before, Mendel then allowed the F₁ plants to self-pollinate. Two possible results could occur in the F₂ generation:

1. If the dominant factors (*TG*) always segregate into the F₁ gametes together, and the recessive factors (*tg*) always stay together, then there would be two phenotypes among the F₂ plants—tall plants with green pods and short plants with yellow pods.
2. If the four factors segregate into the F₁ gametes independently, then there would be four phenotypes among the F₂ plants—tall plants with green pods, tall plants with yellow pods, short plants with green pods, and short plants with yellow pods.

Figure 11.7 shows that Mendel observed four phenotypes among the F₂ plants, supporting the second hypothesis. Therefore, Mendel formulated his second law of heredity—the law of independent assortment.

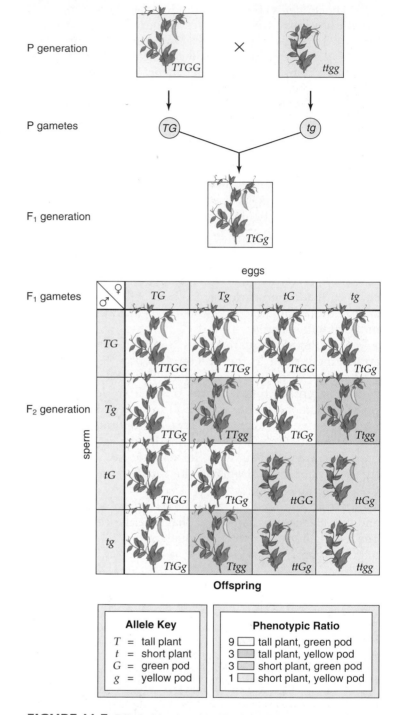

FIGURE 11.7 Dihybrid cross done by Mendel.
P generation plants differ in two regards—length of the stem and color of the pod. The F₁ generation shows only the dominant traits, but all possible phenotypes appear among the F₂ generation. The 9:3:3:1 ratio allowed Mendel to deduce that factors segregate into gametes independently of other factors.

The law of independent assortment states the following:
• Each pair of factors segregates (assorts) independently of the other pairs.
• All possible combinations of factors can occur in the gametes.

Two-Trait Genetics Problems

The fruit fly, *Drosophila melanogaster*, less than one-fifth the size of a housefly, is a favorite subject for genetic research because it has several mutant characteristics that are easily determined. The allele that occurs most frequently in a population or that is designated as normal is known as the **wild type.** For example, a "wild-type" fly has long wings and a gray body, while certain mutant flies may have short (vestigial) wings and black (ebony) bodies. The allele key for a cross involving these traits is *L* = long wing, *l* = short wing, *G* = gray body, and *g* = black body.

Laws of Probability

If two flies heterozygous for both traits are crossed, what are the probable results? Since each characteristic is inherited separately from any other, it is possible to apply again the laws of probability mentioned on page 186. For example, we know the F_2 results for two separate monohybrid crosses are as listed here:

1. The chance of long wings = ¾
   The chance of short wings = ¼
2. The chance of gray body = ¾
   The chance of black body = ¼

Using the multiplicative law, we know that:

The chance of long wings and gray body = ¾ × ¾ = ⁹⁄₁₆
The chance of long wings and black body = ¾ × ¼ = ³⁄₁₆
The chance of short wings and gray body = ¼ × ¾ = ³⁄₁₆
The chance of short wings and black body = ¼ × ¼ = ¹⁄₁₆

Using the additive law, we conclude that the phenotypic ratio is 9:3:3:1. Again, since all genetically different male gametes must have an equal opportunity to fertilize all genetically different female gametes to even approximately achieve these results, a large number of offspring must be counted.

Punnett Square

In Figure 11.8, each P generation fly has only one possible type of gamete because the P_1 fly is either homozygous dominant or homozygous recessive for both traits. All the F_1 flies are heterozygous (*LlGg*) and have the same phenotype (long wings, gray body).

The Punnett square in Figure 11.8 shows the expected results when the F_1 flies are crossed, assuming that all genetically different sperm have an equal opportunity to fertilize all genetically different eggs. Notice that ⁹⁄₁₆ of the offspring have long wings and a gray body, ³⁄₁₆ have long wings and a black body, ³⁄₁₆ have short wings and a gray body, and ¹⁄₁₆ have short wings and a black body. This phenotypic ratio of 9:3:3:1 is expected whenever a heterozygote for two traits is

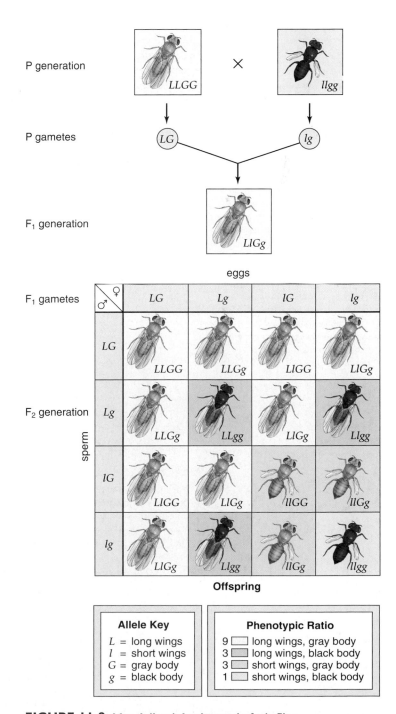

Allele Key

L = long wings
l = short wings
G = gray body
g = black body

Phenotypic Ratio

9 ▢ long wings, gray body
3 ▨ long wings, black body
3 ▢ short wings, gray body
1 ▢ short wings, black body

FIGURE 11.8 Mendelian inheritance in fruit flies.
Each F_1 fly (*LlGg*) produces four types of gametes because all possible combinations of chromosomes can occur in the gametes. Therefore, all possible phenotypes appear among the F_2 offspring.

science focus

Mendel's Laws and Meiosis

Today, we realize that the genes are on the chromosomes and that Mendel's laws hold because of the events of meiosis. Figure 11A assumes a parent cell that has two homologous pairs of chromosomes and that the alleles *A,a* are on one pair and the alleles *B,b* are on the other pair. Following duplication of the chromosomes, the parent cell undergoes meiosis as a first step toward the production of gametes. At metaphase I, the homologous pairs line up independently, and therefore all alignments of homologous chromo- somes can occur at the metaphase plate. Then the pairs of homologous chromosomes separate.

In keeping with Mendel's law of independent assortment and law of segregation, each pair of chromosomes and alleles segregates indepen- dently of the other pairs. It matters not which member of a homologous pair faces which spindle pole. Therefore, the daughter cells from meiosis I have all possible combinations of alleles in a single copy. One daughter cell has both dominant al- leles, namely *A* and *B*. Another daughter cell has both recessive alleles, namely *a* and *b*. The other two are mixed: *A* with *b* and *a* with *B*. Therefore, all possible combinations of alleles occur at meta- phase II and in the gametes.

When you form the gametes for any genetic cross, you are following the dictates of Mendel's laws but also mentally taking the chromosomes and alleles through the process of meiosis. We can also note that fertilization restores both the diploid chromosome number and the paired con- dition of alleles in the zygote.

FIGURE 11A Independent assortment and segregation during meiosis.
Mendel's laws hold because of the events of meiosis. The homologous pairs of chromosomes line up randomly at the metaphase plate during meiosis I. Therefore, the homologous chromosomes, and alleles they carry, segregate independently during gamete formation. All possible combinations of chromosomes and alleles occur in the gametes.

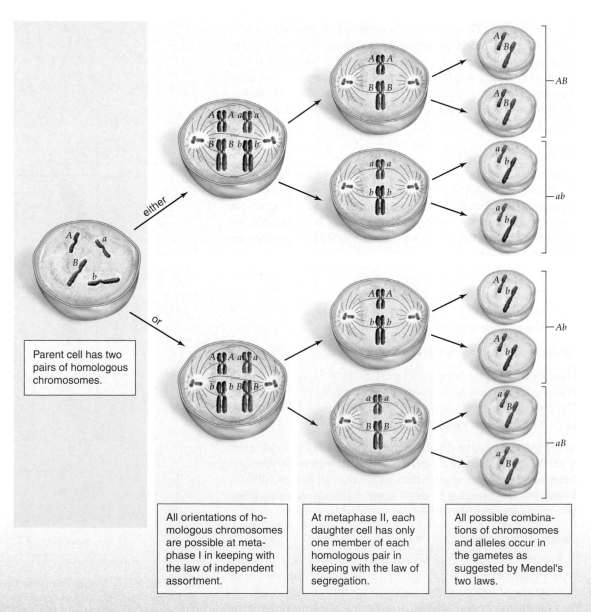

either

or

Parent cell has two pairs of homologous chromosomes.

— AB

— ab

— Ab

— aB

All orientations of ho- mologous chromosomes are possible at meta- phase I in keeping with the law of independent assortment.

At metaphase II, each daughter cell has only one member of each homologous pair in keeping with the law of segregation.

All possible combina- tions of chromosomes and alleles occur in the gametes as suggested by Mendel's two laws.

crossed with another heterozygote for two traits and simple dominance is present in both genes.

A Punnett square can also be used to predict the chances of an offspring having a particular phenotype. What are the chances of an offspring with long wings and gray body? The chances are $^9/_{16}$. What are the chances of an offspring with short wings and gray body? The chances are $^3/_{16}$, and so forth.

Today, we know that these results are obtained and the law of independent assortment holds because of the events of meiosis. The gametes contain one allele for each trait and in all possible combinations because homologues separate independently during meiosis I.

Two-Trait Testcross

A two-trait testcross is used to determine if an individual is homozygous dominant or heterozygous for either of the two traits. Since it is not possible to determine the genotype of a long-winged, gray-bodied fly by inspection, the genotype may be represented as $L\_\_^1\ G\_\_$.

When doing a two-trait testcross, an individual with the dominant phenotype is crossed with an individual with the recessive phenotype. For example, a long-winged, gray-bodied fly is crossed with a short-winged, black-bodied fly. A long-winged, gray-bodied fly heterozygous for both traits will form four different types of gametes. The homozygous fly with short wings and a black body can form only one kind of gamete.

P:	*LlGg*	×	*llgg*
Gametes:	*LG*		*lg*
	Lg		
	lG		
	lg		

Figure 11.9 shows that ¼ of the expected offspring have long wings and a gray body; ¼ have long wings and a black body; ¼ have short wings and a gray body; and ¼ have short wings and a black body. This is a 1:1:1:1 phenotypic ratio. The presence of offspring with short wings and a black body shows that the *L_G_* fly is heterozygous for both traits and has the genotype *LlGg*.

If the *L_G_* fly is homozygous for both traits, then no offspring will have short wings or a black body when this fly is crossed with one that has the recessive phenotype for both traits. If the *L_G_* fly is heterozygous for one trait but not the other, what is the expected phenotypic ratio among the offspring when this fly is crossed with one that has the recessive phenotype for both traits?

In two-trait genetics problems, the individual has four alleles—two for each trait.

---

1. The blank means that a dominant or recessive allele can be present.

Parents

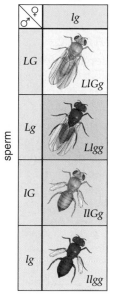

Offspring

Allele Key	**Phenotypic Ratio**
L = long wings	1 ☐ long wings, gray body
l = short wings	1 ▓ long wings, black body
G = gray body	1 ☐ short wings, gray body
g = black body	1 ☐ short wings, black body

FIGURE 11.9 Two-trait testcross.
Testcross to determine if a fly is heterozygous for both traits. If a fly heterozygous for both traits is crossed with a fly that is recessive for both traits, the expected ratio of phenotypes is 1:1:1:1. What would the result be if the test individual were homozygous dominant for both traits? Homozygous dominant for one trait but heterozygous for the other?

Practice Problems 11.5*

1. In horses, B = black coat, b = brown coat, T = trotter, and t = pacer. A black pacer mated to a brown trotter produces a black trotter offspring. Give all possible genotypes for this offspring.
2. In fruit flies, long wings (L) is dominant over short wings (l), and gray body (G) is dominant over black body (g). In each instance, what are the most likely genotypes of the previous generation if a student gets the following phenotypic results?
   a. 1:1:1:1 (all possible combinations in equal number)
   b. 9:3:3:1 (9 dominant; 3 mixed; 3 mixed; 1 recessive)
3. In humans, short fingers and widow's peak are dominant over long fingers and straight hairline. A heterozygote in both regards reproduces with a similar heterozygote. What is the chance of any one child having the same phenotype as the parents?

*Answers to Practice Problems appear in Appendix A.

11.4 HUMAN GENETIC DISORDERS

Many human disorders are genetic in origin. Genetic disorders are medical conditions caused by alleles inherited from parents. Some of these conditions are controlled by dominant or recessive alleles on the autosomal chromosomes. An **autosome** is any chromosome other than a sex (X or Y) chromosome. In humans, the chromosome pairs 1–22 are considered autosomes.

Patterns of Inheritance

When a genetic disorder is autosomal dominant, an individual with the alleles *AA* or *Aa* has the disorder. When a genetic disorder is autosomal recessive, only individuals with the alleles *aa* have the disorder. Genetic counselors often construct pedigrees to determine whether a condition is dominant or recessive. A pedigree shows the pattern of inheritance for a particular condition. Consider these two possible patterns of inheritance:

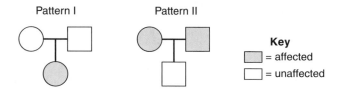

In both patterns, males are designated by squares and females by circles. Shaded circles and squares are affected individuals. A line between a square and a circle represents a union. A vertical line going downward leads, in these patterns, to a single child. (If there are more children, they are placed off a horizontal line.) Which pattern of inheritance (I or II) do you suppose represents an autosomal dominant characteristic, and which represents an autosomal recessive characteristic?

In pattern I, the child is affected, but neither parent is; this can happen if the condition is recessive and the parents are *Aa*. Notice that the parents are **carriers** because they appear normal but are capable of having a child with the genetic disorder. In pattern II, the child is unaffected, but the parents are affected. This can happen if the condition is dominant and the parents are *Aa*.

Figure 11.10 shows other ways to recognize an autosomal recessive pattern of inheritance, and Figure 11.11 shows other ways to recognize an autosomal dominant pattern of inheritance. In these pedigrees, generations are indicated by Roman numerals placed on the left side. Notice in the third generation of Figure 11.10 that two closely related individuals have produced three children, two of which have the affected phenotype. This illustrates that reproduction between closely related persons increases the chances of children inheriting two copies of a potentially harmful recessive allele.

The inheritance pattern of alleles on the X chromosome follows different rules than those on the autosomal chromosomes, and you will learn to recognize this pattern also when we discuss it in Chapter 12.

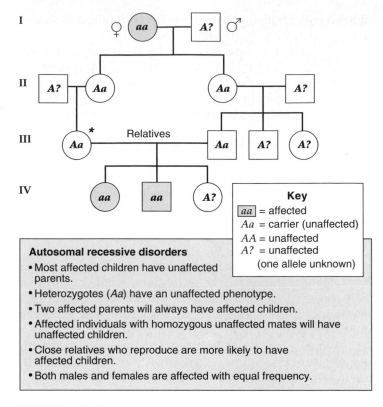

Autosomal recessive disorders
- Most affected children have unaffected parents.
- Heterozygotes (*Aa*) have an unaffected phenotype.
- Two affected parents will always have affected children.
- Affected individuals with homozygous unaffected mates will have unaffected children.
- Close relatives who reproduce are more likely to have affected children.
- Both males and females are affected with equal frequency.

Key
- *aa* = affected
- *Aa* = carrier (unaffected)
- *AA* = unaffected
- *A?* = unaffected (one allele unknown)

FIGURE 11.10 Autosomal recessive pedigree.
The list gives ways to recognize an autosomal recessive disorder. How would you know the individual at the asterisk is heterozygous?[2]

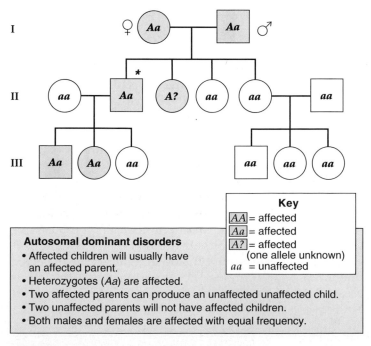

Autosomal dominant disorders
- Affected children will usually have an affected parent.
- Heterozygotes (*Aa*) are affected.
- Two affected parents can produce an unaffected unaffected child.
- Two unaffected parents will not have affected children.
- Both males and females are affected with equal frequency.

Key
- *AA* = affected
- *Aa* = affected
- *A?* = affected (one allele unknown)
- *aa* = unaffected

FIGURE 11.11 Autosomal dominant pedigree.
The list gives ways to recognize an autosomal dominant disorder. How would you know the individual at the asterisk is heterozygous?[2]

2. See Appendix A for answers.

Autosomal Recessive Disorders

In humans, a number of autosomal recessive disorders have been identified. Four of the best known autosomal dominant disorders include Tay-Sachs disease, cystic fibrosis, phenyl-ketonuria, and sickle cell disease.

Tay-Sachs Disease

Tay-Sachs disease is a genetic disease that usually occurs in about 1 in 3,600 Jewish people in the United States, most of whom are of central and eastern European descent. At first, it is not apparent that a baby has Tay-Sachs disease. However, development begins to slow down between four and eight months of age, and neurological impairment and psychomotor difficulties then become apparent. The child gradually becomes blind and helpless, develops uncontrollable seizures, and eventually becomes paralyzed. There is no treatment or cure, and most affected individuals die by the age of three or four years.

Tay-Sachs disease results from a lack of the enzyme hexosaminidase A (Hex A) and the subsequent storage of its substrate, a glycosphingolipid, in lysosomes. As the glycosphingolipid builds up in the lysosomes, it crowds the organelles and impairs their function. The primary sites of the storage problem are the cells of the brain, which accounts for the onset of symptoms and the progressive deterioration of psychomotor functions.

Carriers of Tay-Sachs disease have about half the level of Hex A activity found in homozygous dominant individuals. Prenatal diagnosis of the disease is possible following either amniocentesis or chorionic villi sampling. The gene for Tay-Sachs disease is located on chromosome 15.

Cystic Fibrosis

Cystic fibrosis (CF) is the most common lethal genetic disease among Caucasians in the United States. About 1 in 20 Caucasians is a carrier, and about 1 in 2,000 newborns has the disorder. Abnormal secretions related to the chloride ion channel characterize this disorder. One of the most obvious symptoms of CF patients is having extremely salty sweat. In children with CF, the mucus in the bronchial tubes and pancreatic ducts is particularly thick and viscous, interfering with the function of the lungs and pancreas. To ease breathing, the thick mucus in the lungs has to be loosened periodically, but still the lungs frequently become infected. The clogged pancreatic ducts prevent digestive enzymes from reaching the small intestine, and to improve digestion, patients take digestive enzymes mixed with applesauce before every meal.

Research has demonstrated that chloride ions (Cl^-) fail to pass through plasma membrane channel proteins in the cells of CF patients. Ordinarily, after chloride ions have passed through the membrane, sodium ions (Na^+) and water follow. It is believed that lack of water is the cause of the abnormally thick mucus in the bronchial tubes and pancreatic ducts. In the past few years, new treatments, such as the one shown in Figure 11.12, have raised the average life

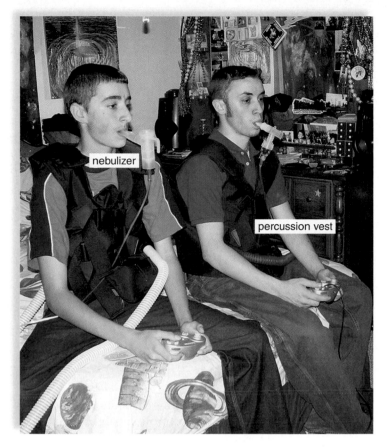

FIGURE 11.12 Cystic fibrosis therapy.
These brothers are undergoing antibiotic and percussion therapy for cystic fibrosis. Antibiotic therapy is used to control lung infections of cystic fibrosis patients. The antibiotic tobramycin can be aerosolized and administered using a nebulizer. It is inhaled twice daily for about 15 minutes. Mucus in the lungs can be loosened automatically by a percussion vest or manually by percussion hand therapy.

expectancy for CF patients to as much as 35 years of age. The CF gene, which is located on chromosome 7, has been identified and isolated. It is hoped that as gene therapy improves, corrected genes can be placed in patients to replace faulty genes. Genetic testing for an allele in adult carriers and in fetuses is possible; if an abnormal allele is present, couples may want to consider that successful gene therapy may be possible some day. To explain the persistence of the CF allele in a population, it has been suggested that those heterozygous for CF are less likely to die from potentially fatal diseases such as cholera.

Phenylketonuria

Phenylketonuria (PKU) occurs once in 5,000 births, so it is not as frequent as the disorders previously discussed. However, it is the most commonly inherited metabolic disorder that affects nervous system development.

Affected individuals lack an enzyme that is needed for the normal metabolism of the amino acid phenylalanine, so an abnormal breakdown product, a phenylketone, accumulates in the urine. The PKU gene is located on chromosome 12, and a prenatal DNA test can determine the presence of this mutation. Newborns are routinely tested in the hospital for elevated levels of phenylalanine in the blood. If elevated levels are detected, newborns are placed on a diet low in phenylalanine, which must be continued until the brain is fully developed (around the age of seven years), or else severe mental retardation develops. Some doctors recommend that the diet continue for life, but in any case, a pregnant woman with phenylketonuria must be on the diet in order to protect her unborn child from harm. Many diet products such as soft drinks have warnings to phenylketonurics that the product contains the amino acid phenylalanine.

Sickle Cell Disease

Sickle cell disease is the most common inherited disorder among blacks, affecting approximately 1 in 500 African Americans. It is estimated that 1 in 12 African Americans are carriers for sickle cell disease. The gene for this disorder is located on chromosome 11. In individuals with sickle cell disease, the red blood cells are shaped like sickles or half-moons instead of biconcave discs. An abnormal hemoglobin molecule (Hb^S) causes the defect. Normal hemoglobin (Hb^A) differs from Hb^S by one amino acid in the protein globin. The single change causes Hb^S to be less soluble than Hb^A.

Sickle cell disease, like cystic fibrosis and several other genetic disorders, is an example of **pleiotropy** [Gk. *pleion,* more or many, *tropos,* turning], a term used to describe a gene that affects more than one characteristic of an individual. Sickling of the red blood cells (see page 35) occurs when the oxygen content of an affected individual's blood is low. The abnormally shaped sickle cells slow down blood flow and clog small blood vessels in affected individuals. In addition, sickled red blood cells have a shorter life span than normal red blood cells.

A person suffering from sickle cell disease (Hb^SHb^S) may exhibit a number of symptoms, including severe anemia, physical weakness, poor circulation, impaired mental function, pain and high fever, rheumatism, paralysis, spleen damage, low resistance to disease, and kidney and heart failure. Individuals who are Hb^AHb^S have sickle cell trait and do not usually have any sickle-shaped cells unless they experience dehydration or mild oxygen deprivation.

Two individuals with sickle cell trait can produce children with three possible phenotypes. The chances of producing an individual with a normal genotype (Hb^AHb^A) are 25%, sickle cell trait (Hb^AHb^S) 50%, and sickle cell disease (Hb^SHb^S) 25%. Because of the three possible phenotypes, some geneticists consider sickle cell disease to represent incomplete dominance. This inheritance pattern will be discussed in Section 11.5.

Present treatments for sickle cell disease include pain management, blood transfusions, and bone marrow transplants. New medications that activate the gene for the production of normal hemoglobin are in the experimental phases. Presently, prenatal diagnosis for sickle cell disease is possible. In the future, gene therapy may be available for these patients. Individuals with sickle cell trait are encouraged to lead a normal, active lifestyle.

Although sickle cell disease is a devastating disorder, it provides heterozygous individuals with a survival advantage. Individuals who have sickle cell trait are resistant to the protozoan parasite that causes malaria. The parasite that spends part of its life cycle in red blood cells cannot complete its life cycle when sickle-shaped cells form and break down earlier than usual.

Practice Problems 11.6*

1. What is the genotype of the children in Figure 11.12? What are the genotypes of their parents if neither parent has cystic fibrosis? (Use this allele key: C = normal; c = cystic fibrosis.)
2. What are the chances that the parents in problem 1 will have a child with cystic fibrosis? (Hint: See Figure 11.5 and use the key in problem 1.)
3. A mother and daughter both have PKU. A son does not have PKU. What is the genotype of the father?

*Answers to Practice Problems appear in Appendix A.

Autosomal Dominant Disorders

A number of autosomal dominant disorders have been identified in humans. Three relatively common autosomal dominant disorders include neurofibromatosis, Huntington disease, and achondroplasia.

Neurofibromatosis

Neurofibromatosis, sometimes called von Recklinghausen disease, is one of the most common genetic disorders. It affects roughly 1 in 3,500 newborns and is seen equally in every racial and ethnic group throughout the world.

At birth or later, the affected individual may have six or more large, tan spots on the skin. Such spots may increase in size and number and get darker. Small, benign tumors (lumps) called neurofibromas that arise from the fibrous coverings of nerves may develop. This genetic disorder shows variable expressivity. In most cases, symptoms are mild and patients live a normal life. In some cases, however, the effects are severe. Some patients have skeletal deformities, including a large head, and eye and ear tumors occur that can lead to blindness and hearing loss. Many children with neurofibromatosis have learning disabilities and are hyperactive.

In 1990, researchers isolated the gene for neurofibromatosis, which was known to be on chromosome 17. The gene controls the production of a protein called neurofibromin, which normally blocks growth signals leading to cell division. Any number of mutations can lead to a neurofibromin that fails to block cell growth, and the result is the formation of tumors. Some mutations are caused by inserted genes that don't belong in their present location. The fact that genes can move from one location in the genome to another was first discovered in other organisms and then in humans also.

a.   b.

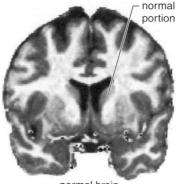

FIGURE 11.13
Huntington disease.
a. Persons with this condition gradually lose psychomotor control of the body. At first, the disturbances are only minor, but the symptoms become worse over time. **b.** Cross section of a normal brain and affected brain. In Huntington disase a portion of the brain involved in motor control atrophies.

normal portion

normal brain

atrophied portion

affected brain

Huntington Disease

Huntington disease is a neurological disorder that leads to progressive degeneration of brain cells, which in turn causes severe muscle spasms and personality disorders (Fig. 11.13). The disease, which occurs in about 1 in 20,000 persons, is caused by a single mutated copy of the gene for a protein called huntingtin. Most patients appear normal until they are of middle age and have already had children, who may then also be stricken. Occasionally, the first sign of the disease in the next generation is seen in teenagers or even younger children. There is no effective treatment, and death comes 10–15 years after the onset of symptoms.

Several years ago, researchers found that the gene for Huntington disease was located on chromosome 4. A test was developed for the presence of the gene, but few people want to know if they have inherited the gene because there is no cure. At least now we know that the disease stems from a mutation that causes huntingtin to have too many copies of the amino acid glutamine. The normal version of the huntingtin protein has stretches of between 10 and 25 glutamines. If the huntingtin protein has more than 36 glutamines, it changes shape and forms large clumps inside neurons. Even worse, it attracts and causes other proteins to clump with it. One of these proteins, called CBP, ordinarily helps nerve cells survive. Researchers hope they may be able to combat the disease by boosting CBP levels.

Achondroplasia

Achondroplasia is a common form of dwarfism associated with a defect in the growth of long bones. Individuals with achondroplasia have short arms and legs, a sway back, and a normal torso and head. About 1 in 25,000 people have achondroplasia. The gene is located on chromosome 4. Individuals who have achondroplasia are heterozygotes (Aa). The homozygous recessive (aa) condition yields normal length limbs, and the homozygous dominant condition (AA) is lethal, and death occurs before the individual can reproduce.

Recessive genetic disorders can be passed on by parents who appear normal. Dominant genetic disorders are passed on by a parent who has or will develop the disorder.

Practice Problems 11.7*

1. What is the genotype of the woman in Figure 11.13 if she is heterozygous? What would be the genotype of a husband who is homozygous recessive? (Use this allele key: H = Huntington; h = unaffected.)
2. If the couple in question 1 reproduces, what is the chance that a child will develop Huntington disease? Will be free of Huntington disease? (Hint: See Figure 11.6b and use the key in problem 1.)
3. Which of the children from problem 2 could pass on Huntington disease? Why?
4. A child has neurofibromatosis. The mother appears unaffected. What is the genotype of the father?

*Answers to Practice Problems appear in Appendix A.

11.5 BEYOND MENDELIAN GENETICS

Mendelian genetics is **unifactorial;** one gene consisting of a single pair of alleles, one dominant and one recessive, controls a certain trait. Mendelian genetics can explain the pattern of inheritance for many traits, but not all. In this section, we will see that some traits are **multifactorial.** They are controlled by the interactions of more than one pair of alleles and also the environment.

Incomplete Dominance

Incomplete dominance is exhibited when the heterozygote has an intermediate phenotype between that of either homozygote. In a cross between a true-breeding, red-flowered four-o'clock strain and a true-breeding, white-flowered strain, the offspring have pink flowers. But this is not an example of the blending theory of inheritance. When the F_1 plants self-pollinate, the F_2 generation has a phenotypic ratio of 1 red-flowered : 2 pink-flowered : 1 white-flowered plant. The reappearance of the parent phenotypes in the F_2 generation makes it clear that we are still dealing with a single pair of alleles (Fig. 11.14).

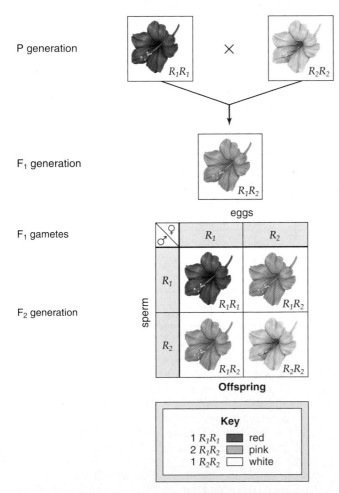

FIGURE 11.14 Incomplete dominance.

Incomplete dominance is illustrated by a cross between red- and white-flowered four-o'clocks. The F_1 plants are pink, a phenotype intermediate between those of the P generation; however, the F_2 results show that neither the red nor the white allele has disappeared.

Incomplete dominance in four-o'clocks can be explained in this manner: Each R_1 allele leads to the production of pigment, but the R_2 allele does not cause the production of pigment. In R_1R_1 individuals, the double dose of pigment results in red flowers, in R_1R_2 individuals, the single dose of pigment results in pink flowers, and because the R_2R_2 individual produces no pigment, the flowers are white.

Human Examples of Incomplete Dominance

In humans, clear-cut examples of incomplete dominance in which the phenotype of the heterozygote is intermediate between the two homozygote phenotypes are relatively rare. One such example is curly versus straight hair in Caucasians. When a curly-haired Caucasian reproduces with a straight-haired Caucasian, their children have wavy hair. When two wavy-haired persons reproduce, the expected phenotypic ratio among the offspring is 1:2:1—that is, 1 curly-haired child : 2 with wavy hair : 1 with straight hair (same as in Figure 11.14).

Sickle cell disease can also be considered an example of incomplete dominance. After all, the expected ratio when two individuals with sickle cell trait reproduce is 1:2:1—that is, 1 child with sickle cell disease : 2 with sickle cell trait : 1 child who is normal.

It can also be argued that the inheritance pattern of other human disorders should be considered one of incomplete dominance. To detect the carriers of cystic fibrosis and Tay-Sachs disease, for example, it is customary to determine the amount of cellular activity of the gene. When the activity is one-half that of the dominant homozygote, the individual is a carrier. In other words, at the level of gene expression, the homozygotes and heterozygotes do differ in the same manner as four-o'clock plants.

Multiple Allelic Traits

When a trait is controlled by **multiple alleles,** the gene exists in several allelic forms. But each person usually has only two of the possible alleles. For example, a person's blood type is determined by multiple alleles.

ABO Blood Types

Three alleles for the same gene control the inheritance of ABO blood types. These alleles determine the presence or absence of antigens on red blood cells.

I^A = A antigen on red blood cells
I^B = B antigen on red blood cells
i = Neither A nor B antigen on red blood cells

Each person has only two of the three possible alleles, and both I^A and I^B are dominant over i. Therefore, there are two possible genotypes for type A blood and two possible genotypes for type B blood. On the other hand, I^A and I^B are fully expressed in the presence of the other. Therefore, if a person inherits one of each of these alleles, that person will have type AB blood. Type O blood can only result from the inheritance of two i alleles.

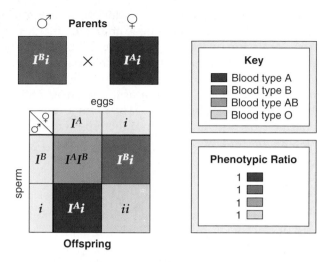

FIGURE 11.15 Inheritance of blood types.
A mating between individuals with type A blood and type B blood can result in any one of the four blood types. Why? Because the parents are $I^A i$ and $I^B i$. If both parents had type AB blood, predict the blood types of their potential offspring.

The possible genotypes and phenotypes for blood type are as follows:

Phenotype	Genotype
A	$I^A I^A$, $I^A i$
B	$I^B I^B$, $I^B i$
AB	$I^A I^B$
O	ii

The inheritance of the ABO blood group in humans is an example of **codominance.** This inheritance pattern differs greatly from Mendel's findings since more than one allele is fully expressed. When an individual has blood type AB, both A and B types of glycolipids appear on the red blood cells. The two different capital letters signify that both alleles are coding for an effective product.

Figure 11.15 shows that matings between certain genotypes can have surprising results in terms of blood type. As a point of interest, the Rh factor is inherited separately from A, B, AB, or O blood types. When you are Rh positive, your red blood cells have a particular antigen, and when you are Rh negative, that antigen is controlled by a single allelic pair in which simple dominance prevails: The Rh-positive allele is dominant while the Rh-negative allele is recessive.

Exceptions to simple Mendelian inheritance include incomplete dominance, multiple alleles, and codominance. In incomplete dominance, the offspring has an intermediate phenotype compared to the parents with two different phenotypes. In multiple alleles, the offspring inherits two of several possible alleles. In codominance, two inherited alleles are expressed equally.

Polygenic Inheritance

Polygenic inheritance [Gk. *poly*, many; L. *genitus*, producing] occurs when a trait is governed by two or more sets of alleles. The individual has a copy of all allelic pairs, possibly located on many different pairs of chromosomes. Each dominant allele has a quantitative effect on the phenotype, and these effects are additive. The result is a continuous variation of phenotypes, resulting in a distribution that resembles a bell-shaped curve. The more genes involved, the more continuous are the variations and distribution of the phenotypes. Pure polygenic traits that are not influenced by the environment are relatively rare. In other words, polygenic traits are multifactorial because they are controlled by more than one pair of alleles and are also influenced by the environment.

Seed Color in Wheat

In wheat, when plants that produce white seeds are crossed with plants that produce dark-red seeds, the F_1 plants are intermediate in color. If the F_1 plants are allowed to self-pollinate, the F_2 plants produce seeds having one of seven degrees of color. These results can be explained by assuming that color of seeds is controlled by genes at three different loci and that each dominant allele contributes a small but equal effect to the phenotype (Fig. 11.16). The dominant alleles are represented by dark dots in this diagram:

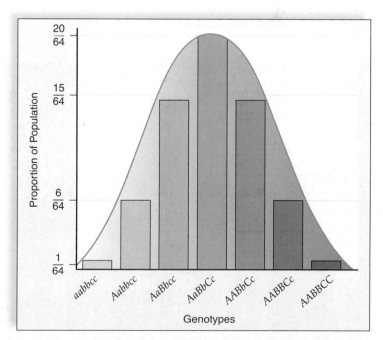

FIGURE 11.16 Polygenic inheritance.
In this example, a trait is controlled by three genes. Only the alleles represented by a capital letter contribute to the phenotype of an individual. When two F_1 individuals with genotypes *AaBbCc* are crossed, seven phenotypes are seen among the F_2 generation as shown in this graph. If environmental effects are considered, a bell-shaped curve results because of many intervening phenotypes.

Human Examples of Polygenic Inheritance

In addition to height, skin color is an example of a polygenic trait. Just how many pairs of alleles control skin color is not known, but a range in colors can be explained on the basis of just two pairs. When a very dark person reproduces with a very light person, the children have medium-brown skin; when two people with medium-brown skin reproduce with one another, the children may range in skin color from very dark to very light. This is possible if skin color is controlled by two pairs of alleles, designated *Aa* and *Bb*, and *each capital letter contributes pigment to the skin.*

Genotypes	Phenotypes
AABB	Very dark
AABb or *AaBB*	Dark
AaBb or *AAbb* or *aaBB*	Medium brown
Aabb or *aaBb*	Light
aabb	Very light

Notice again that a range of phenotypes exists and several possible phenotypes fall between the two extremes. Therefore, the distribution of these phenotypes is expected to follow a bell-shaped curve, meaning that few people have the extreme phenotypes, and most people have the phenotype that lies in the middle.

Eye color is also a polygenic trait. The iris, or colored portion of the eye, is colored by the pigment melanin. The amount of melanin deposited in the iris increases the darker the color of the eye. Different eye colors from the brightest of blue to nearly black eyes are thought to be the result of two genes with alleles each interacting in an additive manner.

Skin and eye color can also be used as an example of **epistasis,** which occurs when a gene at one locus interferes with the expression of a gene at a different locus. Consider, for example, that some individuals are albinos lacking pigment regardless of how many dominant alleles the individual inherits for the production of the pigment melanin, which causes skin color in humans. The problem stems from the inheritance of a gene whose product interferes with the expression of the alleles for skin color.

Polygenic Disorders

Many human disorders, such as cleft lip and/or palate, clubfoot, congenital dislocations of the hip, hypertension, diabetes, schizophrenia, and even allergies and cancers, are most likely due to the combined action of many genes plus environmental influences. Therefore, many investigators are reconsidering the nature versus nurture question—that is, what percentage of the trait is controlled by genes, and what percentage is controlled by the environment? Thus far, it has not been possible to determine precise, generally accepted percentages for any particular trait.

In recent years, reports have surfaced that all sorts of behavioral traits, such as alcoholism, phobias, and even suicide, can be associated with particular genes. No doubt, behavioral traits are to a degree controlled by genes, but again, it is impossible at this time to determine to what degree. And few scientists would support the idea that these behavioral traits are predetermined by our genes.

FIGURE 11.17 Height in human beings.
When you record the heights of a large group of people chosen at random, the values typically follow a bell-shaped curve. Such a continuous distribution is due to control of a trait by several sets of alleles. Environmental effects can also influence gene expression.

FIGURE 11.18 Coat color in Himalayan rabbits.
The influence of the environment on the phenotype has been demonstrated by plucking out the fur from one area and applying an ice pack. The new fur that grew in was black instead of white, showing that the enzyme that produces melanin (a dark pigment) in the rabbit is active only at low temperatures.

Environment and the Phenotype

There is no doubt that both the genotype and the environment affect the phenotype. In the case of human height, nutrition is a major factor (Fig. 11.17). The relative importance of genetic and environmental influences on the phenotype can vary, but in some instances the environment seems to have an extreme effect. In the water buttercup, *Ranunculus peltatus*, the submerged part of the plant has an appearance different from the part above water. Apparently, the presence or absence of an aquatic environment dramatically influences the phenotype.

Temperature can also have a dramatic effect on the phenotypes of plants and animals. Primroses have white flowers when grown above 32°C and red flowers when grown at 24°C. The coats of Siamese cats and Himalayan rabbits are darker in color at the ears, nose, paws, and tail. Himalayan rabbits are known to be homozygous for the allele *ch*, which is involved in the production of melanin. Experimental evidence suggests that the enzyme encoded by this gene is active only at a low temperature and that, therefore, black fur only occurs at the extremities, where body heat is lost to the environment (Fig. 11.18). When the animal is placed in a warmer environment, new fur on these body parts is light in color.

So, we come to the conclusion that matters are a little more complex than Mendel envisioned. Not only do gene interactions contribute to the phenotype, but so does the environment.

An organism's environmental history, in addition to its genotype, can influence the phenotype.

Practice Problems 11.8*

1. What are the chances of an offspring inheriting sickle cell disease when the parents have sickle cell trait?
2. In a paternity suit, the man has blood type AB. He could not be the father if the child has what blood type?
3. A child with type O blood is born to a mother with type A blood. What is the genotype of the child? The mother? What are the possible genotypes of the father?
4. A polygenetic trait is controlled by three different loci. Give seven genotypes among the offspring that will result in seven different phenotypes when *AaBbCc* is crossed with *AaBbCc*.

*Answers to Practice Problems appear in Appendix A.

CONNECTING THE CONCEPTS

Humans have practiced selective plant and animal breeding since agriculture and animal domestication began thousands of years ago. Many early breeders came to realize that their programs were successful because they were manipulating inherited factors for particular characteristics. A good experimental design and a bit of luck allowed Mendel, and not one of these breeders, to work out his rules of inheritance.

The use of a large number of pea crosses and a statistical analysis of the large number of resulting offspring contributed to the success of Mendel's work. Although humans do not usually produce a large number of offspring, it has been possible to conclude that Mendel's laws do apply to humans in many instances. Good historical records of inheritance in large families, such as Mormon families, have allowed researchers to show that a number of human genetic disorders are indeed controlled by a single allelic pair. Such disorders include cystic fibrosis, albinism, achondroplasia, neurofibromatosis, and Huntington disease.

Mendel was lucky in that he chose to study an organism, namely the garden pea, whose observable traits are often determined by a single allelic pair. In the common garden bean, on the other hand, plant height, for example, is determined by several genes. Therefore, crosses of tall and short plants result in F_1 plants of intermediate height rather than all tall plants. This type of inheritance pattern, called polygenic, controls *many* important traits in farm animals, such as weight and milk production in cattle. Other types of inheritance patterns that differ from simple Mendelian inheritance are also discussed in this chapter.

Summary

11.1 GREGOR MENDEL

At the time Mendel began his hybridization experiments, the blending concept of inheritance was popular. This concept stated that whenever the parents are distinctly different, the offspring are intermediate between them. In contrast to preceding plant breeders, Mendel decided to do a statistical study, and his study involved nonblending traits of the garden pea. Therefore, he arrived at a particulate theory of inheritance.

11.2 MENDEL'S LAW OF SEGREGATION

When Mendel crossed heterozygous plants with other heterozygous plants, he found that the recessive phenotype reappeared in about $\frac{1}{4}$ of the F_2 plants. This allowed Mendel to deduce his law of segregation, which states that the individual has two factors for each trait and the factors segregate into the gametes.

The laws of probability can be used to calculate the expected phenotypic ratio of a cross. In practice, a large number of offspring must be counted in order to observe the expected results, and to ensure that all possible types of sperm have fertilized all possible types of eggs.

Because humans do not produce a large number of offspring, it is best to use a predicted ratio as a means of estimating the chances of an individual inheriting a particular characteristic.

Mendel also crossed the F_1 plants having the dominant phenotype with homozygous recessive plants. The results indicated that the recessive factor was present in these F_1 plants (i.e., that they were heterozygous). Today, we call this a testcross, because it is used to test whether an individual showing the dominant characteristic is homozygous dominant or heterozygous.

11.3 MENDEL'S LAW OF INDEPENDENT ASSORTMENT

Mendel conducted two-trait crosses, in which the F_1 individuals showed both dominant characteristics, but there were four phenotypes among the F_2 offspring. This allowed Mendel to deduce the law of independent assortment, which states that the members of one pair of factors separate independently of those of another pair. Therefore, all possible combinations of parental factors can occur in the gametes.

When Mendel crossed individuals heterozygous in two traits, he observed a 9:3:3:1 ratio among the offspring. The two-trait testcross allows an investigator to test whether an individual showing two dominant characteristics is homozygous dominant for both traits or for one trait only, or is heterozygous for both traits.

11.4 HUMAN GENETIC DISORDERS

Studies of human genetics have shown that many genetic disorders can be explained on the basis of simple Mendelian inheritance. When studying human genetic disorders, biologists often construct pedigrees to show the pattern of inheritance of a characteristic within a family. The particular pattern indicates the manner in which a characteristic is inherited. Sample pedigrees for autosomal recessive and autosomal dominant patterns appear in Figures 11.10 and 11.11.

Tay-Sachs disease, cystic fibrosis, sickle cell disease, and PKU are autosomal recessive disorders that have been studied in detail. Sickle cell disease also exemplifies pleiotropy. Neurofibromatosis, achondroplasia, and Huntington disease are autosomal dominant disorders that have been well studied.

11.5 BEYOND MENDELIAN GENETICS

Patterns of inheritance discovered since Mendel's original contribution make it clear that the genotype should be thought of as involving all the genes. For example, different degrees of dominance have been observed. With incomplete dominance, the F_1 individuals are intermediate between the parent phenotypes; this does not support the blending theory because the parent phenotypes reappear in F_2. Some genes have multiple alleles, although each individual organism has only two alleles, as in the inheritance of blood type in human beings. Inheritance of blood type also illustrates codominance.

Polygenic traits are controlled by genes that have an additive effect on the phenotype, resulting in quantitative variations. A bell-shaped curve is seen because environmental influences bring about many intervening phenotypes, as in the inheritance of height in human beings. Skin color and eye color are also examples of polygenic inheritance. Albinism is an example of epistasis. The influence of the environment on the phenotype can vary—there are examples of extreme environmental influence.

Reviewing the Chapter

1. How did Mendel's procedure differ from that of his predecessors? What is his theory of inheritance called? 182
2. How does the F_2 of Mendel's one-trait cross refute the blending concept of inheritance? Using Mendel's one-trait cross as an example, trace his reasoning to arrive at the law of segregation. 184
3. How is a one-trait testcross used today? 187–88
4. Using Mendel's two-trait cross as an example, trace his reasoning to arrive at the law of independent assortment. 188
5. What would be the result of a two-trait testcross if an individual was heterozygous for two traits? Heterozygous for one trait and homozygous dominant for the other? Homozygous dominant for both traits? 191
6. How might you distinguish an autosomal dominant trait from an autosomal recessive trait when viewing a pedigree? 192
7. For autosomal recessive disorders, what are the chances of two carriers having an affected child? 192–194
8. For most autosomal dominant disorders, what are the chances of a heterozygote and a normal individual having an affected child? 194–95
9. Explain the inheritance of incompletely dominant alleles and why this is not an example of blending inheritance. 196
10. Explain inheritance by multiple alleles. List the human blood types, and give the possible genotypes for each. 196–97
11. Explain why traits controlled by polygenes show continuous variation and produce a distribution in the F_2 generation that follows a bell-shaped curve. 197–98

Testing Yourself

Choose the best answer for each question. For questions 1–4, match each item to those in the key.

KEY:

   a. 3:1
   b. 9:3:3:1
   c. 1:1
   d. 1:1:1:1
   e. 3:1:3:1

1. $TtYy \times TtYy$

2. $Tt \times Tt$

3. $Tt \times tt$

4. $TtYy \times ttyy$

5. Which of these could be a normal gamete?
   a. GgRr
   b. GRr
   c. Gr
   d. GgR
   e. None of these are correct.

6. Which of these properly describes a cross between an individual who is homozygous dominant for hairline but heterozygous for finger length and an individual who is recessive for both characteristics? (W = widow's peak, w = straight hairline, S = short fingers, s = long fingers.)
   a. WwSs × WwSs
   b. WWSs × wwSs
   c. Ws × ws
   d. WWSs × wwss

7. In peas, yellow seed (Y) is dominant over green seed (y). In the F_2 generation of a monohybrid cross that begins when a dominant homozygote is crossed with a recessive homozygote, you would expect
   a. three plants with yellow seeds to every plant with green seeds.
   b. plants with one yellow seed for every green seed.
   c. only plants with the genotype Yy.
   d. only plants that produce yellow seeds.
   e. Both c and d are correct.

8. In humans, pointed eyebrows (B) are dominant over smooth eyebrows (b). Mary's father has pointed eyebrows, but she and her mother have smooth. What is the genotype of the father?
   a. BB
   b. Bb
   c. bb
   d. BbBb
   e. Any one of these is correct.

9. In guinea pigs, smooth coat (S) is dominant over rough coat (s), and black coat (B) is dominant over white coat (b). In the cross SsBb × SsBb, how many of the offspring will have a smooth black coat on average?
   a. 9 only
   b. about 9/16
   c. 1/16
   d. 6/16
   e. 2/6

10. In horses, B = black coat, b = brown coat, T = trotter, and t = pacer. A black trotter that has a brown pacer offspring is
   a. BT.
   b. BbTt.
   c. bbtt.
   d. BBtt.
   e. BBTT.

11. In tomatoes, red fruit (R) is dominant over yellow fruit (r), and tallness (T) is dominant over shortness (t). A plant that is RrTT is crossed with a plant that is rrTt. What are the chances of an offspring being heterozygous for both traits?
   a. none
   b. ½
   c. ¼

12. In the cross RrTt × rrtt,
   a. all the offspring will be tall with red fruit.
   b. 75% (¾) will be tall with red fruit.
   c. 50% (½) will be tall with red fruit.
   d. 25% (¼) will be tall with red fruit.

For questions 13–15, match the statements to the items in the key.

KEY:

   a. multiple alleles
   b. polygenes
   c. epistasis
   d. pleiotropic gene

13. People with sickle cell disease have many cardiovascular complications.

14. Although most people have an IQ of about 100, IQ generally ranges from about 50–150.

15. In humans, there are three possible alleles at the chromosomal locus that determine blood type.

16. Alice and Henry are at the opposite extremes for a polygenic trait. Their children will
   a. be bell-shaped.
   b. be a phenotype typical of a 9:3:3:1 ratio.
   c. have the middle phenotype between their two parents.
   d. look like one parent or the other.

17. Determine if the characteristic possessed by the shaded squares (males) and circles (females) below is an autosomal dominant or an autosomal recessive.

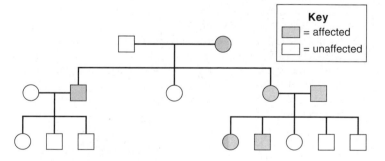

Key
☐ = affected
☐ = unaffected

Additional Genetics Problems*

1. If a man homozygous for widow's peak (dominant) reproduces with a woman homozygous for straight hairline (recessive), what are the chances of their children having a widow's peak? A straight hairline?

2. A son with cystic fibrosis (autosomal recessive) is born to a couple who appear to be normal. What are the chances that any child born to this couple will have cystic fibrosis?

3. John has unattached earlobes (dominant) like his father, but his mother has attached earlobes (recessive). What is John's genotype?

4. In a fruit fly experiment (see key on page 189), two gray-bodied fruit flies produce mostly gray-bodied offspring, but some offspring have black bodies. If there are 280 offspring, how many do you predict will have gray bodies and how many will have black bodies? How many of the 280 offspring do you predict will be heterozygous? If you wanted to test whether a particular gray-bodied fly was homozygous dominant or heterozygous, what cross would you do?

5. In humans, the allele for short fingers is dominant over that for long fingers. If a person with short fingers who had one parent with long fingers reproduces with a person having long fingers, what are the chances of each child having short fingers?

6. A man has type AB blood. What is his genotype? Could this man be the father of a child with type B blood? If so, what blood types could the child's mother have?

7. If blood type in cats were controlled by three codominant and multiple alleles (A, B, C), how many genotypes and phenotypes would occur? Could the cross AC × BC produce a cat with AB blood type?

*Answers to Additional Genetics Problems appear in Appendix A.

Thinking Scientifically

1. In peas, genes C and P are required for pigment production in flowers. Gene C codes for an enzyme that converts a compound into a colorless intermediate product. Gene P codes for an enzyme that converts the colorless intermediate into anthocyanin, a purple pigment. A flower, therefore, will be purple only if it contains at least one dominant allele for each of the two genes. Flowers are white if they do not produce anthocyanin. What phenotypic ratio would you expect in the P_2 generation following a cross between two double heterozygotes? What phenotypic ratio would you expect if a double heterozygote was crossed with a plant that is homozygous recessive for both genes?

2. Mendel wrote a paper reporting his results and conclusions. The first section of his paper, which gives a detailed account of all the crosses he did, indicates that the measured ratios were not exactly 3:1 or 9:3:3:1. The second section, however, reports data that is in keeping with these predicted ratios. What might explain why these two sections are different?

Understanding the Terms

allele 184	monohybrid cross 184
autosome 192	multifactorial 196
carrier 192	multiple alleles 196
codominance 197	phenotype 185
dihybrid cross 188	pleiotropy 194
dominant allele 184	polygenic inheritance 197
epistasis 198	Punnett square 186
gene locus 185	recessive allele 184
genotype 185	testcross 188
heterozygous 185	unifactorial 196
homozygous 185	wild type 189
incomplete dominance 196	

Match the terms to these definitions:

a. _____ Allele that exerts its phenotypic effect only in the homozygote; its expression is masked by a dominant allele.

b. _____ Alternative form of a gene that occurs at the same locus on homologous chromosomes.

c. _____ Allele that exerts its phenotypic effect in the heterozygote; it masks the expression of the recessive allele.

d. _____ Cross between an individual with the dominant phenotype and an individual with the recessive phenotype to see if the individual with the dominant phenotype is homozygous or heterozygous.

e. _____ Genes of an organism for a particular trait or traits; for example, BB or Aa.

ARIS, the *Biology* Website

ARIS, the website for *Biology*, provides a wealth of information organized and integrated by chapter. You will find practice quizzes, interactive activities, labeling exercises, flashcards, and much more that will complement your learning and understanding of general biology.

www.mhhe.com/maderbiology9

12

CHROMOSOMAL PATTERNS OF INHERITANCE

Considering that Mendel knew nothing about chromosomes, genes, or DNA, it is astounding that his laws about the transfer of genetic information from parent to offspring apply to all organisms and many patterns of inheritance. But as brilliant and groundbreaking as Mendel's work was, it was only the first step toward a more thorough understanding of how the genotype functions.

By the early 1900s, researchers knew that the genes are on the chromosomes and that each chromosome must have a sequence of genes. Further, Mendel's laws of segregation and independent assortment only hold if the genes of interest are on separate chromosomes. Mendel would not have been able to formulate his two laws if he had been working with genes on the same chromosome! Determining the sequence of genes on chromosomes began early in the last century with the fruit fly, Drosophila. By now, researchers know the sequence of paired bases in human DNA and the DNA of a great number of other organisms.

Chromosomes can undergo mutations, permanent changes in number and structure. We will observe that such alterations can have profound effects on the phenotype because each chromosome carries many genes.

Chromosomes from a human cell.

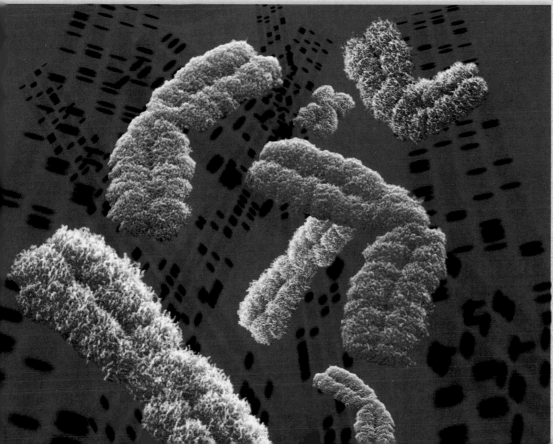

12.1 CHROMOSOMAL INHERITANCE

In 1879, Walther Flemming's discovery of chromosomes in the nuclei of salamander eggs opened the door for scientists of the early twentieth century to rediscover the works of Mendel because it re-created an interest in genetics. In 1920, both Theodor Boveri, a German, and Walter S. Sutton, an American, independently noted the parallel behavior of Mendel's factors and chromosomes. Their observations prompted each of them to propose the **chromosome theory of inheritance,** which states that the chromosomes are carriers of genetic information.

To understand how it was eventually determined that the genes are indeed on the chromosomes, it is first necessary for us to consider the chromosomal makeup of the fruit fly, *Drosophila melanogaster.* In fruit flies, chromosomes differ between the sexes. All but one pair of chromosomes in males and females are the same; these are called **autosomes** because they do not actively determine sex. The pair that is different is called the **sex chromosome.** These chromosomes are active in determining the sex of an individual. In fruit flies and humans, the sex chromosomes in females are XX and those in males are XY (Fig 12.1).

Notice that males produce two different types of gametes—those that contain an X and those that contain a Y—thus, the contribution of the male determines the sex of the new individual:

	X	X
X	XX	XX
Y	XY	XY

In addition to genes that determine sex, the sex chromosomes carry genes for traits that have nothing to do with the sex of the individual. By tradition, the term *sex-linked,* or **X-linked,** is used for genes carried on the X chromosome. The Y chromosome does not carry these genes and indeed carries very few genes.

X-Linked Alleles

A Columbia University group headed by Thomas Hunt Morgan is well known for performing many genetic experiments using *Drosophila* in the early 1900s. Fruit flies are even better subjects for genetic studies than garden peas: They can be easily and inexpensively raised in simple laboratory glassware; females mate and then lay hundreds of eggs during their lifetimes; and the generation time is short, taking only about ten days when conditions are favorable.

Drosophila flies have the same sex chromosome pattern as humans, and this facilitates our understanding of a cross performed by Morgan. Morgan took a newly discovered mutant male with white eyes and crossed it with a red-eyed female:

	♀		♂
P	red-eyed	×	white-eyed
F₁	red-eyed		red-eyed

From these results, he knew that red eyes are the dominant characteristic and white eyes are the recessive characteristic. He then crossed the F_1 flies. In the F_2 generation, there was the expected 3 red-eyed : 1 white-eyed ratio, but it struck him as odd that all of the white-eyed flies were males:

	♀		♂
$F_1 \times F_1$	red-eyed	×	red-eyed
F_2	red-eyed		1 red-eyed : 1 white-eyed

Obviously, a major difference between the male flies and the female flies was their sex chromosomes. Could it be possible that an allele for eye color was on the Y chromosome but not on the X? This idea could be quickly discarded because usually females have red eyes, and they have no Y chromosome. Perhaps an allele for eye color was on the X, but not on the Y, chromosome. Figure 12.2 indicates that this explanation would match the results obtained in the experiment. These results support the chromosome theory of inheritance by showing that the behavior of a specific allele corresponds exactly with that of a specific chromosome—the X chromosome in *Drosophila.*

Notice that X-linked alleles have a different pattern of inheritance than alleles that are on the autosomes because the Y chromosome is lacking for these alleles, and the inheritance of a Y chromosome cannot offset the inheritance of an X-linked recessive allele. For the same

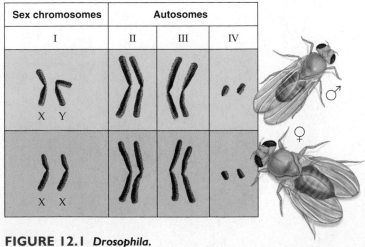

Sex chromosomes	Autosomes		
I	II	III	IV

FIGURE 12.1 *Drosophila.*
Early geneticists knew that the sex chromosomes differ in males and females. For example, in *Drosophila* (fruit flies), both males and females have eight chromosomes: three pairs of autosomes (II–IV), and one pair of sex chromosomes (I). Males are XY, and females are XX.

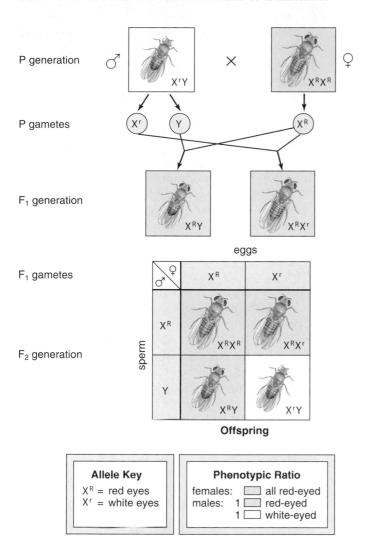

P generation ♂ × ♀
X^rY   X^RX^R

P gametes X^r Y   X^R

F₁ generation
X^RY   X^RX^r

eggs

F₁ gametes

♂ \ ♀	X^R	X^r
X^R	X^RX^R	X^RX^r
Y	X^RY	X^rY

sperm

F₂ generation

Offspring

Allele Key	Phenotypic Ratio
X^R = red eyes X^r = white eyes	females: ▨ all red-eyed males: 1 ▨ red-eyed 1 ▢ white-eyed

FIGURE 12.2 X-linked inheritance.
Once researchers deduced that the alleles for red/white eye color are on the X chromosome in *Drosophila*, they were able to explain their experimental results. Males with white eyes in the F₂ inherit the recessive allele only from the female parent; they receive a Y chromosome lacking the gene for eye color from the male parent.

reason, males always receive an X-linked recessive mutant allele from the female parent—they received the Y chromosome from the male parent.

Practice Problems 12.1*

1. In a cross between a brown-haired female and a black-haired male, all male offspring have brown hair and all female offspring have black hair. What is the genotype of all individuals involved, assuming X-linkage?

2. In *Drosophila*, if a homozygous red-eyed female and a red-eyed male mated, what would be the possible genotypes of their offspring?

3. a. Which *Drosophila* cross would produce white-eyed males: (1) X^RX^R × X^rY or (2) X^RX^r × X^RY? b. In what ratio?

Answers to Practice Problems appear in Appendix A.

X-Linked Problems

Recall that when solving autosomal genetics problems, the allele key and genotypes are represented as follows:

Allele key: Genotypes:

L = long wings LL, Ll, ll
l = short wings

As noted in Figure 12.2, however, the allele key for an X-linked gene shows an allele attached to the X:

Allele key:

X^R = red eyes
X^r = white eyes

The possible genotypes in both males and females are as follows:

X^RX^R = red-eyed female
X^RX^r = red-eyed female
X^rX^r = white-eyed female
X^RY = red-eyed male
X^rY = white-eyed male

Notice that there are three possible genotypes for females but only two for males. Females can be heterozygous X^RX^r, in which case they are carriers. **Carriers** usually do not show a recessive abnormality, but they are capable of passing on a recessive allele for an abnormality. Males cannot be carriers; if the dominant allele is on the single X chromosome, they show the dominant phenotype, and if the recessive allele is on the single X chromosome, they show the recessive phenotype.

We know that males have white eyes when they receive the mutant recessive allele from the female parent. What is the inheritance pattern when females have white eyes? Females can only have white eyes when they receive a recessive allele from both parents.

The question and answer for a genetics problem based on the parental cross in Figure 12.2 would be:

Question: *What would be the expected ratio among the offspring when a white-eyed male fruit fly is crossed with a female homozygous dominant for red eyes?*
Answer: *Both males and females have red eyes.*

The question and answer for a genetics problem based on the F₁ cross in Figure 12.2 would be:

Question: *What would be the expected ratio among the offspring when a red-eyed male fruit fly is crossed with a female heterozygous for red eyes?*
Answer: *Females have red eyes, males are 1:1.*

Or a reverse problem like this is also possible:

Question: *A male fruit fly has white eyes, but its parents have red eyes. What are the genotypes of all flies involved?*
Answer: *Offspring, X^rY; female parent, X^RX^r; male parent, X^RY.*

When a recessive allele is linked to the X chromosome, males will have the characteristic after inheriting only one recessive allele from the female parent.

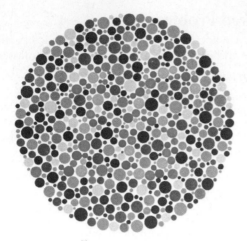

FIGURE 12.3 Red-green color blindness chart.
A person with red-green color blindness cannot see the number 16 within the pattern of circles.
Reproduced from *Ishihara's Tests for Colour Blindness*, published by Kanehara & Co. Ltd., Tokyo, Japan. Tests for color blindness cannot be conducted with this material. For accurate testing, the original plates should be used.

Human X-Linked Disorders

Several X-linked recessive disorders occur in humans including color blindness, Duchenne muscular dystrophy, and hemophilia.

Color Blindness

In humans, the receptors for color vision in the retina of the eyes are three different classes of cone cells. Only one type of pigment protein is present in each class of cone cell; there are blue-sensitive, red-sensitive, and green-sensitive cone cells. The allele for the blue-sensitive protein is autosomal, but the alleles for the red- and green-sensitive proteins are on the X chromosome. About 8% of Caucasian men have red-green color blindness. Most of these see brighter greens as tans, olive greens as browns, and reds as reddish browns. A few cannot tell reds from greens at all. They see only yellows, blues, blacks, whites, and grays (Fig. 12.3).

The pedigree in Figure 12.4 shows the usual pattern of inheritance for color blindness. More males than females have the trait because recessive alleles on the X chromosome are expressed in males. The disorder often passes from grandfather to grandson through a carrier daughter.

Muscular Dystrophy

Muscular dystrophy, as the name implies, is characterized by a wasting away of the muscles. The most common form, Duchenne muscular dystrophy, is X-linked and occurs in about 1 out of every 3,600 male births. Symptoms, such as waddling gait, toe walking, frequent falls, and difficulty in rising, may appear as soon as the child starts to walk. Muscle weakness intensifies until the individual is confined to a wheelchair. Death usually occurs by age 20; therefore, affected males are rarely fathers. The recessive allele remains in the population through passage from carrier mother to carrier daughter.

The allele for Duchenne muscular dystrophy has been isolated, and it was discovered that the absence of a protein now

called dystrophin causes the disorder. Much investigative work determined that dystrophin is involved in the release of calcium from the sarcoplasmic reticulum in muscle fibers. The lack of dystrophin causes calcium to leak into the cell, which promotes the action of an enzyme that dissolves muscle fibers. When the body attempts to repair the tissue, fibrous tissue forms, and this cuts off the blood supply so that more and more cells die.

A test is now available to detect carriers of Duchenne muscular dystrophy. Also, various treatments are being attempted. Immature muscle cells can be injected into muscles, and for every 100,000 cells injected, dystrophin production occurs in 30–40% of muscle fibers. The allele for dystrophin has been inserted into thigh muscle cells, and about 1% of these cells then produced dystrophin.

Hemophilia

About 1 in 10,000 males is a hemophiliac. There are two common types of hemophilia: Hemophilia A is due to the absence or minimal presence of a clotting factor known as factor VIII, and hemophilia B is due to the absence of clotting factor IX. Hemophilia is called the bleeder's disease because the affected person's blood either does not clot or clots very slowly. Although hemophiliacs bleed externally after an injury, they also bleed internally, particularly around joints. Hemorrhages can be stopped with transfusions of fresh blood (or plasma) or concentrates of the clotting protein. Also, clotting factors are now available as biotechnology products.

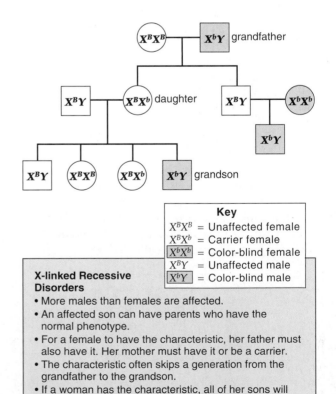

Key

$X^B X^B$ = Unaffected female
$X^B X^b$ = Carrier female
$X^b X^b$ = Color-blind female
$X^B Y$ = Unaffected male
$X^b Y$ = Color-blind male

X-linked Recessive Disorders

- More males than females are affected.
- An affected son can have parents who have the normal phenotype.
- For a female to have the characteristic, her father must also have it. Her mother must have it or be a carrier.
- The characteristic often skips a generation from the grandfather to the grandson.
- If a woman has the characteristic, all of her sons will have it.

FIGURE 12.4 X-linked recessive pedigree.
This pedigree for color blindness exemplifies the inheritance pattern of an X-linked recessive disorder. The list gives various ways of recognizing the X-linked recessive pattern of inheritance.

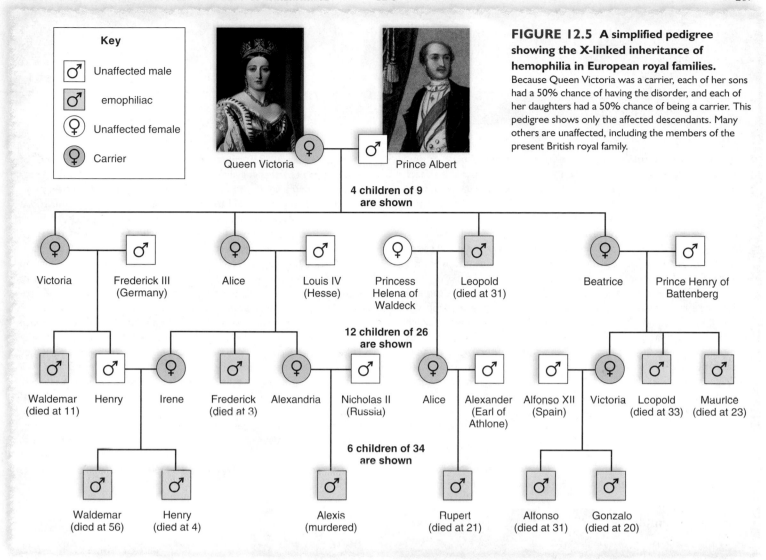

Key

♂ Unaffected male

♂ ☐ emophiliac

♀ Unaffected female

♀ Carrier

Queen Victoria   Prince Albert

FIGURE 12.5 A simplified pedigree showing the X-linked inheritance of hemophilia in European royal families. Because Queen Victoria was a carrier, each of her sons had a 50% chance of having the disorder, and each of her daughters had a 50% chance of being a carrier. This pedigree shows only the affected descendants. Many others are unaffected, including the members of the present British royal family.

4 children of 9 are shown

Victoria — Frederick III (Germany)   Alice — Louis IV (Hesse)   Princess Helena of Waldeck — Leopold (died at 31)   Beatrice — Prince Henry of Battenberg

12 children of 26 are shown

Waldemar (died at 11)   Henry — Irene   Frederick (died at 3)   Alexandria — Nicholas II (Russia)   Alice — Alexander (Earl of Athlone)   Alfonso XII (Spain) — Victoria   Leopold (died at 33)   Maurice (died at 23)

6 children of 34 are shown

Waldemar (died at 56)   Henry (died at 4)   Alexis (murdered)   Rupert (died at 21)   Alfonso (died at 31)   Gonzalo (died at 20)

At the turn of the century, hemophilia was prevalent among the royal families of Europe, and all of the affected males could trace their ancestry to Queen Victoria of England. Figure 12.5 shows that, of Queen Victoria's 26 grandchildren, four grandsons had hemophilia and four granddaughters were carriers. Because none of Queen Victoria's relatives were affected, it seems that the faulty allele she carried arose by mutation either in Victoria or in one of her parents. Her carrier daughters Alice and Beatrice introduced the allele into the ruling houses of Russia and Spain, respectively. Alexis, the last heir to the Russian throne before the Russian Revolution, was a hemophiliac. There are no hemophiliacs in the present British royal family because Victoria's eldest son, King Edward VII, did not receive the allele.

Fragile X Syndrome

Fragile X syndrome is an X-linked genetic disorder with an unusual pattern of inheritance. As discussed in the Health Focus on the next page, fragile X syndrome is due to base triplet repeats in a gene on the X chromosome. We now know that other disorders, such as Huntington disease, are also due to base triplet repeats.

Practice Problems 12.2*

1. Both the mother and the father of a male hemophiliac appear normal. From whom did the son inherit the allele for hemophilia? What are the genotypes of the mother, the father, and the son?

2. A woman is color-blind. What are the chances that her sons will be color-blind? If she is married to a man with normal vision, what are the chances that her daughters will be color-blind? Will be carriers?

3. Both the husband and wife have normal vision. The wife gives birth to a color-blind daughter. What can you deduce about the girl's parentage?

Answers to Practice Problems appear in Appendix A.

Certain traits that have nothing to do with the gender of the individual are controlled by genes on the X chromosomes. Males have only one X chromosome, and therefore X-linked recessive disorders are more likely in males.

health focus

Fragile X Syndrome

Fragile X syndrome is one of the most common genetic causes of mental retardation, second only to Down syndrome. It affects about 1 in 1,500 males and 1 in 2,500 females and is seen in all ethnic groups. As children, fragile X syndrome individuals may be hyperactive or autistic; their speech is delayed in development and often repetitive in nature. As adults, males have large testes and big, usually protruding ears (Fig. 12Aa). They are short in stature, but the jaw is prominent, and the face is long and narrow. Stubby hands, lax joints, and a heart defect may also occur.

In 1991, the DNA sequence at the fragile site was isolated and found to have base triplet repeats: CGG was repeated over and over again. An unaffected person has only 6–50 repeats, while a person with fragile X syndrome has 230–2,000 repeats. This mutation affects a gene located at the site, and the result is mental retardation. The name *fragile X syndrome* originated because diagnosis used to be dependent on observing an X chromosome whose tip is attached to the rest of the chromosome by only a thin thread (Fig. 12Ab).

Fragile X syndrome is an X-linked condition, and therefore you would expect all males having a fragile X to show the condition. However, one-fifth of these males with a fragile X do not have symptoms, and investigators wanted to discover why not. Analysis of DNA in these males shows that they have only an intermediate number of repeats—from 50–230 copies. A female who inherits this number of repeats from her father may have mild symptoms of retardation and may have sons with fragile X syndrome. Therefore, anyone with an intermediate number of repeats is said to have a premutation. And premutations can lead to full-blown mutations in future generations (Fig. 12Ac).

How does the pattern of inheritance differ from that of the usual pattern for an X-linked genetic condition? First, a fragile X can have from 50–2,000 repeats. If the number is below 230, the individual has no symptoms, and if the number is above 230, the individual has symptoms. Second, a male with no symptoms can have grandsons with fragile X syndrome by way of a daughter who has mild symptoms.

This type of mutation—called by some a dynamic mutation because it changes, and by others an expanded trinucleotide repeat because the number of triplet copies increases—is now known to characterize other conditions. Huntington disease is caused by a base triplet repeat of CAG, for example. What might cause repeats to occur in the first place, and why does this cause a syndrome? Repeats might arise during DNA replication prior to cell division, and their presence undoubtedly leads to nonfunctioning or malfunctioning proteins.

Scientists have developed a new technique that can identify repeats in DNA, and they hope that this technique will help them find the genes for other human disorders. They expect expanded trinucleotide repeats to be a common mutation indeed.

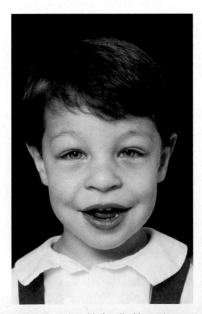
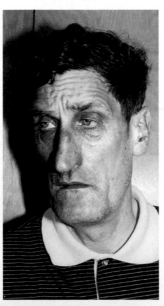

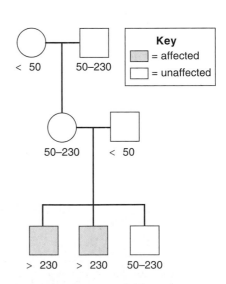

a. Young male with fragile X syndrome   Same individual when mature   b. Fragile X chromosome   c. Inheritance pattern for fragile X syndrome

FIGURE 12A Fragile X syndrome.
*a. A young male with fragile X syndrome appears normal but with age develops an elongated face with a prominent jaw and ears that noticeably protrude. **b.** An arrow points out the fragile site of this fragile X chromosome. **c.** The number of base-pair repeats at the fragile site are given for each person. A grandfather who has a premutation with 50–230 repeats has no symptoms but transmits the condition to his grandsons through his daughter. Two grandsons have full-blown mutations with more than 230 base repeats.*

12.2 GENE LINKAGE

After Thomas Morgan (Fig. 12.6) and his students had done a great number of *Drosophila* crosses, they discovered many more mutants and were able to do various two-trait crosses. They did not always get the expected ratios among the off-spring, however, due to **gene linkage,** the existence of several genes on the same chromosome. The genes on the same chromosome form a **linkage group** because these genes tend to be inherited together.

Drosophila probably has thousands of different genes controlling all aspects of its structure, biochemistry, and behavior. Yet it has only four chromosomes. This paradox alone allows you to reason that each chromosome must carry a large number of genes. For example, it is now known that genes controlling eye color, wing type, body color, leg length, and antennae type are all located on chromosome 2 (Fig. 12.7). The physical location of a gene within a chromosome is called the **locus** (pl., loci).

Figure 12.7 also illustrates a **linkage map,** because it tells you the relative distance between the gene loci on chromosome 2 of *Drosophila*. Geneticists are able to construct linkage maps by doing crosses and observing the number of recombinant gametes due to crossing-over, as will be discussed on the next two pages.

FIGURE 12.6 Thomas Hunt Morgan (1866–1945).
T. H. Morgan received the 1933 Nobel Prize in medicine for his discoveries of the role chromosomes played in heredity. He and his students discovered the white-eyed mutation in fruit flies. His work led to our understanding of sex-linked inheritance and eventually gene mapping.

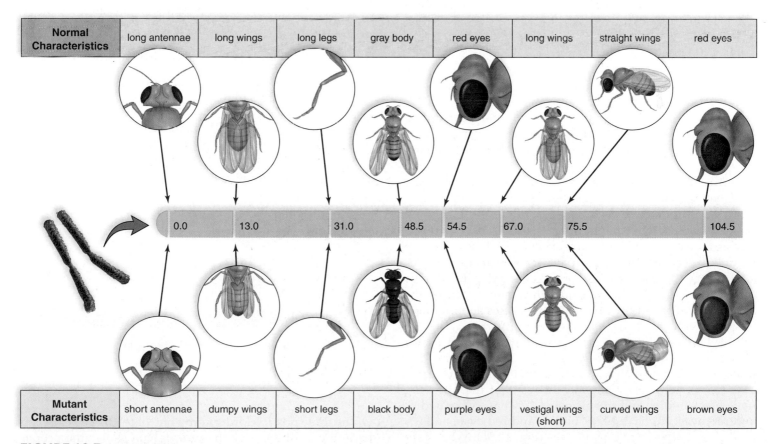

FIGURE 12.7 Linkage group.
Each homologous pair of chromosomes carries a number of genes. The alleles on a chromosome form a linkage group because they tend to go together into the gametes. This simplified chromosome map shows the relative positions of some of the genes on *Drosophila* chromosome 2. The distances between the genes (the numbers = map units) are equivalent to the percentage of crossing-over events that occurs between various alleles. For example, the crossing-over frequency between gray body and long legs would be 48.5 − 31.0 = 17.5%. This means that 17.5% of all gametes would carry recombinant gametes.

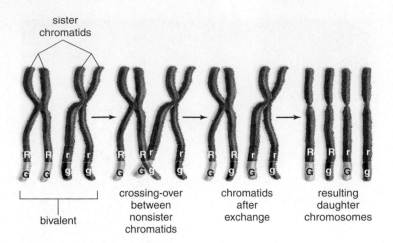

FIGURE 12.8 Crossing-over.
When homologous chromosomes are in synapsis, the nonsister chromatids exchange genetic material. Following crossing-over, recombinant chromosomes occur. Recombinant chromosomes contribute to recombinant gametes.

Constructing a Chromosome Map

A linkage map can also be called a chromosome map because it tells the order of gene loci on chromosomes. To construct a chromosome map, investigators can sometimes rely on crossing-over. Crossing-over, you recall, occurs between nonsister chromatids when homologous pairs of chromosomes pair prior to separation during meiosis. During crossing-over, the nonsister chromatids exchange genetic material and therefore genes. If crossing-over occurs between two linked alleles of interest, a dihybrid produces four types of gametes instead of two (Fig. 12.8). Recombinant means a new combination of alleles. The recombinant gametes occur in reduced number because crossing-over is infrequent. Still, all possible phenotypes will occur among the offspring.

To take an actual example, suppose you are doing a cross between a gray-bodied, red-eyed heterozygous fly and a black-bodied, purple-eyed fly. Since the alleles governing these traits are both on chromosome 2 (see Fig. 12.7), you predict that the results will be 1:1 instead of 1:1:1:1, as is the case for unlinked genes. The reason, of course, is that linked alleles tend to stay together and do not separate independently, as predicted by Mendel's laws. Under these circumstances, the heterozygote forms only two types of gametes and produces offspring with only two phenotypes (Fig. 12.9a).

When you do the cross, however, you find that a very small number of offspring show recombinant phenotypes (i.e., those that are different from the original parents). Specifically, you find that 47% of the offspring have black bodies and purple eyes, 47% have gray bodies and red eyes, 3% have black bodies and red eyes, and 3% have gray bodies and purple eyes (Fig. 12.9b). What happened? In this instance, crossing-over led to a very small

number of recombinant gametes, and when these were fertilized, recombinant phenotypes were observed in the offspring. An examination of chromosome 2 shows that these two sets of alleles are very close together. Doesn't it stand to reason that the closer together two genes are, the less likely they are to cross over? This is exactly what various crosses have repeatedly shown.

All the genes on one chromosome form a linkage group that tends to stay together, except when crossing-over occurs.

Linkage Data

You can use the percentage of recombinant phenotypes to map the chromosomes, because there is a direct relationship between the frequency of crossing-over (the percentage of recombinant phenotypes) and the distance between alleles. In our example (Fig. 12.9), a total of 6% of the offspring are recombinants and this means that the rate of crossing-over is 6%. For the sake of mapping the chromosomes, it is assumed that 1% of crossing-over equals 1 map unit. Therefore, the allele for black body and the allele for purple eyes are 6 map units apart.

Suppose you want to determine the order of any three genes on the chromosomes. To do so, you can perform crosses that tell you the map distance between all three pairs of alleles. If you know, for instance, that:

1. the distance between the black-body and purple-eye alleles = 6 map units;
2. the distance between the purple-eye and vestigial-wing alleles = 12.5 units; and
3. the distance between the black-body and vestigial-wing alleles = 18.5 units, then the order of the alleles must be as shown here:

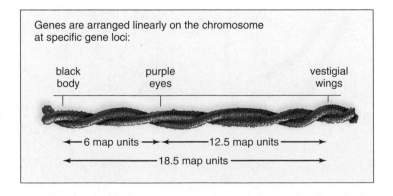

Genes are arranged linearly on the chromosome at specific gene loci:

black body   purple eyes   vestigial wings

← 6 map units → ← 12.5 map units →
← 18.5 map units →

Of the three possible orders (any one of the three genes can be in the middle), only the order given shows the proper distance between all three allelic pairs. Because it is pos-

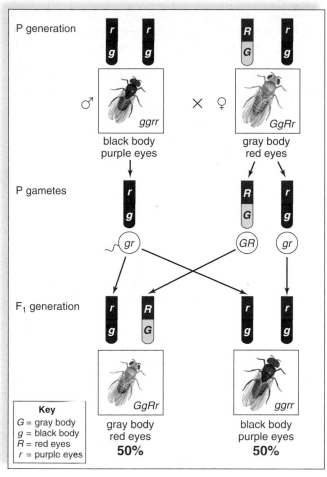

a.

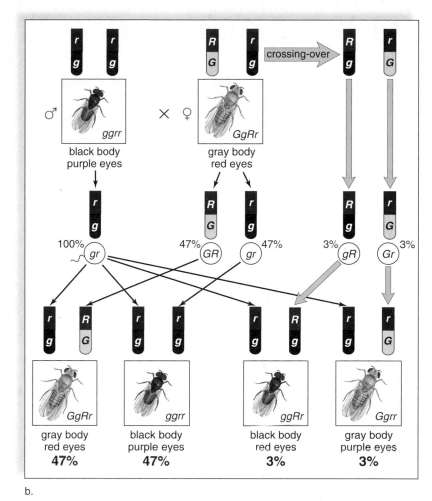

b.

FIGURE 12.9 Complete linkage versus incomplete linkage.
a. Hypothetical cross in which genes for body and eye color on chromosome 2 of *Drosophila* are completely linked. Instead of the expected 1:1:1:1 ratio among the offspring, there would be a 1:1 ratio. **b.** Crossing-over occurs, and 6% of the offspring show recombinant phenotypes. The percentage of recombinant phenotypes is used to map the chromosomes (1% = 1 map unit).

sible to map the chromosomes, the chromosome theory of inheritance includes the concept that alleles are arranged linearly along a chromosome at specific loci.

Linkage data have been used to map the chromosomes of *Drosophila*, but the possibility of using linkage data to map human chromosomes is limited because it would only be possible to work with matings that occur by chance. This, coupled with the fact that humans tend not to have numerous offspring, means that additional methods are used to sequence the genes on human chromosomes. Today, it is customary to also rely on biochemical methods to map the human chromosomes, as will be discussed in Chapter 16.

The frequency of recombinant gametes that occurs due to the process of crossing-over has been used to map chromosomes.

Practice Problems 12.3*

1. When *AaBb* individuals are allowed to self-breed, the phenotypic ratio is just about 3:1. a. What ratio was expected? b. What may have caused the observed ratio?
2. In two sweet pea strains, *B* = blue flowers, *b* = red flowers, *L* = long pollen grains, and *l* = round pollen grains. In a cross between a heterozygous plant with blue flowers and long pollen grains and a plant with red flowers and round pollen grains, 44% of offspring are blue, long; 44% are red, round; 6% are blue, round; and 6% are red, long. How many map units separate these two sets of alleles?
3. Investigators performed crosses that indicated bar-eye and garnet-eye alleles are 13 map units apart, scallop-wing and bar-eye alleles are 6 units apart, and garnet-eye and scallop-wing alleles are 7 units apart. What is the order of these alleles on the chromosome?

* Answers to Practice Problems appear in Appendix A.

12.3 CHANGES IN CHROMOSOME NUMBER

Besides crossing-over, recombination of chromosomes during meiosis, and gamete fusion during fertilization, another factor that increases the amount of variation among offspring is mutation. **Chromosomal mutations** include changes in chromosome number and changes in chromosome structure. The correct number of chromosomes in a species is known as **euploidy.** Changes in the chromosome number include polyploidy and aneuploidy.

Polyploidy

When a eukaryote has three or more complete sets of chromosomes, it is called a **polyploid** [Gk. *polys,* many, and *plo,* fold]. Polyploid organisms are named according to the number of sets of chromosomes they have. Triploids (3n) have three of each kind of chromosome, tetraploids (4n) have four sets, pentaploids (5n) have five sets, and so on.

Although polyploidy is not often seen in animals, it is a major evolutionary mechanism in plants. Some estimate that 47% of all flowering plants are polyploids. Among these are many of our most important crops, such as wheat, corn, cotton, and sugarcane, as well as fruits such as watermelons, strawberries, bananas, and apples. Also, many attractive flowers, including chrysanthemums and daylilies, are polyploids. The flowers and fruits of many polyploid plants are larger than their diploid counterparts.

Polyploidy generally arises following hybridization. When two different species reproduce, the resulting organism, called a hybrid, may have an odd number of chromosomes. If so, the chromosomes cannot pair evenly during meiosis, and the organism will not be fertile. In plants, hybridization, followed by doubling of the chromosome number, will result in an even number of chromosomes, and the chromosomes will be able to undergo synapsis as usual during meiosis:

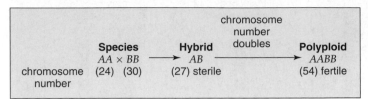

	Species AA × BB	**Hybrid** AB	chromosome number doubles	**Polyploid** AABB
chromosome number	(24) (30)	(27) sterile	→	(54) fertile

Aneuploidy

When an organism has more or less than the normal number of chromosomes, the **aneuploid** condition exists. Monosomy and trisomy are two aneuploid states.

Monosomy (2n − 1) occurs when an individual has only one of a particular type of chromosome, and **trisomy** (2n + 1) occurs when an individual has three of a particular type of chromosome. The usual cause of monosomy and trisomy is nondisjunction during meiosis. **Nondisjunction** occurs during meiosis I when both members of a homologous pair go into the same daughter cell, or during meiosis II when the sister chromatids fail to separate and both daughter chromosomes go into the same gamete (Fig. 12.10).

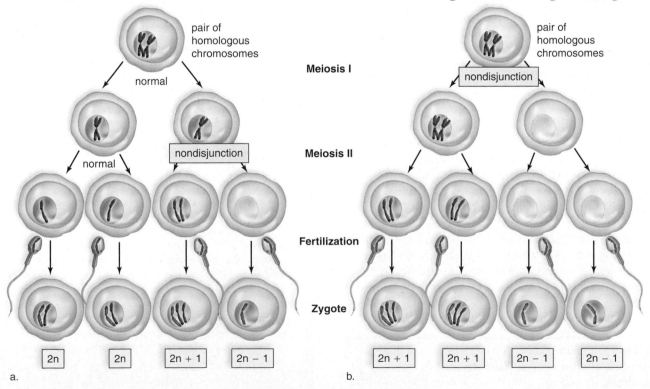

FIGURE 12.10 Nondisjunction of chromosomes during oogenesis, followed by fertilization with normal sperm.
a. Nondisjunction can occur during meiosis II if the sister chromatids separate but the resulting chromosomes go into the same daughter cell. Then the egg will have one more or one less than the usual number of chromosomes. Fertilization of these abnormal eggs with normal sperm produces an abnormal zygote with abnormal chromosome numbers. **b.** Nondisjunction can also occur during meiosis I and result in abnormal eggs that also have one more or one less than the normal number of chromosomes. Fertilization of these abnormal eggs with normal sperm results in a zygote with abnormal chromosome numbers. 2n = diploid number of chromosomes.

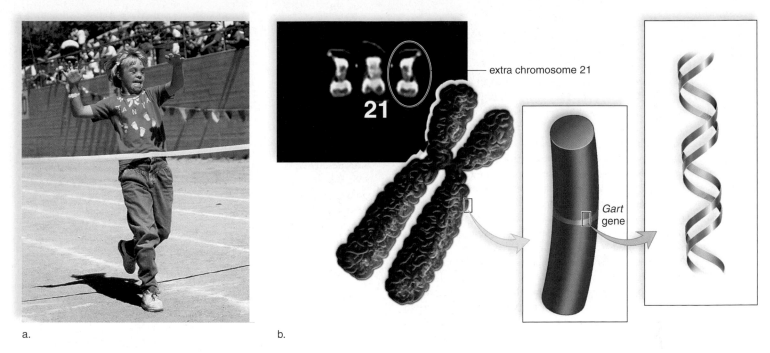

extra chromosome 21

21

Gart
gene

FIGURE 12.11 **Trisomy 21.**

Persons with Down syndrome, or trisomy 21, have an extra chromosome 21. **a.** Common characteristics of the syndrome include a wide, rounded face and a fold on the upper eyelids. Mental retardation, along with an enlarged tongue, makes it difficult for a person with Down syndrome to speak distinctly. **b.** More sophisticated technologies allow investigators to pinpoint the location of specific genes associated with the syndrome. An extra copy of the *Gart* gene, which leads to a high level of purines in the blood, may account for the mental retardation seen in persons with Down syndrome.

Monosomy and trisomy occur in both plants and animals. In animals, autosomal monosomies and trisomies are generally lethal, but a trisomic individual is more likely to survive than a monosomic one. The survivors are characterized by a distinctive set of physical and mental abnormalities, as in the human condition called trisomy 21. Sex chromosome aneuploids have a better chance of producing survivors than autosomal aneuploids.

Trisomy 21

The most common autosomal trisomy seen among humans is trisomy 21, also called Down syndrome (Fig. 12.11). This syndrome is easily recognized by these characteristics: short stature; an eyelid fold; a flat face; stubby fingers; a wide gap between the first and second toes; a large, fissured tongue; a round head; a distinctive palm crease; heart problems; and, unfortunately, mental retardation, which can sometimes be severe. In addition, these individuals have an increased chance of developing Alzheimer disease later in life.

Over 90% of individuals with Down syndrome have three copies of chromosome 21. Usually two copies of chromosome 21 are contributed by the egg; however, recent studies indicate that in 23% of the cases studied, the sperm contributed an extra chromosome. The chances of a woman having a child with Down syndrome increase rapidly with age. In women age 20–30, 1 in 1,400 births have Down syndrome, and in women 30–35, about 1 in 750 births have Down syndrome. It is thought that the longer the oocytes

are stored in the female, the greater the chances of a nondisjunction event.

Although an older woman is more likely to have a Down syndrome child, most babies with Down syndrome are born to women younger than age 40 because this is the age group having the most babies. Karyotyping, as discussed in the Science Focus on the next page, can detect a Down syndrome child. However, young women are not routinely encouraged to undergo the procedures necessary to get a sample of fetal cells (i.e., amniocentesis or chorionic villi sampling) because the risk of complications is greater than the risk of having a Down syndrome child. Fortunately, a test based on substances in maternal blood can help identify fetuses who may need to be karyotyped.

The genes that cause the characteristics of Down syndrome are located on the bottom third of chromosome 21 (Fig. 12.11*b*). Extensive investigative work has been directed toward discovering the specific genes responsible for the characteristics of the syndrome. Thus far, investigators have discovered several genes that may account for various conditions seen in persons with Down syndrome. For example, they have located genes most likely responsible for the increased tendency toward leukemia, cataracts, accelerated rate of aging, and mental retardation. The extra copy of the *Gart* gene causes an increased level of purines in the blood, leading to the mental retardation associated with Down syndrome. One day it may be possible to control the expression of the *Gart* gene even before birth so that at least this symptom of Down syndrome does not appear.

science focus

Viewing the Chromosomes

It is possible to view an individual's chromosomes by constructing a **karyotype,** a visual display of the chromosomes arranged by size, shape, and banding pattern. Usually, both human males and females have 23 pairs of chromosomes; 22 pairs are autosomes, and 1 pair is the sex chromosomes. Males have the sex chromosomes X and Y, and females have two X chromosomes.

Various human disorders result from abnormal chromosome number or structure. Such disorders often cause a **syndrome,** which is a group of symptoms that always occur together. Table 12A lists several syndromes that are due to an abnormal chromosome number. Doing a karyotype will reveal such abnormalities.

Any cell in the body except red blood cells, which lack a nucleus, can be a source of chromosomes for karyotyping. In adults, it is easiest to use white blood cells separated from a blood sample for this purpose. In fetuses, whose chromosomes are often examined in order to be forewarned of a syndrome, cells can be obtained by either amniocentesis or chorionic villi sampling.

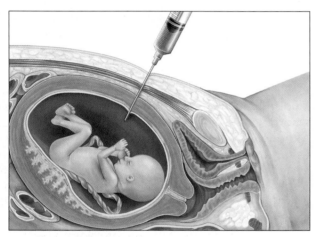

a. During amniocentesis, a long needle is used to withdraw amniotic fluid containing fetal cells.

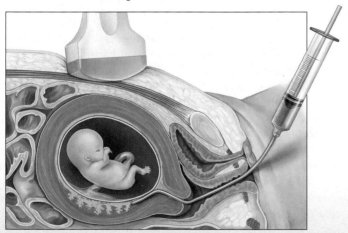

b. During chorionic villi sampling, a suction tube is used to remove cells from the chorion, where the placenta will develop.

FIGURE 12B Human karyotype preparation.

Amniocentesis

Amniocentesis is a procedure for obtaining a sample of amniotic fluid from the uterus of a pregnant woman. Blood tests and the mother's age are used to determine when the procedure should be done. Amniocentesis is not usually performed until about the fourteenth to the seventeenth week of pregnancy. A long needle is passed through the abdominal and uterine walls to withdraw a small amount of fluid, which also contains fetal cells (Fig. 12B*a*). Testing of the cells and karyotyping of the chromosomes may be delayed as long as four weeks so that the cells can be cultured to increase their number. As many as 400 chromosome and biochemical problems can be detected by testing the cells and the amniotic fluid.

The risk of spontaneous abortion increases by about 0.3% due to amniocentesis, and doctors only use the procedure if it is medically warranted.

Chorionic Villi Sampling

Chorionic villi sampling (CVS) is a procedure for obtaining chorionic cells in the region where the placenta will develop. This procedure can be done as early as the fifth week of pregnancy. A long, thin suction tube is inserted through the vagina into the uterus (Fig. 12B*b*). Ultrasound, which gives a picture of the uterine contents, is used to place the tube between the uterine lining and the chorionic villi. Then a sampling of chorionic cells is obtained by suction. The cells do not have to be cultured, and karyotyping can be done immediately. This sampling procedure does not include any amniotic fluid, so the biochemical tests done on the amniotic fluid following amniocentesis are not possible. Also, CVS carries a greater risk of spontaneous abortion than

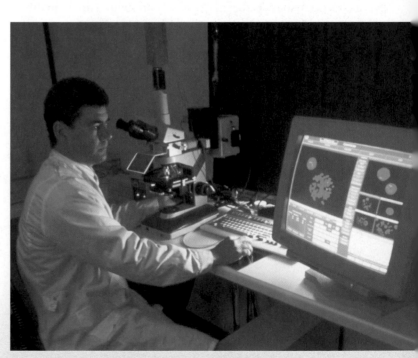

c. Cells are microscopically examined and photographed. A computer is used to arrange chromosomes by pairs.

amniocentesis—0.8% compared with 0.3%. The advantage of CVS is getting the results of karyotyping at an earlier date.

Karyotyping

After a cell sample has been obtained, the cells are stimulated to divide in a culture medium. A chemical is used to stop mitosis during metaphase when chromosomes are the most highly compacted and condensed. The cells are then killed, spread on a microscope slide, and dried. Stains are applied to the slides, and the cells are photographed. Staining produces dark and light crossbands of varying widths, and these can be used in addition to size and shape to help pair up the chromosomes. Today, a computer is used to arrange the chromosomes in pairs (Fig. 12Bc). The karyotype of a person who has Down syndrome usually has three number 21 chromosomes instead of the usual two number 21 chromosomes (Fig. 12Bd).

In about 5% of the cases of Down syndrome, a segment of chromosome 21 is attached to another chromosome, often chromosome 14. This abnormal chromosome arose because a translocation occurred between chromosomes 14 and 21 in one of the parents or even in a relative who lived generations earlier (see page 218). Therefore, when Down syndrome is due to a translocation, it is not related to age of mother, and instead it tends to run in the family of either the father or the mother.

TABLE 12A

Syndromes from Abnormal Chromosome Numbers

Syndrome	Sex	Disorder	Chromosome Number	Frequency	
				Spontaneous Abortions	Live Births
Turner	F	XO	45	1/18	1/2,500
Klinefelter	M	XXY (or XXXY)	47 or 48	1/300	1/800
Poly-X	F	XXX (or XXXX)	47 or 48	0	1/1,500
Jacobs	M	XYY	47	?	1/1,000
Down	M or F	Trisomy 21	47	1/40	1/800

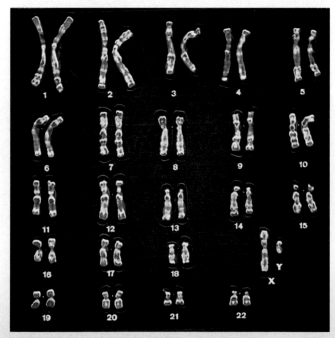

d. Normal male karyotype with 46 chromosomes.

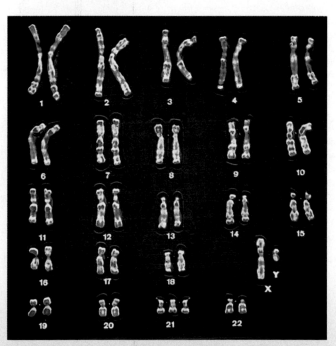

Down syndrome karyotype with an extra chromosome 21.

Changes in Sex Chromosome Number

An abnormal sex chromosome number is the result of inheriting too many or too few X or Y chromosomes. Figure 12.10 on page 212 can be used to illustrate nondisjunction of the sex chromosomes during oogenesis if you assume that the chromosomes shown represent two X chromosomes or X and Y chromosomes. Nondisjunction during oogenesis or spermatogenesis can result in gametes that have too few or too many X or Y chromosomes.

A person with Turner syndrome (XO) is a female, and a person with Klinefelter syndrome (XXY) is a male. This shows that in humans the presence of a Y chromosome, not the number of X chromosomes, determines maleness. The *SRY* gene on the short arm of the Y chromosome produces a hormone called testis-determining factor, which plays a critical role in the development of male genitals.

Why are newborns with an abnormal sex chromosome number more likely to survive than those with an abnormal autosome number (see Table 12A)? Actually, both males and females have only one functioning X chromosome. Any others become an inactive mass called a Barr body after Murray Barr, the person who discovered it (see p. 256).

Turner Syndrome

From birth, an XO individual with Turner syndrome has only one sex chromosome, an X; the O signifies the absence of a second sex chromosome (Fig. 12.12*a*). Therefore, the nucleus does not contain a Barr body. The approximate incidence is 1 in 10,000 females.

Turner females are short, with a broad chest and widely spaced nipples. These individuals also have a low posterior hairline and neck webbing. The ovaries, oviducts, and uterus are very small and underdeveloped. Turner females do not undergo puberty or menstruate, and their breasts do not develop. However, some have given birth following in vitro fertilization using donor eggs. They usually are of normal intelligence and can lead fairly normal lives if they receive hormone supplements.

Klinefelter Syndrome

A male with Klinefelter syndrome has two or more X chromosomes in addition to a Y chromosome (Fig. 12.12*b*). The extra X chromosomes become Barr bodies. The approximate incidence for Klinefelter syndrome is 1 in 10,000 males.

In Klinefelter males, the testes and prostate gland are underdeveloped and there is no facial hair. But there may be some breast development. Affected individuals have large hands and feet and very long arms and legs. They are usually slow to learn but not mentally retarded unless they inherit more than two X chromosomes. No matter how many X chromosomes there are, an individual with a Y chromosome is a male.

FIGURE 12.12

Abnormal sex chromosome number.

People with (**a**) Turner syndrome, who have only one sex chromosome, an X, as shown, and (**b**) Klinefelter syndrome, who have more than one X chromosome plus a Y chromosome, as shown, can look relatively normal (especially as children) and can lead relatively normal lives.

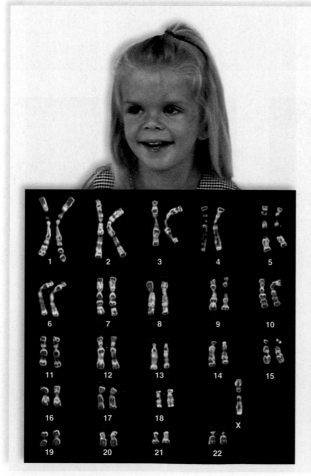

a. Turner syndrome

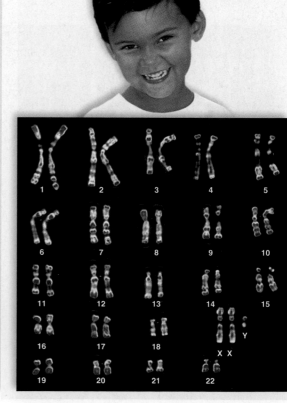

b. Klinefelter syndrome

health focus

Living with Klinefelter Syndrome

In 1996, at the age of 25, I was diagnosed with Klinefelter syndrome (KS). Being diagnosed has changed my life for the better.

I was a happy baby, but when I was still very young, my parents began to believe that there was something wrong with me. I knew something was different about me, too, as early on as five years old. I was very shy and had trouble making friends. One minute I'd be well behaved, and the next I'd be picking fights and flying into a rage. Many psychologists, therapists, and doctors tested me because of school and social problems and severe mood changes. Their only diagnosis was "learning disabilities" in such areas as reading comprehension, abstract thinking, word retrieval, and auditory processing. No one could figure out what the real problem was, and I hated the tutoring sessions I had. In the seventh grade, a psychologist told me that I was stupid and lazy, I would probably live at home for the rest of my life, and I would never amount to anything. For the next five years, he was basically right, and I barely graduated from high school.

I believe, though, that I have succeeded because I was told that I would fail. I quit the tutoring sessions when I enrolled at a community college; I decided I could figure things out on my own. I received an associate degree there, then transferred to a small liberal arts college. I never told anyone about my learning disabilities and never sought special help. However, I never had a semester below a 3.0, and I graduated with two B.S. degrees. I was accepted into a graduate program but decided instead to accept a job as a software engineer even though I did not have an educational background in this field. As I later learned, many KS'ers excel in computer skills. I had been using a computer for many years and had learned everything I needed to know on my own, through trial and error.

Around the time I started the computer job, I went to my physician for a physical. He sent me for blood tests because he noticed that my testes were smaller than usual. The results were conclusive: Klinefelter syndrome with sex chromosomes XXY. I initially felt denial, depression, and anger, even though I now had an explanation for many of the problems I had experienced all my life. But then I decided to learn as much as I could about the condition and treatments available. I now give myself a testosterone injection once every two weeks, and it has made me a different person, with improved learning abilities and stronger thought processes in addition to a more outgoing personality.

I found, though, that the best possible path I could take was to help others live with the condition. I attended my first support group meeting four months after I was diagnosed. By spring 1997, I had developed an interest in KS that was more than just a part-time hobby. I wanted to be able to work with this condition and help people forever. I have been very involved in KS conferences and have helped to start support groups in the United States, Spain, and Australia.

Since my diagnosis, it has been my dream to have a son with KS, although when I was diagnosed, I found out it was unlikely that I could have biological children. Through my work with KS, I had the opportunity to meet my fiancee Chris. She has two wonderful children: a daughter, and a son who has the same condition that I do. There are a lot of similarities between my stepson and me, and I am happy I will be able to help him get the head start in coping with KS that I never had. I also look forward to many more years of helping other people seek diagnosis and live a good life with Klinefelter syndrome.

Stefan Schwarz

The Health Focus on this page tells of the experiences of a person with Klinefelter syndrome. He suggests that it is best for parents to know right away that they have a child with this abnormality because much can be done to help the child lead a normal life.

Poly-X Females

A poly-X female, sometimes called a superfemale, has more than two X chromosomes and, therefore, extra Barr bodies in the nucleus. Females with three X chromosomes have no distinctive phenotype aside from a tendency to be tall and thin. Although some have delayed motor and language development, as well as learning problems, most poly-X females are not mentally retarded. Some may have menstrual difficulties, but many menstruate regularly and are fertile. Children usually have a normal karyotype. The incidence for poly-X females is about 1 in 1,500 females.

Females with more than three X chromosomes occur rarely. Unlike XXX females, XXXX females are usually tall and severely retarded. Various physical abnormalities are seen, but they may menstruate normally.

Jacobs Syndrome

XYY males with Jacobs syndrome can only result from nondisjunction during spermatogenesis. These individuals are sometimes called supermales. Among all live male births, the frequency of the XYY karyotype is about 1 in 1,000. Affected males are usually taller than average, suffer from persistent acne, and tend to have speech and reading problems. Based upon the number of XYY individuals in prisons and mental facilities, at one time it was suggested that these men were likely to be criminally aggressive, but it has since been shown that the incidence of such behavior among them may be no greater than among XY males.

Several syndromes are due to an abnormal number of sex chromosomes.

12.4 Changes in Chromosome Structure

Changes in chromosome structure are another type of chromosomal mutation. Some, but not all, changes in chromosome structure can be detected microscopically. Various agents in the environment, such as radiation, certain organic chemicals, or even viruses, can cause chromosomes to break. Ordinarily, when breaks occur in chromosomes, the two broken ends reunite to give the same sequence of genes. Sometimes, however, the broken ends of one or more chromosomes do not rejoin in the same pattern as before, and the result is various types of chromosomal mutation.

Changes in chromosome structure include deletions, duplications, translocations, and inversions of chromosome segments. A **deletion** occurs when an end of a chromosome breaks off or when two simultaneous breaks lead to the loss of an internal segment (Fig. 12.13a). Even when only one member of a pair of chromosomes is affected, a deletion often causes abnormalities.

A **duplication** is the presence of a chromosomal segment more than once in the same chromosome (Fig. 12.13b). An **inversion** has occurred when a segment of a chromosome is turned around 180° (Fig. 12.13c). This reversed sequence of genes can lead to altered gene activities and to deletions and duplications as described in Figure 12.14.

A **translocation** is the movement of a chromosome segment from one chromosome to another, nonhomologous chromosome (Fig. 12.13d). In 5% of cases, a translocation that occurred in a previous generation between chromosomes 21 and 14 is the cause of Down syndrome. In other words, because a portion of chromosome 21 is now attached to a portion of chromosome 14, the individual has three copies of the alleles that bring about Down syndrome when they are present in triplet copy. In these cases, Down syndrome is not related to the age of the mother but instead tends to run in the family of either the father or the mother.

Human Syndromes

Changes in chromosome structure occur in humans and lead to various syndromes, many of which are just now being discovered. Sometimes changes in chromosome structure can be detected in humans by doing a karyotype. They may also be discovered by studying the inheritance pattern of a disorder in a particular family.

Deletion Syndromes

Williams syndrome occurs when chromosome 7 loses a tiny end piece (Fig. 12.15). Children who have this syndrome look like pixies, with turned-up noses, wide mouths, a small chin, and large ears. Although their academic skills are poor, they exhibit excellent verbal and musical abilities. The gene that governs the production of the protein elastin is missing, and this

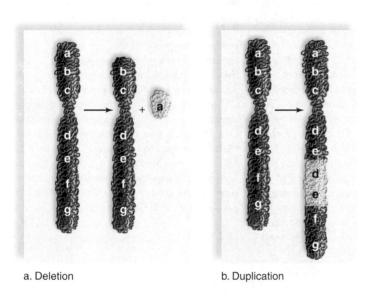

a. Deletion

b. Duplication

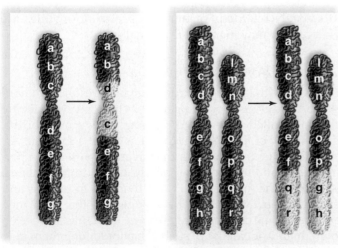

c. Inversion

d. Translocation

FIGURE 12.13 Types of chromosomal mutations.
a. Deletion is the loss of a chromosome piece. **b.** Duplication occurs when the same piece is repeated within the chromosome. **c.** Inversion occurs when a piece of chromosome breaks loose and then rejoins in the reversed direction. **d.** Translocation is the exchange of chromosome pieces between nonhomologous pairs.

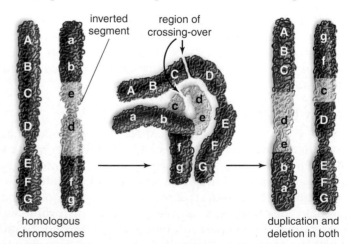

homologous chromosomes

inverted segment

region of crossing-over

duplication and deletion in both

FIGURE 12.14 Inversion.
Left: A segment of one homologue is inverted. Notice that in the shaded segment, *edc* occurs instead of *cde. Middle:* The two homologues can pair only when the inverted sequence forms an internal loop. After crossing-over, a duplication and a deletion can occur. *Right:* The homologue on the left has *AB* and *ab* sequences and neither *fg* nor *FG* genes. The homologue on the right has *gf* and *FG* sequences and neither *AB* nor *ab* genes.

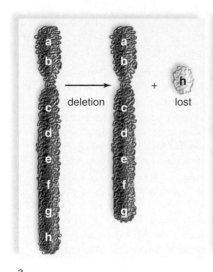

FIGURE 12.15
Deletion.
a. When chromosome 7 loses an end piece, the result is Williams syndrome. **b.** These children, although unrelated, have the same appearance, health, and behavioral problems.

a.   b.

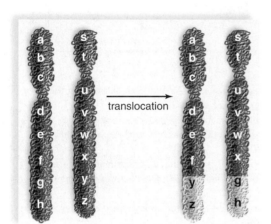

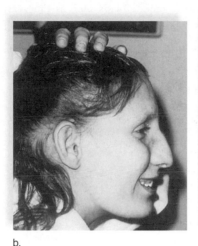

FIGURE 12.16
Translocation.
a. When chromosomes 2 and 20 exchange segments, (**b**) Alagille syndrome, with distinctive facial features, sometimes results because the translocation disrupts an allele on chromosome 20.

a.   b.

affects the health of the cardiovascular system and causes their skin to age prematurely. Such individuals are very friendly but need an ordered life, perhaps because of the loss of a gene for a protein that is normally active in the brain.

Cri du chat (cat's cry) syndrome is seen when chromosome 5 is missing an end piece. The affected individual has a small head, is mentally retarded, and has facial abnormalities. Abnormal development of the glottis and larynx results in the most characteristic symptom—the infant's cry resembles that of a cat.

Translocation Syndromes

A person who has both of the chromosomes involved in a translocation has the normal amount of genetic material and is healthy, unless the chromosome exchange breaks an allele into two pieces. The person who inherits only one of the translocated chromosomes will no doubt have only one copy of certain alleles and three copies of certain other alleles. A genetic counselor begins to suspect a translocation has occurred when spontaneous abortions are commonplace and family members suffer from various syndromes. A special microscopic technique allows a technician to determine that a translocation has occurred.

Figure 12.16 shows a father and son who have a translocation between chromosomes 2 and 20. Although they have the

normal amount of genetic material, they have the distinctive face, abnormalities of the eyes and internal organs, and severe itching characteristic of Alagille syndrome. People with this syndrome ordinarily have a deletion on chromosome 20; therefore, it can be deduced that the translocation disrupted an allele on chromosome 20 in the father. The symptoms of Alagille syndrome range from mild to severe, so some people may not be aware they have the syndrome. This father did not realize it until he had a child with the syndrome.

Translocations can also be responsible for a variety of other disorders including certain types of cancer. In the 1970s, new staining techniques identified that a translocation from a portion of chromosome 22 to chromosome 9 was responsible for chronic myelogenous leukemia. This translocated chromosome was called Philadelphia chromosome. In Burkett lymphoma, a cancer common in children in equatorial Africa, a large tumor develops from lymph glands in the region of the jaw. This disorder involves a translocation from a portion of chromosome 8 to chromosome 14.

Types of chromosomal mutations include deletions, translocations, duplications, and inversions. Chromosomal mutations in humans lead to syndromes.

CONNECTING THE CONCEPTS

By analyzing the outcome of his crosses, Mendel deduced that inheritable factors (now called alleles) contributed by each parent segregate from one another during the production of gametes. Early cytogeneticists (scientists who microscopically observe and study chromosomes) knew nothing of Mendel's work, but they saw how homologous chromosome pairs separate during meiosis. Boveri and Sutton were the first to realize that Mendel's work and the work of cytogeneticists must be interrelated—in other words, that the genes are on the chromosomes! The development of the chromosome theory of inheritance is a fine example of how different avenues of research can re-

sult in a major scientific breakthrough. It also exemplifies why communication between scientists through publications and scientific conferences is so important.

Once it was understood that genes are carried on chromosomes, previously confusing phenomena such as linked genes, sex linkage, and the consequences of chromosomal abnormalities became much easier to interpret correctly. The ability to map genes along chromosomes was an important step toward base sequencing genomes. It is no accident that the first eukaryotic genomes to be completely base sequenced were those that already had excellent genetic maps such as the yeast

Saccharomyces cerevisiae and the roundworm *Caenorhabditis elegans*. By contrast, constructing a base sequence map of the human genome was hampered by the lack of a conventional genetic map based on recombination distances. Humans cannot be asked to procreate in order to construct a genetic map! Chapter 16 discusses how laboratory techniques were combined with computer technology to sequence the human genome. The burgeoning field of genomics has led to other areas of interest such as proteomics, the study of protein activity in cells, and bioinformatics, the use of the computer to assimilate newfound biological information.

Summary

12.1 CHROMOSOMAL INHERITANCE

Sex determination in animals is dependent on the chromosomes. In *Drosophila* and humans, females are XX and males are XY. Solid experimental support for the chromosome theory of inheritance came when Morgan and his group were able to determine that the white-eye allele in *Drosophila* is on the X chromosome.

Alleles on the X chromosome are called X-linked alleles. Therefore, when doing X-linked genetics problems, it is the custom to indicate the sexes by using sex chromosomes and to indicate the alleles by superscripts attached to the X. The Y is blank because it does not carry these genes. Color blindness, Duchenne muscular dystrophy, and hemophilia are X-linked recessive disorders in humans. Fragile X syndrome is an X-linked condition with an unusual mode of inheritance.

12.2 GENE LINKAGE

All the genes on a chromosome form a linkage group. Linked genes do not obey Mendel's law of independent assortment because they tend to go into the same gamete together. Crossing-over can cause recombinant gametes and recombinant phenotypes to occur. The percentage of recombinant phenotypes is used to measure the distance between genes and to map the chromosomes.

12.3 CHANGES IN CHROMOSOME NUMBER

Chromosomal mutations fall into two categories: changes in chromosome number and changes in chromosome structure.

Polyploidy occurs when the eukaryotic individual has more than two complete sets of chromosomes. Aneuploidy is the gain or loss of one or more chromosomes. Monosomy occurs when an individual has only one of a particular type of chromosome $(2n - 1)$; trisomy occurs when an individual has three of a particular type of chromosome $(2n + 1)$. Monosomies and trisomies are due to nondisjunction during meiosis. Down syndrome is a well-known trisomy in human beings.

12.4 CHANGES IN CHROMOSOME STRUCTURE

Changes in chromosome structure include deletions, translocations, duplications, and inversions. Inversions can lead to both deletions and duplications. Syndromes in humans can be due to changes in chromosome structure. For example, cri du chat is due to a deletion on chromosome 5.

Reviewing the Chapter

1. How is sex determined in humans? Which sex determines the sex of the offspring in human beings?   204
2. How did a *Drosophila* cross involving an X-linked gene help investigators show that certain genes are carried on certain chromosomes?   204
3. What are the chances of an affected male offspring if the female parent is a carrier of an X-linked recessive disorder and the male parent is normal?   205
4. How do you recognize a pedigree for an X-linked recessive allele in human beings?   206
5. Discuss the symptoms of color blindness, Duchenne muscular dystrophy, and hemophilia in human beings.   206–7
6. Show that a dihybrid cross between a heterozygote and a recessive homozygote involving linked genes does not produce the expected 1:1:1:1 ratio. What is the significance of the small percentage of recombinants that occurs among the offspring?   210–11
7. What are the two types of chromosomal mutations?   212
8. What is a common way for polyploidy to occur in plants? For a monosomy and a trisomy to occur?   212–213
9. Discuss trisomy 21, a fairly common trisomy in humans. How is trisomy 21 detected before birth?   213–15
10. What four types of changes in chromosome structure were discussed? How might it be possible for a duplication and a deletion to occur at the same time?   218
11. Name some syndromes that occur in humans due to changes in chromosome structure.   218–19

Testing Yourself

Choose the best answer for each question.

1. Which of these would you not find in a pedigree when a male has an X-linked recessive disorder?
   a. Neither parent has the disorder.
   b. Only males in the pedigree have the disorder.
   c. Only females in previous generations have the disorder.
   d. The sons of a female with the disorder will all have the disorder.
   e. Both a and c would not be seen.

2. A boy is color-blind (X-linked recessive) and has a straight hairline (autosomal recessive). Which could be the genotype of his mother?
   a. *bbww*
   b. X^bYWw
   c. bbX^wX^w
   d. X^BX^bWw
   e. X^wX^wBb

3. Investigators found that a cross involving the mutant genes *a* and *b* produced 30% recombinants, a cross involving *a* and *c* produced 5% recombinants, and a cross involving *c* and *b* produced 25% recombinants. Which is the correct order of the genes?
   a. *a, b, c*
   b. *a, c, b*
   c. *b, a, c*
   d. Both a and b are correct.
   e. Both b and c are correct.

4. Nondisjunction during meiosis I of oogenesis will result in eggs that have
   a. the normal number of chromosomes.
   b. one too many chromosomes.
   c. one less than the normal number of chromosomes.
   d. Both b and c are correct.

For questions 5–7, match the conditions in the key with the descriptions.

KEY:
   a. Down syndrome
   b. Turner syndrome
   c. Klinefelter syndrome
   d. XYY
   e. poly-X individual

5. male with underdeveloped testes and some breast development

6. trisomy 21

7. XO female

8. Which two of these chromosomal mutations are most likely to occur when an inverted chromosome is undergoing synapsis?
   a. deletion and translocation
   b. deletion and duplication
   c. duplication and translocation
   d. inversion and duplication

9. Investigators do a dihybrid cross between two heterozygotes and get about a 3:1 ratio among the offspring. The reason must be due to
   a. a chromosomal mutation.
   b. a one-trait cross.
   c. crossing-over.
   d. a linkage group.

10. a. Determine if the characteristic possessed by the shaded squares (males) is autosomal dominant, autosomal recessive, or X-linked recessive. b. Write in the genotype for the starred individual.

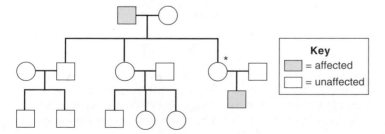

Key
■ = affected
□ = unaffected

11. Which explains why X-linked recessive characteristics often skip a generation?
    a. More males than females are affected.
    b. If a woman has the characteristic all of her sons will have it.
    c. Females can be carriers.
    d. X-linked recessive characteristics are incompletely dominant.

In questions 12–15, indicate whether the statement is true (T) or false (F).

12. As exemplified by color blindness, most X-linked recessive characteristics alter the senses. _____

13. Fragile X syndrome is due to nondisjunction. _____

14. Alleles that are linked can become unlinked due to crossing-over. _____

15. The lower the frequency of recombinant phenotypes, the closer the loci. _____

For questions 16–21, match the chromosomal changes in the key with the descriptions. Answers are used more than once.

KEY:
   a. polyploidy
   b. monosomy
   c. trisomy
   d. Both b and c are correct.
   e. A, b, and c are correct.

16. occurs as a result of nondisjunction

17. an abnormal chromosome number

18. is more common in plants than animals

19. one chromosome of a pair is missing

20. Klinefelter syndrome

21. Turner syndrome

22. Trisomy 21 is due to
    a. a very poor diet.
    b. more than two copies of certain genes.
    c. linked genes.
    d. X-linked inheritance.
    e. Both b and c are correct.

23. Changes in sex chromosome number
    a. are not due to nondisjunction.
    b. do not lead to syndromes.
    c. can be a monosomy or a trisomy.
    d. always involve the Y chromosome.
    e. All of these are correct.

24. Translocation
    a. is a deletion plus a duplication.
    b. always results in a syndrome.
    c. occurs following two deletions on nonhomologous chromosomes.
    d. is exemplified by cri du chat.
    e. None of these are correct.

25. When translocation is the cause of trisomy 21,
    a. the mother is middle-aged.
    b. there are three copies of chromosome 21.
    c. chromosome 14 is joined with a piece of chromosome 21.
    d. all of these are correct.

Additional Genetics Problems*

1. In *Drosophila*, the gene that controls red eye color (dominant) versus white eye color is on the X chromosome. What are the expected phenotypic results if a heterozygous female is crossed with a white-eyed male?

2. In cats, the alleles for black (X^B) and orange (X^b) coat color are carried on the X chromosome. Tortoiseshell cats have both alleles and express both coat colors. (a) If a male black cat mates with a female tortoiseshell cat, predict the possible genotypes of their offspring. (b) Although male tortoiseshell cats are rare, how can their existence be explained?

3. A boy has severe combined immune deficiency syndrome, an autosomal recessive disorder. What are the genotypes of the parents, who have the normal phenotype?

4. If a female who carries an X-linked allele for Lesch-Nyhan syndrome reproduces with an unaffected man, what are the chances that male children will have the condition? That female children will have the condition?

5. An unaffected woman whose father had hemophilia marries a man who has hemophilia. What is the chance their sons will be hemophiliacs? What is the chance their daughters will be hemophiliacs?

6. What is the possible genotype of a man who is color-blind? What is the possible genotype of a woman who is not color-blind but whose father was color-blind?

7. A man who is homozygous for curling the tongue (dominant) is color-blind. He reproduces with a woman who is heterozygous for tongue-curling and homozygous for normal vision. What is the chance this couple will have a color-blind son?

8. How many chromosomes would you expect to find in a person with (a) Turner syndrome? (b) Klinefelter syndrome?

9. It is known that *A* and *B* are 10 map units apart; *A* and *C* are 8 units apart; *A* and *D* are 18 units apart; and *C* and *D* are 10 units apart. What is the order of the genes on the chromosome?

10. In *Drosophila*, S = normal body, s = sable body; W = normal wings, w = miniature wings. In a cross between a heterozygous normal fly and a sable-bodied, miniature-winged fly, the results were 99 normal flies, 99 with a sable body and miniature wings, 11 with a normal body and miniature wings, and 11 with a sable body and normal wings. What inheritance pattern explains these results? How many map units separate the genes for sable body and miniature wing?

* Answers to Additional Genetics Problems appear in Appendix A.

Thinking Scientifically

1. Females with two X chromosomes have two X-linked alleles, but one of the X chromosomes becomes a Barr body. Why do you suppose the body of a woman heterozygous for an X-linked trait would be a mosaic—some cells expressing the recessive and some cells expressing the dominant allele?

Bioethical Issue: Choosing Human Gender

Do you approve of choosing a baby's gender even before it is conceived? As you know, the sex of a child is dependent on whether an X-bearing sperm or a Y-bearing sperm enters the egg. A new technique has been developed that can separate X-bearing sperm from Y-bearing sperm. First, the sperm are dosed with a DNA-staining chemical. Because the X chromosome has slightly more DNA than the Y chromosome, it takes up more dye. When a laser beam shines on the sperm, the X-bearing sperm shine a little more brightly than the Y-bearing sperm. A machine sorts the sperm into two groups on this basis. The results are not perfect. Following artificial insemination, there's about an 85% success rate for a girl and about a 65% success rate for a boy.

Some might argue that it goes against nature to choose gender. But what if the mother is a carrier of an X-linked genetic disorder such as hemophilia or Duchenne muscular dystrophy? Is it acceptable to bring a child into the world with a genetic disorder that may cause an early death? Wouldn't it be better to select sperm for a girl, who at worst would be a carrier like her mother? Some authorities do not find gender selection acceptable even for this reason. Once you separate reproduction from the sex act, they say, it opens the door to children who have been genetically designed in every way. As a society, should we accept certain ways of interfering with nature and not accept other ways? Why or why not?

Understanding the Terms

amniocentesis 214	inversion 218
aneuploid 212	karyotype 214
autosome 204	linkage group 209
carrier 205	linkage map 209
chorionic villi sampling	locus 209
(CVS) 214	monosomy 212
chromosomal mutation 212	nondisjunction 212
chromosome theory of	polyploid 212
inheritance 204	sex chromosome 204
deletion 218	syndrome 214
duplication 218	translocation 218
euploidy 212	trisomy 212
gene linkage 209	X-linked 204

Match the terms to these definitions:

a. _____ Any chromosome other than a sex chromosome.

b. _____ Condition in which an organism has more than two complete sets of chromosomes.

c. _____ Allele located on the X chromosomes that controls a trait unrelated to sex.

ARIS, the *Biology* Website

ARIS, the website for *Biology*, provides a wealth of information organized and integrated by chapter. You will find practice quizzes, interactive activities, labeling exercises, flashcards, and much more that will complement your learning and understanding of general biology.

www.mhhe.com/maderbiology9

13

DNA STRUCTURE AND FUNCTIONS

O ne of the most exciting periods of scientific activity in history occurred during the 30 short years between the 1930s and 1960s. Geneticists knew that chromosomes contain protein and DNA (deoxyribonucleic acid). Of these two organic molecules, proteins are seemingly more complicated; they consist of countless sequences of 20 amino acids, which can coil and fold into complex shapes. DNA, on the other hand, contains only four different nucleotides. Surely, the diversity of life-forms on Earth must be the result of the endless varieties of proteins.

However, due to several elegantly executed experiments, by the mid-1950s researchers realized that DNA, not protein, is the genetic material. But this finding only led to another fundamental question—what exactly is the structure of DNA? The biological community at the time knew that whoever determined the structure of DNA would receive a Nobel Prize and go down in history. Consequently, researchers were racing against time and each other. The story of the discovery of DNA structure resembles a mystery, with each clue adding to the total picture until the breathtaking design of DNA—a double helix—was finally unraveled.

The double-helix structure of DNA.

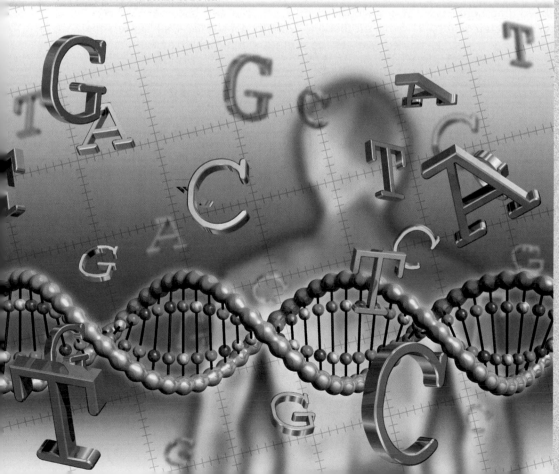

13.1 THE GENETIC MATERIAL

Even though previous investigators had confirmed that the genes are on the chromosomes and had even been able to map the *Drosophila* chromosomes, they did not know the composition of the genes. The search for the genetic material was of utmost importance to biologists at the beginning of the twentieth century. They knew that this material must be:

1. able to *store information* that pertains to the development, structure, and metabolic activities of the cell or organism;
2. stable so that it *can be replicated* with high fidelity during cell division and be transmitted from generation to generation;
3. able to *undergo rare changes* called mutations [L. *muta*, change] that provide the genetic variability required for evolution to occur.

The genetic material must be able to store information, be replicated, and undergo mutations.

Knowledge of the chemistry of DNA was absolutely essential in order to conclude that DNA is the genetic material. In 1869, the Swiss physician Johann Friedrich Miescher removed nuclei from pus cells (these cells have little cytoplasm) and found that they contained a chemical he called *nuclein.* Nuclein, he said, was rich in phosphorus and had no sulfur, properties that distinguished it from protein. Later, other chemists did further work with nuclein and said that it contained an acidic substance they called **nucleic acid.** Soon researchers realized that there are two types of nucleic acids: **DNA (deoxyribonucleic acid)** and **RNA (ribonucleic acid).**

Early in the twentieth century, it was discovered that nucleic acids contain only four types of **nucleotides,** molecules that are composed of a nitrogen-containing base, a phosphate, and a pentose (5-carbon sugar). Perhaps DNA was composed of repeating units, and each unit always had just one of each of the four different nucleotides. In that case, DNA could not vary between genes and could not be the genetic material! Everyone thought that the protein component of chromosomes must be the genetic material because proteins contain 20 different amino acids that can be sequenced in any particular way.

Transformation of Bacteria

During the late 1920s, the bacteriologist Frederick Griffith was attempting to develop a vaccine against *Streptococcus pneumoniae* (pneumococcus), which causes pneumonia in mammals. In 1931, he performed a classic experiment with the bacterium. He noticed that when these bacteria are grown on culture plates, some, called S strain bacteria, produce shiny, smooth colonies, and others, called R strain bacteria, produce colonies that have a rough appearance. Under the microscope, S strain bacteria have a capsule (mucous coat) but R strain bacteria do not. When Griffith injected mice with the S strain of bacteria, the mice died, and when he injected mice with the R strain, the mice did not die (Fig. 13.1). In an effort to determine if the capsule alone was responsible for the virulence (ability to kill) of the S strain bacteria, he injected mice with heat-killed S strain bacteria. The mice did not die.

Finally, Griffith injected the mice with a mixture of heat-killed S strain and live R strain bacteria. Most unexpectedly, the mice died and living S strain bacteria were recovered from the bodies! Griffith concluded that some substance necessary to the synthesis of a capsule and, therefore, viru-

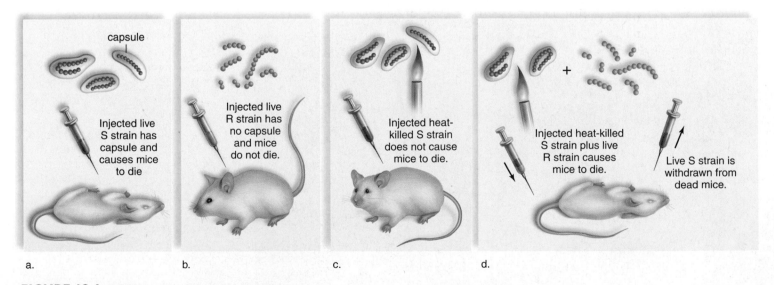

FIGURE 13.1 Griffith's transformation experiment.
a. Encapsulated S strain is virulent and kills mice. **b.** Nonencapsulated R strain is not virulent and does not kill mice. **c.** Heat-killed S strain bacteria do not kill mice. **d.** If heat-killed S strain and R strain are both injected into mice, they die because the R strain bacteria have been transformed into the virulent S strain.

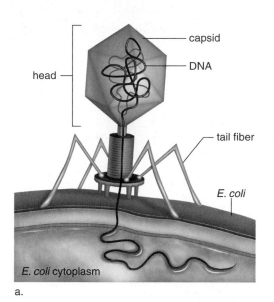

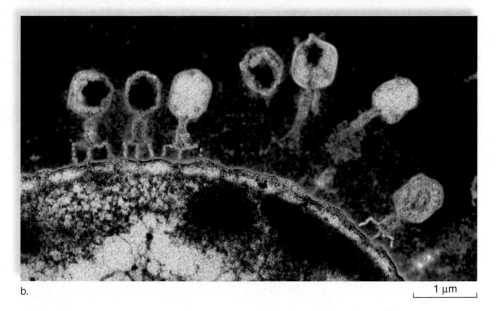

a.   b.   1 μm

FIGURE 13.2 Bacteria and bacteriophages.
a. A T virus consists of a capsid (protein coat) surrounding a DNA core. When a T virus attacks a bacterium, only DNA enters the cell, and later many fully formed viruses emerge from the cell. **b.** An electron micrograph of T viruses attacking an *E. coli* cell.

lence must have passed from the dead S strain bacteria to the living R strain bacteria so that the R strain bacteria were *transformed* (Fig. 13.1*d*). This change in the phenotype of the R strain bacteria must be due to a change in their genotype. Indeed, couldn't the transforming substance that passed from S strain to R strain be genetic material? Reasoning such as this prompted investigators at the time to begin looking for the transforming substance to determine the chemical nature of the genetic material.

DNA: *The Transforming Substance*

Obviously, it is not convenient to look for the transforming substance in mice. It is not surprising, then, that the next group of investigators, led by Oswald Avery, isolated the genetic material in vitro (in laboratory glassware). In 1944, after 16 years of research, Avery and his coinvestigators Colin MacLeod and Maclyn McCarty published a paper demonstrating that the transforming substance is DNA. Their evidence included the following observations:

1. DNA from S strain bacteria causes R strain bacteria to be transformed.
2. Enzymes that degrade proteins cannot prevent transformation, nor does RNase, an enzyme that digests RNA.
3. Enzymatic digestion of the transforming substance with DNase, an enzyme that digests DNA, does prevent transformation.
4. The molecular weight of the transforming substance is so great that it must contain about 1,600 nucleotides! Certainly this is enough for some genetic variability.

These experiments certainly showed that DNA is the transforming substance. Although some remained skeptical, many felt that the evidence for DNA being the hereditary material was overwhelming.

Transformation Experiments Today

Transformation experiments are often done today in various laboratories around the world, from high schools to the most sophisticated research facilities. Transformation occurs when organisms receive foreign DNA and thereby acquire a new characteristic. One typical rudimentary transformation experiment is to allow bacteria to take up DNA that makes them resistant to penicillin. Before they are transformed, the bacteria are unable to grow in the presence of penicillin, but after they are transformed, they are able to grow in the presence of penicillin. The DNA contains a gene that makes bacteria resistant to penicillin!

Reproduction of Viruses

Soon after Avery and his coinvestigators published their results, many geneticists who were interested in determining the chemical nature of the genetic material began to work with bacteriophages [Gk. *bacterion*, rod, and *phagein*, to eat]. **Bacteriophages** are viruses that infect bacteria, such as *Escherichia coli (E. coli)*, which normally lives within the human gut.

Viruses consist only of a protein coat called a capsid surrounding a nucleic acid core (Fig. 13.2). We now know that when a bacteriophage, or simply phage, latches onto a bacterium, only the nucleic acid (i.e., DNA) enters the cell, and the capsid is left behind. Later the bacterium releases many hundreds of new viruses. Why? Because phage DNA contains the genetic information necessary to cause the bacterium to produce new phages.

During the 1950s, biologists were still performing experiments to determine which one—DNA or protein—enters a bacterium. Whichever one does this has to be the genetic material. In 1952, two experimenters, Alfred D. Hershey and Martha Chase, chose a bacteriophage known as T2 to determine which of the phage components—protein

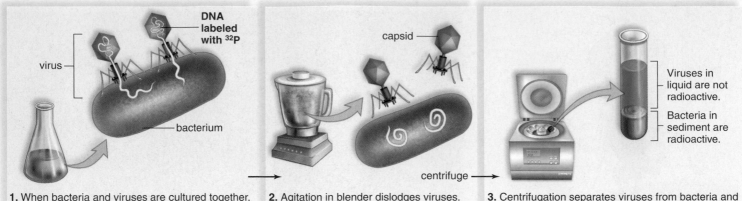

1. When bacteria and viruses are cultured together, radioactive viral DNA enters bacterium.
2. Agitation in blender dislodges viruses. Radioactivity stays inside the bacterium.
3. Centrifugation separates viruses from bacteria and allows investigator to detect location of radioactivity.

a. Viral DNA is labeled (yellow).

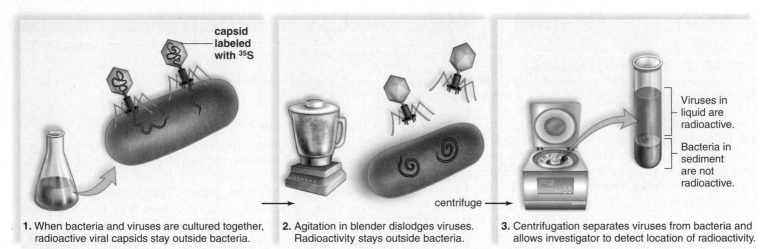

1. When bacteria and viruses are cultured together, radioactive viral capsids stay outside bacteria.
2. Agitation in blender dislodges viruses. Radioactivity stays outside bacteria.
3. Centrifugation separates viruses from bacteria and allows investigator to detect location of radioactivity.

b. Viral capsid is labeled (yellow).

FIGURE 13.3 Hershey and Chase experiments.
The experiment concluded that viral DNA, not protein, was responsible for directing the production of new viruses.

or DNA—entered bacterial cells and directed reproduction of the virus. In their experiment, Hershey and Chase relied on a chemical difference between DNA and protein to solve the mystery. In DNA, phosphorus is present but not sulfur, and in protein, sulfur is present but phosphorus is not. They used radioactive ^{32}P to label the DNA core of the phage and radioactive ^{35}S to label the protein in the capsid of the phage. Recall that radioactive isotopes serve as labels (i.e., tracers) in biological experiments because it is possible to detect the presence of radioactivity by standard laboratory procedures.

Hershey and Chase did two separate experiments (Fig. 13.3). In the first experiment, phage DNA was labeled with radioactive ^{32}P. The phages were allowed to attach to and inject their genetic material into *E. coli* cells. Then the culture was agitated in a kitchen blender to remove whatever remained of the phages on the outside of the bacterial cells. Finally, the culture was centrifuged (spun at high speed) so that the bacterial cells collected as a pellet at the bottom of the centrifuge tube. In this experiment, as you would predict, they found most of the ^{32}P-labeled DNA in

the cells and not in the liquid medium. Why? Because the DNA had entered the cells.

In the second experiment, phage protein in capsids was labeled with radioactive ^{35}S. The phages were allowed to attach to and inject their genetic material into *E. coli* bacterial cells. Then the culture was agitated in a kitchen blender to remove whatever remained of the phages on the outside of the bacterial cells. Finally, the culture was centrifuged so that the bacterial cells collected as a pellet at the bottom of the centrifuge tube. In this experiment, as you would predict, they found ^{35}S-labeled protein in the liquid medium and not in the cells. Why? Because the radioactive capsids remained on the outside of the cells and were removed by the blender.

These results indicated that the DNA (not the protein) of a virus enters the host, where viral reproduction takes place. Therefore, DNA is the genetic material. It transmits all the necessary genetic information needed to produce new viruses.

The work of early investigators confirmed that DNA, and not protein, is the genetic material.

13.2 THE STRUCTURE OF DNA

During the same period that biologists were using viruses to show that DNA is the genetic material, biochemists were trying to determine the molecular configuration of DNA. By understanding the structure of DNA, scientists would be able to understand how DNA stores information and functions in the transmission of the genetic code.

Nucleotide Data

With the development of new chemical techniques in the 1940s, it was possible for Erwin Chargaff to analyze in detail the base content of DNA. It was known that DNA contains four different types of nucleotides: two with **purine** bases, **adenine (A)** and **guanine (G),** which have a double ring, and two with **pyrimidine** bases, **thymine (T)** and **cytosine (C),** which have a single ring (Fig. 13.4*a, b*). It had been hypothesized that DNA has repeating units, each unit having four nucleotides—one for each of the four bases. If so, the DNA of every species would contain 25% of each kind of nucleotide.

A sample of Chargaff's data is seen in Figure 13.4*c.* You can see that while some species—*E. coli* and *Zea mays* (corn), for

example—do have approximately 25% of each type of nucleotide, most do not. Further, the percentage of each type of nucleotide differs from species to species. Therefore, the nucleotide content of DNA is not fixed, and DNA does have the *variability* between species required of the genetic material.

Within each species, however, DNA was found to have the *constancy* required of the genetic material—that is, all members of a species have the same base composition. Also, the percentage of A always equals the percentage of T, and the percentage of G equals the percentage of C. The percentage of A + G equals 50%, and the percentage of T + C equals 50%. These relationships are called Chargaff's rules.

Chargaff's rules:

1. The amount of A, T, G, and C in DNA varies from species to species.
2. In each species, the amount of A = T and the amount of G = C.

At this time, chemists were still uncertain about the molecular configuration of DNA.

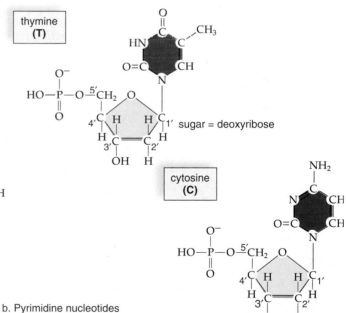

a. Purine nucleotides

b. Pyrimidine nucleotides

Chargaff's DNA Database Composition in Various Species (%)				
Species	A	T	G	C
Homo sapiens (human)	31.0	31.5	19.1	18.4
Drosophila melanogaster (fruit fly)	27.3	27.6	22.5	22.5
Zea mays (corn)	25.6	25.3	24.5	24.6
Neurospora crassa (fungus)	23.0	23.3	27.1	26.6
Escherichia coli (bacterium)	24.6	24.3	25.5	25.6
Bacillus subtilis (bacterium)	28.4	29.0	21.0	21.6

c.

FIGURE 13.4 Nucleotide composition of DNA.
All nucleotides contain phosphate, a 5-carbon sugar, and a nitrogen-containing base. In DNA, the sugar is called deoxyribose because it lacks an oxygen atom in the 2′ position, compared to ribose. The nitrogen-containing bases are (**a**) the purines adenine and guanine, which have a double ring, and (**b**) the pyrimidines thymine and cytosine, which have a single ring. **c.** Chargaff's data show that the DNA of various species differs. For example, in humans the A and T percentages are about 31%, but in fruit flies these percentages are about 27%.

Variation in Base Sequence

Chargaff's data suggest that A is always paired with T and G is always paired with C. Today we know that the paired bases can be in any order:

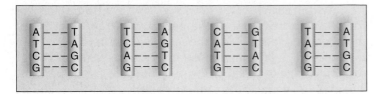

The variability that can be obtained is overwhelming. For example, it has been calculated that an average human chromosome contains on the average about 140 million base pairs. Since any of the four possible nucleotides can be present at each nucleotide position, the total number of possible nucleotide sequences is $4^{140 \times 10^6}$ or $4^{140,000,000}$. No wonder each species has its own base percentages!

Diffraction Data

Rosalind Franklin, a researcher in the laboratory of Maurice H. F. Wilkins at King's College in London, studied the structure of DNA using X-rays. She found that if a concentrated, viscous solution of DNA is made, it can be separated into fibers. Under the right conditions, the fibers are enough like a crystal (a solid substance whose atoms are arranged in a definite manner) that when X-rayed, an X-ray diffraction pattern results (Fig. 13.5a). The X-ray diffraction pattern of DNA shows that DNA is a helix. The helical shape is indicated by the crossed (X) pattern in the center of the photograph in Figure 13.5b. The dark portions at the top and bottom of the photograph indicate that some portion of the helix is repeated.

The Watson and Crick Model

James Watson, an American, was on a postdoctoral fellowship at Cavendish Laboratories in Cambridge, England, and while there he began to work with the biophysicist Francis H. C. Crick. Using the data provided from X-ray diffraction and other sources, they constructed a model of DNA for which they received a Nobel Prize in 1962 (Fig. 13.6).

Watson and Crick knew, of course, that DNA is a polymer of nucleotides, but they did not know how the nucleotides were arranged within the molecule. However, they deduced that DNA is a **double helix** with sugar-phosphate backbones on the outside and paired bases on the inside. This arrangement fits the mathematical measurements provided by the X-ray diffraction data for the spacing between the base pairs (0.34 nm) and for a complete turn of the double helix (3.4 nm).

This model also agreed with Chargaff's rules, which said that A = T and G = C. Figure 13.6 shows that A is hydrogen-bonded to T, and G is hydrogen-bonded to C. This so-called **complementary base pairing** means that a purine is always bonded to a pyrimidine. Only in this way will the molecule have the width (2 nm) dictated by its X-ray diffraction pattern, since two pyrimidines together are too narrow, and two purines together are too wide.

The double-helix model of DNA is like a twisted ladder; sugar-phosphate backbones make up the sides, and hydrogen-bonded bases make up the rungs, or steps, of the ladder.

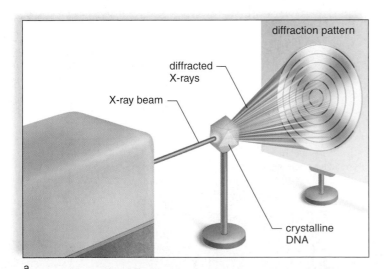

a.

FIGURE 13.5 X-ray diffraction of DNA.

a. When a crystal is X-rayed, the way in which the beam is diffracted reflects the pattern of the molecules in the crystal. The closer together two repeating structures are in the crystal, the farther from the center the beam is diffracted. **b.** The diffraction pattern of DNA produced by Rosalind Franklin. The crossed (X) pattern in the center told investigators that DNA is a helix, and the dark portions at the top and the bottom told them that some feature is repeated over and over. Watson and Crick determined that this feature was the hydrogen-bonded bases.

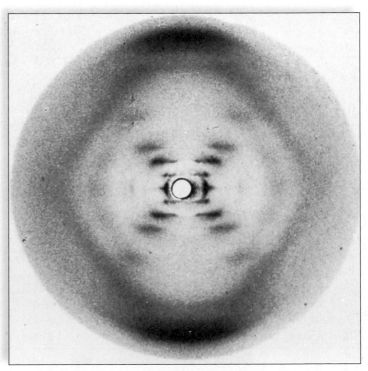

b.

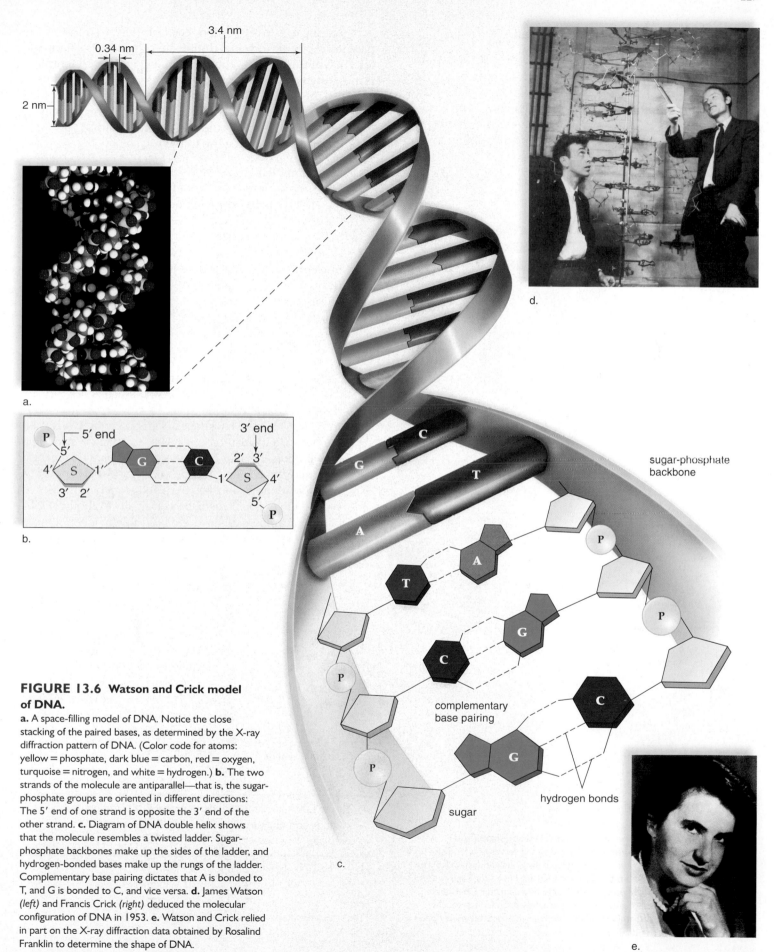

FIGURE 13.6 Watson and Crick model of DNA.

a. A space-filling model of DNA. Notice the close stacking of the paired bases, as determined by the X-ray diffraction pattern of DNA. (Color code for atoms: yellow = phosphate, dark blue = carbon, red = oxygen, turquoise = nitrogen, and white = hydrogen.) **b.** The two strands of the molecule are antiparallel—that is, the sugar-phosphate groups are oriented in different directions: The 5′ end of one strand is opposite the 3′ end of the other strand. **c.** Diagram of DNA double helix shows that the molecule resembles a twisted ladder. Sugar-phosphate backbones make up the sides of the ladder, and hydrogen-bonded bases make up the rungs of the ladder. Complementary base pairing dictates that A is bonded to T, and G is bonded to C, and vice versa. **d.** James Watson (*left*) and Francis Crick (*right*) deduced the molecular configuration of DNA in 1953. **e.** Watson and Crick relied in part on the X-ray diffraction data obtained by Rosalind Franklin to determine the shape of DNA.

13.3 Replication of DNA

The term **DNA replication** refers to the process of copying a DNA molecule. Following replication, there is usually an exact copy of the DNA double helix. As soon as Watson and Crick developed their double-helix model, they commented, "It has not escaped our notice that the specific pairing we have postulated immediately suggests a possible copying mechanism for the genetic material."

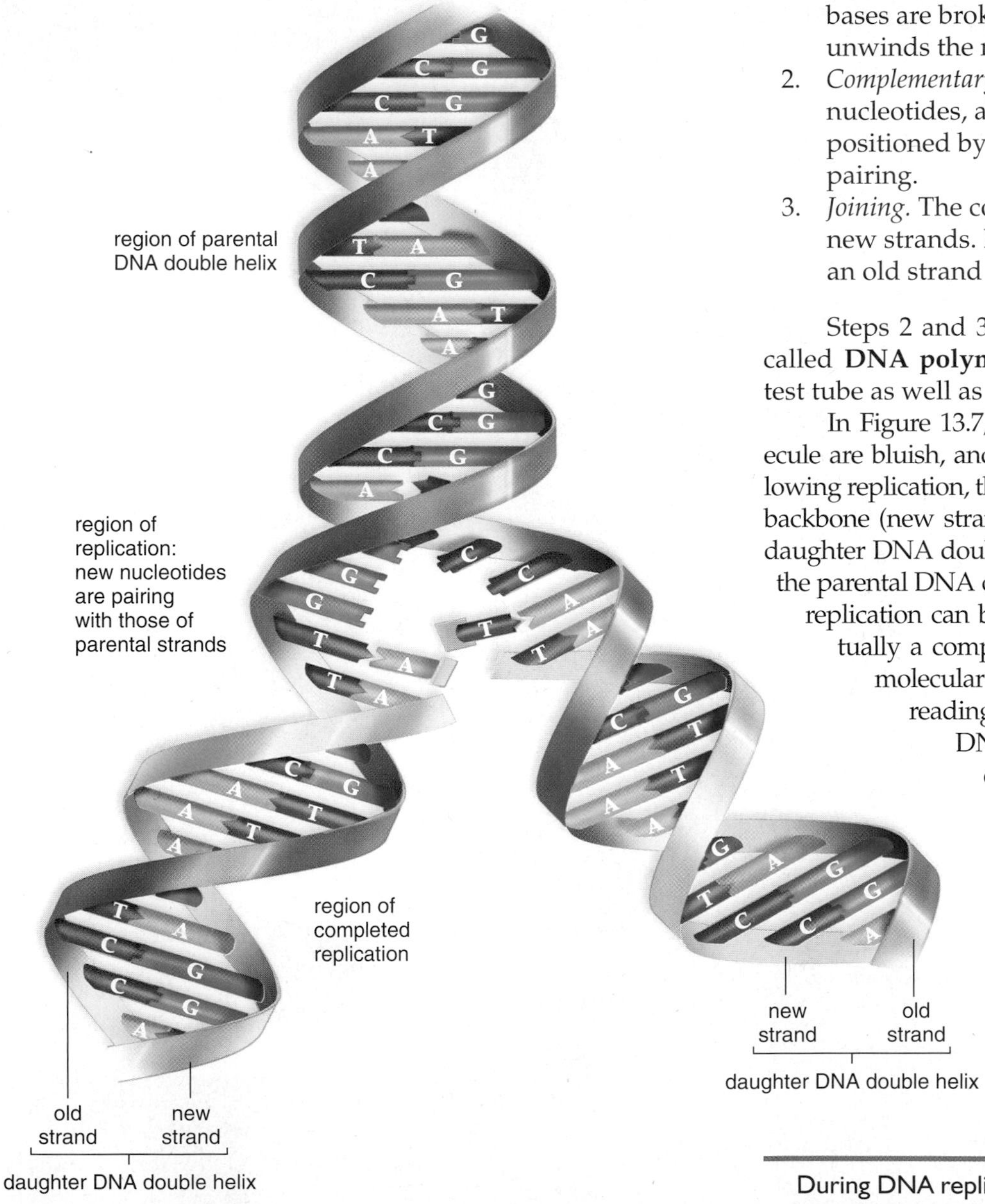

During DNA replication, each old DNA strand of the parental molecule (original double helix) serves as a template for a new strand in a daughter molecule (Fig. 13.7). A **template** is most often a mold used to produce a shape complementary to itself. DNA replication is termed **semiconservative replication** because one of the old strands is conserved, or present, in each daughter DNA molecule (new double helix).

Replication requires the following steps:

1. *Unwinding.* The old strands that make up the parental DNA molecule are unwound and "unzipped" (i.e., the weak hydrogen bonds between the paired bases are broken). A special enzyme called helicase unwinds the molecule.
2. *Complementary base pairing.* New complementary nucleotides, always present in the nucleus, are positioned by the process of complementary base pairing.
3. *Joining.* The complementary nucleotides join to form new strands. Each daughter DNA molecule contains an old strand and a new strand.

Steps 2 and 3 are carried out by an enzyme complex called **DNA polymerase.**[1] DNA polymerase works in the test tube as well as in cells.

In Figure 13.7, the backbones of the parental DNA molecule are bluish, and each base is given a particular color. Following replication, the daughter molecules each have a greenish backbone (new strand) and a bluish backbone (old strand). A daughter DNA double helix has the same sequence of bases as the parental DNA double helix had originally. Although DNA replication can be explained easily in this manner, it is actually a complicated process. Some of the more precise molecular events are discussed in the Science Focus reading on page 232.

DNA replication must occur before a cell can divide. Cancer, which is characterized by rapidly dividing cells, is sometimes treated with chemotherapeutic drugs that are analogs (have a similar, but not identical, structure) to one of the four nucleotides in DNA. When these are mistakenly used by the cancer cells to synthesize DNA, replication stops and the cells die off.

region of parental DNA double helix

region of replication: new nucleotides are pairing with those of parental strands

region of completed replication

old strand   new strand

daughter DNA double helix

new strand   old strand

daughter DNA double helix

FIGURE 13.7 Semiconservative replication (simplified).
After the DNA double helix unwinds, each old strand serves as a template for the formation of the new strand. Complementary nucleotides available in the cell pair with those of the old strand and then are joined together to form a strand. After replication is complete, there are two daughter DNA double helices. Each is composed of an old strand and a new strand. Each daughter double helix has the same sequence of base pairs as the parental double helix had before unwinding occurred.

During DNA replication, the parental DNA molecule unwinds and unzips. Then each old strand serves as a template for a new strand.

[1] The complex contains a number of different DNA polymerases with specific functions.

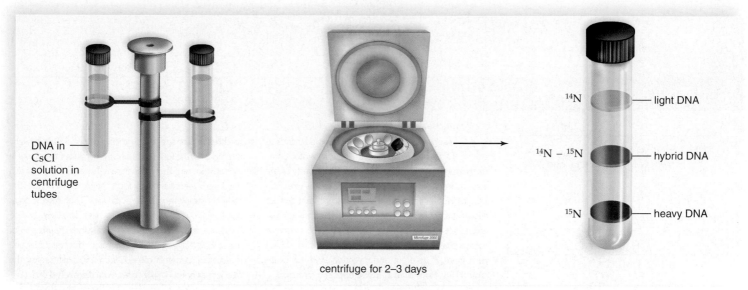

a. Possible results when DNA is centrifuged in CsCl

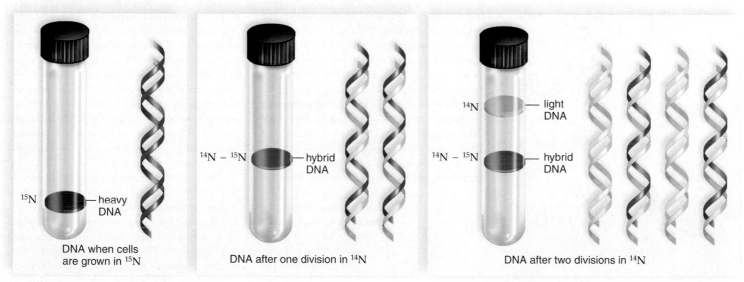

b. Steps in Meselson and Stahl experiment

FIGURE 13.8 Meselson and Stahl's DNA replication experiment.

a. When DNA molecules are centrifuged in a CsCl density gradient, they separate on the basis of density. **b.** Cells grown in heavy nitrogen (^{15}N) have dense (heavy DNA) strands. After one division in light nitrogen (^{14}N), DNA molecules are hybrid and have intermediate density. After two divisions, DNA molecules separate into two bands—one for light DNA and one for hybrid DNA.

Replication Is Semiconservative

As previously noted, DNA replication is termed semiconservative because each daughter double helix contains an old strand and a new strand. Semiconservative replication was experimentally confirmed by Matthew Meselson and Franklin Stahl in 1958.

Centrifuges spin tubes and in that way separate particles from the suspending fluid. Meselson and Stahl knew that it would be possible to centrifuge DNA molecules in a suspending fluid that would separate them on the basis of their different densities (Fig. 13.8a). A DNA molecule in which both strands contain heavy nitrogen (^{15}N, with an atomic weight of 15) is most dense. A DNA molecule in which both strands contain light nitrogen (^{14}N, with an atomic weight of 14) is

least dense. A hybrid DNA molecule in which one strand is heavy and one is light has an intermediate density.

Figure 13.8b shows that Meselson and Stahl first grew bacteria in a medium containing ^{15}N so that only heavy DNA molecules were extracted from the cells. Then they switched the bacteria to a medium containing ^{14}N. After one division, only hybrid DNA molecules were in the cells. After two divisions, half of the DNA molecules were light and half were hybrid. These were exactly the results to be expected if DNA replication is semiconservative.

DNA replication is semiconservative. Each daughter strand contains an old strand and a new strand.

---

science focus

Aspects of DNA Replication

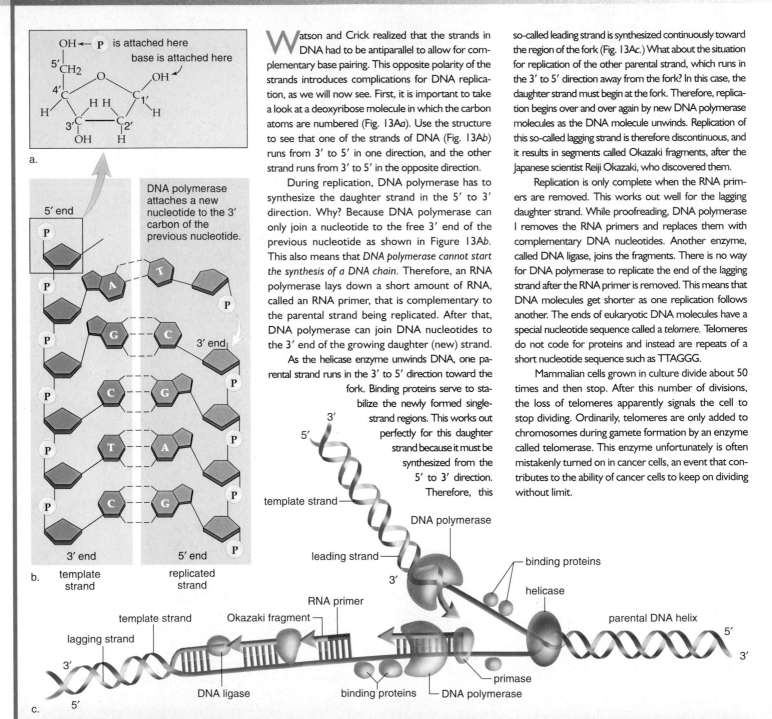

Watson and Crick realized that the strands in DNA had to be antiparallel to allow for complementary base pairing. This opposite polarity of the strands introduces complications for DNA replication, as we will now see. First, it is important to take a look at a deoxyribose molecule in which the carbon atoms are numbered (Fig. 13A*a*). Use the structure to see that one of the strands of DNA (Fig. 13A*b*) runs from 3′ to 5′ in one direction, and the other strand runs from 3′ to 5′ in the opposite direction.

During replication, DNA polymerase has to synthesize the daughter strand in the 5′ to 3′ direction. Why? Because DNA polymerase can only join a nucleotide to the free 3′ end of the previous nucleotide as shown in Figure 13A*b*. This also means that *DNA polymerase cannot start the synthesis of a DNA chain.* Therefore, an RNA polymerase lays down a short amount of RNA, called an RNA primer, that is complementary to the parental strand being replicated. After that, DNA polymerase can join DNA nucleotides to the 3′ end of the growing daughter (new) strand.

As the helicase enzyme unwinds DNA, one parental strand runs in the 3′ to 5′ direction toward the fork. Binding proteins serve to stabilize the newly formed single-strand regions. This works out perfectly for this daughter strand because it must be synthesized from the 5′ to 3′ direction. Therefore, this so-called leading strand is synthesized continuously toward the region of the fork (Fig. 13A*c.*) What about the situation for replication of the other parental strand, which runs in the 3′ to 5′ direction away from the fork? In this case, the daughter strand must begin at the fork. Therefore, replication begins over and over again by new DNA polymerase molecules as the DNA molecule unwinds. Replication of this so-called lagging strand is therefore discontinuous, and it results in segments called Okazaki fragments, after the Japanese scientist Reiji Okazaki, who discovered them.

Replication is only complete when the RNA primers are removed. This works out well for the lagging daughter strand. While proofreading, DNA polymerase I removes the RNA primers and replaces them with complementary DNA nucleotides. Another enzyme, called DNA ligase, joins the fragments. There is no way for DNA polymerase to replicate the end of the lagging strand after the RNA primer is removed. This means that DNA molecules get shorter as one replication follows another. The ends of eukaryotic DNA molecules have a special nucleotide sequence called a *telomere*. Telomeres do not code for proteins and instead are repeats of a short nucleotide sequence such as TTAGGG.

Mammalian cells grown in culture divide about 50 times and then stop. After this number of divisions, the loss of telomeres apparently signals the cell to stop dividing. Ordinarily, telomeres are only added to chromosomes during gamete formation by an enzyme called telomerase. This enzyme unfortunately is often mistakenly turned on in cancer cells, an event that contributes to the ability of cancer cells to keep on dividing without limit.

FIGURE 13A DNA replication (in depth).
a. Structure of deoxyribose showing where nitrogen-containing bases and phosphate groups are attached. b. Template strand and replicated strand. A replicated strand always grows from the 5′ end toward the 3′ end. c. Before replication begins, helicase enzymes separate the DNA parental strands, and binding proteins hold them steady. Replication is a continuous process for the leading daughter strand but a discontinuous process for the lagging daughter strand. In both instances, primase adds a short RNA primer, and DNA polymerase adds nucleotides to the 3′ end of the daughter strand. DNA polymerase replaces the primer with DNA nucleotides. Because replication is discontinuous for the lagging strand, DNA ligase is needed to join the so-called Okazaki fragments to form a completed strand.

Prokaryotic Versus Eukaryotic Replication

There are several differences between prokaryotes and eukaryotes. The process of DNA replication is distinctly different in these cells (Fig. 13.9).

Prokaryotic Replication

Bacteria have a single circular loop of DNA that must be replicated before the cell divides. In some circular DNA molecules, replication moves around the DNA molecule in one direction only. In others, as shown in Figure 13.9*a*, replication occurs in two directions. The process always occurs in the 5′ to 3′ direction.

Bacterial cells are able to replicate their DNA at a rate of about 10^6 base pairs per minute, and about 40 minutes are required to replicate the complete chromosome. Because bacterial cells are able to divide as often as once every 20 minutes, it is possible for a new round of DNA replication to begin even before the previous round is completed!

Eukaryotic Replication

In eukaryotes, DNA replication begins at numerous origins of replication along the length of the chromosome, and the so-called replication bubbles spread bidirectionally until they meet. Notice in Figure 13.9*b* that there is a V shape wherever DNA is being replicated. This is called a **replication fork.**

Although eukaryotes replicate their DNA at a slower rate—500–5,000 base pairs per minute—there are many individual origins of replication. Therefore, eukaryotic cells complete the replication of the diploid amount of DNA (in humans, over 6 billion base pairs) in a matter of hours!

Replication Errors

In molecular genetics, a gene is a section of DNA with a particular sequence of bases. A **genetic mutation** is a permanent change in this sequence of bases.

Some mutations are due to errors in DNA replication. During the replication process, DNA polymerase chooses complementary nucleotide triphosphates from the cellular pool. Then the nucleotide triphosphate is converted to a nucleotide monophosphate and aligned with the template nucleotide. A mismatched nucleotide slips through this selection process only once per 100,000 base pairs at most. The mismatched nucleotide causes a pause in replication, during which it is excised from the daughter strand and replaced with the correct nucleotide. After this so-called **proofreading** has occurred, the error rate is only one mistake per 1 billion base pairs.

Some mutations are a result of DNA damage. Many of these are caused by a variety of environmental factors. Free radicals present in cells, the UV radiation in sunlight, organic chemicals in tobacco smoke, pesticides, and pollutants can all damage DNA. **DNA repair enzymes** are usually available and able to reverse most of the damage to DNA, but some may escape notice.

The effect of a genetic mutation need not necessarily harm an organism. In fact, some genetic mutations are beneficial to the organism and perhaps the species. Other mu-

tations do a great deal of harm, as when they cause cancer in an individual or a genetic disease in offspring. Still, it is important to keep in mind that regardless of their effects, genetic variations are the raw material for the evolutionary process. Without changes in the genetic material, evolution would not be possible.

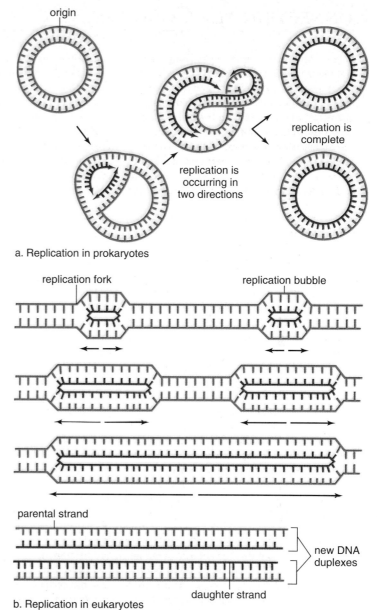

a. Replication in prokaryotes

b. Replication in eukaryotes

FIGURE 13.9 Prokaryotic versus eukaryotic replication.
a. In prokaryotes, replication can occur in two directions at once because the DNA molecule is circular. **b.** In eukaryotes, replication occurs at numerous replication forks. The bubbles thereby created spread out until they meet.

Mutations due to replication errors and environmental factors are rare due to the proofreading function of DNA polymerase and the presence of DNA repair enzymes in cells.

CONNECTING THE CONCEPTS

Many of the Nobel Prizes in physiology, medicine, and chemistry between 1950 and 1990 were awarded to scientists studying the molecular basis of inheritance. Indeed, the biochemistry of genes and the basic structure of chromosomes did not become known until the middle of the twentieth century. The early studies of Griffith, Avery, and their colleagues demonstrated that DNA contains genetic information; Watson and Crick later presented a model of its structure; and finally, many researchers contributed to our knowledge of how DNA is copied.

Even to this date, details of DNA structure and function are still being clarified. Research in such fields as biochemistry, biophysics, bacteriology, and molecular biology are contributing to our understanding of how genetic information is stored and usually faithfully copied for the next generation.

This chapter reviews how DNA fulfills the requirements for the genetic material listed at the start of the chapter. The genetic material must be (1) able to store information, (2) stable and capable of replication, and (3) able occasionally to undergo change. In Chapter 14, we will investigate how the sequence of bases in DNA specifies the blueprint for building a cell and an organism through the processes of transcription and translation.

Summary

13.1 THE GENETIC MATERIAL

Early work on the biochemistry of DNA wrongly suggested that DNA lacks the variability necessary for the genetic material. However, diligent research illustrated that DNA was the hereditary material

Griffith injected strains of pneumococcus into mice and observed that smooth (S) strain bacteria are virulent but rough (R) strain bacteria are not. However, when heat-killed S strain bacteria were injected along with live R strain bacteria, virulent S strain bacteria were recovered from the dead mice. Griffith said that the R strain had been transformed by some substance passing from the dead S strain to the live R strain. Twenty years later, Avery and his colleagues reported that the transforming substance is DNA. Hershey and Chase turned to bacteriophage T2 as their experimental material. In two separate experiments, they labeled the capsid with ^{35}S and the DNA with ^{32}P. They then showed that the radioactive P alone is largely taken up by the bacterial host and that reproduction of viruses proceeds normally. This convinced most researchers that DNA is the genetic material.

13.2 THE STRUCTURE OF DNA

Chargaff performed a chemical analysis of DNA and found that $A = T$ and $G = C$, and that the amount of purine equals the amount of pyrimidine. Franklin prepared an X-ray photograph of DNA that showed it is helical, has repeating structural features, and has certain dimensions. Watson and Crick built a model of DNA in which the sugar-phosphate molecules made up the sides of a twisted ladder, and the complementary-paired bases were the rungs of the ladder.

13.3 REPLICATION OF DNA

The Watson and Crick model immediately suggested a method by which DNA could be replicated. The two strands unwind and unzip, and each parental strand acts as a template for a new (daughter) strand. In the end, each new helix is like the other and like the parental helix.

Meselson and Stahl demonstrated that replication is semiconservative by the following experiment: Bacteria were grown in heavy nitrogen (^{15}N) and then switched to light nitrogen (^{14}N). After one division in light nitrogen, DNA molecules are hybrid and have intermediate density. After two divisions, DNA molecules separate into two bands—one for light DNA and one for hybrid DNA.

The enzyme DNA polymerase joins the nucleotides together and proofreads them to make sure the bases have been paired correctly. Incorrect base pairs that survive the process are a mutation. Replication in prokaryotes proceeds from one point of origin until there are two copies of the circular chromosome. Replication in eukaryotes has many points of origin and many bubbles (places where the DNA strands are separating and replication is occurring). Replication occurs at the ends of the bubbles—at replication forks.

Reviewing the Chapter

1. List and discuss the requirements for genetic material.   224
2. Describe Griffith's experiments with pneumococcus, his surprising results, and his conclusion.   224–25
3. How did Avery and his colleagues demonstrate that the transforming substance is DNA?   225
4. Describe the experiment of Hershey and Chase, and explain how it shows that DNA is the genetic material.   226
5. What are Chargaff's rules?   227
6. Describe the Watson and Crick model of DNA structure. How did it fit the data provided by Chargaff and the X-ray diffraction pattern?   228–29
7. Explain how DNA replicates semiconservatively. What role does DNA polymerase play? What role does helicase play?   230–31
8. How did Meselson and Stahl demonstrate semiconservative replication?   231
9. List and discuss differences between prokaryotic and eukaryotic replication of DNA.   233
10. Explain why the replication process is a source of few mutations.   233

Testing Yourself

Choose the best answer for each question. For questions 1–4, match the statements to the names in the key.

KEY:

   a. Griffith
   b. Chargaff
   c. Meselson and Stahl
   d. Hershey and Chase
   e. Avery, MacLeod, McCarty

1. $A = T$ and $G = C$.

2. Only the DNA from T2 enters the bacteria.

3. R strain bacteria became an S strain through transformation.

4. DNA replication is semiconservative.

5. If 30% of an organism's DNA is thymine, then
   a. 70% is purine.
   b. 20% is guanine.
   c. 30% is adenine.
   d. 70% is pyrimidine.
   e. Both c and d are correct.

6. If you grew bacteria in heavy nitrogen and then switched them to light nitrogen, how many generations after switching would you have some light/light DNA?
   a. never, because replication is semiconservative
   b. the first generation
   c. the second generation
   d. only the third generation
   e. Both b and c are correct.

7. The double-helix model of DNA resembles a twisted ladder in which the rungs of the ladder are
   a. a purine paired with a pyrimidine.
   b. A paired with G and C paired with T.
   c. sugar-phosphate paired with sugar-phosphate.
   d. a 5′ end paired with a 3′ end.
   e. Both a and b are correct.

8. Cell division requires that the genetic material be able to
   a. store information.
   b. undergo replication.
   c. undergo rare mutations.
   d. condense into spindle fibers.
   e. All of these are correct.

9. In a DNA molecule,
   a. the bases are covalently bonded to the sugars.
   b. the sugars are covalently bonded to the phosphates.
   c. the bases are hydrogen-bonded to one another.
   d. the nucleotides are covalently bonded to one another.
   e. All of these are correct.

10. In the following diagram, blue stands for heavy DNA (contains ^{15}N) and green stands for light DNA (does not contain ^{15}N). Label each strand of all three DNA molecules as heavy or light DNA, and explain why the diagram is in keeping with the semiconservative replication of DNA.

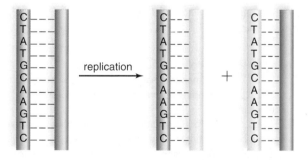

11. The enzyme responsible for adding new nucleotides to a growing DNA chain during DNA replication is
    a. helicase.
    b. RNA polymerase.
    c. DNA polymerase.
    d. ribozymes.

12. If the sequence of bases in one strand of DNA is TAGCCT, then the sequence of bases in the other strand will be
    a. TCCGAT.
    b. ATCGGA.
    c. TAGCCT.
    d. AACGGUA.
    e. Both a and b are correct.

13. Which one of these is not a base?
    a. purine
    b. pyrimidine
    c. adenine
    d. guanine
    e. All of these are bases.

14. Transformation occurs when
    a. R strain bacteria become S strain bacteria.
    b. R strain bacteria have a change of genotype.
    c. bacteria can now grow on penicillin.
    d. organisms receive foreign DNA and thereby acquire a new characteristic.
    e. All of these are correct.

15. In the Hershey and Chase experiment, which one did not occur?
    a. Some viruses were labeled with both radioactive phosphorus and radioactive sulfur.
    b. A blender dislodged viral capsids.
    c. Radioactivity was either in the bacteria or in the liquid medium.
    d. Viruses infected the bacteria.
    e. None of these occurred.

16. Pyrimidines
    a. are always paired with a purine.
    b. are thymine and cytosine.
    c. keep DNA from replicating too often.
    d. are adenine and guanine.
    e. Both a and b are correct.

17. Watson and Crick did not make use of
    a. diffraction data.
    b. Chargaff's rules.
    c. complementary base pairing.
    d. a knowledge of viral structure.
    e. All of these are correct.

18. A nucleotide
    a. is smaller than a base.
    b. is a subunit of nucleic acids.
    c. has a lot of variable parts.
    d. has at least four phosphates.
    e. always joins with other nucleotides.

19. During replication, unwinding requires
    a. backbones to split.
    b. nucleotides to join together.
    c. hydrolysis and synthesis to occur.
    d. hydrogen bonds to unzip.
    e. All of these are correct.

20. Replication is semiconservative because
    a. both day-old and fresh new nucleotides are used.
    b. an old strand becomes a new strand.
    c. Meselson and Stahl did this experiment.
    d. the old strand is a template for the new strand.
    e. All of these are correct.

21. Which of these is mismatched?
    a. DNA replicates—necessary to reproduction of cell and organism
    b. DNA mutates—necessary to the diversity of living things
    c. DNA is constant—necessary to "like begets like"
    d. DNA mutates—is not necessary because it causes abnormalities
    e. All of these are correct.

22. Chargaff arrived at his rules after
    a. doing transformation experiments.
    b. studying viruses.
    c. studying replication.
    d. chemically analyzing DNA.
    e. All of these are correct.

23. In prokaryotes,
    a. replication can occur in two directions at once because their DNA molecule is circular.
    b. bubbles thereby created spread out until they meet.
    c. replication occurs at numerous replication forks.
    d. a new round of DNA replication cannot begin before the previous round is complete.
    e. Both a and b are correct.

24. When DNA replicates,
    a. a single helix becomes a double helix.
    b. a single molecule becomes two molecules.
    c. one nucleus immediately becomes two nuclei.
    d. the nuclear envelope immediately enlarges.
    e. All of these are correct.

25. Complementary base pairing
    a. involves T, A, G, C.
    b. is necessary to replication.
    c. uses hydrogen bonds.
    d. occurs when ribose binds with deoxyribose.
    e. All but d are correct.

26. In DNA, phosphate
    a. is part of the backbone.
    b. joins base to base.
    c. joins with other phosphates.
    d. gives off hydrogen ions because it is an acid.

Thinking Scientifically

1. Replication of DNA in some viruses requires the use of proteins that have no enzymatic activity. Hypothesize how a protein might possibly act as a primer for DNA replication. What would be the advantage of using a protein rather than RNA to get replication started? (See page 232.)
2. Skin cancer is much more common than brain cancer. Why might the frequency of cancer be related to rate of cell division in skin as opposed to rate of cell division in the brain? How might DNA sequencing help test your hypothesis?

Bioethical Issue: Human Cloning

The term *cloning* means making exact multiple copies of genes, a cell, or an organism. Therefore, cloning has been around for some time. Identical twins are clones of a single zygote. When a single bacterium reproduces asexually on a petri dish, a colony of cells results, and each member of the colony is a clone of the original cell. Through biotechnology, bacteria now produce cloned copies of human genes.

Now, for the first time in our history, it is possible to produce a clone of a mammal—and perhaps even one day, a human. The parent need not contribute sperm or an egg to the process. During so-called reproductive cloning, a nucleus (which contains a person's genes) from one adult cell is placed in an enucleated egg, and that egg begins developing. The developing embryo is placed in the uterus of a surrogate mother, and when birth occurs, the clone is an exact copy but, of course, younger than the original parent. The process of cloning whole animals, and certainly humans, has not been perfected. Until it is, therapeutic cloning is more likely to become widespread.

Suppose it were possible to put the nucleus from a cell of a burn victim into an enucleated egg that is cajoled to become skin cells in the laboratory. These cells could be used to provide grafts of new skin that would not be rejected by the recipient. Would this be a proper use of cloning in humans?

Or suppose parents want to produce a child free of genetic disease. Scientists produce an embryo through in vitro fertilization, and then they clone the embryo to produce several embryos. Genetic engineering to correct the defect doesn't work on all the embryos—only a few. They implant just those few in a uterus, where development continues to term. Would this be a proper use of cloning in humans?

What if later scientists were able to produce children with increased intelligence or athletic prowess using this same technique? Would this be an acceptable use of cloning in humans?

Understanding the Terms

adenine (A) 227	nucleic acid 224
bacteriophage 225	nucleotide 224
complementary base	proofreading 233
pairing 228	purine 227
cytosine (C) 227	pyrimidine 227
DNA (deoxyribonucleic	replication fork 233
acid) 224	RNA (ribonucleic
DNA polymerase 230	acid) 224
DNA repair enzyme 233	semiconservative
DNA replication 230	replication 230
double helix 228	template 230
genetic mutation 233	thymine (T) 227
guanine (G) 227	

Match the terms to these definitions:
a. _____ Permanent change in DNA base sequence.
b. _____ Bonding between particular purines and pyrimidines in DNA.
c. _____ During replication, an enzyme that joins the nucleotides complementary to a DNA template.
d. _____ Type of nitrogen-containing base, such as adenine and guanine, having a double-ring structure.
e. _____ Virus that infects bacteria.

ARIS, the *Biology* Website

ARIS, the website for *Biology*, provides a wealth of information organized and integrated by chapter. You will find practice quizzes, interactive activities, labeling exercises, flashcards, and much more that will complement your learning and understanding of general biology.

www.mhhe.com/maderbiology9

14

GENE ACTIVITY: HOW GENES WORK

N*early 1.5 million different species of organisms have been discovered and named. This number represents a small portion of the total number of species on Earth. It certainly represents a small fraction of the total number of species that have ever lived. Yet one gene differs from another only by the sequence of the nucleotide bases in DNA. How does a difference in base sequence determine the uniqueness of a species—for example, whether an individual is a daffodil or a gorilla? Or, for that matter, whether a human has blue, brown, or hazel eyes?*

By studying the activity of genes in cells, geneticists have confirmed that proteins are the link between the genotype and the phenotype. Mendel's peas are smooth or wrinkled according to the presence or absence of a starch-forming enzyme. The allele S in peas dictates the presence of the starch-forming enzyme, whereas the allele s does not.

Through its ability to specify proteins, DNA brings about the development of the unique structures that make up a particular type of organism. It follows that humans may have blue or brown or hazel eyes because of the type of enzymes contained within their cells. When studying gene expression in this chapter, keep in mind this flow diagram: DNA's sequence of nucleotides → sequences of amino acids → specific enzymes → structures in organism. In the next chapter we will consider the control of gene expression, which most likely also plays a powerful role in explaining the diversity of organisms.

The diversity of life is dependent on gene activity.

14.1 THE FUNCTION OF GENES

In the early 1900s, the English physician Sir Archibald Garrod suggested that there is a relationship between inheritance and metabolic diseases. He introduced the phrase *inborn error of metabolism* to dramatize this relationship. Garrod observed that family members often had the same disorder, and he said this inherited defect could be caused by the lack of a particular enzyme in a metabolic pathway. Since it was already known that enzymes are proteins, Garrod was among the first to hypothesize a link between genes and proteins. This hypothesis was later confirmed by elegant experiments carried out by several researchers.

Genes Specify Enzymes

Many years later, in 1940, George Beadle and Edward Tatum performed a series of experiments on *Neurospora crassa,* the red bread mold fungus, which reproduces by means of spores. Normally, the spores become a mold capable of growing on minimal medium (containing only a sugar, mineral salts, and the vitamin biotin) because mold can produce all the enzymes it needs. In their experiments, Beadle and Tatum induced mutations in asexually produced haploid spores by the use of X-rays. Some of the X-rayed spores could no longer become a mold capable of growing on minimal medium; however, growth was possible on me-dium enriched by certain metabolites. In the example given in Figure 14.1, a mold grows only on enriched medium that includes all metabolites or on minimal medium enriched with C and D alone. C and D are part of this hypothetical pathway in which the numbers are enzymes and the letters are metabolites.

$$A \xrightarrow{\ 1\ } B \xrightarrow{\ 2\ } C \xrightarrow{\ 3\ } D$$

Thus, it is concluded that the mold lacks enzyme 2. Beadle and Tatum further found that each of the mutant strains has only one defective gene, leading to one defective enzyme and one additional growth requirement. Therefore, they proposed that each gene specifies the synthesis of one enzyme. This is called the *one gene–one enzyme hypothesis.*

Genes Specify a Polypeptide

The one gene–one enzyme hypothesis suggests that a genetic mutation causes a change in the structure of a protein. To test this idea, in 1949 Linus Pauling and Harvey Itano decided to see if the hemoglobin in the red blood cells of persons with sickle cell disease has a structure different from that of the red blood cells of normal individuals (Fig. 14.2a). Recall that proteins are polymers of amino acids, some of which carry a charge. These investigators decided to see if there was a charge difference between normal hemoglobin (*Hb^A*)

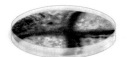

a. Normally, red bread mold grows on minimal medium.

b. Spores (single-celled reproductive bodies) are X-rayed. Mutations are introduced, and spores will produce growth on enriched medium but not minimal medium.

c. Further investigation shows, for example, that a particular mold needs only metabolite C or D to grow on minimal medium.

d. This pathway explains the experimental results. The mold lacks enzyme 2 and cannot change B to C. It can change C to D.

FIGURE 14.1 Beadle and Tatum experiment.

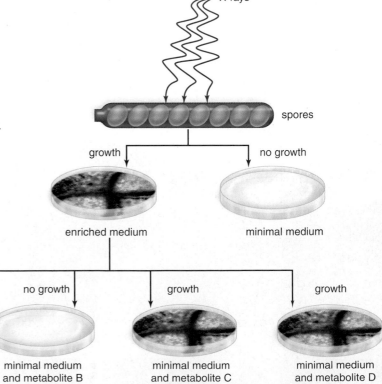

FIGURE 14.1 Beadle and Tatum experiment.

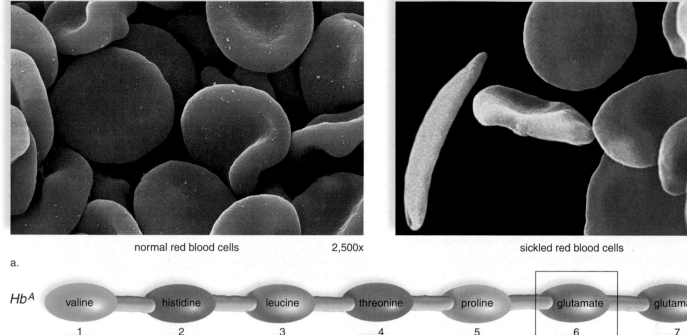

normal red blood cells   2,500x   sickled red blood cells   2,500x

a.

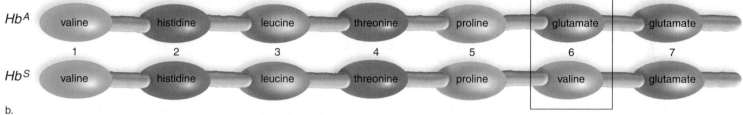

Hb^A | valine | histidine | leucine | threonine | proline | glutamate | glutamate

1   2   3   4   5   6   7

Hb^S | valine | histidine | leucine | threonine | proline | valine | glutamate

b.

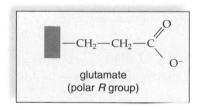

glutamate
(polar *R* group)

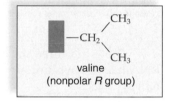

valine
(nonpolar *R* group)

c.

FIGURE 14.2 Sickle cell disease in humans.
a. Scanning electron micrograph of normal *(left)* and sickled *(right)* red blood cells.
b. Portion of the chain in normal hemoglobin *(Hb^A)* and in sickle cell hemoglobin
(Hb^S). Although the chain is 146 amino acids long, the one change from glutamate
to valine in the sixth position results in sickle cell disease. **c.** Glutamate has a polar
R group, while valine has a nonpolar *R* group, and this causes *Hb^S* to be less soluble
and to precipitate out of solution, distorting the red blood cell into the sickle shape.

and sickle cell hemoglobin *(Hb^S)*. To determine this, they subjected hemoglobin collected from normal individuals, sickle cell trait individuals, and sickle cell disease individuals to electrophoresis, a procedure that separates molecules according to their size and charge, whether (+) or (−). Their conclusions were based on the movement of molecules similar to the following illustration:

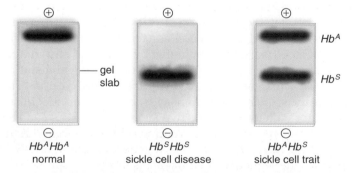

$Hb^A Hb^A$
normal

$Hb^S Hb^S$
sickle cell disease

$Hb^A Hb^S$
sickle cell trait

It is apparent that there is a difference in migration rate toward the positive pole between normal hemoglobin and sickle cell hemoglobin. Further, hemoglobin from individuals with sickle cell trait separates into two distinct bands, one corresponding to that for *Hb^A* hemoglobin and the other

corresponding to that for *Hb^S* hemoglobin. Pauling and Itano therefore demonstrated that a mutation leads to a change in the structure of a protein.

Several years later, Vernon Ingram was able to determine the structural difference between *Hb^A* and *Hb^S*. Normal hemoglobin *(Hb^A)* contains negatively charged glutamate; in sickle cell hemoglobin, the glutamate is replaced by nonpolar valine (Fig. 14.2*b, c*). This causes *Hb^S* to be less soluble and to precipitate out of solution, especially when environmental oxygen is low. At these times, the *Hb^S* molecules stack up into long, semirigid rods that push against the plasma membrane and distort the red blood cell into the sickle shape.

Hemoglobin contains two types of polypeptide chains, designated α (alpha) and β (beta). Only the β chain is affected in persons with sickle cell trait and sickle cell disease; therefore, there must be a gene for each type of chain. A refinement of the one gene–one enzyme hypothesis was needed, so it was replaced by the *one gene–one polypeptide hypothesis*.

Early investigators concluded that a gene is a segment of DNA that specifies the sequence of amino acids in a polypeptide of a protein.

From DNA to RNA to Protein

As with sickle cell disease, geneticists have confirmed many times over that proteins are the link between genotype and phenotype. In Huntington disease, the protein huntingtin is altered and unable to carry out its usual functions. People with cystic fibrosis have a malfunctioning Cl⁻ channel protein in the plasma membrane. Granted there is a connection between genes and proteins, what exactly is it that genes do? A **gene** is a segment of DNA that specifies the amino acid sequence of a protein. Living things, from a prokaryotic *Staphylococcus* bacterium to an eukaryotic elephant, undergo the same basic processes of specifying polypeptides and expressing genes. A gene does not directly control protein synthesis; instead, it passes its genetic information onto **RNA (ribonucleic acid)** molecules, which are more directly involved in protein synthesis.

Types of RNA

Like DNA, RNA is a polymer composed of nucleotides. The nucleotides in RNA, however, contain the sugar ribose and the bases adenine (A), cytosine (C), guanine (G), and uracil (U). In RNA, the base **uracil** replaces the thymine found in DNA. Finally, RNA is single stranded and does not form a double helix in the same manner as DNA (Table 14.1 and Fig. 14.3).

TABLE 14.1

RNA Structure Compared to DNA Structure

	RNA	DNA
Sugar	Ribose	Deoxyribose
Bases	Adenine, guanine, uracil, cytosine	Adenine, guanine, thymine, cytosine
Strands	Single stranded	Double stranded with base pairing
Helix	No	Yes

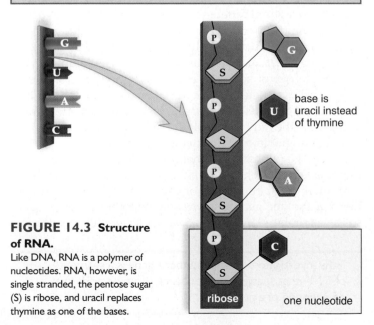

FIGURE 14.3 Structure of RNA.
Like DNA, RNA is a polymer of nucleotides. RNA, however, is single stranded, the pentose sugar (S) is ribose, and uracil replaces thymine as one of the bases.

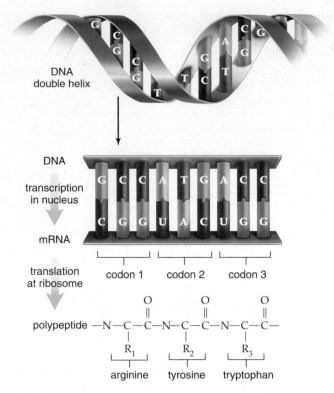

FIGURE 14.4 Overview of gene expression.
One strand of DNA acts as a template for mRNA synthesis, and the sequence of bases in mRNA determines the sequence of amino acids in a polypeptide.

There are three major classes of RNA. Each class of RNA has its own unique size, shape, and function in protein synthesis.

Messenger RNA (mRNA) takes a message from DNA in the nucleus to the ribosomes in the cytoplasm.

Transfer RNA (tRNA) transfers amino acids to the ribosomes.

Ribosomal RNA (rRNA), along with ribosomal proteins, makes up the ribosomes, where polypeptides are synthesized.

Gene Expression

Gene expression (production of a protein) requires two steps (Fig. 14.4). First, during **transcription** [L. *trans,* across, and *scriptio,* a writing], DNA serves as a template for RNA formation. DNA is transcribed monomer by monomer into another type of polynucleotide (RNA). Second, during **translation** [L. *trans,* across, and *latus,* carry bear] the mRNA transcript directs the sequence of amino acids in a polypeptide. Like a translator who understands two languages, the cell changes a nucleotide sequence into an amino acid sequence. With the help of the three types of RNA, a gene (a segment of DNA) specifies the sequence of amino acids in a polypeptide.

During transcription, DNA serves as a template for the formation of RNA. During translation, mRNA is involved in polypeptide synthesis.

14.2 THE GENETIC CODE

Consider that the central dogma of molecular biology tells us that the sequence of nucleotides in DNA specifies the order of amino acids in a polypeptide. It would seem then that there must be a **genetic code** for each of the 20 amino acids found in proteins. But can four nucleotides provide enough combinations to code for 20 amino acids? If each code word, called a **codon,** were made up of two bases, such as AG, there could be only 16 codons. But if each codon were made up of three bases, such as AGC, there would be 64 codons—more than enough to code for 20 amino acids:

number of bases	1	4	number of different
in genetic code	2	16	amino acids specified
	3	64	

The genetic code is a **triplet code.** Each codon consists of three nucleotide bases, such as AUC.

Finding the Genetic Code

In 1961, Marshall Nirenberg and J. Heinrich Matthei performed an experiment that laid the groundwork for cracking the genetic code. First, they found that a cellular enzyme could be used to construct a synthetic RNA (one that does not occur in cells), and then they found that the synthetic RNA polymer could be translated in a test tube that contains the cytoplasmic contents of a cell. Their first synthetic RNA was composed only of uracil, and the protein that resulted was composed only of the amino acid phenylalanine. Therefore, the mRNA codon for phenylalanine was known to be UUU. Later, they were able to translate just three nucleotides at a time; in that way, it was possible to assign an amino acid to each of the mRNA codons (Fig. 14.5).

A number of important properties of the genetic code can be seen by careful inspection of Figure 14.5.

1. The genetic code is degenerate. This means that most amino acids have more than one codon; leucine, serine, and arginine have six different codons, for example. The degeneracy of the code protects against potentially harmful effects of mutations.
2. The genetic code is unambiguous. Each triplet codon has only one meaning.
3. The code has start and stop signals. There is only one start signal, but there are three stop signals.

Figure 14.4 illustrates that it is possible to use the principle of complementary base pairing to determine the original base sequence in DNA from a sequence of codons. Therefore, if the sequence of codons is CGG'UAC'UGG', the sequence of bases in DNA is GCC'ATG'ACC'.

The Code Is Universal

With a few exceptions, the genetic code (Fig. 14.5) is universal to all living things. In 1979, however, researchers

First Base	Second Base				Third Base
	U	**C**	**A**	**G**	
U	UUU phenylalanine	UCU serine	UAU tryosine	UGU cysteine	U
	UUC phenylalanine	UCC serine	UAC tryosine	UGC cysteine	C
	UUA leucine	UCA serine	UAA *stop*	UGA *stop*	A
	UUG leucine	UCG serine	UAG *stop*	UGG tryptophan	G
C	CUU leucine	CCU proline	CAU histidine	CGU arginine	U
	CUC leucine	CCC proline	CAC histidine	CGC arginine	C
	CUA leucine	CCA proline	CAA glutamine	CGA arginine	A
	CUG leucine	CCG proline	CAA glutamine	CGG arginine	G
A	AUU isoleucine	ACU threonine	AAU asparagine	AGU serine	U
	AUC isoleucine	ACC threonine	AAC asparagine	AGC serine	C
	AUA isoleucine	ACA threonine	AAA lysine	AGA arginine	A
	AUG (start) methionine	ACG threonine	AAG lysine	AGG arginine	G
G	GUU valine	GCU alanine	GAU aspartate	GGU glycine	U
	GUC valine	GCC alanine	GAC aspartate	GGC glycine	C
	GUA valine	GCA alanine	GAA glutamate	GGA glycine	A
	GUG valine	GCG alanine	GAG glutamate	GGG glycine	G

FIGURE 14.5 Messenger RNA codons.
Notice that in this chart, each of the codons (in boxes) is composed of three letters representing the first base, second base, and third base. For example, find the box where C for the first base and A for the second base intersect. You will see that U, C, A, or G can be the third base. The bases CAU and CAC are codons for histidine; the bases CAA and CAG are codons for glutamine.

discovered that the genetic code used by mammalian mitochondria and chloroplasts differs slightly from the more familiar genetic code.

The universal nature of the genetic code provides strong evidence that all living things share a common evolutionary heritage. Since the same genetic code is used by all living things, it is possible to transfer genes from one organism to another. Many commercial and medicinal products such as insulin can be produced in this manner. Genetic engineering has also produced some unusual organisms such as mice that literally glow in the dark (see page 267). In this case, the gene responsible for bioluminescence in jellyfish was placed into mouse embryos.

The genetic code is a triplet code. Each codon consists of three DNA nucleotides. Except for the stop codons, all the codons code for amino acids.

14.3 FIRST STEP: TRANSCRIPTION

During *transcription*, a segment of the DNA serves as a template for the production of an RNA molecule. Although all three classes of RNA are formed by transcription, we will focus on transcription to form mRNA, the type of RNA that eventually leads to building a polypeptide as a gene product.

Messenger RNA Is Formed

An mRNA molecule has a sequence of bases complementary to a portion of one DNA strand; wherever A, T, G, or C is present in the DNA template, U, A, C, or G, respectively, is incorporated into the mRNA molecule (Fig. 14.6). A segment of the DNA helix unwinds and unzips, and complementary RNA nucleotides pair with DNA nucleotides of the strand that is being transcribed. This strand is known as the template strand or sense strand. An RNA polymerase joins the nucleotides

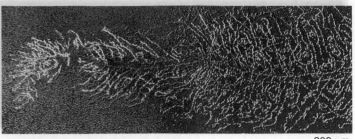

a.   200 μm

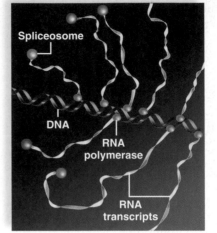

b.

FIGURE 14.7
RNA polymerase.
a. Numerous RNA transcripts extend from a horizontal gene in an amphibian egg cell. **b.** The strands get progressively longer because transcription begins to the left. The dots along the DNA are RNA polymerase molecules. The dots at the end of the strands are spliceosomes involved in RNA processing (see Fig. 14.8).

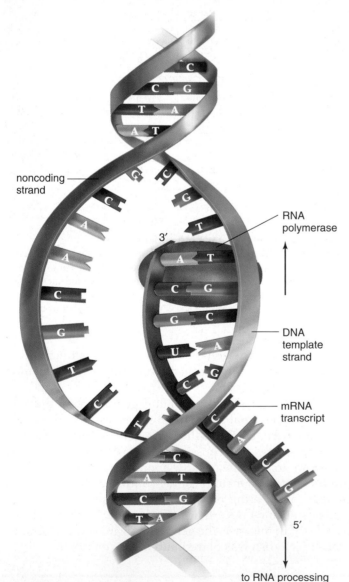

FIGURE 14.6 Transcription.
During transcription, complementary RNA is made from a DNA template. At the point of attachment of RNA polymerase, the DNA helix unwinds and unzips, and complementary RNA nucleotides are joined together. After RNA polymerase has passed by, the DNA strands rejoin and the mRNA transcript dangles to the side.

together in the 5' ⟶ 3' direction. In other words, an **RNA polymerase** only adds a nucleotide to the 3' end of the polymer under construction. The strand of DNA that is not transcribed is called the noncoding strand or antisense strand.

Transcription begins when RNA polymerase attaches to a region of DNA called a promoter. A **promoter** defines the start of a gene, the direction of transcription, and the strand to be transcribed. The RNA-DNA association is not as stable as the DNA helix. Therefore, only the newest portion of an RNA molecule that is associated with RNA polymerase is bound to the DNA, and the rest dangles off to the side. Elongation of the mRNA molecule continues until RNA polymerase comes to a DNA stop sequence. The stop sequence causes RNA polymerase to stop transcribing the DNA and to release the mRNA molecule, now called an **mRNA transcript.**

Many RNA polymerase molecules can be working to produce mRNA transcripts at the same time (Fig. 14.7). This allows the cell to produce many thousands of copies of the same mRNA molecule, and eventually many copies of the same protein, within a shorter period of time than if the single copy of DNA were used to direct protein synthesis.

As a result of transcription, many mRNA molecules direct protein synthesis. Many more copies of a protein are produced within a given time versus using DNA directly.

RNA Molecules Are Processed

Newly formed RNA molecules, called primary mRNA, are modified before leaving the eukaryotic nucleus. For example, messenger RNA molecules receive a cap at the 5'

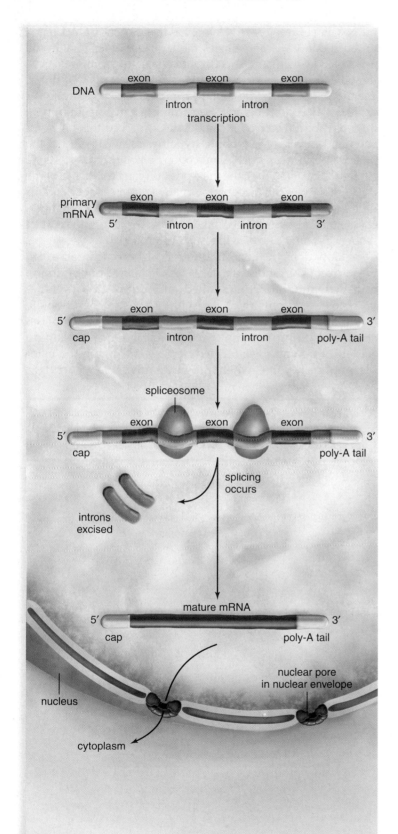

FIGURE 14.8 Messenger RNA (mRNA) processing in eukaryotes.
DNA contains both exons (coding sequences) and introns (noncoding sequences). Both of these are transcribed and are present in primary mRNA. During processing, a cap and a poly-A tail (a series of adenine nucleotides) are added to the molecule. Also, there is excision of the introns and a splicing together of the exons. This is accomplished by complexes called spliceosomes. Then the mature mRNA molecule is ready to leave the nucleus.

end and a tail at the 3′ end (Fig. 14.8). The *cap* is a modified guanine (G) nucleotide that helps tell a ribosome where to attach when translation begins. The tail consists of a chain of 150–200 adenine (A) nucleotides. This so-called *poly-A tail* facilitates the transport of mRNA out of the nucleus and also inhibits degradation of mRNA by hydrolytic enzymes.

Immediately following transcription, primary mRNA, particularly in multicellular eukaryotes, is composed of exons and introns. The exons of mRNA will be expressed, but not the **introns,** which occur in between the **exons.** During RNA processing, the introns are removed by a process called RNA splicing (see Fig. 14.7). In prokaryotes, introns are removed by "self-splicing"—that is, the intron itself has the capability of enzymatically splicing itself out of a primary mRNA. In eukaryotes, the RNA splicing is done by spliceosomes, complexes that contain several kinds of ribonucleoproteins. A spliceosome cuts the primary mRNA and then rejoins the adjacent exons. An mRNA that has been processed and is ready to be translated is called mature mRNA. (RNAs with an enzymatic function are called **ribozymes,** and the presence of ribozymes in both prokaryotes and eukaryotes suggests that RNA could have preceded DNA in the evolutionary history of cells.)

Function of Introns

Introns are far more common in eukaryotes than prokaryotes, perhaps because prokaryotes do not have spliceosomes, and translation precedes directly after transcription. In humans, 95% or more of the average protein-coding gene is introns. In general, as complexity increases so does the proportion of non-protein-coding DNA sequences. This phenomenon has piqued the interest of investigators in what used to be called "junk DNA" but is now thought to probably serve important functions.

The sequencing of the human genome has shown that humans may have only about 25,000 coding genes, far less than the 100,000 formerly projected. How do humans make do with so few genes? It's possible that the presence of introns allows exons to be put together in various sequences so that different mRNAs and proteins can result from a single gene. It's also possible that introns function to regulate gene expression and feedback to help determine which coding genes are to be expressed and how they should be spliced. A heretofore-overlooked RNA-signaling network may be what allows humans, for example, to achieve structural complexity far beyond anything seen in the unicellular world. An entirely different picture of the genome is being developed than that envisioned by the investigators who gave us the central dogma of DNA synthesis. In the next chapter we will see that it has been known for some time that proteins have regulatory functions. But now it seems that the small-sized RNAs are also able to carry out regulatory functions and with more ease at times. This new hypothesis can explain why the structural and developmental complexity of organisms does not parallel the number of protein-coding genes.

Particularly in eukaryotes, the primary mRNA is processed before it becomes a mature mRNA.

14.4 Second Step: Translation

Translation, which takes place in the cytoplasm of eukaryotic cells, is the second step by which gene expression leads to protein synthesis. During translation, the sequence of codons in the mRNA at a ribosome directs the sequence of amino acids in a polypeptide. In other words, one language (nucleic acids) gets translated into another language (protein).

The Role of Transfer RNA

Transfer RNA (tRNA) molecules transfer amino acids to the ribosomes. A tRNA molecule is a single-stranded nucleic acid that doubles back on itself to create regions where complementary bases are hydrogen-bonded to one another. The structure of a tRNA molecule is generally drawn as a flat cloverleaf, but a space-filling model shows the molecule's three-dimensional shape (Fig. 14.9).

There is at least one tRNA molecule for each of the 20 amino acids found in proteins. The amino acid binds to the 3′ end. The opposite end of the molecule contains an **anticodon,** a group of three bases that is complementary to a specific codon of mRNA. For example, a tRNA that has the anticodon GAA binds to the codon CUU and carries the amino acid leucine. In the genetic code, there are 61 codons that encode for amino acids; the other 3 serve as stop sequences. Approximately 40 different tRNA molecules are found in most cells. There are fewer tRNAs than codons because some tRNAs can pair with more than one codon. In 1966, Francis Crick observed this phenomenon and called it the **wobble hypothesis.** He stated that the first two positions in a tRNA anticodon pair obey the A–U/G–C configuration. However, the third position can be variable. Some tRNA molecules can recognize as many as four separate codons differing only in the third nucleotide. The wobble effect helps ensure that despite changes in DNA base sequences, the correct sequence of amino acids will result in a protein.

How does the correct amino acid become attached to the correct tRNA molecule? This task is carried out by amino acid–activating enzymes, called aminoacyl-tRNA synthetases. Just as a key fits a lock, each enzyme has a recognition site for the amino acid to be joined to a particular tRNA. This is an energy-requiring process that uses ATP. Once the amino acid–tRNA complex is formed, it travels through the cytoplasm to a ribosome, where protein synthesis is occurring.

> Transfer RNA (tRNA) molecules bind with one particular amino acid, and they bear an anticodon that is complementary to a particular codon—the codon for that amino acid.

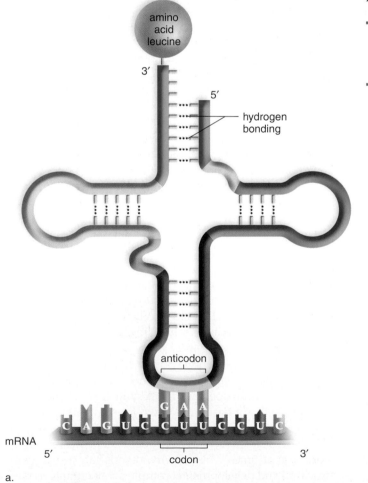

a.

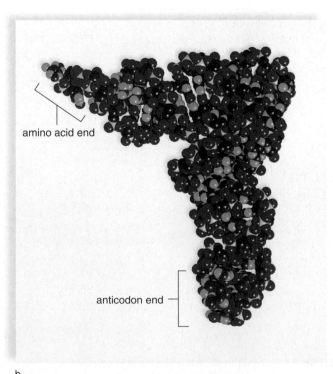

b.

FIGURE 14.9 Structure of a transfer RNA (tRNA) molecule.
a. Complementary base pairing indicated by hydrogen bonding occurs between nucleotides of the molecule, and this causes it to form its characteristic loops. The anticodon that base-pairs with a particular messenger RNA (mRNA) codon occurs at one end of the folded molecule; the other two loops help hold the molecule at the ribosome. An appropriate amino acid is attached at the 3′ end of the molecule. For this mRNA codon and tRNA anticodon, the specific amino acid is leucine. **b.** Space-filling model of tRNA molecule.

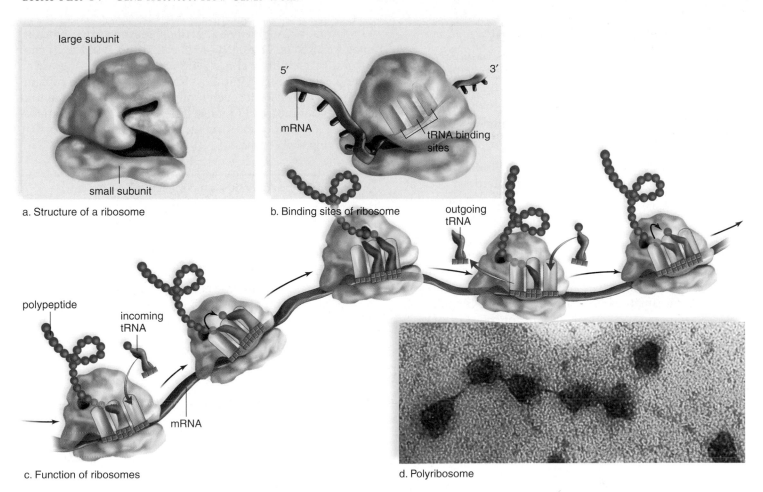

a. Structure of a ribosome

b. Binding sites of ribosome

c. Function of ribosomes

d. Polyribosome

FIGURE 14.10 Ribosome structure and function.
a. Side view of a ribosome shows that it is composed of two subunits, a small subunit and a large subunit. **b.** Frontal view of a ribosome shows its binding sites. mRNA is bound to the small subunit, and the large subunit has three binding sites for tRNAs. **c.** Overview of protein synthesis. A polypeptide increases by one amino acid at a time because a peptide-bearing tRNA passes the peptide to an amino acid-bearing tRNA at a ribosome. Freed of its burden, the "empty" tRNA exits, and the peptide-bearing tRNA moves over one binding site. The polypeptide is formed as this process is repeated. **d.** Electron micrograph of a polyribosome, a number of ribosomes all translating the same mRNA molecule.

The Role of Ribosomal RNA

The structure of a ribosome is suitable to its function.

Structure of a Ribosome

In eukaryotes, ribosomal RNA (rRNA) is produced from a DNA template in the nucleolus of a nucleus. The rRNA is packaged with a variety of proteins into two ribosomal subunits, one of which is larger than the other. Then the subunits move separately through nuclear envelope pores into the cytoplasm, where they combine when translation begins (Fig. 14.10*a*). Ribosomes can remain in the cytoplasm, or they can become attached to endoplasmic reticulum.

Function of a Ribosome

Both prokaryotic and eukaryotic cells contain thousands of ribosomes per cell because they play a significant role in protein synthesis. Ribosomes have a binding site for mRNA and three binding sites for transfer RNA (tRNA) molecules (Fig. 14.10*b*). The tRNA binding sites facilitate complementary base pairing between tRNA anticodons and mRNA codons. A ribosomal RNA (i.e., a ribozyme) is now known to

join one amino acid to another amino acid as a polypeptide is synthesized by the ribosome.

When a ribosome moves down an mRNA molecule, the polypeptide increases by one amino acid at a time (Fig. 14.10*c*). Translation terminates at a stop codon. Once transcription is complete, the polypeptide dissociates from the translation complex and adopts its normal shape. In Chapter 3 we observed that a polypeptide twists and bends into a definite shape. This so-called folding process begins as soon as the polypeptide emerges from a ribosome, and often chaperone molecules are present in the cytoplasm and in the ER to make sure that all goes well. Some proteins contain more than one polypeptide, and if so they join to produce the final three-dimensional structure of a functional protein.

Polyribosome.   As soon as the initial portion of mRNA has been translated by one ribosome, and the ribosome has begun to move down the mRNA, another ribosome attaches to the mRNA. Therefore, several ribosomes are often attached to and translating the same mRNA. The entire complex is called a **polyribosome** (Fig. 14.10*d*).

Translation Requires Three Steps

During translation, the codons of an mRNA base-pair with the anticodons of tRNA molecules carrying specific amino acids. The order of the codons determines the order of the tRNA molecules at a ribosome and the sequence of amino acids in a polypeptide. The process of translation must be extremely orderly so that the amino acids of a polypeptide are sequenced correctly.

Protein synthesis involves three steps: initiation, elongation, and termination. Enzymes are required for each of the three steps to function properly. The first two steps, initiation and elongation, require energy.

Initiation

Initiation is the step that brings all the translation components together. Proteins called initiation factors are required to assemble the small ribosomal subunit, mRNA, initiator tRNA, and the large ribosomal subunit for the start of protein synthesis.

Initiation is shown in Figure 14.11. In prokaryotes, a small ribosomal subunit attaches to the mRNA in the vicinity of the *start codon* (AUG). The first or initiator tRNA pairs with this codon because its anticodon is UAC. Then, a large ribosomal subunit joins to the small subunit (Fig. 14.11). Although similar in many ways, initiation in eukaryotes is much more complex and complicated.

As already discussed, a ribosome has three binding sites for tRNAs. One of these is called the E (for exit) site, second is the P (for peptide) site, and the third is the A (for amino acid) site. The initiator tRNA happens to be capable of binding to the P site, even though it carries only the amino acid methionine (see Fig. 14.5). The A site is for tRNA carrying the next amino acid, and the E site is for any tRNAs that are leaving a ribosome.

Following initiation, mRNA and the first tRNA are at a ribosome.

Elongation

Elongation is the protein synthesis step in which a polypeptide increases in length one amino acid at a time. In addition to the participation of tRNAs, elongation requires elongation factors, which facilitate the binding of tRNA anticodons to mRNA codons at a ribosome.

Elongation is shown in Figure 14.12, where a tRNA with an attached peptide is already at the P site, and a tRNA carrying its appropriate amino acid is just arriving at the A site.① Once the next tRNA is in place at the A site, the peptide will be transferred to this tRNA.② A ribozyme, which is a part of the larger ribosomal subunit, and energy are needed to bring about this transfer. Peptide bond forma-

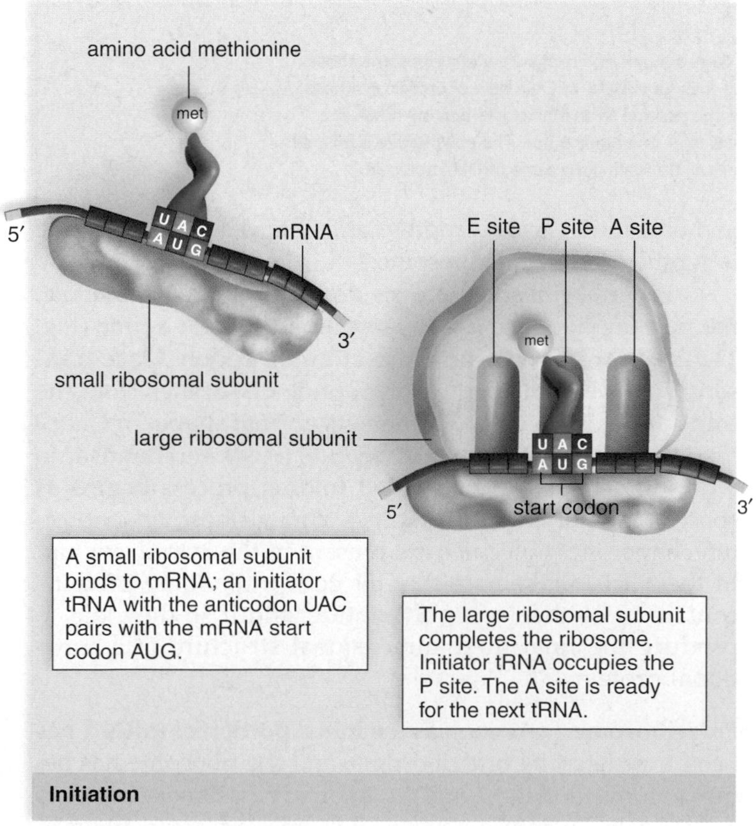

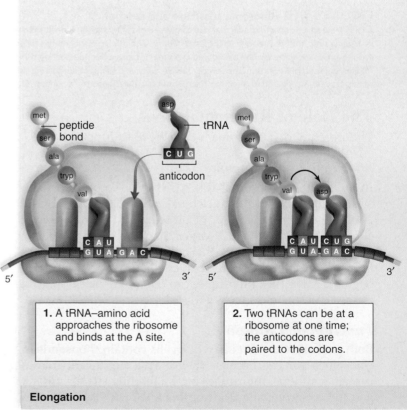

FIGURE 14.11 Initiation.
In prokaryotes, participants in the translation process assemble as shown. The first amino acid is typically methionine.

FIGURE 14.12 Elongation.
Note that a polypeptide is already at the P site. During elongation, polypeptide synthesis occurs as amino acids are added one at a time to the growing chain.

tion means that the peptide is one amino acid longer than it was before.③ Next, translocation occurs: The mRNA moves forward, and the peptide-bearing tRNA is now at the P site of the ribosome. The spent tRNA is now at the E site, and it exits. A new codon is at the A site and is ready to receive another tRNA.④

The complete cycle—complementary base pairing of new tRNA, transfer of peptide chain, and translocation—is repeated at a rapid rate (about 15 times each second in the bacterium *Escherichia coli*).

During elongation, amino acids are added one at a time to the growing polypeptide.

Termination

Termination is the final step in protein synthesis. During termination, as shown in Figure 14.13, the polypeptide and the assembled components that carried out protein synthesis are separated from one another.

Termination of polypeptide synthesis occurs at a *stop codon*—that is, a codon that does not code for an amino acid. Termination requires a protein called a release factor, which cleaves the polypeptide from the last tRNA. After this occurs, the polypeptide is set free and begins to take on its three-dimensional shape. The ribosome dissociates into its two subunits.

During termination, the ribosome separates into its two subunits and the polypeptide is released.

Definition of a Gene and a Genetic Mutation

Scientists who first developed the chromosome theory of inheritance envisioned the gene as a locus on the chromosome. This definition allowed them to solve genetics problems that trace the inheritance of a gene from generation to generation. Geneticists that developed the one gene–one polypeptide concept used the concept of inborn errors in metabolism to say that a gene is a sequence of DNA bases coding for a single polypeptide. This definition is consistent with the diagram for gene expression in Figure 14.4. However, we know today that the definition of a gene should be broadened to include DNA that is transcribed into tRNA and rRNA and perhaps introns. In that case, a **protein-coding gene** is one that is transcribed into mRNA, while a **noncoding gene** is one that is transcribed into any other type of RNA.

Previously we observed that a genetic mutation may result in phenotypic alteration, and we had no direct way to define a mutation. In Chapter 15, we will define a mutation as a permanent change in the sequence of DNA bases. Many mutations do not have an effect on the organism's phenotype, and others provide a survival advantage. We will observe that if an organism inherits a mutated gene, the result can be a genetic disorder or a propensity toward cancer, in which proteins are not fulfilling their usual functions.

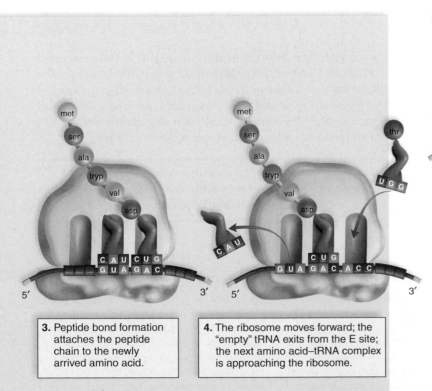

3. Peptide bond formation attaches the peptide chain to the newly arrived amino acid.

4. The ribosome moves forward; the "empty" tRNA exits from the E site; the next amino acid–tRNA complex is approaching the ribosome.

The ribosome comes to a stop codon on the mRNA. A release factor binds to the site.

The release factor hydrolyzes the bond between the last tRNA at the P site and the polypeptide, releasing them. The ribosomal subunits dissociate.

Termination

FIGURE 14.13 Termination.
During termination, the finished polypeptide is released as is the mRNA and the last tRNA.

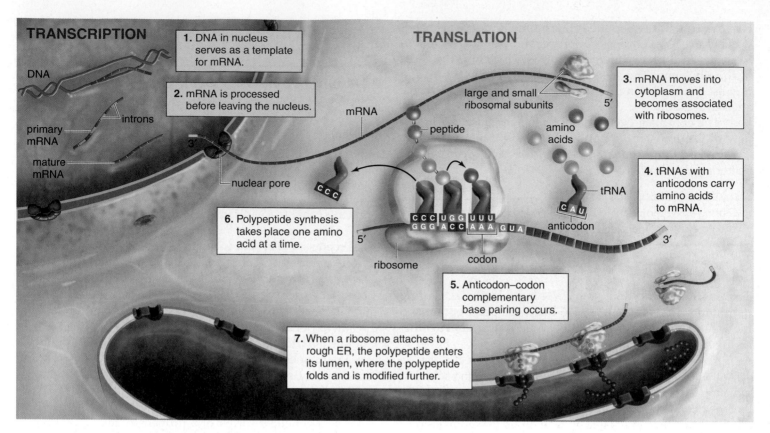

FIGURE 14.14 Summary of gene expression in eukaryotes.
Messenger RNA is produced and processed in the nucleus, and protein synthesis occurs in the cytoplasm. When a ribosome attaches to the rough ER, the polypeptide enters the ER, where it folds and is further modified.

Protein Synthesis and the Eukaryotic Cell

We have already noted that in the eukaryotic cell, mRNA is produced and processed in the nucleus and that protein synthesis occurs at the ribosomes in the cytoplasm. Figure 14.14 reviews transcription and mRNA processing in the nucleus and translation during protein synthesis in the cytoplasm.

As discussed in Chapter 4, some ribosomes (polyribosomes) remain free in the cytoplasm, and some become attached to rough ER. Recall that the first few amino acids of a polypeptide act as a signal peptide that indicates where the polypeptide belongs in the cell or if it is to be secreted from the cell. After the polypeptide enters the lumen of the ER by way of a channel, it is folded and further processed by the addition of sugars, phosphates, or lipids. Transport vesicles carry the proteins between organelles and to the plasma membrane as appropriate for that protein.

Polypeptide synthesis at a ribosome occurs in the cytoplasm. Some ribosomes become attached to the ER. If so, protein folding and modification occur in the ER.

CONNECTING THE CONCEPTS

Early geneticists were amazed to discover that a single nucleotide change in a huge DNA molecule can have profound phenotypic consequences. A well-known example in humans is the single nucleotide change that leads to an amino acid change at one position of the β chain of hemoglobin. The result is the often fatal condition known as sickle cell disease because the red blood cells are sickle-shaped.

The genetic information stored by DNA is realized through the steps of transcription and translation. During transcription, DNA is used as a template for the production of an mRNA molecule. The actions of RNA polymerase, the enzyme that carries out transcription of DNA, and DNA polymerase, the enzyme required for DNA replication, are similar enough to suggest that both enzymes evolved from a common ancestral enzyme. If RNA was the original genetic material, as has been proposed, then it must have specified the synthesis of an enzyme capable of replicating RNA. RNA replicase enzymes are still present in RNA viruses today. Such an enzyme could have been the common ancestor of the DNA and RNA polymerases known today.

Both RNA molecules and proteins are needed for translation. During translation, protein synthesis is carried out according to the genetic information now borne by mRNA. Since a single gene can be transcribed many times, and an mRNA molecule may be repeatedly translated, many copies of a protein can result. In Chapter 15, we will examine the ways cells can regulate the quantity of protein produced per gene.

Summary

14.1 THE FUNCTION OF GENES

Several investigators contributed to our knowledge of what genes do. Garrod is associated with the phrase "inborn error of metabolism" because he suggested that some of his patients had inherited an inability to carry out certain enzymatic reactions.

Beadle and Tatum X-rayed spores of *Neurospora crassa* and found that some of the subsequent cultures lacked a particular enzyme needed for growth on minimal medium. Since they found that the mutation of one gene results in the lack of a single enzyme, they suggested the one gene–one enzyme hypothesis.

Pauling and Itano found that the chemical properties of the β chain of sickle cell hemoglobin differ from those of normal hemoglobin, and therefore the one gene–one polypeptide hypothesis was formulated instead. Later, Ingram showed that the biochemical change is due to the substitution of the amino acid valine (nonpolar) for glutamate (polar).

RNA differs from DNA in these ways: (1) The pentose sugar is ribose, not deoxyribose; (2) the base uracil replaces thymine; and (3) RNA is single stranded.

14.2 THE GENETIC CODE

The central dogma of molecular biology says that (1) DNA is a template for its own replication and also for RNA formation during transcription, and (2) the complementary sequence of nucleotides in mRNA directs the correct sequence of amino acids of a polypeptide during translation.

The sequence of bases in DNA specifies the proper sequence of amino acids in a polypeptide. The genetic code is a triplet code, and each codon (code word) consists of three bases. The code is degenerate—that is, more than one codon exists for most amino acids. There are also one start and three stop codons. The genetic code is considered universal, with a few exceptions.

14.3 FIRST STEP: TRANSCRIPTION

Figure 14.14 provides a summary of transcription and translation. Transcription to form messenger RNA (mRNA) begins when RNA polymerase attaches to a promoter. Elongation occurs until RNA polymerase reaches a stop sequence.

mRNA is processed following transcription. A cap is put onto the 5′ end, and a poly-A tail is put onto the 3′ end; introns are removed in eukaryotes by spliceosomes. It's possible that introns have important regulatory functions.

14.4 SECOND STEP: TRANSLATION

Translation requires mRNA, transfer RNA (tRNA), and ribosomal RNA (rRNA). Each tRNA has an anticodon at one end and an amino acid at the other; amino acid–activating enzymes ensure that the correct amino acid is attached to the correct tRNA. When tRNAs bind with their codon at a ribosome, the amino acids are correctly sequenced in a polypeptide according to the order predetermined by DNA.

In the cytoplasm, many ribosomes move along the same mRNA at a time. Collectively, these are called a polyribosome.

Translation requires these steps: During initiation, mRNA, the first (initiator) tRNA, and the two subunits of a ribosome all come together in the proper orientation at a start codon. During elongation, as the tRNA anticodons bind to their codons, the growing peptide chain is transferred by peptide bonding to the next amino acid in a polypeptide. During termination at a stop codon, the polypeptide is cleaved from the last tRNA. The ribosome now dissociates.

Reviewing the Chapter

1. What made Garrod think there were inborn errors of metabolism?   238
2. Explain Beadle and Tatum's experimental procedure.   238
3. How did Pauling and Itano know that the chemical properties of Hb^S differed from those of Hb^A? What change does a substitution of valine for glutamate cause in the chemical properties of Hb^S compared to Hb^A?   238–39
4. What are the biochemical differences between RNA and DNA?   240
5. What two steps are required for the expression of a gene?   240
6. How did investigators reason that the code must be a triplet code, and in what manner was the code cracked? Why is it said that the code is degenerate, unambiguous, and almost universal?   241
7. What specific steps occur during transcription of RNA off a DNA template?   242
8. How is messenger RNA (mRNA) processed before leaving the eukaryotic nucleus?   242–43
9. Compare the functions of mRNA, transfer RNA (tRNA), and ribosomal RNA (rRNA) during protein synthesis. What are the specific events of translation?   244–47
10. Genes specify the amino acid sequence in polypeptides. Explain with reference to Figure 14.14.   248

Testing Yourself

Choose the best answer for each question. For questions 1–3, match the investigators to the phrase in the key.

KEY:

   a. one gene–one enzyme hypothesis
   b. inborn error of metabolism
   c. one gene–one polypeptide hypothesis
   d. one gene–one mRNA hypothesis

1. Sir Archibald Garrod

2. George Beadle and Edward Tatum

3. Linus Pauling and Harvey Itano

4. Considering the following pathway, if Beadle and Tatum found that *Neurospora* cannot grow if metabolite A is provided, but can grow if B, C, or D is provided, what molecule would be missing?

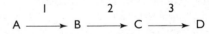

$$A \xrightarrow{\ 1\ } B \xrightarrow{\ 2\ } C \xrightarrow{\ 3\ } D$$

   a. enzyme 1
   b. enzyme 2
   c. enzyme 3
   d. the substrate in addition to the product
   e. All of these are correct.

5. The central dogma of molecular biology
   a. states that DNA is a template for all RNA production.
   b. states that DNA is a template only for DNA replication.
   c. states that translation precedes transcription.
   d. pertains only to prokaryotes because humans are unique.
   e. All of these are correct.

6. mRNA synthesis is called
   a. replication.
   b. translocation.
   c. translation.
   d. transcription.

7. Which of these does not characterize the process of transcription? Choose more than one answer if correct.
   a. RNA is made with one strand of the DNA serving as a template.
   b. In making RNA, the base uracil of RNA pairs with the base thymine of DNA.
   c. The enzyme RNA polymerase synthesizes RNA.
   d. RNA is made in the cytoplasm of eukaryotic cells.

8. Because there are more codons than amino acids,
   a. some amino acids are specified by more than one codon.
   b. some codons specify more than one amino acid.
   c. some codons do not specify any amino acid.
   d. some amino acids do not have codons.

9. If the sequence of bases in DNA is TAGC, then the sequence of bases in RNA will be
   a. ATCG.
   b. TAGC.
   c. AUCG.
   d. GCTA.
   e. Both a and b are correct.

10. RNA processing
    a. is the same as transcription.
    b. is an event that occurs after RNA is transcribed.
    c. is the rejection of old, worn-out RNA.
    d. pertains to the function of transfer RNA during protein synthesis.
    e. Both b and d are correct.

11. Translation can be defined as
    a. the making of protein using mRNA and tRNA.
    b. the making of RNA from a DNA template.
    c. the gathering of amino acids by tRNA molecules in the cytoplasm.
    d. the removal of introns from mRNA.

12. During protein synthesis, an anticodon on transfer RNA (tRNA) pairs with
    a. DNA nucleotide bases.
    b. ribosomal RNA (rRNA) nucleotide bases.
    c. messenger RNA (mRNA) nucleotide bases.
    d. other tRNA nucleotide bases.
    e. Any one of these can occur.

13. This is a segment of a DNA molecule. (Remember that the template strand only is transcribed.) What are (a) the RNA codons, (b) the tRNA anticodons, and (c) the sequence of amino acids in a protein?

template strand

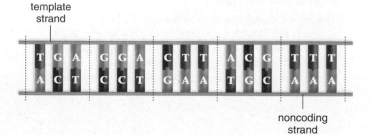

noncoding strand

14. A codon in a fragment of DNA is AAA. By using the table of mRNA codons, you are able to determine that the DNA code is specifying the amino acid
    a. lysine.
    b. glycine.
    c. proline.
    d. phenylalanine.

Thinking Scientifically

1. In one strain of bacteria, you observe an x amount of some enzyme. In a mutant derived from that strain there is 1.53 times the amount of the enzyme. A comparison of the sequence of the two strains shows that there is no change in the coding region of the gene. But there is a substitution of two A–T base pairs in the mutant for two G–C base pairs where DNA polymerase attaches. What hypothesis would explain how such a mutation could produce this change in protein level? What change would be predicted in mRNA level?

2. Changes in DNA sequence—that is, mutations—account for evolutionary changes. Sometimes evolution seems to occur quite rapidly. What type of base change in DNA might bring about the greatest change in a protein and the greatest phenotypic change immediately?

Understanding the Terms

anticodon 244
codon 241
elongation 246
exon 243
gene 240
genetic code 241
initiation 246
intron 243
messenger RNA (mRNA) 240
noncoding gene 247
polyribosome 245
promoter 242
protein-coding gene 247

ribosomal RNA (rRNA) 240
ribozyme 243
RNA (ribonucleic acid) 240
RNA polymerase 242
RNA transcript 242
termination 247
transcription 240
transfer RNA (tRNA) 240
translation 240
triplet code 241
uracil 240
wobble hypothesis 244

Match the terms to these definitions:

a. _____ Enzyme that speeds the formation of mRNA from a DNA template.

b. _____ Noncoding segment of DNA that is transcribed but the transcript is removed before mRNA leaves the nucleus.

c. _____ Process whereby the sequence of codons in mRNA determines (is translated into) the sequence of amino acids in a polypeptide.

d. _____ String of ribosomes, simultaneously translating different regions of the same mRNA strand during protein synthesis.

e. _____ Three mRNA nucleotides that code for a particular amino acid or termination of translation.

f. _____ illustrates that the third position in a tRNA can be variable.

ARIS, the *Biology* Website

ARIS, the website for *Biology*, provides a wealth of information organized and integrated by chapter. You will find practice quizzes, interactive activities, labeling exercises, flashcards, and much more that will complement your learning and understanding of general biology.

www.mhhe.com/maderbiology9

15

REGULATION OF GENE ACTIVITY AND GENE MUTATIONS

Humans begin life as a one-celled zygote that has 46 chromosomes, and these same chromosomes are passed to all the daughter cells during mitosis. Yet, cells become specialized in structure and function. The genes for digestive enzymes are expressed only in digestive tract cells, and the genes for muscle proteins are expressed only in muscle cells. This shows that control of gene transcription and translation does occur within each particular type of cell.

In this chapter, we will see that regulation of gene activity in eukaryotic cells extends from the nucleus (where gene transcription is controlled) to the cytoplasm (where the activity of an enzyme is controlled). Knowledge of gene regulation has many applications. It can help us understand how development progresses and what causes certain disorders. We will see that cancer develops when genes that promote cell division are expressed and when genes that suppress cell division are not expressed. In future chapters, it will become apparent that regulatory gene mutations have evolutionary significance. Such changes might very well cause a population of organisms to evolve rapidly into a new species. Perhaps regulatory mutations contributed to the evolution of humans from an apelike ancestor.

Cell division in cancer cells is no longer regulated.

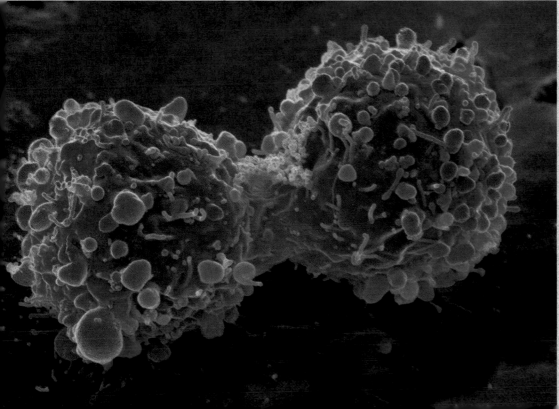

15.1 Prokaryotic Regulation

Because their environment is ever changing, bacteria do not need the same enzymes and possibly other proteins all the time. For example, they need only the enzymes required to break down the nutrients available to them and the enzymes required to synthesize whatever metabolites are absent under the present circumstances.

In 1961, French microbiologists François Jacob and Jacques Monod showed that *Escherichia coli* is capable of regulating the expression of its genes. They proposed the **operon** [L. *opera,* works] model to explain gene regulation in prokaryotes and later received a Nobel Prize for their investigations. A **regulator gene,** located outside the operon, codes for a repressor that controls whether the operon is active or not. An operon includes the following elements:

Promoter—A short sequence of DNA where RNA polymerase first attaches when a gene is to be transcribed. Basically, the promoter signals the start of a gene.

Operator—A short portion of DNA where an active repressor binds. When an active repressor is bound to the operator, RNA polymerase cannot attach to the promoter, and transcription does not occur. In this way, the operator controls mRNA systhesis.

Structural genes—One to several genes coding for the primary structure of enzymes of a metabolic pathway that are transcribed as a unit.

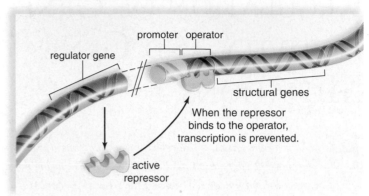

a. **Tryptophan absent.** Enzymes needed to synthesize tryptophan are produced.

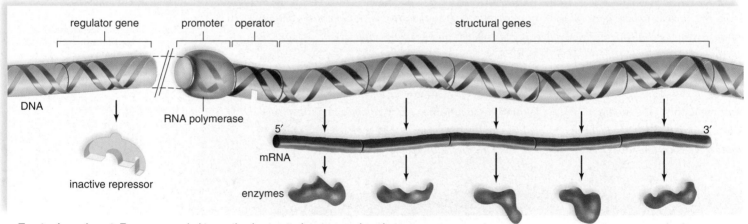

b. **Tryptophan present.** Presence of tryptophan prevents production of enzymes used to synthesize tryptophan.

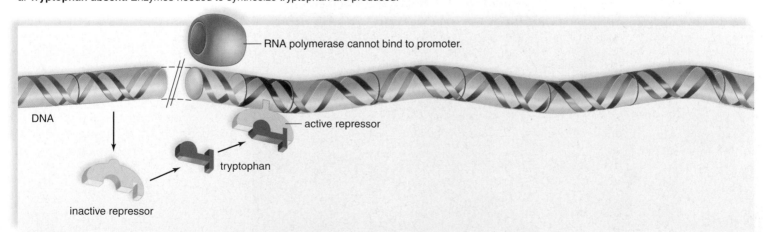

FIGURE 15.1 The *trp* operon.

a. The regulator gene codes for a repressor protein that is normally inactive. RNA polymerase attaches to the promoter, and the structural genes are expressed. **b.** When the nutrient tryptophan is present, it binds to the repressor, changing its shape. Now the repressor is active and can bind to the operator. RNA polymerase cannot attach to the promoter, and the structural genes are not expressed.

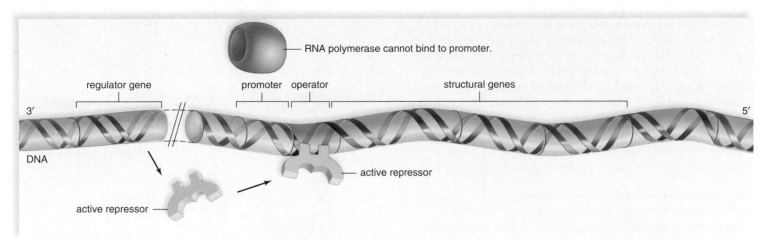

a. **Lactose absent.** Enzymes needed to take up and use lactose are not produced.

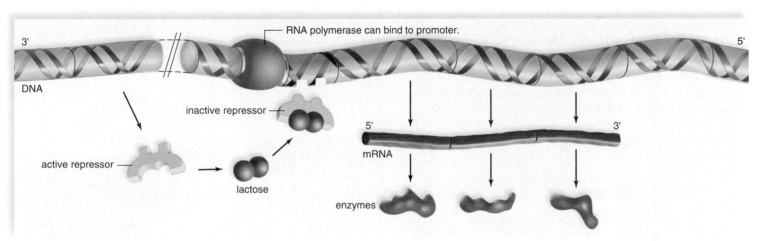

b. **Lactose present.** Enzymes needed to take up and use lactose are produced only when lactose is present.

FIGURE 15.2 The *lac* operon.
a. The regulator gene codes for a repressor that is normally active. When it binds to the operator, RNA polymerase cannot attach to the promoter, and structural genes are not expressed. **b.** When lactose is present, it binds to the repressor, changing its shape so that it is inactive and cannot bind to the operator. Now, RNA polymerase binds to the promoter, and the structural genes are expressed.

The *trp* Operon

Investigators, including Jacob and Monod, found that some operons in *E. coli* usually exist in the "on" rather than "off" condition. For example, in the *trp* operon, the regulator codes for a repressor that ordinarily is unable to attach to the operator. Therefore, RNA polymerase is able to bind to the promoter, and the structural genes of the operon are ordinarily expressed (Fig. 15.1). Their products, five different enzymes, are part of an anabolic pathway for the synthesis of the amino acid tryptophan.

If tryptophan happens to be already present in the medium, these enzymes are not needed by the cell, and the operon is turned off by the following method. Tryptophan binds to the repressor. A change in shape now allows the repressor to bind to the operator, and the structural genes are not expressed. The enzymes are said to be repressible, and the entire unit is called a **repressible operon.** Tryptophan is called the **corepressor.** Repressible operons are usually involved in anabolic pathways that synthesize a substance needed by the cell.

The *lac* Operon

When *E. coli* is denied glucose and is given the milk sugar lactose instead, it immediately begins to make the three enzymes needed for the metabolism of lactose. These enzymes are encoded by three genes: One gene is for an enzyme called β-galactosidase, which breaks down the disaccharide lactose to glucose and galactose; a second gene codes for a permease that facilitates the entry of lactose into the cell; and a third gene codes for an enzyme called transacetylase, which has an accessory function in lactose metabolism.

The three structural genes are adjacent to one another on the chromosome and are under the control of a single promoter and a single operator (Fig. 15.2). The regulator gene codes for a *lac* operon repressor that ordinarily binds to the operator and prevents transcription of the three genes. But when glucose is absent and lactose (or more correctly, allolactose, an isomer formed from lactose) is present, lactose binds to the **repressor,** and the repressor undergoes a change in shape that prevents it from binding to the operator. Because the repressor is unable to bind to the operator, RNA polymerase is better able to bind to the promoter. After RNA polymerase carries out transcription, the three enzymes of lactose metabolism are synthesized.

Because the presence of lactose brings about expression of genes, it is called an **inducer** of the *lac* operon: the enzymes are said to be inducible enzymes, and the entire unit is called an **inducible operon**. Inducible operons are usually necessary to catabolic pathways that break down a nutrient. Why is that beneficial? Because these enzymes need only be active when the nutrient is present.

The *trp* operon is an example of a repressible operon because it is ordinarily turned on. The *lac* operon is an example of an inducible operon because it is ordinarily turned off.

Further Control of the lac Operon

E. coli preferentially breaks down glucose, and the bacterium has a way to ensure that the lactose operon is maximally turned on only when glucose is absent. When glucose is absent, a molecule called *cyclic AMP (cAMP)* accumulates. Cyclic AMP, which is derived from ATP, has only one phosphate group, which is attached to ribose at two locations:

$$^{5'}CH_2 \quad O \quad \text{adenine}$$
$$P \qquad 3' $$
$$OH$$

cyclic AMP
(cAMP)

Cyclic AMP binds to a molecule called a *catabolite activator protein (CAP)*, and the complex attaches to a CAP binding site next to the *lac* promoter. When CAP binds to DNA, DNA bends, exposing the promoter to RNA polymerase. RNA polymerase is now better able to bind to the promoter so that the *lac* operon structural genes are transcribed, leading to their expression (Fig. 15.3).

When glucose is present, there is little cAMP in the cell; CAP is inactive, and the lactose operon does not function maximally. CAP affects other operons as well and takes its name for activating the catabolism of various other metabolites when glucose is absent. A cell's ability to encourage the metabolism of lactose and other metabolites when glucose is absent provides a backup system for survival when the preferred metabolite glucose is absent.

Negative Versus Positive Control

Negative control of an operon is exemplified by the involvement of repressors. Why? Because active repressors shut down the activity of an operon. On the other hand, the CAP molecule is an example of positive control. Why? Because when this molecule is active, it promotes the activity of an

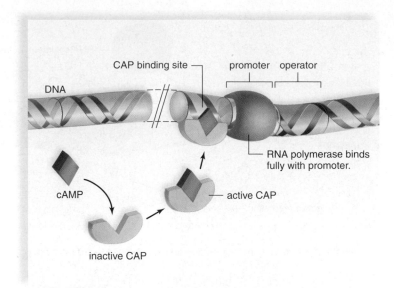

a. **Lactose present, glucose absent (cAMP level high)**

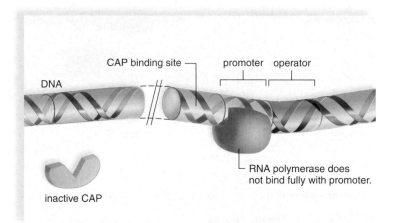

b. **Lactose present, glucose present (cAMP level low)**

FIGURE 15.3 Action of CAP.
When active CAP binds to its site on DNA, the RNA polymerase is better able to bind to the promoter so that the structural genes of the *lac* operon are expressed. **a.** CAP becomes active in the presence of cAMP, a molecule that is prevalent when glucose is absent. Therefore, transcription of lactose enzymes increases, and lactose is metabolized. **b.** If glucose is present, CAP is inactive, and RNA polymerase does not completely bind to the promoter. Therefore, transcription of lactose enzymes decreases, and less metabolism of lactose occurs.

operon. The use of both a negative and a positive control mechanism to control an operon allows the cell to fine-tune its response. In the case of the *lac* operon, the operon is only maximally active when glucose is absent and lactose is present. If both glucose and lactose are present, the cell preferentially metabolizes glucose.

Bacterial DNA contains control regions whose sole purpose is to regulate the transcription of structural genes that code for enzymes or other products.

15.2 EUKARYOTIC REGULATION

The cells in multicellular eukaryotes, including humans, have a copy of all genes; however, different genes are actively expressed in different cells. Muscle cells, for example, have a different set of genes that are turned on in the nucleus and a different set of proteins that are active in the cytoplasm than do nerve cells.

Unlike prokaryotic cells, a variety of mechanisms regulate gene expression in eukaryotic cells. These mechanisms can be grouped under five primary levels of control; three of them pertain to the nucleus, and two pertain to the cytoplasm (Fig. 15.4). In other words, control of gene activity in eukaryotes extends from transcription to protein activity. These are the types of control in eukaryotic cells that can modify the amount of the gene product:

1. *Chromatin structure:* Chromatin packing is used as a way to keep genes turned off. If genes are not accessible to RNA polymerase, they cannot be transcribed.

   In the nucleus, highly condensed chromatin is not available for transcription, while more loosely condensed chromatin is available for transcription. Chromatin structure is a part of **epigenetic** [Gk. *epi*, besides] **inheritance**, the transmission of genetic information outside the coding sequences of a gene.

2. *Transcriptional control:* The degree to which a gene is transcribed into mRNA determines the amount of gene product.

   In the nucleus, transcription factors in particular help initiate transcription, the first step in gene expression. Transposons are DNA sequences that move between chromosomes and shut down genes.

3. *Posttranscriptional control:* Posttranscriptional control involves mRNA processing and how fast mRNA leaves the nucleus. We now know that mRNA processing differences can determine the type of protein product made by a cell.

   Also, mRNA processing differences can affect how fast mRNA leaves the nucleus and the amount of gene product within a given amount of time.

4. *Translational control:* Translational control occurs in the cytoplasm and affects when translation begins and how long it continues. Any influence that can cause the persistence of the 5′ cap and 3′ poly-A tail can affect the length of translation. Excised introns are now believed to be involved in a regulatory system that directly affects the life-span of mRNA.

   Some mRNAs may need further processing before they are translated. The possibility of an RNA-based regulatory system is being investigated.

5. *Posttranslational control:* Posttranslational control, which also takes place in the cytoplasm, occurs after protein synthesis. Only a functional protein is an active gene product.

   The polypeptide product may have to undergo additional changes before it is biologically functional. Also, a functional enzyme is subject to feedback control so that it is no longer able to speed its reaction.

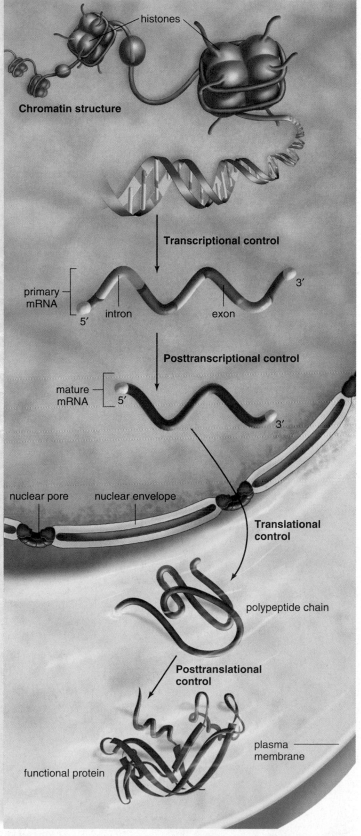

FIGURE 15.4 Levels at which control of gene expression occurs in eukaryotic cells.

The five levels of control are: (1) chromatin structure, (2) transcriptional control, and (3) posttranscriptional control, which occur in the nucleus; and (4) translational and (5) posttranslational control, which occur in the cytoplasm.

Chromatin Structure

Genetic control in eukaryotes is bound to be more complicated than prokaryotes if only because they have a great deal more DNA than prokaryotes. The DNA in eukaryotes is always associated with plentiful proteins, and together they make up the stringy material, called **chromatin,** that can be observed in the interphase nucleus. Previously we discussed how chromatin is compacted to form chromosomes that are visible during cell division (Fig. 15.5*a*). DNA is periodically wound around a core of eight protein molecules so that it looks like beads on a string. The protein molecules are **histones,** and each bead is called a **nucleosome.** The levels of chromatin organization seen in a nucleus can be related to the de-

gree that the nucleosomes coil as chromatin condenses. Just before cell division occurs, chromatin coils tightly into a fiber that has six nucleosomes to a turn. Then the fiber loops back and forth and condenses further to produce highly compacted chromosomes.

Even during interphase, chromatin has to be condensed if a very large amount of DNA is to fit inside a much smaller nucleus. Some portions of chromatin are highly condensed and these portions appear as darkly stained portions, called **heterochromatin** in electron micrographs (Fig. 15.5*b*). A dramatic example of heterochromatin is the Barr body in mammalian females. Females have a small, darkly staining mass of condensed chromatin adhering to the inner edge of the nuclear envelope. This structure, called a **Barr body** after its

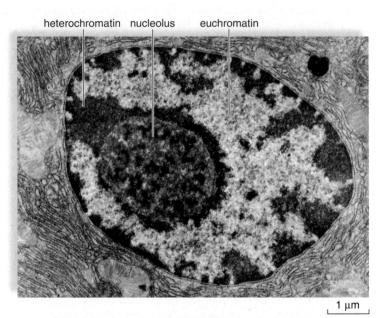

b. Darkly-stained heterochromatin and lightly-stained euchromatin

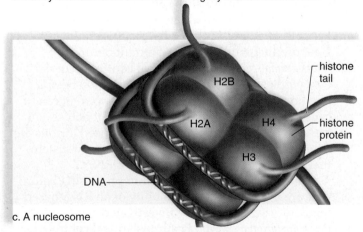

c. A nucleosome

a. Levels of organization of chromatin

FIGURE 15.5 Levels of chromatin structure.
a. Histones, present in nucleosomes, are responsible for the packaging of chromatin so that it fits into the nucleus and so that it becomes highly condensed when cell division occurs. **b.** A eukaryotic nucleus contains heterochromatin (darkly stained, highly condensed chromatin) and euchromatin, which is not as condensed. **c.** Nucleosomes ordinarily prevent access to DNA so that transcription cannot take place. If histone tails are acetylated, access can be achieved; if the tails are methylated, access is more difficult.

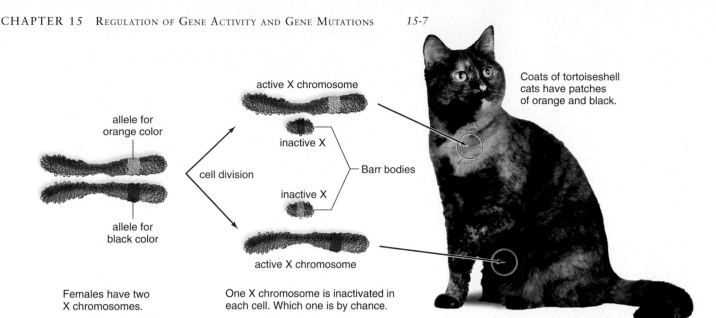

FIGURE 15.6 X-inactivation in mammalian females.
In cats, the alleles for black or orange coat color are carried on the X chromosomes. Random X-inactivation occurs in females. Therefore, in heterozygous females, 50% of the cells have an allele for black coat color and 50% of cells have an allele for orange coat color. The result is tortoiseshell cats that have coats with patches of both black and orange.

discoverer, is an inactive X chromosome. One of the X chromosomes undergoes inactivation in the cells of female embryos. The inactive X chromosome does not produce gene products, and therefore female cells have a reduced amount of product from genes on the X chromosome.

How do we know that Barr bodies are inactive X chromosomes that are not producing gene product? Suppose 50% of the cells have one X chromosome active and 50% have the other X chromosome active. Wouldn't the body of a heterozygous female be a mosaic, with "patches" of genetically different cells? This is exactly what investigators have discovered. Human females who are heterozygous for an X-linked recessive form of ocular albinism have patches of pigmented and nonpigmented cells at the back of the eye. Women heterozygous for Duchenne muscular dystrophy have patches of normal muscle tissue and degenerative muscle tissue (the normal tissue increases in size and strength to make up for the defective tissue). And women who are heterozygous for hereditary absence of sweat glands have patches of skin lacking sweat glands. The female tortoise shell cat also provides dramatic support for a difference in X-inactivation in its cells. In these cats, an allele for black coat color is on one X chromosome, and a corresponding allele for orange coat color is on the other X chromosome. The patches of black and orange in the coat can be related to which X chromosome is in the Barr bodies of the cells found in the patches (Fig. 15.6).

During interphase, most of the chromatin is not highly condensed. Instead, it is in a loosely compacted state called **euchromatin.** Whereas heterochromatin is inactive, euchromatin is potentially active in the sense that the genes located in euchromatin are subject to being expressed. Even now, however, so-called *remodeling proteins* have to push aside the histone portion of a nucleosome so that access to DNA is not barred and transcription can begin. In other words, histones regulate accessibility to DNA,

and euchromatin becomes genetically active when histones no longer bar access to DNA.

Epigenetic Inheritance

What regulates whether chromatin exists as heterochromatin or euchromatin? Histone molecules have *tails*, strings of amino acids that extend beyond the main portion of a nucleosome (see Fig. 15.5c). In euchromatin, the histone tails tend to be acetylated and have attached acetyl groups ($-COCH_3$); in heterochromatin, the histone tails tend to bear methyl groups ($-CH_3$). Methylation of DNA can also occur, and when it does, genes are shut down from generation to generation. Methylation of DNA accounts for a phenomenon called *genomic imprinting* in which gene expression is dependent on whether the chromosome carrying that gene was inherited from a mother or father. If an inherited chromosome is highly methylated, the gene is not expressed, even if it is a normal gene in every other way.

The expression *epigenetic inheritance* is now used for inheritance patterns that do not depend on the genes themselves. In this instance, epigenetic inheritance depends on whether acetyl and methyl groups surround and adhere to DNA. Epigenetic inheritance explains unusual inheritance patterns and also may play an important role in growth, aging, and cancer. Researchers are even hopeful that it will be easier to develop drugs to modify this level of inheritance rather than trying to change DNA itself.

One level of gene expression control in eukaryotes is chromatin structure. If genes are located within highly condensed chromatin (heterochromatin) they are inactive. Methylation appears to be a way to prevent access to DNA. If genes are located within loosely condensed chromatin (euchromatin), they can be activated by outside influences such as remodeling proteins.

Transcriptional Control

Although eukaryotes have various levels of genetic control (see Fig. 15.4), **transcriptional control** remains the most critical of these levels. The first step toward transcription is availability of DNA for transcription. In certain cells it is possible to observe a change in chromatin structure when transcription is occurring. The chromosomes within the developing egg cells of many vertebrates are called lampbrush chromosomes because they have many loops that appear to be bristles. (In olden days, lampbrushes were used to clean the inside glass of kerosene lamps.) Figure 15.7 shows lampbrush chromosomes in which mRNA is being synthesized in great quantity as development proceeds. Lampbrush chromosomes are thought to assist in fulfilling the high demand for mRNA transcripts immediately following fertilization.

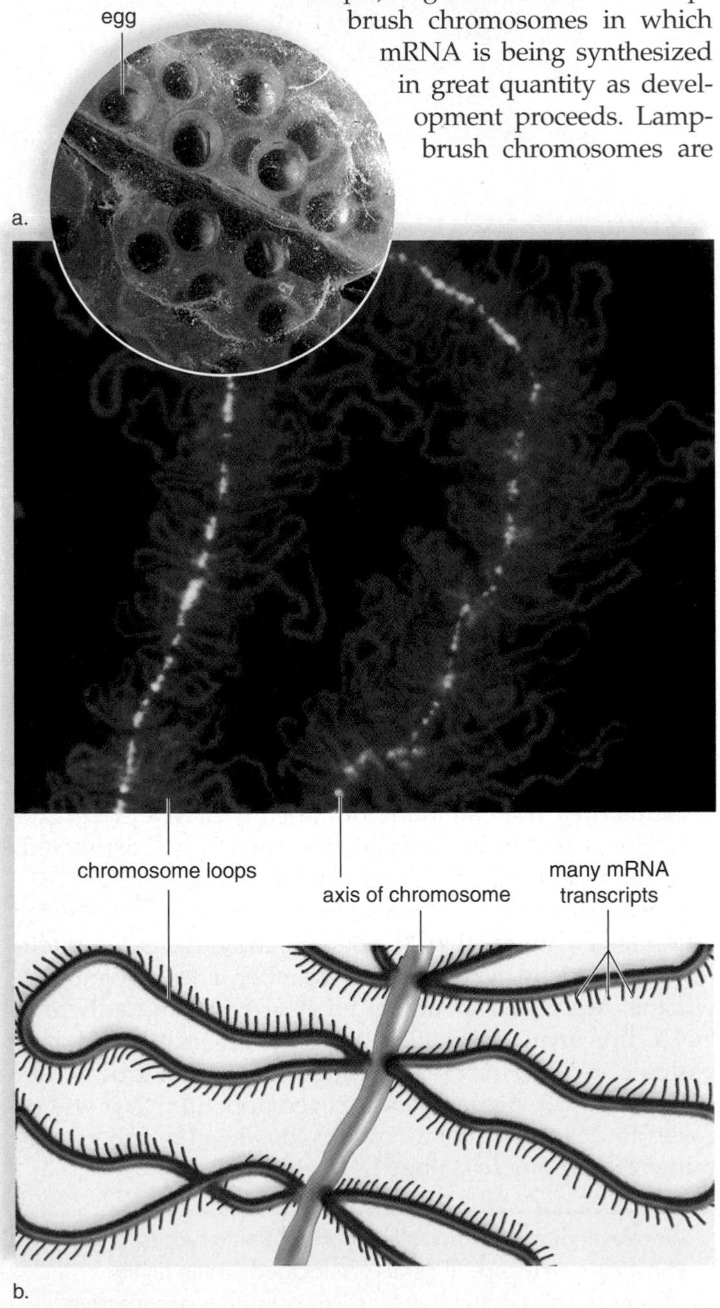

Transcription Factors and Activators

Although no operons like those of prokaryotic cells have been found in eukaryotic cells, transcription is controlled by DNA-binding proteins. Every cell contains many different types of **transcription factors,** proteins that help regulate transcription. A group of transcription factors binds to a promoter adjacent to a gene, and then the complex attracts and binds RNA polymerase, but transcription may still not begin. Quite often, **transcription activators** are also involved in promoting transcription, and a different combination is believed to regulate the activity of any particular gene. Transcription activators bind to regions of DNA called enhancers. Enhancers can be quite a distance from the promoter, but a hairpin loop in the DNA can bring the transcription activators attached to the enhancer into the vicinity of the promoter. Mediator proteins act as a bridge between transcription factors and transcription activators at the promoter. Now RNA polymerase begins the transcription process (Fig. 15.8). Such protein-to-protein interactions are a hallmark of eukaryotic gene regulation.

FIGURE 15.7 Lampbrush chromosomes.
a. An amphibian egg. **b.** These chromosomes are present in maturing amphibian egg cells and give evidence that when mRNA is being synthesized, chromosomes most likely decondense. Each chromosome has many loops extended from its axis (white). Many mRNA transcripts are being made off these DNA loops (red).

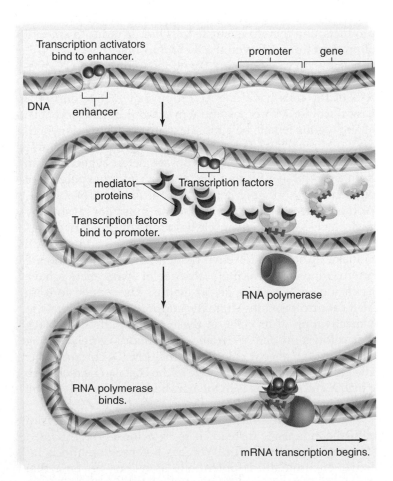

FIGURE 15.8 Initiation of transcription.
Transcription in eukaryotic cells requires that transcription factors bind to the promoter and transcription activators bind to an enhancer. The enhancer is far from the promoter, but the DNA loops and mediator proteins act as a bridge joining activators to factors. Only then does transcription begin.

Transcription factors and activators are always present in the nucleus of a cell, but they most likely have to be activated in some way before they will bind to DNA. Activation probably occurs when they are phosphorylated by a kinase. Kinases, which add a phosphate group to molecules, and phosphatases, which remove a phosphate group, are known to be signaling proteins involved in a growth regulatory network that reaches from receptors in the plasma membrane to the genes in the nucleus.

Transposons

Transposons are specific DNA sequences that have the remarkable ability to move within and between chromosomes. As discussed in the Science Focus on page 260, their movement to a new location sometimes alters neighboring genes, particularly decreasing their expression. In other words, a transposon sometimes acts like a regulator gene. Although "movable elements" in corn were described 40 years ago, their significance was only realized within the past several years. So-called jumping genes now have been discovered in bacteria, fruit flies, humans, and many other organisms.

Transposons are among the 40% of the human genome consisting of repetitive elements—the same short sequence of bases is repeated over and over again. The white-eye mutant in *Drosophila*, first observed by Morgan, is due to a transposon located within a gene coding for a pigment-producing enzyme. In a rare human neurological disorder called Charcot-Marie-Tooth disease, the muscles and nerves of the legs and feet gradually wither away. This syndrome is due to a transposon called *Mariner*. New thinking is that transposons could be considered a part of epigenetic inheritance because they are noncoding DNA sequences that play regulatory functions; namely, they shut down genes.

Posttranscriptional Control

Posttranscriptional control of gene expression occurs in the nucleus and includes mRNA processing and the speed with which mRNA leaves the nucleus.

During mRNA processing, introns (noncoding regions) are excised, and exons (expressed regions) are spliced together (see Fig. 14.8). When introns are removed from mRNA, differential splicing of exons can occur and this effects gene expression. For example, both the hypothalamus and the thyroid gland produce a hormone called calcitonin, but the mRNA that leaves the nucleus is not the same in both types of cells. Radioactive labeling studies show that they vary because of a difference in mRNA splicing (Fig. 15.9). Evidence of different patterns of mRNA splicing is found in other cells, such as those that produce neurotransmitters, muscle regulatory proteins, and antibodies. It is not known how cells specify which form of mature mRNA they will make. Some researchers believe that RNAs regulate the process themselves because artificial RNAs can modify splicing patterns when they are added to cells being grown in petri dishes.

The speed of transport of mRNA from the nucleus into the cytoplasm can ultimately affect the amount of

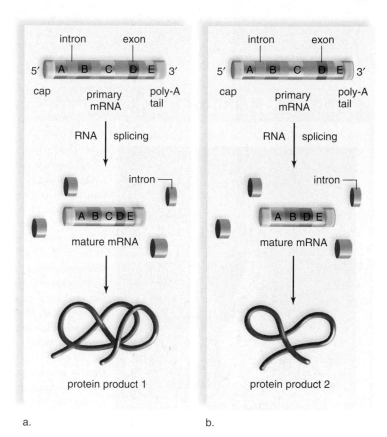

FIGURE 15.9 Processing of mRNA transcripts.
Because the primary mRNAs are processed differently in these two cells (**a** and **b**), distinct proteins result. This is a form of posttranscriptional control of gene expression.

gene product realized per unit time following transcription. Evidence indicates there is a difference in the length of time it takes various mRNA molecules to pass through a nuclear pore.

Translational Control

Translational control begins when the processed mRNA molecule reaches the cytoplasm and before there is a protein product. Presence or absence of the (guanine nucleotide) cap at the 5' end and the poly-A (adenine nucleotide) tail at the 3' end of a mature mRNA transcript can determine whether translation takes place and how long the mRNA is active. The long life of mRNA in mammalian red blood cells, which code for hemoglobin, is attributed to the persistence of their 5' end cap and their 3' poly-A tail. On the other hand, any influence that affects the length of the poly-A tail can lead to removal of the cap and the destruction of mRNA.

Another area of important research today is the role that introns play in regulating which mRNAs are translated. Researchers working with introns suggest that they are processed into smaller signals, called **microRNAs.** After microRNAs are degraded, they combine with a protein and the complex binds to mRNAs. These mRNAs are destroyed instead of being translated. It is even possible that microRNAs might feed back and influence which genes are transcribed in the first place. If so, genetic control in eukaryotes might

science focus

Barbara McClintock and the Discovery of Jumping Genes

FIGURE 15A Barbara McClintock.

When Barbara McClintock (Fig. 15A) first began studying inheritance in corn (maize) plants, geneticists believed that each gene had a fixed locus on a chromosome. Thomas Morgan and his colleagues at Columbia University were busy mapping the chromosomes of *Drosophila*, but McClintock preferred to work with corn. In the course of her studies, she came to the conclusion that "controlling elements" could move from one location to another on the chromosome. If a controlling element landed in the middle of a gene, it prevented the expression of that gene. Today, McClintock's controlling elements are called movable genetic elements, transposons, or (in slang) "jumping genes."

Based on her experiments with maize, McClintock showed that because transposons are capable of suppressing gene expression, they could account for the pigment pattern of the corn strain popularly known as Indian corn. A colorless corn kernel results when cells are unable to produce a purple pigment due to the presence of a transposon within a particular gene needed to synthesize the pigment. While mutations are usually stable, a transposition is very unstable. When the transposon jumps to another chromosomal location, some cells regain the ability to produce the purple pigment, and the result is a corn kernel with a speckled pattern (Fig. 15B).

When McClintock first published her results in the 1950s, the scientific community

paid them little attention. Years later, when molecular genetics was well established, transposons were also discovered in bacteria, yeasts, plants, fruit flies and humans.

Geneticists now believe that transposons:

1. are involved in transcriptional control because they block transcription.
2. can carry a copy of certain host genes with them when they jump. Therefore, they can be a source of chromosome mutations such as translocations, deletions, and inversions.
3. can leave copies of themselves and certain host genes before jumping. Therefore, they can be a source of a duplication, another type of chromosome mutation.
4. can contain one or more genes that make a bacterium resistant to antibiotics.

Considering that transposition has a powerful effect on genotype and phenotype (Fig. 15C), it most likely has played an important role in evolution. For her discovery of transposons, McClintock was, in 1983, finally awarded the Nobel Prize in Physiology or Medicine. In her Nobel Prize acceptance speech, the 81-year-old scientist proclaimed that "it might seem unfair to reward a person for having so much pleasure over the years, asking the maize plant to solve specific problems and then watching its responses."

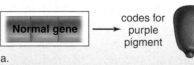 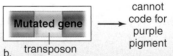

a. b.

FIGURE 15B Indian corn.
Indian corn displays a variety of colors and patterns. Transposons are responsible for the speckled or striped patterns.

FIGURE 15C Transposon.
a. A purple-coding gene ordinarily codes for a purple pigment. b. A transposon "jumps" into the purple-coding gene. This mutated gene is unable to code for purple pigment, and a white kernel results.

be due to an extensive RNA-based regulatory system that could be influenced by pharmaceutical drugs.

Posttranslational Control

Posttranslational control begins once a protein has been synthesized and has become active. Posttranslational control represents the last chance a cell has for influencing gene expression.

Some proteins are not immediately active after synthesis. For example, at first bovine proinsulin is a single, long polypeptide that folds into a three-dimensional structure. Cleavage results in two smaller chains that are bonded together by disulfide (S—S) bonds. Only then is active insulin present.

Some proteins are short-lived in cells because they are degraded or destroyed. For example, the cyclin proteins that control the cell cycle are only temporarily present. The cell has giant protein complexes, called proteasomes that carry out the task of destroying proteins.

Transcription factors, transcription activators, and mediator proteins form a complex that facilitates the binding of RNA polymerase to a promoter so that transcription can begin. Transposons block transcription, and introns may play significant roles at the levels of posttranscriptional and translational regulatory control.

15.3 GENETIC MUTATIONS

A **genetic mutation** is a permanent change in the sequence of bases in DNA. The effect of a DNA base sequence change on protein activity can range from no effect to complete inactivity. Germ-line mutations are those that occur in sex cells and can be passed to subsequent generations. Somatic mutations occur in body cells, and therefore they may affect only a small number of cells in a tissue. Somatic mutations are not passed on to future generations, but they can lead to the development of cancer.

Effect of Mutations on Protein Activity

Point mutations involve a change in a single DNA nucleotide and therefore a change in a specific codon. Figure 15.10 gives an example in which a single base change could have no effect or a drastic effect, depending on the particular base change that occurs. You already know that sickle cell disease is due to a single base change in DNA. Because the β chain of hemoglobin now contains valine instead of glutamate at one location, hemoglobin molecules form semirigid rods. The resulting sickle-shaped cells clog blood vessels and die off more quickly than normal-shaped cells.

Frameshift mutations occur most often because one or more nucleotides are either inserted or deleted from DNA. The result of a frameshift mutation can be a completely new sequence of codons and nonfunctional proteins. Here is how this occurs: The sequence of codons is read from a specific starting point, as in this sentence, THE CAT ATE THE RAT. If the letter C is deleted from this sentence and the reading frame is shifted, we read THE ATA TET HER AT—something that doesn't make sense.

Nonfunctional Proteins

A single nonfunctioning protein can have a dramatic effect on the phenotype. The human transposon *Alu* is responsible for hemophilia when *Alu* inserts into the gene for clotting

factor IX and places a premature stop codon there. One particular metabolic pathway in cells is as follows:

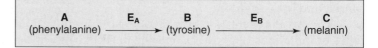

If a faulty code for enzyme E_A is inherited, a person is unable to convert the molecule A to B. Phenylalanine builds up in the system, and the excess causes mental retardation and the other symptoms of the genetic disorder phenylketonuria (PKU). In the same pathway, if a person inherits a faulty code for enzyme E_B, then B cannot be converted to C, and the individual is an albino.

Recall that cystic fibrosis is due to the inheritance of a faulty code for a chloride ion channel protein in the plasma membrane. A rare condition called androgen insensitivity is due to a faulty receptor for androgens, which are male sex hormones such as testosterone. Although there is plenty of testosterone in the blood, the cells are unable to respond to it. Female instead of male external genitals form, and female instead of male secondary sex characteristics occur. The individual, who appears to be a normal female, may be prompted to seek medical advice when menstruation never occurs. The karyotype is that of a male rather than a female, and the individual does not have the internal sexual organs of a female (Fig. 15.11).

Carcinogenesis

It is estimated that one-third of the children born in 1999 will develop cancer at some time in their lives. Of these affected

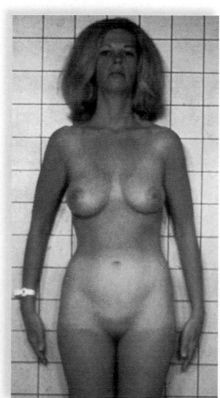

FIGURE 15.11
Androgen insensitivity. This individual has a female appearance but the karyotype of a male, and there are testes instead of ovaries and a uterus in the abdominal cavity. Her cells are unable to respond to the male sex hormones because of a mutation that makes the androgen receptor ineffective.

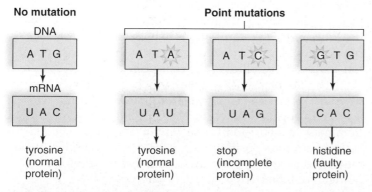

FIGURE 15.10 Point mutation.
The effect of a DNA base alteration can vary. Starting at the *left*, if the DNA codes for the same amino acid after the base change, there is no noticeable effect; if it now codes for a stop codon, the resulting protein will be incomplete; and if it now codes for a different amino acid, a faulty protein is possible.

individuals, one-third of the females and one-fourth of the males will die due to cancer. In the United States, the three deadliest forms of cancer are lung cancer, colon and rectal cancer, and breast cancer.

The development of cancer involves a series of various types of mutations (Fig. 15.12), and the particular series can be different for each type of cancer. As discussed in Chapter 9, tumor suppressor genes ordinarily act as brakes on cell division especially when it begins to occur abnormally. Proto-oncogenes stimulate cell division but are usually turned off in fully differentiated nondividing cells. When proto-oncogenes mutate, they become oncogenes which are active all the time. The growth process of cancer begins with the loss of tumor suppressor gene activity and the gain of oncogene activity.

New mutations arise, and one cell (brown) has the ability to start a tumor.

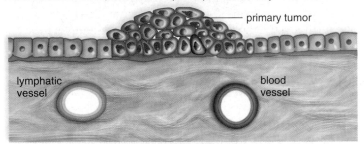

Cancer in situ. The tumor is at its place of origin. One cell (purple) mutates further.

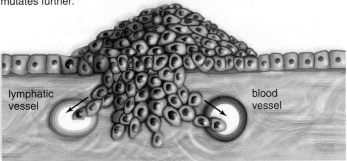

Cancer cells now have the ability to invade lymphatic and blood vessels and travel throughout the body.

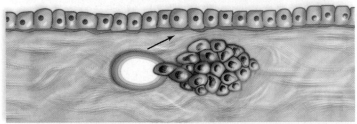

New metastatic tumors are found some distance from the primary tumor.

FIGURE 15.12 Carcinogenesis.
The development of cancer requires a series of mutations leading first to a localized tumor and then to metastatic tumors. With each successive step toward cancer, the most genetically altered and aggressive cell becomes the dominant type of tumor. The cells take on characteristics of embryonic cells; they are not differentiated, they can divide controllably; and they are able to migrate to new locations.

When tumor suppressor genes are inactive and oncogenes are active, cell division occurs uncontrollably.

It often happens that tumor suppressor genes and proto-oncogenes often code for transcription factors or proteins that control transcription factors. As we have seen, transcription factors are a part of the rich and diverse types of mechanisms that control transcription in cells. They are of fundamental importance to DNA replication and repair, cell growth and division, control of apoptosis, and cellular differentiation. Therefore, it is not surprising that inherited or acquired defects in transcription factor structure and function contribute to the development of cancer.

To take an example, a major tumor suppressor gene called *p53* is more frequently mutated in human cancers than any other known gene. It has been found that the p53 protein acts as a transcription factor, and as such is involved in turning on the expression of genes whose products are cell cycle inhibitors. *p53* also promotes apoptosis, programmed cell death, when it is needed. The retinoblastoma protein (RB) controls the activity of a transcription factor for cyclin D and other genes whose products promote entry into the S stage of the cell cycle. When the tumor suppressor gene *p16* mutates, the RB protein is always available and the result is too much active cyclin D in the cell.

Other types of mutations also contribute to the development of cancer. Several proto-oncogenes code for *ras* proteins, which are needed for cells to grow, to make new DNA, and to not grow out of control. A point mutation is sufficient to turn a normally functioning *ras* proto-oncogene into an oncogene. Abnormal growth results.

Inheritance of Cancer

In 1990, a gene called *breast cancer 1* or *BRCA1* (pronounced *brak-uh*) was identified. Later, *BRCA2* was also found. These genes are tumor suppressor genes that are known to behave as if they are autosomal recessive alleles. When one mutated gene is inherited, cancer is more likely wherever the second mutation occurs. If the second mutation is in the ovary, ovarian cancer may develop if additional cancer-causing mutations occur.

Similarly, when a mutated *RB* allele, which causes cancer of the eye, is inherited it takes another mutated allele to increase the chance of cancer. A tumor in one eye is more usual because it takes mutations in both alleles before cancer can develop. The RB protein interacts with transcription factors involved in control of the cell cycle.

The abnormal *RET* gene, which predisposes one to thyroid cancer, can be passed from parent to child. *RET* is a proto-oncogene known to be inherited in an autosomal dominant manner—only one mutated allele is needed to increase a predisposition to cancer. The remainder of the mutations necessary for a thyroid cancer to develop are acquired (not inherited).

Causes of Mutations

Some mutations are spontaneous—they happen for no apparent reason—while others are due to environmental

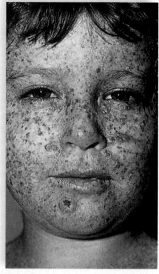

a.   b.

FIGURE 15.13 Achondroplasia and xeroderma pigmentosum.
a. Achondroplasia results from a spontaneous mutation. **b.** In xeroderma pigmentosum, deficient DNA repair enzymes leave the skin cells vulnerable to the mutagenic effects of ultraviolet light, and hundreds of skin cancers (small dark spots) appear on the skin exposed to the sun. This individual also has a tumor on the bridge of the nose.

mutagens. An example of a spontaneous germ-line mutation in humans is achondroplasia, which results in a type of dwarfism, a dominant trait (Fig. 15.13a). Spontaneous mutations due to DNA replication errors are rare. DNA polymerase, the enzyme that carries out replication, proofreads the new strand against the old strand and detects any mismatched nucleotides, and each is usually replaced with a correct nucleotide. In the end, only about one mistake occurs for every 1 billion nucleotide pairs replicated. Also, a cell has DNA repair enzymes that fix any mutations that occur independent of replication. The importance of repair enzymes is exemplified by individuals with the condition known as xeroderma pigmentosum. They lack some of the repair enzymes, and as a consequence, these individuals have a high incidence of skin cancer (Fig. 15.13b).

Environmental Mutagens

A mutagen is an environmental agent that increases the chances of a mutation. Among the best-known mutagens are radiation and organic chemicals. Many mutagens are also **carcinogens** (cancer-causing). Scientists use the Ames test for mutagens to hypothesize that a chemical can be carcinogenic. In the Ames test, a histidine-requiring (His-) strain of bacteria is exposed to a chemical. If the chemical is mutagenic, the bacterium regains the ability to grow without histidine. A large number of chemicals used in agriculture and industry give a positive Ames test. Examples are ethylese dibromide (EDB), which is added to leaded gasoline (to vaporize lead deposits in the engine and send them out the exhaust), and ziram, which is used to prevent fungus disease on crops. Some drugs, such as isoniazed (used to prevent

tuberculosis), are mutagenic according to the Ames test. The mutagenic potency of AF-2, a food additive once widely used in Japan, and safrole, a flavoring agent that used to be added to root beers, caused them to be banned. Although most testing has been done on products of the chemical industry, many naturally occurring substances (like safrole) have been shown to be mutagenic. These include aflatoxin, produced in moldy grain and peanuts (and present in peanut butter at an average level of 2 parts per billion). Traces of nine different substances that give positive Ames tests have been found in fried hamburger.

Tobacco smoke contains a number of organic chemicals that are known carcinogens, and it is estimated that one-third of all cancer deaths can be attributed to smoking. Lung cancer is the most frequent lethal cancer in the United States, and smoking is also implicated in the development of cancers of the mouth, larynx, bladder, kidney, and pancreas. The greater the number of cigarettes smoked per day, the earlier the habit starts, and the higher the tar content, the greater possibility of cancer. When smoking is combined with drinking alcohol, the risk of these cancers increases even more.

Aside from chemicals, certain forms of radiation, such as X-rays and gamma rays, are called ionizing radiation because they create free radicals, ionized atoms with unpaired electrons. Free radicals react with and alter the structure of other molecules, including DNA. Ultraviolet (UV) radiation is easily absorbed by the pyrimidines in DNA. Wherever there are two thymine molecules next to one another, ultraviolet radiation may cause them to bond together, forming thymine dimers. A kink results in the DNA.

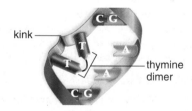

Usually, these dimers are removed from damaged DNA by repair enzymes, which constantly monitor DNA and fix any irregularities. One enzyme excises a portion of DNA that contains the dimer; another makes a new section by using the other strand as a template; and still another seals the new section in place. Because of the carcinogenic effect of X-rays, it is wise to avoid unnecessary exposure. Repair enzymes have failed as when skin cancer develops because of sunbathing.

Genetic mutations are due to a change in base sequences, and when such changes lead to malfunctioning proteins, the effect on the phenotype can be dramatic or can even lead to the development of cancer. Genetic mutations can be due to errors in replication and environmental mutagens including X-rays.

CONNECTING THE CONCEPTS

Gene regulation determines whether a gene is expressed and/or the degree to which a gene is expressed. In differentiated cells such as nerve cells, muscle fibers, reproductive cells, and so forth, only certain genes are actively expressed. The collective work of many researchers was required to determine that the regulation of genes can occur at several different stages during the processes of transcription and translation.

Mutations of regulatory genes may cause structural genes to be abnormally turned off or on, or expressed in abnormal quantity. Such mutations can seriously affect a developing embryo or lead to development of cancer. Cancers can arise when proto-oncogenes, which are normally not expressed, become oncogenes that are expressed. On the other hand, cancers can also arise when tumor suppressor genes fail to be adequately expressed.

Mutations in regulatory genes may account for major evolutionary changes. The proteins in chimpanzees and humans are strikingly similar in amino acid sequence. It's possible that the presence or absence of particular proteins in particular cells may account for their phenotypic differences. In Chapter 16, you will see how the basic knowledge of DNA structure, replication, and expression has contributed to a biotechnology revolution.

Summary

15.1 PROKARYOTIC REGULATION

Regulation in prokaryotes usually occurs at the level of transcription. The operon model developed by Jacob and Monod says that a regulator gene codes for a repressor, which sometimes binds to the operator. When it does, RNA polymerase is unable to bind to the promoter, and transcription of structural genes cannot take place.

The *trp* operon is an example of a repressible operon because when tryptophan, the corepressor, is present, it binds to the repressor. The repressor is then able to bind to the operator, and transcription of structural genes does not take place.

The *lac* operon is an example of an inducible operon because when lactose, the inducer, is present, it binds to the repressor. The repressor is unable to bind to the operator, and transcription of structural genes takes place if glucose is absent.

These are two examples of negative control because a repressor is involved. There are also examples of positive control. The structural genes in the *lac* operon are not maximally expressed unless glucose is absent and lactose is present. At that time, cAMP attaches to a molecule called CAP, and then CAP binds to a site next to the promoter. Now RNA polymerase is better able to bind to the promoter, and transcription occurs.

15.2 EUKARYOTIC REGULATION

The following levels of control of gene expression are possible in eukaryotes: chromatin structure, transcriptional control, posttranscriptional control, translational control, and posttranslational control.

Chromatin structure helps regulate transcription. Highly condensed heterochromatin is genetically inactive, as exemplified by Barr bodies. Less-condensed euchromatin is genetically active, as exemplified by lampbrush chromosomes in vertebrates. Regulatory proteins called transcription factors, as well as DNA sequences called enhancers, play a role in controlling transcription in eukaryotes. Transcription factors bind to the promoter, and transcription activators bind to an enhancer.

Posttranscriptional control refers to variations in messenger RNA (mRNA) processing and the speed with which a particular mRNA molecule leaves the nucleus. Introns are converted to microRNAs that may feed back to regulate translation or even transcription.

Translational control affects mRNA translation and the length of time it is translated. Posttranslational control affects whether or not an enzyme is active and how long it is active.

15.3 GENETIC MUTATIONS

In molecular terms, a gene is a sequence of DNA nucleotide bases, and a genetic mutation is a change in this sequence.

Point mutations can range in effect, depending on the particular codon change. Sickle cell disease is an example of a point mutation. Frameshift mutations result when a base is added or deleted, and the result is usually a nonfunctional protein. Cystic fibrosis is an example of a frameshift mutation. Nonfunctional proteins can affect the phenotype drastically, as in albinism, which is due to a single faulty enzyme, and androgen insensitivity, which is due to a faulty receptor for testosterone.

Cancer is due to an accumulation of genetic mutations among regulator genes. The cell cycle occurs inappropriately when proto-oncogenes become oncogenes and tumor suppressor genes are no longer effective. Mutations can be spontaneous or due to an environmental mutagen. Carcinogens are mutagens such as radiation and organic chemicals that cause cancer.

Reviewing the Chapter

1. Name and state the function of the three components of operons.   252
2. Explain the operation of the *trp* operon, and note why it is considered a repressible operon.   253
3. Explain the operation of the *lac* operon, and note why it is considered an inducible operon.   253–54
4. What are the five levels of genetic regulatory control in eukaryotes?   255
5. Relate heterochromatin and euchromatin to levels of chromatin organization.   256–57
6. With regard to transcriptional control in eukaryotes, explain how Barr bodies show that heterochromatin is genetically inactive.   256–57
7. Explain how lampbrush chromosomes in vertebrates show that euchromatin is genetically active.   258
8. What do transcription factors do in eukaryotic cells? What are enhancers?   258–59
9. Give examples of posttranscriptional, translational, and posttranslational control in eukaryotes.   259–60
10. What are two major types of mutations, and what effect can they have on protein activity?   261
11. Name some causes of mutations.   262–63

Testing Yourself

Choose the best answer for each question.

1. Which type of prokaryotic cell would be more successful as judged by its growth potential?
   a. one that is able to express all its genes all the time
   b. one that is unable to express any of its genes any of the time
   c. one that can alter gene regulation in response to changing environmental conditions
   d. one that limits its growth regardless of the richness of the culture medium
   e. Both a and d are correct.

2. When lactose is present and glucose is absent,
   a. the repressor is able to bind to the operator.
   b. the repressor is unable to bind to the operator.
   c. transcription of structural genes occurs.
   d. transcription of the structural genes, operator, and promoter occur.
   e. Both b and c are correct.

3. When tryptophan is present,
   a. the repressor is able to bind to the operator.
   b. the repressor is unable to bind to the operator.
   c. transcription of structural genes occurs.
   d. transcription of the structural genes, operator, and promoter occurs.
   e. Both b and c are correct.

4. Which of these pairs is mismatched?
   a. posttranslational control—nucleus
   b. transcriptional control—nucleus
   c. translational control—cytoplasm
   d. posttranscriptional control—nucleus
   e. Both b and c are mismatched.

5. When cells become specialized, they
   a. develop similarly.
   b. contain different genes.
   c. transcribe different genes.
   d. must undergo mutations.
   e. Both b and c are correct.

6. Which of these would regulate gene expression?
   a. binding of transcription factors
   b. differential mRNA processing
   c. length of time mRNA is active
   d. modification of protein structure
   e. All of these are correct.

7. RNA processing varies in different cells. This is an example of _____ control of gene expression.
   a. transcriptional   d. posttranslational
   b. posttranscriptional   e. Both b and d are correct.
   c. translational

8. A scientist adds radioactive uridine (label for RNA) to a culture of cells and examines an autoradiograph. Which type of chromatin is apt to show the label?
   a. heterochromatin
   b. euchromatin
   c. the histones, not the DNA
   d. the DNA, not the histones
   e. Both a and d are correct.

9. Barr bodies are
   a. genetically active X chromosomes in males.
   b. genetically inactive X chromosomes in females.
   c. genetically active Y chromosomes in males.
   d. genetically inactive Y chromosomes in females.
   e. Both b and d are correct.

10. Which of these might cause a proto-oncogene to become an oncogene?
    a. exposure of the cell to radiation
    b. exposure of the cell to certain chemicals
    c. viral infection of the cell
    d. exposure of the cell to pollutants
    e. All of these are correct.

11. A cell is cancerous. You might find an abnormality in
    a. a proto-oncogene.
    b. a tumor suppressor gene.
    c. regulation of the cell cycle.
    d. tumor cells.
    e. All of these are correct.

12. A tumor suppressor gene
    a. inhibits cell division.
    b. opposes oncogenes.
    c. prevents cancer.
    d. is subject to mutations.
    e. All of these are correct.

13. Label this diagram of an operon.

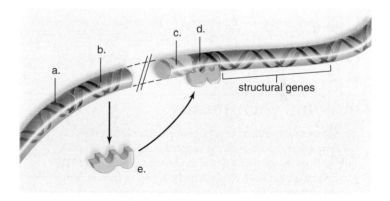

structural genes

14. Which of these is involved in controlling gene expression in eukaryotes?
    a. the occurrence of transcription
    b. activity of the polypeptide product
    c. life expectancy of the mRNA molecule in the cell
    d. All of these are involved.

15. In operon models, the function of the promoter is to
    a. code for the repressor protein.
    b. bind with RNA polymerase.
    c. bind to the repressor.
    d. code for the regulator gene.

16. If the DNA codons are CAT CAT CAT, and a guanine base is added at the beginning, which would result?
    a. CAT CAT CAT G   d. a type of mutation
    b. G CAT CAT CAT   e. Both c and d are correct.
    c. GCA TCA TCA T

17. A mutation in a DNA molecule involving the replacement of one nucleotide base pair with another is called
    a. a frameshift mutation.   d. a point mutation.
    b. a transposon.   e. an insertion mutation.
    c. a deletion mutation.

18. Which of these is characteristic of cancer?
    a. It may involve a lack of mutations over a length of time.
    b. It cannot be tied to particular environmental factors.
    c. Apoptosis is one of the first developmental effects.
    d. Mutations occur in only certain genes.
    e. It typically develops within a short period of time.

19. Which is not evidence that eukaryotes control transcription?
    a. euchromatin/heterochromatin
    b. existence of transcription factors
    c. lampbrush chromosomes
    d. occurrence of mutations
    e. All of these are correct.

For questions 20–27, indicate whether the statement is true (T) or false (F).

20. Operons only occur in prokaryotes. _____

21. Transcription factors only occur in eukaryotes. _____

22. Promoters occur in both prokaryotes and eukaryotes. _____

23. When transposons bind to DNA, transcription begins. _____

24. Tobacco smoke causes genetic mutations. _____

25. Barr bodies are not the cause of mosaicism but a reflection of mosaicism. _____

26. Androgen insensitivity shows that a nonfunctional protein can have a profound effect on the body. _____

27. The CAP protein is an example of positive control of gene expression in prokaryotes. _____

Thinking Scientifically

1. In patients with chronic myelogenous leukemia, an odd chromosome is seen in all the cancerous cells. A small piece of chromosome 9 is connected to chromosome 22. This 9:22 translocation has been termed the Philadelphia chromosome. How could a translocation cause genetic changes that result in cancer?

2. Interferon is a molecule that sets up an "alert" state in cells so that they can respond quickly to viral infection. This alert state actually causes the cell to self-destruct when a virus enters the cell. This is beneficial to the organism since the virus is prevented from producing progeny virus. For this system to work, the cell must be able to respond as rapidly as possible to the incoming virus. What hypothesis would explain how gene expression could produce the most rapid response? How could you use drugs that inhibit transcription to test the hypothesis?

Bioethical Issue: Assuming Responsibility for Environmental Mutagens

One day, your genetic profile will tell physicians which diseases you are likely to develop. Knowledge of your genes might indicate your susceptibility to various types of cancer, for example. This information could be used to develop a prevention program, including avoiding environmental influences associated with diseases. No doubt, you would be less inclined to smoke if you knew your genes make it almost inevitable that smoking will give you lung cancer.

People worry, however, that genetic profiles could be used against them. Perhaps employers will not hire, or insurance companies will not insure, those who have a propensity for a particular disease. About 25 states have passed laws prohibiting genetic discrimination by health insurers, and 11 have passed laws prohibiting genetic discrimination by employers. Is such legislation enough to allay our fears of discrimination?

On the other hand, employers may fear that one day they will be required to provide an environment specific to every employee's need to prevent future illness. Would you approve of this, or should individuals be required to leave an area or job that exposes them to an environmental influence that could be detrimental to their health?

People's medical records are usually considered private. But if scientists could match genetic profiles to environmental conditions that bring on illnesses, they could come up with better prevention guidelines for the next generation. Should genetic profiles and health records become public information under these circumstances? It would particularly help in the study of complex diseases such as cardiovascular disorders, noninsulin-dependent diabetes, and juvenile rheumatoid arthritis.

Understanding the Terms

Barr body 256	operon 252
carcinogen 263	point mutation 261
chromatin 256	posttranscriptional
corepressor 253	control 259
epigenetic inheritance 255	posttranslational control 260
euchromatin 257	promoter 252
frameshift mutation 261	regulator gene 252
genetic mutation 261	repressible operon 253
heterochromatin 256	repressor 253
histone 256	structural gene 252
inducer 254	transcription activator 258
inducible operon 254	transcriptional control 258
microRNA 259	transcription factor 258
nucleosome 256	translational control 259
operator 252	transposon 259

Match the terms to these definitions:

a. _____ A regulation of gene expression that begins once there is an mRNA transcript.

b. _____ Specific DNA sequences that have the ability to move between chromosomes.

c. _____ Dark-staining body in the nuclei of female mammals that contains a condensed, inactive X chromosome.

d. _____ Diffuse chromatin, which is being transcribed.

e. _____ Environmental agent that causes mutations leading to the development of cancer.

ARIS, the *Biology* Website

ARIS, the website for *Biology*, provides a wealth of information organized and integrated by chapter. You will find practice quizzes, interactive activities, labeling exercises, flashcards, and much more that will complement your learning and understanding of general biology.

www.mhhe.com/maderbiology9

16

BIOTECHNOLOGY AND GENOMICS

Since Mendel's work was rediscovered in 1900, geneticists have made startling advances that have led to the present era of DNA technology. This technology allows us to alter the genotype of various organisms and thereby produce improved drugs to treat human illnesses. Insulin is one such product produced by the new biotechnology. Not very long ago, insulin for the treatment of diabetes mellitus was acquired from dead animals. Today, the gene for human insulin is inserted into bacteria, which then produce the hormone as a commercial product. Genetically engineered bacteria have also been used to clean up environmental pollutants, increase the fertility of the soil, and kill insect pests.

Biotechnology also extends beyond unicellular organisms; it is now possible to alter the genotype and subsequently the phenotype of plants and animals. In agriculture, new crops are being developed that are resistant to disease and frost damage. Gene therapy in humans—the insertion of normal human genes to make up for ones that do not function—is already undergoing clinical trials.

Advocacy groups have concerns about the genetic modification of organisms. They worry that modified bacteria and plants might harm the environment. Others are fearful that products produced by altered organisms might not be healthy for humans. Perhaps terrorists could make use of biotechnology to produce weapons of mass destruction. Ethical concerns are also expressed. Is it ethical to give a mouse a bioluminescent gene that makes it glow? To what extent would it be proper to improve the human genome? Everyone should be knowledgeable about genetics and biotechnology so they can participate in deciding these issues.

This mouse received a gene responsible for making jellyfish bioluminescent when it was an embryo.

16.1 DNA CLONING

In biology, **cloning** is the production of genetically identical copies of DNA, cells, or organisms through some asexual means. When an underground stem or root sends up new shoots, the resulting plants are clones of one another. The members of a bacterial colony on a petri dish are clones because they all came from the division of the same original cell. Human identical twins are also considered clones. The first two cells of the embryo separate, and each becomes a complete individual.

DNA cloning can be done to produce many identical copies of the same gene; that is, for the purpose of **gene cloning.** Scientists clone genes for a number of reasons. They might want to determine the difference in base sequence between a normal gene and a mutated gene. Or they might use the genes to genetically modify organisms in a beneficial way. When genes are used to modify a human, the process is called **gene therapy.** Otherwise, the organisms are called **transgenic organisms** [L. *trans,* across, through; Gk. *genic,* producing]. Transgenic organisms are frequently used today to produce a product desired by humans.

Recombinant DNA technology and the polymerase chain reaction are two procedures that scientists can use to clone DNA for various purposes.

Recombinant DNA Technology

Recombinant DNA (rDNA) contains DNA from two or more different sources, such as a human cell and a bacterial cell, as shown in Figure 16.1. To make rDNA, a technician needs a **vector** [L. *vehere,* to carry], by which rDNA will be introduced into a host cell. One common vector is a plasmid. **Plasmids** are small accessory rings of DNA found in bacteria. The ring is not part of the bacterial chromosome and replicates on its own. Plasmids were discovered by investigators studying the bacterium *Escherichia coli* (*E. coli*).

Two enzymes are needed to introduce foreign DNA into vector DNA: (1) a **restriction enzyme,** which cleaves DNA, and (2) an enzyme called **DNA ligase** [L. *ligo,* bind], which seals DNA into an opening created by the restriction enzyme. Hundreds of restriction enzymes occur naturally in bacteria, where they cut up any viral DNA that enters the cell. They are called restriction enzymes because they *restrict* the growth of viruses, but they also act as molecular scissors to cleave any piece of DNA at a specific site. For example, the restriction enzyme called *Eco*RI always cuts double-stranded DNA at this sequence of bases and in this manner:

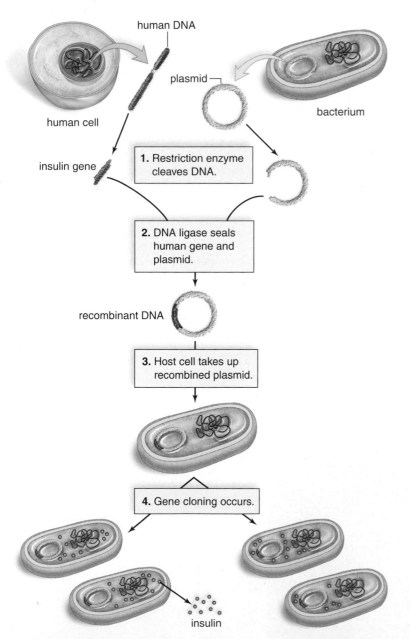

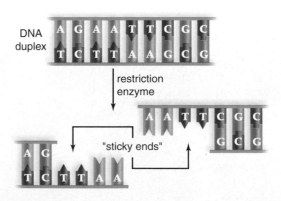

Notice that there is now a gap into which a piece of foreign DNA can be placed if it ends in bases complementary to those exposed by the restriction enzyme. To ensure this, it is only necessary to cleave the foreign DNA with the same type of restriction enzyme. The single-stranded, but complementary, ends of the two DNA molecules are called "sticky

FIGURE 16.1 Cloning a human gene.
Human DNA and plasmid DNA are cleaved by a specific type of restriction enzyme. For example, human DNA containing the insulin gene is spliced into a plasmid by the enzyme DNA ligase. Gene cloning is achieved after a bacterium takes up the plasmid. If the gene functions normally as expected, the product (e.g., insulin) may also be retrieved.

ends" because they can bind a piece of foreign DNA by complementary base pairing. Sticky ends facilitate the insertion of foreign DNA into vector DNA.

Next, genetic engineers use DNA ligase to seal the foreign piece of DNA into the vector. DNA splicing is now complete; an rDNA molecule has been prepared (Fig. 16.1). Bacterial cells take up recombinant plasmids, especially if they are treated to make them more permeable. Thereafter, as the plasmid replicates, DNA is cloned.

For a human gene to express itself in a bacterium, the gene has to be accompanied by regulatory regions unique to bacteria. Also, the gene should not contain introns because bacteria don't have introns. However, it is possible to make a human gene that lacks introns. The enzyme called reverse transcriptase can be used to make a DNA copy of mRNA. The DNA molecule, called **complementary DNA (cDNA),** does not contain introns. Alternatively, it is possible to manufacture small pieces of DNA in the laboratory. A machine called a DNA synthesizer joins together the correct sequence of nucleotides, and the resulting DNA also lacks introns.

The Polymerase Chain Reaction

The **polymerase chain reaction (PCR),** developed by Kary Mullis in 1985, can create copies of a segment of DNA quickly in a test tube. PCR is very specific—it amplifies (makes copies of) a targeted DNA sequence. The targeted sequence can be less than one part in a million of the total DNA sample!

PCR requires the use of DNA polymerase, the enzyme that carries out DNA replication, and a supply of nucleo-

tides for the new DNA strands. PCR is a chain reaction because the targeted DNA is repeatedly replicated as long as the process continues. The colors in Figure 16.2 distinguish old from new DNA. Notice that the amount of DNA doubles with each replication cycle.

PCR has been in use for years, and now almost every laboratory has automated PCR machines to carry out the procedure. Automation became possible after a temperature-insensitive (thermostable) DNA polymerase was extracted from the bacterium *Thermus aquaticus,* which lives in hot springs. The enzyme can withstand the high temperature used to separate double-stranded DNA; therefore, replication does not have to be interrupted by the need to add more enzyme.

Analyzing DNA Segments

DNA amplified by PCR is often analyzed for various purposes. Mitochondrial DNA base sequences in modern living populations were used to decipher the evolutionary history of human populations. Since so little DNA is required for PCR to be effective, it has even been possible to sequence DNA taken from mummified human brains.

Also, following PCR, DNA can be subjected to **DNA fingerprinting** using a method developed by Alec Jefferies and his colleagues in the 1980s. When a piece of DNA is treated with restriction enzymes, the result is a unique collection of different-sized fragments. During a process called gel electrophoresis, the fragments can be separated according to their charge/size ratios, and the result is a pattern of

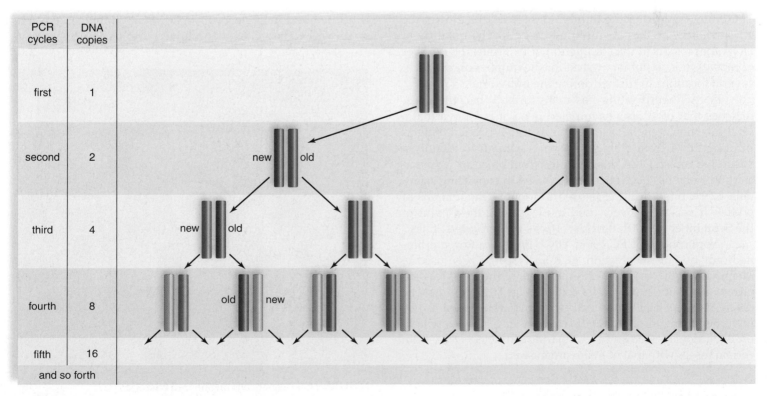

FIGURE 16.2 Polymerase chain reaction (PCR).
PCR allows the production of many identical copies of DNA in a laboratory setting.

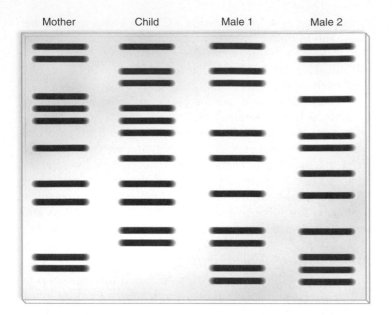

FIGURE 16.3 The use of DNA fingerprints to establish paternity.
In this example, male 1 is the father of the child. His DNA fingerprint is like that of the child except for the portion supplied by the mother.

distinctive bands. If two DNA patterns match, there is a high probability that the DNA came from the same person.

Today, DNA fingerprinting is often carried out by detecting how many times a short sequence (two to five bases) is repeated. People differ by how many repeats they have at particular locations, but how can this be detected? Recall that PCR amplifies only particular portions of the DNA. Therefore, the greater the number of repeats, the greater the amount of DNA that is amplified by PCR. The quantity of DNA that results at the completion of PCR tells the number of repeats. It is customary to test for the number of repeats at several locations to further define the individual.

DNA fingerprinting has many uses. When the DNA matches that of a virus or mutated gene, it is known that a viral infection, genetic disorder, or cancer is present. Fingerprinting DNA from a single sperm is enough to identify a suspected rapist. DNA fingerprinted from blood or tissues at a crime scene has been successfully used in convicting criminals. DNA fingerprinting is used to identify the remains of bodies. It was extensively used in identifying the victims of the September 11, 2001, terrorist attacks in the United States.

Applications of PCR and DNA fingerprinting are limited only by our imagination. PCR analysis has been used to identify unknown soldiers and members of the royal Russian family. Paternity suits can be settled (Fig. 16.3), and genetic disorders and illegally poached ivory and whale meat can be identified using these technologies. They have shed new light on evolutionary studies by comparing extracted DNA from certain fossils with that of living organisms.

Recombinant DNA technology and the polymerase chain reaction are two ways to clone a segment of DNA.

16.2 BIOTECHNOLOGY PRODUCTS

Today, bacteria, plants, and animals are genetically modified in order to have them produce a product desired by humans. The organisms themselves are called transgenic organisms, and the products they produce are called **biotechnology products.** The techniques used to genetically alter organisms are various.

Transgenic Bacteria

Recombinant DNA technology is used to produce transgenic bacteria, which are grown in huge vats called bioreactors. The gene product is collected from the medium. Biotechnology products now on the market that are produced by bacteria include insulin, clotting factor VIII, human growth hormone, t-PA (tissue plasminogen activator), and hepatitis B vaccine. Transgenic bacteria have many other uses as well. Some have been produced to promote the health of plants. For example, bacteria that normally live on plants and encourage the formation of ice crystals have been changed from frost-plus to frost-minus bacteria. As a result, new crops such as frost-resistant strawberries are being developed. Also, a bacterium that normally colonizes the roots of corn plants has now been endowed with genes (from another bacterium) that code for an insect toxin. The toxin protects the roots from insects.

Bacteria can be selected for their ability to degrade a particular substance, and this ability can then be enhanced by **genetic engineering.** For instance, naturally occurring bacteria that eat oil can be genetically engineered to do an even better job of cleaning up beaches after oil spills (Fig. 16.4). Bacteria can also remove sulfur from coal before it is burned and help clean up toxic waste dumps. One such strain was given genes that allowed it to clean up levels of

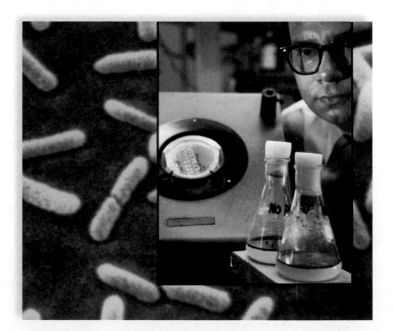

FIGURE 16.4 Genetically engineered bacteria.
Bacteria capable of decomposing oil have been engineered and patented. In the inset, the flask toward the rear contains oil and no bacteria; the flask toward the front contains the bacteria and is almost clear of oil.

toxins that would have killed other strains. Further, these bacteria were given "suicide" genes that caused them to self-destruct when the job had been accomplished.

Organic chemicals are often synthesized by having catalysts act on precursor molecules or by using bacteria to carry out the synthesis. Today, it is possible to go one step further and manipulate the genes that code for these enzymes. For instance, biochemists discovered a strain of bacteria that is especially good at producing phenylalanine, an organic chemical needed to make aspartame, the dipeptide sweetener better known as NutraSweet. They isolated, altered, and formed a vector for the appropriate genes so that various bacteria could be genetically engineered to produce phenylalanine.

Transgenic Plants

Techniques have been developed to introduce foreign genes into immature plant embryos or into plant cells called protoplasts that have had the cell wall removed. It is possible to treat protoplasts with an electric current while they are suspended in a liquid containing foreign DNA. The electric current makes tiny, self-sealing holes in the plasma membrane through which genetic material can enter. Protoplasts go on to develop into mature plants. One altered plant known as the pomato is the result of these technologies. This plant produces potatoes belowground and tomatoes aboveground.

Foreign genes transferred to cotton, corn, and potato strains have made these plants resistant to pests because their cells now produce an insect toxin. Similarly, soybeans have been made resistant to a common herbicide. Some corn and cotton plants are both pest- and herbicide-resistant. These and other genetically engineered crops that are expected to have increased yield are now sold commercially.

Plants are also being engineered to produce human proteins, such as hormones, clotting factors, and antibodies, in their seeds. One type of antibody made by corn can deliver radioisotopes to tumor cells, and another made by soybeans can be used to treat genital herpes.

Transgenic Animals

Techniques have been developed to insert genes into the eggs of animals. It is possible to microinject foreign genes into eggs by hand, but another method uses vortex mixing. The eggs are placed in an agitator with DNA and silicon-carbide needles, and the needles make tiny holes through which the DNA can enter. When these eggs are fertilized, the resulting offspring are transgenic animals. Using this technique, many types of animal eggs have taken up the gene for bovine growth hormone (bGH). The procedure has been used to produce larger fishes, cows, pigs, rabbits, and sheep.

Gene pharming, the use of transgenic farm animals to produce pharmaceuticals, is being pursued by a number of firms. Genes that code for therapeutic and diagnostic proteins are incorporated into an animal's DNA, and the proteins appear in the animal's milk. Plans are under way to produce drugs for the treatment of cystic fibrosis, cancer, blood diseases, and other disorders by this method. Figure 16.5a outlines the procedure for producing transgenic mammals: DNA containing the gene of in-

terest is injected into donor eggs. Following in vitro fertilization, the zygotes are placed in host females, where they develop. After female offspring mature, the product is secreted in their milk.

Cloning Transgenic Animals

For many years, it was believed that adult vertebrate animals could not be cloned because cloning requires that all the genes of an adult cell be turned on if development is to proceed normally. This had long been thought impossible.

In 1997, however, Scottish scientists announced that they had produced a cloned sheep called Dolly. Since then, calves and goats have also been cloned, as described in Figure 16.5b. After

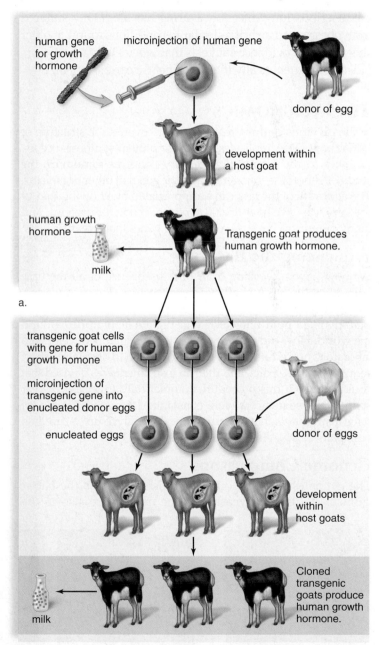

FIGURE 16.5 Transgenic mammals.
a. A genetically engineered egg develops in a host to create a transgenic goat that produces a biotechnology product in its milk. **b.** Nuclei from the transgenic goat are transferred into donor eggs, which develop into cloned transgenic goats.

enucleated eggs from a donor are microinjected with 2n nuclei from the same transgenic animal, they are coaxed to begin development in vitro. Development continues in host females until the clones are born. The offspring are clones because all have the genotype and phenotype of the adult that donated the 2n nuclei.

Now that scientists have a way to clone animals, this procedure will undoubtedly be used routinely to procure biotechnology products. However, animal cloning is a difficult process with a low success rate. The vast majority of cloning attempts are unsuccessful, resulting in the early death of the clone. The clones that survive often have health problems and die prematurely. In 2002, Dolly was euthanized after developing a serious lung infection. During her life, Dolly suffered from obesity and arthritis similar to conditions found in older sheep. Recent evidence suggests that when Dolly was 2 years old, she was genetically the age of her 6-year-old donor.

16.3 GENOMICS

In the previous century, researchers discovered the structure of DNA, how DNA replicates, and how protein synthesis occurs. Genetics in the twenty-first century concerns **genomics,** the study of genomes—our genes, and the genes of other organisms. The enormity of the task can be appreciated by knowing that, at the very least, we have 25,000 genes that code for proteins, and many organisms have more encoding genes than we do.

Sequencing the Bases

We now have a working draft of the sequence of all the base pairs in all the DNA in all our chromosomes. This feat was accomplished by the Human Genome Project, a 13-year effort that involved both university and private laboratories around the world. How did they do it? First, investigators developed a laboratory procedure that would allow them to decipher a short sequence of base pairs, and then an instrument was devised that could carry out this procedure automatically. Over the 13-year span, DNA sequencers were constantly improved. Today, we have instruments that can automatically analyze up to 2 million base pairs of DNA in a 24-hour period.

Genome Comparisons

The genomes of many other organisms, such as the bacterium *E. coli*, baker's yeast, and the mouse, are also in the

TABLE 16.1

Sequenced Genomes

2004	*Rattus norvegicus*, the brown Norway rat
2003	*Homo sapiens*, humans
2002	*Mus musculus*, the mouse; *Fugu rubripes*, the Japanese puffer fish
2000	*Drosophila melanogaster*, the fruit fly
1998	*Caenorhabditis elegans*, a form of roundworm
1997	*Escherichia coli*, a bacterium
1996	*Methanococcus jannaschii*, an archaeon
1995	*Saccharomyces cerevisiae*, baker's yeast

final-draft stage (Table 16.1). There are many similiarities between the sequence of our bases and that of other organisms. From this, we can conclude that we share a large numer of genes with much simpler organisms. Currently it is thought that our uniqueness may be linked to the regulation of our genes rather than the genes themselves.

In one study, researchers compared our genome to that of chromosome 22 in chimpanzees. Among the many genes that differed in sequence were three types of particular interest: a gene for proper speech development, several for hearing, and several for smell. The gene necessary for speech development is thought to have played an important role in human evolution. You can suppose that changes in hearing may also have facilitated using language for communication between people. Changes in smell genes are a little more problematic. The investigators speculated that the olfaction genes may have affected dietary changes or sexual selection. Or, they may have been involved in other traits, rather than just smell (Fig. 16.6).

The researchers who did this study were surprised to find that many of the other genes they located and studied are known to cause human diseases. They wondered if comparing genomes would be a way of finding other genes that are associated with human diseases. Investigators are taking all sorts of avenues to link human base sequence differences to illnesses.

The HapMap Project

The HapMap Project is a new undertaking, whose goal is to catalog common sequence differences that occur in human be-

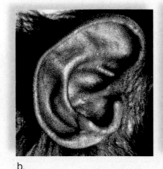
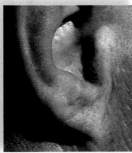
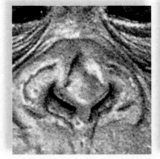
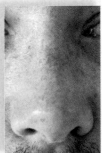

a.   b.   c.

FIGURE 16.6 Studying genomic differences between chimpanzees and humans.
Did changes in the genes for (**a**) speech, (**b**) hearing, and (**c**) smell influence the evolution of humans?

ings. People have been found to inherit patterns of sequence differences, now called haplotypes. For example, if one haplotype of a person has an A rather than a G at a particular location in a chromosome, there are probably other particular base differences near the A. To discover the most common haplotypes, genetic data from African, Asian, and European populations will be analyzed. The goal of the project is to link haplotypes to the risk for specific illnesses, with the hope that it will lead to new methods of preventing, diagnosing, and treating disease.

The Genetic Profile

One important aspect of genomics is to determine how genes work together to control the phenotype. One way to accomplish this is to analyze the genetic profile of many individuals and compare the profiles to the resulting phenotypes. DNA chips (or DNA microarrays) will soon be available that rapidly identify the complete genotype, including all the various mutations in the genome of an individual. This is called the person's genetic profile. To get a genetic profile is easy. The patient need only provide a few cells, even by simply swabbing the inside of a cheek. The DNA is removed from the cells, amplified by PCR, if need be, and then cut into fragments that are tagged by a fluorescent dye. The fragments are applied to a DNA chip, and the results are read (Fig. 16.7).

As discussed in the Health Focus on pages 274–75, knowing the genetic profile is a step toward good health. It is possible that a person has or will have a genetic disorder caused by a single pair of alleles. Polygenic traits are more common, however, and in these instances, the genetic profile will indicate an increased or decreased risk for a disorder. Risk information can be used to design a program of medical surveillance and to foster a lifestyle that will reduce the risk. For example, suppose an indvidual has mutations common to people with colon cancer. It would be helpful to have an annual colonoscopy so that any abnormal growths can be detected and removed before they are invasive.

The genetic profile may indicate what particular drug therapy or what gene therapy might be most appropriate for the individual. Drugs tend to be proteins or small molecules that affect the behavior of proteins. The field of proteomics is especially pertinent because it deals with the development of new drugs for the treatment of genetic disorders.

Proteomics

Proteomics is the study of the structure, function, and interaction of cellular proteins. All our genes—at least 25,000—are translated into proteins at some time, in some of our cells. The translation of all of these genes results in a collection of proteins, called the human proteome.

Computer modeling of the three-dimensional shape of cellular proteins is an important part of proteomics. If the primary structure of these protein is now known, it should be possible to predict their final three-dimensional shape. The study of protein function is also essential to the discovery of better drugs. One day, it may be possible to correlate drug treatment to the particular genome of the individual to increase efficiency and decrease side effects.

Bioinformatics

Bioinformatics is the application of computer technologies to the study of the genome (Fig. 16.8). Genomics and proteomics produce raw data, and these fields depend on computer analysis to find significant patterns in the data. As a result of bioinformatics, scientists are hopeful of finding cause-and-effect relationships between various genetic profiles and genetic disorders caused by multifactorial genes.

Also, the current genome sequence contains 82 gene "deserts," with no functions. Bioinformatics might find these regions have functions by correlating any sequence changes with resulting phenotypes. New computational tools will, most likely, be needed to accomplish these goals.

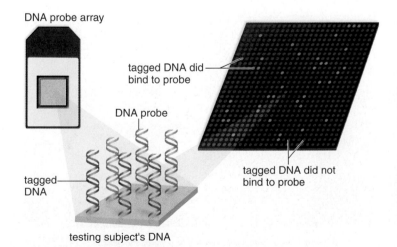

FIGURE 16.7 DNA chip.
The DNA chip contains rows of DNA sequences for mutations that indicate the presence of particular genetic disorders. If DNA fragments derived from an individual's DNA bind to a sequence, the individual has the mutation.

FIGURE 16.8 Bioinformatics.
New computer programs are being developed to make sense out of the raw data generated by genomics and proteomics. Bioinformatics allows researchers to correlate gene activity and protein function so that the phenotype can be better understood in molecular terms.

health focus

New Cures on the Horizon

Back in the 1980s, Leroy Hood could not get funding for the DNA sequencer he was developing. Biologists did not like the idea of just "collecting facts," and it took several years before an entrepreneur decided to fund the project. Without ever-better DNA sequencers, the Human Genome Project would never have completed its monumental task of determining the sequence of bases in our DNA.

Now that we know the sequence of all the bases in the DNA of all our chromosomes, biologists all over the world believe that this knowledge will result in rapid medical advances for ourselves and our children. At least four categories of improvement are expected: (1) Many more medicines will be available to keep us healthy; (2) medicines will be safer due to genome scans; (3) a longer life span, even to over 100 years, may become commonplace; and (4) we will be able to shape the genotypes of our offspring.

First prediction: Many new medicines will be available.

Genome sequence data will allow scientists to determine all the proteins that are active only during development plus all those that are still active in adults.

Most drugs are either proteins or small chemicals that are able to interact with proteins. Many of these small chemicals target proteins that act as signals between cells or within the cytoplasm of cells. Today's drugs were usually discovered in a hit-or-miss fashion, but now we can take a more systematic approach to finding effective drugs. For example, it is known that all receptor and signaling proteins start with the same ten-amino-acid sequence. Now, it is possible to scan the human genome for all genes that code for this sequence of ten amino acids and thereby find all the signaling proteins. Thereafter, they must be tested.

In a recent search for a protein that makes wounds heal, researchers cultured skin cells with 14 proteins (found by chance) that can cause skin cells to grow. Only one of these proteins made skin cells grow and did nothing else. They expect this protein to become an effective drug for conditions such as venous ulcers, which are skin lesions that affect many thousands of people in the United States. Such tests, leading to effective results, can be carried out with all the signaling proteins scientists will discover by scanning the human genome.

People's genotypes differ. We all have mutations that account for our various illnesses. Knowing each patient's mutations will allow physicians to match the right drugs to the particular patient.

Second prediction: Medicines will be safer due to genetic profiles.

Genetic profiles will allow us to discover genetically different subgroups of the population. Physicians will be looking for two types of mutations in particular. One type is called single nucleotide polymorphisms (SNPs, pronounced "snips"), in which individuals differ by only one nucleotide. The other type of mutation is nucleotide repeats, in which the same three bases, repeated over and over again, interrupt a gene and affect its expression. It is not yet clear how many SNPs will be medically significant, but the present estimate is on the order of 300,000.

How will a physician be able to determine which of the 300,000 SNPs and other types of mutations are in your genome? The use of a DNA chip will quickly and efficiently provide knowledge of your genotype. A DNA chip is an array of thousands of genes on one or several glass slides packaged together. After the DNA chip is exposed to an individual's DNA, a technician can note any mutant sequences present in the individual's genes. Soon a chip will be able to hold all the genes carried within the human genome.

Some disorders, such as sickle cell disease, are caused by a single SNP, but many disorders seem to require more than one, and possibly in different combinations. In one study, researchers found that a series of SNPs, numbered 1–12, were associated with the development of asthma. A particular drug, called albuterol, was effective for patients with certain combinations

First prediction: *Many new medicines will be available.*

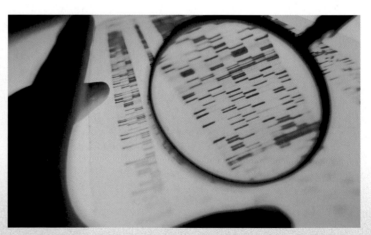

Second prediction: *Medicines will be safer due to genetic profiles.*

and not others. This example and others show that many diseases are polygenic, and that only a genetic profile is able to detect which mutations are causing an individual to have the disease and how it should be properly treated.

Genetic profiles are also expected to make drugs safer to take. As you know, many drugs potentially have unwanted side effects. Why do some people and not others have one or more of the side effects? Most likely, because people have different genotypes. It is expected that a physician will be able to match patients to drugs that are safe for them on the basis of their genotypes.

Biologists also hope that DNA chips will single out the specific oncogenes and mutated tumor suppressor genes that cause the various types of cancer. The protein products of these genes will become targets against which chemists can try to develop drugs. If so, the current methods of treating cancer—surgery, radiation, and chemotherapy to kill all dividing cells—will no longer be necessary.

Third prediction: A longer and healthier life will be yours.

Genome sequence data may allow scientists to determine which genes enable people to live longer. Investigators have already found evidence for genes that extend the life span of animals such as roundworms and fruit flies.

The sequencing of the human genome makes it possible for scientists to find such genes in humans also.

For example, we know that the presence of free radicals causes cellular molecules to become unstable and cells to die. Certain genes are believed to code for antioxidant enzymes that detoxify free radicals. It could be that human beings with particular forms of these genes have more efficient antioxidant enzymes and therefore live longer. If so, researchers will no doubt be able to locate these genes and also others that promote a longer life.

Consider, too, that natural selection favors phenotypes that result in the greatest number of fertile offspring in the next generation. Since children are usually born to younger individuals, natural selection is indifferent to genes that protect the body from the deleterious effects of aging. Researchers can possibly find such genes, however, in individuals who have a long life span. Use of these genes would possibly oppose a destiny determined so far only by evolution.

Possible stem cell therapy has generated much interest of late. Stem cells are embryonic cells and also some adult cells, such as those in red bone marrow, that are nondifferentiated. These cells have the potential to become any type of tissue, depending on which signaling proteins discovered through genomic research

are used. Stem cells can be subjected to gene therapy in order to correct any defective genes before scientists use them to create the tissues or organs of the body. Use of these tissues and organs to repair and/or replace worn-out structures could no doubt expand the human life span.

Fourth prediction: You will be able to design your children.

Genome sequence data will be used to identify many more mutant genes that cause genetic disorders than are presently known. In the future, it may be possible to cure genetic disorders before the child is born by adding a normal gene to any egg that carries a mutant gene. Or an artificial chromosome, constructed to carry a large number of corrective genes, could automatically be placed in eggs. In vitro fertilization would have to be used to take advantage of such measures for curing genetic disorders before birth.

Genome sequence data can also be used to identify polygenic genes for traits such as height, intelligence, or behavioral characteristics. A couple could decide on their own which genes they wish to use to enhance a child's phenotype. In other words, the sequencing of the human genome may bring about a genetically just society, in which all types of genes would be accessible to all parents.

Third prediction: *A longer and healthier life will be yours.*

Fourth prediction: *You will be able to design your children.*

16.4 Gene Therapy

Gene therapy is the insertion of genetic material into human cells for the treatment of a disorder. Gene therapy has been used to cure inborn errors of metabolism and also to treat more generalized disorders such as cardiovascular disease and cancer. Both ex vivo (outside the body) and in vivo (inside the body) gene therapy methods are used.

Ex Vivo Gene Therapy

Figure 16.9 describes the methodology for treating children who have SCID (severe combined immunodeficiency). These children lack the enzyme ADA (adenosine deaminase), which is involved in the maturation of T and B cells. To carry out gene therapy, bone marrow stem cells are removed from bone marrow and infected with an RNA retrovirus that carries a normal gene for the enzyme. Then the cells are returned to the patient. Bone marrow stem cells are preferred for this procedure because they divide to produce more cells with the same genes. Patients who have undergone this procedure show significantly improved immune function associated with a sustained rise in the level of ADA enzyme activity in the blood.

Among the many gene therapy trials, one is for the treatment of familial hypercholesterolemia, a condition that develops when liver cells lack a receptor protein for removing cholesterol from the blood. The high levels of blood cholesterol make the patient subject to fatal heart attacks at a young age. A small portion of the liver is surgically excised and then infected with a retrovirus containing a normal gene for the receptor before being returned to the patient. Several patients have experienced lowered serum cholesterol levels following this procedure.

In Vivo Gene Therapy

Cystic fibrosis patients lack a gene that codes for the transmembrane carrier of the chloride ion. They often suffer from numerous and potentially deadly infections of the respiratory tract. In gene therapy trials, the gene needed to cure cystic fibrosis is sprayed into the nose or delivered to the lower respiratory tract by adenoviruses or by the use of liposomes, microscopic vesicles that spontaneously form when lipoproteins are put into a solution. Investigators are trying to improve uptake, and they are also hypothesizing that a combination of all three vectors might be more successful.

Genes are being used to treat medical conditions such as poor coronary circulation. It has been known for some time that VEGF (vascular endothelial growth factor) can cause the growth of new blood vessels. The gene that codes for this growth factor can be injected alone or within a virus into the heart to stimulate branching of coronary blood vessels. Patients report that they have less chest pain and can run longer on a treadmill.

Gene therapy is increasingly used as a part of cancer therapy. Genes are being used to make healthy cells more tolerant of chemotherapy and to make tumors more vulnerable to chemotherapy. The gene *p53* brings about apoptosis, and there is much interest in introducing it into cancer cells and in that way killing them off.

Both ex vivo and in vivo methods of gene therapy are playing a role in curing illnesses.

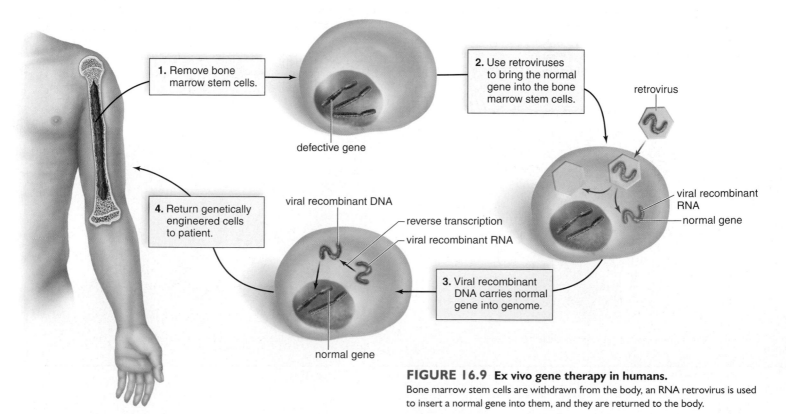

1. Remove bone marrow stem cells.

defective gene

2. Use retroviruses to bring the normal gene into the bone marrow stem cells.

retrovirus

viral recombinant RNA

normal gene

3. Viral recombinant DNA carries normal gene into genome.

viral recombinant DNA

reverse transcription

viral recombinant RNA

4. Return genetically engineered cells to patient.

normal gene

FIGURE 16.9 Ex vivo gene therapy in humans.
Bone marrow stem cells are withdrawn from the body, an RNA retrovirus is used to insert a normal gene into them, and they are returned to the body.

CONNECTING THE CONCEPTS

Basic research into the nature and organization of genes in various organisms allowed geneticists to produce recombinant DNA molecules. A knowledge of transcription and translation makes it possible for scientists to manipulate the expression of genes in foreign organisms. These breakthroughs have spurred a biotechnology revolution. One result is that bacteria and eukaryotic cells are now used to produce vaccines, hormones, and growth factors for use in humans. Today, plants and animals are also engineered to make a product or to possess characteristics desired by humans. Some people are concerned that transgenic organisms may spread out of control and wreak ecological havoc, much as

plants and animals new to an area sometimes do. This is often called the chimera effect. Or they hypothesize that transgenic plants may pass on their traits, such as resistance to pesticides or herbicides, to weeds that will then prosper as never before. It's also possible, as discussed in the bioethical issue on page 279, that transgenic plants may endanger our health.

On the other hand, biotechnology offers the promise of treating and even someday curing human genetic disorders such as muscular dystrophy, cystic fibrosis, hemophilia, and many others. It's possible that genomic research will discover the loci of many other genetic disorders. This information is the first step toward

curing a disorder by the use of gene therapy. It will also allow us to determine people's genetic profiles for the purpose of prescribing medications and preventing future illness. Is it possible, however, that employers or insurance companies might use genetic profiling for discriminatory purposes?

The alteration of organisms as a result of genetic changes is one of the definitions of evolution. Are we changing the course of evolution when we create transgenic organisms or modify our offspring through gene therapy? It will take some time to realize whether evolution, the topic of Part III, will be affected by biotechnology or not.

Summary

16.1 DNA CLONING

DNA cloning can lead to multiple copies of a gene and also a quantity of the gene product. The gene can be studied in the laboratory or inserted into a bacterium, plant, or animal. The gene product, a protein, can become a commercial product for use as a medicine.

Two methods are currently available for making copies of DNA: recombinant DNA technology and the polymerase chain reaction (PCR). Recombinant DNA contains DNA from two different sources. A restriction enzyme is used to cleave plasmid DNA and to cleave foreign DNA. The resulting "sticky ends" facilitate the insertion of foreign DNA into vector DNA. The foreign gene is sealed into the vector DNA by DNA ligase. Both bacterial plasmids and viruses can be used as vectors to carry foreign genes into bacterial host cells.

PCR uses the enzyme DNA polymerase to quickly make multiple copies of a specific piece (target) of DNA. PCR is a chain reaction because the targeted DNA is repeatedly replicated as long as the process continues. Analysis of DNA segments following PCR can be preparatory to determining DNA base sequences or to carry out DNA fingerprinting.

16.2 BIOTECHNOLOGY PRODUCTS

Transgenic organisms have had a foreign gene inserted into them. Genetically engineered bacteria, agricultural plants, and farm animals now produce commercial products of interest to humans, such as hormones and vaccines. Bacteria secrete the product. The seeds of plants and the milk of animals contain the product.

Transgenic bacteria have also been engineered to promote the health of plants, perform bioremediation, extract minerals, and produce chemicals. Transgenic crops, engineered to resist herbicides and pests, are commercially available. Transgenic animals have been given various genes, in particular the one for bovine growth hormone (bGH). Cloning of animals is now possible.

16.3 GENOMICS

Researchers now know the sequence of the base pairs along the length of the human chromosomes. So far, researchers have found only 25,000

genes that code for proteins; the rest of our DNA consists of nucleotide repeats that do not code for a protein. Little difference seems to exist between the DNA sequence of our bases and those of other organisms that have also been studied. It's possible we will discover that our uniqueness is due to the regulation of our genes.

Researchers hope knowing the sequence of the human genome will help them discover the root causes of many illnesses. Also, it may lead to the development of specific drugs based on DNA differences between individuals. Two new fields, proteomics and bioinformatics, will assist in this endeavor.

16.4 GENE THERAPY

Gene therapy, by either ex vivo or in vivo methods, is used to correct the genotype of humans and to cure various human ills. Ex vivo gene therapy has apparently helped children with SCID lead normal lives. In vivo treatment for cystic fibrosis has been less successful. A number of in vivo therapies are being employed in the war against cancer and other human illnesses, such as cardiovascular disease.

Reviewing the Chapter

1. What is the methodology for producing recombinant DNA useful for gene cloning? 268
2. What is the polymerase chain reaction (PCR), and how is it carried out to produce multiple copies of a DNA segment? 269
3. What is DNA fingerprinting? 269
4. What are some practical applications of DNA segment analysis following PCR? 270
5. For what purposes have bacteria, plants, and animals been genetically altered? 270–72
6. How does our genome compare to others that have been sequenced? 272
7. What are the possible benefits of the Human Genome Project? 273–5
8. Explain and give examples of ex vivo and in vivo gene therapies in humans. 276
9. Discuss the pros and cons of genetic engineering. 276

Testing Yourself

Choose the best answer for each question.

1. Using this key, put the phrases in the correct order to form a plasmid-carrying recombinant DNA.

KEY:

(1) use restriction enzymes
(2) use DNA ligase
(3) remove plasmid from parent bacterium
(4) introduce plasmid into new host bacterium

    a. 1, 2, 3, 4   c. 3, 1, 2, 4
    b. 4, 3, 2, 1   d. 2, 3, 1, 4

2. Which is not a clone?
    a. a colony of identical bacterial cells
    b. identical quintuplets
    c. a forest of identical trees
    d. eggs produced by oogenesis
    e. copies of a gene through PCR

3. Restriction enzymes found in bacterial cells are ordinarily used
    a. during DNA replication.
    b. to degrade the bacterial cell's DNA.
    c. to degrade viral DNA that enters the cell.
    d. to attach pieces of DNA together.

4. Which of these would you not expect to be a biotechnology product?
    a. vaccine   c. protein hormone
    b. modified enzyme   d. steroid hormone

5. Recombinant DNA technology is used
    a. for gene therapy.
    b. to clone a gene.
    c. to make a particular protein.
    d. to clone a specific piece of DNA.
    e. All of these are correct.

6. For bacterial cells to express human genes,
    a. the recombinant DNA must not contain introns.
    b. reverse transcriptase is sometimes used to make complementary DNA from an mRNA molecule.
    c. bacterial regulatory genes must be included.
    d. a vector is used to deliver the recombinant DNA.
    e. All of these are correct.

7. The polymerase chain reaction
    a. uses RNA polymerase.
    b. takes place in huge bioreactors.
    c. uses a temperature-insensitive enzyme.
    d. makes lots of nonidentical copies of DNA.
    e. All of these are correct.

8. DNA fingerprinting can be used for which of these?
    a. identifying human remains
    b. identifying infectious diseases
    c. finding evolutionary links between organisms
    d. solving crimes
    e. All of these are correct.

9. DNA amplified by PCR and then used for fingerprinting could come from
    a. any diploid or haploid cell.
    b. only white blood cells that have been karyotyped.
    c. only skin cells after they are dead.
    d. only purified animal cells.
    e. Both b and d are correct.

10. Which of these pairs is incorrectly matched?
    a. DNA ligase—mapping human chromosomes
    b. protoplast—plant cell engineering
    c. DNA fragments—DNA fingerprinting
    d. DNA polymerase—PCR

11. Which is not a correct association with regard to genetic engineering?
    a. plasmid as a vector—bacteria
    b. protoplast as a vector—plants
    c. RNA retrovirus as a vector—human stem cells
    d. All of these are correct.

12. Which of these is an incorrect statement?
    a. Bacteria secrete the biotechnology product into the medium.
    b. Plants are being engineered to have human proteins in their seeds.
    c. Animals are engineered to have a human protein in their milk.
    d. Animals can be cloned, but plants and bacteria cannot.

13. Which of these is not needed to clone an animal?
    a. sperm from a donor animal
    b. nucleus from an adult animal cell
    c. enucleated egg from a donor animal
    d. host female to develop the embryo
    e. All of these are needed.

14. Because of the Human Genome Project, we know or will know
    a. the sequence of the base pairs of our DNA.
    b. the sequence of genes along the human chromosomes.
    c. the mutations that lead to genetic disorders.
    d. All of these are correct.
    e. Only a and c are correct.

15. Gene therapy has been used to treat which of these?
    a. cystic fibrosis
    b. familial hypercholesterolemia
    c. severe combined immunodeficiency
    d. All of these are correct.

16. The restriction enzyme called *Eco*RI has cut double-stranded DNA in the following manner. The piece of foreign DNA to be inserted has what bases from the left and from the right?

17. Which of these is a true statement?
    a. Plasmids can serve as vectors.
    b. Plasmids can carry recombinant DNA, but viruses cannot.
    c. Vectors carry only the foreign gene into the host cell.
    d. Only gene therapy uses vectors.
    e. Both a and d are correct.

18. Which of these is a benefit of having insulin produced by biotechnology?
    a. It is just as effective.
    b. It can be mass-produced.
    c. It is nonallergenic.
    d. It is less expensive.
    e. All of these are correct.

19. What is the benefit of using a retrovirus as a vector in gene therapy?
    a. It is not able to enter cells.
    b. It incorporates the foreign gene into the host chromosome.
    c. It eliminates a lot of unnecessary steps.
    d. It prevents infection by other viruses.
    e. Both b and c are correct.

20. Gene therapy
    a. is still an investigative procedure.
    b. has met with no success.
    c. is only used to cure genetic disorders such as SCID and cystic fibrosis.
    d. makes use of viruses to carry foreign genes into human cells.
    e. Both a and d are correct.

21. These drawings pertain to gene therapy. Label the drawings, using these terms: retrovirus, recombinant RNA (twice), defective gene, recombinant DNA, reverse transcription, and human genome.

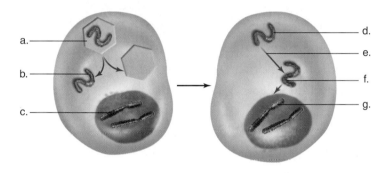

Thinking Scientifically

1. A library is a set of cloned DNA segments that altogether are representative of the genome of an organism. cDNA libraries contain only expressed DNA sequences for a particular cell. Therefore, a cDNA library produced for a liver cell will contain genes unique to a liver cell. Knowing that a probe is a strand of DNA that will bind with a complementary strand, how could a cDNA library for a liver cell be used to acquire a complete copy of a liver-cell gene from a complete DNA library? A complete gene contains the promoter and introns.

2. There has been much popular interest in re-creating extinct animals from DNA obtained from various types of fossils. However, such DNA is always badly degraded, consisting of extremely short pieces. Even if one hypothetically had ten intact genes from a dinosaur, why might it still be impossible to create a dinosaur?

Bioethical Issue: Transgenic Crops

Transgenic plants can possibly allow crop yields to keep up with the ever-increasing worldwide demand for food. And some of these plants have the added benefit of requiring less fertilizer and/or pesticides, which are harmful to human health and the environment.

But some scientists believe transgenic crops pose their own threat to the environment, and many activists believe transgenic plants are themselves dangerous to our health. Studies have shown that wind-carried pollen can cause transgenic crops to hybridize with nearby weedy relatives. Although it has not happened yet, some fear that characteristics acquired in this way might cause weeds to become uncontrollable pests. Or perhaps a toxin produced by transgenic crops could possibly hurt other organisms in the field. Many researchers are conducting tests to see if this might occur. Also, although transgenic crops have not caused any illnesses in humans so far, some scientists concede the possibility that people could be allergic to the transgene's protein product. After unapproved genetically modified corn was detected in Taco Bell taco shells several years ago, a massive recall pulled about 2.8 million boxes of the product from grocery stores.

Already, transgenic plants must be approved by the Food and Drug Administration before they are considered safe for human consumption, and they must meet certain Environmental Protection Administration standards. Some people believe safety standards for transgenic crops should be further strengthened, while others fear stricter standards will result in less food produced. Another possibility is to retain the current standards but require all biotech foods to be clearly labeled so the buyer can choose whether or not to eat them.

Understanding the Terms

bioinformatics   273
biotechnology products   270
cloning   268
complementary DNA (cDNA)   269
DNA fingerprinting   269
DNA ligase   268
gene cloning   268
gene therapy   268
genetic engineering   270
genomics   272
plasmid   268
polymerase chain reaction (PCR)   269
proteomics   273
recombinant DNA (rDNA)   268
restriction enzyme   268
transgenic organism   268
vector   268

Match the terms to these definitions:

a. _____ Bacterial enzyme that stops viral reproduction by cleaving viral DNA; used to cut DNA at specific points during production of recombinant DNA.

b. _____ Free-living organisms in the environment that have had a foreign gene inserted into them.

c. _____ All the genetic information of an individual or a species.

d. _____ Production of identical copies; in genetic engineering, the production of many identical copies of a gene.

e. _____ Self-duplicating ring of accessory DNA in the cytoplasm of bacteria.

ARIS, the *Biology* Website

ARIS, the website for *Biology*, provides a wealth of information organized and integrated by chapter. You will find practice quizzes, interactive activities, labeling exercises, flashcards, and much more that will complement your learning and understanding of general biology.

www.mhhe.com/maderbiology9

PART III

EVOLUTION

Evolution refers to both descent with modification and adaptation to the environment. Descent from a common ancestor explains the unity of life—living things share a common chemistry and cellular structure because they are all descended from the same original source. Each type of living thing has a history that can be traced by way of the fossil record and discerned from a comparative study of other living things.

Adaptation to a particular environment explains the diversity of life. Natural selection is a mechanism that results in adaptation to the environment. Individuals with variations that make them better adapted have more offspring than those who are not as well adapted. In that way, certain characteristics become more common among a population of organisms. Each species has its own unique, evolved solutions to life's challenges, such as how to acquire nutrients, find a mate, and reproduce. The chapters in this part will present the theory of evolution, the evolution of life, and the classification of organisms according to our current knowledge of evolution.

17

DARWIN AND EVOLUTION

Modern geologists believe that the Earth is more than 4 billion years old and that life began about 3.8 billion years ago. From simple, unicellular organisms, new life-forms arose and changed in response to environmental pressures, producing the past and present biodiversity. Prior to the 1800s, however, most people thought the Earth was only a few thousand years old. They also believed that species were specially created and fixed in time. Change was explained by the notion that global catastrophes—mass extinctions—followed by repopulation by new species had occurred periodically throughout history.

Into this prevailing climate came Charles Darwin, Alfred Wallace, and other innovative minds who changed the field of biology forever. On a five-year ocean voyage, Darwin observed many diverse life-forms, and he eventually concluded that (1) species change over time in response to their environment, and (2) all living things share characteristics because they have a single common ancestry. His theory of evolution challenged the widely accepted biblical account of creation, and acceptance of Darwin's ideas required an intellectual revolution of great magnitude.

The modern killer whale, *Orca*, and a fossil ancestor with vestigial limbs.

17.1 HISTORY OF EVOLUTIONARY THOUGHT

In December 1831, a new chapter in the history of biology had its humble origins. A 22-year-old naturalist, Charles Darwin (1809–82), set sail on a journey of a lifetime aboard the British naval vessel the HMS *Beagle* (Fig. 17.1). Darwin's primary mission on his journey around the world was to expand the navy's knowledge of natural resources such as water and food in foreign lands. The captain of the *Beagle*, Robert Fitzroy, also hoped that Darwin would find evidence to support the biblical account of creation. Contrary to Fitzroy's wishes, Darwin amassed observations that would eventually support another way of thinking and change the history of biology and science forever. The pre-Darwinian worldview and the post-Darwinian worldview are contrasted in Table 17.1. Before Darwin, the worldview was forged by deep-seated beliefs that were held to be intractable truths and not by experimentation and observation of the natural word. In contrast, Darwin used all sorts of data to come to the conclusion that the Earth is very old, not young, and that biological evolution is the method by which species arise and change. The acceptance of the Darwinian view of the world was fostered by a scientific and intellectual revolution that occurred in both the scientific and social realms of the mid-1800s.

FIGURE 17.1

Voyage of the HMS *Beagle*.

a. Map shows the journey of the HMS *Beagle* around the world. Notice that the map's encircled colors are keyed to the encircled colors in the photographs, which show us what Charles Darwin may have observed in or near South America.
b. As Darwin traveled along the east coast of South America, he noted that a bird called a rhea looked like the African ostrich.
c. The sparse vegetation of the Patagonian Desert is in the southern part of the continent. **d.** The Andes Mountains of the west coast have strata containing fossilized animals. **e.** The lush vegetation of a rain forest contains its own set of plants and animals. **f.** On the Galápagos Islands, marine iguanas have large claws to help them cling to rocks and blunt snouts for eating algae growing on rocks. **g.** Galápagos finches are specialized to feed in various ways. This finch is using a cactus spine to probe for insects.

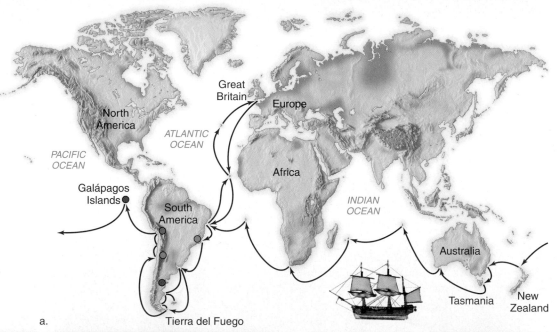

TABLE 17.1

Contrast of Worldviews

Pre-Darwinian View	Post-Darwinian View
1. The Earth is relatively young—age is measured in thousands of years.	1. The Earth is relatively old—age is measured in billions of years.
2. Each species is specially created; species don't change, and the number of species remains the same.	2. Species are related by descent—it is possible to piece together a history of life on Earth.
3. Adaptation to the environment is the work of a creator, who decided the structure and function of each type of organism. Any variations are imperfections.	3. Adaptation to the environment is the interplay of random variations and environmental conditions.
4. Observations are supposed to substantiate the prevailing worldview.	4. Observation and experimentation are used to test hypotheses, including hypotheses about evolution.

Although it is often believed that Darwin (Fig. 17.2) forged this change in worldview by himself, several biologists during the preceding century and some of Darwin's contemporaries slowly began to accept the idea that species change over time. This concept would eventually be known as **evolution** [L. *evolutio*, an unrolling]. Evolution is the unifying principle of the biological sciences. It can be used to explain both the unity and diversity of life on Earth. First, evolution illustrates that living things share common characteristics because they share a common ancestry. Evolution also can explain how species adapt to specific habitats and ways of life.

The history of evolutionary thought is a history of ideas about descent and adaptation. Darwin used the expression "descent with modification," by which he meant that as descent occurs through time, so does diversification. Darwin brilliantly saw the process of adaptation as a means by which the diversity of species arises.

FIGURE 17.2 Charles Darwin.
Portrait of Charles Darwin at the age of 31.

Earth was an imperfect copy of an ideal form, which can be deduced upon reflection and study. To Plato, individual variations were imperfections that only distract the observer. Aristotle saw that organisms were diverse and some were more complex than others. His belief that all organisms could be arranged in order of increasing complexity became the *scala naturae* just described.

Linnaeus and other taxonomists wanted to describe the ideal characteristics of each species and also wanted to discover the proper place for each species in the *scala naturae*. Therefore, for most of his working life, Linnaeus did not even consider the possibility of evolutionary change. There is evidence, however, that he did eventually perform hybridization experiments, which made him think that a species might change with time.

Georges-Louis Leclerc, better known by his title, Count Buffon (1707–88), was a French naturalist who devoted many years of his life to writing a 44-volume natural history that described all known plants and animals. He provided evidence of descent with modification, and he even speculated on various causative mechanisms such as environmental influences, migration, geographic isolation, and the struggle for existence. Buffon seemed to waver, however, as to whether or not he believed in evolutionary descent, and often he professed to believe in special creation and the fixity of species.

Erasmus Darwin (1731–1802), Charles Darwin's grandfather, was a physician and a naturalist. His writings on both botany and zoology contained many comments, although they were mostly in footnotes and asides, that suggested the possibility of common descent. He based his conclusions on changes undergone by animals during development, artificial selection by humans, and the presence of **vestigial organs** (organs that are believed to have been functional in an ancestor but are reduced and nonfunctional in a descendant). Like Buffon, Erasmus Darwin offered no mechanism by which evolutionary descent might occur.

Mid-Eighteenth-Century Contributions

Taxonomy, the science of classifying organisms, was an important endeavor during the mid-eighteenth century. Chief among the taxonomists was Carolus Linnaeus (1707–78), who developed the binomial system of nomenclature (a two-part name for species, such as *Homo sapiens*) and who developed a system of classification for living things. Linnaeus, like other taxonomists of his time, believed in the fixity of species. Each species had an "ideal" structure and function and also a place in the *scala naturae*, a sequential ladder of life. The simplest and most material being was on the lowest rung of the ladder, and the most complex and spiritual being was on the highest rung. In this view, human beings occupied the highest rung of the ladder.

These ideas, which were consistent with Judeo-Christian teachings about special creation, can be traced to the works of the famous Greek philosophers Plato (427–347 BC) and Aristotle (384–322 BC). Plato said that every object on

Late-Eighteenth-/Early-Nineteenth-Century Contributions

Cuvier and Catastrophism

In addition to taxonomy, comparative anatomy was of interest to biologists prior to Darwin. Explorers and collectors traveled the world and brought back not only currently existing species to be classified but also fossils (remains of once-living organisms) to be studied. Baron Georges Cuvier (1769–1832), a distinguished zoologist, was the first to use comparative anatomy to develop a system of classifying animals. He also founded the science of **paleontology** [Gk. *palaios*, old; *ontos*, having existed; *-logy,* study of], the study of fossils, and was quite skilled at using fossil bones to deduce the structure of an animal (Fig. 17.3*a*).

Because Cuvier was a staunch advocate of special creation and the fixity of species, he faced a real problem when a particular region showed a succession of life-forms in the Earth's strata (layers). To explain these observations, he hypothesized that a series of local catastrophes or mass extinctions had occurred whenever a new stratum of that region showed a new mix of fossils. After each catastrophe, the region was repopulated by species from surrounding areas, and this accounted for the appearance of new fossils in the new stratum. The result of all these catastrophes was change appearing over time. Some of Cuvier's followers even suggested that there had been worldwide catastrophes and that after each of these events, God created new sets of species. This explanation of the history of life came to be known as **catastrophism** [Gk. *katastrophe,* calamity, misfortune].

Lamarck and Acquired Characteristics

Jean-Baptiste de Lamarck (1744–1829) was the first biologist to believe that evolution does occur and to link diversity with adaptation to the environment. Lamarck's ideas about descent were entirely different from those of Cuvier, perhaps because Lamarck specialized in the study of invertebrates (animals without backbones), while Cuvier was a vertebrate zoologist, who studied animals with backbones. Lamarck concluded, after studying the succession of life-forms in strata, that more complex organisms are descended from less complex organisms. He mistakenly said, however, that increasing complexity is the result of a natural force—a desire for perfection—that is inherent in all living things.

To explain the process of adaptation to the environment, Lamarck supported the idea of **inheritance of acquired characteristics**—that the environment can bring about inherited change. One example that he gave—and for which he is most famous—is that the long neck of a giraffe developed over time because animals stretched their necks to reach food in tall trees and then passed on a long neck to their offspring (Fig. 17.3*b*). His hypothesis of the inheritance of acquired characteristics has never been substantiated by experimentation. The molecular mechanism of inheritance explains why. Phenotypic changes acquired during an organism's lifetime do not result in genetic changes that can be passed to subsequent generations.

a.

b.

FIGURE 17.3 Evolutionary thought before Darwin.
a. Cuvier reconstructed animals such as extinct mastodons and said that catastrophes followed by repopulations could explain why species change over time. **b.** Lamarck explained the long neck of a giraffe according to his ideas about the inheritance of acquired characteristics.

17.2 DARWIN'S THEORY OF EVOLUTION

When Darwin signed on as the naturalist aboard the HMS *Beagle*, he possessed a suitable background for the position. Since childhood, he was an ardent student of nature and a collector of insects. At 16, Darwin was sent to medical school to follow in the footsteps of his grandfather and father. However, his sensitive nature prevented him from studying medicine, and he enrolled in the school of divinity at Christ College at Cambridge, intent on becoming a clergyman.

While at Christ College, he attended many lectures in biology and geology to satisfy his interest in natural science. During this time, he became the protégé of his teacher and friend, Reverend John Henslow. Darwin gained valuable experience in geology in the summer of 1831, conducting fieldwork with Adam Sedgewick. Shortly after Darwin was awarded his BA, Henslow recommended him to serve as ship's naturalist on the HMS *Beagle*. The trip was to take five years and the ship was to traverse the Southern Hemisphere (see Fig. 17.1), where life is most abundant and varied. Along the way, Darwin encountered forms of life very different from his native England.

Occurrence of Descent

Although it was not his original intent, Darwin began to gather evidence that organisms are related through common descent and that adaptation to various environments results in diversity. Darwin also began contemplating the "mystery of mysteries," the origin of new species.

Geology and Fossils

Darwin took Charles Lyell's *Principles of Geology* on the voyage. This book presented arguments to support a theory of geological change proposed by James Hutton. In contrast to the catastrophists, Hutton believed the Earth was subject to slow but continuous cycles of erosion and uplift. Weather causes erosion; thereafter, dirt and rock debris are washed into the rivers and transported to oceans. These loose sediments are deposited in thick layers, which are converted eventually into sedimentary rocks (Fig. 17.4). Then sedimentary rocks, which often contain fossils, are uplifted from below sea level to form land. Hutton concluded that extreme geological changes can be accounted for by slow, natural processes, given enough time. Lyell went on to propose a theory of **uniformitarianism,** which stated that these slow changes

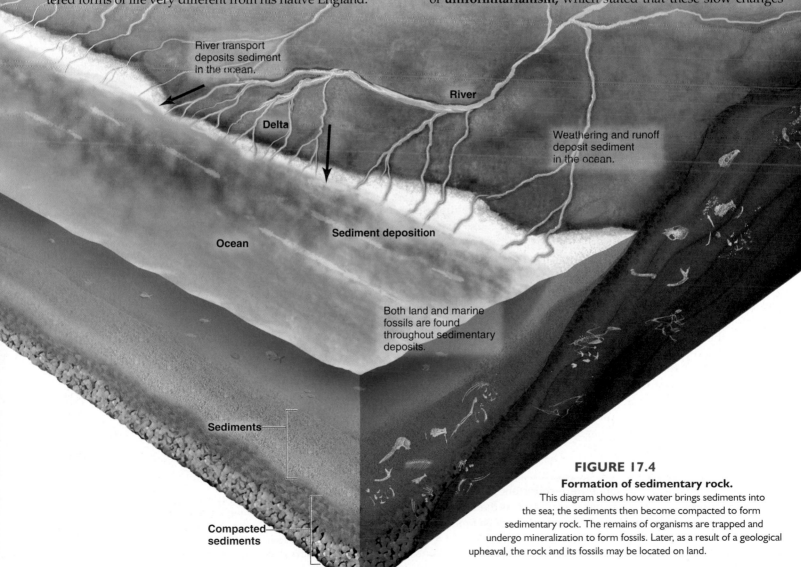

River transport deposits sediment in the ocean.

River

Delta

Weathering and runoff deposit sediment in the ocean.

Ocean

Sediment deposition

Both land and marine fossils are found throughout sedimentary deposits.

Sediments

Compacted sediments

FIGURE 17.4
Formation of sedimentary rock.
This diagram shows how water brings sediments into the sea; the sediments then become compacted to form sedimentary rock. The remains of organisms are trapped and undergo mineralization to form fossils. Later, as a result of a geological upheaval, the rock and its fossils may be located on land.

FIGURE 17.5 A glyptodont and a giant sloth.
a. A giant armadillo-like glyptodont, *Glyptodon*, is known only by the study of its fossil remains. Darwin found such fossils and came to the conclusion that this extinct animal must be related to living armadillos. The glyptodont weighed 2,000 kg. **b.** Darwin also observed the fossil remains of a giant ground sloth, *Mylodon*. These animals exceeded 6 m when standing.

occurred at a uniform rate. Hutton's general ideas about slow and continual geological change are still accepted today, although modern geologists realize that rates of change have not always been uniform. Darwin was not taken by the idea of uniform change, but he was convinced, as was Lyell, that the Earth's massive geological changes are the result of slow processes and that the Earth, therefore, must be very old.

On his trip, Darwin observed massive geological changes firsthand. When he explored what is now Argentina, he saw raised beaches for great distances along the coast. When he got to the Andes Mountains, he was impressed by their great height. In Chile, he found marine shells inland, well above sea level, and witnessed the effects of an earthquake that caused the land to rise several feet. While Darwin was making geological observations, he also collected fossil specimens. For example, on the east coast of South America, he found the fossil remains of an armadillo-like animal and a giant ground sloth (Fig. 17.5). Once Darwin accepted the supposition that the Earth must be very old, he began to think that there would have been enough time for descent with modification to occur. Therefore, living forms could be descended from extinct forms known only from the fossil record. It would seem that species were not fixed; instead, they changed over time.

Darwin's geological observations were consistent with those of Hutton and Lyell. He began to think that the Earth was very old and that there would have been enough time for descent with modification to occur.

Biogeography

Biogeography [Gk. *bios*, life, *geo*, earth, and *grapho*, writing] is the study of the range and geographic distribution of lifeforms on Earth. Darwin could not help but compare the animals of South America to those with which he was familiar. For example, instead of rabbits, he found the Patagonian hare in the grasslands of South America. The Patagonian hare has long legs and ears but the face of a guinea pig, a rodent native to South America (Fig. 17.6). Did the Patagonian hare resemble a rabbit because the two types of animals were adapted to the same type of environment? Both animals ate grass, hid in bushes, and moved rapidly using long hind legs. Did the Patagonian hare have the face of a guinea pig because of common descent with guinea pigs?

As he sailed southward along the eastern coast of the continent of South America, Darwin saw how similar species replaced each other. For example, the greater rhea (an ostrichlike bird) found in the north was replaced by the lesser rhea in the south. Therefore, Darwin reasoned that related species could be modified according to the environment. When he explored the Galápagos Islands, he found further evidence of this phenomenon. The Galápagos Islands are a small group of volcanic islands off the western coast of South America. The few types of plants and animals found there were slightly different from species Darwin had observed on the mainland, and even more important, they also varied from island to island.

Tortoises. Each of the Galápagos Islands seemed to have its own type of tortoise, and Darwin began to wonder if this could be correlated with a difference in vegetation among the islands (Fig. 17.7). Long-necked tortoises seemed to inhabit only dry areas, where food was scarce, most likely because

FIGURE 17.6 The Patagonian hare, *Dolichotis patagonium*.
This animal has the face of a guinea pig and is native to South America, which has no native rabbits. The Patagonian hare has long legs and other adaptations similar to those of rabbits, although they are not closely related to rabbits or hares.

a.   b.

FIGURE 17.7 Galápagos tortoises, *Geochelone*.
Darwin wondered if all of the tortoises of the various islands were descended from a common ancestor. **a.** The tortoises with dome shells and short necks feed at ground level and are from well-watered islands where grass is available. **b.** Those with shells that flare up in front have long necks and are able to feed on tall, treelike cacti. They are from arid islands where prickly pear cactus is the main food source. Only on these islands are the cacti treelike.

a. *Geospiza magnirostris*   b. *Certhidea olivacea*   c. *Cactornis scandens*

FIGURE 17.8 Galápagos finches.
Each of the present-day 13 species of finches has a beak adapted to a particular way of life. **a.** For example, the heavy beak of the large ground-dwelling finch is suited to a diet of seeds. **b.** The beak of the warbler-finch is suited to feeding on insects found among ground vegetation or caught in the air. **c.** The longer beak, somewhat decurved, and the split tongue of the cactus-finch is suited to probing cactus flowers for nectar.

the longer neck was helpful in reaching cacti. In moist regions with relatively abundant ground foliage, short-necked tortoises were found. Had an ancestral tortoise given rise to these different types, each adapted to a different environment?

Finches. Darwin almost overlooked the finches because of their unassuming nature compared with many of the other animals in the Galápagos. However, these birds would eventually play a major role in his thoughts about geographic isolation. The finches of the Galápagos Islands seemed to Darwin like mainland finches, but they exhibited significant variety (Fig. 17.8). Today, there are ground-dwelling finches with different-sized beaks, depending on the size of the seeds they feed on, and a cactus-eating finch with a more pointed beak. The beak size of the tree-dwelling finches also varies but according to the size of their insect prey. The most unusual of the finches is a woodpecker-type finch. This bird has a sharp beak to chisel through tree bark but lacks the woodpecker's long tongue, which probes for insects. To

make up for this, the bird carries a twig or cactus thorn in its beak and uses it to poke into crevices (see Fig. 17.1*g*). Once an insect emerges, the finch drops this tool and seizes the insect with its beak.

Later, Darwin speculated as to whether all the different species of finches he had seen could have descended from a type of mainland finch. In other words, he wondered if a mainland finch was the common ancestor to all the types on the Galápagos Islands. Had speciation occurred because the islands allowed isolated populations of birds to evolve independently? Could the present-day species have resulted from accumulated changes occurring within each of these isolated populations?

Biogeography had a powerful influence on Darwin and made him think that adaptation to the environment accounts for diversification; one species can give rise to many species, each adapted differently.

Natural Selection and Adaptation

Upon returning to England, Darwin began to reflect on the voyage of the *Beagle* and to collect evidence supporting his ideas about how organisms adapt to the environment. Darwin decided that adaptations develop over time (instead of being the instant work of a creator), and he began to think about a mechanism that would allow this to happen. Both Darwin and Alfred Russel Wallace, who is discussed in the Science Focus on page 289, proposed **natural selection** as a mechanism for evolutionary change in the late 1850s. Natural selection is a process in which the following preconditions (1–3) may result in certain consequences (4–5):

1. The members of a population have heritable variations.
2. In a population, many more individuals are produced each generation than the environment can support.
3. Some individuals have adaptive characteristics that enable them to survive and reproduce better than other individuals.
4. An increasing proportion of individuals in succeeding generations have the adaptive characteristics.
5. The result of natural selection is a population adapted to its local environment.

FIGURE 17.9 Variation in a population.
For Darwin, variations, such as those seen in a human population, were highly significant and were required for natural selection to result in adaptation to the environment.

Natural selection uses only variations that happen to be provided by genetic changes; it lacks any directedness or anticipation of future needs. Natural selection is an ongoing process because the environment of living things is constantly changing. Extinction (loss of a species) can occur when previous adaptations are no longer suitable to a changed environment.

Organisms Have Variations

With reference to precondition 1, Darwin emphasized that the members of a population vary in their functional, physical, and behavioral characteristics (Fig. 17.9). Before Darwin, variations were imperfections that should be ignored since they were not important to the description of a species (see Table 17.1). Darwin believed variations were essential to the natural selection process. He suspected—but did not have the evidence we have today—that the occurrence of variations is completely random; they arise by accident and for no particular purpose. New variations are just as likely to be harmful as helpful to the organism.

The variations that make adaptation to the environment possible are those that are passed on from generation to generation. The science of genetics was not yet well established, so Darwin was never able to determine the cause of variations or how they are passed on. Today, we realize that genes, along with the environment, determine the phenotype of an organism, and that mutations and recombination of alleles during sexual reproduction can cause new variations to arise.

Organisms Struggle to Exist

With reference to precondition 2, in Darwin's time, a socioeconomist, Thomas Malthus, stressed the reproductive potential of human beings. He proposed that death and famine were inevitable because the human population tends to increase faster than the supply of food. Darwin applied this concept to all organisms and saw that the available resources were not sufficient for all members of a population to survive. He calculated the reproductive potential of elephants. Assuming a life span of about 100 years and a breeding span of from 30–90 years, a single female probably bears no fewer than six young. If all these young survive and continue to reproduce at the same rate, after only 750 years, the descendants of a single pair of elephants will number about 19 million! This overproduction potential of a species is often referred to as Darwin's geometric ratio of increase.

Each generation has the same reproductive potential as the previous generation. Therefore, there is a constant struggle for existence, and only certain members of a population survive and reproduce each generation.

Organisms Differ in Fitness

With reference to precondition 3, **fitness** is the reproductive success of an individual relative to other members of a population. The most-fit individuals are the ones that capture a disproportionate amount of resources, and that convert these resources into a larger number of viable offspring. Since organisms vary anatomically and physiologically and the challenges of local environments vary, what determines fitness varies for

science focus

Alfred Russel Wallace

Alfred Russel Wallace (1823–1913) is best known as the English naturalist who independently proposed natural selection as a process to explain the origin of species (Fig. 17A). Like Darwin, Wallace was a collector at home and abroad. Even at age 14, while learning the trade of surveying, he became interested in botany and started collecting plants. While he was a schoolteacher at Leicester in 1844–45, he met Henry Walter Bates, an entomologist, who interested him in insects. Together, they went on a collecting trip to the Amazon, which lasted for several years. Wallace's knowledge of the world's extensive flora and fauna was further expanded by a tour he made of the Malay Archipelago from 1854–62. After studying the animals of every important island, he divided the islands into a western group, with animals like those of the Orient, and an eastern group, with animals like those of Australia. The dividing line between the islands of the archipelago is a narrow but deep strait that is now known as the Wallace Line (Fig. 17B).

Like Darwin, Wallace wrote articles and books. As a result of his trip to the Amazon, he wrote two books, entitled *Travels on the Amazon and Rio Negro* and *Palm Trees of the Amazon*. In 1855, during his trip to Malay, he wrote an essay called "On the Law Which Has Regulated the Introduction of New Species." In the essay, he said that "every species has come into existence coincident both in time and space with a preexisting closely allied species." It is clear, then, that by this date Wallace believed in the origin of new species rather than the fixity of species. Later, he said that he had pondered for many years about a mechanism to explain the origin of species. He, too, had read Malthus's treatise on human population increases, and in 1858, while suffering an attack of malaria, the idea of "survival of the fittest" came upon him. He quickly completed an essay discussing a natural selection process, which he chose to send to Darwin for comment. Darwin was stunned upon its receipt. Here before him was the hypothesis he had formulated as early as 1844 but had never dared to publish. In 1856, he had begun to work on a book that would supply copious data to support natural selection as a mechanism for evolutionary change. He told his friend and colleague Charles Lyell that Wallace's ideas were so similar to his own that even Wallace's "terms now stand as heads of my chapters."

Darwin suggested that Wallace's paper be published immediately, even though Darwin himself as yet had nothing in print. Lyell and others who knew of Darwin's detailed work substantiating the process of natural selection suggested that a joint paper be read to the Linnean Society. The title of Wallace's section was "On the Tendency of Varieties to Depart Indefinitely from the Original Type." Darwin presented an abstract of a paper he had written in 1844 and an abstract of his book *On the Origin of Species,* which was published in 1859.

FIGURE 17A Alfred Russel Wallace.

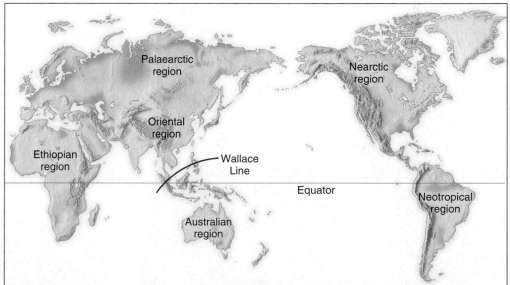

FIGURE 17B Biogeographical regions.
Aside from presenting a hypothesis that natural selection explains the origin of new species, Alfred Wallace is well known for another contribution. He said that the world can be divided into six biogeographical regions separated by impassable barriers. The deep waters between the Oriental and Australian regions are called the Wallace Line.

FIGURE 17.10 Artificial selection of animals.
All dogs, *Canis familiaris,* are descended from the wolf, *Canis lupus,* which began to be domesticated about 14,000 years ago. In evolutionary terms, the process of diversification has been exceptionally rapid. Several factors may have contributed: (1) The wolves under domestication were separated from other wolves because human settlements were separate, and (2) humans in each tribe selected for whatever traits appealed to them. Artificial selection of dogs continues even today.

a.   b.   c.

FIGURE 17.11 Artificial selection of plants.
All these vegetables are derived from one species of *Brassica oleracea*. **a.** Chinese cabbage. **b.** Brussels sprouts. **c.** Kohlrabi. Darwin believed that artificial selection provided a model by which to understand natural selection. With natural selection, however, the environment provides the selective force.

different populations. For example, among western diamond-back rattlesnakes *(Crotalus atrox)* living on lava flows, the most fit are those that are black in color. But among those living on desert soil, the most fit are those with the typical light and dark brown coloring. Background matching helps an animal both capture prey and avoid being captured; therefore, it is expected to lead to survival and increased reproduction.

Darwin noted that when humans help carry out *artificial selection,* the process by which a breeder chooses which traits to perpetuate, they select the animals and plants that will reproduce. For example, prehistoric humans probably noted desirable variations among wolves and selected particular animals for breeding. Therefore, the desired traits increased in frequency in the next generation. This same process was repeated many times over. The result today is the existence of many varieties of dogs, all descended from the wolf (Fig. 17.10). In a similar way, several varieties of vegetables can be traced to a single ancestor. Chinese cabbage, brussels sprouts, and kohlrabi are all derived from a single species, *Brassica oleracea* (Fig. 17.11).

In nature, interactions with the environment determine which members of a population reproduce to a greater degree than other members. In contrast to artificial selection, the result of natural selection is not predesired. Natural selection occurs because certain members of a population happen to have a variation that allows them to survive and reproduce to a greater extent than other members. For example, any variation that increases the speed of a hoofed animal helps it escape predators and live longer; a variation that reduces water loss is beneficial to a desert plant; and one that increases the sense of smell helps a wild dog find its prey. Therefore, we expect organisms with these traits to have increased fitness.

Organisms Become Adapted

With reference to consequences 4 and 5 (see page 288), an **adaptation** [L. *ad*, toward, and *aptus*, fit, suitable] is a trait that helps an organism be more suited to its environment. Adaptations are especially recognizable when unrelated organisms, living in a particular environment, display similar characteristics. For example, manatees, penguins, and sea turtles all have flippers, which help them move through the water. Such adaptations of populations to their specific environments result from natural selection. Because of differential reproduction generation after generation, adaptive traits are increasingly represented in each succeeding generation. There are other processes of evolution aside from natural selection (see pages 304–6), but natural selection is the only process that results in adaptation to the environment.

On the Origin of Species by Darwin

After the HMS *Beagle* returned to England in 1836, Darwin waited more than 20 years to publish his ideas. During the intervening years, he used the scientific process to test his hypothesis that today's diverse life-forms arose by descent from a common ancestor and that natural selection is a mechanism by which species can change and new species can arise. Darwin was prompted to publish his book *On the Origin of Species* after reading a similar hypothesis from Alfred Russel Wallace, as discussed in the Science Focus on page 289.

Darwin became convinced that descent with modification explains the history of life. His theory of natural selection proposes a mechanism by which adaptation to the environment occurs.

17.3 EVIDENCE FOR EVOLUTION

Many different lines of evidence support the hypothesis that organisms are related through common descent. This is significant, because the more varied the evidence supporting a hypothesis, the more certain it becomes. Darwin cited much of the evidence to support his hypothesis. However, he had no knowledge of the genetic and biochemical data that became available after his time.

Fossil Evidence

The **fossil record** represents the history of life on Earth recorded in the remains or traces of organisms that lived in the past. In order to be a fossil, the remains or traces have to be at least 10,000 years old. There are many types of fossilization. Fossil mammoths have been found embedded in ice from Siberia, and a variety of organisms have been encased in fossil tree sap known as amber. Skeletal parts and even wood have undergone petrification, where the living material has been replaced by minerals. Many fern fossils are found as impressions in shale. For the past two centuries, paleontologists have studied fossils from the Earth's strata (layers) in all places of the world and have pieced together the story of life.

The fossil record is rich in information. One of its most striking patterns is a succession of life-forms from the simple to the more complex. Catastrophists offered an explanation for the extinction and subsequent replacement of one group of organisms by another group, but they never could explain successive changes that link groups of organisms historically. Particularly interesting are the fossils that serve as transitional links between groups. For example, famous fossils of *Archaeopteryx* are intermediate between reptiles and birds (Fig. 17.12). The dinosaur-like skeleton of these fossils has reptilian features, including jaws with teeth and a long, jointed tail, but *Archaeopteryx* also had feathers and wings. Other transitional links among fossil vertebrates include the amphibious fish *Eustheopteron*, the reptile-like amphibian *Seymouria*, and the mammal-like reptiles, or therapsids. These fossils allow paleontologists to deduce the order in which vertebrates (animals with backbones) evolved from fishes to amphibians to reptiles to mammals and birds.

Sometimes the fossil record is complete enough to allow us to trace the history of an organism, such as the modern-day horse, *Equus* (Fig. 17.13). *Hyracotherium,* an ancestor of *Equus,* was about the size of a large dog (35 kg). This animal had cusped, low-crowned molars, four toes on each front foot, and three toes on each hind foot. When grasslands replaced the forest home of *Hyracotherium,* the ancestors of *Equus* were subject to selective pressure for the development of strength, intelligence, speed, and durable grinding teeth. A larger size provided the strength needed for combat, a larger skull made room for a larger brain, elongated legs ending in hooves provided greater speed to escape enemies, and the durable grinding teeth enabled the animals to feed efficiently on grasses.

Figure 17.13 shows that the evolutionary history of *Equus* is like a tree, with multiple branchings and rebranchings from one source. A common ancestor is at each fork of the evolutionary tree, and the evolutionary history of *Equus* can be traced back from common

a.

b.

FIGURE 17.12 Transitional fossils.
a. *Archaeopteryx* was a transitional link between reptiles and birds. Fossils indicate it had feathers and wing claws. Most likely, it was a poor flier. Perhaps it ran over the ground on strong legs and climbed up into trees with the assistance of these claws. **b.** *Archaeopteryx* also had a feather-covered, reptilian-type tail that shows up well in this artist's representation. (Red labels = reptilian characteristics; green label = bird characteristic.)

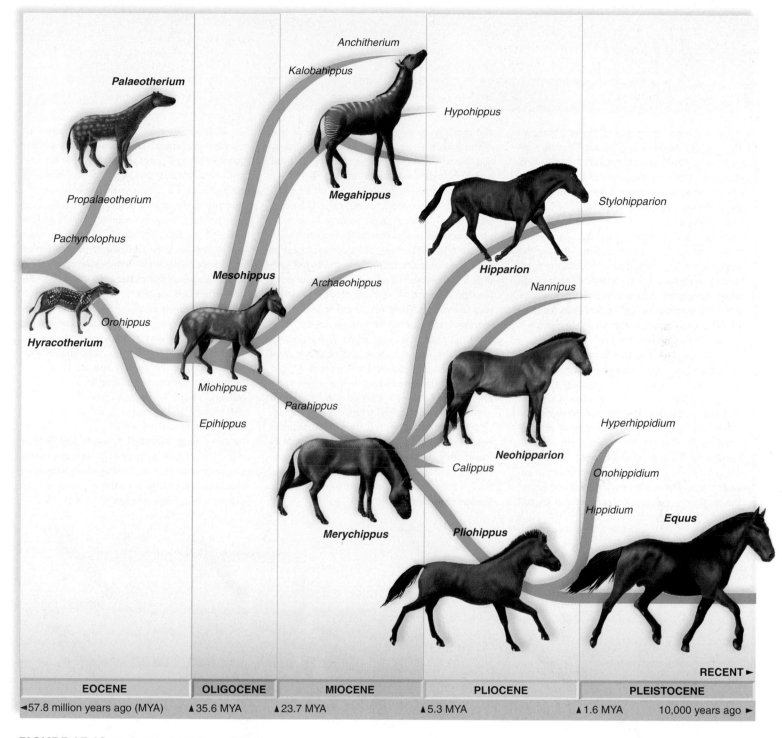

FIGURE 17.13 Evolutionary history of *Equus*.
The evolutionary tree of *Equus* is known to have included many branchings, several of which came to dead ends. By ignoring these, it is possible to trace the history of *Equus* back to *Hyracotherium*. MYA = million years ago.

ancestor to common ancestor through time. There appears to have been a gradual change in form from *Hyracotherium* to *Equus*. Many of the side branches became extinct or died out over time. The paleontologist George Gaylord Simpson estimated that 99.9% of all species eventually became **extinct.** It is reasonable to hypothesize that environments are constantly changing and the ability to adapt to a changing environment is a requirement for long-term survival of a species.

Living organisms closely resemble the most recent fossils in their line of descent. Fossils can be linked over time because there is a similarity in form, despite observed changes; therefore, the fossil record supports common descent.

The fossil record broadly traces the history of life and more specifically allows us to study the history of particular groups.

science focus

The Pace of Evolution

Evolutionists who support phyletic gradualism [Gk. *phyle*, tribe], as did Darwin, suggest that evolutionary change is rather slow and steady. In other words, fossils of the same species designation can show a trend over time—say, a change in plumage color (Fig. 17C). Further divergence, when a common ancestor gives rise to two separate lineages, is not necessarily dependent on speciation—that is, the origination of a new species. Indeed, the fossil record, even if complete, is unlikely to indicate when speciation has occurred. Since evolution occurs gradually, transitional links (see Fig. 17.12) are expected, and most likely more will eventually be found in the fossil record.

Is the phyletic gradualism model applicable to the evolutionary history of *Equus* (see Fig. 17.13)? Is it possible, for example, to show overall trends such as an increase in size, an increase in the grinding surface of the molar teeth, and a reduction in the number of toes? Most agree that phyletic gradualism is applicable only if we pick and choose among the many fossils available. Closer examination reveals that as *Hyracotherium* evolved into *Equus*, the evolution of every character varied greatly,

and there were even times of reversal. If one of the ancestral animals had lived on and *Equus* had become extinct, we no doubt would be discussing a different set of "trends."

Considering such difficulties, other paleontologists—Stephen J. Gould, Niles Eldredge, and Steven Stanley in particular—have proposed another model they call punctuated equilibrium (Fig. 17D). They point out examples of organisms that are called *living fossils* because they are so similar to an ancestor known from the fossil record. A few years ago, investigators found exquisitely preserved specimens of cyanobacteria that have the same sizes, shapes, and organization as living forms. These findings suggest that the cyanobacteria of today have not changed at all in over 3 billion years. Among plants, the dawn redwood was thought to be extinct; then a living specimen was discovered in a small area of China. Horseshoe crabs, crocodiles, and coelacanth fish are animals that still resemble their earliest ancestors. A recently found scaly anteater fossil shows that these animals have changed minimally in 60 million years. Such a time of limited evolutionary change in a lineage is called *stasis*.

In most lineages, however, a period of equilibrium (stasis) is punctuated by evolutionary change—that is, speciation occurs. With reference to the length of the fossil record (about 3.5 billion years), speciation occurs relatively rapidly. Therefore, transitional links are less likely to become fossils, and less likely to be found! Indeed, speciation most likely involves only an isolated population at one locale. Only when a new species evolves and displaces existing species is it apt to show up in the fossil record.

Carlton Brett at the University of Rochester studied the sequence of fossil communities in the Devonian seas that covered the present state of New York 360–408 million years ago. He found that the mix of species in a community did not change as long as the sea level remained high. A low sea level disrupted the community, and a change of species occurred that maintained itself throughout the next interval of a high sea level. It would appear that gradual evolutionary processes are the norm, but environmental disturbances promote rapid evolutionary change and the replacement of old species by new species.

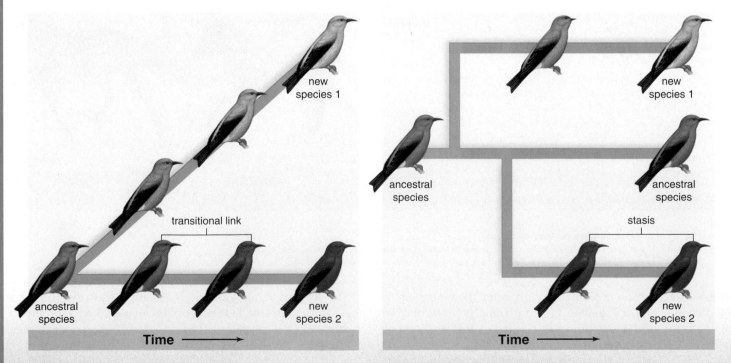

FIGURE 17C Phyletic gradualism.
Speciation occurs gradually and many transitional links occur. Therefore, apparent stasis is not real.

FIGURE 17D Punctuated equilibrium.
Speciation occurs rapidly (transitional links do not occur) and stasis is real.

Biogeographical Evidence

Biogeography is the study of the range and distribution of plants and animals in different places throughout the world. Such distributions are consistent with the hypothesis that, when forms are related, they evolved in one locale and then spread to accessible regions. Therefore, a different mix of plants and animals would be expected whenever geography separates continents, islands, seas, etc. As previously mentioned, Darwin noted that South America lacked rabbits, even though the environment was quite suitable to them. He concluded there are no rabbits in South America because rabbits evolved somewhere else and had no means of reaching South America.

To take another example, both cacti and euphorbia are plants adapted to a hot, dry environment—both are succulent, spiny, flowering plants. Why do cacti grow in North American deserts and euphorbia grow in African deserts when each would do well on the other continent? It seems obvious that they just happened to evolve on their respective continents.

The islands of the world have many unique species of animals and plants that are found no place else, even when the soil and climate are the same. Why do so many species of finches live on the Galápagos Islands when these same species are not on the mainland? The reasonable explanation is that an ancestral finch originally inhabited the different islands. Geographic isolation allowed the ancestral finch to evolve into a different species on each island.

Also, in the history of the Earth, South America, Antarctica, and Australia were originally connected (see Fig. 19.14). Marsupials (pouched mammals) arose at this time and today are found in both South America and Australia. But when Australia separated and drifted away, the marsupials diversified into many different forms suited to various environments of Australia (Fig. 17.14). They were free to do so because there were few, if any, placental mammals in Australia. In South America, where there are placental mammals, marsupials are not as diverse. This supports the hypothesis that evolution is influenced by the mix of plants and animals in a particular continent—that is, by biogeography, not by design.

The distribution of organisms on the Earth is explainable by assuming that related forms evolved in one locale. They then diversified as they spread out into other accessible areas.

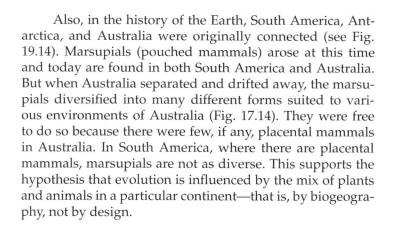

FIGURE 17.14 Biogeography.
Each type of marsupial in Australia is adapted to a different way of life. All of the marsupials in Australia presumably evolved from a common ancestor that entered Australia some 60 million years ago.

Tasmanian wolf, *Thylacinus*, an extinct nocturnal carnivore of deserts and plains

Kangaroo, *Macropus*, an herbivore of plains and forests

Sugar glider, *Petaurus*, a tree dweller

Coarse-haired wombat, *Vombatus*, nocturnal and living in burrows

Australian native cat, *Dasyurus*, a carnivore of forests

Tasmanian devil, *Sarcophilus*, a carnivore of forests

Anatomical Evidence

Darwin was able to show that a common descent hypothesis offers a plausible explanation for anatomical similarities among organisms. Vertebrate forelimbs are used for flight (birds and bats), orientation during swimming (whales and seals), running (horses), climbing (arboreal lizards), or swinging from tree branches (monkeys). Yet all vertebrate forelimbs contain the same sets of bones organized in similar ways, despite their dissimilar functions (Fig. 17.15). The most plausible explanation for this unity is that the basic forelimb plan belonged to a common ancestor, and then the plan was modified in the succeeding groups as each continued along its own evolutionary pathway. Structures that are anatomically similar because they are inherited from a common ancestor are called **homologous structures** [Gk. *homologos*, agreeing, corresponding]. In contrast, **analogous structures** serve the same function, but they are not constructed similarly, nor do they share a common ancestry. The wings of birds and insects and the eyes of octopi and humans are analogous structures. The presence of homology, not analogy, is evidence that organisms are related.

Vestigial structures [L. *vestigium*, trace, footprint] are anatomical features that are fully developed in one group of organisms but that are reduced and may have no function in similar groups. Most birds, for example, have well-developed wings used for flight. Some bird species (e.g., ostrich), however, have greatly reduced wings and do not fly. Similarly, snakes have no use for hindlimbs, and yet some have remnants of a pelvic girdle and legs. Humans have a tailbone but no tail. The presence of vestigial structures can be explained by the common descent hypothesis. Vestigial structures occur

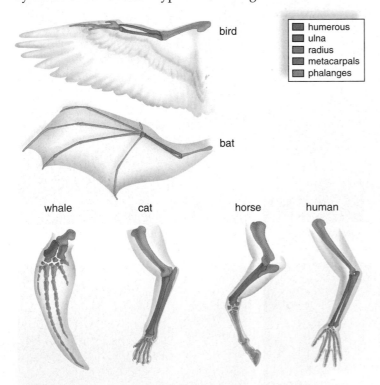

■	humerous
■	ulna
■	radius
■	metacarpals
■	phalanges

FIGURE 17.15 Significance of homologous structures.
Although the specific design details of vertebrate forelimbs are different, the same bones are present (they are color-coded). Homologous structures provide evidence of a common ancestor.

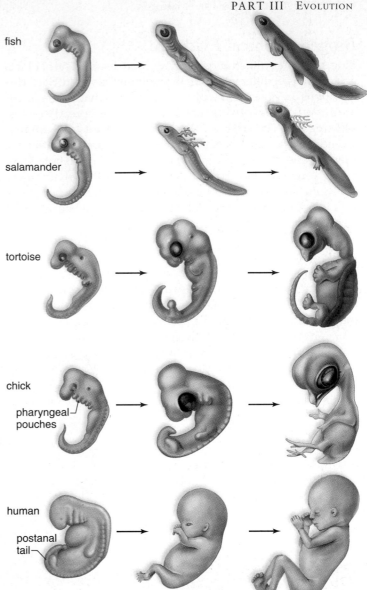

FIGURE 17.16 Significance of developmental similarities.
At these comparable developmental stages, vertebrate embryos have many features in common, which suggests they evolved from a common ancestor. (These embryos are not drawn to scale).

because organisms inherit their anatomy from their ancestors; they are traces of an organism's evolutionary history.

The homology shared by vertebrates extends to their embryological development (Fig. 17.16). At some time during development, all vertebrates have a postanal tail and exhibit paired pharyngeal pouches. In fishes and amphibian larvae, these pouches develop into functioning gills. In humans, the first pair of pouches becomes the cavity of the middle ear and the auditory tube. The second pair becomes the tonsils, while the third and fourth pairs become the thymus and parathyroid glands. Why should terrestrial vertebrates develop and then modify structures like pharyngeal pouches that have lost their original function? The most likely explanation is that fishes are ancestral to other vertebrate groups.

In 1859, Darwin speculated that whales evolved from a land mammal. His hypothesis has now been substantiated. In recent years the fossil record has yielded an incredible parade of fossils that link modern whales and dolphins to

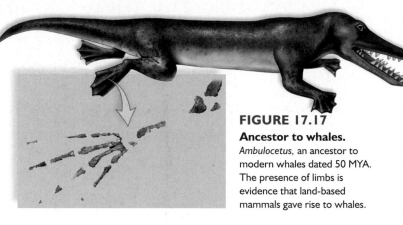

FIGURE 17.17

Ancestor to whales.

Ambulocetus, an ancestor to modern whales dated 50 MYA. The presence of limbs is evidence that land-based mammals gave rise to whales.

land ancestors (Fig. 17.17 and page 281). The presence of a vestigial pelvic girdle and legs in modern whales is also significant evidence to consider.

Organisms that share homologous structures are closely related and have a common ancestry. Studies of comparative anatomy and embryological development reveal homologous structures.

Biochemical Evidence

Almost all living organisms use the same basic biochemical molecules, including DNA (deoxyribonucleic acid), ATP (adenosine triphosphate), and many identical or nearly identical enzymes. Further, organisms use the same DNA triplet code and the same 20 amino acids in their proteins. Since the sequences of DNA bases in the genomes of many organisms are now known, it has become clear that humans share a large number of genes with much simpler organisms. Also of interest, evolutionists who study development have found that many developmental genes are shared in animals ranging from worms to humans. It appears that life's vast diversity has come about by only a slight difference in the same genes. The result has been widely divergent types of bodies. For example, a similar gene in arthropods and vertebrates determines the dorsal-ventral axis. Although the base sequences are similar, the genes have opposite effects. Therefore, in arthropods, such as fruit flies and crayfish, the nerve cord is ventral, whereas in vertebrates, such as chicks and humans, the nerve cord is dorsal. The nerve cord eventually gives rise to the spinal cord and brain.

When the degree of similarity in DNA base sequences or the degree of similarity in amino acid sequences of proteins is examined, the data are as expected, assuming common descent. Cytochrome *c* is a molecule that is used in the electron transport chain of all the organisms appearing in Figure 17.18. Data regarding differences in the amino acid sequence of cytochrome *c* show that the sequence in a human differs from that in a monkey by only one amino acid, from that in a duck by 11 amino acids, and from that in a yeast by 51 amino acids. These data are consistent with other data regarding the anatomical similarities of these organisms and, therefore, their relatedness.

Darwin discovered that many lines of evidence support the hypothesis of common descent. Since his time, biochemical evidence has also been found to support the hypothesis. A hypothesis is strengthened when it is supported by many different lines of evidence.

Evolution is no longer considered a hypothesis. It is one of the great unifying theories of biology. In science, the word *theory* is reserved for those conceptual schemes that are supported by a large number of observations and have not yet been found lacking. The theory of evolution has the same status in biology that the germ theory of disease has in medicine.

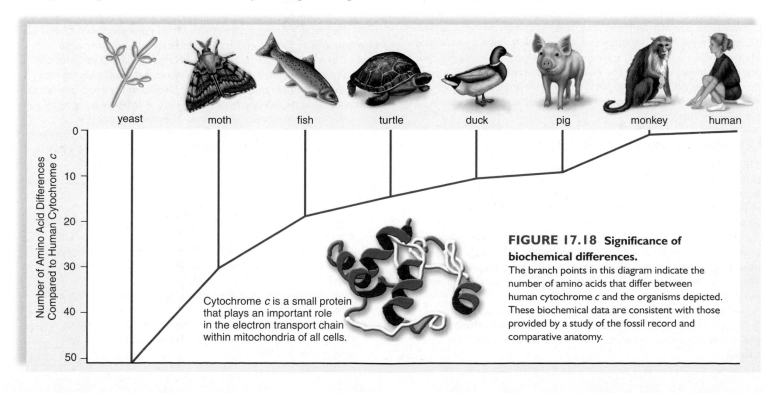

yeast   moth   fish   turtle   duck   pig   monkey   human

Number of Amino Acid Differences Compared to Human Cytochrome *c*

0
10
20
30
40
50

Cytochrome *c* is a small protein that plays an important role in the electron transport chain within mitochondria of all cells.

FIGURE 17.18 Significance of biochemical differences.

The branch points in this diagram indicate the number of amino acids that differ between human cytochrome *c* and the organisms depicted. These biochemical data are consistent with those provided by a study of the fossil record and comparative anatomy.

CONNECTING THE CONCEPTS

Before the 1800s, most people believed that the origin and diversity of life on Earth were due to the work of a supernatural being who created each species at the beginning of the world, and that modern organisms were essentially unchanged descendants of their ancestors.

At the time Charles Darwin boarded the HMS *Beagle,* he had studied the writings of his grandfather Erasmus Darwin, James Hutton, Charles Lyell, Jean-Baptiste de Lamarck, Thomas Malthus, Carolus Linnaeus, and other original thinkers. He had ample time during his five-year voyage to reflect on the ideas of these authors, and from them collectively, he built a framework that helped support his theory of descent with modification.

One aspect of scientific genius is the power of astute observation—to see what others miss or fail to appreciate. In this area, Darwin excelled. By the time he reached the Galápagos Islands, Darwin had already begun to hypothesize that species could be modified according to the environment. In other words, like other scientists, Darwin used a testable hypothesis to explain his observations. His observation of the finches on the isolated Galápagos Islands supported his hypothesis. He concluded that the finches on each island varied from each other and from mainland finches because each species had become adapted to a different habitat; therefore, one species had given rise to many. The fact that Alfred Wallace almost

simultaneously proposed natural selection as an evolutionary mechanism suggests that the world was ready for a revised view of life on Earth.

The theory of evolution has quite rightly been called the grand unifying theory (GUT) of biology. Fossils, comparative anatomy, biogeography, and biochemical data all indicate that living things share common ancestors. Evolutionary principles help us understand why organisms are both different and alike, and why some species flourish and others die out. As the Earth's habitats change over millions of years, those individuals with the traits best adapted to new environments survive and reproduce; thus, populations change over time.

Summary

17.1 HISTORY OF EVOLUTIONARY THOUGHT

In general, the pre-Darwinian worldview was different from the post-Darwinian worldview (see Table 17.1). The scientific community, however, was ready for a new worldview, and it received widespread acceptance.

A century before Darwin's trip, the classification of organisms had been a main concern of biology. Linnaeus thought that each species had a place in the *scala naturae* and that classification should describe the fixed features of species and reveal God's divine plan. Some naturalists, such as Count Buffon and Erasmus Darwin, put forth tentative suggestions that species do change over time.

Georges Cuvier and Jean-Baptiste de Lamarck, contemporaries of Darwin in the late-eighteenth century, differed sharply on evolution. To explain the fossil record of a region, Cuvier proposed that a whole series of catastrophes (extinctions) and repopulations from other regions had occurred. Lamarck said that descent with modification does occur and that organisms do become adapted to their environments; however, he suggested the inheritance of acquired characteristics as a mechanism for evolutionary change.

17.2 DARWIN'S THEORY OF EVOLUTION

Charles Darwin formulated hypotheses concerning evolution after taking a trip around the world as a naturalist aboard the HMS *Beagle* (1831–36). His hypotheses were that common descent does occur and that natural selection results in adaptation to the environment.

Darwin's trip involved two primary types of observations. His study of geology and fossils caused him to concur with Lyell that the observed massive geological changes were caused by slow, continuous changes. Therefore, he concluded that the Earth is old enough for descent with modification to occur.

Darwin's study of biogeography, including the animals of the Galápagos Islands, allowed him to conclude that adaptation to the environment can cause diversification, including the origin of new species.

Natural selection is the mechanism Darwin proposed for how adaptation comes about. Members of a population exhibit random, but inherited, variations. (In contrast to the previous worldview, variations are highly significant.) Relying on Malthus's ideas regarding overpopulation, Darwin stressed that there was a struggle for existence. The most-fit organisms are those possessing characteristics that allow them to acquire more resources and to survive and reproduce more than the less fit. In this way, natural selection results in adaptation to a local environment.

17.3 EVIDENCE FOR EVOLUTION

The hypothesis that organisms share a common descent is supported by many lines of evidence. The fossil record, biogeography, anatomical evidence, and biochemical evidence all support the hypothesis. The fossil record gives us the history of life in general and allows us to trace the descent of a particular group. Biogeography shows that the distribution of organisms on Earth is explainable by assuming organisms evolved in one locale. Comparing the anatomy and the development of organisms reveals a unity of plan among those that are closely related. All organisms have certain biochemical molecules in common, and any differences indicate the degree of relatedness. A hypothesis is greatly strengthened when many different lines of evidence support it.

Today, the theory of evolution is one of the great unifying theories of biology because it has been supported by so many different lines of evidence.

Reviewing the Chapter

1. In general, contrast the pre-Darwinian worldview with the post-Darwinian worldview.   282–83
2. Cite naturalists who made contributions to biology in the mid-eighteenth century, and state their beliefs about evolutionary descent.   283
3. How did Cuvier explain the succession of life-forms in the Earth's strata?   284
4. What is meant by the inheritance of acquired characteristics, a hypothesis that Lamarck used to explain adaptation to the environment?   284

5. What book influenced Darwin's views on geology, and what observations did he make regarding geology?   285–86

6. What observations did Darwin make regarding biogeography? How did these influence his conclusions about the origin of new species?   285–87

7. What are the essential features of the process of natural selection as proposed by Darwin?   288

8. Distinguish between the concepts of fitness and adaptation to the environment.   288, 291

9. How does the fossil record support the concept that organisms are related through common descent? Explain why *Equus* is vastly different from its ancestor *Hyracotherium*, which lived in a forest.   292–93

10. How does biogeography support the concept of common descent? Explain why a diverse assemblage of marsupials evolved in Australia.   295

11. How does anatomical evidence support the concept of common descent? Explain why vertebrate forelimbs are similar despite different functions.   296–97

12. How does biochemical evidence support the concept of common descent? Explain why the sequence of amino acids in cytochrome *c* differs between two organisms.   297

Testing Yourself

Choose the best answer for each question.

1. Which of these pairs is mismatched?
   a. Charles Darwin—natural selection
   b. Linnaeus—classified organisms according to the *scala naturae*
   c. Cuvier—series of catastrophes explains the fossil record
   d. Lamarck—uniformitarianism
   e. All of these are correct.

2. According to the theory of inheritance of acquired characteristics,
   a. if a man loses his hand, then his children will also be missing a hand.
   b. changes in phenotype are passed on by way of the genotype to the next generation.
   c. organisms are able to bring about a change in their phenotype.
   d. evolution is striving toward particular traits.
   e. All of these are correct.

3. Why was it helpful to Darwin to learn that Lyell thought the Earth was very old?
   a. An old Earth has more fossils than a new Earth.
   b. It meant there was enough time for evolution to have occurred slowly.
   c. There was enough time for the same species to spread out into all continents.
   d. Darwin said that artificial selection occurs slowly.
   e. All of these are correct.

4. All the finches on the Galápagos Islands
   a. are unrelated but descended from a common ancestor.
   b. are descended from a common ancestor, and therefore related.
   c. rarely compete for the same food source.
   d. Both a and c are correct.
   e. Both b and c are correct.

5. Organisms
   a. compete with other members of their species.
   b. differ in fitness.
   c. are adapted to their environment.
   d. are related by descent from common ancestors.
   e. All of these are correct.

6. DNA nucleotide differences between organisms
   a. indicate how closely related organisms are.
   b. indicate that evolution occurs.
   c. explain why there are phenotypic differences.
   d. are to be expected.
   e. All of these are correct.

7. If evolution occurs, we would expect different biogeographical regions with similar environments to
   a. all contain the same mix of plants and animals.
   b. each have its own specific mix of plants and animals.
   c. have plants and animals with similar adaptations.
   d. have plants and animals with different adaptations.
   e. Both b and c are correct.

8. The fossil record offers direct evidence for common descent because you can
   a. see that the types of fossils change over time.
   b. sometimes find common ancestors.
   c. trace the ancestry of a particular group.
   d. trace the biological history of living things.
   e. All of these are correct.

9. Organisms such as whales and sea turtles that are adapted to an aquatic way of life
   a. will probably have homologous structures.
   b. will have similar adaptations but not necessarily homologous structures.
   c. may very well have analogous structures.
   d. will have the same degree of fitness.
   e. Both b and c are correct.

For questions 10–17, match the evolutionary evidence in the key to the description. Choose more than one answer if correct.

KEY:
   a. biogeographical evidence
   b. fossil evidence
   c. biochemical evidence
   d. anatomical evidence

10. It's possible to trace the evolutionary ancestry of a species.

11. Rabbits are not found in Patagonia.

12. A group of related species have homologous structures.

13. The same types of molecules are found in all living things.

14. Islands have many unique species not found elsewhere.

15. All vertebrate embryos have pharyngeal pouches.

16. Distantly related species have more amino acid differences in cytochrome *c*.

17. Transitional links have been found between major groups of animals.

18. Which of these is/are necessary to natural selection?
   a. variations
   b. differential reproduction
   c. inheritance of differences
   d. All of these are correct.

19. Which of these is explained incorrectly?
   a. Organisms have variations—mutations and recombination of alleles occur.
   b. Organisms struggle to exist—the environment will support only so many of the same type of species.
   c. Organisms differ in fitness—adaptations enable some members of a species to reproduce more than other members.
   d. Organisms become adapted—species become adapted because of differential fitness.
   e. All of these are correct.

For questions 20–23, offer an explanation for each of these observations based on information in the section indicated. Write out your answer.

20. Transitional fossils serve as links between groups of organisms. See Fossil Evidence (page 292).

21. Cacti and euphorbia exist on different continents, but both have spiny, water-storing, leafless stems. See Biogeographical Evidence (page 295).

22. Amphibians, reptiles, birds, and mammals all have pharyngeal pouches at some time during development. See Anatomical Evidence (page 296).

23. The base sequence of DNA differs from species to species. See Biochemical Evidence (page 297).

Thinking Scientifically

1. Because viruses are rapidly replicated, it is possible to observe hundreds of generations in a single host (infected organism). Some viruses (such as influenza and HIV) evolve rapidly, and others (such as rabies virus and poliovirus) are relatively stable. Two selective forces that influence the speed of viral evolution are the strength of the host immune response that works to destroy the virus, and host behavior in assisting transmission of the virus. How could these two selective forces influence the speed of viral evolution?

2. DNA evidence shows that the closest living relatives of elephants are manatees, aquatic mammals found in the Atlantic Ocean. Two possibilities exist: Manatees evolved from elephants (or their immediate ancestors), or elephants evolved from manatees (or their immediate ancestors). A study of the embryonic and adult anatomy of manatees and elephants might reveal structures that would help to more firmly establish the evolutionary relationship between these two animals. Hypothesize what kind of structures these might be.

Bioethical Issue: Theory of Evolution

The term *theory* in science is reserved for those ideas that scientists have found to be all-encompassing because they are based on data collected in a number of different fields. Evolution is a scientific theory. So is the cell theory, which says that all organisms are composed of cells, and so is the atomic theory, which says that all matter is composed of atoms. No one argues that schools should teach alternatives to the cell theory or the atomic theory, but confusion reigns over the use of the expression "the theory of evolution."

No wonder most scientists in our country are dismayed when state legislatures or school boards rule that teachers must present a variety of "theories" on the origin of life, including one that runs contrary to the mass of data that support the theory of evolution. In California, the Institute for Creation Research advocates that students be taught an "intelligent-design theory," which says that DNA could never have arisen without the involvement of an "intelligent agent," and that gaps in the fossil record mean species arose fully developed with no antecedents.

Since our country forbids the mingling of church and state—that is, no purely religious ideas can be taught in the schools—the advocates for an intelligent-design theory are careful to never mention the Bible or ideas such as "God created the world in seven days." Still, teachers who have a good scientific background do not feel comfortable teaching an intelligent-design theory because it does not meet the test of a scientific theory. Science is based on hypotheses that have been tested by observation and/or experimentation. A scientific theory has stood the test of time—in other words, no hypotheses have been supported by observation and/or experimentation that runs counter to the theory. Indeed, the theory of evolution is supported by data collected in such wide-ranging fields as development, anatomy, geology, and biochemistry.

The polls consistently show that nearly half of all Americans prefer to believe the Old Testament account of how God created the world in seven days. That, of course, is their right, but should schools be required to teach an intelligent-design theory that traces its roots back to the Old Testament and is not supported by observation and experimentation?

Understanding the Terms

Match the terms to these definitions:

adaptation 291	homologous structure 296
analogous structure 296	inheritance of acquired
biogeography 286	characteristics 284
catastrophism 284	natural selection 288
evolution 283	paleontology 284
extinct 293	uniformitarianism 285
fitness 288	vestigial organ 283
fossil record 292	vestigial structure 296

a. _____ Study of the geographic distribution of organisms.

b. _____ Study of fossils that results in knowledge about the history of life.

c. _____ Poorly developed structure that was complete and functional in an ancestor but is no longer functional in a descendant.

d. _____ Organism's modification in structure, function, or behavior suitable to the environment.

e. _____ Lamarckian belief that organisms become adapted to their environment during their lifetime and pass on these adaptations to their offspring.

ARIS, the *Biology* Website

ARIS, the website for *Biology*, provides a wealth of information organized and integrated by chapter. You will find practice quizzes, interactive activities, labeling exercises, flashcards, and much more that will complement your learning and understanding of general biology.

www.mhhe.com/maderbiology9

18

PROCESS OF EVOLUTION

Darwin emphasized that without variation among members of a population, there could be no evolutionary change. The individuals most suited to an environment have a reproductive advantage over those not as suited, and with each succeeding generation the members of a population become more and more adapted to the particular environment. For example, the many different species of finches observed on the Galápagos Islands are the result of this natural selection process in varied environments.

Today we see natural selection at work as bacteria become resistant to antibiotics. When your grandparents were young, infectious diseases such as tuberculosis, pneumonia, and syphilis killed thousands of people every year. Then in the 1940s, penicillin and other antibiotics were developed, and public health officials believed infectious diseases were a thing of the past. Today, however, tuberculosis, pneumonia, and many other ailments are back with a vengeance. What happened? Natural selection occurred.

The bacterium Pscudomonas aeruginosa *causes a great variety of infections, particularly in individuals with burns or preexisting conditions such as AIDS, diabetes, cancer, and cystic fibrosis. In some hospitals, populations of* Pseudomonas aeruginosa *are now resistant to antibiotics. In the soil, where the bacterium lives, it routinely becomes resistant to life-threatening chemicals produced by its neighbors (other bacteria and molds). Therefore, it would seem that the selection process toward resistance began before the bacterium was exposed to antibiotics frequently used in a hospital setting. Exposure to these antibiotics simply continued the selection process. The members of the population that are already resistant to an antibiotic are selected to reproduce, and in this way the entire population becomes resistant. By now most cystic fibrosis patients are infected with a strain of* Pseudomonas *that is so resistant to various antibiotics that their infections cannot be treated. Alarmingly, many other strains of so-called superbugs are now known to cause various illnesses that cannot be killed by antibiotics. One type of bacterium actually feeds on the antibiotic vancomycin.*

The bacterium *Pseudomonas aeruginosa* can become resistant to common antibiotics.

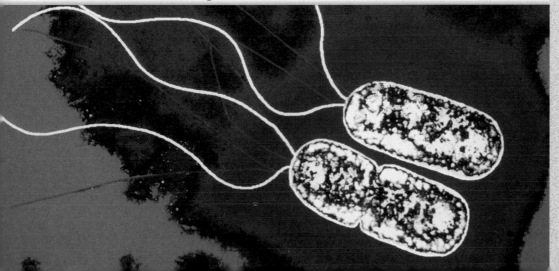

18.1 Microevolution

Darwin stressed that the members of a population vary, but he did not know how variations come about and how they are transmitted. It was not until the 1930s that population geneticists were able to apply the principles of genetics to populations and thereafter develop a way to recognize when evolution had occurred. **Microevolution** pertains to evolutionary changes within a **population,** which is all the members of a single species occupying a particular area.

In **population genetics,** the various alleles at all the gene loci in all individuals make up the **gene pool** of the population. It is customary to describe the gene pool of a population in terms of gene frequencies. Suppose that in a *Drosophila* population, 36% of the flies are homozygous dominant for long wings, 48% are heterozygous, and 16% are homozygous recessive for short wings. Therefore, in a population of 100 individuals, we have

36 *LL,* 48 *Ll,* and 16 *ll*

What is the number of the allele *L* and the allele *l* in the population?

Number of **L** alleles:		Number of *l* alleles:	
LL (2 *L* × 36) =	72	*LL* (0 *l*) =	0
Ll (1 *L* × 48) =	48	*Ll* (1 *l* × 48) =	48
ll (0 *L*) =	0	*ll* (2 *l* × 32) =	32
	120 *L*		80 *l*

To determine the frequency of each allele, calculate its percentage from the total number of alleles in the population. In each case, for the dominant allele *L,* $120/200 = 0.6$; for the recessive allele *l,* $80/200 = 0.4$. The sperm and eggs produced by this population will also contain these alleles in these frequencies. Assuming random mating (all possible gametes have an equal chance to combine with any other), we can calculate the ratio of genotypes in the next generation by using a Punnett square.

There is an important difference between a Punnett square used for a cross between individuals and the following one. Below, the sperm and eggs are those produced by the members of a population—not those produced by a single male and female. As you can see, the frequency of the allele in the next generation is the product of the frequencies of the parental generation. The results of the Punnett square indicate that the frequency for each allele in the next generation is the same as it was in the previous generation.

	eggs		Genotype frequencies:
	0.6 *L*	0.4 *l*	0.36 *LL* + 0.48 *Ll* + 0.16 *ll* = 1
sperm 0.6 *L*	0.36 *LL*	0.24 *Ll*	
0.4 *l*	0.24 *Ll*	0.16 *ll*	

Therefore, sexual reproduction alone cannot bring about a change in allele frequencies. Also, the dominant allele need

$$p^2 + 2\,pq + q^2$$

p^2 = frequency of homozygous dominant individuals *(AA)*

p = frequency of dominant allele *(A)*

q^2 = frequency of homozygous recessive individuals *(aa)*

q = frequency of recessive allele *(a)*

$2\,pq$ = frequency of heterozygous individuals *(Aa)*

Realize that $p + q = 1$ (There are only 2 alleles.)

$p^2 + 2\,pq + q^2 = 1$ (These are the only genotypes.)

Example: An investigator has determined by inspection that 16% of a human population has a recessive trait. Using this information, we can complete all the genotype and allele frequencies for this population.

Given: $q^2 = 16\% = 0.16$ are homozygous recessive individuals

Therefore, $q = \sqrt{0.16} = 0.4$ = frequency of recessive allele
$p = 1.0 - 0.4 = 0.6$ = frequency of dominant allele
$p^2 = (0.6)(0.6) = 0.36 = 36\%$ are homozygous dominant individuals
$2\,pq = 2(0.6)(0.4) = 0.48 = 48\%$ are heterozygous individuals

or

$2\,pq = 1.00 - 0.52 = 0.48$ 84% have the dominant phenotype

FIGURE 18.1 **Calculating gene pool frequencies using the Hardy-Weinberg equation.**

not increase from one generation to the next. Dominance does not cause an allele to become a common allele. The potential constancy, or equilibrium state, of gene pool frequencies was independently recognized in 1908 by G. H. Hardy, an English mathematician, and W. Weinberg, a German physician. They used the binomial expression $(p^2 + 2\,pq + q^2)$ to calculate the genotype and allele frequencies of a population. Figure 18.1 shows you how this is done.

The **Hardy-Weinberg principle** states that an equilibrium of allele frequencies in a gene pool, calculated by using the expression $p^2 + 2\,pq + q^2$, will remain in effect in each succeeding generation of a sexually reproducing population as long as five conditions are met:

1. No mutations: Allele changes do not occur, or changes in one direction are balanced by changes in the opposite direction.
2. No gene flow: Migration of alleles into or out of the population does not occur.
3. Random mating: Individuals pair by chance and not according to their genotypes or phenotypes.

a.　　　　　　　　　　　　　　　　　　　　　　　　　　　　b.

FIGURE 18.2 Industrial melanism and microevolution.
Both dark-colored and light-colored individuals occur in populations of the peppered moth, *Biston betularia*. **a.** When tree trunks are light, dark-colored moths are seen and eaten by predatory birds, and the light-colored moths increase in number. **b.** When tree trunks are dark due to pollution, light-colored moths are seen and eaten by predatory birds, and the dark-colored moths increase in number.

4. No genetic drift: The population is very large, and changes in allele frequencies due to chance alone are insignificant.
5. No selection: No selective agent favors one genotype over another.

In real life, these conditions are rarely, if ever, met, and allele frequencies in the gene pool of a population do change from one generation to the next. Therefore, evolution has occurred. The significance of the Hardy-Weinberg principle is that it tells us what factors cause evolution—those that violate the conditions listed. Evolution can be detected by noting any deviation from a Hardy-Weinberg equilibrium of allele frequencies in the gene pool of a population.

The accumulation of small changes in the gene pool over a relatively short period of time is called microevolution. Microevolution is involved in the origin of species, to be discussed later in this chapter, as well as in the history of life recorded in the fossil record.

A change in allele frequencies may result in a change in phenotype frequencies. Figure 18.2 illustrates a process called **industrial melanism.** Prior to the Industrial Revolution in Great Britain, light-colored pepper moths, *Biston betularia,* were more common living on the light-colored, unpolluted bark of trees than dark-colored peppered moths. When dark-colored moths landed on light tree trunks, they were seen and eaten by predatory birds. It was estimated that only 10% of the moth population was dark at this time. With the advent of industry and an increase in pollution, the bark of the trees be-

came stained darker. On dark-colored bark, light-colored moths were easy prey for predators. Before the Clean Air legislation in the mid-1950s, the numbers of dark-colored moths exceeded 80% of the moth population. After the legislation, a dramatic reversal in the ratio of light-colored moths to dark-colored moths occurred. In 1994, one collecting site recorded a drop in the frequency of dark-colored moths to 19% from a high of 94% in 1960.

A Hardy-Weinberg equilibrium provides a baseline by which to judge whether evolution has occurred. Any change in allele frequencies in the gene pool of a population signifies that evolution has occurred.

Practice Problems 18.1*

1. In a certain population, 21% are homozygous dominant, 49% are heterozygous, and 30% are homozygous recessive. What percentage of the next generation is predicted to be homozygous dominant, assuming a Hardy-Weinberg equilibrium?

2. Of the members of a population of pea plants, 9% are short (recessive). What are the frequencies of the recessive allele *t* and the dominant allele *T*? What are the genotypic frequencies in this population?

   *Answers to Practice Problems appear in Appendix A.

Elaphe obsoleta obsoleta

Elaphe obsoleta quadrivittata

Elaphe obsoleta bairdi

Elaphe obsoleta lindheimeri

Elaphe obsoleta rossalleni

Elaphe obsoleta spiloides

FIGURE 18.3 Gene flow.
Each rat snake represents a separate population of snakes. Because the populations are adjacent to one another, there is interbreeding and therefore gene flow among the populations. This keeps their gene pools somewhat similar, and each of these populations is a subspecies of the species *Elaphe obsoleta*. Therefore, each has a three-part name.

Causes of Microevolution

The list of conditions for allele equilibrium implies that the opposite conditions can cause evolutionary change. The conditions that can cause a deviation from the Hardy-Weinberg equilibrium are mutation, gene flow, nonrandom mating, genetic drift, and natural selection. Only natural selection results in adaptation to the environment.

Genetic Mutations

Mutations are the raw material for evolutionary change; without mutations, there could be no new variations among members of a population. Many traits in organisms are **polymorphic**—that is, two or more distinct phenotypes are present in a population due to mutated genes. For example, in the human population, freckles exist in two forms: freckles or no freckles. On the other hand, four different ABO blood types can be present in a human population. Evidence of mutations in members of a *Drosophila pseudoobscura* population was gathered by R. C. Lewontin and J. L. Hubby in 1966. They extracted various enzymes and subjected them to electrophoresis, a process that separates proteins according to size and charge. These investigators concluded that a fly population has multiple alleles at no less than 30% of all its gene loci. Similar results have been found in other animal studies, demonstrating that high levels of allele variation are the rule in natural populations.

Many mutations do not immediately affect the phenotype, and therefore they may not be detected. In a changing environment, even a seemingly harmful mutation can be a source of an adaptive variation. For example, the water flea *Daphnia* ordinarily thrives at temperatures around 20°C, but there is a mutation that requires *Daphnia* to live at temperatures between 25°C and 30°C. The adaptive value of this mutation is entirely dependent on environmental conditions.

Once alleles have mutated, certain combinations of several alleles might be more adaptive than others in a particular environment. The most favorable phenotype may not occur until just the right alleles are grouped by recombination.

Mutations cause many gene loci to have multiple alleles. Recombination of these alleles increases the possibility of favorable phenotypes.

Gene Flow

Gene flow, also called gene migration, is the movement of alleles between populations by migration of breeding individuals (Fig. 18.3). There can be constant gene flow between adjacent animal populations due to the migration of organisms. Gene flow can increase the variation within a population by introducing novel alleles that were produced by mutation in another population. Continued gene flow makes gene pools similar and reduces the possibility of allele frequency differences among populations due to natural selection and genetic drift. Gene flow among populations can prevent speciation from occurring.

Gene flow tends to decrease the genetic diversity among populations, causing their gene pools to become more similar.

Nonrandom Mating

Random mating occurs when individuals pair by chance and not according to their genotypes or phenotypes. Inbreeding, or mating between relatives to a greater extent than by chance, is an example of **nonrandom mating.** Inbreeding does not change allele frequencies, but it does decrease the proportion of heterozygotes and increase the proportions of both homozygotes at all gene loci. In a human population, inbreeding increases the frequency of recessive abnormalities in the phenotype.

Assortative mating occurs when individuals tend to mate with those that have the same phenotype with respect to a certain characteristic. For example, in fruit flies, those with a high number of bristles tend to mate with each other, and those with a low number of bristles mate with each other. Assortative mating causes the population to subdivide into two phenotypic classes, between which there is reduced gene exchange. Homozygotes for the gene loci that control the trait in question increase in frequency, and heterozygotes for these loci decrease in frequency. Assortive mating can also be observed in human society. Men and women tend to marry individuals with characteristics like themselves in regard to intelligence and height.

Sexual selection occurs when males compete for the right to reproduce, and females choose to mate with males that have a particular phenotype. The elaborate tail of a peacock may have come about because peahens choose to mate with males with such ornate tails.

Nonrandom mating involves inbreeding and assortative mating. The former results in increased frequency of homozygotes at all gene loci, and the latter results in increased frequency of homozygotes at only certain loci.

Genetic Drift

Genetic drift refers to changes in allele frequencies of a gene pool due to chance. Although genetic drift occurs in both large and small populations, a larger population is expected to suffer less of a sampling error than a smaller population. Suppose you had a large bag containing 1,000 green balls and 1,000 blue balls, and you randomly drew 10%, or 200, of the balls. Because there is a large number of balls of each color in the bag, you can reasonably expect to draw 100 green balls and 100 blue balls or at least a ratio close to this. But suppose you had a bag containing only 10 green balls and 10 blue balls and you drew 10%, or only 2 balls. The chances of drawing 1 green ball and 1 blue ball with a single trial are now considerably less.

When a population is small, there is a greater chance that some rare genotype might not participate at all in the production of the next generation. Suppose there is a small population of frogs in which certain frogs for one reason or another do not pass on their traits. Certainly, the next generation will have a change in allele frequencies (Fig. 18.4). When genetic drift leads to a loss of one or more alleles for a gene locus, a particular allele may become fixed in the population over time.

In an experiment involving brown eye color, each of 107 *Drosophila* populations was in its own culture bottle. Every bot-

tle contained eight heterozygous flies of each sex. There were no homozygous recessive or homozygous dominant flies. From the many offspring, the experimenter chose at random eight males and eight females. These were the parents for the next generation, and so forth, for 19 generations. For the first few generations, most populations still contained many heterozygotes. But by the nineteenth generation, 25% of the populations contained only homozygous recessive flies, and 25% contained only homozygous dominant flies for a brown-eye allele.

Genetic drift is a random process, and therefore it is not likely to produce the same results in several populations. In California, there are a number of cypress groves, each a separate population. The phenotypes within each grove are more similar to one another than they are to the phenotypes in the other groves. Some groves have longitudinally shaped trees, and others have pyramidally shaped trees. The bark is rough in some colonies and smooth in others. The leaves are gray to bright green or bluish, and the cones are small or large. Because the environmental conditions are similar for all the groves and no correlation has been found between phenotype and environment across groves, it is hypothesized that these variations among populations are due to genetic drift.

Bottleneck Effect.   Sometimes a species is subjected to near extinction because of a natural disaster (e.g., earthquake or fire) or because of overhunting, overharvesting, and habitat loss. It is as if most of the population has stayed behind and only a few survivors have passed through the neck of a bottle. The **bottleneck effect** prevents the majority of genotypes from participating in the production of the next generation.

The extreme genetic similarity found in cheetahs is believed to be due to a bottleneck. In a study of 47 different enzymes, each of which can come in several different forms, all the cheetahs had exactly the same form. This demonstrates that genetic drift can cause alleles to be lost from a population. Exactly what caused the cheetah bottleneck

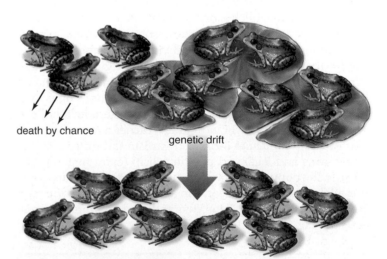

death by chance

genetic drift

FIGURE 18.4 Genetic drift.
Genetic drift occurs when by chance only certain members of a population (in this case, green frogs) reproduce and pass on their alleles to the next generation. The allele frequencies of the next generation's gene pool may be markedly different from those of the previous generation.

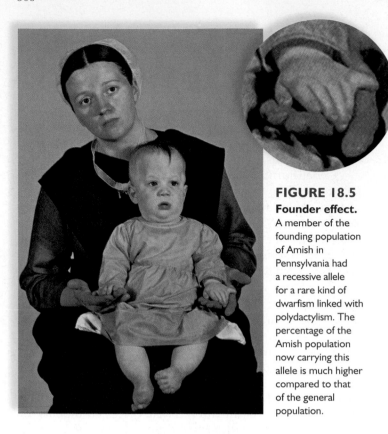

FIGURE 18.5
Founder effect.
A member of the founding population of Amish in Pennsylvania had a recessive allele for a rare kind of dwarfism linked with polydactylism. The percentage of the Amish population now carrying this allele is much higher compared to that of the general population.

18.2 NATURAL SELECTION

Natural selection is the process that results in adaptation of a population to the biotic and abiotic environments. The biotic environment includes organisms that seek resources through competition, predation, and parasitism. The abiotic environment includes weather conditions dependent chiefly on temperatures and precipitation. Charles Darwin, the father of modern evolutionary theory, became convinced that species evolve (change) with time and suggested natural selection as the mechanism for adaptation to the environment (see Chapter 17). Here, we restate Darwin's hypothesis of natural selection in the context of modern evolutionary theory.

Evolution by natural selection requires:

1. Variation. The members of a population differ from one another.
2. Inheritance. Many of these differences are heritable genetic differences.
3. Differential adaptiveness. Some of these differences affect how well an organism is adapted to its environment.
4. Differential reproduction. Individuals that are better adapted to their environment are more likely to reproduce, and their fertile offspring will make up a greater proportion of the next generation.

---

is not known. It is speculated that perhaps cheetahs were slaughtered by nineteenth-century cattle farmers protecting their herds, or were captured by Egyptians as pets 4,000 years ago, or were decimated by a mass extinction tens of thousands of years ago. Today, cheetahs suffer from relative infertility because of the intense inbreeding that occurred after the bottleneck. Other organisms pushed to the brink of extinction suffer a similar plight as the cheetah.

Founder Effect.   The **founder effect** is an example of genetic drift in which rare alleles, or combinations of alleles, occur at a higher frequency in a population isolated from the general population. After all, founding individuals contain only a fraction of the total genetic diversity of the original gene pool. Which particular alleles are carried by the founders is dictated by chance alone. The Amish of Lancaster County, Pennsylvania, are an isolated group that was begun by German founders. Today, as many as 1 in 14 individuals carries a recessive allele that causes an unusual form of dwarfism (affecting only the lower arms and legs) and polydactylism (extra fingers) (Fig. 18.5). In the population at large, only 1 in 1,000 individuals has this allele. The founder effect can also be observed in the high incidence of the genetic neurological disorder Huntington disease (see Fig. 11.13) in Lake Maracaibo, Venezuela.

Genetic drift refers to changes in gene pool frequencies due to the chance reproduction of a few individuals.

Differential reproduction is the measure of an individual's fitness. Population geneticists speak of **relative fitness**—that is, the fitness of one phenotype compared to another.

Types of Selection

Most of the traits on which natural selection acts are polygenic and controlled by more than one pair of alleles located at different gene loci. Such traits have a range of phenotypes, the frequency distribution of which usually resembles a bell-shaped curve.

Three types of natural selection have been described for any particular trait. They are directional selection, stabilizing selection, and disruptive selection.

Directional Selection

Directional selection occurs when an extreme phenotype is favored, and the distribution curve shifts in that direction. Such a shift can occur when a population is adapting to a changing environment.

The resistance to antibiotics by bacteria and insecticides by insects are examples of directional selection. The indiscriminate use of antibiotics and pesticides results in a wide distribution of bacteria and insects that are resistant to these chemicals. When an antibiotic is administered, some bacteria may survive because they are genetically resistant to the antibiotic. These are the bacteria that are likely to pass on their genes to the next generation. As a result, the number of resistant bacteria keeps increasing. Drug-resistant strains of bacteria that cause tuberculosis have now become a serious threat to the health of people worldwide.

Another example of directional selection is the human struggle against malaria, a disease caused by an infection of the liver and the red blood cells. The *Anopheles* mosquito transmits the disease-causing protozoan *Plasmodium vivax* from person to person. In the early 1960s, international health authorities thought that malaria would soon be eradicated. A new drug, chloroquine, seemed effective against *Plasmodium,* and DDT (an insecticide) spraying had reduced the mosquito population. But in the mid-1960s, *Plasmodium* was showing signs of chloroquine resistance, and worse yet, mosquitoes were becoming resistant to DDT. A few drug-resistant parasites and a few DDT-resistant mosquitoes had survived and multiplied, making the fight against malaria more difficult than ever.

The gradual increase in the size of the modern horse, *Equus*, is an example of directional selection that can be correlated with a change in the environment from forest conditions to grassland conditions (Fig. 18.6). Even so, as discussed previously, the evolution of the horse should not be viewed as a straight line of descent because we know of many side branches that became extinct (see Fig. 17.13).

Stabilizing Selection

Stabilizing selection occurs when an intermediate phenotype is favored. It can improve adaptation of the population to those aspects of the environment that remain constant. With stabilizing selection, extreme phenotypes are selected against, and individuals near the average are favored. As an example, consider that when Swiss starlings lay four to five eggs, more young survive than when the female lays more or less than this number (Fig. 18.7). Genes determining physiological characteristics, such as the production of yolk, and behavioral characteristics, such as how long the female will mate, are involved in determining clutch size.

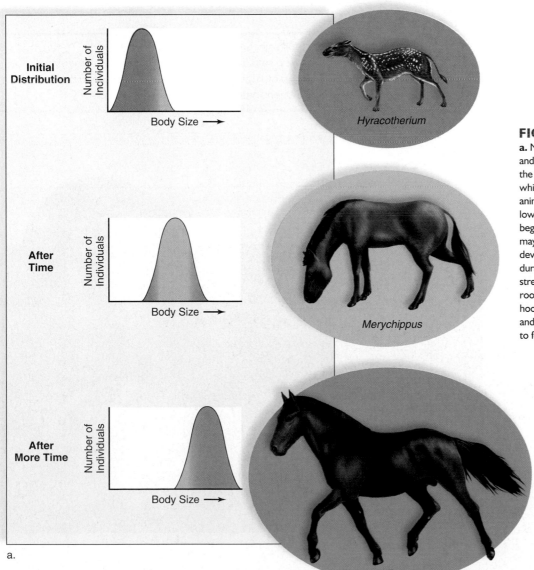

FIGURE 18.6 Directional selection.
a. Natural selection favors one extreme phenotype, and there is a shift in the distribution curve. **b.** *Equus,* the modern-day horse, evolved from *Hyracotherium,* which was about the size of a dog. This small animal could have hidden among trees and had low-crowned teeth for browsing. When grasslands began to replace forests, the ancestors of *Equus* may have been subject to selective pressure for the development of strength, intelligence, speed, and durable grinding teeth. A larger size provided the strength needed for combat, a larger skull made room for a larger brain, elongated legs ending in hooves provided greater speed to escape enemies, and the durable grinding teeth enabled the animals to feed efficiently on grasses.

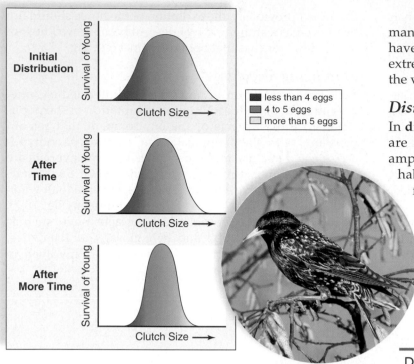

FIGURE 18.7 Stabilizing selection.
Stabilizing selection occurs when natural selection favors the intermediate phenotype over the extremes. For example, Swiss starling birds that lay four to five eggs (usual clutch size) have more young survive than those that lay fewer than four eggs or more than five eggs.

Through the years, hospital data have shown that human infants born with an intermediate birth weight (3–4 kg) have a better chance of survival than those born with an extreme birth weight. Stabilizing selection serves to reduce the variability in birth weight in human populations.

Disruptive Selection

In **disruptive selection,** two or more extreme phenotypes are favored over any intermediate phenotype. For example, British land snails *(Cepaea nemoralis)* have a wide habitat range that includes low-vegetation areas (grass fields and hedgerows) and forests. In forested areas, thrushes feed mainly on light-banded snails, and the snails with dark shells become more prevalent. In low-vegetation areas, thrushes feed mainly on snails with dark shells, and light-banded snails become more prevalent. Therefore, these two distinctly different phenotypes are found in the population (Fig. 18.8). In this instance natural selection has resulted in obvious polymorphism.

Directional selection favors one of the extreme phenotypes; stabilizing selection favors the intermediate phenotype; disruptive selection favors more than one extreme phenotype.

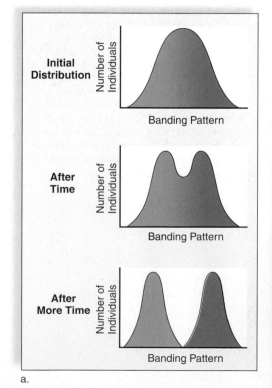

a.

b.

FIGURE 18.8 Disruptive selection.
a. Disruptive selection favors two extreme phenotypes. **b.** Today, it is observed that British land snails comprise mainly two different phenotypes, each adapted to a different habitat. Snails with dark shells are more prevalent in forested areas, and light-banded snails are more prevalent in areas with low-lying vegetation.

Maintenance of Variations

A population always shows some genotypic variation. The maintenance of variation is beneficial because populations with limited variation may not be able to adapt to new conditions and may become extinct. How can variation be maintained in spite of selection constantly working to reduce it?

First, we must remember that the forces that promote variation are still at work: Mutation still creates new alleles, and recombination still recombines these alleles during gametogenesis and fertilization. Second, gene flow might still occur. If the receiving population is small and is mostly homozygous, gene flow can be a significant source of new alleles. Finally, we have seen that natural selection reduces, but does not eliminate, the range of phenotypes. And disruptive selection even promotes polymorphism in a population. There are also other ways variation is maintained.

Diploidy and the Heterozygote

Only alleles that are exposed (cause a phenotypic difference) are subject to natural selection. In diploid organisms, this makes the heterozygote a potential protector of recessive alleles that otherwise would be weeded out of the gene pool. Consider these gene pool frequency distributions:

Frequency of Allele a in Gene Pool	Genotypic Frequencies		
	AA	Aa	aa
0.9	0.01	0.18	0.81
0.1	0.81	0.18	0.01

Notice that even when selection reduces the recessive allele from a frequency of 0.9 to 0.1, the frequency of the heterozygote remains the same. The heterozygote remains a source of the recessive allele for future generations. In a changing environment, the recessive phenotype may then be favored by natural selection.

Sickle Cell Disease

In certain regions of Africa, the importance of the heterozygote in maintaining variation is exemplified by sickle cell disease. The relative fitness of the heterozygote and the two homozygotes will determine their percentages in the population. When the ratio of two or more phenotypes remains the same in each generation, it is called *balanced polymorphism*. The optimum ratio will be the one that correlates with the survival of the most offspring.

Individuals with sickle cell disease have the genotype Hb^SHb^S and tend to die at an early age due to hemorrhaging and organ destruction. Those who are heterozygous and have sickle cell trait (Hb^AHb^S) are better off; their red blood cells usually become sickle-shaped only when the oxygen content of the environment is low. Geneticists studying the distribution of sickle cell disease in Africa have found that the recessive allele (Hb^S) has a higher frequency (0.2 to as high as 0.4 in a few areas) in regions where the disease malaria is also prevalent (Fig. 18.9). What is the connection between the two conditions?

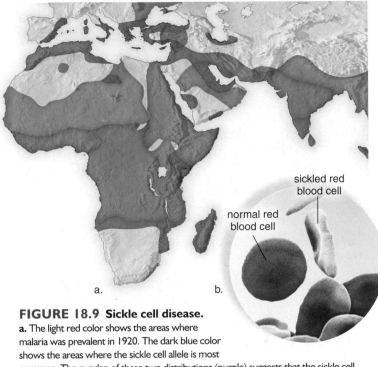

FIGURE 18.9 Sickle cell disease.
a. The light red color shows the areas where malaria was prevalent in 1920. The dark blue color shows the areas where the sickle cell allele is most common. The overlap of these two distributions (purple) suggests that the sickle cell allele is protective against malaria. **b.** Micrograph of a sickled red blood cell.

Malaria is caused by a protozoan parasite that lives in and destroys the red blood cells of the normal homozygote (Hb^AHb^A). The parasite is unable to live in the red blood cells of the heterozygote (Hb^AHb^S) because the infection causes the red blood cells to become sickle-shaped. Sickle-shaped red blood cells lose potassium, and this causes the parasite to die. Therefore, the homozygote (Hb^SHb^S) is maintained at the same level in regions of Africa subject to malaria because the heterozygote is protected both from sickle cell disease and from malaria. Sickle cell disease and perhaps several other genetic disorders such as cystic fibrosis are examples of heterozygote advantage in which the heterozygous condition is favored over the two homozygous conditions. The following table summarizes the effects of the sickle cell genotypes:

Genotype	Phenotype	Result
Hb^AHb^A	Normal	Dies due to malarial infection
Hb^AHb^S	Sickle cell trait	Lives due to protection from both
Hb^SHb^S	Sickle cell disease	Dies due to sickle cell disease

Even after populations become adapted to their environments, variation is promoted and maintained by various mechanisms. In diploid organisms, the heterozygote genotype can help maintain variation.

Least flycatcher, *Empidonax minimus*

Acadian flycatcher, *Empidonax virescens*

Traill's flycatcher, *Empidonax trailli*

FIGURE 18.10 Biological species concept.
Although these three flycatcher species are nearly identical, we would still expect to find some structural feature that distinguishes them. We know they are separate species because they are reproductively isolated—the members of each species reproduce only with one another. Each species has a characteristic song and its own particular habitat during the mating season as well.

18.3 MACROEVOLUTION

In evolutionary biology today the term **macroevolution** is used to refer to any evolutionary change at or above the level of species. **Speciation** is the splitting of one species into two or more species or the transformation of one species into a new species over time. Speciation is the final result of changes in gene pool allele and genotypic frequencies.

What Is a Species?

Sometimes it is difficult to tell one **species** [L. *species,* a kind] from another. Therefore, before we consider the origin of species, it is first necessary to define a species. For Linnaeus, the father of taxonomy, one species was separated from another by morphology—that is, their physical traits differed. The evolutionary biologist Ernst Mayr developed the *biological species concept,* which defines a species as a group of populations that can breed among themselves to produce fertile offspring. Further, the members of a species are reproductively isolated—they are unable to reproduce with members of another species. In terms of population genetics, the members of a species have a shared gene pool, which is isolated from every other species. In other words, gene flow occurs between the populations of a species but not between populations of different species.

The flycatchers in Figure 18.10 meet the criteria for separate species because they do not interbreed. While useful, the biological species concept cannot be applied to asexually reproducing organisms nor to organisms known only by the fossil record nor to species that could possibly interbreed if they lived near one another. In recent years, the biological species concept can be supplemented by our knowledge of biochemical genetics. DNA base sequence data can indicate the relatedness of groups of organisms.

Reproductive Isolating Mechanisms

For two species to be separate, they must be reproductively isolated—that is, gene flow must not occur between them. A *reproductive isolating mechanism* is any structural, functional, or behavioral characteristic that prevents successful reproduction from occurring. Table 18.1 lists the mechanisms by which reproductive isolation is maintained.

Prezygotic (before formation of a zygote) **isolating mechanisms** are those that prevent reproduction attempts and make it unlikely that fertilization will be successful if mating is attempted. Habitat isolation, temporal isolation, behavioral isolation, mechanical isolation, and gamete isolation make it highly unlikely that particular genotypes will contribute to the gene pool of a population.

Habitat isolation When two species occupy different habitats, even within the same geographic range, they are less likely to meet and to attempt to reproduce. This is one of the reasons the flycatchers in Figure 18.10 do not mate, and plants such as the red maple and sugar maple do not exchange pollen. In tropical rain forests, many animal species are restricted to a particular level of the forest canopy, and in this way they are isolated from similar species.

TABLE 18.1

Reproductive Isolating Mechanisms

Isolating Mechanism	Example
Prezygotic	
Habitat isolation	Species at same locale occupy different habitats.
Temporal isolation	Species reproduce at different seasons or different times of day.
Behavioral isolation	In animal species, courtship behavior differs, or individuals respond to different songs, calls, pheromones, or other signals.
Mechanical isolation	Genitalia between species are unsuitable for one another.
Gamete isolation	Sperm cannot reach or fertilize egg.
Postzygotic	
Zygote mortality	Fertilization occurs, but zygote does not survive.
Hybrid sterility	Hybrid survives but is sterile and cannot reproduce.
F_2 fitness	Hybrid is fertile, but F_2 hybrid has reduced fitness.

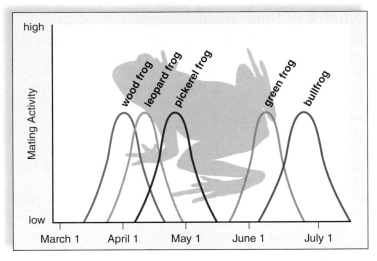

FIGURE 18.11 Temporal isolation.
Five species of frogs of the genus *Rana* are all found at Ithaca, New York. The species remain separate because the period of most active mating is different for each and because whenever there is an overlap, different breeding sites are used. For example, wood frogs are found in woodland ponds or shallow water, leopard frogs in lowland swamps, and pickerel frogs in streams and ponds on high ground.

Temporal isolation Two species can live in the same locale, but if each reproduces at a different time of year, they do not attempt to mate. For example, *Reticulitermes hageni* and *R. virginicus* are two species of termites. The former has mating flights in March through May, whereas the latter mates in the fall and winter months. Similarly, the frogs listed in Figure 18.11 have different periods of most active mating.

Behavioral isolation Many animal species have courtship patterns that allow males and females to recognize one another. Male fireflies are recognized by females of their species by the pattern of their flashings; similarly, male crickets are recognized by females of their species by their chirping. Many males recognize females of their species by sensing chemical signals called pheromones. For example, female gypsy moths secrete chemicals from special abdominal glands. These chemicals are detected downwind by receptors on the antennae of males.

Mechanical isolation When animal genitalia or plant floral structures are incompatible, reproduction cannot occur. Inaccessibility of pollen to certain pollinators can prevent cross-fertilization in plants, and the sexes of different insect species have genitalia that do not match or other characteristics that make mating impossible. For example, male dragonflies have claspers that are suitable for holding only the females of their own species.

Gamete isolation Even if the gametes of two different species meet, they may not fuse to become a zygote. In animals, the sperm of one species may not be able to survive in the reproductive tract of another species, or the egg may have receptors only for sperm of its species. In plants, the stigma prevents hybrid fertilization by controlling the formation of pollen tubes by pollen grains.

Postzygotic (after the formation of a zygote) **isolating mechanisms** prevent hybrid offspring (reproductive product of two different species) from developing or breeding, even if reproduction attempts have been successful.

Zygote mortality A hybrid zygote may not be viable and it dies. A zygote with two different chromosome sets may fail to go through mitosis properly, or the developing embryo may receive incompatible instructions from the maternal and paternal genes so that it cannot continue to exist.

Hybrid sterility The hybrid zygote may develop into a sterile adult. As is well known, a cross between a horse and a donkey produces a mule, which is usually sterile—it cannot reproduce. Sterility of hybrids generally results from complications in meiosis that lead to an inability to produce viable gametes. A cross between a cabbage and a radish produces offspring that cannot form gametes even though the diploid number is even. Most likely, the cabbage chromosomes and radish chromosomes do not pair during meiosis.

F_2 fitness If hybrids can reproduce, their offspring are unable to reproduce. In some cases mules are fertile, but their offspring (the F_2 generation) are not fertile.

The members of a biological species are able to breed and produce fertile offspring only among themselves. Several mechanisms keep species reproductively isolated from one another.

Modes of Speciation

Ernst Mayr also proposed a model of speciation after observing that when a population is geographically isolated from other populations, gene flow stops. Variations due to different mutations, genetic drift, and directional selection build up, causing first postzygotic and then prezygotic reproductive isolation to occur. Mayr called this model **allopatric speciation**.

Figure 18.12 features an example of allopatric speciation, which has been extensively studied in California. Apparently, members of an ancestral population of *Ensatina* salamanders existing in the Pacific Northwest migrated southward, establishing a series of populations. Each population was exposed to its own selective pressures along the coastal mountains and along the Sierra Nevada mountains. Due to the presence of the Central Valley of California, gene flow rarely occurs between the eastern populations and the western populations. Genetic differences increased from north to south, the result being two distinct forms of *Ensatina* salamanders in Southern California that differ dramatically in color and no longer interbreed.

To take another example, the green iguana found in South America is believed to be the common ancestor for both the marine iguana on the Galápagos Islands (to the west) and the rhinoceros iguana on Hispaniola, an island to the north. If so, how could it happen? Green iguanas are strong swimmers, so by chance a few could have migrated to these islands, where they formed populations separate from each other and from the parent population back in South America. Each population continues on its own evolutionary path as new mutations, genetic drift, and different selection pressures occur for each of the populations. Eventually, postzygotic and prezygotic reproductive isolation developed, and the result was three species of iguanas that are reproductively isolated from each other. A speciation model based on geographic isolation of populations is called allopatric speciation because *allo* means different and *patri* loosely means fatherland.

With **sympatric speciation** [Gk. *sym*, together, and *patri*, fatherland], a population develops into two or more reproductively isolated groups without prior geographic isolation. The best example for this type of speciation is found among plants, where it can occur by means of polyploidy, an increase in the number of sets of chromosomes to 3n or more. Figure 18.13 shows that the parents of our present-day bread wheat had 28 and 14 chromosomes, respectively. The hybrid with 21 chromosomes is sterile, but bread wheat with 42 chromosomes is fertile because the chromosomes can pair during

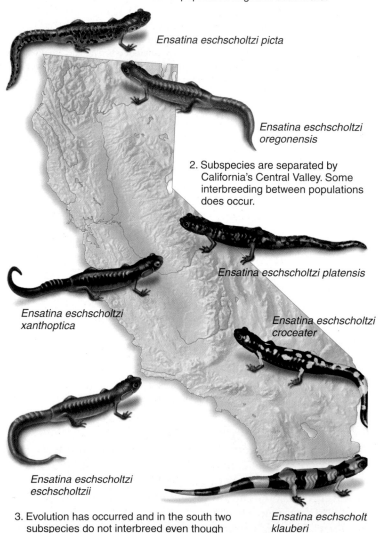

1. Members of a nothern ancestral population migrated southward.

Ensatina eschscholtzi picta

Ensatina eschscholtzi oregonensis

2. Subspecies are separated by California's Central Valley. Some interbreeding between populations does occur.

Ensatina eschscholtzi platensis

Ensatina eschscholtzi xanthoptica

Ensatina eschscholtzi croceater

Ensatina eschscholtzi eschscholtzii

3. Evolution has occurred and in the south two subspecies do not interbreed even though they live in the same environment.

Ensatina eschscholt klauberi

FIGURE 18.12 Allopatric speciation.
In this example of allopatric speciation, the Central Valley of California is separating a range of populations descended from the same northern ancestral species. Those to the east along the coastal mountains and those to the west along the Sierra Nevada mountains experience gene flow, but there is limited gene flow between the eastern populations and the western populations. Members of the most southerly eastern and western populations are by now so genetically different that they do not mate when they meet in Southern California.

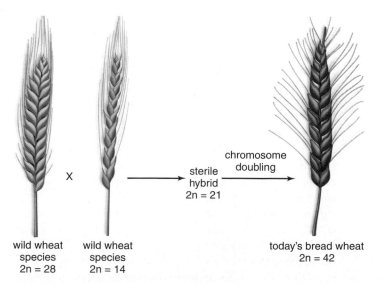

wild wheat species 2n = 28 X wild wheat species 2n = 14 → sterile hybrid 2n = 21 chromosome doubling today's bread wheat 2n = 42

FIGURE 18.13 Sympatric speciation.
In this example of sympatric speciation, two populations of wild wheat hybridized many years ago. The hybrid is sterile, but chromosome doubling allowed some populations to reproduce. These populations are today's bread wheat.

meiosis. Such polyploid plants are isolated by a postzygotic mechanism; they can reproduce successfully only with other similar polyploids, and backcrosses with their parents are sterile. Therefore, there are now three species instead of two species. The presence of sex chromosomes makes it difficult for polyploid speciation to occur in animals.

Adaptive Radiation

Adaptive radiation is an example of allopatric speciation. During adaptive radiation, many new species evolve from a single ancestral species when members of the species become adapted to different environments. The various species of finches that live on the Galápagos Islands are believed to be descended from a single type of ancestral finch from the mainland. The populations on the various islands were subjected to the founder effect and the process of natural selection. Because of natural selection, each population became adapted to a particular habitat on its island. In time, the various populations became so genotypically different that now, when by chance they reside on the same island, they do not interbreed and are therefore separate species (see Fig. 17.18). There is evidence that the finches use beak shape to recognize members of the same species during courtship. Rejection of suitors with the wrong type of beak is a behavioral type of prezygotic isolating mechanism.

FIGURE 18.14

Adaptive radiation in Hawaiian honeycreepers.
More than 20 species, now classified in separate genera, evolved from a single species of a finch-like bird that colonized the Hawaiian Islands. The beaks of honeycreepers are specialized for eating different types of foods.

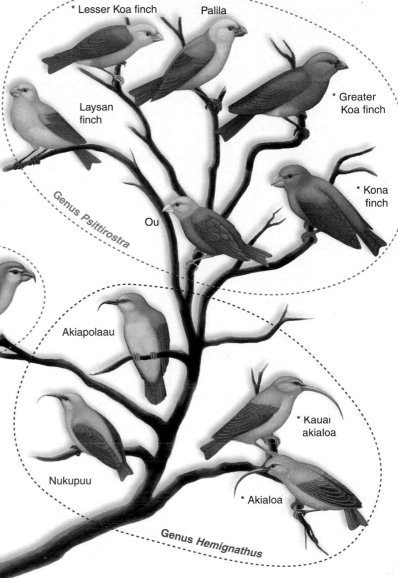

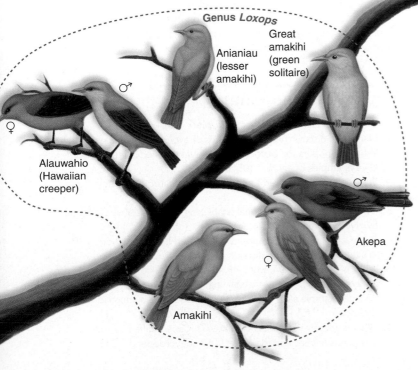

Similarly, on the Hawaiian Islands, a wide variety of honeycreepers are descended from a common goldfinch-like ancestor that arrived from Asia or North America about 5 million years ago. Today, honeycreepers have a range of beak sizes and shapes for feeding on various food sources, including seeds, fruits, flowers, and insects (Fig. 18.14).

Adaptive radiation has occurred throughout the history of life on Earth when a group of organisms exploits a new environment. With the demise of the dinosaurs about 66 million years ago, mammals underwent adaptive radiation as they exploited the environments previously occupied by the dinosaurs.

Allopatric speciation but not sympatric speciation is dependent on a geographic barrier. Adaptive radiation is a particular form of allopatric speciation.

CONNECTING THE CONCEPTS

We have seen that there are variations among the individuals in any population, whether the population is tuberculosis-causing bacteria in a city, the dandelions on a hill, or the squirrels in your neighborhood. Individuals vary because of the presence of mutations and, in sexually reproducing species, because of the recombination of alleles and chromosomes due to the processes of meiosis and fertilization.

The field of population genetics, which uses the Hardy-Weinberg principle, shows us how the study of evolution can be objective rather than subjective. A change in gene pool allele frequencies defines and signifies that evolution has oc-curred. There are various agents of evolutionary change, and natural selection is one of these.

Population genetics has also given us the biological species concept: The members of one species can successfully reproduce with each other but cannot successfully reproduce with members of another species. This criterion allows us to know when a new species has arisen. Species usually come into being after two populations have been geographically isolated. Suppose the present population of squirrels in your neighborhood was divided by a cavern following an earthquake. Then, there would be two genetically similar populations of squirrels that are geographically isolated from each other. Suppose also that the area west of the cavern is slightly drier than that to the east. Over time, each population would become adapted to its particular environment. Eventually, the two populations might become so genetically different that even if members of each population came into contact, they would not be able to produce fertile offspring. Because gene flow between the two populations is no longer possible, the squirrels would be considered separate species. By the same process, multiple species can repeatedly arise, as when a common ancestor led to 13 species of Darwin's finches.

Summary

18.1 MICROEVOLUTION

All the various genes of a population make up its gene pool. The Hardy-Weinberg equilibrium is a constancy of gene pool allele frequencies that remains from generation to generation if certain conditions are met. The conditions are no mutations, no gene flow, random mating, no genetic drift, and no selection. Since these conditions are rarely met, a change in gene pool frequencies is likely. When gene pool frequencies change, microevolution has occurred. Deviations from a Hardy-Weinberg equilibrium allow us to determine when evolution has taken place.

Mutations are the raw material for evolutionary change. Recombinations help bring about adaptive genotypes. Mutations, gene flow, nonrandom mating, genetic drift, and natural selection all cause deviations from a Hardy-Weinberg equilibrium. Certain genotypic variations may be of evolutionary significance only should the environment change. Gene flow occurs when a breeding individual (in animals) migrates to another population or when gametes and seeds (in plants) are carried into another population. Constant gene flow between two populations causes their gene pools to become similar. Nonrandom mating occurs when relatives mate (inbreeding) or assortative mating takes place. Both of these cause an increase in homozygotes. Genetic drift occurs when allele frequencies are altered only because some individuals, by chance, contribute more alleles to the next generation. Genetic drift can cause the gene pools of two isolated populations to become dissimilar as some alleles are lost and others are fixed. Genetic drift is particularly evident after a bottleneck, when severe inbreeding occurs, or when founders start a new population.

18.2 NATURAL SELECTION

The process of natural selection can now be restated in terms of population genetics. A change in gene pool frequencies results in adaptation to the environment.

Most of the traits of evolutionary significance are polygenic; the many variations in population result in a bell-shaped curve. Three types of selection occur: (1) directional—the curve shifts in one direction, as when bacteria become resistant to antibiotics; (2) stabilizing—the peak of the curve increases, as when there is an optimum clutch size for survival of Swiss starling young; and (3) disruptive—the curve has two peaks, as when *Cepaea* snails vary because a wide geographic range causes selection to vary.

Despite constant natural selection, variation is maintained. Mutations and recombination still occur; gene flow among small populations can introduce new alleles; and natural selection itself sometimes results in variation. In sexually reproducing diploid organisms, the heterozygote acts as a repository for recessive alleles whose frequency is low. In regard to sickle cell disease, the heterozygote is more fit in areas where malaria occurs, and therefore both homozygotes are maintained in the population.

18.3 MACROEVOLUTION

The biological species concept recognizes that populations of the same species breed only among themselves and are reproductively isolated from other species. Reproductive isolating mechanisms prevent gene flow between species. Prezygotic isolating mechanisms (habitat, temporal, behavioral, mechanical, and gamete isolation) prevent mating from being attempted. Postzygotic isolating mechanisms (zygote mortality, hybrid sterility, and F_2 fitness) prevent hybrid offspring from surviving and/or reproducing.

Macroevolution refers to evolutionary change at or above the level of a species. During allopatric speciation, geographic isolation occurs before reproductive isolation. A series of salamander subspecies on either side of the Central Valley of California has resulted in at least two subspecies that are unable to successfully reproduce when their ranges overlap. Sympatric speciation does not require geographic isolation for reproductive isolation to develop. The occurrence of polyploidy in plants is an example of this latter type of speciation.

Adaptive radiation is multiple speciation from an ancestral population because varied habitats permit varied adaptations to occur. Adaptive radiation, as exemplified by the Hawaiian honeycreepers, is a form of allopatric speciation.

Reviewing the Chapter

1. What is the Hardy-Weinberg principle? 302–3
2. Name and discuss the five conditions of evolutionary change. 302–3
3. What is a population bottleneck, and what is the founder effect? Give examples of each. 305–6
4. State the steps required for adaptation by natural selection in modern terms. 306
5. Distinguish among directional, stabilizing, and disruptive selection by giving examples. 306–8
6. State ways in which variation is maintained in a population. 309
7. What is the biological definition of a species? 310
8. What is a reproductive isolating mechanism? Give examples of both prezygotic and postzygotic isolating mechanisms. 310–11
9. Give an example each of allopatric speciation and sympatric speciation. 312–13
10. What is adaptive radiation, and how is it exemplified by the Galápagos finches and the Hawaiian honeycreepers? 313

Testing Yourself

Choose the best answer for each question.

1. Assuming a Hardy-Weinberg equilibrium, 21% of a population is homozygous dominant, 50% is heterozygous, and 29% is homozygous recessive. What percentage of the next generation is predicted to be homozygous recessive?
   a. 21%   d. 42%
   b. 50%   e. 58%
   c. 29%

2. A human population has a higher-than-usual percentage of individuals with a genetic disorder. The most likely explanation is
   a. mutations and gene flow.
   b. mutations and natural selection.
   c. nonrandom mating and genetic drift.
   d. nonrandom mating and gene flow.
   e. All of these are correct.

3. The offspring of better-adapted individuals are expected to make up a larger proportion of the next generation. The most likely explanation is
   a. mutations and nonrandom mating.
   b. gene flow and genetic drift.
   c. mutations and natural selection.
   d. mutations and genetic drift.

4. The continued occurrence of sickle cell disease with malaria in parts of Africa is due to
   a. continual mutation.
   b. gene flow between populations.
   c. relative fitness of the heterozygote.
   d. disruptive selection.
   e. protozoan resistance to DDT.

5. Which of these is necessary to natural selection?
   a. variations
   b. differential reproduction
   c. inheritance of differences
   d. differential adaptiveness
   e. All of these are correct.

6. When a population is small, there is a greater chance of
   a. gene flow.
   b. genetic drift.
   c. natural selection.
   d. mutations occurring.
   e. sexual selection.

7. Which of these is an example of stabilizing selection?
   a. Over time, *Equus* developed strength, intelligence, speed, and durable grinding teeth.
   b. British land snails mainly have two different phenotypes.
   c. Swiss starlings usually lay four or five eggs, thereby increasing their chances of more offspring.
   d. Drug resistance increases with each generation; the resistant bacteria survive, and the nonresistant bacteria get killed off.
   e. All of these are correct.

8. The biological definition of a species is simply
   a. the anatomical and developmental differences between two groups of organisms.
   b. the geographic distribution of two groups of organisms.
   c. the differences in the adaptations of two groups of organisms.
   d. the reproductive isolation of a group of organisms.
   e. the difference in mutations between two groups of organisms.

9. Which of these is a prezygotic isolating mechanism?
   a. habitat isolation
   b. temporal isolation
   c. hybrid sterility
   d. zygote mortality
   e. Both a and b are correct.

10. Male moths recognize females of their species by sensing chemical signals called pheromones. This is an example of
    a. gamete isolation.
    b. habitat isolation.
    c. behavioral isolation.
    d. mechanical isolation.
    e. temporal isolation.

11. Allopatric but not sympatric speciation requires
    a. reproductive isolation.
    b. geographic isolation.
    c. prior hybridization.
    d. spontaneous differences in males and females.
    e. changes in gene pool frequencies.

12. The many species of Galápagos finches are each adapted to eating different foods. This is the result of
    a. gene flow.
    b. adaptive radiation.
    c. sympatric speciation.
    d. genetic drift.
    e. All of these are correct.

13. Which of these cannot occur if a population is to maintain an equilibrium of allele frequencies?
    a. People leave one country and relocate in another.
    b. A disease wipes out the majority of a herd of deer.
    c. Members of an Indian tribe only allow the two tallest people in a tribe to marry each spring.
    d. Large black rats are the preferred males in a population of rats.
    e. All of these are correct.

14. Which of these is mechanical isolation?
    a. Sperm cannot reach or fertilize an egg.
    b. Courtship pattern differs.
    c. Living in different locales.
    d. Reproducing at different times of the year.
    e. Genitalia are unsuitable to each other.

15. The homozygote Hb^SHb^S persists because
    a. it offers protection against malaria.
    b. the heterozygote offers protection against malaria.
    c. the genotype Hb^AHb^A offers protection against malaria.
    d. sickle cell disease is worse than sickle cell trait.
    e. Both b and d are correct.

16. The diagrams represent a distribution of phenotypes in a population. Superimpose another diagram on (a) to show that directional selection has occurred, on (b) to show that stabilizing selection has occurred, and on (c) to show that disruptive selection has occurred.

a. Directional selection   b. Stabilizing selection   c. Disruptive selection

Additional Genetics Problems*

1. If $p2 = 0.36$, what percentage of the population has the recessive phenotype, assuming a Hardy-Weinberg equilibrium?
2. If 1% of a human population has the recessive phenotype, what percentage has the dominant phenotype, assuming a Hardy-Weinberg equilibrium?
3. Four percent of the members of a population of pea plants are short (recessive characteristic). What are the frequencies of both the recessive allele and the dominant allele? What are the genotypic frequencies in this population, assuming a Hardy-Weinberg equilibrium?

*Answers to Additional Genetics Problems appear in Appendix A.

Thinking Scientifically

1. You are observing a grouse population in which two feather phenotypes are present in males. One is relatively dark and blends into shadows well, and the other is relatively bright and so is more obvious to predators. The females are uniformly dark-feathered. Observing the frequency of mating between females and the two types of males, you have recorded the following:

   matings with dark-feathered males: 13

   matings with bright-feathered males: 32

   Propose a hypothesis to explain why females apparently prefer bright-feathered males. What selective advantage might there be in choosing a male with alleles that make it more susceptible to predation? What data would help test your hypothesis?
2. A farmer uses a new pesticide. He applies the pesticide as directed by the manufacturer and loses about 15% of his crop to insects. A farmer in the next state learns of these results, uses three times as much pesticide, and loses only 3% of her crop to insects. Each farmer follows this pattern for five years. At the end of five years, the first farmer is still losing about 15% of his crop to insects, but the second farmer is losing 40% of her crop to insects. How could these observations be interpreted on the basis of natural selection?

Understanding the Terms

adaptive radiation   313
allopatric speciation   312
assortative mating   305
bottleneck effect   305
directional selection   306
disruptive selection   308
founder effect   306
gene flow   304
gene pool   302
genetic drift   305
Hardy-Weinberg principle   302
industrial melanism   303
macroevolution   310
microevolution   302

natural selection   306
nonrandom mating   305
polymorphic   304
population   302
population genetics   302
postzygotic isolating mechanism   311
prezygotic isolating mechanism   310
relative fitness   306
sexual selection   305
speciation   310
species   310
stabilizing selection   307
sympatric speciation   312

Match the terms to these definitions:

a. _____Outcome of natural selection in which extreme phenotypes are eliminated and the average phenotype is conserved.
b. _____Anatomical or physiological difference between two species that prevents successful reproduction after mating has taken place.
c. _____Change in the genetic makeup of a population due to chance (random) events; important in small populations or when only a few individuals mate.
d. _____Evolution of a large number of species from a common ancestor.
e. _____Sharing of genes between two populations through interbreeding.

ARIS, the *Biology* Website

ARIS, the website for *Biology*, provides a wealth of information organized and integrated by chapter. You will find practice quizzes, interactive activities, labeling exercises, flashcards, and much more that will complement your learning and understanding of general biology.

www.mhhe.com/maderbiology9

19

ORIGIN AND HISTORY OF LIFE

O ne of the most fascinating and frequently asked questions by laypeople and scientists alike is, Where did life come from? Although various answers have been proposed throughout history, today the most widely accepted hypothesis is that inorganic molecules in the primitive Earth's prebiotic oceans combined to produce organic molecules, and eventually the first cells arose. These earliest cells probably appeared around 3.5 billion years ago. At about that time, oxygen-releasing photosynthesis began, and an atmospheric ozone layer started to form. This shield, which protects terrestrial life from intense ultraviolet radiation, enabled life to move from water onto land. Shortly (at least in geologic time), the number of multicellular life-forms increased dramatically.

If we could trace the lineage of all the millions of species ever to have evolved, the entirety would resemble a dense bush. Some lines of descent are cut off close to the base; some continue in a straight line even to today; others have split, producing two or even several groups. The history of life on Earth has many facets, twists, and turns.

The Earth was once devoid of life.

317

19.1 Origin of Life

Today we do not believe that life arises spontaneously from nonlife, and we say that "life comes only from life." But if this is so, how did the first form of life come about? Since it was the very first living thing, it had to come from nonliving chemicals. Could there have been an increase in the complexity of the chemicals—could a **chemical evolution** have produced the first cell(s) on early Earth?

The Early Earth

The sun and the planets, including Earth, probably formed over a 10-billion-year period from aggregates of dust particles and debris. At 4.6 billion years ago (BYA), the solar system was in place. Intense heat produced by gravitational energy and radioactivity produced several stratified layers. Heavier atoms of iron and nickel became the molten liquid core, and dense silicate minerals became the semiliquid mantle. Upwellings of volcanic lava produced the first crust.

The Earth's mass is such that the gravitational field is strong enough to have an atmosphere. Less mass and atmospheric gases would escape into outer space. The early Earth's atmosphere was not the same as today's atmosphere; it was produced primarily by outgassing from the interior, exemplified by volcanic eruptions. The early atmosphere most likely consisted mainly of these inorganic chemicals: water vapor (H_2O), nitrogen (N_2), and carbon dioxide (CO_2), with only small amounts of hydrogen (H_2), methane (CH_4), ammonia (NH_3), hydrogen sulfide (H_2S), and carbon monoxide (CO). The early atmosphere may have been a reducing atmosphere, with little free oxygen. If so, that would have been fortuitous because oxygen (O_2) attaches to organic molecules, preventing them from joining to form larger molecules.

At first it was so hot that water was present only as a vapor that formed dense, thick clouds. Then, as the early Earth cooled, water vapor condensed to liquid water, and rain began to fall. It rained in such enormous quantity over hundreds of millions of years that the oceans of the world were produced. Our planet is an appropriate distance from the sun: any closer, water would have evaporated; any farther away, water would have frozen.

It is also possible that the oceans were fed by celestial comets that entered the Earth's gravitational field. In 1999, physicist Louis Frank presented images taken by cameras on NASA's *Polar* satellite to substantiate his claim that the Earth is bombarded with 5–30 icy comets the size of a house every minute. The ice becomes water vapor that later comes down as rain, enough rain, says Frank, to raise the oceans' level by an inch in just 10,000 years.

Monomers Evolve

Comets and meteorites have constantly pelted the Earth throughout history. In recent years, scientists have confirmed the presence of organic molecules in some meteorites. Many scientists, championed by Chandra Wickramsinghe, feel that these organic molecules could have seeded the chemical origin of life on early Earth. Others even hypothesize that bacterium-like cells evolved first on another planet and then were carried to Earth. A meteorite from Mars labeled ALH84001 landed on Earth some 13,000 years ago. When examined, experts found tiny rods similar in shape to fossilized bacteria. Therefore this hypothesis is being investigated.

In a letter to a friend in 1871, Charles Darwin accurately speculated about the chemical origin of life. Darwin wrote: "If (and oh, what a big if!) we could conceive in some warm little pond, with all sorts of ammonia and phosphoric salts, light, heat, electricity, etc. . ., present that a protein compound was chemically formed ready to undergo even more complex changes." It would be decades before scientists would revisit this idea.

In the 1920s, A. I. Oparin, a Russian biochemist, and J. B. S. Haldane, a Scottish physiologist and geneticist, independently suggested that the first organic molecules could have been produced from early atmospheric gases in the presence of strong energy sources. The energy sources on early Earth included heat from volcanoes and meteorites, radioactivity from isotopes, powerful electric discharges in lightning, and solar radiation, especially ultraviolet radiation. Oparin's idea is called abiotic synthesis, the formation of simple monomers (sugars, amino acids, nucleotide bases) from inorganic molecules.

In 1953, Stanley Miller and Harold Urey provided support for Oparin's and Haldane's hypotheses through an ingenious experiment (Fig. 19.1). They placed a mixture resembling a strongly reducing atmosphere—methane (CH_4), ammonia (NH_3), hydrogen (H_2), and water (H_2O)—in a closed system, heated the mixture, and circulated it past an electric spark (simulating lightning). After a week's run, Miller and Urey discovered that a variety of amino acids

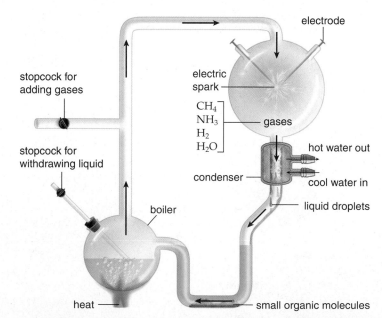

FIGURE 19.1 Stanley Miller's and Harold Urey's apparatus and experiment.

Gases that were thought to be present in the early Earth's atmosphere were admitted to the apparatus, circulated past an energy source (electric spark), and cooled to produce a liquid that could be withdrawn. Upon chemical analysis, the liquid was found to contain various small organic molecules.

plume of hot water
rich in iron-nickel sulfides

hydrothermal
vent

FIGURE 19.2 Chemical evolution at hydrothermal vents.
Minerals that form at deep-sea hydrothermal vents like this one can catalyze the formation of ammonia and even organic molecules.

and organic acids had been produced. Since that time, other investigators have achieved similar results by using other, less-reducing combinations of gases dissolved in water.

These experiments apparently lend support to the hypothesis that the Earth's first atmospheric gases could have reacted with one another to produce small organic compounds. If so, neither oxidation (there was no free oxygen) nor decay (there were no bacteria) would have destroyed these molecules, and rainfall would have washed them into the ocean, where they accumulated for hundreds of millions of years. Therefore, the oceans would have been a thick, warm organic soup.

Other investigators are concerned that Miller and Urey used ammonia as one of the atmospheric gases. They point out that, whereas inert nitrogen gas (N_2) would have been abundant in the primitive atmosphere, ammonia (NH_3) would have been scarce. Where might ammonia have been abundant? A team of researchers at the Carnegie Institution in Washington, D.C., believe they have found the answer: hydrothermal vents on the ocean floor. These vents line huge **ocean ridges,** where molten magma wells up and adds material to the ocean floor (Fig. 19.2). Cool water seeping through the vents is heated to a temperature as high as 350°C, and when it spews back out, it contains various mixed iron-nickel sulfides that can act as catalysts to change N_2 to NH_3. A laboratory test of this hypothesis worked perfectly. Under ventlike conditions, 70% of various nitrogen sources were converted to ammonia within 15 minutes. German organic chemists Gunter Wachtershäuser and Claudia Huber have gone one more step. They have shown that organic molecules will react and amino acids will form peptides in the presence of iron-nickel sulfides under ventlike conditions.

Comets from outer space, atmospheric reactions, and hydrothermal vents could be responsible for the first organic molecules on Earth.

Polymers Evolve

In cells, monomers join to form polymers in the presence of enzymes, which of course are proteins. How did the first organic polymers form if there were no proteins yet? As just mentioned, Wachtershäuser and Huber have managed to achieve the formation of peptides using iron-nickel sulfides as inorganic catalysts under ventlike conditions of high temperature and pressure. These minerals have a charged surface that attracts amino acids and provides electrons so they can bond together.

Sidney Fox has shown that amino acids polymerize abiotically when exposed to dry heat. He suggests that once amino acids were present in the oceans, they could have collected in shallow puddles along the rocky shore. Then the heat of the sun could have caused them to form **proteinoids,** small polypeptides that have some catalytic properties. When he simulates this scenario in the lab and returns proteinoids to water, they form **microspheres** [Gk. *mikros*, small, little, and *sphaera*, ball], structures composed only of protein that have many properties of a cell (Fig. 19.3a). It's possible that even newly formed polypeptides had enzymatic properties, and some proved to be more capable than others. Those that led to the first cell or cells had a selective advantage. Fox's **protein-first hypothesis** assumes that DNA genes came after protein enzymes arose. After all, it is protein enzymes that are needed for DNA replication.

Another hypothesis is put forth by Graham Cairns-Smith. He believes that clay was especially helpful in causing polymerization of monomers to produce both proteins and nucleic acids at the same time. Clay also attracts small organic molecules and contains iron and zinc, which may have served as inorganic catalysts for polypeptide formation. In addition, clay has a tendency to collect energy from radioactive decay and to discharge it when the temperature and/or humidity changes. This could have been a source of energy for polymerization to take place. Cairns-Smith suggests that RNA nucleotides and amino acids became associated in such a way that polypeptides were ordered by and helped synthesize RNA. This hypothesis suggests that both polypeptides and RNA arose at the same time.

There is still another hypothesis concerning this stage in the origin of life. The **RNA-first hypothesis** suggests that only the macromolecule RNA (ribonucleic acid) was needed to progress toward formation of the first cell or cells. Thomas Cech and Sidney Altman shared a Nobel Prize in 1989 because they discovered that RNA can be both a substrate and an enzyme. Some viruses today have RNA genes; therefore, the first genes could have been RNA. It would seem, then, that RNA could have carried out the processes of life commonly associated today with DNA (deoxyribonucleic acid, the genetic material) and proteins (enzymes). Those who support this hypothesis say that it was an "RNA world" some 4 BYA.

Polymerization of monomers to produce proteins and nucleotides is the next step toward the first cell.

A Protocell Evolves

Before the first true cell arose, there would have been a **protocell** or **protobiont** [Gk. *protos*, first], a structure that first and foremost has an outer membrane. After all, life requires chemical reactions that take place within a boundary. The plasma membrane separates the living interior from the nonliving exterior. There are several hypotheses about the origin of the first membrane. Sidney Fox has shown that if lipids are made available to microspheres, lipids tend to become associated with microspheres (Fig. 19.3*a*), producing a lipid-protein membrane. Microspheres have many interesting properties: they resemble bacteria and have an electrical potential difference; they divide and perhaps are subject to selection. Although Fox believes they are a cell, others disagree.

Oparin, who was mentioned previously, showed that under appropriate conditions of temperature, ionic composition, and pH, concentrated mixtures of macromolecules tend to give rise to complex units called **coacervate droplets.** Coacervate droplets have a tendency to absorb and incorporate various substances from the surrounding solution. Eventually, a semipermeable-type boundary may form about the droplet.

In the early 1960s biophysicist Alec Bangham of the Animal Physiology Institute in Cambridge, England, discovered that when he extracted lipids from egg yolks and placed them in water, the lipids would naturally organize themselves into double-layered bubbles roughly the size of a cell. Bangham's bubbles soon became known as **liposomes** [Gk. *lipos*, fat, and *soma*, body] (Fig. 19.3*b*). Later, biophysicist David Deamer of the University of California and Bangham realized that liposomes might have provided life's first boundary. Perhaps liposomes with a phospholipid membrane engulfed early molecules that had enzymatic, even replicative abilities. The liposomes would have protected the molecules from their surroundings and concentrated them so they could react (and evolve) quickly and efficiently. These investigators called this the "membrane-first" hypothesis, meaning that the first cell had to have a plasma membrane before any of its other parts. Perhaps the first membrane formed in this manner, and the protocell contained only RNA, which functioned as both genetic material and enzymes.

The protocell would have had to carry on nutrition so that it could grow. If organic molecules formed in the atmosphere and were carried by rain into the ocean, nutrition would have been no problem because simple organic molecules could have served as food. This hypothesis suggests that the protocell was a heterotroph [Gk. *hetero*, different, and *trophe*, food], an organism that takes in preformed food. On the other hand, if the protocell evolved at hydrothermal vents, it may have carried out chemosynthesis. Chemoautotrophic bacteria obtain energy for synthesizing organic molecules by oxidizing inorganic compounds, such as hydrogen sulfide (H_2S), a molecule that is abundant at the vents. When hydrothermal vents were first discovered in the 1970s, investigators were surprised to discover complex vent ecosystems supported by organic molecules formed by chemosynthesis, a process that does not require the energy of the sun.

At first, the protocell may have used preformed ATP (adenosine triphosphate), but as this supply dwindled, natural selection favored any cells that could extract energy from carbohydrates in order to transform ADP (adenosine diphosphate) to ATP. Glycolysis is a common metabolic pathway in living things, and this testifies to its early evolution in the history of life. Since there was no free oxygen, we can assume that the protocell carried on a form of fermentation.

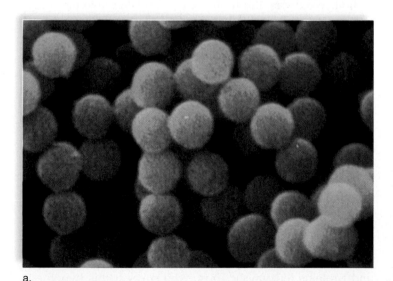

a.

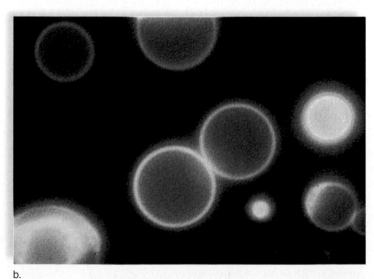

b.

FIGURE 19.3 Origin of plasma membrane.

a. Microspheres have a number of cellular characteristics and could have evolved into the protocell. For example, microspheres are composed only of protein, but if lipids are added, the proteins combine with them. Perhaps this was the step that led to the first plasma membrane. **b.** Liposomes form automatically when lipids are put into water. Phospholipids also have a tendency to form a circle. Perhaps the first plasma membrane was simply a phospholipid bilayer without any proteins being present.

At first the protocell must have had limited ability to break down organic molecules, and scientists speculate that it took millions of years for glycolysis to evolve completely. Interestingly, Fox has shown that microspheres from which protocells may have evolved have some catalytic ability, and Oparin found that coacervates do incorporate enzymes if they are available in the medium.

The protocell is hypothesized to have had a membrane boundary and to have been either a heterotroph or a chemoautotroph.

A Self-Replication System Evolves

Today's cell is able to carry on protein synthesis in order to produce the enzymes that allow DNA to replicate. The central dogma of genetics states that DNA directs protein synthesis and that information flows from DNA to RNA to protein. It is possible that this sequence developed in stages.

According to the RNA-first hypothesis, RNA would have been the first to evolve, and the first true cell would have had RNA genes. These genes would have directed and enzymatically carried out protein synthesis. Today, ribozymes are enzymatic RNA molecules. Also, today we know there are viruses that have RNA genes. These viruses have a protein enzyme called reverse transcriptase that uses RNA as a template to form DNA. Perhaps with time, reverse transcription occurred within the protocell, and this is how DNA genes arose. If so, RNA was responsible for both DNA and protein formation. Once there were DNA genes, protein synthesis would have been carried out in the manner dictated by the central dogma of genetics.

According to the protein-first hypothesis, proteins, or at least polypeptides, were the first of the three (i.e., DNA, RNA, and protein) to arise. Only after the protocell developed a plasma membrane and sophisticated enzymes did it have the ability to synthesize DNA and RNA from small molecules provided by the ocean. Because a nucleic acid is a complicated molecule, the likelihood RNA arose *de novo* (on its own) is minimal. It seems more likely that enzymes were needed to guide the synthesis of nucleotides and then nucleic acids. Again, once there were DNA genes, protein synthesis would have been carried out in the manner dictated by the central dogma of genetics.

Cairns-Smith proposes that polypeptides and RNA evolved simultaneously. Therefore, the first true cell would have contained RNA genes that could have replicated because of the presence of proteins. This eliminates the baffling chicken-and-egg paradox: assuming a plasma membrane, which came first, proteins or RNA? It means, however, that two unlikely events would have had to happen at the same time.

After DNA formed, the genetic code had to evolve before DNA could store genetic information. The present genetic code is subject to fewer errors than a million other possible codes. Also, the present code is among the best at minimizing the effect of mutations. A single-base change in a present codon is likely to result in the substitution of a chemically similar amino acid and, therefore, minimal changes in the final protein. This evidence suggests that the genetic code did undergo a natural selection process before finalizing into today's code.

Once there was a flow of information from DNA to RNA to protein, the protocell became a true cell, and biological evolution began.

Figure 19.4 reviews how most biologists believe life could have evolved on early Earth.

1. An abiotic synthesis process created small organic molecules such as amino acids and nucleotides, perhaps in the atmosphere or at hydrothermal vents.

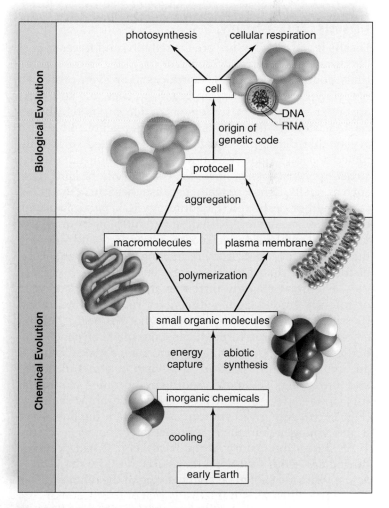

FIGURE 19.4 Origin of the first cell(s).
There was an increase in the complexity of macromolecules, leading to a self-replicating system (DNA ⟶ RNA ⟶ protein) enclosed by a plasma membrane. The protocell, a heterotrophic fermenter, underwent biological evolution, becoming a true cell, which then diversified.

2. These monomers joined together to form polymers along the shoreline (warm seaside rocks or clay) or at the vents. The first polymers could have been proteins or RNA, or they could have evolved together.

3. The aggregation of polymers inside a plasma membrane produced a protocell, which had enzymatic properties such that it could grow. If the protocell developed in the ocean, it was a heterotroph; if it developed at hydrothermal vents, it was a chemoautotroph.

4. Once the protocell contained DNA genes, a true cell had evolved. The first genes may have been RNA molecules, but later DNA became the information storage molecule of heredity. Biological evolution—and the history of life—had begun!

19.2 HISTORY OF LIFE

Macroevolution includes large-scale patterns of change taking place over very long periods of time. The fossil record records such changes.

Fossils Tell a Story

Fossils [L. *fossilis*, dug up] are the remains and traces of past life or any other direct evidence of past life. Traces include trails, footprints, burrows, worm casts, or even preserved droppings. Usually when an organism dies, the soft parts are either consumed by scavengers or decomposed by bacteria. Occasionally, the organism is buried quickly and in such a way that decomposition is never completed or is completed so slowly that the soft parts leave an imprint of their structure. Most fossils, however, consist only of hard parts such as shells, bones, or teeth, because these are usually not consumed or destroyed. **Paleontology** [Gk. *palaios*, ancient, old, and *ontos*, having existed; *-logy*, study of, from *logikos*, rational, sensible] is the science of discovering and studying the fossil record and, from it, making decisions about the history of life, ancient climates, and environments.

The great majority of fossils are found embedded in or recently eroded from sedimentary rock. **Sedimentation** [L. *sedimentum*, a settling], a process that has been going on since the Earth was formed, can take place on land or in bodies of water. Weathering and erosion of rocks produces an accumulation of particles that vary in size and nature and are called sediment. Sediment becomes a **stratum** (pl., strata), a recognizable layer in a stratigraphic sequence (Fig. 19.5*a*). Any given stratum is older than the one above it and younger than the one immediately below it. Figure 19.5*b* shows the history of the Earth as if it had occured during a 24-hour time span that starts at midnight. (The actual years are shown on an inner ring of the diagram.) This figure illustrates dramatically that only unicellular organisms were present during most (about 80%) of the history of the Earth.

If the Earth formed at midnight, prokaryotes do not appear until about 5 A.M., eukaryotes are present at approximately 4 P.M., and multicellular forms do not appear

until around 8 P.M. Invasion of the land doesn't occur until about 10 P.M., and humans don't appear until 30 seconds before the end of the day. This timetable has been worked out by studying the fossil record. In addition to sedimentary fossils, more recent fossils can be found in tar, ice, bogs, and amber. Petrified wood, shells, and bones are also relatively common (Fig. 19.6).

Relative Dating of Fossils

In the early nineteenth century, even before the theory of evolution was formulated, geologists sought to correlate the strata worldwide. The problem was that strata change their character over great distances, and therefore a stratum in England might contain different sediments than one of the same age in Russia. Geologists discovered that each strata of the same age contained certain **index fossils** that serve to identify deposits made at apparently the same time in different parts of the world. These index fossils are used in **relative dating** methods. For example, a particular species of fossil ammonite (an animal related to the chambered nautilus) has been found over a wide range and for a limited time period. Therefore, all strata around the world that contain this fossil must be of the same age.

Absolute Dating of Fossils

An **absolute dating** method that relies on radioactive dating techniques assigns an actual date to a fossil. All radioactive isotopes have a particular half-life, the length of time it takes for half of the radioactive isotope to change into another stable element. If the fossil has organic matter, half of the carbon 14 (^{14}C)

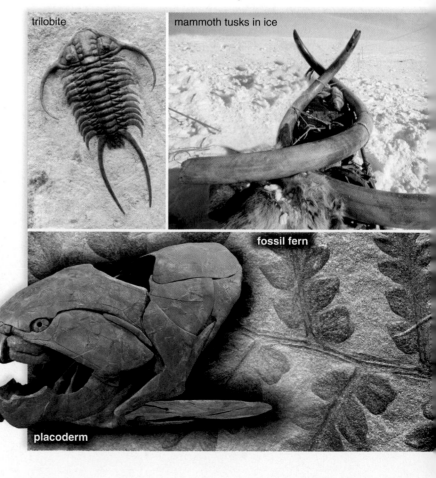

trilobite

mammoth tusks in ice

fossil fern

placoderm

a.

FIGURE 19.5 The history of life.

a. Strata, layers seen in sedimentary rock as exposed by road cuts, are the source of fossils that tell us about the history of life. **b.** The outer ring of this diagram shows the history of life as it would be measured on a 24-hour timescale starting at midnight. (The inner ring shows the actual years starting at 4.6 BYA.) The fossil record suggests that a very large portion of life's history was devoted to the evolution of unicellular organisms. The first multicellular organisms do not appear in the fossil record until just before 8 P.M., and humans are not on the scene until less than a minute before midnight.

first appearance of *Homo sapiens*
(11:59:30)

Age of Dinosaurs

land plants

formation of earth

oldest multicellular fossils

oldest known rocks

oldest fossils (prokaryotes)

first photosynthetic organisms

oldest eukaryotic fossils

free oxygen in atmosphere

1 second = 52,000 years
1 minute = 3,125,000 years
1 hour = 187,500,000 years

b.

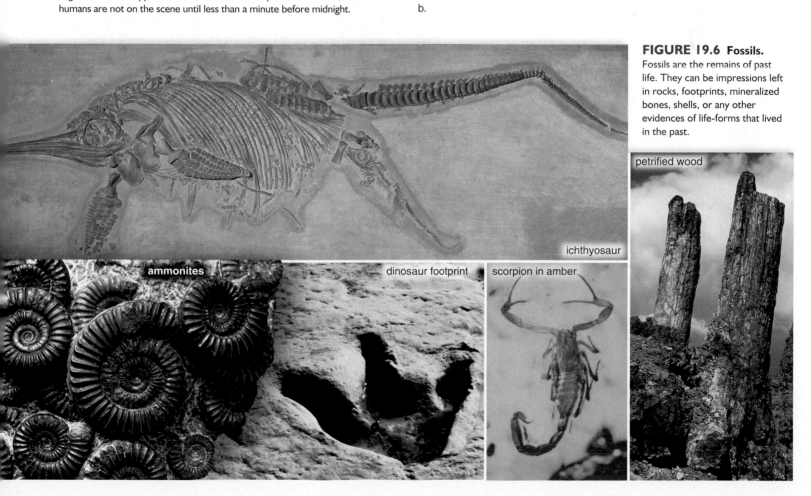

FIGURE 19.6 Fossils.

Fossils are the remains of past life. They can be impressions left in rocks, footprints, mineralized bones, shells, or any other evidences of life-forms that lived in the past.

ichthyosaur

ammonites

dinosaur footprint

scorpion in amber

petrified wood

will have changed to nitrogen 14 (^{14}N) in 5,730 years. To know how much ^{14}C was in the organism to begin with, it is reasoned that organic matter always begins with the same amount of ^{14}C. (In reality, it is known that the ^{14}C levels in the air—and therefore the amount in organisms—can vary from time to time.) Now we need only compare the ^{14}C radioactivity of the fossil to that of a modern sample of organic matter. The amount of radiation left can be converted to the age of the fossil. After 50,000 years, however, the amount of ^{14}C radioactivity is so low that it cannot be used to measure the age of a fossil accurately.

^{14}C is the only radioactive isotope contained within organic matter, but it is possible to use others to date rocks, and from that to infer the age of a fossil contained in the rock. For instance, the ratio of potassium 40 (^{40}K) to argon 40 trapped in rock is often used to date rocks.

Fossils, which can be dated relatively according to their location in strata and absolutely according to their content of radioactive isotopes, give us information about the history of life.

The Precambrian

As a result of their study of fossils in strata, geologists have devised the **geological timescale,** which divides the history of the Earth into eras and then periods and epochs (Table 19.1). We will follow the biologist's tradition of first discussing Precambrian time. The Precambrian is a very long period of time, comprising about 87% of the geological timescale. During this time, life arose and the first cells came into existence. The first cells were probably prokaryotes. Prokaryotes do not have a nucleus or membrane-bounded organelles. Prokaryotes, the archaea and bacteria, can live in the most inhospitable of environments, such as hot springs, salty lakes, and airless swamps—all of which may typify habitats on early Earth. The cell wall, plasma membrane, RNA polymerase, and ribosomes of archaea are more like those of eukaryotes than those of bacteria.

The first identifiable fossils are those of complex prokaryotes. Chemical fingerprints of complex cells are found in sedimentary rocks from southwestern Greenland, dated at 3.8 BYA. But paleobiologist J. William Schopf found the oldest prokaryotic fossils in western Australia. These 3.46-billion-year-old microfossils resemble today's cyanobacteria, prokaryotes that carry on photosynthesis in the same manner as plants (Fig. 19.7a).

At this time, only volcanic rocks jutted above the waves, and there were as yet no continents. Strange-looking boulders, called **stromatolites,** littered beaches and shallow waters (Fig. 19.7b). Living stromatolites can still be found today along Australia's western coast. The outer surface of a

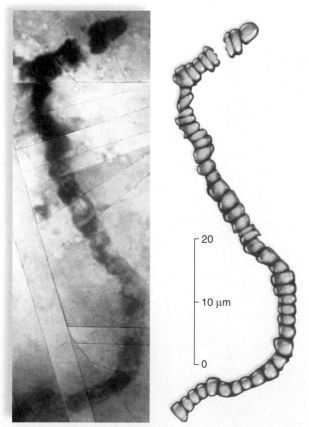

a. *Primaevifilum*

b. Stromatolites

FIGURE 19.7 Prokaryote fossils of the Precambrian.
a. The prokaryotic microorganism *Primaevifilum* (with interpretive drawing) was found in rocks dated 3.46 BYA. **b.** Stromatolites also date back to this time. Living stromatolites are located in shallow waters off the shores of western Australia and also in other tropical seas.

TABLE 19.1

The Geological Timescale: Major Divisions of Geological Time and Some of the Major Evolutionary Events That Occurred

Era	Period	Epoch	Millions of Years Ago	Plant Life	Animal Life
		Holocene	0.01	Human influence on plant life	Age of *Homo sapiens*
		Significant Mammalian Extinction			
	Quaternary	Pleistocene	1.8	Herbaceous plants spread and diversify.	Presence of ice age mammals. Modern humans appear.
Cenozoic*		Pliocene	5.3	Herbaceous angiosperms flourish.	First hominids appear.
		Miocene	2.3	Grasslands spread as forests contract.	Apelike mammals and grazing mammals flourish; insects flourish.
		Oligocene	33.9	Many modern families of flowering plants evolve.	Browsing mammals and monkeylike primates appear.
	Tertiary	Eocene	55.8	Subtropical forests with heavy rainfall thrive.	All modern orders of mammals are represented.
		Paleocene	65.5	Flowering plants continue to diversify.	Primitive primates, herbivores, carnivores, and insectivores appear.
		Mass Extinction: Dinosaurs and Most Reptiles			
	Cretaceous		145.5	Flowering plants spread; conifers persist.	Placental mammals appear; modern insect groups appear.
Mesozoic	Jurassic		199.6	Flowering plants appear.	Dinosaurs flourish; birds appear.
		Mass Extinction			
	Triassic		251	Forests of conifers and cycads dominate.	First mammals appear; first dinosaurs appear; corals and molluscs dominate seas.
		Mass Extinction			
	Permian		299	Gymnosperms diversify.	Reptiles diversify; amphibians decline.
	Carboniferous		359.2	Age of great coal-forming forests: Ferns, club mosses, and horsetails flourish.	Amphibians diversify; first reptiles appear; first great radiation of insects.
		Mass Extinction			
Paleozoic	Devonian		416	First seed plants appear. Seedless vascular plants diversify.	Jawed fishes diversify and dominate the seas; first insects and first amphibians appear.
	Silurian		443.7	Seedless vascular plants appear.	First jawed fishes appear.
		Mass Extinction			
	Ordovician		488.3	Nonvascular land plants appear. Marine algae flourish.	Invertebrates spread and diversify; jawless fishes (first vertebrates) appear.
	Cambrian		542	First plants appear on land. Marine algae flourish.	All invertebrate phyla present; first chordates appear.
Precambrian Time			600	Oldest soft-bodied invertebrate fossils	
			1,400–700	Protists evolve and diversify.	
			2,200	Oldest eukaryotic fossils	
			2,700	O_2 accumulates in atmosphere.	
			3,500	Oldest known fossils (prokaryotes)	
			4,600	Earth forms.	

* Many authorities divide the Cenozoic era into the Paleogene period (contains the Paleocene, Eocene, and Oligocene epochs) and the Neogene period (contains the Miocene, Pliocene, Pleistocene, and Holocene epochs).

stromatolite is alive with cyanobacteria that secrete a mucus. Grains of sand get caught in the mucus and bind with calcium carbonate from the water to form rock. To gain access to sunlight, the photosynthetic organisms move outward toward the surface before they are cemented in. They leave behind a menagerie of aerobic and then anaerobic bacteria caught in the layers of the rock.

The cyanobacteria in ancient stromatolites added oxygen to the atmosphere. By 2 BYA, the presence of oxygen was such that most environments were no longer suitable for anaerobic prokaryotes, and they began to decline in importance. Photosynthetic cyanobacteria and aerobic bacteria proliferated as new metabolic pathways evolved. Due to the presence of oxygen, the atmosphere became an oxidizing one instead of a reducing one. Oxygen in the upper atmosphere forms ozone (O_3), which filters out the ultraviolet (UV) rays of the sun. Before the formation of the **ozone shield,** the amount of ultraviolet radiation reaching the Earth could have helped create organic molecules, but it would have destroyed any land-dwelling organisms. Once the ozone shield was in place, living things were sufficiently protected and able to live on land.

2.2 BYA	Oldest eukaryotic fossils
2.7 BYA	O_2 accumulates in atmosphere
3.5 BYA	Oldest known fossils
4.6 BYA	Formation of the Earth

The evolution of photosynthesizing prokaryotes caused oxygen to enter the atmosphere.

Eukaryotic Cells Arise

The eukaryotic cell, which originated around 2.2 BYA, is nearly always aerobic and contains a nucleus as well as other membranous organelles. Most likely, the eukaryotic cell acquired its organelles gradually. The nucleus may have developed by an invagination of the plasma membrane. The mitochondria of the eukaryotic cell probably were once free-living aerobic prokaryotes, and the chloroplasts probably were free-living photosynthetic prokaryotes. The **endosymbiotic hypothesis** states that a nucleated cell engulfed these prokaryotes, which then became organelles (see Fig. 4.5, p. 64). It's been suggested that flagella (and cilia) also arose by endosymbiosis. First, slender undulating prokaryotes could have attached themselves to a host cell to take advantage of food leaking from the host's plasma membrane. Eventually, these prokaryotes adhered to the host cell and became the flagella and cilia we know today. The first eukaryotes were unicellular, as are prokaryotes.

Multicellularity Arises

Fossils identified as multicellular protists and dated 1.4 BYA have been found in arctic Canada. It's possible that the first multicellular organisms practiced sexual reproduction. Among today's protists (eukaryotes classified in the kingdom Protista), we find colonial forms in which some cells are specialized to produce gametes needed for sexual reproduction. Separation of germ cells, which produce gametes from somatic cells, may have been an important first step toward the development of the Ediacaran invertebrates discussed next, which appeared about 600 MYA and died out about 545 MYA.

In 1946, R. C. Sprigg, a government geologist assessing abandoned lead mines in southern Australia, discovered the first remains of a remarkable biota that has taken its name from the region, the Ediacara Hills. Since then, similar fossils have been discovered on a number of other continents. Many of the fossils, dated 600–545 MYA, are believed to be of soft-bodied invertebrates (animals without a vertebral column) that most likely lived on mudflats in shallow marine waters. Some may have been mobile, but others were large, immobile, bizarre creatures resembling spoked wheels, corrugated ribbons, and lettucelike fronds. All were flat and probably had two tissue layers; few had any type of skeleton (Fig. 19.8). They apparently had no mouths; perhaps they absorbed nutrients from the sea or else had photosynthetic organisms living on their tissues. These soft-bodied animals could flourish because there were no predators to eat them. Their fossils are like footprints—impressions made in the sandy seafloor before their bodies decayed away. Whether the Ediacaran animals were simply a failed evolutionary experiment or whether they are related to animals of the Cambrian period is not known. With few exceptions they disappear from the fossil record at 545 MYA but even so some may have given rise to modern cnidarians and related animals. What caused their demise is not known but they very well could have been eaten by the myriad of animals with mouths that suddenly appear in the Cambrian.

543 MYA	Cambrian animals
600 MYA	Ediacaran animals
1.4 BYA	Protists evolve and diversify
2.2 BYA	Oldest eukaryotic fossils

The Precambrian takes up most of the Earth's history. Prokaryotic cells evolved about 3.5 BYA. After eukaryotic cells arose (about 2.2 BYA), it was nearly another billion years before multicellularity occurred. Ediacaran animal fossils are dated 600 MYA.

a.   b.

FIGURE 19.8 Ediacaran fossils.
The Ediacaran invertebrates lived from about 600–545 MYA. They were all flat, soft-bodied invertebrates. **a.** Classified as *Spriggina,* this bilateral organism had a crescent-shaped head and numerous segments tapering to a posterior end. **b.** Classified as *Dickinsonia,* these fossils are often interpreted to be segmented worms. However, in the opinion of some, they may be cnidarian polyps.

The Paleozoic Era

The Paleozoic era lasted about 300 million years. Even though the era was quite short compared to the length of the Precambrian, many events occurred during this era, including three major mass extinctions (see Table 19.1). An **extinction** is the total disappearance of all the members of a species or higher taxonomic group. **Mass extinctions,** which are the disappearance of a large number of species or a higher taxonomic group within an interval of just a few million years, are discussed on page 336.

Cambrian Animals

The seas of the Cambrian period teemed with invertebrate life as illustrated in Figure 19.9. Life became so abundant that scientists refer to this period in Earth's history as the Cambrian explosion. All of today's groups of animals can trace their ancestry to this time, and perhaps earlier, according to new molecular clock data. A **molecular clock** is based on the principle that DNA differences in certain parts of the genome occur at a fixed rate and are not tied to natural selection. The number of DNA base-pair differences tells how long two species have been evolving separately.

Even if certain of the animals in Figure 19.9 had evolved earlier, no fossil evidence of them occurs until the Cambrian period, perhaps because they lacked a skeleton. Animals that lived during the Cambrian possessed protective outer skeletons known as exoskeletons. These structures are capable of surviving the forces that are apt to destroy soft-bodied organisms. For example, Cambrian seafloors were dominated by now-extinct trilobites, which had thick, jointed armor covering them from head to tail. Trilobites are classified as arthropods, a major phylum of animals today. (Some Cambrian species, with most unusual eating and locomotion appendages, have been classified in phyla that no longer exist today.)

Paleontologists seek an explanation for why animals of the Cambrian period possessed exoskeletons and why this development did not occur during the Precambrian. By this time, not only cyanobacteria but also various algae, which are floating photosynthetic organisms, were pumping oxygen into the atmosphere at a rapid rate. Perhaps the oxygen supply became great enough to permit aquatic animals to acquire oxygen even though they had outer skeletons. The presence of a skeleton reduces possible access to oxygen in seawater. Steven Stanley of Johns Hopkins University suggests that predation may have played a role. Skeletons may have evolved during the Cambrian period because skeletons help protect animals from predators. If so, the evolutionary arms race came of age in the Cambrian seas.

The fossil record is rich during the Cambrian period, but the animals may have evolved earlier. The richness of the Cambrian period may be due to the evolution of outer skeletons.

FIGURE 19.9 Sea life of the Cambrian period.
The animals depicted here are found as fossils in the Burgess Shale, a formation of the Rocky Mountains of British Columbia, Canada. Some lineages represented by these animals are still evolving today; others have become extinct.

Invasion of Land

Plants. During the Ordovician period, algae, which were abundant in the seas, most likely began to take up residence in bodies of fresh water. Eventually, algae invaded damp areas on land. The first land plants were nonvascular (did not possess water-conducting tissues) similar to the mosses and liverworts that survive today. The lack of water-conducting tissues limited the height of these plants to a few centimeters. Although the Ordovician evidence is scarce, fossil spore finds from this time support this hypothesis.

Fossils of seedless vascular plants (those having tissue for water and organic nutrient transport) date back to the Silurian period. They later flourished in the warm swamps of the Carboniferous period. Club mosses, horsetails, and seed ferns were the trees of that time, and they grew to enormous size. A wide variety of smaller ferns and fernlike plants formed an underbrush (Fig. 19.10).

Invertebrates. The jointed appendages and exoskeleton of arthropods are adaptive for living on land. Various arthropods—spiders, centipedes, mites, and millipedes—all pre-

ceded the appearance of insects on land. Insects enter the fossil record in the Carboniferous period. One fossil dragonfly from this period had a wingspan of nearly a meter. The evolution of wings provided advantages that allowed insects to radiate into the most diverse and abundant group of animals today. Flying provides a way to escape enemies and find food.

Vertebrates. Vertebrates are animals with a vertebral column. The vertebrate line of descent began in the early Ordovician period with the evolution of jawless fishes. Jawed fishes appeared later in the Silurian period. Fishes are ectothermic (cold-blooded) aquatic vertebrates that have gills, scales, and fins. The cartilaginous and ray-finned fishes made their appearance in the Devonian period, which is called the Age of Fishes.

At this time, the seas were filled with giant predatory fish covered with protective armor made of external bone. Sharks cruised up deep, wide rivers, and smaller lobe-finned fishes lived at the river's edge in waters too shallow for large predators. Fleshy fins helped the small fishes push aside debris or hold their place in strong currents, and the

a.

b.

FIGURE 19.10 Swamp forests of the Carboniferous period.
a. Vast swamp forest of treelike club mosses and horsetails dominated the land during the Carboniferous period (see Table 19.1). The air contained insects with wide wingspans, such as the predecessors to dragonflies shown here, and amphibians lumbered from pool to pool. **b.** Dragonfly fossil from the Carboniferous period.

fins may also have allowed these fishes to venture onto land and lay their eggs safely in inland pools. Lobe-finned fishes are believed to be ancestral to the amphibians.

Amphibians are thin-skinned vertebrates that are not fully adapted to life on land, particularly because they must return to water to reproduce. The Carboniferous swamp forests provided the water they needed, and amphibians adaptively radiated into many different sizes and shapes. Some superficially resembled alligators and were covered with protective scales; others were small and snakelike; and a few were larger plant eaters. The largest measured 6 m from snout to tail. The Carboniferous period is called the Age of Amphibians.

The process that turned the great Carboniferous forests into the coal we use today to fuel our modern society started during the Carboniferous period. The weather turned cold and dry, and this brought an end to the Age of Amphibians. A major mass extinction event occurred at the end of the Permian period, bringing an end to the Paleozoic era and setting the stage for the Mesozoic era.

Seedless vascular plants and amphibians became larger and more abundant during the Carboniferous period. Insects appeared and flourished, eventually becoming the largest animal group today.

science focus

Real Dinosaurs, Stand Up!

Today's paleontologists are setting the record straight about dinosaurs. Because dinosaurs are classified as reptiles, it is assumed that they must have had the characteristics of today's reptiles. They must have been ectothermic (cold-blooded), slow-moving, and antisocial, right? Wrong!

First of all, not all dinosaurs were great lumbering beasts. Many dinosaurs were less than 1 m long, and their tracks indicate they moved "right along." These dinosaurs stood on two legs that were positioned directly under the body. Perhaps they were as agile as ostriches, which are famous for their great speed.

Dinosaurs may have been endothermic (warm-blooded). Could they have competed successfully with the preevolving mammals otherwise? They must have been able to hunt prey and escape from predators as well as mammals, which are known to be active because of their high rate of metabolism. Some argue that, in contrast, ectothermic animals have little endurance and cannot keep up. They also believe that the bone structure of dinosaurs indicates they were endothermic.

Dinosaurs cared for their young much as birds do today. In Montana, paleontologist Jack Horner has studied fossilized nests complete with eggs, embryos, and nestlings (Fig. 19A). The nests are about 7.5 m apart, the space needed for the length of an adult parent. About 20 eggs are neatly arranged in circles and may have been covered with decaying vegetation to keep them warm. Many contain the bones of juveniles as much as a meter long. It would seem then that baby dinosaurs remained in the nest to be fed by their parents. They must have attained this size within a relatively short period of time, again indicating that dinosaurs were endothermic. Ectothermic animals grow slowly and take a long time to reach this size.

Dinosaurs were also social! The remains of an enormous herd of dinosaurs found by Horner and colleagues is estimated to have nearly 30 million bones, representing 10,000 animals, in an area measuring about 1.6 mi². Most likely, the herd kept on the move in order to be assured of an adequate food supply, which consisted of flowering plants that could be stripped one season and grow back the next season. The fossilized herd is covered by volcanic ash, suggesting that the dinosaurs died following a volcanic eruption.

Some dinosaurs, such as the duck-billed dinosaurs and horned dinosaurs, have a skull crest. How might it have functioned? Perhaps it was a resonating chamber, used when dinosaurs communicated with one another. Or, as with modern horned animals that live in large groups, the males could have used the skull crest in combat to establish dominance.

If dinosaurs were endothermic, fast-moving, and social animals, should they be classified as reptiles? Some say no!

a.

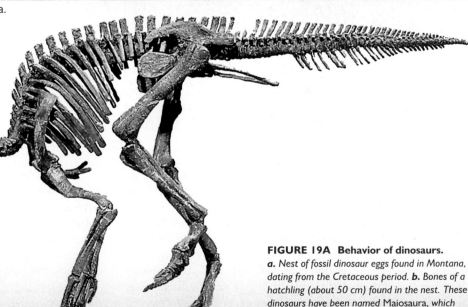

b.

FIGURE 19A Behavior of dinosaurs.
a. Nest of fossil dinosaur eggs found in Montana, dating from the Cretaceous period. b. Bones of a hatchling (about 50 cm) found in the nest. These dinosaurs have been named Maiosaura, *which means "good mother lizard" in Greek.*

FIGURE 19.11 Dinosaurs of the late Cretaceous period.

Parasaurolophus walkeri, although not as large as other dinosaurs, was one of the largest plant-eaters of the late Cretaceous period. Its crest atop the head was about 2 m long and was used to make booming calls. Also living at this time were the rhino-like dinosaurs represented here by *Triceratops* (*left*), another herbivore.

The Mesozoic Era

Although a severe mass extinction occurred at the end of the Paleozoic era, the evolution of certain types of plants and animals continued into the Triassic, the first period of the Mesozoic era. Nonflowering seed plants (collectively called gymnosperms), which had evolved and then spread during the Paleozoic, became dominant. Cycads are short and stout with palmlike leaves, and they produce large cones. Cycads and related plants were so prevalent during the Triassic and Jurassic periods that these periods are sometimes called the Age of Cycads. Reptiles can be traced back to the Permian period of the Paleozoic era. Unlike amphibians, reptiles can thrive in a dry climate because they have scaly skin and lay a shelled egg that hatches on land. Reptiles underwent an adaptive radiation during the Mesozoic era to produce forms that lived in the air, in the sea, and on the land. One group of reptiles, the therapsids, had several mammalian skeletal traits.

During the Jurassic period, large flying reptiles called pterosaurs ruled the air, and giant marine reptiles with paddle-like limbs ate fishes in the sea. But on land, it was dinosaurs that prevented the evolving mammals from taking center stage.

Although the average size of the dinosaurs was about that of a crow, many giant species developed. The gargantuan *Apatosaurus* and the armored, tractor-sized *Stegosaurus* fed on cycad seeds and conifer trees. The size of a dinosaur such as *Apatosaurus* is hard for us to imagine. It was 4.5 m tall at the hips and 27 m long in length and weighed about 40 tons. How might dinosaurs have benefited from being so large? One theory is that, being ectothermic (cold-blooded), the volume-to-surface ratio was favorable for retaining heat. Others believe dinosaurs were endothermic (warm-blooded) for reasons discussed in the Science Focus on page 330.

During the Cretaceous period, great herds of rhino-like dinosaurs, *Triceratops*, roamed the plains, as did the infamous *Tyrannosaurus rex*, which may have been a carnivore, filling the same ecological role as lions do today. *Parasaurolophus* was a unique-looking, long-crested, duck-billed dinosaur (Fig. 19.11). The long, hollow crest was

FIGURE 19.12 Mammals of the Oligocene epoch.
The artist's representation of these mammals and their habitat vegetation is based on fossil remains.

bigger than the rest of its skull and functioned as a resonating chamber for making booming calls, perhaps used during mating or to help members of a herd locate each other. In comparison to *Apatosaurus, Parasaurolophus* was small. It was less than 3 m tall at the hips and weighed only about 3 tons. Still, it was one of the largest plant-eaters of the late Cretaceous period and fed on pine needles, leaves, and twigs. *Parasaurolophus* was easy prey for large predators; its main defense would have been running away in large herds.

At the end of the Cretaceous period, the dinosaurs became victims of a mass extinction, which will be discussed on page 336.

One group of dinosaurs, called theropods, were bipedal and had an elongate, mobile, S-shaped neck. They most likely gave rise to the birds, whose fossil record begins with the famous *Archaeopteryx* (see Fig. 17.12, p. 292). Up until 1999, Mesozoic mammal fossils largely consisted of teeth. This changed when a fossil found in China was dated at 120 MYA and named *Jeholodens*. The animal, identified as a mammal, apparently looked like a long-snouted rat. Surprisingly, *Jeholodens* still had the

sprawling hindlimbs of a reptile, but its forelimbs were under the belly, as in today's mammals.

> During the Mesozoic era, some of the dinosaurs achieved enormous size, while mammals remained small.

The Cenozoic Era

Classically, the Cenozoic era is divided into two periods, the Tertiary period and the Quaternary period. Another theme, dividing the Cenozoic into the Paleogene and the Neogene periods is gaining popularity. This new system divides the epochs differently. In any case, we are living in the Holocene epoch.

Mammalian Diversification

At the end of the Mesozoic era, mammals began an adaptive radiation into the many habitats now left vacant by the demise of the dinosaurs. Mammals are endotherms, and they have hair, which helps keep body heat from escaping. Their name refers to the presence of mammary glands, which produce milk to feed their young. At the start of the

FIGURE 19.13 Woolly mammoth of the Pleistocene epoch.
Woolly mammoths were animals that lived along the borders of continental glaciers.

Paleocene epoch, mammals were small and resembled rats. By the end of the Eocene epoch, mammals had diversified to the point that all of the modern orders were in existence. Mammals adaptively radiated into a number of environments. Several species of mammals, including the bats, conquered the air. Whales, dolphins, manatees, and other mammals returned to the sea from land ancestry. On land, herbivorous hoofed mammals populated the forests and grasslands and were preyed upon by carnivorous mammals. Many of the types of herbivores and carnivores of the Oligocene epoch are extinct today (Fig. 19.12).

Evolution of Primates

Flowering plants (collectively called angiosperms) were already diverse and plentiful by the Cenozoic era. Primates are a type of mammal adapted to living in flowering trees, where there is protection from predators and where food in the form of fruit is plentiful. The ancestors of modern primates appeared during the Eocene epoch about 55 MYA. The first primates were small, squirrel-like animals. Ancestral apes appeared during the Oligocene epoch. These primates were adapted to living in the open grasslands and savannas. Apes diversified during the Miocene and Pliocene epochs and gave rise to the first hominids, a group that includes humans. Many of the skeletal differences between apes and humans relate to the fact that humans walk upright. Exactly what caused humans to adopt bipedalism is still being debated.

The world's climate became progressively colder during the Tertiary period. The Quaternary period begins with the Pleistocene epoch, which is known for multiple ice ages in the Northern Hemisphere. During periods of glaciation, snow and ice covered about one-third of the land surface of the Earth. The Pleistocene epoch was an age of not only humans, but also giant ground sloths, beavers, wolves, bison, woolly rhinoceroses, mastodons, and mammoths (Fig. 19.13). Humans have survived, but what happened to the oversized mammals just mentioned? Some think humans became such skilled hunters that they are at least partially responsible for the extinction of these awe-inspiring animals.

The Cenozoic era is the present era. Only during this time did mammals diversify and human evolution begin.

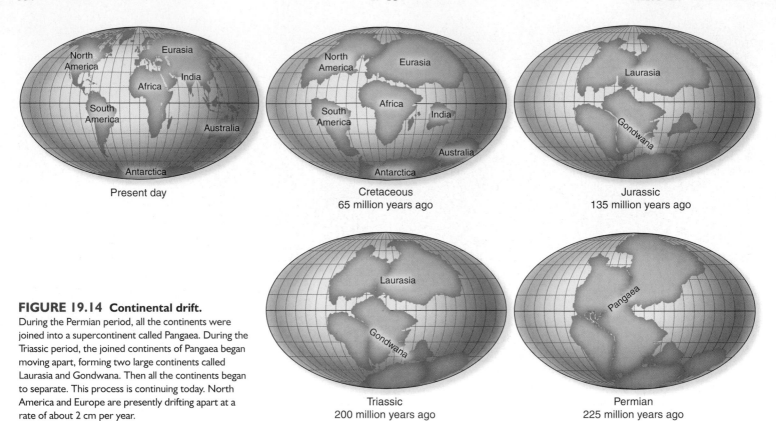

FIGURE 19.14 Continental drift.
During the Permian period, all the continents were joined into a supercontinent called Pangaea. During the Triassic period, the joined continents of Pangaea began moving apart, forming two large continents called Laurasia and Gondwana. Then all the continents began to separate. This process is continuing today. North America and Europe are presently drifting apart at a rate of about 2 cm per year.

19.3 FACTORS THAT INFLUENCE EVOLUTION

In the past, it was thought that the Earth's crust was immobile, that the continents had always been in their present positions, and that the ocean floors were only a catch basin for the debris that washed off the land. But in 1920, Alfred Wegener, a German meteorologist, presented data from a number of disciplines to support his hypothesis of continental drift. His evidence was drawn from several sources, including geology. The rocks in South Africa and southeast Brazil are very similar. If both rocks had formed in the same place at the same time, that would explain why they were the same. Coal only forms under warm, wet conditions and yet coal is found in Britain and also in the Antarctic, both of which could not produce coal today. Perhaps the Antarctic and Britain were at one time nearer the equator. Wegener also noticed similarity of living things, which could only be due to the continents being joined at some time in the past. Since then, the continents have drifted apart.

Continental Drift

Continental drift was finally confirmed in the 1960s, establishing that the continents are not fixed; instead, their positions and the positions of the oceans have changed over time (Fig. 19.14). During the Permean period, the continents joined to form one supercontinent that Wegener called Pangaea [Gk. *pangea,* all lands]. First, Pangaea divided into two large subcontinents, called Gondwana and Laurasia, and then these also split to form the continents of today. Presently, the continents are still drifting in relation to one another.

Continental drift explains why the coastlines of several continents are mirror images of each other—for example, the outline of the west coast of Africa matches that of the east coast of South America. The same geological structures are also found in many of the areas where the continents touched. A single mountain range runs through South America, Antarctica, and Australia. Continental drift also explains the unique distribution patterns of several fossils. Fossils of the same species of seed fern *(Glossopteris)* have been found on all the southern continents. No suitable explanation was possible previously, but now it seems plausible that the plant evolved on one continent and spread to the others while they were still joined as one. Similarly, the fossil reptile *Cynognathus* is found in Africa and South America, and *Lystrosaurus,* a mammal-like reptile, has now been discovered in Antarctica, far from Africa and southeast Asia, where it also occurs. With mammalian fossils, the situation is different: Australia, South America, and Africa all have their own distinctive mammals because mammals evolved after the continents separated. The mammalian biological diversity of today's world is the result of isolated evolution on separate continents. For example, why are marsupials prevalent in Australia but no place else? Most likely marsupials started evolving in the Americas and were able to reach Australia when the southern continents were still joined. Once Australia separated off, marsupials were able to diversify because placental mammals on that continent offered little competition. On the other hand, placental mammals are prevalent in the Americas and few marsupials can be found.

The relationship of the continents to one another has affected the biogeography of the Earth.

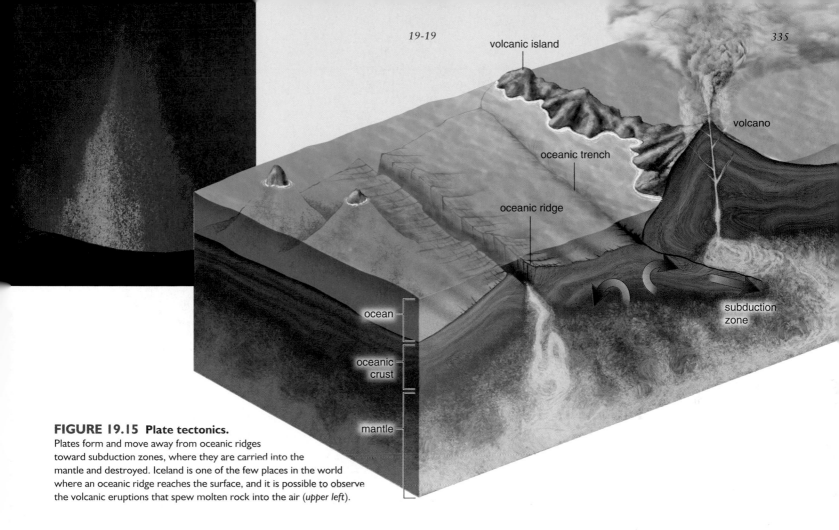

FIGURE 19.15 Plate tectonics.
Plates form and move away from oceanic ridges
toward subduction zones, where they are carried into the
mantle and destroyed. Iceland is one of the few places in the world
where an oceanic ridge reaches the surface, and it is possible to observe
the volcanic eruptions that spew molten rock into the air (*upper left*).

Plate Tectonics

Why do the continents drift? An answer has been suggested
through a branch of geology known as **plate tectonics** [Gk.
tektos, fluid, molten, able to flow]. Tectonics refers to movements of the Earth's crust, which is fragmented into slablike
plates that float on a lower hot mantle layer. The continents
and the ocean basins are a part of these rigid plates, which
move like conveyor belts. At oceanic ridges, seafloor spreading occurs as molten mantle rock rises and material is added
to the ocean floor (Fig. 19.15). Seafloor spreading causes the
continents to move a few centimeters a year on the average.
At *subduction zones,* the forward edge of a moving plate sinks
into the mantle and is destroyed. When the ocean floor is at
the leading edge of a plate, a deep ocean trench forms that is
bordered by volcanoes or volcanic island chains. The Earth
isn't getting bigger or smaller, so the amount of oceanic crust
being formed is as much as that being destroyed. The edges
of continents need not have deep trenches because continental crust (as opposed to oceanic crust) can carry on beyond
the edges of land and end far into the sea.

When two continents collide, the result is often a
mountain range; for example, the Himalayas resulted when
India collided with Eurasia. The place where two plates
meet and scrape past one another is called a *transform
boundary* (Fig. 19.16). The San Andreas fault in Southern
California is at a transform boundary, and the movement
of the two plates is responsible for the many earthquakes

in that region. No one can see the continents moving. The
only visible evidence of movement is an earthquake at
transform boundaries.

The Earth's crust is divided into plates that move because
of seafloor spreading at oceanic ridges.

FIGURE 19.16 San Andreas fault.
A transform boundary occurs where two plates scrape
past each other. The San Andreas fault occurs at a
transform boundary, and earthquakes are
apt to occur there.

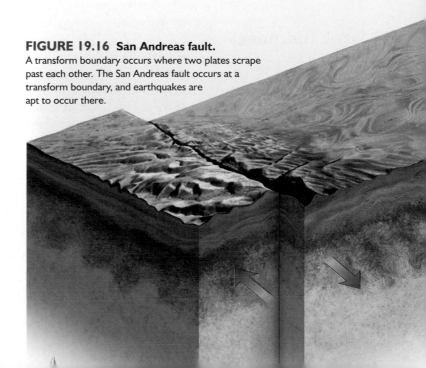

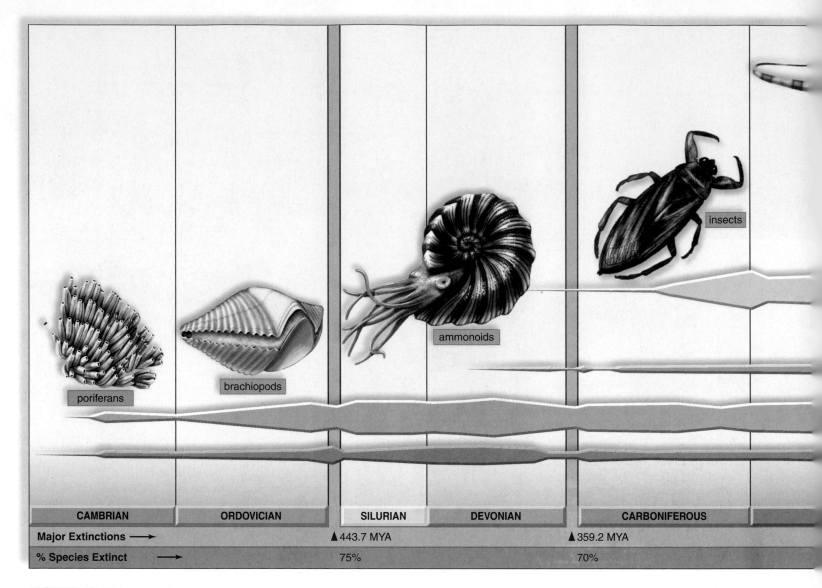

FIGURE 19.17 Mass extinctions.
Five significant mass extinctions and their effects on the abundance of certain forms of marine and terrestrial life.
The width of the horizontal bars indicates the varying abundance of each life-form considered.

Mass Extinctions

At least five mass extinctions have occurred throughout history: at the ends of the Ordovician, Devonian, Permian, Triassic, and Cretaceous periods (Fig. 19.17; see Table 19.1). Is a mass extinction due to some cataclysmic event, or is it a more gradual process brought on by environmental changes, including tectonic, oceanic, and climatic fluctuations? This question was brought to the fore when Walter and Luis Alvarez proposed in 1977 that the Cretaceous extinction was due to a bolide. A bolide is an asteroid (minor planet) that explodes, producing meteorites that fall to Earth. They found that Cretaceous clay contains an abnormally high level of iridium, an element that is rare in the Earth's crust but more common in asteroids and meteorites. The result of a large meteorite striking Earth could have been similar to that of a worldwide atomic bomb explosion: a cloud of dust would have mushroomed into the atmosphere, blocking out the sun and causing plants to freeze and

die. A layer of soot has been identified in the strata alongside the iridium, and a huge crater that could have been caused by a meteorite was found in the Caribbean–Gulf of Mexico region on the Yucatán peninsula.

In 1984, paleontologists David Raup and John Sepkoski suggested, based on the fossil record of marine animals, that mass extinctions have occurred every 26 million years, and surprisingly, astronomers can offer an explanation. Our solar system is in a starry galaxy known as the Milky Way. Because of the vertical movement of our sun, our solar system approaches other members of the Milky Way every 26–33 million years, producing an unstable situation that could lead to a bolide.

Certainly, continental drift contributed to the Ordovician extinction. This extinction occurred after Gondwana arrived at the South Pole. Immense glaciers, which drew water from the oceans, chilled even once-tropical land. Marine invertebrates and coral reefs, which were especially hard hit, didn't recover until Gondwana drifted away from the pole

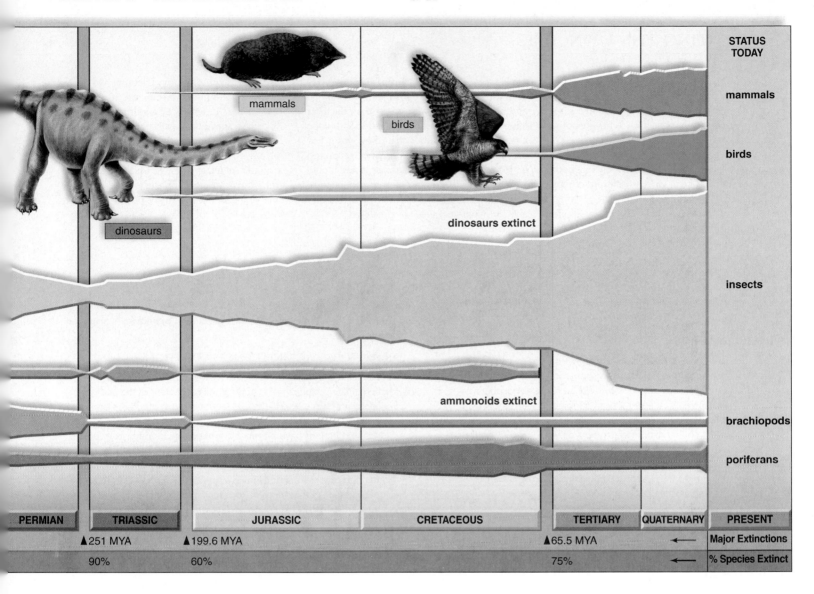

and warmth returned. The mass extinction at the end of the Devonian period saw an end to 70% of marine invertebrates. Helmont Geldsetzer of Canada's Geological Survey notes that iridium has also been found in Devonian rocks in Australia, suggesting that a bolide event was involved in the Devonian extinction. Other scientists believe that this mass extinction could have been due to movement of Gondwana back to the South Pole.

The extinction at the end of the Permian period was quite severe; 90% of species in the ocean and 70% on land disappeared. The latest hypothesis attributes the Permian extinction to excess carbon dioxide. When Pangaea formed, there were no polar ice caps to initiate ocean currents. The lack of ocean currents caused organic matter to stagnate at the bottom of the ocean. Then, as the continents drifted into a new configuration, ocean circulation switched back on. Now, the extra carbon on the seafloor was swept up to the surface where it became carbon dioxide, a deadly gas for

sea life. The trilobites became extinct, and the crinoids (sea lilies) barely survived. Excess carbon dioxide on land led to a global warming that altered the pattern of vegetation. Areas that were wet and rainy became dry and warm, and vice versa. Burrowing animals that could escape land surface changes seemed to have the best chance of survival.

The extinction at the end of the Triassic period is another that has been attributed to the environmental effects of a meteorite collision with Earth. Central Quebec has a crater half the size of Connecticut that some believe is the impact site. The dinosaurs may have benefited from this event because this is when the first of the gigantic dinosaurs took charge of the land. A second wave occurred in the Cretaceous period.

Mass extinctions seem to be due to climatic changes that occur after a meteorite collision and/or after the continents drift into a new and different configuration.

CONNECTING THE CONCEPTS

Does the process of evolution always have to be the same? The traditional view of the evolutionary process proposes that speciation occurs gradually and steadily over time. A new hypothesis suggests that long periods of little or no evolutionary change are punctuated by periods of relatively rapid speciation. Is it possible that both mechanisms may be at work in different groups of organisms and at different times?

Is it also possible that the history of life

could have turned out differently? The species alive today are the end product of the abiotic and biotic changes that occurred on Earth as life evolved. And what if the abiotic and biotic changes had been other than they were? For example, if the continents had not separated 65 MYA, what types of mammals, if any, would be alive today? Given a different sequence of environments, a different mix of plants and animals might very well have resulted.

The history of life on Earth, as we know it, is only one possible scenario. If we could rewind the "tape of life" and let history take its course anew, the result might well be very different, depending on the geologic and biologic events that took place the second time around. As an analogy, consider that if you were born in another time period and in a different country, you might be very different from the "you" of today.

Summary

19.1 ORIGIN OF LIFE

The unique conditions of the early Earth allowed a chemical evolution to occur. An abiotic synthesis of small organic molecules such as amino acids and nucleotides occurred, possibly either in the atmosphere or at hydrothermal vents. These monomers joined together to form polymers either on land (warm seaside rocks or clay) or at the vents. The first polymers could have been proteins or RNA, or they could have evolved together. The aggregation of polymers inside a plasma membrane produced a protocell having some enzymatic properties such that it could grow. If the protocell developed in the ocean, it was a heterotroph; if it developed at hydrothermal vents, it was a chemoautotroph. A true cell had evolved once the protocell contained DNA genes. The first genes may have been RNA molecules, but later DNA became the information storage molecule of heredity. Biological evolution now began.

19.2 HISTORY OF LIFE

The fossil record allows us to trace the history of life. The oldest prokaryotic fossils are cyanobacteria, dated about 3.5 BYA, and they were the first organisms to add oxygen to the atmosphere. The eukaryotic cell evolved about 2.2 BYA, but multicellular animals (the Ediacaran animals) do not occur until 600 MYA.

A rich animal fossil record starts at the Cambrian period of the Paleozoic era. The occurrence of external skeletons, which seems to explain the increased number of fossils at this time, may have been due to the presence of plentiful oxygen in the atmosphere, or perhaps it was due to predation. The fishes were the first vertebrates to diversify and become dominant. Amphibians are descended from lobe-finned fishes.

Plants also invaded land during the Ordovician period. The swamp forests of the Carboniferous period contained seedless vascular plants, insects, and amphibians. This period is sometimes called the Age of Amphibians.

The Mesozoic era was the Age of Cycads and Reptiles. First mammals and then birds evolved from reptilian ancestors. During this era, dinosaurs of enormous size were present. By the end of the Cretaceous period, the dinosaurs were extinct.

The Cenozoic era is divided into the Tertiary period and the Quaternary period. The Tertiary is associated with the adaptive radiation of mammals and flowering plants that formed vast tropical forests. The Quaternary is associated with the evolution of primates; first monkeys appeared, then apes, and then humans. Grasslands were replacing

forests, and this put pressure on primates, who were adapted to living in trees. The result may have been the evolution of humans—primates who left the trees.

19.3 FACTORS THAT INFLUENCE EVOLUTION

The continents are on massive plates that move, carrying the land with them. Plate tectonics is the study of the movement of the plates. Continental drift helps explain the distribution pattern of today's land organisms.

Mass extinctions have played a dramatic role in the history of life. It has been suggested that the extinction at the end of the Cretaceous period was caused by the impact of a large meteorite, and evidence indicates that other extinctions have a similar cause as well. It has also been suggested that tectonic, oceanic, and climatic fluctuations, particularly due to continental drift, can bring about mass extinctions.

Reviewing the Chapter

1. List and describe the various hypotheses concerning the chemical evolution that produced macromolecules. 318–19
2. Trace in general the steps by which the protocell may have evolved from macromolecules. 320
3. List and describe the various hypotheses concerning the origin of a self-replication system. 321
4. Explain how the fossil record develops and how fossils are dated relatively and absolutely. 322, 324
5. When did prokaryotes arise, and what are stromatolites? 324
6. When and how might the eukaryotic cell have arisen? 326
7. Describe the first multicellular animals found in the Ediacara Hills in southern Australia. 326
8. Why might there be so many fossils from the Cambrian period? 327
9. Which plants, invertebrates, and vertebrates were present on land during the Carboniferous period? 328
10. Which type of vertebrate was dominant during the Mesozoic era? Which types began evolving at this time? 331–32
11. Which type of vertebrate underwent an adaptive radiation in the Cenozoic era? 332–33
12. What is continental drift, and how is it related to plate tectonics? Give examples to show how biogeography supports the occurrence of continental drift. 334–35
13. Identify five significant mass extinctions during the history of the Earth. What may have been the cause, and what types of organisms were most affected by each extinction? 336–37

Testing Yourself

Choose the best answer for each question.

For questions 1–6, match the statements with events in the key. Answers may be used more than once.

KEY:

a. early Earth
b. monomers evolve
c. polymers evolve
d. protocell evolves
e. self-replication system evolves

1. The heat of the sun could have caused amino acids to form proteinoids.

2. In a liquid environment, phospholipid molecules automatically form a membrane.

3. As the Earth cooled, water vapor condensed, and subsequent rain produced the oceans.

4. Miller and Urey's experiment shows that under the right conditions, inorganic chemicals can react to form small organic molecules.

5. Some investigators believe that RNA was the first nucleic acid to evolve.

6. An abiotic synthesis may have occurred at hydrothermal vents.

7. Which of these did Stanley Miller and Harold Urey place in their experimental system to show that organic molecules could have arisen from inorganic molecules on the early Earth?
   a. microspheres
   b. purines and pyrimidines
   c. early atmospheric gases
   d. only RNA
   e. All of these are correct.

8. Which of these is not a place where polymers may have arisen?
   a. at hydrothermal vents
   b. on rocks beside the sea
   c. in clay
   d. in the atmosphere
   e. Both b and c are correct.

9. Which of these is the chief reason the protocell was probably a fermenter?
   a. The protocell didn't have any enzymes.
   b. The atmosphere didn't have any oxygen.
   c. Fermentation provides the most energy.
   d. There was no ATP yet.
   e. All of these are correct.

10. Liposomes (phospholipid droplets) are significant because they show that
   a. the first plasma membrane contained protein.
   b. a plasma membrane could have easily evolved.
   c. a biological evolution produced the first cell.
   d. there was water on the early Earth.
   e. the protocell had organelles.

11. Evolution of the DNA ⟶ RNA ⟶ protein system was a milestone because the protocell could now
   a. be a heterotrophic fermenter.
   b. pass on genetic information.
   c. use energy to grow.
   d. take in preformed molecules.
   e. All of these are correct.

12. Fossils
   a. are the remains and traces of past life.
   b. can be dated absolutely according to their location in strata.
   c. are usually found embedded in sedimentary rock.
   d. have been found for all types of animals except humans.
   e. Both a and c are correct.

13. Which of these events did not occur during the Precambrian?
   a. evolution of the prokaryotic cell
   b. evolution of the eukaryotic cell
   c. evolution of multicellularity
   d. evolution of the first animals
   e. All of these occurred during the Precambrian.

14. The organisms with the longest evolutionary history are
   a. prokaryotes that left no fossil record.
   b. eukaryotes that left a fossil record.
   c. prokaryotes that are still evolving today.
   d. animals that had a shell.

For questions 15–19, match the phrases with divisions of geological time in the key. Answers may be used more than once.

KEY:

a. Cenozoic era
b. Mesozoic era
c. Paleozoic era
d. Precambrian

15. dinosaur diversity, evolution of birds and mammals

16. contains the Carboniferous period

17. prokaryotes abound; eukaryotes evolve and become multicellular

18. mammalian diversification

19. invasion of land

20. Which of these occurred during the Carboniferous period?
   a. Dinosaurs evolved twice and became huge.
   b. Human evolution began.
   c. The great swamp forests contained insects and amphibians.
   d. Prokaryotes evolved.
   e. All of these are correct.

21. Continental drift helps explain
   a. mass extinctions.
   b. the distribution of fossils on the Earth.
   c. geological upheavals such as earthquakes.
   d. climatic changes.
   e. All of these are correct.

22. Which of these pairs is mismatched?
   a. Mesozoic—cycads and dinosaurs
   b. Cenozoic—grasses and humans
   c. Paleozoic—rise of prokaryotes and unicellular eukaryotes
   d. Cambrian—marine organisms with external skeletons
   e. Precambrian—origin of the cell at hydrothermal vents

23. Complete the following listings using these phrases: *O$_2$ accumulates in atmosphere, Ediacaran animals, oldest known fossils, Cambrian animals, protists evolve and diversify, oldest eukaryotic fossils*

   2.2 BYA   a. _____   543 MYA   d. _____
   2.7 BYA   b. _____   600 MYA   e. _____
   3.5 BYA   c. _____   1.4 BYA   f. _____
   4.6 BYA   formation of the   2.2 BYA   oldest eukaryotic
             Earth   fossils

24. The protocell is hypothesized to have had a membrane boundary and to have been either a _____ or a _____.

25. Once there was a flow of information from DNA to RNA to protein, the protocell became a _____ cell, and biological evolution began.

26. The evolution of _____ prokaryotes caused oxygen to enter the atmosphere.

27. Primitive vascular plants and amphibians were large and abundant during the _____ period.

28. The mammals diversified and human evolution began during the _____ era.

29. Mass extinctions seem to be due to climatic changes that occur after a _____ bombards the Earth, or after the continents _____ into a new configuration.

30. The longest length of time in geological history is the _____.

31. The earliest _____ fossils are Ediacaran animal fossils dated 600 MYA.

32. The continents are not fixed; instead, they have _____ to their present positions.

Thinking Scientifically

1. From a scientific standpoint, trying to devise an experimental system that mimics the conditions of the early Earth is an inherently frustrating endeavor. While one might make some interesting hypotheses and experimental observations, there is no way to know for sure if experimental conditions are anything like those that really existed billions of years ago. Why do scientists continue in this quest?

2. Many environmentalists are concerned about global warming and ozone depletion. How would we know if current changes in climate are human-made or just part of natural, long-term cycles? Even if they aren't, if life has survived changes in the past, why shouldn't it survive these changes now?

Bioethical Issue: Evolution Research vs. Creation Research

Dr. H. M. Morris, Director of the Institute for Creation Research, lists these contradictions between evolution research and creation research:[1]

Evolution Research	Creation Research
Fishes evolved before fruit trees.	Fruit trees were created before fishes.
Insects evolved before birds.	Birds were created before insects.
The sun was present before land plants.	Land vegetation was created before the sun.
Reptiles evolved before birds.	Birds were created before reptiles.
Reptiles evolved before whales.	Whales were created before reptiles.
Rain was present before humans.	Humans were created before rain.
Evolution is still continuing.	Creation has been completed.

Do we have an obligation to accept one list over the other? Why or why not? On what basis?

[1] Montagu, A. (Ed). 1984. *Science and Creationism.* New York: Oxford University Press, p. 246.

Understanding the Terms

Match the terms to these definitions:

a. _____ Concept that the rate at which mutational changes accumulate in certain types of genes is constant over time.

b. _____ Cell forerunner that possibly developed from cell-like microspheres.

c. _____ Droplet of phospholipid molecules formed in a liquid environment.

d. _____ A region where crust forms and from which it moves laterally in each direction.

e. _____ Formed from oxygen in the upper atmosphere, it protects the Earth from ultraviolet radiation.

ARIS, the *Biology* Website

ARIS, the website for *Biology,* provides a wealth of information organized and integrated by chapter. You will find practice quizzes, interactive activities, labeling exercises, flashcards, and much more that will complement your learning and understanding of general biology.

www.mhhe.com/maderbiology9

20

CLASSIFICATION OF LIVING THINGS

The process of evolution has resulted in a history of life on Earth, called the phylogeny of life. In the past, biologists primarily relied on the fossil record and comparative anatomical data between organisms to decipher evolutionary relationships and to develop a natural as opposed to an artificial system of classification. Modern-day biology is increasingly making use of molecular data to determine evolutionary relationships, and as we shall see in this chapter, the result has been a revolution in the classification of life.

Classification based on molecular data offers several advantages. Specific anatomical differences occur only between certain groups of organisms, but DNA is the genetic material for all known organisms. Therefore, DNA differences can be used to determine the evolutionary relationship between any two species; between a bacterium and a human or between a paramecium and a mushroom. Then, too, an organism's DNA is now readily accessible, and molecular data is almost limitless, because each nucleotide position in a sequence of RNA or DNA nucleotides is a potential difference between species.

It has even been suggested that just as supermarkets use bar codes to identify their products, it would be possible to create a "bar code" based on DNA differences for each species of life on Earth. No longer would a biologist in the field have to make use of just the right anatomical characteristics to identify a species. Only a sample of DNA would be required, for example, to identify the frogs shown in the photo as Agalychnis callidryas. Happily, the new era of biology has reunited those interested in cellular biology with those interested in the entire organism.

Red-eyed tree frog, *Agalychnis callidryas*, Belize tropical rain forest.

20.1 Taxonomy

Suppose you went to Africa on a photo safari and wanted to classify the organisms shown in Figure 20.1 according to your own system. Most likely, you would begin by making a list, and naturally this would require you to give each organism a name. Then you would start assigning the organisms on your list to particular groups. But what criteria would you use—color, shape, size, how the organisms relate to you? Deciding on the number, types, and arrangement of the groups would not be easy, and periodically you might change your mind or even have to start over. Biologists, too, have not had an easy time deciding how living things should be classified and have made changes in their methods throughout history. These changes are often brought about by an increase in fossil, anatomical, or molecular data. Classification is usually based on our understanding of how organisms are related to one another through evolution. A natural system of classification, as opposed to an artificial system, reflects the evolutionary history of organisms.

Taxonomy [Gk. *tasso,* arrange, classify, and *nomos,* usage, law], the branch of biology concerned with identifying, naming, and classifying organisms, began with the ancient Greeks and Romans. The famous Greek philosopher Aristotle was interested in taxonomy, and he identified organisms as belonging to a particular group, such as horses, birds, and oaks. In the Middle Ages, these names were translated into Latin, the language still used for scientific names today. Much later, John Ray (1627–1705), a British naturalist of the seventeenth century, believed that each organism should have a set name. He said, "When men do not know the name and properties of natural objects—they cannot see and record accurately."

The Binomial System

The number of known types of organisms expanded greatly in the mid-eighteenth century as Europeans traveled to distant parts of the world. During this time, Carolus Linnaeus (1707–78) developed **binomial nomenclature,** by which each species receives a two-part name (Fig. 20.2). For example, *Lilium buibiferum* and *Lilium canadense* are two different species of lily. The first word, *Lilium,* is the genus (pl., genera), a classification category that can contain many species. The second word, the **specific epithet,** refers to one species within that genus. The specific epithet sometimes tells us something descriptive about the organism. Notice that the scientific name is in italics; the genus is capitalized, while the specific epithet is not. Both names are separately underlined when handwritten. The species is designated by the

FIGURE 20.1 Classifying organisms.
How would you name and classify these organisms? After naming them, how would you assign each to a particular group? Based on what criteria? An artificial system would not take into account how they might be related through evolution, as would a natural system.

a.

b. *Lilium buibiferum*

c. *Lilium canadense*

FIGURE 20.2 Carolus Linnaeus.
a. Linnaeus was the father of taxonomy and gave us the binomial system of naming and classifying organisms. His original name was Karl von Linne, but he later latinized it because of his fascination with scientific names. Linnaeus was particularly interested in classifying plants. **b, c.** Each of these two lilies are species in the same genus, *Lilium*.

full binomial name—in this case, either *Lilium buibiferum* or *Lilium canadense*. The specific epithet alone gives no clue as to species—just as the house number alone without the street name gives no clue as to which house is specified. The genus name can be used alone, however, to refer to a group of related species. Also, the genus can be abbreviated to a single letter if used with the specific epithet (e.g., *L. buibiferum*) and if the full name has been given previously.

Scientific names are derived in a number of ways. Some scientific names are descriptive in nature, for example, *Acer rubrum* for the red maple. Other scientific names may include geographic descriptions such as *Alligator mississippiensis* for the American alligator. Scientific names can also include eponyms (named after someone), such as the owl mite *Strigophilus garylarsonii* (named after the cartoonist). Many scientific names are derived from mythical characters, such as *Iris versicolor*, named for Iris, the goddess of the rainbow. Some scientific names reflect a humorous slant, such as *Ba humbugi* for a species of snail.

Why do organisms need scientific names? And why do scientists use Latin, rather than common names, to describe organisms? There are several reasons. First, a common name will vary from country to country because different countries use different languages. Second, even people who speak the same language sometimes use different common names to describe the same organism. For example, bowfin, grindle, choupique, and cypress trout describe the same common fish, *Amia calva*. Furthermore, between countries, the same common name is sometimes given to different organisms. A "robin" in England is very different from a "robin" in the United States, for example. Latin, on the other hand, is a universal language that not too long ago was well known by most scholars, many of whom were physicians or clerics. When scientists throughout the world use the same scientific binomial name, they know they are speaking of the same organism.

The task of identifying and naming the species of the world is a daunting one. A new fast and efficient way of identifying species that is based on their DNA is described in the Science Focus on page 348. Of the estimated 3–30 million species now living on Earth, a million species of animals and a half million species of plants and microorganisms have been named. We are further along on some groups than others; we may have finished the birds, but there may be hundreds of thousands of unnamed insects. The Linnaen Society rules on the appropriateness of the binomial name for each species in the world.

The scientific name of an organism consists of its genus and a specific epithet. The complete binomial name indicates the species.

Distinguishing a Species

There are several ways to distinguish a species, and each way has its advantages and disadvantages. For Linnaeus, every species had distinctive structural characteristics not shared by members of a similar species. In birds, structural differences can involve the shape, size, and color of the body, feet, beak, or wings. However, variations do occur among members of a species. Differences between males and females or between juveniles and adults may even make it difficult to tell when an organism belongs to a particular species (Fig. 20.3).

The *biological definition of a species* rests on the recognition that distinctive characteristics are passed from parents to offspring. This definition, which states that members of a species interbreed and share the same gene pool, applies only to sexually reproducing organisms and cannot apply to asexually reproducing organisms. Sexually reproducing organisms are not always as reproductively isolated as we would expect. When a species has a wide geographic range, there may be variant types that tend to interbreed where their populations overlap (see Fig. 18.3). This observation has led to calling these populations subspecies, designated by a three-part name. For example, *Elaphe obsoleta bairdi* and *Elaphe obsoleta obsoleta* are two subspecies within the snake species *Elaphe obsoleta*. It may be that these subspecies are actually distinct species. Even species that seem to be obvi-

FIGURE 20.4 Hybridization between species.
Zebroids are horse-zebra hybrids. Like mules, zebroids are generally infertile, due to differences in the chromosomes inherited from their parents.

ously distinct interbreed on occasion (Fig. 20.4). Therefore, the presence or absence of hybridization may not be indicative of what constitutes a species.

Identifying species on the basis of DNA differences (as described in the Science Focus on page 348) has also been called into question. There are those who feel that nucleotide base differences in a single gene may not yield enough data to distinguish two closely related species or to recognize when hybridization has occured.

Classification Categories

Classification, which begins when an organism is named, includes taxonomy, since genus and species are two classification categories. In the context of this chapter on classification, a species is a taxonomic category below the rank of genus. A **taxon** (pl., taxa) is a group of organisms that fills a particular category of classification; *Rosa* and *Felis* are taxa at the genus level.

The taxonomists mentioned in this chapter contributed to classification. Aristotle divided living things into 14 groups—mammals, birds, fish, and so on. Then he subdivided the groups according to the size of the organisms. Ray used a more natural system, grouping animals and plants according to how he thought they were related. Linnaeus simply used flower part differences to assign plants to the categories species, genus, order, and class. His studies were published in a book called *Systema Naturae* in 1735.

Today, taxonomists use the following major categories of classification: **species, genus, family, order, class, phylum,** and **kingdom.** Recently, a higher taxonomic category, the **domain,** has been added to this list. There can be several species within a genus, several genera within a family, and so forth—the higher the category, the more inclusive it is (Fig. 20.5). Therefore, there is a hierarchy of categories. The organisms that fill a particular classification

FIGURE 20.3 Members of a species.
Identifying the members of a species can be difficult—especially when the male and female members do not look alike, as with these mallards, *Anas platyrhynchos.* The male mallard duck is on the *left* and the female is on the *right*.

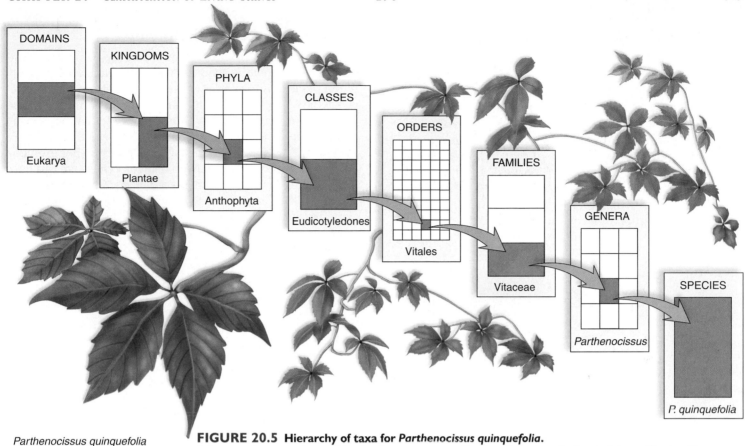

FIGURE 20.5 Hierarchy of taxa for *Parthenocissus quinquefolia*.
A domain is the most inclusive of the classification categories. Kingdom Plantae is in the domain Eukarya. Kingdom Plantae contains several phyla, one of which is Anthophyta. The phylum Anthophyta has two classes (the monocots and eudicots). In the class Eudicotyledones, there are many orders, including Vitales. One of the genera in the family Vitaceae is the genus *Parthenocissus*. This genus includes the species *Parthenocissus quinquefolia*. The specific epithet is due to the plant's whorl of five leaves. This illustration is diagrammatic and is not necessarily representative of the correct number of subcategories.

Parthenocissus quinquefolia
Virginia creeper (five-leaf ivy)

TABLE 20.1

Hierarchy of the Taxa to Which Humans Are Assigned

Domain Eukarya	Organisms whose cells have a membrane-bounded nucleus
Kingdom Animalia	Usually motile, multicellular organisms, without cell walls or chlorophyll; usually, internal cavity for digestion of nutrients
Phylum Chordata	Organisms that at one time in their life history have a dorsal hollow nerve cord, a notochord, pharyngeal pouches, and a postanal tail
Class Mammalia	Warm-blooded vertebrates possessing mammary glands; body more or less covered with hair; well-developed brain
Order Primates	Good brain development, opposable thumb and sometimes big toe; lacking claws, scales, horns, and hoofs
Family Hominidae	Limb anatomy suitable for upright stance and bipedal locomotion
Genus *Homo*	Maximum brain development, especially in regard to particular portions; hand anatomy suitable to the making of tools
Species *Homo sapiens**	Body proportions of modern humans; speech centers of brain well developed

* To specify an organism, you must use the full name, such as Homo sapiens.

category are distinguishable from other organisms by sharing a set of characteristics, or simply characters. A **character** is any structural, chromosomal, or molecular feature that distinguishes one group from another. Organisms in the same domain have general characters in common; those in the same species have quite specific characters in common.

Table 20.1 lists some of the characters that help classify humans into major categories.

In most cases, categories of classification can be subdivided into three additional categories, as in superorder, order, suborder, and infraorder. Considering these, there are more than 30 categories of classification.

20.2 Phylogenetic Trees

Taxonomy and classification are a part of the broader field of **systematics** [Gk. *systema,* an orderly arrangement], which is the study of the diversity of organisms at all levels of organization. One goal of systematics is to determine **phylogeny** [Gk. *phyle,* tribe; L. *genitus,* producing], or the evolutionary history of a group of organisms, which is often represented by a **phylogenetic tree,** a diagram that indicates common ancestors and lines of descent (lineages). Each branching point in an evolutionary tree is a divergence from a **common ancestor,** an organism that gives rise to two new groups. For example, this portion of an evolutionary tree says that monkeys and apes share a common primate ancestor:

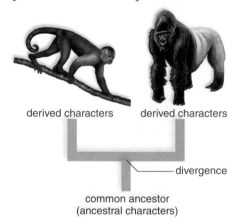

Divergence is presumed because monkeys and apes have their own individual characteristics (often called **derived characters**). For example, skeletal differences allow an ape to swing from limb to limb of a tree while monkeys run along the tops of tree branches. The common primate ancestor to both monkeys and apes has **ancestral characteristics** that are shared by the ancestor and also monkeys and apes. For example, the common primate ancestor must have been able to climb trees, as can monkeys and apes.

A phylogenetic tree has many branch points, and they show that it is possible to trace the ancestry of a group of organisms back farther and farther in the past. For example, reindeer, monkeys, and apes all give birth to live young because they all have a common ancestor that was a placental mammal:

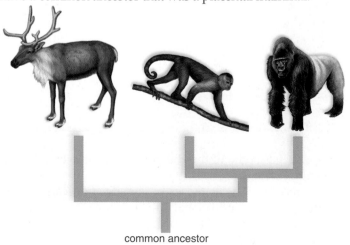

common ancestor

Classification and Phylogeny

Classification is a part of systematics because classification categories list the unique characters of each taxon, which ideally reflect phylogeny. A species is most closely related to other species in the same genus, then to genera in the same family, and so forth, from order to class to phylum to kingdom. When we say that two species (or genera, families, etc.) are closely related, we mean that they share a recent common ancestor. For example, all the animals in Figure 20.6 are related because we can trace their ancestry back to the same order. The animals in the order Artiodactyla all have even-toed hoofs. Animals in the family Cervidae have solid horns, called antlers, but they are highly branched in red deer (genus *Cervus*) and palmate (having the shape of a hand) in reindeer (genus *Rangifer*). In contrast, animals in the family Bovidae have hollow horns and unlike the Cervidae, both males and females have horns, although they are smaller in females.

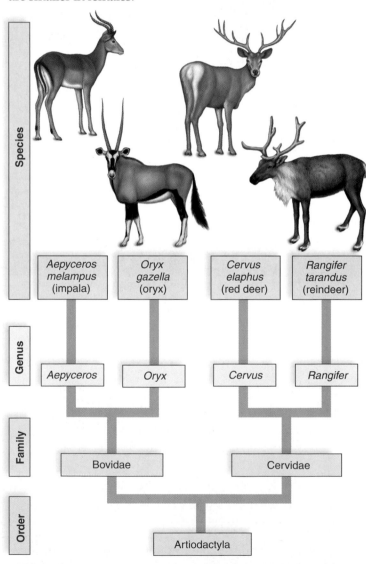

FIGURE 20.6 Classification and phylogeny.
The classification and phylogenetic tree for a group of organisms are ideally constructed to reflect their evolutionary history. A species is most closely related to other species in the same genus, more distantly related to species in other genera of the same family, and so forth, on through order, class, phylum, kingdom, and domain.

Tracing Phylogeny

Systematists gather all sorts of data in order to discover the evolutionary relationship between species. They rely heavily on a combination of data from the fossil record, homology, and molecular data to determine the correct sequence of common ancestors in any particular group of organisms. If you can determine the common ancestors in the history of life, then you know how evolution occurred and can classify organisms accordingly.

Fossil Record

It is possible to use the fossil record to trace the history of life in broad terms, and sometimes to trace the history of lineages. One of the advantages of fossils is that they can be dated, but unfortunately it is not always possible to tell to which group, living or extinct, a fossil is related. For example, at present, paleontologists are discussing whether fossil turtles indicate that turtles are distantly or closely related to crocodiles. On the basis of his interpretation of fossil turtles, Olivier C. Rieppel of the Field Museum of Natural History in Chicago is challenging the conventional interpretation that turtles are ancestral (have traits seen in a common ancestor to all reptiles) and are not closely related to crocodiles, which evolved later. His interpretation is being supported by molecular data that show turtles and crocodiles are closely related.

If the fossil record was more complete, there might be fewer controversies about the interpretation of fossils. One reason the fossil record is incomplete is that most fossils exist as only harder body parts, such as bones and teeth. Soft parts are usually eaten or decayed before they have a chance to be buried. This may be one reason it has been difficult to discover when angiosperms (flowering plants) first evolved. A Jurassic fossil recently found, if accepted as an angiosperm by most botanists, may help pin down the date (Fig. 20.7). As paleontologists continue to explore the world, the sometimes stingy fossil record will reveal some of its secrets.

FIGURE 20.7 Ancestral angiosperm.
The fossil *Archaefructus liaoningensis*, dated from the Jurassic period, may be the earliest angiosperm to be discovered. Without knowing the anatomy of the first flowering plant, it has been difficult to determine the correct classification of these plants.

Prickly pear, *Opuntia*   Spurge, *Euphorbia*

FIGURE 20.8 Convergent evolution.
Cacti and spurges evolved on different continents, and yet they are both succulent flowering plants with spines. Cacti are adapted to living in American deserts, whereas spurges are adapted to living in tropical habitats of Africa. This is an example of convergent evolution.

Homology

Homology [Gk. *homologos*, agreeing, corresponding] is character similarity that stems from having a common ancestor. Comparative anatomy, including embryological evidence, provides information regarding homology. **Homologous structures** are related to each other through common descent. The forelimbs of vertebrates are homologous because they contain the same bones organized in the same general way as in a common ancestor. For example, a horse has but a single digit and toe (the hoof), while a bat has four lengthened digits that provide support for the membranous wings.

Deciphering homology is sometimes difficult because of convergent evolution and parallel evolution. **Convergent evolution** is the acquisition of the same or similar characters in distantly related lines of descent. Similarity due to convergence is termed **analogy**. The wings of an insect and the wings of a bat are analogous. You may recall from Chapter 17 that **analogous structures** have the same function in different groups but do not have a common ancestry. Both cacti and spurges are adapted similarly to a hot, dry environment, and both are succulent (thick, fleshy) with spiny leaves (Fig. 20.8). However, the details of their flower structure indicate that these plants are not closely related. **Parallel evolution** is the acquisition of the same or a similar character in two or more related lineages without it being present in a common ancestor. A similar banding pattern is found in several species of moths, for example. It is sometimes difficult to tell if features are convergent or parallel.

science focus

DNA Bar Coding of Life

Traditionally, taxonomists have often relied on anatomical data to tell species apart. For example, differences in the type of spinning apparatus and the type of web have played a large role in distinguishing one spider from another (Fig. 20A). We can well imagine that if a mother wanted to know if certain spiders in the backyard were dangerous to her children, she might want a faster answer than could be provided by a traditional taxonomist at a university some distance away.

Enter the Consortium for the Barcode of Life (CBOL), which proposes that any scientist, not just taxonomists, will be able to identify a species with the flick of a handheld scanner. Just like the 11-digit Universal Product Code (UPC) used to identify products sold in a supermarket, the consortium believes that a sample of DNA should be able to identify any organism on Earth. The proposed scanner would tap into a bar-code database that contains the bar codes for all species so far identified on planet Earth. Also, a handheld DNA–bar-coding device is expected to provide a fast and inexpensive way for a wide range of researchers, including biology students, to catalog any and all of the world's species that do not yet have a bar code. So far scientists have identified only about 1.5 million species out of a potential 30 million. And there is no central database that keeps track of the known species.

The idea of using bar codes to identify species is not new, but Paul Hebert and his colleagues at the University of Guelph in Canada are the first to believe it would be possible to use the base sequence in DNA to develop a bar code for each living thing. The order of DNA's nucleotides—A, T, C, and G—within a particular gene common to the organisms in each kingdom would fill the role taken by numbers in the UPC used in warehouses and stores. Hebert believes that the gene

- should contain no more than 650 nucleotides so that sequencing can be accomplished speedily with few mistakes;
- should be easy to extract from an organism's complete genome;
- should have mutated to the degree that each species has its own sequence of bases but not so fast that the sequence differs greatly among individuals within the same species.

Hebert's team decided that a mitochondrial gene known as cytochrome *c* oxidase subunit I, or COI, would be a suitable target gene in animals. (This gene codes for one of the carriers in the electron transport chain; see page 110.) Another researcher, John Kress, a plant taxonomist at the Smithsonian Institution in Washington, D. C., has developed a potential method for bar coding plant species. The Consortium for the Barcode of Life is growing by leaps and bounds and now includes various biotech companies, various museums and universities, the U.S. Food and Drug Administration, and also the U.S. Department of Homeland Security. Hebert has received a $3 million grant from the Gordon and Betty Moore Foundation to start the Biodiversity Institute of Ontario, which will be housed on the University of Guelph campus, where he teaches.

Speedy DNA bar coding would not only be a boon to ordinary citizens and taxonomists, but it would also benefit farmers who need to identify a pest attacking their crops, doctors who need to know the correct antivenin for snakebite victims, and college students who are expected to identify the plants, animals, and protists on an ecological field trip.

a.

b.

FIGURE 20A Identifying spiders.
Identification of spiders depends in part on their type of spinning apparatus and the type of web they weave. The orb web of the garden spider Araneus diadematus *(a) differs somewhat from (b) the orb web of the New Zealand spider* Waitkera waitkerensis.

Molecular Data

Speciation occurs when mutations bring about changes in the base-pair sequences of DNA. Systematists, therefore, assume that the more closely species are related, the fewer changes there will be in DNA base-pair sequences. Since DNA codes for amino acid sequences in proteins, it also follows that the more closely species are related, the fewer differences there will be in the amino acid sequences within their proteins.

Because molecular data are straightforward and numerical, they can sometimes sort out relationships obscured by inconsequential anatomical variations or convergence. Software breakthroughs have made it possible to analyze nucleotide sequences or amino acid sequences quickly and accurately using a computer. Also, these analyses are available to anyone doing comparative studies through the Internet, so each investigator doesn't have to start from scratch. The combination of accuracy and availability of past data has made molecular systematics a standard way to study the relatedness of groups of organisms today.

Protein Comparisons.   Before amino acid sequencing became routine, immunological techniques were used to roughly judge the similarity of plasma membrane proteins. In one procedure, antibodies are produced by transfusing a rabbit with the cells of one species. Cells of the second species are exposed to these antibodies, and the degree of the reaction is observed. The stronger the reaction, the more similar the cells from the two species.

Later, it became customary to use amino acid sequencing to determine the number of amino acid differences in a particular protein. Cytochrome *c* is a protein that is found in all aerobic organisms, so its sequence has been determined for a number of different organisms. The amino acid difference in cytochrome *c* between chickens and ducks is only 3, but between chickens and humans there are 13 amino acid

differences. From this data you can conclude that, as expected, chickens and ducks are more closely related than are chickens and humans. Since the number of proteins available for study in all living things at all times is limited, most new studies today study differences in RNA and DNA.

RNA and DNA Comparisons.   All cells have ribosomes, which are essential for protein synthesis. Further, the genes that code for ribosomal RNA (rRNA) have changed very slowly during evolution in comparison to other genes. Therefore, it is believed that comparative rRNA sequencing provides a reliable indicator of the similarity between organisms. Ribosomal RNA sequencing helped investigators conclude that all living things can be divided into the three domains, which will be discussed later in this chapter.

It is possible to determine DNA similarities by **DNA-DNA hybridization.** The DNA double helix of each species is separated into single strands. Then strands from both species are allowed to combine. The more closely related the two species, the better the two strands of DNA will stick together. Some long-standing questions in systematics have been resolved by doing DNA-DNA hybridization. The giant panda, which lives in China, was at one time considered to be a bear, but its bones and teeth resemble those of a raccoon. The giant panda eats only bamboo and has a false thumb by which it grasps bamboo stalks. The red panda, which lives in the same area and has the same raccoonlike features, also feeds on bamboo but lacks the false thumb. The results of DNA hybridization studies suggest that after raccoons and bears diverged from a common lineage 50 million years ago, the giant panda diverged from the bear line and the red panda diverged from the raccoon line (Fig. 20.9). Therefore, it can be seen that some of the characters of the giant panda and the red panda are ancestral (present in a common ancestor), and some are due to parallel evolution.

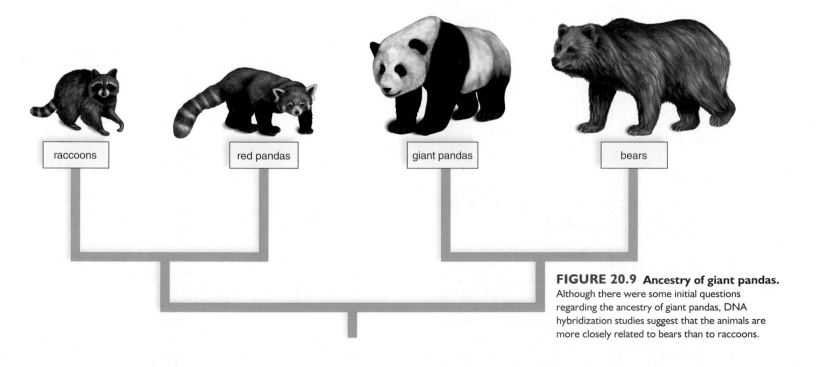

FIGURE 20.9 Ancestry of giant pandas.
Although there were some initial questions regarding the ancestry of giant pandas, DNA hybridization studies suggest that the animals are more closely related to bears than to raccoons.

raccoons   red pandas   giant pandas   bears

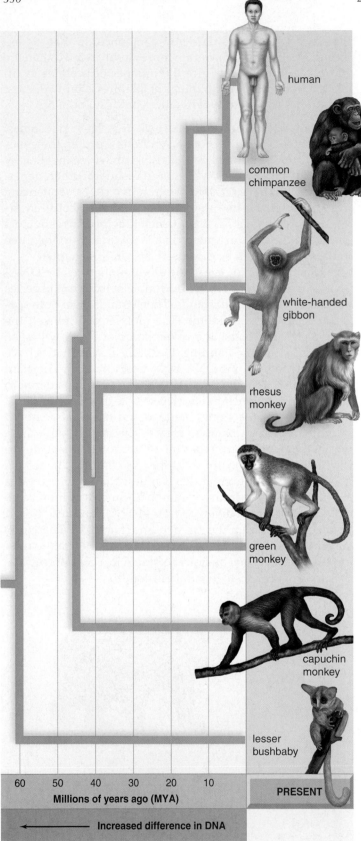

FIGURE 20.10 Molecular data.
The relationship of certain primate species based on a study of their genomes. The length of the branches indicates the relative number of nucleotide pair differences that were found between groups. These data, along with knowledge of the fossil record for one divergence, make it possible to suggest a date for the other divergences in the tree.

Because hybridization studies do not provide numerical data, many researchers prefer to compare the nucleotide sequences of a particular gene or genes. One study involving DNA differences produced the data shown in Figure 20.10. Although the data suggest that chimpanzees are more closely related to humans than to other apes, in most classifications, humans and chimpanzees are placed in different families; humans are in the family Hominidae, and chimpanzees are in the family Pongidae. In contrast, the rhesus monkey and the green monkey, which have more numerous DNA differences, are placed in the same family (Cercopithecidae). To be consistent with the data, shouldn't humans and chimpanzees also be in the same family? Traditional systematists, in particular, believe that since humans are markedly different from chimpanzees because of adaptation to a different environment, it is justifiable to place humans in a separate family.

Mitochondrial DNA (mtDNA) mutates ten times faster than nuclear DNA. Therefore, when determining the phylogeny of closely related species, investigators often choose to sequence mtDNA instead of nuclear DNA. One such study concerned North American songbirds. It had long been suggested that these birds diverged into eastern and western subspecies due to retreating glaciers some 250,000–100,000 years ago. Sequencing of mtDNA allowed investigators to conclude that groups of North American songbirds diverged from one another an average of 2.5 million years ago (MYA). Since the old hypothesis based on glaciation is apparently flawed, a new hypothesis is required to explain why eastern and western subspecies arose among these songbirds.

Molecular Clocks. When nucleic acid changes are neutral (not tied to adaptation) and accumulate at a fairly constant rate, these changes can be used as a kind of **molecular clock** to indicate relatedness and evolutionary time. The researchers doing comparative mtDNA sequencing used their data as a molecular clock when they equated a 5.1% nucleic acid difference among songbird subspecies to 2.5 MYA. In Figure 20.10, the researchers used their DNA sequence data to suggest how long the different types of primates have been separate. The fossil record was used to calibrate the clock: when the fossil record for one divergence is known, it indicates how long it probably takes for each nucleotide pair difference to occur. Even so, the tree drawn from molecular data is usually used as a hypothesis until it is confirmed by the fossil record. When the fossil record and molecular clock data agree, researchers have more confidence that the proposed phylogenetic tree is correct.

The fossil record, homology, and molecular data help systematists decipher phylogeny and construct phylogenetic trees.

20.3 SYSTEMATICS TODAY

There are three main schools of systematics: cladistics, phenetics, and traditional. We will begin by considering cladistics and then compare it to the other methodologies.

Cladistic Systematics

Cladistic systematics, which is based on the work of Willi Hennig, uses shared derived characters to classify organisms and arrange taxa in a type of phylogenetic tree called a **cladogram** [Gk. *klados,* branch, stem, and *gramma,* picture]. A cladogram traces the evolutionary history of the group being studied. Let's see how it works.

The first step when constructing a cladogram is to draw a table that summarizes the characters of the taxa being compared (Fig. 20.11*a*). At least one but preferably several species are studied as an outgroup, a taxon (taxa) that is (are) not part of the study group. In this example, lancelets are the outgroup and selected vertebrates are the study group. Any character found both in the outgroup and the study group is a shared ancestral character (e.g., notochord in embryo) presumed to have been present in a common ancestor to both the outgroup and study group. Any character found in one or scattered taxa (e.g., long cylindrical body) is excluded from the cladogram. The other characters are shared derived characters—that is, they are homologies shared by certain taxa of the study group. In a cladogram, a **clade** is an evolutionary branch that includes a common ancestor, together with all its descendant species. A clade includes the taxa that share homologies.

The cladogram in Figure 20.11*b* has three clades that differ in size because the first includes the other two, and so forth. Notice that the common ancestor at the root of the tree had one ancestral character: notochord in embryo. There follow common ancestors that have vertebrae, lungs and a three-chambered heart, and finally amniotic egg and internal fertilization. Therefore, this is the sequence in which these characters evolved during the evolutionary history of vertebrates. These are also the homologies that show the clade species are closely related to one another. All the taxa in the study group belong to the first clade because they all have vertebrae; newts, snakes, and lizards are in the clade that has lungs and a three-chambered heart; and only snakes and lizards have an amniotic egg and internal fertilization.

A cladogram is objective because it lists the characters that were used to construct the cladogram. Cladists typically use many more characters than appear in our simplified cladogram. They also feel that a cladogram is a hypothesis that can be tested and either corroborated or refuted on the basis of additional data. These are the reasons that cladistics, a relatively young discipline, has now become a respected way to decipher evolutionary history. The terms you need to learn to understand cladistics are given in Table 20.2.

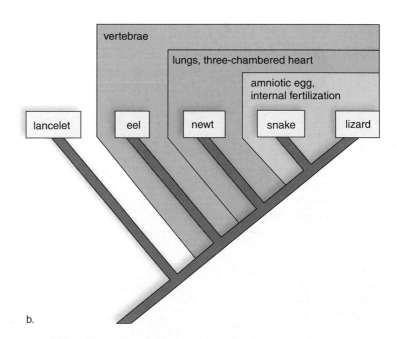

	lancelet	eel	newt	snake	lizard
Notochord in embryo	X	X	X	X	X
Vertebrae		X	X	X	X
Lungs			X	X	X
Three-chambered heart			X	X	X
Internal fertilization				X	X
Amniotic membrane in egg				X	X
Four bony limbs			X		X
Long cylindrical body		X		X	

a.

b.

FIGURE 20.11 Constructing a cladogram.
a. First, a table is drawn that lists characters for all the taxa. An examination of the table shows which characters are ancestral (notochord) and which are derived (blue, purple, and green). The shared derived characters distinguish the taxa. **b.** In a cladogram, the shared derived characters are sequenced in the order they evolved and are used to define clades. A clade contains a common ancestor and all the species that share the same derived characters (homologies). Four bony limbs and a long cylindrical body were not used in constructing the cladogram because they are in scattered taxa.

TABLE 20.2

Terms Used in Cladistics

Outgroup	Taxon (taxa) that define(s) the primitive characters of the study group
Study group	Taxa that will be placed into clades in a cladogram
Ancestral characters	Structures present in the outgroup and also in the study group
Clade	Evolutionary branch of a cladogram; a monophyletic taxon that contains a common ancestor and all its descendant species
Shared derived characters	Homologies present in a particular clade but not in the outgroup
Monophyletic taxon	Contains a single common ancestor and all its descendant species; no descendant species can be in any other taxon
Parsimony	Results in the simplest cladogram possible

Parsimony

Figure 20.12 shows a cladogram in which all three species, represented by X, Y, and Z, belong to a monophyletic taxon, since they all trace their ancestry to the same common ancestor that had the ancestral characters designated by the first arrow. Species Y and Z are placed in the same clade because they share the derived characters designated by the second arrow. How do you know you have done the cladogram correctly, and that the other two patterns shown in Figure 20.12b and c are not likely? In the other two arrangements, the characters represented by the green boxes would have had to evolve twice.

Cladists are always guided by the principle of *parsimony*, which states that the minimum number of assumptions is the most logical. That is, they construct the cladogram that leaves the fewest number of shared derived characters unexplained or that minimizes the number of assumed evolutionary changes. However, cladists must be on the lookout for the possibility that convergent evolution has produced what appears to be common ancestry. Then, too, the reliability of a cladogram is dependent on the knowledge and skill of the particular investigator gathering the data and doing the character analysis.

> Cladistics is based on the premise that shared derived characters (homologies) can be used to define monophyletic taxa and determine the sequence in which evolution occurred.

Phenetic Systematics

In **phenetic systematics,** species are classified according to the number of their similarities. Systematists of this school believe that since it is impossible to construct a classification that truly reflects phylogeny, it is better to rely on a method that does away with personal prejudices. They measure as many traits as possible, count the number of traits the two species share, and then estimate the degree of relatedness. They simply ignore the possibility that some of the shared characters are probably the result of convergence or parallelism, or that some of the characters might depend on one another. For example, a large animal is bound to have larger parts. The results of their analysis are depicted in a phenogram. Figure 20.10 is a phenogram that is based solely on the number of DNA differences among the species shown. (Phenograms have been known to vary for the same group of taxa, depending on how the data are collected and handled.)

Traditional Systematics

Soon after Darwin published his book, *On the Origin of Species*, **traditional systematics** began and still continues today. These systematists mainly use anatomical data to classify organisms and construct phylogenetic trees based on evolutionary principles. Traditionalists differ from today's cladists largely by stressing both common ancestry *and* the degree of structural

FIGURE 20.12 Alternate, simplified cladograms.
a. X, Y, Z share the same characters, designated by the purple shading, and are judged to form a monophyletic taxon. Y and Z are grouped together because they share the same derived character, designated by the green shading. **b, c.** These cladograms are rejected because in each you would have to assume that the same character evolved in different groups. Since this seems unlikely, the first branching pattern is chosen as the hypothesis.

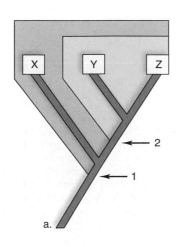
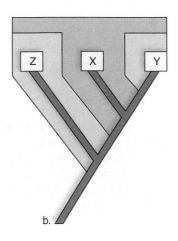
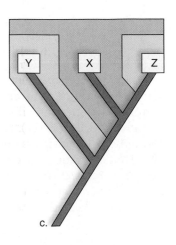

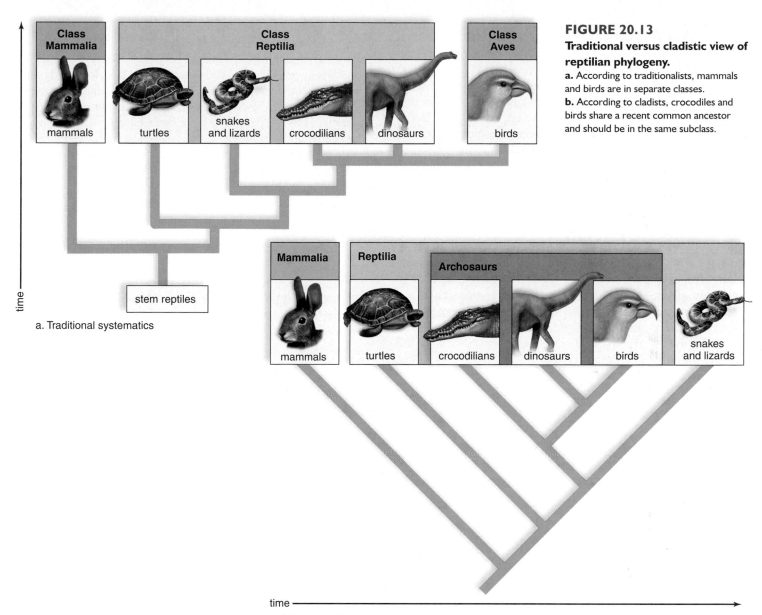

FIGURE 20.13

Traditional versus cladistic view of reptilian phylogeny.

a. According to traditionalists, mammals and birds are in separate classes.
b. According to cladists, crocodiles and birds share a recent common ancestor and should be in the same subclass.

a. Traditional systematics

b. Cladistic systematics

difference among divergent groups. Therefore, a group that has adapted to a new environment and shows a high degree of evolutionary change is not always classified with the common ancestor from which it evolved. In other words, traditionalists are not as strict as cladists are about making sure all taxa are monophyletic.

In the traditional phylogenetic tree shown in Figure 20.13a, birds and mammals are placed in different classes because it is quite obvious to the most casual observer that mammals (having hair and mammary glands) and birds (having feathers) are quite different in appearance from reptiles (having scaly skin). The traditionalist goes on to say that birds and mammals evolved from stem reptiles.

Cladists prefer the cladogram shown in Figure 20.13b. All the animals shown are in one clade because they all evolved from a common ancestor that laid eggs. Mammals can be placed in a class because they all have hair and mammary glands and three middle ear bones. Cladists doubt

there should be a class Reptilia because the only thing that dinosaurs, crocodiles, snakes, lizards, and turtles have in common is that they are not birds or mammals. Crocodiles and birds may share common derived characters not present in snakes and lizards. Then, too, some believe the fossil record indicates that snakes and lizards have a separate common ancestor from crocodiles and birds. Birds just seem different from crocodiles because each is adapted to a different way of life. To indicate that crocodiles and birds, along with the dinosaurs, are closely related, perhaps they should be in a subgroup of their own called Archosaurs. Most biologists today are willing to admit to inconsistencies but still use the classes Aves and Reptilia for the sake of convenience.

Pheneticists and traditionalists are not as strict as cladists about the use of only homologies and monophyletic groups to classify organisms and construct phylogenetic trees.

20.4 CLASSIFICATION SYSTEMS

From Aristotle's time to the middle of the twentieth century, biologists recognized only two kingdoms: kingdom Plantae (plants) and kingdom Animalia (animals). Plants were literally organisms that were planted and immobile, while animals were animated and moved about. After the light microscope was perfected in the late 1600s, unicellular organisms were revealed that didn't fit neatly into the plant or animal kingdoms. In the 1880s, a German scientist, Ernst Haeckel, proposed adding a third kingdom. The kingdom Protista (protists) included unicellular microscopic organisms but not multicellular, largely macroscopic ones.

In 1969, R. H. Whittaker expanded the classification system to five kingdoms: Monera, Protista, Fungi, Plantae, and Animalia. Organisms were placed in these kingdoms based on the type of cell (prokaryotic or eukaryotic), complexity (unicellular or multicellular), and type of nutrition. Kingdom Monera contained all the prokaryotes, which are organisms that lack a membrane-bounded nucleus. These unicellular organisms were collectively called the bacteria. The other four kingdoms contained the eukaryotes, and since they were essentially the same as the four kingdoms recognized as eukaryotes today, they are described later. We can note, however, that Whittaker was the first to give the fungi their own kingdom. He did so because fungi are generally multicellular, yet they are heterotrophic by absorption. Plants, of course, are photosynthesizers while animals are heterotrophic by ingestion.

Three-Domain System

In the late 1970s, Dr. Carl Woese and his colleagues at the University of Illinois were studying relationships among the prokaryotes using rRNA sequences. As mentioned previously, rRNA probably changes only slowly during evolution, and indeed it may change only when there is a major evolutionary event. Woese found that the rRNA sequence of prokaryotes that lived at high temperatures or produced methane was quite different from that of all the other types of prokaryotes and from the eukaryotes. Therefore, he proposed that there are two groups of prokaryotes (rather than one group as in the **five-kingdom system**). Further, Woese said that the rRNA sequences of these two groups, called the bacteria and archaea, are so fundamentally different from each other that they should be assigned to separate domains, a category of classification that is higher than the kingdom category. Originally, Woese used the

term *archaebacteria* but then shortened it to *archaea* because these organisms are not bacteria, they are archaea. The phylogenetic tree shown in Figure 20.14 is based on his rRNA sequencing data. The data suggest that the bacteria diverged first, followed by the archaea and then the eukarya. This means that the archaea and eukarya are more closely related to each other than either is to the bacteria. These relationships have been supported by molecular data gathered by other independent investigators.

Systematists are in the process of sorting out what kingdoms belong within **domain Bacteria** and in **domain Archaea**, but kingdoms Protista, Fungi, Plantae, and Animalia are all in the **domain Eukarya.**

Domain Bacteria

In the three-domain system, most, but not all, prokaryotes are bacteria, a group of organisms that is so diversified and plentiful they are found in large numbers nearly everywhere on Earth. In large part, Chapter 21 will be devoted to discussing the bacteria, which differ from the archaea not structurally but biochemically (Table 20.3).

In the meantime, we can note that the cyanobacteria are large photosynthetic prokaryotes. They carry on photosynthesis in the same manner as plants in that they use solar en-

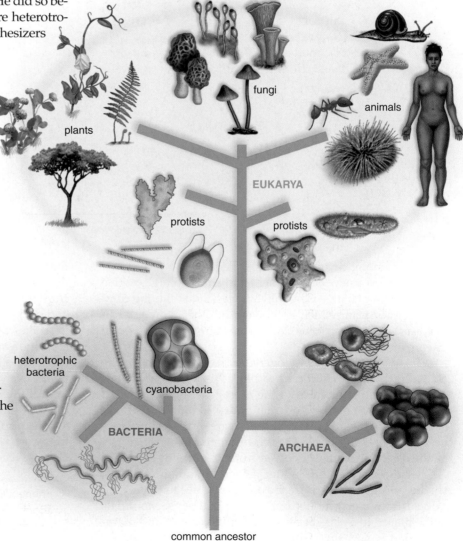

fungi

animals

plants

EUKARYA

protists

protists

heterotrophic bacteria

cyanobacteria

BACTERIA

ARCHAEA

common ancestor

FIGURE 20.14 The three-domain system of classification.
Representatives of each domain are depicted in the ovals, and the phylogenetic tree shows that domain Archaea is more closely related to domain Eukarya than either is to domain Bacteria.

TABLE 20.3

Major Distinctions Among the Three Domains of Life

	Bacteria	Archaea	Eukarya
Unicellularity	Yes	Yes	Some, many multicellular
Membrane lipids	Phospholipids, unbranched	Varied branched lipids	Phospholipids, unbranched
Cell wall	Yes (contains peptidoglycan)	Yes (no peptidoglycan)	Some yes, some no
Nuclear envelope	No	No	Yes
Membrane-bounded organelles	No	No	Yes
Ribosomes	Yes	Yes	Yes
Introns	No	Some	Yes

ergy to convert carbon dioxide and water to a carbohydrate and in the process give off oxygen. Indeed the cyanobacteria may have been the first organisms to contribute oxygen to early Earth's atmosphere, making it hospitable to the evolution of oxygen-using organisms, including animals.

All forms of nutrition are found among the bacteria, but most are heterotrophic. *Escherichia coli,* which lives in the human intestine is heterotropic as are parasitic forms that cause human disease. *Clostridium tetani* (cause of tetanus), *Bacillus anthracis* (cause of anthrax), and *Vibrio cholerae* (cause of cholera) are disease-causing species of bacteria. Heterotrophic bacteria are beneficial in ecosystems because they are organisms of decay that break down organic remains and absorb nutrient molecules. Along with fungi, they keep chemical cycling going so that plants always have a source of inorganic nutrients.

Domain Archaea

Like bacteria, archaea are prokaryotic unicellular organisms that reproduce asexually. Archaea don't look that different from bacteria under the microscope, and the extreme conditions under which many species live has made it difficult to culture them. This may have been the reason that their unique place among the living organisms long went unrecognized.

The archaea are distinguishable from bacteria by a difference in their rRNA base sequences and also by their unique plasma membrane and cell wall chemistry (Table 20.3). The chemical nature of the archaeal cell wall is diverse and never the same as that of the bacterial cell. The archaea also have diverse modes of nutrition.

The archaea live in all sorts of environments, but they are known for thriving in extreme environments thought to be similar to those of the early Earth. For example, the methanogens live in anaerobic environments, such as swamps and marshes and the guts of animals; the halophiles are salt lovers living in bodies of water such as the Great Salt Lake in Utah; and the thermoacidophiles are both high temperature and acid loving. These archaea live in extremely hot acidic environments, such as hot springs and geysers.

At least some of the differences between bacteria and archaea can be attributed to the adaptations of archaea to harsh environments. The branched nature of diverse lipids in the archaeal plasma membrane, for example, could possibly help them live in extreme conditions.

Domain Eukarya

Eukaryotes are unicellular to multicellular organisms whose cells have a membrane-bounded nucleus. Sexual reproduction is common, and various types of life cycles are seen. Later in this text, we will be studying the individual kingdoms that occur within the domain Eukarya (Fig. 20.15 and Table 20.4). In the meantime, we can note that protists are a diverse group of organisms that are hard to classify and define. They are eukaryotes and mainly unicellular, but some are filaments, colonies, or multicellular sheets. Even so, protists do not have true tissues. Nutrition is diverse and some are heterotrophic by ingestion or absorption and some are photosynthetic. Green algae, paramecia, and slime molds are representative protists. There has been considerable debate over the classification of protists, and this debate will most likely continue for some time. Some texts recognize several distinct kingdoms for these organisms instead of a simple kingdom, Protista.

Fungi are eukaryotes that form spores, lack flagella, and have cell walls containing chitin. They are multicellular with a few exceptions. Fungi are heterotrophic by absorption—they secrete digestive enzymes and then absorb nutrients from decaying organic matter. Mushrooms, molds, and yeasts are representative fungi.

Despite appearances, molecular data suggest that fungi and animals are more closely related to each other than either are to plants.

Plants are nonmotile eukaryotic multicellular organisms. They possess true tissues and have the organ system level of organization. Plants are autotrophic and carry on photosynthesis. Examples include cacti, ferns, and cypress trees.

Animals are motile eukaryotic multicellular organisms. They also have true tissues and the organ system level of organization. Animals are heterotrophic by ingestion. Worms, whales, and insects are all examples of animals.

Recently, it has been suggested that there are three evolutionary domains: Bacteria, Archaea, and Eukarya. The domain Eukarya contains four kingdoms.

FIGURE 20.15 The three domains of life.
This pictorial representation of the domains Bacteria, Archaea, and Eukarya includes an example for each of the four kingdoms in the domain Eukarya:
Protista, Fungi, Plantae, and Animalia.

TABLE 20.4

Classification Criteria for the Three Domains

	Domains Bacteria and Archaea	Domain Eukarya			
		Kingdom Protista	*Kingdom Fungi*	*Kingdom Plantae*	*Kingdom Animalia*
Type of cell	Prokaryotic	Eukaryotic	Eukaryotic	Eukaryotic	Eukaryotic
Complexity	Unicellular	Unicellular usual	Multicellular usual	Multicellular	Multicellular
Type of nutrition	Autotrophic or heterotrophic	Photosynthetic or heterotrophic by various means	Heterotrophic by absorption	Autotrophic by photosynthesis	Heterotrophic by ingestion
Motility	Sometimes by flagella	Sometimes by flagella (or cilia)	Nonmotile	Nonmotile	Motile by contractile fibers
Life cycle*	Asexual usual	Various life cycles	Haploid	Alternation of generations	Diploid
Internal protection of zygote	No	No	No	Yes	Yes

*See the Science Focus, Life Cycles Among the Algae, page 386

CONNECTING THE CONCEPTS

We have seen in this chapter that identifying, naming, and classifying living organisms is an ongoing process. Carolus Linnaeus's system of binomial nomenclature is still accepted by virtually all biologists, but many species remain to be found and named (most in the rain forests). For years, Whittaker's five-kingdom concept of life on Earth was widely used. Now new findings suggest that there are three domains of life: Bacteria, Archaea, and Eukarya. The archaea are structurally similar to bacteria, but their rRNA differs from that of bacteria and is instead similar to that of eukaryotes. Also, some archaeal genes are unique only to the archaea. Most likely, several kingdoms will eventually be recognized among the bacteria and archaea just as several are recognized among the eukarya (protists, fungi, plants, and animals).

Most of today's systematists use evolutionary relationships among organisms for classification purposes. The traditional and cladistic schools differ as to how to determine such relationships. The traditionalist is willing to consider both structural similarities and obvious differences due to adaptations to new environments. The cladist believes that only similarities in anatomy and DNA should be used to classify organisms. In the end, it may be the ability to quickly sequence genes that will do away with the need for any subjective analyses and make classification a purely objective science.

Summary

20.1 TAXONOMY

Taxonomy deals with the naming of organisms; each species is given a binomial name consisting of the genus and specific epithet.

Distinguishing species on the basis of structure can be difficult because members of the same species can vary in structure. Distinguishing species on the basis of reproductive isolation runs into problems because some species hybridize and also because reproductive isolation is difficult to observe. In this chapter, the term *species* refers to a taxon occurring below the level of genus. Species in the same genus share a more recent common ancestor than do species in related genera.

Classification involves the assignment of species to categories. When an organism is named, a species has been assigned to a particular genus. Eight obligatory categories of classification are species, genus, family, order, class, phylum, kingdom, and domain. Each higher category is more inclusive; species in the same kingdom share general characters, and species in the same genus share quite specific characters.

20.2 PHYLOGENETIC TREES

Systematics, a very broad field, encompasses both taxonomy and classification. Classification should reflect phylogeny, and one goal of systematics is to create phylogenetic trees, based on ancestral and derived characters.

The fossil record, homology, and molecular data are used to help decipher phylogenies. Because fossils can be dated, available fossils can establish the antiquity of a species. If the fossil record is complete enough, we can sometimes trace a lineage through time. Homology helps indicate when species belong to a monophyletic taxon (share a common ancestor); however, convergent evolution and parallel evolution sometimes make it difficult to distinguish homologous structures from analogous structures. Various molecular data are used to indicate relatedness, but DNA base sequences probably pertain more directly to phylogenetic characters.

20.3 SYSTEMATICS TODAY

Today there are three main schools of systematics: the cladistic, phenetic, and traditional schools. The cladistic school analyzes ancestral and derived characters and constructs cladograms on the basis of shared derived characters. A clade includes a common ancestor and all the species derived from that common ancestor. Cladograms are diagrams based on homologies. The numerical phenetic school clusters species on the basis of the number of shared similarities regardless of whether they might be convergent, parallel, or dependent on one another. The traditional school stresses common ancestry *and* the degree of structural difference among divergent groups in order to construct a phylogenetic tree.

20.4 CLASSIFICATION SYSTEMS

The five-kingdom system of classification recognizes these kingdoms: Plantae, Animalia, Fungi, Protista, and Monera. On the basis of molecular data, three evolutionary domains have been established: Bacteria, Archaea, and Eukarya. The first two domains contain prokaryotes; the domain Eukarya contains the kingdoms Protista, Fungi, Plantae, and Animalia.

Reviewing the Chapter

1. Explain the binomial system of naming organisms. Why must species be designated by a complete name?   342–43
2. Why is it necessary to give organisms scientific names?   343
3. Discuss three ways to distinguish a species. Which way relates to classification?   344
4. What are the eight obligatory classification categories? In what way are they a hierarchy?   344
5. How is it that taxonomy and classification are a part of systematics? What three types of data help systematists construct phylogenetic trees?   346–47, 349–50
6. Discuss the principles of cladistics, and explain how to construct a cladogram.   351
7. In what ways do the cladistic school, the phenetic school, and the traditional school of systematics differ?   351–53
8. Compare the five-kingdom system of classification to the three-domain system.   354
9. Contrast the characteristics of the bacteria, the archaea, and the eukarya.   354–55
10. Contrast the eukaryotic kingdoms Protista, Fungi, Plantae, and Animalia.   355–56

Testing Yourself

Choose the best answer for each question.

1. Which is the scientific name of an organism?
   a. *Rosa rugosa*
   b. *Rosa*
   c. *rugosa*
   d. *Rugosa rugosa*
   e. Both a and d are correct.

2. Which of these best pertains to taxonomy? Species always
   a. have three-part names, such as *Homo sapiens sapiens*.
   b. are reproductively isolated from other species.
   c. share the most recent common ancestor.
   d. look exactly alike.
   e. Both c and d are correct.

3. The classification category below the level of family is
   a. class.
   b. species.
   c. phylum.
   d. genus.
   e. order.

4. Which of these are domains? Choose more than one answer if correct.
   a. bacteria
   b. archaea
   c. eukarya
   d. animals
   e. plants

5. Which of these are eukaryotes? Choose more than one answer if correct.
   a. bacteria
   b. archaea
   c. eukarya
   d. animals
   e. plants

6. Which of these characteristics is shared by bacteria and archaea? Choose more than one answer if correct.
   a. presence of a nucleus
   b. absence of a nucleus
   c. presence of ribosomes
   d. absence of membrane-bounded organelles
   e. presence of a cell wall

7. Which kingdom is mismatched?
   a. Fungi—prokaryotic single cells
   b. Plantae—multicellular only
   c. Plantae—flowers and mosses
   d. Animalia—arthropods and humans
   e. Protista—unicellular eukaryotes

8. Which kingdom is mismatched?
   a. Fungi—usually saprotrophic
   b. Plantae—usually photosynthetic
   c. Animalia—rarely ingestive
   d. Protista—various modes of nutrition
   e. Both c and d are mismatched.

9. Concerning a phylogenetic tree, which is incorrect?
   a. Dates of divergence are always given.
   b. Common ancestors occur at the notches.
   c. The more recently evolved are at the top of the tree.
   d. Ancestors have only primitive characters.

10. Which pair is mismatched?
    a. homology—character similarity due to a common ancestor
    b. molecular data—DNA strands match
    c. fossil record—bones and teeth
    d. homology—functions always differ
    e. molecular data—molecular clock

11. One benefit of the fossil record is
    a. that hard parts are more likely to fossilize.
    b. fossils can be dated.
    c. its completeness.
    d. fossils congregate in one place.
    e. All of these are correct.

12. The discovery of common ancestors in the fossil record, the presence of homologies, and nucleic acid similarities help scientists decide
    a. how to classify organisms.
    b. the proper cladogram.
    c. how to construct phylogenetic trees.
    d. how evolution occurred.
    e. All of these are correct.

13. Molecular clock data are based on
    a. common adaptations among animals.
    b. DNA dissimilarities in living species.
    c. DNA fingerprinting of fossils.
    d. finding homologies among plants.
    e. All of these are correct.

14. In cladistics,
    a. a clade must contain the common ancestor plus all its descendants.
    b. derived characters help construct cladograms.
    c. data for the cladogram are presented.
    d. the species in a clade share homologous structures.
    e. All of these are correct.

15. In the traditional school of systematics, birds are assigned to a different group from reptiles because
    a. they evolved from reptiles and couldn't be a monophyletic taxon.
    b. they are adapted to a different way of life compared to reptiles.
    c. feathers came from scales, and feet came before wings.
    d. all classes of vertebrates are only related by way of a common ancestor.
    e. All of these are correct.

16. Which of these pairs is mismatched?
    a. phylogenetic tree—shows common ancestors
    b. cladogram—shows derived characters
    c. phylogenetic tree—uses names of classification categories
    d. cladogram—based on monophyletic groups
    e. All of these are properly matched.

In questions 17–20, fill in the blanks.

17. The scientific name of an organism consists of its genus and _____.

18. The fossil record, _____, and molecular data help systematists decipher phylogeny and construct phylogenetic trees.

19. The biological definition of a species states that members of two different species cannot successfully _____.

20. Members of a species _____ and share the same gene pool.

21. Label this diagram using these labels: common ancestor, derived characters, ancestral characters divergence.

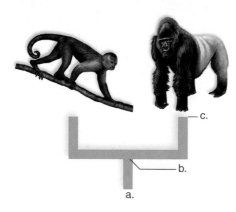

22. Answer these questions about the following cladogram.
    a. This cladogram contains how many clades? How are they designated in the diagram?
    b. What character is shared by all clades in the study group? What characters are shared by only snakes and lizards?
    c. Which clades share a recent common ancestor? How do you know?

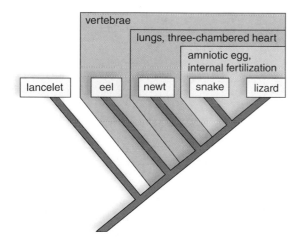

Thinking Scientifically

1. Recent DNA evidence suggests to some plant taxonomists that the traditional way of classifying flowering plants is not correct, and that flowering plants need to be completely reclassified. Other botanists disagree, saying it would be chaotic and unwise to disregard the historical classification groups. Argue for and against keeping traditional classification schemes.

2. Two populations of frogs apparently differ only in skin coloration. What data would you need to determine if both populations belong to the same species? If they are two different species, what data would you need to determine how closely related the two species are?

Bioethical Issue: Classifying Apes

Even though chimpanzees and humans have fewer molecular differences than do any two human beings, chimpanzees and humans are classified in different families by traditional taxonomists. Chimpanzees and humans do have several skeletal and other anatomical differences such as the size of their brains. Humans also have a much more sophisticated culture including use of language.

Do you feel that it is proper to ignore the closeness of their molecular biology and classify chimpanzees in an entirely different family than human beings, or do you feel that prejudice is influencing traditional biologists, who should be more objective? Why do you say so?

Understanding the Terms

analogous structure 347	five-kingdom system 354
analogy 347	genus 344
ancestral characteristic 346	homologous structure 347
binomial nomenclature 342	homology 347
character 345	kingdom 344
clade 351	molecular clock 350
cladistic systematics 351	order 344
cladogram 351	parallel evolution 347
class 344	phenetic systematics 352
common ancestor 346	phylogenetic tree 346
convergent evolution 347	phylogeny 346
derived character 346	phylum 344
DNA-DNA hybridization 349	species 344
domain 344	specific epithet 342
domain Archaea 354	systematics 346
domain Bacteria 354	taxon 344
domain Eukarya 354	taxonomy 342
family 344	traditional systematics 352

Match the terms to these definitions:
a. _____ Branch of biology concerned with identifying, describing, and naming organisms.
b. _____ Diagram that indicates common ancestors and lines of descent.
c. _____ Group of organisms that fills a particular classification category.
d. _____ School of systematics that determines the degree of relatedness by analyzing ancestral and derived characters and constructing cladograms.
e. _____ Similarity in structure due to having a common ancestor.

ARIS, the *Biology* Website

ARIS, the website for *Biology*, provides a wealth of information organized and integrated by chapter. You will find practice quizzes, interactive activities, labeling exercises, flashcards, and much more that will complement your learning and understanding of general biology.

www.mhhe.com/maderbiology9

PART IV

MICROBIOLOGY AND EVOLUTION

Microbiology is the study of viruses, bacteria, archaea, protists (such as algae and protozoans), and fungi. From the fossil record we know that microorganisms have existed on Earth for at least 3.5 billion years. Most (aside from fungi) are unicellular, and from these single cells came multicellular forms such as plants and animals—they are our ancestors. Even though microorganisms are usually too small to be seen without a microscope, they are extremely plentiful, being present on everything we touch, including ourselves.

Although they seem relatively simple, microorganisms are extraordinarily complex. They are diverse in appearance, metabolism, physiology, and genetics. Their metabolic complexity gives them the ability to grow in a wide variety of different environments and to interact with all other forms of life, including human beings. Although they may cause diseases, they also perform services that make life on Earth possible.

21

VIRUSES, BACTERIA, AND ARCHAEA

*V*iruses are noncellular entities responsible for a number of diseases in plants, animals, and humans. Polio, smallpox, rabies, and AIDS are all caused by viruses. Even so, viruses are useful tools in the biotechnology laboratory and are even used as vectors for gene therapy in humans. Bacteria and archaea are prokaryotes, organisms that are simple in structure but metabolically diverse. They can live under conditions that are too hot, too salty, too acidic, or too cold for eukaryotes. They have been found hundreds of meters below ground level and hundreds of meters below Antarctic ice. Through their ability to oxidize sulfides that spew forth from deep-sea vents and to subsequently produce nutrients, they support communities of organisms where the sun never shines. Both on land and in the sea, prokaryotes routinely generate oxygen and recycle the nutrients of dead plants and animals.

Most people are familiar with bacteria because some cause deadly diseases; but others produce antibiotics that are used to cure such illnesses. Due to the ease with which they can be grown and manipulated in the laboratory, bacteria are used to study basic life processes such as protein synthesis. Humans also use bacteria to mine minerals, clean up oil spills, manufacture industrial chemicals, and produce medicines such as vitamins and insulin. All in all, it is safe to say that we could not live without the services of prokaryotes.

Micrograph of anthrax bacteria, *Bacillus anthracis*.

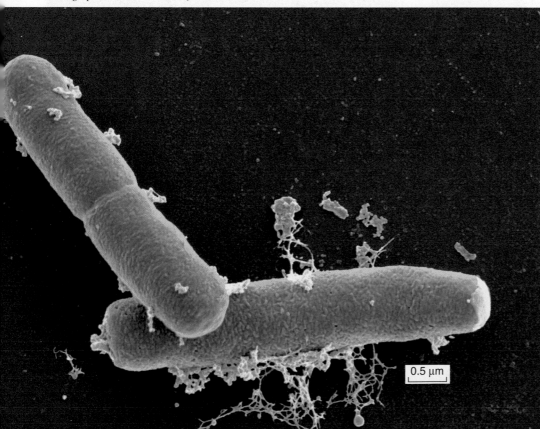

0.5 μm

21.1 VIRUSES, VIROIDS, AND PRIONS

The term **virus** [L. *virus*, poison] is associated with a number of plant, animal, and human diseases (see Table 21A on page 375). The mere mention of the term brings to mind serious illnesses such as polio, rabies, and AIDS (acquired immunodeficiency syndrome), as well as common childhood maladies such as measles, chickenpox, and mumps. Viral diseases are of concern to everyone; it is estimated that the average person catches two or three colds a year.

The viruses are a biological enigma. They have a DNA or RNA genome, but they can only reproduce by using the metabolic machinery of a host cell. Viruses are noncellular, and therefore they do not fit into current classification systems, which are devoted to categorizing the cellular organisms on Earth.

Our knowledge of viruses began in 1884 when the French chemist Louis Pasteur (1822–95) suggested that something smaller than a bacterium was the cause of rabies, and it was he who chose the word *virus* from a Latin word meaning poison. In 1892, Dimitri Ivanowsky (1864–1920), a Russian microbiologist, was studying a disease of tobacco leaves, called tobacco mosaic disease because of the leaves' mottled appearance. He noticed that even when an infective extract was filtered through a fine-pore porcelain filter that retains bacteria, it still caused disease. This substantiated Pasteur's belief because it meant that the disease-causing agent was smaller than any known bacterium. In the next century, electron microscopy was born, and viruses were seen for the first time. By the 1950s, virology was an active field of research; the study of viruses, and now also viroids and prions, has contributed much to our understanding of disease, genetics, and even the characteristics of living things.

Viral Structure

The size of a virus is comparable to that of a large protein macromolecule, and ranges in size from 10–400 nm. Viruses are best studied through electron microscopy. Many viruses can be purified and crystallized, and the crystals can be stored just as chemicals are stored. Still, viral crystals will become infectious when the viral particles they contain are given the opportunity to invade a host cell.

FIGURE 21.1 Viruses.
Despite their diversity, all viruses have an outer capsid composed of protein subunits and a nucleic acid core—composed of either DNA or RNA, but not both. Some types of viruses also have a membranous envelope.

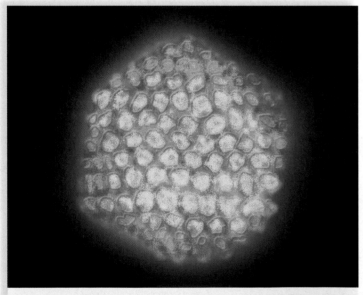

TEM 80,000×

Adenovirus: DNA virus with a polyhedral capsid and a fiber at each corner.

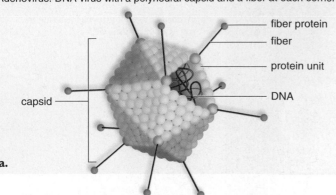

fiber protein
fiber
protein unit
capsid
DNA

a.

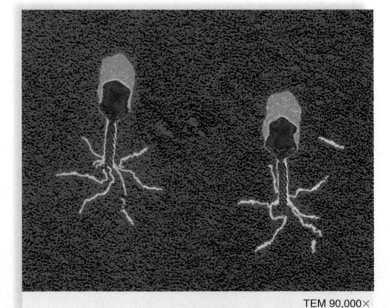

TEM 90,000×

T-even bacteriophage: DNA virus with a polyhedral head and a helical tail.

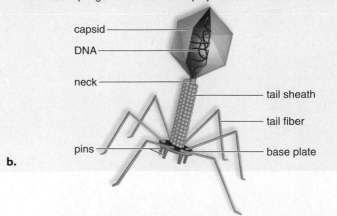

capsid
DNA
neck
tail sheath
tail fiber
pins
base plate

b.

The following diagram summarizes viral structure:

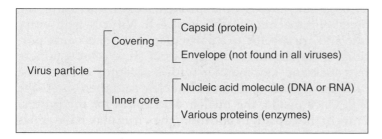

Viruses vary in shape from threadlike to polyhedral (Fig. 21.1). However, all viruses possess the same basic anatomy: an outer **capsid** composed of protein subunits and an inner core of nucleic acid—either DNA (deoxyribonucleic acid) or RNA (ribonucleic acid), but not both. A viral genome has as little as three and as many as 100 genes; a human cell contains tens of thousands of genes. The viral capsid may be surrounded by an outer membranous envelope; if not, the virus is said to be naked. Figure 21.1a, b, c gives examples of naked viruses, while Figure 21.1d is an example of an enveloped virus. The envelope is actually a piece of the host's plasma membrane that also contains viral glycoprotein spikes. Aside from its genome, a viral particle may also contain various proteins, especially enzymes such as the polymerases, needed to produce viral DNA and/or RNA.

TABLE 21.1

Comparison of Viruses and Prokaryotes

Characteristic of Life	Viruses	Prokaryotes
Consist of cell	No	Yes
Metabolize	No	Yes
Respond to stimuli	No	Yes
Multiply	Yes (always inside living cell)	Yes (usually independently)
Evolve	Yes	Yes

Viruses are categorized by (1) their type of nucleic acid, including whether it is single stranded or double stranded, (2) their size and shape, and (3) the presence or absence of an outer envelope. Viruses differ from prokaryotes in the ways stated in Table 21.1. From this comparison, you can see why viruses are considered by some to be nonliving and why they are not in the classification of organisms in Appendix B.

Viruses are noncellular and have at least two parts: an outer capsid composed of protein subunits and an inner core of nucleic acid, either DNA or RNA but not both.

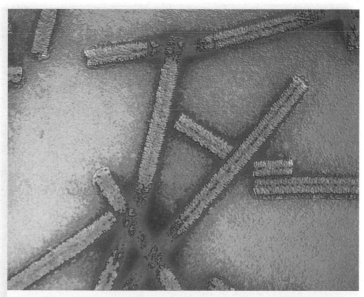

TEM 500,000×

Tobacco mosaic virus: RNA virus with a helical capsid.

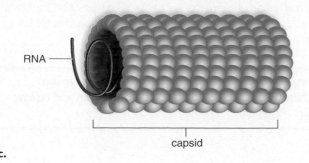

c.

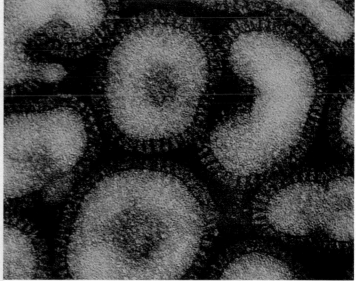

20 nm

Influenza virus: RNA virus with a helical capsid surrounded by an envelope with spikes

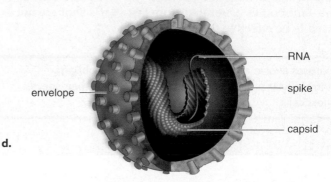

d.

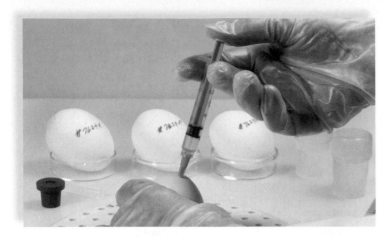

FIGURE 21.2 Growing viruses.
To grow a virus, scientists can inoculate live chicken eggs with viral particles. A virus reproduces only inside a living cell, not because it uses the cell for nutrients, but because it takes over the machinery of the cell.

Parasitic Nature

Viruses are *obligate intracellular parasites,* which means they cannot reproduce outside a living cell. To maintain animal viruses in the laboratory, they are sometimes injected into live chicken embryos (Fig. 21.2). Today, host cells are often maintained in tissue (cell) culture by simply placing a few cells in a glass or plastic container with appropriate medium. The cells can then be infected with the animal virus to be studied. Viruses infect a variety of cells, but they are **host specific.** Bacteriophages infect only bacteria, the tobacco mosaic virus infects only certain species of plants, and the rabies virus infects only mammals, for example. Some human viruses even specialize in a particular tissue. Human immunodeficiency virus (HIV) enters only certain blood cells, the polio virus reproduces in spinal nerve cells, and the hepatitis viruses infect only liver cells. What could cause this remarkable parasite–host cell correlation? Some scientists hypothesize that viruses are derived from the very cell they infect; the nucleic acid of viruses came from their host cell genomes! In that case, viruses evolved after cells came into existence, and new viruses may be evolving even now. Using protein and genetic analysis in 2000, other scientists hypothesize that viruses arose early in the origin of life predating the three domains.

Viruses can also mutate; therefore, it is correct to say that they evolve. Those that mutate often can be quite troublesome because a vaccine that is effective today may not be effective tomorrow. Flu viruses are well known for mutating, and this is why it is necessary to have a flu shot every year—antibodies generated from last year's shot are not expected to be effective this year.

Viruses evolve and reproduce, but they are obligate intracellular parasites. They only grow inside their specific host cells.

Viral Reproduction

Viruses are microscopic pirates, commandeering the metabolic machinery of a host cell. Viruses gain entry into and are specific to a particular host cell because portions of the capsid (or the spikes of an envelope) adhere in a lock-and-key manner with a receptor on the host cell's outer surface. The viral nucleic acid then enters the cell. Once inside, the nucleic acid codes for the protein units in the capsid. In addition, the virus may have genes for special enzymes needed for the virus to reproduce and exit from the host cell. In large measure, however, a virus relies on the host's enzymes, ribosomes, transfer RNA (tRNA), and ATP (adenosine triphosphate) for its own reproduction.

Reproduction of Bacteriophages

Bacteriophages [Gk. *bacterion,* rod, and *phagein,* to eat], or simply phages, are viruses that parasitize bacteria; the bacterium in Figure 21.3 could be *Escherichia coli,* which lives in our intestines, for example. As the figure shows, there are two types of bacteriophage life cycles, termed the lytic cycle and the lysogenic cycle. In the lytic cycle, viral reproduction occurs, and the host cell undergoes *lysis,* a breaking open of the cell to release viral particles. In the lysogenic cycle, viral reproduction does not immediately occur, but reproduction may take place sometime in the future. The following discussion is based on the DNA bacteriophage lambda. This bacteriophage is considered **virulent** because it undergoes both lytic and lysogenic cycles.

Lytic Cycle. The **lytic cycle** [Gk. *lyo,* loose] may be divided into five stages: attachment, penetration, biosynthesis, maturation, and release. During *attachment,* portions of the capsid combine with a receptor on the rigid bacterial cell wall in a lock-and-key manner. During *penetration,* a viral enzyme digests away part of the cell wall, and viral DNA is injected into the bacterial cell. *Biosynthesis* of viral components begins after the virus brings about inactivation of host genes not necessary to viral replication. The virus takes over the machinery of the cell in order to carry out viral DNA replication and production of multiple copies of the capsid protein subunits. During *maturation,* viral DNA and capsids assemble to produce several hundred viral particles. Lysozyme, an enzyme coded for by a viral gene, is produced; this disrupts the cell wall, and the *release* of new viruses occurs. The bacterial cell dies as a result.

During the lytic cycle, a bacteriophage takes over the machinery of the cell so that viral reproduction and release occur.

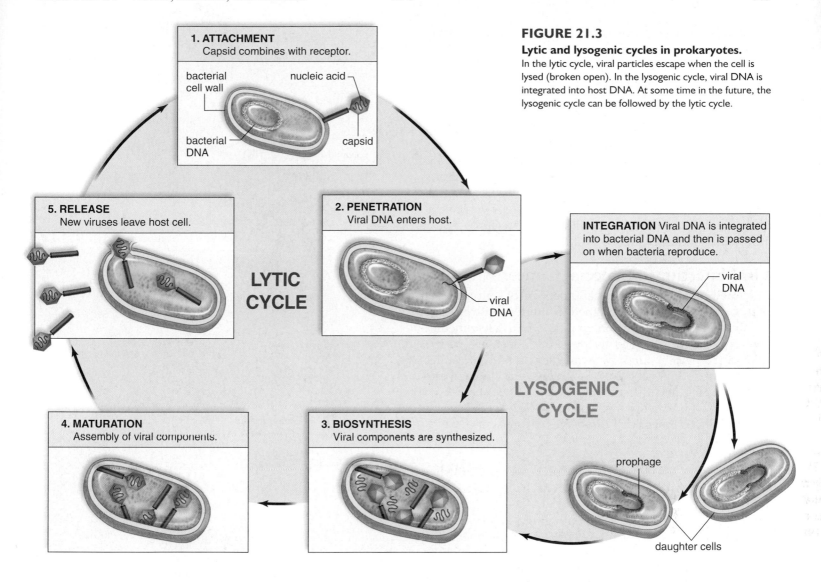

1. ATTACHMENT
Capsid combines with receptor.

bacterial cell wall
nucleic acid
bacterial DNA
capsid

FIGURE 21.3
Lytic and lysogenic cycles in prokaryotes.
In the lytic cycle, viral particles escape when the cell is lysed (broken open). In the lysogenic cycle, viral DNA is integrated into host DNA. At some time in the future, the lysogenic cycle can be followed by the lytic cycle.

5. RELEASE
New viruses leave host cell.

2. PENETRATION
Viral DNA enters host.

viral DNA

INTEGRATION Viral DNA is integrated into bacterial DNA and then is passed on when bacteria reproduce.

viral DNA

LYTIC CYCLE

LYSOGENIC CYCLE

4. MATURATION
Assembly of viral components.

3. BIOSYNTHESIS
Viral components are synthesized.

prophage

daughter cells

Lysogenic Cycle. With the **lysogenic cycle** [Gk. *lyo,* loose, break up, and *genitus,* producing], the infected bacterium does not immediately produce phage but may do so sometime in the future. In the meantime, the phage is *latent*—not actively replicating. Following attachment and penetration, *integration* occurs: viral DNA becomes incorporated into bacterial DNA with no destruction of host DNA. While latent, the viral DNA is called a *prophage.* The prophage is replicated along with the host DNA, and all subsequent cells, called lysogenic cells, carry a copy of the prophage. Certain environmental factors, such as ultraviolet radiation, can induce the prophage to enter the lytic stage of biosynthesis, followed by maturation and release.

During the lysogenic cycle, the phage becomes a prophage that is integrated into the host genome. At a later time, the phage may reenter the lytic cycle. Reproduction and release of the virus then occur.

Reproduction of Animal Viruses

Animal viruses reproduce in a manner similar to that of bacteriophages, but there are modifications. The viral genome covered by the capsid and envelope derived from the host plasma membrane attach and fuse with the host cell. Once inside, the virus is uncoated—that is, the capsid and envelope are removed. The viral genome, either DNA or RNA, is now free of its covering, and biosynthesis plus the other steps then proceed. Viral release occurs by budding. During budding, the virus picks up its envelope consisting of lipids, proteins, and carbohydrates from the plasma membrane. Envelope markers, such as the glycoproteins that allow the virus to enter a host cell, are coded for by viral genes.

After animal viruses enter the host cell, uncoating releases viral DNA or RNA, and reproduction occurs. During release by budding, the viral particle acquires a membranous envelope.

Retroviruses. **Retroviruses** [L. *retro,* backward, and *virus,* poison] are RNA animal viruses that have a DNA stage. Figure 21.4 illustrates the reproduction of a retrovirus— namely, HIV (human immunodeficiency virus), the cause of AIDS. A retrovirus contains a special enzyme called **reverse transcriptase,** which carries out RNA ⟶ cDNA transcription. The DNA is called cDNA because it is a DNA copy of the viral genome. Using host enzymes, a resulting double-stranded DNA is integrated into the host genome. The viral DNA remains in the host genome and is replicated when host DNA is replicated. When and if this DNA is transcribed, new viruses are produced by the steps we have already cited: biosynthesis, maturation, and release—not by destruction of the cell, but by budding.

Viral Infections of Special Concern

All viral infections of humans and pets are of concern, especially because antibiotics designed to interfere with bacterial metabolism have no effect on viral illnesses. A few medications specifically designed to treat viral infections, such as AZT to treat AIDS, are now available. Certain viral infections are of special concern. In humans, papillomaviruses, herpes viruses, hepatitis viruses, adenoviruses, and retroviruses have long been of concern because they lead to specific types of cancer. Emerging viruses are of recent concern.

Emerging Viruses

Some emerging diseases—illness not known previously— are caused by viruses that only recently have caused widespread diseases in humans. These viruses are known as **emerging viruses.** Examples of emerging viral diseases are AIDS, West Nile encephalitis, hantavirus pulmonary syndrome (HPS), severe acute respiratory syndrome (SARS), Ebola hemorrhagic fever, and avian influenza (bird flu).

Several different types of events can cause a viral disease to suddenly "emerge" and start causing a human illness. A virus can extend its range when it is transported from one part of the world to another. West Nile encephalitis is a virus that extended its range after being transported into the United States, where it took hold in bird and mosquito populations. SARS was transported from Southeast Asia to Toronto, Canada. A world in which you can begin your day in Bangkok and end it in Los Angeles is a world in which disease can spread at an unprecedented rate.

Viruses are well known for their high mutation rates. Sometimes viruses that formerly infected other animals and not humans can "jump" species and start infecting humans due to a change in their spikes. For example, AIDS and Ebola hemorrhagic fever are caused by viruses that formerly infected only monkeys and apes. A virus isolated from the palm civet, a catlike carnivore sold for food in China, may be the cause of SARS. Wild ducks are resistant to avian influenza viruses that can spread from them to chickens, which increases the likelihood the disease will spread to humans. Apart from being highly contagious, avian influenza can spread by contaminated equipment and clothing.

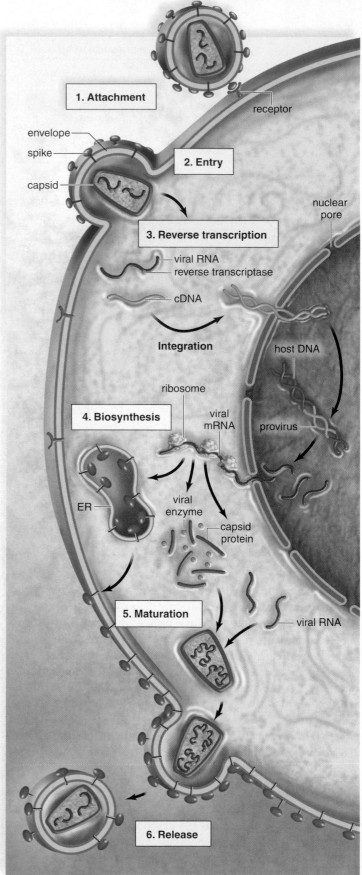

FIGURE 21.4 Reproduction of the retrovirus HIV.
HIV uses reverse transcription to produce cDNA (DNA copy of RNA genes); double-stranded DNA integrates into the cell's chromosomes before the virus reproduces and buds from the cell. The steps in color are unique to retroviruses.

Viroids and Prions

At least a thousand different viruses cause diseases in plants. About a dozen crop diseases have been attributed not to viruses but to **viroids,** which are naked strands of RNA (not covered by a capsid). Like viruses, though, viroids direct the cell to produce more viroids. Since it is difficult to distinguish a plant disease from a mineral deficiency, the policy is to propagate plants known to be healthy and to protect them from infection.

Some diseases in humans have been attributed to **prions,** a term coined for *pro*teinaceous *in*fectious particles. The discovery of prions began when it was observed that members of a primitive tribe in the highlands of Papua New Guinea died from a disease called kuru (meaning trembling with fear) after participating in the cannibalistic practice of eating a deceased person's brain. The causative agent was smaller than a virus—it was a rogue protein. The prion protein is found in healthy brains, although its function is unknown. It appears that illness results when a normal prion protein changes shape so that the polypeptide chain is in a different configuration. The result is a fatal neurodegenerative disorder.

It is believed that a rogue prion can interact with a normal prion protein to change its shape, but the mechanism is unclear. The process has been best studied in a disease called scrapie that attacks sheep. Other prion diseases include the highly publicized mad cow disease, human maladies such as Creutzfeldt-Jakob disease (CJD), and a variety of chronic wasting syndromes in animals.

Viruses, viroids, and prions are all known to cause diseases.

21.2 THE PROKARYOTES

As previously mentioned, the **prokaryotes** include bacteria and archaea, which are fully functioning cells. Because they are microscopic, the prokaryotes were not discovered until the Dutch microscopist Antonie van Leeuwenhoek (1632–1723), better known as the father of the microscope, first described them along with many other microorganisms. Leeuwenhoek and others after him believed that the "little animals" that he observed could arise spontaneously from inanimate matter. For about 200 years, scientists carried out various experiments to determine the origin of microorganisms in laboratory cultures. Finally, in about 1850, Louis Pasteur devised an experiment for the French Academy of Sciences that is described in Figure 21.5. It showed that a previously sterilized broth cannot become cloudy with growth unless it is exposed directly to the air, where bacteria are abundant. Today we know that bacteria are plentiful in air, water, and soil and on most objects. A single spoonful of soil can contain 10^{10} prokaryotes. Clearly, the combined number of all prokaryotes exceeds that of any other type of organism on Earth. In the pages following, the general characteristics of prokaryotes are discussed before those specific to the bacteria (domain Bacteria) and then the archaea (domain Archaea) are considered.

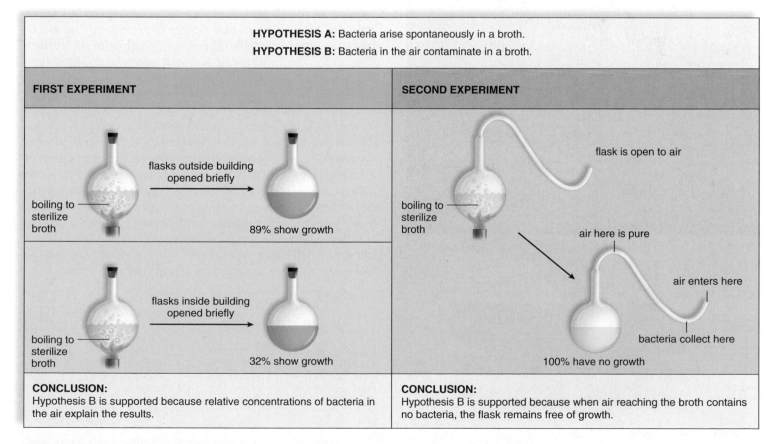

HYPOTHESIS A: Bacteria arise spontaneously in a broth.
HYPOTHESIS B: Bacteria in the air contaminate in a broth.

FIRST EXPERIMENT

flasks outside building opened briefly →
boiling to sterilize broth
89% show growth

flasks inside building opened briefly →
boiling to sterilize broth
32% show growth

CONCLUSION:
Hypothesis B is supported because relative concentrations of bacteria in the air explain the results.

SECOND EXPERIMENT

flask is open to air
boiling to sterilize broth
air here is pure
air enters here
bacteria collect here
100% have no growth

CONCLUSION:
Hypothesis B is supported because when air reaching the broth contains no bacteria, the flask remains free of growth.

FIGURE 21.5 Pasteur's experiment.
Pasteur disproved the theory of spontaneous generation of microbes by performing these types of experiments.

Structure of Prokaryotes

Prokaryotes generally range in size from 1 to 10 μm in length and from 0.7 to 1.5 μm in width. The term *prokaryote* means "before a nucleus," and these organisms lack a eukaryotic nucleus. There are prokaryotic fossils dated as long ago as 3.5 billion years, and the fossil record indicates that the prokaryotes were alone on Earth for at least 1.3 billion years. During that time, they became extremely diverse in structure and especially diverse in metabolic capabilities. Prokaryotes are adapted to living in most environments because the various types differ in the ways they acquire and use energy.

Bacteria have an outer cell wall that is strengthened by the presence of **peptidoglycan,** a complex molecule containing a unique amino disaccharide and peptide fragments. The cell wall prevents them from bursting or collapsing due to osmotic changes. The cell wall may be surrounded by a layer of polysaccharides called a glycocalyx. A well-organized glycocalyx is called a capsule, while a loosely organized one is called a slime layer. In parasitic forms, these outer coverings protect the cell from host defenses.

Some prokaryotes move by means of **flagella** (Fig. 21.6). A flagellum has a filament composed of three strands

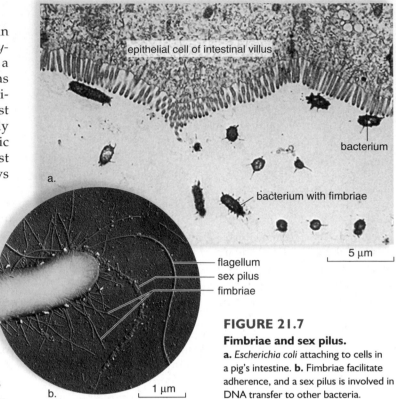

epithelial cell of intestinal villus

bacterium

bacterium with fimbriae

a.

flagellum
sex pilus
fimbriae

5 μm

b.   1 μm

FIGURE 21.7

Fimbriae and sex pilus.

a. *Escherichia coli* attaching to cells in a pig's intestine. **b.** Fimbriae facilitate adherence, and a sex pilus is involved in DNA transfer to other bacteria.

of the protein flagellin wound in a helix. The filament is inserted into a hook that is anchored by a basal body. The 360° rotation of the flagellum causes the cell to spin and move forward. Many prokaryotes adhere to surfaces by means of **fimbriae,** short bristlelike fibers extending from the surface (Fig. 21.7). The fimbriae of *Neisseria gonorrhoeae* allow it to attach to host cells and cause gonorrhea.

A prokaryotic cell lacks the membranous organelles of a eukaryotic cell, and various metabolic pathways are located on the inside of the plasma membrane. Although prokaryotes do not have a nucleus, they do have a dense area called a **nucleoid** where a single chromosome consisting largely of a circular strand of DNA is found. Many prokaryotes also have accessory rings of DNA called **plasmids.** Plasmids can be extracted and used as vectors to carry foreign DNA into host bacteria during genetic engineering processes. Protein synthesis in a prokaryotic cell is carried out by thousands of ribosomes, which are smaller than eukaryotic ribosomes. The following diagram summarizes prokaryotic cell structure:

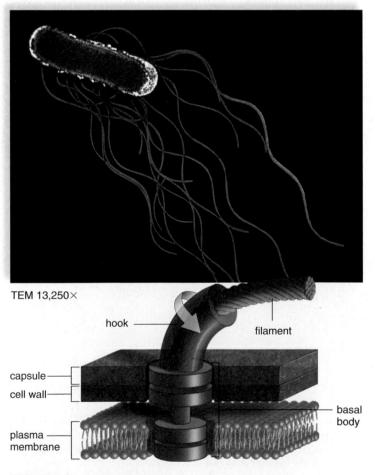

TEM 13,250×

hook

filament

capsule
cell wall

plasma membrane

basal body

FIGURE 21.6 Flagella.

Each flagellum of a bacterium contains a basal body, a hook, and a filament. The arrow indicates that the basal body, hook and filament turn 360°.

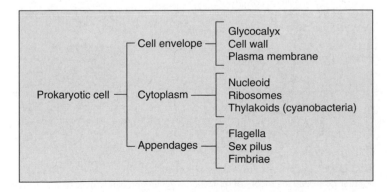

Prokaryotic cell —
- Cell envelope —
  - Glycocalyx
  - Cell wall
  - Plasma membrane
- Cytoplasm —
  - Nucleoid
  - Ribosomes
  - Thylakoids (cyanobacteria)
- Appendages —
  - Flagella
  - Sex pilus
  - Fimbriae

Reproduction in Prokaryotes

Prokaryotes reproduce asexually by means of **binary fission** [L. *binarius*, of two, and *fissura*, cleft, break] (Fig. 21.8), a process that can be diagrammed as follows:

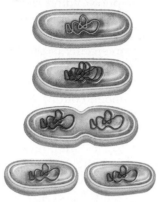

The single circular chromosome replicates, and then two copies separate as the cell enlarges. Newly formed plasma membrane and cell wall separate the cell into two cells. Mitosis, which requires the formation of a spindle apparatus, does not occur in prokaryotes.

Prokaryotes have a generation time as short as 12 minutes under favorable conditions. Mutations are generated and passed on to offspring more quickly than in eukaryotes. Also, prokaryotes are haploid, and so mutations are immediately subjected to natural selection, which determines any possible adaptive benefit.

In eukaryotes, genetic recombination occurs as a result of sexual reproduction. Sexual reproduction does not occur among prokaryotes, but three means of genetic recombination have been observed in bacteria. **Conjugation** occurs between bacteria when the donor cell passes DNA to the recipient cell by way of a sex pilus (see Fig. 21.7*b*), which temporarily joins the two cells. Conjugation takes place only between bacteria in the same or closely related species. **Transformation** occurs when a bacterium picks up (from the surroundings) free pieces of DNA secreted by live prokaryotes or released by dead prokaryotes. During **transduction,** bacteriophages carry portions of bacterial DNA from one cell to another. Plasmids, which sometimes carry genes for resistance to antibiotics, can be transferred between infectious bacteria by any of these ways.

When faced with unfavorable environmental conditions, some bacteria form **endospores** [Gk. *endon,* within, and *spora,* seed] (Fig. 21.9). A portion of the cytoplasm and a copy of the chromosome dehydrate and are then encased by a heavy, protective spore coat. In some bacteria, the rest of the cell deteriorates, and the endospore is released. Spores survive in the harshest of environments—desert heat and dehydration, boiling temperatures, polar ice, and extreme ultraviolet radiation. They also survive for very long periods. When anthrax spores 1,300 years old germinate, they can still cause a severe infection (usually seen in cattle and sheep). Humans also fear a deadly but uncommon type of food poisoning called botulism that is caused by the germination of endospores inside cans of food. To germinate, the endospore absorbs water and grows out of the spore coat. In a few hours' time, it becomes a typical bacterial cell, capable of reproducing once again by binary fission. Spore formation is not a means of reproduction, but it does allow survival and dispersal of bacteria to new places.

Prokaryotes reproduce asexually by binary fission. Mutations are the chief means of achieving genetic variation, but various means of genetic recombination are seen. Spore formation allows survival under unfavorable environmental conditions.

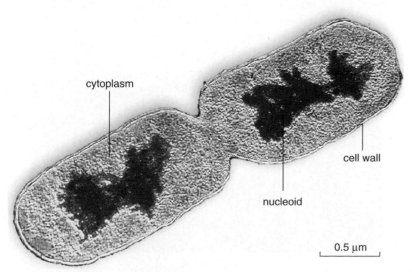

FIGURE 21.8 Binary fission.
When conditions are favorable for growth, prokaryotes divide to reproduce. This is a form of asexual reproduction because the daughter cells have exactly the same genetic material as the parent cell.

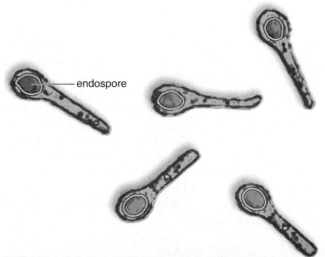

FIGURE 21.9 The endospore of *Clostridium tetani*.
C. tetani produces a terminal spore that causes it to have a drumstick appearance. If spores gain access to a wound, they germinate and release bacteria that produce a neurotoxin. The patient develops tetanus, a progressive rigidity that can result in death; immunization can prevent tetanus.

Prokaryotic Nutrition

With respect to nutrient requirements, prokaryotes are not much different from other organisms. One difference, however, concerns the need for oxygen. Some prokaryotes are **obligate anaerobes** and are unable to grow in the presence of free oxygen. A few serious illnesses—such as botulism, gas gangrene, and tetanus—are caused by anaerobic bacteria. Other prokaryotes, called **facultative anaerobes,** are able to grow in either the presence or the absence of gaseous oxygen. Most prokaryotes, however, are aerobic and, like animals, require a constant supply of oxygen to carry out cellular respiration.

Autotrophic Prokaryotes

Bacteria called **photoautotrophs** [Gk. *photos,* light, *auto,* self, and *trophe,* food] are photosynthetic. They use solar energy to reduce carbon dioxide to organic compounds. There are two types of photoautotrophs: bacteria that evolved first and do not give off oxygen (O_2) and bacteria that evolved later and do give off oxygen. Their characteristics are shown here:

Do Not Give Off O_2	Do Give Off O_2
Photosystem I only	Photosystems I and II
Unique type of chlorophyll called bacteriochlorophyll	Type of chlorophyll *a* found in plants

Green sulfur bacteria and some purple bacteria carry on the first type of photosynthesis. These bacteria do not give off oxygen because they do not use water as an electron donor; instead, they can, for example, use hydrogen sulfide (H_2S).

$$CO_2 + 2\,H_2S \longrightarrow (CH_2O)_n + 2\,S$$

These bacteria usually live in anaerobic conditions, such as the muddy bottom of a marsh, and they cannot photosynthesize in the presence of oxygen. In contrast, the cyanobacteria (see Fig. 21.12) contain chlorophyll *a* and carry on photosynthesis in the second way just as algae and plants do.

$$CO_2 + H_2O \longrightarrow (CH_2O)_n + O_2$$

Prokaryotes called **chemoautotrophs** [Gk. *chemo,* pertaining to chemicals, *auto,* self, and *trophe,* food] carry out chemosynthesis. They oxidize inorganic compounds such as hydrogen gas, hydrogen sulfide, and ammonia to obtain the necessary energy to reduce O_2 to an organic compound. The nitrifying bacteria oxidize ammonia (NH_3) to nitrites (NO_2^-) and nitrites to nitrates (NO_3). Their metabolic abilities keep nitrogen cycling through ecosystems. Other prokaryotes oxidize sulfur compounds at deep-sea vents 2.5 km below sea level. The organic compounds they produce support the growth of a community of organisms found at vents. This discovery lends support to the suggestion that the first cells originated at deep-sea vents.

Heterotrophic Prokaryotes

Prokaryotes called **chemoheterotrophs** [Gk. *chemo,* pertaining to chemicals, *hetero,* different, and *trophe,* food] take in organic nutrients. They are aerobic **saprotrophs** that decompose almost any large organic molecule to smaller ones that can be absorbed. There is probably no natural organic molecule that cannot be digested by at least one prokaryotic species. In ecosystems, saprotrophic bacteria are called decomposers. They play a critical role in recycling matter and making inorganic molecules available to photosynthesizers.

The metabolic capabilities of chemoheterotrophic prokaryotes have long been exploited by human beings. Prokaryotes are used commercially to produce chemicals, such as ethyl alcohol, acetic acid, butyl alcohol, and acetones. Prokaryotic action is also involved in the production of butter, cheese, sauerkraut, rubber, cotton, silk, coffee, and cocoa. Even antibiotics are produced by some bacteria.

Chemoheterotrophs may be free-living or **symbiotic** [Gk. *sym,* together, and *bios,* life], meaning that they form mutualistic, commensalistic, or parasitic relationships. Some mutualistic bacteria are involved in nitrogen cycling. They live in the root nodules of soybean, clover, and alfalfa plants where they reduce atmospheric nitrogen (N_2) to ammonia, a process called nitrogen fixation (Fig. 21.10). Plants are unable to fix atmospheric nitrogen, and those without nodules take up nitrate and ammonia from the soil. Other mutualistic bacteria that live in human intestines release vitamins K and B_{12}, which we can use to help produce blood components. In the stomachs of cows and goats, special mutualistic prokaryotes digest cellulose, enabling these animals to feed on grass.

Commensalism often occurs when one population modifies the environment in such a way that a second population benefits. Obligate anaerobes can live in our intestines only because the bacterium *Escherichia coli* uses up the available oxygen. The parasitic bacteria cause disease,

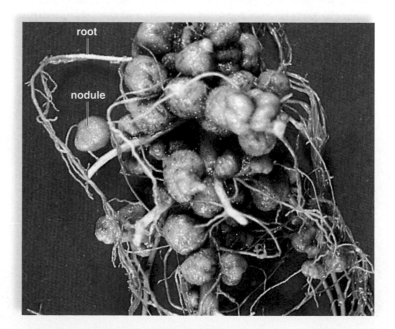

FIGURE 21.10 Nodules of a legume.
While some free-living bacteria carry on nitrogen fixation, those of the genus *Rhizobium* invade the roots of legumes, with the resultant formation of nodules. Here the bacteria convert atmospheric nitrogen to an organic nitrogen that the plant can use. These are nodules on the roots of a soybean plant, *Glycine.*

including human diseases, as discussed in the Health Focus on page 374.

Bacterial Diseases in Humans.   Microbes that can cause disease are called **pathogens.** Pathogens are able to (1) produce a toxin, and/or (2) adhere to surfaces, and sometimes (3) invade organs or cells.

Toxins are small organic molecules, or small pieces of protein or parts of the bacterial cell wall, that are released when bacteria die. Toxins are poisonous, and bacteria that produce a toxin usually cause serious diseases. In almost all cases, the growth of microbes themselves does not cause disease; the toxins they release cause disease. When someone steps on a rusty nail, bacteria can be injected deep into damaged tissue. The damaged area does not have a good blood flow and can become anaerobic. *Clostridium tetani,* the cause of tetanus, proliferates under these conditions. The bacteria never leave the site of the wound, but the tetanus toxin they produce does move throughout the body. This toxin prevents the relaxation of muscles. In time, the body contorts because all the muscles have contracted. Eventually, suffocation occurs.

Adhesion factors allow a pathogen to bind to certain cells, and this determines which organs or cells of the body will be its host. *Shigella dysenteriae* produces a toxin, and it is also able to stick to the intestinal wall, which makes it a more life-threatening cause of dysentery. Also, invasive mechanisms that give a pathogen the ability to move through tissues and into the bloodstream result in a more medically significant disease than if it were localized. Usually a person can recover from food poisoning caused by *Salmonella.* But some strains of *Salmonella* have virulence factors that allow the bacteria to penetrate the lining of the colon and move beyond this organ. Typhoid fever, a life-threatening disease, can then result.

Because bacteria are cells in their own right, a number of antibiotic compounds are active against bacteria and are widely prescribed. Most antibacterial compounds fall within two classes, those that inhibit protein biosynthesis and those that inhibit cell wall biosynthesis. Erythromycin and tetracyclines can inhibit bacterial protein synthesis because bacterial ribosomes function somewhat differently than eukaryotic ribosomes. Cell wall biosynthesis inhibitors generally block the formation of peptidoglycan, required to maintain bacterial integrity. Penicillin, ampicillin, and fluoroquinolone (like Cipro) inhibit bacterial cell wall biosynthesis without harming animal cells.

One problem with antibiotic therapy has been increasing bacterial resistance to antibiotics. When penicillin was first introduced, less than 3% of *Staphylococcus aureus* strains were resistant to it. Now, due to selective advantage, 90% or more are resistant.

21.3 THE BACTERIA

Bacteria (domain Bacteria) are the more common type of prokaryote. Most bacterial cells are protected by a cell wall that contains the unique molecule peptidoglycan. Groups of bacteria are commonly differentiated from one another by using the Gram stain procedure, which was developed in the late 1880s by Hans Christian Gram, a Danish bacteriologist. Gram-positive bacteria retain a dye-iodine complex and appear purple under the light microscope, while Gram-negative bacteria do not retain the complex and appear pink. This difference is dependent on the construction of the cell wall; that is, the Gram-positive bacteria have a thick layer of peptidoglycan in their cell walls, whereas Gram-negative bacteria have only a thin layer. *Clostridium tetani,* which causes tetanus, is an example of a Gram-positive bacterium, and *Vibrio cholerae,* which causes cholera, is an example of a Gram-negative bacterium.

Bacteria (and archaea) can also be classified in terms of their three basic shapes (Fig. 21.11): spirillum (pl., spirilli), spiral-shaped or helical-shaped; bacilli (sing., bacillus), rod-shaped; and cocci (sing., coccus), round or spherical. These three basic shapes may be augmented by particular arrangements or shapes of cells. For example, rod-shaped prokaryotes may appear as very short rods (coccobacilli) or as very long filaments (fusiform). Cocci may form clusters (staphylococci, diplococci) or chains (streptococci).

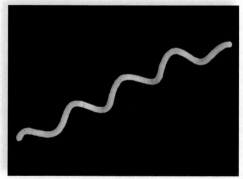
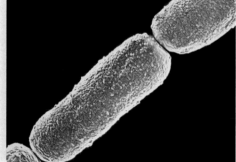

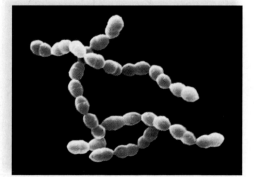

a. Spirillum: SEM 3,520×
   Spirillum volutans

b. Bacilli: SEM 35,000×
   Bacillus anthracis

c. Cocci: SEM 6,250×
   Streptococcus thermophilus

FIGURE 21.11 Diversity of bacteria.
a. Spirillum, a spiral-shaped bacterium. **b.** Bacilli, rod-shaped bacteria. **c.** Cocci, round bacteria.
a. ©Dr. Richard Kessel & Dr. Gene Shih/Visuals Unlimited.

DOMAIN: Bacteria

CHARACTERISTICS
- Prokaryotic
- Unicellular organisms that lack a membrane-bounded nucleus
- Reproduce asexually
- Metabolically diverse; many being heterotrophic by absorption, others being autotrophic by chemosynthesis or by photosynthesis
- Motile forms move by flagella consisting of a single filament

Gram-Negative Bacteria*

KINGDOM: Cyanobacteria** ——————— Prochloron, Oscillatoria, Anabaena

KINGDOM: Proteobacteria** ——————— α: Rhodospirillum, Rickettsia, Caulobacter, Rhizobium, Nitrobacter
β: Neisseria, Burkholderia
γ: Legionella, Pseudomonas, Vibrio, Salmonella, Shigella
δ: Bdellovibrio
ε: Campylobacter

KINGDOM: Spirochetes** ——————— Borrelia, Leptospira

KINGDOM: Bacteriodetes** ——————— Bacteroides, Porphyromonas

Gram-Positive Bacteria*

KINGDOM: Deinococci** ——————— Deinococcus

KINGDOM: Firmicutes** ——————— Clostridium, Mycoplasma, Streptococcus, Enterococcus, Listeria, Staphylococcus

KINGDOM: Actinobacteria** ——————— Actinomyces, Corynebacterium, Mycobacterium, Nocardia

Other*

KINGDOM: Aquificae** ——————— Aquifex

KINGDOM: Chloroflexi** ——————— Chloroflexus, Herpetosiphon, Chlorobium, Pelodictyon

KINGDOM: Planctomycetes** ——————— Planctomyces

*Not a proposed grouping but added here for clarity
**Proposed kingdom.

DOMAIN: Archaea

CHARACTERISTICS
- Prokaryotic
- Unicellular organisms that lack a membrane-bounded nucleus
- Reproduce asexually
- Metabolically diverse; many being autotrophic by chemosynthesis and a few by photosynthesis, some being heterotrophic by absorption
- Distinguish from bacteria by their unique rRNA base sequence and their distinctive plasma membrane and cell wall chemistry

KINGDOM: Euryarchaeota** ——————— Halobacterium, Methanobacterium, Picrophilus, Thermococcus

KINGDOM: Crenarchaeota** ——————— Pyrodictium, Pyrolobus, Pyrobaculum

KINGDOM: Korarchaeota** ——————— Caldisphaerales, Cenarchaeales, Desulfolobales, Thermoproteates

**Proposed kingdom.

Other traditional criteria for the classification of bacteria, aside from Gram stain and shape and arrangement, are presence of endospores, metabolism, growth, nutritional characteristics, and other physiological characteristics. For the past 75 years, bacterial taxonomy has been compiled in *Bergey's Manual of Determinative Bacteriology*. A recent edition of *Bergey's Manual* divided the prokaryotes into 19 major groups, which were further subdivided into orders, families, genera, and species. The names of the groups—for example, "nonmotile Gram-negative curved bacteria" or "nonspore-forming Gram-positive rods"—reflect the phenotypic bias used to group the bacteria. However, the newest edition classifies prokaryotes according to the most recent information on their molecular genetic relatedness.

Since the 1980s, Carl Woese and other researchers have pioneered a new mode of bacterial taxonomy based on the phylogenetic comparisons of bacterial 16S ribosomal RNA sequences. The kingdoms listed in the box above are based on such data. Some of the new kingdoms, such as the spirochetes, are essentially identical to early classification systems. However, other groups contain a diverse assortment of bacteria that appear to be physiologically distant, but nevertheless share common ribosomal RNA sequences. For example, the proteobacteria are a phenotypically diverse group of Gram-negative bacteria with many nutritional types. In spite of their obvious phenotypic differences, these seemingly diverse bacterial types are genetically related to one another. Further genetic studies may explain these phenotypic differences.

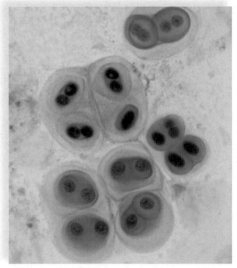

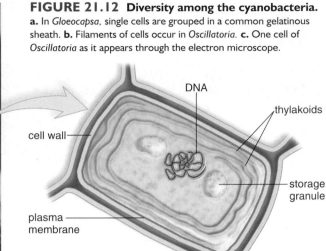

FIGURE 21.12 Diversity among the cyanobacteria.
a. In *Gloeocapsa*, single cells are grouped in a common gelatinous sheath. **b.** Filaments of cells occur in *Oscillatoria*. **c.** One cell of *Oscillatoria* as it appears through the electron microscope.

DNA

thylakoids

cell wall

storage granule

plasma membrane

a. *Gloeocapsa*   LM 250×

b. *Oscillatoria*   LM 100×

c. *Oscillatoria* cell

Cyanobacteria

Cyanobacteria [Gk. *kyanos,* blue, and *bacterion,* rod] are Gram-negative bacteria with a number of unusual traits. They photosynthesize in the same manner as plants and are believed to be responsible for first introducing oxygen into the primitive atmosphere. Formerly, the cyanobacteria were called blue-green algae and were classified with eukaryotic algae, but now they are classified as prokaryotes. Cyanobacteria can have other pigments that mask the color of chlorophyll so that they appear red, yellow, brown, or black, rather than only blue-green.

Cyanobacterial cells are rather large, ranging from 1 to 50 µm in width. They can be unicellular, colonial, or filamentous. Cyanobacteria lack any visible means of locomotion, although some glide when in contact with a solid surface and others oscillate (sway back and forth). Some cyanobacteria have a special advantage because they possess heterocysts, which are thick-walled cells without nuclei, where nitrogen fixation occurs. The ability to photosynthesize and also to fix atmospheric nitrogen (N_2) means that their nutritional requirements are minimal. They can serve as food for heterotrophs in ecosystems.

Cyanobacteria (Fig. 21.12) are common in fresh and marine waters, in soil, and on moist surfaces, but they are also found in harsh habitats, such as hot springs. They are symbiotic with a number of organisms, including liverworts, ferns, and even at times invertebrates such as corals. In association with fungi, they form **lichens** that can grow on rocks. A lichen is a symbiotic relationship in which the cyanobacterium provides organic nutrients to the fungus, while the fungus possibly protects and furnishes inorganic nutrients to the cyanobacterium. It is also possible that the fungus is parasitic on the cyanobacterium. Lichens help transform rocks into soil; other forms of life then may follow. It is hypothesized that cyanobacteria were the first colonizers of land during the course of evolution.

Cyanobacteria are ecologically important in still another way. If care is not taken in disposing of industrial, agricultural, and human wastes, phosphates drain into lakes and ponds, resulting in a "bloom" of these organisms. The surface of the water becomes turbid, and light cannot penetrate to lower levels. When a portion of the cyanobacteria die off, the decomposing prokaryotes use up the available oxygen, causing fish to die from lack of oxygen.

Cyanobacteria are photosynthesizers that sometimes can also fix atmospheric nitrogen. In association with fungi, they form lichens, which contribute to soil formation.

21.4 THE ARCHAEA

Archaea (domain Archaea) are prokaryotes with biochemical characteristics that distinguish them from both bacteria and eukaryotes.

Relationship to Domain Bacteria and Domain Eukarya

Archaea used to be considered bacteria until Carl Woese discovered that their rRNA has a different sequence of bases than the rRNA of bacteria. He chose rRNA because of its involvement in protein synthesis—any changes in rRNA sequence probably occur at a slow, steady pace as evolution occurs. As discussed in Chapter 20, it is proposed that the tree of life contains three domains: Archaea, Bacteria, and Eukarya. Because archaea and some bacteria are found in extreme environments (hot springs, thermal vents, salt basins), they may have diverged from a common ancestor relatively soon after life began. Then later, the eukarya are believed to have split off from the archaeal line of descent. In other words, the eukarya are believed to be more closely related to the archaea than to the bacteria. Archaea and eukarya share some of the same ribosomal proteins (not found in bacteria), initiate transcription in the same manner, and have similar types of tRNA.

health focus

Pathogens as Weapons

Pathogens are viruses and bacteria that cause diseases in humans (Tables 21A and 21B). Some pathogens are more likely than others to be used by terrorists as biological weapons. Such pathogens are cheap and easy to acquire. They disseminate easily—most are airborne—and they have the potential to kill thousands of people within a relatively short period of time. Although anthrax and smallpox are considered the most likely biological agents of mass destruction, other diseases are also possible weapons.

Anthrax, caused by the bacterium *Bacillus anthracis*, is at the top of the list as a possible bioterrorist agent because it is the easiest to acquire and grow. Anthrax occurs in three forms. Inhalation anthrax is far more serious than cutaneous and intestinal anthrax. Inhalation anthrax causes flulike symptoms, and if the correct diagnosis is delayed, the person will most likely die within 24–72 hours from the effect of the toxins the bacterium produces. Thankfully, anthrax is not contagious. The outbreak of anthrax infections in the fall of 2001 in the United States could be controlled because a limited number of people were involved, and enough antibiotic was available to treat all those who were exposed.

Smallpox, caused by the variola virus, is far more dangerous than anthrax because the disease is highly contagious. As soon as a case of smallpox is identified, everyone who has had contact with the patient should be vaccinated immediately. The virus is airborne and causes a fever, headache, and malaise some 12 days after exposure. The rash that develops can be mistaken for chickenpox. Only after the pocks—pus-filled blisters—are full-blown would most physicians be likely to consider a diagnosis of smallpox. Many physicians have never seen smallpox because the disease was eliminated

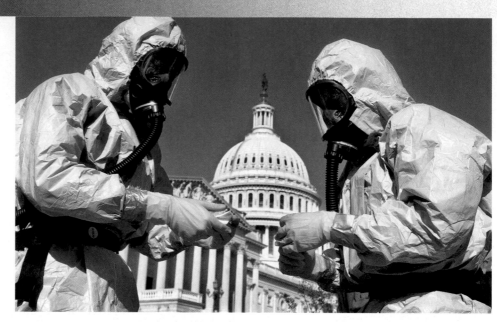

FIGURE 21A Bioterrorism threatens our health safety.

prior to 1980 through consistent vaccination of young people. So vaccination was stopped, and immunity has most likely worn off by this time.

Plague, caused by the bacterium *Yersinia pestis*, occurs in two forms. Bubonic plague is transmitted by the bite of an infected flea, while pneumonic plague is an inhaled form. Terrorists would most likely favor the inhaled form because only immediate treatment can prevent respiratory failure and, most likely, death. Pneumonic plague is not as contagious as smallpox.

Botulism, caused by the toxins of *Clostridium botulinum*, is typically foodborne. Ingesting just a tiny amount can lead to blurred vision and difficulty in swallowing and speaking within 24 hours. Sufficient toxin can cause respiratory failure within 24 hours unless a respirator is immediately available.

Tularemia is caused by a hardy bacterium called *Francisella tularensis*. This disease has several forms, and again terrorists would most likely prefer the inhaled, or typhoidal, variety.

This type causes fever, chills, headache, and general weakness, as well as chest pains, weight loss, and a cough. Survival is dependent on a rapid and accurate diagnosis and the immediate administration of the correct antibiotic.

Hemorrhagic fevers are characterized by high fever and severe bleeding from multiple organs. Infections with the Marburg virus and the Ebola virus are known to cause hemorrhagic fevers, and they, too, might be used as biological weapons.

The bacterial diseases—anthrax, plague, botulism, and tularemia—respond to antibiotics that poison bacterial enzymes. Smallpox, a viral disease, can be cured with the timely administration of a vaccine, but hemorrhagic fevers respond only to antiviral drugs, which are in short supply.

It is important for all of us to promote the work of government agencies and civic groups that are preparing to meet the challenge of a possible bioterrorist attack (Fig. 21A).

Structure and Function

The plasma membranes of archaea contain unusual lipids that allow them to function at high temperatures. The lipids of archaea contain glycerol linked to branched-chain hydrocarbons in contrast to the lipids of bacteria, which contain glycerol linked to fatty acids. The archaea also evolved diverse cell wall types, which facilitate their survival under extreme

conditions. The cell walls of archaea do not contain peptidoglycan as do the cell walls of bacteria. In some archaea, the cell wall is largely composed of polysaccharides, and in others, the wall is pure protein. A few have no cell wall.

Metabolically, the archaea have retained primitive and unique forms of metabolism. Methanogenesis, the ability to form methane, is one type of metabolism that is performed only by some archaea, called methanogens.

TABLE 21A

Viral Diseases in Humans

Category	Disease
Sexually transmitted diseases	AIDS (HIV), genital warts, genital herpes
Childhood diseases	Mumps, measles, chickenpox, German measles
Respiratory diseases	Common cold, influenza, severe acute respiratory infection (SARS)
Skin diseases	Warts, fever blisters, shingles
Digestive tract diseases	Gastroenteritis, diarrhea
Nervous system diseases	Poliomyelitis, rabies, encephalitis
Other diseases	Smallpox, hemorrhagic fevers, cancer, hepatitis, mononucleosis, yellow fever, dengue fever, conjunctivitis, hepatitis C

TABLE 21B

Bacterial Diseases in Humans

Category	Disease
Sexually transmitted diseases	Syphilis, gonorrhea, chlamydia
Respiratory diseases	Strep throat, scarlet fever, tuberculosis, pneumonia, Legionnaires' disease, whooping cough, inhalation anthrax
Skin diseases	Erysipelas, boils, carbuncles, impetigo, acne, infections of surgical or accidental wounds and burns, leprosy (Hansen disease)
Digestive tract diseases	Gastroenteritis, food poisoning, dysentery, cholera, peptic ulcers, dental caries
Nervous system diseases	Botulism, tetanus, leprosy, spinal meningitis
Systemic diseases	Plague, typhoid fever, diphtheria
Other diseases	Tularemia, Lyme disease

Most archaea are chemoautotrophs (see page 370) and few are photosynthetic. This suggests that chemoautotrophy predated photoautotrophy during the evolution of prokaryotes.

Archaea are sometimes mutualistic or even commensalistic, but there are no parasitic archaea—that is, they are not known to cause infectious diseases.

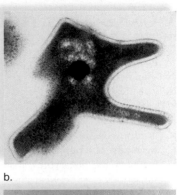

FIGURE 21.13 Thermoacidophile habitat and structure.
a. Boiling springs and geysers in Yellowstone National Park where thermoacidophiles live. **b.** Transmission electron micrograph of *Sulfolobus acidocaldarius*, a thermoacidophile. The dark central region has yet to be identified and could be an artifact of staining. **c.** Transmission electron micrograph of *Thermoproteus tenax*, also a thermoacidophile.

Types of Archaea

Archaea are often discussed in terms of their unique habitats. The **methanogens** (methane makers) are found in anaerobic environments in swamps, marshes, and the intestinal tracts of animals. They use hydrogen gas (H_2) to reduce carbon dioxide (CO_2) to methane and couple the energy released to ATP production. Methane, also called biogas, is released into the atmosphere, where it contributes to the greenhouse effect and global warming. About 65% of the methane found in our atmosphere is produced by these methanogenic archaea.

The **halophiles** require high salt concentrations (usually 12–15%—the ocean is about 3.5% salt) for growth. They have been isolated from highly saline environments such as the Great Salt Lake in Utah, the Dead Sea, solar salt ponds, and hypersaline soils. These archaea have evolved a number of mechanisms to survive in environments that are high in salt. Their proteins have unique chloride pumps that use halorhodopsin (related to the rhodopsin pigment found in our eyes) to pump chloride to the inside of the cell and this prevents water loss. They also use another pigment, bacteriorhodopsin, to synthesize ATP in the presence of light.

A third major type of archaea are the **thermoacidophiles** (Fig. 21.13). These archaea are isolated from extremely hot, acidic environments such as hot springs, geysers, submarine thermal vents, and around volcanoes. They reduce sulfides and survive best at temperatures above 80°C; some can even grow at 105°C (remember that water boils at 100°C)! Metabolism of sulfides results in acidic sulfates; these bacteria grow best at pH 1–2.

Archaea are able to live in extremely stressful environments, perhaps like those of the ancestral Earth. Their biochemical properties suggest that they are more closely related to eukaryotes than to bacteria.

CONNECTING THE CONCEPTS

Microbiology began when Leeuwenhoek first used his microscope to observe microorganisms. Significant advances occurred with the discovery that bacteria and viruses cause disease, and again when microbes were first used in genetic studies. Although there are significant structural differences between prokaryotes and eukaryotes, many biochemical similarities exist between the two. Thus, the details of protein synthesis, first worked out in bacteria, are applicable to all cells, including human cells. Today, transgenic bacteria routinely make products and otherwise serve the needs of human beings.

Many prokaryotes can live in environments that may represent the kinds of habitats available when the Earth first formed. We find prokaryotes in such hostile habitats as swamps, the Dead Sea, and hot sulfur springs. The fossil record suggests that the prokaryotes evolved before the eukaryotes. Not only do all living things trace their ancestry to the prokaryotes, but prokaryotes are believed to have contributed to the evolution of the eukaryotic cell. The mitochondria and chloroplasts of the eukaryotic cell are probably derived from bacteria that took up residence inside a nucleated cell.

Cyanobacteria are believed to have introduced oxygen into the Earth's ancestral atmosphere, and they may have been the first colonizers of the terrestrial environment. Some bacteria are decomposers that recycle nutrients in both aquatic and terrestrial environments. Bacteria play significant roles in the carbon, nitrogen, and phosphorus cycles. Mutualistic bacteria also fix nitrogen in plant nodules, enabling herbivores to digest cellulose, and release certain vitamins in the human intestine. Clearly, humans are dependent on the past and present activities of prokaryotes.

Summary

21.1 VIRUSES, VIROIDS, AND PRIONS

Viruses are noncellular, while prokaryotes are fully functioning organisms. All viruses have at least two parts: an outer capsid composed of protein subunits and an inner core of nucleic acid, either DNA or RNA but not both. Some also have an outer membranous envelope.

Viruses are obligate intracellular parasites that can be maintained only inside living cells, such as those of a chicken egg or those propagated in cell (tissue) culture.

The lytic cycle of a bacteriophage consists of attachment, penetration, biosynthesis, maturation, and release. In the lysogenic cycle of a bacteriophage, viral DNA is integrated into bacterial DNA for an indefinite period of time, but it can undergo the lytic cycle when stimulated.

The reproductive cycle differs for animal viruses. Uncoating is needed to free the genome from the capsid, and budding releases the viral particles from the cell. RNA retroviruses have an enzyme, reverse transcriptase, that carries out reverse transcription. This produces cDNA, which becomes integrated into host DNA. The AIDS virus is a retrovirus.

Viruses cause various diseases in plants and animals, including human beings. Viroids are naked strands of RNA (not covered by a capsid) that can cause disease in plants. Prions are protein molecules that have a misshapen tertiary structure. Prions cause diseases such as CJD in humans and mad cow disease in cattle when they cause other proteins of their own kind also to become misshapen.

21.2 THE PROKARYOTES

The bacteria (domain Bacteria) and archaea (domain Archaea) are prokaryotes. Prokaryotic cells lack a nucleus and most of the other cytoplasmic organelles found in eukaryotic cells. Prokaryotes reproduce asexually by binary fission. Their chief method for achieving genetic variation is mutation, but genetic recombination by means of conjugation, transformation, and transduction has been observed in bacteria. Some bacteria form endospores, which are extremely resistant to destruction; the genetic material can thereby survive unfavorable conditions.

Prokaryotes differ in their need (and tolerance) for oxygen. There are obligate anaerobes, facultative anaerobes, and aerobic prokaryotes.

Some prokaryotes are autotrophic—either photoautotrophs (photosynthetic) or chemoautotrophs (chemosynthetic). Some photosynthetic bacteria (cyanobacteria) give off oxygen, and some (purple and green sulfur bacteria) do not. Chemoautotrophs oxidize inorganic compounds such as hydrogen gas, hydrogen sulfide, and ammonia to acquire energy to make their own food. Surprisingly, chemoautotrophs support communities at deep-sea vents.

Many prokaryotes are chemoheterotrophs (aerobic heterotrophs) and are saprotrophic decomposers that are absolutely essential to the cycling of nutrients in ecosystems. Their metabolic capabilities are so vast that they are used by humans both to dispose of and to produce substances. Many heterotrophic prokaryotes are symbiotic. The mutualistic nitrogen-fixing bacteria live in nodules on the roots of legumes. Some symbionts, however, are parasitic and cause plant and animal, including human, diseases.

21.3 THE BACTERIA

Bacteria (domain Bacteria) are the more prevalent type of prokaryote. The classification of bacteria is still being developed. Of primary importance at this time are the shape of the cell and the structure of the cell wall, which affects Gram staining. Bacteria occur in three basic shapes: spiral-shaped (spirillum), rod-shaped (bacillus), and round (coccus). Of special interest are the cyanobacteria, which were the first organisms to photosynthesize in the same manner as plants. When cyanobacteria are symbionts with fungi, they form lichens.

21.4 THE ARCHAEA

The archaea (domain Archaea) are a second type of prokaryote. On the basis of rRNA sequencing, it is believed that there are three evolutionary domains: Bacteria, Archaea, and Eukarya. In addition, the archaea appear to be more closely related to the eukarya than to the bacteria. Archaea do not have peptidoglycan in their cell walls, as do the bacteria, and they share more biochemical characteristics with the eukarya than do bacteria.

The three types of archaea live under harsh conditions, such as anaerobic marshes (methanogens), salty lakes (halophiles), and hot sulfur springs (thermoacidophiles).

Reviewing the Chapter

1. Describe the general structure of viruses, and describe both the lytic cycle and the lysogenic cycle of bacteriophages.   362–65
2. Contrast viruses with prokaryotes in terms of the characteristics of life.   363
3. How do animal viruses differ in structure and reproductive cycle from bacteriophages?   365
4. How do retroviruses differ from other animal viruses? Describe the reproductive cycle of retroviruses in detail.   366
5. Explain Pasteur's experiment, which showed that bacteria do not arise spontaneously.   367
6. Describe the general structure of prokaryotes, and tell how they reproduce.   368–69
7. How do all prokaryotes introduce variations? How does genetic recombination occur in bacteria?   369
8. How do prokaryotes differ in their tolerance of and need for oxygen?   370
9. Compare photosynthesis between the green sulfur bacteria and the cyanobacteria.   370, 373
10. What are chemoautotrophic prokaryotes, and where have they been found to support whole communities?   370
11. Discuss the importance of cyanobacteria in ecosystems and in the history of the Earth.   373
12. How do the archaea differ from the bacteria?   373–74

Testing Yourself

Choose the best answer for each question.

1. Label this condensed version of bacteriophage reproductive cycles using these terms: viral DNA, penetration, maturation, release, prophage, attachment, biosynthesis, and integration.

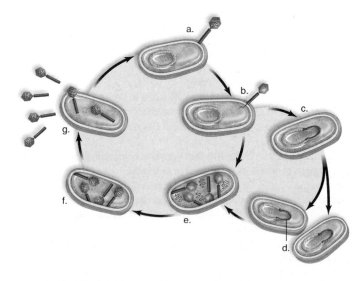

2. Viruses are considered nonliving because
   a. they do not locomote.
   b. they cannot reproduce independently.
   c. their nucleic acid does not code for protein.
   d. they are noncellular.
   e. Both b and d are correct.

3. A virus is a tiny, infectious
   a. cell.
   b. living thing.
   c. particle.
   d. nucleic acid.

4. The envelope of an animal virus is derived from the
   ———————————— of its host cell.
   a. cell wall
   b. plasma membrane
   c. glycocalyx
   d. receptors

5. A prophage occurs during the development of
   a. a lysogenic virus.
   b. a poxvirus.
   c. a lytic virus.
   d. an enveloped virus.

6. Which of these are found in all viruses?
   a. envelope, nucleic acid, capsid
   b. DNA, RNA, and proteins
   c. proteins and a nucleic acid
   d. proteins, nucleic acids, carbohydrates, and lipids
   e. tail fibers, spikes, and rod shape

7. Viruses cannot be cultivated in
   a. tissue culture.
   b. bird embryos.
   c. live mammals.
   d. blood agar.

8. A pathogen would most accurately be described as
   a. a parasite.
   b. a commensal.
   c. a saprobe.
   d. a symbiont.

9. RNA retroviruses have a special enzyme that
   a. disintegrates host DNA.
   b. polymerizes host DNA.
   c. transcribes viral RNA to cDNA.
   d. translates host DNA.
   e. produces capsid proteins.

10. Which is not true of prokaryotes? They
    a. are living cells.
    b. lack a nucleus.
    c. all are parasitic.
    d. are both archaea and bacteria.
    e. evolved early in the history of life.

11. Facultative anaerobes
    a. require a constant supply of oxygen.
    b. are killed in an oxygenated environment.
    c. do not always need oxygen.
    d. are photosynthetic but do not give off oxygen.
    e. All of these are correct.

12. Which of these is most apt to be a bacterial cell wall function?
    a. transport
    b. motility
    c. support
    d. adhesion

13. Cyanobacteria, unlike other types of bacteria that photosynthesize, do
    a. give off oxygen.
    b. not have chlorophyll.
    c. not have a cell wall.
    d. need a fungal partner.

14. Chemoautotrophic prokaryotes
    a. are chemosynthetic.
    b. use the rays of the sun to acquire energy.
    c. oxidize inorganic compounds to acquire energy.
    d. are always bacteria, not archaea.
    e. Both a and c are correct.

15. Archaea differ from bacteria in that they
    a. can form methanogen.
    b. have different rRNA sequences.
    c. do not have peptidoglycan in their cell walls.
    d. rarely photosynthesize.
    e. All of these are correct.

16. Which of these archaea would live at a deep-sea vent?
    a. thermoacidophile
    b. halophile
    c. methanogen
    d. parasitic forms
    e. All of these are correct.

17. Chemosynthetic bacteria
    a. are autotrophic.
    b. use the rays of the sun to acquire energy.
    c. oxidize inorganic compounds to acquire energy.
    d. Both a and c are correct.
    e. Both a and b are correct.

18. A bacterium that can synthesize all its required organic components from CO_2 using energy from the sun is
    a. a photoautotroph.
    b. a photoheterotroph.
    c. a chemoautotroph.
    d. a chemoheterotroph.

19. Bacterial endospores function in
    a. reproduction.
    b. survival.
    c. protein synthesis.
    d. storage.

Thinking Scientifically

1. While a few drugs are effective against some viruses, they often impair the function of body cells and thereby have a number of side effects. Most antibiotics (antibacterial drugs) do not cause side effects. Why would antiviral medications be more likely to produce side effects?
2. *Escherichia coli* is a model organism for modern geneticists. What bacterial characteristics make *E. coli* particularly useful in genetic experiments?

Bioethical Issue: Identifying Carriers

Carriers of disease are persons who do not appear to be ill but can nonetheless pass on an infectious disease. The only way society can protect itself is to identify carriers and remove them from areas or activities where transmission of the pathogen is most likely. Sometimes it's difficult to identify all activities that might pass on a pathogen—for example, HIV. A few people

believe that they have acquired HIV from their dentists, and while this is generally believed to be unlikely, medical personnel are still required to identify themselves when they are carriers of HIV.

Transmission of HIV is believed to be possible in certain sports. In a statistical study, the Centers for Disease Control figured that the odds of acquiring HIV from another football player were 1 in 85 million. But the odds might be higher for boxing, a bloody sport. When two brothers, one of whom had AIDS, got into a vicious fight, the infected brother repeatedly bashed his head against his brother's. Both men bled profusely, and soon after, the previously uninfected brother tested positive for the virus. The possibility of transmission of HIV in the boxing ring has caused several states to require boxers to undergo routine HIV testing. If they are HIV positive, they can't fight.

Should all people who are HIV positive always be required to identify themselves, no matter what the activity? Why or why not? By what method would they identify themselves at school, at work, and in other places?

Understanding the Terms

archaea 373
bacteria 371
bacteriophage 364
binary fission 369
capsid 363
chemoautotroph 370
chemoheterotroph 370
conjugation 369
cyanobacteria 373
emerging virus 366
endospore 369
facultative anaerobe 370
fimbriae 368
flagellum (pl., flagella) 368
halophile 375
host specific 364
lichen 373
lysogenic cycle 365
lytic cycle 364
methanogen 375
nucleoid 368
obligate anaerobe 370
pathogen 371
peptidoglycan 368
photoautotroph 370
plasmid 368
prion 367
prokaryote 367
retrovirus 366
reverse transcriptase 366
saprotroph 370
symbiotic 370
thermoacidophile 375
transduction 369
transformation 369
viroid 367
virulent 364
virus 362

Match the terms to these definitions:
a. _____ Bacteriophage life cycle in which the virus incorporates its DNA into that of the bacterium; only later does it begin a lytic cycle, which ends with the destruction of the bacterium.
b. _____ Organism that contains chlorophyll and uses solar energy to produce its own organic nutrients.
c. _____ Organism that secretes digestive enzymes and absorbs the resulting nutrients back across the plasma membrane.
d. _____ Relationship that could be mutualistic, commensalistic, or parasitic.
e. _____ Type of prokaryote that is most closely related to the Eukarya.

ARIS, the *Biology* Website

ARIS, the website for *Biology*, provides a wealth of information organized and integrated by chapter. You will find practice quizzes, interactive activities, labeling exercises, flashcards, and much more that will complement your learning and understanding of general biology.

www.mhhe.com/maderbiology9

22

THE PROTISTS

*O**f the four kingdoms of domain Eukarya, kingdom Protista is the most biologically diverse. In fact, many systematists consider kingdom Protista to be a "catch-all" kingdom comprised of a number of artificially but conveniently grouped organisms. The protists include an immense assortment of living things that vary in their body form, lifestyle, nutrition, locomotion, and reproduction. Today, nearly 100,000 living and extinct species of protists have been described, including those that are autotrophic as are plants, heterotrophic as are animals, and saprotrophic as are fungi.*

Although algae are often overlooked, they are environmentally and industrially important organisms. Like plants, algae, such as diatoms, are responsible for forming copious amounts of oxygen and serve as the producers in many food chains. Industrially, algin, a product made by seaweeds, contributes to the production of foods, pharmaceuticals, cosmetics, paper, textiles, and even the tungsten filaments of lightbulbs.

At one time botanists classified algae as plants, and on the basis of molecular data, some are again thinking that at least green algae should be classified as plants. Similarly, zoologists used to classify protozoans as animals. Protozoans live in a variety of moist environments, from the ocean to the gut of termites. In aquatic ecosystems, protozoans serve as food for animals, but some are parasitic on animals. Malaria is a medically and economically important protozoan illness of humans.

Like fungi, the slime molds and the water molds play an important role in the recycling of nutrients in ecosystems. The majority of water molds are saprotrophic, as are the fungi, but the slime molds ingest their food by phagocytosis.

Diatoms, photosynthetic protists of the oceans.

300 μm

22.1 GENERAL BIOLOGY OF PROTISTS

Protists (domain Eukarya, kingdom Protista) have eukaryotic cells. The endosymbiotic hypothesis suggests how mitochondria and chloroplasts eventually became part of the eukaryotic cell. Mitochondria may be derived from aerobic bacteria and chloroplasts may be derived from cyanobacteria that were engulfed by a eukaryotic cell on two separate occasions (Fig. 22.1a, b). The protist *Giardia lamblia* possesses two nuclei but no mitochondria, suggesting that a nucleated cell preceded the acquisition of mitochondria (Fig. 22.1c).

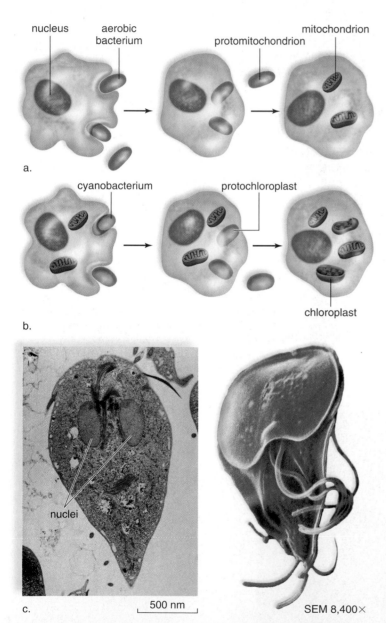

a.

b.

c.

FIGURE 22.1 **Origin of the eukaryotic cell.**
According to the endosymbiotic hypothesis, mitochondria and chloroplasts were once free-living bacteria. **a.** A nucleated cell takes in aerobic bacteria, which become its mitochondria. **b.** A nucleated cell with mitochondria takes in photosynthetic bacteria (cyanobacteria), which become its chloroplasts. **c.** *Giardia* is a protist that has two nuclei but lacks mitochondria.

Protist Overview

Protists vary in size from microscopic algae and protozoans to kelp that can exceed 200 m in length. Kelp, a brown alga, is multicellular, *Volvox*, a green alga, is colonial, while *Spirogyra*, also a green alga, is filamentous. Most protists are unicellular, but despite their small size they have attained a high level of complexity. The amoeboids and ciliates possess unique organelles—their contractile vacuole is an organelle that assists in water regulation.

Protists acquire nutrients in a number of different ways. We will see that the algae are photosynthetic and gather energy from sunlight. The protozoans are heterotrophic, and some ingest food by endocytosis, thereby forming food vacuoles. A slime mold creeps along the forest floor ingesting decaying plant material in the same manner. Other protozoans are parasitic and absorb nutrient molecules meant for their host.

Asexual reproduction by mitosis is the norm in protists. Sexual reproduction involving meiosis and spore formation generally occurs only in a hostile environment. Spores are resistant to adverse conditions and can survive until favorable conditions return once more. Some protozoans form **cysts,** another type of resting stage. In parasites, a cyst often serves as a means of transfer to a new host.

Ecological Importance

While the protists have great medical importance because several cause diseases in humans, they also are of enormous ecological importance. Being aquatic, the photosynthesizers give off oxygen and function as producers in both freshwater and saltwater ecosystems. They are a part of **plankton** [Gr. *plankt,* wandering], organisms that are suspended in the water and serve as food for heterotrophic protists and animals.

Protists enter symbiotic relationships ranging from parasitism to mutualism. Coral reef formation is greatly aided by the presence of a symbiotic photosynthetic protist that lives in the tissues of coral animals, for example.

Classification of Protists

The complexity and diversity of protists make classification difficult. New research in protistan classification is challenging traditional taxonomical schemes. The variety of protists is so great that it has been suggested that they may be split into as many as a dozen kingdoms in the future. Some current taxonomical schemes recognize as many as 60 phyla.

For convenience the protists have been grouped according to modes of nutrition, as noted in the classification table in Figure 22.2. Due to limited space, this text includes 15 protist phyla. Primarily, protists include the algae, which are autotrophic, as are plants; the protozoans and slime molds, which are heterotrophic by ingestion, as are animals; and the water molds, which are heterotrophic by absorption, as are fungi. Some protozoans and some water molds are parasitic.

CLASSIFICATION

DOMAIN: Eukarya
KINGDOM: Protista

CHARACTERISTICS
• Eukaryotes
• Primarily unicellular
• Metabolically diverse
• Structurally complex
• Asexual reproduction usual: sexual reproduction diverse

Photoautotrophs*
PHYLUM: Chlorophyta   green algae
PHYLUM: Rhodophyta   red algae
PHYLUM: Phaeophyta   brown algae
PHYLUM: Chrysophyta   diatoms, golden-brown algae
PHYLUM: Pyrrophyta   dinoflagellates
PHYLUM: Euglenophyta   euglenoids

Heterotrophs by ingestion or parasitic*
PHYLUM: Zoomastigophora zooflagellates
PHYLUM: Rhizopoda   amoeboids
PHYLUM: Foraminifera   foraminiferans
PHYLUM: Actinopoda   radiolarians
PHYLUM: Ciliophora   ciliates
PHYLUM: Apicomplexa   sporozoans
PHYLUM: Myxomycota   plasmodial slime molds
PHYLUM: Acrasiomycota   cellular slime molds

Heterotrophs by absorption (saprotrophs) or parasitic*
PHYLUM: Oomycota   water molds

* Not in the classification of organisms but added here for clarity

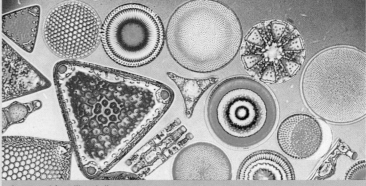

Assorted fossilized diatoms

Onychodromus, a giant ciliate ingesting one of its own kind

Plasmodium, a slime mold

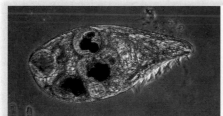
Blepharisma, a ciliate with visible vacuoles   *Licmorpha*, a stalked diatom

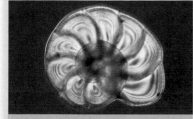
Nonionina, a foraminiferan

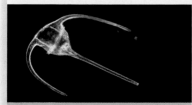
Ceratium, an armored dinoflagellate

Bossiella, a coralline red alga

Acetabularia, a single-celled green alga

Synura, a colony-forming golden alga   *Amoeba proteus*, a protozoan

FIGURE 22.2 Protist diversity.

22.2 Diversity of Protists

Traditionally, the term **algae** [L., seaweed] refers to a diverse group of photosynthetic protists. At one time, botanists classified algae as plants because all have chlorophyll *a* and carry on photosynthesis within a membrane-bounded plastid. However, algae do not develop from an embryo, as do all plants.

The Green Algae

Phylum **Chlorophyta** includes approximately 7,500 species of organisms known as the **green algae.** They inhabit a variety of environments including oceans, freshwater environments, snowbanks, the bark of trees, and the backs of turtles. The green algae also form symbiotic relationships with fungi, plants, and animals. As discussed in Chapter 23, they associate with fungi in lichens. Members of phylum Chlorophyta occur in an abundant variety of forms. The majority of green algae are unicellular; however, filamentous and colonial forms exist. Some multicellular green algae are seaweeds that resemble lettuce leaves.

Chlorophylls *a* and *b* and other pigments found in green algae are similar to those found in plants. However, green algae are not always green; some possess pigments that give them an orange, red, or rust color.

Chlamydomonas, *a Unicellular Green Alga*

Chlamydomonas is a minute, actively moving green alga that inhabits still, freshwater pools. Its fossil ancestors date back over a billion years. Because the alga *Chlamydomonas* is less than 25 μm long, its anatomy is best seen in an electron micrograph (Fig. 22.3). It has a definite cell wall and a single, large, cup-shaped chloroplast that contains a *pyrenoid*, a dense body where starch is synthesized. In many species, a bright

red eyespot, or stigma, exists on the chloroplast, which is sensitive to light and helps bring the organism into the light, where photosynthesis can occur. Two long, whiplike flagella project from the anterior end of this alga and operate with a breaststroke motion.

Chlamydomonas most often reproduces asexually (Fig. 22.4). During asexual reproduction, mitosis produces as many as 16 daughter cells still within the parent cell wall. Each daughter cell then secretes a cell wall and acquires flagella. The daughter cells escape by secreting an enzyme that digests the parent cell wall.

Chlamydomonas occasionally reproduces sexually when growth conditions are unfavorable. Gametes of two different mating types come into contact and join to form a zygote. A heavy wall forms around the zygote, and it becomes a resistant zygospore that undergoes a period of dormancy. When a zygospore germinates, it produces four zoospores by meiosis. **Zoospores,** which are flagellated spores, are typical of aquatic species.

Spirogyra, *a Filamentous Green Alga*

Filaments [L. *filum*, thread] are end-to-end chains of cells that form after cell division occurs in only one plane. In some

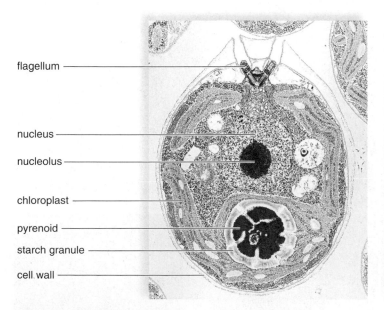

FIGURE 22.3 Electron micrograph of *Chlamydomonas*.
The eyespot is not visible, and only the bases of the flagella can be seen in this micrograph.

flagellum
nucleus
nucleolus
chloroplast
pyrenoid
starch granule
cell wall

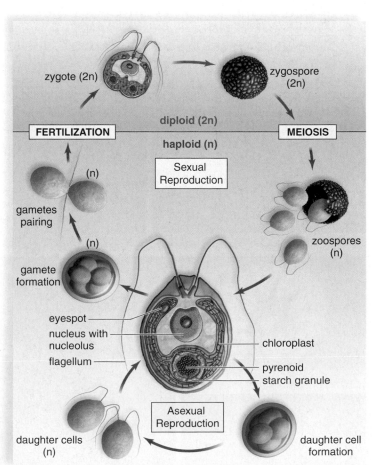

FIGURE 22.4 Reproduction in *Chlamydomonas*.
Chlamydomonas is a motile green alga. During asexual reproduction, all structures are haploid; during sexual reproduction, meiosis follows the zygospore stage, which is the only diploid part of the cycle.

algae, the filaments are branched, and in others the filaments are unbranched. *Spirogyra* is an unbranched, filamentous green alga. Filamentous green algae often grow epiphytically (on but not taking nutrients from) aquatic flowering plants; they also attach to rocks or other objects underwater. Some filaments are suspended in the water.

Spirogyra is found in green masses on the surfaces of ponds and streams. It has ribbonlike, spiralled chloroplasts (Fig. 22.5). During sexual reproduction, *Spirogyra* undergoes **conjugation** [L. *conjugalis*, pertaining to marriage], a temporary union during which the cells exchange genetic material. The two filaments line up parallel to each other, and the cell contents of one filament move into the cells of the other filament, forming diploid zygotes. Resistant zygospores survive the winter, and in the spring they undergo meiosis to produce new haploid filaments.

Multicellular Green Algae

Biologists have long suggested that plants are most closely related to the green algae because, for example, both groups have a cell wall that contains cellulose, possess chlorophylls *a* and *b,* and store reserve food as starch.

Ulva is a multicellular green alga, commonly called sea lettuce because it lives in the sea and has a leafy appearance (Fig. 22.6a). The thallus (body) is two cells thick and can be as much as a meter long. *Ulva* has an alternation of generations life cycle (see Fig. 22A) like that of plants, except that both generations look exactly alike and the gametes all look the same.

Stoneworts are green algae that live in freshwater lakes and ponds. They are called stoneworts because some species, such as *Chara,* are encrusted with calcium carbonate deposits. The main axis of the alga, which can be over a meter long, is a single file of very long cells anchored by rhizoids, which are colorless, hairlike filaments. Only the cell at the upper end of the main axis produces new cells. Whorls of branches occur at multicellular nodes, regions between the giant cells of the main axis. Each of the branches is also a single file of cells (Fig. 22.6b).

Male and female multicellular reproductive structures grow at the nodes, and in some species they occur on separate individuals. The male structure produces flagellated sperm, and the female structure produces a single egg. The zygote is retained until it is enclosed by tough walls. DNA sequencing data suggest that among green algae, the stoneworts are most closely related to plants.

a. *Ulva*, several individuals   One individual

b. *Chara*, several individuals   One individual

FIGURE 22.6 Multicellular green algae.
a. *Ulva* is a green alga known as sea lettuce. **b.** *Chara* is an example of a stonewort, the type of green alga believed to be most closely related to the land plants.

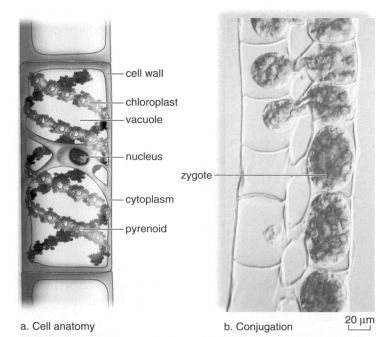

cell wall
chloroplast
vacuole
nucleus
zygote
cytoplasm
pyrenoid

20 μm

a. Cell anatomy   b. Conjugation

FIGURE 22.5 *Spirogyra.*
a. *Spirogyra* is a filamentous green alga in which each cell has a ribbonlike chloroplast. **b.** During conjugation, the cell contents of one filament enter the cells of another filament. Zygote formation follows.

40 µm

daughter colony

vegetative cells

15 µm

FIGURE 22.7 *Volvox.*
Volvox is a colonial green alga. The adult *Volvox* colony often contains daughter colonies, which are asexually produced by special cells.

Volvox, *a Colonial Green Alga*

A number of *colonial* forms occur among the flagellated green algae. *Volvox* is a well-known colonial green alga. A **colony** is a loose association of independent cells. A *Volvox* colony is a hollow sphere with thousands of cells arranged in a single layer surrounding a watery interior. Each cell of a *Volvox* colony resembles a *Chlamydomonas* cell—perhaps it is derived from daughter cells that fail to separate following zoospore formation. In *Volvox,* the cells cooperate in that the flagella beat in a coordinated fashion. Some cells are specialized for reproduction, and each of these can divide asexually to form a new daughter colony (Fig. 22.7). This daughter colony resides for a time within the parent colony, but then it leaves by releasing an enzyme that dissolves away a portion of the parent colony, allowing it to escape.

Green algae share characteristics with plants. DNA analysis suggests that among the green algae, the stoneworts are most closely related to plants.

FIGURE 22.8 Red alga.
Red algae are multicellular marine algae, represented by *Chondrus crispus.* Red algae are smaller and more delicate than brown algae.

The Red Algae

Phylum **Rhodophyta** includes more than 5,000 species of multicellular organisms known as the **red algae.** These algae live primarily in warm seawater, growing in both shallow and deep waters. Some grow attached to rocks in the intertidal zone, where they are exposed at low tide. Red algae are usually much smaller and more delicate than brown algae, although some species can exceed a meter in length. Others can grow at depths exceeding 200 m, where light barely penetrates. A few species of red algae are unicellular.

Some forms of red algae are simple filaments, but more often they are complexly branched, with the branches having a feathery, flat, or expanded, ribbonlike appearance. Coralline algae are red algae that have cell walls impregnated with calcium carbonate. In some instances, they contribute as much to the growth of coral reefs as do coral animals.

Red algae are economically important. Agar is a gelatin-like product made primarily from the algae *Gelidium* and *Gracilaria.* Agar is used commercially to make capsules for vitamins and drugs, as a material for making dental impressions, and as a base for cosmetics. In the laboratory, agar is a solidifying agent for a bacterial culture medium. When purified, it becomes the gel for electrophoresis, a procedure that separates proteins or nucleotides. Agar is also used in food preparation—as an antidrying agent for baked goods and to make jellies and desserts set rapidly. Carrageenin, extracted from *Chondrus crispus* (Fig. 22.8), called an Irish moss but actually a red alga, is an emulsifying agent for the production of chocolate and cosmetics. *Porphyra,* a red alga, is the basis of a billion-dollar aquaculture industry in Japan. The reddish-black wrappings around sushi rolls consist of processed *Porphyra* blades.

Many red algae, which are more delicate than brown algae, have filamentous branches or are multicellular. Several species have commercial importance.

The Brown Algae

Phylum **Phaeophyta** consists of over 1,500 species of sea-weeds and other algae that are called the **brown algae.** No unicellular or colonial brown algae exist. The brown algae range from small forms with simple filaments to large multi-cellular forms that may reach 100 m in length (Fig. 22.9). The vast majority of brown algae live in cold ocean waters. The brown algae have chlorophylls *a* and *c* in their chloroplasts and a type of carotenoid pigment (fucoxanthin) that gives them their characteristic color. The reserve food is stored as a carbohydrate called *laminarin.*

The multicellular forms of green, red, and brown algae are called **seaweeds,** a common term for any large, complex alga. Brown algae are often observed along the rocky coasts in the north temperate zone, where they are pounded by waves as the tide comes in and are exposed to dry air as the tide goes out. They dry out slowly, however, because their cell walls contain a mucilaginous, water-retaining material.

Both *Laminaria,* commonly called *kelp,* and *Fucus,* known as rockweed, are examples of brown algae that grow along the shoreline. In deeper waters, the giant kelps (*Macrocystis* and *Nereocystis*) of-ten grow extensively in vast beds. Individu-als of the genus *Sargassum* sometimes break off from their holdfasts and form floating masses. Brown algae not only provide food and habitat for marine organisms,

they are harvested for human food and for fertilizer in sev-eral parts of the world. *Macrocystis* is the source of algin, a pectinlike material that is added to ice cream, sherbet, cream cheese, and other products to give them a stable, smooth consistency.

Laminaria is unique among the protists because mem-bers of this genus show tissue differentiation—that is, they transport organic nutrients by way of a tissue that resembles phloem in land plants. Most brown algae have the alterna-tion of generations life cycle, but some species of *Fucus* are unique in that meiosis produces gametes, and the adult is always diploid, as in animals.

The multicellular forms of brown algae include the largest of the algae, the giant kelps.

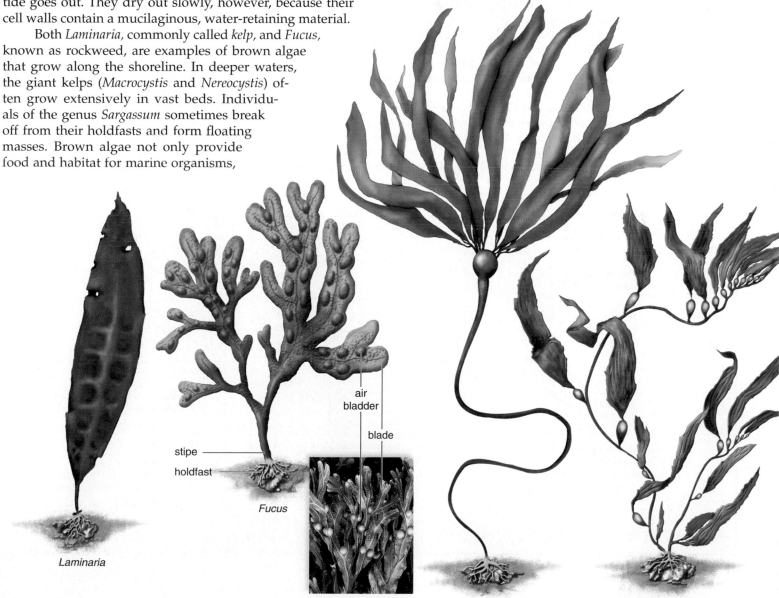

air bladder

blade

stipe

holdfast

Fucus

Laminaria

Rockweed, *Fucus*

Macrocystis

Nereocystis

FIGURE 22.9 Brown algae.
Laminaria and *Fucus* are seaweeds known as kelps. They live along rocky coasts of the north temperate zone. The other brown algae featured, *Nereocystis* and *Macrocystis,* form spectacular underwater "forests" at sea.

science focus

Life Cycles Among the Algae

Both asexual and sexual reproduction occur in algae, depending on species and environmental conditions. The types of life cycles seen in algae occur in other protists, and also in plants or animals.

Asexual reproduction is a frequent mode of reproduction among protists when the environment is favorable to growth. Asexual reproduction requires only one parent. The offspring are identical to this parent because the offspring receive a copy of only this parent's genes. The new individuals are likely to survive and flourish if the environment is steady. Various modes of asexual reproduction occur, but growth alone produces a new individual.

Sexual reproduction, with its genetic recombination due in part to fertilization and independent assortment of chromosomes, is more likely to occur among protists when the environment is changing and is unfavorable to growth. Recombination of genes might produce individuals that are more likely to survive

extremes in the environment—such as high or low temperatures, acidic or basic pH, or the lack of a particular nutrient.

Sexual reproduction requires two parents, each of which contributes chromosomes (genes) to the offspring by way of gametes. The gametes fuse to produce a diploid zygote. A reproductive cycle is isogamous when the gametes look alike—called isogametes—and a cycle is oogamous when the gametes are dissimilar—called heterogametes. Usually a small flagellated sperm fertilizes a large egg with plentiful cytoplasm.

Meiosis occurs during sexual reproduction—just *when* it occurs makes the sexual life cycles diagrammed in Figure 22A differ from one another. In these diagrams, the diploid phase is shown in blue, and the haploid phase is shown in tan. The haploid life cycle (Fig. 22A*a*) most likely evolved first. In the haploid cycle, the zygote divides by meiosis to form haploid spores that develop into haploid individuals. In algae, the spores are typi-

cally zoospores. The zygote is only the diploid stage in this life cycle, and the haploid individual gives rise to gametes. This life cycle is seen in *Chlamydomonas* and a number of other algae.

In alternation of generations, the sporophyte (2n) produces haploid spores by meiosis (Fig. 22A*b*). A spore develops into a haploid gametophyte that produces gametes. The gametes fuse to form a diploid zygote, and the zygote develops into the sporophyte. This life cycle is characteristic of some algae (e.g., *Ulva* and *Laminaria*) and all plants. In *Ulva*, the haploid and diploid generations have the same appearance. In plants, they are noticeably different from each other.

In the diploid life cycle, typical of animals, a diploid individual produces gametes by meiosis (Fig. 22A*c*). Gametes are the only haploid stage in this cycle. They fuse to form a zygote that develops into the diploid individual. This life cycle is rare in algae but does occur in a few species of the brown alga *Fucus*.

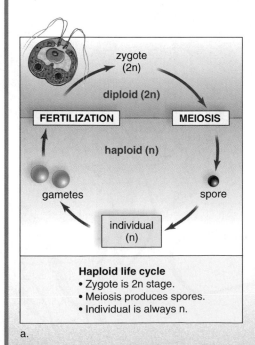

a.

Haploid life cycle
• Zygote is 2n stage.
• Meiosis produces spores.
• Individual is always n.

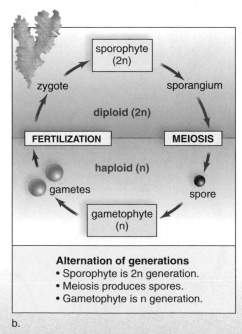

b.

Alternation of generations
• Sporophyte is 2n generation.
• Meiosis produces spores.
• Gametophyte is n generation.

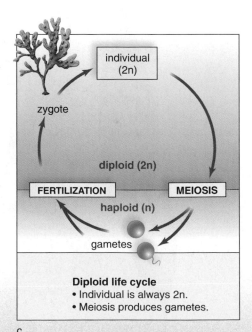

c.

Diploid life cycle
• Individual is always 2n.
• Meiosis produces gametes.

FIGURE 22A Common life cycles in sexual reproduction.

The Diatoms and Golden Brown Algae

Phylum **Chrysophyta** consists of approximately 11,000 species of **diatoms** [Gk. *dia,* through, and *temno,* cut], 500 species of golden-brown algae, and 600 species of yellow-green algae. The majority of members of this phyla are free-living photosynthetic cells that inhabit aquatic and marine environments. Diatoms are the most numerous unicellular algae in the oceans and freshwater environments. Diatoms are a significant part of the **phytoplankton,** photosynthetic organisms that are suspended but sink slowly in the water. Phytoplankton are an important source of food and oxygen for heterotrophs in both freshwater and marine ecosystems.

The structure of a diatom is often compared to a hat box because the cell wall has two halves or valves, with the larger valve acting as a "lid" that fits over the smaller valve (Fig. 22.10*a*). When diatoms reproduce asexually, each receives one old valve. The new valve fits inside the old one; therefore, new diatoms are smaller than the original ones. This continues until diatoms are about 30% of the original size. Then they reproduce sexually. The zygote becomes a structure that grows and then divides mitotically to produce diatoms of normal size.

The cell wall of a diatom has an outer layer of silica, a common ingredient in glass. The valves are covered with a great variety of striations and markings that form beautiful patterns when observed under the microscope. These are actually depressions or pores through which the organism makes contact with the outside environment. The remains of diatoms, called diatomaceous earth, accumulate on the ocean floor and are mined for use as filtering agents, soundproofing materials, and gentle polishing abrasives such as those found in silver polish and toothpaste.

Diatoms are significant producers of food and oxygen in aquatic ecosystems because of their sheer abundance.

The Dinoflagellates

Phylum **Pyrrophyta** consists of approximately 4,000 species of aquatic and marine unicellular organisms known as the **dinoflagellates.** The cell is usually bounded by protective cellulose plates impregnated with silicates (Fig. 22.10*b*). Typically, they have two flagella; one lies in a longitudinal groove with its distal end free, and the other lies in a transverse groove that encircles the organism. The longitudinal flagellum acts as a rudder, and the beating of the transverse flagellum causes the cell to spin as it moves forward.

The chloroplasts of a dinoflagellate vary in color from yellow-green to brown because in addition to chlorophylls *a* and *c*, they also contain carotenoids. Some species such as *Noctiluca* are capable of bioluminescence (producing light). Being a part of the phytoplankton, the dinoflagellates are an important source of food for small animals in the ocean. They also live within the bodies of some invertebrates as symbionts. Symbiotic dinoflagellates lack cellulose plates and flagella and are called zooxanthellae. Corals (see Chapter 29), members of the animal kingdom, usually contain large numbers of zooxanthellae. They provide their hosts with organic nutrients, and the corals provide wastes that fertilize the algae. Some dinoflagellates lack chloroplasts and are heterotrophic; some of these are parasitic.

Dinoflagellates usually reproduce asexually using mitosis and longitudinal division of the cell. A daughter cell gets half the cellulose plates unless they were shed before reproduction began. During sexual reproduction the daughter cells act as gametes. The zygote enters a resting stage broken when meiosis produces only one haploid cell because the others disintegrate.

Like the diatoms, dinoflagellates are one of the most important groups of producers in marine environments. Occasionally, however, particularly in polluted waters in late summer, they undergo a population explosion and become more numerous than usual. At these times, their density can equal 30,000 in a single milliliter. When dinoflagellates such as

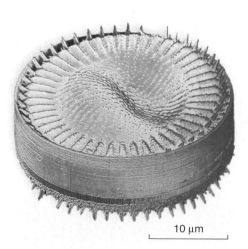

10 μm

a. Diatom, *Cyclotella*

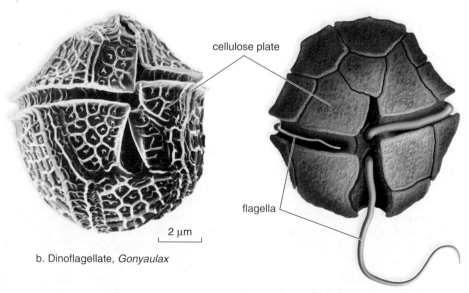

cellulose plate

flagella

2 μm

b. Dinoflagellate, *Gonyaulax*

FIGURE 22.10 Diatoms and dinoflagellates.
a. Diatoms may be variously colored, but even so their chloroplasts contain a unique golden-brown pigment (fucoxanthin), in addition to chlorophylls *a* and *c*. The beautiful pattern results from markings on the silica-embedded wall. **b.** Dinoflagellates such as *Gonyaulax* have cellulose plates.

a.

b.

FIGURE 22.11 Fish kill and dinoflagellate bloom.
a. Fish kills such as this one in University Lake in Baton Rouge, Louisiana, can be the result of a dinoflagellate bloom.
b. This bloom, often called a red tide after the color of the water, appeared near California's central coast.

Gymnodinium brevis increase in number they may cause a phenomenon called "red tide." Massive fish kills can occur as the result of a powerful neurotoxin produced by these dinoflagellates (Fig. 22.11). Humans who consume shellfish that have fed during a *Gymnodinium* outbreak may suffer from paralytic shellfish poisoning, in which the respiratory organs are paralyzed.

> The dinoflagellates, like the diatoms, are significant producers in marine environments.

The Euglenoids

Phylum **Euglenophyta** includes about 1,000 species of small (10–500 μm) freshwater unicellular organisms called **euglenoids** that typify the problem of classifying protists. One-third of all genera have chloroplasts; the rest do not. Those that lack chloroplasts ingest or absorb their food. This may not be surprising when one knows that their chloroplasts are like those of green algae and are probably derived from them through endosymbiosis. The chloroplasts are surrounded by three rather than two membranes. A pyrenoid is a special region of the chloroplast where carbohydrate formation occurs. Euglenoids produce an unusual type of carbohydrate called paramylon.

A common englenoid is *Euglena deces,* an inhabitant of freshwater ditches and ponds. *Euglena's* ability to undergo photosynthesis as well as ingest food has caused the organism to be treated as a plantlike organism in botany texts and an animal-like organism in zoology texts. Euglenoids have two flagella, one of which typically is much longer than the other and projects out of an anterior, vase-shaped invagination (Fig. 22.12). It is called a tinsel flagellum because it has hairs on it. Near the base of this flagellum is an eyespot, which shades a photoreceptor for detecting light. Because euglenoids are bounded by a flexible *pellicle* composed of protein bands lying side by side, they can assume different shapes as the underlying cytoplasm undulates and contracts. As in certain other

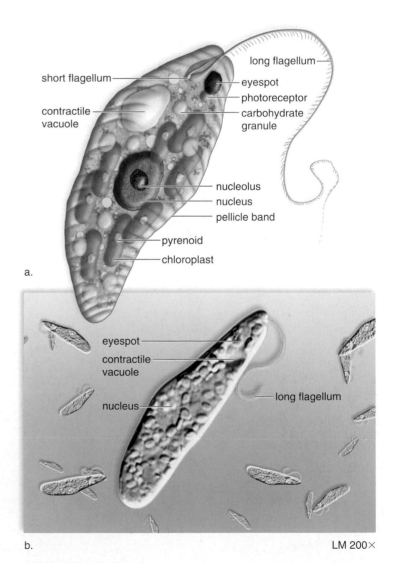

a.

b.   LM 200×

FIGURE 22.12 *Euglena.*
a. In *Euglena,* a very long flagellum propels the body, which is enveloped by a flexible pellicle. **b.** Micrograph of several specimens.

protists, there is a contractile vacuole for ridding the body of excess water. Euglenoids reproduce by longitudinal cell division, and sexual reproduction is not known to occur.

The Zooflagellates

Phylum **Zoomastigophora** includes thousands of species of mostly unicellular, heterotrophic protozoans called **zooflagellates** that are highly variable in form. The phylum includes many symbiotic and some parasitic forms.

Trypanosomes, transmitted by the bite of the tsetse fly, are the cause of African sleeping sickness (Fig. 22.13). The white blood cells in an infected human accumulate around the blood vessels leading to the brain and cut off circulation. The lethargy characteristic of the disease is caused by an inadequate supply of oxygen to the brain. Many thousands of cases of human sleeping sickness are diagnosed each year.

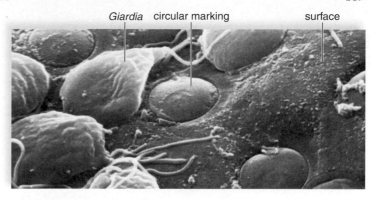

FIGURE 22.14 *Giardia lamblia.*
The protist adheres to any surface including epithelial cells by means of a sucking disk. Characteristic markings can be seen after the disk detaches.

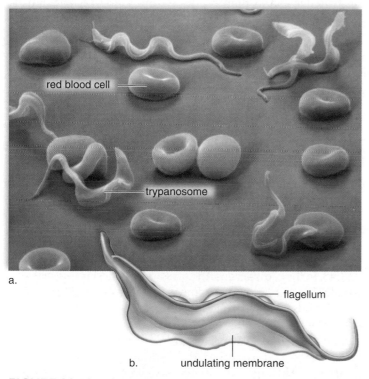

red blood cell

trypanosome

a.

flagellum

b.   undulating membrane

FIGURE 22.13 Zooflagellates.
a. Micrograph of *Trypanosoma brucei*, a causal agent of African sleeping sickness, among red blood cells. **b.** The drawing shows its general structure.

Fatalities or permanent brain damage are common. *Trypanosoma cruzi* causes Chagas disease in humans in Central and South America. Approximately 45,000 people die yearly from this parasite.

Giardia lamblia, whose cysts are transmitted by way of contaminated water, attaches to the human intestinal wall and causes severe diarrhea (Fig. 22.14). *Giardia* is the most common flagellate of the human digestive tract. *Giardia* lives in a variety of mammals in addition to humans. Beavers seem to be an important source of infection in the mountains of the western United States, and many cases of infection have been acquired by hikers who fill their canteens at a beaver pond.

Trichomonas vaginalis, a sexually transmitted organism, infects the vagina and urethra of women and the prostate, seminal vesicles, and urethra of men. Therefore, it is a common culprit in vaginitis.

> The zooflagellates are well known for causing various diseases in humans, ranging from some that are annoying to others that are extremely serious.

Some authorities include all the flagellates among the **protozoans,** a term that can simply mean a unicellular (or colonial) eukaryote. In that case, protozoans include both photosynthetic and heterotrophic organisms. Others prefer to restrict the term *protozoan* to unicellular heterotrophic organisms such as those listed in Table 22.1. Some of these heterotrophic forms ingest their food by endocytosis. Usually, a protozoan has some

TABLE 22.1

Some Protozoans

Name	Unicellular	Locomotion	Examples
Zooflagellates	Yes	Flagella	*Trypanosoma*
Amoeboids	Yes	Pseudopods	*Amoeba, Entamoeba*
Radiolarians	Yes	Pseudopods	—
Foraminiferans	Yes	Pseudopods	*Globigerina*
Ciliates	Yes	Cilia	*Paramecium, Vorticella, Didinia, Stentor*
Sporozoans	Yes	None	*Plasmodium*

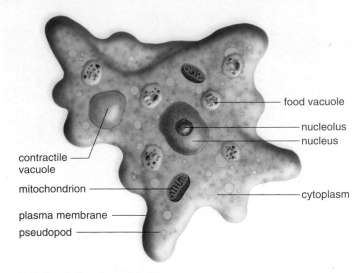

contractile vacuole

food vacuole

nucleolus

nucleus

mitochondrion

cytoplasm

plasma membrane

pseudopod

a. Amoeba, *Amoeba proteus*

250 μm

b. Foraminiferan, *Globigerina*, and the White Cliffs of Dover, England

c. Radiolarian tests   SEM 200×

FIGURE 22.15 Protists with pseudopods.
a. Structure of *Amoeba proteus*, an amoeboid common in freshwater ponds. Bacteria and other microorganisms are digested in food vacuoles, and contractile vacuoles rid the body of excess water. **b.** Pseudopods of a live foraminiferan project through holes in the calcium carbonate shell. Fossilized shells were so numerous they became a large part of the White Cliffs of Dover when a geological upheaval occurred. **c.** Skeletal test of a radiolarian. In life, pseudopods extend outward through the openings of the glassy silicon shell.
©Dr. Richard Kessel & Dr. Gene Shih/Visuals Unlimited.

form of locomotion, either by flagella, pseudopods, or cilia. At one time, zoologists classified protozoans as animals.

Protists with Pseudopods

Pseudopods [Gk. *pseudes*, false, and *podos*, foot] are processes that form when cytoplasm streams forward in a particular direction. Protists that move by pseudopods usually live in aquatic environments. In oceans and freshwater lakes and ponds, they may be a part of the **zooplankton,** microscopic suspended organisms that feed on other organisms.

The **amoeboids** (phylum **Rhizopoda**) are protists that move and also ingest their food with pseudopods. Hundreds of species of amoeboids have been classified. *Amoeba proteus* is a commonly studied freshwater member of this group (Fig. 22.15*a*). When amoeboids feed, the pseudopods surround and **phagocytize** [Gk. *phagein,* eat, and *kytos,* cell] their prey, which may be algae, bacteria, or other protists. Digestion then occurs within a *food vacuole.* Freshwater amoeboids, including *Amoeba proteus,* have *contractile vacuoles* where excess water from the cytoplasm collects before the vacuole appears to "contract," releasing the water through a temporary opening in the plasma membrane.

Entamoeba histolytica is a parasitic amoeboid that lives in the human large intestine and causes amoebic dysentery. The abililty of the organism to form cysts makes amoebic dysentery infectious. Complications arise when this parasite invades the intestinal lining and reproduces there. If the parasites enter the body proper, liver and brain involvement can be fatal.

The **foraminiferans** (phylum **Foraminifera**) and the **radiolarians** (phylum **Actinopoda**) both have a skeleton called a **test.** The tests of foraminiferans and radiolarians are intriguing and beautiful. In the foraminiferans, the calcium carbonate test is often multichambered. The pseudopods extend through openings in the test, which covers the plasma membrane (Fig. 22.15*b*). In the radiolarians, the glassy silicon test is internal and usually has a radial arrangement of spines (Fig. 22.15*c*). The pseudopods are external to the test.

The tests of dead foraminiferans and radiolarians form a deep layer (700–4,000 m) of sediment on the ocean floor. The radiolarians lie deeper than the foraminiferans because their glassy test is insoluble at greater pressures. The presence of either or both is used as an indicator of oil deposits on land and sea. Their fossils date even as far back as to Precambrian times and are evidence of the antiquity of the protists. Because each geological period has a distinctive form of foraminiferan, they can be used as index fossils to date sedimentary rock. Deposits of foraminiferans for millions of years followed by geological upheaval formed the White Cliffs of Dover along the southern coast of England. Also, the great Egyptian pyramids are built of foraminiferan limestone. One foraminiferan test found in the pyramids is about the size of a silver dollar. This species, known as *Nummulites,* has been found in deposits worldwide, including central eastern Mississippi.

Among protists that have pseudopods, *Amoeba proteus* lives in fresh water, but the shelled foraminiferans and radiolarians are abundant in the ocean.

The Ciliates

Phylum **Ciliophora** consists of approximately 8,000 species of unicellular protists that range in size from 10 to 3,000 μm. The **ciliates** are the most structurally complex and specialized of all protozoa (Fig. 22.16*a*). The majority of ciliates are free-living; however, several parasitic, sessile, and colonial forms exist. As the name implies, the ciliates move by means of cilia.

The classic example of a member of phylum Ciliophora belongs to the genus *Paramecium*. These unicellular ciliates are commonly found in ponds and ditches. Hundreds of cilia, which beat in a coordinated rhythmic manner, project through tiny holes in a semirigid outer covering, or pellicle. Numerous oval capsules lying in the cytoplasm just beneath the pellicle contain **trichocysts**. Upon mechanical or chemical stimulation, trichocysts discharge long, barbed threads that are useful for defense and for capturing prey. Toxicysts are similar, but they release a poison that paralyzes prey.

Most ciliates are **holozoic**. For example, when a paramecium feeds, food particles are swept down a gullet, below which food vacuoles form. Following digestion, the soluble nutrients are absorbed by the cytoplasm, and the nondigestible residue is eliminated at the anal pore.

During asexual reproduction, ciliates divide by transverse binary fission. Ciliates have two types of nuclei: a large macronucleus and one or more small micronuclei. The macronucleus controls the normal metabolism of the cell, while the micronuclei are concerned with reproduction. Sexual reproduction involves conjugation (Fig. 22.16*b*). The macronucleus disintegrates and, after the micronuclei undergo meiosis, two ciliates exchange a haploid micronucleus. Then the micronuclei give rise to a new macronucleus, which contains copies of only certain housekeeping genes.

The ciliates are a diverse group of protozoans. The barrel-shaped didiniums expand to consume paramecia much larger than themselves. *Suctoria* have an even more dramatic way of getting food. They rest quietly on a stalk until a hapless victim comes along. Then they promptly paralyze it and use their tentacles like straws to suck it dry. *Stentor* may be the most elaborate ciliate, resembling a giant blue vase decorated with stripes (Fig. 22.16*c*). *Ichthyophthirius* is responsible for a common disease in fishes called "ick." If left untreated, it can be fatal.

Ciliates, which move by cilia, are diverse and very complex. The single cell has regions specialized to carry out various functions.

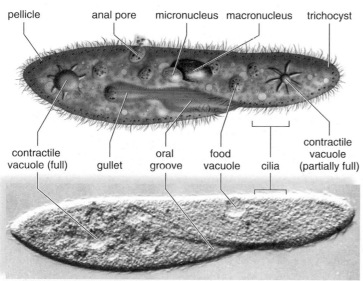

pellicle | anal pore | micronucleus | macronucleus | trichocyst

contractile vacuole (full) | gullet | oral groove | food vacuole | cilia | contractile vacuole (partially full)

a. *Paramecium*

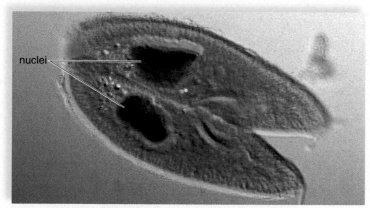

nuclei

b. During conjugation two paramecia first unite at oral areas

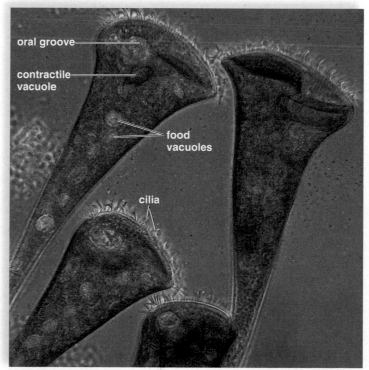

oral groove
contractile vacuole
food vacuoles
cilia

c. *Stentor*

200 μm

FIGURE 22.16 Ciliates.
Ciliates are the most complex of the protists. **a.** Structure of *Paramecium*, adjacent to an electron micrograph. Note the oral groove and the gullet and anal pore. **b.** A form of sexual reproduction called conjugation occurs periodically. **c.** *Stentor*, a large, vase-shaped, freshwater ciliate.

The Sporozoans

Phylum **Apicomplexa** consists of nearly 3,900 species of nonmotile, parasitic, spore-forming protozoans also known as **sporozoans.** Their scientific name refers to a unique collection of organelles at one end of the infective stage. Many apicomplexans have multiple hosts.

Pneumocystis carinii causes the type of pneumonia seen primarily in AIDS patients. During sexual reproduction, thick-walled cysts form in the lining of pulmonary air sacs. The cysts contain spores that successively divide until the cyst bursts and the spores are released. Each spore becomes a new mature organism that can reproduce asexually but may also enter the sexual stage and form cysts.

Today, approximately 3 million people die each year from malaria. In humans, malaria is caused by four distinct members of the genus *Plasmodium*. Despite efforts to control the malaria parasite as well as the mosquito vector, malaria is making a resurgence. International travel coupled with new resistant forms of the parasite and vector is presenting health professionals with formidable problems.

The most widespread human parasite is *Plasmodium vivax*, the cause of one type of malaria. The life cycle alternates between a sexual and an asexual phase in different hosts. Female *Anopheles* mosquitoes acquire protein for production of eggs by biting humans and other animals. When a human is bitten, the parasite eventually invades the red blood cells. The chills and fever of malaria appear when the infected cells burst and release toxic substances into the blood (Fig. 22.17).

Toxoplasma gondii, another sporozoan, causes toxoplasmosis, particularly in cats but also in people. In pregnant women, the parasite can infect the fetus and cause birth defects and mental retardation; in AIDS patients, it can infect the brain and cause neurological symptoms.

The malarial parasite *Plasmodium* is the best known of the sporozoans, but the phylum Apicomplexa contains many examples of human parasites.

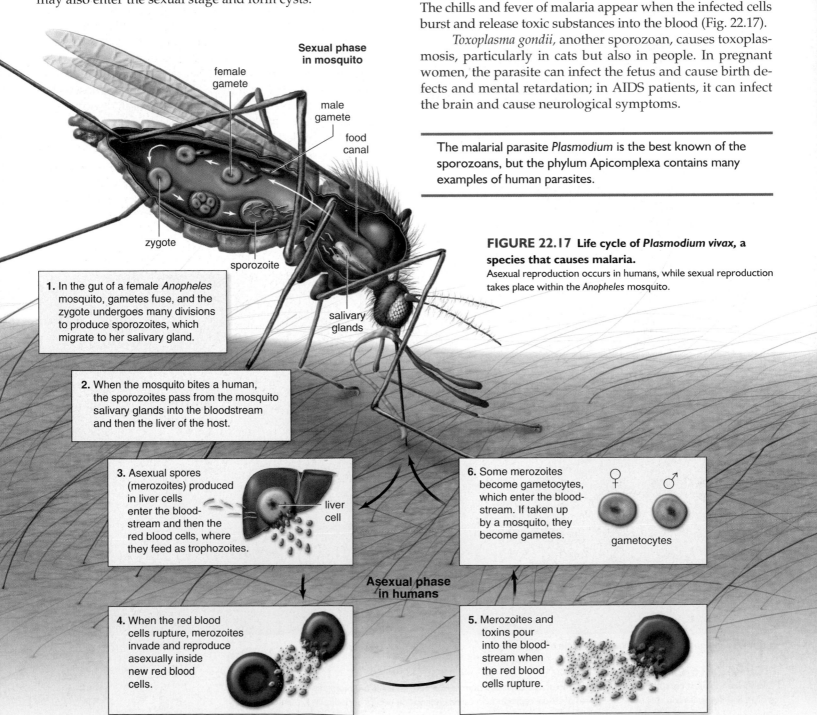

FIGURE 22.17 Life cycle of *Plasmodium vivax*, a species that causes malaria.
Asexual reproduction occurs in humans, while sexual reproduction takes place within the *Anopheles* mosquito.

Sexual phase in mosquito

female gamete

male gamete

food canal

zygote

sporozoite

salivary glands

1. In the gut of a female *Anopheles* mosquito, gametes fuse, and the zygote undergoes many divisions to produce sporozoites, which migrate to her salivary gland.

2. When the mosquito bites a human, the sporozoites pass from the mosquito salivary glands into the bloodstream and then the liver of the host.

3. Asexual spores (merozoites) produced in liver cells enter the bloodstream and then the red blood cells, where they feed as trophozoites. — liver cell

6. Some merozoites become gametocytes, which enter the bloodstream. If taken up by a mosquito, they become gametes. — gametocytes

Asexual phase in humans

4. When the red blood cells rupture, merozoites invade and reproduce asexually inside new red blood cells.

5. Merozoites and toxins pour into the bloodstream when the red blood cells rupture.

The Slime Molds and Water Molds

In forests and woodlands, slime molds phagocytize, and therefore help dispose of, dead plant material. They also feed on bacteria, keeping their population under control. Water molds decompose remains, but they are also significant parasites of plants and animals in ecosystems. These organisms were once classified as fungi, but unlike fungi, both types have flagellated cells at some time during their life cycle. The water molds, but not the slime molds, have a cell wall, and it contains cellulose, not chitin as do the fungi. They form spores; those of the water mold are 2n zoospores. Meiosis produces gametes.

Slime molds and water molds may be more closely related to the amoeboids than to each other. Indeed, the vegetative state of the slime molds is mobile and amoeboid! Like the amoeboids, they ingest their food by phagocytosis.

The Plasmodial Slime Molds

Usually **plasmodial slime molds** (phylum **Myxomycota**) exist as a plasmodium, a diploid, multinucleated, cytoplasmic mass enveloped by a slime sheath that creeps along, phagocytizing decaying plant material in a forest or agricultural field (Fig. 22.18). Approximately 500 species of plasmodial slime molds have been described. Many species are brightly colored. At times unfavorable to growth, such as during a drought, the plasmodium develops many sporangia. A **sporangium** [Gk. *spora*, seed, and *angeion* (dim. of *angos*), vessel] is a reproductive structure that produces spores. An aggregate of sporangia is called a fruiting body.

The spores produced by a sporangium can survive until moisture is sufficient for them to germinate. In plasmodial slime molds, spores release a haploid flagellated cell or an amoeboid cell. Eventually, two of them fuse to form a zygote that feeds and grows, producing a multinucleated plasmodium once again.

The Cellular Slime Molds

Cellular slime molds (phylum **Acrasiomycota**) are called such because they exist as individual amoeboid cells. They are common in soil, where they feed on bacteria and yeasts. Their small size prevents them from being seen. Nearly 70 species of cellular slime molds have been described.

As the food supply runs out or unfavorable environmental conditions develop, the cells release a chemical that causes them to aggregate into a pseudoplasmodium. The pseudoplasmodium stage is temporary and eventually gives rise to a fruiting body in which sporangia produce spores. When favorable conditions return, the spores germinate, releasing haploid amoeboid cells, and the asexual cycle begins again.

Slime molds have an amoeboid stage that phagocytizes organic matter and thereby contributes to ecological balance.

Plasmodium, *Physarum*

Sporangia, *Hemitrichia*   └ 1 mm ┘

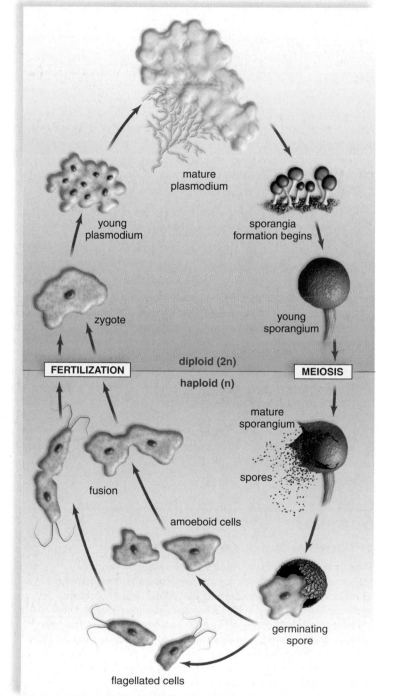

mature plasmodium

young plasmodium

sporangia formation begins

zygote

young sporangium

FERTILIZATION   **diploid (2n)**   **MEIOSIS**

haploid (n)

fusion

mature sporangium

spores

amoeboid cells

germinating spore

flagellated cells

FIGURE 22.18 Plasmodial slime molds.
The diploid adult forms sporangia during sexual reproduction, when conditions are unfavorable to growth. Haploid spores germinate, releasing haploid amoeboid or flagellated cells that fuse.

The Water Molds

The **water molds** (phylum **Oomycota**) usually live in the water, where they form furry growths when they parasitize fishes or insects and decompose remains. In spite of their common name, some water molds live on land and parasit-

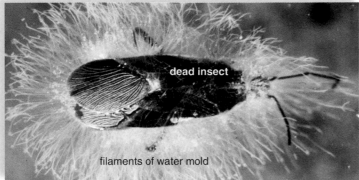

FIGURE 22.19 Water mold.
Saprolegnia, a water mold, feeding on a dead insect.

(image labels: dead insect; filaments of water mold) LM 10×

ize insects and plants. Nearly 500 species of water molds have been described. A water mold, *Phytophthora infestans,* was responsible for the 1840s potato famine in Ireland. However, most water molds are saprotrophic and live off dead organic matter. Another well-known water mold is *Saprolegnia,* which is often seen as a cottonlike white mass on dead organisms (Fig. 22.19).

Water molds have a filamentous body as do fungi, but their cell walls are largely composed of cellulose, whereas fungi have cell walls of chitin. The life cycle of water molds differs from that of fungi. During asexual reproduction, water molds produce motile spores (2n zoospores), which are flagellated. The organism is diploid (not haploid as in the fungi), and meiosis produces gametes. Their phylum name refers to the enlarged tips (called oogonia) where eggs are produced.

> Water molds are filamentous decomposers of organic remains. They are also parasites of animals and plants in ecological settings.

CONNECTING THE CONCEPTS

It's possible that the origin of sexual reproduction played a role in fostering the diversity of protists and their ability to inhabit water, soil, food, and even air. Unicellular *Chlamydomonas* usually reproduces asexually, but when it does reproduce sexually, the gametes look alike. Colonial forms of algae, such as *Volvox,* and multicellular forms, such as the brown alga *Fucus,* have a more specialized form of gametogenesis. They undergo oogamy: One gamete is larger and immobile, and the other is flagellated and motile. The green alga *Ulva* and the foraminiferans have an alternation of generations life cycle in which only the haploid stage produces gametes.

Conjugation is another way to introduce genetic variation among offspring. *Spirogyra* undergoes conjugation, but the ciliates have taken conjugation to another level. In ciliates, one haploid nucleus in each mating cell survives meiosis and divides again by mitosis. An exchange of haploid nuclei through a cytoplasmic channel restores diploidy and ensures diversity of genes.

Reproduction in protists is adaptive in still another way. As a part of their life cycle, some protists form spores or cysts that can survive in a hostile environment. The cell emerging from a spore or cyst is often a re-formed haploid or diploid individual, depending on the protist. Spores and cysts can survive inclement conditions for years on end.

The diversity of reproduction among protists exemplifies why it is difficult to classify them and discover their evolutionary relationships to the fungi, plants, and animals. The protists we study today are distant cousins of these other eukaryotic groups and therefore give only hints of what the ancestry of fungi, plants, and animals may have been.

Summary

22.1 GENERAL BIOLOGY OF PROTISTS

Protists are in the domain Eukarya and the kingdom Protista. Independent endosymbiotic events may account for the presence of mitochondria and chloroplasts in eukaryotic cells. Protists are generally unicellular, but still they are quite complex because they (1) have a variety of characteristics, (2) acquire nutrients in a number of ways, and (3) have complicated life cycles that include the ability to withstand hostile environments.

Protists are of great ecological importance because in largely aquatic environments they are the producers that support communities of organisms. Protists also enter into various types of symbiotic relationships. Their great diversity makes it difficult to classify protists, and as yet, there is no general agreement about their classification. In this chapter, they have been arranged according to mode of nutrition.

22.2 DIVERSITY OF PROTISTS

Green algae possess chlorophylls *a* and *b,* store reserve food as starch, and have cell walls of cellulose as do plants. Green algae are divided into those that are unicellular *(Chlamydomonas),* filamentous *(Spirogyra),* multicellular *(Ulva* and *Chara),* and colonial *(Volvox).* The life cycle varies among the green algae. In most, the zygote undergoes meiosis, and the adult is haploid. *Ulva* has an alternation of generations like plants, but the sporophyte and gametophyte generations are similar in appearance; the gametes look alike.

Red algae are filamentous or multicellular seaweeds that are more delicate than brown algae, and are usually found in warmer waters. Like brown algae, red algae have economic importance. Brown algae have chlorophylls *a* and *c* plus a brownish carotenoid pigment. The large, complex brown algae, commonly called seaweeds, are well known and economically important.

Diatoms, which have an outer layer of silica, are extremely numerous in both marine and freshwater ecosystems. Dinoflagellates usually have cellulose plates and two flagella, one at a right angle to the other. They are extremely numerous in the ocean, and on occasion produce a neurotoxin when they form red tides.

Euglenoids are flagellated cells with a pellicle instead of a cell wall. Their chloroplasts are most likely derived from a green alga through endosymbiosis. Zooflagellates, which also have flagella, are often symbiotic. Trypanosomes are zooflagellates that cause Chagas disease and African sleeping sickness in humans.

Amoeboids move and feed by forming pseudopods. In *Amoeba proteus,* food vacuoles form following phagocytosis of prey. Contractile vacuoles discharge excess water. The tests of foraminiferans and radiolarians form a deep layer of sediment on the ocean floor. The tests of foraminiferans are responsible for the limestone deposits of the White Cliffs of Dover.

The ciliates move by their many cilia. They are remarkably diverse in form, and as exemplified by *Paramecium,* they show how complex a protist can be despite being a single cell.

The sporozoans are nonmotile parasites that form spores; *Plasmodium* causes malaria.

Slime molds, which produce nonmotile spores, are unlike fungi in that they have an amoeboid stage and are heterotrophic by ingestion. Water molds, which are filamentous and hetertrophic by absorption, are unlike fungi in that they produce flagellated 2n zoospores. Meiosis produces specialized gametes: sperm and egg.

Reviewing the Chapter

1. List and discuss three ways that protists are complex.   380
2. Describe the structures of *Chlamydomonas* and *Volvox,* and contrast how they reproduce.   383–84
3. Describe the structures of *Spirogyra* and *Ulva,* and explain how they reproduce.   382–83
4. Describe the structure of *Chara* and the relationship of stoneworts to plants.   383
5. Describe the structure of red algae, and discuss their economic importance.   384
6. Describe the structure of brown algae, and discuss their ecological and economic importance.   385
7. Describe the structures of diatoms and dinoflagellates. What is a red tide?   387–88
8. Describe the unique structure of euglenoids.   388–89
9. Name several illnesses caused by zooflagellates.   389
10. Define the term *protozoan,* and list phyla that are sometimes included among the protozoans.   389
11. Distinguish among the amoeboids, foraminiferans, and radiolarians.   390
12. How are ciliates like and different from amoeboids?   391
13. Describe the life cycle of *Plasmodium vivax,* the causative agent of malaria.   392
14. What features distinguish slime molds and water molds from fungi? Describe the life cycle of a plasmodial slime mold.   393

Testing Yourself

Choose the best answer for each question.
For questions 1–5, match each item to those in the key.

KEY:
a. phylum Chlorophyta
b. phylum Pyrrophyta
c. phylum Rhizopoda
d. phylum Phaeophyta
e. phylum Ciliophora

1. Red algae
2. Ciliates
3. Brown algae
4. Amoeboids
5. Green algae
6. Which of these is not a green alga?
   a. *Volvox*   d. *Chlamydomonas*
   b. *Fucus*   e. *Ulva*
   c. *Spirogyra*
7. Which is not a characteristic of brown algae?
   a. multicellular
   b. chlorophylls *a* and *b*
   c. live along rocky coast
   d. harvested for commercial reasons
   e. contain a brown pigment
8. In *Chlamydomonas,*
   a. the adult is haploid.
   b. the zygospore survives times of stress.
   c. sexual reproduction occurs.
   d. asexual reproduction occurs.
   e. All of these are correct.
9. Which is found in *Ulva* but not in plants?
   a. alternation of generations
   b. generations look alike
   c. sperm and egg are produced
   d. protection of zygote
   e. Both b and c are correct.
10. Which of these algae are not flagellated?
    a. *Volvox*   d. *Chlamydomonas*
    b. *Spirogyra*   e. trypanosomes
    c. dinoflagellates
11. Which pair is mismatched?
    a. diatoms—silica shell, boxlike, golden brown
    b. euglenoids—flagella, pellicle, eyespot
    c. *Fucus*—adult is diploid, seaweed, chlorophylls *a* and *c*
    d. *Paramecium*—cilia, calcium carbonate shell, gullet
    e. foraminiferan—test, pseudopod, digestive vacuole
12. Which is a false statement?
    a. Only heterotrophic and not photosynthetic protists are flagellated.
    b. Among protozoans, sporozoans are symbiotic.
    c. Among protists, the haploid cycle is common.
    d. Ciliates exchange genetic material during conjugation.
    e. Slime molds have an amoeboid stage.
13. Which pair is mismatched?
    a. trypanosome—African sleeping sickness
    b. *Plasmodium vivax*—malaria
    c. amoeboid—severe diarrhea
    d. AIDS—*Giardia lamblia*
    e. dinoflagellates—coral
14. Which is found in slime molds but not in fungi?
    a. nonmotile spores   d. photosynthesis
    b. flagellated cells   e. chitin in cell walls
    c. zygote formation
15. Which is a false statement?
    a. Slime molds and water molds are protists.
    b. There are flagellated algae and flagellated protozoans.
    c. Among protists, some flagellates are photosynthetic.
    d. Among protists, only green algae ever have a sexual life cycle.
    e. Conjugation occurs among green algae.

16. Which pair is properly matched?
    a. water mold—flagellate
    b. trypanosome—protozoan
    c. *Plasmodium vivax*—mold
    d. amoeboid—algae

17. Which is an incorrect statement?
    a. Unicellular protists can be quite complex.
    b. Euglenoids are motile but have chloroplasts.
    c. Plasmodial slime molds are amoeboid but have sporangia.
    d. *Volvox* is colonial but has a boxed shape.
    e. Both b and d are incorrect.

18. All are correct about brown algae except that they
    a. range in size from small to large.
    b. are a type of seaweed.
    c. live on land.
    d. are photosynthetic.
    e. are usually multicellular.

19. In the haploid life cycle (e.g., *Chlamydomonas*),
    a. meiosis occurs following zygote formation.
    b. the adult is diploid.
    c. fertilization is delayed beyond the diploid stage.
    d. the zygote produces sperm and eggs.

20. Dinoflagellates
    a. usually reproduce sexually.
    b. have protective cellulose plates.
    c. are insignificant producers of food and oxygen.
    d. have cilia instead of flagella.
    e. tend to be larger than brown algae.

21. Ciliates
    a. move by pseudopods.
    b. are not as varied as other protists.
    c. have a gullet for food gathering.
    d. does not divide by binary fission.
    e. are closely related to the radiolarians.

22. Label this diagram of the *Chlamydomonas* life cycle.

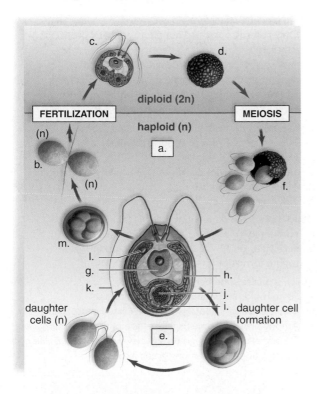

Thinking Scientifically

1. While studying a unicellular alga, you discover a mutant in which the daughter cells do not separate after mitosis. This gives you an idea about how filamentous algae may have evolved. You hypothesize that the mutant alga is missing a protein or making a new form of a protein. How might each possibly lead to a filamentous appearance?
2. You are trying to develop a new anti-termite chemical that will not harm environmentally beneficial insects. Since termites are adapted to eat only wood, they will starve if they cannot digest this food source. What are two possible noninsect targets for your chemical that would result in the death of termites?

Understanding the Terms

algae 382	plankton 380
amoeboid 390	plasmodial slime mold 393
brown algae 385	protist 380
cellular slime mold 393	protozoan 389
ciliate 391	pseudopod 390
colony 384	radiolarian 390
conjugation 383	red algae 384
cyst 380	seaweed 385
diatom 387	sporangium 393
dinoflagellate 387	sporozoan 392
euglenoid 388	test 390
filament 382	trichocyst 391
foraminiferan 390	trypanosome 389
green algae 382	water mold 394
holozoic 391	zooflagellate 389
phagocytize 390	zooplankton 390
phytoplankton 387	zoospore 382

Match the terms to these definitions:

a. _____ Cytoplasmic extension of amoeboid protists; used for locomotion and engulfing food.
b. _____ Flexible freshwater unicellular organism that usually contains chloroplasts and is flagellated.
c. _____ Freshwater or marine unicellular protist with a cell wall consisting of two silica-impregnated valves; extremely numerous in phytoplankton.
d. _____ Causes severe diseases in human beings and domestic animals, including a condition called sleeping sickness.
e. _____ Freshwater and marine organisms that are suspended on or near the surface of the water.

ARIS, the *Biology* Website

ARIS, the website for *Biology*, provides a wealth of information organized and integrated by chapter. You will find practice quizzes, interactive activities, labeling exercises, flashcards, and much more that will complement your learning and understanding of general biology.

www.mhhe.com/maderbiology9

23

THE FUNGI

hat do athlete's foot, homemade bread, and LSD have in common? They all involve species from the kingdom Fungi. The best-known members of this kingdom are the highly visible mushrooms, puffballs, and shelf fungi. Less visible to the casual observer are the mildews, molds, and smuts, which also are fungi.

The fungi used to be classified as plants because they have a cell wall and are nonmotile. Furthermore, certain fungi such as mushrooms superficially resemble plants. Molecular data suggest that fungi are more closely related to animals than they are to plants. However, animals ingest and then digest their food, while fungi first digest their food externally and then absorb the nutrients. In this way and several others, fungi are unique in their structure and function; therefore, they deserve their own kingdom.

Fungi are not usually noted for being extremely large, but researchers have found the body of one fungus in Washington State that spreads, beneath the surface, to over 1,500 acres. Fungi spread by feeding on the organic remains of other organisms in the soil. Dead leaves, tree trunks, and the carcasses of dead animals are their daily fare. Bacterial and also fungal decomposers enrich the environment with nutrients and thereby keep various chemicals cycling in the biosphere. Trees grow, grass turns green, and flowers appear all because fungi, while digesting for themselves, have also digested for others. Not all fungi are decomposers; some are responsible for dreaded diseases in plants and animals, including humans.

Scarlet hood mushroom, *Hygrocybe coccinea*.

23.1 Characteristics of Fungi

Kingdom **Fungi** includes over 80,000 species of mostly multicellular eukaryotes that share a common mode of nutrition. Mycologists, scientists that study fungi, expect this number of species to increase to over 1.5 million in the future. Like animals, they are heterotrophic and consume preformed organic matter. Animals, however, are heterotrophs that ingest food, while fungi are heterotrophs that absorb food. Their cells send out digestive enzymes into the immediate environment and then, when organic matter is broken down, the cells absorb the resulting nutrient molecules.

The majority of fungal species are saprotrophic decomposers that feed on the waste products and dead remains of plants and animals. A dying tree or a cowpat is literally a smorgasbord for millions of fungi. Some fungi are parasitic, living off the tissues of living plants and animals. Fungal diseases such as smuts and rusts account for millions of dollars in crop losses each year. In addition, some plant diseases have greatly reduced the numbers of various types of trees, such as the chestnut. Fungi are also responsible for a number of diseases in humans, including yeast infections, ringworm, aspergillosis, and histoplasmosis.

In addition to causing disease, fungi are also economically important. Approximately 30 species of mushrooms are available in supermarkets throughout the United States. Truffles are gourmet delights that rival the cost of gold per ounce. The unique flavors of many cheeses such as Roquefort, Camembert, and Gorgonzola are the result of fungal metabolism. The baking and alcohol industries use the anaerobic nature of yeast to make bread rise and produce ethyl alcohol. Fungi are also important in the production of antibiotics such as penicillin. One new fungal chemical known as fumagillin has promise as an anticancer agent.

Several types of fungi have a mutualistic relationship with the roots of seed plants; they acquire inorganic nutrients for plants, and in return they are given organic nutrients. Others form an association with a green alga or cyanobacterium within a lichen. Lichens can live on rocks and are important soil formers.

Fungi are multicellular eukaryotes that often function as saprotrophic decomposers. Some are mutualistic and others are disease-causing parasites.

Structure of Fungi

Some fungi, including the yeasts, are unicellular; however, the vast majority of species are multicellular. The thallus or body of most fungi is a multicellular structure known as a mycelium (Fig. 23.1). A **mycelium** [Gk. *mycelium*, fungus filaments] is a network of filaments called **hyphae** [Gk. *hyphe*, web]. Hyphae give the mycelium quite a large surface area per volume of cytoplasm, and this facilitates absorption of nutrients into the body of a fungus. Hyphae grow at their tips, and the mycelium absorbs and then passes nutrients on to the growing tips. When a fungus reproduces, a specific portion of the mycelium becomes a reproductive structure that is then nourished by the rest of the mycelium (Fig. 23.1b).

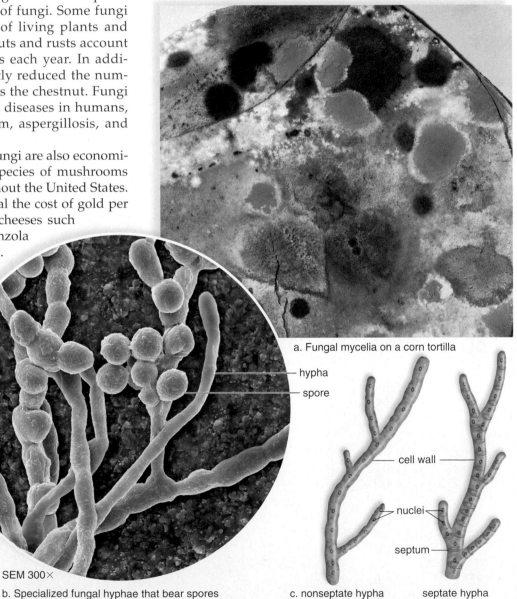

a. Fungal mycelia on a corn tortilla

hypha
spore

SEM 300×
b. Specialized fungal hyphae that bear spores

cell wall

nuclei

septum

c. nonseptate hypha   septate hypha

FIGURE 23.1 Mycelia and hyphae of fungi.
a. Each mycelium grown from a different spore on a corn tortilla is quite symmetrical. **b.** Scanning electron micrograph of specialized aerial fungal hyphae that bear spores. **c.** Hyphae are either nonseptate (do not have cross walls) or septate (have cross walls).

Fungal cells are quite different from plant cells, not only by lacking chloroplasts but also by having a cell wall that contains chitin and not cellulose. Chitin, like cellulose, is a polymer of glucose organized into microfibrils. In chitin, however, each glucose molecule has a nitrogen-containing amino group attached to it. Chitin is also found in the exoskeleton of arthropods, a major phylum of animals that includes the insects and crustaceans. The energy reserve of fungi is not starch but glycogen, as in animals. Fungi are nonmotile; they lack basal bodies and do not have flagella at any stage in their life cycle. They move toward a food source by growing toward it. Hyphae can cover as much as a kilometer a day!

Some fungi have cross walls in their hyphae. These hyphae are called **septate** [L. *septum,* fence, wall]. Actually the presence of septa makes little difference because pores allow cytoplasm and sometimes even organelles to pass freely from one cell to the other. The septa that separate reproductive cells, however, are complete in all fungal groups. **Nonseptate** fungi are multinucleated; they have many nuclei in the cytoplasm of a hypha (Fig. 23.1*c*).

The body of a fungus is a mycelium made up of hyphae. Fungal cells are nonmotile and have a cell wall that contains chitin.

Reproduction of Fungi

Both sexual and asexual reproduction occur in fungi. In general, fungal sexual reproduction involves these stages:

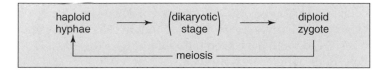

The relative length of time of each phase varies with the species.

During sexual reproduction, hyphae (or a portion thereof) from two different mating types make contact and fuse. It would be expected that the nuclei from the two mating types would also fuse immediately, and they do in some species. In other species, the nuclei pair but do not fuse for days, months, or even years. The nuclei continue to divide in such a way that every cell (in septate hyphae) has at least one of each nucleus. A hypha that contains paired haploid nuclei is said to be n + n or **dikaryotic** [Gk. *dis*, two, and *karyon*, nucleus, kernel]. When the nuclei do eventually fuse, the zygote undergoes meiosis prior to spore formation. Fungal spores germinate directly into haploid hyphae without any noticeable embryological development.

How can a terrestrial and nonmotile organism ensure that the offspring will be dispersed to new locations? As an adaptation to life on land, fungi usually produce nonmo-

a.   20 μm

b.

FIGURE 23.2 Dispersal of spores.
a. Electron micrograph shows the external appearance of spores discharged by (**b**) earth star, *Geastrum*, which releases hordes of spores into the air.
Part *a*: From C. Y. Shih and R. G. Kessel, *Living Images.* Science Books International, Boston, 1982.

tile, but normally windblown, spores during both sexual and asexual reproduction (Fig. 23.2). A **spore** is a reproductive cell that develops into a new organism without the need to fuse with another reproductive cell. A large mushroom may produce billions of spores within a few days. When a spore lands upon an appropriate food source, it germinates and begins to grow.

Asexual reproduction usually involves the production of spores by a specialized part of a single mycelium. Alternately, asexual reproduction can occur by fragmentation—a portion of a mycelium begins a life of its own. Also, unicellular yeasts reproduce asexually by **budding;** a small cell forms and gets pinched off as it grows to full size (see Fig. 23.4).

Fungi produce windblown spores during both asexual and sexual reproduction.

23.2 Evolution of Fungi

The ancestry of fungi, which evolved about 570 million years ago, is under scientific investigation. It is possible that the organisms now classified in the kingdom Fungi do not share a recent common ancestor. In that case, they probably evolved separately from different protists. It's been suggested, though, that fungi evolved from red algae because both fungi and red algae lack flagella in all stages of the life cycle. Fossils of the first vascular plants from the late Silurian period possess petrified fungi called mycorrhizae on their roots. Perhaps fungi were instrumental in the colonization of land by plants.

In the past, fungi have been classified in other kingdoms. Originally they were considered a part of the plant kingdom, and later they were placed in the kingdom Protista. But in 1969, R. H. Whittaker argued that their multicellular nature and mode of nutrition (extracellular digestion and absorption of nutrients) meant they should be in their own kingdom. Table 23.1 contrasts the features of fungi with the features of certain other organisms now included in kingdom Protista by this text. In the absence of detailed knowledge of their evolutionary relationships, fungal groups are classified according to differences in the life cycle and the type of structure that produces spores. Because scientists are now using comparative molecular data to decipher evolutionary relationships, changes in the classification of fungi may be forthcoming.

Zygospore Fungi

Phylum **Zygomycota** includes approximately 1,050 species of fungi known as the **zygospore fungi.** These organisms are saprotrophic, living off plant and animal remains in the soil or in bakery goods in the pantry. Some are parasites of minute soil protists, worms, and insects such as a housefly.

The black bread mold, *Rhizopus stolonifer,* is commonly used as an example of this phylum. The body of this fungus, which is composed of nonseptate hyphae, demonstrates that although there is little cellular differentiation among fungi, the hyphae may be specialized for various purposes. In *Rhizopus,* stolons are horizontal hyphae that exist on the surface of the bread; rhizoids grow into the bread, anchor the mycelium, and carry out digestion; and sporangiophores are aerial hyphae that bear sporangia. A **sporangium** is a capsule that produces spores called sporangiospores. During asexual reproduction, all structures involved are haploid (Fig. 23.3).

The phylum name refers to the zygospore, which is seen during sexual reproduction. The hyphae of opposite mating types, termed plus (+) and minus (−), are chemically attracted, and they grow toward each other until they touch. The ends of the hyphae swell as nuclei enter; then cross walls develop a short distance behind each end, forming gametangia. The gametangia merge, and the result is a large multinucleate cell in which the nuclei of the two mating types pair and then fuse. A thick wall develops around the cell, which is now called a **zygospore.** The zygospore undergoes a period of dormancy before meiosis and germination take place. One or more sporangiophores with sporangia at their tips develop, and many spores are released. The spores, dispersed by air currents, give rise to new haploid mycelia. Spores from black bread mold have been found in the air above the North Pole, in the jungle, and far out at sea.

Zygospore fungi produce spores within sporangia. During sexual reproduction, a zygospore forms prior to meiosis and production of spores.

CLASSIFICATION

DOMAIN: Eukarya
KINGDOM: Fungi

CHARACTERISTICS
- Multicellular eukaryotes
- Heterotrophic by absorption
- Lack flagella
- Nonmotile spores form during both asexual and sexual reproduction

PHYLUM: Zygomycota ⟶ zygospore fungi
PHYLUM: Ascomycota ⟶ sac fungi
PHYLUM: Basidiomycota ⟶ club fungi

TABLE 23.1

Certain Protists Compared to Fungi

	Fungi	*Red Algae*	*Plasmodial Slime Molds*	*Water Molds*
Body form	Filamentous	Filamentous	Multinucleate plasmodium	Filamentous
Mode of nutrition	Heterotrophic by absorption	Autotrophic by photosynthesis	Heterotrophic by ingestion	Heterotrophic by absorption
Basal bodies/flagella	In no stages	In no stages	In one stage	Flagellated zoospores
Cell wall	Contains chitin	Contains cellulose	None	Contains cellulose
Life cycle	Zygotic meiosis (haploid cycle)	Sporic meiosis (alternation of generations)	Unique	Gametic meiosis (diploid cycle)

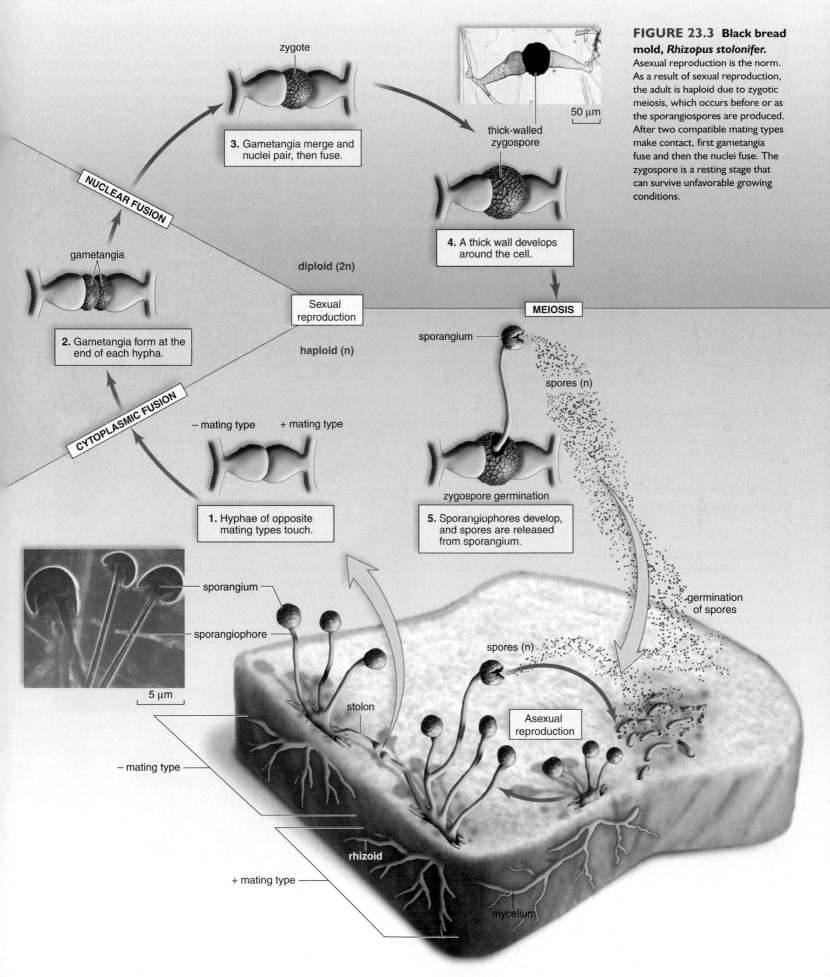

zygote

3. Gametangia merge and nuclei pair, then fuse.

NUCLEAR FUSION

gametangia

2. Gametangia form at the end of each hypha.

CYTOPLASMIC FUSION

− mating type   + mating type

1. Hyphae of opposite mating types touch.

50 μm

thick-walled zygospore

4. A thick wall develops around the cell.

diploid (2n)

Sexual reproduction

haploid (n)

MEIOSIS

sporangium

spores (n)

zygospore germination

5. Sporangiophores develop, and spores are released from sporangium.

germination of spores

spores (n)

Asexual reproduction

sporangium

sporangiophore

5 μm

stolon

− mating type

rhizoid

+ mating type

mycelium

FIGURE 23.3 Black bread mold, *Rhizopus stolonifer.* Asexual reproduction is the norm. As a result of sexual reproduction, the adult is haploid due to zygotic meiosis, which occurs before or as the sporangiospores are produced. After two compatible mating types make contact, first gametangia fuse and then the nuclei fuse. The zygospore is a resting stage that can survive unfavorable growing conditions.

Sac Fungi

Phylum **Ascomycota** consists of about 50,000 species of fungi that are commonly termed the **sac fungi**. Presently, the sac fungi can be thought of as having two main groups: the sexual ascomycetes, in which sexual reproduction has long been known, and the asexual ascomyetes, in which sexual reproduction has not yet been observed.

The sexual ascomycetes include the yeasts, unicellular fungi important in the baking and brewing industries and also in various molecular biological studies. *Neurospora*, the experimental material for the one-gene–one-enzyme studies, and the other red bread molds are sexual ascomycetes. So are the morels and truffles, which are famous gourmet delicacies revered throughout the world. A large number of sexual ascomycetes are parasitic on plants. Powdery mildews grow on leaves, as do leaf curl fungi; chestnut blight and Dutch elm disease destroy the trees named. Ergot, a parasitic sac fungus that infects rye and (less commonly) other grains is discussed in the Health Focus on page 405.

The asexual ascomycetes have only recently been designated as members of the phylum Ascomycota. These fungi used to be in the phylum Deuteromycota, or the imperfect fungi, because their means of sexual reproduction was unknown. However, on the basis of molecular data and structural characteristics, these fungi have now been identified as sac fungi. The asexual ascomycetes include *Aspergillus*, *Candida*, and, formerly, *Penicillium* molds. The asci of *Penicillium* have recently been discovered and, therefore, it has been renamed *Talaromyces*. *Aspergillus* and *Candida* cause serious human infections even though *Aspergillus* is useful to humans in various ways. *Talaromyces* is important for the production of antibiotics and the aging of cheese.

Biology of the Sac Fungi

The body of the ascomyetes can be a single cell, as in yeasts, but more often it is a mycelium composed of septate hyphae.

The ascomycetes play an essential role in recycling by digesting resistant (not easily decomposed) materials containing cellulose, lignin, or collagen. Species are also known that can even consume jet fuel and wall paint. Some are symbiotic with algae, forming lichens, and plant roots, forming mycorrhizae. They also account for most of the known fungal pathogens causing various plant and animal, including human, diseases.

Reproduction. Asexual reproduction is the norm among ascomycetes. The yeasts usually reproduce by budding. A small cell forms and pinches off as it grows to full size (Fig. 23.4*a*). The other ascomycetes produce spores called conidia or **conidiospores** that vary in size and shape and may be multicellular. The conidia usually develop at the tips of specialized aerial hyphae called conidiophores (Fig. 23.4*b*). Conidiophores differ in appearance and this helps mycologists identify the particular ascomycetes. When released, the spores are windblown. The conidia of the allergy-causing mold *Cladosporium* (see Fig. 23.4*b*) are carried easily though the air and transported even over oceans. One researcher found a concentration of more than 35,000 *Cladosporium* conidia/m^3 over Leiden (Germany).

The phylum name, Ascomycota, refers to the **ascus** [Gk. *askos*, bag, sac], a fingerlike sac that develops during sexual re-

FIGURE 23.5 Sexual reproduction in sac fungi.
The ascomycetes reproduce sexually by producing asci, in fruiting bodies called ascocarps. **a.** In the ascocarp of cup fungi, dikaryotic hyphae terminate forming the asci, where meiosis follows nuclear fusion and spore formation takes place. **b.** In morels, the asci are borne on the ridges of pits. **c.** Peach leaf curl, a parasite of leaves, forms asci as shown.

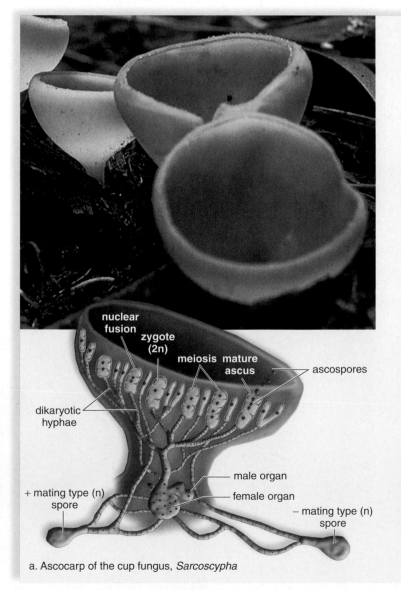

a. Ascocarp of the cup fungus, *Sarcoscypha*

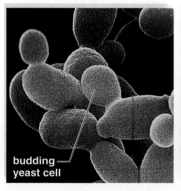

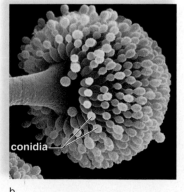

FIGURE 23.4 Asexual reproduction in sac fungi.
a. Yeasts, unique among fungi, reproduce by budding. **b.** The sac fungi usually reproduce asexually by producing spores called conidia or conidiospores.

production. On occasion, the asci are surrounded and protected by sterile hyphae within a fruiting body called an *ascocarp* (Fig. 23.5). A **fruiting body** is a reproductive structure where spores are produced and released. Ascocarps can have different shapes; in cup fungi they are cup shaped and in morels they are stalked and crowned by a pitted bell-shaped ascocarp.

Ascus-producing hyphae remain dikaryotic except in the walled-off portion that becomes the ascus, where nuclear fusion and meiosis take place. Because mitosis follows meiosis, each ascus contains eight haploid nuclei and produces eight spores.

ascospore ———— ascus

In most ascomyetes, the asci become swollen as they mature, and then they burst, expelling the ascopores. If released into the air, the spores are then windblown.

Relationship of Asomycetes to Humans

Many ascomycetes are well known to humans as organisms that produce useful substances such as the drug penicillin, which cures bacterial infections; cyclosporin, which suppresses the immune system leading to the success of transplantation operations; and the steroids, which are present in the birth control pill. They are also used during the production of various foods such as blue cheese and Coke. Ascomycetes cause diseases known as *mycoses*. Many mycoses are acquired from the environment. Ringworm comes from soil fungi, rose gardener's disease from thorns, Chicago's disease from old buildings, and basketweaver's disease from grass cuttings.

Yeasts.   **Yeasts** have proven useful to human beings. In the wild, yeasts grow on fruits, and historically the yeasts already present on grapes were used to produce wine. Today, selected yeasts, such as *Saccharomyces cerevisiae*, are added to relatively sterile grape juice to make wine. Also, this yeast is added to prepared grains to make beer. When *Saccharomyces* ferments, it produces ethanol and also carbon dioxide. Both the ethanol and the carbon dioxide are retained for beers and sparkling wines; carbon dioxide is released for still wines. In baking, the carbon dioxide given off is the leavening agent that causes bread to rise. *Saccharomyces* is serviceable to humans in another way. It is sometimes used in genetic engineering experiments requiring a eukaryote.

Candida albicans is a yeast that causes the widest variety of fungal infections. Disease occurs when the normal balance

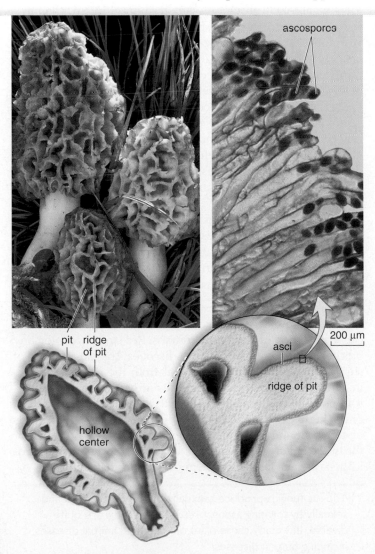

b. Ascocarp of the morel, *Morchella*

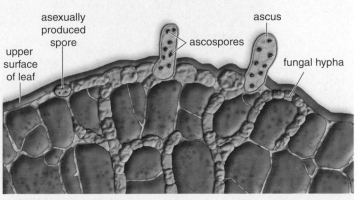

c. Peach leaf curl, *Taphrina*

of microbes in an organ, such as the vagina, is disturbed, particularly by antibiotic therapy. Then *Candida* proliferates and symptoms result. A vaginal infection results in inflammation, itching, and discharge. Oral thrush is a *Candida* infection of the mouth, common in newborns and AIDS patients. In immunocompromised individuals, *Candida* can move through the body, causing a systemic infection that can damage the heart, brain, and other organs.

Molds.   *Aspergillus* is a group of green molds recognized by the bottle-shaped structure that bears their conidiophores. Commonly isolated from soil, plant debris, and house dust, *Aspergillus* is sometimes pathogenic to humans. However, it is used to produce soy sauce by fermentation of soybeans. A Japanese food called miso is made by fermenting soybeans and rice with *Aspergillus*. In the United States, *Aspergillus* is used to produce citric and gallic acids, which serve as additives during the manufacture of a wide variety of products, including foods, inks, medicines, dyes, plastics, toothpaste, soap, and even chewing gum. *Aspergillus flavus*, which grows on moist seeds, secretes a toxin that is the most potent natural carcinogen known. Therefore, in humid climates such as that in the southeastern United States, care must be taken to store grains properly. *Aspergillus* also causes a potentially deadly disease of the respiratory tract that arises after spores have been inhaled.

Various molds can be helpful and/or harmful to humans. The mold *Stachybotrys chartarum* (Fig. 23.6) grows well on building materials. It is known as black mold and is responsible for the "sick-building" syndrome. Individuals with chronic exposure to toxins produced by this fungus have reported cold and flulike symptoms, fatigue, and dermatitis. The toxins may suppress and could destroy the immune system, affecting the lymphoid tissue and the bone marrow.

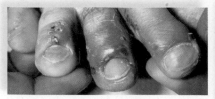
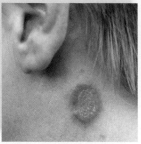

FIGURE 23.7 Tineas.
a. Athlete's foot and (**b**) ringworm are termed tineas.

A species of *Penicillium* (blue molds now classified as *Talaromyces*) is the source of the familiar antibiotic called penicillin, which was first manufactured during World War II and still has many applications today. Since the discovery of penicillin by Sir Alexander Fleming, it has saved countless lives. Other *Penicillium* species give the characteristic flavor and aroma to cheeses such as Roquefort and Camembert. The bluish streaks in blue cheese are patches of conidiospores.

Moldlike fungi in the genus *Tinea* cause infections of the skin called tineas. Athlete's foot is a tinea characterized by itching and peeling of the skin between the toes (Fig. 23.7a). In ringworm, the fungus releases enzymes that degrade keratin and collagen in skin. The area of infection becomes red and inflamed. The fungal colony grows outward, forming a ring of inflammation. The center of the lesion begins to heal, thereby giving ringworm its characteristic appearance, a red ring surrounding an area of healed skin (Fig. 23.7b). Tinea infections of the scalp are rampant among school-age children, affecting as much as 30% of the population, with the possibility of permanent hair loss.

The vast majority of people living in the Midwest have been infected with *Histoplasma capsulatum*, a thermally dimorphic fungus that grows in mold form at 25°C and in yeast form at 37°C. This common soil fungus, often associated with bird droppings, leads in most cases to a mild "fungal flu." Over 40 million people have been infected in the United States with 500,000 new cases annually. Less than half notice any symptoms, with 3,000 showing severe disease and about 50 dying each year from histoplasmosis. The fungal pathogen lives and grows within cells of the immune system and causes systemic illness. Lesions are formed in the lungs that leave calcifications, visible in X-ray images, that resemble those of tuberculosis.

Control of Fungal Infections.   The strong similarities between fungal and human cells make it difficult to design fungal medications that do not also harm humans. Researchers exploit any biochemical differences they can discover. The biosynthesis of steroids in fungi differs somewhat from the same pathways in humans. A variety of fungicides are directed against steroid biosynthesis, including some that are applied to fields of grain. Fungicides based on heavy metals are applied to seeds of sorghum and other crops. Topical agents are available for the treatment of yeast infections and tineas, and systemic medications are available for systemic ascomycete infections.

FIGURE 23.6
Black mold.
Stachybotrys chartarum, or black mold, grows well in moist areas, including the walls of homes. It represents a potential health risk.

—conidiophore

SEM 1,800×

The sac fungi reproduce asexually by forming conidia and sexually by forming ascospores. The sac fungi are quite diverse, but many cause plant diseases and animal diseases, including human diseases.

health focus

Deadly Fungi

It is unwise for amateurs to collect mushrooms in the wild because certain mushroom species are poisonous. The red and yellow *Amanitas* are especially dangerous. This species is also known as fly agaric because it was thought to kill flies (it was gathered and then sprinkled with sugar to attract flies). Its toxins include muscarine and muscaridine, which produce symptoms similar to those of acute alcoholic intoxication. In one to six hours, the victim staggers, loses consciousness, and becomes delirious, sometimes suffering from hallucinations, manic conditions, and stupor. Luckily, it also causes vomiting, which rids the system of the poison, so death occurs in less than 1% of cases. The death angel mushroom (*Amanita phalloides,* Fig. 23A) causes 90% of the fatalities attributed to mushroom poisoning. When this mushroom is eaten, symptoms don't begin until 10–12 hours later. Abdominal pain, vomiting, delirium, and hallucinations are not the real problem; rather, a poison interferes with RNA (ribonucleic acid) transcription by inhibiting RNA polymerase, and the victim dies from liver and kidney damage.

Some hallucinogenic mushrooms are used in religious ceremonies, particularly among Mexican Indians. *Psilocybe mexicana* contains a chemical called psilocybin that is a structural analogue of LSD and mescaline. It produces a dreamlike state in which visions of colorful patterns and objects seem to fill up space and dance past in endless succession. Other senses are also sharpened to produce a feeling of intense reality.

The only reliable way to tell a nonpoisonous mushroom from a poisonous one is to be able to correctly identify the species. Poisonous mushrooms cannot be identified with simple tests, such as whether they peel easily, have a bad odor, or blacken a silver coin during cooking. Only consume mushrooms identified by an expert!

Like club fungi, some sac fungi also contain chemicals that can be dangerous to people. *Claviceps purpurea,* the ergot fungus, infects rye and replaces the grain with ergot—hard, purple-black bodies consisting of tightly cemented hyphae (Fig. 23B). When ground with the rye and made into bread, the fungus releases toxic alkaloids that cause the disease ergotism. In humans, vomiting, feelings of intense heat or cold, muscle pain, a yellow face, and lesions on the hands and feet are accompanied by hysteria and hallucinations. Ergotism was common in Europe during the Middle Ages. During this period, it was known as St. Anthony's Fire and was responsible for 40,000 deaths in an epidemic in 994 A.D. We now know that ergot contains lysergic acid, from which LSD is easily synthesized. Based on recorded symptoms, historians believe that those individuals who were accused of practicing witchcraft in Salem, Massachusetts, during the seventeenth century were actually suffering from ergotism. It is also speculated that ergotism is to blame for supposed demonic possessions throughout the centuries. As recently as 1951, an epidemic of ergotism occurred in Pont-Saint-Esprit, France. Over 150 persons became hysterical, and four died.

Because the alkaloids that cause ergotism stimulate smooth muscle and selectively block the sympathetic nervous system, they can be used in medicine to cause uterine contractions and to treat certain circulatory disorders, including migraine headaches. Although the ergot fungus can be cultured in petri dishes, no one has succeeded in inducing it to form ergot in the laboratory. So far, the only way to obtain ergot, even for medical purposes, is to collect it in an infected field of rye.

FIGURE 23A Poisonous mushroom, *Amanita phalloides.*

FIGURE 23B Ergot infection of rye, caused by *Claviceps purpurea.*

ergot

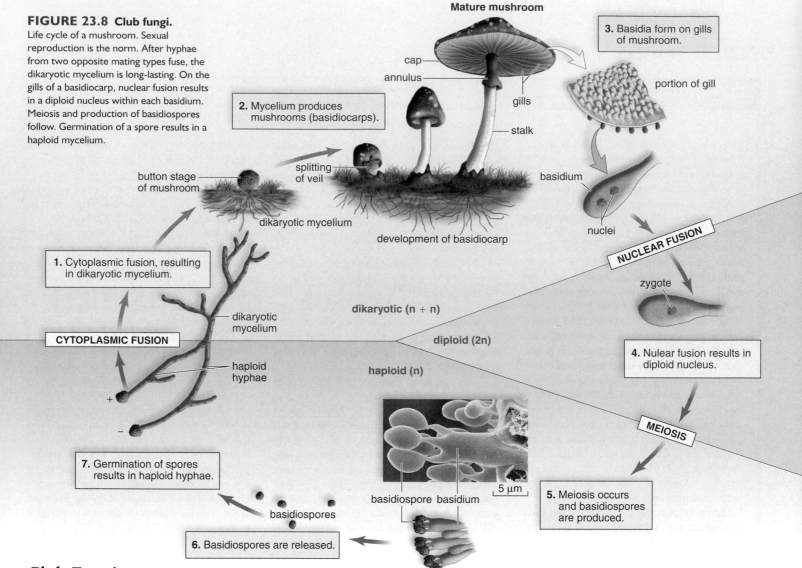

FIGURE 23.8 Club fungi.
Life cycle of a mushroom. Sexual reproduction is the norm. After hyphae from two opposite mating types fuse, the dikaryotic mycelium is long-lasting. On the gills of a basidiocarp, nuclear fusion results in a diploid nucleus within each basidium. Meiosis and production of basidiospores follow. Germination of a spore results in a haploid mycelium.

Mature mushroom

3. Basidia form on gills of mushroom.

cap
annulus
gills
portion of gill
stalk
basidium

2. Mycelium produces mushrooms (basidiocarps).

button stage of mushroom

splitting of veil

dikaryotic mycelium

development of basidiocarp

nuclei

NUCLEAR FUSION

1. Cytoplasmic fusion, resulting in dikaryotic mycelium.

zygote

dikaryotic (n + n)

dikaryotic mycelium

CYTOPLASMIC FUSION

diploid (2n)

4. Nulear fusion results in diploid nucleus.

haploid hyphae

haploid (n)

MEIOSIS

7. Germination of spores results in haploid hyphae.

+

−

5 μm

5. Meiosis occurs and basidiospores are produced.

basidiospore basidium

basidiospores

6. Basidiospores are released.

Club Fungi

Phylum **Basidiomycota** consists of over 22,000 species of fungi known as the **club fungi.** Mushrooms, toadstools, puffballs, shelf fungi, jelly fungi, bird's-nest fungi, and stinkhorns are basidiomycetes. In addition, fungi that cause plant diseases such as the smuts and rusts are placed in this phylum. Several mushrooms such as the *portabella* and *shiitake* mushrooms are savored as foods by humans. Approximately 75 species of basidiomycetes are considered poisonous. The poisonous death angel mushroom is discussed in the Health Focus on page 405.

Biology of Club Fungi

The body of a basidomycete is a mycelium composed of septate hyphae. Most members of this phylum are saprotrophs, although several parasitic species exist.

Reproduction. Although club fungi occasionally do produce conidia asexually, they usually reproduce sexually. The phylum name, Basidiomycota, refers to the **basidium** [L. *basidi*, small pedestal], a club-shaped structure in which spores called basidiospores develop. Basidia are located within a fruiting body called a basidiocarp (Fig. 23.8). Prior to formation of a basidiocarp, haploid hyphae of opposite mating types meet and fuse, producing a dikaryotic (n + n)

mycelium. The dikaryotic mycelium continues its existence year after year, even for hundreds of years on occasion. In many species of mushrooms, the dikaryotic mycelium often radiates out and produces mushrooms in an ever larger, so-called fairy ring (Fig. 23.9*a*).

Mushrooms are composed of nothing but tightly packed hyphae whose walled-off ends become basidia. In gilled mushrooms, the basidia are located on radiating lamellae, the gills. In shelf fungi and pore mushrooms (Fig. 23.9*b, c*), the basidia terminate in tubes. In any case, the extensive surface area of a basidiocarp is lined by basidia, where nuclear fusion, meiosis, and spore production occur. A basidium has four projections in which cytoplasm and a haploid nucleus enter as the basidiospore forms:

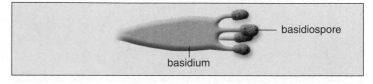

basidiospore
basidium

Basidiospores are often windblown; when they germinate, a new haploid mycelium forms. It is estimated that some large mushrooms can produce up to 40 million spores per hour.

a. Fairy ring

b. Shelf fungus

c. Pore mushroom, *Boletus*

d. Puffball, *Calvatiga gigantea*

FIGURE 23.9 Club Fungi.
a. Fairy ring. Mushrooms develop in a ring on the outer living fringes of a dikaryotic mycelium. The center has used up its nutrients and is no longer living. **b.** A shelf fungus. **c.** Fruiting bodies of *Boletus*. This mushroom is not gilled; instead, it has basidia-lined tubes that open on the undersurface of the cap. **d.** In puffballs, the spores develop inside an enclosed fruiting body. Giant puffballs are estimated to contain 7 trillion spores.

In puffballs, spores are produced inside parchmentlike membranes, and the spores are released through a pore or when the membrane breaks down (Fig. 23.9*d*). In bird's-nest fungi, falling raindrops provide the force that causes the nest's basidiospore-containing "eggs" to fly through the air and land on vegetation that may be eaten by an animal. If so, the spores pass unharmed through the digestive tract. Stinkhorns resemble a mushroom with a spongy stalk and a compact, slimy cap. The long stalk bears the elongated basidiocarp. Stinkhorns emit an incredibly disagreeable odor; flies are attracted by the odor, and when they linger to feed on the sweet jelly, the flies pick up spores that they later distribute.

Smuts and Rusts

Smuts and rusts are club fungi that parasitize cereal crops such as corn, wheat, oats, and rye. They are of great economic

a. Corn smut, *Ustilago*

fungus

leaf

b. Wheat rust, *Puccinia*

FIGURE 23.10 Smuts and rusts.
a. Corn smut. **b.** Micrograph of wheat rust.

importance because of the crop losses they cause every year. Smuts and rusts don't form basidiocarps, and their spores are small and numerous, resembling soot. Some smuts enter seeds and exist inside the plant, becoming visible only near maturity. Other smuts externally infect plants. In corn smut, the mycelium grows between the corn kernels and secretes substances that cause the development of tumors on the ears of corn (Fig. 23.10*a*).

The life cycle of rusts may be particularly complex in that it often requires two different plant host species to complete the cycle. Black stem rust of wheat uses barberry bushes as an alternate host, and blister rust of white pine uses currant and gooseberry bushes. Campaigns to eradicate these bushes in areas where the alternate host grows help keep these rusts in check. Wheat rust (Fig. 23.10*b*) is also controlled by producing new and resistant strains of wheat. The process is continuous, because rust can mutate to cause infection once again.

Club fungi usually reproduce sexually. The dikaryotic stage is prolonged and periodically produces fruiting bodies, where spores are produced in basidia.

23.3 SYMBIOTIC RELATIONSHIPS OF FUNGI

Several instances in which fungi are parasites of plants and animals have been mentioned. Two other symbiotic associations are of interest.

Lichens

Lichens are an association between a fungus, usually a sac fungus, and a cyanobacterium or a green alga. The body of a crustose lichen has three layers: The fungus forms a thin, tough upper layer and a loosely packed lower layer that shield the photosynthetic cells in the middle layer (Fig. 23.11). Specialized fungal hyphae, which penetrate or envelop the photosynthetic cells, transfer nutrients directly to the rest of the fungus. Lichens can reproduce asexually by releasing fragments that contain hyphae and an algal cell. In fruticose lichens, the sac fungus reproduces sexually (Fig. 23.11*b*).

In the past, lichens were assumed to be mutualistic relationships in which the fungus received nutrients from the algal cells, and the algal cells were protected from desiccation by the fungus. Actually, lichens may involve a controlled form of parasitism of the algal cells by the fungus, with the algae not benefiting at all from the association. This is supported by experiments in which the fungal and algal components are removed and grown separately. The algae grow faster when they are alone than when they are part of a lichen. On the other hand, it is difficult to cultivate the fungus, which does not naturally grow alone. The different lichen species are identified according to the fungal partner.

Three types of lichens are recognized. Compact crustose lichens are often seen on bare rocks or on tree bark; fruticose lichens are shrublike; and foliose lichens are leaflike (Fig. 23.11). Lichens are efficient at acquiring nutrients and moisture, and therefore they can survive in areas of low moisture and low temperature as well as in areas with poor or no soil. They produce and improve the soil, thus making it suitable for plants to invade the area. Unfortunately, lichens also take up pollutants and cannot survive where the air is polluted. Therefore, their presence indicates that the air is healthy for humans to breathe.

Lichens, an association between fungal hyphae and algal cells, can live in areas of extreme conditions and contribute to the formation of soil.

Mycorrhizae

Mycorrhizae [Gk. *mykes*, fungus, and *rhizion*, dim. for root] are mutualistic relationships between soil fungi and the roots of most plants. Plants whose roots are invaded by mycorrhizae grow more successfully in poor soils—particularly soils deficient in phosphates—than do plants without mycorrhizae (Fig. 23.12). The fungal partner, usually a sac fungus, may enter the cortex of roots but does not enter the cytoplasm of plant cells. Ectomycorrhizae form a mantle that is exterior to the root, and they grow between cell walls. Endomycorrhizae

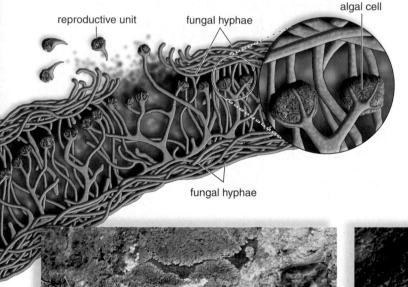

reproductive unit   fungal hyphae   algal cell

fungal hyphae

sac fungi reproductive cups

a. Crustose lichen, *Xanthoria*

b. Fruticose lichen, *Cladonia*

c. Foliose lichen, *Xanthoparmelia*

FIGURE 23.11 Lichen morphology.
a. A section of a compact crustose lichen shows the placement of the algal cells and the fungal hyphae, which encircle and penetrate the algal cells.
b. Fruticose lichens are shrublike. **c.** Foliose lichens are leaflike.

FIGURE 23.12 Plant growth experiment.
A soybean plant (*left*) without mycorrhizae grows poorly compared to two other (*right*) plants infected with different strains of mycorrhizae.

penetrate only the cell walls. In any case, the presence of the fungus gives the plant a greater absorptive surface for the intake of minerals. The fungus also benefits from the association by receiving carbohydrates from the plant.

It is of interest to know that the truffle, a gourmet delight whose ascocarp is somewhat prunelike in appearance, is a mycorrhizal sac fungus living in association with oak and beech tree roots. In the past, the French used pigs (truffle-hounds) to sniff out and dig up truffles, but now they have succeeded in cultivating truffles by inoculating the roots of seedlings with the proper mycelium.

Even the earliest fossil plants have mycorrhizae associated with them. It would appear, then, that mycorrhizae helped plants adapt to and flourish on land.

> Mycorrhizae (fungus roots), a mutualistic association between a fungus and plant roots, help plants acquire mineral nutrients.

CONNECTING THE CONCEPTS

At one time, fungi were considered a part of the plant kingdom, and then later, they were placed in the kingdom Protista. Whittaker argued that, because of their multicellular nature and mode of nutrition, fungi should be in their own kingdom. Like animals, they are heterotrophic, but they do not ingest food—they absorb nutrients. Like plants, they have cell walls, but their cell walls contain chitin instead of cellulose. Fungal cells store energy in the form of glycogen, as do animal cells.

The range of species within the kingdom Fungi is broad. Some fungi are unicellular (such as yeasts), some are multicellular parasites (such as those that cause athlete's foot), and some form mutualistic relationships with other species (such as those found in lichens). Most, however, are saprotrophic decomposers that play a vital role in all ecosystems. The decomposers work with bacteria to break down the waste products and dead remains of plants and animals so that organic materials are recycled.

In addition to their various ways of life, fungi have been successful because they have diverse reproductive strategies. During sexual reproduction, the filaments of different mating types typically fuse. Asexual reproduction via spores, fragmentation, and budding is common.

Fungi have been around for at least 570 million years. It is possible that the species now classified in the kingdom do not share a recent common ancestor, which means they probably evolved from different protists. It is also possible that fungi evolved from red algae because the two groups share similar morphological traits.

Summary

23.1 CHARACTERISTICS OF FUNGI

Fungi are multicellular eukaryotes that are heterotrophic by absorption. After external digestion, they absorb the resulting nutrient molecules. Most fungi act as saprotrophic decomposers that aid the cycling of chemicals in ecosystems by decomposing dead remains. Some fungi are parasitic, especially on plants, and others are mutualistic with plant roots and algae.

The body of a fungus is composed of thin filaments called hyphae, which collectively are termed a mycelium. The cell wall contains chitin, and the energy reserve is glycogen. Fungi do not have flagella at any stage in their life cycle. Nonseptate hyphae have no cross walls; septate hyphae have cross walls, but there are pores that allow the cytoplasm and even organelles to pass through.

Fungi produce nonmotile and often windblown spores during both asexual and sexual reproduction. During sexual reproduction, hyphae tips fuse so that dikaryotic (n + n) hyphae usually result, depending on the type of fungus. Following nuclear fusion, zygotic meiosis occurs during the production of the sexual spores.

23.2 EVOLUTION OF FUNGI

Three significant phyla of fungi are phylum Zygomycota (zygospore fungi), phylum Ascomycota (sac fungi), and phylum Basidiomycota (club fungi).

The zygospore fungi are nonseptate, and during sexual reproduction they have a dormant stage consisting of a thick-walled zygospore. When the zygospore germinates, sporangia produce windblown spores. Asexual reproduction occurs when nutrients are plentiful and sporangia again produce spores. An example of a zygomycete is black bread mold.

The sac fungi are septate, and during sexual reproduction saclike cells called asci produce spores. Asci are sometimes located in fruiting bodies called ascocarps. Asexual reproduction, which is dependent on the production of conidiospores, is more common. Sexual reproduction is unknown in some sac fungi. Sac fungi include *Talaromyces, Aspergillus, Candida,* morels and truffles, and various yeasts and molds, some of which cause disease in plants and animals, including humans.

The club fungi are septate, and during sexual reproduction club-shaped structures called basidia produce spores. Basidia are located in fruiting bodies called basidiocarps. Club fungi have a prolonged dikaryotic stage, and asexual reproduction by conidiospores is rare.

A dikaryotic mycelium periodically produces fruiting bodies. Mushrooms and puffballs are examples of club fungi.

23.3 SYMBIOTIC RELATIONSHIPS OF FUNGI

Lichens are an association between a fungus, usually a sac fungus, and a cyanobacterium or a green alga. Traditionally, this association was considered mutualistic, but experimentation suggests a controlled parasitism by the fungus on the alga. Lichens may live in extreme environments and on bare rocks; they allow other organisms that will eventually form soil to establish.

The term *mycorrhizae* refers to an association between a fungus, usually a sac fungus, and the roots of a plant. The fungus helps the plant absorb minerals, and the plant supplies the fungus with carbohydrates.

Reviewing the Chapter

1. Which characteristics best define fungi? Describe the body of a fungus and how fungi reproduce.   398–99
2. On what basis are fungi classified?   400
3. Explain the term *zygospore fungi*. How does black bread mold reproduce asexually? Sexually?   400–1
4. Explain the terms *sexual ascomycetes* and *asexual ascomycetes*.   402
5. Explain the term *sac fungi*. How do sac fungi reproduce asexually? Describe the structure of an ascocarp.   402–3
6. Describe the structure of yeasts, and explain how they reproduce. How are yeasts useful to humans?   403–4
7. Explain the term *club fungi*. Draw and explain a diagram of the life cycle of a typical mushroom.   406
8. What is the economic importance of smuts and rusts? How can their numbers be controlled?   407
9. Describe the structure of a lichen, and name the three different types. What is the nature of this fungal association?   408
10. Describe the association known as mycorrhizae, and explain how each partner benefits.   408–9

Testing Yourself

Choose the best answer for each question.
For questions 1–3, match the fungi to the phyla in the key.

KEY:
   a. phylum Zygomycota
   b. phylum Ascomycota
   c. phylum Basidiomycota
   d. none of these

1. club fungi
2. zygospore fungi
3. sac fungi

4. During sexual reproduction, the zygospore fungi produce
   a. an acus.
   b. a basidium.
   c. a sporangium.
   d. a conidiophore.

5. An organism that decomposes remains is most likely to use which mode of nutrition?
   a. parasitic   d. chemosynthesis
   b. saprotrophic   e. Both a and b are correct.
   c. ingestion

6. Which feature is best associated with hyphae?
   a. strong, impermeable walls
   b. rapid growth
   c. large surface area
   d. pigmented cells
   e. Both b and c are correct.

7. A fungal spore
   a. contains an embryonic organism.
   b. germinates directly into an organism.
   c. is always windblown.
   d. is most often diploid.
   e. Both b and c are correct.

8. The taxonomy of fungi is based on
   a. sexual reproductive structures.
   b. shape of the sporocarp.
   c. mode of nutrition.
   d. type of cell wall.
   e. level of organization.

9. In the life cycle of black bread mold, the zygospore
   a. undergoes meiosis and produces zoospores.
   b. produces spores as a part of asexual reproduction.
   c. is a thick-walled dormant stage.
   d. is equivalent to asci and basidia.
   e. All of these are correct.

10. In an ascocarp,
    a. there are fertile and sterile hyphae.
    b. hyphae fuse, forming the dikaryotic stage.
    c. a sperm fertilizes an egg.
    d. hyphae do not have chitinous walls.
    e. conidiospores form.

11. In which fungus is the dikaryotic stage longer lasting?
    a. zygospore fungi   c. club fungi
    b. sac fungi   d. fungal partner of a lichen

12. Conidiospores are formed
    a. asexually at the tips of special hyphae.
    b. during sexual reproduction.
    c. by all types of fungi except water molds.
    d. when it is windy and dry.
    e. as a way to survive a harsh environment.

13. The asexual sac fungi are so called because
    a. they have no zygospore.
    b. they cause diseases.
    c. they form conidiospores.
    d. sexual reproduction has not been observed.
    e. All of these are correct.

14. Lichens
    a. cannot reproduce.
    b. need a nitrogen source to live.
    c. are parasitic on trees.
    d. are able to live in extreme environments.

15. Label this diagram of black bread mold structure and asexual reproduction.

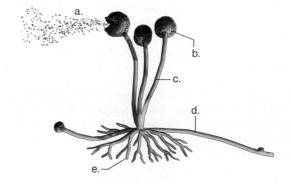

16. The multicellular individual in the life cycle of fungi
    a. is always haploid.
    b. alternates between being haploid and being diploid.
    c. is always diploid.
    d. undergoes meiosis to produce spores.

17. Mycorrhizae
    a. are a type of lichen.
    b. are mutualistic relationships.
    c. help plants gather solar energy.
    d. help plants gather inorganic nutrients.
    e. Both b and d are correct.

18. Classify each of the organisms by stating the kingdom and the phylum.

    Organism   **Classification**
    a. black bread mold
    b. cup fungus
    c. rusts and smuts
    d. truffles

19. Which statement is incorrect?
    a. Some fungi are parasitic, especially on plants, and others are mutualistic with plant roots and algae.
    b. The cell walls of fungi contain chitin.
    c. Following nuclear fusion, zygotic meiosis occurs during the production of spores.
    d. Lichens are an association between a fungus and a bacterium or a protist.
    e. All statements are correct.

20. Symbiotic relationships of fungi include
    a. athlete's foot.
    b. lichens.
    c. mycorrhizae.
    d. Only b and c are correct.
    e. All three examples are correct.

21. Label this diagram of the life cycle of a mushroom.

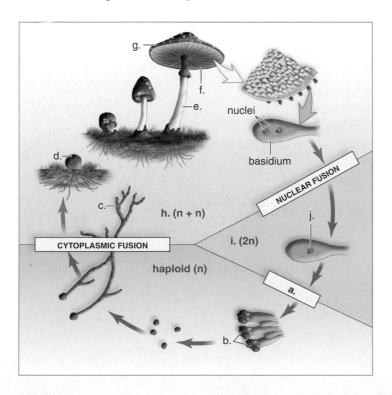

Thinking Scientifically

1. The very earliest bakers observed that dough left in the air would rise. Unknown to them, yeast from the air "contaminated" the bread, began to grow, and produced carbon dioxide. Carbon dioxide caused the bread to rise. Later, cooks began to save some soft dough (before much flour was added) from the previous loaf to use in the next loaf. The saved portion was called the mother. What is in the mother, and why was it important to save it in a cool place?

2. There seems to be a fine line between symbiosis and parasitism when you examine the relationships between fungi and plants. What hypotheses could explain how different selective pressures may have caused particular fungal species to adopt one or the other relationship? Under what circumstances might a mutualistic relationship evolve between fungi and plants? Under what circumstances might a parasitic relationship evolve?

Understanding the Terms

ascus 402	mycelium 398
basidium 406	mycorrhizae 408
budding 399	nonseptate 399
club fungi 406	sac fungi 402
conidiospore 402	septate 399
dikaryotic 399	sporangium 400
fruiting body 403	spore 399
fungus (pl., fungi) 398	yeast 403
hypha 398	zygospore 400
lichen 408	zygospore fungi 400

Match the terms to these definitions:

a. _____ Clublike structure in which nuclear fusion and meiosis occur during sexual reproduction of club fungi.

b. _____ Tangled mass of hyphal filaments composing the vegetative body of a fungus.

c. _____ Spore produced by sac and club fungi during asexual reproduction.

d. _____ Spore-producing and spore-disseminating structure found in sac and club fungi.

e. _____ Symbiotic relationship between fungal hyphae and roots of vascular plants. The fungus allows the plant to absorb more mineral ions and obtain carbohydrates from the plant.

ARIS, the *Biology* Website

ARIS, the website for *Biology*, provides a wealth of information organized and integrated by chapter. You will find practice quizzes, interactive activities, labeling exercises, flashcards, and much more that will complement your learning and understanding of general biology.

www.mhhe.com/maderbiology9

PART V

PLANT EVOLUTION AND BIOLOGY

The organisms we call plants have a long evolutionary history that is still being deciphered by studying the fossil record, the biology of today's representatives, and molecular data. Plants evolved as terrestrial photosynthesizers. Today there are many types of plants with many distinctive features and specialized processes, but the most successful and abundant members of the plant kingdom are the flowering plants. After tracing the evolutionary history of plants, this part explores the basic processes and features of these important organisms.

In flowering plants, the stem supports the leaves, which are the primary organs of photosynthesis. Roots anchor the plant in the soil and absorb water and minerals. A vascular system transports water and minerals up the stem to the leaves and transports the products of photosynthesis to all body parts. A complex network of hormones controls plant growth and therefore a plant's responses to the environment. Plants grow their entire lives and are capable of producing new body parts. Trees can have a much longer life span than animals: some live for thousands of years!

24

EVOLUTION AND DIVERSITY OF PLANTS

Plants are intricate biological entities with a long evolutionary history stretching back hundreds of millions of years. Today's plants are adapted to living in particular terrestrial ecosystems such as a desert, grassland, coniferous forest, temperate deciduous forest, or tropical rain forest. Indeed, terrestrial ecosystems, which differ by their average temperature and annual rainfall, are largely defined by the types of plants that live there. Animals could not exist in the various terrestrial ecosystems were it not for the innumerable services provided by plants.

Plants provide animals with their food and their habitat. Plants are the producers at the base of most ecological pyramids, including those that sustain humans. Food in the form of leaves, stems, roots, and fruits provides all animals with a source of chemical energy, building blocks, minerals, and vitamins. Plants and also algae take up copious amounts of carbon dioxide given off by animals and other organisms in an ecosystem. They then provide the oxygen needed for cellular respiration. Plants supply animals with a place to hide and/or the raw materials needed to build a nest or more elaborate shelter.

The well-being of animals is intricately connected to the well-being of plants. Activities that harm the plants of a region will eventually impact the animals as well.

Grape hyacinth, *Muscaris,* and tulips, *Tulipa,* bloom in spring.

24.1 EVOLUTIONARY HISTORY OF PLANTS

Plants are multicellular photosynthetic eukaryotes placed in kingdom **Plantae.** The 280,000 known species are very diverse, ranging from the diminutive duckweed to the giant coastal redwoods of California. Plants are important ecologically, industrially, and medically. Members of the plant kingdom have an ancient and intriguing evolution.

Plants are believed to have evolved from freshwater green algal species over 500 million years ago. As evidence for a green algal ancestry, scientists have known for some time that both green algae and plants contain chlorophylls *a* and *b* and various accessory pigments, store excess carbohydrates as starch, and have cellulose in their cell walls. In recent years, molecular systematists have compared the sequences of DNA bases coding for ribosomal RNA between organisms. The results suggest that plants are most closely related to a group of green algae known as stoneworts, perhaps those in the genus *Chara.* If

so, stoneworts and plants had a common ancestor sometime in the Paleozoic era.

The evolution of plants is marked by four evolutionary events that can be conveniently associated with the four major groups of plants living today. The nonvascular plants such as mosses and all other groups of plants nourish a multicellular embryo within the body of the female plant (Fig. 24.1*a*). This feature distinguishes plants from green algae. Since green algae do not protect the embryo as all plants do, this may be the first evolved feature that separated plants from green algae.

The seedless vascular plants such as ferns and the other two groups of plants have **vascular tissue** [L. *vasculum,* vessel or duct]. Vascular tissue is specialized for the transport of water and solutes throughout the body of a plant (Fig. 24.1*b*). The fossil record indicates that vascular plants evolved about 430 million years ago (during the Silurian period).

Chara

a. In nonvascular plants (e.g., mosses), multicellular embryos are protected and nourished within the structures that produce an egg.

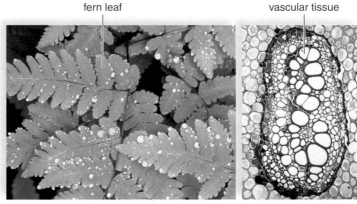

b. In seedless vascular plants (e.g., ferns), vascular tissue conducts water and organic nutrients within its roots, stems, and leaves.

c. In gymnosperms (e.g., conifers), seeds produced in seed cones disperse offspring away from the parent plant.

d. In angiosperms, flowers produce seeds protected by fruits, which aid in the dispersal of offspring.

FIGURE 24.1 Representatives of the four major groups of plants.

The gymnosperms, which are primarily cone-bearing plants such as pine trees, and the angiosperms, the flowering plants such as cherry trees, produce seeds (Fig. 24.1c, d). A **seed** contains an embryo and stored organic nutrients within a protective coat. When a seed is planted and germinates (begins to grow), a plant of the next generation emerges. Seeds ensure the successful dispersal of plants. Seeds are highly resistant structures well suited to protect a plant embryo from drought, and to some extent from predators, until conditions are favorable for germination. Gymnosperms appear in the fossil record about 400 million years ago during the Devonian period.

The fourth evolutionary event was the advent of the **flower,** a reproductive structure (Fig. 24.1d). Flowers attract pollinators such as insects, and they also give rise to fruits that contain the seeds. Fossil angiosperms appear in the fossil record during the late Jurassic period about 135 million years ago.

All the features mentioned here are adaptations to a land existence. Figure 24.2 traces the evolutionary history of these plant adaptations, and the accompanying table illustrates how plants are classified.

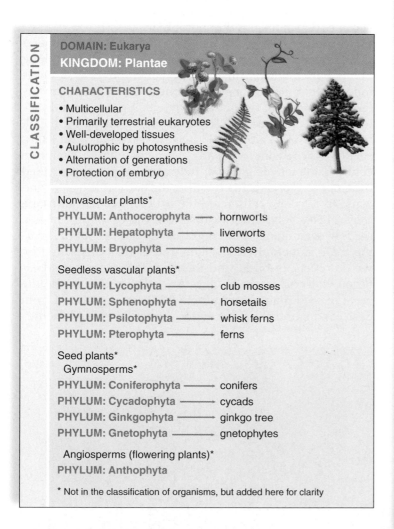

FIGURE 24.2 Evolutionary history of plants.
The evolution of plants involves four significant innovations. Protection of a multicellular embryo was seen in the first plants to live on land. The evolution of vascular tissue was another important adaptation for life on land. The evolution of the seed increased the chance of survival for the next generation. The evolution of the flower fostered the use of animals as pollinators and the use of fruits to aid in the dispersal of seeds.

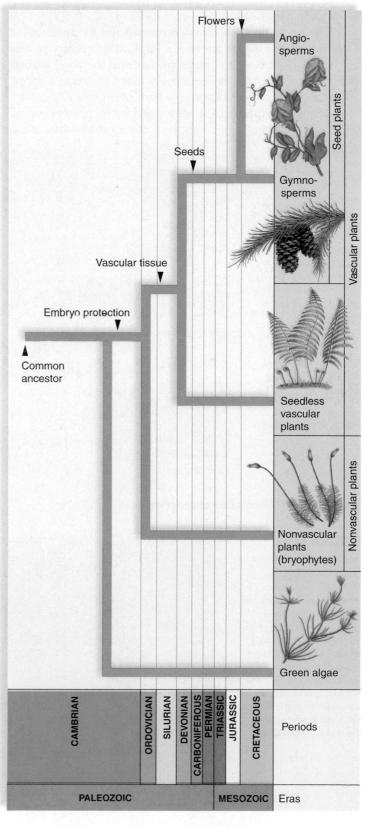

Alternation of Generations

All plants have a life cycle that includes an **alternation of generations.** In this life cycle, two multicellular individuals alternate, each producing the other (Fig. 24.3). The two individuals are (1) a sporophyte, which represents the diploid generation, and (2) a gametophyte, which represents the haploid generation.

The **sporophyte** (2n) is so named for its production of spores by meiosis. A **spore** is a haploid reproductive cell that develops into a new organism without the need to fuse with another reproductive cell. In the plant life cycle, a spore undergoes mitosis and becomes a gametophyte.

The **gametophyte** (n) is so named for its production of gametes. In plants, eggs and sperm are produced by *mitotic*

cell division. A sperm and egg fuse, forming a diploid zygote that undergoes mitosis and becomes the sporophyte.

Two observations are in order. First, meiosis produces haploid spores. This is consistent with the realization that the sporophyte is the diploid generation and spores are haploid reproductive cells. Second, mitosis occurs as a spore becomes a gametophyte, and mitosis occurs as a zygote becomes a sporophyte. Indeed, it is the occurrence of mitosis at these times that results in two generations.

Plants differ as to which generation is dominant—that is, more conspicuous. In nonvascular plants, such as mosses and liverworts, the gametophyte is dominant, but in the other three groups of plants, the sporophyte is dominant (Fig. 24.4). Common examples of the sporophyte generation include ferns, pine trees, and maple trees. In the history of plants, only the sporophyte evolved vascular tissue; therefore, the shift to sporophyte dominance is an adaptation to life on land. Notice that as the sporophyte gains in dominance, the gametophyte becomes microscopic. It also becomes dependent on the sporophyte.

Although all plants have an alternation of generations life cycle, the appearance of the generations among plants varies widely. In ferns, a gametophyte is a small, independent, heart-shaped structure. **Archegonia** are the female portions of the gametophyte that produce and protect the egg (Fig. 24.5a). The eggs are fertilized by flagellated sperm from **antheridia** [Gk. *anthos,* flower, and *-idion,* small], which swim to the archegonia in a film of water. In contrast, the female gametophyte of an angiosperm (called the embryo sac) is retained within the body of the plant and consists of a few cells within a structure called an **ovule** (Fig. 24.5b). Following fertilization, the ovule becomes a seed. In seed plants, **pollen grains** are mature sperm-bearing male gametophytes. Pollen grains are transported by wind, insects, or birds, and therefore they do not need external water to reach the egg. At each juncture in the life cycle of seed plants, reproductive cells are protected from desiccation, or drying out, in the terrestrial environment.

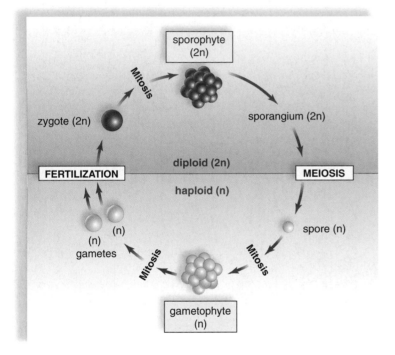

FIGURE 24.3 Alternation of generations.

FIGURE 24.4

Reduction in the size of the gametophyte.

Notice the reduction in the size of the gametophyte and the increase in the size of the sporophyte among these representatives of today's plants. This trend occurred as plants became adapted for life on land. In the moss and fern, spores disperse the gametophyte. In gymnosperms and angiosperms, seeds disperse the sporophyte.

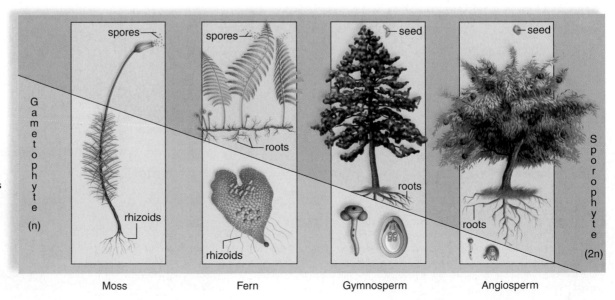

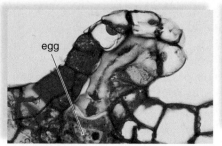
a. Archegonium

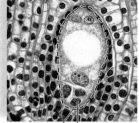
b. Ovule

FIGURE 24.5 Protection of eggs and embryos.
a. The archegonia of a fern gametophyte produce eggs. Flagellated sperm swim in a film of water to the archegonia. **b.** The ovule in a flowering plant contains an egg-bearing female gametophyte called an embryo sac. Following fertilization by sperm from a mature pollen grain, the ovule becomes a seed.

Other Adaptations to a Terrestrial Environment

Sporophyte dominance is accompanied by adaptations for water transport and water conservation. Vascular tissue, which evolved in the sporophyte, transports water and organic nutrients in the body of the plant. The sporophyte is also protected against desiccation. For example, the leaves and other exposed parts of the sporophyte plant are covered by a waxy cuticle (Fig. 24.6*a*). The **cuticle** is relatively impermeable and provides an effective barrier to water loss, but it also limits gas exchange. The thickness of the cuticle varies among different species of plants. Leaves and other photosynthesizing organs have little openings called **stomata** (sing., stoma) (Fig. 24.6*b*). Most stomata occur on the underside of leaves. A stoma is bordered by guard cells that regulate whether it is open or closed. A stoma closes when the weather is hot and dry, and this keeps water loss to a minimum.

Plants evolved from freshwater green algae and have reproductive strategies that adapt them well for life on land. They also evolved conducting systems, cuticles, and stomata that help them transport and conserve water.

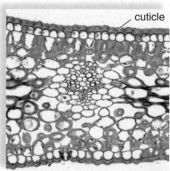
a. Stained photomicrograph of a leaf cross-section.

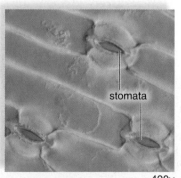
400×
b. Falsely-colored scanning electron micrograph of leaf surface.

FIGURE 24.6 Leaves of vascular plants.
a. A cuticle protects the outer epidermal cells of a plant, and it also keeps the underlying cells and tissues from drying out. **b.** Gas exchange and water loss occur at stomata on surfaces of leaves.

24.2 NONVASCULAR PLANTS

The **nonvascular plants** lack a specialized means of transporting water and organic nutrients. Although they often have a "leafy" appearance, these plants do not have true roots, stems, and leaves—which, by definition, must contain true vascular tissue. Therefore, the nonvascular plants are said to have rootlike, stemlike, and leaflike structures. The term **bryophyte** (lowercase *b*) is a general term for nonvascular plants.

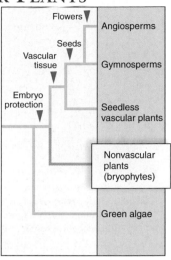

In bryophytes, the gametophyte is the dominant generation, meaning that it is the generation we recognize as the plant. The female gametophyte produces eggs in archegonia and the male gametophyte produces flagellated sperm in antheridia. The sperm swim to the vicinity of the egg in a continuous film of water. Following fertilization, the zygote becomes an embryo that develops into a sporophyte. The sporophyte is attached to, and derives its nourishment from, the photosynthetic gametophyte. The sporophyte produces windblown spores.

The lack of vascular tissue and the need for sperm to swim to archegonia in a film of water largely account for the limited height of bryophytes (usually no bigger than a few centimeters). Nevertheless, some bryophytes compete well in harsh environments because the gametophyte can reproduce asexually, allowing them to spread into stressful and even dry habitats.

Approximately 24,000 species of nonvascular plants have been described. They are classified into three distinct living phyla: phylum Anthocerophyta, the hornworts; phylum Hepatophyta, the liverworts; and phylum Bryophyta, the mosses. Genetic and comparative evidence indicates that hornworts, liverworts, and mosses diverged independently before the origin of vascular plants.

Hornworts

The **hornworts** (phylum **Anthocerophyta**) consist of about a hundred species of inconspicuous nonvascular plants. The suffix *wort* is an Anglo-Saxon term meaning herb. Although some species of hornworts live on trees, the majority of species live in moist, well-shaded areas. They photosynthesize, but they also have a symbiotic relationship with cyanobacteria, which, unlike plants, can fix nitrogen from the air. The cyanobacteria live in special slime chambers provided by the cells of a hornwort gametophyte.

A hornwort can bypass the alternation of generations life cycle by producing asexually through fragmentation. The small sprorophytes of a hornwort resemble tiny, green broom handles rising from a thin gametophyte usually less than 2 cm

FIGURE 24.7 Hornwort, *Anthoceros* sp.
The "horns" of a hornwort are sporophytes that grow continuously from a base anchored in gametophyte tissue.

in diameter (Fig. 24.7). Like the gametophyte, a sporophyte can photosynthesize, although it has only one chloroplast.

Liverworts

Approximately 8,000 species of bryophytes are known as **liverworts** (phylum **Hepatophyta**). Most liverworts are terrestrial, but some are epiphytic, growing on the bark of trees, and others, such as *Riccio carpus,* are aquatic. Liverworts are divided into two groups—the thallose liverworts with flattened bodies known as a thallus, and the leafy liverworts, which superficially resemble mosses. The name *liverwort* refers to the lobes of a thallus, which to some resemble those of the liver. Thallose liverworts grow along creek banks on moist soil after a fire; leafy liverworts are usually found in tropical forests or fog belts. The majority of liverwort species have leafy bodies.

Marchantia is most often used to exemplify a liverwort. Each thallus is thin, being about 30 cells thick in the center. Each branched lobe of the thallus is approximately a centimeter in length; the upper surface is divided into diamond-shaped segments with a small pore, and the lower surface bears numerous hairlike extensions called **rhizoids** [Gk. *rhizion,* dim. of root] that project into the soil (Fig. 24.8). Rhi-

zoids serve in anchorage and limited absorption. *Marchantia* reproduces both asexually and sexually. Gemma cups on the upper surface of the thallus contain *gemmae,* groups of cells that detach from the thallus and can start a new plant. Sexual reproduction depends on disk-headed stalks that bear antheridia, where flagellated sperm are produced, and on umbrella-headed stalks that bear archegonia, where eggs are produced. Following fertilization, tiny sporophytes composed of a foot, a short stalk, and a capsule begin growing within archegonia. Windblown spores are produced within the capsule.

Mosses

Mosses (phylum **Bryophyta**) are the largest phyla of nonvascular plants, with over 15,000 species. The gametophytes of most mosses appear as small leaflike structures arranged around a stemlike axis that sprouts rhizoids. There are three distinct classes of mosses: peat mosses, true mosses, and rock mosses. Mosses can be found from the Antarctic through the tropics to parts of the Arctic. Although most prefer damp, shaded locations in the temperate zone, some survive in deserts, and others inhabit bogs and streams. In forests, they frequently form a mat that covers the ground and rotting logs. In dry environments, they may become shriveled, turn brown, and look completely dead. As soon as it rains, however, the plant becomes green and resumes metabolic activity.

The so-called copper mosses live only in the vicinity of copper and can serve as an indicator plant for copper deposits. Luminous mosses, which glow with a golden-green light, are found in caves, under the roots of trees, and in other dimly lit places. Although mosses do not grow well in polluted areas such as cities, many times they can be seen growing on bricks near moist ground.

The term *moss* is a misnomer for some plants. Many of the common "mosses" are not even nonvascular plants. Irish moss is an edible red alga that grows in leathery tufts along northern seacoasts. Reindeer moss, a lichen, is the dietary mainstay of reindeer and caribou in northern lands. Club mosses, discussed later in this chapter, are vascular plants, and Spanish moss, which hangs in grayish clusters from trees in the southeastern United States, is a flowering plant of the pineapple family.

Most mosses can reproduce asexually by fragmentation, because just about any part of the plant is able to produce

gemma cup
thallus
rhizoids
gemma
a. Gemma cup

FIGURE 24.8 Liverwort, *Marchantia*.
a. Gemmae can detach and start a new plant.
b. Antheridia are present in disk-shaped structures, and archegonia are present in umbrella-shaped structures.

Thallus with gemmae cups

male gametophyte

b. Male gametophytes bear antheridia

female gametophyte

c. Female gametophytes bear archegonia

leafy shoots. Figure 24.9 describes the life cycle of a typical temperate-zone moss. The gametophyte of mosses has two forms. An algalike branching filament of cells, the *protonema,* precedes and produces the upright leafy shoots. The shoots bear antheridia and archegonia. An antheridium has an outer layer of sterile cells and an inner mass of cells that become flagellated sperm. An archegonium, which looks like a vase with a long neck, has an outer layer of sterile cells with a single egg located at the base.

The dependent sporophyte consists of a foot, which is enclosed in female gametophyte tissue; a stalk; and an upper capsule, the **sporangium,** where spores are produced. In some species, the sporangium can produce as many as 50 million spores. At first, the sporophyte is green and photosynthetic; at maturity, it is brown and nonphotosynthetic.

Uses of Mosses

Sphagnum, also called peat moss, has commercial importance. Over 350 species of *sphagnum* have been identified. The cells of this moss have a tremendous ability to absorb water, which is why peat moss is often used in gardening to improve the water-holding capacity of the soil. In some areas where the ground is wet and acidic, such as bogs, dead mosses, especially sphagnum, accumulate and do not decay. This accumulated moss, called **peat,** can be used as fuel. In World War I, peat moss was successfully used as a substitute for bandages.

The three major phyla of nonvascular plants (hornworts, liverworts, and mosses) are all relatively unspecialized, but they are well suited for diverse terrestrial environments.

FIGURE 24.9 Moss life cycle, *Polytrichum* **sp.**

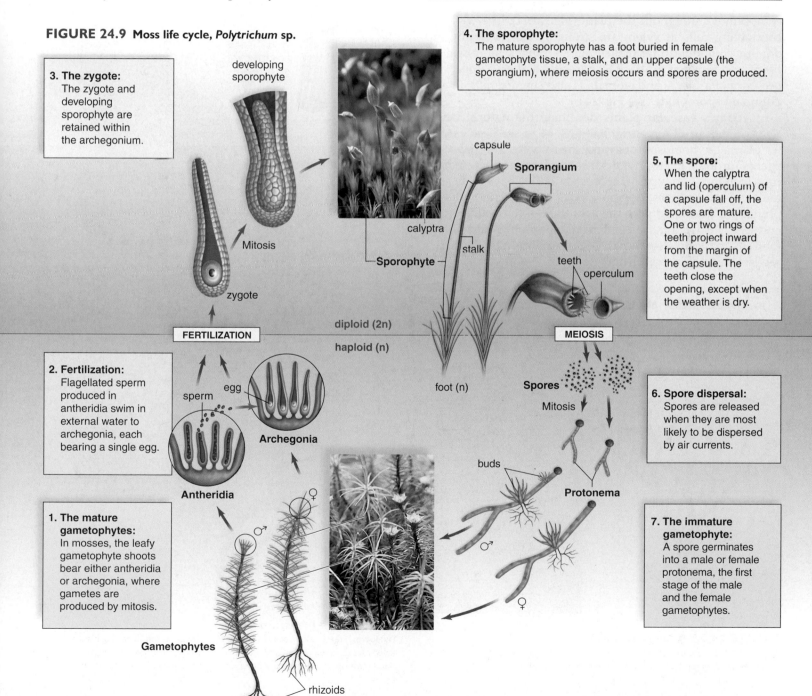

3. The zygote: The zygote and developing sporophyte are retained within the archegonium.

developing sporophyte

Mitosis

zygote

4. The sporophyte: The mature sporophyte has a foot buried in female gametophyte tissue, a stalk, and an upper capsule (the sporangium), where meiosis occurs and spores are produced.

capsule

Sporangium

calyptra

Sporophyte

stalk

teeth

operculum

5. The spore: When the calyptra and lid (operculum) of a capsule fall off, the spores are mature. One or two rings of teeth project inward from the margin of the capsule. The teeth close the opening, except when the weather is dry.

diploid (2n)

haploid (n)

FERTILIZATION

MEIOSIS

2. Fertilization: Flagellated sperm produced in antheridia swim in external water to archegonia, each bearing a single egg.

sperm   egg

Archegonia

Antheridia

foot (n)

Spores

Mitosis

6. Spore dispersal: Spores are released when they are most likely to be dispersed by air currents.

buds

Protonema

1. The mature gametophytes: In mosses, the leafy gametophyte shoots bear either antheridia or archegonia, where gametes are produced by mitosis.

♂

♀

7. The immature gametophyte: A spore germinates into a male or female protonema, the first stage of the male and the female gametophytes.

Gametophytes

rhizoids

24.3 VASCULAR PLANTS

Several phyla of extinct vascular plants are known only from the fossil record. *Cooksonia* is a member of the phylum **Rhyniophyta,** which flourished during the Silurian period but then became extinct by the mid-Devonian period. The rhyniophytes were only about 6.5 cm tall and had no roots or leaves. They consisted simply of a stem that forked evenly to produce branches that ended in sporangia (Fig. 24.10). Like the bryophytes, these plants were **homosporous**—they produced only one type of spore.

Cooksonia and its relatives were successful colonizers of land because of evolved vascular tissues in the dominant sporophyte generation. In today's **vascular plants, xylem** conducts water and dissolved minerals upward from the roots, and **phloem** conducts sucrose and other organic compounds throughout the plant (Fig. 24.11). The walls of conducting cells in xylem are strengthened by **lignin,** an organic compound that makes them stronger, more water-proof, and resistant to attack by parasites and predators. The presence of a cuticle and stomata is also characteristic of the dominant sporophyte (see Fig. 24.6).

Today's vascular plants dominate the natural landscape in nearly all terrestrial habitats. Most seedless vascular plants are homosporous, and their spores are dispersal agents for the gametophyte. All seed plants are **heterosporous** (produce two types of spores), and they have male and female gametophytes. Separate male and female gametophytes are associated with the evolution of the pollen grain and seed. Seeds contain an embryonic sporophyte and stored food in a protective seed coat. Gymnosperms have "naked" seeds, while angiosperms have seeds covered by fruit.

In seedless vascular plants, windblown spores are dispersal agents. In seed plants, seeds disperse offspring.

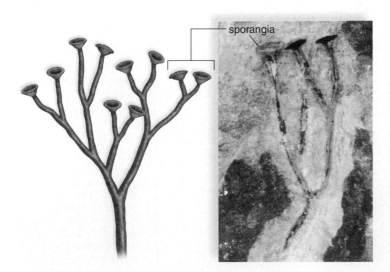

FIGURE 24.10 A *Cooksonia* fossil.
The upright branches of a *Cooksonia* fossil, no more than a few centimeters tall, terminated in sporangia as seen here in the drawing and photo.

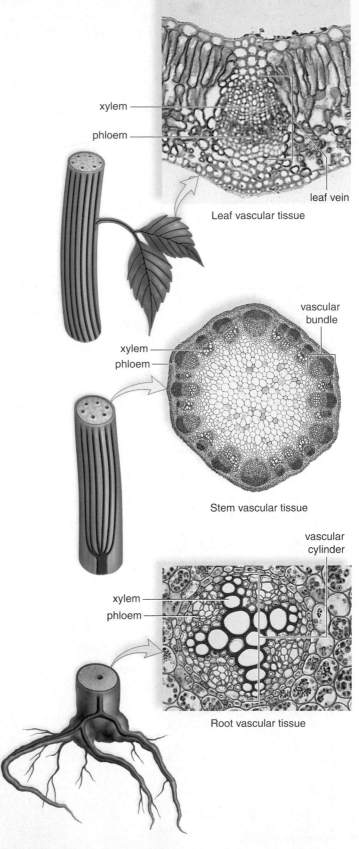

Leaf vascular tissue

Stem vascular tissue

Root vascular tissue

FIGURE 24.11 Vascular tissue.
The leaves, stems, and roots of vascular plants, such as this flowering plant, have vascular tissue (xylem and phloem). Xylem transports water and minerals; phloem transports sucrose and other organic compounds, including hormones.

24.4 SEEDLESS VASCULAR PLANTS

The seedless vascular plants were dominant from the late Devonian period through the Carboniferous period. Club mosses at 35 m, horsetails at 18 m, and ferns at 8 m were much larger than today's specimens, and these plants contributed significantly to the great swamp forests of the time (see the Ecology Focus on page 427).

Club Mosses

The **club mosses** (phylum **Lycophyta**) include 1,150 species of seedless vascular plants grouped in three genera called ground pines (*Lycopodium*), spike mosses (*Selaginella*), and quillworts (*Isoetes*). The leaves of club mosses are microphylls, so called because they have only one strand of vascular tissue. Ferns and seed plants have megaphylls, with many strands of vascular tissue.

The sporophyte is dominant in club mosses. Ground pines are homosporous; the spores germinate into inconspicuous and independent gametophytes. The spike mosses and quillworts are heterosporous; microspores develop into male gametophytes, and megaspores develop into female gametophytes. This means that heterospory evolved at least twice in the history of plants—in club mosses and in seed plants.

In the temperate zone, club mosses inhabit moist woodlands. In ground pines, a branching **rhizome** (horizontal underground stem) produces aerial stems, less than 30 cm tall, and branching underground roots (Fig 24.12). Tightly packed, scalelike leaves cover the body of the plant. The sporangia occur on the surfaces of leaves called **sporophylls,** which are grouped into club-shaped **strobili** (cones). The majority of club mosses live in the tropics and subtropics where many of them are epiphytes—plants that live on, but are not parasitic on, trees.

Some of the extinct relatives of club mosses dominated the Carboniferous swamps (see Ecology Focus on page 427). Today the club mosses are of little economic relevance. In the past, they were used in pills and tablets to keep them from sticking to one another, as a theatrical explosive, and as flash powder before there were photographic flashbulbs. Some ground pines are used for Christmas decorations. Several species have been exploited to the brink of becoming an endangered species.

Ferns and Allies

Horsetails (phylum **Sphenophyta**) consist of one genus, *Equisetum*, and approximately 25 species of distinct seedless vascular plants. Most horsetails inhabit wet, marshy environments around the globe. About 300 millions years ago, horsetails were dominant plants and grew as large as modern trees. Today, horsetails have a rhizome that produces hollow, ribbed aerial stems and reaches a height of 1.3 m (Fig. 24.13). The whorls of slender, green side branches at the joints (nodes) of the stem make the plant bear a resemblance to a horse's tail. The small, scalelike leaves also form whorls at the nodes. Many horsetails have strobili at the tips of all stems; others send up special buff-colored stems that bear the strobili. The spores germinate into inconspicuous and independent gametophytes. The stems are tough and

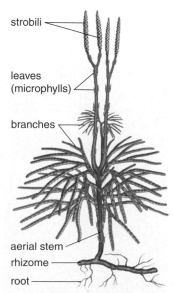

FIGURE 24.12 Ground pine, *Lycopodium*.
Green photosynthetic stems are covered by scalelike leaves, and sporangia are found on sporophylls arranged into strobili (cones).

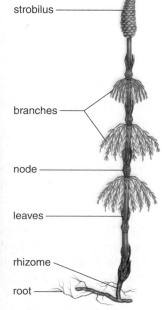

FIGURE 24.13 Horsetail, *Equisetum*.
Whorls of branches and tiny leaves are at the nodes of the stem. Spore-producing sporangia are borne in strobili (cones).

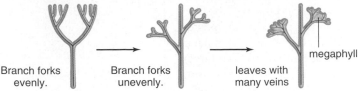

Ferns are the only group of seedless vascular plants to have well-developed megaphylls. Megaphylls are believed to have evolved by the fusion of branched stems:

Branch forks evenly.　Branch forks unevenly.　leaves with many veins　megaphyll

In ferns, the dominant sporophyte produces wind-blown spores. When the spores germinate, a tiny green gametophyte, which lacks vascular tissue, is independent of the sporophyte for its nutrition. In these plants, flagellated sperm are released by antheridia and swim in a film of water to the archegonia, where fertilization occurs. The life cycle of a typical temperate fern is shown in Figure 24.16.

FIGURE 24.14 Whisk fern, *Psilotum*.
Psilotum has no leaves—the branches carry on photosynthesis. The sporangia are yellow.

Cinnamon fern, *Osmunda cinnamomea*

rigid because of silica deposited in cell walls. Early settlers of the American frontier used horsetails for scouring pots and called them "scouring rushes." Other cultures have used horsetails and their extracts in a variety of medicines. Today horsetails are commonly seen in oriental gardens.

Whisk ferns occur as two genera, *Psilotum* and *Tmesipteris*, within the phylum **Psilotophyta.** Both genera live in southern climates as epiphytes, or they can also be found on the ground. The two *Psilotum* species resemble a whisk broom (Fig. 24.14) because they have no leaves. A horizontal rhizome gives rise to an aerial stem that repeatedly forks. The sporangia are borne on short side branches. The two to three species of *Tmesipteris* have appendages that some call leaves, although most botantists consider them extensions of the stem.

Ferns

Ferns (phylum **Pterophyta**) include approximately 11,000 species. Ferns are most abundant in warm, moist, tropical regions, but they can also be found in temperate regions and as far north as the Arctic Circle. Several species live in dry, rocky places and others have adapted to an aquatic life. Ferns range in size from minute aquatic species less than 1 cm in diameter to giant tropical tree ferns that exceed 20 m in length.

The large and conspicuous leaves of ferns, called **fronds,** are commonly divided into leaflets. The royal fern has fronds that stand about 1.8 m tall; those of the hart's tongue fern are straplike and leathery; and those of the maidenhair fern are broad, with subdivided leaflets (Fig. 24.15). In nearly all ferns, the leaves first appear in a curled-up form called a fiddlehead, which unrolls as it grows.

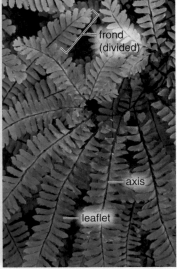

Hart's tongue fern, *Campyloneurum scolopendrium*

Maidenhair fern, *Adiantum pedatum*

FIGURE 24.15 Diversity of ferns.

Uses of Ferns. At first, it may seem that ferns do not have much economic value, but they are much used by florists in decorative bouquets and as ornamental plants in the home and garden. Although not true wood, trunks from tropical tree ferns are often used as a building material because they resist decay. Ferns also have medicinal value; many Native Americans use ferns as an astringent during childbirth to stop bleeding, and the maidenhair fern is the source of an expectorant. Fern extracts were also used to expel intestinal parasites such as tapeworms.

The Environmental Protection Agency has pointed out that Boston fern can substantially remove formaldehyde from the air in closed rooms.

Ferns and other seedless vascular plants have a large, conspicuous sporophyte with vascular tissue. The gametophyte is quite small, but independent. Flagellated sperm swim from antheridia to fertilize eggs within archegonia.

FIGURE 24.16 Fern life cycle.

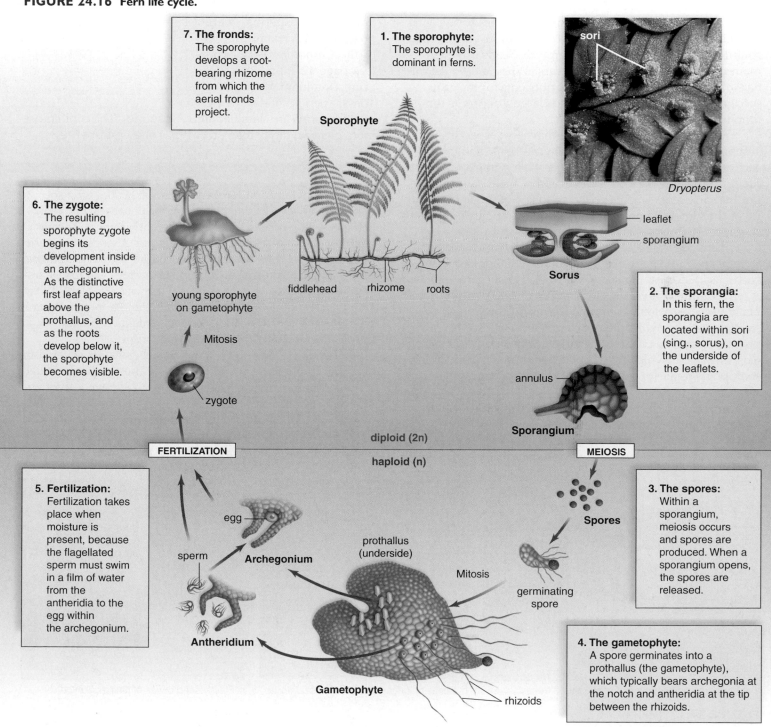

7. The fronds:
The sporophyte develops a root-bearing rhizome from which the aerial fronds project.

1. The sporophyte:
The sporophyte is dominant in ferns.

Dryopterus

6. The zygote:
The resulting sporophyte zygote begins its development inside an archegonium. As the distinctive first leaf appears above the prothallus, and as the roots develop below it, the sporophyte becomes visible.

Sporophyte

young sporophyte on gametophyte

fiddlehead   rhizome   roots

leaflet
sporangium

Sorus

Mitosis

zygote

2. The sporangia:
In this fern, the sporangia are located within sori (sing., sorus), on the underside of the leaflets.

annulus

Sporangium

diploid (2n)
haploid (n)

FERTILIZATION

MEIOSIS

5. Fertilization:
Fertilization takes place when moisture is present, because the flagellated sperm must swim in a film of water from the antheridia to the egg within the archegonium.

egg

sperm   **Archegonium**

Antheridium

prothallus (underside)

Mitosis

Gametophyte

rhizoids

germinating spore

Spores

3. The spores:
Within a sporangium, meiosis occurs and spores are produced. When a sporangium opens, the spores are released.

4. The gametophyte:
A spore germinates into a prothallus (the gametophyte), which typically bears archegonia at the notch and antheridia at the tip between the rhizoids.

24.5 SEED PLANTS

In the first seed plants of the Devonian period, the seeds developed along branches within small, cuplike structures, now called cupules. Following pollination, fertilization of an egg cell occurred within the cupule, prior to seed formation. Were these plants ancestral to both gymnosperms and angiosperms? A fairly good history of gymnosperm evolution is available, but exactly when angiosperms arose is shrouded in mystery. Today, the seed plants—the gymnosperms and angiosperms—are the most plentiful plants in the biosphere.

Seeds contain a sporophyte embryo and stored food within a protective seed coat. The seed coat and stored food allow an embryo to survive harsh conditions during long periods of dormancy (arrested state) until environmental conditions become favorable for growth. Seeds can even remain dormant for hundreds of years. When a seed germinates, the stored food is a source of nutrients for the growing seedling. The survival value of seeds largely accounts for the dominance of seed plants today.

Seed plants are heterosporous (have two types of spores) and produce two kinds of gametophytes—male and female—each of which consists of just a few cells. Pollen grains, which are drought resistant, become a multicellular male gametophyte. **Pollination** occurs when a pollen grain is brought to the vicinity of the female gametophyte by wind or a pollinator. Later, sperm move toward the female gametophyte through a growing pollen tube. *Note that no external water is needed to accomplish fertilization.* The whole male gametophyte, rather than just the sperm as in seedless plants, moves to the female gametophyte. A female gametophyte develops within an ovule, which eventually becomes a seed. In gymnosperms (mostly cone-bearing seed plants), the ovules are not completely enclosed by sporophyte tissue at the time of pollination. In flowering plants (angiosperms), the ovules are completely enclosed within diploid sporophyte tissue (ovaries), which becomes a fruit.

Gymnosperms and angiosperms are the seed plants, which produce pollen grains and seeds. Pollen grains and seeds are well protected from drying out.

24.6 GYMNOSPERMS

The four groups of living **gymnosperms** [Gk. *gymnos,* naked, and *sperma,* seed] are conifers, cycads, ginkgoes, and gnetophytes. All of these plants have ovules and seeds exposed on the surface of sporophylls or analogous structures. (Since the seeds are not enclosed by fruit, gymnosperms have "naked seeds.") Early gymnosperms were present in the swamp forests of the Carboniferous period (see page 427), and they became dominant during the Triassic period. Today, living

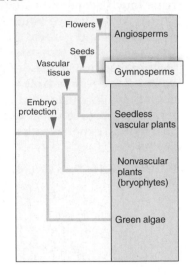

gymnosperms are classified into 780 species. The conifers are more plentiful today than other types of gymnosperms.

Conifers

Conifers (phylum **Coniferophyta**) consist of about 575 species of trees, many evergreen, including pines, spruces, firs, cedars, hemlocks, redwoods, cypresses, yews, and junipers. The name *conifers* signifies plants that bear **cones,** but other gymnosperm phyla are also cone-bearing. The coastal redwood (*Sequoia sempervirens),* a conifer native to northwestern California and southwestern Oregon, is the tallest living vascular plant; it may attain nearly 100 m in height. Another conifer, the bristlecone pine *(Pinus longaeva)* of the White Mountains of California, is the oldest living tree; one is 4,900 years of age.

Vast areas of northern temperate regions are covered in evergreen coniferous forests (Fig. 24.17). The tough, needlelike leaves of pines conserve water because they have a thick cuticle and recessed stomata. Note in the life cycle of the pine (Fig. 24.18) that the sporophyte is dominant, pollen grains are windblown, and the seed is the dispersal stage. Conifers are **monoecious** since a tree produces both pollen and seed cones.

a. A northern coniferous forest of evergreen trees

b. Cones of lodgepole pine, *Pinus contorta*

c. Fleshy seed cones of juniper, *Juniperus*

FIGURE 24.17 Conifers.

Uses of Pines

The wood of pines and other conifers is used extensively in construction. The wood consists primarily of xylem tissue that lacks some of the more rigid cell types found in flowering trees. Therefore, it is considered a "soft" rather than a "hard" wood. Although called soft woods, some soft woods such as yellow pine are actually harder than so-called hard woods. The foundations of the 100-year-old Brooklyn Bridge are made of southern yellow pine. Resin, made by pines as an insect and fungal deterrent, is harvested commercially for a derived product called turpentine.

Cycads

Cycads (phylum **Cycadophyta**) include 10 genera and 140 species of distinctive gymnosperms. The cycads are native to tropical and subtropical forests. *Zamia pumila* found in Florida is the only species of cycad native to North America. Cycads are commonly used in landscaping. One species, *Cycas revoluta*, referred to as the sago palm, is a common landscaping plant. Their large, finely divided leaves grow in clusters at the top of the stem, and therefore they resemble palms or ferns, depending on their height. The trunk of a cycad is unbranched, even if it reaches a height of 15–18 m, as is possible in some species.

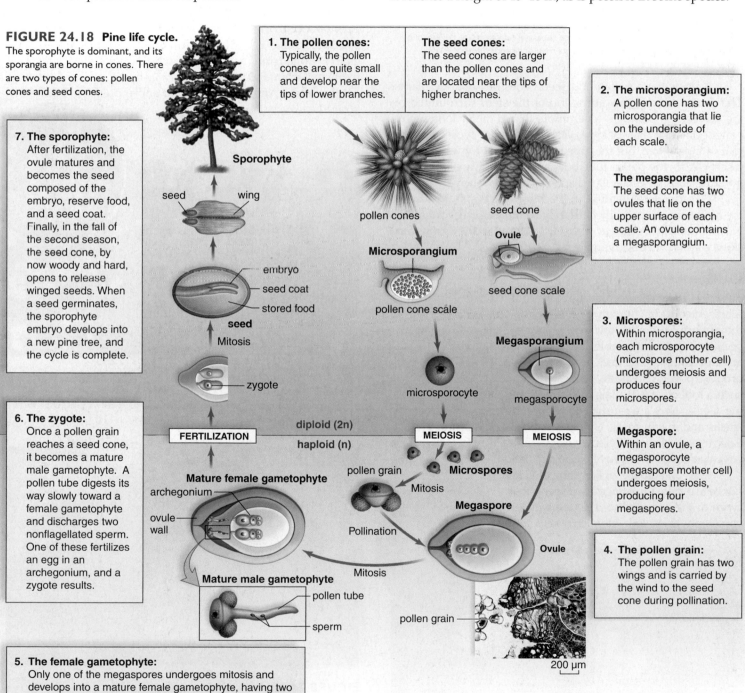

FIGURE 24.18   Pine life cycle.
The sporophyte is dominant, and its sporangia are borne in cones. There are two types of cones: pollen cones and seed cones.

1. The pollen cones:
Typically, the pollen cones are quite small and develop near the tips of lower branches.

The seed cones:
The seed cones are larger than the pollen cones and are located near the tips of higher branches.

2. The microsporangium:
A pollen cone has two microsporangia that lie on the underside of each scale.

The megasporangium:
The seed cone has two ovules that lie on the upper surface of each scale. An ovule contains a megasporangium.

3. Microspores:
Within microsporangia, each microsporocyte (microspore mother cell) undergoes meiosis and produces four microspores.

Megaspore:
Within an ovule, a megasporocyte (megaspore mother cell) undergoes meiosis, producing four megaspores.

4. The pollen grain:
The pollen grain has two wings and is carried by the wind to the seed cone during pollination.

5. The female gametophyte:
Only one of the megaspores undergoes mitosis and develops into a mature female gametophyte, having two to six archegonia. Each archegonium contains a single large egg lying near the ovule opening.

6. The zygote:
Once a pollen grain reaches a seed cone, it becomes a mature male gametophyte. A pollen tube digests its way slowly toward a female gametophyte and discharges two nonflagellated sperm. One of these fertilizes an egg in an archegonium, and a zygote results.

7. The sporophyte:
After fertilization, the ovule matures and becomes the seed composed of the embryo, reserve food, and a seed coat. Finally, in the fall of the second season, the seed cone, by now woody and hard, opens to release winged seeds. When a seed germinates, the sporophyte embryo develops into a new pine tree, and the cycle is complete.

a.   b.

FIGURE 24.19 Cycad cones.
The pollen cones and seed cones occur on separate individuals. **a.** Pollen cone. **b.** Close-up of seed cone, *cycas* sp., and its maturing seeds. Note the finely divided leaves of cycads.

Cycads have pollen and seed cones on separate plants. The cones, which grow at the top of the stem surrounded by the leaves, can be huge—even more than a meter long with a weight of 40 kg (Fig. 24.19). Cycads have a life cycle similar to that of a pine tree, but they are pollinated not by wind but by insects. Also, the pollen tube bursts in the vicinity of the archegonium, and multiflagellated sperm swim to reach an egg.

Cycads were plentiful in the Mesozoic era at the time of the dinosaurs, and it's likely that dinosaurs fed on cycad seeds. Now cycads are in danger of extinction because they grow very slowly, a distinct disadvantage.

Ginkgoes

Phylum **Ginkgophyta,** although plentiful in the fossil record, is represented today by only one surviving species, *Ginkgo biloba,* the **ginkgo** or maidenhair tree. It is called the maidenhair tree because its leaves resemble those of a maidenhair fern. Ginkgoes are **dioecious**—some trees produce seeds (Fig. 24.20) and others produce pollen. The fleshy seeds, which ripen in the fall, give off such a foul odor that male trees are usually preferred for planting. *Ginkgo* trees are resistant to pollution and do well along city streets and in city parks. *Ginkgo* is native to China, and in Asia, *Ginkgo* seeds are considered a delicacy. Extracts from *Ginkgo* trees have been used to improve blood circulation.

Like cycads, the pollen tube of *Gingko* bursts to release multiflagellated sperm that swim to the egg produced by the female gametophyte located within an ovule.

FIGURE 24.20 The *Ginkgo* tree.
Leaves are broad, and the fleshy seeds are borne at the end of stalklike megasporophylls. Inset shows the ovules that become the seeds.

Gnetophytes

Gnetophytes (phylum **Gnetophyta**) are represented by three living genera and 70 species of plants that are very diverse in appearance. In all gnetophytes, xylem is structured similarly, none have archegonia, and their strobili (cones) have a similar construction. Angiosperms don't have archegonia either, and molecular analysis suggests that among gymnosperms, the gnetophytes are most closely related to angiosperms. The reproductive structures of some gnetophyte species produce nectar, and insects play a role in the pollination of these species. *Gnetum*, which occurs in the tropics, consists of trees or climbing vines with broad, leathery leaves arranged in pairs. *Ephedra*, occurring only in southwestern North America and southeast Asia, is a shrub with small, scalelike leaves (Fig. 24.21). Ephedrine, a medicine with serious side effects, is extracted from *Ephedra*. *Welwitschia*, living in the deserts of southwestern Africa, has only two enormous, straplike leaves (Fig. 24.22).

FIGURE 24.21 *Ephedra*.
Ephedra, a type of gnetophyte, is a branched shrub with small, scalelike leaves. This is a pollen-producing plant. The inset shows a strobilus with microsporangia (yellow) that produce pollen.

FIGURE 24.22 *Welwitschia miribilis*.
Welwitschia is a type of gnetophyte. A woody, saucer-shaped disk produces two long, straplike leaves that split one to several times. Pollen-producing strobili (cones) and seed-producing strobili are on different plants. This plant has pollen-producing strobili.

ecology focus

Carboniferous Forests

Our industrial society runs on fossil fuels such as **coal.** The term *fossil fuel* might seem odd at first until one realizes that it refers to the remains of organic material from ancient times. During the Carboniferous period more than 300 million years ago, a great swamp forest (Fig. 24A) encompassed what is now northern Europe, the Ukraine, and the Appalachian Mountains in the United States. The weather was warm and humid, and the trees grew very tall. These are not the trees we know today; instead, they are related to today's seedless vascular plants: the club mosses, horsetails, and ferns! Club mosses today may stand as high as 30 cm, but their ancient relatives were 35 m tall and 1 m wide. The spore-bearing cones were up to 30 cm long, and some had leaves more than 1 m long. Horsetails too—at 18 m tall—were giants compared to today's specimens. The tree ferns were also taller than tree ferns found in the tropics today, and there were two other types of trees: seed

ferns and early gymnosperms. "Seed fern" is a misnomer because it has been shown that these plants, which only resemble ferns, were actually a type of gymnosperm.

The amount of biomass was enormous, and occasionally the swampy water rose and the trees fell. Trees under water do not decompose well, and their partially decayed remains became covered by sediment that sometimes changed to sedimentary rock. Sedimentary rock applied pressure, and the organic material then became coal, a fossil fuel. This process continued for millions of years, resulting in immense deposits of coal. Geological upheavals raised the deposits to the level where they can be mined today.

With a change of climate, the trees of the Carboniferous period became extinct, and only their herbaceous relatives survived to our time. Without these ancient forests, our life today would be far different because they helped bring about our industrialized society.

Fossil seed fern

FIGURE 24A Swamp forest of the Carboniferous period.
Nonvascular plants and early gymnosperms dominated the swamp forests of the Carboniferous period. Among the early gymnosperms were the seed ferns, so named because their leaves looked like fronds, as shown in a micrograph of fossil remains in the upper right.

24.7 ANGIOSPERMS

Angiosperms [Gk. *angion,* dim. of *angos,* vessel, and *sperma,* seed] (phylum **Anthophyta**) are the flowering plants. They are an exceptionally large and successful group of plants, with 240,000 known species—six times the number of all other plant groups combined. Angiosperms live in all sorts of habitats, from fresh water to desert, and from the frigid north to the torrid tropics. They range in size from the tiny, almost microscopic duckweed to *Eucalyptus* trees over 100 m tall. It would be impossible to exaggerate the importance of angiosperms in our everyday lives. As discussed in the Ecology Focus on pages 432–33, they provide us with clothing, food, medicines, and other commercially valuable products.

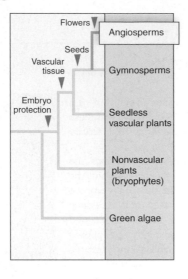

The flowering plants are called angiosperms because their ovules, unlike those of gymnosperms, are always enclosed within diploid tissues. In the Greek derivation of their name, *angio* ("vessel") refers to the ovary, which develops into a fruit, a unique angiosperm feature.

Origin and Radiation of Angiosperms

Although the first fossils of angiosperms are no older than about 135 million years, the angiosperms probably arose much earlier. Indirect evidence suggests the possible ancestors of angiosperms may have originated as long as 200 million years ago. But their exact ancestral past has remained a mystery since Charles Darwin pondered it.

To find the angiosperm of today that might be most closely related to the first angiosperms, botanists have turned to DNA comparisons. Gene-sequencing data single out *Amborella trichopoda* (Fig. 24.23) as having the oldest lineage among today's angiosperms. This small shrub, with small cream-colored flowers, lives only on the island of New Caledonia in the South Pacific.

Although *A. trichopoda* may not be the original angiosperm species, it is sufficiently close that much may be learned from studying its reproductive biology. Botanists hope that this knowledge will help them understand the early adaptive radiation of angiosperms in the late Cretaceous and early Tertiary periods. It could very well be that the rise to dominance of angiosperms is tied to the increasing diversity of flying insects at this time. Insects are significant **pollinators**—animals that carry pollen between flowers.

Monocots and Eudicots

Most flowering plants belong to one of two classes. These classes are the **Monocotyledones (Liliopsida),** often shortened to simply the **monocots** (about 65,000 species), and the **Eudicotyledones,** shortened to **eudicots** (about 175,000 species). The term *eudicot* (meaning true dicot) is more specific than the term *dicot*. It was discovered that some of the plants formerly classified as dicots diverged before the evolutionary split that gave rise to the two major classes of angiosperms. These earlier evolving plants are not included in the designation eudicots.

Monocots are so called because they have only one cotyledon (seed leaf) in their seeds. Several common monocots include corn, tulips, pineapples, bamboos, and sugarcane. Eudicots are so called because they possess two cotyledons in their seeds. Several common eudicots include cactuses, strawberries, dandelions, poplars, and beans. **Cotyledons** are the seed leaves that contain nutrients that nourish the plant embryo. Table 24.1 lists several fundamental differences between monocots and eudicots.

The Flower

Although flowers vary widely in appearance (Fig. 24.24), most have certain structures in common. The **peduncle,** a flower stalk, expands slightly at the tip into a **receptacle,** which bears the other flower parts. These parts, called

FIGURE 24.23 *Amborella trichopoda.*
Genetic data suggest this plant is most closely related to the first flowering plants.

TABLE 24.1	
Monocots and Eudicots	
Monocots	*Eudicots*
One cotyledon	Two cotyledons
Flower parts in threes or multiples of three	Flower parts in fours or fives or multiples of four or five
Usually herbaceous	Woody or herbaceous
Usually parallel venation	Usually net venation
Scattered bundles in stem	Vascular bundles in a ring
Fibrous root system	Taproot system

FIGURE 24.24
Flower diversity.
Regardless of size and shape, flowers share certain features.

Beavertail cactus

Water lily

Louisiana iris

Dogwood flowers

Daisy

Butterfly weed flowers

sepals, petals, stamens, and carpels, are attached to the receptacle in whorls (circles) (Fig. 24.25).

1. The **sepals,** collectively called the calyx, protect the flower bud before it opens. The sepals may drop off or may be colored like the petals. Usually, however, sepals are green and remain attached to the receptacle.
2. The **petals,** collectively called the corolla, are quite diverse in size, shape, and color. The petals often attract a particular pollinator.
3. Next are the **stamens.** Each stamen consists of two parts: the anther, a saclike container, and the filament, a slender stalk. Pollen grains develop from microspores produced in the anther.
4. At the very center of a flower is the **carpel,** a vaselike structure with three major regions: the **stigma,** an enlarged sticky knob; the **style,** a slender stalk; and the **ovary,** an enlarged base that encloses one or more

ovules. The ovule becomes the seed, and the ovary becomes the fruit.

It can be noted that not all flowers have all these parts (Table 24.2). A flower is said to be complete if it has all four parts; otherwise it is incomplete.

TABLE 24.2

Other Flower Terminology

Term	Type of Flower
Complete	All four parts (sepals, petals, stamens, and carpels) present
Incomplete	Lacks one or more of the four parts
Perfect	Has both stamens and (a) carpel(s)
Imperfect	Has stamens or (a) carpel(s), but not both
Inflorescence	A cluster of flowers
Composite	Appears to be a single flower but consists of a group of tiny flowers

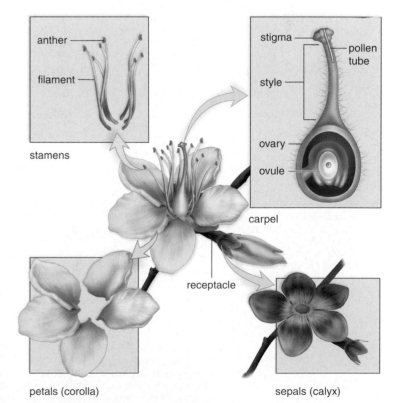

petals (corolla)

sepals (calyx)

anther
filament
stamens

stigma
pollen tube
style
ovary
ovule
carpel

receptacle

FIGURE 24.25 Generalized flower.
A flower has four main parts: sepals, petals, stamens, and carpels. A stamen has an anther and filament. A carpel has a stigma, style, and ovary. An ovary contains ovules.

Flowering Plant Life Cycle

Figure 24.26 depicts the life cycle of a typical flowering plant. Like the gymnosperms, flowering plants are heterosporous, producing two types of spores. A **megaspore** located in an ovule within an ovary of a carpel develops into an egg-bearing female gametophyte called the embryo sac. In most angiosperms, the embryo sac has seven cells; one of these is an egg, and another contains two polar nuclei. (These two nuclei are called the polar nuclei because they came from opposite ends of the embryo sac.)

Microspores, produced within anthers, become pollen grains that, when mature, are sperm-bearing male gametophytes. The full-fledged mature male gametophyte

FIGURE 24.26   Flowering plant life cycle.
The parts of the flower involved in reproduction are the stamens and the carpel. Reproduction has been divided into significant stages of female gametophyte development, male gametophyte development, and also significant stages of sporophyte development.

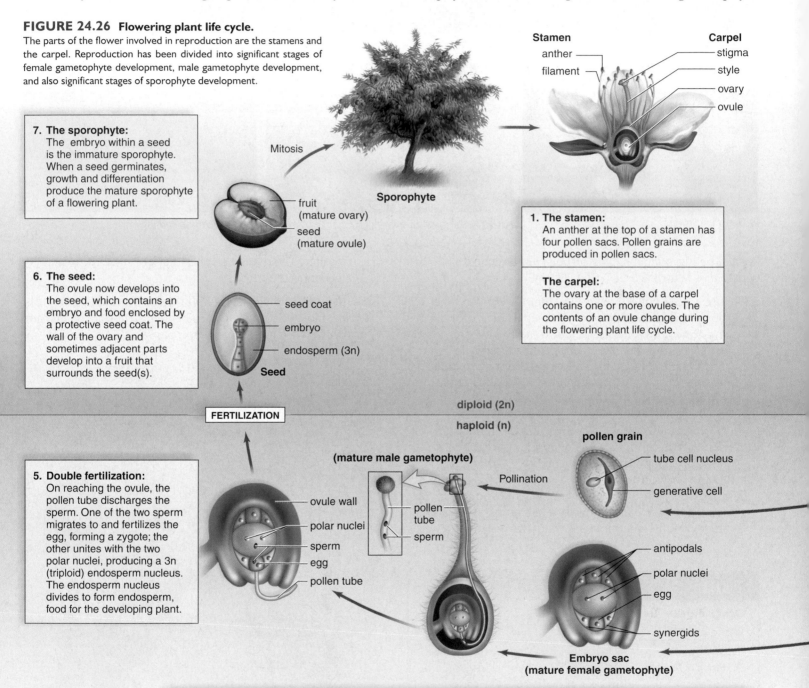

7. The sporophyte:
The embryo within a seed is the immature sporophyte. When a seed germinates, growth and differentiation produce the mature sporophyte of a flowering plant.

6. The seed:
The ovule now develops into the seed, which contains an embryo and food enclosed by a protective seed coat. The wall of the ovary and sometimes adjacent parts develop into a fruit that surrounds the seed(s).

5. Double fertilization:
On reaching the ovule, the pollen tube discharges the sperm. One of the two sperm migrates to and fertilizes the egg, forming a zygote; the other unites with the two polar nuclei, producing a 3n (triploid) endosperm nucleus. The endosperm nucleus divides to form endosperm, food for the developing plant.

1. The stamen:
An anther at the top of a stamen has four pollen sacs. Pollen grains are produced in pollen sacs.

The carpel:
The ovary at the base of a carpel contains one or more ovules. The contents of an ovule change during the flowering plant life cycle.

4. The mature male gametophyte:
A pollen grain that lands on the carpel of the same type of plant germinates and produces a pollen tube, which grows within the style until it reaches an ovule in the ovary. Inside the pollen tube, the generative cell nucleus divides and produces two nonflagellated sperm. A fully germinated pollen grain is the mature male gametophyte.

The mature female gametophyte:
The ovule now contains the mature female gametophyte (embryo sac), which typically consists of eight haploid nuclei embedded in a mass of cytoplasm. The cytoplasm differentiates into cells, one of which is an egg and another of which contains two polar nuclei.

consists of only three cells: the tube cell and two sperm cells. During pollination, a pollen grain is transported by various means from the anther to the stigma of a carpel, where it germinates. During germination, the tube cell produces a pollen tube. The **pollen tube** carries the two sperm to the micropyle (small opening) of an ovule. During double fertilization, one sperm unites with an egg, forming a diploid zygote, and the other unites with polar nuclei, forming a triploid endosperm nucleus.

Ultimately, the ovule becomes a seed that contains the embryo (the sporophyte of the next generation) and stored food enclosed within a seed coat. Endosperm in some seeds is absorbed by the cotyledons, whereas in other seeds endosperm is digested as the seed matures.

A **fruit** is derived from an ovary and possibly accessory parts of the flower. Some fruits such as apples and tomatoes provide a fleshy covering, and other fruits such as pea pods and acorns provide a dry covering for seeds.

Flowers and Diversification

Flowers are involved in the production and development of spores, gametophytes, gametes, and embryos enclosed within seeds. Successful completion of sexual reproduction in angiosperms requires the effective dispersal of pollen and then seeds. The various ways pollen and seeds can be dispersed have resulted in many different types of flowers.

Wind-pollinated flowers are usually not showy, whereas insect-pollinated flowers and bird-pollinated flowers are often colorful. Night-blooming flowers attract nocturnal mammals or insects; these flowers are usually aromatic and white or cream-colored. Although some flowers disperse their pollen by wind, many are adapted to attract specific pollinators such as bees, wasps, flies, butterflies, moths, and even bats, which carry only particular pollen from flower to flower. For example, glands located in the region of the ovary produce nectar, a nutrient that is gathered by pollinators as they go from flower to flower. Bee-pollinated flowers are usually blue or yellow and have ultraviolet shadings that lead the pollinator to the location of nectar. The mouthparts of bees are fused into a long tube that is able to obtain nectar from the base of the flower.

The fruits of flowers protect and aid in the dispersal of seeds. Dispersal occurs when seeds are transported by wind, gravity, water, and animals to another location. Fleshy fruits may be eaten by animals, which transport the seeds to a new location and then deposit them when they defecate. Because animals live in particular habitats and/or have particular migration patterns, they are apt to deliver the fruit-enclosed seeds to a suitable location for seed germination (when the embryo begins to grow again) and development of the plant.

Angiosperms are the flowering plants. The flower attracts animals (e.g., insects), which aid in pollination, and produces seeds enclosed by fruit, which aids in their dispersal.

---

2. The microsporangia: Pollen sacs of the anther are microsporangia, where each microsporocyte undergoes meiosis to produce four microspores.	The megasporangium: First, an ovule within an ovary contains a megasporangium, where a megasporophyte undergoes meiosis to produce four megaspores.

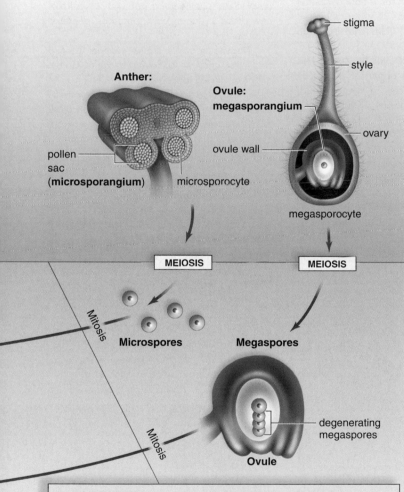

Anther:

pollen sac (**microsporangium**)

microsporocyte

Ovule: megasporangium

stigma

style

ovary

ovule wall

megasporocyte

MEIOSIS

MEIOSIS

Mitosis

Microspores

Megaspores

Mitosis

degenerating megaspores

Ovule

3. **Microspores:**
Each microspore in a pollen sac undergoes mitosis to become an immature pollen grain with two cells: the tube cell and the generative cell. The pollen sacs open, and the pollen grains are windblown or carried by an animal carrier, usually to other flowers. This is pollination.

Megaspores:
Inside the ovule of an ovary, three megaspores disintegrate, and only the remaining one undergoes mitosis to become a female gametophyte.

ecology focus

Plants: Could We Do Without Them?

Plants define and are the producers in most ecosystems. Humans derive most of their sustenance from three flowering plants: wheat, corn, and rice (Fig. 24B). All three of these plants are in the grass family and are collectively, along with other species, called grains. Most of the Earth's 6.4 billion people live a simple way of life, growing their food on family plots. The continued growth of these plants is essential to human existence. A virus or other disease could hit any one of these three plants and cause massive loss of life from starvation.

Wheat, corn, and rice originated and were first cultivated in different parts of the globe. Wheat is commonly used in the United States to produce flour and bread. It was first cultivated in the Near East (Iran, Iraq, and neighboring countries) about 8000 B.C.; hence, it is thought to be one of the earliest cultivated plants. Wheat was brought to North America in 1520 by early settlers; now the United States is one of the world's largest producers of wheat. Corn, or what is properly called maize, was first cultivated in Central America about 7,000 years ago. Maize developed from a plant called teosinte, which grows in the highlands of central Mexico. By the time Europeans were exploring Central America, over 300 varieties were already in existence—growing from Canada to Chile. We now commonly grow six major varieties of corn: sweet, pop, flour, dent, pod, and flint. Rice had its origin in southeastern Asia several thousand years ago, where it grew in swamps. Today we are familiar with white and brown rice, which differ in the extent of processing. Brown rice results when the seeds are threshed to remove the hulls—the seed coat and complete embryo remain. If the seed coat and embryo are removed, leaving only the starchy endosperm, white rice results. Unfortunately, the seed coat and embryo are a good source of vitamin B and fat-soluble vitamins. Today, rice is grown throughout the tropics and subtropics where water is abundant. It is also grown in some parts of western United States by flooding diked fields with irrigation water.

Do you have an "addiction" to sugar? This simple carbohydrate comes almost exclusively from two plants—sugarcane (grown in South America, Africa, Asia, and the Caribbean) and sugar beets (grown mostly in Europe and North America). Each provides about 50% of the world's sugar.

Many foods are bland or tasteless without spices. In the Middle Ages, wealthy Europeans spared no cost to obtain spices from the Near and Far East. In the fifteenth and sixteenth centuries, major expeditions were launched in an attempt to find better and cheaper routes for spice importation. The explorer Columbus convinced the queen of Spain that he would find a shorter route to the Far East by traveling west by ocean rather than east by land. Columbus's idea was sound, but he encountered a little barrier, the New World. This discovery later provided Europe with a wealth of new crops, including corn, potatoes, peppers, and tobacco.

Our most popular drinks—coffee, tea, and cola—also come from flowering plants. Coffee originated in Ethiopia, where it was first used (along with animal fat) during long trips for sustenance and to relieve fatigue. Coffee as a drink was not developed until the thirteenth century in Arabia and Turkey, and it did not catch on in Europe until the seventeenth century. Tea is thought to have been developed somewhere in central Asia. Its earlier uses were almost exclusively medicinal, especially among the Chinese, who still drink tea for medical reasons. The drink as we now know it was not developed until the fourth century. By the mid-seventeenth century it had become popular in Europe. Cola is a common ingredient in tropical drinks and was used around the turn of the century, along with the drug coca (used to make cocaine), in the "original" Coca-Cola.

Plants have been used for centuries for a number of important household items (Fig. 24Ca), including the house itself. We are most familiar with lumber being used as the major structural portion in buildings. This wood

Wheat plants, *Triticum*

FIGURE 24B Cereal grains.
These three cereal grains are the principal source of calories and protein for our civilization.

Corn plants, *Zea*

Rice plants, *Oryza*

a. Dwarf fan palms, *Chamaerops,* for basket weaving

b. Rubber, *Hevea,* for auto tires

d. Tulips, *Tulipa,* for beauty

FIGURE 24C Uses of plants.
a. Dwarf fan palms can be used to make baskets.
b. A rubber plant provides latex for making tires.
c. Cotton can be used for clothing. d. Plants have aesthetic value.

c. Cotton, *Gossypium,* for cloth

comes mostly from a variety of conifers: pine, fir, and spruce, among others. In the tropics, trees and even herbs provide important components for houses. In rural parts of Central and South America, palm leaves are preferable to tin for roofs, since they last as long as ten years and are quieter during a rainstorm. In the Near East, numerous houses along rivers are made entirely of reeds.

Rubber is another plant that has many uses today (Fig. 24C*b*). The product had its origin in Brazil from the thick, white sap (latex) of the rubber tree. Once collected, the sap is placed in a large vat, where acid is added to coagulate the latex. When the water is pressed out, the product is formed into sheets or crumbled and placed into bales. Much stronger rubber, such as that in tires, was made by adding sulfur and heating in a process called vulcanization; this produces a flexible material less sensitive to temperature changes. Today, though, much rubber is synthetically produced.

Before the invention of synthetic fabrics, cotton and other natural fibers were our only source of clothing (Fig. 24C*c*). The 5,200-year-old remains of Ice Man found in the Alps had a cape made of grass. China is now the largest producer of cotton. The cotton fiber

itself comes from filaments that grow on the seed. In sixteenth-century Europe, cotton was a little-understood fiber known only from stories brought back from Asia. Columbus and other explorers were amazed to see the elaborately woven cotton fabrics in the New World. But by 1800, Liverpool was the world's center of cotton trade. (Interestingly, when Levi Strauss wanted to make a tough pair of jeans, he needed a stronger fiber than cotton, so he used hemp. Hemp is now known primarily as a hallucinogenic drug—marijuana—though there has been a resurgence in its use in clothing.) Over 30 species of native cotton now grow around the world, including the United States.

An actively researched area of plant use today is that of medicinal plants. Currently about 50% of all pharmaceutical drugs have their origins from plants. The treatment of cancers appears to rest in the discovery of new

plants. Indeed, the National Cancer Institute (NCI) and most pharmaceutical companies have spent millions (or, more likely, billions) of dollars to send botanists out to collect and test plant samples from around the world. Tribal medicine men, or shamans, of South America and Africa have already been of great importance in developing numerous drugs.

Over the centuries, malaria has caused far more human deaths than any other disease. After European scientists became aware that malaria can be treated by quinine, which comes from the bark of the cinchona tree, a synthetic form of the drug, chloroquine, was developed. But by the late 1960s, it was found that some of the malaria parasites, which live in red blood cells, had become resistant to the synthetically produced drug. Resistant parasites were first seen in Africa but are now showing up in Asia and the Amazon. Today, the only 100%-effective drug for malaria treatment must come directly from the cinchona tree, common to northeastern South America.

Numerous plant extracts continue to be misused for their hallucinogenic or other effects on the human body: coca for cocaine and crack, opium poppy for morphine, and yam for steroids. In addition to all these uses of plants, we should not forget or neglect their aesthetic value (Fig. 24C*d*). Flowers brighten any yard, ornamental plants accent landscaping, and trees provide cooling shade during the summer and break the wind of winter days. Plants also produce oxygen, which is so necessary for all plants and animals.

CONNECTING THE CONCEPTS

The first plants to invade land did not have much competition for space and resources. Through the evolutionary process, they adapted to a dry environment and the threat of water loss (desiccation). Today, the bryophytes are generally small, and they are usually found in moist habitats. However, mosses can store large amounts of water and even become dormant during dry spells. The vascular plants, on the other hand, have specialized tissues to transport water and organic nutrients from one part of the plant to another. Therefore, they can grow tall. Small

pores in their leaves and waxy cuticle open and close to control water loss.

Reproductive strategies in plants are also adapted to a land environment. Mosses and ferns produce flagellated sperm that require external water, but their windblown spores disperse offspring. In seed plants, pollen grains protect sperm until they fertilize an egg. The sporophyte even retains the egg-producing female gametophyte and a seed contains the resulting zygote. The seed protects the sporophyte embryo from drying out until it germinates in a new location. Dispersal of plants

by spores or seeds reduces competition for resources.

Humans use plants for various purposes, including fuel. Massive amounts of biomass were submerged by swamps and covered by sediment during the Carboniferous period. Due to extreme pressure, this organic material became the fossil fuel coal, which, along with the other fuels, makes our way of life today possible. Also, we must not forget that plants produce food and oxygen, two resources that keep our biosphere functioning.

Summary

24.1 EVOLUTIONARY HISTORY OF PLANTS

Plants evolved from multicellular, freshwater green algae over 500 million years ago. During the evolution of plants, protecting the embryo, having vascular tissue for transporting water and organic nutrients, using seeds to disperse offspring, and having flowers were all adaptations to a land existence.

All plants have a life cycle that includes an alternation of generations. In this life cycle, a haploid gametophyte alternates with a diploid sporophyte. During the evolution of plants, the sporophyte gained in dominance, while the gametophyte became microscopic and dependent on the sporophyte.

24.2 NONVASCULAR PLANTS

The bryophytes include three phyla of plants that lack well-developed vascular tissue. Sporophytes of hornworts, liverworts, and mosses are usually nutritionally dependent on the gametophyte, which is more conspicuous and photosynthetic. The life cycle of the moss (see Fig. 24.9) demonstrates reproductive strategies such as flagellated sperm and dispersal of spores.

24.3 VASCULAR PLANTS

The other plant phyla contain vascular plants. The dominant sporophyte has two kinds of well-defined conducting tissues. Xylem is specialized to conduct water and dissolved minerals, and phloem is specialized to conduct organic nutrients and hormones.

24.4 SEEDLESS VASCULAR PLANTS

In the seedless vascular plants, such as the club mosses and the ferns (and fern allies), the sporophyte is dominant, and the gametophyte is usually separate and independent. The life cycle of a fern demonstrates the reproductive strategies of most seedless vascular plants, including flagellated sperm and dispersal by spores.

The seedless vascular plants grew to enormous sizes during the Carboniferous period when the climate was warm and wet. Many of these plants became extinct when the climate became drier and colder. Today, seedless vascular plants that live in the temperate zone use asexual propagation to spread into environments that are not favorable to a water-dependent gametophyte generation.

24.5 SEED PLANTS

Seed plants also have an alternation of generations, but they are heterosporous, producing both microspores and megaspores.

Gametophytes are so reduced that the female gametophyte is retained within an ovule. Microspores become the windblown or animal-transported male gametophytes—the pollen grains. Pollen grains carry sperm to the egg-bearing female gametophyte. Following fertilization, the ovule becomes the seed. A seed contains a sporophyte embryo, and therefore seeds disperse the sporophyte generation. Fertilization no longer uses external water, and sexual reproduction is fully adapted to the terrestrial environment.

24.6 GYMNOSPERMS

The four phyla of gymnosperms are cone-bearing plants, but they do not have an immediate common ancestor. The conifers, represented by the pine tree, exemplify the traits of these plants. Gymnosperms have "naked seeds" because the seeds are not enclosed by fruit, as are those of flowering plants.

24.7 ANGIOSPERMS

In angiosperms, the reproductive organs are found in flowers; the ovules, which become seeds, are located in the ovary, which becomes the fruit. Therefore, angiosperms have "covered seeds." In many angiosperms, pollen is transported from flower to flower by insects and other animals. Both flowers and fruits are found only in angiosperms and may account for the extensive colonization of terrestrial environments by the flowering plants.

Reviewing the Chapter

1. Trace the evolutionary history of plants with regard to four major events.   414–15
2. What is meant when it is said that a plant alternates generations? Distinguish between a sporophyte and a gametophyte.   416
3. Describe the various types of nonvascular plants. How do humans make use of mosses?   417–19
4. Draw a diagram to describe the life cycle of the moss, pointing out significant features.   419
5. When do vascular plants appear in the fossil record, and how do vascular plants differ from nonvascular plants?   420
6. Discuss the various types of seedless vascular plants, pointing out important features.   421–23
7. What are the human uses of ferns? Draw a diagram to describe the life cycle of a fern, pointing out significant features.   423
8. What features do all seed plants have in common?   424
9. What are the human uses of pine trees? List and describe the four phyla of gymnosperms.   424–26

10. Use a diagram of the pine life cycle to point out significant features, including those that distinguish a seed plant's life cycle from that of a seedless vascular plant. 425
11. What is known about the ancestry of flowering plants? 428
12. How do monocots and eudicots differ? What are the parts of a flower? 428–29
13. Use a diagram to explain and point out significant features of the flowering plant life cycle. 430–31
14. Offer an explanation as to why flowering plants are the dominant plants today. 431

Testing Yourself

Choose the best answer for each question.

1. Which of these are characteristics of plants?
   a. multicellular with specialized tissues and organs
   b. photosynthetic and contain chlorophylls *a* and *b*
   c. protect the developing embryo from desiccation
   d. have an alternation of generations life cycle
   e. All of these are correct.

2. In bryophytes, sperm usually move from the antheridium to the archegonium by
   a. swimming.
   b. flying.
   c. insect pollination.
   d. worm pollination.
   e. bird pollination.

3. Seedless vascular plants have
   a. a dominant gametophyte generation.
   b. simple vascular tissue.
   c. flowers.
   d. Both a and b are correct.
   e. Choices a, b, and c are correct.

4. The spore-bearing structure that gives rise to a female gametophyte is called
   a. a microphyll.
   b. a spore.
   c. a megasporangium.
   d. a microsporangium.
   e. a sporophyll.

5. A seedless vascular plant that resembles a tiny upright pine tree and has microphylls spirally arranged belongs to what phylum?
   a. Psilotophyta
   b. Lycophyta
   c. Coniferophyta
   d. Sphenophyta
   e. Pterophyta

6. Trends in the evolution of plants include all of the following except
   a. from homospory to heterospory.
   b. from less to more reliance on water for life cycle.
   c. from nonvascular to vascular.
   d. from nonwoody to woody.

7. Gymnosperms
   a. have flowers.
   b. are eudicots.
   c. are monocots.
   d. reproduce by spores.
   e. reproduce by seeds.

8. In the moss life cycle, the sporophyte
   a. consists of leafy green shoots.
   b. is the heart-shaped prothallus.
   c. consists of a foot, a stalk, and a capsule.
   d. is the dominant generation.
   e. All of these are correct.

9. You are apt to find ferns in a moist location because they have
   a. a water-dependent sporophyte generation.
   b. flagellated spores.
   c. flagellated sperm.
   d. large, leafy fronds.
   e. Both c and d are correct.

10. How are ferns different from mosses?
    a. Only ferns produce spores for reproduction.
    b. Ferns have vascular tissue.
    c. In the fern life cycle, the gametophyte and sporophyte are both independent.
    d. Ferns do not have flagellated sperm.
    e. Both b and c are correct.

11. Which of these pairs is mismatched?
    a. pollen grain—male gametophyte
    b. ovule—female gametophyte
    c. seed—immature sporophyte
    d. pollen tube—spores
    e. tree—mature sporophyte

12. In the life cycle of the pine tree, the ovules are found on
    a. needlelike leaves.   d. root hairs.
    b. seed cones.   e. All of these are correct.
    c. pollen cones.

13. Monocotyledonous plants often have
    a. parallel leaf venation.
    b. flower parts in units of four or five.
    c. a blade and a petiole.
    d. stipules.
    e. Both a and d are correct.
    f. Choices b, c, and d are correct.

14. Which of these pairs is mismatched?
    a. anther—produces microspores
    b. carpel—produces pollen
    c. ovule—becomes seed
    d. ovary—becomes fruit
    e. flower—reproductive structure

15. Which of these plants contributed the most to our present-day supply of coal?
    a. bryophytes
    b. seedless vascular plants
    c. gymnosperms
    d. angiosperms
    e. Both b and c are correct.

16. Which of these is found in seed plants?
    a. complex vascular tissue
    b. pollen grains that are not flagellated
    c. retention of female gametophyte within the ovule
    d. roots, stems, and leaves
    e. All of these are correct.

17. Which of these is a seedless vascular plant?
    a. gymnosperm   d. monocot
    b. angiosperm   e. eudicot
    c. fern

18. Which of these associations is incorrect?
    a. phylum Bryophyta: mosses
    b. phylum Pterophyta: ferns
    c. phylum Coniferophyta: conifers
    d. phylum Sphenophyta: horsetails
    e. All of these are classified correctly.

19. Label this diagram of alternation of generations.

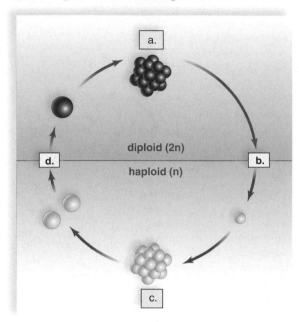

Thinking Scientifically

1. Design an experiment that tests the need for water in order for mosses to undergo alternation of generations instead of reproducing asexually by fragmentation.

2. Before the evolution of the vascular system, the terrestrial landscape was essentially "two-dimensional." By allowing plants to rise above the ground, the vascular system created a three-dimensional world. How does being tall give plants an advantage?

Bioethical Issue: Saving Plant Species

Pollinator populations have been decimated by pollution, pesticide use, and destruction or fragmentation of natural areas. Belatedly, we have come to realize that various types of bees are responsible for pollinating such cash crops as blueberries, cranberries, and squash and are partly responsible for pollinating apple, almond, and cherry trees.

Why are we so short-sighted when it comes to protecting the environment and living creatures like pollinators? Because pollinators are a resource held in common. The term *commons* originally meant a piece of land where all members of a village were allowed to graze their cattle. The farmer who thought only of himself and grazed more cattle than his neighbor was better off. The difficulty is, of course, that eventually the resource is depleted, and everyone loses.

So, a farmer or property owner who uses pesticides is only thinking of his or her field or lawn and not the good of the whole. The commons can only be protected if citizens have the foresight to enact rules and regulations by which all abide. DDT was outlawed in this country in part because it led to the decline of birds of prey. Similarly, we may need legislation to protect pollinators from factors that kill them off. Legislation to protect pollinators would protect the food supply for all of us.

Understanding the Terms

alternation of generations 416	monocot 428
angiosperm 428	monoecious 424
antheridia 416	moss 418
archegonia 416	nonvascular plant 417
bryophyte 417	ovary 429
carpel 429	ovule 416, 424
club moss 421	peat 419
coal 427	peduncle 428
cone 424	petal 429
conifer 424	phloem 420
cotyledon 428	pollen grain 416
cuticle 417	pollen tube 431
cycad 425	pollination 424
dioecious 426	pollinator 428
eudicot 428	receptacle 428
fern 422	rhizoid 418
flower 415	rhizome 421
frond 422	seed 415
fruit 431	sepal 429
gametophyte 416	sporangium 419
ginkgo 426	spore 416
gnetophyte 426	sporophyll 421
gymnosperm 424	sporophyte 416
heterosporous 420	stamen 429
homosporous 420	stigma 429
hornwort 417	stomata (sing., stoma) 417
horsetail 421	strobilus (pl., strobili) 421
kingdom Plantae 414	style 429
lignin 420	vascular plant 420
liverwort 418	vascular tissue 414
megaspore 430	whisk fern 422
microspore 430	xylem 420

Match the terms to these definitions:

a. _____ Diploid generation of the alternation of generations life cycle of a plant; meiosis produces haploid spores that develop into the gametophyte.

b. _____ Flowering plant group; members have one embryonic leaf, parallel-veined leaves, scattered vascular bundles, and other characteristics.

c. _____ Male gametophyte in seed plants.

d. _____ Rootlike hair that anchors a nonvascular plant and absorbs minerals and water from the soil.

e. _____ Vascular tissue that conducts sucrose and organic compounds in plants.

ARIS, the *Biology* Website

ARIS, the website for *Biology*, provides a wealth of information organized and integrated by chapter. You will find practice quizzes, interactive activities, labeling exercises, flashcards, and much more that will complement your learning and understanding of general biology.

www.mhhe.com/maderbiology9

25

STRUCTURE AND ORGANIZATION OF PLANTS

A stunning array of plant life covers the Earth, and over 80% of all living plants are flowering plants, or angiosperms. Therefore, it is fitting that we set aside a chapter of this text to primarily examine the structure and the function of flowering plants. The organization of a flowering plant is suitable to photosynthesizing on land. The elevated leaves have a shape that facilitates absorption of solar energy. Strong stems conduct water up to the leaves and organic food to all parts, including the extensive roots, which not only anchor the plant but also absorb water and minerals. It can be seen, then, that the organs of flowering plants are structured to suit their functions. Roots, stems, and leaves are adapted even further to suit particular environments.

Plants are essential for life on Earth. Animals depend on them for food and oxygen; plants also help regulate the water cycle and the carbon cycle of the biosphere. Plants have many other uses for human beings. For example, they provide fibers for making clothes, wood for construction, and chemicals for commercial and medicinal uses. Plants also give us much pleasure. We like to vacation in natural settings, and we use plants to beautify our parks, lawns, and homes.

Flowering plants are a very important part of the life of humans.

25.1 PLANT ORGANS

From cactuses living in hot deserts to water lilies growing in a nearby pond, the flowering plants, or angiosperms, are extremely diverse. Despite their great diversity in size and shape, flowering plants share many common structural features. Most flowering plants possess a root system and a shoot system (Fig. 25.1). The **root system** simply consists of the roots, while the **shoot system** consists of the stem and leaves. A typical plant features three vegetative **organs** (structures that contain different tissues and perform one or more specific functions) that allow them to live and grow. The roots, stems, and leaves are the vegetative organs common to plants. Flowers, seeds, and fruits are structures involved in reproduction.

Roots

The root system in the majority of plants is located underground. As a rule of thumb, the root system is at least equivalent in size and extent to the shoot system. An apple tree has a much larger root system than a corn plant, for example. A single corn plant may have roots as deep as 2.5 m and spread out over 1.5 m, while a mesquite tree that lives in the desert may have roots that penetrate to a depth of over 20 m.

The extensive root system of a plant anchors it in the soil and gives it support (Fig. 25.2a). The root system absorbs water and minerals from the soil for the entire plant. The cylindrical shape of a root allows it to penetrate the soil as it grows and permits water to be absorbed from all sides. The absorptive capacity of a root is also increased by its many branches and root hairs located in a special zone near a root tip. Root hairs, which are projections from epidermal root-hair cells, are especially responsible for the absorption of water and minerals. Root hairs are so numerous that they increase the absorptive surface of a root tremendously. It has been estimated that a single rye plant has about 14 billion hair cells, and if placed end to end, the root hairs would stretch 10,626 km. Root-hair cells are constantly being replaced, so this same rye plant forms about 100 million new root-hair cells every day. A plant roughly pulled out of the soil will not fare well when transplanted; this is because small lateral roots and root hairs are torn off. Transplantation is more apt to be successful if you take a part of the surrounding soil along with the plant, leaving as much of the lateral roots and the root hairs intact as possible.

Roots have still other functions. Roots produce hormones that stimulate the growth of stems and coordinate their size with the size of the root. It is most efficient for a plant to have root and stem sizes that are appropriate to each other. **Perennial** plants have vegetative structures that live year after year. Herbaceous perennials, which live in temperate areas and die back, store the products of photosynthesis in their roots. Carrots and sweet potatoes are the roots of such plants.

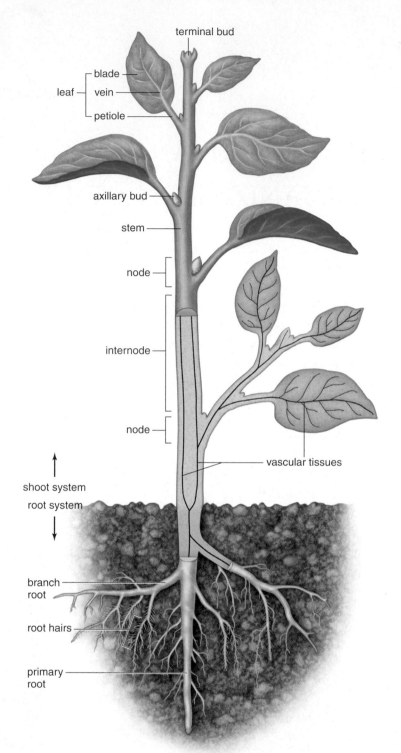

FIGURE 25.1 Organization of a plant body.
The body of a plant consists of a root system and a shoot system. The shoot system contains the stem and leaves, two types of plant vegetative organs. Axillary buds can develop into branches of stems or flowers, the reproductive structures of a plant. The root system is connected to the shoot system by vascular tissue (brown) that extends from the roots to the leaves.

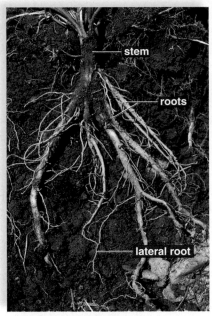

a. Root system, dandelion   b. Shoot system, bean seedling   c. Leaves, pumpkin seedling

FIGURE 25.2 Vegetative organs of several eudicots.
a. The root system anchors the plant and absorbs water and minerals. **b.** The shoot system consists of a stem and its branches, which support the leaves and transport water and organic nutrients. **c.** The leaves, which may be broad and thin, carry on photosynthesis.

Stems

The shoot system of a plant is composed of the stem, the branches, and the leaves. A **stem,** the main axis of a plant, terminates in tissue that allows the stem to elongate and produce leaves (Fig. 25.2*b*). If upright, as most are, stems support leaves in such a way that each leaf is exposed to as much sunlight as possible. A **node** occurs where leaves are attached to the stem and an **internode** is the region between the nodes (see Fig. 25.1). The presence of nodes and internodes is used to identify a stem, even if it happens to be an underground stem. In some plants, the nodes of horizontal stems asexually produce new plants.

In addition to supporting the leaves, a stem has vascular tissue that transports water and minerals from the roots through the stem to the leaves and transports the products of photosynthesis, usually in the opposite direction. Nonliving cells form a continuous pipeline for water and mineral transport, while living cells join end to end for organic nutrient transport. A cylindrical stem can sometimes expand in girth as well as length. As trees grow taller each year, they accumulate woody tissue that adds to the strength of their stems.

Stems may have functions other than transport. In some plants (e.g., cactus), the stem is the primary photosynthetic organ. The stem is a water reservoir in succulent plants. Tubers are horizontal stems that store nutrients.

Leaves

Leaves are the major part of a plant that carries on photosynthesis, a process that requires water, carbon dioxide, and sunlight. Leaves receive water from the root system by way of the stem. Stems and leaves function together to bring about water transport from the roots.

The size, shape, color, and texture of leaves are highly variable. These characterstics are fundamental in plant identification. The leaves of some aquatic duckweeds may be less than 1 mm in diameter while some palms may have leaves that exceed 6 m in length. The shape of leaves can vary from cactus spines to deeply lobed white oak leaves. Leaves can exhibit a variety of colors from various shades of green to deep purple. The texture of leaves varies from smooth and waxy like a magnolia to coarse like a sycamore. Plants that retain their leaves for two to seven years are called **evergreens** and those that lose their leaves every year are called **deciduous.**

Broad and thin plant leaves have the maximum surface area for the absorption of carbon dioxide and the collection of solar energy needed for photosynthesis. Also unlike stems, leaves are almost never woody. With few exceptions, their cells are living, and the bulk of a leaf contains tissue specialized to carry on photosynthesis.

The wide portion of a foliage leaf is called the **blade.** The **petiole** is a stalk that attaches the blade to the stem. The upper acute angle between the petiole and stem is the leaf axil, where an **axillary** (lateral) **bud,** which may become a branch or a flower, originates. Not all leaves are foliage leaves. Some are specialized to protect buds, attach to objects (tendrils), store food (bulbs), or even capture insects.

A flowering plant has three vegetative organs: The root absorbs water and minerals, the stem supports and services leaves, and the leaf carries on photosynthesis.

25.2 MONOCOT VERSUS EUDICOT PLANTS

Flowering plants are divided into two groups, depending on the number of **cotyledons,** or seed leaves, in the embryonic plant (Fig. 25.3). Some plants have one cotyledon, and these plants are known as monocotyledons, or **monocots.** Other embryos have two cotyledons, and these plants are known as eudicotyledons, or **eudicots.** Cotyledons [Gk. *cotyledon,* cup-shaped cavity] of eudicots supply nutrients for seedlings, but the cotyledon of monocots acts as a transfer tissue, and the nutrients are derived from the endosperm before the true leaves begin photosynthesizing.

The vascular (transport) tissue is organized differently in monocots and eudicots. In the monocot root, vascular tissue occurs in a ring. In the eudicot root, phloem, which transports organic nutrients, is located between the arms of xylem, which transports water and minerals, and has a star shape. In the monocot stem, the vascular bundles, which contain vascular tissue surrounded by a bundle sheath, are scattered. In a eudicot stem, the vascular bundles occur in a ring.

Leaf veins are vascular bundles within a leaf. Monocots exhibit parallel venation, and eudicots exhibit netted venation, which may be either pinnate or palmate. Pinnate venation means that major veins originate from points along the centrally placed main vein, and palmate venation means that the major veins all originate at the point of attachment of the blade to the petiole:

Netted venation: pinnately veined   palmately veined

Adult monocots and eudicots have other structural differences, such as the number of flower parts and the number of apertures (thin areas in the wall) of pollen grains. Monocots have their flower parts arranged in multiples of three, and eudicots have their flower parts arranged in multiples of four or five. Eudicot pollen grains usually have three apertures, and monocot pollen grains usually have one aperture.

Although the division between monocots and eudicots may seem of limited importance, it does in fact represent a division that affects many aspects of their structure. The eudicots are the larger group and include some of our most familiar flowering plants—from dandelions to oak trees. The monocots include grasses, lilies, orchids, and palm trees, among others. Some of our most significant food sources are monocots, including rice, wheat, and corn.

FIGURE 25.3 Flowering plants are either monocots or eudicots.
Five features illustrated here are used to distinguish monocots from eudicots: number of cotyledons; the arrangement of vascular tissue in roots, stems, and leaves; and the number of flower parts.

Flowering plants are divided into monocots and eudicots on the basis of structural differences.

	Seed	Root	Stem	Leaf	Flower
Monocots	One cotyledon in seed	Root xylem and phloem in a ring	Vascular bundles scattered in stem	Leaf veins form a parallel pattern	Flower parts in threes and multiples of three
Eudicots	Two cotyledons in seed	Root phloem between arms of xylem	Vascular bundles in a distinct ring	Leaf veins form a net pattern	Flower parts in fours or fives and their multiples

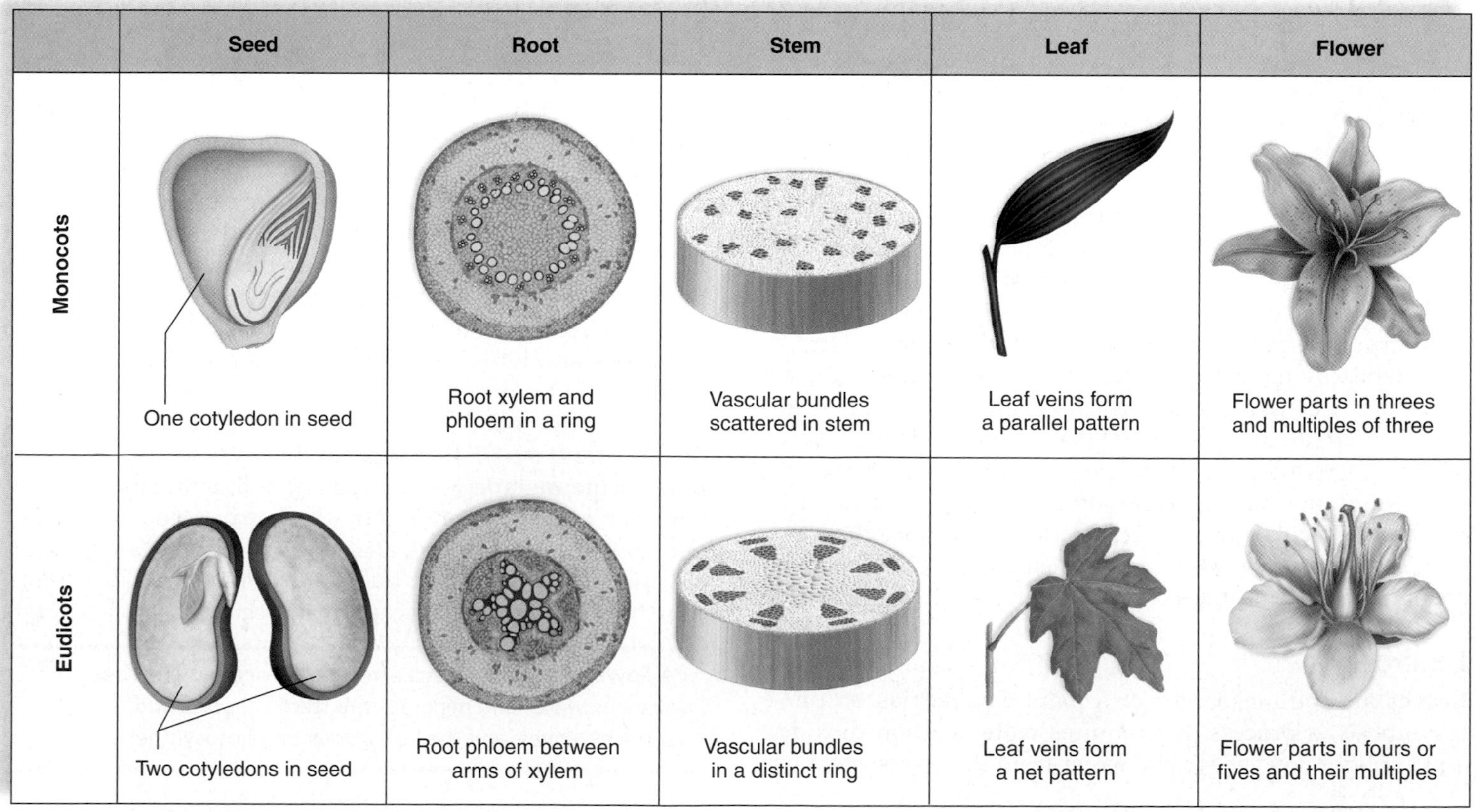

25.3 PLANT TISSUES

A plant has the ability to grow its entire life because it possesses meristematic (embryonic) tissue. Apical **meristems** are located at or near the tips of stems and roots, where they increase the length of these structures. This increase in length is called primary growth. In addition to apical meristems, monocots have a type of meristem called intercalary [L. *intercalare*, to insert] meristem, which allows them to regrow lost parts. Intercalary meristems occur between mature tissues, and they account for why grass can so readily regrow after being grazed by a cow or cut by a lawnmower.

Apical meristem continually produces three types of meristem, and these develop into the three types of specialized primary tissues in the body of a plant: Protoderm gives rise to epidermis; ground meristem produces ground tissue; and procambium produces vascular tissue.

Three specialized tissues include:

1. **Epidermal tissue** forms the outer protective covering of a plant.
2. **Ground tissue** fills the interior of a plant.
3. **Vascular tissue** transports water and nutrients in a plant and provides support.

Epidermal Tissue

The entire body of both nonwoody (herbaceous) and young woody plants is covered by a layer of **epidermis** [Gk. *epi*, over, and *derma*, skin], which contains closely packed epidermal cells. The walls of epidermal cells that are exposed to air are covered with a waxy **cuticle** [L. *cutis*, skin] to minimize water loss. The cuticle also protects against bacteria and other organisms that might cause disease.

In roots, certain epidermal cells have long, slender projections called **root hairs** (Fig. 25.4*a*). As mentioned, the hairs increase the surface area of the root for absorption of water and minerals; they also help anchor the plant firmly in place.

On stems, leaves, and reproductive organs, epidermal cells produce hairs called **trichomes** [Gk. *trichos*, hair] that have two important functions: to protect the plant from too much sun and to conserve moisture. Sometimes trichomes, particularly glandular ones, help protect a plant from herbivores by producing a toxic substance. Under the slightest pressure the stiff trichomes of the stinging nettle lose their tips, forming "hypodermic needles" that inject an intruder with a stinging secretion.

In leaves, the lower epidermis of eudicots and both surfaces of monocots contain specialized cells called guard cells (Fig. 25.4*b*). Guard cells, which are epidermal cells with chloroplasts, surround microscopic pores called **stomata** (sing., stoma). When the stomata are open, gas exchange and water loss occur.

In older woody plants, the epidermis of the stem is replaced by **periderm** [Gk. *peri*, around; *derma*, skin]. The majority component of periderm is boxlike **cork** cells. At maturity, cork cells can be sloughed off (Fig. 25.4*c*). New cork cells are made by a meristem called cork cambium. As the new cork cells mature, they increase slightly in volume, and their walls become encrusted with suberin, a lipid material, so that they are waterproof and chemically inert. These nonliving cells protect the plant and make it resistant to attack by fungi, bacteria, and animals. Some cork tissues are commercially used for bottle corks and other products.

The cork cambium overproduces cork in certain areas of the stem surface; this causes ridges and cracks to appear. These features on the surface are called **lenticels.** Lenticels are important in gas exchange between the interior of a stem and the air.

Epidermal tissue forms the outer protective covering of a herbaceous plant. It is modified in roots, stems, and leaves.

a. Root hairs

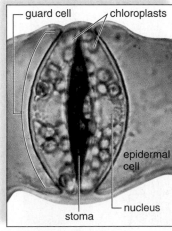

b. Stoma of leaf

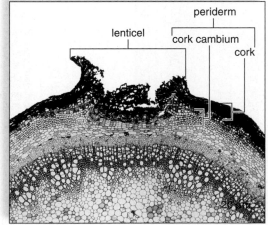

c. Cork of older stem

FIGURE 25.4 Modifications of epidermal tissue.
a. Root epidermis has root hairs to absorb water. **b.** Leaf epidermis contains stomata (sing., stoma) for gas exchange.
c. Periderm includes cork and cork cambium. Lenticels in cork are important in gas exchange.

FIGURE 25.5
Ground tissue cells.
a. Parenchyma cells are the
least specialized of the plant
cells. **b.** Collenchyma cells.
Notice how much thicker
and irregular the walls
are compared to those of
parenchyma cells.
c. Sclerenchyma cells have
very thick walls and are
nonliving—their only function
is to give strong support.

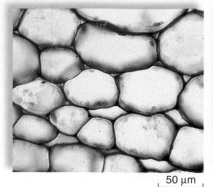
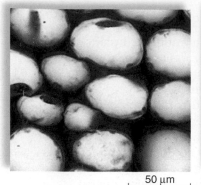
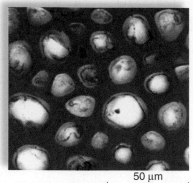

a. Parenchyma cells   b. Collenchyma cells   c. Sclerenchyma cells

Ground Tissue

Ground tissue forms the bulk of a plant and contains parenchyma, collenchyma, and sclerenchyma cells (Fig. 25.5). **Parenchyma** [Gk. *para*, beside, and *enchyma*, infusion] cells are the most abundant and correspond best to the typical plant cell. These are the least specialized of the cell types and are found in all the organs of a plant. They may contain chloroplasts and carry on photosynthesis (chlorenchyma), or they may contain colorless plastids that store the products of photosynthesis. A juicy bite from an apple yields mostly storage parenchyma cells. Parenchyma cells line the connected air spaces of a water lily and other aquatic plants. Parenchyma cells can divide and give rise to more specialized cells, such as when roots develop from stem cuttings placed in water.

Collenchyma cells are like parenchyma cells except they have thicker primary walls. The thickness is uneven and usually involves the corners of the cell. Collenchyma cells often form bundles just beneath the epidermis and give flexible support to immature regions of a plant body. The familiar strands in celery stalks (leaf petioles) are composed mostly of collenchyma cells.

Sclerenchyma cells have thick secondary cell walls impregnated with **lignin,** which is a highly resistant organic substance that makes the walls tough and hard. If we compare a cell wall to reinforced concrete, cellulose fibrils would play the role of steel rods, and lignin would be analogous to the cement. Most sclerenchyma cells are nonliving; their primary function is to support the mature regions of a plant. Two types of sclerenchyma cells are *fibers* and *sclereids*. Although fibers are occasionally found in ground tissue, most

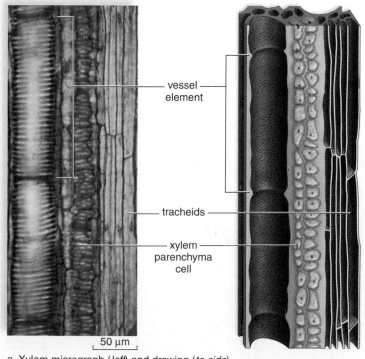
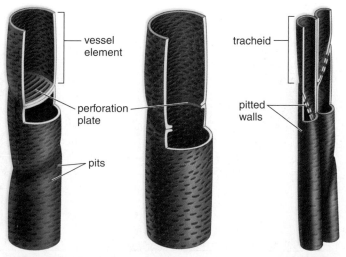

a. Xylem micrograph (*left*) and drawing (*to side*)   b. Two types of vessels   c. Tracheids

FIGURE 25.6 Xylem structure.
a. Photomicrograph of xylem vascular tissue and drawing showing general organization of xylem tissue. **b.** Drawing of two types of vessels (composed of vessel elements)—the perforation plates differ. **c.** Drawing of tracheids.

are in vascular tissue, which is discussed next. Fibers are long and slender and may be grouped in bundles that are sometimes commercially important. Hemp fibers can be used to make rope, and flax fibers can be woven into linen. Flax fibers, however, are not lignified, which is why linen is soft. Sclereids, which are shorter than fibers and more varied in shape, are found in seed coats and nutshells. Sclereids, or "stone cells," are responsible for the gritty texture of pears. The hardness of nuts and peach pits is due to sclereids.

Vascular Tissue

There are two types of vascular (transport) tissue. **Xylem** transports water and minerals from the roots to the leaves, and **phloem** transports sucrose and other organic compounds, including hormones, usually from the leaves to the roots. Both xylem and phloem are considered **complex tissues** because they are composed of two or more kinds of cells. Xylem contains two types of conducting cells: tracheids and vessel elements (VE) (Fig. 25.6). Both types of conducting cells are hollow and nonliving, but the **vessel elements** are larger, may have perforation plates in their end walls, and are arranged to form a continuous vessel for water and mineral transport. The elongated **tracheids,** with tapered ends, form a less obvious means of transport, but water can move across the end walls and side walls because there are **pits,** or depressions, where the secondary wall does not form. In

addition to vessel elements and tracheids, xylem contains parenchyma cells that store various substances. Vascular rays, which are flat ribbons or sheets of parenchyma cells located between rows of tracheids, conduct water and minerals across the width of a plant. Xylem also contains fibers that lend support.

The conducting cells of phloem are **sieve-tube members** arranged to form a continuous sieve tube (Fig. 25.7). Sieve-tube members contain cytoplasm but no nuclei. The term *sieve* refers to a cluster of pores in the end walls, which is known as a sieve plate. Each sieve-tube member has a companion cell, which does have a nucleus. The two are connected by numerous plasmodesmata, and the nucleus of the companion cell may control and maintain the life of both cells. The companion cells are also believed to be involved in the transport function of phloem.

It is important to realize that vascular tissue (xylem and phloem) extends from the root through stems to the leaves and vice versa (see Fig. 25.1). In the roots, the vascular tissue is located in the **vascular cylinder;** in the stem, it forms **vascular bundles;** and in the leaves, it is found in **leaf veins.**

The vascular tissues are xylem and phloem. Xylem transports water and minerals, and phloem transports organic nutrients.

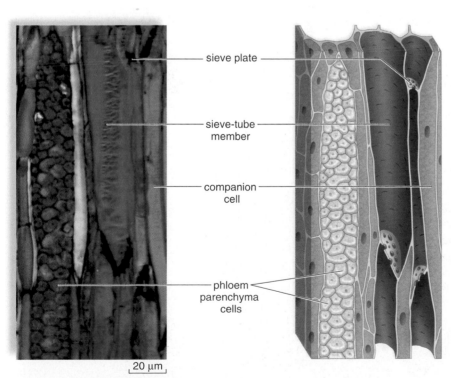
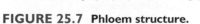

a. Phloem micrograph (*left*) and drawing (*to side*)

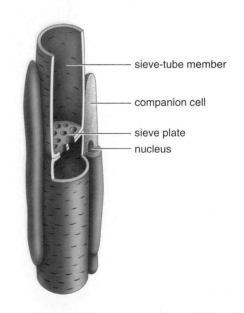

b. Sieve-tube member and companion cells

FIGURE 25.7 Phloem structure.
a. Photomicrograph of phloem vascular tissue and drawing showing general organization of phloem tissue.
b. Drawing of sieve tube (composed of sieve-tube members) and companion cells.

25.4 ORGANIZATION OF ROOTS

Figure 25.8*a*, a longitudinal section of a eudicot root, reveals zones where cells are in various stages of differentiation as primary growth occurs. The root **apical meristem** is in the region protected by the **root cap.** Root cap cells have to be replaced constantly because they get ground off by rough soil particles as the root grows. The primary meristems are in the zone of cell division, which continuously provides cells to the zone of elongation above by mitosis. In the zone of elongation, the cells lengthen as they become specialized. The zone of maturation, which contains fully differentiated cells, is recognizable because here root hairs are borne by many of the epidermal cells.

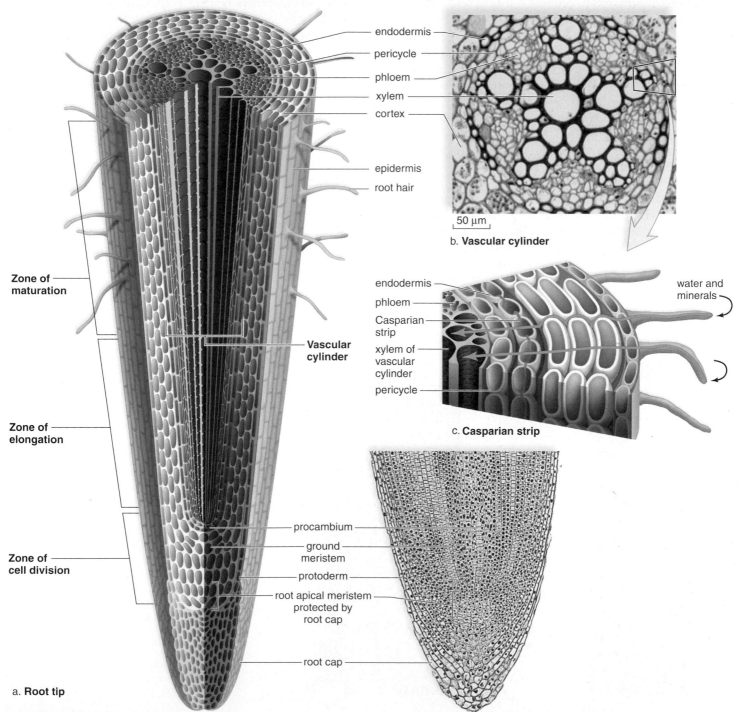

a. **Root tip**

b. **Vascular cylinder**

c. **Casparian strip**

FIGURE 25.8 Eudicot root tip.
a. The root tip is divided into three zones, best seen in a longitudinal section such as this. **b.** The vascular cylinder of a eudicot root contains the vascular tissue. Xylem is typically star-shaped, and phloem lies between the points of the star. **c.** Water and minerals either pass between cells, or they enter root cells directly. In any case, because of the Casparian strip, water and minerals must pass through the cytoplasm of endodermal cells in order to enter the xylem. In this way, endodermal cells regulate the passage of minerals into the vascular cylinder.

Tissues of a Eudicot Root

Figure 25.8*a* also shows a cross section of a root at the region of maturation. These specialized tissues are identifiable:

Epidermis The epidermis, which forms the outer layer of the root, consists of only a single layer of cells. The majority of epidermal cells are thin-walled and rectangular, but in the zone of maturation, many epidermal cells have root hairs. These project as far as 5–8 mm into the soil particles.

Cortex Moving inward, next to the epidermis, large, thin-walled parenchyma cells make up the **cortex** of the root. These irregularly shaped cells are loosely packed, and it is possible for water and minerals to move through the cortex without entering the cells. The cells contain starch granules, and the cortex functions in food storage.

Endodermis The **endodermis** [Gk. *endon*, within, and *derma*, skin] is a single layer of rectangular cells that forms a boundary between the cortex and the inner vascular cylinder. The endodermal cells fit snugly together and are bordered on four sides (but not the two sides that contact the cortex and the vascular cylinder) by a layer of impermeable lignin and suberin known as the **Casparian strip** (Fig. 25.8*c*). This strip prevents the passage of water and mineral ions between adjacent cell walls. Therefore, the only access to the vascular cylinder is through the endodermal cells themselves, as shown by the arrow in Figure 25.8*c*. This arrangement regulates the entrance of minerals into the vascular cylinder.

Vascular tissue The **pericycle,** the first layer of cells within the vascular cylinder, has retained its capacity to divide and can start the development of branch, or lateral, roots (Fig. 25.9). The main portion of the vascular cylinder contains xylem and phloem. The xylem appears star-shaped in eudicots because several arms of tissue radiate from a common center (see Fig. 25.8*b*). The phloem is found in separate regions between the arms of the xylem.

Organization of Monocot Roots

Monocot roots have the same growth zones as eudicot roots, but they do not undergo secondary growth as many eudicot roots go through. Also, the organization of their tissues is slightly different. The ground tissue of a monocot root's **pith,** which is centrally located, is surrounded by a vascular ring composed of alternating xylem and phloem bundles (Fig. 25.10). Monocot roots also have pericycle, endodermis, cortex, and epidermis.

The root system of a plant absorbs water and minerals. These nutrients cross the epidermis and cortex before entering the endodermis, the tissue that regulates their entrance into the vascular cylinder.

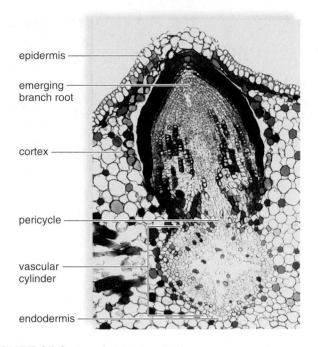

epidermis

emerging branch root

cortex

pericycle

vascular cylinder

endodermis

FIGURE 25.9 Branching of eudicot root.
These cross sections of a willow, *Salix*, show the origination and growth of a branch root from the pericycle.

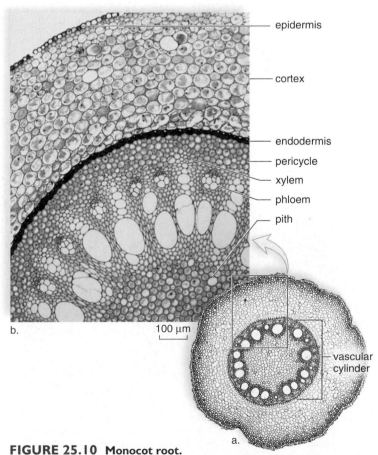

epidermis

cortex

endodermis

pericycle

xylem

phloem

pith

b.   100 µm

vascular cylinder

a.

FIGURE 25.10 Monocot root.
a. In this overall cross section, it is possible to observe that a vascular ring surrounds a central pith. **b.** The enlargement shows the exact placement of various tissues.

a. Taproot

b. Fibrous root system

c. Prop roots, a type of adventitious
   root

d. Pneumatophores of black
   mangrove trees

dodder

e. Dodder

haustorium   dodder

host's vascular tissue

FIGURE 25.11 Root diversity.
a. A taproot may have branch roots in addition to a main root. **b.** A fibrous root has
many slender roots with no main root. **c.** Prop roots are specialized for support.
d. The pneumatophores of a black mangrove tree allow it to acquire oxygen even
though it lives in swampy water. **e.** *Left:* Dodder is a parasitic plant consisting mainly of
orange-brown twining stems. (The green in the photograph is the host plant.) *Right:*
Haustoria are rootlike projections of the stem that tap into the host's vascular system.

Root Diversity

Roots have various adaptations and associations to better
perform their functions: anchorage, absorption of water and
minerals, and storage of carbohydrates.

In some plants, notably eudicots, the first or **primary
root** grows straight down and remains the dominant root
of the plant. This so-called **taproot** is often fleshy and stores
food (Fig. 25.11*a*). Carrots, beets, turnips, and radishes have
taproots that we consume as vegetables. Sweet potato plants
don't have taproots, but they do have roots that expand to
store starch. We call these storage areas sweet potatoes.

In other plants, notably monocots, there is no single,
main root; instead, there are a large number of slender roots.
These grow from the lower nodes of the stem when the first
(primary) root dies. These slender roots and their lateral
branches make up a **fibrous root system** (Fig. 25.11*b*). Many
have observed the fibrous root systems of grasses and have
noted how these roots strongly anchor the plant to the soil.

Root Specializations

When roots develop from organs of the shoot system instead
of the root system, they are known as **adventitious roots.**
Some adventitious roots emerge above the soil line, as they
do in corn plants, in which their main function is to help anchor the plant. If so, they are called prop roots (Fig. 25.11*c*).
Other examples of adventitious roots are those found on
horizontal stems (see Fig. 25.19*a*) or the rootlets at the nodes
of climbing English ivy. As the vines climb, the rootlets attach the plant to any available vertical structure.

Black mangroves live in swampy water and have pneumatophores, root projections that rise above the water and acquire oxygen for cellular respiration (Fig. 25.11*d*).

Some plants, such as dodders and broomrapes, are
parasitic on other plants. Their stems have rootlike projections called haustoria (sing., haustorium) that grow into
the host plant and make contact with vascular tissue from
which they extract water and nutrients (Fig. 25.11*e*). **Mycorrhizae** are associations between roots and fungi that can
extract water and minerals from the soil better than roots
that lack a fungus partner. This is a mutualistic relationship
because the fungus receives sugars and amino acids from
the plant, while the plant receives water and minerals via
the fungus.

Peas, beans, and other legumes have **root nodules**
where nitrogen-fixing bacteria live. Plants cannot extract
nitrogen from the air, but the bacteria within the nodules
can take up and reduce atmospheric nitrogen. This means
that the plant is no longer dependent on a supply of nitrogen
(i.e., nitrate or ammonium) from the soil, and indeed these
plants are often planted just to bolster the nitrogen supply of
the soil.

Roots have various adaptations and associations to enhance
their ability to anchor a plant, absorb water and minerals,
and store the products of photosynthesis.

ecology focus

Paper Comes from Plants

The word *paper* takes its origin from papyrus, the plant Egyptians used to make the first form of paper. The Egyptians manually placed thin sections cut from papyrus at right angles and pressed them together to make a sheet of writing material. From that beginning some 5,500 years ago, the production of paper is now a worldwide industry of major importance (Fig. 25A). The process is fairly simple. Plant material is ground up mechanically to form a pulp that contains "fibers," which biologists know come from vascular tissue. The fibers automatically form a sheet when they are screened from the pulp.

If wood is the source of the fibers, the pulp must be chemically treated to remove lignin. If only a small amount of lignin is removed, the paper is brown, as in paper bags. If more lignin is removed, the paper is white but not very durable, and it crumbles after a few decades. Paper is more durable when it is made from cotton or linen because the fibers from these plants are lignin-free.

Among the other major plants used to make paper are:

Eucalyptus trees. In recent years, Brazil has devoted huge areas of the Amazon region to the growing of cloned eucalyptus seedlings, specially selected and engineered to be ready for harvest after about seven years.

Temperate hardwood trees. Plantation cultivation in Canada provides birch, beech, chestnut, poplar, and particularly aspen wood for paper making. Tropical hardwoods, usually from Southeast Asia, are also used.

Softwood trees. In the United States, several species of pine trees have been genetically improved to have a higher wood density and to be harvestable five years earlier than ordinary pines. Southern Africa, Chile, New Zealand, and Australia also devote thousands of acres to growing pines for paper pulp production.

Bamboo. Several Asian countries, especially India, provide vast quantities of bamboo

FIGURE 25A Paper production.
Today, a revolving wire-screen belt is used to deliver a continuous wet sheet of paper to heavy rollers and heated cylinders, which remove most of the remaining moisture and press the paper flat.

pulp for the making of paper. Because bamboo is harvested without destroying the roots, and the growing cycle is favorable, this plant, which is actually a grass, is expected to be a significant source of paper pulp despite high processing costs to remove impurities.

Flax and cotton rags. Linen and cotton cloth from textile and garment mills are used to produce *rag paper,* whose flexibility and durability are desirable in legal documents, high-grade bond paper, and high-grade stationery.

It has been known for some time that paper largely consists of the cellulose within plant cell walls. It seems reasonable to suppose, then, that paper could be made from synthetic polymers (e.g., rayon). Indeed, synthetic polymers produce a paper that has qualities superior to those of paper made from natural sources, but

the cost thus far is prohibitive. Another consideration, however, is the ecological impact of making paper from trees. Plantations containing stands of uniform trees replace natural ecosystems, and when the trees are clear-cut, the land is laid bare. Paper mill wastes, which include caustic chemicals, add significantly to the pollution of rivers and streams.

The use of paper for packaging and making all sorts of products has increased dramatically in the last century. Each person in the United States uses about 318 kg of paper products per year, and this compares to only 2.3 kg of paper per person in India. It is clear, then, that we should take the initiative in recycling paper. When newspaper, office paper, and photocopies are soaked in water, the fibers are released, and they can be used to make a new batch of paper. It's estimated that recycling the Sunday newspapers alone would save approximately 500,000 trees each week!

25.5 ORGANIZATION OF STEMS

The anatomy of a woody twig ready for next year's growth reviews for us the organization of a stem (Fig. 25.12). The **terminal bud** contains the shoot tip protected by bud scales, which are modified leaves. Leaf scars and bundle scars mark the location of leaves that have dropped. Dormant axillary buds that can give rise to branches or flowers are also found here. Each spring when growth resumes, bud scales fall off and leave a scar. You can tell the age of a stem by counting these bud scale scars because there is one for each year's growth.

When growth resumes, primary growth continues. The apical meristem at the shoot tip produces new cells that elongate and thereby increase the height of the stem. The **shoot apical meristem** is protected within the terminal bud, where leaf primordia (immature leaves) envelop it (Fig. 25.13). The leaf primordia mark the location of a node; the portion of stem in between nodes is an internode. As a stem grows, the internodes increase in length.

In addition to leaf primordia, three specialized types of primary meristem develop from a shoot apical meristem (Fig. 25.13b). These primary meristems contribute to the length of a shoot. As mentioned, the protoderm, the outermost primary meristem, gives rise to epidermis. The ground meristem produces two tissues composed of parenchyma cells. The parenchyma tissue in the center of the stem is the pith, and the parenchyma tissue between the epidermis and the vascular tissue is the cortex.

The procambium, seen as an obvious strand of tissue in Figure 25.13a, produces the first xylem cells, called primary xylem, and the first phloem cells, called primary phloem. Differentiation continues as certain cells become the first tracheids or vessel elements of the xylem within a vascular bundle. The first sieve-tube members of a vascular bundle do not have companion cells and are short-lived (some live only a day before being replaced). Mature vascular bundles contain fully differentiated xylem, phloem, and a lateral meristem called **vascular cambium** [L. *vasculum*, dim. of *vas*, vessel, and *cambio*, exchange]. Vascular cambium is discussed more fully in the next section.

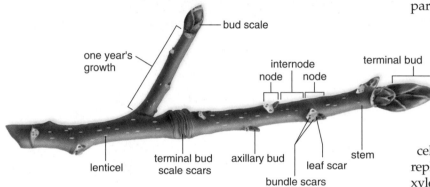

FIGURE 25.12 Woody twig.
The major parts of a stem are illustrated by a woody twig collected in winter.

FIGURE 25.13

Shoot tip and primary meristems.
a. The shoot apical meristem within a terminal bud is surrounded by leaf primordia. **b.** The shoot apical meristem produces the primary meristems: Protoderm gives rise to epidermis; ground meristem gives rise to pith and cortex; and procambium gives rise to vascular tissue, including primary xylem, primary phloem, and vascular cambium.

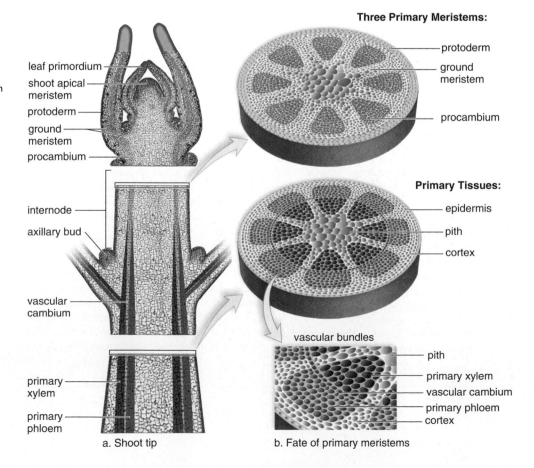

a. Shoot tip　　　　b. Fate of primary meristems

Herbaceous Stems

Mature nonwoody stems, called **herbaceous stems** [L. *herba*, vegetation, plant], exhibit only primary growth. The outermost tissue of herbaceous stems is the epidermis, which is covered by a waxy cuticle to prevent water loss. These stems have distinctive vascular bundles, where xylem and phloem are found. In each bundle, xylem is typically found toward the inside of the stem, and phloem is found toward the outside.

In the herbaceous eudicot stem such as a sunflower, the vascular bundles are arranged in a distinct ring that sep-arates the cortex from the central pith, which stores water and products of photosynthesis (Fig. 25.14). The cortex is sometimes green and carries on photosynthesis. In a mono-cot stem such as corn, the vascular bundles are scattered throughout the stem, and often there is no well-defined cor-tex or well-defined pith (Fig. 25.15).

As a stem grows, the shoot apical meristem produces new leaves and primary meristems. The primary meristems produce the other tissues found in herbaceous stems.

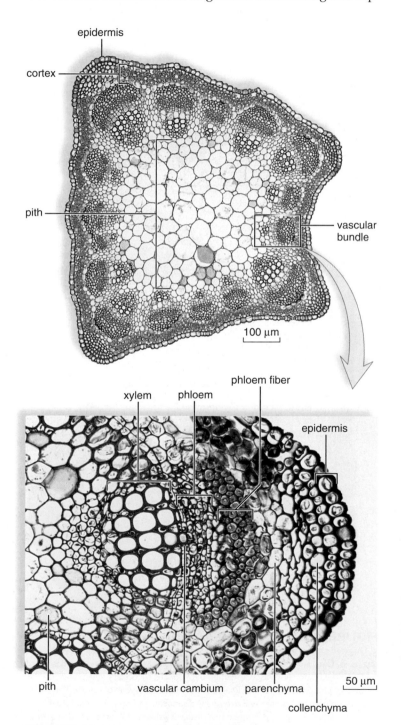

FIGURE 25.14 Herbaceous eudicot stem.

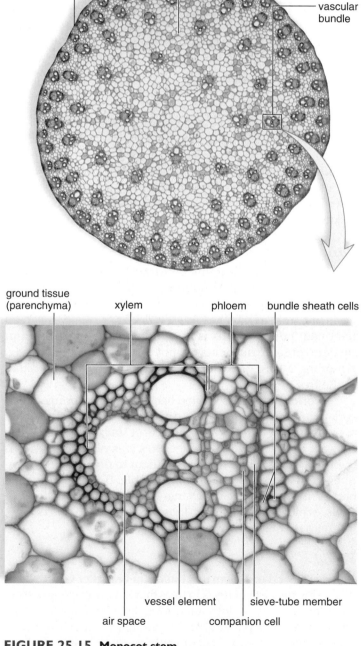

FIGURE 25.15 Monocot stem.

Woody Stems

A woody plant such as an oak tree has both primary and secondary tissues. Primary tissues are those new tissues formed each year from primary meristems right behind the shoot apical meristem. Secondary tissues develop during the first and subsequent years of growth from lateral meristems: vascular cambium and cork cambium. Primary growth, which occurs in all plants, increases the length of a plant, and secondary growth, which occurs only in conifers and woody eudicots, increases the girth of trunks, stems, branches, and roots.

Trees and shrubs undergo secondary growth because of a change in the location and activity of vascular cambium (Fig. 25.16). In herbaceous plants, vascular cambium is present between the xylem and phloem of each vascular bundle. In woody plants, the vascular cambium develops to form a ring of meristem that divides parallel to the surface of the plant, and produces new xylem and phloem each year. Eventually, a woody eudicot stem has an entirely different organization from that of a herbaceous eudicot stem. A woody stem has no distinct vascular bundles and instead has three distinct areas: the bark, the wood, and the pith. Vascular cambium occurs between the bark and the wood.

You will also notice in Figure 25.16 the xylem rays and phloem rays that are visible in the cross section of a woody stem. Rays consist of parenchyma cells that permit lateral conduction of nutrients from the pith to the cortex and some storage of food. A phloem ray is actually a continuation of a xylem ray. Some phloem rays are much broader than other phloem rays.

Bark

The **bark** of a tree contains periderm (cork and cork cambium), and phloem. Although secondary phloem is produced each year by vascular cambium, phloem does not build up from season to season. The bark of a tree can be removed; however, this is very harmful because, without phloem, organic nutrients cannot be transported. Girdling, removing a ring of bark from around a tree, can be lethal to the tree.

Cork cambium is located beneath the epidermis. When cork cambium first begins to divide, it produces tissue that disrupts the epidermis and replaces it with cork cells. Cork cells are impregnated with suberin, a waxy layer that makes them waterproof but also causes them to die. This is protective because now the stem is less edible. But an impervious barrier means that gas exchange is impeded except at lenticels, which are pockets of loosely arranged cork cells not impregnated with suberin.

Wood

Wood is secondary xylem that builds up year after year, thereby increasing the girth of trees. In trees that have a growing season, vascular cambium is dormant during the winter. In the spring, when moisture is plentiful and leaves require much water for growth, the secondary xylem contains wide vessel elements

with thin walls. In this so-called *spring wood*, wide vessels transport sufficient water to the growing leaves. Later in the season, moisture is scarce, and the wood at this time, called *summer wood*, has a lower proportion of vessels (Fig. 25.17). Strength is required because the tree is growing larger and summer wood contains numerous thick-walled tracheids. At the end of the growing season, just before the cambium becomes dormant again, only heavy fibers with especially thick secondary walls may develop. When the trunk of a tree has spring wood followed by summer wood, the two together make up one year's growth, or an **annual ring.** You can tell the age of a tree by counting the annual rings (Fig. 25.18a). The outer annual rings, where transport occurs, are called sapwood.

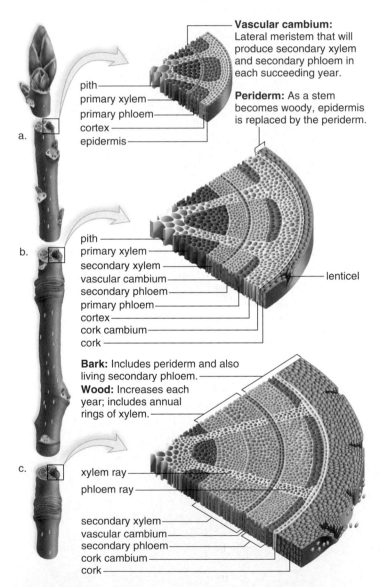

Vascular cambium: Lateral meristem that will produce secondary xylem and secondary phloem in each succeeding year.

Periderm: As a stem becomes woody, epidermis is replaced by the periderm.

a.
pith
primary xylem
primary phloem
cortex
epidermis

b.
pith
primary xylem
secondary xylem
vascular cambium
secondary phloem
primary phloem
cortex
cork cambium
cork

lenticel

Bark: Includes periderm and also living secondary phloem.
Wood: Increases each year; includes annual rings of xylem.

c.
xylem ray
phloem ray

secondary xylem
vascular cambium
secondary phloem
cork cambium
cork

FIGURE 25.16 Diagrams of secondary growth of stems.
a. Diagram showing eudicot herbaceous stem just before secondary growth begins. **b.** Diagram showing that secondary growth has begun. Periderm has replaced the epidermis. Vascular cambium produces secondary xylem and secondary phloem each year. **c.** Diagram showing a two-year-old stem. The primary phloem and cortex will eventually disappear, and only the secondary phloem (within the bark) produced by vascular cambium will be active that year. Secondary xylem builds up to become the annual rings of a woody stem.

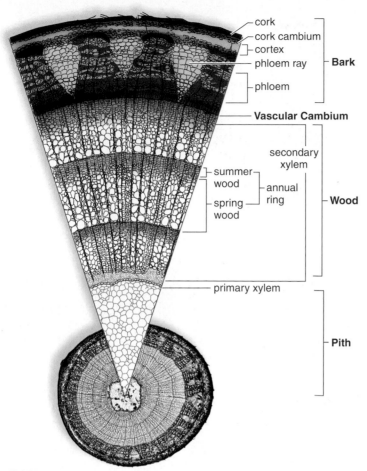

FIGURE 25.17 Three-year-old woody twig.
The buildup of secondary xylem in a woody stem results in annual rings, which tell the age of the stem. The rings can be distinguished because each one begins with spring wood (large vessel elements) and ends with summer wood (smaller and fewer vessel elements).

more defense mechanisms because a long-lasting plant that stays in one spot is likely to be attacked by herbivores and parasites. Then, too, trees don't usually reproduce until they have grown several seasons, by which time they may have succumbed to an accident or disease. In certain habitats, it is more advantageous for a plant to put most of its energy into simply reproducing rather than being woody.

> Woody plants grow in girth due to the presence of vascular cambium and cork cambium. Their bodies have three main parts: bark (which contains cork, cork cambium, and phloem), wood (which contains xylem), and pith.

In older trees, the inner annual rings, called the heartwood, no longer function in water transport. The cells become plugged with deposits, such as resins, gums, and other substances that inhibit the growth of bacteria and fungi. Heartwood may help support a tree, although some trees stand erect and live for many years after the heartwood has rotted away. Figure 25.18*b* shows the layers of a woody stem in relation to one another.

The annual rings are not only important in telling the age of a tree, they can serve as a historical record of tree growth. For example, if rainfall and other conditions were extremely favorable during a season, the annual ring may be wider than usual. If the tree were shaded on one side by another tree or building, the rings may be wider on the favorable side.

Woody Plants. The first flowering plants to evolve may have been woody shrubs; herbaceous plants evolved later. Is it advantageous to be woody? With adequate rainfall, woody plants can grow taller and have more growth because they have adequate vascular tissue to support and service their leaves. However, it takes energy to produce secondary growth and prepare the body for winter if the plant lives in the temperate zone. Also, woody plants need

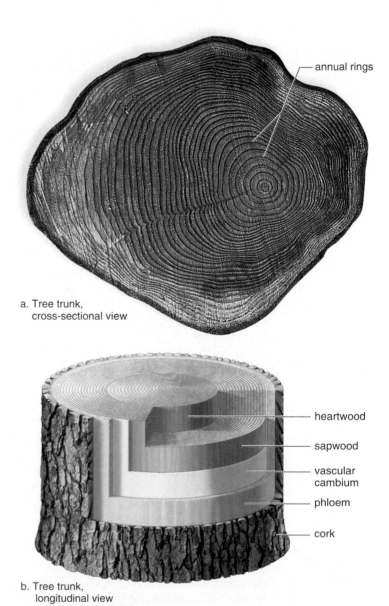

a. Tree trunk,
cross-sectional view

b. Tree trunk,
longitudinal view

FIGURE 25.18 Tree trunk.
a. A cross section of a 39-year-old larch, *Larix decidua*. The xylem within the darker heartwood is inactive; the xylem within the lighter sapwood is active.
b. The relationship of bark, vascular cambium, and wood is retained in a mature stem. The pith has been buried by the growth of layer after layer of new secondary xylem.

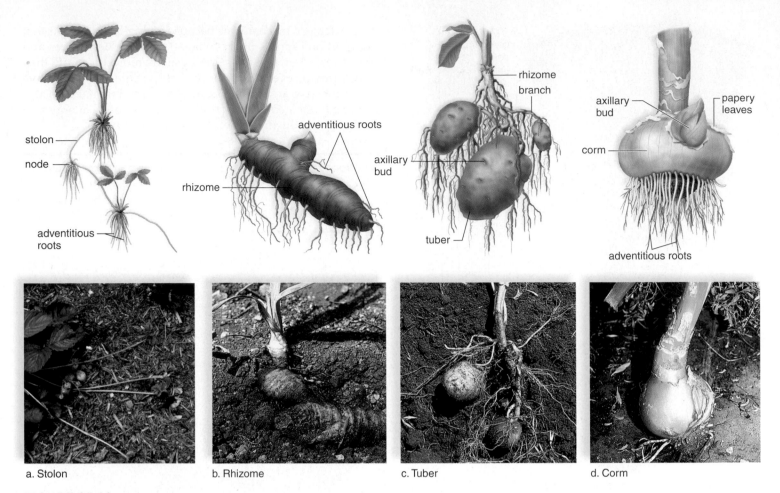

FIGURE 25.19 Stem diversity.
a. A strawberry plant has aboveground horizontal stems called stolons. Every other node produces a new shoot system. **b.** The underground horizontal stem of an iris is a fleshy rhizome. **c.** The underground stem of a potato plant has enlargements called tubers. We call the tubers potatoes. **d.** The corm of a gladiolus is a stem covered by papery leaves.

Stem Diversity

Stem diversity is illustrated in Figure 25.19. Aboveground horizontal stems, called **stolons** [L. *stolo,* shoot] or runners, produce new plants where nodes touch the ground. The strawberry plant is a common example of this type of stem, which functions in vegetative reproduction.

Aboveground vertical stems can also be modified. For example, cacti have succulent stems specialized for water storage, and the tendrils of grape plants (which are stem branches) allow them to climb. Morning glory and relatives have stems that twine around support structures. Such tendrils and twining shoots help plants expose their leaves to the sun.

Underground horizontal stems, **rhizomes** [Gk. *rhiza,* root], may be long and thin, as in sod-forming grasses, or thick and fleshy, as in irises. Rhizomes survive the winter and contribute to asexual reproduction because each node bears a bud. Some rhizomes have enlarged portions called tubers, which function in food storage.

Potatoes are tubers, in which the eyes are buds that mark the nodes.

Corms are bulbous underground stems that lie dormant during the winter, just as rhizomes do. They also produce new plants the next growing season. Gladiolus corms are referred to as bulbs by laypersons, but the botanist reserves the term *bulb* for a structure composed of modified leaves attached to a short vertical stem. An onion is a bulb.

Humans make use of stems in many ways. The stem of the sugarcane plant is a primary source of table sugar. The spice cinnamon and the drug quinine are derived from the bark of *Cinnamomum verum* and various *Cinchona* species, respectively. And wood is necessary for the production of paper as discussed in the Ecology Focus on page 447.

Plants use modified stems for such functions as vegetative reproduction, climbing, survival, and food storage. Modified stems aid adaptation to different environments.

---

science focus

Defense Strategies of Trees

Rainstorms, ice, snow, animals, wind, excess weight, temperature extremes, and chemicals can all injure a tree. So can improper pruning. Pruning, which requires the cutting away of tree parts, can benefit a tree by improving its appearance and helping maintain its balance. But removing the top of a tree, called topping, removing a portion of the roots, and flush-cutting a number of branches at one time is injurious to trees. This is known because a tree reacts to improper pruning in the same manner it reacts to all injuries, no matter what the cause. The wounding of a tree subjects it to disease. Trees, like humans, have defensive strategies against bacterial and fungal invasions that occur when a tree is wounded.

A defense strategy is a mechanism that has arisen through the evolutionary process. In other words, members of the group with the strategy compete better than those that do not. We expect defense strategies to be beneficial—and they are—but the manner in which trees react to disease, called compartmentalization of decay, can still weaken them. Therefore, improper pruning practices should be avoided at all cost if you care about a tree!

Just as with humans, trees have a series of defense strategies against infection. Each one is better than the other at stopping the progress of disease organisms. First, when a tree is injured, the tracheids and vessel elements of xylem immediately plug up with chemicals that block them off above and below the site of the injury. In trees that fail to effectively close off vessel elements, long columns of rot (decay) run up and down the trunk and into branches, which eventually become hollow.

The second defense strategy is a result of tree trunk structure. As you know, a tree trunk has annual rings that tell its age. A dark region at the edge of an annual ring in the cross section of a trunk tells you that this tree was injured, and that disease organisms were unable to advance inward on their way to the pith. It appears, therefore, that disease organisms have a harder time moving across a trunk due to annual ring construction than they do moving through the trunk in vessel elements.

The third defense strategy involves rays. Rays take their name from the fact that they project radially from vascular cambium. Just like the slices of a pie, rays divide the trunk

Turkey oak, *Quercus laevis*

FIGURE 25B Defense strategies.
An oak tree is better at defense against infection than is a weeping willow tree. The oak tree never has to employ the defense strategy (right) that resulted in this dark ring in the trunk, which can lead to cracks and collapse of the tree.

Weeping willow, *Salix bablonica*

reaction zone

Cross section of a damaged tree trunk.

of a tree. Disease organisms can't cross rays either, and this keeps them in a small pie piece of the trunk and prevents them from moving completely around the trunk.

The fourth defense strategy is a so-called reaction zone that develops in the region of the injury along the inner portion of the cambium next to the youngest annual ring. The reaction zone can extend from a few inches to a few feet above and below the injury, and partway or all the way around the trunk. The reaction zone doesn't wall off any annual rings that develop after the injury, but it does wall off any annual rings that were present before the injury occurred. Figure 25B (*right*) shows a cross section of a tree that was topped seven years before it was cut down. The reaction zone is seen as a dark circle that, in this case, extends from the top of the tree to the root system.

Although the fourth defense strategy more effectively retards disease, it has a severe disadvantage. Cracks can develop along the reaction zone, and radial cracks also occur from the reaction zone to and through the bark. Cracks can severely weaken a tree and make it more susceptible to breaking. A closure crack is one that occurs at the site of the wound. Sometimes this crack never actually closes.

Some trees are better defenders against disease than others. Trees that effectively carry out strategies 1–3 need never employ strategy 4, which can lead to cracking. Oak trees, *Quercus* (Fig. 25B), are examples of trees that are good at defending themselves, while willows, *Salix*, are not as good.

25.6 ORGANIZATION OF LEAVES

Leaves are the organs of photosynthesis in vascular plants such as flowering plants. As mentioned earlier, a leaf usually consists of a flattened blade and a petiole connecting the blade to the stem. The blade may be single or composed of several leaflets. Externally, it is possible to see the pattern of the leaf veins, which contain vascular tissue. Leaf veins have a net pattern in eudicot leaves and a parallel pattern in monocot leaves (see Fig. 25.3).

Figure 25.20 shows a cross section of a typical eudicot leaf of a temperate zone plant. At the top and bottom are layers of epidermal tissue that often bear trichomes, protective hairs often modified as glands that secrete irritating substances. These features may prevent the leaf from being eaten by insects. The epidermis characteristically has an outer, waxy cuticle that helps keep the leaf from drying out. The cuticle also prevents gas exchange because it is not gas permeable. However, the lower epidermis of eudicot and both surfaces of monocot leaves contain stomata that allow gases to move into and out of the leaf. Water loss also occurs at stomata, but each stoma has two guard cells that regulate its opening and closing, and stomata close when the weather is hot and dry.

The body of a leaf is composed of **mesophyll** [Gk. *mesos,* middle, and *phyllon,* leaf] tissue. Most eudicot leaves have two distinct regions: **palisade mesophyll,** containing elongated cells, and **spongy mesophyll,** containing irregular cells bounded by air spaces. The parenchyma cells of these layers have many chloroplasts and carry on most of the photosynthesis for the plant. The loosely packed arrangement of the cells in the spongy layer increases the amount of surface area for gas exchange.

Leaf Diversity

The blade of a leaf can be simple or compound (Fig. 25.21). A simple leaf has a single blade in contrast to a compound leaf, which is divided in various ways into leaflets. An example of a plant with simple leaves is a magnolia and a tree with compound leaves is a pecan tree. Pinnately compound leaves have the leaflets occuring in pairs, such as in a black walnut tree, while palmately compound leaves have all of the leaflets attached to a single point, as in a buckeye tree. Some plants such as the mimosa have bipinnately compound leaves. The leaflets of their leaves are subdivided into even smaller leaflets.

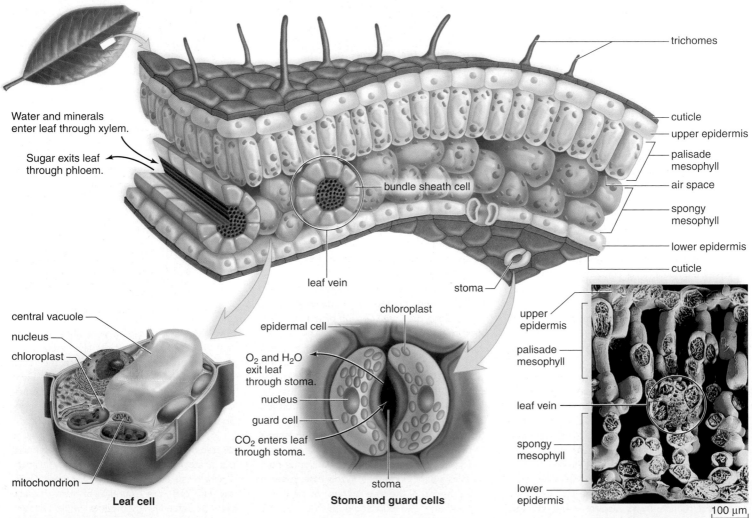

Water and minerals enter leaf through xylem.

Sugar exits leaf through phloem.

trichomes

cuticle

upper epidermis

palisade mesophyll

air space

spongy mesophyll

lower epidermis

cuticle

bundle sheath cell

leaf vein

stoma

central vacuole

nucleus

chloroplast

mitochondrion

Leaf cell

epidermal cell

chloroplast

O_2 and H_2O exit leaf through stoma.

nucleus

guard cell

CO_2 enters leaf through stoma.

stoma

Stoma and guard cells

upper epidermis

palisade mesophyll

leaf vein

spongy mesophyll

lower epidermis

100 μm

SEM of leaf cross section

FIGURE 25.20 Leaf structure.
Photosynthesis takes place in mesophyll tissue of leaves. The leaf is enclosed by epidermal cells covered with a waxy layer, the cuticle. Leaf hairs are also protective. The veins contain xylem and phloem for the transport of water and solutes. A stoma is an opening in the epidermis that permits the exchange of gases.

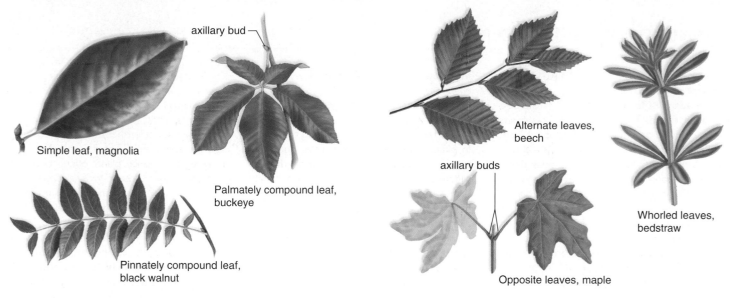

a. Simple versus compound leaves

b. Arrangement of leaves on stem

FIGURE 25.21 Classification of leaves.
a. Leaves are either simple or compound, being either pinnately compound or palmately compound. Note the one axillary bud per compound leaf.
b. Leaf arrangement on stem can be alternate, opposite, or whorled.

Leaves can be arranged on a stem in three ways: alternate, opposite, or whorled. The leaves are alternate in the American beech; in a maple, the leaves are opposite, being attached to the same node. Bedstraw has a whorled leaf arrangement with several leaves originating from the same node.

Leaves are adapted to environmental conditions. Shade plants tend to have broad, wide leaves, and desert plants tend to have reduced leaves with sunken stomata. The leaves of a cactus are the spines attached to the succulent stem (Fig. 25.22a). Other succulents have leaves adapted to hold moisture.

An onion bulb is made up of leaves surrounding a short stem. In a head of cabbage, large leaves overlap one another. The petiole of a leaf can be thick and fleshy, as in celery and rhubarb. Climbing leaves, such as those of peas and cucumbers, are modified into tendrils that can attach to nearby objects (Fig. 25.22b). The leaves of a few plants are specialized for catching insects. A sundew has sticky trichomes that trap insects and others that secrete digestive enzymes. The Venus's flytrap has hinged leaves that snap shut and interlock when an insect triggers its sensitive trichomes that project from inside the leaves (Fig. 25.22c). The leaves of a pitcher plant resemble a pitcher and have downward-pointing hairs that lead insects into a pool of digestive enzymes secreted by trichomes. Insectivorous plants commonly grow in marshy regions, where the supply of soil nitrogen is severely limited. The digested insects provide the plants with a source of organic nitrogen.

a. Cactus, *Opuntia*

b. Cucumber, *Cucumis*

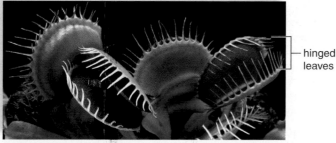

c. Venus's flytrap, *Dionaea*

FIGURE 25.22 Leaf diversity.
a. The spines of a cactus plant are modified leaves that protect the fleshy stem from animal predation. **b.** The tendrils of a cucumber are modified leaves that attach the plant to a physical support. **c.** The modified leaves of the Venus's flytrap serve as a trap for insect prey. When triggered by an insect, the leaf snaps shut. Once shut, the leaf secretes digestive juices that break down the soft parts of the insect's body.

CONNECTING THE CONCEPTS

A multicellular green alga is believed to be ancestral to plants. Multicellularity would have been adaptive for development of the specialized tissues found in plants—organisms that evolved on land. On land, as opposed to an aquatic environment, there is a danger of drying out. Even humid air is drier than a living cell, and the prevention of water loss is critical for plants. The epidermis, the trichomes, and the cuticle it produces help prevent water loss and overheating in sunlight. (The epidermis and glandular trichomes also protect against invasion by bacteria, fungi, and small insects.) The cork of woody plants is especially protective against water loss. Only the presence of lenticels still allows gas exchange to occur.

In an aquatic environment, water is available to all cells, but on land it is adaptive to have a means of water uptake and transport. In plants, the roots absorb water and have special extensions called root hairs that facilitate the uptake of water. Xylem transports water to all plant parts. Chapter 26 discusses how the drying effect of air is used by plants to help move water from the roots to the leaves. Roots buried in soil where they absorb water cannot photosynthesize, and the structure of phloem permits the transport of sugars down to the roots.

In an aquatic environment, water buoys up organisms and keeps them afloat, but on land it is adaptive to have a means to oppose the force of gravity. The stems of plants contain strong-walled sclerenchyma cells, tracheids, and vessel elements. The accumulation of annual rings in woody plants offers more support and allows a tree to grow in diameter.

In an aquatic environment, the surrounding water prevents the gametes and zygote from drying out. As we shall see in Chapter 28, flowering plants do not rely on external water for reproductive purposes and have evolved structures and a means of reproduction that prevent the gametes, zygote, and embryo from drying out.

Summary

25.1 PLANT ORGANS

A flowering plant has three vegetative organs. A root anchors a plant, absorbs water and minerals, and stores the products of photosynthesis. Stems support leaves, conduct materials to and from roots and leaves, and help store plant products. Leaves are specialized for gas exchange and they carry on most of the photosynthesis in the plant.

25.2 MONOCOT VERSUS EUDICOT PLANTS

Flowering plants are divided into the monocots and eudicots according to the number of cotyledons in the seed, the arrangement of vascular tissue in roots, stems, and leaves, and the number of flower parts.

25.3 PLANT TISSUES

Flowering plants have apical meristem plus three types of primary meristem. Protoderm produces epidermal tissue. In the roots, epidermal cells bear root hairs; in the leaves, the epidermis contains guard cells. In a woody stem, epidermis is replaced by periderm.

Ground meristem produces ground tissue. Ground tissue is composed of parenchyma cells, which are thin-walled and capable of photosynthesis when they contain chloroplasts. Collenchyma cells have thicker walls for flexible support. Sclerenchyma cells are hollow, nonliving support cells with secondary walls fortified by lignin.

Procambium produces vascular tissue. Vascular tissue consists of xylem and phloem. Xylem contains two types of conducting cells: vessel elements and tracheids. Vessel elements, which are larger and have perforation plates, form a continuous pipeline from the roots to the leaves. In elongated tracheids with tapered ends, water must move through pits in end walls and side walls. Xylem transports water and minerals. In phloem, sieve tubes are composed of sieve-tube members, each of which has a companion cell. Phloem transports sucrose and other organic compounds including hormones.

25.4 ORGANIZATION OF ROOTS

A root tip has a zone of cell division (containing the primary meristems), a zone of elongation, and a zone of maturation.

A cross section of a herbaceous eudicot root reveals the epidermis, which protects; the cortex, which stores food; the endodermis, which regulates the movement of minerals; and the vascular cylinder, which is composed of vascular tissue. In the vascular cylinder of a eudicot, the xylem appears star-shaped, and the phloem is found in separate regions, between the arms of the xylem. In contrast, a monocot root has a ring of vascular tissue with alternating bundles of xylem and phloem surrounding the pith.

Roots are diversified. Taproots are specialized to store the products of photosynthesis; a fibrous root system covers a wider area. Prop roots are adventitious roots specialized to provide increased anchorage.

25.5 ORGANIZATION OF STEMS

The activity of the shoot apical meristem within a terminal bud accounts for the primary growth of a stem. A terminal bud contains internodes and leaf primordia at the nodes. When stems grow, the internodes lengthen.

In a cross section of a nonwoody eudicot stem, epidermis is followed by cortex tissue, vascular bundles in a ring, and an inner pith. Monocot stems have scattered vascular bundles, and the cortex and pith are not well defined.

Secondary growth of a woody stem is due to vascular cambium, which produces new xylem and phloem every year, and cork cambium, which produces new cork cells when needed. Cork, a part of the bark, replaces epidermis in woody plants. In a cross section of a woody stem, the bark is all the tissues outside the vascular cambium. It consists of secondary phloem, cork cambium, and cork. Wood consists of secondary xylem, which builds up year after year and forms the annual rings.

Stems are diverse. There are horizontal aboveground and underground stems. Corms and some tendrils are also modified stems.

25.6 ORGANIZATION OF LEAVES

The bulk of a leaf is mesophyll tissue bordered by an upper and lower layer of epidermis; the epidermis is covered by a cuticle and may bear trichomes. Stomata tend to be in the lower layer. Vascular tissue is present within leaf veins. Leaves are diverse. The spines of a cactus are leaves. Other succulents have fleshy leaves. An onion is a bulb with fleshy leaves, and the tendrils of peas are leaves. The Venus's flytrap has leaves that trap and digest insects.

Reviewing the Chapter

1. Name and discuss the vegetative organs of a plant. 438
2. List five differences between monocots and eudicots. 440
3. Epidermal cells are found in what type of plant tissue? Explain how epidermis is modified in various organs of a plant. Contrast an epidermal cell with a cork cell. 441
4. Contrast the structure and function of parenchyma, collenchyma, and sclerenchyma cells. These cells occur in what type of plant tissue? 442
5. Contrast the structure and function of xylem and phloem. Xylem and phloem occur in what type of plant tissue? 443
6. Name and discuss the zones of a root tip. Trace the path of water and minerals across a root from the root hairs to xylem. Be sure to mention the Casparian strip. 444
7. Contrast a taproot with a fibrous root system. What are adventitious roots? 446
8. Describe the primary growth of a stem. 448
9. Describe cross sections of a herbaceous eudicot, a monocot, and a woody stem. 449–51
10. Discuss the diversity of stems by giving examples of several adaptations. 452
11. Describe the structure and organization of a typical eudicot leaf. 454
12. Note the diversity of leaves by giving examples of several adaptations. 455

Testing Yourself

Choose the best answer for each question.

1. Which of these is an incorrect contrast between monocots (stated first) and eudicots (stated second)?
   a. one cotyledon—two cotyledons
   b. leaf veins parallel—net veined
   c. vascular bundles in a ring—vascular bundles scattered
   d. flower parts in threes—flower parts in fours or fives
   e. All of these are correct contrasts.
2. Which of these types of cells is most likely to divide?
   a. parenchyma   d. xylem
   b. meristem   e. sclerenchyma
   c. epidermis
3. Which of these cells in a plant is apt to be nonliving?
   a. parenchyma   d. epidermal cells
   b. collenchyma   e. guard cells
   c. sclerenchyma
4. Root hairs are found in the zone of
   a. cell division.   d. apical meristem.
   b. elongation.   e. All of these are correct.
   c. maturation.
5. Cortex is found in
   a. roots, stems, and leaves. d. stems and leaves.
   b. roots and stems.   e. roots only.
   c. roots and leaves.
6. Between the bark and the wood in a woody stem, there is a layer of meristem called
   a. cork cambium.   d. the zone of cell division.
   b. vascular cambium.   e. procambium preceding bark.
   c. apical meristem.

7. Which part of a leaf carries on most of the photosynthesis of a plant?
   a. epidermis
   b. mesophyll
   c. epidermal layer
   d. guard cells
   e. Both a and b are correct.
8. Annual rings are the number of
   a. internodes in a stem.
   b. rings of vascular bundles in a monocot stem.
   c. layers of xylem in a woody stem.
   d. bark layers in a woody stem.
   e. Both b and c are correct.
9. The Casparian strip is found
   a. between all epidermal cells.
   b. between xylem and phloem cells.
   c. on four sides of endodermal cells.
   d. within the secondary wall of parenchyma cells.
   e. in both endodermis and pericycle.
10. Which of these is a stem?
    a. taproot of carrots
    b. stolon of strawberry plants
    c. spine of cactuses
    d. prop roots
    e. Both b and c are correct.
11. Meristem tissue that gives rise to epidermal tissue is called
    a. procambium.
    b. ground meristem.
    c. epiderm.
    d. protoderm.
    e. periderm.
12. New plant cells originate from
    a. the parenchyma.
    b. the collenchyma.
    c. the sclerenchyma.
    d. the base of the shoot.
    e. the apical meristem.
13. Ground tissue does not include
    a. collenchyma cells.
    b. sclerenchyma cells.
    c. parenchyma cells.
    d. chlorenchyma cells.
14. Evenly thickened cells that function to support mature regions of a plant are called
    a. guard cells.
    b. aerenchyma.
    c. parenchyma.
    d. sclerenchyma.
    e. xylem.
15. Roots
    a. are the primary site of photosynthesis.
    b. give rise to new leaves and flowers.
    c. have a thick cuticle to protect the epidermis.
    d. absorb water and nutrients.
    e. contain spores.
16. Monocot stems have
    a. vascular bundles arranged in a ring.
    b. vascular cambium.
    c. scattered vascular bundles.
    d. a cork cambium.
    e. a distinct pith and cortex.

17. Secondary thickening of stems occurs in
   a. all angiosperms.
   b. most monocots.
   c. many eudicots.
   d. few eudicots.

18. All of these may be found in heartwood except
   a. tracheids.
   b. vessel elements.
   c. parenchyma cells.
   d. sclerenchyma cells.
   e. companion cells.

19. How are compound leaves distinguished from simple leaves?
   a. Compound leaves do not have axillary buds at the base of leaflets.
   b. Compound leaves are smaller than simple leaves.
   c. Simple leaves are usually deciduous.
   d. Compound leaves are found only in pine trees.
   e. Simple leaves are found only in gymnosperms.

20. Label this root using these terms: endodermis, phloem, xylem, cortex, and epidermis.

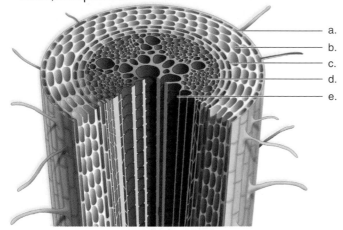

a.
b.
c.
d.
e.

21. Label this leaf using these terms: leaf vein, lower epidermis, palisade mesophyll, spongy mesophyll, and upper epidermis.

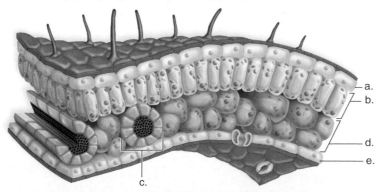

a.
b.
c.
d.
e.

Thinking Scientifically

1. Carrots are taproots. It is somewhat strange that these roots are orange since they are belowground and cannot engage in photosynthesis. Hypothesize how such a pigment could have arisen and its possible advantages.

2. Design an experiment that tests the hypothesis that new plants arise at the nodes of a stolon according to environmental conditions (temperature, water, and sunlight) that affect whether nodes can touch the ground.

Understanding the Terms

adventitious root 446	parenchyma 442
annual ring 450	perennial 438
apical meristem 444	pericycle 445
axillary bud 439	periderm 441
bark 450	petiole 439
blade 439	phloem 443
Casparian strip 445	pith 445
collenchyma 442	pit 443
complex tissue 443	primary root 446
cork 441	rhizome 452
cork cambium 450	root cap 444
cortex 445	root hair 441
cotyledon 440	root nodule 446
cuticle 441	root system 438
deciduous 439	sclerenchyma 442
endodermis 445	shoot apical
epidermal tissue 441	meristem 448
epidermis 441	shoot system 438
eudicot 440	sieve-tube member 443
evergreen 439	spongy mesophyll 454
fibrous root system 446	stem 439
ground tissue 441	stolon 452
herbaceous stem 449	stomata (sing., stoma) 441
internode 439	taproot 446
leaf 439	terminal bud 448
leaf vein 443	tracheid 443
lenticel 441	trichome 441
lignin 442	vascular bundle 443
meristem 441	vascular cambium 448
mesophyll 454	vascular cylinder 443
monocot 440	vascular tissue 441
mycorrhiza 446	vessel element 443
node 439	wood 450
organ 438	xylem 443
palisade mesophyll 454	

Match the terms to these definitions:

a. _____ Inner, thickest layer of a leaf; the site of most photosynthesis.

b. _____ Lateral meristem that produces secondary phloem and secondary xylem.

c. _____ Seed leaf for embryonic plant; provides nutrient molecules before the leaves begin to photosynthesize.

d. _____ Stem that grows horizontally along the ground and establishes plantlets periodically when it contacts the soil (e.g., the runners of a strawberry plant).

e. _____ Vascular tissue that contains vessel elements and tracheids.

ARIS, the *Biology* Website

ARIS, the website for *Biology*, provides a wealth of information organized and integrated by chapter. You will find practice quizzes, interactive activities, labeling exercises, flashcards, and much more that will complement your learning and understanding of general biology.

www.mhhe.com/maderbiology9

26

NUTRITION AND TRANSPORT IN PLANTS

*C*arbon, hydrogen, nitrogen, potassium, calcium, magnesium, and other elements are required in various amounts by all plants. Although more than 60 elements have been identified in plants, less than 20 elements are usually considered essential. These essential elements supply the plant with the raw material to carry out all life functions.

The leaves take up gases such as carbon dioxide and oxygen. The roots absorb water and other essential minerals such as potassium ions and nitrogen in the form of nitrates. Plants, unlike many animals, have the ability to concentrate minerals in their tissues. In fact, plants can concentrate some minerals to a much higher level than they occur in the soil. Some plants, such as those in the legume family (peanuts, clovers, beans), have roots colonized by bacteria that fix atmospheric nitrogen (e.g., NH_4^+) and make it available for the production of proteins. This relationship allows legumes such as beans to be particularly high in protein.

In vertebrates, blood, which contains nutrients, salts, gases, and other substances, is pumped throughout the body by the heart. In plants, there is no central pumping mechanism, yet materials move throughout the body of the plant. The unique properties of water account for the movement of water and minerals in the xylem, while osmosis plays an essential role in phloem transport of sugars. The same mechanisms account for transport in very tall redwood trees and in dwarf gardenias.

Redwoods, *Sequoia sempervirens.*

26.1 PLANT NUTRITION AND SOIL

The ancient Greeks believed that plants were "soil-eaters" and somehow converted soil into plant material. Apparently to test this hypothesis, a seventeenth-century Dutchman named Jean-Baptiste Van Helmont planted a willow tree weighing 5 lb in a large pot containing 200 lb of soil. He watered the tree regularly for five years and then reweighed both the tree and the soil. The tree weighed 170 lb, and the soil weighed only a few ounces less than the original 200 lb. Van Helmont concluded that the increase in weight of the tree was due primarily to the addition of water.

Water is a vitally important nutrient for a plant, but Van Helmont was unaware that water and carbon dioxide (taken in at the leaves) combine in the presence of sunlight to produce carbohydrates, the chief organic matter of plants. Much of the water entering a plant evaporates at the leaves. Roots, like all plant organs, carry on cellular respiration, a process that uses oxygen and gives off carbon dioxide (Fig. 26.1).

FIGURE 26.1
Overview of plant nutrition.
Carbon dioxide, which enters leaves, and water, which enters roots, are combined during photosynthesis to form carbohydrates, with the release of oxygen from the leaves. Root cells, and all other plant cells, carry on cellular respiration, which uses oxygen and gives off carbon dioxide. Aside from the elements carbon, hydrogen, and oxygen, plants require nutrients that are absorbed as minerals by the roots.

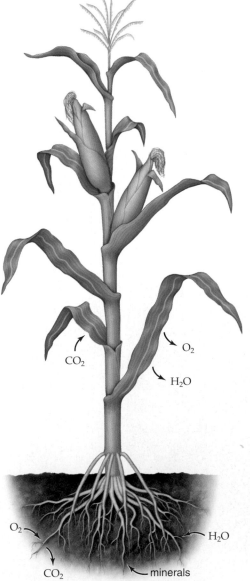

Essential Inorganic Nutrients

Approximately 95% of a typical plant's dry weight (weight excluding free water) is carbon, hydrogen, and oxygen. Why? Because these are the elements that are found in most organic compounds, such as carbohydrates. Carbon dioxide (CO_2) supplies carbon, and water (H_2O) supplies hydrogen and oxygen found in the organic compounds of a plant.

TABLE 26.1

Some Essential Inorganic Nutrients in Plants

Elements	Symbol	Form	Major Functions
Macronutrients			
Carbon	C	CO_2	Major component of organic
Hydrogen	H	H_2O	molecules
Oxygen	O	O_2	
Phosphorus	P	$H_2PO_4^-$	Part of nucleic acids, ATP,
		HPO_4^{2-}	and phospholipids
Potassium	K	K^+	Cofactor for enzymes; water balance and opening of stomata
Nitrogen	N	NO_3^-	Part of nucleic acids,
		NH_4^+	proteins, chlorophyll, and coenzymes
Sulphur	S	SO_4^{2-}	Part of amino acids, some coenzymes
Calcium	Ca	Ca^{2+}	Regulates responses to stimuli and movement of substances through plasma membrane; involved in formation and stability of cell walls
Magnesium	Mg	Mg^{2+}	Part of chlorophyll; activates a number of enzymes
Micronutrients			
Iron	Fe	Fe^{2+}	Part of cytochrome needed
		Fe^{3+}	for cellular respiration; activates some enzymes
Boron	B	BO_3^{3-}	Role in nucleic acid
		$B_4O_7^{2-}$	synthesis, hormone responses, and membrane function
Manganese	Mn	Mn^{2+}	Required for photosynthesis; activates some enzymes such as those of the citric acid cycle
Copper	Cu	Cu^{2+}	Part of certain enzymes, such as redox enzymes
Zinc	Zn	Zn^{2+}	Role in chlorophyll formation; activates some enzymes
Chlorine	Cl	Cl^-	Role in water-splitting step of photosynthesis and water balance
Molybdenum	Mo	MoO_4^{2-}	Cofactor for enzyme used in nitrogen metabolism

In addition to carbon, hydrogen, and oxygen, plants require certain other nutrients that are absorbed as minerals by the roots. A **mineral** is an inorganic substance usually containing two or more elements. Why are minerals from the soil needed by a plant? In plants, nitrogen is a major component of nucleic acids and proteins, magnesium is a component of chlorophyll, and iron is a building block of cytochrome molecules. The major functions of various **essential nutrients** for plants are listed in Table 26.1. A nutrient is essential if (1) it has an identifiable role, (2) no other nutrient can substitute and fulfill the same role, and (3) a deficiency of this nutrient causes a plant to die without completing its life cycle. Essential nutrients are divided into **macronutrients** and **micronutrients** according to their relative concentrations in plant tissue. The following diagram and slogan helps us remember which are the macronutrients and which are the micronutrients for plants:

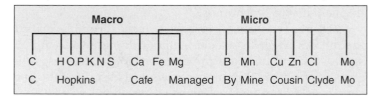

Beneficial nutrients are another category of elements taken up by plants. Beneficial nutrients either are required for or enhance the growth of a particular plant. Horsetails require silicon as a mineral nutrient and sugar beets show enhanced growth in the presence of sodium. Nickel is a beneficial mineral nutrient in soybeans when root nodules are present. Aluminum is used by some ferns, and selenium, which is often fatally poisonous to livestock, is used by locoweeds.

Determination of Essential Nutrients

When a plant is burned, its nitrogen component is given off as ammonia and other gases, but most other essential minerals remain in the ash. The presence of a mineral in the ash, however, does not necessarily mean that the plant normally requires it. The preferred method for determining the mineral requirements of a plant was developed at the end of the nineteenth century by the German plant physiologists Julius von Sachs and Wilhem Knop. This method is called water culture, or **hydroponics** [Gk. *hydrias*, water, and *ponos*, hard work]. Hydroponics allows plants to grow well if they are supplied with all the nutrients they need. The investigator omits a particular mineral and observes the effect on plant growth. If growth suffers, it can be concluded that the omitted mineral is an essential nutrient (Fig. 26.2). This method has been more successful for macronutrients than for micronutrients. For studies involving the latter, the water and the mineral salts used must be absolutely pure, but purity is difficult to attain, because even instruments and glassware can introduce micronutrients. Then, too, the element in question may already be present in the seedling used in the experiment. These factors complicate the determination of essential plant micronutrients by means of hydroponics.

a. Solution lacks nitrogen   Complete nutrient solution

b. Solution lacks phosphorus   Complete nutrient solution

c. Solution lacks calcium   Complete nutrient solution

FIGURE 26.2 Nutrient deficiencies.
The nutrient cause of poor plant growth is diagnosed when plants are grown in a series of complete nutrient solutions except for the elimination of just one nutrient at a time. These experiments show that sunflower plants respond negatively to a deficiency of (**a**) nitrogen, (**b**) phosphorus, and (**c**) calcium.

Certain elements are required by plants for good nutrition. Lack of an essential nutrient causes plants to die. Beneficial nutrients serve particular purposes in some plants.

Soil

Plants acquire carbon when carbon dioxide diffuses into leaves through stomata. Oxygen can enter from the air, but all of the other essential nutrients are absorbed by roots from the soil. It would not be an exaggeration to say that terrestrial life is dependent on the quality of the soil and the ability of soil to provide plants with the nutrients they need.

Soil Formation

Soil formation begins with the weathering of rock in the Earth's crust. Weathering first gradually breaks down rock to rubble and then to soil particles. Some weathering mechanisms, such as the freeze-thaw cycle of ice or the grinding of rock on rock by the action of glaciers or river flow, are purely mechanical. Other forces include a chemical effect, as when acidic rain leaches (washes away) soluble components of rock or when oxygen combines with the iron of rocks.

In addition to these forces, organisms also play a role in the formation of soil. Lichens and mosses grow on pure rock and trap particles that later allow grasses, herbs, and soil animals to follow. When these die, their remains are decomposed, notably by bacteria and fungi. Decaying organic matter, called **humus,** begins to accumulate. Humus supplies nutrients to plants, and its acidity also leaches minerals from rock.

Building soil takes a long time. Under ideal conditions, depending on the type of parent material (the original rock) and the various processes at work, a centimeter of soil may develop within 15 years.

The Nutritional Function of Soil

Soil is defined as a mixture of mineral particles, decaying organic material, living organisms, air, and water, which together support the growth of plants. In a good agricultural soil, the first three components come together in such a way that there are spaces for air and water (Fig. 26.3). It's best if the soil contains particles of different sizes because only then will there be spaces for air. Roots take up oxygen from air spaces. Ideally, water clings to particles by capillary action and does not fill the spaces. That's why you shouldn't overwater your houseplants!

Mineral Particles. Mineral particles vary in size: sand particles are the largest (0.05–2.0 mm in diameter); silt particles have an intermediate size (0.002–0.05 mm); and clay particles are the smallest (less than 0.002 mm). Soils are a mixture of these three types of particles. Because sandy soils have many large particles, they have large spaces, and the water drains readily through the particles. In contrast to sandy soils, a soil composed mostly of clay particles has small spaces that fill completely with water. Most likely, you have experienced the feel of sand and clay in your hand: sand having no moisture flows right through your fingers, while clay clumps together in one large mass because of its water content.

Clay particles have another benefit that sand particles do not have. As Table 26.1 indicates, some minerals are negatively charged and others are positively charged. Clay particles are negative, and they can retain positively charged

minerals such as calcium (Ca^{2+}) and potassium (K^+), preventing these minerals from being washed away by leaching. Plants exchange hydrogen ions for these minerals when they take them up (Fig. 26.3). If rain is acidic, its hydrogen ions displace positive mineral ions and cause them to drain away; this is one reason acid rain kills trees. Because clay particles are unable to retain negatively charged NO_3^-, the nitrogen content of soil is apt to be low. Legumes (see Fig. 1.11) are sometimes planted to replenish the nitrogen in the soil in preference to relying solely on the addition of fertilizer.

The type of soil called loam is composed of roughly one-third sand, silt, and clay particles. This combination sufficiently retains water and nutrients while still allowing the drainage necessary to provide air spaces. Some of the most productive soils are loam.

Humus. Humus, which mixes with the top layer of soil particles, increases the benefits of soil. Plants do well in soils that contain 10–20% humus.

Humus causes soil to have a loose, crumbly texture that allows water to soak in without doing away with air spaces. After a rain, the presence of humus decreases the chances of runoff. Humus swells when it absorbs water and shrinks as it dries. This action helps aerate soil.

Soil that contains humus is nutritious for plants. Humus is acidic; therefore, it retains positively charged minerals until plants take them up. When the organic matter in humus is broken down by bacteria and fungi, inorganic nutrients are returned to plants. Although soil particles are the original source of minerals in soil, recycling of nutrients, as you know, is a major characteristic of ecosystems.

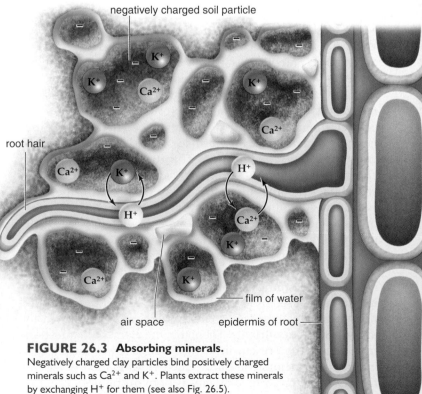

FIGURE 26.3 Absorbing minerals.
Negatively charged clay particles bind positively charged minerals such as Ca^{2+} and K^+. Plants extract these minerals by exchanging H^+ for them (see also Fig. 26.5).

Living Organisms. Small plants play a major role in the formation of soil from bare rock. Due to the process of succession (see Fig. 47.18), larger plants eventually become dominant in certain ecosystems. The roots of larger plants penetrate soil even to the cracks in bedrock. This action slowly opens up soil layers, allowing water, air, and animals to follow.

There are many different types of soil animals. The largest of them, such as toads, snakes, moles, badgers, and rabbits, disturb and mix soil by burrowing. Smaller animals like earthworms ingest fine soil particles and deposit them on the surface as worm casts. Earthworms also loosen and aerate the soil. A range of soil animals, including mites, springtails, and millipedes, help break down leaves and other plant remains by eating them. Soil-dwelling ants construct tremendous colonies with massive chambers and tunnels. These ants also loosen and aerate the soil.

The microorganisms in soil, such as protozoans, fungi, algae, and bacteria, are responsible for the final decomposition of organic remains in humus to inorganic nutrients. Recall that plants are unable to make use of atmospheric nitrogen (N_2) and that soil bacteria play an important nutrient role because they make nitrate available to plants.

Certain soil organisms, such as some roundworms and insects, are parasitic on plants. Insects may improve the properties of soil, but they are also major crop pests when they feed on plant roots.

Soil Profiles

A **soil profile** is a vertical section from the ground surface to the unaltered rock below. Usually, a soil profile has parallel layers known as **horizons**. Mature soil generally has three horizons (Fig. 26.4). The A horizon is the uppermost (or topsoil) layer that contains litter and humus, although most of the soluble chemicals may have been leached away. The B horizon has little or no organic matter but does contain the inorganic nutrients leached from the A horizon. The C horizon is a layer of weathered and shattered rock.

Because the parent material (rock) and climate (e.g., temperature and rainfall) differ in various parts of the biosphere, the soil profile varies according to the particular ecosystem. Soils formed in grasslands tend to have a deep A horizon built up from decaying grasses over many years, but because of limited rain, there has been little leaching into the B horizon. In forest soils, both the A and B horizons have enough inorganic nutrients to allow for root growth. In tropical rain forests, the A horizon is more shallow than the generalized profile, and the B horizon is deeper, signifying that leaching is more extensive. Since the topsoil of a rain forest lacks nutrients, it can only support crops for a few years.

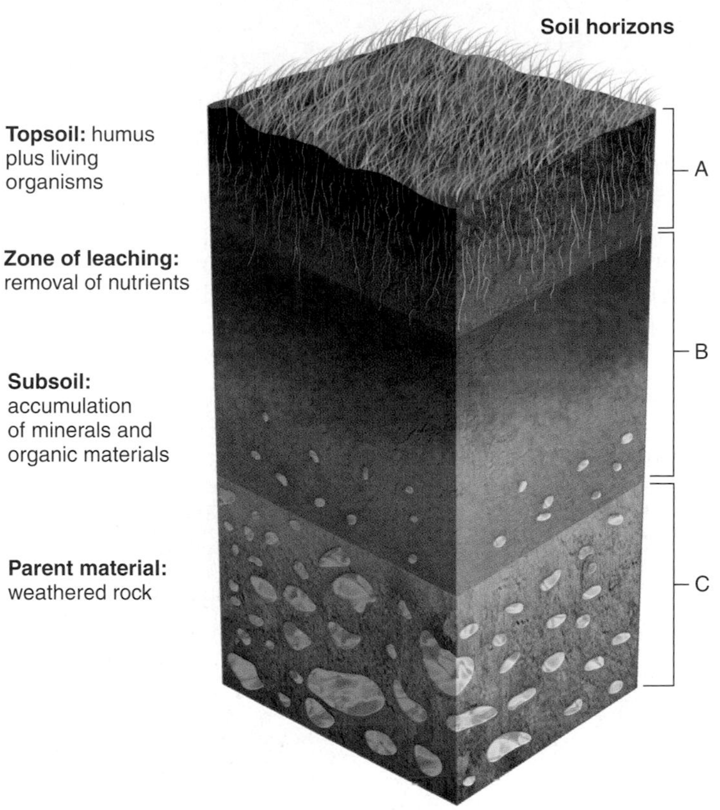

Soil Erosion

Soil erosion occurs when water or wind carry soil away to a new location. Erosion removes about 25 billion tons of topsoil yearly, worldwide. If this rate of loss continues, some scientists predict that the Earth will lose practically all of its topsoil by the middle of the next century. Deforestation (removal of trees) and desertification (increase in deserts due to overgraz-

Soil horizons

Topsoil: humus plus living organisms — A

Zone of leaching: removal of nutrients

Subsoil: accumulation of minerals and organic materials — B

Parent material: weathered rock — C

FIGURE 26.4 Simplified soil profile.
The top layer (A horizon) contains most of the humus; the next layer (B horizon) accumulates materials leached from the A horizon; and the lowest layer (C horizon) is composed of weathered parent material. Erosion removes the A horizon, a primary source of humus and minerals in soil.

ing and overfarming marginal lands) contribute to the occurrence of erosion, and so do poor farming practices in general.

In the United States, soil is eroding faster than it is being formed on about one-third of all cropland. Fertilizers and pesticides, carried by eroding soil into groundwater and rivers, are threatening human health. To make up for the loss of soil due to erosion, more energy is used to apply more fertilizers and pesticides to crops. Instead, it would be best to stop erosion before it occurs by following sound agricultural practices.

The coastal wetlands are losing soil at a tremendous rate. These wetlands are important as nurseries for many species of organisms, such as shrimp and redfish, and as protection against storm surge from hurricanes. In Louisiana, 24 mi^2 of wetlands are lost each year. This equates to one football field being lost every 38 minutes.

Soil formation involves the interplay of physical, chemical, and biological processes over an extended period of time. Soil particles, humus, and living organisms give soil the texture it needs to provide air, water, and minerals to plant roots. Erosion removes topsoil (A horizon) and seriously reduces the ability of soil to provide plants with these requirements.

26.2 WATER AND MINERAL UPTAKE

The pathways for water and mineral uptake and transport in a plant are the same. As Figure 26.5a shows, water along with minerals can enter the root of a flowering plant from the soil simply by passing between the porous cell walls. Eventually, however, the **Casparian strip,** a band of suberin and lignin bordering four sides of root endodermal cells, forces water to enter endodermal cells. Alternatively,

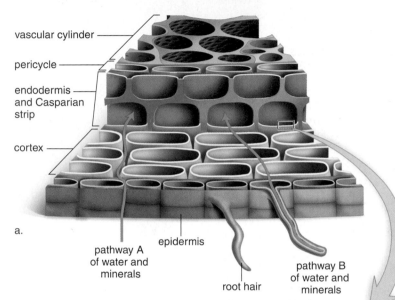

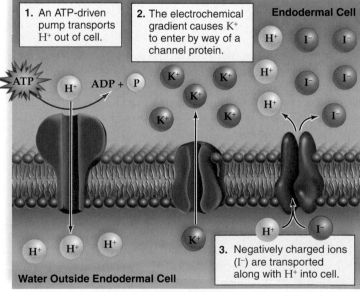

FIGURE 26.5 Water and mineral uptake.
a. Pathways of water and minerals. Water and minerals can travel via porous cell walls but then must enter endodermal cells because of the Casparian strip (pathway A). Alternatively, water and minerals can enter root hairs and move from cell to cell (pathway B). **b.** Transport of minerals across an endodermal plasma membrane. An ATP-driven pump removes hydrogen ions (H^+) from the cell. This establishes an electrochemical gradient that allows potassium (K^+) and other positively charged ions to cross the membrane via a channel protein. Negatively charged mineral ions (I^-) can cross the membrane by way of a carrier when they "hitch a ride" with hydrogen ions (H^+), which are diffusing down their concentration gradient.

water can enter epidermal cells at their **root hairs** and then progress through cells across the cortex and endodermis of a root by means of cytoplasmic strands within plasmodesmata (Fig. 5.16). Regardless of the pathway, water enters root cells when they have a lower osmotic pressure than does the soil solution.

In contrast to water, minerals are actively taken up by plant cells. Plants possess an astonishing ability to concentrate minerals—that is, to take up minerals until they are many times more concentrated in the plant than in the surrounding medium. The concentration of certain minerals in roots is as much as 10,000 times greater than in the surrounding soil. Following their uptake by root cells, minerals move into xylem and are transported into leaves by the upward movement of water. Along the way, minerals can exit xylem and enter those cells that require them. Some eventually reach leaf cells. In any case, minerals must again cross a selectively permeable plasma membrane when they exit xylem and enter living cells. By what mechanism do minerals cross plasma membranes?

Recall that plant cells absorb minerals in the ionic form: nitrogen is absorbed as nitrate (NO_3^-), phosphorus as phosphate (HPO_4^{2-}), potassium as potassium ions (K^+), and so forth. Ions cannot cross the plasma membrane because they are unable to enter the nonpolar phase of the lipid bilayer. It has long been known that plant cells expend energy to actively take up and concentrate mineral ions. If roots are deprived of oxygen or are poisoned so that cellular respiration cannot occur, mineral ion uptake is diminished. The energy of ATP is required for mineral ion transport, but not directly (Fig. 26.5b). A plasma membrane pump, called a proton pump, hydrolyzes ATP and uses the energy released to transport hydrogen ions (H^+) out of the cell. This sets up an electrochemical gradient that drives positively charged ions such as K^+ through a channel protein into the cell. Negatively charged mineral ions are transported, along with H^+, by carrier proteins. Since H^+ is moving down its concentration gradient, no energy is required. Notice that this model of mineral ion transport in plant cells is based on chemiosmosis, the establishment of an electrochemical gradient to perform work.

Adaptations of Roots for Mineral Uptake

Two symbiotic relationships are known to assist roots in taking up mineral nutrients. Some plants, such as legumes, soybeans, and alfalfa, have roots colonized by *Rhizobium* bacteria, which can fix atmospheric nitrogen (N_2). They break the $N \equiv N$ bond and reduce nitrogen to NH_4^+ for incorporation into organic compounds. The bacteria live in **root nodules** [L. *nodulus,* dim. of *nodus,* knot] and are supplied with carbohydrates by the host plant (Fig. 26.6). The bacteria, in turn, furnish their host with nitrogen compounds.

The second type of symbiotic relationship, called a mycorrhizal association, involves fungi and almost all plant roots (Fig. 26.7). Only a small minority of plants do not have **mycorrhizae** [Gk. *mykes,* fungus, and *rhiza,* root], and these plants are most often limited as to the environment in which they can grow. *Ectomycorrhizae* form a mantle that is exterior to the root, and they grow between cell walls. *Endomycorrhizae* can penetrate cell walls. In any case, the fungus increases the surface area available for mineral and water uptake and breaks down organic matter, releasing nutrients that the plant can use. In return, the root furnishes the fungus with sugars and amino acids. Plants are extremely dependent on mycorrhizae. Orchid seeds, which are quite small and contain limited nutrients, do not germinate until a mycorrhizal fungus has invaded their cells. Nonphotosynthetic plants, such as Indian pipe, use their mycorrhizae to extract nutrients from nearby trees.

Some plants have poorly developed roots or no roots at all because minerals and water are supplied by other mechanisms. **Epiphytes** [Gk. *epi*, over, and *phyton*, plant] are "air plants"; they do not grow in soil but on larger plants, which give them support. Spanish moss is an example of an epiphytic organism (Fig. 26.8). Epiphytes do not receive nutrients from their host, however. Some have roots that absorb moisture from the atmosphere, and many catch rain and minerals in special pockets at the base of their leaves. Parasitic plants such as dodders, broomrapes, and pinedrops send out rootlike projections called haustoria that tap into the xylem and phloem of the host stem (see Fig. 25.11*e*). Carnivorous plants such as the Venus's flytrap and sundews obtain some nitrogen and minerals when their leaves capture and digest insects (see Fig. 25.22*c*).

Mineral uptake follows the same pathway as water uptake; however, mineral uptake takes an expenditure of energy. Most plants are assisted in acquiring minerals by a symbiotic relationship with microorganisms and/or fungi.

FIGURE 26.7 Mycorrhizae.
Experimental results show that rough lemon plants with mycorrhizae (*right*) grow much better than plants without mycorrhizae (*left*)

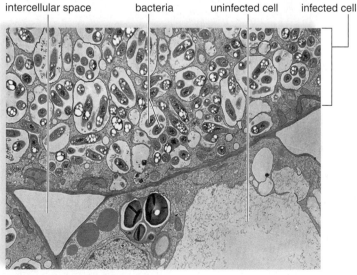

a. Root nodule

b. Cross section of nodule

FIGURE 26.6 Root nodules.
a. Nitrogen-fixing bacteria live in nodules on the roots of plants, particularly legumes. **b.** In infected nodule cells, bacteria fix atmospheric nitrogen and make reduced nitrogen available to a plant. The plant passes carbohydrates to the bacteria.

Roots of epiphyte   Spanish moss

FIGURE 26.8 Spanish moss.
Spanish moss, *Tillandsia usenoides*, is a common epiphyte found in the southern United States. Spanish moss is a member of the pineapple family.

26.3 TRANSPORT MECHANISMS IN PLANTS

Flowering plants are well adapted to living in a terrestrial environment. Their leaves, which carry on photosynthesis, are positioned to catch the rays of the sun because they are held aloft by the stem (Fig. 26.9). Carbon dioxide enters leaves at the stomata, but water, the other main requirement for photosynthesis, is absorbed by the roots. Water must be transported from the roots through the stem to the leaves. Vascular plants have a transport tissue, called **xylem,** that moves water and minerals from the roots to the leaves. Xylem contains two types of conducting cells: tracheids and vessel elements. **Tracheids** are tapered at both ends. The ends overlap with those of adjacent tracheids (see Fig. 25.6). Pits located in adjacent tracheids allow water to pass from cell to cell. **Vessel elements** are long and tubular with perforation plates at each end (see Fig. 25.6). Vessel elements placed end to end form a completely hollow pipeline from the roots to the leaves. Xylem, with its strong-walled, nonliving cells, gives trees much-needed internal support.

The process of photosynthesis results in sugars, which are used as a source of energy and building blocks for other organic molecules throughout a plant. **Phloem** is the type of vascular tissue that transports organic nutrients to all parts of the plant. Roots buried in the soil cannot possibly carry on photosynthesis, but they still require a source of energy in order to carry on cellular metabolism. Vascular plants are able to transport the products of photosynthesis to regions that require them and/or that will store them for future use. In flowering plants, the conducting cells of phloem are **sieve-tube members,** each of which typically has a companion cell (see Fig. 25.7). **Companion cells** can provide proteins to sieve-tube members, which contain cytoplasm but have no nucleus. The end walls of sieve-tube members are called sieve plates because they contain numerous pores. The sieve-tube members are aligned end to end, and strands of cytoplasm within plasmodesmata extend from one cell to the other through the sieve plates. In this way, sieve-tube members form a continuous sieve tube for organic nutrient transport throughout the plant.

Knowing that vascular plants are structured in a way that allows materials to move from one part to another does not tell us the mechanisms by which they move. Plant physiologists have performed numerous experiments to determine how water and minerals rise to the tops of very tall trees in xylem and how organic nutrients move in the opposite direction in phloem. It would be expected that these processes are mechanical in nature and based on the properties of water because water is a large part of both **xylem sap** and **phloem sap,** as the watery contents of these vessels are called. In living systems, water molecules diffuse freely across plasma membranes from the area of higher concentration to the area of lower concentration. Botanists favor describing the movement of water in terms of water potential: water always flows passively from the area of higher water potential to the area of lower water potential. As can be seen in the Science Focus on page 467, the concept

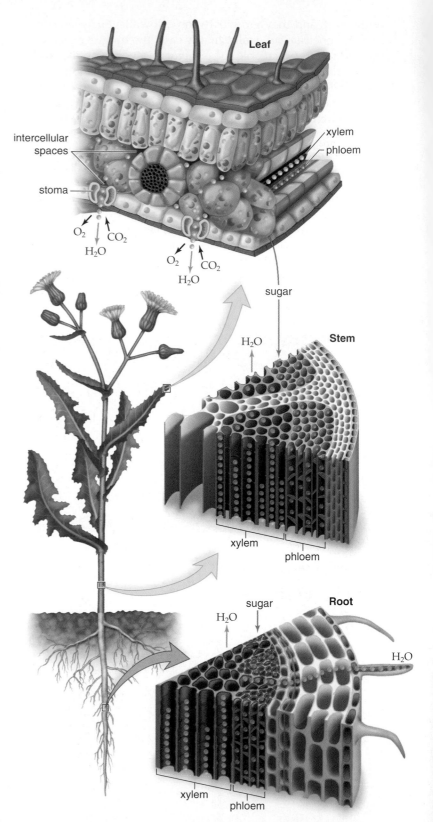

FIGURE 26.9 Plant transport system.
Vascular tissue in plants includes xylem, which transports water and minerals from the roots to the leaves, and phloem, which transports organic nutrients oftentimes in the opposite direction. Notice that xylem and phloem are continuous from the roots through the stem to the leaves, which are the vegetative organs of a plant.

science focus

The Concept of Water Potential

Potential energy is stored energy due to the position of an object. A boulder placed at the top of a hill has potential energy. When pushed, the boulder moves down the hill as potential energy is converted into kinetic (motion) energy. Once it's at the bottom of the hill, the boulder has lost much of its potential energy.

Water potential is defined as the energy of water. Just like the boulder, water at the top of a waterfall has a higher water potential than water at the bottom of the waterfall. As illustrated by this example, water moves from a region of higher potential to a region of lower water potential.

In terms of cells, two factors usually determine water potential, which in turn determines the direction in which water will move across a plasma membrane. These factors concern differences in:

1. Water pressure across a membrane
2. Solute concentration across a membrane

Pressure potential is the effect that pressure has on water potential. With regard to pressure, it is obvious that water will move across a membrane from the area of higher pressure to the area of lower pressure. The higher the water pressure, the higher the water potential. The lower the water pressure, the lower the water potential, and the more likely it is that water will flow in that direction. Pressure potential is the concept that best explains the movement of sap in xylem and phloem.

To fully explain the movement of water into plant cells, the concept of *osmotic potential* is also required. Osmotic potential takes into account the effects of solutes on the movement of water. The presence of solutes restricts the movement of water because water tends to interact with solutes. Indeed, water tends to move across a membrane from the area of lower solute concentration to the area of higher solute concentration. The lower the concentration of solutes (osmotic potential),

the higher the water potential. The higher the concentration of solutes, the lower the water potential and the more likely it is that water will flow in that direction.

Not surprisingly, increasing water pressure will counter the tendency of water to enter a cell because of the presence of solutes. A common situation exists in plant cells. As water enters a plant cell by osmosis, water pressure will increase inside the cell—a plant cell has a strong cell wall that allows water pressure to build up. When will water stop entering the cell? When the pressure potential inside the cell increases and balances the osmotic potential outside the cell.

Pressure potential that increases due to the process of osmosis is often called *turgor pressure*. Turgor pressure is critical, since plants depend on it to maintain the turgidity of their bodies (Fig. 26A). The cells of a wilted plant have insufficient turgor pressure, and the plant droops as a result.

FIGURE 26A
Water potential and turgor pressure.
*Water flows from an area of higher water potential to an area of lower water potential. **a.** The cells of a wilted plant have a lower water potential; therefore, water enters the cells. **b.** Equilibrium is achieved when the water potential is equal inside and outside the cell. Cells are now turgid, and the plant is no longer wilted.*

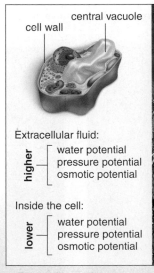

central vacuole | **Wilted**
cell wall

Extracellular fluid:
higher — water potential / pressure potential / osmotic potential

Inside the cell:
lower — water potential / pressure potential / osmotic potential

a. Plant cells need water.

H_2O enters the cell

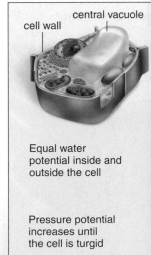

central vacuole | **Turgid**
cell wall

Equal water potential inside and outside the cell

Pressure potential increases until the cell is turgid

b. Plant cells are turgid.

of water potential has the benefit of considering water pressure in addition to osmotic pressure.

Chemical properties of water are also important in movement of xylem sap. The polarity of water molecules and the hydrogen bonding between water molecules allow water to fill xylem cells.

Flowering plants have transport tissues; xylem transports water and minerals from the roots to the leaves, and phloem transports organic nutrients to various parts of the plant, particularly from the leaves to the roots.

Water Transport

Water and minerals enter the root of a plant by the two pathways described in Figure 26.5. Eventually, water and minerals reach xylem, which contains tracheids and vessel elements. The vessel elements are larger than the tracheids, and they are stacked one on top of the other to form vessels that stretch from the roots to the leaves. Vessels constitute an open pipeline because the vessel elements have perforation plates separating one from the other (Fig. 26.10a, b). The tracheids, which are elongated with tapered ends, form a less obvious means of transport, but water can move across the end and side walls of tracheids because of pits, or depressions, where the secondary wall does not form (Fig. 26.10c).

Water entering root cells creates a positive pressure called **root pressure.** Root pressure, which primarily occurs at night, tends to push xylem sap upward. Root pressure may be responsible for **guttation** [L. *gutta*, drops, spots] when drops of water are forced out of vein endings along the edges of leaves (Fig. 26.11). Although root pressure may contribute to the upward movement of water in some instances, it is not believed to be the mechanism by which water can rise to the tops of very tall trees. After an injury or pruning, especially in spring, some plants appear to "bleed" as water exudes from the site. This phenomenon is the result of root pressure.

Cohesion-Tension Model of Xylem Transport

Once water enters xylem, it must be transported upward the entire height of a tree to the leaves, which use it for photosynthesis. This can be a daunting task, given that redwood trees can exceed 90 m in height.

FIGURE 26.11 Guttation.
Drops of guttation water on the edges of a strawberry leaf. Guttation, which occurs at night, may be due to root pressure. Root pressure is a positive pressure potential caused by the entrance of water into root cells. Often guttation is mistaken for early morning dew.

The **cohesion-tension model** of xylem transport outlined in Figure 26.12 suggests a passive mechanism for xylem transport. Stephen Hales (1677–1761), a British plant physiologist, described root pressure as well as the foundations of the cohesion-tension model. The term *cohesion* refers to the tendency of water molecules to cling together. Because of hydrogen bonding, water molecules interact with one an-

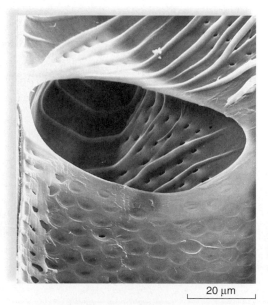

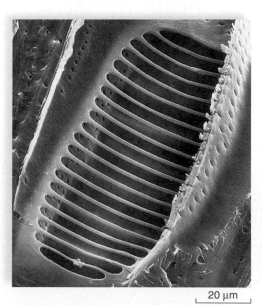

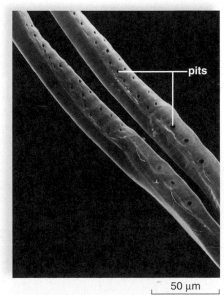

⊢ 20 µm ⊣	⊢ 20 µm ⊣	⊢ 50 µm ⊣
a. Perforation plate with a single, large opening	b. Perforation plate with a series of openings	c. Tracheids

FIGURE 26.10 Conducting cells of xylem.
Water can move from vessel element to vessel element through perforation plates (**a** and **b**). Vessel elements can also exchange water with tracheids through pits. **c.** Tracheids are long, hollow cells with tapered ends. Water can move into and out of tracheids through pits only.

other, and there is a continuous water column in xylem from the leaves to the roots that is not easily broken. *Adhesion* refers to the ability of water, a polar molecule, to interact with the molecules making up the walls of the vessels in xylem. Adhesion is a property of water that gives the water column extra strength and prevents it from slipping back.

Why does the continuous water column move passively upward? Consider the structure of a leaf. When the sun rises, stomata open and carbon dioxide enters a leaf. Within the leaf, the mesophyll cells—particularly the spongy layer—are exposed to the air, which can be quite dry. Water now evaporates from mesophyll cells into the intercellular spaces. Evaporation of water from leaf cells is called **transpiration.** At least 90% of the water taken up by the roots is eventually lost by transpiration. This means that the total amount of water lost by a plant over a long period of time is surprisingly large. A single *Zea mays* (corn) plant loses somewhere between 135 and 200 liters of water through transpiration during a growing season. During the growing season, an average-sized birch tree with over 200,000 leaves will transpire up to 3,700 liters of water a day.

The water molecules that evaporate from cells into the intercellular spaces are replaced by other water molecules from the leaf veins. In this way, transpiration exerts a driving force—that is, a *tension*—that draws the water column up in vessels from the roots to the leaves. As transpiration occurs, the water column is pulled upward, first within the leaf, then from the stem, and finally from the roots.

Tension can reach from the leaves to the root only if the water column is continuous. What happens if the water column within xylem is broken, as by cutting a stem? The water column "snaps back" down the xylem vessel away from the site of breakage, making it more difficult for conduction to occur. This is why it is best to maximize water conduction by cutting flower stems under water. This effect has also allowed investigators to measure the tension in stems. A device called the pressure bomb measures how much pressure it takes to push the xylem sap back to the cut surface of the stem.

There is an important consequence to the way water is transported in plants. When a plant is under water stress, the stomata close. Now the plant loses little water because the leaves are protected against water loss by the waxy **cuticle** of the upper and lower epidermis. When stomata are closed, however, carbon dioxide cannot enter the leaves, and many plants are unable to photosynthesize efficiently. Photosynthesis, therefore, requires an abundant supply of water so that stomata can remain open and allow carbon dioxide to enter.

Water is transported from the roots to the leaves in xylem. The water column remains intact as transpiration pulls water from the roots to the leaves.

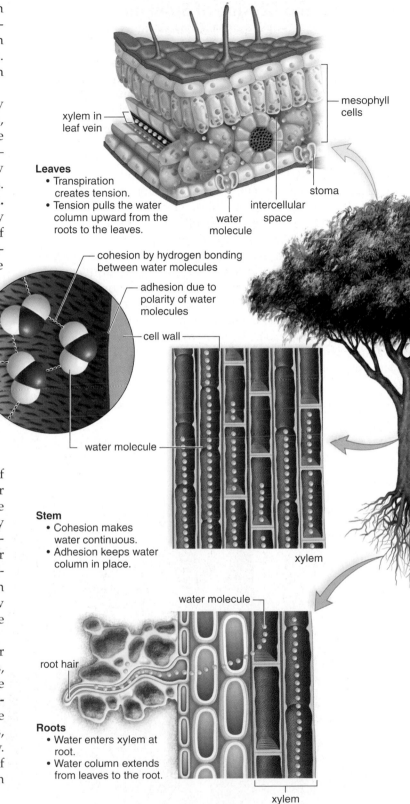

Leaves
- Transpiration creates tension.
- Tension pulls the water column upward from the roots to the leaves.

xylem in leaf vein

mesophyll cells

stoma

intercellular space

water molecule

cohesion by hydrogen bonding between water molecules

adhesion due to polarity of water molecules

cell wall

water molecule

Stem
- Cohesion makes water continuous.
- Adhesion keeps water column in place.

xylem

water molecule

root hair

Roots
- Water enters xylem at root.
- Water column extends from leaves to the root.

xylem

FIGURE 26.12 Cohesion-tension model of xylem transport.
Tension created by evaporation (transpiration) at the leaves pulls water along the length of the xylem—from the root hairs to the leaves.

Opening and Closing of Stomata

Each **stoma,** a small pore in leaf epidermis, is bordered by **guard cells.** When water enters the guard cells and turgor pressure increases, the stoma opens; when water exits the guard cells and turgor pressure decreases, the stoma closes. Notice in Figure 26.13 that the guard cells are attached to each other at their ends and that the inner walls are thicker than the outer walls. When water enters, a guard cell's radial expansion is restricted because of cellulose microfibrils in the walls, but lengthwise expansion of the outer walls is possible. When the outer walls expand lengthwise, they buckle out from the region of their attachment, and the stoma opens.

Since about 1968, it has been clear that potassium ions (K^+) accumulate within guard cells when stomata open. In other words, active transport of K^+ into guard cells causes water to follow by osmosis and stomata to open. Also interesting is the observation that hydrogen ions (H^+) accumulate outside guard cells as K^+ moves into them. A proton pump run by the hydrolysis of ATP transports H^+ to the outside of the cell. This establishes an electrochemical gradient that allows K^+ to enter by way of a channel protein (see Fig. 26.5b).

What regulates the opening and closing of stomata? It appears that the blue-light component of sunlight is a signal that can cause stomata to open. Evidence suggests that a flavin pigment absorbs blue light, and then this pigment sets in motion the cytoplasmic response that leads to activation of the proton pump. Similarly, there could be a receptor in the plasma membrane of guard cells that brings about inactivation of the pump when carbon dioxide (CO_2) concentration rises, as might happen when photosynthesis ceases. Abscisic acid (ABA), which is produced by cells in wilting leaves, can also cause stomata to close (see page 487). Although photosynthesis cannot occur, water is conserved.

If plants are kept in the dark, stomata open and close just about every 24 hours, just as if they were responding to the presence of sunlight in the daytime and the absence of sunlight at night. This means that some sort of internal biological clock must be keeping time. Circadian rhythms (a behavior that occurs nearly every 24 hours) and biological clocks are areas of intense investigation at this time. Other factors that influence the opening and closing of stoma include temperature, humidity, and stress.

When stomata open, first K^+ and then water enter guard cells. Stomata open and close in response to environmental signals, and the exact mechanism is being investigated.

FIGURE 26.13

Opening and closing of stomata.

a. A stoma opens when turgor pressure increases in guard cells due to the entrance of K^+ followed by the entrance of water. **b.** A stoma closes when turgor pressure decreases due to the exit of K^+ followed by the exit of water.

a.   25 μm

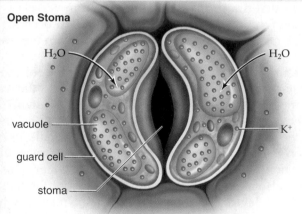

Open Stoma

H_2O H_2O

vacuole

guard cell

stoma

K^+

K^+ enters guard cells, and water follows.

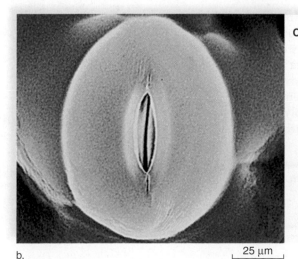

b.   25 μm

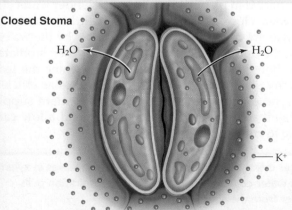

Closed Stoma

H_2O H_2O

K^+

K^+ exits guard cells, and water follows.

ecology focus

Plants Can Clean up Toxic Messes

Most trees planted along the edges of farms are intended to break the wind. But a mile-long stand of spindly poplars outside Amana, Iowa, serves a different purpose. It cleans pollution.

The poplars act like vacuum cleaners, sucking up nitrate-laden runoff from a fertilized cornfield before this runoff reaches a nearby brook—and perhaps other waters. Nitrate runoff into the Mississippi River from Midwest farms, after all, is a major cause of the large "dead zone" of oxygen-depleted water that develops each summer in the Gulf of Mexico.

Before the trees were planted, the brook's nitrate levels were as much as ten times the amount considered safe. But then Louis Licht, a University of Iowa graduate student, had the idea that poplars, which absorb lots of water and tolerate pollutants, could help. In 1991, Licht tested his hunch by planting the trees along a field owned by a corporate farm. The brook's nitrate levels subsequently dropped more than 90%, and the trees have thrived, serving as a prime cleanup method known as **phytoremediation.**

This method uses plants—many of them common species such as poplar, mustard, and mulberry—that have an appetite for lead, uranium, and other pollutants. These plants' genetic makeups allow them to absorb and to store, degrade, or transform substances that kill or harm other plants and animals. "It's an elegantly simple solution," says Licht, who now runs Ecolotree, an Iowa City phytoremediation company.

The idea behind phytoremediation is not new; scientists have long recognized certain plants' abilities to absorb and tolerate toxic substances. But the idea of using these plants on contaminated sites has just gained support in the last decade. The plants clean up sites in two basic ways, depending on the substance involved. If it is an organic contaminant, such as spilled oil, the plants or microbes around their roots break down the substance. The remainders can either be absorbed by the plant or left in the soil or water. For an inorganic contaminant such as cadmium or zinc, the plants absorb the substance and trap it. The plants must then be harvested and disposed of, or processed to reclaim the trapped contaminant.

Different plants work on different contaminants. The mulberry bush, for instance, is ef-

FIGURE 26B Canola plants.
Scientist Gary Bañuelos recommended planting canola to pull selenium out of the soil.

fective on industrial sludge; some grasses attack petroleum wastes; and sunflowers (together with soil additives) remove lead. Canola plants, meanwhile, are grown in California's San Joaquin Valley to soak up excess selenium in the soil to help prevent an environmental catastrophe like the one that occurred there in the 1980s.

Back then, irrigated farming caused naturally occurring selenium to rise to the soil surface. When excess water was pumped onto the fields, some selenium would flow off into drainage ditches, eventually ending up in Kesterson National Wildlife Refuge. The selenium in ponds at the refuge accumulated in plants and fish and subsequently deformed and killed waterfowl, says Gary Bañuelos, a plant scientist with the U.S. Department of Agriculture who helped remedy the problem. He recommended that farmers add selenium-accumulating canola plants to their crop rotations (Fig. 26B). As a result, selenium levels in runoff are being managed. Although the underlying problem of excessive selenium in soils has not been solved, says Bañuelos, "this is a tool to manage mobile selenium and prevent another unlikely selenium-induced disaster."

Phytoremediation has also helped clean up badly polluted sites, in some cases at a fraction of the usual cost. Edenspace Systems Corporation of Reston, Virginia, just concluded a phytoremediation demonstration at a Superfund site on an Army firing range in Aberdeen, Maryland. The company successfully used mustard plants to re-

move uranium from the firing range, at as little as 10% of the cost of traditional cleanup methods. Depending on the contaminant involved, traditional cleanup costs can run as much as $1 million per acre, experts say.

Phytoremediation does have its limitations, however. One of them is its slow pace. Depending on the contaminant, it can take several growing seasons to clean a site—much longer than conventional methods. "We normally look at phytoremediation as a target of one to three years to clean a site," notes Edenspace's Mike Blaylock. "People won't want to wait much longer than that."

Phytoremediation is also only effective at depths that plant roots can reach, making it useless against deep-lying contamination unless the contaminated soils are excavated. Phytoremediation will not work on lead and other metals unless chemicals are added to the soil. In addition, it is possible that animals may ingest pollutants by eating the leaves of plants in some projects.

Despite its shortcomings, experts see a bright future for this technology. David Glass, an independent analyst based in Needham, Massachusetts, predicts the business of phytoremediation will grow to $235 million to $400 million by 2005. It is a promising solution to pollution problems but, says the EPA's Walter W. Kovalick, "it's not a panacea. It's another arrow in the quiver. It takes more than one arrow to solve most problems."

Organic Nutrient Transport

Not only do plants transport water and minerals from the roots to the leaves, but they also transport organic nutrients to the parts of plants that need them. This includes young leaves that have not yet reached their full photosynthetic potential; flowers that are in the process of making seeds and fruits; and the roots, whose location in the soil prohibits them from carrying on photosynthesis

Role of Phloem

As long ago as 1679, Marcello Malpighi suggested that bark is involved in translocating sugars from leaves to roots. He observed the results of removing a strip of bark from around a tree, a procedure called **girdling.** If a tree is girdled below the level of the majority of leaves, the bark swells just above the cut, and sugar accumulates in the swollen tissue. We know today that when a tree is girdled, the phloem is removed, but the xylem is left intact. Therefore, the results of girdling suggest that phloem is the tissue that transports sugars.

Radioactive tracer studies with carbon 14 (^{14}C) have confirmed that phloem transports organic nutrients. When ^{14}C-labeled carbon dioxide (CO_2) is supplied to mature leaves, radioactively labeled sugar is soon found moving down the stem into the roots. It's difficult to get samples of sap from phloem without injuring the phloem, but this problem is solved by using aphids, small insects that are phloem feeders. The aphid drives its stylet, which is a sharp mouthpart that functions like a hypodermic needle, between the epidermal cells, and sap enters its body from a sieve-tube member (Fig. 26.14). If the aphid is anesthetized using ether, its body can be carefully cut away, leaving the stylet. Phloem can then be collected and analyzed by a researcher. By the use of radioactive tracers and aphids, it is known that the movement through phloem can be as fast as 60–100 cm per hour and possibly up to 300 cm per hour.

Pressure-Flow Model of Phloem Transport

The **pressure-flow model** is a current explanation for the movement of organic materials in phloem (Fig. 26.15). Consider the following experiment in which two bulbs are connected by a glass tube. The first bulb contains solute at a higher concentration than the second bulb. Each bulb is bounded by a differentially permeable membrane, and the entire apparatus is submerged in distilled water.

a. An aphid feeding on a plant stem

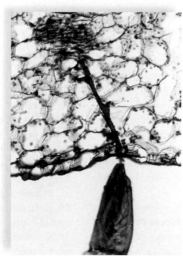

b. Aphid stylet in place

FIGURE 26.14

Acquiring phloem sap.
Aphids are small insects that remove nutrients from phloem by means of a needlelike mouthpart called a stylet. **a.** Excess phloem sap appears as a droplet after passing through the aphid's body. **b.** Micrograph of stylet in plant tissue. When an aphid is cut away from its stylet, phloem sap becomes available for collection and analysis.

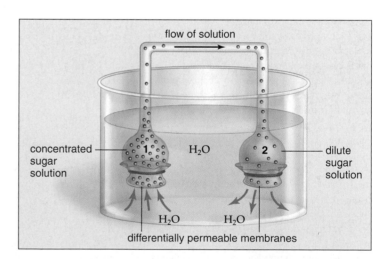

flow of solution

concentrated sugar solution

H_2O

dilute sugar solution

H_2O H_2O

differentially permeable membranes

Distilled water flows into the first bulb because it has the higher solute concentration. The entrance of water creates a positive *pressure*, and water *flows* toward the second bulb. This flow not only drives water toward the second bulb, but it also provides enough force for water to move out through the membrane of the second bulb—even though the second bulb contains a higher concentration of solute than the distilled water.

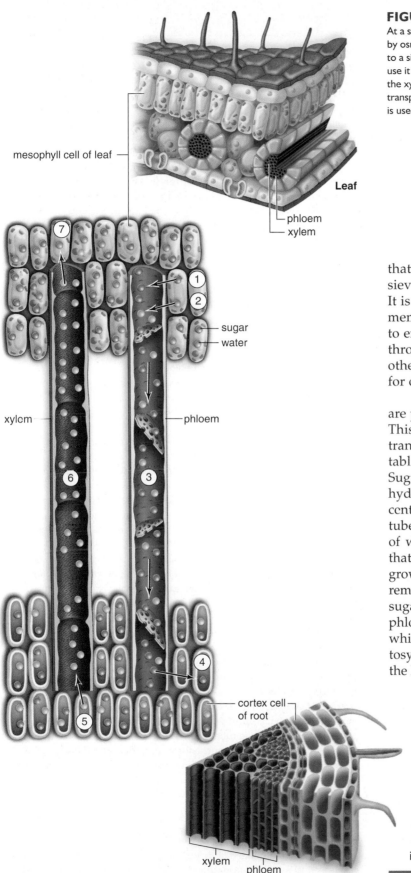

mesophyll cell of leaf

Leaf

phloem
xylem

7

1
2

sugar
water

xylem

phloem

6

3

4

cortex cell
of root

5

xylem
phloem

Root

FIGURE 26.15 Pressure-flow model of phloem transport.
At a source, ① sugar is actively transported into sieve tubes. ② Water follows by osmosis. ③ A positive pressure causes phloem contents to flow from source to a sink. At a sink, ④ sugar is actively transported out of sieve tubes and cells use it for cellular respiration. Water exits by osmosis. ⑤ Some water returns to the xylem, where it mixes with more water absorbed from the soil. ⑥ Xylem transports water to the mesophyll of the leaf. ⑦ Most water is transpired, some is used for photosynthesis, and some reenters phloem by osmosis.

In plants, sieve tubes are analogous to the glass tube that connects the two bulbs. Sieve tubes are composed of sieve-tube members, each of which has a companion cell. It is possible that the companion cells assist the sieve-tube members in some way. The sieve-tube members align end to end, and strands of plasmodesmata (cytoplasm) extend through sieve plates from one sieve-tube member to the other. Sieve tubes, therefore, form a continuous pathway for organic nutrient transport throughout a plant.

During the growing season, photosynthesizing leaves are producing sugar. Therefore, they are a source of sugar. This sugar is actively transported into phloem. Again, transport is dependent on an electrochemical gradient established by a proton pump, a form of active transport. Sugar is carried across the membrane in conjunction with hydrogen ions (H^+), which are moving down their concentration gradient (see Fig. 26.5). After sugar enters sieve tubes, water follows passively by osmosis. The buildup of water within sieve tubes creates the positive pressure that starts a flow of phloem contents. The roots (and other growth areas) are a sink for sugar, meaning that they are removing sugar and using it for cellular respiration. After sugar is actively transported out of sieve tubes, water exits phloem passively by osmosis and is taken up by xylem, which transports water to leaves, where it is used for photosynthesis. Now, phloem contents continue to flow from the leaves (source) to the roots (sink).

The pressure-flow model of phloem transport can account for any direction of flow in sieve tubes if we consider that the direction of flow is always from **source** to **sink**. For example, recently formed leaves can be a sink, and they will receive sucrose until they begin to maximally photosynthesize.

The pressure-flow model of phloem transport suggests that phloem contents can move either up or down as appropriate for the plant at a particular time in its life cycle.

CONNECTING THE CONCEPTS

The land environment offers many advantages for plants, such as greater availability of light and carbon dioxide for photosynthesis. (Water, even if clear, filters out light, and carbon dioxide concentration and rate of diffusion is less in water.) The evolution of a transport system was critical, however, for plants to make full use of these advantages. Only if a transport system is present can plants elevate the leaves so that they are better exposed to solar energy and carbon dioxide in the air. A transport system brings water, a raw material of photosynthesis, from the roots to the leaves and also brings the products of photosynthesis down to the roots. Roots lie beneath the soil, and their cells depend on an input of organic food from the leaves to remain alive. An efficient transport system allows roots to penetrate deeply into the soil to absorb water and minerals.

The presence of a transport system also allows materials to be distributed to those parts of the plant body that are growing most rapidly. New leaves and flower buds would grow rather slowly if they had to depend on their own rate of photosynthesis, for example. Height in vascular plants, due to the presence of a transport system, has other benefits aside from elevation of leaves. It is also adaptive to have reproductive structures located where the wind can better distribute pollen and seeds. Once animal pollination came into existence, it was beneficial for flowers to be located where they are more easily seen by animals.

Clearly, plants with a transport system have a competitive edge in the terrestrial environment.

Summary

26.1 PLANT NUTRITION AND SOIL

Plants need both essential and beneficial inorganic nutrients. Carbon, hydrogen, and oxygen make up 95% of a plant's dry weight. The other necessary nutrients are taken up by the roots as mineral ions. Even nitrogen (N), which is present in the atmosphere, is most often taken up as NO_3^-.

You can determine mineral requirements by hydroponics, in which plants are grown in a solution. The solution is varied by the omission of one mineral. If the plant grows poorly, then the missing mineral must be essential for growth.

Terrestrial life is dependent on soil, which forms by the weathering of rock. Organisms contribute to the formation of humus and soil. Soil is a mixture of mineral particles, humus, living organisms, air, and water. Soil particles are of three types from the largest to the smallest: sand, silt, and clay. Loam, which contains about equal proportions of all three types, retains water but still has air spaces. Humus contributes to the texture of soil and its ability to provide inorganic nutrients to plants. Topsoil (A horizon of a soil profile) contains humus, and this is the layer that is lost by erosion, a worldwide problem.

26.2 WATER AND MINERAL UPTAKE

Water, along with minerals, can enter a root by passing between the porous cell walls (a process called apoplastic transport), until it reaches the Casparian strip, after which it passes through an endodermal cell before entering xylem. Water can also enter root hairs and then pass through the cells of the cortex and endodermis to reach xylem. This is known as symplastic transport.

Mineral ions cross plasma membranes by a chemiosmotic mechanism. A proton pump transports H^+ out of the cell. This establishes an electrochemical gradient that causes positive ions to flow into the cells. Negative ions are carried across in conjunction with H^+, which is moving along its concentration gradient.

Plants have various adaptations that assist them in acquiring nutrients. Legumes have nodules infected with the bacterium *Rhizobium*, which makes nitrogen compounds available to these plants. Many other plants have mycorrhizae, or fungus roots. The fungus gathers nutrients from the soil, and the root provides the fungus with sugars and amino acids. Some plants have poorly developed roots. Most epiphytes live on, but do not parasitize, trees, whereas dodder and some other plants parasitize their hosts.

26.3 TRANSPORT MECHANISMS IN PLANTS

As an adaptation to life on land, plants have a vascular system that transports water and minerals from the roots to the leaves and must also transport the products of photosynthesis in the opposite direction. Vascular tissue includes xylem and phloem.

In xylem, vessels composed of vessel elements aligned end to end form an open pipeline from the roots to the leaves. Particularly at night, root pressure can build in the root. However, this does not contribute significantly to xylem transport.

The cohesion-tension model of xylem transport states that transpiration creates a tension that pulls water upward in xylem. This means of transport works only because water molecules are cohesive with one another and adhesive with xylem walls.

Most of the water taken in by a plant is lost through stomata by transpiration. Only when there is plenty of water do stomata remain open, allowing carbon dioxide to enter the leaf and photosynthesis to occur.

Stomata open when guard cells take up water. The guard cells are anchored at their ends. They can only stretch lengthwise because microfibrils in their walls prevent lateral expansion. Therefore, guard cells buckle out when water enters. Water enters the guard cells after potassium ions (K^+) have entered. Light signals stomata to open, and a high carbon dioxide (CO_2) level signals stomata to close. Abscisic acid produced by wilting leaves also signals for closure.

In phloem, sieve tubes composed of sieve-tube members aligned end to end form a continuous pipeline from the leaves to the roots. Sieve-tube members have sieve plates through which plasmodesmata (strands of cytoplasm) extend from one to the other. The pressure-flow model of phloem transport proposes that a positive pressure drives phloem contents in sieve tubes. Sucrose is actively transported into sieve tubes—by a chemiosmotic mechanism—at a source, and water follows by osmosis. The resulting increase in pressure creates a flow that moves water and sucrose to a sink. A sink can be at the roots or any other part of the plant that requires organic nutrients.

Reviewing the Chapter

1. Name the elements that make up most of a plant's body. What are essential mineral nutrients and beneficial mineral nutrients?   460
2. Briefly describe the use of hydroponics to determine the mineral nutrients of a plant.   461

3. How is soil formed, and how do the components of soil join to provide nutrients to plants? Describe a generalized soil profile and how a profile is affected by erosion. 462–63

4. Give two pathways by which water and minerals can cross the epidermis and cortex of a root. What feature allows endodermal cells to regulate the entrance of molecules into the vascular cylinder? 464

5. Describe the chemiosmotic mechanism by which mineral ions cross plasma membranes. 464

6. Name two symbiotic relationships that assist plants in taking up minerals and two types of plants that tend not to take up minerals by roots from soil. 464–65

7. A vascular system is adaptive for a land existence. Explain. Describe the composition of a plant's vascular system. 466–67

8. What is root pressure, and why can't it account for the transport of water in xylem? 468

9. Describe and give evidence for the cohesion-tension model of water transport. 468–69

10. Describe the structure of stomata and explain how they can open and close. By what mechanism do guard cells take up potassium (K^+) ions? 470

11. What data are available to show that phloem transports organic compounds? Explain the pressure-flow model of phloem transport. 472

Testing Yourself

Choose the best answer for each question.

1. Which of these molecules is not a nutrient for plants?
   a. water
   b. carbon dioxide gas
   c. mineral ions
   d. nitrogen gas
   e. None of these are nutrients.

2. Which is a component of soil?
   a. mineral particles
   b. humus
   c. organisms
   d. air and water
   e. All of these are correct.

3. The Casparian strip affects
   a. how water and minerals move into the vascular cylinder.
   b. vascular tissue composition.
   c. how soil particles function.
   d. how organic nutrients move into the vascular cylinder.
   e. Both a and d are correct.

4. Which of these is not a mineral ion?
   a. NO_3^-
   b. Mg^+
   c. CO_2
   d. Al^{3+}
   e. All of these are correct.

5. What role do cohesion and adhesion play in xylem transport?
   a. Like transpiration, they create a tension.
   b. Like root pressure, they create a positive pressure.
   c. Like sugars, they cause water to enter xylem.
   d. They create a continuous water column in xylem.
   e. All of these are correct.

6. The pressure-flow model of phloem transport states that
   a. phloem contents always flows from the leaves to the root.
   b. phloem contents always flows from the root to the leaves.
   c. water flow brings sucrose from a source to a sink.
   d. water pressure creates a flow of water.
   e. Both c and d are correct.

7. Root hairs do not play a role in
   a. oxygen uptake.
   b. mineral uptake.
   c. water uptake.
   d. carbon dioxide uptake.
   e. the uptake of any of these.

8. Xylem includes all of these except
   a. companion cells.
   b. vessels.
   c. tracheids.
   d. dead tissue.

9. After sucrose enters sieve tubes,
   a. it is removed by the source.
   b. water follows passively by osmosis.
   c. it is driven by active transport to the source, which is usually the roots.
   d. stomata open so that water flows to the leaves.
   e. All of these are correct.

10. An opening in the leaf that allows gas and water exchange is called
    a. the lenticel.
    b. the hole.
    c. the stoma.
    d. the guard cell.
    e. the accessory cell.

11. What main force drives absorption of water, creates tension, and draws water through the plant?
    a. adhesion
    b. cohesion
    c. tension
    d. transpiration
    e. absorption

12. A nutrient element is considered essential if
    a. plant growth increases with a reduction in the concentration of the element.
    b. plant growth suffers in the absence of the element.
    c. plants can substitute a similar element for the missing element with no ill effects.
    d. the element is a positive ion.

13. Humus
    a. supplies nutrients to plants.
    b. is basic in its pH.
    c. is found in the deepest soil horizons.
    d. is inorganic in origin.

14. Which sequence represents the size of soil particles from largest to smallest?
    a. sand, clay, silt
    b. silt, clay, sand
    c. sand, silt, clay
    d. clay, silt, sand
    e. silt, sand, clay

15. Soils rich in which type of soil particle will have a high water-holding capacity?
    a. sand
    b. silt
    c. clay
    d. All soil particles hold water equally well.

16. Stomata are usually open
    a. at night, when the plant requires a supply of oxygen.
    b. during the day, when the plant requires a supply of carbon dioxide.
    c. day or night if there is excess water in the soil.
    d. during the day, when transpiration occurs.
    e. Both b and d are correct.

17. Negatively charged clay particles attract
    a. K⁺.
    a. K^+.
    b. NO_3^-.
    c. Ca^+.
    d. Both a and b are correct.
    e. Both a and c are correct.

18. Explain why this experiment supports the hypothesis that transpiration can cause water to rise to the tops of tall trees.

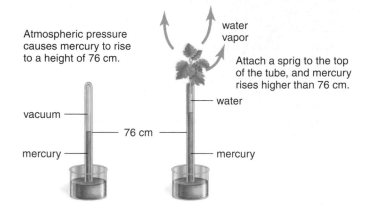

Atmospheric pressure causes mercury to rise to a height of 76 cm.

water vapor

Attach a sprig to the top of the tube, and mercury rises higher than 76 cm.

vacuum

water

76 cm

mercury

mercury

19. Label water (H_2O) and potassium ions (K^+) appropriately in these diagrams. What is the role of K^+ in the opening of stomata?

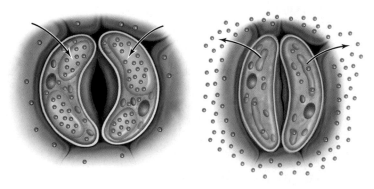

20. Explain why solution flows from the left bulb to the right bulb.

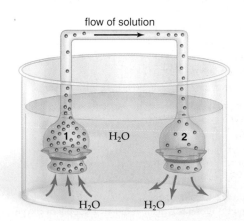

flow of solution

H_2O

1

2

H_2O

H_2O

Thinking Scientifically

1. A field has been watered by irrigation for a number of years. The salts that were in the irrigation water have accumulated in the soil, making the solute concentration much higher than it was before. You notice that while water is still plentiful, plants in the field seem to be wilting. How would a change in the water's solute concentration affect its uptake by plants?

2. The force that pulls water up xylem is not unlike the force that pulls water up into a syringe. The wider the diameter of the syringe, the greater the force on the sides of the syringe. Vessel elements are larger in diameter than tracheids. Which would you expect to have stronger walls—vessel elements or tracheids? The xylem of gymnosperms, such as pine trees, contains only tracheids, while the xylem of angiosperms, such as oak trees, contains both tracheids and vessel elements. Relate this information to the strength of the wood in pine trees and oak trees.

Understanding the Terms

beneficial nutrients 461	phytoremediation 471
Casparian strip 464	pressure-flow model 472
cohesion-tension model 468	root hair 464
companion cell 466	root nodule 464
cuticle 469	root pressure 468
epiphyte 465	sieve-tube member 466
essential nutrient 461	sink 473
girdling 472	soil 462
guard cell 470	soil erosion 463
guttation 468	soil profile 463
horizon 463	source 473
humus 462	stoma 470
hydroponics 461	tracheid 466
macronutrient 461	transpiration 469
micronutrient 461	vessel element 466
mineral 461	water potential 467
mycorrhizae 464	xylem 466
phloem 466	xylem sap 466
phloem sap 466	

Match the terms to these definitions:

a. _____ Model explaining transport in sieve tubes of phloem.

b. _____ Plant that takes its nourishment from the air because its attachment to other plants gives it an aerial position.

c. _____ Plant's loss of water to the atmosphere, mainly through evaporation at leaf stomata.

d. _____ Layer of impermeable lignin and suberin bordering four sides of root endodermal cells; causes water and minerals to enter endodermal cells before entering vascular tissue.

e. _____ Type of plant cell that is found in pairs, with one on each side of a leaf stoma.

ARIS, the *Biology* Website

ARIS, the website for *Biology*, provides a wealth of information organized and integrated by chapter. You will find practice quizzes, interactive activities, labeling exercises, flashcards, and much more that will complement your learning and understanding of general biology.

27

CONTROL OF GROWTH AND RESPONSES IN PLANTS

P lants respond to environmental stimuli such as light, gravity, and seasonal changes. Some plant responses are short term. For example, plants will bend toward the light within a few hours because a hormone produced by the growing tip has moved from the sunny side to the shady side of the stem. Short-term and also long-term plant responses to environmental conditions often involve a change in the pattern of growth. When protected, oak trees tend to be tall with fewer branches, but when exposed to high winds, oak trees are strong and thickly branched like the one in the photograph. Plants could not meet such environmental challenges without a system for controlling growth and responses. Hormones help plants respond to stimuli in a coordinated manner.

Plants, particularly in the temperate zone, undergo seasonal changes. In the spring, seeds germinate and growth begins if the soil is warm enough to contain liquid water. In the fall, when temperatures drop, shoot and root apical growth ceases. Hormones are involved in these responses also. The pigment phytochrome is instrumental in detecting day length, which determines whether a plant flowers or does not flower.

Live oak, *Quercus virginiana.*

27.1 PLANT RESPONSES

All organisms are capable of responding to environmental stimuli, as when you withdraw your hand from a hot stove. It is adaptive for organisms to respond to stimuli because it leads to their longevity and ultimately to the survival of the species. Animals often quickly respond to a stimulus by an appropriate behavior. Presented with a nipple, a newborn automatically begins sucking. Sometimes plants, too, can respond quickly, as when stomata open at sunrise. While animals are apt to change their location, plants, which are rooted in one place, change their growth pattern in response to a stimulus. The events in a tree's life and even the history of the Earth's climate can be determined by studying the growth pattern of tree rings!

Tropisms

Plant growth toward or away from a unidirectional stimulus is called a **tropism** [Gk. *tropos*, turning]. Unidirectional means that the stimulus is coming from only one direction instead of multiple directions. Growth toward a stimulus is called a positive tropism, and growth away from a stimulus is called a negative tropism. Tropisms are due to differential growth—one side of an organ elongates faster than the other,

and the result is a curving toward or away from the stimulus. A number of tropisms have been observed in plants. The three best-known tropisms are phototropism (light), gravitropism (gravity), and thigmotropism (touch).

> Phototropism: a movement in response to a light stimulus
> Gravitropism: a movement in response to gravity
> Thigmotropism: a movement in response to touch

Several other tropisms include chemotropism (chemicals), traumotropism (trauma), skototropism (dark), and aerotropism (oxygen).

What mechanism permits plants to respond to stimuli? When humans respond to light, the stimulus is first received by a pigment in the retina at the back of the eyes, and then nerve impulses are generated that go to the brain. Only then do humans perform an appropriate behavior. Notice that the first step toward a response is *reception* of the stimulus. The next step is *transduction,* meaning that the stimulus has been changed into a form that is meaningful to the organism. (In our example, the light stimulus was changed to nerve impulses.) Finally, there is

FIGURE 27.1 Phototropism.
Time-lapse photograph of a buttercup, *Ranunculus ficaria,* curving toward and tracking a source of light.

tracking light

curving of stem

a *response* by the organism. Animals and plants go through a similar sequence of events when they respond to a stimulus:

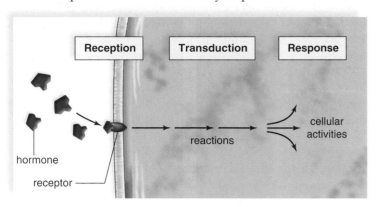

a.

Phototropism

Early researchers, including Charles Darwin and his son Francis, observed that plants curve toward the light. The positive **phototropism** [Gk. *photos*, light, and *tropos*, turning] of stems occurs because the cells on the shady side of the stem elongate (Fig. 27.1). Curving away from light is called negative phototropism. Roots, depending on the species examined, are either insensitive to light or they exhibit negative phototropism.

It is believed that a pigment related to the vitamin riboflavin acts as a photoreceptor for light when phototropism occurs. Following reception, a plant hormone called auxin migrates from the bright side to the shady side of a stem. The cells on that side elongate faster than those on the bright side, causing the stem to curve toward the light. It is not yet known how reception of the stimulus (light) is coupled to the binding of auxin.

As discussed in the next sections, auxin is also involved in gravitropism, apical dominance, root development, and seed development.

Gravitropism

When an upright plant is placed on its side, the stem displays negative **gravitropism** [L. *gravis*, heavy; Gk. *tropos*, turning] because it grows upward, opposite the pull of gravity (Fig. 27.2*a*). Again, Charles Darwin and his son were among the first to say that roots, in contrast to stems, show positive gravitropism (Fig. 27.2*b*). Further, they discovered that if the root cap is removed, roots no longer respond to gravity. Later investigators came up with an explanation. Root cap cells contain sensors called **statoliths,** which are starch grains located within amyloplasts, a type of plastid (Fig. 27.2*c*). Due to gravity, the amyloplasts settle to a lower part of the cell where they come in contact with the endoplasmic reticulum (ER). ER releases stored calcium ions (Ca^{2+}), which lead to the activation of auxin pumps and auxin entering the cell.

Again, the hormone auxin brings about the positive gravitropism of roots and the negative gravitropism of stems. The two types of tissues respond differently to auxin, which appears on the lower side of both stems and roots after gravity has been perceived. Auxin inhibits the growth of root cells; therefore, the cells of the upper surface elongate so that the root curves downward. Auxin stimulates the growth of stem cells; therefore, the cells of the lower surface elongate, and the stem curves upward.

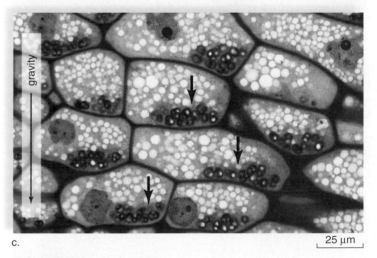
c.   25 μm

FIGURE 27.2 Gravitropism.
a. Negative gravitropism of the stem of a *Coleus* plant 24 hours after the plant was placed on its side. **b.** Positive gravitropism of a root emerging from a corn kernel. **c.** Sedimentation of statoliths (see arrows), which are amyloplasts containing starch granules, is thought to explain how roots perceive gravity.

Thigmotropism

Unequal growth due to contact with solid objects is called **thigmotropism** [Gk. *thigma*, touch, and *tropos*, turning]. An example of this response is the coiling of tendrils or the stems of plants such as morning glory (Fig. 27.3).

The plant grows straight until it touches something. Then the cells in contact with an object, such as a pole, grow less while those on the opposite side elongate. Thigmotropism can be quite rapid; a tendril has been observed to encircle an object within 10 minutes. The response endures; a couple of minutes of touching can bring about a response that lasts for several days. The response can also be delayed; tendrils touched in the dark will respond once they are illuminated. ATP (adenosine triphosphate) rather than light can cause the response; therefore, the need for light may simply be a need for ATP. Also, the hormones auxin and ethylene may be involved since they can induce curvature of tendrils even in the absence of touch.

Thigmomorphogenesis is a touch response related to thigmotropism. In this case, however, the entire plant responds to the presence of environmental stimuli, such as wind or rain. The same type of tree growing in a windy location often has a shorter, thicker trunk than one growing in a more protected location. Even simple mechanical stimulation—rubbing a plant with a stick, for example—can inhibit cellular elongation and produce a sturdier plant with increased amounts of support tissue.

Nastic Movements

In contrast to tropisms, nastic movements do not involve growth and are not dependent on the direction of the stimulus. *Seismonastic* movements result from touch, shaking, or thermal stimulation. If you touch a *Mimosa pudica* leaf, the leaflets fold because the petiole droops (Fig. 27.4). This response, which takes only a second or two, is due to a loss of turgor pressure within cells located in a thickening, called a pulvinus, at the base of each leaflet and at the base of the petiole. Investigation shows that potassium ions (K^+) move out of the cells and then water follows by osmosis. A single stimulus, such as a hot needle, is enough to cause all the leaves to respond. There must be some means of communication for this to occur, and in fact, a nerve impulse-type stimulus has been recorded in these plants!

A Venus's flytrap (see Fig. 25.21c) has three trichomes at the base of the trap, and if these are touched by an insect, a nerve impulse-type stimulus brings about closing of the trap. This, too, is due to turgor pressure changes in the leaf cells that form the trap.

Sleep Movements

A sleep movement is a nastic response that occurs daily in response to light and dark changes. One of the most common examples occurs in a houseplant called the prayer plant because at night the leaves fold upward into a shape resembling hands at prayer (Fig. 27.5). This movement is also due to changes in the turgor pressure of motor cells in a pulvinus located at the base of each leaf.

Circadian Rhythms. Organisms exhibit periodic fluctuations that correspond to environmental changes. For example, body temperature and blood pressure tend to change with the time of day, and people become sleepy at a certain time of night. The prayer plant in Figure 27.5 displays rhythmic "sleep" behavior. A biological rhythm with a 24-hour cycle is called a **circadian rhythm** [L. *circum*, about, and *dies*, day].

Circadian rhythms tend to persist, even if the appropriate environmental cues are no longer present. For example, on a transcontinental flight, one will likely suffer jet lag, and it will take several days to adjust to the time change because

FIGURE 27.3 Coiling response.
The stem of a morning glory plant, *Ipomoea*, coiling around a pole illustrates thigmotropism.

before

before

after

after

FIGURE 27.4 Seismonastic movement.
A leaf of the sensitive plant, *Mimosa pudica*, before and after it is touched.

FIGURE 27.5 Sleep movement.
Prayer plant, *Maranta leuconeura*, before dark and after dark when the leaves fold up.

the body will still be attuned to the day-night pattern of its previous environment. The internal mechanism by which a biological rhythm is maintained in the absence of appropriate environmental stimuli is termed a **biological clock.** Typically, if organisms are sheltered from environmental stimuli, their circadian rhythms continue, but the cycle extends. In prayer plants, for example, the sleep cycle changes to 26 hours. Therefore, it is believed that biological clocks are synchronized by external stimuli to 24-hour rhythms. The length of daylight compared to the length of darkness, called the photoperiod, sets the clock. Temperature has little or no effect. This is adaptive because the photoperiod indicates seasonal changes bet-

ter than temperature changes. Spring and fall, in particular, can have both warm and cold days.

There are other examples of circadian rhythms in plants. Stomata and certain flowers usually open in the morning and close at night, and some plants secrete nectar at the same time of the day or night.

Tropisms are growth responses toward or away from unidirectional stimuli such as light, gravity, or physical contact. Plants may also produce nondirectional nastic movements in response to stimuli such as touch and light.

27.2 PLANT HORMONES

For plants to respond to stimuli, the activities of plant cells and structures have to be coordinated. Almost all communication in a plant is done by **hormones** [Gk. *hormao*, instigate], chemical signals produced in very low concentrations and active in another part of the organism. Some plant responses are probably influenced by several hormones and most likely require a specific ratio of all hormones involved. Hormones are synthesized or stored in one part of the plant, but they travel within phloem or from cell to cell in response to the appropriate stimulus.

Each naturally occurring hormone has a specific chemical structure. Other chemicals, some of which differ only slightly from the natural hormones, also affect the growth of plants. These and the naturally occurring hormones are sometimes grouped together and called plant growth regulators.

Auxins

The most common naturally occurring **auxin** [Gk. *auximos*, promoting growth] is indoleacetic acid (IAA). It is produced in shoot apical meristem and is found in young leaves and in flowers and fruits:

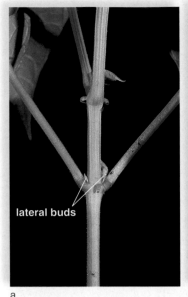 
a.   b.

FIGURE 27.6 Apical dominance.
a. Auxin in the apical bud inhibits lateral bud development, and the plant exhibits apical dominance. **b.** When the apical bud is removed, lateral branches develop.

Structure of
indoleacetic acid (IAA)

CH_2-COOH

Auxins affect many aspects of plant growth and development. Horticulturists often use auxins such as indoleacetic acid and other growth regulators applied as a paste to plant cuttings to stimulate vigorous root formation.

Effects of Auxin

Apically produced auxin prevents the growth of lateral buds, a phenomenon called **apical dominance.** When a terminal bud is removed deliberately or accidentally, the nearest lateral buds begin to grow, and the plant branches. Pruning the top (apical meristem) of a plant generally achieves a fuller look. This removes apical dominance and causes more branching of the main body of the plant (Fig. 27.6).

The application of a weak solution of auxin to a woody cutting causes adventitious roots to develop more quickly than they would otherwise. Auxin production by seeds also promotes the growth of fruit. As long as auxin is concentrated in leaves or fruits rather than in the stem, leaves and fruits do not fall off. Therefore, trees can be sprayed with auxin to keep mature fruit from falling to the ground.

Synthetic auxins are used today in a number of applications. These auxins are sprayed on plants such as tomatoes to induce the development of fruit without pollination. Thus, seedless tomatoes can be commercially developed. Synthetic auxins such as 2,4-D and 2,4,5-T have been used

to control broadleaf weeds such as dandelions and other plants. These substances have little effect on grasses. 2,4-D is still used but 2,4,5-T was banned in 1979 because of its detrimental effects on human and animal life. A mixture of 2,4-D and 2,4,5-T is best known as the defoliant Agent Orange, used in the Vietnam war.

As discussed earlier, auxin is also involved in gravitropism and phototropism. After gravity has been perceived, auxin moves to the lower surface of roots and stems. Thereafter, roots curve downward and stems curve upward (see Fig. 27.2). The role of auxin in the positive phototropism of stems has been studied for quite some time. The experimental material of choice has been oat seedlings with coleoptiles intact. A **coleoptile** is a protective sheath for the young leaves of the seedling. In 1881, the Darwins found that phototropism will not occur if the tip of the seedling is cut off or covered by a black cap. They concluded that some influence that causes curvature is transmitted from the coleoptile tip to the rest of the shoot.

In 1926, Frits W. Went cut off the tips of coleoptiles and placed them on agar (a gelatin-like material). Then he placed an agar block to one side of a tipless coleoptile and found that the shoot would curve away from that side. The bending occurred even though the seedlings were not exposed to light (Fig. 27.7). Went concluded that the agar blocks contained a chemical that had been produced by the coleoptile tips. This chemical, he decided, had caused the shoots to curve. He named the chemical substance auxin after the Greek word *auximos*, which means promoting growth.

How Auxins Work

When a stem is exposed to unidirectional light, auxin moves to the shady side, where it binds to receptors in the plasma

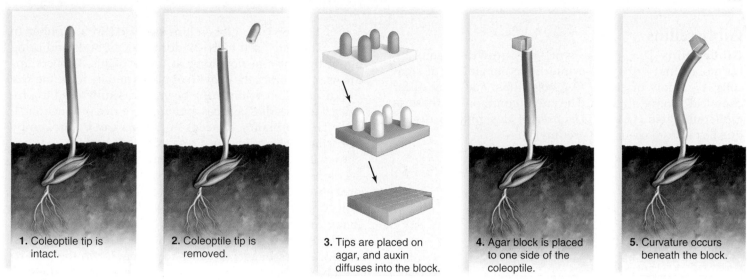

1. Coleoptile tip is intact.

2. Coleoptile tip is removed.

3. Tips are placed on agar, and auxin diffuses into the block.

4. Agar block is placed to one side of the coleoptile.

5. Curvature occurs beneath the block.

FIGURE 27.7 Demonstrating phototropism.
Oat seedlings are protected by a hollow sheath called a coleoptile (pink). After a tip is removed and placed on agar, an agar block placed on one side of the coleoptile can cause it to curve.

membrane (Fig. 27.8). Binding leads to a series of reactions and the generation of at least three specific second messengers, so called because they and not auxin actually bring about the response of the cell. One activates a proton (H⁺) pump, and the resulting acidic conditions loosen the cell wall because hydrogen bonds are broken and cellulose fibrils are weakened. Another activates the Golgi apparatus. The Golgi apparatus then sends out vesicles laden with cell wall materials that will bolster the elongating cell wall. The third second messenger stimulates a DNA-binding protein that enters the nucleus and activates a particular gene. Activation of this gene leads to the production of growth factors. The end result of these activities is elongation of the stem on the shady side so that it bends toward the light.

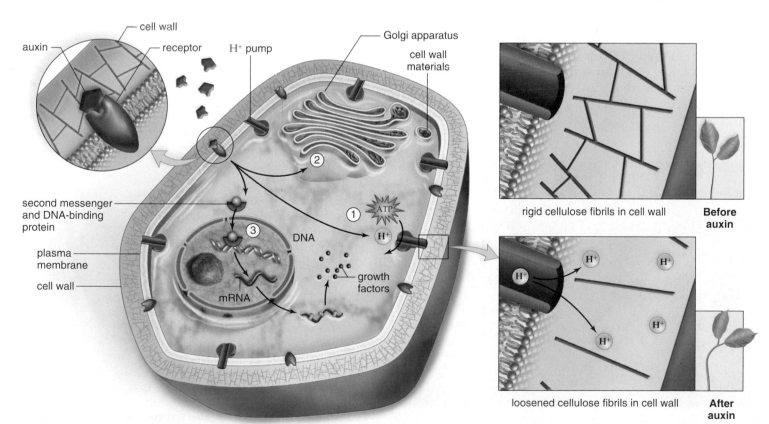

FIGURE 27.8 Auxin mode of action.
After auxin binds to a receptor, a series of reactions results in three second messengers. ① This second messenger activates an ATP-driven H⁺ pump. As hydrogen ions (H⁺) are being pumped out of the cell, the cell wall becomes acidic, loosening the wall and enabling the cell to elongate. ② This second messenger stimulates the Golgi apparatus to discharge transport vesicles carrying cell wall materials. ③ This second messenger stimulates a DNA-binding protein that enters the nucleus and activates a gene, and growth factors are produced.

Gibberellins

Gibberellins [L. *gibbus*, bent] are growth-promoting hormones that bring about internode elongation of stem cells. We know of about 70 gibberellins, and they differ chemically only slightly. The most common of these is gibberellic acid, GA_3 (the subscript designation distinguishes it from other gibberellins):

Structure of
gibberellic acid (GA_3)

When gibberellins are applied externally to plants, the most obvious effect is stem elongation (Fig. 27.9). Gibberellins can cause dwarf plants to grow, cabbage plants to become 2 m tall, and bush beans to become pole beans.

Gibberellins were discovered in 1926, the same year that Went performed his classic experiments with auxin. Ewiti Kurosawa, a Japanese scientist, was investigating a fungal disease of rice plants called "foolish seedling disease." The plants elongated too quickly, causing the stem to weaken and the plant to collapse. Kurosawa found that the fungus infecting the plants produced an excess of a chemical he called gibberellin, named after the fungus *Gibberella fujikuroi*. It wasn't until 1956 that gibberellic acid was isolated from a flowering plant rather than from a fungus. Sources of gibberellin in flowering plant parts are young leaves, roots, embryos, seeds, and fruits.

Commercially, gibberellins are used in a number of ways. They are used to break dormancy of buds and bring about the onset of flowering in many plants. Gibberellins are used to induce the growth of plants and increase the size of flowers. Gibberellins have been successfully used to produce larger seedless grapes and improve rice production.

The **dormancy** of seeds and buds can be broken by applying gibberellins, and research with barley seeds has shown how GA_3 acts as a chemical messenger. Barley seeds have a large, starchy endosperm, which must be broken down into sugars to provide energy for growth. After the embryo produces gibberellins, amylase, an enzyme that breaks down the starch, appears in cells just inside the seed coat. It is hypothesized that GA_3 (the first chemical messenger) attaches to a receptor in the plasma membrane, and then a second messenger inside the cell, namely calcium ions (Ca^{2+}), combines with a DNA-binding protein. This complex is believed to activate the gene that codes for amylase (Fig. 27.10). Amylase then acts on starch to release sugars used as a source of energy by the growing embryo.

The action of hormones on the cellular level is another example of the reception-transduction-response pathway described at the beginning of this chapter. The Science Focus on page 485 describes the work of two investigators in this area.

FIGURE 27.9 Effect of gibberellins.
The *Cyclamen* plant on the right was treated with gibberellins; the plant on the left was not treated. Gibberellins are often used to promote stem elongation in economically important plants, and the mode of action is becoming clear.

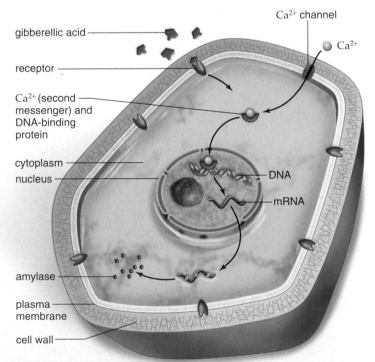

FIGURE 27.10 Gibberellic acid structure and mode of action.
Gibberellic acid is believed to bind to a receptor, and thereafter a second messenger (Ca^{2+}) inside the cell joins with a DNA-binding protein. The complex enters the nucleus and activates a gene. Thereafter, amylase is produced, and it acts on starch to release sugars used by the growing embryo.

science focus

Husband-and-Wife Team Explores Signal Transduction in Plants

Plants perceive and react to a variety of environmental stimuli. Some examples include light intensity and quality, gravity, carbon dioxide levels, pathogen infection, drought, and touch. Plant responses may be short term, such as the rapid localized cell death that occurs at the site of infection by a pathogen. This rapid cell death, termed the hypersensitive response, serves to limit the further progression of infection. Another form of short term response is the opening or closing of leaf stomata in response to light levels. High light causes stomata to open, while low light or darkness causes them to close. In this way, the supply of carbon dioxide to the leaves via stomata is coordinated with light-driven photosynthetic reactions occurring in leaf cells. On the other hand, plant responses to environmental stimuli may be more long term and involve actual plant growth. An example is the plant response to gravity (gravitropism), which results in the downward growth of the root and the upward growth of the stem. In each of these situations, the coupling of a stimulus to an appropriate response represents signal transduction.

Although we know that plants do not have a nervous system and a brain, some system or process must be present in these organisms to allow the perception of a stimulus and its conversion to an appropriate response. While we are both interested in attempting to understand this process, we approach this complex research area from different perspectives related to our training, expertise, and interests.

As a plant biochemist, Donald Briskin focuses on how membrane transport processes are involved in signal transduction. Here, research has shown that membrane transport of calcium ions (Ca^{2+}) can play a fundamental role in the signal transduction process. When a stimulus is perceived by a plant cell, a transient increase in cytoplasmic Ca^{2+} concentration occurs due to the rapid opening of Ca^{2+} channels at plant membranes, which operate not unlike the Ca^{2+} channels in animal membranes. This transient increase in cytoplasmic Ca^{2+} acts as a sort of message linking the stimulus to a response at the biochemical level. In particular, Donald studies how Ca^{2+}-transporting enzymes (called Ca^{2+}-ATPases) and Ca^{2+} channels are coordinated after a stimulus is received.

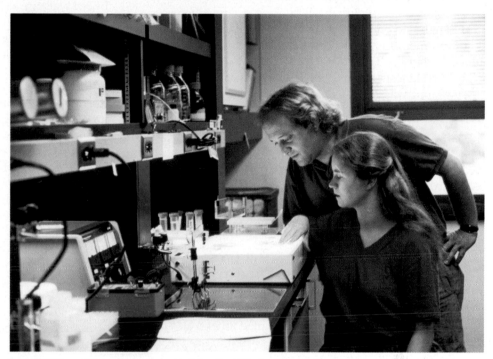

Donald Briskin and Margaret Gawienowski, University of Illinois at Urbana-Champaign.

Just how a single type of message—such as a rapid change in cytoplasmic Ca^{2+}—can elicit so many different types of specific plant responses is Margaret Gawienowski's interest. As a plant molecular biologist, Margaret focuses on how proteins in the plant cytoplasm that bind Ca^{2+} might impart specificity to the response. Of particular interest to her is a Ca^{2+}-binding protein called calmodulin, which controls the activity of a number of other proteins once it binds to Ca^{2+}. Her work has shown that calmodulin in plants is encoded by multiple genes that produce variants of the protein called isoforms. The isoforms are present to a different extent in different plant tissues, have different properties (Ca^{2+} binding, target protein regulation), and may play a role in determining which responses are coupled to transient cytoplasmic Ca^{2+} changes.

Together, we are conducting studies to determine how signal transduction processes are involved in controlling the production of medicinal chemicals generated by medicinal plants such as St. John's wort and *Echinacea*. We have found that the production of medicinal chemi-

cals can be affected by environmental factors during the growth of these plants. The environmental factors detected by the plant can include light levels, nitrogen supply, plant water status, and even the competition for these resources by other plants. Our work will lead to a better understanding of how the growth environment of medicinal plants can signal the plant to modify the quality and quantity of herbal medicines.

Although our approaches toward understanding plant signal transduction are very different, the work of each of us tends to complement that of the other. We view each other's work from a slightly different perspective, and this helps generate ideas for new experiments and strategies. Furthermore, we work together in the same laboratory at the University of Illinois at Urbana-Champaign, and this leads to much discussion and exchange of opinions. For us, research is not only challenging and exciting, but it keeps us wondering what the next day will bring.

Courtesy of Donald Briskin and Margaret Gawienowski
University of Illinois at Urbana-Champaign

Cytokinins

The **cytokinins** [Gk. *kytos,* cell, and *kineo,* move] are a class of plant hormones that promote cell division. These substances are derivatives of adenine, one of the purine bases in DNA and RNA.

The cytokinins were discovered as a result of attempts to grow plant tissue and organs in culture vessels in the 1940s (Fig. 27.11). It was found that cell division occurs when coconut milk (a liquid endosperm) and yeast extract are added to the culture medium. Although the effective agent or agents could not be isolated, they were collectively called cytokinins because cytokinesis means cell division. A naturally occurring cytokinin was not isolated until 1967. Because it came from the kernels of maize (*Zea*), it was called zeatin:

Structure of zeatin

Cytokinins have been isolated from various seed plants, where they occur in the actively dividing tissues of roots and also in seeds and fruits. A synthetic cytokinin called kinetin also promotes cell division. Cytokinins have been used to prolong the life of flower cuttings as well as vegetables in storage.

Plant tissue culturing is now common practice, and researchers are well aware that the ratio of auxin to cytokinin and the acidity of the culture medium determine whether the plant tissue forms an undifferentiated mass, called a callus, or differentiates to form roots, vegetative shoots, leaves, or floral shoots (Fig. 27.11). Researchers have reported that chemicals they call oligosaccharins (chemical fragments released from the cell wall) are also effective in directing differentiation. They hypothesize that auxin and cytokinins are a part of a reception-transduction-response pathway, which leads to the activation of enzymes that release these fragments from the cell wall.

Senescence

When a plant organ, such as a leaf, loses its natural color, it is most likely undergoing an aging process called **senescence.** During senescence, large molecules within the leaf are broken down and transported to other parts of the plant. Senescence does not always affect the entire plant at once; for example, as some plants grow taller, they naturally lose their lower leaves. It has been found that senescence of leaves can be prevented by the application of cytokinins. Not only can cytokinins prevent the death of leaves, but they can also initiate leaf growth. Lateral buds begin to grow despite apical dominance when cytokinin is applied to them.

Cytokinins promote cell division, prevent senescence, and initiate growth. The interaction of hormones is well exemplified by the effect of varying ratios of auxin and cytokinins on differentiation of plant tissues.

a.

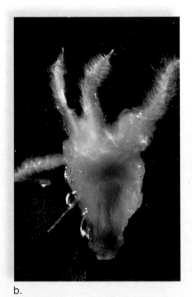

b.

c.

d.

FIGURE 27.11 Interaction of hormones.
Tissue culture experiments have revealed that auxin and cytokinin interact to affect differentiation during development. **a.** In tissue culture that has the usual amounts of these two hormones, tobacco strips develop into a callus of undifferentiated tissue. **b.** If the ratio of auxin to cytokinin is appropriate, the callus produces roots. **c.** Change the ratio, and vegetative shoots and leaves are produced. **d.** Yet another ratio causes floral shoots. It is now clear that each plant hormone rarely acts alone; it is the relative concentrations of hormones that produce an effect. The modern emphasis is to look for an interplay of hormones when a growth response is studied.

Abscisic Acid

Abscisic acid (ABA) is sometimes called the stress hormone because it initiates and maintains seed and bud dormancy and brings about the closure of stomata:

Structure of abscisic acid (ABA)

Dormancy occurs when a plant organ readies itself for adverse conditions by stopping growth (even though conditions at the time are favorable for growth). For example, it is believed that abscisic acid moves from leaves to vegetative buds in the fall, and thereafter these buds are converted to winter buds. A winter bud is covered by thick, hardened scales. A reduction in the level of abscisic acid and an increase in the level of gibberellins are believed to break seed and bud dormancy. Then seeds germinate, and buds send forth leaves.

Abscisic acid brings about the closing of stomata when a plant is under water stress (Fig. 27.12). ABA brings about regulation of channel proteins so that potassium ions (K^+) leave guard cells. Thereafter, the guard cells lose water, and stomata close. It was once believed that abscisic acid functioned in **abscission** [L. *abscissus*, cut off], the dropping of leaves, fruits, and flowers from a plant. Although the external application of abscisic acid promotes abscission, this hormone is no longer believed to function naturally in this process. Instead, the hormone ethylene is thought to bring about abscission (Fig. 27.13). Abscisic acid is produced by any "green tissue" with chloroplasts, monocot endosperm, and roots.

Ethylene

Ethylene is involved in abscission. Low levels of auxin and perhaps gibberellin compared to the stem probably initiate abscission. But once the process of abscission has begun, ethylene stimulates certain enzymes, such as cellulase, which cause leaf, fruit, or flower drop (Fig. 27.13).

Structure of ethylene $H_2C{=}CH_2$

In the early 1900s, it was common practice to prepare citrus fruits for market by placing them in a room with a kerosene stove; only later did researchers realize that an incomplete combustion product of kerosene, namely **ethylene,** ripens fruit. It does so by increasing the activity of enzymes that soften fruits. For example, it stimulates the production of cellulase, an enzyme that hydrolyzes cellulose in plant cell walls.

The use of ethylene in agriculture is extensive. It is used to hasten the ripening of green fruits such as melons and honeydews. It is also applied to citrus fruits to attain pleasing colors before marketing. Ethylene is also used to make members of the pumpkin family produce more female flowers, and thus more fruits.

Because it is a gas, ethylene moves freely through the air—a barrel of ripening apples can induce ripening of a bunch of bananas even some distance away. Ethylene is released at the site of a plant wound due to physical damage or infection (which is why one rotten apple spoils the whole bunch).

Abscisic acid promotes dormancy of buds and seeds; it also brings about the closure of stomata. Ethylene ripens fruits and controls the abscission of leaves, flowers, and fruits.

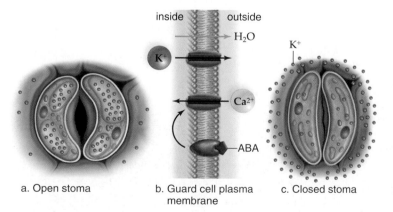

FIGURE 27.12 Abscisic acid control of stoma opening.
a. K^+ is concentrated inside the guard cells and the stoma is open. **b.** Abscisic acid (ABA) binding leads to an influx of Ca^{2+} and the opening of ion channels, including those for K^+. **c.** With the loss of K^+, water (H_2O) exits the guard cells and the stoma closes.

No abscission   Abscission

FIGURE 27.13 Functions of ethylene.
Normally, there is no abscission when a holly twig is placed under a glass jar for a week. When an ethylene-producing ripe apple is also under the jar, abscission of the holly leaves occurs.

27.3 PHOTOPERIODISM

Many physiological changes in plants are related to a seasonal change in day length. Such changes in plants include seed **germination,** the breaking of bud dormancy, and the onset of senescence. A physiological response prompted by changes in the length of day or night is called **photoperiodism** [Gk. *photos*, light, and *periodus*, completed course]. In some plants, photoperiodism influences flowering; for example, violets and tulips flower in the spring, and asters and goldenrod flower in the fall.

In the 1920s, when U.S. Department of Agriculture scientists were trying to improve tobacco, they decided to grow plants in a greenhouse, where they could artificially alter the photoperiod. They came to the conclusion that plants can be divided into three groups:

1. **Short-day plants** flower when the day length is shorter than a *critical length*. (Examples are cocklebur, goldenrod, poinsettia, and chrysanthemum.)
2. **Long-day plants** flower when the day length is longer than a critical length. (Examples are wheat, barley, rose, iris, clover, and spinach.)
3. **Day-neutral plants** are not dependent on day length for flowering. (Examples are tomato and cucumber.)

Further, it should be noted that both a long-day plant and a short-day plant can have the same critical length (Fig. 27.14). Spinach is a long-day plant that has a critical length of 14 hours; ragweed is a short-day plant with the same critical length. Spinach, however, flowers in the summer when the day length increases to 14 hours or more, and ragweed flowers in the fall, when the day length shortens to 14 hours or less. In addition, we now know that some plants may require a specific sequence of day lengths in order to flower.

In 1938, K. C. Hammer and J. Bonner began to experiment with artificial lengths of light and dark that did not necessarily correspond to a normal 24-hour day. They discovered that the cocklebur, a short-day plant, flowers as long as the dark period is continuous for 8.5 hours, regardless of the length of the light period. Further, if this dark period is interrupted by a brief flash of white light, the cocklebur does not flower. (Interrupting the light period with darkness has no effect.) Similar results have been found for long-day plants. They require a dark period that is shorter than a critical length, regardless of the length of the light period. If a slightly longer-than-critical-length night is interrupted by a brief flash of light, however, long-day plants flower. We must conclude, then, that the length of the dark period, not the length of the light period, controls flowering. Of course, in nature, short days always go with long nights, and vice versa. Manipulation of the photoperiod in greenhouses is used to ensure that short-day poinsettias flower in time for the winter holidays.

> In order to flower, short-day plants require a period of darkness that is longer than a critical length, and long-day plants require a period of darkness that is shorter than a critical length.

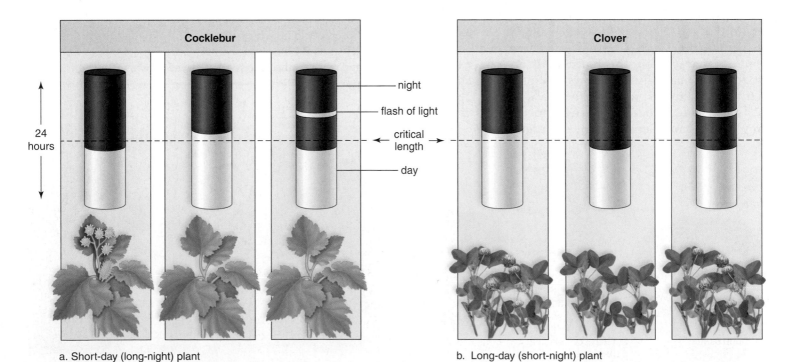

a. Short-day (long-night) plant   b. Long-day (short-night) plant

FIGURE 27.14 Photoperiodism and flowering.
a. Short-day plant. When the day is shorter than a critical length, this type of plant flowers. The plant does not flower when the day is longer than the critical length. It also does not flower if the longer-than-critical-length night is interrupted by a flash of light. **b.** Long-day plant. The plant flowers when the day is longer than a critical length. When the day is shorter than a critical length, this type of plant does not flower. However, it does flower if the slightly longer-than-critical-length night is interrupted by a flash of light.

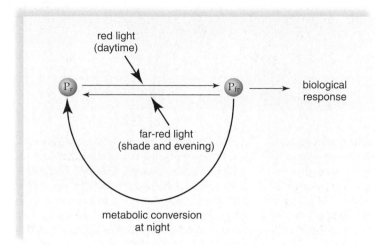

FIGURE 27.15 Phytochrome conversion cycle.
The inactive form P$_r$ is prevalent during the night. At sunset or in the shade, when there is more far-red light, P$_{fr}$ is converted to P$_r$. Also during the night, metabolic processes cause P$_{fr}$ to be converted into P$_r$. P$_{fr}$, the active form of phytochrome, is prevalent during the day because at that time there is more red light than far-red light.

Phytochrome and Plant Flowering

If flowering is dependent on day and night length, plants must have some way to detect these periods. Many years of research by scientists at the U.S. Department of Agriculture led to the discovery of a plant pigment called phytochrome. **Phytochrome** [Gk. *phyton*, plant, and *chroma*, color] is a blue-green leaf pigment that alternately exists in two forms. As Figure 27.15 indicates:

P$_r$ (phytochrome red) absorbs red light (of 660 nm wavelength) and is converted to P$_{fr}$.
P$_{fr}$ (phytochrome far-red) absorbs far-red light (of 730 nm wavelength) and is converted to P$_r$.

Direct sunlight contains more red light than far-red light; therefore, P$_{fr}$ is apt to be present in plant leaves during the day. In the shade and at sunset, there is more far-red light than red light; therefore, P$_{fr}$ is converted to P$_r$ as night approaches. There is a slow metabolic reversion of P$_{fr}$ to P$_r$ during the night.

It is possible that phytochrome conversion is the first step in a reception-transduction-response pathway that results in flowering. At one time, researchers hypothesized that there was a special flowering hormone, called florigen, but such a hormone has never been discovered.

Phytochrome alternates between two forms (P$_{fr}$ during the day and P$_r$ during the night), and this conversion allows a plant to detect photoperiod changes.

Other Functions of Phytochrome

The P$_r$ ⟶ P$_{fr}$ conversion cycle is now known to control other growth functions in plants. P$_{fr}$ promotes seed germi-

nation and inhibits stem elongation, for example. The presence of P$_{fr}$ indicates to some seeds that sunlight is present and conditions are favorable for germination. This is why some seeds must be only partly covered with soil when planted. Germination of other seeds is inhibited by light, so they must be planted deeper. Following germination, the presence of P$_r$ indicates that stem elongation may be needed to reach sunlight. Seedlings that are grown in the dark etiolate—that is, the stem increases in length, and the leaves remain small (Fig. 27.16). Once the seedling is exposed to sunlight and P$_r$ is converted to P$_{fr}$, the seedling begins to grow normally—the leaves expand and the stem branches. It has now been shown that phytochrome in the P$_{fr}$ form leads to the activation of one or more regulatory proteins in the cytoplasm. These DNA-binding proteins migrate to the nucleus where they bind to so-called light-stimulated genes that code for proteins found in chloroplasts.

a. Etiolation

b. Normal growth

FIGURE 27.16 Phytochrome control of growth pattern.
a. If far-red light is prevalent, as it is in the shade, etiolation occurs. **b.** If red light is prevalent, as it is in bright sunlight, normal growth occurs. These effects are due to phytochrome.

science focus

Arabidopsis thaliana, the Valuable Weed

Plants are quite different from animals in many aspects of their development, cell biology, biochemistry, and environmental responses. Plants grow their entire lives and contain meristem tissue that allows them to continuously produce new cells that differentiate into specific tissues. Adult parenchyma cells are totipotent, meaning that each one can easily give rise to a complete plant. Plant biochemistry involves metabolic pathways not seen in animals. Several pathways are needed for various types of photosynthesis and the synthesis of a wide range of chemicals; many of these are used defensively against herbivores and bacterial and fungal pathogens. Plants have their own hormones, which are produced by many tissues and not by organs specialized for this function. Plant hormones, unlike those of animals, allow plants to respond in a unique way to environmental stimuli such as light and gravity. Because plants are unique, a classical and molecular study of their specific genetics is needed.

Until recently, the study of plant genetics was hampered by the lack of a suitable experimental material. Like the fruit fly, *Drosophila,* a suitable material would not take up much laboratory space, but it would produce many offspring within a short time. Enter *Arabidopsis thaliana,* a weed of no food or economic value, even though it is a member of the mustard family, as are cabbages and radishes. Unlike crop plants used formerly, *Arabidopsis* has the characteristics needed to promote the study of both plant classical and molecular genetics. *Arabidopsis* has a short generation time; its entire life cycle takes only about six to eight weeks, and each adult plant may produce 10,000 seeds (potential offspring). Dozens of *Arabidopsis* plants can be grown in a single pot because of its small size (Fig. 27A), and thousands can be grown on a lab bench under fluorescent light, rather than sunlight. A total of 50,000 seeds will fit in a standard 1.5 ml tube. In contrast, crop plants such as corn have generation times of at least several months, and they require a great deal of field space for a large number to grow. In addition to thriving in soil, *Arabidopsis* will grow in a liquid or solid medium whose content can be biochemically controlled.

There are many natural mutants of *Arabidopsis,* and much can be determined by studying them. In one instance, a dwarf plant was found to be deficient in the amount of gibberellin-producing enzymes. Since the plant showed a lack of internodal elongation but showed normal development of leaves and flowers, it was known that gibberellins were affecting only internodal elongation. Working with natural mutagens, researchers discovered three classes of genes that are essential to normal floral pattern formation. These are homeotic genes because they cause sepals, petals, stamens, or carpels to appear in place of one another. Triple mutants that lack all three types of genetic activities have flowers that consist entirely of leaves arranged in whorls (compare Figs. 27B and C). And a mutation of a regulatory gene results in flowers that have three whorls of petals (Fig. 27D). These floral organ-identity genes appear to be regulated by transcription factors that are expressed and required for extended periods.

a.

FIGURE 27A Overall appearance of *Arabidopsis thaliana.*

Many investigators have turned to this weed as an experimental material to study the actions of genes, including those that control growth and development. a. Photograph of actual plant. b. Enlarged drawing.

b.

FIGURE 27B *Arabidopsis thaliana* **flower.**
The structure of the flower is determined in part by three classes of floral organ-identity genes.

FIGURE 27C Mutated flower.
Mutations affecting all three classes of floral organ-identity genes result in flowers that contain only leaves arranged in whorls.

FIGURE 27D Mutated flower.
Mutation of a regulatory gene controlling floral organ-identity genes results in flowers that have three whorls of petals.

Artificial mutagenesis can also be easily accomplished in *Arabidopsis* when the seeds are exposed to chemical mutagens or radiation. *Arabidopsis* cells in tissue culture will take up the plasmid of *Agrobacterium tumefaciens,* the bacterium that can infect certain plants. On occasion, the plasmid inserts itself into the middle of a normal gene, disrupting it and preventing it from functioning as it should. Or the plasmid can be engineered to carry a foreign gene into a host cell. Because the flowers of *Arabidopsis* self-fertilize, a significant number of plants in the next generation will be homozygous following mutagenesis. Many artificially produced mutants have been studied, but we will consider just one example. Following mutagenesis, plants unable to grow in normal levels of CO_2 (they could only grow in high levels of CO_2) were isolated. This classical genetics study became a molecular genetics study when it was discovered that this particular mutation was caused by a gene that codes for a protein

regulating the activity of RuBP carboxylase. It was not previously known that there was such a regulatory protein. By now, the protein has been sequenced, and it is available for other plant cell studies.

A study of the *Arabidopsis* genome will undoubtedly promote plant molecular genetics in general. *Arabidopsis* has just five small chromosomes containing only 70,000 nucleotide base pairs. In contrast, tobacco has 15 times and corn has 30 times the number of nucleotide base pairs as *Arabidopsis*. However, crop plants have about the same number of functional genes as *Arabidopsis*; their excess DNA is noncoding repetitive DNA. Working with the *Arabidopsis* genome, but not the genome of crop plants, investigators can easily find coding genes and determine what these genes do. Remarkably, it has been discovered that many angiosperm plants contain approximately the same coding genes in the same sequence. Therefore, knowledge of

the *Arabidopsis* genome can be used to locate specific genes in the genomes of other plants. Now that the *Arabidopsis* genome has been sequenced, genes of interest can be cloned from the *Arabidopsis* genome and then used as probes for the isolation of the homologous genes from plants of economic value. Also, cellular processes controlled by a family of genes in other plants require only a single gene or fewer genes in *Arabidopsis*. This, too, facilitates molecular biological studies of the plant.

The application of *Arabidopsis* genetics to other plants has been shown. For example, one of the mutant genes that alters the development of flowers has been cloned and reintroduced into tobacco plants where, as expected, it caused sepals and stamens to appear where petals would ordinarily be. The investigators comment that the knowledge about the development of flowers in *Arabidopsis* can have far-ranging applications. It will undoubtedly lead someday to more productive crops.

CONNECTING THE CONCEPTS

Behavior in plants can be understood in terms of three different levels of organization. On the species level, plant responses that promote survival and reproductive success have evolved through natural selection. At the organismal level, hormones coordinate the growth and development of plant parts. And at the cellular level, hormones influence cellular metabolism.

We can illustrate these three perspectives by answering the question, why do plants bend toward the light? On the species level, those plants that bend toward the light will be able to produce more organic food and will have more offspring. On the organismal level, light can cause the movement of auxin in certain plant parts, and when auxin moves from the lit side of a stem to the shady side, elongation occurs; thereafter, the plant bends toward the light. On the cellular level, after auxin is received by a plant cell, cellular activities cause its walls to expand.

The response of both the organism and the cell involve three steps: (1) reception of the stimulus, (2) transduction of the stimulus, and (3) response to the stimulus. Can you designate these steps on the cellular level? Reception of auxin by plasma membrane receptors is the first step, cellular activities is the second step, and stretching of the cell wall is the third step.

If animals were being considered instead of plants, the same type of biological explanations would apply. Plants and animals, and indeed all organisms, share common ancestors, even back to the very first cell(s). Recall that evolution explains both the unity and diversity of living things. Organisms are similar because they share common ancestors; they are different because they are adapted to different ways of life.

Summary

27.1 PLANT RESPONSES

Like animals, plants use a reception-transduction-response pathway when they respond to a stimulus.

When plants respond to stimuli, growth and/or movement occurs. Tropisms are growth responses toward or away from unidirectional stimuli. The positive phototropism of stems results in a bending toward light, and the negative gravitropism of stems results in a bending away from the direction of gravity. Roots that bend toward the direction of gravity show positive gravitropism. Thigmotropism occurs when a plant part makes contact with an object, as when tendrils coil about a pole.

Nastic movements are not directional. Due to turgor pressure changes, some plants respond to touch and some perform sleep movements. Plants exhibit circadian rhythms, which are believed to be controlled by a biological clock. The sleep movements of prayer plants, the closing of stomata, and the daily opening of certain flowers have a 24-hour cycle.

27.2 PLANT HORMONES

Auxin-controlled cell elongation is involved in phototropism and gravitropism. When a plant is exposed to light, auxin moves laterally from the bright to the shady side of a stem. Binding of auxin to a plasma membrane receptor leads to (1) activation of a H^+ pump and a loosening of cellulose fibrils in the cell wall; (2) production of cell wall materials by the Golgi apparatus; and (3) production of growth factors following gene activation.

Gibberellin causes stem elongation between nodes. After this hormone binds to a plasma membrane receptor, a DNA-binding protein activates a gene leading to the production of amylase. Amylase is an enzyme which speeds the break down of amylase.

Cytokinins cause cell division, the effects of which are especially obvious when plant tissues are grown in culture. Abscisic acid (ABA) and ethylene are two plant growth inhibitors. ABA is well known for causing stomata to close, and ethylene is known for causing fruits to ripen.

27.3 PHOTOPERIODISM

Photoperiodism is seen in some plants. For example, short-day plants flower only when the days are shorter than a critical length, and long-day plants flower only when the days are longer than a critical length. Actually, research has shown that it is the length of darkness that is critical. Interrupting the dark period with a flash of white light prevents flowering in a short-day plant and induces flowering in a long-day plant.

Phytochrome is a pigment that responds to both red and far-red light and is involved in flowering. Daylight causes phytochrome to exist as P_{fr}, but during the night, it is reconverted to P_r by metabolic processes. Phytochrome in the P_{fr} form leads to a biological response such as flowering.

P_{fr} also promotes seed germination, leaf expansion, and stem branching. When P_r dominates, the stem elongates and grows toward sunlight.

Reviewing the Chapter

1. Tropisms are responses to stimuli. Describe the pathway that plants use to respond to stimuli leading to tropisms. Why are stems said to exhibit positive phototropism but negative gravitropism? 478–79
2. What are nastic movements, and how do turgor pressure changes bring about movements of plants? 480
3. What is a biological clock, how does it function, and what is its primary usefulness in plants? 481
4. Why does removing a terminal bud cause a plant to get bushier? 482
5. What experiments led to knowledge that a hormone is involved in phototropism? Explain the mechanism by which auxin brings about elongation of cells. 482
6. Gibberellin research supports the hypothesis that plant hormones initiate a reception-transduction-response pathway. Explain. 484–85
7. What is the function of cytokinins? Discuss experimental evidence to suggest that hormones interact when they bring about an effect. 486
8. What are some of the primary effects of abscisic acid and how does it bring about these effects? 487
9. What hormones are involved in abscission? How does ethylene bring about ripening of fruits? 487
10. Define photoperiodism, and discuss its relationship to flowering in certain plants. 488
11. What is the phytochrome conversion cycle, and what are some possible functions of phytochrome in plants? 489

Testing Yourself

Choose the best answer for each question. For questions 1–5, match the statements with the hormones in the key.

KEY:
   a. auxin
   b. gibberellin
   c. cytokinin
   d. ethylene
   e. abscisic acid

 1. One rotten apple can spoil the barrel.

 2. Cabbage plants bolt (grow tall).

 3. Stomata close when a plant is water stressed.

 4. Sunflower plants all point toward the sun.

 5. Coconut milk causes plant tissues to undergo cell division.

 6. After unidirectional light is received, auxin moves to the shady side of a stem, and then the stem bends toward the light. Bending toward light is best associated with
   a. the reception step.
   b. the transduction pathway step.
   c. the response step.
   d. All of these are correct.

 7. Which of these is a correct statement?
   a. Both stems and roots show positive gravitropism.
   b. Both stems and roots show negative gravitropism.
   c. Only stems show positive gravitropism.
   d. Only roots show positive gravitropism.
   e. Only roots show negative gravitropism.

 8. The short-day plant
   a. is the same as the long-day plant.
   b. is apt to flower in the fall.
   c. photoperiod must be longer than a certain length.
   d. photoperiod must be shorter than a certain length.
   e. Both b and d are correct.

 9. A plant requiring a dark period of at least 14 hours will
   a. flower if a 14-hour night is interrupted by a flash of light.
   b. not flower if a 14-hour night is interrupted by a flash of light.
   c. not flower if the days are 14 hours long.
   d. not flower if the nights are longer than 14 hours.
   e. Both b and c are correct.

10. Phytochrome
   a. is a plant pigment.
   b. is present as P_{fr} during the day.
   c. activates DNA-binding proteins.
   d. is a photoreceptor.
   e. All of these are correct.

11. Circadian rhythms
   a. require a biological clock.
   b. do not exist in plants.
   c. are involved in the tropisms.
   d. are involved in sleep movements.
   e. Both a and d are correct.

12. In phototropism, binding of auxin
   a. leads to growth of cells on different sides of the plant.
   b. has no effect on growth of cells.
   c. inhibits growth of cells.
   d. leads to inhibition of cell division on the shady side.
   e. absorbs stimuli and signals the direction of light.

13. The folding of the leaflets of sensitive *Mimosa* plants is classified as a nastic response because
   a. it releases distasteful chemicals.
   b. it involves action potentials.
   c. it occurs the same way no matter which side the plant was touched on.
   d. it is mediated by auxin.
   e. it involves growth on different sides of the plant.

14. The flowering of certain plants only under short- or long-day conditions is due to
   a. high concentrations of ethylene.
   b. differences in temperature.
   c. the onset of dormancy.
   d. photoperiodism.
   e. nastic movements.

15. Phytochrome is normally switched from one form to the other by
   a. red and far-red light in sunlight.
   b. activation of gibberellin or auxin.
   c. different electrical potentials in the plant's plasma membranes and nervous system.
   d. circadian rhythms.

16. The most reliable seasonal cue for plants is usually
   a. temperature.
   b. light.
   c. moisture.
   d. All of the above are correct.

17. A pigment involved in the control of flowering is
   a. phytochrome.
   b. chlorophyll *a*.
   c. lycopene.
   d. beta-carotene.
   e. cytochrome.

18. Nastic movements differ from tropisms in that
   a. tropisms involve growth toward or away from a stimulus, while nastic movements involve growth away from a stimulus only.
   b. nastic movements are independent of the direction of the stimulus.
   c. nastic movements usually take longer than tropisms.
   d. tropisms do not involve any chemical or hormonal changes in the plant.

19. The sensors in the cells of the root cap are called
   a. mitochondria.
   b. central vacuoles.
   c. statoliths.
   d. chloroplasts.
   e. intermediate filaments.

20. A shrub will become fuller in appearance if
   a. the lateral buds are removed.
   b. the root is pruned.
   c. the apical buds are removed from the branches.
   d. the flowers are removed from the lateral branches.

21. Which of these are related to gibberellin activity?
    a. stem elongation
    b. initiation of bud dormancy
    c. repression of amylase production
    d. All of these are correct.
    e. Both b and c are correct.

22. Ethylene
    a. is a gas.
    b. causes fruit to ripen.
    c. is produced by the incomplete combustion of fuels such as kerosene.
    d. All of these are correct.

23. A student places 25 pea seeds in a large pot and allows the seeds to germinate in total darkness. Which of these growth or movement activities would the seedlings exhibit?
    a. gravitropism, as the roots grow down and the shoots grow up
    b. phototropism, as the shoots search for light
    c. thigmotropism, as the tendrils coil around other seedlings
    d. Both a and c are correct.

24. Internode elongation is stimulated by
    a. abscisic acid.
    b. ethylene.
    c. cytokinin.
    d. gibberellin.
    e. auxin.

25. A tissue culture scientist who wants to obtain shoot formation from a callus culture should transfer his callus to a medium that contains
    a. lots of ethylene.
    b. more auxin than cytokinin.
    c. more ABA than auxin.
    d. more cytokinin than auxin.

26. Plants that flower in response to long nights are
    a. day-neutral plants.
    b. long-day plants.
    c. short-day plants.
    d. impossible.

27. Label this diagram. Explain what causes stomata to close when a plant is water stressed.

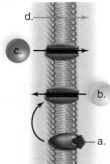

Thinking Scientifically

1. You wish to test the hypothesis that the gravitropic response is stronger than the phototropic response in stems. How could this hypothesis be tested, and what results would be predicted?

2. Gibberellins are responsible for cell elongation between nodes of a stem. Miniature, or dwarf, varieties of plants have short internode cells. Remembering that a hormone only works after it binds to a receptor protein, propose at least two different hypotheses that could explain why a plant could fail to give a hormonal response.

Bioethical Issue: Environmental Activism

The government paid the owner of Pacific Lumber some $500 million so that 3,500 acres of redwood trees along the coast of the Pacific Northwest would be preserved. Activists thought the deal was unfair—they wanted 60,000 acres preserved. Any less and there would not be enough habitat to keep the marbeled murrelets, an endangered seabird, from sinking into oblivion. Besides, it would take hundreds of years for the trees to regrow to their original size.

Activists feel betrayed by their own representatives in the government. After all, the owner of Pacific Lumber is a big contributor to the reelection campaigns of elected officials who approved the deal. Under the circumstances, what would you do to save the trees? Some activists climbed the trees and refused to come down even while the trees were being cut. In September of 1998, David Chain, 24, lost his life when a tree fell on him and crushed his skull. What is the proper action for activist groups when they feel their government is letting them down? Should they defy the law, and if so, what is the proper response of government officials?

Understanding the Terms

abscisic acid (ABA) 487	gibberellin 484
abscission 487	gravitropism 479
apical dominance 482	hormone 482
auxin 482	long-day plant 488
biological clock 481	photoperiodism 488
circadian rhythm 480	phototropism 479
coleoptile 482	phytochrome 489
cytokinin 486	senescence 486
day-neutral plant 488	short-day plant 488
dormancy 484	statolith 479
ethylene 487	thigmotropism 480
germination 488	tropism 478

Match the terms to these definitions:

a. _____ Biological rhythm with a 24-hour cycle.

b. _____ Directional growth of plants in response to the Earth's gravity.

c. _____ Dropping of leaves, fruits, or flowers from a plant.

d. _____ Plant hormone producing increased stem growth between nodes; also involved in flowering and seed germination.

e. _____ Relative lengths of daylight and darkness that affect the physiology and behavior of an organism.

ARIS, the *Biology* Website

ARIS, the website for *Biology*, provides a wealth of information organized and integrated by chapter. You will find practice quizzes, interactive activities, labeling exercises, flashcards, and much more that will complement your learning and understanding of general biology.

www.mhhe.com/maderbiology9

28

REPRODUCTION IN PLANTS

*T*he flowering plants, or angiosperms, are the most diverse and widespread of all the plants, and they have a means of sexual reproduction that is well adapted to life on land. Pollen protects the male gametophyte until fertilization takes place, and seeds protect the embryo until conditions are favorable for germination. The presence of an ovary allows angiosperms to produce seeds within fruits. The evolution of the flower led to pollination not only by wind but also by animals. Flowering plants that rely on animals for pollination have a mutualistic relationship with them. The flower provides nutrients for a pollinator such as a bee, a fly, a beetle, a bird, or even a bat. The animals in turn inadvertently carry pollen from one flower to another, allowing pollination to occur.

Many flowering plants can also reproduce asexually because their cells, especially meristem cells, are totipotent—able to become a complete plant. Totipotency makes it possible for scientists to reproduce plants in laboratory cultures, starting with individual cells or with any organ of a plant. Plants grown from tissue cultures can be endowed with foreign genes that give them new and different characteristics of interest to human beings.

Cherry blossom, *Prunus*, with honeybee, *Apis*.

28.1 REPRODUCTIVE STRATEGIES

In contrast to animals, which have only one multicellular stage in their life cycle, all plants have two. A diploid stage alternates with a haploid one. The diploid plant, called the **sporophyte,** produces haploid spores by meiosis. The spores divide by mitosis to become haploid **gametophytes** that produce gametes by mitosis. In flowering plants, the sporophyte is dominant, and it is the generation that bears flowers (Fig. 28.1).

A **flower,** which is the reproductive structure of angiosperms, produces two types of spores, microspores and megaspores. A **microspore** [Gk. *mikros,* small, little] develops into a male gametophyte; a **megaspore** [Gk. *megas,* great, large] develops into a female gametophyte. A microspore undergoes mitosis and becomes a pollen grain, which is either windblown or carried by an animal to the vicinity of the female gametophyte. In the meantime, the megaspore has undergone mitosis to become the female gametophyte, an embryo sac located within an ovule found within an ovary. At maturity, a pollen grain contains nonflagellated sperm, which travel by way of a pollen tube to the embryo sac. Once a sperm fertilizes an egg, the zygote becomes an embryo, still within an ovule. The ovule develops into a **seed,** which contains the embryo and stored food surrounded by a seed coat. The ovary becomes a fruit, which aids in dispersing the seeds. When a seed germinates, a new sporophyte emerges and develops into a mature organism.

The life cycle of flowering plants is adapted to a land existence. The microscopic gametophytes develop within the sporophyte and are thereby protected from desiccation. Pollen grains are not released until they develop a thick wall. No external water is needed to bring about fertilization in flowering plants. Instead, the pollen tube provides passage for a sperm to reach an egg. Following fertilization, the embryo and its stored food are enclosed within a protective seed coat until external conditions are favorable for germination.

The flower is unique to angiosperms. Aside from producing the spores and protecting the gametophytes, flowers often attract pollinators, which aid in transporting pollen from plant to plant. Flowers also produce the fruits that enclose the seeds. The evolution of the flower was a major factor leading to the success of angiosperms, with over 240,000 species.

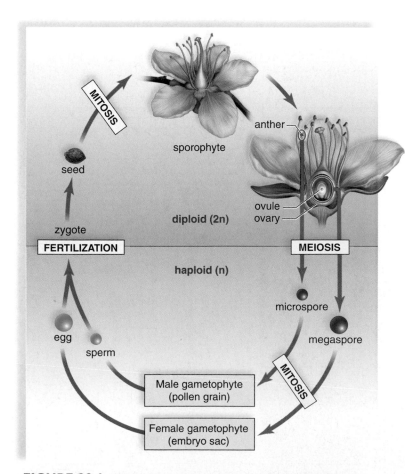

FIGURE 28.1 Alternation of generations in flowering plants.
The sporophyte bears flowers. The flower produces microspores within anthers and megaspores within ovules by meiosis. A megaspore becomes a female gametophyte, which produces an egg within an embryo sac, and a microspore becomes a male gametophyte (pollen grain), which produces sperm. Fertilization results in a seed-enclosed zygote and stored food.

Flowers

A flower (Fig. 28.2) develops in response to environmental signals such as the length of the day (see page 488). In many plants, shoot apical meristem that previously formed leaves suddenly stops producing leaves and starts producing a flower enclosed within a bud. In other plants, axillary buds develop directly into flowers. In monocots, flower parts occur in threes and multiples of three; in eudicots, flower parts are in fours or fives and multiples of four or five (Fig. 28.3).

A typical flower has four whorls of modified leaves attached to a receptacle at the end of a flower stalk. The receptacle is attached to a peduncle if it bears a single flower but is called a pedicel if it bears one of several flowers.

1. The **sepals,** which are the most leaflike of all the flower parts, are usually green, and they protect the bud as the flower develops within.
2. An open flower next has a whorl of **petals,** whose color accounts for the attractiveness of many flowers. The size, the shape, and the color of petals are attractive to a specific pollinator. Wind-pollinated flowers may have no petals at all.
3. **Stamens** are the "male" portion of the flower. Each stamen has two parts: the **anther,** a saclike container, and the **filament** [L. *filum,* thread], a slender stalk. Pollen grains develop from the microspores produced in the anther.
4. At the very center of a flower is the **carpel,** a vaselike structure that represents the "female" portion of the flower. A carpel usually has three parts: the **stigma,** an enlarged sticky knob; the **style,** a slender stalk; and the **ovary,** an enlarged base that encloses one or more ovules.

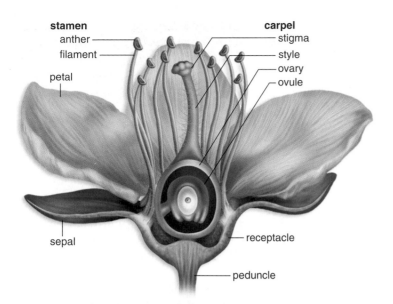

FIGURE 28.2 Anatomy of a flower.
A complete flower has all flower parts: sepals, petals, stamens, and at least one carpel.

A flower can have a single carpel or multiple carpels. Sometimes several carpels are fused into a single structure, in which case the ovary is compound and has several chambers, each of which contains ovules. For example, an orange develops from an ovary, and every section of the orange is a chamber.

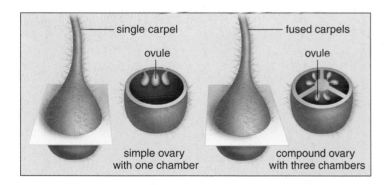

An ovary usually contains a number of **ovules** [L. *ovulum,* little egg], which play a significant role in the production of megaspores and, therefore, female gametophytes.

Not all flowers have sepals, petals, stamens, or carpels. Those that do are said to be complete (see Fig. 28.2), and those that do not are said to be incomplete. Flowers that have both stamens and carpels are called perfect (bisexual) flowers; those with only stamens and those that have only carpels are imperfect (unisexual) flowers. If staminate flowers and carpellate flowers are on one plant, the plant is monoecious [Gk. *monos,* one, and *oikos,* home, house] (Fig. 28.4). If staminate and carpellate flowers are on separate plants, the plant is dioecious. Holly trees are dioecious, and if red berries are a priority, it is necessary to acquire a plant with staminate flowers and another plant with carpellate flowers. Corn is an example of a plant that is monoecious.

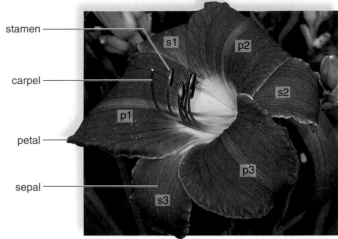

a. Daylily, *Hemerocallis* sp.

b. Festive azalea, *Rhododendron* sp.

FIGURE 28.3 Monocot versus eudicot flowers.
a. Monocots, such as daylilies, have flower parts usually in threes. In particular, note the three petals and three sepals. **b.** Azaleas are eudicots. They have flower parts in fours or fives; note the five petals of this flower. p = petal, s = sepal

a. Staminate flowers   b. Carpellate flowers

FIGURE 28.4 Corn plants are monoecious.
A corn plant has clusters of staminate flowers (**a**) and carpellate flowers (**b**). Staminate flowers produce the pollen that is carried by wind to the carpellate flowers, where an ear of corn develops.

Life Cycle of Flowering Plants

In plants, the sporophyte produces haploid spores by meiosis. The haploid spores grow and develop into haploid gametophytes, which produce gametes by mitotic division.

Flowering plants are heterosporous—they produce microspores and megaspores. Microspores become mature male gametophytes (sperm-bearing pollen grains), and megaspores become mature female gametophytes (egg-bearing embryo sacs).

Development of Male Gametophyte

Microspores are produced in the anthers of flowers (Fig. 28.5). An anther has four pollen sacs, each containing many **microsporocytes** (microspore mother cells). A microsporocyte undergoes meiosis to produce four haploid microspores. In each, the haploid nucleus divides mitotically, followed by unequal cytokinesis, and the result is two cells enclosed by a finely sculptured wall. This structure, called the **pollen grain,** is at first an immature **male gametophyte**

that consists of a tube cell and a generative cell. The larger tube cell will eventually produce a *pollen tube.* The smaller generative cell divides mitotically either now or later to produce two sperm. Once these events take place, the pollen grain has become the mature male gametophyte.

Development of Female Gametophyte

The ovary contains one or more ovules. An ovule has a central mass of parenchyma cells almost completely covered by layers of tissue called integuments except where there is an opening, the micropyle. One parenchyma cell enlarges to become a **megasporocyte** (megaspore mother cell), which undergoes meiosis, producing four haploid megaspores (Fig. 28.5). Three of these megaspores are nonfunctional, and one is functional. In a typical pattern, the nucleus of the functional megaspore divides mitotically until there are eight nuclei in the **female gametophyte** [Gk. *megas,* great, large]. When cell walls form later, there are seven cells, one

FIGURE 28.5 Life cycle of flowering plants.
Development of gametophytes: A pollen sac in the anther contains microsporocytes, which produce microspores by meiosis. A microspore develops into a pollen grain which germinates and has two sperm. An ovule in an ovary contains a megasporocyte, which produces a megaspore by meiosis. A megaspore develops into an embryo sac containing seven cells, one of which is an egg. *Development of sporophyte:* A pollen grain contains two sperm by the time it germinates and forms a pollen tube. During double fertilization, one sperm fertilizes the egg to form a diploid zygote, and the other fuses with the polar nuclei to form a triploid (3n) endosperm cell. A seed contains the developing embryo plus stored food.

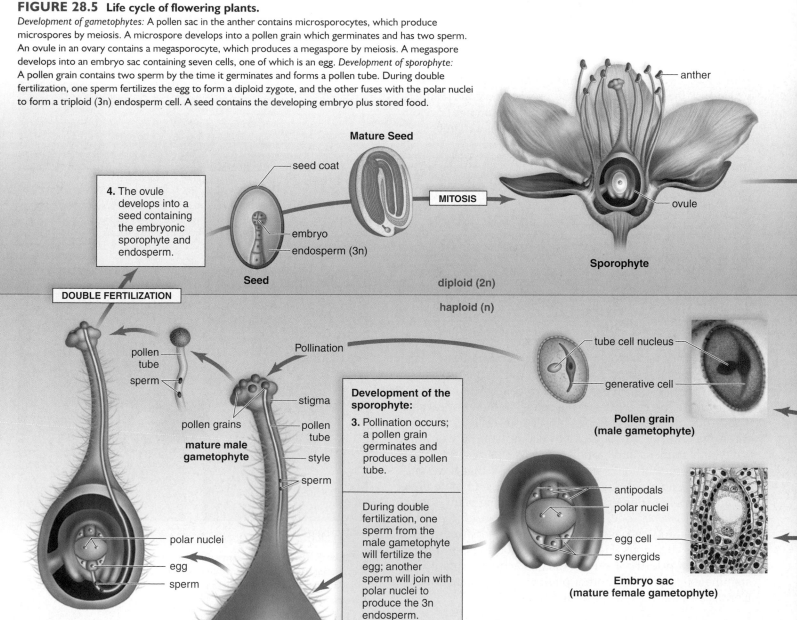

of which is binucleate. The female gametophyte, also called the **embryo sac,** consists of these seven cells:

> one egg cell, associated with
> two synergid cells;
> one central cell, with two polar nuclei;
> three antipodal cells

Development of Sporophyte

The walls separating the pollen sacs in the anther break down when the pollen grains are ready to be released (Fig. 28.6). **Pollination** is simply the transfer of pollen from an anther to the stigma of a carpel. Self-pollination occurs if the pollen is from the same plant, and cross-pollination occurs if the pollen is from a different plant of the same species.

Angiosperms often have adaptations to foster cross-pollination; for example, the carpels may mature only after the anthers have released their pollen. As discussed in the Science Focus on pages 500–501, cross-pollination may also be brought about with the assistance of a particular pollinator. If a pollinator goes from flower to flower of only one type of plant, cross-pollination is more likely to occur in an efficient

FIGURE 28.6 Pollination.

a. Cocksfoot grass, *Dactylus glomerata*, releasing pollen. **b.** Pollen grains of Canadian goldenrod, *Solidago canadensis*. **c.** Pollen grains of pussy willow, *Salix discolor*. The shape and pattern of pollen grain walls are quite distinctive, and experts can use them to identify the genus, and even sometimes the species, that produced a particular pollen grain. Pollen grains have strong walls resistant to chemical and mechanical damage; therefore, they frequently become fossils.

manner. Color, odor, and secretion of nectar are ways that pollinators are attracted to plants. Over time, certain pollinators have become adapted to reach the nectar of only one type of flower. In the process, pollen is inadvertently picked up and taken to another plant of the same type.

When a pollen grain lands on the stigma of the same species, it germinates, forming a pollen tube (see Fig. 28.5). The germinated pollen grain, containing a tube cell and two sperm, is the mature male gametophyte. As it grows, the pollen tube passes between the cells of the stigma and the style to reach the micropyle, a pore of the ovule. Now **double fertilization** occurs. One sperm nucleus unites with the egg nucleus, forming a 2n zygote, and the other sperm nucleus migrates and unites with the polar nuclei of the central cell, forming a 3n endosperm cell. The zygote divides mitotically to become the **embryo,** a young sporophyte, and the endosperm cell divides mitotically to become the endosperm. **Endosperm** [Gk. *endon*, within, and *sperma*, seed] is the tissue that will nourish the embryo and seedling as they undergo development.

Development of the male gametophyte:	Development of the female gametophyte:
1. In pollen sacs of the anther, a microsporocyte undergoes meiosis to produce 4 microspores each.	In an ovule with an ovary, a megasporocyte undergoes meiosis to produce 4 megaspores.

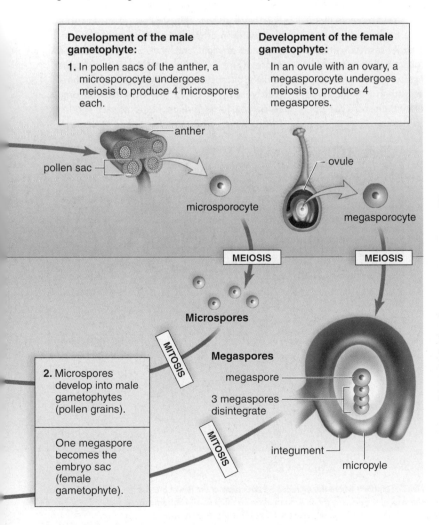

anther

pollen sac

ovule

microsporocyte

megasporocyte

MEIOSIS   MEIOSIS

Microspores

MITOSIS

Megaspores

megaspore

3 megaspores disintegrate

integument

micropyle

MITOSIS

2. Microspores develop into male gametophytes (pollen grains).

One megaspore becomes the embryo sac (female gametophyte).

Flowering plants are heterosporous. Microspores develop into sperm-bearing mature male gametophytes. A megaspore develops into an egg-bearing mature female gametophyte. During double fertilization, one sperm nucleus unites with the egg nucleus, producing a zygote, and the other unites with the polar nuclei, forming a 3n endosperm cell.

science focus

Plants and Their Pollinators

A plant and its pollinator(s) are adapted to one another. They have a mutualistic relationship in which each benefits—the plant uses its pollinator to ensure that cross-pollination takes place, and the pollinator uses the plant as a source of food. This mutualistic relationship came about through the process of **coevolution**—that is, the codependency of the plant and the pollinator is the result of suitable changes in the structure and function of each. The evidence for coevolution is observational. For example, floral coloring and odor are suited to the sense perceptions of the pollinator; the mouthparts of the pollinator are suited to the structure of the flower; the type of food provided is suited to the nutritional needs of the pollinator; and the pollinator forages at the time of day that specific flowers are open. The following are examples of such coevolution.

Bee-Pollinated Flowers

There are 20,000 known species of bees that pollinate flowers. The best-known pollinators are the honeybees (Fig. 28Aa). Bee eyes see a spectrum of light that is different from the spectrum seen by humans. The bees' visible spectrum is shifted so that they do not see red wavelengths but do see ultraviolet wavelengths. Bee-pollinated flowers are usually brightly colored and are predominantly blue or yellow; they are not entirely red. They may also have ultraviolet shadings called nectar guides, which highlight the portion of the flower that contains the reproductive structures. The mouthparts of bees are fused into a long tube that contains a tongue. This tube is an adaptation for sucking up nectar provided by the plant, usually at the base of the flower.

Bee flowers are delicately sweet and fragrant to advertise that nectar is present. The nectar guides often point to a narrow floral tube large enough for the bee's feeding apparatus but too small for other insects to reach the nectar. Bees also collect pollen as food for their larvae. Pollen clings to the hairy body of a bee, and the bees also gather it by means of bristles on their legs. They then store the pollen in pollen baskets on the third pair of legs.

Bee-pollinated flowers are sturdy and irregular in shape because they often have a landing platform where the bee can alight. The landing platform requires the bee to brush up against the anther and stigma as it moves toward the floral tube to feed. One type of orchid, *Ophrys*, has evolved a unique adaptation. The flower resembles a female wasp, and when the male of that species attempts to copulate with the flower, the wasp receives pollen.

Moth- and Butterfly-Pollinated Flowers

Contrasting moth- and butterfly-pollinated flowers emphasizes the close adaptation between pollinator and flower. Both moths and butterflies have a long, thin, hollow proboscis, but they differ in other characteristics. Moths usually feed at night and have a well-developed sense of smell. The flowers they visit are visible at night because they are lightly shaded (white, pale yellow, or pink), and they have strong, sweet perfume, which helps attract moths. Moths hover when they feed, and their flow-

a.

b.

FIGURE 28A Pollinators.
*a. A bee-pollinated flower is a color other than red (bees cannot detect this color) and has a landing platform where the reproductive structures of the flower brush up against the bee's body. **b.** A butterfly-pollinated flower is often a composite, containing many individual flowers. The broad expanse provides room for the butterfly to land, after which it lowers its proboscis into each flower in turn. **c.** Hummingbird-pollinated flowers are curved back, allowing the bird to insert its beak to reach the rich supply of nectar. While doing this, the bird's forehead and other body parts touch the reproductive structures. **d.** Bat-pollinated flowers are large, sturdy flowers that can take rough treatment. Here the head of the bat is positioned so that its bristly tongue can lap up nectar.*

ers have deep tubes with open margins that allow the hovering moths to reach the nectar with their long proboscis. Butterflies are active in the daytime and have good vision but a weak sense of smell. Their flowers have bright colors—even red because butterflies can see the color red—but the flowers tend to be odorless. Unable to hover, butterflies need a place to land. Flowers that are visited by butterflies often have flat landing platforms (Fig. 28A*b*). Composite flowers (composed of a compact head of numerous individual flowers) are especially favored by butterflies. Each flower has a long, slender floral tube, accessible to the long, thin butterfly proboscis.

Bird- and Bat-Pollinated Flowers

In North America, the most well-known bird pollinators are the hummingbirds. These tiny animals have good eyesight but do not have a well-developed sense of smell. Like moths, they hover when they feed. Typical flowers pollinated by hummingbirds are red, with a slender floral tube and margins that are curved back and out of the way. And although they produce copious amounts of nectar, the flowers have little odor. As a hummingbird feeds on nectar with its long, thin beak, its head comes into contact with the stamens and pistil (Fig. 28A*c*).

Bats are adapted to gathering food in various ways, including feeding on the nectar and pollen of plants. Bats are nocturnal and have an acute sense of smell. Those that are pollinators also have keen vision and a long, extensible, bristly tongue. Typically, bat-pollinated flowers open only at night and are light-colored or white. They have a strong, musty smell similar to the odor that bats produce to attract one another. The flowers are generally large and sturdy and are able to hold up when a bat inserts part of its head to reach the nectar. While the bat is at the flower, its head is dusted with pollen (Fig. 28A*d*).

Coevolution

There are many examples of plant and animal coevolution, but how did this coevolution come about? Some 200 million years ago, when seed plants were just beginning to evolve and insects were not as diverse as they are today, wind alone was used to carry pollen. Wind pollination, however, is a hit-or-miss affair. Perhaps beetles feeding on vegetative leaves were the first insects to carry pollen directly from plant to plant by chance. This use of animal motility to achieve cross-fertilization no doubt resulted in the evolution of flowers, which have features, such as the production of nectar, to attract pollinators. Then, if beetles developed the habit of feeding on flowers, other features, such as the protection of ovules within ovaries, may have evolved.

As cross-fertilization continued, more and more flower variations likely developed, and pollinators became increasingly adapted to specific angiosperm species. Today, there are some 240,000 species of flowering plants and over 900,000 species of insects. This diversity suggests that the success of angiosperms has contributed to the success of insects, and vice versa.

c.

d.

28.2 SEED DEVELOPMENT

Development of the embryo within the seed is the next event in the life cycle of the angiosperm. Plant growth and development involves cell division, cell elongation, and differentiation of cells into tissues and then organs. Development is a programmed series of stages from a simple to a more complex form. Cellular differentiation, or specialization of structure and function, occurs as development proceeds.

Development of the Eudicot Embryo

Immediately after double fertilization, the endosperm cell begins to divide, forming endosperm tissue (Fig. 28.7). The zygote divides, but asymmetrically. One of the resulting cells is small with dense cytoplasm. This cell is destined to become the embryo, and it divides repeatedly in different planes, forming several cells called a proembryo. The other larger cell also divides repeatedly, but it forms an elongated structure called a suspensor, which has a basal cell. The suspensor, which anchors the embryo and transfers nutrients to it from the sporophyte plant, will disintegrate later. Suspensor development and function are independent of endosperm development.

Globular Stage

During the globular stage, the proembryo is largely a ball of cells. The root-shoot axis of the embryo is already established at this stage because the embryonic cells near the suspensor will become a root, while those at the other end will ultimately become a shoot.

The outermost cells of the plant embryo will become dermal tissue. These cells divide with their cell plate perpendicular to the surface; therefore, they produce a single outer layer of cells. Recall that dermal tissue protects the plant from desiccation and includes the stomata, which open and close to facilitate gas exchange and minimize water loss.

The Heart Stage and Torpedo Stage Embryos

The embryo has a heart shape when the **cotyledons** [Gk. *cotyledon*, cup-shaped cavity], or seed leaves, appear because of local, rapid cell division. As the embryo continues to enlarge and elongate, it takes on a torpedo shape. Now the root and shoot apical meristems are distinguishable. The shoot apical meristem is responsible for aboveground growth, and the root apical meristem is responsible for underground growth. Ground meristem, which gives rise to the bulk of the embryonic interior, is also now present.

The Mature Embryo

In the mature embryo, the epicotyl is the portion between the cotyledon(s) that contributes to shoot development. The plumule is found at the tip of the epicotyl and consists of the shoot tip and a pair of small leaves. The hypocotyl is the portion below the cotyledon(s). It contributes to stem development and terminates in the radicle or embryonic root.

The cotyledons are quite noticeable in a eudicot embryo and may fold over. Procambium at the core of the embryo is destined to form the future vascular tissue responsible for water and nutrient transport.

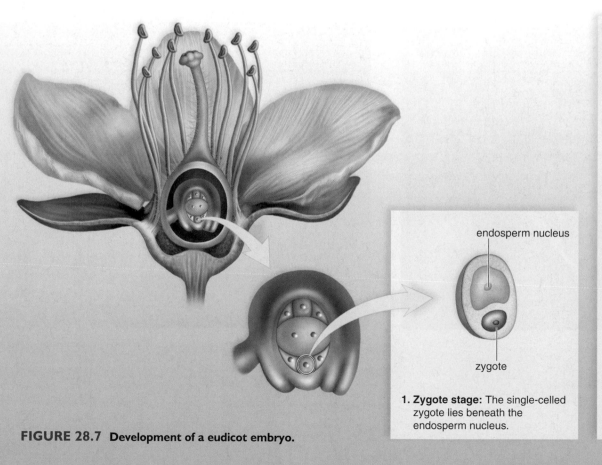

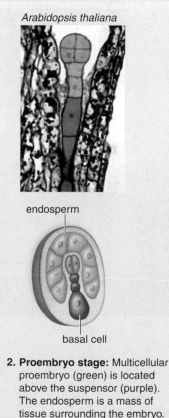

Arabidopsis thaliana

endosperm nucleus

zygote

1. Zygote stage: The single-celled zygote lies beneath the endosperm nucleus.

endosperm

basal cell

2. Proembryo stage: Multicellular proembryo (green) is located above the suspensor (purple). The endosperm is a mass of tissue surrounding the embryo.

FIGURE 28.7 Development of a eudicot embryo.

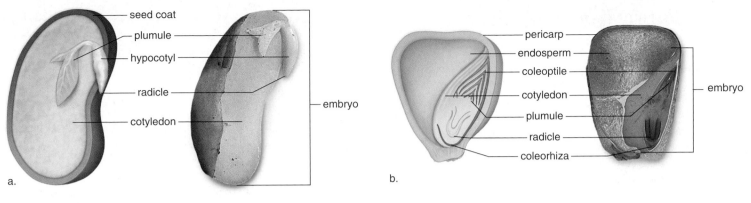

FIGURE 28.8 Monocot versus eudicot.
a. In a bean seed (eudicot), the endosperm has disappeared; the bean embryo's cotyledons take over food storage functions. **b.** The corn kernel (monocot) has endosperm that is still present at maturity.

As the embryo develops, the integuments of the ovule become the seed coat. The seed coat encloses and protects the embryo and its food supply.

Monocots Versus Eudicots

Monocots, unlike eudicots, have only one cotyledon. Another important difference between monocots and eudicots is the manner in which nutrient molecules are stored in the seed. In monocots, the cotyledon, in addition to storing certain nutrients, absorbs other nutrient molecules from the endosperm and passes them to the embryo. In eudicots, the cotyledons usually store all the nutrient molecules that the embryo uses. Therefore, in Figure 28.7 we can see that the endosperm seemingly disappears. Actually, it has been taken up by the two cotyledons.

Figure 28.8 contrasts the structure of a bean seed (eudicot) and a corn kernel (monocot). The size of seeds may vary from the dust-sized seeds of orchids to the 27 kg seed of the double coconut.

The embryo plus its stored food is contained within a seed coat.

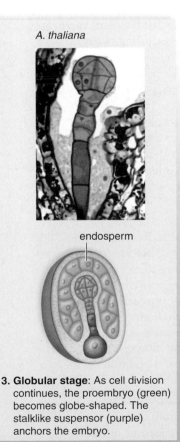

A. thaliana

endosperm

3. Globular stage: As cell division continues, the proembryo (green) becomes globe-shaped. The stalklike suspensor (purple) anchors the embryo.

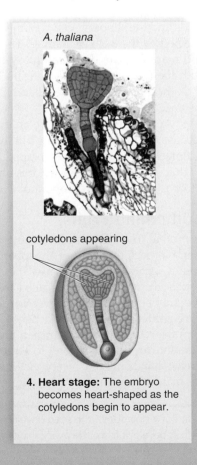

A. thaliana

cotyledons appearing

4. Heart stage: The embryo becomes heart-shaped as the cotyledons begin to appear.

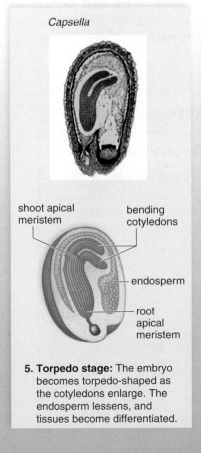

Capsella

shoot apical meristem

bending cotyledons

endosperm

root apical meristem

5. Torpedo stage: The embryo becomes torpedo-shaped as the cotyledons enlarge. The endosperm lessens, and tissues become differentiated.

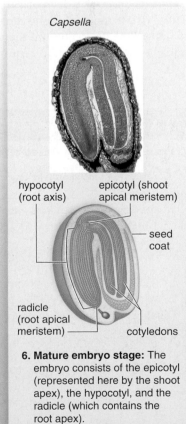

Capsella

hypocotyl (root axis)

epicotyl (shoot apical meristem)

seed coat

radicle (root apical meristem)

cotyledons

6. Mature embryo stage: The embryo consists of the epicotyl (represented here by the shoot apex), the hypocotyl, and the radicle (which contains the root apex).

28.3 FRUIT TYPES AND SEED DISPERSAL

Fruits can be classified based on composition and texture (Table 28.1). A **fruit** is derived from an ovary and sometimes other flower parts (Table 28.2). As a fruit develops, the ovary wall thickens to become the pericarp, which can have as many as three layers: exocarp, mesocarp, and endocarp. The exocarp forms the outermost skin of a fruit. The mesocarp is often the fleshy tissue between the exocarp and endocarp of the fruit. The endocarp serves as the boundary around the seed(s). The endocarp may be hard as in peach pits or papery as in apples.

Fruits protect and help disperse the sporophyte. Some fruits are better at one of these functions than the other. The fruit of a peach protects the seed well, but the pit may make it difficult for germination to occur. Peas easily escape from pea pods, but once they are free, they are protected only by the seed coat.

Simple Fruits

Simple fruits are derived from a simple ovary of a single carpel or from a compound ovary of several fused carpels. Figure 28.9 compares the parts of a pea flower to the parts of the pea pod. A pea pod is dehiscent; at maturity, the pea pod breaks open on both sides of the pod to release the seeds. Peas and beans, you will recall, are legumes. The definition of a *legume* is a fruit that splits along two sides when mature. Legumes play a significant role in human nutrition, and so do the cereal grains.

Like legumes, cereal grains of wheat, rice, and corn are *dry fruits.* Sometimes, fruits like those of grains are mistaken for seeds because a dry pericarp adheres to the seed within. These dry fruits are indehiscent—they don't split open. Humans gather grains before they are released from the plant and then process them to acquire their nutrients.

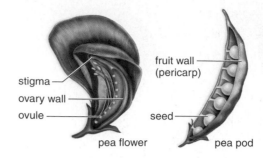

FIGURE 28.9
Pea flower and pea pod.
As a fruit (e.g., pea pod) develops, the ovary wall becomes the fruit wall (pericarp), and the ovule becomes the seed.

Some other dry fruits have special adaptations for dispersal. The hooks and spines of clover, bur, and cocklebur attach to the fur of animals and the clothing of humans. Some plants have fruits with trapped air, or seeds with inflated sacs, that help them float in water. Many seeds are dispersed by wind. Woolly hairs, plumes, and wings are all adaptations for this type of dispersal. The seeds of an orchid are so small and light that they need no special adaptation to carry them far away. The somewhat heavier dandelion fruit uses a tiny "parachute" for dispersal. The winged fruit of a maple tree, which contains two seeds, has been known to travel up to 10 km from its parent (Fig. 28.10a).

In some simple fruits, mesocarp becomes fleshy. When ripe, *fleshy fruits* often attract animals and provide them with food (Fig. 28.10b). Peach and cherry are examples of fleshy fruits that have a hard endocarp. This type of endocarp protects the seed so it can pass through the digestive system of an animal and remain unharmed. In a tomato, the entire pericarp is fleshy. A tomato consists of several chambers because the flower's carpel is composed of several fused carpels. The small size of the seeds and the slippery seed coat means that the seeds rarely get crushed by the teeth of animals. The seeds swallowed by birds and mammals are defecated (passed out of the digestive tract with the feces) some distance from the parent plant. Squirrels and other animals gather seeds and fruits, which they bury some distance away.

Among fruits, apples are an example of an *accessory fruit* because the bulk of the fruit is not from the ovary but from the receptacle. Only the core of an apple is derived from the ovary. If you cut an apple crosswise, it is obvious that an apple, like a tomato, comes from a compound ovary with several chambers.

Compound Fruits

Compound fruits develop from several individual ovaries. For example, each little part of a raspberry or blackberry is derived from a separate ovary. Because the flower had many separate carpels, the resulting fruit is called an *aggregate fruit* (Fig. 28.10c). The strawberry is also an aggregate fruit, but each ovary becomes a one-seeded fruit called an achene. The flesh of a strawberry is from the receptacle. In contrast, a pineapple comes from many different carpels that belong to separate flowers. However, the flowers have only one receptacle (called a pedicel). As the ovaries mature, they fuse to form a large, *multiple fruit* (Fig. 28.10d).

In flowering plants, the seed develops from the ovule, and the fruit develops from the ovary. Fruits aid dispersal of seeds.

TABLE 28.1

Fruit Classification Based on Composition and Texture

Composition (based on type and arrangement of ovaries and flowers)

Simple: develops from a simple ovary or compound ovary

Compound: develops from a group of individual ovaries

Aggregate: ovaries are from a single flower on a common receptacle

Multiple: ovaries are from separate flowers on a common receptacle

Texture (based on mature pericarp)

Fleshy: the entire pericarp or portions of it are soft and fleshy at maturity

Dry: the pericarp is papery, leathery, or woody when the fruit is mature

Dehiscent: the fruit splits open when ripe

Indehiscent: the fruit does not split open when ripe

TABLE 28.2

Kinds of Fruit

Category	Fruit Type	Description	Example
Simple fruits		**Develop from a Flower with a Single Ovary**	
Fleshy			
	Berry	compound ovary, exocarp becomes skin	grape, muscadine, tomato, bell pepper
	Hesperidium	berrylike, leathery rind, juice sacs	lemon, orange
	Pepo	berrylike accessory fruit, receptacle fused with ovary	squash, pumpkin, watermelon, cucumber
	Drupe	simple ovary, mesocarp thick and fleshy, endocarp is stony (pit)	plum, cherry, peach, olive, (walnut, pecan, coconut)
	Pome	accessory fruit, receptacle plus true fruit (ovary tissue)	apple, rose, pear
Dry			
Dehiscent	Legume	simple ovary, splits on two sides	pea, bean, peanut
	Follicle	simple ovary, splits on one side	milkweed, peony
	Capsule	compound ovary, each carpel splits on one side	iris, poppy, cotton, okra
	Silique	compound ovary with two chambers that separate	mustard, radish
Indehiscent	Caryopsis (grain)	simple ovary, one seed, pericarp fused to seed coat	rice, oats, barley, wheat, corn kernel
	Achene	simple ovary, one seed, pericarp and seed coat not fused	dandelion, sunflower
	Samara	simple ovary, pericarp winglike	ash, elm
	Nut	thick, hard pericarp surrounded by bracts and/or receptacle	hickory, oak (acorn), beech, hazelnut
	Schizocarp	fruit divides into two parts at maturity, each with one seed	maple (double samara), dill
Compound fruits		**Develop from a Group of Individual Ovaries**	
Aggregate fruits		Ovaries are from a single flower	
Fleshy			
	Achenes		strawberry
	Drupes		blackberry, dewberry
Dry			
	Follicles		southern magnolia
	Samaras		tulip, poplar
Multiple fruits		Ovaries are from separate flowers.	
Fleshy			
	Achenes		fig
	Drupes		mulberry
	Fused berries		pineapple
Dry			
	Capsules		sweetgum
	Achenes		sycamore
	Caryopses		corn (cob and kernels)

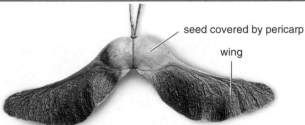

a.

seed covered by pericarp

wing

one fruit

b.

flesh is from receptacle

one fruit

fruits from ovaries of one flower

c.

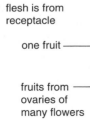

one fruit

fruits from ovaries of many flowers

d.

FIGURE 28.10 Structure and function of fruits.
a. Maple trees produce a dry, indehiscent fruit. The wings rotate in the wind and keep the fruit aloft. **b.** Strawberry plants produce an accessory fruit; each "seed" is actually a fruit on a fleshy, expanded receptacle. **c.** Like strawberries, raspberries are aggregate fruits. Each "berry" is derived from an ovary within the same flower. **d.** A pineapple is a multiple fruit derived from the ovaries of many flowers.

Seed Germination

When **seed germination** occurs, the embryo resumes growth and metabolic activity. We have seen how the embryo forms and that it already has both shoot apical meristem and root apical meristem when enclosed in a seed. Further, the embryo has its primary tissues, which are also meristematic: Protoderm gives rise to the epidermis; ground meristem produces the cells of the cortex and pith; and procambium produces vascular tissue (see page 441).

The length of time seeds retain their viability—their ability to germinate—is quite variable. For some maple seeds, the length of time is only a week. For other seeds, such as lotus seeds, viability is retained for hundreds of years. Most cereal plants retain viability for about ten years.

Some seeds do not germinate until they have been dormant for a period of time. Dormancy is a length of time in which no growth occurs, even though conditions may be favorable for growth. In the temperate zone, seeds often have to be exposed to a period of cold weather before stimulators bring about germination. In deserts, dormancy usually lasts until there is a rain. This requirement helps ensure that seeds do not germinate

until the most favorable growing season has arrived. It is known that fleshy fruits (e.g., apples, pears, oranges, and tomatoes) contain inhibitors that prevent germination until the seeds are released and washed. For some seeds, water, bacterial action, mechanical abrasion, scarification, and even fire may be needed before the seed coat is permeable to water.

The environmental requirements for seed germination are:

- availability of oxygen for increased metabolic needs
- adequate temperature for enzymes to act
- adequate moisture for hydration of cells
- light (in some cases)

Respiration and metabolism continue throughout dormancy, but at a reduced level. The cells of most seeds are dry; for germination to occur, the cells must be hydrated. If water

FIGURE 28.11 Common garden bean seed structure and germination.
a. Seed structure. **b.** Germination and development of the seedling.

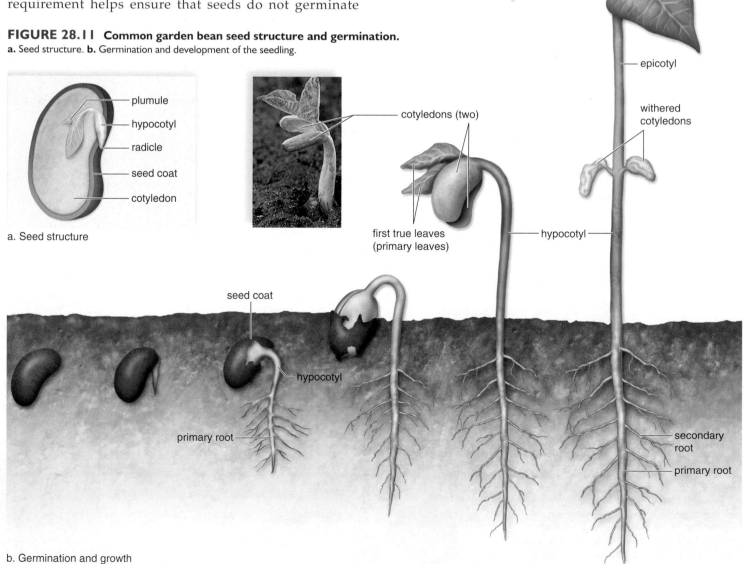

plumule

hypocotyl

radicle

seed coat

cotyledon

a. Seed structure

cotyledons (two)

epicotyl

withered cotyledons

first true leaves (primary leaves)

hypocotyl

seed coat

hypocotyl

primary root

secondary root

primary root

b. Germination and growth

is available, an uptake of water, called imbibition, occurs. For certain seeds, the amount of moisture needed to initiate germination can be minimal, depending on the adaptations of the species. Some seeds have a surface coating that attracts water, for example. Imbibing plant cells can swell dramatically. Dry seeds have been encased in plaster, and when water was added, the seeds swelled and cracked open the plaster.

Most seeds germinate in the dark, but those that must be planted near the surface probably require light. Lettuce seeds, for example, will germinate only if exposed to light. When a seedling, as opposed to a seed, grows in the dark, it etiolates—the stem is elongated, the roots and leaves are small, and the plant is pale yellow and appears spindly. Phytochrome, a pigment that is sensitive to red and far-red light, regulates this response and induces normal growth once proper lighting is available (see page 489).

Germination in Eudicots and Monocots

As mentioned, the embryo of a eudicot, such as a bean plant, has two seed leaves, called cotyledons. The cotyledons, which supply nutrients to the embryo and seedling, eventually shrivel and disappear. If the two cotyledons of a bean seed are parted, you can see a rudimentary plant (Fig. 28.11). The epicotyl bears young leaves and is called a **plumule** [L. *plumulla*, small feather]. As the eudicot seedling emerges from the soil, the shoot is hook-shaped to protect the delicate plumule. The hypocotyl becomes the stem, and the radicle develops into the roots.

A corn plant is a monocot that contains a single cotyledon. The single cotyledon has a storage role, and it also absorbs nutrients from the endosperm. A corn kernel is a type of fruit called a caryopsis or grain, and the outer covering of the kernel is the pericarp (Fig. 28.12). The plumule and radicle are enclosed in protective sheaths called the coleoptile and the coleorhiza, respectively. The plumule and the radicle burst through these coverings when germination occurs.

Germination is a complex event regulated by many factors. The embryo breaks out of the seed coat and becomes a seedling with leaves, stem, and roots.

FIGURE 28.12 Corn kernel structure and germination.
a. Grain structure. **b.** Germination and development of the seedling.

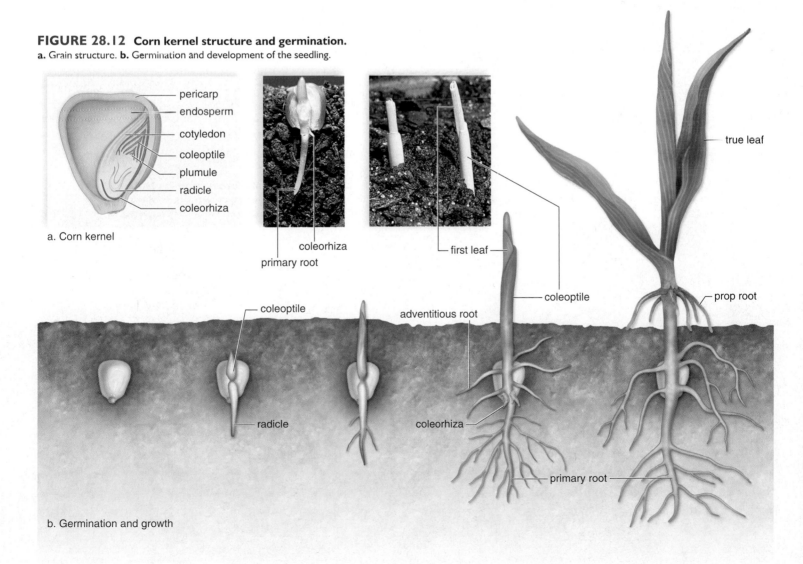

a. Corn kernel

- pericarp
- endosperm
- cotyledon
- coleoptile
- plumule
- radicle
- coleorhiza

coleorhiza

primary root

first leaf

true leaf

coleoptile

prop root

coleoptile

adventitious root

radicle

coleorhiza

primary root

b. Germination and growth

28.4 Asexual Reproduction in Plants

Because plants contain nondifferentiated meristem tissue, they routinely reproduce asexually by vegetative propagation. In asexual reproduction, there is only one parent, instead of two as in sexual reproduction. Violets will grow from the nodes of rhizomes (underground horizontal stems), and complete strawberry plants will grow from the nodes of stolons (aboveground horizontal stems) (Fig. 28.13). White potatoes are actually portions of underground stems, and each eye is a bud that will produce a new potato plant if it is planted with a portion of the swollen tuber. Sweet potatoes are modified roots; they can be propagated by planting sections of the root. You may have noticed that the roots of some fruit trees, such as cherry and apple trees, produce "suckers," small plants that can be used to grow new trees.

In addition to the plants mentioned, pineapple, sugarcane, azalea, gardenia, and many other food and ornamental plants have been propagated from stem cuttings. In these plants, pieces of stem will automatically produce roots. The discovery that the plant hormone auxin can cause roots to develop has expanded the list of plants that can be propagated from stem cuttings.

Tissue Culture of Plants

Hydroponics, the growth of plants in aqueous solutions, had begun by the 1860s. This practice, along with the ability of plants to reproduce asexually, led the German botanist Gottlieb Haberlandt to speculate in 1902 that entire plants could be produced by tissue culture. **Tissue culture** is the growth of a *tissue* in an artificial liquid or solid *culture* medium. Haberlandt said that plant cells are **totipotent** [L. *totus*, all, whole, and *potens*, powerful], meaning that each cell has the full genetic potential of the organism—and therefore, a single cell could become a complete plant. But it wasn't until 1958 that Cornell botanist F. C. Steward grew a complete carrot plant from a tiny piece of phloem. Like former investigators, he provided the cells with sugars, minerals, and vitamins, but he also added coconut milk. (Later, it was discovered that coconut milk contains the plant hormone cytokinin.) When the cultured cells began dividing, they produced a callus, an undifferentiated group of cells. When properly stimulated, the callus differentiated into shoot and roots and eventually developed into complete plants.

Tissue culture techniques have by now led to micropropagation, a commercial method of producing thousands, even millions, of identical seedlings in a limited amount of

Parent plant

FIGURE 28.13 Asexual reproduction in plants.
Meristem tissue at nodes can generate new plants, as when the stolons of strawberry plants, *Fragaria*, give rise to new plants.

Asexually produced offspring

stolon

space. One favorite micropropagation method is meristem culture. If the correct proportions of auxin and cytokinin are added to a liquid medium, many new shoots will develop from a single shoot tip. When these are removed, more shoots form. Since the shoots are genetically identical, the adult plants that develop from them, called clonal plants, all have the same traits. Another advantage to meristem culture is that meristem, unlike other portions of a plant, is virus-free; therefore, the plants produced are also virus-free. (The presence of plant viruses weakens plants and makes them less productive.)

Because plant cells are totipotent, it should be possible to grow an entire plant from a single cell. This, too, has been done. Enzymes are used to digest the cell walls of a small piece of tissue, usually mesophyll tissue from a leaf, and the result is naked cells without walls, called **protoplasts** [Gk. *protos,* first, and *plastos,* formed, molded] (Fig. 28.14*a*). The protoplasts regenerate a new cell wall (Fig. 28.14*b*) and begin to divide (Fig. 28.14*c, d*). These clumps of cells can be manipulated to produce **somatic** (asexually produced) **embryos** (Fig. 28.14*e*). Somatic embryos that are encapsulated in a protective hydrated gel (and sometimes called artificial seeds) can be shipped anywhere. It's possible to produce millions of somatic embryos at once in large tanks called bioreactors. This is done for certain vegetables, such as tomato, celery, and asparagus, and for ornamental plants, such as lilies, begonias, and African violets. A mature plant develops from each somatic embryo (Fig. 28.14*f*). Plants generated from the somatic embryos vary somewhat because of mutations that arise during the production process. These so-called somaclonal variations are another way to produce new plants with desirable traits.

Anther culture is a technique in which mature anthers are cultured in a medium containing vitamins and growth regulators. The haploid tube cells within the pollen grains divide, producing proembryos consisting of as many as 20–40 cells. Finally, the pollen grains rupture, releasing haploid embryos. The experimenter can now generate a haploid plant, or chemical agents can be added that encourage chromosomal doubling. After chromosomal doubling, the resulting plants are diploid but homozygous for all their alleles. Anther culture is a direct way to produce plants that express recessive alleles. If the recessive alleles govern desirable traits, the plants have these traits.

The culturing of plant tissues has led to a technique called cell suspension culture. Rapidly growing calluses are cut into small pieces and shaken in a liquid nutrient medium so that single cells or small clumps of cells break off and form a suspension. These cells will produce the same chemicals as the entire plant. For example, cell suspension cultures of *Cinchona ledgeriana* produce quinine, and those of *Digitalis lanata* produce digitoxin. Scientists envision that it will also be possible to maintain cell suspension cultures in bioreactors for the purpose of producing chemicals used to make drugs, cosmetics, and agricultural chemicals. If so, it will no longer be necessary to farm plants simply for the purpose of acquiring the chemicals they produce.

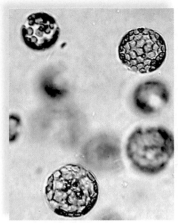
a. Protoplasts, naked cells

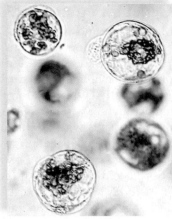
b. Cell wall regeneration

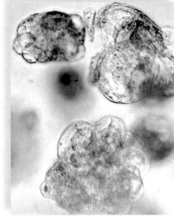
c. Aggregates of cells

d. Callus, undifferentiated mass

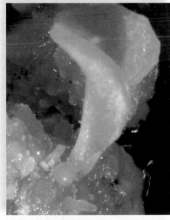
e. Somatic embryo

f. Plantlet

FIGURE 28.14 Tissue culture.
a. When plant cell walls are removed by digestive enzyme action, the result is naked cells, or protoplasts. **b.** Regeneration of cell walls and the beginning of cell division. **c.** Cell division produces aggregates of cells. **d.** An undifferentiated mass, called a callus. **e.** Somatic cell embryos such as this one appear. **f.** The embryos develop into plantlets that can be transferred to soil for growth into adult plants.

Plant tissue culture is now well established. The starting material can be meristem tissue from almost any part of a plant, or it can be adult cells, because plant cells are totipotent if provided with the correct hormonal/nutrient solution.

Genetic Engineering of Plants

Many of our most important crops were developed through artificial selection and hybridization over the past 10,000 years by our ancestors. Modern maize, or corn, is one such plant with a heritage that can be traced to teosinte (Fig. 28.15). Traditionally, **hybridization** [L. *hybrida,* mongrel], the crossing of different varieties of plants or even species, was used to produce plants with desirable traits. Hybridization, followed by vegetative propagation of the mature plants, generated a large number of identical plants with these traits. Today, it is possible to directly alter the genes of organisms.

Today we have the capability of working directly with an organism's genes. The genomes of *Arabidopsis* (see page 490) and rice can be used as models because all flowering eudicots and all flowering monocots are closely related. In other words, when scientists know the location and function of genes in *Arabidopsis* and rice, they will have a fair idea of their location and function in the genomes of crop plants such as wheat, corn, sorghum, millet, and other cereals also.

Tissue Culture and Genetic Engineering

Since a whole plant will grow from a protoplast, it is necessary only to place the foreign gene into a living protoplast. A foreign gene isolated from any type of organism is placed in the tissue culture medium. High-voltage electric pulses can then be used to create pores in the plasma membrane so that the DNA enters. In one of the first procedures, a gene for the production of the firefly enzyme luciferase was inserted into tobacco protoplasts, and the adult plants glowed when sprayed with the substrate luciferin.

Unfortunately, the regeneration of cereal grains from protoplasts has been difficult. Corn and wheat protoplasts produce infertile plants. As a result, other methods are used to introduce DNA into plant cells with intact cell walls. In one technique, foreign DNA is inserted into the plasmid of the bacterium *Agrobacterium,* which normally infects plant cells. When the bacterium infects the plant, the recombinant plasmid is introduced into the plant cells. In 1987, John C. Sanford and Theodore M. Klein of Cornell University developed another method of introducing DNA into a plant tissue culture callus. They constructed a device, called a gene gun, that bombards a callus with DNA-coated microscopic metal particles. Then, genetically altered somatic embryos develop into genetically altered adult plants. Many plants, including corn and wheat varieties, have been genetically engineered by this method. Such plants are called **transgenic plants** because they carry a foreign gene and have new and different traits.

Agricultural Plants with Improved Traits

As discussed in the Health Focus on page 512, corn, potato, soybean, and cotton plants have been engineered to be resistant to either insect predation or herbicides that are widely used (Fig. 28.16). Some corn and cotton plants have been developed that are both insect- and herbicide-resistant. In 2001, transgenic crops were planted on more than 72 million acres worldwide, and the acreage is expected to triple in about five years. If crops are resistant to a broad-spectrum herbicide and weeds are not, then the herbicide can be used to kill the weeds. When herbicide-resistant plants were planted, weeds were easily controlled, less tillage was needed, and soil erosion was minimized.

Crops with other improved agricultural and food quality traits are desirable (Fig. 28.17a). A salt-tolerant tomato has already been developed (Fig. 28.17b). First, scientists identified a gene coding for a protein that transports Na^+ across the vacuole membrane. Sequestering the Na^+ in a vacuole prevents it from interfering with plant metabolism. Then, the scientists cloned the gene and used it to genetically engineer plants that overproduce the channel protein. The modified plants thrived when watered with a salty solution. Today, crop production is limited by the effects of salinization on about 50% of irrigated lands. Salt-tolerant crops would increase yield on this land. Salt- and also drought- and cold-tolerant crops might help provide enough food for a world population that may nearly double by 2050.

Potato blight is the most serious potato disease in the world. About 150 years ago, it was responsible for the Irish potato famine, which caused the death of millions of people. By placing a gene from a naturally blight-resistant wild potato into a farmed variety, researchers have now made potato plants that are no longer vulnerable to a range of blight strains. Some progress has also been made to increase the food quality of crops. Soybeans have been developed that mainly produce the monounsaturated fatty acid oleic acid, a change that may improve human health. These altered plants also produce vernolic acid and ricinoleic acids, derivatives of oleic acid that can be used as hardeners in paints and plastics. The necessary genes were derived from *Vernonia* (ironweed) and castor bean seeds and were transferred into the soybean genomes.

Other types of genetically engineered plants are also expected to increase productivity. Stomata might be altered to take in more carbon dioxide or lose less water. The efficiency of the enzyme RuBP carboxylase, which captures carbon dioxide in plants, could be improved. A team of Japanese scientists is working on introducing the C_4 photosynthetic cycle into rice. Unlike C_3 plants, C_4 plants do well in hot, dry weather. These modifications would require a more complete reengineering of plant cells than the single-gene transfers that have been done so far.

a. Ear of maize

maize kernel

b. Ear of teosinte with maize kernel

FIGURE 28.15 Maize.
Our ancestors used artificial selection to develop maize *(left)* from teosinte *(right)*.

a. Herbicide-resistant soybean plants

b. Nonresistant potato plant

c. Pest-resistant potato plant

FIGURE 28.16 Genetically engineered plants.
a. Genetically engineered herbicide-resistant soybean plants. **b.** Potato plant that has not been genetically engineered to be resistant to pests. **c.** Potato plant that is pest resistant.

Commercial Products

Single-gene transfers have allowed plants to produce various products, including human hormones, clotting factors, and antibodies. One type of antibody made by corn can deliver radioisotopes to tumor cells, and another made by soybeans may be developed to treat genital herpes.

The tobacco mosaic virus has been used as a vector to introduce a human gene into adult tobacco plants in the field. (Note that this technology bypasses the need for tissue culture completely.) Tens of grams of α-galactosidase, an enzyme needed for the treatment of a human lysosome storage disease, were harvested per acre of tobacco plants. And it took only 30 days to get tobacco plants to produce antibodies to treat non-Hodgkin's lymphoma after being sprayed with a genetically engineered virus.

> Genetic engineering of plants is now a reality. The next generation of transgenic crops is expected to have improved agricultural traits and food qualities and to result in higher yields.

Transgenic Crops of the Future	
Improved Agricultural Traits	
Disease-protected	Wheat, corn, potatoes
Herbicide-resistant	Wheat, rice, sugar beets, canola
Salt-tolerant	Cereals, rice, sugarcane
Drought-tolerant	Cereals, rice, sugarcane
Cold-tolerant	Cereals, rice, sugarcane
Improved yield	Cereals, rice, corn, cotton
Modified wood pulp	Trees
Improved Food Quality Traits	
Fatty acid/oil content	Corn, soybeans
Protein/starch content	Cereals, potatoes, soybeans, rice, corn
Amino acid content	Corn, soybeans
Disease protected	Wheat, corn, potatoes

a. Desirable traits

b. Salt-intolerant   Salt-tolerant

FIGURE 28.17
Transgenic crops of the future.
a. Transgenic crops of the future include those with improved agricultural or food quality traits such as those listed. **b.** A salt-tolerant plant has been engineered. The tomato plant to the left does poorly when watered with a salty solution, but the engineered plant to the right is tolerant of the solution. The development of salt-tolerant crops would increase food production in the future.

health focus

Are Genetically Engineered Foods Safe?

A series of focus groups conducted by the Food and Drug Administration (FDA) in 2000 showed that although most participants believed that genetically engineered foods might offer benefits, they also feared unknown long-term health consequences that might be associated with the technology. In Canada, Conrad G. Brunk, a bioethicist at the University of Waterloo in Ontario, has said, "When it comes to human and environmental safety, there should be clear evidence of the absence of risks. The mere absence of evidence is not enough."

The discovery by activists that a genetically engineered corn called StarLink had in-

advertently made it into the food supply triggered the recall of taco shells, tortillas, and many other corn-based foodstuffs from supermarkets. Further, the makers of StarLink were forced to buy back StarLink from farmers and to compensate food producers at an estimated cost of several hundred million dollars. StarLink is a type of "BT" corn. It contains a foreign gene taken from a common soil organism, *Bacillus thuringiensis,* whose insecticidal properties have been long known. About a dozen BT varieties, including corn, potato, and even a tomato, have now been approved for human consumption. These strains contain a gene for an insecticidal protein called CrylA. The makers of StarLink decided to use a gene for a related protein called Cry9C. They thought that using this molecule might slow down the chances of pest resistance to BT corn. To get FDA approval for use in foods, the makers of StarLink performed the required tests.

Like the other now-approved strains, StarLink wasn't poisonous to rodents, and its biochemical structure is not similar to those of most food allergens. But the Cry9C protein resisted digestion longer than the other BT proteins when it was put in simulated stomach acid and subjected to heat. Because most food allergens are stable like this, StarLink was not approved for human consumption.

The scientific community is now trying to devise more tests for allergens because it has not been possible to determine conclusively whether Cry9C is or is not an allergen. Also, at this point, it is unclear how resistant to digestion a protein must be in order to be an allergen, and it is also unclear what degree of sequence similarity a potential allergen must have to a known allergen to raise concern. Dean D. Metcalfe, chief of the Laboratory of Allergic Diseases at the National Institute of Allergy and Infectious Diseases, said, "We need to understand thresholds for sensitization to food allergens and thresholds for elicitation of a reaction with food allergens."

Other scientists are concerned about the following potential drawbacks to the planting of BT corn: (1) resistance among populations of the target pest, (2) exchange of genetic material between the transgenic crop and related plant species, and (3) BT crops' impact on nontarget species. They feel that many more studies are needed before it can be said for certain that BT corn has no ecological drawbacks.

Despite controversies, the planting of genetically engineered corn increased in 2004. The USDA reports that U.S. farmers planted genetically engineered corn on 45% of all corn acres, 5% more than in 2003. In all, U.S. farmers planted at least 150 million acres with mostly genetically engineered corn, soybeans, and cotton (Fig. 28B). The public wants all genetically engineered foods to be labeled as such, but this may not be easy to accomplish because, for example, most cornmeal is derived from both conventional and genetically engineered corn. So far, there has been no attempt to sort out one type of food product from the other.

a.

b.

c.

FIGURE 28B Genetically engineered crops.
Genetically engineered (a) corn, (b) soybeans, and (c) cotton crops are increasingly being planted by today's farmers.

CONNECTING THE CONCEPTS

Life as we know it would not be possible without vascular plants and specifically flowering plants, which now dominate the biosphere. *Homo sapiens* evolved in a world dominated by flowering plants and therefore does not know a world without them. The earliest humans were mostly herbivores; they relied on foods they could gather for survival—fruits, nuts, seeds, tubers, roots, and so forth. Plants also provided protection from the environment, offering shelter from heavy rains and noonday sun. Later on, human civilizations could not have begun without the development of agriculture. The majority of the world's population still relies primarily on three flowering plants—corn, wheat, and rice—for the majority of its sustenance. Sugar, coffee, spices of all kinds, cotton, rubber, and tea are plants that have even promoted wars because of their importance to a country's economy. Although we now live in an industrialized society, we are still dependent on plants and have put them to even more uses. To take a couple of examples, they produce substances needed to lubricate the engines of supersonic jets and to make cellulose acetate for films. For millions of urban dwellers, plants are their major contact with the natural world. We grow them not only for food and shelter, but also for their simple beauty.

Most people fail to appreciate the importance of plants, but plants may be even more critical to our lives today than they were to our early ancestors on the African plains. Currently, half of all pharmaceutical drugs have their origin in plants. The world's major drug companies are engaged in a frantic rush to collect and test plants from the rain forests for their drug-producing potential. Why the rush? Because the rain forests may be gone before all the possible cures for cancer, AIDS, and other killers have been found. Wild plants cannot only help cure human ills, but they can also serve as a source of genes for improving the quality of the plants that support our way of life.

Summary

28.1 REPRODUCTIVE STRATEGIES

Flowering plants exhibit an alternation of generations life cycle. Flowers borne by the sporophyte produce microspores and megaspores by meiosis. Microspores develop into a male gametophyte, and megaspores develop into the female gametophyte. The gametophytes produce gametes by mitotic cell division. Following fertilization, the sporophyte is enclosed within a seed covered by fruit.

The flowering plant life cycle is adapted to a land existence. The microscopic gametophytes are protected from desiccation by the sporophyte; the pollen grain has a protective wall and fertilization does not require external water. The seed has a protective seed coat, and seed germination does not occur until conditions are favorable.

A typical flower has several parts: sepals, which are usually green in color, form an outer whorl; petals, often colored, are the next whorl; and stamens, each having a filament and anther, form a whorl around the base of at least one carpel. The carpel, in the center of a flower, consists of a stigma, style, and ovary. The ovary contains ovules.

Each ovule contains a megasporocyte, which divides meiotically to produce four haploid megaspores, only one of which survives. This megaspore divides mitotically to produce the female gametophyte (embryo sac), which usually has seven cells. One is an egg cell and another is a central cell with two polar nuclei.

The anthers contain microsporocytes, each of which divides meiotically to produce four haploid microspores. Each of these divides mitotically to produce a two-celled pollen grain. One cell is the tube cell, and the other is the generative cell. The generative cell later divides mitotically to produce two sperm cells. The pollen grain is the male gametophyte. After pollination, the pollen grain germinates, and as the pollen tube grows, the sperm cells travel to the embryo sac. Pollination is simply the transfer of pollen from anther to stigma.

Flowering plants experience double fertilization. One sperm nucleus unites with the egg nucleus, forming a 2n zygote, and the other unites with the polar nuclei of the central cell, forming a 3n endosperm cell.

After fertilization, the endosperm cell divides to form multicellular endosperm. The zygote becomes the sporophyte embryo. The ovule matures into the seed (its integuments become the seed coat). The ovary becomes the fruit.

28.2 SEED DEVELOPMENT

As the ovule is becoming a seed, the zygote is becoming an embryo. After the first several divisions, it is possible to discern the embryo and the suspensor. The suspensor attaches the embryo to the ovule and supplies it with nutrients. The eudicot embryo becomes first heart-shaped and then torpedo-shaped. Once you can see the two cotyledons, it is possible to distinguish the shoot tip and the root tip, which contain the apical meristems. In eudicot seeds, the cotyledons frequently take up the endosperm.

28.3 FRUIT TYPES AND SEED DISPERSAL

The seeds of flowering plants are enclosed by fruits. There are different types of fruits. Simple fruits are derived from a single ovary (which can be simple or compound). Some simple fruits are fleshy, such as a peach or an apple. Others are dry, such as peas, nuts, and grains. Compound fruits consist of aggregate fruits, which develop from a number of ovaries of a single flower, and multiple fruits develop from a number of ovaries of separate flowers.

Flowering plants have several ways to disperse seeds. Seeds may be blown by the wind, attached to animals that carry them away, eaten by animals that defecate them some distance away, or adapted to water transport.

Prior to germination, you can distinguish a bean (eudicot) seed's two cotyledons and plumule, which is the shoot that bears leaves. Also present are the epicotyl, the hypocotyl, and the radicle. In a corn kernel (monocot), the endosperm, the cotyledon, the plumule, and the radicle are visible.

28.4 ASEXUAL REPRODUCTION IN PLANTS

Many flowering plants reproduce asexually, as when the nodes of stems (either aboveground or underground) give rise to entire plants, or when roots produce new shoots.

The practice of hydroponics—and the recognition that plant cells can be totipotent—led to plant tissue culture, a technique that now has many applications.

Micropropagation, the production of clonal plants as a result of meristem culture in particular, is now a commercial venture. Flower meristem culture results in somatic embryos that can be packaged in gel for worldwide distribution. Anther culture results in homozygous plants that express recessive genes. Leaf, stem, and root culture can result in cell suspensions that may eventually allow plant chemicals to be produced in large tanks. Development of adult plants from protoplasts results in somaclonal variations, a new source of plant varieties.

Protoplasts in particular lend themselves to direct genetic engineering in tissue culture. Otherwise, the *Agrobacterium* technique or the particle-gun technique allow foreign genes to be introduced into plant cells, which then develop into adult plants with particular traits. Some crops (e.g., soybean, corn, and cotton) have been engineered to be herbicide- and/or pest-resistant. In the future, crops that have these and other improved agricultural traits, improved food quality, and higher productivity are expected.

Reviewing the Chapter

1. Draw a diagram of alternation of generations in flowering plants, and indicate which structures are protected by the sporophyte. Explain.   496
2. Draw a diagram of a flower, and name the parts.   496–97
3. Describe the development of a male gametophyte, from the microsporocyte to the production of sperm.   498
4. Describe the development of a female gametophyte, from the megasporocyte to the production of an egg.   498–99
5. What is the difference between pollination and fertilization? Why doesn't fertilization require any external water? What is double fertilization?   499
6. Describe the sequence of events as a eudicot zygote becomes an embryo enclosed within a seed.   502
7. Distinguish between simple dry fruits and simple fleshy fruits. Give an example of each type. What is an aggregate fruit? A multiple fruit?   504–5
8. Name several mechanisms of seed and/or fruit dispersal.   504
9. What are the requirements for seed germination? Contrast the germination of a bean seed with that of a corn kernel.   506
10. In what ways do plants ordinarily reproduce asexually? What is the importance of totipotency in regard to tissue culture?   508
11. How is plant tissue culture carried out, and what are the benefits of plant tissue culture?   509
12. How are transgenic plants produced? What types of plants have been produced, and for what purposes have they been genetically engineered?   510

Testing Yourself

Choose the best answer for each question.

1. In plants,
    a. gametes become a gametophyte.
    b. spores become a sporophyte.
    c. both sporophyte and gametophyte produce spores.
    d. only a sporophyte produces spores.
    e. Both a and b are correct.

2. The flower part that contains ovules is the
    a. carpel.   d. petal.
    b. stamen.   e. seed.
    c. sepal.

3. The megasporocyte and the microsporocyte
    a. both produce pollen grains.
    b. both divide meiotically.
    c. both divide mitotically.
    d. produce pollen grains and embryo sacs, respectively.
    e. All of these are correct.

4. A pollen grain is
    a. a haploid structure.
    b. a diploid structure.
    c. first a diploid and then a haploid structure.
    d. first a haploid and then a diploid structure.
    e. the mature gametophyte.

5. Which of these pairs is incorrectly matched?
    a. polar nuclei—plumule   d. ovary—fruit
    b. egg and sperm—zygote   e. stigma—carpel
    c. ovule—seed

6. Which of these is not a fruit?
    a. walnut   d. peach
    b. pea   e. All of these are fruits.
    c. green bean

7. Animals assist with
    a. pollination and seed dispersal.
    b. control of plant growth and response.
    c. translocation of organic nutrients.
    d. asexual propagation of plants.
    e. germination of seeds.

8. A seed contains
    a. a seed coat.   d. cotyledon(s).
    b. an embryo.   e. All of these are correct.
    c. stored food.

9. Which of these is mismatched?
    a. plumule—leaves   d. pericarp—corn kernel
    b. cotyledon—seed leaf   e. carpel—ovule
    c. epicotyl—root

10. Which of these is not a common procedure in the tissue culture of plants?
    a. shoot tip culture for the purpose of micropropagation
    b. flower meristem culture for the purpose of somatic embryos
    c. leaf, stem, and root culture for the purpose of cell suspension cultures
    d. protoplast culture for the purpose of genetic engineering of plants
    e. culture of hybridized mature plant cells

11. In the life cycle of flowering plants, a microspore develops into
    a. a megaspore.   d. an ovule.
    b. a male gametophyte.   e. an embryo.
    c. a female gametophyte.

12. Carpels
    a. are the female part of a flower.
    b. contain ovules.
    c. are the innermost part of a flower.
    d. may be absent in a flower.
    e. All of these are correct.

13. Which of these is part of a male gametophyte?
    a. synergid cells
    b. the central cell
    c. polar nuclei
    d. a tube nucleus
    e. antipodal cells

14. Bat-pollinated flowers
    a. are colorful.
    b. are open throughout the day.
    c. are strongly scented.
    d. have little scent.
    e. Both b and c are correct.

15. Heart, torpedo, and globular refer to
    a. embryo development.
    b. sperm development.
    c. female gametophyte development.
    d. seed development.
    e. Both b and d are correct.

16. Fruits
    a. nourish embryo development.
    b. help with seed dispersal.
    c. signal gametophyte maturity.
    d. attract pollinators.
    e. signal when they are ripe.

17. Asexual reproduction in flowering plants
    a. is unknown.
    b. is a rare event.
    c. is common.
    d. occurs in all plants.
    e. is no fun.

18. Plant tissue culture takes advantage of
    a. a difference in flower structure.
    b. sexual reproduction.
    c. gravitropism.
    d. phototropism.
    e. totipotency.

19. In plants, meiosis directly produces
    a. new xylem.   d. egg.
    b. phloem.   e. sperm.
    c. spores.

20. Label this diagram of alternation of generations in flowering plants.

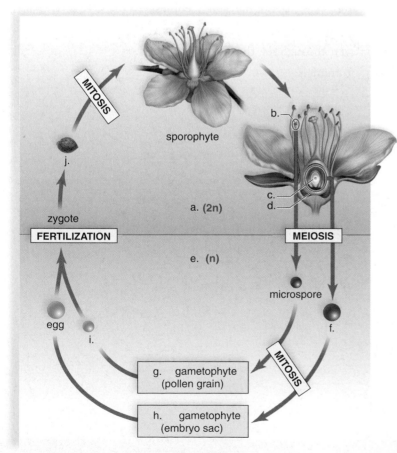

Thinking Scientifically

1. Peanuts are fruits that develop underground. The stems bend toward the ground, and the flowers push into the soil like roots. Since this unusual gravitropic response only happens after flowering, it must be linked to reproduction. What hypothesis would explain the bending? What steps in reproduction would you examine to test the hypothesis?

2. You have discovered an unusual lettuce variant with an orange leaf. Realizing the commercial potential of such a plant, you would like to propagate it. However, by the time of your discovery, the growing season is almost over—the flowers are withered, and the seeds gone. How would you propagate this plant? Why might it be beneficial to cross the propagated plant with green-leafed lettuce?

Understanding the Terms

anther 496	microsporocyte 498
carpel 496	ovary 496
coevolution 500	ovule 497
cotyledon 502	petal 496
double fertilization 499	plumule 507
embryo 499	pollen grain 498
embryo sac 499	pollination 499
endosperm 499	protoplast 509
female	seed 496
gametophyte 498	seed germination 506
filament 496	sepal 496
flower 496	somatic embryo 509
fruit 504	sporophyte 496
gametophyte 496	stamen 496
hybridization 510	stigma 496
male gametophyte 498	style 496
megaspore 496	tissue culture 508
megasporocyte 498	totipotent 508
microspore 496	transgenic plant 510

Match the terms to these definitions:

a. _____ Flower structure consisting of an ovary, a style, and a stigma.

b. _____ Flowering plant structure consisting of one or more ripened ovaries that usually contain seeds.

c. _____ The gametophyte that produces an egg; an embryo sac in flowering plants.

d. _____ Mature ovule that contains an embryo, with stored food enclosed in a protective coat.

e. _____ Mature male gametophyte in seed plants.

ARIS, the *Biology* Website

ARIS, the website for *Biology*, provides a wealth of information organized and integrated by chapter. You will find practice quizzes, interactive activities, labeling exercises, flashcards, and much more that will complement your learning and understanding of general biology.

www.mhhe.com/maderbiology9

PART VI

ANIMAL EVOLUTION

The members of the animal kingdom exhibit tremendous diversity, from microscopic shrimp to the great blue whale. Despite their diversity, evolution from a common ancestor has provided animal life with an unbroken thread of unity. At the molecular level, animals share a common chemistry including a versatile genetic code. At the cellular level, animals are eukaryotic, multicellular organisms with organs and tissues specialized for particular functions. At the organismic level, all animals are heterotrophs requiring a constant supply of organic food.

The great diversity of form and function found throughout the animal world is the result of millions of years of evolutionary forces in action. The search for food, shelter, and mates under a variety of environmental conditions has shaped the diversity of the animal world. A journey through the fossil record reveals a myriad of extinct animals, including the dinosaurs, which possibly were unable to adapt to new conditions.

In his book *On the Origin of Species*, Charles Darwin speaks of the grandeur of evolution. While planet Earth cycles year after year round the sun, "endless forms most beautiful and most wonderful" have appeared and will keep on appearing. Indeed, evolution is the unifying theme woven throughout the history of life that accounts for both the unity and diversity of animals.

29

INTRODUCTION TO INVERTEBRATES

Jeremy Jackson of the Smithsonian Tropical Research Institute wonders if he is doing enough to warn the public that we may lose 60% of the Earth's coral reefs by the year 2050. A coral reef is formed of limestone deposited by invertebrate animals called stony corals. At a reef, many different types of aquatic protists and animals find a home and interact with one another in a complex ecosystem.

Reefs around the globe are in danger. Tons of soil from deforested tracts bring nutrients that stimulate the overgrowth of algae. Off the coast of Australia this has caused a population explosion of the crown-of-thorns sea star, which is ravaging the Great Barrier Reef of Australia. So-called coral bleaching occurs when pollutants and unusually warm seawater make corals expel their symbiotic dinoflagellates.

Overfishing of the reefs is commonplace. The methods are sinister, including the use of dynamite to kill fish, cyanide to stun them, and satellite navigation systems to home in on areas where mature fish spawn. Coral is also being destroyed at an alarming rate by people who collect coral for aquariums and jewelry. Yet, intact coral reefs are storm barriers that protect the shoreline and provide safe harbors for ships. And, like tropical rain forests, reefs are most likely sources of medicines yet to be discovered.

Yellow coral polyps, *Parazoanthus gracilis.*

29.1 EVOLUTION OF ANIMALS

From earthworms to toads, members of kingdom **Animalia,** domain Eukarya, are multicellular heterotrophs that ingest their food (Fig. 29.1). Although biologists have described over 1.5 million species of animals, this number may represent only 10–20% of the animals living today.

The more than 35 living animal phyla are believed to have evolved from a colonial protistan ancestor over 600 million years ago. All but one of these phyla contain only **invertebrates,** which are animals without a backbone. One phylum, Chordata, is mainly composed of **vertebrates,** which are animals with a backbone of bone or cartilage. In this text, only the 12 animal phyla depicted in Figure 29.2 will be considered in depth.

Criteria for Classification

Biologists use a variety of characters to classify animals. Some of the main characters used in animal taxonomy are organization of tissues, type of symmetry, type of body plan, type of **coelom,** and the presence or absence of segmentation. A new criterion for classification is the use of molecular data.

Organization of Tissues

Three levels of organization exist in the animal kingdom. The simplest level, called the cellular level, is found in organisms such as sponges, which have no true tissues. Beyond the cellular level of organization, animals have embryological *germ layers.* Animals such as jellyfishes, which have only two germ layers (ectoderm and endoderm), are diploblastic with the tissue level of organization. Those animals that have all three germ layers (ectoderm, mesoderm, and endoderm), such as octopuses, are triploblastic and have the organ level of organization.

Type of Symmetry

Three types of symmetry exist in the animal world. **Asymmetrical** organisms, such as some sponges, have no particular body shape. In **radially symmetrical** organisms such as sea stars, the animal is organized circularly, similar to a wheel, and two identical halves are obtained no matter how the animal is sliced longitudinally. **Bilaterally symmetrical** organisms such as cats have a definite left and right half, and only a longitudinal cut down the center of the animal will produce two equal halves.

Radially symmetrical animals are sometimes attached to a substrate—that is, they are **sessile.** This type of symmetry is useful because it allows these animals to reach out in all directions from one center. Bilaterally symmetrical animals tend to be active and to move forward with an anterior end. During the evolution of animals, bilateral symmetry is accompanied by **cephalization,** localization of a brain and specialized sensory organs at the anterior end of an animal.

Type of Body Plan

Two body plans are observed in the animal kingdom: the sac plan and the tube-within-a-tube plan. Animals with the sac plan, such as jellyfishes and tapeworms, have an incomplete digestive system. It has only one opening, which is used both as an entrance for food and as an exit for undigested material. Animals with the tube-within-a-tube plan, such as roundworms and alligators, have a complete digestive system, with a separate entrance for food and an exit for undigested material. Having two openings allows specialization of parts to occur along the length of the tube.

Type of Coelom

The space that contains the internal organs is called a body cavity. Some animals such as tapeworms are **acoelomates,** meaning that they have no body cavity. They have mesoderm but no coelom. **Pseudocoelomates** such as roundworms have a coelom incompletely lined by mesoderm because the coelom develops between the mesoderm and the endoderm. A layer of mesoderm exists beneath the body wall but not around the gut. Most animals are **coelomates,** because they have a coelom that is completely lined with mesoderm. Coelomate animals are either **protostomes** or **deuterostomes.** When the first embryonic opening or blastopore becomes the mouth, as in spiders, the animal is a protostome. When the blastopore becomes the anus, as in monkeys, the animal is a deuterostome. Several other features of embryonic development also distinguish protostomes and deuterostomes, as we shall be discussing.

Segmentation

Segmentation is the repetition of body parts along the length of the body. Acoelomates and pseudocoelomates are nonsegmented. Some coelomates are nonsegmented and some are segmented. Molluscs, represented by snails, and echinoderms, represented by sea stars, are nonsegmented. Annelids, represented by earthworms, arthropods, represented by grasshoppers, and chordates, represented by hawks, are segmented (see Fig. 29.2). Segmentation leads to specialization of parts because the various segments can become differentiated for specific purposes.

Molecular Data

Modern phylogenetic investigations take into account molecular data, primarily nucleotide sequences, when classifying animals. It is assumed that the more closely related two organisms are, the more nucleotide sequences they will have in common. As discussed in the Science Focus on page 520, the animal phylogenetic tree based on molecular data is quite different from the one based only on the anatomical characteristics we have been discussing.

Classification of animals is based on organization of tissues, type of symmetry, type of body plan, type of coelom, segmentation, and also molecular data.

FIGURE 29.1

Animals—multicellular, heterotrophic eukaryotes.
The toad is a vertebrate, and the earthworm that it is eating is an invertebrate. All animals are multicellular heterotrophic organisms that must take in preformed food.

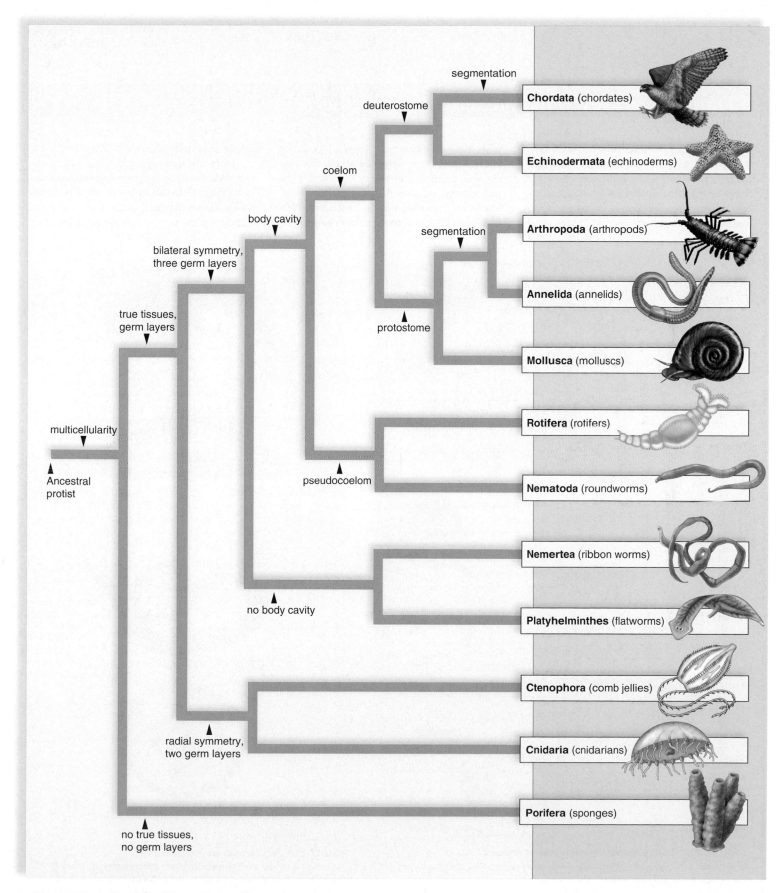

FIGURE 29.2 Traditional phylogenetic tree of animals.

All animal phyla living today are most likely descended from a colonial flagellated protist living about 600 million years ago. This traditional phylogenetic tree uses only anatomical and embryological data to determine which phyla are most closely related to one another. For a tree based on molecular data see Fig. 29C.

science focus

Data for a New Phylogenetic Tree of the Animals

For many years biologists have based the phylogenetic tree of the animals on anatomical and embryological data. In recent years, new trees based on ribosomal RNA (rRNA) nucleotide sequences have been proposed. In the traditional tree, the protostomes are restricted to the three phyla that have a true coelom (Fig. 29A). So, for example, flatworms, which are acoelomates (do not have a coelom), are not included among the protostomes, whereas the annelids (segmented worms, e.g., earthworms) are protostomes because they have a coelom, and their development follows a particular pattern (Fig. 29B).

Molecular data suggest that many more animal phyla should be designated protostomes because their rRNA sequences are so similar. Also their early development preceding formation of a coelom is the same. Note in Figure 29C, that

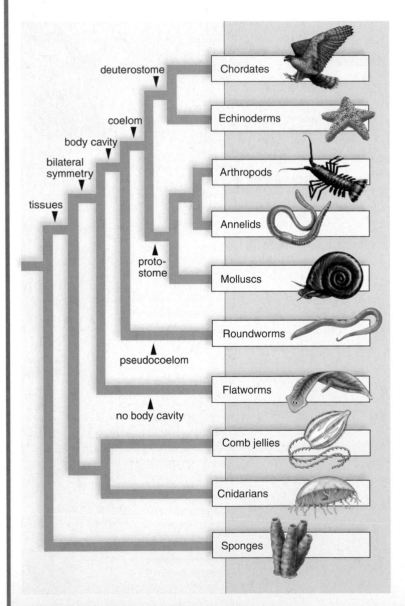

FIGURE 29A Traditional phylogenetic tree of animals.
In the traditional tree, based on anatomical and embryological data, the protostomes comprise only the molluscs, annelids, and arthropods.

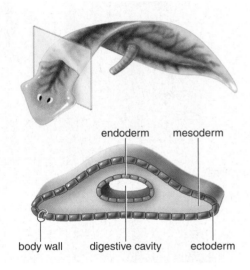

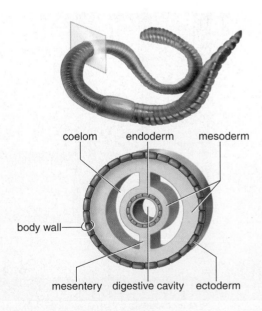

FIGURE 29B Acoelomate versus coelomate.
In the traditional tree, the flatworms (top) are not protostomes, because they are acoelomates (do not have a coelom). The annelids, which are the segmented worms (bottom), are protostomes because they do have a coelom, and it develops in a particular way.

the protostomes are divided into two groups. One group has a particular type of immature stage, the *trocophore* larva, as reflected in their name, lophotrochozoa. (The *lopho* stands for a feeding apparatus, the lophophore, seen in animal phyla we are not studying.)

The other group, called ecdysozoans, contains the roundworms and the arthropods. Both of these types of animals shed their outer covering as they grow. Ecdysozoan means molting animal (Fig. 29D).

Notice, too, that segmentation doesn't play a defining role in the phylogenetic tree based on molecular data (Fig. 29C). In the traditional tree, the annelids are grouped with the arthropods because both phyla contain segmented animals. In the new tree, the segmented worms are in one group and the arthropods are in the other.

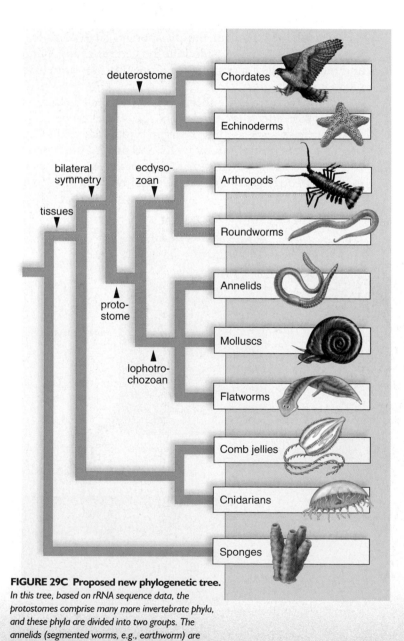

FIGURE 29C Proposed new phylogenetic tree.

In this tree, based on rRNA sequence data, the protostomes comprise many more invertebrate phyla, and these phyla are divided into two groups. The annelids (segmented worms, e.g., earthworm) are placed in one group, and the arthropods are placed in another, even though both are segmented animals.

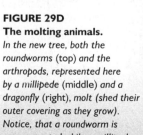

FIGURE 29D
The molting animals.

In the new tree, both the roundworms (top) and the arthropods, represented here by a millipede (middle) and a dragonfly (right), molt (shed their outer covering as they grow). Notice, that a roundworm is nonsegmented while a millipede and dragonfly are segmented.

29.2 MULTICELLULARITY

While all animals are multi-cellular, sponges are the only animals to lack true tissues and have the cellular level of organization. Sponges exhibit various levels of symmetry. Evolutionary biologists believe that sponges are out of the mainstream of animal evolution and represent an evolutionary dead end.

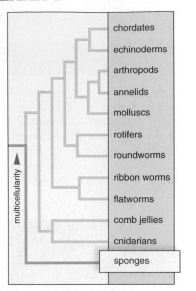

Sponges

Sponges are placed in phylum **Porifera** [L. *porus*, pore, and *-fera*, bearing], because their saclike bodies are perforated by many pores (Fig. 29.3*a*). Presently, over 5,150 species of sponges, most of which are marine, have been described. These organisms are highly variable in size, shape, and color.

The cellular organization of sponges is demonstrated by experimentation. After a sponge is broken into separate cells, the cells exist individually until they spontaneously reorganize into a sponge once again. Sponges have several types of cells. The outer layer of the wall contains flattened epidermal cells, some of which have contractile fibers; the middle layer is a semifluid matrix with wandering amoebocytes; and the inner layer is composed of flagellated cells called collar cells (or choanocytes) that look like protozoans (Fig. 29.3*b*). There are no nerve cells or other means of coordination between the cells. To some, a sponge can be thought of as a colony of protozoans.

The beating of collar-cell flagella produces water currents that flow through the pores into the central cavity and out through the osculum, the upper opening of the body. Although it may seem that sponges can't do much, even a simple one only 10 cm tall is estimated to filter as much as 100 liters of water each day. It takes this much water to meet the needs of the organism. Simple sponges have pores leading directly from the outside into the central cavity. Larger and more complex sponges have canals leading from external to internal pores.

Sponges are **sessile filter feeders,** organisms that stay in one place and filter food from the water. Microscopic food particles such as bits of organic material and microorganisms are brought in by water and are engulfed by the collar cells and then digested in food vacuoles or are passed to the amoebocytes for digestion. The amoebocytes also act

a. Yellow tube sponge, *Aplysina fistularis*

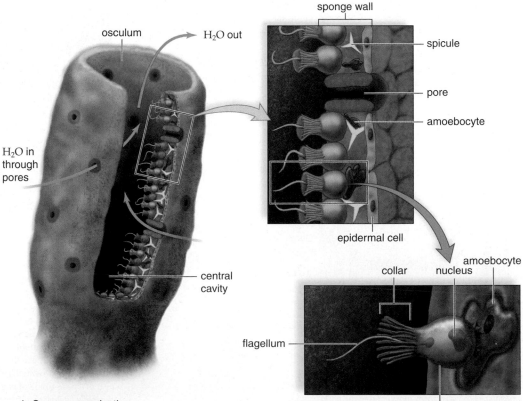

b. Sponge organization

FIGURE 29.3 Simple sponge anatomy.
a. Photograph of simple sponges. **b.** The wall contains two layers of cells: the outer epidermal cells and the inner collar cells. The collar cells (enlarged) have flagella that beat, moving the water through pores as indicated by the arrows. Food particles in the water are trapped by the collar cells and digested within their food vacuoles. Amoebocytes transport nutrients from cell to cell; spicules form an internal skeleton in some sponges.

DOMAIN: Eukarya	
KINGDOM: Animalia	

CHARACTERISTICS
- Multicellular
- Well-developed tissues (except sponges)
- Usually motile
- Heterotrophic by ingestion or absorption, generally a digestive cavity
- Diploid life cycle

CLASSIFICATION

Invertebrates*
PHYLUM: Porifera — sponges
PHYLUM: Cnidaria — jellyfishes, sea anemones, corals
PHYLUM: Ctenophora — comb jellies, sea walnuts
PHYLUM: Platyhelminthes — flatworms (e.g., planarians, flukes, tapeworms)
PHYLUM: Nemertea — ribbon worms
PHYLUM: Nematoda — roundworms
PHYLUM: Rotifera — rotifers
PHYLUM: Mollusca — chitons, snails, slugs, mussels, clams, oysters, squids, octopuses
PHYLUM: Annelida — segmented worms (e.g., clam worms, earthworms, leeches)
PHYLUM: Arthropoda — spiders, scorpions, horseshoe crabs, lobsters, crayfish, shrimps, crabs, millipedes, centipedes, insects
PHYLUM: Echinodermata — sea lilies, sea stars, brittle stars, sea urchins, sand dollars, sea cucumbers, sea daisies
PHYLUM: Chordata
  SUBPHYLUM: Urochordata — sea squirts
  SUBPHYLUM: Cephalochordata — lancelets
Vertebrates*
  SUBPHYLUM: Vertebrata
    SUPERCLASS: Agnatha — jawless fishes (e.g., lampreys, hagfishes)
    SUPERCLASS: Gnathostomata — jawed fishes; all tetrapods
      CLASS: Chondrichthyes — cartilaginous fishes (e.g., sharks, skates, rays)
      CLASS: Osteichthyes — bony fishes (e.g., herring, salmon, cod, eel, flounder)
      CLASS: Amphibia — frogs, toads, salamanders, newts, caecilians
      CLASS: Reptilia — snakes, lizards, turtles, crocodiles
      CLASS: Aves — birds (e.g., sparrows, penguins, ostriches)
      CLASS: Mammalia — mammals (e.g., cats, dogs, horses, rats, humans)
* Not in the classification of organisms, but added here for clarity

as a circulatory device to transport nutrients from cell to cell. The constant flow of water passing through a sponge also carries oxygen and allows for the removal of wastes.

Sponges can reproduce asexually by fragmentation or by budding. During budding, a small protuberance appears and gradually increases in size until a complete organism forms. Budding produces colonies of sponges that can become quite large. During sexual reproduction, cross-fertilization is the rule, and the zygote develops internally into a ciliated larva, which is released into the central cavity. Such a larva ensures dispersal of offspring for sessile sponges because it may swim to a new location. Like all less specialized organisms, sponges are capable of regeneration, or growth of a whole from a small part. Thus, if a sponge is chopped up and returned to the water, each piece may grow into a complete sponge.

Generally, sponges are classified according to their skeletons. Sponges maintain their body shape by pos-

sessing **spicules** or collagenous fibers known as **spongin** throughout their body wall. Spicules are needle-shaped structures with one to six rays. In class Calcarea, a group of marine sponges, spicules are composed of calcium carbonate. In class Hexactinellida, the glass sponges primarily found in deep marine environments, spicules are composed of silica. Class Demospongiae, the largest class of sponges, are found in marine, brackish, and freshwater environments. Members of this class contain both spicules and spongin. Some sponges contain only spongin fibers; a bath sponge is the dried spongin skeleton from which all living tissue has been removed. Today, however, most commercial "sponges" are synthetic.

Sponges have a cellular level of organization and most likely evolved independently from protozoans. They are the only animals in which digestion occurs within cells.

29.3 TRUE TISSUE LAYERS

The animals in the remaining phyla to be studied have true tissues derived from the embryonic germ layers. As mentioned, during animal development, a total of three germ layers are possible: ectoderm, endoderm, and mesoderm.

Comb jellies and cnidarians develop only ectoderm and endoderm. Therefore, they are said to be diploblasts (Fig. 29.4). Animals in these phyla are radially symmetrical, meaning that any longitudinal cut produces two identical halves. If an animal is bilaterally symmetrical, only one longitudinal cut yields two roughly identical halves:

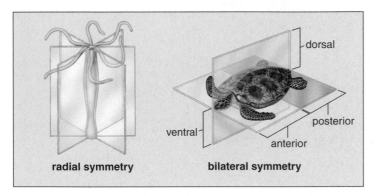

radial symmetry   **bilateral symmetry**

Comb Jellies

Comb jellies (phylum **Ctenophora**) consist of approximately 90 species of solitary, mostly free-swimming marine invertebrates that are usually found in warm waters. Ctenophores represent the largest animals propelled by beating cilia and range in size from a few centimeters to 1.5 m in length. Their body is made up of a transparent jellylike substance called **mesoglea**. Most ctenophores do not have stinging cells and capture their prey by using sticky adhesive cells called colloblasts. Some ctenophores are bioluminescent, capable of producing their own light.

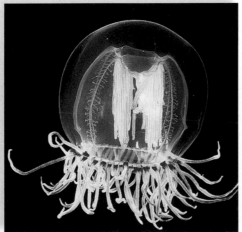

a.   b.

FIGURE 29.4 Comb jelly compared to cnidarian.
a. *Pleurobrachia pileus*, a comb jelly. Despite similar symmetry, diploblastic organization, and gastrovascular cavities, the close relationship of these animals is now in dispute. **b.** *Polyorchis penicillatus*, medusan form of a cnidarian.

Cnidarians

Cnidarians (phylum **Cnidaria**) are a large (10,000 species) and ancient group of invertebrates with one of the oldest fossil records of any animal. This phylum contains some of nature's most bizarre and beautiful creatures, including sea anemones, jellyfishes, sea fans, and corals.

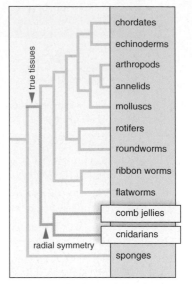

Cnidarians are tubular or bell-shaped animals that reside mainly in shallow coastal waters; however, there are some freshwater, brackish, and oceanic forms. The term *cnidaria* is derived from the presence of specialized stinging cells called cnidocytes. Each cnidocyte has a fluid-filled capsule called a **nematocyst** [Gk. *nema*, thread, and *kystis*, bladder] that contains a long, spirally coiled hollow thread. When the trigger of the cnidocyte is touched, the nematocyst is discharged. Some threads merely trap a prey, and others have spines that penetrate and inject paralyzing toxins.

The body of a cnidarian is a two-layered sac. The outer tissue layer is a protective epidermis derived from ectoderm. The inner tissue layer, which is derived from endoderm, secretes digestive juices into the internal cavity, called the **gastrovascular cavity** [Gk. *gastros*, stomach; L. *vasculum*, dim. of *vas*, vessel] because it serves for digestion of food and circulation of nutrients. The gastrovascular cavity also serves as a supportive hydrostatic skeleton. The two tissue layers are separated by mesoglea. There are muscle fibers at the base of the epidermal and gastrodermal cells. Nerve cells located below the epidermis near the mesoglea interconnect and form a **nerve net** throughout the body. Cnidarians lack a centralized nervous system. In contrast to highly organized nervous systems, the nerve net allows transmission of impulses in several directions at once. Having both muscle fibers and nerve fibers, these animals are capable of directional movement; the body can contract or extend, and the tentacles that ring the mouth can reach out and grasp prey.

Two basic body forms are seen among cnidarians. The mouth of a **polyp** is directed upward, while the mouth of a jellyfish or **medusa** is directed downward (Fig. 29.5a). The bell-shaped medusa has more mesoglea than a polyp, and the tentacles are concentrated on the margin of the bell. At one time, both body forms may have been a part of the life cycle of all cnidarians. When both are present, the animal is **dimorphic:** The sessile polyp stage produces medusae by asexual budding, and the motile medusan stage produces egg and sperm. In some cnidarians, one stage is dominant and the other is reduced; in other species, one form is absent altogether.

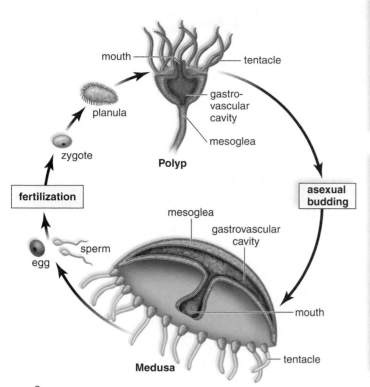

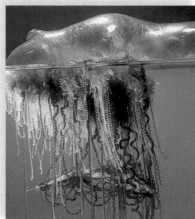

b. Sea anemone, *Corynactis*

c. Cup coral, *Tubastrea*

d. Portuguese man-of-war, *Physalia*

e. Jellyfish, *Crambione*

FIGURE 29.5 Cnidarian diversity.
a. The life cycle of a cnidarian. Some cnidarians have both a polyp stage and a medusa stage; in others, one stage may be dominant or absent altogether. **b.** The anemone, which is sometimes called the flower of the sea, is a solitary polyp. **c.** Corals are colonial polyps residing in a calcium carbonate or proteinaceous skeleton. **d.** The Portuguese man-of-war is a colony of modified polyps and medusae. **e.** True jellyfishes undergo the complete life cycle; this is the medusa stage. The polyp is small.

Cnidarian Diversity

Class Anthozoa are known as the flower animals because their bodies are made of flower-shaped polyps. This class includes solitary sea anemones and colonial corals (Fig. 29.5b, c). Anthozoans can be found in both deep and shallow marine waters worldwide.

Sea anemones are sessile and live attached to submerged rocks, timbers, or other substrate. These animals have an upwardly turned oral disk that contains the mouth surrounded by a large number of hollow tentacles containing nematocysts. The tentacles are used in capturing various marine invertebrates and a variety of fishes. Most sea anemones range in size from 5–200 mm in length and 5–100 mm in diameter and are often colorful. A number of species of sea anemones form mutualistic (both species benefit) relationships with hermit crabs and live attached to the shell of the crab. The anemone provides protection and camouflage while the crab provides locomotion and food.

Corals resemble sea anemones encased in a calcareous house. The coral polyp can extend into the water to feed on microorganisms and retreat into the house for safety. Some corals are solitary, but the vast majority live in colonies that vary in shape from rounded to branching. Many corals exhibit elaborate geometric designs and stunning colors.

Many species of coral are responsible for building reefs. Through the years, the slow accumulation of calcium carbonate, i.e. limestone, from their skeletons can result in massive structures such as the Great Barrier Reef along the eastern coast of Australia. Coral reef ecosystems are very productive, and a diverse group of organisms are dependent on the reef for food and shelter.

In class Hydrozoa, the polyp stage is dominant. One of the most unusual hydrozoans is the Portuguese man-of-war, *Physalia*, which looks as if it might be an odd-shaped medusa (Fig. 29.5d) but actually is a colony of polyps. The original polyp becomes a gas-filled float that provides buoyancy, keeping the colony afloat. Other polyps, which bud from this one, are specialized for feeding or for reproduction. A long, single tentacle armed with numerous nematocysts arises from the base of each feeding polyp. Swimmers who accidentally come upon a Portuguese man-of-war can receive painful, even serious, injuries from these stinging tentacles.

Class Scyphozoa includes the true jellyfishes, such as *Crambione* (Fig. 29.5e). One species, *Cyanea capillata*, may attain a bell diameter of over 2 m with tentacles of up to 70 m in length. In true jellyfishes, the medusa is the primary stage, and the polyp remains small and insignificant. Jellyfishes are zooplankton and depend on tides and currents for their primary means of movement. They feed on a variety of invertebrates and fishes and are food themselves for marine animals.

Cnidarians have a tissue level of organization and are radially symmetrical. They have a sac body plan and exist as polyps and/or medusae.

Hydra and Obelia

Hydra and *Obelia* are two common hydrozoans of interest. *Hydra* is an example of a solitary form, and *Obelia* is an example of a colonial form that has both a polyp and medusa stage in its life cycle.

Hydra. **Hydras** [Gk. *hydra,* a many-headed serpent] are freshwater cnidarians. Hydras are likely to be found attached to underwater plants or rocks in most lakes and ponds. The body is a small tubular polyp about one-quarter inch in length. The only opening (the mouth) is in a raised area surrounded by four to six tentacles that contain a large number of nematocysts. The central cavity of the animal is the gastrovascular cavity.

Hydras usually remain in one location. However, they are capable of limited movement by gliding along on their base or pedal disc, secreting a bubble and floating, or by somersaulting into a new location:

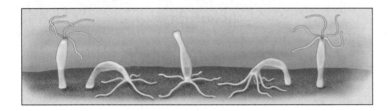

Hydras have both muscle and nerve fibers that enable them to respond to a stimulus. For example, if a tentacle is touched with a needle, all the tentacles and the body contract, only to extend later.

Figure 29.6 shows the microscopic anatomy of *Hydra.* The cells of the epidermis are termed epitheliomuscular cells because they contain muscle fibers. Also present in the epidermis are cnidocytes and sensory cells. The latter have long extensions that make contact with the nerve cells within the nerve net. The interstitial, or embryonic, cells also seen in this layer are capable of becoming other types of cells. For example, they can produce an ovary and/or a testis and probably also account for the animal's great regenerative powers. Like the sponges, cnidarians can grow an entire organism from a small piece.

Gland cells of the gastrodermis secrete digestive juices that pour into the gastrovascular cavity. Hydras feed on small prey such as crustaceans, insect larvae, and worms that they capture with their tentacles and nematocysts. Prey is placed in an elevated mouth, and from there it travels into the gastrovascular cavity. Digestion, which begins in the cavity, is completed within food vacuoles of nutritive-muscular cells, the main type of gastrodermal cell. Nutrient molecules are passed by diffusion to the rest of the cells of the body. Nutritive-muscular cells contain contractile fibers that run circularly about the body; when these contract, the animal lengthens.

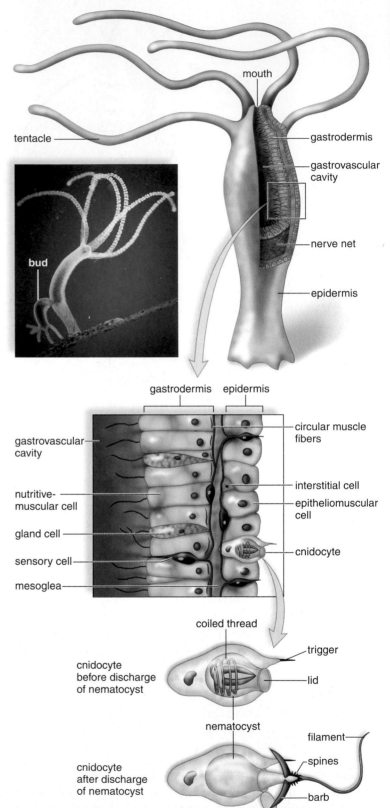

FIGURE 29.6 Anatomy of *Hydra.*

Top: The body of *Hydra* is a small tubular polyp whose wall contains two tissue layers. *Middle:* Various types of cells in the body wall. *Bottom:* Cnidocytes are cells that contain nematocysts. *Top left:* Hydra reproduces asexually by forming outgrowths called buds that develop into a complete animal.

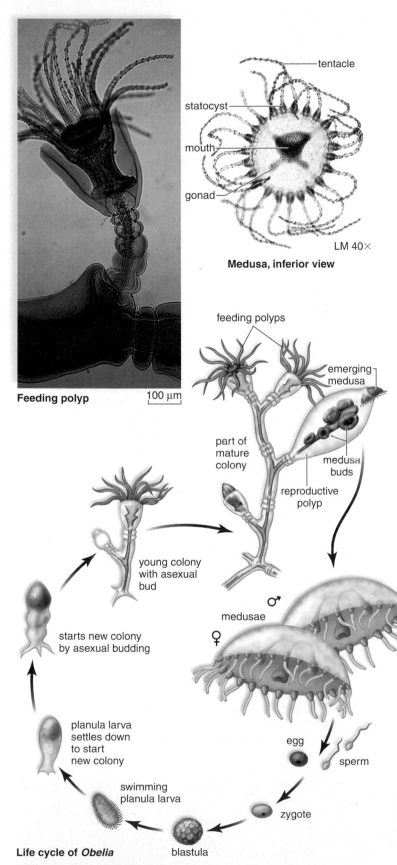

Feeding polyp   100 μm

tentacle

statocyst

mouth

gonad

LM 40×

Medusa, inferior view

feeding polyps

emerging medusa

part of mature colony

medusa buds

reproductive polyp

young colony with asexual bud

medusae

♂

♀

starts new colony by asexual budding

egg

sperm

planula larva settles down to start new colony

zygote

swimming planula larva

blastula

Life cycle of *Obelia*

FIGURE 29.7 *Obelia* life cycle.
Obelia, a colony of feeding polyps and reproductive polyps, undergoes an alternation of generations life cycle.

Hydras can reproduce both asexually and sexually. They reproduce asexually by forming buds, small outgrowths that develop into a complete animal and then detach. When hydras reproduce sexually, sperm from a testis swim to an egg within an ovary. Fertilization and early development occur within the ovary, after which the embryo is encased within a hard, protective shell that allows it to survive until conditions are optimum for it to emerge and develop into a new polyp.

Obelia. *Obelia* (Fig. 29.7) is a common brackish water and marine colonial hydrozoan. It often appears as a small brushy growth on submerged pilings. *Obelia* is dimorphic. The life cycle includes an asexual sessile colonial polyp form and a sexual free-swimming medusa form.

The sessile form of *Obelia* is a branching colony of polyps that is enclosed by a hard, chitinous covering. There are two types of polyps, feeding and reproductive. The feeding polyps extend beyond the covering and can withdraw for protection. They have nematocyst-bearing tentacles that can capture and bring prey, such as tiny crustaceans, worms, and larvae, into the gastrovascular cavity. The polyps are connected, and the partially digested food is distributed to the rest of the colony. The colony increases in size asexually by the budding of new polyps.

Sexual reproduction involves the production of male and female medusae, which bud from reproductive polyps. These medusae tend to be smaller than those of the true jellyfishes. The tentacles attached to the bell margin have numerous nematocysts. Food is brought into the gastrovascular cavity, which extends even into the tentacles. The nerve net is concentrated into two nerve rings; the bell margin is supplied with sensory organs such as statocysts, which function to maintain equilibrium, and ocelli, which are light sensitive. After being released into the environment, medusae swim freely, and produce either eggs or sperm. The fertilized egg, or zygote, eventually develops into a free-swimming planula larva that attaches to a substrate and starts a new colony.

Obelia is an example of a colonial hydroid consisting of feeding and reproductive polyps. Aside from this polyp stage, its life cycle also has a medusa stage.

The relationship of radially symmetrical cnidarians to bilaterally symmetrical animal groups has not been easy to determine. However, some scientists believe that a planuloid-type organism could have given rise to both the cnidarians and the flatworms, which are discussed next. The cnidarians have two germ layers and radial symmetry. The flatworms have three germ layers and bilateral symmetry.

29.4 BILATERAL SYMMETRY

The majority of the animal phyla are bilaterally symmetrical at least in some stage of their development. As embryos, they have three germ layers and are known as triploblastic. In the adult stage, these animals reflect the organ level of organization. Flatworms and ribbon worms have these features. Although flatworms may have evolved from ancestors that had a coelom, and ribbon worms have the suggestion of a coelom, traditional phylogeny considers them to be acoelomates. Flatworms, like the cnidarians, have a sac body plan. Animals with a sac body plan are said to have an incomplete digestive tract, while those with the **tube-within-a-tube body plan** have a complete digestive tract:

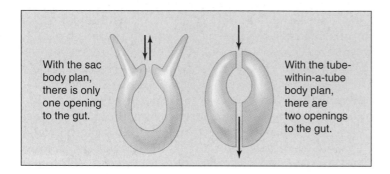

With the sac body plan, there is only one opening to the gut.

With the tube-within-a-tube body plan, there are two openings to the gut.

In the tube-within-a-tube body plan, there is the possibility of specialization of parts along the length of the tube.

Ribbon Worms

Members of phylum **Nemertea** are commonly called **ribbon worms** because of their fragile thread-shaped or ribbon-shaped bodies. This phylum consists of approximately 650 species of primarily bottom-dwelling marine worms that have a distinctive proboscis apparatus. The proboscis can be used to capture prey, but also can be used in defense, burrowing, and locomotion. These worms also possess a long, hollow muscular tube above the digestive tract. It provides muscular pressure to evert the proboscis. Nemerteans are generally less than 20 cm in length. While some of these animals are brightly colored, most are dull in appearance. Figure 29.8 shows a typical ribbon worm. Like roundworms (p. 533), ribbon worms have a tube-within-a-tube body plan.

Flatworms

The organisms in phylum **Platyhelminthes** are known as **flatworms** because of their flat bodies. Over 20,000 species of free-living and parasitic flatworms have been described. Three classes of flatworms are traditionally recognized. Turbellarians are the free-living aquatic planarians and their relatives. Trematodes are the ectoparasitic or endoparasitic flukes. Cestodes are intestinal parasites of vertebrate hosts and are called the tapeworms.

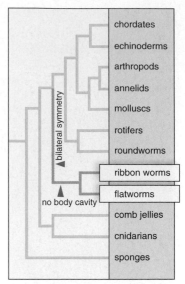

Flatworms are complex animals. In addition to an endodermis derived from endoderm and an epidermis derived from ectoderm, an embryonic mesoderm layer gives rise to muscles and reproductive organs. Flatworms do not have a coelom; instead, mesoderm fills the space between the ectoderm and mesoderm.

Although flatworms exhibit the organ level of organization, there are no circulatory or respiratory structures. The digestive system is incomplete. Nutrients are distributed throughout the body by means of a highly branched gastrovascular cavity. Gas exchange in platyhelminths occurs as a result of diffusion through the thin-walled integument. The excretory system often functions as a osmotic-regulating system. The majority of platyhelminths have complex reproductive systems with both sexes in the same body.

Flatworms are bilaterally symmetrical, and free-living forms do undergo cephalization [Gk. *kaphale*, head], the development of a head region. There is a ladder-type nervous system, so called because the two lateral nerve cords plus the connecting nerves look like a ladder. Paired ganglia (collections of nerve cells) function as a brain; sensory cells are located in the body wall, and the animal is able to respond to various stimuli. In several species, eyespots, chemoreceptors, and tactile receptors are present.

Flatworms have a sac body plan but three germ layers and the organ level of organization. They are bilaterally symmetrical, and cephalization does occur.

FIGURE 29.8

Ribbon worm, *Lineus.*
Ribbon worms (blue), like flatworms, are bilaterally symmetrical and have three germ layers and the organ level of organization. Unlike flatworms, ribbon worms have a complete digestive tract.

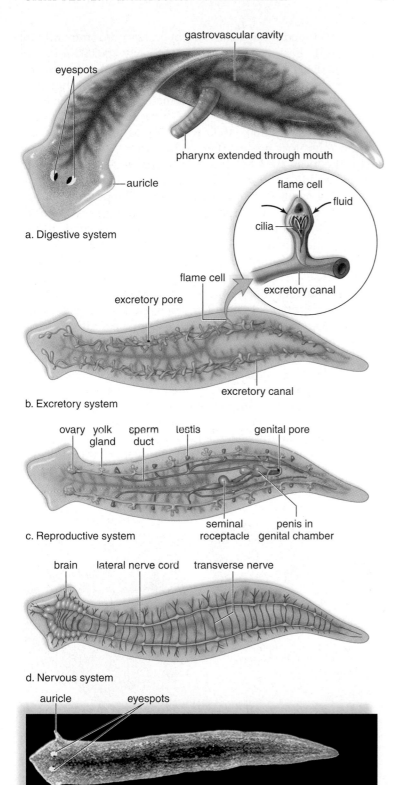

Free-living Flatworms

In class Turbellaria are nearly 3,000 flatworm species, most of which are free-living in marine, freshwater, or terrestrial environments. Their size ranges from 5 mm to less than 50 cm. The best-known turbellarians are the free-living planarians with a very flat body colored brown or black (Fig. 29.9e). *Dugesia* is a common planarian that resides in freshwater lakes, streams, and ponds, where it feeds on small living or dead organisms.

A planarian captures food by wrapping itself around the prey, entangling it in slime, and pinning it down. Then a muscular pharynx is extended, and by a sucking motion, the food is torn up and swallowed. The pharynx leads into a three-branched gastrovascular cavity in which digestion is both extracellular and intracellular (Fig. 29.9a).

Living in fresh water, *Dugesia* has a well-developed excretory system (Fig. 29.9b). The excretory organ functions in osmotic regulation as well as in water excretion. The organ consists of a series of interconnecting canals that run the length of the body on each side. Bulblike structures containing cilia are at the ends of the side branches of the canals. The cilia move back and forth, bringing water into the canals that empty at pores. The beating of the cilia reminded an early investigator of the flickering of a flame, and so the excretory organ of the flatworm is called a flame cell.

Planarians can reproduce asexually. They constrict beneath the pharynx, and each part grows into a whole animal again. Because planarians have the ability to regenerate, they have been the subject of much research. Planarians also reproduce sexually. They are **hermaphroditic,** which means that they possess both male and female sex organs (Fig. 29.9c). The worms practice cross-fertilization when the penis of one is inserted into the genital pore of the other. The fertilized eggs are enclosed in a cocoon and hatch in two or three weeks as tiny worms.

The head of a planarian is bluntly arrow shaped, with lateral extensions called auricles that function as sense organs to detect potential food sources and enemies. Inside, the brain is connected to a ladder-type nervous system (Fig. 29.9d). There are two light-sensitive eyespots whose pigmentation causes the worm to look cross-eyed. Three kinds of muscle layers—an outer circular layer, an inner longitudinal layer, and a diagonal layer—allow for quite varied movement. In larger forms, locomotion is accomplished by the movement of cilia on the ventral and lateral surfaces. Numerous gland cells secrete a mucous material upon which the animal moves.

Free-living planarians best exhibit the bilateral symmetry and organ development—including the nervous system and muscles—of a flatworm.

FIGURE 29.9 Planarian anatomy.
a. When a planarian extends the pharynx, food is sucked up into a gastrovascular cavity that branches throughout the body. **b.** The excretory system with flame cells is shown in detail. **c.** The reproductive system (shown in pink and blue) has both male and female organs. **d.** The nervous system has a ladderlike appearance. **e.** The photograph shows that a flatworm, *Dugesia*, is bilaterally symmetrical and has a head region with eyespots.

Parasitic Flatworms

Flukes (trematodes) and tapeworms (cestodes) are parasitic flatworms. Both are highly modified for the parasitic mode of life. Flukes and tapeworms are covered by a protective tegument, which is a specialized body covering resistant to host digestive juices. Concomitant with the loss of predation is an absence of cephalization; the anterior end notably carries hooks and/or suckers for attachment to the host. Parasites no longer seek out and digest prey, and the nervous system is not well developed. On the other hand, a well-developed reproductive system helps ensure transmission to a new host.

Both flukes and tapeworms utilize a secondary, or intermediate, host to transport the species from primary host to primary host. The primary host is infected with the sexually mature adult; the secondary host contains the larval stage or stages.

Flukes. The members of class Trematoda are all parasitic flukes. Most are vertebrate parasites named for the organ they inhabit; for example, there are blood, liver, and lung flukes. The almost 11,000 species have an oval to more elongated flattened body. At the anterior end is an oral sucker surrounded by sensory papilla and at least one other sucker used for attachment to a host. Trematodes have a reduced digestive, nervous, and sensory system. Their excretory and muscular systems are similar to those in the turbellarians. Most flukes have one set of reproductive organs, either male or female.

The blood fluke (*Schistosoma*) occurs predominantly in the Middle East, Asia, and Africa. Nearly 800,000 infected persons die each year from **schistosomiasis.** Adults are approximately 2.5 cm long and may live for years in their human hosts. In this disease, the female flukes deposit their eggs in small blood vessels close to the lumen of the intestine, and the eggs make their way into the digestive tract by a slow migratory process (Fig. 29.10). After the eggs pass out with the feces, they hatch into tiny larvae that swim about in rice paddies and elsewhere until they enter a particular species of snail. Within the snail, asexual reproduction occurs; sporocysts, which are spore-containing sacs, eventually produce new larval forms that leave the snail. If the larvae penetrate the skin of a human, they begin to mature in the liver and implant themselves in the small intestinal blood vessels. The flukes and their eggs can cause dysentery, anemia, bladder inflammation, brain damage, and severe liver complications. Infected persons usually die of secondary diseases brought on by their weakened condition.

The Chinese liver fluke, *Clonorchis sinensis,* is a parasite of cats, dogs, pigs, and humans and requires two secondary hosts: a snail and a fish. The adults reside in the liver and deposit their eggs in bile ducts, which carry them to the intestines for elimination in feces. Nonhuman species generally become infected through the fecal route, but humans usually become infected by eating raw fish. A heavy *Clonorchis* infection can cause severe cirrhosis of the liver and death.

FIGURE 29.10 Life cycle of a blood fluke, *Schistosoma*.
a. Micrograph of *Schistosoma*. **b.** This infection of humans, caused by blood flukes, *Schistosoma*, is an extremely prevalent disease in Egypt—especially since the building of the Aswan High Dam. Standing water in irrigation ditches, combined with unsanitary practices, has created the conditions for widespread infection.

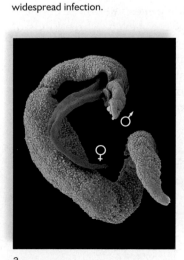

a.

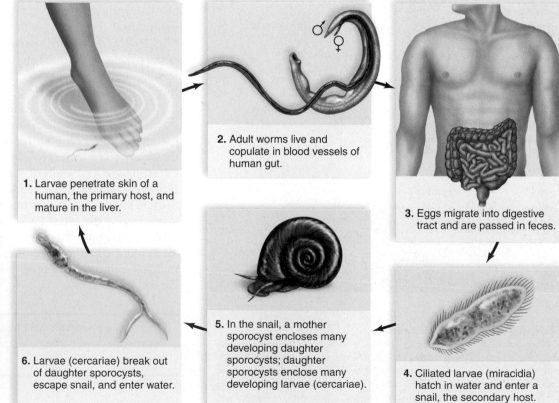

1. Larvae penetrate skin of a human, the primary host, and mature in the liver.

2. Adult worms live and copulate in blood vessels of human gut.

3. Eggs migrate into digestive tract and are passed in feces.

4. Ciliated larvae (miracidia) hatch in water and enter a snail, the secondary host.

5. In the snail, a mother sporocyst encloses many developing daughter sporocysts; daughter sporocysts enclose many developing larvae (cercariae).

6. Larvae (cercariae) break out of daughter sporocysts, escape snail, and enter water.

b.

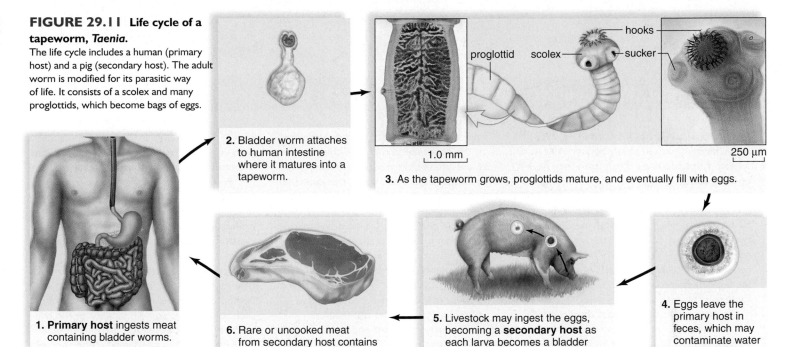

FIGURE 29.11 Life cycle of a tapeworm, *Taenia*.
The life cycle includes a human (primary host) and a pig (secondary host). The adult worm is modified for its parasitic way of life. It consists of a scolex and many proglottids, which become bags of eggs.

2. Bladder worm attaches to human intestine where it matures into a tapeworm.

3. As the tapeworm grows, proglottids mature, and eventually fill with eggs.

1. Primary host ingests meat containing bladder worms.

6. Rare or uncooked meat from secondary host contains many bladder worms.

5. Livestock may ingest the eggs, becoming a **secondary host** as each larva becomes a bladder worm encysted in muscle.

4. Eggs leave the primary host in feces, which may contaminate water or vegetation.

Tapeworms. Class Cestoda consists of nearly 4,000 species of vertebrate parasites known as tapeworms. Tapeworms vary in length from a few millimeters to nearly 20 m. They have a highly modified head region called the **scolex** that contains hooks for attachment to the intestinal wall of the host and suckers for feeding. Behind the scolex, a series of reproductive units called **proglottids** are found that contain a full set of female and male sex organs. The number of proglottids may vary depending on the species. After fertilization, the organs within a proglottid disintegrate and become filled with mature eggs. These egg-filled proglottids are called gravid, and in some species may contain 100,000 eggs. Once mature, depending on the species, the eggs are released through a pore into the host's intestine, where they exit with the feces. In other species, the gravid proglottids break off and are eliminated with the feces.

Most tapeworms have complicated life cycles that usually involve several hosts. Figure 29.11 illustrates the life cycle of the pork tapeworm, *Taenia solium*, which involves the human as the primary host and the pig as the secondary host. After a pig feeds on feces-contaminated food, the larvae are released. They burrow through the intestinal wall and travel in the bloodstream to finally lodge and encyst in muscle. This **cyst** is a small, hard-walled structure that contains a larva called a bladder worm. When humans eat infected meat that has not been thoroughly cooked, the bladder worms break out of the cysts, attach themselves to the intestinal wall, and grow to adulthood. Then the cycle begins again. Generally tapeworm infections cause diarrhea, weight loss, and fatigue in the primary host.

Flukes and tapeworms illustrate the modifications that occur when animals adapt to the parasitic way of life.

Table 29.1 contrasts the features of planarians, flukes, and tapeworms to illustrate the anatomical modifications associated with the parasitic way of life.

TABLE 29.1

Free-Living Flatworms Versus Parasitic Flatworms

	Free-Living	Parasitic	
	Planarians	*Flukes*	*Tapeworms*
Body wall	Ciliated epidermis	Tegument	Tegument
Cephalization	Yes, eyespots and auricles	No, oral suckers	No, scolex with hooks and suckers
Nervous organization	Nerves and brain	Reduced	Reduced
Gastrovascular cavity	Branched	Reduced	Absent
Reproductive organs	Hermaphroditic	Separate sexes	Hermaphroditic proglottids
Larva	Absent	Present	Present

science focus

Acoelomate, Pseudocoelomate, and Coelomate Animals

In acoelomates, such as flatworms (Fig. 29E*a*), mesoderm is packed solidly between the ectoderm and endoderm. Therefore, they have no body cavity.

Roundworms, on the other hand, are pseudocoelomates (Fig. 29E*b*). They have a body cavity but it is incompletely lined with mesoderm. A layer of mesoderm lies beneath the body wall but not around the gut. The internal organs are suspended within their pseudocoelom. Apparently, this is disadvantageous to an increase in size, because most of these worms are small.

In coelomates, the coelom develops as a cavity within the mesoderm; therefore, the coelom is completely lined with mesoderm (Fig. 29E*c*). Such a coelom is sometimes called a "true coelom." A layer of mesoderm lies beneath the body wall and around the gut. The inner and outer layers of mesoderm connect to form tissues called mesenteries that surround the internal organs. Mesenteries ensure a more stable arrangement of internal organs with less crowding. In coelomates (all the other animals we will study), the internal organs are more complex than in animals without a coelom, the gut wall is muscular (muscles are derived from mesoderm), and the gut shows specialization of parts not seen in the pseudocoelomates.

A coelom also serves other functions. It allows the internal organs and the body wall to move independently. This means an animal can stretch and bend without putting a strain on the internal organs. The coelom is filled with fluid, and this fluid protects and cushions the internal organs. In some animals, coelomic fluid aids in the movement of materials, such as metabolic wastes; in others, this function is taken over by blood vessels. Not only metabolic wastes but also sex cells may be deposited in the coelomic cavity before they are transported away by ducts.

The gastrovascular cavity of acoelomates (flatworms) and the fluid-filled coelom of soft-bodied coelomates can act as a hydrostatic skeleton. It offers some resistance to the contraction of muscles and yet permits flexibility, so that the animal can change shape and perform a variety of movements.

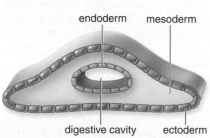
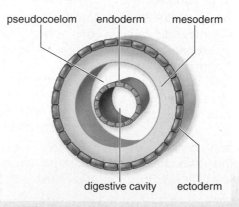
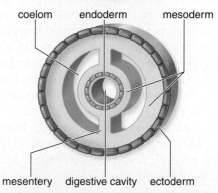

a. **Acoelomate** (flatworms)   b. **Pseudocoelomate** (roundworms)   c. **Coelomate** (molluscs, annelids, arthropods, echinoderms, chordates)

FIGURE 29E Acoelomate, pseudocoelomate, and coelomate comparison.

29.5 BODY CAVITY

The animals in the remaining phyla to be studied have a body cavity.

Roundworms (phylum Nematoda) and rotifers (phylum Rotifera) are pseudocoelomates with a complete digestive tract (tube-within-a-tube body plan). A pseudocoelom [Gk. *pseudes*, false, and *koiloma*, cavity] is a body cavity that is incompletely lined by mesoderm. In other words, mesoderm occurs inside the body wall but not around the digestive cavity (gut). Traditionally, acoelomates and pseudocoelomates are not consid-ered protostomes because they don't have a coelom completely lined by mesoderm. The phylogenetic tree based on molecular data considers them to be protostomes. The accompanying Science Focus compares animals on the basis of coelom type and describes how a coelom can serve as a *hydrostatic skeleton.*

> Pseudocoelomate animals (e.g., roundworms and rotifers) have a coelom that is incompletely lined by mesoderm. Their coelom provides a space for internal organs and can serve as a hydrostatic skeleton.

Roundworms

Phylum Nematoda consists of over 90,000 species of smooth, nonsegmented **roundworms.** Nematodes are prevalent in almost any environment. Generally, nematodes are colorless and range in size from microscopic to exceeding 1 m in length. The internal organs, including the tubular reproductive organs, lie within the pseudocoelom.

Nematodes have developed a variety of lifestyles from free-living to parasitic. Some nematodes cause great agricultural damage. One species, *Caenorhabditis elegans*, a free-living nematode, is a model animal used in genetics and developmental biology. Several parasitic nematodes, including pinworms, hookworms, *Trichinella*, and *Ascaris*, infect humans and other animals.

Ascaris

Ascaris lumbricoides is one of the most common intestinal parasites in humans. Other species of *Ascaris* can be found in cats, dogs, pigs, and a number of other vertebrates. As in other nematodes, *Ascaris* males tend to be smaller (15–31 cm long) than females (20–49 cm long)(Fig. 29.12*a*). In males, the posterior end is curved and comes to a point. Both sexes move by means of a characteristic whiplike motion because only longitudinal muscles lie next to the body wall.

Because mating produces eggs that mature in the soil, the parasite is limited to warmer environments. A typical female *Ascaris* is very prolific, producing over 200,000 eggs daily. The eggs are eliminated with the host's feces and can remain viable in the soil for many months. Proper sanitation is the best means to prevent infection with nematodes such as *Ascaris*. Eggs enter the body via uncooked vegetables, soiled fingers, or ingested fecal material and hatch in the intestines. The juveniles make their way into the veins and lymphatic vessels and are carried to the heart and lungs. From the lungs, the larvae travel up the trachea, where they are swallowed and eventually reach the intestines. There, the larvae mature and begin feeding on intestinal contents.

The symptoms of an *Ascaris* infection depend on the stage of the infection. Larval *Ascaris* in the lungs can cause pneumonia-like symptoms. In the intestines, *Ascaris* can cause malnutrition; blockage of the bile duct, pancreatic duct, and appendix; and poor health.

Other Roundworm Parasites

Trichinosis, caused by *Trichinella spiralis*, is a serious infection that humans can contract when they eat rare pork containing encysted *Trichinella* larvae. After maturation, the female adult burrows into the wall of the small intestine and produces living offspring that are carried by the bloodstream to the skeletal muscles, where they encyst (Fig. 29.12*c*). Heavy infections can be painful and lethal.

Filarial worms, a type of roundworm, cause various diseases. In the United States, a filarial worm is a parasite of dogs. Because the worms live in the heart and the arteries that serve the lungs, the infection is called **heartworm disease.** The condition can be fatal; therefore, heartworm medicine is recommended as a preventive measure for all dogs.

Elephantiasis is a disease of humans caused by the filarial worm, *Wuchereria bancrofti*. Restricted to tropical areas of Africa, the parasite uses a mosquito as a secondary host. Because the adult worms reside in lymphatic vessels, collection

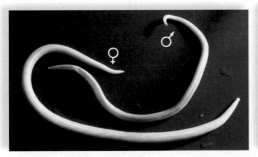
a. *Ascaris*

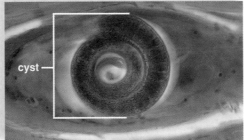
c. *Trichinella*

SEM 400×

FIGURE 29.12 Roundworm anatomy.
a. The roundworm *Ascaris*. **b.** Roundworms such as *Ascaris* have a pseudocoelom and a complete digestive tract with a mouth and an anus. **c.** The larvae of the roundworm *Trichinella* penetrate striated muscle fibers, where they coil in a sheath formed from the muscle fiber.

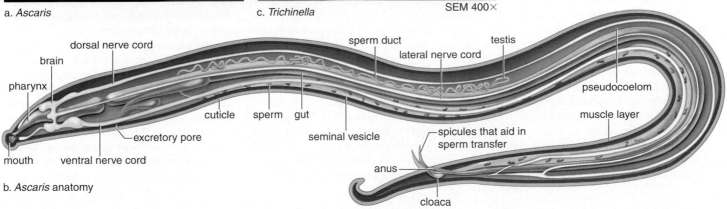
b. *Ascaris* anatomy

FIGURE 29.13

Filarial worm.

An infection from a filarial worm, *Wuchereria bancrofti*, causes elephantiasis, a condition in which the individual experiences extreme swelling in regions where the worms have blocked the lymphatic vessels.

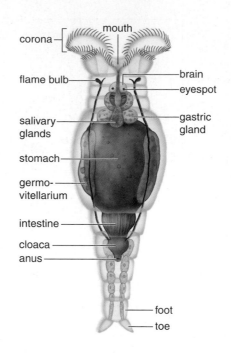

FIGURE 29.14

Rotifer.

Rotifers are microscopic animals only 0.1–3 mm in length. The beating of cilia on two lobes at the anterior end of the animal gives the impression of a pair of spinning wheels

of fluid is impeded, and the limbs of an infected person may swell to a monstrous size (Fig. 29.13). Elephantiasis is treatable in its early stages but usually not after scar tissue has blocked lymphatic vessels.

Pinworms are the most common nematode parasite in the United States. The adult parasites live in the cecum and large intestine. Females migrate to the anal region at night and lay their eggs. Scratching the resultant itch can contaminate hands, clothes, and bedding. The eggs are swallowed and the life cycle begins again.

Rotifers

Anton von Leeuwenhoek viewed rotifers through his microscope and called them the "wheel animacules." **Rotifers,** phylum **Rotifera,** have a crown of cilia, known as the corona, on their heads (Fig. 29.14). When in motion the corona, which looks like a spinning wheel, serves as an organ of locomotion and also directs food into the mouth.

The approximately 2,000 species primarily live in fresh water; however, some marine and terrestrial forms exist. The majority of rotifers are transparent, but some are very colorful. Many species of rotifers can desiccate during harsh conditions and remain dormant for lengthy periods of time. This characteristic has earned them the title "resurrection animacules."

Both roundworms and rotifers have a pseudocoelom. Roundworms are extremely plentiful and cause various diseases in humans.

CONNECTING THE CONCEPTS

Animals are a diverse group with certain features in common. Traditionally, zoologists have relied on such features as level of organization, body plan, symmetry, type of body cavity, and presence of segmentation in order to classify animals. The type of body cavity was also used to help categorize animals into certain groups. While we still rely on these criteria in this chapter, we have to keep in mind that a new type of classification based on molecular data is available. Still in its infancy, it may become the preferred way in the future.

So far there is little controversy about the early evolving groups. Sponges, the simplest multicellular animals, are significantly different from other phyla because they have no true tissues. Apparently, no other groups evolved from them and many biologists consider sponges to be outside the mainstream of animal evolution. The sea anemones, jellyfishes, and related species (cnidarians and ctenophores) are radially symmetrical and have only two germ layers as embryos.

The classification of animals more complex than the cnidarian group is currently being reexamined. Based on rRNA sequence data, should the flatworms and roundworms be considered protostomes even though flatworms have no coelom and roundworms have a pseudocoelom? Some biologists are including these groups among the protostomes and some are not. In the next chapter we consider the groups of animals that are traditionally considered protostomes: molluscs, annelids, and arthropods. The proposed new method of classification places the annelids and arthropods in different groups even though both are segmented. In this instance molecular data can be bolstered by anatomical characteristics shared only by roundworms and arthropods. Still, not all biologists are ready at this time to use the new system, and they await further findings before adopting the classification system described in the Science Focus on pages 520–21.

Summary

29.1 EVOLUTION OF ANIMALS

Animals are multicellular organisms that are heterotrophic and ingest their food. They have the diploid life cycle. Typically, they have the power to move by means of contracting fibers.

It's possible to construct a phylogenetic tree for animals, but this is largely based on a study of today's forms. Level of organization, type of body plan, type of symmetry, type of body cavity, and presence or absence of segmentation are criteria used in classification. The use of molecular data is a new criterion of classification.

29.2 MULTICELLULARITY

All animals are multicellular. Sponges may have evolved separately from other animals, since they have features that set them apart. They have the cellular level of organization, lack tissues, and have various symmetries. Sponges are sessile and depend on a flow of water through the body to acquire food, which is digested in vacuoles within collar cells that line a central cavity.

29.3 TRUE TISSUE LAYERS

Animals in the remaining phyla to be studied have true tissue layers. Comb jellies and cnidarians are diploblastic and have tissue layers derived from the germ layers ectoderm and endoderm. They are radially symmetrical.

Cnidarians have a sac body plan. They exist as either polyps or medusae, or they can alternate between the two. Hydras and their relatives—sea anemones and corals—are polyps; in jellyfishes, the medusan stage is dominant. In *Hydra* and other cnidarians, an outer epidermis is separated from an inner gastrodermis by mesoglea. They possess tentacles to capture prey and nematocysts to stun it. A nerve net coordinates movements. Digestion of prey begins in the gastrovascular cavity and is finished within gastrodermal cells.

29.4 BILATERAL SYMMETRY

Animals in the remaining phyla to be studied have bilateral symmetry. They also are triploblastic and have organs derived from mesoderm. Ribbon worms and flatworms have these features. They are also acoelomates.

Flatworms may be free-living or parasitic. Freshwater planarians exemplify the features of flatworms in general and free-living forms in particular. They have muscles and a ladder-type nervous system, and they show cephalization. They take in food through an extended pharynx leading to a gastrovascular cavity, which extends throughout the body. There is an osmotic-regulating organ that contains flame cells.

Flukes and tapeworms are parasitic. Flukes have two suckers by which they attach to and feed from their hosts. Tapeworms have a scolex with hooks and suckers for attaching to the host intestinal wall. The body of a tapeworm is made up of proglottids, which, when mature, contain thousands of eggs. If these eggs are taken up by pigs or cattle, larvae become encysted in their muscles. If humans eat this meat, they too may become infected with a tapeworm.

29.5 BODY CAVITY

Animals in the remaining phyla to be studied have a body cavity. They also have a complete digestive tract. Roundworms and rotifers have a pseudocoelom. A coelom provides a space for internal organs and can serve as a hydrostatic skeleton. Roundworms are usually small and very diverse; they are present almost everywhere in great numbers. The parasite *Ascaris* is representative of the group. Infections can also be caused by *Trichinella*, whose larval stage encysts in the muscles of humans. Elephantiasis is caused by a filarial worm that blocks lymphatic vessels.

Reviewing the Chapter

1. List the criteria used to classify animals. 518
2. What does the traditional phylogenetic tree (see Fig. 29.2) tell you about the evolution of the animals studied in this chapter? 519
3. What features make sponges different from the other organisms placed in the animal kingdom? 522–23
4. List the types of cells found in a sponge, and describe their functions. 522
5. What are the two body forms found in cnidarians? Explain how they function in the life cycle of various types of cnidarians. 524
6. Describe the anatomy of *Hydra*, pointing out those features that typify cnidarians. 526–27
7. Describe the anatomy of a free-living planarian, pointing out those features that typify nonparasitic flatworms. 528–29
8. Describe the parasitic flatworms, and give the life cycle of both the blood fluke that causes schistosomiasis and the pork tapeworm. 530
9. What is a pseudocoelom? What are the advantages of a coelom? What two groups of animals have a pseudocoelom? 532
10. Describe the anatomy of *Ascaris*, pointing out those features that typify roundworms. 533

Testing Yourself

Choose the best answer for each question.

1. Which of these is not a characteristic of animals?
   a. heterotrophic
   b. diploid life cycle
   c. have contracting fibers
   d. single cells or colonial
   e. lack of chlorophyll

2. The traditional phylogenetic tree of animals shows that
   a. three germ layers evolved before a coelom.
   b. both molluscs and annelids are protostomes.
   c. some animals have radial symmetry.
   d. sponges were the first to evolve from an ancestral protist.
   e. All of these are correct.

3. Which of these sponge characteristics is not typical of animals?
   a. They practice sexual reproduction.
   b. They have the cellular level of organization.
   c. They have various symmetries.
   d. They have flagellated cells.
   e. Both b and c are not typical.

4. Which of these pairs is mismatched?
   a. sponges—spicules
   b. tapeworms—proglottids
   c. cnidarians—nematocysts
   d. roundworms—cilia
   e. cnidarians—polyp and medusa

5. Flukes and tapeworms
   a. show cephalization.
   b. have well-developed reproductive systems.
   c. have well-developed nervous systems.
   d. are free-living.

6. The presence of mesoderm
   a. restricts the development of a coelom.
   b. is associated with the organ level of organization.
   c. is associated with the development of muscles.
   d. means the animal is diploblastic.
   e. Both b and c are correct.
7. *Ascaris* is a parasitic
   a. roundworm.   d. sponge.
   b. flatworm.   e. comb jelly.
   c. hydra.
8. The traditional phylogenetic tree of animals shows that
   a. cnidarians evolved directly from sponges.
   b. flatworms evolved directly from roundworms.
   c. ribbon worms are closely related to flatworms.
   d. coelomates gave rise to the acoelomates.
   e. All of these are correct.
9. Comb jellies are most closely related to
   a. cnidarians.   d. roundworms.
   b. sponges.   e. Both a and b are correct.
   c. flatworms.
10. Label the following diagram of the cnidarian polyp.

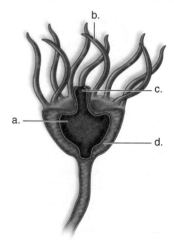

11. Write a correct phylum name beside each of the following terms.
    a. radial symmetry:
    b. pseudocoelom:
    c. tissue level of organization:
    d. tube-within-a-tube body plan:
    e. cephalization:
    f. two body forms:
12. Write the correct type of animal beside each of the following terms.
    a. proglottids:
    b. eyespots on head:
    c. collar cells:
    d. cnidocytes:
    e. crown of cilia:
    f. branched gastrovascular cavity:
    g. nerve net:

Thinking Scientifically

1. Think of the animals in this chapter that are radially symmetrical (cnidarians, comb jellies). How is the lifestyle of many radially symmetrical animals different from that of bilaterally symmetrical animals? How does their body plan complement their lifestyle?
2. Roundworms are tubular and have a complete digestive tract. What advantages can you associate with these anatomical features that might account for roundworms being more plentiful than flatworms?

Understanding the Terms

acoelomate 518
Animalia 518
asymmetrical 518
bilateral symmetry 518
cephalization 518
cestode 528
cnidarian 524
coelom 518
coelomate 518
comb jelly 524
cyst 531
deuterostome 518
dimorphic 524
elephantiasis 533
flatworm 528
gastrovascular cavity 524
heartworm disease 533
hermaphroditic 529
hydra 526
invertebrate 518
medusa 524
mesoglea 524
nematocyst 524
nerve net 524
Obelia 527
polyp 524
proglottid 531
protostome 518
pseudocoelomate 518
radial symmetry 518
ribbon worm 528
rotifer 534
roundworm 533
schistosomiasis 530
scolex 531
segmentation 518
sessile 518
sessile filter feeder 522
spicule 523
sponge 522
spongin 523
trematode 528
trichinosis 533
tube-within-a-tube body plan 528
turbellarian 528
vertebrate 518

Match the terms to these definitions:

a. _____ Blind digestive cavity that also serves a circulatory (transport) function in animals lacking a circulatory system.
b. _____ Body cavity lying between the digestive tract and body wall that is completely lined by mesoderm.
c. _____ Body cavity lying between the digestive tract and body wall that is incompletely lined by mesoderm.
d. _____ Body plan having two corresponding or complementary halves.
e. _____ Body with a digestive tract that has both a mouth and an anus.

ARIS, the *Biology* Website

ARIS, the website for *Biology*, provides a wealth of information organized and integrated by chapter. You will find practice quizzes, interactive activities, labeling exercises, flashcards, and much more that will complement your learning and understanding of general biology.

www.mhhe.com/maderbiology9

30

More Invertebrates

Ordinarily, mussels are relatively simple animals with two shells connected by a ligament. Not so the female of the species Lampsilis reeveiana, *which lives in the streams of Missouri. This female mussel develops and displays an outer flap that looks like a minnow, specifically a darter. Why might that be? Unlike oysters and marine clams, the larvae of freshwater mussels need a fish to complete their life cycle. The fake fish acts as a lure to attract a bass, a fish that likes to eat darters. When the bass chomps down on the "darter," it breaks open, releasing thousands of tiny larvae called glochidia. The larvae clamp onto the gills of the bass, where they encyst and continue developing. In three weeks they drop off the gills, which appear to be unharmed, and settle to the bottom as juveniles. When the juveniles become adults the reproductive cycle will begin again: Males release sperm that fertilize eggs retained by the female. The fertilized eggs develop into larvae, which are enclosed within the lure. Animals, as illustrated by Lampsilis reeveiana, have developed all sorts of ways to ensure their genes are passed to a new generation.*

Mussels belong to a group of animals known as molluscs. In addition to molluscs, this chapter also discusses annelids, arthropods, and echinoderms. Many of these animals have their own incredible life history.

Female mussel, *Lampsilis reeveiana.*

lure

30.1 Advantages of Coelom in Protostomes and Deuterostomes

The remaining phyla to be studied are either **protostomes** [Gk. *protos*, first; L. *stoma*, mouth] or **deuterostomes** [Gk. *deuteros*, second; L. *stoma*, mouth]. Traditionally, the protostomes include only molluscs, annelids, and arthropods because they are also coelomate animals. As discussed in the Science Focus on pages 520–21, molecular data suggest that other invertebrate phyla, including, in particular, the flatworms and roundworms, are also protostomes. The flatworms do not have a coelem, and roundworms do not have a true coelom, but both of these groups exhibit the first two protostome features discussed below. The designation of echinoderms and chordates as deuterostomes is supported by both traditional and molecular data. Protostomes and deuterostomes are differentiated by three major events in their embryological development (Fig. 30.1):

1. Cleavage, the first event of development, is cell division without an increase in the size of the cells. In protostomes, spiral cleavage occurs, and daughter cells sit in grooves formed by the previous cleavages. The fate of these cells is fixed and determinate in protostomes; each can contribute to development in only one particular way. In deuterostomes, radial cleavage occurs, and the daughter cells sit right on top of the previous cells. The fate of these cells is indeterminate—that is, if they are separated from one another, each cell can go on to become a complete organism.

2. As development proceeds, a hollow sphere of cells, or blastula, forms and the indentation that follows produces an opening called the blastopore. In protostomes, the mouth appears at or near the blastopore, hence the origin of their name; in deuterostomes, the anus appears at or near the blastopore, and only later does a new opening form the mouth, hence the origin of their name.

3. Traditionally both protostomes and deuterostomes have a body cavity completely lined by mesoderm, called a true **coelom** [Gk. *koiloma*, cavity]. However, the coelom develops differently in the two groups. In protostomes, the mesoderm arises from cells located near the embryonic blastopore, and a splitting occurs that produces the coelom, which is thus called a **schizocoelom**. In deuterostomes, the coelom arises as a pair of mesodermal pouches from the wall of the primitive gut. The pouches enlarge until they meet and fuse, forming an **enterocoelom.**

As mentioned in Chapter 29, the presence of a coelom has many advantages. Body movements are freer because the outer wall can move independently of the enclosed organs. Also, the ample space of a coelom allows complex organs and organ systems to develop. For example, the digestive tract can coil and provide a greater surface area for absorption

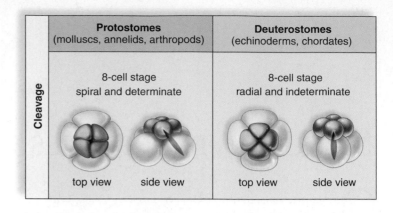

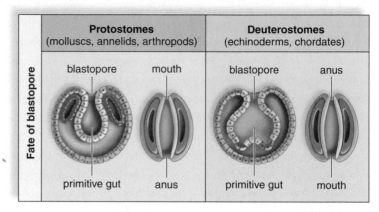

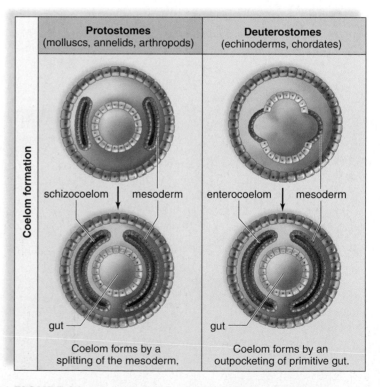

FIGURE 30.1 Protostomes compared to deuterostomes.
In the embryo of protostomes (e.g., molluscs, annelids, and arthropods), cleavage is spiral (*orange*)—new cells are at an angle to old cells—and each cell has limited potential and cannot develop into a complete embryo; the blastopore is associated with the mouth, and the coelom is a schizocoelom. In deuterostomes (echinoderms and chordates), cleavage is radial (*orange*)—new cells sit on top of old cells—and each one can develop into a complete embryo; the blastopore is associated with the anus, and the coelom is an enterocoelom.

DOMAIN: Eukarya
KINGDOM: Animalia

CHARACTERISTICS
• Multicellular
• Well-developed tissues (except sponges)
• Usually motile
• Heterotrophic by ingestion or absorption, generally a digestive cavity
• Diploid life cycle

Invertebrates*
PHYLUM: Porifera ——————————— sponges
PHYLUM: Cnidaria ——————————— jellyfishes, sea anemones, corals
PHYLUM: Ctenophora ——————————— comb jellies, sea walnuts
PHYLUM: Platyhelminthes ——————— flatworms (e.g., planarians, flukes, tapeworms)
PHYLUM: Nemertea ——————————— ribbon worms
PHYLUM: Nematoda ——————————— roundworms
PHYLUM: Rotifera ——————————— rotifers
PHYLUM: Mollusca ——————————— chitons, snails, slugs, clams, oysters, mussels, squids, octopuses
PHYLUM: Annelida ——————————— segmented worms (e.g., clam worms, earthworms, leeches)
PHYLUM: Arthropoda ——————————— spiders, scorpions, horseshoe crabs, lobsters, crayfish, shrimps, crabs, millipedes, centipedes, insects
PHYLUM: Echinodermata ——————— sea lilies, sea stars, brittle stars, sea urchins, sand dollars, sea cucumbers, sea daisies
PHYLUM: Chordata
   SUBPHYLUM: Urochordata ——————— sea squirts
   SUBPHYLUM: Cephalochordata ——— lancelets
Vertebrates*
   SUBPHYLUM: Vertebrata
      SUPERCLASS: Agnatha ——————— jawless fishes (e.g., lampreys, hagfishes)
      SUPERCLASS: Gnathostomata — jawed fishes; all tetrapods
         CLASS: Chondrichthyes ——————— cartilaginous fishes (e.g., sharks, skates, rays)
         CLASS: Osteichthyes ——————— bony fishes (e.g., herring, salmon, cod, eel, flounder)
         CLASS: Amphibia ——————————— frogs, toads, salamanders, newts, caecilians
         CLASS: Reptilia ——————————— snakes, lizards, turtles, crocodiles
         CLASS: Aves ——————————— birds (e.g., sparrows, penguins, ostriches)
         CLASS: Mammalia ——————————— mammals (e.g., cats, dogs, horses, rats, humans)

* Not in the classification of organisms, but added here for clarity

of nutrients. And the coelomic cavity can serve as a storage area for eggs and sperm before they are released into the environment.

Coelomic fluid protects internal organs against damage and against marked temperature changes. It can also assist in respiration and circulation by providing oxygen and nutrients to nearby cells. Metabolic wastes can accumulate in the cavity prior to being taken away by the excretory system. Finally, fluid within the cavity protects internal organs and can provide a hydrostatic skeleton—that is, muscular contraction pushes against the fluid and allows the animal to move.

Advanced Characteristics

In addition to having a true coelom, the animals discussed in this chapter have three germ layers, the organ level of organization, and a complete digestive tract. Like other groups of animals, they evolved in the sea, but the molluscs, annelids, and arthropods have successful terrestrial representatives. In this chapter, we will contrast animal adaptations suitable to living in water with adaptations suitable to living on land. Terrestrial existence requires breathing air, preventing desiccation, and having a means of locomotion and reproduction that are not dependent on external water. The excretory system may be modified for the excretion of a solid nitrogenous waste to help conserve water.

The presence of a coelom enables the digestive system and the body wall to move independently and allows organs to become more complex. Coelomic fluid can assist respiration, circulation, and excretion, and also serves as a hydrostatic skeleton.

30.2 MOLLUSCS

Almost everyone enjoys looking at the intricate patterns and beauty of seashells; however, few people are aware of the tremendous diversity found in the phylum **Mollusca** [L. *mollusc*, soft]. At present, over 110,000 living species and 35,000 fossil species dating back to the Cambrian period have been classified in what is considered the second largest animal phylum. The **molluscs** inhabit a variety of environments including marine, freshwater, and terrestrial habitats. This diverse

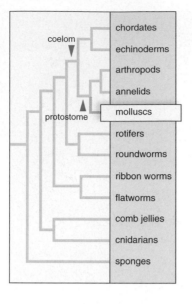

phylum includes chitons, limpets, slugs, snails, abalones, conchs, nudibranchs, clams, scallops, and octopuses. Molluscs vary in size from microscopic to the giant squid, which can attain lengths of over 20 m and weigh over 450 kg. Phylum Mollusca includes herbivores, carnivores, filter feeders, and parasites.

Although diverse, molluscs share a three-part body plan consisting of the visceral mass, mantle, and foot (Fig. 30.2a). The visceral mass contains the internal organs, including a highly specialized digestive tract, paired kidneys, and reproductive organs. The **mantle** is a covering that lies to either side of but does not completely enclose the visceral mass. It may secrete a shell and/or contribute to the development of gills or lungs. The space between the folds of the mantle is called the mantle cavity. The foot is a muscular organ that may be adapted for locomotion, attachment, food capture, or a combination of functions. Another feature often present in molluscs is a rasping, tonguelike **radula,** an organ that bears many rows of teeth and is used to obtain food (Fig. 30.3b).

The coelem is reduced and largely limited to the region around the heart in molluscs. Most molluscs have an open circulatory system. The heart pumps blood, more properly called hemolymph, through vessels into sinuses (cavities) collectively called a hemocoel. Blue hemocyanin, rather than red hemoglobin, is the respiratory pigment.

The nervous system of molluscs consists of several ganglia connected by nerve cords. The amount of cephalization and sensory organs varies from nonexistent in clams to complex in squid and octopuses. The molluscs also exhibit variations in mobility. Oysters are sessile, snails are extremely slow moving, and squid are fast-moving, active predators.

Three of the most commonly studied classes of molluscs include Bivalvia (clams), Gastropoda (snails), and Cephalopoda (octopuses). Each has a unique anatomy and role in the environment. Class Polyplacophora (chitons) and class Scaphopoda (toothshells) also include some common molluscs. Chitons have a shell that consists of a row of eight overlapping plates. Their flat foot is used for creeping along or clinging to rocks. The chiton scrapes algae and other plant food from rocks with its well-developed radula. Toothshells, or tuskshells, are bottom-dwelling marine molluscs. They have a slender, mantle-covered, tube-shaped body that is open at both ends. They use mucus-covered knobs on long tentacles to gather food.

Bivalves

Clams, oysters, shipworms, mussels, and scallops are all **bivalves** (class Bivalvia) with a two-part shell that is hinged and closed by powerful muscles (Fig. 30.3). They have no head, no radula, and very little cephalization. Clams use their hatchet-shaped foot for burrowing in sandy or muddy soil, and mussels use their foot to produce threads that attach them to nearby objects. Scallops both burrow and swim; rapid clapping of the valves releases water in spurts and causes the animal to move forward in a jerky fashion for a few feet.

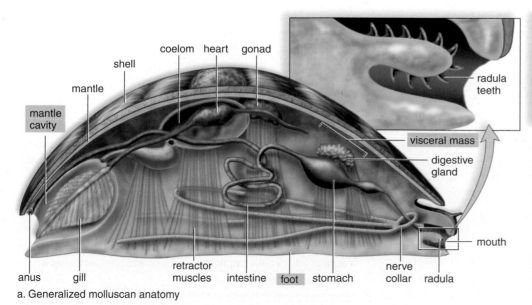

a. Generalized molluscan anatomy

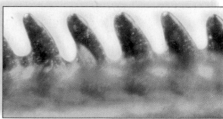

b. Radula

FIGURE 30.2 Body plan of molluscs.
a. Molluscs have a three-part body consisting of a ventral, muscular foot that is specialized for various means of locomotion; a visceral mass that includes the internal organs; and a mantle that covers the visceral mass and may secrete a shell. Ciliated gills may lie in the mantle cavity and direct food toward the mouth. **b.** In the mouth of many molluscs, such as snails, the radula is a tonguelike organ that bears rows of tiny teeth that point backward, shown here in a drawing and a micrograph.

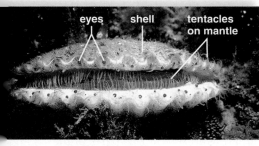

eyes   shell   tentacles on mantle

a. Scallop, *Pecten* sp.

growth lines of shell

b. Mussels, *Mytilus edulis*

FIGURE 30.3 Bivalve diversity.
Bivalves have a two-part shell. **a.** Scallops clap their valves and swim by jet propulsion. This scallop has sensory organs consisting of blue eyes and tentacles along the mantle edges. **b.** Mussels form dense beds in the intertidal zone of northern shores. **c.** In this drawing of a clam, the mantle has been removed from one side. Follow the path of food from the incurrent siphon to the gills, the mouth, the stomach, the intestine, the anus, and the excurrent siphon. Locate the three ganglia: anterior, foot, and posterior. The heart lies in the reduced coelom

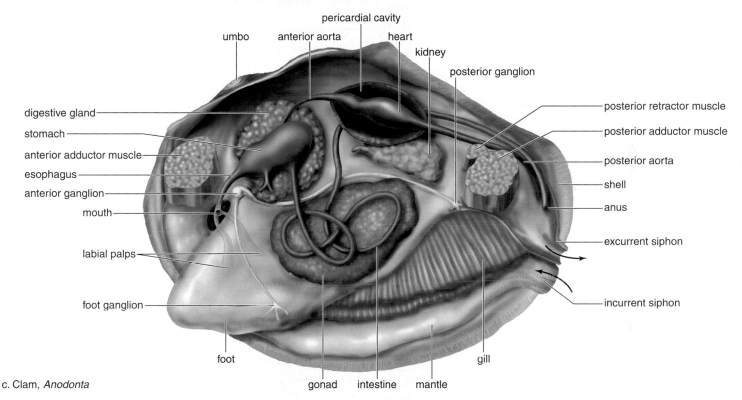

c. Clam, *Anodonta*

In freshwater clams such as *Anodonta* (Fig. 30.3*c*), the shell, secreted by the mantle, is composed of protein and calcium carbonate with an inner layer, called mother of pearl. If a foreign body is placed between the mantle and the shell, pearls form as concentric layers of shell are deposited about the particle. The compressed muscular foot of a clam projects ventrally from the shell; by expanding the tip of the foot and pulling the body after it, the clam moves forward.

Within the mantle cavity, the ciliated gills hang down on either side of the visceral mass. The beating of the cilia causes water to enter the mantle cavity by way of the incurrent siphon and to exit by way of the excurrent siphon. The clam is a filter feeder; small particles in this constant stream of water adhere to the gills, and ciliary action sweeps them toward the mouth.

The mouth leads to a stomach and then to an intestine, which coils about in the visceral mass before going right through the heart and ending in an anus. The anus empties at the excurrent siphon. There is also an accessory organ of digestion called a digestive gland. The heart lies just below the hump of the shell within the pericardial cavity, the only remains of the coelom. The circulatory system is open;

the heart pumps hemolymph into vessels that open into the hemocoel. The nervous system is composed of three pairs of ganglia (located anteriorly, posteriorly, and in the foot), which are connected by nerves.

There are two excretory kidneys, which lie just below the heart and remove waste from the pericardial cavity for excretion into the mantle cavity. The clam excretes ammonia (NH_3), a toxic substance that requires the concomitant excretion of water.

In freshwater clams, the sexes are separate and fertilization is internal. Fertilized eggs develop into specialized larvae and are released from the clam. Some larvae attach to the gills of a fish and become a parasite before they sink to the bottom and develop into a clam. Certain clams and annelids have the same type of larva, and this indicates a possible evolutionary relationship between molluscs and annelids.

In clams, which are bivalves, the body is protected by a heavy shell. They are filter feeders with a hatchet-shaped foot that allows them to burrow slowly in sand and mud.

Cephalopods

Nautiluses, cuttlefish, squids, and octopuses are members of class Cephalopoda [Gk. *kaphale*, head, and *podos*, foot]. Many of the 650 members of this class are fast-swimming, intelligent marine predators (Fig. 30.4). Zoologists consider them the most complex of all invertebrates. **Cephalopods** range in length from 2 cm to 20 m as in the giant squid, *Architeuthis.* Cephalopod means head footed; both squids and octopuses can squeeze their mantle cavity so that water is forced out through a funnel, propelling them by jet propulsion. Also, the tentacles and arms that circle the head capture prey by adhesive secretions or by suckers. A powerful, parrotlike beak is used to tear prey apart. They have well-developed sense organs, including eyes that are similar to those of vertebrates and focus like a camera. Cephalopods, particularly octopuses, have well-developed brains and show a remarkable capacity for learning. Nautiluses are enclosed in shells, but squids and cuttlefishes have a shell that is reduced and internal. The internal shell of cuttlefish is sold in pet stores for birds to sharpen their beaks. Octopuses lack shells entirely. For protection, squids and octopuses possess ink sacs, from which they can squirt a cloud of brown or black ink. This action often leaves a potential predator completely confused.

In squids, such as *Loligo* (Fig. 30.4c), a tough, muscular mantle, containing a vestigial skeleton called the pen, surrounds the visceral mass. In a squid, the funnel can be directed anteriorly or posteriorly, resulting in either forward or backward movement. A squid has an effective means of seizing and eating food. For example, *Loligo* darts backward rapidly into a school of young mackerel and seizes a fish with its tentacles, quickly biting the neck and severing the nerve cord with its jaws.

Unlike other molluscs and in keeping with its active life, the squid has a closed circulatory system, meaning that blood is always enclosed within blood vessels or a heart. The squid has three hearts—one pumps blood to all the internal organs, while the other two pump blood to the gills located in the mantle cavity. This efficient closed system effectively circulates oxygen and nutrients to body parts. The brain is formed from a fusion of the three molluscan ganglia. Nerves leave the brain and supply various parts of the body; an especially large pair of nerves controls the rapid contraction of the mantle. The gonads take up a large part of the visceral mass, and the sexes are separate. Packets called spermatophores contain sperm, which the male passes to the female mantle cavity by means of its specialized tentacle. Once the eggs are fertilized, they are attached to the substratum in elongated strings, each containing as many as 100 eggs. After hatching, the juvenile is free-swimming and ready to join the ocean realm.

> Squids, which are cephalopods, have a closed circulatory system and a well-developed nervous system with cephalization. They are active predators in the deep ocean.

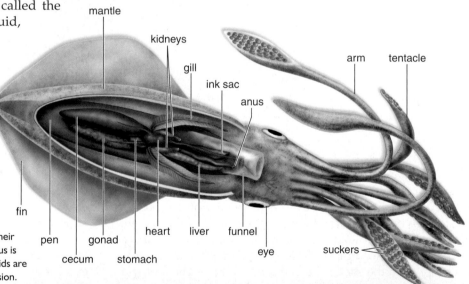

FIGURE 30.4 Cephalopod diversity.
Cephalopods have tentacles and/or arms. **a.** Octopuses use their eight arms for both swimming and crawling. **b.** When a nautilus is active, some 30 tentacles protrude from a coiled shell. **c.** Squids are torpedo-shaped and adapted for fast swimming by jet propulsion.

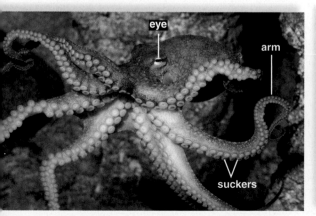
a. Two-spotted octopus, *Octopus bimaculatus*

b. Chambered nautilus, *Nautilus belauensis*

c. Bigfin reef squid, *Sepioteuthis lessoniana*

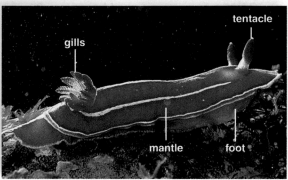

a. Flamingo tongue shell,
   Cyphoma gibbosum

b. Nudibranch,
   Glossodoris macfarlandi

c. Land snail,
   Helix aspersa

FIGURE 30.5 Gastropod diversity.
Gastropods have an elongated, flattened, muscular foot. **a.** Extensions of the mantle, having orange, brown, and white markings, cover the shell of this marine snail when it is active. **b.** Nudibranchs are also marine and, as their name suggests, have no shell. **c.** Photograph of the land snail, *Helix aspersa*. **d.** Diagram of *Helix* anatomy.

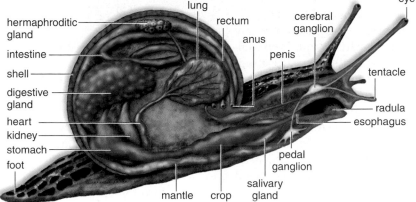

d.

Gastropods

Class Gastropoda [Gk. *gastros,* stomach, and *podos,* foot], the largest and most diverse of the mollusc classes, consists of slugs, snails, whelks, conchs, limpets, and nudibranchs. **Gastropods** are usually found in marine habitats (Fig. 30.5*a, b*), but slugs and garden snails are adapted to terrestrial environments (Fig. 30.5*c*). Many gastropods are herbivorous; some are scavengers, and a few species are carnivorous, even feeding on other molluscs. Molluscs known as cone shells deliver a painful sting that can be lethal even to humans.

Gastropods have an elongated, flattened foot and most, except for slugs and nudibranchs, have a one-piece coiled shell that protects the visceral mass. The anterior end bears a well-developed head region with a cerebral ganglion and eyes on the ends of tentacles. During development, gastropods undergo a torsion, or twisting, that brings the anus and mantle cavity downward, then forward and around to a position above the head. Torsion positions the visceral mass squarely above the foot:

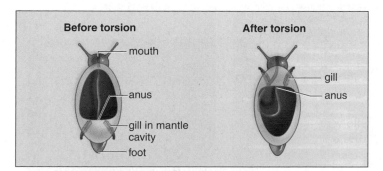

Aquatic gastropods have gills in their mantle cavity, but in those adapted to land, the mantle, richly supplied with blood vessels, functions as a lung when air is moved in and out through respiratory pores. Also, terrestrial gas-

tropod development does not include the swimming larval stages of aquatic species.

Land snails, such as *Helix aspersa,* have three obvious divisions of the body: a head; the flat long, muscular foot; and a visceral mass surrounded by the shell (Fig. 30.5*d*). The shell not only offers protection but also prevents desiccation (drying out). Waves of contraction run from the anterior to the posterior portion of the foot, and a lubricating mucus is secreted to facilitate movement.

Land snails are hermaphroditic; when two snails meet, they shoot calcareous darts into each other's body wall as a part of premating behavior. Then each inserts a penis into the vagina of the other to provide sperm for the future fertilization of eggs, which are deposited in the soil. Development proceeds directly without the formation of larvae.

The presence of a copulatory organ such as the penis, and even hermaphroditism, are adaptations to life on land. The penis allows easy transfer of sperm from one animal to another; hermaphroditism ensures that any two animals can mate. This is especially useful in slow-moving animals that have large ranges.

Land snails have a coiled shell; a flat, long, muscular foot; and a head region. In addition, the mantle in a garden snail becomes a lung.

30.3 ANNELIDS

Approximately 15,000 species of segmented worms have been placed into phylum **Annelida** [L. *annelus*, dim. of *annulus*, ring]. The most familiar members of this phylum include marine worms, leeches, and earthworms. The majority of **annelids** live in the marine environment. Annelids vary in size from microscopic to tropical earthworms that can be over 4 m long. The body of an annelid is segmented; internally the segments are partitioned by septa. The well-developed coelom is fluid-filled and serves as a supportive hydroskeleton. A hydrostatic skeleton along with partitioning of the coelom permit each body segment independence of movement. Instead of just burrowing in the mud, an annelid can crawl on a surface.

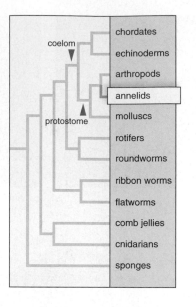

Segmentation and the tube-within-a-tube body plan have led to increased specialization of the digestive tract in annelids. For example, the digestive system may include a pharynx, a stomach, and accessory glands. Annelids have an extensive closed circulatory system with blood vessels that run the length of the body and branch to every segment. They also possess muscular aortic arches called "hearts" that propel blood through the vessels. The nervous system consists of a brain connected to a ventral solid nerve cord, with a ganglion in each segment. Cephalization is apparent in some annelids and in these tentacles, palps and eyespots are common. The excretory system consists of paired **nephridia** [Gk. *nephros*, kidney], which are coiled tubules in each segment that collect waste material from the coelom and excrete it through openings in the body wall.

Polychaetes

Most annelids are marine **polychaetes** (class Polychaeta); their name refers to the presence of many setae. **Setae** [L. *seta*, bristle] are bristles that anchor the worm or help it move. In polychaetes, the setae are in bundles on parapodia [Gk. *para*, beside, and *podos*, foot], which are paddlelike appendages found on most segments. These are used not only in swimming but also as respiratory organs, where the expanded surface area allows for exchange of gases. Some polychaetes are free-swimming, but the majority live in crevices or burrow into the ocean bottom. Clam worms, such as *Nereis* (Fig. 30.6*a*), are predators. They prey on crustaceans and other small animals, which are captured by a pair of strong chitinous jaws that extend with a part of the pharynx when the animal is feeding. Associated with its way of life, *Nereis* undergoes cephalization, having a head region with eyes and other sense organs.

Other polychaetes are sedentary (sessile) tube worms, with tentacles that form a funnel-shaped fan (Fig 30.6*b*). Food particles are directed toward the mouth by the action of cilia. Fanworms are sessile filter feeders; a sorting mechanism rejects large particles, and only the smaller ones are accepted for consumption. Featherduster worms are beautiful marine tube worms with ornate radioles, or feathery arms, that extend into the water for feeding.

Polychaetes have breeding seasons, and only during these times do the worms have sex organs. In *Nereis*, many worms concurrently shed a portion of their bodies containing either eggs or sperm, and these float to the surface where fertilization takes place. The zygote rapidly develops into a type of larva that is similar to that of a marine clam. The existence of this larva in both the annelids and molluscs shows that these two groups of animals are related.

Polychaetes are marine worms with bundles of setae attached to parapodia.

FIGURE 30.6 Polychaete diversity.

a. Clam worms are predaceous polychaetes that undergo cephalization. Note also the parapodia, which are used for swimming and as respiratory organs.
b. Christmas tree worms (a type of tube worm) are sessile feeders whose ciliated tentacles spiral in this example.

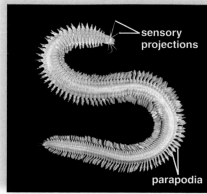
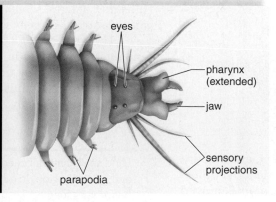

a. Ragworm, *Nereis diversicolor*

b. Christmas tree worms, *Spirobranchus giganteus*

Oligochaetes

Members of class Oligochaeta [Gk. *oligos*, few, and *chaete*, long hair] are primarily found in terrestrial and freshwater habitats. About 3,500 species of oligochaetes have been described. **Oligochaetes** can vary in size from several millimeters to 4 m in length. The oligochaetes have neither well-developed heads nor parapodia (Fig. 30.7). The common earthworm, *Lumbricus terrestris,* is a typical representative of this class. Earthworm setae protrude in pairs directly from the surface of the body. Locomotion, which is accomplished section by section, uses muscle contraction and the setae. When longitudinal muscles contract, segments bulge and their setae protrude into the soil; then, when circular muscles contract, the setae are withdrawn, and these segments move forward.

Earthworms reside in soil where there is adequate moisture to keep the body wall moist for gas exchange. They are scavengers and feed on leaves or any other organic matter that can conveniently be taken into the mouth along with dirt. Food drawn into the mouth by the action of the muscular pharynx is stored in a crop and ground up in a thick, muscular gizzard. Digestion and absorption occur in a long intestine whose dorsal surface has an expanded region called a **typhlosole** that increases the surface for absorption. Waste is eliminated through the anus.

Segmentation

Earthworm segmentation, which is obvious externally, is also internally evidenced by septa. The long, ventral nerve cord leading from the brain has ganglionic swellings and lateral nerves in each segment. The paired nephridia in most segments have two openings: One is a ciliated funnel that collects coelomic fluid, and the other is an exit in the body wall. Between the two openings is a convoluted region where waste material is removed from the blood vessels about the tubule. Red blood moves anteriorly in the dorsal blood vessel, which connects to the ventral blood vessel by five pairs of connectives called "hearts." Pulsations of the dorsal blood vessel and the five pairs of hearts are responsible for blood flow. As the ventral vessel takes the blood toward the posterior regions of the worm's body, it gives off branches in every segment. Altogether, segmentation is evidenced by:

- body rings
- coelom divided by septa
- setae on most segments
- ganglia and lateral nerves in each segment
- nephridia in most segments
- branch blood vessels in each segment

Reproduction

Earthworms are hermaphroditic; the male organs are the testes, the seminal vesicles, and the sperm ducts, and the female organs are the ovaries, the oviducts, and the seminal receptacles. During mating, two worms lie parallel to each other facing in opposite directions. The fused midbody segment, called a clitellum, secretes mucus, protecting the sperm from drying out as they pass between the worms. After the worms separate, the clitellum of each produces a slime tube, which is moved along over

the anterior end by muscular contractions. As it passes, eggs and the sperm received earlier are deposited, and fertilization occurs. The slime tube then forms a cocoon to protect the worms as they develop. There is no larval stage in earthworms.

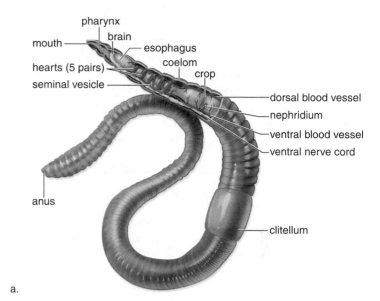

a.

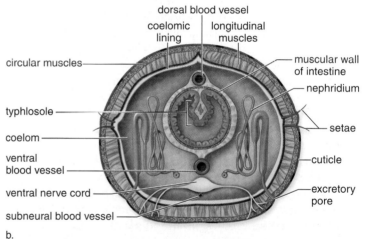

b.

c.

FIGURE 30.7 Earthworm, *Lumbricus terrestris*.
a. Internal anatomy of the anterior part of an earthworm. Notice that each body segment bears a pair of setae and that internal septa divide the coelom into compartments. **b.** Cross section of an earthworm. **c.** When earthworms mate, they are held in place by a mucus secreted by the clitellum. The worms are hermaphroditic, and when mating, sperm pass from the seminal vesicles of each to the seminal receptacles of the other.

Comparison with Clam Worm

Comparing the anatomy of marine clam worms to that of terrestrial earthworms highlights the manner in which earthworms are adapted to life on land. A lack of cephalization is seen in the nonpredatory earthworms that extract organic remains from the soil they eat. The lack of parapodia helps reduce the possibility of water loss and facilitates burrowing in soil. The clam worm uses external water, while the earthworm provides a mucous secretion to aid fertilization. The aquatic form, not the land form, has the swimming, or trochophore, larva.

Earthworms, which burrow in the soil, lack obvious cephalization and parapodia. They are hermaphroditic and have no larval stage.

Leeches

Leeches are placed in class Hirudinea. The majority of the 500 species of described leeches live in freshwater habitats. Leeches range in size from less than 2 cm to the medicinal leech, which can be 20 cm in length. They exhibit a variety of patterns and colors, but most are brown or olive green. The body of leeches is flattened dorsoventrally. They have the same body plan as other annelids, but they have no setae and each body ring has several transverse grooves.

Among their modifications are two suckers, a small oral one around the mouth and a large posterior one. While some leeches are free-living, feeding on plant material or invertebrates, most are fluid feeders that attach themselves to open wounds. Some bloodsuckers, such as the medicinal leech, can cut through tissue (Fig. 30.8). Leeches are able to keep blood flowing and prevent clotting by means of a substance in their saliva known as hirudin, a powerful anticoagulant. Medicinal leeches have been used for centuries in blood-letting and other procedures. Today, the medicinal leech *Hirudo medicinalis* is used in a number of applications, including reconstructive surgery for severed digits and in plastic surgery.

Leeches are modified in a way that lends them to the parasitic way of life. Some are external parasites known as bloodsuckers.

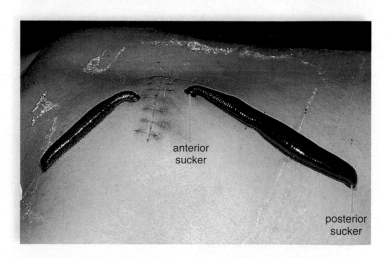

anterior
sucker

posterior
sucker

30.4 ARTHROPODS

Phylum **Arthropoda** [Gk. *arthron*, joint, and *podos*, foot] is the largest and most diverse phylum in the animal kingdom, consisting of over 1 million species to date. Living **arthropods** include barnacles, spiders, beetles, millipedes, and shrimp. Arthropods have successfully adapted to every habitat and all modes of life. The arthropods vary in size from a parasitic mite, which is less than 0.1 mm in length to the Japanese crab, which measures up to 4 m in length. The fossil record of the arthropods is quite extensive.

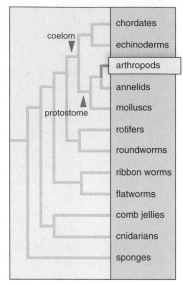

The arthropods are considered the most successful animal phylum. The success of this remarkable group of animals is dependent on five characteristics. First, arthropods have a rigid but jointed **exoskeleton** (Fig. 30.9*a, b*) composed primarily of **chitin** [Gk. *chiton*, tunic], a strong, flexible, nitrogenous polysaccharide. The exoskeleton serves many functions, including protection, attachment for muscles, locomotion, and prevention of desiccation. However, because it is hard and nonexpandable, arthropods must **molt,** or shed, the exoskeleton as they grow larger. Before molting, the body secretes a new, larger exoskeleton, which is soft and wrinkled, underneath the old one. After enzymes partially dissolve and weaken the old exoskeleton, the animal breaks it open and wriggles out. The new exoskeleton then quickly expands and hardens.

Second, the segmentation of arthropods is often modified for specialization of body regions. Segmentation is readily apparent when each segment has a pair of jointed appendages but not when certain segments are fused into a head, thorax, and abdomen. The jointed appendages of arthropods are basically hollow tubes moved by muscles. Typically, the appendages are highly adapted for a particular function, such as food gathering, reproduction, and locomotion. In addition, many appendages are associated with sensory structures and used for tactile purposes.

Third, arthropods have a well-developed nervous system. They possess a brain and a ventral nerve cord. The head bears various types of sense organs, including eyes

FIGURE 30.8 Medicinal leeches, *Hirudo medicinalis.*
On occasion, leeches are used to remove blood from bruised skin. The anterior sucker of the medicinal leech has three jaws that saw through the skin, producing a Y-shaped incision through which blood is drawn up by the sucking action of the leech's muscular pharynx. The salivary glands secrete an anticoagulant called hirudin that keeps the blood flowing while the leech is feeding. Other salivary ingredients dilate the patient's blood vessels and act as an anesthetic.

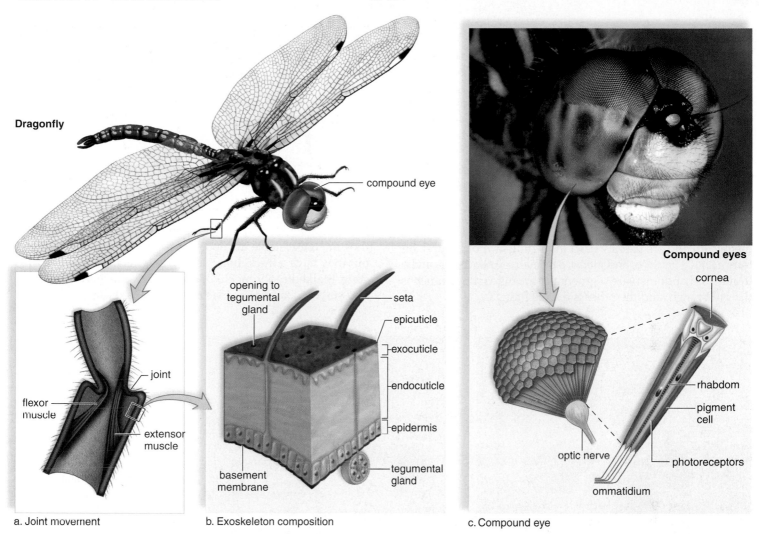

a. Joint movement

b. Exoskeleton composition

c. Compound eye

FIGURE 30.9 Arthropod skeleton and eye.

a. The joint in an arthropod skeleton is a region where the exocuticle is thinner and not as hard as the rest of the exocuticle. The direction of movement is toward the flexor muscle or the extensor muscle, whichever one has contracted. **b.** The exoskeleton is secreted by the epidermis and consists of the endocuticle; the exocuticle, hardened by the deposition of calcium carbonate; and the epicuticle, a waxy layer. Chitin makes up the bulk of the exo- and endocuticles. **c.** Arthropods have a compound eye that contains many individual units, each with its own lens and photoreceptors.

of two types—simple and compound. The compound eye is composed of many complete visual units, each of which operates independently (Fig 30.9c). The lens of each visual unit focuses an image on the light-sensitive membranes of a small number of photoreceptors within that unit. The simple eye, like that of vertebrates, has a single lens that brings the image to focus onto many receptors, each of which receives only a portion of the image. In addition to sight, many arthropods have well-developed touch, smell, taste, balance, and hearing. They constantly monitor their environment for food, mates, and danger. Arthropods display many complex behaviors and communication skills.

Fourth, arthropods have a variety of respiratory organs. Marine forms use gills, which are vascularized, highly convoluted, thin-walled tissue specialized for gas exchange. Terrestrial forms have book lungs (e.g., spiders) or air tubes called **tracheae** [L. *trachia,* windpipe]. Tracheae serve as a rapid way to transport oxygen directly to the cells.

Finally, the occurrence of metamorphosis has contributed to the success of arthropods. **Metamorphosis** [Gk. *meta,* implying change, and *morphe,* shape, form] is a drastic change in form and physiology that occurs as an immature stage, called a larva, becomes an adult. Among arthropods, the larva eats different food and lives in a different environment than the adult. This reduces competition and allows more members of a species to exist at one time. For example, larval crabs live among and feed on plankton, while adult crabs are bottom dwellers that catch live prey or scavenge dead organic matter. Among insects, such as butterflies, the caterpillar feeds on leafy vegetation, while the adult feeds on nectar.

Because phylum Arthropoda is so large, variations in classification may occur but most authorities recognize these three subphyla: Crustacea (crayfish, shrimp, and crabs), Uniramia (centipedes, millipedes, and insects), and Chelicerata (horseshoe crabs, spiders, ticks, and mites).

Crustaceans

Approximately 40,000 species of **crustaceans** (subphylum Crustacea) have been classified. This large and diverse subphylum also contains lobsters, crabs, water fleas, pillbugs, and barnacles. The name *crustacean* is derived from the hard, crusty exoskeleton made from chitin. The majority of crustaceans live in marine and aquatic environments; however, a few species are terrestrial (Fig. 30.10).

Although crustaceans are extremely diverse, the head usually bears a pair of compound eyes and five pairs of appendages. The first two pairs, called antennae and antennules, lie in front of the mouth and have sensory functions. The other three pairs (mandibles, first and second maxillae) lie behind the mouth and are usually used in feeding as mouthparts. Biramous [Gk. *bis*, two, and *ramus*, a branch] appendages on the thorax and abdomen are segmentally arranged; one branch is the gill branch, and the other is the leg branch.

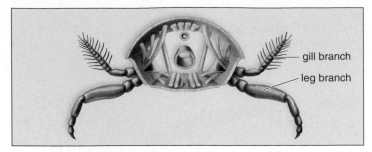

Copepods and krill are small crustaceans that live in the water, where they feed on algae. In the marine environment, they serve as food for fishes, sharks, and whales. They are so numerous that, despite their small size, some believe they are harvestable as food. Barnacles are also crustaceans, but they have a thick, heavy shell as befits their inactive lifestyle. Stalked (gooseneck) barnacles are attached by a stalk, and stalkless (acorn) barnacles are attached directly by their

a. Sally lightfoot crab, *Grapsus grapsus*

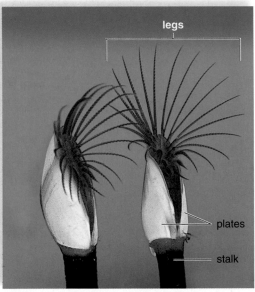

b. Gooseneck barnacles, *Lepas anatifera*

FIGURE 30.10

Crustacean diversity.
The crayfish on the next page, crabs (**a**), and shrimp (**c**) are decapods—they have five pairs of walking legs. Shrimp resemble crayfish more closely than crabs, which have a reduced abdomen. Pelagic shrimp feed on copepods, such as the one seen from below in (**e**). A copepod has long antennae used for floating, and feathery maxillae used for filter feeding. Barnacles have no abdomen and a reduced head; the thoracic legs project through a shell to filter feed. Barnacles often live on human-made objects such as ships, buoys, and cables. The gooseneck barnacle (**b**) is attached to an object by a long stalk. Pillbugs (**d**) are common crustaceans found under wood, bricks, and other objects.

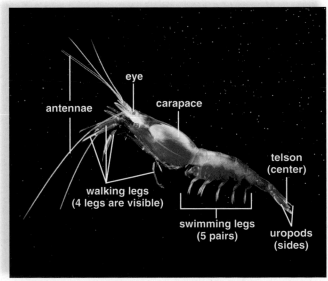

c. Red-backed cleaning shrimp, *Lysmata grasbhami*

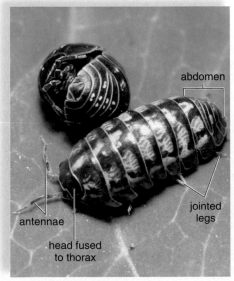

d. Pillbug, *Armadillidium vulgare*

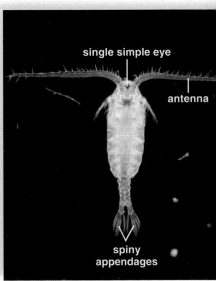

e. Copepod, *Diaptomus*

shells. Barnacles can live on wharf pilings, ship hulls, seaside rocks, and even the bodies of whales. They begin life as free-swimming larvae, but they undergo a metamorphosis that transforms their swimming appendages to cirri, feathery structures that are extended and allow them to filter feed when they are submerged. The pillbug, also known as a roly-poly bug for its habit of rolling up into a ball, is a terrestrial crustacean often found in shady spots as under rocks.

Decapods

Members of the order Decapoda are the most familiar and numerous crustaceans. **Decapods** include lobsters, crabs, crayfish, hermit crabs, and shrimp. These animals have a thorax that bears five pairs of walking appendages. The first pair may be modified as claws. Typically, there are gills situated above the walking legs. Figure 30.11a gives a view of the external anatomy of the crayfish. The head and thorax are fused into a cephalothorax, which is covered on the top and sides by a nonsegmented carapace. The abdominal segments are equipped with swimmerets, small paddlelike structures. The first two pairs of swimmerets in the male are quite strong and are used to pass sperm to the female.

The last two segments bear the uropods and the telson, which make up a fan-shaped tail. Ordinarily, a crayfish lies in wait for prey. It faces out from an enclosed spot with the claws extended, and the antennae moving about. The claws seize any small animal, either dead ones or live ones that happen by, and carry them to the mouth. When a crayfish moves about, it generally crawls slowly, but may swim rapidly by using its heavy abdominal muscles and tail.

The respiratory system consists of gills that lie above the walking legs protected by the carapace. As shown in Figure 30.11b, the digestive system includes a stomach, which is divided into two main regions: an anterior portion called the gastric mill, equipped with chitinous teeth to grind coarse food, and a posterior region, which acts as a filter to prevent coarse particles from entering the digestive glands, where absorption takes place. Green glands lying in the head region, anterior to the esophagus, excrete metabolic wastes through a duct that opens externally at the base of the antennae. The coelom, which is so well developed in the annelids, is reduced in the arthropods and is composed chiefly of the space about the reproductive system. A heart pumps hemolymph containing the respiratory pigment hemocyanin into a **hemocoel** [Gk. *haima*, blood, and *koiloma*, cavity] consisting of sinuses (open spaces) where the hemolymph flows about the organs. (Whereas hemoglobin is a red pigment, hemocyanin is a blue pigment.) This is an open circulatory system because blood is not contained within blood vessels.

The crayfish nervous system is well developed. Crayfish have a brain and a ventral nerve cord that passes posteriorly. Along the length of the nerve cord, periodic ganglia give off lateral nerves. Sensory organs are well developed. The compound eyes are found on the ends of movable eyestalks. These eyes are accurate and can detect motion and respond to polarized light. Other sensory organs include tactile antennae and chemosensitive setae. Crayfish also have statocysts that serve as organs of equilibrium.

The sexes are separate in the crayfish, and the gonads are located just ventral to the pericardial cavity. In the male, a coiled sperm duct opens to the outside at the base of the fifth walking leg. Sperm transfer is accomplished by the modified first two swimmerets of the abdomen. In the female, the ovaries open at the bases of the third walking legs. A stiff fold between the bases of the fourth and fifth pairs serves as a seminal receptacle. Following fertilization, the eggs are attached to the swimmerets of the female. Young hatchlings are miniature adults, and no metamorphosis occurs

> Crustaceans are mainly marine arthropods in which the head bears five pairs of appendages, including mandibles and maxillae. Typically, there are biramous appendages.

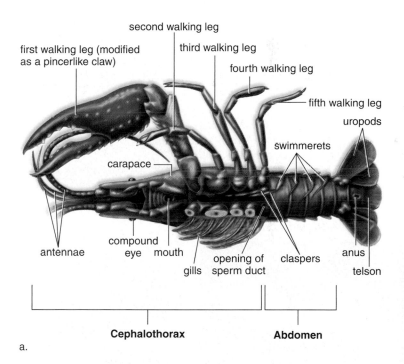

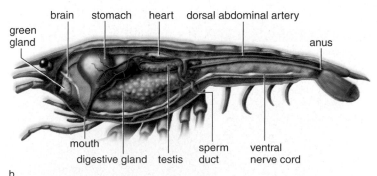

FIGURE 30.11 Male crayfish, *Cambarus*.
a. Externally, it is possible to observe the jointed appendages, including the swimmerets, and the walking legs, which include the claws. These appendages, plus a portion of the carapace, have been removed from the right side so that the gills are visible. **b.** Internally, the parts of the digestive system are particularly visible. The circulatory system can also be clearly seen. Note the ventral nerve cord.

Uniramians

Subphylum Uniramia includes insects, millipedes, and centipedes. **Uniramians** live in every environment on the Earth except for the open sea. In this subphylum, the appendages that attach to the thorax and abdomen have only one branch, the leg branch; thus these animals are called uniramous [Gk. *uni,* one, and *ramus,* a branch]. The head appendages include antennae, mandibles, and maxillae. Most uniramians live on land and breathe by means of a system of air tubes called tracheae.

Insects

Insects (superclass Insecta) number over 900,000 species. Traditionally, insects have been placed into orders. A few examples of insect orders are represented in Figure 30.12. Insects are adapted for an active life on land, although some have secondarily invaded aquatic habitats. The body of an insect is divided into a head, a thorax, and an abdomen. The head bears the sense organs and mouthparts (Fig. 30.13); the thorax bears three pairs of legs and one or two pairs of wings; and the abdomen contains most of the internal organs. Wings enhance an insect's ability to survive by providing a way of escaping enemies, finding food, facilitating mating, and dispersing the species. The exoskeleton of an insect is lighter and contains less chitin than that of many other arthropods.

In the grasshopper (Fig. 30.14), the third pair of legs is suited to jumping. There are two pairs of wings. The forewings are tough and leathery, and when folded back at rest, they protect the broad, thin hindwings. On the lateral surface, the first abdominal segment bears a large tympanum on each side for the reception of sound waves. The posterior region of the exoskeleton in the female has two pairs of projections that form an ovipositor, used to dig a hole in which eggs are laid.

The digestive system is suitable for a herbivorous diet. In the mouth, food is broken down mechanically by mouthparts and enzymatically by salivary secretions. Food is temporarily stored in the crop before passing into the gizzard, where it is finely ground. Digestion is completed in the stomach, and nutrients are absorbed into the hemocoel from outpockets called gastric ceca (*cecum,* a cavity open at one end only). The excretory system consists of **Malpighian tubules,** which extend into a hemocoel and collect nitrogenous wastes that are concentrated and excreted into the digestive tract. The formation of a solid nitrogenous waste, namely uric acid, conserves water.

Grasshopper, order *Orthoptera*

Mealybug, order *Homoptera*

Beetle, order *Coleoptera*

Leafhopper, order *Homoptera*

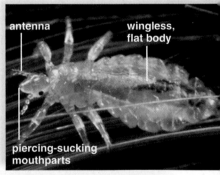

Head louse, order *Anoplura*

Wasp, order *Hymenoptera*

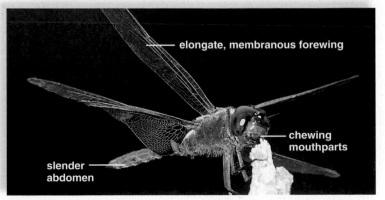

Dragonfly, order *Odonata*

FIGURE 30.12 **Insect diversity.**

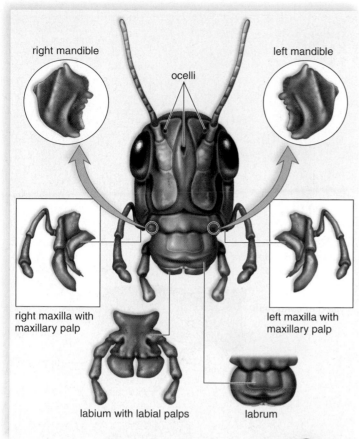

Chewing (grasshopper)

Monarch butterfly, *Danaus*

Housefly, *Musca*

Siphoning tube (butterfly and housefly)

FIGURE 30.13 Two types of insect mouthparts.

The respiratory system begins with openings in the exoskeleton called spiracles. From here, the air enters small tubules called tracheae (Fig. 30.14*a*). The tracheae branch and rebranch, finally ending in moist areas where the actual exchange of gases takes place. The movement of air through this complex of tubules is not a passive process; air is pumped through by a series of bladderlike structures (air sacs), attached to the tracheae near the spiracles. Air enters the anterior four spiracles and exits by the posterior six spiracles. Breathing by tracheae may account for the small size of insects (most are less than 60 mm in length) since the tracheae are so tiny and fragile that they would be crushed by any amount of weight.

The circulatory system contains a slender, tubular heart that lies against the dorsal wall of the abdominal exoskeleton and pumps hemolymph into the hemocoel, where it circulates before returning to the heart again. The hemolymph is colorless and lacks a respiratory pigment, and the tracheal system transports gases.

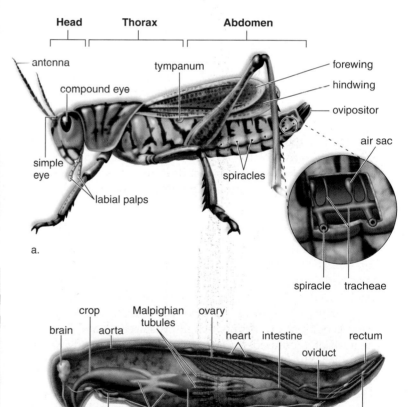

FIGURE 30.14 Female grasshopper, *Romalea*.

a. Externally, the body of a grasshopper is divided into three sections and has three pairs of legs. The tympanum receives sound waves, and the jumping legs and the wings are for locomotion. **b.** Internally, the digestive system is specialized. The Malpighian tubules excrete a solid nitrogenous waste (uric acid). A seminal receptacle receives sperm from the male, which has a penis.

Reproduction is adapted to life on land. The male has a penis, which passes sperm to the female. Internal fertilization protects both gametes and zygotes from drying out. The female deposits the fertilized eggs in the ground with her ovipositor.

Grasshoppers undergo *incomplete metamorphosis,* a gradual change in form as the animal matures. The immature grasshopper, called a nymph, is recognizable as a grasshopper, even though it differs somewhat in shape and form from the adult. Other insects, such as butterflies, undergo *complete metamorphosis,* involving drastic changes in form. At first, the animal is a wormlike larva (caterpillar) with chewing mouthparts. It then forms a case, or cocoon, about itself and becomes a pupa. During this stage, the body parts are completely reorganized; the adult then emerges from the cocoon. This life cycle allows the larvae and adults to use different food sources.

Insects show remarkable behavior adaptations. Bees, wasps, ants, termites, and other colonial insects have complex societies.

Comparison with Crayfish.   The grasshopper is adapted to a terrestrial environment, while the crayfish is adapted to an aquatic environment. In crayfish, gills take up oxygen from water, while in the grasshopper, tracheae allow oxygen-laden air to enter the body. Appropriately, the crayfish has an oxygen-carrying pigment in its blood and a grasshopper has no such pigment. A liquid waste (ammonia) is excreted by a crayfish, while a solid waste (uric acid) is excreted by a grasshopper. Only in grasshoppers are there (1) a tympanum for the reception of sound waves, and (2) a penis in males and an ovipositor in females. To swim, crayfish use tail (telson and uropods) and abdominal muscles; a grasshopper has legs for jumping and wings for flying.

Centipedes and Millipedes

The centipedes and millipedes are known for their many legs (Fig. 30.15). In **centipedes** ("hundred-leggers"), each of their many body segments has a pair of walking legs. The approximately 3,000 species prefer to live in moist environments such as under logs, in crevices, and in leaf litter, where they are active predators on worms, small crustaceans, and insects. The head of a centipede includes paired antennae and jawlike mandibles. Maxillipeds (poison claws) kill or immobilize prey while mandibles chew.

In **millipedes** ("thousand leggers"), each of four thoracic segments bears one pair of legs while abdominal segments have two pairs of legs. Millipedes are harmless animals that live under stones or burrow in the soil as they feed on leaf litter. Their cylindrical bodies have a tough chitinous exoskeleton.

Uniramia—in which the legs have only one branch—include insects, centipedes, and millipedes. Insects, which are adapted to life on land, comprise more species than any other group of animals.

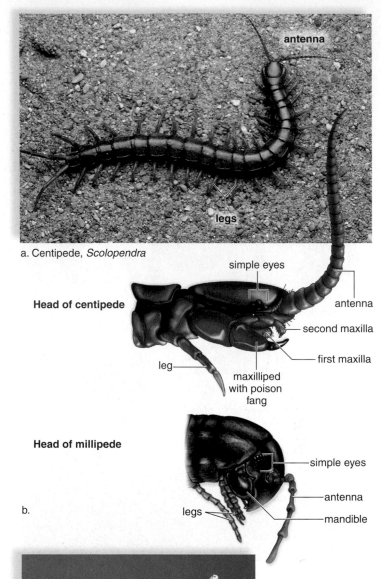

a. Centipede, *Scolopendra*

b.

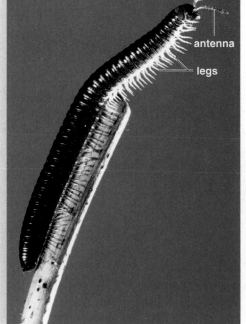

c. Millipede, *Tachypodoiulus*

FIGURE 30.15
Centipede and millipede.
a. A centipede has a pair of appendages on almost every segment. **b.** Head of centipede compared to head of millipede. A centipede is a carnivorous animal and kills its prey with a poison fang. Most millipedes are herbivorous and usually feed on decayed plant matter. **c.** A millipede has two pairs of legs on most segments.

Chelicerates

The 60,000 species of arthropods in subphylum Chelicerata includes ancient lineages such as horseshoe crabs, scorpions, and spiders (Fig. 30.16). The **chelicerates** live in terrestrial, aquatic, and marine environments. The first pair of appendages is the pincerlike chelicerae, used in feeding and defense. The second pair, the pedipalps, is used in feeding but also has a sensory function. Four pairs of walking legs follow the pedipalps. All of these appendages are attached to a **cephalothorax** [Gk. *kephale,* head, and *thorax,* breastplate] (fused head and thorax), which is followed by an abdomen that contains internal organs. No appendages are found on the heads of these animals (antennae, mandibles, or maxillae), as in the other subphyla.

Horseshoe crabs of the genus *Limulus* are familiar along the east coast of North America. They scavenge sandy and muddy substrates for annelids, small molluscs, and other invertebrates. The body is covered by exoskeletal shields. The anterior shield is a horseshoe-shaped carapace, which bears two prominent compound eyes. A long, unsegmented telson projects to the rear. These marine animals have book gills, named for their resemblance to the pages of a closed book. Despite their ferocious appearance, they are harmless (Fig. 30.16*a*).

Ticks, mites, scorpions, spiders, and harvestmen are all arachnids. Over 25,000 species of mites and ticks have been classified, some of which are parasitic on a variety of other animals. Chiggers (red bugs) are the larvae of certain mites and cause irritation and itching when they burrow under the skin. Mange in dogs is due to a parasitic chigger. Ticks are ectoparasites of various vertebrates, and they are carriers for such diseases as Rocky Mountain spotted fever and Lyme disease. When not attached to a host, ticks hide on plants and in the soil.

The tropics, subtropics, and temperate regions of North America are home to approximately 1,500 species of scorpions (Fig. 30.16*b*). Scorpions are nocturnal and spend most of the day hidden under a log or a rock. Their pedipalps are large pincers, and their long abdomen ends with a stinger that contains venom.

Presently, over 35,000 species of spiders have been classified (Fig. 30.16*c*). Spiders, the most familiar arachnids, have a narrow waist that separates the cephalothorax from the abdomen. Spiders do not have compound eyes; instead, they have numerous simple eyes that perform a similar function. The chelicerae are modified as fangs, with ducts from poison glands, and the pedipalps are used to hold, taste, and chew food. The abdomen often contains silk glands, and spiders spin a web in which to trap their prey. Invaginations of the body wall form lamellae ("pages") of their so-called book lungs. In the United States, two species of spiders are dangerous to humans: the brown recluse and the black widow (Fig. 30.16*d*).

Harvestmen are a conspicuous group of arachnids that are often confused with spiders. Although there are nearly 5,000 species, most people are only familiar with the "daddy longlegs" that live in large groups under eaves of houses (Fig. 30.16*e*).

The chelicerates include horseshoe crabs, scorpions, ticks, mites, spiders, and harvestmen. These animals have pincerlike appendages called chelicerae, which are modified as fangs in spiders.

FIGURE 30.16

Chelicerate diversity.
a. Horseshoe crabs are common along the east coast. **b.** Scorpions are more common in tropical areas. **c.** Ventral view of a spider. **d.** The black widow spider is a poisonous spider that spins a web. **e.** Harvestmen are common backyard chelicerates.

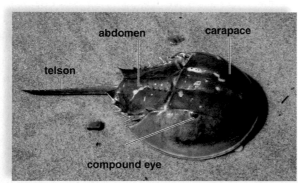

a. Horseshoe crab, *Limulus*

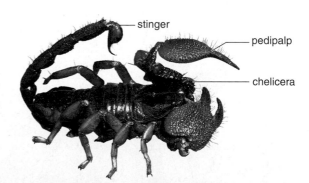

b. Kenyan giant scorpion, *Pandinus*

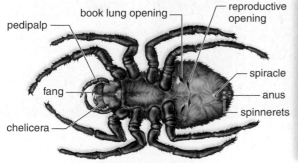

c. Spider external anatomy, dorsal view

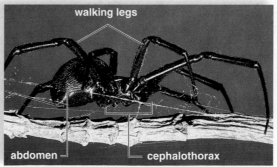

d. Black widow spider, *Latrodectus*

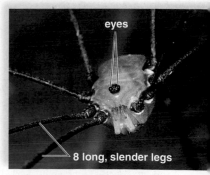

e. Harvestmen (daddy longlegs)

30.5 ECHINODERMS

Echinoderms [Gk. *echinos,* spiny, and *derma,* skin] (phylum Echinodermata) number 6,000 living species of animals including sea urchins, sand dollars, and sea stars (starfishes) (Fig. 30.17). Echinoderms are primarily bottom-dwelling marine animals. They range in size from brittle stars less than 1 cm in length to giant sea cucumbers over 2 m long.

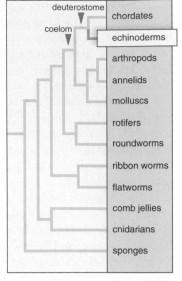

The most striking feature of echinoderms is their 5-pointed radial symmetry, as illustrated by a sand dollar. Although echinoderms are radially symmetrical as adults, their larvae are free-swimming filter feeders with bilateral symmetry. Echinoderms have an endoskeleton of spiny calcium-rich plates called ossicles. The spines protruding from their skin account for the phylum name Echinodermata. Another innovation is their unique **water vascular system** consisting of canals and appendages that function in locomotion, feeding, gas exchange, and sensory reception.

Zoologists recognize six classes of living echinoderms. Class Asteroidea consists of sea stars or starfishes. Class Concentricycloidea includes the sea daisies. Class Crinoidea is the oldest of the classes and includes the stalked sea lilies and the motile feather stars. Class Holothuroidea includes the sea cucumbers, which have long, leathery bodies that resemble a cucumber. Class Echinoidea includes sea urchins and sand dollars, both of which use their spines for locomotion, defense, and burrowing. Class Ophiuroidea includes the brittle stars, which have a central disk surrounded by radially flexible arms.

Sea Stars

Sea stars (class Asteroidea) number about 1,600 species. They are commonly found along rocky coasts, where they feed on clams, oysters, and other bivalve molluscs. Various structures project through the body wall: (1) spines from the endoskeletal plates offer some protection; (2) pincerlike structures around the bases of spines keep the surface free of small particles; and (3) skin gills, tiny fingerlike extensions of the skin, are used for respiration. On the oral surface, each arm has a groove lined by little **tube feet** (Fig. 30.17).

To feed, a sea star positions itself over a bivalve and attaches some of its tube feet to each side of the shell. By working its tube feet in alternation, it pulls the shell open. A very small crack is enough for the sea star to evert its cardiac stomach and push it through the crack, so that it contacts the soft parts of the bivalve. The stomach secretes enzymes, and digestion begins even while the bivalve is attempting to close its shell. Later, partly digested food is taken into the sea star's body, where digestion continues in the pyloric stomach using enzymes from the digestive glands found in each arm. A short intestine opens at the anus on the aboral side (side opposite the mouth).

In each arm, the well-developed coelom contains a pair of digestive glands and gonads (either male or female) that open on the aboral surface by very small pores. The nervous system consists of a central nerve ring that gives off radial nerves in each arm. A light-sensitive eyespot is at the tip of each arm.

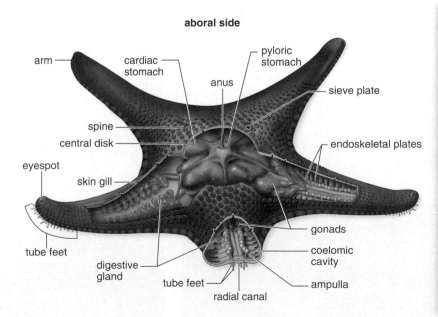

a.

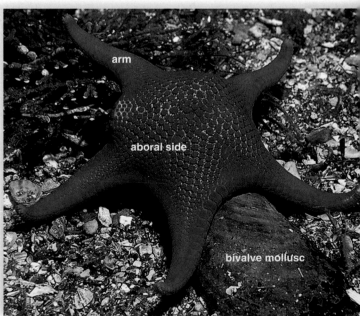

b. Red sea star, *Mediastar*

FIGURE 30.17 Echinoderms

a. Sea star (starfish) anatomy. Like other echinoderms, sea stars have a water vascular system that begins with the sieve plate and ends with expandable tube feet. **b.** The red sea star, *Mediastar,* uses the suction of its tube feet to open a clam, a primary source of food. **c.** Sea cucumber. **d.** Sea urchin.

Locomotion depends on the water vascular system. Water enters this system through a structure on the aboral side called the sieve plate, or madreporite. From there it passes through a stone canal to a ring canal, which surrounds the mouth, and then to a radial canal in each arm. From the radial canals, many lateral canals extend into the tube feet, each of which has an ampulla. Contraction of an ampulla forces water into the tube foot, expanding it. When the foot touches a surface, the center is withdrawn, giving it suction so that it can adhere to the surface. By alternating the expansion and contraction of the tube feet, a sea star moves slowly along.

Echinoderms do not have a complex respiratory, excretory, or circulatory system. Fluids within the coelomic cavity and the water vascular system carry out many of these functions. For example, gas exchange occurs across the skin gills and the tube feet. Nitrogenous wastes diffuse through the coelomic fluid and the body wall. Cilia on the peritoneum lining the coelom keep the coelomic fluid moving.

Sea stars reproduce asexually and sexually. If the body is fragmented, each fragment can regenerate a whole animal. Sea stars spawn and release either eggs or sperm at the same time. The bilateral larva will undergo a metamorphosis to become the radially symmetrical adult.

Echinoderms have a well-developed coelom and internal organs despite being radially symmetrical. Spines project from their internal skeleton, and there is a unique water vascular system.

CONNECTING THE CONCEPTS

About 90% of all animal species have a coelom and are in the phyla Mollusca, Annelida, Arthropoda, Echinodermata, and Chordata. (The chordates, which include the vertebrates, will be studied in Chapter 31.) This indicates that there are some adaptive advantages to having a coelom, or perhaps a combination of bilateral symmetry and the complexity made possible by the presence of a coelom. Most animals with a coelom have both circulatory and respiratory systems, which permit them to be active. The echinoderms are an anomaly because they lack these systems and have radial symmetry.

Animals with a coelom are divided into the protostomes (molluscs, annelids, and arthropods) and the deuterostomes (echinoderms and chordates) on the basis of embryological development. However, it is interesting that both lines of descent have representatives that are segmented. Segmentation of the annelids and many arthropods is obvious because the body has demarcations that can be seen externally. The repeating vertebrae of the backbone in vertebrates signal that they, too, are segmented. Was a coelom present before the evolution of segmentation? Perhaps it was. Later, a partitioned coelom may have provided a hydrostatic skeleton that enabled a worm to burrow more efficiently in the soil. In other words, there was a selective advantage to having a segmented coelom in organisms that lacked limbs. The later evolution of limbs in jointed arthropods (e.g., insects) and vertebrates (e.g., frogs, lizards, horses) was even more adaptive for locomotion on land.

Animals that live on land must also have a means of reproduction that allows fertilization and development to take place without requiring external water.

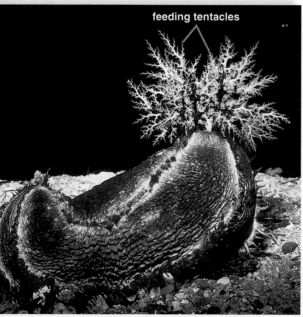

c. Sea cucumber, *Pseudocolochirus*

d. Purple sea urchin, *Strongylocentrotus*

Summary

30.1 ADVANTAGES OF COELOM IN PROTOSTOMES AND DEUTEROSTOMES

There are two groups of coelomate animals. In the protostomes, the mouth appears at or near the blastopore, and in the deuterostomes, the anus appears at or near the blastopore. Other differences between the two are spiral cleavage and a schizocoelom in protostomes versus radial cleavage and an enterocoelom in deuterostomes.

The presence of a coelom has many advantages. The digestive system and body wall can move independently, and internal organs can become more complex. Coelomic fluids can assist respiration, circulation, and excretion and can serve as a hydrostatic skeleton.

Annelids, arthropods, and molluscs are protostomial coelomates. As adults, they all have bilateral symmetry, a complete digestive tract, and organs derived from the three germ layers.

30.2 MOLLUSCS

The body of a mollusc typically contains a visceral mass, a mantle, and a foot. Many also have a head and a radula. The nervous system consists of several ganglia connected by nerve cords. There is a reduced coelom and an open circulatory system. Clams (bivalves) are adapted to a sedentary coastal life, squids (cephalopods) to an active life in the sea, and snails (gastropods) to life on land.

Bivalves are filter feeders. Water enters and exits by siphons, and food trapped on the gills is swept toward the mouth. The nervous system has three major ganglia.

Predaceous squids undergo cephalization, move rapidly by jet propulsion, and have a closed circulatory system.

30.3 ANNELIDS

Annelids are segmented worms. They have a well-developed coelom divided by septa, a closed circulatory system, a ventral solid nerve cord, and paired nephridia. Polychaetes ("many bristles") are marine worms that have parapodia. They may be predators, with a definite head region, or they may be filter feeders, with ciliated tentacles to filter food from the water. Earthworms are oligochaetes ("few bristles") that use the body wall for gas exchange. Leeches (class Hirudinea) are also part of this phylum.

30.4 ARTHROPODS

Arthropods are the most varied and numerous of animals. Their success is largely attributable to a flexible exoskeleton, specialized body regions, and jointed appendages. Also important are a high degree of cephalization, a variety of respiratory organs, and reduced competition through metamorphosis.

Most authorities recognize three subphyla of arthropods: Crustacea, Uniramia, and Chelicerata. Animals in the subphylum Crustacea (crayfish, lobsters, shrimps, copepods, krill, and barnacles) have a head that bears compound eyes, antennae, antennules, mandibles, and maxillae. There are biramous (two-branched) appendages on the thorax and abdomen. The crayfish illustrates other features, such as an open circulatory system, respiration by gills, and a ventral solid nerve cord.

Arthropods in the subphylum Uniramia (insects, centipedes, and millipedes) have uniramous (one-branched) appendages on the thorax and abdomen. The head has one pair of antennae, one pair of mandibles, and one or two pairs of maxillae. These terrestrial animals breathe by means of tracheae. Insects include butterflies, grasshoppers, bees, and beetles. The anatomy of the grasshopper illustrates insect anatomy and the ways they are adapted to life on land. Like other insects, grasshoppers have wings and three pairs of legs attached to the thorax. Grasshoppers have a tympanum for sound reception, a digestive system specialized for a grass diet, Malpighian tubules for excretion of solid nitrogenous waste, tracheae for respiration, internal fertilization, and incomplete metamorphosis.

Arthropods in subphylum Chelicerata (horseshoe crabs, spiders, scorpions, ticks, and mites) have chelicerae, pedipalps, and four pairs of walking legs attached to a cephalothorax.

30.5 ECHINODERMS

Echinoderms (e.g., sea stars, sea urchins, sea cucumbers, and sea lilies) have radial symmetry as adults (not as larvae) and internal calcium-rich plates with spines. Typical of echinoderms, sea stars have tiny skin gills, a central nerve ring with branches, and a water vascular system for locomotion. Each arm of a sea star contains branches from the nervous, digestive, and reproductive systems.

Reviewing the Chapter

1. List and discuss several advantages of the coelom.   538–39
2. Contrast the development of protostomes and deuterostomes in three ways. List the phyla that belong to each group.   538
3. Name at least three members of each phylum studied in this chapter.   539
4. What are the general characteristics of molluscs and the specific features of bivalves, cephalopods, and gastropods?   540–43
5. Contrast the anatomy of the clam, the squid, and the snail, indicating how each is adapted to its way of life.   540–43
6. What are the general characteristics of annelids and the specific features of the three major classes?   544
7. Contrast the anatomy of the clam worm and the earthworm, indicating how each is adapted to its way of life.   546
8. What are the general characteristics of arthropods and the specific features of the three major subphyla?   546–53
9. Describe the specific features of insects using the grasshopper as an example.   550–52
10. Contrast the anatomy of the crayfish and the grasshopper, indicating how each is adapted to its way of life.   552
11. In what ways are spiders adapted to living on land?   553
12. What are the general characteristics of echinoderms? Explain how the water vascular system works in sea stars.   554–55

Testing Yourself

Choose the best answer for each question.

1. Which of these does not pertain to a protostome?
   a. spiral cleavage
   b. blastopore is associated with the anus
   c. schizocoelom
   d. annelids, arthropods, and molluscs
   e. mouth is associated with first opening
2. Which of these best shows that annelids and arthropods are closely related?
   a. Both have a complete digestive tract.
   b. Both have a ventral nerve cord.
   c. Both are segmented.
   d. Both have a tube-within-a-tube body plan.
   e. All of these are correct.

3. Which of these best shows that snails are not closely related to crayfish?
   a. Snails are terrestrial, and crayfish are aquatic.
   b. Snails have a broad foot, and crayfish have jointed appendages.
   c. Snails are hermaphroditic, and crayfish have separate sexes.
   d. Snails are insects, but crayfish are fishes.
   e. Snails are bivalves, and crayfish are chelicerates.

4. Which of these pairs is mismatched?
   a. clam—gills
   b. lobster—gills
   c. grasshopper—book lungs
   d. polychaete—parapodia
   e. chelicerates—pedipalps

5. Which of these is an incorrect statement?
   a. Spiders are carnivores in the phylum Arthropoda.
   b. Clams are filter feeders in the phylum Mollusca.
   c. Earthworms are scavengers in the phylum Arthropoda.
   d. Squids are predators in the phylum Mollusca.
   e. Snails are gastropods in the phylum Mollusca.

6. A radula is a unique organ for feeding found in
   a. molluscs.
   b. annelids.
   c. arthropods.
   d. only insects.
   e. All of these are correct.

7. Which of these pairs is mismatched?
   a. crayfish—walking legs
   b. clam—hatchet foot
   c. grasshopper—wings
   d. earthworm—many cilia
   e. squid—jet propulsion

8. Which of these pairs is mismatched?
   a. molluscs—schizocoelom
   b. insects—hemocoel
   c. crayfish—coelom divided by septa
   d. clam worms—true coelom
   e. vertebrates—enterocoelom

9. Which of these is an adaptation to a terrestrial way of life in the snail?
   a. using the mantle instead of gills for respiration
   b. cephalization with antennae and eyes
   c. hatchet foot for crawling
   d. torsion so that anus is above the head
   e. incomplete metamorphosis

10. Associate each of these terms to annelids, to arthropods, and/or to molluscs.

Terms:
   a. organ system level of organization _____
   b. segmentation _____
   c. true coelom _____
   d. cephalization in some representatives _____
   e. bilateral symmetry _____
   f. complete gut _____
   g. jointed appendages _____
   h. three-part body plan _____

11. Associate each of these terms to clams and/or to earthworms.

Terms:
   a. annelid _____
   b. mollusc _____
   c. three ganglia _____
   d. gills _____
   e. closed circulatory system _____
   f. setae _____
   g. open circulatory system _____
   h. hatchet foot _____
   i. hydrostatic skeleton _____
   j. ventral nerve cord _____

12. Label this diagram.

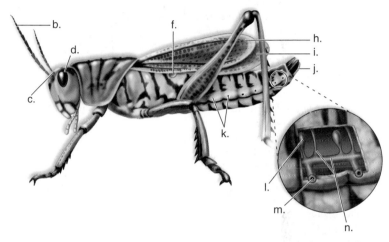

For questions 13–17, match the description to a body system in the key.

KEY:
   a. respiratory system
   b. digestive system
   c. cardiovascular system
   d. nervous system

13. particles adhere to labial palps

14. passes through the heart

15. has three pairs of ganglia

16. hemocoel

17. gills

For questions 18–21, match the classification to an animal in the key.

KEY:
   a. shrimp
   b. squid
   c. earthworm
   d. spider

18. cephalopod in the phylum Mollusca

19. oligochaete in the phylum Annelida

20. crustacean in the phylum Arthropoda
21. chelicerate in the phylum Arthropoda
22. Which of these is an incorrect statement?
    a. The body of a mollusc typically has a visceral mass, a mantle, and a foot.
    b. Arthropods are the most varied and numerous of animals.
    c. Insects include butterflies, grasshoppers, bees, and beetles.
    d. Polychaetes have chelicerae for eating their prey.
    e. The uniramians have one-branched appendages.
23. Segmentation in the earthworm is not exemplified by
    a. body rings.
    b. coelom divided by septa.
    c. setae on most segments.
    d. nephridia interior in most segments.
    e. tympanum exterior to segments.
24. Which of these is an incorrect difference between clam and squid?
    Clam   Squid
    a. filter feeder—active predator
    b. hatchet foot—jet propulsion
    c. brain and nerves—three separate ganglia
    d open circulation—closed circulation
    e. no cephalization—marked cephalization
25. Which pair is incorrectly matched?
    a. cephalopod—octopus
    b. gastropod—leech
    c. crustacean—copepod
    d. uniramians—millipedes
    e. arachnid—spider
26. Which characteristic accounts for the success of arthropods?
    a. jointed exoskeleton
    b. well-developed nervous system
    c. segmentation
    d. respiration adapted to environment
    e. All of these are correct.
27. Which phylum is paired incorrectly with a terrestrial representative?
    a. molluscs—snail
    b. annelids—earthworm
    c. arthropods—grasshopper
    d. annelids—millipede
    e. All of these are correct.
28. Sea stars
    a. have no respiratory, excretory, or circulatory systems.
    b. usually reproduce asexually.
    c. are protostomes.
    d. can evert their stomach to digest food.
    e. Both a and d are correct.
29. The tube feet of echinoderms
    a. are their head.
    b. are a part of the water vascular system.
    c. are found in the coelom.
    d. help pass sperm to females during reproduction.
    e. All of these are correct.

Thinking Scientifically

1. All life-forms are believed to have first evolved in the ocean. What made water a better nursery than land for the evolution of new lineages?
2. A farmer sprays a field of soybeans with pesticide to kill an insect that has been eating his crop. Later in the season, his plants are tall, but there are very few soybeans on the plants. What hypothesis could explain the connection between these two observations?

Understanding the Terms

annelid 544	Malpighian tubule 550
arthropod 546	mantle 540
bivalve 540	metamorphosis 547
centipede 552	millipede 552
cephalopod 542	mollusc 540
cephalothorax 553	molt 546
chelicerate 553	nephridium (pl., nephridia) 544
chitin 546	oligochaete 545
coelom 538	polychaete 544
crustacean 548	protostome 538
decapod 549	radula 540
deuterostome 538	schizocoelom 538
echinoderm 554	sea star 554
enterocoelom 538	seta (pl., setae) 544
exoskeleton 546	trachea (pl., tracheae) 547
gastropod 543	tube foot 554
hemocoel 549	typhlosole 545
insect 550	uniramians 550
leech 546	water vascular system 554

Match the terms to these definitions:

a. _____ Air tube in insects that opens at spiracles.
b. _____ Change in shape and form that some animals, such as insects, undergo during development.
c. _____ Coelom arises as a pair of mesodermal pouches from the wall of the primitive gut.
d. _____ Paired excretory tubules found in the earthworm and other invertebrates.
e. _____ Strong but flexible nitrogenous polysaccharide found in the exoskeleton of arthropods.

ARIS, the *Biology* Website

ARIS, the website for *Biology*, provides a wealth of information organized and integrated by chapter. You will find practice quizzes, interactive activities, labeling exercises, flashcards, and much more that will complement your learning and understanding of general biology.

www.mhhe.com/maderbiology9

31

VERTEBRATES

The New Guinea (NG) singing dog lives on the island of New Guinea, but it doesn't actually sing. It yelps, whines, and howls at different pitches unlike other dogs. The howl is eerie to the point of causing goose bumps when one tone blends with the next.

How the NG singing dog should be classified is a matter of debate. Is it a unique species, a type of gray wolf, or is it related to the dingo, the nonbarking canine that lives in Australia? Some suggest that the NG singing dog might be the most primitive known "breed" of domestic dog because Stone Age people brought it to New Guinea 6,000 years ago! Living on an island, it has remained geographically and reproductively isolated ever since. Therefore, it could be a living fossil, an organism with the same genes and characteristics as its original ancestor. If so, the NG singing dog offers an opportunity to study not only why it "sings" but also what the first domesticated dog was like.

Like so many other unique animals in today's world, the NG singing dog is threatened with extinction. Its present known population is small and inbred, and so far no members of this species have been located in the wild. Expeditions into the highlands of New Guinea have only yielded a few droppings, tracks, and haunting howls in the distance.

New Guinea singing dog, *Canis familiaris hallstromi.*

31.1 CHORDATES

Phylum Chordata consists of approximately 45,000 species of animals. The **chordates** represent less than 5% of the total number of species alive today. Phylum Chordata consists of such familiar animals as sharks, goldfish, toads, turtles, sparrows, antelopes, and humans. An animal must have four basic characteristics during its life history to be considered a chordate:

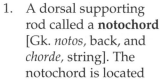

1. A dorsal supporting rod called a **notochord** [Gk. *notos*, back, and *chorde*, string]. The notochord is located just below the nerve cord. The majority of vertebrates have an embryonic notochord that is replaced by the vertebral column during development.

2. A dorsal tubular **nerve cord.** The anterior portion becomes the brain in most chordates. In vertebrates, the nerve cord, often called the spinal cord, is protected by vertebrae.

3. Pharyngeal pouches. These are seen only during embryonic development in most vertebrates. In the nonvertebrate chordates, the fishes, and amphibian larvae, the pharyngeal pouches become functioning **gills** (respiratory organs of aquatic vertebrates). Water passing into the mouth and the pharynx goes through the gill slits, which are supported by gill arches. In terrestrial vertebrates, the pouches are modified for various purposes. In humans, the first pair of pouches become the auditory tubes. The second pair become the tonsils, while the third and fourth pairs become the thymus gland and the parathyroid glands.

4. Postanal tail. A tail—in the embryo if not in the adult—extends beyond the anus:

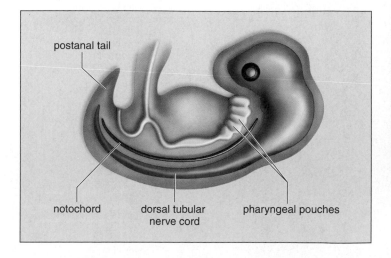

Nonvertebrate Chordates

In the nonvertebrate chordates, the notochord persists in the adult form and is never replaced by the vertebral column. These animals represent the vestiges of early chordate evolution. Two subphyla of lower chordates include subphylum Cephalochordata and subphylum Urochordata.

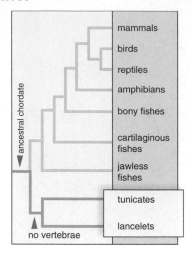

Approximately 25 species of nonvertebrate chordates have been placed in subphylum Cephalochordata. These are commonly called lancelets in the single genus *Branchiostoma*. The former genus for this organism, *Amphioxus*, is now used as a common name. These marine chordates, which are only a few centimeters long, are named for their resemblance to a lancet—a small, two-edged surgical knife (Fig. 31.1). Lancelets

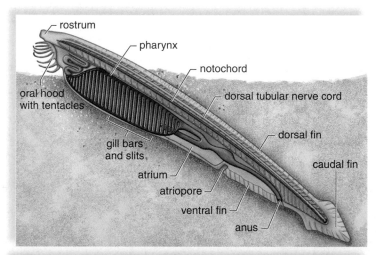

FIGURE 31.1 Lancelet, *Branchiostoma*.
Lancelets are filter feeders. Water enters the mouth and exits at the atriopore after passing through the gill slits.

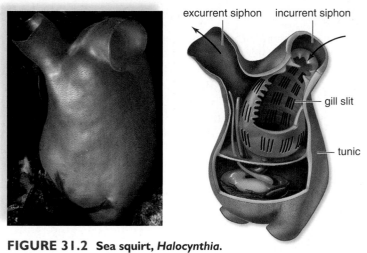

FIGURE 31.2 Sea squirt, *Halocynthia*.
Note that the only chordate characteristic remaining in the adult is gill slits.

are found in the shallow water along most coasts, where they usually lie partly buried in sandy or muddy substrates with only their anterior mouth and gill apparatus exposed. They feed on microscopic particles filtered out of the constant stream of water that enters the mouth and passes through the gill slits into an atrium that opens at the atriopore.

Lancelets retain the four chordate characteristics as an adult. The notochord extends from the tail to the head, and this accounts for their subphylum name, Cephalochordata. In addition, segmentation is present, as witnessed by the fact that the muscles are segmentally arranged and the dorsal tubular nerve cord has periodic branches. Lancelets are important in comparative anatomy and evolutionary studies.

The sea squirts, members of subphylum Urochordata, live on the ocean floor and are so called because they squirt out water from their excurrent siphon when disturbed. They are also called tunicates because they have a tunic that makes them look like thick-walled, squat sacs. Presently, about 2,000 species of urochordates have been described. The sea squirt larva is bilaterally symmetrical and has the four chordate characteristics. Metamorphosis produces the sessile adult with an incurrent and excurrent siphon (Fig. 31.2).

The pharynx is lined by numerous cilia whose beating creates a current of water that moves into the pharynx and out the numerous gill slits, the only chordate characteristic that remains in the adult. Microscopic particles adhere to a mucous secretion and are eaten.

Many evolutionists believe that the sea squirts are directly related to the vertebrates. It has been suggested that a larva with the four chordate characteristics may have become sexually mature without developing the other adult sea squirt characteristics. Then it may have evolved into a fishlike vertebrate. Figure 31.3 shows how the main groups of chordates may have evolved.

The nonvertebrate chordates include the lancelets and the sea squirts. A lancelet has the four chordate characteristics as an adult.

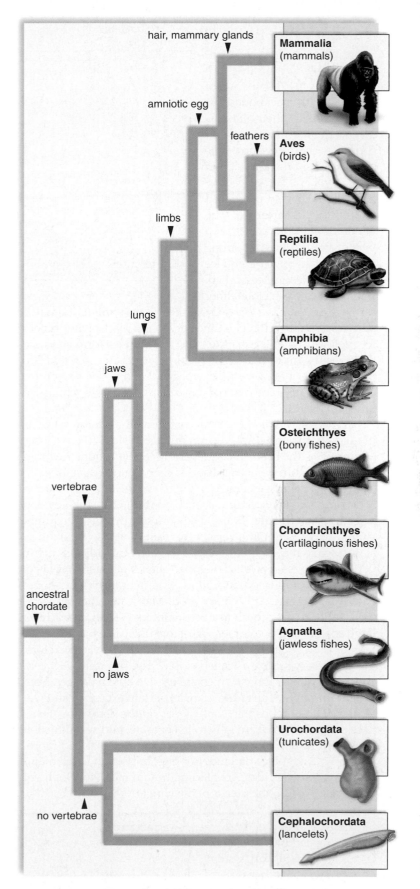

FIGURE 31.3 Phylogenetic tree of the chordates.
Each of the stated features is an evolved characteristic that is shared by the classes beyond this point.

31.2 Vertebrates

Subphylum Vertebrata consists of the **vertebrates,** approximately 43,700 species of animals with backbones. At some stage in their life history, vertebrates have all four chordate characteristics. The embryonic notochord, however, is generally replaced by a vertebral column composed of individual vertebrae. The vertebral column, which is a part of the flexible but strong endoskeleton, gives evidence that vertebrates are segmented. The vertebrate skeleton (either cartilage or bone) is a living tissue that grows with the animal. It also protects internal organs and serves as a place of attachment for muscles. Together, the skeleton and muscles form a system that permits rapid and efficient movement. Two pairs of appendages are characteristic. The pectoral and pelvic fins of fish evolved into the jointed appendages that allowed vertebrates to move onto land.

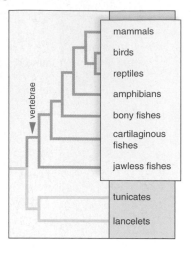

The main axis of the internal jointed skeleton consists of not only the vertebral column but also a skull that encloses and protects the brain. During vertebrate evolution, the brain increased in complexity, and specialized regions developed to carry out specific functions. The high degree of cephalization is accompanied by complex sense organs. The eyes develop as outgrowths of the brain. The ears are primarily equilibrium devices in aquatic vertebrates, but they also function as sound-wave receivers in land vertebrates. In addition, many vertebrates possess well-developed senses of smell and taste.

The evolution of jaws allowed vertebrates to develop a variety of strategies to acquire food. Many are efficient herbivores, and others are voracious predators. Vertebrates have a large coelom and a complete digestive tract. The blood in vertebrates is contained entirely within blood vessels, therefore, the circulatory system is called closed. The respiratory system consists of gills or lungs, which are used to obtain oxygen from the environment. The kidneys are important excretory and water-regulating organs that conserve or rid the body of water as necessary. The sexes are generally separate, and reproduction is usually sexual. The evolution of the amnion allowed reproduction to take place on land. Reptiles, birds, and some mammals lay a shelled egg. In placental mammals, development takes place within the uterus of the female.

These features are usually seen in vertebrates:
- living endoskeleton with vertebral column
- closed circulatory system
- paired appendages
- efficient respiration and excretion
- high degree of cephalization

In short, vertebrates are adapted for an active lifestyle.

Fishes

Fishes are water-dwelling, gill-breathing vertebrates that usually have fins and skin covered with **scales.** Fishes are **ectotherms** [Gk. *ekto,* outer, and *therme,* heat], meaning that they depend on the environment to regulate their body temperature. The fishes are the largest group of vertebrates with nearly 25,000 recognized species. They range in size from a few millimeters in length to the whale shark, which may reach lengths of 12 m.

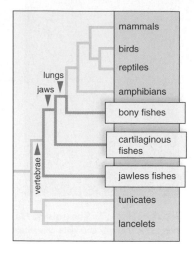

The fossil record of the fishes is extensive. The earliest fossils of Cambrian origin were the small, filter-feeding, jawless and finless **ostracoderms.** Several groups of ostracoderms developed heavy dermal armor for protection. The ostracoderms are the ancient relatives of modern vertebrates.

Jawless Fishes

Hagfishes and lampreys are modern-day **jawless fishes,** or **agnathans,** that lack a bony skeleton. They are cylindrical, up to a meter long, and have smooth, nonscaly skin (Fig. 31.4). The hagfishes are exclusively marine scavengers that feed on softbodied invertebrates and dead fishes. Approximately 43 species of hagfishes have been identified. The lampreys are represented by about 22 species of marine and freshwater jawless fishes. Many species are filter feeders like their ancestors. Parasitic lampreys have a round, muscular mouth used to at-

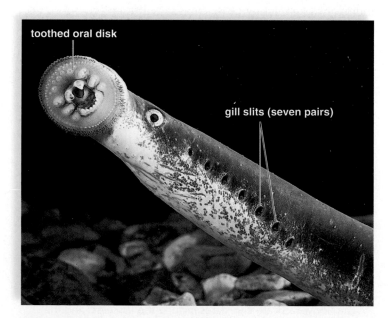

FIGURE 31.4 Lamprey, *Petromyzon.*
Lampreys, which are agnathans, have an elongated, rounded body and nonscaly skin. Note the lamprey's toothed oral disk, which is attached to the aquarium glass.

tach themselves to another fish and suck nutrients from the host's cardiovascular system. The parasitic sea lamprey gained access to the Great Lakes in 1829 and by 1950 had almost demolished the resident trout and whitefish populations.

Fishes with Jaws

All other vertebrates possess **jaws,** tooth-bearing bones of the head. Jaws are believed to have evolved from the first pair of gill arches present in ancestral agnathans. The second pair of gill arches became support structures for the jaws:

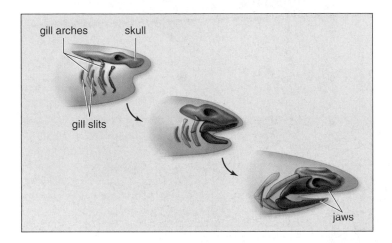

The **placoderms** were extinct jawed fishes of the Devonian period and are probably the ancestors of early sharks and bony fishes. Approximately 200 genera of placoderms of various shapes and sizes have been recorded in the fossil record. One species reached a length of 30 m. Placoderms were armored with heavy, bony plates and strong jaws. Like modern-day fishes, they had paired pectoral and pelvic fins (see Fig. 31.6).

Cartilaginous Fishes.   Sharks, rays, skates, and chimaeras are members of class Chondrichthyes, which consists of about 850 species of primarily marine **cartilaginous fishes.** Members of this class have a skeleton composed of cartilage instead of bone; have five to seven gill slits on both sides of the pharynx; and lack the gill cover of bony fishes. Their body is covered with epidermal placoid (toothlike) scales that project posteriorly, which is why a shark's skin feels like sandpaper. The menacing teeth of sharks and their relatives are simply larger, specialized versions of these scales. At any one time, a shark such as the great white shark may have up to 3,000 teeth in its mouth, arranged in 6 to 20 rows. Only the first row or two are actively used for feeding; the other rows are replacement teeth.

Three well-developed senses enable sharks and rays to detect their prey. They have the ability to sense electric currents in water—even those generated by the muscle movements of animals. They have a lateral line system, a series of pressure-sensitive cells that lie within canals along both sides of the body, which can sense pressure caused by a fish or other animal swimming nearby. They also have a very keen sense of smell; the part of the brain associated with this sense is very well developed. Sharks can detect about one drop of blood in 115 liters of water.

The largest sharks are filter feeders, not predators. The basking sharks and whale sharks ingest tons of small crustaceans, collectively called krill. Many sharks are fast-swimming predators in the open sea (Fig. 31.5a). The great white shark, about 7 m in length, feeds regularly on dolphins, sea lions, and seals. Humans are normally not attacked except when mistaken for sharks' usual prey. Tiger sharks, so named because the young have dark bands, reach 6 m in length and are unquestionably one of the most predaceous sharks. As it swims through the water, a tiger shark will swallow anything, including rolls of tar paper, shoes, gasoline cans, paint cans, and even human parts. The number of bull shark attacks has increased in the Gulf of Mexico in recent years. This is due to more swimmers in the shallow waters and loss of the sharks' natural food supply.

In rays and skates (Fig. 31.5b), the pectoral fins are greatly enlarged into a pair of large, winglike **fins,** and the bodies are dorsoventrally flattened. These bottom-dwelling animals feed on organisms such as crustaceans, small fishes, and molluscs.

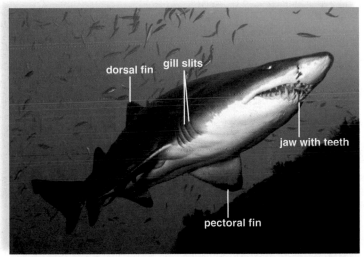

a. Sand tiger shark, *Carcharias taurus*

b. Blue-spotted stingray, *Taeniura lymma*

FIGURE 31.5 Cartilaginous fishes.
a. Sharks are predators or scavengers that move gracefully through open ocean waters. **b.** Most stingrays grovel in the sand, feeding on bottom-dwelling invertebrates. This blue-spotted stingray is protected by the spine on its whiplike tail.

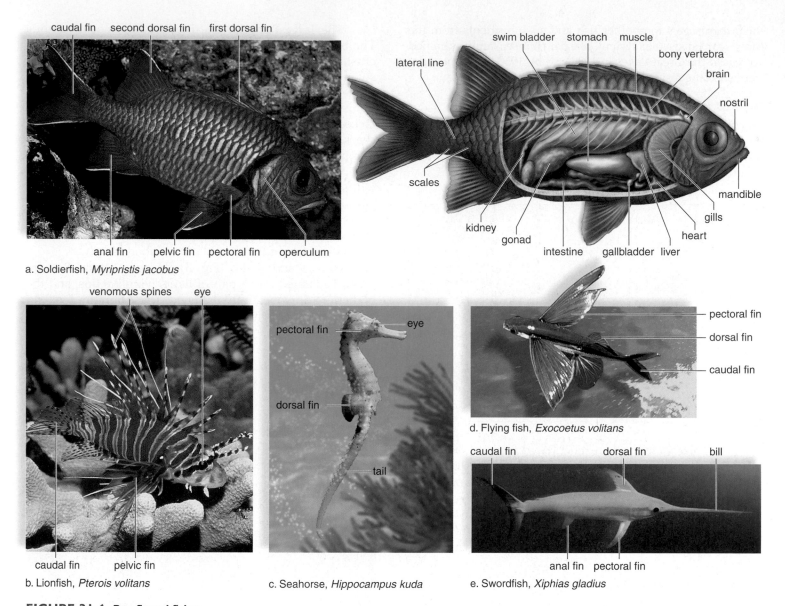

FIGURE 31.6 Ray-finned fishes.
a. A soldierfish has the typical appearance and anatomy of a ray-finned fish. A lionfish (**b**), a seahorse (**c**), a flying fish (**d**), and a swordfish
(**e**) show how diverse ray-finned fishes can be.

Stingrays have a whiplike tail that has serrated spines with venom glands at the base. These animals can deliver a painful "sting." Manta rays are harmless oceanic filter-feeding giants with a fin span of up to 6 m and a weight of 2,000 kg. Sawfish rays are named for their large, protruding anterior "saw." Skates resemble stingrays but possess a muscular tail, two dorsal fins, and a caudal fin. Some species of skate can deliver an electric shock. Their large electric organs, located at the base of their pectoral fins, can discharge over 300 volts.

The chimaeras, or ratfishes, are a group of cartilaginous fishes that live in cold marine waters. They are known for their unusual shape and iridescent colors.

Bony Fishes.   The majority of living vertebrates, approximately 25,000 species, are **bony fishes.** Originally all the

bony fishes belonged to class Osteichthyes. However, many zoologists now recognize two classes of bony fishes: class Actinopterygii (the ray-finned fishes) and class Sarcopterygii (the lobe-finned fishes). The bony fishes range in size from gobies, which are less than 7.5 mm long, to the giant sturgeons, which can obtain a length of 4 m.

The majority of fish species are **ray-finned bony fishes** with fan-shaped fins supported by a thin, bony ray. These fishes are the most successful and diverse of all the vertebrates (Fig. 31.6). Some, like herrings, are filter feeders; others, like trout, are opportunists; and still others are predaceous carnivores, such as piranhas and barracudas.

Despite their diversity, bony fishes have many features in common. The gills do not open separately and are covered by an operculum. Many bony fishes have a **swim bladder,** a gas-filled sac whose pressure can be altered to change buoyancy and therefore the fishes' depth in the water. Bony fishes have a single-loop, closed cardiovascular system (see Fig. 31.10a). The nervous system and brain in bony fishes is well-developed, and complex behaviors are common. Bony fishes have separate sexes, and the majority undergo external fertilization after females deposit eggs and males deposit sperm into the water.

The **lobe-finned fishes** possess fleshy fins supported by central bones. Lobe-finned fishes are abundant in the fossil record. However, today only six species of lungfish and two species of coelacanths survive (Fig. 31.7a). Lungfishes live in Africa, South America, and Australia, either in stagnant fresh water or in ponds that dry up annually. These fishes use a lung for respiration. In 1938, a coelacanth was caught from the deep waters off the eastern coast of South Africa. It took the scientific world by surprise because these animals were thought to be extinct for 70 million years. Approximately 200 coelacanths have been captured in recent years. The lobe-finned fishes are thought to have given rise to the amphibians.

These characteristics are usually seen in fishes:
- aquatic ectotherms
- skin covered with scales
- fins for swimming
- single-loop cardiovascular pathway
- breathe with gills
- lay eggs

Amphibians

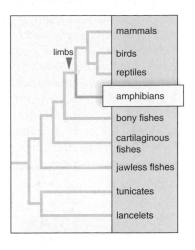

Amphibians [Gk. *amphibios*, living both on land and in water], as well as reptiles, birds, and mammals, are **tetrapods** [Gk. *tetra*, four, and *podos*, foot], meaning that they have four limbs. Class Amphibia consists of approximately 4,200 species of living tetrapods. The lobe-finned fishes of the Devonian period are ancestral to the amphibians, the first tetrapods. Figure 31.7b compares the fins of a lobe-finned fish to the limbs of an early amphibian and shows that the same bones are present in both animals. Animals that live on land use limbs to support their bodies, especially since air is less buoyant than water. Internal nares and lungs allowed lobe-finned fishes and early amphibians to breathe air.

Two hypotheses have been suggested to account for the evolution of amphibians from lobe-finned fishes. Perhaps lobe-finned fishes had an advantage over others because they could use their lobed fins to move from pond to pond. Or perhaps the supply of food on land in the form of plants and insects—and the absence of predators—promoted further adaptations to the land environment.

a. Coelacanth, *Latimeria chalumnae*

FIGURE 31.7 Lobe-finned fish versus amphibian.
a. A coelacanth is a lobe-finned fish once thought to be extinct. **b.** These drawings compare the skeletal structure of the appendages of a lobe-finned fish and an ancestral amphibian. A shift in the position of the bones in the forelimbs and hindlimbs lifts and supports the body.

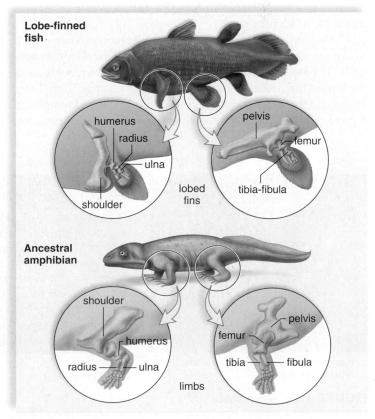

b. Fin to limb

Diversity of Amphibians

The amphibians of today occur in three groups: salamanders and newts; frogs and toads; and caecilians. Salamanders and newts have elongated bodies, long tails, and usually two pairs of limbs (Fig. 31.8*a*). There are about 360 species of these unique animals. Salamanders and newts range in size from less than 15 cm to the giant Japanese salamander, which exceeds 1.5 m in length. Most have limbs that are set at right angles to the body and resemble the earliest fossil amphibians. They move like a fish, with side-to-side, sinusoidal (S-shaped) movements:

Both salamanders and newts are carnivorous, feeding on small invertebrates such as insects, slugs, snails, and worms. Salamanders practice internal fertilization; in most, males produce a sperm-containing spermatophore that females pick up with their **cloaca** (the common receptacle for the urinary, genital, and digestive canals). Then the eggs are laid in water or on land, depending on the species. Some amphibians, such as the mudpuppy of eastern North America, remain in the water and retain the gills of the larva.

Frogs and toads are the tailless amphibians (Fig. 31.8*b*). There are approximately 3,450 species of these common animals. The smallest frog measures less than 1 cm in length, and the largest (the West African Goliath frog) measures 30 cm long. Frogs and toads are common in the subtropical and temperate regions of the world. In these animals, the head and trunk are fused, and the long hindlimbs are specialized for jumping. Numerous muscles are located in the limbs as well as on the trunk. Frogs have smooth skin and and they live in or near fresh water; toads have stout bodies and warty skin, and they live in dark, damp places away from the water.

Caecilians are legless, often sightless, worm-shaped amphibians that range in length from about 10 cm to more than 1 m (Fig. 31.8*c*). Most of the 160 species burrow in moist soil, feeding on worms and other soil invertebrates. Some species have folds of skin that make them look like a segmented earthworm.

The class name of amphibians is appropriate because most return to water for the purpose of reproduction. They shed their eggs and sperm into the water, where external fertilization takes place. Generally, the eggs are protected only by a jelly coat and not by a shell. When the young hatch, they are tadpoles (aquatic larvae with gills) that feed and grow in the water. After they undergo a **metamorphosis** (change in form), amphibians emerge from the water as adults that breathe air (Fig. 31.9). Some amphibians, however, have evolved mechanisms that allow them to reproduce on land.

Anatomy and Physiology of Amphibians

The majority of amphibians have general adaptations for living in a terrestrial environment. The skeletal system, particularly the pelvic and pectoral girdles, is well-developed to promote locomotion. Special senses such as sight, hearing, and smell are fine-tuned for life on land. Amphibian brains are larger than those of fish, and the cerebral cortex is more developed. These animals also have a specialized tongue for catching prey, eyelids for keeping their eyes moist, and a sound-producing larynx.

Most adult amphibians have **lungs,** respiratory organs typical of terrestrial vertebrates, but they are relatively small, and respiration is supplemented by exchange of gases across the porous skin. Amphibians have a double-loop circulatory pathway (Fig. 31.10*b, c*). Amphibians have a three-chambered heart with a single ventricle and two atria, but there is little mixing of O_2-rich and O_2-poor blood.

moist, smooth skin   hindlimb (to side)

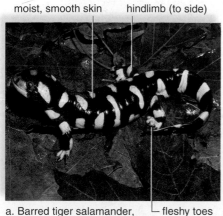

a. Barred tiger salamander,
   Ambystoma tigrinum   └ fleshy toes

tympanum ── eye

b. Tree frog,
   Hyla andersoni   └ forelimb

sightless head   smooth skin

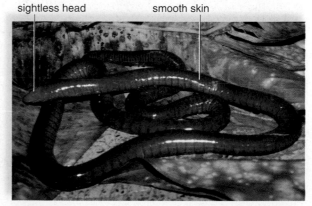

c. Caecilian,
   Caecilia nigricans

FIGURE 31.8 Amphibians.
Living amphibians are divided into three orders: **a.** Salamanders and newts. Members of this order have a tail throughout their lives and, if present, unspecialized limbs. **b.** Frogs and toads. Like this frog, members of this order are tailless and have limbs specialized for jumping. **c.** The caecilians are wormlike burrowers.

1. Tadpoles hatch.

2. Tadpole respires with external gills.

3. Front and hind legs are present.

4. Frog respires with lungs.

FIGURE 31.9 Metamorphosis.
Frogs undergo metamorphosis. During frog metamorphosis, the animal changes from a form specialized for an aquatic life to one that is specialized for a terrestrial one.

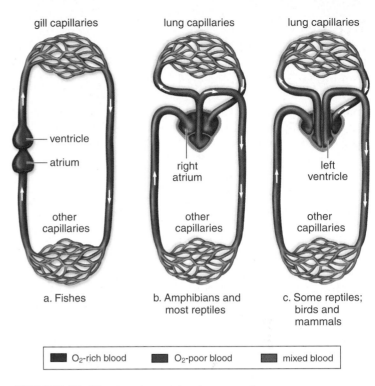

| gill capillaries | lung capillaries | lung capillaries |

ventricle
atrium
right atrium
left ventricle

other capillaries
other capillaries
other capillaries

a. Fishes   b. Amphibians and most reptiles   c. Some reptiles; birds and mammals

■ O₂-rich blood   ■ O₂-poor blood   ■ mixed blood

FIGURE 31.10 Vertebrate circulatory pathways.
a. The single-loop pathway of fishes has a two-chambered heart. **b.** The double-loop pathway of other vertebrates sends blood to the lungs and to the body. In amphibians and most reptiles, there is limited mixing of O_2-rich and O_2-poor blood in the single ventricle of their three-chambered heart. **c.** The four-chambered heart of some reptiles, birds, and mammals sends only O_2-poor blood to the lungs and O_2-rich blood to the body.

The smooth, nonscaly skin of an amphibian is kept moist by numerous mucous glands. The skin plays an active role in water balance and respiration and can also help in temperature regulation when on land, through evaporative cooling. A thin, moist skin does mean, however, that most amphibians stay close to water or else risk drying out. Glands in the skin secrete poisons that make the animal distasteful to eat. Some tropical species with brilliant fluorescent green and red coloration are particularly poisonous. Colombian Indians dip their darts in the deadly secretions of these frogs, aptly called poison-dart frogs.

Like fishes, amphibians are ectotherms but are able to live in environments where the temperature fluctuates greatly. During winters in the temperate zone, they become inactive and enter torpor. The European common frog can survive in temperatures dropping to as low as −6°C.

These features are seen in amphibians:
- usually tetrapods
- usually lungs in adult
- metamorphosis
- smooth and moist skin
- three-chambered heart
- ectothermy
- most lay eggs in water

Reptiles

Class Reptilia [L. *reptile,* snake] includes some of the most intriguing animals that have ever lived. Turtles, lizards, snakes, alligators, and tuataras are among the nearly 8,000 species of living reptiles.

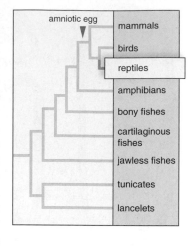

To conquer the land, reptiles developed a means of reproduction that is not dependent on external water. Reptiles practice internal fertilization and lay eggs that are protected by a leathery shell (see Fig. 31.12*a*). The **amniotic egg** contains extraembryonic membranes, which protect the embryo, remove nitrogenous wastes, and provide the embryo with oxygen, food, and water. These membranes are not part of the embryo itself and are disposed of after development is complete. One of the membranes, the amnion, is a sac that fills with fluid and provides a "private pond" within which the embryo develops.

Reptiles are believed to have evolved from amphibian ancestors and were present by the Permian period (Fig. 31.11). The first reptiles, known as the stem reptiles, gave rise to several other lineages, each adapted to a different way of life. Of interest to us are the pelycosaurs, or sail lizards, so named

because of a sail-like web of skin held above the body by slender spines. They are related to the **therapsids,** mammal-like reptiles that came later. Other lines of descent returned to the aquatic environment; one marine reptile (ichthyosaurs) of the Jurassic period was fishlike, while another (plesiosaurs) had a long neck and large rowing paddles for limbs.

The flying reptiles (pterosaurs) of the Jurassic period had a keel for the attachment of large flight muscles and air spaces in their bones to reduce weight. Their wings were membranous and supported by elongated bones of the fourth finger. *Quetzalcoatlus,* the largest flying animal ever to live, had an estimated wingspan of nearly 13.5 m.

Dinosaurs varied in size and behavior. The average size of a dinosaur was about the size of a chicken. Some of the dinosaurs, however, were the largest land animals ever to live. *Brachiosaurus,* a herbivore, was about 23 m long and about 17 m tall. *Tyrannosaurus rex,* a carnivore, was 5 m tall when standing on its hind legs. A bipedal stance freed the forelimbs for seizing prey or fighting off predators. It was also preadaptive for the evolution of wings since birds are descended from dinosaurs; in fact, some hypothesize that birds are actually living dinosaurs.

Reptiles dominated the Earth for about 170 million years throughout the Mesozoic era. Suddenly, the dinosaurs, as well as much of the flora and fauna of the period, died out. The sudden mass extinction of the Cretaceous period is known as the time of dying. One of the major hypotheses addressing the mass extinction is that a massive meteorite struck the Earth

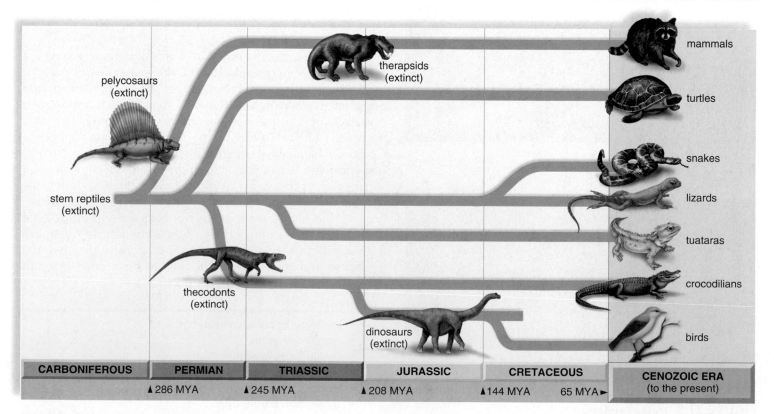

FIGURE 31.11 Phylogenetic tree of reptiles.
This phylogenetic tree shows the presumed evolutionary relationships among major groups of reptiles, starting with the stem reptiles in the Paleozoic era (including the Carboniferous and Permian periods). The Mesozoic era includes the Triassic, Jurassic, and Cretaceous periods.

near the Yucatán Peninsula. The resultant cataclysmic events disrupted existing ecosystems, destroying many living things. This hypothesis is supported by the presence of a layer of the mineral iridium, which is rare on Earth but common in meteorites in the late Cretaceous strata. Luis and Walter Alvarez and others proposed the meteorite hypothesis in the 1980s.

Diversity of Reptiles

Modern reptiles are diverse in size, shape, and mode of life. They have adapted to live in a variety of environments, from the ocean to parching deserts. Living reptiles are represented by crocodiles and alligators, turtles, lizards, snakes, amphisbaenians, and tuataras (Fig. 31.12).

There are approximately 22 species of *crocodiles and alligators* (Fig. 31.12*a*). The majority of these animals live in fresh water, feeding on fishes, turtles, and terrestrial animals that venture too close to the water. They have long, powerful jaws with numerous teeth and a muscular tail that serves as both a weapon and a paddle. Although other reptiles are voiceless, male crocodiles and alligators bellow to attract mates. In some species, the male protects the eggs and cares for the young.

The 270 species of *turtles* (Fig. 31.12*b*) living today have one of the oldest reptilian lineages. Along with tortoises, turtles can be found in marine, freshwater, and terrestrial environments. Most turtles have a heavy shell to which the ribs and thoracic vertebrae are fused. They lack teeth but have a sharp beak. The legs of sea turtles are flattened and paddle-like, while terrestrial tortoises have strong limbs for walking.

Approximately 3,700 species of *lizards* (Fig. 31.12*c*) have been classified. Lizards have four clawed feet and resemble their prehistoric ancestors in appearance. They are carnivorous and feed on insects and small animals, including other lizards. Marine iguanas of the Galápagos Islands are adapted to spend long periods of time at sea, where they feed on sea lettuce and other marine vegetation. Chameleons are adapted to live in trees and have long, sticky tongues for catching insects some distance away. They can change color to blend in with their background. Geckos are primarily nocturnal lizards with adhesive pads on their toes. The glass lizards do not have limbs and resemble a snake. Skinks are common elongated lizards with reduced limbs and shiny scales. The bites of monitor lizards and Gila monsters represent a deadly threat to humans.

a. American crocodile, *Crocodylus acutus*

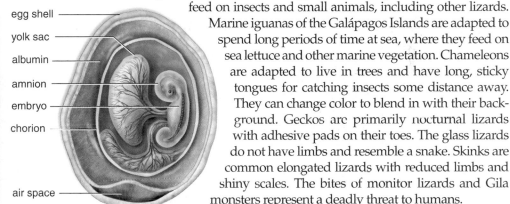

b. Green sea turtle, *Chelonia mydas*

c. Gila monster, *Heloderma suspectum*

d. Diamondback rattlesnake, *Crotalus atrox*

e. Worm lizard, *Amphisbaena* sp.

f. Tuatara, *Sphenodon punctatus*

FIGURE 31.12 Reptilian diversity.
Representative living reptiles include (**a**) a young crocodile hatching from an egg. The egg shell is leathery and flexible, not brittle like birds' eggs. Inside the egg, the embryo is surrounded by three membranes. The chorion aids in gas exchange, the allantois stores waste, and the amnion encloses a fluid that prevents drying out and provides protection. The yolk sac provides nutrients for the embryo. **b.** Green sea turtles. **c.** The venomous Gila monster. **d.** The diamondback rattlesnake. **e.** The worm lizard. **f.** The tuatara.

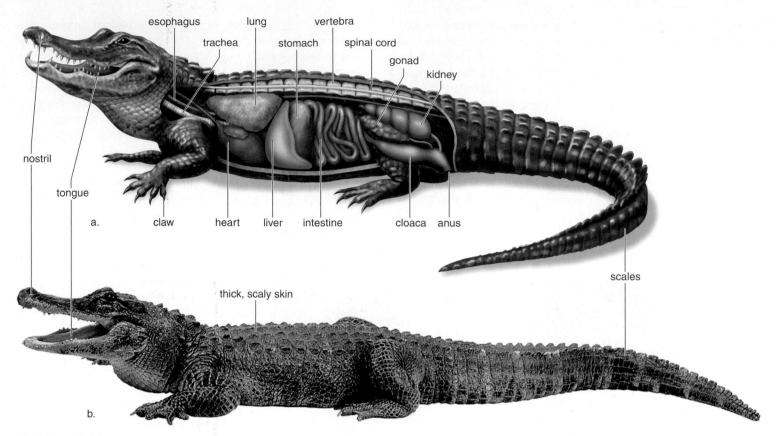

FIGURE 31.13 Reptilian anatomy.
Internal (**a**) and external (**b**) anatomy of an alligator, *Alligator mississippiensis.*

Although the majority of the 2,300 species of *snakes* (Fig. 31.12*d*) are harmless, several venomous species including rattlesnakes, cobras, mambas, and copperheads have given the whole group a bad reputation. Snakes evolved from lizards and have lost their limbs as an adaptation to burrowing. A few species such as pythons and boas still possess the vestiges of pelvic girdles. Snakes are carnivorous and have a jaw that is loosely attached to the skull; therefore, they can eat prey that is much larger than their head size. When a snake flicks out its tongue, it is collecting airborne molecules and transferring them to a Jacobson's organ at the roof of the mouth. A Jacobson's organ is an olfactory organ for the analysis of airborne chemicals. A transparent membrane protects the cornea of snakes, thus snakes do not blink their eyes. Snakes possess internal ears that are capable of detecting low-frequency sounds and vibrations.

The *amphisbaenians* are commonly called worm lizards (Fig. 31.12*e*). There are approximately 140 species worldwide. Worm lizards are highly specialized for burrowing in the soil. They have elongated bodies that are divided into numerous rings and lack visible ears, eyes, and external limbs.

There are two living species of *tuataras* found in New Zealand (Fig. 31.12*f*). They are lizardlike animals that can attain a length of 66 cm and can live for nearly 80 years. These animals possess a well-developed eye known as a third eye, which is light sensitive and buried beneath the skin in the upper part of the head.

Anatomy and Physiology of Reptiles

The alligator in Figure 31.13 is representative of the anatomy of most reptiles. Reptiles have a thick, scaly skin that is keratinized and impermeable to water. Keratin is the protein found in hair, fingernails, and feathers. The protective skin prevents water loss but requires several molts a year. The reptilian lung is more developed than in amphibians, and air rhythmically moves in and out due to the presence of an expandable rib cage, except in turtles. Although most reptiles have a single ventricle, it is partially divided by a septum. The heart is completely four-chambered in crocodiles, therefore, O_2-poor blood is completely separate from O_2-rich blood (see Fig. 31.10*c*). The well-developed kidneys excrete uric acid, and therefore less water is required to rid the body of nitrogenous wastes.

Reptiles, like fishes and amphibians, are ectothermic. This feature allows them to survive on a fraction of the food per body weight required by birds and mammals. Still, they are adapted behaviorally to maintain a warm body temperature by warming themselves in the sun.

These features are seen in reptiles:
- usually tetrapods
- lungs with expandable rib cage
- leathery-shelled amniotic egg
- dry, scaly skin
- endothermy

science focus

Vertebrates and Human Medicine

Hundreds of pharmaceutical products come from other vertebrates, and even those that produce poisons and toxins give us medicines that benefit us. The Thailand cobra paralyzes its victim's nerves and muscles with a potent venom that eventually leads to respiratory arrest. However, that venom is also the source of the drug Immunokine, which has been used for ten years in multiple sclerosis patients. Immunokine, which is almost without side effects, actually protects the patient's nerve cells from destruction by their immune system. A compound known as ABT-594, derived from the skin of the poison-dart frog, is approximately 50 times more powerful than morphine in relieving chronic and acute pain without the addictive properties. The southern copperhead snake, the cone snail, and the fer-de-lance pit viper are some of the unlikely vertebrates that either serve as the source of pharmaceuticals or provide a chemical model for the synthesis of effective drugs in the laboratory. These drugs include anticoagulants ("clot busters"), pain killers, antibiotics, and anticancer drugs.

A variety of friendlier vertebrates produce proteins that are similar enough to human proteins to be used for medical treatment. Until 1978, when recombinant DNA human insulin was produced, diabetics injected insulin purified from pigs. Currently, the flu vaccine is produced in fertilized chicken eggs. The production of these drugs, however, is often time-consuming, labor intensive, and expensive. In 2003, pharmaceutical companies used 90 million chicken eggs and took 9 months to produce the flu vaccine.

Some of the most powerful applications of genetic engineering can be found in the development of drugs and therapies for human diseases. In fact, this new biotechnology has actually led to a new industry: animal pharming. Animal pharming uses genetically altered vertebrates, such as mice, sheep, goats, cows, pigs, and chickens, to produce medically useful pharmaceutical products. The human gene for some useful product is inserted into the embryo of the vertebrate. That embryo is implanted into a foster mother, which gives birth to the transgenic animal, so called because it contains genes from two sources. An adult transgenic vertebrate produces large quantities of the pharmed product in its blood, eggs, or milk, from which the product can be easily harvested and purified. The first such product, alpha-1-antitrypsin for the treatment of emphysema and cystic fibrosis, is now undergoing clinical trials. It is not expected to be marketed until 2007, because some trial patients experienced some wheezing while taking the medication.

Xenotransplantation, the transplantation of vertebrate tissues and organs into human beings, is another benefit of genetically altered animals. There is an alarming shortage of human donor organs to fill the need for hearts, kidneys, and livers. The first animal–human transplant occurred in 1984 when a team of surgeons im-planted a baboon heart into an infant, who unfortunately lived only a short while before dying of circulatory complications. In the late 1990s, two patients were kept alive using a pig liver outside of their body to filter their blood until a human organ was available for transplantation. Although baboons are genetically closer to humans than pigs, pigs are generally healthier, produce more offspring in a shorter time, and are already farmed for food. Despite the fears of some, scientists think that viruses unique to pigs are unlikely to cross the species barrier and infect the human recipient. Currently, pig heart valves and skin are routinely used for treatment of humans. Miniature pigs, whose heart size is appropriate to humans, are being genetically engineered to make them less foreign to the human body, in order to avoid rejection.

The use of transgenic vertebrates for medical purposes does raise health and ethical concerns. What viral AIDS-like epidemic might be unleashed by cross-species transplantation? What other unseen health consequences might there be? Is it ethical to change the genetic makeup of vertebrates, in order to use them as drug or organ factories? Are we redefining the relationship between humans and other vertebrates to the detriment of both? These questions will continue to be debated as the research goes forward. Meanwhile, several U.S. regulatory bodies, including the FDA, have adopted voluntary guidelines for this new technology.

a. Poison-dart frogs, source of a medicine

b. Pigs, source of organs

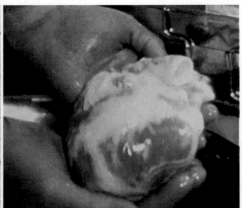

c. Heart for transplantation

FIGURE 31A Use of other vertebrates for medical purposes.
a. The poison-dart frog is the source of a pain medication. b. Pigs are now being genetically altered to provide a supply of (c) hearts for heart transplant operations.

Birds

To many people, the **birds** (class Aves) are the most conspicuous, melodic, beautiful, and fascinating group of vertebrates. Over 9,000 species of birds have been described. Birds range in size from the tiny "bee" hummingbird at 1.8 g to the ostrich at a maximum weight of 160 kg and a height of 2.7 m. The fossil record of birds is unraveling evolutionary mysteries because *Archaeopteryx* and newly described fossils are providing data that birds evolved from bipedal dinosaurs. Birds may be classified as dinosaurs in the future; their legs have scales and their **feathers** (Fig. 31.14*a*) are modified reptilian scales. However, birds lay a hard-shelled amniotic egg rather than the leathery egg of reptiles.

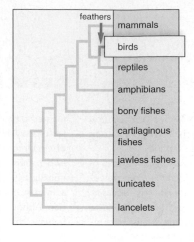

Diversity of Birds

The majority of birds, including eagles, geese, and mockingbirds, have the ability to fly. However, some birds, such as emus, penguins, and ostriches, are flightless. Traditionally birds have been classified according to beak and foot type (Fig. 31.15) and, to some extent, on their habitat and behavior. The various orders include birds of prey with notched beaks and sharp talons; shorebirds with long, slender, probing beaks and long, stiltlike legs; woodpeckers with sharp, chisel-like beaks and grasping feet; waterfowl with broad beaks and webbed toes; penguins with wings modified as paddles; and songbirds with perching feet.

Anatomy and Physiology of Birds

Nearly every anatomical feature of a bird can be related to its ability to fly (see Fig. 31.14*b*). The forelimbs are modified as wings. Respiration is efficient since the lobular lungs connect to anterior and posterior air sacs. The presence of these sacs means the air circulates one way through the lungs, and gases are continuously exchanged across respiratory tissues. Another benefit of air sacs is that they lighten the body and

FIGURE 31.14 Bird anatomy and flight.
a. Bird anatomy. *Top:* In feathers, a hollow central shaft gives off barbs and barbules, which interlock in a latticelike array. *Bottom:* The anatomy of a pigeon is representative of bird anatomy. **b.** Bird flight. *Left:* The skeleton of an eagle shows that birds have a large, keeled sternum to which flight muscles attach. The bones of the forelimb help support the wings. *Right:* Birds fly by flapping their wings. Bird flight requires an airstream and a powerful wing downstroke for lift, a force at right angles to the airstream.

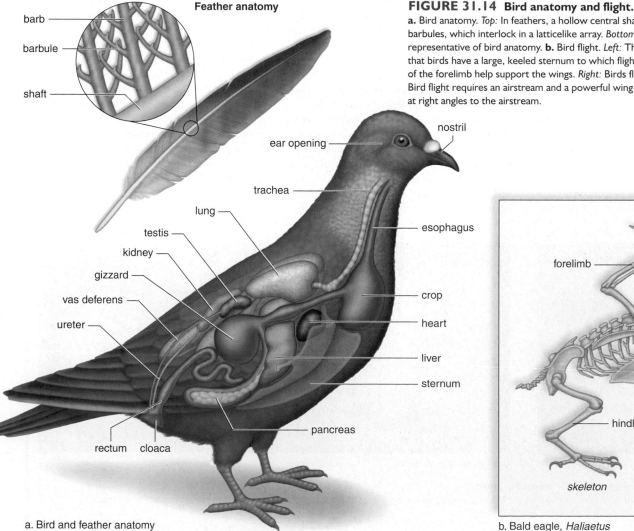

a. Bird and feather anatomy

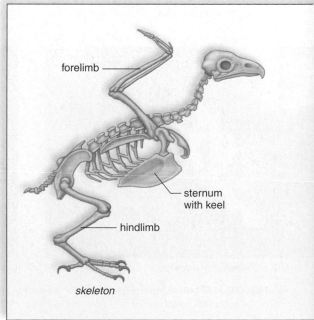

b. Bald eagle, *Haliaetus*

a. Bald eagle, *Haliaetus leucocephalus*

b. Pileated woodpecker, *Dryocopus pileatus*

FIGURE 31.15
Bird beaks.
a. A bald eagle's beak allows it to tear apart prey. **b.** A woodpecker's beak is used to drill in wood. **c.** A flamingo's beak strains food from the water with bristles that fringe the mandibles. **d.** A vulture's beak is modified to eat flesh. **e.** A cardinal's beak allows it to crack tough seeds.

c. Flamingo, *Phoenicopterus ruber*

d. Turkey vulture, *Cathartes aura*

e. Cardinal, *Cardinalis cardinalis*

bones for flying. The air sacs are present in cavities within the bones. A horny beak has replaced jaws equipped with teeth, and a slender neck connects the head to a rounded, compact torso. The sternum is enlarged and has a keel to which strong muscles are attached for flying.

Birds have a four-chambered heart that completely separates O_2-rich blood from O_2-poor blood. The left ventricle pumps O_2-rich blood under pressure to the muscles. Birds are **endotherms** and generate internal heat. Many endotherms can use metabolic heat to maintain a constant internal temperature. This may be associated with their efficient nervous, respiratory, and circulatory systems. Also, their feathers provide insulation. Birds have no bladder and excrete uric acid in a semidry state.

Flight requires that the sense organs and nervous system be well developed. Birds have particularly acute vision and well-developed brains. Their muscle reflexes are excellent. An enlarged portion of the brain seems to be the area responsible for instinctive behavior. A ritualized courtship often precedes mating. Many newly hatched birds require parental care before they are able to fly away and seek food for themselves. A remarkable aspect of bird behavior is the seasonal migration of many species over long distances. Birds navigate by day and night, whether it's sunny or cloudy, by using the sun and stars and even the Earth's magnetic field to guide them. Birds are very vocal animals. Bird vocalizations are distinctive and convey an abundance of information.

downstroke

upstroke

These features are seen in birds:
- feathers
- hard-shelled amniotic egg
- four-chambered heart
- usually wings for flying
- air sacs
- endothermy

Mammals

Class Mammalia [L. *mamma*, breast, teat] includes approximately 4,800 species. Examples of this class include the largest animal ever to live, the blue whale (130 metric tons); the smallest mammal, the Kitti's bat (1.5 g); and the fastest land animal, the cheetah (110 km/hr).

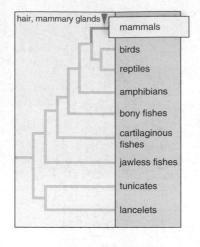

Mammals evolved during the Mesozoic era from mammal-like reptiles called therapsids. True mammals appeared during the Jurassic period, about the same time as the first dinosaurs. These first mammals were similar in size to mice. During the reign of the dinosaurs (165 million years), mammals were a minor group that changed little. Some of the earliest mammalian groups, represented today by the monotremes and marsupials, are not abundant today. The placental mammals that evolved later eventually assumed many habitats previously occupied by the dinosaurs.

One of the major differences between reptiles and mammals is the skull. The mammalian skull accommodates a larger brain relative to body size than does the reptilian skull. Also, mammalian cheek teeth are differentiated as premolars and molars. The vertebrae of mammals are highly differentiated, and the middle region of the vertebral column is arched, which provides more effective movement on land. Mammals have efficient respiratory and circulatory systems, which ensure a ready oxygen supply to muscles whose contraction produces body heat. Like birds, mammals have a double-loop circulatory pathway and a four-chambered heart. The kidneys of mammals have terrestrial origins and are adapted to conserve water. The nervous system of mammals is highly developed. Special senses in mammals are well developed, and mammals exhibit complex behavior.

The most distinguishing characteristics of mammals are the presence of hair and milk-producing mammary glands. Being endothermic, hair provides insulation against heat loss and allows mammals to be active even in cold weather. In addition, hair can be ornamental and can serve sensory functions. Mammary glands enable females to feed (nurse) their young without leaving them to find food. Nursing also creates a bond between mother and offspring that helps ensure parental care while the young are helpless. In most mammals, the young are born alive after a period of development in the uterus, a part of the female reproductive tract. Internal development shelters the young and allows the female to move actively about while the young are maturing. Although mammals are quite diverse, they can be divided into three major groups: mammals that lay eggs, mammals that have pouches, and mammals that have placentas.

Mammals That Lay Eggs

Monotremes [Gk. *monos*, one, and *trema*, hole] are egg-laying mammals that include only the duckbill platypus and the spiny anteater. Both of these unusual animals, found in Australia (Fig. 31.16*a*), possess a cloaca and lay hard-shelled amniotic eggs. The female duckbill platypus lays her eggs in a burrow in the ground. She incubates the eggs, and after hatching, the young lick up milk that seeps from modified sweat glands on the abdomen. The spiny anteater, which actually feeds mainly on termites and not ants, has pores that seep milk in a shallow belly pouch formed by skin folds on each side. The egg moves from the cloaca to this pouch, where hatching takes place and the young remain for about 53 days. Then they stay in a burrow, where the mother periodically visits and nurses them.

Mammals That Have Pouches

The **marsupials** [Gk. *marsupium*, pouch] are also known as the pouched mammals. Example marsupials include kangaroos, koala bears, Tasmanian devils, wombats, sugar gliders, and the opossum. The young of marsupials begin their development inside the female's body, but they are born in a very immature condition. Newborns crawl up into a pouch on their mother's abdomen. Inside the pouch, they attach to nipples of mammary glands and continue to develop. Frequently, more are born than can be accommodated by the number of nipples, and it's "first come, first served."

a. Duckbill platypus, *Ornithorhynchus anatinus*

b. Virginia opossum,
   Didelphis virginianus

c. Koala,
   Phascolarctos cinereus

FIGURE 31.16 Monotremes and marsupials.
a. The duckbill platypus is a monotreme that inhabits Australian streams. **b.** The opossum is the only marsupial in North America. The Virginia opossum is found in a variety of habitats. **c.** The koala is an Australian marsupial that lives in trees.

a. White-tailed deer, *Odocoileus virginianus*

b. African lioness, *Panthera leo*

c. Squirrel monkey, *Saimiri sciureus*

d. Killer whale, *Orcinus orca*

FIGURE 31.17
Placental mammals.
Placental mammals have adapted to various ways of life. **a.** Deer are herbivores that live in forests. **b.** Lions are carnivores on the African plain. **c.** Monkeys inhabit tropical forests. **d.** Whales are sea-dwelling placental mammals.

Today, marsupial mammals are most abundant in Australia and New Guinea. A significant number of marsupial species are found in South and Central America. The opossum is the only North American marsupial (Fig. 31.16*b*). Among herbivorous marsupials in Australia today, koalas are tree-climbing browsers (Fig. 31.16*c*), and kangaroos are grazers.

Mammals That Have Placentas

The placental mammals are the dominant group of mammals on Earth. Shrews, elephants, bats, dolphins, and humans represent this diverse group. Developing placental mammals are dependent on the **placenta,** an organ of exchange between maternal blood and fetal blood. Nutrients are supplied to the growing offspring, and wastes are passed to the mother for excretion. While the fetus is clearly parasitic on the female, in exchange, she is able to freely move about while the fetus develops.

Placental mammals lead an active life. The senses are acute, and the brain is enlarged due to the convolution and expansion of the foremost part—the cerebral hemispheres. The brain is not fully developed for some time after birth, and there is a long period of dependency on the parents, during which the young learn to take care of themselves.

Placental mammals populate all continents except Antarctica. Most mammals live on land, but some (e.g., whales, dolphins, seals, sea lions, and manatees) are secondarily adapted to live in water, and bats are able to fly. The following are some of the major orders of mammals:

The hoofed mammals include the orders Perissodactyla (e.g., horses, zebras, tapirs, rhinoceroses; 17 species) and the Artiodactyla (e.g., pigs, cattle, deer, hippopotamuses, buffaloes, giraffes; 185 species) whose elongated limbs are adapted for running, often across open grasslands (Fig. 31.17*a*). Both groups of animals are herbivorous and have large, grinding teeth.

Order Carnivora (270 species) includes dogs, cats, bears, raccoons, and skunks (Fig. 31.17*b*). The canines of meat eaters are large and conical. Some carnivores are aquatic—namely, seals, sea lions, and walruses—and must return to land to reproduce.

Order Primates (180 species) includes lemurs, monkeys, gibbons, chimpanzees, gorillas, and humans (Fig. 31.17*c*). Typically, primates are tree-dwelling fruit eaters, although some, like humans, are ground dwellers. Among the primates, humans are well known for their opposable thumb and well-developed brain.

Order Cetacea (80 species) includes the whales and dolphins (Fig. 31.17*d*), which are mammals despite their lack of hair or fur. Blue whales, the largest animal ever to have lived on this planet, are baleen whales that feed by straining large quantities of water containing plankton. Toothed whales feed mainly on fish and squid.

Order Chiroptera (925 species) includes the nocturnal bats, whose wings consist of two layers of skin and connective tissue stretched between the elongated bones of all fingers but the first. Many species use echolocation to locate their usual insect prey. But there are also bird-, fish-, frog-, and plant-eating bats.

Order Rodentia (1,760 species), the largest order, includes mice, rats, squirrels, beavers, and porcupines. Rodents have incisors that grow continuously. They usually feed on seeds, but some are omnivorous and others eat mainly insects.

Order Proboscidea (2 species) includes the elephants, the largest living land mammals, whose upper lip and nose have become elongated and muscularized to form a trunk.

Order Lagomorpha (65 species) includes the herbivorous rabbits, hares, and pikas—animals that superficially resemble rodents. They also have two pairs of continually growing incisors, and their hind legs are longer than their front legs.

Order Insectivora (419 species) includes the shrews and moles, which are mammals with short snouts that live primarily underground.

Order Sirenia (4 species) includes the manatees and seacows, which are large aquatic mammals with large heads and forelimbs modified into flippers.

Order Pinnipedia (34 species) includes walruses, seals, and sea lions, which are aquatic mammals with both forelimbs and hindlimbs modified into flippers.

Order Xenarthra (29 species) includes anteaters, sloths, and armadillos, which are toothless or have peglike teeth.

Advanced mammal classification is primarily based on skull anatomy.

These features are seen in mammals:

- body hair
- differentiated teeth
- well-developed brain
- usually live births and newborn dependency
- mammary glands
- endothermy
- internal development

CONNECTING THE CONCEPTS

How do you measure success? As human beings, we may assume that vertebrate chordates, such as ourselves, are the most successful organisms. But, depending on the criteria used, organisms that are in some ways less complex may come out on top!

For example, vertebrates are eukaryotes, which have been assigned to one domain, while the prokaryotes are now divided into two domains. In fact, the total number of prokaryotes is greater than the number of eukaryotes, and there are possibly more types of prokaryotes than any other living form. Thus, the unseen world is much larger than the seen world. Furthermore, prokaryotes are adapted to use most types of energy sources and to live in most any type of environment.

Human beings, being terrestrial mammals, might also assume that terrestrial species are more successful than aquatic ones. However, if not for the myriad types of terrestrial insects, there would be more aquatic species than terrestrial ones on Earth. The adaptive radiation of mammals has taken place on land, and this might seem impressive to some. But actually, the number of mammalian species (4,800) is small compared to, say, the molluscs (110,000 species), which radiated in the sea.

On land, vertebrate species are far outnumbered by insect species. Out of about 3 million species of organisms that have been discovered, at least 900,000 are insects, and although insects are very small, that too can be an advantage. To a large animal like a cow, grass is a more or less uniform carpet, but to small insects it is a highly varied habitat. Specific insects are able to feed on roots, stems, leaves, flowers, pollen, and seeds.

In terms of size alone, vertebrates might seem the most successful when you consider that a blue whale can weigh as much as 130 metric tons and have a length of 30 m. But when plants are brought into the comparison, there are redwood trees over 100 m tall and eucalyptus trees measuring as much as 140 m. Furthermore, metabolically speaking, plants are much more capable than animals. They can make their own food and change glucose into all the other organic molecules they require. An ecosystem only needs plants and decomposers to sustain itself.

Of course, the size and complexity of the brain is cited as a criterion by which vertebrates are more successful than other living things. However, it has been suggested that this very characteristic is associated with others that make an animal prone to extinction. Studies have indicated that animals that are large, have a long life span, are slow to mature, have few offspring, and expend much energy caring for their offspring tend to become extinct if their normal way of life is destroyed. And finally, vertebrates in general are more threatened than other types of organisms by our present biodiversity crisis—a crisis brought on by the activities of the vertebrate with the most complex brain of all, *Homo sapiens*.

Summary

31.1 CHORDATES

Chordates (sea squirts, lancelets, and vertebrates) have a notochord, a dorsal tubular nerve cord, and pharyngeal pouches at some time in their life history. Also, there is a postanal tail.

Lancelets and sea squirts are the nonvertebrate chordates. Lancelets are the only chordate to have the four characteristics in the adult stage. Sea squirts lack chordate characteristics (except gill slits) as adults, but they have a larva that could be ancestral to the vertebrates.

31.2 VERTEBRATES

Vertebrates are in the phylum Chordata and have the four chordate characteristics as embryos. As adults, the notochord is replaced by the vertebral column. Vertebrates undergo cephalization, and have an endoskeleton, paired appendages, and well-developed internal organs.

The first vertebrates lacked jaws and paired appendages. They are represented today by the hagfishes and lampreys. Ancestral bony fishes, which had jaws and paired appendages, gave rise during the Devonian period to two groups: today's cartilaginous fishes (skates, rays, and sharks) and the bony fishes, including the ray-finned fishes and the lobe-finned fishes. The ray-finned fishes became the most diverse group among the vertebrates. The lobe-finned fishes, represented today by six species of lungfishes and two species of coelacanths, gave rise to the amphibians.

Amphibians are tetrapods represented primarily today by frogs and salamanders. Most frogs and some salamanders return to the water to reproduce and then metamorphose into terrestrial adults.

Reptiles (today's alligators and crocodiles, turtles, lizards, snakes, amphisbaenians, and tuataras) lay a shelled egg, which allows them to reproduce on land. One main group of ancient reptiles, the stem reptiles that presumably evolved from amphibian ancestors, produced a line of descent that evolved into both dinosaurs and birds during the Mesozoic era. A different line of descent from stem reptiles evolved into mammals.

Birds are feathered, which helps them maintain a constant body temperature. They are adapted for flight: Their bones are hollow, their shape is compact, their breastbone is keeled, and they have well-developed sense organs.

Mammals remained small and insignificant while the dinosaurs existed, but when dinosaurs became extinct at the end of the Cretaceous period, mammals became the dominant land organisms.

Mammals are vertebrates with hair and mammary glands. Hair helps them maintain a constant body temperature, and the mammary glands allow them to nurse their young. Monotremes lay eggs, while marsupials have a pouch in which the newborn crawls and continues to develop. The placental mammals, which are the most varied and numerous, retain offspring inside the uterus until birth.

Reviewing the Chapter

1. What four characteristics do all chordates have at some time in their life history? 560
2. Describe the two groups of nonvertebrate chordates, and explain how the sea squirts might be ancestral to vertebrates. 560–61
3. What is the vertebrate body plan? Discuss the distinguishing characteristics of vertebrates. 562
4. Describe the jawless fishes, including ancient ostracoderms. 562
5. What is the significance of having jaws? Describe the ancient placoderms and today's cartilaginous and bony fishes. The amphibians evolved from what type of fish? 563–65
6. What is the significance of being a tetrapod? Discuss the characteristics of amphibians, stating which ones are especially adaptive to a land existence. Explain how their class name (Amphibia) characterizes these animals. 565
7. What is the significance of the amniotic egg? What other characteristics make reptiles less dependent on a source of external water? 568
8. Draw a simplified phylogenetic tree for reptiles showing their relationship to birds and mammals. Your tree should also include stem reptiles and dinosaurs. 568
9. What is the significance of wings? In what other ways are birds adapted to flying? 572–73
10. What is the significance of endothermy? 573
11. What are the three major groups of mammals, and what are their primary characteristics? 574–75

Testing Yourself

Choose the best answer for each question.

1. Which of these is not a chordate characteristic?
   a. dorsal supporting rod, the notochord
   b. dorsal tubular nerve cord
   c. pharyngeal pouches
   d. postanal tail
   e. vertebral column

2. Adult sea squirts
   a. do not have all four chordate characteristics.
   b. are also called tunicates.
   c. are fishlike in appearance.
   d. are the first chordates to be terrestrial.
   e. All of these are correct.

3. Cartilaginous fishes and bony fishes are different in that only
   a. bony fishes have paired fins.
   b. bony fishes have a keen sense of smell.
   c. bony fishes have an operculum.
   d. cartilaginous fishes have a complete skeleton.
   e. cartilaginous fishes are predaceous.

4. Amphibians arose from
   a. sea squirts and lancelets.
   b. cartilaginous fishes.
   c. jawless fishes.
   d. ray-finned fishes.
   e. lobe-finned fishes.

5. Which of these is not a feature of amphibians?
   a. dry skin that resists desiccation
   b. metamorphosis from a swimming form to a land form
   c. small lungs and a supplemental means of gas exchange
   d. reproduction in the water
   e. a single ventricle

6. Reptiles
   a. were dominant during the Mesozoic era.
   b. are closely related to the birds.
   c. lay shelled eggs.
   d. are ectotherms.
   e. All of these are correct.

7. Which of these is a true statement? Choose more than one answer if correct.
   a. In all mammals, offspring develop completely within the female.
   b. All mammals have hair and mammary glands.
   c. All mammals have one birth at a time.
   d. All mammals are land-dwelling forms.
   e. All of these are true.

8. Which of these is not an invertebrate? Choose more than one answer if correct.
   a. a tunicate
   b. a frog
   c. a lancelet
   d. a squid
   e. a roundworm

9. Which of these is not a characteristic of vertebrates? Choose more than one answer if correct.
   a. All vertebrates have a complete digestive system.
   b. Vertebrates have a closed circulatory system.
   c. Sexes are usually separate in vertebrates.
   d. Vertebrates have a jointed endoskeleton.
   e. Most vertebrates never have a notochord.

10. Bony fishes are divided into which two groups?
    a. hagfishes and lampreys
    b. sharks and ray-finned fishes
    c. ray-finned fishes and lobe-finned fishes
    d. jawless fishes and cartilaginous fishes

11. Which of these is an incorrect difference between reptiles and birds?

Reptiles	Birds
a. shelled egg	partial internal development
b. scales	feathers
c. tetrapods	wings
d. ectothermy	endothermy
e. no air sacs	air sacs

12. Which of these does not produce an amniotic egg? Choose more than one answer if correct.
    a. bony fishes
    b. duckbill platypus
    c. snake
    d. robin
    e. frog

13. Label the following diagram of a chordate embryo.

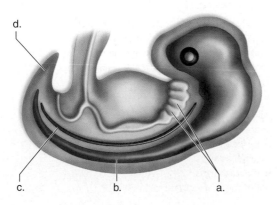

Thinking Scientifically

1. *Archaeopteryx* was a birdlike reptile that had a toothed beak. Modern birds have no teeth in their beaks. Even though modern birds are not thought to have evolved directly from *Archaeopteryx*, presumably teeth were present in the ancestors of birds. Give an evolutionary explanation for the elimination of teeth in a bird's beak.
2. While amphibians have rudimentary lungs, skin is also a respiratory organ. Because of this, they are much more sensitive to air pollution, especially particulate matter, than are higher vertebrates. Why would a thin skin be more sensitive to pollution than lungs?

Understanding the Terms

agnathan 562	lung 566
amniotic egg 568	mammal 574
amphibian 565	marsupial 574
bird 572	metamorphosis 566
bony fish 564	monotreme 574
cartilaginous fish 563	nerve cord 560
chordate 560	notochord 560
cloaca 566	ostracoderm 562
ectotherm 562	placenta 575
endotherm 562	placoderm 563
feather 572	ray-finned bony fishes 564
fin 563	reptile 568
fishes 562	scale 562
gills 560	swim bladder 564
jaw 563	tetrapod 565
jawless fishes 562	therapsid 568
lobe-finned fishes 563	vertebrate 562

Match the terms to these definitions:

a. _____ Animal (bird or mammal) that maintains a uniform body temperature independent of the environmental temperature.
b. _____ Egg-laying mammal—for example, duckbill platypus and spiny anteater.
c. _____ Member of a class of terrestrial vertebrates with internal fertilization, scaly skin, and an egg with a leathery shell; includes crocodiles, turtles, lizards, and snakes.
d. _____ Dorsal supporting rod that exists in all chordates sometime in their life history; replaced by the vertebral column in vertebrates.
e. _____ Mammals that have pouches—for example, kangaroos and koala bears.

ARIS, the *Biology* Website

ARIS, the website for *Biology,* provides a wealth of information organized and integrated by chapter. You will find practice quizzes, interactive activities, labeling exercises, flashcards, and much more that will complement your learning and understanding of general biology.

www.mhhe.com/maderbiology9

32

HUMAN EVOLUTION

Some people mistakenly believe that Darwin, Wallace, and others said that humans evolved from apes. On the contrary, it is known that modern humans and apes evolved from a common apelike ancestor. Today's apes are our cousins, and we couldn't have evolved from our cousins because we are all contemporaries—living on Earth at the same time. Our relationship to apes is analogous to you and your first cousin both being descended from your grandparents.

Scientists study human evolution in the same objective way they approach any other research topic. They have found that while human evolution follows the same patterns of evolutionary descent as other groups, it has more complexity. Various prehuman groups died out, migrated, or interbred with other groups in a short time; nevertheless, recently found fossils such as that of Sahelanthropus tchadensis have brought knowledge of our evolutionary history closer to an apelike ancestor.

Sahelanthropus tchadensis, a possible human ancestor that lived 7 million years aog (MYA).

32.1 EVOLUTION OF PRIMATES

Primates [L. *primus*, first] include prosimians, monkeys, apes, and humans (Fig. 32.1). In contrast to other types of mammals, primates are adapted for an **arboreal** life—that is, for living in trees. The evolution of primates is characterized by trends toward mobile limbs; grasping hands; a flattened face; binocular vision; a large, complex brain; and a reduced reproductive rate. These traits are particularly useful for living in trees.

Mobile Forelimbs and Hindlimbs

In primates, the limbs are mobile, and the hands and feet have five digits each. In most primates, flat nails have replaced the claws of ancestral primates, and sensitive pads on the undersides of fingers and toes assist the grasping of objects. All primates have a thumb, but it is only truly opposable in Old World monkeys, great apes, and humans. Because an opposable thumb can touch each of the other fingers, the grip is both powerful and precise (Fig. 32.2). In all but humans, primates with an opposable thumb also have an opposable toe.

The evolution of the primate limb was a very important adaptation for their life in trees. Mobile limbs with clawless opposable digits allow primates to freely grasp and release tree limbs. They also allow primates to easily reach out and bring food such as fruit to the mouth.

Binocular Vision

A foreshortened snout and a relatively flat face are also evolutionary trends in primates. These may be associated with a general decline in the importance of smell and an increased reliance on vision, leading eventually to binocular vision. In most primates, the eyes are located in the front, where they can focus on the same object from slightly different angles (Fig. 32.3). The stereoscopic (three-dimensional) vision and good depth perception that result permit primates to make accurate judgments about the distance and position of adjoining tree limbs.

Some primates, humans in particular, have color vision and greater visual acuity because the retina contains cone cells in addition to rod cells. Rod cells are activated in dim light, but the blurry image is in shades of gray. Cone cells require bright light, but the image is sharp and in color. The lens of the eye focuses light directly on the fovea, a region of the retina where cone cells are concentrated.

PROSIMIANS

Ring-tailed lemur, *Lemus catta*

Tarsier, *Tarsius bancanus*

a.

NEW WORLD MONKEY

White-faced monkey, *Cebus capucinus*

OLD WORLD MONKEY

Anubis baboon, *Papio anubis*

b.

ASIAN APES

Orangutan, *Pongo pygmaeus*

White-handed gibbon, *Hylobates lar*

c.

Large, Complex Brain

Sense organs are only as beneficial as the brain that processes their input. The evolutionary trend among primates is toward larger and more complex brains. This is evident when comparing the brains of prosimians, such as lemurs and tarsiers, with that of apes and humans. The portion of the brain devoted to smell is smaller, and the portions devoted to sight have increased in size and complexity. Also, more of the brain is devoted to controlling and processing information received from the hands and the thumb. The result is good hand-eye coordination. A larger portion of the brain is devoted to communication skills, and primates tend to live in social groups.

Reduced Reproductive Rate

One other trend in primate evolution is a general reduction in the rate of reproduction, associated with increased age of sexual maturity and extended life spans. Gestation is lengthy, allowing time for forebrain development. One birth at a time is the norm in primates; it is difficult to care for several offspring while moving from limb to limb. The juvenile period of dependency is extended, and there is an emphasis on learned behavior and complex social interactions.

These characteristics especially distinguish primates from other mammals:

- Opposable thumb (and, in some cases, big toe)
- Nails (not claws)
- Single births
- Binocular vision
- Large, complex brain
- Emphasis on learned behavior

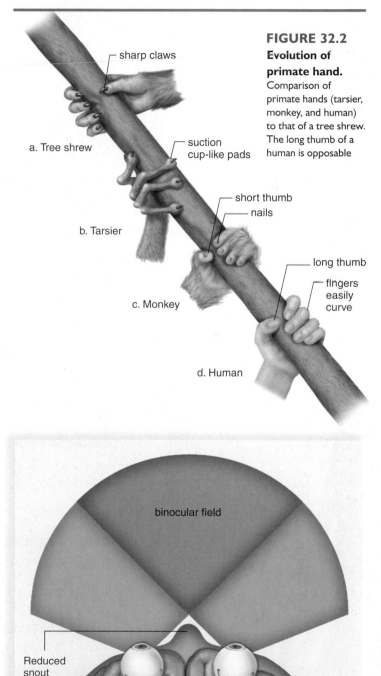

FIGURE 32.2
Evolution of primate hand.
Comparison of primate hands (tarsier, monkey, and human) to that of a tree shrew. The long thumb of a human is opposable

sharp claws

a. Tree shrew

suction cup-like pads

b. Tarsier

short thumb

nails

c. Monkey

long thumb

fingers easily curve

d. Human

AFRICAN APES

Chimpanzee, *Pan troglodytes*

Western lowland gorilla, *Gorilla gorilla*

Humans, *Homo sapiens*

d.

FIGURE 32.1
Primate diversity.
a. Today's prosimians are related to the first group of primates to evolve. **b.** Today's monkeys are divided into the New World monkeys and the Old World monkeys. **c.** Today's apes can be divided into the Asian apes (orangutans and gibbons) and the African apes (chimpanzees and gorillas). **d.** Humans are also classified as primates.

binocular field

Reduced snout does not block vision.

FIGURE 32.3 Binocular vision.
In primates, the snout is reduced, and the eyes are at the front of the head. The result is a binocular field that aids depth perception.

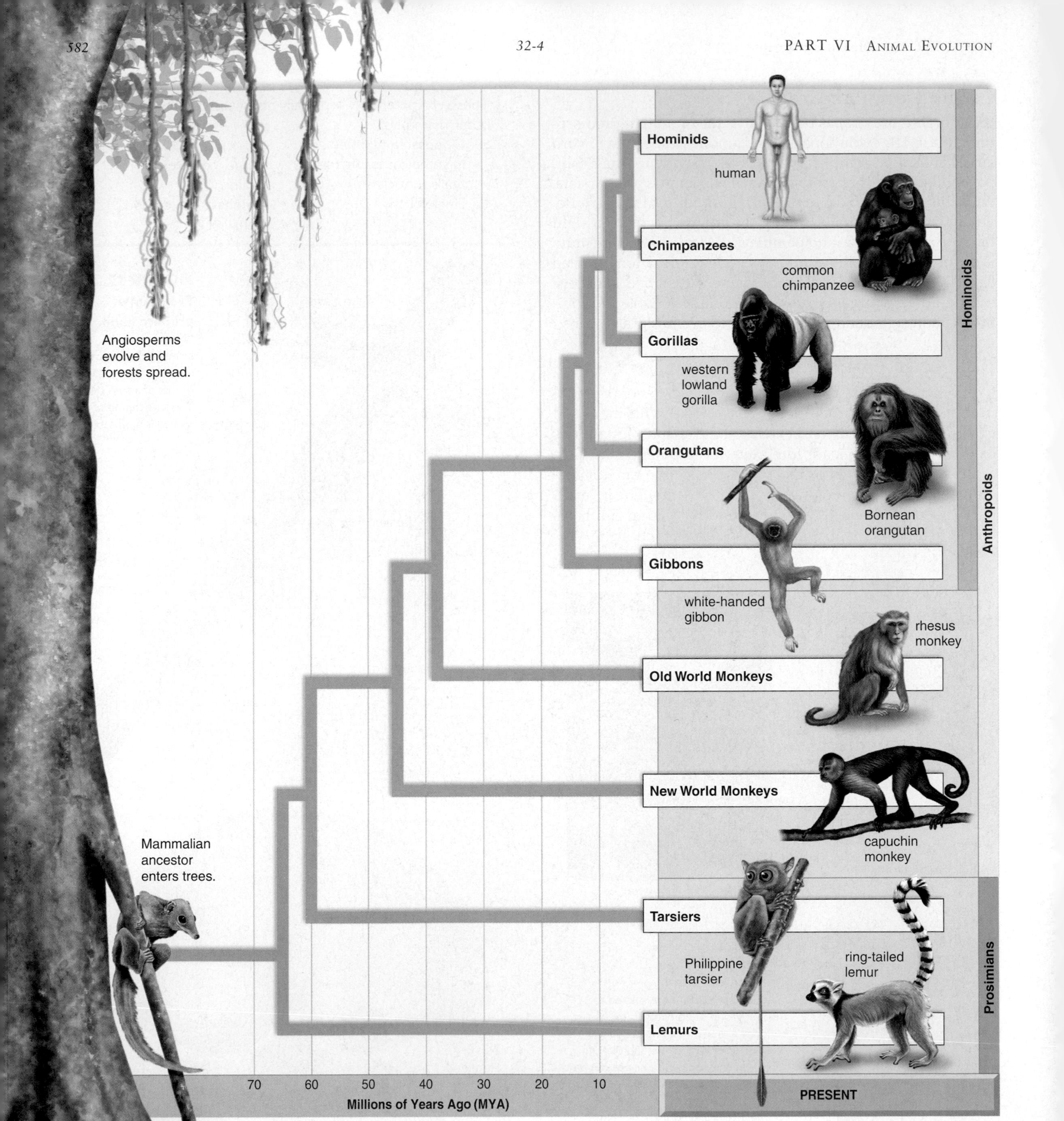

Angiosperms evolve and forests spread.

Mammalian ancestor enters trees.

Hominids

human

Chimpanzees

common chimpanzee

Gorillas

western lowland gorilla

Orangutans

Bornean orangutan

Gibbons

white-handed gibbon

rhesus monkey

Old World Monkeys

New World Monkeys

capuchin monkey

Tarsiers

Philippine tarsier

ring-tailed lemur

Lemurs

Hominoids

Anthropoids

Prosimians

70   60   50   40   30   20   10   PRESENT

Millions of Years Ago (MYA)

FIGURE 32.4 Evolution of primates.
Primates are descended from an ancestor that may have resembled a tree shrew. The descendants of this ancestor adapted to the new way of life and developed traits such as a shortened snout and nails instead of claws. The time when each type of primate diverged from the main line of descent is known from the fossil record. A common ancestor was living at each point of divergence; for example, there was a common ancestor for monkeys, apes, and hominids about 45 MYA; one for all apes and hominids about 15 MYA; and one for just African apes and hominids about 7 MYA. The split between the ape and human lineage may have occurred about this time.

Sequence of Primate Evolution

Figure 32.4 illustrates the sequence of primate evolution during the Cenozoic era. This phylogenetic tree shows that all primates share one common ancestor and that the other types of primates diverged from the human line of descent (called a lineage) over time. Notice that **prosimians** [L. *pro*, before, and *simia*, ape, monkey], represented by lemurs and tarsiers, were the first type of primate to diverge from the human line of descent, and African apes were the last group to diverge.

The surviving **anthropoids** [Gk. *anthropos*, man, and *eides*, like] are classified into three superfamilies: New World monkeys, Old World monkeys, and the **hominoids** (the apes and humans). The New World monkeys often have long prehensile (grasping) tails and flat noses, and Old World monkeys, which lack such tails, have protruding noses. Two of the well-known New World monkeys are the spider monkey and the capuchin, the "organ grinder's monkey." Some of the better-known Old World monkeys are the baboon, a ground dweller, and the rhesus monkey, which has been used in medical research.

Primate fossils similar to monkeys are first found in Africa, but not until 45 MYA. At that time, the Atlantic Ocean would have been too expansive for some of them to have easily made their way to South America, where the New World monkeys live today. It is hypothesized that a common ancestor to both the New World and Old World monkeys arose much earlier when a narrower Atlantic would have made crossing much more reasonable. The New World monkeys evolved in South America, and the Old World monkeys evolved in Africa.

Hominoid Evolution

About 15 MYA, there were dozens of hominoid [L. *homo*, man; Gk. *eides*, like] species, but the anatomy of *Proconsul* makes it a prime candidate to be the ancestral ape. *Proconsul* was about the size of a baboon, and the size of its brain (165 cc) was also comparable. This fossil species didn't have a tail or the expansive pelvis of a monkey (Fig. 32.5). Although its elbow is similar to that of modern apes, its limb proportions suggest that it walked as a quadraped on top of tree limbs as monkeys do. Although primarily a tree dweller, *Proconsul* may have also spent time exploring nearby grasslands for food.

About 10 MYA, Africarabia (Africa plus the Arabian Peninsula) joined with Asia, and the apes migrated into Europe and Asia. Two groups of primates can be distinguished: dryomorphs and ramamorphs. At one time, it was believed that ramamorphs were ancestral to the human lineage. Now, ramamorphs are classified as an ancestral orangutan group. In 1966, Spanish paleontologists announced the discovery of a specimen of *Dryopithecus* dated at 9.5 MYA near Barcelona. The anatomy of these bones clearly indicates that dryomorphs were tree dwellers and locomoted by swinging from branch to branch as orangutans do today. They did not walk along the top of tree limbs as *Proconsul* did.

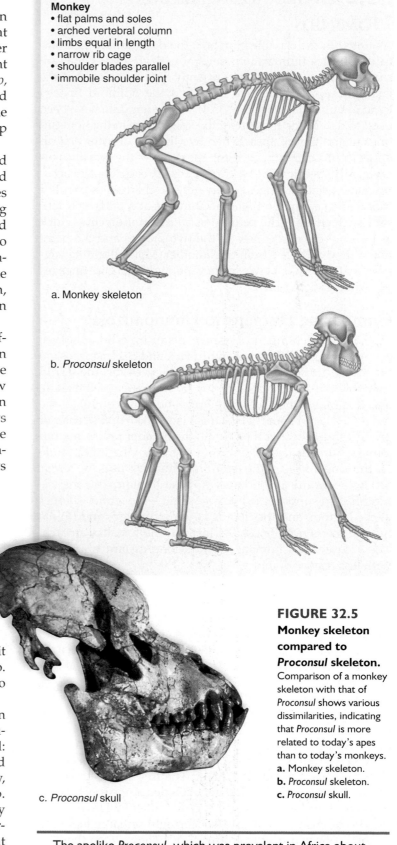

Monkey
- flat palms and soles
- arched vertebral column
- limbs equal in length
- narrow rib cage
- shoulder blades parallel
- immobile shoulder joint

a. Monkey skeleton

b. *Proconsul* skeleton

c. *Proconsul* skull

FIGURE 32.5
Monkey skeleton compared to *Proconsul* skeleton.
Comparison of a monkey skeleton with that of *Proconsul* shows various dissimilarities, indicating that *Proconsul* is more related to today's apes than to today's monkeys.
a. Monkey skeleton.
b. *Proconsul* skeleton.
c. *Proconsul* skull.

The apelike *Proconsul*, which was prevalent in Africa about 15 MYA, is believed to be ancestral to today's hominoids—apes and humans.

32.2 EVOLUTION OF EARLY HOMINIDS

As noted in the classification box, the designation the **hom-inid** includes humans and several extinct species that are closely related to humans. Presently, it is believed that the last common ancestor for apes and the hominid lineage lived about 7 MYA (see Fig. 32.4). Molecular data have been used to determine the date of the split between the hominid lineage and that of apes. When two lines of descent first diverge from a common ancestor, the genes of the two lineages are nearly identical. But as time goes by, each lineage accumulates genetic changes. Many genetic changes are neutral (not tied to adaptation) and accumulate at a fairly constant rate; such changes can be used as a kind of **molecular clock** to indicate the relatedness of the two groups and the point when they diverged from one another. Molecular data suggest that the split between the ape and hominid lineages occurred about 6 MYA.

Comparing Humans to Chimpanzees

Humans and chimpanzees share many traits in common. Several distinct differences in humans and chimpanzees are illustrated in Figure 32.6. These differences include: (1) In humans, the spine exits inferior to the center of the skull, and this places the skull in the midline of the body; (2) the longer S-shaped spine of humans places the trunk's center of gravity squarely over the feet; (3) the broader pelvis and hip joint of humans keep them from swaying when they walk; (4) the longer neck of the femur in humans causes the femur to angle inward at the knees; (5) the human knee joint is modified to support the body's weight—the femur is larger at the bottom, and the tibia is larger at the top; and (6) the human toe is not opposable; instead, the foot has an arch. The arch enables humans to walk long distances and run with less chance of injury.

The Early Hominids

To be a hominid, a fossil must have an anatomy suitable for standing erect and walking on two feet (called **bipedalism**). Human anatomy differs from that of an ape largely because humans are bipedal while apes are quadrupedal (walk on all fours).

Until recently, many scientists thought that hominids evolved in response to a dramatic change in climate that caused forests to be replaced by grassland. Now, some biologists suggest that the first hominid evolved even while it lived in trees because they see no evidence of a dramatic shift in vegetation about 7 MYA. The first hominid's environment is now thought to have included some forest, some woodland, and some grassland. While still living in trees, the first hominids may have walked upright on large branches as they collected fruit from overhead. Then, when they began to forage on the ground, an upright stance would have made it easier for them to travel from woodland to woodland and/or to forage among bushes. Bipedalism may have

had the added advantage of making it easier for males to carry food back to females. If so, bipedal males would have mated more and had more offspring.

Early Hominid Fossils

In Figure 32.7, early hominids are represented by orange-colored bars. The bars show the date of a species' appearance in the fossil record and the date it became extinct. Paleontologists have now found several fossils dated around the time the ape lineage and the human lineage are believed to have split, and one of these is *Sahelanthropus tchadensis* (see page 579). Only the braincase has been found and dated at 7 MYA. Although the braincase is very apelike, a point at the back of the skull where neck muscles attach suggests bipedalism. Also the canines are smaller and the tooth enamel is thicker than that of an ape.

Another early hominid, *Ardipithecus ramidus*, is representative of the ardipithecines of 4.5 MYA. So far only skull fragments of *A. ramidus* have been described. Indirect evidence only suggests that the species was possibly bipedal, and that some individuals may have been 122 cm tall. The teeth seem intermediate between those of earlier apes and later hominids, which are discussed next.

At this point it is not possible to determine if these and other early hominids not mentioned are related to the later hominids and indeed whether they should be included in the human lineage. More remains need to be found to decide these issues.

The early hominids appear in the fossil record soon after the human and ape lineages split. Classification as hominids is based on evidence of bipedalism and their dentition.

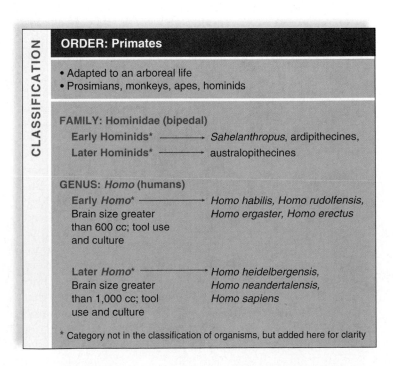

CLASSIFICATION

ORDER: Primates
- Adapted to an arboreal life
- Prosimians, monkeys, apes, hominids

FAMILY: Hominidae (bipedal)
Early Hominids* ——→ *Sahelanthropus*, ardipithecines,
Later Hominids* ——→ australopithecines

GENUS: *Homo* (humans)
Early *Homo** ——→ Homo habilis, Homo rudolfensis,
Brain size greater   Homo ergaster, Homo erectus
than 600 cc; tool use
and culture

Later *Homo** ——→ Homo heidelbergensis,
Brain size greater   Homo neandertalensis,
than 1,000 cc; tool   Homo sapiens
use and culture

* Category not in the classification of organisms, but added here for clarity

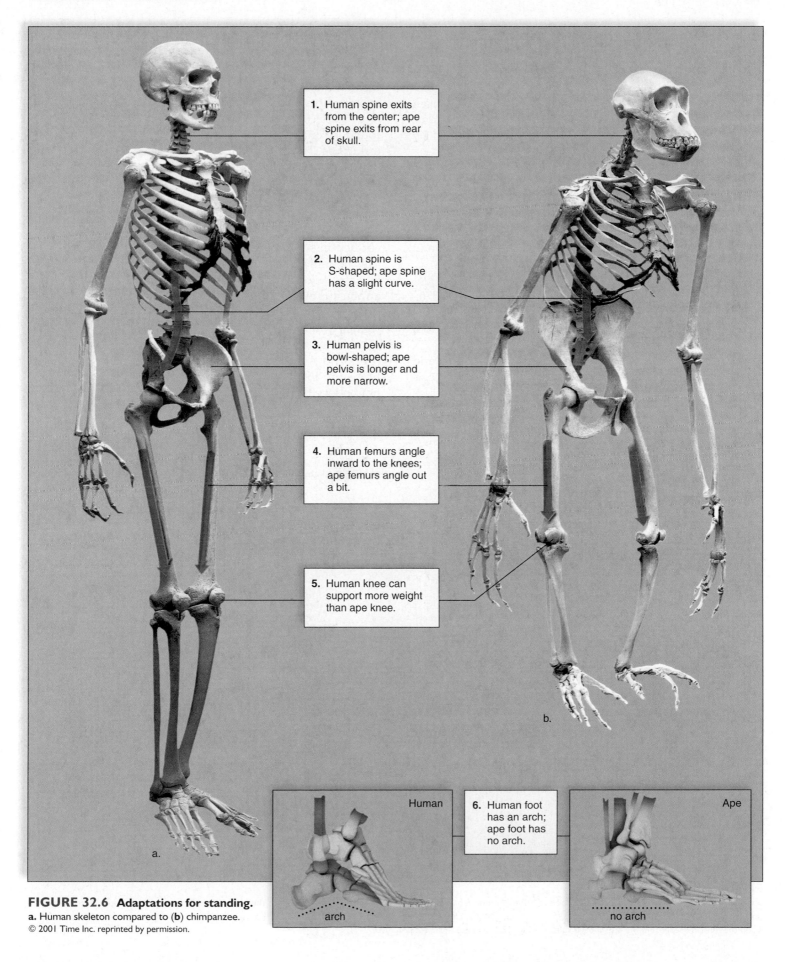

1. Human spine exits from the center; ape spine exits from rear of skull.

2. Human spine is S-shaped; ape spine has a slight curve.

3. Human pelvis is bowl-shaped; ape pelvis is longer and more narrow.

4. Human femurs angle inward to the knees; ape femurs angle out a bit.

5. Human knee can support more weight than ape knee.

Human

6. Human foot has an arch; ape foot has no arch.

Ape

arch

no arch

a.

b.

FIGURE 32.6 Adaptations for standing.
a. Human skeleton compared to (**b**) chimpanzee.
© 2001 Time Inc. reprinted by permission.

32.3 EVOLUTION OF LATER HOMINIDS

It is possible that one of the **australopithecines,** a group of hominids that evolved and diversified in Africa from 4 MYA until about 1 MYA, is a direct ancestor of humans. In Figure 32.7, the australopithecines are represented by green-colored bars.

The australopithecines had a small brain (an apelike characteristic) and walked erect (a humanlike characteristic). Therefore, it seems that human characteristics did not evolve all together at the same time. Australopithecines give evidence of **mosaic evolution,** meaning that different body parts change at different rates and therefore at different times.

Australopithecines stood about 100–115 cm in height and had relatively small brains averaging from about 370–515 cc—slightly larger than that of a chimpanzee. The forehead was low and the face projected forward (Fig. 32.8). Tool use is not in evidence.

Some australopithecines were slight of frame and termed *gracile* (slender). Others were robust (powerful) and tended to have massive jaws because of their large grinding teeth. Their well-developed chewing muscles were an-

chored to a prominent bony crest along the top of the skull. The gracile types most likely fed on soft fruits and leaves, while the robust types had a more fibrous diet that may have included hard nuts. Therefore, the australopithecines show an adaptation to different ways of life.

Some fossil remains of australopithecines have been found in southern Africa, and some have been found in eastern Africa. The exact relationship between these two groups is not known, and it is uncertain how the australopithecines are related to the next group of fossils we will be discussing, namely, the early *Homo* species, represented in Figure 32.7 by lavender-colored bars.

The first australopithecine to be discovered was unearthed in southern Africa by Raymond Dart in the 1920s. This hominid, named ***Australopithecus africanus,*** is a gracile type. A second southern African specimen, *A. robustus,* is a robust type. Both *A. africanus* and *A. robustus* had a brain size of about 500 cc; variations in their skull anatomy are essentially due to their different diets.

These hominids walked upright. Nevertheless, the proportions of their limbs are apelike—that is, the forelimbs are longer than the hindlimbs. Therefore, most do not believe that *A. africanus* is ancestral to early *Homo,* discussed on page 587.

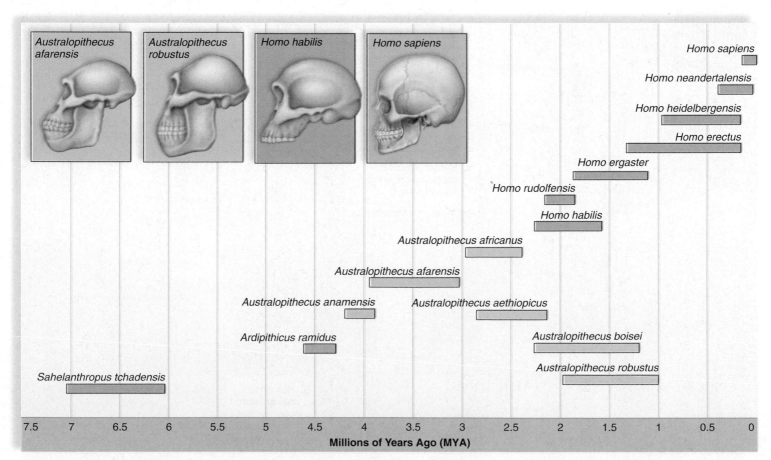

FIGURE 32.7 Human Evolution.
Several groups of extinct hominids preceded the evolution of modern humans. These groups have been divided into the early hominids (orange), later hominids (green), early *Homo* species (lavendar), and later *Homo* species (blue). Only modern humans are classified as *Homo sapiens*. The early hominid fossils are dated closest to the time when the human lineage and the ape lineage split (see Fig. 32.4) and are the most primitive.

More than 20 years ago, a team led by Donald Johanson unearthed nearly 250 fossils of a hominid called ***Australopithecus afarensis*** in eastern Africa. A now-famous female skeleton is known worldwide by its field name, Lucy. (The name derives from the Beatles song "Lucy in the Sky with Diamonds.") Although her brain was quite small (400 cc), the shapes and relative proportions of Lucy's limbs indicate that she stood upright and walked bipedally (Fig. 32.8a). Even better evidence of bipedal locomotion comes from a trail of footprints in Laetoli dated about 3.7 MYA. The larger prints are double, as though a smaller-sized being was stepping in the footprints of another—and there are additional small prints off to the side, within hand-holding distance (Fig. 32.8b). *A. afarensis* is usually favored as being related more directly to early *Homo*.

A. afarensis, a gracile type, is believed to be ancestral to the robust types found in eastern Africa, *A. aethiopicus* and *A. boisei*. *A. boisei* had a powerful upper body and the largest molars of any hominid.

The australopithecines were the later hominids. The brain was small and the face projected forward. However, good evidence suggests they were bipedal. It is unknown which australopithecine is ancestral to early *Homo*.

b.

FIGURE 32.8
Australopithecus afarensis.
a. A reconstruction of Lucy on display at the St. Louis Zoo.
b. These fossilized footprints occur in ash from a volcanic eruption some 3.7 MYA. The larger footprints are double, and a third, smaller individual was walking to the side. (A female holding the hand of a youngster may have been walking in the footprints of a male.) The footprints suggest that *A. afarensis* walked bipedally.

a.

32.4 EVOLUTION OF EARLY *HOMO*

Fossils designated as early *Homo* species are represented in Figure 32.7 by a blue-colored bar. These fossils appear in the fossil record somewhat earlier or later than 2 MYA. They all have a brain size that is 600 cc or greater, their jaw and teeth resemble those of humans, and tool use is in evidence.

In general, *Homo habilis* and *Homo rudolfensis* have a more primitive anatomy than *Homo ergaster* and *Homo erectus*, which are all grouped together as early *Homo* species. The latter two species have a higher forehead and a flatter face and a larger brain size than the first two species.

Homo habilis and *Homo rudolfensis*

Homo habilis and *Homo rudolfensis* are closely related and will be considered together. *H. rudolfensis* is the larger of the two species. Although the height of *H. rudolfensis* did not exceed those of the australopithecines, some of this species' fossils have a brain size as large as 800 cc, which is considerably larger than that of *A. afarensis*. The cheek teeth of these hominids tend to be smaller than even those of the gracile australopithecines. Therefore, it is likely that they were omnivorous and ate meat in addition to plant material. Bones at the campsites of *H. habilis* bear cut marks, indicating that these hominids used tools to strip off the meat.

The stone tools made by *H. habilis* and *H. rudolfensis* are rather crude. Those found may be the cores from which these hominids took flakes sharp enough to scrape away hide, cut tendons, and easily remove meat from bones.

The skulls of these two species suggest that the portions of the brain associated with speech areas were enlarged. We can speculate that the ability to speak may have led to hunting cooperatively. Some members of the group may have remained plant gatherers, and if so, both hunters and gatherers most likely ate together and shared their food. In this way, society and culture could have begun. **Culture,** which encompasses human behavior and products (such as technology and the arts), is dependent on the capacity to speak and transmit knowledge. We can further speculate that the advantages of a culture to these hominids may have hastened the extinction of the australopithecines.

H. habilis and *H. rudolfensis* warrant classification as *Homo* because of brain size, dentition, and tool use. These early *Homo* species may have had the rudiments of a culture.

---

science focus

Origins of the Genus *Homo*

Fossil evidence shows that the earliest *Homo* species evolved in Africa from the *Australopithecus* line of descent about 2 MYA. Remains of australopithecines indicate that they spent part of their time climbing trees and that they retained many apelike traits. Australopithecine arms, like those of an ape, were long compared to the length of the legs. *A. afarensis* also had strong wrists and long, curved fingers and toes. These traits would have served well for climbing, and the australopithecines probably climbed trees for the same reason that chimpanzees do today: to gather fruits and nuts in trees and to sleep aboveground at night so as to avoid predatory animals, such as lions and hyenas.

Whereas our brain is about the size of a grapefruit, that of the australopithecines was about the size of an orange. Their brain was only slightly larger than that of a chimpanzee. There is no evidence that the australopithecines manufactured stone tools; presumably, they were not smart enough to do so.

We know that the genus *Homo* evolved from the genus *Australopithecus*, but several years ago I [Stephen Stanley] concluded that this could not have happened as long as the australopithecines climbed trees every day. The obstacle relates to the way we, members of *Homo*, develop our large brain. Unlike other primates, we retain the high rate of fetal brain growth through the first year after birth. (That is why a one-year-old child has a very large head.) The brain of other primates, including monkeys and apes, grows rapidly before birth, but immediately after birth their brain grows more slowly. An adult human brain is more than three times as large as that of an adult chimpanzee.

A continuation of the high rate of fetal brain growth eventually allowed the genus *Homo* to evolve from the genus *Australopithecus*. But there was a problem in that continued brain growth is linked to underdevelopment of the entire body. Although the human brain becomes much larger, human babies are remarkably weak and uncoordinated. Such helpless infants must be carried about and tended. Human babies are unable to cling to their mothers the way chimpanzee babies can (Fig. 32A).

The origin of the *Homo* genus entailed a great evolutionary compromise. Humans gained a large brain, but they were saddled with the largest interval of infantile helplessness in the entire class Mammalia. The positive value of a large brain must have outweighed the negative aspects of infantile helplessness, however, or natural selection would not have produced the *Homo* genus. Having a larger brain meant that humans were able to outsmart or ward off predators with weapons they were clever enough to manufacture.

Probably very few genetic changes were required to delay the maturation of *Australopithecus* and produce the large brain of *Homo*. The mutation of a regulatory gene that controls one or more other genes most likely could have delayed early maturation. As we learn more about the human genome, we will eventually uncover the particular gene or gene combinations that cause us to have a large brain, and this will be a very exciting discovery.

Steven Stanley
Johns Hopkins University

FIGURE 32A Human infant.
A human infant is often cradled and has no means to cling to its mother when she goes about her daily routine.

Homo ergaster and *Homo erectus*

Homo ergaster evolved in Africa perhaps from *H. rudolfensis.* Similar fossils found in Asia are different enough to be classified as **Homo erectus** [L. *homo,* man, and *erectus,* upright]. These fossils span the dates between 1.9 and 0.3 MYA. A Dutch anatomist named Eugene Dubois was the first to unearth *H. erectus* bones in Java in 1891, and since that time many other fossils belonging to both species have been found in Africa and Asia.

Compared to *H. habilis, H. ergaster* had a larger brain (about 1,000 cc) and a flatter face with a nose that projected. This type of nose is adaptive for a hot, dry climate because it permits water to be removed before air leaves the body. The recovery of an almost complete skeleton of a 10-year-old boy indicates that *H. ergaster* was much taller than the hominids discussed thus far (Fig. 32.9). Males were 1.8 m tall, and females were 1.55 m. Indeed, these hominids stood erect and most likely had a *striding gait* like that of modern humans. The robust and most likely heavily muscled skeleton still retained some australopithecine features. Even so, the size of the birth canal indicates that infants were born in an immature state that required an extended period of care.

H. ergaster first appeared in Africa but then migrated into Europe and Asia. At one time, the migration was thought to have occurred about 1 MYA, but recently *H. erectus* fossils in Java and the Republic of Georgia have been dated at 1.9 and 1.6 MYA, respectively. These remains push the evolution of *H. ergaster* in Africa to an earlier date than has yet been determined because most likely *H. erectus* evolved from *H. ergaster* after *H. ergaster* arrived in Asia. In any case, such an extensive population movement is a first in the history of humankind and a tribute to the intellectual and physical skills of the species.

These hominids were the first to use fire, and both types of hominids fashioned more advanced tools than the other early *Homo* species. They used heavy, teardrop-shaped axes and cleavers as well as flakes, which were probably used for cutting and scraping. Some investigators believe that these hominids were systematic hunters and brought kills to the same site over and over again. In one location, researchers have found over 40,000 bones and 2,647 stones. These sites could have been "home bases" where social interaction occurred and a prolonged childhood allowed time for learning. Perhaps a language evolved and a culture more like our own developed.

Homo Floresiensis. In 2004, scientists announced the discovery of the fossil remains of *Homo floresiensis.* The 18,000-year-old fossil of a 1 m tall, 25 kg adult female was discovered on the island of Flores in the South Pacific. The specimen was the size of a three-year-old *Homo sapiens* but possessed a braincase only one-third the size of modern humans. Researchers suspect that this diminutive hominid and her peers evolved from normal-sized, island hopping *Homo erectus* populations that reached Flores about 840,000 years ago. Apparently *H. floresiensis* used tools and fire. Some scientists think that its small size is due to island dwarfing, a phenomenon seen in other island populations.

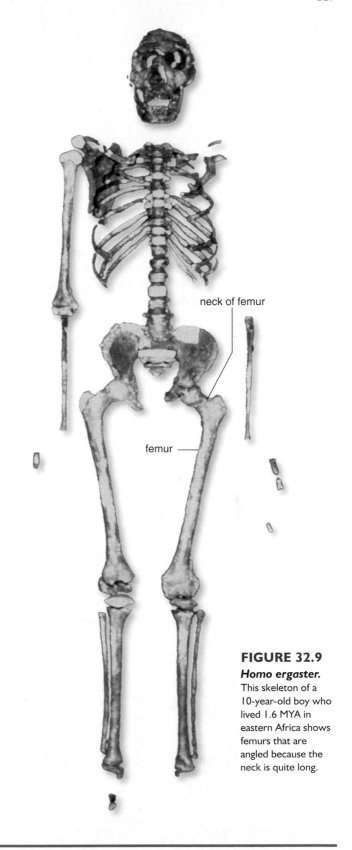

neck of femur

femur

FIGURE 32.9
Homo ergaster.
This skeleton of a 10-year-old boy who lived 1.6 MYA in eastern Africa shows femurs that are angled because the neck is quite long.

H. ergaster could have evolved from *H. rudolfensis* in Africa. This hominid had a striding gait, made well-fashioned tools (perhaps for hunting), and migrated out of Africa between 2 and 1 MYA. A related form, *H. erectus,* is also found in Asia.

32.5 EVOLUTION OF LATER *HOMO*

Most researchers believe that modern humans (*Homo sapiens*) evolved from *H. ergaster,* but they differ as to the details. Many disparate early *Homo* species in Europe are now classified as *Homo heidelbergensis.* Just as *H. erectus* is believed to have evolved from *H. ergaster* in Asia, so *H. heidelbergensis* is believed to have evolved from *H. ergaster* in Europe. Further, for the sake of discussion, *H. ergaster* in Africa, *H. erectus* in Asia, and *H. heidelbergensis* in Europe can be grouped together as archaic humans who lived as long as a million years ago. The presence of archaic humans at these different locations suggests to some that modern humans evolved from archaic humans separately in all three places (Fig. 32.10*a*). This hypothesis, called the **multiregional continuity hypothesis,** proposes that modern humans arose from archaic humans in essentially the same manner. Even so, the humans of each region should show a continuity of unique anatomical characteristics from about the time of the archaic species.

Opponents argue that it is unlikely that evolution would have produced essentially the same result in these different places. They suggest, instead, the **out-of-Africa hypothesis,** which proposes that modern humans evolved from *H. ergaster* only in Africa, and then *H. sapiens* migrated to Europe and Asia, where it replaced the archaic species about 100,000 years BP (before the present) (Fig. 32.10*b*). If so, fossils dated 200,000 BP and 100,000 BP are expected to be markedly different from each other.

According to which hypothesis would modern humans be most genetically alike? The multiregional hypothesis states that human populations have been evolving separately for a long time; therefore, genetic differences are expected between groups. According to the out-of-Africa hypothesis, we are all descended from a few individuals from about 100,000 years BP. Therefore, the out-of-Africa hypothesis suggests that we are more genetically similar.

Several years ago, a study attempted to show that all the people of Europe (and the world for that matter) have essentially the same mitochondrial DNA. Called the "mitochondrial Eve" hypothesis by the press (note that this is a misnomer because no single ancestor is proposed), the statistics that calculated the date of the African migration were found to be flawed. Still, the raw data—which indicate a close genetic relationship among all Europeans—support the out-of-Africa hypothesis.

These opposing hypotheses have sparked many other innovative studies to test them. The final conclusions are still being determined, but evidence is leaning toward the out-of-Africa hypothesis.

Investigators are currently testing two hypotheses: (1) Modern humans evolved separately in Asia, Africa, and Europe, or (2) modern humans evolved in Africa and then migrated to Asia and Europe.

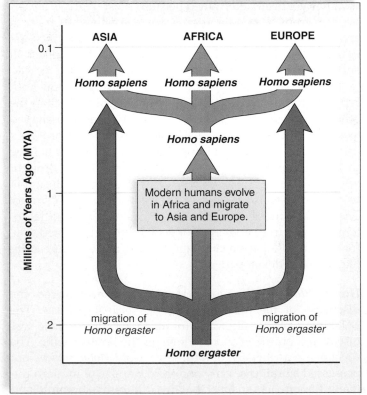

a. Multiregional continuity

b. Out of Africa

FIGURE 32.10 Evolution of modern humans.
a. The multiregional continuity hypothesis proposes that *Homo sapiens* evolved separately in at least three different places: Asia, Africa, and Europe. Therefore, continuity of genotypes and phenotypes is expected in each region but not between regions. **b.** The out-of-Africa hypothesis proposes that *Homo sapiens* evolved only in Africa and then migrated out of Africa, as *H. ergaster* did many years before. *H. sapiens* would have supplanted populations of archaic *Homo* in Asia and Europe about 100,000 years ago.

Neandertals

The **Neandertals,** sometimes classified as *Homo neandertalensis,* are an intriguing species of archaic humans that lived between 200,000 and 28,000 years ago. Neandertal fossils are known from the Middle East and throughout Europe. Neandertals take their name from Germany's Neander Valley, where one of the first Neandertal skeletons, dated some 200,000 years ago, was discovered.

According to the out-of-Africa hypothesis (see page 590), archaic human species, including Neandertals, were supplanted by modern humans. Surprisingly, however, the Neandertal brain was, on the average, slightly larger than that of *Homo sapiens* (1,400 cc, compared with 1,360 cc in most modern humans). The Neandertals had massive brow ridges and wide, flat noses. They also had a forward-sloping forehead and a receding lower jaw. Their nose, jaws, and teeth protruded far forward. Physically, the Neandertals were powerful and heavily muscled, especially in the shoulders and neck (Fig. 32.11). The bones of Neandertals were shorter and thicker than those of modern humans. New fossils show that the pubic bone was long compared to that of modern humans. The Neandertals lived in Europe and Asia during the last Ice Age, and their sturdy build could have helped conserve heat.

Archaeological evidence suggests that Neandertals were culturally advanced. Some Neandertals lived in caves; however, others probably constructed shelters. They manufactured a variety of stone tools, including spear points, which could have been used for hunting, and scrapers and knives, which would have helped in food preparation. They most likely successfully hunted bears, woolly mammoths, rhinoceroses, reindeer, and other contemporary animals. They used and could control fire, which probably helped in cooking frozen meat and in keeping warm. They even buried their dead with flowers and tools and may have had a religion.

FIGURE 32.12 Cro-Magnons.
Cro-Magnon people are the first to be designated *Homo sapiens.* Their toolmaking ability and other cultural attributes, such as their artistic talents, are legendary.

Cro-Magnons

Cro-Magnons are the oldest fossils to be designated *Homo sapiens.* In keeping with the out-of-Africa hypothesis, the Cro-Magnons, who are named after a fossil location in France, were the modern humans who entered Asia and Europe from Africa 100,000 years ago or even earlier. They probably reached western Europe about 40,000 years ago. Cro-Magnons had a thoroughly modern appearance (Fig. 32.12). They had lighter bones, flat high foreheads, domed skulls housing brains of 1,350 cc, small teeth, and a distinct chin. They made advanced stone tools, including compound tools, as when stone flakes were fitted to a wooden handle. They may have been the first to make knifelike blades and to throw spears, enabling them to kill animals from a distance. They were such accomplished hunters that some researchers believe they may have been responsible for the extinction of many larger mammals, such as the giant sloth, the mammoth, the saber-toothed tiger, and the giant ox, during the late Pleistocene epoch. This event is known as the Pleistocene overkill.

Cro-Magnons hunted cooperatively, and perhaps they were the first to have a language. They are believed to have lived in small groups, with the men hunting by day while the women remained at home with the children. It's quite possible that this hunting way of life among prehistoric people influences our behavior even today. The Cro-Magnon culture included art. They sculpted small figurines out of reindeer bones and antlers. They also painted beautiful drawings of animals on cave walls in Spain and France (Fig. 32.12).

FIGURE 32.11 Neandertals.
This drawing shows that the nose and the mouth of the Neandertals protruded from their faces, and their muscles were massive. They made stone tools and were most likely excellent hunters.

Most likely, modern humans migrated out of Africa and supplanted other *Homo* species in Asia and Europe. Today, there is only one *Homo* species in existence—namely, *Homo sapiens.*

Human Variation

Human beings have been widely distributed about the globe ever since they evolved. As with any other species that has a wide geographic distribution, phenotypic and genotypic variations are noticeable between populations. Today, we say that people have different ethnicities (Fig. 32.13*a*).

It has been hypothesized that human variations evolved as adaptations to local environmental conditions. One obvious difference among people is skin color. A darker skin is protective against the high UV intensity of bright sunlight. On the other hand, a white skin ensures vitamin D production in the skin when the UV intensity is low. Harvard University geneticist Richard Lewontin points out, however, that this hypothesis concerning the survival value of dark and light skin has never been tested.

a.

b.

c.

FIGURE 32.13 Ethnic groups.
a. Some of the differences between the various prevalent ethnic groups in the United States may be due to adaptations to the original environment. **b.** The Massai live in East Africa. **c.** Eskimos live near the Arctic Circle.

Two correlations between body shape and environmental conditions have been noted since the nineteenth century. The first, known as Bergmann's rule, states that animals in colder regions of their range have a bulkier body build. The second, known as Allen's rule, states that animals in colder regions of their range have shorter limbs, digits, and ears. Both of these effects help regulate body temperature by increasing the surface-area-to-volume ratio in hot climates and decreasing the ratio in cold climates. For example, Figure 32.13*b, c* shows that the Massai of East Africa tend to be slightly built with elongated limbs, while the Eskimos, who live in northern regions, are bulky and have short limbs.

Other anatomical differences among ethnic groups, such as hair texture, a fold on the upper eyelid (common in Asian peoples), or the shape of lips, cannot be explained as adaptations to the environment. Perhaps these features became fixed in different populations due simply to genetic drift. As far as intelligence is concerned, no significant disparities have been found among different ethnic groups.

Genetic Evidence for a Common Ancestry

The two hypotheses regarding the evolution of humans, discussed on page 590, pertain to the origin of ethnic groups. The multiregional continuity hypothesis suggests that different human populations came into existence as long as a million years ago, giving time for significant ethnic differences to accumulate despite some gene flow. The out-of-Africa hypothesis, on the other hand, proposes that all modern humans have a relatively recent common ancestor, that is, Cro-Magnon, who evolved in Africa and then spread into other regions. Paleontologists tell us that the variation among modern populations is considerably less than among archaic human populations some 250,000 years ago. This would mean that all ethnic groups evolved from the same single, ancestral population.

A comparative study of mitochondrial DNA shows that the differences among human populations are consistent with their having a common ancestor no more than a million years ago. Lewontin has also found that the genotypes of different modern populations are extremely similar. He examined variations in 17 genes, including blood groups and various enzymes, among seven major geographic groups: Caucasians, black Africans, mongoloids, south Asian Aborigines, Amerinds, Oceanians, and Australian Aborigines. He found that the great majority of genetic variation—85%—occurs within ethnic groups, not among them. In other words, the amount of genetic variation between individuals of the same ethnic group is greater than the variation between ethnic groups.

Certain variations among ethnic groups can be attributed to adaptations to local environments. The genotypes of all ethnic groups are extremely similar, suggesting that all groups evolved from the same fairly recent ancestral population.

CONNECTING THE CONCEPTS

Aside from various anatomical differences related to human bipedalism and intelligence, a cultural evolution separates us from the apes. A hunter-gatherer society evolved when humans became able to make and use tools. That society then gave way to an agricultural economy about 12,000 to 15,000 years ago, perhaps because we were too efficient at killing big game so that a food shortage arose. The agricultural period extended from that time to about 200 years ago, when the Industrial Revolution began. Now most people live in urban areas. Perhaps as a result, modern humans are for the most part divorced from nature and often endowed with the philosophy of exploiting and controlling nature.

Our cultural evolution has had far-reaching effects on the biosphere, especially since the human population has expanded to the point that it is crowding out many other species. Our degradation and disruption of the environment threaten the continued existence of many species, including our own. As discussed in Part VIII of this text, however, we have recently begun to realize that we must work with, rather than against, nature if biodiversity is to be maintained and our own species is to continue to exist.

Before we examine the environment and the role of humans in ecosystems, we will study the various organ systems of the human body. Humans need to keep themselves and the environment fit so that they and their species can endure.

Summary

32.1 EVOLUTION OF PRIMATES

Primates, in contrast to other types of mammals, are adapted for an arboreal life. The evolution of primates is characterized by trends toward mobile limbs; grasping hands; a flattened face; binocular vision; a large, complex brain; and one birth at a time. These traits are particularly useful for living in trees.

During the evolution of primates, various groups diverged from the main line of descent in a particular sequence. Prosimians (tarsiers and lemurs) diverged first. They were followed by the monkeys and then the apes followed by humans. *Proconsul* is representative of the first type of ape.

32.2 EVOLUTION OF EARLY HOMINIDS

Fossil and molecular data tell us we shared a common ancestor with the apes about 7 MYA and the split between the ape and human lineage occurred somewhat later.

Human anatomy differs from ape anatomy. In humans, the spinal cord curves and exits from the center of the skull rather than from the rear of the skull. The human pelvis is broader and more bowl-shaped to place the weight of the body over the legs. Humans use only the longer, heavier lower limbs for walking; in apes, all four limbs are used for walking, and the upper limbs are longer than the lower limbs.

Only humans and their closest relatives are hominids. To be a hominid, a fossil must have an anatomy suitable to standing erect. Perhaps bipedalism developed when hominids stood on branches to reach fruit overhead, and then they continued to use this stance when foraging among bushes. An upright posture reduces exposure of the body to the sun's rays, and leaves the hands free to carry food, perhaps as a gift to receptive females.

Several early hominid fossils, such as *Sahelanthropus tchadensis,* have been dated around the time of a shared ancestor for apes and humans (7 MYA). The ardipithecines appeared about 4.5 MYA. All the early hominids have a chimp-sized braincase but are believed to have walked erect.

32.3 EVOLUTION OF LATER HOMINIDS

It is possible that an australopithecine (4 MYA–1 MYA) is a direct ancestor for humans. These hominids walked upright and had a brain size of 370–515 cc. In southern Africa, hominids classified as australopithecines include *Australopithecus africanus*, a gracile form, and *A. robustus*, a robust form. In eastern Africa, hominids classified as australopithecines include, *A. afarensis* (Lucy), a gracile form, and also robust forms. Many of the australopithecines coexisted, and it is difficult to tell who is ancestral to whom. It is not known whether *A. africanus* or *A. afarensis* is directly ancestral to humans. *A. africanus* had a larger brain than *A. afarensis,* but the proportion of its limbs was more apelike.

32.4 EVOLUTION OF EARLY HOMO

Early *Homo,* such as *Homo habilis* and *Homo rudolfensis,* dated around 2 MYA, is characterized by a brain size of at least 600 cc, a jaw with teeth that resembled those of modern humans, and the use of tools.

Homo ergaster and *Homo erectus* (1.9–0.3 MYA) had a striding gait, made well-fashioned tools, and could control fire. *Homo ergaster* migrated into Asia and Europe from Africa between 2 and 1 MYA. *Homo erectus* evolved in Asia.

32.5 EVOLUTION OF LATER HOMO

Two contradicting hypotheses have been suggested about the origin of modern humans. The multiregional continuity hypothesis says that modern humans originated separately in Asia, Europe, and Africa as much as a million years ago. If so, a difference in the genes is expected between human populations at different locations. The out-of-Africa hypothesis says that modern humans originated only in Africa and, after migrating into Europe and Asia, replaced the archaic *Homo* species found there. Many studies are being done to determine which hypothesis is supported by data.

The Neandertals, a group of archaic humans, lived in Europe and Asia. Their chinless faces, squat frames, and heavy muscles are apparently adaptations to the cold. Cro-Magnon is a name often given to modern humans. Their tools were sophisticated, and they definitely had a culture, as witnessed by the paintings on the walls of caves. The human ethnic groups of today differ in ways that can be explained in part by adaptation to the environment. Genetic studies tell us that there are more genetic differences between people of the same ethnic group than between ethnic groups. We are one species.

Reviewing the Chapter

1. List and discuss various evolutionary trends among primates, and state how they would be beneficial to animals with an arboreal life. 580
2. What is the significance of the fossils classified as *Proconsul?* 583
3. How do modern humans differ anatomically from chimpanzees? 584–85
4. Discuss the possible benefits of bipedalism in early hominids. 584
5. Why does the term *mosaic evolution* apply to the australopithecines? 586

6. Why are the early *Homo* species classified as humans? If these hominids did make tools, what does this say about their probable way of life?   587

7. What role(s) might *H. ergaster* have played in the evolution of modern humans according to the multiregional continuity hypothesis? The out-of-Africa hypothesis?   590

8. Who were the Neandertals and the Cro-Magnons, and what is their place in the evolution of humans according to the out-of-Africa hypothesis mentioned in question 7?   591

Testing Yourself

Choose the best answer for each question.

1. Which of these gives the correct order of divergence from the main primate line of descent?
   a. prosimians, monkeys, gibbons, orangutans, African apes, humans
   b. gibbons, orangutans, prosimians, monkeys, African apes, humans
   c. monkeys, gibbons, prosimians, African apes, orangutans, humans
   d. African apes, gibbons, monkeys, orangutans, prosimians, humans
   e. *H. habilis, H. ergaster, H. neandertalensis,* Cro-Magnon

2. Lucy is a(n)
   a. early *Homo.*
   b. australopithecine.
   c. ardipithecine.
   d. modern human.

3. What possibly influenced the evolution of bipedalism?
   a. Humans wanted to stand erect in order to use tools.
   b. With bipedalism, it's possible to reach food overhead.
   c. With bipedalism, sexual intercourse is facilitated.
   d. An upright stance exposes more of the body to the sun, and vitamin D production requires sunlight.
   e. All of these are correct.

4. Which of these is an incorrect association with robust types?
   a. massive chewing muscles attached to bony skull crest
   b. some australopithecines
   c. a fibrous diet
   d. lived during an Ice Age
   e. Both a and c are incorrect associations.

5. *H. ergaster* could have been the first to
   a. use and control fire.
   b. migrate out of Africa.
   c. make axes and cleavers.
   d. have a brain of at least 850 cc.
   e. All of these are correct.

6. Which of these characteristics is not consistent with the others?
   a. brow ridges   d. projecting face
   b. small cheek teeth (molars)   e. binocular vision
   c. high forehead

7. Which of these statements is correct? The last common ancestor for African apes and hominids
   a. has been found, and it resembles a gibbon.
   b. was probably alive around 7 MYA.
   c. has been found, and it has been dated at 30 MYA.
   d. is not expected to be found because there was no such common ancestor.
   e. is now believed to have lived in Asia, not Africa.

8. Which of these pairs is incorrectly matched?
   a. gibbon—hominoid
   b. *A. africanus*—hominid
   c. tarsier—anthropoid
   d. *H. erectus—H. ergaster*
   e. early *Homo—H. habilis*

9. If the multiregional continuity hypothesis is correct, then
   a. hominid fossils in China after 100,000 BP would not be expected to resemble earlier fossils.
   b. hominid fossils in China after 100,000 BP would be expected to resemble earlier fossils.
   c. modern humans did not migrate out of Africa.
   d. Both b and c are correct.
   e. Both a and c are correct.

10. Which of these pairs is incorrectly matched?
    a. *H. erectus*—made tools
    b. Neandertal—good hunter
    c. *H. habilis*—controlled fire
    d. Cro-Magnon—good artist
    e. *A. robustus*—fibrous diet

11. Classify humans by filling in the missing lines.
    Kingdom: Animalia
    Phylum: a. _____
    Subphylum: b. _____
    c. _____   Mammalia
    d. _____   Primates
    Family: e. _____
    f. _____   *Homo*
    Species: g. _____

12. Which hominids could have inhabited the Earth at the same time?
    a. australopithecines and Cro-Magnons
    b. *Australopithecus robustus* and *Homo habilis*
    c. *Homo habilis* and *Homo sapiens*
    d. apes and humans

For questions 13–17, indicate whether the statement is true (T) or false (F).

13. Australopithecines were adapted to different diets. _____

14. *Homo habilis* made stone tools. _____

15. The human pelvis is bowl-shaped, and the ape pelvis is long and narrow. _____

16. The gibbon is an Asian ape, while the chimpanzee is an African ape. _____

17. Mitochondrial DNA differences are inconsistent with the existence of a recent human common ancestor for all ethnic groups. _____

For questions 18–23, fill in the blanks.

18. Along with monkeys and apes, humans are classified as _____.

19. The out-of-Africa hypothesis proposes that modern humans evolved in _____ only.

20. The australopithecines could probably walk _____, but they had a _____ brain.

21. The only fossil rightly called *Homo sapiens* is that of _____.

22. Modern humans evolved _____ (choose billions, millions, thousands) of years ago.

23. To describe the out-of-Africa hypothesis, where in the following diagram would you place:
   a. *Homo sapiens?* _____
   b. *Homo ergaster?* _____

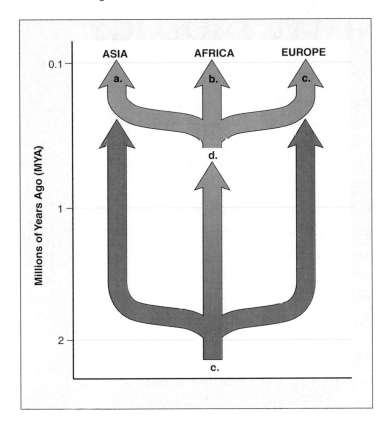

Thinking Scientifically

1. Recently, three fossil skulls of *Homo sapiens,* dating to about 160,000 years BP, were discovered in eastern Africa. These skulls fill a gap between the 100,000-year-old *H. sapiens* skulls found in Africa and Israel and the 500,000-year-old skulls of archaic *H. sapiens* found in Ethiopia. Supporters of the out-of-Africa hypothesis argue that their position is strengthened by the documentation of the succession of human ancestors from 7 MYA through the group in this recent African find. Does this recent find support the out-of-Africa hypothesis? If not, what fossil evidence might yet be found to support the multiregional continuity hypothesis?

2. Bipedalism has many selective advantages. However, there is one particular disadvantage to walking on two feet: Giving birth to an offspring with a large head through the smaller pelvic opening that is necessitated by upright posture is very difficult. This situation results in a high percentage of deaths (of both mother and child) during birth compared to other primates. How do you explain the selection of a trait that is both positive and negative?

Bioethical Issue: Manipulation of Evolution

Since the dawn of civilization, humans have carried out cross-breeding programs to develop plants and animals of use to them. With the advent of DNA technology, we have entered a new era in which even greater control can be exerted over the evolutionary process. We can manipulate genes and give organisms traits that they would not ordinarily possess. Some plants today produce human proteins that can be extracted from their seeds, and some animals grow larger because we have supplied them with an extra gene for growth hormone. Does this type of manipulation seem justifiable?

What about the possibility that we are manipulating our own evolution? Should doctors increase the fitness of certain couples by providing them with a means to reproduce that they cannot achieve on their own? Is the use of alternate means of reproduction bioethically justifiable? In the near future, it may be possible for parents to choose the phenotypic traits of their offspring; in effect, this might enable humans to ensure that their offspring are stronger and brighter than their parents. Does this choosing of "designer babies" seem ethical to you?

Understanding the Terms

anthropoid 583	*Homo erectus* 589
arboreal 580	*Homo ergaster* 589
australopithecine 586	molecular clock 584
Australopithecus afarensis 587	mosaic evolution 586
Australopithecus africanus 586	multiregional continuity hypothesis 590
bipedalism 584	Neandertal 591
Cro-Magnon 591	out-of-Africa hypothesis 590
culture 587	
hominid 584	primate 580
hominoid 583	prosimian 583

Match the terms to these definitions:
a. _____ Group of primates that includes only monkeys, apes, and humans.
b. _____ The common name for the first fossils generally accepted as being modern humans.
c. _____ Hominid with a sturdy build who lived during the last Ice Age in Eurasia; hunted large game and lived together in a kind of society.
d. _____ Type of early human to first have a striding gait similar to that of modern humans.
e. _____ Member of a group containing humans and apes.

ARIS, the *Biology* Website

ARIS, the website for *Biology,* provides a wealth of information organized and integrated by chapter. You will find practice quizzes, interactive activities, labeling exercises, flashcards, and much more that will complement your learning and understanding of general biology.

www.mhhe.com/maderbiology9

PART VII

COMPARATIVE ANIMAL BIOLOGY

In contrast to plants, which are autotrophic and make their own organic food, animals are heterotrophic and feed on other organisms. Their mobility, which is dependent upon nerve fibers and muscle fibers, is essential in escaping predators, finding a mate, and acquiring food. The food that is acquired is digested, and the nutrients are distributed to cells. Finally, waste products are expelled.

In complex animals, a distinct division of labor exists in that the body contains organ systems specialized to carry out specific functions. A circulatory system moves materials from one body part to another; a respiratory system carries out gas exchange; and an excretory system filters the blood and removes wastes. Coordination of the systems is accomplished by a nervous system and an endocrine system. The lymphatic system, along with the immune system, protects the body from infectious diseases.

Certain small, microscopic animals do not possess these systems or organs of any kind. In Part VII, the development of organ systems within the animal kingdom, as well as how these systems function in humans and other animals, will be discussed.

33

ANIMAL ORGANIZATION AND HOMEOSTASIS

T*he organization of complex animals includes organ systems, such as the nervous, digestive, respiratory, and circulatory systems. In turn, the organ systems consist of organs including the brain, stomach, lungs, and heart. Organs are composed of tissues, and each type of tissue has like cells that perform specific duties within the organism. This chapter concerns the tissue level of organization and will examine the various types of tissues found in humans, a complex animal.*

Organs and organ systems function best if the internal environment stays within normal limits. For example, a warm temperature speeds enzymatic reactions, a moderate blood pressure helps blood circulate, and sufficient oxygen concentration facilitates ATP production. Working in harmony and under the coordination of the nervous and endocrine systems, healthy organisms are capable of maintaining homeostasis, a dynamic equilibrium of the internal environment. Swim the English Channel, cross the Sahara Desert by camel, hike to the South Pole, or take a space walk—your body temperature will stay at just about 37°C as long as you take proper precautions. An astronaut must depend on artificial systems in addition to natural systems to maintain homeostasis.

An astronaut takes a walk outside his spaceship.

33.1 TYPES OF TISSUES

Like all living things, animals are highly organized. Animals begin life as a single cell, the fertilized egg or zygote. The zygote undergoes cell division, and the cells differentiate into a variety of tissues that go on to become parts of organs. Several organs are found in an organ system. In this chapter, we consider the tissue, organ, and organ system levels of organization.

A **tissue** is composed of specialized cells of the same type that perform a common function in the body. The tissues of the human body can be categorized into four major types:

1. *Epithelial tissue* covers body surfaces, lines body cavities, and forms glands.

2. *Connective tissue* binds and supports body parts.

3. *Muscular tissue* moves the body and its parts.

4. *Nervous tissue* receives stimuli and transmits nerve impulses.

Epithelial Tissue

Epithelial tissue, also called epithelium (pl., epithelia), consists of tightly packed cells that form a continuous layer. Epithelial tissue covers surfaces and lines body cavities. Usually, it has a protective function, but it can also be modified to carry out secretion, absorption, excretion, and filtration.

Epithelial cells may be connected to one another by three types of junctions composed of proteins (see Fig. 5.14). Regions where proteins join them together are called tight junctions. In the intestine, the gastric juices stay out of the body, and in the kidneys, the urine stays within kidney tubules because epithelial cells are joined by tight junctions. For example, adhesion junctions in the skin allow epithelial cells to stretch and bend, while gap junctions are protein channels that permit the passage of molecules between two adjacent cells.

Epithelial cells are exposed to the environment on one side, but on the other side they have a **basement membrane.** The basement membrane should not be confused with the plasma membrane or the body membranes we will be discussing. It is simply a thin layer of various types of proteins that anchors the epithelium to underlying connective tissue.

Simple Epithelia

Epithelial tissue is either simple or complex. Simple epithelia have only a single layer of cells (Fig. 33.1) and are classified according to cell type. **Squamous epithelium,** which is composed of flattened cells, is found lining blood vessels and the air sacs of lungs. **Cuboidal epithelium** contains cube-shaped cells and is found lining the kidney tubules and various glands. **Columnar epithelium** has cells resembling rectangular pillars or columns, with nuclei usually located near the bottom of each cell. This epithelium is found lining the digestive tract, where it efficiently absorbs nutrients from the small intestine because of minute cellular extensions called microvilli. Ciliated columnar epithelium is found lining the oviducts, where it propels the egg toward the uterus.

When an epithelium is pseudostratified, it appears to be layered, but true layers do not exist because each cell touches the basement membrane. The lining of the windpipe, or trachea, is pseudostratified ciliated columnar epithelium. A secreted covering of mucus traps foreign particles, and the upward motion of the cilia carries the mucus to the back of the throat, where it may be either swallowed or expectorated.

FIGURE 33.1

Types of epithelial tissues in vertebrates.

Basic epithelial tissues found in vertebrates are shown, along with locations of the tissue and the primary function of the tissue at these locations.

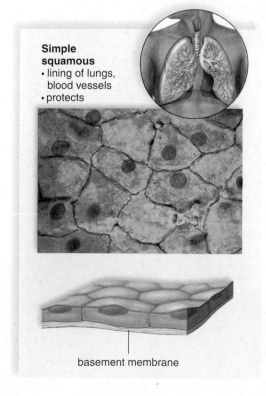

Simple squamous
• lining of lungs, blood vessels
• protects

basement membrane

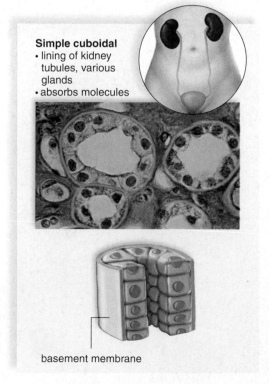

Simple cuboidal
• lining of kidney tubules, various glands
• absorbs molecules

basement membrane

Smoking can cause a change in mucus secretion and inhibit ciliary action, resulting in a chronic inflammatory condition called bronchitis.

Stratified Epithelia

Stratified epithelia has layers of cells piled one on top of the other. Only the bottom layer touches the basement membrane. The nose, mouth, esophagus, anal canal, and vagina are all lined with stratified squamous epithelium. As we shall see, the outer layer of skin is also stratified squamous epithelium, but the cells have been reinforced by keratin, a protein that provides strength. Stratified cuboidal and stratified columnar epithelia also occur in the body.

Glandular Epithelia

When an epithelium secretes a product, it is said to be glandular. A **gland** can be a single epithelial cell, as in the case of mucus-secreting goblet cells within the columnar epithelium lining the digestive tract, or a gland can contain many cells. Glands that secrete their product into ducts are called **exocrine glands,** and those that secrete their product directly into the bloodstream are called **endocrine glands.** The pancreas is both an exocrine and an endocrine gland; it secretes digestive juices into the small intestine via ducts, and it secretes insulin into the bloodstream.

Epithelial tissue is named according to the shape of the cell. These tightly packed protective cells can occur in more than one layer, and the cells lining a cavity can be ciliated and/or glandular.

Connective Tissue

Connective tissue is the most abundant and widely distributed tissue in complex animals. It is quite diverse in structure and function, but, even so, all types have three components: specialized cells, ground substance, and protein fibers (Fig. 33.2). The ground substance is a noncellular material that separates the cells and varies in consistency from solid to semifluid to fluid. The fibers are of three possible types. White **collagen fibers** contain collagen, a protein that gives them flexibility and strength. **Reticular fibers** are very thin collagen fibers that are highly branched and form delicate supporting networks. Yellow **elastic fibers** contain elastin, a protein that is not as strong as collagen but is more elastic.

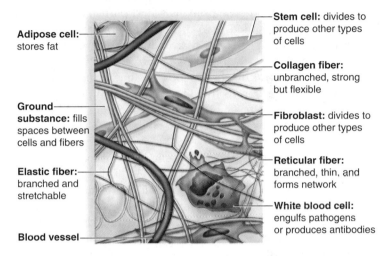

Adipose cell: stores fat

Stem cell: divides to produce other types of cells

Collagen fiber: unbranched, strong but flexible

Fibroblast: divides to produce other types of cells

Ground substance: fills spaces between cells and fibers

Reticular fiber: branched, thin, and forms network

Elastic fiber: branched and stretchable

White blood cell: engulfs pathogens or produces antibodies

Blood vessel

FIGURE 33.2 Diagram of fibrous connective tissue.

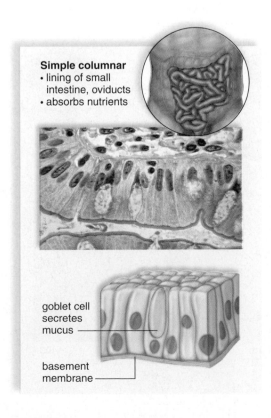

Simple columnar
• lining of small intestine, oviducts
• absorbs nutrients

goblet cell secretes mucus

basement membrane

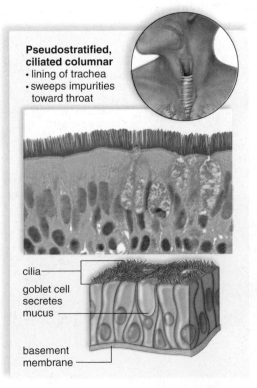

Pseudostratified, ciliated columnar
• lining of trachea
• sweeps impurities toward throat

cilia

goblet cell secretes mucus

basement membrane

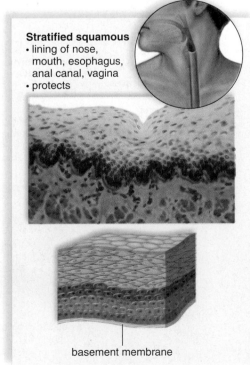

Stratified squamous
• lining of nose, mouth, esophagus, anal canal, vagina
• protects

basement membrane

Fibrous Connective Tissue

Both loose fibrous and dense fibrous connective tissues have cells called **fibroblasts** [L. *fibra*, thread, and Gk. *blastos*, bud] that are located some distance from one another and are separated by a jellylike matrix containing white collagen fibers and yellow elastic fibers.

Loose fibrous connective tissue supports epithelium and also many internal organs (Fig. 33.3*a*). Its presence in lungs, arteries, and the urinary bladder allows these organs to expand. It forms a protective covering enclosing many internal organs, such as muscles, blood vessels, and nerves.

Adipose tissue [L. *adipalis*, fatty] serves as the body's primary energy reservoir (Fig. 33.3*b*). Adipose tissue also insulates the body, contributes to body contours, and provides cushioning. In mammals, adipose tissue is found particularly beneath the skin, around the kidneys, and on the surface of the heart. The number of adipose cells in an individual is fixed. When a person gains weight, the cells become larger, and when weight is lost, the cells shrink. In obese people, the individual cells called adipocytes may be up to five times larger than normal. Most adipose tissue is white fat, but fetuses, infants, and children also have brown fat, which is rich in mitochondria. The brown fat is better at heat generation.

Dense fibrous connective tissue contains many collagen fibers that are packed together (Fig. 33.3*c*). This type of tissue has more specific functions than does loose connective tissue. For example, dense fibrous connective tissue is found in **tendons** [L. *tendo*, stretch], which connect muscles to bones, and in **ligaments** [L. *ligamentum*, band], which connect bones to other bones at joints.

Supportive Connective Tissue

In **cartilage,** the cells lie in small chambers called lacunae (sing., **lacuna**), separated by a matrix that is solid yet flexible. Unfortunately, because this tissue lacks a direct blood supply, it heals very slowly. There are three types of cartilage, distinguished by the type of fiber in the matrix.

Hyaline cartilage (Fig. 33.3*d*), the most common type of cartilage, contains only very fine collagen fibers. The matrix has a white, translucent appearance. Hyaline cartilage is found in the nose and at the ends of the long bones and the ribs, and it forms rings in the walls of respiratory passages. The fetal skeleton also is made of this type of cartilage. Later, the cartilaginous fetal skeleton is replaced by bone.

Elastic cartilage has more elastic fibers than hyaline cartilage. For this reason, it is more flexible and is found, for example, in the framework of the outer ear.

Fibrocartilage has a matrix containing strong collagen fibers. Fibrocartilage is found in structures that withstand tension and pressure, such as the pads between the vertebrae in the backbone and the wedges in the knee joint.

Bone

Bone is the most rigid connective tissue. It consists of an extremely hard matrix of inorganic salts, notably calcium salts, deposited around protein fibers, especially collagen fibers. The inorganic salts give bones rigidity, and the protein fibers provide elasticity and strength, much as steel rods do in reinforced concrete.

Compact bone makes up the shaft of a long bone (Fig. 33.3*e*). It consists of cylindrical structural units called osteons (Haversian systems). The central canal of each osteon is surrounded by rings of hard matrix. Bone cells are located in spaces called lacunae between the rings of matrix. Blood vessels in the central canal carry nutrients that allow bone to renew itself. Thin extensions of bone cells within canaliculi (minute canals) connect the cells to each other and to the central canal.

FIGURE 33.3

Types of connective tissue in vertebrates. Pertinent information about each type of connective tissue is given.

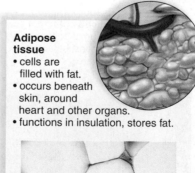
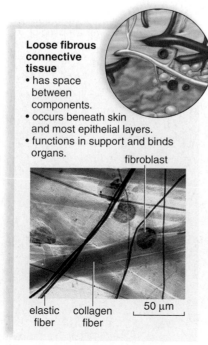

Loose fibrous connective tissue
• has space between components.
• occurs beneath skin and most epithelial layers.
• functions in support and binds organs.

fibroblast

elastic   collagen   50 µm
fiber   fiber

a.

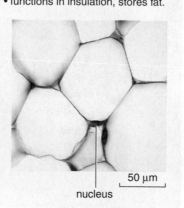

Adipose tissue
• cells are filled with fat.
• occurs beneath skin, around heart and other organs.
• functions in insulation, stores fat.

50 µm

nucleus

b.

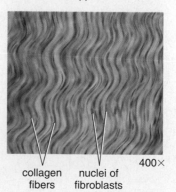

Dense fibrous connective tissue
• has collagenous fibers closely packed.
• in dermis of skin, tendons, ligaments.
• functions in support.

400×

collagen   nuclei of
fibers   fibroblasts

c.

The ends of a long bone contain spongy bone, which has an entirely different structure. **Spongy bone** contains numerous bony bars and plates, separated by irregular spaces. Although lighter than compact bone, spongy bone is still designed for strength. Just as braces are used for support in buildings, the solid portions of spongy bone follow lines of stress.

Fluid Connective Tissues

Blood, which consists of formed elements and plasma, is a fluid connective tissue located in blood vessels (Fig. 33.4). In adults, the production of blood cells, known as hematopoiesis, occurs in the red bone marrow.

The internal environment of the body consists of blood and **tissue fluid.** The systems of the body help keep blood composition and chemistry within normal limits, and blood in turn creates tissue fluid. Blood transports nutrients and oxygen to tissue fluid and removes carbon dioxide and other wastes. It helps distribute heat and also plays a role in fluid, ion, and pH balance. The formed elements, discussed following, each have specific functions.

The **red blood cells** are small, biconcave, disk-shaped cells without nuclei. The presence of the red pigment hemoglobin makes the cells red and, in turn, makes the blood red. Hemoglobin is composed of four units; each unit is composed of the protein globin and a complex iron-containing structure called heme. The iron forms a loose association with oxygen, and in this way red blood cells transport oxygen.

White blood cells may be distinguished from red blood cells by the fact that they are usually larger, have a nucleus, and without staining would appear translucent. White blood cells characteristically look bluish because they have been stained that color. White blood cells fight infection, primarily in two ways. Some white blood cells are phagocytic and

FIGURE 33.4
Blood, a liquid tissue.
a. Blood is classified as connective tissue because the cells are separated by a matrix—plasma. Plasma, the liquid portion of blood, usually contains several types of cells. **b.** Drawing of the components of blood: red blood cells, white blood cells, and platelets (which are actually fragments of a larger cell).

plasma

white blood cells (leukocytes)

red blood cells (erythrocytes)

a. Blood sample after centrifugation

white blood cell

platelets

red blood cell

plasma

b. Blood smear

engulf infectious **pathogens,** while other white blood cells either produce antibodies, molecules that combine with foreign substances to inactivate them, or they kill cells outright.

Platelets are not complete cells; rather, they are fragments of giant cells present only in bone marrow. When a blood vessel is damaged, platelets form a plug that seals the vessel, and injured tissues release molecules that help the clotting process.

Lymph is a fluid connective tissue located in lymphatic vessels. Lymphatic vessels absorb excess tissue fluid and various dissolved solutes in the tissues and transport them to particular vessels of the cardiovascular system. Special lymphatic capillaries, called lacteals, absorb fat molecules from the small intestine. Lymph nodes, composed of fibrous connective tissue, occur along the length of lymphatic vessels. In particular, lymph is cleansed as it passes through lymph nodes because white blood cells congregate there. Lymp nodes enlarge when you have an infection.

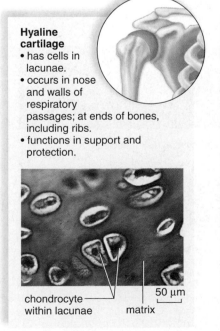

Hyaline cartilage
• has cells in lacunae.
• occurs in nose and walls of respiratory passages; at ends of bones, including ribs.
• functions in support and protection.

chondrocyte within lacunae   matrix   50 μm

d.

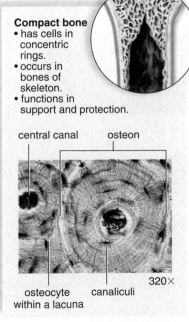

Compact bone
• has cells in concentric rings.
• occurs in bones of skeleton.
• functions in support and protection.

central canal   osteon

osteocyte within a lacuna   canaliculi   320×

e.

Connective tissue is classified into three types: fibrous, supportive, and fluid connective tissue (Fig. 33.3). Fibrous connective tissue forms the structural framework, binds organs, and stores fat. Supportive connective tissue supports, protects, and is involved in movement of the body. Fluid connective tissue transports materials and fights infections.

Muscular Tissue

Muscular (contractile) tissue is composed of cells called muscle fibers. Muscle fibers contain actin filaments and myosin filaments, whose interaction accounts for movement. The muscles are also important in the generation of body heat. There are three distinct types of muscle tissue: skeletal, smooth, and cardiac. Each type differs in appearance, physiology, and function.

Skeletal muscle, also called voluntary muscle (Fig. 33.5a), is attached by tendons to the bones of the skeleton, and when it contracts, body parts move. Contraction of skeletal muscle is under voluntary control and occurs faster than in the other muscle types. Skeletal muscle fibers are cylindrical and quite long—sometimes they run the length of the muscle. They arise during development when several cells fuse, resulting in one fiber with multiple nuclei. The nuclei are located at the periphery of the cell, just inside the plasma membrane. The fibers have alternating light and dark bands that give them a **striated** appearance. These bands are due to the placement of actin filaments and myosin filaments in the cell.

Smooth (visceral) muscle is so named because the cells lack striations. The spindle-shaped cells form layers in which the thick middle portion of one cell is opposite the thin ends of adjacent cells. Consequently, the nuclei form an irregular pattern in the tissue (Fig. 33.5b). Smooth muscle is not under voluntary control and therefore is said to be involuntary. Smooth muscle, found in the walls of viscera (intestine, stomach, and other internal organs) and blood vessels, contracts more slowly than skeletal muscle but can remain contracted for a longer time. When the smooth muscle of the intestine contracts, food moves along its lumen (central cavity). When the smooth muscle of the blood vessels contracts, blood vessels constrict, helping to raise blood pressure. Small amounts of smooth muscle are also found in the iris of the eye and in the skin.

Cardiac muscle (Fig. 33.5c) makes up the walls of the heart. Its contraction pumps blood and accounts for the heartbeat. Cardiac muscle combines features of both smooth muscle and skeletal muscle. Like skeletal muscle, it has striations, but the contraction of the heart is involuntary for the most part. Cardiac muscle cells also differ from skeletal muscle cells in that they have a single, centrally placed nucleus. The cells are branched and seemingly fused one with the other, and the heart appears to be composed of one large interconnecting mass of muscle cells. Actually, cardiac muscle cells are separate and individual, but they are bound end to end at **intercalated disks,** areas where folded plasma membranes between two cells contain adhesion junctions and gap junctions.

All muscular tissue contains actin filaments and myosin filaments; these form a striated pattern in skeletal and cardiac muscle, but not in smooth muscle.

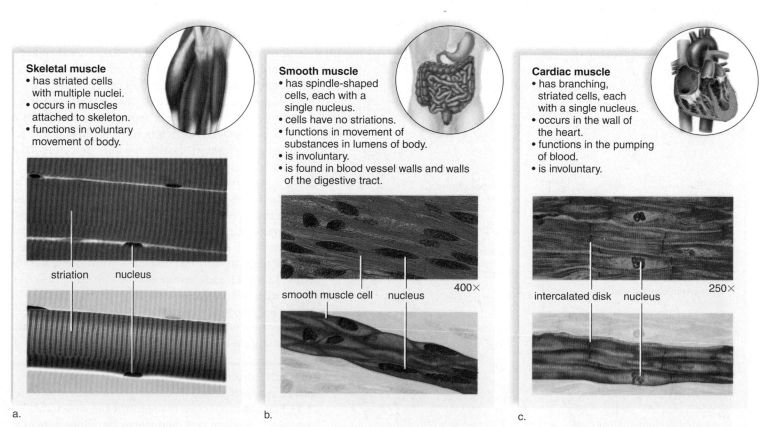

Skeletal muscle
- has striated cells with multiple nuclei.
- occurs in muscles attached to skeleton.
- functions in voluntary movement of body.

striation   nucleus

Smooth muscle
- has spindle-shaped cells, each with a single nucleus.
- cells have no striations.
- functions in movement of substances in lumens of body.
- is involuntary.
- is found in blood vessel walls and walls of the digestive tract.

smooth muscle cell   nucleus   400×

Cardiac muscle
- has branching, striated cells, each with a single nucleus.
- occurs in the wall of the heart.
- functions in the pumping of blood.
- is involuntary.

intercalated disk   nucleus   250×

a.   b.   c.

FIGURE 33.5 Muscular tissue.
a. Skeletal muscle is voluntary and striated. **b.** Smooth muscle is involuntary and nonstriated. **c.** Cardiac muscle is involuntary and striated. Cardiac muscle cells branch and fit together at intercalated disks.

Nervous Tissue

Nervous tissue contains nerve cells called neurons. An average person has about 1 trillion neurons. A **neuron** is a specialized cell that has three parts: dendrites, a cell body, and an axon (Fig. 33.6, *top*). A dendrite is a process that conducts signals toward the cell body. The cell body contains the major concentration of the cytoplasm and the nucleus of the neuron. An axon is a process that typically conducts nerve impulses away from the cell body. Long axons are covered by myelin, a white, fatty substance. The term *fiber*[1] is used here to refer to an axon along with its myelin sheath if it has one. Outside the brain and spinal cord, fibers bound by connective tissue form **nerves.**

The nervous system has just three functions: sensory input, integration of data, and motor output. Nerves conduct impulses from sensory receptors to the spinal cord and the brain, where integration occurs. The phenomenon called sensation occurs only in the brain, however. Nerves also conduct nerve impulses away from the spinal cord and brain to the muscles and glands, causing them to contract and secrete, respectively. In this way, a coordinated response to the stimulus is achieved.

In addition to neurons, nervous tissue contains neuroglia.

Neuroglia

Neuroglia are cells that outnumber neurons as much as 50 to 1, and take up more than half the volume of the brain. Although the primary function of neuroglia is to support and nourish neurons, research is currently being conducted to determine how much they directly contribute to brain function. Various types of neuroglia are found in the brain. Microglia, astrocytes, and oligodendrocytes are shown in Figure 33.6, *bottom*. Microglia, in addition to supporting neurons, engulf bacterial and cellular debris. Astrocytes provide nutrients to neurons and produce a hormone known as glia-derived growth factor, which someday might be used as a cure for Parkinson disease and other diseases caused by neuron degeneration. Oligodendrocytes form myelin. Neuroglia do not have a long process, but even so, researchers are now beginning to gather evidence that they do communicate among themselves and with neurons!

Mature neurons have little capacity for cell division and seldom form tumors. The majority of brain tumors in adults involve actively dividing neuroglia cells. Most brain tumors have to be treated with surgery or radiation therapy because of a blood-brain barrier.

Nerve cells, called neurons, have fibers (processes) called axons and dendrites. In general, neuroglia support and service neurons.

---

[1] In connective tissue, a fiber is a component of the matrix; in muscular tissue, a fiber is a muscle cell; in nervous tissue, a nerve fiber is an axon and its myelin sheath.

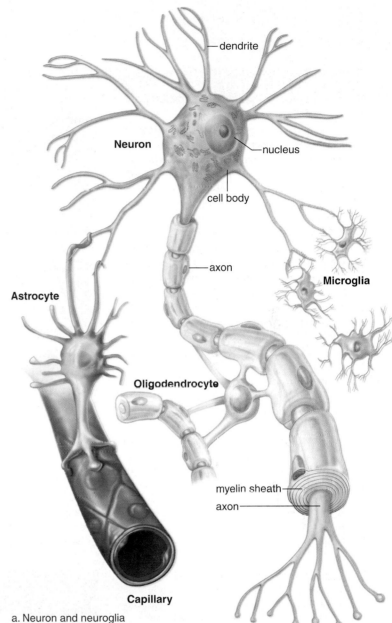

a. Neuron and neuroglia

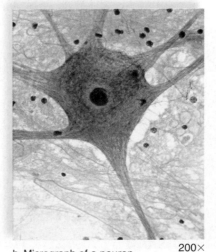

b. Micrograph of a neuron   200×

FIGURE 33.6

Neurons and neuroglia.
Neurons conduct nerve impulses. Neuroglia consist of cells that support and service neurons and have various functions: Microglia are a type of neuroglia that become mobile in response to inflammation and phagocytize debris. Astrocytes lie between neurons and a capillary; therefore, nutrients entering neurons from the blood must first pass through astrocytes. Oligodendrocytes form the myelin sheaths around fibers in the brain and spinal cord.

health focus

Nerve Regeneration

In humans, axons outside the brain and spinal cord can regenerate, but not those inside these organs. After injury, axons in the human central nervous system (CNS) degenerate, resulting in permanent loss of nervous function. Not so in cold-water fishes and amphibians, where axon regeneration in the CNS does occur. So far, investigators have identified several proteins that seem to be necessary to axon regeneration in the CNS of these animals (Fig. 33A), but it will be a long time before biochemistry can offer a way to bring about axon regeneration in the human CNS. It's possible, though, that one day these proteins will become drugs or that gene therapy might be used to cause humans to produce the same proteins when CNS injuries occur.

In the meantime, some accident victims are trying other ways to bring about a cure. In 1995, Christopher Reeve, best known for his acting role as "Superman," was thrown headfirst from his horse, crushing the spinal cord just below the neck's top two vertebrae. Immediately, his brain lost almost all communication with the portion of his body below the site of damage and he could not move his arms and legs. Many years later, Reeve could move his left index finger slightly and could take tiny steps while being held upright in a

a.

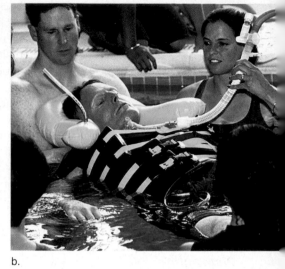
b.

FIGURE 33B Treatment today for spinal cord injuries.
a. Reeve suffered a spinal cord injury when horseback riding in 1995. *b.* He exercised many hours a day. Here he receives aqua therapy. Reeve died in 2004.

FIGURE 33A Researchers at work.
a. Some researchers are studying the activity of proteins that allow cold-bodied animals to regenerate axons in the CNS. *b.* Others are doing stem cell research. Stem cells might one day be used to cure people with spinal cord injuries.

pool. He had sensation throughout his body and could feel his wife's touch.

Reeve's improvement was not the result of cutting-edge drugs or gene therapy—it was due to exercise (Fig. 33B)! Reeve exercised as much as five hours a day, especially using a recumbent

bike outfitted with electrodes that made his leg muscles contract and relax. The bike cost him $16,000. It could cost less if commonly used by spinal cord injury patients in their own homes. Reeve, who was an activist for the disabled, was pleased that insurance would pay for the bike about 50% of the time.

It's possible that Reeve's advances were the result of improved strength and bone density, which lead to stronger nerve signals. Normally, nerve cells are constantly signaling one another, but after a spinal cord injury, the signals cease. Perhaps Reeve's intensive exercise brought back some of the normal communication between nerve cells. Reeve's physician, John McDonald, a neurologist at Washington University in St. Louis, is convinced that his axons were regenerating. The neuroscientist Fred Gage at the Salk Institute in La Jolla, California, has shown that exercise does enhance the growth of new cells in adult brains.

For himself, Reeve was convinced that stem cell therapy would one day allow him to be off his ventilator and functioning normally; however, Reeve died in 2004. So far, researchers have shown that both embryonic stem cells and bone marrow stem cells can differentiate into neurons in the laboratory. Bone marrow stem cells apparently can also become neurons when injected into the body.

a.

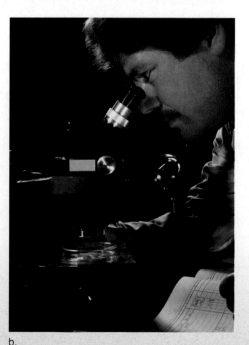
b.

33.2 ORGANS AND ORGAN SYSTEMS

Specific tissues are associated with particular organs. For example, nervous tissue is associated with the brain. In actuality, an **organ** is composed of two or more types of tissues working together to perform particular functions. An **organ system** contains many different organs that cooperate to carry out a process, such as the digestion of food.

The integumentary system, consisting of the skin and its derivatives (hair, nails, and cutaneous glands), is the most conspicuous system in the body. Being the largest organ, the skin covers an area of 1.5–2 m² and accounts for nearly 15% of the weight of an average human.

Derivatives of the skin differ throughout the vertebrate world. Fishes possess a number of bony scales. Amphibians have smooth skin covered with mucous glands. Reptiles possess epidermal scales that vary in color and shape. Birds have scales on their legs, but most of the body is covered with feathers. Mammals are characterized by hair, nails, glands, and a number of sensory detectors.

Skin as an Organ

Human skin covers the body, protecting underlying parts from physical trauma, pathogen invasion, and water loss. The skin is also important in thermoregulation or regulating body temperature. Skin is equipped with a variety of sensory structures that monitor touch, pressure, temperature, and pain. In addition, skin cells manufacture precursor molecules that are converted to vitamin D after exposure to UV (ultraviolet) light.

Regions of Skin

The **skin** has two regions: the epidermis and the dermis (Fig. 33.7). A **subcutaneous layer** known as the hypodermis is found between the skin and any underlying structures, such as muscle or bone.

The Epidermis

The **epidermis** [Gk. *epi*, over, and *derma*, skin] is made up of stratified squamous epithelium. Skin can be described as thin skin or thick skin based on the thickness of the epidermis. Thin skin covers most of the body and is associated with hair follicles, sebaceous (oil) glands, and sweat glands. Thick skin appears in regions of wear and tear, such as the palms of the hands and soles of the feet. Thick skin has sweat glands but no sebaceous glands or hair follicles. In both types of skin, new cells derived from stem (basal) cells become flattened and hardened as they push to the surface (Fig. 33.8*a*). Hardening takes place because the cells produce

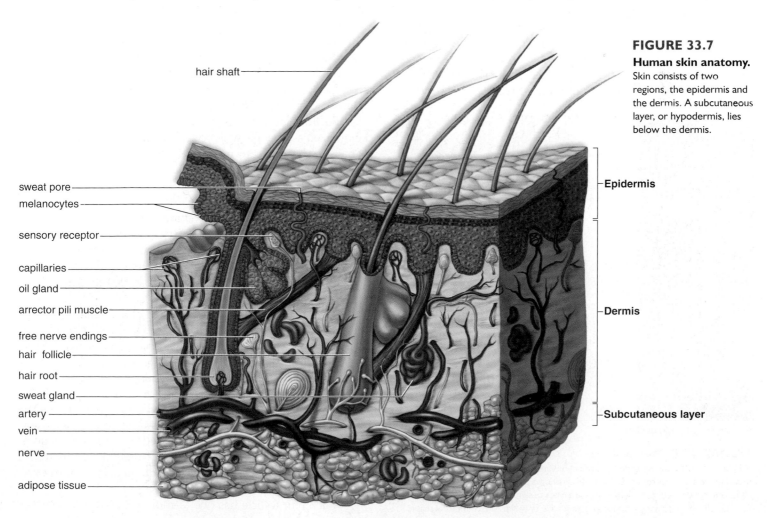

FIGURE 33.7

Human skin anatomy. Skin consists of two regions, the epidermis and the dermis. A subcutaneous layer, or hypodermis, lies below the dermis.

hair shaft

sweat pore
melanocytes
sensory receptor
capillaries
oil gland
arrector pili muscle
free nerve endings
hair follicle
hair root
sweat gland
artery
vein
nerve
adipose tissue

Epidermis
Dermis
Subcutaneous layer

keratin, a waterproof protein. Dandruff occurs when the rate of keratinization in the skin of the scalp is two or three times the normal rate. A thick layer of dead keratinized cells, arranged in spiral and concentric patterns, forms fingerprints and footprints.

Specialized cells in the epidermis called **melanocytes** produce melanin, the pigment responsible for skin color. The amount of melanin varies throughout the body. It is concentrated in freckles and moles. Tanning occurs after a light-skinned person is exposed to sunlight because melanocytes produce more melanin, which is distributed to epidermal cells before they rise to the surface. While we tend to associate a tan with health, actually it signifies that the body is trying to protect itself from the dangerous rays of the sun. Some ultraviolet radiation does serve a purpose, however. As mentioned, certain cells in the epidermis convert a steroid related to cholesterol into **vitamin D** only with the aid of ultraviolet radiation. Vitamin D is required for proper bone growth.

Too much ultraviolet radiation is dangerous and can lead to skin cancer. Basal cell carcinoma (Fig. 33.8b) derived from stem cells gone awry is the more common type of skin cancer and the most curable. Melanoma (Fig. 33.8c), the type of skin cancer derived from melanocytes, is extremely serious.

The Dermis

The **dermis** [Gk. *derma*, skin] is a region of dense fibrous connective tissue beneath the epidermis. The dermis contains collagen and elastic fibers. The collagen fibers are flexible but offer great resistance to overstretching; they prevent the skin from being torn. Stretching of the dermis, as occurs in obesity and pregnancy, can produce stretch marks, or striae. The elastic fibers maintain normal skin tension but also stretch to allow movement of underlying muscles and joints. (The number of collagen and elastic fibers decreases with age and with exposure to the sun, causing the skin to become less supple and more prone to wrinkling.) The dermis also contains blood vessels that nourish the skin. When blood rushes into these vessels, a person blushes, and when blood is minimal in them, a person turns "blue."

Sensory receptors are specialized nerve endings in the dermis that respond to external stimuli. There are sensory receptors for touch, pressure, pain, and temperature. The fingertips contain the most touch receptors, and these add to our ability to use our fingers for delicate tasks.

The Subcutaneous Layer

Technically speaking, the subcutaneous layer (the hypodermis) beneath the dermis is not a part of skin. It is composed of loose connective tissue and adipose tissue, which stores fat. Fat is a stored source of energy in the body. Adipose tissue helps to thermally insulate the body from either gaining heat from the outside or losing heat from the inside. A well-developed subcutaneous layer gives the body a rounded appearance and provides protective padding against external assaults. Excessive development of the subcutaneous layer accompanies obesity.

Accessory Organs of the Skin

Nails, hair, and glands are structures of epidermal origin, even though some parts of hair and glands are largely found in the dermis.

Nails are a protective covering of the distal part of fingers and toes, collectively called digits. Nails grow from special epithelial cells at the base of the nail in the portion called the nail root. The cuticle is a fold of skin that hides the nail root. The whitish color of the half-moon-shaped base, or lunula, results from the thick layer of cells in this area. The cells of a nail become keratinized as they grow out over the nail bed. The appearance of nails can be medically important. For example, clubbing of the nails and fingertips is associated with a deficiency of oxygen in the blood.

Hair follicles begin in the dermis and continue through the epidermis, where the hair shaft extends beyond the skin. Contraction of the arrector pili muscles attached to hair follicles causes the hairs to "stand on end" and goose bumps to develop. Epidermal cells form the root of hair, and their division causes a hair to grow. The cells become keratinized and die as they are pushed farther from the root.

Hair, except for the root, is formed of dead, hardened epidermal cells; the root is alive and resides at the base of a follicle in the dermis. An average person has about 100,000 hair follicles on the scalp. The number of follicles varies from one body region to another. The texture of hair is dependent on the shape of the shaft. In wavy hair, the shaft is oval, and in straight hair, the shaft is round. Hair color is determined by pigmentation. Dark hair is due to melanin concentration, and blond hair has scanty amounts of melanin. Red hair is caused by an iron-containing pigment called trichosiderin. Gray or white hair results from a lack of pigment. A hair on the scalp grows about 1 mm every three days.

Each hair follicle has one or more **oil glands,** also called sebaceous glands, which secrete sebum, an oily substance that

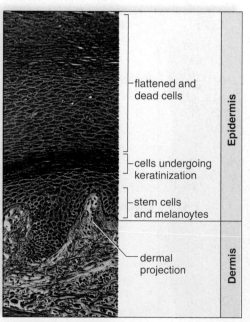

a. Photomicrograph of skin

- flattened and dead cells

Epidermis

- cells undergoing keratinization

- stem cells and melanoytes

- dermal projection

Dermis

b. Basal cell carcinoma

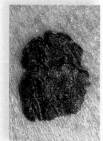

c. Melanoma

FIGURE 33.8 The epidermis.
a. Epidermal ridges following dermal projections are clearly visible. Stem cells and melanocytes are in this region. **b.** Basal cell carcinoma derived from stem cells and melanoma (**c**) derived from melanocytes are types of skin cancer. Remember the A, B, C, D rule when examining a questionable freckle or mole: A–asymmetrical shape; B–border irregularity; C–color change; D–diameter or sudden change in size.

lubricates the hair within the follicle and the skin itself. If the sebaceous glands fail to discharge, the secretions collect and form "whiteheads" or "blackheads." The color of blackheads is due to oxidized sebum. Acne is an inflammation of the sebaceous glands that most often occurs during adolescence due to hormonal changes.

Sweat glands, also called sudoriferous glands, are quite numerous and are present in all regions of skin. A sweat gland is a tubule that begins in the dermis and either opens into a hair follicle, or more often opens onto the surface of the skin. Sweat glands located all over the body play a role in modifying body temperature. When the body temperature starts to rise, sweat glands become active. Sweat absorbs body heat as it evaporates. Once the body temperature lowers, sweat glands are no longer active. Other sweat glands occur in the groin and axillary regions and are associated with distinct scents.

Skin has two regions: the epidermis and the dermis. A subcutaneous layer lies beneath the skin. The accessory organs of the skin—nails, hair, and glands—are of epidermal origin, even those portions located in the dermis.

Organ Systems

In most animals, individual organs function as part of an organ system, the next higher level of animal organization. These same systems are found in all vertebrate animals. The organ systems of vertebrates carry out the life processes that are common to all animals, and indeed to all organisms.

Life Processes	Human Systems
Coordinate body activities	Nervous system Endocrine system
Acquire materials and energy (food)	Skeletal system Muscular system Digestive system
Maintain body shape	Skeletal system Muscular system
Exchange gases	Respiratory system
Transport materials	Cardiovascular system
Excrete wastes	Urinary system
Protect the body from disease	Lymphatic system Immune system
Produce offspring	Reproductive system

Body Cavities

Each organ system has a particular distribution within the human body. There are two main body cavities: the smaller dorsal cavity and the larger ventral cavity (Fig. 33.9a). The brain and the spinal cord are in the dorsal cavity.

During development, the ventral cavity develops from the coelom. In humans and other mammals, the coelom is divided by a muscular diaphragm that assists breathing. The heart (a

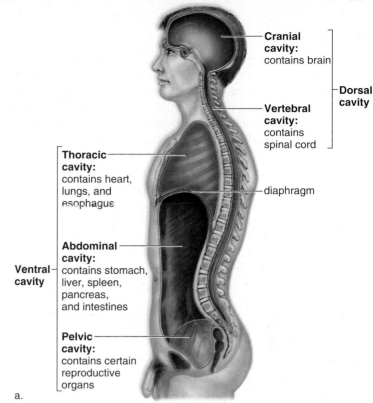

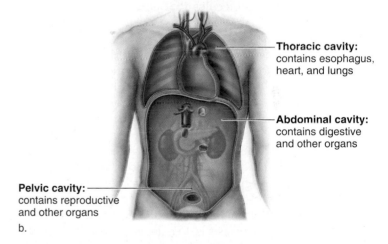

FIGURE 33.9 Mammalian body cavities.
a. Side view. The dorsal (toward the back) cavity contains the cranial cavity and the vertebral canal. The brain is in the cranial cavity, and the spinal cord is in the vertebral canal. The well-developed ventral (toward the front) cavity is divided by the diaphragm into the thoracic cavity and the abdominopelvic cavity (abdominal cavity and pelvic cavity). The heart and lungs are in the thoracic cavity, and most other internal organs are in the abdominal cavity. b. Frontal view of the thoracic cavity.

pump for the cardiovascular system) and the lungs are located in the upper (thoracic or chest) cavity (Fig. 33.9b). The major portions of the digestive system, including the accessory organs (e.g., the liver and pancreas) are located in the abdominal cavity, as are the kidneys of the urinary system. The urinary bladder, the female reproductive organs, or certain of the male reproductive organs, are located in the pelvic cavity.

The animal body is organized; the organs have a specific structure, function, and location. The performances of the organs of each system are coordinated.

33.3 HOMEOSTASIS

Claude Bernard, a famous French physiologist, suggested in 1859 that while an animal lives in an external environment, the cells of the body lie within an internal environment. The internal environment is tissue fluid, which bathes the cells of the body and blood, which services the cells. Bernard concluded that the relative stability of the internal environment allows animals to live in an external environment that can vary considerably. Later, Walter Cannon, an American physiologist, introduced the term **homeostasis** [Gk. *homoios*, like, resembling, and *stasis*, standing]. He said there is a dynamic interplay between events that tend to change the internal environment and events that act against this possibility. For example, to achieve homeostasis, the composition of blood and tissue fluid, the body temperature, and the blood pressure must all stay within a normal range.

The internal environment of an animal's body consists of tissue fluid, which bathes the cells.

The organ systems of the human body contribute to homeostasis. The digestive system takes in and digests food, providing nutrient molecules that enter the blood and replace the nutrients that are constantly being used by the body cells. The respiratory system adds oxygen to the blood and removes carbon dioxide. The amount of oxygen taken in and carbon dioxide given off can be increased to meet body needs. The liver and the kidneys contribute greatly to homeostasis. For example, immediately after glucose enters the blood, it can be removed by the liver and stored as glycogen. Later, glycogen is broken down to replace the glucose used by the body cells; in this way, the glucose composition of the blood remains constant. The hormone insulin, secreted by the pancreas, regulates glycogen storage. The kidneys are also under hormonal control as they excrete wastes and salts, substances that can affect the pH level of the blood.

Although homeostasis is, to a degree, controlled by hormones, it is ultimately controlled by the nervous system. In humans, the brain contains regulatory centers that control the function of other organs, maintaining homeostasis. These regulatory centers are often a part of a negative feedback system.

Negative Feedback

Negative feedback is the primary homeostatic mechanism that keeps a variable, such as the blood glucose level, close to a particular value, or set point. A homeostatic mechanism has at least two components: a sensor and a control center (Fig. 33.10). The sensor detects a change in the internal environment; the control center then brings about an effect to bring conditions back to normal again. Now, the sensor is no longer activated. In other words, a negative feedback mechanism is present when the output of the system dampens the original stimulus. For example, when the pancreas detects that the blood glucose level is too high, it secretes insulin, the hormone that causes cells to take up glucose. Now, the

blood sugar level returns to normal, and the pancreas is no longer stimulated to secrete insulin.

Mechanical Example

A home heating system is often used to illustrate how a more complicated negative feedback mechanism works (Fig. 33.11). You set the thermostat at, say, 68°F. This is the *set point*. The thermostat contains a thermometer, a sensor that detects when the room temperature is above or below the set point. The thermostat also contains a control center; it turns the furnace off when the room is warm and turns it on when the room is cool. When the furnace is off, the room cools a bit, and when the furnace is on, the room warms a bit. In other words, typical of negative feedback mechanisms, there is a fluctuation above and below normal.

Human Example: Regulation of Body Temperature

The sensor and control center for body temperature is located in the brain. Notice in Figure 33.12 that a negative feedback mechanism prevents change in the same direction; body temperature does not get warmer and warmer because warmth brings about a change toward a lower body temperature. Also, body temperature does not get colder and colder because a body temperature below normal brings about a change toward a warmer body temperature.

Above Normal Temperature. When the body temperature is above normal, the control center directs the blood vessels of the skin to dilate. This allows more blood to flow near the surface of the body, where heat can be lost to the environment. In addition, the nervous system activates the sweat

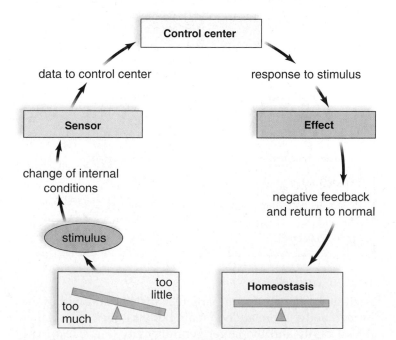

FIGURE 33.10 Negative feedback mechanism.
This diagram shows how the basic elements of a feedback mechanism work. A sensor detects the stimulus and a control center brings about an effect that dampens the stimulus.

glands, and the evaporation of sweat helps lower body temperature. Gradually, body temperature decreases to 37°C.

Below Normal Temperature. When the body temperature falls below normal, the control center directs (via nerve impulses) the blood vessels of the skin to constrict (Fig. 33.12). This conserves heat. If body temperature falls even lower, the control center sends nerve impulses to the skeletal muscles, and shivering occurs. Shivering generates heat, and gradually body temperature rises to 37°C. When the temperature rises to normal, the control center is inactivated.

Positive Feedback

Positive feedback is a mechanism that brings about an ever greater change in the same direction. When a woman is giving birth, the head of the baby begins to press against the cervix, stimulating sensory receptors there. When nerve impulses reach the brain, the brain causes the pituitary gland to secrete the hormone oxytocin. Oxytocin travels in the blood and causes the uterus to contract. As labor continues, the cervix is ever more stimulated and uterine contractions become ever stronger until birth occurs.

A positive feedback mechanism can be harmful, as when a fever causes metabolic changes that push the fever still higher. Death occurs at a body temperature of 45°C because cellular proteins denature at this temperature and metabolism stops. Still, positive feedback loops such as those involved in childbirth, blood clotting, and the stomach's digestion of protein assist the body in completing a process that has a definite cut-off point.

Negative feedback mechanisms keep conditions within the range of normality; in contrast, positive feedback allows rapid change in one direction and does not achieve relative stability.

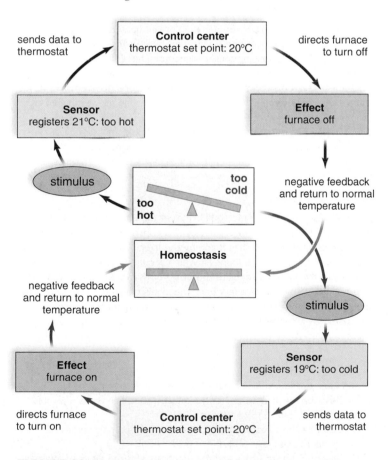

FIGURE 33.11 Regulation of room temperature.
This diagram shows how room temperature is returned to normal when the room becomes too hot (*above*) or too cold (*below*). The thermostat contains both the sensor and the control center. *Above:* The sensor detects that the room is too hot, and the control center turns the furnace off. The stimulus is no longer present when the temperature returns to normal. *Below:* The sensor detects that the room is too cold, and the control center turns the furnace on. The stimulus is no longer present when the temperature returns to normal.

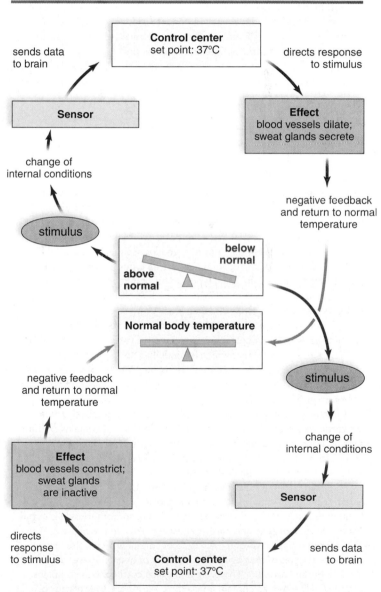

FIGURE 33.12 Regulation of body temperature.
Above: When body temperature rises above normal, the hypothalamus senses the change and causes blood vessels to dilate and sweat glands to secrete so that body temperature returns to normal. *Below:* When body temperature falls below normal, the hypothalamus senses the change and causes blood vessels to constrict. In addition, shivering may occur to bring body temperature back to normal. In this way, the original stimulus was removed.

Connecting the Concepts

In this chapter, we have concentrated on complex coelomate animals, but homeostasis occurs even in the simplest of animals. Homeostasis in unicellular organisms and thin invertebrate animals occurs only at the cellular level, and each cell must carry out its own exchanges with the external environment to maintain a relative constancy of cytoplasm.

In complex animals, there are localized boundaries where materials are exchanged with the external environment. In terrestrial animals, gas exchange usually occurs within lungs, food is digested within a digestive tract, and kidneys collect and excrete metabolic wastes. Exchange boundaries are an effective way to regulate the internal environment if there is a transport system to carry materials from one body part to an-

other. Circulation in complex animals carries out this function.

Regulating mechanisms occur at all levels of organization. At the cellular level, the actions of enzymes are often controlled by feedback mechanisms. However, in animals with organ systems, the nervous and endocrine systems regulate the actions of organs. In addition, an animal's nervous system gathers and processes information about the external environment. Sensory receptors act as specialized boundaries through which external stimuli are received and converted into a form that can be processed by the nervous system. The information received and processed by the nervous system may then influence the organism's behavior in a way that contributes to homeostasis.

Homeostasis is so critical that without it the organism dies. Let us take a familiar example in humans. After eating, when the hormone insulin is present, glucose is removed from blood and stored in the liver as glycogen. In between eating, glycogen breakdown keeps the blood glucose level at just about 0.1%. When a person has type I diabetes, the pancreas fails to secrete insulin, and glucose is not stored in the liver. Worse yet, cells are unable to take up glucose even after eating, when there is a plentiful supply in the blood. Lacking glucose for cellular respiration, cells begin to break down fats with the result that acids are released in cells and enter the bloodstream. Now the person has acidosis—a low pH, which may hinder enzymatic activity to the point that cellular metabolism falters and the person dies.

Summary

33.1 TYPES OF TISSUES

During development, the zygote divides to produce cells that go on to become tissues composed of similar cells specialized for a particular function. Tissues make up organs, and organ systems make up the organism. This sequence describes the levels of organization within an organism.

Tissues are categorized into four groups. Epithelial tissue, which covers the body and lines cavities, is of three types: squamous, cuboidal, and columnar epithelium. Each type can be simple or stratified; it can also be glandular or have modifications, such as cilia. Epithelial tissue protects, absorbs, secretes, and excretes.

Connective tissue has a matrix between cells. Loose fibrous connective tissue and dense fibrous connective tissue contain fibroblasts and fibers. Loose fibrous connective tissue has both collagen and elastic fibers. Dense fibrous connective tissue, like that of tendons and ligaments, contains closely packed collagen fibers. In adipose tissue, the cells enlarge and store fat.

Both cartilage and bone have cells within lacunae, but the matrix for cartilage is more flexible than that for bone, which contains calcium salts. In bone, the lacunae lie in concentric circles within an osteon (or Haversian system) about a central canal. Blood is a connective tissue in which the matrix is a liquid called plasma.

Muscular (contractile) tissue can be smooth or striated (skeletal and cardiac), and involuntary (smooth and cardiac) or voluntary (skeletal). In humans, skeletal muscle is attached to bone, smooth muscle is in the wall of internal organs, and cardiac muscle makes up the heart.

Nervous tissue has one main type of conducting cell, the neuron, and several types of neuroglia. The majority of neurons have dendrites, a cell body, and an axon. The brain and spinal cord contain complete neurons, while nerves contain only axons. Axons are specialized to conduct nerve impulses.

33.2 ORGANS AND ORGAN SYSTEMS

Organs contain various tissues. Skin is an organ that has two regions. Epidermis (stratified squamous epithelium) overlies the dermis (fibrous connective tissue containing sensory receptors, hair follicles, blood vessels, and nerves). A subcutaneous layer is composed of loose connective tissue.

Organ systems contain several organs. The organ systems of humans have specific functions and carry out the life processes that are common to all organisms.

The human body has two main cavities. The dorsal cavity contains the brain and spinal cord. The ventral cavity is divided into the thoracic cavity (heart and lungs) and the abdominal cavity (most other internal organs).

33.3 HOMEOSTASIS

Homeostasis is the dynamic equilibrium of the internal environment (blood and tissue fluid). All organ systems contribute to homeostasis, but special contributions are made by the liver, which keeps the blood glucose constant, and the kidneys, which regulate the pH. The nervous and hormonal systems regulate the other body systems. Both of these are controlled by negative feedback mechanisms, which result in slight fluctuations above and below desired levels. Body temperature is regulated by a center in the hypothalamus.

Reviewing the Chapter

1. Name the four major types of tissues.   598
2. Describe the structure and the functions of three types of simple epithelial tissue.   598–99
3. Describe the structure and the functions of six major types of connective tissue.   600–1
4. Describe the structure and the functions of three types of muscular tissue.   602

5. Nervous tissue contains what types of cells?   603
6. Describe the structure of skin, and state at least two functions of this organ.   605–7
7. In general terms, describe the locations of the human organ systems.   607
8. Tell how the various systems of the body contribute to homeostasis.   608
9. What is the function of sensors, the regulatory center, and effectors in a negative feedback mechanism? Why is it called negative feedback?   608–9

Testing Yourself

Choose the best answer for each question.

1. Which of these pairs is incorrectly matched?
   a. tissues—like cells
   b. epithelial tissue—protection and absorption
   c. muscular tissue—contraction and conduction
   d. connective tissue—binding and support
   e. nervous tissue—conduction and message sending

2. Which of these is not a type of epithelial tissue?
   a. simple cuboidal and stratified columnar
   b. bone and cartilage
   c. stratified squamous and simple squamous
   d. pseudostratified and ciliated
   e. All of these are epithelial tissue.

3. Which tissue is more apt to line a lumen?
   a. epithelial tissue
   b. connective tissue
   c. nervous tissue
   d. muscular tissue
   e. only smooth muscle

4. Tendons and ligaments
   a. are connective tissue.
   b. are associated with the bones.
   c. are found in vertebrates.
   d. contain collagen.
   e. All of these are correct.

5. Which tissue has cells in lacunae?
   a. epithelial tissue
   b. cartilage
   c. bone
   d. smooth muscle
   e. Both b and c are correct.

6. Cardiac muscle is
   a. striated.
   b. involuntary.
   c. smooth.
   d. many fibers fused together.
   e. Both a and b are correct.

7. Which of these components of blood fights infection?
   a. red blood cells
   b. white blood cells
   c. platelets
   d. hydrogen ions
   e. All of these are correct.

8. Which of these body systems contribute to homeostasis?
   a. digestive and urinary systems
   b. respiratory and nervous systems
   c. nervous and endocrine systems
   d. immune and cardiovascular systems
   e. All of these are correct.

9. In a negative feedback mechanism,
   a. the output cancels the input.
   b. there is a fluctuation above and below the average.
   c. there is self-regulation.
   d. a regulatory center communicates with other body parts.
   e. All of these are correct.

10. When a human being is cold, the superficial blood vessels
    a. dilate, and the sweat glands are inactive.
    b. dilate, and the sweat glands are active.
    c. constrict, and the sweat glands are inactive.
    d. constrict, and the sweat glands are active.
    e. contract so that shivering occurs.

11. Give the name, the location, and the function for each of these tissues in the human body.
    a. type of epithelial tissue
    b. type of muscular tissue
    c. type of connective tissue

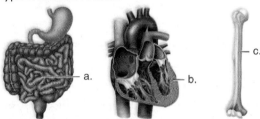

12. Which of these is a function of skin?
    a. temperature regulation
    b. manufacture of vitamin D
    c. collection of sensory input
    d. protection from invading pathogens
    e. All of these are correct.

13. Which of these is an example of negative feedback?
    a. Air conditioning goes off when room temperature lowers.
    b. Insulin decreases blood sugar levels after eating a meal.
    c. Heart rate increases when blood pressure drops.
    d. All of these are examples of negative feedback.

14. Which of these correctly describes a layer of the skin?
    a. The epidermis is simple squamous epithelium in which hair follicles develop and blood vessels expand when we are hot.
    b. The subcutaneous layer lies between the epidermis and the dermis. It contains adipose tissue, which keeps us warm.
    c. The dermis is a region of connective tissue that contains sensory receptors, nerve endings, and blood vessels.
    d. The skin has a special layer, still unnamed, in which there are all the accessory structures such as nails, hair, and various glands.

15. The _____ separates the thoracic cavity from the abdominal cavity.
    a. liver
    b. pancreas
    c. diaphragm
    d. pleural membrane
    e. intestines

In questions 16–18, match each type of muscle tissue to as many terms in the key as possible.

KEY:

   a. voluntary
   b. involuntary
   c. striated
   d. nonstriated
   e. spindle-shaped cells
   f. branched cells
   g. long, cylindrical cells

16. skeletal muscle

17. smooth muscle

18. cardiac muscle

In questions 19–22, match each description to the tissues in the key.

KEY:

   a. loose fibrous connective tissue
   b. hyaline cartilage
   c. adipose tissue
   d. compact bone

19. occurs in nose and walls of respiratory passages

20. occurs only within bones of skeleton

21. occurs beneath most epithelial layers

22. occurs beneath skin, and around organs, including the heart

Thinking Scientifically

1. Many cancers develop from epithelial tissue. These include lung, colon, and skin cancers. What are two attributes of this tissue type that make cancer more likely to develop?

2. When infected with certain bacteria or viruses, our bodies produce a febrile response, or fever. Fevers occur when the hypothalamus changes its temperature set point. Signaling of the hypothalamus could be direct (from the infectious agent itself) or indirect (from the immune system). Which of these would enable the hypothalamus to respond to the greatest variety of infectious agents? Is there any disadvantage to such a signaling system?

Bioethical Issue: Organ Transplants

Despite widespread efforts to convince people of the need for organ donors, supply continues to lag far behind demand. One proposed strategy to bring supply and demand into better balance is to develop an "insurance" program for organs. In this program, participants would pay the "premium" by promising to donate their organs at death. They, in turn, would receive priority for transplants as their "benefit." To avoid the problem of too many high-risk people applying for this type of insurance, a medical exam would be required so that only people with a normal risk of requiring a transplant would be accepted. Do you suppose this system would be an improvement over the current system in which organ donation is voluntary, with no tangible benefit? Can you think of any other strategy that would be more effective for increasing organ donations?

Understanding the Terms

adipose tissue 600
basement membrane 598
blood 601
bone 600
cardiac muscle 602
cartilage 600
collagen fiber 599
columnar epithelium 598
compact bone 600
connective tissue 599
cuboidal epithelium 598
dense fibrous connective tissue 600
dermis 606
elastic cartilage 600
elastic fiber 599
endocrine gland 599
epidermis 605
epithelial tissue 598
exocrine gland 599
fibroblast 600
fibrocartilage 600
gland 599
hair follicle 608
homeostasis 608
hyaline cartilage 600
intercalated disk 602
lacuna 600
ligament 600
loose fibrous connective tissue 600
lymph 601
melanocyte 606
muscular (contractile) tissue 602
nail 606
negative feedback 608
nerve 603
nervous tissue 603
neuroglia 603
neuron 603
oil gland 606
organ 605
organ system 605
pathogen 601
platelet 601
positive feedback 609
red blood cell 601
reticular fiber 599
skeletal muscle 602
skin 605
smooth (visceral) muscle 602
spongy bone 601
squamous epithelium 598
striated 602
subcutaneous layer 605
sweat gland 607
tendon 600
tissue 598
tissue fluid 601
vitamin D 606
white blood cell 601

Match the terms to these definitions:

a. _____ Fibrous connective tissue that joins bone to bone at a joint.
b. _____ Outer region of the skin composed of stratified squamous epithelium.
c. _____ Having striations, such as in cardiac and skeletal muscle.
d. _____ Self-regulatory state in which imbalances result in a fluctuation above and below a mean.
e. _____ Porous bone found at the ends of long bones where blood cells are formed.

ARIS, the *Biology* Website

ARIS, the website for *Biology*, provides a wealth of information organized and integrated by chapter. You will find practice quizzes, interactive activities, labeling exercises, flashcards, and much more that will complement your learning and understanding of general biology.

www.mhhe.com/maderbiology9

34

CIRCULATION AND CARDIOVASCULAR SYSTEMS

*A*ll *animal cells acquire nutrients and oxygen from the environment and give off carbon dioxide and other wastes to the environment. In small aquatic animals, each cell directly exchanges materials with the external environment by using diffusion and plasma membrane transport mechanisms. These animals have no need for a circulatory system.*

Larger, more active animals have a cardiovascular system in which the heart pumps a fluid about the body to all organs. In humans, an 11-ounce heart keeps blood flowing into a system of vessels 60,000 miles long. The pumping of the heart allows the brain to think, the lungs to breathe, and the muscles to move. The heart is one of the first organs to form during development, and when it stops, death occurs. The blood transports gases, nutrients, and metabolic wastes to parts of the body including organs, which have exchange boundaries with the external environment. As an example, think of the lungs, which carry on gas exchange with the external environment.

The blood cells featured in the micrograph have specific functions. The red blood cells assist in the transport of gases. Collectively, the white blood cells defend the body from infection. Platelets, which are fragments of a much larger cell, are essential to the clotting of blood.

Scanning electron micrograph of mammalian red blood cells.

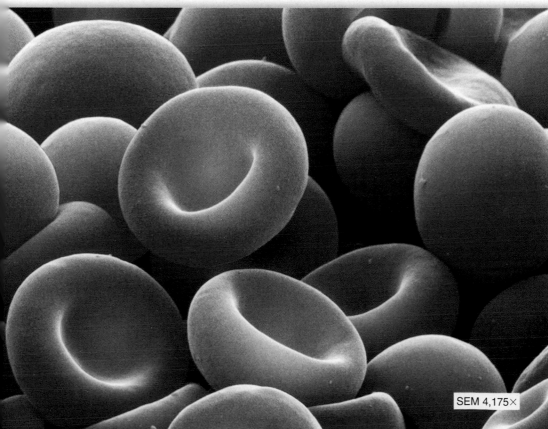

SEM 4,175×

34.1 TRANSPORT IN INVERTEBRATES

Unicellular protists with a high surface-area-to-volume rely on exchanges with the surrounding external environment to acquire nutrients and oxygen and rid their bodies of waste molecules (Fig. 34.1a). Even some small multicellular animals do not have an internal transport system because their cells can be serviced without one. Larger invertebrates (e.g., octopuses and earthworms) and vertebrates have a circulatory system. These animals have either a closed circulatory system or an open circulatory system.

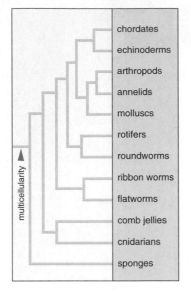

Invertebrates Without a Circulatory System

Cnidarians (e.g., sea anemones) and platyhelminthes (e.g., flatworms such as planarians) have a saclike body (Fig. 34.1b, c). This type of body makes a circulatory system unnecessary. In a sea anemone, cells are either part of an external layer, or they line the gastrovascular cavity. In either case, each cell is exposed to water and can independently exchange gases and rid itself of wastes. The cells that line the gastrovascular cavity are specialized to carry out digestion. They pass nutrient molecules to other cells by diffusion. In a planarian, a trilobed gastrovascular cavity branches throughout the small, flattened body. No cell is very far from one of the three digestive branches, so nutrient molecules can diffuse from cell to cell. Similarly, diffusion meets the respiratory and excretory needs of the cells.

Pseudocoelomate invertebrates, such as nematodes, use the coelomic fluid of their body cavity for transport purposes. The coelomate echinoderms also rely on movement of coelomic fluid within a body cavity as a circulatory system.

Invertebrates with an Open or a Closed Circulatory System

All other animals have a **circulatory system** in which a pumping heart moves a fluid into blood vessels. There are two types of circulatory fluids: **blood**, which is always contained within blood vessels, and **hemolymph**, which flows into a body cavity called a hemocoel. Hemolymph is a mixture of blood and tissue fluid.

Hemolymph is seen in animals that have an **open circulatory system.** In most molluscs and arthropods, the heart pumps hemolymph via vessels into tissue spaces that are sometimes enlarged into saclike sinuses (Fig. 34.2a). Eventually, hemolymph drains back to the heart. In the grasshopper, an arthropod, the dorsal tubular heart pumps hemolymph into a dorsal aorta, which empties into the

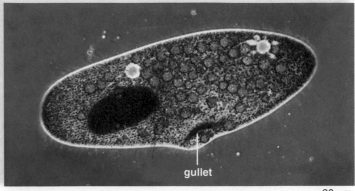

a. Paramecium, *Paramecium*   20 μm

b. Sea anemone, *Apitasia*

gastrovascular cavity

7×

c. Flatworm, *Dugesia*

FIGURE 34.1 Aquatic organisms without a circulatory system.
a. A paramecium is a unicellular organism that carries on gas exchange across its cell surface. Food particles that flow into a specialized region called a gullet are enclosed within food vacuoles (green), where digestion occurs. Molecules leave these vacuoles as they are distributed about the cell by movement of the cytoplasm. **b.** In a sea anemone, a cnidarian, digestion takes place inside the gastrovascular cavity, so named because it (like a vascular system) makes digested material available to the cells that line the cavity. These cells can also acquire oxygen from the watery contents of the cavity and discharge their wastes there. **c.** In a planarian, a flatworm, the gastrovascular cavity branches throughout the body, bringing nutrients to body cells. Diffusion is sufficient to pass molecules to every cell from either the cavity or the exterior surface.

hemocoel. When the heart contracts, openings called ostia (sing., ostium) are closed; when the heart relaxes, the hemolymph is sucked back into the heart by way of the ostia. The hemolymph of a grasshopper is colorless because it does not contain hemoglobin or any other respiratory pigment. It carries nutrients but no oxygen. Oxygen is taken to cells, and carbon dioxide is removed from them by way of air tubes, called tracheae, which are found throughout the body. The tracheae provide efficient transport and delivery of respiratory gases, while at the same time restricting water loss.

A **closed circulatory system** exists in annelids such as earthworms and in some molluscs such as squid and octopuses. Blood, which usually consists of cells and plasma, is pumped by the heart into a system of blood vessels (Fig. 34.2*b*). Valves prevent the backward flow of blood. In the segmented earthworm, five pairs of anterior hearts (aortic arches) pump blood into the ventral blood vessel (an artery), which has a branch

called a lateral vessel in every segment of the worm's body. Blood moves through these branches into capillaries, where exchanges with tissue fluid take place. Blood then moves from small veins into the dorsal blood vessel (a vein). This dorsal blood vessel returns blood to the heart for repumping.

The earthworm has red blood that contains the respiratory pigment hemoglobin. Hemoglobin is dissolved in the blood and is not contained within cells. The earthworm has no specialized boundary, such as lungs, for gas exchange with the external environment. Gas exchange takes place across the body wall, which must always remain moist for this purpose.

Animals with a gastrovascular cavity use this cavity for transport purposes. Other animals have an open or a closed circulatory system.

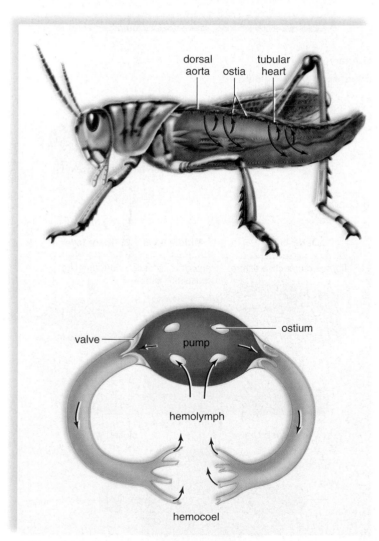

a. Open circulatory system

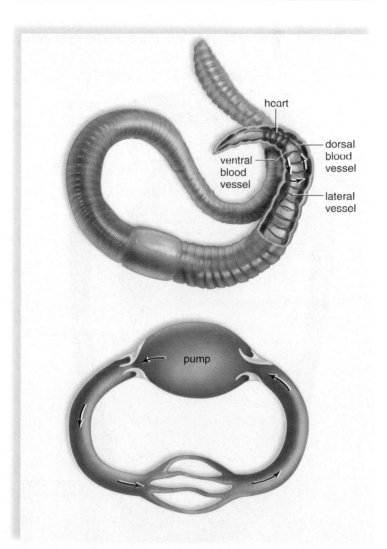

b. Closed circulatory system

FIGURE 34.2 Open versus closed circulatory systems.
a. The grasshopper, an arthropod, has an open circulatory system. A hemocoel is a body cavity filled with hemolymph, which freely bathes the internal organs. The heart, a pump, keeps the hemolymph moving, but this open system probably could not supply oxygen to wing muscles rapidly enough. These muscles receive oxygen directly from tracheae (air tubes). **b.** The earthworm, an annelid, has a closed circulatory system. The dorsal and ventral blood vessels are joined by five pairs of anterior hearts, which pump blood. The lateral vessels distribute blood to the rest of the worm.

34.2 TRANSPORT IN VERTEBRATES

All vertebrate animals have a closed circulatory system, which is called a **cardiovascular system** [Gk. *kardia*, heart; L. *vascular*, vessel]. It consists of a strong, muscular heart in which the atria (sing., atrium) receive blood and the muscular ventricles pump blood through the blood vessels. There are three kinds of blood vessels: **arteries,** which carry blood away from the heart; **capillaries** [L. *capillus*, hair], which exchange materials with tissue fluid; and **veins** [L. *vena*, blood vessel], which return blood to the heart (Fig. 34.3).

An artery or a vein has three distinct layers (Fig. 34.3b, d). The outer layer consists of fibrous connective tissue, which is rich in elastic and collagen fibers. The middle layer is composed of smooth muscle and elastic tissue. The innermost layer, called the endothelium, is similar to squamous epithelium.

Arteries have thick walls, and those attached to the heart are resilient, meaning that they are able to expand and accommodate the sudden increase in blood volume that results after each heartbeat. **Arterioles** are small arteries whose diameter

can be regulated by the nervous system. Arteriole constriction and dilation affect blood pressure in general. The greater the number of vessels dilated, the lower the blood pressure.

Arterioles branch into capillaries, which are extremely narrow, microscopic tubes with a wall composed of only one layer of cells. Capillary beds (many capillaries interconnected)

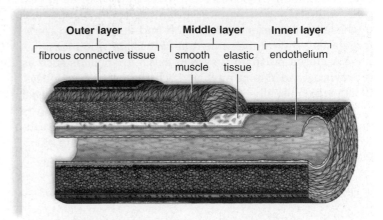

b. Artery

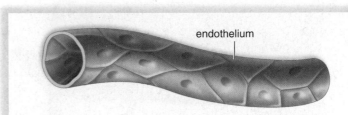

c. Capillary

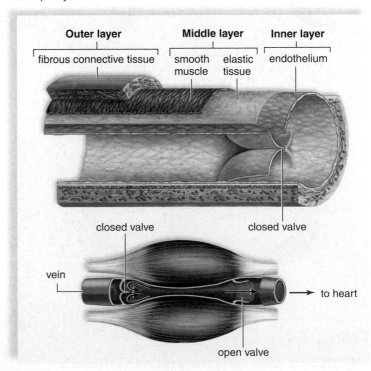

d. Vein

FIGURE 34.3 Transport in birds and mammals.

a. Blood leaving the heart moves from an artery to arterioles to capillaries to venules and then returns to the heart by way of a vein. **b.** Arteries have well-developed walls with a thick middle layer of elastic tissue and smooth muscle. **c.** Capillary walls are only one cell thick. **d.** Veins have flabby walls, particularly because the middle layer is not as thick as in arteries. Veins have valves, which ensure one-way flow of blood back to the heart.

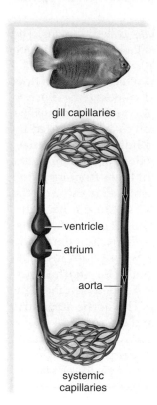

gill capillaries

ventricle

atrium

aorta

systemic
capillaries

a. Fishes

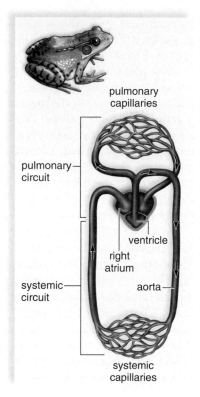

pulmonary
capillaries

pulmonary
circuit

ventricle

right
atrium

systemic
circuit

aorta

systemic
capillaries

b. Amphibians and most reptiles

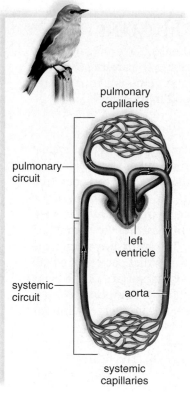

pulmonary
capillaries

pulmonary
circuit

left
ventricle

systemic
circuit

aorta

systemic
capillaries

c. Some reptiles; birds and mammals

FIGURE 34.4

Comparison of circulatory circuits in vertebrates.

a. In fishes, the blood moves in a single circuit. The heart has a single atrium and ventricle and pumps the blood into the gill region, where gas exchange takes place. Blood pressure created by the pumping of the heart is dissipated after the blood passes through the gill capillaries. This is a disadvantage of this one-circuit system. **b.** Amphibians and most reptiles have a two-circuit system in which the heart pumps blood to both the pulmonary capillaries in the lungs and the systemic capillaries in the body itself. Although there is a single ventricle, there is little mixing of O_2-rich and O_2-poor blood. **c.** The pulmonary and systemic circuits are completely separate in crocodiles (a reptile) and in birds and mammals, because the heart is divided by a septum into right and left halves. The right side pumps blood to the lungs, and the left side pumps blood to the body proper.

■ O_2-rich blood
■ O_2-poor blood
■ mixed blood

are so prevalent that in humans, all cells are within 60–80 µm of a capillary. But only about 5% of the capillary beds are open at the same time. After an animal has eaten, precapillary sphincters relax, and the capillary beds in the digestive tract are usually open. During muscular exercise, the capillary beds of the muscles are open. Capillaries, which are usually so narrow that red blood cells pass through in single file, allow exchange of nutrient and waste molecules across their thin walls.

Venules and veins collect blood from the capillary beds and take it to the heart. First, the venules drain the blood from the capillaries, and then they join to form a vein. The wall of a vein is much thinner than that of an artery, and this may be associated with a lower blood pressure in the veins. Valves within the veins point, or open, toward the heart, preventing a backflow of blood when they close (Fig. 34.3d).

Comparison of Circulatory Pathways

Two different types of circulatory pathways are seen among vertebrate animals. In fishes, blood follows a one-circuit (single-loop) circulatory pathway through the body. The heart has a single atrium and a single ventricle (Fig. 34.4a). The pumping action of the ventricle sends blood under pressure to the gills, where gas exchange occurs. After passing through the gills, blood returns to the dorsal aorta, which distributes blood throughout the body. Veins return O_2-poor blood to an enlarged chamber called the sinus venosus that leads to the atrium. The atrium pumps blood back to the ventricle. This single circulatory loop has an advantage in that the gill capillaries receive O_2-poor blood and the capillaries of the body, called systemic capillaries, receive fully O_2-rich blood. It is disadvantageous in that after leaving the gills the blood is under reduced pressure.

As a result of evolutionary changes, the other vertebrates have a two-circuit (double-loop) circulatory pathway. The heart pumps blood to the tissues, called a **systemic circuit,** and also pumps blood to the lungs, called a **pulmonary circuit** [L. *pulmonarius*, of the lungs]. This double pumping action is an adaptation to breathing air on land.

In amphibians, the heart has two atria and a single ventricle (Fig. 34.4b). The sinus venosus collects O_2-poor blood returning via the veins and pumps it to the right atrium. O_2-rich blood returning from the lungs passes to the left atrium. Both of the atria empty into the single ventricle. O_2-rich and O_2-poor blood are kept somewhat separate because O_2-poor blood is pumped out of the ventricle before O_2-rich blood enters. The O_2-rich blood is pumped out of the ventricle for distribution to the body. The O_2-poor blood is delivered to the lungs and, perhaps, to the skin for recharging with oxygen.

In most reptiles, a septum partially divides the ventricle. In these animals, mixing of O_2-rich and O_2-poor blood is kept to a minimum. In crocodilians (alligators and crocodiles), the septum completely separates the ventricle. These reptiles have a four-chambered heart.

In all birds and mammals, as well as crocodilians, the heart is divided into left and right halves (Fig. 34.4c). The right ventricle pumps blood to the lungs, and the larger left ventricle pumps blood to the rest of the body. This arrangement provides adequate blood pressure for both the pulmonary and systemic circuits.

In mammals, birds, and crocodiles, the heart is divided into a right side and a left side; this ensures adequate blood pressure for both the pulmonary and systemic circuits.

34.3 TRANSPORT IN HUMANS

In the cardiovascular system of humans, the pumping of the heart keeps blood moving primarily in the arteries. Skeletal muscle contraction pressing against veins is primarily responsible for the movement of blood in the veins.

The Human Heart

The **heart** is a cone-shaped, muscular organ about the size of a fist (Fig. 34.5). It is located between the lungs directly behind the sternum (breastbone) and is tilted so that the apex (the pointed end) is oriented to the left. The major portion of the heart, called the myocardium, consists largely of cardiac muscle tissue. The muscle fibers of the myocardium are branched and tightly joined to one another. The heart lies within the pericardium, a thick, membranous sac that secretes a small quantity of lubricating liquid. The inner surface of the heart is lined with endocardium, a membrane composed of connective tissue and endothelial tissue. The lining is continuous with the endothelium lining of the blood vessels.

Internally, a wall called the **septum** separates the heart into a right side and a left side (Fig. 34.6). The heart has four chambers. The two upper, thin-walled atria (sing., **atrium**) have wrinkled, protruding appendages called auricles. The two lower chambers are the thick-walled **ventricles,** which pump the blood.

The heart also has four valves, which direct the flow of blood and prevent its backward movement. The two valves that lie between the atria and the ventricles are called the **atrioventricular valves.** These valves are supported by strong fibrous strings called chordae tendineae. The chordae, which are attached to muscular projections of the ventricular walls, support the valves and prevent them from inverting when the heart contracts. The atrioventricular valve on the right side is called the tricuspid valve because it has three flaps, or cusps. The valve on the left side is called the bicuspid (or the mitral) because it has two flaps. The remaining two valves are the **semilunar valves,** whose flaps resemble half-moons, between the ventricles and their attached vessels. The pulmonary semilunar valve lies between the right ventricle and the pulmonary trunk. The aortic semilunar valve lies between the left ventricle and the aorta.

Humans have a four-chambered heart (two atria and two ventricles). A septum separates the right side from the left side.

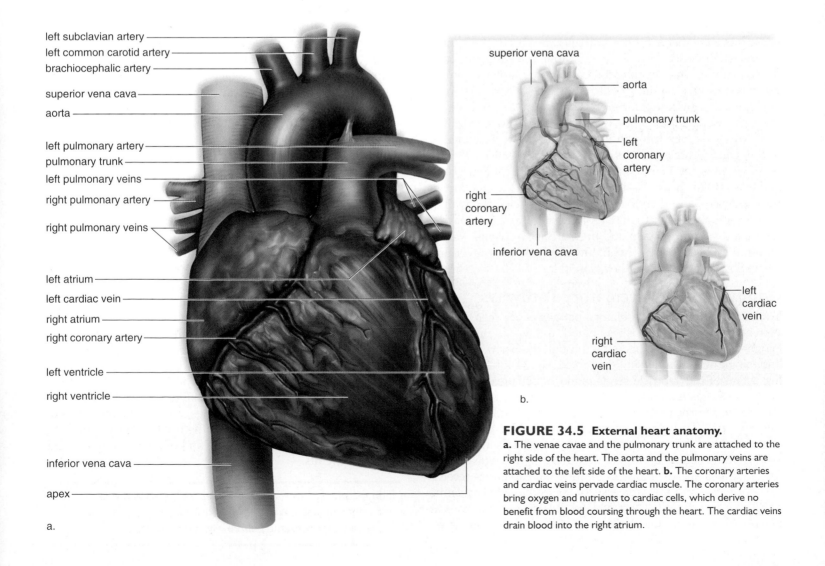

a.

FIGURE 34.5 External heart anatomy.
a. The venae cavae and the pulmonary trunk are attached to the right side of the heart. The aorta and the pulmonary veins are attached to the left side of the heart. **b.** The coronary arteries and cardiac veins pervade cardiac muscle. The coronary arteries bring oxygen and nutrients to cardiac cells, which derive no benefit from blood coursing through the heart. The cardiac veins drain blood into the right atrium.

Path of Blood Through the Heart

By referring to Figure 34.6b, we can trace the path of blood through the heart (Fig. 34.6a) in the following manner:

- The superior vena cava and the inferior vena cava, which carry O_2-poor blood that is relatively high in carbon dioxide, enter the right atrium.
- The right atrium sends blood through an atrioventricular valve (the tricuspid valve) to the right ventricle.
- The right ventricle sends blood through the pulmonary semilunar valve into the pulmonary trunk and the two pulmonary arteries to the lungs.
- Four pulmonary veins, which carry O_2-rich blood, enter the left atrium.
- The left atrium sends blood through an atrioventricular valve (the bicuspid or mitral valve) to the left ventricle.
- The left ventricle sends blood through the aortic semilunar valve into the aorta to the body proper.

From this description, it is obvious that O_2-poor blood never mixes with O_2-rich blood and that blood must go through the lungs in order to pass from the right side to the left side of the heart. In fact, the heart is a double pump because the right ventricle of the heart sends blood into the pulmonary circuit, and the left ventricle sends blood into the systemic circuit. Since the left ventricle has the harder job of pumping blood to the entire body, its walls are thicker than those of the right ventricle, which pumps blood a relatively short distance to the lungs.

The pumping of the heart sends blood out under pressure into the arteries. Because the left side of the heart is the stronger pump, blood pressure is greatest in the aorta. Blood pressure then decreases as the cross-sectional area of arteries and then arterioles increases.

The right side of the heart pumps blood into the pulmonary circuit, and the left side of the heart pumps blood into the systemic circuit.

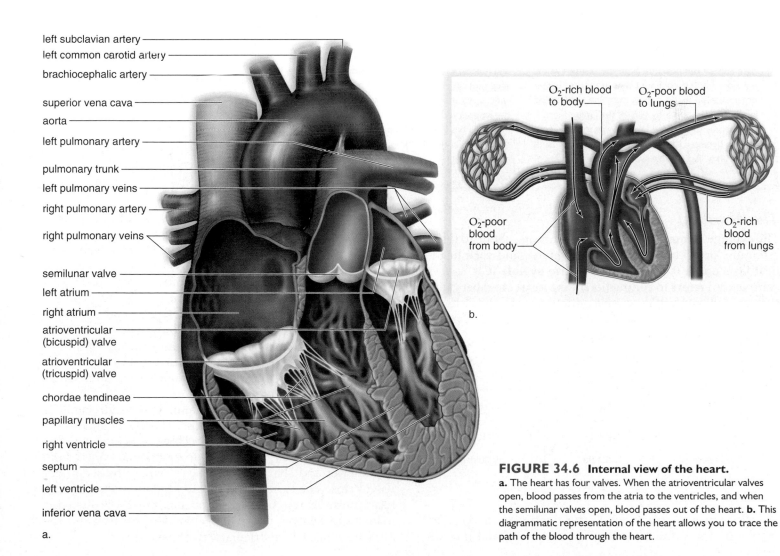

left subclavian artery
left common carotid artery
brachiocephalic artery
superior vena cava
aorta
left pulmonary artery
pulmonary trunk
left pulmonary veins
right pulmonary artery
right pulmonary veins
semilunar valve
left atrium
right atrium
atrioventricular (bicuspid) valve
atrioventricular (tricuspid) valve
chordae tendineae
papillary muscles
right ventricle
septum
left ventricle
inferior vena cava

a.

O_2-rich blood to body
O_2-poor blood to lungs
O_2-poor blood from body
O_2-rich blood from lungs

b.

FIGURE 34.6 Internal view of the heart.
a. The heart has four valves. When the atrioventricular valves open, blood passes from the atria to the ventricles, and when the semilunar valves open, blood passes out of the heart. **b.** This diagrammatic representation of the heart allows you to trace the path of the blood through the heart.

science focus

William Harvey and Circulation of the Blood

William Harvey (1578–1657) was the first to offer data that the blood circulates in the body of humans and other animals. Harvey was an English scientist of the seventeenth century, a time of renewed interest in the collection of facts and the use of the hypothesis, experimentation, and mathematics to arrive at a conclusion. This time period is known as the Scientific Revolution.

After many years of research and study, Harvey hypothesized that the heart is a pump for the entire circulatory system and that blood flows in a circuit. In contrast to former anatomists, Harvey dissected not only dead but also live organisms and observed that when the heart beats, it contracts, forcing blood into the aorta. Had this blood come from the right side of the heart? To do away with the complication of the lungs (pulmonary circulation), Harvey turned to fishes and noticed that the heart first receives and then pumps blood forward. He observed that blood in the mammalian fetus passes directly from the right side of the heart through the septum to the left side. He felt confident that in mature higher organisms, all blood moves from the right to the left side of the heart by way of the lungs.

Harvey then wanted to show an intimate connection between the arteries and veins in the tissues of the body. Again using live organisms, he demonstrated that if an artery is slit, the whole blood system empties, including arteries and veins. He measured the capacity of the left ventricle in humans and found it to be 2 oz. Since the heart beats 70 times a minute, in one hour the left ventricle forces into the aorta no less than $70 \times 60 \times 2 = 8{,}400$ oz $=$ 525 lb of blood, or about three times a man's weight! Since so much blood could not be created and consumed every hour, he reasoned that the same blood must return again and again to the heart.

Harvey also studied the valves in the veins and suggested their true purpose. Using ligatures, he demonstrated that a tight ligature on the arm causes the artery to swell on the side of the heart, and a slack ligature causes the vein to swell on the opposite side (Fig. 34A). He said, "This is an obvious indication the blood passes from the arteries into the veins . . . and there is an anastomosis of the two orders of vessels."

Harvey's methods showed how fruitful research might be done. He established that physical and mechanical evidence could provide data for a theory of circulation. He erred, however, when he speculated that the heart heated the blood, and the lung served to cool it or control the degree of heat. His basic method, however, contributed to the Scientific Revolution and set an example for others to follow.

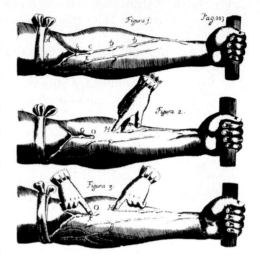

FIGURE 34A Harvey's drawings.
Harvey tied a ligature (Figure 1) above the elbow to observe the accumulation of blood in the veins. Blood can be forced past a valve from H to O (Figure 2), but not in the opposite direction (Figure 3). Therefore, Harvey concluded that venous blood ordinarily moves toward the heart.

The Heartbeat

The average human heart contracts, or beats, about 70 times a minute, or 2.5 billion times in a lifetime, and each heartbeat lasts about 0.85 second. The term **systole** [Gk. *systole,* contraction] refers to contraction of the heart chambers, and the word **diastole** [Gk. *diastole,* dilation, spreading] refers to relaxation of these chambers. Each heartbeat, or **cardiac cycle,** consists of the following phases:

Cardiac Cycle		
Time	Atria	Ventricles
0.15 sec	Systole	Diastole
0.30 sec	Diastole	Systole
0.40 sec	Diastole	Diastole

First the atria contract (while the ventricles relax), then the ventricles contract (while the atria relax), and then all chambers rest. Note that the heart is in diastole about 50% of the time. The short systole of the atria is appropriate since the atria send blood only into the ventricles. It is the muscular ventricles that actually pump blood out into the cardiovascular system proper. When the word *systole* is used alone, it usually refers to the left ventricular systole. The volume of blood that the left ventricle pumps per minute into the systemic circuit is called the **cardiac output.** A person with a heartbeat of 70 beats per minute has a cardiac output of 5.25 liters a minute. This is almost equivalent to the amount of blood in the body. During heavy exercise, the cardiac output can increase manyfold.

When the heart beats, the familiar lub-dub sound is heard as the valves of the heart close. The longer and lower-pitched *lub* is caused by vibrations of the heart when the atrioventricular valves close due to ventricular contraction. The shorter and sharper *dub* is heard when the semilunar valves close due to back pressure of blood in the arteries. A heart murmur, a slight slush sound after the lub, is often due to ineffective valves, which allow blood to pass back into the atria after the atrioventricular valves have closed.

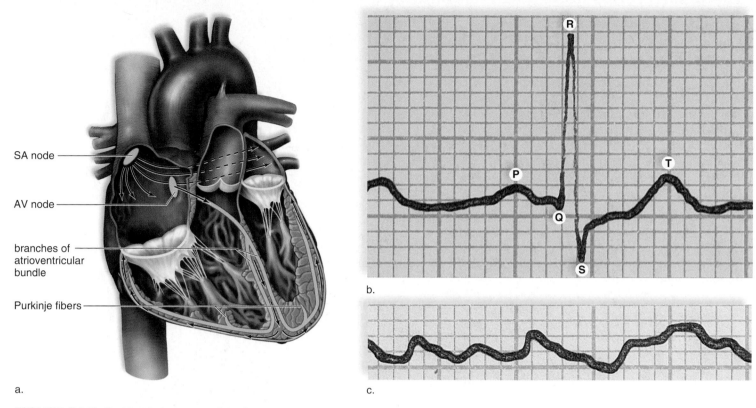

SA node

AV node

branches of atrioventricular bundle

Purkinje fibers

a.

b.

c.

FIGURE 34.7   Conduction system of the heart.
a. The SA node sends out a stimulus (black arrows), which causes the atria to contract. When this stimulus reaches the AV node, it signals the ventricles to contract. Impulses pass down the two branches of the atrioventricular bundle to the Purkinje fibers, and thereafter the ventricles contract. **b.** A normal ECG usually indicates that the heart is functioning properly. The P wave occurs just prior to atrial contraction; the QRS complex occurs just prior to ventricular contraction; and the T wave occurs when the ventricles are recovering from contraction. **c.** Ventricular fibrillation produces an irregular electrocardiogram due to irregular stimulation of the ventricles.

The **pulse** is a wave effect that passes down the walls of the arterial blood vessels when the aorta expands and then recoils following ventricular systole. Because there is one arterial pulse per ventricular systole, the arterial pulse rate can be used to determine the heart rate.

The rhythmic contraction of the heart is due to the **cardiac conduction system.** Nodal tissue, which has both muscular and nervous characteristics, is a unique type of cardiac muscle located in two regions of the heart. The SA (sinoatrial) node is found in the upper dorsal wall of the right atrium; the AV (atrioventricular) node is found in the base of the right atrium very near the septum (Fig. 34.7a). The SA node initiates the heartbeat and every 0.85 second automatically sends out an excitation impulse, which causes the atria to contract. Therefore, the SA node is called the **cardiac pacemaker** because it usually keeps the heartbeat regular. When the impulse reaches the AV node, the AV node signals the ventricles to contract by way of large fibers terminating in the more numerous and smaller Purkinje fibers. Although the beat of the heart is intrinsic, it is regulated by the nervous system, which can increase or decrease the heartbeat rate.

An **electrocardiogram (ECG)** is a recording of the electrical changes that occur in myocardium during a cardiac cycle. Body fluids contain ions that conduct electrical currents, and therefore the electrical changes in myocardium can be detected on the skin's surface. When an electrocardiogram is being taken, electrodes placed on the skin are connected by wires to an instrument that detects the myocardium's electrical changes. Thereafter, a pattern appears that reflects the contractions of the heart. Figure 34.7b depicts the pattern that results from a normal cardiac cycle.

When the SA node triggers an impulse, the atrial fibers produce an electrical change called the P wave. The P wave indicates that the atria are about to contract. After that, the QRS complex signals that the ventricles are about to contract and the atria are relaxing. The electrical changes that occur as the ventricular muscle fibers recover produce the T wave.

Various types of abnormalities can be detected by an electrocardiogram. One of these, called ventricular fibrillation, is caused by uncoordinated contraction of the ventricles (Fig. 34.7c). Ventricular fibrillation is of special interest because it can be caused by an injury or drug overdose. It is the most common cause of sudden cardiac death in a seemingly healthy person. Once the ventricles are fibrillating, they have to be defibrillated by applying a strong electric current for a short period of time. Then the SA node may be able to reestablish a coordinated beat.

The conduction system of the heart includes the SA node, the AV node, and the Purkinje fibers. An ECG helps determine if the conduction system, and therefore the heartbeat, is regular.

Vascular Pathways

The human cardiovascular system includes two major circular pathways, the pulmonary circuit and the systemic circuit (Fig. 34.8).

The Pulmonary Circuit

In the pulmonary circuit, the path of blood can be traced as follows. O_2-poor blood from all regions of the body collects in the right atrium and then passes into the right ventricle, which pumps it into the pulmonary trunk. The pulmonary trunk divides into the right and left pulmonary arteries, which carry blood to the lungs. As blood passes through pulmonary capillaries, carbon dioxide is given off and oxygen is picked up. O_2-rich blood returns to the left atrium of the heart, through pulmonary venules that join to form pulmonary veins.

The Systemic Circuit

The **aorta** [L. *aorte*, great artery] and the **venae cavae** (sing., vena cava) [L. *vena*, blood vessel, and *cavus*, hollow] are the major blood vessels in the systemic circuit. To trace the path of blood to any organ in the body, you need only start with the left ventricle, mention the aorta, the proper branch of the aorta, the organ, and the vein returning blood to the vena cava, which enters the right atrium. In the systemic circuit, arteries contain O_2-rich blood and have a bright red color, but veins contain O_2-poor blood and appear dull red or, when viewed through the skin, blue.

The coronary arteries are extremely important because they serve the heart muscle itself (see Fig. 34.5). (The heart is not nourished by the blood in its chambers.) The coronary arteries arise from the aorta just above the aortic semilunar valve. They lie on the exterior surface of the heart, where they branch into arterioles and then capillaries. In the capillary beds, nutrients, wastes, and gases are exchanged between the blood and the tissues. The capillary beds enter venules, which join to form the cardiac veins, and these empty into the right atrium.

A **portal system** [L. *porto*, carry, transport] is one that begins and ends in capillaries. The hepatic portal system takes blood from the intestines to the liver. The liver, an organ of homeostasis, modifies substances absorbed by the intestines, removes toxins and bacteria picked up from the intestines, and monitors the normal composition of the blood. Blood leaves the liver by way of the hepatic vein, which enters the inferior vena cava.

The pulmonary circuit takes O_2-poor blood to the lungs and returns O_2-rich blood to the heart. The systemic circuit takes blood throughout the body from the heart via the aorta and back to the heart via the venae cavae.

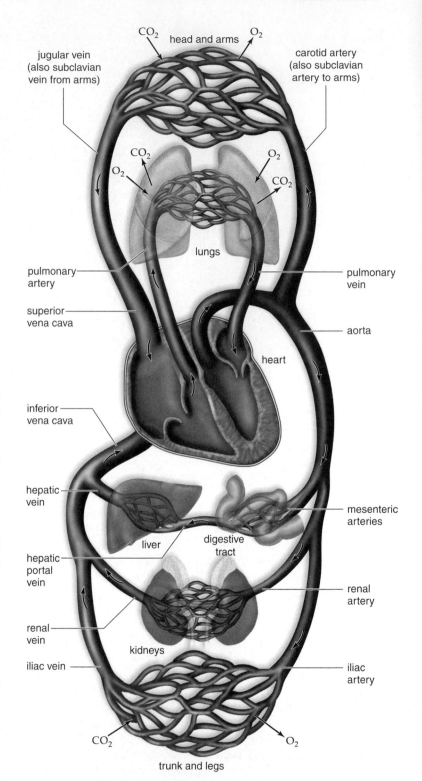

FIGURE 34.8 Path of blood.
When tracing blood from the right to the left side of the heart in the pulmonary circuit, you must mention the pulmonary vessels. When tracing blood from the digestive tract to the right atrium in the systemic circuit, you must mention the hepatic portal vein, the hepatic vein, and the inferior vena cava. The blue-colored vessels carry O_2-poor blood, and the red-colored vessels carry O_2-rich blood; the arrows indicate the flow of blood.

Blood Pressure

When the left ventricle contracts, blood is forced into the aorta and then other systemic arteries under pressure. Systolic pressure results from blood being forced into the arteries during ventricular systole, and diastolic pressure is the pressure in the arteries during ventricular diastole. Human **blood pressure** can be measured with a **sphygmomanometer,** which has a pressure cuff that determines the amount of pressure required to stop the flow of blood through an artery. Blood pressure is normally measured on the brachial artery, an artery in the upper arm. Today, digital manometers are often used to take one's blood pressure instead. Blood pressure is given in millimeters of mercury (mm Hg). A blood pressure reading consists of two numbers—for example, 120/80—that represent systolic and diastolic pressures, respectively.

As blood flows from the aorta into the various arteries and arterioles, blood pressure falls. Also, the difference between systolic and diastolic pressure gradually diminishes. In the capillaries, there is a slow, fairly even flow of blood. This may be related to the very high total cross-sectional area of the capillaries (Fig. 34.9). It has been calculated that if all the blood vessels in a human were

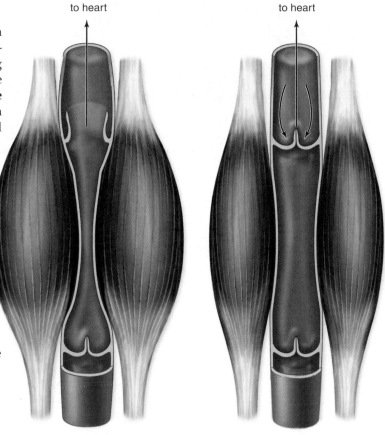

a. Contracted skeletal muscle pushes blood past open valve.

b. Closed valve prevents backward flow of blood.

FIGURE 34.10 Cross section of a valve in a vein.
a. Pressure on the walls of a vein, exerted by skeletal muscles, increases blood pressure within the vein and forces a valve open. **b.** When external pressure is no longer applied to the vein, blood pressure decreases, and back pressure forces the valve closed. Closure of the valves prevents the blood from flowing in the opposite direction.

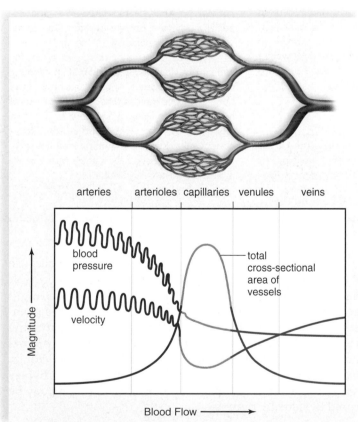

FIGURE 34.9 Velocity and blood pressure related to vascular cross-sectional area.
In capillaries, blood is under minimal pressure and has the least velocity. Blood pressure and velocity drop off because capillaries have a greater total cross-sectional area than arterioles.

connected end to end, the total distance would reach around the Earth at the equator two times. A large portion of this distance would be due to the quantity of capillaries.

Blood pressure in the veins is low and cannot move blood back to the heart, especially from the limbs of the body. When skeletal muscles near veins contract, they put pressure on the collapsible walls of the veins and on blood contained in these vessels. Veins, however, have valves that prevent the backward flow of blood, and therefore pressure from muscle contraction is sufficient to move blood through veins toward the heart (Fig. 34.10). Varicose veins, abnormal dilations in superficial veins, develop when the valves of the veins become weak and ineffective due to a backward pressure of the blood. Varicose veins of the anal canal are known as hemorrhoids.

The beat of the heart supplies the pressure that keeps blood moving in the arteries, and skeletal muscle contraction pushes blood in the veins toward the heart. Veins have valves that prevent the backward flow of blood.

health focus

Prevention of Cardiovascular Disease

All of us can take steps to prevent cardio-vascular disease, the most frequent cause of death in the United States. Certain genetic factors predispose an individual to cardiovascular disease, such as family history of heart attack under age 55, male gender, and ethnicity (African Americans are at greater risk). People with one or more of these risk factors need not despair, however. It means only that they should pay particular attention to the following guidelines for a heart-healthy lifestyle.

The Don'ts

Smoking

Hypertension is well recognized as a major contributor to cardiovascular disease. When a person smokes, the drug nicotine, present in cigarette smoke, enters the bloodstream. Nicotine causes the arterioles to constrict and the blood pressure to rise. Restricted blood flow and cold hands are associated with smoking in most people. More serious is the need for the heart to pump harder to propel the blood through the lungs at a time when the oxygen-carrying capacity of the blood is reduced.

Drug Abuse

Stimulants, such as cocaine and amphetamines, can cause an irregular heartbeat and lead to heart attacks and strokes in people who are using drugs even for the first time. Intravenous drug use may result in a cerebral embolism.

Too much alcohol can destroy just about every organ in the body, the heart included. But investigators have discovered that people who take an occasional drink have a 20% lower risk of heart disease than do teetotalers. Two to four drinks a week is the recommended limit for men; one to three drinks for women.

Weight Gain

Hypertension is prevalent in persons who are more than 20% above the recommended weight for their height. In those who are overweight, more tissues require servicing, and the heart sends the extra blood out under greater pressure. It may be harder to lose weight once it is gained, and therefore it is recommended that weight control be a lifelong endeavor. Even a slight decrease in weight can bring with it a reduction in hypertension. A 4.5 kg weight loss doubles the chance that blood pressure can be normalized without drugs.

The Dos

Healthy Diet

Diet influences the amount of cholesterol in the blood. Cholesterol is ferried by two types of plasma proteins, called LDL (low-density lipo-protein) and HDL (high-density lipoprotein). LDL (called "bad" lipoprotein) takes cholesterol from the liver to the tissues, and HDL (called "good" lipoprotein) transports cholesterol out of the tissues to the liver. When the LDL level in blood is high or the HDL level is abnormally low, plaque, which interferes with circulation, accumulates on arterial walls (Fig. 34B).

Eating foods high in saturated fat (red meat, cream, and butter) and foods containing so-called trans-fats (most margarines, commercially baked goods, and deep-fried foods) raises the LDL-cholesterol level. Replacement of these harmful fats with healthier ones, such as monounsaturated fats (olive and canola oils) and polyunsaturated fats (corn, safflower, and soybean oils), is recommended. Cold water fish (e.g., halibut, sardines, tuna, and salmon) contain polyunsaturated fatty acids and especially omega-3 polyunsaturated fatty acids, which can reduce plaque.

Evidence is mounting to suggest a role for antioxidant vitamins (A, E, and C) in preventing cardiovascular disease. Antioxidants protect the body from free radicals that oxidize cholesterol and damage the lining of an artery, leading to a blood clot that can block blood vessels. Nutritionists believe that consuming at least five servings of fruits and vegetables a day may protect against cardiovascular disease.

34.4 CARDIOVASCULAR DISORDERS

Cardiovascular disease (CVD) is the leading cause of untimely death in the Western countries. Modern research efforts have resulted in improved diagnosis, treatment, and prevention. This section discusses the advances that have been made in these areas. The Health Focus on this page emphasizes how to prevent CVD from developing in the first place.

Hypertension

It is estimated that about 20% of all Americans suffer from hypertension, which is high blood pressure. Women of any age are considered to have hypertension if their blood pressure reading is 160/95 or above. For a man under age 45, a reading above 130/90 is hypertensive, and beyond age 45, a reading above 140/95 is hypertensive. While both systolic and diastolic pressures are important, it is the diastolic pressure that is emphasized when medical treatment is being considered.

Hypertension is sometimes called a silent killer because it may not be detected until a stroke or heart attack occurs. It has long been thought that a certain genetic makeup might account for the development of hypertension. Now researchers have discovered two genes that may be involved in some individuals. One gene codes for angiotensinogen, a plasma protein that is converted to a powerful vasoconstrictor in part by the product of the second gene. Persons with hypertension due to overactivity of these genes might one day be cured by gene therapy.

Atherosclerosis

Hypertension is also seen in individuals who have atherosclerosis (formerly called arteriosclerosis), an accumulation of soft masses of fatty materials, particularly cholesterol, beneath the inner linings of arteries (see Fig. 34B). Such deposits are called plaque. As deposits occur, plaque tends

Cholesterol Profile

Starting at age 20, all adults are advised to have their cholesterol levels tested at least every five years. Even in healthy individuals, an LDL level above 160 mg/100 ml and an HDL level below 40 mg/100 ml are matters of concern. If a person has heart disease or is at risk for heart disease, an LDL level below 100 mg/100 ml is now recommended. Medications will most likely be prescribed for individuals who do not meet these minimum guidelines.

Exercise

People who exercise are less apt to have cardiovascular disease. One study found that moderately active men who spent an average of 48 minutes a day on a leisure-time activity such as gardening, bowling, or dancing had one-third fewer heart attacks than peers who spent an average of only 16 minutes each day being active. Exercise helps keep weight under control, may help minimize stress, and reduces hypertension.

The heart beats faster when exercising, but exercise slowly increases its capacity. This means that the heart can beat slower when we are at rest and still do the same amount of work. One physician recommends that his cardiovascular patients walk for one hour, three times a week, and, in addition, practice meditation and yogalike stretching and breathing exercises to reduce stress.

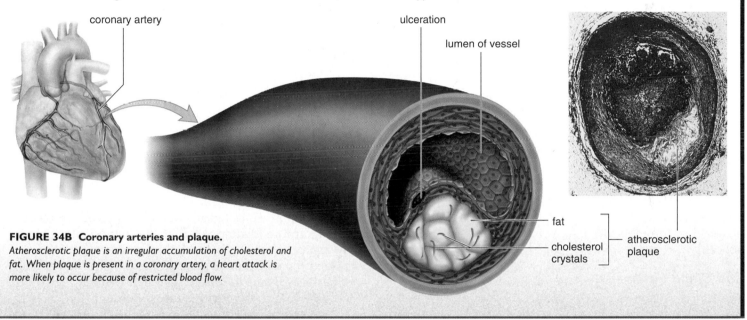

FIGURE 34B   Coronary arteries and plaque.
Atherosclerotic plaque is an irregular accumulation of cholesterol and fat. When plaque is present in a coronary artery, a heart attack is more likely to occur because of restricted blood flow.

to protrude into the lumen of the vessel, interfering with the flow of blood. Atherosclerosis begins in early adulthood and develops progressively through middle age, but symptoms may not appear until an individual is 50 or older. To prevent its onset and development, the American Heart Association and other organizations recommend a diet low in saturated fat and cholesterol and rich in fruits and vegetables.

Plaque can cause a clot to form on the irregular arterial wall. As long as the clot remains stationary, it is called a thrombus, but when and if it dislodges and moves along with the blood, it is called an embolus. If thromboembolism is not treated, complications can arise (see the following section).

In certain families, atherosclerosis is due to an inherited condition such as familial hypercholesterolemia. The presence of the disease-associated mutation can be detected, and this information is helpful if measures are taken to prevent the occurrence of the disease.

Stroke and Heart Attack

Stroke, heart attack, and aneurysm are associated with hypertension and atherosclerosis. A cardiovascular accident, also called a **stroke,** often results when a small cranial arteriole bursts or is blocked by an embolus. A lack of oxygen causes a portion of the brain to die, and paralysis or death can result. A person is sometimes forewarned of a stroke by a feeling of numbness in the hands or the face, difficulty in speaking, or temporary blindness in one eye.

When a coronary artery is completely blocked, perhaps because of thromboembolism, a portion of the heart muscle dies due to a lack of oxygen. This is a myocardial infarction, also called a **heart attack.** If a coronary artery becomes partially blocked, the individual may suffer from **angina pectoris,** characterized as a squeezing sensation or a flash of burning. Nitroglycerin or related drugs dilate blood vessels and help relieve the pain.

34.5 Blood, a Transport Medium

The blood of mammals has numerous functions that help maintain homeostasis. Blood (1) transports gases, nutrients, waste products, and hormones throughout the body; (2) helps destroy pathogenic microorganisms; (3) distributes antibodies that are important in immunity; (4) aids in maintaining water balance and pH; (5) helps regulate body temperature; and (6) carries platelets and factors that ensure clotting and prevent blood loss.

In humans, blood has two main portions: the liquid portion, called plasma, and the formed elements, consisting of various cells and platelets (Fig. 34.11). **Plasma** [Gk. *plasma*, something molded] contains many types of molecules, including nutrients, wastes, salts, and proteins. The salts and proteins are involved in buffering the blood, effectively keeping the pH near 7.4. They also maintain the blood's osmotic pressure so that water has an automatic tendency to enter blood capillaries. Several plasma proteins are involved in blood clotting, and others transport large organic molecules in the blood. Albumin, the most plentiful of the plasma proteins, transports bilirubin, a breakdown product of hemoglobin. Globulins have various functions; among the globulins are the lipoproteins, which transport cholesterol.

Formed Elements

The formed elements are of three types: red blood cells, or erythrocytes [Gk. *erythros*, red, and *kytos*, cell]; white blood cells, or leukocytes [Gk. *leukos*, white, and *kytos*, cell]; and platelets, or thrombocytes [Gk. *thrombos*, blood clot, and *kytos*, cell].

Red Blood Cells

Red blood cells are small, biconcave disks that at maturity lack a nucleus and contain the respiratory pigment hemoglobin. Approximately 25 trillion red blood cells exist in an average adult. There are 6 million red blood cells per cubic millimeter (mm^3) of whole blood, and each one of these cells contains about 250 million hemoglobin molecules. **Hemoglobin** [Gk. *haima*, blood; L. *globus*, ball] contains four globin protein chains, each associated with heme, an iron-containing group. Iron combines loosely with oxygen, and in this way oxygen is carried in the blood. If there is an insufficient number of red blood cells, or if the cells do not have enough hemoglobin, the individual suffers from anemia and has a tired, run-down feeling.

Red blood cells are manufactured continuously in the red bone marrow of the skull, the ribs, the vertebrae, and the ends of the long bones. The hormone erythropoietin is produced by the kidneys when it acts on a precursor made by the liver. Erythropoietin stimulates the production of red blood cells. Now available as a drug, erythropoietin is helpful to persons with anemia and is also sometimes abused by athletes who want to enhance their performance.

Before they are released from the bone marrow into blood, red blood cells lose their nuclei and synthesize hemoglobin. After living about 120 days, they are destroyed chiefly in the liver and the spleen, where they are engulfed by large phagocytic cells. When red blood cells are destroyed, hemoglobin is released. The iron is recovered and returned to the red bone marrow for reuse. The heme portions of the molecules undergo chemical degradation and are excreted by the liver as bile pigments in the bile. The bile pigments are primarily responsible for the color of feces.

White Blood Cells

White blood cells differ from red blood cells in that they are usually larger and have a nucleus, they lack hemoglobin, and, without staining, they appear translucent. With staining, white blood cells appear light blue unless they have granules that bind with certain stains. The following are granular leukocytes with a lobed nucleus: neutrophils, which have granules that stain slightly pink; eosinophils, which have granules that take up the red dye eosin; and basophils, which have granules that take up a basic dye, staining them a deep blue. The agranular leukocytes with no granules and a circular or indented nucleus are the larger monocytes and the smaller lymphocytes. There are approximately 5,000–11,000 white blood cells per mm^3. Stem cell growth factor can be used to increase the production of all white blood cells, and various other factors are also available to stimulate the production of specific stem cells. These growth factors are helpful to people with low immunity, such as AIDS patients.

When microorganisms enter the body due to an injury, the response is called an inflammatory reaction because swelling and reddening occur at the injured site. Cells in the vicinity release substances that cause vasodilation (increase in the diameter of a vessel), and increased capillary permeability. **Neutrophils** [Gk. *neuter*, neither, and *phileo*, love], which are amoeboid, squeeze through the capillary wall and enter the tissue fluid, where they phagocytize foreign material. **Monocytes** appear and are transformed into **macrophages** [Gk. *makros*, long, and *phagein*, to eat], which are large phagocytizing cells that release white blood cell growth factors. Soon there is an explosive increase in the number of leukocytes. The thick, yellowish fluid called pus contains a large proportion of dead white blood cells that have fought the infection.

Lymphocytes [L. *lympha*, clear water; Gk. *kytos*, cell] also play an important role in fighting infection. Certain lymphocytes called T cells attack infected cells that contain viruses. Other lymphocytes called B cells produce antibodies. Each B cell produces just one type of antibody, which is specific for one type of antigen. An **antigen** [Gk. *anti*, against; L. *genitus*, forming, causing], which is most often a protein but sometimes a polysaccharide, causes the body to produce an antibody because the antigen doesn't belong to the body. Antigens are present in the outer covering of parasites or in their toxins. When **antibodies** [Gk. *anti*, against] combine with antigens, the complex is often phagocytized

FIGURE 34.11
Composition of blood.
a. When blood is transferred to a test tube containing an anticoagulant to prevent clotting and then centrifuged, it consists of three layers. The transparent straw-colored or yellow top layer is the plasma, the liquid portion of the blood. The thin, middle buffy coat layer consists of leukocytes and platelets. The bottom layer contains the erythrocytes.
b. Micrograph of the formed elements, which are also listed and drawn in (**c**).

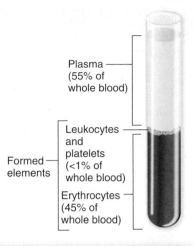

Plasma
(55% of
whole blood)

Leukocytes
and
platelets
(<1% of
whole blood)

Formed
elements

Erythrocytes
(45% of
whole blood)

Plasma		
Type	**Function**	**Source**
Water (90–92% of plasma)	Maintains blood volume; transports molecules	Absorbed from intestine
Plasma proteins (7–8% of plasma)	Maintain blood osmotic pressure and pH	
Albumin	Maintains blood volume and pressure	Liver
Globulins	Transport; fight infection	
Fibrinogen	Clotting	
Salts (less than 1% of plasma)	Maintain blood osmotic pressure and pH; aid metabolism	Absorbed from intestine
Gases		
Oxygen	Cellular respiration	Lungs
Carbon dioxide	End product of metabolism	Tissues
Nutrients		
Lipids Glucose Amino acids	Food for cells	Absorbed from intestine
Nitrogenous wastes	Excretion by kidneys	Liver
Urea Uric acid		
Other		
Hormones, vitamins, etc.	Aid metabolism	Varied

a.

platelets

neutrophils

red blood
cell

eosinophil

monocyte

basophil

lymphocyte

b.   250x

Formed Elements		
Type	**Function and Description**	**Source**
Red blood cells (erythrocytes)	Transport O_2 and help transport CO_2	Red bone marrow
4 million–6 million per mm³ blood	7–8 µm in diameter Bright-red to dark-purple biconcave disks without nuclei	
White blood cells (leukocytes) 5,000–11,000 per mm³ blood	Fight infection	Red bone marrow
Neutrophils ★ 40–70%	10–14 µm in diameter Spherical cells with multilobed nuclei; fine, pink granules in cytoplasm; phagocytize pathogens	
Eosinophils ★ 1–4%	10–14 µm in diameter Spherical cells with bilobed nuclei; coarse, deep-red, uniformly sized granules in cytoplasm; phagocytize antigen-antibody complexes and allergens	
Basophils ★ 0–1%	10–12 µm in diameter Spherical cells with lobed nuclei; large, irregularly shaped, deep-blue granules in cytoplasm; release histamine, which promotes blood flow to injured tissues	
Lymphocytes ★ 20–45%	5–17 µm in diameter (average 9–10 µm) Spherical cells with large, round nuclei; responsible for specific immunity	
Monocytes ★ 4–8%	10–24 µm in diameter Large spherical cells with kidney-shaped, round, or lobed nuclei; become macrophages that phagocytize pathogens and cellular debris	
Platelets (thrombocytes)	Aid clotting	Red bone marrow
150,000–300,000 per mm³ blood	2–4 µm in diameter Disk-shaped cell fragments with no nuclei; purple granules in cytoplasm	

Granular leukocytes

Agranular leukocytes

c.   ★ Appearance with Wright's stain.

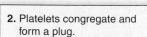

1. Blood vessel is punctured.

2. Platelets congregate and form a plug.

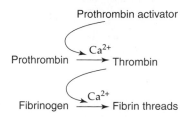

3. Platelets and damaged tissue cells release prothrombin activator, which initiates a cascade of enzymatic reactions.

Prothrombin activator

$$\text{Prothrombin} \xrightarrow{\text{Ca}^{2+}} \text{Thrombin}$$

$$\text{Fibrinogen} \xrightarrow{\text{Ca}^{2+}} \text{Fibrin threads}$$

4. Fibrin threads form and trap red blood cells.

a. Blood-clotting process

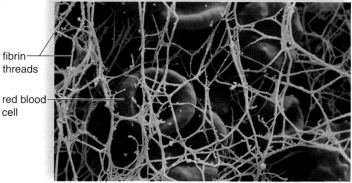

fibrin threads

red blood cell

b. Blood clot   4,400×

FIGURE 34.12 Blood clotting.

a. Platelets and damaged tissue cells release prothrombin activator, which acts on prothrombin in the presence of Ca^{2+} (calcium ions) to produce thrombin. Thrombin acts on fibrinogen in the presence of Ca^{2+} to form fibrin threads. **b.** A scanning electron micrograph of a blood clot shows red blood cells caught in the fibrin threads.

by a macrophage. An individual is actively immune when a large number of B cells are all producing the specific antibody needed for a particular infection.

The **eosinophils** are granular leukocytes that are important in releasing enzymes used in fighting parasites and destroying allergens. **Basophils** are the least common leukocyte. They contain the anticoagulant heparin, which prevents blood from clotting too quickly. Basophils also contain the vasodilator histamine, which promotes blood flow to tissues.

Platelets

Platelets (thrombocytes) result from fragmentation of certain large cells, called megakaryocytes, in the red bone mar-

TABLE 34.1	
Body Fluids	
Name	*Composition*
Blood	Formed elements and plasma
Plasma	Liquid portion of blood
Serum	Plasma minus fibrinogen
Tissue fluid	Plasma minus most proteins
Lymph	Tissue fluid within lymphatic vessels

row. Platelets are produced at a rate of 200 billion a day, and the blood contains 150,000–300,000 per mm³. These formed elements are involved in blood clotting, or coagulation.

There are at least 12 clotting factors in the blood that participate in the formation of a blood clot. Hemophilia is an inherited clotting disorder in which the liver is unable to produce one of the clotting factors. The slightest bump can cause the affected person to bleed into the joints, and this leads to degeneration of the joints. Bleeding into muscles can lead to nerve damage and muscular atrophy. The most frequent cause of death due to hemophilia is bleeding into the brain.

We will discuss the roles played by platelets, prothrombin, and fibrinogen in blood clotting. Fibrinogen and prothrombin are proteins manufactured and deposited in blood by the liver. Vitamin K, found in green vegetables and also formed by intestinal bacteria, is necessary for the production of prothrombin, and if by chance this vitamin is missing from the diet, hemorrhagic disorders can also develop.

Blood Clotting. When a blood vessel in the body is damaged, platelets clump at the site of the puncture and partially seal the leak. Platelets and the injured tissues release a clotting factor called prothrombin activator that converts prothrombin to thrombin. This reaction requires calcium ions (Ca^{2+}). **Thrombin,** in turn, acts as an enzyme that severs two short amino acid chains from each fibrinogen molecule. These activated fragments then join end to end, forming long threads of fibrin. Fibrin threads wind around the platelet plug in the damaged area of the blood vessel and provide the framework for the clot. Red blood cells also are trapped within the fibrin threads; these cells make a clot appear red (Fig. 34.12). A fibrin clot is present only temporarily. As soon as blood vessel repair is initiated, an enzyme called plasmin destroys the fibrin network and restores the fluidity of plasma.

If blood is allowed to clot in a test tube, a yellowish fluid develops above the clotted material. This fluid is called **serum,** and it contains all the components of plasma except fibrinogen. Table 34.1 lists the terms used to refer to various body fluids related to blood.

A blood clot consists of platelets and red blood cells entangled within fibrin threads.

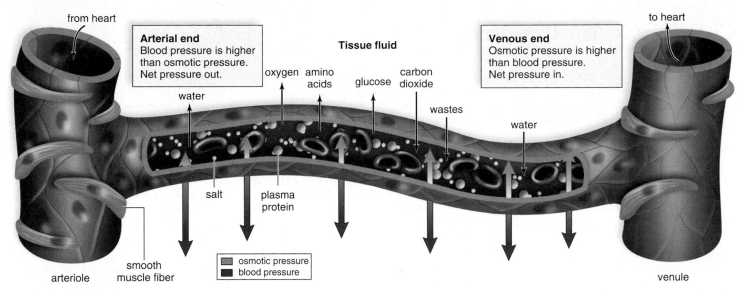

FIGURE 34.13 Capillary exchange.
A capillary, illustrating the exchanges that take place and the forces that aid the process. At the arterial end of a capillary, the blood pressure is higher than the osmotic pressure; therefore, water (H_2O) tends to leave the bloodstream. In the midsection, molecules, including oxygen (O_2) and carbon dioxide (CO_2), follow their concentration gradients. At the venous end of a capillary, the osmotic pressure is higher than the blood pressure; therefore, water tends to enter the bloodstream. Notice that the red blood cells and the plasma proteins are too large to exit a capillary.

Capillary Exchange

Two forces primarily control movement of fluid through the capillary wall: osmotic pressure, which tends to cause water to move from tissue fluid to blood, and blood pressure, which tends to cause water to move in the opposite direction. At the arterial end of a capillary, blood pressure (30 mm Hg) is higher than the osmotic pressure of blood (21 mm Hg) (Fig. 34.13). Osmotic pressure is created by the presence of salts and the plasma proteins. Because blood pressure is higher than osmotic pressure at the arterial end of a capillary, water exits a capillary at this end.

Midway along the capillary, where blood pressure is lower, the two forces essentially cancel each other, and there is no net movement of water. Solutes now diffuse according to their concentration gradient: Oxygen and nutrients (glucose and amino acids) diffuse out of the capillary; carbon dioxide and wastes diffuse into the capillary. Red blood cells and almost all plasma proteins remain in the capillaries. The substances that leave a capillary contribute to **tissue fluid,** the fluid between the body's cells. Since plasma proteins are too large to readily pass out of the capillary, tissue fluid tends to contain all components of plasma except much lesser amounts of protein.

At the venule end of a capillary, where blood pressure has fallen even more, osmotic pressure is greater than blood pressure, and water tends to move into the capillary. Almost the same amount of fluid that left the capillary returns to it, although some excess tissue fluid is always collected by the lymphatic capillaries (Fig. 34.14). Tissue fluid contained within lymphatic vessels is called **lymph.** Lymph is returned to the systemic venous blood when the major lymphatic vessels enter the subclavian veins in the shoulder region.

Not all capillary beds are open at the same time. When the precapillary sphincters (circular muscles) shown in Figure 34.14 are relaxed, the capillary bed is open and blood flows through

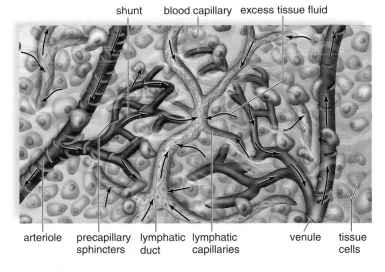

FIGURE 34.14 Capillary bed.
A lymphatic capillary bed lies near a blood capillary bed. When lymphatic capillaries take up excess tissue fluid, it becomes lymph. Precapillary sphincters can shut down a blood capillary, and blood then flows through the shunt.

the capillaries. When precapillary sphincters are contracted, blood flows through a shunt that carries blood directly from an arteriole to a venule. In addition to nutrients and wastes, the blood distributes heat to body parts. When you are warm, many capillaries that serve the skin are open, and your face is flushed. This helps rid the body of excess heat. When you are cold, skin capillaries close, conserving heat, and your skin turns blue.

Oxygen and nutrients exit a capillary and carbon dioxide and waste molecules enter a capillary midway between the arterial and venous ends.

CONNECTING THE CONCEPTS

Small aquatic animals with no circulatory system may rely on the passage of external water through a gastrovascular cavity to service cells. Roundworms and other pseudocoelomates use a fluid-filled body cavity as a means of transporting substances. A fluid-filled cavity, whether it is a gastrovascular cavity, a pseudocoelom, or a coelom as in earthworms, can act as a hydrostatic skeleton (see Chapter 41). Unlike a hard endo- or exoskeleton, a hydrostatic skeleton doesn't take much energy to carry around. Even animals that have a rigid skeleton may still rely on body fluids for the purpose of locomotion. Bivalves, for example, pump hemo-

lymph into spaces within the foot, thereby expanding it for digging into the mud.

Most animals have a circulatory system that uses either hemolymph or blood as a circulatory fluid for moving gases, nutrients, and wastes from one part of the body to another. The respiratory function of a circulatory fluid is often augmented by the presence of a pigment that combines with and carries oxygen. Body fluids make ideal culture media for the growth of infectious parasites, and these fluids often have some way to ward off an invasion. In mammals, the white blood cells perform this function by phagocytizing microbes and producing antibodies.

In birds and mammals, which are endothermic, blood plays a role in temperature regulation. In animals that live in cold climates, blood vessels are arranged so that heat can pass from the arteries to the veins within exposed limbs. This conserves heat and also keeps the limbs from freezing. In vertebrates that dive under water, most of the blood pumped by the heart flows through the capillaries of the brain and heart. Closure of other possible pathways keeps these organs active despite the cessation of breathing. These extreme examples show how the circulatory system can be modified as an adaptation to the environment.

Summary

34.1 TRANSPORT IN INVERTEBRATES

Some invertebrates do not have a transport system. The presence of a gastrovascular cavity allows diffusion alone to supply the needs of cells in cnidarians and flatworms. Roundworms make use of their pseudocoelom in the same way that echinoderms use their coelom.

Other invertebrates do have a transport system. Insects have an open circulatory system, and earthworms have a closed one.

34.2 TRANSPORT IN VERTEBRATES

Vertebrates have a closed system in which arteries carry blood away from the heart to capillaries, where exchange takes place, and veins carry blood to the heart.

Fishes have a one-circuit circulatory pathway because the heart, with the single atrium and ventricle, pumps blood only to the gills. The other vertebrates have both pulmonary and systemic circuits. Amphibians have two atria but a single ventricle. Birds and mammals, including humans, have a heart with two atria and two ventricles, in which O_2-rich blood is always separate from O_2-poor blood.

34.3 TRANSPORT IN HUMANS

The heartbeat in humans begins when the SA (sinoatrial) node (pacemaker) causes the two atria to contract, and blood moves through the atrioventricular valves to the two ventricles. The SA node also stimulates the AV (atrioventricular) node, which in turn causes the two ventricles to contract. Ventricular contraction sends blood through the semilunar valves to the pulmonary trunk and the aorta. Now all chambers rest. The heart sounds, lub-dub, are caused by the closing of the valves.

In the pulmonary circuit, blood can be traced to and from the lungs. In the systemic circuit, the aorta divides into blood vessels that serve the body's cells. The venae cavae return O_2-poor blood to the heart.

Blood pressure created by the beat of the heart accounts for the flow of blood in the arteries, but skeletal muscle contraction is largely responsible for the flow of blood in the veins, which have valves preventing a backward flow.

34.4 CARDIOVASCULAR DISORDERS

Hypertension and atherosclerosis are two circulatory disorders that lead to heart attack and to stroke. Following a heart-healthy diet, getting regular exercise, maintaining a proper weight, and not smoking cigarettes are protective against the development of these conditions.

34.5 BLOOD, A TRANSPORT MEDIUM

Blood has two main parts: plasma and formed elements. Plasma contains mostly water (90–92%) and proteins (7–8%) but also contains nutrients and wastes.

The red blood cells contain hemoglobin and function in oxygen transport. Defense against disease depends on the various types of white blood cells. Neutrophils and monocytes are phagocytic and are very active during the inflammatory reaction. Lymphocytes are involved in the development of immunity to disease. Eosinophils are important in allergic reactions and parasitic infections, and basophils contain the anticoagulant heparin.

The platelets and two plasma proteins, prothrombin and fibrinogen, function in blood clotting, an enzymatic process that results in fibrin threads. Blood clotting is a complex process that includes three major events: (1) Platelets and injured tissue release prothrombin activator, which (2) enzymatically changes prothrombin to thrombin; (3) thrombin is an enzyme that causes fibrinogen to be converted to fibrin threads.

When blood reaches a capillary, water moves out at the arterial end, due to blood pressure. At the venous end, water moves in, due to osmotic pressure. In between, nutrients diffuse out of and wastes diffuse into the capillary.

Reviewing the Chapter

1. Describe transport in invertebrates that have no circulatory system; in those that have an open circulatory system; and in those that have a closed circulatory system. 614–15
2. Compare the circulatory systems of a fish, an amphibian, and a mammal. 617
3. Trace the path of blood in humans from the right ventricle to the left atrium; from the left ventricle to the kidneys and to the right atrium; from the left ventricle to the small intestine and to the right atrium. 619

4. Describe the mechanism of a heartbeat, mentioning all the factors that account for this repetitive process. Describe how the heartbeat affects blood flow. What other factors are involved in blood flow? 620–21

5. Define these terms: pulmonary circuit, systemic circuit, and portal system. 622

6. Discuss the life cycle and function of red blood cells. 626–27

7. How are white blood cells classified? What are the functions of neutrophils, monocytes, and lymphocytes? 626–27

8. Name the steps that take place when blood clots. Which substances are present in blood at all times, and which appear during the clotting process? 628

9. What forces facilitate exchange of molecules across the capillary wall? 629

Testing Yourself

Choose the best answer for each question.

1. Which one of these would you expect to be part of a closed, but not an open, circulatory system?
   a. ostia
   b. capillary beds
   c. hemocoel
   d. heart
   e. All of these are correct.

2. In a one-circuit circulatory pathway, blood pressure
   a. is constant throughout the system.
   b. drops significantly after gas exchange has taken place.
   c. is higher at the intestinal capillaries than at the gill capillaries.
   d. brings O_2-rich blood directly to the heart.
   e. does not occur in the animal kingdom.

3. In which animal does aortic blood have less oxygen than blood in the pulmonary vein?
   a. frog
   b. chicken
   c. monkey
   d. fish
   e. All of these are correct.

4. Which of these factors has little effect on blood flow in arteries?
   a. heartbeat
   b. blood pressure
   c. total cross-sectional area of vessels
   d. skeletal muscle contraction
   e. the amount of blood leaving the heart

5. In humans, blood returning to the heart from the lungs returns to
   a. the right ventricle.
   b. the right atrium.
   c. the left ventricle.
   d. the left atrium.
   e. both the right and left sides of the heart.

6. Systole refers to the contraction of the
   a. major arteries.
   b. SA node.
   c. atria and ventricles.
   d. major veins.
   e. All of these are correct.

7. Which of these is an incorrect association?
   a. white blood cells—infection fighting
   b. red blood cells—blood clotting
   c. plasma—water, nutrients, and wastes
   d. red blood cells—hemoglobin
   e. platelets—blood clotting

8. Water enters capillaries on the venous end as a result of
   a. active transport from tissue fluid.
   b. an osmotic pressure gradient.
   c. higher blood pressure on the venous end.
   d. higher blood pressure on the arterial side.
   e. higher red blood cell concentration on the venous end.

9. The last step in blood clotting
   a. is the only step that requires calcium ions.
   b. occurs outside the bloodstream.
   c. is the same as the first step.
   d. converts prothrombin to thrombin.
   e. converts fibrinogen to fibrin.

10. Macrophages are derived from
    a. basophils.
    b. eosinophils.
    c. neutrophils.
    d. lymphocytes.
    e. monocytes.

11. Which of the following is not a formed element of blood?
    a. leukocyte
    b. eosinophil
    c. fibrinogen
    d. platelet

12. Which of these is an incorrect statement concerning the heartbeat?
    a. The atria contract at the same time.
    b. The ventricles relax at the same time.
    c. The atrioventricular valves open at the same time.
    d. The semilunar valves open at the same time.
    e. First, the right side contracts, and then the left side contracts.

13. All arteries in the body contain O_2-rich blood, with the exception of the
    a. aorta.
    b. pulmonary arteries.
    c. renal arteries.
    d. coronary arteries.

14. The cardiac veins directly enter
    a. the inferior vena cava.
    b. the superior vena cava.
    c. the right atrium.
    d. the left atrium.

15. The "lub," the first heart sound, is produced by the closing of
    a. the aortic semilunar valve.
    b. the pulmonary semilunar valve.
    c. the right (atrioventricular) tricuspid valve.
    d. the left (atrioventricular) bicuspid valve, or mitral valve.
    e. both atrioventricular valves.

For questions 16–19, indicate whether the statement is true (T) or false (F).

16. Carbon dioxide exits the arterial end of the capillary, and oxygen enters the venous end of the capillary. _____

17. Platelets form a plug by sticking to each other. _____

18. Another term for blood clotting is agglutination. _____

19. SA node impulses cause the atria to contract. _____

20. Label arrows a–d as either blood pressure or osmotic pressure.

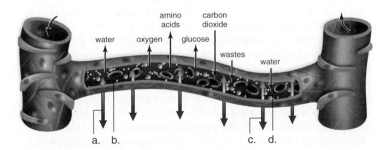

21. Label this diagram of the heart.

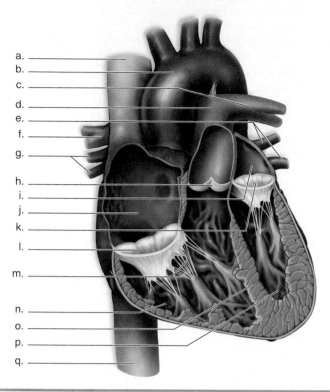

a. _____
b. _____
c. _____
d. _____
e. _____
f. _____
g. _____
h. _____
i. _____
j. _____
k. _____
l. _____
m. _____
n. _____
o. _____
p. _____
q. _____

Thinking Scientifically

1. For several years, researchers have attempted to produce an artificial blood for transfusions. Artificial blood would most likely be safer and more readily available than human blood. While artificial blood might not have all the characteristics of human blood, it would be useful on the battlefield and in emergency situations. Which characteristics must artificial blood have to be useful, and which would probably be too difficult to reproduce?

2. You have to stand in front of the class to give a report. You are nervous, and your heart is pounding. How would your ECG appear?

Bioethical Issue: A Healthy Lifestyle

According to a 1993 study, about one million deaths a year in the United States could be prevented if people adopted the healthy lifestyle described in the Health Focus on page 624. Tobacco, lack of exercise, and a high-fat diet probably cost the nation about $200 billion per year in health-care costs. To what lengths should we go to prevent these deaths and reduce health-care costs?

E. A. Miller, a meatpacking entity of ConAgra in Hyrum, Utah, charges extra for medical coverage of employees who smoke. Eric Falk, Miller's director of human resources, says, "We want to teach employees to be responsible for their behavior." Anthem Blue Cross–Blue Shield of Cincinnati, Ohio, takes a more positive approach. They give insurance plan participants $240 a year in extra benefits, such as additional vacation days, if they get good scores in five out of seven health-related categories. The University of Alabama, Birmingham, School of Nursing has a health-and-wellness program that counsels employees about how to get into shape in order to keep their insurance coverage. Audrey Brantley, who is in the program, has mixed feelings.

She says, "It seems like they are trying to control us, but then, on the other hand, I know of folks who found out they had high blood pressure or were borderline diabetics and didn't know it."

Does it really work? Turner Broadcasting System in Atlanta has a policy that affects all employees hired after 1986. They will be fired if caught smoking—whether at work or at home—but some admit they still manage to sneak a smoke. What steps do you think are ethical to encourage people to adopt a healthy lifestyle?

Understanding the Terms

angina pectoris 625
antibody 626
antigen 626
aorta 622
arteriole 616
artery 616
atrioventricular valve 618
atrium 618
basophil 628
blood 614
blood pressure 623
capillary 616
cardiac conduction
    system 621
cardiac cycle 620
cardiac output 620
cardiac pacemaker 621
circulatory (or cardiovascular)
    system 614, 616
closed circulatory system 615
diastole 620
electrocardiogram
    (ECG) 621
eosinophil 628
heart 618
heart attack 625
hemoglobin 626
hemolymph 614

lymph 629
lymphocyte 626
macrophage 626
monocyte 626
neutrophil 626
open circulatory
    system 614
plasma 626
platelet 628
portal system 622
pulmonary circuit 617
pulse 621
red blood cell 626
semilunar valve 618
septum 618
serum 628
sphygmomanometer 623
stroke 625
systemic circuit 617
systole 620
thrombin 628
tissue fluid 629
vein 616
vena cava 622
ventricle 618
venule 617
white blood cell 626

Match the terms to these definitions:

a. _____ Blood vessel that transports blood away from the heart.

b. _____ Cell fragment that is necessary to blood clotting.

c. _____ The liquid portion of blood; contains nutrients, wastes, salts, and proteins.

d. _____ The major systemic veins that take blood to the heart from the tissues.

e. _____ Iron-containing respiratory pigment occurring in vertebrate red blood cells and in the blood plasma of many invertebrates.

ARIS, the *Biology* Website

ARIS, the website for *Biology*, provides a wealth of information organized and integrated by chapter. You will find practice quizzes, interactive activities, labeling exercises, flashcards, and much more that will complement your learning and understanding of general biology.

www.mhhe.com/maderbiology9

35

LYMPH TRANSPORT AND IMMUNITY

W hen HIV arrived on the scene in the 1970s, the general public was awakened to the function of the immune system. Human immunodefiency virus, or HIV, is transmitted to a new host by bodily fluids, such as semen and blood. The HIV virus does not kill a host directly; it kills by destroying cells of the host's immune system, thereby enabling other pathogens known as opportunistic organisms (bacteria and fungi) to invade the body. In contrast, individuals with an intact immune system can usually defend themselves against opportunistic infections.

The immune system consists of a variety of cells, tissues, and organs that defend the body against pathogens, cancer cells, and foreign proteins. The first line of defense is nonspecific and consists of mechanisms that occur quickly and against any pathogen, giving time for specific defenses to be mounted. Working in harmony, the various branches of the immune system help maintain homeostasis.

The lymphatic organs (red bone marrow, thymus gland, spleen, and lymph nodes) can also be classified as immune system organs because they produce and store immune system cells. The lymphatic system has other functions as well; lymphatic vessels collect excess tissue fluid and return it to the bloodstream, thereby helping to maintain blood pressure.

Scanning electron micrograph of a T lymphocyte infected with HIV (purple).

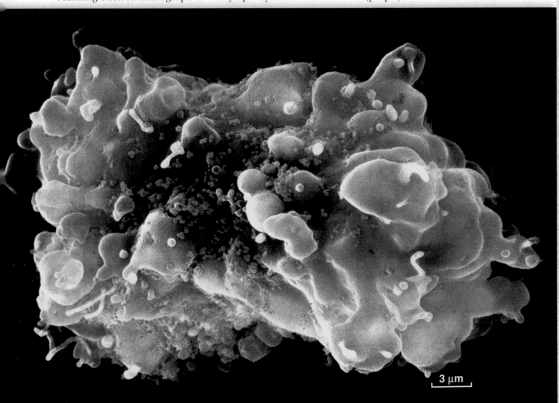

3 μm

35.1 THE LYMPHATIC SYSTEM

The **lymphatic system** [L. *lympha,* clear water] consists of lymphatic vessels and the lymphatic organs. This system, which is closely associated with the cardiovascular system, has four main functions that contribute to homeostasis: (1) Lymphatic capillaries absorb excess tissue fluid and return it to the bloodstream; (2) in the small intestines, lymphatic capillaries called lacteals absorb fats in the form of lipoproteins and transport them to the bloodstream; (3) the lymphatic system is responsible for the production, maintenance, and distribution of lymphocytes; and (4) the lymphatic system helps defend the body against pathogens.

Lymphatic Vessels

Lymphatic vessels form a one-way system that begins with lymphatic capillaries. Most regions of the body are richly supplied with lymphatic capillaries—tiny, closed-ended vessels whose walls consist of simple squamous epithelium (Fig. 35.1). Lymphatic capillaries take up excess tissue fluid. Tissue fluid is mostly water, but it also contains solutes (i.e., nutrients, electrolytes, and oxygen) derived from plasma and cellular products (i.e., hormones, enzymes, and wastes) secreted by cells. The fluid inside lymphatic vessels is called **lymph.** Lymph is usually a colorless liquid but after a meal, it appears creamy because of its lipid content.

The lymphatic capillaries join to form lymphatic vessels that merge before entering one of two ducts: the thoracic duct or

Right lymphatic duct: empties lymph into the right subclavian vein

Right subclavian vein: transports blood away from the right arm and the right ventral chest wall toward the heart

Axillary lymph nodes: located in the underarm region

Thoracic duct: empties lymph into the left subclavian vein

Inguinal lymph nodes: located in the groin region

Lymph nodes cleanse lymph and alert the immune system to pathogens.

Tonsil: patches of lymphatic tissue that help to prevent entrance of pathogens by way of the nose and mouth

Left subclavian vein: transports blood away from the left arm and the left ventral chest wall toward the heart

Red bone marrow: site for the origin of all types of blood cells

Thymus gland: lymphoid tissue where T lymphocytes mature and learn to tell "self" from "nonself"

Spleen: cleanses the blood of cellular debris and bacteria while resident T and B cells respond to the presence of antigens

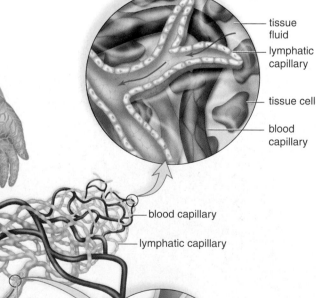

tissue fluid

lymphatic capillary

tissue cell

blood capillary

blood capillary

lymphatic capillary

valve

lymphatic vessel

Lymph node

FIGURE 35.1

Lymphatic system.
Lymphatic vessels drain excess fluid from the tissues and return it to the cardiovascular system. The enlargement shows that lymphatic vessels, like cardiovascular veins, have valves to prevent backward flow. The lymph nodes, spleen, thymus gland, and red bone marrow are the main lymphatic organs that assist immunity.

the right lymphatic duct. The larger thoracic duct returns lymph collected from the body below the thorax and the left arm and left side of the head and neck into the left subclavian vein. The right lymphatic duct returns lymph from the right arm and right side of the head and neck into the right subclavian vein.

The construction of the larger lymphatic vessels is similar to that of cardiovascular veins, including the presence of valves. The movement of lymph within lymphatic capillaries is largely dependent on skeletal muscle contraction. Lymph forced through lymphatic vessels as a result of muscular compression is prevented from flowing backwards by one-way valves.

Edema is localized swelling caused by the accumulation of tissue fluid that has not been collected by the lymphatic system. This can happen if too much tissue fluid is made and/or if not enough of it is drained away. Edema can lead to tissue damage and eventual death, illustrating the importance of the function of the lymphatic system. The fat absorption and defense functions of the lymphatic system are equally important. Unfortunately, cancer cells sometimes enter lymphatic vessels and move undetected to other regions of the body, where they produce secondary tumors. In this way, the lymphatic system sometimes assists metastasis, the spread of cancer far from its place of origin.

Lymphatic Organs

Lymphatic (lymphoid) organs contain large numbers of lymphocytes, the type of white blood cell that plays a pivotal role in immunity. Lymphocytes exist as **B lymphocytes (B cells)** and **T lymphocytes (T cells).** They are produced and mature in the primary lymphatic organs: red bone marrow and the thymus gland (Fig. 35.2).

Primary Lymphatic Organs

Red bone marrow is the site of stem cells, which are ever capable of dividing and producing blood cells. Some of these cells become the various types of white blood cells: neutrophils, eosinophils, basophils, lymphocytes, and monocytes (see Fig. 34.11).

In a child, most bones have red bone marrow, but in an adult, it is present only in the bones of the skull, the sternum (breastbone), the ribs, the clavicle (collarbone), the pelvic bones, the vertebral column, and the proximal heads of the femur and humerus.

The red bone marrow consists of a network of connective tissue fibers that supports the stem cells and their progeny. They are packed around thin-walled sinuses filled with venous blood. Differentiated blood cells enter the bloodstream at these sinuses.

Bone marrow is not only the source of B cells but also the place where B cells mature. T cells mature in the thymus.

The soft, bilobed **thymus gland** is located in the thoracic cavity between the trachea and the sternum ventral to the heart. The thymus varies in size, but it is largest in children and shrinks as we get older. In the elderly, it is barely detectable. Connective tissue divides the thymus into lobules, which are filled with T cells. Immature T cells migrate from the bone marrow through the bloodstream to the thymus, where they mature. Only about 5% of these cells ever leave the thymus. It is said that the T cells remaining are capable of telling "self" from "nonself." Those T cells that are capable of reacting to the body's own normal cells have undergone apoptosis, and only those that have the capability to react to specific molecules leave the thymus. **Apoptosis** is a process of programmed cell death (PCD) involving a cascade of specific cellular events leading to the death and destruction of the cell (see Fig. 9.2, p. 151).

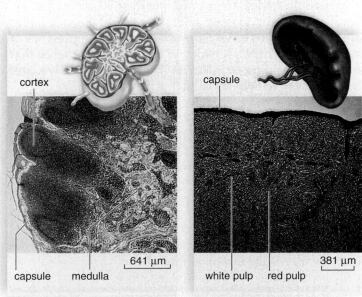

FIGURE 35.2 The lymphatic organs.
Left: Red bone marrow and the thymus gland are the primary lymphatic organs. Blood cells, including lymphocytes, are produced in red bone marrow. B cells mature in the bone marrow, but T cells mature in the thymus. *Right:* Lymph nodes and the spleen are secondary lymphatic organs. Lymph is cleansed in lymph nodes, while blood is cleansed in the spleen.

The thymus produces thymic hormones, such as thymosin, that are thought to aid in the maturation of T cells. Thymosin may also have other functions in immunity. The thymus is absolutely critical to immunity because of its role in the maturation of T cells.

Secondary Lymphatic Organs

In the secondary lymphatic organs, lymphocytes encounter and bind with specific molecules, after which they proliferate and become active.

Lymph nodes, which cleanse lymph, are small (about 1–25 mm in diameter), ovoid structures occurring along lymphatic vessels. Connective tissue divides the organ into nodules; each of these is packed with B cells and contains a sinus. As lymph courses through the many sinuses, it is filtered by macrophages, which engulf debris and **pathogens,** disease-causing agents such as viruses and bacteria. T cells, also present in sinuses, fight infections and attack cancer cells.

The **spleen** is located in the upper left side of the abdominal cavity posterior to the stomach. Most of the spleen is red pulp that filters the blood. Red pulp consists of blood vessels and sinuses where macrophages remove old and defective blood cells and lymphocytes cleanse the blood of foreign particles. The spleen also has white pulp that is inside the red pulp and consists of little lumps of lymphatic tissue.

The spleen's outer capsule is relatively thin, and an infection or a blow can cause the spleen to burst. Although the spleen's functions are largely replaced by other organs, a person without a spleen is often slightly more susceptible to infections and may have to receive antibiotic therapy indefinitely.

Lymph nodes are named for their location. For example, inguinal nodes are in the groin, and axillary nodes are in the armpits. Physicians often feel for the presence of swollen, tender lymph nodes in the neck as evidence that the body is fighting an infection. This is a noninvasive, preliminary way to help make such a diagnosis.

Other Lymphatic Organs

The **tonsils** are patches of lymphatic tissue located in the pharynx (see Fig. 36.5). The tonsils perform the same functions as lymph nodes, but because of their location, they may be the first to encounter pathogens and antigens that enter the body by way of the mouth.

Peyer's patches located in the intestinal wall and the **vermiform appendix** attached to the cecum, a blind pouch of the large intestine, encounter pathogens that enter the body by way of the intestinal tract.

Lymphocytes and other white blood cells are produced by the red bone marrow, which is a primary lymphatic organ, as is the thymus, where T cells mature. Lymphocytes congregate in the secondary lymphatic organs: the spleen, the lymph nodes, and other organs, such as the tonsils.

35.2 NONSPECIFIC AND SPECIFIC DEFENSES

Immunity is the body's capability to remove foreign substances and to kill pathogens and cancer cells. Immunity involves both nonspecific defenses and specific defenses.

Nonspecific Defenses

Nonspecific defenses occur automatically because they are innate; however, no memory is involved—there is no recognition that this same intruder has been attacked before. Barriers to entry, the inflammatory response, phagocytes and natural killer cells, and protective proteins are types of nonspecific defenses.

Barriers to Entry

Barriers to entry include nonchemical barriers such as the skin and the mucous membranes lining the respiratory, digestive, and urinary tracts, which serve as mechanical barriers to entry by pathogens. Then, too, the upper respiratory tract is lined by ciliated cells that sweep mucus and trapped particles up into the throat, where they can be swallowed or expectorated (coughed out). Or consider that the various bacteria that normally reside in the intestine and other areas, such as the vagina, prevent pathogens from taking up residence.

Barriers to entry also include antimicrobial molecules. As an example, oil gland secretions contain chemicals that weaken or kill certain bacteria on the skin. Mucous membranes secrete lysozyme, an enzyme that can lyse bacteria. The stomach has an acidic pH, which inhibits the growth of or kills many types of bacteria.

Inflammatory Response

Whenever tissue is damaged by physical or chemical agents or by pathogens, a series of events occur known as the **inflammatory response.** Figure 35.3 illustrates the participants in the inflammatory response.

An inflamed area has four outward signs: redness, heat, swelling, and pain. All of these signs are due to capillary changes in the damaged area. Chemical signals (mediators) such as **histamine,** released by damaged tissue cells and **mast cells,** a type of white blood cell in tissues, cause the capillaries to dilate and become more permeable. Increased blood flow, due to enlarged capillaries, causes the skin to redden and become warm. Increased permeability allows proteins and fluids to escape into tissues. The swollen area stimulates free nerve endings, causing the sensation of pain.

Some inflammatory responses also trigger fever, whose onset is controlled by the brain. Fever serves to inhibit the growth of some microorganisms, promotes accelerated tissue repair, facilitates phagocytosis of pathogens, stimulates immune cells to divide more rapidly, and increases the production of viral-fighting interferon, discussed on page 638. Although high fevers are dangerous, moderate fevers of a short duration are considered beneficial.

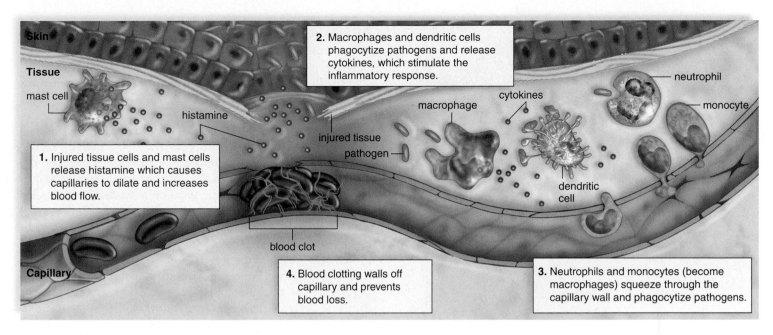

FIGURE 35.3 Inflammatory response.
Due to capillary changes in a damaged area and the release of chemical mediators, such as histamine by mast cells, an inflamed area exhibits redness, heat, swelling, and pain. The inflammatory response can be accompanied by other reactions to the injury. Macrophages and dendritic cells, present in the tissues, phagocytize pathogens, as do neutrophils, which squeeze through capillary walls from the blood. Macrophages and dendritic cells release cytokines, which stimulate the inflammatory and other immune response. A blood clot can form to seal a break in a blood vessel.

Chronic Inflammation. Regardless of the specific cause, a **chronic inflammation** is one that persists for weeks or longer. Chronic inflammations are often treated by administering anti-inflammatory agents, such as aspirin, ibuprofen, or cortisone. These medications act against the chemical signals, such as histamines, that bring about an inflammation.

Chronic inflammation is now thought to play a role in many human ills. It may destabilize cholesterol deposits in the coronary arteries, leading to heart attacks, even when the patient's cholesterol level is within a normal range. Inflammation may destroy nerve cells in the brains of Alzheimer victims, and it could even be involved in the proliferation of abnormal cells and their transformation into cancer cells. Evidence is even growing that inflammation could be involved in the development of diabetes in obese individuals because fat cells produce inflammatory chemical signals. At every turn investigators are finding that chronic inflammation precipitates various illnesses. Autoimmune diseases in which the body attacks itself (see page 649) also involve inflammation, but in these instances specific immunity seems to be the leading cause. The good news is that a healthy diet (low in fat and rich in fruits and vegetables), exercise, and good dental hygiene can reduce the occurrence of inflammation and help keep us well by reducing the occurrence of chronic inflammation.

Phagocytes and Natural Killer Cells

Migration of phagocytes, namely neutrophils and monocytes, occurs during the inflammatory response. Neutrophils and monocytes are amoeboid and can change shape to squeeze through capillary walls and enter tissue fluid. Also present are **dendritic cells,** notably in the skin and mucous membranes, and **macrophages** in other tissues, which are able to devour many pathogens and still survive. Macrophages and dendritic cells have receptors that allow them to recognize the presence of pathogens. Then, they release **cytokines,** chemical signals that stimulate other white cells, such as neutrophils and monocytes, that then mature into macrophages. Neutrophils, dendritic cells, and macrophages engulf pathogens, which are destroyed by enzymes when endocytic vesicles combine with lysosomes. As the infection is being overcome, a blood clot can form to seal a break in the blood vessel. Some phagocytes die. These—along with dead tissue cells, dead bacteria, and living white blood cells—form pus, a whitish material. The presence of pus indicates that the body is trying to overcome an infection. Dendritic cells and macrophages move through the tissue fluid and lymph to the lymph nodes and spleen, where they activate B cells and T cells to mount a specific defense to an infection.

Natural killer (NK) cells are large, granular lymphocytes that kill virus-infected cells and cancer cells by cell-to-cell contact. NK cells do their work while specific defenses are still mobilizing, and they produce cytokines that stimulate these cells.

What makes an NK cell attack and kill a cell? First, they normally congregate in the tonsils, lymph nodes, and spleen, where they are stimulated by dendritic cells before they travel forth. Then, NK cells look for a self-protein on the body's cells. As may happen, if a virus-infected cell or a cancer cell has lost its self-proteins, the NK cell kills it in the same manner used by cytotoxic T cells (see Fig. 35.8). Unlike cytotoxic T cells, NK cells are not specific; they have no memory; and their numbers do not increase after stimulation occurs.

FIGURE 35.4
Action of the complement system against a bacterium. When complement proteins in the blood plasma are activated by an immune response, they form a membrane attack complex that makes holes in bacterial cell walls and plasma membranes, allowing fluids and salts to enter until the cell eventually bursts.

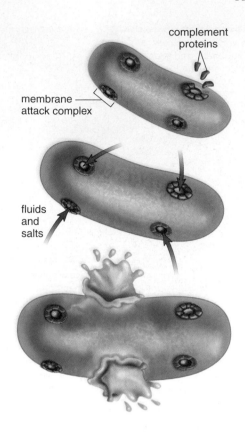

complement proteins

membrane attack complex

fluids and salts

Protective Proteins

The complement system, often simply called complement, and the interferons are considered protective proteins.

Complement is composed of a number of blood plasma proteins. The complement proteins "complement" certain immune responses, which accounts for their name. For example, they are involved in and amplify the inflammatory response because certain complement proteins can bind to mast cells and trigger histamine release, and others can attract phagocytes to the scene. Some complement proteins bind to the surface of pathogens already coated with antibodies, which ensures that the pathogens will be phagocytized by a neutrophil, dendritic cell, or macrophage.

Certain other complement proteins join to form a **membrane attack complex** that produces holes in the surface of bacteria and some viruses. Fluids and salts then enter the bacterial cell or virus to the point that it bursts (Fig. 35.4).

Interferons are proteins produced by virus-infected cells as a warning to noninfected cells in the area. Interferon binds to receptors of noninfected cells, causing them to prepare for possible attack by producing substances that interfere with viral replication. Interferons are used as treatment in certain viral infections, such as hepatitis C.

> Nonspecific defenses provide a varied and rapid means for the body to respond to a threat such as an invading pathogen. Specific defense mechanisms do not occur unless macrophages and dendritic cells have responded to the pathogen and started producing chemical signals called cytokines.

Specific Defenses

When nonspecific defenses have been inadequate, specific defenses come into play. *First*, a specific defense requires that the immune system be able to recognize a specific molecule, called an **antigen**. Some antigens are termed **foreign antigens** because the body does not produce them. Pathogens, such as bacteria and viruses and transplanted tissues and organs, bear foreign antigens the body usually recognizes. Other antigens are termed **self-antigens** because the body itself produces them. It is unfortunate when the immune system reacts to pancreatic cells (results in diabetes mellitus) or nerve fiber sheaths (results in multiple sclerosis) as if they were foreign, but fortunate when the body can destroy the cancerous cells of a tumor. *Second*, the immune system can respond to the antigen. Unlike the nonspecific defenses which occur immediately, it usually takes five to seven days to mount a specific defense. However, *third*, the immune system can remember an antigen. This is the reason, for example, that once we recover from measles, we usually do not get measles a second time.

Specific defenses primarily depend on the action of B cells and T cells. These cells are capable of recognizing antigens because they have specific **antigen receptors** that combine with particular antigens. Each lymphocyte has only one type of receptor. It is often said that the receptor and the antigen fit together like a lock and a key. Remarkably, diversification occurs to such an extent during maturation that there are specific B cells and/or T cells for any possible antigen we are likely to encounter during a lifetime.

B cells give rise to plasma cells, which produce antibodies capable of combining with and neutralizing a particular antigen. In contrast, T cells do not produce antibodies. Instead, they differentiate into either helper T cells, which release cytokines, or cytotoxic T cells that attack and kill virus-infected cells and cancer cells. Macrophages

TABLE 35.1

Immune Cells for Specific Defenses

Cell	Function
Lymphocytes	Responsible for specific immunity
B cells	Produce plasma cells and memory B cells
Plasma cells	Produce specific antibodies
Memory cells	Ready to produce antibodies in the future
T cells	Regulate immune response; produce cytotoxic T cells, helper T cells, and memory T cells
Cytotoxic T cells	Kill virus-infected and tumor cells
Helper T cells	Regulate immunity
Memory T cells	Ready to react to antigens in the future
Macrophages	Phagocytize pathogens; inflammatory reponse and specific immunity
Dendritic cells	Phagocytize pathogens; inflammatory response and specific immunity

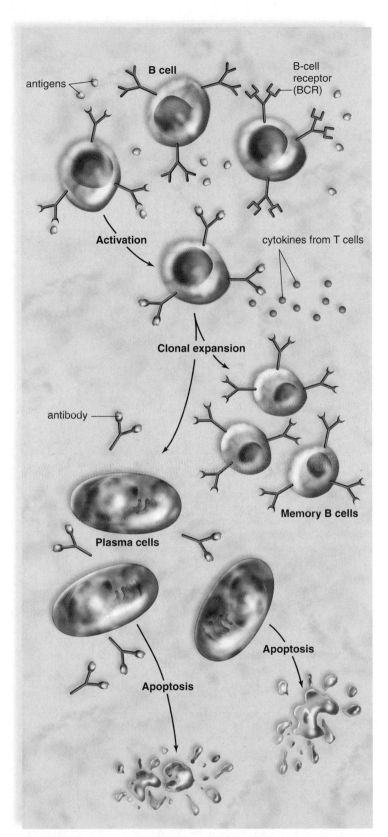

FIGURE 35.5 Clonal selection model as it applies to B cells.
Each B cell has a B-cell receptor (BCR) designated by shape that will combine with
a specific antigen. Activation of a B cell occurs when its BCR can combine with an
antigen (colored green). In the presence of cytokines, the B cell undergoes clonal
expansion, producing many plasma cells that secrete antibodies specific to the antigen
and memory B cells that immediately recognize the antigen in the future. After the
infection passes, plasma cells undergo apoptosis.

and dendritic cells help T cells learn to recognize an antigen.
All these cells are listed in Table 35.1 for easy reference.

B Cells and Antibody-Mediated Immunity

The receptor on a B cell is called a *B-cell receptor (BCR)*. A
B cell is activated in a lymph node or the spleen, when its
BCRs bind to a specific antigen. Thereafter, the B cell divides
by mitosis many times. In other words, it makes many clones
of itself. Most of the resulting cells (clones) become **plasma
cells,** which circulate in the blood and lymph. Plasma cells are
larger than regular B cells because they have extensive rough
endoplasmic reticulum for the mass production and secretion
of antibodies to a specific antigen. Antibodies are identical to
the BCR of the B cell that was activated.

The **clonal selection model** states that an antigen selects, then binds to the BCR of only one type of B cell or T cell,
and then this B cell or T cell clones. Note in Figure 35.5 that
each B cell has a specific BCR represented by shape. Only the
B cell with a BCR that has a shape that fits the antigen (colored
green) undergoes clonal expansion. During clonal expansion,
cytokines secreted by helper T cells (see page 642) stimulate
B cells to clone. Some cloned B cells become memory cells,
which are the means by which long-term immunity is possible. If the same antigen enters the system again, **memory
B cells** quickly divide and give rise to more plasma cells capable of producing the correct type antibody. Once the threat
of an infection has passed, the development of new plasma
cells ceases, and those present undergo apoptosis.

Defense by B cells is called **antibody-mediated immunity** because the various types of activated B cells become plasma cells that produce antibodies. It is also called
humoral immunity because these antibodies are present
in blood and lymph. (A humor is any fluid normally occurring in the body.) Collectively, plasma cells probably
produce as many as two million different antibodies. A
human being doesn't have two million genes, so there
cannot be a separate gene for each type of antibody. As
discussed in the Science Focus on page 641, it has been
found that the genome contains scattered DNA segments
that can be shuffled and combined in various ways to
produce the DNA sequence coding for the BCR unique
to each type of B cell. The shuffling of DNA segments is
called somatic recombination because it occurs when the
B cells are produced in red bone marrow, and it does not
occur when the gametes are produced.

Characteristics of B Cells
- Antibody-mediated immunity against pathogens
- Produced and mature in bone marrow
- Reside in lymph nodes and spleen; circulate in blood
  and lymph
- Directly recognize antigen and then undergo clonal
  selection
- Clonal expansion produces antibody-secreting plasma
  cells as well as memory B cells

Structure of an Antibody. Antibodies are also called **immunoglobulins (Igs).** They are typically Y-shaped molecules with two arms. Each arm has a "heavy" (long) polypeptide chain and a "light" (short) polypeptide chain. These chains have constant regions, where the sequence of amino acids is set, and variable regions, where the sequence of amino acids varies between antibodies (Fig. 35.6). The constant regions are not identical among all the antibodies. Instead, they are almost the same within a particular antibody class. The variable regions become hypervariable at their tips and form antigen-binding sites, and their shape is specific for a particular antigen. It is the variable and hypervariable regions that change to be specific for a particular antigen. The antigen combines with the antibody at the antigen-binding site in a lock-and-key manner.

The antigen-antibody reaction can take several forms, but quite often the reaction produces complexes of antigens combined with antibodies. Such antigen-antibody complexes, sometimes called immune complexes, mark the antigens for destruction. For example, an antigen-antibody complex may be engulfed by neutrophils or macrophages, or it may activate complement. Complement makes pathogens more susceptible to phagocytosis, as discussed previously.

Types of Antibodies. There are five different classes of circulating antibody proteins, or immunoglobulins (Igs) (Table 35.2). IgG antibodies are the major type in blood, lymph, and tissue fluid. IgG antibodies bind to pathogens and their toxins. They can activate the complement system. IgGs are the only antibodies that can cross the placenta and, along with IgAs, are in breast milk. IgM antibodies are pentamers, meaning that they contain five of the Y-shaped structures shown in Figure 35.6a. These antibodies appear in blood soon after a vaccination or infection and disappear before it ends. They are good activators of the complement system. IgMs are the antibodies in plasma that can cause red blood cells to clump. IgA antibodies are monomers in blood and lymph or dimers in tears, saliva, gastric juice, and mucous secretions. They are the main type of antibody found in body secretions. IgAs bind to pathogens before they reach the blood-

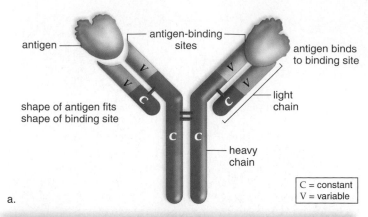

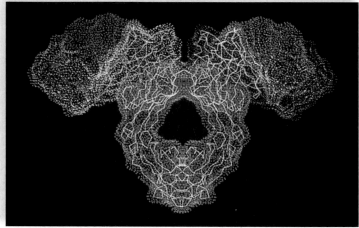

b.

FIGURE 35.6 Structure of an antibody.
a. An antibody contains two heavy (long) polypeptide chains and two light (short) chains arranged so there are two variable regions, where a particular antigen is capable of binding with an antibody (V = variable region, C = constant region). **b.** Computer model of an antibody molecule. The antigen combines with the two side branches.

stream. IgD molecules appear in the surface of B cells when they are ready to be activated. IgE antibodies are important in the immune system's response to parasites (e.g., parasitic worms). They are best known, however, for immediate allergic responses, including anaphylactic shock and asthma (see page 649).

TABLE 35.2

Antibodies

Class	Presence	Function
IgG	Main antibody type in circulation	Binds to pathogens, activates complement, and enhances phagocytosis
IgM	Antibody type found in circulation; largest antibody	Activates complement; clumps cells
IgA	Main antibody type in secretions such as saliva and milk	Prevents pathogens from attaching to epithelial cells in digestive and respiratory tracts
IgD	Antibody type found on surface of immature B cells	Presence signifies readiness of B cell
IgE	Antibody type found as antigen receptors on basophils in blood and on mast cells in tissues	Responsible for immediate allergic response; protection against certain parasitic worms

Antibody Diversity

In 1987, Susumu Tonegawa (Fig. 35A*a*) became the first Japanese scientist to win the Nobel Prize in Physiology or Medicine, after dedicating himself to finding the solution to an engrossing puzzle. Immunologists and geneticists knew that each B cell makes an antibody especially equipped to recognize the specific shape of a particular antigen. But they did not know how the human genome contained enough genetic information to permit the production of up to 2 million different antibody types needed to combat all of the pathogens we are likely to encounter during our lives.

An antibody is composed of two light and two heavy polypeptide chains, which are divided into constant and variable regions. The constant region determines the antibody class, and the variable region determines the specificity of the antibody, because this is where an antigen binds to a specific antibody (see Fig. 35.6). Each B cell must have a genetic way to code for the variable regions of both the light and heavy chains.

Tonegawa's colleagues say that he is a creative genius who intuitively knows how to design experiments to answer specific questions. In this instance, he examined the DNA sequences of immature lymphoblasts and compared them to mature B cells. He found that the DNA segments coding for the variable and constant regions were scattered throughout the genome in lymphoblasts, and that only certain of these segments were present in each mature antibody-secreting B cell, where they randomly came together and coded for a specific variable region. Later, the variable and constant regions are joined to give a specific antibody (Fig. 35A*b*). As an analogy, consider that each person entering a supermarket chooses various items for purchase, and that the possible combination of items in any particular grocery bag is astronomical. Tonegawa also found that mutations occur as the variable segments are undergoing rearrangements. Such mutations are another source of antibody diversity.

Invariably, some B cells with receptors that could bind to the body's own cell surface molecules arise. It is believed that these cells undergo apoptosis, or programmed cell death.

Tonegawa received his BS in chemistry in 1963 at Kyoto University and earned his PhD in biology from the University of California at San Diego (UCSD) in 1969. After that, he worked as a research fellow at UCSD and the Salk Institute. In 1971, he moved to the Basel Institute for Immunology and began the experiments that eventually led to his Nobel Prize–winning discovery. Tonegawa also contributed to the effort to decipher the receptors of T cells. This was an even more challenging area of research than the diversity of antibodies produced by B cells. Since 1981, he has been a full professor at Massachusetts Institute of Technology (MIT), where he has a reputation for being an "aggressive, determined researcher" who often works late into the night.

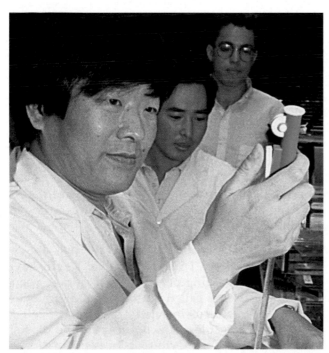

a. Susumu Tonegawa in the laboratory

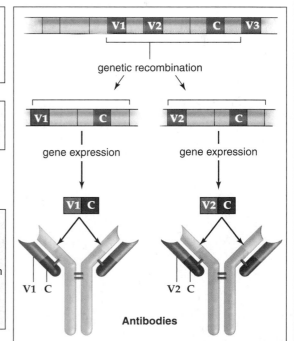

b. Antibody variable regions

FIGURE 35A Antibody diversity.
a. Susumu Tonegawa received a Nobel Prize for his findings regarding antibody diversity. b. Different genes for the variable regions of heavy and light chains are brought together during the production of B lymphocytes so that the antigen receptors of each one can combine with only a particular antigen.

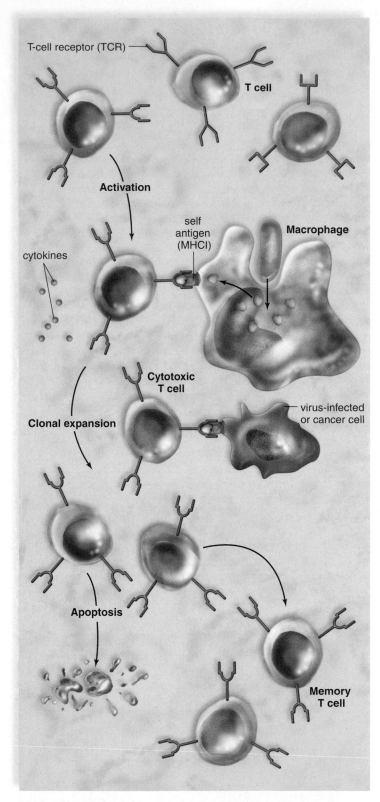

FIGURE 35.7 Clonal selection model as it applies to T cells.
Each T cell has a T-cell receptor (TCR) designated by a shape that will combine only with a specific antigen. Activation of a T cell occurs when its TCR can combine with an antigen. A macrophage presents the antigen (colored green) in the groove of an MHCI. Thereafter, the T cell undergoes clonal expansion, and many copies of the same type T cell are produced. After the immune response has been successful, the majority of T cells undergo apoptosis, but a small number are memory T cells. Memory T cells provide protection should the same antigen enter the body again at a future time.

T Cells and Cell-Mediated Immunity

Cell-mediated immunity is named for the action of T cells that directly attack diseased cells and cancer cells. Other T cells, however, release cytokines that stimulate both nonspecific and specific defenses.

T cells are formed in red bone marrow before they migrate to the thymus, a gland that secretes thymic hormones. These hormones stimulate T cells to develop *T-cell receptors (TCR)*. When a T cell leaves the thymus, it has a unique T-cell receptor (TCR) just as B cells have. Unlike B cells, however, T cells are unable to recognize an antigen without help. The antigen must be displayed to them by an **antigen-presenting cell (APC)**, such as a dendritic cell or a macrophage. After phagocytizing a pathogen, APCs travel to a lymph node or spleen, where T cells also congregate. In the meantime, the APC has broken the pathogen apart in a lysosome. A piece of the pathogen is then displayed in the groove of an MHC (major histocompatibility complex) protein on the cell's surface. MHC proteins are self-antigens because they mark cells as belonging to a particular individual and make transplantation of organs difficult.

In Figure 35.7, the different types of T cells have specific TCRs represented by their different shapes. A macrophage is presenting an antigen to a T cell that has a TCR capable of combining with this particular antigen (colored green). Now, the T cell is activated and undergoes clonal expansion. Many copies of the activated T cell are produced during clonal expansion. There are two classes of MHC proteins (called MHC I and MHC II). If an APC displays an antigen within the groove of an MHC I protein, the activated T cell will form cytotoxic T cells. If an APC displays an antigen within the groove of an MHC II protein, the activated T cell will form helper T cells.

As the illness disappears, the immune reaction wanes, and activated T cells become susceptible to apoptosis. As mentioned previously, apoptosis contributes to homeostasis by regulating the number of cells present in an organ, or in this case, in the immune system. When apoptosis does not occur as it should, T cell cancers (i.e., lymphomas and leukemias) can result. Also, in the thymus, any T cell that has the potential to destroy the body's own cells undergoes apoptosis.

Types of T Cells.   The two main types of T cells are cytotoxic T cells and helper T cells. **Cytotoxic T cells** have storage vacuoles containing perforins and storage vacuoles containing enzymes called granzymes. After a cytotoxic T cell binds to a virus-infected cell or cancer cell, it releases perforin molecules, which perforate the plasma membrane, forming a pore. Cytotoxic T cells then deliver granzymes into the pore, and these cause the cell to undergo apoptosis and die. Once cytotoxic T cells have released the perforins and granzymes, they move on to the next target cell. Cytotoxic T cells are responsible for so-called **cell-mediated immunity** (Fig. 35.8).

Helper T cells play a critical role in coordinating nonspecific defenses and specific defenses, including both cell-mediated immunity and also antibody-mediated immunity. How do helper T cells perform this function? When a helper T cell recognizes an antigen attached to an MHC protein, it secretes cytokines. Cytokines are chemical mediators that attract neutrophils, natural killer cells, and macrophages to where they are needed. Cytokines stimulate phagocytosis of pathogens, and they stimulate the clonal expansion of T and B cells.

Because more and more immune cells are recruited by helper T cells, the number of pathogens eventually begins to wane. But by now the body is fully prepared to deal with this pathogen again. Therefore, it is said that the immune system possesses memory—it can often remember former antigens. Notice in Figure 35.7 that following clonal expansion, some cloned cells undergo apoptosis, and some become memory cells. **Memory T cells,** like memory B cells, are long-lived, and their number is far greater than the original number of T cells that could recognize a specific antigen. Therefore, when the same antigen enters the body later on, the immune response may occur so rapidly that no detectable illness occurs.

HIV Infections. The primary host for an HIV (human immunodeficiency virus) is a helper T cell, but macrophages are also under attack. After HIV enters a host cell, it reproduces as described in Figure 21.4, and many progeny bud from the cell. In other words, the host cell produces the viruses that go on to destroy more helper T cells. Figure 35B in the Health Focus on page 644 shows the progression of an HIV infection over time. At first an individual is able to stay ahead of the virus by producing enough helper T cells to keep their number within the normal range. But gradually as the HIV count rises in plasma, the helper T cell count drops to way below normal. Then the person comes down with what are called opportunistic infections—infections that would be unable to take hold in a person with a healthy immune system. Now the individual has AIDS (acquired immunodeficiency syndrome). An HIV infection is presently a treatable disease, but the regimen is very difficult to maintain, and viral resistance to these medications is becoming apparent. Therefore, it is much wiser for individuals to prevent becoming infected by following the recommendations given on page 793.

Characteristics of T Cells
- Cell-mediated immunity against virus-infected cells and cancer cells
- Produced in bone marrow; mature in thymus
- Antigen must be presented in groove of an MHC protein
- Cytotoxic T cells destroy nonself antigen-bearing cells
- Helper T cells secrete cytokines, which control the immune response

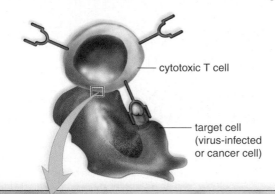
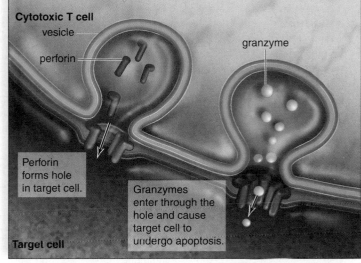

Cytotoxic T cell
vesicle
perforin
granzyme

Perforin forms hole in target cell.

Granzymes enter through the hole and cause target cell to undergo apoptosis.

Target cell

a.

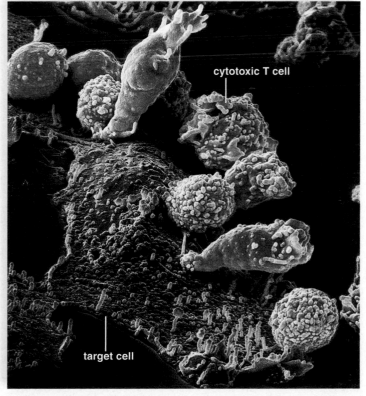

cytotoxic T cell

target cell

b.

FIGURE 35.8 Cell-mediated immunity.
a. How a T cell destroys a virus-infected cell or cancer cell. **b.** The scanning electron micrograph shows cytotoxic T cells attacking and destroying a cancer cell (target cell).

health focus

Opportunistic Infections and HIV

AIDS (acquired immunodeficiency syndrome) is caused by the destruction of the immune system, following an HIV (human immunodeficiency virus) infection. An HIV infection leads to the eventual destruction of immune system cells, known as helper T lymphocytes or, simply, helper T cells. Then, the individual succumbs to many unusual types of infections that would not cause disease in a person with a healthy immune system. Such infections are known as "opportunistic infections" (OIs).

HIV not only kills helper T cells by directly infecting them; it also causes many uninfected T cells to die by a variety of mechanisms, including apoptosis (programmed cell death or "cell suicide"). Many helper T cells are also killed by the person's own immune system as it tries to overcome the HIV infection. After initial infection with HIV, it may take up to ten years for an individual's helper T cells to become so depleted that the immune

system can no longer organize a specific response to OIs (Fig. 35B). In a healthy individual, the number of helper T cells typically ranges from 800–1,000 cells per mm^3 of blood. The appearance of specific OIs can be associated with the helper T-cell count.

- Shingles. Painful infection with varicella-zoster (chickenpox) virus. Helper T-cell count of less than 500/mm^3.
- Candidiasis. Fungal infection of the mouth, throat, or vagina. Helper T-cell count of about 350/mm^3.
- Pneumocystis pneumonia. Fungal infection causing the lungs to become useless as they fill with fluid and debris. Helper T-cell count of less than 200/mm^3.
- Kaposi's sarcoma. Cancer of blood vessels due to human herpesvirus 8 gives rise to reddish purple, coin-sized spots and lesions on the skin. Helper T-cell count of

less than 200/mm^3.
- Toxoplasmic encephalitis. Protozoan infection characterized by severe headaches, fever, seizures, and coma. Helper T-cell count of less than 100/mm^3.
- *Mycobacterium avium* complex (MAC). Bacterial infection resulting in persistent fever, night sweats, fatigue, weight loss, and anemia. Helper T-cell count of less than 75/mm^3.
- Cytomegalovirus. Viral infection that leads to blindness; inflammation of the brain, throat ulcerations. Helper T-cell count of less than 50/mm^3.

Due to development of powerful drug therapies that slow the progression of AIDS, people infected with HIV in the United States are suffering lower incidence of OIs than in the 1980s and 1990s. It is hoped that a vaccine will be developed for AIDS one day.

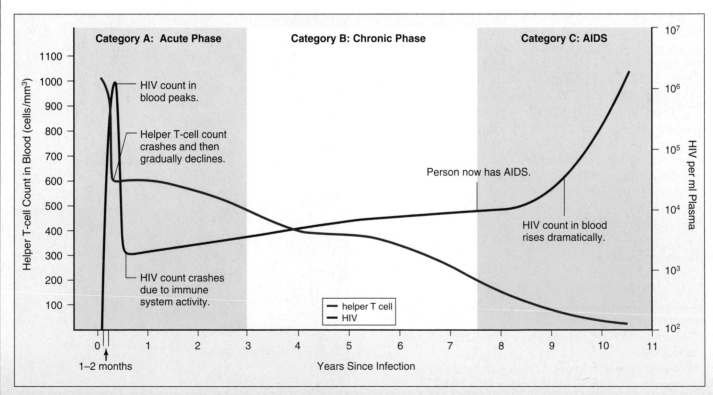

FIGURE 35B Progression of HIV infection during its three stages, called categories A, B, and C.
In category A, the individual may have no symptoms or very mild symptoms associated with the infection. By category B, opportunistic infections have begun to occur, such as candidiasis, shingles, and diarrhea. Category C is characterized by more severe opportunistic infections and is clinically described as AIDS.

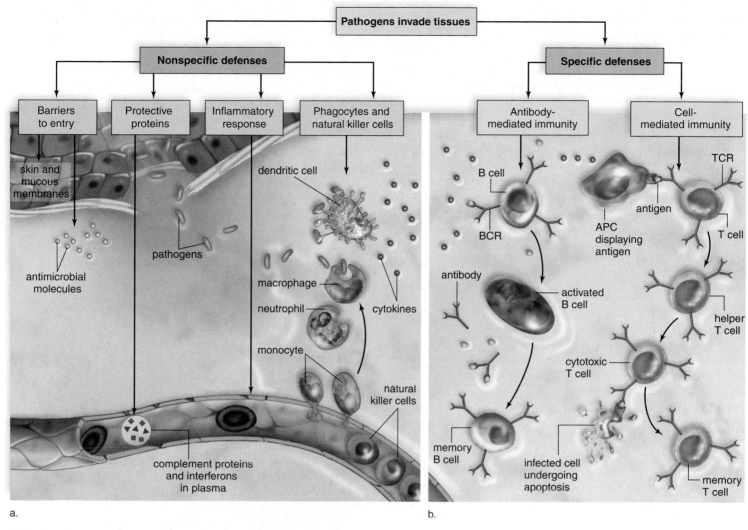

FIGURE 35.9 Overview of nonspecific and specific defenses.
Immunity involves two types of defense mechanisms. Nonspecific defenses act rapidly to detect and respond to an infection by any and all pathogens and cancer cells. If these defenses prove inadequate, specific defenses come into play, and the attack is then directed against a particular pathogen or cancer cell. **a.** Nonspecific defenses include barriers to entry, protective proteins, the inflammatory response, phagocytes and natural killer cells. **b.** Specific defenses include B cells, which produce antibodies, and T cells, which kill virus-infected or cancer cells on contact.

Overview of Immunity

The body has the ability to defend itself against pathogens and cancer cells. As summarized in Figure 35.9, immunity involves nonspecific defenses and specific defenses. The nonspecific defenses are especially geared to defend the body against a pathogen invasion. Barriers to entry such as the skin and mucous membranes help prevent pathogen invasion in the first place. But if tissue damage permits the entry of pathogens, the other nonspecific defenses—protective proteins, the inflammatory response, plus phagocytes and natural killer cells—may be able to prevent an infection. If not, the specific defenses come into play. Specific defenses rely on lymphocytes that have the ability to recognize antigens and to respond to them. Recognition is dependent on B and T cell receptors, abbreviated as BCR and TCR respectively. Each antigen can bind with only one type of receptor, but after binding takes place, the lymphocyte undergoes clonal expansion. When a B cell undergoes clonal expansion, the result is antibody-producing plasma cells. Therefore, the response of B cells is called antibody-mediated immunity. Antigen-presenting cells, either macrophages or dendritic cells, present antigens to T cells. Only a T cell with a receptor capable of binding the antigen presented in a MHC protein can be activated. Depending on the MHC protein, either cytotoxic T cells result or helper T cells result. Cytotoxic T cells kill virus-infected or cancer cells outright. Helper T cells coordinate the response of nonspecific and specific defenses by producing cytokines. When the infection has passed, memory B and T cells remain to jump-start a response to the same antigen again. Humans are dependent on an adequate immune response to continue their existence.

Immunity in Other Animals

The abundance of invertebrates suggests that they must have some way to defend themselves against pathogens. In 1882, the Russian zoologist Elie Metchnikoff stuck a rose thorn into a sea star and noted that phagocytes gathered in an attempt to remove the foreign object. Chemowarfare has also been found among invertebrates. In 1979, Swedish scientist Hans G. Boman discovered a class of silk moth antibacterial peptides, which he named cecropins. Like complement, they perforate bacteria, causing them to burst. (These peptides are currently being developed as antibacterial agents for use in humans.) Similarly, many invertebrates possess molecules related to vertebrate cytokines. In the sea star, cells that resemble macrophages release a cytokine-like chemical. Invertebrates possess proteins called lectins that may be the ancestor to antibodies. Lectins serve to mark invading microorganisms for phagocytic destruction.

Apparently, specific defense mechanisms evolved only among the vertebrates. Gary W. Litman has studied specific immunity in the horned shark. He found that antibody diversity has a genetic basis in this shark, just as it does in humans. But the horned shark relies more heavily on "inherited" diversity. This shark more quickly produces antibodies directed against pathogens it is apt to encounter than antibodies against new and different antigens. This characteristic of antibody-mediated immunity may be an advantage if the environment remains relatively constant. Investigators have hypothesized that cell-mediated immunity based on the presence of T cells predates antibody-mediated immunity based on B cells. Therefore, Litman decided to test the horned shark for evidence of T cell activity. He used the polymerase chain reaction (PCR) to replicate human T-cell receptor genes prior to sequencing them. Thereafter, he found evidence of corresponding proteins in the horned shark.

> There is evidence of nonspecific defenses in invertebrates, but specific defenses seem to be unique to vertebrates.

35.3 INDUCED IMMUNITY

Immunity occurs naturally through infection or is brought about artificially by medical intervention. There are two types of induced immunity: active and passive. In active immunity, the individual alone produces antibodies against an antigen; in passive immunity, the individual is given prepared antibodies.

Active Immunity

Active immunity sometimes develops naturally after a person is infected with a pathogen such as measles or chickenpox. However, active immunity is often induced when a person is well so that possible future infection will not take place. Individuals can receive artificial immunization to prevent infection (Fig. 35.10). The United States is committed to immunizing all children against the common types of childhood disease.

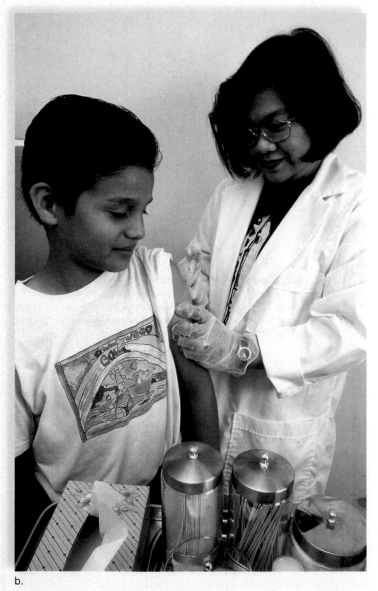

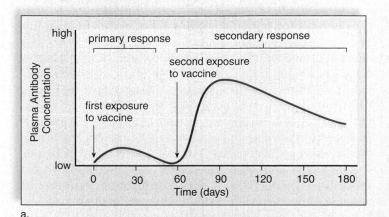

a.

b.

FIGURE 35.10 Active immunity due to immunizations.
a. The primary response, after the first exposure to a vaccine, is minimal, but the secondary response, which may occur after the second exposure, shows a dramatic rise in the amount of antibody present in plasma. **b.** Vaccines are used to immunize children against various childhood diseases.

Immunization involves the use of **vaccines,** substances that contain an antigen to which the immune system responds. Traditionally, vaccines are the pathogens themselves, or their products, that have been treated so they are no longer virulent (able to cause disease). Vaccines against smallpox, polio, and tetanus have been successfully used worldwide. Today, it is possible to genetically engineer bacteria to mass-produce a protein from pathogens, and this protein can be used as a vaccine. This method has helped produce a vaccine against hepatitis B, a viral disease, and is being used to prepare a vaccine against malaria, a disease caused by a protozoan parasite.

After a vaccine is given, it is possible to follow an immune response by examining the antibody titer (the amount of antibody present in a sample of plasma). After the first exposure to a vaccine, a primary response occurs. For a period of several days, no antibodies are present; then, there is a slow rise in the titer, followed by first a plateau and then a gradual decline as the antibodies bind to the antigen or simply break down (Fig. 35.10). After a second exposure, a secondary response is expected. The titer rises rapidly to a plateau level much greater than before. The second exposure is called a "booster" because it boosts the antibody titer to a high level. The high antibody titer now is expected to help prevent disease symptoms even if the individual is exposed to the disease-causing antigen.

Active (long-lasting) immunity can be induced by the use of vaccines. Active immunity depends upon the presence of memory B cells and memory T cells in the body.

Passive Immunity

Passive immunity occurs when an individual is given prepared antibodies (immunoglobulins) to combat a disease (Fig. 35.11). Since these antibodies are not produced by the individual's B cells, passive immunity is short-lived. For example, newborn infants are passively immune to some diseases because antibodies have crossed the placenta from the mother's blood. These antibodies soon disappear, however, so that within a few months, infants become more susceptible to infections. Breast-feeding prolongs the natural passive immunity an infant receives from the mother because antibodies are present in the mother's milk.

Even though passive immunity does not last, it sometimes is used to prevent illness in a patient who has been unexpectedly exposed to an infectious disease. Artificial passive immunity is used in the emergency treatment of rabies, measles, tetanus, diptheria, botulism, hepatitis A, and snakebites. Usually, the patient receives a gamma globulin injection (serum that contains antibodies), perhaps taken from individuals who have recovered from the illness. In the past, horses were immunized, and serum was taken from them to provide the needed antibodies against diseases. A patient who received these antibodies became ill about 50% of the time because the serum contained proteins that

the individual's immune system recognized as foreign. This was called serum sickness. But problems can still occur with products produced in other ways. An immunoglobulin intravenous product called Gammagard was withdrawn from the market because of its possible implication in the transmission of hepatitis.

Many times both passive and active measures are taken in combating a pathogen. In the majority of cases, if a rabid animal bites a person, both passive and active immunizations are administered. The passively injected antibodies immediately combat the rabies virus for a few weeks and then these are aided by the actively induced immune response.

Active (long-lasting) immunity can be induced by the use of vaccines. Passive immunity is temporary because the antibodies are used up by the individual.

FIGURE 35.11 Passive immunity.
Breast-feeding is believed to prolong the passive immunity an infant receives from the mother during pregnancy because antibodies are present in the mother's milk.

Cytokines and Immunity

Cytokines are signaling molecules produced by lymphocytes, monocytes, or other cells. Because cytokines regulate white blood cell formation and/or function, they are being investigated as possible adjunct therapy for cancer and AIDS. Both interferon and **interleukins,** which are cytokines produced by various white blood cells, have been used as immunotherapeutic drugs, particularly to enhance the ability of the individual's own T cells (and possibly B cells) to fight cancer.

Interferon, discussed on page 638, is a substance produced by leukocytes, fibroblasts, and probably most cells in response to a viral infection. Interferon is still being investigated as a possible cancer drug, but so far it has proven to be effective only in certain patients, and the exact reasons for this as yet cannot be discerned.

When and if cancer cells carry an altered protein on their cell surface, they should be attacked and destroyed by cytotoxic T cells. Whenever cancer does develop, it is possible that the cytotoxic T cells have not been activated. In that case, cytokines might awaken the immune system and lead to the destruction of the cancer. In one technique being investigated, researchers first withdraw T cells from the patient and activate the cells by culturing them in the presence of an interleukin. The cells then are reinjected into the patient, who is given doses of interleukin to maintain the killer activity of the T cells.

Scientists who are actively engaged in interleukin research believe that interleukins soon will be used as adjuncts for vaccines, for the treatment of chronic infectious diseases, and perhaps for the treatment of cancer. Interleukin antagonists may also prove helpful in preventing skin and organ rejection, autoimmune diseases, and allergies.

Various cytokines show some promise for potentiating the individual's own immune system.

Monoclonal Antibodies

Every plasma cell derived from the same B cell secretes antibodies against a specific antigen. These are **monoclonal antibodies** because all of them are the same type and because they are produced by plasma cells derived from the same B cell. One method of producing monoclonal antibodies in vitro (outside the body in glassware) is depicted in Figure 35.12. B cells are removed from an animal (today, usually mice are used) and are exposed to a particular antigen. The resulting *plasma cells* are fused with myeloma cells (malignant plasma cells that live and divide indefinitely). The fused cells are called hybridomas—*hybrid-* because they result from the fusion of two different cells, and *-oma* because one of the cells is a cancer cell.

At present, monoclonal antibodies are being used for quick and certain diagnosis of various conditions. For example, a particular hormone is present in the urine of a pregnant woman. A monoclonal antibody can be used to detect this hormone; if it is present, the woman knows she is pregnant. Monoclonal antibodies are also used to identify infections. And because they can distinguish between cancer and normal tissue cells, they are used to carry radioactive isotopes or toxic drugs to tumors so that they can be selectively destroyed.

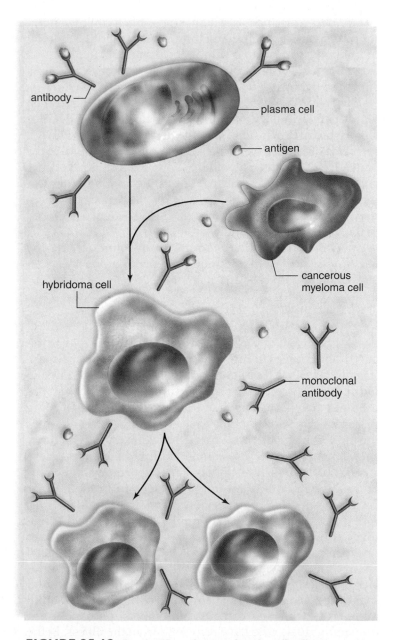

FIGURE 35.12 Production of monoclonal antibodies.
Plasma cells of the same type (derived from immunized mice) are fused with myeloma (cancerous) cells, producing hybridoma cells that are "immortal." Hybridoma cells divide and continue to produce the same type of antibody, called monoclonal antibodies.

35.4 IMMUNITY SIDE EFFECTS

Sometimes the immune system responds in a manner that harms the body, as when individuals develop allergies, suffer tissue rejection, have an autoimmune disease, or receive an incompatible blood type.

Allergies

Allergies are hypersensitivities to substances, such as pollen, food, or animal hair, that ordinarily would do no harm to the body. The response to these antigens, called **allergens,** usually includes some degree of tissue damage.

An **immediate allergic response** can occur within seconds of contact with the antigen. The response is caused by antibodies known as IgE (see Table 35.2). IgE antibodies are attached to the plasma membrane of mast cells in the tissues and also to basophils in the blood. When an allergen attaches to the IgE antibodies on these cells, mast cells release histamine and other substances that bring about the allergic symptoms. When pollen is an allergen, histamine stimulates the mucous membranes of the nose and eyes to release fluid, causing the runny nose and watery eyes typical of **hay fever.** If a person has **asthma,** the airways leading to the lungs constrict, resulting in difficult breathing accompanied by wheezing. When food contains an allergen, nausea, vomiting, and diarrhea result.

Anaphylactic shock is an immediate allergic response that occurs because the allergen has entered the bloodstream. Bee stings and penicillin shots are known to cause this reaction because both inject the allergen into the blood. Anaphylactic shock is characterized by a sudden and life-threatening drop in blood pressure due to increased permeability of the capillaries by histamine.

People with allergies produce ten times more IgE than those people without allergies. A new treatment using injections of monoclonal IgG antibodies for IgEs is currently being tested in individuals with severe food allergies. More routinely, injections of the allergen are given so that the body will build up high quantities of IgG antibodies. The hope is that these will combine with allergens received from the environment before they have a chance to reach the IgE antibodies located in the membrane of mast cells and basophils.

A **delayed allergic response** is initiated by memory T cells at the site of allergen contact in the body. The allergic response is regulated by the cytokines secreted by both T cells and macrophages. A classic example of a delayed allergic response is the skin test for tuberculosis (TB). When the test result is positive, the tissue where the antigen was injected becomes red and hardened. This shows that there was prior exposure to tubercle bacilli, the cause of TB. Contact dermatitis, which occurs when a person is allergic to poison ivy, jewelry, cosmetics, and many other substances that touch the skin, is also an example of a delayed allergic response.

Tissue Rejection

Certain organs, such as skin, the heart, and the kidneys, could be transplanted easily from one person to another if the body did not attempt to reject them. Rejection occurs because antibodies and cytotoxic T cells bring about destruction of foreign tissues in the body. When rejection occurs, the immune system is correctly distinguishing between self and nonself.

Organ rejection can be controlled by carefully selecting the organ to be transplanted and administering immunosuppressive drugs. It is best if the transplanted organ has the same type of antigens as those of the recipient, because cytotoxic T cells recognize foreign antigens. Two well-known immunosuppressive drugs, cyclosporine and tacrolimus, both act by inhibiting the response of T cells to cytokines.

The hope is that tissue engineering, the production of organs that lack antigens or that can be protected in some way from the immune system, will one day do away with the problem of rejection. Xenotransplantation, the transplantation of animal tissues and organs into human beings, is another way to solve the problem of supply. The first animal-human heart transplant occurred in 1984 when a baboon heart was given to an infant, who unfortunately lived only a short while. Although baboons are genetically closer to humans, pigs are generally healthier, produce more offspring in a shorter time, and are already farmed for food. Currently, pig heart valves and skin are routinely used for treatment of humans. Miniature pigs, whose heart size is appropriate to humans, are also being genetically engineered to make them less foreign to the human body, in order to avoid rejection.

Autoimmune Diseases

In an **autoimmune disease,** chronic inflammation is present, and cytotoxic T cells or antibodies mistakenly attack the body's own cells as if they bear foreign antigens. Sometimes autoimmune diseases set in following a noticeable infection; otherwise they probably start after an undetectable inflammatory response.

In the autoimmune disease myasthenia gravis, neuromuscular junctions do not work properly, and muscular weakness results. In multiple sclerosis, the myelin sheath of nerve fibers breaks down, and this causes various neuromuscular disorders. A person with systemic lupus erythematosus has various symptoms due to kidney damage. In rheumatoid arthritis, the joints are affected. Researchers suggest that heart damage following rheumatic fever and type 1 diabetes are also autoimmune illnesses. As yet, there are no cures for autoimmune diseases, but they can be controlled with medications.

Blood-Type Reactions

Many early blood transfusions resulted in the illness and even the death of some recipients. Eventually, it was discovered that only certain types of blood are compatible because red blood cell membranes carry specific proteins or carbohydrates that are antigens to blood recipients. Several groups of red blood cell antigens exist, the most significant being the ABO system. Clinically, it is very important that the blood groups are properly cross-matched to avoid a potentially deadly transfusion reaction. In these reactions, the recipient may die of kidney failure within a week.

ABO System

In the ABO system, the presence or absence of type A and type B antigens on red blood cells determines a person's blood type. For example, if a person has type A blood, the A antigen is on his or her red blood cells. This molecule is not an antigen to this individual, although it can be an antigen to a recipient who does not have type A blood.

In the simplified ABO system, there are four types of blood: A, B, AB, and O. Within the plasma, there are antibodies to the antigens that are *not* present on the person's red blood cells. These antibodies are called anti-A and anti-B. This chart tells you what antibodies are present in the plasma of each type blood:

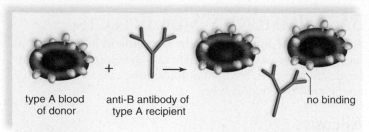

a. No agglutination

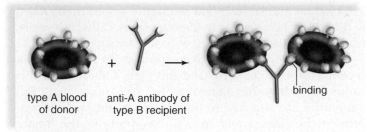

b. Agglutination

FIGURE 35.13 Blood transfusions.
No agglutination (**a**) versus agglutination (**b**) is determined by whether the recipient has antibodies in the plasma that can combine with antigens on the donor's red blood cells.

Blood Type	Antigen on Red Blood Cells	Antibody in Plasma
A	A	Anti-B
B	B	Anti-A
AB	A,B	None
AB	None	Anti-A and anti-B

Because type A blood has anti-B and not anti-A antibodies in the plasma, a donor with type A blood can give blood to a recipient with type A blood (Fig. 35.13a). Try, however, giving type A blood to a type B recipient and agglutination occurs (Fig. 35.13b). **Agglutination,** or clumping of red blood cells, can cause blood to stop circulating in small blood vessels, and this leads to organ damage. It is also followed by *hemolysis,* or bursting of red blood cells, which, if extensive, can cause the death of the individual.

Theoretically, which type blood would be accepted by all other blood types? Type O blood has no antigens on the red blood cells and is sometimes called the universal donor. Which type blood could receive blood from any other blood type? Type AB blood has no anti-A or anti-B antibodies in the plasma and is sometimes called the universal recipient. In practice, however, it is not safe

to rely solely on the ABO system when matching blood. Instead, samples of the two types of blood are physically mixed, and the result is microscopically examined before blood transfusions are done.

Today, blood transfusions are a matter of concern not only because blood types should match, but also because each person wants to receive blood that is of good quality and free of infectious agents. Blood is tested for the more serious agents, such as those that cause AIDS, hepatitis, and syphilis. Donors can help protect the nation's blood supply by knowing when not to give blood.

Rh System

Another important antigen in matching blood types is the Rh factor. Eighty-five percent of the U.S. population have this particular antigen on the red blood cells and are Rh positive. Fifteen percent do not have this antigen and are Rh negative. Rh-negative individuals normally do not have antibodies to the Rh factor, but they may make them when exposed to the Rh factor. The designation of blood type usually also includes whether the person has or does not have the Rh factor on the red blood cells.

During pregnancy, if the mother is Rh-negative and the father is Rh-positive, the child may be Rh-positive (Fig. 35.14). The Rh-positive red blood cells may begin leaking across the placenta into the mother's cardiovascular system, as placental tissues normally break down before and at birth. Now, the mother produces anti-Rh antibodies. In this or a subsequent pregnancy with another Rh-positive baby, these antibodies may cross the placenta and destroy

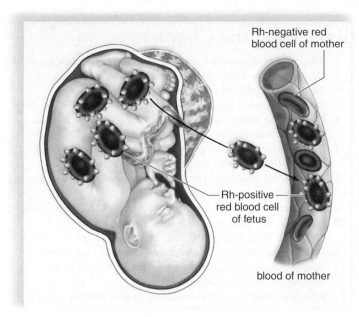

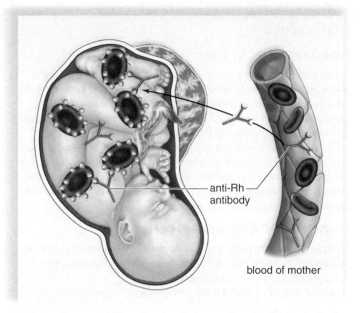

a. Fetal Rh-positive red blood cells leak across placenta into mother's bloodstream.

b. Mother forms anti-Rh antibodies that cross the placenta and attack fetal Rh-positive red blood cells.

FIGURE 35.14 Hemolytic disease of the newborn.
Due to a pregnancy in which the child is Rh positive, an Rh-negative mother can begin to produce antibodies against Rh-positive red blood cells. In the same, but more likely another pregnancy, these antibodies can cross the placenta and cause hemolysis of an Rh-positive child's red blood cells.

the child's red blood cells. As the red blood cells break down, the liver produces a substance called bilirubin in such excess that bilirubin ends up in the blood and other tissues and fluids of the baby's body. Bilirubin is a pigment that causes the baby's blood and tissue to turn yellow. This is called jaundice. The presence of jaundice helps a physician to diagnose hemolytic disease of the newborn (HDN) and order a blood transfusion. Without the blood transfusion, the baby will make more red blood cells in the bone marrow, but they are immature and cannot function as mature red blood cells do. HDN can lead to brain damage and mental retardation or even death due to excess bilirubin in the blood.

The Rh problem is prevented by giving Rh-negative women an Rh immunoglobulin injection midway through the first pregnancy and no later than 72 hours after giving birth to an Rh-positive child. This injection contains anti-Rh antibodies that attack any of the baby's red blood cells in the mother's blood before these cells can stimulate her immune system to produce her own antibodies. This injection must be given before the woman starts producing her own anti-Rh antibodies, so timing is important.

When agglutination or HDN occurs, antibodies in the plasma have reacted to antigens on red blood cells.

CONNECTING THE CONCEPTS

The role of the lymphatic system in homeostasis cannot be overemphasized. The internal environment of cells consists of tissue fluid and lymph. If the composition of these fluids stays relatively constant, homeostasis is maintained. The lymphatic system is also intimately involved in immunity.

The defense systems of humans have been extensively studied, but little is known about these same systems in other animals. In humans, the levels of defense against invasion of the body by pathogens can be compared to how we protect our homes. Homes usually have external defenses such as a fence, a dog, or locked doors. Similarly, the body has barriers, such as the skin and mucous membranes, that prevent pathogens from entering the blood and lymph. Like a home alarm system, if invasion does occur, a signal goes off. First, nonspecific defense mechanisms such as the complement system and phagocytosis by white blood cells come into play. Finally, specific defense, which is dependent on the activities of B and T cells, occurs.

In humans, a strong connection exists between the immune, nervous, and endocrine systems. Lymphocytes have receptors for a wide variety of hormones, and the thymus gland produces hormones that influence the immune response. Cytokines help the body recover from disease by affecting the brain's temperature control center. A fever is thought to create an unfavorable environment for foreign invaders. Also, cytokines bring about a feeling of sluggishness, sleepiness, and loss of appetite. These behaviors tend to make us take care of ourselves until we feel better. A close connection between the immune and endocrine systems is illustrated by the ability of cortisone to mollify the inflammatory response in joints.

Summary

35.1 THE LYMPHATIC SYSTEM

The lymphatic system consists of lymphatic vessels and organs. The lymphatic vessels receive lipoproteins at intestinal villi and excess tissue fluid at blood capillaries, and they carry these to the bloodstream.

Lymphocytes are produced and accumulate in the lymphatic organs (lymph nodes, tonsils, spleen, thymus gland, and red bone marrow). Lymph is cleansed of pathogens and/or their toxins in lymph nodes, and blood is cleansed of pathogens and/or their toxins in the spleen. T lymphocytes mature in the thymus, while B lymphocytes mature in the red bone marrow where all blood cells are produced. White blood cells are necessary for nonspecific and specific defenses.

35.2 NONSPECIFIC AND SPECIFIC DEFENSES

Immunity involves nonspecific and specific defenses. Nonspecific defenses include barriers to entry, the inflammatory response, phagocytes and natural killer cells, and protective proteins.

Specific defenses require B lymphocytes and T lymphocytes, also called B cells and T cells. B cells undergo clonal selection with production of plasma cells and memory B cells after their B-cell receptor combines with a specific antigen. Plasma cells secrete antibodies and eventually undergo apoptosis. Plasma cells are responsible for antibody-mediated immunity. An antibody is usually a Y-shaped molecule that has two binding sites for a specific antigen. Memory B cells remain in the body and produce antibodies if the same antigen enters the body at a later date.

T cells have T-cell receptors and are responsible for cell-mediated immunity. For a T cell to recognize an antigen, the antigen must be presented by an antigen-presenting cell (APC), a dendritic cell, or a macrophage, along with an MHC (major histocompatibility complex) protein. Therefore, the activated T cell undergoes clonal expansion until the illness has been stemmed. Then, most of the activated T cells undergo apoptosis. A few cells remain, however, as memory T cells.

The two main types of T cells are cytotoxic T cells and helper T cells. Cytotoxic T cells kill virus-infected or cancer cells on contact because they bear a nonself protein. Helper T cells produce cytokines and stimulate other immune cells.

35.3 INDUCED IMMUNITY

Active (long-lived) immunity can be induced by vaccines when a person is well and in no immediate danger of contracting an infectious disease. Active immunity is dependent upon the presence of memory cells in the body.

Passive immunity is needed when an individual is in immediate danger of succumbing to an infectious disease. Passive immunity is short-lived because the antibodies are administered to and not made by the individual.

Cytokines, including interferon, are used in attempts to treat AIDS and to promote the body's ability to recover from cancer.

Monoclonal antibodies, which are produced by the same plasma cell, have various functions, from detecting infections to treating cancer.

35.4 IMMUNITY SIDE EFFECTS

Allergic responses occur when the immune system reacts vigorously to substances not normally recognized as foreign. Immediate allergic responses, usually consisting of coldlike symptoms, are due to the activity of antibodies. Delayed allergic responses, such as contact dermatitis, are due to the activity of T cells. Immune side effects also include blood-type reactions, tissue rejection, and autoimmune diseases.

Reviewing the Chapter

1. What is the lymphatic system, and what are its four functions?   634
2. Describe the structure and the function of lymph nodes, the spleen, the thymus gland, and red bone marrow.   635–36
3. Discuss the body's nonspecific defense mechanisms.   636–38
4. Describe the inflammatory response, and give a role for each type of cell and molecule that participates in the response.   636–37
5. Describe the clonal selection model as it applies to B cells. B cells are responsible for which type of immunity?   639
6. Describe the structure of an antibody, and define the terms variable regions and constant regions.   640
7. Discuss the clonal selection model as it applies to T cells.   642
8. Name the two main types of T cells, and state their functions.   638, 642–43
9. Describe the evidence for immunity in other animals.   646
10. How is active immunity artificially achieved? How is passive immunity achieved?   646–47
11. What are cytokines, and how are they used in immunotherapy?   648
12. How are monoclonal antibodies produced, and what are their applications?   648
13. Discuss allergies, blood typing, tissue rejection, and autoimmune diseases as they relate to the immune system.   648–51

Testing Yourself

Choose the best answer for each question.

1. Both veins and lymphatic vessels
   a. have thick walls of smooth muscle.
   b. contain valves for one-way flow of fluids.
   c. empty directly into the heart.
   d. are fed fluids from arterioles.

2. Complement
   a. is a nonspecific defense mechanism.
   b. is involved in the inflammatory response.
   c. is a series of proteins present in the plasma.
   d. plays a role in destroying bacteria.
   e. All of these are correct.

3. Which of these pertain(s) to T cells?
   a. have specific receptors
   b. are more than one type
   c. are responsible for cell-mediated immunity
   d. stimulate antibody production by B cells
   e. All of these are correct.

4. Which one of these does not pertain to B cells?
   a. have passed through the thymus
   b. have specific receptors
   c. are responsible for antibody-mediated immunity
   d. become plasma cells which synthesize and liberate antibodies

5. The clonal selection model says that
   a. an antigen selects certain B cells and suppresses them.
   b. an antigen stimulates the multiplication of B cells that produce antibodies against it.
   c. T cells select those B cells that should produce antibodies, regardless of antigens present.
   d. T cells suppress all B cells except the ones that should multiply and divide.
   e. Both b and c are correct.

6. Plasma cells are
   a. the same as memory cells.
   b. formed from blood plasma.
   c. B cells that are actively secreting antibody.
   d. inactive T cells carried in the plasma.
   e. a type of red blood cell.

7. Which of these pairs is incorrectly matched?
   a. helper T cells—help complement react
   b. cytotoxic T cells—active in tissue rejection
   c. macrophages—activate T cells
   d. memory T cells—long-living line of T cells
   e. T cells—mature in thymus

8. Vaccines are
   a. the same as monoclonal antibodies.
   b. treated bacteria or viruses, or one of their proteins.
   c. short-lived.
   d. MHC proteins.
   e. All of these are correct.

9. During blood typing, agglutination indicates that the
   a. plasma contains certain antibodies.
   b. red blood cells carry certain antigens.
   c. plasma contains certain antigens.
   d. red blood cells carry certain antibodies.
   e. white blood cells fight infection.

10. Label a–c on this IgG molecule using these terms: antigen-binding sites, light chain, heavy chain.
    d. What do V and C stand for in the diagram?

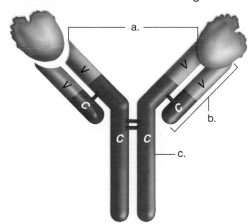

11. A person with AB type blood could receive which of the following blood types?
    a. AB
    b. B
    c. A
    d. O
    e. All are correct.

12. The lymphatic system does not
    a. transport interstitial fluid back to the blood.
    b. transport absorbed lipoproteins from the small intestine to the blood.
    c. play a role in immunological defense.
    d. filter metabolic wastes, such as urea.

13. Which is a nonspecific defense against a pathogen?
    a. skin
    b. gastric juice
    c. complement
    d. interferons
    e. All of these are correct.

14. Which cell does not phagocytize?
    a. neutrophil
    b. lymphocyte
    c. monocyte
    d. macrophage

15. B cells mature within
    a. the lymph nodes.
    b. the spleen.
    c. the thymus.
    d. the bone marrow.

16. Plasma cells secrete
    a. antibodies.
    b. perforins.
    c. lysosomal enzymes.
    d. histamine.
    e. lymphokines.

17. Mast cell secretion occurs after an allergen combines with
    a. IgG antibodies.
    b. IgE antibodies.
    c. IgM antibodies.
    d. IgA antibodies.

18. During a secondary immune response,
    a. antibodies are made quickly and in great amounts.
    b. antibody production lasts longer than in a primary response.
    c. antibodies of the IgG class are produced.
    d. lymphocyte cloning occurs.
    e. All of these are correct.

19. Active immunity may be produced by
    a. having a disease.
    b. receiving a vaccine.
    c. receiving gamma globulin injections.
    d. Both a and b are correct.
    e. Both b and c are correct.

20. T cells do not
    a. promote the activity of B cells.
    b. undergo apoptosis.
    c. secrete cytokines.
    d. produce antibodies.

21. MHC proteins
    a. are present only on the surface of certain cells.
    b. help present the antigen to T cells.
    c. are unnecessary to the immune response.
    d. Both b and c are correct.

Thinking Scientifically

1. The transplantation of organs from one person to another was impossible until the discovery of immunosuppressant drugs. Now, with the use of drugs such as cyclosporine, organs can be transplanted without rejection. Transplant patients must take immunosuppressant drugs for the remainder of their lives. How can a person do this and not eventually succumb to disease?

2. Laboratory mice are immunized with a measles vaccine. When the mice are challenged with measles virus to test the strength of their immunity, the memory cells do not completely prevent replication of the measles virus. The virus undergoes a few rounds of replication before the immune response is observed. You have developed a strain of mice with a much faster response to a viral challenge, but these mice often develop an autoimmune disease. Speculate on the connection between speed of response and an autoimmune disease.

654   35-22   PART VII COMPARATIVE ANIMAL BIOLOGY

Bioethical Issue: Cost of Drugs to Treat AIDS

Over 36 million people worldwide are living with AIDS. This disease is deadly without proper medical care but a chronic disease if treated. Drug companies typically charge a high price for AIDS medications because Americans and their insurance companies can afford them. However, these drugs are out of reach in many countries, such as those in Africa, where AIDS is a widespread problem. Some people argue that drug companies should use the profits from other drugs (such as those for heart disease, depression, and impotence) to make AIDS drugs affordable to those who need them. This has not happened yet. In some countries, governments have allowed companies to infringe on foreign patents held by major drug companies so that they can produce affordable AIDS drugs. Do drug companies have a moral obligation to provide low-cost AIDS drugs, even if they have to do so at a loss of revenue? Is it right for governments to ignore patent laws in order to provide their citizens with affordable drugs?

Understanding the Terms

agglutination 650
allergen 649
allergy 649
anaphylactic shock 649
antibody-mediated
   immunity 639
antigen 638
antigen-presenting cell
   (APC) 642
antigen receptor 638
apoptosis 635
asthma 649
autoimmune disease 649
B lymphocyte (B cell) 635
cell-mediated immunity 642
chronic inflammation 637
clonal selection model 639
complement 638
cytokine 637
cytotoxic T cell 642
delayed allergic response 649
dendritic cell 637
edema 635
foreign antigen 638
hay fever 649
helper T cell 642
histamine 636
immediate allergic
   response 649

immunity 636
immunization 647
immunoglobulin (Ig) 640
inflammatory response 636
interferon 638
interleukin 648
lymph 634
lymphatic (lymphoid) organ 635
lymphatic system 634
lymphatic vessel 634
lymph node 636
macrophage 637
mast cell 636
membrane attack complex 638
memory B cell 639
memory T cell 643
monoclonal antibody 648
natural killer (NK) cell 637
pathogen 636
Peyer's patches 636
plasma cell 639
red bone marrow 635
self-antigen 638
spleen 636
T lymphocyte (T cell) 635
thymus gland 635
tonsils 636
vaccine 647
vermiform appendix 636

Match the terms to these definitions:

a. _____ Antigens prepared in such a way that they can promote active immunity without causing disease.

b. _____ Fluid, derived from tissue fluid, that is carried in lymphatic vessels.

c. _____ Foreign substance, usually a protein or a polysaccharide, that stimulates the immune system to react, such as by producing antibodies.

d. _____ Process of programmed cell death involving a cascade of specific cellular events leading to the death and destruction of the cell.

e. _____ Lymphocyte that matures in the thymus and exists in three varieties, one of which kills antigen-bearing cells outright.

ARIS, the *Biology* Website

ARIS, the website for *Biology*, provides a wealth of information organized and integrated by chapter. You will find practice quizzes, interactive activities, labeling exercises, flashcards, and much more that will complement your learning and understanding of general biology.

www.mhhe.com/maderbiology9

36

DIGESTIVE SYSTEMS AND NUTRITION

Animals are heterotrophic organisms that use food as a source of energy and building blocks for their own organic molecules. The variety of diets found in the animal kingdom is astounding—from beetles eating rotting flesh to killer whales and great blue herons consuming live prey. The majority of animals have a complete digestive tract with one opening that serves as an entrance and another that serves as an exit. Modifications of this type of tract are innumerable, and animals have a great range of adaptations to acquire, manipulate, and digest food.

A great blue heron stands on long legs along the water's edge or even in the water. Periodically, the long neck is cocked into the familiar S shape, and then the swordlike bill is launched to grasp a fish. Most birds, like the heron, are discontinuous feeders and feed periodically. Some birds, like a hummingbird, are continuous feeders and feed almost every waking moment. In keeping with feeding discontinuously, the esophagus of a great blue heron has a crop, a storage area that allows the bird to quickly ingest a large fish and digest it at leisure. The bird's stomach has a gizzard that crushes hard materials with its muscular walls and abrades them with sand the bird swallowed sometime in the past. Similar to humans, the bulk of enzymatic digestion and also absorption occur in the small intestine.

In a study of mantled howling monkeys, it was concluded that howlers choose foods that give them the greatest nutrient return. In contrast, humans often must be taught good nutrition. Eating correctly and exercising regularly is essential to staying healthy.

This great blue heron has caught a fish.

36.1 DIGESTIVE TRACTS

The majority of animals have some sort of gut, or digestive tract, where food is digested into small nutrient molecules that can cross plasma membranes. Digestion contributes to homeostasis by providing the body with the nutrients needed to sustain the life of cells. A digestive tract (1) ingests food, (2) breaks food down into small molecules that can cross plasma membranes, (3) absorbs these nutrient molecules, and (4) eliminates nondigestible remains.

Incomplete Versus Complete Tracts

An **incomplete digestive tract** has a single opening, usually called a mouth. However, the single opening is used both as an entrance for food and an exit for wastes. Planarians, which are flatworms, have an incomplete tract (Fig. 36.1). It begins with a mouth and muscular pharynx and then the tract, a gastrovascular cavity, branches throughout the body. Planarians are primarily carnivorous and feed largely on smaller aquatic animals, as well as bits of organic debris. When a planarian is feeding, the pharynx actually extends beyond the

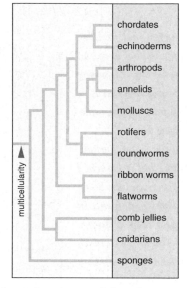

mouth. The body is wrapped about the prey and the pharynx sucks up minute quantities at a time. Digestive enzymes present in the tract allow some extracellular digestion to occur. Digestion is finished intracellularly by the cells that line the tract. No cell in the body is far from the digestive tract; therefore, diffusion alone is sufficient to distribute nutrient molecules.

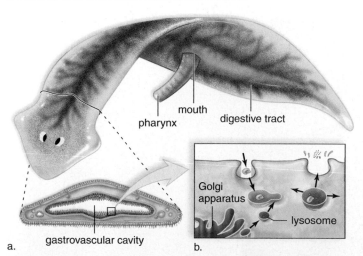

FIGURE 36.1 Incomplete digestive tract of a planarian.
a. Planarians, which are flatworms, have a gastrovascular cavity with a single opening that acts as both an entrance and an exit. Planarians rely on intracellular digestion to complete the digestive process. **b.** Phagocytosis produces a vacuole, which joins with an enzyme-containing lysosome. The digested products pass from the vacuole into the cytoplasm before any nondigestible material is eliminated at the plasma membrane.

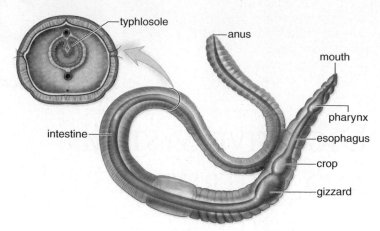

FIGURE 36.2 Complete digestive tract of an earthworm.
Complete digestive tracts have both a mouth and an anus and can have many specialized parts, such as those labeled in this drawing. Also in earthworms, which are annelids, the absorptive surface of the intestine is increased by an internal fold called the typhlosole.

The digestive tract of a planarian is notable for its lack of specialized parts. It is saclike because the pharynx serves not only as an entrance for food but also as an exit for nondigestible material. Specialization of parts does not occur under these circumstances.

Planarians have some modified parasitic relatives. Tapeworms, which are parasitic flatworms, lack a digestive system. Nutrient molecules are absorbed by the tapeworm from the intestinal juices of the host, which surround the tapeworm's body. The integument and body wall of the tapeworm are highly modified for this purpose. They have millions of microscopic, fingerlike projections that increase the surface area for absorption.

In contrast to planarians, earthworms, which are annelids, have a **complete digestive tract,** meaning that the tract has a mouth and an anus (Fig. 36.2). Earthworms feed mainly on decayed organic matter in soil. The muscular pharynx draws in food with a sucking action. Food then enters the crop, which is a storage area with thin, expansive walls. From there, food goes to the gizzard, where thick, muscular walls crush and sand grinds the food. Digestion is extracellular within an intestine. The surface area of digestive tracts is often increased for absorption of nutrient molecules, and in earthworms, this is accomplished by an intestinal fold called the **typhlosole.** Undigested remains pass out of the body at the anus. Specialization of parts is obvious in the earthworm because the pharynx, the crop, the gizzard, and the intestine each has a particular function in the digestive process.

In contrast to an incomplete, saclike tract, the complete digestive tract, with both a mouth and an anus, can have many specialized parts depending on the way of life of the animal.

Continuous Versus Discontinuous Feeders

Clams, which are molluscs, are continuous feeders, called filter feeders (Fig. 36.3a). Water is always moving into the mantle cavity by way of the incurrent siphon (slitlike opening) and depositing particles, including algae, protozoans, and minute invertebrates, on the gills. The size of the incurrent siphon permits the entrance of only small particles, which adhere to the gills. Ciliary action moves suitably sized particles to the labial palps, which force them through the mouth into the stomach. Digestive enzymes are secreted by a large digestive gland, but amoeboid cells present throughout the tract are believed to complete the digestive process by intracellular digestion.

Marine fanworms, which are annelids, are also filter feeders. The feathery tentacles of these worms are specialized for gathering fine particles and microscopic plankton from the water. They allow only small particles to enter their digestive tract. Larger particles are rejected. A baleen whale, such as a blue whale, is an active filter feeder. Baleen, a keratinized curtain-like fringe, hangs from the roof of the mouth and filters small shrimp called krill from the water. A baleen whale filters up to a ton of krill every few minutes.

Squids, which are molluscs, are discontinuous feeders (Fig. 36.3b). The body of a squid is streamlined, and the animal moves rapidly through the water using jet propulsion (forceful expulsion of water from a tubular funnel). The head of a squid is surrounded by ten arms, two of which have developed into long, slender tentacles whose suckers have toothed, horny rings. These tentacles seize prey (fishes, shrimps, and worms) and bring it to the squid's beaklike jaws, which bite off pieces pulled into the mouth by the action of a **radula,** a tonguelike structure. An esophagus leads to a stomach and a cecum (blind sac), where digestion occurs. The stomach, supplemented by the cecum, retains food until digestion is complete. Discontinuous feeders, whether they are carnivores, like owls, or herbivores, like elephants, require a storage area for food, which can be a stomach and cecum or just a stomach.

Continuous feeders, such as filter feeders, do not need a storage area for food; discontinuous feeders do need a storage area for food.

Adaptation to Diet

Some animals are omnivores; they eat both plants and animals. Others are herbivores; they feed only on plants. Still others are carnivores; they eat only other animals. Among invertebrates, filter feeders such as clams and tube worms are omnivores. Land snails, which are terrestrial molluscs, and some insects, such as grasshoppers and locusts, are herbivores. Spiders (arthropods) are carnivores, as are sea stars (echinoderms), which feed on clams. A sea star positions itself above a clam and uses its tube feet to pull the valves of the shell apart (see Fig. 30.17). Then, it everts a part of its stomach to start the digestive process, even while the clam is trying to close its shell. Some invertebrates are cannibalistic. A female praying mantis (an insect), if starved, will feed on her mate as the reproductive act is taking place!

Among mammals, the dentition differs according to mode of nutrition. Among herbivores, the koala of Australia is famous for its diet of only eucalyptus leaves, and likewise many other mammals are browsers, feeding off bushes and trees. Grazers, like the horse, feed off grasses. The horse has sharp, even incisors for neatly clipping off blades of grass and large, flat premolars and molars for grinding and crushing the grass (Fig. 36.4a). Extensive grinding and crushing disrupts plant cell walls, allowing

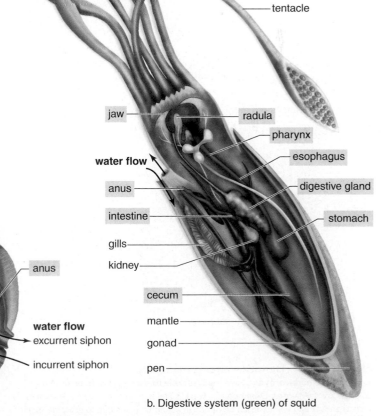

b. Digestive system (green) of squid

a. Digestive system (green) of clam

FIGURE 36.3 Nutritional mode of a clam compared to a squid.
Clams and squids are molluscs. A clam burrows in the sand or mud, where it filter feeds, whereas a squid swims freely in open waters and captures prey. In keeping with their lifestyles, a clam (**a**) is a continuous feeder and a squid (**b**) is a discontinuous feeder.

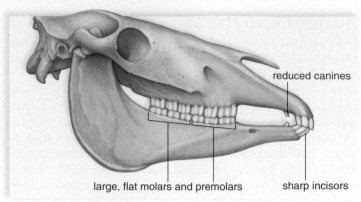

a. Horses are herbivores.

reduced canines

large, flat molars and premolars

sharp incisors

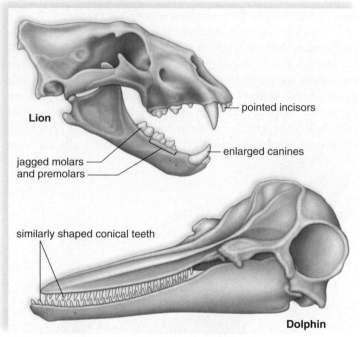

Lion

pointed incisors

enlarged canines

jagged molars
and premolars

similarly shaped conical teeth

Dolphin

b. Lions and dolphins are carnivores.

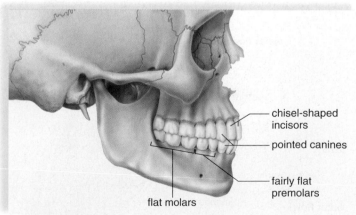

chisel-shaped
incisors

pointed canines

fairly flat
premolars

flat molars

c. Humans are omnivores.

FIGURE 36.4 Dentition among mammals.
a. Horses are herbivores and have teeth suitable to eating grass. **b.** Lions and dolphins are carnivores. Dentition in a lion is suitable to chewing up large animals such as zebras and wildebeests. Dentition in a dolphin is suitable to grasping small animals like fishes, which are swallowed whole. **c.** Humans are omnivores and have teeth suitable to a mixed diet of vegetables and meat.

bacteria located in a part of the digestive tract called the cecum to digest cellulose. Other mammalian grazers, such as cattle, sheep, and deer, are ruminants with a large, four-chambered stomach. In contrast to horses, they graze quickly and swallow partially chewed grasses into a special part of the stomach called a rumen. Here, microorganisms start the digestive process, and the result, called cud, is regurgitated at a later time when the animal is no longer feeding. The cud is chewed again before being swallowed for complete digestion.

Many mammals, including dogs, lions, toothed whales, and dolphins, are carnivores. Dogs and lions use pointed incisors and enlarged canine teeth to tear off pieces small enough to be quickly swallowed (Fig. 36.4b, top). Dolphins and toothed whales swallow food whole without chewing it first; they are equipped with many identical, conical teeth that are used to catch and grasp their slippery prey before swallowing (Fig. 36.4b, bottom). Meat is rich in protein and fat and is easier to digest than plant material. The intestine of a rabbit, a herbivore, is much longer than that of a similarly sized cat, a carnivore.

Humans, as well as raccoons, rats, and brown bears, are omnivores. Therefore, the dentition has a variety of specializations to accommodate both a vegetable diet and a meat diet. An adult human has 32 teeth. One-half of each jaw has teeth of four different types: two chisel-shaped incisors for shearing; one pointed canine for tearing; two fairly flat premolars for grinding; and three molars, well flattened for crushing (Fig. 36.4c).

Herbivores have large, flat molars that grind food; some carnivores have incisors and canines to tear off chunks of meat. Omnivores have a variety of teeth.

36.2 HUMAN DIGESTIVE TRACT

Humans have a tube-within-a-tube body plan and possess a complete digestive tract, which begins with a mouth and ends in an anus. The major structures found in the human digestive tract are illustrated in Figure 36.5, and their features and major functions are summarized in Table 36.1.

The digestion of food in humans is an extracellular event. Digestion requires a cooperative effort between different parts

TABLE 36.1

Path of Food

Organ	Special Features	Function
Mouth	Teeth, tongue, salivary glands	Chewing of food; digestion of starch
Esophagus		Movement of food by peristalsis
Stomach	Gastric glands	Storage of food; acidity kills some bacteria; digestion of protein
Small intestine	Villi	Digestion of all foods; absorption of nutrients
Large intestine		Absorption of water; storage of nondigestible remains
Anus		Defecation

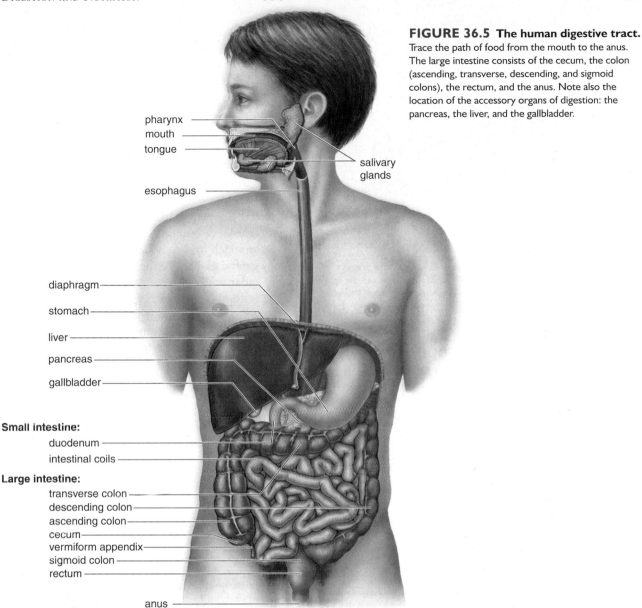

FIGURE 36.5 The human digestive tract.
Trace the path of food from the mouth to the anus.
The large intestine consists of the cecum, the colon
(ascending, transverse, descending, and sigmoid
colons), the rectum, and the anus. Note also the
location of the accessory organs of digestion: the
pancreas, the liver, and the gallbladder.

of the body. Digestion consists of two major stages: mechanical digestion and chemical digestion. Mechanical digestion involves the physical breakdown of food into smaller particles. This is accomplished through the chewing of food in the mouth and the physical churning and mixing of food in the stomach and small intestine. Chemical digestion requires enzymes that are secreted by the digestive tract or by accessory glands that lie nearby. Specific enzymes break down particular macromolecules into smaller molecules that can be absorbed.

Mouth

The **mouth,** or oral cavity, serves as the beginning of the digestive tract. The palate, or roof of the mouth, separates the oral cavity from the nasal cavity. It consists of the anterior hard palate and the posterior soft palate. The fleshy uvula is the posterior extension of the soft palate (see Fig. 36.6). The cheeks and lips retain food while it is chewed by the teeth and mixed with saliva.

There are three major pairs of **salivary glands** that send their juices by way of ducts to the mouth. Saliva contains

the enzyme **salivary amylase,** which begins the process of starch digestion. The disaccharide maltose is a typical end product of salivary amylase digestion.

$$starch \quad + \quad H_2O \xrightarrow{\text{salivary amylase}} maltose$$

While in the mouth, food is manipulated by a muscular tongue, which has touch and pressure receptors similar to those in the skin. Taste buds, sensory receptors that are stimulated by the chemical composition of food, are also found primarily on the tongue as well as on the surface of the mouth. The tongue, which is composed of striated muscle and an outer layer of mucous membrane, mixes the chewed food with saliva. It then forms this mixture into a mass called a bolus in preparation for swallowing.

The salivary glands send saliva into the mouth, where food is chewed and formed into a bolus for swallowing.

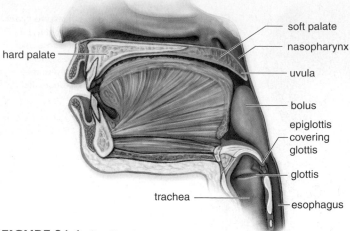

FIGURE 36.6 Swallowing.
Respiratory and digestive passages converge and diverge in the pharynx. When food is swallowed, the soft palate closes off the nasopharynx, and the epiglottis covers the glottis, forcing the bolus to pass down the esophagus. Therefore, a person does not breathe when swallowing.

The Pharynx and the Esophagus

The digestive and respiratory passages come together in the **pharynx** and then separate (Fig. 36.6). When food is swallowed, the soft palate, the rear portion of the mouth's roof, moves back to close off the nasopharynx. A flap of tissue called the epiglottis covers the glottis, an opening into the trachea. Now the bolus must move through the pharynx into the esophagus because the air passages are blocked.

The **esophagus** [Gk. *eso,* within, and *phagein,* eat] is a tubular structure, of about 25 cm in length, that takes food to the stomach. When food enters the esophagus, peristalsis begins (Fig. 36.7). **Peristalsis** [Gk. *peri,* around, and *stalsis,* compression] is a rhythmical contraction that serves to move the contents along in tubular organs, such as the digestive tract.

During swallowing, the air passages are blocked, and the bolus enters the esophagus, which conducts it to the stomach.

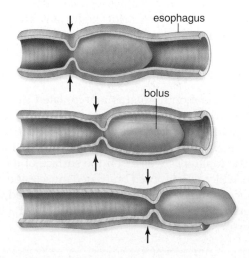

FIGURE 36.7 Peristalsis in the digestive tract.
These three drawings show how a peristaltic wave moves through a single section of the esophagus over time. The arrows point to areas of contraction.

Stomach

The **stomach** (Fig. 36.8) is a thick-walled, J-shaped organ that lies on the left side of the body beneath the diaphragm. The wall of the stomach has deep folds (rugae), that disappear as the stomach fills to an approximate capacity of 1 liter. Therefore, humans can periodically eat relatively large meals and spend the rest of their time at other activities.

The stomach is much more than a storage organ, as was discovered by William Beaumont in the mid-nineteenth century. Beaumont, an American doctor, had a French Canadian patient, Alexis St. Martin. St. Martin had been shot in the stomach, and when the wound healed, he was left with a fistula, or opening, that allowed Beaumont to look inside the stomach and to collect gastric (stomach) juices produced by gastric glands. Beaumont was able to determine that the muscular walls of the stomach contract vigorously and mix food with juices that are secreted whenever food enters the stomach. He found that gastric juice contains hydrochloric acid (HCl) and a substance, now called pepsin, that is active in digestion. He also found that the gastric juices are produced independently of the protective mucous secretions of the stomach. Beaumont's work, which was carefully and painstakingly done, pioneered the study of digestive physiology.

The epithelial lining of the stomach has millions of gastric pits, which lead into gastric glands. The gastric glands produce gastric juice. So much hydrochloric acid is secreted by the gastric glands that the stomach routinely has a pH of

FIGURE 36.8 Anatomy of the stomach.
a. The stomach, which has thick walls, expands as it fills with food. **b.** The mucous membrane layer of its walls secretes mucus and contains gastric glands, which secrete a gastric juice active in the digestion of protein.

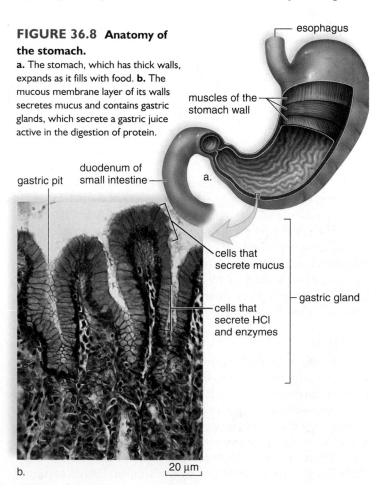

about 2. Such a high acidity usually is sufficient to kill bacteria and other microorganisms that might be in food. This low pH also stops the activity of salivary amylase, which functions optimally at the near-neutral pH of saliva.

As with the rest of the digestive tract, a thick layer of mucus protects the wall of the stomach from enzymatic action. Still, an ulcer, which is an open sore in the wall caused by the gradual destruction of tissues, does occur in some individuals. Ulcers can be caused by an infection by an acid-resistant bacterium, *Helicobacter pylori,* which is able to attach to the epithelial lining. Wherever the bacterium attaches, the lining stops producing mucus, and the area becomes exposed to digestive action. As a result, an ulcer develops.

Eventually, food mixing with gastric juice in the stomach contents become **chyme,** which has a thick, creamy consistency. At the base of the stomach is a narrow opening controlled by a sphincter. A sphincter is a muscle that surrounds a tube and closes or opens the tube by contracting and relaxing. Whenever the sphincter relaxes, a small quantity of chyme passes through the opening into the small intestine. When chyme enters the small intestine, it sets off a neural reflex that causes the muscles of the sphincter to contract vigorously and to close the opening temporarily. Then the sphincter relaxes again and allows more chyme to enter. The slow manner in which chyme enters the small intestine allows for thorough digestion.

> The stomach can expand to accommodate large amounts of food. When food is present, the stomach churns, mixing food with acidic gastric juice, resulting in chyme.

The Small Intestine

The **small intestine** is named for its small diameter (compared to that of the large intestine), but perhaps it should be called the long intestine. The small intestine averages about 6 m in length, compared to the large intestine, which is about 1.5 m in length.

The first 25 cm of the small intestine is called the **duodenum.** A duct brings bile from the liver and gallbladder, and pancreatic juice from the pancreas, into the small intestine (see Fig. 36.10*a*). **Bile** emulsifies fat—emulsification causes fat droplets to disperse in water. The intestine has a slightly basic pH because pancreatic juice contains sodium bicarbonate ($NaHCO_3$), which neutralizes chyme. The enzymes in pancreatic juice and enzymes produced by the intestinal wall complete the process of food digestion.

It has been suggested that the surface area of the small intestine is approximately that of a tennis court. What factors contribute to increasing its surface area? The wall of the small intestine contains fingerlike projections called villi (sing. **villus**), which give the intestinal wall a soft, velvety appearance (Fig. 36.9). A villus has an outer layer of columnar epithelial cells, and each of these cells has thousands of microscopic extensions called microvilli. Collectively, in electron micrographs, microvilli give the villi a fuzzy border, known as a "brush border." Since the microvilli bear the intestinal enzymes, these enzymes are called brush-border enzymes. The microvilli greatly increase the surface area of the villus for the absorption of nutrients.

Nutrients are absorbed into the vessels of a villus, which contains blood capillaries and a lymphatic capillary, called a **lacteal.** Sugars (digested from carbohydrates) and amino acids (digested from proteins) enter the blood capillaries of a villus. Glycerol and fatty acids (digested from fats) enter the epithelial cells of the villi, and within these cells they are joined and packaged as lipoprotein droplets, which enter a lacteal. After nutrients are absorbed, they are eventually carried to all the cells of the body by the bloodstream.

> The large surface area of the small intestine facilitates absorption of nutrients into the cardiovascular system (glucose and amino acids) and the lymphatic system (fats).

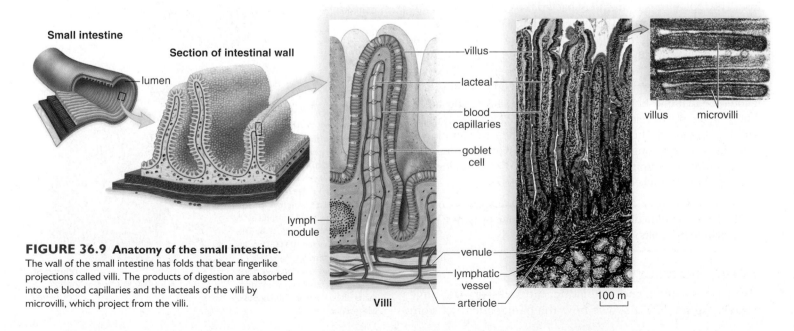

FIGURE 36.9 Anatomy of the small intestine.
The wall of the small intestine has folds that bear fingerlike projections called villi. The products of digestion are absorbed into the blood capillaries and the lacteals of the villi by microvilli, which project from the villi.

Small intestine
lumen
Section of intestinal wall
lymph nodule
Villi
villus
lacteal
blood capillaries
goblet cell
venule
lymphatic vessel
arteriole
villus   microvilli
100 m

Large Intestine

The **large intestine,** which includes the cecum, the colon, the rectum, and the anus, is larger in diameter (6.5 cm) but shorter in length (1.5 m) than the small intestine. The large intestine absorbs water, salts, and some vitamins. It also stores nondigestible material until it is eliminated at the anus. No digestion takes place in the large intestine.

The cecum, which lies below the junction with the small intestine, is the blind end of the large intestine. The cecum has a small projection called the vermiform **appendix** [L. *verm,* worm, and *form,* shape, and *append,* an addition]. In humans, the appendix may play a role in fighting infections. In the case of appendicitis, the appendix becomes infected and so filled with fluid that it may burst. If an infected appendix bursts before it can be removed, it can lead to a serious, generalized infection of the abdominal lining, called peritonitis.

The **colon** is subdivided into the ascending, transverse, descending, and sigmoid colons (see Fig. 36.5). The sigmoid colon enters the rectum, the last 20 cm of the large intestine. About 1.5 liters of water enters the digestive tract daily as a result of eating and drinking. An additional 8.5 liters enter the digestive tract each day carrying the various substances secreted by the digestive glands. About 95% of this water is absorbed by the small intestine, and much of the remaining portion is absorbed into the cells of the colon. If this water is not reabsorbed, **diarrhea** can lead to serious dehydration and ion loss, especially in children.

The large intestine has a large population of bacteria, notably *Escherichia coli.* The bacteria break down nondigestible material, and they also produce some vitamins, such as vitamin K. Vitamin K is necessary to blood clotting. Digestive wastes (feces) eventually leave the body through the **anus,** the opening of the anal canal. Feces are about 75% water and 25% solid matter. Almost one-third of this solid matter is made up of intestinal bacteria. The remainder is undigested plant material, fats, waste products (such as bile pigments), inorganic material, mucus, and dead cells from the intestinal lining. The color of feces is the result of the breakdown of bilirubin and the presence of oxidized iron. The foul odor is the result of bacterial action.

The colon is subject to the development of **polyps,** which are small growths arising from the epithelial lining. Polyps, whether they are benign or cancerous, can be removed surgically. Some investigators believe that dietary fat increases the likelihood of colon cancer. Dietary fat causes an increase in bile secretion, and it could be that intestinal bacteria convert bile salts to substances that promote the development of colon cancer. Dietary fibers absorb water and add bulk, thereby diluting the concentration of bile salts and facilitating the movement of substances through the intestine. Regular elimination reduces the time that the colon wall is exposed to any cancer-promoting agents in feces.

The large intestine does not produce digestive enzymes; it absorbs water, salts, and some vitamins.

Three Accessory Organs

The pancreas, liver, and gallbladder are accessory digestive organs. Figure 36.10*a* shows how the pancreatic duct from the pancreas and the common bile duct from the liver and gallbladder enter the duodenum.

The Pancreas

The **pancreas** lies deep in the abdominal cavity, resting on the posterior abdominal wall. It is an elongated and somewhat flattened organ that has both an endocrine and an exocrine function. As an endocrine gland, it secretes insulin and glucagon, hormones that help keep the blood glucose level within normal limits. In this chapter, however, we are interested in its exocrine function. Most pancreatic cells produce pancreatic juice, which contains sodium bicarbonate ($NaHCO_3$) and digestive enzymes for all types of food. Sodium bicarbonate neutralizes acid chyme from the stomach. Pancreatic amylase digests starch, trypsin digests protein, and lipase digests fat.

The Liver

The **liver,** which is the largest gland in the body, lies mainly in the upper right section of the abdominal cavity, under the diaphragm (see Fig. 36.5). The liver contains approximately 100,000 lobules that serve as its structural and functional units (Fig. 36.10*b*). Triads, located between the lobules, consist of a bile duct, which takes bile away from the liver; a branch of the hepatic artery, which brings O_2-rich blood to the liver; and a branch of the hepatic portal vein, which transports nutrients

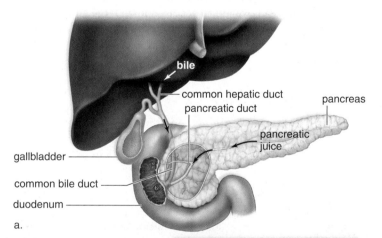

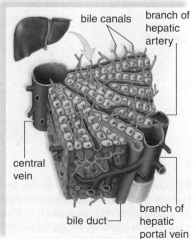

FIGURE 36.10 Liver, gallbladder, and pancreas.
a. The liver makes bile, which is stored in the gallbladder and sent (black arrow) to the small intestine by way of the common bile duct. The pancreas produces digestive enzymes that are sent (black arrows) to the small intestine by way of the pancreatic duct. **b.** The liver contains over 100,000 lobules. Each lobule contains many cells that perform the various functions of the liver. They remove and add materials to the blood and deposit bile in a duct.

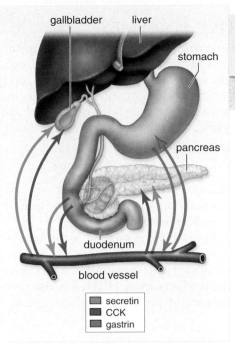

FIGURE 36A Hormonal control of digestive gland secretions.

Gastrin (blue), produced by the lower part of the stomach, enters the bloodstream and thereafter stimulates the upper part of the stomach to produce more digestive juices. Secretin (green) and CCK (purple), produced by the duodenal wall, stimulate the pancreas to secrete its digestive juices and the gallbladder to release bile.

Labels in figure: gallbladder, liver, stomach, pancreas, duodenum, blood vessel

Legend: secretin, CCK, gastrin

science focus

Control of Digestive Juices

The study of the control of digestive gland secretion began in the late 1800s. At that time, Ivan Pavlov showed that dogs would begin to salivate at the ringing of a bell because they had learned to associate the sound of the bell with being fed. Pavlov's experiments demonstrated that even the thought of food can cause the nervous system to order the secretion of digestive juices. If food is present in the mouth, the stomach, and the small intestine, digestive secretion occurs because of simple reflex action. The presence of food sets off nerve impulses that travel to the brain. Thereafter, the brain stimulates the digestive glands to secrete.

In the twentieth century, investigators discovered that specific control of digestive secretions is achieved by hormones. Hormones are chemical messengers often transported in the bloodstream from one set of cells to another. When investigators cut the nerves leading to the digestive tract, the stomach and pancreas continued to secrete their enzymes. The gallbladder continued to secrete bile. They concluded that hormones, not nerves, controlled these secretions.

When a person has eaten a meal particularly rich in protein, the gastric glands of the lower stomach wall produce the hormone gastrin (Fig. 36A). Gastrin enters the bloodstream, and soon stomach churning and the secretory activity of gastric glands increase.

Cells of the duodenal wall also produce hormones, two of which are of particular interest—secretin and CCK (cholecystokinin). Acid, especially hydrochloric acid (HCl) present in chyme, stimulates the release of secretin, while partially digested protein and fat stimulate the release of CCK. Soon after these hormones enter the bloodstream, the pancreas increases its output of pancreatic juice, and the liver increases its output of bile. The gallbladder contracts to release bile.

Another hormone produced by the duodenal wall, GIP (gastric inhibitory peptide), works opposite to gastrin—it inhibits gastric gland secretion and stomach motility. This is not surprising, because the body's hormones often have opposite effects.

---

from the intestines (see Fig. 34.8). The central veins of lobules enter a hepatic vein. Blood moves from the intestines to the liver via the hepatic portal vein and from the liver to the inferior vena cava via the hepatic veins.

In some ways, the liver acts as the gatekeeper to the blood (Table 36.2). As blood in the hepatic portal vein passes through the liver, it removes poisonous substances and detoxifies them. The liver also removes and stores iron and the vitamins A, B_{12}, D, E, and K. The liver makes the plasma proteins and helps regulate the quantity of cholesterol in the blood.

The liver maintains the blood glucose level at about 100 mg/100 ml (0.1%), even though a person eats intermittently. When insulin is present, any excess glucose present in blood is removed and stored by the liver as glycogen. Between meals, glycogen is broken down to glucose, which enters the hepatic veins, and in this way, the blood glucose level remains constant.

If the supply of glycogen is depleted, the liver converts glycerol (from fats) and amino acids to glucose molecules. The conversion of amino acids to glucose necessitates deamination, the removal of amino groups. By a complex metabolic pathway, the liver then combines ammonia with carbon dioxide to form urea. Urea is the usual nitrogenous waste product from amino acid breakdown in humans.

The liver produces bile, which is stored in the gallbladder. Bile has a yellowish green color because it contains the bile pigment *bilirubin*, derived from the breakdown of hemoglobin, the red pigment of red blood cells. Bile also contains bile salts. Bile salts are derived from cholesterol, and they emulsify fat in the small intestine. When fat is emulsified, it breaks up into

TABLE 36.2

Functions of the Liver

1. Detoxifies blood by removing and metabolizing poisonous substances

2. Stores iron and the vitamins A, B_{12}, D, E, and K

3. Makes plasma proteins, such as albumins and fibrinogen, from amino acids

4. Stores glucose as glycogen after a meal, and breaks down glycogen to glucose to maintain the glucose level of blood between eating periods

5. Produces urea after breaking down amino acids

6. Removes bilirubin, a breakdown product of hemoglobin from the blood, and excretes it in bile, a liver product

7. Helps regulate blood cholesterol level, converting some to bile salts

droplets, providing a much larger surface area, which can be acted upon by a digestive enzyme from the pancreas.

Liver Disorders.　Hepatitis and cirrhosis are two serious diseases that affect the entire liver and hinder its ability to repair itself. Therefore, they are life-threatening diseases. When a person has a liver ailment, jaundice may occur. **Jaundice** is present when the skin and the whites of the eyes have a yellowish tinge. Jaundice occurs because bilirubin is deposited in the skin, due to an abnormally large amount in the blood. Jaundice can also result from **hepatitis,** inflammation of the liver. Viral hepatitis occurs in several forms. Hepatitis A is usually acquired from sewage-contaminated drinking water. Hepatitis B, which is usually spread by sexual contact, can also be spread by blood transfusions or contaminated needles. The hepatitis B virus is more contagious than the AIDS virus, which is spread in the same way. Thankfully, however, a vaccine is now available for hepatitis B. Hepatitis C, which is usually acquired by contact with infected blood and for which there is no vaccine, can lead to chronic hepatitis, liver cancer, and death.

Cirrhosis is another chronic disease of the liver. First, the organ becomes fatty, and then liver tissue is replaced by inactive fibrous scar tissue. Cirrhosis of the liver is often seen in alcoholics, due to malnutrition and to the excessive amounts of alcohol (a toxin) the liver is forced to break down.

The liver has amazing regenerative powers and can recover if the rate of regeneration exceeds the rate of damage. During liver failure, however, there may not be enough time to let the liver heal itself. Liver transplantation is usually the preferred treatment for liver failure, but artificial livers have been developed and tried in a few cases. One type is a cartridge that contains liver cells. The patient's blood passes through the cellulose acetate tubing of the cartridge and is serviced in the same manner as with a normal liver.

The Gallbladder

The **gallbladder** is a pear-shaped, muscular sac attached to the surface of the liver (see Fig. 36.5). About 1,000 ml of bile are produced by the liver each day, and any excess is stored in the gallbladder. Water is reabsorbed by the gallbladder so that bile becomes a thick, mucuslike material. When needed, bile leaves the gallbladder and proceeds to the duodenum via the common bile duct.

The cholesterol content of bile can come out of solution and form crystals. If the crystals grow in size, they form gallstones. The passage of the stones from the gallbladder may block the common bile duct and cause obstructive jaundice. Then, the gallbladder must be removed.

The pancreas produces pancreatic juice, which contains enzymes for the digestion of food. Among the liver's many functions is the production of bile, which is stored in the gallbladder.

36.3 Digestive Enzymes

The various digestive enzymes present in the digestive juices, mentioned previously, help break down carbohydrates, proteins, nucleic acids, and fats, the major components of food. Starch is a polysaccharide, and its digestion begins in the mouth. Saliva from the salivary glands has a neutral pH and contains **salivary amylase,** the first enzyme to act on starch.

$$\text{starch} + H_2O \xrightarrow{\text{salivary amylase}} \text{maltose}$$

Maltose molecules cannot be absorbed by the intestine; additional digestive action in the small intestine converts maltose to glucose, which can be absorbed.

Protein digestion begins in the stomach. Gastric juice secreted by gastric glands has a very low pH—about 2—because it contains hydrochloric acid (HCl). Pepsinogen, a precursor that is converted to **pepsin** when exposed to HCl, is also present in gastric juice. Pepsin acts on protein to produce peptides.

$$\text{protein} + H_2O \xrightarrow{\text{pepsin}} \text{peptides}$$

Peptides are usually too large to be absorbed by the intestinal lining, but later they are broken down to amino acids in the small intestine.

Starch, proteins, nucleic acids, and fats are all enzymatically broken down in the small intestine. Pancreatic juice, which enters the duodenum, has a basic pH because it contains sodium bicarbonate (NaHCO₃). One pancreatic enzyme, **pancreatic amylase,** digests starch (Fig. 36.11a).

$$\text{starch} + H_2O \xrightarrow{\text{pancreatic amylase}} \text{maltose}$$

Another pancreatic enzyme, **trypsin,** digests protein (Fig. 36.11b).

$$\text{protein} + H_2O \xrightarrow{\text{trypsin}} \text{peptides}$$

Trypsin is secreted as trypsinogen, which is converted to trypsin in the duodenum.

Maltase and peptidases, enzymes produced by the small intestine, complete the digestion of starch to glucose and protein to amino acids, respectively. Glucose and amino acids are small molecules that cross into the cells of the villi and enter the blood (Fig. 36.11a, b).

Maltose, a disaccharide that results from the first step in starch digestion, is digested to glucose by **maltase.**

$$\text{maltose} + H_2O \xrightarrow{\text{maltase}} \text{glucose} + \text{glucose}$$

Other disaccharides have their own enzyme and are digested in the small intestine. The absence of any one of these enzymes can cause illness.

Peptides, which result from the first step in protein digestion, are digested to amino acids by **peptidases.**

$$\text{peptides} + H_2O \xrightarrow{\text{peptidases}} \text{amino acids}$$

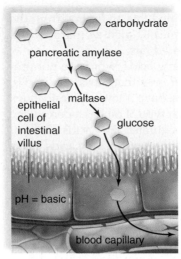

a. Carbohydrate digestion

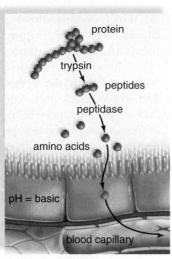

b. Protein digestion

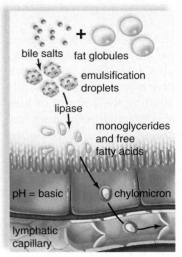

c. Fat digestion

FIGURE 36.11 Digestion and absorption of nutrients. a. Starch is digested to glucose, which is actively transported into the epithelial cells of intestinal villi. From there, glucose moves into the bloodstream. **b.** Proteins are digested to amino acids, which are actively transported into the epithelial cells of intestinal villi. From there, amino acids move into the bloodstream. **c.** Fats are emulsified by bile and digested to monoglycerides and fatty acids. These diffuse into epithelial cells, where they recombine and join with proteins to form lipoproteins, called chylomicrons. Chylomicrons enter a lacteal.

Lipase, a third pancreatic enzyme, digests fat molecules in fat droplets after they have been emulsified by bile salts.

$$fat \xrightarrow{\text{bile salts}} fat\ droplets$$

$$fat\ droplets + H_2O \xrightarrow{\text{lipase}} glycerol + 3\ fatty\ acids$$

Specifically, the end products of lipase digestion are monoglycerides (glycerol + one fatty acid) and fatty acids. These enter the cells of the villi, and within these cells, they are rejoined and packaged as lipoprotein droplets, called chylomicrons. Chylomicrons enter the lacteals (Fig. 36.11c).

Digestive enzymes break down food to the nutrient molecules: glucose, amino acids, fatty acids, and glycerol. The first two are absorbed into the blood capillaries, and the last two re-form within epithelial cells before entering the lacteals.

36.4 NUTRITION

In 2005, the U.S. Department of Agriculture (USDA) released new dietary recommendations. Figure 36.12 gives the recommended proportions in pyramid form for a person on a 2,000 Calorie[1] diet. The daily diet should include about 6 oz of whole-grain breads, crackers, pasta, cereals, or rice. Fruits and vegetables, collectively, should be consumed in even greater quantity than grains. The USDA recommends limiting solid fats such as butter, stick margarine, lard, and shortening. However, the USDA does not shy away from dairy: They recommend that fat-free or low-fat milk or equivalent milk products be a part of the

diet every day. Interestingly, the USDA says people should fulfill their protein needs by consuming lean meat such as poultry. Fish, beans, peas, nuts, and seeds are alternative sources of protein.

USDA recognizes that not only diet but also exercise is a vital part of staying fit. The department suggests that you go to www.mypyramid.gov on the Internet for an individualized diet and exercise program to meet your particular needs.

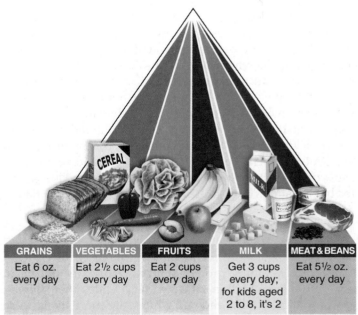

Source: U.S. Dept. of Agriculture

GRAINS	VEGETABLES	FRUITS	MILK	MEAT & BEANS
Eat 6 oz. every day	Eat 2½ cups every day	Eat 2 cups every day	Get 3 cups every day; for kids aged 2 to 8, it's 2	Eat 5½ oz. every day

FIGURE 36.12 Food guide pyramid.
The United States Department of Agriculture (USDA) developed this pyramid as a guide to better health. The different widths of the food group bands suggest how much food a person should choose from each group. The six different colors illustrate that foods from all groups are needed each day for good health. The wider base is supposed to encourage the selection of foods with little or no solid fats or added sugars.

---

[1] A calorie is a unit of energy. Specifically, a calorie is the amount of heat needed to raise one gram of pure water one °C. A Calorie is 1000 calories.

Carbohydrates

Complex sources of carbohydrates, such as whole-grain foods, are recommended because they are digested to sugars gradually and contain fiber. Insoluble fiber, such as that found in wheat bran, has a laxative effect and may possibly guard against colon cancer by limiting the amount of time cancer-causing substances are in contact with the intestinal wall. Soluble fiber, such as that found in oat bran, combines with bile acids and cholesterol in the intestine and prevents them from being absorbed. A diet too high in fiber can be detrimental, possibly impairing the body's ability to absorb iron, zinc, and calcium.

Simple sugars in foods such as candy and ice cream immediately enter the bloodstream as do those from the digestion of starch within white bread and potatoes. These foods are said to have a high **glycemic index (GI),** because the blood glucose response to these foods is high. When the blood glucose level rises rapidly, the pancreas produces an overload of insulin to bring the level under control. (Cells take up glucose in response to insulin.) Investigators hypothesize that a chronically high insulin level may lead to insulin resistance and type 2 diabetes and increased fat metabolism.

Proteins

Adequate protein formation requires 20 different types of amino acids. Of these, 8 are required from the diet in adults (9 in children) because the body is unable to produce them. These are termed the **essential amino acids.** Some protein sources such as meat, milk, and eggs are complete; they provide all 20 types of amino acids. Legumes (beans and peas), other types of vegetables, seeds and nuts, and also grains supply us with amino acids, but each of these alone is an incomplete protein source because of a deficiency in at least one of the essential amino acids. Therefore, vegetarians should combine two or more plant products to acquire all the essential amino acids.

While the body is harmed if the amount of protein in the diet is severely limited, it is also likely harmed if the diet contains an overabundance of protein. Deamination of excess amino acids in the liver results in urea, our main nitrogen excretion product. The water needed for excretion of urea can cause dehydration when a person is exercising and losing water by sweating. High-protein diets, especially those rich in animal proteins, can also increase calcium loss in urine. Excretion of calcium can lead to kidney stones. The same comments apply to protein and amino acid supplements, which are generally not recommended by nutritionists.

Certain animal protein sources, especially red meat, are known to be high in saturated fats, while other sources, such as chicken, fish, and egg whites, are more likely to be low in saturated fats. As discussed in the next section, an intake of saturated fats leads to cardiovascular disease.

Lipids

Fat, oils, and cholesterol are lipids. Saturated fats, which are solids at room temperature, usually have an animal origin. Two well-known exceptions are palm oil and coconut oil, which contain mostly saturated fats and come from the plants mentioned. Butter and meats, such as marbled red meats and bacon, contain saturated fats.

Oils contain unsaturated fatty acids, which do not promote cardiovascular disease. Not only that, unsaturated oils, like whole grains and vegetables, have a low glycemic index and are actually more filling. Each type of oil has a percentage of monounsaturated and polyunsaturated fatty acids. Corn oil and safflower oil are high in polyunsaturated fatty acids. Polyunsaturated oils are nutritionally essential because they are the only type of fat that contains linoleic acid and linolenic acid, two fatty acids the body cannot make. The body needs these two polyunsaturated fatty acids to produce various hormones and the plasma membrane of cells. Since these fatty acids must be supplied by the diet, they are called essential fatty acids.

Olive or canola oil are well known to contain a larger percentage of monounsaturated fatty acids than other types of cooking oils. Omega-3 fatty acids—a double bond in the third position—are believed to be especially protective against heart disease. Some cold-water fishes, such as salmon, sardines, and trout, are rich sources of omega-3, but they contain only about half that of flaxseed oil, the best source from plants.

Fats That Cause Disease

Cardiovascular disease is often due to arteries blocked by **plaque.** An intake of saturated fats and the presence of cholesterol transporterd by LDL (see page 624) leads to plaque. Of new concern is an intake of **trans fatty acids** found in commercially processed foods. Cookies, crackers, fried foods, packaged snacks, as well as some margarines contain trans fats. Any packaged food that contains partially hydrogenated vegetable oils or "shortening," is likely to contain trans fats.

Vitamins

Vitamins are organic compounds (other than carbohydrates, fats, and protein) that the body is unable to produce but requires for metabolic purposes. Many vitamins are portions of coenzymes, which are enzyme helpers. For example, niacin is part of the coenzyme NAD^+, and riboflavin is part of FAD. Coenzymes are needed in only small amounts because each can be used over and over again. Not all vitamins are coenzymes; vitamin A, for example, is a precursor for the visual pigment that prevents night blindness. It has been known for several years that if vitamins are lacking in the diet, various symptoms develop. Altogether, there are 13 vitamins, which are divided into those that are fat soluble and those that are water soluble (Table 36.3).

Minerals

In addition to vitamins, various **minerals** are also required by the body (Table 36.4). Some minerals (calcium, phosphorus, potassium, sulfur, sodium, chlorine, and magnesium) are recommended in amounts of more than 100 mg per day. These major minerals serve as constituents of cells and body fluids, and as structural components of tissues. Calcium is needed for the construction of bones and teeth and also for nerve conduction and muscle contraction.

Other minerals (zinc, iron, fluorine, copper, and iodine) are recommended in amounts of less than 20 mg per day. These trace minerals are more likely to have very specific functions. Iron is needed for the production of hemoglobin, and iodine is used in the production of the hormone thyroxine by the thyroid glands. As research continues, more and more elements have

been added to the list of those considered essential. During the past three decades, very small amounts of molybdenum, selenium, chromium, nickel, vanadium, silicon, and even arsenic have been found to be essential to good health.

Some individuals do not receive enough iron (especially women), calcium, magnesium, or zinc in their diets. Adult females need more iron in their diet than males (18 mg compared to 10 mg) because they lose hemoglobin each month during

TABLE 36.3

Vitamins: Their Role in the Body and Their Food Sources

Vitamin	Major Role in Body	Good Food Sources
Fat Soluble		
Vitamin A	Vision; health of skin, hair, bones, and sex organs	Deep green or yellow vegetables, dairy products
Vitamin D	Health of bones and teeth	Dairy products, tuna, eggs
Vitamin E	Strengthening of red blood cell membrane	Green leafy vegetables, whole grains
Vitamin K	Clotting of blood, bone metabolism	Green leafy vegetables, cabbage, cauliflower
Water Soluble		
Thiamine (B_1)	Carbohydrate metabolism	Pork, whole grains
Riboflavin (B_2)	Energy metabolism	Whole grains, milk, green vegetables
Niacin (B_3)	Energy metabolism	Organ meats, whole grains
Pyridoxine (B_6)	Amino acid metabolism	Meats, fish, whole grains
Vitamin B_{12}	Red blood cell formation	Meats, dairy foods
Biotin	Carbohydrate metabolism	Eggs, most foods
Folic acid	Formation of red blood cells, DNA, and RNA	Green leafy vegetables, nuts, whole grains
Pantothenic acid	Energy metabolism	Most foods
Vitamin C	Collagen formation	Citrus fruits, tomatoes

TABLE 36 .4

Minerals: Their Role in the Body and Their Food Sources

Mineral	Major Role in Body	Good Food Sources
Major Minerals		
Calcium (Ca)	Strong bones and teeth, nerve conduction, muscle contraction	Dairy products, green leafy vegetables
Phosphorus (P)	Strong bones and teeth	Meat, dairy products, whole grains
Potassium (K)	Nerve conduction, muscle contraction	Many fruits and vegetables
Sulfur	Constituent of proteins	Meat and dairy products
Sodium (Na)	Nerve conduction, pH balance	Table salt
Chlorine (Cl)	Water balance	Table salt
Magnesium (Mg)	Protein synthesis	Whole grains, green leafy vegetables
Trace Minerals		
Zinc (Zn)	Wound healing, tissue growth	Whole grains, legumes, meats
Iron (Fe)	Hemoglobin synthesis	Whole grains, legumes, eggs, green leafy vegetables
Fluorine (F)	Strong bones and teeth	Fluoridated drinking water, tea
Copper (Cu)	Hemoglobin synthesis	Seafood, whole grains, legumes
Iodine (I)	Thyroid hormone synthesis	Iodized table salt, seafood

menstruation. Stress can bring on a magnesium deficiency, and a vegetarian diet may lack zinc, which is usually obtained from meat. A varied and complete diet, however, usually supplies the recommended daily allowances for minerals.

Calcium

There is much interest in calcium supplements to counteract the development of osteoporosis [Gk. *osteon*, bone, and *poros*, hole, passage], a degenerative bone disease that afflicts an estimated one-fourth of older men and one-half of older women in the United States. Osteoporosis develops because bone-eating cells, called osteoclasts, are more active than bone-forming cells, called osteoblasts. Therefore, the bones are porous, and they break easily because they lack sufficient calcium. Due to recent studies that show consuming more calcium does slow bone loss in adults, the guidelines have been revised. A calcium intake of 1,000–1,500 mg a day is recommended. To achieve this amount, supplemental calcium is most likely necessary.

Exercise is also effective in building bone mass. In addition, there are medications that slow bone loss while increasing skeletal mass, but these should be taken only when under a physician's care.

Sodium

The recommended amount of sodium intake per day is 400–3,300 mg, and the average intake in the United States is 4,000–4,700 mg, mostly as salt (sodium chloride). In recent years, this imbalance has caused concern because high sodium intake has been linked to hypertension in some people. About one-third of the sodium we consume occurs naturally in foods; another third is added during commercial processing; and we add the last third either during home cooking or at the table in the form of table salt.

Clearly, it is possible to cut down on the amount of sodium in the diet by reducing the amount of salt we eat.

Proper nutrition is extremely important to our health. Eating disorders interfere with proper nutrition, as do poor food choices.

CONNECTING THE CONCEPTS

In humans, digestive enzymes are produced by the salivary glands, gastric glands, and intestinal glands. Two other organs (the pancreas and the liver) also contribute secretions that help break down food. The liver produces bile (stored by the gallbladder), which emulsifies fat. The pancreas produces enzymes for the digestion of carbohydrates, proteins, and fat. Secretions from these glands, which are sent by ducts into the small intestine, are regulated by hormones, such as secretin, produced by the digestive tract.

Assigning any organ to a particular body system seems arbitrary. Granted, a good argument can be made for declaring that the mouth, esophagus, stomach, and intestines are in the digestive system. But actually, even these organs assist other body systems. The stomach and small intestine contribute to the endocrine system, for example, by producing hormones (see the Science Focus on page 663). And how could muscles and nerves function without a supply of calcium from the digestive tract? Or for that matter, how could any other system in the body function without a supply of nutrients absorbed by the digestive tract and distributed by the cardiovascular system?

The liver has so many functions it really belongs to many systems of the body. Doesn't it belong to the cardiovascular system because it produces plasma proteins and to the urinary system because it produces urea, as well as to the digestive system because it produces bile? The liver is a vital organ, meaning that we cannot live without it. Among its many functions, it detoxifies blood by removing and metabolizing poisonous substances. The liver has amazing regenerative powers and in some instances can recover if the rate of regeneration exceeds the rate of damage. Otherwise, liver transplantation is usually the preferred treatment, but artificial livers have been developed and tried in a few cases. As mentioned, one type of artificial liver consists of a cartridge that contains liver cells. The patient's blood passes through a cellulose acetate tube of the cartridge and is serviced in the same manner as by a normal liver. In the meantime, the patient's liver has a chance to recover.

Summary

36.1 DIGESTIVE TRACTS

Some animals (e.g., planarians) have an incomplete digestive tract, so called because there is only one opening. An incomplete tract has little specialization of parts. Other animals (e.g., earthworms) have a complete digestive tract, so called because the tract has both a mouth and an anus. A complete tract tends to have specialization of parts.

Some animals are continuous feeders (e.g., clams, which are filter feeders); others are discontinuous feeders (e.g., squid). Discontinuous feeders need a storage area for food.

Most mammals have teeth. Herbivores need teeth that can clip off plant material and grind it up. Also, the herbivore's stomach contains bacteria that can digest cellulose.

Carnivores need teeth that can tear and rip meat into pieces. Meat is easier to digest than plant material, so the digestive system of carnivores has fewer specialized regions and the intestine is shorter than that of herbivores.

36.2 HUMAN DIGESTIVE TRACT

In the human digestive tract, food is chewed and manipulated in the mouth, where salivary glands secrete saliva. Saliva contains salivary amylase, which begins carbohydrate digestion.

Food then passes to the pharynx and down the esophagus by peristalsis to the stomach. The stomach stores and mixes food with mucus and gastric juices to produce chyme. Pepsin begins protein digestion in the stomach.

The duodenum of the small intestine receives bile from the liver and pancreatic juice from the pancreas. Bile emulsifies fat and readies it for digestion by an enzyme produced by the pancreas. The pancreas also produces enzymes that digest starch and protein. The intestinal enzymes finish the process of chemical digestion.

The walls of the small intestine have fingerlike projections called villi where small nutrient molecules are absorbed. Amino acids and glucose enter the blood vessels of a villus. Glycerol and fatty acids are joined and packaged as lipoproteins before entering lymphatic vessels called lacteals in a villus.

The large intestine consists of the cecum, the colon (including the ascending, transverse, descending, and sigmoid colons), and the rectum, which ends at the anus. The large intestine does not produce digestive enzymes; it does absorb water, salts, and some vitamins. Reduced water absorption results in diarrhea. The intake of water and fiber help prevent constipation.

Three accessory organs of digestion—the pancreas, liver, and gallbladder—send secretions to the duodenum via ducts. The pancreas produces pancreatic juice, which contains digestive enzymes for carbohydrates, protein, and fat.

The liver produces bile, which is stored in the gallbladder. The liver receives blood from the small intestine by way of the hepatic portal vein. It has numerous important functions, and any malfuncton of the liver is a matter of considerable concern.

36.3 DIGESTIVE ENZYMES

Digestive enzymes are present in digestive juices and break down food into the nutrient molecules glucose, amino acids, fatty acids, and glycerol. Salivary amylase and pancreatic amylase begin the digestion of starch. Pepsin and trypsin digest protein to peptides. Following emulsification by bile, lipase digests fat to glycerol and fatty acids. Intestinal enzymes finish the digestion of starch and protein.

Each digestive enzyme is present in a particular part of the digestive tract. Salivary amylase functions in the mouth; pepsin functions in the stomach; trypsin, lipase, and pancreatic amylase occur in the intestine along with the various enzymes that digest disaccharides and peptides.

36.4 NUTRITION

The nutrients released by the digestive process should provide us with an adequate amount of energy, essential amino acids and fatty acids, and all necessary vitamins and minerals.

Carbohydrates are necessary in the diet, but high glycemic index carbohydrates (simple sugars and refined starches) are not helpful because they cause a rapid release of insulin, which can lead to health problems. Proteins supply us with essential amino acids, but it is wise to avoid meats that are fatty because fats from animal sources are saturated. While unsaturated fatty acids, particularly the omega-3 fatty acids, are protective against cardiovascular disease, the saturated fatty acids lead to plaque, which blocks blood vessels. Aside from carbohydrates, proteins, and fats, the body requires vitamins and minerals. The vitamins C, E, and A are antioxidants that protect cell contents from damage due to free radicals. The mineral calcium is needed for strong bones.

Reviewing the Chapter

1. Contrast the incomplete digestive tract with the complete digestive tract, using the planarian and earthworm as examples.   656
2. Contrast a continuous feeder with a discontinuous feeder, using the clam and squid as examples.   656–57
3. Contrast the dentition of the mammalian herbivore with that of the mammalian carnivore, using the horse and lion as examples.   658
4. List the parts of the human digestive tract, anatomically describe them, and state the contribution of each to the digestive process.   658–64
5. Discuss the absorption of the products of digestion into the lymphatic and cardiovascular systems.   661
6. Describe the structure and function of the large intestine. Name two medical conditions associated with the large intestine.   662
7. State the location and describe the functions of the pancreas, the liver, and the gallbladder.   662–64
8. Name and discuss three serious illnesses of the liver.   664
9. Assume that you have just eaten a ham sandwich. Discuss the digestion of the contents of the sandwich. Mention all the necessary enzymes.   664–65
10. Explain why good nutrition is important to human health. Give examples.   665–66

Testing Yourself

Choose the best answer for each question.

1. Animals that feed discontinuously
   a. have digestive tracts that permit storage.
   b. are always filter feeders.
   c. exhibit extremely rapid digestion.
   d. have a nonspecialized digestive tract.
   e. usually eat only meat.

2. In which of these types of animals would you expect the digestive tract to be more complex?
   a. those with a single opening for the entrance of food and exit of wastes
   b. those with two openings, one serving as an entrance and the other as an exit
   c. only those complex animals that also have a respiratory system
   d. those with two openings that use the digestive tract to help fight infections
   e. All but a are correct.

3. The typhlosole within the gut of an earthworm compares best to which of these organs in humans?
   a. teeth in the mouth
   b. esophagus in the thoracic cavity
   c. folds in the stomach
   d. villi in the small intestine
   e. the large intestine because it absorbs water

4. Which of these animals is a continuous feeder with a complete digestive tract?
   a. planarian   d. lion
   b. clam   e. human
   c. squid

5. Tracing the path of food in humans, which step is out of order first?
   a. mouth
   b. pharynx
   c. esophagus
   d. small intestine
   e. stomach
   f. large intestine

6. The products of digestion are
   a. large macromolecules needed by the body.
   b. enzymes needed to digest food.
   c. small nutrient molecules that can be absorbed.
   d. regulatory hormones of various kinds.
   e. the food we eat.

7. Why can a person not swallow food and talk at the same time?
   a. To swallow, the epiglottis must close off the trachea.
   b. The brain cannot control two activities at once.
   c. To speak, air must come through the larynx to form sounds.
   d. A swallowing reflex is only initiated when the mouth is closed.
   e. Both a and c are correct.

8. Which of these could be absorbed directly without need of digestion?
   a. glucose   d. nucleic acid
   b. fat   e. All of these are correct.
   c. protein

9. Which association is incorrect?
   a. protein—trypsin   d. starch—amylase
   b. fat—lipase   e. protein—pepsin
   c. maltose—pepsin

10. Most of the absorption of the products of digestion takes place in humans across
    a. the squamous epithelium of the esophagus.
    b. the convoluted walls of the stomach.
    c. the fingerlike villi of the small intestine.
    d. the smooth wall of the large intestine.
    e. the lacteals of the lymphatic system.

11. The hepatic portal vein is located between
    a. the hepatic vein and the vena cava.
    b. two capillary beds.
    c. the pancreas and the small intestine.
    d. the small intestine and the liver.
    e. Both b and d are correct.

12. Bile in humans
    a. is an important enzyme for the digestion of fats.
    b. is made by the gallbladder.
    c. emulsifies fat.
    d. must be activated during the first few weeks of life.
    e. All of these are correct.

13. Which of these is not a function of the liver in adults?
    a. produces bile
    b. stores glucose
    c. produces urea
    d. makes red blood cells
    e. produces proteins needed for blood clotting

14. The large intestine in humans
    a. digests all types of food.
    b. is the longest part of the intestinal tract.
    c. absorbs water.
    d. is connected to the stomach.
    e. All of these are correct.

15. The appendix connects to
    a. the cecum.
    b. the small intestine.
    c. the esophagus.
    d. the large intestine.
    e. the liver.
    f. All of these are correct.

16. Which association is incorrect?
    a. mouth—starch digestion
    b. esophagus—protein digestion
    c. small intestine—starch, lipid, protein digestion
    d. stomach—food storage
    e. liver—production of bile

17. Predict and explain the expected digestive results per test tube for this experiment.

Incubator

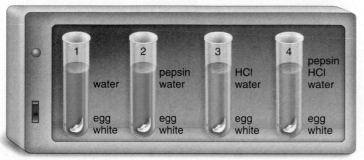

Thinking Scientifically

1. A drug for leukemia is not broken down in the stomach and is well absorbed by the intestine. However, the molecular form of the drug collected from the blood is not the same as the form that was swallowed by the patient. What explanation is most likely?

2. Snakes often swallow whole animals, a process that takes a long time. Then snakes spend some time digesting their food. What structural modifications would allow slow swallowing and storage of a whole animal to occur? What chemical modifications would be necessary to digest a whole animal?

Understanding the Terms

Match the terms to these definitions:

a. _____ Essential requirement in the diet, needed in small amounts. They are often part of coenzymes.
b. _____ Fat-digesting enzyme secreted by the pancreas.
c. _____ Lymphatic vessel in an intestinal villus; it aids in the absorption of fats.
d. _____ Muscular tube for moving swallowed food from the pharynx to the stomach.
e. _____ Organ attached to the liver that serves to store and concentrate bile.

ARIS, the *Biology* Website

ARIS, the website for *Biology*, provides a wealth of information organized and integrated by chapter. You will find practice quizzes, interactive activities, labeling exercises, flashcards, and much more that will complement your learning and understanding of general biology.

www.mhhe.com/maderbiology9

37

RESPIRATORY SYSTEMS

Many animals have a storage area for food that allows them to eat first and digest later. But when it comes to respiratory gases, animals have no such storage area; oxygen and carbon dioxide are continually exchanged with the environment.

In many small aquatic invertebrates, gas exchange occurs directly between the animal's cells and the environment. Dissolved oxygen in water diffuses into the cells, while carbon dioxide diffuses from the cells into the water. In these animals, no specialized respiratory structure is present. Other aquatic invertebrates have highly branched gills filled with blood vessels. Water moving over the gills supplies oxygen to and removes carbon dioxide from the blood.

Among vertebrate animals, fishes and some amphibians also rely on gills for respiration. Most adult amphibians have lungs but also use their thin skin for gas exchange. Reptiles, birds, and mammals have efficient vascularized lungs for respiration. Air tubes bring oxygen into the lungs, where gas exchange with the blood occurs. Diffusion is the basic mechanism for gas exchange in all animals, whether the animal has gills, lungs, or no respiratory organ.

Gas exchange is essential for survival of all animals.

37.1 GAS EXCHANGE SURFACES

Respiration is the sequence of events that results in gas exchange between the body's cells and the environment. In terrestrial vertebrates, respiration includes these steps:

- **Ventilation** (i.e., breathing): includes inspiration (entrance of air into lungs) and expiration (exit of air from the lungs).
- **External respiration:** gas exchange between the air and the blood within the lungs. Blood then transports oxygen from the lungs to the tissues.
- **Internal respiration:** gas exchange between blood and tissue fluid. The body's cells exchange gases with tissue fluid. The blood then transports carbon dioxide to the lungs.

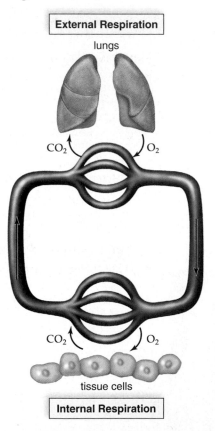

Gas exchange takes place by the physical process of diffusion. For diffusion to be effective, the gas-exchange region must be (1) moist, (2) thin, and (3) large in relation to the size of the body. Some animals, such as planarians, are small and shaped in a way that allows the surface of the animal to be the gas-exchange surface. Most complex animals have specialized gas-exchange surfaces, such as gills in aquatic animals and lungs in terrestrial animals. The effectiveness of diffusion is enhanced by vascularization, and delivery to cells is promoted when the blood contains a respiratory pigment such as hemoglobin.

Regardless of the particular gas-exchange surface and the manner in which gases are delivered to the cells, in the end oxygen enters mitochondria, where cellular respiration takes place. Without the delivery of oxygen to the body's cells, ATP production does not take place, and life ceases.

Water Environments

It is more difficult for animals to obtain oxygen from water than from air. Water fully saturated with air contains only a fraction of the amount of oxygen that would be present in the same volume of air. Also, water is more dense than air. Therefore, aquatic animals expend more energy to respire than do terrestrial animals. Fishes use up to 25% of their energy output to respire, while terrestrial mammals use only 1–2% of their energy output.

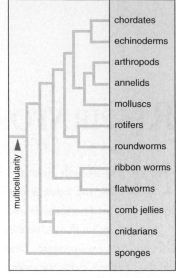

Hydras, which are cnidarians, and planarians, which are flatworms, have a large surface area in comparison to their size (Fig. 37.1). This makes it possible for most of their cells to exchange gases directly with the environment. In hydras, the outer layer

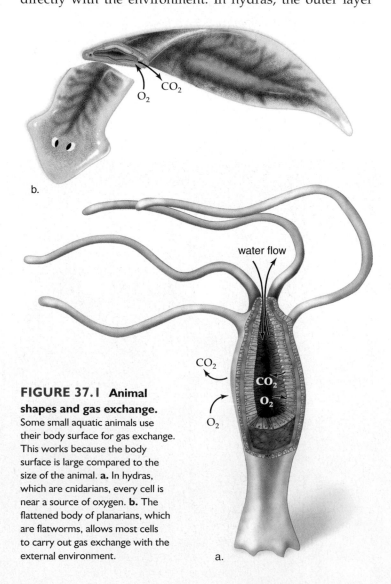

FIGURE 37.1 Animal shapes and gas exchange.
Some small aquatic animals use their body surface for gas exchange. This works because the body surface is large compared to the size of the animal. **a.** In hydras, which are cnidarians, every cell is near a source of oxygen. **b.** The flattened body of planarians, which are flatworms, allows most cells to carry out gas exchange with the external environment.

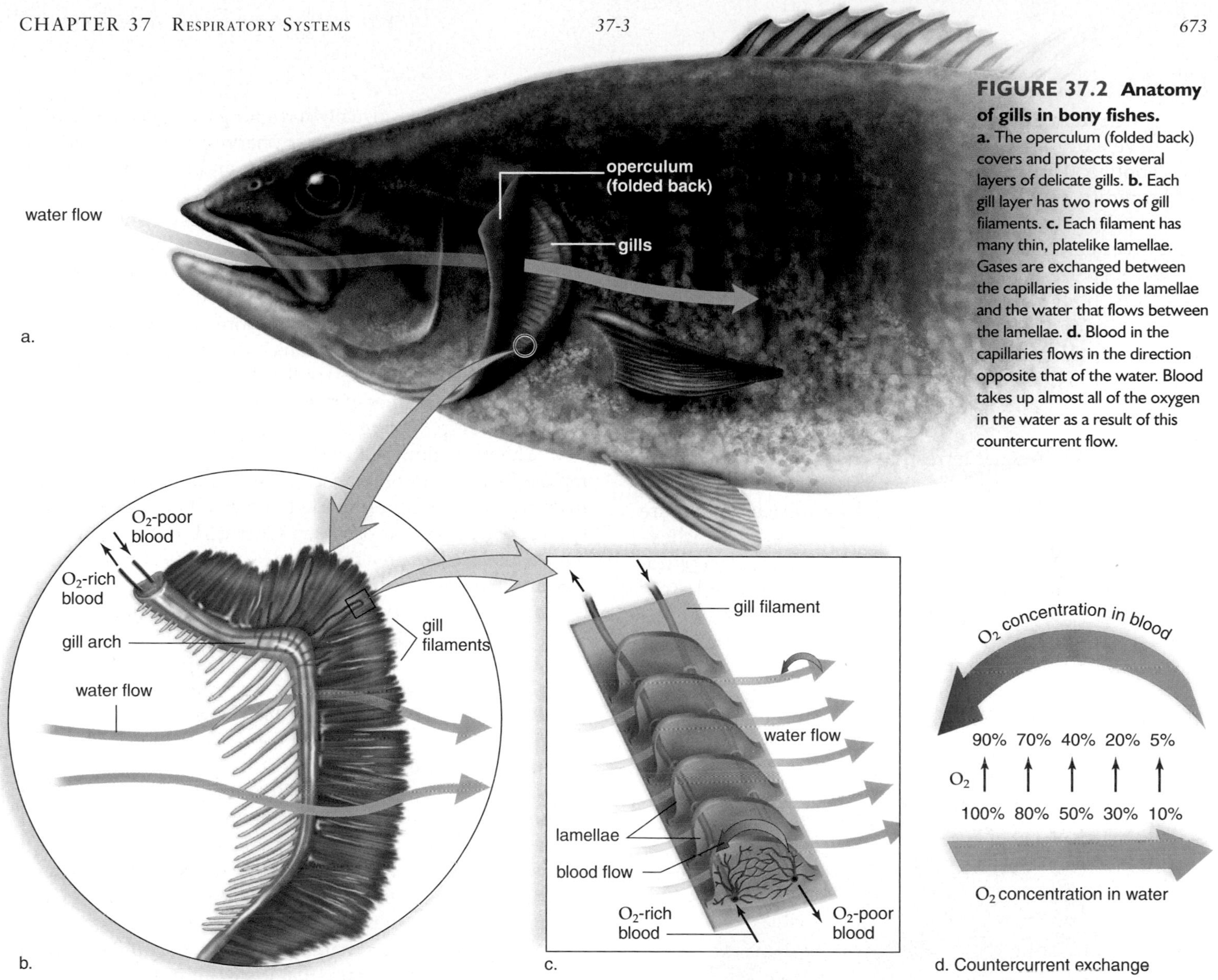

FIGURE 37.2 Anatomy of gills in bony fishes.
a. The operculum (folded back) covers and protects several layers of delicate gills. **b.** Each gill layer has two rows of gill filaments. **c.** Each filament has many thin, platelike lamellae. Gases are exchanged between the capillaries inside the lamellae and the water that flows between the lamellae. **d.** Blood in the capillaries flows in the direction opposite that of the water. Blood takes up almost all of the oxygen in the water as a result of this countercurrent flow.

of cells is in contact with the external environment, and the inner layer can exchange gases with the water in the gastrovascular cavity. In planarians, the flattened body permits cells to exchange gases with the external environment.

The tubular shape of annelids (segmented worms) also provides a surface area adequate for gas-exchange purposes. In addition to a tubular shape, polychaete worms have extensions of the body wall called parapodia, which are vascularized (see Fig. 30.6). Quite often, aquatic animals have **gills,** which are finely divided and vascularized outgrowths of either the outer or the inner body surface. Among molluscs, such as a clam or squid, water is drawn into the mantle cavity, where it passes through the gills. In decapod crustaceans, such as crabs and shrimps, which are arthropods, the gills are located in thoracic chambers covered by the exoskeleton. Water is kept moving by the action of specialized appendages located near the mouth.

Among aquatic vertebrates, the gills of bony fishes are outward extensions of the pharynx (Fig. 37.2). Ventilation, which brings oxygen to respiratory surfaces, is brought about by the combined action of the mouth and gill covers, or opercula (sing., operculum). When the mouth is open, the

opercula are closed and water is drawn in. Then the mouth closes and the opercula open, drawing the water from the pharynx through the gill slits located between the gill arches. On the outside of the gill arches, the gills are composed of filaments that are folded into platelike lamellae.

In the capillaries of each lamella, the blood flows in a direction opposite to the movement of water across the gills. With a countercurrent flow as blood gains oxygen, it always encounters water having an even higher oxygen content, and no equilibrium point is reached. If the flowing were concurrent (oxygen-rich water and oxygen-poor blood flow in the same direction), an equilibrium point would occur, and only about half the oxygen in the water would be captured. With a countercurrent mechanism, an equilibrium point is not reached, and about 80–90% of the initial dissolved oxygen in water is extracted.

Small aquatic animals sometimes use the body surface for gas exchange, but many larger ones have localized gas-exchange surfaces known as gills.

Land Environments

Air is a rich source of oxygen compared to water; however, it does have a drying effect on respiratory surfaces. A human loses about 350 ml of water per day when the air has a relative humidity of only 50%.

The earthworm, which is an annelid, is an example of an invertebrate terrestrial animal that uses its body surface for respiration. An earthworm expends much energy to keep its body surface moist by secreting mucus and by releasing fluids from excretory pores. Further, the worm is behaviorally adapted to remain in damp soil during the day, when the air is driest.

Tracheae

Insects and certain other terrestrial arthropods have a respiratory system that consists of air tubes called **tracheae** (Fig. 37.3). Oxygen enters the tracheae at spiracles, which are valvelike openings on each side of the body. The tracheae branch and then rebranch, ending in tiny channels, the tracheoles, that are in direct contact with the body cells. Larger insects have a mechanism to ventilate—keep the air moving in and out of—the tracheae. Many have air sacs located near major muscles; contraction of these muscles causes the air sacs to empty, and relaxation causes the air sacs to expand and draw in air. The tracheal system of insects is very effective in delivering oxygen to the cells, so that the circulatory system plays no role in gas transport.

Lungs

Terrestrial vertebrates usually have **lungs,** which are vascularized outgrowths from the lower pharyngeal region. The tadpoles of amphibians live in the water and have gills as respiratory organs, but adult amphibians possess simple, saclike lungs (Fig. 37.4a). Many amphibians possess a short trachea, which divides into two bronchi that open into the lungs. Most amphibians respire to some extent through the skin, which is kept moist by the presence of mucus produced by numerous glands on the surface of the body. During the winter in temperate climates, many amphibians burrow in the mud, and all gas exchange occurs by way of the skin. A few species of salamanders possess neither gills nor lungs and depend entirely on exchange of respiratory gases across the skin.

The inner lining of the lungs is more finely divided in reptiles than in amphibians. The lungs of birds and mammals are elaborately subdivided into small passageways and spaces, respectively. It has been estimated that human lungs have a total surface area that is at least 50 times the skin's surface area. To keep the lungs from drying out, air is moistened as it moves through passageways leading to the lungs.

Terrestrial vertebrates ventilate the lungs by moving air into and out of the respiratory tract. Amphibians use positive pressure to force air into the respiratory tract. With the mouth and nostrils firmly shut, the floor of the mouth rises and pushes the air into the lungs (Fig. 37.4b). Reptiles, birds, and mammals use negative pressure to move air into the lungs. Reptiles have jointed ribs that can be raised to expand the lungs. Mammals have a rib cage that is lifted up and out and a muscular **diaphragm** [Gk. *diaphragma*, partition wall] that is flattened. As the thoracic cavity (chest cavity) expands and lung volume increases, air will flow into the lungs due to differences in air pressure. Following **inspiration** (or inhalation), **expiration** (or exhalation) occurs. In reptiles, lowering the ribs exerts a pressure that forces air out. In mammals, when the rib cage is lowered and the diaphragm rises, the thoracic pressure increases, forcing air out of the lungs.

All terrestrial vertebrates except birds use a *tidal ventilation mechanism,* so called because the air moves in and out by the same route. This means that the lungs of amphibians, reptiles, and mammals are not completely emptied and refilled during each breathing cycle. Because of this, the air entering mixes with used air remaining in the lungs. While this does help conserve water, it also decreases gas-exchange efficiency. In contrast, birds use a *one-way ventilation mechanism.* Incoming air is carried past the lungs by a trachea, which takes it to a set of posterior air sacs (Fig. 37.5). The air then passes

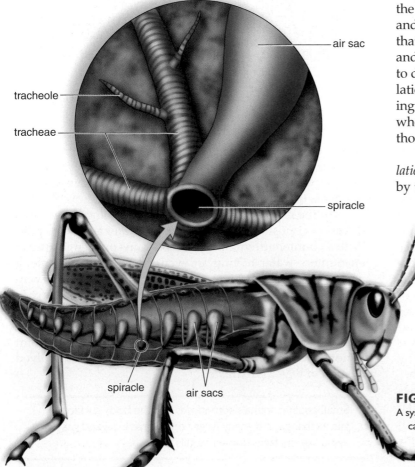

FIGURE 37.3 Tracheae of insects.
A system of air tubes extends throughout the body of an insect, and the tubes carry oxygen to the cells. Air enters the tracheae at openings called spiracles. From here, the air moves to the smaller tracheoles, which take it to the cells, where gas exchange takes place.

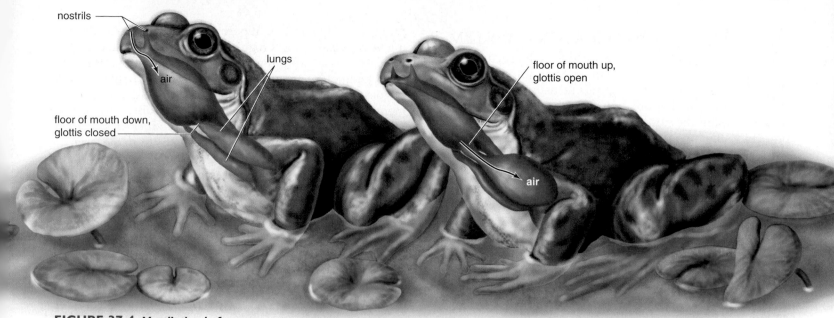

FIGURE 37.4 Ventilation in frogs.
Amphibians have simple saclike lungs that they fill by using positive pressure. **a.** First, frogs bring air into the mouth when the glottis is closed. **b.** Then, with the mouth and nostrils closed, the floor of the mouth rises to push air through an open glottis into the lungs.

forward through the lungs into a set of anterior air sacs. From here, it is finally expelled. Notice that fresh, oxygen-rich air never mixes with used air in the lungs of birds, and thereby gas-exchange efficiency is greatly improved.

The **partial pressure** of a gas is the pressure a gas exerts in a mixture of gases. Because the ventilation mechanism of birds results in a higher partial pressure of oxygen in the lungs, the uptake of oxygen with each breath is greater than in other vertebrates, even though bird lungs may be smaller.

Most terrestrial animals have a specialized gas-exchange region. Insects use tracheae, while terrestrial vertebrates depend on lungs, which are ventilated variously.

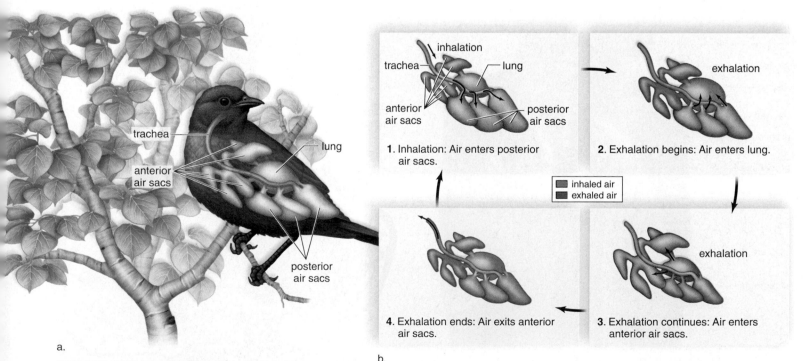

a.

b.

FIGURE 37.5 Respiratory system in birds.
a. Air sacs are attached to the lungs of birds. These allow birds to have a one-way mechanism of ventilating their lungs. **b.** When a bird inhales, air enters the posterior air sacs, and when a bird exhales, air moves through the lungs to the anterior air sacs before exiting the trachea. This one-way flow of air through the lungs allows more fresh air to be present in the lungs with each breath, and this leads to greater uptake of oxygen from one breath of air.

37.2 HUMAN RESPIRATORY SYSTEM

The human respiratory system includes all structures that conduct air in a continuous pathway to and from the lungs (Fig. 37.6 and Table 37.1). The lungs lie deep within the thoracic cavity, where they are protected from drying out. As air moves through the nose, the pharynx, the trachea, and the bronchi to the lungs, it is filtered so that it is free of debris, warmed, and humidified. By the time the air reaches the lungs, it is at body temperature and is saturated with water. In the nose, hairs and cilia act as a screening device. In the trachea and the bronchi, cilia beat upward, carrying mucus, dust, and occasional small bits of food that "went down the wrong way" into the throat, where the accumulation may be swallowed or expectorated.

The hard and soft palates separate the nasal cavities from the mouth, but the air and food passages cross in the **pharynx** [Gk. *pharynx*, throat]. This may seem inefficient, and there is danger of choking if food accidentally enters the trachea, but this arrangement does have the advantage of letting you breathe through your mouth in case your nose is plugged up. In addition, it permits greater intake of air during heavy exercise, when greater gas exchange is required.

Air passes from the pharynx through the **glottis** [Gk. *glotti*, tongue], an opening into the **larynx** [Gk. *larynx*, gullet], or voice box. At the edges of the glottis, embedded in mucous membrane, are the **vocal cords.** These flexible and pliable bands of connective tissue vibrate and produce sound when air is expelled past them through the glottis from the larynx.

The larynx and the **trachea** [L. *trachia*, windpipe] are permanently held open to receive air. The larynx is held open by a complex of nine cartilages; among them is the cartilage of the Adam's apple. Easily seen in many men, the Adam's apple resembles a small, rounded apple just under the skin in the front of the neck. The trachea is held open by a series of C-shaped, cartilaginous rings that do not completely meet in the rear. When food is being swallowed, the larynx rises, and the glottis is closed by a flap of tissue called the **epiglottis** [Gk. *epi*, over, and *glotta*, tongue]. A backward movement of the soft palate covers the entrance of the nasal passages into the pharynx. The food then enters the esophagus, which lies behind the larynx.

The trachea divides into two primary **bronchi** [Gk. *bronchos*, windpipe], which enter the right and left lungs. Branching continues until there are a great number of smaller passages called **bronchioles** [Gk. *bronchos*, windpipe]. The two bronchi resemble the trachea in structure, but as the bronchial tubes divide and subdivide, their walls become thinner,

nasal cavity

nostril

pharynx

epiglottis

glottis

larynx

trachea

bronchus

bronchiole

lung

diaphragm

FIGURE 37.6 The human respiratory tract.
The respiratory tract extends from the nose to the lungs, which are composed of air sacs called alveoli. Gas exchange occurs between air in the alveoli and blood within a capillary network that surrounds the alveoli.

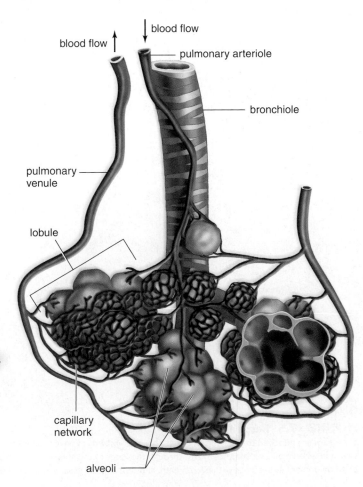

blood flow

blood flow

pulmonary arteriole

bronchiole

pulmonary venule

lobule

capillary network

alveoli

TABLE 37.1

Path of Air

Structure	Function
Nasal cavities	Filter, warm, and moisten
Pharynx (throat)	Connection to larynx
Glottis	Permits passage of air
Larynx (voice box)	Sound production
Trachea (windpipe)	Passage of air to bronchi
Bronchi	Passage of air to lung
Bronchioles	Passage of air to alveoli
Alveoli	Air sacs for gas exchange

and rings of cartilage are no longer present. Each bronchiole terminates in an elongated space enclosed by a multitude of air pockets, or sacs, called **alveoli** [L. *alveolus*, cavity], which make up the lungs. Gas exchange occurs between air in the alveoli and blood in the capillaries.

Ventilation

Humans use a tidal ventilation mechanism the same as other terrestrial vertebrates except birds. As in other mammals, the volume of the thoracic cavity and lungs is increased by muscle contractions that lower the diaphragm and raise the ribs (Fig. 37.7). These movements create a negative pressure in the thoracic cavity and lungs, and air then flows into the lungs during inspiration. When rib and diaphragm muscles relax, air moves out during expiration as a result of increased pressure in the thoracic cavity and lungs.

At rest, the average person takes a breath about 14 times per minute. But the respiratory control center located in the pons and medulla oblongata can change the normal rate according to the circumstances. When the control center detects a drop in pH (increase in CO_2), it increases the rate and depth of breathing. Also, the chemical content of the blood is detected by the aortic and carotid bodies, which are located in the walls of the aorta and the carotid arteries. Information from the chemoreceptors goes to the respiratory control centers. These receptors are sensitive to O_2 in addition to pH and CO_2. Normally, O_2 concentration in blood has little effect on the breathing rate. However, when the O_2 level is particularly depressed, the aortic and carotid bodies convey an alarm to the breathing control centers to increase the breathing rate. Temperature, touch, and pain receptors, as well as higher brain functions such as speech and emotions, can also affect the rate of breathing.

In humans, air moves through passages that repeatedly divide until the smallest passages terminate at the alveoli of the lungs. Inspiration and expiration are due to movements of the rib cage and diaphragm that affect the size of the lungs.

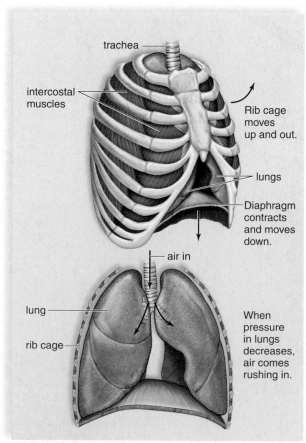

a. Inspiration

trachea

intercostal muscles

Rib cage moves up and out.

lungs

Diaphragm contracts and moves down.

air in

lung

rib cage

When pressure in lungs decreases, air comes rushing in.

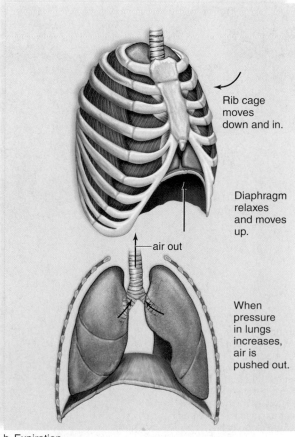

b. Expiration

Rib cage moves down and in.

Diaphragm relaxes and moves up.

air out

When pressure in lungs increases, air is pushed out.

FIGURE 37.7

Inspiration versus expiration.

a. During inspiration, the thoracic cavity and lungs expand so that air is drawn in. **b.** During expiration, the thoracic cavity and lungs resume their original positions and pressures. Now, air is forced out.

Gas Exchange and Transport

Diffusion primarily accounts for the exchange of gases between the air in the alveoli and the blood in the pulmonary capillaries, called external respiration (Fig. 37.8). Blood flowing into the pulmonary capillaries has a higher carbon dioxide concentration than alveolar air. Therefore, carbon dioxide diffuses out of the blood and into the alveoli. The pattern is the reverse for oxygen. Blood coming into the pulmonary capillaries has a lower concentration of oxygen than alveolar air. Therefore, oxygen diffuses from the alveoli into the capillaries.

Transport of Oxygen and Carbon Dioxide

Most oxygen entering the pulmonary capillaries from the alveoli combines with **hemoglobin (Hb)** [Gk. *haima*, blood; L. *globus*, ball] in red blood cells to form **oxyhemoglobin**:

$$\underset{\text{deoxyhemoglobin}}{Hb} + \underset{\text{oxygen}}{O_2} \longrightarrow \underset{\text{oxyhemoglobin}}{HbO_2}$$

Each hemoglobin molecule contains four polypeptide chains, and each chain is folded around an iron-containing group called **heme** (Fig. 37.9). It is actually the iron that forms a loose association with oxygen. Since there are about

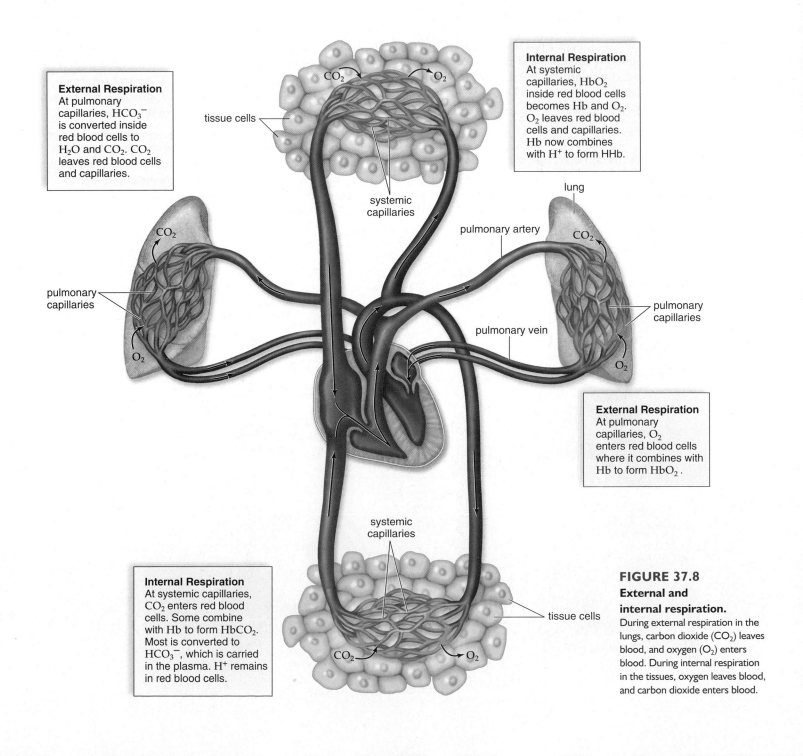

External Respiration
At pulmonary capillaries, HCO_3^- is converted inside red blood cells to H_2O and CO_2. CO_2 leaves red blood cells and capillaries.

Internal Respiration
At systemic capillaries, HbO_2 inside red blood cells becomes Hb and O_2. O_2 leaves red blood cells and capillaries. Hb now combines with H^+ to form HHb.

External Respiration
At pulmonary capillaries, O_2 enters red blood cells where it combines with Hb to form HbO_2.

Internal Respiration
At systemic capillaries, CO_2 enters red blood cells. Some combine with Hb to form $HbCO_2$. Most is converted to HCO_3^-, which is carried in the plasma. H^+ remains in red blood cells.

tissue cells

systemic capillaries

lung

pulmonary artery

pulmonary capillaries

pulmonary vein

pulmonary capillaries

systemic capillaries

tissue cells

FIGURE 37.8

External and internal respiration.
During external respiration in the lungs, carbon dioxide (CO_2) leaves blood, and oxygen (O_2) enters blood. During internal respiration in the tissues, oxygen leaves blood, and carbon dioxide enters blood.

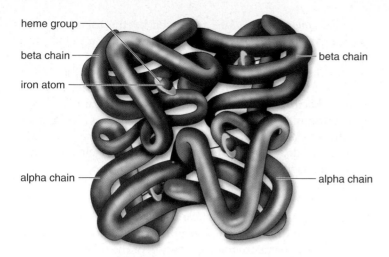

FIGURE 37.9 Hemoglobin.
Hemoglobin consists of four polypeptide chains (two alpha—red; two beta—purple), each associated with a heme group. Each heme group contains an iron atom, which can bind to O_2.

250 million hemoglobin molecules in each red blood cell, each red blood cell is capable of carrying at least 1 billion molecules of oxygen.

Figure 37.10 shows the percent saturation of hemoglobin at various partial pressures, depending on the temperature and acidity. At the normal partial pressure of oxygen in the lungs, hemoglobin is practically saturated with oxygen, but at the partial pressure in the tissues, oxyhemoglobin quickly gives up much of its oxygen during internal respiration:

$$HbO_2 \longrightarrow Hb + O_2$$

The acid pH and warmer temperature of the tissues also promote the release of oxygen by hemoglobin. In tissues, oxygen leaves the blood and enters tissue fluid, where it is taken up by cells.

Carbon dioxide enters blood from the tissues during internal respiration. Some carbon dioxide combines with hemoglobin to form **carbaminohemoglobin.** Most of the carbon dioxide, however, is transported in the form of the **bicarbonate ion** (HCO_3^-). First carbon dioxide combines with water, forming carbonic acid, and then this dissociates to a hydrogen ion (H^+) and HCO_3^-:

$$CO_2 + H_2O \longrightarrow H_2CO_3 \longrightarrow H^+ + HCO_3^-$$

carbon dioxide   water   carbonic acid   hydrogen ion   bicarbonate ion

Carbonic anhydrase [Gk. *an*, without, and *hydrias*, water], an enzyme in red blood cells, speeds this reaction. The release of H^+ could drastically change the pH of the blood. However, the H^+ is absorbed by the globin portions of hemoglobin, and the HCO_3^- diffuses out of the red blood cells to be carried in the plasma. Hemoglobin that combines with a H^+ is called reduced hemoglobin and can be symbolized as HHb. HHb plays a vital role in maintaining the normal pH of the blood.

As blood enters the pulmonary capillaries, most of the carbon dioxide is present in plasma as HCO_3^-. Reduced hemoglobin (HHb) gives up the H^+ it has been carrying, as carbonic anhydrase speeds this reaction:

$$H^+ + HCO_3^- \longrightarrow H_2CO_3 \longrightarrow H_2O + CO_2$$

Now free carbon dioxide diffuses out of blood into the alveoli of the lungs.

Oxygen is transported in the blood by hemoglobin, and carbon dioxide is largely carried in plasma as the bicarbonate ion.

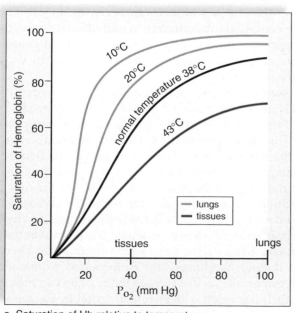

a. Saturation of Hb relative to temperature

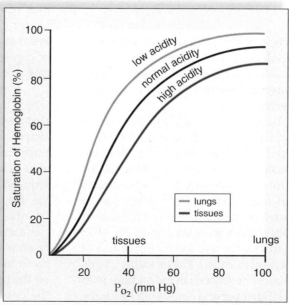

b. Saturation of Hb relative to pH

FIGURE 37.10

Hemoglobin saturation in relation to temperature and acidity.

The P_{O_2} (partial pressure of oxygen) in lungs is close to 100 mm Hg. Because of this and (**a**) the lower than normal temperature and (**b**) the lower than normal acidity in the lungs, hemoglobin is about 98% saturated. The P_{O_2} in tissues is only about 40 mm Hg. Because of this, and also because of (**a**) the higher than normal temperature and (**b**) the higher than normal acidity in the tissues, hemoglobin is less than 60% saturated. This change in saturation releases O_2 to the tissue cells.

37.3 RESPIRATION AND HEALTH

We have seen that the entire respiratory tract has a warm, wet, mucous membrane lining that is constantly exposed to environmental air. The quality of this air, determined by the pollutants and the pathogens therein, can affect our health.

Upper Respiratory Tract Infections

The upper respiratory tract consists of the nose, the pharynx, and the larynx. Upper respiratory infections can spread from the nasal cavities to the sinuses, to the middle ears, and to the larynx. Viral infections sometimes lead to secondary bacterial infections. What we call "strep throat" is a primary bacterial infection caused by *Streptococcus pyogenes* that can lead to a generalized upper respiratory infection and even a systemic (affecting the body as a whole) infection. The symptoms of strep throat are severe sore throat, high fever, and white patches on a dark red throat.

Sinusitis is an infection of the sinuses, which are cavities within the facial skeleton that drain into the nasal cavities. Only about 1–3% of upper respiratory infections are accompanied by sinusitis. Sinusitis develops when nasal congestion blocks the tiny openings leading to the sinuses. *Tonsillitis* occurs when tonsils become inflamed and enlarged. **Tonsils** are masses of lymphatic tissue that occur in the pharynx. The tonsils in the dorsal wall of the nasopharynx are often called adenoids. The tonsils remove many of the pathogens that enter the pharynx; therefore, they are a first line of defense against invasion of the body. *Laryngitis* is an infection of the larynx with an accompanying hoarseness, leading to the inability to talk in an audible voice. Usually laryngitis disappears with treatment of the upper respiratory infection. Persistent hoarseness without the presence of an upper respiratory infection is one of the warning signs of cancer and therefore should be looked into by a physician.

Lower Respiratory Tract Disorders

Lower respiratory tract disorders are shown in Figure 37.11.

Infections

Acute bronchitis is an infection of the primary and secondary bronchi. Usually it is preceded by a viral upper respiratory infection that has led to a secondary bacterial infection. Most likely, a nonproductive cough has become a deep cough that expectorates mucus and perhaps pus.

Pneumonia is a viral, bacterial, or fungal infection of the lungs in which bronchi and alveoli fill with pus and fluid (Fig. 37.11). Most often it is preceded by influenza. Rather than being a generalized lung infection, pneumonia may be localized in specific lobules of the lungs. Obviously the more lobules involved, the more serious the infection. Pneumonia can be caused by a microbe that is usually held in check but has gained the upper hand due to stress and/or reduced immunity. AIDS patients are subject to a particularly rare form of pneumonia caused by a fungal-like organism called *Pneumocystis carinii*. Pneumonia of this type is almost never seen in individuals with a healthy immune system. High fever and chills with headache and chest pain are symptoms of pneumonia.

Pulmonary tuberculosis is caused by the tubercle bacillus, a type of bacterium. It is possible to tell if a person has ever been exposed to tuberculosis with a skin test in which a highly diluted extract of the bacillus is injected into the skin of the patient. A person who has never been in contact with the tubercle bacillus shows no reaction, but one who has developed immunity to the organism shows an area of inflammation that peaks in about 48 hours. When tubercle bacilli invade the lung tissue, the cells build a protective capsule about the foreigners, isolating them from the rest of the body. This tiny capsule is called a tubercle. If the resistance of the body is high, the imprisoned organisms die, but if the resistance is low, the organisms eventually can be liberated. If a chest X ray detects active tubercles, the individual is put on appropriate drug therapy to ensure the localization of the disease and the eventual destruction of any live bacterial organisms.

Tuberculosis was a major killer in the United States before the middle of the twentieth century, after which antibiotic therapy brought it largely under control. In recent years, the incidence of tuberculosis has been on the rise, particularly among AIDS patients, the homeless, and the rural poor. Worse, the new strains are resistant to the usual antibiotic therapy.

Disorders

Inhaling particles such as silica (sand), coal dust, asbestos, and, now it seems, fiberglass can lead to *pulmonary fibrosis*, a condition in which fibrous connective tissue builds up in the lungs. The lungs cannot inflate properly and are always tending toward deflation. Breathing asbestos is also associated with the development of cancer. Since asbestos has been used so widely as a fireproofing and insulating agent, unwarranted exposure has occurred. It is projected that 2 million deaths could be caused by asbestos exposure—mostly in the workplace—between 1990 and 2020.

In *chronic bronchitis*, the airways are inflamed and filled with mucus. A cough that brings up mucus is common. The bronchi have undergone degenerative changes, including the loss of cilia and their normal cleansing action. Under these conditions, an infection is more likely to occur. Smoking cigarettes and cigars is the most frequent cause of chronic bronchitis. Exposure to other pollutants can also cause chronic bronchitis.

Emphysema is a chronic and incurable lung disorder in which the alveoli are distended and their walls damaged so that the surface area available for gas exchange is reduced. Emphysema is often preceded by chronic bronchitis. Air trapped in the lungs leads to alveolar damage and a noticeable ballooning of the chest. The elastic recoil of the lungs is reduced, so not only are the airways narrowed, but the driving force behind expiration is also reduced. The patient is breathless and may have a cough. Because the surface area for gas exchange is reduced, oxygen reaching the heart and the brain is reduced. Even so, the heart works furiously to force more blood through the lungs, and an increased workload on the heart can result. Lack of oxygen to the brain can make the person feel depressed, sluggish, and irritable. Exercise, drug therapy, and supplemental oxygen, along with giving

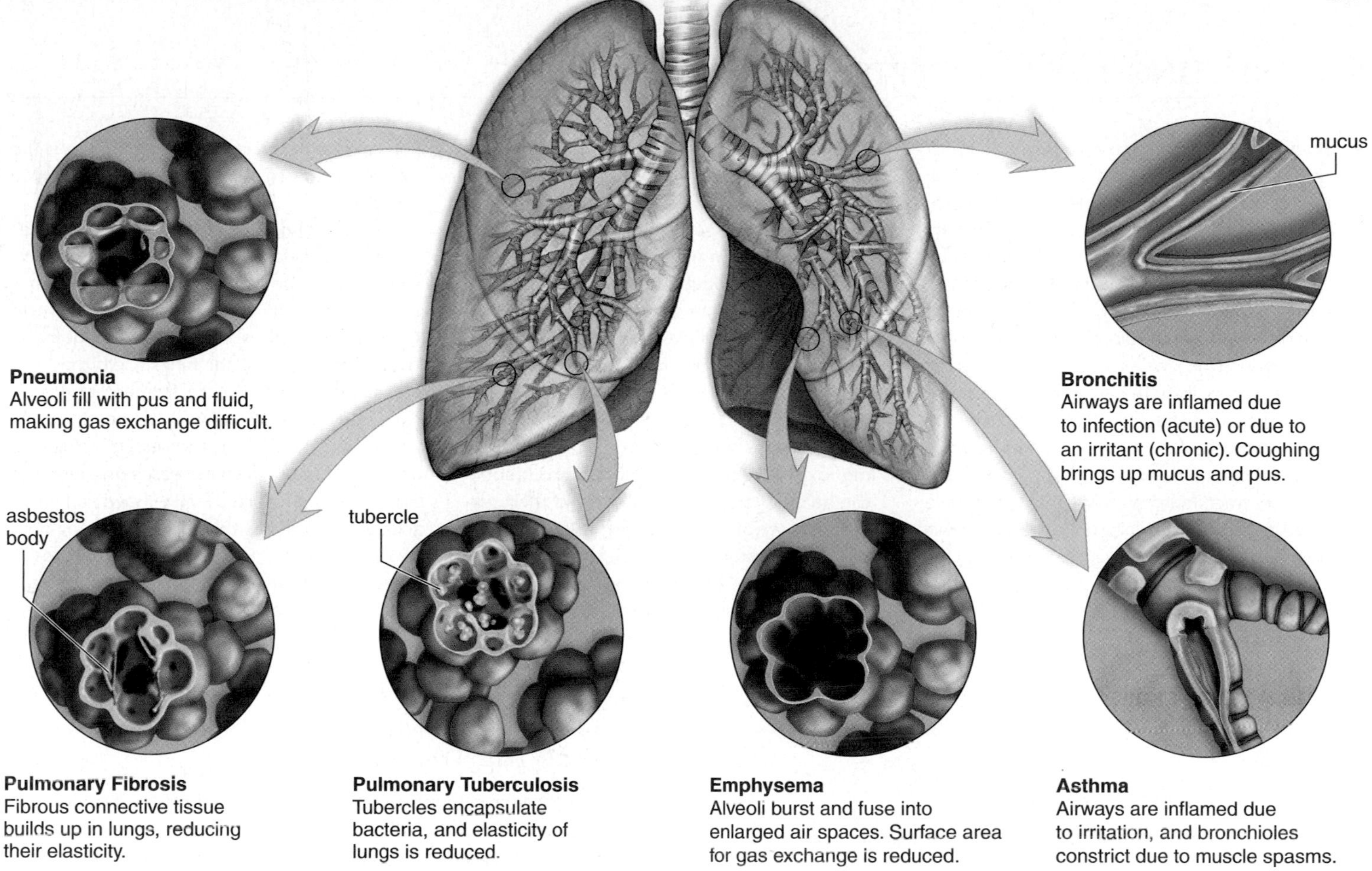

Pneumonia
Alveoli fill with pus and fluid, making gas exchange difficult.

Bronchitis
Airways are inflamed due to infection (acute) or due to an irritant (chronic). Coughing brings up mucus and pus.

mucus

asbestos body

tubercle

Pulmonary Fibrosis
Fibrous connective tissue builds up in lungs, reducing their elasticity.

Pulmonary Tuberculosis
Tubercles encapsulate bacteria, and elasticity of lungs is reduced.

Emphysema
Alveoli burst and fuse into enlarged air spaces. Surface area for gas exchange is reduced.

Asthma
Airways are inflamed due to irritation, and bronchioles constrict due to muscle spasms.

FIGURE 37.11 Common bronchial and pulmonary diseases.
Exposure to infectious pathogens and/or polluted air, including tobacco smoke, causes the diseases and disorders shown here.

up smoking, may relieve the symptoms and possibly slow the progression of emphysema.

Asthma is a disease of the bronchi and bronchioles that is marked by wheezing, breathlessness, and sometimes cough and expectoration of mucus. The airways are unusually sensitive to specific irritants, which can include a wide range of allergens such as pollen, animal dander, dust, cigarette smoke, and industrial fumes. Even cold air can be an irritant. When exposed to the irritant, the smooth muscle in the bronchioles undergoes spasms. It now appears that chemical mediators given off by immune cells in the bronchioles result in the spasms. Most asthma patients have some degree of bronchial inflammation that reduces the diameter of the airways and contributes to the seriousness of an attack. Asthma is not curable, but it is treatable. Special inhalers can control the inflammation and possibly prevent an attack, while other types of inhalers can stop the muscle spasms should an attack occur.

Lung Cancer

Lung cancer used to be more prevalent in men than in women, but recently it has surpassed breast cancer as a cause of death in women. This can be linked to an increase in the number of women who smoke today. Nearly 150,000 people in the United States die of lung cancer each year. The American Cancer Society links over 85% of these deaths to smoking. Autopsies on smokers have revealed the progressive steps by which the most common form of lung cancer develops. The first event appears to be thickening and callusing of the cells lining the airways. (Callusing occurs whenever cells are exposed to irritants.) Then there is a loss of cilia so that it is impossible to prevent dust and dirt from settling in the lungs. Following this, cells with atypical nuclei appear in the callused lining. A tumor consisting of disordered cells with atypical nuclei is considered to be cancer in situ (at one location). A final step occurs when some of these cells break loose and penetrate other tissues, a process called metastasis. Now the cancer has spread. The original tumor may grow until a bronchus is blocked, cutting off the supply of air to that lung. The entire lung then collapses, the secretions trapped in the lung spaces become infected, and pneumonia or a lung abscess (localized area of pus) results. The only treatment that offers a possibility of cure is to remove a lobe or the whole lung before metastasis has had time to occur. This operation is called a *pneumonectomy*.

Lung infections and disorders occur at a higher incidence among smokers than nonsmokers.

CONNECTING THE CONCEPTS

Both the human intestine and the lungs are exchange boundaries with the external environment. The lungs are composed of alveoli that number about 300 million, and each alveolus may have as many as 1,800 blood capillary contacts. The lungs are adapted to facilitate gas exchange, freely allowing oxygen to enter and carbon dioxide to exit the blood. The characteristics that enhance the function of the lungs for gas exchange also facilitate the entrance of substances and pathogens that could be harmful to the body. Defense mechanisms are in place that limit possible pathogen invasion of the body along the respiratory tract. Numerous lymphoid nodules located in the bronchial lining are a source of white blood cells, which protect against infection.

The body, however, has no means by which to inhibit the entrance of dangerous chemicals across the alveolar wall. Breathing air pollutants can result in respiratory distress, headache, and exhaustion. On occasion, when smog gets trapped above a city by a blanket of warm air, the results can be disastrous. In 1966, about 168 people died in New York City due to an accumulation of air pollutants over several days. We know the manner in which one air pollutant causes death. Carbon monoxide (CO) is an air pollutant that comes from the incomplete combustion of natural gas and gasoline. Because carbon monoxide is a colorless, odorless gas, people can be unaware that it has entered their system by way of the lungs. But once CO is in the bloodstream, it combines with the iron of hemoglobin 200 times more tightly than oxygen. The result is that the delivery of oxygen to mitochondria is impaired, and so is the functioning of mitochondria. The result can be death.

Those who smoke cigars and cigarettes should be aware that CO is in tobacco smoke. The soot in cigarette smoke impairs the ability of the lungs to breathe, and the CO impairs the ability of hemoglobin to transport oxygen to the lungs. Cells that lack sufficient oxygen turn to fermentation as a way to keep producing ATP. Fermentation, as you know, is inefficient and results in the end product lactate, which increases the acidity of the blood. As Figure 37.10b shows, a higher acidity also reduces the ability of hemoglobin to carry oxygen. Clearly, smoking makes it more difficult for blood concentrations to remain normal and threatens homeostasis.

Summary

37.1 GAS EXCHANGE SURFACES

Some aquatic animals, such as hydras and planarians, use their entire body surface for gas exchange. Most animals have a specialized gas-exchange area. Large aquatic animals usually pass water through gills. On land, insects use tracheal systems, and vertebrates have lungs.

Lungs are found inside the body, where water loss is reduced. To ventilate the lungs, some vertebrates use positive pressure, but most inhale, using muscular contraction to produce a negative pressure that causes air to rush into the lungs. When the breathing muscles relax, air is exhaled.

Birds have a series of air sacs that allow a one-way flow of air over the gas-exchange area. This results in greater gas-exchange efficiency than other vertebrates.

37.2 HUMAN RESPIRATORY SYSTEM

Table 37.1 lists the structures in the human respiratory system.

Humans breathe by negative pressure, as do other mammals. During inspiration, the rib cage goes up and out, and the diaphragm lowers. The lungs expand and air comes rushing in. During expiration, the rib cage goes down and in, and the diaphragm rises. Therefore, air rushes out.

The rate of breathing increases when the amount of H^+ and carbon dioxide in the blood rises, as detected by chemoreceptors such as the aortic and carotid bodies.

Gas exchange in the lungs and tissues is brought about by diffusion. Hemoglobin transports oxygen in the blood; carbon dioxide is mainly transported in plasma as the bicarbonate ion. The enzyme carbonic anhydrase found in red blood cells speeds the formation of the bicarbonate ion.

37.3 RESPIRATION AND HEALTH

The respiratory tract is subject to infections. Two major lung disorders, emphysema and cancer, are usually due to cigarette smoking.

Reviewing the Chapter

1. Compare the respiratory organs of aquatic animals to those of terrestrial animals.   673–75
2. How does the countercurrent flow of blood within gill capillaries and water passing across the gills assist respiration in fishes?   673
3. Why is it beneficial for the body wall of earthworms to be moist? Why don't insects require circulatory system involvement in air transport?   674
4. Explain the phrase "breathing by using negative pressure."   674
5. Contrast the tidal ventilation mechanism in humans with the one-way ventilation mechanism in birds, and explain the benefits of the ventilation mechanism in birds.   674–75
6. Name the parts of the human respiratory system, and list a function for each part.   676–77
7. The concentration of what substances in blood controls the breathing rate in humans? Explain.   677
8. How are oxygen and carbon dioxide transported in blood? What does carbonic anhydrase do?   678–79
9. Which conditions depicted in Figure 37.11 are due to infection? Which are due to behavioral or environmental factors? Explain.   680–81

Testing Yourself

Choose the best answer for each question.

1. One problem faced by terrestrial animals with lungs, but not by freshwater aquatic animals with gills, is that
   a. gas exchange involves water loss.
   b. breathing requires considerable energy.
   c. oxygen diffuses very slowly in air.
   d. the concentration of oxygen in water is greater than that in air.
   e. All of these are correct.

2. In which animal is the circulatory system not involved in gas transport?
   a. mouse
   b. dragonfly
   c. trout
   d. sparrow
   e. human

3. Birds have more efficient lungs than humans because the flow of air
   a. is the same during both inspiration and expiration.
   b. travels in only one direction through the lungs.
   c. never backs up as it does in human lungs.
   d. is not hindered by a larynx.
   e. enters their bones.

4. Which animal breathes by positive pressure?
   a. fish
   b. human
   c. bird
   d. frog
   e. planarian

5. Which of these is a true statement?
   a. In lung capillaries, carbon dioxide combines with water to produce carbonic acid.
   b. In tissue capillaries, carbonic acid breaks down to carbon dioxide and water.
   c. In lung capillaries, carbonic acid breaks down to carbon dioxide and water.
   d. In tissue capillaries, carbonic acid combines with hydrogen ions to form the carbonate ion.
   e. All of these statements are true.

6. Air enters the human lungs because
   a. atmospheric pressure is less than the pressure inside the lungs.
   b. atmospheric pressure is greater than the pressure inside the lungs.
   c. although the pressures are the same inside and outside, the partial pressure of oxygen is lower within the lungs.
   d. the residual air in the lungs causes the partial pressure of oxygen to be less than it is outside.
   e. the process of breathing pushes air into the lungs.

7. If the digestive and respiratory tracts were completely separate in humans, there would be no need for
   a. swallowing.
   b. a nose.
   c. an epiglottis.
   d. a diaphragm.
   e. All of these are correct.

8. In tracing the path of air in humans, you would list the trachea
   a. directly after the nose.
   b. directly before the bronchi.
   c. before the pharynx.
   d. directly before the lungs.
   e. Both a and c are correct.

9. In humans, the respiratory control center
   a. is stimulated by carbon dioxide.
   b. is located in the medulla oblongata.
   c. controls the rate of breathing.
   d. is stimulated by hydrogen ion concentration.
   e. All of these are correct.

10. Carbon dioxide is carried in the plasma
    a. in combination with hemoglobin.
    b. as the bicarbonate ion.
    c. combined with carbonic anhydrase.
    d. only as a part of tissue fluid.
    e. All of these are correct.

11. Which of these is anatomically incorrect?
    a. The nose has two nasal cavities.
    b. The pharynx connects the nasal and oral cavities to the larynx.
    c. The larynx contains the vocal cords.
    d. The trachea enters the lungs.
    e. The lungs contain many alveoli.

12. How is inhaled air modified before it reaches the lungs?
    a. It must be humidified.
    b. It must be warmed.
    c. It must be filtered.
    d. All of these are correct.

13. Internal respiration refers to
    a. the exchange of gases between the air and the blood in the lungs.
    b. the movement of air into the lungs.
    c. the exchange of gases between the blood and tissue fluid.
    d. cellular respiration, resulting in the production of ATP.

14. The chemical reaction that converts carbon dioxide to a bicarbonate ion takes place in
    a. the blood plasma.
    b. red blood cells.
    c. the alveolus.
    d. the hemoglobin molecule.

15. Which of these would affect hemoglobin's O_2-binding capacity?
    a. pH
    b. partial pressure of oxygen
    c. blood pressure
    d. temperature
    e. All of these except c are correct.

16. The enzyme carbonic anhydrase
    a. causes the blood to be more basic in the tissues.
    b. speeds the conversion of carbonic acid to carbon dioxide and water.
    c. actively transports carbon dioxide out of capillaries.
    d. is active only at high altitudes.
    e. All of these are correct.

17. Which of these is incorrect concerning inspiration?
    a. Rib cage moves up and out.
    b. Diaphragm contracts and moves down.
    c. Pressure in lungs decreases, and air comes rushing in.
    d. The lungs expand because air comes rushing in.

18. Label this diagram of the human respiratory system.

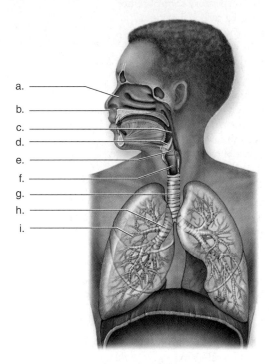

a. _____
b. _____
c. _____
d. _____
e. _____
f. _____
g. _____
h. _____
i. _____

Thinking Scientifically

1. A roadside museum advertises that it has fossils of meter-long insects. Explain why the respiratory system of insects would not support a large size and therefore the existence of such fossils is highly unlikely.
2. Fetal hemoglobin picks up oxygen from the maternal blood. If the oxygen-binding characteristics of hemoglobin in the fetus were identical to the hemoglobin of the mother, oxygen could never be transferred at the placenta to fetal circulation. What hypothesis about the oxygen-binding characteristics of fetal hemoglobin would explain how fetuses get the oxygen they need?

Bioethical Issue: Antibiotic Therapy

Antibiotics cure respiratory infections, but there are problems associated with antibiotic therapy. Aside from a possible allergic reaction, antibiotics not only kill off disease-causing bacteria, but they also reduce the number of beneficial bacteria in the intestinal tract and other locations. These beneficial bacteria hold in check the growth of other pathogens that now begin to flourish. Diarrhea can result, as can a vaginal yeast infection. The use of antibiotics can also prevent natural immunity from occurring, leading to the need for recurring antibiotic therapy. Especially alarming at this time is the occurrence of resistance. Resistance takes place when vulnerable bacteria are killed off by an antibiotic, and this allows resistant bacteria to become prevalent. The bacteria that cause ear, nose, and throat infections as well as scarlet fever and pneumonia are becoming widely resistant because we have

not been using antibiotics properly. Tuberculosis is on the rise, and the new strains are resistant to the usual combined antibiotic therapy.

Every citizen needs to be aware of our present crisis situation. Stuart Levy, a Tufts University School of Medicine microbiologist, says that we should do what is ethical for society and ourselves. What is needed? Antibiotics kill bacteria, not viruses—therefore, we shouldn't take antibiotics unless we know for sure we have a bacterial infection. And we shouldn't take them prophylactically—that is, just in case we might need one. If antibiotics are taken in low dosages and intermittently, resistant strains are bound to take over. Animal and agricultural use should be pared down, and household disinfectants should no longer be spiked with antibacterial agents. Perhaps then, Levy says, vulnerable bacteria will begin to supplant the resistant ones in the population. Are you doing all you can to prevent bacteria from becoming resistant?

Understanding the Terms

alveolus (pl., alveoli)   677
bicarbonate ion   679
bronchiole   676
bronchus (pl., bronchi)   676
carbaminohemoglobin   679
carbonic anhydrase   679
diaphragm   674
epiglottis   676
expiration   674
external respiration   672
gills   673
glottis   676
heme   678

hemoglobin (Hb)   678
inspiration   674
internal respiration   672
larynx   676
lungs   674
oxyhemoglobin   678
partial pressure   675
pharynx   676
respiration   672
tonsils   680
trachea (pl., tracheae)   674, 676
ventilation   672
vocal cord   676

Match the terms to these definitions:

a. _____ In terrestrial vertebrates, the mechanical act of moving air in and out of the lungs; breathing.
b. _____ Dome-shaped muscularized sheet separating the thoracic cavity from the abdominal cavity in mammals.
c. _____ Fold of tissue within the larynx; creates vocal sounds when it vibrates.
d. _____ Respiratory organ in most aquatic animals; in fish, an outward extension of the pharynx.
e. _____ Stage during breathing when air is pushed out of the lungs.

ARIS, the *Biology* Website

ARIS, the website for *Biology*, provides a wealth of information organized and integrated by chapter. You will find practice quizzes, interactive activities, labeling exercises, flashcards, and much more that will complement your learning and understanding of general biology.

www.mhhe.com/maderbiology9

38

BODY FLUID REGULATION AND EXCRETORY SYSTEMS

*H*omeostasis involves the distribution of oxygen and nutrients to cells, and also getting rid of their wastes. The bodies of animals must have some way of eliminating metabolic wastes, a process known as **excretion.**

Some aquatic invertebrates, like the hydra, do not have an excretory organ. The body wall has only two layers of cells, and water enters and exits a large, central, fluid-filled cavity. Each cell excretes metabolic wastes directly into the cavity, and water washes them away. In more complex animals, nitrogenous wastes and excess fluids or salts are discharged into the external environment by an excretory organ. A marine turtle may appear to be crying because it has glands near the eyes that are used to eliminate excess salt.

An abnormal osmolarity of internal fluids can cause an animal to lose or gain fluids to or from the external environment. Excretory organs play a major role in maintaining homeostasis because, in addition to excreting wastes, they regulate the salt balance, and therefore the water balance, of the body. As in humans, excretory organs also play an important role in maintaining the normal pH of body fluids.

Excretion, which has these numerous functions, should not be confused with defecation, which is the elimination of nondigested material from the digestive tract.

Marine organisms, such as marine turtles, must rid the body of excess salt.

38.1 BODY FLUID REGULATION

An excretory system is involved in regulating body fluid concentrations. It does this by retaining or eliminating certain ions and water. The regulation of body fluids is dependent on the concentration of mineral ions such as sodium (Na^+), chloride (Cl^-), potassium (K^+), and the bicarbonate ion (HCO_3^-). Body fluids gain mineral ions as a result of eating foods and drinking fluids. Excretion is the primary way the body loses ions.

In many animals, water can enter the body through metabolism (e.g., cellular respiration produces water), by eating foods that contain water, and by drinking water. Water is lost from the body through evaporation (e.g., from skin and lungs), feces formation, and excretion (Fig. 38.1). To be in fluid balance, the amount of water exiting the body must equal the amount entering.

When there are differences in osmolarity (i.e., solute concentration) between two regions, water tends to move from regions of greater concentration to regions of lesser concentration. Thus, water moves to the region with the highest amount of ions. A marine environment, which is high in salts, tends to promote the osmotic loss of water and the gain of ions by such means as drinking water. Fresh water tends to promote a gain of water by osmosis and a loss of ions as excess water is excreted. Terrestrial animals tend to lose both water and ions to the environment.

Aquatic Animals

Among aquatic animals, only marine invertebrates, such as molluscs and arthropods, and cartilaginous fishes, such as sharks and rays, have body fluids that are nearly isotonic to seawater—that is, having nearly equal concentrations of solutes. These organisms have little difficulty maintaining their normal salt and water balance. It is surprising, though, that while the body fluids of cartilaginous fishes are isotonic, they do not contain the same amount of mineral ions as seawater. The answer to this paradox is that their blood contains a concentration of urea high enough to match the tonicity of the sea! For some unknown reason, this amount of urea is not toxic to cartilaginous fishes.

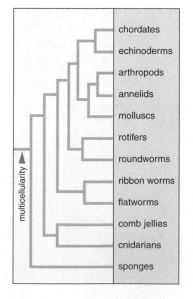

Bony Fishes

The body fluids of all bony fishes normally have only a moderate amount of salt. Apparently, their common ancestor evolved in fresh water, and only later did some groups

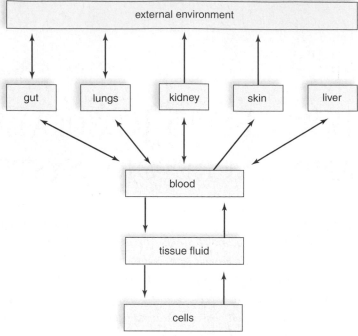

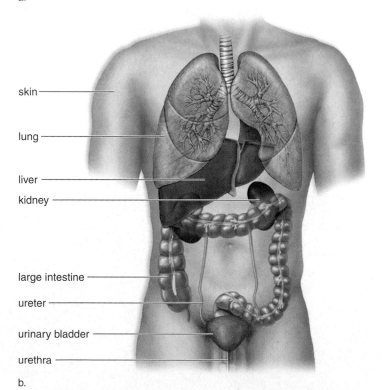

FIGURE 38.1 Overview of excretion.
a. The internal environment of cells (blood and tissue fluid) in humans and other animals stays relatively constant because the blood is continually exchanging substances with the external environment. Fluids and solutes enter the blood and then the tissue fluid by way of the digestive tract. Water is lost by the lungs which carry on gas exchange; mineral ions are lost by the skin at sweat glands but the kidneys are primarily responsible for water and ion balance of the internal environment. **b.** Excretion rids the body of metabolic wastes. Urea produced by the liver exits the body by way of the sweat glands in the skin and by way of the kidneys, ureters, urinary bladder, and urethra. The lungs excrete carbon dioxide derived from the bicarbonate ion in blood. Bile produced by the liver exits the body by way of the large intestine.

invade the sea. Marine bony fishes (Fig. 38.2*a*) are therefore prone to water loss and could become dehydrated. To counteract this, they drink seawater almost constantly. On the average, marine bony fishes swallow an amount of water estimated to be equal to 1% of their body weight every hour. This is equivalent to a human drinking about 700 ml of water every hour around the clock. While they get water by drinking, this habit also causes these fishes to acquire salt. To rid the body of excess salt, they actively transport sodium (Na^+) and chloride (Cl^-) ions into the surrounding seawater at the gills. This causes a passive loss of water through gills.

The osmotic problems of freshwater bony fishes (Fig. 38.2*b*) are exactly opposite to those of marine bony fishes. The body fluids of freshwater bony fishes are hypertonic to fresh water, and they are prone to passively gain water. These fishes never drink water but instead eliminate excess water by producing large quantities of dilute (hypotonic) urine. They discharge a quantity of urine equal to one-third their body weight each day. Because this causes them to lose salts, they actively transport salts into the blood across the membranes of their gills.

The difference in adaptation between marine and freshwater bony fishes makes it remarkable that some fishes actually can move between the two environments during their life cycle. Salmon, for example, begin their lives in freshwater streams and rivers, move to the ocean for a period of time, and finally return to fresh water to breed. These fishes alter their behavior and their gill and kidney functions in response to the osmotic changes they encounter when moving from one environment to the other. Fishes living in water that is subject to changes in salinity have the ability to osmoregulate to survive.

Terrestrial Animals

Most terrestrial animals need to drink water occasionally to make up for the water lost by excretion and respiration, as well as in sweat and feces. Like marine bony fishes, birds and reptiles that live near the sea are able to drink seawater despite its high osmolarity because they have salt glands located above each eye that can excrete large volumes of concentrated salt solution.

To prevent water loss, some animals excrete a nitrogenous waste that is rather insoluble. An impermeable outer covering also helps. Compare the moist, thin, permeable skin of a frog to the dry, horny, thick skin of a lizard, and you know immediately which one is adapted to a dry terrestrial environment. Aside from these measures to prevent water loss, unique adaptations abound. To prevent loss of water during the process of breathing, certain animals, such as the camel and the kangaroo rat, have a nasal passage that has a highly convoluted mucous membrane surface. This surface captures condensed water from exhaled air. The water is reabsorbed into the bloodstream, and respiratory water loss is reduced. Exhaled air is usually always full of moisture, which is why you can see it on cold winter mornings—the moisture in exhaled air is

condensing. As we shall see, humans mainly conserve water by producing a hypertonic urine. The kangaroo rat also forms a very concentrated urine, and its fecal material is almost completely dry. This animal is so adapted to conserving water that it can survive by using metabolic water derived from cellular respiration. Some invertebrates, such as several species of nematodes, literally have the ability to dry up under adverse conditions.

To maintain body fluids within normal limits, animals regulate the excretion of salts and water.

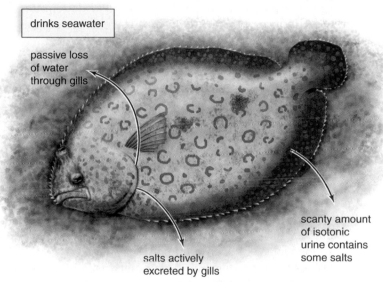

drinks seawater

passive loss of water through gills

scanty amount of isotonic urine contains some salts

salts actively excreted by gills

a. Marine fish

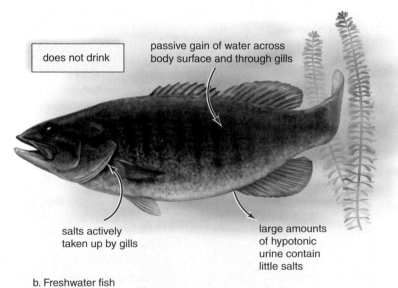

does not drink

passive gain of water across body surface and through gills

salts actively taken up by gills

large amounts of hypotonic urine contain little salts

b. Freshwater fish

FIGURE 38.2 Body fluid regulation in bony fishes.
a. Marine bony fishes employ different mechanisms compared to (**b**) freshwater fishes in order to osmoregulate their body fluids.

38.2 NITROGENOUS WASTE PRODUCTS

The breakdown of various molecules, including nucleic acids and amino acids, results in nitrogenous wastes. For simplicity's sake, however, we will limit our discussion to amino acid metabolism. Amino acids not used for protein synthesis are broken down by the body to generate energy, or are converted to fats or carbohydrates that can be stored. In either case, the amino groups ($—NH_2$) must be removed because only the carbon chains are used when amino acids are broken down or converted to forms of stored energy. Once the amino groups have been removed from amino acids, they may be excreted from the body in the form of ammonia, urea, or uric acid, depending on the species (Table 38.1). Removal of amino groups from amino acids requires a fairly set amount of energy. However, the energy requirement for the conversion of amino groups to ammonia, urea, or uric acid differs, as indicated in Figure 38.3.

Excretion of Nitrogenous Wastes

Amino groups removed from amino acids immediately form **ammonia** (NH_3) by the addition of a third hydrogen ion. Little or no energy is required to convert an amino group to ammonia by the addition of a hydrogen ion. Ammonia is quite toxic and can be used as a nitrogenous excretory product only if a good deal of water is available to wash it from the body. The high solubility of ammonia permits this means of excretion in bony fishes, aquatic invertebrates, and amphibians, whose gills and skin surfaces are in direct contact with the water of the environment.

Terrestrial amphibians and mammals usually excrete **urea** [Gk. *urina*, urine] as their main nitrogenous waste. Urea is much less toxic than ammonia and can be excreted in a moderately concentrated solution. This allows body water to be conserved, an important advantage for terrestrial animals with limited access to water. Production of urea, however, requires the expenditure of energy. Urea is produced in the liver by a set of energy-requiring enzymatic reactions known as the urea cycle. In this cycle, carrier molecules take up carbon dioxide and two molecules of ammonia, finally releasing urea.

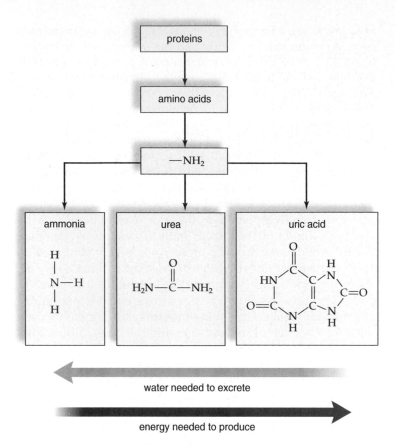

water needed to excrete

energy needed to produce

FIGURE 38.3 Nitrogenous wastes.
Proteins are hydrolyzed to amino acids, whose breakdown results in carbon chains and amino groups ($—NH_2$). The carbon chains can be used as an energy source, but the amino groups must be excreted as ammonia, urea, or uric acid.

Uric acid is routinely excreted by insects, reptiles, and birds. Uric acid is not very toxic, and it is poorly soluble in water. Poor solubility is an advantage if water conservation is needed, because uric acid can be concentrated even more readily than can urea. In reptiles and birds, a dilute solution of uric acid passes from the kidneys to the *cloaca*, a common reservoir for the products of the digestive, urinary, and reproductive systems. The cloacal contents are refluxed into the large intestine, where water is reabsorbed. The white substance in bird feces is uric acid. Embryos of reptiles and birds develop inside completely enclosed shelled eggs. The production of insoluble, relatively nontoxic uric acid is advantageous for shelled embryos because all nitrogenous wastes are stored inside the shell until hatching takes place.

Uric acid is synthesized by a long, complex series of enzymatic reactions that requires expenditure of even more ATP than does urea synthesis. Here again, there seems to be a trade-off between the advantage of water conservation and the disadvantage of energy expenditure for synthesis of an excretory molecule.

Animals excrete nitrogenous wastes as ammonia, urea, or uric acid. Ammonia requires the most water to excrete; uric acid requires the most energy to produce.

TABLE 38.1

Nitrogenous Waste Excretion

Product	Habitat	Animals
Ammonia	Water	Aquatic invertebrates Bony fishes Amphibian larvae
Urea	Land	Adult amphibians Mammals
Uric acid	Land	Insects Birds Reptiles

38.3 ORGANS OF EXCRETION

A number of mechanisms have evolved to cope with water balance in animals. Most animals have tubular excretory organs that regulate the water-salt balance and excrete metabolic wastes into the environment.

Flame Cells in Planarians

Planarians, which are flatworms, live in fresh water and have two strands of branching excretory tubules that open to the outside of the body through excretory pores (Fig. 38.4a). Located along the tubules are bulblike **flame cells** (protonephridia), each of which contains a cluster of beating cilia that looks like a flickering flame under the microscope. The beating of flame-cell cilia propels fluid through the excretory tubules and out of the body. The system is believed to function in ridding the body of excess water and also in excreting wastes.

Nephridia in Earthworms

The body of an earthworm, an annelid, is divided into segments, and nearly every body segment has a pair of excretory structures called **nephridia** [Gk. *nephros*, kidney] (metanephridia). Each nephridium is a tubule with a ciliated opening (the nephridiostome) and an excretory pore (the nephridiopore) (Fig. 38.4b). As fluid from the coelom is propelled through the tubule by beating cilia, nutrient substances are reabsorbed and carried away by a network of capillaries surrounding the tubule. The urine of an earthworm contains only metabolic wastes, salts, and water.

Each day, an earthworm excretes a lot of water and may produce a volume of urine equal to 60% of its body weight. The excretion of ammonia is consistent with these data.

Malpighian Tubules in Insects

Insects have a unique excretory system consisting of long, thin tubules called **Malpighian tubules,** attached to the gut. Uric acid simply flows from the surrounding hemolymph into these tubules, and water follows a salt gradient established by active transport of K^+. Water and other useful substances are reabsorbed at the rectum, but the uric acid leaves the body at the anus. Insects that live in water or that eat large quantities of moist food reabsorb little water. But insects in dry environments reabsorb most of the water and excrete a dry, semisolid mass of uric acid.

A number of other excretory structures have evolved in arthropods. Crayfish possess antennal glands (green glands) located in the ventral portion of the head region. Selective filtration and secretion of salts regulate the amount of urine excreted. In shrimp and pillbugs, the excretory organs are located in the maxillary segments and are called maxillary glands. Spiders, scorpions, and other arachnids possess coxal glands used in excretion near the joint of one or more appendages.

Most animals have a primary excretory organ. Planarians have flame cells, earthworms have nephridia, and insects have Malpighian tubules.

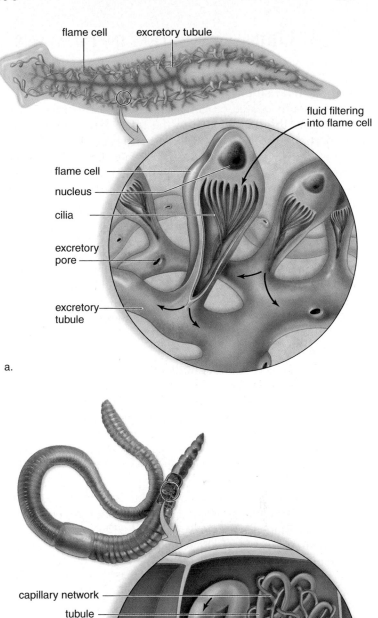

FIGURE 38.4 Excretory organs in animals.
a. The flame-cell excretory system in planarians. Two or more tracts of branching tubules run the length of the body and open to the outside by pores. At the ends of side branches are small bulblike cells called flame cells. The beating of cilia causes fluid to enter tubules that remove excess water from the body.
b. The earthworm nephridium. The nephridium has a ciliated opening, the nephridiostome, that leads to a coiled tubule surrounded by a capillary network. Urine can be temporarily stored in the bladder before being released to the outside via a pore termed a nephridiopore. Most segments contain a pair of nephridia, one on each side.

38.4 URINARY SYSTEM IN HUMANS

The urinary system of humans contains excretory organs called the kidneys (Fig. 38.5). The human **kidneys** are bean-shaped, reddish-brown organs, each about the size of a fist. They are located on either side of the vertebral column just below the diaphragm, in the lower back, where they are partially protected by the lower rib cage. The right kidney is slightly lower than the left kidney. **Urine** [Gk. *urina*, urine] made by the kidneys is conducted from the body by the other organs in the urinary system. Each kidney is connected to a **ureter,** a duct that takes urine from the kidney to the **urinary bladder,** where it is stored until it is voided from the body through the single **urethra.** In males, the urethra passes through the penis, and in females, the opening of the urethra is ventral to that of the vagina. There is no connection between the genital (reproductive) and urinary systems in females, but there is a connection in males. In males, the urethra also carries sperm during ejaculation.

Kidneys

If a kidney is sectioned longitudinally, three major parts can be distinguished (Fig. 38.6). The **renal cortex,** which is the outer region of a kidney, has a somewhat granular appearance. The **renal medulla** consists of six to ten cone-shaped renal pyramids that lie on the inner side of the renal cortex. The innermost part of the kidney is a hollow chamber called the **renal pelvis.** Urine collects in the renal pelvis and then is carried to the bladder by a ureter.

A kidney stone or renal calculus is a hard granule of phosphate, calcium, protein, or uric acid that forms in the renal pelvis. Many are passed unnoticed. However, larger and jagged stones can block the renal pelvis or ureter causing intense pain and damage.

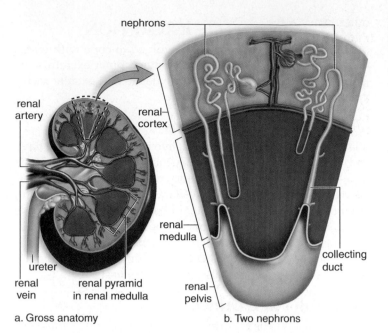

a. Gross anatomy   b. Two nephrons

FIGURE 38.6 Macroscopic and microscopic anatomy of the kidney.
a. Longitudinal section of a kidney, showing the location of the renal cortex, the renal medulla, and the renal pelvis. **b.** An enlargement of one renal lobe, showing the placement of nephrons.

Nephrons

Microscopically, each kidney is composed of over 1 million tiny tubules called **nephrons** [Gk. *nephros*, kidney]. The nephrons of a kidney produce urine. Some nephrons are located primarily in the renal cortex, but others dip down into the renal medulla, as shown in Figure 38.6*b*. Each nephron is made of several parts (Fig. 38.7). The blind end of a nephron is pushed in on itself to form a cuplike structure called the **glomerular capsule** [L. *glomeris*, ball] (Bowman's capsule). The

Fig 38.0
Fig 38.0

FIGURE 38.5 The human urinary system.
Urine is found only within the kidneys, the ureters, the urinary bladder, and the urethra.

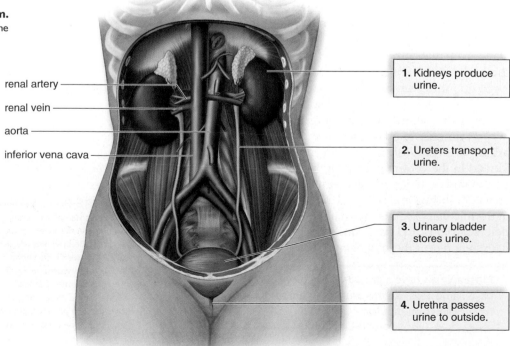

renal artery

renal vein

aorta

inferior vena cava

1. Kidneys produce urine.

2. Ureters transport urine.

3. Urinary bladder stores urine.

4. Urethra passes urine to outside.

outer layer of the glomerular capsule is composed of squamous epithelial cells; the inner layer is composed of specialized cells that allow easy passage of molecules. Leading from the glomerular capsule is a portion of the nephron known as the **proximal convoluted tubule** [L. *proximus,* nearest], which is lined by cells with many mitochondria and tightly packed microvilli. Then, simple squamous epithelium appears in the **loop of the nephron** (loop of Henle), which has a descending limb and an ascending limb. This is followed by the **distal convoluted tubule** [L. *distantia,* far]. Several distal convoluted tubules enter one **collecting duct.** The collecting duct transports urine down through the renal medulla and delivers it to the renal pelvis. The loop of the nephron and the collecting duct give the pyramids of the renal medulla their striped appearance (see Fig. 38.6).

Each nephron has its own blood supply (Fig. 38.7). The renal artery branches into numerous small arteries, which branch into arterioles, one for each nephron. Each arteriole, called an afferent arteriole, divides to form a capillary bed, the **glomerulus** [L. *glomeris,* ball], which is surrounded by the glomerular capsule. The glomerulus drains into an efferent arteriole, which subsequently branches into a second capillary bed around the tubular parts of the nephron. These capillaries, called peritubular capillaries, lead to venules that join to form veins leading to the renal vein, a vessel that enters the inferior vena cava.

Microscopically, the human kidney is composed of nephrons, tubules with specific parts that are richly supplied with capillaries.

Urine Formation

An average human produces between 1 and 2 liters of urine daily. Urine production requires three distinct processes:

1. glomerular filtration at the glomerular capsule;
2. tubular reabsorption at the convoluted tubules; and
3. tubular secretion at the convoluted tubules.

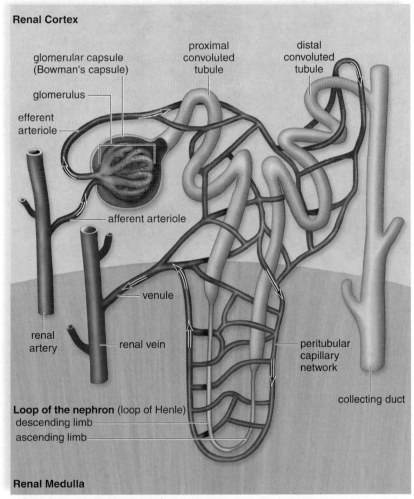

a. A nephron and its blood supply

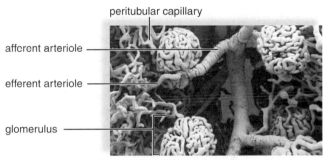

b. Surface view of glomerulus and its blood supply

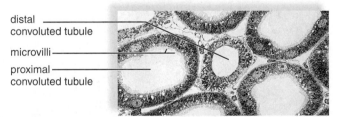

c. Cross sections of proximal and distal convoluted tubules   20 μm

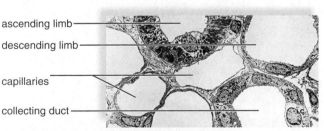

d. Cross sections of a loop of nephron limbs and collecting duct. (The other cross sections are those of capillaries.)   10 μm

FIGURE 38.7 Nephron anatomy.
a. You can trace the path of blood about a nephron by following the arrows. A nephron is made up of a glomerular capsule, the proximal convoluted tubule, the loop of the nephron, the distal convoluted tubule, and the collecting duct. The micrographs in (**b**), (**c**), and (**d**) show these structures.
© R. G. Kessel and R. H. Kardon, *Tissues and Organs: A Text-Atlas of Scanning Electron Microscopy,* 1979.

Glomerular Filtration

Glomerular filtration is the movement of small molecules across the glomerular wall into the glomerular capsule as a result of blood pressure. When blood enters the glomerulus, blood pressure is sufficient to cause small molecules, such as water, nutrients, salts, and wastes, to move from the glomerulus to the inside of the glomerular capsule, especially since the glomerular walls are 100 times more permeable than the walls of most capillaries elsewhere in the body. The molecules that leave the blood and enter the glomerular capsule are called the glomerular filtrate. Plasma proteins and blood cells are too large to be part of this filtrate, so they remain in the blood as it flows into the efferent arteriole.

Glomerular filtrate is essentially protein free, but otherwise it has the same composition as blood plasma. If this composition were not altered in other parts of the nephron, death from loss of nutrients (starvation) and loss of water (dehydration) would quickly follow. The total blood volume averages about 5 liters, and this amount of fluid is filtered every 40 minutes. Thus 180 liters of filtrate is produced daily, some 60 times the amount of blood plasma in the body. Most of the filtered water is obviously quickly returned to the blood, or a person would actually die from urination. Tubular reabsorption prevents this from happening.

Tubular Reabsorption

Tubular reabsorption takes place when substances move across the walls of the tubules into the associated peritubular capillary network (Fig. 38.8). The osmolarity of the blood is essentially the same as that of the filtrate within the glomerular capsule, and therefore osmosis of water from the filtrate into the blood cannot yet occur. However, sodium ions (Na^+) are actively pumped into the peritubular capillary, and then chloride ions (Cl^-) follow passively. Now the osmolarity of the blood is such that water moves passively from the tubule into the blood. About 60–70% of salt and water are reabsorbed at the proximal convoluted tubule.

Nutrients, such as glucose and amino acids, also return to the blood at the proximal convoluted tubule. This is a selective process, because only molecules recognized by carrier proteins in plasma membranes are actively reabsorbed. The cells of the proximal convoluted tubule have numerous microvilli, which increase the surface area, and numerous mitochondria, which supply the energy needed for active transport (Fig. 38.8). Glucose is an example of a molecule that ordinarily is reabsorbed completely because there is a plentiful supply of carrier molecules for it. However, if there is more glucose in the filtrate than there are carriers to handle it, glucose will exceed its renal threshold, or transport maximum. When this happens, the excess glucose in the filtrate will appear in the urine. In diabetes mellitus, there is an abnormally large amount of glucose in the filtrate because the liver fails to store glucose as glycogen. The presence of glucose in the filtrate results in less water being absorbed; the increased thirst and frequent urination in untreated diabetics are due to less water being reabsorbed into the peritubular capillary network.

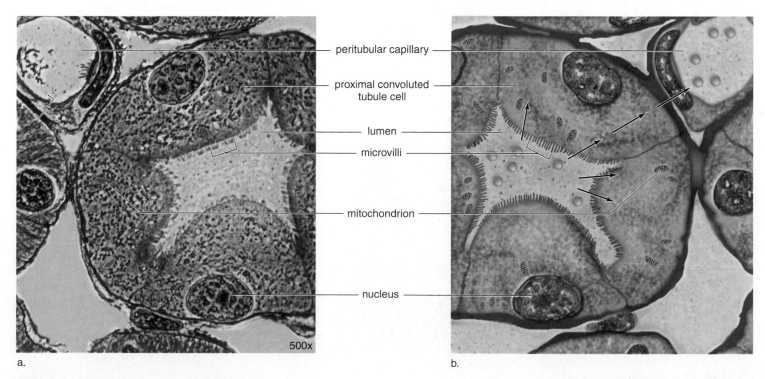

peritubular capillary

proximal convoluted tubule cell

lumen

microvilli

mitochondrion

nucleus

500x

a.

b.

FIGURE 38.8 Proximal convoluted tubule.
a. This photomicrograph shows that the cells lining the proximal convoluted tubule have a brush border composed of microvilli, which greatly increases the surface area exposed to the lumen. The peritubular capillary adjoins the cells. **b.** Diagrammatic representation of (**a**) shows that each cell has many mitochondria, which supply the energy needed for active transport, the process that moves molecules (green) from the lumen of the tubule to the capillary, as indicated by the arrows.

Urea is an example of a substance that is passively reabsorbed from the filtrate. At first, the concentration of urea within the filtrate is the same as that in blood plasma. But after water is reabsorbed, the urea concentration is greater than that of peritubular plasma. In the end, about 50% of the filtered urea is reabsorbed.

Tubular Secretion

Tubular secretion is the second way substances are removed from blood and added to tubular fluid (Fig. 38.9). Substances such as uric acid, hydrogen ions, ammonia, creatinine, histamine, and penicillin are eliminated by tubular secretion. The process of tubular secretion may be viewed as helping to rid the body of potentially harmful compounds that were not filtered into the glomerulus.

Glomerular filtration and tubular secretion add molecules to urine; tubular reabsorption removes them.

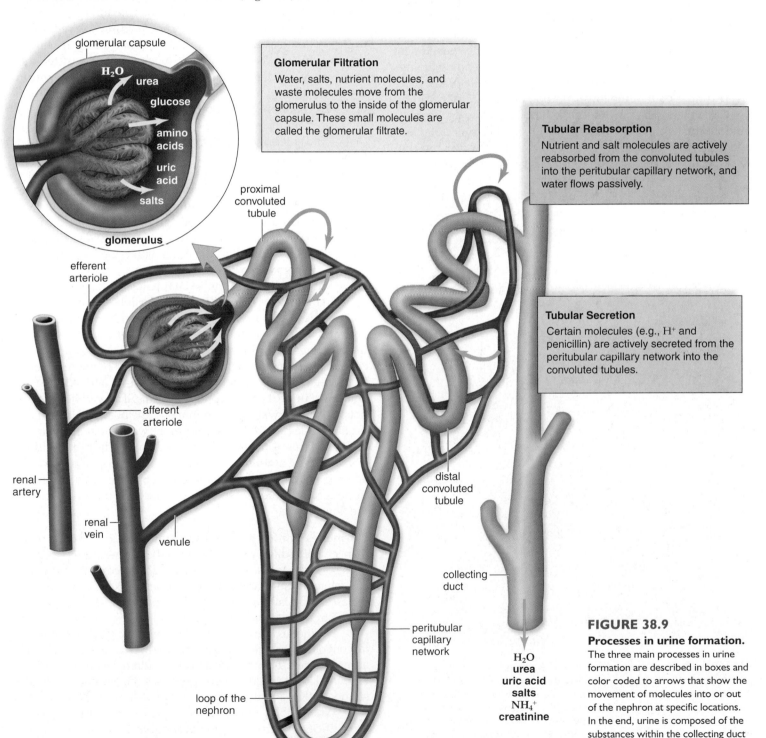

Glomerular Filtration
Water, salts, nutrient molecules, and waste molecules move from the glomerulus to the inside of the glomerular capsule. These small molecules are called the glomerular filtrate.

Tubular Reabsorption
Nutrient and salt molecules are actively reabsorbed from the convoluted tubules into the peritubular capillary network, and water flows passively.

Tubular Secretion
Certain molecules (e.g., H^+ and penicillin) are actively secreted from the peritubular capillary network into the convoluted tubules.

FIGURE 38.9

Processes in urine formation.
The three main processes in urine formation are described in boxes and color coded to arrows that show the movement of molecules into or out of the nephron at specific locations. In the end, urine is composed of the substances within the collecting duct (see blue arrow).

Urine Formation and Homeostasis

The kidneys regulate the water balance of the blood. In this way, they also maintain the blood volume and blood pressure. Most of the water and salt (NaCl) present in the filtrate is reabsorbed across the wall of the proximal convoluted tubule. Reabsorption also occurs along the remainder of the nephron.

Maintaining the Salt-Water Balance

The excretion of a hypertonic urine (one that is more concentrated than blood) is dependent on the reabsorption of water from the loop of the nephron and the collecting duct.

A long loop of the nephron, which typically penetrates deep into the renal medulla, is made up of a descending (going down) limb and an ascending (going up) limb. Salt (NaCl) passively diffuses out of the lower portion of the ascending limb, but the upper, thick portion of the limb actively extrudes salt out into the tissue of the outer renal medulla (Fig. 38.10). Less and less salt is available for transport as fluid moves up the thick portion of the ascending limb. Because of these circumstances, there is an osmotic gradient within the tissues of the renal medulla: The concentration of salt is greater in the direction of the inner medulla. (Note that water cannot leave the ascending limb because the limb is impermeable to water.)

The thick arrow in Figure 38.10 indicates that the innermost portion of the inner medulla has the highest concentration of solutes. This cannot be due to salt because active transport of salt does not start until fluid reaches the thick portion of the ascending limb. Urea is believed to leak from the lower portion of the collecting duct, and it is this molecule that contributes to the high solute concentration of the inner medulla.

Because of the osmotic gradient within the renal medulla, water leaves the descending limb along its entire length. This is a countercurrent mechanism: As water diffuses out of the descending limb, the remaining fluid within the limb encounters an even greater osmotic concentration of solute; therefore, water will continue to leave the descending limb from the top to the bottom.

Fluid enters the collecting duct from the distal convoluted tubule. This fluid is isotonic to the cells of the renal cortex. This means that to this point, the net effect of reabsorption of water and salt is the production of a fluid that has the same tonicity as blood plasma. However, the filtrate within the collecting duct also encounters the same osmotic gradient mentioned earlier (Fig. 38.10). Therefore, water diffuses out of the collecting duct into the renal medulla, and the urine within the collecting duct becomes hypertonic to blood plasma.

Antidiuretic hormone (ADH) [Gk. *anti*, against; L. *ouresis*, urination] released by the posterior lobe of the pituitary plays a role in water reabsorption at the collecting duct. To understand the action of this hormone, consider its name. *Diuresis* means increased amount of urine, and *antidiuresis* means decreased amount of urine. When ADH is present, more water is reabsorbed (blood volume and pressure rise), and a decreased amount of urine results. In practical terms, if an individual does not drink much water on a certain day, the posterior lobe of the pituitary releases ADH, causing more water to be reabsorbed and less urine to form. On the other hand, if an individual drinks a large amount of water and does not perspire much, ADH is not released. Now more water is excreted, and more urine forms. Diuretics, such as caffeine and alcohol, increase the flow of urine.

Hormones Control the Reabsorption of Salt.   Usually, more than 99% of sodium (Na^+) filtered at the glomerulus is returned to the blood. Most sodium (67%) is reabsorbed at the proximal convoluted tubule, and a sizable amount (25%) is extruded by the ascending limb of the loop of the nephron. The rest is reabsorbed from the distal convoluted tubule and collecting duct.

Hormones regulate the reabsorption of sodium at the distal convoluted tubule. **Aldosterone** is a hormone secreted by the adrenal cortex, the outer portion of the adrenal glands, which lie atop the kidneys. Aldosterone promotes the excretion of potassium ions (K^+) and the reabsorption of sodium ions (Na^+).

When blood volume, and therefore blood pressure, is not sufficient to promote glomerular filtration, the kidneys secrete renin. **Renin** is an enzyme that changes angiotensinogen

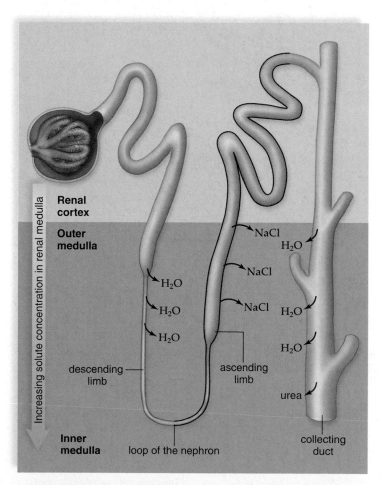

FIGURE 38.10 Countercurrent mechanism.
Salt (NaCl) diffuses and is actively transported out of the ascending limb of the loop of the nephron into the renal medulla; also, urea is believed to leak from the collecting duct and to enter the tissues of the renal medulla. This creates a hypertonic environment, which draws water out of the descending limb and the collecting duct. This water is returned to the cardiovascular system. (The thick black outline of the ascending limb means that it is impermeable to water.)

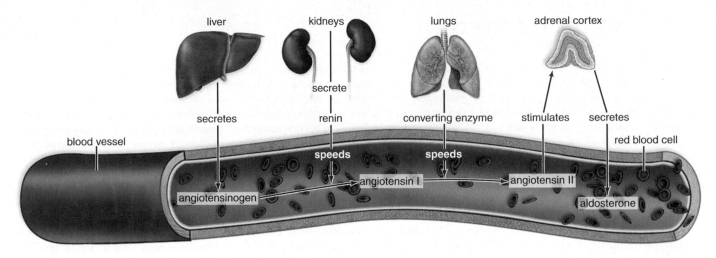

FIGURE 38.11 The renin-angiotensin-aldosterone system.
The liver secretes angiotensinogen into the bloodstream. Due to the action of renin from the kidneys and converting enzyme in lung capillaries, angiotensin II results. Angiotensin II acts on the adrenal cortex to secrete aldosterone, which causes reabsorption of sodium ions by the kidneys and a subsequent rise in blood pressure.

(a large plasma protein produced by the liver) into angiotensin I. Later, angiotensin I is converted to angiotensin II, a powerful vasoconstrictor that also stimulates the adrenal cortex to release aldosterone (Fig. 38.11). The reabsorption of sodium ions is followed by the reabsorption of water. Therefore, blood volume and blood pressure increase.

Atrial natriuretic hormone (ANH) is a hormone secreted by the atria of the heart when cardiac cells are stretched due to increased blood volume. ANH inhibits the secretion of renin by the juxtaglomerular apparatus and the secretion of aldosterone by the adrenal cortex. Its effect, therefore, is to promote the excretion of Na^+—that is, natriuresis. When Na^+ is excreted, so is water, and therefore blood volume and blood pressure decrease.

These examples show that the body regulates the water balance in blood by controlling the excretion and the reabsorption of various ions. Sodium (Na^+) is an important ion in plasma that must be regulated, but the kidneys also excrete or reabsorb other ions, such as potassium ions (K^+), bicarbonate ions (HCO_3^-), and magnesium ions (Mg^{2+}), as needed.

Maintaining the Acid-Base Balance

The bicarbonate (HCO_3^-) buffer system and breathing work together to maintain the pH of the blood. Central to the mechanism is this reaction, which you have seen before:

$$H^+ \; + \; HCO_3^- \; \rightleftharpoons \; H_2CO_3 \; \rightleftharpoons \; H_2O \; + \; CO_2$$

The excretion of carbon dioxide (CO_2) by the lungs helps keep the pH within normal limits, because when carbon dioxide is exhaled, this reaction is pushed to the right, and hydrogen ions are tied up in water. Indeed, when blood pH decreases, chemoreceptors in the carotid bodies (located in the carotid arteries) and in aortic bodies (located in the aorta) stimulate the respiratory control center, and the rate and depth of breathing increase. On the other hand, when blood pH begins to rise, the respiratory control center is depressed, and the amount of bicarbonate ion increases in the blood.

As powerful as this system is, only the kidneys can rid the body of a wide range of acidic and basic substances. The kidneys are slower acting than the buffer/breathing mechanism, but they have a more powerful effect on pH. For the sake of simplicity, we can think of the kidneys as reabsorbing bicarbonate ions and excreting hydrogen ions as needed to maintain the normal pH of the blood:

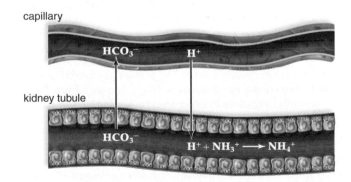

If the blood is acidic, hydrogen ions are excreted and bicarbonate ions are reabsorbed. If the blood is basic, hydrogen ions are not excreted and bicarbonate ions are not reabsorbed. The fact that urine is usually acidic (pH about 6) shows that usually an excess of hydrogen ions are excreted. Ammonia (NH_3) provides a means for buffering these hydrogen ions in urine: ($NH_3 + H^+ \longrightarrow NH_4^+$). Ammonia (whose presence is quite obvious in the diaper pail or kitty litter box) is produced in tubule cells by the deamination of amino acids. Phosphate provides another means of buffering hydrogen ions in urine.

The reabsorption of salt, which is under the control of hormones, determines the reabsorption of water, which is also regulated by ADH. The reabsorption of HCO_3^- and the excretion of H^+ help regulate the acid-base balance of the blood.

CONNECTING THE CONCEPTS

An artificial kidney employs the principle of dialysis to cleanse the blood. Blood from a patient's artery is passed through a semi-membranous tubing, which is surrounded by a dialysis solution that contains salt and various other small molecules. The concentration gradient between blood and the dialysis solution is such that waste molecules and excess salts diffuse through the membrane from the blood into the dialysis solution, while blood cells and plasma proteins stay behind. Because the dialysis solution is continually refreshed, wastes continually diffuse out of the blood. In the course of a six-hour treatment, 50–250 g of urea can be removed from a patient, an amount that greatly exceeds the urea clearance rate of normal kidneys. On the other hand, the dialysis solution contains salts at a level that maintains their proper concentration in the blood before it is returned to the patient.

How does a kidney machine function differently from an animal's excretory organs? In the human body, blood is also passing through a tube, namely the glomerulus. The solution that forms on the other side of the glomerulus (the glomerular filtrate) is free of proteins but otherwise has the same composition as blood plasma. Why? Because blood pressure forces water and small molecules to cross a highly permeable membrane. There is no opposing solution that maintains the normal nutrient concentration and osmolarity of the blood, nor an extreme concentration gradient that draws urea out of the blood.

In the human kidney, energy must be expended to retrieve the water, salts, and nutrients that are on their way out of the body. Following simple diffusion, pumps are employed to actively return Na^+ to the blood, and then water follows passively. Nutrients, such as glucose and amino acids, are reabsorbed by carrier-mediated active transport processes. Therefore, we see that in contrast to an artificial kidney, an animal's kidney is selective about what small molecules are returned to the blood, rather than what small molecules leave the blood. Finally, hormones regulate the osmolarity of the blood and fine-tune just how much water and salts are excreted.

Summary

38.1 BODY FLUID REGULATION

Osmotic regulation is important to animals. Most have to balance their water and salt intake and excretion to maintain normal solute and water concentration in body fluids. Marine fishes constantly drink water, excrete salts at the gills, and pass an isotonic urine. Freshwater fishes never drink water; they take in salts at the gills and excrete a hypotonic urine.

38.2 NITROGENOUS WASTE PRODUCTS

Animals excrete nitrogenous wastes. The amount of water and energy required to excrete nitrogenous wastes differs. Aquatic animals usually excrete ammonia (needs much water to excrete), and land animals excrete either urea or uric acid (needs much energy to produce).

38.3 ORGANS OF EXCRETION

Animals often have an excretory organ. The flame cells of planarians rid the body of excess water. Earthworm nephridia exchange molecules with the blood in a manner similar to that of vertebrate kidneys. Malpighian tubules in insects take up metabolic wastes and water from the hemolymph. Later, the water is absorbed by the gut.

38.4 URINARY SYSTEM IN HUMANS

The kidneys, excretory organs, are part of the human urinary system. Microscopically, each kidney is made up of nephrons, each of which has several parts and its own blood supply.

Urine formation by a nephron requires three steps: glomerular filtration, when nutrients, water, and wastes enter the nephron's glomerular capsule; tubular reabsorption, when nutrients and most water are reabsorbed into the peritubular capillary network; and tubular secretion, when additional wastes are added to the convoluted tubules.

Humans excrete a hypertonic urine. The ascending limb of the loop of the nephron actively extrudes salt so that the renal medulla is increasingly hypertonic relative to the contents of the descending limb and the collecting duct. Since urea leaks from the lower end of the collecting duct, the inner renal medulla has the highest concentration of solute. Therefore, a countercurrent mechanism ensures that water will diffuse out of the descending limb and the collecting duct.

Three hormones are involved in maintaining the water content of the blood. The hormone ADH (antidiuretic hormone), which makes the collecting duct more permeable, is secreted by the posterior pituitary in response to an increase in the osmotic pressure of the blood. The hormone aldosterone is secreted by the adrenal cortex after low blood pressure has caused the kidneys to release renin. The presence of renin leads to the formation of angiotensin II, which causes the adrenal cortex to release aldosterone. Aldosterone acts on the kidneys to retain Na^+; therefore, water is reabsorbed and blood pressure rises. The atrial natriuretic hormone prevents the secretion of renin and aldosterone.

The kidneys keep blood pH within normal limits. They reabsorb HCO_3^- and excrete H^+ as needed to maintain the pH at about 7.4.

Reviewing the Chapter

1. Contrast the osmotic regulation of a marine bony fish with that of a freshwater bony fish.   686–87
2. Give examples of how other types of animals regulate their water and salt balance.   687
3. Relate the three primary nitrogenous wastes to the habitat of animals.   688
4. Describe how the excretory organs of the earthworm and the insect function.   689
5. Describe the path of urine in humans, and give a function for each structure mentioned.   690
6. Describe the macroscopic anatomy of a human kidney, and relate it to the placement of nephrons.   690
7. List the parts of a nephron, and give a function for each structure mentioned.   690–91
8. Describe how urine is made by outlining what happens at each part of the nephron.   691–93
9. Describe the reabsorption of water and salt along the length of the nephron. Include the contribution of the loop of the nephron.   694
10. Name and describe the action of antidiuretic hormone (ADH), the renin-aldosterone connection, and the atrial natriuretic hormone (ANH).   694–95
11. How does the nephron regulate the pH of the blood?   695

Testing Yourself

Choose the best answer for each question.

1. Which of these pairs is mismatched?
   a. insects—excrete uric acid
   b. humans—excrete urea
   c. fishes—excrete ammonia
   d. birds—excrete ammonia
   e. All of these are correct.

2. One advantage of urea excretion over uric acid excretion is that urea
   a. requires less energy than uric acid to form.
   b. can be concentrated to a greater extent.
   c. is not a toxic substance.
   d. requires no water to excrete.
   e. is a larger molecule.

3. Freshwater bony fishes maintain water balance by
   a. excreting salt across their gills.
   b. periodically drinking small amounts of water.
   c. excreting a hypotonic urine.
   d. excreting wastes in the form of uric acid.
   e. Both a and c are correct.

4. Animals with which of these are most likely to excrete a semisolid nitrogenous waste?
   a. nephridia
   b. Malpighian tubules
   c. human kidneys
   d. flame cells
   e. All of these are correct.

5. In which of these human structures are you least apt to find urine?
   a. large intestine
   b. urethra
   c. collecting duct
   d. bladder
   e. Both a and c are correct.

6. Excretion of a hypertonic urine in humans is associated best with
   a. the glomerular capsule.
   b. the proximal convoluted tubule.
   c. the loop of the nephron.
   d. the collecting duct.
   e. Both c and d are correct.

7. The presence of ADH (antidiuretic hormone) causes an individual to excrete
   a. less salt.
   b. less water.
   c. more water.
   d. more salt.
   e. Both a and c are correct.

8. In humans, water is
   a. found in the glomerular filtrate.
   b. reabsorbed from the nephron.
   c. in the urine.
   d. reabsorbed from the collecting duct.
   e. All of these are correct.

9. Which of these is out of order first?
   a. glomerular capsule
   b. proximal convoluted tubule
   c. distal convoluted tubule
   d. loop of the nephron
   e. collecting duct

10. Normally in humans, glucose
    a. is always in the filtrate and urine.
    b. is always in the filtrate with little or none in urine.
    c. undergoes tubular secretion and is in urine.
    d. undergoes tubular secretion and is not in urine.
    e. is not in the filtrate and is not in the urine.

11. Which of these is not likely to cause a rise in blood pressure?
    a. aldosterone
    b. antidiuretic hormone (ADH)
    c. renin
    d. atrial natriuretic hormone (ANH)

12. If a drug inhibits the kidneys' ability to reabsorb bicarbonate so that bicarbonate is excreted in the urine, the blood will become
    a. acidic.
    b. alkaline.
    c. first acidic and then alkaline.
    d. first alkaline and then acidic.

13. Which of these materials is not filtered from the blood at the glomerulus?
    a. water
    b. urea
    c. protein
    d. glucose
    e. sodium ions

14. The renal medulla has a striped appearance due to the presence of which structures?
    a. loop of the nephron
    b. collecting ducts
    c. peritubular capillaries
    d. Both a and b are correct.

15. By what process are most molecules secreted from the blood into the distal convoluted tubule?
    a. osmosis
    b. diffusion
    c. active transport
    d. facilitated diffusion

16. Which of these is not correct?
    a. Uric acid is produced from the breakdown of amino acids.
    b. Urea is produced from the breakdown of proteins.
    c. Ammonia results from the deamination of amino acids.
    d. All of these are correct.

17. When tracing the path of blood, the blood vessel that follows the renal artery is
    a. the peritubular capillary.
    b. the efferent arteriole.
    c. the afferent arteriole.
    d. the renal vein.
    e. the glomerulus.

18. Absorption of the glomerular filtrate occurs at
   a. the convoluted tubules.
   b. only the distal convoluted tubule.
   c. the loop of the nephron.
   d. the collecting duct.

19. Label this diagram of a nephron.

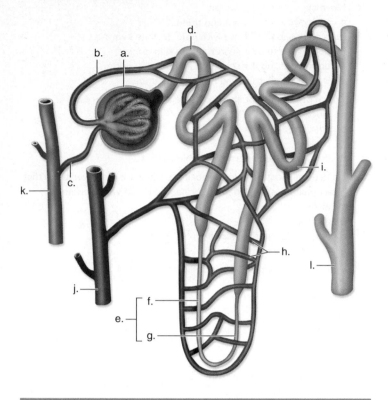

Thinking Scientifically

1. High blood pressure often is accompanied by kidney damage. In some people, the kidney damage is subsequent to the high blood pressure, but in others the kidney damage is what caused the high blood pressure. Would a low-salt diet enable you to determine whether the high blood pressure or the kidney damage came first?

2. The renin-angiotensin-aldosterone system can be inhibited in order to reduce high blood pressure. Usually, the angiotensin-converting enzyme is inhibited by drug therapy. Why would this enzyme be the most effective point to disrupt the system?

Bioethical Issue: Increasing Life Span

As a society, we are accustomed to thinking that as we grow older, diseases such as urinary disorders will begin to occur. Almost everyone is aware that most males are subject to enlargement of the prostate as they age, and that cancer of the prostate is not uncommon among older men. However, as with many illnesses associated with aging, medical science now knows how to treat or even cure prostate problems. Because of these successes, our life span has lengthened. A child born in the United States in 1900 lived to, say, the age of 47. If that same child were born today, he or she would probably live to at least 76. Even more exciting is the probability that scientists will improve the life-span still further. People could live beyond 100 years and have the same vigor and vitality they had when they were young.

Most people are appreciative of living longer, especially if they can expect to be free of the illnesses and inconveniences associated with aging. But have we examined how we feel about longevity as a society? We are accustomed to considering that if the birthrate increases, so does the size of a population. But what about the death rate? If the birthrate stays constant and the death rate decreases, obviously population size also increases. Most experts agree that population growth depletes resources and increases environmental degradation. Having more people in the older population can also put a strain on the economy if they are unable to meet their financial, including medical, needs without government assistance.

What is the ethical solution to this problem? Should we just allow the population to increase as older people live longer? Should we decrease the birthrate? Should we reduce government assistance to older people so they realize that they must be able to take care of themselves? Or should we call a halt to increasing the life-span through advancements in medical science?

Understanding the Terms

aldosterone 694	nephridium 689
ammonia 688	nephron 690
antidiuretic hormone	proximal convoluted
(ADH) 694	tubule 691
atrial natriuretic hormone	renal cortex 690
(ANH) 695	renal medulla 690
collecting duct 691	renal pelvis 690
distal convoluted tubule 691	renin 694
excretion 685	tubular reabsorption 692
flame cell 689	tubular secretion 693
glomerular capsule 690	urea 688
glomerular filtration 692	ureter 690
glomerulus 691	urethra 690
kidneys 690	uric acid 688
loop of the nephron 691	urinary bladder 690
Malpighian tubule 689	urine 690

Match the terms to these definitions:

a. _____ Blind, threadlike excretory tubule near the anterior end of an insect hindgut.

b. _____ Cuplike structure that is the initial portion of a nephron; where glomerular filtration occurs.

c. _____ Main nitrogenous waste of terrestrial amphibians and most mammals.

d. _____ Hormone secreted by the adrenal cortex that regulates the sodium and potassium ion balance of the blood.

e. _____ Main nitrogenous waste of insects, reptiles, and birds.

ARIS, the *Biology* Website

ARIS, the website for *Biology*, provides a wealth of information organized and integrated by chapter. You will find practice quizzes, interactive activities, labeling exercises, flashcards, and much more that will complement your learning and understanding of general biology.

www.mhhe.com/maderbiology9

39

NEURONS AND NERVOUS SYSTEMS

The human nervous system consists of the brain, spinal cord, and nerves. **Sensory receptors** detect changes in stimuli, and nerve impulses race through sensory fibers to the brain and spinal cord. The brain and spinal cord sum up the data before sending impulses via motor fibers to **effectors** (muscles and glands) so that a response to stimuli is possible. Disease can affect the operation of the nervous system, as when someone such as Stephen Hawking, a well-known British mathematician, has ALS (amyotrophic lateral sclerosis), also called Lou Gehrig disease. Lou Gehrig disease, named after a famous baseball player who had ALS, damages spinal cord pathways and motor neurons.

   The human brain carries out all sorts of higher mental functions, and these are not affected by Lou Gehrig disease. An affected person can still think, remember, and learn all sorts of things. Even someone with an advanced stage of the disease can still feel hot, cold, pain, and pressure. But with time, the muscles become smaller and weaker. Eventually, paralysis sets in.

   Besides sending commands to our muscles, the nervous and also the endocrine systems coordinate the other systems to achieve homeostasis, the relative stability of the internal environment. Digestion of food, breathing, and transport of nutrients and more are regulated by the nervous system, with the help of the endocrine system. The survival of all animals, even the minuscule rotifer or flatworm, depends on sensing the environment and responding to changes appropriately.

Stephen Hawking suffers from amyotrophic lateral sclerosis, or Lou Gehrig disease.

39.1 EVOLUTION OF THE NERVOUS SYSTEM

In complex animals, the ability to survive is dependent on a nervous system that monitors internal and external conditions and makes appropriate changes to maintain homeostasis. A comparative study of animal nervous system organization indicates the evolutionary trends that may have led to the nervous system of vertebrates.

Invertebrate Nervous Organization

Simple animals, such as sponges, which have the cellular level of organization, can respond to stimuli; the most common observable response is closure of the osculum (central opening). Hydras, which are cnidarians with the tissue level of organization, can contract and extend their bodies, move their tentacles to capture prey, and even turn somersaults. They have a **nerve net** that is composed of neurons in contact with one another and with contractile cells in the body wall (Fig. 39.1a). Sea

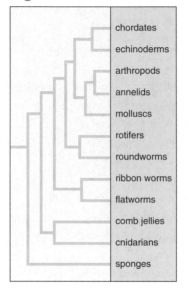

chordates
echinoderms
arthropods
annelids
molluscs
rotifers
roundworms
ribbon worms
flatworms
comb jellies
cnidarians
sponges

anemones and jellyfishes, which are also cnidarians, seem to have two nerve nets. A fast-acting one allows major responses, particularly in times of danger, and the other coordinates slower and more delicate movements.

Planarians, which are flatworms, have a nervous organization that reflects their bilateral symmetry. They possess two ventrally located lateral or longitudinal nerve cords (bundles of nerves) that extend from the cerebral ganglia to the posterior end of their body. Transverse nerves connect the nerve cords, as well as the cerebral ganglia, to the eyespots. The entire arrangement is a **ladderlike nervous system. Cephalization** has occurred, as evidenced by a concentration of ganglia and sensory receptors in a head region. Anterior cerebral ganglia receive sensory information from photoreceptors in the eyespots and sensory cells in the auricles (Fig. 39.1b). The two lateral nerve cords allow a rapid transfer of information from the cerebral ganglia to the posterior end, and the transverse nerves between the nerve cords keep the movement of the two sides coordinated. Bilateral symmetry plus cephalization are two significant trends in the development of a nervous organization that is adaptive for an active way of life. Also, the nervous organization in planarians is a foreshadowing of the central nervous system and peripheral nervous system seen in vertebrates.

Annelids (e.g., earthworm, Fig. 39.1c), arthropods (e.g., crab, Fig. 39.1d), and molluscs (e.g., squid, Fig. 39.1e) are complex animals with true nervous systems. The annelids and arthropods have the typical invertebrate nervous

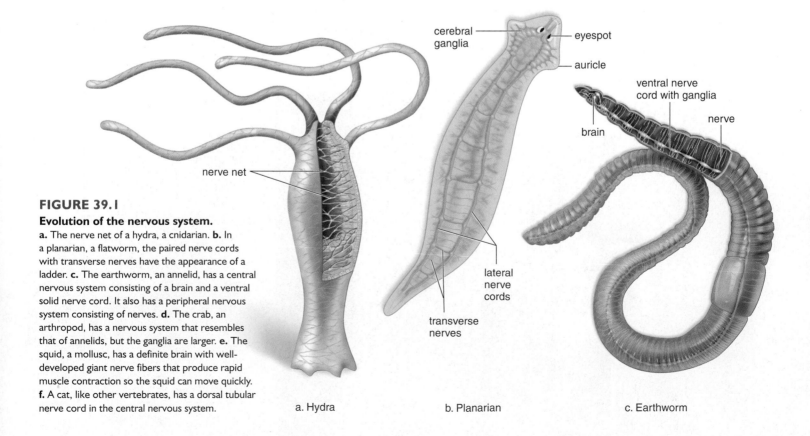

FIGURE 39.1

Evolution of the nervous system.

a. The nerve net of a hydra, a cnidarian. **b.** In a planarian, a flatworm, the paired nerve cords with transverse nerves have the appearance of a ladder. **c.** The earthworm, an annelid, has a central nervous system consisting of a brain and a ventral solid nerve cord. It also has a peripheral nervous system consisting of nerves. **d.** The crab, an arthropod, has a nervous system that resembles that of annelids, but the ganglia are larger. **e.** The squid, a mollusc, has a definite brain with well-developed giant nerve fibers that produce rapid muscle contraction so the squid can move quickly. **f.** A cat, like other vertebrates, has a dorsal tubular nerve cord in the central nervous system.

nerve net

cerebral ganglia — eyespot
— auricle

ventral nerve cord with ganglia
brain — nerve

lateral nerve cords

transverse nerves

a. Hydra   b. Planarian   c. Earthworm

system. There is a brain and a ventral nerve cord having a ganglion in each segment. The brain, which normally receives sensory information, controls the activity of the ganglia and assorted nerves so that the muscle activity of the entire animal is coordinated. The crab and squid show marked cephalization—the anterior end has a well-defined brain, and there are well-developed sense organs, such as eyes. The presence of a brain and other ganglia in the body of all these animals indicates an increase in the number of neurons (nerve cells) among more complex invertebrates.

> Bilateral symmetry, cephalization, and an increase in the number of neurons are evolutionary trends that are observable among the invertebrates.

Vertebrate Nervous Organization

In vertebrates (e.g., cat, Fig. 39.1f), the central nervous system, consisting of a brain and spinal cord, develops from an embryonic dorsal tubular nerve cord. Cephalization, coupled with bilateral symmetry, results in several types of paired sensory receptors, including the paired eyes, ears, and olfactory structures that allow the animal to gather information from the environment. Paired cranial and spinal nerves contain numerous nerve fibers. In vertebrates, there is a vast increase in the number of neurons. For example, an insect's entire nervous system may contain a total of about 1 million neurons, while a vertebrate nervous system may contain many thousand to several billion times that number.

The Vertebrate Brain

The vertebrate **brain** is the enlarged anterior end of the dorsal tubular nerve cord. It is customary to divide the brain into the hindbrain, midbrain, and forebrain. Nearly all vertebrates have a well-developed hindbrain that regulates organs below the level of consciousness. In humans, for example, the lungs and heart function even when we are sleeping. The hindbrain also functions to coordinate motor activity associated with limb movement, posture, and balance.

The optic lobes are part of the midbrain, which was originally a center for coordinating reflex responses to visual input. The forebrain receives sensory input from the midbrain and the hindbrain and regulates their output. The cerebrum is the foremost part of the brain in vertebrates. The cerebrum, which is highly developed in mammals, integrates sensory and motor input and is particularly associated with higher mental capabilities. In humans, the outer layer of the cerebrum, called the cerebral cortex, is especially large and complex.

> In vertebrates, the brain is organized into three areas: the hindbrain, the midbrain, and the forebrain. The forebrain is highly developed in mammals, particularly humans.

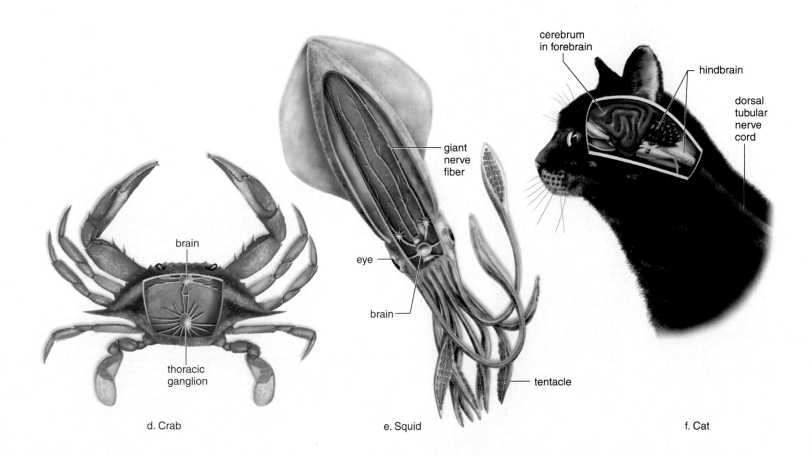

d. Crab   e. Squid   f. Cat

The Human Nervous System

The human nervous system has three specific functions: (1) it receives sensory input—sensory receptors in skin and other organs respond to external and internal stimuli by generating nerve impulses that travel to the central nervous system (CNS); (2) it performs integration—the CNS sums up the input it receives from all over the body; and (3) it generates motor output—nerve impulses from the CNS go to the muscles and glands. Muscle contractions and gland secretions are responses to stimuli received by sensory receptors. As an example, consider the events that occur as a person raises a glass to the lips. Continual sensory input to the CNS from the eyes and hand informs the CNS of the position of the glass, and the CNS continually sums up the incoming data before commanding the hand to proceed. At any time, integration with other sensory data might cause the CNS to command a different motion instead.

In humans, the **central nervous system (CNS)** consists of the brain and spinal cord (Fig. 39.2). The brain is housed in the skull and the spinal cord is housed in the vertebral column. The **peripheral nervous system (PNS)** [Gk. *periphereia*, circumference] consists of all the nerves and ganglia that lie outside the central nervous system. The paired cranial and spinal nerves are part of the PNS. In the PNS, the somatic nervous system has sensory and motor functions that control the skeletal muscles. The autonomic nervous system controls smooth muscle, cardiac muscle, and the glands. It is further divided into the sympathetic and parasympathetic divisions.

The components and functions of the central and peripheral nervous systems are complex. For an organism to maintain homeostasis, both systems have to work in harmony.

The CNS and PNS work together to perform the functions of the human nervous system.

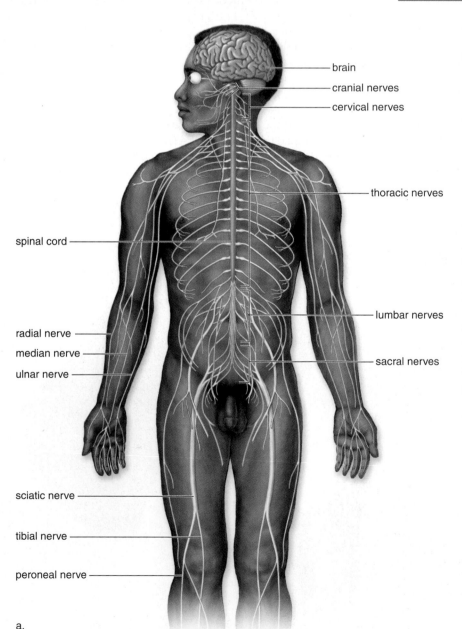

a.

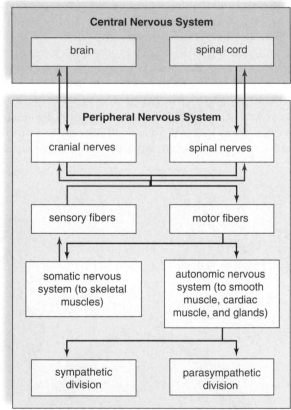

b.

FIGURE 39.2 Organization of the nervous system in humans.

a. This pictorial representation shows the central nervous system (CNS, composed of brain and spinal cord) and some of the nerves of the peripheral nervous system (PNS). **b.** The CNS communicates with the PNS, which contains nerves. In the somatic system, nerves conduct impulses from sensory receptors located in the skin and internal organs to the CNS and motor impulses from the CNS to the skeletal muscles. In the autonomic system, consisting of the sympathetic and parasympathetic divisions, motor impulses travel to smooth muscle, cardiac muscle, and glands.

39.2 NERVOUS TISSUE

Although complex, nervous tissue is composed of two principal types of cells. **Neurons,** also known as nerve cells, are the functional units of the nervous system. They receive sensory information, convey the information to an integration center such as the brain, and conduct signals from the integration center to effector structures such as the glands and muscles. **Neuroglia** serve as supporting cells, providing support and nourishment to the neurons.

Neurons and Neuroglia

Neurons [Gk. *neuron,* nerve] vary in appearance depending on their function and location. They consist of three major parts: a cell body, dendrites, and an axon (Fig. 39.3). The **cell body** contains a nucleus and a variety of organelles. The **dendrites** [Gk. *dendron,* tree] are short, highly branched processes that receive signals from the sensory receptors or other neurons and transmit them to the cell body. The **axon** [Gk. *axon,* axis] is the portion of the neuron that conveys information to another neuron or to other cells. Axons can be bundled together to form nerves. For this reason, axons are often called **nerve fibers.** Many axons are covered by a white insulating layer called the **myelin sheath** [Gk. *myelos,* spinal cord].

Neuroglia, or glial cells, which were discussed on page 601, greatly outnumber neurons in the brain. There are several different types in the CNS, each with specific functions. Some (microglia) help remove bacteria and debris, some (astrocytes) provide metabolic and structural support directly to the neurons. The myelin sheath is formed from the membranes of tightly spiraled neuroglia. In the PNS, **Schwann cells,** or neurolemmocytes, perform this function, leaving gaps called **nodes of Ranvier,** or neurofibril nodes. In the CNS another type of neuroglial cell (oligodendrocyte) performs this function.

Types of Neurons

Neurons can be classified according to their function and shape. **Motor (efferent) neurons** take nerve impulses from the CNS to muscles or glands. Motor neurons are said to be multipolar because they have many dendrites and a single axon (Fig. 39.3a). Motor neurons cause muscle fibers to contract or glands to secrete, and therefore they are said to innervate these structures.

Sensory (afferent) neurons take nerve impulses from sensory receptors to the CNS. The sensory receptor, which is the distal end of the long axon of a sensory neuron, may be as simple as a naked nerve ending (a pain receptor), or may be built into a highly complex organ, such as the eye or ear. Almost all sensory neurons have a structure that is termed unipolar (Fig. 39.3b). In unipolar neurons, the process that extends from the cell body divides into a branch that extends to the periphery and another that extends to the CNS. Since both of these extensions are long and myelinated and transmit nerve impulses, it is now generally accepted to refer to them as an axon.

Interneurons [L. *inter,* between; Gk. *neuron,* nerve], also known as association neurons, occur entirely within the CNS. Interneurons, which are typically multipolar (Fig. 39.3c), convey nerve impulses between various parts of the CNS. Some lie between sensory neurons and motor neurons, and some take messages from one side of the spinal cord to the other or from the brain to the cord, and vice versa. They also form complex pathways in the brain where processes accounting for thinking, memory, and language occur.

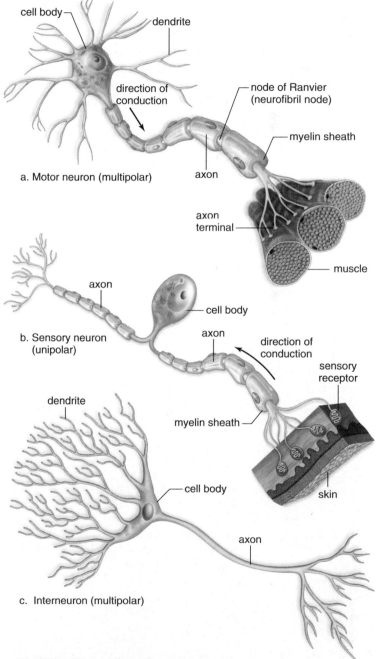

FIGURE 39.3 Neuron anatomy.
a. Motor neuron. Note the branched dendrites and the single, long axon, which branches only near its tip. **b.** Sensory neuron with dendritelike structures projecting from the peripheral end of the axon. **c.** Interneuron (from the cortex of the cerebellum) with very highly branched dendrites.

Transmission of the Nerve Impulses

Italian investigator Luigi Galvani discovered in 1786 that a nerve can be stimulated by an electric current. But it was realized later that the speed of the **nerve impulse,** a signal that travels down the axon, is too slow to be simply an electric current traveling within an axon. In the early 1900s, Julius Bernstein at the University of Halle, Germany, suggested that the nerve impulse is an electrochemical phenomenon involving the movement of unequally distributed ions on either side of an axonal membrane, the plasma membrane of an axon. It was not until later, however, that the investigators developed a technique that enabled them to support this hypothesis. A. L. Hodgkin and A. F. Huxley, English neurophysiologists, received the Nobel Prize in 1963 for their work in this field. They and a group of researchers, headed by K. S. Cole and J. J. Curtis at Woods Hole, Massachusetts, managed to insert a tiny electrode into the giant axon of the squid *Loligo.* This internal electrode was then connected to a voltmeter, an instrument with a screen that shows voltage differences over time (Fig. 39.4). Voltage is a measure of the electrical potential difference between two points, which in this case is the difference between two electrodes—one placed inside and another placed outside the axon. (An electrical potential difference across a membrane is called the membrane potential.) When a membrane potential exists, we can say that a plus pole and a minus pole exist; therefore, the voltmeter indicates the existence of polarity and records polarity changes.

Resting Potential

When the axon is not conducting an impulse, the voltmeter records a membrane potential equal to about -65 mV (millivolts), indicating that the inside of the neuron is more negative than the outside (Fig. 39.4a). This is called the **resting potential** because the axon is not conducting an impulse.

The existence of this polarity can be correlated with a difference in ion distribution on either side of the axonal membrane. As Figure 39.4a shows, there is a higher concentration of sodium ions (Na^+) outside the axon and a higher concentration of potassium ions (K^+) inside the axon. The unequal distribution of these ions is in part due to the activity of the sodium-potassium pump. This pump is an active transport system in the plasma membrane that pumps three sodium ions out of and two potassium ions into the axon. The pump is always working because the membrane is somewhat permeable to these ions and they tend to diffuse toward their lesser concentration. Since the membrane is more permeable to potassium ions than to sodium ions, there are always more positive ions outside the membrane than inside; this accounts for some of the polarity recorded by the voltmeter. There are also large, negatively charged proteins in the cytoplasm of the axon; altogether, then, the voltmeter records that the inside is -65 mV compared to the outside. This is the resting potential.

FIGURE 39.4 Resting and action potential of the axonal membrane.

a. Resting potential. A voltmeter that records voltage changes indicates the axonal membrane has a resting potential of -65 mV. There is a preponderance of Na^+ outside the axon and a preponderance of K^+ inside the axon. The permeability of the membrane to K^+ compared to Na^+ causes the inside to be negative compared to the outside. **b.** Action potential. Depolarization occurs when Na^+ gates open and Na^+ moves inside the axon, and (**c**) repolarization occurs when K^+ gates open and K^+ moves outside the axon. **d.** Graph of the action potential.

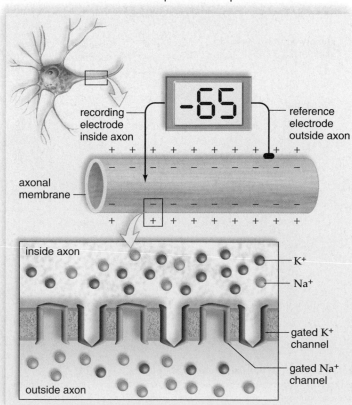

a. Resting potential: more Na^+ outside the axon and more K^+ inside the axon causes polarization.

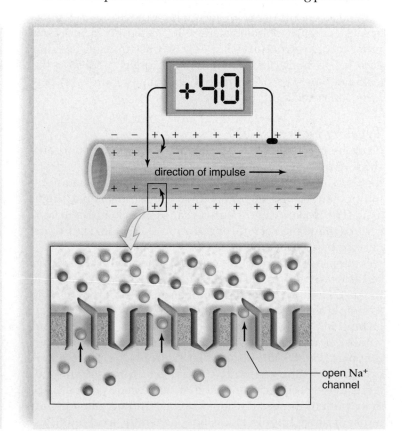

b. Action potential begins: depolarization occurs when Na^+ gates open and Na^+ moves to inside the axon.

Action Potential

An **action potential** is a rapid change in polarity across a portion of an axonal membrane as the nerve impulse occurs. An action potential uses two types of gated ion channels in the axonal membrane. In the axonal membrane, a gated ion channel allows sodium (Na^+) to pass through the membrane, and another allows potassium (K^+) to pass through the membrane. In contrast to ungated ion channels, which constantly allow ions across the membrane, gated ion channels open and close in response to a stimulus such as a signal from another neuron.

The action potential is generated only after the occurrence of a threshold value. Threshold is the minimum change in polarity across the axonal membrane that is required to generate an action potential. During a depolarization, the inside of a neuron becomes positive because of the sudden entrance of sodium ions. If threshold depolarization occurs, many more sodium channels open, and the action potential begins. As sodium ions rapidly move across the membrane to the inside of the axon, the action potential swings up from −65 mV to +40 mV (Fig. 39.4*b*). This reversal in polarity causes the sodium channels to close and the potassium channels to open. Now potassium ions move from inside the axon to outside the axon. As potassium ions leave, the action potential swings down from +40 mV to −65 mV. In other words, a repolarization occurs (Fig. 39.4*c*). An action potential only takes two milliseconds. In order to visualize such rapid fluctuations in voltage across the axonal membrane, researchers generally find it useful to plot the voltage changes over time (Fig. 34.4*d*).

Propagation of Action Potentials

A stimulus causes an action potential to begin. In the laboratory, electrical currents are used to generate a nerve impulse. In your body, nerve impulses begin due to all sorts of stimuli. For instance, a stimulus could consist of someone pinching your arm, thereby activating sensory receptors in your skin. If an axon is unmyelinated, an action potential at one locale stimulates an adjacent part of the axonal membrane to produce an action potential. In myelinated fibers, an action potential at one neurofibril node causes an action potential at the next node. This type of conduction, called **saltatory conduction** [L. *saltator*, hopper], is much faster than otherwise. In thin, unmyelinated axons, the action potential travels about 1.0 m/second, and in thick, myelinated axons, the rate is more than 100 m/second. In any case, action potentials are self-propagating; each action potential generates another along the length of an axon.

The conduction of a nerve impulse (action potential) is an all-or-none event; that is, either a fiber conducts a nerve impulse or it does not. The intensity of a message is determined by how many nerve impulses are generated within a given time span. A fiber can conduct a volley of nerve impulses because only a small number of ions are exchanged with each impulse. As soon as an impulse has passed by each successive portion of a fiber, it undergoes a short refractory period during which it is unable to conduct an impulse. This ensures a one-way direction of the impulse. During a refractory period, the sodium gates cannot yet open.

All neurons transmit the same type of nerve impulse—a change in potential across a portion of an axonal membrane that is self-propagating.

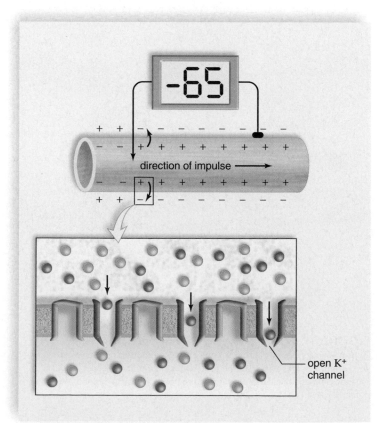

c. Action potential ends: repolarization occurs when K^+ gates open and K^+ moves to outside the axon.

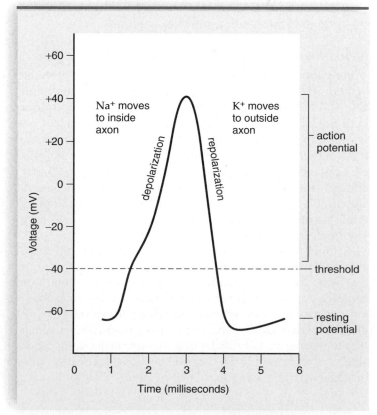

d. An action potential can be visualized if voltage changes are graphed over time.

Transmission Across a Synapse

Every axon branches into many fine endings, each tipped by a small swelling, called an axon terminal (Fig. 39.5). Each terminal lies very close to the dendrite (or the cell body) of another neuron. This region of close proximity is called a **synapse.** At a synapse, the membrane of the first neuron is called the *pre*synaptic membrane, and the membrane of the next neuron is called the *post*synaptic membrane. The small gap between the neurons is called the **synaptic cleft.**

A nerve impulse cannot cross a synaptic cleft. Transmission across a synapse is carried out by molecules called **neurotransmitters,** which are stored in synaptic vesicles. When nerve impulses traveling along an axon reach an axon terminal, gated channels for calcium ions (Ca^{2+}) open, and calcium enters the terminal. This sudden rise in Ca^{2+} stimulates synaptic vesicles to merge with the presynaptic membrane, and neurotransmitter molecules are released into the synaptic cleft. They diffuse across the cleft to the postsynaptic membrane, where they bind with specific receptor proteins.

Depending on the type of neurotransmitter and/or the type of receptor, the response of the postsynaptic neuron can be toward excitation or toward inhibition. Excitatory neurotransmitters that use gated ion channels are fast acting. Other neurotransmitters affect the metabolism of the postsynaptic cell and therefore are slower acting.

Neurotransmitters and Neuromodulators

Among the more than 100 substances known or suspected to be neurotransmitters are **acetylcholine (ACh), norepinephrine (NE), dopamine,** and **serotonin,** which are present in both the CNS and PNS. The effect of ACh on muscle tissue varies. It excites skeletal muscle but inhibits cardiac muscle. It has either an excitory or inhibitory effect on smooth muscle or glands, depending on their location. In the CNS, norepinephrine is important to dreaming, waking, and mood. Dopamine is involved in emotions, learning, and attention while serotonin is involved in thermoregulation, sleeping, emotions, and perception.

Once a neurotransmitter has been released into a synaptic cleft and has initiated a response, it is removed from the cleft. In some synapses, the postsynaptic membrane contains enzymes that rapidly inactivate the neurotransmitter. For example, the enzyme **acetylcholinesterase (AChE)** breaks down acetylcholine. In other synapses, the presynaptic membrane rapidly reabsorbs the neurotransmitter, possibly for repackaging in synaptic vesicles or for molecular breakdown. The short existence of neurotransmitters at a

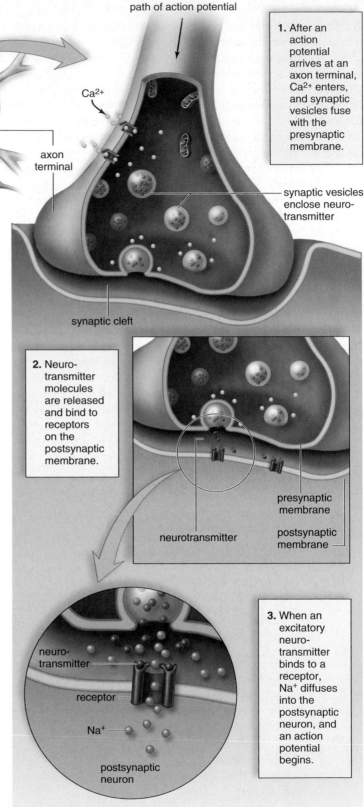

path of action potential

Ca^{2+}

axon terminal

cell body of postsynaptic neuron

1. After an action potential arrives at an axon terminal, Ca^{2+} enters, and synaptic vesicles fuse with the presynaptic membrane.

synaptic vesicles enclose neurotransmitter

synaptic cleft

2. Neurotransmitter molecules are released and bind to receptors on the postsynaptic membrane.

neurotransmitter

presynaptic membrane

postsynaptic membrane

neurotransmitter

receptor

Na^+

postsynaptic neuron

3. When an excitatory neurotransmitter binds to a receptor, Na^+ diffuses into the postsynaptic neuron, and an action potential begins.

FIGURE 39.5 Synapse structure and function.
Transmission across a synapse from one neuron to another occurs when a neurotransmitter is released at the presynaptic membrane, diffuses across a synaptic cleft, and binds to a receptor in the postsynaptic membrane. An action potential may begin.

synapse prevents continuous stimulation (or inhibition) of postsynaptic membranes.

It is of interest to note here that many drugs that affect the nervous system act by interfering with or potentiating the action of neurotransmitters. As described in the Science Focus on pages 716–17, drugs can enhance or block the release of a neurotransmitter, mimic the action of a neurotransmitter or block the receptor, or interfere with the removal of a neurotransmitter from a synaptic cleft. Neurotransmitter imbalances are associated with a number of disorders. Parkinson disease is associated with a lack of dopamine in the brain. Alzheimer disease is associated with a deficiency of ACh.

Neuromodulators are molecules that block the release of a neurotransmitter or modify a neuron's response to a neurotransmitter. Two well-known neuromodulators are substance P and endorphins. Substance P is released by sensory neurons when pain is present. Endorphins block the release of substance P and therefore serve as natural painkillers. They are associated with the "runner's high" of joggers and are produced by the brain not only when physical stress but also when emotional stress is present.

Transmission across a synapse is dependent on the release of neurotransmitters, which diffuse across the synaptic cleft from one neuron to the next.

Synaptic Integration

A single neuron has many dendrites plus the cell body, and both can have synapses with many other neurons. One thousand to 10,000 synapses per a single neuron is not uncommon. Therefore, a neuron is on the receiving end of many excitatory and inhibitory signals. An excitatory neurotransmitter produces a potential change called a signal that drives the neuron closer to an action potential, and an inhibitory neurotransmitter produces a signal that drives the neuron further from an action potential. Excitatory signals have a depolarizing effect, and inhibitory signals have a hyperpolarizing effect.

Neurons integrate these incoming signals. **Integration** is the summing up of excitatory and inhibitory signals (Fig. 39.6). If a neuron receives many excitatory signals (either from different synapses or at a rapid rate from one synapse), chances are the axon will transmit a nerve impulse. On the other hand, if a neuron receives both inhibitory and excitatory signals, the summing up of these signals may prohibit the axon from firing.

Integration is the summing up of inhibitory and excitatory signals received by a postsynaptic neuron.

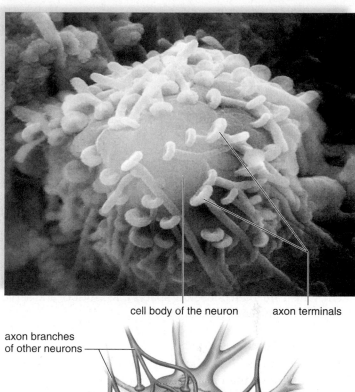

cell body of the neuron   axon terminals

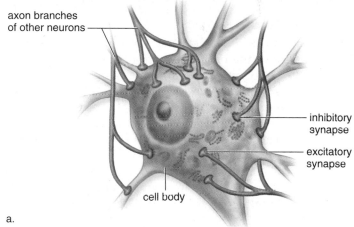

axon branches of other neurons

inhibitory synapse

excitatory synapse

cell body

a.

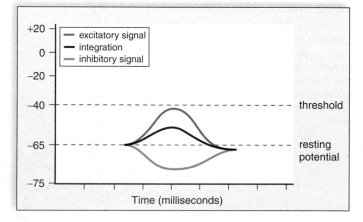

b.

FIGURE 39.6 Synaptic integration.
a. Inhibitory signals and excitatory signals are summed up in the dendrite and cell body of the postsynaptic neuron. Only if the combined signals cause the membrane potential to rise above threshold does an action potential occur. **b.** In this example, threshold was not reached.

39.3 CENTRAL NERVOUS SYSTEM: BRAIN AND SPINAL CORD

The central nervous system (CNS) consists of the spinal cord and the brain, where sensory information is received and motor control is initiated. The spinal cord and the brain are both protected by bone; the spinal cord is surrounded by vertebrae (see Fig. 39.10*b*), and the brain is enclosed by the skull (Fig. 39.7). Both the spinal cord and brain are wrapped in three protective membranes known as **meninges** [Gk. *meninga*, membranes covering the brain]. **Meningitis** (inflammation of the meninges) is a serious disorder caused by a number of bacteria or viruses that invade the meninges. The spaces between the meninges are filled with **cerebrospinal fluid** [L. *cerebrum*, brain, and *spina*, backbone], which cushions and protects the central nervous system. Cerebrospinal fluid is contained in the central canal of the spinal cord and within the **ventricles** of the brain, which are interconnecting spaces that produce and serve as reservoirs for cerebrospinal fluid.

The Spinal Cord

The **spinal cord** is a bundle of nervous tissue that extends from the base of the brain through a large opening in the skull called the foramen magnum through the vertebral canal to the first lumbar vertebrae (see Fig. 41.5). The spinal cord has two main functions: (1) it is the center for many reflex actions, which are discussed on page 713; and (2) it provides a means of communication between the brain and the spinal nerves, which leave the spinal cord.

A cross section of the spinal cord reveals that it is composed of a central portion of **gray matter** and a peripheral region of white matter (see Fig. 39.11). The gray matter

consists of cell bodies and unmyelinated fibers. It is shaped like a butterfly or the letter H with two dorsal (posterior) horns and two ventral (anterior) horns surrounding a central canal. Portions of sensory neurons and motor neurons are found in the gray matter, as well as short interneurons that serve to connect sensory and motor neurons.

Myelinated long fibers of interneurons that run together in bundles called **tracts** give **white matter** its color. These tracts connect the spinal cord to the brain. Dorsally, there are primarily ascending tracts taking information to the brain, and ventrally, there are primarily descending tracts carrying information from the brain. Because the tracts at one point cross over, the left side of the brain controls the right side of the body, and the right side of the brain controls the left side of the body.

If the spinal cord is severed as the result of an injury, paralysis will result. If the injury occurs in the cervical region (neck), all four limbs are usually paralyzed, a condition known as quadriplegia. If the injury occurs in the thoracic region, the lower body may be paralyzed, a condition called paraplegia.

The Brain

Embryologically, the human brain begins as a simple neural tube. Eventually, three distinct regions, the forebrain, the midbrain, and the hindbrain, become evident as the neural tube differentiates. The forebrain consists of the cerebrum and diencephalon, and the hindbrain consists of the cerebellum, pons, and medulla oblongata. Four internal chambers exist in the brain called the ventricles. Two lateral ventricles are associated with the cerebrum, the third ventricle is associated with the diencephalon, and the fourth ventricle lies between the cerebellum and pons. Cerebrospinal fluid circulates through the ventricles.

FIGURE 39.7

The human brain.
The cerebrum is divided into the right and left cerebral hemispheres. The right cerebral hemisphere is shown here. The hemispheres are connected by the corpus callosum.

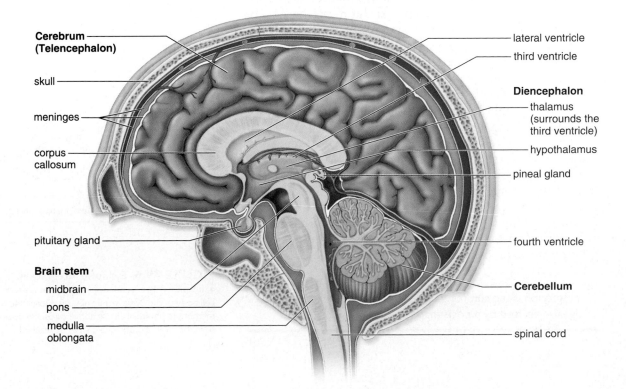

The Cerebrum

The **cerebrum,** also called the telencephalon, is the largest portion of the brain in humans. The cerebrum is the last center to receive sensory input and carry out integration before commanding voluntary motor responses. It communicates with and coordinates the activities of the other parts of the brain. The cerebrum carries out higher thought processes required for learning and memory and for language and speech.

The cerebrum is divided into two halves called **cerebral hemispheres** (see Fig. 39.7). A deep groove called the longitudinal fissure divides the cerebrum into the right and left hemispheres. Each hemisphere receives information from, and controls, the opposite side of the body. Although the hemispheres appear the same, the right hemisphere is associated with artistic and musical ability, emotion, spatial relationships, and pattern recognition. The left hemisphere is more adept at mathematics, language, and analytical reasoning. The two cerebral hemispheres are connected by a bridge of tracts within the corpus callosum.

Shallow grooves called sulci (sing., sulcus) divide each hemisphere into lobes (Fig. 39.8). The *frontal lobe* lies toward the front of the hemispheres and is associated with memory, emotion, planning, judgment, and aggression. The *parietal lobes* lie posterior to the frontal lobe and are concerned with sensory reception and integration, as well as taste. The *temporal lobe* is located laterally and is associated with learning, memory, hearing, smell, visual recognition, and emotional behavior. The *occipital lobe* is the most posterior lobe and serves as the visual center.

The Cerebral Cortex. The **cerebral cortex** is a thin (less than 5 mm thick) but highly convoluted outer layer of gray matter that covers the cerebral hemispheres. The convolutions, or gyri,

increase the surface area of the cerebral cortex. The cerebral cortex contains tens of billions of neurons and is the region of the brain that accounts for sensation, voluntary movement, and all the thought processes we associate with consciousness.

The cerebral cortex contains motor areas and sensory areas as well as association areas. The **primary motor area** is in the frontal lobe just ventral to (before) the central sulcus. Voluntary commands to skeletal muscles begin in the primary motor area, and each part of the body is controlled by a certain section. For example, the versatile human hand takes up an especially large portion of the primary motor area. Ventral to the primary motor area is a premotor area. The *premotor area* organizes motor functions for skilled motor activities, and then the primary motor area sends signals to the cerebellum, which integrates them. The unique ability of humans to speak is partially dependent on *Broca's area*, a motor speech area in the left frontal lobe. Signals originating here pass to the premotor area before reaching the primary motor area.

The **primary somatosensory area** is just dorsal to the central sulcus in the parietal lobe. Sensory information from the skin and skeletal muscles arrives here, where each part of the body is sequentially represented. A *primary taste area*, also in the parietal lobe, accounts for taste sensations. A *primary visual area* in the occipital lobe receives information from our eyes, and a *primary auditory area* in the temporal lobe receives information from our ears.

Association areas are places where integration occurs. For example, the *somatosensory association area*, located just dorsal to the primary somatosensory area, processes and analyzes sensory information from the skin and muscles. The *visual association area* in the occipital lobe associates new visual information with previously received visual information. It might

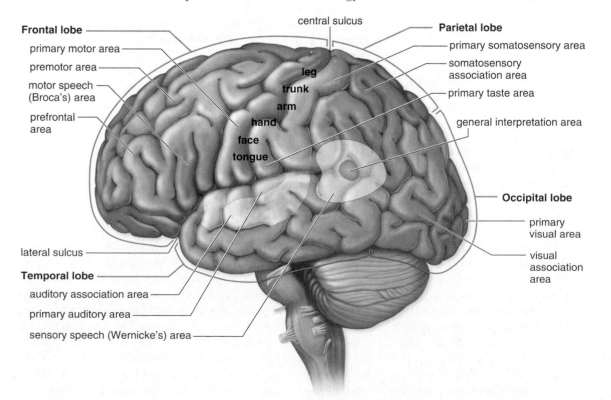

Frontal lobe
- primary motor area
- premotor area
- motor speech (Broca's) area
- prefrontal area

central sulcus

Parietal lobe
- primary somatosensory area
- somatosensory association area
- primary taste area
- general interpretation area

leg
trunk
arm
hand
face
tongue

lateral sulcus

Temporal lobe
- auditory association area
- primary auditory area
- sensory speech (Wernicke's) area

Occipital lobe
- primary visual area
- visual association area

FIGURE 39.8

The lobes of a cerebral hemisphere.

Each cerebral hemisphere is divided into four lobes: frontal, parietal, temporal, and occipital. These lobes contain centers for reasoning and movement, somatic sensing, hearing, and vision, respectively.

"decide," for example, if we have seen this face or tool or whatever before. The *auditory association area* in the temporal lobe performs the same functions with regard to sounds. These association areas meet near the dorsal end of the lateral sulcus. This region is called the general interpretation area because it receives information from all the sensory association areas and allows us to quickly integrate incoming signals and send them on to the prefrontal area so that an immediate response is possible. The general interpretation area is the part of the brain in operation when people are able to quickly assess a situation and take actions that save themselves or others from danger. The **prefrontal area,** an association area in the frontal lobe, receives information from the other association areas and uses this information to reason and plan our actions. Integration in this area accounts for our most cherished human abilities to think critically and to formulate appropriate behaviors.

White Matter. Much of the rest of the cerebrum is composed of white matter. As you know, white matter in the CNS consists of long, myelinated axons organized into tracts. Descending tracts from the primary motor area communicate with lower brain centers, and ascending tracts from lower brain centers send sensory information up to the primary somatosensory area. Because the tracts cross over in the medulla, the left side of the cerebrum controls the right side of the body, and vice versa. Tracts within the cerebrum also take information between the different sensory, motor, and association areas pictured in Figure 39.8. As previously mentioned, the corpus callosum contains tracts that join the two cerebral hemispheres.

Basal Nuclei.   While the bulk of the cerebrum is composed of tracts, there are masses of gray matter located deep within the white matter. These so-called **basal nuclei** (formerly termed basal ganglia) integrate motor commands, ensuring that proper muscle groups are activated or inhibited. Huntington disease and Parkinson disease, which are both characterized by uncontrollable movements, are believed to be due to malfunctioning of the basal nuclei.

The Diencephalon

The hypothalamus and the thalamus are in the **diencephalon,** a region that encircles the third ventricle. The **hypothalamus** forms the floor of the third ventricle. It is an integrating center that helps maintain homeostasis by regulating hunger, sleep, thirst, body temperature, and water balance. The hypothalamus controls the pituitary gland and thereby serves as a link between the nervous and endocrine systems.

The **thalamus** consists of two masses of gray matter located in the sides and roof of the third ventricle. It is on the receiving end for all sensory input except smell. Visual, auditory, and somatosensory information arrives at the thalamus via the cranial nerves and tracts from the spinal cord. The thalamus integrates this information and sends it on to the appropriate portions of the cerebrum. The thalamus is involved in arousal of the cerebrum, and it also participates in higher mental functions such as memory and emotions.

The pineal gland, which secretes the hormone melatonin, is located in the diencephalon. Presently there is much popular interest in the role of melatonin in our daily rhythms; some researchers believe it may be involved in jet lag and insomnia. Scientists are also interested in the possibility that the hormone may regulate the onset of puberty.

The Cerebellum

The **cerebellum** is separated from the brain stem by the fourth ventricle. It is the largest part of the hindbrain. The cerebellum has two portions that are joined by a narrow median portion. Each portion is primarily composed of white matter, which in longitudinal section has a treelike pattern. Overlying the white matter is a thin layer of gray matter that forms a series of complex folds.

The cerebellum receives sensory input from the eyes, ears, joints, and muscles about the present position of body parts, and it also receives motor output from the cerebral cortex about where these parts should be located. After integrating this information, the cerebellum sends motor impulses by way of the brain stem to the skeletal muscles. In this way, the cerebellum maintains posture and balance. It also ensures that all of the muscles work together to produce smooth, coordinated voluntary movements. The cerebellum assists the learning of new motor skills such as playing the piano or hitting a baseball. New evidence indicates that the cerebellum is important in judging the passage of time.

The Brain Stem

The **brain stem** contains the midbrain, the pons, and the medulla oblongata (see Fig. 39.7). The **midbrain** acts as a relay station for tracts passing between the cerebrum and the spinal cord or cerebellum. It also has reflex centers for visual, auditory, and tactile responses. The word **pons** means "bridge" in Latin, and true to its name, the pons contains bundles of axons traveling between the cerebellum and the rest of the CNS. In addition, the pons functions with the medulla oblongata to regulate breathing rate and has reflex centers concerned with head movements in response to visual and auditory stimuli.

The **medulla oblongata** contains a number of reflex centers for regulating heartbeat, breathing, and vasoconstriction (blood pressure). It also contains the reflex centers for vomiting, coughing, sneezing, hiccuping, and swallowing. The medulla oblongata lies just superior to the spinal cord, and it contains tracts that ascend or descend between the spinal cord and higher brain centers.

The reticular formation, a complex network of nuclei (masses of gray matter) and nerve tracts in the brain stem, is a part of the reticular activating system (RAS). The RAS receives sensory signals and sends them up to higher centers. It arouses the cerebrum via the thalamus and causes a person to be alert. Apparently, the RAS can filter out unnecessary sensory stimuli, explaining why you can study with the TV on. If you want to awaken the RAS, surprise it with sudden stimuli like splashing your face with cold water; if you want to deactivate it, remove visual and auditory stimuli. General anesthetics function by artificially suppressing the RAS.

The Limbic System

The **limbic system** is a complex network of tracts and nuclei that incorporates medial portions of the cerebral lobes, the basal nuclei, and the diencephalon (Fig. 39.9). The limbic system blends higher mental functions and primitive emotions into a united whole. It accounts for why activities like sexual behavior and eating seem pleasurable and also why, say, mental stress can cause high blood pressure.

Two significant structures within the limbic system are the hippocampus and the amygdala, which are essential for learning and memory. The hippocampus, a seahorse-shaped structure that lies deep in the temporal lobe, is well situated in the brain to make the prefrontal area aware of past experiences stored in sensory association areas. The amygdala, in particular, can cause these experiences to have emotional overtones. For example, the smell of smoke may serve as an alarm to search for fire in the house. The inclusion of the frontal lobe in the limbic system means that reason can keep us from acting out strong feelings.

Learning and Memory. **Memory** is the ability to hold a thought in mind or recall events from the past, ranging from a word we learned only yesterday to an early emotional experience that has shaped our lives. Learning takes place when we retain and use past memories.

The prefrontal area in the frontal lobe is active during short-term memory as when we temporarily recall a telephone number. Some telephone numbers go into long-term memory. Think of a telephone number you know by heart, and see if you can bring it to mind without also thinking about the place or person associated with that number. Most likely you cannot, because typically long-term memory is a mixture of what is called semantic memory (numbers, words, etc.) and episodic memory (persons, events, etc.). Skill memory is a type of memory that can exist independent of episodic memory. Skill memory is being able to perform motor activities like riding a bike or playing ice hockey.

What parts of the brain are functioning when you remember something from long ago? As mentioned, our long-term memories are stored in bits and pieces throughout the sensory association areas of the cerebral cortex. The hippocampus gathers this information together for use by the prefrontal area of the frontal lobe when we remember Uncle Frank or our summer holiday. Why are some memories so emotionally charged? The amygdala is responsible for fear conditioning and associating danger with sensory information received from the thalamus and the cortical sensory areas.

Long-term potentiation (LTP) is an enhanced response at synapses seen particularly within the hippocampus. LTP is most likely essential to memory storage, but unfortunately, it sometimes causes a postsynaptic neuron to become so excited that it undergoes apoptosis, a form of cell death. This phenomenon, called excitotoxicity, is due to the action of glutamate, a neurotransmitter. When glutamate binds to the postsynaptic membrane, calcium may rush in too fast because of a receptor that is malformed due to a mutation. A gradual extinction of brain cells, particularly in the hippocampus, appears to be the underlying cause of Alzheimer disease (AD), a condition characterized by a gradual loss of memory.

AD neurons have neurofibrillary tangles (bundles of fibrous protein) surrounding the nucleus and protein-rich accumulations called amyloid plaques enveloping the axon branches. Although it is not yet known how excitotoxicity is related to structural abnormalities of AD neurons, some researchers are trying to develop neuroprotective drugs that can possibly guard brain cells against damage due to glutamate.

Each portion of the brain has particular functions; some portions are particularly concerned with sensory input or motor control or homeostasis, while the cerebral cortex of the cerebrum has integrative activities. The limbic system is involved in memory and learning.

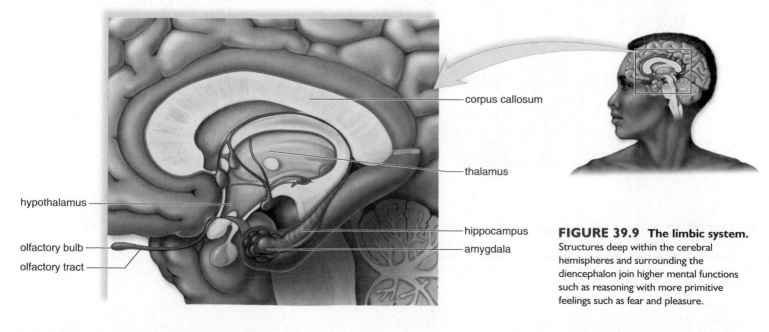

corpus callosum

thalamus

hippocampus

amygdala

hypothalamus

olfactory bulb

olfactory tract

FIGURE 39.9 The limbic system. Structures deep within the cerebral hemispheres and surrounding the diencephalon join higher mental functions such as reasoning with more primitive feelings such as fear and pleasure.

39.4 PERIPHERAL NERVOUS SYSTEM

The peripheral nervous system (PNS) lies outside the central nervous system and contains **nerves,** which are bundles of axons. Axons that occur in nerves are also called nerve fibers.

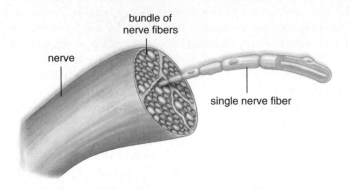

The cell bodies of neurons are found in the CNS and in ganglia. Ganglia (sing., **ganglion**) are collections of cell bodies within the PNS.

Humans have 12 pairs of **cranial nerves** attached to the brain (Fig. 39.10*a*). Some of these are sensory nerves; that is, they contain only sensory nerve fibers. Some are motor nerves that contain only motor fibers, and others are mixed nerves that contain both sensory and motor fibers. Cranial nerves are largely concerned with the head, neck, and facial regions of the body. However, the vagus nerve has branches not only to the pharynx and larynx but also to most of the internal organs.

Humans have 31 pairs of **spinal nerves** (Figs. 39.10*b* and 39.11). The paired spinal nerves emerge from the spinal cord by two short branches, or roots. The dorsal root contains the axons of sensory neurons, which conduct impulses to the spinal cord from sensory receptors. The cell body of a sensory neuron is in the **dorsal root ganglion.** The ventral root contains the axons of motor neurons, which conduct impulses away from the spinal cord to effectors. These two roots join to form a spinal nerve. All spinal nerves are mixed nerves that contain many sensory and motor fibers. Each spinal nerve serves the particular region of the body in which it is located.

In the PNS, cranial nerves take impulses to and/or from the brain, and spinal nerves take impulses to and from the spinal cord.

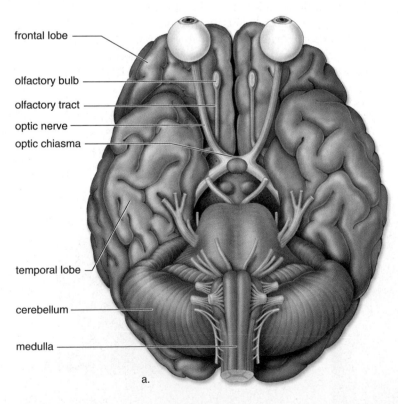

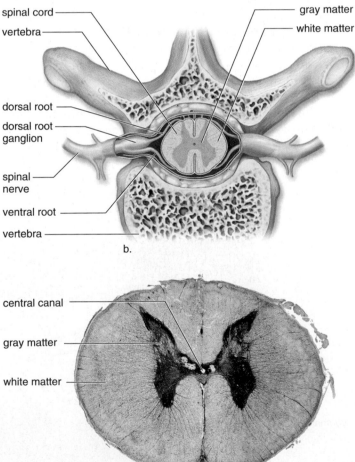

FIGURE 39.10 Cranial and spinal nerves.
a. Ventral surface of the brain, showing the attachment of the cranial nerves.
b. Cross section of the vertebral column and spinal cord, showing a spinal nerve. Each spinal nerve has a dorsal root and a ventral root attached to the spinal cord.
c. Photomicrograph of spinal cord cross section.

Somatic System

The **somatic system** includes the nerves that take sensory information from external sensory receptors to the CNS and motor commands away from the CNS to skeletal muscles. The neurotransmitter acetylcholine is active in the somatic system (see Table 39.1). Voluntary control of skeletal muscles always originates in the brain. Involuntary responses to stimuli, called **reflexes,** can involve either the brain or just the spinal cord. Reflexes enable the body to react swiftly to stimuli that could disrupt homeostasis. Flying objects cause eyes to blink, and sharp pins cause hands to jerk away even without us having to think about it.

The Reflex Arc

Figure 39.11 illustrates the path of a reflex that involves only the spinal cord. If your hand touches a sharp pin, sensory receptors in the skin generate nerve impulses that move along sensory axons toward the spinal cord. Sensory neurons that enter the cord dorsally pass signals on to many interneurons. Some of these interneurons synapse with motor neurons. The short dendrites and the cell bodies of motor neurons are in the spinal cord, but their axons leave the cord ventrally. Nerve impulses travel along motor axons to an effector, which brings about a response to the stimulus. In this case, a muscle contracts so that you withdraw your hand from the pin. Various other reactions are possible—you will most likely look at the pin, wince, and cry out in pain. This whole series of responses is explained by the fact that some of the interneurons involved carry nerve impulses to the brain. The brain makes you aware of the stimulus and directs subsequent reactions to the situation.

In the somatic system, nerves take information from external sensory receptors to the CNS and motor commands to skeletal muscles. Involuntary reflexes allow us to respond rapidly to external stimuli.

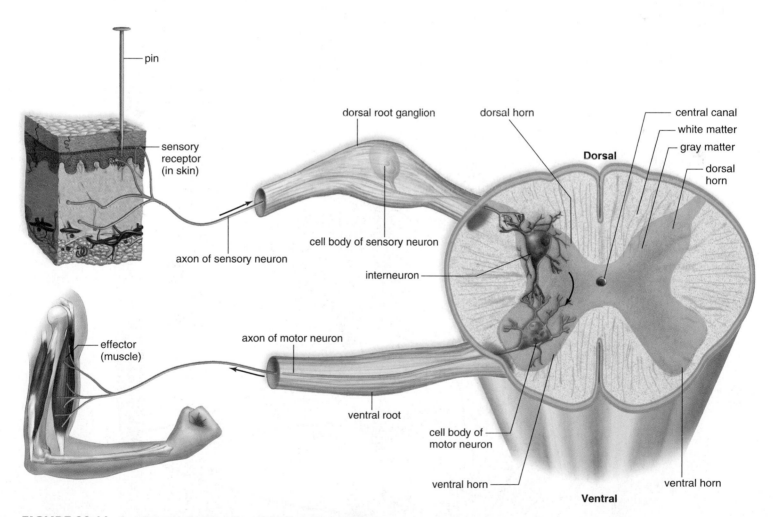

FIGURE 39.11 A reflex arc showing the path of a spinal reflex.
A stimulus (e.g., sharp pin) causes sensory receptors in the skin to generate nerve impulses that travel in sensory axons to the spinal cord. Interneurons integrate data from sensory neurons and then relay signals to motor axons. Motor axons convey nerve impulses from the spinal cord to a skeletal muscle, which contracts. Movement of the hand away from the pin is the response to the stimulus.

FIGURE 39.12

Autonomic system structure and function.
Sympathetic preganglionic fibers *(left)* arise from the
cervical, thoracic, and lumbar portions of the spinal cord;
parasympathetic preganglionic fibers *(right)* arise from the
cranial and sacral portions of the spinal cord. Each system
innervates the same organs but has contrary effects.

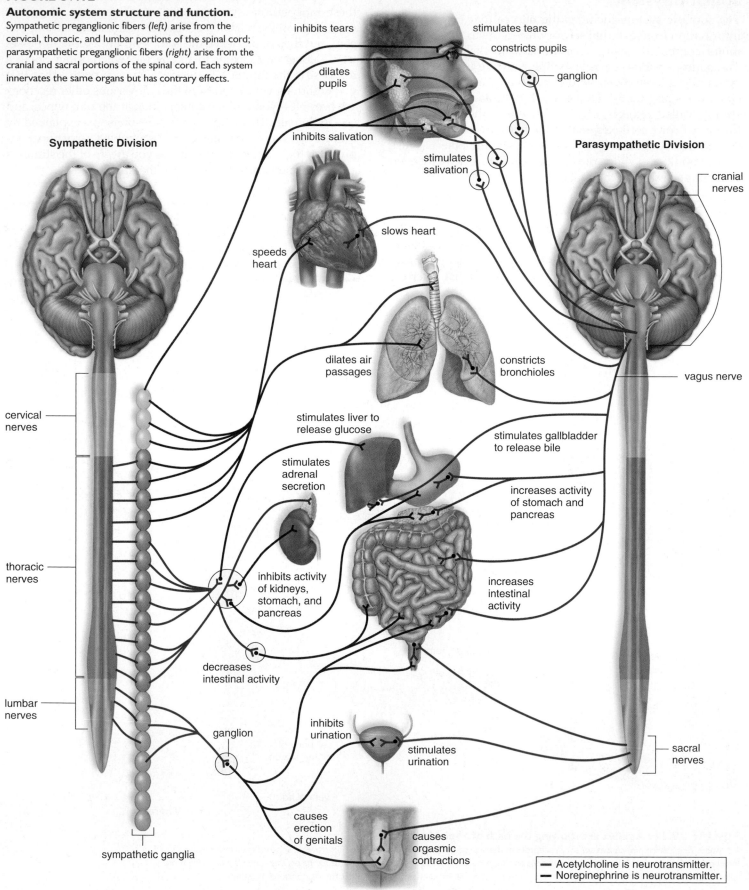

Sympathetic Division

Parasympathetic Division

inhibits tears

stimulates tears

constricts pupils

dilates pupils

ganglion

inhibits salivation

stimulates salivation

cranial nerves

speeds heart

slows heart

cervical nerves

vagus nerve

dilates air passages

constricts bronchioles

stimulates liver to release glucose

stimulates gallbladder to release bile

stimulates adrenal secretion

increases activity of stomach and pancreas

thoracic nerves

inhibits activity of kidneys, stomach, and pancreas

increases intestinal activity

lumbar nerves

decreases intestinal activity

ganglion

inhibits urination

stimulates urination

sacral nerves

sympathetic ganglia

causes erection of genitals

causes orgasmic contractions

— Acetylcholine is neurotransmitter.
— Norepinephrine is neurotransmitter.

Autonomic System

The **autonomic system** of the PNS regulates the activity of cardiac and smooth muscle and glands. It carries out its duties without our awareness or intent. The system is divided into the sympathetic and parasympathetic divisions (Fig. 39.12 and Table 39.1). Both of these divisions (1) function automatically and usually in an involuntary manner; (2) innervate all internal organs; and (3) use two neurons and one ganglion for each impulse. The first neuron has a cell body within the CNS and a preganglionic fiber. The second neuron has a cell body within the ganglion and a postganglionic fiber.

Reflex actions, such as those that regulate the blood pressure and breathing rate, are especially important to the maintenance of homeostasis. These reflexes begin when the sensory neurons in contact with internal organs send information to the CNS. They are completed by motor neurons within the autonomic system.

Sympathetic Division

Most preganglionic fibers of the **sympathetic division** arise from the middle, or thoracolumbar, portion of the spinal cord and almost immediately terminate in ganglia that lie near the cord. Therefore, in this division, the preganglionic fiber is short, but the postganglionic fiber that makes contact with an organ is long.

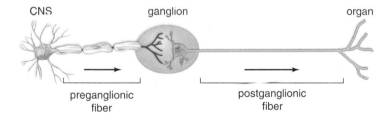

The sympathetic division is especially important during emergency situations and is associated with "fight or flight." If you need to fend off a foe or flee from danger, active muscles require a ready supply of glucose and oxy- gen. The sympathetic division accelerates the heartbeat and dilates the bronchi. On the other hand, the sympathetic division inhibits the digestive tract, since digestion is not an immediate necessity if you are under attack. The neurotransmitter released by the postganglionic axon is primarily norepinephrine (NE). The structure of NE is like that of epinephrine (adrenaline), an adrenal medulla hormone that usually increases heart rate and contractility.

The sympathetic division brings about those responses we associate with "fight or flight."

Parasympathetic Division

The **parasympathetic division** includes a few cranial nerves (e.g., the vagus nerve) and also fibers that arise from the sacral (bottom) portion of the spinal cord. Therefore, this division often is referred to as the craniosacral portion of the autonomic system. In the parasympathetic division, the preganglionic fiber is long, and the postganglionic fiber is short because the ganglia lie near or within the organ.

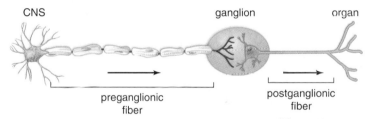

The parasympathetic division, sometimes called the "housekeeper division," promotes all the internal responses we associate with a relaxed state; for example, it causes the pupil of the eye to contract, promotes digestion of food, and slows the heartbeat. The neurotransmitter used by the parasympathetic division is acetylcholine (ACh).

The parasympathetic division brings about the responses we associate with a relaxed state.

TABLE 39.1

Comparison of Somatic Motor and Autonomic Motor Pathways

	Somatic Motor Pathway	Autonomic Motor Pathways	
		Sympathetic	*Parasympathetic*
Type of control	Voluntary/involuntary	Involuntary	Involuntary
Number of neurons per message	One	Two (preganglionic shorter than postganglionic)	Two (preganglionic longer than postganglionic)
Location of motor fiber	Most cranial nerves and all spinal nerves	Thoracolumbar spinal nerves	Cranial (e.g., vagus) and sacral spinal nerves
Neurotransmitter	Acetylcholine	Norepinephrine	Acetylcholine
Effectors	Skeletal muscles	Smooth and cardiac muscle, glands	Smooth and cardiac muscle, glands

science focus

Five Drugs of Abuse

Drugs that people take to alter the mood and/or emotional state affect normal body functions, often by interfering with neurotransmitter release or uptake in the brain.

Alcohol

It is possible to drink alcoholic beverages in moderation, but alcohol is often abused. Alcohol use becomes "abuse," or an illness, when alcohol ingestion impairs an individual's social relationships, health, job efficiency, or judgment. While it is general knowledge that alcoholics are prone to drink until they become intoxicated, there is much debate as to what causes alcoholism. Some believe that alcoholism is due to an underlying psychological disorder, while others blame an inherited physiological disorder.

Alcohol is primarily metabolized in the liver, where it disrupts the normal workings of glycolysis and the citric acid cycle. The liver contains dehydrogenase enzymes, which carry out the following reactions, reducing NAD in the process:

$$NAD \longrightarrow NADH$$
$$alcohol \longrightarrow \longrightarrow acetyl\text{-}CoA$$

The supply of NAD in liver cells is used up by these reactions, and there is not enough free NAD left to keep glycolysis and the citric acid cycle running. The cell begins to ferment, and lactic acid builds up. The pH of the blood decreases and becomes acidic.

Since the citric acid cycle is not working, excess active acetate cannot be broken down, and it is converted to fat—that is, the liver turns fatty. Fat accumulation, the first stage in liver deterioration, begins after only a single night of heavy drinking. If heavy drinking continues, fibrous scar tissue appears during a second stage of deterioration. If heavy drinking stops, the liver can still recover and become normal once again. If not, the final and irrevocable stage, cirrhosis of the liver, occurs: liver cells die, harden, and turn orange ("cirrhosis" means "orange").

The U.S. surgeon general recommends that pregnant women drink no alcohol at all. Alcohol crosses the placenta freely and can cause fetal alcohol syndrome, which is characterized by mental retardation and various physical defects.

Another problem is that heavy drinking interferes with good nutrition. Alcohol is energy intensive—the NADH molecules that result from its breakdown can be used to produce ATP molecules. However, these calories are empty because they do not supply any amino acids, vitamins, and minerals as other energy sources do. Without adequate vitamins, red and white blood cells cannot be formed in the bone marrow. The immune system becomes depressed, and the chances of stomach, liver, lung, pancreas, colon, and tongue cancers increase. Protein digestion and amino acid metabolism are so upset that even adequate protein intake will not prevent amino acid deficiencies. Muscles atrophy, and weakness results. Fat deposits accumulate in the heart wall, and hypertension develops. There is an increased risk of cardiac arrhythmias and stroke.

Nicotine

Nicotine, an alkaloid derived from tobacco, is a widely used neurological agent. When a person smokes a cigarette, nicotine is quickly distributed to all body organs, including the central and peripheral nervous systems. In the central nervous system, nicotine causes neurons to release dopamine, a neurotransmitter associated with behavioral states. The excess dopamine has a reinforcing effect that leads to dependence on the drug. In the peripheral nervous system, nicotine stimulates the same postsynaptic receptors as acetylcholine and leads to increased skeletal muscular activity. It also increases the heartbeat rate and blood pressure as well as digestive tract mobility. Nicotine may even occasionally induce vomiting and/or diarrhea. It also causes water retention by the kidneys.

Many cigarette smokers find it difficult to give up the habit because nicotine induces both physiological and psychological dependence. Withdrawal symptoms include headache, stomach pain, irritability, and insomnia. Tobacco not only contains nicotine, but it also contains many other harmful substances. Cigarette smoking contributes to early death from cancer, including not only lung cancer but also cancer of the larynx, mouth, throat, pancreas, and urinary bladder: chronic diseases such as

bronchitis and emphysema are likely to develop, and the risk of heart attack due to cardiovascular disease increases.

Now that women are as apt to smoke as men, lung cancer has surpassed breast cancer as a cause of death in women. Cigarette smoking in young women who are sexually active is most unfortunate because nicotine, like other psychoactive drugs, adversely affects a developing embryo and fetus.

Marijuana

The dried flowering tops, leaves, and stems of the Indian hemp plant *Cannabis sativa* contain and are covered by a resin that is rich in THC (tetrahydrocannabinol). The names *Cannabis* and marijuana apply to either the plant or THC.

The effects of marijuana differ depending on the strength and the amount consumed, the expertise of the user, and the setting in which it is taken. Usually, the user reports experiencing a mild euphoria along with alterations in vision and judgment, which result in distortions of space and time. Motor incoordination occurs, as well as the inability to concentrate and to speak coherently.

Intermittent use of low-potency marijuana generally is not associated with obvious symptoms of toxicity, but heavy use can produce chronic intoxication. Intoxication is recognized by the presence of hallucinations, anxiety, depression, rapid flow of ideas, body image distortions, paranoid reactions, and similar psychotic symptoms. The terms *cannabis psychosis* and *cannabis delirium* refer to such reactions.

Marijuana is classified as a hallucinogen. It is possible that, like LSD (lysergic acid diethylamide), it has an effect on the action of serotonin, an excitatory neurotransmitter in the brain.

Marijuana use does not seem to produce physical dependence, but a psychological dependence on the euphoric and sedative effects can develop. Craving can also occur as a part of regular heavy use.

Usually marijuana is smoked in a cigarette form called a joint. Since this allows toxic substances, including carcinogens, to enter the lungs, chronic respiratory disease and lung cancer are considered dangers of long-term, heavy use. Some researchers claim that marijuana use leads to long-term brain impairment as well. Others report that males and females suffer reproductive dysfunctions. Fetal cannabis syndrome, which resembles fetal alcohol syndrome, has also been reported. In addition, marijuana has been called a gateway drug because adolescents who have used it also tend to try other drugs. For example, in a study of 100 cocaine abusers, 60% had smoked marijuana for more than ten years.

Some psychologists are very concerned about the use of marijuana among adolescents. Marijuana can be used to avoid dealing with the personal problems that often develop during this maturational phase.

Cocaine

Cocaine is an alkaloid derived from the shrub *Erythroxylon coca*. Cocaine is sold in powder form and as crack, a more potent extract. Users often describe the feeling of euphoria that follows intake of the drug as a rush. Snorting (inhaling) produces this effect in a few minutes; injection, within 30 seconds; and smoking, in less than 10 seconds. Persons dependent on the drug are, therefore, most likely to smoke cocaine. The rush lasts only a few seconds and then is replaced by a state of arousal, which lasts from 5–30 minutes. Then the user begins to feel restless, irritable, and depressed. To overcome these symptoms, the user is apt to take more of the drug, repeating the cycle again and again. A binge of this sort can go on for days, after which the individual suffers a crash. During the binge period, the user is hyperactive and has little desire for food or sleep but has an increased sex drive. During the crash period, the user is fatigued, depressed, and irritable, has memory and concentration problems, and displays no interest in sex. Indeed, men are often impotent. Other drugs, such as marijuana, alcohol, or heroin, often are taken to ease the symptoms of the crash.

Cocaine affects the concentration of dopamine, a neurotransmitter associated with the limbic system and a sense of pleasure. After release into a synapse, dopamine ordinarily is withdrawn into the presynaptic cell for recycling and reuse. Cocaine prevents the reuptake of dopamine by the presynaptic membrane; this causes an excess of dopamine in the synaptic cleft so that the user experi-

FIGURE 39A Drug use.
Blood-borne diseases such as AIDS and hepatitis B pass from one drug abuser to another when they share needles.

ences the sensation of a rush. The epinephrine-like effects of dopamine account for the state of arousal that lasts for some minutes after the rush experience.

With continued cocaine use, the body begins to make less dopamine to compensate for a seemingly excess supply. The user then experiences tolerance, withdrawal symptoms, and an intense craving for the drug. Cocaine, thus, is extremely addictive.

The number of deaths from cocaine use and the number of emergency-room admissions for drug reactions involving cocaine have increased greatly. High doses can cause seizures and cardiac and respiratory arrest.

Individuals who snort the drug can suffer damage to the nasal tissues and even perforation of the septum between the nostrils. Whether or not long-term cocaine abuse causes brain damage is not yet known. It is known, however, that babies born to addicts suffer withdrawal symptoms and may experience neurological and developmental problems.

Heroin

Heroin is derived from morphine, an alkaloid of opium. Heroin is usually injected. Diseases pass between addicts sharing needles (Fig. 39A). After intravenous injection, the onset of action is noticeable within 1 minute and reaches its peak in about 5 minutes. There is a feeling of euphoria along with relief of pain. Side effects

can include nausea, vomiting, hoarseness, and respiratory and circulatory depression leading to death.

Heroin increases dopamine release and binds to receptors meant for the endorphins, the special neurotransmitters that kill pain and produce a feeling of tranquility. They are believed to alleviate pain by preventing the release of a neurotransmitter termed substance P from certain sensory neurons in the region of the spinal cord. When substance P is released, pain is felt, and when substance P is not released, pain is not felt. Endorphins and heroin also bind to receptors on neurons that travel from the spinal cord to the limbic system. Stimulation of these can cause a feeling of pleasure.

Individuals who inject heroin become physically dependent on the drug. With time, the body's production of endorphins decreases. Tolerance develops so that the user needs to take more of the drug just to prevent withdrawal symptoms. The euphoria originally experienced upon injection is no longer felt.

Heroin withdrawal symptoms include perspiration, dilation of pupils, tremors, restlessness, abdominal cramps, gooseflesh, defecation, vomiting, and increase in systolic pressure and respiratory rate. Those who are excessively dependent may experience convulsions, respiratory failure, and death. Infants born to women who are physically dependent also experience these withdrawal symptoms.

CONNECTING THE CONCEPTS

Like the wiring of a modern office building, the peripheral nervous system of humans contains nerves that reach to all parts of the body. There is a division of labor among the nerves. The cranial nerves serve the head region, with the exception of the vagus nerve. All body movements are controlled by spinal nerves, and this is why paralysis may follow a spinal injury. Except for the vagus nerve, only spinal nerves make up the autonomic system, which controls the internal organs. As in most other animals, much of the work of the nervous system in humans is below the level of consciousness.

The nervous system has just three functions: sensory input, integration, and motor output. Sensory input would be impossible without sensory receptors, which are sensitive to external and internal stimuli. You might even argue that sense organs like the eyes and ears should be considered a part of the nervous system, since there would be no sensory nerve impulses without their ability to generate them.

Nerve impulses are the same in all neurons, so how is it that stimulation of eyes causes us to see, and stimulation of ears causes us to hear? Essentially, the central nervous system carries out the function of integrating incoming data. The brain allows us to perceive our environment, reason, and remember. After sensory data have been processed by the CNS, motor output occurs. Muscles and glands are the effectors that allow us to respond to the original stimuli. Without the musculoskeletal system, we would never be able to respond to a danger detected by our eyes and ears.

Summary

39.1 EVOLUTION OF THE NERVOUS SYSTEM

A comparative study of the invertebrates shows a gradual increase in the complexity of the nervous system. The vertebrate nervous system, like that of the earthworm, is divided into the central and peripheral nervous systems.

39.2 NERVOUS TISSUE

The anatomical unit of the nervous system is the neuron, of which there are three types: sensory, motor, and interneuron. Each of these is made up of a cell body, an axon, and dendrites.

When an axon is not conducting an action potential (nerve impulse), the resting potential indicates that the inside of the fiber is negative compared to the outside. The sodium-potassium pump helps maintain a concentration of Na^+ outside the fiber and K^+ inside the fiber. When the axon is conducting a nerve impulse, an action potential (i.e., a change in membrane potential) travels along the fiber. Depolarization occurs (inside becomes positive) due to the movement of Na^+ to the inside, and then repolarization occurs (inside becomes negative again) due to the movement of K^+ to the outside of the fiber.

Transmission of the nerve impulse from one neuron to another takes place across a synapse. In humans, synaptic vesicles release a chemical, known as a neurotransmitter, into the synaptic cleft. The binding of neurotransmitters to receptors in the postsynaptic membrane can either increase the chance of an action potential (stimulation) or decrease the chance of an action potential (inhibition) in the next neuron. A neuron usually transmits several nerve impulses, one after the other.

39.3 CENTRAL NERVOUS SYSTEM: BRAIN AND SPINAL CORD

The CNS consists of the spinal cord and brain, which are both protected by bone. The CNS receives and integrates sensory input and formulates motor output. The gray matter of the spinal cord contains neuron cell bodies; the white matter consists of myelinated axons that occur in bundles called tracts. The spinal cord sends sensory information to the brain, receives motor output from the brain, and carries out reflex actions.

In the brain, the cerebrum has two cerebral hemispheres connected by the corpus callosum. Sensation, reasoning, learning and memory, and language and speech take place in the cerebrum. The cerebral cortex is a thin layer of gray matter covering the cerebrum. The cerebral cortex of each cerebral hemisphere has four lobes: a frontal, parietal, occipital, and temporal lobe. The primary motor area in the frontal lobe sends out motor commands to lower brain centers, which pass them on to motor neurons. The primary somatosensory area in the parietal lobe receives sensory information from lower brain centers in communication with sensory neurons. Association areas for vision are in the occipital lobe, and those for hearing are in the temporal lobe.

The brain has a number of other regions. The hypothalamus controls homeostasis, and the thalamus specializes in sending sensory input on to the cerebrum. The cerebellum primarily coordinates skeletal muscle contractions. The medulla oblongata and the pons have centers for vital functions such as breathing and the heartbeat.

39.4 PERIPHERAL NERVOUS SYSTEM

The peripheral nervous system contains the somatic system and the autonomic system. Reflexes are automatic, and some do not require involvement of the brain. A simple reflex requires the use of neurons that make up a reflex arc. In the somatic system, a sensory neuron conducts nerve impulses from a sensory receptor to an interneuron, which in turn transmits impulses to a motor neuron, which stimulates an effector to react.

While the motor portion of the somatic system of the PNS controls skeletal muscle, the motor portion of the autonomic system controls smooth muscle of the internal organs and glands. The sympathetic division, which is often associated with reactions that occur during times of stress, and the parasympathetic division, which is often associated with activities that occur during times of relaxation, are both parts of the autonomic system.

Reviewing the Chapter

1. Trace the evolution of the nervous system by contrasting its organization in hydras, planarians, earthworms, and humans.   700–702
2. Describe the structure of a neuron, and give a function for each part mentioned. Name three types of neurons, and give a function for each.   703
3. What are the major events of an action potential, and what ion changes are associated with each event?   705
4. Describe the mode of action of a neurotransmitter at a synapse, including how it is stored and how it is destroyed.   706–7
5. Name the major parts of the human brain, and give a principal function for each part.   708–10

6. Describe the limbic system, and discuss its possible involvement in learning and memory.   711
7. Discuss the structure and function of the peripheral nervous system.   712
8. Trace the path of a spinal reflex.   713
9. Contrast the sympathetic and parasympathetic divisions of the autonomic system.   715

Testing Yourself

Choose the best answer for each question.

1. Which is the most complete list of animals that have a central nervous system (CNS) and a peripheral nervous system (PNS)?
   a. hydra, planarian, earthworm, rabbit, human
   b. planarian, earthworm, rabbit, human
   c. earthworm, rabbit, human
   d. rabbit, human

2. Which of these are the first and last elements in a spinal reflex?
   a. axon and dendrite
   b. sense organ and muscle effector
   c. ventral horn and dorsal horn
   d. motor neuron and sensory neuron
   e. sensory receptor and the brain

3. A spinal nerve takes nerve impulses
   a. to the CNS.
   b. away from the CNS.
   c. both to and away from the CNS.
   d. only inside the CNS.
   e. only from the cerebrum.

4. Which of these correctly describes the distribution of ions on either side of an axon when it is not conducting a nerve impulse?
   a. more sodium ions (Na^+) outside and fewer potassium ions (K^+) inside
   b. K^+ outside and Na^+ inside
   c. charged protein outside; Na^+ and K^+ inside
   d. Na^+ and K^+ outside and water only inside
   e. Ca^{2+} inside and outside

5. When the action potential begins, sodium gates open, allowing Na^+ to cross the membrane. Now the polarity changes to
   a. negative outside and positive inside.
   b. positive outside and negative inside.
   c. There is no difference in charge between outside and inside.
   d. Any one of these could be correct.

6. Transmission of the nerve impulse across a synapse is accomplished by
   a. the release of Na^+ at the presynaptic membrane.
   b. the release of neurotransmitters at the postsynaptic membrane.
   c. the reception of neurotransmitters at the postsynaptic membrane.
   d. Only a and c are correct.

7. The autonomic system has two divisions, called
   a. the CNS and PNS.
   b. the somatic and skeletal systems.
   c. the efferent and afferent systems.
   d. the sympathetic and parasympathetic divisions.

8. Synaptic vesicles are
   a. at the ends of dendrites and axons.
   b. at the ends of axons only.
   c. along the length of all long fibers.
   d. at the ends of interneurons only.
   e. Both b and d are correct.

9. Which of these pairs is mismatched?
   a. cerebrum—thinking and memory
   b. thalamus—motor and sensory centers
   c. hypothalamus—internal environment regulator
   d. cerebellum—motor coordination
   e. medulla oblongata—fourth ventricle

10. Repolarization of an axon during an action potential is produced by
    a. inward diffusion of Na^+.
    b. active extrusion of K^+.
    c. outward diffusion of K^+.
    d. inward active transport of Na^+.

11. Which two parts of the brain are least likely to work together?
    a. thalamus and cerebrum
    b. cerebrum and cerebellum
    c. hypothalamus and medulla oblongata
    d. cerebellum and medulla oblongata

12. The spinal cord does not contain or is not attached to
    a. the central canal.   d. the dorsal root.
    b. the white matter area.   e. the tracts.
    c. the association areas.

13. A drug that inactivates acetylcholinesterase
    a. stops the release of ACh from presynaptic endings.
    b. prevents the attachment of ACh to its receptor.
    c. increases the ability of ACh to stimulate muscle contraction.
    d. All of these are correct.

14. Which of these statements about autonomic neurons is correct?
    a. They are motor neurons.
    b. Preganglionic neurons have cell bodies in the CNS.
    c. Postganglionic neurons innervate smooth muscles, cardiac muscle, and glands.
    d. All of these are correct.

15. Which of these fibers release norepinephrine?
    a. preganglionic sympathetic axons
    b. postganglionic sympathetic axons
    c. preganglionic parasympathetic axons
    d. postganglionic parasympathetic axons

16. Sympathetic nerve stimulation does not cause
    a. the liver to release glycogen.
    b. dilation of bronchioles.
    c. the gastrointestinal tract to digest food.
    d. an increase in the heart rate.

17. The limbic system
    a. involves portions of the cerebral lobes, basal nuclei, and the diencephalon.
    b. is responsible for our deepest emotions, including pleasure, rage, and fear.
    c. is not responsible for reason and self-control.
    d. All of these are correct.

18. Which of these would be covered by a myelin sheath?
   a. short dendrites
   b. globular cell bodies
   c. long axons
   d. interneurons
   e. All of these are correct.
19. Label this diagram of a reflex arc.

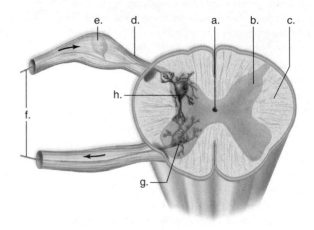

Thinking Scientifically

1. In individuals with panic disorder, the fight-or-flight response is activated by inappropriate stimuli. How might it be possible to directly control this response in order to treat panic disorder? Why is such control often impractical?
2. A man who lost his leg several years ago continues to experience pain as though it were coming from the missing limb. What hypothesis could explain the neurological basis of this pain?

Bioethical Issue: Declaration of Death

An electroencephalogram is a record of the brain's activity picked up from an array of electrodes on the forehead and scalp. The complete and persistent absence of brain activity is often used as a clinical and legal criterion for brain death. Brain death can be present even though the body is still warm and moist. Indeed, the heart has been known to continue beating for more than a month after a declaration of brain death. Even so, should it be permissible to remove organs for transplantation as soon as the person is declared legally dead on the basis of no brain activity? (Keep in mind that organs that have never been deprived of oxygen and nutrients are more likely to be successfully transplanted.)

   As a safeguard, should we have to sign a card that permits organ removal if we have been declared legally dead on the basis of brain activity? Also, should organs only be removed if family members confirm that the person is legally dead and consent to their removal? What evidence should the family require?

   Perhaps it would be best if society simply disallowed the removal of organs unless the person is no longer breathing and the heart has stopped beating. In that case, society would be denying some of its citizens the gift of life. On the other hand, does the end justify the means—do the potential benefits to recipients outweigh the remote possibility that an organ will be removed from a still-living person?

Understanding the Terms

acetylcholine (ACh)   706
acetylcholinesterase
   (AChE)   706
action potential   705
association areas   709
autonomic system   715
axon   703
basal nuclei   710
brain   701
brain stem   710
cell body   703
central nervous system
   (CNS)   702
cephalization   700
cerebellum   710
cerebral cortex   709
cerebral hemisphere   709
cerebrospinal fluid   708
cerebrum   709
cranial nerve   712
dendrite   703
diencephalon   710
dopamine   706
dorsal root ganglion   712
effector   699
ganglion   712
gray matter   708
hypothalamus   710
integration   707
interneuron   703
ladderlike nervous
   system   700
limbic system   711
medulla oblongata   710
memory   711
meninges   708
meningitis   708

midbrain   710
motor (efferent) neuron   703
myelin sheath   703
nerve   712
nerve fiber   703
nerve impulse   704
nerve net   700
neuroglia   703
neuromodulator   707
neuron   703
neurotransmitter   706
nodes of Ranvier   703
norepinephrine (NE)   706
parasympathetic division   715
peripheral nervous system
   (PNS)   702
pons   710
prefrontal area   710
primary motor area   709
primary somatosensory area   709
reflex   713
resting potential   704
saltatory conduction   705
Schwann cell   703
sensory (afferent) neuron   703
sensory receptor   699
serotonin   706
somatic system   713
spinal cord   708
spinal nerve   712
sympathetic division   715
synapse   706
synaptic cleft   706
thalamus   710
tract   708
ventricle   708
white matter   708

Match the terms to these definitions:

a. _____ Automatic, involuntary response of an organism to a stimulus.
b. _____ Chemical stored at the ends of axons that is responsible for transmission across a synapse.
c. _____ System within the peripheral nervous system that regulates internal organs.
d. _____ Collection of neuron cell bodies usually outside the central nervous system.
e. _____ Neurotransmitter active in the somatic system of the peripheral nervous system.

ARIS, the *Biology* Website

ARIS, the website for *Biology*, provides a wealth of information organized and integrated by chapter. You will find practice quizzes, interactive activities, labeling exercises, flashcards, and much more that will complement your learning and understanding of general biology.

www.mhhe.com/maderbiology9

40

SENSE ORGANS

W hen you smell a flower, molecules in the air bind to olfactory cells, causing them to generate nerve impulses that travel to your brain. Interpretation of these impulses is the function of the brain, which has a specific region for receiving information from each of the sense organs. Impulses arriving at a particular sensory area of the brain can be interpreted in only one way; for example, those arriving at the olfactory area result in smell sensation, and those arriving at the visual area result in sight sensation. The brain integrates data from various sensory receptors in order to perceive—for example, a flower—what caused the sight and smell sensations.

Humans have a variety of sensory receptors. Chemoreceptors are sensitive to the presence of particular molecules, photoreceptors are sensitive to light, and mechanoreceptors detect touch, vibration, and pressure. Our sensory receptors form an exchange area with the external environment, just as our digestive tract and lungs are exchange areas. They gather the information that allows the brain to make decisions about finding prey, escaping a predator, and any number of other adaptive behaviors. Therefore, sensory receptors play a significant role in maintaining homeostasis. Learning also helps survival, and sensations can rekindle memories of meaningful experiences that occurred years before.

Smelling fireweed flowers, *Epilobium angustifolium*

40.1 CHEMICAL SENSES

The sensory receptors responsible for taste and smell are termed **chemoreceptors** [Gk. *chemo*, pertaining to chemicals; L. *receptor*, receiver] because they are sensitive to certain chemical substances in food, including liquids, and air. Chemoreception is found almost universally in animals and is therefore believed to be the most primitive sense.

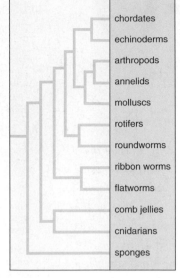

The location and sensitivity of chemoreceptors vary throughout the animal kingdom. They are important in finding food, locating a mate, and detecting potentially dangerous chemicals in the environment. Although chemoreceptors are present throughout the body of planarians, they are concentrated in the auricles located on the sides of the head. Insects, crustaceans, and other arthropods possess a number of chemoreceptors. In the housefly, chemoreceptors are located primarily on the feet. A fly literally tastes with its feet instead of its mouth. Insects also detect airborne pheromones, which are chemical messages passed between individuals. In crustaceans such as lobsters and crabs, chemoreceptors are widely distributed in their appendages and antennae. In vertebrates such as amphibians, chemoreceptors are located in the nose, mouth, and skin. Snakes possess Jacobsen's organs, a pair of sensory pitlike organs located in the roof of the mouth. When a snake flicks its forked tongue, scent molecules are carried to the Jacobsen's organs and sensory information is transmitted to the brain for interpretation. In mammals, the receptors for taste are located in the mouth, and the receptors for smell are located in the nose.

Sense of Taste

In adult humans, approximately 3,000 **taste buds** are located primarily on the tongue (Fig. 40.1). Many taste buds lie along the walls of the papillae, the small elevations on the tongue that are visible to the unaided eye. Isolated taste buds are also present on the hard palate, the pharynx, and the epiglottis.

Taste buds open at a taste pore. Taste buds have supporting cells and a number of elongated taste cells that end in microvilli. The microvilli, which project into the taste pore, bear receptor proteins for certain molecules. When molecules bind to receptor proteins, nerve impulses are generated in associated sensory nerve fibers. These nerve impulses go to the brain, including cortical areas that interpret them as tastes.

There are at least four primary types of taste (sweet, sour, salty, and bitter). A fifth taste, called umami, may exist for certain flavors of cheese, beef broth, and some seafood. Taste buds for each of these tastes are located throughout the tongue, althouth certain regions may be most sensitive to particular tastes: the tip of the tongue is most sensitive to sweet tastes; the margins to salty and sour tastes; and the rear of the tongue to bitter tastes. A particular food can stimulate more than one of these types of taste buds. In this way, the response of taste buds can result in a range of sweet, sour, salty, and bitter tastes. The brain appears to survey the overall pattern of incoming sensory impulses and to take a "weighted average" of their taste messages as the perceived taste.

The taste cells of humans are located in taste buds. The microvilli of taste cells have receptor proteins for molecules that cause the brain to perceive sweet, sour, salty, bitter, and perhaps umami tastes.

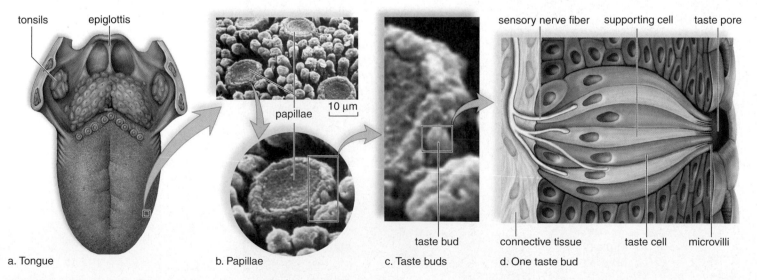

a. Tongue   b. Papillae   c. Taste buds   d. One taste bud

tonsils   epiglottis   papillae   10 μm   sensory nerve fiber   supporting cell   taste pore   taste bud   connective tissue   taste cell   microvilli

FIGURE 40.1 Taste buds in humans.
a. Papillae on the tongue contain taste buds that are sensitive to sweet, sour, salty, bitter, and perhaps umami. **b.** Photomicrograph and enlargement of papillae. **c.** Taste buds occur along the walls of the papillae. **d.** Taste cells end in microvilli that bear receptor proteins for certain molecules. When molecules bind to the receptor proteins, nerve impulses are generated and go to the brain, where the sensation of taste occurs.

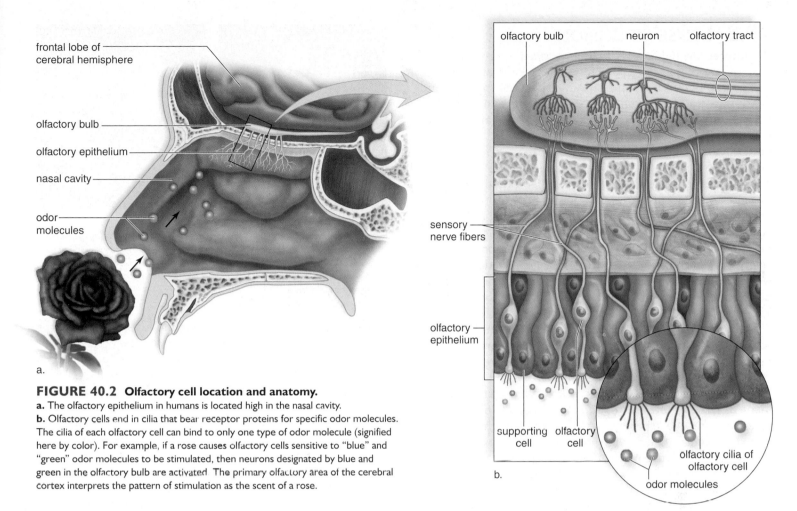

FIGURE 40.2 Olfactory cell location and anatomy.
a. The olfactory epithelium in humans is located high in the nasal cavity.
b. Olfactory cells end in cilia that bear receptor proteins for specific odor molecules. The cilia of each olfactory cell can bind to only one type of odor molecule (signified here by color). For example, if a rose causes olfactory cells sensitive to "blue" and "green" odor molecules to be stimulated, then neurons designated by blue and green in the olfactory bulb are activated. The primary olfactory area of the cerebral cortex interprets the pattern of stimulation as the scent of a rose.

Sense of Smell

In humans, the sense of smell, or olfaction, is dependent on between 10 and 20 million **olfactory cells.** These structures are located within olfactory epithelium high in the roof of the nasal cavity (Fig. 40.2). Olfactory cells are modified neurons. Each cell ends in a tuft of about five olfactory cilia that bear receptor proteins for odor molecules. Each olfactory cell has only 1 out of 1,000 different types of receptor proteins. Nerve fibers from like olfactory cells lead to the same neuron in the olfactory bulb, an extension of the brain. An odor contains many odor molecules that activate a characteristic combination of receptor proteins. A rose might stimulate olfactory cells, designated by blue and green in Figure 40.2, while a gardenia might stimulate a different combination. An odor's signature in the olfactory bulb is determined by which neurons are stimulated. When the neurons communicate this information via the olfactory tract to the olfactory areas of the cerebral cortex, we know we have smelled a rose or a gardenia.

Have you ever noticed that a certain aroma vividly brings to mind a certain person or place? A whiff of perfume may remind you of a specific person, or the smell of boxwood may remind you of your grandfather's farm. The olfactory bulbs have direct connections with the lim-

bic system and its centers for emotions and memory. One investigator showed that when subjects smelled an orange while viewing a painting, they not only remembered the painting when asked about it later, they had many deep feelings about the painting.

The number of olfactory cells declines with age and the remaining population of receptors becomes less sensitive. Thus, older people tend to apply excessive amounts of perfume or aftershave to detect its smell. This can become dangerous if these individuals cannot smell smoke or a gas leak.

Usually, the sense of taste and the sense of smell work together to create a combined effect when interpreted by the cerebral cortex. For example, when you have a cold, you think food has lost its taste, but most likely you have lost the ability to sense its smell. This method works in reverse also. When you smell something, some of the molecules move from the nose down into the mouth region and stimulate the taste buds there. Therefore, part of what we refer to as smell may in fact be taste.

Olfactory epithelium contains olfactory cells. The cilia of olfactory cells have receptor proteins for odor molecules that cause the brain to distinguish odors.

40.2 SENSE OF VISION

Photoreceptors [Gk. *photos*, light; L. *receptor*, receiver] are sensory receptors that are sensitive to light. Some animals lack photoreceptors and depend on senses such as smelling and hearing instead; other animals have photoreceptors but live in environments that do not require them. For example, moles live underground and use their sense of smell and touch rather than eyesight.

Not all photoreceptors form images. The "eyespots" of planarians allow these animals to determine the direction of light. Image-forming eyes are found among four invertebrate groups: cnidarians, annelids, molluscs, and arthropods. Arthropods have **compound eyes** composed of many independent visual units called ommatidia [Gk. *ommation*, dim. of *omma*, eye], each possessing all the elements needed for light reception (Fig. 40.3). Both the cornea and crystalline cone function as lenses to direct light rays toward the photoreceptors. The photoreceptors generate nerve impulses, which pass to the brain by way of optic nerve fibers. The outer pigment cells absorb stray light rays so that the rays do not pass from one visual unit to the other. The image that results from all the stimulated visual units is crude because the small size of compound eyes limits the number of visual units, which still might number as many as 28,000. How arthropod brains integrate images from the compound eye to perceive objects is not known.

Insects have color vision, but they make use of a slightly shorter range of the electromagnetic spectrum compared to humans. However, they can see the longest of the ultraviolet rays, and this enables them to be especially sensitive to the particular parts of flowers such as nectar guides that have particular ultraviolet patterns (Fig. 40.4). Some fishes, all reptiles, and most birds are believed to have color vision, but among mammals, only humans and other primates have color vision. It would seem, then, that this trait was adaptive for a diurnal habit (active during the day), which accounts for its retention in a few mammals.

Vertebrates (including humans) and certain molluscs, such as the squid and the octopus, have a **camera-type eye.** Since molluscs and vertebrates are not closely related, this similarity is an example of convergent evolution. A single lens focuses an image of the visual field on photoreceptors, which are closely packed together. In vertebrates, the lens

FIGURE 40.4 Nectar guides.
Evening primrose, *Oenothera*, as seen by humans (*left*) and insects (*right*). Humans see no markings, but insects see distinct blotches because their eyes respond to ultraviolet rays. These types of markings, known as nectar guides, often highlight the reproductive parts of flowers, where insects feed on nectar and pick up pollen at the same time.

changes shape to aid focusing, but in molluscs the lens moves back and forth. All of the photoreceptors taken together can be compared to a piece of film in a camera. The human eye is more complex than a camera, however, as we shall see.

Animals with two eyes facing forward have three-dimensional, or **stereoscopic, vision.** The visual fields overlap and each eye is able to view an object from a different angle. Predators tend to have stereoscopic vision and so do humans. Animals with eyes facing sideways, such as rabbits, don't have stereoscopic vision, but they do have **panoramic vision,** meaning that the visual field is very wide. Panoramic vision is useful to prey animals because it makes it more difficult for a predator to sneak up on them.

The Human Eye

The most important parts of the human eye and their functions are listed in Table 40.1. The human eye, which is an elongated sphere about 2.5 cm in diameter, has three layers, or coats: the sclera, the choroid, and the retina (Fig. 40.5). The outer layer, the **sclera** [Gk. *skleros*, hard], is an opaque, white, fibrous layer that covers most of the eye; in front of the eye, the sclera becomes the transparent **cornea,** the window of the eye. A thin layer of epithelial cells forms a mucous membrane called the **conjunctiva** that covers the surface of the sclera and keeps the

FIGURE 40.3
Compound eye.
Each visual unit of a compound eye has a cornea and a lens that focus light onto photoreceptors. The photoreceptors generate nerve impulses that are transmitted to the brain, where interpretation produces a mosaic image.

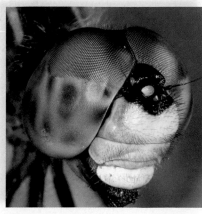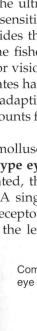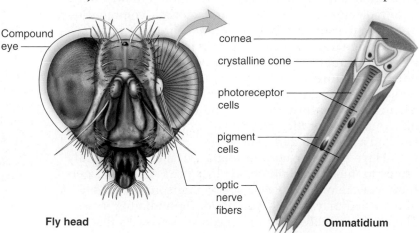

Compound eye — cornea — crystalline cone — photoreceptor cells — pigment cells — optic nerve fibers

Fly head   **Ommatidium**

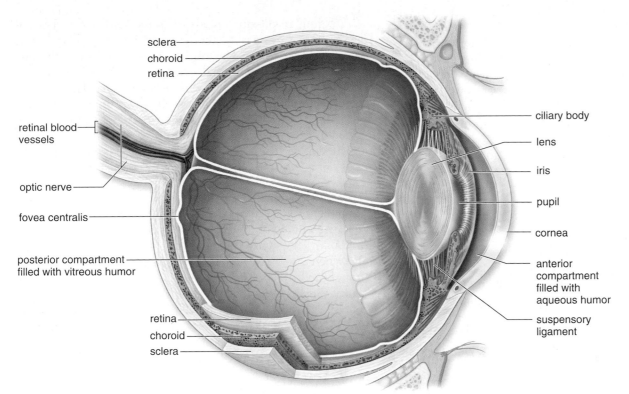

sclera
choroid
retina

retinal blood
vessels

optic nerve

fovea centralis

posterior compartment
filled with vitreous humor

retina
choroid
sclera

ciliary body

lens

iris

pupil

cornea

anterior
compartment
filled with
aqueous humor

suspensory
ligament

FIGURE 40.5
**Anatomy of
the human eye.**
Notice that the sclera,
the outer layer of the
eye, becomes the cornea
and that the choroid, the
middle layer, is continuous
with the ciliary body and
the iris. The retina, the
inner layer, contains the
photoreceptors for vision.
The fovea centralis is the
region where vision is
most acute.

eyes moist. The middle, thin, dark-brown layer, the **choroid** [Gk. *chorion*, membrane], contains many blood vessels and a brown pigment that absorbs stray light rays. Toward the front of the eye, the choroid thickens and forms the ring-shaped ciliary body and a thin, circular, muscular diaphragm, the iris. The **iris** is the colored portion of the eye and regulates the size of an opening called the pupil. The **pupil,** like the aperture on a camera lens, regulates light entering the eye. The **lens,** which is attached to the ciliary body by ligaments, divides the cavity of the eye into two portions and helps form images. A basic, watery solution called aqueous humor fills the anterior compartment between the cornea and the lens. The aqueous humor provides a fluid cushion and nutrient and waste transport for the eye. Glaucoma results when aqueous humor builds up and increases intraocular pressure. A viscous, gelatinous material, the vitreous humor, fills the large posterior compartment behind the lens. The vitreous humor helps stabilize the shape of the eye and supports the retina.

The inner layer of the eye, the **retina** [L. *rete,* net], is located in the posterior compartment. The retina contains photoreceptors called rod cells and cone cells. The rods are very sensitive to light, but they do not see color; therefore, at night or in a darkened room, we see only shades of gray. The cones, which require bright light, are sensitive to different wavelengths of light, and therefore, we have the ability to distinguish colors. The retina has a very special region called the **fovea centralis,** where cone cells are densely packed. Light is normally focused on the fovea when we look directly at an object. This is helpful because vision is most acute in the fovea centralis. Rods are distributed in the peripheral regions of the retina. Sensory fibers form the optic nerve, which takes nerve impulses to the brain.

TABLE 40.1

Functions of the Parts of the Eye

Part	Function
Sclera	Protects and supports eyeball
Conjunctiva	Moistens eye surface
Cornea	Refracts light rays
Pupil	Admits light
Choroid	Absorbs stray light
Ciliary body	Holds lens in place, accommodation
Iris	Regulates light entrance
Retina	Contains sensory receptors for sight
Rods	Make black-and-white vision possible
Cones	Make color vision possible
Fovea centralis	Makes acute vision possible
Other	
Lens	Refracts and focuses light rays
Humors	Transmit light rays and support eyeball
Optic nerve	Transmits impulse to brain

The human eye has three layers: the outer sclera, the middle choroid, and the retina. Only the retina contains receptors for sight.

Focusing of the Eye

When we look at an object, light rays pass through the pupil and are focused on the retina. The image produced is much smaller than the object because light rays are bent (refracted) when they are brought into focus. Focusing starts at the cornea and continues as the rays pass through the lens and the humors. The image on the retina is inverted (it is upside down) and reversed from left to right.

The lens provides additional focusing power as **visual accommodation** occurs for close vision. The shape of the lens is controlled by the **ciliary muscle** within the ciliary body. When we view a distant object, the ciliary muscle is relaxed, causing the suspensory ligaments attached to the ciliary body to be taut; therefore, the lens remains relatively flat (Fig. 40.6a). When we view a near object, the ciliary muscle contracts, releasing the tension on the suspensory ligaments, and the lens becomes more round due to its natural elasticity (Fig. 40.6b). Because close work requires contraction of the ciliary muscle, it very often causes muscle fatigue known as eyestrain. With normal aging, the lens loses its ability to accommodate for near objects (Fig. 40.6b); therefore, people frequently need reading glasses once they reach middle age.

Aging, or possibly exposure to the sun, also makes the lens subject to cataracts; the lens can become opaque and therefore incapable of transmitting light rays. Currently, surgery is the only viable treatment for cataracts. First, a surgeon opens the eye near the rim of the cornea. The enzyme zonulysin may be used to digest away the ligaments holding the lens in place. Most surgeons then use a cryoprobe, which freezes the lens for easy removal. An intraocular lens attached to the iris can then be implanted so that the patient does not need to wear thick glasses or contact lenses.

Distance Vision. Those who can easily see a near object but have trouble seeing what is designated as a size 20 letter 20 ft away on an optometrist's chart are said to be *nearsighted* (myopia). These individuals often have an elongated eyeball, and when they attempt to look at a distant object, the image is brought to focus in front of the retina. These people can wear concave lenses, which diverge the light rays so that the image can be focused on the retina (Fig. 40.7a). Rather than wear glasses or contact lenses, many nearsighted people are now choosing to undergo laser surgery. First, specialists determine how much the cornea needs to be flattened to achieve visual

---

The lens, assisted by the cornea and the humors, focuses images on the retina.

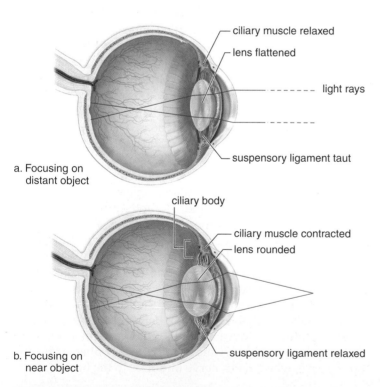

a. Focusing on distant object

- ciliary muscle relaxed
- lens flattened
- light rays
- suspensory ligament taut

b. Focusing on near object

- ciliary body
- ciliary muscle contracted
- lens rounded
- suspensory ligament relaxed

FIGURE 40.6 Focusing of the human eye.

Light rays from each point on an object are bent by the cornea and the lens in such a way that an inverted and reversed image of the object forms on the retina. **a.** When focusing on a distant object, the lens is flat because the ciliary muscle is relaxed and the suspensory ligament is taut. **b.** When focusing on a near object, the lens accommodates; that is, it becomes rounded because the ciliary muscle contracts, causing the suspensory ligament to relax.

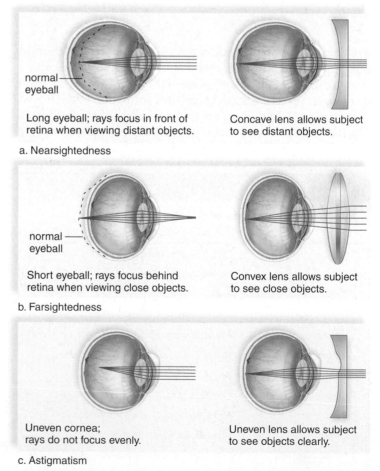

a. Nearsightedness

normal eyeball

Long eyeball; rays focus in front of retina when viewing distant objects.

Concave lens allows subject to see distant objects.

b. Farsightedness

normal eyeball

Short eyeball; rays focus behind retina when viewing close objects.

Convex lens allows subject to see close objects.

c. Astigmatism

Uneven cornea; rays do not focus evenly.

Uneven lens allows subject to see objects clearly.

FIGURE 40.7 Common abnormalities of the eye with possible corrective lenses.

a. A concave lens in nearsighted persons focuses light rays on the retina. **b.** A convex lens in farsighted persons focuses light rays on the retina. **c.** An uneven lens in persons with astigmatism focuses light rays on the retina.

acuity. Controlled by a computer, the laser then removes this amount of the cornea. Most patients achieve at least 20/40 vision, but a few complain of glare and varying visual acuity.

Those who can easily see the optometrist's chart but cannot easily see near objects are *farsighted* (hyperopia). They often have a shortened eyeball, and when they try to see near objects, the image is focused behind the retina. These individuals must wear a convex lens to increase the bending of light rays so that the image can be focused on the retina (Fig. 40.7*b*). When the cornea or lens is uneven, the image is fuzzy. This condition, called astigmatism, can be corrected by an unevenly ground lens to compensate for the uneven cornea (Fig. 40.7*c*).

> The inability of the lens to accommodate as we age requires corrective lenses for close vision. The shape of the eyeball determines the need for corrective lenses for distance vision.

Photoreceptors of the Eye

Vision begins once light has been focused on the photoreceptors in the retina. Figure 40.8 illustrates the structure of the photoreceptors called **rod cells** and **cone cells.** Both rods and cones have an outer segment joined to an inner segment by a stalk. Pigment molecules are embedded in the membrane of the many disks present in the outer segment. Synaptic vesicles are located at the synaptic endings of the inner segment.

The visual pigment in rods is a deep-purple pigment called rhodopsin. **Rhodopsin** is a complex molecule made up of the protein opsin and a light-absorbing molecule called

retinal, which is a derivative of vitamin A. When a rod absorbs light, rhodopsin splits into opsin and retinal, leading to a cascade of reactions and the closure of ion channels in the rod cell's plasma membrane. The release of inhibitory transmitter molecules from the rod's synaptic vesicles ceases. Thereafter, nerve impulses go to the visual areas of the cerebral cortex. Rods are very sensitive to light and therefore are suited to night vision. (Since carrots are rich in vitamin A, it is true that eating carrots can improve your night vision.) Rod cells are plentiful in the peripheral region of the retina; therefore, they also provide us with peripheral vision and perception of motion.

The cones, on the other hand, are located primarily in the fovea centralis and are activated by bright light. They allow us to detect the fine detail and the color of an object. Color vision depends on three different kinds of cones, which contain pigments called the B (blue), G (green), and R (red) pigments. Each pigment is made up of retinal and opsin, but there is a slight difference in the opsin structure of each, which accounts for their individual absorption patterns. Various combinations of cones are believed to be stimulated by in-between shades of color. For example, the color yellow is perceived when green cones are highly stimulated, red cones are partially stimulated, and blue cones are not stimulated. In color blindness, an individual lacks certain visual pigments. As indicated in Chapter 12, color blindness is a hereditary disorder.

> The receptors for sight are the rods and the cones. The rods permit vision in dim light, and the cones permit vision in the bright light needed for color vision.

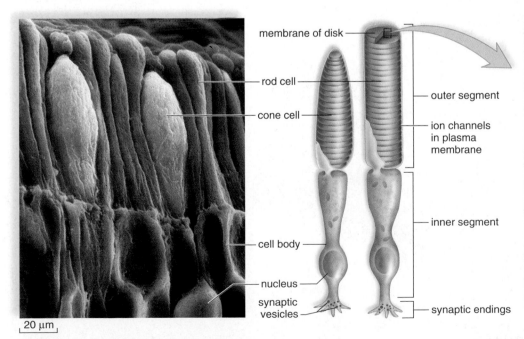
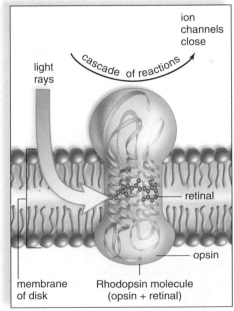

FIGURE 40.8 Photoreceptors in the eye.
The outer segment of rods and cones contains stacks of membranous disks, which contain visual pigments. In rods, the membrane of each disk contains rhodopsin, a complex molecule containing the protein opsin and the pigment retinal. When rhodopsin absorbs light energy, it splits, releasing opsin, which sets in motion a cascade of reactions that cause ion channels in the plasma membrane to close. Thereafter, nerve impulses go to the brain.

Integration of Visual Signals in the Retina

The retina has three layers of neurons (Fig. 40.9). The layer closest to the choroid contains the rod cells and cone cells; the middle layer contains bipolar cells; and the innermost layer contains ganglion cells, whose sensory fibers become the optic nerve. Only the rod cells and the cone cells are sensitive to light, and therefore light must penetrate to the back of the retina before they are stimulated.

The rod cells and the cone cells synapse with the bipolar cells, which in turn synapse with ganglion cells that initiate nerve impulses. Notice in Figure 40.9 that there are many more rod cells and cone cells than ganglion cells. In fact, the retina has as many as 150 million rod cells and 6 million cone cells but only 1 million ganglion cells. The sensitivity of cones versus rods is mirrored by how directly they connect to ganglion cells. As many as 150 rods may excite the same ganglion cell. Stimulation of rods results in vision that is blurred and indistinct. In contrast, some cone cells in the fovea centralis excite only one ganglion cell. This explains why cones, especially in the fovea, provide us with a sharper, more detailed image of an object.

As signals pass to bipolar cells and ganglion cells, integration occurs. Each ganglion cell receives signals from rod cells covering about 1 mm^2 of retina (about the size of a thumbtack hole). This region is the ganglion cell's receptive field. Some time ago, scientists discovered that a ganglion cell is stimulated only by messages received from the center of its receptive field; otherwise, it is inhibited. If all the rod cells in the receptive field receive light, the ganglion cell responds in a neutral way—that is, it reacts only weakly or perhaps not at all. This supports the hypothesis that considerable processing occurs in the retina before nerve impulses are sent to the brain. Additional integration occurs in the visual areas of the cerebral cortex.

> Synaptic integration and processing begin in the retina before nerve impulses are sent to the brain.

Blind Spot

Figure 40.9 provides an opportunity to point out that there are no rods and cones where the optic nerve exits the retina. Therefore, no vision is possible in this area. You can prove this to yourself by putting a dot to the right of center on a piece of paper. Use your right hand to move the paper slowly toward your right eye while you look straight ahead. The dot will disappear at one point—this is your **blind spot.**

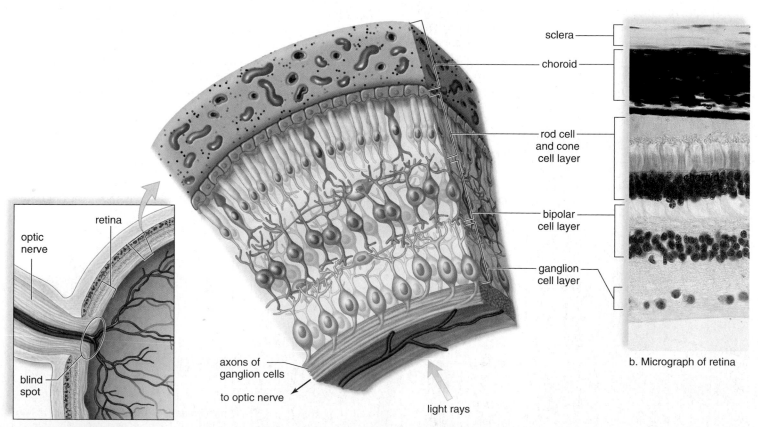

sclera

choroid

rod cell and cone cell layer

bipolar cell layer

ganglion cell layer

b. Micrograph of retina

optic nerve

retina

blind spot

a. Location of retina

axons of ganglion cells

to optic nerve

light rays

FIGURE 40.9 Structure and function of the retina.

a. The retina is the inner layer of the eyeball. Rod cells and cone cells, located at the back of the retina nearest the choroid, synapse with bipolar cells, which synapse with ganglion cells. Integration of signals occurs at these synapses; therefore, much processing occurs in bipolar and ganglion cells. Further, notice that many rod cells share one bipolar cell, but cone cells do not. Certain cone cells synapse with only one ganglion cell. Cone cells, in general, distinguish more detail than do rod cells. **b.** Micrograph shows that the sclera and choroid are relatively thin compared to the retina, which has several layers of cells.

health focus

Protecting Vision and Hearing

Age can be accompanied by a serious loss of vision and hearing. The time to start preventive measures for such problems, however, is when we are younger.

Preventing a Loss of Vision

The eye is subject to both injuries and disorders. Although flying objects sometimes penetrate the cornea and damage the iris, lens, or retina, careless use of contact lenses is the most common cause of injuries to the eye. Injuries cause only 4% of all cases of blindness; the most frequent causes are retinal disorders, glaucoma, and cataracts, in that order. Retinal disorders are varied. In diabetic retinopathy, which blinds many people between the ages of 20 and 74, capillaries to the retina burst, and blood spills into the vitreous humor. Careful regulation of blood glucose levels in these patients may be protective. In macular degeneration, the cones are destroyed because thickened choroid vessels no longer function as they should. Glaucoma occurs when the drainage system of the eyes fails, so that fluid builds up and destroys nerve fibers responsible for peripheral vision. Eye doctors always check for glaucoma, but it is advisable to be aware of the disorder in case it comes on quickly. Those who have experienced acute glaucoma report that the eyeball feels as heavy as a stone. In cataracts, cloudy spots on the lens of the eye eventually pervade the whole lens. The milky yellow-white lens scatters incoming light and blocks vision.

There are preventive measures that we can take to reduce the chance of defective vision as we age. Accumulating evidence suggests that both macular degeneration and cataracts, which tend to occur in older people, are caused by long-term exposure to the ultraviolet rays of the sun. It is recommended, therefore, that everyone, especially those who live in sunny climates or work outdoors, wear glass, not plastic, sunglasses to absorb ultraviolet light. Large lenses worn close to the eyes offer further protection. Special-purpose lenses that block at least 99% of UV-B, 60% of UV-A, and 20–97% of visible light are good for bright sun combined with sand, snow, or water. Healthcare providers have found an increased incidence of cataracts in heavy cigarette smokers. In men, smoking 20 cigarettes or more a day, and in women, smoking more than 35 cigarettes a day, doubles the risk of cataracts. It is possible that smoking reduces the delivery of blood and therefore nutrients to the lens.

Preventing a Loss of Hearing

Especially when we are young, the middle ear is subject to infections that can lead to hearing impairment if they are not treated promptly by a physician. The mobility of ossicles decreases with age, and in the condition called otosclerosis, new filamentous bone grows over the stirrup, impeding its movement. Surgical treatment is the only remedy for this type of conduction deafness. However, age-associated nerve deafness due to stereocilia damage from exposure to loud noises is preventable. Hospitals are now aware that even the ears of the newborn need to be protected from noise and are taking steps to make sure neonatal intensive care units and nurseries are as quiet as possible.

In today's society, exposure to excessive noise is often possible. Noise is measured in decibels, and any noise above a level of 80 decibels could result in damage to the hair cells of the organ of Corti. Eventually, the stereocilia and then the hair cells disappear completely (Fig. 40A). If listening to city traffic for extended periods can damage hearing, it stands to reason that frequent attendance at rock concerts, constantly playing a stereo loudly, or using earphones at high volume are also damaging to hearing. The first hint of danger could be temporary hearing loss, a "full" feeling in the ears, muffled hearing, or tinnitus (e.g., ringing in the ears). If you have any of these symptoms, modify your listening habits immediately to prevent further damage. If exposure to noise is unavoidable, specially designed noise-reduction earmuffs are available, and it is also possible to purchase earplugs made from a compressible, spongelike material at the drugstore or sporting-goods store. These earplugs are not the same as those worn for swimming, and they should not be used interchangeably.

Aside from loud music, noisy indoor or outdoor equipment, such as a rug-cleaning machine or a chain saw, is also troublesome. Even motorcycles and recreational vehicles such as snowmobiles and motocross bikes can contribute to a gradual loss of hearing. Exposure to intense sounds of short duration, such as a burst of gunfire, can result in an immediate hearing loss. Hunters may have a significant hearing reduction in the ear opposite the shoulder where the gun is carried. The butt of the rifle offers some protection to the ear nearest the gun when it is shot.

Finally, people need to be aware that some medicines are ototoxic. Anticancer drugs, most notably cisplatin, and certain antibiotics (e.g., streptomycin, kanamycin, gentamicin) make the ears especially susceptible to hearing loss. Anyone taking such medications needs to be especially careful to protect his or her ears from any loud noises.

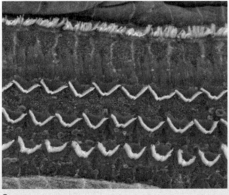
a.

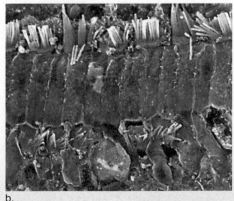
b.

FIGURE 40A Hearing loss.
a. Normal hair cells in the organ of Corti of a guinea pig. b. Damaged hair cells in the organ of Corti of a guinea pig. This damage occurred after 24-hour exposure to a noise level typical of a rock concert.

40.3 SENSES OF HEARING AND BALANCE

The ear has two sensory functions: hearing and balance (equilibrium). The sensory receptors for both of these are located in the inner ear, and each consists of *hair cells* with stereocilia (long microvilli) that are sensitive to mechanical stimulation. They are **mechanoreceptors.**

Anatomy of the Ear

The ear has three distinct divisions: the outer, inner, and middle ear (Figure 40.10). The **outer ear** consists of the pinna (external flap) and the auditory canal. The opening of the auditory canal is lined with fine hairs and glands. Glands that secrete earwax are located in the upper wall of the auditory canal. Earwax helps guard the ear against the entrance of foreign materials, such as air pollutants and microorganisms.

The **middle ear** begins at the **tympanic membrane** (eardrum) and ends at a bony wall containing two small openings covered by membranes. These openings are called the *oval window* and the *round window*. Three small bones are found between the tympanic membrane and the oval window. Collectively called the **ossicles,** individually they are the *malleus* (hammer), the *incus* (anvil), and the *stapes* (stirrup) because their shapes resemble these objects. The malleus adheres to the tympanic membrane, and the stapes touches the oval window. An *auditory tube* (eustachian tube), which extends from each middle ear to the nasopharynx, permits equalization of air pressure. Chewing gum, yawning, and swallowing in elevators and airplanes help move air through the auditory tubes upon ascent and descent. As this occurs, we often hear the ears "pop."

Whereas the outer ear and the middle ear contain air, the inner ear is filled with fluid. Anatomically speaking, the **inner ear** has three areas: the **semicircular canals** and the **vestibule** are both concerned with equilibrium; the **cochlea** is concerned with hearing. The cochlea resembles the shell of a snail because it spirals.

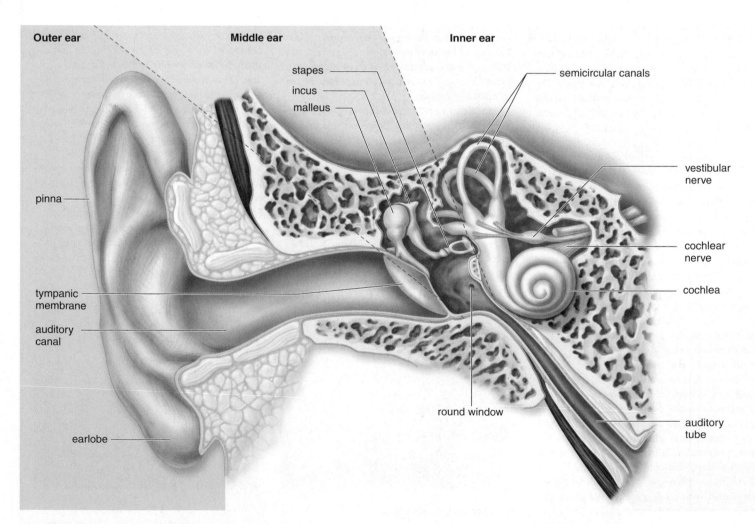

FIGURE 40.10 Anatomy of the human ear.
In the middle ear, the malleus (hammer), the incus (anvil), and the stapes (stirrup) amplify sound waves. In the inner ear, the mechanoreceptors for equilibrium are in the semicircular canals and the vestibule, and the mechanoreceptors for hearing are in the cochlea.

Process of Hearing

The process of hearing begins when sound waves enter the auditory canal. Just as ripples travel across the surface of a pond, sound waves travel by the successive vibrations of molecules. Ordinarily, sound waves do not carry much energy, but when a large number of waves strike the tympanic membrane, it moves back and forth (vibrates) ever so slightly. The malleus then takes the pressure from the inner surface of the tympanic membrane and passes it by means of the incus to the stapes in such a way that the pressure is multiplied about 20 times as it moves. The stapes strikes the membrane of the oval window, causing it to vibrate, and in this way, the pressure is passed to the fluid (endolymph) within the cochlea.

When the snail-shaped cochlea is examined in cross section (Fig. 40.11), the vestibular canal, the cochlear canal, and the tympanic canal become apparent. Along the length of the basilar membrane, which forms the lower wall of the *cochlear canal*, are little hair cells whose stereocilia are embedded within a gelatinous material called the *tectorial membrane*. The hair cells of the cochlear canal, called the **organ of Corti,** or spiral organ, synapse with nerve fibers of the *cochlear nerve* (auditory nerve).

When the stapes strikes the membrane of the oval window, pressure waves move from the vestibular canal to the tympanic canal across the basilar membrane, and the round window membrane bulges. The basilar membrane moves up and down, and the stereocilia of the hair cells embedded in the tectorial membrane bend. Then nerve impulses begin in the cochlear nerve and travel to the brain stem. When they reach the auditory areas of the cerebral cortex, they are interpreted as a sound.

Each part of the organ of Corti is sensitive to different wave frequencies, or pitch. Near the tip, the organ of Corti responds to low pitches, such as a tuba, and near the base, it responds to higher pitches, such as a bell or a whistle. The nerve fibers from each region along the length of the organ of Corti lead to slightly different areas in the brain. The pitch sensation we experience depends on which region of the basilar membrane vibrates and which area of the brain is stimulated.

Volume is a function of the amplitude of sound waves. Loud noises cause the fluid within the vestibular canal to exert more pressure and the basilar membrane to vibrate to a greater extent. The resulting increased stimulation is interpreted by the brain as volume. It is believed that the brain interprets the tone of a sound based on the distribution of the hair cells stimulated.

The mechanoreceptors for sound are hair cells on the basilar membrane (the organ of Corti). When the basilar membrane vibrates, the stereocilia of the hair cells bend, and nerve impulses are transmitted to the brain.

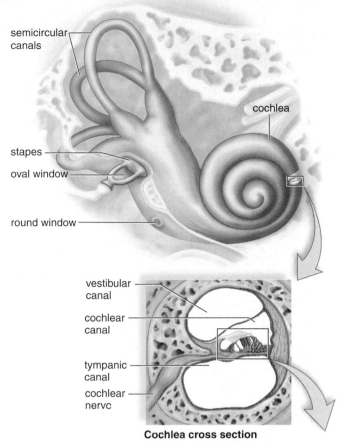

Cochlea cross section

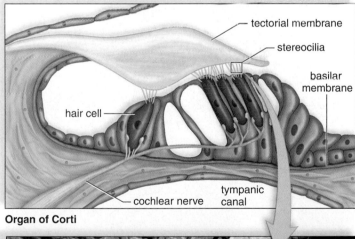

Organ of Corti

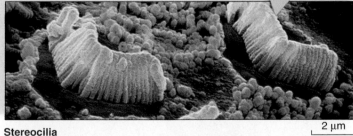

Stereocilia

2 μm

FIGURE 40.11 Mechanoreceptors for hearing.
The organ of Corti is located within the cochlea. In the uncoiled cochlea, note that the organ consists of hair cells resting on the basilar membrane, with the tectorial membrane above. Pressure waves move from the vestibular canal to the tympanic canal, causing the basilar membrane to vibrate. This causes the stereocilia (or at least a portion of the more than 20,000 hair cells) embedded in the tectorial membrane to bend. Nerve impulses traveling in the cochlear nerve result in hearing.

Sense of Balance

Mechanoreceptors in the semicircular canals detect rotational and/or angular movement of the head **(rotational equilibrium)**, while mechanoreceptors in the utricle and saccule detect straight-line movement of the head in any direction **(gravitational equilibrium)** (Table 40.2 and Fig. 40.12).

Rotational Equilibrium

Rotational equilibrium involves the semicircular canals, which are arranged so that there is one in each dimension of space. The base of each of the three canals, called the ampulla, is slightly enlarged. Little hair cells, whose stereocilia are embedded within a gelatinous material called a cupula, are found within the ampullae. Because there are three semicircular ca-

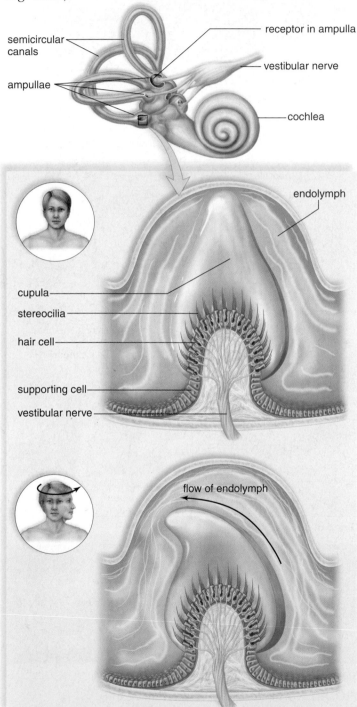

a. Rotational equilibrium: receptors in ampullae of semicircular canal

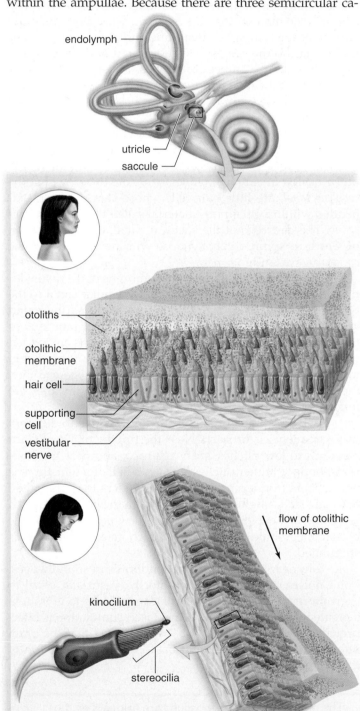

b. Gravitational equilibrium: receptors in utricle and saccule of vestibule

FIGURE 40.12 Mechanoreceptors for equilibrium.

a. Rotational equilibrium. The ampullae of the semicircular canals contain hair cells with stereocilia embedded in a cupula. When the head rotates, the cupula is displaced, bending the stereocilia. Thereafter, nerve impulses travel in the vestibular nerve to the brain. **b.** Gravitational equilibrium. The utricle and the saccule contain hair cells with stereocilia embedded in an otolithic membrane. When the head bends, otoliths are displaced, causing the membrane to sag and the stereocilia to bend. If the stereocilia bend toward the kinocilium, the longest of the stereocilia, nerve impulses increase in the vestibular nerve. If the stereocilia bend away from the kinocilium, nerve impulses decrease in the vestibular nerve. This difference tells the brain in which direction the head moved.

TABLE 40.2

Functions of the Parts of the Ear

Outer Ear	Pinna collects sound waves; auditory canal filters air.
Middle Ear	Tympanic membrane and ossicles amplify sound waves; auditory tube equalizes air pressure.
Inner Ear	Cochlea transmits pressure waves that cause the organ of Corti to generate nerve impulses, resulting in hearing. Semicircular canals function in balance (rotational equilibrium). Utricle and saccule function in balance (gravitational equilibrium).

nals, each ampulla responds to head movement in a different plane of space. As fluid (endolymph) within a semicircular canal flows over and displaces a cupula, the stereocilia of the hair cells bend, and the pattern of impulses carried by the vestibular nerve to the brain changes. Continuous movement of fluid in the semicircular canals causes one form of motion sickness.

Vertigo is dizziness and a sensation of rotation. It is possible to simulate a feeling of vertigo by spinning rapidly and stopping suddenly. When the eyes are rapidly jerked back to a midline position, the person feels like the room is spinning. This shows that the eyes are also involved in our sense of equilibrium.

Gravitational Equilibrium

Gravitational equilibrium depends on the **utricle** and **saccule,** two membranous sacs located in the vestibule. Both of these sacs contain little hair cells, whose stereocilia are embedded within a gelatinous material called an otolithic membrane. Calcium carbonate ($CaCO_3$) granules, or **otoliths,** rest on this membrane. The utricle is especially sensitive to horizontal (back-forth) movements of the head, while the saccule responds best to vertical (up-down) movements.

When the head is still, the otoliths in the utricle and the saccule rest on the otolithic membrane above the hair cells. When the head moves in a straight line, the otoliths are displaced and the otolithic membrane sags, bending the stereocilia of the hair cells beneath. If the stereocilia move toward the largest stereocilium, called the kinocilium, nerve impulses increase in the vestibular nerve. If the stereocilia move away from the kinocilium, nerve impulses decrease in the vestibular nerve. If you are upside down, nerve impulses in the vestibular nerve cease. These data tell the brain the direction of the movement of the head.

Movement of a cupula within the semicircular canals contributes to the sense of rotational equilibrium. Movement of the otolithic membrane within the utricle and the saccule accounts for gravitational equilibrium.

Sensory Receptors in Other Animals

The **lateral line** system of fishes guides them in their movements and in locating other fish, including predators and prey and mates. The system detects water currents and pressure waves from nearby objects in the same manner as the sensory receptors in the human ear. In bony fishes, the sensory receptors are located within a canal that has openings to the outside. A lateral line receptor is a collection of hair cells with cilia embedded in a gelatinous cupula. When the cupula bends due to pressure waves, the hair cells initiate nerve impulses.

Gravitational equilibrium organs, called statocysts, are found in cnidarians, molluscs, and crustaceans, which are arthropods. These organs give information only about the position of the head; they are not involved in the sensation of movement (Fig. 40.13). When the head stops moving, a small particle called a statolith stimulates the cilia of the closest hair cells, and these cilia generate impulses, indicating the position of the head.

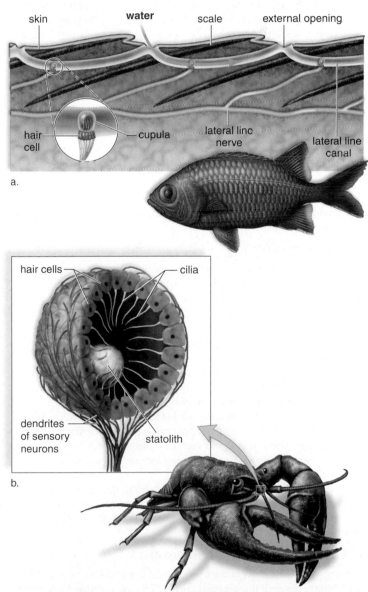

FIGURE 40.13 Sensory receptors in other animals.
a. Hairs located within cupulas of a lateral line detect wave vibrations and currents helping to guide fish movements in order to locate predators and prey and mates. **b.** Within a statocysts, a small particle (the statolith) comes to rest on hair cells and tells a crustacean the position of its head.

CONNECTING THE CONCEPTS

An animal's information exchange with the internal and external environment is dependent on just a few types of sensory receptors. We have examined chemoreceptors, such as taste cells and olfactory cells; photoreceptors, such as eyes; and mechanoreceptors, such as the hair cells for hearing and balance. The senses are not equally developed in all animals. Male moths have chemoreceptors on the filaments of their antennae to detect minute amounts of an airborne sex attractant released by a female. This is certainly a more efficient method than searching for a mate by sight.

Birds that live in forested areas signal that a territory is occupied by singing because it is difficult to see a bird in a tree, as most birders know. On the other hand, hawks have such a keen sense of sight that they are able to locate a small mouse far below them. Insectivorous bats have an unusual adaptation for finding prey in the dark. They send out a series of sound pulses and listen for the echoes that come back. The time it takes for an echo to return indicates the location of an insect. A unique adaptation is found among the so-called electric fishes of Africa and Australia. They have electroreceptors that can detect disturbances in an electrical current they emit into the water. These disturbances indicate the location of obstacles and prey.

Animals that migrate use various senses to find their way. Salmon hatch in a freshwater stream but drift to the ocean as larvae. By the end of the third or fourth year, they migrate back to where they were hatched. Like other migrating animals, they apparently can use the sun as a compass to find their way back to the vicinity of their home river, but then salmon switch to a sense of smell to find the exact location of their hatching. Perhaps the odor of the plants and soil in this stream was imprinted on the nervous system of the larval fish.

Through the evolutionary process, animals tend to rely on those stimuli and senses that are adaptive to their particular environment and way of life. In all cases, sensory receptors generate nerve impulses that travel to the brain, where sensation occurs. In mammals, and particularly human beings, integration of the data received from various sensory receptors results in perception of events occurring in the external environment.

Summary

40.1 CHEMICAL SENSES

Chemoreception is found universally in animals and is, therefore, believed to be the most primitive sense.

Human olfactory cells and taste buds are chemoreceptors. They are sensitive to chemicals in water and air.

40.2 SENSE OF VISION

Vision is dependent on the eye, the optic nerves, and the visual areas of the cerebral cortex. The eye has three layers. The outer layer, the sclera, can be seen as the white of the eye; it also becomes the transparent bulge in the front of the eye called the cornea. The middle pigmented layer, called the choroid, absorbs stray light rays. The rod cells (sensory receptors for dim light) and the cone cells (sensory receptors for bright light and color) are located in the retina, the inner layer of the eyeball. The cornea, the humors, and especially the lens bring the light rays to focus on the retina. To see a close object, accommodation occurs as the lens rounds up.

When light strikes rhodopsin within the membranous disks of rod cells, rhodopsin splits into opsin and retinal. A cascade of reactions leads to the closing of ion channels in a rod cell's plasma membrane. Inhibitory transmitter molecules are no longer released, and nerve impulses are carried in the optic nerve to the brain.

Integration occurs in the retina, which is composed of three layers of cells: the rod and cone layer, the bipolar cell layer, and the ganglion cell layer. Integration also occurs in the visual areas of the cerebral cortex.

40.3 SENSES OF HEARING AND BALANCE

Hearing in humans is dependent on the ear, the cochlear nerve, and the auditory areas of the cerebral cortex. The ear is divided into three parts. The outer ear consists of the pinna and the auditory canal, which direct sound waves to the middle ear. The middle ear begins with the tympanic membrane and contains the ossicles (malleus, incus, and stapes). The malleus is attached to the tympanic membrane, and the stapes is attached to the oval window, which is covered by membrane. The inner ear contains the cochlea and the semicircular canals, plus the utricle and saccule.

Hearing begins when the outer and middle portions of the ear convey and amplify the sound waves that strike the oval window. Its vibrations set up pr essure waves within the cochlea, which contains the organ of Corti, consisting of hair cells whose stereocilia are embedded within the tectorial membrane. When the stereocilia of the hair cells bend, nerve impulses begin in the cochlear nerve and are carried to the brain.

The ear also contains receptors for our sense of equilibrium. Rotational equilibrium is dependent on the stimulation of hair cells within the ampullae of the semicircular canals. Gravitational equilibrium relies on the stimulation of hair cells within the utricle and the saccule.

Reviewing the Chapter

1. Discuss the structure and the function of human chemoreceptors.   722–23
2. In general, how does the eye in arthropods differ from that in humans?   724
3. What types of animals have eyes that are constructed like the human eye?   724
4. Name the parts of the human eye, and give a function for each part.   724–25
5. Explain focusing and accommodation in terms of the anatomy of the human eye.   726–27
6. Contrast the location and the function of rod cells to those of cone cells.   727
7. Explain the process of integration in the retina and the brain.   728
8. Describe the structure of the human ear.   730
9. Describe how we hear.   731
10. Describe the role of the semicircular canals, utricle, and saccule in balance.   732–33

Testing Yourself

Choose the best answer for each question.

1. A sensory receptor
   a. is the first portion of a reflex arc.
   b. initiates nerve impulses.
   c. responds to only one type of stimulus.
   d. is associated with a sensory neuron.
   e. All of these are correct.

2. Which of these gives the correct path for light rays entering the human eye?
   a. sclera, retina, choroid, lens, cornea
   b. fovea centralis, pupil, aqueous humor, lens
   c. cornea, pupil, lens, vitreous humor, retina
   d. optic nerve, sclera, choroid, retina, humors
   e. All of these are correct.

3. Which gives an incorrect function for the structure?
   a. lens—focusing
   b. iris—regulation of amount of light
   c. choroid—location of cones
   d. sclera—protection
   e. fovea centralis—acute vision

4. Retinal is
   a. a derivative of vitamin A.
   b. sensitive to light energy.
   c. a part of rhodopsin.
   d. found in rod cells.
   e. All of these are correct.

5. Which association is incorrect?
   a. taste buds—humans
   b. compound eye—arthropods
   c. camera-type eye—squid
   d. statocysts—sea stars
   e. chemoreceptors—planarians

6. Which one of these wouldn't you mention if you were tracing the path of sound vibrations?
   a. auditory canal
   b. tympanic membrane
   c. semicircular canals
   d. cochlea
   e. ossicles

7. Which one of these correctly describes the location of the organ of Corti?
   a. between the tympanic membrane and the oval window in the inner ear
   b. in the utricle and saccule within the vestibule
   c. between the tectorial membrane and the basilar membrane in the cochlear canal
   d. between the outer and inner ear within the semicircular canals

8. Which of these pairs is mismatched?
   a. semicircular canals—inner ear
   b. utricle and saccule—outer ear
   c. auditory canal—outer ear
   d. ossicles—middle ear
   e. cochlear nerve—inner ear

9. Both olfactory receptors and sound receptors have cilia, and they both
   a. are chemoreceptors.
   b. are a part of the brain.
   c. are mechanoreceptors.
   d. initiate nerve impulses.
   e. All of these are correct.

10. Which of these is an incorrect difference between olfactory receptors and equilibrium receptors?

Olfactory receptors	Equilibrium receptors
a. located in nasal cavities	located in the inner ear
b. chemoreceptors	mechanoreceptors
c. respond to molecules in air	respond to movements of the body
d. communicate with brain via a tract	communicate with brain via vestibular nerve
e. All of these contrasts are correct.	

11. Stimulation of hair cells in the semicircular canals results from the movement of
    a. endolymph.
    b. aqueous humor.
    c. basilar membrane.
    d. otoliths.

12. To focus on objects that are close to the viewer,
    a. the suspensory ligaments must be pulled tight.
    b. the lens needs to become more rounded.
    c. the ciliary muscle will be relaxed.
    d. the image must focus on the area of the optic nerve.

13. Which abnormality of the eye is incorrectly matched?
    a. astigmatism—either the lens or cornea is not even
    b. farsightedness—eyeball is shorter than usual
    c. nearsightedness—image focuses behind the retina
    d. color blindness—genetic disorder in which certain types of cones may be missing

14. Which of these would allow you to know that you were upside down, even if you were in total darkness?
    a. utricle and saccule
    b. cochlea
    c. semicircular canals
    d. tectorial membrane

15. The thin, darkly pigmented layer that underlies most of the sclera is
    a. the conjunctiva.
    b. the cornea.
    c. the retina.
    d. the choroid.

16. Adjustment of the lens to focus on objects close to the viewer is called
    a. convergence.
    b. accommodation.
    c. focusing.
    d. constriction.

17. The middle ear is separated from the inner ear by
    a. the oval window.
    b. the tympanic membrane.
    c. the round window.
    d. Both a and c are correct.

18. Label this diagram of the human eye. State a function for each structure labeled.

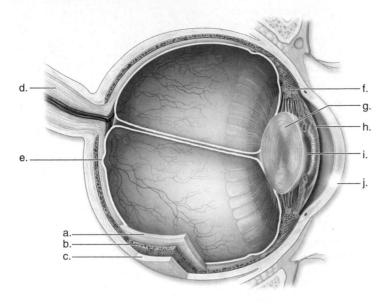

Thinking Scientifically

1. The density of taste buds on the tongue can vary. Some obese individuals have a lower density of taste buds than those who are not obese. Assume that taste perception is related to taste bud density. If so, what hypothesis would you test to see if there is a relationship between taste bud density and obesity?

2. A man who has spent many years serving on submarines complains of hearing loss, particularly the inability to hear high tones. When a submarine submerges, the inside air pressure intensifies. What hypothesis might explain hearing loss in this individual?

Bioethical Issue: Cataract Prevention

A cataract is a cloudiness of the lens that occurs in 50% of people between the ages 65 and 74, and in 70% of those age 75 or older. The extent of visual impairment depends on the size and density of the cataract, and where it is located in the lens. A dense, centrally placed cataract causes severe blurring of vision.

Are cataracts preventable? For most people, the answer is yes. These factors have been identified as contributing to the chances of having a cataract:

- Smoking 20 or more cigarettes a day doubles the risk of cataracts in men. Women have to smoke more than 35 cigarettes a day to increase their chances of cataracts.
- Exposure to the ultraviolet radiation in sunlight can more than double the risk of a cataract. In addition, this relationship is dose-dependent—the more sunlight, the higher the risk.
- As many as one-third of cataracts may be caused by being overweight. Diet, rather than exercise, seems to reduce cataract formation, perhaps through lower blood sugar levels or improved antioxidant properties of the blood.

It's clear, then, that our own behavior contributes to the occurrence of cataracts for most of us. Should we be responsible in our actions and take all possible steps to prevent developing a cataract, such as not smoking, wearing sunglasses and a wide-brim hat, and watching our weight? Or should we simply rely on medical science to restore our eyesight? Most cataract operations today are performed on an outpatient basis with minimal postoperative discomfort and with a high expectation of restoration of sight. Perhaps it's better, though, to take all possible steps to prevent the occurrence of cataracts, just in case our experience is atypical.

Understanding the Terms

blind spot 728	organ of Corti 731
camera-type eye 724	ossicle 730
chemoreceptor 722	otolith 733
choroid 725	outer ear 730
ciliary muscle 726	panoramic vision 724
cochlea 730	photoreceptor 724
compound eye 724	pupil 725
cone cell 727	retina 725
conjunctiva 724	rhodopsin 727
cornea 724	rod cell 727
fovea centralis 725	rotational equilibrium 732
gravitational	saccule 733
equilibrium 732	sclera 724
inner ear 730	semicircular canal 730
iris 725	stereoscopic vision 724
lateral line 733	taste bud 722
lens 725	tympanic membrane 730
mechanoreceptor 730	utricle 733
middle ear 730	vestibule 730
olfactory cell 723	visual accomodation 726

Match the terms to these definitions:

a. _____ Photoreceptor, composed of many independent units, which is typical of arthropods.

b. _____ Inner layer of the eyeball containing the photoreceptors—rod cells and cone cells.

c. _____ Outer, white, fibrous layer of the eye that surrounds the eye except for the transparent cornea.

d. _____ Receptor that is sensitive to chemical stimulation—for example, receptors for taste and smell.

e. _____ Specialized region of the cochlea containing the hair cells for sound detection and discrimination.

ARIS, the *Biology* Website

ARIS, the website for *Biology*, provides a wealth of information organized and integrated by chapter. You will find practice quizzes, interactive activities, labeling exercises, flashcards, and much more that will complement your learning and understanding of general biology.

www.mhhe.com/maderbiology9

41

LOCOMOTION AND SUPPORT SYSTEMS

Not all animals have muscles and bones, but they all use contractile fibers for locomotion. In vertebrates, the muscular system and the skeletal system work together to provide movement, whether it be a pitcher throwing a baseball, a fish swimming, or an eagle flying. Most animals also have a nervous system, which integrates data received from sensory receptors and coordinates muscular activity so that movement achieves goals such as feeding, escaping enemies, reproducing, or simply playing.

To create movement, muscle contraction must be directed against some sort of medium. In planarians, hydras, and earthworms, muscles push against body fluids located inside either a gastrovascular cavity or a coelom. In vertebrates, the muscles are attached to a bony endoskeleton. Both the skeletal system and the muscular system contribute to homeostasis. Aside from giving the body shape and protecting internal organs, the skeleton serves as a storage area for inorganic calcium and produces blood cells. The skeleton also supports the body and its organs against the pull of gravity. Aside from moving the body and its parts, the skeletal muscles give off heat, which warms the body.

Pitching requires coordination between the nervous and support systems.

41.1 DIVERSITY OF SKELETONS

Skeletons serve as support systems for animals, providing rigidity, protection, and surfaces for muscle attachment. Several different kinds of skeletons occur in the animal kingdom. Cnidarians, flatworms, roundworms, and annelids have a hydrostatic skeleton. Typically, molluscs and arthropods have an **exoskeleton** (external skeleton) composed of calcium carbonate or chitin, respectively. Sponges, echinoderms, and vertebrates possess an **endoskeleton** (internal skeleton). The endoskeleton of sponges consists of mineralized spicules and spongin. In echinoderms, the endoskeleton is composed of calcareous plates and in vertebrates, the endoskeleton is comprised of cartilage, bone, or both.

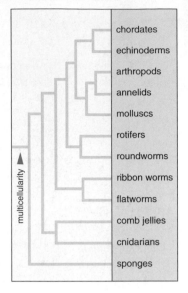

(multicellularity)

chordates
echinoderms
arthropods
annelids
molluscs
rotifers
roundworms
ribbon worms
flatworms
comb jellies
cnidarians
sponges

Hydrostatic Skeleton

In animals that lack a hard skeleton, a fluid-filled gastrovascular cavity or a fluid-filled coelom can act as a hydrostatic skeleton. A **hydrostatic skeleton** [Gk. *hydrias*, water, and *stasis*, standing] offers support and resistance to the contraction of muscles so that mobility results. As analogies, consider that a garden hose stiffens when filled with water, and that a water-filled balloon changes shape when squeezed at one end. Similarly, an animal with a hydrostatic skeleton can change shape and perform a variety of movements.

Hydras and planarians use their fluid-filled gastrovascular cavity as a hydrostatic skeleton. When muscle fibers at the base of epidermal cells in a hydra contract, the body or tentacles shorten rapidly. Planarians usually glide over a substrate with the help of muscular contractions that control the body wall and cilia. Roundworms have a fluid-filled pseudocoelom and move in a whiplike manner when their longitudinal muscles contract. Annelids, such as earthworms, are segmented and have septa that divide the coelom into compartments (Fig. 41.1). Each segment has its own set of longitudinal and circular muscles and its own nerve supply, so each segment or group of segments may function independently. When circular muscles contract, the segments become thinner and elongate. When longitudinal muscles contract, the segments become thicker and shorten. By alternating circular muscle contraction and longitudinal muscle contraction and by using its setae to hold its position during contractions, the animal moves forward.

Muscular Hydrostat

Even animals that have an exoskeleton or an endoskeleton move selected body parts by means of muscular hydrostats, meaning that fluid contained within certain muscle fibers assists movement of that part. Muscular hydrostats are used by clams to extend their muscular foot and by sea stars to extend their tube feet. Spiders depend on them to move their legs, and moths depend on them to extend their proboscis. In vertebrates, movement of an elephant's trunk involves a muscular hydrostat that allows the trunk to reach as high as 23 feet, or pick up a morsel of food, or pull down a tree.

Exoskeletons and Endoskeletons

Molluscs, arthropods, and vertebrates have rigid skeletons. The exoskeleton of molluscs and arthropods protects and supports these animals and provides a location for muscle attachment. Strength of an exoskeleton can be improved by increasing its thickness and weight, but this leaves less room for internal organs.

In molluscs, such as snails and clams, a thick and nonmobile calcium carbonate shell is primarily used for protection against the environment and predators. A mollusc's shell can grow as the animal grows.

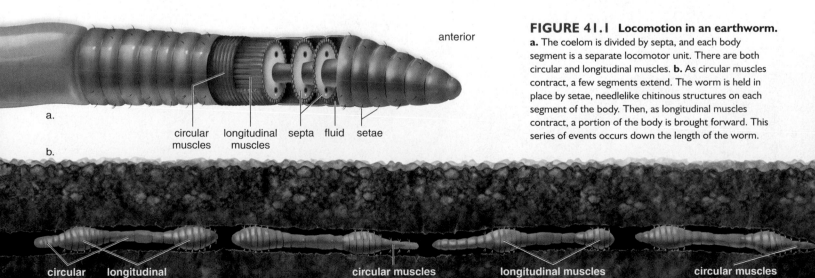

anterior

circular muscles   longitudinal muscles   septa   fluid   setae

a.
b.

FIGURE 41.1 Locomotion in an earthworm.
a. The coelom is divided by septa, and each body segment is a separate locomotor unit. There are both circular and longitudinal muscles. **b.** As circular muscles contract, a few segments extend. The worm is held in place by setae, needlelike chitinous structures on each segment of the body. Then, as longitudinal muscles contract, a portion of the body is brought forward. This series of events occurs down the length of the worm.

circular muscles contracted   longitudinal muscles contracted   circular muscles contract, and anterior end moves forward   longitudinal muscles contract, and segments catch up   circular muscles contract, and anterior end moves forward

FIGURE 41.2 Exoskeleton.
Exoskeletons support muscle contraction and prevent drying out. The chitinous exoskeleton of an arthropod is shed as the animal molts; until the new skeleton dries and hardens, the animal is vulnerable to predators, and muscle contractions may not translate into body movements. In this photo, a dog-day cicada, *Tibicen*, has just finished molting.

The exoskeleton of arthropods, such as insects and crustaceans, is composed of chitin, a strong, flexible nitrogenous polysaccharide. Their exoskeleton protects them against wear and tear, predators, and drying out—an important feature for arthropods that live on land. The jointed and movable exoskeleton of arthropods is particularly suitable for terrestrial life in another way. The jointed and movable appendages allow flexible movements. To grow, however, arthropods must molt to rid themselves of an exoskeleton that has become too small (Fig. 41.2). When molting, arthropods are vulnerable to predators.

Both echinoderms and vertebrates have an endoskeleton. The skeleton of echinoderms consists of spicules and plates of calcium carbonate embedded in the living tissue of the body wall. In contrast, the vertebrate endoskeleton is living tissue. Sharks and rays have skeletons composed only of cartilage. Other vertebrates, such as bony fishes, amphibians, reptiles, birds, and mammals, have endoskeletons composed of bone and cartilage. The advantages of the jointed vertebrate endoskeleton are listed in Figure 41.3. An endoskeleton grows with the animal, and molting is not required. It supports the weight of a large animal without limiting the space for internal organs. An endoskeleton also offers protection to vital internal organs, but it is protected by the soft tissues around it. Injuries to soft tissue are easier to repair than injuries to a hard skeleton. The vertebrate endoskeleton is also jointed, allowing for complex movements such as swimming, jumping, flying, and running.

Arthropods have a flexible exoskeleton and vertebrates have a flexible endoskeleton suitable to life on land. The exoskeleton of arthropods requires molting, but the endoskeleton of vertebrates grows with the animal.

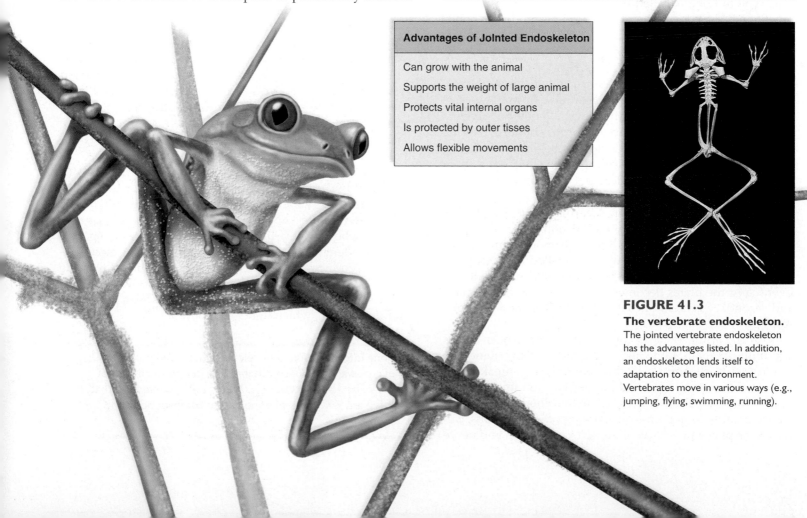

Advantages of Jointed Endoskeleton
Can grow with the animal
Supports the weight of large animal
Protects vital internal organs
Is protected by outer tisses
Allows flexible movements

FIGURE 41.3
The vertebrate endoskeleton.
The jointed vertebrate endoskeleton has the advantages listed. In addition, an endoskeleton lends itself to adaptation to the environment. Vertebrates move in various ways (e.g., jumping, flying, swimming, running).

41.2 THE HUMAN SKELETAL SYSTEM

The human skeletal system has many functions that contribute to homeostasis. The skeleton protects vital internal organs, such as the brain, heart, and lungs. The bones of the skeleton are sites of muscle attachment, making movement possible. The rigid skeleton supports the body and aids in locomotion. The bones serve as an important storage reservoir for ions, such as calcium and phosphorous. Additionally, blood cells and other blood elements are produced within the red bone marrow of the skull, ribs, sternum, pelvis, and long bones.

The skeleton supports and protects the body while at the same time permitting flexible movement. All bones serve as a storehouse for calcium and phosphate ions, and certain ones produce blood cells and platelets.

Bone Growth and Renewal

Most of the bones of the human skeleton are composed of cartilage during prenatal development. Because the cartilaginous structures are shaped like the future bones, they provide "models" of these bones. The cartilaginous models are converted to bones when calcium salts are deposited in the matrix, first by the cartilage cells and later by bone-forming cells called **osteoblasts** [Gk. *osteon*, bone, and *blastos*, bud]. This type of bone is called endochondral bone, or replacement bone. The conversion of cartilaginous models to bones is called endochondral ossification.

There are also examples of ossification that have no previous cartilaginous model. This type of ossification occurs in the dermis and forms bones called dermal bones. Several dermal bones include the mandible (lower jaw), certain bones of the skull, and the clavicle (collarbone). During intramembranous ossification, fibrous connective tissue membranes give support as ossification begins.

During endochondral ossification of a long bone, there is at first only a primary ossification center in the middle of the cartilaginous model. Later, secondary ossification centers form at the ends of the model. A cartilaginous *growth plate* remains between the primary ossification center and each secondary center. As long as these plates remain, growth is possible. The rate of growth is controlled by hormones, particularly growth hormone (GH) and the sex hormones. Eventually, the plates become ossified, causing the primary and secondary centers of ossification to fuse, and the bone stops growing.

In the adult, bone is continually being broken down and built up again. Bone-absorbing cells, called **osteoclasts** [Gk. *osteon*, bone, and *klastos*, broken in pieces], break down bone, remove worn cells, and deposit calcium in the blood. In this way, osteoclasts help maintain the blood calcium level and contribute to homeostasis. Among other functions, calcium ions play a major role in muscle contraction and nerve conduction. The blood calcium level is closely regulated by the antagonistic hormones parathyroid hormone (PTH) and calcitonin. PTH promotes the activity of osteoclasts, and calcitonin inhibits their activity to keep the blood calcium level within normal limits.

Assuming that the blood calcium level is normal, bone destruction caused by the work of osteoclasts is repaired by osteoblasts. As they form bone, osteoblasts take calcium from the blood. Eventually, some of these cells get caught in the matrix (nonliving material) they secrete and are converted to **osteocytes** [Gk. *osteon*, bone, and *kytos*, cell], the cells found within the lacunae of osteons. Adults are thought to require more calcium in the diet than do children to promote the work of osteoblasts.

Through this process of remodeling, old bone tissue is replaced by new bone tissue. Osteoclasts and osteoblasts work together to heal broken bones by breaking down and building bone at the site of the damage. Therefore, the thickness of bones can change, depending on exercise and hormone balances. As discussed in the Health Focus on page 746, a thinning of the bones called osteoporosis can occur as we age if proper precautions are not taken.

Anatomy of a Long Bone

A long bone, such as the humerus, illustrates principles of bone anatomy. When the bone is split open, as in Figure 41.4, the longitudinal section shows that it is not solid but has a cavity called the medullary cavity bounded at the sides by compact bone and at the ends by spongy bone. Beyond the spongy bone, there is a thin shell of compact bone and finally a layer of hyaline cartilage. The cavity of a long bone usually contains yellow bone marrow, which is a fat-storage tissue.

Compact bone contains many osteons (Haversian systems), where osteocytes lie in tiny chambers called lacunae. The lacunae are arranged in concentric circles around central canals that contain blood vessels and nerves. The lacunae are separated by a matrix of collagen fibers and mineral deposits, primarily calcium and phosphorous salts.

Spongy bone has numerous bony bars and plates separated by irregular spaces. Although lighter than compact bone, spongy bone is still designed for strength. Just as braces are used for support in buildings, the solid portions of spongy bone follow lines of stress. The spaces in spongy bone are often filled with **red bone marrow,** a specialized tissue that produces blood cells. This is an additional way the skeletal system assists homeostasis. As you know, red blood cells transport oxygen, and white blood cells are a part of the immune system, which fights infection.

Bone is formed during development, but the process of renewal continues even in adults due to the action of osteoclasts and osteoblasts. A long bone exemplifies that there are two types of bone tissue: compact bone and spongy bone.

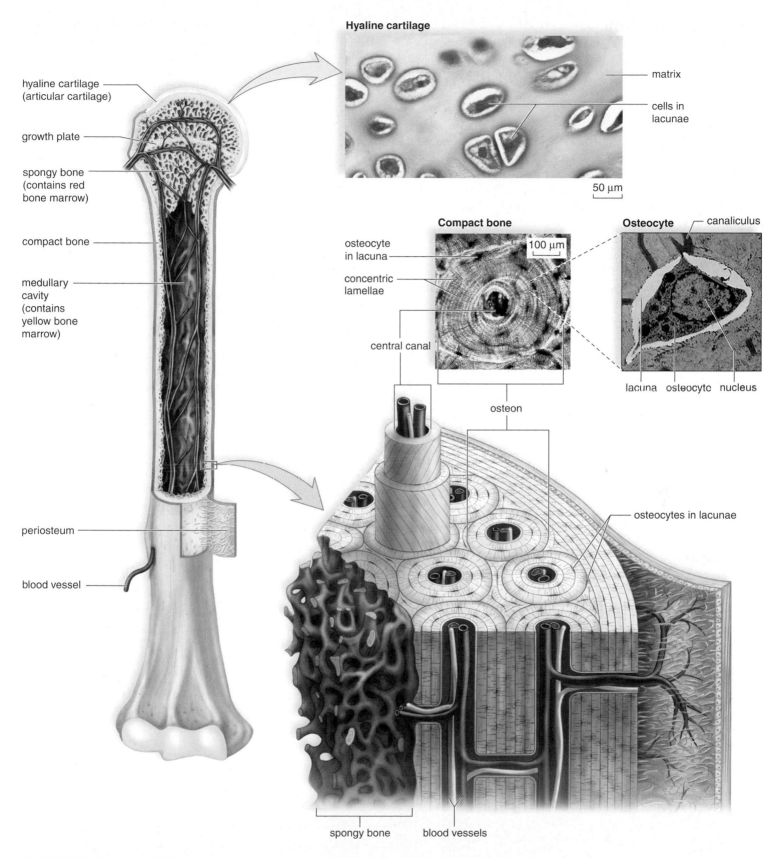

Hyaline cartilage

matrix

cells in lacunae

50 μm

hyaline cartilage (articular cartilage)

growth plate

spongy bone (contains red bone marrow)

compact bone

medullary cavity (contains yellow bone marrow)

periosteum

blood vessel

Compact bone

Osteocyte

canaliculus

osteocyte in lacuna

100 μm

concentric lamellae

central canal

lacuna   osteocyte   nucleus

osteon

osteocytes in lacunae

spongy bone   blood vessels

FIGURE 41.4 Anatomy of a long bone.

Left: A long bone is encased by fibrous membrane (periosteum) except where it is covered at the ends by hyaline cartilage (see micrograph). Spongy bone located beneath the cartilage may contain red bone marrow. The central shaft contains yellow bone marrow and is bordered by compact bone, which is shown in the enlargement and micrograph *(right).*

The Axial Skeleton

The **axial skeleton** [L. *axis*, axis, hinge; Gk. *skeleton*, dried body] lies in the midline of the body and consists of the skull, the vertebral column, the thoracic cage, the sacrum, and the coccyx (blue labels in Fig. 41.5). A total of 80 bones make up the axial skeleton.

The Skull

The skull, which protects the brain, is formed by the cranium and the facial bones (Fig. 41.6). In newborns, certain bones of the cranium are joined by membranous regions called **fontanels** (or "soft spots"), all of which usually close and become **sutures** by the age of two years. The bones of the cranium contain the sinuses [L. *sinus*, hollow], air spaces lined by mucous membrane that reduce the weight of the skull and give a resonant sound to the voice. Two sinuses, called the mastoid sinuses, drain into the middle ear. Mastoiditis, a condition that can lead to deafness, is an inflammation of these sinuses.

The major bones of the cranium have the same names as the lobes of the brain. On the top of the cranium, the frontal bone forms the forehead, and the parietal bones extend to the sides. Below the much larger parietal bones, each temporal bone has an opening that leads to the middle ear. In the rear of the skull, the occipital bone curves to form the base of the skull. At the base of the skull, the spinal cord passes upwards through a large opening called the **foramen magnum** [L. *foramen*, hole, and *magnus*, great, large] and becomes the brain stem.

The temporal and frontal bones are cranial bones that contribute to the face. The sphenoid bones account for the flattened areas on each side of the forehead, which we call the temples. The frontal bone not only forms the forehead, but it also has supraorbital ridges where the eyebrows are located. Glasses sit where the frontal bone joins the nasal bones.

The most prominent of the facial bones are the mandible [L. *mandibula*, jaw], the maxillae, the zygomatic bones, and the nasal bones. The mandible, or lower jaw, is the only freely movable portion of the skull (Fig. 41.6), and its action permits us to chew our food. It also forms the "chin." Tooth sockets are located on the mandible and on the maxillae, which form the upper jaw and a portion of the hard palate. The zygomatic bones are the cheekbone prominences, and the nasal bones form the bridge of the nose. Other bones make up the nasal septum, which divides the nose cavity into two regions.

Whereas the ears are formed only by cartilage and not by bone, the nose is a mixture of bones, cartilage, and connective tissues. The lips and cheeks have a core of skeletal muscle.

The skull, whose major bones are named after the lobes of the brain, is formed by the cranium and the facial bones. The most prominent facial bones are the mandible, the maxillae, the zygomatic bones, and the nasal bones.

FIGURE 41.5

The human skeleton.
a. Anterior view.
b. Posterior view.
The bones of the axial skeleton are in blue and the rest is the appendicular skeleton.

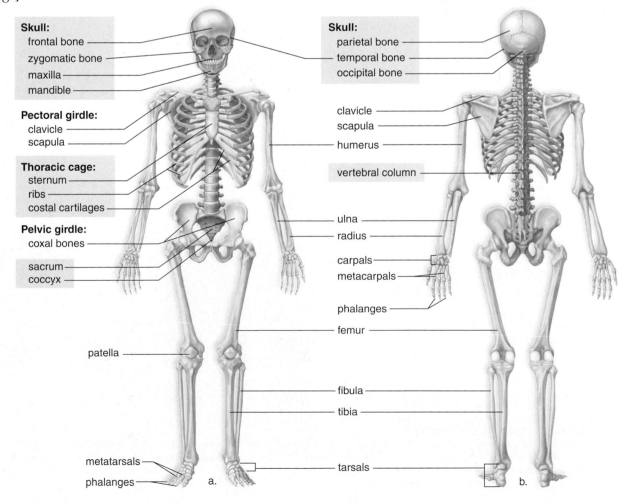

Skull:
  frontal bone
  zygomatic bone
  maxilla
  mandible

Pectoral girdle:
  clavicle
  scapula

Thoracic cage:
  sternum
  ribs
  costal cartilages

Pelvic girdle:
  coxal bones

  sacrum
  coccyx

patella

metatarsals
phalanges   a.

Skull:
  parietal bone
  temporal bone
  occipital bone

clavicle
scapula
humerus

vertebral column

ulna
radius

carpals
metacarpals

phalanges

femur

fibula
tibia

tarsals   b.

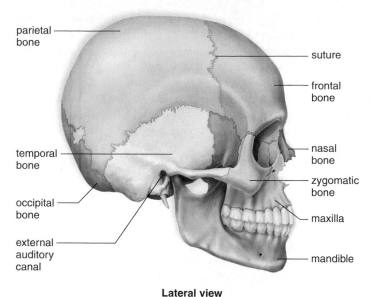

parietal bone

suture

frontal bone

temporal bone

nasal bone

zygomatic bone

occipital bone

maxilla

external auditory canal

mandible

Lateral view

frontal bone

parietal bone

temporal bone

nasal bone

zygomatic bone

maxilla

mandible

Frontal view

FIGURE 41.6
The skull.
The skull consists of the cranium and the facial bones. The frontal bone is the forehead; the zygomatic arches form the cheekbones, and the maxillae form the upper jaw. The mandible has a projection we call the chin.

The Vertebral Column and Rib Cage

The **vertebral column** [L. *vertebra*, bones of backbone] supports the head and trunk and protects the spinal cord and the roots of the spinal nerves. It is a longitudinal axis that serves either directly or indirectly as an anchor for all the other bones of the skeleton.

Twenty-four vertebrae make up the vertebral column (see Fig. 41.5). The 7 cervical vertebrae are located in the neck, and the 12 thoracic vertebrae are in the thorax. The 5 lumbar vertebrae are found in the small of the back. The 5 sacral vertebrae are fused to form a single sacrum. The coccyx, or tailbone, is also composed of several fused vertebrae. Normally, the vertebral column has four curvatures that provide more resilience and strength for an upright posture than could a straight column. Scoliosis is an abnormal lateral (sideways) curvature of the spine. Another well-known abnormal curvature results in a hunchback (kyphosis), and still another results in a swayback (lordosis).

Intervertebral disks, composed of fibrocartilage between the vertebrae, act as a kind of padding. They prevent the vertebrae from grinding against one another and absorb shock caused by movements such as running, jumping, and even walking. The presence of the disks allows the vertebrae to move as we bend forward, backward, and from side to side. Unfortunately, these disks become weakened with age and can herniate and rupture. Pain results if a disk presses against the spinal cord and/or spinal nerves. The body may heal itself, or the disk can be removed surgically. If so, the vertebrae can be fused together, but this limits the flexibility of the body.

Rib Cage.   The thoracic vertebrae are a part of the rib cage. The rib cage also contains the ribs, the costal cartilages, and the sternum, or breastbone (Fig. 41.7).

There are 12 pairs of ribs. The upper 7 pairs are "true ribs" because they attach to the sternum. The lower 5 pairs do not connect directly to the sternum and are called the

"false ribs." Three pairs of false ribs attach by means of a common cartilage, and 2 pairs are "floating ribs" because they do not attach to the sternum at all.

The rib cage demonstrates how the skeleton is protective but also flexible. The rib cage protects the heart and lungs; yet it swings outward and upward upon inspiration and then downward and inward upon expiration.

The vertebral column contains the vertebrae and forms the longitudinal axis of the body. The rib cage protects the heart and lungs and assists breathing.

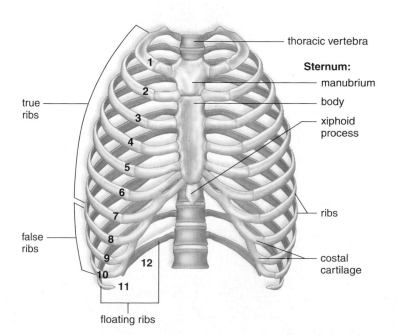

thoracic vertebra

Sternum:
manubrium

body

xiphoid process

true ribs

false ribs

ribs

costal cartilage

floating ribs

FIGURE 41.7   The rib cage.
The rib cage consists of the thoracic vertebrae, the 12 pairs of ribs, the costal cartilages, and the sternum, or breastbone.

The Appendicular Skeleton

The **appendicular skeleton** [L. *appendicula,* dim. of *appendix,* appendage] consists of the bones within the pectoral and pelvic girdles and the attached limbs (see Fig. 41.5). The pectoral (shoulder) girdle and upper limbs are specialized for flexibility, but the pelvic girdle (hipbones) and lower limbs are specialized for strength. A total of 126 bones make up the appendicular skeleton.

The Pectoral Girdle and Upper Limb

The components of the **pectoral girdle** [Gk. *pechys,* forearm] are only loosely linked together by ligaments (Fig. 41.8). Each clavicle (collarbone) connects with the sternum in front and the scapula (shoulder blade) behind, but the scapula is freely movable and held in place only by muscles. This allows it to freely follow the movements of the arm. The single long bone in the arm, the humerus, has a smoothly rounded head that fits into a socket of the scapula. The socket, however, is very shallow and much smaller than the head. Although this means that the arm can move in almost any direction, there is little stability. Therefore, this is the joint that is most apt to

dislocate. The opposite end of the humerus meets the two bones of the lower arm, the ulna and the radius, at the elbow. (The prominent bone in the elbow is the topmost part of the ulna.) When the upper limb is held so that the palm is turned frontward, the radius and ulna are about parallel to one another. When the upper limb is turned so that the palm is next to the body, the radius crosses in front of the ulna, a feature that contributes to the easy twisting motion of the forearm.

The many bones of the hand increase its flexibility. The wrist has eight carpal bones, which look like small pebbles. From these, five metacarpal bones fan out to form a framework for the palm. The metacarpal bone that leads to the thumb is placed in such a way that the thumb can reach out and touch the other digits. (Digits is a term that refers to either fingers or toes.) Beyond the metacarpals are the phalanges, the bones of the fingers and the thumb. The phalanges of the hand are long, slender, and lightweight.

The Pelvic Girdle and Lower Limb

The **pelvic girdle** [L. *pelvis,* basin] (Fig. 41.9) consists of two heavy, large coxal bones (hipbones). The coxal bones are

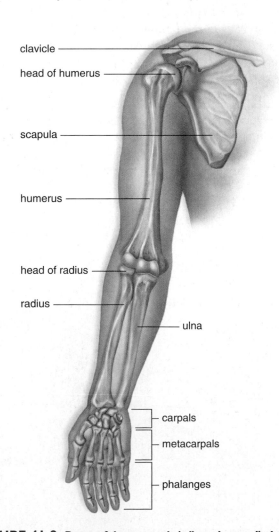

clavicle
head of humerus
scapula
humerus
head of radius
radius
ulna
carpals
metacarpals
phalanges

FIGURE 41.8 Bones of the pectoral girdle and upper limb.
The humerus is known as the "funny bone" of the elbow. The sensation upon bumping it is due to the activation of a nerve that passes across its end.

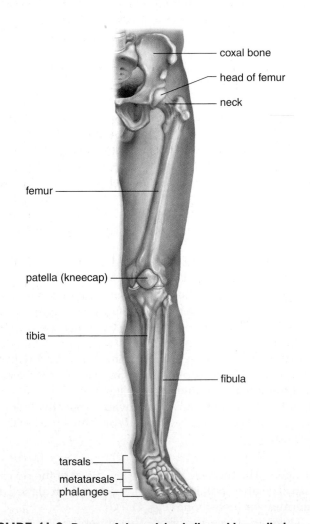

coxal bone
head of femur
neck
femur
patella (kneecap)
tibia
fibula
tarsals
metatarsals
phalanges

FIGURE 41.9 Bones of the pelvic girdle and lower limb.
The femur is our strongest bone—it withstands a pressure of 540 kg per 2.5 cm³ when we walk.

anchored to the sacrum, and together these bones form a hollow cavity called the pelvic cavity. The wider pelvic cavity in females than males accommodates childbearing. The weight of the body is transmitted through the pelvis to the lower limbs and then onto the ground. The largest bone in the body is the femur, or thighbone.

In the leg, the larger of the two bones, the tibia, has a ridge we call the shin. Both of the bones of the leg have a prominence that contributes to the ankle—the tibia on the inside of the ankle and the fibula on the outside of the ankle. Although there are seven tarsal bones in the ankle, only one receives the weight and passes it on to the heel and the ball of the foot. If you wear high-heeled shoes, the weight is thrown toward the front of your foot. The metatarsal bones participate in forming the arches of the foot. There is a longitudinal arch from the heel to the toes and a transverse arch across the foot. These provide a stable, springy base for the body. If the tissues that bind the metatarsals together become weakened, "flat feet" are apt to result. The bones of the toes are called phalanges, just as are those of the fingers, but in the foot the phalanges are stout and extremely sturdy.

The appendicular skeleton contains the bones of the pectoral and pelvic girdles and the limbs.

Classification of Joints

Bones are connected at the **joints**, which are classified as fibrous, cartilaginous, and synovial. Fibrous joints, such as the sutures that exist between the cranial bones, are immovable. Cartilaginous joints, such as those between the vertebrae, are slightly movable. The vertebrae are also separated by disks, which increase their flexibility. The two hipbones are slightly movable because they are ventrally joined by cartilage. Owing to hormonal changes, this joint becomes more flexible during late pregnancy, allowing the pelvis to expand during childbirth.

In freely movable **synovial joints,** the two bones are separated by a cavity. **Ligaments,** composed of fibrous connective tissue, bind the two bones to each other, holding them in place as they form a capsule. In a "double-jointed" individual, the ligaments are unusually loose. The joint capsule is lined by synovial membrane, which produces synovial fluid, a lubricant for the joint.

The knee is an example of a synovial joint (Fig. 41.10). In the knee, as in other freely movable joints, the bones are capped by a layer of articular cartilage. In addition, there are crescent-shaped pieces of cartilage between the bones called menisci [Gk. *meniscus*, crescent]. These give added stability, helping to support the weight placed on the knee joint. Unfortunately, athletes often suffer injury of the meniscus, known as torn cartilage. Thirteen fluid-filled sacs called bursae (sing., **bursa**) occur around the knee joint. Bursae ease friction between tendons and ligaments and between tendons and bones.

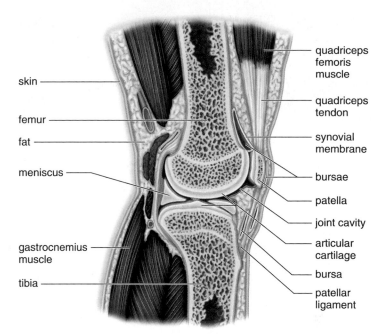

FIGURE 41.10 Knee joint.
The knee joint is an example of a synovial joint. The cavity between the bones is encased by ligaments and lined by synovial membrane. The patella (kneecap) serves to guide the quadriceps tendon over the joint when flexion or extension occurs.

Inflammation of the bursae is called bursitis. Tennis elbow is a form of bursitis.

There are different types of movable joints. The knee and elbow joints are hinge joints because, like a hinged door, they largely permit movement in one direction only. In a pivot joint, a small, cylindrical projection of one bone pivots within the ring formed of bone and ligament of another bone, making rotation possible. The joint between the first two cervical vertebrae, which permits side-to-side movement of the head, is an example of a pivot joint. More movable are the ball-and-socket joints; for example, the ball of the femur fits into a socket on the hipbone. Ball-and-socket joints allow movement in all planes and even rotational movement.

Synovial Joints

Synovial joints are subject to arthritis. In rheumatoid arthritis, the synovial membrane becomes inflamed and thickened. Degenerative changes take place that make the joint almost immovable and painful to use. There is evidence that these effects are brought on by an autoimmune reaction. In old-age arthritis, or osteoarthritis, the cartilage at the ends of the bones disintegrates so that the two bones become rough and irregular. This type of arthritis is apt to affect the joints that have received the greatest use over the years.

Bones are connected at joints. Synovial joints are freely movable and allow particular types of movements.

health focus

You Can Avoid Osteoporosis

Osteoporosis is a condition in which the bones are weakened due to a decrease in the bone mass that makes up the skeleton. Throughout life, bones are continuously remodeled. While a child is growing, the rate of bone formation is greater than the rate of bone breakdown. The skeletal mass continues to increase until ages 20 to 30. After that, there is an equal rate of formation and breakdown of bone mass until ages 40 to 50. Then, reabsorption begins to exceed formation, and the total bone mass slowly decreases.

Over time, men are apt to lose 25% and women to lose 35% of their bone mass. But we have to consider that men tend to have denser bones than women anyway, and their testosterone (male sex hormone) level generally does not begin to decline significantly until after age 65. In contrast, the estrogen (female sex hormone) level in women begins to decline at about age 45. Since sex hormones play an important role in maintaining bone strength, this difference means that women are more likely than men to suffer a higher incidence of fractures, involving especially the hip vertebrae, long bones, and pelvis. Although osteoporosis may at times be the result of various disease processes, it is essentially a disease that occurs as we age.

Everyone can take measures to avoid having osteoporosis when they get older. Adequate dietary calcium throughout life is an important protection against osteoporosis. The U.S. National Institutes of Health recommend a calcium intake of 1,200–1,500 mg per day during puberty. Males and females require 1,000 mg per day until the age of 65 and 1,500 mg per day after age 65. In older women 1,500 mg per day is especially desirable.

A small daily amount of vitamin D is also necessary for the body to use calcium correctly. Exposure to sunlight is required to allow skin to synthesize a precursor to vitamin D. If you reside on or north of a "line" drawn from Boston to Milwaukee, to Minneapolis, to Boise, chances are you're not getting enough vitamin D during the winter months. Therefore, you should avail yourself of vitamin D present in fortified foods such as low-fat milk and cereal.

Very inactive people, such as those confined to bed, lose bone mass 25 times faster than people who are moderately active. On the other hand, a regular, moderate, weight-bearing exercise like walking or jogging is another good way to maintain bone strength (Fig. 41A).

Postmenopausal women with any of the following risk factors should have an evaluation of their bone density:

- white or Asian race
- thin body type
- family history of osteoporosis
- early menopause (before age 45)
- smoking
- a diet low in calcium, or excessive alcohol consumption and caffeine intake
- sedentary lifestyle

Presently bone density is measured by a method called dual energy X-ray absorptiometry (DEXA). This test measures bone density based on the absorption of photons generated by an X-ray tube. Soon there may be blood and urine tests to detect the biochemical markers of bone loss. Then it will be made easier for physicians to screen older women and at-risk men for osteoporosis.

If the bones are thin, it is worthwhile to take all possible measures to gain bone density because even a slight increase can significantly reduce fracture risk. Hormone therapy includes black cohosh, which is a phytoestrogen (estrogen made by a plant as opposed to an animal). Calcitonin is a naturally occurring hormone whose main site of action is the skeleton, where it inhibits the action of osteoclasts, the cells that break down bone. Also, alendronate is a drug that acts similarly to calcitonin. After three years of alendronate therapy, an increase in spinal density by about 8% and hip density by about 7% is obtained. Promising new drugs include slow-release fluoride therapy and certain growth hormones. These medications stimulate the formation of new bone.

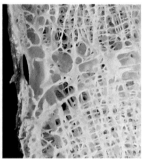

a. normal bone

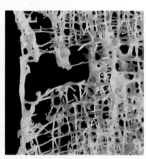

b. osteoporosis

c.

FIGURE 41A Osteoporosis.
*Exercise can help prevent osteoporosis, so when playing golf you should carry your own clubs and walk instead of using a golf cart. **a.** Normal bone compared to bone from a person with (**b**) osteoporosis. **c.** A patient with osteoporosis.*

41.3 THE HUMAN MUSCULAR SYSTEM

Three types of muscle tissue can be found in humans: smooth, cardiac, and skeletal muscle. Skeletal muscle, or striated muscle, is important in maintaining posture, providing support, and allowing for movement. In addition, skeletal muscle is important in homeostasis by helping maintain a constant body temperature. The contraction of skeletal muscle causes ATP breakdown, releasing heat that is distributed about the body.

Macroscopic Anatomy and Physiology

The nearly 700 skeletal muscles and their associated tissues make up approximately 40% of the weight of an average human. Several of the major superficial muscles are illustrated in Figure 41.11. Skeletal muscles are attached to the skeleton by bands of fibrous connective tissue called **tendons** [L. *tendo,* stretch]. When muscles contract, they shorten. Therefore, muscles can only pull; they cannot push. Because of this, skeletal muscles must work in antagonistic pairs. If one muscle of an antagonistic pair flexes the joint and bends the limb, the other one extends the joint and straightens the limb. Figure 41.12 illustrates this principle.

In the laboratory, if a muscle is given a rapid series of threshold stimuli, it can respond to the next stimulus without relaxing completely. In this way, muscle contraction summates until maximal sustained contraction, called **tetanus,** is achieved. Tetanic contractions ordinarily occur in the body's muscles whenever skeletal muscles are actively used. Even when muscles appear to be at rest, they exhibit **tone,** in which some of their fibers are contracting. Muscle tone is particularly important in maintaining posture. If all the fibers within the muscles of the neck, trunk, and legs were to suddenly relax, the body would collapse.

> Muscles work in antagonistic pairs. A muscle at rest exhibits tone dependent on tetanic contractions.

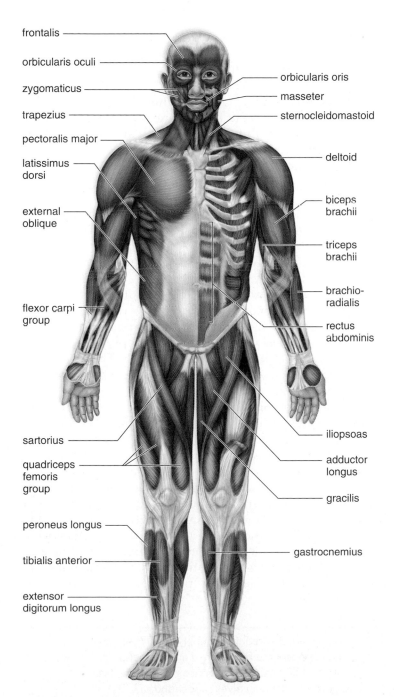

FIGURE 41.11 Human musculature.
Anterior view of some of the major superficial skeletal muscles.

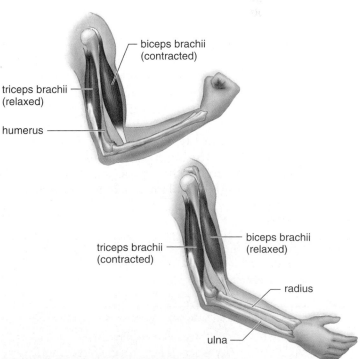

FIGURE 41.12 Antagonistic muscles.
Muscles can exert force only by shortening; therefore, they often work as antagonistic pairs. The biceps and triceps brachii exemplify an antagonistic pair of muscles that act opposite to one another. The biceps brachii raises the forearm, and the triceps brachii lowers the forearm.

Microscopic Anatomy and Physiology

A whole skeletal muscle is composed of a number of muscle fibers in bundles. Each muscle fiber is a cell containing the usual cellular components, but some components have special features (Fig. 41.13). The **sarcolemma,** or plasma membrane [Gk. *sarkos,* flesh, and *lemma,* sheath], forms a T (for transverse) system. The T tubules penetrate, or dip down, into the cell so that they come into contact—but do not fuse—with expanded portions of modified endoplasmic reticulum, called the **sarcoplasmic reticulum.** These expanded portions serve as storage sites for calcium ions (Ca^{2+}), which are essential for muscle contraction. The sarcoplasmic reticulum encases hundreds and sometimes even thousands of **myofibrils** [Gk. *myos,* muscle; L. *fibra,* thread], which are the contractile portions of muscle fiber.

Myofibrils are cylindrical in shape and run the length of the muscle fiber. The light microscope shows that a myofibril has light and dark bands called striations. It is these bands that cause skeletal muscle to appear striated. The electron microscope reveals that the striations of myofibrils are formed by the placement of protein filaments within contractile units called **sarcomeres** [Gk. *sarkos,* flesh, and *meros,* part]. A sarcomere extends between two dark lines called the Z lines. There are two types of protein filaments: thick filaments made up of **myosin,** and thin filaments made up of **actin.** The I band is light colored because it contains only actin filaments attached to a Z line. The dark regions of the A band contain overlapping actin and myosin filaments, and its H zone has only myosin filaments.

Sliding Filament Model

As a muscle fiber contracts, the sarcomeres within the myofibrils shorten. When a sarcomere shortens, the actin (thin) filaments slide past the myosin (thick) filaments and approach one another. This causes the I band to shorten and the H zone to nearly or completely disappear. The movement of actin filaments in relation to myosin filaments is called the **sliding filament model** of muscle contraction. During the sliding process, the sarcomere shortens, even though the filaments themselves remain the same length.

TABLE 41.1

Muscle Contraction

Name	Function
Actin filaments	Slide past myosin, causing contraction
Ca^{2+}	Needed for myosin to bind to actin
Myosin filaments	Pull actin filaments by means of cross-bridges; are enzymatic and split ATP
ATP	Supplies energy for muscle contraction

The participants in muscle contraction have the functions listed in Table 41.1. ATP supplies the energy for muscle contraction. Although the actin filaments slide past the myosin filaments, it is the myosin filaments that do the work. Myosin filaments break down ATP and form cross-bridges that attach to and pull the actin filaments toward the center of the sarcomere.

The sliding filament model states that actin filaments slide past myosin filaments because myosin has cross-bridges that pull the actin filaments inward.

ATP. ATP provides the energy for muscle contraction. Although muscle cells contain **myoglobin,** a molecule that stores oxygen, cellular respiration does not immediately supply all the ATP that is needed. In the meantime, muscle fibers rely on **creatine phosphate** (phosphocreatine), a storage form of high-energy phosphate. Creatine phosphate cannot directly participate in muscle contraction. Instead, it anaerobically regenerates ATP by the following reaction:

$$creatine—P \ + \ ADP \longrightarrow ATP \ + \ creatine$$

This reaction occurs in the midst of sliding filaments, and therefore, this method of supplying ATP is the speediest energy source available to muscles.

When all of the creatine phosphate is depleted, mitochondria may by then be producing enough ATP for muscle contraction to continue. If not, fermentation is a second way for muscles to supply ATP without consuming oxygen. Fermentation, which is apt to occur when strenuous exercise first begins, can supply ATP for only a short time because of lactate buildup. The buildup is noticeable when it produces muscle aches and fatigue upon exercising.

We all have had the experience of needing to continue deep breathing following strenuous exercise. This continued intake of oxygen, which is required to complete the metabolism of lactate and restore cells to their original energy state, represents an **oxygen debt.** The lactate is transported to the liver, where 20% of it is completely broken down to carbon dioxide (CO_2) and water (H_2O). The ATP gained by this respiration is then used to reconvert 80% of the lactate to glucose. In persons who train, the number of mitochondria increases, and there is a greater reliance on them rather than on fermentation to produce ATP. Less lactate is produced, and there is less oxygen debt.

Working muscles require a supply of ATP. Anaerobic creatine phosphate breakdown and fermentation quickly generate ATP. Sustained exercise requires cellular respiration, which involves mitochondria.

FIGURE 41.13 Skeletal muscle fiber structure and function.
A muscle fiber contains many myofibrils, divided into sarcomeres, which are contractile. When the myofibrils of a muscle fiber contract, the sarcomeres shorten: the actin (thin) filaments slide past the myosin (thick) filaments toward the center so that the H zone gets smaller, to the point of disappearing.

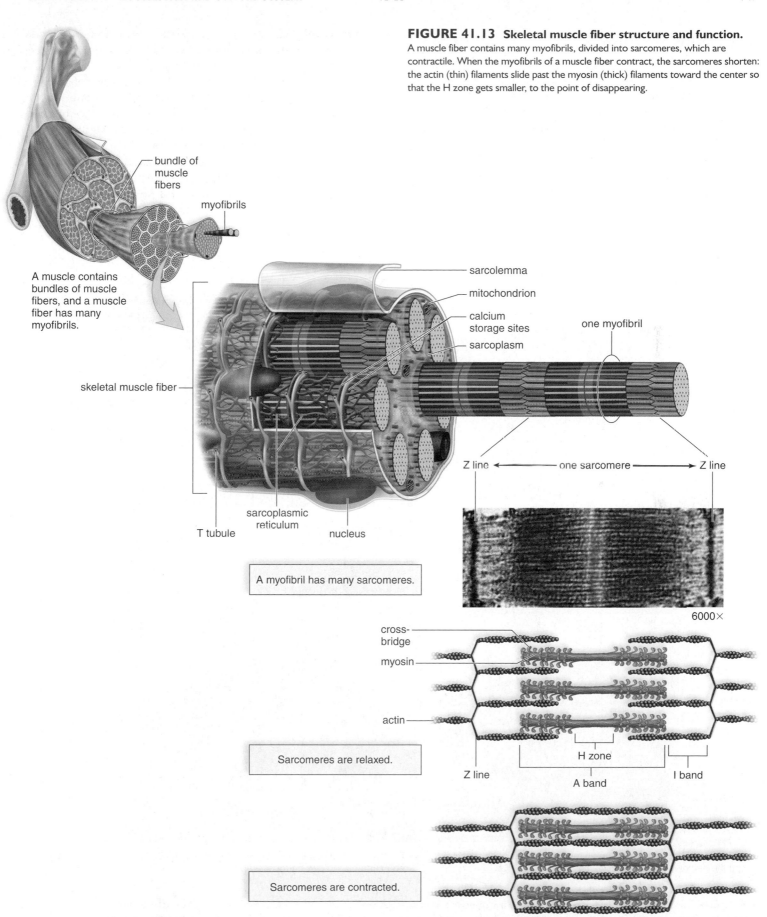

bundle of muscle fibers

myofibrils

A muscle contains bundles of muscle fibers, and a muscle fiber has many myofibrils.

skeletal muscle fiber

T tubule

sarcoplasmic reticulum

nucleus

sarcolemma

mitochondrion

calcium storage sites

sarcoplasm

one myofibril

Z line ← one sarcomere → Z line

A myofibril has many sarcomeres.

6000×

cross-bridge

myosin

actin

Sarcomeres are relaxed.

Z line

A band

H zone

I band

Sarcomeres are contracted.

Muscle Innervation

Muscles are stimulated to contract by motor nerve fibers. Nerve fibers have several branches, each of which ends at an axon terminal that lies in close proximity to the sarcolemma of a muscle fiber. A small gap, called a synaptic cleft, separates the axon terminal from the sarcolemma. This entire region is called a **neuromuscular junction** (Fig. 41.14).

Axon terminals contain synaptic vesicles that are filled with the neurotransmitter acetylcholine (ACh). When nerve impulses traveling down a motor neuron arrive at an axon terminal, the synaptic vesicles release ACh into the synaptic cleft. ACh quickly diffuses across the cleft and binds to receptors in the sarcolemma. Now the sarcolemma generates impulses that spread over the sarcolemma and down T tubules to the sarcoplasmic reticulum. The release of calcium from the sarcoplasmic reticulum causes the filaments within sarcomeres to slide past one another. Sarcomere contraction results in myofibril contraction, which in turn results in muscle fiber, and finally muscle, contraction.

At a neuromuscular junction, nerve impulses bring about the release of a neurotransmitter that signals a muscle fiber to contract.

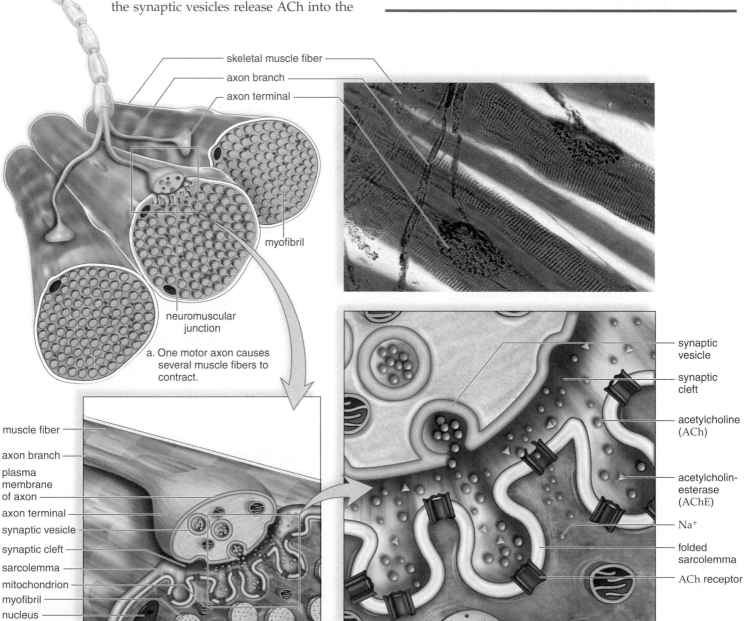

a. One motor axon causes several muscle fibers to contract.

b. A neuromuscular junction is the juxtaposition of an axon terminal and the sarcolemma of a muscle fiber.

c. The release of a neurotransmitter (ACh) causes receptors to open and Na^+ to enter a muscle fiber.

FIGURE 41.14 Neuromuscular junction.
The branch of a motor nerve fiber ends in an axon terminal that meets but does not touch a muscle fiber. A synaptic cleft separates the axon terminal from the sarcolemma of the muscle fiber. Nerve impulses traveling down a motor fiber cause synaptic vesicles to discharge a neurotransmitter that diffuses across the synaptic cleft. When the neurotransmitter is received by the sarcolemma of a muscle fiber, impulses begin that lead to muscle fiber contraction.

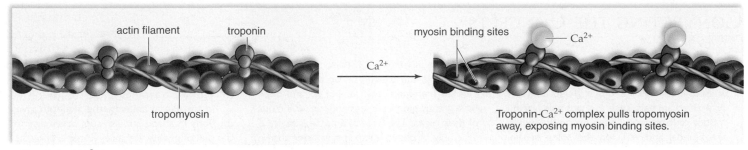

a. Function of Ca²⁺ in muscle contraction

Troponin-Ca²⁺ complex pulls tropomyosin away, exposing myosin binding sites.

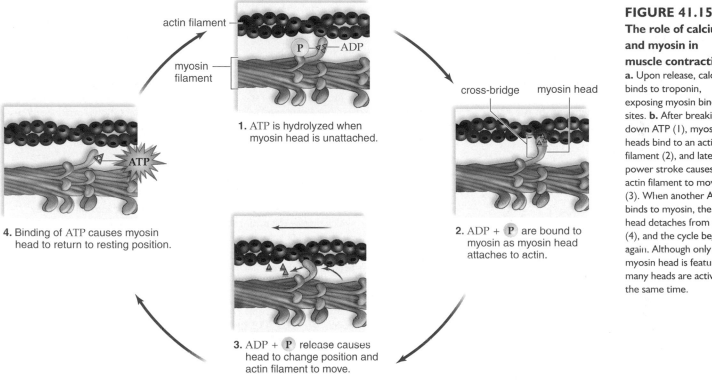

1. ATP is hydrolyzed when myosin head is unattached.

2. ADP + P are bound to myosin as myosin head attaches to actin.

3. ADP + P release causes head to change position and actin filament to move.

4. Binding of ATP causes myosin head to return to resting position.

b. Function of cross-bridges in muscle contraction

FIGURE 41.15
The role of calcium and myosin in muscle contraction.
a. Upon release, calcium binds to troponin, exposing myosin binding sites. **b.** After breaking down ATP (1), myosin heads bind to an actin filament (2), and later, a power stroke causes the actin filament to move (3). When another ATP binds to myosin, the head detaches from actin (4), and the cycle begins again. Although only one myosin head is featured, many heads are active at the same time.

Figure 41.15 illustrates the placement of two other proteins associated with a thin filament, which is composed of a double row of twisted actin molecules. Threads of tropomyosin wind about an actin filament, and troponin occurs at intervals along the threads. Calcium ions (Ca²⁺) that have been released from the sarcoplasmic reticulum combine with troponin. After binding occurs, the tropomyosin threads shift their position, and myosin binding sites are exposed.

Thick filaments are bundles of myosin molecules with double globular heads. Myosin heads function as ATPase enzymes, splitting ATP into ADP and P. This reaction activates the heads so that they bind to actin. The ADP and P remain on the myosin heads until the heads attach to actin, forming cross-bridges. Now, ADP and P are released, and this causes the cross-bridges to change their positions. This

is the power stroke that pulls the thin filaments toward the middle of the sarcomere. When more ATP molecules bind to myosin heads, the cross-bridges are broken as the heads detach from actin. The cycle begins again; the actin filaments move nearer the center of the sarcomere each time the cycle is repeated.

Contraction continues until nerve impulses cease and calcium ions are returned to their storage sites. The membranes of the sarcoplasmic reticulum contain active transport proteins that pump calcium ions back into the calcium storage sites, and muscle relaxation occurs.

Myosin filament heads break down ATP and then attach to an actin filament, forming cross-bridges that pull the actin filament to the center of a sarcomere.

CONNECTING THE CONCEPTS

The adage that structure fits function is evident when one observes how different animals locomote. Among a group of widely diversified animals, such as mammals, various modes of locomotion have evolved. Humans are bipedal and walk on the soles of their feet formed by the tarsal bones. This form of locomotion allows the hands to be free and may have evolved from the habit of monkeys and apes to use only forelimbs as they swing through the branches of trees. Dexterity of hands and feet is actually the ancestral mammalian condition. In humans and apes, the bones of the hands and feet are not fused, and the wrist and ankle can rotate in three dimensions.

Carnivores, such as members of the cat family, walk on their toes. This is an obvious adaptation to running—we also raise the heel and engage the toes in order to run faster. Hoofed mammals, such as horses and deer, have greatly elongated legs and run on the tips of their digits. The hoof of a horse is its third digit only. A cheetah can sprint faster than a horse, but a horse will eventually outdistance the cheetah because it has more endurance.

Mammals that jump, such as kangaroos and rabbits, have a squat shape and elongated hindlimbs that propel them forward. Several groups of arboreal mammals, including flying squirrels and sugar gliders, have membranes attached to their bodies that permit them to glide through the air from tree to tree. Among mammals, only bats truly fly. Their membranous wings are stretched between greatly elongated forelimbs and fingers. In both birds and bats, the wing is moved downward and forward in one motion, and then backward and upward in another.

In terrestrial animals, the skeleton gives the body its shape but also supports it against the pull of gravity. Because water is buoyant, gravity is not much of a problem for aquatic animals, but a shape that reduces friction is quite helpful. Seals, sea lions, whales, and dolphins have a streamlined torpedo shape that facilitates movement through water. A whale has few protruding parts; a male's penis is completely hidden within muscular folds, and the teats of the female lie behind slits on either side of the genital area. Thus, we see that while locomotion chiefly involves musculoskeletal adaptations, other body systems are involved as well.

Summary

41.1 DIVERSITY OF SKELETONS

Three types of skeletons are found in the animal kingdom: hydrostatic skeleton (cnidarians, flatworms, and segmented worms); exoskeleton (certain molluscs and arthropods); and endoskeleton (vertebrates). The rigid but jointed skeleton of arthropods and vertebrates helped them colonize the terrestrial environment.

41.2 THE HUMAN SKELETAL SYSTEM

The human skeleton gives support to the body, helps protect internal organs, provides sites for muscle attachment, and is a storage area for calcium and phosphorous salts, as well as a site for blood cell formation.

Most bones are cartilaginous in the fetus but are converted to bone during development. A long bone undergoes endochondral ossification in which a cartilaginous growth plate remains between the primary ossification center in the middle and the secondary centers at the ends of the bones. Growth of the bone is possible as long as the growth plates are present, but eventually they too are converted to bone. Bone is constantly being renewed; osteoclasts break down bone, and osteoblasts build new bone. Osteocytes are in the lacunae of osteons; a long bone has a shaft of compact bone and two ends that contain spongy bone. The shaft contains a medullary cavity with yellow marrow, and the ends contain red marrow.

The human skeleton is divided into two parts: (1) the axial skeleton, which is made up of the skull, the vertebral column, the sternum, and the ribs, and (2) the appendicular skeleton, which is composed of the girdles and their appendages.

Joints are classified as immovable, like those of the cranium; slightly movable, like those between the vertebrae; and freely movable (synovial joints), like those in the knee and hip. In synovial joints, ligaments bind the two bones together, forming a capsule containing synovial fluid.

41.3 THE HUMAN MUSCULAR SYSTEM

Whole skeletal muscles can only shorten when they contract; therefore, they work in antagonistic pairs. For example, if one muscle flexes the joint and brings the limb toward the body, the other one extends the joint and straightens the limb. A muscle at rest exhibits tone, which is dependent on tetanic contractions.

A whole skeletal muscle is composed of muscle fibers. Each muscle fiber is a cell that contains myofibrils in addition to the usual cellular components. Longitudinally, myofibrils are divided into sarcomeres, which display the arrangement of actin and myosin filaments.

The sliding filament model of muscle contraction says that myosin filaments have cross-bridges, which attach to and detach from actin filaments, causing actin filaments to slide and the sarcomere to shorten. (The H zone disappears as actin filaments approach one another.) Myosin breaks down ATP, and this supplies the energy for muscle contraction. Anaerobic creatine phosphate breakdown and fermentation quickly generate ATP. Sustained exercise requires cellular respiration for the generation of ATP.

Nerves innervate muscles. Nerve impulses traveling down motor neurons to neuromuscular junctions cause the release of ACh, which binds to receptors on the sarcolemma (plasma membrane of a muscle fiber). Impulses begin and move down T tubules that approach the sarcoplasmic reticulum (endoplasmic reticulum of muscle fibers), where calcium is stored. Thereafter, calcium ions are released and bind to troponin. The troponin-Ca^{2+} complex causes tropomyosin threads winding around actin filaments to shift their position, revealing myosin binding sites. Myosin filaments are composed of many myosin molecules with double globular heads. When myosin heads break down ATP, they are ready to attach to actin. The release of ADP + P causes myosin heads to change their position. This is the power stroke that causes the actin filament to slide toward the center of a sarcomere. When more ATP molecules bind to myosin, the heads detach from actin, and the cycle begins again.

Reviewing the Chapter

1. What are the three types of skeletons found in the animal kingdom and how do they differ? Cite animals that have these types of skeletons.   738–39
2. Give several functions of the skeletal system in humans. How does the skeletal system contribute to homeostasis?   740
3. Contrast compact bone with spongy bone. Explain how bone grows and is renewed.   740
4. Distinguish between the axial and appendicular skeletons.   742, 744

5. List the bones that form the pectoral girdle and upper limb; the pelvic girdle and lower limb. 744–45

6. How are joints classified? Describe the anatomy of a freely movable joint. 745

7. Give several functions of the muscular system in humans. How does the muscular system contribute to homeostasis? 747

8. Describe how muscles are attached to bones. What is accomplished by muscles acting in antagonistic pairs? 747

9. Discuss the microscopic structural features of a muscle fiber and a sarcomere. What is the sliding filament model? 748–49

10. Discuss the availability and the specific role of ATP during muscle contraction. What is oxygen debt, and how is it repaid? 748

11. Describe the structure and function of a neuromuscular junction. 750

12. Describe the cyclical events as myosin pulls actin toward the center of a sarcomere. 751

Testing Yourself

Choose the best answer for each question. For questions 1–4, match each bone to the location in the key.

KEY:

a. arm
b. forearm
c. pectoral girdle
d. pelvic girdle
e. thigh
f. leg

1. ulna

2. tibia

3. clavicle

4. femur

5. Spongy bone
   a. contains osteons.
   b. contains red bone marrow, where blood cells are formed.
   c. lends no strength to bones.
   d. contributes to homeostasis.
   e. Both b and d are correct.

6. Which of these pairs is mismatched?
   a. slightly movable joint—vertebrae
   b. hinge joint—hip
   c. synovial joint—elbow
   d. immovable joint—sutures in cranium
   e. ball-and-socket joint—hip

7. The skeletal system does not
   a. produce blood cells.
   b. store minerals.
   c. help produce movement.
   d. store fat.
   e. produce body heat.

8. All blood cells—red, white, and platelets—are produced by which of the following?
   a. yellow bone marrow
   b. red bone marrow
   c. periosteum
   d. medullary cavity

9. Which of the following is not a bone of the appendicular skeleton?
   a. the scapula
   b. a rib
   c. a metatarsal bone
   d. the patella

10. The vertebrae that articulate with the ribs are
    a. the lumbar vertebrae.
    b. the sacral vertebrae.
    c. the thoracic vertebrae.
    d. the cervical vertebrae.
    e. the coccyx.

11. In a muscle fiber,
    a. the sarcolemma is connective tissue holding the myofibrils together.
    b. the sarcoplasmic reticulum stores calcium.
    c. both myosin and actin filaments have cross-bridges.
    d. there is a T system but no endoplasmic reticulum.
    e. All of these are correct.

12. When muscles contract,
    a. sarcomeres shorten.
    b. myosin breaks down ATP.
    c. actin slides past myosin.
    d. the H zone disappears.
    e. All of these are correct.

13. Which of these is the direct source of energy for muscle contraction?
    a. ATP
    b. creatine phosphate
    c. lactic acid
    d. glycogen
    e. Both a and b are correct.

14. Nervous stimulation of muscles
    a. occurs at a neuromuscular junction.
    b. involves the release of ACh.
    c. results in impulses that travel down the T system.
    d. causes calcium to be released from the sarcoplasmic reticulum.
    e. All of these are correct.

15. A neuromuscular junction occurs between an axon terminal and
    a. a muscle fiber.
    b. a myofibril.
    c. a myosin filament.
    d. a sarcomere only.
    e. Both a and d are correct.

16. When calcium is released from the sarcoplasmic reticulum, it binds to
    a. myosin.
    b. actin.
    c. troponin.
    d. sarcomeres.
    e. Both b and d are correct.

17. At what point is ATP hydrolyzed?
    a. Just as myosin attaches to troponin.
    b. Just before myosin attaches to actin.
    c. Just when myosin pulls on actin.
    d. Just when impulses move down a T tubule.

18. ACh
   a. is active at somatic synapses but not at neuromuscular junctions.
   b. binds to receptors in the sarcolemma.
   c. preceded the buildup of ATP in mitochondria.
   d. is stored in the sarcoplasmic reticulum.
   e. Both b and d are correct.

19. Label this diagram of a muscle fiber, using these terms: myofibril, Z line, T tubule, sarcomere, sarcolemma, sarcoplasmic reticulum.

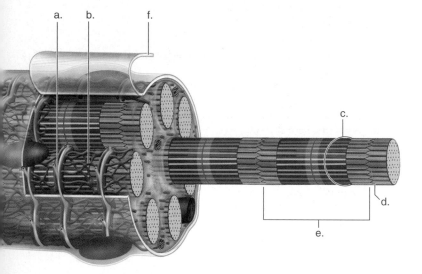

Thinking Scientifically

1. It is observed that some motor neurons innervate only a few muscle fibers in the biceps brachii. Other motor neurons each innervate many muscle fibers. How might this observation correlate with our ability to pick up a pencil or a 2-liter soda bottle? On what basis would the brain bring about the correct level of contraction?

2. Some athletes believe that taking oral creatine will increase their endurance because it will increase the amount of phosphate available to their muscles for ATP synthesis. This statement can be regarded as two hypotheses: (1) oral creatine increases endurance; (2) oral creatine increases the amount of creatine available in muscles for ATP synthesis. How could these two hypotheses be tested?

Bioethical Issue: Support Systems and Locomotion

A natural advantage does not bar an athlete from participating in and winning a medal in a particular sport at the Olympic Games. Nor are athletes restricted to a certain amount of practice or required to eliminate certain foods from their diets.

Athletes are, however, prevented from participating in the Olympic Games if they have taken certain performance-enhancing drugs. There is no doubt that regular use of drugs such as anabolic steroids leads to kidney disease, liver dysfunction, hypertension, and a myriad of other undesirable side effects. Even so, shouldn't the individual be allowed to take these drugs if he or she wants to? Anabolic steroids are synthetic forms of the male sex hormone testosterone. Taking large doses, along with strength training, leads to much larger muscles than otherwise. Extra strength and endurance can give an athlete an advantage in certain sports, such as racing, swimming, and weight lifting.

Should the Olympic committee outlaw the taking of anabolic steroids, and if so, on what basis? The basis can't be an unfair advantage, because some athletes naturally have an unfair advantage over other athletes. Should these drugs be outlawed on the basis of health reasons? Excessive practice or a purposeful decrease or increase in weight to better perform in a sport can also be injurious to one's health. In other words, how can you justify allowing some behaviors that enhance performance and not others?

Understanding the Terms

actin	748	osteoclast	740
appendicular skeleton	744	osteocyte	740
axial skeleton	742	oxygen debt	748
bursa	745	pectoral girdle	744
compact bone	740	pelvic girdle	744
creatine phosphate	748	red bone marrow	740
endoskeleton	738	sarcolemma	748
exoskeleton	738	sarcomere	748
fontanel	742	sarcoplasmic reticulum	748
foramen magnum	742	sliding filament model	748
hydrostatic skeleton	738	spongy bone	740
joint	745	suture	742
ligament	745	synovial joint	745
myofibril	748	tendon	747
myoglobin	748	tetanus	747
myosin	748	tone	747
neuromuscular junction	750	vertebral column	743
osteoblast	740		

Match the terms to these definitions:

a. _____ Bone-forming cell.
b. _____ Movement of actin filaments in relation to myosin filaments, which accounts for muscle contraction.
c. _____ Muscle protein making up the thin filaments in a sarcomere; its movement shortens the sarcomere, yielding muscle contraction.
d. _____ Part of the skeleton that consists of the pectoral and pelvic girdles and the bones of the arms and legs.
e. _____ Portion of the skeleton that provides support and attachment for the arms.

ARIS, the *Biology* Website

ARIS, the website for *Biology*, provides a wealth of information organized and integrated by chapter. You will find practice quizzes, interactive activities, labeling exercises, flashcards, and much more that will complement your learning and understanding of general biology.

www.mhhe.com/maderbiology9

42

HORMONES AND ENDOCRINE SYSTEMS

The "fight-or-flight" reaction illustrates that the nervous system and endocrine system are intimately linked. Sympathetic nerve fibers control the release of epinephrine and norepinephrine from the adrenal medulla (an endocrine gland) when a gazelle is trying to escape a cheetah. Continuing to work together, hormones and direct nervous stimulation cause senses to sharpen as pupils dilate, heart rate and breathing rate increase, sweat glands become active, and the mouth becomes dry. Undetected, blood supply to the muscles increases and, due to hormonal stimulation alone, glucose pours out of the liver.

The nervous system is well known for bringing about an immediate response to environmental stimuli as in the fight-and-flight reaction. Most often the chemical signals released by the nervous system, called neurotransmitters, help maintain homeostasis. Heart rate, breathing rate, and blood pressure are all regulated to stay relatively constant. Hormones released by endocrine glands help maintain homeostasis by keeping the level of calcium, sodium, glucose, and other blood constituents within normal limits.

The endocrine system is slower acting than the nervous system and often regulates processes that occur over days or even months. Hormones secreted into the bloodstream control whole-body processes, such as growth and reproduction, and complex behaviors, including courtship and migration. Hormones influence the metabolism after they arrive at cells with appropriate receptor proteins. It can take a while to metabolically change a cell, but the effect is longer lasting.

A cheetah, *Acinonyx jubatus*, capturing a Thomson's gazelle, *Gazella thomsoni*.

42.1 ENDOCRINE GLANDS

The nervous system and the endocrine system both regulate the other organ systems. Control of the other systems permits coordination of their functions and brings about homeostasis. As we have seen, a nervous system is composed of neurons. In this system, sensory receptors (specialized dendrites) detect changes in the internal and external environment; the CNS integrates the information and can respond by stimulating muscles and glands. Communication in the nervous system depends on nerve impulses conducted in axons and neurotransmitters, which cross synapses. Axon conduction occurs rapidly and so does diffusion of a neurotransmitter across the short distance of a synapse. In other words, the nervous system is organized to respond rapidly to stimuli. This is particularly useful if the stimulus is an external event that endangers our safety—we can move quickly to avoid being hurt.

An endocrine system functions differently (Table 42.1). An **endocrine system** is notably composed of glands. These glands secrete **hormones,** which are carried by the bloodstream to target cells throughout the body (Fig. 42.1 and Table 42.2). The blood concentration of a substance often prompts an endocrine gland to secrete its hormone. For example, the parathyroid glands secrete a hormone when the

TABLE 42.1

Comparison of Nervous and Endocrine Systems

	Nervous System	Endocrine System
Composed of	Neurons	Usually glands
Delivery	Nerve impulse and neurotransmitter	Hormone
How delivered	Axon and synapse	Usually bloodstream
Target	Muscles and glands	Cells throughout body
Response	Rapid, short-lived	Slow, long-lasting
Controlled by	Negative feedback	Negative feedback

blood Ca^{2+} level falls below normal. Hormones influence the metabolism of cells and the growth and development of body parts. In our example, parathyroid hormone (PTH) causes bone to break down and release Ca^{2+}. It takes time to deliver hormones and it takes time for cells to respond, but the effect is longer lasting. In other words, the endocrine system is organized for a slow but prolonged response.

Endocrine glands can be contrasted with exocrine glands. Exocrine glands have ducts and secrete their products into these

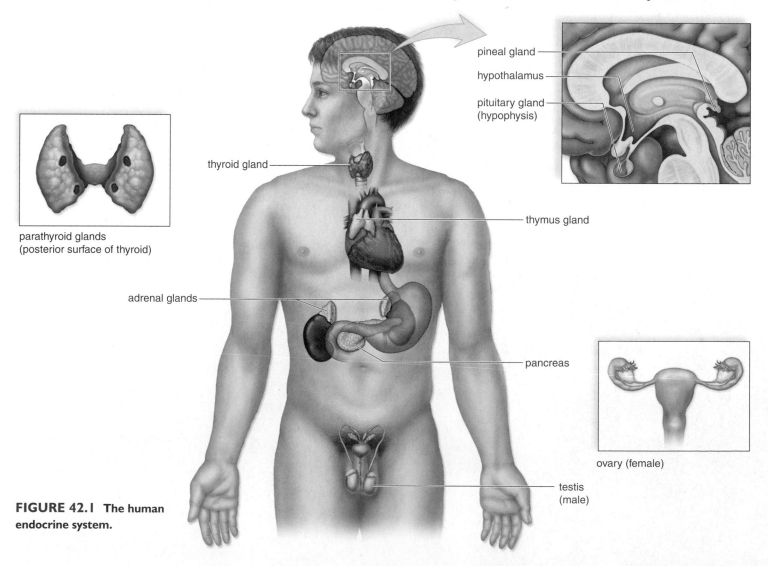

parathyroid glands (posterior surface of thyroid)

pineal gland
hypothalamus
pituitary gland (hypophysis)

thyroid gland

thymus gland

adrenal glands

pancreas

ovary (female)

testis (male)

FIGURE 42.1 The human endocrine system.

TABLE 42.2

Principal Endocrine Glands and Hormones

Endocrine Gland	Hormone Released	Chemical Class	Target Tissues/Organs	Chief Function(s) of Hormone
Hypothalamus	Hypothalamic-releasing and -inhibiting hormones	Peptide	Anterior pituitary	Regulate anterior pituitary hormones
Pituitary gland				
Posterior pituitary	Antidiuretic (ADH)	Peptide	Kidneys	Stimulates water reabsorption by kidneys
	Oxytocin	Peptide	Uterus, mammary glands	Stimulates uterine muscle contraction; release of milk by mammary glands
Anterior pituitary	Thyroid-stimulating (TSH)	Glycoprotein	Thyroid	Stimulates thyroid
	Adrenocorticotropic (ACTH)	Peptide	Adrenal cortex	Stimulates adrenal cortex
	Gonadotropic (FSH, LH)	Glycoprotein	Gonads	Egg and sperm production; sex hormone production
	Prolactin (PRL)	Protein	Mammary glands	Milk production
	Growth (GH)	Protein	Soft tissues, bones	Cell division, protein synthesis, and bone growth
	Melanocyte-stimulating (MSH)	Peptide	Melanocytes in skin	Unknown function in humans; regulates skin color in lower vertebrates
Thyroid	Thyroxine (T_4) and triiodothyronine (T_3)	Iodinated amino acid	All tissues	Increases metabolic rate; regulates growth and development
	Calcitonin	Peptide	Bones, kidneys, intestine	Lowers blood calcium level
Parathyroids	Parathyroid (PTH)	Peptide	Bones, kidneys, intestine	Raises blood calcium level
Adrenal gland				
Adrenal cortex	Glucocorticoids (cortisol)	Steroid	All tissues	Raise blood glucose level; stimulate breakdown of protein
	Mineralocorticoids (aldosterone)	Steroid	Kidneys	Reabsorb sodium and excrete potassium
	Sex hormones	Steroid	Gonads, skin, muscles, bones	Stimulate reproductive organs and bring about sex characteristics
Adrenal medulla	Epinephrine and norepinephrine	Modified amino acid	Cardiac and other muscles	Released in emergency situations; raise blood glucose level
Pancreas	Insulin	Protein	Liver, muscles, adipose tissue	Lowers blood glucose level; promotes formation of glycogen
	Glucagon	Protein	Liver, muscles, adipose tissue	Raises blood glucose level
Gonads				
Testes	Androgens (testosterone)	Steroid	Gonads, skin, muscles, bones	Stimulate male sex characteristics
Ovaries	Estrogens and progesterone	Steroid	Gonads, skin, muscles, bones	Stimulate female sex characteristics
Thymus	Thymosins	Peptide	T lymphocytes	Stimulate production and maturation of T lymphocytes
Pineal gland	Melatonin	Modified amino acid	Brain	Controls circadian and circannual rhythms; possibly involved in maturation of sexual organs

ducts, which take them to the lumens of other organs or outside the body. For example, in humans the salivary glands send saliva into the mouth by way of the salivary ducts. **Endocrine glands,** as stated, secrete their products into the bloodstream, which delivers it throughout the body. It must be stressed that only certain cells, called target cells, can respond to certain hormones. If a cell can respond to a hormone, the hormone and receptor proteins in the plasma membrane bind together as a key fits a lock.

It is of interest to note that both the nervous system and the endocrine system make use of negative feedback mechanisms. If the blood pressure falls, sensory receptors signal a control center in the brain. This center sends out nerve impulses to the arterial walls, so that they constrict and blood pressure rises. Now, the sensory receptors are no longer stimulated and the system is inactivated. Similarly, to continue our example, when the blood Ca^{2+} level rises, the parathyroid gland no longer secretes PTH.

Hormones Are Chemical Signals

Hormones are a type of chemical signal. **Chemical signals** are a means of communication between cells, between body parts, and even between individuals. They typically affect the metabolism of cells that have receptors to receive them (Fig. 42.2). In a condition called androgen insensitivity, an individual has X and Y sex chromosomes and the testes, which remain in the abdominal cavity, produce the sex hormone testosterone. However, the body cells lack receptors that are able to combine with testosterone and the individual appears to be a normal female.

Like testosterone, most hormones act at a distance between body parts. They travel in the bloodstream from the gland that produced them to their target cells. Also counted as hormones are the secretions produced by neurosecretory cells in the hypothalamus, a part of the brain. They travel in the capillary network that runs between the hypothalamus and the pituitary gland. Some of these secretions stimulate the pituitary to secrete its hormones, and others prevent it from doing so.

FIGURE 42.2 Target cell concept.
Most hormones are distributed by the bloodstream to target cells. Target cells have receptors for the hormone, and the hormone combines with the receptor as a key fits a lock.

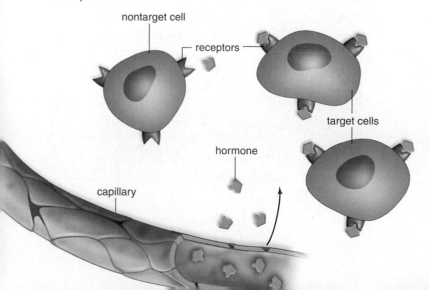

Not all hormones act between body parts. As we shall see, prostaglandins are a good example of a *local hormone.* After prostaglandins are produced, they are not carried in the bloodstream; instead, they affect neighboring cells, sometimes promoting pain and inflammation. Also, growth factors are local hormones that promote cell division and mitosis.

Chemical signals that affect the metabolism and influence the behavior of other individuals are called **pheromones.** Pheromones are active between other animals but not between humans. Still, studies suggest that women prefer the axillary odors of men who are a different HLA type from themselves. (HLA are plasma membrane proteins involved in immunity.) Choosing a mate of a different HLA type could conceivably improve the immune response of offspring. Several studies indicate that axillary secretions can affect the menstrual cycle. Women who live in the same household often have menstrual cycles in synchrony. Researchers have found that a woman's axillary extract can alter another woman's cycle by a few days.

The Action of Hormones

Hormones have a wide range of effects on cells. Some of these effects induce a target cell to increase its uptake of particular substances, such as glucose, or ions, such as calcium. Some bring about an alteration of the target cell's structure in some way. Here, we will concentrate on two ways hormones can influence cell metabolism. One typifies the action of a peptide hormone and the other typifies the action of a steroid hormone. The term **peptide hormone** is used to include hormones that are peptides, proteins, glycoproteins, and modified amino acids. **Steroid hormones** have the same complex of four carbon rings because they are all derived from cholesterol.

The Action of Peptide Hormones. Most hormonal glands secrete peptide hormones (see Table 42.2). The actions of peptide hormones can vary, and we will concentrate on what happens in muscle cells after the hormone epinephrine binds to a receptor in the plasma membrane (Fig. 42.3). In muscle cells, the reception of epinephrine leads to the breakdown of glycogen to glucose, which ends up in the blood. The immediate result of binding is the formation of **cyclic adenosine monophosphate (cAMP).** Cyclic AMP contains one phosphate group attached to adenosine at two locations. Therefore, the molecule is cyclic. Cyclic AMP activates a protein kinase enzyme in the cell, and this enzyme, in turn, activates another enzyme, and so forth. The series of enzymatic reactions that follows cAMP formation is called an enzyme cascade. Because each enzyme can be used over and over again at every step of the cascade, more enzymes are involved. Finally, many molecules of glycogen are broken down to glucose, which enters the bloodstream.

Typical of epinephrine, a peptide hormone never enters the cell. Therefore, the hormone is called the **first messenger,** while cAMP, which sets the metabolic machinery in motion, is called the **second messenger.** To understand this terminology, let's imagine that the adrenal medulla, which produces epinephrine, is like the home office that sends out a courier (i.e., the hormone epinephrine is the first messenger) to a

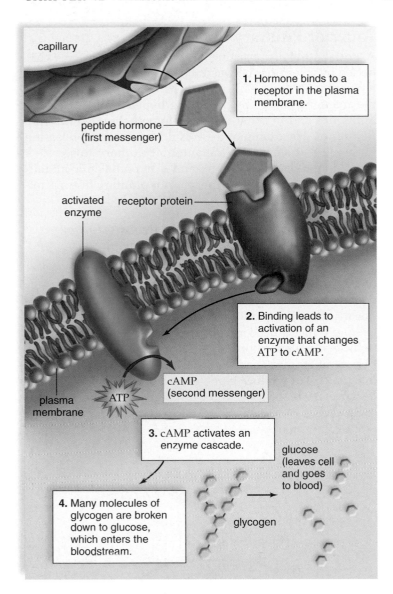

FIGURE 42.3 Peptide hormone.
The peptide hormone (first messenger) binds to a receptor in the plasma membrane. Thereafter cyclic AMP (second messenger) forms and activates an enzyme cascade.

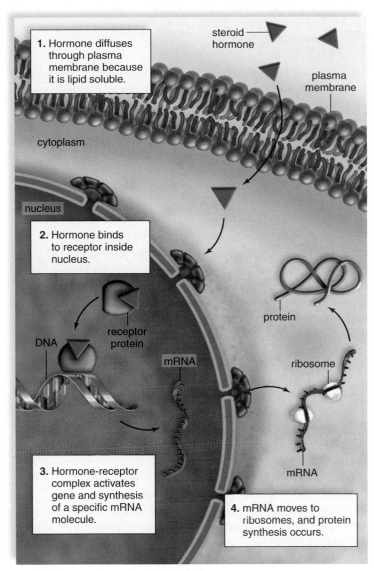

FIGURE 42.4 Steroid hormone.
A steroid hormone passes directly through the target cell's plasma membrane before binding to a receptor in the nucleus or cytoplasm. The hormone-receptor complex binds to DNA and gene expression follows.

factory (the cell). The courier doesn't have a pass to enter the factory, so when he arrives at the factory, he tells a supervisor through the screen door that the home office wants the factory to produce a particular product. The supervisor (i.e., cAMP, the second messenger) walks over and flips a switch that starts the machinery (the enzymatic pathway), and a product is made.

The Action of Steroid Hormones. The number of glands that produce steroid hormones is limited to the adrenal cortex, the ovaries, and the testes (see Table 42.2). Thyroid hormones act similarly to steroid hormones, even though they have a different structure.

Steroid hormones do not bind to plasma membrane receptors, and instead they are able to enter the cell because they are lipids (Fig. 42.4). Once inside, steroid hormones bind to receptors usually in the nucleus but sometimes in the

cytoplasm. Inside the nucleus, the hormone-receptor complex binds with DNA and activates transcription of certain portions of DNA. Translation of messenger RNA (mRNA) transcripts results in enzymes and other proteins that can carry out a response to the hormonal signal.

Steroids act more slowly than peptides because it takes more time to synthesize new proteins than to activate enzymes already present in cells. Their action lasts longer, however.

Hormones bind to receptor proteins in target cells. Peptide hormones (e.g., epinephrine) bind to a receptor in the plasma membrane and cause the formation of cAMP, which activates an enzyme cascade. Steroid hormones diffuse through the plasma membrane and enter the nucleus to bind with a receptor. The complex affects gene activity and protein synthesis.

42.2 HYPOTHALAMUS AND PITUITARY GLAND

The **hypothalamus** regulates the internal environment through the autonomic system. For example, it helps control heartbeat, body temperature, and water balance. The hypothalamus also controls the glandular secretions of the **pituitary gland** (hypophysis). The pituitary, a small gland about 1 cm in diameter, is connected to the hypothalamus by a stalklike structure. The pituitary has two portions: the posterior pituitary and the anterior pituitary.

Posterior Pituitary

Neurons in the hypothalamus called neurosecretory cells produce the hormones **antidiuretic hormone (ADH)** [Gk. *anti*, against; L. *ouresis*, urination] and oxytocin (Fig. 42.5, *left*). These hormones pass through axons into the **posterior pituitary,** where they are stored in axon endings. Certain neurons in the hypothalamus are sensitive to the water-salt balance of the blood. When these cells determine that the blood is too concentrated, ADH is released from the posterior pituitary. Upon reaching the kidneys, ADH causes water to be reabsorbed. As the blood becomes dilute, ADH is no longer released. This is an example of control by **negative feedback** because the *effect* of the hormone (to dilute blood) acts to shut down the *release* of the hormone. Negative feedback maintains stable conditions and homeostasis.

Inability to produce ADH causes diabetes insipidus (watery urine), in which a person produces copious amounts of urine with a resultant loss of ions from the blood. The condition can be corrected by the administration of ADH. The release of ADH is inhibited by the consumption of alcohol, which explains the frequent urination associated with drinking alcohol.

Oxytocin [Gk. *oxys*, quick, and *tokos*, birth], the other hormone made in the hypothalamus, causes uterine contraction during childbirth and milk letdown when a baby is nursing. The more the uterus contracts during labor, the more nerve impulses reach the hypothalamus, causing oxytocin to be released. Similarly, the more a baby suckles, the more oxytocin is released. In both instances, the release of oxytocin from the posterior pituitary is controlled by **positive feedback**—that is, the stimulus continues to bring about an effect that ever increases in intensity. Positive feedback is not a way to maintain stable conditions and homeostasis. Oxytocin may also play a role in the propulsion of semen through the male reproductive tract and may affect feelings of sexual satisfaction and emotional bonding.

The posterior pituitary releases ADH and oxytocin, both of which are produced by the hypothalamus.

Anterior Pituitary

A portal system, consisting of two capillary systems connected by a vein, lies between the hypothalamus and the anterior pituitary (Fig. 42.5, *right*). The hypothalamus controls the anterior pituitary by producing **hypothalamic-releasing hormones** and in some instances **hypothalamic-inhibiting hormones.** For example, there is a hypothalamic-releasing hormone that stimulates the anterior pituitary to secrete a thyroid-stimulating hormone and a hypothalamic-inhibiting hormone that prevents the anterior pituitary from secreting prolactin.

Anterior Pituitary Hormones Affecting Other Glands

Three hormones produced by the **anterior pituitary** affect other glands: **thyroid-stimulating hormone (TSH)** stimulates the thyroid to produce thyroxine and triiodothyronine; **adrenocorticotropic hormone (ACTH)** stimulates the adrenal cortex to produce glucocorticoid; and **gonadotropic hormones** (follicle-stimulating hormone, FSH; and luteinizing hormone, LH) stimulate the gonads—the testes in males and the ovaries in females—to produce gametes and sex hormones. In each instance, the blood level of the last hormone in the sequence exerts negative feedback control over the secretion of the first two hormones.

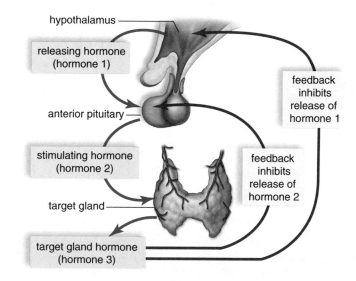

Anterior Pituitary Hormones Not Affecting Other Glands

Other types of hormones produced by the anterior pituitary are under the control of the hypothalamus, but they do not affect other endocrine glands. **Prolactin (PRL)** [L. *pro*, before, and *lactis*, milk] is produced in quantity only after childbirth. It causes the mammary glands in the breasts to develop and produce milk. It also plays a role in carbohydrate and fat metabolism.

Growth hormone (GH), or somatotropic hormone, promotes skeletal and muscular growth. It stimulates the rate at which amino acids enter cells and protein synthesis occurs. It also promotes fat metabolism as opposed to glucose metabolism.

Melanocyte-stimulating hormone (MSH) [Gk. *melanos*, black, and *kytos*, cell] causes skin-color changes in many fishes, amphibians, and reptiles having melanophores, special skin cells that produce color variations. The concentration of this hormone in humans is very low.

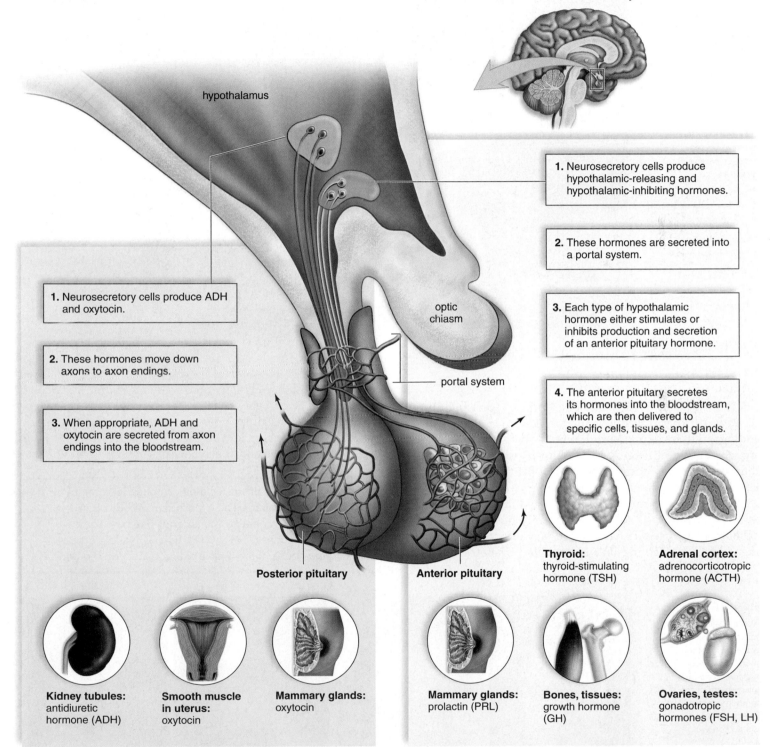

FIGURE 42.5 Hypothalamus and the pituitary.
Left: The hypothalamus produces two hormones, ADH and oxytocin, which are stored and secreted by the posterior pituitary. *Right:* The hypothalamus controls the secretions of the anterior pituitary, and the anterior pituitary controls the secretions of the thyroid, adrenal cortex, and gonads, which are also endocrine glands.

FIGURE 42.6
Effect of growth hormone.
a. The amount of growth hormone produced by the anterior pituitary during childhood affects the height of an individual. Plentiful growth hormone produces very tall basketball players. **b.** Too much growth hormone can lead to giantism, while an insufficient amount results in limited stature and even pituitary dwarfism.

a.

b.

Effects of Growth Hormone

GH is produced by the anterior pituitary. The quantity is greatest during childhood and adolescence, when most body growth is occurring. If too little GH is produced during childhood, the individual has **pituitary dwarfism,** characterized by normal proportions but small stature. Through the administration of GH, growth patterns can be restored. If too much GH is secreted, a person can become a giant (Fig. 42.6b). Giants usually have poor health, primarily because GH has a secondary effect on the blood sugar level, promoting an illness called diabetes mellitus (see page 768).

On occasion, GH is overproduced in the adult, and a condition called **acromegaly** results. Since long bone growth is no longer possible in adults, only the feet, hands, and face (particularly the chin, nose, and eyebrow ridges) can respond, and these portions of the body become overly large (Fig. 42.7).

The hypothalamus produces the hormones secreted by the posterior pituitary. The hypothalamus controls the secretion of hormones by the anterior pituitary. Some of these hormones (TSH, ACTH, FSH, LH) affect other endocrine glands. Others have a direct effect on the body.

FIGURE 42.7
Acromegaly.
Acromegaly is caused by overproduction of GH in the adult. It is characterized by enlargement of the bones in the face, the fingers, and the toes as a person ages.

Age 9

Age 16

Age 33

Age 52

42.3 OTHER ENDOCRINE GLANDS AND HORMONES

Thyroid and Parathyroid Glands

The **thyroid gland** [Gk. *thyreos*, large, door-shaped shield] is a large gland located in the neck, where it is attached to the trachea just below the larynx (see Fig. 42.1). The parathyroid glands are embedded in the posterior surface of the thyroid gland.

Thyroid Gland

The thyroid gland is the largest endocrine gland, weighing approximately 20 g. It is excessively red in appearance because of its high blood volume. It consists of two distinct lobes connected by a slender isthmus. The thyroid gland is composed of a large number of follicles, each a small spherical structure made of thyroid cells filled with the thyroid hormones triiodothyronine (T_3), which contains three iodine atoms, and **thyroxine (T_4)**, which contains four iodine atoms.

Effects of Thyroid Hormones.   To produce triiodothyronine and thyroxine, the thyroid gland actively acquires iodine. The concentration of iodine in the thyroid gland can increase to as much as 25 times that of the blood. If iodine is lacking in the diet, the thyroid gland is unable to produce the thyroid hormones. In response to constant stimulation by the anterior pituitary, the thyroid enlarges, resulting in a **simple goiter** (Fig. 42.8). Some years ago, it was discovered that the use of iodized salt allows the thy-

FIGURE 42.9 Cretinism.
Individuals who develop hypothyroidism during infancy or childhood do not grow and develop as others do.   Unless medical treatment is begun, the body is short and stocky; mental retardation is also likely.

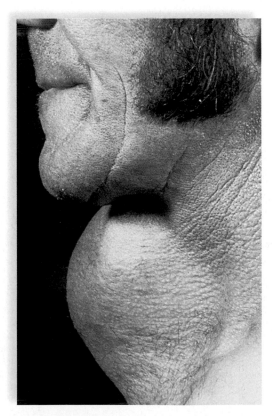

FIGURE 42.8
Simple goiter.
An enlarged thyroid gland is often caused by a lack of iodine in the diet. Without iodine, the thyroid is unable to produce its hormones but still receives stimulation from the anterior pituitary, causing the gland to enlarge.

roid to produce the thyroid hormones and therefore helps prevent simple goiter.

Thyroid hormones increase the metabolic rate. They do not have a target organ; instead, they stimulate all the cells of the body to metabolize at a faster rate. More glucose is broken down, and more energy is used.

If the thyroid fails to develop properly, a condition called **cretinism** (congenital hypothyroidism) results (Fig. 42.9). Individuals with this condition are short and stocky and have had extreme hypothyroidism (undersecretion of thyroid hormone) since infancy or childhood. Thyroid hormone therapy can initiate growth, but unless treatment is begun within the first two months of life, mental retardation results. The occurrence of hypothyroidism in adults produces the condition known as **myxedema,** which is characterized by lethargy, weight gain, loss of hair, slower pulse rate, lowered body temperature, and thickness and puffiness of the skin. The administration of adequate doses of thyroid hormones restores normal function and appearance.

In the case of hyperthyroidism (oversecretion of thyroid hormone), or Graves disease, the thyroid gland is overactive, and a goiter forms. This type of goiter is called **exophthalmic goiter.** The eyes protrude (exophthalmos) because of edema in eye socket tissues and swelling of the muscles that move the eyes. The patient usually becomes hyperactive, nervous, and irritable and suffers from insomnia. In addition, the individual may have unusual sweating and heat sensitivity. Removal or destruction of a portion of the thyroid by means of radioactive iodine is sometimes effective in curing the condition. Hyperthyroidism can also be caused by a thyroid tumor, which is usually detected as a lump during physical examination. Again, the treatment is surgery in combination with administration of radioactive iodine. The prognosis for most patients is excellent.

Calcitonin. Calcium (Ca^{2+}) plays a significant role in both nervous conduction and muscle contraction. It is also necessary to blood clotting. The blood calcium level is regulated in part by **calcitonin,** a hormone secreted by the thyroid gland when the blood calcium level rises (Fig. 42.10). The primary effect of calcitonin is to bring about the deposit of calcium in the bones. It does this by temporarily reducing the activity and number of osteoclasts. When the blood calcium lowers to normal, the release of calcitonin by the thyroid is inhibited, but a low level stimulates the release of **parathyroid hormone (PTH)** by the parathyroid glands.

Calcitonin is important in growing children but seems to have little effect in adults. It may be helpful in reducing bone loss in osteoporosis and in pregnant women. The deficiency of calcitonin is not linked with any specific disorder.

Parathyroid Glands

Many years ago, the four **parathyroid glands** were sometimes mistakenly removed during thyroid surgery because of their size and location. Parathyroid hormone causes the blood phosphate (HPO_4^{2-}) level to decrease and the blood calcium level to increase.

As previously mentioned, a low blood calcium level stimulates the release of PTH. PTH promotes the activity of osteoclasts and the release of calcium from the bones. PTH also promotes the reabsorption of calcium by the kidneys, where it activates vitamin D. Vitamin D, in turn, stimulates the absorption of calcium from the intestine. These effects bring the blood calcium level back to the normal range so that the parathyroid glands no longer secrete PTH.

When insufficient parathyroid hormone production (hypoparathyroidism) leads to a dramatic drop in the blood calcium level, tetany results. In **tetany,** the body shakes from continuous muscle contraction. This effect is brought about by increased excitability of the nerves, which initiate nerve impulses spontaneously and without rest.

In hyperparathyroidism, the calcium levels become abnormally high. This disorder can cause the bones to become abnormally soft and fragile. In addition, it can cause the individual to become unusually irritable and prone to kidney stones.

The antagonistic actions of calcitonin from the thyroid gland and parathyroid hormone from the parathyroid glands maintain the blood calcium level within normal limits.

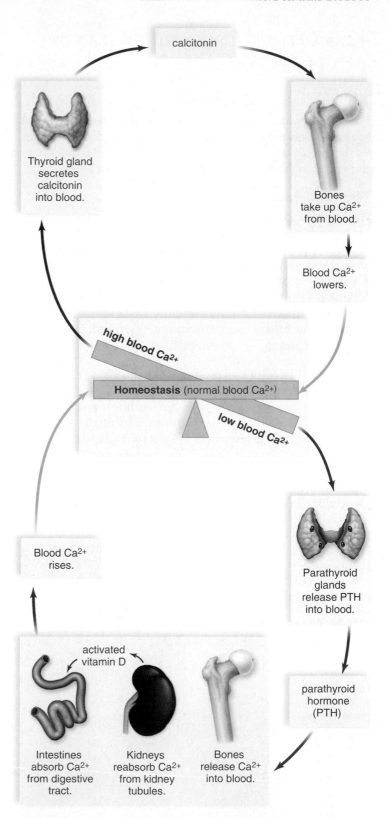

FIGURE 42.10 Regulation of blood calcium level.
Top: When the blood calcium (Ca^{2+}) level is high, the thyroid gland secretes calcitonin. Calcitonin promotes the uptake of Ca^{2+} by the bones, and therefore the blood Ca^{2+} level returns to normal. *Bottom:* When the blood Ca^{2+} level is low, the parathyroid glands release parathyroid hormone (PTH). PTH causes the bones to release Ca^{2+} and the kidneys to reabsorb Ca^{2+} and activate vitamin D; thereafter, the intestines absorb Ca^{2+}. Therefore, the blood Ca^{2+} level returns to normal.

Adrenal Glands

Two **adrenal glands** [L. *ad*, toward, and *renis*, kidney] sit atop the kidneys (see Fig. 42.1). The glands are about 5 cm long and 3 cm wide and weigh about 5 g. Each adrenal gland consists of an inner portion called the **adrenal medulla** and an outer portion called the **adrenal cortex.** These portions, like the anterior pituitary and the posterior pituitary, have no physiological connection with one another.

The hypothalamus exerts control over the activity of both portions of the adrenal glands. It initiates nerve impulses that travel by way of the brain stem, spinal cord, and sympathetic nerve fibers to the adrenal medulla, which then secretes its hormones. The hypothalamus, by means of ACTH-releasing hormone, controls the anterior pituitary's secretion of ACTH, which in turn stimulates the adrenal cortex to secrete glucocorticoids. Stress of all types, including both emotional and physical trauma, prompts the hypothalamus to stimulate the adrenal glands.

Epinephrine (adrenaline) and **norepinephrine** (noradrenaline) produced by the adrenal medulla rapidly bring about all the bodily changes that occur when an individual reacts to an emergency situation. The effects of these hormones are short-term (Fig. 42.11). Epinephrine and norepinephrine accelerate the breakdown of glucose to form ATP, trigger the mobilization of glycogen reserves in skeletal muscle, and increase the cardiac rate and force of contraction.

In contrast, the hormones produced by the adrenal cortex provide a long-term response to stress. The two major types of hormones produced by the adrenal cortex are the mineralocorticoids and the glucocorticoids. The **mineralocorticoids** regulate salt and water balance, leading to increases in blood volume and blood pressure. The **glucocorticoids** regulate carbohydrate, protein, and fat metabolism, leading to an increase in blood glucose level.

The adrenal cortex also secretes a small amount of male sex hormones and a small amount of female sex hormones in both sexes—that is, in the male, both male and female sex hormones are produced by the adrenal cortex, and in the female, both male and female sex hormones are also produced by the adrenal cortex.

The adrenal medulla is under nervous control, and the adrenal cortex is under the control of ACTH, an anterior pituitary hormone. The adrenal hormones help us respond to stress.

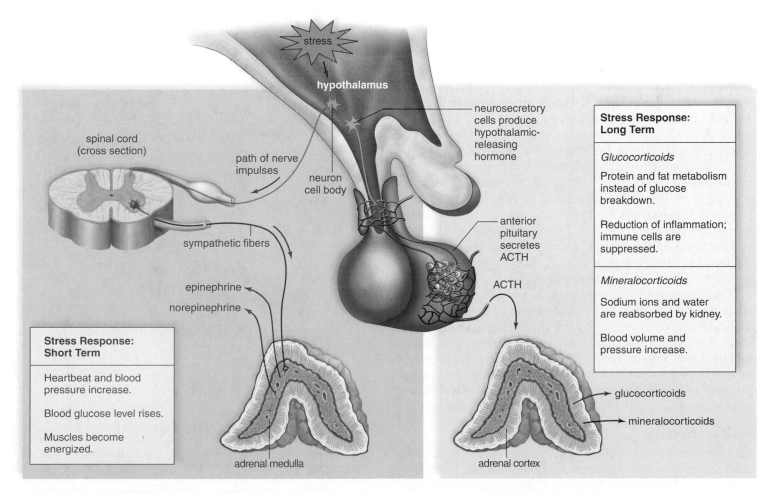

FIGURE 42.11 Adrenal glands.
Both the adrenal cortex and the adrenal medulla are under the control of the hypothalamus when they help us respond to stress. *Left:* Nervous stimulation causes the adrenal medulla to provide a rapid, but short-term, stress response. *Right:* The adrenal cortex provides a slower, but long-term, stress response. ACTH causes the adrenal cortex to release glucocorticoids. Independently, the adrenal cortex releases mineralocorticoids.

Glucocorticoids

Cortisol is a biologically significant glucocorticoid produced by the adrenal cortex. Cortisol raises the blood glucose level in at least two ways: (1) It promotes the breakdown of muscle proteins to amino acids, which are taken up by the liver from the bloodstream. The liver then breaks down these excess amino acids to glucose, which enters the blood. (2) Cortisol promotes the metabolism of fatty acids rather than carbohydrates, and this spares glucose.

Cortisol also counteracts the inflammatory response, which leads to the pain and swelling of joints in arthritis and bursitis. The administration of cortisol in the form of cortisone aids these conditions because it reduces inflammation. Very high levels of glucocorticoids in the blood can suppress the body's defense system, including the inflammatory response that occurs at infection sites. Cortisone and other glucocorticoids can relieve swelling and pain from inflammation, but by suppressing pain and immunity, they can also make a person highly susceptible to injury and infection.

Mineralocorticoids

Aldosterone is the most important of the mineralocorticoids. Aldosterone primarily targets the kidney, where it promotes renal absorption of sodium (Na^+) and renal excretion of potassium (K^+).

The secretion of mineralocorticoids is not controlled by the anterior pituitary. When the blood sodium level and therefore blood pressure are low, the kidneys secrete **renin** (Fig. 42.12). Renin is an enzyme that converts the plasma protein angiotensinogen to angiotensin I, which is changed to angiotensin II by a converting enzyme found in lung capillaries. Angiotensin II stimulates the adrenal cortex to release aldosterone. The effect of this process, called the renin-angiotensin-aldosterone system, is to raise blood pressure in two ways: (1) angiotensin II constricts the arterioles, and (2) aldosterone causes the kidneys to reabsorb sodium. When the blood sodium level rises, water is reabsorbed, in part because the hypothalamus secretes ADH (see page 694). Then blood pressure increases to normal.

As you might suspect, there is an antagonistic hormone to aldosterone. When the atria of the heart are stretched due to increased blood volume, cardiac cells release a hormone called **atrial natriuretic hormone (ANH),** which inhibits the secretion of aldosterone from the adrenal cortex. The effect of this hormone is to cause the excretion of sodium—that is, *natriuresis.* When sodium is excreted, so is water, and therefore blood pressure lowers to normal.

Malfunction of the Adrenal Cortex

When the level of adrenal cortex hormones is low due to hyposecretion, a person develops **Addison disease.** The presence of excessive but ineffective ACTH causes bronzing of the skin because ACTH, like MSH, can lead to a buildup of melanin (Fig. 42.13). Other symptoms include weight loss,

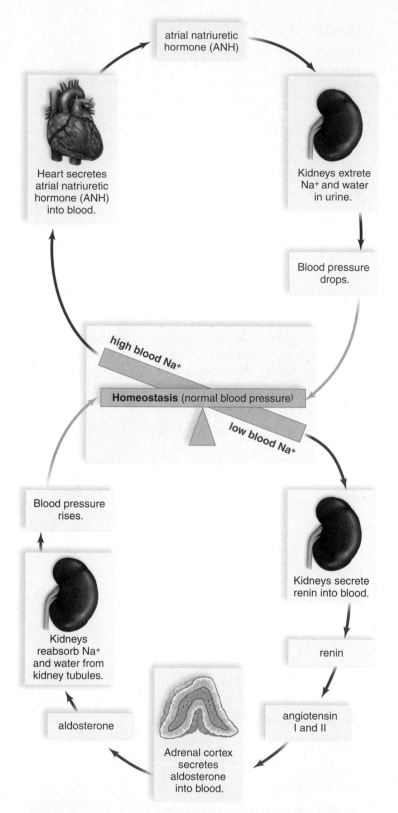

FIGURE 42.12 Regulation of blood pressure and volume.
Bottom: When the blood sodium (Na^+) level is low, a low blood pressure causes the kidneys to secrete renin. Renin leads to the secretion of aldosterone from the adrenal cortex. Aldosterone causes the kidneys to reabsorb Na^+, and water follows, so that blood volume and pressure return to normal. *Top:* When the blood Na^+ is high, a high blood volume causes the heart to secrete atrial natriuretic hormone (ANH). ANH causes the kidneys to excrete Na^+, and water follows. The blood volume and pressure return to normal.

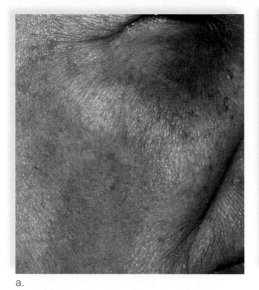 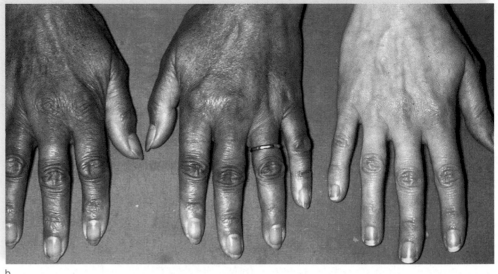

a.   b.

FIGURE 42.13 Addison disease.
Addison disease is characterized by a peculiar bronzing of the skin, particularly noticeable in light-skinned individuals.
Note the color of (**a**) the face and (**b**) the hands compared to the hand of an individual without the disease.

weakness, and hypotension (low blood pressure). Without cortisol, glucose cannot be replenished when a stressful situation arises. Even a mild infection can lead to death. The lack of aldosterone results in the loss of sodium and water, the development of low blood pressure, and possibly severe dehydration. Left untreated, Addison disease can be fatal.

When the level of adrenal cortex hormones is high due to hypersecretion, a person develops **Cushing syndrome** (Fig. 42.14). The excess cortisol results in a tendency toward diabetes mellitus as muscle protein is metabolized and subcutaneous fat is deposited in the midsection. The trunk is obese, while the arms and legs remain a normal size. An excess of aldosterone and reabsorption of sodium and water by the kidneys lead to a basic blood pH and hypertension. The face is moon-shaped due to edema. Hypertension (high blood pressure) is common in patients with Cushing syndrome. Masculinization may occur in women because of excess adrenal male sex hormones. This is referred to as adrenogenital syndrome (AGS). In women with AGS, an increase in body hair, deepening of the voice, and beard growth may occur.

The adrenal cortex hormones are essential to homeostasis. Addison disease is due to adrenal cortex hyposecretion, and Cushing syndrome is due to adrenal cortex hypersecretion.

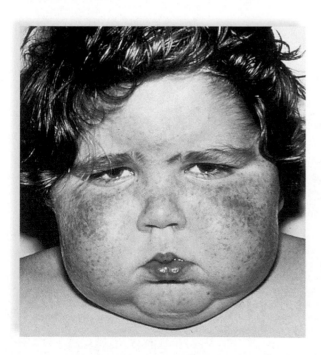 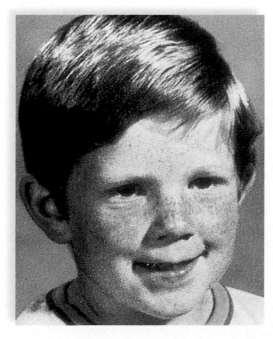

FIGURE 42.14
Cushing syndrome.
Cushing syndrome results from hypersecretion of hormones due to an adrenal cortex tumor. *Left:* Patient first diagnosed with Cushing syndrome. *Right:* Four months later, after therapy.

Pancreas

The **pancreas** is a slender, pale-colored organ that lies transversely in the abdomen between the kidneys and near the duodenum of the small intestine. It is about 20 cm long and 3 cm thick and weighs about 80 g. The pancreas is rather lumpy in consistency and is composed of two types of tissue. Exocrine tissue produces and secretes digestive juices that go by way of ducts to the small intestine. Endocrine tissue, called the **pancreatic islets** (islets of Langerhans), produces and secretes the hormones insulin and glucagon directly into the blood (Fig. 42.15). The majority of pancreas tissues are exocrine in nature.

Insulin is secreted when there is a high blood glucose level, which usually occurs just after eating. Insulin stimulates the uptake of glucose by cells, especially liver cells, muscle cells, and adipose tissue cells. In liver and muscle cells, glucose is then stored as glycogen. In muscle cells, the breakdown of glucose supplies energy for protein metabolism, and in fat cells the breakdown of glucose supplies glycerol for the formation of fat. In these ways, insulin lowers the blood glucose level.

Glucagon is secreted from the pancreas, usually in between eating, when there is a low blood glucose level. The major target tissues of glucagon are the liver and adipose tissue. Glucagon stimulates the liver to break down glycogen to glucose and to use fat and protein in preference to glucose as energy sources. Adipose tissue cells break down fat to glycerol and fatty acids. The liver takes these up and uses them as substrates for glucose formation. In these ways, glucagon raises the blood glucose level.

The antagonistic hormones insulin and glucagon, both produced by the pancreas, maintain the normal level of glucose in the blood.

Diabetes Mellitus

Diabetes mellitus is a fairly common hormonal disease in which liver cells, and indeed most body cells, are unable to take up glucose as they should. Therefore, cellular famine exists in the midst of plenty, and the person becomes extremely hungry. As the blood glucose level rises, glucose, along with water, is excreted in the urine. Urination is frequent and the loss of water in this way causes the diabetic to be extremely thirsty.

The glucose tolerance test assists in the diagnosis of diabetes mellitus. After the patient is given 100 g of glucose, the blood glucose concentration is measured at intervals. In a diabetic, the blood glucose level rises greatly and remains elevated for several hours. In the meantime, glucose appears in the urine. In a nondiabetic, the blood glucose level rises somewhat and then reutrns to normal after about two hours.

Types of Diabetes.   There are two types of diabetes mellitus. In *type 1 diabetes*, the pancreas is not producing insulin. This

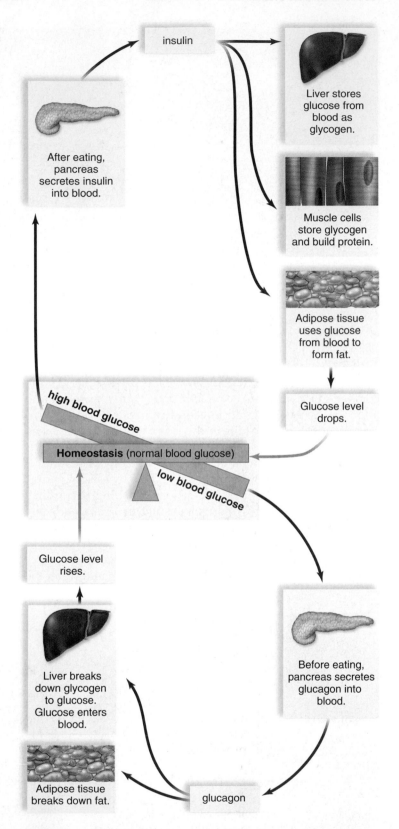

FIGURE 42.15 Regulation of blood glucose level.
Top: When the blood glucose level is high, the pancreas secretes insulin. Insulin promotes the storage of glucose as glycogen and the synthesis of proteins and fats (as opposed to their use as energy sources). Therefore, insulin lowers the blood glucose level to normal. *Bottom:* When the blood glucose level is low, the pancreas secretes glucagon. Glucagon acts opposite to insulin; therefore, glucagon raises the blood glucose level to normal.

Isolation of Insulin

The pancreas is both an exocrine gland and an endocrine gland. It sends digestive juices to the duodenum by way of the pancreatic duct, and it secretes the hormones insulin and glucagon into the bloodstream. In 1920, physician Frederick Banting decided to try to isolate insulin. Previous investigators had been unable to do this because the enzymes in the digestive juices destroyed insulin (a protein) during the isolation procedure. Banting hit upon the idea of tying off the pancreatic duct, which he knew from previous research would lead to the degeneration only of the cells that produce digestive juices and not of the pancreatic islets (of Langerhans), where insulin is made. His professor, J. J. Macleod, made a laboratory available to him at the University of Toronto and also assigned a graduate student, Charles Best, to assist him. Banting and Best had limited funds and spent that summer working,

sleeping, and eating in the lab. By the end of the summer, they had obtained pancreatic extracts that did lower the blood glucose level in diabetic dogs. Macleod then brought in biochemists, who purified the extract. Insulin therapy for the first human patient began in 1922, and large-scale production of purified insulin from pigs and cattle

followed. Banting and Macleod received a Nobel Prize for their work in 1923. The amino acid sequence of insulin was determined in 1953. Insulin is presently synthesized using recombinant DNA technology. Banting and Best performed the required steps given in the following chart to identify a chemical messenger.

Steps	Example
Identify the source of the chemical	Pancreatic islets are source
Identify the effect to be studied	Presence of pancreas in body lowers blood sugar
Isolate the chemical	Insulin isolated from pancreatic secretions
Show that the chemical alone has the effect	Insulin alone lowers blood sugar

condition is believed to be brought on by exposure to an environmental agent, most likely a virus, whose presence causes cytotoxic T cells to destroy the pancreatic islets. The body turns to the metabolism of fat, which leads to the buildup of ketones in the blood, called ketonuria, and, in turn, to acidosis (acid blood), which can lead to coma and death. As a result, the individual must have daily insulin injections. These injections control the diabetic symptoms but still can cause inconveniences, since either an overdose of insulin or missing a meal can bring on the symptoms of hypoglycemia (low blood sugar). These symptoms include perspiration, pale skin, shallow breathing, and anxiety. Because the brain requires a constant supply of glucose, unconsciousness can result. The cure is quite simple: Immediate ingestion of a sugar cube or fruit juice can very quickly counteract hypoglycemia.

It is possible to transplant a working pancreas into patients with type 1 diabetes. To do away with the necessity of taking immunosuppressive drugs after the transplant, fetal pancreatic islet cells have been injected into patients. Another experimental procedure is to place pancreatic islet cells in a capsule that allows insulin to get out but prevents antibodies and T lymphocytes from getting in. This artificial organ is implanted in the abdominal cavity.

Of the over 18 million people who now have diabetes in the United States, most have *type 2 diabetes*. Often, the patient is obese—adipose tissue produces a substance that impairs insulin receptor function. Normally, but not in the type 2 diabetic, the binding of insulin to a receptor

causes the number of glucose transporters to increase in the plasma membrane. Also, the blood insulin level is low and cells do not have enough insulin receptors.

It is possible to prevent or at least control type 2 diabetes by adhering to a low-fat, low-sugar diet and exercising regularly. If this fails, oral drugs that stimulate the pancreas to secrete more insulin and enhance the metabolism of glucose in the liver and muscle cells are available. It is projected that as many as 5 million Americans may have type 2 diabetes without being aware of it. Yet, the effects of untreated type 2 diabetes are as serious as those of type 1 diabetes.

Long-term complications of both types of diabetes are blindness, kidney disease, and cardiovascular disorders, including atherosclerosis, heart disease, stroke, and reduced circulation. The latter can lead to gangrene in the arms and legs. Pregnancy carries an increased risk of diabetic coma, and the child of a diabetic is somewhat more likely to be stillborn or to die shortly after birth. These complications of diabetes are not expected to appear if the mother's blood glucose level is carefully regulated and kept within normal limits.

Diabetes mellitus is caused by the lack of insulin or by the inability of cells to respond to the presence of insulin, a hormone that lowers the blood glucose level by causing cells to take it up. Following uptake, the liver and muscle cells store glucose as glycogen.

Testes and Ovaries

The **testes** are located in the scrotum, and the **ovaries** are located in the pelvic cavity. The testes produce **androgens** (e.g., **testosterone**), which are the male sex hormones, and the ovaries produce **estrogens** and **progesterone**, the female sex hormones. The hypothalamus and the pituitary gland control the hormonal secretions of these organs in the same manner they regulate the secretions of the thyroid gland (see page 760).

Greatly increased testosterone secretion at the time of puberty stimulates the growth of the penis and the testes. Testosterone also brings about and maintains the male secondary sex characteristics that develop during puberty. Testosterone causes growth of a beard, axillary (underarm) hair, and pubic hair. It prompts the larynx and the vocal cords to enlarge, causing the voice to change. It is partially responsible for the muscular strength of males, and this is the reason some athletes take supplemental amounts of **anabolic steroids,** which are either testosterone or related chemicals. The contraindications of taking anabolic steroids are listed in Figure 42.16. Testosterone also stimulates oil and sweat glands in the skin; therefore, it is largely responsible for acne and body odor. Another side effect of testosterone is baldness. Genes for baldness are probably inherited by both sexes, but baldness is seen more often in males because of the presence of testosterone.

The female sex hormones, estrogens and progesterone, have many effects on the body. In particular, estrogens secreted at the time of puberty stimulate the growth of the uterus and the vagina. Estrogen [Gk. *oistros*, sexual heat; L. *genitus*, producing] is necessary for egg maturation and is largely responsible for the secondary sex characteristics in females, including female body hair and fat distribution. In general, females have a more rounded appearance than males because of a greater accumulation of fat beneath the skin. Also, the pelvic girdle is wider in females than in males, resulting in a larger pelvic cavity. Both estrogen and progesterone are required for breast development and regulation of the uterine cycle, which includes monthly menstruation (discharge of blood and mucosal tissues from the uterus).

Pineal Gland

The **pineal gland,** which is located in the brain (see Fig. 42.1), produces the hormone **melatonin,** primarily at night. Melatonin is involved in our daily sleep-wake cycle; normally we grow sleepy at night when melatonin levels increase and awaken once daylight returns and melatonin levels are low (Fig. 42.17). Daily 24-hour cycles such as this are called **circadian rhythms** [L. *circum,* around, and *dies,* day], and circadian rhythms are controlled by an internal timing mechanism called a biological clock.

Based on animal research, it appears that melatonin also regulates sexual development. It has been noted that children whose pineal gland has been destroyed due to a brain tumor experience early puberty.

Thymus Gland

The **thymus gland** is a lobular gland that lies just beneath the sternum (see Fig. 42.1). This organ reaches its largest size and is most active during childhood. With aging, the organ gets smaller and becomes fatty. Lymphocytes that originate in the bone marrow and then pass through the thymus are transformed into T lymphocytes. The lobules of the thymus are lined by epithelial cells that secrete hormones called thymosins. These hormones aid in the differentiation of T lymphocytes packed inside the lobules.

balding in men and women; hair on face and chest in women

deepening of voice in women

breast enlargement in men and breast reduction in women

liver dysfunction and cancer

kidney disease and retention of fluids, called "steroid bloat"

reduced testicular size, low sperm count, and impotency

'roid mania– delusions and hallucinations; depression upon withdrawal

severe acne

high blood cholesterol and atherosclerosis; high blood pressure and damage to heart

in women, increased size of ovaries; cessation of ovulation and menstruation

stunted growth in youngsters by prematurely halting fusion of the growth plates

FIGURE 42.16 **The effects of anabolic steroid use.**

Although the hormones secreted by the thymus ordinarily work in the thymus, there is hope that these hormones could be injected into AIDS or cancer patients, where they would enhance T lymphocyte function.

Other Hormones

Some hormones in the body, as exemplified by the ones discussed here, are produced by tissues/cells rather than by one of the glands listed in Table 42.2.

Leptin

Leptin is a peptide hormone secreted by adipose (fat) tissue throughout the body. It was first described in the 1990s. One of its most interesting functions is its role in the feedback control of appetite. Leptin binds to neurons in the CNS that are concerned with the control of appetite. It can bring about feelings of satiation and can suppress appetite. Early researchers hoped that leptin could be used to control obesity in humans. Unfortunatley, the trials have not yielded satisfactory results.

Erythropoietin

Erythropoietin (EPO) is a peptide hormone produced by the kidneys. It is released in response to low oxygen levels in kidney tissues. EPO serves to stimulate the production of red blood cells (erythropoiesis) and speed up the maturation of red blood cells. Under the influence of EPO, bone marrow can increase the rate of red blood cell production upward to 30 million per second. In recent years, some athletes have practiced blood doping, in which EPO is used to improve performance. The potential dangers of blood doping far outweigh the temporary advantages.

Local Hormones

Local hormones are produced by cells, and they act on neighboring cells. Examples include growth factors, cytokines, and prostaglandins. **Prostaglandins** are potent chemical signals

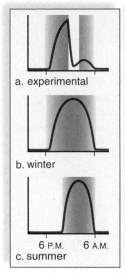

a. experimental

b. winter

6 P.M.   6 A.M.

c. summer

FIGURE 42.17 Melatonin production.
Melatonin production is greatest at night when we are sleeping. Light suppresses melatonin production (**a**), so its duration is longer in the winter (**b**) than in the summer (**c**).

produced within cells from arachidonate, a fatty acid. In the uterus, prostaglandins cause muscles to contract; therefore, they are implicated in the pain and discomfort of menstruation in some women. Also, prostaglandins mediate the effects of pyrogens, chemicals that are believed to reset the temperature regulatory center in the brain. Aspirin reduces body temperature and controls pain because of its effect on prostaglandins.

Certain prostaglandins reduce gastric secretion and have been used to treat ulcers; others lower blood pressure and have been used to treat hypertension; and yet others inhibit platelet aggregation and have been used to prevent thrombosis. However, different prostaglandins have contrary effects, and it has been very difficult to successfully standardize their use. Therefore, prostaglandin therapy is still considered experimental.

CONNECTING THE CONCEPTS

The nervous system and the endocrine system are structurally and functionally related. The hypothalamus, a portion of the brain, controls the pituitary, an endocrine gland. The hypothalamus even produces the hormones that are released by the posterior pituitary. Neurosecretory cells in the hypothalamus produce chemical signals that control the activity of the anterior pituitary. Indeed, they are hormones carried by blood vessels (a portal system) that act at a distance between organs.

A survey of the animal kingdom also shows an overlap of the two systems. In the snail *Aplysia*, a hormone called egg-laying hormone (ELH) stimulates the laying of long strings of eggs. ELH is an excitatory neurotransmitter that also diffuses into the circulatory system and excites the smooth

muscle cells of the reproductive duct, causing them to contract and expel strings of eggs. Similarly in mammals, norepinephrine is both a neurotransmitter in the sympathetic division of the autonomic system and a hormone released by the adrenal medulla. Sympathetic nerve endings stimulate the adrenal medulla to release norepinephrine, which promotes the "fight-or-flight" reaction.

Molting in insects is controlled by the balance of three different hormones, one of which is called brain hormone because it is produced by neurosecretory cells. Brain hormone controls a gland that releases ecdysone, which promotes molting. As long as another gland is producing a hormone called juvenile hormone, molting produces a more mature larva. With time, the level of juvenile hormone

falls off, and molting produces a pupa in which metamorphosis leads to the adult insect.

The close connection between the nervous and endocrine systems suggests that the categorization of organs in any one human system is somewhat arbitrary. Recall that the digestive tract produces hormones that control the secretion of digestive juices (see page 661); that the kidneys produce renin (see page 694), which leads to the release of aldosterone; and that the heart produces atrial natriuretic hormone, which opposes the action of aldosterone (see page 695). Local hormones that affect only their neighbors are produced by most cells. It has even been discovered that brain cells produce insulin that is used locally to influence the metabolism of adjacent cells.

Summary

42.1 ENDOCRINE GLANDS

The nervous system and the endocrine system can be contrasted as in Table 42.1. Endocrine glands secrete hormones into the bloodstream, and from there they are distributed to target organs or tissues.

Hormones are a type of chemical signal that usually act at a distance between body parts. Hormones are either peptides or steroids. Reception of a peptide hormone at the plasma membrane activates an enzyme cascade inside the cell. Steroid hormones combine with a receptor in the cell, and the complex attaches to and activates DNA. Protein synthesis follows.

42.2 HYPOTHALAMUS AND PITUITARY GLAND

Neurosecretory cells in the hypothalamus produce antidiuretic hormone (ADH) and oxytocin, which are stored in axon endings in the posterior pituitary until they are released.

The hypothalamus produces hypothalamic-releasing and hypothalamic-inhibiting hormones, which pass to the anterior pituitary by way of a portal system. The anterior pituitary produces several types of hormones, and some of these stimulate other hormonal glands to secrete hormones.

42.3 OTHER ENDOCRINE GLANDS AND HORMONES

The thyroid gland requires iodine to produce thyroxine and triiodothyronine, which increase the metabolic rate. If iodine is available in insufficient quantities, a simple goiter develops; if the thyroid is overactive, an exophthalmic goiter develops. The thyroid gland also produces calcitonin, which helps lower the blood calcium level. The parathyroid glands secrete parathyroid hormone, which raises the blood calcium and decreases the blood phosphate levels.

The adrenal glands respond to stress: immediately, the adrenal medulla secretes epinephrine and norepinephrine, which bring about responses we associate with emergency situations. On a long-term basis, the adrenal cortex produces the glucocorticoids (e.g., cortisol) and the mineralocorticoids (e.g., aldosterone). Cortisol stimulates hydrolysis of proteins to amino acids that are converted to glucose; in this way, it raises the blood glucose level. Aldosterone causes the kidneys to reabsorb sodium ions (Na^+) and to excrete potassium ions (K^+). Addison disease develops when the adrenal cortex is underactive, and Cushing syndrome develops when the adrenal cortex is overactive.

The pancreatic islets secrete insulin, which lowers the blood glucose level, and glucagon, which has the opposite effect. The most common illness caused by hormonal imbalance is diabetes mellitus, which is due to the failure of the pancreas to produce insulin or the failure of the cells to take it up.

The gonads produce the sex hormones; the pineal gland produces melatonin, which may be involved in circadian rhythms and the development of the reproductive organs; and the thymus secretes thymosins, which stimulate T-lymphocyte production and maturation.

Tissue and organs having other functions also produce hormones. Leptin is a newly described hormone that regulates appetite, and erythropoietin stimulates the production of red blood cells. Cells produce local hormones, for example, prostaglandins are produced and act locally.

Reviewing the Chapter

1. Categorize chemical signals into three groups based on the distance between site of secretion and site of reception, and give examples for each group. 758
2. Explain how steroid hormones and peptide hormones affect the metabolism of the cell. 758–59
3. Explain the relationship of the hypothalamus to the posterior pituitary gland and to the anterior pituitary gland. List the hormones secreted by the posterior and anterior pituitary. 760–61
4. Give an example of the negative feedback relationship among the hypothalamus, the anterior pituitary, and other endocrine glands. 760–61
5. Discuss the effect of there being too much or too little growth hormone when a young person is growing. What is the result if the anterior pituitary produces growth hormone in an adult? 762
6. What two types of goiters are associated with a malfunctioning thyroid? Explain each type. 763–64
7. How do the thyroid and the parathyroid work together to control the blood calcium level? 764
8. How do the adrenal glands respond to stress? What hormones are secreted by the adrenal medulla, and what effects do these hormones have? 765
9. Name the most significant glucocorticoid and mineralocorticoid, and discuss the function of each. Explain the symptoms of Addison disease and Cushing syndrome. 766–67
10. Draw a diagram to explain how insulin and glucagon maintain the blood glucose level. Use your diagram to explain three major symptoms of type I diabetes mellitus. 768–69
11. Name the other endocrine glands cited in this chapter, and discuss the functions of the hormones they secrete. Also discuss the hormones not produced by endocrine glands. 770–71

Testing Yourself

Choose the best answer for each question. For questions 1–5, match each hormone to a gland in the key.

KEY:

   a. pancreas
   b. anterior pituitary
   c. posterior pituitary
   d. thyroid
   e. adrenal medulla
   f. adrenal cortex

1. cortisol
2. growth hormone (GH)
3. oxytocin storage
4. insulin
5. epinephrine
6. The blood cortisol level controls the secretion of
   a. hypothalamic-releasing hormone from the hypothalamus.
   b. adrenocorticotropic hormone (ACTH) from the anterior pituitary.
   c. cortisol from the adrenal cortex.
   d. All of these are correct.

7. The anterior pituitary controls the secretion(s) of both
   a. the adrenal medulla and the adrenal cortex.
   b. the thyroid and the adrenal cortex.
   c. the ovaries and the testes.
   d. Both b and c are correct.

8. Diabetes mellitus is associated with
   a. too much insulin in the blood.
   b. too high a blood glucose level.
   c. blood that is too dilute.
   d. All of these are correct.

9. Which of these is not a pair of antagonistic hormones?
   a. insulin—glucagon
   b. calcitonin—parathyroid hormone
   c. cortisol—epinephrine
   d. aldosterone—atrial natriuretic hormone (ANH)
   e. thyroxine—growth hormone

10. Which hormone and condition are mismatched?
    a. growth hormone—acromegaly
    b. thyroxine—goiter
    c. parathyroid hormone—tetany
    d. cortisol—myxedema
    e. insulin—diabetes

11. Which of the following hormones could affect fat metabolism?
    a. growth hormone   d. glucagon
    b. thyroxine   e. All of these are correct.
    c. insulin

12. The difference between type I and type II diabetes is that
    a. for type II diabetes, insulin is produced but not used; type I results from lack of insulin production.
    b. treatment for type II involves insulin injections, while type I can be controlled, usually by diet.
    c. only type I can result in complications such as kidney disease, reduced circulation, or stroke.
    d. type I can be a result of lifestyle, and type II is thought to be caused by a virus or other agent.

13. Which of the following hormones is/are found in females?
    a. estrogen
    b. testosterone
    c. follicle-stimulating hormone
    d. Both a and c are correct.
    e. All of these are correct.

14. Parathyroid hormone causes
    a. the kidneys to excrete more calcium ions.
    b. bone tissue to break down and release calcium into the bloodstream.
    c. fewer calcium ions to be absorbed by the intestines.
    d. more calcium ions to be deposited in bone tissue.

15. Hormones from all but which of the following glands can affect glucose levels in the body?
    a. pancreas   d. hypothalamus
    b. adrenal glands   e. thymus
    c. pituitary

16. Tropic hormones are hormones that affect other endocrine tissues. Which of the following would be considered a tropic hormone?
    a. calcitonin   d. melatonin
    b. oxytocin   e. follicle-stimulating
    c. glucagon   hormone

17. One of the chief differences between endocrine hormones and local hormones is
    a. the distance over which they act.
    b. that one is a chemical signal and the other is not.
    c. only endocrine hormones are made by humans.
    d. All of these are correct.

18. Peptide hormones
    a. are received by a receptor located in the plasma membrane.
    b. are received by a receptor located in the cytoplasm.
    c. bring about the transcription of DNA.
    d. Both b and c are correct.

19. Complete this diagram by filling in blanks a–e.

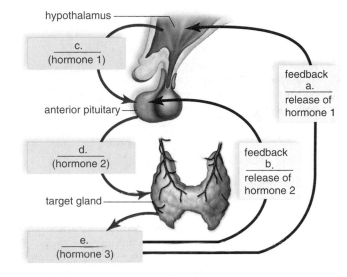

For questions 20–25, match the function to a hormone in the key. Choose more than one answer if correct.

KEY:
   a. antidiuretic
   b. oxytocin
   c. glucocorticoids
   d. glucagon
   e. parathyroid

20. raises blood glucose level

21. stimulates uterine muscle contraction

22. stimulates water reabsorption by kidneys

23. stimulates release of milk by mammary glands

24. raise blood glucose and stimulate breakdown of protein

25. raises blood calcium level

26. Steroid hormones are secreted by
    a. the adrenal cortex.
    b. the gonads.
    c. the thyroid.
    d. Both a and b are correct.
    e. Both b and c are correct.

27. Steroid hormones
    a. bind to a receptor located in the plasma membrane.
    b. cause the production of cAMP.
    c. activate protein kinase.
    d. stimulate the production of mRNA.

28. Which of the following statements about the pituitary gland is incorrect?
    a. The pituitary lies inferior to the hypothalamus.
    b. Growth hormone and prolactin are secreted by the anterior pituitary.
    c. The anterior pituitary and posterior pituitary communicate with each other.
    d. Axons run between the hypothalamus and the posterior pituitary.
29. Prostaglandins
    a. have a consistent effect.
    b. are useful in the treatment of cancer.
    c. are carried in the blood.
    d. stimulate other glands.
    e. act locally.
30. Erythropoietin
    a. stimulates platelet production.
    b. stimulates leukocyte production.
    c. inhibits red blood cell production.
    d. inhibits leukocyte production.
    e. stimulates red blood cell production.

Thinking Scientifically

1. Caffeine inhibits the breakdown of cAMP in the cell. Referring to Figure 42.3, how would this influence a stress response brought about by epinephrine?
2. Both males and females can develop secondary sex characteristics of the opposite sex if they take enough of the appropriate sex hormone. Hypothesize the pattern of gene inheritance for these characteristics, and explain why certain of these genes and not others are expressed.

Bioethical Issue: Fertility Drugs

Higher-order multiple births (triplets or more) in the United States increased 19% between 1980 and 1994. During these years, it became customary to use fertility drugs (gonadotropic hormones) to stimulate the ovaries. The risks for premature delivery, low birth weight, and developmental abnormalities rise sharply for higher-order multiple births. And the physical and emotional burden placed on the parents is extraordinary. They face endless everyday chores and find it difficult to maintain normal social relationships, if only because they get insufficient sleep. Finances are strained to provide for the children's needs, including housing and child-care assistance. About one-third report that they received no help from relatives, friends, or neighbors in the first year after the birth. Trips to the hospital for accidental injury are more frequent because parents with only two arms and two legs cannot keep so many children safe at one time.

Many clinicians are now urging that all possible steps be taken to ensure that the chance of higher-order multiple births be reduced. However, none of the ethical choices to bring this about are attractive. If fertility drugs are outlawed, some couples might be denied the possibility of ever having a child. A higher-order multiple pregnancy can be terminated, or selective reduction can be done. During selective reduction, one or more of the fetuses is killed by an injection of potassium chloride. Selective reduction could very well result in psychological and social complications for the mother and surviving children. The parents could opt to use in vitro fertilization (in which the eggs are fertilized in the lab), with the intent that only one or two zygotes will be placed in the woman's womb. But then any leftover zygotes may never have an opportunity to continue development.

Understanding the Terms

acromegaly 762
Addison disease 766
adrenal cortex 765
adrenal gland 765
adrenal medulla 765
adrenocorticotropic hormone (ACTH) 760
aldosterone 766
anabolic steroid 770
androgen 770
anterior pituitary 760
antidiuretic hormone (ADH) 760
atrial natriuretic hormone (ANH) 766
calcitonin 764
chemical signal 758
circadian rhythm 770
cortisol 766
cretinism 763
Cushing syndrome 767
cyclic adenosine monophosphate (cAMP) 758
diabetes mellitus 768
endocrine gland 758
endocrine system 756
epinephrine 765
erythropoietin (EPO) 771
estrogen 770
exophthalmic goiter 763
first messenger 758
glucocorticoid 765
gonadotropic hormone 760
growth hormone (GH) 761
hormone 756
hypothalamic-inhibiting hormone 760
hypothalamic-releasing hormone 760
hypothalamus 760
leptin 771
melanocyte-stimulating hormone (MSH) 761
melatonin 770
mineralocorticoid 765
myxedema 763
negative feedback 760
norepinephrine 765
ovary 770
oxytocin 760
pancreas 768
pancreatic islet 768
parathyroid gland 764
parathyroid hormone (PTH) 764
peptide hormone 758
pheromone 758
pineal gland 770
pituitary dwarfism 762
pituitary gland 760
positive feedback 760
posterior pituitary 760
progesterone 770
prolactin (PRL) 760
prostaglandin 771
renin 766
second messenger 758
simple goiter 763
steroid hormone 758
testes 770
testosterone 770
tetany 764
thymus gland 770
thyroid gland 763
thyroid-stimulating hormone (TSH) 760
thyroxine (T_4) 763

Match the terms to these definitions:
a. _____ Organ in the neck; secretes several important hormones, including thyroxine and calcitonin.
b. _____ Common homeostatic control mechanism in which the output of a system shuts off or reduces the intensity of the original stimulus.
c. _____ Organ that produces melatonin.
d. _____ Type of hormone that binds to a plasma membrane receptor; results in activation of enzyme cascade.
e. _____ Chemical substance secreted by one organism that influences the behavior of another.

ARIS, the *Biology* Website

ARIS, the website for *Biology*, provides a wealth of information organized and integrated by chapter. You will find practice quizzes, interactive activities, labeling exercises, flashcards, and much more that will complement your learning and understanding of general biology.

www.mhhe.com/maderbiology9

43

REPRODUCTIVE SYSTEMS

To maintain homeostasis and survive, an individual organism must digest food, exchange gases, maintain salt and water balance, and coordinate body systems. To perpetuate the species, organisms must reproduce. Reproduction serves as the link to generations past, present, and future.

The many different ways that animals reproduce can be categorized as asexual or sexual. Asexual reproduction does not involve sex cells (sperm and egg)—offspring have exactly the same traits as the single parent unless mutations occur. Usually, a parent's adaptations for survival are passed on unchanged to the offspring, and this can be an advantage if the environment is not changing.

Because meiosis and fertilization occur during sexual reproduction, this form of reproduction usually has the advantage of producing offspring that are not exactly like either parent. The introduction of genetic variation among offspring may help to ensure the survival of the species if environmental conditions are changing. Roseate spoonbills, named for their coloring and spoon-shaped bill, unfortunately went through a bottleneck in the 1800s that may have reduced genetic variations. They were hunted to extinction for their feathers, which were used in the making of ladies' hats. After receiving legal protection in the 1940s, their numbers recovered, but today scientists are worried because broad use of pesticides to control mosquitoes and changes made to the spoonbills' wetland habitat may be causing their numbers to decline once again. Roseate spoonbills keep the same mate for the breeding season, and both sexes help build a nest located in trees. Both take turns sitting on the eggs and feeding the nestlings until they are proficient flyers.

Roseate spoonbill, *Ajaia ajaja*, nestlings

43.1 HOW ANIMALS REPRODUCE

Although the majority of animals reproduce sexually, several species are capable of asexual reproduction on occasion. In asexual reproduction, a single parent gives rise to offspring that are identical to the parent, unless mutations have occurred. In sexual reproduction, sex cells, or gametes, produced by the parents unite forming a genetically unique individual.

Asexual Reproduction

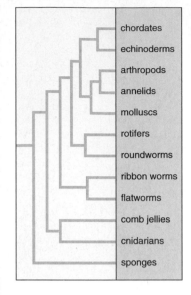

Asexual reproduction can occur in several phyla of invertebrates, including sponges, cnidarians, flatworms, annelids, and echinoderms. Sponges produce asexual gemmules that develop into new individuals. Cnidarians, such as *Hydra,* can reproduce asexually by budding (Fig. 43.1). A new individual arises as an outgrowth (bud) of the parent. *Obelia* is dimorphic and has an asexual colonial stage and a sexually reproducing medusa stage (see Fig. 29.7). Many flatworms can constrict into two halves; each half regenerates to become a new individual. Several annelids, such as earthworms and sandworms, have the ability to regenerate from fragments. Fragmentation followed by regeneration is also seen among sponges and echinoderms. If a sea star is chopped up, it has the potential to regenerate into several new individuals.

Several types of flatworms, roundworms, crustaceans, annelids, insects, fishes, lizards, and even some turkeys have the ability to reproduce parthenogenetically. **Parthenogenesis** [Gk. *parthenos,* virgin, and *genitus,* producing] is a modification of sexual reproduction in which an unfertilized egg develops into a complete individual. In honeybees, the queen bee can fertilize eggs or allow eggs to pass unfertilized as she lays them. The fertilized eggs become diploid females called workers, and the unfertilized eggs become haploid males called drones.

Sexual Reproduction

Usually during sexual reproduction, the egg of one parent is fertilized by the sperm of another. The majority of animals are dioecious, having separate sexes. Monoecious, or hermaphroditic, organisms have both male and female sex organs in the same body. Some hermaphroditic organisms, such as tapeworms, are capable of self-fertilization, but the majority of hermaphroditic species, such as earthworms, practice cross-fertilization. Sequential hermaphroditism, or sex reversal, also occurs. In coral reef fishes called wrasses, a male has a harem of several females. If the male dies, the largest female becomes a male.

Animals usually produce gametes in specialized organs called **gonads** [Gk. *gone,* seed]. Sponges are an exception to this rule because the collar cells lining the central cavity of a sponge give rise to sperm and eggs. Hydras and other cnidarians produce only temporary gonads in the fall, when sexual reproduction occurs (Fig. 43.1). Animals in other phyla have permanent reproductive organs. The gonads are **testes,** which produce sperm, and **ovaries** [L. *ovaris,* egg-keeper], which produce eggs. Eggs or sperm are derived from germ cells, which become specialized for this purpose during early development. Other cells in a gonad support and nourish the developing gametes or produce hormones necessary to the reproductive process. The reproductive system also consists of a number of accessory structures, such as ducts and storage areas, that aid in bringing the gametes together.

A number of reproductive strategies occur in the animal kingdom to ensure that gametes unite. Aquatic animals that practice external fertilization are programmed to release their eggs in the water only at certain times. One environmental signal that seems to influence the release of eggs is the lunar cycle. Each month the moon moves closer to the Earth, and the tides become somewhat higher. Aquatic animals able to sense this change can release their gametes at the same time. Hundreds of thousands of palolo worms, an annelid, rise to the surface of the sea and release their eggs during a two- to four-hour period on two or three specific successive days of the year. Most likely they are under the control of a biological clock that can sense the passage of time so that their reproductive behavior is synchronized.

Copulation [L. *copulatus,* join] is sexual union to facilitate the reception of sperm by a partner, usually a female. In terrestrial animals, males typically have a penis for depositing sperm into the vagina of females. Aquatic animals also have other types of copulatory organs. Lobsters and crayfish, which are arthropods, have modified swimmerets. Cuttlefish and octopuses, which are molluscs, use an arm; and sharks, which are vertebrates, have a modified pelvic fin that passes packets of sperm to the female. Among terrestrial animals, most birds lack a penis or a vagina. They

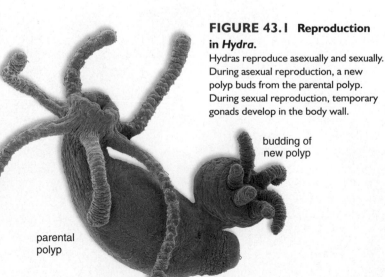

FIGURE 43.1 Reproduction in *Hydra.*
Hydras reproduce asexually and sexually. During asexual reproduction, a new polyp buds from the parental polyp. During sexual reproduction, temporary gonads develop in the body wall.

budding of new polyp

parental polyp

have a cloaca, a chamber that receives products from the digestive, urinary, and reproductive tracts. A male transfers sperm to a female after placing his cloacal opening adjacent to hers.

Life History Strategies

Many aquatic animals practice external fertilization; that is, eggs and sperm join outside the body in the water. Terrestrial animals tend to practice internal fertilization, in which egg and sperm join inside the female's body. Both types of animals are usually oviparous, meaning that they deposit eggs in the external environment. In insects, the eggs are produced in the ovaries, and as they mature, they increase in size because yolk has been added to them. **Yolk** is stored food to be used by the developing embryo. Then, to prevent the eggs from drying out, they are covered by a shell consisting of several layers of protein- and wax-containing material. Small holes are left at one end of the egg for the entry of sperm. Some insects have a special internal organ for storing sperm for some time after copulation so that the eggs can be fertilized internally before they are deposited in the environment.

A **larva** is an independent form of an animal that is often quite different in appearance and way of life from the adult. A larva is able to seek its own food and to sustain itself until it becomes an adult. Some terrestrial insects undergo several larval stages, and then the animal pupates. A pupa is enclosed by a hardened cuticle, often within a cocoon. Here **metamorphosis** [Gk. *meta*, implying change, and *morphe*, shape, form], which is a dramatic change in shape, takes place. Then the adult insect emerges and flies off to find a mate and reproduce. Other insects (e.g., grasshoppers) undergo incomplete metamorphosis; pupation does not occur, and there are a number of nymph stages, each one looking more like the adult.

Many aquatic animals also have a larval stage. Since the larva has a different lifestyle, it is able to use a different food source than the adult. In sea stars, which are echinoderms, the bilaterally symmetrical larva simply attaches itself to a substratum and undergoes metamorphosis to become a radially symmetrical juvenile. Among barnacles, which are arthropods, the free-swimming larva metamorphoses into the sessile adult with calcareous plates. Crayfish, on the other hand, do not have a larval stage; the egg hatches into a tiny juvenile with the same form as the adult.

Reptiles, and particularly birds, provide their eggs with plentiful yolk; there is no larval stage. Complete development takes place within a shelled egg containing **extraembryonic membranes** [L. *extra*, on the outside] to serve the needs of the embryo. The outermost membrane, the chorion, lies next to the shell and functions in gas exchange. The amnion forms a water-filled sac around the embryo, ensuring that it will not dry out. The yolk sac holds the yolk, which nourishes the embryo, and the allantois holds nitrogen waste products (see Fig. 44.11). The shelled egg frees these animals from the need to reproduce in the water and is a significant adaptation to the terrestrial environment.

FIGURE 43.2 Parenting in birds.
Birds, such as the American goldfinch, *Carduelis tristis*, are oviparous and lay hard-shelled eggs. They are well known for incubating their eggs and caring for their offspring after they hatch.

Birds in particular tend their eggs, and newly hatched birds usually have to be fed before they are able to fly away and seek food for themselves. Complex hormones and neural regulation are involved in the reproductive behavior of parental birds (Fig. 43.2). Some animals are ovoviviparous. In these animals, such as a dogfish shark, the eggs are retained in the body until they hatch. Then, fully developed offspring, which have a way of life like the parent, are released. Oysters, which are molluscs, retain their eggs in the mantle cavity, and male sea horses, which are vertebrates, have a special brood pouch in which the eggs develop. Garter snakes, water snakes, and pit vipers retain their eggs in their bodies until they hatch and give birth to living young.

With the exception of the oviparous duckbill platypus and spiny anteater, mammals are **viviparous,** producing living young. Viviparity represents the ultimate in caring for the zygote and embryo. Viviparous mammals are also called placental mammals. The **placenta** is a complex structure derived in part from the chorion, which appeared first in shelled eggs. The placenta allows exchanges to occur between the blood of the developing offspring and the blood of the mother. During the exchange the offspring receives nutrient molecules and oxygen from the mother, and the mother receives carbon dioxide and waste molecules from the offspring. After birth, the nutrients needed for further growth and development are usually supplied by the mother. Marsupial offspring such as kangaroos are born in an extremely immature state; they complete their development within a pouch or marsupium, where they are nourished with milk. The uterus of most placental mammals has two horns, where several embryos can develop. Only among primates, including humans, is there a simplex uterus where, usually, a single embryo develops.

Most animals have gonads in which gametes are produced. The details concerning gamete union and care of the eggs and the young vary greatly.

43.2 MALE REPRODUCTIVE SYSTEM

The human male reproductive system includes the organs pictured in Figure 43.3 and listed in Table 43.1. The male gonads are paired testes, which are suspended within the scrotal sacs of the scrotum. The testes begin their development inside the abdominal cavity, but they descend into the scrotal sacs as development proceeds. If the testes do not descend—and the male does not receive hormone therapy or undergo surgery to place the testes in the scrotum—sterility (the inability to produce offspring) results. Sterility occurs because undescended testes developing in the body cavity are subject to higher body temperatures than those in the scrotum. A cooler temperature is critical for the normal development of sperm.

Sperm produced in the seminiferous tubules of the testes mature within the epididymides (sing., epididymis), which are tightly coiled tubules lying just outside the testes. Maturation seems to be required for the sperm to swim to the egg. Once the sperm have matured, they are propelled into the vasa deferentia (sing., vas deferens) by muscular contractions. Sperm are stored in both the epididymides and the vasa deferentia. When a male becomes sexually aroused, sperm enter the ejaculatory ducts and then the urethra, part of which is located within the penis.

The **penis** is a cylindrical organ that usually hangs in front of the scrotum. Three cylindrical columns of spongy, erectile tissue containing distensible blood spaces extend

TABLE 43.1

Male Reproductive System

Organ	Function
Testes	Produce sperm and sex hormones
Epididymides	Sites of maturation and some storage of sperm
Vasa deferentia	Conduct and store sperm
Seminal vesicles	Contribute fluid to semen
Prostate gland	Contributes fluid to semen
Urethra	Conducts sperm (and urine)
Bulbourethral glands	Contribute fluid to semen
Penis	Organ of copulation

through the shaft of the penis (Fig. 43.4). During sexual arousal, nervous reflexes cause an increase in arterial blood flow to the penis. This increased blood flow fills the blood space in the erectile tissue, and the penis, which is normally limp (flaccid), stiffens and increases in size. These changes are called erection. If the penis fails to become erect, the condition is called erectile dysfunction (formerly called impotency).

Semen (seminal fluid) [L. *semen*, seed] is a thick, whitish fluid that contains sperm and secretions from three glands (Table 43.1). The seminal vesicles lie at the

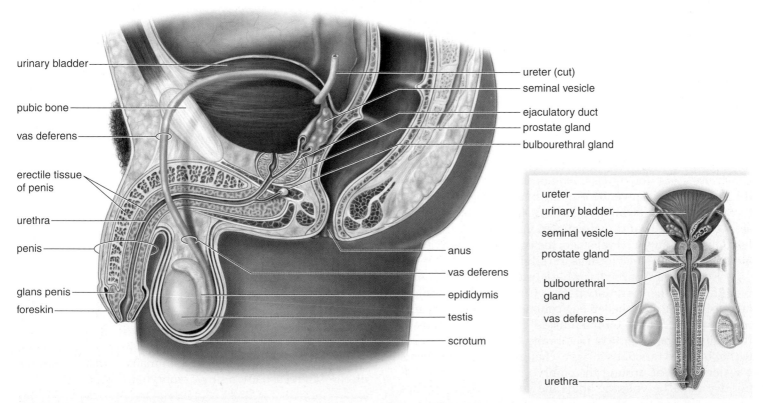

FIGURE 43.3 The male reproductive system.
The testes produce sperm. The seminal vesicles, the prostate gland, and the bulbourethral glands provide a fluid medium for the sperm, which move from the vas deferens through the ejaculatory duct to the urethra in the penis. The foreskin (prepuce) is removed when a penis is circumcised.

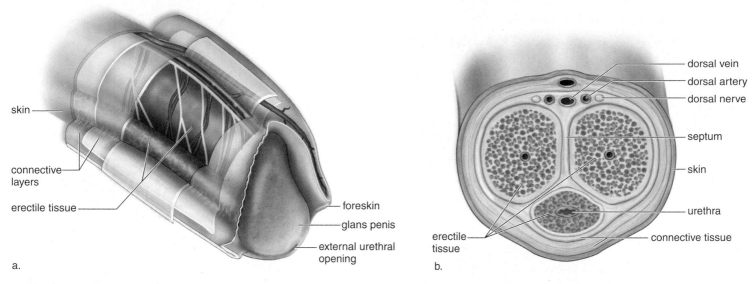

FIGURE 43.4 Penis anatomy.
a. Beneath the skin and the connective tissue lies the urethra, surrounded by erectile tissue. This tissue expands to form the glans penis, which in uncircumcised males is partially covered by the foreskin (prepuce). **b.** Two other columns of erectile tissue in the penis are located dorsally.

base of the bladder. As sperm pass from the vasa deferentia into the ejaculatory ducts, these vesicles secrete a thick, viscous fluid containing nutrients (fructose) for possible use by the sperm. The fluid also contains prostaglandins that stimulate smooth muscle contraction along the male and female reproductive tracts. Just below the bladder lies the prostate gland, which secretes a milky alkaline fluid believed to activate or increase the motility of the sperm. Sperm are more viable in a basic solution, and semen, which is milky in appearance, has a slightly basic pH. Prostatic secretions also include an antibiotic called seminalplasmin that helps prevent urinary tract infections in males. Slightly below the prostate gland, on either side of the urethra, is a pair of small glands called bulbourethral glands that produce mucous secretions with a lubricating effect. An average male releases 2–5 ml of semen during orgasm. Typically semen is composed of 60% seminal vesicle secretions, 30% prostatic fluids, less than 10% sperm and spermatic duct secretions, and a trace amount of bulbourethral fluid.

Notice from Figure 43.3 that the urethra also carries urine from the bladder during urination. In older men, the prostate gland may become enlarged, thereby constricting the urethra and making urination difficult. Also, prostate cancer is the most common form of cancer in men.

Sperm produced by the testes mature in the epididymides and pass from the vasa deferentia to the ejaculatory ducts and then the urethra, where certain glands add fluid to semen.

Ejaculation

If sexual arousal reaches its peak, ejaculation follows an erection. The first phase of ejaculation is called emission. During emission, the spinal cord sends nerve impulses via appropriate nerve fibers to the epididymides and vasa deferentia. Their subsequent motility causes sperm to enter the ejaculatory duct, whereupon the seminal vesicles, prostate gland, and bulbourethral glands release their secretions. Secretions from the bulbourethral glands are the first to enter the urethra, and they function to cleanse the urethra of acidic residue from urine. This fluid does not normally contain sperm, but considering that the young human adult male produces up to 1 billion sperm a day, it may. It is therefore possible, though not probable, for fertilization of an egg and pregnancy of a female to take place even though ejaculation has not occurred.

During the second phase of ejaculation, called expulsion, rhythmical contractions of muscles at the base of the penis and within the urethral wall expel semen in spurts from the opening of the urethra. These rhythmical contractions are an example of release from myotonia, or muscle tenseness. Myotonia is another important sexual response. An erection lasts for only a limited amount of time. The penis now returns to its normal flaccid state. Following ejaculation, a male may typically experience a period of time, called the refractory period, during which stimulation does not bring about an erection. The contractions that expel semen from the penis are a part of male **orgasm** [Gk. *orgasmos*, sexual excitement], the physiological and psychological sensations that occur at the climax of sexual stimulation.

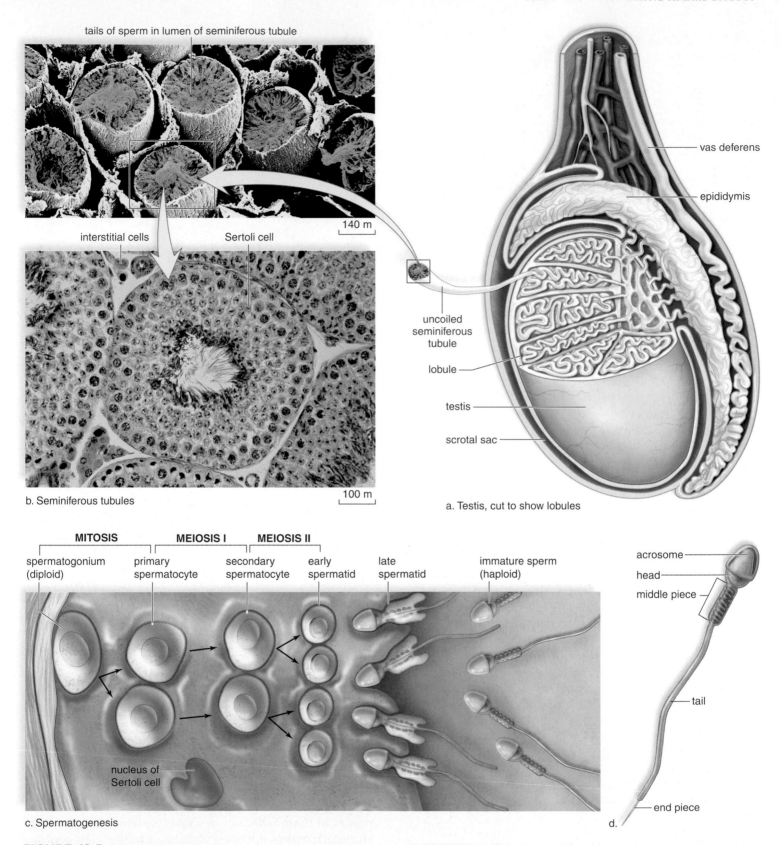

tails of sperm in lumen of seminiferous tubule

140 m

interstitial cells   Sertoli cell

100 m

b. Seminiferous tubules

vas deferens

epididymis

uncoiled seminiferous tubule

lobule

testis

scrotal sac

a. Testis, cut to show lobules

MITOSIS   **MEIOSIS I**   **MEIOSIS II**

spermatogonium (diploid)   primary spermatocyte   secondary spermatocyte   early spermatid   late spermatid   immature sperm (haploid)

nucleus of Sertoli cell

c. Spermatogenesis

acrosome

head

middle piece

tail

end piece

d.

FIGURE 43.5 Testis and sperm.
a. The lobules of a testis contain seminiferous tubules. **b.** Light micrograph of a cross section of the seminiferous tubules, where spermatogenesis occurs. Note the location of interstitial cells in clumps among the seminiferous tubules. **c.** Diagrammatic representation of spermatogenesis, which occurs in wall of tubules. **d.** A sperm has a head, a middle piece, and a tail. The nucleus is in the head, which is capped by the enzyme-containing acrosome.

The Testes

A longitudinal section of a testis shows that it is composed of compartments called lobules, each of which contains one to three tightly coiled **seminiferous tubules** (Fig. 43.5a). Altogether, these tubules have a combined length of approximately 250 m. A microscopic cross section of a seminiferous tubule shows that it is packed with cells undergoing spermatogenesis (Fig. 43.5b, c), the production of sperm. Newly formed cells move away from the outer wall, increase in size, and undergo meiosis to become spermatids, which contain 23 chromosomes. Spermatids then differentiate into sperm. Also present are Sertoli (sustentacular) cells, which support, nourish, and regulate the spermatogenic cells. It takes approximately 74 days for sperm to undergo development from spermatogonia to spermatozoa.

Mature **sperm,** or spermatozoa, have three distinct parts: a head, a middle piece, and a tail (Fig. 43.5d). The middle piece and the tail contain microtubules, in the characteristic 9 + 2 pattern of cilia and flagella. In the middle piece, mitochondria are wrapped around the microtubules and provide the energy for movement. The head contains a nucleus covered by a cap called the acrosome [Gk. akros, at the tip, and soma, body], that stores enzymes needed to penetrate the egg. (Because the human egg is surrounded by several layers of cells and a thick membrane, the acrosomal enzymes play a role in allowing a sperm to reach the surface of the egg.) The ejaculated semen of a normal male contains between 50 and 150 million sperm per milliliter. If less than 25 million sperm per milliliter are released, infertility may result. Fewer than 100 ever reach the vicinity of the egg, however, and only one sperm normally enters an egg. With contractions of the female uterus, it takes spermatozoa about two hours to reach the egg traveling approximately 12.5 cm per hour.

Hormonal Regulation in Males

The hypothalamus has ultimate control of the testes' sexual function because it secretes a hormone called gonadotropin-releasing hormone, or GnRH, that stimulates the anterior pituitary to produce the gonadotropic hormones. There are two gonadotropic hormones—follicle-stimulating hormone (FSH) and luteinizing hormone (LH)—in both males and females. In males, FSH promotes spermatogenesis in the seminiferous tubules.

LH in males controls the production of the androgen testosterone by the interstitial cells (Leydig cells), which are scattered in the spaces between the seminiferous tubules (Fig. 43.5b). All these hormones, including inhibin, a hormone released by the seminiferous tubules, are involved in a negative feedback relationship that maintains the fairly constant production of sperm and testosterone (Fig. 43.6).

Functions of Testosterone

Testosterone is the main sex hormone in males. It is essential for the normal development and functioning of the organs listed in Table 43.1. Testosterone is also necessary for the maturation of sperm.

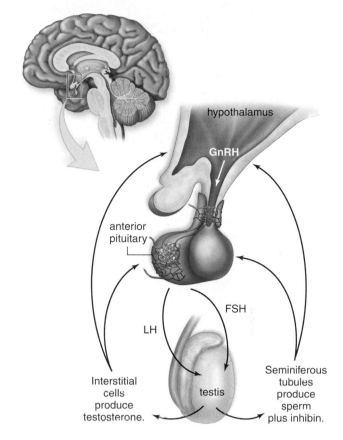

FIGURE 43.6 Hormonal control of testes.
GnRH (gonadotropin-releasing hormone) stimulates the anterior pituitary to produce FSH and LH. FSH stimulates the testes to produce sperm, and LH stimulates the testes to produce testosterone. Testosterone from interstitial cells and inhibin from the seminiferous tubules exert negative feedback control over the hypothalamus and the anterior pituitary, and this ultimately regulates the level of testosterone in the blood.

In addition, testosterone brings about and maintains the male **secondary sex characteristics** that develop at the time of **puberty.** Males are generally taller than females and have broader shoulders and longer legs relative to trunk length. The deeper voice of males compared to females is due to the fact they have a larger larynx with longer vocal cords. Since the so-called Adam's apple is a part of the larynx, it is usually more prominent in males than in females.

Testosterone is responsible for the greater muscular development in males. Because of this effect, males and females sometimes take anabolic steroids (either the natural or synthetic form of testosterone) to build up their muscles (see page 770). Testosterone causes males to develop noticeable hair on the face, chest, and occasionally other regions of the body, such as the back. Testosterone also leads to the receding hairline and pattern baldness that occur in males. Testosterone levels peak at about 20 years of age and then decline as a male ages.

The gonads in males are the testes, which produce sperm and testosterone, the most significant male sex hormone.

43.3 FEMALE REPRODUCTIVE SYSTEM

The human female reproductive system includes the ovaries, the oviducts, the uterus, and the vagina (Fig. 43.7 and Table 43.2). The ovaries, which produce a secondary **oocyte** each month, lie in shallow depressions, one on each side of the upper pelvic cavity. The oviducts [L. *ovum,* egg, and *duco,* lead out], also called uterine or fallopian tubes, extend from the ovaries to the uterus; however, the oviducts are not attached to the ovaries. Instead, they have fingerlike projections called fimbriae (sing., fimbria) that sweep over the ovaries. When an oocyte bursts from an ovary during ovulation, it usually is swept into an oviduct by the combined action of the fimbriae and the beating of cilia that line the oviducts. Fertilization, if it occurs, normally takes place in an oviduct, and the developing embryo is propelled slowly by ciliary movement and tubular muscle contraction to the uterus. The **uterus** [L. *uterus,* womb] is a thick-walled muscular organ about the size and shape of an inverted pear. The narrow end of the uterus is called the **cervix.** The embryo completes its development after embedding itself in the uterine lining, called the **endometrium** [Gk. *endon,* within, and *metra,* womb]. If, by chance, the embryo should embed itself in another location, such as an oviduct, a so-called ectopic pregnancy results.

TABLE 43.2	
Female Reproductive Organs	
Organ	*Function*
Ovaries	Produce egg and sex hormones
Oviducts (fallopian tubes)	Conduct egg; location of fertilization
Uterus (womb)	Houses developing embryo and fetus
Vagina	Receives penis during copulation and serves as birth canal

A small opening in the cervix leads to the vaginal canal. The vagina [L. *vagina,* sheath] is a tube at a 45-degree angle to the small of the back. The mucosal lining of the vagina lies in folds and the vagina can distend. This is especially important when the vagina serves as the birth canal, and it also can facilitate intercourse, when the vagina receives the penis during copulation. Several different types of bacteria normally reside in the vagina and create an acidic environment. This environment is protective against the possible growth of pathogenic bacteria, but sperm prefer the basic environment provided by semen.

The external genital organs of a female are known collectively as the vulva (Fig. 43.7b). The mons pubis and two

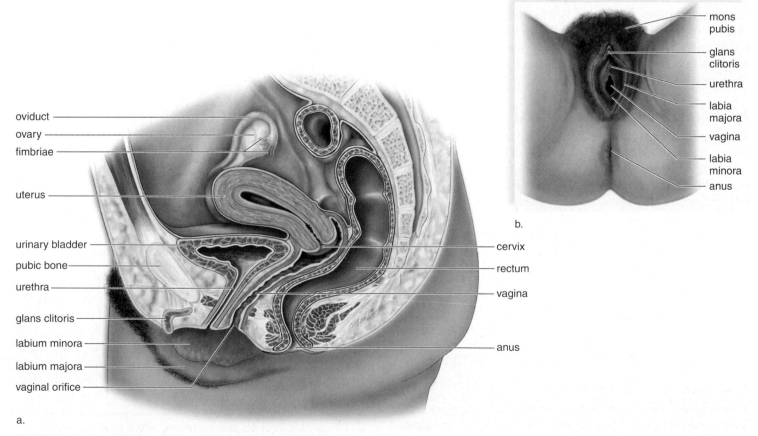

a.

b.

FIGURE 43.7 Female reproductive system.
a. The ovaries produce one oocyte (egg) per month. Fertilization occurs in the oviduct, and development occurs in the uterus. The vagina is the birth canal and organ of sexual intercourse. **b.** Vulva. At birth, the opening of the vagina is partially occluded by a membrane called the hymen. Physical activities and sexual intercourse disrupt the hymen.

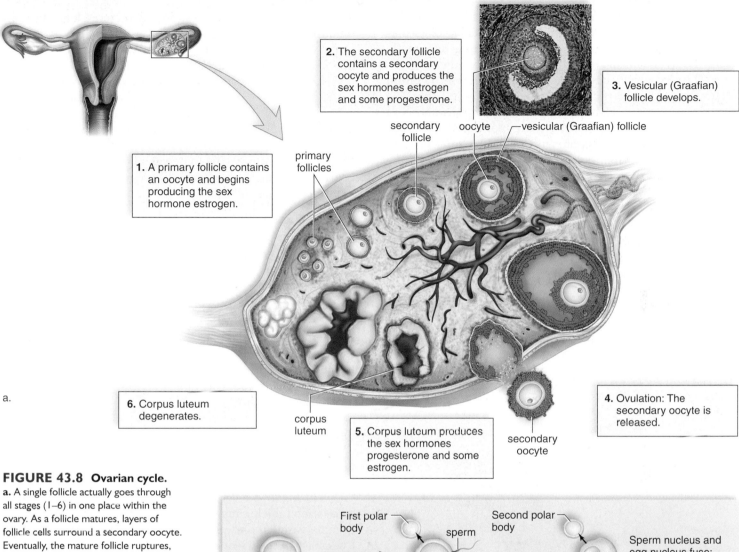

2. The secondary follicle contains a secondary oocyte and produces the sex hormones estrogen and some progesterone.

3. Vesicular (Graafian) follicle develops.

secondary follicle   oocyte   vesicular (Graafian) follicle

primary follicles

1. A primary follicle contains an oocyte and begins producing the sex hormone estrogen.

a.

6. Corpus luteum degenerates.

corpus luteum

5. Corpus luteum produces the sex hormones progesterone and some estrogen.

secondary oocyte

4. Ovulation: The secondary oocyte is released.

FIGURE 43.8 Ovarian cycle.
a. A single follicle actually goes through all stages (1–6) in one place within the ovary. As a follicle matures, layers of follicle cells surround a secondary oocyte. Eventually, the mature follicle ruptures, and the secondary oocyte is released. The follicle then becomes the corpus luteum, which eventually disintegrates. **b.** During oogenesis, the chromosome number is reduced from 46 to 23. Fertilization restores the full number of chromosomes.

First polar body

Second polar body

sperm

Sperm nucleus and egg nucleus fuse: Zygote with 46 chromosomes results.

primary oocyte (46 chromosomes) → meiosis I → secondary oocyte (23 chromosomes) → meiosis II → zygote

b.

folds of skin called labia minora and labia majora are on either side of the urethral and vaginal openings. Beneath the labia majora, pea-sized greater vestibular glands (Bartholin glands) open on either side of the vagina. They keep the vulva moist and lubricated during intercourse.

At the juncture of the labia minora is the clitoris, which is homologous to the penis in males. The clitoris has a shaft of erectile tissue and is capped by a pea-shaped glans. The many sensory receptors of the clitoris allow it to function as a sexually sensitive organ. The clitoris has twice as many nerve endings as the penis. Orgasm in the female is a release of neuromuscular tension in the muscles of the genital area, vagina, and uterus.

The Ovaries

Normally, the ovaries alternate in producing one oocyte each month. For the sake of convenience, the released oocyte

is often called an **ovum,** or egg. The ovaries also produce the female sex hormones, **estrogens,** collectively called estrogen, and **progesterone,** during the ovarian cycle.

The Ovarian Cycle

The **ovarian cycle** occurs as a **follicle** [L. dim. of *folliculus,* bag] changes from a primary to a secondary to a vesicular (Graafian) follicle (Fig. 43.8*a*). Epithelial cells of a primary follicle surround a primary oocyte. Pools of follicular fluid surround the oocyte in a secondary follicle. In a vesicular follicle, a fluid-filled cavity increases to the point that the follicle wall balloons out on the surface of the ovary.

As a follicle matures, oogenesis, a form of meiosis depicted in Figure 43.8*b*, is initiated and continues. The primary oocyte divides, producing two haploid cells. One cell is a secondary oocyte, and the other is a polar body. The vesicular follicle bursts, releasing the secondary oocyte (often called an egg for convenience) surrounded by a clear membrane. This process is referred to as **ovulation.** Once a vesicular follicle has lost the secondary oocyte, it develops into a **corpus luteum,** [L. *corpus,* body, and *luteus,* yellow] a glandlike structure.

The secondary oocyte enters an oviduct. If fertilization occurs, a sperm enters the secondary oocyte and then the oocyte completes meiosis. An egg with 23 chromosomes and a second polar body results. When the sperm nucleus unites with the egg nucleus, a zygote with 46 chromosomes is produced. If zygote formation and pregnancy do not occur, the corpus luteum begins to degenerate after about ten days.

Phases of the Ovarian Cycle

The ovarian cycle is controlled by the gonadotropic hormones, follicle-stimulating hormone (FSH) and luteinizing hormone (LH) (Fig. 43.9). The gonadotropic hormones are not present in constant amounts and instead are secreted at different rates during the cycle. For simplicity's sake, it is convenient to emphasize that during the first half, or **follicular phase,** of the cycle, FSH promotes the development of a follicle that primarily secretes estrogen. As the estrogen level in the blood rises, it exerts feedback control over the anterior pituitary secretion of FSH so that the follicular phase comes to an end.

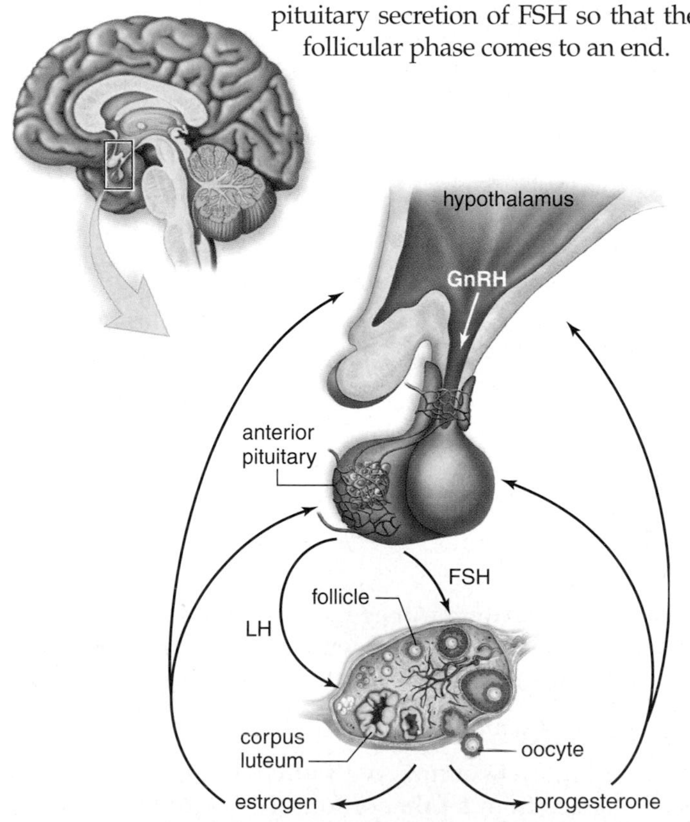

FIGURE 43.9 Hormonal control of ovaries.
The hypothalamus produces GnRH (gonadotropin-releasing hormone). GnRH stimulates the anterior pituitary to produce FSH (follicle-stimulating hormone) and LH (luteinizing hormone). FSH stimulates the follicle to produce primarily estrogen, and LH stimulates the corpus luteum to produce primarily progesterone. Estrogen and progesterone maintain the sex organs (e.g., uterus) and the secondary sex characteristics, and they exert feedback control over the hypothalamus and the anterior pituitary. Feedback control regulates the relative amounts of estrogen and progesterone in the blood.

Presumably, the high level of estrogen in the blood causes the hypothalamus to suddenly secrete a large amount of GnRH. This leads to a surge of LH production by the anterior pituitary and to ovulation at about the fourteenth day of a 28-day cycle (Fig. 43.10, *top*).

During the second half, or **luteal phase,** of the ovarian cycle, it is convenient to emphasize that LH promotes the development of the corpus luteum, which primarily secretes progesterone. As the blood level of progesterone rises, it exerts feedback control over anterior pituitary secretion of LH so that the corpus luteum begins to degenerate. As the luteal phase comes to an end, the low levels of progesterone and estrogen in the body cause menstruation to begin, as discussed next.

One ovarian follicle per month produces a secondary oocyte. Following ovulation, the follicle develops into the corpus luteum.

The Uterine Cycle

The female sex hormones produced in the ovarian cycle (estrogen and progesterone) affect the endometrium of the uterus, causing the cyclical series of events known as the **uterine cycle** (Fig. 43.10, *bottom*). Twenty-eight-day cycles are divided as follows.

During *days 1–5,* there is a low level of female sex hormones in the body, causing the endometrium to disintegrate and its blood vessels to rupture. A flow of blood passes out of the vagina during **menstruation** [L. *menstrualis,* happening monthly], also known as the menstrual period.

During *days 6–13,* increased production of estrogen by an ovarian follicle causes the endometrium to thicken and to become vascular and glandular. This is called the proliferative phase of the uterine cycle.

Ovulation usually occurs on the fourteenth day of the 28-day cycle.

During *days 15–28,* increased production of progesterone by the corpus luteum causes the endometrium to double in thickness and the uterine glands to mature, producing a thick mucoid secretion. This is called the secretory phase of the uterine cycle. The endometrium now is prepared to receive the developing embryo. If pregnancy does not occur, the corpus luteum degenerates, and the low level of sex hormones in the female body causes the endometrium to break down as menstruation occurs. Table 43.3 compares the stages of the uterine cycle with those of ovarian cycle.

Menstruation

Seven to ten days before the start of menstruation, some women suffer from premenstrual syndrome (PMS). During this time a woman may exhibit breast enlargement, achiness, headache, and irritability. The exact cause for PMS has yet to be discovered.

During menstruation, arteries that supply the lining constrict and the capillaries weaken. Blood spilling from the damaged vessels detaches layers of the lining, not all at once but in random patches. Endometrium, mucus, and blood

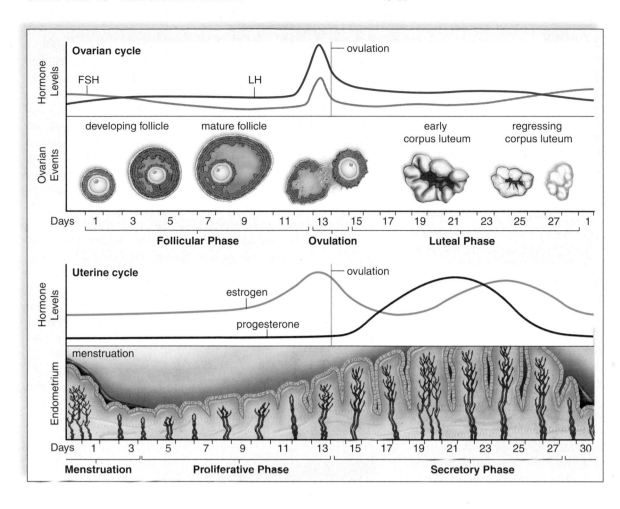

FIGURE 43.10

Female hormone levels during the ovarian and uterine cycles.
During the follicular phase of the ovarian cycle *(top)*, FSH released by the anterior pituitary promotes the maturation of a follicle in the ovary. The ovarian follicle produces increasing levels of estrogen, which causes the endometrium to thicken during the proliferative phase of the uterine cycle *(bottom)*. After ovulation and during the luteal phase of the ovarian cycle, LH promotes the development of the corpus luteum. This structure produces increasing levels of progesterone, which causes the endometrium to become secretory. Menstruation and the proliferative phase begin when progesterone production declines to a low level.

descend from the uterus, through the vagina, creating menstrual flow. Fibrinolysin, an enzyme released by dying cells, prevents the blood from clotting. Menstruation lasts from three to five days, as the uterus sloughs off the thick lining that was three weeks in the making.

The first menstrual period, called **menarche,** typically occurs between the ages of 11 and 12. Menarche signifies that the ovarian and uterine cycles have begun. If menarche does not occur by age 16, or normal uterine cycles are interrupted for six months or more, amenorrhea exists. Primary amenorrhea is usually caused by nonfunctional ovaries or developmental abnormalities. Secondary amenorrhea may be caused by weight loss and/or excessive exercise.

Menopause, which usually occurs between ages 45 and 55, is the time in a woman's life when menstruation ceases because the ovaries are no longer functioning. Menopause is not complete until menstruation is absent for a year.

The ovarian cycle regulates the uterine cycle, which consists of a proliferative phase and a secretory phase, which ends with menstruation.

TABLE 43.3

Ovarian and Uterine Cycles (Simplified)

Ovarian Cycle	Events	Uterine Cycle	Events
Follicular phase—Days 1–13	FSH Follicle maturation Estrogen	Menstruation—Days 1–5 Proliferative phase—Days 6–13	Endometrium breaks down Endometrium rebuilds
Ovulation—Day 14*	LH spike		
Luteal phase—Days 15–28	LH Corpus luteum Progesterone	Secretory phase—Days 15–28	Endometrium thickens and glands are secretory

* Assuming a 28-day cycle

Fertilization and Pregnancy

If fertilization does occur, an embryo begins development even as it travels down the oviduct to the uterus. The endometrium is now prepared to receive the developing embryo, which becomes embedded in the lining several days following fertilization. The placenta originates from both maternal and embryonic tissues. It is shaped like a large, thick pancake and is the site of the exchange of gases and nutrients between fetal and maternal blood, although the two rarely mix. At first, the placenta produces **human chorionic gonadotropin (HCG),** which maintains the corpus luteum until the placenta begins its own production of progesterone and estrogen. Progesterone and estrogen have two effects. They shut down the anterior pituitary so that no new follicles mature, and they maintain the lining of the uterus so that the corpus luteum is not needed. No menstruation occurs during pregnancy.

Estrogen and Progesterone

Estrogen in particular is essential for the normal development and functioning of the female reproductive organs listed in Table 43.2. Estrogen is also largely responsible for the secondary sex characteristics in females, including body hair and fat distribution. In general, females have a more rounded appearance than males because of a greater accumulation of fat beneath their skin. Also, the pelvic girdle enlarges so that females have wider hips than males, and the thighs converge at a greater angle toward the knees. Both estrogen and progesterone are required for breast development as well.

The Female Breast

A female breast contains between 15 and 24 lobules, each with its own mammary duct (Fig. 43.11). A duct begins at the nipple and divides into numerous other ducts that end in blind sacs called alveoli. **Lactation,** the production of milk by the cells of the alveoli, is caused by the hormone prolactin. Milk is not produced during pregnancy because production of prolactin is suppressed by the feedback inhibition effect of estrogen and progesterone on the anterior pituitary. A couple of days after delivery of a baby, milk production begins. In the meantime, the breasts produce a watery, yellowish-white fluid called colostrum, which has a similar composition to milk but contains more protein and less fat. Colostrum is rich in IgA antibodies that may provide some degree of immunity to the newborn. Milk contains water, proteins, amino acids, sugars, and lysozymes (enzymes with antibiotic properties). Milk contains about 750 calories per liter.

Women should regularly check their breasts for lumps and have mammograms (X-ray photographs) taken as recommended by their physician. Although breast cancer genes have been described, most forms of breast cancer are nonhereditary.

43.4 Control of Reproduction

Several means are available to dampen or enhance the human reproductive potential. Contraceptives are medications and devices that reduce the chance of pregnancy.

Birth Control Methods

The most reliable method of birth control is abstinence—that is, not engaging in sexual intercourse. This form of birth control has the added advantage of preventing transmission of a sexually transmitted disease. The male and female condoms also offer some protection against sexually transmitted diseases. These and other common means of birth control used in the United States are listed in Table 43.4. The use of a condom, unlike most of the other methods listed in the table, does not require the assistance of a physician.

Table 43.4 lists the types of birth control methods in the order of their effectiveness. The effectiveness of the method refers to the number of women per year who will not get pregnant even though they are regularly engaging in sexual intercourse. For example, with natural family planning, one of the least effective methods given in the table, we expect that within a year, 70 out of 100, or 70%, of sexually active women will not get pregnant, while 30 out of 100, or 30%, of women will get pregnant.

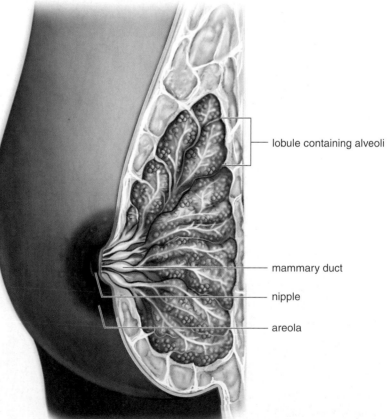

lobule containing alveoli

mammary duct

nipple

areola

FIGURE 43.11 Anatomy of the breast.
The female breast contains lobules consisting of ducts and alveoli. In the lactating breast, cells lining the alveoli have been stimulated to produce milk by the hormone prolactin.

TABLE 43.4

Common Birth Control Methods

Name	Procedure	Methodology	Effectiveness	Risk
Abstinence	Refrain from sexual intercourse	No sperm in vagina	100%	None
Vasectomy	Vasa deferentia cut and tied	No sperm in seminal fluid	Almost 100%	Irreversible sterility
Tubal ligation	Oviducts cut and tied	No eggs in oviduct	Almost 100%	Irreversible sterility
Oral contraception	Hormone medication taken daily	Anterior pituitary does not release FSH and LH	Almost 100%	Thromboembolism, especially in smokers
Contraceptive injections	Injections of hormones	Anterior pituitary does not release FSH and LH	About 99%	Presently none known
Contraceptive implants	Tubes of progestin (form of progesterone) implanted under skin	Anterior pituitary does not release FSH and LH	More than 90%	Presently none known
Intrauterine device (IUD)	Plastic coil inserted into uterus by physician	Prevents implantation	More than 90%	Infection (pelvic inflammatory disease, PID)
Diaphragm	Latex cup inserted into vagina to cover cervix before intercourse	Blocks entrance of sperm to uterus	With jelly, about 90%	Presently none known
Cervical cap	Latex cap held by suction over cervix	Delivers spermicide near cervix	Almost 85%	Cancer of cervix
Male condom	Latex sheath fitted over erect penis	Traps sperm and helps prevent STDs	About 85%	Presently none known
Female condom	Polyurethane liner fitted inside vagina	Blocks entrance of sperm to uterus and prevents STDs	About 85%	Presently none known
Coitus interruptus	Penis withdrawn before ejaculation	Prevents sperm from entering vagina	75%	Presently none known
Jellies, creams, foams	These spermicidal products inserted before intercourse	Kills a large number of sperm	About 75%	Presently none known
Natural family planning	Day of ovulation determined by record keeping, various methods of testing	Intercourse avoided on certain days of the month	About 70%	Presently none known
Douche	Vagina and uterus cleansed after intercourse	Washes out sperm	Less than 70%	Presently none known

Less Common Birth Control Methods

The expression "morning-after pill" refers to a medication that will prevent pregnancy after unprotected intercourse. The expression is a misnomer, in that medication can begin one to several days after unprotected intercourse.

A kit called Preven is made up of four synthetic progesterone pills; two are taken up to 72 hours after unprotected intercourse, and two more are taken 12 hours later. The medication upsets the normal uterine cycle, making it difficult for the embryo to implant itself in the endometrium. A recent study estimated that the medication was 85% effective in preventing unintended pregnancies.

Mifepristone, better known as RU-486, is a pill that is presently used to cause the loss of an implanted embryo by blocking the progesterone receptors of endometrial cells. Without functioning receptors for progesterone, the endometrium sloughs off, carrying the embryo with it. When taken in conjunction with a prostaglandin to induce uterine contractions, RU-486 is 95% effective. It is possible that some day this medication will also be a "morning-after pill," taken when menstruation is late without evidence that pregnancy has occurred.

Contraceptive vaccines are now being developed. For example, a vaccine intended to immunize women against HCG, the hormone so necessary to maintaining the implantation of the embryo, was successful in a limited clinical trial. Since HCG is not normally present in the body, no autoimmune reaction is expected, but the immunization does wear off with time. Others believe that it would also be possible to develop a safe antisperm vaccine that could be used in women.

Numerous well-known birth control methods and devices are available to those who wish to prevent pregnancy. Their effectiveness varies. In addition, new methods are expected to be developed.

Infertility

Infertility is the inability of a couple to achieve pregnancy after one year of regular, unprotected intercourse. The American Medical Association estimates that 15% of all couples are infertile. The cause of infertility can be attributed to the male (40%), the female (40%), or both (20%).

Causes of Infertility

The most frequent cause of infertility in males is low sperm count and/or a large proportion of abnormal sperm, which can be due to environmental influences. The public is particularly concerned about endocrine-disrupting contaminants, which are discussed in the Ecology Focus on page 789. But thus far, it appears that a sedentary lifestyle coupled with smoking and alcohol consumption most often leads to male infertility. When males spend most of the day sitting in front of a computer or the TV or driving, the testes temperature remains too high for adequate sperm production.

Body weight appears to be the most significant factor in causing female infertility. In women of normal weight, fat cells produce a hormone called leptin that stimulates the hypothalamus to release GnRH. In overweight women, the ovaries often contain many small follicles, and the woman fails to ovulate. Other causes of infertility in females are blocked oviducts due to pelvic inflammatory disease (see page 792) and endometriosis. **Endometriosis** is the presence of uterine tissue outside the uterus, particularly in the uterine tubes and on the abdominal organs. Backward flow of menstrual fluid allows living uterine cells to establish themselves in the abdominal cavity, where they go through the usual uterine cycle, causing pain and structural abnormalities that make it more difficult for a woman to conceive.

Sometimes the causes of infertility can be corrected by medical intervention so that couples can have children. If no obstruction is apparent and body weight is normal, it is possible for females to take fertility drugs, which are gonadotropic hormones that stimulate the ovaries and bring about ovulation. Such hormone treatments may cause multiple ovulations and multiple births.

When reproduction does not occur in the usual manner, many couples adopt a child. Others sometimes try one of the assisted technologies discussed in the following paragraphs.

Assisted Reproductive Technologies

Assisted reproductive technologies (ART) consist of techniques used to increase the chances of pregnancy. Often, sperm and/or eggs are retrieved from the testes and ovaries, and fertilization takes place in a clinical or laboratory setting.

Artificial Insemination by Donor (AID). During artificial insemination, sperm are placed in the vagina by a physician. Sometimes a woman is artificially inseminated by her partner's sperm. This is especially helpful if the partner has a low sperm count, because the sperm can be collected over a period of time and concentrated so that the sperm count is sufficient to result in fertilization. Often, however, a woman is inseminated by sperm acquired from a donor. At times, a combination of partner and donor sperm is used.

A variation of AID is *intrauterine insemination (IUI)*. In IUI, fertility drugs are given to stimulate the ovaries, and then the donor's sperm is placed in the uterus rather than in the vagina.

If the prospective parents wish, sperm can be sorted into those that are believed to be X-bearing or Y-bearing to increase the chances of having a child of the desired sex.

In Vitro Fertilization (IVF). During IVF, conception occurs in laboratory glassware. Ultrasound machines can now spot follicles in the ovaries that hold immature eggs; therefore, the latest method is to forgo the administration of fertility drugs and retrieve immature eggs by using a needle. The immature eggs are then brought to maturity in glassware before concentrated sperm are added. After about two to four days, the embryos are ready to be transferred to the uterus of the woman, who is now in the secretory phase of her uterine cycle. If desired, the embryos can be tested for a genetic disease, and only those found to be free of disease will be used. If implantation is successful, development is normal and continues to term.

Gamete Intrafallopian Transfer (GIFT). Recall that the term **gamete** refers to a sex cell, either a sperm or an egg. Gamete intrafallopian transfer was devised to overcome the low success rate (15–20%) of in vitro fertilization. The method is exactly the same as in vitro fertilization, except the eggs and the sperm are placed in the oviducts immediately after they have been brought together. GIFT has the advantage of being a one-step procedure for the woman—the eggs are removed and reintroduced all in the same time period. A variation on this procedure is to fertilize the eggs in the laboratory and then place the zygotes in the oviducts.

Surrogate Mothers. In some instances, women are contracted and paid to have babies. These women are called surrogate mothers. The sperm and even the egg can be contributed by the contracting parents.

Intracytoplasmic Sperm Injection (ICSI). In this highly sophisticated procedure, a single sperm is injected into an egg. It is used effectively when a man has severe infertility problems.

If all the assisted reproductive technologies discussed were employed simultaneously, it would be possible for a baby to have five parents: (1) sperm donor, (2) egg donor, (3) surrogate mother, and (4) and (5) contracting mother and father.

When corrective procedures fail to reverse infertility, assisted reproductive technologies may be considered.

ecology focus

Endocrine-Disrupting Contaminants

Rachel Carson's book *Silent Spring*, published in 1962, predicted that pesticides would have a deleterious effect on animal life. Soon thereafter, it was found that pesticides caused the thinning of eggshells in bald eagles to the point that their eggs broke and the chicks died. Additionally, populations of terns, gulls, cormorants, and lake trout declined after they ate fish contaminated by high levels of environmental toxins. The concern was so great that the United States Environmental Protection Agency (EPA) came into existence. This agency and civilian environmental groups have brought about a reduction in pollution release and a cleaning up of emissions. Even so, we are now aware of some more subtle effects of pollutants.

Hormones influence nearly all aspects of physiology and behavior in animals, including tissue differentiation, growth, and reproduction. Therefore, when wildlife in contaminated areas began to exhibit certain types of abnormalities, researchers began to think that certain pollutants can affect the endocrine system. In England, male fish exposed to sewage developed ovarian tissue and produced a metabolite normally found only in females during egg formation. In California, western gulls displayed abnormalities in gonad structure and nesting behaviors. Hatchling alligators in Florida possessed abnormal gonads and hormone concentrations linked to nesting.

At first, such effects seemed to indicate only the involvement of the female hormone estrogen, and researchers therefore called the contaminants ecoestrogens. Many of the contaminants interact with hormone receptors, and in that way cause developmental effects. Others bind directly with sex hormones such as testosterone and estradiol. Still others alter the physiology of growth hormones and neurotransmitters responsible for brain development and behavior. Therefore, the preferred term today for these pollutants is endocrine-disrupting contaminants (EDCs).

Many EDCs are chemicals used as pesticides and herbicides in agriculture, and some are associated with the manufacture of various other organic molecules such as PCBs (polychlorinated biphenyls). Some chemicals shown to influence hormones are found in plastics, food additives, and personal hygiene products. In mice, phthalate esters, which are plastic components, affect neonatal development when present in the part-per-trillion range. It is, therefore, of great concern that EDCs have been found to exist at levels a thousand times greater than this— even in amounts comparable to functional hormone levels in the human body. Also, it is not surprising that EDCs are affecting the endocrine systems of a wide range of organisms (Fig. 43A).

Scientists and representatives for industrial manufacturers continue to debate whether EDCs pose a health risk to humans. Some suspect that EDCs lower sperm counts, reduce male and female fertility, and increase rates of certain cancers (breast, ovary, testis, and prostate). Additionally, some studies suggest that EDCs contribute to learning deficits and behavioral problems in children. Laboratory and field research continues to identify chemicals that have the ability to influence the endocrine system. Millions of tons of potential EDCs are produced annually in the United States, and the EPA is under pressure to certify these compounds as safe. The European Economic Community has already restricted the use of certain EDCs, and has banned the production of specific plastic components found in items intended for use by children, specifically toys. Only through continued scientific research and the cooperation of industry can we identify the risks that EDCs pose to the environment, wildlife, and humans.

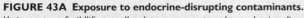

FIGURE 43A Exposure to endocrine-disrupting contaminants.
Various types of wildlife as well as humans are exposed to endocrine-disrupting contaminants that can seriously affect their health and reproductive abilities.

43.5 SEXUALLY TRANSMITTED DISEASES

Sexually transmitted diseases (STDs) are caused by organisms ranging from viruses to arthropods; however, we will discuss only certain STDs caused by viruses and bacteria. Unfortunately, for unknown reasons, humans cannot develop good immunity to any STD. Therefore, prompt medical treatment should be sought when exposed to an STD. To prevent the spread of STDs, a latex condom can be used.

Among those STDs caused by viruses, treatment is available for AIDS and genital herpes. Only STDs caused by bacteria (e.g., chlamydia, gonorrhea, and syphilis) are treatable with antibiotics.

AIDS

The organism that causes acquired immunodeficiency syndrome (AIDS) is a virus called **human immunodeficiency virus (HIV).** HIV attacks the type of lymphocyte known as helper T cells (Fig. 43.12). Helper T cells, you will recall, stimulate the activities of B lymphocytes, which produce antibodies. After an HIV infection sets in, helper T cells begin to decline in number, and the person becomes more susceptible to other types of infections.

Symptoms

AIDS has three stages of infection, called categories A, B, and C. During the category A stage, which may last about a year, the individual is an asymptomatic carrier. He or she may exhibit no symptoms but can pass on the infection. Immediately after infection and before the blood test becomes positive, a large number of infectious viruses are present in the blood, and these could be passed on to another person. Even after the blood test becomes positive, the person remains well as long as the body produces sufficient helper T lymphocytes to keep the count higher than 500 per mm^3. During the category B stage, or AIDS-related complex (ARC), which may last six to eight years, the lymph nodes swell, and the person may experience weight loss, night sweats,

fatigue, fever, and diarrhea. Infections such as thrush (white sores on the tongue and in the mouth) and herpes recur. Finally, the person may progress to category C, which is full-blown AIDS, characterized by nervous disorders and the development of an opportunistic disease, such as an unusual type of pneumonia or skin cancer. Opportunistic diseases are those that occur only in individuals who have little or no capability of fighting an infection. Without intensive medical treatment, the AIDS patient dies about seven to nine years after infection. Now, with a combination therapy of several drugs, AIDS patients are beginning to live longer in the United States.

Transmission

AIDS is transmitted by sexual contact with an infected person, including vaginal or rectal intercourse and oral/genital contact. Also, needle-sharing among intravenous drug users is high-risk behavior. A less common mode of transmission (now occurring only rarely in countries where donated blood is screened for HIV) is through transfusions of infected blood or blood-clotting factors.

HIV first spread through the homosexual community, and male-to-male sexual contact still accounts for the largest percentage of new AIDS cases in the United States. But the largest increases in HIV infections are occurring through heterosexual contact or by intravenous drug use. Now, women account for 19% of all newly diagnosed cases of AIDS.

Treatment

There is no cure for AIDS, but a treatment called highly active antiretroviral therapy (HAART) is usually able to stop HIV replication to such an extent that the viral load becomes undetectable. HAART uses a combination of drugs that interfere with the life cycle of HIV. Entry inhibitors stop HIV from entering a cell by, for example, preventing the virus from binding to a receptor in the plasma membrane. Reverse transcriptase inhibitors, such as AZT, interfere with the operation of the reverse transcriptase enzyme. Integrase inhibitors prevent HIV from inserting its own genetic material into that of the host cells. Protease inhibitors prevent protease from processing newly created polypeptides. Assembly and budding inhibitors are in the experimental stage, and none are available as yet. The hope is that by giving a patient a combination of these drugs, the virus is less likely to replicate, successfully mutate to make drug therapy ineffective, and/or become resistent to drug therapy.

An HIV-positive pregnant woman who takes reverse transcriptase inhibitors during her pregnancy reduces the chances of HIV transmission to her newborn. If possible, drug therapy should be delayed until the tenth to twelfth week of pregnancy to minimize any adverse effects of AZT on fetal development. But if treatment begins at this time the chance of transmission is reduced by 66%.

There is a general consensus that control of the AIDS epidemic will not occur until a vaccine is available. Many different approaches have thus far been tried with limited success. Perhaps a combination of these vaccines will work best, as in drug therapy.

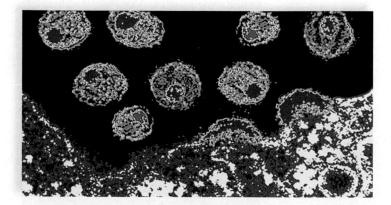

FIGURE 43.12 HIV, the AIDS virus.
False-colored micrograph showing HIV particles budding from an infected helper T cell. These viruses can infect other helper T cells and also macrophages, which work with helper T cells to stem infection.

Genital Warts

Genital warts are caused by the human papillomaviruses (HPVs) (Fig. 43.13). Many times, carriers either do not have any sign of warts or merely have flat lesions. When present, flat or raised warts can be found on the penis and the foreskin of males and at the vaginal orifice and cervix in females. Warts on the cervix are always flat and can be hard to detect.

Presently, there is no cure for an HPV infection, but it can be treated effectively by surgery, freezing, application of an acid, or laser burning, depending on severity. If visible warts are removed, they may recur. Genital warts are associated with cancer of the cervix, as well as tumors of the vulva, the vagina, the anus, and the penis. A vaccine against the most common HPV has been developed, and children should be vaccinated before they become sexually active.

Genital Herpes

Genital herpes is caused by herpes simplex virus (Fig. 43.14). Type 1 usually causes cold sores and fever blisters, while type 2 more often causes genital herpes. Crossover infections do occur, however. That is, type 1 has been known to cause a genital infection, while type 2 has been known to cause cold sores and fever blisters.

Genital herpes is one of the more prevalent sexually transmitted diseases today. At any one time, millions of persons may be having recurring symptoms. After infection, some people exhibit no symptoms; others may experience a tingling or itching sensation before blisters appear on the genitals. Once the blisters rupture, they leave painful ulcers that may take as long as three weeks or as little as five days to heal. The blisters may be accompanied by fever, pain on urination, swollen lymph nodes in the groin, and in women, a copious discharge. At this time, the individual has an increased risk of acquiring an AIDS infection.

After the ulcers heal, the disease is only latent, and blisters can recur, although usually at less frequent intervals and with milder symptoms. Fever, stress, sunlight, and menstruation are associated with recurrence of symptoms. Exposure to herpes in the birth canal can cause an infection in the new-

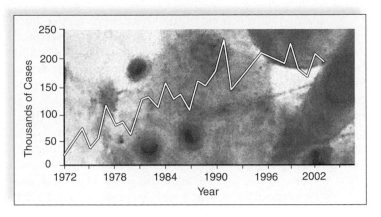

FIGURE 43.14 Genital herpes.
A graph depicting the incidence of new reported cases of genital herpes in the United States from 1972 to 2003 is superimposed on a photomicrograph of cells infected with the herpes simplex virus.

born, which leads to neurological disorders and even death. Birth by cesarean section prevents this possibility.

Hepatitis

There are several types of hepatitis. The type of hepatitis and the virus that causes it are designated by the same letter. Hepatitis A is usually acquired from sewage-contaminated drinking water, but this infection can also be sexually transmitted through oral/anal contact. Hepatitis B, which can be spread by blood transfusions and bodily fluids, can also be spread in the same manner as AIDS and is more infectious. Hepatitis C is called posttransfusion hepatitis but can be transmitted through sexual contact. Hepatitis C has the potential to become epidemic in the future. Hepatitis infects the liver and can lead to liver failure, cancer, and death. Fortunately, a vaccine is now available for hepatitis B. No vaccines are available for hepatitis C.

Chlamydia

Chlamydia is named for the tiny bacterium that causes it (*Chlamydia trachomatis*). The incidence of new chlamydial infections has steadily increased since 1984 (Fig. 43.15).

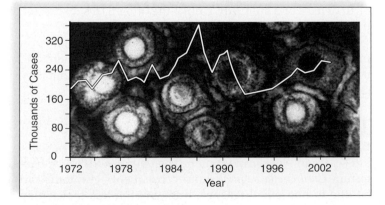

FIGURE 43.13 Genital warts.
A graph depicting the incidence of new cases of genital warts reported in the United States from 1972–2003 is superimposed on a photomicrograph of human papillomaviruses.

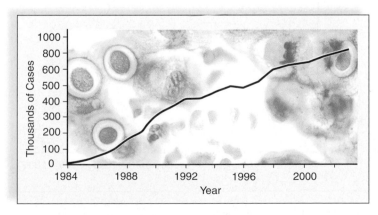

FIGURE 43.15 Chlamydial infection.
A graph depicting the incidence of reported cases of chlamydia in the United States from 1984 to 2003 is superimposed on a photomicrograph of a cell containing different stages of the organism.

Chlamydial infections of the lower reproductive tract are usually mild or asymptomatic, especially in women. About 8–21 days after infection, men may experience a mild burning sensation on urination and a mucoid discharge. Women may have a vaginal discharge along with the symptoms of a urinary tract infection. Chlamydia also causes cervical ulcerations, which increase the risk of acquiring AIDS.

If the infection is misdiagnosed or if a woman does not seek medical help, there is a particular risk of the infection spreading from the cervix to the oviducts so that pelvic inflammatory disease (PID) results. This painful condition can result in blockage of the oviducts with the possibility of sterility and infertility. If a baby comes in contact with chlamydia during birth, inflammation of the eyes or pneumonia can result.

Gonorrhea

Gonorrhea is caused by the bacterium *Neisseria gonorrhoeae* (Fig. 43.16). Diagnosis in the male is not difficult, since typical symptoms are pain upon urination and a thick, greenish yellow urethral discharge. In males and females, a latent infection leads to pelvic inflammatory disease (PID), which affects the vasa deferentia or oviducts. As the inflamed tubes heal, they may become partially or completely blocked by scar tissue, resulting in sterility or infertility. If a baby is exposed during birth, an eye infection leading to blindness can result. All newborns are given eyedrops to prevent this possibility.

Gonorrhea proctitis, an infection of the anus characterized by anal pain and blood or pus in the feces, also occurs in patients. Oral/genital contact can cause infection of the mouth, throat, and tonsils. Gonorrhea can spread to internal parts of the body, causing heart damage or arthritis. If, by chance, the person touches infected genitals and then touches his or her eyes, a severe eye infection can result. Up to now, gonorrhea was curable by antibiotic therapy, but resistance to antibiotics is becoming more and more common, and 40% of all strains are now known to be resistant to therapy.

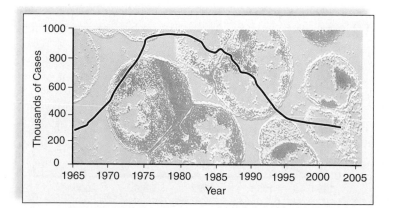

FIGURE 43.16 Gonorrhea.
A graph depicting the incidence of new cases of gonorrhea in the United States from 1965 to 2003 is superimposed on a photomicrograph of a urethral discharge from an infected male. Gonorrheal bacteria *(Neisseria gonorrhoeae)* occur in pairs; for this reason, they are called diplococci.

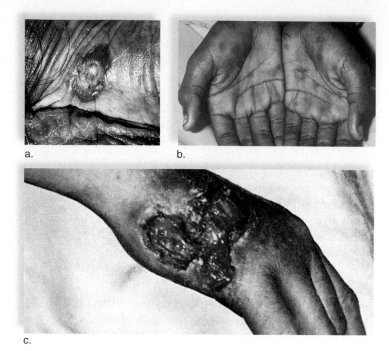

a.   b.

c.

FIGURE 43.17 Syphilis.
a. The primary stage of syphilis is a chancre at the site where the bacterium enters the body. **b.** The secondary stage is a body rash that occurs even on the palms of the hands and soles of the feet. **c.** In the tertiary stage, gummas may appear on the skin or internal organs.

Syphilis

Syphilis, which is caused by the bacterium *Treponema pallidum*, has three stages that are typically separated by latent periods. In the primary stage, a hard chancre (ulcerated sore with hard edges) appears (Fig. 43.17*a*). In the secondary stage, a rash appears all over the body—even on the palms of the hands and the soles of the feet (Fig. 43.17*b*). During the tertiary stage, syphilis may affect the cardiovascular and/or nervous system. An infected person may become mentally retarded, become blind, walk with a shuffle, or show signs of insanity. Gummas, which are large, destructive ulcers, may develop on the skin or within the internal organs (Fig. 43.17*c*). Syphilitic bacteria can cross the placenta, causing birth defects or a stillbirth. Unlike the other STDs discussed, there is a blood test to diagnose syphilis.

Syphilis is a devastating disease. Control depends on prompt and adequate treatment of all new cases; therefore, it is important for all sexual contacts to be traced so that they can be treated with antibiotic therapy.

Two Other Infections

Bacterial vaginosis is believed to account for 50% of vaginitis cases in American women. The overgrowth of the bacterium *Gardnerella vaginalis* and consequent symptoms can be due to nonsexual reasons, but symptomless males may pass on the bacteria to women, who do exhibit symptoms. Also females very often have vaginitis, or infection of the vagina, caused by either the flagellated protozoan *Trichomonas vaginalis* or the yeast *Candida albicans*. The protozoan infec-

health focus

Preventing Transmission of STDs

It is wise to protect yourself from getting a sexually transmitted disease (STD). Some of the STDs, such as gonorrhea, syphilis, and chlamydia, can be cured by taking an antibiotic, but medication for the ones transmitted by viruses is much more problematic. In any case, it is best to prevent the passage of STDs from person to person so that treatment becomes unnecessary.

Sexual Activities Transmit STDs

Abstain from sexual intercourse or develop a long-term monogamous (always the same partner) sexual relationship with a partner who is free of STDs.

Refrain from multiple sex partners or having relations with someone who has multiple sex partners. If you have sex with two other people and each of these has sex with two people and so forth, the number of people who are relating is quite large.

Remember that, although the prevalence of AIDS is presently higher among homosexuals and bisexuals, the highest rate of increase is now occurring among heterosexuals.

The lining of the uterus is generally only one cell thick, and it does allow infected cells from a sexual partner to enter.

Be aware that having relations with an intravenous drug user is risky because the behavior of this group risks hepatitis and an HIV infection. Be aware that anyone who already has another sexually transmitted disease is more susceptible to an HIV infection.

Uncircumcised males are more likely to become infected with an STD than circumcised males because vaginal secretions can remain under the foreskin for a long period of time.

Avoid anal-rectal intercourse (in which the penis is inserted into the rectum) because the lining of the rectum is thin and cells infected with HIV can easily enter the body there.

Unsafe Sexual Practices Transmit STDs

Always use a latex condom during sexual intercourse if you do not know for certain that your partner has been free of STDs for some time. Be sure to follow the directions supplied by the manufacturer.

Avoid fellatio (kissing and insertion of the penis into a partner's mouth) *and cunnilingus* (kissing and insertion of the tongue into the vagina) because they may be a means of transmission. The mouth and gums often have cuts and sores that facilitate the entrance of infected cells.

Be cautious about the use of alcohol or any drug that may prevent you from being able to control your behavior. Females have to be particularly aware of the "date-rape" drug GHB (gamma-hydroxy-butyramine), a drug that puts a person in an uninhibited state with no memory of what transpired.

Drug Use Transmits Hepatitis and HIV

Stop, if necessary, or do not start the habit of injecting drugs into your veins. Be aware that hepatitis and HIV can be spread by blood-to-blood contact.

Always use a new sterile needle for injection or one that has been cleaned in bleach if you are a drug user and cannot stop your behavior.

FIGURE 43B Sexual activities transmit STDs.

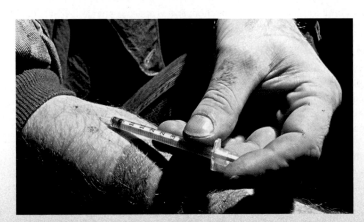

FIGURE 43C Sharing needles transmits STDs.

tion causes a frothy white or yellow, foul-smelling discharge accompanied by itching, and the yeast infection causes a thick, white, curdy discharge, also accompanied by itching. **Trichomoniasis** is most often acquired through sexual intercourse, and the asymptomatic male is usually the reservoir of infection. *Candida albicans,* however, is a normal organism found in the vagina; its growth simply increases beyond normal under certain circumstances. For example, women taking birth control pills are sometimes prone to yeast infections. Also, the indiscriminate use of antibiotics can alter the normal balance of organisms in the vagina so that a yeast infection flares up.

CONNECTING THE CONCEPTS

The dizzying array of reproductive technologies has resulted in many legal complications. Questions range from which mother has first claim to the child—the surrogate mother, the woman who donated the egg, or the primary caregiver—to which partner has first claim to frozen embryos following a divorce. Legal decisions about who has the right to use what techniques have rarely been discussed, much less decided upon. Some clinics will help anyone, male or female, no questions asked, as long as they have the ability to pay. And most clinics are heading toward doing any type of procedure, including guaranteeing the sex of the child or making sure the child will be free from a particular genetic disorder. It would not be surprising if, in the future, zygotes could be engineered to have any particular trait desired by the parents.

Even today, eugenic (good gene) goals are evidenced by the fact that reproductive clinics advertise for egg and sperm donors, primarily in elite college newspapers. Is it too late for us as a society to make ethical decisions about reproductive issues? Should we come to a consensus about what techniques should be allowed and who should be able to use them? We all want to avoid, if possible, what happened to Jonathan Alan Austin. Jonathan, who was born to a surrogate mother, later died from injuries inflicted by his father. Should background checks be legally required? Should surrogate mothers only make themselves available to individuals or couples who possess certain psychological characteristics?

Summary

43.1 HOW ANIMALS REPRODUCE

Ordinarily, asexual reproduction may quickly produce a large number of offspring genetically identical to the parent. Sexual reproduction involves gametes and produces offspring that are genetically slightly different from the parents. The gonads are the primary sex organs, but there are also accessory organs. The accessory organs consist of storage areas for sperm and ducts that conduct the gametes. They also contribute to formation of the semen.

Animals typically protect their eggs and embryos. The egg of oviparous animals contains yolk, and in terrestrial animals a shelled egg prevents drying out. The amount of yolk is dependent on whether there is a larval stage.

Reptiles and birds have extraembryonic membranes that allow them to develop on land; these same membranes are modified for internal development in mammals. Ovoviviparous animals retain their eggs until the offspring have hatched, and viviparous animals retain the embryo. Placental mammals exemplify viviparous animals.

43.2 MALE REPRODUCTIVE SYSTEM

In human males, sperm are produced in the testes, mature in the epididymides, and may be stored in the vasa deferentia before entering the urethra, along with seminal fluid (produced by seminal vesicles, the prostate gland, and bulbourethral glands). Sperm are ejaculated during male orgasm, when the penis is erect.

Spermatogenesis occurs in the seminiferous tubules of the testes, which also produce testosterone in interstitial cells. Testosterone brings about the maturation of the primary sex organs during puberty and promotes the secondary sex characteristics of males, such as low voice, facial hair, and increased muscle strength.

Follicle-stimulating hormone (FSH) from the anterior pituitary stimulates spermatogenesis, and luteinizing hormone (LH) stimulates testosterone production. A hypothalamic-releasing hormone, gonadotropic-releasing hormone (GnRH), controls anterior pituitary production and FSH and LH release. The level of testosterone in the blood controls the secretion of GnRH and the anterior pituitary hormones by a negative feedback system.

43.3 FEMALE REPRODUCTIVE SYSTEM

In females, an oocyte produced by an ovary enters an oviduct, which leads to the uterus. The uterus opens into the vagina. The external genital area of women includes the vaginal opening, the clitoris, the labia minora, and the labia majora.

In either ovary, one follicle a month matures, produces a secondary oocyte, and becomes a corpus luteum. This is called the ovarian cycle. The follicle and the corpus luteum produce estrogens, collectively called estrogen, and progesterone, the female sex hormones.

The uterine cycle occurs concurrently with the ovarian cycle. In the first half of these cycles (days 1–13, before ovulation), the anterior pituitary produces FSH and the follicle produces estrogen. Estrogen causes the endometrium to increase in thickness. In the second half of these cycles (days 15–28, after ovulation), the anterior pituitary produces LH and the follicle produces progesterone. Progesterone causes the endometrium to become secretory. Feedback control of the hypothalamus and anterior pituitary causes the levels of estrogen and progesterone to fluctuate. When they are at a low level, menstruation begins.

If fertilization occurs, a zygote is formed, and development begins. The resulting embryo travels down the oviduct and implants itself in the prepared endometrium. A placenta, which is the region of exchange between the fetal blood and the mother's blood, forms. At first, the placenta produces HCG, which maintains the corpus luteum; later, it produces progesterone and estrogen.

The female sex hormones, estrogen and progesterone, also affect other traits of the body. Primarily, estrogen brings about the maturation of the primary sex organs during puberty and promotes the secondary sex characteristics of females, including less body hair than males, a wider pelvic girdle, a more rounded appearance, and development of breasts.

43.4 CONTROL OF REPRODUCTION

Numerous birth control methods and devices, such as the birth control pill, diaphragm, and condom, are available for those who wish to prevent pregnancy. A morning-after pill, RU-486, is now available. Some couples are infertile, and if so, they may use assisted reproductive technologies to have a child. Artificial insemination and in vitro fertilization have been followed by more sophisticated techniques, such as intracytoplasmic sperm injection.

43.5 SEXUALLY TRANSMITTED DISEASES

Sexually transmitted diseases include AIDS, an epidemic disease; genital warts, which lead to cancer of the cervix; genital herpes, which repeatedly flares up; hepatitis, especially types A and B; chlamydia and gonorrhea, which cause pelvic inflammatory disease (PID); and syphilis, which has cardiovascular and neurological complications if untreated.

Reviewing the Chapter

1. Contrast asexual reproduction with sexual reproduction, reproduction in water with reproduction on land, and the life history of an insect with that of a bird.   776–77
2. Trace the path of sperm in a human male. What glands contribute fluids to semen?   778–79
3. Discuss the anatomy and physiology of the testes. Describe the structure of sperm.   781
4. Name the endocrine glands involved in maintaining the sex characteristics of males and the hormones produced by each.   781
5. Trace the path of an oocyte in a human female. Where do fertilization and implantation occur? Name two functions of the vagina.   782
6. Describe the external genital organs in females.   782–83
7. Discuss the anatomy and physiology of the ovaries. Describe the ovarian cycle and ovulation.   783–84
8. Describe the uterine cycle, and relate it to the ovarian cycle. In what way is menstruation prevented if pregnancy occurs?   784
9. What events occur at fertilization? Name three functions of the female sex hormones, aside from their involvement in the uterine cycle.   786
10. Describe the anatomy and physiology of the breast.   786
11. Which means of birth control require surgery, use hormones, use barrier methods, or are dependent on none of these?   786–87
12. If couples are infertile, what assisted reproductive technologies are available?   788
13. List the cause, symptoms, and treatment for the most common types of sexually transmitted diseases.   790–93

Testing Yourself

1. Label this diagram of the male reproductive system and trace the path of sperm.

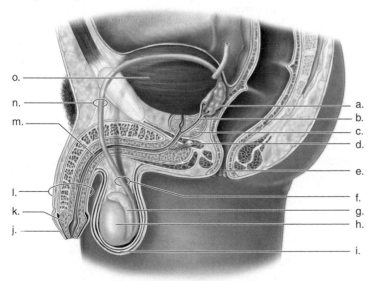

Choose the best answer for each question.

2. Which of these is a requirement for sexual reproduction?
   a. male and female parents
   b. production of gametes
   c. optimal environmental conditions
   d. aquatic habitat
   e. All of these are correct.

3. Internal fertilization
   a. can prevent the drying out of gametes and zygotes.
   b. must take place on land.
   c. is practiced by humans.
   d. requires that males have a penis.
   e. Both a and c are correct.

4. Which of these pairs is mismatched?
   a. interstitial cells—testosterone
   b. seminiferous tubules—sperm production
   c. vasa deferentia—seminal fluid production
   d. urethra—conducts sperm
   e. Both c and d are mismatched.

5. Follicle-stimulating hormone (FSH)
   a. is secreted by females but not males.
   b. stimulates the seminiferous tubules to produce sperm.
   c. secretion is controlled by gonadotropin-releasing hormone (GnRH).
   d. is the same as luteinizing hormone.
   e. Both b and c are correct.

6. Which of these combinations is most likely to be present before ovulation occurs?
   a. FSH, corpus luteum, estrogen, secretory uterine lining
   b. luteinizing hormone (LH), follicle, progesterone, thick endometrium
   c. FSH, follicle, estrogen, endometrium becoming thick
   d. LH, corpus luteum, progesterone, secretory endometrium
   e. Both c and d are correct.

7. In tracing the path of sperm, you would mention vasa deferentia before
   a. testes.
   b. epididymides.
   c. urethra.
   d. uterus.
   e. All of these are correct.

8. An oocyte is fertilized in
   a. the vagina.
   b. the uterus.
   c. the oviduct.
   d. the ovary.
   e. All of these are correct.

9. During pregnancy,
   a. the ovarian and uterine cycles occur more quickly than before.
   b. GnRH is produced at a higher level than before.
   c. the ovarian and uterine cycles do not occur.
   d. the female secondary sex characteristics are not maintained.
   e. Both b and c are correct.

10. Which means of birth control is most effective in preventing sexually transmitted diseases?
    a. condom
    b. pill
    c. diaphragm
    d. spermicidal jelly
    e. vasectomy

11. Which of these is a sexually transmitted disease caused by a bacterium?
    a. gonorrhea
    b. hepatitis B
    c. genital warts
    d. genital herpes
    e. HIV

12. The HIV virus has a preference for binding to
   a. B lymphocytes.
   b. cytotoxic T lymphocytes.
   c. helper T lymphocytes.
   d. All of these are correct.

For questions 13–15, match the descriptions with the sexually transmitted diseases in the key.

KEY:

   a. AIDS
   b. hepatitis B
   c. genital herpes
   d. genital warts
   e. gonorrhea
   f. chlamydia
   g. syphilis

13. blisters, ulcers, pain on urination, swollen lymph nodes

14. flulike symptoms, jaundice; eventual liver failure possible

15. males have a thick, greenish-yellow discharge; no symptoms in female; can lead to PID

16. Which of these is the primary sex organ of the male?
   a. penis
   b. scrotum
   c. testis
   d. prostate
   e. vasectomy

17. The luteal phase of the uterine cycle is characterized by
   a. high levels of LH and progesterone.
   b. low levels of estrogen and progesterone.
   c. increasing estrogen and little or no progesterone.
   d. high levels of LH only.

18. The secretory phase of the uterine cycle occurs during which ovarian phase?
   a. follicular phase
   b. ovulation
   c. luteal phase
   d. menstrual phase

19. Which of these is the correct path of an oocyte?
   a. oviduct, fimbriae, uterus, vagina
   b. ovary, fimbriae, oviduct, uterine cavity
   c. oviduct, fimbriae, abdominal cavity
   d. fimbriae, uterine tube, uterine cavity, ovary

20. Following ovulation, the corpus luteum develops under the influence of
   a. progesterone.   c. LH.
   b. FSH.   d. estradiol.

Thinking Scientifically

1. Female athletes who train intensively often stop menstruating. The important factor appears to be the reduction of body fat below a certain level. Give a possible evolutionary explanation for a relationship between body fat in females and reproductive cycles.

2. The average sperm count in males is now lower than it was several decades ago. The reasons for the lower sperm count usually seen today are not known. What data might be helpful in order to formulate a testable hypothesis?

Understanding the Terms

bacterial vaginosis 792	metamorphosis 777
cervix 782	oocyte 782
contraceptive vaccine 787	orgasm 779
copulation 776	ovarian cycle 783
corpus luteum 784	ovary 776
endometriosis 788	ovulation 784
endometrium 782	ovum 783
estrogen 783	parthenogenesis 776
extraembryonic membrane 777	penis 778
follicle 783	placenta 777
follicular phase 784	progesterone 783
gamete 788	puberty 781
gonad 776	secondary sex
human chorionic gonadotropin	characteristic 781
(HCG) 786	semen (seminal fluid) 778
human immunodeficiency virus	seminiferous tubule 781
(HIV) 790	sperm 781
infertility 788	testes 776
lactation 786	testosterone 781
larva 777	trichomoniasis 793
luteal phase 784	uterine cycle 784
menarche 785	uterus 782
menopause 785	viviparous 777
menstruation 784	yolk 777

Match the terms to these definitions:

a. _____ Release of an oocyte from the ovary.

b. _____ Development of an egg into a whole organism without fertilization.

c. _____ Female sex hormone that causes the endometrium of the uterus to become secretory during the uterine cycle; along with estrogen, it maintains secondary sex characteristics in females.

d. _____ Thick, whitish fluid consisting of sperm and secretions from several glands of the male reproductive tract.

e. _____ Organ that produces gametes; the ovary, which produces eggs, and the testis, which produces sperm.

ARIS, the *Biology* Website

ARIS, the website for *Biology,* provides a wealth of information organized and integrated by chapter. You will find practice quizzes, interactive activities, labeling exercises, flashcards, and much more that will complement your learning and understanding of general biology.

www.mhhe.com/maderbiology9

44

ANIMAL DEVELOPMENT

A s development proceeds, specialization of cells becomes apparent. How does this occur, considering that all cells contain the same genes? Scientists do not have all the answers to this question, but they do know that specific molecules signal specific genes to become active at different times. In the end, nerve cells have axons and dendrites, not myofibrils as are in muscle cells, and bone cells deposit calcium salts, not skin pigment. Furthermore, a baby has eyes and ears, fingers and toes in the right number and place. Unfortunately when development does not proceed normally, serious problems and complications may emerge.

The same processes observed during embryological development are also seen as a newborn matures, as lost parts regenerate, as a wound heals, and even as organisms age. Therefore, it has become increasingly clear that the study of development encompasses not only embryology but these other events as well.

The goal of developmental genetics is to discover which signaling molecules turn on which genes to make specialization come about. It's not legal to perform experiments on human embryos, but luckily scientists have discovered that concepts gained from studying frog, roundworm, and fly embryos apply to humans too.

As development occurs, complexity increases.

four days

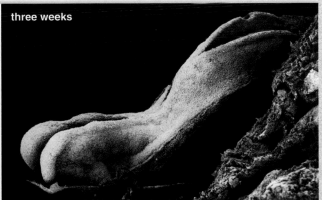

three weeks

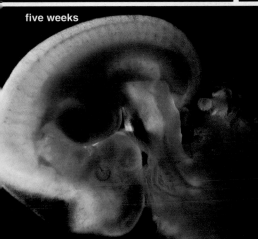

five weeks

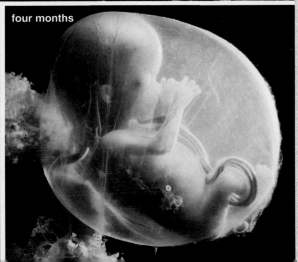

four months

44.1 EARLY DEVELOPMENTAL STAGES

Fertilization is the union of a sperm and an egg to form a zygote. Figure 44.1 illustrates how fertilization occurs in mammals, including humans. Human sperm have three distinct parts; a tail, a middle piece, and a head. The tail is a flagellum, which allows the sperm to swim toward the egg, and the middle piece contains energy-producing mitochondria. The head contains the sperm nucleus and is capped by a membrane-bounded acrosome. The egg of a mammal (actually the secondary oocyte) is surrounded by a few layers of adhering follicular cells, collectively called the *corona radiata*. These cells nourished the egg when it was in a follicle of the ovary. Next the egg has an extracellular matrix termed the *zona pellucida* just outside the plasma membrane.

Fertilization requires three series of events that will result in the diploid zygote. First, a sperm must reach the egg plasma membrane. Several sperm begin the journey. They squeeze through the corona radiata and bind to the zona pellucida. After a sperm head binds tightly to the zona pellucida, the acrosome releases digestive enzymes that forge a pathway for the sperm through the zona pellucida to the egg plasma membrane.

When the first sperm binds to the egg plasma membrane the next series of events prevent *polyspermy* (entrance of more than one sperm). As soon as a sperm touches the plasma membrane of an egg, the egg's plasma membrane depolarizes and this change in charge, known as the "fast block," serves to repel sperm only for a few seconds. Then vesicles in the egg called *cortical granules* secrete enzymes that turn the zona pellucida into an impenetrable *fertilization membrane*. The longer-lasting cortical reaction is known as the "slow block."

The last series of events includes formation of the diploid zygote. Microvilli exending from the plasma membrane of the egg (Fig. 44.1) bring the entire sperm into the egg. The sperm nucleus releases its chromatin, which re-forms into chromosomes enclosed within the co-called sperm pronucleus. In the meantime the secondary oocyte completes ovulation and its chromosomes are also enclosed in a pronucleus. A single nuclear envelope soon surrounds both sperm and egg pronuclei. Cell division is imminent, and the centrosomes that give rise to a spindle apparatus are derived from the basal body of the sperm's flagellum! The two haploid sets of chromosomes share the first spindle apparatus of the fertilized egg, also called a zygote.

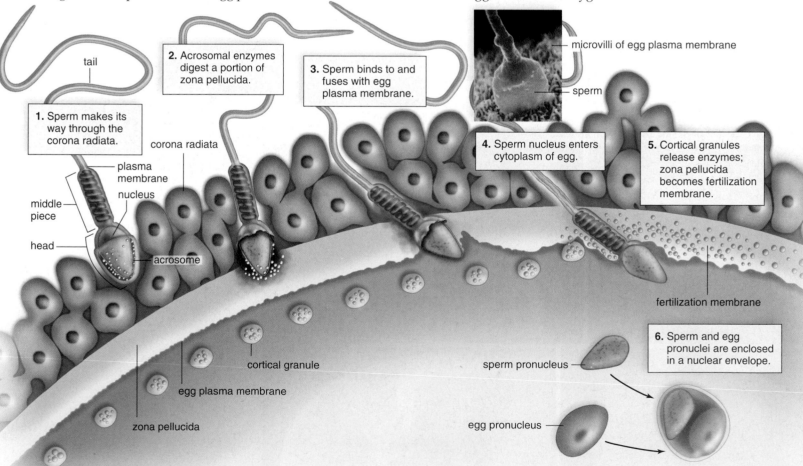

1. Sperm makes its way through the corona radiata.

2. Acrosomal enzymes digest a portion of zona pellucida.

3. Sperm binds to and fuses with egg plasma membrane.

microvilli of egg plasma membrane

sperm

4. Sperm nucleus enters cytoplasm of egg.

5. Cortical granules release enzymes; zona pellucida becomes fertilization membrane.

tail

plasma membrane

nucleus

middle piece

head

acrosome

corona radiata

fertilization membrane

cortical granule

egg plasma membrane

zona pellucida

6. Sperm and egg pronuclei are enclosed in a nuclear envelope.

sperm pronucleus

egg pronucleus

FIGURE 44.1   Fertilization.
During fertilization, a single sperm is drawn into the egg by microvilli of its plasma membrane (micrograph). The head of a sperm has a membrane-bounded acrosome filled with enzymes. When released, these enzymes digest a pathway for the sperm through the zona pellucida. After a sperm binds to the plasma membrane of the egg, changes occur that prevent other sperm from entering the egg. Fertilization is complete when the sperm pronucleus and the egg pronucleus contribute chromosomes to the zygote.

Embryonic Development

Development is all the changes that occur during the life cycle of an organism. During the first stages of development, an organism is called an **embryo.**

Cellular Stages of Development

The cellular stages of development are (1) cleavage resulting in a multicellular embryo and (2) formation of the blastula. **Cleavage** is cell division without growth. DNA replication and mitotic cell division occur repeatedly, and the cells get smaller with each division. In other words, cleavage increases only the number of cells; it does not change the original volume of the egg cytoplasm.

As shown in Figure 44.2, cleavage in a lancelet is equal and results in uniform cells that form a **morula,** which is a ball of cells. The 16-cell morula resembles a mulberry and continues to divide forming a blastula. A **blastula** is a hollow ball of cells having a fluid-filled cavity called a **blastocoel.** The blastocoel forms when the cells of the morula extrude Na$^+$ into extracellular spaces and water follows by osmosis. The water collects in the center, and the result is a hollow ball of cells.

The zygotes of other animals, such as a frog, chick, or human, which are vertebrates, also undergo cleavage and form a morula. In frogs, cleavage is not equal because of the presence of **yolk,** a dense nutrient material. When yolk is present, the zygote and embryo exhibit polarity, and the embryo has an animal pole and a vegetal pole. The animal pole of a frog embryo has a deep gray color because the cells contain melanin granules, and the vegetal pole has a yellow color because the cells contain yolk.

Similarly, all vertebrates have a blastula stage, but the appearance of the blastula can be different from that of a lancelet. A chick is a vertebrate animal that develops on land and lays a hard-shelled egg containing plentiful yolk. Because yolk-filled eggs do not participate in cleavage, the blastula is a layer of cells that spreads out over the yolk. The blastocoel is a space that separates these cells from the yolk:

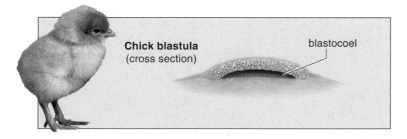
Chick blastula
(cross section)   blastocoel

The blastula of humans resembles that of the chick embryo, yet this resemblance cannot be related to the amount of yolk because the human egg contains little yolk. But the evolutionary history of these two animals can provide an explanation for this similarity. Both birds and mammals are related to reptiles. This explains why all three groups develop similarly, despite a difference in the amount of yolk in their eggs.

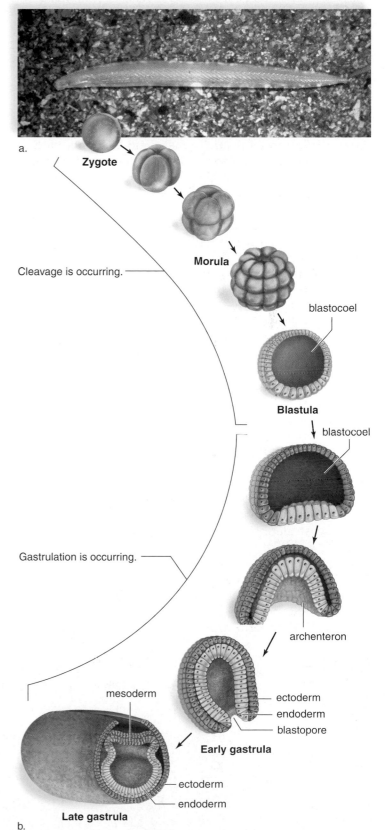

a.

Cleavage is occurring. —

Zygote

Morula

blastocoel

Blastula

blastocoel

Gastrulation is occurring. —

archenteron

mesoderm

ectoderm
endoderm
blastopore

Early gastrula

ectoderm

endoderm

Late gastrula

b.

FIGURE 44.2 Lancelet early development.
a. A lancelet. **b.** The early stages of development are exemplified in the lancelet. Cleavage produces a number of cells that form a cavity, the blastocoel. Invagination during gastrulation produces the germ layers ectoderm and endoderm. Then the mesoderm arises.

Tissue Stages of Development

The tissue stages of development are (1) the early gastrula and (2) the late gastrula. The early **gastrula** stage begins when certain cells begin to push, or invaginate, into the blastocoel, creating a double layer of cells (see Fig. 44.2). Cells migrate during this and other stages of development, sometimes traveling quite a distance before reaching a destination, where they continue developing. Extracellular proteins and cytoskeletal elements participate in allowing migration to occur. As cells migrate, they "feel their way" by changing their pattern of adhering to extracellular proteins.

An early gastrula has two layers of cells. The outer layer of cells is called the **ectoderm,** and the inner layer is called the **endoderm.** The endoderm borders the gut, but at this point, it is termed either the archenteron or the primitive gut. The pore, or hole, created by invagination is the **blastopore,** and in a lancelet, the blastopore eventually becomes the anus. **Gastrulation** is not complete until three layers of cells that will develop into adult organs are produced. In addition to ectoderm and endoderm, the late gastrula has a middle layer of cells called the **mesoderm.**

Figure 44.2 illustrates gastrulation in a lancelet and Figure 44.3 compares the lancelet, frog, and chick late gastrula stages. In the lancelet, mesoderm formation begins as outpocketings from the primitive gut (Fig. 44.3). These outpocketings will grow in size until they meet and fuse forming two layers of mesoderm. The space between them is the coelom. The coelom is a body cavity lined by mesoderm that contains internal organs. (In humans, the coelom becomes the thoracic and abdominal cavities of the body.)

In the frog, the cells containing yolk do not participate in gastrulation, and therefore, they do not invaginate. Instead, a slitlike blastopore is formed when the animal pole cells begin to invaginate from above, forming endoderm. Animal pole cells also move down over the yolk, to invaginate from below. Some yolk cells, which remain temporarily in the region of the blastopore, are called the yolk plug. Mesoderm forms when cells migrate between

TABLE 44.1

Embryonic Germ Layers

Embryonic Germ Layer	Vertebrate Adult Structures
Ectoderm (outer layer)	Nervous system; epidermis of skin and derivatives of the epidermis (hair, nails, glands); tooth enamel, dentin, and pulp; epithelial lining of oral cavity and rectum
Mesoderm (middle layer)	Musculoskeletal system; dermis of skin; cardiovascular system; urinary system; lymphatic system; reproductive system—including most epithelial linings; outer layers of respiratory and digestive systems
Endoderm (inner layer)	Epithelial lining of digestive tract and respiratory tract, associated glands of these systems; epithelial lining of urinary bladder; thyroid and parathyroid glands

the ectoderm and endoderm. Later, a splitting of the mesoderm creates the coelom.

The chick egg contains so much yolk that endoderm formation does not occur by invagination. Instead, an upper layer of cells becomes ectoderm, and a lower layer becomes endoderm. Mesoderm arises by an invagination of cells along the edges of a longitudinal furrow in the midline of the embryo. Because of its appearance, this furrow is called the *primitive streak.* Later, the newly formed mesoderm splits to produce a coelomic cavity.

Ectoderm, mesoderm, and endoderm are called the embryonic **germ layers.** No matter how gastrulation takes place, the result is the same: three germ layers are formed. It is possible to relate the development of future organs to these germ layers (Table 44.1).

A mature gastrula has three germ layers: ectoderm, endoderm, and mesoderm. Each germ layer develops into specific organs.

FIGURE 44.3

Comparative development of mesoderm.
a. In the lancelet, mesoderm forms by an outpocketing of the archenteron. **b.** In the frog, mesoderm forms by migration of cells between the ectoderm and endoderm. **c.** In the chick, mesoderm also forms by invagination of cells.

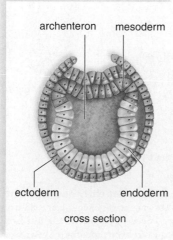

cross section

a. Lancelet late gastrula

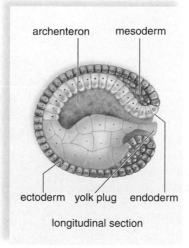

longitudinal section

b. Frog late gastrula

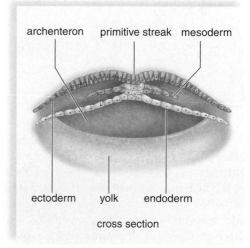

cross section

c. Chick late gastrula

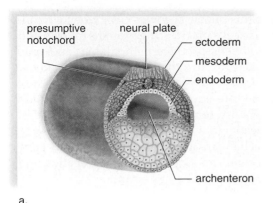

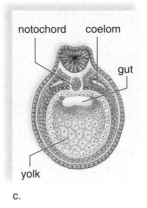

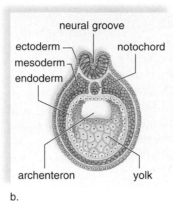

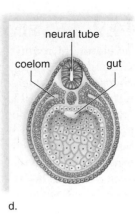

a.

b.

c.

d.

FIGURE 44.4 Development of neural tube and coelom in a frog embryo.
a. Ectodermal cells that lie above the future notochord (called presumptive notochord) thicken to form a neural plate. **b.** The neural groove and folds are noticeable as the neural tube begins to form. **c.** A splitting of the mesoderm produces a coelom, which is completely lined by mesoderm. **d.** A neural tube and a coelom have now developed.

Organ Stages of Development

The organs of an animal's body develop from the three embryonic germ layers. Much study has been devoted to how the nervous system develops.

The newly formed mesoderm cells lie along the main longitudinal axis of the animal and coalesce to form a dorsal supporting rod called the **notochord.** The notochord persists in lancelets, but in frogs, chicks, and humans, it is later replaced by the vertebral column. Therefore, these animals are called vertebrates.

The nervous system develops from midline ectoderm located just above the notochord. At first, a thickening of cells, called the **neural plate,** is seen along the dorsal surface of the embryo. Then, neural folds develop on either side of a neural groove, which becomes the **neural tube** when these folds fuse. Figure 44.4 shows cross sections of frog development to illustrate the formation of the neural tube. At this point, the embryo is called a **neurula.** Later, the anterior end of the neural tube develops into the *brain,* and the rest becomes the *spinal cord.* In addition, the neural crest is a band of cells that develops where the neural tube pinches off from the ectoderm. Neural crest cells migrate to various locations, where they contribute to formation of skin and muscles, in addition to the adrenal medulla and the ganglia of the peripheral nervous system.

Midline mesoderm cells that did not contribute to the formation of the notochord now become two longitudinal masses of tissue. These two masses become blocked off into somites, which are serially arranged along both sides along the length of the notochord. Somites give rise to muscles associated with the axial skeleton and to the vertebrae. The serial origin of axial muscles and the vertebrae testify that vertebrates are segmented animals. Lateral to the somites, the mesoderm splits, forming the mesodermal lining of the coelom.

A primitive gut tube is formed by endoderm as the body itself folds into a tube. The heart, too, begins as a simple tubular pump. Organ formation continues until the germ layers have given rise to the specific organs listed in Table 44.1. Figure 44.5 will help you relate the formation of vertebrate structures and organs to the three embryonic layers of cells: the ectoderm, the mesoderm, and the endoderm.

> During neurulation, the neural tube develops just above the notochord. At the neurula stage of development, cross sections of all chordate embryos are similar in appearance.

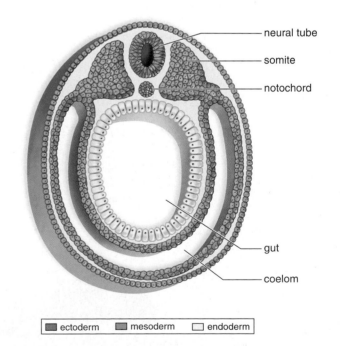

ectoderm   mesoderm   endoderm

FIGURE 44.5 Vertebrate embryo, cross section.
At the neurula stage, each of the germ layers, indicated by color (see key), can be associated with the later development of particular parts. The somites give rise to the muscles of each segment and to the vertebrae, which replace the notochord in vertebrates.

44.2 DEVELOPMENTAL PROCESSES

Development requires (1) growth, (2) cellular differentiation, and (3) morphogenesis. **Cellular differentiation** occurs when cells become specialized in structure and function; that is, a muscle cell looks different and acts differently than a nerve cell. **Morphogenesis** produces the shape and form of the body. One of the earliest indications of morphogenesis is cell movement. Later, morphogenesis includes **pattern formation**, which means how tissues and organs are arranged in the body. **Apoptosis,** or programmed cell death, which was first discussed on page 151, is an important part of pattern formation.

Developmental genetics has benefited from research using the roundworm, *Caenorhabditis elegans,* and the fruit fly, *Drosophila melanogaster.* These organisms are referred to as model organisms because the study of their development produced concepts that help us understand development in general.

Cellular Differentiation

At one time, investigators mistakenly believed that irreversible genetic changes must account for differentiation and morphogenesis. Perhaps the genes are parceled out as development occurs and that is why cells of the body have a different structure and function. Our ability today to clone mammals such as sheep, goats, and mice from specialized adult cells shows that every cell in an organism's body contains a full complement of genes. It's said that cells in the adult body are **totipotent;** each one contains all the instructions needed by any other specialized cell in the body.

The answer to this puzzle becomes clear when we consider that only muscle cells produce the proteins myosin and actin; only red blood cells produce hemoglobin; and only skin cells produce keratin. In other words, we now know that specialization is not due to a parceling out of genes; rather, it is due to differential gene expression. Certain genes and not others are turned on in differentiated cells. In recent years, investigators have turned their attention to discovering the mechanisms that lead to differential gene expression. Two mechanisms—cytoplasmic segregation and induction—seem to be especially important.

Cytoplasmic Segregation

Differentiation must begin long before we can recognize specialized types of cells. Ectodermal, endodermal, and mesodermal cells in the gastrula look quite similar, but they must be different because they develop into different organs. The egg is now known to contain substances called maternal determinants, which influence the course of development. Cytoplasmic segregation is the parceling out of maternal determinants as mitosis occurs:

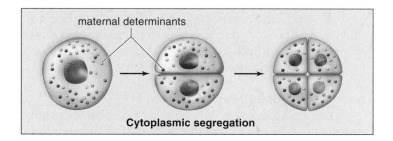

maternal determinants

Cytoplasmic segregation

FIGURE 44.6

Cytoplasmic influence on development.

a. A frog's egg has anterior/posterior and dorsal/ventral axes that correlate with the position of the gray crescent. **b.** The first cleavage normally divides the gray crescent in half, and each daughter cell is capable of developing into a complete tadpole. **c.** But if only one daughter cell receives the gray crescent, then only that cell can become a complete embryo. This shows that maternal determinants are present in the cytoplasm of a frog's egg.

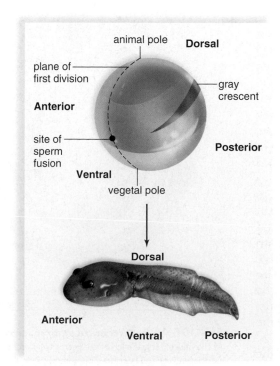

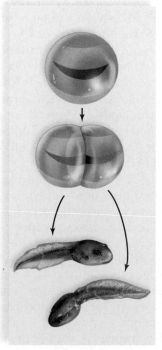

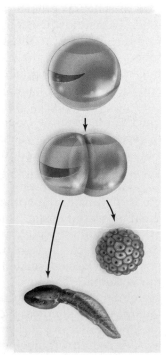

a. Frog's egg is polar and has axes.

b. Each cell receives a part of the gray crescent.

c. Only the cell on the left receives the gray crescent.

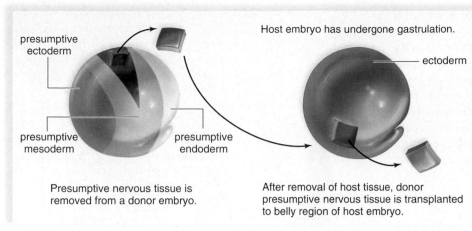
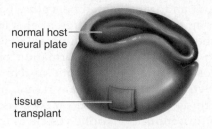

Host embryo has undergone gastrulation.

ectoderm

Host embryo undergoes neurulation.

normal host neural plate

tissue transplant

Presumptive nervous tissue is removed from a donor embryo.

After removal of host tissue, donor presumptive nervous tissue is transplanted to belly region of host embryo.

Due to normal induction process, a host neural plate develops. But donated tissue is not induced to develop into a neural plate.

a.

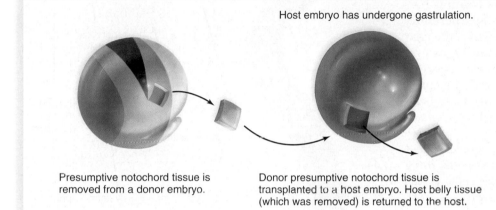
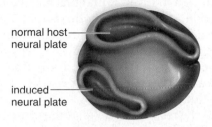

Host embryo has undergone gastrulation.

Host embryo undergoes neurulation.

normal host neural plate

induced neural plate

Presumptive notochord tissue is removed from a donor embryo.

Donor presumptive notochord tissue is transplanted to a host embryo. Host belly tissue (which was removed) is returned to the host.

Host develops two neural plates—one induced by host notochord tissue, the second induced by transplanted notochord tissue.

b.

FIGURE 44.7 Control of nervous system development.
a. In this experiment, the presumptive nervous system (blue) does not develop into the neural plate if moved from its normal location. **b.** In this experiment, the presumptive notochord (pink) can cause even belly ectoderm to develop into the neural plate (blue). This shows that the notochord induces ectoderm to become a neural plate, most likely by sending out chemical signals.

Cytoplasmic segregation helps determine how the various cells of the morula will develop.

An early experiment showed that the cytoplasm of a frog's egg is not uniform. It is polar and has both an anterior/posterior axis and a dorsal/ventral axis, which can be correlated with the **gray crescent,** a gray area that appears after the sperm fertilizes the egg (Fig. 44.6a). Hans Spemann, who received a Nobel Prize in 1935 for his extensive work in embryology, showed that if the gray crescent is divided equally by the first cleavage, each experimentally separated daughter cell develops into a complete embryo (Fig. 44.6b). However, if the egg divides so that only one daughter cell receives the gray crescent, only that cell becomes a complete embryo (Fig. 44.6c). This experiment allows us to speculate that the gray crescent must contain particular chemical signals that are needed for development to proceed normally.

Induction and Frog Experiments

As development proceeds, specialization of cells and formation of organs are influenced not only by maternal determinants but also by signals given off by neighboring cells. **Induction** [L. *in,* into, and *duco,* lead] is the ability of one embryonic tissue to influence the development of another tissue.

Spemann showed that a frog embryo's gray crescent becomes the dorsal lip of the blastopore, where gastrulation begins. Since this region is necessary for complete development, he called the dorsal lip of the blastopore the primary organizer. The cells closest to Spemann's primary organizer become endoderm, those farther away become mesoderm, and those farthest away become ectoderm. This suggests that there may be a molecular concentration gradient that acts as a chemical signal to induce germ layer differentiation.

The gray crescent of a frog's egg marks the dorsal side of the embryo where the mesoderm becomes notochord and ectoderm becomes nervous system. In a classic experiment Spemann and his colleague Hilde Mangold showed that presumptive (potential) notochord tissue induces the formation of the nervous system (Fig. 44.7). If presumptive nervous system tissue, located just above the presumptive notochord, is cut out and transplanted to the belly region of the embryo, it does not form a neural tube. On the other hand, if presumptive notochord tissue is cut out and transplanted beneath what would be belly ectoderm, this ectoderm differentiates into neural tissue. Still other examples of induction are now known. In 1905, Warren Lewis studied the formation of the eye in frog embryos. He found that an optic vesicle, which is a lateral outgrowth of developing brain tissue, induces overlying ectoderm to thicken and become a lens. The developing lens in turn induces an optic vesicle to form an optic cup, where the retina develops.

Induction in Caenorhabditis elegans

The minute nematode, *Caenorhabditis elegans,* is only 1 mm long, and vast numbers can be raised in the laboratory either in petri dishes or a liquid medium. The worm is hermaphroditic, and self-fertilization is the rule. Therefore, even though induced mutations may be recessive, the next generation will yield individuals that are homozygous recessive and will show the mutation. Many modern genetic studies have been performed on *C. elegans,* and the entire genome has been sequenced. Individual genes have been altered and cloned and their products injected into cells or extracellular fluid.

As the result of genetic studies, much has been learned about *C. elegans.* Development of *C. elegans* takes only three days, and the adult worm contains only 959 cells. It has been possible for investigators to watch the process from beginning to end, especially since the worm is transparent. **Fate maps** have been developed that show the destiny of each cell as it arises following successive cell divisions (Fig. 44.8*a*). Some investigators have studied in detail the development of the vulva, a pore through which eggs are laid. A cell called the anchor cell induces the vulva to form. The cell closest to the anchor cell receives the most inducer and becomes the inner vulva. This cell in turn produces another inducer, which acts on its two neighboring cells, and they become the outer vulva. The inducers are growthlike factors that alter the metabolism of the receiving cell and activate particular genes. Work with *C. elegans* has shown that induction requires the transcriptional regulation of genes in a particular sequence. This diagram shows how induction can occur sequentially:

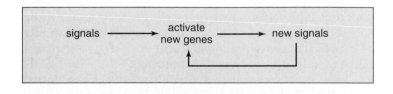

Cytoplasmic segregation and induction are two mechanisms that help explain the developmental processes of cellular differentiation.

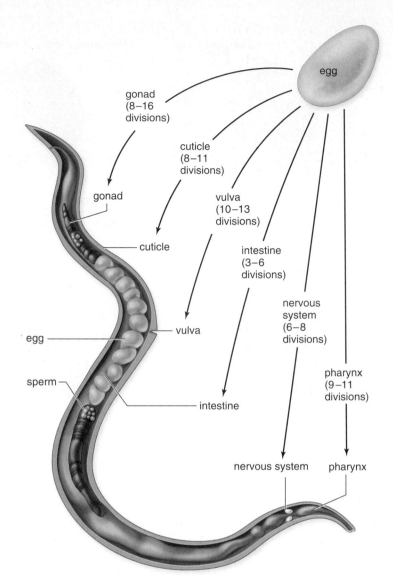

FIGURE 44.8 Development of *C. elegans,* a nematode.
A fate map of the worm showing that as cells arise by cell division, they are destined to become particular structures.

Morphogenesis

Pattern formation is the ultimate in morphogenesis. To understand the concept of pattern formation, think about the pattern of your own body. Your vertebral column runs along the main axis, and your arms and legs occur in certain locations. Pattern formation refers to how this pattern comes about during development. *Drosophila melanogaster* (fruit fly) experiments, in particular, have contributed to our knowledge of pattern formation. A fruit fly may be larger than a roundworm, but a few pairs can produce hundreds of offspring in a couple of weeks, all within a small bottle kept on a laboratory bench.

Morphogenesis in Drosophila melanogaster

Investigators studying morphogenesis in the fruit fly have discovered that some genes determine the animal's anterior/posterior and dorsal/ventral axes, others determine the fly's segmentation pattern, and still others, called homeotic genes, determine the body parts on each segment.

The Anterior/Posterior Axes. One of the first events toward successful development is the establishment of the body axes: location of the head versus the tail and location of the back versus the belly. In the *Drosophila* egg, there is a greater concentration of a protein called bicoid at one end. This end becomes the anterior region, where the head develops. (*Bicoid* means "two tailed," and a *bicoid* gene mutation can cause the embryo to have only two posterior ends.) Researchers have cloned the *bicoid* gene and used it as a probe to establish that mRNA for the bicoid protein is present in a concentration gradient from the anterior end to the posterior end of the embryo.

Many other protein gradients in the *Drosophila* egg help determine the axes of the body. Proteins that influence morphogenesis are called **morphogens.** The bicoid gradient is a morphogen gradient. The bicoid gradient switches on the expression of segmentation genes in *Drosophila*. What is the advantage of a morphogen gradient? A morphogen gradient can have a range of effects depending on its concentration in a particular portion of the embryo.

The Segmentation Pattern. The next event in the development of *Drosophila* is the establishment of its segments. The bicoid gradient initiates a cascade in which a series of segmentation gene sets are turned on, one after the other.

Christiane Nusslein-Vollard and Eric Wieschaus received a Nobel Prize for their work in discovering the segmentation genes of *Drosophila*. They exposed the flies to mutagenic chemicals and then performed innumerable crosses to map the mutated genes that caused segmental abnormalities. They went on to clone many of these genes. The first set of segmental genes to be activated are called *gap* genes (Fig. 44.9a). (If one of these genes mutates, there are *gaps*—that is, large blocks of segments are missing.) Then the *pair-rule* genes become active, and the embryo has precisely 14 segments (Fig. 44.9b). If one of these mutates, the animal has half the number of segments. Next, the *segment-polarity* genes are expressed, and each segment has an anterior and posterior half (Fig. 44.9c).

Work with *Drosophila* has suggested how morphogenesis comes about. Sequential sets of master genes code for morphogen gradients that activate the next set of master genes in turn. How do morphogen gradients turn on genes? They are transcription factors that regulate which genes are active in which parts of the embryo in what order.

Homeotic Genes. **Homeotic genes** act as main switches and control pattern formation, the organization of differentiated cells into specific three-dimensional structures. During normal development of *Drosophila*, the homeo-

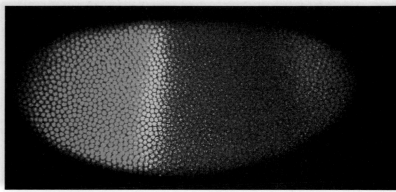

a. Protein products of *gap* genes

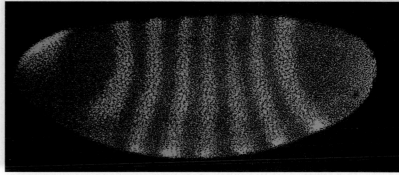

b. Protein products of *pair-rule* genes

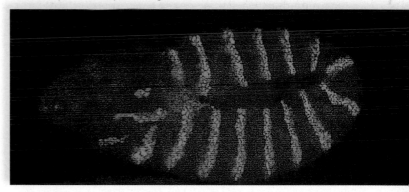

c. Protein products of *segment-polarity* genes

FIGURE 44.9 Development in *Drosophila*, a fruit fly.
a. The different colors show that two different *gap* gene proteins are present from the anterior to the posterior end of an egg. **b.** The green stripes show that a *pair-rule* gene is being expressed as segmentation of the fly occurs. **c.** Now *segment-polarity* genes help bring about division of each segment into an anterior and posterior end.

tic genes are activated after the segmentation genes. If a mutant is missing the set of homeotic genes called the *bithorax complex*, the fly still has the normal number of posterior segments.

As early as the 1940s, Edward B. Lewis had discovered the homeotic genes in *Drosophila*. He was able to determine that certain genes controlled whether a particular segment would bear antennae, legs, or wings. A homeotic mutation caused these appendages to be misplaced—a mutant fly could have extra legs where antennae should be or two pairs of wings (Fig. 44.10a).

Homeotic genes have now been found in many other organisms, and surprisingly, they all contain the same particular sequence of nucleotides, called a **homeobox.** (Because homeotic genes contain a homeobox in mammals, they are called *Hox* genes.) The homeobox codes for a particular sequence of 60 amino acids called a homeodomain:

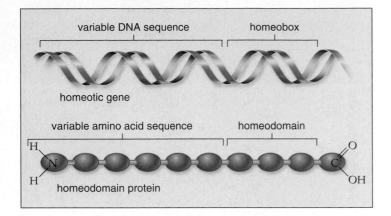

Homeotic genes, like many other developmental genes, code for transcription factors. The homeodomain protein is that part of a transcription factor that binds to DNA, but the other more variable sequences of a transcription factor determine which particular genes are turned on. Researchers envision that a homeodomain protein produced by one homeotic gene binds to and turns on the next homeotic gene, and so forth. This orderly process in the end determines the morphology of particular segments.

Mice and humans have the same four clusters of homeotic genes located on four different chromosomes (Fig. 44.10*b*). In *Drosophila,* homeotic genes are located on a single chromosome. In all three types of animals, homeotic genes are expressed from anterior to posterior in the same order. The first clusters determine the final development of anterior segments of the animal, while those later in the sequence determine the final development of posterior segments of the animal.

Since the homeotic genes of so many different organisms contain the same homeodomain, we know that this nucleotide sequence arose early in the history of life and has been largely conserved as evolution occurred. In general, it has been very surprising to learn how similar developmental genetics is in organisms ranging from yeasts to plants to a wide variety of animals. Certainly, the genetic mechanisms of development appear to be quite similar in all animals.

Apoptosis. We have already discussed the importance of apoptosis (programmed cell death) in the normal day-to-day operation of the immune system and in preventing the occurrence of cancer. Apoptosis is also an important part of morphogenesis. During development of humans, we know that apoptosis is necessary to the shaping of the hands and feet; if it does not occur, the child is born with webbing between its fingers and toes.

The fate maps of *C. elegans* (see Fig. 44.8) indicate that apoptosis occurs in 131 cells as development takes place.

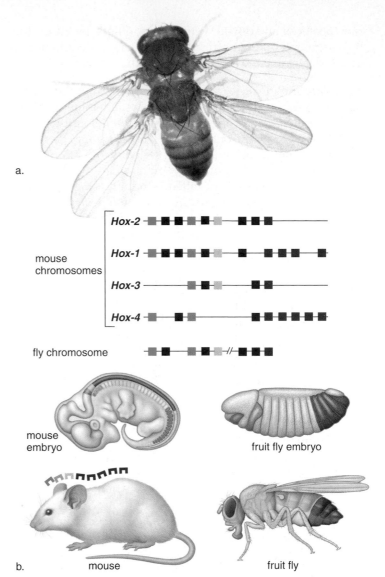

FIGURE 44.10 Pattern formation in *Drosophila.*
Homeotic genes control pattern formation, an aspect of morphogenesis. **a.** If homeotic genes are activated at inappropriate times, abnormalities such as a fly with four wings occur. **b.** The green, blue, yellow, and red colors show that homologous homeotic genes occur on four mouse chromosomes and on a fly chromosome in the same order. These genes are color coded to the region of the embryo, and therefore the adult, where they regulate pattern formation. The black boxes are homeotic genes that are not identical between the two animals. In mammals, homeotic genes are called *Hox* genes.

When a cell-death signal is received, an inhibiting protein becomes inactive, allowing a cell-death cascade to proceed that ends in enzymes destroying the cell.

A hierarchy of gene activity causes morphogenesis to be orderly. Morphogen gradients contain transcription factors that cause cells to produce yet other morphogen gradients until finally specific structures are formed. Apoptosis is also a part of morphogenesis.

44.3 HUMAN EMBRYONIC AND FETAL DEVELOPMENT

In humans, the length of time from conception (fertilization followed by **implantation**) to birth (parturition) is approximately nine months (266 days). It is customary to calculate the time of birth by adding 280 days to the start of the last menstruation, because this date is usually known, whereas the day of fertilization is usually unknown. Because the time of birth is influenced by so many variables, only about 5% of babies actually arrive on the forecasted date.

In humans, pregnancy, or gestation, is the time in which the developing embryo is carried by the mother. Human development is often divided into embryonic development (months 1 and 2) and fetal development (months 3–9). During the **embryonic period,** the major organs are formed, and during fetal development, these structures are refined.

Development can also be divided into trimesters. Each trimester can be characterized by specific developmental accomplishments. During the first trimester, embryonic and early fetal development occur. The second trimester is characterized by the development of organs and organ systems. By the end of the second trimester, the fetus appears distinctly human. In the third trimester, the fetus grows rapidly and the major organ systems become functional. An infant born one or perhaps two months premature has a reasonable chance of survival.

Before we consider human development chronologically, we must understand the placement of **extraembryonic membranes** [L. *extra,* on the outside]. Extraembryonic membranes are best understood by considering their function in reptiles and birds. In reptiles, these membranes made development on land first possible. If an embryo develops in the water, the water supplies oxygen for the embryo and takes away waste products. The surrounding water prevents desiccation, or drying out, and provides a protective cushion. For an embryo that develops on land, all these functions are performed by the extraembryonic membranes.

In the chick, the extraembryonic membranes develop from extensions of the germ layers, which spread out over the yolk. Figure 44.11 shows the chick surrounded by the membranes. The **chorion** [Gk. *chorion,* membrane] lies next to the shell and carries on gas exchange. The **amnion** [Gk. *amnion,* membrane around fetus] contains the protective amniotic fluid, which bathes the developing embryo. The **allantois** [Gk. *allantos,* sausage] collects nitrogenous wastes, and the **yolk sac** surrounds the remaining yolk, which provides nourishment.

Humans (and other mammals) also have these extraembryonic membranes. The chorion develops into the fetal half of the placenta; the yolk sac, which lacks yolk, is the first site of blood cell formation; the allantoic blood vessels become the umbilical blood vessels; and the amnion contains fluid to cushion and protect the embryo, which develops into a fetus. Therefore, the function of the membranes in humans has been modified to suit internal development, but their very presence indicates our relationship to birds and to reptiles. It is interesting to note that all chordate animals develop in water—either in bodies of water or surrounded by amniotic fluid within a shell or uterus.

The presence of extraembryonic membranes in reptiles made development on land possible. Humans also have these membranes, but their function has been modified for internal development.

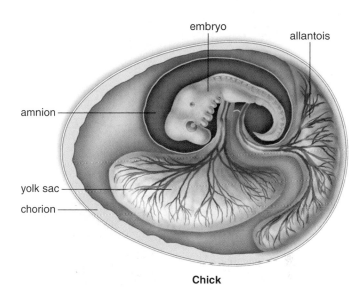

Chick

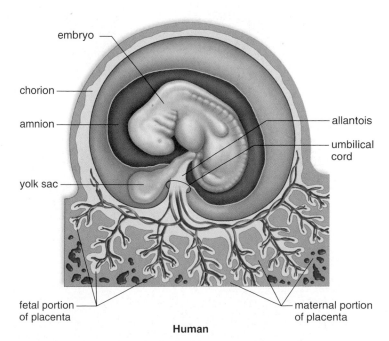

Human

FIGURE 44.11 Extraembryonic membranes.
Extraembryonic membranes, which are not part of the embryo, are found during the development of chicks and humans. Each has a specific function.

Embryonic Development

Embryonic development includes the first two months of development.

The First Week

Fertilization occurs in the upper third of the oviduct (Fig. 44.12). Cleavage begins 30 hours after fertilization and continues as the embryo passes through the oviduct to the uterus. By the time the embryo reaches the uterus on the third day, it is a morula. The morula is not much larger than the zygote because, even though multiple cell divisions have occurred, there has been no growth of these newly formed cells. By about the fifth day, the morula is transformed into the blastocyst. The **blastocyst** has a fluid-filled cavity, a single layer of outer cells called the **trophoblast** [Gk. *trophe*, food, and *blastos*, bud], and an inner cell mass. The early function of the trophoblast is to provide nourishment for the embryo. Later, the trophoblast, reinforced by a layer of mesoderm, gives rise to the chorion, one of the extraembryonic membranes (see Fig. 44.11). The inner cell mass eventually becomes the embryo, which develops into a fetus.

The Second Week

At the end of the first week, the embryo begins the process of implanting in the wall of the uterus. The trophoblast secretes enzymes to digest away some of the tissue

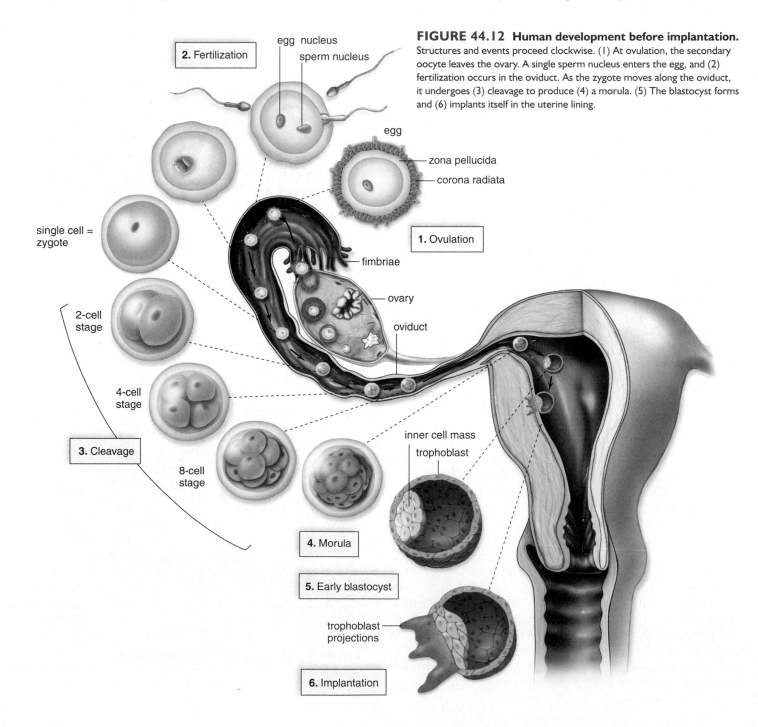

FIGURE 44.12 Human development before implantation. Structures and events proceed clockwise. (1) At ovulation, the secondary oocyte leaves the ovary. A single sperm nucleus enters the egg, and (2) fertilization occurs in the oviduct. As the zygote moves along the oviduct, it undergoes (3) cleavage to produce (4) a morula. (5) The blastocyst forms and (6) implants itself in the uterine lining.

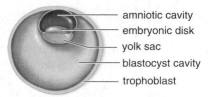

- amniotic cavity
- embryonic disk
- yolk sac
- blastocyst cavity
- trophoblast

a. 14 days

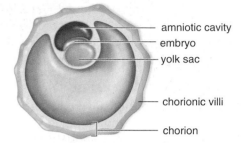

- amniotic cavity
- embryo
- yolk sac
- chorionic villi
- chorion

b. 18 days

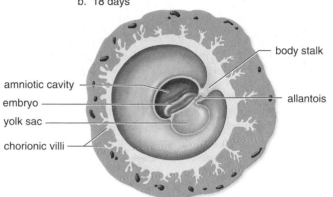

- amniotic cavity
- embryo
- yolk sac
- chorionic villi
- body stalk
- allantois

c. 21 days

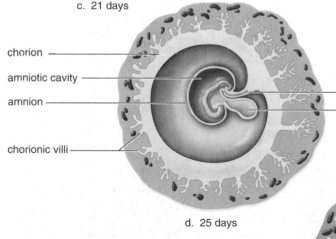

- chorion
- amniotic cavity
- amnion
- chorionic villi
- allantois
- yolk sac

d. 25 days

and blood vessels of the endometrium of the uterus (Fig. 44.12). The embryo is now about the size of the period at the end of this sentence. The trophoblast begins to secrete **human chorionic gonadotropin (HCG),** the hormone that is the basis for the pregnancy test and that serves to maintain the corpus luteum past the time it normally disintegrates. (Recall that the corpus luteum is a yellow body formed in the ovary from a follicle that has discharged its secondary oocyte.) Because of this, the endometrium is maintained, and menstruation does not occur.

As the week progresses, the inner cell mass detaches itself from the trophoblast, and two more extraembryonic membranes form (Fig. 44.13a). The yolk sac, which forms below the embryonic disk, has no nutritive function as in chicks, but it is the first site of blood cell formation. However, the amnion and its cavity are where the embryo (and then the fetus) develops. In humans, amniotic fluid acts as an insulator against cold and heat and also absorbs shock, such as that caused by the mother exercising.

Gastrulation occurs during the second week. The inner cell mass now has flattened into the **embryonic disk,** composed of two layers of cells: ectoderm above and endoderm below. Once the embryonic disk elongates to form the primitive streak, the third germ layer, mesoderm, forms by invagination of cells along the streak. The trophoblast is reinforced by mesoderm and becomes the chorion (Fig. 44.13b). It is possible to relate the development of future organs to these germ layers (see Table 44.1).

FIGURE 44.13

Human embryonic development.

a. At first, the embryo contains no organs, only tissues. The amniotic cavity is above the embryo, and the yolk sac is below. **b.** The chorion develops villi, the structures so important to the exchange between mother and child. **c, d.** The allantois and yolk sac, two more extraembryonic membranes, are positioned inside the body stalk as it becomes the umbilical cord. **e.** At 35+ days, the embryo has a head region and a tail region. The umbilical cord takes blood vessels between the embryo and the chorion (placenta).

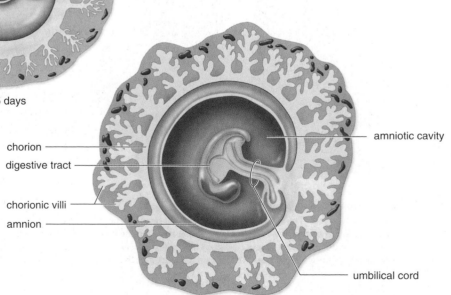

- chorion
- digestive tract
- chorionic villi
- amnion
- amniotic cavity
- umbilical cord

e. 35+ days

The Third Week

Two important organ systems make their appearance during the third week. The nervous system is the first organ system to be visually evident. At first, a thickening appears along the entire dorsal length of the embryo; then the neural folds appear. When the neural folds meet at the midline, the neural tube, which later develops into the brain and the nerve cord, is formed (see Fig. 44.4). After the notochord is replaced by the vertebral column, the nerve cord is called the spinal cord.

Development of the heart begins in the third week and continues into the fourth week. At first, there are right and left heart tubes; when these fuse, the heart begins pumping blood, even though the chambers of the heart are not fully formed. The veins enter posteriorly, and the arteries exit anteriorly from this largely tubular heart, but later the heart twists so that all major blood vessels are located anteriorly.

The Fourth and Fifth Weeks

At four weeks, the embryo is barely larger than the height of this print. A bridge of mesoderm called the body stalk connects the caudal (tail) end of the embryo with the chorion, which has treelike projections called **chorionic villi** [Gk. *chorion,* membrane; L. *villus,* shaggy hair] (Fig. 44.13*c, d*). The chorionic villi eventually form the placental sinus. The fourth extraembryonic membrane, the allantois, is contained within this stalk, and its

blood vessels become the umbilical blood vessels. The head and the tail then lift up, and the body stalk moves anteriorly by constriction. Once this process is complete, the **umbilical cord** [L. *umbilicus,* navel], which connects the developing embryo to the placenta, is fully formed (Fig. 44.13*e*).

Little flippers called limb buds appear (Fig. 44.14); later, the arms and the legs develop from the limb buds, and even the hands and the feet become apparent. At the same time—during the fifth week—the head enlarges, and the sense organs become more prominent. It is possible to make out the developing eyes, ears, and even the nose.

The Sixth Through Eighth Weeks

During the sixth to eighth weeks of development, the embryo becomes easily recognizable as human. Concurrent with brain development, the head achieves its normal relationship with the body as a neck region develops. The nervous system is developed well enough to permit reflex actions, such as a startle response to touch. At the end of this period, the embryo is about 38 mm long and weighs no more than an aspirin tablet, even though all organ systems are established.

During the embryonic period of development, the extraembryonic membranes appear and serve important functions; the embryo acquires organ systems.

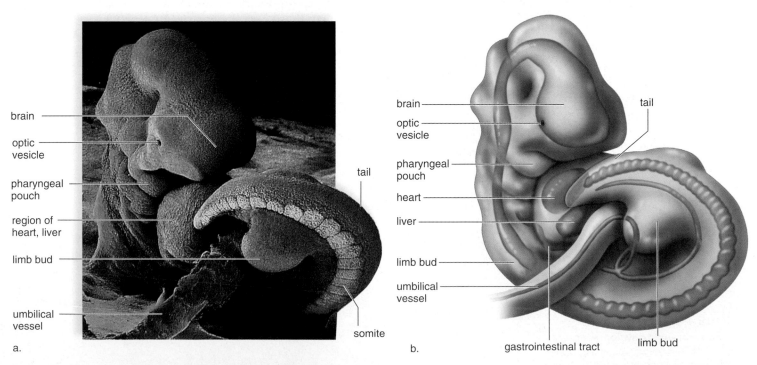

a.

b.

FIGURE 44.14 Human embryo at beginning of fifth week.

a. Scanning electron micrograph. **b.** The embryo is curled so that the head touches the heart, the two organs whose development is farther along than the rest of the body. The organs of the gastrointestinal tract are forming, and the arms and the legs develop from the bulges that are called limb buds. The tail is an evolutionary remnant; its bones regress and become those of the coccyx (tailbone). The pharyngeal arches become functioning gills only in fishes and amphibian larvae; in humans, the first pair of pharyngeal pouches becomes the auditory tubes. The second pair becomes the tonsils, while the third and fourth become the thymus gland and the parathyroid glands.

The Structure and Function of the Placenta

The **placenta** is a mammalian structure that functions in gas, nutrient, and waste exchange between embryonic (later fetal) and maternal cardiovascular systems. The placenta begins formation once the embryo is fully implanted. At first, the entire chorion has chorionic villi that project into endometrium. Later, these disappear in all areas except where the placenta develops. By the tenth week, the placenta (Fig. 44.15) is fully formed and is producing progesterone and estrogen. These hormones have two effects: (1) due to their negative feedback control of the hypothalamus and the anterior pituitary, they prevent any new follicles from maturing, and (2) they maintain the lining of the uterus, so now the corpus luteum is not needed. No menstruation occurs during pregnancy.

The placenta has a fetal side contributed by the chorion and a maternal side consisting of uterine tissues. Notice in Figure 44.15 how the chorionic villi are surrounded by maternal blood; yet maternal and fetal blood do not mix under normal conditions because exchange always takes place across plasma membranes. Carbon dioxide and other wastes move from the fetal side to the maternal side of the placenta and nutrients and oxygen move from the maternal side to the fetal side. The umbilical cord stretches between the placenta and the fetus. Although it may seem that the umbilical cord travels from the placenta to the intestine, actually the umbilical cord is simply taking fetal blood to and from the placenta. The umbilical cord is the lifeline of the fetus because it contains the umbilical arteries and vein, which transport waste molecules (carbon dioxide and urea) to the placenta for disposal into the maternal blood and take oxygen and nutrient molecules from the placenta to the rest of the fetal circulatory system. If the placenta prematurely tears from the uterine wall, the life of the fetus and mother are endangered.

Harmful chemicals can also cross the placenta as discussed in the Health Focus on pages 812–13. This is of particular concern during the embryonic period, when various structures are first forming. Each organ or part seems to have a sensitive period during which a substance can alter its normal development. For example, if a woman takes the drug thalidomide, a tranquilizer, between days 27 and 40 of her pregnancy, the infant is likely to be born with deformed limbs. After day 40, however, the infant is born with normal limbs.

During mammalian development, the embryo and later the fetus are dependent on the placenta for gas exchange and also for acquiring nutrients and ridding the body of wastes.

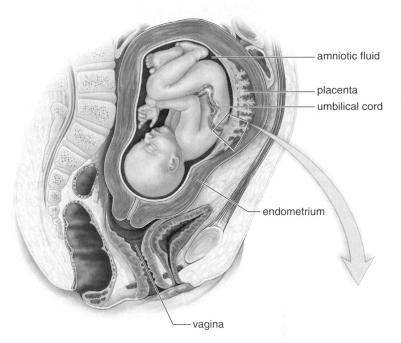

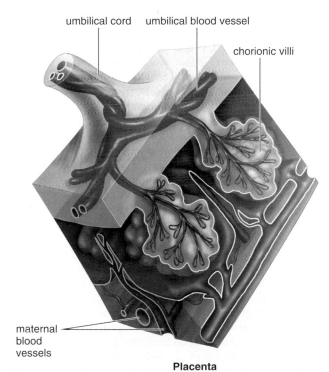

FIGURE 44.15 Anatomy of the placenta in a fetus at six to seven months.
The placenta is composed of both fetal and maternal tissues. Chorionic villi penetrate the uterine lining and are surrounded by maternal blood. Exchange of molecules between fetal and maternal blood takes place across the walls of the chorionic villi.

health focus

Preventing Birth Defects

It is believed that at least 1 in 16 newborns has a birth defect, either minor or serious, and the actual percentage may be even higher. Most likely, only 20% of all birth defects are due to heredity. Those that are hereditary can sometimes be detected before birth. Amniocentesis allows the fetus to be tested for abnormalities of development; chorionic villi sampling allows the embryo to be tested; and a new method has been developed for screening eggs to be used for in vitro fertilization (Fig. 44A).

It is recommended that all females take everyday precautions to protect any future and/or presently developing embryos and fetuses from defects. Proper nutrition is a must because deficiency in folic acid causes neural tube defects, such as spina bifida. X-ray diagnostic therapy should be avoided during pregnancy because X-rays cause mutations in the developing embryo or fetus. Children born to women who received X-ray treatment are apt to have birth defects and/or to develop leukemia later. Toxic chemicals, such as pesticides and many organic industrial chemicals, are also mutagenic and can cross the placenta. Cigarette smoke not only contains carbon monoxide but also other fetotoxic chemicals. Babies born to smokers are often underweight and subject to convulsions.

Pregnant Rh$^-$ women should receive an Rh immunoglobulin injection to prevent the production of Rh antibodies. These antibodies can cause nervous system and heart defects.

Sometimes, birth defects are caused by microbes. Females can be immunized before the childbearing years for rubella (German measles), which in particular causes birth defects such as deafness. Unfortunately, immunization for sexually transmitted diseases is not possible. The AIDS virus can cross the placenta and cause mental retardation. Proper medication can greatly reduce the chance of this happening. When a mother has herpes, gonorrhea, or chlamydia, newborns can become infected as they pass through the birth canal. Blindness and other physical and mental defects may develop. Birth by cesarean section could prevent these occurrences.

Pregnant women should not take any type of drug without a doctor's permission. Certainly, illegal drugs, such as marijuana, cocaine, and heroin, should be completely avoided. "Cocaine babies" now make up a large percentage of drug-affected babies. Severe fluctuations in blood pressure produced by the use of cocaine temporarily deprive the developing brain of oxygen. Cocaine babies have visual problems, lack coordination, and are mentally retarded. Intake of the drugs aspirin, caffeine (present in coffee, tea, and cola), and alcohol should be severely limited. It is not unusual for babies of drug addicts and alcoholics to display withdrawal symptoms and to have various abnormalities. Babies born to women who have about 45 drinks a month and as many as 5 drinks on one occasion are apt to have fetal alcohol syndrome (FAS). These babies have decreased weight, height, and head size, with malformation of the head and face. Mental retardation is common in FAS infants.

Medications can also cause problems. When the synthetic hormone DES was given to pregnant women to prevent miscarriage, their daughters showed various abnormalities of the reproductive organs and an increased tendency toward cervical cancer. Other sex hormones, such as birth control pills, can possibly cause abnormal fetal development, including abnormalities of the sex organs. The tranquilizer thalidomide is well known for having caused deformities of the arms and legs in children born to women who took the drug. Therefore, a woman has to be very careful about taking medications while pregnant.

Now that physicians and laypeople are aware of the various ways birth defects can be prevented, it is hoped that the incidence of birth defects will decrease in the future.

FIGURE 44A   Three methods for genetic defect testing before birth.
a. Amniocentesis is usually performed from the fifteenth to the seventeenth week of pregnancy. A long needle is passed through the abdominal wall to withdraw a small amount of amniotic fluid, along with fetal cells. Since there are only a few cells in the amniotic fluid, testing may be delayed as long as four weeks until cell culture produces enough cells for testing purposes. About 40 tests are available for different defects. b. Chorionic villi sampling is usually performed from the eighth to the twelfth week of pregnancy. The doctor inserts a long, thin tube through the vagina into the uterus. With the help of ultrasound, which gives a picture of the uterine contents, the tube is placed between the lining of the uterus and the chorion. Then a sampling of the chorionic villi cells is obtained by suction. Chromosome analysis and biochemical tests for genetic defects can be done immediately on these cells. c. Screening eggs for genetic defects is a new technique. Preovulatory eggs are removed by aspiration after a laparoscope (optical telescope) is inserted into the abdominal cavity through a small incision in the region of the navel. The first polar body is tested. If the woman is heterozygous (Aa) and the defective gene (a) is found in the polar body, then the egg must have received the normal gene (A). Normal eggs then undergo in vitro fertilization and are placed in the prepared uterus.

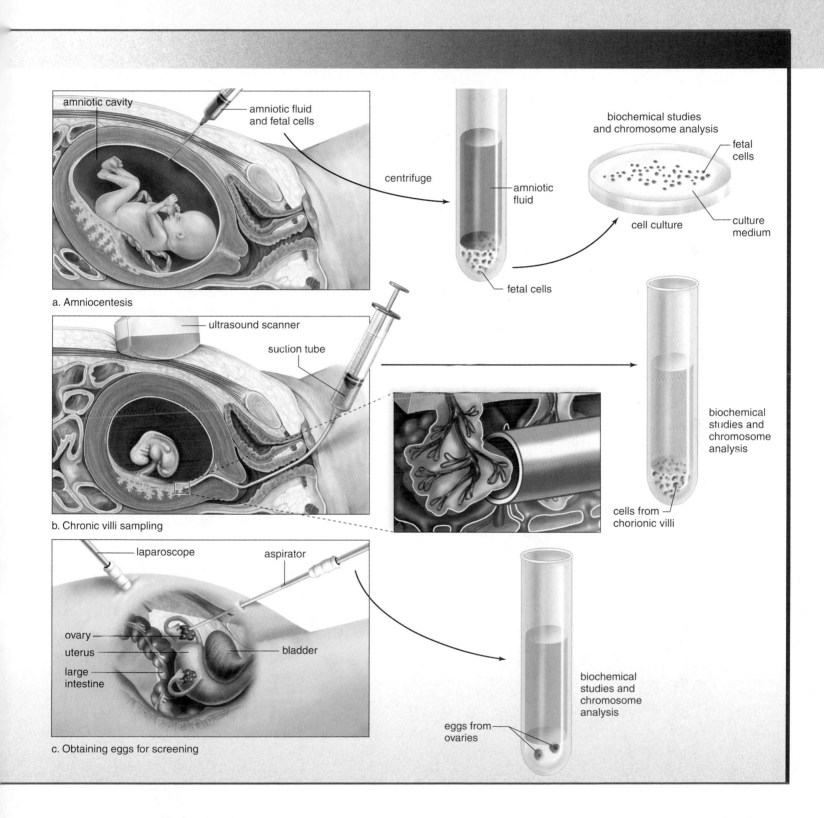

a. Amniocentesis

b. Chronic villi sampling

c. Obtaining eggs for screening

Fetal Development and Birth

Fetal development (months 3–9) is marked by an extreme increase in size. Weight multiplies 600 times, going from less than 28 g to 3 kg. During this time, too, the fetus grows to about 50 cm in length. The genitalia appear in the third month, so it is possible to tell if the fetus is male or female.

Soon, hair, eyebrows, and eyelashes add finishing touches to the face and head. In the same way, fingernails and toenails complete the hands and feet. A fine, downy hair (lanugo) covers the limbs and trunk, only to later disappear. The fetus looks very old because the skin is growing so fast that it wrinkles. A waxy, almost cheese-like substance (vernix caseosa) [L. *vernix*, varnish, and *caseus*, cheese] protects the wrinkly skin from the watery amniotic fluid.

The fetus at first only flexes its limbs and nods its head, but later it can move its limbs vigorously to avoid discomfort. The mother feels these movements from about the fourth month on. The other systems of the body also begin to function. After 16 weeks, the fetal heartbeat is heard through a stethoscope. A fetus born at 24 weeks has a chance of surviving, although the lungs are still immature and often cannot capture oxygen adequately. Weight gain during the last couple of months increases the likelihood of survival.

The Stages of Birth

The latest findings suggest that when the fetal brain is sufficiently mature, the hypothalamus causes the pituitary to stimulate the adrenal cortex so that androgens are released into the bloodstream. The placenta uses androgens as a precursor for estrogens, hormones that stimulate the production of prostaglandin (a molecule produced by many cells that acts as a local hormone) and oxytocin. All three of these molecules cause the uterus to contract and expel the fetus.

The process of birth (parturition) includes three stages. During the first stage, the cervix dilates to allow passage of the baby's head and body. The amnion usually bursts about this time. During the second stage, the baby is born and the umbilical cord is cut. During the third stage, the placenta is delivered (Fig. 44.16).

During the fetal period, organ systems are refined, and the fetus gains weight. Finally, birth occurs.

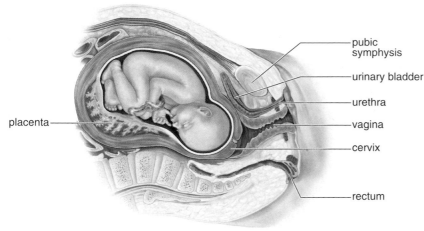

a. 9-month-old fetus

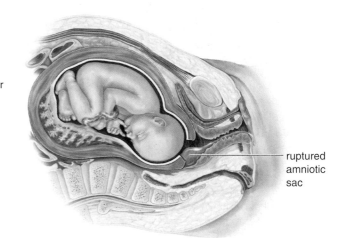

b. First stage of birth: cervix dilates

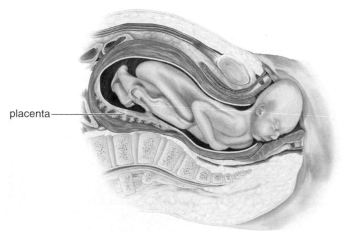

c. Second stage of birth: baby emerges

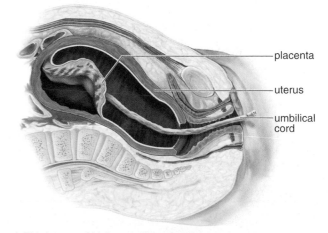

d. Third stage of birth: expelling afterbirth

FIGURE 44.16 Three stages of parturition.
a. Position of fetus just before birth begins. **b.** Dilation of cervix. **c.** Birth of baby. **d.** Expulsion of afterbirth.

CONNECTING THE CONCEPTS

We have come full circle. We began our study of biology by considering the structure of the cell and its genetic machinery, including how the expression of genes is regulated. In this chapter, we have observed that animals go through the same early embryonic stages of morula, blastula, gastrula, and so forth. The set sequence of these stages is due to the expression of genes that bring about cellular changes. Therefore, once again, we are called upon to study organisms at the cellular level of organization.

We have seen that hormones are signals that affect cellular metabolism. Like gibberellins in plants, steroid hormones in animals turn on the expression of genes.

When a steroid hormone binds to a specific hormone receptor, a gene is transcribed, and then translation produces the corresponding protein. This same type of signal transduction pathway occurs during development. Transduction means that the signal has been transformed into an event that has an effect on the organism.

A set sequence of signaling molecules is produced as development occurs. Each new signal in the sequence turns on a specific gene, or more likely, a sequence of genes. Gene expression cascades are common during development. Homeotic genes are arranged in sets on the chromosomes, and a protein product from one set of genes acts as a transcription factor to turn on another set, and so forth. During the development of flies, it is possible to observe that first one region of a chromosome and then another puffs out, indicating that genes are transcribed in sequence as development occurs.

Developmental biology is now making a significant contribution to the field of evolution. Homeotic genes with the same homeoboxes (sequence of about 180 base pairs) have been discovered in many different types of organisms. This suggests that homeotic genes arose early in the history of life, and mutations in these genes could possibly account for macroevolution—the appearance of new species or even higher taxons.

Summary

44.1 EARLY DEVELOPMENTAL STAGES

Development begins at fertilization. Only one sperm actually enters the egg. Both sperm and egg contribute chromosomes to the diploid zygote.

The early developmental stages in animals proceed from cellular stages to tissue stages to organ stages. During the cellular stage, cleavage (cell division) occurs, but there is no overall growth. The result is a morula, which becomes the blastula when an internal cavity (the blastocoel) appears.

During the tissue stage, gastrulation (invagination of cells into the blastocoel) results in formation of the germ layers: ectoderm, mesoderm, and endoderm. Both the cellular and tissue stages can be afftected by the amount of yolk. The human blastula stage resembles that of the chick whose egg has a large amount of yolk.

Organ formation can be related to germ layers. For example, during neurulation, the nervous system develops from midline ectoderm, just above the notochord. At this point, it is possible to draw a typical cross section of a vertebrate embryo in which the notochord has not been replaced by the vertebral column (see Fig. 44.5).

44.2 DEVELOPMENTAL PROCESSES

Cellular differentiation begins with cytoplasmic segregation in the egg. After the first cleavage of a frog embryo, only a daughter cell that receives a portion of the gray crescent is able to develop into a complete embryo. Therefore, cytoplasmic segregation of maternal determinants occurs during early development of a frog. Induction is also part of cellular differentiation. For example, the notochord induces the formation of the neural tube in frog embryos. The reciprocal induction that occurs between the lens and the optic vesicle is another good example of induction. In *C. elegans*, investigators have shown that induction is an ongoing process in which one tissue after the other regulates the development of another, through chemical signals coded for by particular genes.

Some morphogen genes determine the axes of the body, and others regulate the development of segments. An important concept has emerged: during development, sequential sets of master genes code for morphogen gradients that activate the next set of master genes, in turn. Morphogens are transcription factors that bind to DNA.

Homeotic genes control pattern formation such as the presence of antennae, wings, and limbs on the segments of *Drosophila*. Homeotic genes code for proteins that contain a homeodomain, a particular sequence of 60 amino acids. These proteins are also transcription factors, and the homeodomain is the portion of the protein that binds to DNA. Homologous homeotic genes have been found in a wide variety of organisms, and therefore they must have arisen early in the history of life and been conserved.

44.3 HUMAN EMBRYONIC AND FETAL DEVELOPMENT

Human development can be divided into embryonic development (months 1 and 2) and fetal development (months 3-9). The early stages in human development resemble those of the chick. The similarities are probably due to their evolutionary relationship, not to the amount of yolk the eggs contain, because the human egg has little yolk.

The extraembryonic membranes appear early in human development. The trophoblast of the blastocyst is the first sign of the chorion, which goes on to become the fetal part of the placenta. Exchange occurs between fetal and maternal blood at the placenta. The amnion contains amniotic fluid, which cushions and protects the embryo. The yolk sac and allantois are also present.

Fertilization occurs in the oviduct, and cleavage occurs as the embryo moves toward the uterus. The morula becomes the blastocyst before implanting in the endometrium of the uterus. Organ development begins with neural tube and heart formation. There follows a steady progression of organ formation during embryonic development. During fetal development, refinement of organ systems occurs, and the fetus adds weight.

Reviewing the Chapter

1. Describe the events of fertilization that (a) allow the sperm to reach the plasma membrane of the egg, (b) prevent polyspermy, and (c) result in a diploid zygote.   798

2. What happens during the cellular stages, the tissue stages, and the organ stages of early development in animals!   799–801
3. What are the germ layers, and which organs are derived froom each of the germ layers?   800
4. Draw a cross section of a typical vertebrate embryo at the neurula stage, and label your drawing.   801
5. Describe two mechanisms that are known to be involved in the processes of cellular differentiation.   802–3
6. Describe an experiment performed by Spemann suggesting that the notochord induces formation of the neural tube. Give another well-known example of induction between tissues.   803–4
7. With regard to *C. elegans*, what is a fate map? How does induction occur?   804
8. With regard to *Drosophila*, what is a morphogen gradient, and what does such a gradient do to bring about morphogenesis?   805
9. What is the function of homeotic genes, and what is the significance of the homeobox within these genes?   805–6
10. List the human extraembryonic membranes, give a function for each, and compare their functions to those in the chick.   807
11. Tell where fertilization, cleavage, the morula stage, and the blastocyst stage occur in humans. What happens to the embryo in the uterus?   808–9
12. Describe the structure and the function of the placenta in humans.   811
13. List and describe the stages of birth.   814

Testing Yourself

Choose the best answer for each question.

1. Which of these stages is the first one out of sequence?
   a. cleavage
   b. blastula
   c. morula
   d. gastrula
   e. neurula

2. Which of these stages is mismatched?
   a. cleavage—cell division
   b. blastula—gut formation
   c. gastrula—three germ layers
   d. neurula—nervous system
   e. Both b and c are mismatched.

3. In many embryos, differentiation begins at what stage?
   a. cleavage
   b. blastula
   c. gastrula
   d. neurula
   e. after the completion of these stages

4. Morphogenesis is associated with
   a. protein gradients.
   b. induction.
   c. transcription factors.
   d. homeotic genes.
   e. All of these are correct.

5. In humans, the placenta develops from the chorion. This indicates that human development
   a. resembles that of the chick.
   b. is dependent on extraembryonic membranes.
   c. cannot be compared to that of lower animals.
   d. begins only upon implantation.
   e. Both a and b are correct.

6. In humans, the fetus
   a. is surrounded by four extraembryonic membranes.
   b. has developed organs and is recognizably human.
   c. is dependent on the placenta for excretion of wastes and acquisition of nutrients.
   d. is embedded in the endometrium of the uterus.
   e. Both b and c are correct.

7. Developmental changes
   a. require growth, differentiation, and morphogenesis.
   b. stop occurring when one is grown.
   c. are dependent on a parceling out of genes into daughter cells.
   d. are dependent on activation of master genes in an orderly sequence.
   e. Both a and d are correct.

8. Which of these pairs is mismatched?
   a. brain—ectoderm
   b. gut—endoderm
   c. bone—mesoderm
   d. lens—endoderm
   e. heart—mesoderm

9. Label this diagram illustrating the placement of the extraembryonic membranes, and give a function for each membrane in humans.

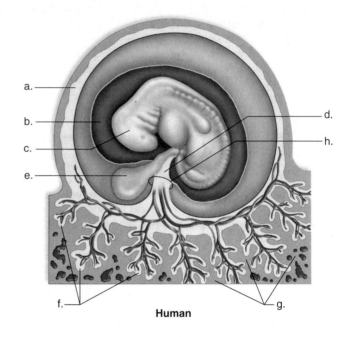

Human

For questions 10–13, match the statement with the terms in the key.

KEY:
   a. apoptosis
   b. homeotic genes
   c. fate maps
   d. morphogen gradients
   e. segment-polarity genes

10. have been developed that show the destiny of each cell that arises through cell division during the development of *C. elegans*
11. occurs when a cell-death cascade is activated
12. can have a range of effects, depending on its concentration in a particular portion of the embryo
13. genes that control pattern formation, the organization of differentiated cells into specific three-dimensional structures
14. The ability of one embryonic tissue to influence the growth and development of another tissue is termed
   a. morphogenesis.
   b. pattern formation.
   c. apoptosis.
   d. cellular differentiation.
   e. induction.
15. Which hormone is administered to begin the process of birth?
   a. estrogen
   b. oxytocin
   c. prolactin
   d. testosterone
   e. Both b and d are correct.

16. Only one sperm enters a human egg because
    a. sperm have an acrosome.
    b. the corona radiata gets larger.
    c. the zona pellucida lifts up.
    d. the plasma membrane hardens.
    e. All of these are correct.
17. Which is a correct sequence in humans that ends with the stage that implants?
    a. morula, blastocyst, embryonic disk, gastrula
    b. ovulation, fertilization, cleavage, morula, early blastocyst
    c. embryonic disk, gastrula, primitive streak, neurula
    d. primitive streak, neurula, extraembryonic membranes, chorion
    e. cleavage, neurula, early blastocyst, morula

Thinking Scientifically

1. A mutant gene is known to disrupt the earliest stages of development in sea urchins. Individuals with two copies of the mutant gene seem to develop normally. However, when normal males are crossed with mutant (normal-appearing) females, none of the offspring develop at all. How could this pattern of expression be explained?
2. Babies that were malnourished in utero are at a higher risk for many adult-onset diseases than those that were well nourished. To determine how well babies were nourished in utero, physicians often determine the ratio of head circumference to abdominal circumference. If malnourished, the head circumference to abdominal circumference is larger than usual because blood in the malnourished fetuses is preferentially directed toward the brain. Why is it better to use the circumference data rather than birth weight data to detect malnourished babies?

Bioethical Issue: Prosecution of Mothers

Because we are now aware of the need for maternal responsibility before a child is born, there has been a growing acceptance of prosecuting women when a newborn has a condition such as fetal alcohol syndrome, which can only be caused by the drinking habits of the mother. Employers have also become aware that they might be subject to prosecution. To protect themselves, Johnson Controls, a U.S. battery manufacturer, developed a fetal protection policy. To be hired for a job that might harm a fetus, a woman had to show that she had been sterilized or was otherwise incapable of having children. In 1991, the U.S. Supreme Court declared this policy unconstitutional on the basis of sexual discrimination. The decision was hailed as a victory for women, but was it? The decision was written in such a way that women alone, and not an employer, are responsible for any harm done to the fetus by workplace toxins.

Some have noted that prosecuting women for causing prenatal harm can itself have a detrimental effect. The women may tend to avoid prenatal treatment, thereby increasing the risk to their children. Or they may opt for an abortion to avoid the possibility of prosecution. The women feel they are in a no-win situation. If they have a child that has been harmed due to their behavior, they feel they are bad mothers, but if they abort, they also feel they are bad mothers.

Understanding the Terms

allantois 807	gray crescent 803
amnion 807	homeobox 806
apoptosis 802	homeotic genes 805
blastocoel 799	human chorionic gonadotropin (HCG) 809
blastocyst 808	
blastopore 800	implantation 807
blastula 799	induction 803
cellular differentiation 802	mesoderm 800
chorion 807	morphogen 805
chorionic villus 810	morphogenesis 802
cleavage 799	morula 799
ectoderm 800	neural plate 801
embryo 799	neural tube 801
embryonic disk 809	neurula 801
embryonic period 807	notochord 801
endoderm 800	pattern formation 802
extraembryonic membrane 807	placenta 811
	totipotent 802
fate map 804	trophoblast 808
fertilization 798	umbilical cord 810
gastrula 800	yolk 799
gastrulation 800	yolk sac 807
germ layer 800	

Match the terms to these definitions:

a. _____ Ability of a chemical or a tissue to influence the development of another tissue.
b. _____ Primary tissue layer of a vertebrate embryo—namely, ectoderm, mesoderm, or endoderm.
c. _____ Extraembryonic membrane of birds, reptiles, and mammals that forms an enclosing, fluid-filled sac.
d. _____ A 180-nucleotide sequence located in nearly all homeotic genes.
e. _____ Stage of early animal development during which the germ layers form, at least in part, by invagination.

ARIS, the *Biology* Website

ARIS, the website for *Biology*, provides a wealth of information organized and integrated by chapter. You will find practice quizzes, interactive activities, labeling exercises, flashcards, and much more that will complement your learning and understanding of general biology.

www.mhhe.com/maderbiology9

PART VIII

BEHAVIOR AND ECOLOGY

An organism's behavior is vitally important to its survival in the natural world and to the perpetuation of the species. The behavior of an organism is regulated primarily by the nervous and endocrine systems, whose development and function are controlled by genetics and the environment. Behavioral characteristics are subject to natural selection and are adaptive in nature. The forces governing the evolution of behavior are the same as those that apply to any other trait.

An animal's behavior is particularly pertinent to ecology, the study of the interrelationships between organisms and their environment. Organisms do not exist alone; rather they are part of a population that makes up a community. A community is an assemblage of populations of organisms in a specific area. Ecosystems include the abiotic (nonliving) and biotic (living) factors in a particular environment. The various ecosystems make up the biosphere. Ultimately, the biosphere includes the zones of air, land, and water occupied by organisms. Ecology considers the distribution, abundance, and interactions of organisms living in the biosphere.

Human populations have a major impact on natural ecosystems. Loss of habitat, pollution, and interference with natural biogeochemical cycles eventually affect all life on Earth. It is a goal of conservation biology, an active field of current research, to help preserve species and manage ecosystems for sustainable human welfare.

45

ANIMAL BEHAVIOR

I n 1973, the Nobel Prize in Physiology and Medicine was awarded to Konrad Lorenz, Karl von Frisch, and Niko Tinbergen for their contributions to the science of ethology, the study of animal behavior under natural conditions. Today, ethology is one of the most fascinating and complex disciplines of the biological sciences.

The photograph shows goslings swimming in the wake of Konrad Lorenz, one of the scientists who helped shape our modern understanding of behavior. Lorenz realized that organisms are members of a population in an ecological setting. Therefore, he reasoned that behavior can be explained in terms of its evolutionary history, and that it must have served to improve the reproductive success of the organism.

For example, Lorenz found that birds such as goslings become attached to and follow the first moving object they see. He termed this learning process imprinting. Ordinarily, the object followed is the mother; however, the goslings could also become imprinted on Lorenz, and in that case, they chose him over their own mother. Therefore, the behavior was at least in part due to social experience. In the wild, does imprinting lead to reproductive success? Indeed, following the mother when young increases the chances of survival, and imprinting also leads to being able to recognize one's species and, therefore, an appropriate mate.

Goslings learned to follow Konrad Lorenz, a famous behaviorist.

45.1 NATURE VERSUS NURTURE: GENETIC INFLUENCES

The nature versus nurture question asks to what extent both our genes (nature) and environmental influences (nurture) affect behavior. **Behavior** encompasses any action that can be observed and described. The anatomy and physiology of an animal determine what types of behavior are possible for that animal. Therefore, genes, which control anatomy and physiology, must also control behavior. Many experiments have been done to discover the degree to which genetics controls behavior.

Experiments with Lovebirds, Snakes, Snails, and Humans

Lovebirds are small, green and pink African parrots that nest in tree hollows. There are several closely related species of lovebirds in the genus *Agapornis* that build their nests differently. Fischer lovebirds, *Agapornis fischeri,* cut large leaves (or in the laboratory, pieces of paper) into long strips with their bills. They use their bills to carry the strips (Fig. 45.1*a*) to the nest where they weave them in with others to make a deep cup. Peach-faced lovebirds, *Agapornis roseicollis,* cut somewhat shorter strips and they carry them to the nest in a very unusual manner. They pick up the strips in their bills and then insert them into their feathers (Fig. 45.1*b*). In this way, they can carry several of these short strips with each trip to the nest, while Fischer lovebirds can carry only one of the longer strips at a time.

Researchers hypothesized that if the behavior for obtaining and carrying nesting material is inherited, then hybrids might show intermediate behavior. When the two species of birds were mated, it was observed that the hybrid birds have difficulty carrying nesting materials. They cut strips of intermediate length and then attempt to tuck the strips into their rump feathers. They do not push the strips far enough into the feathers, however, and when they walk or fly, the strips always come out. Hybrid birds eventually (about three years in this study) learn to carry the cut strips in their beak, but they still briefly turn their head toward their rump before flying off. These studies support the hypothesis that behavior has a genetic basis.

Several experiments using the garter snake, *Thamnophis elegans,* to determine if food preference has a genetic basis have been conducted. There are two different types of garter snake populations in California. Inland populations are aquatic and commonly feed underwater on frogs and fish. Coastal populations are terrestrial and feed mainly on slugs. In the laboratory, inland adult snakes refused to eat slugs, while coastal snakes readily did so. The experimental results of matings between snakes from the two populations (inland and coastal) show their newborns have an overall intermediate incidence of slug acceptance.

a. Fischer lovebird with nesting material in its beak.

b. Peach-faced lovebird with nesting material in its rump feathers.

FIGURE 45.1 Nest building behavior in lovebirds.
a. Fischer lovebirds carry strips of nesting material in the bill, as do most other birds. **b.** Peach-faced lovebirds tuck strips of nesting material into their rump feathers before flying back to the nest.

Differences between slug acceptors and slug rejecters appear to be inherited, but what physiological difference is there between the two populations? A clever experiment answered this question. When snakes eat, their tongues carry chemicals to an odor receptor in the roof of the mouth. They use tongue flicks to recognize their prey. Even newborns will flick their tongues at cotton swabs dipped in fluids of their prey. Swabs were dipped in slug extract, and the number of tongue flicks were counted for newborn inland snakes and coastal snakes. Coastal snakes had a higher number of tongue flicks than inland snakes (Fig. 45.2). Apparently, inland snakes do not eat slugs because they are not sensitive to their smell. A genetic difference between the two populations of snakes results in a physiological difference in their nervous systems. Although hybrids showed a great deal of variation in the number of tongue flicks, they were generally intermediate, as predicted by the genetic hypothesis.

The nervous and endocrine systems are both responsible for the coordination of body systems. Is the endocrine system also involved in behavior? Research studies answer this question in the affirmative. For example, the egg-laying behavior in the marine snail *Aplysia* involves a set sequence of movements. Following copulation, the animal extrudes long strings of more than a million egg cases. It takes the egg case string in its mouth, covers it with mucus, waves its head back and forth to wind the string into an irregular mass, and attaches the mass to a solid object, such as a rock. Several years ago, scientists isolated and analyzed an egg-laying hormone (ELH) that causes the snail to lay eggs even if it has not mated. ELH was found to be a small protein of 36 amino acids that diffuses into the circulatory system and excites the smooth muscle cells of the reproductive duct, causing them to contract and expel the egg string. Using recombinant DNA techniques, the investigators isolated the ELH gene. The gene's product turned out to be a protein with 271 amino acids. The protein can be cleaved into as many as 11 possible products, and ELH is one of these. ELH alone, or in conjunction with these other products, is thought to control all the components of egg-laying behavior in *Aplysia*.

Experiments with Humans

Human twins, on occasion, have been separated at birth and raised under different environmental conditions. Studies of such twins show that they have similar food preferences, activity patterns, and even select mates with similar characteristics. These twin studies lend support to the hypothesis that at least certain types of behavior are primarily influenced by nature (i.e., the genes).

The results of many types of studies support the hypothesis that many types of behavior have a genetic basis. Apparently, genes influence the development of neural and hormonal mechanisms that control behavior.

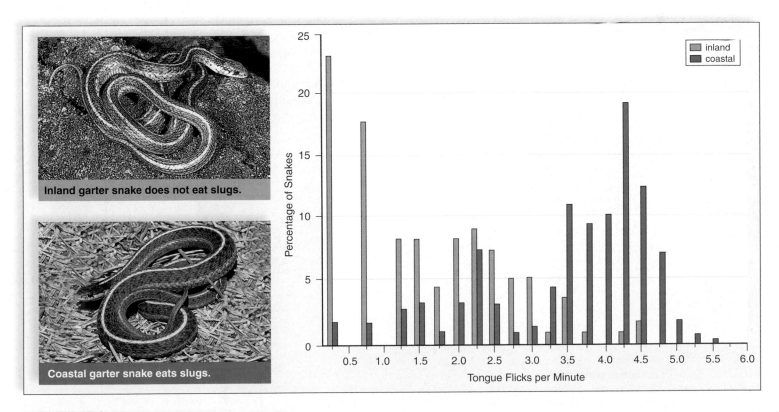

Inland garter snake does not eat slugs.

Coastal garter snake eats slugs.

FIGURE 45.2 Feeding behavior in garter snakes.
The number of tongue flicks by inland and coastal garter snakes is measured in terms of their response to slug extract on cotton swabs. Coastal snakes tongue-flicked more than inland snakes.

45.2 NATURE VERSUS NURTURE: ENVIRONMENTAL INFLUENCES

Even though genetic inheritance serves as a basis for behavior, it is possible that environmental influences (nurture) also affect behavior. For example, behaviorists originally believed that some behaviors were **fixed action patterns (FAP)** elicited by a sign stimulus. But, then they found that many behaviors, formerly thought to be FAPs, improve with practice, for example, by learning. In this context, **learning** is defined as a durable change in behavior brought about by experience.

Learning in Birds

Laughing gull chicks' begging behavior appears to be a FAP, because it is always performed the same way in response to the parent's red bill (the sign stimulus). A chick directs a pecking motion toward the parent's bill, grasps it, and strokes it downward (Fig. 45.3). Parents can bring about the begging behavior by swinging their bill gently from side to side. After the chick responds, the parent regurgitates food onto the floor of the nest. If need be, the parent then encourages the chick to eat. This interaction between the chicks and their parents suggests that the begging behavior involves learning. To test this hypothesis, diagrammatic pictures of gull heads were painted on small cards, and then eggs were collected in the field. The eggs were hatched in a dark incubator to eliminate visual stimuli before the test. On the day of hatching, each chick was allowed to make about a dozen pecks at the model. The chicks were returned to the nest, and then each was retested. The tests showed that on the average, only one-third of the pecks by a newly hatched chick strike the model. But one day after hatch-

ing, more than half of the pecks are accurate, and two days after hatching, the accuracy reaches a level of more than 75%. Investigators concluded that improvement in motor skills, as well as visual experience, strongly affect development of chick begging behavior.

Imprinting

Imprinting is considered a form of learning. Imprinting was first observed in birds when chicks, ducklings, and goslings followed the first moving object they saw after hatching. This object is ordinarily their mother, but investigators found that birds can seemingly be imprinted on any object—a human or a red ball—if it is the first moving object they see during a sensitive period of two to three days after hatching. The term *sensitive period* means that the behavior develops only during this time.

A chick imprinted on a red ball follows it around and chirps whenever the ball is moved out of sight. Social interactions between parent and offspring, during the sensitive period, seem key to the normal imprinting. For example, female mallards cluck during the entire time imprinting is occurring, and it could be that vocalization before and after hatching is necessary to normal imprinting.

Song Learning

White-crowned sparrows sing a species-specific song, but males of a particular region have their own dialect. Birds were caged in order to test the hypothesis that young white-crowned sparrows learn how to sing from older members of their species.

Three groups of birds were tested. Birds in the first group *heard no songs at all.* When grown, these birds sang a song, but it was not fully developed. Birds in the second group *heard tapes of white-crowns singing.* When grown, they sang in that dialect, as long as the tapes had been played during a sensitive period from about age 10–50 days. White-crowned sparrows' dialects (or other species' songs)

FIGURE 45.3 Pecking behavior in laughing gulls.
At about three days, a laughing gull chick grasps the red bill of a parent, stroking it downward, and the parent then regurgitates food. *Right:* The top diagrams show the accuracy of a chick when striking a test probe, painted red. The bottom diagram shows chick-pecking accuracy graphically. Note from these diagrams that a chick markedly improves its ability (within only two days) to peck a bill, a behavior that normally causes a parent to regurgitate food.

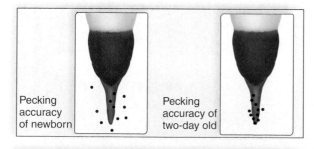

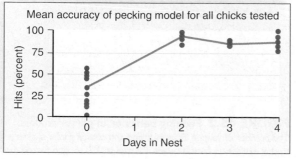

played before or after this sensitive period had no effect on the birds. Birds in a third group did not hear tapes and instead were *given an adult tutor.* These birds sang a song of even a different species—no matter when the tutoring began—showing that social interactions apparently assist learning in birds.

Associative Learning

A change in behavior that involves an association between two events is termed **associative learning.** For example, birds that get sick after eating a monarch butterfly no longer prey on monarch butterflies, even though they may be readily available. Or the smell of fresh baked bread may entice you, even though you may have just eaten. If so, perhaps you associate the taste of bread with a pleasant memory, such as being at home. Both classical conditioning and operant conditioning are examples of associative learning.

Classical Conditioning

In **classical conditioning,** the paired presentation of two different types of stimuli (at the same time) causes an animal to form an association between them. The best-known laboratory example of classical conditioning is that of an experiment done by the Russian psychologist Ivan Pavlov. First, Pavlov observed that dogs salivate when presented with food. Then, he rang a bell whenever the dogs were fed. Eventually, the dogs would salivate whenever the bell was rung, regardless of whether food was present (Fig. 45.4).

Classical conditioning suggests that an organism can be trained—that is, conditioned—to associate any response to any stimulus. Unconditioned responses are those that occur naturally, as when salivation follows the presentation of food. Conditioned responses are those that are learned, as when a dog learns to salivate when it hears a bell. Advertisements attempt to use classical conditioning to sell products. Why do commercials pair attractive people with a product being advertised? The hope is that viewers will associate attractiveness with the product. This pleasant association may cause them to buy the product.

Some types of classical conditioning can be helpful. For example, it's been suggested that you hold a child on your lap when reading to them. Why? Because the child will associate a pleasant feeling with reading.

Operant Conditioning

During **operant conditioning,** a stimulus-response connection is strengthened. Most people know that it is helpful to give an animal an award, such as food or affection, when teaching it a trick. When we go to an animal show, it is quite obvious that trainers use operant conditioning. They present a stimulus, say, a hoop, and then give a reward (food) for the proper response (jumping through the hoop). Sometimes the reward need not be immediate. In latent operant conditioning, an animal makes an association without the immediate reward, as when squirrels make a mental map of where they have hidden nuts.

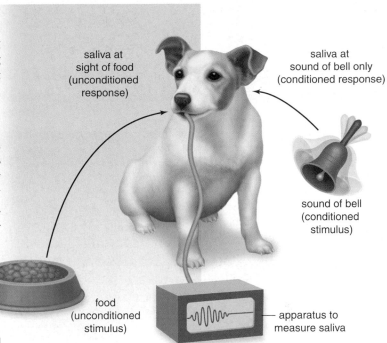

FIGURE 45.4 Classical conditioning.
Ivan Pavlov discovered classical conditioning by performing this experiment with dogs. A bell is rung when a dog is fed food. Salivation is noted. Eventually, the dog salivates when the bell is rung even though no food is presented. Food is an unconditioned stimulus, and the sound of the bell is a conditioned stimulus that brings about the response—that is, salivation.

B. F. Skinner is well known for studying this type of learning in the laboratory. In the simplest type of experiment performed by Skinner, a caged rat happens to press a lever and is rewarded with sugar pellets, which it avidly consumes. Thereafter, the rat regularly presses the lever whenever it wants a sugar pellet. In more sophisticated experiments, Skinner even taught pigeons to play ping-pong by reinforcing desired responses to stimuli.

As an example in child rearing again, it's been suggested that parents who give a positive reinforcement for good behavior will be more successful than parents who punish behaviors they believe are undesirable.

Other Means of Learning

In addition to the modes of learning already discussed, animals may learn through insight, imitation, and habituation. Insight learning occurs when an animal suddenly solves a problem without any prior experience with the problem. The animal appears to adapt prior experience to solving the problem. For example, chimpanzees have been observed stacking boxes to reach bananas in laboratory settings.

Many organisms learn through observation and imitation. Japanese macaques learn to wash sweet potatoes before eating them by imitating others. Habituation occurs when an animal no longer responds to a repeated stimulus. Deer grazing on the side of a busy highway, ignoring traffic, is an example of habituation.

science focus

Courtship Display of Male Bowerbirds

At the start of the breeding season, male bowerbirds use small sticks and twigs to build elaborate display areas called bowers (Fig. 45A). They clear the space around the bower and decorate the area with fresh flowers, fruits, pebbles, shells, bits of glass, tinfoil, and any bright baubles they can find. The Satin Bowerbird, *Ptilonorhynchus violaceus*, of eastern Australia prefers blue objects, a color that harmonizes with the male's glossy blue-back plumage. Males collect blue parrot feathers, flowers, berries, ballpoint pens, clothespins, and even toothbrushes from researchers' cabins.

After the bower is complete, a male bowerbird spends most of his time near this bower, calling to females, renewing his decorations, and guarding his work against possible raids by other males. After inspecting many bowers and their owners, a female approaches one, and the male begins a display. He faces her, fluffs up his feathers, and flaps his wings to the beat of a call. The female enters the bower, and if she crouches, the two mate.

Female bowerbirds build their own nests and raise the young without help from their mates, so attractive males can mate with multiple females. The reproductive advantage gained by attractive male bowerbirds is quite large; the most attractive males may mate with up to 25 females per year, but most males mate rarely or not at all. As already discussed, Dr. [Gerald] Borgia found that the males most often chosen by females have well-built bowers with well-decorated platforms. In addition, it is possible that the ability of a male bowerbird to respond appropriately to the female during courtship might influence his success. I [Gail Patricelli] and my colleagues studied this interactive component of Satin Bowerbird mating behavior as part of my doctoral dissertation research at the University of Maryland, in collaboration with Dr. Borgia, my graduate advisor.

Male bowerbirds are not gaudy in appearance, but their displays are highly intense and aggressive. Their courting displays are similar to those used by males to intimidate each other in aggressive encounters—with males puffing their feathers, rapidly extending their wings, and running, while making a loud, buzzing vocalization. Analysis of natural

FIGURE 45A Male and female bowerbird.
The male bowerbird (right) has prepared this bower and decorated its platform with particularly blue objects. A female bowerbird (left) has entered the bower and a male courtship display will now begin.

courtships has shown that males must display intensely to be attractive, *but males that are too intense too soon can startle females.* Females may benefit from preferentially mating with the most intensely displaying males (e.g., if these displays indicate male health or vigor), but when females are startled repeatedly by male displays, they may not be able to efficiently assess male traits. Thus both sexes can maximize the potential benefits of intense male courtship displays—and minimize the potential costs—by communicating. Indeed, a female behavior (degree of crouching) reflects the level of display intensity that the female will tolerate without being startled.

By giving higher-intensity displays only when females increase their crouching, males could increase their courtship success by displaying intensely enough to be attractive without threatening females with displays more intense than they are ready to tolerate.

With this information in mind, we specifically tested the hypothesis that males respond to female crouching signals by adjusting their intensity, and that a particular male's ability to respond to female signals is related to his success in courtship. A male's ability to modify his courtship display according to the rapidity with which the female crouches was difficult to measure in natural court-

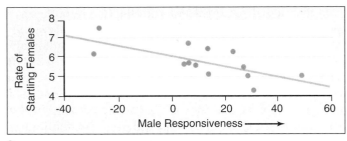

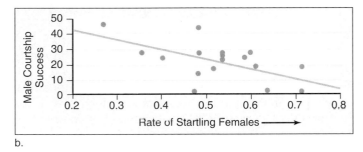

FIGURE 45B Courtship success of males.
*Some males have more courtship success than others. This experiment may explain why. **a.** An experimenter using a remote control (photograph) can regulate the crouch rate of a robotic female. Some males are better able than others to vary the intensity of their courtship depending on the crouch rate of the robotic female (top graph). In other words, only if the robotic female crouches more do they respond more. Therefore, they are predicted to startle females less. **b.** Experimenters found that males who respond best under experimental conditions do startle live females less and do have better courtship success.*

ships, since it was not clear whether males were responding to females, or vice versa. To solve this problem, we collaborated with an engineer to develop robotic female bowerbirds, which used tiny servo motors to mimic female movements (Fig. 45B *right*).

Using these "fembots," we were able to control female signals and measure male response in experimental courtships. During the bowerbird mating season at our field site in Wallaby Creek, Australia, we worked with student volunteers to test each male in our population with robots that crouched at four different rates: no crouch, slow, moderate, and fast. These experiments showed that male Satin Bowerbirds in general modulate their displays in response to robotic female crouching (Fig. 45B*a*). This supports the prediction that males are able to respond to female signals by giving their highest-intensity courtship displays for females who crouched the fastest and are least likely to be startled.

Utilizing automatically triggered video cameras that monitor behaviors at bowers, it was possible to measure each male's courtship/mating success with no difficulty. We found that males who modulate their displays more effectively in response to robotic female signals startle real females less often in natural courtships, and are thus more-successful in courting females (Fig. 45B*b*).

Our results suggest that females prefer intensely displaying males as mates, but that successful males do not always display at maximum intensity; they modulate their intensity in response to female signals, thus producing displays attractive to females without threatening them.

Male responsiveness to female signals may be an important part of successful courtship in many species—even if males do not dance aggressively during courtship like male bowerbirds. For instance, when females choose their mates based on bright coloration, successful males may respond to female signals by altering their position relative to the sun, their distance from the female, or the way they shake their tail when displaying their colors. So, along with extreme male traits—such as gaudy colors and aggressive dances—sexual selection may favor the ability of males to read female signals and adjust courtship displays accordingly.

*Courtesy of Gail Patricelli
University of Maryland*

45.3 ADAPTIVE MATING BEHAVIOR

Sexual selection refers to adaptive changes in males and females that lead to an increased ability to secure a mate. Sexual selection in males may result in an increased ability to compete with other males for a mate, while females may select a male with the best **fitness** (ability to produce surviving offspring). In that way, the female increases her own fitness.

Female Choice

Females produce few eggs, so the choice of a mate becomes a serious consideration. In a study of satin bowerbirds (see Fig. 45A), two opposing hypotheses regarding female choice were tested:

1. *Good genes hypothesis:* Females choose mates on the basis of traits that improve the chance of survival.
2. *Runaway hypothesis:* Females choose mates on the basis of traits that improve male appearance. The term *runaway* pertains to the possibility that the trait will be exaggerated in the male until its mating benefit is checked by the trait's unfavorable survival cost.

As investigators observed the behavior of satin bowerbirds, they discovered that aggressive males were usually chosen as mates by females. It could be that inherited aggressiveness does improve the chance of survival, or it could be aggressive males are good at stealing blue feathers from other males. Females prefer blue feathers as bower decorations. Therefore, the data did not clearly support either hypothesis.

The Raggiana Bird of Paradise is remarkably *dimorphic*, meaning that males and females differ in size and other traits. The males are larger than the females and have beautiful orange flank plumes. In contrast, the females are drab (Fig. 45.5). Female choice can explain why male birds are more ornate than females. Consistent with the two hypotheses, it is possible that the remarkable plumes of the male signify health and vigor to the female. Or, it's possible that females choose the flamboyant males on the basis that their sons will have an increased chance of being selected by females. Some investigators have hypothesized that extravagant male features could indicate that they are relatively parasite-free. In barn swallows, females also chose those with the longest tails, and investigators have shown that males that are relatively free of parasites have longer tails than otherwise.

Male Competition

Males can father many offspring because they continuously produce sperm in great quantity. We expect males to compete in order to inseminate as many females as possible. **Cost-benefit analyses** have been done to determine if the *benefit* of access to mating is worth the *cost* of competition among males.

Baboons, a type of Old World monkey, live together in a troop. Males and females have separate **dominance hierarchies** in which a higher-ranking animal has greater access to resources than a lower-ranking animal. Dominance is decided by confrontations, resulting in one animal giving way to the other.

Baboons are dimorphic; the males are larger than the females, and they can threaten other members of the troop with their long, sharp canines (Fig. 45.6). One or more males become dominant by frightening the other males. However, the male baboon pays a cost for his dominant position. Being larger means that he needs more food, and being willing and able to fight predators means that he may get hurt, and so forth. Is there a reproductive benefit to his behavior? Yes, in that dominant males do indeed monopolize females when they are most fertile. Nevertheless, there may be other ways to father offspring. A male may act as a helper to a female and her offspring; then, the next time she is in estrus, she may mate preferentially with him instead of a dominant male. Or subordinate males may form a friendship group that opposes a dominant male, making him give up a receptive female.

A **territory** is an area that is defended against competitors. Scientists are able to track an animal in the wild in order to determine its home range or territory. **Territoriality** includes the type of defensive behavior needed to defend a territory. Baboons travel within a home range, foraging for food each day and sleeping in trees at

FIGURE 45.5
Raggiana Bird of Paradise.
Raggiana males have brilliantly colored plumage brought about by sexual selection. The drab females tend to choose flamboyant males as mates.

FIGURE 45.6

A male olive baboon displaying full threat. In olive baboons, males are larger than females and have enlarged canines. Competition between males establishes a dominance hierarchy for the distribution of resources.

night. Dominant males decide where and when the troop will move. If the troop is threatened, dominant males protect the troop as it retreats and attack intruders when necessary. Vocalization and displays, rather than outright fighting, may be sufficient to defend a territory. In songbirds, for example, males use singing to announce their willingness to defend a territory. Other males of the species become reluctant to make use of the same area.

Red deer stags (males) on the Scottish island of Rhum compete to be the harem master of a group of hinds (females) that mate only with them. The reproductive group occupies a territory that the harem master defends against other stags. Harem masters first attempt to repel challengers by roaring. If the challenger remains, the two lock antlers and push against one another (Fig. 45.7). If the challenger then withdraws, the master pursues him for a short distance, roaring the whole time. If the challenger wins, he becomes the harem master.

A harem master can father two dozen offspring at most, because he is at the peak of his fighting ability for only a short time. And there is a cost to being a harem master. Stags must be large and powerful in order to fight; therefore, they grow faster and have less body fat. During bad times, they are more likely to die of starvation, and in general, they have shorter lives. Harem master behavior will persist in the population only if its cost (reduction in the potential number of offspring because of a shorter life) is less than its benefit (increased number of offspring due to harem access).

Evolution by sexual selection occurs when females have the opportunity to select among potential mates and/or when males compete among themselves for access to reproductive females.

a.

FIGURE 45.7

Competition between male red deer. Male red deer compete for a harem within a particular territory. **a.** Roaring alone may frighten off a challenger, but (**b**) outright fighting may be necessary, and the victor is most likely the stronger of the two animals.

b.

828 45-10 PART VIII BEHAVIOR AND ECOLOGY

Mating in Humans

A study of human mating behavior shows that the concepts of female choice and male competition apply to humans as well as to the animals we have been discussing. Increased fitness (ability to produce surviving offspring) again seems to influence behavior, this time mating between men and women, in most human cultures.

Human Males Compete

Consider that women, by nature, must invest more in having a child than men. After all, it takes nine months to have a child, and pregnancy is followed by lactation, when a woman may nurse her infant. Men, on the other hand, need only contribute sperm during a sex act that may require only a few minutes. The result is that men are generally more available for mating than are women. Because more men are available, they necessarily have to compete with others for the privilege of mating.

Like many other animals, humans are dimorphic. Males tend to be larger and more aggressive than females, perhaps as a result of past sexual selection by females. As in other animals, males pay a price for their physical attractiveness to females. Male humans live on the average seven years less than females do.

Females Choose

A study in modern Quebec sampled a large number of respondents on how often they had copulated with different sexual partners in the preceding year. Male mating success correlated best with income—those males who had both wealth and status were much more successful in acquiring mates than those who lacked these attributes. In this study, it would appear that females prefer to mate with a male who is wealthy and has a successful career because these men are more likely to be able to provide them with the resources they need to raise their children (Fig. 45.8).

The desire of women for just certain types of men has led to the practice of polygamy in many primitive human societies and even in some modern societies. Women would rather share a husband who can provide resources than to have a one-on-one relationship with a poor man, because the resources provided by the wealthy man make it all the more certain that her children will live to reproduce. On the other hand, polygamy works for wealthy men because having more than one wife will undoubtedly increase his fitness as well. As an alternative to polygamy, modern societies stress monogamy in which the male plays a prominent role in helping to raise the children. This is another way males can raise their fitness.

Men Also Have a Choice

Just as women choose men who can provide resources, men prefer women who are most likely to present them

with children. It has been shown that the "hour-glass figure" so touted by men actually correlates with the best distribution of body fat for reproductive purposes! Men responding to questionnaires about their preferences in women list attributes that biologists associate with a strong immune system, good health, high estrogen levels, and especially with youthfulness. Young males prefer partners who are their own age, give or take five years, but as men age, they prefer women who are many years younger than themselves. Men can reproduce for many more years than women can. Therefore, by choosing younger women, older men increase their fitness as judged by the number of children they have (Fig. 45.8).

Men, unlike women, do not have the same assurance that a child is their own. Therefore, men put a strong emphasis on having a wife who is faithful to them. Both men and women respondents to questionnaires view adultery in women as more offensive than adultery in men.

FIGURE 45.8 King Hussein and family.
The tendency of men to mate with fertile younger women is exemplified by King Hussein of Jordan, who was about 16 years older than his wife, Queen Noor. This photo shows some of their children, one of whom is now King Abdullah of Jordan.

45.4 SOCIOBIOLOGY AND ANIMAL BEHAVIOR

Sociobiology applies the principles of evolutionary biology to the study of social behavior in animals. Sociobiologists hypothesize that living in a society has a greater reproductive benefit than reproductive cost. A cost-benefit analysis can help determine if this hypothesis is supported.

Group living does have its benefits. It can help an animal avoid predators, rear offspring, and find food. A group of impalas is more likely to hear an approaching predator than a solitary one. Many fish moving rapidly in many directions might distract a would-be predator. Weaver birds form giant colonies that help protect them from predators, but the birds may also share information about food sources. Primate members of the same troop signal to one another when they have found an especially bountiful fruit tree. Lions working together are able to capture large prey, such as zebra and buffalo. Pair bonding of trumpet manucodes helps the birds raise their young. Due to their particular food source, the female cannot rear as many offspring alone as she can with the male's help.

Group living also has its disadvantages. When animals are crowded together into a small area, disputes can arise over access to the best feeding places and sleeping sites. Dominance hierarchies are one way to apportion resources, but this puts subordinates at a disadvantage. Among red deer, sons are preferable because, as a harem master, sons will result in a greater number of grandchildren. However, sons, being larger than daughters, need to be nursed more frequently and for a longer period of time. Subordinate females do not have access to enough food resources to adequately nurse sons, and therefore, they tend to rear daughters, not sons. Still, like the subordinate males in a baboon troop, subordinate females in a red deer harem may be better off in terms of fitness if they stay with a group, despite the cost involved.

Living in close quarters exposes individuals to illness and parasites that can easily pass from one animal to another. Social behavior helps to offset some of the proximity disadvantages. For example, baboons and other types of social primates invest much time in grooming one another, and this most likely helps them remain healthy. Humans use extensive medical care to help offset the health problems that arise from living in the densely populated cities of the world.

Social living has both advantages and disadvantages. Only if the benefits, in terms of individual reproductive success, outweigh the cost will societies come into existence.

Altruism Versus Self-Interest

Altruism [L. *alter*, the other] is a behavior that has the potential to decrease the lifetime reproductive success of the altruist, while benefiting the reproductive success of another member of the society. In insect societies, especially, reproduction is limited to only one pair, the queen and her mate. For example, among army ants, the queen is inseminated only during her nuptial flight, and thereafter she spends her time reproducing (Fig. 45.9). The society has three different sizes of sterile female workers. The smallest workers (3 mm), called the nurses, take care of the queen and larvae, feeding them and keeping them clean. The intermediate-sized workers, constituting most of the population, go out on raids to collect food. The soldiers (14 mm), with huge heads and powerful jaws, run along the sides and rear of raiding parties protecting the column of ants from attack by intruders.

Can the altruistic behavior of sterile workers be explained in terms of fitness, which is judged by reproductive

FIGURE 45.9 The queen ant, *Solenopsis geminata*.
A queen ant has a large abdomen for egg production and is cared for by small ants, called nurses. The idea of inclusive fitness suggests that relatives in addition to offspring increase an individual's reproductive success. Therefore, sterile nurses are being altruistic when they help the queen produce offspring to whom they are closely related.

success? Genes are passed from one generation to the next in two quite different ways. The first way is direct: a parent can pass a gene directly to an offspring. The second way is indirect: a relative that reproduces can pass the gene to the next generation. *Direct selection* is adaptation to the environment due to the reproductive success of an individual. *Indirect selection,* called **kin selection,** is adaptation to the environment due to the reproductive success of the individual's relatives. The **inclusive fitness** of an individual includes personal reproductive success and the reproductive success of relatives.

Among social bees, social wasps, and ants, the queen is diploid (2n), but her mate is haploid (n). If the queen has had only one mate, sister workers are more closely related to each other. They share, on average, 75% of their genes because they inherit 100% of their father's alleles. Their potential offspring would share, on average, only 50% of their genes with the queen. Therefore, a worker can achieve a greater inclusive fitness by helping her mother (the queen) produce additional sisters than by directly reproducing. Under these circumstances, behavior that appears altruistic is more likely to evolve.

Indirect selection can also occur among animals whose offspring receive only a half set of genes from both parents. Consider that your brother or sister shares 50% of your genes, your niece or nephew shares 25%, and so on. Therefore, the survival of two nieces (or nephews) is worth the survival of one sibling, assuming they both go on to reproduce.

Among chimpanzees in Africa, a female in estrus frequently copulates with several members of the same group, and the males make no attempt to interfere with each other's matings. How can they be acting in their own self-interest? Genetic relatedness appears to underlie their apparent altruism. Members of a group share more than 50% of their genes in common because members never leave the territory in which they are born.

Reciprocal Altruism

In some bird species, offspring from a previous clutch of eggs may stay at the nest to help parents rear the next batch of offspring. In a study of Florida scrub jays, the number of fledglings produced by an adult pair doubled when they had helpers. Mammalian offspring are also observed to help their parents (Fig. 45.10). Among jackals in Africa, solitary pairs managed to rear an average of 1.4 pups, whereas pairs with helpers reared 3.6 pups. What are the benefits of staying behind to help? First, a helper is contributing to the survival of its own kin. Therefore, the helper actually gains a fitness benefit. Second, a helper is more likely than a nonhelper to inherit a parental territory—including other helpers. Helping, then, involves making a minimal, short-term reproductive sacrifice in order to maximize future reproductive potential. Therefore, helpers at the nest are also practicing a form of **reciprocal altruism.** Reciprocal altruism also occurs in animals that are not necessarily closely related. In this event, an animal helps or cooperates with another animal

FIGURE 45.10 Inclusive fitness.
A meerkat is acting as a babysitter for its young sisters and brothers while their mother is away. Researchers point out that the helpful behavior of the older meerkat can lead to increased inclusive fitness.

with no immediate benefit. However, the animal that was helped will repay the debt at some later time. Reciprocal altruism usually occurs in groups of animals that are mutually dependent. Cheaters in reciprocal altruism are recognized and not reciprocated in future events. Reciprocal altruism occurs in vampire bats that live in the tropics. Bats returning to the roost after a feeding activity share their blood meal with other bats in the roost. If a bat fails to share blood with one that had previously shared blood with it, the cheater bat will be excluded from future blood sharing.

Inclusive fitness is measured by the genes an individual contributes to the next generation, either directly by offspring or indirectly by way of relatives. Many of the behaviors once thought to be altruistic turn out, on closer examination, to be examples of indirect selection and are, therefore, adaptive.

45.5 ANIMAL COMMUNICATION

Animals exhibit a wide diversity of social behaviors. Some animals are largely solitary and join with a member of the opposite sex only for the purpose of reproduction. Others pair, bond, and cooperate in raising offspring. Still others form a **society** in which members of species are organized in a cooperative manner, extending beyond sexual and parental behavior. We have already mentioned the social groups of baboons and red deer. Social behavior in these and other animals requires that they communicate with one another.

Communicative Behavior

Communication is an action by a sender that may influence the behavior of a receiver. The communication can be purposeful, but it does not have to be. Bats send out a series of sound pulses and listen for the corresponding echoes to find their way through dark caves and locate food at night. Some moths have an ability to hear these sound pulses, and they begin evasive tactics when they sense that a bat is near. Are the bats purposefully communicating with the moths? No, bat sounds are simply a cue to the moths that danger is near.

Communication is an action by a sender that affects the behavior of a receiver.

Chemical Communication

Chemical signals have the advantage of being effective both night and day. The term **pheromone** [Gk. *phero,* bear, carry, and *monos,* alone] designates chemical signals in low concentration that are passed between members of the same species. Some animals are capable of secreting different pheromones, each with a different meaning. Female moths secrete

FIGURE 45.11   Use of a pheromone.
This male cheetah is spraying urine onto a tree to mark its territory.

chemicals from special abdominal glands, which are detected downwind by receptors on male antennae. The antennae are especially sensitive, and this ensures that only male moths of the correct species (not predators) will be able to detect them.

Ants and termites mark their trails with pheromones. Cheetahs and other cats mark their territories by depositing urine, feces, and anal gland secretions at the boundaries (Fig. 45.11). Klipspringers (small antelope) use secretions from a gland below the eye to mark twigs and grasses of their territory. Whether humans secrete pheromones is discussed on page 758.

Auditory Communication

Auditory (sound) communication has some advantages over other kinds of communication (Fig. 45.12). It is faster

**FIGURE 45.12
A chimpanzee with a researcher.**
Chimpanzees are unable to speak but can learn to use a visual language consisting of symbols. Some believe chimps only mimic their teachers and never understand the cognitive use of a language. Here the researcher holds a sweater with the animal's name on it and it has learned to sign the word "me."

than chemical communication, and it too is effective both night and day. Further, auditory communication can be modified not only by loudness but also by pattern, duration, and repetition. In an experiment with rats, a researcher discovered that an intruder can avoid attack by increasing the frequency with which it makes an appeasement sound.

Male crickets have calls, and male birds have songs for a number of different occasions. For example, birds may have one song for distress, another for courting, and still another for marking territories. Sailors have long heard the songs of humpback whales transmitted through the hull of a ship. But only recently has it been shown that the song has six basic themes, each with its own phrases, that can vary in length and be interspersed with sundry cries and chirps. The purpose of the song is probably sexual, serving to advertise the availability of the singer. Bottlenose dolphins have one of the most complex languages in the animal kingdom.

Language is the ultimate auditory communication. Only humans have the biological ability to produce a large number of different sounds and to put them together in many different ways. Nonhuman primates have at most only 40 different vocalizations, each having a definite meaning, such as the one meaning "baby on the ground," which is uttered by a baboon when a baby baboon falls out of a tree. Although chimpanzees can be taught to use an artificial language, they never progress beyond the capability level of a two-year-old child. It has also been difficult to prove that chimps understand the concept of grammar or can use their language to reason. It still seems as if humans possess a communication ability unparalleled by other animals.

Visual Communication

Visual signals are most often used by species that are active during the day. Contests between males make use of threat postures and possibly prevent outright fighting, a behavior that might result in reduced fitness. A male baboon displaying full threat is an awesome sight that establishes his dominance and keeps peace within the baboon troop (see Fig. 45.6). Hippopotamuses perform territorial displays that include mouth opening.

Many animals use complex courtship behaviors and displays. The plumage of a male Raggiana Bird of Paradise allows him to put on a spectacular courtship dance to attract a female, giving her a basis on which to select a mate (see Fig. 45.5). Defense and courtship displays are exaggerated and always performed in the same way so that their meaning is clear.

Visual communication allows animals to signal others of their intentions without the need to provide any auditory or chemical messages. The body language of students during a lecture provides an example. Some students lean forward in their seats and make eye contact with the instructor. They want the instructor to know they are interested and find the material of value. Others lean back in their chairs and look at the floor or doodle. These students indicate they are not interested in the material. Teachers can use students' body language to determine if they are effectively presenting the material and make changes accordingly.

Other human behaviors also send visual clues to others. The hairstyle and dress of a person or the way he or she walks and talks are ways to send messages to others. Some studies have suggested that women are apt to dress in an appealing manner and be sexually inviting when they are ovulating. People who dress in black, move slowly, fail to make eye contact, and sit alone may be telling others that they are unhappy. Psychologists have long tried to understand how visual clues can be used to better understand human emotions and behavior. Similarly, body language in animals is being used by researchers to suggest that they, as well as humans, have emotions.

Tactile Communication

Tactile communication occurs when one animal touches another. For example, laughing gull chicks peck at the parent's bill to induce the parent to feed them (see Fig. 45.3). A male leopard nuzzles the female's neck to calm her and to stimulate her willingness to mate. In primates, grooming—one animal cleaning the coat and skin of another—helps cement social bonds within a group (Fig. 45.13).

Honeybees use a combination of communication methods, but especially tactile ones, to impart information about the environment. When a foraging bee returns to the hive, it performs a waggle dance that indicates the distance and the

FIGURE 45.13 Grooming among baboons.
Grooming occurs when one member of the troop picks through the hair of another and removes dirt, parasites, and bits of debris with fingers or teeth. The dominant males are groomed by other members of the troop and, in general, an animal is groomed by a less dominant one.

a.

FIGURE 45.14 Communication among bees.
a. Honeybees do a waggle dance to indicate the direction of food. **b.** If the dance is done outside the hive on a horizontal surface, the straight run of the dance will point to the food source. If the dance is done inside the hive on a vertical surface, the angle of the straightaway to that of the direction of gravity is the same as the angle of the food source to the sun.

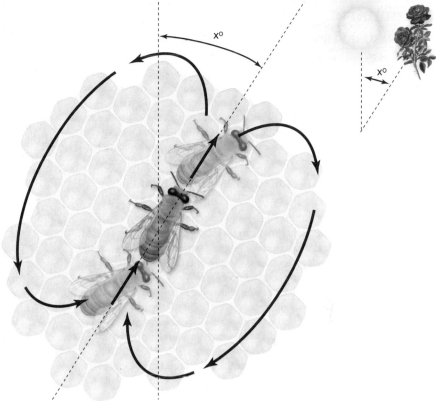

b.

direction of a food source (Fig. 45.14). As the bee moves between the two loops of a figure 8, it buzzes noisily and shakes its entire body in so-called waggles. Outside the hive, the dance is done on a horizontal surface, and the straight run indicates the direction of the food. Inside the hive, the angle of the straight run to that of the direction of gravity is the same as the angle of the food source with the sun. In other words, a 40° angle to the left of vertical means that food is 40° to the left of the sun.

Bees can use the sun as a compass to locate food because they have a biological clock, which allows them to compensate for the movement of the sun in the sky. A biological clock is an internal means of telling time. Today, we know that the ticking of the clock, in both insects and mammals (including humans), requires alterations in the expression of a gene called *period*.

Animals use a number of different ways to communicate, and communication facilitates cooperation.

CONNECTING THE CONCEPTS

Birds build nests, dogs bury bones, cats chase quick-moving objects, and snakes bask in the sun. All animals, including humans, behave—they respond to stimuli, both from the physical environment and from other individuals of the same or different species. The work of some behaviorists centers around the nature-nurture question. It can be shown that behaviors have a genetic basis and yet behaviors can be modified by experience. Regardless, behaviorists use an evolutionary approach to generate hypotheses that can be tested to better understand how the behavior increases individual fitness (i.e., the capacity to produce surviving offspring).

Behaviorists concede that both nature and nurture determine behavior. Genetics determines, for example, that hawks hunt by using vision rather than smell, and that cats, not dogs, climb trees. Songbirds are born with the ability to sing, but which song or dialect they sing is strongly dependent on the songs they hear from their parents and siblings. A new chimpanzee mother naturally cares for her young, but she is a better mother if she has observed other females in her troop raising young.

Inheritance produces the behavioral variations that are subject to natural selection, resulting in the most adaptive responses to stimuli. Sexual selection explains much about mating behavior. Males are apt to compete and females to choose because this type of behavior is most apt to lead to reproductive success. Also, there is survival value, after all, in the ability of male baboons to react ag-gressively when the troop is under attack. There is a cost to certain behaviors. Why should subordinate members of a baboon troop or subordinate females in a red deer harem remain in a situation that seemingly leads to reduced fitness?

Why should older offspring help younger offspring? The evolutionary answer is that, in the end, there are benefits that outweigh the costs. Otherwise, the behavior would not continue. Similarly, group living in which animals communicate must have some benefits. The evolutionary approach to studying behavior has proved fruitful in helping us understand why birds sing their melodious songs, dolphins frolic in groups, and wondrous male Raggiana Birds of Paradise display to females.

Summary

45.1 NATURE VERSUS NURTURE: GENETIC INFLUENCES

Investigators have long been interested in the degree to which nature (genetics) or nurture (environment) influences behavior. Studies with birds, snakes, snails, and humans have been done, among many others. Hybrid studies with lovebirds produce results consistent with the hypothesis that behavior has a genetic basis. Garter snake experiments indicate that the nervous system controls behavior. *Aplysia* DNA studies indicate that the endocrine system also controls behavior. Twin studies in humans show that certain types of behavior are apparently inherited.

45.2 NATURE VERSUS NURTURE: ENVIRONMENTAL INFLUENCES

Even behaviors formerly thought to be fixed action patterns (FAPs), or otherwise inflexible, sometimes can be modified by learning. The red bill of laughing gulls initiates chick begging behavior. However, with experience, chick begging behavior improves and they demonstrate an increased ability of chicks to recognize parents.

Other studies suggest that learning is involved in behaviors. Imprinting in birds, during a sensitive period, causes them to follow the first moving object they see. Song learning in birds involves various elements—including the existence of a sensitive period, during which an animal is primed to learn—and the positive benefit of social interactions.

Associative learning includes classical conditioning and operant conditioning. In classical conditioning, the pairing of two different types of stimuli causes an animal to form an association between them. In this way, dogs will salivate at the sound of a bell. In operant conditioning, animals learn behaviors because they are rewarded when they perform them.

45.3 ADAPTIVE MATING BEHAVIOR

Traits that promote reproductive success are expected to be advantageous overall, despite any possible disadvantage. Males produce many sperm and are expected to compete to inseminate females. Females produce few eggs and are expected to be selective about their mates. Studies of satin bowerbirds and birds of paradise have been done to test hypotheses regarding female choice. A cost-benefit analysis can be applied to competition between males for mates, in reference to a dominance hierarchy (e.g., baboons) and territoriality (e.g., red deer).

It is possible that male competition and female choice also occur among humans. Biological differences between the sexes may promote certain mating behaviors because they increase fitness.

45.4 SOCIOBIOLOGY AND ANIMAL BEHAVIOR

Living in a social group can have its advantages (e.g., ability to avoid predators, raise young, and find food). It also has disadvantages (e.g., tension between members, spread of illness and parasites, and reduced reproductive potential). When animals live in groups, the benefits must outweigh the costs or the behavior would not exist.

In most instances, the individuals of a society act to increase their own fitness (ability to produce surviving offspring). In this context, it is necessary to consider inclusive fitness, which includes personal reproductive success and the reproductive success of relatives, also. Sometimes, animals perform altruistic acts, as when individuals help their parents rear siblings. Social insects help their mother reproduce, but this behavior seems reasonable when we consider that siblings share 75% of their genes. Among mammals, a parental helper may be likely to inherit the parent's territory. In reciprocal altruism, animals aid one another for future benefits.

These examples show that altruistic behavior, most likely, does result in increased inclusive fitness.

45.5 ANIMAL COMMUNICATION

Animals that form social groups communicate with one another. In this context, communication is an action by a sender that affects the behavior of a receiver. Chemical, auditory, visual, and tactile signals are forms of communication that foster cooperation that benefits both the sender and the receiver. Pheromones are chemical signals that are passed between members of the same species. Auditory communication includes language, which may occur between other types of animals and not just humans. Visual communication allows animals to signal others without the need of auditory or chemical messages. Tactile communication is especially associated with sexual behavior.

Reviewing the Chapter

1. In what way do studies of hybrid lovebirds support the hypothesis that genetics (nature) influences behavior? 820–21
2. What body system is involved in the behavior of garter snakes toward slugs? Explain an experiment what supports your answer. 821
3. Studies of *Aplysia* DNA show that the endocrine system is also involved in behavior. Explain. 821
4. How does an experiment with laughing gull chicks support the hypothesis that environment (nurture) influences behavior? 822
5. Describe two types of associative learning. 823
6. What is sexual selection, and why does it foster female choice and male competition during mating? 826–28
7. What is a cost-benefit analysis, and how does it apply to a dominance hierarchy and territoriality? Give examples. 826
8. Give examples of behaviors that appear to be altruistic but actually increase the inclusive fitness of an individual. 829–30
9. Give examples of the different types of communication among members of a social group. 831–33

Testing Yourself

Choose the best answer for each question.

1. Behavior is
   a. any action that is learned.
   b. all responses to the environment.
   c. any action that can be observed and described.
   d. all activity that is controlled by hormones.
   e. unique to birds and humans.
2. A behaviorist would most likely study which one of the following items?
   a. the flow of energy through an ecosystem
   b. the way a bird digests seeds
   c. the number of times a fiddler crab flashes its claw to attract a mate
   d. the structure of a horse's leg
   e. All of the above are correct.
3. Which one of the following is considered, by behaviorists, to control (in part or in whole) animal behavior?
   a. circulatory and respiratory systems
   b. respiratory and digestive systems
   c. digestive and nervous systems
   d. nervous and endocrine systems
   e. All systems of the body control behavior.

4. In lovebirds, if carrying strips in the bill is controlled by a single dominant gene, then hybrids between peach-faced lovebirds and Fischer lovebirds would
   a. carry strips in their bills.
   b. carry strips in their rump feathers.
   c. not carry strips.
   d. carry strips in both their bills and rump feathers.
   e. have decreased fitness.

5. Which of the following is not an example of a genetically based behavior?
   a. Inland garter snakes do not eat slugs, while coastal populations do.
   b. One species of lovebird carries nesting strips one at a time, while another carries several.
   c. One species of warbler migrates, while another one does not.
   d. Snails lay eggs in response to egg-laying hormone.
   e. Wild foxes raised in captivity are not capable of hunting for food.

6. How would the following graph differ if pecking behavior in laughing gulls was a fixed action pattern?
   a. It would be a diagonal line with an upward incline.
   b. It would be a diagonal line with a downward incline.
   c. It would be a horizontal line.
   d. It would be a vertical line.
   e. None of these is correct.

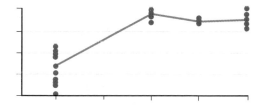

7. Egg-laying hormone causes snails to lay eggs, implicating which system in this behavior?
   a. digestive   d. lymphatic
   b. endocrine   e. respiratory
   c. nervous

8. Using treats to train a dog to do a trick is an example of
   a. imprinting.   c. vocalization.
   b. tutoring.   d. operant conditioning.

9. The benefits of imprinting (following and obeying the mother) generally outweigh the costs (following the wrong object instead of the mother) because
   a. an animal that has been imprinted on the wrong object can be reimprinted on the mother.
   b. imprinting behavior never lasts more than a few months.
   c. animals in the wild rarely imprint on anything other than their mother.
   d. animals that imprint on the wrong object generally die before they pass their genes on.

10. In white-crowned sparrows, social experience exhibits a very strong influence over the development of singing patterns. What observation led to this conclusion?
    a. Birds only learned to sing when they were trained by other birds.
    b. The window in which birds learn from other birds is wider than that when birds learn from tape recordings.

    c. Birds could learn different dialects only from other birds.
    d. Birds that learned to sing from a tape recorder could change their song when they listened to another bird.

11. Which of the following best describes classical conditioning?
    a. The gradual strengthening of stimulus-response connections that seemingly are unrelated.
    b. A type of associative learning in which there is no contingency between response and reinforcer.
    c. The learning behavior in which an organism follows the first moving object it encounters.
    d. The learning behavior in which an organism exhibits a fixed-action pattern from the time of birth.

12. The observation that the most fit male bowerbirds are the ones that can keep their nests intact supports which hypothesis?
    a. good genes hypothesis—females choose mates based on their improved chances of survival
    b. runaway hypothesis—females choose mates based on their appearance
    c. Either hypothesis could be true.
    d. Neither hypothesis is true.

13. In some bird species, the female bird chooses a mate that is most similar to her in size. This supports
    a. the good genes hypothesis.
    b. the runaway hypothesis.
    c. Either hypothesis could be true.
    d. Neither hypothesis is true.

14. Which of the following are costs that a dominant male baboon must pay in order to gain a reproductive benefit?
    a. He requires more food and must travel larger distances.
    b. He requires more food and must care for his young.
    c. He is more prone to injury and requires more food.
    d. He is more prone to injury and must care for his young.
    e. He must care for his young and travel larger distances.

15. A red deer harem master typically dies earlier than other males because he is
    a. likely to get expelled from the herd and cannot survive alone.
    b. more prone to disease because he interacts with so many animals.
    c. likely to starve to death.
    d. apt to place himself between a predator and the herd to protect the herd.

For problems 16–20, match the type of communication in the key with its description. Answers may be used more than once.

KEY:
   a. chemical communication
   b. auditory communication
   c. visual communication
   d. tactile communication

16. Aphids (insects) release an alarm pheromone when they sense they are in danger.

17. Male peacocks exhibit an elaborate display of feathers to attract females.

18. Ground squirrels give an alarm call to warn others of the approach of a predator.

19. Male silk moths are attracted to females by a sex attractant released by the female moth.

20. Sage grouses perform an elaborate courtship dance.

Thinking Scientifically

1. Meerkats are said to exhibit altruistic behavior because certain members of a population act as sentries. These sentries stand on rocks or other high places and serve as lookouts while others feed. However, recent observations have shown that these sentries are the first ones to reach safety when a predator is spotted, and that sentries only serve after they have eaten. Their behavior is still beneficial to the group since without the sentries there would be no warning of approaching predators, but can it still be termed altruistic? How would you test the hypothesis that sentries are engaged in altruistic behavior?

2. You are testing the hypothesis that human infants instinctively respond to higher-pitched voices. Your design is to record head turns toward speakers placed on opposite sides of a month-old infant. The speakers would play voices (all making the same noises) in different pitches and you would see if the infants turned toward some voices more often than others. When you do the experiment using several different infants, your data support your hypothesis. However, prior learning by infants is still a serious criticism. What is the basis of this criticism?

Bioethical Issue: Putting Animals in Zoos

Is it ethical to keep animals in zoos, where they are not free to behave as they would in the wild? If we keep animals in zoos, are we depriving them of their freedom? Some point out that freedom is never absolute. Even an animal in the wild is restricted in various ways by its abiotic and biotic environment. Animals are said to have five freedoms: freedom from starvation, cold, injury, and fear, as well as freedom to wander and express their natural behavior. Perhaps it's worth giving up a bit of the last freedom to achieve the first four? Many modern zoos keep animals in habitats that nearly match their natural ones so that they have some freedom to roam and behave naturally. Perhaps, too, we should consider the education and enjoyment of the many thousands of human visitors to a zoo compared to the freedom lost by a much smaller number of animals kept in a zoo.

Today, reputable zoos rarely go out and capture animals in the wild—they usually get their animals from other zoos. Most people feel it is not a good idea to take animals from the wild except for very serious reasons. Certainly, zoos should not be involved in the commercial and often illegal trade of wild animals that still goes on today. When animals are captured, it should be done by skilled biologists or naturalists who know how to care for and transport the animals.

Many zoos today are involved in the conservation of animals. They provide the best home possible while animals are recovering from injury or increasing their numbers until they can be released to the wild. Can we perhaps look at zoos favorably if they show that they are keeping animals under good conditions and are also involved in preserving animals?

Understanding the Terms

altruism 829
associative learning 823
behavior 820
classical conditioning 823
communication 831
cost-benefit analysis 826
dominance hierarchy 826
fitness 826
fixed action pattern
　(FAP) 822
imprinting 822
inclusive fitness 830
kin selection 830
learning 822
operant conditioning 823
pheromone 831
reciprocal altruism 830
sexual selection 826
society 831
sociobiology 829
territoriality 826
territory 826

Match the terms to these definitions:
a. _____ Behavior related to defending a particular area, which is often used for the purpose of feeding, mating, and caring for young.
b. _____ Social interaction that benefits others but has the potential to decrease the lifetime reproductive success of the member exhibiting the behavior.
c. _____ Signal by a sender that may influence the behavior of a receiver.
d. _____ Chemical substance secreted into the environment by one organism that may influence the behavior of another.

ARIS, the *Biology* Website

ARIS, the website for *Biology*, provides a wealth of information organized and integrated by chapter. You will find practice quizzes, interactive activities, labeling exercises, flashcards, and much more that will complement your learning and understanding of general biology.

www.mhhe.com/maderbiology9

46

ECOLOGY OF POPULATIONS

E *lephants attain a large size, are social, live a long time, even up to 70 years, and produce few offspring. Females live in social family units, and the much larger males visit them only during breeding season. Females give birth about every five years to a single calf that is well cared for and has a good chance of meeting the challenges of its lifestyle. Normally, an elephant population exists at the carrying capacity of the environment. Carrying capacity is the largest number of organisms of a given species that can be maintained indefinitely within a given environment.*

A population ecologist studies the distribution and abundance of organisms and relates the population statistics to a species' life history in order to determine what causes population growth or decline. Elephants are threatened because of human population growth, and also because males, in preference to females, are killed for their ivory tusks. Males tend not to breed until they have reached their largest size—the size that makes them prized by humans. Now, even a moratorium on killing male elephants may not help their plight. So few breeding males are left that elephant populations are expected to continue to decline for quite some time. The study of population ecology is important in the preservation of the species and the maintenance of the diversity of life on Earth.

Social unit of female elephants, *Loxodonta africana.*

46.1 Scope of Ecology

In 1866, the German zoologist Ernst Haeckel coined the word *ecology* from two Greek roots [Gk. *oikos,* home, house, and *-logy,* "study of" from *logikos,* rational, sensible]. He said that **ecology** is the study of the interactions of organisms with other organisms and with the physical environment. Haeckel also pointed out that ecology and evolution are intertwined because ecological interactions are selection pressures that result in evolutionary change, which in turn affects ecological interactions.

Ecology, like so many biological disciplines, is wide-ranging. At one of its lowest levels, ecologists study how the individual organism is adapted to its environment. For example, they study how a fish is adapted to and survives in its **habitat** (the place where the organism lives) (Fig. 46.1). Most organisms do not exist singly; rather, they are part of a population, a functional unit that interacts with the environment. A **population** is defined as all the organisms within an area belonging to the same species. At this level of study, ecologists are interested in factors that affect the growth and regulation of population size.

A **community** consists of all the various populations interacting at a locale. In a coral reef, there are numerous populations of algae, corals, crustaceans, fishes, and so forth. At this level, ecologists want to know how interactions such as predation and competition affect the organization of a community. An **ecosystem** contains a community of populations and also the abiotic environment (e.g., the availability of sunlight for plants). Energy flow and chemical cycling are significant aspects of understanding how an ecosystem functions. Ecosystems rarely have distinct boundaries and are not totally self-sustaining. Usually, a transition zone called an ecotone, which has a mixture of organisms from adjacent ecosystems, exists between ecosystems. The **biosphere** encompasses the zones of the Earth's soil, water, and air where living organisms are found. Table 46.1 summarizes the levels of biological study.

Modern ecology is not just descriptive, it is predictive. It analyzes levels of organization and develops models and hypotheses that can be tested. A central goal of modern ecology is to develop models that explain and predict the distribution and abundance of organisms. Ultimately, ecology considers not one particular area, but the distribution and abundance

TABLE 46.1

Ecological Terms

Term	Definition
Ecology	Study of the interactions of organisms with each other and with the physical environment
Population	All the members of the same species that inhabit a particular area
Community	All the populations found in a particular area
Ecosystem	A community and its physical environment, including both nonliving (abiotic) and living (biotic) components
Biosphere	All the communities on Earth whose members exist in air and water and on land

FIGURE 46.1 Ecological levels.
The study of ecology encompasses levels of organization that proceed from the individual organism to the population, to the community, and finally to an ecosystem.

Organism ⟶ Population ⟶ Community ⟶ Ecosystem

of populations in the biosphere. For example, what factors have brought about the mix of plants and animals in a tropical rain forest at one latitude and in a desert at another? While modern ecology is useful in and of itself, it also has unlimited application possibilities, including the proper management of plants and wildlife, the identification of and efficient use of renewable and nonrenewable resources, the preservation of habitats and natural cycles, the maintenance of food resources, and the ability to predict the impact and course of a disease such as malaria or AIDS.

> Ecology is the study of the interactions of organisms with other organisms and with the physical environment. These interactions determine the distribution and abundance of organisms at a particular locale and over the Earth's surface.

46.2 DEMOGRAPHICS OF POPULATIONS

Demography is the statistical study of a population, such as its density, its distribution, and its rate of growth, which is dependent on such factors as its mortality pattern and age distribution.

Density and Distribution

Population density is the number of individuals per unit area, such as there are 73 persons per square mile in the United States. Population density figures make it seem as if individuals are uniformly distributed, but this often is not the case. For example, we know full well that most people in the United States live in cities, where the number of people per unit area is dramatically higher than in the country. And even within a city, more people live in particular neighborhoods than others, and such distributions can change over time. Therefore, basing ecological models solely on population density, as has often been done in the past, can lead to misleading results.

Population distribution is the pattern of dispersal of individuals across an area of interest. The availability of resources can affect where populations live. **Resources** are nonliving (abiotic) and living (biotic) components of an environment that support living organisms. Light, water, space, mates, and food are some important resources for populations. **Limiting factors** are those environmental aspects that particularly determine where an organism lives. For example, trout live only in cool mountain streams, where the oxygen content is high, but carp and catfish are found in rivers near the coast because they can tolerate warm waters, which have a low concentration of oxygen. The timberline is the limit of tree growth in mountainous regions or in high latitudes. Trees cannot grow above the high timberline because of low temperatures and the fact that water remains frozen most of the year. The distribution of organisms can also be due to biotic factors. In Australia, the red kangaroo does not live outside arid inland areas because it is adapted to feeding on the grasses that grow there.

Three descriptions—*clumped, random,* and *uniform*—are often used to characterize observed patterns of distribution. Suppose you considered the distribution of a species across its full range. A range is that portion of the globe where the species can be found. For example, red kangaroos live in Australia. On that scale, you would expect to find a clumped distribution because organisms are located in areas suitable to their adaptations. For example, as mentioned, red kangaroos live in grasslands, and catfish live in warm river water near the coast.

Within a smaller area such as a single body of water or a single forest, the availability of resources again influences which of the patterns of distribution is common for a particular population. For example, as discussed in the Ecology Focus on page 842, a study of the distribution of hard clams in a bay on the south shore of Long Island, New York, showed that clam abundance is associated with sediment shell content. Indeed, it might be possible to use this information to transform areas that have few clams into high-abundance areas. Distribution patterns need not be constant. In a study of desert shrubs, it was found that the distribution changed from clumped to random to a uniform distribution pattern as the plants matured. As time passed, it was found that competition for belowground resources caused the distribution pattern to become uniform (Fig. 46.2).

a. Clumped

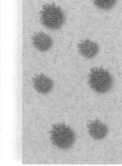
b. Random

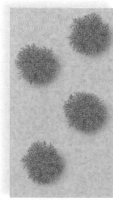
c. Uniform

d. Mature desert shrubs

FIGURE 46.2 Distribution patterns of the creosote bush.
a. Young, small desert shrubs are clumped. **b.** Medium shrubs are randomly distributed. **c.** Mature shrubs are uniformly distributed. **d.** Photograph of mature shrub distribution.

a.

b.

FIGURE 46.3 Biotic potential.
A population's maximum growth rate under ideal conditions—that is, its biotic potential—is greatly influenced by the number of offspring produced in each reproductive event. **a.** Pigs, which produce many offspring that quickly mature to produce more offspring, have a much higher biotic potential than (**b**) the rhinoceros, which produces only one or two offspring per infrequent reproductive event.

Other factors besides resource availability can influence distribution patterns. Breeding golden eagles, like many other birds, exhibit territoriality, and this behavioral characteristic discourages a clumped distribution at this time. On the other hand, cedar trees tend to be clumped near the parent plant because seeds are not widely dispersed.

Population Growth

The **rate of natural increase (r)**, which for our purposes is the same as the growth rate, is dependent on the number of individuals born each year and the number of individuals that die each year. Usually it is possible to assume that immigration and emigration are equal and need not be considered in the calculation of the growth rate. Populations grow when the number of births exceeds the number of deaths. If the number of births is 30 per year and the number of deaths is 10 per year per 1,000 individuals, the growth rate would be:

$$(30 - 10)/1{,}000 = 0.02 = 2.0\%$$

The highest possible rate of natural increase for a population when resources are unlimited is called its **biotic potential** (Fig. 46.3). Whether the biotic potential is high or low depends on demographic characteristics of the population, such as the following:

- Usual number of offspring per reproductive event
- Chances of survival until age of reproduction
- How often each individual reproduces
- Age at which reproduction begins

Mortality Patterns

Population growth patterns assume that populations are made up of identical individuals. Actually the individuals of a population are in different stages of their life span. A **cohort** is all the members of a population born at the same time. Some investigators study population dynamics and construct life tables that show how many members of a cohort are still alive after certain intervals of time. For example, Table 46.2 is a life table for a bluegrass cohort. The cohort contains 843 individuals. The table tells us that after three months, 121 individuals have died, and therefore the mortality rate is 0.143 per capita. Another way to express this same statistic, however, is to consider that 722 individuals are still alive—have survived—after three months. **Survivorship** is the probability of newborn individuals of a cohort surviving to particular ages. If we plot the number surviving at each age, a survivorship curve is produced (see Fig. 46.4b). The results of such investigations show that each species has a typical survivorship curve.

TABLE 46.2

A Life Table for a Bluegrass Cohort

Age (months)	Number Observed Alive	Number Dying	Mortality Rate per Capita	Avg. Number of Seeds per Individual
0–3	843	121	0.143	0
3–6	722	195	0.271	300
6–9	527	211	0.400	620
9–12	316	172	0.544	430
12–15	144	95	0.626	210
15–18	54	39	0.722	60
18–21	15	12	0.800	30
21–24	3	3	1.000	10
24	0	—	—	—

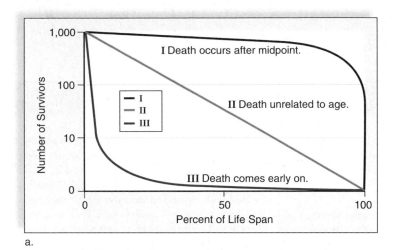

a.

b. Bluegrasses

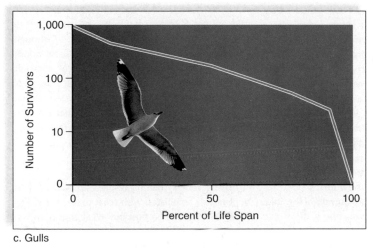

c. Gulls

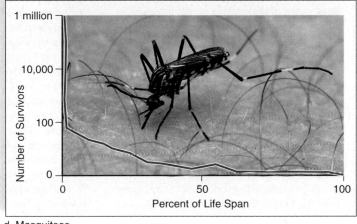

d. Mosquitoes

FIGURE 46.4 Survivorship curves.
Survivorship curves show the number of individuals of a cohort that are still living over time. **a.** Three generalized survivorship curves. **b.** The survivorship curve for bluegrasses seems to be a combination of the type I and type II curves. **c.** The survivorship curve for gulls fits the type II curve somewhat. **d.** The survivorship curve for mosquitoes is a type III curve.

Three types of idealized survivorship curves, numbered I, II, and III, are typically recognized (Fig. 46.4*a*). The type I curve is characteristic of a population in which most individuals survive well past the midpoint of the life span, and death does not come until near the end of the life span. Animals that have this type of survivorship curve include large mammals and humans in more-developed countries. On the other hand, the type III curve is typical of a population in which most individuals die very young. This type of survivorship curve occurs in many invertebrates, fishes, and humans in less-developed countries. In the type II curve, survivorship decreases at a constant rate throughout the life span. In many songbirds and small mammals, death is usually unrelated to age; thus they represent a type II survivorship curve.

The survivorship curves of natural populations do not fit these three idealized curves exactly. In a bluegrass cohort, as shown in Table 46.2 for example, most individuals survive till six to nine months, and then the chances of survivorship diminish at an increasing rate (Fig. 46.4*b*). Statistics for a gull cohort are close enough to classify the survivorship curve in the type II category (Fig. 46.4*c*), while a mosquito cohort has a type III curve (Fig. 46.4*d*).

Much can be learned about the life history of a species by studying its life table and the survivorship curve that can be constructed based on this table. Would you predict that most or few members of a population with a type III survivorship curve are contributing offspring to the next generation? Obviously, since death comes early for most members, only a few are living long enough to reproduce. What about the other two types of survivorship curves?

Other types of information are also available from studying life tables. Look again at Table 46.2, the bluegrass life table. It tells us that per capita seed production increases as plants mature, and then seed production drops off. How do you predict this would compare to a cohort of human beings?

Age Distribution

When the individuals in a population reproduce repeatedly, several generations may be alive at any given time. From the perspective of population growth, a population contains three major age groups: prereproductive, reproductive, and

ecology focus

Increasing a Clam Population by Using Oyster Shells

With the help of my students, I [Jeffrey Kassner] study the distribution and abundance of hard clams in the Great South Bay on the south shore of Long Island, New York. We have discovered that hard clam abundance is strongly associated with sediment oyster shell content, and we now want to explore the feasibility of placing oyster shells in low-abundance areas in order to change them to high-abundance areas. If this strategy works, it would be a tremendous boon to the shellfish industry because nearly three-quarters of the study area is low abundance.

Purpose of the Study

The Great South Bay has often been referred to as a "hard clam factory" because in the 1970s, over half of the hard clams harvested in the United States came from its waters. For the past 20 years, I have been studying the eastern third of the bay for the Town of Brookhaven which, by virtue of a seventeenth-century grant from the King of England, owns 6,000 hectares of prime hard clam habitat.

Knowing the distribution and abundance of hard clams is important to Brookhaven because ownership carries with it an obligation to manage a hard clam fishing industry that has an annual value of $10 million and employs 300 fishermen. Of particular interest to Brookhaven is the possibility of using this information to increase

hard clam abundance, and hence the hard clam harvest, for the economic benefits the hard clam resource provides to Brookhaven.

The Study

My study began by censusing the hard clam population. For several weeks during the summers, a barge-mounted crane with a one-square-meter clamshell bucket was used to take bottom grabs at 462 stations located throughout the study area. Each bottom sample was placed in a one-square-meter wire sieve and washed with a high-pressure water hose to separate the hard clams from the sediment so that the hard clams could be counted and measured in order to calculate various demographic parameters. The fieldwork was messy and physically hard, and by the end of the day, my crew of students and biologists were tired, wet, and covered with mud. The results, however, have proven to be well worth the effort.

Working with Dr. Robert Cerrato of the State University of New York, I was able to draw up a composite census map showing the distribution of hard clams. I was surprised to find that hard clams were not distributed uniformly throughout the study area, but occurred in distinct patches of high and low abundance. I termed the high-abundance areas "beds" (a dense assemblage of clams is traditionally referred to as a "clam bed"), and identified six such areas.

When I overlaid the census map on a sediment map that I had generated from my field notes, nearly all of the beds coincided with areas of high-shell-content sediment associated with formerly productive oyster reefs, or what I call "relict" oyster reefs. This observation is of historical note as well as of biological interest because up through the early years of this century, the Great South Bay was a major producer of oysters, and was the source of the world-famous "Blue Point Oyster." Although oysters are no longer found in the Great South Bay because of environmental changes, they left behind a legacy in the sediment that now supports high abundances of hard clams. As stated, I want to use this information to increase the abundance of clams where they are now in low abundance.

Further Study

Having found that hard clam abundance was positively associated with relict oyster reefs (Fig. 46A), I wanted to find out what aspect of the sediment of the relict oyster reefs makes them so productive, and conversely, why low-abundance areas have so few hard clams. This information would be useful to the Town of Brookhaven because it might lead to additional ways in which the sediment in low-abundance areas can be transformed into the type found in high-abundance areas. I therefore needed a sedimentary "portrait" of high- and low-abundance areas, and to develop one, I borrowed techniques generally associated with other marine science disciplines: From shipwreck hunting, I used side-scan sonar to map the topography of the bottom; from deep-sea research, an ROV (Remotely Operated Vehicle) to photograph the bottom; from commercial fishing, a fathometer to map the bottom of the bay; and from pollution studies, a sediment profile camera to photograph the sediment-water interface where hard clams live and feed.

I am now putting all this different information on a single map. It will take time and perhaps more studies to develop the portrait. My research project is not only scientifically interesting, but it is also personally rewarding. Over the years, I have become friends with many fishermen, and I know that if I am successful, I will be helping them continue in an occupation that, in some cases, goes back generations.

Jeffrey Kassner

Hard clams in study

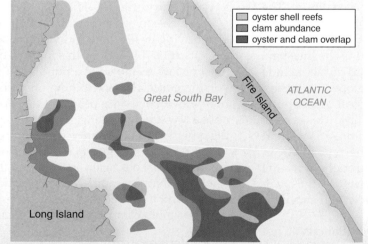

oyster shell reefs
clam abundance
oyster and clam overlap

Great South Bay

Fire Island

ATLANTIC OCEAN

Long Island

FIGURE 46A Map of Great South Bay.
Kassner mapped areas of clam abundance (medium pink) and oyster shell reefs (light pink). He found that, in general, the two coincided (dark pink).

FIGURE 46.5 Age structure diagrams. Typical age structure diagrams for hypothetical populations that are increasing, stable, or decreasing. Different numbers of individuals in each age class create these distinctive shapes. In each diagram, the left half represents males while the right half represents females.

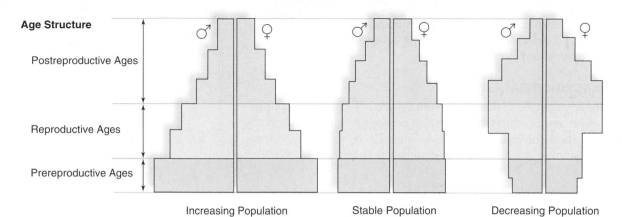

Age Structure

Postreproductive Ages

Reproductive Ages

Prereproductive Ages

Increasing Population   Stable Population   Decreasing Population

postreproductive. Populations differ according to what proportion of the population falls in each age group. At least three **age structure diagrams** are possible (Fig. 46.5).

When the prereproductive group is the largest of the three groups, the birthrate is higher than the death rate, and a pyramid-shaped diagram is expected. Under such conditions, even if the growth for that year were matched by the deaths for that year, the population would continue to grow in the following years. Why? Because there are more individuals entering than leaving the reproductive years. Eventually, as the size of the reproductive group equals the size of the prereproductive group, a bell-shaped diagram will result. The postreproductive group will still be the smallest, however, because of mortality. If the birthrate falls below the death rate, the prereproductive group will become smaller than the reproductive group. The age structure diagram will then be urn-shaped, because the postreproductive group is now the largest.

The age distribution reflects the past and future history of a population. Because a postwar baby boom occurred in the United States between 1946 and 1964, the postreproductive group will soon be the largest group.

Demographics includes statistical data about a population's density and distribution and its growth rate, which is dependent on resource availability and on the population's mortality pattern and age distribution.

46.3 POPULATION GROWTH MODELS

We shall assume that there are two patterns of population growth. In the pattern called **discrete breeding,** the members of the population have only a single reproductive event in their lifetime. When this event draws near, the mature adults largely cease to grow, expend all their energy in reproduction, and then die. Many insects, such as mayflies, and annual plants, such as zinnias, reproduce in this manner. They produce offspring or seeds that can survive dryness and/or cold and resume growth the next favorable season. In the pattern called **continuous breeding,** members experience many reproductive events throughout their lifetime. During their entire lives, they continue to invest energy in their future survival, increasing their chances of reproducing again. Most vertebrates, shrubs, and trees have this pattern of reproduction.

Ecologists have developed mathematical models of population growth based on these two very different patterns of reproduction. Do these models have value even if the manner in which organisms reproduce does not always fit either of these two patterns, as exemplified in Figure 46.6? Although the mathematical models we will be describing are simplifications, they still may predict how best to control the distribution and abundance of organisms, or how to predict the responses of populations when their environment

a.

b.

FIGURE 46.6

Patterns of reproduction. Although we assume that members of populations either have a single suicidal reproductive event or reproduce repeatedly, actually there are exceptions. **a.** Aphids reproduce repeatedly by asexual reproduction during the summer and then reproduce sexually only once, right before the onset of winter. Therefore, aphids use both patterns of reproduction. **b.** The offspring of annual plants can germinate several seasons later. Under these circumstances, population size of these organisms could fluctuate according to environmental conditions.

is altered in some way. Testing predictions permits the development of new hypotheses that can then be tested.

Exponential Growth

As an example of discrete breeding, we will consider a population of insects in which females reproduce only once a year and then the adult population dies. Each female produces on the average 2.4 eggs per generation that will survive the winter and become offspring the next year. In the next generation, each female will again produce 2.4 eggs. In the case of discrete breeding, R = net reproductive rate.[1] Why net reproductive rate? Because it is the observed rate of natural increase after deaths have occurred.

Figure 46.7a shows how the population would grow year after year for ten years, assuming that R stays constant from generation to generation. This growth is equal to the size of the population because all members of the previous generation have died. Figure 46.7b shows the growth curve for this population. This growth curve, which has a J shape, depicts exponential growth. With **exponential growth,** the number of individuals added each generation increases as the total number of females increases.

Notice that the curve has these phases:

Lag phase During this phase, growth is slow because the population is small.
Exponential growth phase During this phase, growth is accelerating.

Figure 46.7c gives the mathematical equation that allows you to calculate growth and size for any population that has discrete (nonoverlapping) generations. In other words, as discussed, all members of the previous generation die off before the new generation appears. To use this equation to determine future population size, it is necessary to know R, which is the net reproductive rate determined after gathering mathematical data regarding past population increases. Notice that even though R remains constant, growth is exponential because the number of individuals added each year is increasing. Therefore, the growth of the population is accelerating.

For exponential growth to continue unchecked, each member of the population has to have unlimited resources. Plenty of room, food, shelter, and any other requirements necessary to sustain growth have to be available. But in reality, environmental conditions prevent exponential growth. Eventually any further growth is impossible because the food supply runs out and waste products begin to accumulate. Also, as the population increases in size so do the effects of competition between members, predation, parasites, and disease.

[1] The change of r to R is simply customary in discrete breeding calculations; both coefficients deal with the same thing (birth minus death).

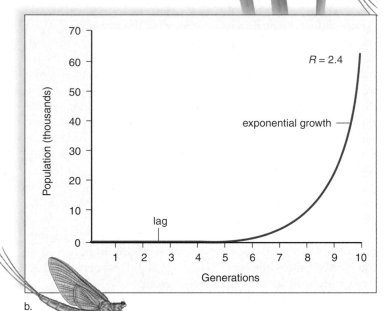

FIGURE 46.7
Model for exponential growth. When the data for discrete reproduction in (a) are plotted, the exponential growth curve in (b) results. **c.** This formula produces the same results as (a) and generates the same graph.

Generation	Population Size
0	10.0
1	24.0
2	57.6
3	138.2
4	331.7
5	796.1
6	1,910.6
7	4,585.4
8	11,005.0
9	26,412.0
10	63,388.8

a.

b.

To calculate population size from year to year, use this formula:

$$N_{t+1} = RN_t$$

N_t = number of females already present
R = net reproductive rate
N_{t+1} = population size the following year

c.

Exponential growth produces a characteristic J-shaped curve. Because growth accelerates over time, the size of the population can increase dramatically. Eventually, environmental conditions will prevent any further growth.

Logistic Growth

What type of growth curve results when environmental resistance comes into play? In 1930, Raymond Pearl developed a method for estimating the number of yeast cells accruing every two hours in a laboratory culture vessel. His data are shown in Figure 46.8a. When the data are plotted, the growth curve has the appearance shown in Figure 46.8b. This type of growth curve is a sigmoidal (S) or S-shaped curve.

Notice that this so-called **logistic growth** has these phases:

Lag phase During this phase, growth is slow because the population is small.
Exponential growth phase During this phase, growth is accelerating.
Deceleration phase During this phase, growth slows down.
Stable equilibrium phase During this phase, there is little if any growth because births and deaths are about equal.

Figure 46.8c gives the mathematical equation that allows us to calculate logistic growth (so called because the exponential portion of the curve would produce a straight line if the log of N were plotted). The entire equation for logistic growth is:

$$\frac{dN}{dt} = rN\,\frac{(K-N)}{K}$$

but let's consider each portion of the equation separately.

Because the population has repeated reproductive events, we need to consider growth as a function of change in time (Δ):

$$\frac{\Delta N}{\Delta t} = rN$$

If the change in time is very small, then we can turn to differential calculus, and the instantaneous population growth (d) is given by:

$$\frac{dN}{dt} = rN$$

This portion of the equation applies to the first two phases of growth—the lag phase and the exponential growth phase. Here, also, we do not expect exponential growth to continue. Charles Darwin calculated that a single pair of elephants could have over 19 million live descendants after 750 years. Others have calculated that a single female housefly could produce over 5 trillion flies in one year! Such explosive growth does not occur because environmental conditions, both abiotic and biotic, cause population growth to slow. The yeast population was grown in a vessel in which food could run short and waste products could accumulate. These environmental conditions oppose exponential growth particularly when a population is displaying its biotic potential, the highest rate of increase possible for a population. Look again at Figure 46.8. Following

exponential growth, a population is expected to enter a deceleration phase and then a stable equilibrium phase of the logistic growth curve. Now the population is at the carrying capacity of the environment.

Growth of Yeast Cells in Laboratory Culture

Time (t) (hours)	Number of individuals (N)	Number of individuals added per 2-hour period $\left(\dfrac{\Delta N}{\Delta t}\right)$
0	9.6	0
2	29.0	19.4
4	71.1	42.1
6	174.6	103.5
8	350.7	176.1
10	513.3	162.6
12	594.4	81.1
14	640.8	46.4
16	655.9	15.1
18	661.8	5.9

a.

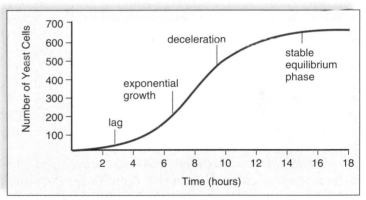

b.

To calculate population growth as time passes, use this formula:

$$\frac{dN}{dt} = rN\left(\frac{K-N}{K}\right)$$

N = population size

dN/dt = change in population size

r = rate of natural increase

K = carrying capacity

$\dfrac{K-N}{K}$ = effect of carrying capacity on population growth

c.

FIGURE 46.8 Model for logistic growth.
When the data for repeated reproduction (**a**) are plotted, the logistic growth curve in (**b**) results. **c.** This formula produces the same results as (a) and generates the same graph.

Carrying Capacity

The **carrying capacity** of any environment is the maximum number of individuals of a given species the environment can support. The closer population size nears the carrying capacity of the environment, the more likely resources will become scarce and biotic effects such as competition and predation will become evident. The birthrate is expected to decline and the death rate is expected to increase. This will result in a decrease in population growth; eventually, the population stops growing and its size remains stable.

How does our mathematical model for logistic growth take this process into account? To our equation for growth under conditions of exponential growth we add the term:

$$\frac{(K-N)}{K}$$

In this expression, K is the carrying capacity of the environment. The easiest way to understand the effects of this term is to consider two extreme possibilities. First, consider a time at which the population size is well below carrying capacity. Resources are relatively unlimited, and we expect rapid, nearly exponential growth to take place. Does the model predict this? Yes, it does. When N is very small relative to K, the term $(K-N)/K$ is very nearly $(K-0)/K$, or approximately 1. Therefore, $dN/dt =$ approximately rN.

Similarly, consider what happens when the population reaches carrying capacity. Here, we predict that growth will stop and the population will stabilize. What happens in the model? When $N=K$, the term $(K-N)/K$ declines from nearly 1 to 0, and the population growth slows to zero.

As mentioned, the model we have developed predicts that exponential growth will occur only when population size is much lower than the carrying capacity. So, as a practical matter, if we are using a fish population as a continuous food source, it would be best to maintain the population size in the exponential phase of the logistic growth curve. Biotic potential can have its full effect, and the birthrate is the highest it can be during this phase. If we overfish, the population will sink into the lag phase, and it will be years before exponential growth recurs. On the other hand, if we are trying to limit the growth of a pest, it is best, if possible, to reduce the carrying capacity rather than reduce the population size. Reducing the population size only encourages exponential growth to begin once again. Farmers can reduce the carrying capacity for a pest by alternating rows of different crops rather than growing one type of crop throughout the entire field.

Logistic growth produces a characteristic S-shaped curve. Population size stabilizes when the carrying capacity of the environment has been reached.

46.4 REGULATION OF POPULATION SIZE

We have developed two models of population growth: exponential growth with its J-shaped curve and logistic growth with its S-shaped curve. In a study of winter moth population dynamics, it was discovered that a large proportion of eggs did not survive the winter and exponential growth never occurred. Perhaps the low number of individuals at the start of each season helps prevent the occurrence of exponential growth. This observation raises the question, "How well do the models for exponential and logistic growth predict population growth in natural populations?"

Is it possible, for example, that exponential growth may cause population size to rise above the carrying capacity of the environment, and as a consequence a population crash may occur? For example, in 1911, 4 male and 21 female reindeer *(Rangifer)* were released on St. Paul Island in the Bering Sea off Alaska. St. Paul Island had a completely undisturbed environment—there was little hunting pressure, and there were no predators. The herd grew exponentially to about 2,000 reindeer in 1938, overgrazed the habitat, and then abruptly declined to only 8 animals in 1950 (Fig. 46.9). This pattern of a population explosion eventually followed by a population crash is called irruptive, or Malthusian, growth. It is named in honor of the eighteenth-century economist

FIGURE 46.9 Density-dependent effect.
On St. Paul Island, Alaska, reindeer grew exponentially for several seasons and then underwent a sharp decline as a result of overgrazing the available range.

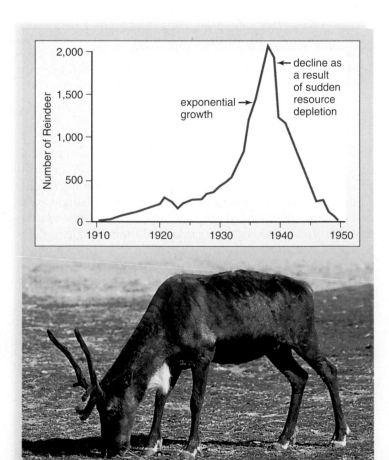

a.

b.

FIGURE 46.10 Density-independent effects. The impact of a density-independent factor, such as weather or natural disasters, is not influenced by population density. The impact of a flash flood (indicated by the blue-shaded area) on (**a**) a low-density population (mortality rate of 3/5, or 60%) is similar to the impact on (**b**) a high-density population (mortality rate of 12/20, or 64%).

who had a great influence on Charles Darwin. Populations do not ordinarily undergo Malthusian growth because of factors that regulate population growth.

Factors That Regulate Population Growth

Ecologists have long recognized that both abiotic conditions and biotic conditions play an important role in regulating population size in nature.

Density-Independent Factors

Abiotic factors, such as weather or natural disasters, are typically **density-independent factors.** Abiotic factors can have drastic effects on a population, causing sudden and catastrophic reductions in population size. However, density-independent factors cannot in and of themselves regulate population size because the effect is not influenced by the number of individuals in the population (Fig. 46.10). In other words, the intensity of the effect does not increase with increased population size. For example, the proportion of a population killed in a flash flood event is independent of density—floods do not necessarily kill a larger percentage of a dense population than of a less dense population. Of course, the larger the population, the greater the number of individuals killed, but the percentage killed does not depend on the density.

Density-Dependent Factors

Biotic factors are considered to be **density-dependent factors** because the percentage of the population affected does increase as the density of the population increases. Competition, predation, and parasitism are all biotic factors that increase in intensity as the density increases. We will discuss these interactions between populations again in the next chapter because they influence community composition and diversity.

Competition can occur when members of the same species attempt to use needed resources (such as light, food, space) that are in limited supply. As a result, not all members of the population can have access to the resource to the degree necessary to ensure survival or reproduction or some other aspect of their life history. As an example, let's consider a woodpecker population in which members have to compete for nesting sites. Each pair of birds requires a tree hole to raise offspring. If there are more holes than breeding pairs, each pair can have a hole in which to lay eggs and rear young birds (Fig. 46.11a). But if there are fewer holes than there are breeding pairs, then each pair must compete to acquire a nesting site (Fig. 46.11b). Pairs that fail to gain access to holes will be unable to contribute new members to the population.

a.

b.

FIGURE 46.11 Density-dependent effects—competition. The impact of competition on a population is directly proportional to the density of the population. When density is low (**a**), every member of the population has access to the resource. But when the density is high (**b**), there is competition between members of the population to gain access to available resources.

Competition for food as among the reindeer on St. Paul Island also controls population growth. However, resource partitioning among different age groups is a way to reduce competition for food. The life cycle of butterflies includes caterpillars, which require a different food from the adults. The caterpillars graze on leaves, while the adults feed on nectar produced by flowers. Therefore, parents are less apt to compete with their offspring for food.

Predation occurs when one living organism, the predator, eats another, the prey. In the broadest sense, predation can include not only animals such as lions, which kill zebras, but also filter-feeding blue whales, which strain krill from the ocean waters; parasitic ticks, which suck blood from their hosts; and even herbivorous deer, which browse on trees and bushes. The effect of predation on a prey population generally increases as the population grows more dense, because prey are easier to find when hiding places are limited. Consider a field inhabited by a population of mice (Fig. 46.12). Each mouse must have a hole in which to hide to avoid being eaten by a hawk. If there are 102 mice, but only 100 holes, 2 mice will be left out in the open. It might be hard for the hawk to find only 2 mice in the field. If neither mouse is caught, then the predation rate is 0/2 = 0%. However, if there are 200 mice and only 100 holes, there is a greater chance that the hawk will be able to find some of these 100 mice without holes. If half of the exposed mice are caught, the predation rate is 50/100 = 50%. Therefore, increasing the density of the available prey has increased the proportion of the population preyed upon.

Other Considerations

Density-independent and density-dependent factors are extrinsic to the organism. Is it possible that intrinsic factors—those based on the anatomy, physiology, or behavior of the organism—might affect population size and growth rates? Territoriality in birds and dominance hierarchies in mammals are behaviors that affect population size and growth rates. Recruitment, immigration, and emigration are other intrinsic social means by which the population size of more complex organisms is regulated.

Do populations ever show extreme fluctuations in size and growth rates in spite of extrinsic and intrinsic regulating mechanisms? Yes, they do. Outside of any density-dependent and density-independent factors, it could be that populations have an innate instability. Ecologists have developed models that predict complex, erratic changes in even simple systems. For example, a computer model of Dungeness crab populations assumed that adults produce many larvae and then die. Most of the larvae do not survive, and those that do stay close to home. Under these circumstances, the model predicted wild fluctuations in population size with no recurring pattern, which are now termed *chaos*.

Regulation of population size is a complex issue. Density-dependent factors are expected to play a bigger role than density-independent factors. Behavioral characteristics assist also. Even so, population size fluctuates widely in some types of populations.

FIGURE 46.12

Density-dependent effects—predation.

The impact of predation on a population is directly proportional to the density of the population. In a low-density population (**a**), the chances of a predator finding the prey are low, resulting in little predation. But in the higher density population (**b**), there is a greater likelihood of the predator locating potential prey, resulting in a greater predation rate.

a.

b.

46.5 LIFE HISTORY PATTERNS

Populations vary on such particulars as the number of births per reproduction, the age of reproduction, the life span, and the probability of living the entire life span. Such particulars are part of a species' life history. Life histories contain characteristics that can be thought of as trade-offs. Each population is able to capture only so much of the available energy, and how this energy is distributed between its life span (short versus long), reproduction events (few versus many), care of offspring (little versus much), and so forth has evolved over the years. Natural selection shapes the final life history of individual species, and therefore it is not surprising that even related species, such as frogs and toads, may have different life history patterns, if they occupy different types of environments (Fig. 46.13).

The logistic population growth model has been used to suggest that members of some populations are subject to *r*-selection and members of other populations are subject to *K*-selection. In fluctuating and/or unpredictable environments, density-independent factors will keep populations in the lag or exponential phase of population growth. Population size is low relative to *K*, and **r-selection** favors *r*-strategists, which produce large numbers of offspring when young. As a consequence of this pattern of energy allocation, small individuals that mature early and have a short life span are favored. They will tend to produce many relatively small offspring and to forgo parental care in favor of a greater number of offspring. The more offspring, the more likely it is that some of them will survive a population crash. Because of low population densities, density-dependent mechanisms such as predation and intraspecific competition are unlikely to play a major role in regulating population size and growth rates most of the time. Such organisms are often very good dispersers and colonizers of new habitats.

a. Mouth-brooding frog, *Rhinoderma darwinii*

b. Strawberry poison arrow frog, *Phyllobates lugubris*

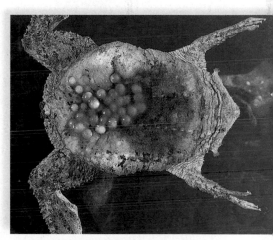

c. Surinam toad, *Pipa pipa*

d. Wood frog, *Rana sylvatica*

e. Midwife toad, *Alyces obstetricans*

FIGURE 46.13 Parental care among frogs and toads.
a. In mouth-brooding frogs of South America, the male carries the larvae in a vocal pouch (brown area), which elongates the full length of his body before the froglets are released. **b.** In poison arrow frogs of Costa Rica, after the eggs hatch, the tadpoles wiggle onto the parent's back (at arrow) and are then carried to water. **c.** Most amazing, in the Surinam toads of South America, males fertilize the eggs during a somersaulting bout because the eggs are placed on the female's back. Each egg develops in a separate pocket, where the tail of the tadpole acts as a placenta to take nourishment from the female's circulatory system. **d.** Wood frogs live mainly in wooded areas, but they breed in temporary ponds arising from spring snow melt. Toads and any frogs that lay their eggs on land exhibit various forms of parental care. **e.** The midwife toad of Europe carries strings of eggs entwined around his hind legs and takes them to water when they are ready to hatch.

Opportunistic Pattern

- Small individuals
- Short life span
- Fast to mature
- Many offspring
- Little or no care of offspring
- Many offspring die before reproducing
- Early reproductive age

Equilibrium Pattern

- Large individuals
- Long life span
- Slow to mature
- Few and large offspring
- Much care of offspring
- Most young survive to reproductive age
- Adapted to stable environment

FIGURE 46.14 Life history strategies.
Are dandelions *r*-strategists with the characteristics noted, and are bears *K*-strategists with the characteristics noted? Most often the distinctions between these two possible life strategies are not as clear-cut as they may seem.

Classic examples of such *opportunistic species* are bacteria, some fungi, many insects, rodents, and annual plants (Fig. 46.14, *left*).

In contrast, we can imagine environments that are relatively stable and/or predictable, where populations tend to be near *K,* with minimal fluctuations in population size. Resources such as food and shelter will be relatively scarce for these individuals, and those that are best able to compete will have the largest number of offspring. **K-selection** favors *K*-strategists, which allocate energy to their own growth and survival and to the growth and survival of their offspring. Therefore, they are fairly large, are slow to mature, and have a fairly long life span. Because these organisms, termed *equilibrium species,* are strong competitors, they can become established and exclude opportunistic species. They are specialists rather than colonizers and tend to become extinct when their normal way of life is destroyed. The best possible examples of *K*-strategists include long-lived plants (saguaro cactus, oaks, cypress, and pine), birds of prey (hawks and eagles), and large mammals (whales, elephants, bears, and humans) (Fig. 46.14, *right*). Another example of a *K*-strategist is the Florida panther, the largest animal in the Florida Everglades. It requires a very large range and produces few offspring that must be cared for. Currently, the Florida panther is unable to compensate for a reduction in its range and is therefore on the verge of extinction.

Nature is actually more complex than these two possible life history patterns suggest. It now appears that our descriptions of *r*-strategist and *K*-strategist populations are at the ends of a continuum, and most populations lie somewhere in between these two extremes. For example, recall that plants have a two-generation life cycle, the sporophyte and the gametophyte. Ferns, which could be classified as *r*-strategists, distribute many spores and leave the gametophyte to fend for itself, but gymnosperms (e.g., pine trees) and angiosperms (e.g., oak trees), which could be classified as *K*-strategists, retain and protect the gametophyte. They produce seeds that contain the next sporophyte generation plus stored food. The added investment is significant, but these plants still release copious numbers of seeds.

A cod is a rather large fish weighing up to 12 kg and measuring nearly 2 m in length—but the cod releases gametes in vast numbers, the zygotes form in the sea, and the parents make no further investment in developing offspring. Of the 6–7 million eggs released by a single female cod, only a few will become adult fish. Cod are considered *r*-strategists.

Differences in the environment result in different selection pressures and a range of life history patterns.

46.6 HUMAN POPULATION GROWTH

The human population has gone through a period of rapid exponential growth. The position of 2002 on the growth curve in Figure 46.15c shows that growth is still quite substantial despite a slight deceleration in growth. Approximately 216,000 people are added to the world population each day. The number of people added annually to the world population peaked at about 87 million around 1990, and currently it is a little over 79 million. That is roughly the population of Germany, the Philippines, or Vietnam.

The potential for future population growth can be appreciated by considering the doubling time. The **doubling time**—the length of time it takes for the population size to double—is now estimated to be 53 years. Such an increase in population size would put extreme demands on our ability to produce and distribute resources. In 53 years, the world would need double the amount of food, jobs, water, energy, and so on just to maintain the present standard of living.

Many people are gravely concerned that the amount of time needed to add each additional billion persons to the world population has become shorter and shorter. The first billion didn't occur until 1800; the second billion was attained in 1930; the third billion in 1960; and today there are over 6 billion. Only when the number of young women entering the reproductive years is the same as those leaving these years behind can there be zero population growth, when the birthrate equals the death rate, and population size remains steady. The world's population may level off at 8, 10.5, or 14.2 billion, depending on the speed with which the growth rate declines.

More-Developed Versus Less-Developed Countries

The countries of the world can be divided into two groups. In the **more-developed countries (MDCs),** typified by countries in North America and Europe, Japan, and Australia, population growth is low, and the people enjoy a good standard of living (Fig. 46.15a). In the **less-developed countries (LDCs),** such as countries in Latin America, Africa, and Asia, population growth is expanding rapidly, and the majority of people live in poverty (Fig. 46.15b). (Sometimes the term *third-world countries* is used to mean the less-developed countries. This term was introduced by those who thought of the United States and Europe as the first world and the former USSR as the second world.)

The MDCs doubled their populations between 1850 and 1950. This was largely due to a decline in the death rate, the development of modern medicine, and improved socioeconomic conditions. The decline in the death rate was followed shortly thereafter by a decline in the birthrate, so that populations in the MDCs experienced only modest growth between 1950 and 1975. This sequence of events (i.e., decreased death rate followed by decreased birthrate) is termed a **demographic transition.**

Yearly growth of the MDCs as a whole has now stabilized at about 0.1%. The populations of several of the MDCs, including Germany, Greece, Italy, Hungary, and Sweden, are not growing or are actually decreasing in size. In contrast, there is no leveling off and no end in sight to U.S. population growth (see the Science Focus on pages 854–55). Although yearly growth of the U.S. population is only 0.6%, many people immigrate to the United States each year. In

a.

b.

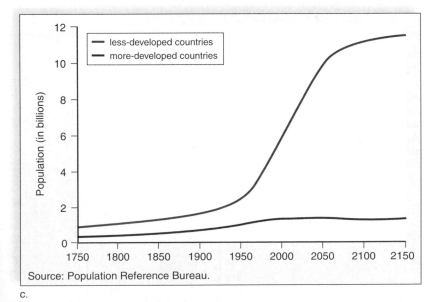

c.

FIGURE 46.15 World population growth.
People in the (**a**) more-developed countries have a high standard of living and will contribute the least to world population growth, while people in the (**b**) less-developed countries have a low standard of living and will contribute the most to world population growth. **c.** The graph shows world population in the past, with estimates to 2150.

addition, an unusually large number of babies were born between 1946 and 1964 (called a baby boom). Therefore, a large number of women are still of reproductive age.

Although the death rate began to decline steeply in the LDCs following World War II with the importation of modern medicine from the MDCs, the birthrate remained high. The yearly growth of the LDCs peaked at 2.5% between 1960 and 1965. Since that time, a demographic transition has occurred: The decline in the death rate slowed, and the birthrate fell. The yearly growth rate is now 1.6%. Still, because of past exponential growth, the population of the LDCs may explode from 5 billion today to 8 billion in 2050. Some of this 3 billion increase will occur in Africa, but most will occur in Asia, India, and Latin America due to many deaths from AIDS in the African countries. Today, 82% of the world's population already lives in Africa, Latin America, and Asia. Ways to greatly reduce the expected increase have been suggested:

1. Establish and/or strengthen family planning programs. A decline in growth is seen in countries with good family planning programs supported by community leaders. Currently, 25% of women in sub-Saharan Africa say they would like to delay or stop childbearing, yet they are not practicing birth control; likewise, 15% of women in Asia and Latin America have an unmet need for birth control.

2. Use social progress to reduce the desire for large families. Some couples in the LDCs still desire as many as four to six children. But providing education, raising the status of women, and reducing child mortality are desirable social improvements that could cause them to think differently.

3. Delay the onset of childbearing. A delay in the onset of childbearing and wider spacing of births could cause a temporary decline in the birthrate and reduce the present reproductive rate.

Age Distributions

The age structure diagrams of MDCs and LDCs in Figure 46.16 divide the population into three age groups: prereproductive, reproductive, and postreproductive. The LDCs are experiencing a population momentum because they have more women entering the reproductive years than older women leaving them.

Laypeople are sometimes under the impression that if each couple has two children, **zero population growth** (no increase in population size) will take place immediately. However, **replacement reproduction,** as it is called, will still cause most countries today to continue growing due to the age structure of the population. If there are more young women entering the reproductive years than there are older women leaving them, replacement reproduction will still result in population growth.

Many MDCs have a stable age structure, but most LDCs have a youthful profile—a large proportion of the population younger than 15. For example, in Nigeria, 44% of the population is under age 15. On average, Nigerian women have 5.8 children. As a result of these rates, the population of Nigeria is expected to increase from 130 million presently to 300 million in 2050. The population of the continent of Africa is projected to increase from 840 million to 1.8 billion between 2002 and 2050. This means that the LDC populations will still expand, even after replacement reproduction is attained. The more quickly replacement reproduction is achieved, however, the sooner zero population growth will result.

Currently, the less-developed countries are expanding dramatically because of past exponential growth.

FIGURE 46.16

Age structure diagrams (2002). The diagrams illustrate that (**a**) the MDCs are approaching stabilization, whereas (**b**) the LCDs will expand rapidly due to their age distributions.

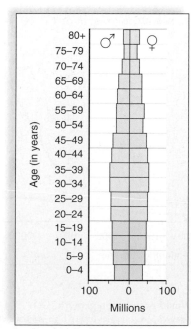

a. More-developed countries (MDCs)

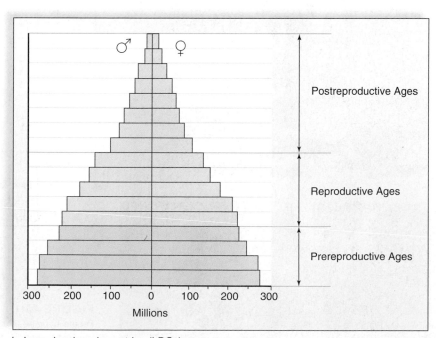

b. Less-developed countries (LDCs)

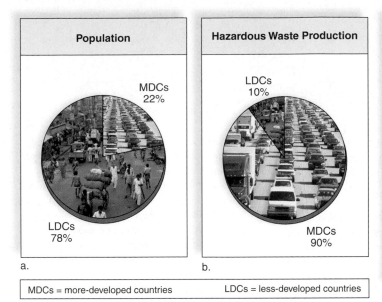

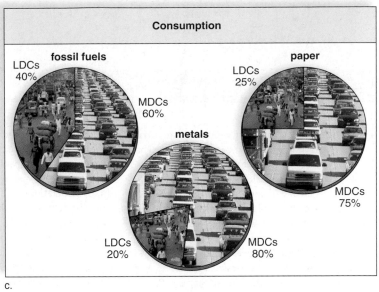

MDCs = more-developed countries	LDCs = less-developed countries

FIGURE 46.17 Environmental impact caused by MDCs and LDCs.
a. The combined population of MDCs is smaller than that of LDCs: The MDCs account for 22% of the world's population, and the LDCs account for 78%. **b.** MDCs produce most of the world's hazardous wastes—90% for MDCs compared with 10% for LDCs. The production of hazardous waste is tied to (**c**), the consumption of fossil fuels, metals, and paper, among other resources. The MDCs consume 60% of fossil fuels compared with 40% for the LDCs, 80% of metals compared with 20% for the LDCs, and 75% of paper compared with 25% for the LDCs.

Population Growth and Environmental Impact

Population growth is putting extreme pressure on each country's social organization, the Earth's resources, and the biosphere. Since the population of the LDCs is still growing at a significant rate, it might seem that their population increase will be the greater cause of future environmental degradation. But this is not necessarily the case because the MDCs consume a much larger proportion of the Earth's resources than do the LDCs. This consumption leads to environmental degradation, which is also of the utmost concern.

Environmental Impact

The environmental impact (E.I.) of a population is measured not only in terms of population size, but also in terms of resource consumption per capita and the pollution that results because of this consumption. In other words:

> E.I. = population size × resource consumption per capita
>   = pollution per unit of resource used

Therefore, there are two possible types of overpopulation: The first is simply due to population growth, and the second is due to increased resource consumption caused by population growth. The first type of overpopulation is more obvious in LDCs, and the second type is more obvious in MDCs because the per capita consumption is so much higher in the MDCs. For example, an average family in the United States, in terms of per capita resource consumption and waste production, is the equivalent of 30 people in India. We need to realize, therefore, that only a limited number of people can be sustained anywhere near the standard of living in the MDCs.

The comparative environmental impact of MDCs and LDCs is shown in Figure 46.17. The MDCs account for only about one-fourth of the world population (Fig. 46.17a). But the MDCs account for 90% of the hazardous waste production (pollution) (Fig. 46.17b). Why should that be? MDCs account for much more pollution than LDCs because they consume much greater amounts of fossil fuel, metal, and paper than do LDCs, as shown in Figure 46.17c.

Results of Environmental Impact

In the chapters that follow, we will examine the results of environmental impact on the biosphere. In Chapter 48, Ecosystems and Human Interferences, you will learn how resource consumption affects the cycling of chemicals in ecosystems, leading to the various pollution problems that have a negative effect on the biosphere's ability to support the world's population.

In Chapter 50, Conservation Biology, you will learn how our increasing environmental impact may cause a mass extinction of wildlife greater than any other since the evolution of life began. This chapter also reviews the direct and indirect value of other species to humans and suggests ways to conserve these species.

science focus

The United States Population

The U.S. Constitution requires that a national census be taken every ten years. The latest census concluded that the U.S. population numbered 281,421,906 persons in April 2000. This is a gain of 32.7 million since the last census in 1990 and the largest increase ever between censuses. It is clear that the United States population, unlike other developed countries, is still increasing. Most industrialized countries now have a growth rate at or lower than 0.1%, but the growth rate for the United States is 0.6%. This positive difference results in the addition of many persons within a short period of time because the population is very large. Also, many persons immigrate to the United States each year. By 2001, the United States population was believed to be around 285 million.

Minorities

Most people in the United States are Caucasian, but that group's percentage of the total population is decreasing because the major minorities—the Hispanic, Black, and Asian populations—are increasing at a faster rate than the Caucasian population. The Hispanic population grew the most over the past ten years, followed by the Asian and then the Black populations.

The American Indian population increased only slightly, from 0.6 to 0.7% of the population. For the first time, Americans were allowed to identify with more than one ethnic group, and about 2% did so. A significant number of these chose American Indian as one of the ethnic groups, and then they could not be counted as belonging to this single minority.

California is now a "minority majority" state, meaning that those Californians who considered themselves Caucasian make up less than one-half of the population. Other statistics regarding minorities are also of interest.

Hispanics

Hispanics who indicate they are from Mexico account for nearly 60% of this particular segment of the total population. Whereas ten years ago Hispanics were 9% of the total population, now they are 12.5%. This is the largest percent increase of all the minorities.

Altogether, the percent of persons identified as Hispanic is now slightly larger than that of those who consider themselves Black. The Hispanic population overtook the Black population as the largest minority by having a higher birthrate and higher immigration rate than do Blacks.

Hispanics are now more dispersed than formerly. Hispanics no longer tend to live in the Southwest and the West and in cities such as Miami, New York, New Jersey, and Chicago. They are dispersing to smaller cities and even to rural areas in the Midwest, South, and Northeast.

Asians

The Asian population includes people from countries such as China, the Philippines, Japan, Korea, India, Pakistan, and so forth. The Asian population increased by two-thirds, but even

FIGURE 46B Percent changes in U.S. population.
Over the past ten years, the Caucasian population dropped in percent of total population, while the Hispanic, Asian, and Black populations increased in percent of total population. Not shown is the American Indian population, which rose modestly from 0.6 to 0.7%. About 2% of the total population identified with more than one ethnicity, according to the 2000 census.

Ethnicity: Caucasian
2000 69.1%
1990 75.6%
  −6.5%

Ethnicity: Hispanic
2000 12.5%
1990 9.0%
  +3.5%

so it accounts for only 3.7% of the year 2000 population. About 60% of Asian Americans are foreign born; therefore, immigration accounts for most of the increase in the Asian population.

The Asian population is also more widely dispersed than before. As many as 36% live in California, but there are also large Asian-American populations living in Georgia, Pennsylvania, Minnesota, and several other states. Many have dispersed to where they can find jobs rather than where they can find others of the same ethnic background.

Blacks

Most Blacks in the United States identify themselves as African Americans. The Black population increased at a faster rate than the Caucasian population, but it had a much slower rate of increase than the Hispanics and also a slower rate than the Asians. Unlike these other minority populations, there is little immigration from other countries, and this accounts for

why the Black population did not increase as fast as other minorities.

African Americans remain the predominant minority group in the South, however. While they make up about 12% of the U.S. population as a whole, they are now about 19% of the population in the South. During the 1990s, many Blacks left the North and decided to return to the South because of the region's booming economy, attractive lifestyle, improved racial climate, and preexisting family ties there.

What's Ahead?

The 2000 census found that nearly 40% of the population under age 18 (i.e., children) belong to a minority group. This means that minorities are sure to increase their percent of the total population during the next decade. More than one-half of the children in California, Arizona, Hawaii, New Mexico, and Texas are members of a minority. By the next census, all these states will likely be minority majorities as California is now.

As mentioned, this was the first time the census allowed individuals to identify with more than one ethnicity. While 2% of the total population said they were multiracial, 4% of the children indicated they were multiracial. Most were a mixture of any two of these three ethnicities: Caucasian, Black, and Asian. It would appear then that we can safely predict that the percent of people identifying themselves as multiracial will increase over the next ten years.

The number of people living in the western and southern states continued to increase faster than those living in the northeastern states. At present, the South, at 36% has the largest share of the population. But Nevada, Arizona, Colorado, Utah, and Idaho—all in the West—were the five fastest growing states over the decade. The other states mentioned grew at a more modest rate. The West (now 22%) will most likely overtake the Midwest (now 23%) before the next census, just as it overtook the Northeast (now 19%) after the 1990 census.

Ethnicity: Asian
2000 3.7%
1990 2.8%
    +0.9%

Ethnicity: Black
2000 12.1%
1990 11.7%
    +0.4%

Connecting the Concepts

Modern ecology began with descriptive studies by nineteenth-century naturalists. In fact, an early definition of the field was "scientific natural history." However, modern ecology has grown to be much more than a simple descriptive field. Ecology is now very much an experimental, predictive science.

Much of the success in the development of ecology as a predictive science has come from studies of populations and the creation of models that examine how populations change over time. The simplest models are based on population growth when there are unlimited resources. This results in exponential growth, a type of population growth that is only rarely seen in nature. Pest species may exhibit expo-

nential growth until they run out of resources. Because so few natural populations exhibit exponential growth, population ecologists realized they must incorporate resource limitation into their models. The simplest models that account for limited resources result in logistic growth. Populations that exhibit logistic growth will cease growth when they reach the environmental carrying capacity.

Many modern ecological studies are concerned with identifying the factors that place limits on population growth and that set the environmental carrying capacity. A combination of careful, descriptive studies, experiments done in nature, and sophisticated models has allowed ecologists to make good

predictions about which factors have the greatest influence on population growth.

The next step in the development of modern ecology has been to try to understand how populations of different species affect each other. This is known as community ecology. Because each population in a community responds to environmental changes in slightly different ways, developing predictive models that explain how communities change has been challenging. However, ecologists are beginning to be able to predict how communities will change through time and to understand what factors influence community properties such as species number, abundance of individuals, and species interactions.

Summary

46.1 SCOPE OF ECOLOGY

Ecology is the study of the interactions of organisms with other organisms and with the physical environment. Ecology encompasses several levels of study: organism, population, community, ecosystem, and finally the biosphere. Ecologists are particularly interested in how interactions affect the distribution and abundance of organisms.

46.2 DEMOGRAPHICS OF POPULATIONS

Demographics includes statistical data about a population. For example, population density is the number of individuals per unit area or volume. Distribution of individuals can be uniform, random, or clumped. A population's distribution is often determined by resources, that is, abiotic factors such as water, temperature, and availability of nutrients.

Population growth is dependent on number of births and number of deaths, immigration, and emigration. The number of births minus the number of deaths results in the rate of natural increase (growth rate). Mortality (deaths per capita) within a population can be recorded in a life table and illustrated by a survivorship curve. The pattern of population growth is reflected in the age distribution of a population, which consists of prereproductive, reproductive, and postreproductive segments. Populations that are growing exponentially have a pyramid-shaped age distribution pattern.

46.3 POPULATION GROWTH MODELS

One model for population growth assumes that the environment offers unlimited resources. In the example given, the members of the population have discrete reproductive events, and therefore the size of next year's population is given by the equation $N_{t+1} = RN_t$. Under these conditions, exponential growth results in a J-shaped curve.

Most environments restrict growth, and exponential growth cannot continue indefinitely. Under these circumstances, an S-shaped, or logistic, growth curve results. The growth of the population is given by

the equation $dN/dt = rN (K−N)/K$ for populations in which individuals have repeated reproductive events. The term $(K−N)/K$ represents the unused portion of the carrying capacity (K). When the population reaches carrying capacity, the population stops growing because environmental conditions oppose biotic potential, the maximum rate of natural increase (growth rate) for a population.

46.4 REGULATION OF POPULATION SIZE

Population growth is limited by density-independent factors (e.g., abiotic factors such as weather) and density-dependent factors (biotic factors such as predation and competition). Are there other means of regulating population growth? For example, territoriality is possibly a means of population regulation. Other populations seem to not be regulated, and their population size fluctuates widely.

46.5 LIFE HISTORY PATTERNS

The logistic growth model has been used to suggest that the environment promotes either r-selection or K-selection. So-called r-selection occurs in unpredictable environments where density-independent factors affect population size. Energy is allocated to producing as many small offspring as possible. Adults remain small and do not invest in parental care of offspring. K-selection occurs in environments that remain relatively stable, where density-dependent factors affect population size. Energy is allocated to survival and repeated reproductive events. The adults are large and invest in parental care of offspring. Actual life histories contain trade-offs between these two patterns.

46.6 HUMAN POPULATION GROWTH

The human population is still expanding, but deceleration has begun. It is unknown when the population size will level off, but it may occur by 2050. Substantial increases are expected in certain LDCs (less-developed countries) of Asia and also Africa. Support for family planning, human development, and delayed childbearing could help prevent the increase.

Reviewing the Chapter

1. What are the various levels of ecological study? 838
2. What largely determines a population's density and distribution and growth rate? 839–40
3. How does the mortality pattern and age distribution affect the growth rate? 840–41, 843
4. What type of growth curve indicates that exponential growth is occurring? What are the environmental conditions for exponential growth? 844
5. What type of growth curve indicates that biotic potential is being opposed by environmental conditions? What environmental conditions are involved? 845
6. What is the carrying capacity of an area? 846
7. Are density-independent or density-dependent factors more likely to regulate population size? Why? 846–47
8. What other types of factors are involved in regulating population size? 848
9. Why would you expect the life histories of natural populations to vary and contain some characteristics that are so-called *r*-selected and some that are so-called *K*-selected? 849–50
10. What type of growth curve presently describes the growth of the human population? In what types of countries is most of this growth occurring, and how might it be curtailed? 851–52

Testing Yourself

Choose the best answer for each question.

1. Which of these levels of ecological study involves an interaction between abiotic and biotic components?
   a. organisms
   b. populations
   c. communities
   d. ecosystem
   e. All of these are correct.

2. When phosphorus is made available to an aquatic community, the algal populations suddenly bloom. This indicates that phosphorus is
   a. a density-dependent regulating factor.
   b. a reproductive factor.
   c. a limiting factor.
   d. an *r*-selection factor.
   e. All of these are correct.

3. A J-shaped growth curve can be associated with
   a. exponential growth.
   b. biotic potential.
   c. unlimited resources.
   d. rapid population growth.
   e. All of these are correct.

4. An S-shaped growth curve
   a. occurs when resources become limited.
   b. includes an exponential growth phase.
   c. occurs in natural populations but not in laboratory ones.
   d. is subject to a sharp decline.
   e. All of these are correct.

5. If a population has a type I survivorship curve (most live the entire life span), which of these would you also expect?
   a. a single reproductive event per adult
   b. overlapping generations
   c. reproduction occurring near the end of the life span
   d. a very low birthrate
   e. None of these are correct.

6. A pyramid-shaped age distribution means that
   a. the prereproductive group is the largest group.
   b. the population will grow for some time in the future.
   c. the country is more likely an LDC than an MDC.
   d. fewer women are leaving the reproductive years than entering them.
   e. All of these are correct.

7. Which of these is a density-independent regulating factor?
   a. competition
   b. predation
   c. weather
   d. resource availability
   e. the average age when childbearing begins

8. Fluctuations in population growth can correlate to changes in
   a. predation.
   b. weather.
   c. resource availability.
   d. parasitism.
   e. All of these are correct.

9. A species that has repeated reproductive events, lives a long time, but suffers a crash due to the weather is exemplifying
   a. *r*-selection.
   b. *K*-selection.
   c. a mixture of both *r*-selection and *K*-selection.
   d. density-dependent and density-independent regulation.
   e. a pyramid-shaped age structure diagram.

10. In which pair is the first included in the second?
    a. habitat—population
    b. population—community
    c. community—ecosystem
    d. ecosystem—biosphere
    e. All of these except a are correct.

11. How does population density differ from population distribution?
    a. Actually, population density is the same as population distribution.
    b. The greater the population density, the more likely that the population distribution will be clumped.
    c. Population density has nothing to do with population distribution.
    d. The less dense the population, the more likely that the population distribution will be equally spaced.
    e. Both b and c are correct.

12. Which one of these has nothing to do with a population's growth curve?
    a. exponential growth
    b. biotic potential
    c. environmental conditions
    d. rate of natural increase
    e. All of these pertain to a population's growth curve.

13. When a population is undergoing logistic growth, environmental conditions determine
    a. whether to expect exponential growth.
    b. the carrying capacity.
    c. the length of the lag phase.
    d. whether the population is subject to density-independent regulation.
    e. All of these are correct.

14. Which of these statements is correct?
    a. A life table is based on whether a population has cohorts or not.
    b. The life table supplies the information for constructing the survivorship curve.
    c. A life table, like an ideal weight table, supplies information that can keep members of a population healthy.
    d. A life table tells which ideal survivorship curve is most appropriate to a particular population.
    e. Both b and d are correct.

15. Because of the postwar baby boom, the U.S. population
    a. can never level off.
    b. presently has a pyramid-type age distribution.
    c. is subject to regulation by density-dependent factors.
    d. is more conservation-minded than before.
    e. None of these is correct.

Thinking Scientifically

1. In the winter moth life cycle, parasites were found to be a less important cause of mortality than cold winter weather and predators. Give an evolutionary explanation for the inefficiency of parasites in controlling population size.

2. Many people enjoy backyard bird feeders and feel that they are helping preserve birds by providing them with food. Many more people unintentionally feed animals by leaving garbage cans and recycling bins where animals have access to the contents. This selects for what animal behaviors (and thus species with that behavior)? Which are selected against? How can feeding animals decrease the diversity of local wildlife?

Bioethical Issue: Population Control

The answer to how to curb the expected increase in the world's population lies in discovering how to curb the rapid population growth of the less-developed countries. In these countries, population experts have discovered what they call the "virtuous cycle." Family planning leads to healthier women, and healthier women have healthier children, and the cycle continues. Women no longer need to have many babies for a few to survive. More education is also helpful because better-educated people are more interested in postponing childbearing and promoting women's rights. Women who have equal rights with men tend to have fewer children.

"There isn't any place where women have had the choice that they haven't chosen to have fewer children," says Beverly Winikoff at the Population Council in New York City. "Governments don't need to resort to force." Bangladesh is a case in point. Bangladesh is one of the densest and poorest countries in the world. In 1990, the birthrate was 4.9 children per woman, and now it is 3.3. This achievement was due in part to the Dhaka-based Grameen Bank, which loans small amounts of money mostly to destitute women to start a business. The

bank discovered that when women start making decisions about their lives, they also start making decisions about the size of their families. Family planning within Grameen families is twice as common as the national average; in fact, those women who get a loan promise to keep their families small! Also helpful has been the network of village clinics that counsel women who want to use contraceptives. The expression "contraceptives are the best contraceptives" refers to the fact that you don't have to wait for social changes to get people to use contraceptives—the two feed back on each other.

Recently, some of the less-developed countries, faced with economic crises, have cut back on their family planning programs, and the more-developed countries have not taken up the slack. Indeed, some foreign donors have also cut back on aid—the United States by one-third. Are you in favor of foreign aid to help countries develop family planning programs? Why or why not?

Understanding the Terms

age structure diagram   843
biosphere   838
biotic potential   840
carrying capacity   846
cohort   840
community   838
continuous breeding   843
demographic transition   851
demography   839
density-dependent factor   847
density-independent factor   847
discrete breeding   843
doubling time   851
ecology   838
ecosystem   838
exponential growth   844
habitat   838

K-selection   850
less-developed country
   (LDC)   851
limiting factor   839
logistic growth   845
more-developed country
   (MDC)   851
population   838
population density   839
population distribution   839
rate of natural increase (r)   840
replacement
   reproduction   852
resource   839
r-selection   849
survivorship   840
zero population growth   852

Match the terms to these definitions:

a. _____ Due to industrialization, a decline in the birthrate following a reduction in the death rate so that the population growth rate is lowered.

b. _____ Group of organisms of the same species occupying a certain area and sharing a common gene pool.

c. _____ Growth, particularly of a population, in which the increase occurs in the same manner as compound interest.

d. _____ Largest number of organisms of a particular species that can be maintained indefinitely in a given environment.

e. _____ Maximum population growth rate under ideal conditions.

ARIS, the *Biology* Website

ARIS, the website for *Biology*, provides a wealth of information organized and integrated by chapter. You will find practice quizzes, interactive activities, labeling exercises, flashcards, and much more that will complement your learning and understanding of general biology.

www.mhhe.com/maderbiology9

47

COMMUNITY ECOLOGY

oyotes and badgers both prey on prairie dogs, and sometimes they improve their chances of catching them by working together. When coyotes are around, prairie dogs tend to stay in their burrows where they are safe. But badgers can enter a prarie dog burrow and flush them out to waiting coyotes. For the coyote, cooperation with badgers prevents wasting time and energy stalking prairie dogs aboveground. For badgers, they are sure to catch a prairie dog if coyotes are waiting outside the burrow. The relationship is so beneficial that coyotes have been known to perform various antics to encourage badgers to move on to a new and different hunting ground. The behavior of coyotes and badgers illustrates how two different species living in the same community can develop a cooperative relationship.

Within a community, a number of relationships and interactions occur between different species. One species may prey on or parasitize another species or may even assist another species. Often, interactions can be complex, as when coyotes sometimes kill young badgers. This chapter examines specific interactions and how they determine the distribution and abundance of species in a community. We will see that the structure of communities is subject to changes over time. Although an intermediate amount of disturbance seems to increase diversity, extreme disturbances can be damaging to the structure of a community.

Coyotes, *Canis latrans*, and badgers, *Taxidea taxus*, interact within a prairie community.

47.1 CONCEPT OF THE COMMUNITY

Populations do not occur as single entities; they are part of a community. A **community** is an assemblage of populations interacting with one another within the same environment. Communities come in different sizes, and it is sometimes difficult to decide where one community ends and another begins. A fallen log can be considered a community because the various populations living on and within a fallen log, such as plants, fungi, worms, and insects, interact with one another and form a community. The fungi break down the log and provide food for the earthworms and insects living in and on the log. Those insects may feed on one another, too. If birds flying throughout the entire forest feed on the insects and worms living in and on the log, then they are also part of the larger forest community.

squirrel
moose
snowshoe hare
bear
red fox
wolf

a.

kinkajou
monkey
anteater
jaguar
tapir
bat
sloth

b.

FIGURE 47.1 Community structure.
Communities differ in their composition, as witnessed by their predominant plants and animals. The diversity of communities is described by the richness of species and their relative abundance.
a. A coniferous forest. Some mammals found here are listed to the *right*. **b.** A tropical rain forest. Some mammals found here are listed to the *left*.
c. Species richness increases with decreasing latitude.

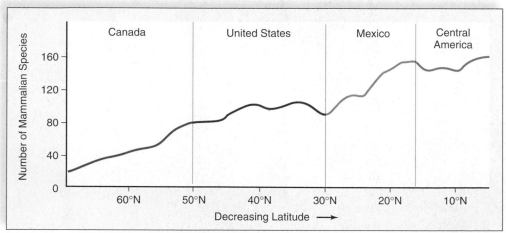

c.

This chapter examines the various types of community interactions and their importance to the structure of a community. Such interactions illustrate some of the most important selection pressures impinging on individuals. They also help us develop an understanding of how biodiversity can be preserved.

Community Composition and Diversity

The composition and the diversity of a community are two fundamental community characteristics that are examined when comparing various communities. The *composition* of a community is a thorough listing of the various species in a particular community. Just glancing at the photographs in Figure 47.1 makes it apparent that a coniferous forest has a composition different from a tropical rain forest. Pictorially, it is evident that the narrow-leaved evergreen trees are prominent in a coniferous forest, and broad-leaved evergreen trees are numerous in a tropical rain forest. Mammals also differ between the two communities, as those listed demonstrate.

The *diversity* of a community goes beyond composition because it includes not only a listing of species but also the abundance of each species. To take an extreme example: A forest in West Virginia has, among other species, 76 yellow poplar trees but only one American elm. If we were simply walking through this forest, we might miss seeing the American elm. If, instead, the forest had 36 poplar trees and 41 American elms, the forest would seem more diverse to us, and indeed it would be more diverse. The greater the diversity, the greater the number and the more even the distribution of species.

Models Regarding Composition and Diversity

What causes the populations within a community to assemble in the same place at the same time? The graphical information in Figure 47.1c illustrates that the number of species in a community increases as we move from the northern latitudes to the equator. We know that at the equator the weather is warmer and there is more precipitation. Could it be that community composition is controlled by abiotic factors? A species' range is based on its tolerance for such abiotic factors in the environment as temperature, light, water availability, salinity, and so forth. To determine a species' tolerance range, we plot the species' ability to survive and reproduce under a particular environmental condition. The result is a bell-shaped curve that shows the species' range for that abiotic factor (Fig. 47.2a). It is possible that species may assemble because their tolerance ranges simply overlap (Fig. 47.2b).

This view, advanced by H. L. Gleason in the early 1900s, is called the *individualistic model* because it suggests that each population in a community is there because its own particular abiotic requirements are met by a particular habitat. The individualistic model predicts that species will have independent distributions, and that the boundaries between communities will not be distinct from one another.

For many years, the majority of ecologists supported the *interactive model* of community structure suggested by F. E.

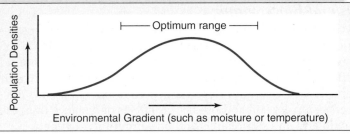

a. One species

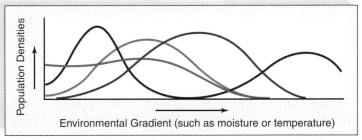

b. Several species

FIGURE 47.2 Species richness of communities.
According to the individualistic model, (**a**) each species is distributed along environmental gradients according to its own tolerance for abiotic factors, and (**b**) a community is an assemblage of species that happen to occupy the same area because of similar tolerances.

Clements, also in the early 1900s. Clements saw a community as the highest level of organization—that is, proceeding from cell to tissue, to organism, to populations, and finally, to a community. Just as the parts of an organism are dependent on one another, so a species is dependent on biotic interactions such as its food source. Further, like an organism, a community remains stable because of homeostatic mechanisms. This model predicts that the same species will recur in communities whose boundaries are distinct from one another.

The data collected by modern ecologists lend support to the individualistic model. For example, F. H. Talbot and colleagues built artificial reefs of a constant size out of cement building blocks and put them in a warm tropical lagoon a few meters apart from one another. Altogether, 42 species of fish colonized the artificial reefs, but the similarity of species between the reefs was only 32%. Moreover, Talbot observed a very high turnover from month to month. From one month to the next, 20–40% of species had been replaced by other species. Therefore, species composition seemed to depend on chance migrations. However, we know that certain animals are more likely to live where they find their usual food source. In Australia, koala bears feed on eucalyptus leaves and are found only in regions where these plants are found. Plants, in turn, are found where temperature, moisture, and soil are most favorable to them. Most likely, community structure is dependent on both abiotic and biotic factors.

Community composition is probably dependent on both abiotic gradients (e.g., climate, inorganic nutrients) and biotic interactions (e.g., organic food source).

Island Biogeography

Would you expect larger coral reefs to have greater species richness (number of species) than smaller coral reefs? The area (space) occupied by a community can have a profound effect on its diversity. American ecologists Robert MacArthur and E. O. Wilson developed a general model of *island biogeography* to explain and predict the effects of distance from the mainland and size of an island on community diversity.

Imagine two new islands that as yet contain no species at all. One of these islands is near the mainland, and one is far from the mainland. Which island will receive more immigrants from the mainland? Most likely, the near one because it's easier for immigrants to get there (Fig. 47.3a).

Similarly, imagine two islands that differ in size. Which island will be able to support a greater number of species? The large one, because a greater amount of resources can support more populations (Fig. 47.3b). MacArthur and Wilson studied the diversity on many island chains, including the West Indies, and discovered that species richness does correlate positively with island size. Conservationists note that the model of island biogeography they developed applies to their work because conserved land is like an island surrounded by farms, towns, and cities. The model suggests that the larger the conserved area, the better the chance of preserving more species.

Is it possible to increase the amount of space without using more area? The *spatial heterogeneity model* holds that if the environment has *patches,* the greater the number of habitats, the greater the diversity. As gardeners, we are urged to create patches in our yards if we wish to attract more butterflies and birds! Stratification is one way to introduce patchiness. Just as a high-rise apartment building allows more human families to live in an area, so stratification within a community provides more and different types of living space for different species. Stratification of its canopy (treetops) is another reason a tropical rain forest has many more species than a coniferous forest.

The fact that space is limited suggests that there must be an end point beyond which community richness cannot increase. Figure 47.3c suggests there will be an equilibrium point for four types of islands (i.e., near and large, near and small, far and large, far and small). Notice that the equilibrium point is highest for a large island that is near the mainland. An equilibrium is reached when the rate of species immigration matches the rate of species extinction. The equilibrium could be dynamic (new species keep on arriving, and new extinctions keep on occurring), or the composition of the community could remain steady unless disturbed. This fundamental concept is used throughout ecology.

> The model of island biogeography suggests that the size of an island and its distance from a population source affect species diversity. When immigration and extinction rates are equal, an equilibrium in species diversity develops.

47.2 STRUCTURE OF THE COMMUNITY

In the living world, competition for resources and predator-prey, parasite-host, and other types of interactions integrate species into a system of dynamically interacting populations (Table 47.1). Competition for limited resources between two species has a negative effect on the abundance of both species. In predation, one animal, known as the predator, eats another called the prey, and in parasitism one individual obtains nutrients from another, called the host. Predation and parasitism are expected to increase the abundance of the predator or parasite at the expense of the abundance of the prey or host. In commensalism, one species is benefited, while the other is not harmed. As an example, commensalism often occurs when one species provides a home or transportation for another. In mutualism, two species help one another, and both species benefit. In Table 47.1, a plus sign (+) is used whenever the fitness of the individual in the interaction is expected to increase, and a minus sign (−) is used whenever the fitness of the individual is expected to decrease. These effects are manifested by a change in the abundance of the species as indicated by the table.

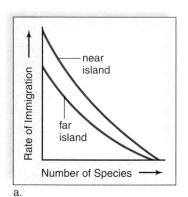

a.

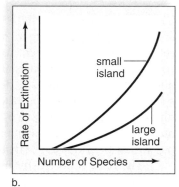

b.

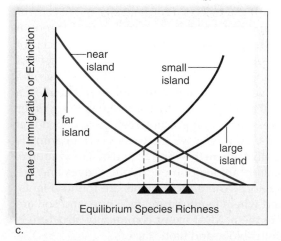

c.

FIGURE 47.3 Model of island biogeography.
a. An island near the mainland will have a higher immigration rate than an island far from the mainland. **b.** A small island will have a higher extinction rate than a large island. **c.** The balance between immigration and extinction for four possible types of islands. A large island that is near the mainland will have a higher equilibrium species richness than the other types of islands.

TABLE 47.1

Species Interactions

Interaction	Expected Outcome
Competition (− −)	Abundance of both species decreases.
Predation (+ −)	Abundance of predator increases, and abundance of prey decreases.
Symbiotic Relationships	
Parasitism (+ −)	Abundance of parasite increases, and abundance of host decreases.
Commensalism (+ 0)	Abundance of one species increases, and the other is not affected.
Mutualism (+ +)	Abundance of both species increases.

Habitat and Ecological Niche

Each species occupies a particular position in the community, both in a spatial sense (where it lives) and in a functional sense (what role it plays). A particular place where an organism lives and reproduces is its **habitat.** The habitat of an organism might be the forest floor, a swift stream, or the ocean's edge. The **ecological niche** of an organism is the role it plays in its community, including its habitat and its interactions with other organisms. The niche includes the resources an organism uses to meet its energy, nutrient, and survival demands. For a dragonfly larva, home is a pond or lake where it eats other insects. The pond must contain vegetation where the dragonfly larva can hide from its predators, such as fish and birds. On the other hand, the water must be clear enough for it to see its prey and warm enough for it to be in active pursuit. Since it is difficult to study the total niche

of an organism, some observations focus on a certain aspect of an organism's niche, as with the birds featured in Figure 47.4.

Because an organism's niche is affected by both abiotic factors (such as climate and habitat) and biotic factors (such as competitors, parasites, and predators in the habitat), ecologists distinguish between the fundamental and realized niches. An organism's *fundamental niche* comprises all conditions under which the organism can potentially survive and reproduce; the *realized niche* is the set of conditions under which it exists in nature. Provided that there is no great competition pressure from other species, an organism can occupy its fundamental niche. However, competition forces an organism to occupy its realized niche. The realized niche may be much smaller than the fundamental niche.

In nature, **generalist species,** such as raccoons, roaches, and humans, have a broad range of niches. These organisms have a diversified diet, tolerate a wide range of environmental conditions, and can live in a variety of places. **Specialist species,** such as pandas, spotted owls, and freshwater dolphins, have a narrow range of niches. These organisms have a limited diet, tolerate only small changes in environmental conditions, and live in a specific habitat. When environmental conditions are apt to change, generalist species have a survival advantage. Specialist species have a survival advantage in stable environments.

A habitat is where an organism lives, and an ecological niche is the role an organism plays in its community, including its habitat and its interactions with other organisms.

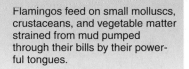
Flamingos feed on small molluscs, crustaceans, and vegetable matter strained from mud pumped through their bills by their powerful tongues.

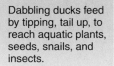
Dabbling ducks feed by tipping, tail up, to reach aquatic plants, seeds, snails, and insects.

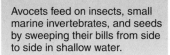
Avocets feed on insects, small marine invertebrates, and seeds by sweeping their bills from side to side in shallow water.

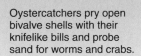
Oystercatchers pry open bivalve shells with their knifelike bills and probe sand for worms and crabs.

Plovers dart around on beaches and grasslands hunting for insects and small invertebrates.

FIGURE 47.4 Feeding niches for wading birds.
Flamingos feed in deeper water by filter feeding; dabbling ducks feed in shallower areas by upending; avocets feed by sifting. Oystercatchers and plovers are restricted to the shallows by their shorter legs.

Competition Between Populations

Interspecific competition occurs when members of different species try to use a resource (such as light, space, or nutrients) that is in limited supply. If the resource is not in limited supply, there is no competition. In the 1920s, Lotka and Volterra developed mathematical formulas that predicted competition can lead to the predominance of one species and the virtual elimination of the other. Later, this was demonstrated in the laboratory by the Russian ecologist G. F. Gause, who grew two species of *Paramecium* in one test tube containing a fixed amount of bacterial food. Although each population survived when grown separately, only one survived when they were grown together (Fig. 47.5). The successful *Paramecium* population had a higher biotic potential than the unsuccessful population. After observing the outcome of other similar experiments in the laboratory, ecologists formulated the **competitive exclusion principle,** which states that no two species can indefinitely occupy the same niche at the same time.

What does it take to have different ecological niches so that extinction of one species is avoided? In another laboratory experiment, Gause found that two species of *Paramecium* did continue to occupy the same tube when one species fed on bacteria at the bottom of the tube and the other fed on bacteria suspended in solution. **Resource partitioning** decreased competition between the two species leading to increased niche specialization. An example of resource partitioning involves owl and hawk populations. Owls and hawks feed on similar prey (small rodents), but owls are nocturnal hunters and hawks are diurnal hunters. What could have been one niche became two more specialized niches because of a divergence of behavior.

It is possible to observe the process of niche specialization in nature. When three species of ground finches of the Galápagos Islands occur on separate islands, their beaks tend to be the same intermediate size, enabling each to feed on a wider range of seeds (Fig. 47.6). Where they co-occur, selection has favored divergence in beak size because the size of the beak affects the kinds of seeds that can be eaten. In other words, competition has led to resource partitioning and therefore niche specialization. The tendency for characteristics to be more divergent when populations belong to the same community than when they are isolated is termed **character displacement.** Character displacement is often used as evidence that competition and resource partitioning have taken place.

Niche specialization can be subtle. Five different species of warblers that occur in North American forests are all about the same size, and all feed on budworms, a type of caterpillar found on spruce trees. In a very famous study, ecologist Robert MacArthur recorded the length of time each of five species of warblers spent in different regions of spruce canopies to determine where each species did most of its feeding (Fig. 47.7). He discovered that each type of bird primarily used different parts of the tree canopy and in that way had a more specialized niche.

As another example, consider that swallows, swifts, and martins all eat flying insects and parachuting spiders. These birds even frequently fly in mixed flocks. But each type of bird has a different nesting site and migrates at a slightly different time of year. Therefore, they are not competing for the same food source when they are feeding their young.

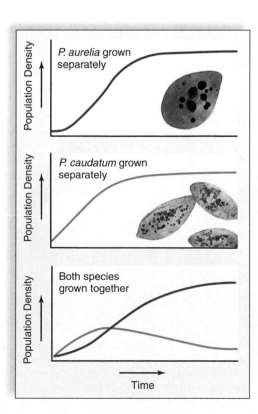

FIGURE 47.5
Competition between two laboratory populations of *Paramecium*.
When grown alone in pure culture, *Paramecium caudatum* and *Paramecium aurelia* exhibit sigmoidal growth. When the two species are grown together in mixed culture, *P. aurelia* is the better competitor, and *P. caudatum* dies out. Both attempted to exploit the same resources, which led to competitive exclusion.
Source: Data from G. F. Gause, *The Struggle for Existence,* 1934, Williams & Wilkins Company, Baltimore, MD.

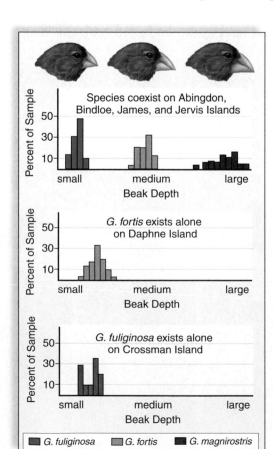

FIGURE 47.6
Character displacement in finches on the Galápagos Islands.
When *Geospiza fuliginosa, G. fortis,* and *G. magnirostris* are on the same island, their beak sizes are appropriate to eating small-, medium-, and large-sized seeds. When *G. fortis* and *G. fuliginosa* are on separate islands, their beaks have the same intermediate size, which allows them to eat seeds that vary in size.

In all these cases of niche specialization, we have merely assumed that what we observe today is due to competition in the past. Some ecologists are fond of saying that in doing so we have invoked the "ghosts of competition past." Are there any instances in which competition has actually been observed? Joseph Connell has studied the distribution of barnacles on the Scottish coast, where a small barnacle (*Chthamalus stellatus*) lives on the high part of the intertidal zone, and a large barnacle (*Balanus balanoides*) lives on the lower part (Fig. 47.8). Free-swimming larvae of both species attach themselves to rocks at any point in the intertidal zone, where they develop into the sessile adult forms. In the lower zone, the large *Balanus* barnacles seem to either force the smaller *Chthamalus* individuals off the rocks or grow over them. To test his observation, Connell removed the larger barnacles and found that the smaller barnacles grew equally well on all parts of the rock. The entire intertidal zone is the fundamental niche for *Chthamalus*, but competition for space is restricting the range of *Chthamalus* on the rocks. *Chthamalus* is more resistant to drying out than is *Balanus*; therefore, it has an advantage that permits it to grow in the upper intertidal zone. The upper intertidal zone becomes the realized niche for *Chthamalus*.

Interspecific competition may result in resource partitioning leading to niche specialization. When similar species seem to be occupying the same ecological niche, it is usually possible to find differences that indicate niche specialization has occurred.

FIGURE 47.7 Niche specialization among five species of coexisting warblers.

The diagrams represent spruce trees. The time each species spent in various portions of the trees was determined; each species spent more than half its time in the blue regions.

Cape May warbler

Black-throated green warbler

Bay-breasted warbler

Blackburnian warbler

Yellow-rumped warbler

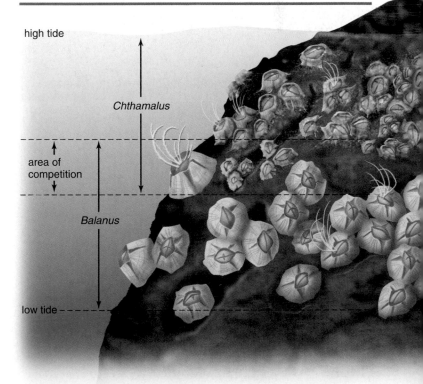

high tide

Chthamalus

area of competition

Balanus

low tide

FIGURE 47.8 Competition between two species of barnacles.

Competition prevents two species of barnacles from occupying as much of the intertidal zone as possible. Both exist in the area of competition between *Chthamalus* and *Balanus*. Above this area only *Chthamalus* survives, and below it only *Balanus* survives.

Predator-Prey Interactions

Predation occurs when one living organism, called the **predator,** feeds on another, called the **prey.** In its broadest sense, predaceous consumers include not only animals like lions, which kill zebras, but also filter-feeding blue whales, which strain krill from ocean waters; parasitic ticks, which suck blood from their victims, and *parasitoids,* which are wasps that lay their eggs inside the body of a host. The resulting larvae feed on the host, sometimes causing death. In the broadest sense, predation also includes herbivorous deer, which browse on trees and shrubs.

Predator-Prey Population Dynamics

Do predators reduce the population density of prey? In a classic experiment, G. F. Gause illustrated that predators do reduce the population density of their prey. When Gause reared the ciliated protozoans *Paramecium caudatum* (prey) and *Didinium nasutum* (predator) together in a culture medium, *Didinium* ate all the *Paramecium* and then died of starvation (Fig. 47.9). In nature, we can find a similar exam-

ple. When a gardener brought prickly-pear cactus to Australia from South America, the cactus spread out of control until millions of acres were covered with nothing but cacti. The cacti were brought under control when a moth from South America, whose caterpillar feeds only on the cactus, was introduced. The caterpillar was a voracius predator on the cactus, efficiently reducing the cactus population. Now both cactus and moth are found at greatly reduced densities in Australia.

This raises an interesting point: The population density of the predator can be affected by the prevalence of the prey. In other words, the predator-prey relationship is actually a two-way street. In that context, consider that at first the biotic potential (maximum reproductive rate) of the prickly-pear cactus was maximized, but environmental resistance (factors that oppose biotic potential) increased after the moth was introduced. And the biotic potential of the moth was maximized when it was first introduced, but the carrying capacity decreased after its food supply was diminished.

In nature, the presence of predators can decrease prey densities, and vice versa.

Instead of remaining in a steady state, predator and prey populations cycle in a predictable pattern:

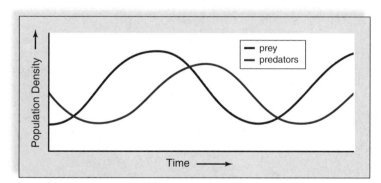

Notice that the predator population is smaller than the prey population, and that it lags behind the prey population. What might cause a cycling like this? We can certainly observe that as the prey population increases, the predator population also increases, most likely because more food has become available. At this point, at least two possibilities account for the reduction in population densities that follow: (1) Perhaps the biotic potential (reproductive rate) of the predator is so great that its increased numbers overconsume the prey, and then as the prey population declines, so does the predator population; or (2) perhaps the biotic potential of the predator is unable to keep pace with the prey, and the prey population overshoots the carrying capacity and suffers a crash. Now the predator population follows suit because of a lack of food. In either case (1 or 2), the result will be a series of peaks and valleys, with the predator population lagging slightly behind the prey.

a. *Didinium* is attacking *Paramecium*　　SEM 600×

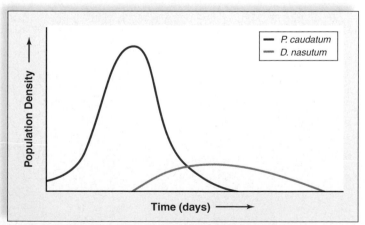

b. Population density of *Paramecium* and *Didinium* over time

FIGURE 47.9 Predator-prey interaction between *Paramecium caudatum* and *Didinium nasutum*.

a. Scanning electron micrograph of *Didinium* preying on *Paramecium*. **b.** In an oat medium, *Didinium* ate all the *Paramecium* and then died out.

Source: Data from G. F. Gause, *The Struggle for Existence,* 1934, Williams & Wilkins Company, Baltimore, MD.

A famous example of predator-prey cycles occurs between the snowshoe hare and the Canadian lynx, a type of small predatory cat (Fig. 47.10). The snowshoe hare is a common herbivore in the coniferous forests of North America, where it feeds on terminal twigs of various shrubs and small trees. The Canadian lynx feeds on snowshoe hares but also on ruffed grouse and spruce grouse, two types of birds. Investigators at first assumed that the lynx brings about a decline in the hare population and that this accounts for the cycling seen in Figure 47.10. But others noted that the decline in snowshoe hare abundance was accompanied by low growth and reproductive rates, which could be signs of a food shortage.

To test whether a limited food supply for the predator (lynx) or the prey (hare) was causing the observed cycling between the two populations, three experiments were done:

1. To do away with both possibilities, a hare population was given a constant supply of food, and predators were excluded. Under these circumstances, the cycling of the hare population ceased.

2. To test the predator effect alone, a hare population was given a constant supply of food, but predators were not excluded. Under these circumstances, the cycling of both populations continued.

3. To test the food effect alone, predators were excluded, but no food was added to the environment of the hare population. Under these circumstances, the cycling of the hare population continued.

These results suggest that a hare-food cycle and a predator-hare cycle have combined to produce the cycling observed in Figure 47.10. It is interesting to note that the densities of the grouse populations also cycle, perhaps because the lynx switches to this food source when the hare population declines. Predators and prey do not normally exist as simple, two-species systems, and therefore abundance patterns should be viewed with the complete community in mind.

Interactions between predator and prey and between prey and its own food source can produce complex cycles.

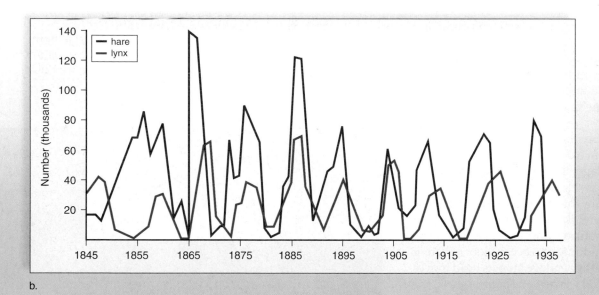

b.

FIGURE 47.10 Predator-prey interaction between a lynx and a snowshoe hare. **a.** A Canadian lynx, *Felis canadensis*, is a solitary predator. A long, strong forelimb with sharp claws grabs its main prey, the snowshoe hare, *Lepus americanus*. The lynx lives in northern forests. Its brownish gray coat blends in well against tree trunks, and its long legs enable it to walk through deep snow. **b.** The number of pelts received yearly by the Hudson Bay Company for almost 100 years shows a pattern of ten-year cycles in population densities. The snowshoe hare population reaches a peak abundance before that of the lynx by a year or more.

a.

Prey Defenses and Other Interactions

Prey defenses are mechanisms that thwart the possibility of being eaten by a predator. Prey species have evolved a variety of mechanisms that enable them to avoid predators, including heightened senses, speed, protective armor, protective spines or thorns, tails and appendages that break off, and chemical defenses.

Plants have defenses that discourage herbivores. The sharp spines of a cactus, pointed leaves of holly, irritating prickles of raspberry, and the tough, leathery leaves of oak serve to deter herbivores. Many plant species discourage herbivores by producing irritating or poisonous chemicals. Over 10,000 defensive chemical compounds have been identified from plants, including caffeine, cocaine, nicotine, peyote, and strychnine. Even the unique taste of jalapeño pepper is a chemical defense. Many of the chemical defenses taste bad to a predator, and others interfere with the normal metabolism of the predator. Several chemical defenses act as hormone analogues that interfere with the development of insect larvae. Some insects can still inhabit plants that produce toxins because the insects have evolved either detoxifying enzymes or a method of holding the toxin within their bodies so that it is not harmful to them.

Animals, like plants, exhibit a number of physical and chemical defenses. Bees, wasps, spiders, and scorpions use venom to defend themselves as well as kill their prey. In addition, venoms have evolved in sea urchins, fishes, salamanders, and reptiles. Some animals, such as skunks and spittle bugs, produce foul-smelling or irritating chemicals to discourage predation.

Among animals, there are number of ways that a predator deceives its prey and prey avoids capture by predators. One common strategy is **camouflage**, or the ability to blend into a background. Animals that are camouflaged possess **cryptic coloration,** which blends them into their surroundings. The anglerfish puts camouflage to work both as a prey defense and a predator mechanism. The mottled skin, grotesque jaw fringe, and flat body allow it to blend into the ocean floor where it waits, while a fleshy, wormlike appendage atop its head moves slowly back and forth (Fig. 47.11). When a fish draws near, the anglerfish opens its mouth and sucks in its prey. This anglerfish ranges from Newfoundland to North Carolina and appears in seafood stores and menus as "monkfish."

Many examples of protective camouflage are known: Walking sticks look like twigs; katydids look like sprouting green leaves; some caterpillars resemble bird droppings; and some insects and moths blend into the bark of trees. Normally, pepper moths are a light color. But in industrial areas where the tree trunks have turned dark due to pollution, the pepper moth populations have mostly dark-colored members. This change in population statistics is often used as an example of natural selection, with predatory birds being the selective agent. The birds capture the more visible light-colored moths on dark tree trunks, and the dark-colored ones remain to contribute a greater number of offspring to the next generation.

Many animals use warning coloration to advertise themselves as potentially dangerous. As a warning to possible predators, poison arrow frogs are brightly colored (Fig. 47.12a). Once stung by a black and yellow insect makes one wary of that color pattern. Many organisms, including caterpillars, moths, and fishes, possess false eyespots that serve to confuse or startle another animal (Fig. 47.12b). Other animals have elaborate anatomical structures that aid in defense. The South American lantern fly has a large false head that resembles that of an alligator (Fig. 47.12c). The porcupine sends out arrowlike quills with barbs that dig into the predator's flesh and penetrate even deeper as the enemy struggles after being struck. In the meantime, the porcupine runs away.

Association with other prey may sometimes help avoid capture. Flocks of birds, schools of fish, and herds of mammals stick together as protection against predators. Baboons that detect predators visually, and antelopes that detect predators by smell, sometimes forage together, gaining double protection against stealthy predators. The gazellelike springboks of southern Africa jump stiff-legged 2–4 m into the air when alarmed. Such a jumble of shapes and motions might confuse an attacking lion, and the herd can escape.

Many prey defenses involve not only structural but also behavioral adaptations. A blowfish has the ability to puff its body up to 300 times its usual volume and extend spikes for protection. Animals that use camouflage usually stay very still so as not to be detected; those that use fright must spring into action at the right time; and those that use other members as shields must stay together.

FIGURE 47.11
Camouflage in the anglerfish.
Top: A wormlike appendage on the head of an anglerfish, *Lophius americanus,* lures possible prey. *Bottom:* Then it opens its cavernous mouth and devours its prey—in this case, a black seabass.

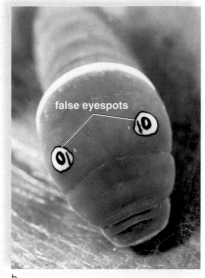

a. b.

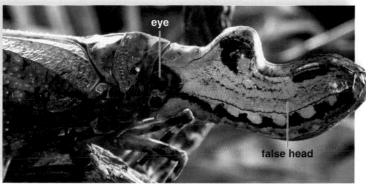

c.

FIGURE 47.12 Antipredator defenses.
a. The skin secretions of poison arrow frogs are so poisonous that they were used by natives to make their arrows instant lethal weapons. The coloration of these frogs warns others to beware. **b.** The caterpillar of the eastern tiger swallowtail butterfly has false eyespots used to confuse a predator. **c.** The South American lantern fly has a large false head that resembles that of an alligator. This may frighten a predator into thinking it is facing a dangerous animal.

Mimicry

Warning coloration is a mechanism used by poisonous animals to prevent attack in the first place. The skin secretions of poison arrow frogs, such as the one shown in Figure 47.12a, are so poisonous that South American Indians used these secretions to turn their arrows into instant lethal weapons. **Mimicry** occurs when one species resembles another that possesses an overt antipredator defense. A mimic that lacks the defense of the organism it resembles is called a Batesian mimic (named for Henry Bates, who discovered it). Once an animal experiences the defense of the model, it remembers the coloration and avoids all animals that look similar. Figure 47.13a–c shows several insects that resemble a yellow jacket wasp but lack the ability to sting as does the wasp. Classical examples of Batesian mimicry include the scarlet kingsnake mimicking the venomous coral snake and the viceroy butterfly mimicking the foul-tasting monarch butterfly.

There are also examples of species that have the same defense and resemble each other. Many stinging insects—

bees, wasps, and hornets—all have the familiar black and yellow color bands. Mimics that share the same protective defense are called Müllerian mimics, after Fritz Müller, who suggested that this, too, is a form of mimicry. Each of these Müllerian mimics reinforces the message that insects with this coloration can sting (Fig. 47.13d, e). In recent years, some researchers have suggested that perhaps the viceroy/monarch mimicry example cited earlier may actually be Müllerian mimicry because some birds apparently do not like the taste of viceroy butterflies either.

Just as with other prey defenses, behavior plays a role in mimicry. Mimicry works better if the mimic acts like the model. For example, beetles that resemble a wasp actively fly from place to place and spend most of their time in the same habitat as the wasp model, and ants that resemble spiders move about like spiders. Their behavior makes them resemble the model to an even greater degree.

Prey escape predation by using camouflage, fright, flocking together, warning coloration, and mimicry.

a. b.

c. d.

e.

FIGURE 47.13 Mimicry among insects with yellow and black stripes.
a. A flower fly, *Chrysotoxum*, (**b**) a longhorn beetle, *Strophiona*, and (**c**) a moth, *Aegeria*, are Batesian mimics because they are incapable of stinging another animal, yet they have the same appearance as the yellow jacket wasp. **d.** A yellow jacket, *Vespula*, and (**e**) a bumblebee, *Bombus*, are Müllerian mimics. They have a similar appearance, and they both use stinging as a defense.

Symbiotic Relationships

Symbiosis, an intimate relationship between two or more species, is commonly found in any community. Table 47.1 (see page 863) lists the three types of symbiosis—parasitism, commensalism, and mutualism—and tells how each type of relationship affects the fitness of the species involved.

Parasitism

Parasitism is similar to predation in that an organism called a **parasite** derives nourishment from another, called the **host** (just as the predator derives nourishment from its prey). Viruses, such as HIV, that reproduce inside human lymphocytes are always parasitic. Parasites occur in all kingdoms of life also. Bacteria (e.g., strep infection), protists (e.g., malaria), fungi (e.g., rusts and smuts), plants (e.g., mistletoe), and animals (e.g., tapeworms and fleas) all contain parasitic members. While small parasites can be endoparasites (pinworms), larger ones are more likely to be ectoparasites (leeches), which remain attached to the exterior of the body by means of specialized organs and appendages. The effects of parasites on the health of the host can range from slightly weakening them to actually killing them over time. When host populations are at a high density, parasites readily spread from one host to the next, causing intense infestations and a subsequent decline in host density. Parasites that do not kill their host can still play a role in reducing the host's population density because an infected host is less fertile and becomes more susceptible to another cause of death.

In addition to nourishment, host organisms also provide their parasites with a place to live and reproduce, as well as a mechanism for dispersing offspring to new hosts. Many parasites have both a primary and secondary host. The secondary host may be a vector that transmits the parasite to the next primary host. Usually both hosts are required to complete the life cycle. The association between parasite and host is so intimate that parasites are usually specific and require only certain species as hosts.

As an example, we will consider the deer tick, called *Ixodes dammini* in the eastern and *I. ricinus* in the western United States. Deer ticks are arthropods that go through a number of stages (egg, larva, nymph, adult). They are so named because adults feed and mate on white-tailed deer in the fall. The female lays her eggs on the ground, and when the eggs hatch in the spring, they become larvae that feed primarily on white-footed mice. If a mouse is infected with the bacterium *Borrelia burgdorferi*, the larvae become infected also. The fed larvae overwinter and molt the next spring to become nymphs that can, by chance, take a blood meal from a human. At this time, the tick may pass the bacterium on to a human, who subsequently develops Lyme disease, characterized by arthritis-like symptoms. The fed nymphs develop into adults, and the cycle begins again (Fig. 47.14).

Parasites are very different from predators. They take nourishment from their host but also use the host as a habitat and a way to transmit themselves to the next host. Many complicated associations have evolved.

1. Adult ticks feed and mate on white-tailed deer.

2. Female adult ticks lay eggs in soil.

3. Tick larvae feed mainly on white-footed mice. If the mouse is infected with *B. burgdorferi* the larvae can become infected.

B. burgdorferi

4. Larvae are dormant through the winter.

5. The following summer, tick nymphs feed on a variety of animals, including humans.

a.

FIGURE 47.14

The life cycle of a deer tick.

a. Adult ticks feed and mate on white-tailed deer, which accounts for the common name of the ticks. Female adult ticks lay their eggs in soil and then die. In the spring and summer, tick larvae feed mainly on white-footed mice, which may be infected with the bacterium *Borrelia burgdorferi*, which is responsible for Lyme disease. Larvae then overwinter. During the next summer, nymphs feed on white-footed mice or other animals, including humans. If bitten by an infected nymph, humans get Lyme disease. **b.** Deer tick before feeding and after feeding. Actual size is shown along with enlarged size.

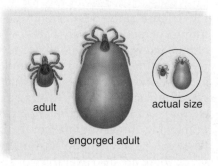

adult

actual size

engorged adult

b.

ecology focus

Interactions and Coevolution

Coevolution is present when two species adapt in response to selective pressure imposed by the other. Symbiosis (close association between two species), which includes parasitism, commensalism, and mutualism, is especially prone to the process of coevolution. Flowers pollinated by animals have features that attract them (see the Science Focus on page 498). As an example of this type of coevolution, a butterfly-pollinated flower is often a composite, containing many individual flowers; the broad expanse provides room for the butterfly to land, and the butterfly has a proboscis that it inserts into each flower in turn. In this case, coevolution is highly beneficial to the flower because the insect will carry pollen to other flowers of the same type.

Coevolution also occurs between predators and prey. For example, a cheetah sprints forward to catch its prey, and this behavior might be selective for those gazelles that are able to run away or jump high in the air (such as a springbok) to avoid capture. The adaptation of the prey may very well put selective pressure on the predator for an adaptation to the prey's defense mechanism. Hence, an "arms race" can develop. The process of coevolution has been studied in the cuckoo, a social parasite that reproduces at the expense of other birds by laying its eggs in their nests. It is a strange sight to see a small bird feeding a cuckoo nestling several times its size. How did this strange happening develop? Investigators discovered that in order to "trick" a host bird, the adult cuckoo has to (1) lay an egg that mimics the host's egg, (2) lay its egg very rapidly (only 10 seconds are required) in the afternoon while the host is away from the nest, and (3) leave most of the eggs in the nest because hosts will desert a nest that has only one egg in it. (The cuckoo chick hatches first and is adapted to removing any other eggs in the nest [Fig.47A].) It seems that the host birds may very well next evolve a way to distinguish the cuckoo from their own young.

Coevolution can take many forms. In the case of *Plasmodium,* a cause of malaria, the sexual portion of the life cycle occurs within mosquitoes (the vector), and the asexual portion occurs in humans. The human immune system uses surface proteins to detect pathogens, and *Plasmodium* has numerous genes for surface proteins. But it is capable of changing its surface

proteins repeatedly, and in this way it stays one step ahead of the host's immune system. A similar capability of HIV has added to the difficulty of producing an AIDS vaccine.

The relationship between parasite and host can even include the ability of parasites to seemingly manipulate the behavior of their hosts in self-serving ways. Ants infected with the lance fluke (but not those uninfected) mysteriously cling to blades of grass with their mouthparts. There, the infested ants are eaten by grazing sheep, and the flukes are transmitted to the next host in their life cycle. Similarly, when snails of the genus *Succinea* are parasitized by worms of the genus *Leucochloridium,* they are eaten by birds. As the worms mature, they invade the snail's eyestalks, making them resemble edible caterpillars. Now the birds eat the snails, and the parasites release their eggs, which complete development inside the urinary tracts of birds.

It used to be thought that as host and parasite coevolved, each would become more tolerant of the other since, if the opposite occurred, the parasite would soon run out of hosts. Parasites could first become com-

mensal, or harmless to the host. Then, given enough time, the parasite and host might even become mutualists. In fact, the evolution of the eukaryotic cell by endosymbiosis is predicated on the supposition that bacteria took up residence inside a larger cell, and then the parasite and cell became mutualists.

However, this argument is too teleological for some; after all, no organism is capable of "looking ahead" at its evolutionary fate. Rather, if an aggressive parasite could transmit more of itself in less time than a benign one, aggressiveness would be favored by natural selection. On the other hand, other factors, such as the life cycle of the host, can determine whether aggressiveness is beneficial or not. For example, a benign parasite of newts will do better than an aggressive one. Why? Because newts take up solitary residence outside ponds in the forest for six years, and parasites have to wait that long before they are likely to meet up with another possible host.

b.

a.

FIGURE 47A Social parasitism.
a. The cuckoo, Cuculus, *is a social parasite of the reed warbler,* Acrocephalus. *A cuckoo chick is heaving the eggs of its host out of the nest.* **b.** *Its own egg (see inset) mimics and is accepted by the host as its own.*

Commensalism

Commensalism is a symbiotic relationship between two species in which one species is benefited and the other is neither benefited nor harmed.

Instances are known in which one species provides a home and/or transportation for the other species. Barnacles that attach themselves to the backs of whales and the shells of horseshoe crabs are provided with both a home and transportation. It is possible though that the movement of the host is impeded by the presence of the attached animals, and therefore some are reluctant to use these as instances of commensalism.

Epiphytes, such as Spanish moss and some species of orchids and ferns, grow in the branches of trees, where they receive light, but they take no nourishment from the trees. Instead, their roots obtain nutrients and water from the air. Clownfishes live within the waving mass of tentacles of sea anemones (Fig. 47.15). Because most fishes avoid the poisonous tentacles of the anemones, clownfishes are protected from predators. Perhaps this relationship borders on mutualism, because the clownfishes actually may attract other fishes on which the anemone can feed.

Commensalism often turns out, on closer examination, to be an instance of either mutualism or parasitism. Cattle egrets are so named because these birds stay near cattle, which flush out their prey—insects and other animals—from vegetation. The relationship becomes mutualistic when egrets remove ectoparasites from the cattle. Remoras are fishes that attach themselves to the bellies of sharks by means of a modified dorsal fin acting as a suction cup. Remoras benefit by getting a free ride and they also feed on a shark's leftovers. However, the shark benefits when remoras remove its ectoparasites. To some, it seems like wasted effort to try to classify symbiotic relationships into the three categories of parasitism, commensalism, and mutualism. The amount of harm or good two species seem to do one another is dependent on what the investigator chooses to measure.

FIGURE 47.15 Clownfish among sea anemone's tentacles.
If the clownfish, *Premnasa biaculeztus*, performs no service for the sea anemone, this association is a case of commensalism. If the clownfish lures other fish to be eaten by the sea anemone, this is a case of mutualism.

Mutualism

Mutualism is a symbiotic relationship in which both members of the association benefit. Mutualistic relationships often help organisms obtain food or avoid predation.

As with parasitism, it is possible to find examples of mutualism in all kingdoms of life. Bacteria that reside in the human intestinal tract are provided with food, but they also provide us with vitamins, molecules we are unable to synthesize for ourselves. Termites would not even be able to digest wood if it were not for the protozoans that inhabit their intestinal tract and digest cellulose, which termites cannot. Mycorrhizae are symbiotic associations between the roots of plants and fungal hyphae. It is now known that the roots of most plants form these relationships. The hyphae increase the solubility of minerals in the soil, improve the uptake of nutrients for the plant, protect the plant's roots against pathogens, and produce plant growth hormones. In return, the fungus obtains carbohydrates from the plant. Some sea anemones make their home on the backs of crabs. The crab uses the stinging tentacles of the sea anemone to gather food and for protection; the sea anemone gets a free ride that allows it greater access to food than other anemones.

Most mutualistic relationships need not be equally beneficial to both species. The mutualistic relationship between plants and their animal pollinators has already been mentioned. We can imagine that the relationship began when herbivores feasted on pollen. The provision of nectar may have spared the pollen and at the same time allowed the animal to become an instrument of pollination. Lichens can grow on rocks because their fungal member conserves water and leaches minerals that are provided to the algal partner, which photosynthesizes and provides organic food for both populations. If this were a true mutualistic relationship, each partner would do poorly when grown alone. The algae seem to do fine without the fungus, which suffers when grown alone. For that reason, it's been suggested that the fungus is parasitic at least to a degree on the algae.

Ants form mutualistic relationships with both plants and insects. In tropical America, the bullhorn acacia tree is adapted to provide a home for ants of the species *Pseudomyrmex ferruginea* (Fig. 47.16). Unlike other acacias, this species has swollen thorns with a hollow interior, where ant larvae can grow and develop. In addition to housing the ants, acacias provide them with food. The ants feed from nectaries at the base of the leaves and eat fat- and protein-containing nodules called Beltian bodies, which are found at the tips of the leaves. The ants constantly protect the plant from herbivores and other plants that might shade it because, unlike other ants, they are active 24 hours a day. Indeed, when the ants on experimental trees were poisoned, the trees died. Other types of trees, such as those in the genus *Croton*, also have nectaries for ants—the very same ants that form a mutualistic relationship with the butterfly *Thisbe irenea*, whose caterpillars

a.

b.

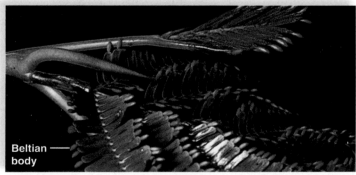

Beltian body

c.

FIGURE 47.16 Mutualism between the bullhorn acacia tree and ants.

The bullhorn acacia tree, *Acacia,* is adapted to provide nourishment for ants, *Pseudomyrmex ferruginea.* **a.** The thorns are hollow, and the ants live inside. **b.** The bases of the leaves have nectaries (openings) where ants can feed. **c.** The tips of leaves of the bullhorn acacia have Beltian bodies that ants harvest for larval food.

FIGURE 47.17 Cleaning symbiosis.

A cleaner wrasse, *Labroides dimidiatus,* in the mouth of a spotted sweetlip, *Plectorhincus chaetodontoides,* is feeding off parasites. Does this association improve the health of the sweetlip, or is the sweetlip being exploited? Investigation is under way.

feed on *Croton* saplings. *Thisbe* caterpillars have nectary organs that offer nourishment to ants, keeping them nearby. The caterpillar also exploits the social behavior of ants by releasing chemicals that normally cause ants to defend the ant colony against predators. The result is that caterpillars are protected while they feast on the trees. In fact, the caterpillars, besides eating the leaves, feed from the ant nectaries on the trees!

Cleaning symbiosis is a symbiotic relationship in which crustaceans, fish, and birds act as cleaners for a variety of vertebrate clients (Fig. 47.17). Large fish in coral reefs line up at cleaning stations and wait their turn to be cleaned by small fish that even enter the mouths of the large fish. Whether cleaning symbiosis is an example of mutualism has been questioned because of the lack of experimental data. If clients respond to tactile stimuli by remaining immobile while cleaners pick at them, then cleaners may be exploiting this response by feeding on host tissues as well as on ectoparasites. Still, it could be that, while stimulation may be the immediate cause of client behavior, cleaning could ultimately lead to net gains in client fitness.

Symbiotic relationships do occur between species, but it may be too simplistic to divide them into parasitic, commensalistic, and mutualistic relationships.

FIGURE 47.18

Secondary succession.
This example of secondary succession occurred in a former cornfield in New Jersey on the east coast of the United States. **a.** During the first year, only the remains of corn plants are seen. **b.** During the second year, wild grasses have invaded the area. **c.** By the fifth year, the grasses look more mature, and sedges have joined them. **d.** During the tenth year, there are goldenrod plants, shrubs (blackberry), and juniper trees. **e.** After 20 years, the juniper trees are mature and there are also birch and maple trees in addition to the blackberry shrubs.

a.

b.

47.3 COMMUNITY DEVELOPMENT

Each community has a history that can be surveyed over a short time period or even over geological time. We know that the distribution of life has been influenced by dynamic changes occurring during the history of Earth. We have previously discussed how continental drifting contributed to various mass extinctions that have occurred in the past. For example, when the continents joined to form the supercontinent Pangaea, many forms of marine life became extinct. Or when the continents drifted toward the poles, immense glaciers drew water from the oceans and even chilled once-tropical lands. During an ice age, glaciers moved southward, and then in between ice ages, glaciers retreated, changing the environment and allowing life to colonize the land once again. After time, complex communities came into being. Many ecologists, however, try to observe changes as they occur during their own lifetime.

Ecological Succession

Communities are subject to disturbances that can range in severity from a storm blowing down a patch of trees, to a beaver damming a pond, to a volcanic eruption. We know from observation that following these disturbances, we'll see changes in the plant and animal communities over time; often, we'll wind up with the same kind of community with which we started.

Ecological succession is a change involving a series of species replacements in a community following a disturbance. *Primary succession* occurs in areas where there is no soil formation, such as following a volcanic eruption or a glacial retreat. *Secondary succession* begins in areas where soil is present, as when a cultivated field, such as a cornfield in New Jersey, returns to a natural state (Fig. 47.18). Notice that the progression changes from grasses to shrubs to a mixture of shrubs and trees.

The first species to begin secondary succession are called **pioneer species**—that is, plants that are invaders of disturbed areas—and then the area progresses through the series of stages described in Figure 47.19. Again we observe a series that begins with grasses and proceeds from shrub stages to a mixture of shrubs and trees, until finally there are only trees. Ecologists have tried to determine the processes and mechanisms by which the changes described in Figures 47.18 and 47.19 take place—and whether these processes always have the same "end point" of community composition and diversity.

Models About Succession

In 1916, F. E. Clements proposed the *climax-pattern model* of succession and said that succession in a particular area will always lead to the same type of community, which he called a **climax community.** He believed that climate, in particular, determined whether a desert, a type of grassland, or a particular type of forest results. This is the reason, he said, that coniferous forests occur in northern latitudes, deciduous forests in temperate zones, and tropical rain forests in the tropics. Secondarily, he believed that soil conditions might also affect the results. Shallow, dry soil might produce a grassland where a forest is expected, or the rich soil of a riverbank might produce a woodland where a prairie is expected.

Further, Clements believed that each stage facilitated the invasion and replacement by organisms of the next stage. Shrubs can't grow on dunes until dune grass has caused soil to develop. Similarly, in the example given in Figure 47.19, shrubs can't arrive until grasses have made the soil suitable for them. Each successive community prepares the way for the next, so that grass-shrub-forest development occurs in a sequential way. Therefore, this is known as the *facilitation model* of succession.

Aside from this facilitation model, there is also an *inhibition model.* That model predicts that colonists hold onto

c.   d.   e.

their space and inhibit the growth of other plants until the colonists die or are damaged. Still another model, called the *tolerance model*, predicts that different types of plants can colonize an area at the same time. Sheer chance determines which seeds arrive first, and successional stages may simply reflect the length of time it takes species to mature. This alone could account for the herb-shrub-forest development that is often seen (Fig. 47.19). The length of time it takes for trees to develop might give the impression that there is a recognizable series of plant communities, from the simple to the complex. But in reality, the models we have mentioned are not mutually exclusive, and succession is probably a multiple, complex process.

Although it may not have been apparent to early ecologists, we now recognize that the most outstanding characteristic of natural communities is their dynamic nature. Also, it seems obvious to us now that the most complex communities most likely consist of habitat patches that

are at various stages of succession. Each successional stage has its own mix of plants and animals, and if a sample of all stages is present, community diversity is greatest. Further, we do not know if succession continues to certain end points, because the process may not be complete anywhere on the face of the Earth.

Ecological succession that occurs after a disturbance probably involves complex processes, and the end result cannot be foretold.

grass → low shrub → high shrub → shrub-tree → low tree → high tree

FIGURE 47.19 Secondary succession in a forest.

In secondary succession in a large conifer plantation in central New York State, certain species are common to particular stages. However, the process of regrowth shows approximately the same stages as secondary succession from a cornfield (see Fig. 47.18).

47.4 COMMUNITY BIODIVERSITY

Community stability can be recognized in three ways: persistence through time, resistance to change, and recovery once a disturbance has occurred. Persistence occurs when a forest remains just about the same year after year. A forest resists change when its trees are able to regrow their leaves after an insect infestation. A chaparral community shows resilience when it quickly returns to its normal state after a fire.

The Intermediate Disturbance Hypothesis

The *intermediate disturbance hypothesis* states that moderate amounts of disturbances at moderate frequency are required for a high degree of community diversity (Fig. 47.20). Fire, wind, moving water, and severe weather changes are possible abiotic factors that cause a disturbance. For example, fires promote understory plant diversity by preventing longer-lived shrub species from out-competing annuals. A tropical rain forest has a great diversity of types of trees, each with their own specialized ecological niche requirements. This suggests that the environment consists of many small patches, each slightly different from the other. As long as disturbances are small enough to affect one type of patch and not another, overall stability might still be maintained.

If widespread disturbances occur frequently, diversity is expected to be limited. The community will be dominated by species with a rapid growth rate, a short life span, and a strong ability to colonize disturbed areas. (Refer to Figure 46.14, and note that these attributes characterize *r*-strategists.) If disturbances are less widespread and less frequent, other types of species with slower growth rates and a longer life span will also be able to compete. (Refer to Figure 46.14, and note these are the attributes of *K*-strategists.) Under these circumstances, diversity will be the greatest.

It is possible for a disturbance to be so extensive that the community never returns to its original state. Archaeological remains suggest that hundreds of square kilometers were cleared and cultivated by the Maya at Tikal from A.D. 300 to 900. This civilization collapsed for some unknown reason, and then the forest began to regrow. Even today, plant diversity in the community is low, and many of the common tree species are ones known to have been prized by the Maya for wood or for edible fruits or nuts. The animal wildlife is largely restricted to animals that use these food resources. Even after 1,200 years, the site does not have the expected composition for a tropical rain forest in that area.

The intermediate disturbance hypothesis states that moderate amounts of disturbances at moderate frequency are required for increased species richness.

a.

FIGURE 47.20 The intermediate disturbance hypothesis.
a. A disturbance, such as the Yellowstone fire in 1988, increases the diversity of a community by providing patches where annual/perennial plants can grow. **b.** This graph suggests that diversity is greatest when disturbances are intermediate in frequency and size.

Predation, Competition, and Biodiversity

In certain communities, predation by a particular species reduces competition and increases diversity. Robert Paine was among the first to show this possible effect of predation. He removed the starfish *Pisaster* from test areas along the rocky intertidal zone on the west coast of North America but not from control areas. In the control areas, the number of species did not change but in the test areas, the mussel *Mytilus* increased in number and excluded other invertebrates and algae from attachment sites on the rocks. The species richness declined drastically (Fig. 47.21). Today, we call a species such as *Pisaster* a keystone species. **Keystone species** are organisms that play a greater role in maintaining the function and diversity of an ecosystem than would be predicted by their abundance. The sea otter is a keystone species of a kelp forest ecosystem off the coast of California. Kelp forests, created by large brown seaweeds, provide a home for a vast assortment of organisms. The kelp forests occur just off the coast and protect coastline ecosystems from damaging wave action. Among other species, sea otters eat sea urchins, keeping their population size in check. Otherwise, sea urchins feed on the kelp, causing the kelp forest and its associated species to severely decline. Fishermen don't like sea otters because they also prey on abalone, a mollusc prized for its commercial value. They do not realize that without the otters, abalone and many other species would not be around because their natural habitat, a kelp forest, would no longer exist. The elephant is a keystone species in Africa. Elephants feed on shrubs and small trees, causing woodland habitats to become open grassland. This is not beneficial to the elephant,

which needs woody species in its diet, but it is beneficial to other ungulates that graze on grasses.

Island Biogeography and Biodiversity

Recall that MacArthur and Wilson have proposed a general model of island biogeography that states that a large island close to the mainland (the source of dispersing species) will have greater diversity than a small island distant from the mainland (see Fig. 47.3). Of late, humans have created many islands; some of these are not true islands surrounded by water—they are terrestrial patches surrounded by human cities, towns, or farms.

In Panama, Barro Colorado Island (BCI) was created in the 1910s when a river was dammed to form a lake. As predicted by the model of island biogeography, BCI lost species because it was a small island now cut off from the mainland. Among the species that became extinct were the top predators on the island, namely the jaguar, puma, and ocelot. Thereafter, medium-sized terrestrial mammals, such as the coatimundi, increased in number. Because the coatimundi is an avid predator of bird eggs, nestlings, and other small vertebrates, it is not surprising that fewer bird species now inhabit BCI, even though the island is large enough to support them. Therefore, it appears that the loss of certain predators from a community can sometimes result in chain reactions that lead to decreased diversity at various levels of community structure.

Such changes as this can also be expected in isolated terrestrial areas. We are finding that preserves need to be very large in order to conserve K-strategists, especially top predators.

Exotic Species and Biodiversity

The previous examples illustrate that unbridled competition and predation can reduce biodiversity. The introduction of exotic (alien) species into new areas also provides many good examples. African bees were introduced into Brazil to help increase honey production. Instead, they displaced domestic honeybees and started moving northward. These so-called killer bees are now established as far north as Texas and New Mexico. The endemic bird populations of many Pacific islands have been devastated due to the accidental introduction of the brown tree snake, which preys on birds. The red fox was deliberately imported into Australia to prey on the introduced European rabbit, but it has now reduced the populations of native small mammals. The nutria, a South American rodent, was accidently introduced into south Louisiana. It has disturbed the natural population of muskrats and is responsible for many acres of wetland loss.

a.

b.

FIGURE 47.21 Effect of a keystone species.
a. The sea star *Pisaster ochracheous* is eating a mussel, *Mytilus*, its favorite food.
b. Along the west coast, *Pisaster* was removed from experimental areas (tidepools) but not from control areas. Diversity decreased in experimental areas because the mussel population increased and crowded out other invertebrates and algae.

Predation and competition also influence biodiversity. K-strategists, which are sometimes top predators, are maintained best on large preserves. Exotic species outcompete and/or prey on native plants and animals.

CONNECTING THE CONCEPTS

Community ecology is concerned with how populations of different species interact with each other. Historically, some ecologists felt that communities functioned as "super-organisms," with each species a vital and integral part of the community. At the other extreme, other ecologists felt that each population simply occurred where conditions were best for it and that communities were chance aggregations of species. Most likely, the truth lies somewhere in between these two extreme views. We know that a population of any species is distributed where conditions are best for it. However, communities do exhibit emergent properties that result from certain groupings of species co-occurring.

The number of individuals in many populations within ecological communities is determined by negative interactions such as interspecific competition, predation, and parasitism. The sizes of all populations are negatively affected by at least one of those interactions. Positive interactions such as mutualisms are fairly common in nature (especially for plants), but their effect on population size is not well understood yet.

Perhaps one of the most important recent discoveries about communities is that they are highly dynamic. The number of species, kinds of species, and sizes of populations within most communities are constantly changing due to disturbances and climatic variability.

Because the species composition of communities can be so variable, many ecologists study the movement of energy and nutrients through communities. The physical environment has a large influence on energy and nutrient flow, and thus the nonliving world must be incorporated into our studies of communities, as discussed in Chapter 48.

Summary

47.1 CONCEPT OF THE COMMUNITY

A community is an assemblage of populations interacting with one another within the same environment. Communities differ in their composition (species found there) and their diversity (species richness and relative abundance). Abiotic factors, such as latitude and environmental gradients, seem to largely control community composition, but biotic interactions also play a role. As suggested by the model of island biogeography, space limitations affect species diversity.

47.2 STRUCTURE OF THE COMMUNITY

An organism's habitat is where it lives in the community. An ecological niche is defined by the role an organism plays in its community, including its habitat and how it interacts with other species in the community. Competition, predator-prey, parasite-host, commensalistic, and mutualistic relationships help organize populations into an intricate dynamic system.

According to the competitive exclusion principle, no two species can indefinitely occupy the same niche at the same time. When resources are partitioned between two or more species, increased niche specialization occurs. Character displacement is a structural change that gives evidence of resource partitioning and niche specialization. But the difference between species can be more subtle, as when warblers feed at different parts of the tree canopy. Usually we have to assume that we are seeing the results of competition. Barnacles competing on the Scottish coast may be an example of present ongoing competition.

Predator-prey interactions between two species are influenced by the biotic potential of the two populations, the carrying capacity of the environment, and the presence of other food sources in the community. Predation can cause prey populations to decline and remain at relatively low densities, or a cycling of population densities may occur.

Prey defenses take many forms: Camouflage, use of fright, and warning coloration are three possible mechanisms among many others. Batesian mimicry occurs when one species has the warning coloration but lacks the defense. Müllerian mimicry occurs when two species with the same warning coloration have the same defense.

We would expect coevolution to occur within a community. For example, the better the predator becomes at catching prey, the better the prey becomes at escaping the predator. Flowers are adapted to attract specific pollinators, and hosts are adapted to cope with potential parasites.

Like predators, parasites take nourishment from their host. However, the host usually provides a home and a place where parasites reproduce. Parasites escape detection and manipulate their hosts in various ways. Whether they are aggressive (kill their host) or benign probably depends on which results in the highest fitness, considering the life cycles of both organisms in the relationship.

Symbiotic relationships may be too complicated to classify as commensalistic, parasitic, or mutualistic. For example, it is questionable whether certain relationships, such as clownfishes that live among sea anemone tentacles or lichens that contain fungal and algal members, are commensalistic or mutualistic, respectively. *Croton* saplings provide food for ants, but so do the caterpillars that feed on *Croton* saplings! In this case, the possible mutualistic relationship between tree and ant has been undermined by a predator.

47.3 COMMUNITY DEVELOPMENT

Directional patterns of community change are called ecological succession. Ecologists have recognized for some time that communities undergo succession in terms of both a geological timescale and a present-day timescale. There is now much discussion about the cause of succession and whether it results in so-called climax communities or not. In the past, the term *climax community* has implied that a diverse community is stable in its composition.

47.4 COMMUNITY BIODIVERSITY

It is perfectly obvious now that the presence of patches in different stages of succession results in the most diversity. The intermediate disturbance hypothesis states that moderate amounts of disturbances at moderate frequency are required for a high degree of community diversity.

Predation and competition are examples of interactions that affect biodiversity. Predation by keystone predators increases the diversity of a community. If a preserve is so small that top predators are eliminated, diversity will decrease because other predators are not held in check. Exotic species often out-compete and/or prey on native species so that they are reduced in numbers.

Reviewing the Chapter

1. Contrast the individualistic model of community composition with the interactive model.   861
2. What do experiments with artificial reefs and the model of island biogeography tell us about species richness?   861–62
3. Describe the habitat and ecological niche of a particular organism.   863
4. What is the competitive exclusion principle? How does the principle relate to character displacement and niche specialization?   864–65
5. What is the special significance of the barnacle experiment off the coast of Scotland?   865
6. Would you expect all predator and prey interactions to produce similar results with regard to population densities? Why or why not?   866–67
7. What is mimicry, and why does it work as a prey defense?   869
8. Give examples of parasitism, commensalism, and mutualism, and show that it is not always easy to distinguish between these symbiotic relationships.   870–73
9. Describe an example of coevolution between a predator and its prey, between a parasite and a host, and between mutualists.   871
10. What is ecological succession? What is the present controversy surrounding the concept?   874–75
11. What is the intermediate disturbance hypothesis, and how does it relate to community diversity?   876
12. What interactions affect biodiversity? In what way?   876–77
13. What is the danger of introducing species from other parts of the world into a region?   877

Testing Yourself

Choose the best answer for each question.

1. Six species of monkeys are found in a tropical forest. Most likely, they
   a. occupy the same ecological niche.
   b. eat different foods and occupy different ranges.
   c. spend much time fighting each other.
   d. are from different stages of succession.
   e. All of these are correct.

2. Leaf-cutter ants keep fungal gardens. The ants provide food for the fungus but also feed on the fungus. This is an example of
   a. competition.
   b. predation.
   c. commensalism.
   d. parasitism.
   e. mutualism.

3. Clownfishes live among sea anemone tentacles, where they are protected. If the clownfish provides no service to the anemone, this is an example of
   a. competition.
   b. predation.
   c. parasitism.
   d. commensalism.
   e. mutualism.

4. Two species of barnacles vie for space in the intertidal zone. The one that remains is
   a. the better competitor.
   b. better adapted to the area.
   c. the better predator on the other.
   d. the better parasite on the other.
   e. Both a and b are correct.

5. A bullhorn acacia provides a home and nutrients for ants. Which statement is likely?
   a. The plant is under the control of pheromones produced by the ants.
   b. The ants protect the plant.
   c. The plant and the ants compete with one another.
   d. The plant and the ants have coevolved to occupy different ecological niches.
   e. All of these are correct.

6. The ecological niche of an organism
   a. is the same as its habitat.
   b. includes how it competes and acquires food.
   c. is specific to the organism.
   d. is usually occupied by another species.
   e. Both b and c are correct.

7. The frilled lizard of Australia suddenly opened its mouth wide and unfurled folds of skin around its neck. Most likely, this was a way to
   a. conceal itself.
   b. warn that it was noxious to eat.
   c. scare a predator.
   d. scare its prey.
   e. All of these are correct.

8. When one species mimics another species, the mimic sometimes
   a. lacks the defense of the model.
   b. possesses the defense of the model.
   c. is brightly colored.
   d. competes with a mimic.
   e. All of these are correct.

9. Which of these most likely would help account for diversity in a coral reef?
   a. warm temperatures
   b. presence of predation
   c. moderate disturbances
   d. prey defenses
   e. All of these are correct.

10. The presence of patches in an environment may be associated with
    a. spatial and temporal heterogeneity.
    b. a greater degree of diversity.
    c. past disturbances.
    d. community instability.
    e. All of these are correct.

11. The species within a community are
    a. used to compare communities.
    b. present due to their abiotic requirements.
    c. more diverse as the size of the area increases.
    d. present due to their biotic interactions.
    e. All of these are correct.

12. Which of these pairs is incorrectly matched?
   a. interactive model of community structure—abiotic requirements determine the species present
   b. model of island biogeography—size of area determines number of species present
   c. climax-pattern model of succession—colonists inhibit the growth of other plants
   d. intermediate disturbance hypothesis—species richness increases with disturbances
   e. Both a and d are correct.

13. The model of island biogeography is pertinent to
   a. only islands surrounded by water.
   b. explaining community composition and diversity.
   c. explaining the intermediate disturbance hypothesis.
   d. why exotics are such a problem today.
   e. All of these are correct.

14. Competition between species most likely
   a. increases the number of niches.
   b. accounts for a decrease in the number of species today.
   c. is a type of symbiotic relationship.
   d. leads to coevolution.
   e. All of these are correct.

15. Predator-prey interactions
   a. increase the number of niches.
   b. account for the decrease in the number of species today.
   c. are a type of symbiotic relationship.
   d. lead to coevolution.
   e. Both c and d are correct.

16. Which one of these pertains to ecological succession?
   a. facilitation model
   b. model of island biogeography
   c. competition between species
   d. Both b and c are correct.

17. Complex cycles in species abundance can occur due to
   a. competition.
   b. predator-prey interactions.
   c. mutualistic relationships.
   d. commensalism only.
   e. All of these are correct.

18. Resource partitioning pertains to
   a. niche specialization.
   b. character displacement.
   c. increased species diversity.
   d. the development of mutualism.
   e. All but d are correct.

19. Label blanks a and b on this diagram, and tell how it applies to the individualistic concept of community composition.

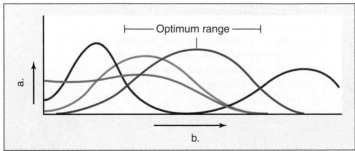

Several species

Thinking Scientifically

1. You suspect that a particular animal's coloration pattern is mimicking that of another, poisonous species. What data would support this hypothesis?

2. For a long time, fire was considered destructive and was suppressed, without regard to the role it might be playing in the environment. Now, fire is recognized as a tool to help maintain healthy ecosystems. However, people living on the edge of lands susceptible to fires are nervous about allowing minimally controlled fires to burn out. Imagine that you are a government ecologist in charge of determining which wild areas would benefit by allowing naturally occurring fires to burn with minimal control and which have no need of fire as a source of ecological disturbance. What criteria would help you make a decision?

Understanding the Terms

camouflage 868	host 870
character displacement 864	interspecific competition 864
climax community 874	keystone species 876
coevolution 871	mimicry 869
commensalism 872	mutualism 872
community 860	parasite 870
community stability 876	parasitism 870
competitive exclusion principle 864	pioneer species 874
cryptic coloration 868	predation 866
ecological niche 863	predator 866
ecological succession 874	prey 866
generalist species 863	resource partitioning 864
habitat 863	specialist species 863
	symbiosis 870

Match the terms to these definitions:

a. _____ Assemblage of populations interacting with one another within the same environment.

b. _____ Place where an organism lives and is able to survive and reproduce.

c. _____ Directional pattern of change in which one community replaces another until a community typical of the area results.

d. _____ Role an organism plays in its community, including its habitat and its interactions with other organisms.

e. _____ Symbiotic relationship in which both species benefit in terms of growth and reproduction.

ARIS, the *Biology* Website

ARIS, the website for *Biology*, provides a wealth of information organized and integrated by chapter. You will find practice quizzes, interactive activities, labeling exercises, flashcards, and much more that will complement your learning and understanding of general biology.

www.mhhe.com/maderbiology9

48

ECOSYSTEMS AND HUMAN INTERFERENCES

*A*n ecosystem is characterized by energy flow and chemical cycling. Both of these begin when algae and green plants capture a small percentage of the sun's energy and use it to transform inorganic chemicals such as carbon dioxide and water into organic compounds. These compounds serve as organic nutrients for plants and also either directly or indirectly for all the other populations in an ecosystem. Eventually, when decomposers break down organic matter, inorganic chemicals are liberated once again, but the energy has dissipated as heat.

Ecosystems are not self-contained; they have inputs and outputs to the other ecosystems of the biosphere. Therefore, some ecologists think of the biosphere as a global ecosystem. Most persons now realize that humans are a part of the biosphere and that our activities affect all of its ecosystems. One activity of great concern is the destruction of forests and woodlands. Approximately 120,000 km² of forests are cleared annually, an amount of forest that equals nearly 20 football fields a minute. Forests are cleared for a number of purposes, including the logging of trees, the building of subdivisions, and the development of agricultural lands. Deforestation results in soil erosion, silting in of waterways, extinction of species, and disruption of natural cycles. The mistreatment of forests and other ecosystems impacts the biosphere and threatens the existence of all living things, including ourselves.

Deforestation due to clear-cutting of trees.

48.1 THE NATURE OF ECOSYSTEMS

Our planet is unique in many ways. Unlike the other planets in our solar system, Earth has water and abundant life. Perhaps our planet should have been called "water" instead of Earth. Water is present in the *hydrosphere* [Gk. *hydrias*, of water; L. *sphaera*, ball], which covers over three-quarters of the Earth's surface. The oceans moderate the temperature of the Earth; as surface temperatures rise, the oceans take up a great deal of heat, and then as temperatures cool, they return heat slowly to the atmosphere. This helps keep the temperature on Earth suitable for life.

The *atmosphere* [Gk. *atmos*, vapor] is concentrated in the lowest 10 km near Earth but extends out at least 1,000 km. Among other gases, the atmosphere contains carbon dioxide, nitrogen, and oxygen, which are each used and released by living things. Carbon dioxide is necessary for photosynthesis. Oxygen is necessary for cellular respiration, and in the upper atmosphere it becomes ozone, a substance that shields Earth from damaging ultraviolet radiation and makes life on land possible.

A rocky substratum called the *lithosphere* [Gk. *lithos*, stone] extends from Earth's surface to about 100 km deep. The weathering of rocks supplies minerals to plants, which take root in weathered rocks and slowly form soil. Besides minerals, soil contains decaying organic material known as humus. The organisms of decay play a vital role in breaking down organic matter, returning inorganic nutrients to plants so that photosynthesis can continue.

The **biosphere** is that part of the atmosphere, hydrosphere, and lithosphere that contains living things. Taking the global view, the entire biosphere is an **ecosystem,** a place where organisms interact among themselves and with the physical and chemical environment. These interactions help maintain ecosystems and in turn the biosphere. Human activities can alter the interactions between organisms and their environments in ways that reduce the abundance and diversity of life that the environments can support. It is important to understand how ecosystems function so that we can repair past damage and predict how human activities might change current conditions.

Usually, ecologists study ecosystems on a smaller scale: a tundra, a desert, or a wooded area are all ecosystems (Fig. 48.1). Regardless of their size, ecosystems are characterized by (1) one-way flow of energy through the biotic community of an ecosystem, and (2) cycling of materials from the abiotic environment through the biotic community and back to the abiotic environment.

FIGURE 48.1 Ecosystems.
The biosphere can be considered one giant ecosystem, or it can be divided up into smaller sections, each considered to be an ecosystem. Aquatic ecosystems include ponds, rivers, or lakes. Terrestrial ecosystems include forests, grasslands, or deserts.

Biotic Components of an Ecosystem

The abiotic components of an ecosystem are the nonliving components. The biotic components are living things that can be categorized according to their food source (Fig. 48.2). Some living things are autotrophs and some are heterotrophs.

Autotrophs

Autotrophs require only inorganic nutrients and an outside energy source to produce organic nutrients for their own use and for all the other members of a community. They are called **producers** because they produce food. Photoautotrophs, often called photosynthetic organisms, produce most of the organic nutrients for the biosphere. Algae of all types possess chlorophyll and carry on photosynthesis in freshwater and marine habitats. Algae make up the phytoplankton, which are photosynthesizing organisms suspended in water. Green plants are the dominant photosynthesizers on land.

Some autotrophic bacteria are chemosynthetic. They obtain energy by oxidizing inorganic compounds such as ammonia, nitrites, and sulfides, and they use this energy to synthesize organic compounds. Chemoautotrophs have been found to support communities in some caves and also at hydrothermal vents along deep-sea oceanic ridges.

Heterotrophs

Heterotrophs need a preformed source of organic nutrients. They are called **consumers** because they consume food. **Herbivores** are animals that graze directly on plants or algae. In terrestrial habitats, insects are small herbivores; antelopes and bison are large herbivores. In aquatic habitats, zooplankton are small herbivores; fishes and manatees are large herbivores. **Carnivores** feed on other animals; birds that feed on insects are carnivores, and so are hawks that feed on birds. **Omnivores** are animals that feed on both plants and animals. Chickens, raccoons, and humans are omnivores. Some animals are scavengers, such as vultures, and jackals, which eat the carcasses of dead animals.

Detritus feeders are organisms that feed on detritus, which is decomposing particles of organic matter. Marine fan worms take detritus from the water, while clams take it from the substratum. Earthworms and some beetles, termites, and ants are all terrestrial detritus feeders. Bacteria and fungi, including mushrooms, are decomposers; they acquire nutrients by breaking down dead organic matter, including animal wastes. **Decomposers** perform a valuable service because they release inorganic substances that are taken up by plants once more. Otherwise, plants would be completely dependent only on physical processes, such as the release of minerals from rocks, to supply them with inorganic nutrients.

The populations in the biotic community of an ecosystem are either autotrophs or heterotrophs. Autotrophs produce organic nutrients, and heterotrophs consume organic nutrients.

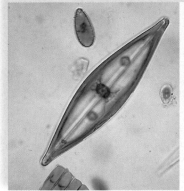 

a. Producers

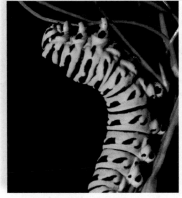 

b. Herbivores

c. Carnivores

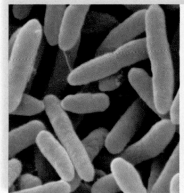 

d. Decomposers

FIGURE 48.2 Biotic components.
a. Diatoms and green plants are photoautrophs. **b.** Caterpillars and rabbits are herbivores. **c.** Spiders and osprey are carnivores. **d.** Bacteria and mushrooms are decomposers.

science focus

Productivity of Ecosystems

Producers provide organic nutrients for an ecosystem when they photosynthesize and produce organic nutrients. One way to compare ecosystems is based on their productivity (Fig. 48A*a*). The gross primary productivity (GPP) of an ecosystem is the rate at which producers convert solar energy into their own biomass. A portion of the GPP is used by producers for respiration (RS) in order to carry on life's activities. The remainder that is available for use by other organisms is called the net primary productivity (NPP). Temperature and moisture, and secondarily the nature of the soil, influence the primary productivity, as does the assemblage of species in an ecosystem. In terrestrial ecosystems, primary productivity is generally lowest in high-latitude tundras and deserts and highest at the equator, where tropical rain forests occur. The high productivity of tropical rain forests provides varied niches and much food for consumers. The number and diversity of species in tropical rain forests is the highest of all the terrestrial ecosystems. Therefore, conservation biologists are interested in preserving as much of this ecosystem as possible.

The primary productivity of aquatic communities is largely dependent on the availability of inorganic nutrients. Estuaries, swamps, and marshes are rich in organic nutrients and in decomposers that convert those organic nutrients into their inorganic chemical components. Estuaries, swamps, and marshes also contain a large number of varied species, particularly in the early stages of their development before they venture forth into the sea. Therefore, all of these coastal regions are in great need of preservation. The open ocean has a productivity somewhere between that of a desert and the tundra because it lacks a concentrated supply of inorganic nutrients. Coral reefs exist near the coasts in warm tropical waters, where currents and waves bring nutrients and where sunlight penetrates to the ocean floor. Coral reefs are areas of remarkable biological abundance, equivalent to that of tropical rain forests.

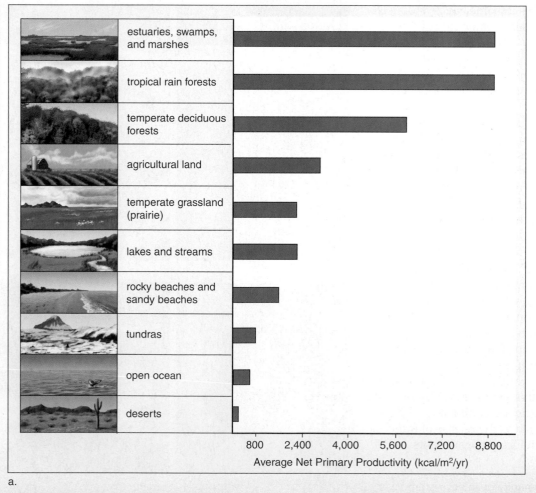

Gross primary productivity (GPP): Total rate of photosynthesis in a specified area, usually expressed as kilocalories produced per square meter per year ($kcal/m^2/yr$).

Autotrophs (mostly photosynthetic plants and algae) absorb the energy of the sun and produce food for all organisms in a community.

Net primary productivity (NPP): The rate at which plants produce usable food, which contains chemical energy (also expressed as kilocalories per square meter per year).

$$NPP = GPP - RS$$

where RS is the energy used by the autotrophs for respiration.

Because plants respire, they use some of the food they produce (GPP) for respiration, and what is left (NPP) is available for growth and storage and can be used as food by heterotrophs.

Usually, between 50% and 90% of GPP remains as NPP.

a.   b.

FIGURE 48A Gross and net primary productivity.
a. *Average net primary productivity (NPP) for several communities. NPP is influenced by such factors as temperature and rainfall. Also important is the availability of sunlight and nutrients and the length of the growing season.* ***b.*** *The mathematical relationship between gross and net primary productivity.*

Energy Flow and Chemical Cycling

A diagram of all the biotic components of an ecosystem illustrates that every ecosystem is characterized by two fundamental phenomena: energy flow and chemical cycling (Fig. 48.3). Energy flow begins when producers absorb solar energy, and chemical cycling begins when producers take in inorganic nutrients from the physical environment. Thereafter, via photosynthesis, producers make organic nutrients (food) directly for themselves and indirectly for the other populations of the ecosystem. Energy flows through an ecosystem via photosynthesis because as organic nutrients pass from one component of the ecosystem to another, such as when an herbivore eats a plant or a carnivore eats an herbivore, a portion is used as an energy source. Eventually, the energy dissipates into the environment as heat. Therefore, the vast majority of ecosystems cannot exist without a continual supply of solar energy.

Only a portion of the organic nutrients made by producers is passed on to consumers because plants use organic molecules to fuel their own cellular respiration. Similarly, only a small percentage of nutrients consumed by lower-level consumers, such as herbivores, is available to higher-level consumers, or carnivores. As Figure 48.4 demonstrates, a certain amount of the food eaten by an herbivore is never digested and is eliminated as feces. Metabolic wastes are excreted as urine. Of the assimilated energy, a large portion

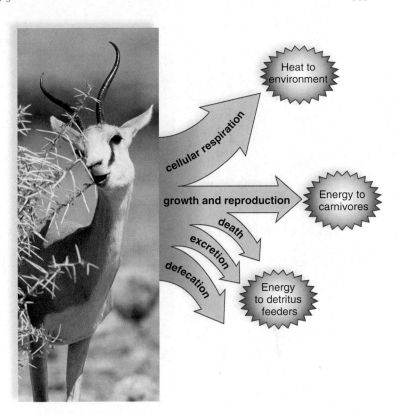

FIGURE 48.4 Energy balances.
Only about 10% of the food energy taken in by an herbivore is passed on to carnivores. A large portion goes to detritus feeders via defecation, excretion, and death, and another large portion is used for cellular respiration.

is used during cellular respiration for the production of ATP and thereafter becomes heat. Only the remaining energy, which is converted into increased body weight or additional offspring, becomes available to carnivores.

The elimination of feces and urine by a heterotroph, and indeed the death of all organisms, does not mean that organic nutrients are lost to an ecosystem. Instead, they represent the organic nutrients made available to decomposers. Decomposers convert the organic nutrients, such as glucose, back into inorganic chemicals, such as carbon dioxide and water, and release them to the soil or atmosphere. Chemicals complete their cycle within an ecosystem when inorganic chemicals are absorbed by the producers from the atmosphere or from soil.

The laws of thermodynamics support the concept that energy flows through an ecosystem. The first law states that energy cannot be created (or destroyed). This explains why ecosystems are dependent on a continual outside source of energy, usually solar energy, which is used by photosynthesizers to produce organic nutrients. The second law states that, with every transformation, some energy is degraded into a less available form such as heat. Because plants carry on cellular respiration, for example, only about 55% of the original energy absorbed by plants is available to an ecosystem.

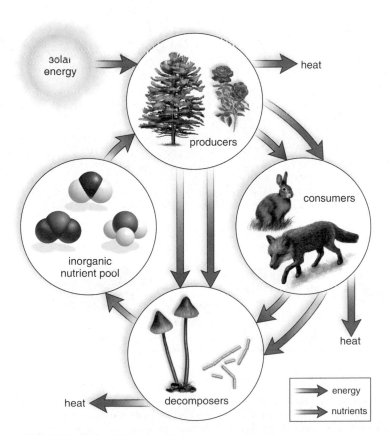

FIGURE 48.3 Nature of an ecosystem.
Chemicals cycle, but energy flows through an ecosystem. As energy transformations repeatedly occur, all the energy derived from the sun eventually dissipates as heat.

Energy flows through the populations of an ecosystem, while chemicals cycle within and between ecosystems.

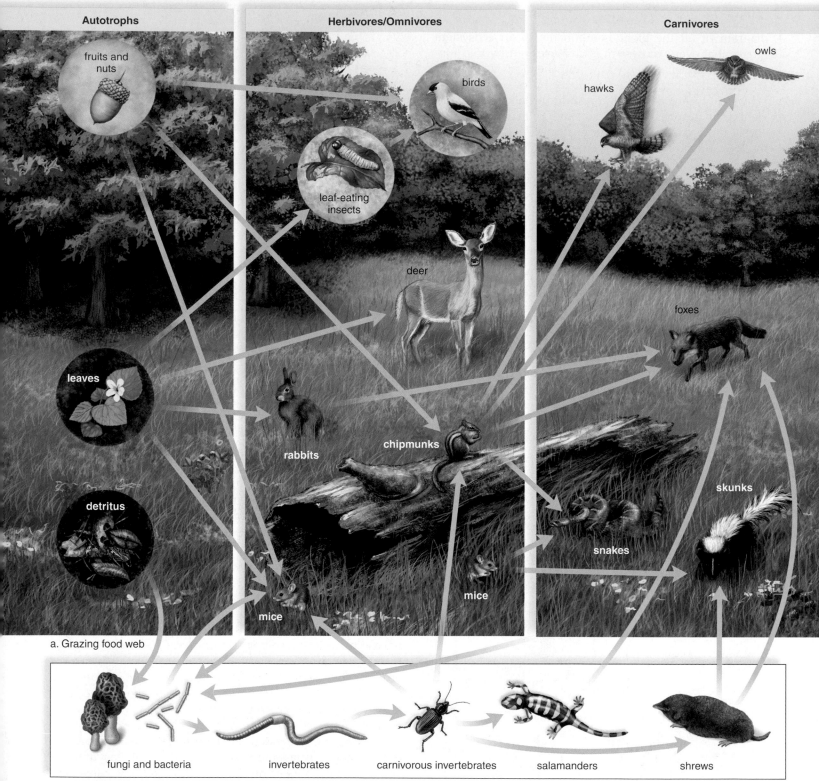

a. Grazing food web

b. Detritus food web

FIGURE 48.5 Grazing and detrital food web.

Food webs are descriptions of who eats whom. **a.** Tan arrows illustrate possible grazing food webs. For example, birds, which feed on nuts, may be eaten by a hawk. Autotrophs such as the tree are producers (first trophic, or feeding, level), the first series of animals are primary consumers (second trophic level), and the next group of animals are secondary consumers (third trophic level). **b.** Green arrows illustrate possible detrital food webs, which begin with detritus—the bacteria and fungi of decay and the remains of dead organisms. A large portion of these remains are from the grazing food web illustrated in (**a**). The organisms in the detrital food web are sometimes fed on by animals in the grazing food web, as when robins feed on earthworms. Thus, the grazing food web and the detrital food web are connected to one another.

48.2 ENERGY FLOW

The principles discussed in the previous section can now be applied to an actual ecosystem—a forest of 132,000 m² in New Hampshire. The various interconnecting paths of energy flow are represented by a **food web,** a diagram that describes **trophic (feeding) relationships.** Figure 48.5a is a **grazing food web** because it begins with a producer, specifically the oak tree depicted. Insects in the form of caterpillars feed on leaves, while mice, rabbits, and deer feed on leaf tissue at or near the ground. Birds, chipmunks, and mice feed on fruits and nuts, but they are in fact omnivores because they also feed on caterpillars. These herbivores and omnivores all provide food for a number of different carnivores.

Figure 48.5b is a **detrital food web,** which begins with detritus. Detritus is food for soil organisms such as earthworms. Earthworms are in turn fed on by carnivorous invertebrates, and they may be fed on by shrews or salamanders. Because the members of a detrital food web may become food for aboveground carnivores, the detrital and grazing food webs are joined.

We naturally tend to think that aboveground plants such as trees are the largest storage form of organic matter and energy, but this is not necessarily the case. In this particular forest, the organic matter lying on the forest floor and mixed into the soil contains over twice as much energy as the leaf matter of living trees. Therefore, more energy in a forest may be funneling through the detrital food web than through the grazing food web.

Trophic Levels

The arrangement of the species in Figure 48.5 suggests that organisms are linked to one another in a straight line, according to feeding relationships, or who eats whom. Diagrams that show a single path of energy flow in an ecosystem are called **food chains.** For example, in the grazing food web, we could find this **grazing food chain:**

leaves ⟶ caterpillars ⟶ birds ⟶ hawks

And in the detrital food web, we could find this **detrital food chain:**

detritus ⟶ earthworms ⟶ salamanders

A **trophic level** is a level of nourishment within a food web or chain. In the grazing food web in Figure 48.5a, going from left to right, the trees are producers (first trophic level), the first series of animals are primary consumers (second trophic level), and the next group of animals are secondary consumers (third trophic level).

Ecological Pyramids

The shortness of food chains can be attributed to the loss of energy between trophic levels. In general, only about 10% of the energy of one trophic level is available to the next trophic level. Therefore, if an herbivore population consumes 1,000 kg of plant material, only about 100 kg is converted to herbivore tissue, 10 kg to first-level carnivores,

and 1 kg to second-level carnivores. The so-called 10% rule of thumb explains why few carnivores can be supported in a food web. The flow of energy with large losses between successive trophic levels is sometimes depicted as an **ecological pyramid** (Fig. 48.6).

Energy losses between trophic levels also result in pyramids based on the number of organisms or the amount of biomass at each trophic level. When constructing such pyramids, problems arise, however. For example, in Figure 48.5, each tree would contain numerous caterpillars; therefore, there would be more herbivores than autotrophs! The explanation, of course, has to do with size. An autotroph can be as tiny as a microscopic alga or as big as a beech tree; similarly, an herbivore can be as small as a caterpillar or as large as an elephant.

Pyramids of biomass eliminate size as a factor because **biomass** is the number of organisms multiplied by the dry weight of the organic matter within one organism. The biomass of the producers is expected to be greater than the biomass of the herbivores, and that of the herbivores is expected to be greater than that of the carnivores. In aquatic ecosystems, such as some lakes and open seas where algae are the only producers, the herbivores may have a greater

FIGURE 48.6 Ecological pyramid.
The biomass, or dry weight (g/m²), for trophic levels in a grazing food web in a bog at Silver Springs, Florida. There is a sharp drop in biomass between the producer level and herbivore level, which is consistent with the common knowledge that the detrital food web plays a significant role in bogs.

biomass than the producers when their measurements are taken because the algae are consumed at a high rate. Such pyramids, which have more herbivores than producers, are called inverted pyramids:

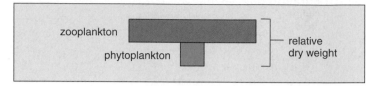

These kinds of problems are making some ecologists hesitant about using pyramids to describe ecological relationships. One more problem is what to do with the decomposers, which are rarely included in pyramids, even though a large portion of energy becomes detritus in many ecosystems.

The flow of energy through the populations explains in large part the organization of an ecosystem as depicted in food webs, food chains, and ecological pyramids.

48.3 GLOBAL BIOGEOCHEMICAL CYCLES

All organisms require a variety of organic and/or inorganic nutrients. For example, carbon dioxide and water are necessary nutrients for photosynthesizers. Nitrogen is a component of all the structural and functional proteins and nucleic acids that sustain living tissues. Phosphorus is essential for ATP and nucleotide production.

Since the pathways by which chemicals circulate through ecosystems involve both living (biotic) and nonliving (geological) components, they are known as **biogeochemical cycles.** For each element, chemical cycling may involve (1) a reservoir—a source normally unavailable to producers, such as fossilized remains, rocks, and deep-sea sediments; (2) an exchange pool—a source from which organisms do generally take chemicals, such as the atmosphere, soil, or water; and (3) the biotic community—through which chemicals move along food chains, perhaps never entering a pool (Fig. 48.7).

With the exception of water, which exists as a gas, a liquid, and a solid, there are two types of biogeochemical cycles. In a gaseous cycle, exemplified by the carbon and nitrogen cycles, the element returns to and is withdrawn from the atmosphere as a gas. In a sedimentary cycle, exemplified by the phosphorus cycle, the element is absorbed from the sediment by plant roots, passed to heterotrophs, and eventually returned to the soil by decomposers, usually in the same general area.

The diagrams on the next several pages make it clear that nutrients flow between ecosystems. In the nitrogen and phosphorus cycles, nutrients run off from a terrestrial to an aquatic ecosystem and in that way enrich the aquatic ecosystem. Decaying organic matter in aquatic ecosystems can be a source of nutrients for intertidal inhabitants such as fiddler crabs. Seabirds feed on fish but deposit guano (droppings) on land,

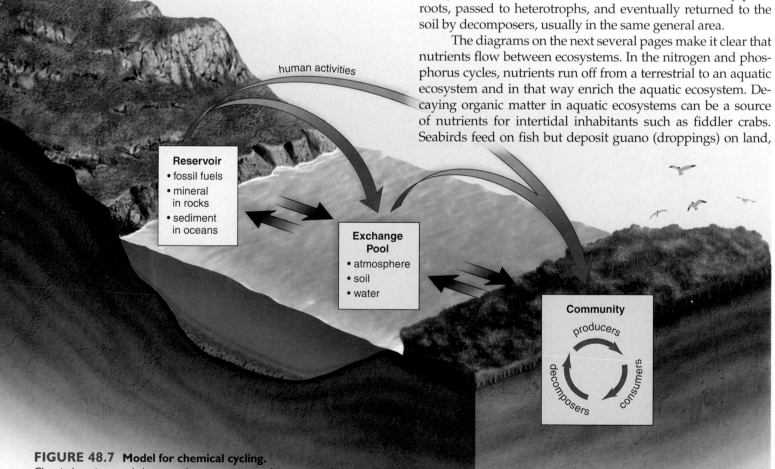

FIGURE 48.7 Model for chemical cycling.
Chemical nutrients cycle between these components of ecosystems. Reservoirs, such as fossil fuels, minerals in rocks, and sediments in oceans, are normally relatively unavailable sources, but exchange pools, such as those in the atmosphere, soil, and water, are available sources of chemicals for the biotic community. When human activities (purple arrows) remove chemicals from reservoirs and pools and make them available to the biotic community, pollution can result.

and in that way phosphorus from the water is deposited on land. It seems that anything put into the environment in one ecosystem could find its way to another ecosystem. As proof, scientists find the soot from urban areas and pesticides from agricultural fields in the snow and animals of the Arctic.

The Water Cycle

The **water (hydrologic) cycle** is described in Figure 48.8. A **transfer rate** is defined as the amount of a substance that moves from one component of the environment to another within a specified period of time. The width of the arrows in Figure 48.8 indicates the transfer rate of water.

During the water cycle, fresh water is distilled from salt water. First, evaporation occurs. During **evaporation,** a liquid, in this case water, changes from a liquid state to a gaseous state. The sun's rays cause fresh water to evaporate from the seawater, and the salts are left behind. Next, condensation occurs. During **condensation,** a gas is changed into a liquid. For example, vaporized fresh water rises into the atmosphere, is stored in clouds, cools, and falls as rain over the oceans and the land.

Water evaporates from land and from plants (evaporation from plants is called transpiration) and also from bodies of fresh water. Because land lies above sea level, gravity eventually returns all fresh water to the sea. In the meantime, water is contained within standing waters (lakes and ponds), flowing water (streams and rivers), and groundwater.

Some of the water from **precipitation** (e.g.. rain, snow, sleet, hail, and fog) sinks, or percolates, into the ground and saturates the Earth to a certain level. The top of the satu-

ration zone is called the groundwater table, or simply, the water table. Because water infiltrates through the soil and rock layers, sometimes groundwater is also located in **aquifers,** rock layers that contain water and release it in appreciable quantities to wells or springs. Aquifers are recharged when rainfall and melted snow percolate into the soil.

Human Activities

In some parts of the United States, especially the arid West and southern Florida, withdrawals from aquifers exceed any possibility of recharge. This is called "groundwater mining." In these locations, the groundwater is dropping, and residents may run out of groundwater, at least for irrigation purposes, within a few short years.

Fresh water, which makes up only about 3% of the world's supply of water, is called a renewable resource because a new supply is always being produced because of the water cycle. But it is possible to run out of fresh water when the available supply is not adequate or is polluted so that it is not usable. This topic is discussed further later in this chapter.

In the water cycle, fresh water evaporates from bodies of water. Precipitation on land enters the ground, surface waters, or aquifers. Water ultimately returns to the ocean—even the quantity that remains in aquifers for some time.

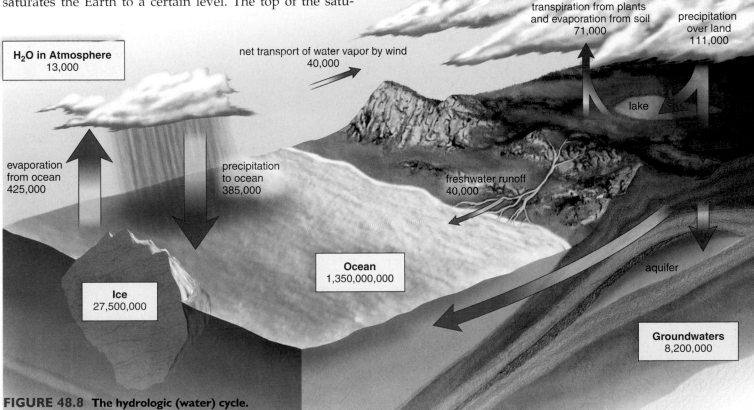

FIGURE 48.8 The hydrologic (water) cycle.
Evaporation from the ocean exceeds precipitation, so there is a net movement of water vapor onto land, where precipitation results in surface water and groundwater that flow back to the sea. On land, transpiration by plants contributes to evaporation. The numbers in this diagram indicate water flow in cubic kilometers per year.

The Carbon Cycle

In the **carbon cycle,** organisms exchange carbon dioxide with the atmosphere (Fig. 48.9). On land, plants take up carbon dioxide from the air, and through photosynthesis they incorporate carbon into organic nutrients that are food for other living things. When organisms (e.g., plants, animals, and decomposers) respire, a portion of this carbon is returned to the atmosphere as carbon dioxide. In aquatic ecosystems, carbon dioxide from the air combines with water to produce bicarbonate ion (HCO_3^-), a source of carbon for photosynthetic protists. When aquatic organisms respire, the carbon dioxide they give off becomes bicarbonate ion.

Living and dead organisms contain organic carbon and serve as one of the reservoirs for the carbon cycle. The world's biotic components, particularly trees, contain 800 billion tons of organic carbon and an additional 1,000–3,000 billion metric tons are estimated to be held in the remains of plants and animals in the soil. Before decomposition can occur, some of these remains are subjected to physical processes that transform them into coal, oil, and natural gas. We call these materials the **fossil fuels.** Most of the fossil fuels were formed during the Carboniferous period, 299–360 million years ago, when an exceptionally large amount of organic matter was buried before decomposing. Another reservoir is the inorganic carbonate

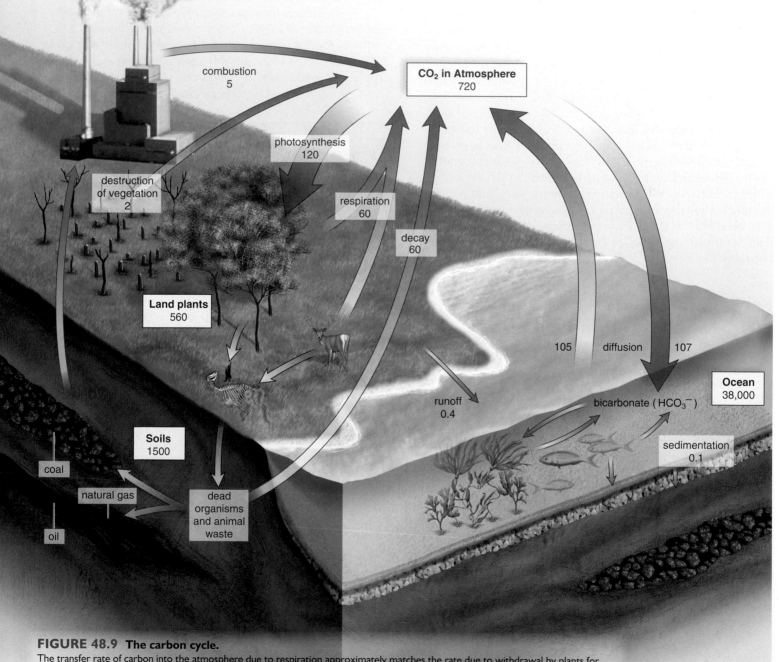

FIGURE 48.9 The carbon cycle.
The transfer rate of carbon into the atmosphere due to respiration approximately matches the rate due to withdrawal by plants for photosynthesis. However, due to the burning of fossil fuels and destruction of vegetation by human activities (purple arrows), more carbon dioxide is added to the atmosphere than is withdrawn. The numbers in this diagram indicate 10^{15} g C/yr.

that accumulates in limestone and calcium carbonate ($CaCO_3$) shells. Many marine organisms have calcium carbonate shells that remain in bottom sediments long after the organisms have died. Geological forces change these sediments into limestone.

Carbon Dioxide and Global Warming

The width of the arrows in Figure 48.9 indicates the transfer rate of carbon dioxide (CO_2). The transfer rates due to photosynthesis and respiration, which includes decay, are just about even. However, more carbon dioxide is now being deposited in the atmosphere than is being removed. In 1850, atmospheric carbon dioxide was about 280 parts per million (ppm), and today it is about 350 ppm. This increase is largely due to the burning of wood and fossil fuels and the destruction of forests to make way for farmland and pasture.

Other gases are also being emitted due to human activities, including nitrous oxide (N_2O) from fertilizers and animal wastes and methane (CH_4) from bacterial decomposition, particularly in the guts of grazing animals, in sediments, and in flooded rice paddies. These gases are known as **greenhouse gases** because, just like the panes of a greenhouse, they allow solar radiation to pass through but hinder the escape of infrared rays (heat) back into space. The greenhouse gases are contributing significantly to an overall rise in the Earth's ambient temperature, a phenomenon called **global warming**, because of their **greenhouse effect.**

Earth's radiation balances are illustrated in Figure 48.10. One thing to be noted in this diagram is that water vapor is a greenhouse gas, so clouds (which are composed of water vapor) also reradiate heat back to Earth. If the Earth's temperature rises, more water will evaporate, forming more clouds and setting up a positive feedback effect that could increase global warming still more.

Today, data collected around the world show a steady rise in the concentration of greenhouse gases. For example, methane, which is a more effective greenhouse gas than carbon dioxide, is increasing by about 1% per year. Such data have been used to generate computer models that predict the environmental temperature may become warmer than ever before. The global climate has already warmed about 0.6°C since the Industrial Revolution. Computer models are unable to consider all possible variables, but the Earth's temperature may rise 1.5–4.5°C by 2100 if greenhouse emissions continue at the current rates.

Global warming will bring about other effects, which computer models attempt to forecast. It is predicted that, as the oceans warm, temperatures in the polar regions will rise to a greater degree than in other regions. If so, glaciers will melt, and sea levels will rise, not only due to this melting but also because water expands as it warms. Water evaporation will increase, and most likely there will be increased rainfall along the coasts and dryer conditions inland. The occurrence of droughts will reduce agricultural yields and also cause trees to die off. Expansion of forests into arctic areas might not offset the loss of forests in the temperate zones. Coastal agricultural lands, such as the deltas of Bangladesh and China, will be inundated, and billions of dollars will have to be spent to keep coastal cities such as New Orleans, New York, Boston, Miami, and Galveston from disappearing into the sea. In addition, many species of flora and fauna will become extinct as ecosystems change and habitat is destroyed.

The atmosphere is an exchange pool for carbon dioxide. Fossil fuel combustion in particular has increased the amount of carbon dioxide in the atmosphere. Global warming is predicted because carbon dioxide and other gases impede the escape of infrared radiation from the surface of the Earth.

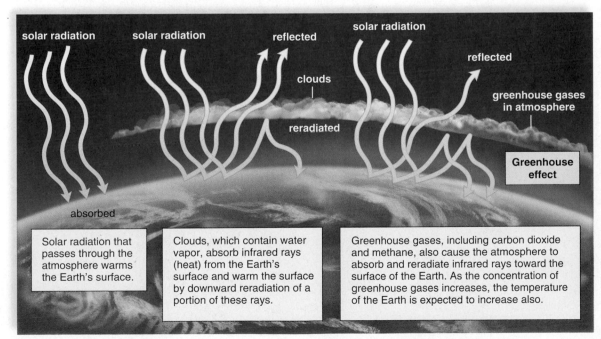

FIGURE 48.10

Earth's radiation balances. The effect of the greenhouse gases *(far right)* is contributing to global warming. A positive feedback cycle is predicted. As the Earth's temperature rises, more water will evaporate, causing clouds to thicken and absorb more solar radiation. Water vapor in clouds and the other greenhouse gases prevent the escape of heat (infrared rays) and it is reradiated back to the Earth, causing even more evaporation, and so forth.

solar radiation   solar radiation   reflected   solar radiation

clouds   reflected

reradiated   greenhouse gases in atmosphere

Greenhouse effect

absorbed

Solar radiation that passes through the atmosphere warms the Earth's surface.

Clouds, which contain water vapor, absorb infrared rays (heat) from the Earth's surface and warm the surface by downward reradiation of a portion of these rays.

Greenhouse gases, including carbon dioxide and methane, also cause the atmosphere to absorb and reradiate infrared rays toward the surface of the Earth. As the concentration of greenhouse gases increases, the temperature of the Earth is expected to increase also.

The Nitrogen Cycle

Nitrogen serves as an important building block of amino acids, proteins, and nucleic acids. It makes up about 78% of the atmosphere by volume, but even so, nitrogen deficiency sometimes limits plant growth. Plants cannot incorporate nitrogen gas into organic compounds, and therefore they depend on various types of bacteria and physical processes to make nitrogen available to them through the **nitrogen cycle** (Fig. 48.11).

Nitrogen fixation occurs when nitrogen (N_2) is converted to a form that plants can use. Some nitrogen-fixing bacteria live in nodules on the roots of legumes (plants of the pea family, such as peas, beans, and alfalfa). They make nitrogen-containing organic compounds available to a host plant. Free-living bacteria in bodies of water and in the soil are able to fix nitrogen gas as ammonium (NH_4^+). Plants can

use NH_4^+ and also nitrate (NO_3^-) from the soil to produce amino acids and nucleic acids.

Nitrification is the production of nitrates. Nitrogen gas (N_2) is converted to nitrate (NO_3^-) in the atmosphere when cosmic radiation, meteor trails, and lightning provide the high energy needed for nitrogen to react with oxygen. Ammonium (NH_4^+) in the soil is converted to nitrate by chemoautotrophic soil bacteria in a two-step process. First, nitrite-producing bacteria convert ammonium to nitrite (NO_2^-), and then nitrate-producing bacteria convert nitrite to nitrate. Notice the subcycle in Figure 48.11 that involves dead organisms and animal wastes, ammonium, nitrites, nitrates, and plants. This subcycle does not necessarily depend on nitrogen gas at all.

Denitrification is the conversion of nitrate back to nitrogen gas, which enters the atmosphere. Denitrifying bacteria living in the anaerobic mud of lakes, bogs, and estuaries carry

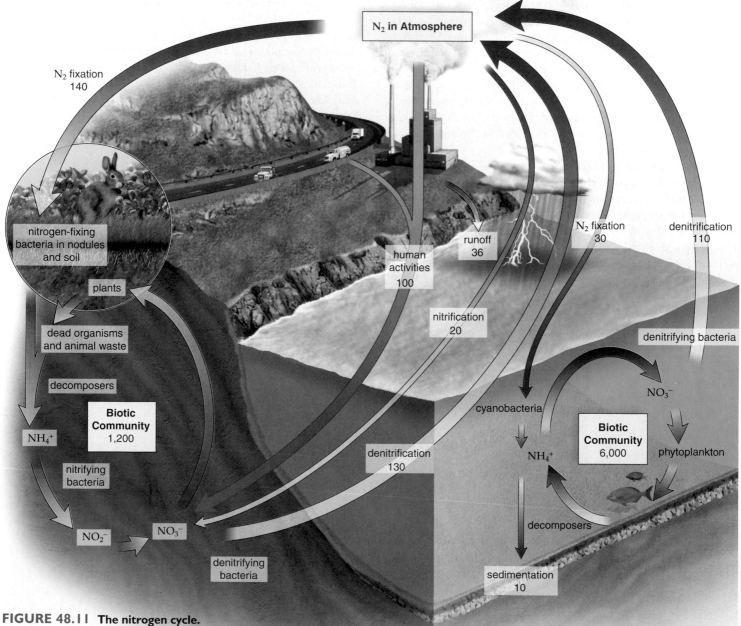

FIGURE 48.11 The nitrogen cycle.
Nitrogen is primarily made available to biotic communities by internal cycling of the element. Without human activities the amount of nitrogen returned to the atmosphere (denitrification in terrestrial and aquatic communities) exceeds withdrawal from the atmosphere (N_2 fixation and nitrification). Human activities (purple arrows) result in an increased amount of NO_3^- in terrestrial communities with resultant runoff to aquatic biotic communities. The numbers in this diagram indicate 10^{12} g N/yr.

out this process as a part of their own metabolism. In the nitrogen cycle, denitrification would counterbalance nitrogen fixation except for human activities.

Nitrogen and Air Pollution

Fertilizer production (purple arrow in Fig. 48.11) and use results in the release of nitrous oxide (N_2O), a greenhouse gas and a contributor to ozone shield depletion. The ozone shield is a layer of ozone high in the atmosphere that protects the Earth from dangerous levels of solar radiation (see the Ecology Focus on page 896).

Due to the burning of fossil fuels, the atmosphere contains three times more nitrogen oxides (NO_x) than it would otherwise. Fossil fuel combustion also pumps much sulfur dioxide (SO_2) into the atmosphere. Both nitrogen oxides and sulfur dioxide are converted to acids when they combine with water vapor in the atmosphere. These acids eventually return to Earth.

Acid deposition has drastically affected forests and lakes in northern Europe, Canada, and the northeastern United States because their soils are naturally acidic and their surface waters are only mildly alkaline (basic) to begin with. The forests in these areas are dying (Fig. 48.12a), and their waters cannot support normal fish populations. Acid deposition also reduces agricultural yields and corrodes marble, metal, and stonework, an effect that is noticeable in cities.

Nitrogen oxides (NO_x) and hydrocarbons (HC) react with one another in the presence of sunlight to produce **photochemical smog,** which contains ozone (O_3) and **PAN**

FIGURE 48.12 Acid deposition.
a. Many forests in higher elevations of northeastern North America and northern Europe are dying due to acid deposition. **b.** Air pollution due to fossil fuel burning in factories and modes of transportation is the major cause of acid deposition.

a.

b.

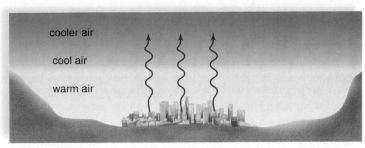

a. Normal pattern

b. Thermal inversion

c.

FIGURE 48.13 Thermal inversion.
a. Normally, pollutants escape into the atmosphere when warm air rises. **b.** During a thermal inversion, a layer of warm air (warm inversion layer) overlies and traps pollutants in cool air below. **c.** Los Angeles is particularly susceptible to thermal inversions, and this accounts for why this city is the "air pollution capital" of the United States.

(peroxyacetylnitrate). Hydrocarbons come from fossil fuel combustion (Fig. 48.12b), but additional amounts come from various other sources, including paint solvents and pesticides. Breathing ozone affects the respiratory and nervous systems, resulting in respiratory distress, headache, and exhaustion. These symptoms are particularly apt to appear in young people. Ozone is especially damaging to plants, resulting in leaf mottling and reduced growth.

Warm air near the Earth usually escapes into the atmosphere, taking pollutants with it. However, during a **thermal inversion,** pollutants are trapped near the Earth beneath a layer of warm, stagnant air. Because the air does not circulate, pollutants can build up to dangerous levels. Areas surrounded by hills are particularly susceptible to the effects of a thermal inversion because the air tends to stagnate, and little turbulent mixing can occur (Fig. 48.13).

Nitrogen cycles within the biotic community, and only a limited amount is fixed by bacteria. Environmental problems are associated with the release of excess nitrous oxide (N_2O) due to fertilizer production and the release of nitrogen oxides (NO_x) due to fossil fuel combustion.

The Phosphorus Cycle

As you can verify by following the appropriate arrows in Figure 48.14, in the **phosphorus cycle**, phosphorus moves from rocks on land to the oceans, where it gets trapped in sediments; then, phorphorus moves back into land again following a geological upheaval. However, on land, the very slow weathering of rocks makes phosphate ions (PO_4^{3-} and HPO_4^{2-}) available to plants, which take up phosphate from the soil.

Producers use phosphate in a variety of molecules, including phospholipids, ATP, and the nucleotides that become a part of DNA and RNA. Animals eat producers and incorporate

some of the phosphate into teeth, bones, and shells, which take many years to decompose. Death and decay of all organisms and also decomposition of animal wastes do, however, make phosphate ions available to producers once again. Because the available amount of phosphate is already being used within food chains, phosphate is usually a limiting inorganic nutrient for plants—that is, the lack of it limits their growth.

Some phosphate runs off into aquatic ecosystems, where algae acquire phosphate from the water before it becomes trapped in sediments. Phosphate in marine sediments does not become available to producers on land again until a geological upheaval exposes sedimentary rocks to weathering once more. Phosphorus does not enter the atmosphere; therefore, the phosphorus cycle is called a sedimentary cycle.

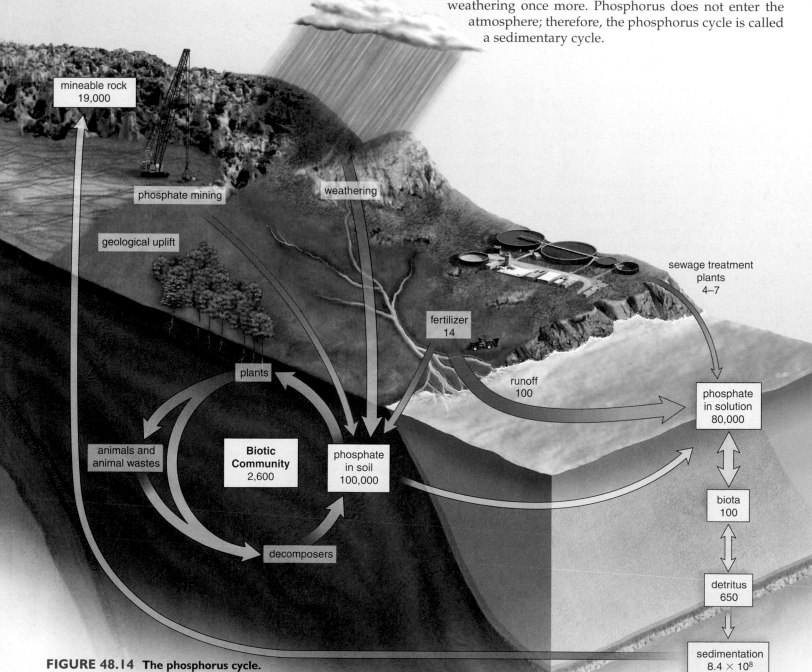

FIGURE 48.14 The phosphorus cycle.
The weathering of rocks provides phosphorus, which cycles locally in both terrestrial and aquatic biota. Human activities (purple arrows) produce fertilizers, which add to the amount of phosphorus available to biotic communities—eventually, fertilizers become a part of the runoff that enriches waters. Sewage treatment plants directly add phosphorus to local waters. When phosphorus becomes a part of oceanic sediments, it is lost to biotic communities for many years. The numbers in this diagram indicate 10^{12} g P/yr.

Phosphorus and Water Pollution

Human beings boost the supply of phosphate by mining phosphate ores for fertilizer and detergent production. Runoff of phosphate and nitrogen due to fertilizer use, animal wastes from livestock feedlots, and discharge from sewage treatment plants results in **eutrophication** (over-enrichment) of waterways. Eutrophication can lead to an algal bloom, apparent when green scum floats on the water. When the algae die off, decomposers use up all the available oxygen during cellular respiration. The result is a massive fish kill. An abundance of phosphate in the water can also be linked to algal blooms, such as red tide, which can produce deadly toxins.

Figure 48.15 lists the various sources of water pollution. Point sources of pollution are specific, and nonpoint sources are those caused by runoff from the land. Industrial wastes can include heavy metals and organochlorides, such as those in some pesticides. These materials are not readily degraded under natural conditions or in conventional sewage treatment plants. **Biological magnification** occurs as they pass along a food chain and become more and more concentrated because they remain in the body and are not excreted. Biological magnification occurs more readily in aquatic food chains, which have more links than terrestrial food chains. In any case, humans are the final consumers in food chains, and in some areas, human milk contains detectable amounts of DDT and PCBs, which are organochlorides.

Coastal regions are the immediate receptors for local pollutants and the final receptors for pollutants carried by rivers that empty at a coast. Waste dumping occurs at sea, but ocean currents sometimes transport both trash and pollutants back to shore. Offshore mining and shipping add pollutants to the oceans. Some 5 million metric tons of oil a year—or more than 1 g per 100 m^2 of the oceans' surfaces—end up in the oceans. Large oil spills kill plankton, fish fry, and shellfishes, as well as birds and marine mammals.

In the last 50 years, humans have polluted the seas and exploited their resources to the point that many species are at the brink of extinction. Fisheries once rich and diverse, such as George's Bank off the coast of New England, are in severe decline. Haddock was once the most abundant species in this fishery, but now it accounts for less than 2% of the total catch. Cod and bluefin tuna have suffered a 90% reduction in population size. In warm, tropical regions, many areas of coral reefs are now overgrown with algae because the fish that normally keep the algae under control have been killed off.

Sedimentary rock is a reservoir for phosphorus; for the most part, producers are dependent on decomposers to make phosphate available to them. Fertilizer production and other human activities add phosphate to aquatic ecosystems, contributing to water pollution.

Sources of Water Pollution	
Leading to Cultural Eutrophication	
oxygen-demanding waste	Biodegradable organic compounds (e.g., sewage, wastes from food-processing plants, paper mills, and tanneries)
plant nutrients	Nitrates and phosphates from detergents, fertilizers, and sewage treatment plants
sediments	Enriched soil in water due to soil erosion
thermal discharges	Heated water from power plants
Health Hazards	
disease-causing agents	Bacteria and viruses from sewage and barnyard waste (causing, for example, cholera, food poisoning, and hepatitis)
synthetic organic compounds	Pesticides, industrial chemicals (e.g., PCBs)
inorganic chemicals and minerals	Acids from mines and air pollution; dissolved salts; heavy metals (e.g., mercury) from industry
radiation	Radioactive substances from nuclear power plants, medical and research facilities

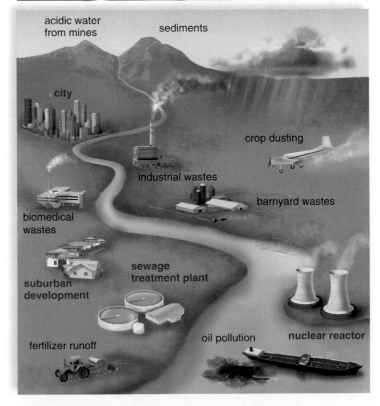

FIGURE 48.15 Sources of surface water pollution.
Many bodies of water are dying due to the introduction of pollutants from point sources, which are easily identifiable (red), and nonpoint sources, which cannot be specifically identified.

ecology focus

Ozone Shield Depletion

The Earth's atmosphere is divided into layers. The troposphere is the layer that envelops us as we go about our day-to-day lives. When ozone is present in the troposphere (called ground-level ozone), it is considered a pollutant because it adversely affects a plant's ability to grow and our ability to breathe oxygen (O_2). In the stratosphere, some 50 km above the Earth, ozone forms the **ozone shield,** a layer of ozone that absorbs most of the ultraviolet (UV) rays of the sun so that fewer rays strike the Earth. Ozone forms when ultraviolet radiation from the sun splits oxygen molecules (O_2), and then the oxygen atoms (O) combine with other oxygen molecules to produce ozone (O_3).

The absorption of UV radiation by the ozone shield is critical for living things. In humans, UV radiation causes mutations that can lead to skin cancer and can make the lens of the eye develop cataracts. In addition, it adversely affects the immune system and our ability to resist infectious diseases. UV radiation also impairs crop and tree growth and kills off algae (phytoplankton) and tiny shrimplike animals (krill) that sustain oceanic life. Without an adequate ozone shield, therefore, our health and food sources are threatened.

It became apparent in the 1980s that depletion of ozone had occurred worldwide and that the depletion was most severe above the Antarctic every spring. There ozone depletion became so great that it covered an area two and a half times the size of Europe, and it exposed not only Antarctica but also the southern tip of South America and vast areas of the Pacific and Atlantic Oceans to harmful UV rays. In the popular press, severe depletions of the ozone layer are called **ozone holes** (Fig. 48B*a*). Of even greater concern, an ozone hole has now appeared above the Arctic as well, and ozone holes were also detected within northern and southern latitudes, where many people live. Whether or not these holes develop in the spring depends on prevailing winds, weather conditions, and the type of particles in the atmosphere. A United Nations Environmental Program report predicts a 26% rise in cataracts and nonmelanoma skin cancers for every 10% drop in the ozone level. A 26% increase translates into 1.75 million additional cases of cataracts and 300,000 more skin cancers every year, worldwide.

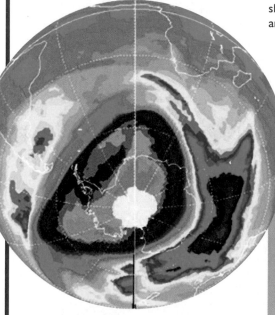

Ozone layer

thickest   thinnest

a.

FIGURE 48B Ozone shield depletion.
a. Map of ozone levels in the atmosphere of the Southern Hemisphere, September 2000. The ozone depletion, often called an ozone hole, is larger than the size of Europe. *b.* The release of chlorine atoms from chlorofluorocarbons (CFCs) contributes to the occurrence of ozone holes.

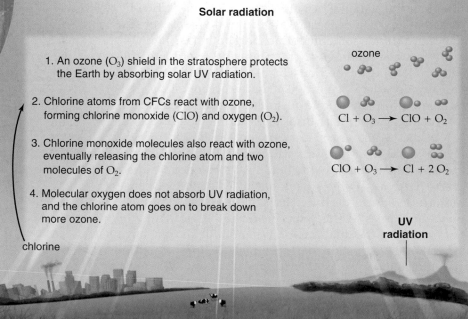

Solar radiation

1. An ozone (O_3) shield in the stratosphere protects the Earth by absorbing solar UV radiation.

2. Chlorine atoms from CFCs react with ozone, forming chlorine monoxide (ClO) and oxygen (O_2).

3. Chlorine monoxide molecules also react with ozone, eventually releasing the chlorine atom and two molecules of O_2.

4. Molecular oxygen does not absorb UV radiation, and the chlorine atom goes on to break down more ozone.

chlorine

ozone

$Cl + O_3 \longrightarrow ClO + O_2$

$ClO + O_3 \longrightarrow Cl + 2\,O_2$

UV radiation

b.

The seriousness of the situation caused scientists around the globe to begin studying the cause of ozone depletion. A vortex of cold wind (a whirlpool in the atmosphere) circles the poles during the winter months, creating polar stratospheric clouds. These clouds contain ice crystals in which chemical reactions occur that break down ozone. The cause of ozone depletion can be traced to the release of chlorine atoms (Cl) into the stratosphere. Chlorine atoms combine with ozone and form chlorine monoxide (ClO), which also breaks down ozone before releasing the same chlorine atom once again (Fig. 48Bb). One atom of chlorine can destroy up to 100,000 molecules of ozone before settling to the Earth's surface as chloride years later.

The chlorine atoms that enter the troposphere and eventually reach the stratosphere come primarily from the breakdown of **chlorofluorocarbons (CFCs),** chemicals much in use by humans. The best-known CFC is Freon, a coolant found in refrigerators and air conditioners. CFCs are also used as cleaning agents and as foaming agents during the production of Styrofoam coffee cups, egg cartons, insulation, and paddings. Formerly, CFCs were used as propellants in spray cans, but this application is now banned in the United States and several European countries. Other molecules, such as the cleaning solvent methyl chloroform, are also sources of harmful chlorine atoms.

Most of the countries of the world have stopped using CFCs, and the United States halted production in 1995. Since that time, satellite measurements indicate that the amount of harmful chlorine pollution in the stratosphere has started to decline. Scientists determined that chlorine concentrations peaked between 1992 and 1994. But because it takes several years for air from the troposphere to leak up into the stratosphere, chlorine concentrations didn't peak in the stratosphere until 1997.

The decline in chlorine pollution should lead to an increase in the ozone layer, but other factors are involved. Researchers report that during the winter of 2000, there were more and longer-lasting polar clouds

FIGURE 48C Polar stratospheric clouds.
Global warming is expected to make the stratosphere (as opposed to the troposphere) cool. As the stratosphere cools, polar clouds will increase and last longer. Cloud cover contributes to the breakdown of the ozone shield by chlorine pollution.

than previously. Why might that be? As the Earth's surface warms due to global warming (see Fig. 48.10), less heat reradiates into the stratosphere, and the stratosphere is becoming cooler than usual. Mathematical modeling suggests that polar clouds could last twice as long over the Arctic by the year 2010, when the coldest winter ever is expected.

Lingering polar clouds may inhibit the possible recovery of the ozone shield, despite lower amounts of chlorine pollution (Fig. 48C). Chlorine monoxide (ClO) ordinarily leads to ozone depletion, but it can also react with nitric oxide (NO), forming nitrogen dioxide (NO_2), which then breaks down, releasing an oxygen atom. The oxygen atom reacts with molecular oxygen to form more ozone, doing away with at least one part of the damaging ozone breakdown cycle caused by chlorine. Unfortunately, polar clouds drip nitrogen, and thereby lower the nitrogen concentration in the stratosphere. Scientists speculate that once polar clouds become twice as persistent, there could be an ozone loss of 30% due to reduction in nitric oxide levels. It is clear, then, that recovery of the ozone shield may take several more years and involve other pollution-fighting approaches, aside from lowering chlorine pollution.

CONNECTING THE CONCEPTS

One of the fundamental ways to study ecosystem function is to study energy flow within ecosystems. By measuring the amount of carbon entering an ecosystem via photosynthesis, we can estimate how much solar energy is being fixed by primary producers. We can then determine how much carbon is moving from one trophic level to the next, and finally how much carbon is released through respiration. By determining which factors govern different levels of energy flow (factors such as precipitation, nutrient availability, and seasonality), we can develop a good model for each specific ecosystem.

Many ecologists feel that understanding energy flow is necessary to understanding nutrient cycling and availability. Nutrient availability does not just influence plant growth; it also determines the distribution of animals. Many large grazing mammals preferentially consume plants with high nutrient content and thus tend to occur in areas with high nutrient availability.

Ecosystem studies have become increasingly sophisticated and have moved beyond studying energy flow and chemical cycling in general. Much current research is directed toward learning how particular species influence the function of an ecosystem. Thus, we are now becoming able to predict which species are most important and to determine which species must be preserved in order to have natural ecosystem functioning.

Human activities are affecting the transfer rates of biogeochemical cycles and therefore the functioning of the biosphere, the largest ecosystem of all. For example, pumping water from aquifers is not normally a part of the water cycle, and burning fossil fuels and trees is altering the carbon cycle so that the end result may be global warming. Carbon dioxide and other greenhouse gases allow the sun's rays to pass through, but they absorb and reradiate heat back to the Earth. Transfer rates in both the phosphorus and nitrogen cycles are affected when we produce fertilizers and detergents. Nitrogen and phosphorus runoff in aquatic ecosystems cause eutrophication. Pollution can be defined as a change in transfer rates that leads directly or indirectly to degradation of human health or plant and animal life.

Summary

48.1 THE NATURE OF ECOSYSTEMS

The biosphere encompasses those portions of the atmosphere, lithosphere, and hydrosphere where living things live in ecosystems. An ecosystem is composed of populations of organisms plus the chemical and physical environment. Some populations are producers, and some are consumers. Producers are autotrophs that produce their own organic food. Consumers are heterotrophs that take in organic food. Consumers may be herbivores, carnivores, omnivores, or decomposers.

Ecosystems are characterized by energy flow and chemical cycling. Energy is lost from the biosphere, but inorganic nutrients are not. They recycle within and between ecosystems. Decomposers return some proportion of inorganic nutrients to autotrophs, and other portions are imported or exported between ecosystems in global cycles.

Producers transform solar energy into food for themselves and all consumers. As herbivores feed on plants (or algae) and carnivores feed on herbivores, energy is converted to heat. Feces, urine, and dead bodies become food for decomposers. Eventually, all the solar energy that enters an ecosystem is converted to heat, and thus ecosystems require a continual supply of solar energy.

48.2 ENERGY FLOW

Ecosystems contain food webs in which the various organisms are connected by trophic relationships. In a grazing food web, food chains begin with a producer. In a detrital food web, food chains begin with detritus. The two food webs are joined when the same consumer is a link in both a grazing and a detrital food chain. A trophic level is all the organisms that feed at a particular link in a food chain. Ecological pyramids depict trophic levels stacked one on top of the other like building blocks. Generally they show that energy content, and therefore numbers of organisms and biomass, decreases from one trophic level to the next.

48.3 GLOBAL BIOGEOCHEMICAL CYCLES

Biogeochemical cycles contain reservoirs such as fossil fuels, sediments, and rocks, which are ecosystem components that make elements available on a limited basis to living things. Exchange pools are components of ecosystems, such as the atmosphere, soil, and water, which are ready sources of nutrients for living things.

In the water cycle, evaporation over the ocean is not compensated for by precipitation. Precipitation over land results in bodies of fresh water plus groundwater, including aquifers. Eventually, all water returns to the oceans.

In the carbon cycle, organisms add as much carbon dioxide to the atmosphere as they remove. Shells in ocean sediments, organic compounds in living and dead organisms, and fossil fuels are reservoirs for carbon. Human activities such as burning fossil fuels and trees are adding carbon dioxide to the atmosphere. It is predicted that a buildup of carbon dioxide and other greenhouse gases will lead to global warming and a rise in sea level. A change in climate patterns could follow.

In the nitrogen cycle, the biotic community recycles nitrogen back to the producers, and only limited quantities are made available by nitrifying bacteria in water, soil, and root nodules. Other bacteria return nitrogen to the atmosphere. Humans convert atmospheric nitrogen to fertilizer, which releases nitrous oxide (N_2O), a greenhouse gas that also contributes to the breakdown of the ozone shield. When humans burn fossil fuels, large quantities of nitrogen oxide (NO_x) and sulfur dioxide (SO_2) enter the atmosphere, where the oxides contribute to acid deposition. Acid deposition kills lakes and forests and corrodes stonework, etc. Nitrogen oxides and hydrocarbons (HC) react to form smog, which contains ozone and PAN (peroxyacetylnitrate), which are harmful oxidants.

In the phosphorus cycle, the biotic community recycles phosphorus back to the producers, and only limited quantities are made available by the weathering of rocks. Phosphates are mined for fertilizer production; when phosphates and nitrates enter lakes and ponds, overenrichment occurs.

Reviewing the Chapter

1. Distinguish between autotrophs and heterotrophs, and describe four different types of heterotrophs found in natural ecosystems. 883
2. Discuss energy flow in an ecosystem and chemical cycling within and between ecosystems. Why are chemicals able to cycle between ecosystems? 885
3. Describe the two types of food webs found in ecosystems. Which of these typically moves more energy through an ecosystem? 886–87
4. With reference to food chains, what is a trophic level? What is an ecological pyramid? 887–88
5. Give examples of reservoirs and pools in biogeochemical cycles. Which of these is less accessible to living things? 888
6. Draw diagrams to illustrate the water cycle and the carbon cycle. 889–90
7. How and why is the global temperature expected to change, and what are the predicted consequences of this change? 891
8. Draw a diagram of the nitrogen cycle. What types of bacteria are involved in this cycle? 892
9. What causes acid deposition, and what are its effects? 893
10. How does photochemical smog develop, and what is a thermal inversion? 893
11. Draw a diagram of the phosphorus cycle. 894
12. Why is the use of phosphate-based detergents dangerous to aquatic ecosystems? 894–95
13. What are several ways in which fresh water and marine waters can be polluted? What is biological magnification? 895
14. Of what benefit is the ozone shield? What pollutant in particular is associated with ozone shield depletion, and what are the consequences of this depletion? 896

Testing Yourself

Choose the best answer for each question.

1. Of the total amount of energy that passes from one trophic level to another, about 10% is
   a. respired and becomes heat.
   b. passed out as feces or urine.
   c. stored as body tissue.
   d. recycled to autotrophs.
   e. All of these are correct.
2. Compare this food chain:
       algae ⟶ water fleas ⟶ fish ⟶ green herons
   to this food chain:
       trees ⟶ tent caterpillars ⟶ red-eyed vireos ⟶ hawks

   Both water fleas and tent caterpillars are
   a. carnivores.
   b. primary consumers.
   c. detritus feeders.
   d. present in grazing and detrital food webs.
   e. Both a and b are correct.
3. Which of these contribute(s) to the carbon cycle?
   a. respiration
   b. photosynthesis
   c. fossil fuel combustion
   d. decomposition of dead organisms
   e. All of these are correct.

4. How do nitrogen-fixing bacteria contribute to the nitrogen cycle?
   a. They return nitrogen (N_2) to the atmosphere.
   b. They change ammonium to nitrate.
   c. They change N_2 to ammonium.
   d. They decompose and return nitrogen to autotrophs.
5. In what way are decomposers like producers?
   a. Either may be the first member of a grazing or a detrital food chain.
   b. Both produce oxygen for other forms of life.
   c. Both require a source of nutrient molecules and energy.
   d. Both produce organic nutrients for other members of ecosystems.
6. Choose the statement that is true concerning this food chain:
       grass ⟶ rabbits ⟶ snakes ⟶ hawks
   a. Each predator population has a greater biomass than its prey population.
   b. Each prey population has a greater biomass than its predator population.
   c. Each population returns inorganic nutrients and energy to the producer.
   d. Both a and c are correct.

For questions 7–9, match each term to those in the key.

KEY:
   a. sulfate salts
   b. ozone
   c. carbon dioxide
   d. pesticides

7. acid deposition
8. greenhouse effect
9. photochemical smog

10. Because of biological magnification,
    a. aquatic environments are overenriched.
    b. pollutants are trapped below a stagnant layer of warm air.
    c. certain pollutants are more concentrated in final consumers.
    d. nutrients move from one environment to another in a specific time frame.

11. Acid deposition causes
    a. lakes and forests to die.
    b. acid indigestion in humans.
    c. the greenhouse effect to lessen.
    d. pests to increase decomposition.
    e. All of these are correct.
12. Water is a renewable resource, and
    a. there will always be a plentiful supply.
    b. the oceans can never become polluted.
    c. it is still subject to pollution.
    d. primary sewage treatment plants ensure clean drinking water.
    e. Both a and c are correct.

13. Which of these statements are true? Choose more than one if correct.
   a. Energy flows through populations except for the amount that becomes urine and feces.
   b. While it might seem otherwise, more energy flows through grazing food webs than detrital food webs.
   c. Ecological pyramids typically pertain to grazing food webs only.
   d. Because humans produce fertilizers, global warming will continue to worsen.
   e. Unlike the other biogeochemical cycles, the phosphorus cycle is a gaseous cycle.

14. In this diagram, label the trophic levels (blanks a–d), using two of these terms for each level: producers, top carnivores, secondary consumers, autotrophs, primary consumers, tertiary consumers, carnivores, herbivores.
   e. Tell what the numbers on the left refer to.

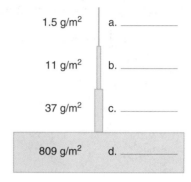

1.5 g/m²   a. _____

11 g/m²   b. _____

37 g/m²   c. _____

809 g/m²   d. _____

15. Label this diagram of an ecosystem:

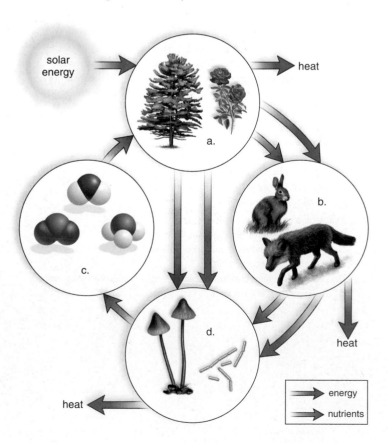

Thinking Scientifically

1. A large forest has been removed by clear-cutting, and the land has not been replanted. After several years, humidity in the area seems to have decreased. How can knowledge of the water cycle be used to interpret these observations?

2. In mountainous regions of the western United States, there is interest in reintroducing large mammalian predators that were previously driven out of the area. For these projects to work, predator populations must have enough wild food, or else domestic livestock may be preyed upon. What type of data is needed concerning food webs and ecological pyramids to successfully reintroduce these predators?

Understanding the Terms

acid deposition 893	fossil fuel 890
aquifer 889	global warming 891
autotroph 883	grazing food chain 887
biogeochemical cycle 888	grazing food web 887
biological magnification 895	greenhouse effect 891
biomass 887	greenhouse gases 891
biosphere 882	herbivore 883
carbon cycle 890	heterotroph 883
carnivore 883	nitrification 892
chlorofluorocarbons (CFCs) 897	nitrogen cycle 892
	nitrogen fixation 892
condensation 889	omnivore 883
consumer 883	ozone hole 896
decomposer 883	ozone shield 896
denitrification 892	PAN (peroxyacetylnitrate) 893
detrital food chain 887	phosphorus cycle 894
detrital food web 887	photochemical smog 893
detritus feeders 883	precipitation 889
ecological pyramid 887	producer 883
ecosystem 882	thermal inversion 893
eutrophication 895	transfer rate 889
evaporation 889	trophic level 887
food chain 887	trophic relationship 887
food web 887	water (hydrologic) cycle 889

Match the terms to these definitions:

a. _____ Partially decomposed remains of plants and animals.

b. _____ Formed from oxygen in the upper atmosphere, it protects the Earth from ultraviolet radiation.

c. _____ Remains of once-living organisms that are burned to release energy, such as coal, oil, and natural gas.

d. _____ Process by which atmospheric nitrogen gas is changed to forms that plants can use.

e. _____ Complex pattern of interlocking and crisscrossing food chains.

ARIS, the *Biology* Website

ARIS, the website for *Biology,* provides a wealth of information organized and integrated by chapter. You will find practice quizzes, interactive activities, labeling exercises, flashcards, and much more that will complement your learning and understanding of general biology.

49

THE BIOSPHERE

From space, the Earth is a pristine aqua globe, hovering against a vast backdrop of darkness. Get somewhat closer and the Earth's churning atmosphere and immense water systems come into view. Not until the Earth's surface is visible can you make out our many farms, towns, and cities, which have reduced Earth's natural ecosystems to mere remnants of their original size. Most of the ecosystems of the world have been greatly impacted by human activities so that they no longer function as they once did. Pollutants abound in all parts of the biosphere, and significant numbers of plants and animals have become extinct only within the past several hundred years.

In this chapter, we see how ecosystems were originally distributed over the globe and how climate determines the characteristics of the major biomes, such as forests, deserts, grasslands, and oceans. Each biome has its own mix of species, which are adapted to living under particular environmental conditions. Through biogeochemical cycles driven by solar energy, natural ecosystems transformed the Earth's crust, its waters, and the atmosphere into a life-supporting environment. It is critical that we learn to value the services of ecosystems and work toward preserving the remaining portions of the original biomes. In this way, we can help preserve species, including our own.

Satellite image of Earth.

49.1 CLIMATE AND THE BIOSPHERE

Climate refers to the prevailing weather conditions in a particular region. Climate is dictated by temperature and rainfall, which are influenced by the following factors: (1) variations in solar radiation distribution due to the tilt of the spherical Earth as it orbits about the sun, and (2) other effects, such as topography and whether a body of water is nearby.

Effect of Solar Radiation

Because the Earth is a sphere, the sun's rays are more direct at the equator and more spread out at the polar regions (Fig. 49.1*a*). Therefore, the tropics are warmer than the temperate regions. The tilt of the Earth as it orbits around the sun causes one pole or the other to be closer to the sun (except at the spring and fall equinoxes, when the sun aims directly at the equator), and this accounts for the seasons that occur in all parts of the Earth except at the equator (Fig. 49.1*b*). When the Northern Hemisphere is having winter, the Southern Hemisphere is having summer, and vice versa.

If the Earth were standing still and were a solid, uniform ball, all air movements—which we call winds—would be in two directions. Warm equatorial air would rise and move directly to the poles, creating a zone of lower pressure that would be filled by cold polar air moving equatorward.

However, because the Earth rotates on its axis daily and its surface consists of continents and oceans, the flows of warm and cold air are modified into three large circulation cells in each hemisphere (Fig. 49.2). At the equator, the sun heats the air and evaporates water. The warm, moist air rises, cools, and loses most of its moisture as rain. The greatest amounts of rainfall on Earth are near the equator. The rising air flows toward the poles, but at about 30° north and south latitude, it sinks toward the Earth's surface and reheats. As the air descends and warms, it becomes very dry, creating zones of low rainfall. The great deserts of Africa, Australia, and the Americas occur at these latitudes. At the Earth's surface, the air flows both poleward and equatorward. At about 60° north and south latitude, the air rises and cools, producing additional zones of high rainfall. This moisture supports the great forests of the temperate zone. Part of this rising air flows equatorward, and part continues poleward, descending near the poles, which are zones of low precipitation.

Besides affecting precipitation, the spinning of the Earth also affects the winds (Fig. 49.2). In the Northern Hemisphere, large-scale winds generally bend clockwise, and in the Southern Hemisphere, they bend counterclockwise. The curving pattern of the winds, ocean currents, and cyclones is the result of the fact that the Earth rotates in an eastward direction. At about 30° north latitude and 30° south latitude, the winds blow from the east-southeast in the Southern Hemisphere and from the east-northeast in the Northern Hemisphere (the east coasts of continents at these latitudes are wet). The doldrums, periods of calm, occur at the equator. These are called trade winds because sailors depended on them to fill the sails of their trading ships. Between 30° and 60° north and south latitude, strong winds, called the prevailing westerlies, blow from west to east. The west coasts of the continents at these latitudes are wet, as is the Pacific Northwest, where a massive evergreen forest is located. Weaker winds, called the polar easterlies, blow from east to west at still higher latitudes of their respective hemispheres.

The distribution of solar energy caused by a spherical Earth, and the rotation and path of the Earth around the sun, affect how the winds blow and the amount of rainfall various regions receive.

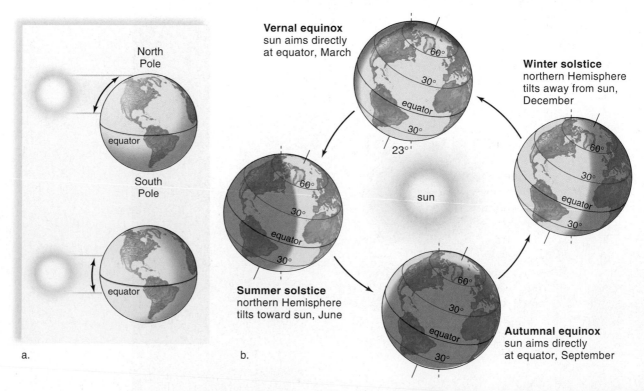

FIGURE 49.1

Distribution of solar energy.

a. Since the Earth is a sphere, beams of solar energy striking the Earth near one of the poles are spread over a wider area than similar beams striking the Earth at the equator. **b.** The seasons of the Northern and Southern Hemispheres are due to the tilt of the Earth on its axis as it rotates about the sun.

North Pole

equator

South Pole

equator

a.

Vernal equinox
sun aims directly at equator, March

60°
30°
equator
30°
23°

Winter solstice
northern Hemisphere tilts away from sun, December

60°
30°
equator
30°

sun

60°
30°
equator
30°

Summer solstice
northern Hemisphere tilts toward sun, June

60°
30°
equator
30°

Autumnal equinox
sun aims directly at equator, September

b.

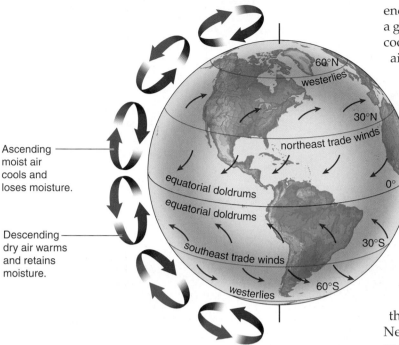

FIGURE 49.2 Global wind circulation.
Air ascends and descends as shown because the Earth rotates on its axis.
Also, the trade winds move from the northeast to the west in the Northern
Hemisphere, and from the southeast to the west in the Southern Hemisphere.
The westerlies move toward the east.

ence in temperature between the land and the ocean causes a gigantic circulation of air: Warm air rises over the land, and cooler air comes in off the ocean to replace it. As the warm air rises, it loses its moisture, and the monsoon season begins. As just discussed, rainfall is particularly heavy on the windward side of hills. Cherrapunji in northern India receives an annual average of 1,090 cm of rain a year because of its high altitude. The weather pattern has reversed by November. The land is now cooler than the ocean; therefore, dry winds blow from the Asian continent across the Indian Ocean. In the winter, the air over the land is dry, the skies cloudless, and temperatures pleasant. The chief crop of India is rice, which starts to grow when the monsoon rains begin.

In the United States, people often speak of the "lake effect," meaning that in the winter, arctic winds blowing over the Great Lakes become warm and moisture-laden. When these winds rise and lose their moisture, snow begins to fall. Places such as Buffalo, New York, get heavy snowfalls due to the lake effect, and snow is on the ground there for an average of 90–140 days every year.

Atmospheric circulations between the ocean and the landmasses influence regional climate conditions.

Other Effects

Topography means the physical features, or "the lay," of the land. One physical feature that affects climate is the presence of mountains. As air blows up and over a coastal mountain range, it rises and cools. One side of the mountain, called the windward side, receives more rainfall than the other side, called the leeward side. On the leeward side, the air descends, picks up moisture, and produces clear weather (Fig. 49.3). The difference between the windward side and the leeward side can be quite dramatic. In the Hawaiian Islands, for example, the windward side of the mountains receives more than 750 cm of rain a year, while the leeward side, which is in a **rain shadow,** gets on the average only 50 cm of rain and is generally sunny. In the United States, the western side of the Sierra Nevada Mountains is lush, while the eastern side is a semidesert.

The temperature of the oceans is more stable than that of landmasses. Oceanic water gains or loses heat more slowly than terrestrial environments. This causes coasts to have a unique weather pattern that is not observed inland. During the day, the land warms more quickly than the ocean, and the air above the land rises. Then a cool sea breeze blows in from the ocean. At night, the reverse happens; the breeze blows from the land toward the sea.

India and some other countries in southern Asia have a **monsoon** climate, in which wet ocean winds blow onshore for almost half the year. The land heats more rapidly than the waters of the Indian Ocean during spring. The differ-

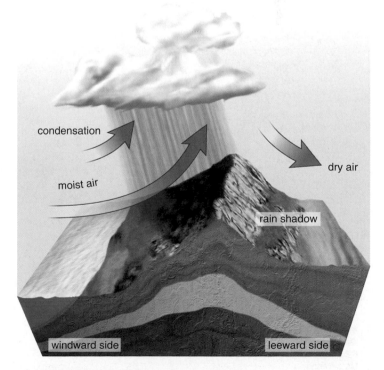

FIGURE 49.3 Formation of a rain shadow.
When winds from the sea cross a coastal mountain range, they rise and release their moisture as they cool this side of a mountain, called the windward side. The leeward side of a mountain receives relatively little rain and is therefore said to lie in a "rain shadow."

FIGURE 49.4 Pattern of biome distribution.
a. Pattern of world biomes in relation to temperature and moisture. The dashed line encloses a wide range of environments in which either grasses or woody plants can dominate the area, depending on the soil type.
b. The same type of biome can occur in different regions of the world, as shown on this global map.

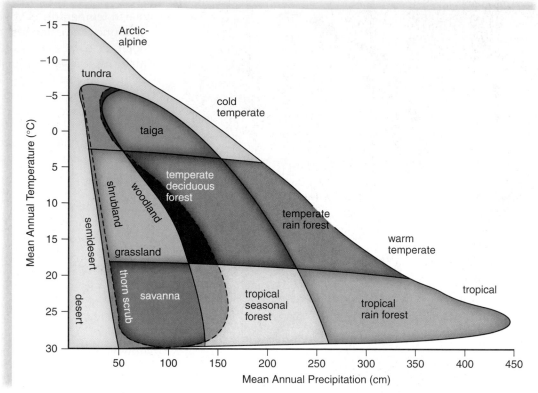

a.

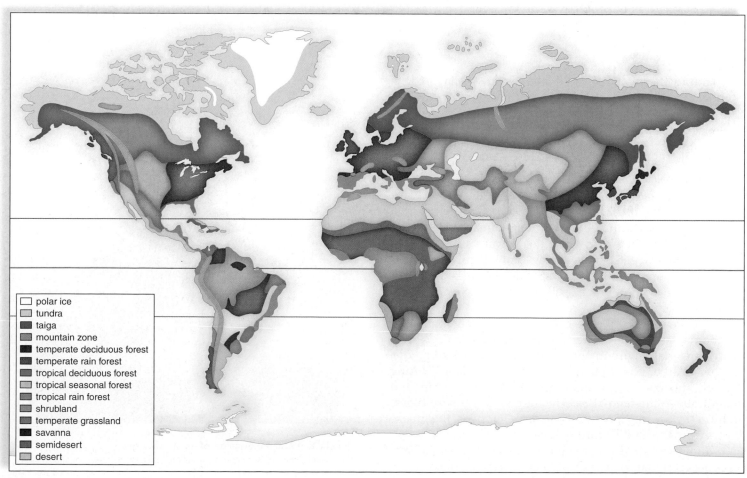

b.

49.2 TERRESTRIAL ECOSYSTEMS

A major type of terrestrial ecosystem is called a **biome.** A biome has a particular mix of plants and animals that are adapted to living under certain environmental conditions, of which climate is an overriding influence. For example, when terrestrial biomes are plotted according to their mean annual temperature and mean annual precipitation, a particular pattern results (Fig. 49.4*a*). The distribution of biomes is shown in Figure 49.4*b*. Even though Figure 49.4 shows definite demarcations, keep in mind that the biomes gradually change from one type to the other. Also, although we will be discussing each type of biome separately, we should remember that each biome has inputs from and outputs to all the other terrestrial and aquatic ecosystems of the biosphere.

The distribution of the biomes and their corresponding organismal populations are determined principally by differences in climate due to the distribution of solar radiation and defining topographical features. Both latitude and altitude are responsible for temperature gradients. If one travels from the equator to the North Pole, it is possible to observe first a tropical rain forest, followed by a temperate deciduous forest, a coniferous forest, and tundra, in that order, and this sequence is also seen when ascending a mountain (Fig. 49.5). The coniferous forest of a mountain is called a **montane coniferous forest,** and the tundra near the peak of a mountain is called an **alpine tundra.** When going from the equator to the South

TABLE 49.1	
Selected Biomes	
Name	*Characteristics*
Tundra	Around North Pole; average annual temperature is $-12C°$ to $-6C°$; low annual precipitation (less than 25 cm); permafrost (permanent ice) year-round within a meter of surface.
Taiga (coniferous forest)	Large northern biome that circles just below the Arctic Circle; temperature is below freezing for half the year; moderate annual precipitation (30–85 cm); long nights in winter and long days in summer.
Temperate deciduous forest	Eastern half of United States, Canada, Europe, and parts of Russia; four seasons of the year with hot summers and cold winters; goodly annual precipitation (75–150 cm)
Grasslands	Called prairies in North America, savannas in Africa, pampas in South America, steppes in Europe; hot in summer and cold in winter (United States); moderate annual precipitation; good soil for agriculture.
Tropical rain forests	Located near the equator in Latin America, Southeast Asia, and West Africa; warm (20–25C°) and wet (190 cm/year); has wet/dry season.
Deserts	Northern and Southern Hemispheres at 30° latitude; hot (38C°) days and cold (7C°) nights; low annual precipitation (less than 25 cm).

Pole, one would not reach a region corresponding to a coniferous forest and tundra of the Northern Hemisphere because the distribution of the landmasses is shifted toward the north.

> The distribution of biomes is determined by physical factors such as climate (principally temperature and rainfall), which varies according to latitude and altitude (Table 49.1).

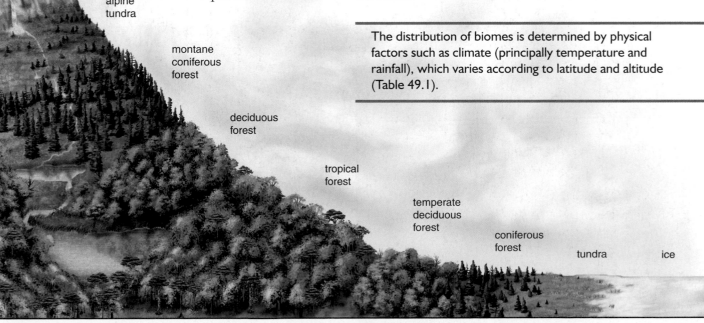

FIGURE 49.5 Climate and biomes.
Biomes change with altitude just as they do with latitude because vegetation is partly determined by temperature. Precipitation also plays a significant role, which is one reason grasslands, instead of tropical or deciduous forests, are sometimes found at the base of mountains.

Tundra

The **Arctic tundra** biome, which encircles the Earth just south of ice-covered polar seas in the Northern Hemisphere, covers about 20% of the Earth's land surface (Fig. 49.6). (A similar ecosystem, called the alpine tundra, occurs above the timberline on mountain ranges.) The Arctic tundra is cold and dark much of the year. Arctic tundra has extremely long, cold, harsh winters and short summers (6–8 weeks). Because rainfall amounts to only about 20 cm a year, the tundra could possibly be considered a desert, but melting snow creates a landscape of pools and mires in the summer, especially because so little evaporates. Only the topmost layer of soil thaws; the **permafrost** beneath this layer is always frozen, and therefore, drainage is minimal. The available soil in the tundra is nutrient-poor.

Trees are not found in the tundra because the growing season is too short, their roots cannot penetrate the permafrost, and they cannot become anchored in the boggy soil of summer. In the summer, the ground is covered with short grasses and sedges, as well as numerous patches of lichens and mosses. Dwarf woody shrubs, such as dwarf birch, flower and seed quickly while there is plentiful sun for photosynthesis.

A few animals live in the tundra year-round. For example, the mouselike lemming stays beneath the snow; the ptarmigan, a grouse, burrows in the snow during storms; and the musk ox conserves heat because of its thick coat and short, squat body. Other animals that live in the tundra include snowy owls, lynx, voles, Arctic foxes, and snowshoe hares. In the summer, the tundra is alive with numerous insects and birds, particularly shorebirds and waterfowl that migrate inland. Caribou in North America and reindeer in Asia and Europe also migrate to and from the tundra, as do the wolves that prey upon them. Polar bears are common near the coastal regions.

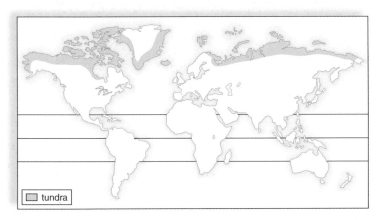
a.

b.

c. Caribou, *Rangifer taradanus*

FIGURE 49.6 The tundra.
a. In this biome, which is nearest the polar regions, the vegetation consists principally of lichens, mosses, grasses, and low-growing shrubs. **b.** Pools of water that do not evaporate or drain into the permanently frozen ground attract many birds. **c.** Caribou, more plentiful in the summer than in the winter, feed on lichens, grasses, and shrubs.

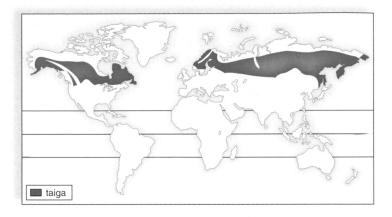

bull moose, *Alces americanus*

FIGURE 49.7 The taiga.
The taiga, which means swampland, spans northern Europe, Asia, and North America. The appellation "spruce-moose" refers to the dominant presence of spruce trees and moose, which frequent the ponds.

Coniferous Forests

Coniferous forests are found in three locations: in the **taiga**, which extends around the world in the northern part of North America and Eurasia; near mountaintops (where it is called a montane coniferous forest); and along the Pacific coast of North America, as far south as northern California.

The taiga, or boreal forest, exists south of the tundra and covers approximately 11% of the Earth's landmasses (Fig. 49.7). There are no comparable biomes in the Southern Hemisphere because no large landmasses exist at that latitude. The taiga typifies the coniferous forest with its cone-bearing trees, such as spruce, fir, and pine. These trees are well adapted to the cold because both the leaves and bark have thick coverings. Also, the needlelike leaves can withstand the weight of heavy snow. There is a limited understory of plants, but the floor is covered by low-lying mosses and lichens beneath a layer of needles. Birds harvest the seeds of the conifers, and bears, deer, moose, beaver, and muskrat live around the cool lakes and along the streams. Wolves prey on these larger mammals. A montane coniferous forest also harbors the wolverine and the mountain lion.

The coniferous forest that runs along the west coast of Canada and the United States is sometimes called a **temperate rain forest.** The prevailing winds moving in off the Pacific Ocean lose their moisture when they meet the coastal mountain range. The plentiful rainfall and rich soil have produced some of the tallest conifer trees ever in existence, including the coastal redwoods. This forest is also called an old-growth forest because some trees are as old as 800 years. It truly is an evergreen forest because mosses, ferns, and other plants grow on all the tree trunks. Squirrels, lynx, and numerous species of amphibians, reptiles, and birds inhabit the temperate rain forest. Whether the limited portion of the forest that remains should be preserved from logging has been quite controversial. Unfortunately, the controversy has centered around the Northern Spotted Owl, which is endemic to this area, rather than around the larger issue, the conservation of this particular ecosystem.

Temperate Deciduous Forests

Temperate deciduous forests are found south of the taiga in eastern North America, eastern Asia, and much of Europe (Fig. 49.8). The climate in these areas is moderate, with relatively high rainfall (75–150 cm per year). The seasons are well defined, and the growing season ranges between 140 and 300 days. The trees, such as oak, beech, sycamore, and maple, have broad leaves and are termed deciduous trees; they lose their leaves in the fall and grow them in the spring. In the southern temperate deciduous forests, evergreen magnolia trees can be found.

The tallest trees form a canopy, an upper layer of leaves that are the first to receive sunlight. Even so, enough sunlight penetrates to provide energy for another layer of trees, called understory trees. Beneath these trees are shrubs that may flower in the spring before the trees have put forth their leaves. Still another layer of plant growth—mosses, lichens, and ferns—resides beneath the shrub layer. This stratification provides a variety of habitats for insects and birds. Ground life is also plentiful. Squirrels, cottontail rabbits, shrews, skunks, woodchucks, and chipmunks are small herbivores. These and ground birds such as turkeys, pheasants, and grouse are preyed on by red foxes. White-tailed deer and black bears have increased in number in recent years. In contrast to the taiga, amphibians and reptiles occur in this biome because the winters are not as cold. Frogs and turtles prefer an aquatic existence, as do the beaver and muskrat, which are mammals.

Autumn fruits, nuts, and berries provide a supply of food for the winter, and the leaves, after turning brilliant colors and falling to the ground, contribute to the rich layer of humus. The minerals within the rich soil are washed far into the ground by spring rains, but the deep tree roots capture these and bring them back up into the forest system again.

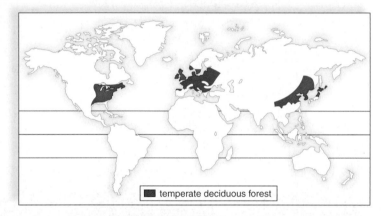

temperate deciduous forest

millipede, *Marceus* sp.

eastern chipmunk, *Tamias striatus*

marsh marigolds, *Caltha howellii*

bobcat, *Felis rufus*

FIGURE 49.8 Temperate deciduous forest.
A temperate deciduous forest is home to many varied plants and animals. Millipedes can be found among leaf litter, chipmunks feed on acorns, and bobcats prey on these and other small mammals.

ecology focus

Wildlife Conservation and DNA

After DNA analysis, scientists were amazed to find that some 60% of loggerhead turtles drowning in the nets and hooks of fisheries in the Mediterranean Sea were from beaches in the southeastern United States. Since the unlucky creatures were a good representative sample of the turtles in the area, that meant more than half of the young turtles living in the Mediterranean Sea had hatched from nests on beaches in Florida, Georgia, and South Carolina (Fig. 49Aa). Some 20,000–50,000 loggerheads die each year due to the Mediterranean fisheries, which may partly explain the decline in loggerheads nesting on southeastern U.S. beaches for the last 25 years.

The sequencing of DNA from Alaskan brown bears allowed Sandra Talbot (a graduate student at the University of Alaska's Institute of Arctic Biology) and wildlife geneticist Gerald Shields to conclude that there are two types of brown bears in Alaska. One type resides only on southeastern Alaska's Admiralty, Baranof, and Chichagof islands, known as the ABC Islands. The other brown bear in Alaska is found throughout the rest of the state, as well as in Siberia and western Asia (Fig. 49Ab).

A third distinct type of brown bear, known as the Montana grizzly, resides in other parts of North America. These three types comprise all of the known brown bears in the New World.

The ABC bears' uniqueness may be bad news for the timber industry, which has expressed interest in logging parts of the ABC Islands. Says Shields, "Studies show that when roads are built and the habitat is fragmented, the population of brown bears declines. Our genetic observations suggest they are truly unique, and we should consider their heritage. They could never be replaced by transplants."

In what will become a classic example of how DNA analysis might be used to protect endangered species from future ruin, scientists from the United States and New Zealand carried out discreet experiments in a Japanese hotel room on whale sushi bought in local markets. Sushi, a staple of the Japanese diet, is a rice and meat concoction wrapped in seaweed. Armed with a miniature DNA sampling machine, the scientists found that, of the 16 pieces of whale sushi they examined, many were from whales that are endangered or protected under an international moratorium on whaling. "Their findings demonstrated the true power of DNA studies," says David Woodruff, a conservation biologist at the University of California, San Diego.

One sample was from an endangered humpback, four were from fin whales, one was from a northern minke, and another from a beaked whale. Stephen Palumbi of the University of Hawaii says the technique could be used for monitoring and verifying catches. Until then, he says, "no species of whale can be considered safe."

Meanwhile, Ken Goddard, director of the unique U.S. Fish and Wildlife Service Forensics Laboratory in Ashland, Oregon, is already on the watch for wildlife crimes in the United States and 122 other countries that send samples to him for analysis. "DNA is one of the most powerful tools we've got," says Goddard, a former California police crime-lab director.

The lab has blood samples, for example, for all of the wolves being released into Yellowstone Park—"for the obvious reason that we can match those samples to a crime scene," says Goddard. The lab has many cases currently pending in court that he cannot discuss. But he likes to tell the story of the lab's first DNA-matching case. Shortly after the lab opened in 1989, California wildlife authorities contacted Goddard. They had seized the carcass of a trophy-sized deer from a hunter. They believed the deer had been shot illegally on a 3,000-acre preserve owned by actor Clint Eastwood. The agents found a gut pile on the property but had no way to match it to the carcass. The hunter had two witnesses to deny the deer had been shot on the preserve.

Goddard's lab analysis made a perfect match between tissue from the gut pile and tissue from the carcass. Says Goddard: "We now have a cardboard cutout of Clint Eastwood at the lab saying 'Go ahead: Make my DNA.'"

FIGURE 49A DNA Studies.
a. Many loggerhead turtles found in the Mediterranean Sea are from the southeastern United States. *b.* These two brown bears appear similar, but one type, known as an ABC bear, resides only on southeastern Alaska's Admiralty, Baranof, and Chichagof islands.

a. Loggerhead turtle, *Caretta caretta*

b. Brown bears, *Ursus arctos*

Tropical Forests

In the **tropical rain forests** of South America, Africa, and the Indo-Malayan region near the equator, the weather is always warm (between 20° and 25°C), and rainfall is plentiful (with a minimum of 190 cm per year). This may be the richest biome, in terms of both number of different kinds of species and their abundance.

A tropical rain forest has a complex structure, with many levels of life, including the forest floor, understory, and canopy. The vegetation of the forest floor is very sparse because much of the sunlight is filtered out by the canopy. The understory consists of smaller plants that are specialized for life in the shade. The canopy, topped by the crowns of tall trees, is the most productive level of the tropical rain forest (Fig. 49.9). Some of the broadleaf evergreen trees grow from 15–50 m or more. These tall trees often have trunks buttressed at ground level to prevent their toppling over. Lianas, or woody vines, which encircle the tree as it grows, also help strengthen the trunk. The diversity of species is enormous—a 10 km² area of tropical rain forest may contain 750 species of trees and 1,500 species of flowering plants.

Although some animals live on the forest floor (e.g., pacas, agoutis, peccaries, and armadillos), most live in the trees (Fig. 49.10). Insect life is so abundant that the majority of species have not been identified yet. Termites play a vital role in the decomposition of woody plant material, and ants are found everywhere, particularly in the trees. The various birds, such as hummingbirds, parakeets, parrots, and toucans, are often beautifully colored. Amphibians and reptiles are well represented by many types of frogs, snakes, and lizards. Lemurs, sloths, and monkeys are well-known primates that feed on the fruits of the trees. The largest carnivores are the big cats—the jaguars in South America and the leopards in Africa and Asia.

Many animals spend their entire life in the canopy, as do some plants. **Epiphytes** are plants that grow on other plants but usually have roots of their own that absorb moisture and minerals leached from the canopy; others catch rain and debris in hollows produced by overlapping leaf bases. The most common epiphytes are related to pineapples, orchids, and ferns.

While we usually think of tropical forests as being nonseasonal rain forests, tropical forests that have wet and dry seasons are found in India, Southeast Asia, West Africa, South and Central America, the West Indies, and northern

canopy

understory

forest floor

lianas

epiphyte

fern

FIGURE 49.9 Levels of life in a tropical rain forest.
The primary levels within a tropical rain forest are the canopy, the understory, and the forest floor. But the canopy (solid layer of leaves) contains levels as well, and some organisms spend their entire life in one particular level. Long lianas (hanging vines) climb into the canopy, where they produce leaves. Epiphytes are air plants that grow on the trees but do not parasitize them.

Australia. Here, there are deciduous trees, with many layers of growth beneath them. In addition to the animals just mentioned, some of these forests also contain elephants, tigers, and hippopotamuses.

Whereas the soil of a temperate deciduous forest biome is rich enough for agricultural purposes, the soil of a tropical rain forest biome is not. Nutrients are cycled directly from the litter to the plants again. Productivity is high because of high temperatures, a yearlong growing season, and the rapid recycling of nutrients from the litter. (In humid tropical forests, iron and aluminum oxides occur at the surface, causing a reddish residue known as laterite. When the trees are cleared, laterite bakes in the hot sun to a bricklike consistency that will not support crops.) Swidden agriculture, often called slash-and-burn agriculture, has been successful, but also destructive, in the tropics. Trees are felled and burned, and the ashes provide enough nutrients for several harvests. Thereafter, the forest must be allowed to regrow, and a new section must be cut and burned.

It is estimated that 2.4 acres of rain forest are destroyed per second. This rate equates to nearly 78 million acres annually. Unless conservation strategies are employed soon, rain forests will be destroyed beyond recovery, taking with them unique and interesting life-forms. Ecologists estimate that an average of 137 species are driven to extinction every day in rain forests.

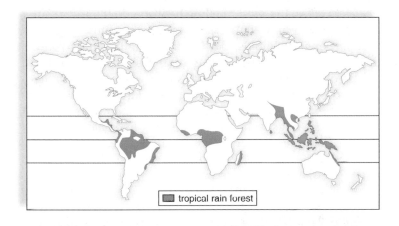

tropical rain forest

poison arrow frog,
Dendrobates histrionicus

cone-headed katydid,
Panacanthus cuspidatus

ocelot,
Felis pardalis

blue and gold macaw,
Ara ararauna

brush-footed butterfly,
Anartia amalthea linnaeus

lemur,
Propithecus verreauxi

arboreal lizard,
Calotes calotes

FIGURE 49.10 Representative animals of the tropical rain forests of the world.

Shrublands

It is difficult to define a shrub, but in general, shrubs are shorter than trees (4.5–6 m) with a woody, persistent stem and no central trunk. Shrubs have small but thick evergreen leaves, which are often coated with a waxy material that prevents loss of moisture from the leaves. Their thick underground roots can survive dry summers and frequent fires and take deep moisture from the soil. Shrubs are adapted to withstand arid conditions and can also quickly sprout new growth after a fire. As a point of interest, you will recall from Chapter 47 that a shrub stage is part of the process of both primary and secondary succession.

Shrublands tend to occur along coasts that have dry summers and receive most of their rainfall in the winter. Shrublands are found along the cape of South Africa, the western coast of North America, and the southwestern and southern shores of Australia, as well as around the Mediterranean Sea and in central Chile. The dense shrubland that occurs in California is known as **chaparral** (Fig. 49.11). This type of shrubland, called Mediterranean, lacks an understory and ground litter and is highly flammable. The seeds of many species require the heat and scarring action of fire to induce germination. Other shrubs sprout from the roots after a fire. Typical animals of the chaparral include mule deer, rodents, lizards, and scrub jays.

There is also a northern shrub area that lies west of the Rocky Mountains. This area is sometimes classified as a cold desert, but the region is dominated by sagebrush and other hardy plants. Some of the birds found there are dependent on sagebrush for their existence.

Grasslands

Grasslands occur where annual rainfall is greater than 25 cm but generally insufficient to support trees. For example, in temperate areas, where rainfall is between 25 and 75 cm, it is too dry for forests and too wet for deserts to form.

Grasses are well adapted to a changing environment and can tolerate a high degree of grazing, flooding, drought, and sometimes fire. Where rainfall is high, tall grasses that reach more than 2 m in height (e.g., pampas grass) can flourish. In drier areas, shorter grasses (between 5 and 10 cm) are dominant. Low-growing bunch grasses (e.g., grama grass) grow in the United States near deserts. The growth of grasses is seasonal. As a result, grassland animals such as bison migrate, and others such as ground squirrels hibernate, when there is little grass for them to eat.

Temperate Grasslands

The temperate grasslands include the Russian steppes, the South American pampas, and the North American prairies (Fig. 49.12). In these grasslands, winters are bitterly cold and summers are hot and dry. When traveling across the United States from east to west, the line between the temperate deciduous forest and a tall-grass prairie is roughly along the border between Illinois and Indiana. The tall-grass prairie requires more rainfall than does the short-grass prairie, which occurs near deserts. Large herds of bison—estimated at hundreds of thousands—once roamed the prairies, as did herds of pronghorn antelope. Now, small mammals, such as mice, prairie dogs, and rabbits, typically live below ground, but usually feed aboveground. Hawks, snakes, badgers, coyotes, and foxes feed on these mammals. Virtually all of these grasslands, however, have been converted to agricultural lands because of their fertile soils.

Savannas

Savannas occur in regions where a relatively cool dry season is followed by a hot rainy season (Fig. 49.13). The largest savannas are in central and southern Africa. Other savannas exist in Australia, southeast Asia, and South America. The savanna is characterized by large expanses of grasses with sparse populations of trees. The plants of the savanna have extensive and deep root systems that enable them to survive drought and fire. One tree that can survive the severe dry season is the thorny flat-topped acacia, which sheds its leaves during a drought. The African savanna supports

Scrub jay, *Aphelocoma coerulescens*

Chemise, *Adenostema faciculatum*

FIGURE 49.11 Shrubland.
Shrublands, such as chaparral in California, are subject to raging fires, but the shrubs are adapted to quickly regrow. Scrub jays find a home here as does a plant commonly called chemise.

the greatest variety and number of large herbivores of all the biomes. Elephants and giraffes are browsers that feed on tree vegetation. Antelopes, zebras, wildebeests, water buffalo, and rhinoceroses are grazers that feed on grasses. Any plant litter that is not consumed by grazers is attacked by a variety of small organisms, among them termites. Termites build towering nests in which they tend fungal gardens, their source of food. The herbivores support a large population of carnivores. Lions and hyenas sometimes hunt in packs, cheetahs hunt singly by day, and leopards hunt singly by night.

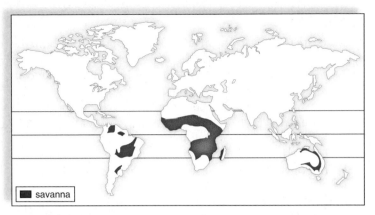
savanna

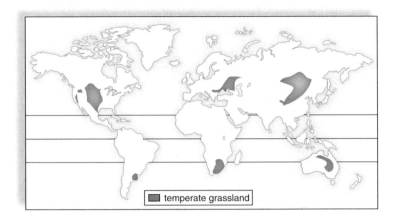
temperate grassland

cheetah, *Acinonyx jubatus*

giraffe, *Giraffa camelopardalis*

wildebeast, *Connochaetes* sp.

American bison, *Bison bison*

FIGURE 49.12 Temperate grassland.
Tall-grass prairies are seas of grasses dotted by pines and junipers. Bison, once abundant, are now being reintroduced into certain areas.

zebra (in foreground), *Equus quagga*

FIGURE 49.13 The savanna.
The African savanna varies from grassland to widely spaced shrubs and trees. This biome supports a large assemblage of herbivores (e.g., zebras, wildebeests, and giraffes). Carnivores (e.g., cheetahs) prey on these.

Deserts

Deserts are usually found at latitudes of about 30°, in both the Northern and Southern Hemispheres. Deserts cover nearly 30% of the Earth's land surface. The winds that descend in these regions lack moisture. Therefore, the annual rainfall is less than 25 cm. Days are hot because a lack of cloud cover allows the sun's rays to penetrate easily, but nights are cold because heat escapes easily into the atmosphere.

The Sahara, which stretches all the way from the Atlantic coast of Africa to the Arabian peninsula, and a few other deserts have little or no vegetation. But most have a variety of plants (Fig. 49.14, *left*). Desert plants are highly adapted to survive long droughts, extreme heat, and extreme cold. Adaptations to these conditions include thick epidermal layers, water-storing stems and leaves, and the ability to set seeds quickly in the spring. The best-known desert perennials in North America are the succulent, spiny-leafed cactuses, which have stems that store water and carry on photosynthesis. Also common are nonsucculent shrubs, such as the many-branched sagebrush with silvery gray leaves and the spiny-branched ocotillo, which produces leaves during wet periods and sheds them during dry periods.

Some animals are adapted to the desert environment. To conserve water, many desert animals such as reptiles and insects are nocturnal or burrowing and have a protective outer body covering. A desert has numerous insects, which pass through the stages of development from pupa to pupa again when there is rain. Reptiles, especially lizards and snakes, are perhaps the most characteristic group of vertebrates found in deserts, but running birds (e.g., the roadrunner) and rodents (e.g., the kangaroo rat) are also well known (Fig. 49.14, *right*). Larger mammals, such as the kit fox, prey on the rodents, as do hawks.

FIGURE 49.14 The desert.
Plants and animals that live in a desert are adapted to arid conditions. The plants are either succulents, which retain moisture, or shrubs with woody stems and small leaves, which lose little moisture. The kangaroo rat feeds on seeds and other vegetation; the roadrunner preys on insects, lizards, and snakes. The kit fox is a desert carnivore.

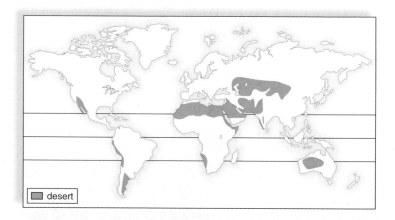

desert

bannertail kangaroo rat,
Dipodomys spectabilis

greater roadrunner,
Geococcyx californianus

kit fox, *Vulpes velox*

49.3 AQUATIC ECOSYSTEMS

Aquatic ecosystems are classified as two types: freshwater (inland) or saltwater (usually marine). Brackish water, however, is a mixture of fresh and salt water. Figure 49.15 shows how these ecosystems are joined physically and discusses some of the organisms that are adapted to live in them.

In the water cycle, the sun's rays cause seawater to evaporate, and the salts are left behind. As discussed in Chapter 48, the vaporized fresh water rises into the atmosphere, cools, and falls as rain either over the ocean or over the land. A lesser amount of water also evaporates from and returns to the land. When rain falls, some of the water sinks, or percolates, into the ground and saturates the Earth to a certain level. The top of the saturation zone is called the groundwater table, or simply the water table.

Since land lies above sea level, gravity eventually returns all fresh water to the sea, but in the meantime, it is contained as standing water within basins, called lakes and ponds, or as flowing water within channels, called streams or rivers. Sometimes groundwater is also located in underground rivers called aquifers. Whenever the Earth contains basins or channels, water will appear to the level of the water table.

Wetlands are areas that are wet for at least part of the year. Generally, wetlands are classified by their vegetation. **Marshes** are wetlands that are frequently or continually inundated by water and characterized by the presence of rushes, reeds, and other grasses. They provide excellent habitat for waterfowl and small mammals. Marshes are one of the most productive ecosystems on Earth. **Swamps** are wetlands that are dominated by either woody plants or shrubs. Common swamp trees include cypress, red maple, and tupelo. The American alligator is a top predator in many swamp ecosystems. **Bogs** are wetlands that are characterized by acidic waters, peat deposits, and sphagnum moss. Bogs receive most of their water from precipitation and are nutrient-poor. Several species of plants thrive in bogs, including cranberries, orchids, and insectivorous plants such as the pitcher plant. Moose and a number of other animals are inhabitants of bogs in the northern United States and Canada.

Humans have the habit of channeling aboveground rivers and filling in wetlands. These activities degrade ecosystems and eventually cause seasonal flooding. Wetlands provide food and habitats for fish, waterfowl, and other wildlife. They also purify waters by filtering them and by diluting and breaking down toxic wastes and excess nutrients. Wetlands directly absorb storm waters and also absorb overflows from lakes and rivers. In this way, they protect farms, cities, and towns from the devastating effects of floods. Federal and local laws have been enacted for the protection of wetlands, but they are not always enforced.

Between the 1780s and 1980s, approximately 53% of the original wetlands in the lower 48 states of the United States has been lost. On average, about 60 acres of wetlands have been lost hourly during this 200-year time span.

Aquatic ecosystems can be classified as freshwater or saltwater. The two sets of communities interact and are joined by the water cycle.

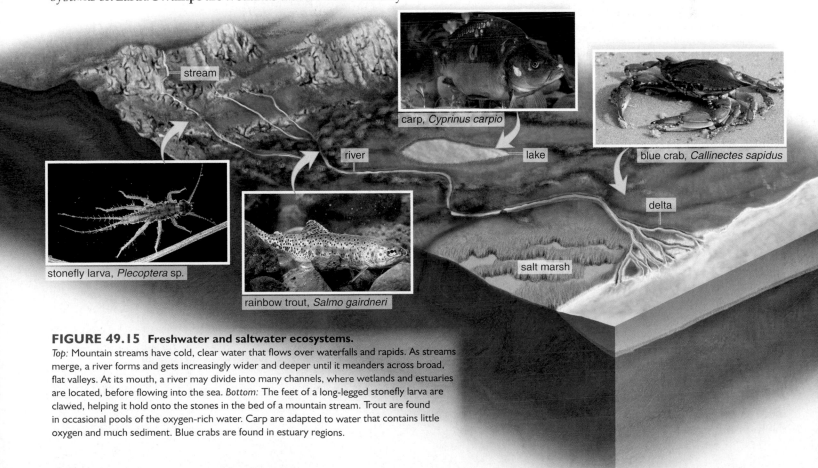

FIGURE 49.15 Freshwater and saltwater ecosystems.
Top: Mountain streams have cold, clear water that flows over waterfalls and rapids. As streams merge, a river forms and gets increasingly wider and deeper until it meanders across broad, flat valleys. At its mouth, a river may divide into many channels, where wetlands and estuaries are located, before flowing into the sea. *Bottom:* The feet of a long-legged stonefly larva are clawed, helping it hold onto the stones in the bed of a mountain stream. Trout are found in occasional pools of the oxygen-rich water. Carp are adapted to water that contains little oxygen and much sediment. Blue crabs are found in estuary regions.

a.

b.

FIGURE 49.16 Types of lakes.
Lakes can be classified according to whether they are (**a**) oligotrophic (nutrient-poor) or (**b**) eutrophic (nutrient-rich). Eutrophic lakes tend to have large populations of algae and rooted plants, resulting in a large population of decomposers that use up much of the oxygen and leave little oxygen for fishes.

Lakes

Lakes are bodies of fresh water often classified by their nutrient status. Oligotrophic (nutrient-poor) lakes are characterized by a small amount of organic matter and low productivity (Fig. 49.16a). Eutrophic (nutrient-rich) lakes are characterized by plentiful organic matter and high productivity (Fig. 49.16b). Such lakes are usually situated in naturally nutrient-rich regions or are enriched by agricultural or urban and suburban runoff. Oligotrophic lakes can become eutrophic through large inputs of nutrients. This process is called **eutrophication.**

In the temperate zone, deep lakes are stratified in the summer and winter and have distinct vertical zones. In summer, lakes in the temperate zone have three layers of water that differ in temperature (Fig. 49.17). The surface layer, the epilimnion, is warm from solar radiation; the middle layer, the thermocline, experiences an abrupt drop in temperature; and the lowest layer, the hypolimnion, is cold. These differences in temperature prevent mixing. The warmer, less dense water of the epilimnion "floats" on top of the colder, more dense water of the thermocline, which floats on top of the hypolimnion.

As the season progresses, the epilimnion becomes nutrient-poor, while the hypolimnion begins to be depleted of oxygen. The phytoplankton found in the sunlit epilimnion use up nutrients as they photosynthesize. Photosynthesis releases oxygen, giving this layer a ready supply. Detritus naturally falls by gravity to the bottom of the lake, and there oxygen is used up as decomposition occurs. Decomposition releases nutrients, however.

In the fall, as the epilimnion cools, and in the spring, as it warms, an overturn occurs. In the fall, the upper epilimnion waters become cooler than the hypolimnion waters. This causes the surface water to sink and the deep water to rise. This **fall overturn** continues until the temperature is uniform throughout the lake. At this point, wind aids in the circulation of water so that mixing occurs. Eventually, oxygen and nutrients become evenly distributed.

As winter approaches, the water cools. Ice formation begins at the top, and the ice remains there because ice is less dense than cool water. Ice has an insulating effect, preventing further cooling of the water below. This permits aquatic organisms to live through the winter in the water beneath the surface of the ice.

In the spring, as the ice melts, the cooler water on top sinks below the warmer water on the bottom. This

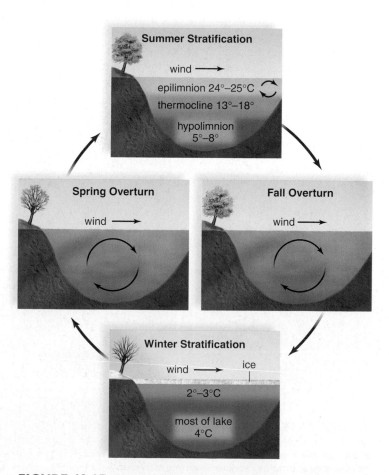

FIGURE 49.17 Lake stratification in a temperate region.
Temperature profiles of a large oligotrophic lake in a temperate region vary with the season. During the spring and fall overturns, the deep waters receive oxygen from surface waters, and surface waters receive inorganic nutrients from deep waters.

spring overturn continues until the temperature is uniform throughout the lake. At this point, wind aids in the circulation of water as before. When the surface waters absorb solar radiation, thermal stratification occurs once more.

The vertical stratification and seasonal change of temperatures in a lake influence the seasonal distribution of fish and other aquatic life in the lake basin. For example, coldwater fish move to the deeper water in summer and inhabit the upper water in winter. In the fall and spring just after mixing occurs, phytoplankton growth at the surface is most abundant.

Life Zones

In both fresh and salt water, free-drifting microscopic organisms, called *plankton* [Gk. *planktos*, wandering], are important components of the ecosystem. **Phytoplankton** [Gk. *phyton*, plant, and *planktos*, wandering] are photosynthesizing algae that become noticeable when a green scum or red tide appears on the water. **Zooplankton** [Gk. *zoon*, animal, and *planktos*, wandering] are minute animals that feed on the phytoplankton. Lakes and ponds can be divided into several life zones (Fig. 49.18). The *littoral zone* is closest to the shore, the *limnetic zone* forms the sunlit body

of the lake, and the *profundal zone* is below the level of light penetration. The *benthic zone* includes the sediment at the soil-water interface. Aquatic plants are rooted in the shallow littoral zone of a lake, providing habitat for numerous protozoans, invertebrates, fishes, and some reptiles. Pike species are "lurking perdators." They wait among vegetation around the margins of lakes and surge out to capture passing prey. Wading birds are commonly seen feeding in the littoral zone. Some organisms, such as the water strider, live at the water-air interface and can literally walk on water. In the limnetic zone, small fishes, such as minnows and killifish, feed on plankton and also serve as food for larger fish, such as bass. In the profundal zone, zooplankton, invertebrates, and fishes such as catfish and whitefish feed on debris that falls from higher zones.

The bottom of the lake is known as the benthic zone. Primarily, the benthic zone is composed of silt, sand, inorganic sediment, and dead organic material (detritus). Bottom-dwelling organisms are known as the benthos and include worms, snails, clams, crayfish, and insect larvae. Decomposers, such as bacteria, are also found in the benthic zone and serve to break down wastes and dead organisms into nutrients that are eventually used by producers.

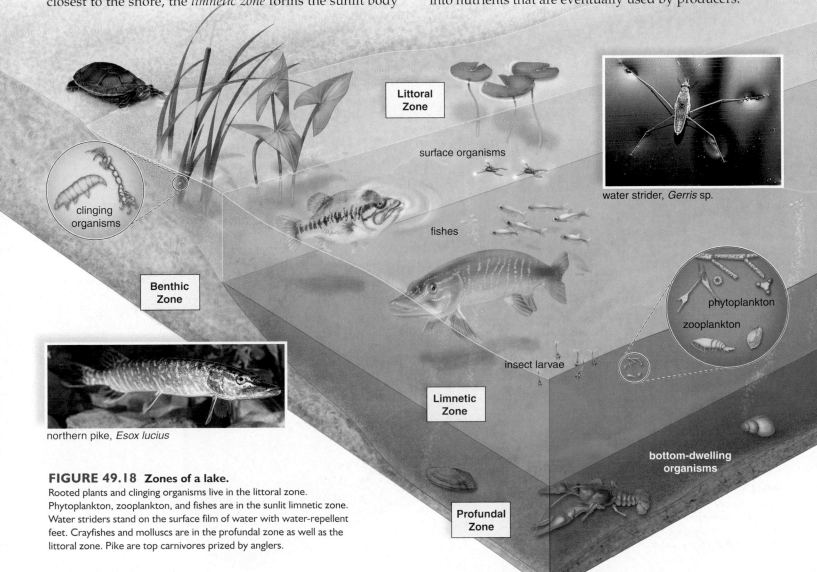

water strider, *Gerris* sp.

northern pike, *Esox lucius*

FIGURE 49.18 Zones of a lake.
Rooted plants and clinging organisms live in the littoral zone. Phytoplankton, zooplankton, and fishes are in the sunlit limnetic zone. Water striders stand on the surface film of water with water-repellent feet. Crayfishes and molluscs are in the profundal zone as well as the littoral zone. Pike are top carnivores prized by anglers.

Coastal Ecosystems

Marine ecosystems are divided into coastal ecosystems and the oceans. Coastal ecosystems, which are more productive than the oceans, include estuaries and seashores.

Estuaries

An **estuary** is a partially enclosed body of water where fresh water and seawater meet and mix (Fig. 49.19). A river brings fresh water into the estuary, and the sea, because of the tides, brings salt water. Coastal bays, tidal marshes, fjords (an inlet of water between high cliffs), some deltas (a triangular-shaped area of land at the mouth of a river), and lagoons (a body of water separated from the sea by a narrow strip of land) are all examples of estuaries.

Mudflats, mangrove swamps, and salt marshes dominated by salt marsh cordgrass are often associated with estuaries. All occur at the mouth of a river, where sediment and nutrients from the land collect (Fig. 49.20). Mangrove swamps develop in subtropical and tropical zones, while marshes occur in temperate zones.

Organisms living in an estuary must be able to withstand constant mixing of waters and rapid changes in salinity. Those organisms adapted to an estuarine environment find an abundance of nutrients. An estuary acts as a nutrient trap because the sea prevents the rapid escape of nutrients brought by a river. As the result of usually warm,

a. Black skimmers, *Rynchops niger*

b. Mangrove, *Avicennia* sp.

FIGURE 49.20 Types of estuaries.
Some of the many types of regions that are associated with estuaries include the salt marsh depicted in Figure 49.19; mudflats (**a**), which are frequented by migrant birds; and mangrove swamps (**b**) skirting the coastlines of many tropical and subtropical lands. The tangled roots of mangrove trees trap sediments and nutrients that sustain many immature forms of sea life.

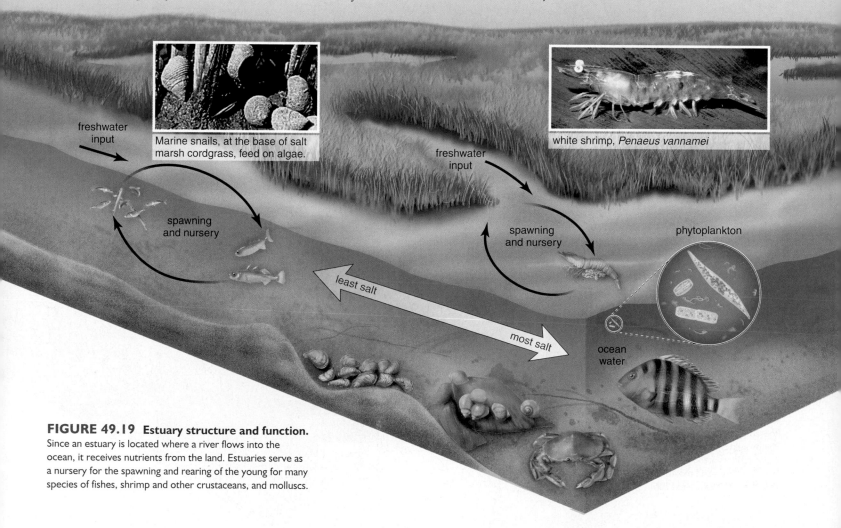

freshwater input

Marine snails, at the base of salt marsh cordgrass, feed on algae.

spawning and nursery

white shrimp, *Penaeus vannamei*

freshwater input

spawning and nursery

phytoplankton

least salt

most salt

ocean water

FIGURE 49.19 Estuary structure and function.
Since an estuary is located where a river flows into the ocean, it receives nutrients from the land. Estuaries serve as a nursery for the spawning and rearing of the young for many species of fishes, shrimp and other crustaceans, and molluscs.

calm waters and plentiful nutrients, estuaries are biologically diverse and highly productive.

Phytoplankton and shore plants thrive in the nutrient-rich estuaries, providing an abundance of food and habitat for animals. It is estimated that nearly two-thirds of marine fishes and shellfish spawn and develop in the protective and rich environment of estuaries. The estuarine environment is known as the nursery of the sea. An abundance of larval, juvenile, and mature shellfish and fish attract a number of predators such as reptiles, birds, and fishers.

Seashores

Both rocky and sandy shores are constantly bombarded by the sea as the tides roll in and out. The **littoral zone** lies between the high and low water marks. The littoral zone of a rocky beach is divided into subzones (Fig. 49.21a). In the upper portion of the littoral zone, barnacles are glued so tightly to the stone by their own secretions that their calcareous outer plates remain in place even after the enclosed shrimplike animal dies. In the midportion of the littoral zone, brown algae known as rockweed may overlie the barnacles. In the lower portions of the littoral

zone, oysters and mussels attach themselves to the rocks by filaments called *byssal threads*. Also present are sea stars and also snails called limpets and periwinkles (Fig. 49.21b). Periwinkles have a coiled shell and secure themselves by hiding in crevices or under seaweeds, while limpets press their single, flattened cone tightly to a rock. Below the littoral zone, macroscopic seaweeds, which are the main photosynthesizers, anchor themselves to the rocks by holdfasts. Sandy shores are usually gently sloping beaches associated with the mainland or a barrier island. Undisturbed sandy beaches may feature natural sand dunes that are held in place by specialized plants.

Organisms cannot attach themselves to shifting, unstable sands on a sandy beach; therefore, nearly all the permanent residents dwell underground. They either burrow during the day and surface to feed at night, or they remain permanently within their burrows and tubes. Ghost crabs and sandhoppers (amphipods) burrow themselves above the high tide mark and feed at night when the tide is out. Sandworms and sand (ghost) shrimp remain within their burrows in the littoral zone and feed on detritus whenever possible. Still lower in the sand, clams, cockles, and sand dollars are found. A variety of shorebirds visit the beaches and feed on various invertebrates and fishes (Fig. 49.21c).

b. Sea stars at low tide

c. Shorebirds

FIGURE 49.21 Seacoasts.
a. The littoral zone of a rocky coast, where the tide comes in and out, has different types of shelled and algal organisms at its upper, middle, and lower portions. **b.** Some organisms of a rocky coast live in tidal pools. **c.** A sandy shore looks devoid of life except for the birds that feed there.

Oceans

Climate is driven by the sun, but the oceans play a major role in redistributing heat in the biosphere. Water tends to be warm at the equator and much cooler at the poles because of the distribution of the sun's rays, as we have discussed before (see Fig. 49.1a). Air takes on the temperature of the water below, and warm air moves from the equator to the poles. In other words, the oceans make the winds blow. (The landmasses also play a role, but the oceans hold heat longer and remain cool longer during periods of changing temperature than do solid continents.)

When the wind blows strongly and steadily across a great expanse of ocean for a long time, friction from the moving air begins to drag the water along with it. Once the water has been set in motion, its momentum, aided by the wind, keeps it moving in a steady flow called a current. Because the ocean currents eventually strike land, they move in a circular path—clockwise in the Northern Hemisphere and counterclockwise in the Southern Hemisphere (Fig. 49.22). As the currents flow, they take warm water from the equator to the poles. One such current, called the Gulf Stream, brings tropical Caribbean water to the east coast of North America and the higher latitudes of western Europe. Without the Gulf Stream, Great Britain, which has a relatively

warm temperature, would be as cold as Greenland. In the Southern Hemisphere, another major ocean current warms the eastern coast of South America.

Also in the Southern Hemisphere, a current called the Humboldt Current flows toward the equator. The Humboldt Current carries phosphorus-rich cold water northward along the west coast of South America. During a process called **upwelling,** cold offshore winds cause cold nutrient-rich waters to rise and take the place of warm nutrient-poor waters. In South America, the enriched waters cause an abundance of marine life that supports the fisheries of Peru and northern Chile. Birds feeding on these organisms deposit their droppings on land, where they are mined as guano, a commercial source of phosphorus. When the Humboldt Current is not as cool as usual, upwelling does not occur, stagnation results, the fisheries decline, and climate patterns change globally. This phenomenon, which is discussed in the Ecology Focus on page 921, is called an **El Niño–Southern Oscillation.**

Major ocean currents move heat from the equator to cooler parts of the biosphere. The Gulf Stream warms the east coast of North America and parts of Europe.

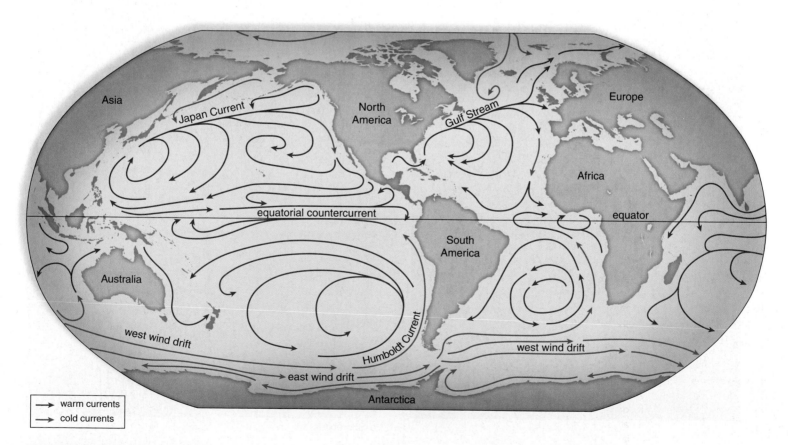

FIGURE 49.22 Ocean currents.
The arrows on this map indicate the locations and directions of the major ocean currents set in motion by the global wind circulation. By carrying warm water to cool latitudes (e.g., the Gulf Stream) and cool water to warm latitudes (e.g., the Humboldt Current), these currents have a major effect on the world's climates.

ecology focus

El Niño–Southern Oscillation

Climate largely determines the distribution of life on Earth. Short-term variations in climate, which we call weather, also have a pronounced effect on living things. There is no better example than an El Niño. Originally, El Niño referred to a warming of the seas off the coast of Peru at Christmastime—hence, the name El Niño, "the boy child," for the Christ child Jesus.

Now scientists prefer the term El Niño–Southern Oscillation (ENSO) for a severe weather change brought on by an interaction between the atmosphere and ocean currents. Ordinarily, the southeast trade winds move along the coast of South America and turn west because of the Earth's daily rotation on its axis. As the winds drag warm ocean waters from east to west, there is an upwelling of nutrient-rich cold water from the ocean's depths, resulting in a bountiful Peruvian harvest of anchovies. When the warm ocean waters reach their western destination, the monsoons bring rain to India and Indonesia. Scientists have noted that these events correlate with a difference in the barometric pressure over the Indian Ocean and the southeastern Pacific—that is, the barometric pressure is low over the Indian Ocean and high over the southeastern Pacific. But when a "southern oscillation" occurs and the barometric pressures switch, an El Niño begins.

During an El Niño, both the northeast and the southeast trade winds slacken. Upwelling no longer occurs, and the anchovy catch off the coast of Peru plummets. During a severe El Niño, waters from the east never reach the west, and the winds lose their moisture in the middle of the Pacific instead of over the Indian Ocean. The monsoons fail, and drought occurs in India, Indonesia, Africa, and Australia. Harvests decline, cattle must be slaughtered, and famine is likely in highly populated India and Africa, where funds to import replacement supplies of food are limited.

A backward movement of winds and ocean currents may even occur so that the waters warm to more than 14° above normal along the west coast of the Americas. This is a sign that a severe El Niño has occurred, and the weather changes are dramatic in the Americas also. Southern California is hit by storms and even hurricanes, and the deserts of Peru

and Chile receive so much rain that flooding occurs. A jet stream (strong wind currents) can carry moisture into Texas, Louisiana, and Florida, with flooding a near certainty. Or the winds can turn northward and deposit snow in the mountains along the west coast so that flooding occurs in the spring. Some parts of the United States, however, benefit from an El Niño. The Northeast is warmer than usual, few if any hurricanes hit the east coast, and there is a lull in tornadoes throughout the Midwest. Altogether, a severe El Niño affects the weather over three-quarters of the globe.

Eventually, an El Niño dies out, and normal conditions return. The normal cold-water state off the coast of Peru is known as La Niña

(the girl). Figure 49B contrasts the weather conditions of a La Niña with those of an El Niño. Since 1991, the sea surface has been almost continuously warm, and two record-breaking El Niños have occurred. What could be causing more of the El Niño state than the La Niña state? Some scientists are seeking data to relate this environmental change to global warming, a rise in environmental temperature due to greenhouse gases in the atmosphere. Like the glass of a greenhouse, the gases allow the sun's rays to pass through, but they trap the heat. Greenhouse gases, including carbon dioxide, are pollutants that human beings have been pumping into the atmosphere since the Industrial Revolution.

La Niña

- Upwelling off the west coast of South America brings cold waters to the surface.
- Barometric pressure is high over the southeastern Pacific.
- Monsoons associated with the Indian Ocean occur.
- Hurricanes occur off the east coast of the United States.

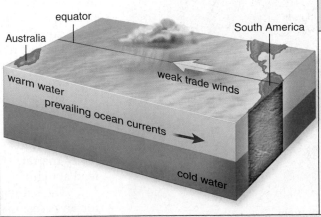

El Niño

- Great ocean warming occurs off the west coast of the Americas.
- Barometric pressure is low over the southeastern Pacific.
- Monsoons associated with the Indian Ocean fail.
- Hurricanes occur off the west coast of the United States.

FIGURE 49B La Niña and El Niño.

Pelagic Division

Oceans cover approximately three-quarters of our planet. It is customary to place the organisms of the oceans into either the pelagic division (open waters) or the benthic division (ocean floor). The geographic areas and zones of an ocean are shown in Figure 49.23. The **pelagic division** includes the neritic province and the oceanic province.

Neritic Province. The neritic province has a greater concentration of organisms than the oceanic province because the neritic province is sunlit and has a supply of inorganic nutrients for photosynthesizers. Phytoplankton, consisting of suspended algae, is food not only for zooplankton but also for small fishes. These small fishes in turn attract a number of predatory and commercially valuable fishes.

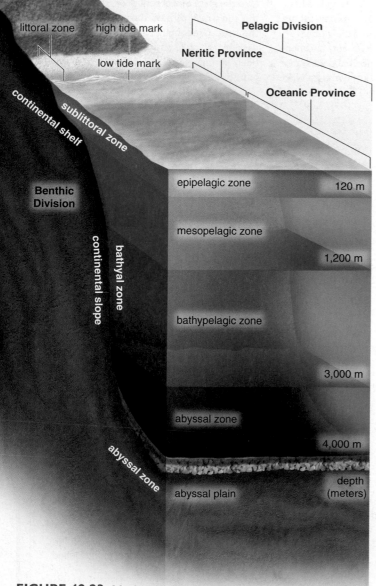

littoral zone   high tide mark

low tide mark

continental shelf
sublittoral zone
continental slope
bathyal zone

Benthic Division

abyssal zone

abyssal plain

Pelagic Division

Neritic Province

Oceanic Province

epipelagic zone   120 m

mesopelagic zone

1,200 m

bathypelagic zone

3,000 m

abyssal zone

4,000 m

depth (meters)

FIGURE 49.23 Marine environment.
Organisms reside in the pelagic division (blue), where waters are divided as indicated. Organisms also reside in the benthic division (brown) in the surfaces and zones indicated.

Coral reefs are areas of biological abundance found in shallow, warm, tropical waters just below the surface. Their chief constituents are stony corals, animals that have a calcium carbonate (limestone) exoskeleton, and calcareous red and green algae. Corals do not usually occur individually; rather, they form colonies derived from an individual coral that has reproduced by means of budding. Corals provide a home for a microscopic alga called zooxanthella. The corals, which feed at night, and the algae, which photosynthesize during the day, are mutualistic and share materials and nutrients. The close relationship of corals and zooxanthellae may be the reason coral reefs form only in shallow, sunlit water—the algae use sunlight for photosynthesis.

A reef is densely populated with life. The large number of crevices and caves provide shelter for filter feeders (sponges, sea squirts, and fanworms) and for scavengers (crabs and sea urchins). The barracuda, moray eel, and shark are top predators in coral reefs. There are many types of small, beautifully colored fishes. Parrot fishes feed directly on corals, and others feed on plankton or detritus. Small fishes become food for larger fishes, including snappers that are caught for human consumption.

There is much concern about the future survival of coral reefs. They are vulnerable to environmental changes such as shifts in temperature, salinity, and light availability. In addition, coral is slow growing, and physical destruction of coral (including overcollecting) may cause irreparable damage.

Oceanic Province. The oceanic province lacks the inorganic nutrients of the neritic province, and therefore it does not have as high a concentration of phytoplankton, even though the epipelagic zone is sunlit. Still, the photosynthesizers are food for a large assembly of zooplankton, which then become food for herrings and bluefishes. These, in turn, are eaten by larger mackerels, tunas, and sharks. Flying fishes, which glide above the surface, are commonly found in the epipelagic zone. A number of porpoise and dolphin species visit and feed in this region of the oceanic province. Whales are other mammals found in the epipelagic zone (Fig. 49.24). Baleen whales strain krill (small crustaceans) from the water, and toothed sperm whales feed primarily on the common squid.

Animals in the mesopelagic zone are carnivores, are adapted to the absence of light, and tend to be translucent, red colored, or even luminescent. There are luminescent shrimps, squids, and fishes, including lantern and hatchet fishes. Various species of zooplankton, invertebrates, and fishes migrate from the mesopelagic zone to the surface to feed at night.

The bathypelagic zone is in complete darkness except for an occasional flash of bioluminescent light. Carnivores and scavengers are found in this zone. Strange-looking fishes with distensible mouths and abdomens and small, tubular eyes feed on infrequent prey.

Benthic Division

The **benthic division** includes organisms that live on or in the soil of the continental shelf (sublittoral zone), the continental slope (bathyal zone), and the abyssal plain (abyssal zone) (see Fig. 49.23).

Seaweed grows in the sublittoral zone, and it can be found in batches on outcroppings as the water gets deeper. More diversity of life exists in the sublittoral and bathyal zones than in the abyssal zone. In the first two zones, clams, worms, and sea urchins are preyed upon by sea stars, lobsters, crabs, and brittle stars. Photosynthesizing algae occur in the sunlit sublittoral zone, but benthic organisms in the bathyal zone are dependent on the slow rain of detritus from the waters above.

The abyssal zone is inhabited by animals that live at the soil-water interface of the abyssal plain (Fig. 49.24). It once was thought that few animals exist in this zone because of the intense pressure and the extreme cold. Yet many invertebrates live here by feeding on debris floating down from the mesopelagic zone. Sea lilies (crinoids) rise above the seafloor; sea cucumbers and sea urchins crawl around on the sea bottom; and tube worms burrow in the mud.

The flat abyssal plain is interrupted by enormous underwater mountain chains called oceanic ridges. Along the axes of the ridges, crustal plates spread apart, and molten magma rises to fill the gap. At **hydrothermal vents,** seawater percolates through cracks and is heated to about 350°C, causing sulfate to react with water and form hydrogen sulfide (H_2S). Chemosynthetic bacteria that obtain energy from oxidizing hydrogen sulfide exist freely or mutualistically within the tissues of organisms. They are the start of food chains for an ecosystem that includes huge tubeworms, clams, crustaceans, echinoderms, and fishes. It may seem surprising to find ecosystems of organisms living so deep in the ocean, where light never penetrates, until we consider that, unlike photosynthesis, chemosynthesis does not require light energy.

The neritic province and the epipelagic zone of the oceanic province receive sunlight and contain the organisms with which we are most familiar. The organisms of the benthic division are dependent on debris that floats down from above.

FIGURE 49.24 Ocean inhabitants.
Different organisms are characteristic of the epipelagic, mesopelagic, and bathypelagic zones of the pelagic division compared to the abyssal zone of the benthic division.

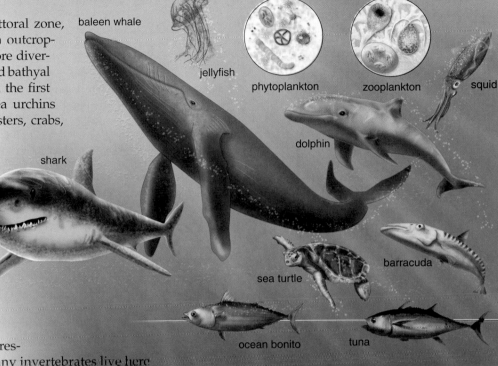

baleen whale

jellyfish   phytoplankton   zooplankton   squid

dolphin

shark

barracuda

sea turtle

ocean bonito   tuna

mackerel   prawn   lantern fish

giant squid   midshipman

sperm whale

viperfish

hagfish   anglerfish

deep-sea shrimp

tripod fish   gulper   glass sponges

brittle stars   sea cucumber

Epipelagic Zone (0 – 120 m)

Mesopelagic Zone (120 – 1,200 m)

Bathypelagic Zone (1,200 – 3,000 m)

Abyssal Zone (3,000 m – bottom)

CONNECTING THE CONCEPTS

The biosphere is the product of interactions of living organisms with the Earth's physical and chemical environment over billions of years. Life has affected the atmosphere, modifying its composition and influencing global climate. Organisms have helped create the chemical and physical conditions of streams, lakes, and oceans. The soils of terrestrial ecosystems and the sediments of aquatic ecosystems are structured largely by the activities of organisms. Reef-building plants and animals have helped build the islands on which millions of humans live. The Earth's diverse biomes also result from interactions of the biotic communities and the abiotic environment.

Over geological time, the biosphere has been changing constantly. The biomes of the age of dinosaurs were strikingly different from those of today. Astrophysical events have triggered some of this change. Changes in the sun's radiation output and in the tilt of the Earth's axis have altered the pattern of solar energy reaching the Earth's surface. Geological processes have also modified conditions for life. The drifting of continents has changed the arrangement of continents and oceans. Mountain ranges have been thrust up and eroded down. Through these changing conditions, life has evolved, and the structure of the Earth's biomes has evolved as well. In the last few million years, humans appeared and learned to exploit the Earth's biomes.

Humans have transformed vast areas of many of the terrestrial biomes into farmland, cities, highways, and other developments. Through our resource use and release of pollutants, we have now become an agent of global importance. Yet we still depend on the biodiversity that exists in the Earth's biomes, and on the interactions of other organisms within the biosphere. These interactions still influence climate, patterns of nutrient cycling and waste processing, and basic biological productivity. The Earth's biotic diversity also provides enjoyment and inspiration to millions of people, who spend billions of dollars to visit coral reefs, deserts, rain forests, and even the Arctic tundra. Unfortunately, as we shall see in Chapter 50, many of the Earth's biomes may not survive in their present form for our descendants' benefit. Although most of us value Earth's biodiversity, human activities are threatening many species with extinction.

Summary

49.1 CLIMATE AND THE BIOSPHERE

Because the Earth is a sphere, the sun's rays at the poles are spread out over a larger area than the vertical rays at the equator. The temperature at the surface of the Earth therefore decreases from the equator to each pole. The Earth is tilted on its axis, and the seasons change as the Earth rotates annually around the sun.

Warm air rises near the equator, loses its moisture, and then descends at about 30° north and south latitude, and so forth, to the poles. When the air descends, it retains moisture, and therefore the great deserts of the world are formed at 30° latitudes. Because the Earth rotates on its axis daily, the winds blow in opposite directions above and below the equator. Topography also plays a role in the distribution of moisture. Air rising over coastal ranges loses its moisture on the windward side, making the leeward side arid.

49.2 TERRESTRIAL ECOSYSTEMS

A biome is a major type of terrestrial community. Biomes are distributed according to climate—that is, temperature and rainfall influence the pattern of biomes about the world. The effect of temperature causes the same sequence of biomes when traveling to northern latitudes as when traveling up a mountain.

The Arctic tundra is the northernmost biome and consists largely of short grasses, sedges, and dwarf woody plants. Because of cold winters and short summers, most of the water in the soil is frozen year-round. This is called the permafrost.

The taiga, a coniferous forest, has less rainfall than other types of forests. The temperate deciduous forest has trees that gain and lose their leaves because of the alternating seasons of summer and winter. Tropical rain forests are the most complex and productive of all biomes.

Shrublands usually occur along coasts that have dry summers and receive most of their rainfall in the winter. Among grasslands, the savanna, a tropical grassland, supports the greatest number of different types of large herbivores. Temperate grasslands, such as that found in the central United States, have a limited variety of vegetation and animal life.

Deserts are characterized by a lack of water—they are usually found in places with less than 25 cm of precipitation per year. Some desert plants, such as cactuses, are succulents, and others are shrubs with thick leaves that often fall off during dry periods.

49.3 AQUATIC ECOSYSTEMS

Streams, rivers, lakes, and wetlands are different freshwater ecosystems. In deep lakes of the temperate zone, the temperature and the concentration of nutrients and gases in the water vary with depth. The entire body of water is cycled twice a year, distributing nutrients from the bottom layers. Lakes and ponds have three life zones. Rooted plants and clinging organisms live in the littoral zone, plankton and fishes live in the sunlit limnetic zone, and bottom-dwelling organisms such as crayfishes and molluscs live in the profundal zone.

Marine ecosystems are divided into coastal ecosystems and the oceans. The coastal ecosystems, especially estuaries, are more productive than the oceans. Estuaries (and associated salt marshes, mudflats, and mangrove forests) are near the mouth of a river. Estuaries are considered the nurseries of the sea.

An ocean is divided into the pelagic division and the benthic division. The oceanic province of the pelagic division (open waters) has three zones. The epipelagic zone receives adequate sunlight and supports the most life. The mesopelagic zone contains organisms adapted to minimum or no light. The benthic division (ocean floor) includes organisms living on the continental shelf in the sublittoral zone, the continental slope in the bathyal zone, and the abyssal plain in the abyssal zone.

Reviewing the Chapter

1. Describe how a spherical Earth and the path of the Earth about the sun affect climate.   902
2. Describe the air circulation about the Earth, and tell why deserts are apt to occur at 30° north and south of the equator. Why does the Pacific coast of California get plentiful rainfall?   902–3
3. How does a coastal mountain range affect climate? What causes a monsoon climate?   903

4. Name the terrestrial biomes you would expect to find when going from the base to the top of a mountain.   905
5. Describe the location, the climate, and the populations of the Arctic tundra, coniferous forests (both taiga and temperate rain forest), temperate deciduous forests, tropical rain forests, shrublands, grasslands (both temperate grasslands and savanna), and deserts.   906–14
6. Describe the importance of wetlands and the major types of wetlands.   915
7. Describe the overturn of a temperate lake, the life zones of a lake, and the organisms you would expect to find in each life zone.   916–17
8. Describe the coastal communities, and discuss the importance of estuaries to the productivity of the ocean.   918–19
9. Describe the ocean currents and how the Gulf Stream accounts for Great Britain having a mild temperature.   920
10. Describe the zones of the open ocean and the organisms you would expect to find in each zone.   922–23

Testing Yourself

Choose the best answer for each question.

1. The seasons are best explained by
   a. the distribution of temperature and rainfall in biomes.
   b. the tilt of the Earth as it orbits about the sun.
   c. the daily rotation of the Earth on its axis.
   d. the fact that the equator is warm and the poles are cold.

2. The mild climate of Great Britain is best explained by
   a. the winds called the westerlies.
   b. the spinning of the Earth on its axis.
   c. Great Britain being a mountainous country.
   d. the flow of ocean currents.

3. Which of these pairs is mismatched?
   a. tundra—permafrost
   b. savanna—acacia trees
   c. prairie—epiphytes
   d. coniferous forest—evergreen trees

4. All of these phrases describe the tundra except
   a. low-lying vegetation.
   b. northernmost biome.
   c. short growing season.
   d. many different types of species.

5. The forest with a multilevel understory is
   a. the tropical rain forest.
   b. the coniferous forest.
   c. the tundra.
   d. the temperate deciduous forest.

6. Why are lush evergreen forests present in the Pacific Northwest of the United States?
   a. The rotation of the Earth causes this.
   b. Winds blow from the ocean, bringing moisture.
   c. They are located on the leeward side of a mountain range.
   d. Both b and c are correct.
   e. All of these are correct.

7. Which of these influences the location of a particular biome?
   a. latitude
   b. average annual rainfall
   c. average annual temperature
   d. altitude
   e. All of these are correct.

8. Which of these is a function of a wetland?
   a. purifies water
   b. is an area where toxic wastes can be broken down
   c. helps absorb overflow and prevents flooding
   d. is a home for organisms that are links in food chains
   e. All of these are correct.

9. The area of a lake closest to shore and where rooted plants are found is
   a. the littoral zone.
   b. the limnetic zone.
   c. the profundal zone.
   d. the benthic zone.

10. An estuary acts as a nutrient trap because of
    a. the action of rivers and tides.
    b. the depth at which photosynthesis can occur.
    c. the amount of rainfall received.
    d. the height of the water table.

11. Which area of an ocean has the greatest concentration of nutrients?
    a. epipelagic zone and benthic zone
    b. epipelagic zone only
    c. benthic zone only
    d. neritic province

12. Plankton, both zooplankton and phytoplankton, would be found in which division of the ocean?
    a. oceanic province
    b. neritic province
    c. benthic division
    d. epipelagic zone
    e. Both b and d are correct.

13. Which zone of the oceanic province is completely dark?
    a. epipelagic zone
    b. mesopelagic zone
    c. bathypelagic zone
    d. abyssal zone

14. Runoff of fertilizer and animal wastes from a large farm that drains into a lake would be an example of which process?
    a. fall overturn
    b. eutrophication
    c. spring overturn
    d. upwelling

15. Phytoplankton are more likely to be found in which life zone of a lake?
    a. limnetic zone
    b. profundal zone
    c. benthic zone
    d. All of these are correct.

16. Which area of the seashore is only exposed during low tide?
    a. upper littoral zone
    b. mid littoral zone
    c. lower littoral zone
    d. None of these are correct.

17. Which of these would normally be found in the benthic zone of a lake?
    a. minnows
    b. clams
    c. zooplankton
    d. plants
    e. large fish

Thinking Scientifically

1. Pharmaceutical companies are interested in "bioprospecting" in tropical rain forests. These companies are looking for naturally occurring compounds in plants or animals that can be used as drugs for a variety of diseases. The most promising compounds act as antibacterial or antifungal agents. Even discounting the fact that the higher density of species in tropical rain forests would produce a wider array of compounds than another biome, why would antibacterial and antifungal compounds be more likely to evolve in this particular biome?

2. Many scientists predict that global warming will lead to a partial melting of glaciers at the north and south poles and a rise in sea level. Among the many places affected by such a rise would be the estuaries that now serve as breeding grounds for many birds and fish. While it is impossible to predict exactly how different estuaries would change in response to an increase in sea level, what possibilities can be predicted based on the general structure of estuaries?

Bioethical Issue: Water Pollution

Agricultural fertilizers are the chief cause of nitrate contamination of drinking-water wells. Excessive nitrates in a baby's bloodstream can lead to slow suffocation, known as blue-baby syndrome. Agricultural herbicides are suspected carcinogens in the tap water of scattered ecosystems coast to coast. What can be done?

Some farmers are already using irrigation methods that deliver water directly to plant roots, no-till agriculture that reduces the loss of topsoil and cuts back on herbicide use, and integrated pest management, which relies heavily on good bugs to kill bad bugs. Perhaps more should do so. Encouraged—in some cases, compelled—by state and federal agents, dairy farmers have built sheds, concrete containments, and underground liquid storage tanks to hold wastes when it rains. Later, the manure is trucked to fields and spread as fertilizer.

Homeowners, like business golf clubs and ski resorts, also contribute to the problem. The manicuring of lawns, the use of motor vehicles, and the construction and use of roads and buildings all add contaminants to streams, lakes, and aquifers. Citizens around Grand Traverse Bay on the eastern shore of Lake Michigan have also gotten the message, especially because they want to keep on enjoying water-dependent activities such as boating, swimming, and fishing. James Haverman, a concerned member of the Traverse Bay Watershed Initiative says, "If we can't change the way people live their everyday lives, we are not going to be able to make a difference." Builders in Traverse County are already required to control soil erosion with filter fences, steer rainwater away from exposed soil, build sediment basins, and plant protective buffers. Presently, homeowners must have a 25-foot setback from wetlands and a 50-foot setback from lakes and creeks. They are also encouraged to pump out their septic systems every two years. Do you approve of legislation that requires farmers and homeowners to protect freshwater supplies? Why or why not?

Understanding the Terms

alpine tundra 905
Arctic tundra 906
benthic division 923
biome 905
bog 915
chaparral 912
climate 902
coral reef 922
desert 914
El Niño–Southern
   Oscillation 920
epiphyte 910
estuary 918
eutrophication 916
fall overturn 916
grassland 912
hydrothermal vent 923
lake 916
littoral zone 919
marsh 915
monsoon 903
montane coniferous forest 905
pelagic division 922
permafrost 906
phytoplankton 917
rain shadow 903
savanna 912
shrubland 912
spring overturn 917
swamp 915
taiga 907
temperate deciduous
   forest 907
temperate rain forest 908
tropical rain forest 910
upwelling 920
wetland 915
zooplankton 917

Match the terms to these definitions:

a. _____ End of a river where fresh water and salt water mix as they meet.

b. _____ Ocean floor, which supports a unique set of organisms in contrast to the pelagic division.

c. _____ Terrestrial biome that is a coniferous forest extending in a broad belt across northern Eurasia and North America.

d. _____ Terrestrial biome that is a grassland in Africa, characterized by few trees and a severe dry season.

e. _____ Oxygen-rich top waters mix with nutrient-rich bottom waters in stratified lakes.

ARIS, the *Biology* Website

ARIS, the website for *Biology*, provides a wealth of information organized and integrated by chapter. You will find practice quizzes, interactive activities, labeling exercises, flashcards, and much more that will complement your learning and understanding of general biology.

www.mhhe.com/maderbiology9

50

CONSERVATION BIOLOGY

The growth of the human population and our economic activities have brought us face to face with the limits set by the natural environment. The danger of permanent environmental damage has successively increased, and the biosphere's biodiversity is being challenged. The biodiversity crisis demands a twofold response. We must take quick action to protect biodiversity, but we must also address the human forces that are threatening this valuable resource.

The 1990s saw the emergence of conservation biology as a branch of environmental science directed at managing our living heritage—biodiversity. Conservation biology is in part a basic science and in part an applied science. As a basic science, for example, conservation biology examines the nature and evolutionary origin of biodiversity. It also considers how changing environmental conditions, especially those due to human activities, are reducing biodiversity. As an applied science, however, its ultimate role is goal-oriented: the preservation and management of ecosystems for human welfare.

To illustrate that conservation measures are effective, consider the past plight of pelicans. In the previous century, the pesticide DDT easily entered various food chains after being used to control mosquitoes and other insect pests worldwide. Unfortunately, the presence of DDT in their food caused the eggs of birds to be extremely thin and easily crushed by incubating parents. Therefore, the size of bird populations, including those of pelicans, declined dramatically. The Environmental Protection Agency banned the use of DDT in the United States in 1972, and since then pelican populations have recovered nicely. Even so, DDT lingers and is still found in every body of water on Earth and in the tissues of organisms, including humans.

The brown pelican, *Pelecanus occidentalis*, is making a remarkable recovery from drastically reduced populations caused by DDT.

50.1 CONSERVATION BIOLOGY AND BIODIVERSITY

Conservation biology [L. *conservatio*, keep, save] is a relatively new discipline of biology that studies all aspects of biodiversity with the goal of conserving natural resources for this generation and all future generations. Conservation biology is unique in that it is concerned with both the development of scientific concepts and the application of these concepts to the everyday world. A primary goal is the management of biodiversity for sustainable use by humans. To achieve this goal, conservation biologists are interested in, and come from, many subfields of biology that only now have been brought together into a cohesive whole.

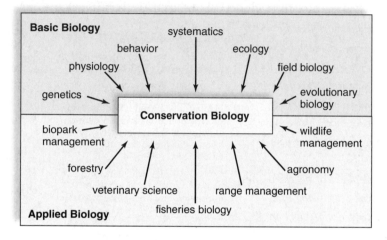

Like a physician, a conservation biologist must be aware of the latest findings, both theoretical and practical, and be able to use this knowledge to diagnose the source of trouble and suggest a suitable treatment. Often, it is necessary for conservation biologists to work with government officials at both the local and federal levels. Public education is another important duty of conservation biologists.

Conservation biology is a unique science in another way. It unabashedly supports the following ethical principles: (1) Biodiversity is desirable for the biosphere and therefore for humans; (2) extinctions, due to human actions, are therefore undesirable; (3) the complex interactions in ecosystems support biodiversity and are desirable; and (4) biodiversity brought about by evolutionary change has value in and of itself, regardless of any practical benefit.

Conservation biology has emerged in response to a crisis—never before in the history of the Earth are so many extinctions expected in such a short period of time. Estimates vary, but at least 10–20% of all species now living most likely will become extinct in the next 20–50 years unless immediate action is taken. It is urgently important, then, that all citizens understand the concept of biodiversity, the value of biodiversity, the likely causes of present-day extinctions, and what could be done to prevent extinctions from occurring.

To protect biodiversity, **bioinformatics,** the science of collecting and analyzing biological information, is applied.

Throughout the world, molecular, descriptive, and biogeographical information on organisms is being collected. Eventually this information will be used in understanding and protecting biodiversity.

Biodiversity

At its simplest level, **biodiversity** [Gk. *bio*, life; L. *diversus*, various] is the variety of life on Earth. It is common practice to describe biodiversity in terms of the number of species among various groups of organisms. Figure 50.1 only accounts for the species that have so far been described. It has been estimated that there may be between 10 and 50 million species in all; if so, many species are still to be found and described.

Of these, nearly 1,200 species in the United States and 30,000 species worldwide are in danger of extinction. An **endangered species** is one that is in peril of immediate extinction throughout all or most of its range. Examples of endangered species include the black lace cactus, armored snail, hawksbill sea turtle, California condor, West Indian manatee, and snow leopard. **Threatened species** are organisms that are likely to become endangered species in the foreseeable future. Examples of threatened species include the Navaho sedge, puritan tiger beetle, gopher tortoise, bald eagle, gray wolf, and Louisiana black bear.

To develop a meaningful understanding of life on Earth, we need to know more about species than their total number. Ecologists describe biodiversity as an attribute of three other levels of biological organization: genetic diversity, community diversity, and landscape diversity.

Genetic diversity refers to variations among the members of a population. Populations with high genetic diversity

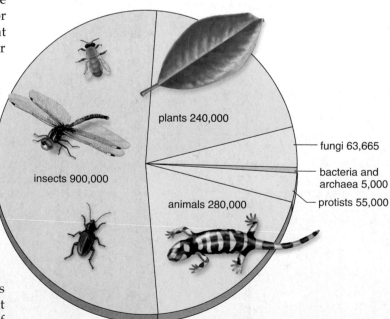

FIGURE 50.1 Number of described species.
There are only about 1.75 million described species; insects are far more prevalent than organisms in other groups. Undescribed species probably number far more than those species that have been described.

are more likely to have some individuals that can survive a change in the structure of their ecosystem. For example, the 1846 potato blight in Ireland, the 1922 wheat failure in the Soviet Union, and the 1984 outbreak of citrus canker in Florida were all made worse by limited genetic variation among these crops. If a species' populations are quite small and isolated, it is more likely to eventually become extinct because of a loss of genetic diversity. As organisms become endangered and threatened, they lose their genetic diversity.

Community (and therefore ecosystem) diversity is dependent on the interactions of species at a particular locale. One community's species composition can be completely different from that of other communities. Community composition, therefore, increases the amount of biodiversity in the biosphere. Although past conservation efforts frequently concentrated on saving particular charismatic species, such as the California condor, the black-footed ferret, or the spotted owl, this is a shortsighted approach. A better approach is to conserve species that have a critical role to play in an ecosystem. Saving an entire community can save many species, and the contrary is also true—disrupting a community threatens the existence of more than one species. Opossum shrimp, *Mysis relicta*, were introduced into Flathead Lake in Montana and its tributaries as food for salmon. The shrimp ate so much zooplankton that there was in the end far less food for the fish and ultimately for the grizzly bears and bald eagles as well (Fig. 50.2).

Landscape diversity involves a group of interacting ecosystems; within one **landscape,** for example, there may be plains, mountains, and rivers. Any of these ecosystems can be so fragmented that they are connected by only patches (remnants) or strips of land that allow organisms to move from one ecosystem to the other. Fragmentation of the landscape reduces reproductive capacity and food availability and can disrupt seasonal behaviors.

Distribution of Diversity

Biodiversity is not evenly distributed throughout the biosphere; therefore, protecting some areas will save more species than protecting other areas. Biodiversity is highest at the tropics, and it declines toward each pole on land, in fresh water, and in the ocean. Also, more species are found in the coral reefs of the Indonesian archipelago than in other coral reefs as one moves westward across the Pacific.

Some regions of the world are called **biodiversity hotspots** because they contain unusually large concentrations of species. Hotspots contain about 44% of all known higher plant species and 35% of all terrestrial vertebrate species but cover only about half of 1.4% of the Earth's land area. The island of Madagascar, the Cape region of South Africa, Indonesia, the coast of California, and the Great Barrier Reef of Australia are all biodiversity hotspots.

One surprise of late has been the discovery that rain forest canopies and the deep-sea benthos have many more species than formerly thought. Some conservationists refer to these two areas as biodiversity frontiers.

> Conservation biology is the study and sustainable management of biodiversity for the benefit of human beings. Biodiversity is often defined as a variety of life at the species level, but biodiversity is also important at the genetic, community (ecosystem), and landscape levels.

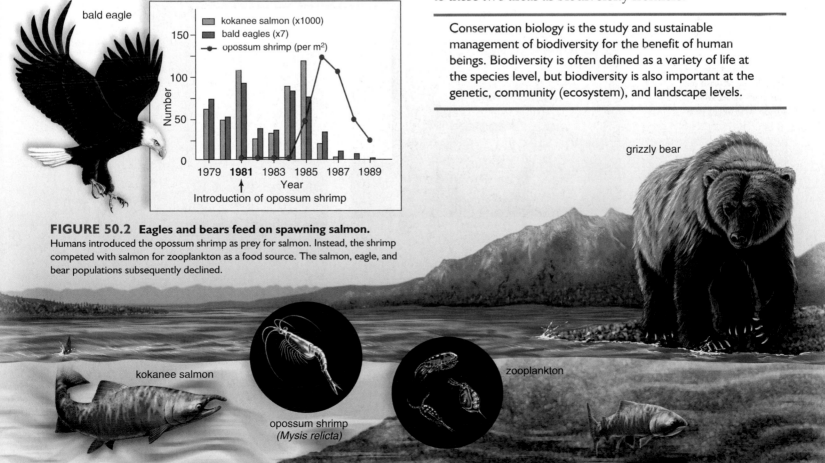

FIGURE 50.2 Eagles and bears feed on spawning salmon. Humans introduced the opossum shrimp as prey for salmon. Instead, the shrimp competed with salmon for zooplankton as a food source. The salmon, eagle, and bear populations subsequently declined.

50.2 VALUE OF BIODIVERSITY

Conservation biology strives to reverse the trend toward the possible extinction of tens of thousands of living things. To bring this about, it is necessary to make all people aware that biodiversity is a resource of immense value.

Direct Value

Various individual species perform services for human beings and contribute greatly to the value we should place on biodiversity. Only some of the most obvious values are discussed here and illustrated in Figure 50.3.

Medicinal Value

Most of the prescription drugs used in the United States, valued at over $200 billion, were originally derived from living organisms. The rosy periwinkle from Madagascar is an excellent example of a tropical plant that has provided us with useful medicines. Potent chemicals from this plant are now used to treat two forms of cancer: leukemia and Hodgkin disease. Because of these drugs, the survival rate for childhood leukemia has gone from 10% to 90%, and Hodgkin disease is usually curable. Although the value of saving a life cannot be calculated, it is still sometimes easier for us to appreciate the worth of a resource if it is explained in monetary terms. Thus, researchers tell us that, judging from the success rate in the past, an additional 328 types of drugs are yet to be found in tropical rain forests, and the value of this resource to society is probably $147 billion.

You may already know that the antibiotic penicillin is derived from a fungus and that certain species of bacteria produce the antibiotics tetracycline and streptomycin. These drugs have proven to be indispensable in the treatment of diseases, including certain sexually transmitted diseases.

Leprosy is among those diseases for which there is as yet no cure. The bacterium that causes leprosy will not grow in the laboratory, but scientists discovered that it grows naturally in the nine-banded armadillo. Having a source for the bacterium may make it possible to find a cure for leprosy. The blood of horseshoe crabs contains a substance called limulus amoebocyte lysate, which is used to ensure that medical devices such as pacemakers, surgical implants, and prosthetic devices are free of bacteria. Blood is taken from 250,000 crabs a year, and then they are returned to the sea unharmed.

Agricultural Value

Crops such as wheat, corn, and rice are derived from wild plants that have been modified to be high producers. The same high-yield, genetically similar strains tend to be grown worldwide. When rice crops in Africa were being devastated by a virus, researchers grew wild rice plants from thousands of seed samples until they found one that contained a gene for resistance to the virus. These wild plants were then used in a breeding program to transfer the gene into high-yield rice plants. If this variety of wild rice had become extinct be-

FIGURE 50.3

Direct value of wildlife.
The direct services of wild species benefit human beings immensely, and it is sometimes possible to calculate the monetary value, which is always surprisingly large.

Wild species, like the rosy periwinkle, *Catharanthus roseus,* are sources of many medicines.

Wild species, like many marine species, provide us with food.

Wild species, like the nine-banded armadillo, *Dasypus novemcinctus,* play a role in medical research.

fore it could be discovered, rice cultivation in Africa might have collapsed.

Biological pest controls—natural predators and parasites—are often preferable to using chemical pesticides. When a rice pest called the brown planthopper became resistant to pesticides, farmers began to use natural brown planthopper enemies instead. The economic savings were calculated at well over $1 billion. Similarly, cotton growers in Cañete Valley, Peru, found that pesticides were no longer working against the cotton aphid because of resistance. Research identified natural predators that are now being used to an ever greater degree by cotton farmers. Again, savings have been enormous.

Most flowering plants are pollinated by animals, such as bees, wasps, butterflies, beetles, birds, and bats. The honeybee, *Apis mellifera*, has been domesticated, and it pollinates almost $10 billion worth of food crops annually in the United States. The danger of this dependency on a single species is exemplified by mites, which have now wiped out more than 20% of the commercial honeybee population in the United States. Where can we get resistant bees? From the wild, of course. The value of wild pollinators to the U.S. agricultural economy has been calculated at $4.1–$6.7 billion a year.

Consumptive Use Value

Humans have had much success cultivating crops, keeping domesticated animals, growing trees in plantations, and so forth. But so far, aquaculture, the growing of fish and shellfish for human consumption, has contributed only minimally to human welfare—instead, most freshwater and marine harvests depend on the catching of wild animals, such as fishes (e.g., trout, cod, tuna, and flounder), crustaceans (e.g., lobsters, shrimps, and crabs), and mammals (e.g., whales). Obviously, these aquatic organisms are an invaluable biodiversity resource.

The environment provides a variety of other products that are sold in the marketplace worldwide, including wild fruits and vegetables, skins, fibers, beeswax, and seaweed. Also, by hunting and fishing, some people obtain their meat directly from the environment. In one study, researchers calculated that the economic value of wild pig in the diet of native hunters in Sarawak, East Malaysia, was approximately $40 million per year.

Similarly, many trees are still felled in the natural environment for their wood. Researchers have calculated that a species-rich forest in the Peruvian Amazon is worth far more if the forest is used for fruit and rubber production than for timber production. Fruit and the latex needed to produce rubber can be brought to market for an unlimited number of years, whereas once the trees are gone, no more timber can be harvested.

Wild species directly provide us with all sorts of goods and services whose monetary value can sometimes be calculated.

Wild species, like the lesser long-nosed bat, *Leptonycteris curasoae*, are pollinators of agricultural and other plants.

Wild species, like ladybugs, *Coccinella*, play a role in biological control of agricultural pests.

Wild species, like rubber trees, *Hevea*, can provide a product indefinitely if the forest is not destroyed.

Indirect Value

The wild species we have been discussing live in ecosystems. If we want to preserve them, it is more economical to save the ecosystems than the individual species. Ecosystems perform many services for modern humans, who increasingly live in cities. These services are said to be indirect because they are pervasive and not easily discernible (Fig. 50.4). Even so, our very survival depends on the functions that ecosystems perform for us.

Biogeochemical Cycles

Ecosystems are characterized by energy flow and chemical cycling. The biodiversity within ecosystems contributes to the workings of the water, carbon, nitrogen, phosphorus, and other biogeochemical cycles. We are dependent on these cycles for fresh water, removal of carbon dioxide from the atmosphere, uptake of excess soil nitrogen, and provision of phosphate. When human activities upset the usual workings of biogeochemical cycles, the dire environmental consequences include the release of excess pollutants that are harmful to us. Technology is unable to artificially contribute to or create any of the biogeochemical cycles.

Waste Disposal

Decomposers break down dead organic matter and other types of wastes to inorganic nutrients that are used by the producers within ecosystems. This function aids humans immensely because we dump millions of tons of waste material into natural ecosystems each year. If it were not for decomposition, waste would soon cover the entire surface of our planet. We can build sewage treatment plants, but they are expensive, and few of them break down solid wastes completely to inorganic nutrients. It is less expensive and more efficient to water plants and

a.

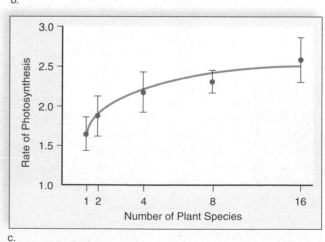

b.

c.

FIGURE 50.4 Indirect value of ecosystems.
a. Natural ecosystems provide (**b**) human-impacted ecosystems with many ecological services. **c.** Research results show that the higher the biodiversity (measured by number of plant species), the greater the rate of photosynthesis of an experimental community.

trees with partially treated wastewater and let soil bacteria cleanse it completely.

Biological communities are also capable of breaking down and immobilizing pollutants, such as heavy metals and pesticides, that humans release into the environment. A review of wetland functions in Canada assigned a value of $50,000 per hectare (100 acres, or 10,000 m^2) per year to the ability of natural areas to purify water and take up pollutants.

Provision of Fresh Water

Few terrestrial organisms are adapted to living in a salty environment—they need fresh water. The water cycle continually supplies fresh water to terrestrial ecosystems. Humans use fresh water in innumerable ways, including drinking it and irrigating their crops. Freshwater ecosystems such as rivers and lakes also provide us with fish and other types of organisms for food.

Unlike other commodities, there is no substitute for fresh water. We can remove salt from seawater to obtain fresh water, but the cost of desalination is about four to eight times the average cost of fresh water acquired via the water cycle.

Forests and other natural ecosystems exert a "sponge effect." They soak up water and then release it at a regular rate. When rain falls in a natural area, plant foliage and dead leaves lessen its impact, and the soil slowly absorbs it, especially if the soil has been aerated by organisms. The water-holding capacity of forests reduces the possibility of flooding. The value of a marshland outside Boston, Massachusetts, has been estimated at $72,000 per hectare per year solely on its ability to reduce floods. Forests release water slowly for days or weeks after the rains have ceased. Rivers flowing through forests in West Africa release twice as much water halfway through the dry season, and between three and five times as much at the end of the dry season, as do rivers from coffee plantations.

Prevention of Soil Erosion

Intact ecosystems naturally retain soil and prevent soil erosion. The importance of this ecosystem attribute is especially observed following deforestation. In Pakistan, the world's largest dam, the Tarbela Dam, is losing its storage capacity of 12 billion m^3 many years sooner than expected because silt is building up behind the dam due to deforestation. At one time, the Philippines were exporting $100 million worth of oysters, mussels, clams, and cockles each year. Now silt carried down rivers following deforestation is smothering the mangrove ecosystem that serves as a nursery for the sea. Most coastal ecosystems are not as bountiful as they once were because of deforestation and a myriad of other assaults.

Regulation of Climate

At the local level, trees provide shade and reduce the need for fans and air conditioners during the summer. Proper placement of shade trees near a home can reduce energy bills.

Globally, forests ameliorate the climate because they take up carbon dioxide. The leaves of trees use carbon dioxide when they photosynthesize, and the bodies of the trees store carbon. When trees are cut and burned, carbon dioxide is released into the atmosphere. Carbon dioxide makes a significant contribution to global warming, which is expected to be stressful for many plants and animals. Only a small percentage of wildlife will be able to move northward, where the weather will be suitable for them.

Ecotourism

Almost everyone prefers to vacation in the natural beauty of an ecosystem. In the United States, nearly 100 million people enjoy vacationing in a natural setting. To do so, they spend $4 billion each year on fees, travel, lodging, and food. Many tourists want to go sport fishing, whale watching, boat riding, hiking, birdwatching, and the like. Others want to merely immerse themselves in the beauty and serenity of a natural environment.

Biodiversity and Natural Ecosystems

Massive changes in biodiversity, such as deforestation, have a significant impact on ecosystem services such as those we have been discussing. Researchers are interested in determining whether a high degree of biodiversity also helps ecosystems function more efficiently. To test the benefits of biodiversity in a Minnesota grassland habitat, researchers sowed plots with seven levels of plant diversity. Their study found that ecosystem performance improves with increasing species richness. The more diverse the plots, the lower the concentrations of inorganic soil nitrogen, indicating a higher level of nitrate uptake. A similar study in California also showed greater overall resource use in more diverse plots.

Another group of experimenters tested the effects of an increase in diversity at four levels: producers, herbivores, parasites, and decomposers. They found that the rate of photosynthesis increased as diversity increased (Fig. 50.4c). A computer simulation has shown that the response of a deciduous forest to elevated carbon dioxide is a function of species diversity. The more complex community, composed of nine tree species, exhibited a 30% greater amount of photosynthesis than a community composed of a single species.

More studies are needed to test whether biodiversity maximizes resource acquisition and retention within an ecosystem. Also, are more diverse ecosystems better able to withstand environmental changes and invasions by other species, including pathogens? Then, too, how does fragmentation affect the distribution of organisms within an ecosystem and the functioning of an ecosystem?

Ecosystems perform services that most likely depend on a high degree of biodiversity.

50.3 CAUSES OF EXTINCTION

To stem the tide of extinction due to human activities, it is first necessary to identify its causes. Researchers examined the records of 1,880 threatened and endangered wild species in the United States and found that habitat loss was involved in 85% of the cases (Fig. 50.5a). Exotic species had a hand in nearly 50%, pollution was a factor in 24%, overexploitation in 17%, and disease in 3%. The percentages add up to more than 100% because most of these species are imperiled for more than one reason. Macaws are a good example that a combination of factors can lead to a species decline (Fig. 50.5b). Not only has their habitat been reduced by encroaching timber and mining companies, but macaws are also hunted for food and collected for the pet trade.

Habitat Loss

Habitat loss has occurred in all ecosystems, but concern has now centered on tropical rain forests and coral reefs because they are particularly rich in species. A sequence of events in Brazil offers a fairly typical example of the manner in which rain forest is converted to land uninhabitable for wildlife. The construction of a major highway into the forest first provided a way to reach the interior of the forest (Fig. 50.5c). Small towns and industries sprang up along the highway, and roads branching off the main highway gave rise to even more roads. The result was fragmentation of the once immense forest. The government offered subsidies to anyone willing to take up residence in the forest, and the people who came cut and burned trees in patches (Fig. 50.5d). Tropical soils contain limited nutrients, but when the trees are burned, nutrients are released that support a lush growth for the grazing of cattle for about three years. However, once the land was degraded (Fig. 50.5e), the farmers moved on to another portion of the forest to start over again.

Loss of habitat also affects freshwater and marine biodiversity. Coastal degradation is mainly due to the large concentration of people living on or near the coast. Already, 60% of coral reefs have been destroyed or are on the verge of destruction; it is possible that all coral reefs may disappear during the next 40 years. Mangrove forest destruction is also a problem; Indonesia, with the most mangrove acreage, has lost 45% of its mangroves, and the percentage is even higher for other tropical countries. Wetland areas, estuaries, and seagrass beds are also being rapidly destroyed by the actions of humans.

FIGURE 50.5 Habitat loss.
a. In a study examining records of imperiled U.S. plants and animals, habitat loss emerged as the greatest threat to wildlife. **b.** Macaws that reside in South American tropical rain forests are endangered for some of the reasons listed in the graph in (a). **c.** The construction of roads in an area in Brazil opened up the rain forest and subjected it to fragmentation. **d.** The result was patches of forest and degraded land. **e.** Wildlife could not live in destroyed portions of the forest.

ecology focus

Cyanide Fishing and Coral Reefs

Coral reefs are areas of biological abundance found in shallow, warm tropical waters just below the surface of the water. They occur in the Caribbean and off the coasts of the Southern Hemisphere continents. The Great Barrier Reef off the coast of Australia is the largest coral reef in the world. The chief constituents of a coral reef are stony corals, animals that exist as small polyps with a calcium carbonate (limestone) exterior (Fig. 50A). Corals do not usually occur individually; rather they form colonies derived from just one individual. When the corals die off, they leave behind a stony, branching limestone structure. The large number of crevices and caves of a reef provide shelter for many animals, including sponges, nudibranchs, small fish (groupers, clownfish, parrotfish, snapper, and scorpion fish), jellyfish, anemones, sea stars (including the destructive Crown of Thorns), crustaceans (like crabs, shrimp, and lobsters), turtles, sea snakes, snails, and molluscs (like octopuses, nautiluses, and clams). Barracudas, moray eels, and sharks are top predators in coral reefs.

Coral reefs are such incredibly diverse ecosystems that they are often referred to as the "rain forests of the sea." They are home to thousands of known species, with perhaps millions more yet to be documented. Coral reef degradation is a worldwide problem that involves overfishing for the food and aquarium trades, careless tourist divers, impacts from boat anchors and propellers, oil spills, nutrient pollution, global climate change, and an increase in coral diseases. It has been estimated that 58% of all coral reefs are being harmed by human activities. Even so, many aquarium hobbyists enjoy collecting and caring for the beautiful fishes, sea stars, sea anemones, and other organisms that call the reef home.

One of the most damaging methods of collecting aquarium fish from coral reefs is known as cyanide fishing (Fig. 50A). Cyanide is a poison that interferes with the mitochondrial electron transport chain and effectively shuts down cellular respiration. Reef fish can be challenging for divers to capture with handheld nets alone, since they are able to seek refuge in hard-to-reach nooks and crannies. How-

Figure 50A Coral reef ecosystem.
Gathering tropical fish in coral reefs with cyanide.

ever, spewing a cyanide solution over fish stuns them and makes them much easier to catch. Unfortunately, cyanide dosing in open water is a tricky proposition. Many fish die immediately, and even more expire later from aftereffects of the poison. To make matters worse, the coral itself, which takes thousands of years to grow and form a reef, can succumb to cyanide. Even the divers do not always escape unscathed; they can inadvertently poison themselves.

The United States imports almost half of all marine aquarium organisms. Approximately two-thirds of them come from the Philippines, where cyanide fishing began, and Indonesia, throughout which the practice has spread. Even though cyanide fishing is officially illegal, many of the people in this region live in poverty. As a result, numerous individuals feel that they must assign a lower priority to protection of the coral reefs than the effort to earn a living.

There is hope that cyanide fishing may become a less popular way of obtaining organisms for the aquarium trade, as hobbyists and retailers become better educated about the harmful effects of this practice. The Marine Aquarium Council (MAC) is a nonprofit, international organization that certifies marine organisms as having been harvested using environmentally responsible methods. However, some scientists contend that removing fish by any means may be detrimental to the reef community, since some of the most popular aquarium fish are herbivores that keep the plants of the reef from growing too much and overwhelming the coral.

Exotic Species

Exotic species, sometimes called alien species, are non-native members of an ecosystem. Ecosystems around the globe are characterized by unique assemblages of organisms that have evolved together in one location. Migrating to a new location is not usually possible because of barriers such as oceans, deserts, mountains, and rivers. Humans, however, have introduced exotic species into new ecosystems chiefly due to:

Colonization Europeans, in particular, brought various familiar species with them when they colonized new places. For example, the pilgrims brought the dandelion to the United States as a familiar salad green. In addition, they introduced pigs to North America that have become feral, reverting to their wild state. In some parts of the United States, feral pigs are very destructive.

Horticulture and agriculture Some exotics now taking over vast tracts of land have escaped from cultivated areas. Kudzu is a vine from Japan that the U.S. Department of Agriculture thought would help prevent soil erosion. The plant now covers much landscape in the South, including even walnut, magnolia, and sweet gum trees (Fig. 50.6*a*). The water hyacinth was introduced to the United States from South America because of its beautiful flowers. Today, it clogs up waterways and diminishes natural diversity.

Accidental transport Global trade and travel accidentally bring many new species from one country to another. Researchers found that the ballast water released from ships into Coos Bay, Oregon, contained 367 marine species from Japan. The zebra mussel from the Caspian Sea was accidentally introduced into the Great Lakes in 1988. It now forms dense beds that squeeze out native mussels. Other organisms accidentally introduced into the United States include the Formosan termite, the Argentinian fire ant, and the nutria, a type of rodent.

Exotic species can disrupt food webs. As mentioned earlier, opossum shrimp introduced into a lake in Montana added a trophic level that in the end meant less food for bald eagles and grizzly bears (see Fig. 50.2).

Exotics on Islands

Islands are particularly susceptible to environmental discord caused by the introduction of exotic species. Islands have unique assemblages of native species that are closely adapted to one another and cannot compete well against exotics. Myrtle trees, *Myrica faya,* introduced into the Hawaiian Islands from the Canary Islands, are symbiotic with a type of bacterium that is capable of nitrogen fixation. This feature allows the species to establish itself on nutrient-poor volcanic soil, a distinct advantage in Hawaii. Once established, myrtle trees call a halt to the normal succession of native plants on volcanic soil.

The brown tree snake has been introduced onto a number of islands in the Pacific Ocean. The snake eats eggs, nestlings, and adult birds. On Guam, it has reduced ten native bird species to the point of extinction. On the Galápagos Islands, black rats have reduced populations of giant tortoise, while goats and feral pigs have changed the vegetation from highland forest to pampaslike grasslands and destroyed stands of cactus. In Australia, mice and rabbits have stressed native marsupial populations. Mongooses introduced into the Hawaiian Islands to control rats also prey on native birds (Fig. 50.6*b*).

Pollution

In the present context, **pollution** can be defined as any environmental change that adversely affects the lives and health of living things. Pollution has been identified as the third main cause of extinction. Pollution can also weaken organisms and lead to disease, the fifth main cause of extinction. Biodiversity is particularly threatened by the following types of environmental pollution:

Acid deposition Both sulfur dioxide from power plants and nitrogen oxides in automobile exhaust are converted to acids when they combine with water vapor in the atmosphere. These acids return to Earth as either wet deposition (acid rain or snow) or dry deposition (sulfate and nitrate salts). Sulfur dioxide and nitrogen oxides are emitted in one locale, but deposition occurs across state and national boundaries. Acid deposition causes trees to weaken and increases their susceptibility to disease and insects. It also kills small invertebrates and decomposers so that the entire ecosystem is threatened. Many lakes in the northern United States are now lifeless because of the effects of acid deposition.

Eutrophication Lakes are also under stress due to over-enrichment. When lakes receive excess nutrients due to runoff from agricultural fields and wastewater from sewage treatment, algae begin to grow in

a.   b.

FIGURE 50.6 Exotic species.
a. Kudzu, a vine from Japan, was introduced in several southern states to control erosion. Today, kudzu has taken over and displaced many native plants.
b. Mongooses were introduced into Hawaii to control rats, but they also prey on native birds.

abundance. An algal bloom is apparent as a green scum or excessive mats of filamentous algae. Upon death, the decomposers break down the algae, but in so doing, they use up oxygen. A decreased amount of oxygen is available to fish, leading sometimes to a massive fish kill.

Ozone depletion The ozone shield is a layer of ozone (O_3) in the stratosphere, some 50 km above the Earth. The ozone shield absorbs most of the wavelengths of harmful ultraviolet (UV) radiation so that they do not strike the Earth. The cause of ozone depletion can be traced to chlorine atoms (Cl^-) that come from the breakdown of chlorofluorocarbons (CFCs). The best-known CFC is Freon, a heat transfer agent still found in refrigerators and air conditioners today. Severe ozone shield depletion can impair crop and tree growth and also kill plankton (microscopic plant and animal life) that sustain oceanic life. The immune system and the ability of all organisms to resist infectious diseases will most likely be weakened.

Organic chemicals Our modern society uses organic chemicals in all sorts of ways. Organic chemicals called nonylphenols are used in products ranging from pesticides to dishwashing detergents, cosmetics, plastics, and spermicides. These chemicals mimic the effects of hormones, and in that way most likely harm wildlife. Salmon are born in fresh water but mature in salt water. After investigators exposed young fish to nonylphenol, they found that 20–30% were unable to make the transition between fresh and salt water. Nonylphenols cause the pituitary to produce prolactin, a hormone that may prevent saltwater adaptation.

Global warming The expression **global warming** refers to an expected increase in average temperature during the twenty-first century. You may recall from Chapter 48 that carbon dioxide is a gas that comes from the burning of fossil fuels, and methane is a gas that comes from oil and gas wells, rice paddies, and animals. These gases are known as greenhouse gases because, just like the panes of a greenhouse, they allow solar radiation to pass through but hinder the escape of its heat back into space. Data collected around the world show a steady rise in the concentration of the various greenhouse gases. These data are used to generate computer models that predict the Earth may warm to temperatures never before experienced by living things (Fig. 50.7*a*).

As the oceans warm, temperatures in the polar regions will rise to a greater degree than in other regions. The sea level will then rise because glaciers will melt and water expands as it warms. A 1 m rise in sea level could inundate 25–50% of U.S. coastal wetlands. This loss of habitat could be higher if wetlands cannot move inward because of coastal development and levees.

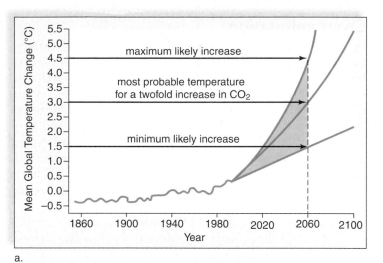

a.

b.

FIGURE 50.7
Global warming.
a. Mean global temperature change is expected to rise due to the introduction of greenhouse gases into the atmosphere. Global warming has the potential to significantly affect the world's biodiversity.
b. A temperature rise of only a few degrees causes coral reefs to "bleach" and become lifeless.

The tropics will also feel the effects of global warming. The growth of corals is very dependent on mutualistic algae living in their walls. When the temperature rises by 4 degrees, corals expel their algae and are said to be "bleached" (Fig. 50.7*b*). Almost no growth or reproduction occurs until the algae return. Also, coral reefs prefer shallow waters, and if sea levels rise, they may "drown." Multiple assaults on coral are even now causing them to be stricken with various diseases.

Global warming could very well cause many extinctions on land also. As temperatures rise, regions of suitable climate for various species will shift toward the poles and higher elevations. The present assemblages of species in ecosystems will be disrupted as some species migrate northward, leaving others behind. Plants migrate when seeds disperse and growth occurs in a new locale. For example, to remain in a favorable habitat, it's been calculated that the rate of beech tree migration would have to be 40 times faster than has ever been observed. It seems unlikely that beech or any other type of tree would be able to meet the pace required. Then, too, many species of organisms are confined to relatively small habitat patches that are surrounded by agricultural or urban areas they would not be able to cross. And even if they have the capacity to disperse to new sites, suitable habitats may not be available.

Overexploitation

Overexploitation occurs when the number of individuals taken from a wild population is so great that the population becomes severely reduced in numbers. A positive feedback cycle explains overexploitation: The smaller the population, the more valuable its members, and the greater the incentive to capture the few remaining organisms. Poachers and participants in organized crime are very active in the collecting and sale of endangered and threatened species because it has become so lucrative. The overall international value of trading wildlife species is $20 billion, of which $8 billion is attributed to the illegal sale of rare species.

Markets for decorative plants and exotic pets support both legal and illegal trade in wild species. Rustlers dig up rare cactuses such as the single-crested saguaro and sell them to gardeners for as much as $15,000 each. Parakeets and macaws are among the birds taken from the wild for sale to pet owners. For every bird delivered alive, many more have died in the process. The same holds true for tropical fish, which often come from the coral reefs of Indonesia and the Philippines. Divers dynamite reefs or use plastic squeeze-bottles of cyanide to stun them; in the process, many fish and valuable corals die.

Declining species of mammals, such as the Siberian tiger, are still hunted for their hides, tusks, horns, or bones. Because of its rarity, a single Siberian tiger is now worth more than $500,000—its bones are pulverized and used as a medicinal powder. The horns of rhinoceroses become ornate carved daggers, and their bones are ground up to sell as a medicine. The ivory of an elephant's tusk is used to make art objects, jewelry, or piano keys. The fur of a Bengal tiger sells for as much as $100,000 in Tokyo.

The U.N. Food and Agricultural organization tells us that we have now overexploited 11 of 15 major oceanic fishing areas. Fish are a renewable resource if harvesting does not exceed the ability of the fish to reproduce. Our society uses larger and more efficient fishing fleets to decimate fishing stocks. Pelagic species such as tuna are captured by purse-seine fishing, in which a very large net surrounds a school of fish, and then the net is closed in the same manner as a drawstring purse. Up to thousands of dolphins that swim above schools of tuna are often captured and then killed in this type of net. Other fishing boats drag huge trawling nets, large enough to accommodate 12 jumbo jets, along the seafloor to capture bottom-dwelling fish (Fig. 50.8a). Only large fish are kept; undesirable small fish and sea turtles are discarded, dying, back into the ocean. Trawling has been called the marine equivalent of clear-cutting trees because after the net goes by, the sea bottom is devastated (Fig. 50.8b). Today's fishing practices don't allow fisheries to recover. Cod and haddock, once the most abundant bottom-dwelling fish along the northeast coast of the United States, are now often outnumbered by dogfish and skate.

A marine ecosystem can be disrupted by overfishing, as exemplified on the U.S. west coast. When sea otters began to decline in numbers, investigators found that they were being eaten by orcas (killer whales). Usually orcas prefer seals and sea lions to sea otters, but they began eating sea otters when few seals and sea lions could be found. What caused a decline in seals and sea lions? Their preferred food sources—perch and herring—were no longer plentiful due to overfishing. Ordinarily, sea otters keep the population of sea urchins, which feed on kelp, under control. But with fewer sea otters around, the sea urchin population exploded and decimated the kelp beds. Thus, overfishing set in motion a chain of events that detrimentally altered the food web of an ecosystem.

The five main causes of extinction are disease, habitat loss, introduction of exotic species, pollution, and overexploitation.

FIGURE 50.8

Trawling.

a. These Alaskan pollock were caught by dragging a net along the seafloor. **b.** Appearance of the seabed before *(top)* and after *(bottom)* the net passed.

a.

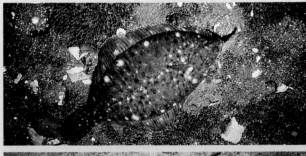

b.

50.4 CONSERVATION TECHNIQUES

Despite the value of biodiversity to our very survival, human activities are causing the extinction of thousands of species a year. Clearly, we need to reverse this trend and preserve as many species as possible. Habitat preservation and habitat restoration are important in preserving biodiversity.

Habitat Preservation

Preservation of a species' habitat is of primary concern, but first we must decide which species to preserve. As mentioned previously, the biosphere contains biodiversity hotspots, relatively small areas having a concentration of endemic (native) species not found anyplace else. In the tropical rain forests of Madagascar, 93% of the primate species, 99% of the frog species, and over 80% of the plant species are endemic to Madagascar. Preserving these forests and other hotspots will save a wide variety of organisms.

Keystone species are species that influence the viability of a community, although their numbers may not be excessively high. The extinction of a keystone species can lead to other extinctions and a loss of biodiversity. For example, bats are designated a keystone species in tropical forests of the Old World. They are pollinators that also disperse the seeds of trees. When bats are killed off and their roosts destroyed, the trees fail to reproduce. The grizzly bear is a keystone species in the northwestern United States and Canada (Fig. 50.9a). Bears disperse the seeds of berries; as many as 7,000 seeds may be in one dung pile. Grizzlies kill the young of many hoofed animals and thereby keep their populations under control. Grizzlies are also a principal mover of soil when they dig up roots and prey upon hibernating ground squirrels and marmots. Other keystone species are beavers in wetlands, bison in grasslands, alligators in swamps, and elephants in grasslands and forests.

Keystone species should not be confused with flagship species, which evoke a strong emotional response in humans. Flagship species are considered charismatic and are treasured for their beauty, cuteness, and regal nature. These species can motivate the public to preserve biodiversity. Flagship species include lions, tigers, dolphins, and the giant panda.

Metapopulations

The grizzly bear population is actually a metapopulation [Gk. *meta*, between; L. *populus*, people], a population subdivided into several small, isolated populations due to habitat fragmentation. Originally there were probably 50,000–100,000 grizzlies south of Canada, but this number has been reduced because communities have encroached on their home range and bears have been killed by frightened homeowners. Now there are six virtually isolated subpopulations totaling about 1,000 individuals. The Yellowstone National Park population numbers 200, but the others are even smaller.

Saving metapopulations sometimes requires determining which of the populations is a source and which are sinks. A source population is one that most likely lives in a

a. Grizzly bear, *Ursus arctos horribilis*

b. Old-growth forest; northern spotted owl, *Strix occidentalis caurina* (inset)

FIGURE 50.9

Habitat preservation.
When particular species are protected, other wildlife benefits. **a.** The Greater Yellowstone Ecosystem has been delineated in an effort to save grizzly bears, which need a very large habitat. **b.** Currently, the remaining portions of old-growth forests in the Pacific Northwest are not being logged in order to save the northern spotted owl (inset).

favorable area, and its birthrate is most likely higher than its death rate. Individuals from source populations move into sink populations, where the environment is not as favorable and where the birthrate equals the death rate at best. When trying to save the northern spotted owl, conservationists determined that it was best to avoid having owls move into sink habitats. The northern spotted owl reproduces successfully in old-growth rain forests of the Pacific Northwest (Fig. 50.9b) but not in nearby immature forests that are in the process of recovering from logging. Distinct boundaries that hindered the movement of owls into these sink habitats proved to be beneficial in maintaining source populations.

Landscape Preservation

Grizzly bears inhabit a number of different types of ecosystems, including plains, mountains, and rivers. Saving any one of these types of ecosystems alone would not be sufficient to preserve grizzly bears. Instead, it is necessary to save diverse ecosystems that are at least connected by corridors. You will recall that a landscape encompasses different types of ecosystems. An area called the Greater Yellowstone Ecosystem, where bears are free to roam, has now been defined. It contains millions of acres in Yellowstone National Park; state lands in Montana, Idaho, and Wyoming; five different national forests; various wildlife refuges; and even private lands.

Landscape protection for one species is often beneficial for other wildlife that share the same space. The last of the contiguous 48 states' harlequin ducks, bull trout, westslope cutthroat trout, lynx, pine martens, wolverines, mountain caribou, and great gray owls are found in areas occupied by grizzlies. The recent return of gray wolves has occurred in this territory also. Then, too, grizzly range overlaps with 40% of Montana's vascular plants of special conservation concern.

The Edge Effect.   When preserving landscapes, it is necessary to consider the **edge effect.** An edge reduces the amount of habitat typical of an ecosystem because the edges around a patch have a habitat slightly different from the interior of the patch. For example, forest edges are brighter, warmer, drier, and windier, with more vines, shrubs, and weeds than the forest interior. Also, Figure 50.10a shows that a small and a large patch of habitat have the same amount of edge; therefore, the effective habitat shrinks as a patch gets smaller.

Many popular game animals, such as turkeys and white-tailed deer, are more plentiful in the edge region of a particular area. However, today it is known that creating edges can be detrimental to wildlife because of fragmentation.

The edge effect can have a serious impact on population size. Songbird populations west of the Mississippi have been declining of late, and ornithologists have noticed that the nesting success of songbirds is quite low at the edge of a forest. The cause turns out to be the brown-headed cowbird, a social parasite of songbirds. Adult cowbirds prefer to feed in open agricultural areas, and they only briefly enter the forest when searching for a host nest in which to lay their eggs (Fig. 50.10b). Cowbirds are therefore benefited, while songbirds are disadvantaged, by the edge effect.

Computer Analyses

Two types of computer analyses, in particular, are now available to help conservationists plan how best to protect a species.

Gap analysis is used to find gaps in preservation—places where biodiversity is high outside of preserved areas. First, computerized maps are drawn up showing the topography, vegetation, hydrology, and land ownership of a region. Then computer maps are done showing the geographic distribution of a region's animal and plant species. Once the distribution maps are superimposed onto the land-use maps, it is obvious where preserved habitats still need to be and/or could be located.

A **population viability analysis** can help researchers determine how much habitat a species requires to maintain itself. First it is necessary to calculate the minimum population size needed to prevent extinction. This size should protect the species from unforeseen events such as natural catastrophes or chance swings in the birth and death rates. Another component to consider is the size needed to protect genetic diversity. This number varies according to the species. For example, analysis of red-cockaded woodpecker populations showed that an adult population of about 1,323 birds is needed to result in a genetically effective population of 500 offspring because of the breeding system of the species. After you know the minimum population size, you can determine how much total acreage is needed for that population.

All life history characteristics of organisms must be taken into account when a population viability analysis is done. For example, female grizzlies do not give birth until

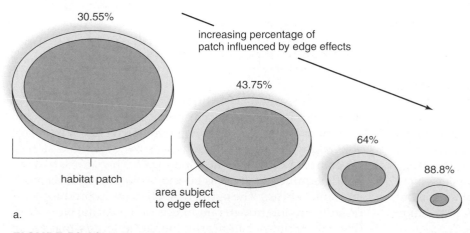

30.55%

increasing percentage of patch influenced by edge effects

43.75%

64%

88.8%

habitat patch

area subject to edge effect

a.

brown-headed cowbird chick

yellow warbler chick

b.

FIGURE 50.10 Edge effect.
a. The smaller the patch, the greater the proportion that is subject to the edge effect. **b.** Cowbirds lay their eggs in the nests of songbirds (yellow warblers). A cowbird is bigger than a warbler nestling and will be able to acquire most of the food brought by the warbler parent.

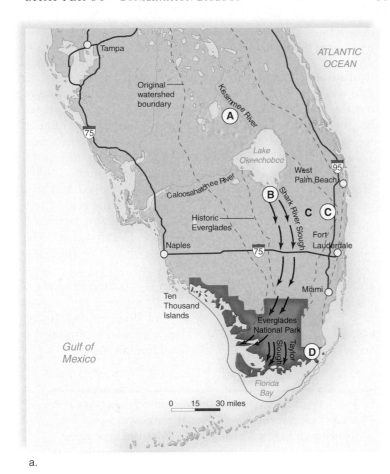

a.

Florida panther, *Puma concolor coryi*

American alligator, *Alligator mississippiensis*

White Ibis, *Eudocimus albus*   Roseate spoonbill, *Ajaia ajaia*
b.

FIGURE 50.11 Restoration of the Everglades.
a. Restoration plans call for adding curves and habitat back to the Kissimmee River **(A)**; creating large marshes and making the Shark River Slough free-flowing again **(B)**; creating a buffer zone of wetlands between urban development along Florida's eastern coast **(C)**; and reducing salinity by letting fresh water flow into and through Taylor Slough **(D)**. **b.** Wildlife of the Florida Everglades.

they are five or six, and then they typically wait three years before reproducing again. After doing one of the first population viability analyses, Mark Shaffer of the Wilderness Society predicted that a total grizzly bear population of 70–90 individuals, each with a suitable home range of 9,600 km, will have about a 95% chance of surviving for 100 years. But Fred Allendorf pointed out that because only a few dominant males breed, the population needs to be larger than this to protect genetic diversity. Also, to prevent inbreeding, he recommended the introduction of one or two unrelated bears each decade into populations of 100 individuals. The bottom line is that dispersal among subpopulations is needed to prevent inbreeding and extinction.

Habitat Restoration

Restoration ecology is a new subdiscipline of conservation biology that seeks scientific ways to return ecosystems to their former state. Three principles have so far emerged. First, it is best to begin as soon as possible before remaining fragments of the original habitat are lost. These fragments are sources of wildlife and seeds from which to restock the restored habitat. Second, once the natural history of the habitat is understood, it is best to use biological techniques

that mimic natural processes to bring about restoration. This might take the form of using controlled burns to bring back grassland habitats, biological pest controls to rid the area of exotic species, or bioremediation techniques to clean up pollutants. Third, the goal is **sustainable development,** the ability of an ecosystem to maintain itself while providing services to human beings. We will use the Everglades ecosystem to illustrate these principles. Although habitat restoration is good, there is some concern that the restored areas may not be functionally equivalent to the natural regions.

The Everglades

Originally, the Everglades encompassed the whole of southern Florida from Lake Okeechobee down to Florida Bay (Fig. 50.11). This ecosystem is a vast sawgrass prairie, interrupted occasionally by a cypress dome or hardwood tree island. Within these islands, both temperate and tropical evergreen trees grow amongst dense and tangled vegetation. Mangroves are found along sloughs (creeks) and at the shoreline. The prop roots of red mangroves protect over 40 different types of juvenile fishes as they grow to maturity. During a wet season, from May to November, animals disperse throughout the region, but in the dry season, from December to April, they congregate wherever pools of water are found. Alligators are famous for making "gator holes," where water collects and fish, shrimp, crabs, birds, and a host of living things survive until the rains come again. The Everglades once supported millions of large and beautiful birds, including herons, egrets, the white ibis, and the roseate spoonbill (Fig. 50.11b).

At the turn of the century, settlers began to drain the land just south of Lake Okeechobee to grow crops in the newly established Everglades Agricultural Area (EAA). A large dike now rings Lake Okeechobee and prevents water from overflowing its banks and moving slowly southward. To provide flood protection for urban development, water is shunted through the St. Lucie Canal to the Atlantic Ocean or through the canalized Caloosahatchee River to the Gulf of Mexico. In times of drought, water is contained not only in the lake but also in three so-called conservation areas established to the south of the lake. Water must be conserved to irrigate the farmland and to recharge the Biscayne aquifer (underground river), which supplies drinking water for the cities on the east coast of Florida. The Central and Southern Florida Flood Control Project (C&SF) included the construction of over 2,250 km of canals, 125 water control stations, and 18 large pumping stations. Now the Everglades National Park receives water only when it is discharged artificially from a conservation area, and the discharge is according to the convenience of the C&SF rather than according to the natural wet/dry season of southern Florida. Largely because of this, the Everglades are now dying, as witnessed by declining bird populations. The birds, which used to number in the millions, now number in the thousands.

Restoration Plan. A restoration plan has been developed that will sustain the Everglades ecosystem while maintaining the services society requires. The U.S. Army Corps of Engineers is to redesign the C&SF so that the Everglades receive a more natural flow of water from Lake Okeechobee. This will require flooding the EAA and growing only crops such as sugarcane and rice that can tolerate these conditions. This has the benefit of stopping the loss of topsoil and preventing possible residential development in the area. There will also be an extended buffer zone between an expanded Everglades and the urban areas on Florida's east coast. The buffer zone will contain a contiguous system of interconnected marsh areas, detention reservoirs, seepage barriers, and water treatment areas. This plan is expected to stop the decline of the Everglades, while still allowing agriculture to continue and providing water and flood control to the eastern coast. Sustainable development will maintain the ecosystem indefinitely and still meet human needs.

Today, landscape preservation is commonly needed to protect metapopulations. Often the preserved area must be restored before sustainable development is possible.

CONNECTING THE CONCEPTS

Our industrial societies are overusing the environment to the point of exhaustion. Forests throughout tropical, temperate, and subarctic regions are being harvested and cut for timber at unsustainable rates. Urban sprawl is completely replacing natural ecosystems in highly populated regions. Fresh waters are being diverted for agricultural and urban uses to the extent that riverbeds and lake beds are becoming dry in places. Dams are being constructed for hydropower and irrigation with little consideration for their impact on aquatic life. Exotic species of plants, animals, and microbes are being released into new environments with little or no restraint. Marine fisheries are being exploited by major fishing nations at unsustainable levels.

All these actions, and others, are reducing biodiversity, which we now realize is a resource of enormous economic value. If properly managed, sustainable yields of food and fiber can be obtained from many natural lands and waters. Modern genetic engineering technologies make the genes of millions of wild species available for use in breeding improved crops, domestic animals, and biological control agents. Enjoyment of nature can also enrich human life enormously.

As natural forests, grasslands, streams, lakes, and seas are degraded, human society must expend greater amounts of nonrenewable energy and materials to substitute for benefits that biodiversity provides at no cost. Lost species, and ultimately, lost ecosystems, cannot be replaced. Biodiversity is therefore a nonrenewable resource. The goal of conservation biology is to protect, restore, and use this resource wisely. To that end, the vision of conservation biology is:

A world where leaders are committed to long-term environmental protection and to international leadership and cooperation in addressing the world's environmental problems.

A world with an environmentally literate citizenry that has the knowledge, skills, and ethical values needed to achieve sustainable development.

A world in which market prices and economic indicators reflect the full environmental and social costs of human activities.

A world in which a new generation of technologies contributes to the conservation of resources and the protection of the environment.

A world landscape that sustains natural systems, maximizes biological diversity, and uplifts the human spirit.

A world in which human numbers are stabilized, all people enjoy a decent standard of living through sustainable development, and the global environment is protected for future generations.

Modified from *The Report of the National Commission on the Environment*, 1993.

Summary

50.1 CONSERVATION BIOLOGY AND BIODIVERSITY

Conservation biology is the scientific study of biodiversity and its management for sustainable human welfare. The unequaled present rate of extinctions has drawn together scientists and environmentalists in basic and applied fields to address the problem.

Biodiversity is the variety of life on Earth; the exact number of species is not known, but there are many more insects than other types of organisms. Biodiversity must also be preserved at the genetic, community (ecosystem), and landscape levels of organization.

Conservationists have discovered that biodiversity is not evenly distributed in the biosphere, and therefore saving particular areas may protect more species than saving other areas.

50.2 VALUE OF BIODIVERSITY

The direct value of biodiversity is seen in the observable services of individual wild species. Wild species are our best source of new medicines to treat human ills, and they help meet other medical needs. For example, the bacterium that causes leprosy grows naturally in armadillos, and horseshoe crab blood contains a bacteria-fighting substance.

Wild species have agricultural value. Domesticated plants and animals are derived from wild species, and they use wild species as a source of genes for the improvement of their phenotypes. Instead of pesticides, wild species can be used as biological controls, and most flowering plants benefit from animal pollinators. Much of our food, particularly fish and shellfish, is still caught in the wild. Hardwood trees from natural forests supply us with lumber for various purposes, such as making furniture.

The indirect services provided by ecosystems is largely unseen but absolutely necessary to our well-being. These services include the workings of biogeochemical cycles, waste disposal, provision of fresh water, prevention of soil erosion, and regulation of climate. Many people enjoy vacationing in natural settings. Various studies show that more diverse ecosystems function better than less diverse systems.

50.3 CAUSES OF EXTINCTION

Researchers have identified the major causes of extinction. Habitat loss is the most frequent cause, followed by introduction of exotic species, pollution, overexploitation, and disease. (Pollution often leads to disease, so these were discussed at the same time.) Habitat loss has occurred in all parts of the biosphere, but concern has now centered on tropical rain forests and coral reefs, where biodiversity is especially high. Exotic species have been introduced into foreign ecosystems through colonization, horticulture and agriculture, and accidental transport. Among the various causes of pollution (acid deposition, eutrophication, ozone depletion, and organic chemicals), global warming is expected to cause the most instances of extinction. Overexploitation is exemplified by commercial fishing, which is so efficient that fisheries of the world are collapsing.

50.4 CONSERVATION TECHNIQUES

To preserve species, it is necessary to preserve their habitat. Some emphasize the need to preserve biodiversity hotspots because of their richness. Often today it is necessary to save metapopulations because of past habitat fragmentation. If so, it is best to determine the source populations and save those instead of the sink populations. A keystone species such as the grizzly bear requires the preservation of a landscape consisting of several types of ecosystems over millions of acres of

territory. Obviously, in the process, many other species will also be preserved.

Conservation today is assisted by two types of computer analysis. A gap analysis tries for a fit between biodiversity concentrations and land still available to be preserved. A population viability analysis indicates the minimum size of a population needed to prevent extinction from happening.

Since many ecosystems have been degraded, habitat restoration may be necessary before sustainable development is possible. Three principles of restoration are: (1) start before sources of wildlife and seeds are lost; (2) use simple biological techniques that mimic natural processes; and (3) aim for sustainable development so that the ecosystem fulfills the needs of humans.

Reviewing the Chapter

1. Explain these attributes of conservation biology: (1) both academic and applied; (2) supports ethical principles; and (3) is responding to a biodiversity crisis. 928
2. Discuss the conservation of biodiversity at the species, genetic, community, and landscape levels. 928–29
3. Describe the uneven distribution of diversity in the biosphere. What is the implication of uneven distribution for conservation biologists? 929
4. List various ways in which individual wild species provide us with valuable services. 930–31
5. List various ways in which ecosystems provide us with indispensable services. 932–33
6. List and discuss the five major causes of extinction, starting with the most frequent cause. 934
7. Introduction of exotic species is usually due to what events? 936
8. List and discuss five major types of pollution that particularly affect biodiversity. Of these, why is global warming considered the most significant? 936–37
9. Use the positive feedback cycle to explain why overexploitation occurs. 938
10. Using the grizzly bear population as an example, explain keystone species, metapopulations, landscape preservation, and a population viability analysis. 939–40
11. Explain the three principles of habitat restoration with reference to the Everglades. 941

Testing Yourself

Choose the best answer for each question.

1. Which of these would not be within the realm of conservation biology?
   a. helping to manage a national park
   b. a government board charged with restoring an ecosystem
   c. writing textbooks and/or popular books on the value of biodiversity
   d. introducing endangered species back into the wild
   e. All of these are concerns of conservation biology.

2. Which of these pairs does not show a contrast in the number of species?
   a. temperate zone—tropical zone
   b. hotspots—cold spots
   c. rain forest canopy—rain forest floor
   d. pelagic zone—deep-sea benthos

3. The value of wild pollinators to the U.S. agricultural economy has been calculated to be $4.1–$6.7 billion a year. What is the implication?
   a. Society could easily replace wild pollinators by domesticating various types of pollinators.
   b. Pollinators may be valuable, but that doesn't mean any other species also provide us with valuable services.
   c. If we did away with all natural ecosystems, we wouldn't be dependent on wild pollinators.
   d. Society doesn't always appreciate the services that wild species provide naturally and without any fanfare.
   e. All of these statements are correct.

4. The services provided to us by ecosystems are unseen. This means
   a. they are not valuable.
   b. they are noticed particularly when the service is disrupted.
   c. biodiversity is not needed for ecosystems to keep functioning as before.
   d. we should be knowledgeable about them and protect them.
   e. Both b and d are correct.

5. Which of these is a true statement?
   a. Habitat loss is the most frequent cause of extinctions today.
   b. Exotic species are often introduced into ecosystems by accidental transport.
   c. Global warming is expected to cause many extinctions in the twenty-first century.
   d. Overexploitation of fisheries could very well lead to a complete collapse of the fishing industry.
   e. All of these statements are true.

6. Which of these is not expected because of global warming?
   a. the inability of species to migrate to cooler climates as environmental temperatures rise
   b. the bleaching and drowning of coral reefs
   c. rise in sea levels and loss of wetlands
   d. preservation of species because cold weather causes hardships
   e. All of these are expected.

7. Why is a grizzly bear a keystone species existing as a metapopulation?
   a. Grizzly bears require many thousands of miles of preserved land because they are large animals.
   b. Grizzly bears have functions that increase biodiversity, but presently the population is subdivided into isolated subpopulations.
   c. When grizzly bears are preserved, so are many other types of species within a diverse landscape.
   d. Grizzly bears are a source population for many other types of organisms across several population types.
   e. All of these statements are correct.

8. Sustainable development of the Everglades will mean that
   a. the various populations that make up the Everglades will continue to exist indefinitely.
   b. human needs will also be met while successfully managing the ecosystem.
   c. the means used to maintain the Everglades will mimic the processes that naturally maintain the Everglades.
   d. the restoration plan is a workable plan.
   e. All of these statements are correct.

9. Which statement accepted by conservation biologists best shows that they support ethical principles?
   a. Biodiversity is the variety of life observed at various levels of biological organization.
   b. Wild species directly provide us with all sorts of goods and services.
   c. Reduction in the burning of fossil fuels would help reduce the effects of global warming.
   d. There are three principles of restoration biology that need to be adhered to in order to restore ecosystems.
   e. Biodiversity is desirable and has value in and of itself, regardless of any practical benefit.

10. A population in an unfavorable area with a high infant mortality rate would be
    a. a metapopulation.　　　　c. a sink population.
    b. a source population.　　　d. a new population.

11. What is the edge effect?
    a. More species live near the edge of an ecosystem, where more resources are available to them.
    b. New species originate at the edge of ecosystems due to interactions with other species.
    c. The edge of an ecosystem is not a typical habitat and may be an area where survival is more difficult.
    d. More species are found at the edge of a rain forest due to deforestation of the forest interior.

12. Eagles and bears feed on spawning salmon. If shrimp are introduced that compete with salmon for food,
    a. the salmon population will decline.
    b. the eagle and bear populations will decline.
    c. only the shrimp population will decline.
    d. all populations will increase in size.
    e. Both a and b are correct.

13. Biodiversity hotspots
    a. have few populations because the temperature is too hot.
    b. contain about 20% of the Earth's species even though their area is small.
    c. are always found in tropical rain forests and coral reefs.
    d. are sources of species for the ecosystems of the world.
    e. All except a are correct.

14. Consumptive use value
    a. means we should think of conservation in terms of the long run.
    b. means we are placing too much emphasis on living things that are useful to us.
    c. means some organisms, other than crops and farm animals, are valuable as products.
    d. is a type of direct value.
    e. Both c and d are correct.

15. Which of these is not an indirect value of species?
    a. participates in biogeochemical cycles
    b. participates in waste disposal
    c. helps provide fresh water
    d. prevents soil erosion
    e. All of these are indirect values.

16. Most likely, ecosystem performance improves
    a. the more diverse the ecosystem.
    b. as long as selected species are maintained.
    c. as long as species have both direct and indirect value.
    d. if extinctions are diverse.
    e. Both b and c are correct.

17. Global warming has nothing to do with
    a. habitat loss.
    b. introduction of exotic species into new environments.
    c. pollution.
    d. overexploitation.
    e. Global warming pertains to all of these.

18. Sea urchins feed on kelp beds, and sea otters feed on sea urchins. If sea otters are killed off, which of these statement(s) is true? Choose more than one answer if correct.
    a. Kelp beds will increase.   c. Sea urchins will increase.
    b. Kelp beds will decrease.   d. Sea urchins will decrease.

19. Complete the following graph by labeling each bar (a–e) with a cause of extinction, from the most influential to the least influential.

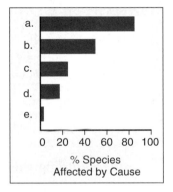

% Species Affected by Cause

Thinking Scientifically

1. The scale at which conservation biologists work often makes direct experimentation difficult. But computer models can assist in predicting the fate of populations or ecosystems. Some scientists feel that these models are inadequate because they cannot reproduce all the variables found in the real world. If you were trying to predict the impact on songbirds of clear-cutting a portion of a forest, what information would you need to develop a good model?

2. Bioprospecting is the search for medically useful molecules derived from living things. The desire for monetary gain from such discoveries provides an impetus to preserve endangered habitats. Bioprospecting protects ecosystems and in this way saves many species rather than individual species. What types of living things would bioprospectors be most interested in?

Bioethical Issue: Protecting Bighorn Sheep

To protect the state's declining bighorn sheep populations, the New Mexico Game Commission approved a plan that calls for killing scores of mountain lions over several years. The commission pointed out that the lions have killed 36 of 43 radio-collared bighorn sheep released into the wild since 1996 and have increasingly turned to killing sheep as the state's deer herd has declined.

Lisa Jennings, director of Animal Protection of New Mexico, said that mountain lions and bighorns have long coexisted and that, rather than killing mountain lions, the best way to increase the bighorn sheep population would be to consider a restoration management plan that would improve the ecosystem. Then, too, the state recently engaged the Hornocker Wildlife Institute of Idaho to study the situation. The institute's report concluded that the number of lions does not necessarily affect the size of the sheep population. Rather, diseases from domestic livestock grazing in the area are responsible for more than 50% of the deaths in some bighorn sheep populations.

In reply, Bill Dunn, a state game department biologist who specializes in sheep, said that the plan to kill mountain lions was solidly based in science. He said that killing lions in selected areas will give the sheep population a chance to rebound because there are 2,000 lions and only 760 bighorn sheep in two separate populations. Do you approve of taking steps to protect bighorn sheep from mountain lions? Why or why not? If you do approve, how would you proceed?

Understanding the Terms

biodiversity   928
biodiversity hotspot   929
bioinformatics   928
conservation biology   928
edge effect   940
endangered species   928
exotic species   936
flagship species   939
gap analysis   940
global warming   937
keystone species   939

landscape   929
metapopulation   939
overexploitation   938
pollution   936
population viability
    analysis   940
restoration ecology   941
sink population   939
source population   939
sustainable development   941
threatened species   928

Match the terms to these definitions:

a. _____ A rather small area with an unusually large concentration of species.
b. _____ A subdivided population in isolated patches of habitat.
c. _____ A population that has a positive growth rate and net emigration rate to other locations.
d. _____ Determination of whether a population of a given size can survive indefinitely in a particular region.
e. _____ Regional assemblage of different, interacting ecosystem units, such as a forest, open fields, wetlands, and streams.

ARIS, the *Biology* Website

ARIS, the website for *Biology*, provides a wealth of information organized and integrated by chapter. You will find practice quizzes, interactive activities, labeling exercises, flashcards, and much more that will complement your learning and understanding of general biology.

www.mhhe.com/maderbiology9

APPENDIX A

ANSWER KEY

CHAPTER 1

Testing Yourself

1. d; **2.** c; **3.** e; **4.** a; **5.** b; **6.** a; **7.** d; **8.** d; **9.** c; **10.** b; **11.** a; **12.** c; **13.** d; **14.** b; **15.** c; **16.** d; **17.** b; **18.** copy; **19.** Everyone; **20.** human; **21.** archaea; **22. a.** After dye is spilled on culture plate, investigator notices that bacteria live despite exposure to sunlight; **b.** Dye protects bacteria against death by UV light; **c.** Expose all culture plates to UV light: one set of plates contains bacteria and dye, and the other set contains only bacteria. The bacteria in both sets die; **d.** Dye does not protect bacteria against death by UV light. Rejects hypothesis.

Understanding the Terms

a. metabolism; **b.** evolution; **c.** experimental variable; **d.** photosynthesis; **e.** control group

CHAPTER 2

Testing Yourself

1. d; **2.** e; **3.** c; **4.** c; **5.** a; **6.** d; **7.** e; **8.** d; **9.** b; **10.** c; **11.** c; **12.** b; **13.** a; **14.** a; **15.** 7p and 7n in nucleus; two electrons in first shell, and five electrons in outer shell. This means that nitrogen needs three more electrons in the outer shell to be stable; because each hydrogen contributes one electron, the formula for ammonia is NH_3. **16.** b; **17.** e; **18.** a; **19.** F; **20.** F; **21.** T; **22.**

Understanding the Terms

a. polar covalent bond; **b.** ion; **c.** acid; **d.** molecule; **e.** buffer

CHAPTER 3

Testing Yourself

1. c; **2.** a; **3.** e; **4.** b; **5.** c; **6.** b; **7.** c; **8.** c; **9.** e; **10.** c; **11.** e; **12.** c; **13.** See Figure 3.5, page 38; **14.** c; **15.** d; **16.** c; **17.** d; **18.** a; **19.** a; **20.** d;

21. a; **22.** c; **23.** b; **24.** c; **25.** b; **26.** d; **27.** a; **28.** F; **29.** F; **30.** F; **31.** T; **32.** T; **33.** F; **34.** T

Understanding the Terms

a. carbohydrate; **b.** lipid; **c.** polymer; **d.** isomer; **e.** peptide

CHAPTER 4

Testing Yourself

1. c; **2.** c; **3.** d; **4.** a; **5.** c; **6.** c; **7.** a; **8.** d; **9. a.** rough ER—produces proteins; **b.** chromatin—DNA specifies the order of amino acids in proteins; **c.** nucleolus—forms ribosomal RNA, which participates in protein synthesis; **d.** smooth ER—forms transport vesicles; **e.** Golgi apparatus—processes and packages proteins for distribution. **10.** b; **11.** c; **12.** e; **13.** c; **14.** a; **15. a.** example; **b.** Mitochondria and chloroplasts are a pair because they are both membranous structures involved in energy metabolism; **c.** Centrioles and flagella are a pair because they both contain microtubules; centrioles give rise to the basal bodies of flagella; **d.** ER and ribosomes are a pair because together they are rough ER, which produces proteins.

Understanding the Terms

a. Golgi apparatus; **b.** peroxisome; **c.** nucleolus; **d.** cytoskeleton; **e.** fimbria

CHAPTER 5

Testing Yourself

1. a. hypertonic—cell shrinks due to loss of water; **b.** hypotonic—central vacuole expands due to gain of water; **2.** b; **3.** b; **4.** b; **5.** c; **6.** c; **7.** e; **8.** e; **9.** b; **10.** b; **11.** See Figure 5.2, page 85; **12.** b; **13.** e; **14.** d; **15.** a

Understanding the Terms

a. differentially permeable; **b.** osmosis; **c.** hypertonic solution; **d.** glycoprotein; **e.** phagocytosis

CHAPTER 6

Testing Yourself

1. e; **2.** e; **3.** e; **4.** e; **5.** d; **6.** d; **7.** b; **8.** e; **9.** c; **10. a.** active site; **b.** substrates; **c.** product; **d.** enzyme; **e.** enzyme-substrate complex;

f. enzyme. The shape of an enzyme is important to its activity because it allows an enzyme-substrate complex to form. **11.** e; **12.** See Figure 6.13, page 111. **13.** c; **14.** a; **15.** a; **16.** c; **17.** a; **18.** b; **19.** d; **20.** d; **21.** e

Understanding the Terms

a. metabolism; **b.** potential energy; **c.** vitamin; **d.** entropy; **e.** coenzyme; **f.** oxidation

CHAPTER 7

Testing Yourself

1. e; **2.** d; **3.** a; **4.** a, b; **5.** c; **6.** a, b, c; **7.** b; **8.** d; **9.** T; **10.** F; **11.** F; **12.** F; **13.** T; **14.** e; **15.** e; **16.** b; **17.** c; **18.** See Figure 7.2, page 117 and Figure 7.4, page 119; **19. a.** water; **b.** oxygen; **c.** carbon dioxide; **d.** carbohydrate; **e.** ADP + P; **f.** ATP; **g.** $NADP^+$; **h.** NADPH; **20.** a; **21.** e; **22.** e; **23.** c; **24.** c; **25.** e

Understanding the Terms

a. light reaction; **b.** photosystem; **c.** electron transport chain; **d.** photosynthesis; **e.** Calvin cycle

CHAPTER 8

Testing Yourself

1. b; **2.** c; **3.** a; **4.** c; **5.** c; **6.** c; **7.** a; **8.** b; **9.** e; **10.** c; **11.** a; **12.** c; **13.** b; **14.** b; **15.** c; **16.** d; **17.** d; **18.** c; **19.** a; **20.** b; **21.** b; **22.** b; **23.** d; **24.** a, c, d; **25.** b, d; **26.** a, b, d; **27.** d; **28. a.** cristae, contains electron transport chain and ATP synthase complex; **b.** matrix, location of preparatory reaction and citric acid cycle; **c.** outer membrane, defines the boundary of the mitochondrion; **d.** intermembrane space, accumulation of H^+; **e.** inner membrane, partitions the mitochondrion into the intermembrane space and the matrix.

Understanding the Terms

a. glycolysis; **b.** oxygen debt; **c.** catabolism; **d.** oxidative phosphorylation; **e.** chemiosmosis; **f.** cellular respiration

CHAPTER 9

Testing Yourself

1. c; **2.** b; **3.** e; **4.** d; **5.** a; **6.** c; **7.** b; **8.** d; **9.** e; **10.** b; **11.** b; **12.** e; **13.** c; **14.** e; **15.** a; **16.** c; **17.** d;

18. b; **19.** d; **20.** a; **21.** c; **22.** d; **23.** a; **24. a.** chromatid of chromosome; **b.** centriole; **c.** spindle fiber or aster; **d.** nuclear envelope (fragment)

Understanding the Terms

a. centrosome; **b.** centromere; **c.** spindle; **d.** sister chromatid; **e.** apoptosis

CHAPTER 10

Testing Yourself

1. b; **2.** d; **3.** e; **4.** b; **5.** a; **6.** d; **7.** d; **8.** e; **9.** c; **10.** T; **11.** T; **12.** F; **13.** F; **14.** T; **15.** F; **16.** 24, 12; **17.** spermatogenesis, oogenesis; **18.** fertilization; **19.** gametes, spore; **20.** diploid, haploid; **21.** d; **22.** b; **23.** d; **24.** c; **25.** b, c; **26.** e; **27.** a; **28.** The cell on the right represents metaphase I because bivalents are present at the metaphase plate.

Understanding the Terms

a. spermatogenesis; **b.** bivalent; **c.** polar body; **d.** secondary oocyte; **e.** homologue

CHAPTER 11

Practice Problems 11.1

1. a. all *W*; **b.** 1/2 *W*, 1/2 *w*; **c.** 1/2 *T*, 1/2 *t*; **d.** all *T*; **2. a.** gamete; **b.** genotype; **c.** gamete

Practice Problems 11.2

1. *bb*; **2.** 75% yellow; **25%** green; **3.** 75%; **4.** *Tt x tt, tt*

Practice Problems 11.3

1. Testcross black-coated horse x brown-coated horse; **2.** *Ll x ll*; **3.** 200 have wrinkled seeds

Practice Problems 11.4

1. a. 1/2 *TG*, 1/2 *tG*; **b.** 1/4 *TG*, 1/4 *Tg*, 1/4 *tG*, 1/4 *tg*; **c.** 1/2 *TG*, 1/2 *Tg*; **2. a.** gamete; **b.** genotype; **c.** gamete

Practice Problems 11.5

1. *BbTt*; **2. a.** *LlGg x llgg*; **b.** *LlGg x LlGg*; **3.** 9/16

Practice Problems 11.6

1. *cc*; *Cc and Cc*; **2.** 25%; **3.** heterozygous

Practice Problems 11.7

1. a. Woman: *Hh*; husband *hh*; **2.** 50%, 50%; **3.** Children with the genotype *Hh, because the gene (H) that causes the condition is dominant.* **4. homozygous dominant or heterozygous**

Practice Problems 11.8

1. 25% **2.** Type O; **3.** *Child ii, mother: I^Ai, father:*

I^Ai, ii, I^Bi. **4.** See Figure 11.6, Page 197

Testing Yourself

1. b; **2.** a; **3.** c; **4.** d; **5.** c; **6.** d; **7.** a; **8.** b; **9.** b; **10.** b; **11.** c; **12.** d; **13.** d; **14.** b; **15.** a; **16.** c; **17.** autosomal dominant

Additional Genetics Problems

1. 100% chance for widow's peak and 0% chance for straight hairline; **2.** 25%; **3.** *Ee* **4.** 210 gray bodies and 70 black bodies; 140 = heterozygous; cross fly with recessive (black body); **5.** 50%; **6.** I^AI^B; *yes, mother could be AO, BO, OO*; **7.** 6 genotypes and 6 phenotypes; yes

Understanding the Terms

a. recessive allele; **b.** allele; **c.** dominant allele; **d.** testcross; **e.** genotype

CHAPTER 12

Practice Problems 12.1

1. Mother: X^bX^b; *father:* X^BY; *female offspring are* X^BX^b, *and males are* X^bY; **2.** X^RX^R: X^RY **3. a. cross (2); b.** 1:1

Practice Problems 12.2

1. His mother; X^HX^h, X^HY, X^hY; **2. a.** 100%; **b.** none; **c.** 100%; **3.** RrX^BX^h x RrX^BY; rrX^bY; **4.** The husband is not the father.

Practice Problems 12.3

1. a. 9:3:3:1; **b.** linkage; **2.** 12 map units; **3.** bar eye, scalloped wings, garnet eye

Testing Yourself

1. c; **2.** d; **3.** b; **4.** d; **5.** c; **6.** a; **7.** b; **8.** b; **9.** d; **10. a.** X-linked recessive; **b.** X^AX^a **11.** c; **12.** F; **13.** F; **14.** T; **15.** T; **16.** d; **17.** e; **18.** a; **19.** b; **20.** c; **21.** b; **22.** b; **23.** c; **24.** e; **25.** c

Additional Genetics Problems

1. Both males and females are 1:1; **2. a.** Males: X^BY, X^bY; *females:* X^BX^B: X^BX^b; **b.** nondisjunction during oogenesis; **3.** heterozygous; **4.** males 50%, females 0%; **5.** sons 50%, daughters 50%; **6.** Man: X^bY, *Woman:* X^BX^b; **7.** 0% **8. a.** 45; **b.** 47 **9.** A, C, B, D; **10.** linkage, 11 map units

Understanding the Terms

a. autosome; **b.** polyploid; **c.** X-linked

CHAPTER 13

Testing Yourself

1. b; **2.** d; **3.** a; **4.** c; **5.** c; **6.** c; **7.** a; **8.** b; **9.** e; **10.** The parental helix is heavy-heavy, and each daughter helix is heavy-light. This shows that each daughter helix is composed of one old strand and one new strand, which is consistent with semiconservative replication. **11.** c; **12.** b; **13.** e; **14.** e; **15.** a; **16.** e; **17.** d; **18.** b; **19.** d; **20.** d; **21.** d; **22.** d; **23.** a; **24.** b; **25.** e; **26.** a

Understanding the Terms

a. genetic mutation; **b.** complementary base pairing; **c.** DNA polymerase; **d.** purine; **e.** bacteriophage

CHAPTER 14

Testing Yourself

1. b; **2.** a; **3.** c; **4.** a; **5.** a; **6.** d; **7.** d; **8.** a; **9.** c; **10.** b; **11.** a; **12.** c; **13. a.** ACU CCU GAA UGC AAA; **b.** UGA GGA CUU ACG UUU; **c.** threonine–proline–glutamate–cysteine–lysine; **14.** d

Understanding the Terms

a. RNA polymerase; **b.** intron; **c.** translation; **d.** polyribosome; **e.** codon; **f.** wobble hypothesis

CHAPTER 15

Testing Yourself

1. c; **2.** e; **3.** a; **4.** a; **5.** c; **6.** e; **7.** b; **8.** b; **9.** b; **10.** e; **11.** e; **12.** e; **13. a.** DNA; **b.** regulator gene; **c.** promoter; **d.** operator; **e.** active repressor; **14.** d; **15.** b; **16.** e; **17.** d; **18.** d; **19.** d; **20.** T; **21.** T; **22.** T; **23.** F; **24.** T; **25.** T; **26.** T; **27.** T

Understanding the Terms

a. posttranscriptional control; **b.** transposons; **c.** Barr body; **d.** euchromatin; **e.** carcinogen

CHAPTER 16

Testing Yourself

1. c; **2.** d; **3.** c; **4.** d; **5.** e; **6.** e; **7.** c; **8.** e; **9.** a; **10.** a; **11.** b; **12.** d; **13.** a; **14.** d; **15.** d; **16.** AATTTTAA; **17.** a; **18.** e; **19.** b; **20.** e; **21. a.** retrovirus **b.** recombinant RNA **c.** defective gene **d.** recombinant RNA **e.** reverse transcription **f.** recombinant DNA **g.** human genome

Understanding the Terms

a. restriction enzyme; **b.** transgenic organism; **c.** genome; **d.** clone; **e.** plasmid

CHAPTER 17

Testing Yourself

1. d; **2.** e; **3.** b; **4.** e; **5.** e; **6.** e; **7.** e; **8.** e; **9.** e; **10.** b; **11.** a; **12.** b, d; **13.** c; **14.** a; **15.** d; **16.** c; **17.** b, d; **18.** d; **19.** e; **20.** Life has a history, and it's possible to trace the history of individual organisms. **21.** Two different continents can have similar environments, and therefore unrelated organisms that are similarly adapted. **22.** All vertebrates share a common ancestor, who had pharyngeal pouches during development. **23.** Genetic differences account for speciation, and therefore evolution.

Understanding the Terms

a. biogeography; **b.** paleontology; **c.** vestigial structure; **d.** adaptation; **e.** inheritance of acquired characteristics

CHAPTER 18

Practice Problems 18.1

1. 21%; **2.** $t = q = 0.3$, $T = p = 0.7$; *homozygous recessive* $= q^2 = 9\%$; *homozygous dominant* $= p^2 = 49\%$; *heterozygous* $= 2pq = 42\%$

Testing Yourself

1. c; **2.** c; **3.** c; **4.** c; **5.** e; **6.** b; **7.** c; **8.** d; **9.** e; **10.** c; **11.** b; **12.** b; **13.** e; **14.** e; **15.** b; **16. a.** See Figure 18.6, page 307; **b.** See Figure 18.7, page 308; **c.** See Figure 18.8, page 308.

Additional Genetics Problems

1. 16%; **2.** 99%; **3.** recessive allele = 0.2, dominant allele = 0.8; homozygous recessive = 0.04, homozygous dominant = 0.64, heterozygous = 0.32

Understanding the Terms

a. stabilizing selection; **b.** postzygotic isolating mechanism; **c.** genetic drift; **d.** adaptive radiation; **e.** gene flow

CHAPTER 19

Testing Yourself

1. c; **2.** d; **3.** a; **4.** b; **5.** e; **6.** a; **7.** c; **8.** d; **9.** b; **10.** b; **11.** b; **12.** e; **13.** e; **14.** a; **15.** b; **16.** c; **17.** d; **18.** a; **19.** c; **20.** c; **21.** e; **22.** c; **23. a.** oldest eukaryotic fossils; **b.** O_2 accumulates; **c.** oldest known fossils; **d.** Cambrian animals; **e.** Ediacaran animals; **f.** protists evolve and diversify; **24.** heterotroph, chemoautotroph; **25.** true; **26.** photosynthesizing; **27.** Carboniferous; **28.** Cenozoic; **29.** meteorite, drift; **30.** Precambrian; **31.** invertebrate; **32.** drifted

Understanding the Terms

a. molecular clock; **b.** protocell; **c.** liposome; **d.** ocean ridge; **e.** ozone shield

CHAPTER 20

Testing Yourself

1. e; **2.** c; **3.** d; **4.** a, b, c; **5.** c, d, e; **6.** b, c, d, e;

7. a; **8.** c; **9.** b; **10.** d; **11.** b; **12.** e; **13.** b; **14.** e; **15.** b; **16.** e; **17.** specific epithet; **18.** homology; **19.** interbreed; **20.** interbreed; **21. a.** common ancestor, ancestral characters; **b.** divergence; **c.** derived characteristics; **22. a.** three; by color; **b.** vertebrae; amniotic egg and internal fertilization; **c.** all of them; they share the same derived characters.

Understanding the Terms

a. taxonomy; **b.** phylogenetic tree; **c.** taxon; **d.** cladistics; **e.** homology

CHAPTER 21

Testing Yourself

1. a. attachment; **b.** penetration; **c.** integration; **d.** prophage; **e.** biosynthesis; **f.** maturation; **g.** release; See also Figure 21.3, page 365; **2.** e; **3.** c; **4.** b; **5.** a; **6.** c; **7.** d; **8.** a; **9.** c; **10.** c; **11.** c; **12.** c; **13.** a; **14.** e; **15.** e; **16.** a; **17.** d; **18.** a; **19.** b

Understanding the Terms

a. lysogenic; **b.** photoautotroph; **c.** saprotroph; **d.** symbiotic; **e.** archaea

CHAPTER 22

Testing Yourself

1. b; **2.** e; **3.** d; **4.** c; **5.** a; **6.** b; **7.** b; **8.** e; **9.** b; **10.** b; **11.** d; **12.** a; **13.** d; **14.** b; **15.** d; **16.** b; **17.** d; **18.** c; **19.** a; **20.** b; **21.** c; **22. a.** sexual reproduction; **b.** gametes pairing; **c.** zygote (2n); **d.** zygospore (2n); **e.** asexual reproduction; **f.** zoospores (n); **g.** nucleus with nucleolus; **h.** chloroplast; **i.** starch granule; **j.** pyrenoid; **k.** flagellum; l. eyespot; **m.** gamete formation. See also Figure 22.4, page 382.

Understanding the Terms

a. pseudopod; **b.** euglenoid; **c.** diatom; **d.** trypanosome; **e.** plankton

CHAPTER 23

Testing Yourself

1. c; **2.** a; **3.** b; **4.** c; **5.** b; **6.** c; **7.** e; **8.** a; **9.** c; **10.** a; **11.** d; **12.** a; **13.** d; **14.** d; **15. a.** spores; **b.** sporangium; **c.** sporangiophore; **d.** stolon; **e.** rhizoid; **16.** a; **17.** e; **18.** All are in kingdom Fungi **a.** phylum Zygomycota; **b.** phylum Ascomycota; **c.** phylum Basidiomycota; **d.** phylum Ascomycota; **19.** d; **20.** d; **21. a.** meiosis; **b.** basidiospores; **c.** dikaryotic mycelium; **d.** button stage of the mushroom (basidiocarp); **e.** stalk; **f.** gill; **g.** cap; **h.** dikaryotic; **i.** diploid; **j.** zygote. See also Figure 23.8, page 406.

Understanding the Terms

a. basidium; **b.** mycelium; **c.** conidiospore; **d.** fruiting body; **e.** mycorrhizae

CHAPTER 24

Testing Yourself

1. e; **2.** a; **3.** b; **4.** c; **5.** b; **6.** b; **7.** e; **8.** c; **9.** c; **10.** e; **11.** d; **12.** b; **13.** a; **14.** b; **15.** b; **16.** e; **17.** c; **18.** e; **19. a.** sporophyte (2n); **b.** meiosis; **c.** gametophyte (n); **d.** fertilization. See also Figure 24.3, page 416.

Understanding the Terms

a. sporophyte; **b.** monocotyledon; **c.** pollen grain; **d.** rhizoid; **e.** phloem

CHAPTER 25

Testing Yourself

1. c; **2.** b; **3.** c; **4.** c; **5.** b; **6.** b; **7.** b; **8.** c; **9.** c; **10.** b; **11.** d; **12.** e; **13.** d; **14.** d; **15.** d; **16.** c; **17.** c; **18.** e; **19.** a; **20. a.** epidermis; **b.** cortex; **c.** endodermis; **d.** phloem; **e.** xylem. See also Figure 25.8, page 444; **21. a.** upper epidermis; **b.** palisade mesophyll; **c.** leaf vein; **d.** spongy mesophyll; **e.** lower epidermis. See also Figure 25.20, page 454.

Understanding the Terms

a. mesophyll; **b.** vascular cambium; **c.** cotyledon; **d.** stolon; **e.** xylem

CHAPTER 26

Testing Yourself

1. d; **2.** e; **3.** a; **4.** c; **5.** d; **6.** c; **7.** d; **8.** a; **9.** b; **10.** c; **11.** d; **12.** b; **13.** a; **14.** c; **15.** c; **16.** e; **17.** e; **18.** The diagram shows that air pressure pushing down on mercury in the pan can raise a column of mercury only to 76 cm. When water above the column is transpired, it pulls on the mercury and raises it higher. This suggests that transpiration would be able to raise water to the tops of trees. **19.** See Figure 26.13, page 470. After K^+ enters guard cells, water follows by osmosis and the stoma opens. **20.** There is more solute in bulb 1 than in bulb 2, therefore water enters bulb 1. This creates a positive pressure that causes water, along with solute, to flow toward bulb 2. See also illustration, page 472.

Understanding the Terms

a. pressure-flow model; **b.** epiphyte; **c.** transpiration; **d.** Casparian strip; **e.** guard cell

CHAPTER 27

Testing Yourself

1. d; **2.** b; **3.** e; **4.** a; **5.** c; **6.** b; **7.** d; **8.** e; **9.** e;
10. e; **11.** e; **12.** a; **13.** c; **14.** d; **15.** a; **16.** b; **17.** a;
18. b; **19.** c; **20.** c; **21.** a; **22.** d; **23.** d; **24.** d;
25. d; **26.** c; **27.** See Figure 27.12, page 487.

Understanding the Terms

a. circadian rhythm; **b.** gravitropism;
c. abscission; **d.** gibberellin;
e. photoperiodism

CHAPTER 28

Testing Yourself

1. d; **2.** a; **3.** b; **4.** a; **5.** a; **6.** e; **7.** a; **8.** e; **9.** c;
10. e; **11.** b; **12.** e; **13.** d; **14.** c; **15.** a; **16.** b; **17.** c;
18. e; **19.** c; **20. a.** diploid ; **b.** anther; **c.** ovule
d. ovary; **e.** haploid; **f.** megaspore; **g.** male;
h. female; **i.** sperm; **j.** seed. See also Figure
28.1, page 496.

Understanding the Terms

a. carpel; **b.** fruit; **c.** female gametophyte;
d. seed; **e.** pollen grain

CHAPTER 29

Testing Yourself

1. d; **2.** e; **3.** e; **4.** d; **5.** b; **6.** e; **7.** a; **8.** c; **9.** a;
10. a. gastrovascular cavity; **b.** tentacle;
c. mouth; **d.** mesoglea; **11. a.** Cnidaria and
Ctenophora; **b.** Nematoda, Rotifera;
c. Cnidaria and Ctenophora; **d.** Nematoda,
Nemertea; **e.** Platyhelminthes; **f.** Cnidaria;
12. a. tapeworm; **b.** planarian; **c.** sponge;
d. *Hydra*; **e.** rotifer; **f.** planarian; **g.** *Hydra*

Understanding the Terms

a. gastrovascular cavity; **b.** coelom;
c. pseudocoelom; **d.** bilateral symmetry;
e. tube-within-a-tube body plan

CHAPTER 30

Testing Yourself

1. b; **2.** c; **3.** b; **4.** c; **5.** c; **6.** a; **7.** d; **8.** c; **9.** a;
10. a. all three; **b.** annelids, arthropods;
c. all three; **d.** all three; **e.** all three; **f.** all
three; **g.** arthropods; **h.** molluscs; **11.**
a. earthworms; **b.** clams; **c.** clams; **d.** clams;
e. earthworms; **f.** earthworms; **g.** clams;
h. clams; **i.** earthworms; **j.** earthworms; **12.**
a. head; **b.** antenna; **c.** simple eye; **d.** compound
eye; **e.** thorax; **f.** tympanum; **g.** abdomen;
h. forewing; **i.** hindwing; **j.** ovipositor;
k. spiracles; **l.** air sac; **m.** spiracle. **n.** tracheae.
See also Figure 30.14*a*, page 551. **13.** b; **14.** b;
15. d; **16.** c; **17.** a; **18.** b; **19.** c; **20.** a; **21.** d; **22.** d;

23. e; **24.** c; **25.** b; **26.** e; **27.** d; **28.** e; **29.** b.

Understanding the Terms

a. trachea; **b.** metamorphosis;
c. enterocoelom; **d.** nephridia; **e.** chitin

CHAPTER 31

Testing Yourself

1. e; **2.** b; **3.** c; **4.** e; **5.** a; **6.** e; **7.** b; **8.** b; **9.** e;
10. c; **11.** a; **12.** a; **13. a.** pharyngeal pouches;
b. dorsal tubular nerve cord; **c.** notochord;
d. postanal tail

Understanding the Terms

a. endothermic; **b.** monotreme; **c.** reptile;
d. notochord; **e.** marsupial

CHAPTER 32

Testing Yourself

1. a; **2.** b; **3.** b; **4.** d; **5.** e; **6.** d; **7.** b; **8.** c; **9.** d;
10. c; **11. a.** Chordata; **b.** Vertebrata; **c.** Class;
d. Order; **e.** Hominidae; **f.** Genus; **g.** *Homo
sapiens*; **12.** b; **13.** T; **14.** T; **15.** T; **16.** T; **17.** F;
18. anthropoids; **19.** Africa; **20.** erect, small;
21. Cro-Magnon; **22.** thousands; **23. a.** a, b, c,
d; **b.** e. See also Figure 32.10*b*, page 588.

Understanding the Terms

a. anthropoid; **b.** Cro-Magnon; **c.** Neandertal;
d. *Homo ergaster*; **e.** hominoid

CHAPTER 33

Testing Yourself

1. c; **2.** b; **3.** a; **4.** e; **5.** e; **6.** e; **7.** b; **8.** e; **9.** e;
10. c; **11. a.** columnar epithelium, lining of
intestine (digestive tract), protection and
absorption; **b.** cardiac muscle, wall of heart,
pumps blood; **c.** compact bone, skeleton,
support and protection. **12.** e; **13.** d; **14.** c;
15. c; **16.** c, a, g; **17.** e, d, b; **18.** b, c, f; **19.** b;
20. d; **21.** a; **22.** c

Understanding the Terms

a. ligament; **b.** epidermis; **c.** striated;
d. homeostasis; **e.** spongy bone

CHAPTER 34

Testing Yourself

1. b; **2.** b; **3.** a; **4.** d; **5.** d; **6.** c; **7.** b; **8.** b; **9.** e;
10. e; **11.** c; **12.** e; **13.** b; **14.** c; **15.** e; **16.** F; **17.** T;
18. F; **19.** T; **20. a.** blood pressure; **b.** osmotic
pressure; **c.** blood pressure; **d.** osmotic
pressure; **21.** See Figure 34.6, page 617.

Understanding the Terms

a. artery; **b.** platelet; **c.** plasma; **d.** venae
cavae; **e.** hemoglobin

CHAPTER 35

Testing Yourself

1. b; **2.** e; **3.** e; **4.** a; **5.** b; **6.** c; **7.** a; **8.** b; **9.** b;
10. a. antigen-binding sites; **b.** light chain;
c. heavy chain; **d.** *V stands for variable region;
C stands for constant region;* **11.** e; **12.** d; **13.** e;
14. b; **15.** d; **16.** a; **17.** b; **18.** e; **19.** d; **20.** d; **21.** b

Understanding the Terms

a. vaccine; **b.** lymph; **c.** antigen; **d.** apoptosis;
e. T lymphocyte

CHAPTER 36

Testing Yourself

1. a; **2.** b; **3.** d; **4.** b; **5.** d; **6.** c; **7.** e; **8.** a; **9.** c;
10. c; **11.** e; **12.** c; **13.** d; **14.** c; **15.** a; **16.** b;
17. Test tube 1: no digestion—no enzyme
and no HCl; Test tube 2: some digestion—no
HCl; Test tube 3: no digestion—no enzyme;
Test tube 4: digestion—both enzyme and
HCl are present

Understanding the Terms

a. vitamins; **b.** lipase; **c.** lacteal; **d.** esophagus;
e. gallbladder

CHAPTER 37

Testing Yourself

1. a; **2.** b; **3.** b; **4.** d; **5.** c; **6.** b; **7.** c; **8.** b; **9.** e;
10. b; **11.** d; **12.** d; **13.** c; **14.** b; **15.** e; **16.** b; **17.** d;
18. a. nasal cavity; **b.** nostril; **c.** pharynx;
d. epiglottis; **e.** glottis; **f.** larynx; **g.** trachea;
h. bronchus; **i.** bronchiole. See also Figure
37.6, page 676.

Understanding the Terms

a. ventilation; **b.** diaphragm; **c.** vocal cord;
d. gill; **e.** expiration

CHAPTER 38

Testing Yourself

1. d; **2.** a; **3.** c; **4.** b; **5.** a; **6.** e; **7.** b; **8.** e; **9.** c; **10.** b;
11. a; **12.** a; **13.** c; **14.** d; **15.** c; **16.** d; **17.** c; **18.** a;
19. a. glomerulus; **b.** efferent arteriole; **c.**
afferent arteriole; **d.** proximal convoluted
tubule; **e.** loop of the nephron; **f.** descending
limb; **g.** ascending limb; **h.** peritubular
capillary network; **i.** distal convoluted tubule;
j. renal vein; **k.** renal artery; **l.** collecting duct

Understanding the Terms

a. Malpighian tubule; **b.** glomerular capsule;
c. urea; **d.** aldosterone; **e.** uric acid

CHAPTER 39

Testing Yourself

1. b; **2.** b; **3.** c; **4.** a; **5.** a; **6.** c; **7.** d; **8.** b; **9.** b; **10.** c; **11.** d; **12.** c; **13.** c; **14.** d; **15.** b; **16.** c; **17.** d; **18.** c; **19. a.** central canal; **b.** gray matter; **c.** white matter; **d.** dorsal root ganglion; **e.** cell body of sensory neuron; **f.** spinal nerve; **g.** cell body of motor neuron; **h.** interneuron

Understanding the Terms

a. reflex; **b.** neurotransmitter; **c.** autonomic system; **d.** ganglion; **e.** acetylcholine

CHAPTER 40

Testing Yourself

1. e; **2.** c; **3.** c; **4.** e; **5.** d; **6.** c; **7.** c; **8.** b; **9.** d; **10.** e; **11.** a; **12.** b; **13.** c; **14.** a; **15.** d; **16.** b; **17.** d; **18. a.** retina—contains sensory receptors; **b.** choroid—absorbs stray light; **c.** sclera—protects and supports eyeball; **d.** optic nerve—transmits impulses to brain; **e.** fovea centralis—makes acute vision possible; **f.** muscle in ciliary body—holds lens in place, accommodation; **g.** lens—refracts and focuses light rays; **h.** iris—regulates light entrance; **i.** pupil—admits light; **j.** cornea—refracts light rays

Understanding the Terms

a. compound eye; **b.** retina; **c.** sclera; **d.** chemoreceptor; **e.** spiral organ

CHAPTER 41

Testing Yourself

1. b; **2.** f; **3.** c; **4.** e; **5.** e; **6.** b; **7.** e; **8.** b; **9.** b; **10.** c; **11.** b; **12.** e; **13.** a; **14.** e; **15.** a; **16.** c; **17.** b; **18.** b; **19. a.** T tubule; **b.** sarcoplasmic reticulum; **c.** myofibril; **d.** Z line; **e.** sarcomere; **f.** sarcolemma of muscle fiber

Understanding the Terms

a. osteoblast; **b.** sliding filament model; **c.** actin; **d.** appendicular skeleton; **e.** pectoral girdle

CHAPTER 42

Testing Yourself

1. f; **2.** b; **3.** c; **4.** a; **5.** e; **6.** d; **7.** d; **8.** b; **9.** e; **10.** d; **11.** e; **12.** a; **13.** e; **14.** b; **15.** e; **16.** e; **17.** a; **18.** a; **19. a.** inhibits; **b.** inhibits; **c.** releasing hormone; **d.** stimulating hormone; **e.** target gland hormone; **20.** d; **21.** b; **22.** a; **23.** b; **24.** c; **25.** e; **26.** d; **27.** d; **28.** c; **29.** e **30.** e

Understanding the Terms

a. thyroid gland; **b.** negative feedback; **c.** pineal gland; **d.** peptide hormone; **e.** pheromone

CHAPTER 43

Testing Yourself

1. a. seminal vesicle; **b.** ejaculatory duct; **c.** prostate gland; **d.** bulbourethral gland; **e.** anus; **f.** vas deferens; **g.** epididymis; **h.** testis; **i.** scrotum; **j.** foreskin; **k.** glans penis; **l.** penis; **m.** urethra; **n.** vas deferens; **o.** urinary bladder. Path of sperm: testis, epididymis, vas deferens, urethra (in penis). See also Figure 43.3, page 778. **2.** b; **3.** e; **4.** c; **5.** e; **6.** c; **7.** c; **8.** c; **9.** c; **10.** a; **11.** a; **12.** c; **13.** c; **14.** b; **15.** e; **16.** c; **17.** a; **18.** c; **19.** b; **20.** c

Understanding the Terms

a. ovulation; **b.** parthenogenesis; **c.** progesterone; **d.** semen; **e.** gonad

CHAPTER 44

Testing Yourself

1. b; **2.** b; **3.** a; **4.** e; **5.** b; **6.** e; **7.** e; **8.** d; **9. a.** chorion (contributes to forming placenta where wastes are exchanged for nutrients and oxygen); **b.** amnion (protects and prevents desiccation); **c.** embryo; **d.** allantois (blood vessels become umbilical blood vessels); **e.** yolk sac (first site of blood cell formation); **f.** chorionic villi (embryonic portion of placenta); **g.** maternal portion of placenta; **h.** umbilical cord (connects developing embryo to the placenta). See also Figure 44.11, page 807. **10.** c; **11.** a; **12.** d; **13.** b; **14.** e; **15.** b; **16.** c; **17.** b

Understanding the Terms

a. induction; **b.** germ layer; **c.** amnion; **d.** homeobox; **e.** gastrula

CHAPTER 45

Testing Yourself

1. c; **2.** c; **3.** d; **4.** d; **5.** e; **6.** c; **7.** b; **8.** d; **9.** c; **10.** c; **11.** a; **12.** d; **13.** b; **14.** c; **15.** c; **16.** a; **17.** c; **18.** b; **19.** a; **20.** c

Understanding the Terms

a. territoriality; **b.** altruism; **c.** communication; **d.** pheromone

CHAPTER 46

Testing Yourself

1. d; **2.** c; **3.** e; **4.** b; **5.** b; **6.** e; **7.** c; **8.** e; **9.** c; **10.** e; **11.** c; **12.** e; **13.** b; **14.** e; **15.** e

Understanding the Terms

a. demographic transition; **b.** population; **c.** exponential growth; **d.** carrying capacity; **e.** biotic potential

CHAPTER 47

Testing Yourself

1. b; **2.** e; **3.** d; **4.** e; **5.** b; **6.** e; **7.** c; **8.** e; **9.** e; **10.** e; **11.** e; **12.** e; **13.** b; **14.** a; **15.** d; **16.** a; **17.** b; **18.** e; **19. a.** population densities; **b.** environmental gradient. A community contains species with overlapping tolerances to environmental factors (see also Figure 47.2, page 861).

Understanding the Terms

a. community; **b.** habitat; **c.** ecological succession; **d.** ecological niche; **e.** mutualism

CHAPTER 48

Testing Yourself

1. c; **2.** b; **3.** e; **4.** c; **5.** c; **6.** b; **7.** a; **8.** c; **9.** b; **10.** c; **11.** a; **12.** c; **13.** c; **14. a.** top carnivore, tertiary consumers; **b.** carnivore, secondary consumers; **c.** herbivore, primary consumers; **d.** producer, autotrophs; **e.** The numbers are the dry biomass weights of the organisms at each level. **15. a.** producers; **b.** consumers; **c.** inorganic nutrient pool; **d.** decomposers

Understanding the Terms

a. detritus; **b.** ozone shield; **c.** fossil fuel; **d.** nitrogen fixation; **e.** food web

CHAPTER 49

Testing Yourself

1. b; **2.** d; **3.** c; **4.** d; **5.** d; **6.** d; **7.** e; **8.** e; **9.** a; **10.** a; **11.** d; **12.** e; **13.** c; **14.** b; **15.** a; **16.** c; **17.** b

Understanding the Terms

a. estuary; **b.** benthic division; **c.** taiga; **d.** savanna; **e.** spring overturn

CHAPTER 50

Testing Yourself

1. e; **2.** b; **3.** d; **4.** e; **5.** e; **6.** d; **7.** b; **8.** e; **9.** e; **10.** c; **11.** c; **12.** e; **13.** b; **14.** e; **15.** e; **16.** a; **17.** e; **18.** b, c; **19. a.** habitat loss; **b.** exotic species; **c.** pollution; **d.** overexploitation; **e.** disease

Understanding the Terms

a. biodiversity hotspot; **b.** metapopulation; **c.** source population; **d.** population viability analysis; **e.** landscape

APPENDIX B

Classification of Organisms

DOMAIN BACTERIA

Prokaryotic, unicellular organisms that lack a membrane-bounded nucleus and reproduce asexually. Metabolically diverse, being heterotrophic by absorption; autotrophic by chemosynthesis or by photosynthesis. Motile forms move by flagella consisting of a single filament. (371)

DOMAIN ARCHAEA

Prokaryotic, unicellular organisms that lack a membrane-bounded nucleus and reproduce asexually. Metabolically diverse, many being autotrophic by chemosynthesis and a few by photosynthesis; some are heterotrophic by absorption. Most live in extreme or anaerobic environments. Archaea are distinguishable from bacteria by their unique rRNA base sequence and their distinctive plasma membrane and cell wall chemistry. (373)

DOMAIN EUKARYA

Eukaryotic, unicellular to multicellular organisms that have a membrane-bounded nucleus containing several chromosomes. Sexual reproduction is common. Phenotypes and nutrition are diverse; each kingdom has specializations that distinguish it from the other kingdoms. Flagella, if present, have a 9 + 2 organization. (373)

Kingdom Protista

Eukaryotic, unicellular organisms and their immediate multicellular descendan Asexual reproduction is common, but sexual reproduction as a part of various life cycles does occur. Metabolically diverse, being either photosynthetic or heterotrophic by various means. Locomotion, if present, utilizes flagella, cilia, or pseudopods. (380)

Algae*

- Phylum Chlorophyta: green algae (382)
- Phylum Rhodophyta: red algae (384)
- Phylum Phaeophyta: brown algae (385)
- Phylum Chrysophyta: diatoms, golden-brown algae (387)
- Phylum Pyrrophyta: dinoflagellates (387)
- Phylum Euglenophyta: euglenoids (388)

Protozoans*

- Phylum Zoomastigophora: zooflagellates (389)
- Phylum Rhizopoda: amoeboids (390)
- Phylum Foraminifera: foraminiferans (390)
- Phylum Actinopoda: radiolarians (390)
- Phylum Ciliophora: ciliates (391)
- Phylum Apicomplexa: sporozoans (392)

Slime Molds*

- Phylum Myxomycota: plasmodial slime molds (393)
- Phylum Acrasiomycota: cellular slime molds (393)

Water Molds*

- Phylum Oomycota: water molds (394)

Kingdom Fungi

Multicellular eukaryotes that form nonmotile spores during both asexual and sexual reproduction as a part of the haploid life cycle. The only multicellular forms of life to be heterotrophic by absorption. They lack flagella in all life-cycle stages. (400)

- Phylum Zygomycota: zygospore fungi (400)
- Phylum Ascomycota: sac fungi (402)
- Phylum Basidiomycota: club fungi (406)

*Not a classification category, but added for clarity

Kingdom Plantae

Multicellular, primarily terrestrial eukaryotes with well-developed tissues. Plants have an alternation of generations life cycle and are usually autotrophic by photosynthesis. Like green algae, they contain chlorophylls *a* and *b,* carotenoids; store starch in chloroplasts; and have a cell wall that contains cellulose. (414)

Nonvascular Plants*

> Phylum Anthocerophyta: hornworts (417)
>
> Phylum Hepatophyta: liverworts (418)
>
> Phylum Bryophyta: mosses (418)

Vascular Plants*

> Phylum Rhyniophya: Cooksonia (extinct) (420)

Seedless Vascular Plants*

> Phylum Lycophyta: club mosses, spike mosses, quillworts (421)
>
> Phylum Sphenophyta: horsetails (421)
>
> Phylum Psilotophyta: whisk ferns (422)
>
> Phylum Pterophyta: ferns (422)

Gymnosperms*

> Phylum Coniferophyta: conifers, such as pines, firs, yews, redwoods, spruces (424)
>
> Phylum Cycadophyta: cycads (425)
>
> Phylum Ginkgophyta: maidenhair tree (426)
>
> Phylum Gnetophyta: gnetophytes (426)

Angiosperms*

> Phylum Anthophyta: flowering plants (428)
>
>> Class Monocotyledones: monocots (428)
>>
>> Class Eudicotyledones: eudicots (428)

Kingdom Animalia

Multicellular organisms with well-developed tissues that have a diploid life cycle. Animals tend to be mobile and are heterotrophic by ingestion or absorption, generally in a digestive cavity. Complexity varies; the more complex forms have well-developed organ systems. More than 1.5 million species have been described. (518)

Invertebrates*

> Phylum Porifera: sponges (522)
>
> Phylum Ctenophora: comb jellies, sea walnuts (524)
>
> Phylum Cnidaria: hydras, jellyfishes (524)
>
>> Class Anthozoa: sea anemones, corals (525)
>>
>> Class Hydrozoa: *Hydra, Obelia* (525)
>>
>> Class Scyphozoa: jellyfishes (525)
>
> Phylum Platyhelminthes: flatworms (528)
>
>> Class Turbellaria: planarians (529)

> Class Trematoda: flukes (530)
>
> Class Cestoda: tapeworms (531)

Phylum Nemertea: ribbon worms (528)

Phylum Nematoda: roundworms (533)

Phylum Rotifera: rotifers (534)

Phylum Mollusca: molluscs (540)

> Class Polyplacophora: chitons (540)
>
> Class Scaphopoda: toothshells (540)
>
> Class Bivalvia: clams, scallops, oysters, mussels (540)
>
> Class Cephalopoda: squid, chambered nautilus, octopus (542)
>
> Class Gastropoda: snails, slugs, nudibranchs (543)

Phylum Annelida: annelids (544)

> Class Polychaeta: clam worms, tube worms (544)
>
> Class Oligochaeta: earthworms (545)
>
> Class Hirudinea: leeches (546)

Phylum Arthropoda: arthropods (546)

> Subphylum Crustacea: crustaceans (shrimps, crabs, lobsters, barnacles) (548)
>
> Subphylum Uniramia: insects, millipedes, centipedes (550)
>
> Subphylum Chelicerata: spiders, scorpions, horseshoe crabs (553)

Phylum Echinodermata: echinoderms (554)

> Class Concentricycloidea: sea daisies (554)
>
> Class Crinoidea: sea lilies, feather stars (554)
>
> Class Holothuroidea: sea cucumbers (554)
>
> Class Ophiuroidea: brittle stars (554)
>
> Class Echinoidea: sea urchins, sand dollars (554)
>
> Class Asteroidea: sea stars (554)

Phylum Chordata: chordates (560)

> Subphylum Urochordata: tunicates (560)
>
> Subphylum Cephalochordata: lancelets (560)

Vertebrates*

> Subphylum Vertebrata: Vertebrates (562)
>
> Superclass Agnatha: jawless fishes (hagfishes, lampreys) (562)
>
> Superclass Gnathostomata: jawed fishes, tetrapods (561)
>
>> Class Chondrichthyes: cartilaginous fishes (sharks, skates) (563)
>>
>> Class Osteichthyes: bony fishes (lobe-finned fishes and ray-finned fishes) (564)
>>
>> Class Amphibia: amphibians (frogs, toads) (565)
>>
>> Class Reptilia: reptiles (turtles, snakes, lizards, crocodiles) (568)
>>
>> Class Aves: birds (owls, woodpeckers, kingfishers, cuckoos) (572)
>>
>> Class Mammalia: mammals (monotremes— duckbill platypus, spiny anteater; marsupials— opossums, kangaroos, koalas; placental mammals—shrews, whales, rats, rabbits, dogs, cats) (574)
>>
>>> Order Primates: primates (prosimians, monkeys, apes, and humans) (578)
>>>
>>>> Family Hominidae (582)
>>>>
>>>>> Genus *Homo:* humans (585)

*Not a classification category, but added for clarity

APPENDIX C

Metric System

Unit and Abbreviation	Metric Equivalent	Approximate English-to-Metric Equivalents	Units of Temperature
Length			
nanometer (nm)	$= 10^{-9}\,m\ (10^{-3}\,\mu m)$		
micrometer (μm)	$= 10^{-6}\,m\ (10^{-3}\,mm)$		
millimeter (mm)	$= 0.001\ (10^{-3})\,m$		
centimeter (cm)	$= 0.01\ (10^{-2})\,m$	1 inch = 2.54 cm 1 foot = 30.5 cm	
meter (m)	$= 100\ (10^{2})\,cm$ = 1,000 mm	1 foot = 0.30 m 1 yard = 0.91 m	
kilometer (km)	$= 1,000\ (10^{3})\,m$	1 mi = 1.6 km	
Weight (mass)			
nanogram (ng)	$- 10^{-9}\,g$		
microgram (μg)	$= 10^{-6}\,g$		
milligram (mg)	$= 10^{-3}\,g$		
gram (g)	$- 1,000\,mg$	1 ounce = 28.3 g 1 pound = 454 g	
kilogram (kg)	$= 1,000\ (10^{3})\,g$	= 0.45 kg	
metric ton (t)	= 1,000 kg	1 ton = 0.91 t	
Volume			
microliter (μl)	$= 10^{-6}\,l\ (10^{-3}\,ml)$		
milliliter (ml)	$= 10^{-3}$ liter $= 1\,cm^{3}\ (cc)$ $= 1,000\,mm^{3}$	1 tsp = 5 ml 1 fl oz = 30 ml	
liter (l)	= 1,000 ml	1 pint = 0.47 liter 1 quart = 0.95 liter 1 gallon = 3.79 liter	
kiloliter (kl)	= 1,000 liter		

°C	°F	
100	212	Water boils at standard temperature and pressure.
71	160	Flash pasteurization of milk
57	134	Highest recorded temperature in the United States, Death Valley, July 10, 1913
41	105.8	Average body temperature of a marathon runner in hot weather
37	98.6	Human body temperature
13.7	56.66	Human survival is still possible at this temperature.
0	32.0	Water freezes at standard temperature and pressure.

To convert temperature scales:

$$°C = \frac{5(°F - 32)}{1.8}$$

$$°F = 1.8\ (°C) + 32$$

APPENDIX D

PERIODIC TABLE OF THE ELEMENTS

Atomic number → **1** ¹ ← Atomic mass
H ← Atomic symbol
hydrogen

group Ia																	VIIIa
1 ¹ **H** hydrogen	IIa											IIIa	IVa	Va	VIa	VIIa	**2** ⁴ **He** helium
3 ⁷ **Li** lithium	**4** ⁹ **Be** beryllium											**5** ¹¹ **B** boron	**6** ¹² **C** carbon	**7** ¹⁴ **N** nitrogen	**8** ¹⁶ **O** oxygen	**9** ¹⁹ **F** fluorine	**10** ²⁰ **Ne** neon
11 ²³ **Na** sodium	**12** ²⁴ **Mg** magnesium	IIIb	IVb	Vb	VIb	VIIb	—— VIIIb ——			Ib	IIb	**13** ²⁷ **Al** aluminum	**14** ²⁸ **Si** silicon	**15** ³¹ **P** phosphorus	**16** ³² **S** sulfur	**17** ³⁵ **Cl** chlorine	**18** ⁴⁰ **Ar** argon
19 ³⁹ **K** potassium	**20** ⁴⁰ **Ca** calcium	**21** ⁴⁵ **Sc** scandium	**22** ⁴⁸ **Ti** titanium	**23** ⁵¹ **V** vanadium	**24** ⁵² **Cr** chromium	**25** ⁵⁵ **Mn** manganese	**26** ⁵⁶ **Fe** iron	**27** ⁵⁹ **Co** cobalt	**28** ⁵⁹ **Ni** nickel	**29** ⁶⁴ **Cu** copper	**30** ⁶⁵ **Zn** zinc	**31** ⁷⁰ **Ga** gallium	**32** ⁷³ **Ge** germanium	**33** ⁷⁵ **As** arsenic	**34** ⁷⁹ **Se** selenium	**35** ⁸⁰ **Br** bromine	**36** ⁸⁴ **Kr** krypton
37 ⁸⁵ **Rb** rubidium	**38** ⁸⁸ **Sr** strontium	**39** ⁸⁹ **Y** yttrium	**40** ⁹¹ **Zr** zirconium	**41** ⁹³ **Nb** niobium	**42** ⁹⁶ **Mo** molybdenum	**43** ⁹⁸ **Tc** technetium	**44** ¹⁰¹ **Ru** ruthenium	**45** ¹⁰³ **Rh** rhodium	**46** ¹⁰⁶ **Pd** palladium	**47** ¹⁰⁸ **Ag** silver	**48** ¹¹² **Cd** cadmium	**49** ¹¹⁵ **In** indium	**50** ¹¹⁹ **Sn** tin	**51** ¹²² **Sb** antimony	**52** ¹²⁸ **Te** tellurium	**53** ¹²⁷ **I** iodine	**54** ¹³¹ **Xe** xenon
55 ¹³³ **Cs** cesium	**56** ¹³⁷ **Ba** barium	**57** ¹³⁹ **La** lanthanum	**72** ¹⁷⁸ **Hf** hafnium	**73** ¹⁸¹ **Ta** tantalum	**74** ¹⁸⁴ **W** tungsten	**75** ¹⁸⁶ **Re** rhenium	**76** ¹⁹⁰ **Os** osmium	**77** ¹⁹² **Ir** iridium	**78** ¹⁹⁵ **Pt** platinum	**79** ¹⁹⁷ **Au** gold	**80** ²⁰¹ **Hg** mercury	**81** ²⁰⁴ **Tl** thallium	**82** ²⁰⁷ **Pb** lead	**83** ²⁰⁹ **Bi** bismuth	**84** ²¹⁰ **Po** polonium	**85** ²¹⁰ **At** astatine	**86** ²²² **Rn** radon
87 ²²³ **Fr** francium	**88** ²²⁶ **Ra** radium	**89** ²²⁷ **Ac** actinium	**104** ²⁶¹ **Rf** rutherfordium	**105** ²⁶⁰ **Db** dubnium	**106** ²⁶³ **Sg** seaborgium	**107** ²⁶¹ **Bh** bohrium	**108** ²⁶⁵ **Hs** hassium	**109** ²⁶⁶ **Mt** meitnerium	**110** ²⁸¹ **Ds** darmstadtium	**111** ²⁷² **Rg** Roentgenium	**112** ²⁷⁷ *******		**114** ²⁸⁵ *******		**116** ²⁹² *******		

58 ¹⁴⁰ **Ce** cerium	**59** ¹⁴¹ **Pr** praseodymium	**60** ¹⁴⁴ **Nd** neodymium	**61** ¹⁴⁷ **Pm** promethium	**62** ¹⁵⁰ **Sm** samarium	**63** ¹⁵² **Eu** europium	**64** ¹⁵⁷ **Gd** gadolinium	**65** ¹⁵⁹ **Tb** terbium	**66** ¹⁶³ **Dy** dysprosium	**67** ¹⁶⁵ **Ho** holmium	**68** ¹⁶⁷ **Er** erbium	**69** ¹⁶⁹ **Tm** thulium	**70** ¹⁷³ **Yb** ytterbium	**71** ¹⁷⁵ **Lu** lutetium
90 ²³² **Th** thorium	**91** ²³¹ **Pa** protactinium	**92** ²³⁸ **U** uranium	**93** ²³⁷ **Np** neptunium	**94** ²⁴² **Pu** plutonium	**95** ²⁴³ **Am** americium	**96** ²⁴⁷ **Cm** curium	**97** ²⁴⁷ **Bk** berkelium	**98** ²⁴⁹ **Cf** californium	**99** ²⁵⁴ **Es** einsteinium	**100** ²⁵³ **Fm** fermium	**101** ²⁵⁶ **Md** mendelevium	**102** ²⁵⁴ **No** nobelium	**103** ²⁵⁷ **Lr** lawrencium

GLOSSARY

A

abscisic acid (ABA) (ab SIH sick) Plant hormone that causes stomata to close and initiates and maintains dormancy. 487

abscission (ab SIH shun) Dropping of leaves, fruits, or flowers from a plant. 487

absolute dating (of fossils) Determining the age of a fossil by direct measurement, usually involving radioisotope decay. 322

absorption spectrum Spectrum produced when atoms absorb specific wavelengths of incoming light as they become excited from lower to higher energy levels. 118

acetylcholine (ACh) (uh see tuhl KOH lean) Neurotransmitter active in both the peripheral and central nervous systems. 706

acetylcholinesterase (AChE) (uh see tuhl KOH lean ESS turr raze) Enzyme that breaks down acetylcholine within a synapse. 706

acetyl CoA Molecule made up of a 2-carbon acetyl group attached to coenzyme A. During cellular respiration, the acetyl group enters the citric acid cycle for further breakdown. 136

acid Molecules tending to raise the hydrogen ion concentration in a solution and to lower its pH numerically. 30

acid deposition The return to Earth in rain or snow of sulfate or nitrate salts of acids produced by commercial and industrial activities. 893

acoelomate Animal that has no body cavity (i.e., tapeworm). 518

acquired immunodeficiency syndrome (AIDS) Disease caused by a retrovirus and transmitted via body fluids; characterized by failure of the immune system. 790

acromegaly (ack row MEG uh lee) Condition resulting from an increase in growth hormone production after adult height has been achieved. 762

actin (ACK tin) One of two major proteins of muscle; makes up thin filaments in myofibrils of muscle fibers. *See also* myosin. 748

actin filament Muscle protein filament in a sarcomere; its movement shortens the sarcomere, yielding muscle contraction. Actin filaments play a role in the movement of the cell and its organelles. 76

action potential Electrochemical changes that take place across the axomembrane; the nerve impulse. 705

action spectrum Spectrum of light that elicits a particular response. 118

active site Region on the surface of an enzyme where the substrate binds and where the reaction occurs. 106

active transport Use of a plasma membrane carrier protein to move a molecule or ion from a region of lower concentration to one of higher concentration; it opposes equilibrium and requires energy. 92

adaptation Organism's modification in structure, function, or behavior suitable to the environment. 5, 291

adaptive radiation Evolution of several species from a common ancestor into new ecological or geographical zones. 313

Addison disease (ADD dih sun) Condition resulting from a deficiency of adrenal cortex hormones; characterized by low blood glucose, weight loss, and weakness. 766

adenine (A) (AD duh neen) One of four nitrogen-containing bases in nucleotides composing the structure of DNA and RNA. 227

adenosine (ad DEN oh sign) Portion of ATP and ADP that is composed of the base adenine and the sugar ribose. 52

adhesion junction (ad HE shun) Junction between cells in which the adjacent plasma membranes do not touch but are held together by intercellular filaments attached to buttonlike thickenings. 96

adipose tissue (AD dip pose) Connective tissue in which fat is stored. 600

ADP (adenosine diphosphate) (ad DEN oh seen dye FOSS fate) Nucleotide with two phosphate groups that can accept another phosphate group and become ATP. 52, 104

adrenal cortex (uh DREEN ul) Outer portion of the adrenal gland; secretes mineralocorticoids, such as aldosterone, and glucocorticoids, such as cortisol. 765

adrenal gland Gland that lies atop a kidney; the *adrenal medulla* produces the hormones epinephrine and norepinephrine, and the *adrenal cortex* produces the glucocorticoid and mineralocorticoid hormones. 765

adrenal medulla Inner portion of the adrenal gland; secretes the hormones epinephrine and norepinephrine. 765

adrenocorticotropic hormone (ACTH) (uh DREEN oh core tic oh TROH pick) Hormone secreted by the anterior lobe of the pituitary gland that stimulates activity in the adrenal cortex. 760

adventitious roots (ad vin TIH shus) Fibrous roots that develop from stems or leaves, such as the prop roots of corn or the holdfast roots of ivy. 446

age structure diagram In demographics, a display of the age groups of a population; a growing population has a pyramid-shaped diagram. 843

agglutination (ag glue tin NAY shun) Clumping of red blood cells due to a reaction between antigens on red blood cell plasma membranes and antibodies in the plasma. 650

agnathan (ag NATH uhn) Vertebrate that lacks jaws; superclass Agnatha includes only the jawless fishes—i.e., the lampreys and hagfishes. 562

alcoholic fermentation Fermentation process that produces ethanol and CO_2 from sugars. 142

aldosterone (al DOSS turr own) Hormone secreted by the adrenal cortex that regulates the sodium and potassium ion balance of the blood. 694, 766

algae (sing., alga) (AL jee, AL guh) Type of protist that carries on photosynthesis; unicellular forms are a part of phytoplankton, and multicellular forms are called seaweed. 382

allantois (uh LANN toys) Extraembryonic membrane that accumulates nitrogenous wastes in birds and reptiles and contributes to the formation of umbilical blood vessels in mammals. 807

allele (uh LEEL) Alternative form of a gene—alleles occur at the same locus on homologous chromosomes. 168, 184

allergen (AL lurr jen) Foreign substance capable of stimulating an allergic response. 648

allergy Immune response to substances that usually are not recognized as foreign. 648

allopatric speciation (al low PAT trick spee see AY shun) Origin of new species between populations that are separated geographically. 312

alpine tundra Tundra near the peak of a mountain. 905

alternation of generations Life cycle, typical of plants, in which a diploid sporophyte alternates with a haploid gametophyte. 174, 416

altruism (AL true is uhm) Social interaction that has the potential to decrease the lifetime reproductive success of the member exhibiting the behavior. 829

alveolus (pl., alveoli) (al VEE oh luss) In humans, terminal, microscopic, grapelike air sac found in lungs. 677

amino acid (uh MEAN no) Organic molecule composed of an amino group and an acid group; covalently bonds to produce peptide molecules. 46

ammonia Colorless gas, which has a penetrating odor and is soluble in water. 688

amniocentesis (am nee oh cent TEE sis) Procedure for removing amniotic fluid surrounding the developing fetus for testing of the fluid or cells within the fluid. 214

amnion (AM nee ahn) Extraembryonic membrane of birds, reptiles, and mammals that forms an enclosing, fluid-filled sac. 807

amniotic egg (am nee AH tick) Egg that has an amnion, as seen during the development of reptiles, birds, and mammals. 568

amoeboid (uh ME boid) Cell that moves and engulfs debris with pseudopods. 390

amphibian Member of a class of vertebrates that includes frogs, toads, and salamanders; they are still tied to a watery environment for reproduction. 565

anabolic steroid (ann uh BAH lick STARE oid) Synthetic steroid that mimics the effect of testosterone. 770

anabolism (ann AB oh liz uhm) Metabolic process by which larger molecules are synthesized from smaller ones; anabolic metabolism. 144

anaerobic (ann air OH bick) Growing or metabolizing in the absence of oxygen. 133, 141

analogous structure (ann AL oh gus) Structure that has a similar function in separate lineages but differs in anatomy and ancestry. 296, 347

analogy (ann AL oh gee) Similarity of function but not of origin. 347

anaphase (ANN uh faze) Mitotic phase during which daughter chromosomes move toward the poles of the spindle. 156

anaphylactic shock (ann uh fuh LACK tick) Severe systemic form of anaphylaxis involving bronchiolar constriction, impaired breathing, vasodilation, and a rapid drop in blood pressure with a threat of circulatory failure. 648

ancestral character Structural, physiological, or behavioral trait that is present in a common ancestor and all members of a group. 346

anchoring junction Junctions that mechanically attach adjoining cells (e.g., adhesion, tight, and gap junctions). 96

androgen (ANN droh jen) Male sex hormone (e.g., testosterone). 770

aneuploid (ANN you ploid) Individual whose chromosome number is not an exact multiple of the haploid number for the species. 212

angina pectoris (ann JEYE nuh peck TORE iss) Condition characterized by thoracic pain resulting from occluded coronary arteries; precedes a heart attack. 625

angiogenesis (ann jee oh JEN uh sis) Formation of new blood vessels; one mechanism by which cancer spreads. 159

angiosperm (ANN jee oh sperm) Flowering plant; the seeds are borne within a fruit. 428

animal Multicellular, heterotrophic organism belonging to the kingdom Animalia. 10

Animalia Kingdom of animals. 518

annelid (ANN uh lid) Member of a phylum of invertebrates (phylum Annelida) that contains segmented worms, such as the earthworm and the clam worm. 544

annual ring Layer of wood (secondary xylem) usually produced during one growing season. 450

anterior pituitary (an TEER rih urr pit TWO uh tare ree) Portion of the pituitary gland that is controlled by the hypothalamus and produces six types of hormones, some of which control other endocrine glands. 760

anther (ANN thurr) In flowering plants, pollen-bearing portion of stamen. 496

antheridia Sperm-producing structures, as in the moss life cycle. 416

anthropoid (ANN throw poid) Group of primates that includes monkeys, apes, and humans. 583

antibody Protein produced in response to the presence of an antigen; each antibody combines with a specific antigen. 626

antibody-mediated immunity Specific mechanism of defense in which plasma cells derived from B cells produce antibodies that combine with antigens. 637

anticodon (ann tie COH don) Three-base sequence in a transfer RNA molecule base that pairs with a complementary codon in mRNA. 244

antidiuretic hormone (ADH) (an tee die you REH tick) Hormone secreted by the posterior pituitary that increases the permeability of the collecting ducts in a kidney. 694, 760

antigen (ANN tih jen) Foreign substance, usually a protein or a polysaccharide, that stimulates the immune system to react, such as to produce antibodies. 626, 638

antigen-presenting cell (APC) Cell that displays the antigen to certain cells of the immune system so they can defend the body against that particular antigen. 642

antigen receptor Receptor proteins in the plasma membrane of immune system cells whose shape allows them to combine with a specific antigen. 638

anus (AY nuss) Outlet of the digestive tube. 662

aorta (ay OR tuh) In humans, the major systemic artery that takes blood from the heart to the tissues. 622

apical dominance (AY pick uhl) Influence of a terminal bud in suppressing the growth of lateral buds. 482

apical meristem (AY pick uhl MARE uh stem) In vascular plants, masses of cells in the root and shoot that reproduce and elongate as primary growth occurs. 444

apoptosis (ay pop TOE sis) Programmed cell death involving a cascade of specific cellular events leading to death and destruction of the cell. 71, 151, 635, 802

appendicular skeleton (app pen DICK you lurr) Part of the vertebrate skeleton forming the appendages, shoulder girdle, and hip girdle. 744

appendix (app PEN dicks) In humans, small, tubular appendage that extends outward from the cecum of the large intestine. 662

aquifer (ACK kwih fur) Rock layers that contain water and will release it in appreciable quantities to wells or springs. 889

arboreal (are BORE ree uhl) Living in trees. 580

archaea Member of the domain Archaea. 373

archegonia Egg-producing structures, as in the moss life cycle. 416

Arctic tundra Biome that encircles the Earth just south of ice-covered polar seas in the Northern Hemisphere. 906

arteriole (are TEER ree ohl) Vessel that takes blood from an artery to capillaries. 616

artery Blood vessel that transports blood away from the heart. 616

arthropod (ARTH throw pod) Member of a phylum of invertebrates (phylum Arthropoda) that contains, among other groups, crustaceans and insects, which have an exoskeleton and jointed appendages. 546

ascus (ASK us) Fingerlike sac in which nuclear fusion, meiosis, and ascospore production occur during sexual reproduction of sac fungi. 402

asexual reproduction Reproduction that requires only one parent and does not involve gametes. 162

association areas Regions of the cerebral cortex related to memory, reasoning, judgment, and emotional feelings. 709

associative learning Acquired ability to associate two stimuli or between a stimulus and a response. 823

assortative mating (ah Sor tah tive) Mating of individuals with similar phenotypes. 305

aster (ASS turr) Short, radiating fibers produced by the centrosomes in animal cells. 155

asthma Condition in which bronchioles constrict and cause difficulty in breathing. 649

asymmetrical Dissimilar in corresponding parts or organs on opposite sides of the body that are normally alike in symmetrical animals. 518

atom Smallest particle of an element that displays the properties of the element. 20

atomic mass Mass of an atom equal to the number of protons plus the number of neutrons within the nucleus. 20

atomic number Number of protons within the nucleus of an atom. 21

atomic symbol One or two letters that represent the name of an element—e.g., H stands for a hydrogen atom, and Na stands for a sodium atom. 20

ATP (adenosine triphosphate) (ad DEN no seen try FOSS fate) Nucleotide with three phosphate groups. The breakdown of ATP into ADP + P makes energy available for energy-requiring processes in cells. 50, 104

ATP synthase (SIN thaze) Complex formed of enzymes and their carrier proteins; functions in the production of ATP in chloroplasts and mitochondria. 111, 122

atrial natriuretic hormone (ANH) (AY tree uhl nat tree you RETT tick) Hormone secreted by the heart that increases sodium excretion. 695, 766

atrioventricular valve (ay tree oh vinn TRICK you lurr) Heart valve located between an atrium and a ventricle. 618

atrium (AY tree uhm) Chamber; particularly an upper chamber of the heart lying above a ventricle. 618

australopithecine (oss stray loh PITH ih seen) One of several species of *Australopithecus*, a genus that contains the first generally recognized hominids. 586

autoimmune disease (ah toe ih MUNE) Disease that results when the immune system mistakenly attacks the body's own tissues. 649

autonomic system (ah toe NAHM mick) Portion of the peripheral nervous system that regulates internal organs. 715

autosome (AH toe sohm) Any chromosome other than the sex-determining pair. 192

autotroph (AH toe trofe) Organism that can capture energy and synthesize organic molecules from inorganic nutrients. 116, 883

auxin (OX sin) Plant hormone regulating growth, particularly cell elongation; also called indoleacetic acid (IAA). 482

axial skeleton (AXE ee uhl) Part of the vertebrate skeleton forming the vertical support or axis, including the skull, the rib cage, and the vertebral column. 742

axillary bud (AXE ill air ree) Bud located in the axil of a leaf. 439

axon (AXE ahn) Elongated portion of a neuron that conducts nerve impulses, typically from the cell body to the synapse. 703

B

bacillus (buh SILL us) A rod-shaped bacterium. 62

bacterial vaginosis Sexually transmitted disease caused by *Gardnerella vaginalis*, *Mobiluncus* spp., *Mycoplasma hominis*, and various anaerobic bacteria. Although a mild disease, it is a risk factor for obstetric infections and pelvic inflammatory disease. 792

bacteriophage (back TEER ree oh fahj) Virus that infects bacteria. 225, 364

bacterium (pl., bacteria) Member of the domain Bacteria. 373

bark External part of a tree, containing cork, cork cambium, and phloem. 450

Barr body Dark-staining body (discovered by M. Barr) in the nuclei of female mammals that contains a condensed, inactive X chromosome. 256

basal body (BAY zull) A cytoplasmic structure that is located at the base of—and may organize—cilia or flagella. 78

basal nuclei (BAY zull NEW clee eye) Subcortical nuclei deep within the white matter that serve as relay stations for motor impulses and produce dopamine to help control skeletal muscle activities. 710

base Molecules tending to lower the hydrogen ion concentration in a solution and raise the pH numerically. 30

basement membrane Layer of nonliving material that anchors epithelial tissue to underlying connective tissue. 598

basidium (buh SIH dee uhm) Clublike structure in which nuclear fusion, meiosis, and basidiospore production occur during sexual reproduction of club fungi. 406

basophil (BASE oh fill) White blood cell with a granular cytoplasm; able to be stained with a basic dye. 628

behavior Observable, coordinated responses to environmental stimuli. 820

benign (buh NINE) Mass of cells derived from a single mutated cell that has repeatedly undergone cell division but has remained at the site of origin. 159

benthic division (BEN thick) Ocean or lake floor, from the high tidemark to the deepest depths, which supports a unique set of organisms. 923

bicarbonate ion (by CAR boh nate EYE ahn) Ion that participates in buffering the blood, and the form in which carbon dioxide is transported in the bloodstream. 679

bilateral symmetry (by LATT turr uhl SIMM met tree) Body plan having two corresponding or complementary halves. 518

bile Secretion of the liver that is temporarily stored and concentrated in the gallbladder before being released into the small intestine, where it emulsifies fat. 661

binary fission (BY nuh ree FISH uhn) Splitting of a parent cell into two daughter cells; serves as an asexual form of reproduction in bacteria. 162, 369

binomial nomenclature (by NO mee uhl) Scientific name of an organism, the first part of which designates the genus and the second part of which designates the specific epithet. 10, 342

binomial system Assignment of two names to each organism; a system developed by Carolus Linnaeus in the mid-eighteenth century. 342

biodiversity (by oh die VERSE sit tee) Total number of species, the variability of their genes, and the communities in which they live. 7, 928

biodiversity hotspots Regions of the world that contain unusually large concentrations of species. 929

biogeochemical cycle (by oh jee oh KEM ick cull) Circulating pathway of elements such as carbon and nitrogen involving exchange pools, storage areas, and biotic communities. 888

biogeography (by oh jee AH gruh fee) Study of the geographical distribution of organisms. 286

bioinformatics (by oh in for MAT ticks) Computer technologies used to study the genome. 273, 928

biological clock Internal mechanism that maintains a biological rhythm in the absence of environmental stimuli. 481

biological magnification Process by which substances become more concentrated in organisms in the higher trophic levels of a food web. 895

biology Scientific study of life. 10

biomass (BY oh mass) The number of organisms multiplied by their weight. 887

biome (BY ohm) One of the biosphere's major communities, characterized in particular by certain climatic conditions and particular types of plants. 905

biosphere (BY ohs fear) Zone of air, land, and water at the surface of the Earth in which living organisms are found. 5, 838, 882

biotic potential (by AH tick) Maximum population growth rate under ideal conditions. 840

bipedalism (by PED uh liz uhm) Walking erect on two feet. 584

bird Endothermic vertebrate that has feathers and wings, is often adapted for flight, and lays hard-shelled eggs. 572

bivalent (by VAY lent) Homologous chromosomes, each having sister chromatids that are joined by a nucleoprotein lattice during meiosis; also called a tetrad. 169

bivalve (BY valve) Type of mollusc with a shell composed of two valves; includes clams, oysters, and scallops. 540

blade Broad, expanded portion of a plant leaf that may be single or compound leaflets. 439

blastocoel (BLAST toe seal) Fluid-filled cavity of a blastula. 799

blastocyst (BLAST toe sist) Early stage of human embryonic development that consists of a hollow, fluid-filled ball of cells. 808

blastopore Opening into the primitive gut formed at gastrulation. 800

blastula (BLAST you luh) Hollow, fluid-filled ball of cells occurring during animal development prior to gastrula formation. 799

blind spot Region of the retina, lacking rods or cones, where the optic nerve leaves the eye. 728

blood Fluid circulated by the heart through a closed system of vessels. 601, 614

blood pressure Force of blood pushing against the inside wall of blood vessels. 623

B lymphocyte (B cell) (LIMM foe site) Lymphocyte that matures in the bone marrow and, when stimulated by the presence of a specific antigen, gives rise to antibody-producing plasma cells. 635

bog Wet, spongy ground in a low-lying area. 915

bone Connective tissue having protein fibers and a hard matrix of inorganic salts, notably calcium salts. 600

bony fishes Members of a class of vertebrates (class Osteichthyes) containing numerous diverse fishes, with a bony rather than cartilaginous skeleton. 564

bottleneck effect Cause of genetic drift; occurs when a majority of genotypes are prevented from participating in the production of the next generation as a result of a natural disaster or human interference. 305

brain Ganglionic mass at the anterior end of the nerve cord; in vertebrates, the brain is located in the cranial cavity of the skull. 701

brain stem In mammals; portion of the brain consisting of the medulla oblongata, pons, and midbrain. 710

bronchiole (BRAHN key ohl) In terrestrial vertebrates, small tube that conducts air from a bronchus to the alveoli. 676

bronchus (pl., bronchi) (BRAHN cuss, BRAHN kie) In terrestrial vertebrates, branch of the trachea that leads to the lungs. 676

brown algae Marine photosynthetic protist with a notable abundance of xanthophyll pigments; this group includes well-known seaweeds of northern rocky shores. 385

bryophyte (BRY oh fite) Member of one of three phyla of nonvascular plants—the mosses, liverworts, and hornworts. 417

budding Asexual form of reproduction whereby a new organism develops as an outgrowth of the body of the parent. 399

buffer Substance or group of substances that tend to resist pH changes of a solution, thus stabilizing its relative acidity and basicity. 30

bursa (BURR suh) Saclike, fluid-filled structure, lined with synovial membrane, that occurs near a synovial joint. 745

C

C$_3$ plant Plant that fixes carbon dioxide via the Calvin cycle; the first stable product of C$_3$ photosynthesis is a 3-carbon compound. 126

C$_4$ plant Plant that fixes carbon dioxide to produce a C$_4$ molecule that releases carbon dioxide to the Calvin cycle. 126

calcitonin (cal sit ON in) Hormone secreted by the thyroid gland that increases the blood calcium level. 764

calorie Amount of heat energy required to raise the temperature of one gram of water 1°C. 27

Calvin cycle reaction Portion of photosynthesis that takes place in the stroma of chloroplasts and can occur in the dark; it uses the products of the light reactions to reduce CO$_2$ to a carbohydrate. 119

calyx The sepals collectively; the outermost flower whorl. 428

CAM Plant that fixes carbon dioxide at night to produce a C$_4$ molecule that releases carbon dioxide to the Calvin cycle during the day; CAM stands for crassulacean-acid metabolism. 127

camera-type eye Type of eye found in vertebrates and certain molluscs; a single lens focuses an image on closely packed photoreceptors. 724

camouflage (CAM oh flaj) Method of hiding from predators in which the organism's behavior, form, and pattern of coloration allow it to blend into the background and prevent detection. 868

cancer Malignant tumor whose nondifferentiated cells exhibit loss of contact inhibition, uncontrolled growth, and the ability to invade tissue and metastasize. 159

capillary (CAP pill air ree) Microscopic blood vessel; gases and other substances are exchanged across the walls of a capillary between blood and tissue fluid. 616

capsid (CAP sid) Protective protein containing the genetic material of a virus. 363

capsule Gelatinous layer surrounding the cells of blue-green algae and certain bacteria. 62

carbaminohemoglobin (car buh meen oh HEE muh glow bin) Hemoglobin carrying carbon dioxide. 679

carbohydrate (car boh HI drate) Class of organic compounds that includes monosaccharides, disaccharides, and polysaccharides. 38

carbon cycle Continuous process by which carbon circulates in the air, water, and organisms of the biosphere. 890

carbon dioxide (CO_2) fixation Photosynthetic reaction in which carbon dioxide is attached to an organic compound. 124

carbonic anhydrase (car BAH nick ann HI draze) Enzyme in red blood cells that speeds the formation of carbonic acid from water and carbon dioxide. 679

carcinogen (car SIN uh jen) Environmental agent that causes mutations leading to the development of cancer. 263

carcinogenesis (car sin oh JEN uh sis) Development of cancer. 159

cardiac conduction system System of specialized cardiac muscle fibers that conducts impulses from the SA node to the chambers of the heart, causing them to contract. 621

cardiac cycle One complete cycle of systole and diastole for all heart chambers. 620

cardiac muscle Striated, involuntary muscle tissue found only in the heart. 602

cardiac output Blood volume pumped by each ventricle per minute (not total output pumped by both ventricles). 620

cardiac pacemaker Mass of specialized cardiac muscle tissue that controls the rhythm of the heartbeat; the SA node. 621

cardiovascular system (car dee oh VASS cue lurr) Organ system in which blood vessels distribute blood under the pumping action of the heart. 616

carnivore (CAR nih vore) Consumer in a food chain that eats other animals. 883

carotenoid (car RAH ten oid) Yellow or orange pigment that serves as an accessory to chlorophyll in photosynthesis. 118

carpel (CAR pull) Ovule-bearing unit that is a part of a pistil. 429, 496

carrier Heterozygous individual who has no apparent abnormality but can pass on an allele for a recessively inherited genetic disorder. 192, 205

carrier protein Protein that combines with and transports a molecule or ion across the plasma membrane. 87

carrying capacity Largest number of organisms of a particular species that can be maintained indefinitely by a given environment. 846

cartilage (CAR tih ledge) Connective tissue in which the cells lie within lacunae embedded in a flexible, proteinaceous matrix. 600

cartilaginous fishes (car tih LAJJ jen us) Members of a class of vertebrates (class Chondrichthyes) with a cartilaginous rather than bony skeleton; includes sharks, rays, and skates. 563

Casparian strip (cass PAIR ree uhn) Layer of impermeable lignin and suberin bordering four sides of root endodermal cells; prevents

water and solute transport between adjacent cells. 445, 464

catabolism (cuh TAB uh liz uhm) Metabolic process that breaks down large molecules into smaller ones; catabolic metabolism. 144

catastrophism (cuh TASS troh fizz uhm) Belief espoused by Georges Cuvier that periods of catastrophic extinctions occurred, after which repopulation of surviving species took place, giving the appearance of change through time. 284

cell Smallest unit that displays the properties of life; composed of cytoplasm surrounded by a plasma membrane. 2, 58

cell body Portion of a neuron that contains a nucleus and from which dendrites and an axon extend. 703

cell cycle Repeating sequence of events in eukaryotes that involves cell growth and nuclear division; consists of the stages G_1, S, G_2, and M. 150

cell envelope In a prokaryotic cell, the portion composed of the plasma membrane, the cell wall, and the glycocalyx. 62

cell-mediated immunity Specific mechanism of defense in which T cells destroy antigen-bearing cells. 642

cell plate Structure across a dividing plant cell that signals the location of new plasma membranes and cell walls. 157

cell recognition protein Glycoprotein that helps the body defend itself against pathogens. 87

cell theory One of the major theories of biology, which states that all organisms are made up of cells; cells are capable of self-reproduction and come only from preexisting cells. 58

cellular differentiation Process and developmental stages by which a cell becomes specialized for a particular function. 802

cellular respiration Metabolic reactions that use the energy from carbohydrate, fatty acid, or amino acid breakdown to produce ATP molecules. 132

cellular slime mold Free-living amoeboid cells that feed on bacteria and yeasts by phagocytosis and aggregate to form a plasmodium that produces spores. 393

cellulose (SELL you lohs) Polysaccharide that is the major complex carbohydrate in plant cell walls. 41

cell wall Structure that surrounds a plant, protistan, fungal, or bacterial cell and maintains the cell's shape and rigidity. 62, 97

centipede (SEN tih peed) Elongated arthropod characterized by having one pair of legs to each body segment; they may have 15 to 173 pairs of legs. 552

central nervous system (CNS) Portion of the nervous system consisting of the brain and spinal cord. 702

central vacuole (VACK you ohl) In a plant cell, a large, fluid-filled sac that stores metabolites. During growth, it enlarges, forcing the primary cell wall to expand and the cell surface-area-to-volume ratio to increase. 73

centriole (SENT tree ohl) Cell organelle, existing in pairs, that occurs in the centrosome and may help organize a mitotic spindle for chromosome movement during animal cell division. 78, 154

centromere (SENT troh meer) Constriction where sister chromatids of a chromosome are held together. 153

centrosome (SENT troh sohm) Central microtubule organizing center of cells. In animal cells, it contains two centrioles. 76, 154

cephalization (seff full lih ZAY shun) Having a well-recognized anterior head with a brain and sensory receptors. 518, 700

cephalopod (SEFF full oh pod) Type of mollusc in which a modified foot develops into the head region; includes squids, cuttlefish, octopuses, and nautiluses. 542

cephalothorax (seff full oh THORE axe) Fused head and thorax found in decapods (shrimps, lobsters, crayfish, and crabs). 553

cerebellum (sair uh BELL uhm) In terrestrial vertebrates, portion of the brain that coordinates skeletal muscles to produce smooth, graceful motions. 710

cerebral cortex (sir REE brull CORE tex) Outer layer of cerebral hemispheres; receives sensory information and controls motor activities. 709

cerebral hemisphere Either of the two lobes of the cerebrum in vertebrates. 709

cerebrospinal fluid (sir ree broh SPY null) Fluid found in the ventricles of the brain, in the central canal of the spinal cord, and in association with the meninges. 708

cerebrum (sir REE brumm) Largest part of the brain in mammals. 709

cervix Narrow end of the uterus, which leads into the vagina. 782

cestode Member of the phylum Platyhelminthes (flatworm); tapeworms that are intestinal parasites of vertebrate hosts. 528

channel protein Protein that forms a channel to allow a particular molecule or ion to cross the plasma membrane. 87

chaparral (shapp purr AL) Biome characterized by broad-leafed evergreen shrubs forming dense thickets. 912

chaperone protein (shapp purr OHN) Molecule that binds to a protein during synthesis and keeps it from making incorrect interactions. 49

character Any structural, chromosomal, or molecular feature that distinguishes one group from another. 345

character displacement Tendency for characteristics to be more divergent when similar species belong to the same community than when they are isolated from one another. 864

chelicerate (shell lih sir AH tuh) Group of arthropods (e.g., horseshoe crabs, sea spiders, arachnids), which have a pair of appendages in the form of pinchers or fangs. 553

chemical energy Energy associated with the interaction of atoms in a molecule. 102

chemical evolution Increase in the complexity of chemicals over time that could have led to the first cells. 318

chemical signal Molecule that brings about a change in a cell, tissue, organ, or individual when it binds to a specific receptor. 758

chemiosmosis (kim mee oz MOW sis) Ability of certain membranes to use a hydrogen ion gradient to drive ATP formation. 111, 122, 139

chemoautotroph (key mow AH toe trofe) Organism able to synthesize organic molecules by using carbon dioxide as the carbon source and the oxidation of an inorganic substance (such as hydrogen sulfide) as the energy source. 370

chemoheterotroph (key mow HETT turr row trofe) Organism that is unable to produce its own organic molecules, and therefore requires organic nutrients in its diet. 370

chemoreceptor (key mow ree SEPP turr) Sensory receptor that is sensitive to chemical stimulation—for example, receptors for taste and smell. 722

chitin (KITE in) Strong but flexible nitrogenous polysaccharide found in the exoskeleton of arthropods. 41, 546

chlorofluorocarbons (CFCs) (klore oh flur oh CAR buns) Organic compounds containing carbon, chlorine, and fluorine atoms. CFCs such as Freon can deplete the ozone shield by releasing chlorine atoms in the upper atmosphere. 897

chlorophyll (KLORE uh fill) Green pigment that absorbs solar energy and is important in algal and plant photosynthesis; occurs as chlorophyll *a* and chlorophyll *b*. 117

chloroplast (KLORE oh plast) Membrane-bounded organelle in algae and plants with chlorophyll-containing membranous thylakoids; where photosynthesis takes place. 74, 116

cholesterol (koh LESS turr all) One of the major lipids found in animal plasma membranes; makes the membrane impermeable to many molecules. 85

chordate (CORE date) Animals in the phylum Chordata that have a dorsal tubular nerve cord, a notochord, pharyngeal gill pouches, and a postanal tail at some point in their life cycle. 560

chorion (CORE ree ahn) Extraembryonic membrane functioning for respiratory exchange in birds and reptiles; contributes to placenta formation in mammals. 807

chorionic villi sampling (CVS) (core ree AH nick VILL eye) Prenatal test in which a sample of chorionic villi cells is removed for diagnostic purposes. 214

chorionic villus (core ree AH nick VILL us) Treelike extension of the chorion of the embryo, projecting into the maternal tissues at the placenta. 810

choroid (CORE oid) Vascular, pigmented middle layer of the eyeball. 725

chromatin (CROW muh tin) Network of fibrils consisting of DNA and associated proteins observed within a nucleus that is not dividing. 68, 153, 256

chromosomal mutation (crow mow SO mull) Alteration in the chromosome structure or number typical of the species. 212

chromosome (CROW muh sohm) Structure consisting of DNA complexed with proteins that transmits genetic information from the previous generation of cells and organisms to the next generation. 68

chromosome theory of inheritance The idea that chromosomes are the carriers of genes. 204

chronic inflammation Slow, progressing inflammation of connective tissue, usually causes permanent tissue damage. 637

chyme (KIME) Thick, semiliquid food material that passes from the stomach to the small intestine. 661

ciliary muscle (SILL lee air ree) Within the ciliary body of the vertebrate eye, the ciliary muscle controls the shape of the lens. 726

ciliate (SILL lee ate) Complex unicellular protist that moves by means of cilia and digests food in food vacuoles. 391

cilium (SILL lee uhm) Short, hairlike projection from the plasma membrane, occurring usually in larger numbers (cilia). 78

circadian rhythm (sir KAY dee uhn) Biological rhythm with a 24-hour cycle. 480, 770

circulatory system In animals other than humans, an organ system that moves substances to and from cells, usually via a heart, blood, and blood vessels. 614

cirrhosis Chronic, irreversible injury to liver tissue; commonly caused by frequent alcohol consumption. 664

citric acid cycle Cycle of reactions in mitochondria that begins with citric acid. It breaks down an acetyl group and produces CO_2, ATP, NADH, and $FADH_2$; also called the Krebs cycle. 133, 137

clade Taxon or other group consisting of an ancestral species and all of its descendants, forming a distinct branch on a phylogenetic tree. 351

cladistic systematics (kluh DISS tick) School of systematics that uses derived characters to determine monophyletic groups and construct cladograms. 351

cladogram (CLADD doe gram) In cladistics, a branching diagram that shows the relationship among species in regard to their shared derived characters. 351

class One of the categories, or taxa, used by taxonomists to group species; class is the taxon above the order level. 8, 344

classical conditioning Type of learning whereby an unconditioned stimulus that elicits a specific response is paired with a neutral stimulus so that the response becomes conditioned. 823

cleavage (CLEAVE edge) Cell division without cytoplasmic addition or enlargement; occurs during the first stage of animal development. 799

climate Weather condition of an area, including especially prevailing temperature and average daily/yearly rainfall. 902

climax community In ecology, community that results when succession has come to an end. 874

cloaca (kloh AY cuh) Posterior portion of the digestive tract in certain vertebrates that receives feces and urogenital products. 566

clonal selection theory (KLOH null) States that the antigen selects which lymphocyte will undergo clonal expansion and produce more lymphocytes bearing the same type of receptor. 639

cloning Production of identical copies. In organisms, the production of organisms with the same genes; in genetic engineering, the production of many identical copies of a gene. 268

closed circulatory system Blood is confined to vessels and is kept separate from the interstitial fluid. 615

club fungi Members of the phylum Basidiomycota. 406

club moss Type of seedless vascular plant that is also called ground pine because it has the appearance of a miniature pine tree. 421

cnidarian (neye DARE ree uhn) Invertebrate in the phylum Cnidaria existing as either a polyp or medusa with two tissue layers and radial symmetry. 524

coacervate droplet (coh AY sir vate) An aggregate of colloidal droplets held together by electrostatic forces. 320

coal Fossil fuel formed millions of years ago from plant material that did not decay. 427

coccus (COCK us) A spherical-shaped bacterium. 62

cochlea (COKE lee uh) Spiral-shaped structure of the vertebrate inner ear containing the sensory receptors for hearing. 730

codominance (koh DAH men unce) Inheritance pattern in which both alleles of a gene are equally expressed. 197

codon (KOH dahn) Three-base sequence in messenger RNA that causes the insertion of a particular amino acid into a protein, or termination of translation. 241

coelom (SEE lumm) Body cavity lying between the digestive tract and body wall that is completely lined by mesoderm. 518, 538

coelomate Animal possessing a coelom (body cavity) completely lined by mesoderm (e.g., protostomes and deuterostomes). 518

coenzyme (koh IN zime) Nonprotein organic molecule that aids the action of the enzyme to which it is loosely bound. 50, 108

coevolution Joint evolution in which one species exerts selective pressure on the other species. 500, 871

cofactor Nonprotein adjunct required by an enzyme in order to function; many cofactors are metal ions, others are coenzymes. 108

cohesion-tension model Explanation for upward transport of water in xylem based upon transpiration-created tension and the cohesive properties of water molecules. 468

cohort (KOH hort) Group of individuals having a statistical factor in common, such as year of birth, in a population study. 840

coleoptile (koh lee OPP tile) Protective sheath that covers the young leaves of a seedling. 482

collagen fiber White fiber in the matrix of connective tissue, giving flexibility and strength. 599

collecting duct Duct within the kidney that receives fluid from several nephrons; the reabsorption of water occurs here. 691

collenchyma (kuh LENN kih muh) Plant tissue composed of cells with unevenly thickened walls; supports growth of stems and petioles. 442

colon (KOH lunn) In humans, the major portion of the large intestine. 662

colony Loose association of cells that remain independent for most functions. 384

columnar epithelium (kuh LUM nurr epp pih THEE lee uhm) Type of epithelial tissue with cylindrical cells. 598

comb jelly Member of phylum Ctenophora; free-swimming marine invertebrates. 524

commensalism (kuh MENS suh liz uhm) Symbiotic relationship in which one species is benefited, and the other is neither harmed nor benefited. 872

common ancestor Ancestor held in common by at least two lines of descent. 346

communication Signal by a sender that influences the behavior of a receiver. 831

community Assemblage of populations interacting with one another within the same environment. 5, 838, 860

community stability A community is said to be stable when it can persist through time, resist changes, and recover after a disturbance. 876

compact bone Type of bone that contains osteons consisting of concentric layers of matrix and osteocytes in lacunae. 600, 740

companion cell Cell associated with sieve-tube members in phloem of vascular plants. 466

competitive exclusion principle Theory that no two species can occupy the same niche. 864

competitive inhibition Form of enzyme inhibition where the substrate and inhibitor are both able to bind to the enzyme's active site; each complexes with the enyzme. Only when the substrate is at the active site will product form. 109

complement Collective name for a series of enzymes and activators in the blood, some of which may bind to antibody and may lead to rupture of a foreign cell. 638

complementary base pairing Hydrogen bonding between particular purines and pyrimidines in DNA. 51, 228

complementary DNA (cDNA) DNA that has been synthesized from mRNA by the action of reverse transcriptase. 269

complete digestive tract Digestive tract that has both a mouth and an anus. 656

complex tissues In plants, tissues composed of two or more kinds of cells (e.g., xylem, containing tracheids and vessel elements; phloem, containing sieve-tube members and companion cells). 443

compound eye Type of eye found in arthropods; it is composed of many independent visual units. 724

concentration gradient Gradual change in chemical concentration from one point to another. 88

conclusion Statement made following an experiment as to whether or not the results support the hypothesis. 11

condensation Conversion from a gaseous state to a liquid or solid. 889

cone Structure comprised of scales bearing sporangia; pollen cones bear microsporangia, and seed cones bear megasporangia. 424

cone cell Photoreceptor in vertebrate eyes that responds to bright light and makes color vision possible. 727

conidiospore (koh NIDD dee uh spore) Spore produced by sac and club fungi during asexual reproduction. 402

conifer (KAH nih fur) Member of a group of cone-bearing gymnosperm plants that includes pine, cedar, and spruce trees. 424

conjugation (kahn jew GAY shun) Transfer of genetic material from one cell to another. 369, 383

conjunctiva (kahn junk TY vuh) Delicate membrane that lines the eyelid protecting the sclera. 724

connective tissue Type of animal tissue that binds structures together, provides support and protection, fills spaces, stores fat, and forms blood cells; adipose tissue, cartilage, bone, and blood are types of connective tissue. 599

conservation biology Scientific discipline that seeks to understand the effects of human activities on species, communities, and ecosystems and to develop practical approaches to preventing the extinction of species and the destruction of ecosystems. 928

consumer Organism that feeds on another organism in a food chain; primary consumers eat plants, and secondary consumers eat animals. 883

continental drift Movement of continents with respect to one another over the Earth's surfaces. 334

continuous breeding Population growth pattern where members experience many reproductive events throughout their lifetime. 843

contraceptive vaccine Under development, this birth control method immunizes against the hormone HCG, crucial to maintaining implantation of the embryo. 787

control Sample that goes through all the steps of an experiment but does not contain the variable being tested; a standard against which the results of an experiment are checked. 11

convergent evolution (kahn VERGE gent) Similarity in structure in distantly related groups due to adaptation to the environment. 347

copulation (cop you LAY shun) Sexual union between a male and a female. 776

coral reef Area of biological abundance found in shallow, warm, tropical waters on and around coral formations. 922

corepressor (koh ree PRESS her) Molecule that binds to a repressor, allowing the repressor to bind to an operator in a repressible operon. 253

cork Outer covering of the bark of trees; made of dead cells that may be sloughed off. 441

cork cambium Lateral meristem that produces cork. 450

cornea (CORE nee uh) Transparent, anterior portion of the outer layer of the eyeball. 724

corolla The petals, collectively; usually the conspicuously colored flower whorl. 429

corpus luteum (CORE pus LU tee uhm) Follicle that has released an egg and increases its secretion of progesterone. 784

cortex (CORE tex) In plants, ground tissue bounded by the epidermis and vascular tissue in stems and roots; in animals, outer layer of an organ, such as the cortex of the kidney or adrenal gland. 445

cortisol (CORE tih zahl) Glucocorticoid secreted by the adrenal cortex that responds to stress on a long-term basis; reduces inflammation and promotes protein and fat metabolism. 766

cotyledon (cot tih LEE dunn) Seed leaf for embryo of a flowering plant; provides nutrient molecules for the developing plant before photosynthesis begins. 428, 440, 502

coupled reactions Reactions that occur simultaneously; one is an exergonic reaction that releases energy, and the other is an endergonic reaction that requires an input of energy in order to occur. 105

covalent bond (koh VALE lunt) Chemical bond in which atoms share one pair of electrons. 25

cranial nerve (CRANE nee uhl) Nerve that arises from the brain. 712

creatine phosphate (KREE uh teen FOSS fate) High-energy phosphate molecule found in vertebrate muscles; used to generate ATP molecules for muscle contraction. 748

crenation (krin AY shun) In animal cells, shriveling of the cell due to water leaving the cell when the environment is hypertonic. 91

cretinism (KREE tin iz uhm) Condition resulting from improper development of the thyroid in an infant; characterized by stunted growth and mental retardation. 763

cristae (sing., crista) (KRISS tee, KRISS tuh) Short, fingerlike projections formed by the folding of the inner membrane of mitochondria. 75

Cro-Magnon (crow MAG nahn) Common name for the first fossils to be designated *Homo sapiens*. 591

crossing-over Exchange of segments between nonsister chromatids of a bivalent during meiosis. 170

crustacean (crust TAY shun) Member of a group of marine arthropods that contains, among others, shrimps, crabs, crayfish, and lobsters. 548

cryptic coloration (KRIPP tick) Coloration of an animal that helps it to conceal itself in its surroundings. 868

cuboidal epithelium (cube OID uhl epp pih THEE lee uhm) Type of epithelial tissue with cube-shaped cells. 598

culture Total pattern of human behavior; includes technology and the arts, and is dependent upon the capacity to speak and transmit knowledge. 587

Cushing syndrome Condition resulting from hypersecretion of glucocorticoids; characterized by thin arms and legs and a "moon face," and accompanied by high blood glucose and sodium levels. 767

cuticle Waxy layer covering the epidermis of plants that protects the plant against water loss and disease-causing organisms. 417, 441, 469

cyanobacterium (pl., cyanobacteria) (SIGH uhn no back TEER ree uhm) Photosynthetic bacterium that contains chlorophyll and releases oxygen; formerly called a blue-green alga. 62, 373

cycad (SIGH cad) Type of gymnosperm with palmate leaves and massive cones; cycads are most often found in the tropics and subtropics. 425

cyclic electron pathway (SICK lick) Portion of the light reaction that involves only photosystem I and generates ATP. 121

cyclic monophosphate (cAMP) ATP-related compound that acts as the second messenger in peptide hormone transduction; it initiates activity of the metabolic machinery. 758

cyclin (SIGH klin) Protein that cycles in quantity as the cell cycle progresses; combines with and activates the kinases that function to promote the events of the cycle. 151

cyst (SIST) In protists and invertebrates, resting stage that contains reproductive bodies or embryos. 380, 531

cytochrome (SIGH toe krome) Any of several iron-containing protein molecules that serve as electron carriers in photosynthesis and cellular respiration. 138

cytokine (SIGH toe kine) Type of protein secreted by a T lymphocyte that attacks viruses, virally infected cells, and cancer cells. 637

cytokinesis (sigh toe kin NEE sis) Division of the cytoplasm following mitosis and meiosis. 150

cytokinin (sigh toe KINE ninn) Plant hormone that promotes cell division; often works in combination with auxin during organ development in plant embryos. 486

cytoplasm (SIGH toe plaz uhm) Contents of a cell between the nucleus (nucleoid) region of bacteria and the plasma membrane. 62

cytosine (C) (SIGH toe zeen) One of four nitrogen-containing bases in the nucleotides composing the structure of DNA and RNA; pairs with guanine. 227

cytoskeleton (sigh toe SKELL luh ton) Internal framework of the cell, consisting of microtubules, actin filaments, and intermediate filaments. 76

cytotoxic T cell (sigh toe TOX ick) T lymphocyte that attacks and kills antigen-bearing cells. 642

D

data (sing., datum) (DAY tuh, DAY tum) Facts or information collected through observation and/or experimentation. 11

day-neutral plant Plant whose flowering is not dependent on day length—e.g., tomato and cucumber. 488

deamination (dee am in AY shun) Removal of an amino group (—NH₂) from an amino acid or other organic compound. 144

decapod (DECK uh pod) Type of crustacean in which the thorax bears five pairs of walking legs; includes shrimps, lobsters, crayfish, and crabs. 549

deciduous (dih SIDD you us) Plant which sheds its leaves annually. 439

decomposer Organism, usually a bacterium or fungus, that breaks down organic matter into inorganic nutrients that can be recycled in the environment. 883

deductive reasoning Process of logic and reasoning, using "if . . . then" statements. 11

dehydration reaction Chemical reaction resulting in a covalent bond with the accompanying loss of a water molecule. 38

delayed allergic response Allergic response initiated at the site of the allergen by sensitized T cells, involving macrophages and regulated by cytokines. 649

deletion (duh LEE shun) Change in chromosome structure in which the end of a chromosome breaks off or two simultaneous breaks lead to the loss of an internal segment; often causes abnormalities—e.g., cri du chat syndrome. 218

demographic transition (dem oh GRAFF ick) Due to industrialization, a decline in the birthrate following a reduction in the death rate so that the population growth rate is lowered. 851

demography Properties of the rate of growth and the age structure of populations. 839

denatured (dee NATE churd) Loss of an enzyme's normal shape so that it no longer functions; caused by a less than optimal pH and temperature. 49, 108

dendrite (DEN drite) Part of a neuron that sends signals toward the cell body. 703

dendritic cell Antigen-presenting cell of the epidermis and mucous membranes. 637

denitrification (dee nite trih fih KAY shun) Conversion of nitrate or nitrite to nitrogen gas by bacteria in soil. 892

dense fibrous connective tissue Type of connective tissue containing many collagen fibers packed together; found in tendons and ligaments, for example. 600

density-dependent factor Biotic factor, such as disease or competition, that affects population size according to the population's density. 847

density-independent factor Abiotic factor, such as fire or flood, that affects population size independent of the population's density. 847

deoxyribose (dee ox ee RYE bohs) Pentose sugar found in DNA. 39

derived character Structural, physiological, or behavioral trait that is present in a specific lineage and is not present in the common ancestor for several lineages. 346

dermis (DER miss) In mammals, thick layer of the skin underlying the epidermis. 606

desert Ecological biome characterized by a limited amount of rainfall; deserts have hot days and cool nights. 914

desmosome (DEZ moh sohm) Intercellular junction that connects cytoskeletons of adjacent cells. 96

detrital food chain (dee TRY tull) Straight-line linking of organisms according to who eats whom, beginning with detritus. 887

detrital food web Complex pattern of interlocking and crisscrossing food chains that begins with detritus. 887

detritus feeder Any organism that obtains most of its nutrients from the detritus in an ecosystem. 883

deuterostome (DEW turr row stome) Group of coelomate animals in which the second embryonic opening is associated with the mouth; the first embryonic opening, the blastopore, is associated with the anus. 518, 538, 556

diabetes mellitus Condition characterized by a high blood glucose level and the appearance of glucose in the urine due to a deficiency of insulin production and failure of cells to take up glucose. 768

diaphragm (DIE uh framm) In mammals, dome-shaped muscularized sheet separating the thoracic cavity from the abdominal cavity. 674

diarrhea Excessively frequent and watery bowel movements. 662

diastole (die ASS tuh lee) Relaxation period of a heart chamber during the cardiac cycle. 620

diatom (DIE uh tom) Golden-brown alga with a cell wall in two parts, or valves; significant part of phytoplankton. 387

diencephalon (die in SEF uh lahn) In vertebrates, portion of the brain in the region of the third ventricle that includes the thalamus and hypothalamus. 710

differentially permeable Ability of plasma membranes to regulate the passage of substances into and out of the cell, allowing some to pass through and preventing the passage of others. 88

diffusion Movement of molecules or ions from a region of higher to lower concentration; it requires no energy and tends to lead to an equal distribution. 89

dihybrid cross (die HIGH brid) Cross between parents that differ in two traits. 188

dikaryotic (die care ree AH tick) Having two haploid nuclei that stem from different parent cells; during sexual reproduction, sac and club fungi have dikaryotic cells. 399

dimorphic In cnidarians, when both body forms (polyp and medusa) are present. 524

dinoflagellate (dine no FLAJ jell late) Photosynthetic unicellular protist with two flagella, one whiplash and the other located within a groove between protective cellulose plates; significant part of phytoplankton. 387

dioecious Having unisexual flowers or cones, with the male flowers or cones confined to certain plants and the female flowers or cones of the same species confined to other different plants. 426

diploid (2n) number (DIP loid) Cell condition in which two of each type of chromosome are present. 153, 168

directional selection Outcome of natural selection in which an extreme phenotype is favored, usually in a changing environment. 306

disaccharide (die SACK uh ride) Sugar that contains two units of a monosaccharide; e.g., maltose. 39

discrete breeding Population growth pattern where members have only one reproductive event in their lifetime. 843

disruptive selection Outcome of natural selection in which the two extreme phenotypes are favored over the average phenotype, leading to more than one distinct form. 308

distal convoluted tubule (DISS tull KAHN vole loot ted TUBE yule) Final portion of a nephron that joins with a collecting duct; associated with tubular secretion. 691

DNA (deoxyribonucleic acid) (dee OX ee RYE bow new CLAY ick) Nucleic acid polymer produced from covalent bonding of nucleotide monomers that contain the sugar deoxyribose; the genetic material of nearly all organisms. 50, 224

DNA-DNA hybridization Method to determine relatedness by allowing single DNA strands from two different species to join, and thereafter observing how well they joined. 349

DNA fingerprinting The use of DNA fragment lengths resulting from restriction enzyme cleavage to identify particular individuals. 269

DNA ligase (LIE gaze) Enzyme that links DNA fragments; used during production of recombinant DNA to join foreign DNA to vector DNA. 268

DNA polymerase (pah LIMM murr race) During replication, an enzyme that joins the nucleotides complementary to a DNA template. 230

DNA repair enzyme One of several enzymes that restore the original base sequence in an altered DNA strand. 233

DNA replication Synthesis of a new DNA double helix prior to mitosis and meiosis in eukaryotic cells and during prokaryotic fission in prokaryotic cells. 230

domain Largest of the categories, or taxa, used by taxonomists to group species; the three domains are Archaea, Bacteria, and Eukarya. 8, 344

domain Archaea One of the three domains of life; contains prokaryotic cells that often live in extreme habitats and have unique genetic, biochemical, and physiological characteristics; its members are sometimes referred to as *archaea*. 8, 354

domain Bacteria One of the three domains of life; contains prokaryotic cells that differ from archaea because they have their own unique genetic, biochemical, and physiological characteristics. 8, 354

domain Eukarya One of the three domains of life, consisting of organisms with eukaryotic cells and further classified into the kingdoms Protista, Fungi, Plantae, and Animalia. 8, 354

dominance hierarchy Organization of animals in a group that determines the order in which the animals have access to resources. 826

dominant allele (uh LEEL) Allele that exerts its phenotypic effect in the heterozygote; it masks the expression of the recessive allele. 184

dopamine Neurotransmitter in the central nervous system. 706

dormancy In plants, a cessation of growth under conditions that seem appropriate for growth. 484

dorsal root ganglion (GANG lee uhn) Mass of sensory neuron cell bodies located in the dorsal root of a spinal nerve. 712

double fertilization In flowering plants, one sperm nucleus unites with the egg nucleus, and a second sperm nucleus unites with the polar nuclei of an embryo sac. 499

double helix Double spiral; describes the three-dimensional shape of DNA. 41, 228

doubling time Number of years it takes for a population to double in size. 851

duodenum (dew ODD duh num) First part of the small intestine, where chyme enters from the stomach. 661

duplication Change in chromosome structure in which a particular segment is present more than once in the same chromosome. 218

E

echinoderm (ee KINE oh derm) Phylum of marine animals that includes sea stars, sea urchins, and sand dollars; characterized by radial symmetry and a water vascular system. 554

ecological niche Role an organism plays in its community, including its habitat and its interactions with other organisms. 863

ecological pyramid Pictorial graph based on the biomass, number of organisms, or energy content of various trophic levels in a food web—from the producer to the final consumer populations. 887

ecological succession The gradual replacement of communities in an area following a disturbance (secondary succession) or the creation of new soil (primary succession). 874

ecology Study of the interactions of organisms with other organisms and with the physical and chemical environment. 838

ecosystem Biological community together with the associated abiotic environment; characterized by a flow of energy and a cycling of inorganic nutrients. 5, 838, 882

ectoderm (EK toe derm) Outermost primary tissue layer of an animal embryo; gives rise

to the nervous system and the outer layer of the integument. 800

ectotherm (ek toe THERM) Organism having a body temperature that varies according to the environmental temperature. 562

edema (eh DEE muh) Swelling due to tissue fluid accumulation in the intercellular spaces. 635

edge effect Phenomenon in which the edges around a landscape patch provide a slightly different habitat than the favorable habitat in the interior of the patch. 940

effector Muscle or gland that receives signals from motor fibers and thereby allows an organism to respond to environmental stimuli. 699

elastic cartilage Type of cartilage composed of elastic fibers, allowing greater flexibility. 600

elastic fiber Yellow fiber in the matrix of connective tissue, providing flexibility. 599

electrocardiogram (ECG) (ee leck troh CARD dee oh gram) Recording of the electrical activity associated with the heartbeat. 621

electron Negative subatomic particle, moving about in an energy level around the nucleus of an atom. 20

electronegativity The ability of an atom to attract electrons toward itself in a chemical bond. 26

electron shell Concentric energy levels in which electrons orbit. 20

electron transport chain Passage of electrons along a series of membrane-bound electron carrier molecules from a higher to lower energy level; the energy released is used for the synthesis of ATP. 110, 119, 133, 138

element Substance that cannot be broken down into substances with different properties; composed of only one type atom. 20

elephantiasis (ell uh fun TIE uh sis) A swelling of the limbs due to blockage of lymphatic vessels by parasitic filarial roundworms. 533

El Nino–Southern Oscillation Warming of water in the Eastern Pacific equatorial region such that the Humboldt Current is displaced, with possible negative results such as reduction in marine life. 920

embryo Stage of a multicellular organism that develops from a zygote before it becomes free-living; in seed plants, the embryo is part of the seed. 499, 799

embryonic disk (em bree AHN ick) During human development, flattened area during gastrulation from which the embryo arises. 809

embryonic period Period that spans from approximately the second to the eighth week of human development, during which the major organ systems are organized. 807

embryo sac Female gametophyte (megagametophyte) of flowering plants. 499

emergent property Quality that appears as biological complexity increases. 2

emerging virus Newly identified viruses that are becoming more prominent. 366

endangered species A species that is in peril of immediate extinction throughout all or most of its range (e.g., California condor, snow leopard). 928

endergonic reaction (en der GONN ick) Chemical reaction that requires an input of energy; opposite of exergonic reaction. 104

endocrine gland (EN doe crinn) Ductless organ that secretes hormone(s) into the bloodstream. 599, 758

endocrine system Organ system involved in the coordination of body activities; uses hormones as chemical signals secreted into the bloodstream. 756

endocytosis (en doe site TOE sis) Process by which substances are moved into the cell from the environment by phagocytosis (cellular eating) or pinocytosis (cellular drinking); includes receptor-mediated endocytosis. 94

endoderm (EN doe derm) Innermost primary tissue layer of an animal embryo that gives rise to the linings of the digestive tract and associated structures. 800

endodermis (en doe DERM miss) Internal plant root tissue forming a boundary between the cortex and the vascular cylinder. 445

endomembrane system (en doe MEM brain) Cellular system that consists of the nuclear envelope, endoplasmic reticulum, Golgi apparatus, and vesicles. 70

endometriosis (en doe meet tree OH sis) Presence of uterine tissue outside the uterus, which can contribute to infertility; possibly the result of irregular menstrual flow. 788

endometrium (en doe MEET tree uhm) Mucous membrane lining the interior surface of the uterus. 782

endoplasmic reticulum (ER) (en doe PLAZ mick ruh TICK you lumm) System of membranous saccules and channels in the cytoplasm, often with attached ribosomes. 70

endoskeleton (en doe SKELL uh ton) Protective internal skeleton, as in vertebrates. 738

endosperm (EN doe sperm) In flowering plants, nutritive storage tissue that is derived from the union of a sperm nucleus and polar nuclei in the embryo sac. 499

endospore (EN doe spore) Spore formed within a cell; certain bacteria form endospores. 369

endosymbiotic hypothesis (en doe simm bee AH tick) Possible explanation of the evolution of eukaryotic organelles by phagocytosis of prokaryotes. 326

endotherm (en doe THERM) Organism in which maintenance of a constant body temperature is independent of the environmental temperature. 562

energy Capacity to do work and bring about change; occurs in a variety of forms. 4, 102

energy of activation Energy that must be added in order for molecules to react with one another. 106

enterocoelom (en ter oh SEE lumm) Body cavity that forms by the fusion of a pair of mesodermal pouches from the wall of the primitive gut. 538

entropy (EN truh pee) Measure of disorder or randomness. 103

enzymatic protein (en zih MATT tick) Protein that catalyzes a specific reaction. 87

enzyme (EN zime) Organic catalyst, usually a protein, that speeds a reaction in cells due to its particular shape. 38, 106

enzyme inhibition Means by which cells regulate enzyme activity; may be competitive or noncompetitive inhibition. 109

eosinophil (ee oh SIN uh fill) White blood cell containing cytoplasmic granules that stain with acidic dye. 628

epidermal tissue Exterior tissue, usually one cell thick, of leaves, young stems, roots, and other parts of plants. 441, 605

epidermis (eh pih DERM miss) In mammals, the outer, protective layer of the skin; in plants, tissue that covers roots, leaves, and stems of nonwoody organisms. 441, 605

epigenetic inheritance State of gene functionality that is not encoded within the DNA sequence but that is still inheritable from one generation to the next. 255

epiglottis (eh pih GLOTT tiss) Structure that covers the glottis, the air-tract opening, during the process of swallowing. 676

epinephrine (eh pih NEFF rinn) Hormone secreted by the adrenal medulla in times of stress; adrenaline. 765

epiphyte (EPP pih fite) Plant that takes its nourishment from the air because its placement in other plants gives it an aerial position. 465, 910

epistasis (eh PISS stuh sis) Inheritance pattern in which one gene masks the expression of another gene that is at a different locus and is independently inherited. 198

epithelial tissue (eh pih THEE lee uhl) Tissue that lines hollow organs and covers surfaces. 598

erythropoietin (EPO) (eh rih throw poe EE tin) Hormone produced by the kidneys that speeds red blood cell formation. 771

esophagus (eh SOFF uh gus) Muscular tube for moving swallowed food from the pharynx to the stomach. 660

essential amino acids Amino acids required in the human diet because the body cannot make them. 666

essential nutrient In plants, substance required for normal growth, development, or reproduction. 461

estrogen (ESS truh jen) Female sex hormone that helps maintain sexual organs and secondary sex characteristics. 770, 783

estuary (ESST you air ree) Portion of the ocean located where a river enters and fresh water mixes with salt water. 918

ethylene (ETH uh leen) Plant hormone that causes ripening of fruit and is also involved in abscission. 487

euchromatin (you CROW muh tin) Chromatin that is extended and accessible for transcription. 257

eudicot (you DIE cot) Abbreviation of eudicotyledon. Flowering plant group; members have two embryonic leaves (cotyledons), net-veined leaves, vascular bundles in a ring, flower parts in fours or fives and their multiples, and other characteristics. 428, 440

euglenoid (YOU glen oid) Flagellated and flexible freshwater unicellular protist that usually contains chloroplasts and has a semirigid cell wall. 388

eukaryotic cell (you care ree AH tick) Type of cell that has a membrane-bounded nucleus and membranous organelles; found in organisms within the domain Eukarya. 62

euploidy (you PLOY dee) Cells containing only complete sets of chromosomes. 212

eutrophication (you troh fih KAY shun) Enrichment of water by inorganic nutrients used by phytoplankton. Often, overenrichment caused by human activities leads to excessive bacterial growth and oxygen depletion. 895, 916

evaporation Conversion of a liquid or a solid into a gas. 889

evolution Descent of organisms from common ancestors with the development of genetic and phenotypic changes over time that make them more suited to the environment. 5, 283

excretion Elimination of metabolic wastes by an organism at exchange boundaries such as the plasma membrane of unicellular organisms and excretory tubules of multicellular animals. 685

exergonic reaction (ex urr GONN ick) Chemical reaction that releases energy; opposite of endergonic reaction. 104

exocrine gland (EX oh krinn) Gland that secretes its product to an epithelial surface directly or through ducts. 599

exocytosis (ex oh sigh TOE sis) Process in which an intracellular vesicle fuses with the plasma membrane so that the vesicle's contents are released outside the cell. 94

exophthalmic goiter (ex opp THOWL mick GOI turr) Enlargment of the thyroid gland accompanied by an abnormal protrusion of the eyes. 763

exoskeleton (ex oh SKELL uh ton) Protective external skeleton, as in arthropods. 546, 738

exotic species Nonnative species that migrate or are introduced by humans into a new ecosystem; also called alien species. 936

experiment Artificial situation devised to test a hypothesis. 11

experimental design Methodology by which an experiment will seek to support the hypothesis. 11

experimental variable Factor of the experiment being tested. 12

expiration Act of expelling air from the lungs; exhalation. 674

exponential growth Growth, particularly of a population, in which the increase occurs in the same manner as compound interest. 844

external respiration Exchange of oxygen and carbon dioxide between alveoli and blood. 672

extinct; extinction Total disappearance of a species or higher group. 7, 293, 327

extraembryonic membrane (ex truh em bree AH nick) Membrane that is not a part of the embryo but is necessary to the continued existence and health of the embryo. 777, 807

F

facilitated transport Passive transfer of a substance into or out of a cell along a concentration gradient by a process that requires a carrier. 92

facultative anaerobe (fac ull TAY tihv ANN air robe) Prokaryote that is able to grow in either the presence or the absence of gaseous oxygen. 370

FAD Flavin adenine dinucleotide; a coenzyme of oxidation-reduction that becomes $FADH_2$ as oxidation of substrates occurs, and then delivers electrons to the electron transport chan in mitochondria during cellular respiration. 132

fall overturn Mixing process that occurs in fall in stratified lakes, whereby oxygen-rich top waters mix with nutrient-rich bottom waters. 916

family One of the categories, or taxa, used by taxonomists to group species; the taxon above the genus level. 8, 344

fat Organic molecule that contains glycerol and fatty acids and is found in adipose tissue of vertebrates. 42

fate map Diagram that traces the differentiation of cells during development from their origin to their final structure and function. 804

fatty acid Molecule that contains a hydrocarbon chain and ends with an acid group. 42

feather One of the light, horny, epidermal outgrowths that form the external covering of the bodies of birds and the greater part of the surface of their wings. 572

feedback inhibition Mechanism for regulating metabolic pathways in which the concentration of the product is kept within a certain range until binding at an allosteric site shuts down the pathway, and no more product is produced. 109

female gametophyte (guh MEET oh fite) In seed plants, the gametophyte that produces an egg; in flowering plants, an embryo sac. Sometimes called a megagametophyte. 498

fermentation Anaerobic breakdown of glucose that results in a gain of two ATP and end products such as alcohol and lactate. 133, 141

fern Member of a group of plants that have large fronds; in the sexual life cycle, the independent gametophyte produces flagellated sperm, and the vascular sporophyte produces windblown spores. 422

fertilization Fusion of sperm and egg nuclei, producing a zygote that develops into a new individual. 170, 798

fibroblast (FIF broh blast) Connective tissue cell that synthesizes fibers and ground substance. 600

fibrocartilage Cartilage with a matrix of strong collagenous fibers. 600

fibrous protein Principle structural proteins of the body; generally insoluble; include collagens, elastins, keratins, and actin and myosin. 49

fibrous root system In most monocots, a mass of similarly sized roots that cling to the soil. 446

filament (FILL uh mint) End-to-end chains of cells that form as cell division occurs in only one plane; in plants, the elongated stalk of a stamen. 382, 496

fimbria (pl., fimbriae) (FIMM bree uh, FIMM bree ee) Small, bristlelike fiber on the surface of a bacterial cell, which attaches bacteria to a surface; also fingerlike extension from the oviduct near the ovary. 63, 368

fin In fish and other aquatic animals, membranous, winglike, or paddlelike process used to propel, balance, or guide the body. 563

first messenger Chemical signal such as a peptide hormone that binds to a plasma membrane receptor protein and alters the metabolism of a cell because a second messenger is activated. 758

fishes Aquatic, gill-breathing vertebrate that usually has fins and skin covered with scales; fishes were among the earliest vertebrates that evolved. 562

fitness Ability of an organism to reproduce and pass its genes to the next fertile generation; measured against the ability of other organisms to reproduce in the same environment. 288, 826

five-kingdom system System of classification that contains the kingdoms Monera, Protista, Plantae, Animalia, and Fungi. 354

fixed action pattern (FAP) Innate behavior pattern that is stereotyped, spontaneous, independent of immediate control, genetically encoded, and independent of individual learning. 822

flagellum (pl., flagella) (fluh JELL uhm) Long, slender extension used for locomotion by some bacteria, protozoans, and sperm. 62, 78, 368

flagship species Species that evoke a strong emotional response in humans; charismatic, cute, regal (e.g., lions, tigers, dolphin, panda). 939

flame cell Found along excretory tubules of planarians; functions in propulsion of fluid through the excretory canals and out of the body. 689

flatworm Member of the phylum Platyhelminthes (e.g., tapeworms, planarians). 528

flower Reproductive organ of a flowering plant, consisting of several kinds of modified leaves arranged in concentric rings and attached to a modified stem called the receptacle. 415, 496

fluid-mosaic model Model for the plasma membrane based on the changing location and pattern of protein molecules in a fluid phospholipid bilayer. 84

follicle (FOLL lick cull) Structure in the ovary of animals that contains an oocyte; site of oocyte production. 783

follicular phase (foe LICK you lurr) First half of the ovarian cycle, during which the follicle matures and much estrogen (and some progesterone) is produced. 784

fontanel (fahn tuh NELL) Membranous region located between certain cranial bones in the skull of a vertebrate fetus or infant. 742

food chain The order in which one population feeds on another in an ecosystem, thereby showing the flow of energy from a detrivore (detrital food chain) or a producer (grazing food chain) to the final consumer. 887

food web In ecosystems, a complex pattern of interlocking and crisscrossing food chains. 887

foramen magnum (for AY men MAG numm) Opening in the occipital bone of the vertebrate skull through which the spinal cord passes. 742

foraminiferan (for am men IF furr uhn) Member of the phylum Foraminifera bearing a calcium carbonate test with many openings through which pseudopods extend. 390

fossil Any past evidence of an organism that has been preserved in the Earth's crust. 322

fossil fuel Fuels such as oil, coal, and natural gas that are the result of partial decomposition of plants and animals coupled with exposure to heat and pressure for millions of years. 890

fossil record History of life recorded from remains from the past. 292

founder effect Cause of genetic drift due to colonization by a limited number of individuals who, by chance, have different gene frequencies than the parent population. 306

fovea centralis (FOE vee uh sen TRAHL liss) Region of the retina consisting of densely packed cones; responsible for the greatest visual acuity. 725

frameshift mutation Insertion or deletion of at least one base so that the reading frame of the corresponding mRNA changes. 261

free energy Useful energy in a system that is capable of performing work. 104

frond Leaf of a fern. 422

fruit Flowering plant structure consisting of one or more ripened ovaries that usually contain seeds. 431, 504

fruiting body Spore-producing and spore-disseminating structure found in sac and club fungi. 403

functional group Specific cluster of atoms attached to the carbon skeleton of organic molecules that enters into reactions and behaves in a predictable way. 37

fungus (pl., fungi) Saprotrophic decomposer; the body is made up of filaments called hyphae that form a mass called a mycelium. 9, 398

G

gallbladder Organ attached to the liver that serves to store and concentrate bile. 664

gamete (GAMM meet) Haploid sex cell; e.g., egg and sperm. 168, 788

gametogenesis (gamm meet oh JEN uh sis) Development of the male and female sex gametes. 176

gametophyte (guh MEET uh fite) Haploid generation of the alternation of generations life cycle of a plant; produces gametes that unite to form a diploid zygote. 176, 416, 496

ganglion (GANG lee ahn) Collection or bundle of neuron cell bodies usually outside the central nervous system. 712

gap analysis Use of computers to discover places where biodiversity is high outside of preserved areas. 940

gap junction Junction between cells formed by the joining of two adjacent plasma membranes; it lends strength and allows ions, sugars, and small molecules to pass between cells. 96

gastropod (gas trah POD) Mollusc with a broad, flat foot for crawling (e.g., snails and slugs). 543

gastrovascular cavity (gas troh VASS cue lurr) Blind digestive cavity in animals that have a sac body plan. 524

gastrula (GAS true luh) Stage of animal development during which the germ layers form, at least in part, by invagination. 800

gastrulation (gas true LAY shun) Formation of a gastrula from a blastula; characterized by an invagination of the cell layers to form a caplike structure. 800

gene (JEEN) Unit of heredity existing as alleles on the chromosomes; in diploid organisms, typically two alleles are inherited—one from each parent. 5, 240

gene cloning DNA cloning to produce many identical copies of the same gene. 268

gene flow Sharing of genes between two populations through interbreeding. 304

gene linkage Relationship between genes on the same chromosome. 209

gene locus Specific location of a particular gene on homologous chromosomes. 185

gene pool Total of all the genes of all the individuals in a population. 302

generalist species Species that have a broad range of niches, such as diversified diet,

wide range of environmental tolerances, and diverse habitat (e.g., racoons, roaches, and humans). 863

gene therapy Correction of a detrimental mutation by the addition of new DNA and its insertion in a genome. 268

genetic code Universal code that has existed for eons; specifies protein synthesis in the cells of all living things. Each codon consists of three letters standing for the DNA nucleotides that make up one of the 20 amino acids found in proteins. 241

genetic drift Mechanism of evolution due to random changes in the allelic frequencies of a population; more likely to occur in small populations or when only a few individuals of a large population reproduce. 305

genetic engineering Alteration of genomes for medical or industrial purposes. 270

genetic mutation Altered gene whose sequence of bases differs from the previous sequence. 233, 261

genetic recombination Process in which new genetic information is incorporated into a chromosome or DNA fragment. 170

genomics Study of whole genomes. 272

genotype (JEEN oh type) Genes of an organism for a particular trait or traits; often designated by letters—for example, *BB* or *Aa*. 185

genus (JEEN us) One of the categories, or taxa, used by taxonomists to group species; contains those species that are most closely related through evolution. 8, 344

geological timescale History of the Earth based on correlations between rocks (or the fossils contained in them) and time periods of the past. 324

germination Beginning of growth of a seed, spore, or zygote, especially after a period of dormancy. 488

germ layer Primary tissue layer of a vertebrate embryo—namely, ectoderm, mesoderm, or endoderm. 800

gibberellin (jib urr ELL uhn) Plant hormone promoting increased stem growth; also involved in flowering and seed germination. 484

gills Respiratory organ in most aquatic animals; in fish, an outward extension of the pharynx. 560, 673

ginkgo Member of phylum Ginkgophyte; maidenhair tree. 426

girdling Removing a strip of bark from around a tree. 472

gland Epithelial cell or group of epithelial cells that are specialized to secrete a substance. 599

global warming Predicted increase in the Earth's temperature due to human activities that promote the greenhouse effect. 891, 937

globular protein Most of the proteins in the body; soluble in water or salt solution; includes albumins, globulins, histones. 49

glomerular capsule (glow MARE you lurr) Cuplike structure that is the initial portion of a nephron. 690

glomerular filtration Movement of small molecules from the glomerulus into the glomerular capsule due to the action of blood pressure. 692

glomerulus (glow MARE you luss) Capillary network within the glomerular capsule of a nephron. 691

glottis (GLAH tiss) Opening for airflow in the larynx. 676

glucocorticoid (glue koh CORE tih coid) Type of hormone secreted by the adrenal cortex that influences carbohydrate, fat, and protein metabolism; *See* also cortisol. 765

glucose (GLUE kohs) Six-carbon sugar that organisms degrade as a source of energy during cellular respiration. 39

glycemic index (GI) The blood glucose response of a given food, compared to, for example, white bread. 666

glycerol (GLISS sir all) Three-carbon carbohydrate with three hydroxyl groups attached; a component of fats and oils. 42

glycocalyx (glie koh KAY licks) Gel-like coating outside the cell wall of a bacterium. If compact, it is called a capsule; if diffuse, it is called a slime layer. 62

glycogen (GLIE kuh jen) Storage polysaccharide found in animals; composed of glucose molecules joined in a linear fashion but having numerous branches. 40

glycolipid (glie koh LIP pidd) Lipid in plasma membranes that bears a carbohydrate chain attached to a hydrophobic tail. 86

glycolysis (glie KAH lih sis) Anaerobic breakdown of glucose that results in a gain of two ATP and the end product pyruvate. 133, 134

glycoprotein (glie koh PRO teen) Protein in plasma membranes that bears a carbohydrate chain. 86

gnetophyte Member of one of the four phyla of gymnosperms; Gnetophyta has only three living genera, which differ greatly from one another—e.g., *Welwitschia* and *Ephedra*. 426

Golgi apparatus (GOAL ghee app uh RAT us) Organelle consisting of saccules and vesicles that processes, packages, and distributes molecules about or from the cell. 70

gonad (GO nadd) Organ that produces gametes; the ovary produces eggs, and the testis produces sperm. 776

gonadotropic hormone (go nadd oh TROH pick) Substance secreted by the anterior pituitary that regulates the activity of the ovaries and testes; principally, follicle-stimulating hormone (FSH) and luteinizing hormone (LH). 760

granum (GRAY numm) Stack of chlorophyll-containing thylakoids in a chloroplast. 74, 117

grassland Biome characterized by rainfall greater than 25 cm/yr, grazing animals, and warm summers; includes the prairie of the U.S. midwest and the African savanna. 912

gravitational equilibrium Maintenance of balance when the head and body are motionless. 732

gravitropism (grav ih TROPE is uhm) Growth response of roots and stems of plants to the Earth's gravity; roots demonstrate positive gravitropism, and stems demonstrate negative gravitropism. 479

gray crescent Gray area that appears in an amphibian egg after being fertilized by the sperm; thought to contain chemical signals that turn on the genes that control development. 803

gray matter Nonmyelinated axons and cell bodies in the central nervous system. 708

grazing food chain Flow of energy to a straight-line linking of organisms according to who eats whom. 887

grazing food web Complex pattern of interlocking and crisscrossing food chains that begins with a population of photosynthesizers serving as producers. 887

green algae Members of a diverse group of photosynthetic protists; contain chlorophylls *a* and *b* and have other biochemical characteristics like those of plants. 382

greenhouse effect Reradiation of solar heat toward the Earth, caused by gases such as carbon dioxide, methane, nitrous oxide, water vapor, ozone, and nitrous oxide in the atmosphere. 891

greenhouse gases Gases in the atmosphere such as carbon dioxide, methane, nitrous oxide, water vapor, ozone, and nitrous oxide that are involved in the greenhouse effect. 891

ground tissue Tissue that constitutes most of the body of a plant; consists of parenchyma, collenchyma, and sclerenchyma cells that function in storage, basic metabolism, and support. 441

growth hormone (GH) Substance secreted by the anterior pituitary; controls size of an individual by promoting cell division, protein synthesis, and bone growth. 761

guanine (G) (GWAH neen) One of four nitrogen-containing bases in nucleotides composing the structure of DNA and RNA; pairs with cytosine. 227

guard cell One of two cells that surround a leaf stoma; changes in the turgor pressure of these cells cause the stoma to open or close. 470

guttation (gutt TAY shun) Liberation of water droplets from the edges and tips of leaves. 468

gymnosperm (JIM no sperm) Type of woody seed plant in which the seeds are not enclosed by fruit and are usually borne in cones, such as those of the conifers. 424

H

habitat Place where an organism lives and is able to survive and reproduce. 838, 863

hair follicle Tubelike depression in the skin in which a hair develops. 606

halophile (HAL uh file) Type of archaea that lives in extremely salty habitats. 375

haploid (n) number (HAP loid) Cell condition in which only one of each type of chromosome is present. 153, 168

Hardy-Weinberg principle Law stating that the gene frequencies in a population remain stable if evolution does not occur due to nonrandom mating, selection, migration, and genetic drift. 302

hay fever Seasonal variety of allergic reaction to a specific allergen. Characterized by sudden attacks of sneezing, swelling of nasal mucosa, and often asthmatic symptoms. 649

heart Muscular organ whose contraction causes blood to circulate in the body of an animal. 618

heart attack Damage to the myocardium due to blocked circulation in the coronary arteries; myocardial infarction. 625

heartworm disease Disease in dogs caused by the filarial worm, a roundworm; worms live in the heart and arteries that serve the lungs. 533

heat Type of kinetic energy; captured solar energy eventually dissipates as heat in the environment. 102

helper T cell Secretes lymphokines, which stimulate all kinds of immune cells. 642

heme (HEEM) Iron-containing group found in hemoglobin. 678

hemocoel (HEEM uh seel) Residual coelom found in arthropods, which is filled with hemolymph. 549

hemoglobin (Hb) (HEEM uh globe in) Iron-containing respiratory pigment occurring in vertebrate red blood cells and in the blood plasma of some invertebrates. 46, 626, 674

hemolymph (HEEM uh limf) Circulatory fluid that is a mixture of blood and interstitial fluid; seen in animals that have an open circulatory system, such as molluscs and arthropods. 614

hepatitis Inflammation of the liver. Viral hepatitis occurs in several forms. 664

herbaceous stem (her BAY shus) Nonwoody stem. 449

herbivore (HER bih vore) Primary consumer in a grazing food chain; a plant eater. 883

hermaphroditic Type of animal that has both male and female sex organs. 529

heterochromatin (hett turr oh CROW muh tin) Highly compacted chromatin that is not accessible for transcription. 256

heterosporous Seed plant that produces two types of spores—microspores and megaspores. A plant that produces only one type of spore is *homosporous*. 420

heterotroph (HETT turr uh trofe) Organism that cannot synthesize organic compounds from inorganic substances and therefore must take in organic food. 116, 883

heterozygous (hett turr oh ZYE guss) Possessing unlike alleles for a particular trait. 185

hexose (HEX ohs) Six-carbon sugar. 39

histamine (HISS tuh mean) Substance, produced by basophils in blood and mast cells in connective tissue, that causes capillaries to dilate. 636

histone Small basic protein with large amounts of lysine and arginine that is associated with eukaryotic DNA in chromatin. 256

holozoic (hoe low ZOE ick) Obtaining nourishment by ingesting solid food particles. 391

homeobox (HOME me oh box) 180-nucleotide sequence located in all homeotic genes. 806

homeostasis (home me oh STAY sis) Maintenance of normal internal conditions in a cell or an organism by means of self-regulating mechanisms. 4, 608

homeotic genes (home me AH tick) Genes that control the overall body plan by controlling the fate of groups of cells during development. 805

hominid (HAH men idd) Member of the family Hominidae, which contains australopithecines and humans. 584

hominoid (HAH men oid) Member of the superfamily Hominoidea, which includes apes, humans, and their recent ancestors. 583

Homo erectus (HOE mow eh RECK tuss) Hominid who used fire and migrated out of Africa to Europe and Asia. 589

homologous chromosome (hoe MOLL uh gus) Member of a pair of chromosomes that are alike and come together in synapsis during prophase of the first meiotic division; a *homologue*. 168

homologous structure In evolution, a structure that is similar in different types of organisms because these organisms are derived from a common ancestor. 296, 347

homologue (HOE mow log) Member of a homologous pair of chromosomes. 168

homology (hoe MAH low jee) Similarity of parts or ograns of different organisms caused by evolutionary derivation from a corresponding part or organ in a remote ancestor, and usually having a similar embryonic origin. 347

homosporous In some plants, production of only one type of spore rather than differentiated types. 420

homozygous (hoe mow ZYE guss) Possessing two identical alleles for a particular trait. 185

horizon Major layer of soil visible in vertical profile; for example, topsoil is the A horizon. 463

hormone Chemical messenger produced in one part of the body that controls the activity of other parts. 482, 756

hornwort Member of phylum Anthocerophyta. 417

horsetails Division of seedless vascular plants having only one genus (*Equisetum*) in existence today; characterized by rhizomes, scale-like leaves, strobili, and tough, rigid stems. 421

host Organism that provides nourishment and/or shelter for a parasite. 870

host specific Organisms that can be infected by a virus, specifically. 364

human chorionic gonadotropin (HCG) (core ree AH nick go nadd uh TROPE in) Gonadotropic hormone produced by the chorion that functions to maintain the uterine lining. 786, 809

human immunodeficiency virus (HIV) (im you no duh FISH ens see) Virus responsible for AIDS. 790

humus (HUE muss) Decomposing organic matter in the soil. 462

hyaline cartilage Cartilage whose cells lie in lacunae separated by a white translucent matrix containing very fine collagen fibers. 600

hybridization (high brih dih ZAY shun) Crossing of different species. 510

hydra Freshwater member of phylum Cnidaria. 526

hydrogen bond Weak bond that arises between a slightly positive hydrogen atom of one molecule and a slightly negative atom of another molecule or between parts of the same molecule. 26

hydrogen ion (H⁺) Hydrogen atom that has lost its electron and therefore bears a positive charge. 30

hydrolysis reaction (high DRAH lih sis) Splitting of a compound by the addition of water, with the H⁺ being incorporated in one fragment and the OH⁻ in the other. 38

hydrophilic (high droh FILL ick) Type of molecule that interacts with water by dissolving in water and/or by forming hydrogen bonds with water molecules. 28, 37

hydrophobic (high droh FOE bick) Type of molecule that does not interact with water because it is nonpolar. 28, 37

hydroponics (high droh PAH nicks) Technique for growing plants by suspending them with their roots in a nutrient solution. 461

hydrostatic skeleton (high droh STAT ick) Fluid-filled body compartment that provides

support for muscle contraction resulting in movement; seen in cnidarians, flatworms, roundworms, and segmented worms. 738

hydrothermal vent (high droh THERM mull) Hot springs in the seafloor along ocean ridges where heated seawater and sulfate react to produce hydrogen sulfide; here, chemosynthetic bacteria support a community of varied organisms. 923

hydroxide ion (OH⁻) (high DROX side EYE ahn) One of two ions that results when a water molecule dissociates; it has gained an electron and therefore bears a negative charge. 30

hypertonic solution (high purr TAH nick) Higher solute concentration (less water) than the cytoplasm of a cell; causes cell to lose water by osmosis. 91

hypha (HIGH fuh) Filament of the vegetative body of a fungus. 398

hypothalamic-inhibiting hormone (high poh THOWL mick) One of many hormones produced by the hypothalamus that inhibits the secretion of an anterior pituitary hormone. 760

hypothalamic-releasing hormone One of many hormones produced by the hypothalamus that stimulates the secretion of an anterior pituitary hormone. 760

hypothalamus (high poh THOWL uh muss) In vertebrates, part of the brain that helps regulate the internal environment of the body— for example, heart rate, body temperature, and water balance. 710, 760

hypothesis (high PAH thuh sis) Supposition established by reasoning after consideration of available evidence; it can be tested by obtaining more data, often by experimentation. 10

hypotonic solution (high poh TAH nick) Lower solute (more water) concentration than the cytoplasm of a cell; causes cell to gain water by osmosis. 90

I

immediate allergic response Allergic response that occurs within seconds of contact with an allergen; caused by the attachment of the allergen to IgE antibodies. 649

immunity Ability of the body to protect itself from foreign substances and cells, including disease-causing agents. 636

immunization Strategy for achieving immunity to the effects of specific disease-causing agents. 647

immunoglobulin (Ig) (imm you no GLOB you linn) Globular plasma protein that functions as an antibody. 640

implantation In placental mammals, the embedding of an embryo at the blastocyst stage into the endometrium of the uterus. 807

imprinting Learning to make a particular response to only one type of animal or object. 822

inclusion body In a bacterium, stored nutrients for later use. 62

inclusive fitness Fitness that results from personal reproduction and from helping nondescendant relatives reproduce. 830

incomplete digestive tract Digestive tract that has a single opening, usually called a mouth. 656

incomplete dominance Inheritance pattern in which the offspring has an intermediate phenotype, as when a red-flowered plant and a white-flowered plant produce pink-flowered offspring. 196

independent assortment Alleles of unlinked genes segregate independently of each other during meiosis so that the gametes contain all possible combinations of alleles. 170

index fossil Deposits found in certain layers of strata; similar fossils can be found in the same strata around the world. 322

induced fit model Change in the shape of an enzyme's active site that enhances the fit between the active site and its substrate(s). 106

inducer Molecule that brings about activity of an operon by joining with a repressor and preventing it from binding to the operator. 254

inducible operon (in DO sih bull AH purr ahn) In a catabolic pathway, an operon causes transcription of the genes controlling a group of enzymes. 254

induction Ability of a chemical or a tissue to influence the development of another tissue. 803

inductive reasoning Using specific observations and the process of logic and reasoning to arrive at a hypothesis. 10

industrial melanism (MELL uh nizz uhm) Increased frequency of darkly pigmented (melanic) forms in a population when soot and pollution make lightly pigmented forms easier for predators to see against a pigmented background. 303

infertility Inability to have as many children as desired. 788

inflammatory response Tissue response to injury that is characterized by redness, swelling, pain, and heat. 636

inheritance of acquired characteristics Lamarckian belief that characteristics acquired during the lifetime of an organism can be passed on to offspring. 284

inner ear Portion of the ear consisting of a vestibule, semicircular canals, and the cochlea where equilibrium is maintained and sound is transmitted. 730

inorganic chemistry Branch of science which deals with compounds that do not occur in the plant or animal worlds. 36

insect Type of arthropod. The head has antennae, compound eyes, and simple eyes; the thorax has three pairs of legs and often wings; and the abdomen has internal organs. 550

inspiration Act of taking air into the lungs; inhalation. 674

integration Summing up of excitatory and inhibitory signals by a neuron or by some part of the brain. 707

intercalated disk (in TURK uh lay tidd) Region that holds adjacent cardiac muscle cells together; disks appear as dense bands at right angles to the muscle striations. 602

interferon (in turr FEAR ron) Antiviral agent produced by an infected cell that blocks the infection of another cell. 638

interkinesis (in turr kuh NEE sis) Period of time between meiosis I and meiosis II during which no DNA replication takes place. 174

interleukin (in turr LOO kin) Cytokine produced by macrophages and T lymphocytes that functions as a metabolic regulator of the immune response. 648

intermediate filaments Ropelike assemblies of fibrous polypeptides in the cytoskeleton that provide support and strength to cells; so called because they are intermediate in size between actin filaments and microtubules. 76

internal respiration Exchange of oxygen and carbon dioxide between blood and tissue fluid. 672

interneuron (in turr NURE ron) Neuron located within the central nervous system that conveys messages between parts of the central nervous system. 703

internode (IN turr node) In vascular plants, the region of a stem between two successive nodes. 439

interphase (IN turr faze) Stages of the cell cycle (G_1, S, G_2) during which growth and DNA synthesis occur when the nucleus is not actively dividing. 150

interspecific competition Similar species trying to occupy the same niche in an ecosystem compete with one another for a share of resources, and in this way the number of niches increases. 864

inversion Change in chromosome structure in which a segment of a chromosome is turned around 180°; this reversed sequence of genes can lead to altered gene activity and abnormalities. 218

invertebrate (in VURR tuh brate) Animal without endoskeleton of bone or cartilage. 518

ion (EYE ahn) Charged particle that carries a negative or positive charge. 24

ionic bond (eye AH nick) Chemical bond in which ions are attracted to one another by opposite charges. 24

iris Muscular ring that surrounds the pupil and regulates the passage of light through this opening. 725

isomer (EYE so murr) Molecules with the same molecular formula but a different structure, and therefore a different shape. 37

isotonic solution (eye so TAH nick) Solution that is equal in solute concentration to that of the cytoplasm of a cell; causes cell to neither lose nor gain water by osmosis. 90

isotope (EYE so tope) Atom of the same element having the same atomic number but a different mass number due to the number of neutrons. 22

J

jaundice Yellowish tint to the skin caused by an abnormal amount of bilirubin (bile pigment) in the blood, indicating liver malfunction. 664

jaw Tooth-bearing bone of the head. 563

jawless fishes Type of fish that has no jaws; includes today's hagfishes and lampreys. 562

joint Articulation between two bones of a skeleton. 745

K

karyotype (CARE ree oh type) Chromosomes arranged by pairs according to their size, shape, and general appearance in mitotic metaphase. 214

keystone species Species whose activities significantly affect community structure. 876, 939

kidneys Paired organs of the vertebrate urinary system that regulate the chemical composition of the blood and produce a waste product called urine. 690

kinetic energy (kin NET tick) Energy associated with motion. 102

kinetochore (kin NET uh core) Disk-shaped structure within the centromere of a chromosome to which spindle microtubules become attached during mitosis and meiosis. 153, 171

kingdom One of the categories, or taxa, used by taxonomists to group species; the taxon above phylum. 8, 344

kin selection Indirect selection; adaptation to the environment due to the reproductive success of an individuals relatives. 830

K-selection Favorable life-history strategy under stable environmental conditions characterized by the production of a few offspring with much attention given to offspring survival. 850

L

lactation (lack TAY shun) Secretion of milk by mammary glands, usually for the nourishment of an infant. 786

lacteal (LACK tee uhl) Lymphatic vessel in an intestinal villus; aids in the absorption of fats. 661

lactic acid fermentation Fermentation that produces lactic acid as the sole or primary product. 142

lacuna (luh COON uh) Small pit or hollow cavity, as in bone or cartilage, where a cell or cells are located. 600

ladderlike nervous system In planarians, two lateral nerve cords joined by transverse nerves. 700

lake Body of fresh water, often classified by nutrient status, such as oligotrophic (nutrient-poor) or eutrophic (nutrient-rich). 916

landscape A number of interacting ecosystems. 929

large intestine In vertebrates, portion of the digestive tract that follows the small intestine; in humans, consists of the cecum, colon, rectum, and anal canal. 662

larva (LARR vuh) Immature form in the life cycle of some animals; it sometimes undergoes metamorphosis to become the adult form. 777

larynx (LAIR inks) Cartilaginous organ located between the pharynx and the trachea; in humans, contains the vocal cords; sometimes called the voice box. 676

lateral line Canal system containing sensory receptors that allow fishes and amphibians to detect water currents and pressure waves from nearby objects. 733

law See *principle*

laws of thermodynamics Two laws explaining energy and its relationships and exchanges. The first, also called the "law of conservation," says that energy cannot be created or destroyed but can only be changed from one form to another; the second says that energy cannot be changed from one form to another without a loss of usable energy. 102

leaf Lateral appendage of a stem, highly variable in structure, often containing cells that carry out photosynthesis. 439

leaf vein Vascular tissue within a leaf. 443

learning Relatively permanent change in an animal's behavior that results from practice and experience. 822

leech Blood-sucking annelid, usually found in fresh water, with a sucker at each end of a segmented body. 546

lens Clear, membranelike structure found in the vertebrate eye behind the iris; brings objects into focus. 725

lenticel (LENN tiss uhl) Frond of usually numerous, lightly rasied, somewhat spongy, groups of cells in the bark of woody plants. Permits gas exchange between the interior of a plant and the external atmosphere. 441

leptin Hormone produced by adipose tissue that acts on the hypothalamus to signal satiety (fullness). 771

less-developed country (LDC) Country that is becoming industrialized; typically, population growth is expanding rapidly, and the majority of people live in poverty. 851

leucoplast (LOO coh plast) Plastid, generally colorless, that synthesizes and stores starch and oils. 74

leukemia Cancer of the blood-forming tissues leading to the overproduction of abnormal white blood cells. 160

lichen (LIKE in) Symbiotic relationship between certain fungi and algae, in which the fungi possibly provide inorganic food or water and the algae provide organic food. 373, 408

life cycle Recurring pattern of genetically programmed events by which individuals grow, develop, maintain themselves, and reproduce. 176

ligament Tough cord or band of dense fibrous tissue that binds bone to bone at a joint. 600, 745

light reaction Portion of photosynthesis that captures solar energy and takes place in thylakoid membranes of chloroplasts; it produces ATP and NADPH. 119

lignin (LIGG nihn) Chemical that hardens the cell walls of plants. 420, 442

limbic system (LIMM bick) In humans, functional association of various brain centers, including the amygdala and hippocampus; governs learning and memory and various emotions such as pleasure, fear, and happiness. 711

limiting factor Resource or environmental condition that restricts the abundance and distribution of an organism. 839

linkage group Alleles of different genes that are located on the same chromosome and tend to be inherited together. 209

linkage map Depicts the distances between loci as well as the order in which they occur on the organism. 209

lipase (LIE pace) Fat-digesting enzyme secreted by the pancreas. 665

lipid (LIP pid) Class of organic compounds that tends to be soluble in nonpolar solvents; includes fats and oils. 42

liposome (LIP uh sohm) Droplet of phospholipid molecules formed in a liquid environment. 320

littoral zone (LIT turr uhl) Shore zone between high tidemark and low tidemark; also, shallow water of a lake where light penetrates to the bottom. 919

liver Large, dark red internal organ that produces urea and bile, detoxifies the blood, stores glycogen, and produces the plasma proteins, among other functions. 662

liverwort Type of bryophyte. 418

lobe-finned fishes Type of fishes with limblike fins. 565

logistic growth (luh JISS tick) Population increase that results in an S-shaped curve; growth is slow at first, steepens, and then levels off due to environmental resistance. 845

long-day plant Plant that flowers when day length is longer than a critical length; e.g., wheat, barley, clover, and spinach. 488

loop of the nephron (NEFF ron) Portion of a nephron between the proximal and distal convoluted tubules; functions in water reabsorption. 691

loose fibrous connective tissue Tissue composed mainly of fibroblasts widely separated by a matrix containing collagen and elastic fibers. 600

lungs Internal respiratory organ containing moist surfaces for gas exchange. 566, 674

luteal phase (LOO tee uhl) Second half of the ovarian cycle, during which the corpus luteum develops and much progesterone (and some estrogen) is produced. 784

lymph (LIMF) Fluid, derived from tissue fluid, that is carried in lymphatic vessels. 601, 629, 634

lymphatic (lymphoid) organ (LIMM foid) Organ other than a lymphatic vessel that is part of the lymphatic system; the lymphoid organs are the lymph nodes, tonsils, spleen, thymus gland, and bone marrow. 635

lymphatic system (limm FAT ick) Organ system consisting of lymphatic vessels and lymphoid organs; transports lymph and lipids, and aids the immune system. 634

lymphatic vessel Vessel that carries lymph. 634

lymph node Mass of lymphoid tissue located along the course of a lymphatic vessel. 636

lymphocyte (LIMM foe site) Specialized white blood cell that functions in specific defense; occurs in two forms—T lymphocytes and B lymphocytes. 626

lysogenic cycle (lie so JEN ick) Bacteriophage life cycle in which the virus incorporates its DNA into that of a bacterium; occurs preliminary to the lytic cycle. 365

lysosome (LIE so sohm) Membrane-bounded vesicle that contains hydrolytic enzymes for digesting macromolecules. 71

lytic cycle (LIH tick) Bacteriophage life cycle in which the virus takes over the operation of the bacterium immediately upon entering it and subsequently destroys the bacterium. 364

M

macroevolution (mac crow evv oh LOO shun) Large-scale evolutionary change, such as the formation of new species. 310

macronutrient Essential element needed in large amounts for plant growth, such as nitrogen, calcium, or sulfur. 461

macrophage (MAC crow fahj) In vertebrates, large phagocytic cell derived from a monocyte that ingests microbes and debris. 637

male gametophyte (guh MEET toe fite) In seed plants, the gametophyte that produces sperm; a pollen grain. Sometimes called a microgametophyte. 498

malignant (muh LIGG nunt) The power to threaten life; cancerous. 159

Malpighian tubule (mal PIG ee uhn TUBE yule) Blind, threadlike excretory tubule near the anterior end of an insect's hindgut. 550, 689

maltase Enzyme produced in small intestine that breaks down maltose to two glucose molecules. 664

mammal Endothermic vertebrate characterized especially by the presence of hair and mammary glands. 574

mantle In molluscs, an extension of the body wall that may secrete a shell. 540

marsh Soft, wet land, which is treeless. 915

marsupial Member of a group of mammals bearing immature young nursed in a marsupium, or pouch—for example, kangaroo and opossum. 574

mass extinction Episode of large-scale extinction in which large numbers of species disappear in a few million years or less. 327

mast cell Connective tissue cell that releases histamine in allergic reactions. 636

matrix (MAY tricks) Unstructured semifluid substance that fills the space between cells in connective tissues or inside organelles. 75

matter Anything that takes up space and has mass. 20

mechanical energy A type of kinetic energy, such as walking or running. 102

mechanoreceptor (muh can oh ree SEPP turr) Sensory receptor that responds to mechanical stimuli, such as pressure, sound waves, or gravity. 730

medulla oblongata (muh DULE uh ahb long AH tuh) In vertebrates, part of the brain stem that is continuous with the spinal cord; controls heartbeat, blood pressure, breathing, and other vital functions. 710

medusa Among cnidarians, bell-shaped body form that is directed downward and contains much mesoglea. 524

megaspore (MEG uh spore) One of the two types of spores produced by seed plants; develops into a female gametophyte (embryo sac). 430, 496

megasporocyte (meg uh SPORE uh site) Megaspore mother cell; produces megaspores by meiosis; only one megaspore persists. 498

meiosis (my OH sis) Type of nuclear division that occurs as part of sexual reproduction, in which the daughter cells receive the haploid number of chromosomes in varied combinations. 168

melanocyte (mell ANN oh site) Specialized cell in the epidermis that produces melanin, the pigment responsible for skin color. 606

melanocyte-stimulating hormone (MSH) Substance that causes melanocytes to secrete melanin in lower vertebrates. 761

melatonin (mell uh TONE in) Hormone, secreted by the pineal gland, that is involved in biorhythms. 770

membrane attack complex Group of complement proteins that form channels in microbe surface and destroy microbes. 638

memory Capacity of the brain to store and retrieve information about past sensations and perceptions; essential to learning. 711

memory B cell Forms during a primary immune response but enters a resting phase until a secondary immune response occurs. 639

menarche Onset of menstruation. 785

meninges (men IN jeez) Protective membranous coverings around the central nervous system. 708

meningitis (men in JIE tuss) A condition that refers to inflammation of the brain or spinal cord meninges (membranes). 708

menopause Termination of the ovarian and uterine cycles in older women. 785

menstruation (men strew AY shun) Periodic shedding of tissue and blood from the inner lining of the uterus in primates. 784

meristem (MARE uh stem) Undifferentiated embryonic tissue in the active growth regions of plants. 441

mesoderm (MESS oh derm) Middle primary tissue layer of an animal embryo that gives rise to muscle, several internal organs, and connective tissue layers. 800

mesoglea Transparent jellylike substance. 524

mesophyll (MESS oh fill) Inner, thickest layer of a leaf consisting of palisade and spongy mesophyll; the site of most of photosynthesis. 454

mesosome (MESS oh sohm) In a bacterium, plasma membrane that folds into the cytoplasm and increases surface area. 62

messenger RNA (mRNA) Type of RNA formed from a DNA template and bearing coded information for the amino acid sequence of a polypeptide. 240

metabolic pathway (met uh BAH lick) Series of linked reactions, beginning with a particular reactant and terminating with an end product. 106

metabolic pool Metabolites that are the products of and/or the substrates for key reactions in cells, allowing one type of molecule to be changed into another type, such as carbohydrates converted to fats. 144

metabolism (met TAB uh liz uhm) All of the chemical reactions that occur in a cell during growth and repair. 4, 104

metamorphosis (met uh MORE foh sis) Change in shape and form that some animals, such as insects, undergo during development. 547, 566, 777

metaphase (MET uh faze) Mitotic phase during which chromosomes are aligned at the metaphase plate. 155

metaphase plate A disk formed during metaphase in which all of a cell's chromosomes lie in a single plane at right angles to the spindle fibers. 156

metapopulation Population subdivided into several small and isolated populations due to habitat fragmentation. 939

metastasis (muh TASS tuh sis) Spread of cancer from the place of origin throughout the body; caused by the ability of cancer cells to migrate and invade tissues. 159

methanogen (meth THANN uh jen) Type of archaea that lives in oxygen-free habitats, such as swamps, and releases methane gas. 375

microevolution Change in gene frequencies between populations of a species over time. 302

micronutrient Essential element needed in small amounts for plant growth, such as boron, copper, and zinc. 461

microRNA Introns that are processed into smaller signals; after being degraded, they combine with a protein, and the complex binds to mRNAs. These are then destroyed instead of being translated. 259

microsphere Formed from proteinoids exposed to water; has properties similar to those of today's cells. 319

microspore (MY crow spore) One of the two types of spores produced by seed plants; develops into a male gametophyte (pollen grain). 430, 496

microsporocyte (my crow SPORE oh site) Microspore mother cell; produces microspores by meiosis. 498

microtubule (my crow TUBE yule) Small, cylindrical organelle composed of tubulin protein around an empty central core; present in the cytoplasm, centrioles, cilia, and flagella. 76

midbrain In mammals, the part of the brain located below the thalamus and above the pons. 710

middle ear Portion of the ear consisting of the tympanic membrane, the oval and round windows, and the ossicles, where sound is amplified. 730

millipede (MILL ih peed) More or less cylindrical arthropod characterized by having two pairs of short legs on most of its body segments; may have 13 to almost 200 pairs of legs. 552

mimicry (MIMM ick kree) Superficial resemblance of two or more species; a mechanism that avoids predation by appearing to be noxious. 869

mineral Naturally occurring inorganic substance containing two or more elements; certain minerals are needed in the diet. 461, 667

mineralocorticoid (men urr ull oh CORE tih coid) Hormones secreted by the adrenal cortex that regulate salt and water balance, leading to increases in blood volume and blood pressure. 765

mitochondrion (mite oh KAHN dree uhn) Membrane-bounded organelle in which ATP molecules are produced during the process of cellular respiration. 74, 136

mitosis (my TOE sis) Process in which a parent nucleus produces two daughter nuclei, each having the same number and kinds of chromosomes as the parent nucleus. 150

model Simulation of a process that aids conceptual understanding until the process can be studied firsthand; a hypothesis that describes how a particular process could possibly be carried out. 11

molecular clock Idea that the rate at which mutational changes accumulate in certain genes is constant over time and is not involved in adaptation to the environment. 327, 350, 584

molecule Union of two or more atoms of the same element; also, the smallest part of a compound that retains the properties of the compound. 24

mollusc Member of the phylum Mollusca, which includes squids, clams, snails, and chitons; characterized by a visceral mass, a mantle, and a foot. 540

molt Periodic shedding of the exoskeleton in arthropods. 546

monoclonal antibody (mah no CLONE uhl) One of many antibodies produced by a clone of hybridoma cells that all bind to the same antigen. 648

monocot (MAH no cot) Abbreviation of monocotyledon. Flowering plant group; members have one embryonic leaf (cotyledon), parallel-veined leaves, scattered vascular bundles, flower parts in threes or multiples of three, and other characteristics. 428, 440

monocyte (MAH no site) Type of a granular leukocyte that functions as a phagocyte, particularly after it becomes a macrophage, which is also an antigen-presenting cell. 626

monoecious Having unisexual male flowers or cones and unisexual female flowers or cones both on the same plant. 424

monohybrid cross Cross between parents that differ in only one trait. 184

monomer (MAH nuh murr) Small molecule that is a subunit of a polymer—e.g., glucose is a monomer of starch. 38

monosaccharide (mah no SACK uh ride) Simple sugar; a carbohydrate that cannot be decomposed by hydrolysis—e.g., glucose. 39

monosomy (MAH no sohm mee) One less chromosome than usual. 212

monotreme (MAH no treem) Egg-laying mammal—e.g., duckbill platypus and spiny anteater. 574

monsoon (mahn SOON) Climate in India and southern Asia caused by wet ocean winds that blow onshore for almost half the year. 903

montane coniferous forest (MAHN tane cuh NIFF urr us) Coniferous forest of a mountain. 905

more-developed country (MDC) Country that is industrialized; typically, population growth is low, and the people enjoy a good standard of living. 851

morphogen (MORF uh jen) Protein that is part of a gradient that influences morphogenesis. 805

morphogenesis (morf oh JEN uh sis) Emergence of shape in tissues, organs, or entire embryo during development. 802

morula (MORE you luh) Spherical mass of cells resulting from cleavage during animal development prior to the blastula stage. 799

mosaic evolution Concept that human characteristics did not evolve at the same rate; for example, some body parts are more humanlike than others in early hominids. 586

moss Type of bryophyte. 418

motor molecule Protein that moves along either actin filaments or microtubules and translocates organelles. 76

motor (efferent) neuron Nerve cell that conducts nerve impulses away from the central nervous system and innervates effectors (muscle and glands). 703

mouth In humans, organ of the digestive tract where food is chewed and mixed with saliva. 659

multicellular Organism composed of many cells; usually has organized tissues, organs, and organ systems. 2

multifactorial The result of the interaction of several genes. 196

multiple alleles (uh LEEL) Inheritance pattern in which there are more than two alleles for a particular trait; each individual has only two of all possible alleles. 196

multiregional continuity hypothesis Proposal that modern humans evolved separately in at least three different places: Asia, Africa, and Europe. 590

muscular (contractile) tissue (cunn TRACK tile) Type of animal tissue composed of fibers that shorten and lengthen to produce movements. 602

mutualism (mute you uh LIZ uhm) Symbiotic relationship in which both species benefit in terms of growth and reproduction. 872

mycelium (my SEE lee uhm) Tangled mass of hyphal filaments composing the vegetative body of a fungus. 398

mycorrhizae (sing., mycorhiza) (my coh RIZE ee) Mutualistic relationship between fungal hyphae and roots of vascular plants. 408, 446, 464

myelin sheath (MY uh linn) White, fatty material—derived from the membrane of neurolemmocytes—that forms a covering for nerve fibers. 703

myofibril (my oh FIBE rull) Specific muscle cell organelle containing a linear arrangement of sarcomeres, which shorten to produce muscle contraction. 748

myoglobin (MY oh globe in) Pigmented molecule in muscle tissue that stores oxygen. 748

myosin (MY oh sin) Muscle protein making up the thick filaments in a sarcomere; it pulls actin to shorten the sarcomere, yielding muscle contraction. 748

myxedema (mikes uh DEEM uh) Condition resulting from a deficiency of thyroid hormone in an adult. 763

N

NAD⁺ (nicotinamide adenine dinucleotide) (nick coh TIN uh mide ADD uh neen die NUKE klee oh tide) Coenzyme of oxidation-reduction that accepts electrons and hydrogen ions to become NADH + H⁺ as oxidation of substrates occurs. During cellular respiration, NADH carries electrons to the electron transport chain in mitochondria. 110, 132

NADP⁺ (nicotinamide adenine dinucleotide phosphate) (nick coh TIN uh mide ADD uh neen die NUKE klee oh tide FOSS fate) Coenzyme of oxidation-reduction that accepts electrons and hydrogen ions to become NADPH + H⁺. During photosynthesis, NADPH participates in the reduction of carbon dioxide to glucose. 110

nail Flattened epithelial tissue from the stratum lucidum of the skin; located on the tips of fingers and toes. 606

natural killer (NK) cell Lymphocyte that causes an infected or cancerous cell to burst. 637

natural selection Mechanism of evolution caused by environmental selection of organisms most fit to reproduce; results in adaptation to the environment. 5, 288, 306

Neandertal (nee AND urr tall) Hominid with a sturdy build that lived during the last Ice Age in Europe and the Middle East; hunted large game and left evidence of being culturally advanced. 591

negative feedback Mechanism of homeostatic response by which the output of a system suppresses or inhibits activity of the system. 608, 760

nematocyst (nuh MAT uh sist) In cnidarians, a capsule that contains a threadlike fiber, the release of which aids in the capture of prey. 524

neoplasm (NEE oh plazz uhm) Any new and abnormal growth of tissue in which the growth is uncontrolled and progressive. 159

nephridium (pl., nephridia) (nuh FRIDD ee uhm, nuh FRIDD ee uh) Segmentally arranged, paired excretory tubules of many invertebrates, as in the earthworm. 544, 689

nephron (NEFF rahn) Microscopic kidney unit that regulates blood composition by glomerular filtration, tubular reabsorption, and tubular secretion. 690

nerve Bundle of long axons outside the central nervous system. 603, 712

nerve cord In many complex animals, a centrally placed cord of nervous tissue that receives sensory information and exercises motor control. 560

nerve fiber Axon; conducts nerve impulses away from the cell. They are classified as either myelinated or unmyelinated based on the presence or absence of a myelin sheath. 703

nerve impulse Action potential (electrochemical change) traveling along a neuron. 704

nerve net Diffuse, noncentralized arrangement of nerve cells in cnidarians. 524, 700

nervous tissue Tissue that contains nerve cells (neurons), which conduct impulses, and neuroglia, which support, protect, and provide nutrients to neurons. 603

neural plate (NURE uhl) Region of the dorsal surface of the chordate embryo that marks the future location of the neural tube. 801

neural tube Tube formed by closure of the neural groove during development. In vertebrates, the neural tube develops into the spinal cord and brain. 801

neuroglia (nure RAH glee uh) Nonconducting nerve cells that are intimately associated with neurons and function in a supportive capacity. 603, 703

neuromodulator (nure oh MAH dew lay turr) Electrical stimulant of a peripheral nerve, the spinal cord, or the brain; used to ease pain. 707

neuromuscular junction (nure oh MUSS cue lurr) Region where an axon bulb approaches a muscle fiber; contains a presynaptic membrane, a synaptic cleft, and a postsynaptic membrane. 750

neuron (NURE ahn) Nerve cell that characteristically has three parts: dendrites, cell body, and an axon. 603, 703

neurotransmitter (nure oh trans MITT urr) Chemical stored at the ends of axons that is responsible for transmission across a synapse. 706

neurula The early embryo during the development of the neural tube from the neural plate, marking the first appearance of the nervous system; the next stage after the gastrula. 801

neutron (NEW trahn) Neutral subatomic particle, located in the nucleus and assigned one atomic mass unit. 20

neutrophil (NEW troh fill) Granular leukocyte that is the most abundant of the white blood cells; first to respond to infection. 626

nitrification (nite trih fih KAY shun) Process by which nitrogen in ammonia and organic compounds is oxidized to nitrites and nitrates by soil bacteria. 892

nitrogen cycle Continuous process by which nitrogen circulates in the air, soil, water, and organisms of the biosphere. 892

nitrogen fixation Process whereby free atmospheric nitrogen is converted into compounds, such as ammonium and nitrates, usually by bacteria. 892

node In plants, the place where one or more leaves attach to a stem. 439

nodes of Ranvier (RAN veer) Gap in the myelin sheath around a nerve fiber. 703

noncoding gene Transcribes into any other types of RNA other than mRNA. 247

noncompetitive inhibition Form of enzyme inhibition where the inhibitor binds to an enzyme at a location other than the active site; while at this site, the enzyme shape changes, the inhibitor is unable to bind to its substrate, and no product forms. 109

noncyclic electron pathway Portion of the light reactions of photosynthesis that involves both photosystem I and photosystem II. It generates both ATP and NADPH. 120

nondisjunction Failure of homologous chromosomes or daughter chromosomes to separate during meiosis I and meiosis II, respectively. 212

nonpolar covalent bond (nahn POH lurr coh VALE lent) Bond in which the sharing of electrons between atoms is fairly equal. 26

nonrandom mating Mating among individuals on the basis of their phenotypic similarities or differences, rather than mating on a random basis. 305

nonseptate (nahn SEPP tate) Lacking cell walls; some fungal species have hyphae that are nonseptate. 109

nonvascular plants Bryophytes, such as mosses and liverworts, that have no vascular tissue and either occur in moist locations or have special adaptations for living in dry locations. 417

norepinephrine (NE) (nor epp pin EFF renn) Neurotransmitter of the postganglionic fibers in the sympathetic division of the autonomic system; also, a hormone produced by the adrenal medulla. 706, 765

notochord (NO toh cord) Cartilaginous-like supportive dorsal rod in all chordates sometime in their life cycle; replaced by vertebrae in vertebrates. 560, 801

nuclear envelope Double membrane that surrounds the nucleus in eukaryotic cells and is connected to the endoplasmic reticulum; has pores that allow substances to pass between the nucleus and the cytoplasm. 68

nuclear pore Opening in the nuclear envelope that permits the passage of proteins into the nucleus and ribosomal subunits out of the nucleus. 68

nucleic acid (new CLAY ick) Polymer of nucleotides; both DNA and RNA are nucleic acids. 50, 224

nucleoid (NEW klee oid) Region of prokaryotic cells where DNA is located; it is not bounded by a nuclear envelope. 62, 162, 368

nucleolus (new KLEE uh luss) Dark-staining, spherical body in the nucleus that produces ribosomal subunits. 68

nucleoplasm (NEW klee oh plazz uhm) Semifluid medium of the nucleus containing chromatin. 68

nucleosome (NEW klee oh sohm) In the nucleus of a eukaryotic cell, a unit composed of DNA wound around a core of eight histone proteins, giving the appearance of a string of beads. 256

nucleotide (NEW klee oh tide) Monomer of DNA and RNA consisting of a 5-carbon sugar bonded to a nitrogenous base and a phosphate group. 50, 224

nucleus Membrane-bounded organelle within a eukaryotic cell that contains chromosomes and controls the structure and function of the cell. 64

O

Obelia Member of phylum Cnidaria; common colonial hydrozoan found in brackish water or the ocean. 527

obligate anaerobe (AHB lih gate ANN urr robe) Prokaryote unable to grow in the presence of free oxygen. 370

observation Step in the scientific method by which data are collected before a conclusion is drawn. 10

ocean ridge Ridge on the ocean floor where oceanic crust forms and from which it moves laterally in each direction. 319

octet rule States that an atom other than hydrogen tends to form bonds until it has eight electrons in its outer shell; an atom that already has eight electrons in its outer shell does not react and is inert. 23

oil Triglyceride, usually of plant origin, that is composed of glycerol and three fatty acids and is liquid in consistency due to many unsaturated bonds in the hydrocarbon chains of the fatty acids. 42

oil gland Gland of the skin, associated with hair follicle, that secretes sebum; sebaceous gland. 606

olfactory cell (ohl FACT toh ree) Modified neuron that is a sensory receptor for the sense of smell. 723

oligochaete Invertebrate member of the phylum Annelida; characterized by body segmentation and the presence of setae (e.g., earthworms). 545

omnivore (AHM nih vore) Organism in a food chain that feeds on both plants and animals. 883

oncogene (AHN coh jeen) Cancer-causing gene. 160

oocyte (OH oh site) Immature egg that is undergoing meiosis; upon completion of meiosis, the oocyte becomes an egg. 782

oogenesis (oh oh JENN us sis) Production of eggs in females by the process of meiosis and maturation. 177

open circulatory system Arrangement of internal transport in which blood bathes the organs directly, and there is no distinction between blood and interstitial fluid. 614

operant conditioning (AH purr unt) Learning that results from rewarding or reinforcing a particular behavior. 823

operator In an operon, the sequence of DNA that binds tightly to a repressor, and thereby regulates the expression of structural genes. 252

operon (AH purr rahn) Group of structural and regulating genes that function as a single unit. 252

orbital (OR bit uhl) Volume of space around a nucleus where electrons can be found most of the time. 23

order One of the categories, or taxa, used by taxonomists to group species; the taxon above the family level. 8, 344

organ Combination of two or more different tissues performing a common function. 438, 605

organelle Small, often membranous structure in the cytoplasm having a specific structure and function. 64

organic chemistry Branch of science which deals with compounds that contain carbon. 36

organic molecule Molecule that always contains carbon and hydrogen, and often contains oxygen as well; organic molecules are associated with living things. 36

organism Individual living thing. 2

organ of Corti (CORE tie) Structure in the vertebrate inner ear that contains auditory receptors (also called spiral organ). 731

organ system Group of related organs working together. 605

orgasm (OR gazz uhm) Physiological and psychological sensations that occur at the climax of sexual stimulation. 779

osmosis (oz MOH sis) Diffusion of water through a differentially permeable membrane. 90

osmotic pressure (oz MAH tick) Measure of the tendency of water to move across a differentially permeable membrane; visible as an increase in liquid on the side of the membrane with higher solute concentration. 90

ossicle (AH sick cull) One of the small bones of the vertebrate middle ear—malleus, incus, and stapes. 730

osteoblast (AH stee oh blast) Bone-forming cell. 740

osteoclast (AH stee oh clast) Cell that causes erosion of bone. 740

osteocyte (AH stee oh site) Mature bone cell located within the lacunae of bone. 740

ostracoderm (ah STRAH cuh derms) Earliest vertebrate fossils of the Cambrian and Devonian periods; these fishes were small, jawless, and finless. 562

otolith (OH toe lith) Calcium carbonate granule associated with sensory receptors for detecting movement of the head; in vertebrates, located in the utricle and saccule. 733

outer ear Portion of the ear consisting of the pinna and the auditory canal. 730

out-of-Africa hypothesis Proposal that modern humans originated only in Africa; then they migrated and supplanted populations of *Homo* in Asia and Europe about 100,000 years ago. 590

ovarian cycle (oh VAIR ree uhn) Monthly changes occurring in the ovary that determine the level of sex hormones in the blood. 783

ovary In flowering plants, the enlarged, ovule-bearing portion of the carpel that develops into a fruit; female gonad in animals that produces an egg and female sex hormones. 429, 496, 770, 776

ovulation (ah view LAY shun) Bursting of a follicle when a secondary oocyte is released from the ovary; if fertilization occurs, the secondary oocyte becomes an egg. 784

ovule (OH vule) In seed plants, a structure that contains the female gametophyte and has the potential to develop into a seed. 424, 497

ovum (OH vuhm) Haploid egg cell that is usually fertilized by a sperm to form a diploid zygote. 783

oxidation Loss of one or more electrons from an atom or molecule; in biological systems, generally the loss of hydrogen atoms. 110

oxidative phosphorylation (ox ih DAY tiv foss for ill LAY shun) Process by which ATP production is tied to an electron transport system that uses oxygen as the final acceptor; occurs in mitochondria. 138

oxygen debt Amount of oxygen required to oxidize lactic acid produced anaerobically during strenuous muscle activity. 142, 748

oxyhemoglobin (ox zee HEEM uh glow bin) Compound formed when oxygen combines with hemoglobin. 678

oxytocin (ox zee TOE sin) Hormone released by the posterior pituitary that causes contraction of the uterus and milk letdown. 760

ozone hole Seasonal thinning of the ozone shield in the lower stratosphere at the North and South Poles. 896

ozone shield Accumulation of O_3, formed from oxygen in the upper atmosphere; a filtering layer that protects the Earth from ultraviolet radiation. 326, 896

P

p53 **gene** For control of cell division, the *p53* gene halts the cell cycle when DNA mutates and is in need of repair. 160

paleontology (pale lee uhn TAH loh jee) Study of fossils that results in knowledge about the history of life. 284, 322

palisade mesophyll (PAL uh sade MESS oh fill) Layer of tissue in a plant leaf containing elongated cells with many chloroplasts. 454

PAN (peroxyacetylnitrate) (purr ox zee uh see tull NITE rate) Type of noxious chemical found in photochemical smog. 893

pancreas (PAN kree us) Internal organ that produces digestive enzymes and the hormones insulin and glucagon. 662, 768

pancreatic amylase (pan kree AT tick AM uh laze) Enzyme that digests starch to maltose. 664

pancreatic islet (pan kree AT tick EYE lit) Masses of cells that constitute the endocrine portion of the pancreas. 768

panoramic vision Vision characterized by having a wide field of vision; found in animals with eyes to the side. 724

parallel evolution Similarity in structure in related groups that cannot be traced to a common ancestor. 347

parasite Species that is dependent on a host species for survival, usually to the detriment of the host species. 870

parasitism (PAIR uh sit tiz uhm) Symbiotic relationship in which one species (the *parasite*) benefits in terms of growth and reproduction to the detriment of the other species (the *host*). 870

parasympathetic division (pair uh simm puh THETT ick) Division of the autonomic system that is active under normal conditions; uses acetylcholine as a neurotransmitter. 715

parathyroid gland (pair uh THIGH roid) Gland embedded in the posterior surface of the thyroid gland; it produces parathyroid hormone. 764

parathyroid hormone (PTH) Hormone secreted by the four parathyroid glands that increases the blood calcium level and decreases the phosphate level. 764

parenchyma (puh RENN kih muh) Plant tissue composed of the least-specialized of all plant cells; found in all organs of a plant. 442

parthenogenesis (par thin oh JENN uh sis) Development of an egg cell into a whole organism without fertilization. 776

partial pressure Pressure exerted by each gas in a mixture of gases. 675

pathogen Disease-causing agent such as viruses, parasitic bacteria, fungi, and animals. 371, 601, 636

pattern formation Positioning of cells during development that determines the final shape of an organism. 802

peat Organic fuel consisting of the partially decomposed remains of peat mosses that accumulate in bogs. 419

pectoral girdle (PECK tore uhl) Portion of the vertebrate skeleton that provides support and attachment for the upper (fore) limbs; consists of the scapula and clavicle on each side of the body. 744

peduncle Flower stalk; expands into the receptacle. 428

pelagic division (puh LAJJ ick) Open portion of the sea. 922

pelvic girdle Portion of the vertebrate skeleton to which the lower (hind) limbs are attached; consists of the coxal bones. 744

penis Male copulatory organ; in humans, the male organ of sexual intercourse. 778

pentose (PEN toes) Five-carbon sugar. Deoxyribose is the pentose sugar found in DNA; ribose is the pentose sugar found in RNA. 39

pepsin (PEP sin) Enzyme secreted by gastric glands that digests proteins to peptides. 664

peptidase Intestinal enzyme that breaks down short chains of amino acids to individual amino acids that are absorbed across the intestinal wall. 664

peptide (PEP tide) Two or more amino acids joined together by covalent bonding. 46

peptide bond Type of covalent bond that joins two amino acids. 46

peptide hormone Type of hormone that is a protein, a peptide, or derived from an amino acid. 758

peptidoglycan (pep tih doe GLIKE can) Unique molecule found in bacterial cell walls. 41, 368

perennial (purr IN nee uhl) Flowering plant that lives more than one growing season because the underground parts regrow each season. 438

pericycle (pair ih SIGH cull) Layer of cells surrounding the vascular tissue of roots; produces branch roots. 445

periderm (PAIR ih derm) Protective tissue that replaces epidermis; includes cork, cork cambium. 441

peripheral nervous system (PNS) (purr IF fur uhl) Nerves and ganglia that lie outside the central nervous system. 702

peristalsis (pair iss STALL sis) Wavelike contractions that propel substances along a tubular structure such as the esophagus. 660

permafrost Permanently frozen ground, usually occurring in the tundra, a biome of Arctic regions. 906

peroxisome (purr OX ih sohm) Enzyme-filled vesicle in which fatty acids and amino acids are metabolized to hydrogen peroxide that is broken down to harmless products. 73

petal A flower part that occurs just inside the sepals; often conspicuously colored to attract pollinators. 429, 496

petiole (PET tee ohl) The part of a plant leaf that connects the blade to the stem. 439

Peyer's patches Lymphatic organs located in small intestine. 636

phagocytize (fag OSS sit tize) To ingest extracellular particles by engulfing them, as do amoeboid cells. 390

phagocytosis (fag oh site OH sis) Process by which amoeboid-type cells engulf large substances, forming an intracellular vacuole. 94

pharynx (FAIR inks) In vertebrates, common passageway for both food intake and air movement; located between the mouth and the esophagus. 660, 676

phenetic systematics (fin ETT tick sis tim MAT ticks) School of systematics that determines the degree of relatedness between species by counting the number of their similarities. 352

phenomenon (fin NAH men ahn) Observable event. 10

phenotype (FEE no type) Visible expression of a genotype—e.g., brown eyes or attached earlobes. 185

pheromone (FAIR oh moan) Chemical messenger that works at a distance and alters the behavior of another member of the same species. 758, 831

phloem (FLOW emm) Vascular tissue that conducts organic solutes in plants; contains sieve-tube members and companion cells. 420, 443, 466

phospholipid (foss foe LIP id) Molecule that forms the bilayer of the cell's membranes; has a polar, hydrophilic head bonded to two nonpolar, hydrophobic tails. 44

phospholipid bilayer Comprises the plasma membrane; each polar, hydrophilic head is bonded to two nonpolar, hydrophobic tails; contains embedded proteins. 85

phosphorus cycle Continuous process by which phosphorus circulates in the soil, water, and organisms of the biosphere. 894

phosphorylation (foss for ill LAY shun) In metabolic processes, a way to activate an enzyme in which the enzyme either attaches an inorganic phosphate to a molecule or mediates the transfer of a phosphate group from one molecule to another. 109

photoautotroph (foe toe AH toe trofe) Organism able to synthesize organic molecules by using carbon dioxide as the carbon source and sunlight as the energy source. 370

photochemical smog Air pollution that contains nitrogen oxides and hydrocarbons, which react to produce ozone and PAN (peroxyacetylnitrate). 893

photoperiodism Relative lengths of daylight and darkness that affect the physiology and behavior of an organism. 488

photoreceptor Sensory receptor that responds to light stimuli. 724

photorespiration Series of reactions that occurs in plants when carbon dioxide levels are depleted but oxygen continues to accumulate, and the enzyme RuBP carboxylase fixes oxygen instead of carbon dioxide. 126

photosynthesis (foe toe SIN thuh sis) Process occurring usually within chloroplasts whereby chlorophyll-containing organelles trap solar energy to reduce carbon dioxide to carbohydrate. 4, 116

photosystem Photosynthetic unit where solar energy is absorbed and high-energy electrons are generated; contains a pigment complex and an electron acceptor; occurs as PS (photosystem) I and PS II. 120

phototropism (foe toe TROH piz uhm) Growth response of plant stems to light; stems demonstrate positive phototropism. 479

pH scale Measurement scale for hydrogen ion concentration. 30

phylogenetic tree (file oh jenn ETT ick) Diagram that indicates common ancestors and lines of descent among a group of organisms. 346

phylogeny (file AH jenn ee) Evolutionary history of a group of organisms. 346

phylum (FILE uhm) One of the categories, or taxa, used by taxonomists to group species; the taxon above the class level. 8, 344

phytochrome (FITE toe chrome) Photoreversible plant pigment that is involved in photoperiodism and other responses of plants, such as etiolation. 489

phytoplankton (fite oh PLANK ton) Part of plankton containing organisms that photosynthesize, releasing oxygen to the atmosphere and serving as food producers in aquatic ecosystems. 387, 917

phytoremediation (FITE toe ruh mee dee AY shun) The use of plants to restore a natural area to its original condition. 471

pineal gland (PIN nee uhl) Gland—either at the skin surface (fish, amphibians) or in the third ventricle of the brain (mammals)—that produces melatonin. 770

pinocytosis (pie no site OH sis) Process by which vesicle formation brings macromolecules into the cell. 94

pioneer species Early colonizer of barren or disturbed habitats that usually has rapid growth and a high dispersal rate. 874

pistil Femal reproductive structure of a flower; composed of one or more carpels and consisting of stigma, style, and ovary. 429

pit Any depression or opening; usually in reference to the small openings in the cell walls of xylem cells that function in providing a continuum between adjacent xylem cells. 442

pith Parenchyma tissue in the center of some stems and roots. 445

pituitary dwarfism (pit TWO it air ree) Condition caused by inadequate growth hormone in which affected individual has normal proportions but small stature. 762

pituitary gland Small gland that lies just inferior to the hypothalamus; consists of the anterior and posterior pituitary, both of which produce hormones. 760

placenta (pluh SENT uh) Organ formed during the development of placental mammals from the chorion and the uterine wall; allows the embryo, and then the fetus, to acquire nutrients and rid itself of wastes; produces hormones that regulate pregnancy. 575, 777, 811

placoderm (PLACK uh derm) First jawed vertebrates; heavily armored fishes of the Devonian period. 563

plankton (PLANK ton) Freshwater and marine organisms that are suspended on or near the surface of the water; includes phytoplankton and zooplankton. 380

plant Multicellular, usually photosynthetic, organism belonging to the plant kingdom. 10

plaque Accumulation of soft masses of fatty material, particularly cholesterol, beneath the inner linings of the arteries. 666

plasma (PLAZZ muh) In vertebrates, the liquid portion of blood; contains nutrients, wastes, salts, and proteins. 626

plasma cell Mature B cell that mass-produces antibodies. 639

plasma membrane Membrane surrounding the cytoplasm that consists of a phospholipid bilayer with embedded proteins; functions to regulate the entrance and exit of molecules from cell. 62

plasmid (PLAZZ mid) Self-duplicating ring of accessory DNA in the cytoplasm of bacteria. 62, 268, 368

plasmodesmata (plazz moh dezz MAH tuh) In plants, cytoplasmic strands that extend through pores in the cell wall and connect the cytoplasm of two adjacent cells. 97

plasmodial slime mold (plazz MOH dee uhl) Free-living mass of cytoplasm that moves by pseudopods on a forest floor or in a field, feeding on decaying plant material by phagocytosis; reproduces by spore formation. 393

plasmolysis (plazz MOLL ih sis) Contraction of the cell contents due to the loss of water. 91

plastid (PLASS tidd) Organelles of plants and algae that are bounded by a double membrane and contain internal membranes and/or vesicles (i.e., chloroplasts, chromoplasts, leucoplasts). 74

platelet (PLATE let) Component of blood that is necessary to blood clotting. 601, 628

plate tectonics (tec TAH nicks) Concept that the Earth's crust is divided into a number of fairly rigid plates whose movements account for continental drift. 335

pleiotropy (ply AH troh pee) Inheritance pattern in which one gene affects many phenotypic characteristics of the individual. 194

plumule (PLOO mule) In flowering plants, the embryonic plant shoot that bears young leaves. 507

point mutation Change of one base only in the sequence of bases in a gene. 261

polar body In oogenesis, a nonfunctional product; two to three meiotic products are of this type. 177

polar covalent bond Bond in which the sharing of electrons between atoms is unequal. 26

pollen grain In seed plants, structure that is derived from a microspore and develops into a male gametophyte. 416, 498

pollen tube In seed plants, a tube that forms when a pollen grain lands on the stigma and germinates. The tube grows, passing between the cells of the stigma and the style to reach the egg inside an ovule, where fertilization occurs. 431

pollination In gymnosperms, the transfer of pollen from pollen cone to seed cone; in angiosperms, the transfer of pollen from anther to stigma. 424, 499

pollinator Animal (e.g., a bee) that inadvertently transfers pollen from anther to stigma. 424

pollution Any environmental change that adversely affects the lives and health of living things. 936

polychaete Invertebrate member of the phylum Annelida; marine organisms characterized by the presence of many setae (e.g., tube worms and clam worms). 544

polygenic inheritance (pah lee JENN ick) Pattern of inheritance in which a trait is controlled by several allelic pairs; each dominant allele contributes to the phenotype in an additive and like manner. 197

polymer (PAH lee murr) Macromolecule consisting of covalently bonded monomers; for example, a polypeptide is a polymer of monomers called amino acids. 38

polymerase chain reaction (PCR) (pah LIMM mare raze) Technique that uses the enzyme DNA polymerase to produce millions of copies of a particular piece of DNA. 269

polymorphic (pah lee MORE fick) Genes that have more than one wild-type allele. 304

polyp (PAH lip) Among cnidarians, body form that is directed upward and contains much mesoglea; in anatomy; small, abnormal growth that arises from the epithelial lining. 524, 662

polypeptide (pah lee PEP tide) Polymer of many amino acids linked by peptide bonds. 46

polyploid (PAH lee ploid) Having a chromosome number that is a multiple greater than twice that of the monoploid number. 212

polyribosome (pah lee RIBE uh sohm) String of ribosomes simultaneously translating regions of the same mRNA strand during protein synthesis. 69, 245

polysaccharide (pah lee SACK uh ride) Polymer made from sugar monomers; the polysaccharides starch and glycogen are polymers of glucose monomers. 39

pons (PAHNS) Portion of the brain stem above the medulla oblongata and below the midbrain; assists the medulla oblongata in regulating the breathing rate. 710

population Group of organisms of the same species occupying a certain area and sharing a common gene pool. 5, 302, 838

population density The number of individuals per unit area or volume living in a particular habitat. 839

population distribution The pattern of dispersal of individuals living within a certain area. 839

population genetics The study of gene frequencies and their changes within a population. 302

population viability analysis Calculation of the minimum population size needed to prevent extinction. 940

portal system Pathway of blood flow that begins and ends in capillaries, such as the portal system located between the small intestine and liver. 622

positive feedback Mechanism of homeostatic response in which the output of the system intensifies and increases the activity of the system. 609, 760

posterior pituitary (pit YOU ih tare rree) Portion of the pituitary gland that stores and secretes oxytocin and antidiuretic hormone produced by the hypothalamus. 760

posttranscriptional control Gene expression following translation regulated by the way mRNA transcripts are processed. 259

posttranslational control Gene expression following translation regulated by the activity of the newly synthesized protein. 260

postzygotic isolating mechanism (post zie GAH tick) Anatomical or physiological difference between two species that prevents successful reproduction after mating has taken place. 311

potential energy Stored energy as a result of location or spatial arrangement. 102

precipitation (prih sip ih TAY shun) Water deposited on the Earth in the form of rain, snow, sleet, hail, or fog. 889

predation (preh DAY shun) Interaction in which one organism (the *predator*) uses another (the *prey*) as a food source. 866

predator Organism that practices predation. 866

prediction Step of the scientific process that follows the formulation of a hypothesis and assists in creating the experimental design. 11

prefrontal area Association area in the frontal lobe that receives information from other association areas and uses it to reason and plan actions. 710

preparatory (prep) reaction Reaction that oxidizes pyruvate with the release of carbon dioxide; results in acetyl CoA and connects glycolysis to the citric acid cycle. 133, 136

pressure-flow model Explanation for phloem transport; osmotic pressure following active transport of sugar into phloem brings a flow of sap from a source to a sink. 472

prey Organism that provides nourishment for a predator. 866

prezygotic isolating mechanism (pree zie GAH tick) Anatomical or behavioral difference between two species that prevents the possibility of mating. 310

primary motor area Area in the frontal lobe where voluntary commands begin; each section controls a part of the body. 709

primary root Original root that grows straight down and remains the dominant root of the plant; contrasts with fibrous root system. 446

primary somatosensory area (so mat oh SENSE uh ree) Area dorsal to the central sulcus where sensory information arrives from the skin and skeletal muscles. 709

primate Member of the order Primate; includes prosimians, monkeys, apes, and hominids, all of whom have adaptations for living in trees. 580

principle Theory that is generally accepted by an overwhelming number of scientists; also called a law. 12

prion (PRY ahn) Infectious particle consisting of protein only and no nucleic acid. 49, 367

producer Photosynthetic organism at the start of a grazing food chain that makes its own food—e.g., green plants on land and algae in water. 883

product Substance that forms as a result of a reaction. 104

progesterone (pro JEST turr ohn) Female sex hormone that helps maintain sexual organs and secondary sex characteristics. 770, 783

proglottid (pro GLAH tid) Segment of a tapeworm that contains both male and female sex organs and becomes a bag of eggs. 531

prokaryote (pro CARE ree oat) Organism that lacks the membrane-bounded nucleus and membranous organelles typical of eukaryotes. 367

prokaryotic cell (pro care ree AH tick) Lacking a membrane-bounded nucleus and organelles; the cell type within the domains Bacteria and Archaea. 62

prolactin (PRL) (pro LACK tin) Hormone secreted by the anterior pituitary that stimulates the production of milk from the mammary glands. 760

prometaphase (pro MET uh faze) Phase of mitosis which generally begins with the disintegration of the nuclear membrane. 155

promoter In an operon, a sequence of DNA where RNA polymerase binds prior to transcription. 242, 252

proofreading Process used to check the accuracy of DNA replication as it occurs and to replace a mispaired base with the right one. 233

prophase (PRO faze) Mitotic phase during which chromatin condenses so that chromosomes appear; chromosomes are scattered. 154

prosimian (pro SIMM me uhn) Group of primates that includes lemurs and tarsiers, and may resemble the first primates to have evolved. 583

prostaglandin (pro stah GLAN din) Hormone that has various and powerful local effects. 771

protein (PRO teen) Molecule consisting of one or more polypeptides. 46

protein-coding gene Transcribes into mRNA. 247

protein-first hypothesis In chemical evolution, the proposal that protein originated before other macromolecules and made possible the formation of protocells. 319

proteinoid (PRO tin oid) Abiotically polymerized amino acids that, when exposed to water, become microspheres having cellular characteristics. 319

proteomics Study of the complete collection of proteins that an organism produces. 273

protist (PRO teest) Member of the kingdom Protista. 9, 380

protobiont Also called protocell, possible first cell. 320

protocell (PRO toe cell) In biological evolution, a possible cell forerunner that became a cell once it acquired genes. 320

proton (PRO tahn) Positive subatomic particle located in the nucleus and assigned one atomic mass unit. 20

proto-oncogene (pro toe AHN coh jeen) Normal gene that can become an oncogene through mutation. 160

protoplast (PRO toe plast) Plant cell from which the cell wall has been removed. 509

protostome (PRO toe stome) Group of coelomate animals in which the first embryonic opening (the blastopore) is associated with the mouth. 518, 538

protozoan (pro toe ZOH uhn) Heterotrophic, unicellular protist that moves by flagella, cilia, or pseudopodia, or is immobile. 389

proximal convoluted tubule Portion of a nephron following the glomerular capsule where tubular reabsorption of filtrate occurs. 691

pseudocoelomate (sue doe SEE lumm ate) Animal possessing a coelom (body cavity) incompletely lined by mesoderm (i.e., roundworms). 518

pseudopod (SUE doe pod) Cytoplasmic extension of amoeboid protists; used for locomotion and engulfing food. 76, 390

puberty Period of life when secondary sex changes occur in humans; marked by the onset of menses in females and sperm production in males. 781

pulmonary circuit (PULL moh nair ree) Circulatory pathway between the lungs and the heart. 617

pulse Vibration felt in arterial walls due to expansion of the aorta following ventricle contraction. 621

Punnett square (PUN net) Grid used to calculate the expected results of simple genetic crosses. 186

pupil Opening in the center of the iris of the vertebrate eye. 725

purine (PURE reen) Type of nitrogen-containing base, such as adenine and guanine, having a double-ring structure. 227

pyrimidine (pih RIM uh dean) Type of nitrogen-containing base, such as cytosine, thymine, and uracil, having a single-ring structure. 227

pyruvate (pie ROO vate) End product of glycolysis; its further fate, involving fermentation or entry into a mitochondrion, depends on oxygen availability. 133

R

radial symmetry (RAY dee uhl SIM meh tree) Body plan in which similar parts are arranged around a central axis, like spokes of a wheel. 518

radiolarian (ray dee oh LAIR ree uhn) Member of the phylum Actinopoda bearing a glassy silicon test, usually with a radial arrangement of spines; pseudopods are external to the test. 390

radula (RADD you luh) Tonguelike organ found in molluscs that bears rows of tiny teeth, which point backward; used to obtain food. 540, 657

rain shadow Leeward side (side sheltered from the wind) of a mountainous barrier, which

receives much less precipitation than the windward side. 903

rate of natural increase (r) Growth rate dependent on the number of individuals that are born each year and the number of individuals that die each year. 840

ray-finned bony fishes Group of bony fishes with fins supported by parallel bony rays connected by webs of thin tissue. 564

reactant (ree ACT unt) Substance that participates in a reaction. 104

receptacle Area where a flower attaches to a floral stalk. 428

receptor-mediated endocytosis (en doe site TOE sis) Selective uptake of molecules into a cell by vacuole formation after they bind to specific receptor proteins in the plasma membrane. 94

receptor protein Protein located in the plasma membrane or within the cell; binds to a substance that alters some metabolic aspect of the cell. 87

recessive allele (re SESS ihv uh LEEL) Allele that exerts its phenotypic effect only in the homozygote; its expression is masked by a dominant allele. 184

reciprocal altruism The trading of helpful or cooperative acts, such as helping at the nest, by individuals—the animal that was helped will repay the debt at some later time. 830

recombinant DNA (rDNA) (ree CAHM bih nunt) DNA that contains genes from more than one source. 268

red algae Marine photosynthetic protists with a notable abundance of phycobilin pigments; include coralline algae of coral reefs. 384

red blood cell Erythrocyte; contains hemoglobin and carries oxygen from the lungs or gills to the tissues in vertebrates. 601, 626

red bone marrow Vascularized, modified connective tissue that is sometimes found in the cavities of spongy bone; site of blood cell formation. 635, 740

reduction Gain of electrons by an atom or molecule with a concurrent storage of energy; in biological systems, the electrons are accompanied by hydrogen ions. 110

reflex Automatic, involuntary response of an organism to a stimulus. 713

regulator gene In an operon, a gene that codes for a protein that regulates the expression of other genes. 252

relative dating (of fossils) Determining the age of fossils by noting their sequential relationships in strata; *absolute dating* relies on radioactive dating techniques to assign an actual date. 322

relative fitness Reproductive success of a genotype as measured by survival, fecundity, or other life-history parameters. 306

renal cortex (REE null CORE tex) Outer portion of the kidney that appears granular. 690

renal medulla (REE null muh DOO luh) Inner portion of the kidney that consists of renal pyramids. 690

renal pelvis Hollow chamber in the kidney that lies inside the renal medulla and receives freshly prepared urine from the collecting ducts. 690

renin (REN ninn) Enzyme released by the kidneys that leads to the secretion of aldosterone and a rise in blood pressure. 694, 766

replacement reproduction Population in which each person is replaced by only one child. 852

replication fork In eukaryotes, the point where the two parental DNA strands separate to allow replication. 233

repressible operon (AH purr ahn) Operon that is normally active because the repressor is normally inactive. 253

repressor In an operon, protein molecule that binds to an operator, preventing transcription of structural genes. 253

reproduce To produce a new individual of the same kind. 5

reproductive cloning Genetically identical to the original individual. 157

reptile Member of a class of terrestrial invertebrates with internal fertilization, scaly skin, and an egg with a leathery shell; includes snakes, lizards, turtles, and crocodiles. 568

resource In economic terms, anything with potential use in creating wealth or giving satisfaction. 839

resource partitioning Mechanism that increases the number of niches by apportioning the supply of a resource such as food or living space between species. 864

respiration Sequence of events that results in gas exchange between the cells of the body and the environment. 672

responding variable Result or change that occurs when an experimental variable is utilized in an experiment. 12

resting potential Membrane potential of an inactive neuron. 704

restoration ecology Subdiscipline of conservation biology that seeks ways to return ecosystems to their former state. 941

restriction enzyme Bacterial enzyme that stops viral reproduction by cleaving viral DNA; used to cut DNA at specific points during production of recombinant DNA. 268

reticular fiber (reh TICK cue lurr) Very thin collagen fibers in the matrix of connective tissue, highly branched and forming delicate supporting networks. 599

retina (RETT tih nuh) Innermost layer of the vertebrate eyeball containing the photoreceptors—rod cells and cone cells. 725

retrovirus (rett troh VIE russ) RNA virus containing the enzyme reverse transcriptase that carries out RNA/DNA transcription. 366

rhizoid (RYE zoid) Rootlike hair that anchors a plant and absorbs minerals and water from the soil. 418

rhizome (RYE zohm) Rootlike underground stem. 421, 452

rhodopsin (rode AHP sin) Light-absorbing molecule in rod cells and cone cells that contains a pigment and the protein opsin. 727

ribbon worm Marine invertebrate of the phylum Nemertea having a distinctive proboscis apparatus. 528

ribose (RYE bohs) Pentose sugar found in RNA. 39

ribosomal RNA (rRNA) (rye boh SOHM uhl) Type of RNA found in ribosomes that translate messenger RNAs to produce proteins. 240

ribosome (RYE boh sohm) RNA and protein in two subunits; site of protein synthesis in the cytoplasm. 62, 69

ribozyme (RYE boh zime) Enzyme that carries out mRNA processing. 243

RNA (ribonucleic acid) (rye boh new CLAY ick) Nucleic acid produced from covalent bonding of nucleotide monomers that contain the sugar ribose; occurs in three forms: messenger RNA, ribosomal RNA, and transfer RNA. 50, 224, 240

RNA-first hypothesis In chemical evolution, the proposal that RNA originated before other macromolecules and allowed the formation of the first cell(s). 319

RNA polymerase (pah LIMM mare raze) During transcription, an enzyme that joins nucleotides complementary to a DNA template. 242

RNA transcript mRNA molecule formed during transcription that has a sequence of bases complementary to a gene. 242

rod cell Photoreceptor in vertebrate eyes that responds to dim light. 727

root cap Protective cover of the root tip, whose cells are constantly replaced as they are ground off when the root pushes through rough soil particles. 444

root hair Extension of a root epidermal cell that collectively increases the surface area for the absorption of water and minerals. 441, 464

root nodule (NOD yule) Structure on plant root that contains nitrogen-fixing bacteria. 446, 464

root pressure Osmotic pressure caused by active movement of mineral into root cells; serves to elevate water in xylem for a short distance. 468

root system Includes the main root and any and all of its lateral (side) branches. 438

rotational equilibrium Maintenance of balance when the head and body are suddenly moved or rotated. 732

rotifer Member of phylum Rotifera; rotifers are primarily freshwater organisms. 534

rough ER (endoplasmic reticulum) (in doe PLAZZ mick ruh TICK you lumm) Membranous system of tubules, vesicles, and sacs in cells; has attached ribosomes. 70

roundworm Member of the phylum Nematoda with a cylindrical body that has a complete digestive tract and a pseudocoelom; some forms are free-living in water and soil, and many are parasitic. 533

r-selection Favorable life history strategy under certain environmental conditions; characterized by a high reproductive rate with little or no attention given to offspring survival. 849

RuBP carboxylase (car BOX ill laze) An enzyme that starts the Calvin cycle reactions by catalyzing attachment of the carbon atom from CO_2 to RuBP. 124

S

saccule (SACK yule) Saclike cavity in the vestibule of the vertebrate inner ear; contains sensory receptors for gravitational equilibrium. 733

sac fungi Members of the phylum Ascomycota. 402

salivary amylase (SAL lih vair ree AM uh laze) In humans, enzyme in saliva that digests starch to maltose. 659, 664

salivary gland In humans, gland associated with the mouth that secretes saliva. 659

salt Ionic compound that results from a classical acid-base reaction. 24

saltatory conduction (SALT tuh tore ree) Movement of nerve impulses from one neurolemmal node to another along a myelinated axon. 705

saprotroph (SAP pro trofe) Organism that secretes digestive enzymes and absorbs the resulting nutrients back across the plasma membrane. 370

sarcolemma (sark oh LIMM uh) Plasma membrane of a muscle fiber; also forms the tubules of the T system involved in muscular contraction. 748

sarcomere (SARK oh meer) One of many units, arranged linearly within a myofibril, whose contraction produces muscle contraction. 748

sarcoplasmic reticulum (sark oh PLAZZ mick ruh TICK you lumm) Smooth endoplasmic reticulum of skeletal muscle cells; surrounds the myofibrils and stores calcium ions. 748

saturated fatty acid Fatty acid molecule that lacks double bonds between the carbons of its hydrocarbon chain. The chain bears the maximum number of hydrogens possible. 42

savanna (suh VANN uh) Terrestrial biome that is a grassland in Africa, characterized by few trees and a severe dry season. 912

scale In fishes and reptiles, a thin flake; scales cover the body and offer protection. 562

schistosomiasis (skiss toe so MY uh sis) Disease caused by the blood fluke, a parasitic flatworm of the phylum Platyhelminthes. 530

schizocoelom (skits oh SEE lumm) In protostomes, coelom formed by splitting of the embryonic mesoderm. 538

Schwann cell Cell that surrounds a fiber of a peripheral nerve and forms the myelin sheath. 703

scientific process Process by which scientists formulate a hypothesis, gather data by observation and experimentation, and come to a conclusion. 10

scientific theory Concept supported by a broad range of observations, experiments, and data. 11

sclera (SKLARE uh) White, fibrous, outer layer of the eyeball. 724

sclerenchyma (skluh RINK ih muh) Plant tissue composed of cells with heavily lignified cell walls; functions in support. 442

scolex (SCOLE lex) Tapeworm head region; contains hooks and suckers for attachment to host. 531

sea star Invertebrate member of the phylum Echinodermata; characterized by water vascular system and tube feet (also called starfish). 554

seaweed Multicellular forms of red, green, and brown algae found in marine habitats. 385

secondary oocyte (OH oh site) In oogenesis, the functional product of meiosis I; becomes the egg. 177

secondary sex characteristic Trait that is sometimes helpful but not absolutely necessary for reproduction and is maintained by the sex hormones in males and females. 781

second messenger Chemical signal such as cyclic AMP that causes the cell to respond to the first messenger—a hormone bound to plasma membrane receptor protein. 758

secretion (suh KREE shun) Release of a substance by exocytosis from a cell that may be a gland or part of a gland. 70

sedimentation (sed ih men TAY shun) Process by which particulate material accumulates and forms a stratum. 322

seed Mature ovule that contains an embryo, with stored food enclosed in a protective coat. 415, 496

seed germination Beginning of plant growth. 506

segmentation (seg men TAY shun) Repetition of body units as seen in the earthworm. 518

semen (seminal fluid) (SEE men, SIMM in uhl) Thick, whitish fluid consisting of sperm and secretions from several glands of the male reproductive tract. 778

semicircular canal One of three half-circle-shaped canals of the vertebrate inner ear; contains sensory receptors for rotational equilibrium. 730

semiconservative replication Duplication of DNA resulting in two double helix molecules, each having one parental and one new strand. 230

semilunar valve Valve resembling a half moon located between the ventricles and their attached vessels. 618

seminiferous tubule (seh men IF furr us TUBE yule) Long, coiled structure contained within chambers of the testis where sperm are produced. 781

senescence (seh NESS sense) Sum of the processes involving aging, decline, and eventual death of a plant or plant part. 486

sensory (afferent) neuron Nerve cell that transmits nerve impulses to the central nervous system after a sensory receptor has been stimulated. 703

sensory receptor Structure that receives either external or internal environmental stimuli and is a part of a sensory neuron or transmits signals to a sensory neuron. 699

sepal (SEE pull) Outermost, sterile, leaflike covering of the flower; usually green in color. 429, 496

septate (SEPP tate) Having cell walls; some fungal species have hyphae that are septate. 399

septum (SEPP tum) Partition or wall that divides two areas; the septum in the heart separates the right half from the left half. 618

serotonin A neurotransmitter. 706

serum Light yellow liquid left after clotting of blood. 628

sessile (SESS isle) Tending to stay in one place. 518

sessile filter feeder Animal that stays in one place and filters small food particles from the water (e.g., sponge). 522

seta (pl., setae) (SEE tuh, SEE tee) A needlelike, chitinous bristle in annelids, arthropods, and others. 544

sex chromosome Chromosome that determines the sex of an individual; in humans, females have two X chromosomes, and males have an X and a Y chromosome. 204

sex pilus (pl., sex pili) (PIE luss, PIE lie) In a bacterium, elongated, hollow appendage used to transfer DNA to other cells. 63

sexual reproduction Reproduction involving meiosis, gamete formation, and fertilization; produces offspring with chromosomes inherited from each parent with a unique combination of genes. 168

sexual selection Changes in males and females, often due to male competition and female selectivity, leading to increased fitness. 305, 826

shoot apical meristem (AY pick uhl MARE ih stem) Group of actively dividing embryonic cells at the tips of plant shoots. 448

shoot system Aboveground portion of a plant consisting of the stem, leaves, and flowers. 438

short-day plant Plant that flowers when day length is shorter than a critical length—e.g., cocklebur, poinsettia, and chrysanthemum. 488

shrubland Arid terrestrial biome characterized by shrubs and tending to occur along coasts that have dry summers and receive most of their rainfall in the winter. 912

sieve-tube member Member that joins with others in the phloem tissue of plants as a means of transport for nutrient sap. 443, 466

signal A molecule that stimulates or inhibits a metabolic event. 151

simple goiter Condition in which an enlarged thyroid produces low levels of thyroxine. 763

sink population Population that is found in an unfavorable area where at best the birthrate equals the death rate; sink populations receive new members from source populations. 939

sister chromatid (CROW muh tid) One of two genetically identical chromosomal units that are the result of DNA replication and are attached to each other at the centromere. 153

skeletal muscle Striated, voluntary muscle tissue that comprises skeletal muscles; also called striated muscle. 602

skin Outer covering of the body; can be called the integumentary system because it contains organs such as sense organs. 605

sliding filament model An explanation for muscle contraction based on the movement of actin filaments in relation to myosin filaments. 748

small intestine In vertebrates, the portion of the digestive tract that precedes the large intestine; in humans, consists of the duodenum, jejunum, and ileum. 661

smooth (visceral) muscle Nonstriated, involuntary muscles found in the walls of internal organs. 602

smooth ER (endoplasmic reticulum) (in doe PLAZZ mick ruh TICK cue lumm) Membranous system of tubules, vesicles, and sacs in eukaryotic cells; lacks attached ribosomes. 70

society Group in which members of species are organized in a cooperative manner, extending beyond sexual and parental behavior. 831

sociobiology (so see oh by AH low jee) Application of evolutionary principles to the study of social behavior of animals, including humans. 829

sodium-potassium pump Carrier protein in the plasma membrane that moves sodium ions out of and potassium ions into animal cells; important in nerve and muscle cells. 92

soil Accumulation of inorganic rock material and organic matter that is capable of supporting the growth of vegetation. 462

soil erosion Movement of topsoil to a new location due to the action of wind or running water. 463

soil profile Vertical section of soil from the ground surface to the unaltered rock below. 463

solute (SAHL yute) Substance that is dissolved in a solvent, forming a solution. 27, 89

solution Fluid (the solvent) that contains a dissolved solid (the solute). 27, 89

solvent (SAHL vent) Liquid portion of a solution that serves to dissolve a solute. 89

somatic cell (so MAT tick) Body cell; excludes cells that undergo meiosis and become sperm or egg. 151

somatic embryo Plant cell embryo that is asexually produced through tissue culture techniques. 509

somatic system Portion of the peripheral nervous system containing motor neurons that control skeletal muscles. 713

source population Population that can provide members to other populations of the species because it lives in a favorable area, and the birthrate is most likely higher than the death rate. 939

specialist species Species that have a narrow range of niches, such as limited diet, narrow environmental tolerances, and specific habitat (e.g., pandas and spotted owls). 863

speciation (spee see AY shun) Origin of new species due to the evolutionary process of descent with modification. 310

species Group of similarly constructed organisms capable of interbreeding and producing fertile offspring; organisms that share a common gene pool; the taxon at the lowest level of classification. 5, 8, 344

specific epithet (spuh SIFF ick EPP pih thett) In the binomial system of taxonomy, the second part of an organism's name; it may be descriptive. 342

sperm Male gamete having a haploid number of chromosomes and the ability to fertilize an egg, the female gamete. 781

spermatogenesis (sperm mat oh JENN uh sis) Production of sperm in males by the process of meiosis and maturation. 177

sphygmomanomter (sfig moh mah NAHM met turr) Device consisting of inflatable cuff and pressure gauge for measuring arterial blood pressure. 623

spicule (SPICK yule) Skeletal structure of sponges composed of calcium carbonate or silicate. 523

spinal cord In vertebrates, the nerve cord that is continuous with the base of the brain and housed within the vertebral column. 708

spinal nerve Nerve that arises from the spinal cord. 712

spindle Microtubule structure that brings about chromosomal movement during nuclear division. 154

spirillum (pl., spirilla) (spy RILL lumm) Long, rod-shaped bacterium that is twisted into a rigid spiral; if the spiral is flexible rather than rigid, it is called a spirochete. 62

spirochete (SPY roe keet) Long, rod-shaped bacterium that is twisted into a flexible spiral; if the spiral is rigid rather than flexible, it is called a spirillum. 62

spleen Large, glandular organ located in the upper left region of the abdomen; stores and purifies blood. 636

sponge Invertebrate animal of the phylum Porifera; pore-bearing filter feeder whose inner body wall is lined by collar cells. 522

spongin Collagenous fibers found in the body wall of sponges. 523

spongy bone Type of bone that has an irregular, meshlike arrangement of thin plates of bone. 601, 740

spongy mesophyll (MESS oh fill) Layer of tissue in a plant leaf containing loosely packed cells, increasing the amount of surface area for gas exchange. 454

sporangium (pl., sporangia) (spore RAN jee uhm) Structure that produces spores. 393, 400, 419

spore Asexual reproductive or resting cell capable of developing into a new organism without fusion with another cell, in contrast to a gamete. 174, 399, 416

sporophyll Modified leaf that bears a sporangium or sporangia. 421

sporophyte (SPORE oh fite) Diploid generation of the alternation of generations life cycle of a plant; produces haploid spores that develop into the haploid generation. 176, 416, 496

sporozoan (spore oh ZOH uhn) Spore-forming protist that has no means of locomotion and is typically a parasite with a complex life cycle having both sexual and asexual phases. 392

spring overturn Mixing process that occurs in spring in stratified lakes whereby oxygen-rich top waters mix with nutrient-rich bottom waters. 917

squamous epithelium (SQUAY muss epp pih THEE lee uhm) Type of epithelial tissue that contains flat cells. 598

stabilizing selection Outcome of natural selection in which extreme phenotypes are eliminated and the average phenotype is conserved. 307

stamen (STAY men) In flowering plants, the portion of the flower that consists of a filament and an anther containing pollen sacs where pollen is produced. 429, 496

starch Storage polysaccharide found in plants that is composed of glucose molecules joined in a linear fashion with few side chains. 40

statolith (STAT oh lith) Sensors found in root cap cells that cause a plant to demonstrate gravitropism. 479

stem Usually the upright, vertical portion of a plant that transports substances to and from the leaves. 439

stereoscopic vision Vision characterized by depth perception and three-dimensionality. 724

steroid (STARE oid) Type of lipid molecule having a complex of four carbon rings—e.g., cholesterol, estrogen, progesterone, and testosterone. 44

steroid hormone Type of hormone that has the same complex of four carbon rings, but each one has different side chains. 758

stigma (STIG muh) In flowering plants, portion of the carpel where pollen grains adhere and germinate before fertilization can occur. 429, 496

stolon (STOLE uhn) Stem that grows horizontally along the ground and may give rise to new plants where it contacts the soil—e.g., the runners of a strawberry plant. 452

stoma (pl., stomata) (STOME muh, stoh MAH tuh) Small opening between two guard cells on the underside of leaf epidermis through which gases pass. 116, 417, 441

stomach In vertebrates, muscular sac that mixes food with gastric juices to form chyme, which enters the small intestine. 660

stratum (STRAY tum) Ancient layer of sedimentary rock; results from slow deposition of silt, volcanic ash, and other materials. 322

striated (STRY ate ted) Having bands; in cardiac and skeletal muscle, alternating light and dark bands produced by the distribution of contractile proteins. 602

strobilus (stroh BILL us) In club mosses, terminal clusters of leaves that bear sporangia. 421

stroke Condition resulting when an arteriole in the brain bursts or becomes blocked by an embolism; cerebrovascular accident. 625

stroma (STROH muh) Fluid within a chloroplast that contains enzymes involved in the synthesis of carbohydrates during photosynthesis. 74, 116

stromatolite (stroh MAT oh lite) Domed structure found in shallow seas consisting of cyanobacteria bound to calcium carbonate. 324

structural gene Gene that codes for an enzyme in a metabolic pathway. 252

style Elongated, central portion of the carpel between the ovary and stigma. 429, 496

subcutaneous layer A sheet that lies just beneath the skin and consists of loose connective and adipose tissue. 605

substrate Reactant in a reaction controlled by an enzyme. 106

substrate-level phosphorylation (foss for ill LAY shun) Process in which ATP is formed by transferring a phosphate from a metabolic substrate to ADP. 134

surface-area-to-volume ratio Ratio of a cell's outside area to its internal volume. 59

survivorship Probability of newborn individuals of a cohort surviving to particular ages. 840

sustainable development Management of an ecosystem so that it maintains itself while providing services to human beings. 941

swamp Wet, spongy land that is saturated and sometimes partially or intermittently covered with water. 915

sweat gland Skin gland that secretes a fluid substance for evaporative cooling; sudoriferous gland. 607

swim bladder In fishes, a gas-filled sac whose pressure can be altered to change buoyancy. 565

symbiosis Relationship that occurs when two different species live together in a unique way; it may be beneficial, neutral, or detrimental to one and/or the other species. 870

symbiotic *See symbiosis.* 370

sympathetic division Division of the autonomic system that is active when an organism is under stress; uses norepinephrine as a neurotransmitter. 715

sympatric speciation (simm PAT trick spee see AY shun) Origin of new species in populations that overlap geographically. 312

synapse (SIN naps) Junction between neurons consisting of the presynaptic (axon) membrane, the synaptic cleft, and the postsynaptic (usually dendrite) membrane. 706

synapsis (sin NAP sis) Pairing of homologous chromosomes during meiosis I. 169

synaptic cleft (sin NAP tick) Small gap between presynaptic and postsynaptic membranes of a synapse. 706

syndrome (SIN drome) Group of symptoms that appear together and tend to indicate the presence of a particular disorder. 214

synovial joint (sin OH vee uhl) Freely moving joint in which two bones are separated by a cavity. 745

systematics (sis tim MAT ticks) Study of the diversity of organisms to classify them and determine their evolutionary relationships. 346

systemic circuit (sis TIM mick SIR kit) Circulatory pathway of blood flow between the tissues and the heart. 617

systole (SIS toe lee) Contraction period of the heart during the cardiac cycle. 620

T

taiga (TIE guh) Terrestrial biome that is a coniferous forest extending in a broad belt across northern Eurasia and North America. 907

taproot Main axis of a root that penetrates deeply and is used by certain plants (such as carrots) for food storage. 446

taste bud Structure in the vertebrate mouth containing sensory receptors for taste; in humans, most taste buds are on the tongue. 722

taxon (pl., taxa) (TAX ahn, TAX uh) Group of organisms that fills a particular classification category. 344

taxonomy (tax AH no mee) Branch of biology concerned with identifying, describing, and naming organisms. 8, 342

telomere (TELL oh meer) Tip of the end of a chromosome that shortens with each cell division and may thereby regulate the number of times a cell can divide. 160

telophase (TELL oh faze) Mitotic phase during which daughter cells are located at each pole. 156

temperate deciduous forest (TIM purr utt duh SIDD you us) Forest found south of the taiga; characterized by deciduous trees such as oak, beech, and maple, moderate climate, relatively high rainfall, stratified plant growth, and plentiful ground life. 907

temperate rain forest Coniferous forest—e.g., that running along the west coast of Canada and the United States—characterized by plentiful rainfall and rich soil. 908

template (TEM plate) Parental strand of DNA that serves as a guide for the complementary daughter strand produced during DNA replication. 230

tendon Strap of fibrous connective tissue that connects skeletal muscle to bone. 600, 747

terminal bud Bud that develops at the apex of a shoot. 448

territoriality Marking and/or defending a particular area against invasion by another species member; area often used for the purpose of feeding, mating, and caring for young. 826

territory Area occupied and defended exclusively by an animal or group of animals. 826

test Loose-fitting shell of a foraminiferan or a radiolarian; made of calcium carbonate or silicon, respectively. 390

testcross Cross between an individual with the dominant phenotype and an individual with the recessive phenotype. The resulting phenotypic ratio indicates whether the dominant phenotype is homozygous or heterozygous. 188

testes (sing., testis) (TEST tiss, TEST teez) Male gonad that produces sperm and the male sex hormones. 770, 776

testosterone (test TOSS turr ohn) Male sex hormone that helps maintain sexual organs and secondary sex characteristics. 770, 781

tetanus (TETT uh nuss) Sustained muscle contraction without relaxation. 747

tetany (TETT uh nee) Severe twitching caused by involuntary contraction of the skeletal muscles due to a calcium imbalance. 764

tetrapod (TETT truh pod) Four-footed vertebrate; includes amphibians, reptiles, birds, and mammals. 565

thalamus (THAL uh muss) In vertebrates, the portion of the diencephalon that passes on selected sensory information to the cerebrum. 710

therapeutic cloning Used to create mature cells of various cell types. Also, used to learn about specialization of cells and provide cells and tissue to treat human illnesses. 157

therapsid (thurr RAP sid) Mammal like reptiles appearing in the middle Permian period; ancestral to mammals. 568

thermal inversion Temperature inversion that traps cold air and its pollutants near the Earth, with the warm air above it. 893

thermoacidophile (therm moh uh SIDD oh file) Type of archaea that lives in hot, acidic, aquatic habitats, such as hot springs or near hydrothermal vents. 375

thigmotropism (thig MAH troh piz uhm) In plants, unequal growth due to contact with solid objects, as the coiling of tendrils around a pole. 480

threatened species Species that is likely to become an endangered species in the foreseeable future (e.g., bald eagle, gray wolf, Louisiana black bear). 928

thrombin (THROMM bin) Enzyme that converts fibrinogen to fibrin threads during blood clotting. 628

thylakoid (THIGH luh koid) Flattened sac within a granum whose membrane contains chlorophyll and where the light reactions of photosynthesis occur. 62, 74, 117

thymine (T) (THIGH men) One of four nitrogen-containing bases in nucleotides composing the structure of DNA; pairs with adenine. 227

thymus gland (THIGH muss) Lymphoid organ involved in the development and functioning of the immune system; T lymphocytes mature in the thymus gland. 635, 770

thyroid gland (THIGH roid) Large gland in the neck that produces several important hormones, including thyroxine, triiodothyronine, and calcitonin. 763

thyroid-stimulating hormone (TSH) Substance produced by the anterior pituitary that causes the thyroid to secrete thyroxine and triiodothyronine. 760

thyroxine (T_4) (thigh ROCKS sin) Hormone secreted from the thyroid gland that promotes growth and development; in general, it increases the metabolic rate in cells. 763

tight junction Junction between cells when adjacent plasma membrane proteins join to form an impermeable barrier. 96

tissue Group of similar cells combined to perform a common function. 598

tissue culture Process of growing tissue artificially, usually in a liquid medium in laboratory glassware. 508

tissue fluid Fluid that surrounds the body's cells; consists of dissolved substances that leave the blood capillaries by filtration and diffusion. 601, 629

T lymphocyte (T cell) (LIMM foe site) Lymphocyte that matures in the thymus and exists in four varieties, one of which kills antigen-bearing cells outright. 635

tone Continuous, partial contraction of muscle. 747

tonicity (tone ISS ih tee) Osmolarity of a solution compared to that of a cell. If the solution is isotonic to the cell, there is no net movement of water; if the solution is hypotonic, the cell gains water; and if the solution is hypertonic, the cell loses water. 90

tonsils Partially encapsulated lymph nodules located in the pharynx. 636, 680

totipotent (toe TIP uh tent) Cell that has the full genetic potential of the organism, including the potential to develop into a complete organism. 508, 802

tracer Substance having an attached radioactive isotope that allows a researcher to track its whereabouts in a biological system. 22

trachea (pl., tracheae) (TRAY kee uh, TRAY kee ee) In insects, air tubes located between the spiracles and the tracheoles. In tetrapod vertebrates, air tube (windpipe) that runs between the larynx and the bronchi. 547, 674, 676

tracheid (TRAY kee id) In vascular plants, type of cell in xylem that has tapered ends and pits through which water and minerals flow. 443, 466

tract Bundle of myelinated axons in the central nervous system. 708

traditional systematics School of systematics that takes into consideration the degree of difference between derived characters to construct phylogenetic trees. 352

transcription Process whereby a DNA strand serves as a template for the formation of mRNA. 240

transcription activator Protein that speeds transcription. 258

transcriptional control Control of gene expression during the transcriptional phase determined by mechanisms that control whether transcription occurs or the rate at which it occurs. 258

transcription factor In eukaryotes, protein required for the initiation of transcription by RNA polymerase. 258

transduction (trans DUCK shun) Exchange of DNA between bacteria by means of a bacteriophage. 369

trans fatty acid Form of an unsaturated fatty acid, usually a monounsaturated one when found in food. Stick margarine, shortenings, and deep-fat fried foods in general are rich sources. 666

transfer rate Amount of a substance that moves from one component of the environment to another within a specified period of time. 889

transfer RNA (tRNA) Type of RNA that transfers a particular amino acid to a ribosome during protein synthesis; at one end, it binds to the amino acid, and at the other end it has an anticodon that binds to an mRNA codon. 240

transformation Taking up of extraneous genetic material from the environment by bacteria. 369

transgenic organism (trans JENN ick) Free-living organism in the environment that has had a foreign gene inserted into it. 268

transgenic plant Plant that carries the genes of another organism as a result of DNA technology; also genetically modified plant. 510

translation Process whereby ribosomes use the sequence of codons in mRNA to produce a polypeptide with a particular sequence of amino acids. 240

translational control Gene expression regulated by the activity of mRNA transcripts. 259

translocation (trans low KAY shun) Movement of a chromosomal segment from one chromosome to another nonhomologous chromosome, leading to abnormalities—e.g., Down syndrome. 218

transpiration Plant's loss of water to the atmosphere, mainly through evaporation at leaf stomata. 469

transposon (trans POSE ahn) DNA sequence capable of randomly moving from one site to another in the genome. 259

trematode Member of the phylum Platyhelminthes (flatworm); ectoparasitic or endoparasitic flukes. 528

trichinosis (trick in OH sis) Serious infection caused by parasitic roundworm of the phylum Nematoda whose larvae encyst in muscles. 533

trichocyst (TRICK oh sist) Found in ciliates; contains long, barbed threads useful for defense and capturing prey. 391

trichomes (TRY cohmz) In plants, specialized outgrowth of the epidermis (e.g, root hairs). 441

trichomoniasis (trih coh moh NIE uh sis) Sexually transmitted disease caused by the parasitic protozoan *Trichomonas vaginalis*. 793

triglyceride (try GLISS suh ride) Neutral fat composed of glycerol and three fatty acids. 42

triplet code During gene expression, each sequence of three nucleotide bases stands for a particular amino acid. 241

trisomy (try SO mee) Having three of a particular type of chromosome (2n + 1). 212

trophic level (TROFE ick) Feeding level of one or more populations in a food web. 887

trophic relationship In ecosystems, feeding relationships such as grazing food webs or detrital food webs. 887

trophoblast (TROFE oh blast) Outer membrane surrounding the embryo in mammals; when thickened by a layer of mesoderm, it becomes the chorion, an extraembryonic membrane. 808

tropical rain forest Biome near the equator in South America, Africa, and the Indo-Malay regions; characterized by warm weather, plentiful rainfall, a diversity of species, and mainly tree-living animal life. 910

tropism (TROPE iz uhm) In plants, a growth response toward or away from a directional stimulus. 478

trypanosome (try PAN uh sohm) Parasitic zooflagellate that causes severe disease in human beings and domestic animals, including a condition called sleeping sickness. 389

trypsin (TRIP sin) Protein-digesting enzyme secreted by the pancreas. 664

tube foot Part of the water vascular system in sea stars, located on the oral surface of each arm; functions in locomotion. 554

tube-within-a-tube body plan Body with a digestive tract that has both a mouth and an anus. 528

tubular reabsorption (TUBE yule lurr ree ab SORP shun) Movement of primarily nutrient molecules and water from the contents of the nephron into blood at the proximal convoluted tubule. 692

tubular secretion Movement of certain molecules from blood into the distal convoluted tubule of a nephron so that they are added to urine. 693

tumor Cells derived from a single mutated cell that has repeatedly undergone cell division; benign tumors remain at the site of origin, while malignant tumors metastasize. 159

tumor suppressor gene Gene that codes for a protein that ordinarily suppresses cell division; inactivity can lead to a tumor. 160

turbellarian Member of the phylum Platyhelminthes (flatworm); free-living aquatic planarians and their relatives. 528

turgor pressure (TURR gurr) Pressure of the cell contents against the cell wall; in plant cells, determined by the water content of the vacuole and provides internal support. 91

tympanic membrane (tim PAN ick) Membranous region that receives air vibrations in an auditory organ; in humans, the eardrum. 730

typhlosole (TIFE low sole) Expanded dorsal surface of long intestine of earthworms, allowing additional surface for absorption. 545, 656

U

umbilical cord (uhm BILL lick cull) Cord connecting the fetus to the placenta through which blood vessels pass. 810

unicellular (you nih SELL you lurr) Made up of but a single cell, as in the bacteria. 2

unifactorial One gene consisting of a single pair of alleles, one dominant and one recessive. 196

uniformitarianism (you nih form ih TARE ree uhn iz uhm) Belief espoused by James Hutton that geological forces act at a continuous, uniform rate. 285

uniramians (you nih RAM me un) Members of animal phylum Arthropoda that includes centipedes, millipeds, and insects. 550

unsaturated fatty acid Fatty acid molecule that has one or more double bonds between the carbons of its hydrocarbon chain. The chain bears fewer hydrogens than the maximum number possible. 42

upwelling Upward movement of deep, nutrient-rich water along coasts; it replaces surface waters that move away from shore when the direction of prevailing wind shifts. 920

urea (you REE uh) Main nitrogenous waste of terrestrial amphibians and most mammals. 688

ureter (you REE turr) Tubular structure conducting urine from the kidney to the urinary bladder. 690

urethra (you REE thruh) Tubular structure that receives urine from the bladder and carries it to the outside of the body. 690

uric acid (YOUR rick) Main nitrogenous waste of insects, reptiles, and birds. 688

urinary bladder (YOUR rinn air ree) Organ where urine is stored. 690

urine Liquid waste product made by the nephrons of the vertebrate kidney through the processes of glomerular filtration, tubular reabsorption, and tubular secretion. 690

uterine cycle (YOU turr rinn) Cycle that runs concurrently with the ovarian cycle; it prepares the uterus to receive a developing zygote. 784

uterus (YOU turr us) In mammals, expanded portion of the female reproductive tract through which eggs pass to the environment or in which an embryo develops and is nourished before birth. 782

utricle (YOU trick cull) Cavity in the vestibule of the vertebrate inner ear; contains sensory receptors for gravitational equilibrium. 733

V

vaccine (vax SEEN) Antigens prepared in such a way that they can promote active immunity without causing disease. 647, 945

vacuole (VAC you ohl) Membrane-bounded sac, larger than a vesicle; usually functions in storage and can contain a variety of substances. In plants, the central vacuole fills much of the interior of the cell. 73

vascular bundle (VASS cue lurr) In plants, primary phloem and primary xylem enclosed by a bundle sheath. 443

vascular cambium (VASS cue lurr CAMM bee uhm) In plants, lateral meristem that produces secondary phloem and secondary xylem. 448

vascular cylinder In eudicots, the tissues in the middle of a root, consisting of the pericycle and vascular tissues. 443

vascular tissue Transport tissue in plants, consisting of xylem and phloem. 414, 441

vector (VECK turr) In genetic engineering, a means to transfer foreign genetic material into a cell—e.g., a plasmid. 268

vein Blood vessel that arises from venules and transports blood toward the heart. 616

vena cava (VEE nuh CAVE uh) Large systemic vein that returns blood to the right atrium of the heart in tetrapods; either the superior or inferior vena cava. 622

ventilation (venn tih LAY shun) Process of moving air into and out of the lungs; breathing. 672

ventricle (VENT trih cull) Cavity in an organ, such as a lower chamber of the heart or the ventricles of the brain. 618, 708

venule (VENN yule) Vessel that takes blood from capillaries to a vein. 617

vermiform appendix Small, tubular appendage that extends outward from the cecum of the large intestine. 636

vertebral column (VERT tih brull) Portion of the vertebrate endoskeleton that houses the spinal cord; consists of many vertebrae separated by intervertebral disks. 743

vertebrate (VERT tih brate) Chordate in which the notochord is replaced by a vertebral column. 518, 562

vesicle (VESS sick cull) Small, membrane-bounded sac that stores substances within a cell. 70

vessel element Cell that joins with others to form a major conducting tube found in xylem. 443, 466

vestibule (VESS tibb yule) Space or cavity at the entrance to a canal, such as the cavity that lies between the semicircular canals and the cochlea. 730

vestigial structure (vest TIH jee uhl) Remains of a structure that was functional in some ancestor but is no longer functional in the organism in question. 296

villus (VILL us) Small, fingerlike projection of the inner small intestinal wall. 661

viroid (VYE roid) Infectious strand of RNA devoid of a capsid and much smaller than a virus. 367

virulent (VEER uh lent) Pathogenicity of an organism as indicated by ability to invade host tissues and cause disease. 364

virus Noncellular parasitic agent consisting of an outer capsid and an inner core of nucleic acid. 362

visible light Portion of the electromagnetic spectrum that is visible to the human eye. 118

visual accommodation Ability of the eye to focus at different distances by changing the curvature of the lens. 726

vitamin Essential requirement in the diet, needed in small amounts. Vitamins are often part of coenzymes. 109, 666

vitamin D Fat-soluble compound; deficiency tends to cause rickets in children. 606

viviparous (vie VIP purr us) Animal that gives birth after partial development of offspring within mother. 777

vocal cord In humans, fold of tissue within the larynx; creates vocal sounds when it vibrates. 676

W

water (hydrologic) cycle (high droh LAH jick) Interdependent and continuous circulation of water from the ocean, to the atmosphere, to the land, and back to the ocean. 889

water mold Filamentous organisms having cell walls made of cellulose; typically decomposers of dead freshwater organisms, but some are parasites of aquatic or terrestrial organisms. 394

water potential Potential energy of water; a measure of the capability to release or take up water relative to another substance. 467

water vascular system Series of canals that takes water to the tube feet of an echinoderm, allowing them to expand. 554

wax Sticky, solid, waterproof lipid consisting of many long-chain fatty acids usually linked to long-chain alcohols. 45

wetland Wet area. (*See also bog or swamp.*) 915

whisk fern Common name for seedless vascular plant that consists only of stems and has no leaves or roots. 422

white blood cell Leukocyte, of which there are several types, each having a specific function in protecting the body from invasion by foreign substances and organisms. 601, 626

white matter Myelinated axons in the central nervous system. 708

wild type Phenotype or genotype that is characteristic of the majority of individuals of a species in a natural environment. 189

wobble hypothesis Ability of the 5'-most nucleotide of an anticodon to interact with more than one nucleotide at the 3'-end of codons; helps explain the degeneracy of the genetic code. 244

wood Secondary xylem that builds up year after year in woody plants and becomes the annual rings. 450

X

X-linked Allele that is located on an X chromosome but may control a trait that has nothing to do with the sexual characteristics of an animal. 204

xylem (ZIE lumm) Vascular tissue that transports water and mineral solutes upward through the plant body; it contains vessel elements and tracheids. 420, 442, 466

Y

yeast Unicellular fungus that has a single nucleus and reproduces asexually by budding or fission, or sexually through spore formation. 403

yolk Dense nutrient material in the egg of a bird or reptile. 777, 799

yolk sac One of the extraembryonic membranes that, in shelled vertebrates, contains yolk for the nourishment of the embryo, and in placental mammals is the first site for blood cell formation. 807

Z

zero population growth No growth in population size. 852

zooflagellate (zoh oh FLAJ jell ate) Nonphotosynthetic protist that moves by flagella; typically zooflagellates enter into symbiotic relationships, and some are parasitic. 389

zooplankton (zoe oh PLANK ton) Part of plankton containing protozoans and other types of microscopic animals. 390, 917

zygospore (ZIE go spore) Thick-walled resting cell formed during sexual reproduction of zygospore fungi. 382, 400

zygospore fungi Members of the phylum Zygomycota. 400

zygote (ZIE goat) Diploid cell formed by the union of two gametes; the product of fertilization. 168

CREDITS

Associates/Photo Researchers, Inc.; **9.9a-c:** © Stanley C. Holt/Biological Photo Service.

Chapter 10

Opener: © Y. Nikas/Photo Researchers, Inc.; **10.1:** © L. Willatt/Photo Researchers, Inc.; **10.3a:** Courtesy of Dr. D. Von Wettstein; **10.5:** © American Images, Inc./Getty Images; **10.6a-j:** © Ed Reschke.

Chapter 11

Opener: © Jorg & Petra Wegner/Animals Animals/Earth Scenes; **11.1:** © National Geographic Image Sales; **11.12:** © Pat Pendarvis; **11.13a(both):** © Steve Uzzell; **11.13b(both):** Courtesy Jane S. Paulsen/ University of Iowa; **11.16:** Courtesy University of Connecticut, Peter Morenus, photographer; **11.18:** © Jane Burton/Bruce Coleman, Inc.

Chapter 12

Opener: © Andrew Syred/Photo Researchers, Inc.; **12.5(right):** © Huton Archive/Getty Images; **12.5(left):** © Stapleton Collection/ Corbis; **12A:** From R. Simensen and R. Curtis Rogers, "Fragile X Syndrome," AMERICAN FAMILY PHYSICIAN 39(5):186, May 1989. © American Academy of Family Physicians; **12Ab:** © David M. Phillips/Visuals Unlimited; **12Ad:** © CNRI/SPL/Photo Researchers, Inc.; **12.6:** © Ned Seidler National Geographic Image Collection; **12.11a:** © Jose Carrilo/ PhotoEdit; **12.11b:** © CNRI/SPL/Photo Researchers, Inc.; **12Bc:** © James King-Holmes/SPL/Photo Researchers, Inc.; **12.12a(background):** UNC Medical Illustration and Photography; **12.12a(foreground):** © CNRI/SPL/Photo Researchers, Inc.; **12.12b(background):** Courtesy Robert H. Shelton/Klinefelter Syndrome & Associates; **12.12b(foreground):** © CNRI/Photo Researchers, Inc.; **12.15:** Courtesy The Williams Syndrome Association; **12.16b:** From N.B. Spinner et al., AMERICAN JOURNAL OF HUMAN GENETICS 55 (1994):p. 239. The University of Chicago Press.

Chapter 13

Opener: © Digital Art/Corbis; **13.2b:** © Eye of Science/Photo Researchers, Inc.; **13.5b:** © Science Source/Photo Researchers, Inc.; **13.6a:** © Kenneth Eward/Photo Researchers, Inc.; **13.6b:** © A. Barrington Brown/ Photo Researchers, Inc.; **13.6e:** © Photo Researchers, Inc.

Chapter 14

Opener(Leopard): © James Martin/Stone/Getty Images; **(Hibiscus):** © Rosemary Calvert/Stone/ Getty Images; **(Vorticella):** © A.M. Siegelman/ Visuals Unlimited; **(Crab):** © Tui DeRoy/Bruce Coleman; **14.2a(left):** © SPL/Photo Researchers, Inc.; **14.2a(right):** © Dr. Gopal Murti/Photo Researchers, Inc.; **14.7a:** © Oscar L. Miller/Photo Researchers, Inc.; **14.9b:** Courtesy University of California Lawrence Livermore National Library and the U.S. Department of Energy; **14.10d:** Courtesy Alexander Rich.

Chapter 15

Opener: © SPL/Photo Researchers, Inc.; **15.5b:** Courtesy Stephen Wolfe; **15.6:** © Chanan Photo

2004; **15.7a:** © J. L. Stone/Photo Researchers, Inc.; **15.7b:** From M.B. Roth and J.G. Gall /it/ Journal of Cell Biology, /xit/ 105:1047-1054, 1987. Reproduced by copyright permission of The Rockefeller University Press; **15A:** Courtesy Cold Spring Harbor Archive; **15B(left):** © David Young-Wolf/PhotoEdit; **15B(right):** © John N.A. Lott/Biological Photo Service; **15.11:** Courtesy Dr. Howard Jones, Eastern VA Medical School; **15.13a:** © AP Photo/The Republic, Darron Cummings; **15.13b:** © Ken Greer/Visuals Unlimited.

Chapter 16

Opener: © Eye of Science/Photo Researchers, Inc.; **16.4(both):** Courtesy General Electric Research & Development.; **16.6a(top):** © Digital Vision/Getty Images; **16.6a(bottom):** © Royalty Free/Getty Images; **16.6b(right):** © Digital Vision/Getty Images; **16.6b(left):** © The McGraw-Hill Companies, Inc./Bob Coyle, Photographer; **16.6c(left):** © Digital Vision/Getty Images; **16.6c(right):** © Royal Free/Getty Images; **16.7:** © Affymetrix; **16.8:** © Cindy Charles/Photo Edit; **p. 274(left):** © A. Ramey/Photo Edit; **p. 274(right):** © Tek Image/Photo Researchers, Inc.; **p. 275(left):** © Ron Chapple/FPG/Getty; **p. 275(right):** © Camille Tokerud/Stone/Getty Images.

Chapter 17

Opener (Fossil skeleton): Courtesy Philip D. Gingerich/University of Michigan, Museum of Paleontology Exhibit Museum; **(Whale):** © Kelli Jaunsen/Eccoblue; **17.1b:** © Wolfgang Kaehler/ Corbis; **17.1c:** © C. Luiz Claudio Marigo/Peter Arnold; **17.1d:** © Gary J. James/Biological Photo Service; **17.1e:** © Charles Benes/Index Stock Imagery; **17.1f:** © Galen Rowell/Corbis; **17.1g:** © D. Parer & E. Parer-Cook/Ardea; **17.2:** © Carolina Biological/Visuals Unlimited; **17.3a:** © James H. Bailey/Field Museum of Natural History; **17.3b:** © Daryl Balfour/Photo Researchers, Inc.; **17.6:** © Juan & Carmecita Munoz/Photo Researchers, Inc.; **17.7a:** © Kevin Schafer/Corbis; **17.7b:** © Michael Dick/Animals Animals/Earth Scenes; **17.8a:** © Adrienne T. Gibson/Animals Animals/ Earth Scenes; **17.8b:** © Joe McDonald/Animals Animals/Earth Scenes; **17.8c:** © Leonard Lee Rue/Animals Animals/Earth Scenes; **17.9:** © Lisette Le Bon/SuperStock; **p. 289:** © Stock Montage; **17.10(Timber wolf):** © Gary Milburn/Tom Stack & Assoc.; **17.10(Red Chow):** © Jeanne White/Photo Researchers, Inc.; **17.10(Bloodhound):** © Mary Bloom/Peter Arnold, Inc.; **17.10(Dalmatian):** © Alexander Lowry/Photo Researchers, Inc.; **17.10(Boston Terrier):** © Robert Dowling/ Corbis; **17.10(Sheltie):** © Ralph Reinhold/Index Stock Imagery; **17.10(Beagle):** © Tim Davis/ Photo Researchers, Inc.; **17.10(Chihuahua):** © Kent & Donna Dannen/Photo Researchers, Inc.; **17.10(Scottie):** © Carolyn A. McKeone/ Photo Researchers, Inc.; **17.10(Sheepdog):** © Yann Arthus-Bertrand/Corbis; **17.10(Shih Tzu):** © Bob Shirtz/SuperStock; **17.10(Newfoundland):** © Robert Dowling/ Corbis; **17.10(Irish Wolfhound):** © Ralph Reinhold/Index Stock Imagery; **17.11a-c,b:** Courtesy W. Atlee Burpee Company; **17.12a:** © Jean-Claude Carton/Bruce Coleman Inc.; **17.12b:** © Joe Tucciarone; **17.14(Tasmanian wolf):** © Tom McHugh/Photo Researchers,

Inc.; **17.14(Kangaroo):** © George Holton/Photo Researchers, Inc.; **17.14(Sugar Glider):** © Ken Lucas/Visuals Unlimited; **17.14(Tasmanian):** © PhotoDisc Green/Getty; **17.14(Wombat):** © PhotoDisc Blue/Getty; **17.14(Dasyurus):** © Tom McHugh/Photo Researchers, Inc.; **17.17:** © J.G.M. Thewissen, http://darla.neoucom. edu/DEPTS/ANAT/Thewissen/.

Chapter 18

Opener: © Dr. Linda Stannard, Uct/Photo Researchers, Inc.; **18.2a,b:** © Michael Wilmer Forbes Tweedie/Photo Researchers, Inc., **18.3(Bardi):** © Visuals Unlimited; **18.3(Lindheimeri):** © Zig Leszczynski/Animals Animals/Earth Scenes;**18.3(Spiloides):** © Joseph Collins/Photo Researchers, Inc.; **18.3(Rossaleni):** © Dale Jackson/Visuals Unlimited; **18.3(Quadrivittate):** © Zig Leszczynski/Animals Animals/Earth Scenes; **18.3(Obsoleta):** © William Weber/Visuals Unlimited; **18.5:** Courtesy Victor McKusick; **18.7:** © Maier, Robert/Animals Animals/Earth Scenes; **18.8b:** © Bob Evans/Peter Arnold, Inc.; **18A:** Courtesy Gerald D. Carr; **18.8(both):** © Farley Bridges; **18.9:** © Dr. Gopal Murti/ Photo Researchers, Inc.; **18.10(Least):** © Stanley Maslowski/Visuals Unlimited; **18.10(Acadian):** © Karl Maslowski/Visuals Unlimited; **18.10(Traill's):** © Ralph Reinhold/Animals Animals/Earth Scenes; **18.11:** © Doug Wechsler/ Animals Animals/Earth Scenes.

Chapter 19

Opener: © George Ranalli/Photo Researchers, Inc.; **19.2:** © Ralph White/ Corbis; **19.3a:** © Science VU/Visuals Unlimited; **19.3b:** Courtesy Dr. David Deamer; **19.5:** © Henry W. Robinson/ Visuals Unlimited; **19.6(Trilobite):** © Francois Gohier/Photo Researchers, Inc.; **19.6(Tusks):** © AP Photo/Francis Latreille/ Nova Productions; **19.6(Placoderm):** © The Cleveland Museum of Natural History; **19.6(Fern):** © George Bernard/Natural History Photo Agency; **19.6(Petrified wood):** Courtesy National Park Service; **19.6(Ammonites):** © Sinclair Stammers/ SPL/Photo Researchers; **19.6(Scorpion):** © George O. Poinar; **19.6(Dinosaur track):** © Scott Berner/Visuals Unlimited; **19.6(Ichthyosaur):** © The Natural History Museum, London; **19.7a:** Courtesy J. William Schopf; **19.7b:** © Francois Gohier/Photo Researchers, Inc.; **19.8a:** Courtesy James G. Gehling, South Australian Museum; **19.8b:** Courtesy Dr. Bruce N. Runnegar; **19.9:** © D.W. Miller; **19.10a:** © The Field Museum; **19.10b:** © John Cancalosi/Peter Arnold, Inc.; **19A(both):** Courtesy Museum of the Rockies; **19.11:** © Chase Studio/Photo Researchers, Inc.; **19.12:** © Chase Studio/Photo Researchers, Inc.; **19.13:** © Gianni Dagli Orti/Corbis; **19.15:** © Matthew Shipp/Photo Researchers, Inc.

Chapter 20

Opener: © Michael Sewell/Peter Arnold, Inc.; **20.1(all):** © Sylvia S. Mader; **20.2a:** Courtesy Uppsala University Library, Sweden; **20.2(Bubil lily):** © Arthur Gurmankin/Visuals Unlimited; **20.2(Canada lily):** © Dick Poe/Visuals Unlimited; **20.3:** © Tim Davis/Photo Researchers, Inc.; **20.4:** © Jen & Des Bartlett/Bruce Coleman,

Inc.; **20.7:** Courtesy Dr. David Dilcher, Florida Museum of Natural History, University of Florida; **20.8(left):** © John D. Cunningham/Visuals Unlimited; **20.8(right):** © John Shaw/Tom Stack & Assoc.; **20Ab:** © Kjell B. Sandved/Visuals Unlimited; **20Ac:** © Brent Opell; **20.15(E. coli):** © David M. Phillips/Visuals Unlimited; **20.1(Methanosarcina):** © Ralph Robinson/Visuals Unlimited; **20.15(Black-eyed Susan):** © Ed Reschke/Peter Arnold, Inc.; **20.15(Paramecium):** © M. Abbey/Visuals Unlimited; **20.15(Mushroom):** © S. Gerig/Tom Stack & Associates; **20.15(Wolf):** © Art Wolf/Stone/Getty Images.

Chapter 21

Opener: © Dr. Gary Gaugler/Photo Researchers, Inc.; **21.1a:** © Dr. Hans Gelderblom/Visuals Unlimited; **21.1b:** © Eye of Science/Photo Researchers, Inc.; **21.1c:** © Dr. O. Bradfute/Peter Arnold, Inc.; 21.1d: © K.G. Murti/Visuals Unlimited; **21.2:** © Ed Degginger/Color Pic Inc.; 21.6(top): © RDF/Visuals Unlimited; **21.7a:** Harley W. Moon, U.S. Dept. of Agriculture; **21.7b:** Courtesy of C. Brinton, Jr.; **21.8:** © CNRI/SPL/Photo Researchers, Inc.; **21.9:** © Alfred Pasieka/SPL/Photo Researchers, Inc.; **21.10:** Courtesy Nitragin Company, Inc.; **21.11a:** © Dr. Richard Kessel & Dr. Gene Shih/Visuals Unlimited; 21.11b: © Gary Gaugler/Visuals Unlimited; 21.11c: © SciMAT/Photo Researchers, Inc.; 21.12a: © Michael Abbey/Photo Researchers, Inc.; 21.12b: © Tom Adams/Visuals Unlimited; 21A: © AP Photo/Kenneth Lambert; **21.13a:** © Jeff Lepore/Photo Researchers, Inc.; 21.13b: Courtesy Dennis W. Grogan, Univ. of Cincinnati; 21.13c: Courtesy Prof. Dr. Karl O. Stetter, Univ. Regensburg, Germany.

Chapter 22

Opener: © CBS/Phototake; **22.1c(left):** © E. White/Visuals Unlimited; **22.1c(right):** © Jerome Paulin/Visuals Unlimited; **22.2(Diatoms):** © M.I. Walker/Photo Researchers, Inc.; **22.2(Nonionina):** © Astrid & Hanns-Frieder Michler/Photo Researchers, Inc.; **22.2(Synura):** Courtesy Dr. Ronald W. Hoham; **22.2(Plasmodium):** © Patrick W. Grace/Photo Researchers, Inc.; **22.2(Blepharisma):** © Eric Grave/Photo Researchers, Inc.; **22.2(Onychodromus):** Courtesy Dr. Barry Wicklow; **22.2(Ceratium):** © D.P. Wilson/Photo Researchers, Inc.; **22.2(Licmortha):** © Biophoto Associates/Photo Researchers, Inc.; **22.2(Acetabularia):** © Linda L. Sims/Visuals Unlimited; **22.2(Amoeba):** © Michael Abbey/Visuals Unlimited; **22.2(Bossiella):** © Daniel V. Gotschall/Visuals Unlimited; **22.3:** © W.L. Dentler/Biological Photo Service; **22.5b:** © M.I. Walker/Science Source/Photo Researchers, Inc.; **22.6a:** © William E. Ferguson; **22.6b(left):** © Dr. John D. Cunningham/Visuals Unlimited; **22.6b(right):** © Kingsley Stern; **22.7(top):** © John D. Cunningham/Visuals Unlimited; **22.7(bottom):** © Cabisco/Visuals Unlimited; **22.8:** © Walter Hodge/Peter Arnold, Inc.; **22.9:** © D.P Wilson/Eric & David Hosking/Photo Researchers, Inc.; **22.10a:** © Dr. Ann Smith/Photo Researchers, Inc.; **22.10b:** © Biophoto Assoc./Photo Researchers, Inc.; **22.11a:** © C.C. Lockwood/Cactus Clyde Productions;

22.11b: © Sanford Berry/Visuals Unlimited; **22.12b:** © Michael Abbey/Visuals Unlimited; **22.13a:** Image acquired by Prof. Michael Duszenko, University of Tubingen and Eye of Science, Reutlingen (Germany); **22.14:** © Stanley Erlandsen; **22.15b(background):** © Rick Ergenbright/Corbis; **22.15b(inset):** © Manfred Kage/Peter Arnold, Inc.; **22.15c:** © Dr. Richard Kessel & Dr. Gene Shih/Visuals Unlimited; **22.16a:** © CABISCO/Phototake; **22.16b:** © Manfred Kage/Peter Arnold, Inc.; **22.16c:** © Eric Grave/Photo Researchers, Inc.; **22.18:** © CABISCO/Visuals Unlimited; **22.18:** © V. Duran/Visuals Unlimited; 22.19: © James Richardson/Visuals Unlimited.

Chapter 23

Opener: © Biophoto Assoc./Photo Researchers, Inc.; **23.1a:** © Gary R. Robinson/Visuals Unlimited; **23.1b:** © Dennis Kunkel/Visuals Unlimited; **23.2a:** From C.Y. Shih and R.G. Kessel, /it/LIVING IMAGES/xit/ Science Books International, Boston, 1982; **23.2b:** © Jeffrey Lepore/Photo Researchers, Inc.; **23.3(top):** © James W. Richardson/Visuals Unlimited; **23.3(Bottom):** © David M. Phillips/Visuals Unlimited; **23.4a,b:** © David Philips/Visuals Unlimited; **23.5a:** © Walter H. Hodge/Peter Arnold, Inc.; **23.5b(left):** © Michael Viard/Peter Arnold, Inc.; **23.5b(right):** © James Richardson/Visuals Unlimited; **23.5c:** © Kingsley Stern; **23.6:** © SciMAT/Photo Researchers, Inc.; **23.7a:** © Dr. P. Marazzi/SPL/Photo Researchers, Inc.; **23.7b:** © John Hadfield/SPL/Photo Researchers, Inc.; **23A:** © Patrick Endres/Visuals Unlimited; **23B:** © R. Calentine/Visuals Unlimited; **23.8** © Biophoto Associates; **23.9a:** © Glenn Oliver/Visuals Unlimited; **23.9b:** © Larry Lefever/Jane Grushow/Grant Heilman Photography; **23.9c:** © M. Eichelberger/Visuals Unlimited; **23.9d:** © L. West/Photo Researchers, Inc.; **23.10a:** © Leonard L. Rue/Photo Researchers, Inc.; **23.10b:** © Arthur M. Siegelman/Visuals Unlimited; **23.11a:** © Dr. Jeremy Burgess/SPL/Photo Researchers, Inc.; **23.11b:** © Stephen Sharnoff/Visuals Unlimited; **23.11c:** © Kerry T. Givens/Tom Stack & Assoc.; **23.12:** © R. Roncadori/Visuals Unlimited.

Chapter 24

Opener: © J.J. Alcalay/Peter Arnold, Inc.; **24.1a(left):** © Dr. Jeremy Burgess/SPL/Photo Researchers, Inc.; **24.1a(right):** © Ed Reschke; **24.1b(left):** © Simon Fraser/SPL/Photo Researchers, Inc.; **24.1b(right):** Courtesy Graham Kent; **24.1c(left):** © Kent Dannen/Photo Researchers, Inc.; **24.1c(right):** Courtesy Graham Kent; **24.1d(left):** © E. Webber/Visuals Unlimited; **24.1d(right):** © Richard Shiell/Animals Animals/Earth Scenes; **24.5a:** © Triarch/Visuals Unlimited; **24.5b:** © Ed Reschke; **24.6a:** © Kingsley Stern; **24.6b:** © Andrew Syred/SPL/Photo Researchers, Inc.; **24.7:** © Barry Runk/Stan/Grant Heilman Photography; **24.8a:** © Ed Reschke/Peter Arnold, Inc.; **24.8 b:** © J.M. Conrarder/Nat'l Audubon Society/Photo Researchers, Inc.; **24.8c:** © R. Calentine/Visuals Unlimited; **24.9(top):** © Heather Angel/Biofotos; **24.9(lower):** © Bruce Iverson; **24.10:** Courtesy of Hans Steur, the Netherlands; **24.11(top):** © Kingsley Stern; **24.11(center):** © Runk/Schoenberger/Grant Heilman Photography, Inc.; **24.11(bottom):** © Ed Reschke; **24.12:** © Steve Solum/Bruce Coleman, Inc.; **24.13:** © Robert P. Carr/Bruce

Coleman, Inc.; **24.14:** © CABISCO/Phototake; **24.15(top):** © James Strawser/Grant Heilman Photography; **24.15(bottom right):** © Larry Lefever/Jane Grushow/Grant Heilman Photography; **24.15(bottom left):** © Walter H. Hodge/Peter Arnold, Inc.; **24.16:** © Matt Meadows/Peter Arnold, Inc.; **24.17a:** © Grant Heilman/Grant Heilman Photography, Inc.; **24.17b:** © Walt Anderson/Visuals Unlimited; **24.17c:** © James Mauseth; **24.18:** © Phototake; **24.19a,b:** Pat Pendarvis; **24.20(main):** © Runk/Schoenburger/Grant Heilman Photography; **24.20(right):** Courtesy Ken Robertson, University of Illinois/INHS; **24.21(left):** Courtesy Dan Nickrent; **24.21(right):** Courtesy K.J. Niklas; **24.22:** © NHPA/Steve Robinson; **24A:** © Sinclair Stammers/SPL/Photo Researchers, Inc.; **24.23:** Courtesy Stephen McCabe/Arboretum at University of California Santa Cruz; **24.24(Daisy):** © Ed Reschke; **24.24(Cactus):** © Christi Carter/Grant Heilman Photography, Inc.; **24.24(Butterfly weed):** © Evelyn Jo Hebert; **24.24(Waterlily):** © Pat Pendarvis; **24.24(Dogwood):** © Adam Jones/Photo Researchers, Inc.; **24.24(Iris):** © David Cavanaugh/Peter Arnold, Inc.; **24B(Wheat):** © Earl Roberge/Photo Researchers, Inc.; **24B(Corn ear):** © Doug Wilson/Corbis; **24B(Corn plants):** © Adam Hart-Davis/SPL/Photo Researchers, Inc.; **24B(Rice plant):** © CORBIS RF; **24B(Rice grains):** © Dex Image/Getty RF; **24C(Palm):** © Heather Angel; **24C(Rubber):** © Steven King/Peter Arnold, Inc.; **24C(Cotton):** © Dale Jackson/Visuals Unlimited; **24C(Tulip):** © PhotoDisc Blue/Getty RF.

Chapter 25

Opener (Prop root): © David Newman/Visuals Unlimited; **(Root hairs):** © Runk-Schoenberger/Grant Heilman Photography; **(Trunk cross section):** © Ardea London Ltd.; **(Onion plant):** © Dwight Kuhn; **(Micrograph):** Courtesy of George Ellmore, Tufts University; **25.2a-c:** © Dwight Kuhn; **25.4a:** © Runk Schoenberger/Grant Heilman Photography; **25.4b:** © J.R. Waaland/Biological Photo Service; **25.4c:** © Kingsley Stern; **25.5a-c:** © Biophoto Associates/Photo Researchers, Inc.; **25.6a:** © J. Robert Waaland/Biological Photo Service; **25.7a:** © George Wilder/Visuals Unlimited; **25.8(top):** © CABISCO/Phototake; **25.8(bottom):** Courtesy Ray F. Evert/University of Wisconsin Madison; **25.9a:** © Dwight Kuhn; **25.10a:** © John D. Cunningham/Visuals Unlimited; **25.10b:** Courtesy of George Ellmore, Tufts University; **25.11a:** © Dr. Robert Calentine/Visuals Unlimited; **25.11b:** © Ed Degginger/Color Pic; **25.11c:** © David Newman/Visuals Unlimited; **25.11d:** © Terry Whittaker/Photo Researchers, Inc.; **25.11e-1:** © Pat Pendarvis; **25.11e-2:** © Biophot; **25A:** © James Schnepf Photography, Inc.; **25.13a:** © J. Robert Waaland/Biological Photo Service; **25.14(bottom):** Courtesy Ray F. Evert/University of Wisconsin Madison; **25.14(top):** © Ed Reschke; **25.15(top):** © CABISCO/Visuals Unlimited; **25.15(bottom):** © Kingsley Stern; **25.17:** © Ed Reschke/Peter Arnold, Inc.; **25.18a:** © Ardea London Limited; **25.19a:** © Stanley Schoenberger/Grant Heilman Photography; **25.19b:** © William E. Ferguson; **25.19c,d:** © The McGraw Hill Companies, Inc./Carlyn Iverson, photographer; **25B(top left):** © Jim Strawser/Grant Heilman Photogorapy, Inc.; **25.B(top right):** © Stephen Owens/SPL/Photo Researchers; **25B(Center):** Courtesy Edward F. Gilman/U. Florida; **25.20(right):**

© Jeremy Burgess/SPL/Photo Researchers, Inc.; **25.22a:** © Patti Murray Animals Animals/Earth Scenes; **25.22b:** © Gerald & Buff Corsi/Visuals Unlimited; **25.22c:** © P. Goetgheluck/Peter Arnold, Inc.

Chapter 26

Opener: © Tim Davis/Photo Researchers, Inc.; **26.2a-c:** Courtesy Mary E. Doohan; **26.6a:** © Dwight Kuhn; **26.6b:** © E.H. Newcomb & S.R. Tardon/Biological Photo Service; **26.7:** © Runk Schoenberger/Grant Heilman; **26.8:** © Pat Pendarvis; **26.8(inset):** © Gary Retherford/Photo Researchers, Inc.; **26A(both):** © Dwight Kuhn; **26.10a,b:** Courtesy Wilfred A. Cote, from H.A. Core, W.A. Cote, and A.C. Day, Wood: Structure and Identification, 2/e; **26.10c:** Courtesy W.A. Cote, Jr., N.C. Brown Center for Ultrastructure Studies, SUNY-ESF; **26.11:** © Ed Reschke/Peter Arnold, Inc.; **26.13a,b:** © Jeremy Burgess/SPL/Photo Researchers, Inc.; **26B:** Courtesy Gary Banuelos/Agriculture Research Service/USDA; **26.14a:** From M.H. Zimmerman "Movement of Organic Substances in Trees" in SCIENCE 133 (13) January 1961 page 667, Fig. 3 page 73, © 1961 AAAS; **26.14b:** © Bruce Iverson/SPL/Photo Researchers, Inc.

Chapter 27

Opener: © Lightwave Photography, Inc./Animals/Animals/Earth Scenes; **27.1:** © Kim Taylor/Bruce Coleman, Inc.; **27.2a:** © Kingsley Stern; **27.2b:** Courtesy Malcom Wilkins, Botany Department, Glascow University; **27.2c:** © Biophot; **27.3:** © John D. Cunningham/Visuals Unlimited; **27.4(both):** © John Kaprielian/Photo Researchers, Inc.; **27.5(top):** © Tom McHugh/Photo Researchers, Inc.; **27.5(Bottom):** © Tom McIugh/Photo Researchers, Inc.; **27.6a:** Courtesy Prof. Malcolm B. Wilkins; **27.6b:** Courtesy Prof. Malcolm B. Wilkins; **27.9:** © Robert E. Lyons/Visuals Unlimited; **p. 485:** Courtesy Donald Briskin; **27.11a-d:** Courtesy Alan Darvill and Stefan Eberhard, Complex Carbohydrate Research Center, University of Georgia; **27.13(both):** © Kingsley Stern; **27.16a,b:** © Grant Heilman Photography; **27A-D:** Courtesy Elliot Meyerowitz/California Institute of Technology.

Chapter 28

Opener: © Runk/Schoenberger/Grant Heilman Photography, Inc.; **28.3a:** © Farley Bridges; **28.3b:** © Pat Pendarvis; **28.4a:** © Arthur C. Smith, III/Grant Heilman Photography, Inc.; **28.4b:** © Larry Lefever/Grant Heilman Photography, Inc.; **28.5(top):** Courtesy Graham Kent; **28.5(Bottom):** © Ed Reschke; **28.6a:** © George Bernard/Animals Animals/Earth Scenes; **28.6b:** © Simko/Visuals Unlimited; **28.6c:** © Dwight Kuhn; **28E:** © Elliot Meyerowitz/California Institute of Technology, Biology Dept. of Pasadena; **28Aa:** © Comstock; **28Ab:** © Robert Maier/Animals/Animals/Earth Scenes; **28Ac:** © Anthony Mercieca/Photo Researchers, Inc.; **28Ad:** © Merlin D. Tuttle/Bat Conservation International; **28.7(Arabidopsis):** Courtesy Dr. Chun-Ming Liu; **28.7(Torpedo):** © Biology Media/Photo Researchers, Inc.; **28.7(Embryo):** Jack Bostrack/Visuals Unlimited; **28.8a:** © Dwight Kuhn; **28.10a:** © James Mauseth; **28.8b:** Courtesy Ray F. Evert/University of Wisconsin Madison; **28.10b:** © Corbis Royalty-Free; **28.10c:** © Runk/Schoenberger/Grant Heilman Photography, Inc.;

28.10d: © BJ Miller/Biological Photo Service; **28.11b:** © Ed Reschke; **28.12(left):** © James Mauseth; **28.12(right):** © Barry L. Runk/Grant Heilman, Inc.; **28.13:** © G.I. Bernard/Animals Animals/Earth Scenes; **28.14a-f:** Courtesy Prof. Dr. Hans-Ulrich Koop, from Plant Cell Reports, 17:601-604; **28.15a:** © Wally Eberhart/Visuals Unlimited; **28.15b:** © Science VU/Visuals Unlimited; **28.16a,b:** Courtesy Monsanto; **28.17b:** Courtesy Eduardo Blumwald; **28Ba:** © Larry Lefever/Grant Heilman Photography, Inc.; **28Bb:** © James Shaffer/PhotoEdit; **28Bc:** Courtesy Carolyn Merchant/ UC Berkeley.

Chapter 29

Opener: © James Castner; **29.1:** © Joe McDonald/Visuals Unlimited; **29D(top):** © Arthur Siegelman/Visuals Unlimited; **29D(center):** © John MacGregor/Peter Arnold, Inc.; **29D(bottom):** © OSF/London Scientific Films/Animals Animals/Earth Scenes; **29.3a:** © Andrew J. Martinez/Photo Researchers, Inc.; **29.4a:** © Jeff Rotman; **29.4b:** © J. McCollugh/Visuals Unlimited; **29.5b:** © Azure Computer & Photo Services/Animals Animals/Earth Scenes; **29.5c:** © Ron Taylor/Bruce Coleman; **29.5d:** © Runk/Schoenberger/Grant Heilman Photography; **29.5e:** © Under Watercolours; **29.6:** © CABISCO/Visuals Unlimited; **29.7(left):** © Runk/Schoenberger/Grant Heilman Photography; **29.7(right):** © Biodisc/Visuals Unlimited; **29.8:** © Fred Bavendam/Peter Arnold, Inc.; **29.9e:** © Tom E. Adams/Peter Arnold, Inc.; **29.10a:** © SPL/Photo Researchers Inc.; **29.11(left):** © John D. Cunningham/Visuals Unlimited; **29.11(right):** © James Webb/Phototake, NYC; **29.12a:** © Lauritz Jensen/Visuals Unlimited; **29.12c:** © James Solliday/Biological Photo Service; **29.13:** From E.K. Markell and M. Voge, *Medical Parasitology*, **7/e**, 1992 W.B. Saunders Co.

Chapter 30

Opener: Courtesy of M. Christopher Barnhart/Southwest Missouri State University; **30.2b:** © Kjell Sandved/Butterfly Alphabet; **30.3a:** Courtesy of Larry S. Roberts; **30.3b:** © Fred Whitehead/Animals/Animals/Earth Scenes; **30.4a:** © Ken Lucas/Visuals Unlimited; **30.4b:** © Douglas Faulkner/Photo Researchers, Inc.; **30.4c:** © Georgette Douwma/Photo Researchers, Inc.; **30.5a:** © M. Gibbs/OSF/Animals Animals/Earth Scenes; **30.5b:** © Kenneth W. Fink/Bruce Coleman, Inc.; **30.5b:** © Farley Bridges; **30.6a:** © Heather Angel/Natural Visions; **30.6b:** © James H. Carmichael; **30.7c:** © Roger K. Burnard/Biological Photo Service; **30.8:** © St. Bartholomews Hospital/SPL/Photo Researchers, Inc.; **30.9c:** © Farley Bridges; **30.10a:** © Michael Lustbader/Photo Researchers, Inc.; **30.10b:** © Kjell Sandved/Butterfly Alphabet; **30.10c:** © Bruce Robinson/Corbis; **30.10d:** © James Robinson/Animals Animals/Earth Scenes; **30.10e:** © Kim Taylor/Bruce Coleman, Inc.; **30.12(Wasp):** © Johnathan Smith Cordaiy Photo Library/Corbis; **30.12(Leafhopper, Dragonfly, Mealybug):** © Farley Bridges; **30.12(Beetle):** © Wolfgang Kaehler/Corbis; **30.12(Louse):** © Darlyne A. Murawski/Peter Arnold, Inc.; **30.12(Grasshopper):** © Chris Mattison;Frank Lane Picture Agency/Corbis; **30.13(top):** © McDonald Wildlife Photography/Animals Animals/Earth Scenes; **30.13(Bottom):** © L. West/Bruce Coleman, Inc.; **30.15a:** © David M. Dennis/Animals Animals/Earth Scenes;

30.15c: Geof du Feu/Imagestate; **30.16a:** © Jana R. Jirak/Visuals Unlimited; **30.16b:** © Tom McHugh/Photo Researchers, Inc.; **30.16d:** © Ken Lucas; **30.16e:** © Farley Bridges; **30.17b:** © Randy Morse/Tom Stack & Assoc.; **30.17c:** © Alex Kerstitch/Visuals Unlimited; **30.17d:** © Randy Morse/Animals Animals/Earth Scenes.

Chapter 31

Opener: © 2003 Monty Sloan; **31.1:** © Heather Angel; **31.2:** © Rick Harbo; **31.4:** © Heather Angel; **31.5a:** © James Watt/Animals Animals/Earth Scenes; **31.5b:** © Fred Bavendam/Minden Pictures; **31.6a:** © Ron & Valarie Taylor/Bruce Coleman, Inc.; **31.6b:** © Hal Beral/Visuals Unlimited; **31.6c:** © Jane Burton/Bruce Coleman, Inc.; **31.6d:** © Claus Qvist Jessen; **31.6e:** © Franco Banfi/SeaPics.com; **31.7:** © Peter Scoones/SPL/Photo Researchers, Inc.; **31.8a:** © Suzanne L. Collins & Joseph T. Collins/Photo Researchers, Inc.; **31.8b:** © Joe McDonald/Visuals Unlimited; **31.8c:** © Juan Manuel Renjifo/Animals Animals/Earth Scenes; **31.9a-c:** © Michael Redmer/Visuals Unlimited; **31.9d:** © Joel McDonald/Visuals Unlimited; **31.12a:** © Martin Harvey/Gallo Images/Corbis; **31.12b:** © H. Hall/OSF/Animals Animals/Earth Scenes; **31.12c:** © Joe McDonald/Visuals Unlimited; **31.12d:** © Joel Sartorie/National Geographic/Getty Images; **31.12e:** © Fabio Colombini Mederios/Animals Animals/Earth Scenes; **31.12f:** © Nathan W. Cohen/Visuals Unlimited; **31.13:** © OS21/PhotoDisc/Getty Images; **31Aa:** © Mark Smith/Photo Researchers, Inc.; **31Ab:** © Allan Friedlander/SuperStock; **31Ac:** © Account Phototake/Phototake; **31.15b:** © Joel McDonald/Corbis; **31.15e:** © Kirtley Perkins/Visuals Unlimited; **31.16a:** © Thomas Kitchin/Tom Stack & Associates;**31.16c:** © Brian Parker/Tom Stack & Associates; **31.16d:** © Robert Comport/Animals Animals/Earth Scenes; **p. 571(both):** © Daniel J. Cox; **31.16a:** © D. Parer & E. Parer-Cook/Ardea; **31.16b:** © Leonard Lee Rue/Photo Researchers, Inc.; **31.16c:** © Fritz Prenzel/Animals Animals/Earth Scenes; **31.17a,b:** © Stephen J. Krasemann/Photo Researchers, Inc.; **31.17b:** © Stephen J. Krasemann/DRK Photo; **31.17c:** © Gerald Lacz/Animals Animals/Earth Scenes; **31.17d:** © Mike Bacon/Tom Stack & Associates.

Chapter 32

Opener: © Kazuhiko Sano; **32.1a(Lemur):** © Frans Lanting/Minden Pictures; **32.1a(Tarsier):** © Doug Wechsler; **32.1b(White-faced):** © C.C. Lockwod/DRK Photo; **32.1b(Anubis):** © St. Meyers/Okapia/Photo Researchers, Inc.; **32.1c:** © Tim Davis/Photo Researchers, Inc.; **32.1(Gibbon):** © Hans & Judy Beste/Animals Animals/Earth Scenes; **32.1(Chimp, Gorilla):** © Martin Harvey/Peter Arnold, Inc.; **32.1d:** © Tim Davis/Photo Researchers, Inc.; **32.5c:** © National Museum of Kenya; **32.8a:** © Dan Dreyfus and Associates; **32.8b:** © John Reader/Photo Researchers, Inc.; **32A:** © Margaret Miller/Photo Researchers, Inc.; **32.11:** The Field Museum, #A102513c; **32.12:** Transp. #608 Courtesy Dept. of Library Services, American Museum of Natural History; **32.13a:** © PhotoDisc/Getty Images; **32.13b:** Sylvia Mader; **32.13c:** © B & C Alexander/Photo Researchers, Inc.

47.18a–e: © Breck P. Kent/Animals Animals/ Earth Scenes; **47.20a:** © Jeff Foott/Bruce Coleman, Inc.; **47.21a:** © William E. Townsend, Jr./Photo Researchers, Inc.

Chapter 48

Opener: © Stephen Ferry/Liaison/Getty Images; **48.1(African Grassland):** © Gregory G. Dimijian/Photo Researchers, Inc.; **48.1(rest of images):** © Corbis RF; **Fig 48.2(Cottontail):** © Gerald C. Kelley/Photo Researchers, Inc.; **48.2(Diatom):** © Ed Reschke/Peter Arnold, Inc.; **48.2(Tree):** © Hermann Eisenbeiss/ Photo Researchers, Inc.; **48.2(Caterpillar):** © Royalty-free/Corbis; **48.2(Spider):** © Bill Beatty/Visuals Unlimited; **48.2(Mushroom):** © Michael Beug; **48.2(Osprey):** © Joe McDonald/Visuals Unlimited; **48.2(Decomposer):** © SciMAT/Photo Researchers, Inc.; **48.4:** © George D. Lepp/ Photo Researchers, Inc.; **48.12a,b:** © John Shaw/Tom Stack & Assoc.; **48.13c:** © Bill Aron/Photo Edit; **48Ba:** Courtesy GSFC/ NASA; **48C:** © Vol. 9/Getty Images.

Chapter 49

Opener: NASA; **49.6a:** © John Shaw/Tom Stack & Assoc.; **49.6b:** © John Eastcott/Animals Animals/Earth Scenes; **49.6c:** © John Shaw/ Bruce Coleman, Inc.; **49.7(right):** © Mack Henly/ Visuals Unlimited; **49.7(left):** © Bill Silliker, Jr./ Animals Animals/Earth Scenes; **49.8(Forest):** © E. R. Degginger/Animals Animals/ Earth Scenes; **49.8(Chipmunk):** © Carmela Lesczynski/Animals Animals/Earth Scenes; **49.8(Millipede):** © OSF/Animals Animals/Earth Scenes; **49.8(Bobcat):** © Tom McHugh/Photo Researchers, Inc.; **49.8(Marigolds):** © Virginia Neefus/Animals Animals/Earth Scenes; **49Aa:** © Porterfield/Chickering/Photo Researchers, Inc.; **49Ab:** © Michio Hoshino/Minden Pictures; **49.10(Arboreal lizard):** © Kjell Sandved/Butterfly Alphabet; **49.10(Katydid):** © James Castner; **49.10(Butterfly):** © Kjell Sandved/Butterfly Alphabet; **49.10(Ocelot):** © Martin Wendler/Peter Arnold, Inc.; **49.10(Lemur):** © Erwin & Peggy Bauer/Bruce Coleman, Inc.; **49.10(Macaw):** © Tony Craddock/SPL/Photo Researchers, Inc.; **49.10(Frog):** © Art Wolfe/Photo Researchers, Inc.; **49.11(Chaparral):** © Bruce Iverson; **49.11(Jay):** © H.P. Smith, Jr./VIREO; **49.11(Chaparral):** © Kathy Merrifield/Photo Researchers, Inc.; **49.12(Prairie):** © Jim Steinberg/Photo Researchers, Inc.; **49.12(Bison):** © Steven Fuller/ Animals Animals/Earth Scenes; **49.13(Zebra, Wildebeest):** © Darla G. Cox; **49.13(Giraffe):** © George W. Cox; **49.13(Cheetah):** © Digital Vision/Getty Images; **49.14(Kangaroo rat):** © Bob Calhoun/Bruce Coleman, Inc.; **49.14(Roadrunner):** © Jack Wilburn/Animals Animals/Earth Scenes; **49.14(Desert):** © John Shaw/Bruce Coleman; **49.14(Kit fox):** © Jeri Gleiter/Peter Arnold, Inc.; **49.15(Stonefly):** © Kim Taylor/Bruce Coleman, Inc.; **49.15(Trout):** © William H. Mullins/Photo Researchers, Inc.; **49.15(Carp):** © Robert Maier/Animals Animals/ Earth Scenes; **49.15(Blue crab):** © Gerlach Nature Photography/Animals Animlas/Earth Scenes; **49.16a:** © Roger Evans/Photo Researchers, Inc.; **49.16b:** © Michael Gadomski/Animals Animals/ Earth Scenes; **49.18(Pike):** © Robert Maier/ Animals Animals/Earth Scenes; **49.18(Pond skater):** © G.I. Bernard/Animals Animals/Earth Scenes; **49.19(Shrimp):** © Ken Lucas/Ardea; **49.19(Snails):** © Heather Angel; **49.20a:** © John

Eastcott/Yva Momatiuk/Animals Animals/ Earth Scenes; **49.20b:** © James Castner; **49.21b:** © Brandon Cole/Visuals Unlimited; **49.21c:** © Jeff Greenburg/Photo Researchers, Inc.

Chapter 50

Opener: © Annie Griffiths-Belt/Corbis; **50.3(Rose):** © Kevin Schaefer/Peter Arnold, Inc.; **50.3(Man):** © Bryn Campbell/Stone/ Getty; **50.3(Armadillo):** © John Cancalosi/ Peter Arnold, Inc.; **50.3(Fishermen):** © Herve Donnezan/Photo Researchers, Inc.; **50.3(Bat):** © Merlin D. Tuttle/Bat Conservation International; **50.3(Ladybug):** © Anthony Mercieca/Photo Researchers, Inc.; **50.4a:** © William M. Smithy/Getty Images; **50.4b:** © Don and Pat Valenti/DRK; **50.5b:** © Gunter Ziesler/Peter Arnold, Inc.; **50.5c:** Courtesy Woods Hole Research Center; **50.5d:** Courtesy R.O. Bierregaard; **50.5e:** Courtesy Thomas Stone, Woods Hole Research Center; **p. 935:** © Marine Aquarium Council Headquarters; **50.6a:** © Chris Johns/National Geographic Society Image Collection; **50.6b:** © Chuck Pratt/Bruce Coleman, Inc.; **50.7b:** Courtesy Walter C. Jaap/Florida Fish & Wildlife Conservation Commission; **50.8a:** © Shane Moore/Animals Animals/Earth Scenes; **50.8b(both):** © Peter Auster/ University of Connecticut; **50.9a:** © Gerard Lacz/Peter Arnold, Inc.; **50.9a(top):** © Art Wolfe; **50.9b(bottom):** © Pat & Tom Leeson/ Photo Researchers, Inc.; **50.10b:** © Jeff Foott Productions; **50.11(Panther):** © Tom & Pat Leeson/Photo Researchers, Inc.; **50.11(Alligator):** © Fritz Polking/Visuals Unlimited; **50.11(Ibis):** © Stephen G. Maka; **50.11(Spoonbill):** © Kim Heacox/Peter Arnold, Inc.

Line Art and Text

Chapter 1

Opener: Courtesy of J. William Schopf, Director, UCLA Center for the Study of Evolution and the Origin of Life. p. 1.

Chapter 2

Ecology Focus: Data from G. Tyler Miller, *Living in the Environment,* 1983, Wadsworth Publishing Company, Belmont, CA; and Lester R. Brown, *State of the World,* 1992, W.W. Norton & Company, Inc., New York, NY. p. 31

Chapter 8

Health Focus: From Scott K. Powers and Edward T. Howley, *Exercise Physiology,* 2/e. Copyright 1994 The McGraw-Hill Companies. All Rights Reserved. p. 143.

Chapter 15

Science Focus: Courtesy of Joyce Haines. p. 263.

Chapter 18

Science Focus: Courtesy of Gerald D. Carr, University of Hawaii at Manoa. p. 314.

Chapter 19

19.16: Data supplied by J. John Sepkoski, Jr., Professor of Paleontology, University of Chicago. p. 328.

Chapter 21

Table 21.1: Data Courtesy of Lansing M. Prescott. p. 363.

Chapter 24

Ecology Focus: Courtesy of Charles Horn. p. 432-33.

Chapter 26

Science Focus: From Joe Bower, *National Wildlife Magazine,* June/July 2000, Vol. 38, No. 4, Reprinted with permission of the author. p. 471.

Chapter 27

Science Focus: Courtesy of Donald Briskin and Margaret Gawienowski, University of Illinois at Urbana-Champaign. p. 485.

Chapter 32

Science Focus: Courtesy of Steven Stanley, The John Hopkins University. p. 586.

Chapter 36

36.12: Data from U. S. Department of Agriculture. p. 663.

Chapter 43

43.12: Data from Division of STD Prevention, Sexually Transmitted Disease Surveillance, 2003. U.S. Department of Health and Human Services, Public Health Service, Atlanta: Centers for Disease Control and Prevention, September 2003. p. 788; **43.13:** Data from Division of STD Prevention, Sexually Transmitted Disease Surveillance, 2003. U.S. Department of Health and Human Services, Public Health Service, Atlanta: Centers for Disease Control and Prevention, September 2003. p. 789; **43.14:** Data from Division of STD Prevention, Sexually Transmitted Disease Surveillance, 2003. U.S. Department of Health and Human Services, Public Health Service, Atlanta: Centers for Disease Control and Prevention, September 2003. p. 789; **43.15:** Data from Division of STD Prevention, Sexually Transmitted Disease Surveillance, 2003. U.S. Department of Health and Human Services, Public Health Service, Atlanta: Centers for Disease Control and Prevention, September 2003. p. 789; **43.16:** Data from Division of STD Prevention, Sexually Transmitted Disease Surveillance, 2003. U.S. Department of Health and Human Services, Public Health Service, Atlanta: Centers for Disease Control and Prevention, September 2003. p. 790.

Chapter 45

45.2b: Data from S.J. Arnold, "The Microevolution of Feeding Behavior" in *Foraging Behavior: Ecology, Ethological, and Psychology Approaches,* edited by A. Kamil and T. Sargent, 1980, Garland Publishing Company, New York, NY. p. 819; **Science Focus:** Courtesy of Gail Patricelli, University of Maryland. p. 824.

Chapter 46

46.4b: Data from W.K. Purves, et al., Life: *The*

Science of Biology, 4/e, Sinaeur & Associates. p. 839.; **46.4c:** Data from E.J. Kormondy, 1984, *Concepts of Ecology,* 3/e, Prentice-Hall, Inc., Figure 4.6, p. 107. p. 839; **46.4d:** Data from A.K. Hegazy, 1990, "Population Ecology & Implications for Conservation of Cleome Droserifolia: A Threatened Xerophyte," *Journal of Arid Environments,* 19:269-82. p. 839; **Ecology Focus:** Courtesy of Jeffrey Kassner. p. 840; **46.8:** From Raymond Pearl, *The Biology of Population Growth.* Copyright 1925 The McGraw-Hill Companies. All Rights Reserved. p. 843; **46.9b:** Data from Charles J. Krebs, *Ecology,* 3/e, 1984, Harper & Row; after Scheffer, 1951. p. 844; **46.16:** United Nations Population Division, 2002. p. 850.

Chapter 47

47.1c: Data from G.G. Simpson, "Species Density of North America Recent Mammals" in *Systemic Zoology,* Vol. 13:57-73, 1964. p. 858; **47.2:** Data from Charles J. Krebs, *Ecology,* 3/e, 1984, Harper & Row. p. 859; **47.5:** Data from G.F. Gause, *The Struggle for Existance,* 1934, Williams & Wilkins Company, Baltimore, MD. p. 862; **47.9:** Data from G.F. Gause, *The Struggle for Existance,* 1934, Williams & Wilkins Company, Baltimore, MD. p. 864; **47.10b:** Data from D.A. MacLulich, *Fluctuations in the Numbers of the Varying Hare (Lepus americanus),* University of Toronoto Press, Toronto, 1937, reprinted 1974. p. 865.

Chapter 49

Ecology Focus: From T. Friend, "DNA Fingerprinting: Power Rool," *National Wildlife Magazine,* October/November 1995, Vol. 33, No. 6. Reprinted with permission of the author. p. 907.

Chapter 50

50.2: Redrawn from "Shrimp Stocking, Salmon Collapse, and Eagle Displacement" by C.N. Spencer, B.R. McClelland and J.A. Stanford, Bioscience, 41(1):14-21. Copyright © 1991 American Institute of Biological Sciences. p. 927; **Ecology Focus:** Courtesy of Stephanie Songer, North Georgia College and State University. p. 933; **50.7:** Data from David M. Gates, *Climate Change and Its Biological Consequences,* 1993, Sinauer & Associates, Inc., Sunderland, MA. p. 935.

Line Art

Electronic Publishing Services Inc. Illustration Team

Lead Illustrator: Erin Daniel
Lead Illustrator: Matthew McAdams
Lead Illustrator: Tommy Moorman
Art Coordinator: Haydee Martinez

Amadeo Bachar: Figures 25.03, 25.19, 27.8, 27.11, 27.12

Rachel Bedno Robinson: Figures 12B, 20.13

Leigh Campbell: Figures 1.2, 1.5, 2.11, 4C, 4.2, uf 4.1, uf 4.3, uf 4.4, 4.18, 5.6, 5.7, 6.1, uf 6.2, 6.4, 31.11, 48.3

Raychel Ciemma: Figures 18.14, 20.6, 20.13, uf20.02, 25.1, 26.1, 26.9, 27.14, 28.7, 31.3, 31.11, 31.14, 32.5, 32.7, 36.4, 37.5, 44.10, 45.1, 45.3, 45.4, 45.5, 45.14, 46.1, 46.2, 46.7, 46.10, 46.11, 46.12, 47.4, 47.6, 47.7, 47.14, 50.1

Emily Damstra Marinovic: Figures 18.4, 19.17, 20.13, 29A, 29.2, 29.3, 29.5, uf 29.3, uf 29.5, 29.6, 29.7, 29.10, 29.14, 30.2, 30.3, 30.4, 30.5, 30.10, 30.14, uf 30.5, 31.3, 31.6, 31.11, 31.12, 31.13, uf 31.8, 34.2, 36.3, 37.2, 37.3, 37.4, 38.2, 39.1, 40.3, 40.13, uf 40.2, 41.3, 47.19, 48A, 48.3, 48.5, 49.5, 49.9, 49.17, 50.1

Erin Daniel: Figures 8.3, 9.9, 13.1, 13.2, 13.3, 13.5, 15C, 15.12, 16.1, 18.12, 20.5, 20.9, 20.10, 20.13, uf 20.01, uf20.2, 21.3, uf21.3, 24.2, 24.4, 24.7, 24.8, 24.9, 24.10, 24.11, 24.12, 24.13, 24.14, 24.16, 24.18, 24A, 24.25, 24.26, uf24.4, 25.8c, 25.03, 25.13, 25.16, 25.17, 25.20, 26.3, 26.5, 26.13, uf 26.4, 27.12, 28.7, 28.1, 28.2, 28.5, 28.8, 28.11, 28.12, uf28.1, uf28.2, 30.3, 30.4, 31.13, 30.17, 32.4, 35.3, 35.9, 36A, 36.3, 37.5, 38.7, 38.9, 40.2, 40.11, 44.1, 48.3, 48.7, 49.18, 49.19, 49.22

Jennifer Nicole Gentry: Figures 33.1, 33.6, 35.1, 35.3, 35.4, 35.5, 35.6, 35.7, 35.8, 35.9, 35.12, 35.13, 35.14, 35A, uf 35.1, 37.6, uf 38.3, 38.7, 38.9, 42.2, 42.3, 42.4, 42.5, 42.11, 42.16, 43.4, 43.8, 44.1, 44.13, 44.14, 44.15, 44.16, 44A

Jonathan P. Higgins: Figures 2.11, 4.2, 10.1, 10.2, 10.3, 10.4, 10.7, 10.8, 10.9, 10.6, uf10.1, 11.4, 11A, uf12.4, 12.8, 12.10, 12.11, 12.13, 12.14, 12.15, 12.16, 16.5, 17.5, 17.13, 18.6, 18.8, 19.7, 19.15, 19.16, 19.17, uf 20.1, uf20.2, 25.3, 25.6, 25.7, 25.8a, 25.16, 25.17, 25.18, uf 25.2, 26.3, 26.4, 26.5a, 26.9, 27.8, 27.9, 27.01A, uf27.1a, 31.3, 32.4, 36.10, 36.11, 36.12, 39.1, 44.1, 44.13, 44.14, 44.15, 44.16, 44A, 48.7, 48.8, 48.9, 48.11, 48.14, 49.3, 49.15, 49.23, 50.2

Kellie Marsh Holoski: Figures 2.11, 3.12, 4.2, 4.6, 4.7, 4.8, 4.9, 4.10, 4.11, 4.13, 4.16, 4.17, 4.19, 4.20, uf 4.2, 5.1, 5.2, 5.4, 5.5, 5.7, 5.9, 5.10, 5.11, 5.12, 5.13, 5.14, 5.15, 5.16, uf 5.1, uf 5.2, uf 5.3, uf 5.4, uf 5.5, 6.13, 7.2, 7.4, 7.5, 7.6, 7.7, 7.8, uf 7.7, 8.1, 8.2, 8.5, 8.6, 8.7, 8.8, 8.9, 9.1, 9.2, 9.3, 9.4, 9.5, 9.8, 9A, uf 9.1, uf9.2, uf 9.3, 10.1, 10.2, 10.3, 10.4, 10.6, 10.7, 10.8, 10.9, uf 10.1, 11A, 12.9, 12.10, 12.11, 13.1, 13.3, 14.7, 14.8, 14.9, 14.10, 14.11, 14.12, 14.13, 14.14, 15.4, 16.1, 16.5, 16.9, uf16.7, 17.18, 21.3, 21.4, 21.12, 25.16, 25.20, uf 25.3, 29.10, 29B, 30.9, 30.10, 30.14, 30.13, 30.16, 31.3, 31.10, 31.11, 32.2, 33.9, 34B, 34.3, 34.4, 34.6b, 34.8, 34.9, 34.12, 34.13, 34.14, 35.1, 37.2, uf37.1, 38.8,

39.3, 39.4, 39.5, 39.6, uf 39.2, 40.8, 40.2, 40.11, 40.12, 41.13, 41.14, 43.5

Eliza Jewett: Figures 25.12, 25.17, 25.21, uf 25.1, 27B, 50.1

Alison Kendall: Figure 22.9

Matthew McAdams: Figures 1.5, 1.6, 1.13, 2.12, 2.13, 3.11, 3.17, 3.18, 4A, 4C, 5.3, 5.6, 5.8, 6.12, 8.8, 10.9, 14.14, 15.4, 17C, 18.12, 19.1, 19.17, 21.1, 21.5, 21.6, 21.12, 23.1, 23.3, 23.5, 23.8, 23.11, 29A, 29.2, 29.3, 30.2, 30.5, 30.14, 31.3, 33.5, 36.12, 37.09, 38.4, 38.6, 39.1, 39.4, 46.10, 46.12, 48.3, 48.9, 48.11, 48.15, 49.1, 49.2, 49.5, 49.14, 49.21, 49.24, 49B

Tommy Moorman: Figures 4.5, 5.6, 5.7, 6.1, 6.12, 13.6, 13.7, 14.3, 14.4, 15.1, 15.2, 15.3, 15.8, uf15.1, uf 15.5, 16.3, 17.4, 19.14, 19.15, 19.17, 22.1, 22.4, 22.5, 22.10, 22.12, 22.13, 22.15, 22.16, 22.17, 22.18, uf 22.1, 31.1, 31.2, 31.3, 31.11, uf31.2, 41.3, 44.1

Bob Morreale: Figures 4.6, 4.7

Kim Moss: Figures 3.21, 4.4, 4.6

Evelyn Pence: Figures 10.1, 10.2, 10.4, 10.8, 10.9, uf10.1, 11A, 12.9, 13.6, 13.7, 13.8, 13.9, 31.11, 13A, uf 13.2, 14.2, 14.3, 14.4, 14.5, 14.6, 14.7, 14.8, 14.9, 14.10, 14.11, 14.12, 14.13, 15.1, 15.2, 15.3, 15.4, 15.5, 15.6, 15.7, 15.8, 15.9, 15.10, 15.11, 15.12, 15.13, uf15.1, uf 15.4, uf15.5, 17.15, 17.16, 17.17, 20.5, 23.1, 24A, 24.25, 29A, 29B, 29E, 29.2, uf29.6, 29.9, 29.11, 29.12, 30.1, 30.6, 30.7, 30.15, 30.17, uf30.2, 31.1, 31.2, 31.3, 31.7b, 31.11, uf31.2, uf31.6, 34.02, 36.1, 36.2, 37.1, 38.4, 39.1, 41.1, uf 44.1, 44.2, 44.3, 44.4, 44.5, 44.6, 44.7, 44.8, 44.11, uf 44.2, 47.8, 48.6, 48.10, 48.13, 48.15, 48B, 49.18, 49.19, 49B, 49.24

Tara Russo: Figures 18.12, 18.13, 20.13, 29.5, uf29.9, 47.8

Melissa Thomas: Figure 49.21

Staci Washington: Figures 11.2, 11.3, 11.6, 11.7, 11.8, 11.9, 11.14, 11.18, 12.1, 12.7, 12.9, 20.9, 20.10, 25.03, 26A, 26.5b, 26.13, uf26.02, uf26.03, 27.7, 28.09, uf28.01, 31.11

Craig Zuckerman: Figure 10.8

Contributing Illustrators:

The following illustrators worked with the illustration team to enhance the artwork that appears throughout this book. Their talents and contributions were invaluable to the production of this art program.

Jen Christiansen
Bob Morreale
Fiona Morris
Kim Moss

INDEX

History of Biology

Antonie van Leeuwenhoek

Charles Darwin

Louis Pasteur

Robert Koch

Ivan Pavlov

Year	Name	Country	Contribution
1628	William Harvey	Britain	Demonstrates that the blood circulates and the heart is a pump.
1665	Robert Hooke	Britain	Uses the word *cell* to describe compartments he sees in cork under the microscope.
1668	Francesco Redi	Italy	Shows that decaying meat protected from flies does not spontaneously produce maggots.
1673	Antonie van Leeuwenhoek	Holland	Uses microscope to view living microorganisms.
1735	Carolus Linnaeus	Sweden	Initiates the binomial system of naming organisms.
1809	Jean B. Lamarck	France	Supports the idea of evolution but thinks there is inheritance of acquired characteristics.
1825	Georges Cuvier	France	Founds the science of paleontology and shows that fossils are related to living forms.
1828	Karl E. von Baer	Germany	Establishes the germ layer theory of development.
1838	Matthias Schleiden	Germany	States that plants are multicellular organisms.
1839	Theodor Schwann	Germany	States that animals are multicellular organisms.
1851	Claude Bernard	France	Concludes that a relatively constant internal environment allows organisms to survive under varying conditions.
1858	Rudolf Virchow	Germany	States that cells come only from preexisting cells.
1858	Charles Darwin	Britain	Presents evidence that natural selection guides the evolutionary process.
1858	Alfred R. Wallace	Britain	Independently comes to same conclusions as Darwin.
1865	Louis Pasteur	France	Disproves the theory of spontaneous generation for bacteria; shows that infections are caused by bacteria, and develops vaccines against rabies and anthrax.
1866	Gregor Mendel	Austria	Proposes basic laws of genetics based on his experiments with garden peas.
1882	Robert Koch	Germany	Establishes the germ theory of disease and develops many techniques used in bacteriology.
1900	Walter Reed	United States	Discovers that the yellow fever virus is transmitted by a mosquito.
1902	Walter S. Sutton Theodor Boveri	United States Germany	Suggest that genes are on the chromosomes, after noting the similar behavior of genes and chromosomes.
1903	Karl Landsteiner	Austria	Discovers ABO blood types.
1904	Ivan Pavlov	Russia	Shows that conditioned reflexes affect behavior, based on experiments with dogs.
1910	Thomas H. Morgan	United States	States that each gene has a locus on a particular chromosome, based on experiments with *Drosophila*.
1922	Sir Frederick Banting Charles Best	Canada	Isolate insulin from the pancreas.
1924	Hans Spemann Hilde Mangold	Germany	Show that induction occurs during development, based on experiments with frog embryos.
1927	Hermann J. Muller	United States	Proves that X rays cause mutations.
1929	Sir Alexander Fleming	Britain	Discovers the toxic effect of a mold product he called penicillin on certain bacteria.